Some Physical Properties

Air (dry, at 20°C and 1 atm)

Density	1.20 kg/m^3
Specific heat at constant pressure	$1.00 \times 10^3 \text{ J/kg·K}$
	0.240 cal/g·K
Ratio of specific heats (γ)	1.40
Speed of sound	343 m/s
	1130 ft/s

Water (20°C and 1 atm)

Density	$1.00 \times 10^3 \text{ kg/m}^3$
	1.00 g/cm^3
Speed of sound	1460 m/s
	4790 ft/s
Index of refraction ($\lambda = 589$ nm)	1.33
Specific heat at constant pressure	4180 J/kg·K
	1.00 cal/g·K
Heat of fusion (0°C)	$3.33 \times 10^5 \text{ J/kg}$
	79.7 cal/g
Heat of vaporization (100°C)	$2.26 \times 10^6 \text{ J/kg}$
	539 cal/g

The Earth*

Mass	$5.98 \times 10^{24} \text{ kg}$
Mean radius	$6.37 \times 10^6 \text{ m}$
	3960 mi
Mean earth-sun distance	$1.50 \times 10^8 \text{ km}$
	$9.30 \times 10^7 \text{ mi}$
Mean earth-moon distance	$3.84 \times 10^5 \text{ km}$
	$2.39 \times 10^5 \text{ mi}$
Standard gravity	9.81 m/s^2
	32.2 ft/s^2
Standard atmosphere	$1.01 \times 10^5 \text{ Pa}$
	14.7 lb/in^2
	760 mm-Hg
	29.9 in-Hg
Magnetic dipole moment	$8.0 \times 10^{22} \text{ A·m}^2$

* See also Appendix C.

FUNDAMENTALS OF
PHYSICS

Second Edition   Extended Version

REVISED PRINTING

David Halliday
University of Pittsburgh

Robert Resnick
Rensselaer Polytechnic Institute

with the assistance of
W. Farrell Edwards
Utah State University

John Merrill
Brigham Young University

JOHN WILEY & SONS, New York • Chichester • Brisbane • Toronto • Singapore

Library of Congress Cataloging-in-Publication Data:

Halliday, David, 1916–
    Fundamentals of physics.

    Includes index.
    1. Physics. I.   Resnick, Robert, 1923–
II. Title.
QC21.2.H35   1986b   530   85-22622
ISBN 0-471-82770-3

Printed in the United States of America
10 9 8 7 6 5 4 3 2 1

Preface

To the Revised Printing
of the Second Edition

We have prepared this revised printing of *Fundamentals of Physics, Second Edition,* and of *Fundamentals of Physics, Second Edition, Extended Version,* in response to long-time users of the text who asked for still more new problems in the book. We have gone beyond meeting that request to provide changes and additions in the Questions. We have also made revisions at a number of specific places in the text itself where we have learned from instructors that students can particularly benefit from improvements in the clarity of presentation. Overall this provides a significant improvement in the utility of the book for homework and class discussion purposes.

The new set of problems, about 2900 overall, represents a net increase of almost 40 percent over the second edition, giving us the largest set by far of any comparable textbook. To assist students and instructors in organizing and evaluating this large number of problems, we have done several things. First we have separated out the easiest problems, those that typically involve one step or formula or represent a single application, and we have called them Exercises. These Exercises, which have been increased substantially to almost 1000, are for building student confidence and constitute about one third of the total set of problems. They are numbered separately and are grouped in each chapter according to section number, the first section needed to be covered in order to be able to work out the problem.

The remaining two thirds (about 2000) of the problems have been split into two categories of roughly equal size, namely Problems Grouped by Section and Additional Problems, and numbered consecutively. These differ from each other only in the manner of listing and are of equal difficulty overall. The latter set, Additional Problems, are put in the order of difficulty regardless of section and should appeal to professors who prefer not to hint at section location. Problems Grouped by Section are put in the order of difficulty within each section of the chapter, that is, the first section needed to study to be able to solve the problem. In all cases a small number of particularly difficult problems are identified by stars* next to the number. Hence, there is much guidance provided in evaluating this generous and flexible set of problems.

Perhaps of equal importance is the fact that a considerable number of the new problems fit the "real world" category of student interest, involving situations familiar to students from their daily experiences and concerns. Of course, many new problems serve other pedagogic functions, as well, and interest and clarity have been heightened by the addition of more than 200 new figures. Most of the new problems were chosen from the classroom-tested material of the authors or of experienced users of the text.

In preparing this new set of problems we have been careful not to discard the many tried and true problems that have survived the test of the classroom for many years. Many of the old problems have been edited for greater clarity, and an occasional figure has been added or altered to make a point. Long-time users of our text will not find their favorites missing.

The Questions have served many professors as excellent sources of classroom discussion and clarification of homework concepts, so we have correspondingly enlarged and improved them. There are now more than 1200 thought questions, an increase of 25 percent over the second edition, with many new figures added as well. All have been edited for greater student appeal and relevance to the text. Many relate to everyday phenomena — atmospheric, kitchen, athletic, occupational, and the like — or otherwise serve to arouse curiosity and interest. All these changes and additions in the questions and the problems are designed to increase the appeal and relevance of the subject to the students.

As for the changes in the text itself, we have been responsive to the suggestions of users and have made changes at those places where, we are told, student understanding could benefit from improved clarity. Many of these changes relate to the sign conventions that plague a student's life — the Doppler effect, heat and work, and geometrical optics being particularly troublesome areas. In addition, we have rewritten a few sections, such as the rocket problem and the treatment of nonconservative forces, where improvements in clarity seemed particularly called for. We have introduced the new definition of length, based on the speed of light, being careful to include its implications for electromagnetism. Finally, we have reorganized the Appendices somewhat, adding new material that we have learned is needed. Hence, this revised printing — though not a new edition, strictly speaking — is a significant step towards greater utility and effectiveness of the text pending completion of a new edition now in progress.

We gratefully acknowledge the active assistance and important role of Edward Derringh (Wentworth Institute), J. Richard Christman (U.S. Coast Guard Academy), and Kenneth Brownstein (University of Maine) in the preparation and evaluation of the problems and the generation of new ones, and also of Albert Altman (University of Lowell), Robert Bowden (Virginia Polytechnic Institute and State University), W. Farrell Edwards (Utah State University), George W. Ficken, Jr. (Cleveland State University), Andrew L. Gardner (Brigham Young University), Richard Guglielmino (Glendale College), William J. Kernan, Jr. (Iowa State University), Brij M. Khorana (Rose-Hulman Institute of Technology), Kenneth Krane (Oregon State University), James H. Stith (U. S. Military Academy), M. J. Stott (Queen's University), and Gordon A. Wolfe (Southern Oregon State College).

We are also grateful to Benjamin Chi (SUNY Albany), who has been a special consultant for figures, and to John Balbalis, our outstanding chief illustrator for the past twenty-five years, for their valuable contributions. We thank Farrell Edwards (Utah State University), John Merrill (Brigham Young University), and Edward Derringh (Wentworth Institute) for detailed production assistance and Kay Guyette for so ably providing the wide range of secretarial services required. Finally, Robert McConnin, Physics Editor, and the entire professional staff at John Wiley & Sons, which he organized so effectively, were of invaluable help to us.

We hope the final product proves worthy of the effort.

David Halliday
3 Clement Road
Hanover, New Hampshire 03755

Robert Resnick
Rensselaer Polytechnic Institute
Troy, New York 12181

January, 1986

Preface

This second edition of *Fundamentals of Physics* is a major revision of the first edition of that text. Originally a careful condensation of *Physics* (1966), a longer and more sophisticated text, *Fundamentals of Physics* would have been substantially improved merely by incorporation of the relevant changes made in preparing the recent (1978) new edition of *Physics*. However, in addition to including those significant improvements, we also made use of extensive and detailed comments from a large number of users of the first edition of *Fundamentals* itself to reevaluate the effectiveness of virtually every aspect of that text. Moreover, major improvements in pedagogic devices, careful addition of suitable new material, expansion and clarifications of the questions, problems and worked-examples, and reorganization of the presentation of material in a great many chapters — much of this arrived at independently — were carried out. Thereafter we subjected the resulting changes to an in-depth evaluation by several reviewers. All this has led to a qualitatively new textbook, one that has now matured, grown somewhat independent of the parent text *Physics,* and acquired a special character of its own.

Users of the first edition of *Fundamentals* can appreciate the changes better if we list them in some detail, as follows:

(a) We have restyled the physical layout of the book to give it a less crowded appearance than formerly, making it easier now for the student to read the material, to make notations and to differentiate between the various components of each chapter (text, figures, examples, tables, quotes, references, questions, problems, and so forth).

(b) The art work is totally new: We now use color when it can help the pedagogy; all numerical graphs have been rechecked and updated and computer plots are used to increase accuracy; a significant number of new photographs has been added; sizing, lettering, and placement have been reevaluated for greater effectiveness of figures.

(c) Marginal notes are provided as a running outline of principal developments in each chapter and serve as a "locator" for students. Moreover, a "Review Guide and Summary," at the end of each chapter, clearly identifies the main points, repeats key equations, and refers to pertinent worked-examples for review purposes.

(d) The thought questions, the problems, and the worked-examples have been edited and revised and the range, the spread in level, and the choices have been enhanced by significant additions. The thought questions now number 844 (a 33% increase); the problems now number 1872 (a 45% increase over the first edition and a 14% increase over the 1974 revised printing); and the worked-out examples now number 256 (a 25% increase).

(e) All tables have been carefully reviewed, updated, and improved. The text itself has been updated in terms of recent measurements and discoveries, and the Appendices are correspondingly modernized.

(f) We have taken a major step toward a complete SI treatment of units, retaining the British system in parallel at selected places in early mechanics chapters only for valid pedagogic reasons.

(g) New sections and new material have been added selectively where user demand was greatest. Examples are internal work and kinetic energy, combined rotational and translational motion, damped harmonic motion, forced oscillations and resonance, satellite motions, sound intensity, heat transfer including convection and radiation, temperature dependence of resistivity, electrical measuring instruments, mutual induction, geomagnetism, laser light, virtual objects in optics, optical instruments, and an entirely new chapter on alternating currents.

(h) There has been much rewriting and reorganization of the text proper, to improve clarity and learning and to permit greater flexibility in coverage. Some examples are: particle dynamics, formerly in one chapter, is now split into two, more manageable chapters; the Biot-Savart law now precedes the presentation of Ampere's law in the appropriate chapter; the material in "Capacitors and Dielectrics" has been rearranged so that all topics involving dielectrics are treated together in the latter part of the chapter; the sequence of topics in the chapters on electromagnetic oscillations, alternating currents, and Maxwell's equations is changed for greater coherence and understanding and for more direct analogy to mechanics; treatments of mass-energy, cross sections, rotational dynamics, traveling waves, thermal expansion, internal energy, entropy, flux, current density, magnetic properties of matter, radiation sources, sign convention in mirrors and lenses, Doppler effect, and wave mechanics have been significantly rewritten.

In making all these changes—and others—we have been guided by the objective situation in physics courses today, a decade after the first edition was published. At the end of this interval we find that the student composition has changed, students' preparation has changed, the length of courses is different, applied material is in greater favor, topics considered relevant are of greater diversity, motivating factors are different, and the need for new teaching and learning devices is enhanced. All this can be for the better in revealing the richness of the subject to diverse audiences and in raising the quality of instruction. We hope that this second edition of *Fundamentals of Physics* will widen the appeal of physics and contribute to the improvement of physics education.

A textbook contains far more contributions to the elucidation of a subject than those made by the authors alone. We have benefited from the active assistance of Farrell Edwards (Utah State University) and John Merrill (Brigham Young University), as in the first edition, and from constructive review comments of Lawrence E. Evans (Duke University), Russell K. Hobbie (University of Minnesota), and B. A. McInnes (University of Sydney). Benjamin Chi (SUNY Albany) has been a special consultant for figures. We are grateful to these physicists for their substantial contributions to the book. The professional staff at John Wiley and Sons, organized with great effectiveness by Robert McConnin, Physics Editor, has been outstanding in every respect and we acknowledge with pleasure our debt to the members of that firm. John Balbalis, our chief illustrator for over twenty years, deserves special mention in this connection. We also wish to thank Edward Derringh (Wentworth Institute) for assistance with the problem sets and Carolyn Clemente for providing the wide range of secretarial services required. And, finally, to our many user co-respondents, who showed interest in, concern for, and support of effective teaching, we express our genuine appreciation. May the students be the beneficiaries.

David Halliday
3 Clement Road
Hanover, New Hampshire 03755

Robert Resnick
Rensselaer Polytechnic Institute
Troy, New York 12181

January 1, 1981

Preface

To the Extended Version

Many colleges and universities have reported a need for a survey of modern physics in the introductory physics course. At some universities, optics and modern physics constitute the subject matter of a separate semester of that course; at others, selected topics in modern physics are surveyed at the end of the course. In each of these cases there is not enough time for a full development of quantum theory or of relativity theory, but there is a desire to apply the basic concepts of these theories to topics in atomic, nuclear, and solid state physics. We have responded to this need in our extended version of *Fundamentals of Physics,* second edition. Its seven concluding chapters, and a special supplementary topic, provide the necessary material to meet these modern physics requirements.

Chapters 42 and 43 develop the basic concepts of quantum physics, describing fully the experimental bases on which they rest. These concepts—which include the wave-particle duality, the uncertainty principle, and the wave function, among other topics—constitute an introduction to quantum theory in the context of modern physical phenomena. These two chapters are the same as those that conclude the second edition of *Fundamentals of Physics.* * In that book, also, concepts and results of special relativity theory are presented at appropriate places in the text. In the extended version we have added a supplementary topic in special relativity that gathers in one place those concepts and results, with coherent new writing and with a

large set of new worked-examples, new questions, and new problems. This constitutes a review and an extension of the text material, much like a study guide, that instructors can use when, where, and how they wish.

The final five chapters, on atomic, solid state, and nuclear physics, are entirely new. Taken in conjunction with the modern physics material already in the main text and with the special relativity supplement, they constitute a fairly substantial introduction to modern physics. There is a balance in these chapters between basic principles and selected applications, as a look at the table of contents will reveal, and between qualitative and quantitative treatment of phenomena, as perusing the sections will confirm. The treatment is in no way superficial and lends itself to substantial problem-solving exercises and thought discussions as is true of the earlier classical physics chapters. In the new material, for example, we have included 37 worked examples, 126 thought questions, and 215 problems. However, the selection of topics and the depth of treatment have been limited by the time available in typical introductory course. Furthermore, the instructor has many valid options in selecting a menu to suit her or his course.

We welcome feedback on the classroom experience with this extended version of our text and hope that it satisfies the expressed needs that brought it about.

David Halliday

Robert Resnick

* See the Preface to the second edition of *Fundamentals of Physics* on the preceding pages.

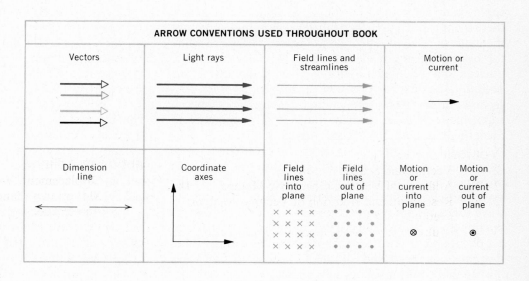

ARROW CONVENTIONS USED THROUGHOUT BOOK			
Vectors	Light rays	Field lines and streamlines	Motion or current
Dimension line	Coordinate axes	Field lines into plane / Field lines out of plane	Motion or current into plane / Motion or current out of plane

John Balbalis, Art Program Designer
John Wiley & Sons, Inc.

Benjamin E. Chi, Illustration Consultant
Department of Physics
State University of New York, Albany

Contents

SUPPLEMENTS

STUDENT STUDY GUIDE REVISED
to accompany PHYSICS, 3RD
and FUNDAMENTALS OF PHYSICS

Stanley A. Williams, Iowa State University
Kenneth L. Brownstein, University of Maine
Robert L. Gray, University of Massachusetts

Developed for student use with this text and also PHYSICS, 3RD by Halliday and Resnick. Includes learning objectives carefully and thoroughly developed, plus exercises in problem solving for each chapter of the textbooks.

SELECTED SOLUTIONS FOR
FUNDAMENTALS OF PHYSICS, 2ND

David Halliday, University of Pittsburgh
Robert Resnick, Rensselaer Polytechnic Institute
Edward Derringh, Wentworth Institute

Contains solutions to selected problems for student use with FUNDAMENTALS OF PHYSICS, 2ND, including the extended version.

LABORATORY PHYSICS, 2ND

Harry P. Meiners, Rensselaer Polytechnic Institute
Walter Eppenstein, Rensselaer Polytechnic Institute
Kenneth Moore, Rensselaer Polytechnic Institute
Ralph A. Oliva, Texas Instruments, Inc.

A laboratory manual designed to be used in the calculus-based general physics course. The new edition includes material on calculators, integrated circuits, and microprocessors.

PHYSICS PROBLEMS FOR
PROGRAMMABLE CALCULATORS

J. Richard Christman, U. S. Coast Guard Academy
Robert Resnick, Rensselaer Polytechnic Institute
David Halliday, University of Pittsburgh

Designed for use with Halliday and Resnick PHYSICS, 3RD and FUNDAMENTALS OF PHYSICS, 2ND, as a supplement demonstrating to students the solution of physics problems using the calculator.

1 Measurement

1–1 The Physical Quantities, Standards, and Units

The building blocks of physics are the physical quantities that we use to express the laws of physics. Among these are length, mass, time, force, velocity, density, resistivity, temperature, luminous intensity, magnetic field strength, and many more. Many of these words, such as length and force, are part of our everyday vocabulary. You might say, for example: "I will go to any *length* to satisfy you as long as you do not *force* me to do so." In physics, however, we must define words that we associate with physical quantities, such as force and length, clearly and precisely, and we must not confuse them with their everyday meanings. In this example the precise scientific definitions of length and force have no connection at all with the uses of these words in the quoted sentence.

We say that we have defined a physical quantity such as mass, for example, when we have laid down a set of procedures, a recipe if you will, for measuring that quantity and assigning a unit, such as the kilogram, to it. That is, we set up a *standard*. The procedures are arbitrary. We can define the kilogram in any way we want. The important thing is to define it in a useful and practical way, and to obtain international acceptance of the definition.

The need for standards

There are so many physical quantities that it becomes a problem to organize them. They are not independent of each other. As a simple example, a speed is the ratio of a length to a time. What we do is select from all possible physical quantities a certain small number that we choose to call basic, all others being derived from them. We then assign standards to each of these basic quantities and to no others. If, for example, we select length as a basic quantity, we choose a standard called the meter (see Section 1–3) and we define it in terms of precise laboratory operations.

Several questions arise: (*a*) How many basic quantities should be selected? (*b*) Which quantities should they be? (*c*) Who is going to do the selecting?

The answers to the first two questions are that we select the smallest number of physical quantities that will lead to a complete description of physics in the simplest terms. Many choices are possible. In one system, for example, force is a basic quantity. In the system used in this book it is a derived quantity.

The answer to the third question has been decided by international agreement. The International Bureau of Weights and Measures, located near Paris and established in 1875, is the fountainhead for these matters. It maintains contact with standardizing laboratories throughout the world, our own National Bureau of Standards being an example. Periodically the General Conference of Weights and Measures, an international body, meets and makes resolutions and recommendations. Its first meeting was in 1889 and its fifteenth meeting was in 1975.

Standards must be accessible and invariable

Once we have set up a basic standard, for length say, we must also set up procedures that allow us to measure the length of any object by comparing it with the standard. This means that the standard must be accessible. Also, we want within acceptable limits to get the same answer every time we compare the standard with a given object. This means that the standard must be invariable. These two requirements are often incompatible. If we choose length as a basic quantity, define its standard as the distance between a person's nose and the fingertips of the outstretched arm, and assign the yard as the unit, we have a standard that is certainly accessible but is not invariable. The demands of science and technology steer us just the other way. We achieve accessibility by creating more readily available secondary, tertiary, etc., standards, and we strongly stress invariability.

We often make comparisons with a basic standard in very indirect ways. In the case of length, consider the problems of measuring: (*a*) the distance to the Andromeda galaxy, (*b*) your height, and (*c*) the distances between nuclei in the molecule NH_3. It is clear that the comparison techniques vary greatly. For example, we cannot use a ruler directly in either the first or the third problem.

1–2 The International System of Units*

Base units and derived units

The 14th General Conference on Weights and Measures (1971), building on the work of preceding conferences and international committees, selected as *base units* the seven quantities displayed in Table 1–1. This is the basis of the International System of Units, abbreviated SI from the French "Le Système International d'Unités."

Table 1–1 SI base units

Quantity	Name	Symbol
Length	meter	m
Mass	kilogram	kg
Time	second	s
Electric current	ampere	A
Thermodynamic temperature	kelvin	K
Amount of substance	mole	mol
Luminous intensity	candela	cd

* See Appendix A and also NBS Special Publication 330, 1972, "The International System of Units: SI." Write to the Superintendent of Documents, U.S. Government Printing Office, Washington, DC 20402.

Throughout the book we shall give many examples of SI *derived units* such as velocity, force, electric resistance, and so on, that follow from Table 1–1. For example, the SI unit of force, called the *newton* (abbreviation N), is defined in terms of the SI base units as

$$1 \text{ N} = 1 \text{ m} \cdot \text{kg/s}^2,$$

as we shall make clear in Chapter 5.

Often if we express physical quantities such as the time interval between two events in a nuclear physics experiment, or the power output of a particular electric power generating plant, in the appropriate SI units we end up with very small or very large numbers. For convenience, the 14th General Conference of Weights and Measures recommended the prefixes shown in Table 1–2. Thus we can write an electric power output of 1.2×10^9 watts as 1.2 gigawatts ($= 1.2$ GW) and a time interval of, say, 2.35×10^{-9} seconds as 2.35 nanoseconds ($= 2.35$ ns).

Table 1–2 SI prefixes[a]

Factor	Prefix	Symbol
10^{18}	exa	E
10^{15}	peta	P
10^{12}	tera	T
10^{9}	giga	G
10^{6}	mega	M
10^{3}	kilo	k
10^{2}	hecto	h
10^{1}	deka	da
10^{-1}	deci	d
10^{-2}	centi	c
10^{-3}	milli	m
10^{-6}	micro	μ
10^{-9}	nano	n
10^{-12}	pico	p
10^{-15}	femto	f
10^{-18}	atto	a

[a] In all cases, the first syllable is accented, as in ná-no-mé-ter.

To fortify Table 1–1 we need seven sets of procedures that tell us how to produce in the laboratory the seven SI base units. We shall explore those for length, mass, and time in the next three sections.

Gaussian and British units

Two other major systems of units compete with the International System (SI). One is the Gaussian system, in terms of which much of the literature of physics is expressed. We shall not use this system in this book. Appendix E gives conversion factors to SI units.

The second is the British system, in daily use in this country and elsewhere. The basic units, in mechanics, are length (the foot), force (the pound), and time (the second). Again Appendix E gives conversion factors to SI units. We shall use SI units in this book except that in mechanics we shall sometimes use the British system, especially in the early chapters. The British system is being phased out in Britain in favor of the officially adopted International System. In fact, as of 1978, the only countries that had neither adopted the metric system (which emerged as SI) nor officially indicated that they intended to do so were Borneo, Brunei, Liberia, Southern Yemen, and the United States.* The Metric Conversion Act of

---

* See "Conversion to the Metric System," by Lord Ritchie-Calder, *Scientific American*, July 1970.

1975, however, established a United States Metric Board, whose function is to urge and facilitate the voluntary adoption of the metric system; several major corporations and large government departments have done so.

Example 1 *Changing units*   Ocean depths are often measured in fathoms (1 fathom = 6 ft). A submarine is diving at 36 fathoms/minute. Express this speed in meters per second, miles per hour, and light-years per year. A light-year, defined as the distance that light travels in 1 year, is 9.46×10^{12} km.

To change these units we use what we call *chain-link conversions*. In this method the conversion factor is written as a ratio that equals unity. Thus, in place of 1 min = 60 s we write

$$\left(\frac{1 \text{ min}}{60 \text{ s}}\right) = 1 \quad \text{or} \quad \left(\frac{60 \text{ s}}{1 \text{ min}}\right) = 1.$$

Since multiplication by unity leaves a quantity unchanged, this form for the conversion factor is not only convenient but makes errors less likely. We use the factor in such a way that the unwanted units divide out. For instance,

$$2.0 \text{ min} = (2.0 \text{ min})(1) = (2.0 \text{ min})\left(\frac{60 \text{ s}}{1 \text{ min}}\right) = 120 \text{ s}$$

We could also write

$$2.0 \text{ min} = (2.0 \text{ min})(1) = (2.0 \text{ min})\left(\frac{1 \text{ min}}{60 \text{ s}}\right) = 0.033 \text{ min}^2/\text{s},$$

which is true but not particularly useful. Use the conversion factor in the form that causes the appropriate units to cancel

out. If it is not right when you first write it down, just interchange the numerator and denominator (turn it over).

Now to solve the problem: We shall use conversion factors found in Appendix E.

$$36 \frac{\text{fathoms}}{\text{min}} = \left(36 \frac{\text{fathoms}}{\text{min}}\right)\left(\frac{1 \text{ min}}{60 \text{ s}}\right)\left(\frac{6 \text{ ft}}{1 \text{ fathom}}\right)\left(\frac{1 \text{ m}}{3.28 \text{ ft}}\right)$$
$$= 1.1 \text{ m/s}.$$

Going on, we find that

$$36 \frac{\text{fathoms}}{\text{min}} = \left(36 \frac{\text{fathoms}}{\text{min}}\right)\left(\frac{60 \text{ min}}{1 \text{ hr}}\right)\left(\frac{6 \text{ ft}}{1 \text{ fathom}}\right)\left(\frac{1 \text{ mi}}{5280 \text{ ft}}\right)$$
$$= 2.5 \text{ mi/h}$$

To find the speed in light-years per year we start from the first result and find

$$1.1 \frac{\text{m}}{\text{s}} = \left(1.1 \frac{\text{m}}{\text{s}}\right)\left(\frac{1 \text{ ly}}{9.46 \times 10^{12} \text{ km}}\right)\left(\frac{1 \text{ km}}{10^3 \text{ m}}\right)\left(\frac{3.16 \times 10^7 \text{ s}}{1 \text{ y}}\right)$$
$$= 3.7 \times 10^{-9} \text{ ly/y}.$$

We can write this in the even more unlikely form of 3.7 nly/y, where nly is an abbreviation for nano-light-year.

1–3   The Standard of Length

The first international standard of length was a platinum-iridium bar kept at the International Bureau of Weights and Measures near Paris. The distance between two fine lines engraved near the ends of the bar, when the bar was maintained at 0°C and supported in a prescribed way, was defined to be one meter. Historically, the meter was intended to be one ten-millionth of the distance from the north pole to the equator along the meridian line through Paris. Later measurements, however, showed that the standard meter bar differed slightly (about 0.023%) from this intended value.

The meter bar

Because the standard meter was not very accessible, accurate copies of it were made and sent to standardizing laboratories throughout the world. These secondary standards were used to calibrate other, still more accessible, standards so that ultimately every measuring rod derived its authority from the standard meter through a complicated chain of comparisons, using microscopes and dividing engines. In 1959 this also became true of the yard, whose legal definition in this country was adopted in that year to be

A legal definition

$$1 \text{ yard} = 0.9144 \text{ m (exactly)}$$

which is equivalent to

$$1 \text{ in.} = 2.54 \text{ cm (exactly)}.$$

As time went on, however, the accuracy with which intercomparisons of length could be made by measuring fine scratches under a microscope became no longer good enough for the demands of science and industry. In 1960, then, the 11th General Conference of Weights and Measures adopted a new length standard, based on the wavelength of light. The wavelength in vacuum of a particular orange-red radiation emitted by atoms of krypton-86 was chosen. Specifically, one meter was then defined to be 1,650,763.73 wavelengths of this light. This unwieldy number was chosen so that the new standard, based on the wavelength of light, would be as consistent as possible with the old standard, based on the meter bar. This new standard was only feasible because techniques, based on the Michelson interferometer, had been developed that made it possible to measure the length of a meter bar in terms of light waves. The American physicist A. A. Michelson received the Nobel prize for his pioneering work in this field.

The choice of an atomic standard has advantages other than increased precision in length measurements. Krypton-86 atoms are available everywhere, are identical, and emit light of the same wavelength. The standard cannot be destroyed by fire or war. These are not idle threats. When the British Houses of Parliament burned in 1834, the British standard yard and standard pound were destroyed. As Philip Morrison has written, ". . . every atom is a storehouse of natural units safer than the Bureau International des Poids et Mésures in Sèvres."

Figure 1–1 shows how the length of a master machinist's gage block, used in industry as a precise secondary standard, is compared with that of a standard reference block at the National Bureau of Standards. The comparison is carried out using fringes generated by the interference of light. If the fringe patterns crossing the two rectangular blocks outlined in the figure match, the gage blocks are of equal length; if the fringe patterns fail to match by, say, one-tenth of a fringe, the blocks differ in length by one-twentieth of the wavelength of light, or $\sim 3 \times 10^{-8}$ m. The horizontal background fringes come from the polished plate on which the two blocks rest.

By 1983 the demands of science and industry for increased precision had reached such a point that even the krypton-86 standard could not meet them and in that year a very bold step was taken. The meter was redefined as the distance traveled by a light wave in a specified time interval. Doing so required that the speed of light be regarded as an *exactly defined quantity*, no longer capable of independent measurement.

Specifically, the 17th General Conference of Weights and Measures, meeting near Paris in 1983, defined the meter to be ". . . the length of the path traveled by light in vacuum in 1/299,792,458 of a second." This is equivalent to saying that the speed of light is now *defined* to be (exactly) 299,792,458 m/s. This new definition of the meter was only possible because measurements of the speed of light had become so precise that the reproducibility of the krypton-86 meter itself became the limiting factor. In view of this it then made sense to adopt the speed of light as a defined quantity and use it to define the meter.

Given the speed-of-light meter as basic, we still need convenient secondary standards calibrated against it for practical use. Often, as in measuring intramolecular or interstellar distances, we cannot make a direct comparison to a physical standard such as a ruler. However indirect the method, though, there must always be a clear traceable link to the speed-of-light meter as the ultimate standard.

The krypton-86 meter

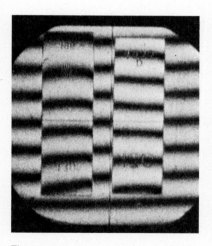

Figure 1–1 The length of a master machinist's gage block being compared with a reference standard by means of light waves. (Courtesy National Bureau of Standards.)

The speed-of-light meter

Table 1-3 shows some measured lengths. Note that they vary by a factor of about 10^{41}.

Table 1-3 Some measured lengths

Length	Meters
Distance to the farthest quasar (1984)	$\sim 10^{26}$
Distance to the nearest galaxy (in Andromeda)	2×10^{22}
Radius of our galaxy	6×10^{18}
Distance to the nearest star (Alpha Centauri)	4.3×10^{16}
Mean orbit radius for our most distant planet (Pluto)	5.9×10^{12}
Radius of the earth	6.4×10^{6}
Height of Mt. Everest	8.9×10^{3}
Height of a typical person	1.7×10^{0}
Thickness of a page in this book	1×10^{-4}
Size of a poliomyelitis virus	1.2×10^{-8}
Radius of a hydrogen atom	5.0×10^{-11}
Effective radius of a proton	1.2×10^{-15}

1-4 The Standard of Mass

The SI standard of mass is a platinum-iridium cylinder kept at the International Bureau of Weights and Measures and assigned by international agreement a mass of 1 kg. Secondary standards are sent to standardizing laboratories in other countries and the masses of other bodies can be found by an equal-arm balance technique to a precision of two parts in 10^8.

The standard kilogram

The U.S. copy of the international standard of mass, known as Prototype Kilogram No. 20, is housed in a vault at the National Bureau of Standards (see Fig. 1-2). It is removed no more than once a year for checking the values of tertiary standards. Since 1889 Prototype No. 20 has been taken to France twice for recomparison with the master kilogram. When it is removed from the vault two people are always present, one to carry the kilogram in a pair of forceps, the second to catch the kilogram if the first person should fall.

Table 1-4 shows some measured masses. Note that they vary by a factor of about 10^{71}. Most masses have been measured in terms of the standard kilogram by

Figure 1-2 The National Standard Kilogram No. 20, which is kept under a double bell jar in a vault at the National Bureau of Standards. It is an accurate copy of the International Standard Kilogram kept at the International Bureau of Weights and Measures in Sèvres, France. (Courtesy National Bureau of Standards.)

Table 1-4 Some measured masses

Object	Kilograms
Our galaxy	2.2×10^{41}
The sun	2.0×10^{30}
The earth	6.0×10^{24}
The moon	7.4×10^{22}
The waters of the oceans	1.4×10^{21}
An ocean liner	7.2×10^{7}
An elephant	4.5×10^{3}
A person	5.9×10^{1}
A grape	3.0×10^{-3}
A speck of dust	6.7×10^{-10}
A tobacco mosaic virus	2.3×10^{-13}
A penicillin molecule	5.0×10^{-17}
A uranium atom	4.0×10^{-26}
A proton	1.7×10^{-27}
An electron	9.1×10^{-31}

indirect methods. For example, we can measure the mass of the earth (see Section 15–2) by measuring in the laboratory the gravitational force of attraction between two lead spheres. Their masses must be known by direct comparison with the standard kilogram, using, say, an equal-arm balance.

On the atomic scale we have a second standard of mass, not an SI unit. It is the mass of the carbon-12 atom, which by international agreement is assigned an atomic mass of 12 *unified atomic mass units* (abbreviation u), exactly and by definition. We can find the masses of other atoms to considerable accuracy by using a mass spectrometer. Table 1–5 shows some selected atomic masses, including the probable errors of measurement. We need a second standard of mass because present laboratory techniques permit us to compare atomic masses to each other with greater precision than we can compare them to the standard kilogram. The relationship is approximately

The unified atomic mass unit

$$1 \text{ u} = 1.661 \times 10^{-27} \text{ kg}.$$

Table 1–5 Some measured atomic masses

Atom	Mass in atomic mass units
^{1}H	$1.00782522 \pm 0.00000002$
^{12}C	12.00000000 (exactly)
^{64}Cu	63.9297568 ± 0.0000035
^{102}Ag	101.911576 ± 0.000024
^{137}Cs	136.907074 ± 0.000005
^{190}Pt	189.959965 ± 0.000026
^{238}Pu	238.049582 ± 0.000011

1–5 The Standard of Time

The measurement of time has two aspects. For civil and for some scientific purposes we want to know the time of day so that we can order events in sequence. In most scientific work we want to know how long an event lasts (the time interval). Thus any time standard must be able to answer the questions "At what time does it occur?" and "How long does it last?" Table 1–6 shows the range of time intervals that can be measured. They vary by a factor of about 10^{40}.

We can use any phenomenon that repeats itself as a measure of time. The measurement consists of counting the repetitions. We could use an oscillating pendulum, a mass-spring system, or a quartz crystal, for example. Of the many repetitive phenomena in nature, the rotation of the earth on its axis, which determines the length of the day, has been used as a time standard for centuries. It is still the basis of our civil time standard, one (mean solar) second being defined to be 1/86,400 of a (mean solar) day. Time defined in terms of the rotation of the earth is called universal time (UT).

Universal time must be measured by astronomical observations extended over several weeks. Thus we need a good terrestrial clock, calibrated by the astronomical observations. Quartz crystal clocks based on the electrically sustained periodic vibrations of a quartz crystal serve well as secondary time standards. The best of these have kept time for a year with a maximum error of 0.02 s.

One of the most common uses of a time standard is to measure frequencies. In the radio range, frequency comparisons to a quartz clock can be made electronically to a precision of at least 1 part in 10^{10} and, indeed, we often need such preci-

Table 1–6 Some measured time intervals

Time interval	Seconds
Age of the earth	1.3×10^{17}
Age of the pyramid of Cheops	1.2×10^{11}
Human life expectancy at birth (USA)	2×10^9
Time of earth's orbit around the sun (1 year)	3.1×10^7
Time of earth's rotation about its axis (1 day)	8.6×10^4
Period of a typical satellite	5.1×10^3
Half-life of the free neutron	7.0×10^2
Time between normal heartbeats	8.0×10^{-1}
Period of concert-A tuning fork	2.3×10^{-3}
Half-life of the muon	2.2×10^{-6}
Period of oscillation of 3-cm microwaves	1.0×10^{-10}
Typical period of rotation of a molecule	1×10^{-12}
Half-life of the neutral pion	2.2×10^{-16}
Period of oscillation of a 1-MeV gamma ray (calculated)	4×10^{-21}
Time for a fast neutron to pass through a medium-sized nucleus (calculated)	2×10^{-23}

sion. However this precision is about 100 times greater than that with which a quartz clock itself can be calibrated by astronomical observations. To meet the need for a better time standard, atomic clocks have been developed in several countries, using periodic atomic vibrations as a standard. Figure 1–3 shows such a clock, based on a characteristic frequency of the atom ^{133}Cs, at the U.S. National Bureau of Standards.

Atomic time unit

In 1967 the second based on the cesium clock was adopted as an international standard by the 13th General Conference on Weights and Measures. The second

Figure 1–3 A cesium atomic frequency standard (No. NBS-6) at the National Bureau of Standards in Boulder, Colorado. It is the heart of the NBS Atomic Clock System. (Courtesy National Bureau of Standards.)

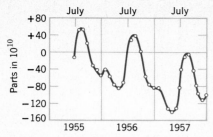

Figure 1–4 Variation in the rate of rotation of the earth as revealed by comparison with a cesium clock. (From L. Essen, *Physics Today,* July 1960.)

Radio time signals

was defined as 9,192,631,770 periods of the particular ^{133}Cs transition selected. This action increased the accuracy of time measurements to 1 part in 10^{12}, an improvement over the accuracy associated with astronomical methods of about 10^3. If two cesium clocks are operated at this precision they will differ by no more than 1 s after running for 6000 years. Hydrogen maser clocks, not (yet) adopted as time standards, have achieved the incredible precision of 1 s in 50,000,000 y!

Figure 1–4 shows, by comparison with the cesium clock, variations in the rate of rotation of the earth over nearly a 3-year period. Note that the earth's rotation rate is high in summer and low in winter (northern hemisphere) and decreases steadily from year to year. You may ask how we can be sure that the rotating earth and not the cesium clock is at fault. There are two answers. (1) The relative simplicity of the atom compared to the earth leads us to attribute any difference between these two timekeepers to the earth. Tidal friction between the water and the land, for example, causes a slowing down of the earth's rotation. Also, the seasonal motion of the winds introduces a seasonal variation in the rotation. Other variations may be associated with the melting and refreezing of polar icecaps. (2) The solar system contains other timekeepers such as the orbiting planets and the orbiting moons of the planets. The rotation of the earth shows variations with respect to these, too, which are similar to but less accurately observable than the variations shown in Fig. 1–4.

The time standard is available at remote locations by radio transmission.* Stations WWV in Colorado and WWVH in Hawaii, operated by the National Bureau of Standards, are examples of such stations. They broadcast signals indicating one second intervals and also, once a minute, voice time announcements of Greenwich Mean Time. These signals are superimposed on carrier frequencies of 2.5, 5, 10, 15, 20, and 25 MHz, stabilized to one part in 10^{11} by comparison with a cesium clock standard. (1 MHz = 10^6 hertz = 10^6 Hz = 10^6 cycles/s.) Time signals are also broadcast using a binary coded system.

REVIEW GUIDE AND SUMMARY

SI units

The unit system used in this book is the International System of Units (SI), for which the seven physical quantities displayed in Table 1–1 serve as a base. *Standards,* which must be both accessible and invariable, define units for these base quantities and are established by international agreement. These standards underlie all physical measurement, for both the base quantities and for quantities derived from them, such as force or electric conductivity. The prefixes of Table 1–2 simplify notation in many cases.

The meter

The unit of length—the meter—is defined in terms of the wavelength of light emitted by an atomic (^{86}Kr) source. The yard, together with its multiples and submultiples, is legally defined in this country in terms of the meter.

The kilogram

The unit of mass—the kilogram—is defined in terms of a particular platinum-iridium prototype kept in Paris. For atomic problems the unified atomic mass unit, defined in terms of the atom ^{12}C, is also used.

The second

The unit of time—the second—was formerly defined in terms of the rotation of the earth. It is now defined in terms of the period of oscillation of the light emitted by an atomic (^{133}Cs) source. Accurate time signals are disseminated worldwide by radio signals keyed to atomic clocks in various standardizing laboratories.

---

* See "NBS Time and Frequency Dissemination Services; Special Publication 432," National Bureau of Standards, January 1976 (write to the Superintendent of Documents, U.S. Government Printing Office, Washington, DC 20402).

QUESTIONS

1. How would you criticize this statement: "Once you have picked a standard, by the very meaning of 'standard' it is invariable"?

2. List characteristics other than accessibility and invariability that you would consider desirable for a physical standard.

3. Can you imagine a system of base units (Table 1–1) in which time was not included?

4. Of the seven base units listed in Table 1–1 only one, the kilogram, has a prefix (see Table 1–2). Would it be wise to redefine the mass of that platinum-iridium cylinder at the International Bureau of Weights and Measures as 1 g rather than 1 kg?

5. The meter was originally intended to be one ten-millionth of the meridian line from the north pole to the equator that passes through Paris. In Section 1–3 we learned that this definition disagrees with the standard meter bar by 0.023%. Does this mean that the standard meter bar is inaccurate to this extent?

6. In defining the meter bar as the standard of length, why specify its temperature? Can length be called a fundamental quantity if another physical quantity, such as temperature, must be specified in choosing a standard?

7. In the seventeenth century the followers of Newton believed that the earth was flattened at the poles and bulged at the equator and the followers of Descartes believed just the opposite. The matter can be settled by measuring the length of a degree of latitude, both at or near the poles and at or near the equator. Measurement showed that Newton was right. Where is a degree of latitude longer, at the poles or at the equator?

8. In redefining the meter in terms of the speed of light, why did not the delegates to the 1983 General Conference on Weights and Measures simplify matters by defining the speed of light to be 3×10^8 m/s, exactly? For that matter, why did they not define it to be 1 m/s, exactly? Were both of these possibilities open to them? If so, why did they reject them?

9. If someone told you that every dimension of every object had shrunk to half its former value overnight, how could you refute this statement?

10. Can you suggest a way to measure (a) the radius of the earth,

(b) the distance between the sun and the earth, (c) the radius of the sun?

11. Can you suggest a way to measure (a) the thickness of a sheet of paper, (b) the thickness of a soap bubble film, (c) the diameter of an atom?

12. Why do we find it useful to have two standards of mass, the kilogram and the carbon-12 atom?

13. Is the current standard kilogram of mass accessible, invariable, reproducible, indestructible? Does it have simplicity for comparison purposes? Would an atomic standard be better in any respect? Why don't we adopt an atomic standard, as we do for length and time?

14. Suggest practical ways by which one could determine the mass of the various objects listed in Table 1–4.

15. Suggest objects whose mass would fall in the wide range in Table 1–4 between that of an ocean liner and all the water in the oceans. Estimate their mass.

16. Name several repetitive phenomena occurring in nature that could serve as reasonable time standards.

17. You could define "one second" to be one pulse beat of the current president of the American Physical Society. Galileo used his pulse as a timing device in some of his work. Why is a definition based on the atomic clock better?

18. What criteria should a good clock satisfy?

19. From what you know about pendulums, cite the drawbacks to using the period of a pendulum as a time standard.

20. On June 30, 1981, the "minute" extending from 10:59 AM to 11:00 AM was arbitrarily lengthened to contain 61 s. This "leap second" was introduced to compensate for the fact that, as measured by our atomic time standard, the earth's rotation rate is slowly decreasing. Why is it desirable to readjust our clocks in this way?

21. Critics of the metric system often cloud the issue by saying things such as "Instead of buying one pound of butter you will have to ask for 0.454 kg of butter." The implication is that life would be more complicated. How would you refute this?

EXERCISES

Section 1–2   The International System of Units

1. Use the prefixes in Table 1–2 and express (a) 10^6 phones; (b) 10^{-6} phones; (c) 10^1 cards; (d) 10^9 los; (e) 10^{12} bulls; (f) 10^{-1} mates; (g) 10^{-2} pedes; (h) 10^{-9} Nannettes; (i) 10^{-12} boos; (j) 10^{-18} boys; (k) 2×10^2 withits; (l) 2×10^3 mockingbirds. Now that you have the idea, invent a few more similar expressions. (See *A Random Walk in Science,* p. 61, compiled by R. L. Weber, Crane, Russak & Co. Inc., New York, 1974.)

2. Some of the prefixes of the SI units have crept into everyday language. (a) What is the weekly equivalent of an annual salary of $36K (= 36 kilobucks)? (b) A lottery awards 10 megabucks as the top prize, payable over 20 years. How much is received in each monthly check?

Section 1–3   The Standard of Length

3. A space shuttle orbits the earth at an altitude of 300 km. What is this distance in miles?

4. What is your height in meters?

5. The micrometer (10^{-6} m = 1 μm) is often called the *micron.* (a) How many microns make up 1 kilometer? (b) What fraction of a centimeter equals 1 micron? (c) How many microns are in 1 yard?

Section 1–4   The Standard of Mass

6. Using appropriate conversions and data in the chapter, determine the number of hydrogen atoms required to obtain 1 kg of mass.

7. One molecule of water (H_2O) contains two atoms of hydrogen and one atom of oxygen. An atom of oxygen has a mass of 16 u, approximately. (a) What is the mass in kilograms of one molecule of water? (b) How many molecules of water are in the oceans of the world (see Table 1–4)?

Section 1–5   The Standard of Time
8. Express the speed of light, 3.0×10^8 m/s, in (a) feet per nanosecond and (b) in millimeters per picosecond.

9. Enrico Fermi once pointed out that a standard lecture period (50 min) is close to 1 microcentury. How long is a microcentury in minutes, and what is the percent difference from Fermi's approximation?

10. A convenient substitution for the number of seconds in a year is $\pi \times 10^7$. To within what percentage error is this correct?

11. A certain pendulum clock (with a 12-hour dial) happens to gain one minute per day. After setting the clock to the correct time, how long must one wait until it again indicates the correct time?

PROBLEMS GROUPED BY SECTION

Section 1–3   The Standard of Length
1. Calculate the number of kilometers in 20 mi using only the following conversion factors: 1 mi = 5280 ft, 1 ft = 12 in., 1 in. = 2.54 cm, 1 m = 100 cm, and 1 km = 1000 m.

2. The *cord* is a volume of cut wood equal to a stack 8 ft long, 4 ft wide, and 4 ft high. How many cords of wood are in 1 m^3?

3. A room is 20 ft, 2 in. long and 12 ft, 5 in. wide. (a) What is the floor area in square feet? (b) In square meters? (c) If the ceiling is 12 ft, $2\frac{1}{2}$ in. above the floor, what is the volume of the room in cubic feet? (d) In cubic meters?

4. Hydraulic engineers often use, as a unit of volume of water, the acre · foot: the volume of water that will cover 1 acre of land to a depth of 1 ft. A severe thunderstorm dumps 2 in. of rain in 30 min on a town of area 26 km^2. What volume of water, in acre·feet, fell on the town?

5. A certain brand of house paint claims a "coverage" of 460 ft^2/gal. (a) Express this quantity in square meters per liter. (b) Express this quantity in SI base units (see Appendix A).

6. Master machinists would like to have master gauges (1 in. long, for example) good to 0.0000001 in. Show that the platinum-iridium meter is not measurable to this accuracy but that the ^{86}Kr meter is. Use data given in this chapter.

7. Astronomical distances are so large compared to terrestrial ones that much larger units of length are used for easy comprehension of the relative distances of astronomical objects. An *astronomical unit* (AU) is equal to the average distance from the earth to the sun, about 92.9×10^6 mi. A *parsec* is the distance at which one astronomical unit would subtend an angle of 1″. A *light-year* is the distance that light, traveling through a vacuum with a speed of 186,000 mi/s, would cover in 1 year. (a) Express the distance from the earth to the sun in parsecs and in light-years. (b) Express a light-year and a parsec in miles. Although the light-year is much used in popular writing, the parsec (abbr. pc) is the unit used professionally by astronomers.

Section 1–4   The Standard of Mass
8. If you remember the Avogadro constant (see Appendix B), you can think of the mass of the earth as being 10 moles of kilograms. What does this statement mean, and how accurate is it? The actual mass of the earth is 5.98×10^{24} kg.

9. What mass of water fell on the town in Problem 4 during the thunderstorm? 1 m^3 of water has a mass of 10^3 kg.

10. (a) Assuming that the density (mass/volume) of water is exactly one gram per cubic centimeter (1 g/cm^3), express the density of water in kilograms per liter. (b) Suppose that it takes exactly 10 h to drain a container of 1 liter of water. What is the average mass flow rate, in kilograms per second, of water from the container?

Section 1–5   The Standard of Time
11. An astronomical unit (AU) is the average distance of the earth from the sun, approximately 149,000,000 km. The speed of light is about 3×10^8 m/s. Express the speed of light in terms of astronomical units per minute.

12. Isaac Asimov has proposed a unit of time involving the highest known speed and the smallest known measureable distance. It is the *light-fermi*, the time taken by light to move a distance of 1 fermi (= 1 femtometer = 10^{-15} m). The radius of the proton is about 1 fermi. How many seconds are there in a light-fermi?

13. From Fig. 1–4, calculate by what length of time the earth's rotation period in midsummer differs from that in the following spring.

14. Until 1883, every city and town in the United States kept its own local time. Today, travelers reset their watches only when the time change equals one hour. How far must you travel in longitude until your watch must be reset by one hour?

*15. The time it takes the moon to return to a given position as seen against the background of the fixed stars is called a sidereal month. The time interval between identical phases of the moon is called a lunar month. The lunar month is longer than a sidereal month. Why, and by how much?

ADDITIONAL PROBLEMS

16. The maximum speeds of various animals are given roughly as follows in miles per hour: (a) snail, 3×10^{-2}; (b) spider, 1.2; (c) squirrel, 12; (d) man, 23; (e) rabbit, 35; (f) fox, 42; (g) lion, 50; and (h) cheetah, 70. Convert these data to meters per second.

17. (*a*) In track meets both 100 yd and 100 m are used as distances for dashes. Which is longer? By how many meters is it longer? By how many feet? (*b*) Track and field records are kept for the mile and the so-called metric mile (1500 m). Compare these distances.

18. (*a*) A unit of time sometimes used in microscopic physics is the *shake*. One shake equals 10^{-8} s. Are there more shakes in a second than there are seconds in a year? (*b*) Mankind has existed for about 10^6 years, whereas the universe is about 10^{10} years old. If the age of the universe is taken to be one day, for how many seconds has mankind existed?

19. A man on a diet loses 2.3 kg (corresponding to about 5 lb) per week. Express his mass loss rate in milligrams per second.

20. In two *different* track meets, the winners of the mile race ran their races in 3 min 58.05 s and 3 min 58.20 s. In order to conclude that the runner with the shorter time was indeed faster, what is the maximum tolerable error, in feet, in laying out the distances?

21. Give the relation between (*a*) a square inch and a square centimeter; (*b*) a square mile and a square kilometer; (*c*) a cubic meter and a cubic centimeter; (*d*) a square foot and a square yard.

22. The distance between neighboring atoms, or molecules, in a solid substance can be estimated by calculating twice the radius of a sphere with volume equal to the volume per atom of the material. Calculate the distance between neighboring atoms in (*a*) iron and (*b*) sodium. The densities of iron and sodium are 7.87 g/cm^3 and 1.013 g/cm^3, respectively; the mass of an iron atom is 9.27×10^{-26} kg, and the mass of a sodium atom is 3.82×10^{-26} kg.

23. Five clocks are being tested in a laboratory. Exactly at noon, as determined by the WWV time signal, on the successive days of a week the clocks read as follows:

Clock	Sun.	Mon.	Tues.	Wed.	Thurs.	Fri.	Sat.
A	12:36:40	12:36:56	12:37:12	12:37:27	12:37:44	12:37:59	12:38:14
B	11:59:59	12:00:02	11:59:57	12:00:07	12:00:02	11:59:56	12:00:03
C	15:50:45	15:51:43	15:52:41	15:53:39	15:54:37	15:55:35	15:56:33
D	12:03:59	12:02:52	12:01:45	12:00:38	11:59:31	11:58:24	11:57:17
E	12:03:59	12:02:49	12:01:54	12:01:52	12:01:32	12:01:22	12:01:12

How would you arrange these five clocks in the order of their relative value as good timekeepers? Justify your choice.

24. Assume that the average distance of the sun from the earth is 400 times the average distance of the moon from the earth. Now consider a total eclipse of the sun and state conclusions that can be drawn about (*a*) the relation between the sun's diameter and the moon's diameter; (*b*) the relative volumes of the sun and the moon. (*c*) Hold up a dime (or another small coin) so that it just eclipses the full moon and measure the angle it intercepts at the eye. From this experimental result and the given distance between the moon and the earth ($= 3.80 \times 10^5$ km), estimate the diameter of the moon.

*25. Assuming that the length of the day uniformly increases by 0.001 s in a century, calculate the cumulative effect on the measure of time over 20 centuries. Such a slowing down of the earth's rotation is indicated by observations of the occurrences of solar eclipses during this period.

*26. The navigator of the oil tanker *Gulf Supernox* uses the satellites of the Global Positioning System (GPS/NAVSTAR) to find his latitude and longitude; see Fig. 1–5. These are 43°36′25.3″ N and 77°31′48.2″W. If the accuracy of these determinations is ±0.5″, what is the uncertainty in the tanker's position measured along (*a*) a north-south line and (*b*) an east-west line? (*c*) Where is the tanker?

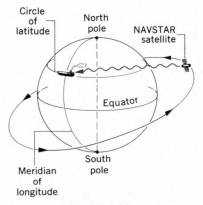

Figure 1–5 Problem 26.

2 Vectors

2-1 Vectors and Scalars

A change of position of a particle is called a *displacement*. If a particle moves from position A to position B (Fig. 2–1a), we can represent its displacement by drawing a line from A to B; the direction of displacement can be shown by putting an arrowhead at B indicating that the displacement was from A to B. The path of the particle need not necessarily be a straight line from A to B; the arrow represents only the net effect of the motion, not the actual motion.

In Fig. 2–1b, for example, we plot an actual path followed by a particle from A to B. The path is not the same as the displacement AB. If we were to take snapshots of the particle when it was at A and, later, when it was at some intermediate position P, we could obtain the displacement vector AP, representing the net effect of the motion during this interval, even though we would not know the actual path taken between these points. Furthermore, a displacement such as $A'B'$ (Fig. 2–1a), which is parallel to AB, similarly directed and equal in length to AB, represents the same change in position as AB. We make no distinction between these two displacements. A displacement is therefore characterized by a length and a direction.

In a similar way, we can represent a subsequent displacement from B to C (Fig. 2–1c). The net effect of the two displacements will be the same as a displacement from A to C. We speak then of AC as the *sum* or *resultant* of the displacements AB and BC. Notice that this sum is not an algebraic sum and that a number alone cannot uniquely specify it.

Quantities that behave like displacements are called *vectors*. Vectors, then, are quantities that have both magnitude and direction and combine according to certain rules of addition. These rules are stated below. The displacement vector is a convenient prototype. Some other physical quantities that are vectors are force,

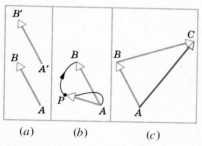

Figure 2–1 Displacement vectors. (a) Vectors AB and $A'B'$ are identical since they have the same length and point in the same direction. (b) The actual *path* of the particle in moving from A to B may be the curve shown; the *displacement* remains the vector AB. At some intermediate point P the displacement from A is the vector AP. (c) After displacement AB the particle undergoes another displacement BC. The net effect of the two displacements is represented by the vector AC.

velocity, acceleration, momentum, the electric field, and the magnetic field. There are two reasons that vectors are important in physics. (1) Vector notation is compact. If we can express a law of physics in vector form we usually find it easier to understand and to manipulate mathematically. (2) The laws of physics have the same content no matter what coordinate system we pick to express them in. Our ability to put these laws in vector form concisely expresses that fact because vector equations are the same in all coordinate systems.

Quantities that can be completely specified by a number and unit and that therefore have magnitude only are called *scalars*. Some physical quantities that are scalars are mass, length, time, density, energy, and temperature. Scalars are manipulated by the rules of ordinary algebra.

Figure 2–2 The vector sum **a** + **b** = **r**. Compare with Fig. 2–1c.

2–2 Addition of Vectors, Geometric Method

To represent a vector on a diagram we draw an arrow. We choose the length of the arrow proportional to the magnitude of the vector (that is, we choose a scale), and we choose the direction of the arrow to be the direction of the vector, with the arrowhead giving the sense of the direction. A vector such as this is represented conveniently in printing by a boldface symbol such as **d**. In handwriting it is convenient to put an arrow above the symbol to denote a vector quantity, such as $\vec{d}$.

Often we shall be interested only in the *magnitude* of the vector and not in its direction. The magnitude of **d** may be written as |**d**|, called the *absolute value* of **d**; more frequently we represent the magnitude alone by the italic letter symbol d. The boldface symbol is meant to signify both properties of the vector, magnitude and direction.

Consider now Fig. 2–2, in which we have redrawn and relabeled the vectors of Fig. 2–1c. The relation among these displacements (vectors) can be written as

Adding vectors geometrically

$$\mathbf{a} + \mathbf{b} = \mathbf{r}. \qquad (2–1)$$

The rules to be followed in performing this (vector) addition geometrically are these: On a diagram drawn to scale lay out the displacement vector **a**; then draw **b** with its tail at the head of **a**, and draw a line from the tail of **a** to the head of **b** to construct the vector sum **r**. This is a displacement equivalent in length and direction to the successive displacements **a** and **b**. This procedure can be generalized to obtain the sum of any number of successive displacements.

Since vectors are new quantities, we must expect new rules for their manipulation. The symbol "+" in Eq. 2–1 simply has a different meaning than it has in arithmetic or scalar algebra. It tells us to carry out a different set of operations.

Using Fig. 2–3 we can prove two important properties of vector addition:

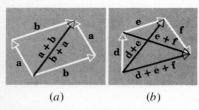

(a)   *(b)*

Figure 2–3 *(a)* The commutative law for vector sums, which states that **a** + **b** = **b** + **a**. *(b)* The associative law, which states that **d** + (**e** + **f**) = (**d** + **e**) + **f**.

$$\mathbf{a} + \mathbf{b} = \mathbf{b} + \mathbf{a}, \qquad \text{(commutative law)} \qquad (2–2)$$

and

$$\mathbf{d} + (\mathbf{e} + \mathbf{f}) = (\mathbf{d} + \mathbf{e}) + \mathbf{f}. \qquad \text{(associative law)} \qquad (2–3)$$

These laws assert that it makes no difference in what order or in what grouping we add vectors; the sum is the same. In this respect, vector addition and scalar addition follow the same rules.

Subtracting vectors

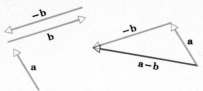

Figure 2–4 The vector difference $\mathbf{a} - \mathbf{b} = \mathbf{a} + (-\mathbf{b})$.

The operation of subtraction can be included in our vector algebra by defining the negative of a vector to be another vector of equal magnitude but opposite direction. Then

$$\mathbf{a} - \mathbf{b} = \mathbf{a} + (-\mathbf{b}), \qquad (2-4)$$

as shown in Fig. 2–4.

Remember that, although we have used displacements to illustrate these operations, the rules apply to all vector quantities.

2–3 Resolution and Addition of Vectors, Analytic Method

The geometric method of adding vectors is not very useful for vectors in three dimensions; often it is even inconvenient for the two-dimensional case. Another way of adding vectors is the analytic method, involving the resolution of a vector into components with respect to a particular coordinate system.

Finding vector components

Figure 2–5a shows a vector **a** whose tail has been placed at the origin of a rectangular coordinate system. If we drop perpendicular lines from the head of **a** to the axes, the quantities a_x and a_y so formed are called the *components* of the vector **a**. The process is called resolving a vector into its components. Figure 2–5 shows a two-dimensional case; the extension of our conclusions to three dimensions should be clear.

A vector may have many sets of components. For example, if we rotated the x-axis and y-axis in Fig. 2–5a by 10° counterclockwise, the components of **a** would be different. Thus the components of a vector are uniquely specified only if we specify the particular coordinate system being used. The vector need not be drawn with its tail at the origin of the coordinate system to find its components—although we have done so for convenience; the vector may be moved anywhere in the coordinate space and, as long as its angles with the coordinate directions are maintained, its components will be unchanged.

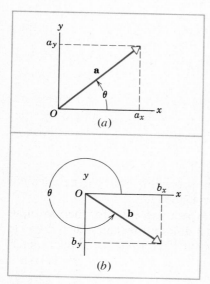

Figure 2–5 Two examples of the resolution of a vector into its scalar components in a particular coordinate system.

The components a_x and a_y in Fig. 2–5a are found readily from

$$a_x = a \cos \theta \qquad \text{and} \qquad a_y = a \sin \theta, \qquad (2-5)$$

where θ is the angle that the vector **a** makes with the positive x-axis, measured counterclockwise from this axis. Note that, depending on the angle θ, a_x and a_y can be positive or negative. For example, in Fig. 2–5b, b_y is negative and b_x is positive. The components of a vector behave like scalar quantities because, in any particular coordinate system, only a number, with an algebraic sign, is needed to specify them.

Once a vector is resolved into its components, the components themselves can be used to specify the vector. Instead of the two numbers a (magnitude of the vector) and θ (direction of the vector relative to the x-axis), we now have two other numbers, a_x and a_y. We can pass back and forth between the description of a vector in terms of its components a_x, a_y and the equivalent description in terms of magnitude and direction a and θ. To obtain a and θ from a_x and a_y, we note from Fig. 2–5a that

Reconstructing the vector from its components

$$a = \sqrt{a_x{}^2 + a_y{}^2} \qquad \text{and} \qquad \tan \theta = \frac{a_y}{a_x}. \qquad (2-6)$$

The quadrant in which the vector lies is determined from the signs of a_x and a_y.

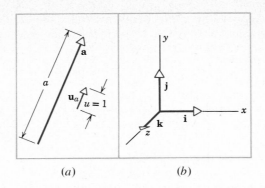

Figure 2–6 (a) The vector **a** may be written as $\mathbf{u}_a a$, in which $\mathbf{u}_a$ is a unit vector in the direction of **a**. (b) The unit vectors **i**, **j**, and **k**, used to specify the positive x-, y-, and z-directions, respectively.

When resolving a vector into components it is sometimes useful to introduce a vector of unit length in a given direction. Thus vector **a** in Fig. 2–6a may be written, for example, as

$$\mathbf{a} = \mathbf{u}_a a, \tag{2-7}$$

Unit vectors

where $\mathbf{u}_a$ is a *unit vector* in the direction of **a**. Often it is convenient to draw unit vectors along the particular coordinate axes chosen. In the rectangular coordinate system the special symbols **i**, **j**, and **k** are usually used for unit vectors in the positive x-, y-, and z-directions, respectively; see Fig. 2–6b. Note that **i**, **j**, and **k** need not be located at the origin. Like all vectors, they can be translated anywhere in the coordinate space as long as their directions with respect to the coordinate axes are not changed.

The vectors **a** and **b** of Fig. 2–5 may be written in terms of their components and the unit vectors as

$$\mathbf{a} = \mathbf{i}a_x + \mathbf{j}a_y \tag{2-8a}$$

and

$$\mathbf{b} = \mathbf{i}b_x + \mathbf{j}b_y; \tag{2-8b}$$

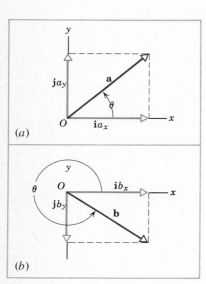

Figure 2–7 Two examples of the resolution of a vector into its vector components in a particular coordinate system; compare with Fig. 2–5.

see Fig. 2–7. The vector relation Eq. 2–8a is equivalent to the scalar relations Eqs. 2–6; each equation relates the vector (**a**, or a and θ) to its components (a_x and a_y). Sometimes we shall call quantities such as $\mathbf{i}a_x$ and $\mathbf{j}a_y$ in Eq. 2–8a the *vector components* of **a**; they are drawn as vectors in Fig. 2–7a. The word *component* alone will continue to refer to the scalar quantities a_x and a_y.

We now consider the addition of vectors by the analytic method. Let **r** be the sum of the two vectors **a** and **b** lying in the x-y plane, so that

$$\mathbf{r} = \mathbf{a} + \mathbf{b}. \tag{2-9}$$

Adding vectors analytically

In a given coordinate system, two vectors such as **r** and $\mathbf{a} + \mathbf{b}$ can be equal only if their corresponding components are equal, or

$$r_x = a_x + b_x \tag{2-10a}$$

and

$$r_y = a_y + b_y. \tag{2-10b}$$

These two algebraic equations, taken together, are equivalent to the single vector relation Eq. 2–9. From Eqs. 2–6 we may find r and the angle θ that **r** makes with the x-axis; that is,

$$r = \sqrt{r_x{}^2 + r_y{}^2} \quad \text{and} \quad \tan \theta = \frac{r_y}{r_x}$$

Thus we have the following analytic rule for adding vectors: Resolve each vector into its components in a given coordinate system; the algebraic sum of the individual components along a particular axis is the component of the sum vector along that same axis; the sum vector can be reconstructed once its components are known. This method for adding vectors may be generalized to many vectors and to three dimensions.

The advantage of the method of breaking up vectors into components, rather than adding directly with the use of suitable trigonometric relations, is that we always deal with right triangles and thus simplify the calculations.

When you are adding vectors by the analytic method, your choice of coordinate axes may simplify or complicate the process. Sometimes the components of the vectors with respect to a particular set of axes are known to begin with, so that the choice of axes is obvious. Other times a judicious choice of axes can greatly simplify the job of resolution of the vectors into components. For example, the axes can be oriented so that at least one of the vectors lies parallel to an axis.

Example 1 An airplane traveled 130 mi (=210 km) on a straight course making an angle of 22° east of due north. How far north and how far east did the plane travel from its starting point?

We choose the positive x-direction to be east and the positive y-direction to be north. Next (Fig. 2–8) we draw a displacement vector from the origin (starting point), making an angle of 22° with the y-axis (north) inclined along the positive x-direction (east). The length of the vector is chosen to represent a magnitude of 130 mi. If we call this vector **d**, then d_x gives the distance traveled east of the starting point and d_y gives the distance traveled north of the starting point. We have

$$\theta = 90° - 22° = 68°,$$

so that (see Eqs. 2–5)

$$d_x = d \cos \theta = (130 \text{ mi}) \cos 68° = 49 \text{ mi } (=78 \text{ km})$$

and

$$d_y = d \sin \theta = (130 \text{ mi}) \sin 68° = 120 \text{ mi } (=190 \text{ km}).$$

Figure 2–8 Example 1.

Example 2 An automobile travels due east on a level road for 30 km. It then turns due north at an intersection and travels 40 km before stopping. Find the resultant displacement of the car.

We choose a coordinate system fixed with respect to the earth, with the positive x-direction pointing east and the positive y-direction pointing north. The two successive displacements, **a** and **b**, are then drawn as shown in Fig. 2–9. The resultant displacement **r** is obtained from $\mathbf{r} = \mathbf{a} + \mathbf{b}$. Since **b** has no x-component and **a** has no y-component, we obtain (see Eqs. 2–10)

$$r_x = a_x + b_x = 30 \text{ km} + 0 = 30 \text{ km};$$

$$r_y = a_y + b_y = 0 + 40 \text{ km} = 40 \text{ km}.$$

The magnitude and direction of **r** are then (see Eqs. 2–6)

$$r = \sqrt{r_x^2 + r_y^2} = \sqrt{(30 \text{ km})^2 + (40 \text{ km})^2} = 50 \text{ km}.$$

$$\tan \theta = \frac{r_y}{r_x} = \frac{40 \text{ km}}{30 \text{ km}} = 1.33 \qquad \theta = \tan^{-1}(1.33) = 53°.$$

The resultant vector displacement **r** has a magnitude of 50 km and makes an angle of 53° north of east.

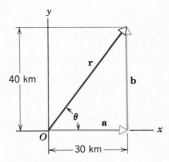

Figure 2–9 Example 2.

Example 3 Three coplanar vectors are expressed, with respect to a certain rectangular coordinate system, as

$$\mathbf{a} = 4\mathbf{i} - \mathbf{j},$$

$$\mathbf{b} = -3\mathbf{i} + 2\mathbf{j},$$

and

$$\mathbf{c} = -3\mathbf{j},$$

in which the components are given in arbitrary units. Find the vector **r** that is the sum of these vectors.

From Eqs. 2–10 we have

$$r_x = a_x + b_x + c_x = 4 - 3 + 0 = 1,$$

and

$$r_y = a_y + b_y + c_y = -1 + 2 - 3 = -2.$$

Thus

$$\mathbf{r} = \mathbf{i}r_x + \mathbf{j}r_y = \mathbf{i} - 2\mathbf{j}.$$

Figure 2–10 shows the four vectors. From Eqs. 2–6 we can calculate that the magnitude of **r** is $\sqrt{5}$ and that the angle that **r** makes with the positive x-axis, measured counterclockwise from that axis, is

$$\theta = \tan^{-1}(-2/1) = \tan^{-1}(-2) = 297°.$$

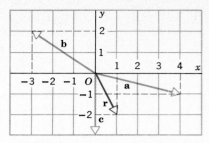

Figure 2–10 Three vectors **a**, **b**, and **c**, and their vector sum **r**.

Note that $\theta = 117°$ also has a tangent of -2 but we reject this solution because it is not consistent with Fig. 2–10.

2–4 Multiplication of Vectors*

We have assumed in the previous discussion that the vectors being added together are of like kind; that is, displacement vectors are added to displacement vectors, or velocity vectors are added to velocity vectors. Just as it would be meaningless to add together scalar quantities of different kinds, such as mass and temperature, so it would be meaningless to add together vector quantities of different kinds.

Like scalars, vectors of different kinds can be multiplied by one another to generate quantities of new physical dimensions. Because vectors have direction as well as magnitude, vector multiplication cannot follow exactly the same rules as the algebraic rules of scalar multiplication. We must establish new rules of multiplication for vectors.

We find it useful to define three kinds of multiplication operations for vectors: (1) multiplication of a vector by a scalar, (2) multiplication of two vectors in such a way as to yield a scalar, and (3) multiplication of two vectors in such a way as to yield another vector. There are still other possibilities, but we shall not consider them here.

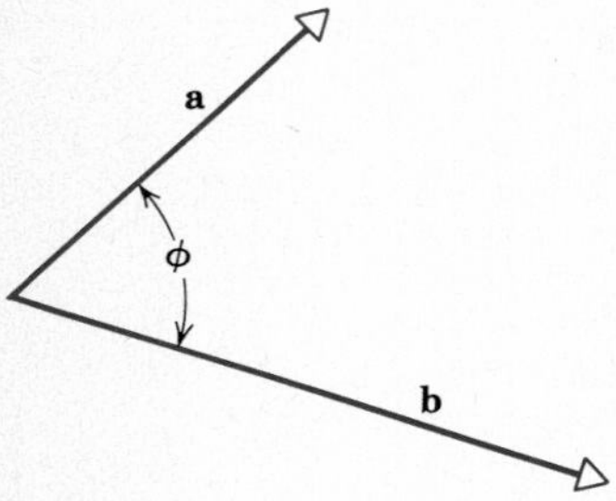

The multiplication of a vector by a scalar has a simple meaning: The product of a scalar k and a vector **a**, written $k\mathbf{a}$, is defined to be a new vector whose magnitude is k times the magnitude of **a**. The new vector has the same direction as **a** if k is positive and the opposite direction if k is negative. To divide a vector by a scalar we simply multiply the vector by the reciprocal of the scalar.

When we multiply a vector quantity by another vector quantity, we must distinguish between the *scalar* (or *dot*) *product* and the *vector* (or *cross*) *product*. The *scalar product* of two vectors **a** and **b**, written as **a** · **b**, is defined to be

$$\mathbf{a} \cdot \mathbf{b} = ab \cos \phi, \tag{2–11}$$

where a is the magnitude of vector **a**, b is the magnitude of vector **b**, and $\cos \phi$ is the cosine of the angle ϕ between the two vectors (see Fig. 2–11).

Since a and b are scalars and $\cos \phi$ is a pure number, the scalar product of two vectors is a scalar. The scalar product of two vectors can be regarded as the

Figure 2–11 Illustrating Eq. 2–11.

The scalar product

---

* The material of this section will be used later in the text. The scalar product is used first in Chapter 7 and the vector product in Chapter 12. The instructor may wish to postpone this section accordingly. Its presentation here gives a unified treatment of vector algebra and serves as a convenient reference for later work.

product of the magnitude of one vector and the component of the other vector in the direction of the first (see Fig. 2–12). Because of the notation, **a · b** is also called the dot product of **a** and **b** and is spoken as "**a** dot **b**."

We could have defined **a · b** to be any operation we want, for example, to be $a^{1/3}b^{1/3} \tan(\phi/2)$, but this would turn out to be of no use to us in physics. With our definition of the scalar product, a number of important physical quantities can be described as the scalar product of two vectors. Some of them are mechanical work, gravitational potential energy, electrical potential, electric power, and electromagnetic energy density.

The *vector product* of two vectors **a** and **b** is written as **a × b** and is another vector **c**, where **c = a × b**. The magnitude of **c** is defined by

$$c = ab \sin \phi, \qquad (2–12)$$

where ϕ is the (smaller) angle between **a** and **b**.

The *direction of* **c**, the vector product of **a** and **b**, is defined to be perpendicular to the plane formed by **a** and **b**. To specify the sense of the vector **c** refer to Fig. 2–13. Imagine rotating a right-handed screw whose axis is perpendicular to the plane formed by **a** and **b** so as to turn it from **a** to **b** through the angle ϕ between them. Then the direction of advance of the screw gives the direction of the vector product **a × b** (Fig. 2–13a). Another convenient way to obtain the direction is to imagine an axis perpendicular to the plane of **a** and **b** through their origin. Now wrap the fingers of the right hand around this axis and push the vector **a** into the vector **b** through the smaller angle between them with the fingertips, keeping the thumb erect; the direction of the erect thumb then gives the direction of the vector product **a × b** (Fig. 2–13b). Because of the notation, **a × b** is also called the cross product of **a** and **b** and is spoken as "**a** cross **b**."

The vector product

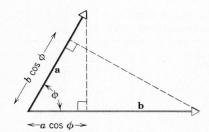

Figure 2–12 The scalar product **a · b** ($=ab \cos \phi$) is the product of the magnitude of either vector (a, say) and the component of the other vector in the direction of the first vector ($b \cos \phi$, say).

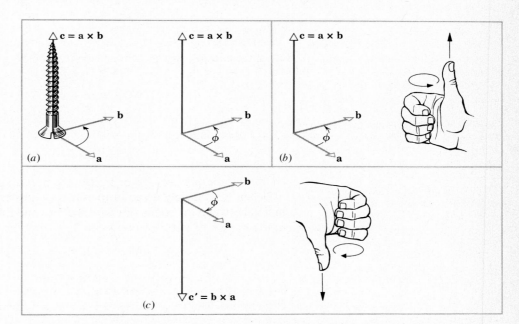

Figure 2–13 The vector product. (*a*) In **c = a × b**, the direction of **c** is that in which a right-handed screw advances when turned from **a** to **b** through the smaller angle. (*b*) The direction of **c** can also be obtained from the "right-hand rule": If the right hand is held so that the curled fingers follow the rotation of **a** into **b**, the extended right thumb will point in the direction of **c**. (*c*) The vector product changes sign when the order of the factors is reversed: **a × b = −b × a**. Apply the right-hand rule or the rule for the advance of a right-handed screw to show that **c** and **c′** have opposite directions.

Notice that **b** × **a** is not the same vector as **a** × **b**, so that the order of factors in a vector product is important. This is not true for scalars because the order of factors in algebra or arithmetic does not affect the resulting product. Actually, **a** × **b** = −(**b** × **a**) (Fig. 2–13c). This can be deduced from the fact that the magnitude $ab \sin \phi$ equals the magnitude $ba \sin \phi$, but the direction of **a** × **b** is opposite to that of **b** × **a;** this is so because the right-handed screw advances in one direction when rotated from **a** to **b** through ϕ but advances in the opposite direction when rotated from **b** to **a** through ϕ. You can obtain the same result by applying the right-hand rule.

If ϕ is 90°, **a**, **b**, and **c** (= **a** × **b**) are all at right angles to one another and give the directions of a three-dimensional right-handed coordinate system.

The reason for defining the vector product in this way is that it proves to be useful in physics. We often encounter physical quantities that are vectors whose product, defined as above, is a vector quantity having important physical meaning. Some examples of physical quantities that are vector products are torque, angular momentum, the force on a moving charge in a magnetic field, and the flow of electromagnetic energy. When such quantities are discussed later, your attention will be drawn to their connection with the vector product.

Example 4   A certain vector **a** in the x-y plane is 250° counter-clockwise from the positive x-axis and has a magnitude 1.8 units. Vector **b** has magnitude 1.2 units and is directed parallel to the z-axis. Calculate (a) the scalar product **a · b** and (b) the vector product **a** × **b**.

The two vectors are shown in Fig. 2–14.

(a) Since **a** and **b** are perpendicular to one another, the angle ϕ between them is 90° and cos ϕ = cos 90° = 0. Therefore, from Eq. 2–11, the scalar product is

$$\mathbf{a} \cdot \mathbf{b} = ab \cos \phi = ab \cos 90° = (1.8)(1.2)0 = 0,$$

consistent with the fact that neither vector has a component in the direction of the other.

(b) The magnitude of the vector product is, from Eq. 2–12,

$$|\mathbf{a} \times \mathbf{b}| = ab \sin \phi = (1.8)(1.2) \sin 90° = 2.2.$$

The direction of the vector product is perpendicular to the plane formed by **a** and **b**. Therefore, as shown in Fig. 2–14, it lies in

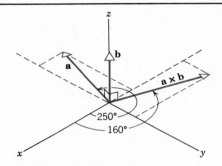

Figure 2–14   Example 4.

the x-y plane (perpendicular to **b**) at an angle of 250° − 90° = 160° from the +x-axis (perpendicular to **a**) in accordance with the right-hand rule.

REVIEW GUIDE AND SUMMARY

Scalars and vectors

Scalars, such as temperature, have magnitude only and are specified by a number with a unit (82°F) and obey the rules of ordinary algebra. Vectors, such as displacement, have both magnitude and direction (5.0 m, north) and obey the special rules of vector algebra.

Adding vectors geometrically

Two vectors **a** and **b** may be added geometrically by drawing them to a common scale and placing them head to tail. The vector connecting the tail of **a** to the head of **b** is the sum vector **r**, as Fig. 2–2 shows. To subtract **b** from **a**, reverse the direction of **b** and then add to **a** as above; see Fig. 2–4.

The components of a vector

The components a_x and a_y of any vector **a** are the perpendicular projections of **a** on the coordinate axes, as Fig. 2–5 and Example 1 show. Analytically, the components are given by*

$$a_x = a \cos \theta \quad \text{and} \quad a_y = a \sin \theta. \qquad [2\text{–}5]$$

---

* Equation references previously given are enclosed in square brackets.

Given the components, we can reconstruct the vector from

$$a = \sqrt{a_x{}^2 + a_y{}^2} \quad \text{and} \quad \tan \theta = \frac{a_y}{a_x}. \qquad [2\text{-}6]$$

Adding vectors analytically

To add vectors analytically use this rule: Each component of the sum vector is the sum of the corresponding components of the vectors being added; see Examples 1 and 2. Once we have the components of the sum vector we can reconstruct that vector analytically from Eq. 2–6.

Unit vectors

Often we find it useful to introduce unit vectors **i**, **j**, and **k**, whose magnitudes are unity and whose directions are the x, y, and z-axes, respectively; see Fig. 2–6. We can write any vector **a** in terms of unit vectors as

$$\mathbf{a} = \mathbf{i}a_x + \mathbf{j}a_y,$$

in which $\mathbf{i}a_x$ and $\mathbf{j}a_y$ are the vector components of **a**. Example 3 shows how to add vectors using vector components.

The scalar product

The scalar (or dot) product of two vectors is written as $\mathbf{a} \cdot \mathbf{b}$ and is a scalar quantity given by $ab \cos \phi$, where ϕ is the angle between a and b; see Fig. 2–11. The scalar product may be positive, zero, or negative, depending on the value of ϕ. Figure 2–12 shows that the scalar product can be viewed as the magnitude of either vector (**a**, say) multiplied by the component of the other vector ($b \cos \phi$) in the direction of the first vector.

The vector product

The vector (or cross) product is written as $\mathbf{a} \times \mathbf{b}$ and is a vector **c** whose magnitude is given by $ab \sin \phi$. The direction of **c** is at right angles to the plane defined by **a** and **b**. The sense in which **c** points along this direction is given by a right-hand rule described in Fig. 2–13. Note that (1) $\mathbf{a} \times \mathbf{b} = 0$ if **a** and **b** are either parallel or antiparallel, (2) $\mathbf{a} \times \mathbf{b}$ has its maximum value ($= ab$) if **a** and **b** are at right angles, and (3) $\mathbf{a} \times \mathbf{b} = -\mathbf{b} \times \mathbf{a}$; in contrast, $\mathbf{a} \cdot \mathbf{b} = \mathbf{b} \cdot \mathbf{a}$. Example 4 illustrates scalar and vector multiplication.

QUESTIONS

1. In 1969 three Apollo astronauts left Cape Canaveral, went to the moon and back, and splashed down in the Pacific Ocean. An admiral bid them goodby at the Cape and then sailed to the Pacific Ocean in an aircraft carrier where he picked them up. For their respective journeys, did the astronauts or the admiral have the larger displacement?

2. Can two vectors of different magnitude be combined to give a zero resultant? Can three vectors?

3. Can a vector have zero magnitude if one of its components is not zero?

4. Can the sum of the magnitudes of two vectors ever be equal to the magnitude of the sum of these two vectors?

5. Can the magnitude of the difference between two vectors ever be greater than the magnitude of either vector? Can it be greater than the magnitude of their sum? Give examples.

6. If three vectors add up to zero, they must all be in the same plane. Make this plausible.

7. Does a unit vector have units?

8. Name several scalar quantities. Does the value of a scalar quantity depend on the coordinate system you choose?

9. You can order events in time. For example, event b may precede event c but follow event a, giving a time order of events a, b, c. Hence there is a sense of time, distinguishing past, present, and future. Is time a vector therefore? If not, why not?

10. Do the commutative and associative laws apply to vector subtraction?

11. Can a scalar product be a negative quantity?

12. (a) If $\mathbf{a} \cdot \mathbf{b} = 0$, does it follow that **a** and **b** are perpendicular to one another? (b) If $\mathbf{a} \cdot \mathbf{b} = \mathbf{a} \cdot \mathbf{c}$, does it follow that **b** equals **c**?

13. If $\mathbf{a} \times \mathbf{b} = 0$, must **a** and **b** be parallel to each other? Is the converse true?

14. Must you specify a coordinate system when you (a) add two vectors, (b) form their scalar product, (c) form their vector product, (d) find their components?

EXERCISES

Section 2-2   Addition of Vectors, Geometric Method

1. Consider two displacements, one of magnitude 3 m and another of magnitude 4 m. Show how the displacement vectors may be combined to get a resultant displacement of magnitude (a) 7 m, (b) 1 m, and (c) 5 m.

2. What are the properties of two vectors **a** and **b** such that
(a) $\mathbf{a} + \mathbf{b} = \mathbf{c}$ and $a + b = c$,
(b) $\mathbf{a} + \mathbf{b} = \mathbf{a} - \mathbf{b}$,
(c) $\mathbf{a} + \mathbf{b} = \mathbf{c}$ and $a^2 + b^2 = c^2$.

3. A woman walks 250 m in the direction 30° east of north, then 175 m directly east. (a) Using graphical methods, find her final displacement from the starting point. (b) Compare the magnitude of her displacement with the distance she walked.

Section 2–3   Resolution and Addition of Vectors, Analytic Method

4. What are the components of a vector **A** in the x-y plane if its direction is 250° counterclockwise from the positive x-axis and its magnitude is 7.3 units?

5. The x-component of a certain vector is -25 units and the y-component is $+40$ units. (a) What is the magnitude of the vector? (b) What is the angle between the direction of the vector and the positive x-axis?

6. Two vectors **a** and **b** have equal magnitudes of 10 units. They are oriented as shown in Fig. 2–15 and their vector sum is **r**. Find (a) the x- and y-components of **r**, (b) the magnitude of **r**, and (c) the angle **r** makes with the x-axis.

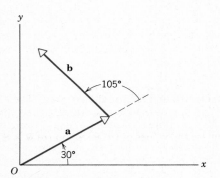

Figure 2–15   Exercise 6.

7. A displacement vector in the x-y plane is 15 m long and directed as shown in Fig. 2–16. For each of the three coordinate systems shown in the figure, determine the x- and y-components of the vector.

8. A heavy piece of machinery is raised by sliding it along a 12.5-m plank oriented at 20° to the horizontal, as shown in Fig. 2–17. (a) How high above its original position is it raised? (b) How far is it moved horizontally?

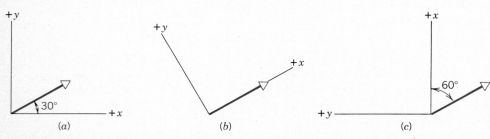

Figure 2–16   Exercise 7.

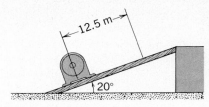

Figure 2–17   Exercise 8.

9. (a) What is the sum in unit vector notation of the two vectors $\mathbf{a} = 4\mathbf{i} + 3\mathbf{j}$ and $\mathbf{b} = -3\mathbf{i} + 7\mathbf{j}$? (b) What are the magnitude and direction of $\mathbf{a} + \mathbf{b}$?

10. Calculate the components, magnitudes, and directions of (a) $\mathbf{A} + \mathbf{B}$ and (b) $\mathbf{B} - \mathbf{A}$ if $\mathbf{A} = 3\mathbf{i} + 4\mathbf{j}$ and $\mathbf{B} = 5\mathbf{i} - 2\mathbf{j}$.

Section 2–4   Multiplication of Vectors

11. A vector **d** has a magnitude of 2.5 m and points north. What are the magnitudes and directions of the vectors (a) $4.0\mathbf{d}$ and (b) $-3.0\mathbf{d}$?

12. A vector **a** of magnitude 10 units and another vector **b** of magnitude 6 units point in directions differing by 60°. Find (a) the scalar product of the two vectors and (b) the vector product of the two vectors.

13. Given vector **a** in the $+x$-direction, a vector **b** in the $+y$-direction, and the scalar quantity d: (a) What is the direction of $\mathbf{a} \times \mathbf{b}$? (b) What is the direction of $\mathbf{b} \times \mathbf{a}$? (c) What is the direction of $\mathbf{b}/d$? (d) What is the magnitude of $\mathbf{a} \cdot \mathbf{b}$?

14. Show for any vector **a** that $\mathbf{a} \cdot \mathbf{a} = a^2$ and that $\mathbf{a} \times \mathbf{a} = 0$.

15. In the coordinate system of Fig. 2–6b, show that

$$\mathbf{i} \cdot \mathbf{i} = \mathbf{j} \cdot \mathbf{j} = \mathbf{k} \cdot \mathbf{k} = 1$$

and

$$\mathbf{i} \cdot \mathbf{j} = \mathbf{j} \cdot \mathbf{k} = \mathbf{k} \cdot \mathbf{i} = 0.$$

16. In the right-handed coordinate system of Fig. 2–6b, show that

$$\mathbf{i} \times \mathbf{i} = \mathbf{j} \times \mathbf{j} = \mathbf{k} \times \mathbf{k} = 0$$
$$\mathbf{i} \times \mathbf{j} = \mathbf{k}; \quad \mathbf{k} \times \mathbf{i} = \mathbf{j}; \quad \mathbf{j} \times \mathbf{k} = \mathbf{i}.$$

17. Find (a) "north" cross "west," (b) "down" dot "south," (c) "east" cross "up," (d) "west" dot "west," and (e) "south" cross "south." Let each vector have unit magnitude.

PROBLEMS GROUPED BY SECTION

Section 2–2   Addition of Vectors, Geometric Method

1.   A car is driven east for a distance of 50 km, then north for 30 km, and then in a direction 30° east of north for 25 km. Draw the vector diagram and determine the total displacement of the car from its starting point.

2.   A person walks in the following pattern: 3.1 km north, then 2.4 km west, and finally 5.2 km south. (a) Construct the vector diagram that represents this motion. (b) How far and in what direction would a bird fly in a straight line to arrive at the same final point?

3.   Three vectors **A**, **B**, and **C**, each having a magnitude of 50 units, lie in the x-y plane and make angles of 30°, 195°, and 315° with the positive x-axis, respectively. Find graphically the magnitudes and directions of the vectors (a) $\mathbf{A} + \mathbf{B} + \mathbf{C}$, (b) $\mathbf{A} - \mathbf{B} + \mathbf{C}$, (c) a vector **D** such that $(\mathbf{A} + \mathbf{B}) - (\mathbf{C} + \mathbf{D}) = 0$.

Section 2–3   Resolution and Addition of Vectors, Analytic Method

4.   Two vectors are given by $\mathbf{a} = 4\mathbf{i} - 3\mathbf{j} + \mathbf{k}$ and $\mathbf{b} = -\mathbf{i} + \mathbf{j} + 4\mathbf{k}$. Find (a) $\mathbf{a} + \mathbf{b}$, (b) $\mathbf{a} - \mathbf{b}$, and (c) a vector **c** such that $\mathbf{a} - \mathbf{b} + \mathbf{c} = 0$.

5.   A particle undergoes three successive displacements in a plane, as follows: 4.0 m southwest, 5.0 m east, 6.0 m in a direction 60° north of east. Choose the y-axis pointing north and the x-axis pointing east and find (a) the components of each displacement, (b) the components of the resultant displacement, (c) the magnitude and direction of the resultant displacement, and (d) the displacement that would be required to bring the particle back to the starting point.

6.   Find the vector components of the sum **r** of the vector displacements **c** and **d** whose components in meters along three perpendicular directions are

$$c_x = 7.4,\ c_y = -3.8,\ c_z = -6.1;\ d_x = 4.4,\ d_y = -2.0,\ d_z = 3.3.$$

7.   A person desires to reach a point that is 3.4 km from her present location and in a direction that is 35° north of east. However, she must travel along streets that go either north-south or east-west. What is the minimum distance she could travel to reach her destination?

8.   A particle moves in the x-y plane as shown in the table:

t (s)	0.0	0.1	0.2	0.3	0.4
x (m)	+1.0	+0.707	0.0	−0.707	−1.0
y (m)	0.0	+0.707	+1.0	+0.707	0.0

What is the vector displacement from $t = 0$ to $t = 0.4$ s? Express in unit vector notation.

9.   Two vectors of lengths a and b make an angle θ with each other when placed tail to tail. Prove, by taking components along two perpendicular axes, that the length of their sum is

$$r = \sqrt{a^2 + b^2 + 2ab\cos\theta}.$$

10.   Two vectors **A** and **B** have components, in arbitrary units, $A_x = 3.2,\ A_y = 1.6;\ B_x = 0.50,\ B_y = 4.5$. (a) Find the angle between **A** and **B**. (b) Find the components of a vector **C** that is perpendicular to **A**, is in the x-y plane, and has a magnitude of 5.0 units.

11.   A wheel with a radius of 45 cm rolls without slipping along a horizontal floor, as shown in Fig. 2–18. P is a dot painted on the rim of the wheel. At time t_1, P is at the point of contact between the wheel and the floor. At a later time t_2, the wheel has rolled through one-half of a revolution. What is the displacement of P during this interval?

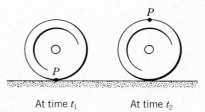

At time t_1 At time t_2

Figure 2–18   Problem 11.

*12.   A room has the dimensions 10 ft × 12 ft × 14 ft. A fly starting at one corner ends up at a diametrically opposite corner. (a) What is the magnitude of its displacement? (b) Could the length of its path be less than this distance? Greater than this distance? Equal to this distance? (c) Choose a suitable coordinate system and find the components of the displacement vector in this frame. (d) If the fly walks rather than flies, what is the length of the shortest path it can take?

Section 2–4   Multiplication of Vectors

13.   Two vectors, **R** and **S**, lie in the x-y plane. Their magnitudes are 4.5 and 7.3 units, respectively, whereas their directions are 320° and 85° measured counterclockwise from the positive x-axis. What are the values of (a) $\mathbf{R}\cdot\mathbf{S}$ and (b) $\mathbf{R}\times\mathbf{S}$?

14.   (a) Using unit vectors, express the diagonals of a cube in terms of its edges, which have length a. (b) Determine the angles made by the diagonals with the adjacent edges. (c) Determine the length of the diagonals. (d) Find the angles between the diagonals.

15.   Show that $\mathbf{a} \cdot (\mathbf{b} \times \mathbf{c})$ is equal in magnitude to the volume of the parallelepiped formed on the three vectors **a**, **b**, and **c**, as shown in Fig. 2–19.

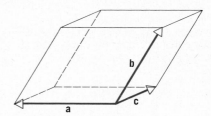

Figure 2–19   Problem 15.

16.   (a) We have seen that the commutative law does *not* apply to vector products; that is, $\mathbf{a} \times \mathbf{b}$ does not equal $\mathbf{b} \times \mathbf{a}$. Show that the commutative law *does* apply to scalar products; that is, $\mathbf{a} \cdot \mathbf{b} = \mathbf{b} \cdot \mathbf{a}$. (b) Show that the distributive law applies to both scalar products and vector products; that is, show that

$$\mathbf{a} \cdot (\mathbf{b} + \mathbf{c}) = \mathbf{a} \cdot \mathbf{b} + \mathbf{a} \cdot \mathbf{c}$$

and that $\mathbf{a} \times (\mathbf{b} + \mathbf{c}) = \mathbf{a} \times \mathbf{b} + \mathbf{a} \times \mathbf{c}$.

(c) Does the associative law apply to vector products; that is, does $\mathbf{a} \times (\mathbf{b} \times \mathbf{c})$ equal $(\mathbf{a} \times \mathbf{b}) \times \mathbf{c}$? (d) Does it make any sense to talk about an associative law for scalar products?

17. *Scalar product in unit vector notation.* Let two vectors be represented in terms of their coordinates as

$$\mathbf{a} = \mathbf{i}a_x + \mathbf{j}a_y + \mathbf{k}a_z$$

and

$$\mathbf{b} = \mathbf{i}b_x + \mathbf{j}b_y + \mathbf{k}b_z.$$

Show analytically that

$$\mathbf{a} \cdot \mathbf{b} = a_x b_x + a_y b_y + a_z b_z.$$

(Hint: See Exercise 15.)

18. *Vector product in unit vector notation.* Show that $\mathbf{a} \times \mathbf{b} = \mathbf{i}(a_y b_z - a_z b_y) + \mathbf{j}(a_z b_x - a_x b_z) + \mathbf{k}(a_x b_y - a_y b_x)$. (Hint: See Exercise 16.)

19. Vector **A** lies in the *y-z* plane 63° counterclockwise from the +*y*-axis and has magnitude 3.2. Vector **B** lies in the *x-z* plane 48° counterclockwise from the +*x*-axis and has magnitude 1.4. Find (a) $\mathbf{A} \cdot \mathbf{B}$, (b) $\mathbf{A} \times \mathbf{B}$, and (c) the angle between **A** and **B**.

20. Two vectors are given by $\mathbf{A} = 3\mathbf{i} + 5\mathbf{j}$ and $\mathbf{B} = 2\mathbf{i} + 4\mathbf{j}$. Find (a) $\mathbf{A} \times \mathbf{B}$, (b) $\mathbf{A} \cdot \mathbf{B}$, and (c) $(\mathbf{A} + \mathbf{B}) \cdot \mathbf{B}$.

21. A golfer takes three putts to get his ball into the hole once he is on the green. The first putt displaces the ball 12 ft north, the second 6.0 ft southeast, and the third 3.0 ft southwest. What displacement was needed to get the ball into the hole on the first putt?

22. (a) A man leaves his front door, walks 1000 m east, 2000 m north, and then takes a penny from his pocket and drops it from a cliff 500 m high. Set up a coordinate system and write down an expression, using unit vectors, for the displacement of the penny. (b) The man then returns to his front door, following a different path on the return trip. What is his resultant displacement for the round trip?

23. Given two vectors, $\mathbf{a} = 4\mathbf{i} - 3\mathbf{j}$ and $\mathbf{b} = 6\mathbf{i} + 8\mathbf{j}$, find the magnitude and direction of **a**, of **b**, of $\mathbf{a} + \mathbf{b}$, of $\mathbf{b} - \mathbf{a}$, and of $\mathbf{a} - \mathbf{b}$.

24. (a) Determine the components and magnitude of $\mathbf{R} = \mathbf{A} - \mathbf{B} + \mathbf{C}$ if $\mathbf{A} = 5\mathbf{i} + 4\mathbf{j} - 6\mathbf{k}$, $\mathbf{B} = -2\mathbf{i} + 2\mathbf{j} + 3\mathbf{k}$, and $\mathbf{C} = 4\mathbf{i} + 3\mathbf{j} + 2\mathbf{k}$. (b) Calculate the angle between **R** and the positive *z*-axis.

25. A missile early-warning radar station detects a Cruise missile approaching from the east. At first observation, the range to the missile is 1200 ft at 40° above the horizon. The missile is tracked for another 123° in the east-west plane, the range at final contact being 2580 ft. See Fig. 2–20. Find the displacement of the Cruise missile during the period of observation.

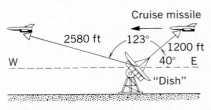

Figure 2–20 Problem 25.

26. Three vectors are given by $\mathbf{A} = 3\mathbf{i} + 3\mathbf{j} - 2\mathbf{k}$, $\mathbf{B} = -\mathbf{i} - 4\mathbf{j} + 2\mathbf{k}$, and $\mathbf{C} = 2\mathbf{i} + 2\mathbf{j} + \mathbf{k}$. Find (a) $\mathbf{A} \cdot (\mathbf{B} \times \mathbf{C})$. (b) $\mathbf{A} \cdot (\mathbf{B} + \mathbf{C})$ and (c) $\mathbf{A} \times (\mathbf{B} + \mathbf{C})$.

ADDITIONAL PROBLEMS

27. Vector **A** has a magnitude of 5.0 units and is directed east. Vector **B** is directed 35° west of north and has a magnitude of 4.0 units. Construct vector diagrams for calculating $(\mathbf{A} + \mathbf{B})$ and $(\mathbf{B} - \mathbf{A})$. Estimate the magnitudes and directions of $(\mathbf{A} + \mathbf{B})$ and $(\mathbf{B} - \mathbf{A})$ from your diagrams.

28. Rock *faults* are ruptures along which opposite faces of rock have moved across each other, parallel to the fracture surface. Earthquakes often accompany this movement. In Fig. 2–21, points *A* and *B* coincided before faulting. The component of the net displacement *AB* parallel to the horizontal surface fault line is called the *strike-slip* (*AC*). The component of the net displacement along the steepest line of the fault plane is the *dip-slip* (*AD*).

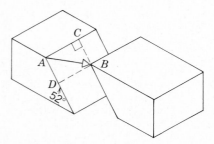

Figure 2–21 Problem 28.

(a) What is the net shift if the strike-slip is 22 m and the dip-slip is 17 m? (b) If the fault plane is inclined 52° to the horizontal, what is the net *vertical* displacement of *B* as a result of the faulting in (a)?

29. Two vectors **a** and **b** have components, in arbitrary units, $a_x = 3.2$, $a_y = 1.6$; $b_x = 0.50$, $b_y = 4.5$. (a) Find the angle between **a** and **b**. (b) Find the components of a vector **c** that is perpendicular to **a**, is in the *x-y* plane, and has a magnitude of 5.0 units.

30. Show that the area of the triangle contained between the vectors **a** and **b** in Fig. 2–22 is $\frac{1}{2}|\mathbf{a} \times \mathbf{b}|$, where the vertical bars signify magnitude.

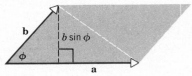

Figure 2–22 Problem 30.

31. Use the definition of scalar product $\mathbf{a} \cdot \mathbf{b} = ab \cos \phi$ and the fact that $\mathbf{a} \cdot \mathbf{b} = a_x b_x + a_y b_y + a_z b_z$ (see Problem 17) to obtain the angle between the two vectors given by $\mathbf{a} = 3\mathbf{i} + 3\mathbf{j} + 3\mathbf{k}$ and $\mathbf{b} = 2\mathbf{i} + \mathbf{j} + 3\mathbf{k}$.

32. (a) Show that $\mathbf{A} \cdot (\mathbf{B} \times \mathbf{A})$ is zero for all vectors **A** and **B**. (b) What is the magnitude of $\mathbf{A} \times (\mathbf{B} \times \mathbf{A})$ if there is an angle ϕ between the directions of **A** and **B**?

33. By examining the triangle formed by any vector and its x- and y-components, prove the inequality

$$|\sin \theta| + |\cos \theta| \geq 1.$$

34. Prove that two vectors must have equal magnitudes if their sum is perpendicular to their difference.

35. In Fig. 2–23, show that the vector $\mathbf{T} = \mathbf{i} \sin \theta - \mathbf{j} \cos \theta$ is in the direction of the tangent to the circle at P. The radius of the circle is 1 unit.

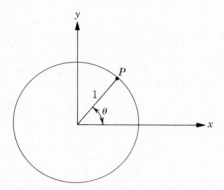

Figure 2–23   Problem 35.

36. The three vectors shown in Fig. 2–24 have magnitudes $A = 3$, $B = 4$, $C = 10$. (a) Calculate the x- and y-components of these vectors. (b) Find the numbers a and b such that $\mathbf{C} = a\mathbf{A} + b\mathbf{B}$.

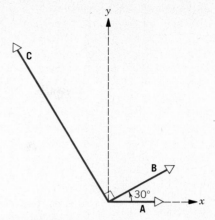

Figure 2–24   Problem 36.

37. Two places A and B on the surface of the earth in South America differ by $1°$ in latitude and $1°$ in longitude. Show that the magnitude of the displacement vector from A to B is approximately $(111)(1 + \cos^2 L)^{1/2}$ in kilometers, where L is the latitude of A.

*38. Let $\mathbf{r}$ be the displacement of a point from the origin and $\mathbf{K}$ be a vector perpendicular to a plane. (a) Show that $\mathbf{r} \cdot \mathbf{K}$ has the same value for all points on the plane. (b) If $\mathbf{K} = 3.50\mathbf{i} - 7.20\mathbf{j}$ (the units of $\mathbf{K}$ being m^{-1}) and if $\mathbf{r} \cdot \mathbf{K} = 5$, what is the perpendicular distance from the origin to the plane?

*39. A person flies from Washington, D.C., to Manila. (a) Describe the displacement vector. (b) What is its magnitude if the latitude and longitude of the two cities are $39°$N, $77°$W and $15°$N, $121°$E?

3 Motion in One Dimension

3–1 Mechanics

Mechanics, the oldest of the physical sciences, is the study of the motion of objects. The calculation of the path of a baseball or of a space probe sent from Earth to Mars are among its problems. So too is the analysis of tracks formed in bubble chambers, representing the decay and interactions of elementary particles.

When we describe motion we are dealing with that part of mechanics called *kinematics*. When we relate motion to the forces associated with it and to the properties of the moving objects, we are dealing with *dynamics*. In this chapter we shall define some kinematical quantities and study them in detail for the special case of motion in one dimension. In Chapter 4 we discuss some cases of two- and three-dimensional motion. Chapters 5 and 6 deal with the more general case of dynamics.

3–2 Particle Kinematics

An object can rotate as it moves. For example, a satellite may tumble as it orbits the earth. Also, a body may vibrate as it moves, as, for example, a falling water droplet. These complications can be avoided by considering the motion of an idealized body called a *particle*. Mathematically, a particle is treated as a point, an object without extent, so that rotational and vibrational considerations are not involved.

What is a particle?

Actually, there is probably no such thing in nature as an object without extent. The concept of "particle" is nevertheless useful because real objects often behave to a very good approximation as though they were particles. A body need not be "small" in the usual sense of the word in order to be treated as a particle. For example, if we consider the distance from the earth to the sun, with respect to this distance the earth and the sun can usually be considered to be particles. We can find out a great deal about the motion of the sun and planets, without appreciable error, by treating these bodies as particles. Baseballs, atoms, and protons can often be treated as particles. Even if a body is too large to be considered a particle for a particular problem, it can always be thought of as made up of a number of particles, and the results of particle motion may be useful in analyzing the problem. As a simplification, therefore, we confine our present treatment to the motion of a particle.

3-3 Average Velocity

In the present chapter we shall emphasize one-dimensional motion. Nevertheless we must define quantities, such as velocity, that are basically vectors and later will be used to describe two- and three-dimensional situations. For that reason we shall first give a short general discussion of such quantities before moving to the one-dimensional special cases, which will be developed in some detail.

The velocity of a particle is the rate at which its position changes with time. The position of a particle in a particular reference frame is given by a *position vector* drawn from the origin of that frame to the particle. At time t, let a particle be at point P in Fig. 3-1a, its position in the x-y plane being described by position vector $\mathbf{r}$. For simplicity we treat motion in two dimensions only; the extension to three dimensions will not be difficult.

At a later time t_1 let the particle be at point Q, described by position vector $\mathbf{r}_1$. The *displacement vector* describing the *change* in position of the particle as it moves from P to Q is $\Delta\mathbf{r}$ ($=\mathbf{r}_1 - \mathbf{r}$) and the elapsed time for the motion between these points is Δt ($=t_1 - t$). The *average velocity* for the particle during this interval is defined by

Average velocity

$$\overline{\mathbf{v}} = \frac{\Delta\mathbf{r}}{\Delta t} = \frac{\text{displacement (a vector)}}{\text{elapsed time (a scalar)}}. \qquad (3-1)$$

A bar above a symbol indicates an average value for the quantity in question.

The quantity $\overline{\mathbf{v}}$ is a vector, for it is obtained by dividing the vector $\Delta\mathbf{r}$ by the scalar Δt. Velocity, therefore, involves both direction and magnitude. Its direction is the direction of $\Delta\mathbf{r}$ and its magnitude is $|\Delta\mathbf{r}/\Delta t|$. The magnitude is expressed in distance units divided by time units, as, for example, meters per second.

Notice that the *average* velocity defined by Eq. 3-1 involves only the total displacement and the total elapsed time and does not tell us anything about the de-

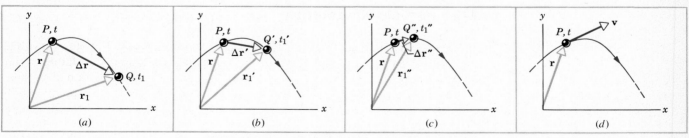

Figure 3-1 (a) A particle moves from P to Q in a time Δt ($=t_1 - t$), undergoing a displacement $\Delta\mathbf{r}$ ($=\mathbf{r}_1 - \mathbf{r}$). The *average velocity* $\overline{\mathbf{v}}$ for this motion is in the direction of $\Delta\mathbf{r}$. (b) The same, at an earlier time t_1'. (c) At a still earlier time t_1''. As Q moves closer to P, the average velocity approaches the *instantaneous velocity* $\mathbf{v}$ at point P, as shown in (d); $\mathbf{v}$ is tangent to the path at P.

tails of the motion between P and Q. For example, suppose that a man leaves his house for work and returns nine hours later. His average velocity for this trip is zero because the displacement corresponding to the nine-hour time interval is zero; he starts from his house and he also ends there.

3–4   Instantaneous Velocity

Suppose that a particle is moving in such a way that its average velocity, measured for a number of different time intervals, does *not* turn out to be constant. This particle is said to move with variable velocity. Then we must seek to determine a velocity of the particle at any given instant of time, called the *instantaneous velocity*.

Velocity can vary by a change in magnitude, by a change in direction, or both. For the motion portrayed in Fig. 3–1a, the average velocity during the time interval $t_1 - t$ may differ both in magnitude and direction from the average velocity obtained during another time interval $t_1' - t$. In Fig. 3–1b we illustrate this by choosing the point Q to be successively closer to point P. Points Q' and Q'' show two intermediate positions of the particle corresponding to the times t_1' and t_1'' and described by position vectors $\mathbf{r}_1'$ and $\mathbf{r}_1''$, respectively. The vector displacements $\Delta\mathbf{r}$, $\Delta\mathbf{r}'$, and $\Delta\mathbf{r}''$ differ in direction and become successively smaller. Likewise, the corresponding time intervals $\Delta t \ (= t_1 - t)$, $\Delta t' \ (= t_1' - t)$, and $\Delta t'' \ (= t_1'' - t)$ become successively smaller.

As we continue this process, letting Q approach P, we find that the ratio of displacement to elapsed time approaches a definite limiting value. Although the displacement in this process becomes extremely small, the time interval by which we divide it becomes small also and the ratio is not necessarily a small quantity. Similarly, while growing smaller, the displacement vector approaches a limiting direction, that of the tangent to the path of the particle at P. This limiting value of $\Delta\mathbf{r}/\Delta t$ is called the *instantaneous* velocity at the point P, or the velocity of the particle at the instant t.

If $\Delta\mathbf{r}$ is the displacement in a small interval of time Δt, following the time t, the velocity at the time t is the limiting value approached by $\Delta\mathbf{r}/\Delta t$ as both $\Delta\mathbf{r}$ and Δt approach zero. That is, if we let $\mathbf{v}$ represent the instantaneous velocity at point P,

$$\mathbf{v} = \lim_{\Delta t \to 0} \frac{\Delta\mathbf{r}}{\Delta t}.$$

The direction of $\mathbf{v}$ is the limiting direction that $\Delta\mathbf{r}$ takes as Q approaches P or as Δt approaches zero. As we have seen, this limiting direction is that of the tangent to the path of the particle at point P.

In the notation of the calculus, the limiting value of $\Delta\mathbf{r}/\Delta t$ as Δt approaches zero is written $d\mathbf{r}/dt$ and is called the *derivative* of $\mathbf{r}$ with respect to t. We have then

Instantaneous velocity

$$\mathbf{v} = \lim_{\Delta t \to 0} \frac{\Delta\mathbf{r}}{\Delta t} = \frac{d\mathbf{r}}{dt}. \qquad (3\text{–}2)$$

The magnitude v of the instantaneous velocity is called the *speed* and is simply the absolute value of $\mathbf{v}$. That is,

$$v = |\mathbf{v}| = \left|\frac{d\mathbf{r}}{dt}\right|. \qquad (3\text{–}3)$$

Speed, being the magnitude of a vector, is intrinsically positive.

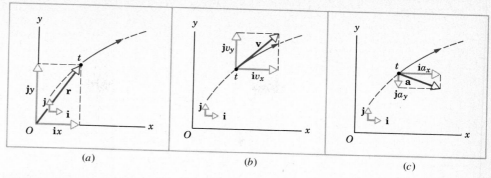

Figure 3–2 A particle at time t has (a) a position described by **r**, (b) an instantaneous velocity **v**, and (c) an instantaneous acceleration **a**. The vector components ix and jy of Eq. 3–4, iv_x and jv_y of Eq. 3–5, and ia_x and ja_y of Eq. 3–10 are also shown, as are the unit vectors **i** and **j**.

3–5 One-Dimensional Motion —Variable Velocity

Figure 3–2 shows a particle moving along a path in the x-y plane. At time t its position with respect to the origin is described by position vector **r** (see Fig. 3–2a) and it has a velocity **v** (see Fig. 3–2b) tangent to its path as shown. We can write

$$\mathbf{r} = \mathbf{i}x + \mathbf{j}y, \qquad (3\text{–}4)$$

where **i** and **j** are unit vectors in the positive x- and y-directions, respectively, and x and y are the (scalar) components of vector **r**. Because **i** and **j** are constant vectors, we have, on combining Eqs. 3–2 and 3–4,

$$\mathbf{v} = \frac{d\mathbf{r}}{dt} = \mathbf{i}\frac{dx}{dt} + \mathbf{j}\frac{dy}{dt},$$

which we can express as

$$\mathbf{v} = \mathbf{i}v_x + \mathbf{j}v_y \qquad \text{(two-dimensional motion)}, \qquad (3\text{–}5)$$

where $v_x (= dx/dt)$ and $v_y (= dy/dt)$ are the (scalar) components of the vector **v**.

We now consider motion in one dimension only, chosen for convenience to be the x-axis. We must then have $v_y = 0$, so that Eq. 3–5 reduces to

$$\mathbf{v} = \mathbf{i}v_x \qquad \text{(one-dimensional motion)}. \qquad (3\text{–}6)$$

Since **i** points in the positive x-direction, v_x will be positive (and equal to $+v$) when **v** points in that direction, and negative (and equal to $-v$) when it points in the other direction. Since, in one-dimensional motion, there are only two choices as to the direction of **v**, the full power of the vector method is not needed, as we have pointed out; we can work with the (scalar) velocity component v_x alone.

Example 1 *The limiting process* As an illustration of the limiting process in one dimension, consider the table shown on p. 30 describing motion along the x-axis. The first four columns are experimental data. The symbols refer to Fig. 3–3, in which the particle is moving from left to right, that is, in the positive x-direction. The particle was at position x_1 (100 cm from the origin) at time t_1 (1.00 s). It was at position x_2 at time t_2. As we con-

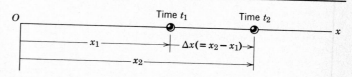

Figure 3–3 A particle is moving to the right along the x-axis.

sider different values for x_2, and different corresponding times t_2, we find

x_1, cm	t_1, s	x_2, cm	t_2, s	$x_2 - x_1$ $= \Delta x$, cm	$t_2 - t_1$ $= \Delta t$, s	$\Delta x/\Delta t$, cm/s
100.0	1.00	200.0	11.00	100.0	10.00	+10.0
100.0	1.00	180.0	9.60	80.0	8.60	+9.3
100.0	1.00	160.0	7.90	60.0	6.90	+8.7
100.0	1.00	140.0	5.90	40.0	4.90	+8.2
100.0	1.00	120.0	3.56	20.0	2.56	+7.8
100.0	1.00	110.0	2.33	10.0	1.33	+7.5
100.0	1.00	105.0	1.69	5.0	0.69	+7.3
100.0	1.00	103.0	1.42	3.0	0.42	+7.1
100.0	1.00	101.0	1.14	1.0	0.14	+7.1

Equation 3–2, which holds for the general case of motion in three dimensions, is

$$\mathbf{v} = \lim_{\Delta t \to 0} \frac{\Delta \mathbf{r}}{\Delta t} = \frac{d\mathbf{r}}{dt}.$$

For one-dimensional motion along the x-axis we have a similar relation, scalar in character, in which each vector quantity is replaced by its corresponding component or

$$v_x = \lim_{\Delta t \to 0} \frac{\Delta x}{\Delta t} = \frac{dx}{dt}. \qquad (3\text{–}7)$$

It is clear from the table that as we select values of x_2 closer to x_1, Δt approaches zero and the ratio $\Delta x/\Delta t$ approaches the apparent limiting value $+7.1$ cm/s. At time t_1, therefore, $v_x = +7.1$ cm/s, as closely as we are able to determine from the data. Since v_x is positive, the velocity $\mathbf{v}$ ($= \mathbf{i}v_x$; see Eq. 3–6) points to the right in Fig. 3–3. This is tangent to the path in the direction of motion, as it must be.

Example 2 Figure 3–4a shows six successive "snapshots" of a particle moving along the x-axis with variable velocity. At $t = 0$ it is at position $x = +1.00$ m measured from the origin; at $t = 2.5$ s it has come to rest at $x = +5.00$ m; at $t = 4.0$ s it has returned to $x = +1.40$ m. Figure 3–4b is a plot of position x versus time t for this motion. The average velocity for the entire 4.0-s interval is the net displacement or change of position ($+0.40$ m) divided by the elapsed time (4.0 s) or $\overline{v_x} = +0.10$ m/s. (We call $\overline{v_x}$ average velocity and v_x velocity, for one-dimensional motion, even though velocity is a vector and not a scalar. This conforms to common usage and should cause no misunderstandings. These quantities are not speeds because they may be negative whereas speed is intrinsically positive.) The average velocity vector $\bar{\mathbf{v}}$ points in the positive x-direction (that is, upward in Fig. 3–4a) because the net displacement points in this direction. The quantity $\overline{v_x}$ can be obtained directly from the slope of the dashed line af in Fig. 3–4b, where by slope we mean the ratio of the net displacement gf to the elapsed time ga. (The slope is *not* the tangent of the angle fag measured on the graph with a protractor. This angle is arbitrary because it depends on the scales we choose for x and t.)

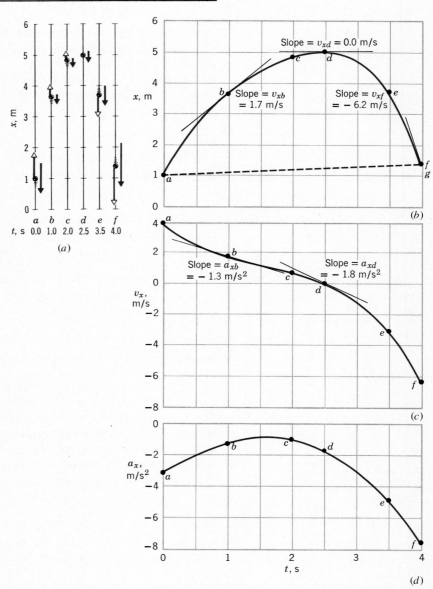

Figure 3–4 (a) Six consecutive "snapshots" of a particle moving along the x-axis. The vector joined to the particle is its instantaneous velocity; that to the right of the particle is its instantaneous acceleration. (b) A plot of x versus t for the motion of the particle. (c) A plot of v_x versus t. (d) A plot of a_x versus t.

The velocity v_x at any instant is found from the slope of the curve of Fig. 3–4b at that instant. Equation 3–7 is in fact the relation by which the slope of the curve is defined in the calculus. In our example the slope at b, which is the value of v_x at b, is $+1.7$ m/s; the slope at d is zero and the slope at f is -6.2 m/s.

When we determine the slope dx/dt at each instant t, we can make a plot of v_x versus t, as in Fig. 3–4c. Note that for the interval $0 < t < 2.5$ s, v_x is positive so that the velocity vector **v** points upward in Fig. 3–4a; for the interval $2.5\,s < t < 4.0\,s\ v_x$ is negative so that **v** points down in Fig. 3–4a.

3–6 Acceleration

Often the velocity of a moving body changes either in magnitude, in direction, or both as the motion proceeds. The body is then said to have an acceleration. *The acceleration of a particle is the rate of change of its velocity with time.* Suppose that at the instant t_1 a particle, as in Fig. 3–5, is at point A and is moving in the x-y plane with the instantaneous velocity $\mathbf{v}_1$, and at a later instant t_2 it is at point B and moving with the instantaneous velocity $\mathbf{v}_2$. The *average acceleration* $\bar{\mathbf{a}}$ during the motion from A to B is defined to be the *change of velocity* divided by the corresponding time interval, or

$$\bar{\mathbf{a}} = \frac{\mathbf{v}_2 - \mathbf{v}_1}{t_2 - t_1} = \frac{\Delta \mathbf{v}}{\Delta t}. \tag{3–8}$$

The quantity $\bar{\mathbf{a}}$ is a vector, for it is obtained by dividing a vector $\Delta \mathbf{v}$ by a scalar Δt. Acceleration is therefore characterized by magnitude and direction. Its direction is the direction of $\Delta \mathbf{v}$ and its magnitude is $|\Delta \mathbf{v}/\Delta t|$. The magnitude of the acceleration is expressed in velocity units divided by time units, as, for example, meters per second per second (written m/s^2 and read "meters per second squared"), cm/sec^2, and ft/sec^2.

We call $\bar{\mathbf{a}}$ of Eq. 3–8 the *average* acceleration because nothing has been said about the time variation of velocity during the interval Δt. We know only the net change in velocity and the total elapsed time. If the change in velocity (a vector) divided by the corresponding elapsed time, $\Delta \mathbf{v}/\Delta t$, were to remain constant, regardless of the time intervals over which we measured the acceleration, we would have *constant* acceleration. Constant acceleration, therefore, implies that the *change* in velocity is uniform with time in direction and magnitude. If there is no change in velocity, that is, if the velocity remains constant both in magnitude and direction, then $\Delta \mathbf{v}$ is zero for all time intervals and the acceleration is zero.

If a particle is moving in such a way that its average acceleration, measured for a number of different time intervals, does not turn out to be constant, the particle is said to have a variable acceleration. The acceleration can vary in magnitude, or in direction, or both. In such cases we seek to determine the acceleration of the particle at any given time, called the instantaneous acceleration.

The *instantaneous acceleration* is defined by

$$\mathbf{a} = \lim_{\Delta t \to 0} \frac{\Delta \mathbf{v}}{\Delta t} = \frac{d\mathbf{v}}{dt}. \tag{3–9}$$

That is, the acceleration of a particle at time t is the limiting value of $\Delta \mathbf{v}/\Delta t$ at time t as both $\Delta \mathbf{v}$ and Δt approach zero. The direction of the instantaneous acceleration **a** is the limiting direction of the vector change in velocity $\Delta \mathbf{v}$. The magnitude a of the instantaneous acceleration is simply $a = |\mathbf{a}| = |d\mathbf{v}/dt|$. When the acceleration is constant, the instantaneous acceleration equals the average acceleration. Note that the relation of **a** to **v** in Eq. 3–9 is the same as that of **v** to **r** in Eq. 3–2.

Two special cases illustrate that acceleration can arise from a change in either the magnitude or the direction of the velocity. In one case we have motion along a straight line with uniformly changing speed (as in Section 3–8). Here the velocity

Average acceleration

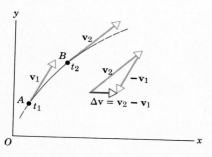

Figure 3–5 A particle has velocity $\mathbf{v}_1$ at point A and moves to point B, where its velocity is $\mathbf{v}_2$. The triangle shows the (vector) change in velocity $\Delta \mathbf{v}$ $(= \mathbf{v}_2 - \mathbf{v}_1)$ experienced by the particle as it moves from A to B.

Instantaneous acceleration

does not change in direction, but its magnitude changes uniformly with time. This is a case of constant acceleration. In the second case we have motion in a circle at constant speed (Section 4–4). Here the velocity vector changes continuously in direction but its magnitude remains constant. This, too, is accelerated motion, though the direction of the acceleration vector is not constant. Later we shall encounter other important cases of accelerated motion.

3–7 One-Dimensional Motion —Variable Acceleration

From Eqs. 3–5 and 3–9 we can write, for motion in two dimensions as in Fig. 3–2c,

$$\mathbf{a} = \frac{d\mathbf{v}}{dt} = \mathbf{i}\frac{dv_x}{dt} + \mathbf{j}\frac{dv_y}{dt}$$

or

$$\mathbf{a} = \mathbf{i}a_x + \mathbf{j}a_y, \tag{3–10}$$

where $a_x\,(=dv_x/dt)$ and $a_y\,(=dv_y/dt)$ are the (scalar) components of the acceleration vector $\mathbf{a}$.

We again restrict ourselves to motion in one dimension only, chosen for convenience to be the x-axis. Since v_y for such motion does not change with time (and is, in fact, zero), a_y, which is dv_y/dt, must also be zero, so that

Acceleration in one dimension

$$\mathbf{a} = \mathbf{i}a_x. \tag{3–11}$$

Since $\mathbf{i}$ points in the positive x-direction, a_x will be positive when $\mathbf{a}$ points in this direction and negative when it points in the other direction.

Example 3 The motion of Fig. 3–4a is one of variable acceleration along the x-axis. To find the acceleration a_x at each instant we must determine dv_x/dt at each instant. This is simply the slope of the curve of v_x versus t at that instant. The slope of Fig. 3–4c at point b is -1.3 m/s² and that at point d is -1.8 m/s², as shown in the figure. The result of calculating the slope for all points is shown in Fig. 3–4d. Notice that a_x is negative at all instants, which means that the acceleration vector $\mathbf{a}$ points in the negative x-direction. This means that v_x is always decreasing with time, as is clearly seen from Fig. 3–4c. The motion is one in which the acceleration vector has a constant direction but varies in magnitude (see Fig. 3–4a).

3–8 One-Dimensional Motion —Constant Acceleration

Let us now further restrict our considerations to motion that not only occurs in one dimension (the x-axis) but for which a_x = a constant. For such constant acceleration the average acceleration for any time interval is equal to the (constant) instantaneous acceleration a_x. Let $t_1 = 0$ and let t_2 be any arbitrary time t. Let v_{x0} be the value of v_x at $t = 0$ and let v_x be its value at the arbitrary time t. With this notation we find a_x (see Eq. 3–8) from

$$a_x = \frac{\Delta v_x}{\Delta t} = \frac{v_x - v_{x0}}{t - 0}$$

or

$$v_x = v_{x0} + a_x t. \tag{3–12}$$

This equation states that the velocity v_x at time t is the sum of its value v_{x0} at time $t = 0$ plus the change in velocity during time t, which is $a_x t$.

Figure 3–6c shows a graph of v_x versus t for constant acceleration; it is a graph of Eq. 3–12. Notice that the slope of the velocity curve is constant, as it must be because the acceleration a_x ($= dv_x/dt$) has been assumed to be constant, as Fig. 3–6d shows.

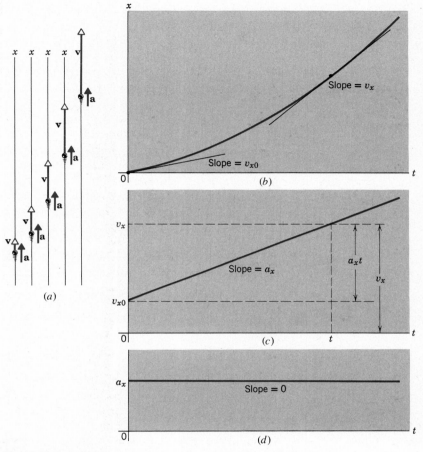

Figure 3–6 (a) Five successive "snapshots" of rectilinear motion with constant acceleration. The arrows on the spheres represent **v**; those to the right represent **a**. (b) The displacement increases according to $x = v_{x0}t + \frac{1}{2}a_x t^2$. Its slope increase uniformly and at each instant has the value v_x, the velocity. (c) The velocity v_x increases uniformly according to $v_x = v_{x0} + a_x t$. Its slope is constant and at each instant has the value a_x, the acceleration. (d) The acceleration a_x has a constant value; its slope is zero. Figure 3–4 shows similar plots for one-dimensional motion in which the acceleration is *not* constant.

When the velocity v_x changes uniformly with time, its average value over any time interval equals one-half the sum of the values of v_x at the beginning and at the end of the interval. That is, the average velocity $\overline{v_x}$ between $t = 0$ and $t = t$ is

$$\overline{v_x} = \tfrac{1}{2}(v_{x0} + v_x). \tag{3–13}$$

This relation would not be true if the acceleration were not constant, for then the curve of v_x versus t would not be a straight line.

If the position of the particle at $t = 0$ is x_0, the position x at $t = t$ can be found from

$$x = x_0 + \bar{v}_x t.$$

which can be combined with Eq. 3–13 to yield

$$x = x_0 + \tfrac{1}{2}(v_{x0} + v_x)t. \qquad (3\text{–}14)$$

The displacement due to the motion in time t is $x - x_0$. Often the origin is chosen so that $x_0 = 0$.

Notice that aside from initial conditions of the motion, that is, the values of x and v_x at $t = 0$ (taken here as $x = x_0$ and $v_x = v_{x0}$), there are four parameters of the motion. These are x, the displacement; v_x, the velocity; a_x, the acceleration; and t, the elapsed time. If we know only that the acceleration is constant, but not necessarily its value, from any two of these parameters we can obtain the other two. For example, if a_x and t are known, Eq. 3–12 gives v_x, and having obtained v_x, we find x from Eq. 3–14.

In most problems in uniformly accelerated motion, two parameters are known and a third is sought. It is convenient, therefore, to obtain relations among any three of the four parameters. Equation 3–12 contains v_x, a_x, and t, but *not* x; Eq. 3–14 contains x, v_x, and t, but *not* a_x. To complete our system of equations we need two more relations, one containing x, a_x, and t but *not* v_x and another containing x, v_x, and a_x but *not* t. These are easily obtained by combining Eqs. 3–12 and 3–14.

Thus, if we substitute into Eq. 3–14 the value of v_x from Eq. 3–12, we thereby eliminate v_x and obtain

$$x = x_0 + v_{x0}t + \tfrac{1}{2}a_x t^2. \qquad (3\text{–}15)$$

When Eq. 3–12 is solved for t and this value for t is substituted into Eq. 3–14, we obtain

$$v_x{}^2 = v_{x0}{}^2 + 2a_x(x - x_0). \qquad (3\text{–}16)$$

Equations 3–12, 3–14, 3–15, and 3–16 (see Table 3–1) are the complete set of equations for motion along a straight line with constant acceleration.

Constant acceleration equations

A special case of motion with constant acceleration is one in which the acceleration is zero, that is, $a_x = 0$. In this case the four equations in Table 3–1 reduce to the expected results $v_x = v_{x0}$ (the velocity does not change) and $x = x_0 + v_{x0}t$ (the displacement changes linearly with time).

Table 3–1 Kinematic equations for straight-line motion with constant acceleration

(The position x_0 and the velocity v_{x0} at the initial instant $t = 0$ are the given initial conditions)

Equation number	Equation	x	v_x	a_x	t
3-12	$v_x = v_{x0} + a_x t$	×	✓	✓	✓
3-14	$x = x_0 + \tfrac{1}{2}(v_{x0} + v_x)t$	✓	✓	×	✓
3-15	$x = x_0 + v_{x0}t + \tfrac{1}{2}a_x t^2$	✓	×	✓	✓
3-16	$v_x{}^2 = v_{x0}{}^2 + 2a_x(x - x_0)$	✓	✓	✓	×

Example 4 Suppose that we wish to find the speed of a particle that has a uniform acceleration of 5.00 cm/s² for an interval of 0.50 h if the particle has a speed of 10.0 ft/s at the beginning of this interval. We decide to choose the foot as our length unit and the second as our time unit. Then

$$a_x = 5.00 \text{ cm/s}^2 = 5.00 \text{ cm/s}^2 \times \left(\frac{1 \text{ in.}}{2.54 \text{ cm}}\right)$$

$$\times \left(\frac{1 \text{ ft}}{12 \text{ in.}}\right) = 0.164 \text{ ft/s}^2.$$

The time interval

$$\Delta t = t - t_0 = 0.50\,\cancel{hr} \times \left(\frac{60\,\cancel{min}}{1\,\cancel{hr}}\right) \times \left(\frac{60\,s}{1\,\cancel{min}}\right) = 1800\,s.$$

Note that, as in Example 1 of Chapter 1, the conversion factors in large parentheses are equal to unity. Taking the initial time $t_0 = 0$, as in Eq. 3–12, we then have

$$v_x = v_{x0} + a_x t = 10.0\,ft/s + (0.164\,ft/s^2)(1800\,s) = 305\,ft/s.$$

Example 5   The nucleus of a helium atom (an alpha particle) travels along the inside of a straight hollow tube 2.0 m long which forms part of a particle accelerator. (a) If one assumes uniform acceleration, how long is the particle in the tube if it enters at a speed of 1.0×10^4 m/s and leaves at 5.0×10^6 m/s? (b) What is its acceleration during this interval?

(a) We choose an x-axis parallel to the tube, its positive direction being that in which the particle is moving and its origin at the tube entrance. We are given x and v_x and we seek t. The acceleration a_x is not involved. Hence we use Eq. 3–14 of Table 3–1, $x = x_0 + \frac{1}{2}(v_{x0} + v_x)\,t$, with $x_0 = 0$, or

$$t = \frac{2x}{v_{x0} + v_x}$$

$$= \frac{(2)(2.0\,m)}{(500 + 1) \times 10^4\,m/s} = 8.0 \times 10^{-7}\,s,$$

or 0.80 microseconds ($= 0.80\,\mu s$).

(b) The acceleration follows from Eq. 3–12, $v_x = v_{x0} + a_x t$, or

$$a_x = \frac{v_x - v_{x0}}{t} = \frac{(500 - 1) \times 10^4\,m/s}{8.0 \times 10^{-7}\,s} = +6.2 \times 10^{12}\,m/s^2.$$

Although this acceleration is enormous by ordinary standards, it occurs over an extremely short time. The acceleration **a** is in the positive x-direction, that is, in the direction in which the particle is moving, because a_x is positive.

Example 6   The speed of an automobile traveling due east is uniformly reduced from 75.0 km/h ($= 46.6$ mi/h) to 45.0 km/h ($= 28.0$ mi/h) in a distance of 88.5 m ($= 290$ ft).

(a) What is the magnitude and direction of the constant acceleration?

We choose, arbitrarily, the direction from west to east to be the positive x-direction. We are given x and v_x and we seek a_x. The time is not involved. Equation 3–16 is therefore appropriate (see Table 3–1). We have $v_x = +45.0$ km/h, $v_{x0} = +75.0$ km/h, and $x - x_0 = +88.5$ m $= 0.0885$ km. From Eq. 3–16 we obtain

$$a_x = \frac{v_x^2 - v_{x0}^2}{2(x - x_0)}$$

or

$$a_x = \frac{(45.0\,km/h)^2 - (75.0\,km/h)^2}{2(0.0885\,km)}$$

$$= -2.03 \times 10^4\,km/h^2 = -1.57\,m/s^2\ (= -5.15\,ft/s^2).$$

The direction of the acceleration **a** is due west, that is, the negative x-direction, because a_x is negative. The car is slowing down as it moves eastward, as it must do if it is being accelerated toward the west. When the speed of a body is decreasing, we often say that the body is *decelerating*.

(b) How much time elapses during this deceleration?

If we use only the original data, Table 3–1 shows that Eq. 3–14 is appropriate. This yields

$$t = \frac{2(x - x_0)}{v_{x0} + v_x}$$

or

$$t = \frac{(2)(0.0885\,km)}{(75.0 + 45.0)\,km/h} = 1.48 \times 10^{-3}\,h = 5.31\,s.$$

If we use the derived data of part (a), Eq. 3–12 is appropriate. This gives us a check on our answer. From Eq. 3–12 we have

$$t = \frac{v_x - v_{x0}}{a_x}$$

or

$$t = \frac{(45.0 - 75.0)\,km/h}{-2.03 \times 10^4\,km/h^2} = 1.48 \times 10^{-3}\,h = 5.31\,s.$$

(c) If we assume that the car continues to decelerate at the rate calculated in (a), how much time would elapse in bringing it to rest from 75.0 km/h?

Equation 3–12 is useful here. We have $v_{x0} = +75.0$ km/h, $a_x = -2.03 \times 10^4$ km/h², and the final velocity $v_x = 0$. Thus

$$t = \frac{v_x - v_{x0}}{a_x}$$

$$= \frac{0 - (75.0\,km/h)}{(-2.03 \times 10^4\,km/h^2)} = 3.69 \times 10^{-3}\,h = 13.3\,s.$$

Then, from Eq. 3–15, we can find the corresponding displacement, or

$$x - x_0 = v_{x0}\,t + \frac{1}{2}a_x t^2$$

$$= (75.0\,km/h)(3.69 \times 10^{-3}\,h)$$

$$+ \frac{1}{2}(-2.03 \times 10^4\,km/h^2)(3.69 \times 10^{-3}\,h)^2$$

$$= 0.139\,km = 139\,m\ (= 454\,ft).$$

3-9 Consistency of Units and Dimensions

You should not feel compelled to memorize relations such as those of Table 3–1. The important thing is to be able to follow the line of reasoning used to obtain the results. These relations will be recalled automatically after you have used them repeatedly to solve problems, partly as a result of the familiarity acquired, but chiefly as a result of the better understanding obtained through application.

We can use any convenient units of time and distance in these equations. If we choose to express time in seconds and distance in meters, for self-consistency we must express velocity in meters per second (m/s) and acceleration in meters per second per second (m/s²). If we are given data in which the units of one quantity, such as velocity, are not consistent with the units of another quantity, such as acceleration, then before using the data in our equations we should transform both quantities to units that are consistent with one another. Having chosen the units of our fundamental quantities, we automatically determine the units of our derived quantities consistent with them. In carrying out any calculation, always remember to attach the proper units to the final result, for the result is meaningless without this label.

Check the dimensions

One way to spot an erroneous equation is to check the *dimensions* of all its terms. The dimensions of any physical quantity can always be expressed as some combination of the fundamental quantities, such as mass, length, and time, from which they are derived. The dimensions of velocity are length (L) divided by time (T); the dimensions of acceleration are length divided by time squared, and so on. *In any legitimate physical equation the dimensions of all the terms must be the same.* This means, for example, that we cannot equate a term whose total dimension is a velocity to one whose total dimension is an acceleration. The dimensional labels attached to various quantities may be treated just like algebraic quantities and may be combined, canceled, and so on, just as if they were factors in the equation. For example, to check Eq. 3.15, $x = x_0 + v_{x0}t + \frac{1}{2}a_x t^2$, dimensionally, we note that x and x_0 have the dimension of a length. Therefore the two remaining terms must also have the dimension of a length. The dimension of the term $v_{x0}t$ is

$$\frac{\text{length}}{\text{time}} \times \text{time} = \text{length} \qquad \text{or} \qquad \frac{L}{T} \times T = L,$$

and that of $\frac{1}{2}a_x t^2$ is

$$\frac{\text{length}}{\text{time}^2} \times \text{time}^2 = \text{length} \qquad \text{or} \qquad \frac{L}{T^2} \times T^2 = L.$$

The equation is therefore dimensionally correct. You should check the dimensions of all the equations you use.

3–10 Freely Falling Bodies*

The most common example of motion with (nearly) constant acceleration is that of a body falling toward the earth. In the absence of air resistance we find that all bodies, regardless of their size, weight, or composition, fall with the same acceleration at the same point on the earth's surface, and if the distance covered is not too great, the acceleration remains constant throughout the fall. This ideal motion, in which air resistance and the small change in acceleration with altitude are neglected, is called "free fall."

The acceleration of a freely falling body is called the acceleration due to gravity and is denoted by the symbol **g**. Near the earth's surface its magnitude is approximately 9.8 m/s², which is 980 cm/s², or 32 ft/s², and it is directed down

---

* See "Galileo's Discovery of the Law of Free Fall," by Stillman Drake, *Scientific American*, May 1973.

toward the center of the earth. The variation of the exact value with latitude and altitude will be discussed later (Chapter 15).

We shall select a reference frame rigidly attached to the earth. The y-axis will be taken as positive vertically upward. Then the acceleration due to gravity **g** will be a vector pointing vertically down (toward the center of the earth) in the negative y-direction. (This choice is arbitrary. In other problems it may be convenient to choose down as positive.) Our equations for constant acceleration are applicable here. We simply replace x by y and set $y_0 = 0$ in Eqs. 3-12, 3-14, 3-15, and 3-16, obtaining

Free-fall equations

$$v_y = v_{y0} + a_y t,$$
$$y = \tfrac{1}{2}(v_{y0} + v_y)t,$$
$$y = v_{y0}t + \tfrac{1}{2}a_y t^2,$$
$$v_y{}^2 = v_{y0}{}^2 + 2a_y y,$$

(3-17)

and, for problems in free fall, we set $a_y = -g$. Notice that we have chosen the initial position as the origin, that is, we have chosen $y_0 = 0$ at $t = 0$. Note also that g is the magnitude of the free-fall acceleration.

Example 7 A body is dropped from rest and falls freely. Determine the position and speed of the body after 1.0, 2.0, 3.0, and 4.0 s have elapsed.

We choose the starting point as the origin. We know the initial speed and the acceleration and we are given the time. To find the position we use

$$y = v_{y0}t - \tfrac{1}{2}gt^2.$$

Then, $v_{y0} = 0$ and $g = 32$ ft/s^2, and with $t = 1.0$ s we obtain

$$y = 0 - \tfrac{1}{2}(32 \text{ ft/s}^2)(1.0 \text{ s})^2 = -16 \text{ ft}.$$

To find the speed with $t = 1.0$ s, we use

$$v_y = v_{y0} - gt$$

and obtain

$$v_y = 0 - (32 \text{ ft/s}^2)(1.0 \text{ s}) = -32 \text{ ft/s}.$$

After 1.0 s of falling from rest, the body is 16 ft (=4.9 m) below its starting point and has a velocity directed downward whose magnitude is 32 ft/s (=9.8 m/s); the negative signs for y and v_y show that the associated vectors each point in the negative y-direction, that is, downward.

Show that the values of y, v_y, and a_y obtained at times $t = 2.0$, 3.0, and 4.0 s are those shown in Fig. 3-7 and determine the metric equivalents.

Example 8 A ball is thrown vertically upward from the ground with a speed of 80 ft/s (=24.4 m/s).

(a) How long does it take to reach its highest point?

At its highest point, $v_y = 0$, and we have $v_{y0} = +80$ ft/s. To obtain the time t we use $v_y = v_{y0} - gt$, or

$$t = \frac{v_{y0} - v_y}{g}$$
$$= \frac{(80 - 0) \text{ ft/s}}{32 \text{ ft/s}^2} = 2.5 \text{ s}.$$

(b) How high does the ball rise?

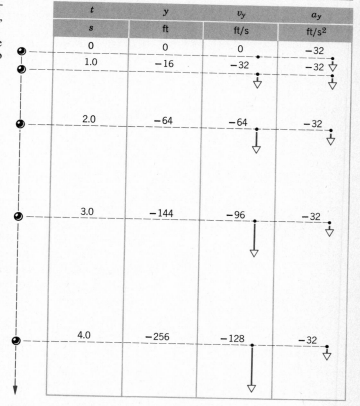

t	y	v_y	a_y
s	ft	ft/s	ft/s^2
0	0	0	-32
1.0	-16	-32	-32
2.0	-64	-64	-32
3.0	-144	-96	-32
4.0	-256	-128	-32

Figure 3-7 A body in free fall; showing y, v_y, and a_y at particular times t.

Using only the original data, we choose the relation $v_y{}^2 = v_{y0}{}^2 - 2gy$, or

$$y = \frac{v_{y0}{}^2 - v_y{}^2}{2g}$$

$$= \frac{(80 \text{ ft/s})^2 - 0}{2 \times 32 \text{ ft/s}^2} = +100 \text{ ft } (=30.5 \text{ m}).$$

(c) At what times will the ball be 96 ft ($=29$ m) above the ground?

Using $y = v_{y0}t - \frac{1}{2}gt^2$, we have

$$\frac{1}{2}gt^2 - v_{y0}t + y = 0,$$

$$\frac{1}{2}(32 \text{ ft/s}^2)t^2 - (80 \text{ ft/s})t + 96 \text{ ft} = 0,$$

or

$$t^2 - 5.0t + 6.0 = 0,$$

which yields $t = 2.0$ s and $t = 3.0$ s.

At $t = 2.0$ s, the ball is moving upward with a speed of 16 ft/s ($=4.9$ m/s), for

$$v_y = v_{y0} - gt = 80 \text{ ft/s} - (32 \text{ ft/s}^2)(2.0 \text{ s}) = +16 \text{ ft/s}.$$

At $t = 3.0$ s, the ball is moving downward with the same speed, for

$$v_y = v_{y0} - gt = 80 \text{ ft/s} - (32 \text{ ft/s}^2)(3.0 \text{ s}) = -16 \text{ ft/s}.$$

Notice that in this 1.0-s interval the velocity changed by -32 ft/s ($=-9.8$ m/s), corresponding to an acceleration of -32 ft/s^2 ($=-9.8$ m/s^2).

You should be able to convince yourself that in the absence of air resistance the ball will take as long to rise as to fall the same distance, and that it will have the same speed going down at each point as it had going up.

REVIEW GUIDE AND SUMMARY

The position vector r

The average velocity v̄

The instantaneous velocity v

To describe the motion of a particle we first give its *position vector* **r**, which locates it with respect to the origin of a convenient coordinate system. During any interval Δt this vector will *change* by $\Delta \mathbf{r}$ and we define the *average velocity* **v̄** of the particle for this time interval as $\Delta \mathbf{r}/\Delta t$. If, as in Fig. 3–1, we allow Δt to approach zero, then $\Delta \mathbf{r}$ will also approach zero but their ratio, which is **v̄**, will approach a definite limiting value **v**, the *instantaneous velocity* of the particle at the time in question, or

$$\mathbf{v} = \lim_{\Delta t \to 0} \frac{\Delta \mathbf{r}}{\Delta t} = \frac{d\mathbf{r}}{dt}. \qquad [3-2]$$

Note that **v** is always tangent to the path of the particle.

In one dimension we replace the position vector **r** by its scalar component x and the velocity **v** by *its* scalar component v_x. Equation 3–2 becomes

Velocity in one dimension

$$v_x = \lim_{\Delta t \to 0} \frac{\Delta x}{\Delta t} = \frac{dx}{dt}. \qquad [3-7]$$

Example 1 illustrates this limiting process. Figure 3–4b shows $x(t)$ for a particular assumed motion of the particle, and Fig. 3–4c shows $v_x(t)$ for that same motion. The latter curve is formed by plotting, point by point, the slope (that is, the time derivative) of the former curve.

We now define the *instantaneous acceleration* **a** of a particle as the rate of change of its velocity **v** with time, or (compare Eq. 3–2 above)

The acceleration a

$$\mathbf{a} = \lim_{\Delta t \to 0} \frac{\Delta \mathbf{v}}{\Delta t} = \frac{d\mathbf{v}}{dt}. \qquad [3-9]$$

For one-dimensional motion we describe the acceleration **a** by its scalar component a_x, and Eq. 3–9 becomes $a_x = dv_x/dt$. Figure 3–4d shows $a_x(t)$ for the one-dimensional motion described by Fig. 3–4a. The curve $a_x(t)$ is formed by plotting, point by point, the slope (or time derivative) of the curve of Fig. 3–4c.

Figure 3–6 shows $x(t)$, $v_x(t)$, and $a_x(t)$ for the important special case in which a_x is a constant. In this circumstance the following four equations (see Table 3–1) relate a_x to the variables x, v_x, and t:

Case of constant acceleration

$$v_x = v_{x0} + a_x t,$$

$$x = x_0 + \frac{1}{2}(v_{x0} + v_x)t,$$

$$x = x_0 + v_{x0}t + \frac{1}{2}a_x t^2,$$

and

$$v_x{}^2 = v_{x0}{}^2 + 2a_x(x - x_0).$$

These equations do *not* hold for the more general motion of Fig. 3–4 because here a_x is *not* constant. Examples 5 and 6 show applications.

Falling bodies

An important example of one-dimensional motion with constant acceleration is that of a body falling freely (in a vacuum) near the earth's surface. The above equations hold, but we make these changes in notation: (1) We refer the motion to the y-axis rather than the x-axis, and we chose the $+ y$-direction to be vertically up; (2) we replace a_y by $-g$, where g is the acceleration due to gravity. Near the earth's surface $g = +9.8 \ \mathrm{m/s^2} = +32 \ \mathrm{ft/s^2}$. Examples 7 and 8 illustrate typical falling (or rising) body situations.

QUESTIONS

1. Can you think of physical phenomena involving the earth in which the earth cannot be treated as a particle?

2. Give examples of bodies that can be treated as particles. Can you find some counter-examples?

3. If the average velocity of the particle shown in Fig. 3–1 turned out to be the same (in magnitude and direction) between any two points along the path, what would you conclude about the instantaneous velocity and the shape of the path?

4. Can the speed of a particle ever be negative?

5. Each second a rabbit moves half the remaining distance from his nose to a head of lettuce. Does he ever get to the lettuce? What is the limiting value of his average velocity? Draw graphs showing his velocity and position as time increases.

6. Average speed can mean the magnitude of the average velocity vector. Another meaning given to it is that average speed is the total length of path traveled divided by the elapsed time. Are these meanings different? If so, given an example.

7. A racing driver, in a qualifying two-lap heat, covers the first lap with an average speed of 90 mi/h. He wants to speed up during the second lap so that his average speed for the two laps together will be 180 mi/h. Can you show that he can't do it?

8. Bob beats Judy by 10 m in a 100-m dash. Bob, claiming to give Judy an equal chance, agrees to race her again but to start from 10 m behind the starting line. Does this really give Judy an equal chance?

9. When the velocity is constant, does the average velocity over any time interval differ from the instantaneous velocity at any instant?

10. Is the average velocity of a particle moving along the x-axis $\frac{1}{2}(v_{x0} + v_x)$ when the acceleration is not uniform? Prove your answer with the use of graphs.

11. Does the speedometer on an automobile register speed as we defined it?

12. (*a*) Can a body have zero velocity and still be accelerating? (*b*) Can a body have a constant speed and still have a varying velocity? (*c*) Can a body have a constant velocity and still have a varying speed?

13. Can an object have an eastward velocity while experiencing a westward acceleration?

14. Can the direction of the velocity of a body change when its acceleration is constant?

15. Can a body be increasing in speed as its acceleration decreases? Explain.

Figure 3–8 Question 22. Courtesy Baltimore Office of Promotion and Tourism.

16. Of the following situations, which one is impossible? A body having: (a) velocity east and acceleration east, (b) velocity east and acceleration west, (c) zero velocity but acceleration not zero, (d) constant acceleration and variable velocity, (e) constant velocity and variable acceleration.

17. If a particle is released from rest ($v_{y0} = 0$) at $y_0 = 0$ at the time $t = 0$, Eq. 3–17 for constant acceleration says that it is at position y at two different times, namely, $+\sqrt{2y/a_y}$ and $-\sqrt{2y/a_y}$. What is the meaning of the negative root of this quadratic equation?

18. What are some examples of falling bodies in which it would be unreasonable to neglect air resistance?

19. As an experiment, a skydiver agrees to weigh a pound of butter as she falls, using a spring scale. What will the scale read when she is in free fall, just after clearing the plane? What will it read after her parachute has opened and she is falling at constant terminal speed?

20. On a windless day you drop a pebble over the railing of the Golden Gate bridge and, a moment later, you drop a second pebble. Will the two pebbles maintain a constant vertical separa-

tion as they fall? Consider the cases in which you neglect air resistance and also those in which you take it fully into account.

21. Consider a ball thrown vertically up. Taking air resistance into account, would you expect the time during which the ball rises to be longer or shorter than the time during which it falls?

22. Figure 3–8 shows a shot tower in Baltimore, Maryland. It was built in 1829 and used to manufacture lead shot pellets by pouring molten lead through a sieve at the top of the tower. The lead pellets solidify as they fall into a tank of water at the bottom of the tower, 230 ft below. What is the advantage of manufacturing shot in this way?

23. Can there be motion in two dimensions with an acceleration in only one dimension?

24. A man standing on the edge of a cliff at some height above the ground below throws one ball straight up with initial speed u and then throws another ball straight down with the same initial speed. Which ball, if either, has the larger speed when it hits the ground? Neglect air resistance.

EXERCISES

Section 3–3 Average Velocity
1. How far does a car, moving at 55 mi/h (=88 km/h), travel forward during the 1 s of time that the driver takes to look at an accident on the side on the road?

2. Carl Lewis runs the 100-m dash in about 10.0 s, and Bill Rodgers runs the marathon (26 mi, 385 yd) in about 2 h 10 min. (a) What are their average speeds? (b) If Carl Lewis could maintain his sprint speed during a marathon, how long would it take him to finish?

3. The legal speed limit on a thruway is changed from 65 mi/h (=105 km/h) to 55 mi/h (=88.6 km/h) to conserve fuel. How much time is thereby added to the trip from the Buffalo entrance to the New York City exit of the New York Thruway for someone traveling at the legal speed limit over this 435-mile (=700 km) stretch of highway?

4. Raindrops falling past a mercury-arc lamp at night appear, by a stroboscopic effect, to be "frozen" into a string of stationary streaks 6 cm apart. At what speed are the raindrops falling? Note that raindrops, because of air resistance, quickly reach a constant, terminal, speed. The street lamp operates from a 60-Hz power line, which means that the lamp is at maximum brightness 120 times each second.

Section 3–4 Instantaneous Velocity
5. The coordinate of a particle moving along the x-axis is given in meters by $x = 3.5t - 5.7t^2$, where t is the time in seconds. By calculating its average velocity for each of the intervals, 0 to 1 s, 0 to 0.1 s, 0 to 0.01 s, and 0 to 0.001 s, estimate the particle's instantaneous velocity at $t = 0$.

Section 3–6 Acceleration
6. A particle had a velocity of 18 m/s and 2.4 s later its velocity was 30 m/s in the opposite direction. What was the average acceleration of the particle during this 2.4-s interval?

7. A man stands still from $t = 0$ to $t = 5$ min; from $t = 5$ min to $t = 10$ min he walks briskly in a straight line at a constant speed of

2 m/s. What are his average velocity and average acceleration during the time intervals (a) 2 min to 8 min, and (b) 3 min to 9 min?

Section 3–7 One-Dimensional Motion — Variable Acceleration
8. The graph of x versus t (see Fig. 3–9a) is for a particle in straight line motion. (a) State for each interval whether the velocity v_x is $+$, $-$, or 0, and whether the acceleration a_x is $+$, $-$, or 0. The intervals are OA, AB, BC, and CD. (b) From the curve, is there any interval over which the acceleration is obviously not constant? (Ignore the behavior at the end points of the intervals.)

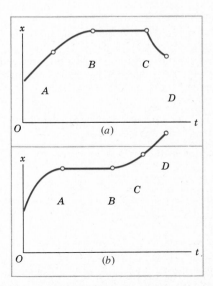

Figure 3–9 Problems 8 and 9.

9. Answer the previous questions for the motion described by the graph of Fig. 3–9b.

10. For each of the following situations, sketch a graph that is a possible description of position as a function of time for a particle that moves along the x-axis. At $t = 1$ s, the particle has (a) zero velocity and positive acceleration; (b) zero velocity and negative acceleration; (c) negative velocity and positive acceleration; (d) negative velocity and negative acceleration. (e) For which of these situations is the speed of the particle increasing at $t = 1$ s?

Section 3–8 One-Dimensional Motion—Constant Acceleration

11. An automobile increases its speed uniformly from 25 to 55 km/h in one-half minute. A bicycle rider uniformly speeds up to 30 km/h from rest in one-half minute. Compare the accelerations.

12. The brakes on a certain automobile are capable of creating a deceleration of 17 ft/s². If a driver going 85 mi/h suddenly sees a state trooper, what is the minimum time in which he can get his car under the 55-mi/h speed limit?

13. A rocket-driven sled running on a straight level track is used to investigate the physiological effects of large accelerations on humans. One such sled can attain a speed of 1600 km/h in 1.8 s starting from rest. (a) Assume the acceleration is constant and compare it to g. (b) What is the distance traveled in this time?

14. On a dry road a car with good tires may be able to brake with a deceleration of 11.0 mph/s (=4.92 m/s²). (a) How long does such a car, initially traveling at 55 mi/h (=24.6 m/s) take to come to rest? (b) How far does it travel in this time?

15. A rocketship in free space moves with constant acceleration equal to 9.8 m/s². (a) If it starts from rest, how long will it take to acquire a speed one-tenth that of light? (b) How far will it travel in so doing?

16. The world's land speed record is held by Col. John P. Stapp. On March 19, 1954, he rode a rocket-propelled sled that moved down a track at 632 mi/h. He and the sled were brought to a stop in 1.4 s. See Fig. 3–10. What acceleration did he experience? Express your answer in terms of g, the acceleration due to gravity.

17. The Owner's Manual for a 1985 Toyota Cressida says that, if lightly loaded, the car can be braked to a halt from 60 mi/h on dry roads in 43 m. (a) What acceleration does this imply? Express both in SI units and in "g" units. (b) What is the stopping time? If your reaction time T for braking is 400 ms, how many "reaction times" does this correspond to?

18. The head of a rattlesnake can accelerate 50 m/s² in striking a victim. If a car could do as well, how long would it take for it to reach a speed of 60 mi/h from rest?

Section 3–9 Consistency of Units and Dimensions

19. Consider the two quantities $(dx/dt)^2$ and d^2x/dt^2. (a) Are these merely two equivalent expressions for the same thing? (b) What are the SI units of these two quantities?

20. Using the tables in Appendix E and following the dimensions carefully, find the speed of light (=3×10^8 m/s) in miles per hour, feet per second, and light-years per year.

Figure 3–10 Exercise 16. U.S. Air Force Photos/UPI-Bettmann.

Section 3–10 Freely Falling Bodies

21. A diver steps off a cliff in Acapulco into the Pacific Ocean. He falls through the air for 3.12 s. (a) How high is the cliff? (b) With what speed does he enter the water?

22. Raindrops fall to earth from a cloud 5000 ft above the earth's surface. If they were not slowed by air resistance, how fast would the drops be moving when they struck the ground? Would it be safe to walk outside during a rainstorm?

23. A shell is fired directly up from a gun and a rocket, propelled by burning fuel, takes off vertically from a launching area. Plot qualitatively (numbers not required) possible graphs of a_y versus t, of v_y versus t, and of y versus t for each. Take $t = 0$ at the instant the shell leaves the gun barrel or the rocket leaves the ground. Continue the plots until the rocket and the shell fall back to earth; neglect air resistance; assume that up is positive and down is negative.

24. A ball thrown straight up takes 2.25 s to reach a height of 36.8 m. (a) What was its initial speed? (b) What is its speed at this height? (c) How much higher will the ball go?

PROBLEMS GROUPED BY SECTION

Section 3-3   Average Velocity

1. An automobile travels on a straight road for 40 km at 30 km/h. It then continues in the same direction for another 40 km at 60 km/h. What is the average speed of the car during this 80-km trip?

2. Compare your average speed in the following two cases. (*a*) You walk 240 ft at a speed of 4.0 ft/s and then run 240 ft at a speed of 10 ft/s along a straight track. (*b*) You walk for 1.0 min at a speed of 4.0 ft/s and then run for 1.0 min at 10 ft/s along a straight track.

3. In 3.5 h, a balloon drifts 21.5 km north, 9.70 km east, and 2.88 km in elevation from its release point on the ground. Find (*a*) the magnitude of its average velocity and (*b*) the angle its average velocity makes with the horizontal.

4. A train moving at an essentially constant speed of 60 km/h moves east for 40 min, then in a direction 45° east of north for 20 min, and finally west for 50 min. What is the average velocity of the train during this run?

5. The airport terminal in Geneva has a "moving sidewalk" to speed passengers through a long corridor. Peter, who walks through the corridor but does not use the sidewalk, takes 150 s to do so. Paul, who simply stands on the moving sidewalk, covers the same distance in 70 s. Mary not only uses the sidewalk but walks along it. How long does Mary take? Assume that Peter and Mary walk at the same speed.

6. The position of an object moving in a straight line is given by $x = 3t - 4t^2 + t^3$, where x is in meters and t in seconds. (*a*) What is the position of the object at $t = 1, 2, 3,$ and 4 s? (*b*) What is the object's displacement between $t = 0$ and $t = 4$ s? (*c*) What is the average velocity for the time interval from $t = 2$ to $t = 4$ s?

Section 3-4   Instantaneous Velocity

7. Construct a graph of x versus t for the object of Problem 6 during the time interval from $t = 0$ to $t = 4$ s. From your graph, estimate (*a*) the initial velocity of the object and (*b*) the instantaneous velocities at $t = 1, 2, 3,$ and 4 s.

8. Two trains, each having a speed of 30 km/h, are headed at each other on the same straight track. A bird that can fly 60 km/h flies off the front of one train when they are 60 km apart and heads directly for the other train. On reaching the other train it flies directly back to the first train, and so forth. (*a*) How many trips can the bird make from one train to the other before they crash? (*b*) What is the total distance the bird travels?

Section 3-6   Acceleration

9. A particle moving along the positive x-axis has the following positions at various times:

x(m)	0.080	0.040	0.010	0.050	0.080	0.13	0.20
t(s)	0.0	1.0	2.0	3.0	4.0	5.0	6.0

(*a*) Plot displacement (not position) versus time. (*b*) Find the average velocity of the particle in the intervals 0.0 to 1.0 s, 0.0 to 2.0 s, 0.0 to 3.0 s, 0.0 to 4.0 s. (*c*) Find the slope of the curve drawn in part *a* at the points $t = 0.0, 1.0, 2.0, 3.0, 4.0,$ and 5.0 s. (*d*) Plot the slope (units?) versus time. (*e*) From the curve of part (*d*), determine the acceleration of the particle at times $t = 2.0, 3.0$ and 4.0 s.

Section 3-7   One-Dimensional Motion—Variable Acceleration

10. A particle moves along the x-axis with a displacement versus time as shown in Fig. 3-11. Sketch roughly curves of velocity versus time and acceleration versus time for this motion.

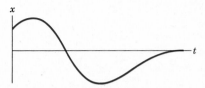

Figure 3-11   Problem 10.

11. A particle moves along the x-axis according to the equation $x = 50t + 10t^2$ where x is in meters and t is in seconds. Calculate (*a*) the average velocity of the particle during the first 3 s of its motion, (*b*) the instantaneous velocity of the particle at $t = 3.0$ s, and (*c*) the instantaneous acceleration of the particle at $t = 3.0$ s.

12. If the position of an object is given by $x = 2t^3$, where x is measured in meters and t in seconds, find: (*a*) the average velocity and the average acceleration between $t = 1$ s and $t = 2$ s, (*b*) the instantaneous velocities and the instantaneous accelerations at $t = 1$ s and $t = 2$ s. (*c*) Compare the average and instantaneous quantities and in each case explain why the larger one is larger.

Section 3-8   One-Dimensional Motion—Constant Acceleration

13. A muon (an elementary particle) is shot with constant speed 5.00×10^6 m/s into a region where an electric field produces an acceleration of the muon of 1.25×10^{14} m/s^2 directed opposite to the initial velocity. How far does the muon travel before coming to rest?

14. A certain drag racer can accelerate from 0 to 60 mi/h in 5.0 s. (*a*) What is its average acceleration, in g's, during this time? (*b*) How far will it travel during the 5.0 s, assuming its acceleration to be constant? (*c*) How much time would be required to go a distance of 0.25 mi if the acceleration could be maintained at the same value?

15. An electron with initial velocity $v_0 = 1.0 \times 10^4$ m/s enters a region 1.0 cm long where it is electrically accelerated (Fig. 3-12). It emerges with a velocity $v = 4.0 \times 10^6$ m/s. What was its acceleration, assumed constant? (Such a process occurs in the electron

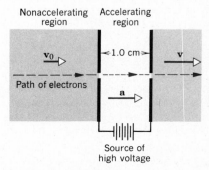

Figure 3-12   Problem 15.

gun in a cathode-ray tube, used in television receivers and computer terminals.

16. A subway train accelerates from rest at one station at a rate of 1.2 m/s² for half of the distance to the next station, then decelerates at this same rate for the final half. If the stations are 1100 m apart, find (a) the time of travel between stations and (b) the maximum speed of the train.

17. A car moving with constant acceleration covers the distance between two points 60 m apart in 6.0 s. Its speed as it passes the second point is 15 m/s. (a) What is its speed at the first point? (b) What is its acceleration? (c) At what prior distance from the first point was the car at rest?

18. At the instant the traffic light turns green, an automobile starts with a constant acceleration a_x of 6.0 ft/s². At the same instant a truck, traveling with a constant speed of 30 ft/s, overtakes and passes the automobile. (a) How far beyond the starting point will the automobile overtake the truck? (b) How fast will the car be traveling at that instant? (It is instructive to plot a qualitative graph of x versus t for each vehicle.)

19. A particle is moving in the direction of the positive x-axis with a constant speed of 3.0 m/s. As it passes the origin a second particle, initially at rest at x = 10 m, begins accelerating toward the origin with a constant acceleration whose magnitude is 2.0 m/s². (a) Where and when do the two particles meet? (b) What are the velocities of the two particles just before they collide?

20. A fast train running at 60 mi/h (= 97 km/h) rounds a curve onto a straightaway. The engineer observes, 200 ft ahead, a slower train running in the same direction on the same track at 30 mi/h (= 48 km/h). The engineer instantly applies his brakes. What must be the resulting constant acceleration if a collision is to be just avoided?

Section 3-9 Consistency of Units and Dimensions

21. The position of a particle moving along the x-axis depends on the time according to the equation

$$x = at^2 - bt^3,$$

where x is in feet and t in seconds. (a) What dimensions and units must a and b have? For the following, let their numerical values be 3 and 1 respectively. (b) At what time does the particle reach its maximum positive x-position? (c) What total length of path does the particle cover in the first 4 s? (d) What is its displacement during the first 4 s? (e) What is the particle's speed at the end of each of the first four seconds? (f) What is the particle's acceleration at the end of each of the first four seconds?

Section 3-10 Freely Falling Bodies

22. (a) With what speed must a ball be thrown vertically up in order to rise to a height of 50 ft? (b) How long will it be in the air?

23. A rock is dropped from a 100-ft high cliff. How long does it take to fall (a) the first 50 ft, and (b) the second 50 ft?

24. A rocket is fired vertically and ascends with a constant vertical acceleration of 20 m/s² (= 64 ft/s²) for 1 min. Its fuel is then all used and it continues as a free-fall particle. (a) What is the maximum altitude reached? (b) What is the total time elapsed from take-off until the rocket strikes the earth?

25. An object falls from a bridge that is 45 m above the water. It falls directly into a boat moving with constant velocity that was 12 m from the point of impact when the object was released. What was the speed of the boat?

26. An arrow is shot straight up in the air with an initial speed of 250 ft/s. If on striking the ground it imbeds itself 6.0 in. into the ground, find (a) the acceleration (assumed constant) required to stop the arrow and (b) the time required for it to come to rest.

27. A ball of clay falls to the ground from a height of 15 m. It is in contact with the ground for 0.02 s before coming to rest. What is the average acceleration of the clay during the time it is in contact with the ground?

28. A ball is thrown down vertically with an initial speed of 20 m/s from a height of 60 m. (a) What will be its speed just before it strikes the ground? (b) How long will it take for the ball to reach the ground? (c) What would be the answers to (a) and (b) if the ball were thrown directly up from the same height and with the same initial speed?

29. A balloon is ascending at the rate of 12 m/s at a height 80 m above the ground when a package is dropped. (a) How long does it take the package to reach the ground? (b) With what speed does it hit the ground?

ADDITIONAL PROBLEMS

30. A jumbo jet needs to reach a speed of 360 km/h (= 225 mi/h) on the runway for takeoff. Assuming a constant acceleration and a runway 1.8 km (= 1.1 mi) long, what minimum acceleration from rest is required?

31. You drive on Interstate 10 from San Antonio to Houston, half the time at 35.0 mi/h (= 56.3 km/h) and the other half at 55.0 mi/h (= 88.5 km/h). On the way back you travel half the *distance* at 35.0 mi/h and the other half at 55.0 mi/h. (a) What is your average speed from San Antonio to Houston? (b) from Houston back to San Antonio? (c) for the entire trip?

32. Figure 3–13 shows a simple device for measuring your reaction time. It consists of a strip of cardboard marked with a scale and two large dots. A friend holds the strip with his thumb and forefinger at the upper dot and you position your thumb and forefinger at the lower dot, being careful not to touch the strip.

Your friend releases the strip, and you try to pinch it as soon as possible after you see it begin to fall. The mark at the place where you pinch the strip gives your reaction time. How far from the lower dot should you place the 50-ms, 100-ms, 150-ms, 200-ms, and 250-ms marks?

33. An automobile traveling 35 mi/h (= 56 km/h) is 110 ft (= 35 m) from a barrier when the driver slams on the brakes. Four seconds later the car hits the barrier. (a) What was the automobile's deceleration before impact? (b) How fast was the car traveling at impact?

34. A juggler tosses balls vertically 1.00 m into the air. How much higher must he toss them if they are to spend twice as much time in the air?

35. A stone is thrown vertically upward. On its way up it passes

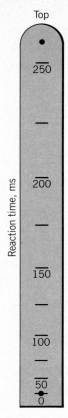

Top

Reaction time, ms

250

200

150

100

50

0

Figure 3–13   Problem 32.

point A with speed V, and point B, 3 m higher than A, with speed $\frac{1}{2}V$. Calculate (a) the speed V and (b) the maximum height reached by the stone above point B.

36. A tennis ball is dropped onto the floor from a height of 4.0 ft. It rebounds to a height of 3.0 ft. If the ball was in contact with the floor for 0.010 s, what was its average acceleration during contact?

37. (a) If the maximum acceleration that would be tolerable for passengers in a subway train is 3.0 mph/s and subway stations are located one-half mile apart, what is the maximum speed a subway train could attain in this distance? (b) What would be the time between stations? (c) If the subway train stops for a 20-s interval at each station, what is the maximum average speed of a subway train?

38. Water drips from the nozzle of a shower onto the floor 200 cm below. The drops fall at regular intervals of time, the first drop striking the floor at the instant the fourth drop begins to fall. Find the location of the individual drops when a drop strikes the floor.

39. The position of a particle moving along the x-axis is given in centimeters by $x = 9.75 + 1.50t^3$, where t is in seconds. Consider the time interval $t = 2$ s to $t = 3$ s and calculate (a) the average velocity; (b) the instantaneous velocity at $t = 2$ s; (c) the instantaneous velocity at $t = 3$ s; (d) the instantaneous velocity at $t = 2.5$ s; (e) the instantaneous velocity when the particle is midway between its positions at $t = 2$ s and $t = 3$ s.

40. In an arcade video game a spot is programmed to move across the screen according to $x = 9.00t - 0.750t^3$, where x is in centimeters measured from the left edge of the screen and t is time

in seconds. When the spot reaches a screen edge, at either $x = 0$ or $x = 15.0$ cm, it starts over. (a) At what time after starting is the spot instantaneously at rest? (b) Where does this occur? (c) What is its acceleration when this occurs? (d) In which direction does it move in the next instant after coming to rest? (e) When does it move off the screen?

41. A lead ball is dropped into a lake from a diving board 5.0 m ($= 16$ ft) above the water. It hits the water with a certain velocity and then sinks to the bottom with this same constant velocity. It reaches the bottom 5.0 s after it is dropped. (a) How deep is the lake? (b) What is the average velocity of the ball? (c) Suppose that all the water is drained from the lake. The ball is thrown from the diving board so that it again reaches the bottom in 5.0 s. What is the initial velocity of the ball?

42. A train started from rest and moved with constant acceleration. At one time it was traveling 30 m/s and 160 m farther on it was traveling 50 m/s. Calculate (a) the acceleration, (b) the time required to travel the 160 m mentioned, (c) the time required to attain the speed of 30 m/s, (d) the distance moved from rest to the time the train had a speed of 30 m/s.

43. A stone is thrown vertically down with an initial speed of 12 m/s from the roof of a building, 30 m above the ground. (a) How long does it take the stone to reach the ground? (b) What is the speed of the stone at impact?

44. If a body travels half its total path in the last second of its fall from rest, find (a) the time and (b) height of its fall.

45. A woman fell 144 ft from the top of a building, landing on the top of a metal ventilator box, which she crushed to a depth of 18 in. She survived without serious injury. What acceleration (assumed uniform) did she experience during the collision? Express your answer in terms of g, the acceleration due to gravity.

46. Two trains, one traveling at 95 km/h ($= 59$ mi/h) and the other at 130 km/h ($= 81$ mi/h), are headed toward one another along a straight level track. When they are 3.2 km ($= 2.0$ mi) apart, both engineers simultaneously see the other's train and apply their brakes. If the brakes decelerate each train at the rate of 1.0 m/s² ($= 3.3$ ft/s²), determine if there is a collision.

47. A stone is dropped into the water from a bridge 144 ft above the water. Another stone is thrown vertically down 1 s after the first is dropped. Both stones strike the water at the same time. (a) What was the initial speed of the second stone? (b) Plot speed versus time on a graph for each stone, taking zero time as the instant the first stone was released.

48. A parachutist after bailing out falls 50 m without friction. When the parachute opens, he decelerates at 2 m/s². He reaches the ground with a speed of 3 m/s. (a) How long is the parachutist in the air? (b) At what height did he bail out?

49. A driver's handbook states that an automobile with good brakes and going 50 mi/h can stop in a distance of 186 ft. The corresponding distance for 30 mi/h is 80 ft. Assume that the driver reaction time, during which the acceleration is zero, and the accelerations after the brakes are applied are both the same for the two speeds. Calculate (a) the driver reaction time and (b) the acceleration.

50. Two bodies begin a free fall from rest from the same height 1 s apart. How long after the first body begins to fall will the two bodies be 10 m apart?

51. An open elevator is ascending with a constant speed of 10

m/s. A ball is thrown straight up by a boy on the elevator when it is a height of 30 m above the ground. The initial speed of the ball with respect to the elevator is 20 m/s. (*a*) What is the maximum height obtained by the ball? (*b*) How long does it take for the ball to return to the elevator?

52. As Fig. 3–14 shows, Clara jumps from a bridge, followed closely by Jim. How long did Jim wait after Clara jumped? Assume that Jim is 170 cm tall and that the jumping-off level is at the top of the figure. Make scale measurements directly on the figure.

*53. A dog sees a flowerpot sail up and then back past a window 5.0 ft (= 1.5 m) high. If the total time the pot is in sight is 1 s, find the height above the top of the window that the pot rises.

*54. A steel ball bearing is dropped from the roof of a building (the initial velocity of the ball is zero). An observer standing in front of a window 120 cm high notes that the ball takes 0.125 s to fall from the top to the bottom of the window. The ball bearing continues to fall, makes a completely elastic collision with a horizontal sidewalk, and reappears at the bottom of the window 2 s after passing it on the way down. How tall is the building? (The ball will have the same speed at a point going up as it had going down after a completely elastic collision.)

*55. An elevator ascends with an upward acceleration of 4 ft/s². At the instant its upward speed is 8 ft/s, a loose bolt drops from the ceiling of the elevator 9 ft from the floor. Calculate (*a*) the time of flight of the bolt from ceiling to floor and (*b*) the distance it has fallen relative to the elevator shaft.

Figure 3–14   Problem 52. Globe Staff Photo by John Tlumaki.

4 Motion in a Plane

4–1 Displacement, Velocity, and Acceleration

In this chapter we return to a consideration of motion in two dimensions, taken, for convenience, to be the *x-y* plane. Figure 4–1 shows a particle at time *t* moving along a curved path in this plane. Its *position,* or displacement from the origin, is measured by the vector **r**; its *velocity* is indicated by the vector **v**, which, as we have seen in Section 3–4, is tangent to the path of the particle. The *acceleration* is indicated by the vector **a**; the direction of **a**, as we shall see more explicitly later, does not bear any unique relationship to the path of the particle but depends rather on the rate at which the velocity **v** changes with time as the particle moves along its path.

The vectors **r**, **v**, and **a** are interrelated (see Eqs. 3–4, 3–5, and 3–10) and can be expressed in terms of their components, using unit vector notation, as

$$\mathbf{r} = \mathbf{i}x + \mathbf{j}y, \tag{4–1}$$

$$\mathbf{v} = \frac{d\mathbf{r}}{dt} = \mathbf{i}v_x + \mathbf{j}v_y, \tag{4–2}$$

Figure 4–1 A particle moves along a curved path in the *x-y* plane. (*a*) Its position **r**, (*b*) its velocity **v**, and (*c*) its acceleration **a** are shown at time *t*, along with the vector components of those vectors. Note that x, y, v_x, v_y, and a_x are positive but that a_y is negative. Compare to Fig. 3–2.

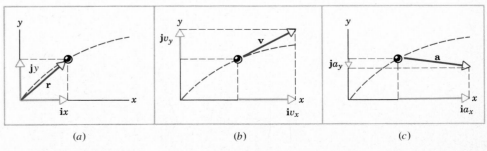

(*a*)   (*b*)   (*c*)

and

$$\mathbf{a} = \frac{d\mathbf{v}}{dt} = \mathbf{i}a_x + \mathbf{j}a_y. \tag{4–3}$$

These equations can easily be extended to three dimensions by adding to them the terms $\mathbf{k}z$, $\mathbf{k}v_z$, and $\mathbf{k}a_z$, respectively, in which $\mathbf{k}$ is a unit vector in the z-direction.

In Chapter 3 we considered the special case in which the particle moved in one dimension only, say, along the x-axis, where the vectors $\mathbf{r}$, $\mathbf{v}$, and $\mathbf{a}$ were directed along this axis, either in the positive x-direction or the negative x-direction. The components y, v_y, and a_y were zero and we described the motion in terms of equations relating the scalar quantities x, v_x, and a_x. Or, when the particle moved along the y-axis, the components x, v_x, and a_x were zero and the motion was described in terms of equations relating the scalar quantities y, v_y, and a_y. In this chapter we consider motion in the x-y plane so that, in general, both sets of components have nonzero values.

4–2   Motion in a Plane with Constant Acceleration

Let us consider first the special case of motion in a plane with constant acceleration. Here, as the particle moves, the acceleration $\mathbf{a}$ does not vary either in magnitude or in direction. Hence the components of $\mathbf{a}$ also will not vary; that is, $a_x = $ constant and $a_y = $ constant. We then have a situation that can be described as the sum of two component motions occurring simultaneously with constant acceleration along each of two perpendicular directions. The particle will move, in general, along a curved path in the plane. This may be so even if one component of the acceleration—say, a_x—is zero, for then the corresponding component of the velocity—say, v_x—may have a constant, nonzero value. An example of this latter situation is the motion of a projectile that follows a curved path in a vertical plane and, neglecting the effects of air resistance, is subject to a constant acceleration $\mathbf{g}$ directed down along the y-axis only.

We can obtain the general equations for plane motion with constant $\mathbf{a}$ simply by setting

$$a_x = \text{constant} \qquad \text{and} \qquad a_y = \text{constant}.$$

The equations for constant acceleration, summarized in Table 3–1, then apply to both the x- and y-components of the position vector $\mathbf{r}$, the velocity vector $\mathbf{v}$, and the acceleration vector $\mathbf{a}$ (see Table 4–1).

Constant acceleration

The two sets of equations in Table 4–1 are related in that the time parameter t is the same for each, since t represents the time at which the particle, moving in a curved path in the x-y plane, occupied a position described by the position components x and y.

Table 4–1   Motion with constant acceleration in the x-y plane

Equation number	x-Motion equations	Equation number	y-Motion equations
4-4a	$v_x = v_{x0} + a_x t$	4-4a$'$	$v_y = v_{y0} + a_y t$
4-4b	$x = x_0 + \frac{1}{2}(v_{x0} + v_x)t$	4-4b$'$	$y = y_0 + \frac{1}{2}(v_{y0} + v_y)t$
4-4c	$x = x_0 + v_{x0}t + \frac{1}{2}a_x t^2$	4-4c$'$	$y = y_0 + v_{y0}t + \frac{1}{2}a_y t^2$
4-4d	$v_x{}^2 = v_{x0}{}^2 + 2a_x(x - x_0)$	4-4d$'$	$v_y{}^2 = v_{y0}{}^2 + 2a_y(y - y_0)$

The equations of motion in Table 4–1 may also be expressed in vector form. For example, substituting Eqs. 4–4a, 4a′ into Eq. 4–2 yields

$$\begin{aligned}
\mathbf{v} &= \mathbf{i}v_x + \mathbf{j}v_y \\
&= \mathbf{i}(v_{x0} + a_x t) + \mathbf{j}(v_{y0} + a_y t) \\
&= (\mathbf{i}v_{x0} + \mathbf{j}v_{y0}) + (\mathbf{i}a_x + \mathbf{j}a_y)t.
\end{aligned}$$

The first quantity in parentheses is the initial velocity vector $\mathbf{v}_0$ (see Eq. 4–2) and the second is the (constant) acceleration vector $\mathbf{a}$ (see Eq. 4–3). Thus the vector relation

Velocity as a function of time

$$\mathbf{v} = \mathbf{v}_0 + \mathbf{a}t \qquad (4\text{–}5)$$

is equivalent to the two scalar relations Eqs. 4–4a, a′ in Table 4–1. It shows simply and compactly that the velocity $\mathbf{v}$ at time t is the sum of the initial velocity $\mathbf{v}_0$ that the particle would have in the absence of acceleration plus the (vector) change in velocity, $\mathbf{a}t$, acquired during the time t under the constant acceleration $\mathbf{a}$. Similarly, the scalar equations 4–4c, c′ are equivalent to the single vector equation

Position as a function of time

$$\mathbf{r} = \mathbf{r}_0 + \mathbf{v}_0 t + \tfrac{1}{2}\mathbf{a}t^2, \qquad (4\text{–}6)$$

which is also easily interpreted.

4–3 Projectile Motion

An example of motion in a plane with constant acceleration is projectile motion. This is the two-dimensional motion of a particle thrown obliquely into the air. The ideal motion of a baseball or a golf ball is an example of projectile motion. We assume that we can neglect the effect of the air on this motion.

   The motion of a projectile is one of constant acceleration $\mathbf{g}$, directed downward, and thus should be described by the equations in Table 4–1. There is no horizontal component of acceleration. If we choose a coordinate system with the positive y-axis vertically upward, we may put $a_y = -g$ and $a_x = 0$ in these equations.

   Let us further choose the origin of our coordinate system to be the point at which the projectile begins its flight (see Fig. 4–2). Hence the origin will be the point at which the ball leaves the thrower's hand or the fuel in the rocket burns out, for example. In Table 4–1 this choice of origin implies that $x_0 = y_0 = 0$. The velocity at $t = 0$, the instant the projectile begins its flight, is $\mathbf{v}_0$, which makes an angle θ_0 with the positive x-direction. The x- and y-components of $\mathbf{v}_0$ (see Fig. 4–2) are then

$$v_{x0} = v_0 \cos\theta_0 \qquad \text{and} \qquad v_{y0} = v_0 \sin\theta_0.$$

   Because there is no horizontal component of acceleration, the horizontal component of the velocity will be constant. In Eq. 4–4a we set $a_x = 0$ and $v_{x0} = v_0 \cos\theta_0$, so that

Horizontal velocity component

$$v_x = v_0 \cos\theta_0. \qquad (4\text{–}7)$$

The horizontal velocity component retains its initial value throughout the flight.

   The vertical component of the velocity will change with time just as for a body in free fall. In Eq. 4–4a′ (see Table 4–1) we set

$$a_y = -g \qquad \text{and} \qquad v_{y0} = v_0 \sin\theta_0,$$

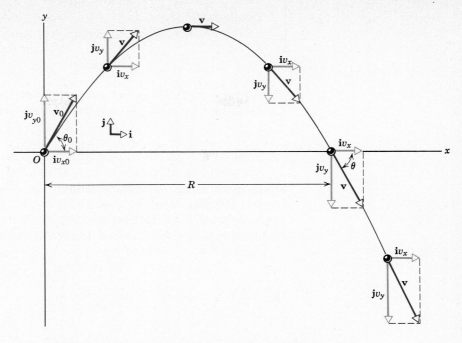

Figure 4-2 The trajectory of a projectile showing the initial velocity $\mathbf{v}_0$ and its vector components and also the velocity $\mathbf{v}$ and its vector components at five later times. Note that $v_x = v_{x0}$ throughout the flight. The distance R is called the *horizontal range*.

so that

Vertical velocity component

$$v_y = v_0 \sin \theta_0 - gt. \tag{4-8}$$

The vertical motion is independent of whether or not there is any horizontal motion. Indeed, if we view the motion of Fig. 4-2 from a reference frame that moves to the right with a speed v_{x0}, the motion will be that of an object thrown vertically upward with an initial speed $v_0 \sin \theta_0$.

The magnitude of the resultant velocity vector at any instant is

$$v = \sqrt{v_x^2 + v_y^2}. \tag{4-9a}$$

The angle θ that the velocity vector makes with the horizontal at that instant is given by

$$\tan \theta = \frac{v_y}{v_x}. \tag{4-9b}$$

As with any type of motion, the velocity vector is tangent to the path of the particle at every point, as shown in Fig. 4-2.

The x-coordinate of the particle's position at any time obtained from Eq. 4-4c with $x_0 = 0$, $a_x = 0$, and $v_{x0} = v_0 \cos \theta_0$, is

Horizontal position component

$$x = (v_0 \cos \theta_0)t. \tag{4-10}$$

The y-coordinate, obtained from Eq. 4-4c′ with $y_0 = 0$, $a_y = -g$, and $v_{y0} = v_0 \sin \theta_0$, is

Vertical position component

$$y = (v_0 \sin \theta_0)t - \tfrac{1}{2}gt^2. \tag{4-11}$$

Equations 4–10 and 4–11 give us x and y as functions of the common parameter t, the time of flight. By combining and eliminating t from them, we obtain

The trajectory of a projectile

$$y = (\tan \theta_0)x - \left(\frac{g}{2v_0{}^2 \cos^2 \theta_0}\right) x^2, \qquad (4\text{--}12)$$

which relates y to x and is the equation of the trajectory of the projectile. Since v_0, θ_0, and g are constants, this equation has the form

$$y = a + bx + cx^2,$$

the equation of a parabola. Hence the trajectory of a projectile is parabolic.

We wish now to find the *horizontal range R* of the projectile. This is the horizontal distance $x\,(=R)$ from the origin to the point where the projectile strikes the ground; see Fig. 4–2. We may find R by setting $y = 0$, which corresponds to ground level, in Eq. 4–12 and solving for x. We expect two solutions because y is zero both at $x = 0$ and at $x = R$. Applying the condition $y = 0$, we obtain

$$x \left[\tan \theta_0 - \frac{gx}{2(v_0 \cos \theta_0)^2}\right] = 0.$$

This has two solutions, as we expected,

$$x = 0 \qquad \text{and} \qquad x = R = \frac{2v_0{}^2}{g} \sin \theta_0 \cos \theta_0,$$

where we have used the relation $\tan \theta_0 = \sin \theta_0/\cos \theta_0$. We now use the identity $\sin 2\theta_0 = 2 \sin \theta_0 \cos \theta_0$ to obtain

The horizontal range

$$R = \frac{v_0{}^2}{g} \sin 2\theta_0. \qquad (4\text{--}13)$$

Note that the range is a maximum when $\sin 2\theta_0 = 1$, which corresponds to $2\theta_0 = 90°$ or a launching angle θ_0 of 45°.

Figure 4–3 is a multiflash photograph showing a modern reproduction of an experiment done by Galileo in 1608 in which he first proved that a projectile

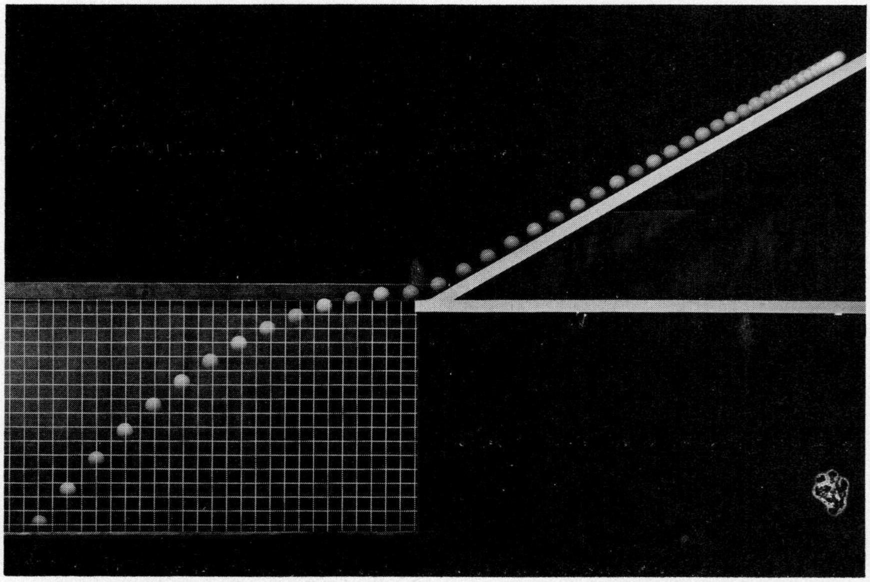

Figure 4–3 A multi-flash photograph of a ball rolling down an inclined plane and following a parabolic path after being projected horizontally from the foot of the plane. (*Scientific American*, March 1975. Photo by Ben Rose.)

follows a parabolic trajectory.* He did so by rolling a ball down a groove in an inclined plane, starting from various heights h, and measuring the corresponding distances r at which the ball struck the floor. A short horizontal extension of the grooved plane ensures that the ball is projected horizontally as it leaves the plane. Note how the ball speeds up as it rolls down the plane. Note also that, after it leaves the plane, its horizontal velocity component does indeed seem to remain constant while its vertical velocity component increases with time. When we take into account effects due to the fact that the object under investigation is not a true particle but is a ball rolling in a groove, Galileo's experimental results agree with modern theory to about 1%.

Example 1 A plane is flying at a constant horizontal velocity of 500 km/h ($= 310$ mi/h) at an elevation of 1.0 km ($= 3300$ ft) toward a point directly above a man struggling in the water. At what angle of sight ϕ should a survival package be released to strike very close to the man in the water (Fig. 4–4)?

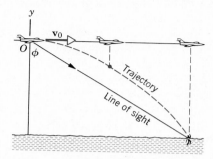

Figure 4–4 Example 1. A survival package is released from an airplane with horizontal velocity $\mathbf{v}_0$.

We choose a reference frame fixed with respect to the earth, its origin O being the release point. The motion of the package at the moment of release is the same as that of the plane. Hence the initial package velocity $\mathbf{v}_0$ is horizontal and its magnitude is 500 km/h. The angle of projection θ_0 is zero.

We find the time of fall from Eq. 4–11. With $\theta_0 = 0$ and $y \doteq -1.0$ km, this gives

$$t = \sqrt{-\frac{2y}{g}} = \sqrt{-\frac{(2)(-1.0 \times 10^3 \text{ m})}{(9.8 \text{ m/s}^2)}} = 14 \text{ s}.$$

Note that the time of fall does not depend on the speed of the plane for a horizontal projection. (See, however, Problem 14.)

The horizontal distance traveled by the package in this time is given by Eq. 4–10, $x = (v_0 \cos \theta_0)t$, or

$$x = (500 \text{ km/h}) \times (10^3 \text{m/km}) \times (1 \text{ h}/3600 \text{ s}) \times (14 \text{ s}) = 1900 \text{ m}.$$

so that the angle of sight (Fig. 4–4) should be

$$\phi = \tan^{-1}\frac{x}{|y|} = \tan^{-1}\frac{1900 \text{ m}}{1000 \text{ m}} = 62°.$$

Does the motion of the package appear to be parabolic when viewed from a reference frame fixed with respect to the plane?

Example 2 A soccer player kicks a ball at an angle of 37° from the horizontal with an initial speed of 50 ft/s. (A right triangle, one of whose angles is 37°, has sides in the ratio 3:4:5, or 6:8:10.) Assume that the ball moves in a vertical plane and that air resistance is negligible.

(a) Find the time t_1 at which the ball reaches the highest point of its trajectory.

At the highest point, the vertical component of velocity v_y is zero. Solving Eq. 4–8 for t, we obtain

$$t = \frac{v_0 \sin \theta_0 - v_y}{g}.$$

With

$$v_y = 0, \qquad v_0 = 50 \text{ ft/s}, \qquad \theta_0 = 37°, \qquad g = 32 \text{ ft/s}^2,$$

we have

$$t_1 = \frac{[50(\frac{6}{10}) - 0] \text{ ft/s}}{32 \text{ ft/s}^2} = \frac{15}{16} \text{ s}.$$

(b) How high does the ball go?

The maximum height is reached at $t = 15/16$ s. By using Eq. 4–11,

$$y = (v_0 \sin \theta_0)t - \tfrac{1}{2}gt^2,$$

we have

$$y_{\max} = (50 \text{ ft/s})(\tfrac{6}{10})(\tfrac{15}{16} \text{ s}) - \tfrac{1}{2}(32 \text{ ft/s}^2)(\tfrac{15}{16})^2 \text{ s}^2 = 14 \text{ ft}.$$

(c) What is the range of the ball and how long is it in the air?

The horizontal distance from the starting point at which the ball returns to its original elevation (ground level) is the *horizontal range* R. We set $y = 0$ in Eq. 4–11 and find the time t_2 required to transverse this range. We obtain

$$t_2 = \frac{2v_0 \sin \theta_0}{g} = \frac{2(50 \text{ ft/s})(\frac{6}{10})}{32 \text{ ft/s}^2} = \frac{15}{8} \text{ s}.$$

Notice that $t_2 = 2t_1$. This corresponds to the fact that the same time is required for the ball to go up (reach its maximum height from ground) as is required for the ball to come down (reach the ground from its maximum height).

The range R can then be obtained by inserting this value t_2 for t in Eq. 4–10. We obtain, from $x = (v_0 \cos \theta_0)t$,

$$R = (v_0 \cos \theta_0)t_2 = (50 \text{ ft/s})(\tfrac{8}{10})(\tfrac{15}{8} \text{ s}) = 75 \text{ ft}.$$

---

* See "Galileo's Discovery of the Parabolic Trajectory," by Stillman Drake and James MacLachlan, *Scientific American*, March 1975.

Verify that the same value follows by direct application of Eq. 4–13.

(d) What is the velocity of the ball as it strikes the ground?
From Eq. 4–7 we obtain

$$v_x = v_0 \cos \theta_0 = (50 \text{ ft/s})(\tfrac{8}{10}) = 40 \text{ ft/s}.$$

From Eq. 4–8 we obtain for $t = t_2 = \tfrac{15}{8}$ s,

$$v_y = v_0 \sin \theta_0 - gt = (50 \text{ ft/s})(\tfrac{6}{10}) - (32 \text{ ft/s}^2)(\tfrac{15}{8} \text{ s}) = -30 \text{ ft/s}.$$

Hence, from Eq. 4–9a,

$$v = \sqrt{v_x^2 + v_y^2} = \sqrt{(40 \text{ ft/s})^2 + (-30 \text{ ft/s})^2} = 50 \text{ ft/s},$$

and

$$\tan \theta = \frac{v_y}{v_x} = -\frac{30}{40},$$

so that $\theta = -37°$, or 37° *clockwise* from the x-axis. Notice that $\theta = -\theta_0$ and $v = v_0$, as we expect from symmetry (Fig. 4–2).

4–4   Uniform Circular Motion

In Section 3–6 we saw that acceleration arises from a change in velocity. In the simple case of free fall the velocity changed in magnitude only, but not in direction. For a particle moving in a circle with constant speed, called *uniform circular motion*, the velocity vector changes continuously in direction but not in magnitude. We seek now to obtain the acceleration in uniform circular motion.

The situation is shown in Fig. 4–5a. Let P be the position of the particle at the time t and P' its position at the time $t + \Delta t$. The velocity at P is $\mathbf{v}$, a vector tangent to the curve at P. The velocity at P' is $\mathbf{v}'$, a vector tangent to the curve at P'. Vectors $\mathbf{v}$ and $\mathbf{v}'$ are equal in magnitude, the speed being constant, but their directions are different. The length of path traversed during Δt is the arc length PP', which is equal to $v \Delta t$, v being the constant speed.

Now redraw the vectors $\mathbf{v}$ and $\mathbf{v}'$, as in Fig. 4–5b, so that they originate at a common point. We are free to do this as long as the magnitude and direction of each vector are the same as in Fig. 4–5a. This diagram (Fig. 4–5b) enables us to see clearly the *change in velocity* as the particle moved from P to P'. This change, $\mathbf{v}' - \mathbf{v} = \Delta\mathbf{v}$, is the vector that must be added to $\mathbf{v}$ to get $\mathbf{v}'$. Notice that it points inward, approximately toward the center of the circle.

Now the triangle OQQ' formed by $\mathbf{v}$, $\mathbf{v}'$, and $\Delta\mathbf{v}$ is similar to the triangle CPP' (Fig. 4–5c) formed by the chord PP' and the radii CP and CP'. This is so because both are isosceles triangles having the same vertex angle; the angle θ between $\mathbf{v}$ and $\mathbf{v}'$ is the same as the angle PCP' because $\mathbf{v}$ is perpendicular to CP and $\mathbf{v}'$ is perpendicular to CP'. We can therefore write

$$\frac{\Delta v}{v} = \frac{v \, \Delta t}{r}, \qquad \text{approximately,}$$

the chord PP' being taken equal to the arc length PP'. This relation becomes more nearly exact as Δt is diminished, since the chord and the arc then approach each other. Notice also that $\Delta\mathbf{v}$ approaches closer and closer to a direction perpendic-

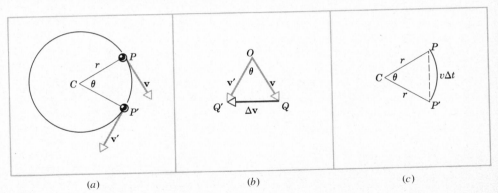

Figure 4–5 Uniform circular motion. The particle travels around a circle at constant speed. Its velocity at two points P and P' is shown. Its change in velocity in going from P to P' is $\Delta\mathbf{v}$.

(a)   (b)   (c)

ular to **v** and **v′** as Δt is diminished and therefore approaches closer and closer to a direction pointing to the exact center of the circle. It follows from this relation that

$$\frac{\Delta v}{\Delta t} = \frac{v^2}{r}, \qquad \text{approximately,}$$

and in the limit when $\Delta t \to 0$ this expression becomes exact. We therefore obtain

$$a = \lim_{\Delta t \to 0} \frac{\Delta v}{\Delta t} = \frac{v^2}{r} \qquad (4-14)$$

as the magnitude of the acceleration. The direction of **a** is instantaneously along a radius inward toward the center of the circle.

Figure 4–6 shows the instantaneous relation between **v** and **a** at various points of the motion. The magnitude of **v** is constant, but its direction changes continuously. This gives rise to an acceleration **a** that is also constant in magnitude (but not zero) but continuously changing in direction. The velocity **v** is always tangent to the circle in the direction of motion; the acceleration **a** is always directed radially inward. Because of this, **a** is called a *centripetal* (that is, "center-seeking") acceleration.

Both in free fall and in projectile motion, **a** is constant in direction and magnitude and we can use the equations developed for constant acceleration (see Table 4–1). We cannot use these equations for uniform circular motion because **a** varies in direction and is therefore not constant.

The units of centripetal acceleration are the same as those of an acceleration resulting from a change in the magnitude of a velocity. Dimensionally, we have

$$\frac{v^2}{r} = \left(\frac{\text{length}}{\text{time}}\right)^2 \Big/ \text{length} = \frac{\text{length}}{\text{time}^2} \qquad \text{or} \qquad \frac{L}{T^2},$$

which are the dimensions of acceleration. The units therefore may be feet per second per second (ft/s^2) or meters per second per second (m/s^2), among others.

The acceleration resulting from a change in direction of a velocity is just as real and just as much an acceleration in every sense as that arising from a change in magnitude of a velocity. By definition, acceleration is the time rate of change of velocity, and velocity, being a vector, can change in direction as well as magnitude. If a physical quantity is a vector, its directional aspects cannot be ignored, for their effects will prove to be every bit as important and real as those produced by changes in magnitude.

It is worth emphasizing at this point that there need not be any motion in the direction of an acceleration and that there is no fixed relation in general between the directions of **a** and **v**. In Fig. 4–7 we give examples in which the angle between **v** and **a** varies from 0 to 180°. Only in one case, $\theta = 0°$, is the motion in the direction of **a**.

Centripetal acceleration

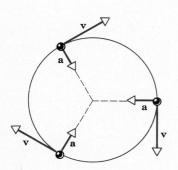

Figure 4–6   In uniform circular motion the acceleration **a** is always directed toward the center of the circle and hence is perpendicular to **v**.

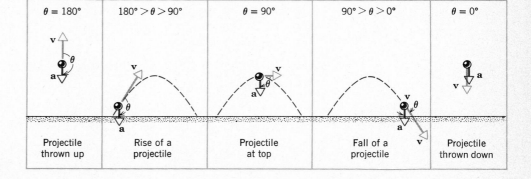

Figure 4–7   Showing the relation between **v** and **a** for various motions.

Example 3   The moon revolves about the earth, making a complete revolution in 27.3 days. Assume that the orbit is circular and has a radius of 3.85×10^5 km ($=2.39 \times 10^5$ mi). What is the magnitude of the acceleration of the moon toward the earth?

We have $r = 3.85 \times 10^8$ m. The time for one complete revolution, called the period, is $T = 27.3$ d $= 2.36 \times 10^6$ s. The speed of the moon (assumed constant) is therefore

$$v = \frac{2\pi r}{T} = 1020 \text{ m/s.}$$

The centripetal acceleration is

$$a = \frac{v^2}{r} = \frac{(1020 \text{ m/s})^2}{3.85 \times 10^8 \text{ m}} = 0.00273 \text{ m/s}^2.$$

This is only $2.78 \times 10^{-4} g$.

Example 4   Calculate the speed of an earth satellite, assuming that it is traveling at an altitude h of 200 km ($= 124$ mi) above the surface of the earth where $g = 9.2$ m/s^2. The mean radius R of the earth is 6.37×10^6 m ($=3960$ mi).

Like any free object near the earth's surface, the satellite has an acceleration g toward the earth's center. It is this acceleration that causes it to follow the circular path. Hence the centripetal acceleration is g, and from Eq. 4–14, $a = v^2/r$, we have

$$g = \frac{v^2}{R + h},$$

or

$$v = \sqrt{(R + h)g} = \sqrt{(6.37 \times 10^6 \text{ m} + 200 \times 10^3 \text{ m})(9.20 \text{ m/s}^2)}$$
$$= 7770 \text{ m/s } (=4.82 \text{ mi/s or } 17,400 \text{ mi/h}).$$

4–5 Relative Velocity and Acceleration

In earlier sections we considered the addition of velocities in a particular reference frame. Let us now consider the relation between the velocity of an object as determined by one observer S (in reference frame S) and the velocity of the same object as determined by another observer S' (in reference frame S') who is moving with respect to the first.

Consider observer S fixed to the earth, so that his reference frame is the earth. The other observer S' is moving on the earth—for example, a passenger sitting on a moving train—so that his reference frame is the train. They each follow the motion of the same object, say an automobile on a road or a man walking through the train. Each observer will record a displacement, a velocity, and an acceleration for this object measured *relative to his own reference frame*. How will these measurements compare? In this section we will assume that frame S' is moving with respect to frame S with a constant velocity $\mathbf{v}_{S'S}$.

In Fig. 4–8 the reference frame S represented by the x- and y-axes can be thought of as fixed to the earth. The shaded region indicates another reference frame S', represented by x'- and y'-axes, which moves along the x-axis with a constant velocity $\mathbf{v}_{S'S}$, as measured in the S-system; it can be thought of as drawn on the floor of a railroad flatcar.

Suppose that a bowling ball, which we will refer to as particle P, is rolling diagonally across the floor of the flatcar. Figure 4–8 shows its position both at an initial time $t = 0$ and at a later time t. During this time frame S' has moved to the right with respect to frame S by a distance $\mathbf{v}_{S'S} t$. If we ask each observer what displacement of the particle (ball) occurred during time t, what will they say? Observer S' will report the displacement to be $\mathbf{r}_{PS'}$ and observer S will report it to be $\mathbf{r}_{PS}$, as shown in Fig. 4–8b. From that figure we see that

$$\mathbf{r}_{PS} = \mathbf{r}_{PS'} + \mathbf{v}_{S'S} t. \tag{4–15}$$

Differentiating yields

$$\frac{d\mathbf{r}_{PS}}{dt} = \frac{d\mathbf{r}_{PS'}}{dt} + \mathbf{v}_{S'S}.$$

But $d\mathbf{r}_{PS}/dt$ is just $\mathbf{v}_{PS}$, the velocity of particle P as seen by S, and $d\mathbf{r}_{PS'}/dt$ is just $\mathbf{v}_{PS'}$, the velocity of the particle as seen by S'. Thus

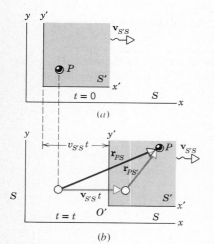

Figure 4–8   (a) A particle P at time $t = 0$. (b) The particle at a later time t. The open circles show the initial positions of the particle.

Velocities in different
reference frames

$$\mathbf{v}_{PS} = \mathbf{v}_{PS'} + \mathbf{v}_{S'S}. \qquad (4-16)$$

Hence the velocity of a particle in the S-frame ($\mathbf{v}_{PS}$) is the vector sum of its velocity in the S'-frame ($\mathbf{v}_{PS'}$) and the velocity $\mathbf{v}_{S'S}$ of the S'-frame relative to the S-frame.

Example 5. (a) The compass of an airplane indicates that it is heading due east. Ground information indicates a wind blowing due north. Show on a diagram the velocity of the plane with respect to the ground.

The moving "particle" in this case is the plane P. The ground is one reference frame (G) and the air (A) is another, moving with respect to it. We rewrite Eq. 4–16 as

$$\mathbf{v}_{PG} = \mathbf{v}_{PA} + \mathbf{v}_{AG}.$$

These three terms are, in sequence, the velocity of the plane with respect to the ground, the velocity of the plane with respect to the air, and the velocity of the air with respect to the ground, that is, the wind velocity. In this case $\mathbf{v}_{AG}$ points north and $\mathbf{v}_{PA}$ points east, as Fig. 4–9a shows.

The angle α is the angle north of east of the plane's course with respect to the ground and is

$$\tan \alpha = v_{AG}/v_{PA}.$$

The plane's ground speed is given by

$$v_{PG} = \sqrt{v_{PA}{}^2 + v_{AG}{}^2}.$$

For example, if the air-speed indicator shows that the plane is moving relative to the air at a speed of 315 km/h (=196 mi/h), and if the speed of the wind with respect to the ground is 65.0 km/h (=40.4 mi/h), then

$$v_{PG} = \sqrt{(315)^2 + (65.0)^2}\ \text{km/h} = 322\ \text{km/h}\ (=200\ \text{mi/h})$$

is the ground speed of the plane and

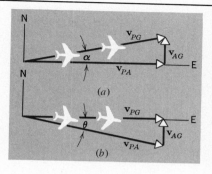

Figure 4–9 Example 5.

$$\alpha = \tan^{-1}\frac{65.0}{315} = 11.7°$$

gives the course of the plane north of east.

(b) Now draw the vector diagram showing the direction the pilot must steer the plane through the air for the plane to travel due east with respect to the ground.

He would naturally head partly into the wind. His speed relative to the earth will therefore be less than before. The vector diagram is shown in Fig. 4–9b. Calculate θ and v_{PG}, using the previous data for $\mathbf{v}_{AG}$ and the air-speed v_{PA}.

We have seen that different velocities are assigned to a particle by different observers when the observers are in relative motion. These velocities always differ by the relative velocity of the two observers, which here is a constant velocity. It follows that when the particle velocity changes, the change will be the same for both observers. Hence they each measure the same acceleration for the particle. *The acceleration of a particle is the same in all reference frames moving relative to one another with constant velocity;* that is, $\mathbf{a}_{PS} = \mathbf{a}_{PS'}$. This result follows readily in a formal way if we differentiate Eq. 4–16.

REVIEW GUIDE AND SUMMARY

In two dimensions the x- and y-motions are independent

Consider a particle moving in two dimensions with a constant acceleration $\mathbf{a}$. This means that its acceleration components a_x and a_y are each separately constant. The four kinematic equations of Table 3–1 then apply separately and independently to the x- and y-motions, as Table 4–1 shows.

Figure 4–2 shows a particle moving in projectile motion, for which $a_x = 0$ and $a_y = -g$. The initial velocity $\mathbf{v}_0$ has components v_{x0} and v_{y0} and a launching angle of θ_0. From the equations of Table 4–1 we find for the velocity components

The velocity components

$$v_x(t) = v_0 \cos \theta_0 \qquad \text{(a constant)} \qquad [4-7]$$

and

$$v_y(t) = v_0 \sin \theta_0 - gt. \qquad [4-8]$$

For the position components we find

The position components

$$x(t) = (v_0 \cos \theta_0)t \qquad [4-10]$$

and

$$y(t) = (v_0 \sin \theta_0)t - \tfrac{1}{2}gt^2. \qquad [4-11]$$

By eliminating t between Eqs. 4–10 and 4–11 we can find the equation of the trajectory followed by the particle. It is a parabola and is

The trajectory

$$y = (\tan \theta_0)x - \left(\frac{g}{2v_0^2 \cos^2 \theta_0}\right) x^2. \qquad [4-12]$$

Putting $y = 0$ here corresponds to the return of the particle to its horizontal launching plane. The corresponding value of x is the horizontal range R, and we find that

The horizontal range

$$R = \frac{v_0^2}{g} \sin 2\theta_0. \qquad [4-13]$$

Examples 1 and 2 illustrate the aspects of projectile motion.

Uniform circular motion

Another example of motion in a plane is uniform circular motion, in which a particle moves at constant speed v in a circle of radius r. This motion is accelerated because even though the magnitude of the particle velocity $\mathbf{v}$ is constant, its direction is changing continuously. The acceleration $\mathbf{a}$ always points toward the center of the circle and has a constant magnitude given by

$$a = \frac{v^2}{r}. \qquad [4-14]$$

Note that in uniform circular motion all three vectors $\mathbf{r}$, $\mathbf{v}$, and $\mathbf{a}$ have constant magnitudes but all vary continuously in direction; see Fig. 4–6. Study Examples 3 and 4.

Motion viewed from different reference frames

The nature of the motion of a particle in a plane depends on the reference frame of the observer. Suppose that two observers, S and S', are in different reference frames that move at a constant velocity $\mathbf{v}_{S'S}$ with respect to each other, as in Fig. 4–8. If S' measures a velocity $\mathbf{v}_{PS'}$ for a moving particle P, S will measure a different velocity $\mathbf{v}_{PS}$, the two being related by

$$\mathbf{v}_{PS} = \mathbf{v}_{PS'} + \mathbf{v}_{S'S}. \qquad [4-16]$$

Even though the velocities differ, the acceleration is the same in both frames; that is, $\mathbf{a} = \mathbf{a}'$.

QUESTIONS

1. Can the acceleration of a body change its direction without its displacement changing direction? . . . without its velocity changing direction?

2. If the acceleration of a body is constant in a given reference frame, is it necessarily constant in all other reference frames?

3. In broad jumping, sometimes called long jumping, does it matter how high you jump? What factors determine the span of the jump?

4. Why doesn't the electron in the beam from an electron gun fall as much because of gravity as a water molecule in the stream from a hose? Assume horizontal motion initially in each case.

5. At what point in its path does a projectile have its minimum speed? Its maximum?

6. Figure 4–10 shows the path followed by a NASA Learjet in a run designed to simulate low-gravity conditions for a short period of time. Make an argument to show that, if the plane follows a parabolic path, the passengers will experience weightlessness.

7. Is it possible to be accelerating if you are traveling at constant speed? Is it possible to round a curve with zero acceleration? . . . with constant acceleration?

8. In his book, *Sport Science,* Peter Brancazio, with such projectiles as baseballs and golfballs in mind, writes: "Everything else being equal, a projectile will travel farther on a hot day than on a cold day, farther at high altitude than at sea level, farther in humid than in dry air." How can you explain these claims?

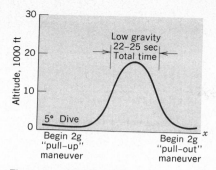

Figure 4-10 Question 6.

9. Long-range artillery pieces are not set at the "maximum range" angle of 45° but at larger elevation angles, in the range 55° to 65°. What's wrong with 45°?

10. Show that, taking the earth's rotation and revolution into account, a book resting on your table moves faster at night than it does during the daytime. In what reference frame is this statement true?

11. An aviator, pulling out of a dive, follows the arc of a circle. He was said to have "experienced 3 g's" in pulling out of the dive. Explain what this statement means.

12. A boy sitting in a railroad car moving at constant velocity throws a ball straight up into the air. Will the ball fall behind him? In front of him? Into his hand? What happens if the car accelerates forward or goes around a curve while the ball is in the air?

13. A man on the observation platform of a train moving with constant velocity drops a coin while leaning over the rail. Describe the path of the coin as seen by (a) the man on the train, (b) a person standing on the ground near the track, and (c) a person in a second train moving in the opposite direction to the first train on a parallel track.

14. Water is collecting in a bucket during a steady downpour. Will the rate at which the bucket is filling change if a steady horizontal wind starts to blow?

15. A bus with a vertical windshield moves along in a rainstorm at speed v_b. The raindrops fall vertically with a terminal speed v_r. At what angle do the raindrops strike the windshield?

16. Drops are falling vertically in a steady rain. In order to go through the rain from one place to another in such a way as to encounter the least number of raindrops, should you move with the greatest possible speed, the least possible speed, or some intermediate speed?

17. What's wrong with Fig. 4-11? The boat is sailing with the wind.

Figure 4-11 Question 17.

18. An elevator is descending at a constant speed. A passenger drops a coin to the floor. What accelerations would (a) the passenger and (b) a person at rest with respect to the elevator shaft observe for the falling coin?

EXERCISES

Section 4-1   Displacement, Velocity, and Acceleration

1. A particle moving in a plane has a position vector $\mathbf{r} = 2t\mathbf{i} + 8t^3\mathbf{j} + \mathbf{k}$ in SI units. By finding the acceleration, prove that the particle does not move in a parabola.

2. At time t_0 the velocity of an object is given in feet per second, by $\mathbf{v}_0 = 125\mathbf{i} + 25\mathbf{j}$. At 3.0 s later the velocity is $\mathbf{v} = 100\mathbf{i} - 75\mathbf{j}$. What was the average acceleration of the object during this time interval?

Section 4-2   Motion in a Plane with Constant Acceleration

3. A particle moves so that its position as a function of time in SI units is

$$\mathbf{r}(t) = \mathbf{i} + 4t^2\mathbf{j} + t\mathbf{k}.$$

(a) Write expressions for its velocity and acceleration as functions of time. (b) What is the shape of the particle's trajectory?

4. Show (a) that Eqs. 4-4b, b' can be expressed in vector form as

$$\mathbf{r} = \mathbf{r}_0 + \tfrac{1}{2}(\mathbf{v}_0 + \mathbf{v})t$$

and (b) Eqs. 4-4c, c' as

$$\mathbf{r} = \mathbf{r}_0 + \mathbf{v}_0 t + \tfrac{1}{2}\mathbf{a}t^2.$$

Also, show (c) that Eqs. 4-4d, d' can be combined to give

$$\mathbf{v} \cdot \mathbf{v} = \mathbf{v}_0 \cdot \mathbf{v}_0 + 2\mathbf{a} \cdot (\mathbf{r} - \mathbf{r}_0).$$

Section 4-3   Projectile Motion

5. A ball rolls off the edge of a horizontal table top 4.0 ft high. If it strikes the floor at a point 5.0 ft horizontally away from the edge of the table, what was its speed at the instant it left the table?

6. A baseball leaves the pitcher's hand horizontally at a speed of 100 mi/h. The distance to the batter is 60 ft. (a) How far does the ball fall under gravity during the first 30 ft of its horizontal travel? (b) During the second 30 ft? (c) Why aren't these quantities equal? Ignore the effects of air resistance.

7. A rifle is aimed horizontally at a target 100 ft away. The bullet hits the target 3.0 in. below the aiming point. What is the muzzle velocity of the rifle?

8. A ball is thrown horizontally from a height of 20 m and hits the ground with a speed that is three times its initial speed. What was the initial speed?

9. A projectile is fired horizontally from a gun located 45 m above a horizontal plane with a muzzle speed of 250 m/s. (a) How long does the projectile remain in the air? (b) At what horizontal distance does it strike the ground? (c) What is the magnitude of the vertical component of its velocity as it strikes the ground?

10. Electrons, like all forms of matter, fall under the influence of gravity. If an electron is projected horizontally with a speed of 3.0×10^7 m/s (one-tenth the speed of light), how far will it fall in traversing 1.0 m?

11. A rifle with a muzzle velocity of 1500 ft/s shoots a bullet at a target 150 ft away. How high above the target must the rifle barrel be pointed so that the bullet will hit the target?

12. A ball is thrown from a cliff with an initial velocity of 15 m/s at an angle of 20° below the horizontal. Find (a) its horizontal displacement and (b) its vertical displacement 2.3 s later.

13. A stone is thrown with an initial velocity of 20 m/s at an angle of 40° above the horizontal. Find its horizontal and vertical displacements (a) 1.1 s and (b) 1.8 s after launch.

14. A projectile is launched with an initial velocity of 30 m/s at an angle of 60° above the horizontal. Calculate the magnitude and direction of its velocity (a) 2 s and (b) 5 s after launch.

15. A ball is thrown with a speed of 25 m/s at an angle of 40° above the horizontal directly toward a wall, as shown in Fig. 4–12. The wall is 22 m from the release point of the ball. (a) How long is the ball in the air? (b) How far above the release point does the ball hit the wall? (c) What are the horizontal and vertical components of its velocity as it hits the wall? (d) Has it passed the highest point on its trajectory when it hits?

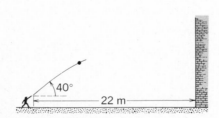

Figure 4–12   Exercise 15.

Section 4–4   Uniform Circular Motion
16. An earth satellite moves in a circular orbit 640 km above the earth's surface. The time for one revolution is 98 min. What is the acceleration of gravity at the orbit?

17. Find the magnitude of the centripetal acceleration of a particle on the tip of a fan blade, 0.30 m in diameter, rotating at 1200 rev/min.

18. The fast train known as the TGV (Train Grand Vitesse) that runs between Paris and Macon in France has a scheduled average speed of 216 km/h. (a) If the train goes around a curve at this speed and the acceleration experienced by the passengers is to be limited to $0.05g$, what is the smallest radius of curvature for the track that can be tolerated? (b) If there is a curve with a 1.0-km radius, to what speed must the train be slowed?

19. Certain neutron stars (extremely dense stars) are believed to be rotating at about 1 rev/s. If such a star has a radius of 20 km, what is the centripetal acceleration of an object on the equator of the star?

20. A magnetic field can force a charged particle to move in a circular path. An electron experiences a radial acceleration of 3.0×10^{14} m/s² in a particular magnetic field. What is its speed if the radius of its circular path is 15 cm?

Section 4–5   Relative Velocity and Acceleration
21. A person walks up a stalled escalator in 90 s. When standing on the same escalator, now moving, he is carried up in 60 s. How much time would it take him to walk up the moving escalator?

22. A transcontinental flight of 2700 mi is scheduled to take 50 min longer westward than eastward. The airspeed of the jet is 600 mi/h. What assumptions about the jet stream wind velocity, presumed to be east or west, are made in preparing the schedule?

23. Two highways intersect, as shown in Fig. 4–13. At the instant shown, a police car P is 800 m from the intersection and moving at 80 km/h. Motorist M is 600 m from the intersection and moving at 60 km/h. At this moment, what is the velocity, magnitude and direction, of the motorist with respect to the police car?

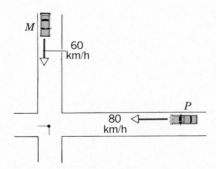

Figure 4–13   Exercise 23.

24. Snow is falling vertically at a constant speed of 8.0 m/s. At what angle from the vertical do the snowflakes appear to be falling as viewed by the driver of a car traveling on a straight road with a speed of 50 km/h?

PROBLEMS GROUPED BY SECTION

Section 4–1   Displacement, Velocity, and Acceleration
1. A plane flies 300 mi east from city A to city B in 45 min and then 600 mi south from city B to city C in 1.5 h. (a) What are the magnitude and direction of the displacement vector that repre-

sents the total trip? What are (b) the average velocity vector, and (c) the average speed for the trip?

2. The position vector **r** of a particle moving in the x-y plane is given by $\mathbf{r} = (2t^3 - 5t)\mathbf{i} + (6 - 7t^4)\mathbf{j}$. Here $\mathbf{r}$ is in meters and t is in seconds. Calculate (a) $\mathbf{r}$, (b) $\mathbf{v}$, and (c) $\mathbf{a}$ when $t = 2$ s.

Section 4-2   Motion in a Plane with Constant Acceleration

3. An ice boat sails across the surface of a frozen lake with constant acceleration produced by the wind. At a certain instant its velocity is $6.30\mathbf{i} - 8.42\mathbf{j}$ in meters per second. Three seconds later the boat is instantaneously at rest. What is its acceleration during this interval?

4. A particle moves in the x-y plane with constant acceleration. At time $t = 0$ it is at the origin and has velocity $3.6\mathbf{i} - 2.9\mathbf{j}$ in meters per second, but 4.6 s later its position is $5.6\mathbf{i} + 7.1\mathbf{j}$ in meters. What is its acceleration?

Section 4-3   Projectile Motion

5. Show that the horizontal range of a projectile having an initial speed v_0 and angle of projection θ_0 is $R = (v_0^2/g) \sin 2\theta_0$. Then show that a projection angle of $45°$ gives the maximum horizontal range (Fig. 4-14).

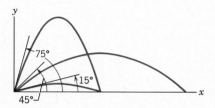

Figure 4-14   Problems 5 and 12.

6. Show that the maximum height reached by a projectile is $y_{max} = (v_0 \sin \theta_0)^2/2g$.

7. In a detective story a body is found 15 ft from the base of a building and beneath an open window 80 ft above. Would you guess the death to be accidental or not? Why?

8. A ball is thrown from the ground into the air. At a height of 9.1 m the velocity is observed to be $\mathbf{v} = 7.6\mathbf{i} + 6.1\mathbf{j}$ in meters per second (x-axis horizontal). (a) To what maximum height will the ball rise? (b) What will be the total horizontal distance traveled by the ball? (c) What is the velocity of the ball (magnitude and direction) the instant before it hits the ground?

9. A third baseman wishes to throw to first base, 127 ft distant. His best throwing speed is 85 mi/h. (a) If the ball leaves his hand, 3 ft above the ground, in a horizontal direction, what will happen to it? (b) At what upward angle must the third baseman launch the ball if the first baseman is to catch it? Assume that the first baseman's glove is also 3 ft above the ground. (c) What will be the time of flight?

10. At what initial speed must the basketball player throw the ball, at $55°$ above the horizontal, to make the foul shot, as shown in Fig. 4-15. The basket rim is 18 in. in diameter. See the figure for other dimensions.

11. A football player punts the football so that it will have a "hang time" (time of flight) of 4.5 s and land 50 yd ($= 45.7$ m) away. If the ball leaves the player's foot 5.0 ft ($= 1.52$ m) above the ground, what is its initial velocity, magnitude and direction?

12. In Galileo's *Two New Sciences* the author states that "for elevations (angles of projection) which exceed or fall short of $45°$

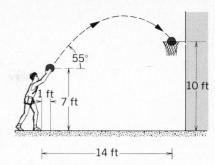

Figure 4-15   Problem 10.

by equal amounts, the ranges are equal. . . ." Prove this statement. See Fig. 4-14.

13. The B-52 shown in Fig. 4-16 is 49 m long and is traveling at an air speed of 820 km/h over a bombing range. How far apart will the bomb craters be? Make any measurements you need directly from the figure. Assume that there is no wind and ignore air resistance. How would air resistance affect your answer?

Figure 4-16   Problem 13. U.S. Air Force photo.

14. A plane, diving at an angle of $53°$ with the vertical, releases a projectile at an altitude of 730 m. The projectile hits the ground 5.0 s after being released. (a) What is the speed of the plane? (b) How far did the projectile travel horizontally during its flight? (c) What were the horizontal and vertical components of its velocity just before striking the ground?

15. (*a*) Show that if the acceleration due to gravity changes by an amount dg, the range of a projectile (see Problem 5) of given initial speed v_0 and angle of projection θ_o changes by dR where $dR/R = -dg/g$. (*b*) If the acceleration due to gravity changes by a small amount Δg (say, by going from one place to another), the range for a given projectile system will change as well. Let the change in range be ΔR. If Δg and ΔR are small enough, we may write $\Delta R/R = -\Delta g/g$. In 1936, Jesse Owens established a world's running broad jump record of 8.09 m at the Olympic Games at Berlin ($g = 9.8128$ m/s²). By how much would his record have differed if he had competed instead in 1956 at Melbourne ($g = 9.7999$ m/s²)? (In this connection see "The Earth's Gravity," by Weikko A. Heiskanen, *Scientific American,* September 1955.)

16. A projectile is fired with an initial speed v_0 from the ground at a target on the ground a distance R away. (*a*) Show that there are two possible paths, a "high" and a "low" trajectory, to the target; see Fig. 4–17. (*b*) For $v_0 = 30$ m/s, $R = 20$ m, find the two possible elevation angles of fire.

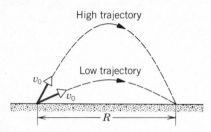

High trajectory

Low trajectory

v_0
v_0

R

Figure 4–17 Problem 16.

Section 4–4 Uniform Circular Motion

17. An astronaut is rotated in a centrifuge of radius 5 m. (*a*) What is his speed if his acceleration is $7g$? (*b*) How many revolutions per minute are required to produce this acceleration?

18. If a spaceship can withstand the stresses of a $20g$ acceleration, (*a*) what is the minimum turning radius of such a craft moving at a speed of one-tenth the speed of light? (*b*) How long would it take to complete a $90°$ turn at this speed?

19. Calculate the acceleration of a person at latitude $40°$ due to the rotation of the earth.

20. (*a*) What is the centripetal acceleration of an object on the earth's equator due to the rotation of the earth? (*b*) By what factor would the speed of the earth's rotation have to increase for a body on the equator to require a centripetal acceleration of g to keep it on the earth?

21. (*a*) Use the data of Appendix C to calculate the ratio of the centripetal accelerations of the Earth and Saturn due to their revolution about the sun. Assume that both planets move in circular orbits with constant speed. (*b*) What is the ratio of the distances of these two planets from the sun? (*c*) Compare your answers in parts (*a*) and (*b*) and suggest a simple relationship between centripetal acceleration and distance from the sun. Check your hypothesis by calculating the same ratios for another pair of planets.

22. (*a*) Write an expression for the position vector **r** for a particle describing uniform circular motion, using rectangular coordi-

nates and the unit vectors **i** and **j**. (*b*) From (*a*) derive vector expressions for the velocity **v** and the acceleration **a**. (*c*) Prove that the acceleration is directed toward the center of the circular motion.

Section 4–5 Relative Velocity and Acceleration

23. In a large department store a shopper is standing on the "up" escalator, which is traveling in a northerly direction at an angle of $40°$ above the horizontal and at a speed of 0.75 m/s. She passes her husband, who is standing on the identical, parallel "down" escalator. Find the velocity of the shopper relative to her husband.

24. A helicopter is flying in a straight line over a level field at a constant speed of 6.2 m/s and at a constant altitude of 9.5 m. A package is ejected horizontally from the helicopter with an initial velocity of 12 m/s relative to the helicopter, and in a direction opposite to the helicopter's motion. (*a*) Find the initial velocity of the package relative to the ground. (*b*) What is the horizontal distance between the helicopter and the package at the instant the package strikes the ground? (*c*) What angle does the velocity vector of the package make with the ground at the instant before impact?

25. A light plane attains an air speed of 500 km/h. The pilot sets out for his destination 800 km to the north, but discovers he has to head his plane $20°$ east of north to fly there directly. He arrives in exactly 2 h. What was the vector wind velocity?

26. A woman can row a boat 4.0 mi/h in still water. (*a*) If she is crossing a river where the current is 2.0 mi/h, in what direction will her boat be headed if she wants to reach a point directly opposite from her starting point? (*b*) If the river is 4.0 mi wide, how long will it take her to cross the river? (*c*) How long will it take her to row 2.0 mi *down* the river and then back to her starting point? (*d*) How long will it take her to row 2.0 mi *up* the river and then back to her starting point? (*e*) In what direction should she head the boat if she wants to cross in the shortest possible time?

27. A battleship steams due east at 24 km/h. A submarine 4.0 km away fires a torpedo that has a speed of 50 km/h; see Fig. 4–18. If the bearing of the ship as seen from the submarine is $20°$ east of north, (*a*) in what direction should the torpedo be fired to hit the ship, and (*b*) what will be the running time for the torpedo to reach the battleship?

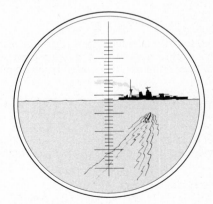

Figure 4–18 Problem 27.

ADDITIONAL PROBLEMS

28. In Bohr's model of the hydrogen atom an electron revolves around a proton in a circular orbit of radius 5.28×10^{-11} m with a speed of 2.18×10^6 m/s. What is the acceleration of the electron in this model of the hydrogen atom?

29. A particle A moves along the line $y = d$ (30 m) with a constant velocity $\mathbf{v}$ ($v = 3.0$ m/s) directed parallel to the positive x-axis (Fig. 4–19). A second particle B starts at the origin with zero speed and constant acceleration $\mathbf{a}$ ($a = 0.40$ m/s^2) at the same instant that particle A passes the y-axis. What angle θ between $\mathbf{a}$ and the positive y-axis would result in a collision between these two particles?

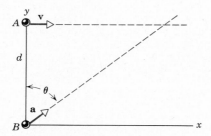

Figure 4–19 Problem 29.

30. A carnival Ferris wheel has a 15-m radius and completes five turns about its horizontal axis every minute. (a) What is the acceleration of a passenger at his highest point? (b) What is his acceleration at the lowest point?

31. In a cathode-ray tube a beam of electrons is projected horizontally with a speed of 1.0×10^9 cm/s into the region between a pair of horizontal plates 2.0 cm long. An electric field between the plates exerts a constant downward acceleration on the electrons of magnitude 1.0×10^{17} cm/s^2. Find (a) the vertical displacement of the beam in passing through the plates and (b) the components of the velocity of the beam as it emerges from the plates.

32. Find the angle of projection at which the horizontal range and the maximum height of a projectile are equal.

33. If the coordinates of a particle moving in a plane are given by $x = 3.0t - 4.0t^2$ and $y = -6.0t^2 + t^3$, where x and y are in meters and t is in seconds, find (a) the coordinates and displacement vector at $t = 3.0$ s, (b) the average velocity during the first 3.0 s, (c) the instantaneous velocity at $t = 3.0$ s, (d) the average acceleration during the first three seconds, and (e) the instantaneous acceleration at $t = 3.0$ s.

34. A woman 1.6 m tall stands upright at latitude 50° for 24 h. (a) During this interval, how much farther does the top of her head move than the soles of her feet? (b) How much greater is the acceleration of the top of her head than the acceleration of the soles of her feet? Consider only effects associated with the rotation of the earth.

35. The velocity $\mathbf{v}$ of a particle moving in the x-y plane is given by $\mathbf{v} = (6t - 4t^2)\mathbf{i} + 8\mathbf{j}$. Here $\mathbf{v}$ is in meters per second and t (>0) is in seconds. (a) What is the acceleration when $t = 3$ s? (b) When (if ever) is the acceleration zero? (c) When (if ever) is the velocity zero? (d) When (if ever) does the speed equal 10 m/s?

36. A boy whirls a stone in a horizontal circle 2.0 m above the ground by means of a string 1.5 m long. The string breaks, and the stone flies off horizontally, striking the ground 10 m away. What was the centripetal acceleration while in circular motion?

37. A particle leaves the origin with an initial velocity $\mathbf{v}$ ($= 3.0$ m/s) in the direction of the positive x-axis. It experiences a constant acceleration $\mathbf{a} = -1.0\mathbf{i} - 0.5\mathbf{j}$, in meters per second squared. (a) What is the velocity of the particle when it reaches its maximum x-coordinate? (b) Where is the particle at this time?

38. What is the maximum vertical height to which a baseball player can throw a ball if he can throw it a maximum distance of 60 m? Hint: see Problems 5 and 6.

39. A football is kicked off with an initial speed of 64 ft/s at a projection angle of 45°. A receiver on the goal line 60 yd away in the direction of the kick starts running to meet the ball at that instant. What must his speed be if he is to catch the ball before it hits the ground? Neglect air resistance. (See, in this connection, "Catching a Baseball," by Seville Chapman, *American Journal of Physics,* October 1968.)

40. A ball rolls off the top of a stairway with a horizontal velocity of magnitude 5.0 ft/s. The steps are 8.0 in. high and 8.0 in. wide. Which step will the ball hit first?

41. A train travels due south at 30 m/s (relative to ground) in a rain that is blown toward the south by the wind. The path of each raindrop makes the angle 22° with the vertical, as measured by an observer stationary on the earth. An observer seated in the train, however, sees perfectly vertical tracks of rain on the windowpane. Determine the speed of each raindrop relative to the earth.

42. A policeman is chasing a burglar across a rooftop; both are running at 15 ft/s. Before the burglar reaches the edge of the roof he has to decide whether or not to jump to the roof of the next building, which is 20 ft away but 16 ft lower; see Fig. 4–20. If he jumps horizontally, can he make it?

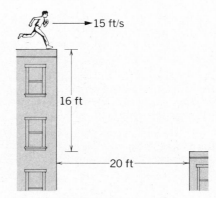

Figure 4–20 Problem 42.

43. A particle P travels with constant speed on a circle of radius 3.0 m and completes 1 rev in 20 s (Fig. 4–21). The particle passes through O at $t = 0$. Starting from the origin O, find (a) the magnitude and direction of the position vectors 5.0 s, 7.5 s, and 10 s later; (b) the magnitude and direction of the displacement in the 5.0-s interval from the fifth to the tenth second; (c) the average velocity vector in this interval; (d) the instantaneous velocity vector at the beginning and at the end of this interval; (e) the average acceleration vector in this interval; and (*f*) the instantaneous acceleration vector at the beginning and at the end of this interval.

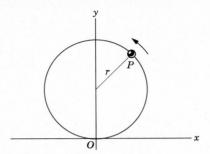

Figure 4–21   Problem 43.

44. A certain airplane has a speed of 180 mi/h and is diving at an angle of 30° below the horizontal when an object is dropped. The horizontal distance between the release point and the point where the object first strikes the ground is 2300 ft. (*a*) How high was the plane when the object was released? (*b*) How long was the object in the air?

45. A particle moves in a clockwise circular path of radius 0.60 m centered at the origin of the *x-y* plane with a period of revolution of 2.0 s. (*a*) What are the *x*- and *y*-components of the velocity when $x = 0.48$ m in the first quadrant? (*b*) What will be the *x*- and *y*-components of the radial acceleration at this same instant?

46. An airplane has a speed of 135 mi/h in still air. It is flying straight north so that it is at all times directly above a north-south highway. A ground observer tells the pilot by radio that a 70-mi/h wind is blowing; but neglects to tell him the wind direction. The pilot observes that in spite of the wind he can travel 135 mi along the highway in 1 h. In other words, his ground speed is the same as if there were no wind. (*a*) What is the direction of the wind? (*b*) What is the heading of the plane, that is, the angle between its axis and the highway?

47. The launching speed of a certain projectile is five times the speed it has at its maximum height. Calculate the elevation angle at launching.

48. A batter hits a pitched ball at a height 4.0 ft above ground so that its angle of projection is 45° and its horizontal range is 350 ft. The ball is fair down the left field line where a 24-ft-high fence is located 320 ft from home plate. Will the ball clear the fence?

49. The kicker on a football team can give the ball an initial speed of 25 m/s. Within what angular range must he kick the ball if he is to just score a field goal from a point 50 m in front of the goalposts whose horizontal bar is 3.44 m above the ground?

50. A wooden boxcar is moving along a straight railroad track at a speed v_1. A sniper fires a bullet (initial speed v_2) at it from a high-powered rifle. The bullet passes through both walls of the car,

its entrance and exit holes being exactly opposite to each other as viewed from within the car. From what direction, relative to the track, was the bullet fired? Assume that the bullet was not deflected upon entering the car, but that its speed decreased by 20%. Take $v_1 = 85$ km/h and $v_2 = 650$ m/s. (Are you surprised that you don't need to know the width of the boxcar?)

51. Two ships, *A* and *B*, leave port at the same time. *A* travels northwest at 24 knots and ship *B* travels at 28 knots in a direction 40° west of south. (1 knot = 1 nautical mile per hour.) (*a*) What are the magnitude and direction of the velocity of ship *A* relative to *B*? (*b*) After what time will they be 160 nautical miles apart? (*c*) What will be the bearing of *B* from *A* at that time?

52. A man wants to cross a river 500 m wide. His rowing speed (relative to the water) is 3000 m/h. The river flows at a speed of 2000 m/h. The man's walking speed on shore is 5000 m/h. (*a*) Find the path (combined rowing and walking) he should take to get to the point directly opposite his starting point in the shortest time. (*b*) How long does it take?

53. An anti-tank gun is located on the edge of a plateau that is 60 m above the surrounding plain. The gun crew sights an enemy tank stationary on the plain at a horizontal distance of 2.2 km from the gun. At the same moment, the tank crew sees the gun and starts to move directly away from it with an acceleration of 0.90 m/s². If the anti-tank gun fires a shell with a muzzle speed of 240 m/s at an elevation angle of 10° above the horizontal, how long should the gun crew wait before firing if they are to hit the tank? See Fig. 4–22.

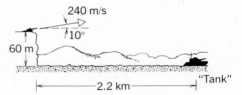

Figure 4–22   Problem 53.

***54.** A rocket is launched from rest and moves in a straight line at 70° above the horizontal with an acceleration of 46 m/s². After 30 s of powered flight, the engines shut off and the rocket follows a parabolic path back to earth; see Fig. 4–23. (*a*) Find the time of flight from launching to impact. (*b*) What is the maximum altitude reached? (*c*) What is the distance from launch pad to impact point?

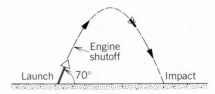

Figure 4–23   Problem 54.

5 Particle Dynamics—I

5–1 Introduction

In Chapters 3 and 4 we studied the motion of a particle, with emphasis on motion along a straight line or in a plane. We did not ask what "caused" the motion; we simply described it in terms of the vectors **r**, **v**, and **a**. Our discussion was thus largely geometrical. In this chapter we discuss the causes of motion, an aspect of mechanics called *Dynamics*. As before, bodies will be treated as though they were single particles. Later in the book we shall treat groups of particles and rigid bodies as well.

Dynamics

The concept of force and Newton's laws of motion will first be discussed. Then the particle dynamics for bodies subject to a force that is constant in both magnitude and direction will be presented. Gravitational and elastic forces, exerted respectively by the earth or by taut cords, will be introduced as illustrations.

5–2 Classical Mechanics

The motion of a given particle is determined by the nature and the arrangement of the other bodies that form its *environment*. Table 5–1 shows some "particles" and possible environments for them.

Classical mechanics

In what follows we limit ourselves to the study of gross objects moving at speeds that are small compared with the speed of light. This is the realm of *classical mechanics*. Specifically, we shall not inquire here into the motions of electrons in atoms or the collisions of high-energy protons. The first inquiry would involve us with the quantum theory and the second with the theory of relativity.

Table 5-1

	System	The Particle	The Environment
1.		A block	The spring; the rough surface
2.		A golf ball	The earth
3.		An orbiting satellite	The earth
4.		An electron	A large uniformly charged sphere
5.		A bar magnet	A second bar magnet

Although we now view classical mechanics as a special case of these more general theories we stress that: (1) It is an extremely important special case, encompassing as it does the motions of objects that range in size from molecules to galaxies. (2) Within its realm of applicability it is extremely accurate, as the highly successful celestial navigation of our space probes attests.

The central problem of classical particle mechanics is this: (1) We are given a particle whose characteristics (mass, charge, magnetic dipole moment, etc.) we know. (2) We place this particle, with a known initial velocity, in an environment of which we have a complete description. (3) Problem: What is the subsequent motion of the particle?

This problem was solved, at least for a large variety of environments, by Isaac Newton (1642–1727) when he put forward his laws of motion and formulated his law of universal gravitation. The program for solving this problem, in terms of our present understanding of classical mechanics, is this: (1) We introduce the concept of *force* **F** and define it in terms of the acceleration **a** experienced by a particular standard body. (2) We develop a procedure for assigning a *mass m* to a body so that we may understand the fact that different particles of the same kind experience different accelerations in the same environment. (3) Finally, we try to find ways of calculating the forces that act on particles from the properties of the particle and of its environment; that is, we look for *force laws*. Force, which is at root a technique for relating the environment to the motion of the particle, appears both in the laws of motion (which tell us what acceleration a given body will experience under the action of a given force) and in the force laws (which tell us how to calculate the force that will act on a given body in a given environment). The laws of motion and the force laws, taken together, constitute the laws of mechanics.

A program for mechanics

The program of mechanics cannot be tested piecemeal. We must view it as a unit and we shall judge it to be successful if we can say "yes" to these two questions. (1) Does the program yield results that agree with experiment? (2) Are the force laws simple in form? It is the crowning glory of Newtonian mechanics that for a fantastic variety of phenomena we can indeed answer each of these questions in the affirmative. The exceptions can usually be handled using the two extensions of Newtonian mechanics, namely, quantum mechanics and Einstein's special theory of relativity.

In this section we have used the terms force and mass rather unprecisely, having identified force with the influence of the environment, and mass with the resistance of a body to being accelerated when a force acts on it, a property often called inertia. In later sections we shall refine these primitive ideas about force and mass.

5–3 Newton's First Law

For centuries the problem of motion and its causes was a central theme of natural philosophy, an early name for what we now call physics. It was not until the time of Galileo and Newton, however, that dramatic progress was made. Isaac Newton, born in England in the year of Galileo's death, was the principal architect of classical mechanics. He carried to full fruition the ideas of Galileo and others who preceded him. His three laws of motion were first presented (in 1686) in his *Philosophiae Naturalis Principia Mathematica*, usually called the *Principia*.

Before Galileo's time most philosophers thought that some influence or "force" was needed to keep a body moving. To them, a body was in its "natural state" when it was at rest. For a body to move in a straight line at constant speed, for example, they believed that some external agent had continually to propel it; otherwise it would "naturally" stop moving.

If we wanted to test these ideas experimentally, we would first have to find a way to free a body from all influences of its environment or from all forces. This is hard to do, but in certain cases we can make the forces very small. If we study the motions as we make the forces smaller and smaller, we shall have some idea of what the motion would be like if the external forces were truly zero.

Let us place a test body—say, a block—on a rigid horizontal plane. If we let the block slide along this plane, we notice that it gradually slows down and stops. This observation was used, in fact, to support the idea that motion stopped when the external force, in this case the hand initially pushing the block, was removed. We can argue against this idea, however, reasoning somewhat as follows: Repeat the experiment using a smoother block and a smoother plane and providing a lubricant. You will notice that the velocity decreases more slowly than before. Now use still smoother blocks and surfaces and better lubricants; a block floating on a film of air on an "air track" is an easily realizeable near limit. As you might expect, you will find that the block decreases in velocity at a slower and slower rate and travels farther each time before coming to rest. We can now extrapolate and say that if all friction could be eliminated, the body would continue indefinitely in a straight line with constant speed. Some external force is necessary to *change* the velocity of a body but no external force is necessary to *maintain* the velocity of a body. Our hand, for example, exerts a force on the block when it sets it in motion. The rough plane exerts a force on it when it slows it down. Both of these forces produce a change in the velocity, that is, they produce an acceleration.

The idea that objects maintain their motion without outside influence was adopted by Newton as the first of his three laws of motion. Newton stated his first law in these words: "*Every body persists in its state of rest or of uniform motion in a straight line unless it is compelled to change that state by forces impressed on it.*"

Newton's first law

Newton's first law is really a statement about reference frames. For, in general, the acceleration of a body depends on the reference frame relative to which it is measured. The first law tells us that, if there are no nearby objects (and by this we mean that there are no forces because every force must be associated with an object in the environment) then it is possible to find a family of reference frames in which a particle has no acceleration. The fact that bodies stay at rest or retain their uniform linear motion in the absence of applied forces is often described by assigning a property to matter called *inertia*. Newton's first law is often called the law of inertia, and the reference frames to which it applies are therefore called *inertial frames*. Such frames are assumed to be unaccelerated with respect to the distant matter that establishes the large-scale structure of the universe.

Notice that there is no distinction in the first law between a body at rest and one moving with a constant velocity. Both motions are "natural" in the absence of forces. That this is so becomes clear when a body at rest in one inertial frame is viewed from a second inertial frame, that is, a frame moving with constant velocity with respect to the first. An observer in the first frame finds the body to be at rest; an observer in the second frame finds the same body to be moving with uniform velocity. Both observers find the body to have no acceleration, that is, no change in velocity with time, and both may conclude from the first law that no force acts on the body.

Notice, too, that by implication there is no distinction in the first law between the absence of all forces and the presence of forces whose resultant is zero. For example, if the push of our hand on a block exactly counteracts the force of friction on it, the block will move with uniform velocity. Hence another way of stating the first law is this: *If no net force acts on a body, its acceleration* **a** *is zero.*

If there is an interaction between the body and objects present in the environment, the effect may be to change the body's motion from its "natural" state. To investigate this we must now examine carefully the concept of force.

5–4 Force

Let us refine our concept of force by defining it operationally. In our everyday language force is associated with a push or a pull, perhaps exerted by our muscles or bones. In physics, however, we need a more precise definition. We define force here in terms of the acceleration that a given standard body experiences when placed in a suitable environment.

As a standard body we find it convenient to use (or rather to imagine that we use!) the standard kilogram; see Section 1–4. This has been assigned, by definition, a mass m_0 of exactly 1 kg. Later we shall describe how masses are assigned to other bodies.

For an environment we place the standard body on a horizontal table having negligible friction and we attach a spring to one end. We hold the other end of the spring in our hand, as in Fig. 5–1a. Now we pull the spring horizontally to the right so that by trial and error the standard body experiences a measured uniform acceleration of 1.00 m/s². We then declare, as a matter of definition, that the spring (which is the significant body in the environment) is exerting a constant force, whose magnitude we shall call "1.00 newton" (abbreviated N), on the standard body. We note that, in imparting this force, the spring is kept stretched an amount Δl beyond its normal unextended length, as Fig. 5–1b shows.

We can repeat the experiment, either stretching the spring more or using a stiffer spring, so that we measure an acceleration of 2.00 m/s² for the standard body. We now declare that the spring is exerting a force of 2.00 N on the standard

Inertia

The newton defined

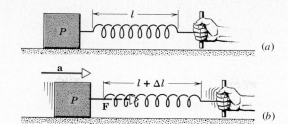

Figure 5-1 (*a*) A "particle" *P* (the standard kilogram) at rest on a horizontal frictionless surface. (*b*) The body is accelerated by pulling the spring to the right.

body. In general, if we observe this particular standard body to have an acceleration *a* in a particular environment, we then say that the environment is exerting a force *F* on the standard body, where *F* (in newtons) is numerically equal to *a* (in meters per second²).

Now let us see whether force, as we have defined it, is a vector quantity. In Fig. 5-1*b* we assigned a magnitude to the force *F*, and it is a simple matter to assign a direction to it as well, namely, the direction of the acceleration that the force produces. However, to be a vector it is not enough for a quantity to have magnitude and direction; it must also obey the laws of vector addition described in Chapter 2. We can learn only from experiment if forces, as we defined them, do indeed obey these laws.

We could arrange to exert a 4.00-N force along the *x*-axis and a 3.00-N force along the *y*-axis. Let us apply these forces simultaneously to the standard body placed, as before, on a horizontal, frictionless surface; see Fig. 5-2. What will be the acceleration of the standard body? We would find by experiment that it is 5.00 m/s², directed along a line that makes an angle of 36.9° with the *x*-axis. In other words, we would say that the standard body is experiencing a force of 5.00 N in this same direction. This same result can be obtained by adding the 4.00-N and 3.00-N forces vectorially according to the method of Section 2-2. Experiments of this kind show conclusively that forces are vectors; they have magnitude; they have direction; they add according to the laws of vector addition.

The result of experiments of this general type is often stated as follows: When several forces act on a body, each can produce its own acceleration independently. The resulting acceleration is the vector sum of the several independent accelerations.

Force is a vector

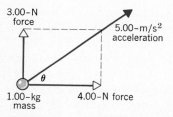

Figure 5-2 An object acted upon by two or more forces accelerates as if a single force equal to their vector sum were acting. In this example $\theta = \tan^{-1} \frac{3}{4} = 36.9°$.

5-5 Mass; Newton's Second Law

In Section 5-4 we considered only the accelerations given to one particular object, the standard kilogram. We were able thereby to define forces quantitatively. What effect would these forces have on other objects? Since our standard body was chosen arbitrarily in the first place, we know that for any given object the acceleration will be directly proportional to the force applied. The significant question remaining then is this: What effect will the same force have on different objects? Everyday experience gives us a qualitative answer. The same force will produce different accelerations on different bodies. A baseball will be accelerated more by a given force than will an automobile. In order to obtain a quantitative answer to this question we need a method to measure mass, the property of a body that determines its resistance to a change in its motion.

Let us attach a spring to our standard body (the standard kilogram, to which we have arbitrarily assigned a mass $m_0 = 1$ kg, exactly) and arrange to give it an acceleration a_0 of 2.00 m/s² using the method of Fig. 5-1*b*. Let us measure carefully the extension Δl of the spring associated with the force that the spring is exerting on the block.

Now we remove the standard kilogram and substitute an arbitrary body, whose mass we label m_1. We apply the same force (the one that accelerated the standard kilogram 2.00 m/s²) to the arbitrary body (by stretching the spring by the same amount) and we measure an acceleration a_1 of 0.50 m/s².

We define the ratio of the masses of the two bodies to be the inverse ratio of the acceleration given to these bodies by the same force, or

$$\frac{m_1}{m_0} = \frac{a_0}{a_1} \quad \text{(same force acting)}.$$

Measurement of mass

In this example we have, numerically,

$$m_1 = m_0 \left(\frac{a_0}{a_1}\right) = 1.00 \text{ kg} \left(\frac{2.00 \text{ m/s}^2}{0.50 \text{ m/s}^2}\right) = 4.00 \text{ kg}.$$

The second body, which has only one-fourth the acceleration of the first body when the same force acts on it, has, by definition, four times the mass of the first body. Hence mass may be regarded as a quantitative measure of inertia.

If we repeat the preceding experiment with a different common force acting, we find the ratio of the accelerations, a_0'/a_1', to be the same as in the previous experiment, or

$$\frac{m_1}{m_0} = \frac{a_0}{a_1} = \frac{a_0'}{a_1'}.$$

The ratio of the masses of two bodies is thus independent of the common force used.

Furthermore, experiment shows that we can consistently assign masses to any body by this procedure. For example, let us compare a second arbitrary body with the standard body, and thus determine its mass, say, m_2. We can now compare the two arbitrary bodies, m_2 and m_1, directly, obtaining accelerations a_2'' and a_1'' when the same force is applied. The mass ratio, defined as usual from

$$\frac{m_2}{m_1} = \frac{a_1''}{a_2''} \quad \text{(same force acting)},$$

turns out to have the same value that we obtain by using the masses m_2 and m_1 determined previously by direct comparison with the standard.

We can show, in still another experiment of this type, that if objects of mass m_1 and m_2 are fastened together, they behave mechanically as a single object of mass $(m_1 + m_2)$. In other words, masses add like (and are) scalar quantities.

We can now summarize all the experiments and definitions described above in one equation of classical mechanics,

Newton's second law

$$\mathbf{F} = m\mathbf{a}. \tag{5-1}$$

In this equation $\mathbf{F}$ is the (vector) sum of all the forces acting on the body, m is the mass of the body, and $\mathbf{a}$ is its acceleration. Equation 5–1 may be taken as a statement of *Newton's second law*. If we write it in the form $\mathbf{a} = \mathbf{F}/m$, we can see easily that the acceleration of the body is directly proportional to the resultant force acting on it and parallel in direction to this force and that the acceleration, for a given force, is inversely proportional to the mass of the body.

Notice that the first law of motion is contained in the second law as a special case, for if $\mathbf{F} = 0$, then $\mathbf{a} = 0$. In other words, if the resultant force on a body is zero, the acceleration of the body is zero. Therefore, in the absence of a net applied force, a body will move with constant velocity or be at rest (zero velocity), which is what the first law of motion says. The division of translational particle dynamics that includes only systems for which the resultant force **F** is zero is called *statics*.

Equation 5–1 is a vector equation. We can write this single vector equation as three scalar equations,

Newton's second law—scalar form

$$F_x = ma_x, \qquad F_y = ma_y, \qquad \text{and} \qquad F_z = ma_z, \qquad (5\text{–}2)$$

relating the x-, y-, and z-components of the resultant force (F_x, F_y, and F_z) to the x-, y-, and z-components of acceleration (a_x, a_y, and a_z) for the mass m.

Equation 5–2 is almost universally referred to as Newton's second law of motion. Newton himself, however, expressed this law in a different form, which we shall explore in Section 9-4. The form **F** = m**a** was first given by the Swiss mathematician and physicist Leonhard Euler in 1752, 65 years after the publication of Newton's *Principia*.*

Example 1 *Linear motion* A 360-kg box rests on the bed of a truck that is going down a slope on a straight section of road as in Fig. 5–3. The truck starts down the slope with a speed of 100 km/h and slows to 60 km/h in 20 s. Assume that the acceleration is constant during this interval and calculate the strength and direction of the resultant force that must act on the box during this period of time.

Figure 5–3 Example 1. A truck slows as it goes down an incline. Note the directions of its velocity and acceleration. In what direction is the resultant force on the box?

We can use Newton's second law to calculate the required force if we first calculate the acceleration of the box. From Eq. 3–12 we have

$$a = \frac{v - v_0}{t}$$

$$= \frac{60 \text{ km/h} - 100 \text{ km/h}}{20 \text{ s}}$$

$$= -2.0 \frac{\text{km}}{\text{h} \cdot \text{s}} \left(\frac{1 \text{ h}}{3600 \text{ s}}\right)\left(\frac{10^3 \text{ m}}{1 \text{ km}}\right) = -0.56 \text{ m/s}^2.$$

The negative sign reminds us that the box is slowing down, so that its acceleration is in the opposite direction to its velocity. The magnitude of the resultant force follows from

$$F = ma$$
$$= (360 \text{ kg})(0.56 \text{ m/s}^2) = 200 \text{ N}.$$

This force acts in the same direction as the acceleration: parallel to the incline and opposite the direction of the truck's velocity. If the cable holding the box to the truck bed were to snap, which way would the box move with respect to the truck?

Example 2 *Circular motion* Calculate the magnitude and direction of the force acting on the moon in its circular orbit around the earth.

We calculated the centripetal acceleration of the moon to be $a = 2.73 \times 10^{-3} \text{ m/s}^2$ in Example 3 of Chapter 4. Using the moon's mass of 7.36×10^{22} kg from Appendix C, we can calculate the magnitude of the required force to be

$$F = ma$$
$$= (7.36 \times 10^{22} \text{ kg})(2.73 \times 10^{-3} \text{ m/s}^2) = 2.01 \times 10^{20} \text{ N}.$$

This force must be directed in the same direction as the acceleration: toward the center of the orbit (toward the earth), perpendicular to the moon's velocity.

Newton's second law applies to centripetal acceleration the same as it does to any other acceleration. Whenever an object with mass m is accelerating, we may be sure that its acceleration is caused by one or more forces whose vector sum is **F** = m**a**. Note that it is appropriate to treat both the box of Example 1 and the moon in this example as particles. Why?

5–6 Systems of Mechanical Units

Unit force is defined as a force that causes a unit of acceleration when applied to a unit mass. In SI terms unit force is the force that will accelerate a 1-kg mass at 1 m/s²; we have seen that this unit is called the newton. In the cgs (centimeter, gram, second) system unit force is the force that will accelerate a 1-g mass at

* See "The Second Law of Motion and Newton's Equations," by V. V. Raman, *The Physics Teacher*, March 1972.

1 cm/s²; this unit is called the *dyne*. Since 1 kg = 10^3 g and 1 m/s² = 10^2 cm/s², it follows that 1 N = 10^5 dynes.

In each of our systems of units we have chosen mass, length, and time as our fundamental quantities. Standards were adopted for these fundamental quantities and units defined in terms of these standards. Force appears as a derived quantity, determined from the relation $\mathbf{F} = m\mathbf{a}$.

In the British system of units, however, *force,* length, and time are chosen as the fundamental quantities and mass is a derived quantity. In this system, mass is determined from the relation $m = F/a$. The standard and unit of force in this system is the *pound*. Actually, the pound of force was originally defined to be the pull of the earth on a certain standard body at a certain place on the earth. We can get this force in an operational way by hanging the standard body from a spring at the particular point where the earth's pull on it is defined to be 1 lb of force (exactly). If the body is at rest, the earth's pull on the body, its weight W, is balanced by the tension in the spring. Therefore $T = W = 1$ lb, in this instance. We can now use this spring (or any other one thus calibrated) to exert a force of 1 lb on any other body; to do this we simply attach the spring to another body and stretch it the same amount as the pound force had stretched it. The standard body has been compared to the kilogram and found to have the mass 0.45359237 kg. The acceleration due to gravity at the certain place on the earth is found to be 32.1740 ft/s². The pound of force can therefore be defined from $F = ma$ as the force that accelerates a mass of 0.45359237 kg at the rate of 32.1740 ft/s².

This procedure enables us to compare the pound-force with the newton. Using the fact that 32.1740 ft/s² equals 9.8066 m/s², we find that

$$
\begin{aligned}
1 \text{ lb} &= (0.45359237 \text{ kg})(32.1740 \text{ ft/s}^2) \\
&= (0.45359237 \text{ kg})(9.8066 \text{ m/s}^2) \\
&\cong 4.45 \text{ N}
\end{aligned}
$$

The unit of mass in the British system can now be derived. It is defined as the mass of a body whose acceleration is 1 ft/s² when the force on it is 1 lb; this mass is called the *slug*. Thus, in this system

$$F[\text{lb}] = m \text{ [slugs]} \times a \text{ [ft/s}^2\text{]}.$$

The unit of mass in the British system is the slug. The units of force, mass, and acceleration in the three systems are summarized in Table 5–2.

The slug

Three systems of units

Table 5–2 Units in $F = ma$

System of units	Force	Mass	Acceleration
SI	newton (N)	kilogram (kg)	m/s²
Cgs	dyne	gram (g)	cm/s²
British	pound (lb)	slug	ft/s²

The dimensions of force are the same as those of mass times acceleration. In a system in which mass, length, and time are the fundamental quantities, the dimensions of force are, therefore, mass × length/time² or MLT^{-2}.

5–7 Newton's Third Law

Forces acting on a body originate in other bodies that make up its environment. Any single force is only one aspect of a mutual interaction between two bodies.

We find by experiment that when one body exerts a force on a second body, the second body always exerts a force on the first. Furthermore, we find that these forces are equal in magnitude but opposite in direction. A single isolated force is therefore an impossibility.

If one of the two forces involved in the interaction between two bodies is called an "action" force, the other is called the "reaction" force. Either force may be considered the "action" and the other the "reaction." Cause and effect is not implied here, but a mutual simultaneous interaction is implied.

Newton's third law

This property of forces was first stated by Newton in his third law of motion: "*To every action there is always opposed an equal reaction; or, the mutual actions of two bodies upon each other are always equal, and directed to contrary parts.*"

In other words, if body A exerts a force on body B, body B exerts an equal but oppositely directed force on body A; and furthermore the forces lie along the line joining the bodies. Notice that the action and reaction forces, which always occur in pairs, act on different bodies. If they were to act on the same body, we could never have accelerated motion because the resultant force on every body would always be zero.

Imagine a boy kicking open a door. The force exerted by the boy B on the door D accelerates the door (it flies open); at the same time, the door D exerts an equal but opposite force on the boy B, which decelerates the boy (his foot loses forward velocity). The boy will be painfully aware of the "reaction" force to his "action."

The following examples illustrate the application of the third law and clarify its meaning.

Example 3 *A dynamic case* Consider a man pulling horizontally on a rope attached to a block on a horizontal table as in Fig. 5–4. The man pulls on the rope with a force $\mathbf{F}_{MR}$. The rope exerts a reaction force $\mathbf{F}_{RM}$ on the man. According to Newton's third law, $\mathbf{F}_{MR} = -\mathbf{F}_{RM}$. Also, the rope exerts a force $\mathbf{F}_{RB}$ on the block, and the block exerts a reaction force $\mathbf{F}_{BR}$ on the rope. Again according to the third law, $\mathbf{F}_{RB} = -\mathbf{F}_{BR}$.

Suppose that the rope has a mass m_R. Then, in order to start the block and rope moving from rest, we must have an acceleration, say, **a**. The only forces acting on the rope are $\mathbf{F}_{MR}$ and $\mathbf{F}_{BR}$, so that the resultant force on it is $\mathbf{F}_{MR} + \mathbf{F}_{BR}$, and this must be different from zero if the rope is to accelerate. In fact, from the second law we have

$$\mathbf{F}_{MR} + \mathbf{F}_{BR} = m_R\mathbf{a}.$$

Since the forces and the acceleration are along the same line, we can drop the vector notation and write the relation between the magnitudes of the vectors, namely,

$$F_{MR} - F_{BR} = m_R a.$$

We see therefore that in general $\mathbf{F}_{MR}$ does not have the same magnitude as $\mathbf{F}_{BR}$ (Fig. 5–4*b*). These two forces act on the same body and are not action and reaction pairs.

According to Newton's third law the magnitude of $\mathbf{F}_{MR}$ always equals the magnitude of $\mathbf{F}_{RM}$, and the magnitude of $\mathbf{F}_{RB}$ always equals the magnitude of $\mathbf{F}_{BR}$. However, only if the acceleration **a** of the system is zero will we have the pair of forces $\mathbf{F}_{MR}$ and $\mathbf{F}_{RM}$ equal in magnitude to the pair of forces $\mathbf{F}_{RB}$ and $\mathbf{F}_{BR}$ (Fig. 5–4*a*). In this special case only, we could imagine that the rope merely transmits the force exerted by the man to the block without change. This same result holds in principle if $m_R = 0$.

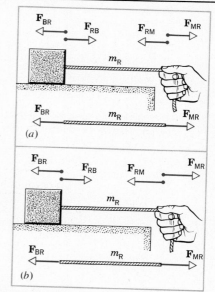

Figure 5–4 Example 3. A man pulls on a rope attached to a block. (*a*) The forces exerted on the rope by the block and by the man are equal and opposite. Thus the resultant horizontal force on the rope is zero, as is shown in the free-body diagram. The rope does not accelerate. (*b*) The force exerted on the rope by the man exceeds that exerted by the block. The net horizontal force has magnitude $F_{MR} - F_{BR}$ and points to the right. Thus the rope is accelerated to the right. The block is also acted upon by a frictional force not shown here.

In practice, we never find a massless rope. However, we can often neglect the mass of a rope, in which case the rope is assumed to transmit a force unchanged. Speaking loosely, the force transmitted by a stretched rope at any point is called the *tension* at that point. We can measure this tension by cutting the rope at the point in question and inserting a spring scale; the tension is the reading of the scale. The tension is the same at all points in the rope only if the rope is unaccelerated or assumed to be massless.

Example 4 *A static case*   Consider a spring attached to the ceiling and at the other end holding a block at rest (Fig. 5–5a). Since no body is accelerating, all the forces on any body add vectorially to zero. For example, the forces on the suspended block are **T**, the tension in the stretched spring, pulling vertically up on the mass, and **W**, the pull of the earth acting vertically down on the body, called its weight. These are drawn in Fig. 5–5b, where we show only the block for clarity. There are no other forces on the block.

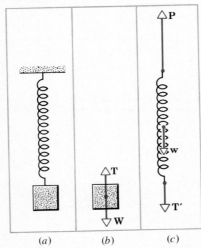

Figure 5–5   Example 4. (a) A block is suspended by a spring. (b) A free-body diagram showing all the vertical forces exerted on the block. (c) A similar diagram for the vertical forces on the spring.

In Newton's second law, **F** represents the (vector) sum of all the forces acting on a body, so that for the block

$$\mathbf{F} = \mathbf{T} + \mathbf{W}.$$

The block is at rest so that its acceleration is zero, or **a** = 0. Hence, from the relation $\mathbf{F} = m\mathbf{a}$, we obtain $\mathbf{T} + \mathbf{W} = 0$, or

$$\mathbf{T} = -\mathbf{W}.$$

The forces act along the same line, so that their magnitudes are equal, or

$$T = W.$$

Therefore the tension in the spring is an exact measure of the weight of the block. We shall use this result later in presenting a static procedure for measuring forces.

It is instructive to examine the forces exerted on the spring; they are shown in Fig. 5–5c. **T**′ is the pull of the block on the spring and is the reaction force of the action force **T**. **T**′ therefore has the same magnitude as **T**, which is W. **P** is the upward pull of the ceiling on the spring, and **w** is the weight of the spring, that is, the pull of the earth on it. Since the spring is at rest and all forces act along the same line, we have

$$\mathbf{P} + \mathbf{T}' + \mathbf{w} = 0,$$

or

$$P = W + w.$$

The ceiling therefore pulls up on the spring with a force whose magnitude is the sum of the weights of the block and spring.

From the third law of motion, the force exerted by the spring on the ceiling, **P**′, must be equal in magnitude to **P**, which is the reaction force to the action force **P**′. Therefore **P**′ has a magnitude $W + w$.

In general, the spring exerts different forces on the bodies attached at its different ends, for $P' \neq T$. In the special case in which the weight of the spring is negligible, $w = 0$ and $P' = W = T$. Therefore a weightless spring (or cord) may be considered to transmit a force from one end to the other without change.

It is instructive to classify all the forces in this problem according to action and reaction pairs. The reaction to **W**, a force exerted by the earth on the block, must be a force exerted by the block on the earth. Similarly, the reaction to **w** is a force exerted by the spring on the earth. Because the earth is so massive, we do not expect these forces to impart a noticeable acceleration to the earth. Since the earth is not shown in our diagrams, these forces have not been shown. The forces **T** and **T**′ are action-reaction pairs, as are **P** and **P**′. Notice that although $\mathbf{T} = -\mathbf{W}$ in our problem, these forces are not an action-reaction pair because they act on the same body.

5–8   The Force Laws

The three laws of motion that we have described are only part of the program of mechanics that we outlined in Section 5–2. It remains to investigate the force laws, which are the procedures by which we calculate the force acting on a given body in terms of the properties of the body and its environment. Newton's second law,

$$\mathbf{F} = m\mathbf{a}, \tag{5-3}$$

is essentially not a law of nature but a definition of force. We need to identify various functions of the type

$\mathbf{F}$ = a function of the properties of the particle and of the environment (5-4)

so that we can, in effect, eliminate $\mathbf{F}$ between Eqs. 5-3 and 5-4, thus obtaining an equation that will let us calculate the acceleration of a particle in terms of the properties of the particle and its environment. Force is a concept that connects the acceleration of the particle on the one hand with the properties of the particle and its environment on the other. We indicated earlier that one criterion for declaring the program of mechanics to be successful would be the discovery that *simple* laws of the type of Eq. 5-4 do indeed exist. This turns out to be the case, and this fact constitutes the essential reason that we "believe" the laws of classical mechanics. If the force laws had turned out to be very complicated, we would not be left with the feeling that we had gained much insight into the workings of nature.

The number of possible environments for an accelerated particle is so great that a detailed discussion of all the force laws is not feasible in this chapter. We shall, however, indicate in Table 5-3 the force laws that apply to the five particle-plus-environment situations of Table 5-1. At appropriate places throughout the text we discuss these and other force laws in detail; several of the laws in Table 5-3 are approximations or special cases.

Table 5-3   The force laws for the systems of Table 5-1

System	Force
1. A block propelled by a stretched spring over a rough horizontal surface	(a) Spring force: $F = -kx$, where x is the extension of the spring and k is a constant that describes the spring; $\mathbf{F}$ points to the right; see Chapter 14. (b) Friction force: $F = \mu mg$, where μ is the coefficient of friction and mg is the weight of the block; $\mathbf{F}$ points to the left; see Section 6-1.
2. A golf ball in flight	$F = mg$; $\mathbf{F}$ points down (see Section 5-9).
3. An orbiting satellite	$F = GmM/r^2$, where G is the gravitational constant, M the mass of the earth, and r the orbit radius; $\mathbf{F}$ points toward the center of the earth; see Chapter 15. This is *Newton's law of universal gravitation.*
4. An electron near a charged sphere	$F = (1/4\pi\epsilon_0)eQ/r^2$, where ϵ_0 is a constant, e is the electron charge, Q is the charge on the sphere, and r is the distance from the electron to the center of the sphere; $\mathbf{F}$ points to the right; see Chapter 23. This is *Coulomb's law of electrostatics.*
5. Two bar magnets	$F = (3\mu_0/2\pi)\mu^2/r^4$, where μ_0 is a constant, μ is the magnetic-dipole moment of each magnet, and r is the center-to-center separation of the magnets; we assume that $r \gg l$, where l is the length of each magnet; $\mathbf{F}$ points to the right.

5-9   Weight and Mass

The *weight* of a body is the gravitational force exerted on it by the earth. Weight, being a force, is a vector quantity. The direction of this vector is the direction of the gravitational force, that is, toward the center of the earth. The magnitude of the weight is expressed in force units, such as pounds or newtons.

When a body of mass m is allowed to fall freely, its acceleration is that of gravity **g** and the force acting on it is its weight **W**. Newton's second law, $\mathbf{F} = m\mathbf{a}$, when applied to a freely falling body, gives us $\mathbf{W} = m\mathbf{g}$. Both **W** and **g** are vectors directed toward the center of the earth. We can therefore write

$$W = mg, \tag{5-5}$$

where W and g are the magnitudes of the weight and acceleration vectors, respectively. To keep an object from falling we have to exert on it an upward force equal in magnitude to W, so as to make the net force zero. In Fig. 5–5a the tension in the spring supplies this force.

We stated previously that g is found experimentally to have the same value for all objects at the same place. From this it follows that the ratio of the weights of two objects must be equal to the ratio of their masses. Therefore an equal-arm balance, which actually is an instrument for comparing two downward forces, can be used to compare masses. If a sample of salt in one pan of a balance is pushing down on that pan with the same force as is a standard 1-g mass on the other pan, we infer that the mass of salt is equal to 1 g. We are likely to say that the salt "weighs" 1 g, although a gram is a unit of mass, not weight. However, it is always important to distinguish carefully between weight and mass.

We have seen that the weight of a body, the downward pull of the earth on that body, is a vector quantity. The mass of a body is a scalar quantity. The quantitative relation between weight and mass is given by $\mathbf{W} = m\mathbf{g}$. Because **g** varies from point to point on the earth, **W**, the weight of a body of mass m, is different in different localities. Thus, the weight of a one kg mass in a locality where g is 9.80 m/s² is 9.80 N; in a locality where g is 9.78 m/s², the same one kg mass weighs 9.78 N. If these weights were determined by measuring the amount of stretch required in a spring to balance them, the difference in weight of the same one kg mass at the two different localities would be evident in the slightly different stretch of the spring at these two localities. Hence, unlike the mass of a body, which is an intrinsic property of the body, the weight of a body depends on its location relative to the center of the earth. Spring scales read differently, balances the same, at different parts of the earth.

We shall generalize the concept of weight in Chapter 15 in which we discuss universal gravitation. There we shall see that the weight of a body is zero in regions of space where the gravitational effects are nil, although the inertial effects, and hence the mass of the body, remain unchanged from those on earth. In a space ship free from the influence of gravity it is a simple matter to "lift" a large block of lead ($\mathbf{W} = 0$), but the astronaut would still stub his toe if he were to kick the block ($m \neq 0$). It takes the same force to accelerate a body in gravity-free space as it does to accelerate it along a horizontal frictionless surface on earth, for its mass is the same in each place.

Often, instead of being given the mass, we are given the weight of a body on which forces are exerted. The acceleration **a** produced by the force **F** acting on a body whose weight has a magnitude W can be obtained by combining Eq. 5–3 and Eq. 5–5. Thus, from $\mathbf{F} = m\mathbf{a}$ and $W = mg$, we obtain

$$m = \frac{W}{g}, \qquad \text{so that} \qquad \mathbf{F} = \left(\frac{W}{g}\right)\mathbf{a}. \tag{5-6}$$

The quantity W/g plays the role of m in the equation $F = ma$ and is, in fact, the mass of a body whose weight has the magnitude W. For example, a man whose weight is 160 lb at a point where $g = 32.0$ ft/s² has a mass $m = W/g = (160 \text{ lb})/(32.0 \text{ ft/s}^2) = 5.00$ slugs. Notice that his weight at another point where $g = 32.2$ ft/s² is $W = mg = (5.00 \text{ slugs})(32.2 \text{ ft/s}^2) = 161$ lb.

Weight vs. mass

5–10   A Static
Procedure for Measuring Forces

In Section 5–4 we defined force by measuring the acceleration imparted to a standard body by pulling on it with a stretched spring. That may be called a dynamic method for measuring force. Although convenient for the purposes of definition, it is not a particularly practical procedure for the measurement of forces. Another method for measuring forces is based on measuring the change in shape or size of a body (a spring, say) on which the force is applied when the body is unaccelerated. This may be called the static method of measuring forces.

The idea of the static method is to use the fact that when a body, under the action of several forces, has zero acceleration, the vector sum of all the forces acting on the body must be zero. This is, of course, just the content of the first law of motion.

The instrument most commonly used to measure forces in this way is the *spring balance*. It consists of a coiled spring having a pointer at one end that moves over a scale. A force exerted on the balance changes the length of the spring. If a body weighing 1.00 N is hung from the spring, the spring stretches until the pull of the spring on the body is equal in magnitude but opposite in direction to its weight. A mark can be made on the scale next to the pointer and labeled "1.00-N force." Similarly, 2.00-N, 3.00-N, and so on, weights may be hung from the spring and corresponding marks can be made on the scale next to the pointer in each case. In this way the spring is calibrated. We assume that the force exerted on the spring is always the same when the pointer stands at the same position. The calibrated balance can now be used to measure any suitable unknown force, not merely the pull of the earth on some body.

The third law is tacitly used in our static procedure because we assume that the force exerted by the spring on the body is the same in magnitude as the force exerted by the body on the spring. This latter force is the force we wish to measure.

If a spring balance is kept in a fixed location, so that the value of g remains unchanged, the balance can be calibrated to read masses because of the relation (see Eq. 5–6) $m = W/g$. In chemistry laboratories spring balances, suitably cased and with a digital read-out, are commonly so used.

5–11   Some Applications
of Newton's Laws of Motion

Steps in problem solving

It will be helpful to write down some procedures for solving problems in classical mechanics and to illustrate them by several examples. Newton's second law states that the vector sum of all the forces acting on a body is equal to its mass times its acceleration. The first step in problem solving is therefore: (1) Identify the body to whose motion the problem refers. Lack of clarity on the point as to what has been or should be picked as "the body" is a major source of mistakes. (2) Having selected "the body," we next turn our attention to the objects in "the environment" because these objects (inclined planes, springs, cords, the earth, etc.) exert forces on the body. We must be clear as to the nature of these forces. (3) The next step is to select a suitable (inertial) reference frame. We should position the origin and orient the coordinate axes so as to simplify the task of our next step as much as possible. (4) We now make a separate diagram of the body alone, showing the reference frame and all of the forces acting on the body. This is called a *free-body* diagram. (5) Finally apply Newton's second law, in the form of Eq. 5–2, to each component of force and acceleration.

The following examples illustrate the method of analysis used in applying Newton's laws of motion. Each body is treated as if it were a particle of definite mass, so that the forces acting on it may be assumed to act at a point. Strings and pulleys are considered to have negligible mass. Although some of the situations picked for analysis may seem simple and artificial, they are the prototypes for many interesting real situations; but, more important, the method of analysis—which is the chief thing to understand—is applicable to all the modern and sophisticated situations of classical mechanics, even sending a spaceship to Mars.

Example 5   Figure 5–6a shows an object of weight W hung by strings. Consider the knot at the junction of the three strings to be "the body." The body remains at rest under the action of the three forces shown in Fig. 5–6b. Suppose we are given the magnitude of one of these forces. How can we find the magnitude of the other forces?

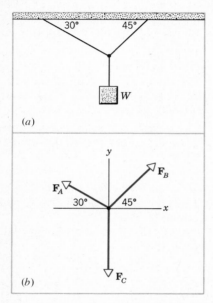

(a)

(b) $\mathbf{F}_C$

Figure 5–6   Example 5. (a) A block of weight W is suspended by strings. (b) A free-body diagram showing all the forces acting on the knot. The strings are assumed to be weightless.

$\mathbf{F}_A$, $\mathbf{F}_B$, and $\mathbf{F}_C$ are *all* the forces acting *on* the body. Since the body is unaccelerated, $\mathbf{F}_A + \mathbf{F}_B + \mathbf{F}_C = 0$. Choosing the x- and y-axes as shown, we can write this vector equation as three scalar equations:

$$F_{Ax} + F_{Bx} = 0,$$
$$F_{Ay} + F_{By} + F_{Cy} = 0,$$

using Eq. 5–2. The third scalar equation for the z-axis is simply

$$F_{Az} = F_{Bz} = F_{Cz} = 0.$$

That is, the vectors all lie in the x-y plane so that they have no z-components.

From the figure we see that

$$F_{Ax} = -F_A \cos 30° = -0.866F_A,$$
$$F_{Ay} = F_A \sin 30° = 0.500F_A,$$

and

$$F_{Bx} = F_B \cos 45° = 0.707F_B,$$
$$F_{By} = F_B \sin 45° = 0.707F_B.$$

Also,

$$F_{Cy} = -F_C = -W,$$

because the string C merely serves to transmit the force on one end to the junction at its other end. Substituting these results into our original equations, we obtain

$$-0.866F_A + 0.707F_B = 0,$$
$$0.500F_A + 0.707F_B - W = 0.$$

If we are given the magnitude of any one of these three forces, we can solve these equations for the other two. For example, if $W = 100$ N, we obtain $F_A = 73.3$ N and $F_B = 89.6$ N.

Example 6   We wish to analyze the motion of a block on a smooth incline.

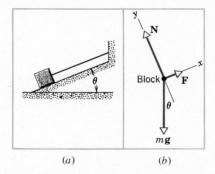

(a)   (b)

Figure 5–7   Example 6. (a) A block is held on a smooth inclined plane by a string. (b) A free-body diagram showing all the forces acting on the block.

(a) *Static case.* Figure 5–7a shows a block of mass m kept at rest on a smooth plane, inclined at an angle θ with the horizontal, by means of a string attached to the vertical wall. The forces acting *on* the block, which we choose as "the body," are shown in Fig. 5–7b. $\mathbf{F}$ is the force exerted *on* the block by the string; $m\mathbf{g}$ is the force exerted *on* the block by the earth, that is, its weight; and $\mathbf{N}$ is the force exerted *on* the block by the inclined surface. $\mathbf{N}$, called the *normal force*, is perpendicular (that is, normal) to the surface of contact because there is no frictional force between the surfaces. If there were a frictional force, $\mathbf{N}$ would have a component parallel to the incline. Because we wish to analyze the motion of *the block,* we choose *ALL the forces acting ON the block.* Note that the block will exert forces on other bodies in its environment (the string, the earth, the surface of the incline) in accordance with the action-reaction principle; these forces, however, are not needed to determine the motion of the block because they do not act on the block.

Suppose that θ and m are given. How do we find F and N? Since the block is unaccelerated, we obtain

$$\mathbf{F} + \mathbf{N} + m\mathbf{g} = 0.$$

It is convenient to choose the x-axis of our reference frame to be along the incline and the y-axis to be normal to the incline (Fig. 5–7b). With this choice of coordinates, only one force, $m\mathbf{g}$, must be resolved into components in solving the problem. The two scalar equations obtained by resolving $m\mathbf{g}$ along the x- and y-axes are

$$F - mg \sin \theta = 0 \quad \text{and} \quad N - mg \cos \theta = 0,$$

from which F and N can be obtained if θ and m are given. Note that these equations reduce to expected results for the special cases of θ = zero and θ = 90°.

(*b*) *Dynamic case.* Now suppose that we cut the string. Then the force $\mathbf{F}$, the pull of the string on the block, will be removed. The resultant force on the block will no longer be zero, and the block will accelerate. What is its acceleration?

From Eq. 5–2 we have $F_x = ma_x$ and $F_y = ma_y$. Using these relations we obtain

$$N - mg \cos \theta = ma_y = 0$$

and

$$-mg \sin \theta = ma_x,$$

which yield

$$a_y = 0, \qquad a_x = -g \sin \theta.$$

The acceleration is directed down the incline with a magnitude of $g \sin \theta$.

(*c*) *Constant speed.* With what force would we need to pull on the string so that the block could move up the incline at constant speed?

In this case, the block has zero acceleration, just as it did in static situation discussed above. The analysis proceeds exactly as in part (*a*), with the result that the required force has a magnitude

$$F = mg \sin \theta.$$

Notice that the same force would be required if the block were to slide down the incline with constant speed.

Example 7 Figure 5–8a shows a block of mass m_1 on a smooth horizontal surface pulled by a massless string attached to a block of mass m_2 hanging over a pulley. We assume that the pulley has no mass and is frictionless and that it merely serves to change the direction of the tension in the string at that point. The magnitude of the tension is the same throughout a massless string (see Example 4). Find the acceleration of the system and the tension in the string.

The blocks m_1 and m_2 shown in Fig. 5–8a are arranged and connected so that, if either block suffers a displacement, the other block will suffer a simultaneous displacement of the same magnitude. The directions of these displacements differ, however, those for m_1 being horizontal and those for m_2 being vertical. The instantaneous velocities—and also the instan-

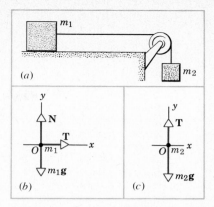

Figure 5–8 Example 7. (*a*) Two masses are connected by a string; m_1 lies on a smooth table, m_2 hangs freely. (*b*) A free-body diagram showing all the forces acting on m_1. (*c*) A similar diagram for m_2.

taneous accelerations—of the two blocks are related in these same ways.

Suppose we choose the block of mass m_1 as the body whose motion we investigate. The forces on this block, taken to be a particle, are shown in Fig. 5–8b. $\mathbf{T}$, the tension in the string, pulls on the block to the right; $m_1\mathbf{g}$ is the downward pull of the earth on the block and $\mathbf{N}$ is the vertical force exerted on the block by the smooth table. The block will accelerate in the x-direction only, so that $a_{1y} = 0$. We, therefore, can write

$$N - m_1 g = 0 = m_1 a_{1y},$$

and

$$T = m_1 a_{1x}. \tag{5-7}$$

From these equations we conclude that $N = m_1 g$. We do not know T, so we cannot solve for a_{1x}.

To determine T we must consider the motion of the block m_2. The forces acting on m_2 are shown in Fig. 5–8c. Because the string and block are accelerating, we cannot conclude that T equals $m_2 g$. In fact, if T were to equal $m_2 g$, the resultant force on m_2 would be zero, a condition holding only if the system is not accelerated.

The equation of motion for the suspended block is

$$T - m_2 g = m_2 a_{2y}. \tag{5-8}$$

If a is the common magnitude of the two accelerations we have, taking the directions of the two acceleration vectors properly into account,

$$a_{1x} = a \quad \text{and} \quad a_{2y} = -a.$$

We then obtain from Eqs. 5–7 and 5–8

$$T = m_1 a,$$

and $\tag{5-9}$

$$m_2 g - T = m_2 a.$$

These yield

$$a = \frac{m_2}{m_1 + m_2} g, \tag{5-10}$$

and

$$T = \frac{m_1 m_2}{m_1 + m_2} g, \tag{5-11}$$

which gives us the acceleration of the system a and the tension in the string T.

Notice that the tension in the string is always less than m_2g. This is clear from Eq. 5–11, which can be written

$$T = m_2g \frac{m_1}{m_1 + m_2}.$$

Notice also that a is always less than g, the acceleration due to gravity. Only when m_1 equals zero, which means that there is no block at all on the table, do we obtain $a = g$ (from Eq. 5–10). In this case $T = 0$ (from Eq. 5–11).

We can interpret Eq. 5–10 in a simple way. If we take as "the body" the two blocks together, of mass $m_1 + m_2$, the net unbalanced force on this system is represented by m_2g. Hence, from $F = ma$, we obtain Eq. 5–10 directly.

To make the example specific, suppose that $m_1 = 2.0$ kg and $m_2 = 1.0$ kg. Then

$$a = \frac{m_2}{m_1 + m_2} g = \tfrac{1}{3}g = 3.3 \text{ m/s}^2,$$

and

$$T = \frac{m_1m_2}{m_1 + m_2} g = (0.67 \text{ kg})(9.8 \text{ m/s}^2) = 6.5 \text{ N}.$$

Example 8 Consider two unequal masses connected by a string that passes over a frictionless and massless pulley, as shown in Fig. 5–9a. Let m_2 be greater than m_1. Find the tension in the string and the acceleration of the masses.

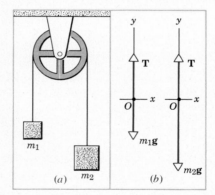

Figure 5–9 Example 8. (*a*) Two unequal masses are suspended by a string from a pulley (Atwood's machine). (*b*) Free-body diagrams for m_1 and m_2.

We consider an *upward* acceleration *positive*. If the acceleration of m_1 is a, the acceleration of m_2 must be $-a$. The forces acting on m_1 and on m_2 are shown in Fig. 5–9b, in which T represents the tension in the string.

The equation of motion for m_1 is

$$T - m_1g = m_1a$$

and for m_2 is

$$T - m_2g = -m_2a.$$

Combining these equations, we obtain

$$a = \frac{m_2 - m_1}{m_2 + m_1} g,$$

and

$$T = \frac{2m_1m_2}{m_1 + m_2} g. \tag{5–12}$$

For example, if $m_2 = 3.0$ slugs ($W_2 = m_2g = 96$ lb) and $m_1 = 1.0$ slug ($W_1 = m_1g = 32$ lb),

$$a = \left(\frac{3.0 - 1.0}{3.0 + 1.0}\right) g = \tfrac{1}{2}g$$

and

$$T = \frac{2(1.0 \text{ slug})(3.0 \text{ slug})}{(1.0 + 3.0) \text{ slug}} (32 \text{ ft/s}^2) = 48 \text{ lb}.$$

Notice that the magnitude of T is always intermediate between the weight of the mass m_1 (32 lb in our example) and the weight of the mass m_2 (96 lb in our example). This is to be expected, since T must exceed m_1g to give m_1 an upward acceleration, and m_2g must exceed T to give m_2 a downward acceleration. In the special case when $m_1 = m_2$, we obtain $a = 0$ and $T = m_1g = m_2g$, which is the static result to be expected.

Example 9 Consider an elevator moving vertically with an acceleration **a**. We wish to find the force exerted by a passenger on the floor of the elevator.

Acceleration will be taken *positive upward* and *negative downward*. Thus positive acceleration in this case means that the elevator is either moving upward with increasing speed or is moving downward with decreasing speed. Negative acceleration means that the elevator is moving upward with decreasing speed or downward with increasing speed.

From Newton's third law the force exerted by the passenger on the floor will always be equal in magnitude but opposite in direction to the force exerted by the floor on the passenger. We can therefore calculate either the action force or the reaction force. (When the forces acting on the passenger are used, we solve for the latter force. When the forces acting on the floor are used, we solve for the former force.)

The situation is shown in Fig. 5–10. The passenger's true weight is **W** and the force exerted on him by the floor, called **P**, is his *apparent* weight in the accelerating elevator. The resultant force acting on him is **P** + **W**. Forces will be taken as positive when directed upward. From the second law of motion we have

$$F = ma,$$

or

$$P - W = ma, \tag{5–13}$$

where m is the mass of the passenger and a is his (and the elevator's) acceleration.

Suppose, for example, that the passenger weighs 160 lb and the acceleration is 2.0 ft/s^2 upward. We have

$$m = \frac{W}{g} = \frac{160 \text{ lb}}{32 \text{ ft/s}^2} = 5.0 \text{ slugs}$$

and from Eq. 5–13,

$$P - 160 \text{ lb} = (5.0 \text{ slugs})(2.0 \text{ ft/s}^2)$$

or

$$P = \text{apparent weight} = 170 \text{ lb}.$$

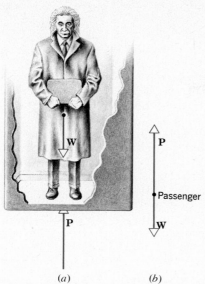

If we were to measure this force directly by having the passenger stand on a spring scale fixed to the elevator floor (or suspended from the ceiling), we would find the scale reading to be 170 lb for a man whose weight is 160 lb. The passenger feels himself pressing down on the floor with greater force (the floor is pressing upward on him with greater force) than when he and the elevator are at rest. Everyone experiences this feeling when an elevator starts upward from rest.

If the acceleration were taken as 2.0 ft/s² downward, then $a = -2.0$ ft/s² and $P = 150$ lb for the passenger. The passenger who weighs 160 lb feels himself pressing down on the floor with less force than when he and the elevator are at rest.

If the elevator cable were to break and the elevator were to fall freely with an acceleration $a = -g$, then P would equal $W + (W/g)(-g) = 0$. Then the passenger and floor would exert no forces on each other. The passenger's apparent weight as indicated by the spring scale on the floor would be zero. Such a situation is popularly referred to as "weightlessness." The passenger's weight (the pull of gravity on him) has not changed, of course, but the force he exerts on the floor and the reaction force of the floor on him are zero.

Figure 5–10 Example 9. (*a*) A passenger stands on the floor of an elevator. (*b*) A free-body diagram for the passenger.

REVIEW GUIDE AND SUMMARY

A program for mechanics

If we see an accelerating particle, we say that the particle'e environment is exerting a force **F** on it. We need to know: (1) How do we calculate **F** from the properties of the particle and its surroundings? The relations that allow us to do this are called the *force laws*. (2) How will the particle move when the force **F** acts on it? The relations that tell us are called the *laws of motion*.

Newton's first law

In the simplest case the environment exerts no force on the particle. Newton's first law then tells us that the particle moves with zero acceleration; that is, it is either at rest or moving in a straight line at constant speed.

Measuring force

If a force **F** *does* act, we must know how to measure it. Attach a spring to a body of standard mass ($m_0 = 1$ kg, by definition) and pull it over a frictionless surface so that it has a constant acceleration **a** whose magnitude a is 1 m/s². We then say that the spring exerts a force **F** on the standard body, its magnitude being defined as one *newton* (N) and its direction being that of **a**. If the standard body m_0 experiences a different acceleration, say 3.7 m/s², the force will then be 3.7 N, and so on. Experiment shows that force is a vector in that it obeys the laws of vector addition; see Fig. 5–2.

Measuring mass

We must also know how to assign masses to bodies other than the standard kilogram. Let a constant force **F** act separately on an arbitrary body (mass m) and on our standard body (mass $m_0 = 1$ kg). Measure the accelerations a and a_0 that result. We *define* the mass m of the arbitrary body from

$$m = m_0 \left(\frac{a_0}{a} \right) \quad \text{(same force } \mathbf{F} \text{ acting).}$$

Experiment shows that masses so assigned to various bodies are internally consistent and add like scalars.

Newton's second law

Newton's second law is summed up by the relation

$$\mathbf{F} = m\mathbf{a}, \qquad [5-1]$$

in which **F** is the vector sum of *all* the forces that act *on* the body. In scalar form this law becomes

$$F_x = ma_x; \qquad F_y = ma_y; \qquad F_z = ma_z. \qquad [5-2]$$

Examples 1 and 2 show the application of this law in two simple cases. Section 5–11 gives five more examples.

Units for F = ma

The appropriate SI units for $\mathbf{F} = m\mathbf{a}$ are, respectively, the newton (N), the kilogram (kg), and the meter per second^2 (m/s^2). The British and the centimeter-gram-second (cgs) systems are also used. Table 5–2 shows the units for all three systems.

Newton's third law

Newton's third law tells us that forces always come in equal but opposite action-reaction pairs. If you push on a wall with a force $\mathbf{F}$, the wall pushes back on you with a force $-\mathbf{F}$. Forces that form such pairs *always* act on different bodies. Study Examples 3 and 4 with this in mind.

Weight and mass

An important force acting on a body is its weight $\mathbf{W}$, which is the earth's gravitational pull on it. Acting alone on a body it produces an acceleration $\mathbf{g}$ and we can write Newton's second law as $\mathbf{W} = m\mathbf{g}$.

In SI units we prefer to describe an object (a car, say) by giving its mass in kilograms (1500 kg) rather than its weight in newtons. In the British system we prefer to describe an object by giving its weight in pounds (3200 lb) rather than its mass in slugs. To avoid dealing directly with masses—and thus with slugs—in the British system we often write Newton's second law in the form $\mathbf{F} = (W/g)\mathbf{a}$.

QUESTIONS

1. Why do you fall forward when a moving bus decelerates to a stop and fall backward when it accelerates from rest? What would happen if the bus rounded a curve at constant speed?

2. Subway standees often find it convenient to face the side of the car when the train is starting or stopping and to face the front or rear when it is running at constant speed. Why?

3. A block of mass m is supported by a cord C from the ceiling, and another cord D is attached to the bottom of the block (Fig. 5–11). Explain this: If you give a sudden jerk to D, it will break, but if you pull on D steadily, C will break.

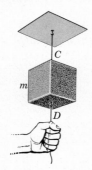

Figure 5–11   Question 3.

4. Criticize the statement, often made, that the mass of a body is a measure of the "quantity of matter" in it.

5. Suppose that a body that is acted upon by exactly two forces is accelerated. Does it then follow that (*a*) the body cannot move with constant speed? (*b*) the velocity can never be zero? (*c*) the sum of the two forces cannot be zero? (*d*) the two forces must act in the same line?

6. You drop two bodies of different masses simultaneously from the top of a tower. Show that, if you assume the air resistance to have the same constant value for each body, the one of larger mass will strike the ground first. How good is this assumption?

7. The owner's manual of a car suggests that your seat belt should be adjusted "to fit snugly" and that the front seat head rest should be adjusted "to fit snugly" and that the front seat head rest should *not* be adjusted so that it fits comfortably at the back of your neck but so that "the top of the head rest is level with the top of your ears." How do Newton's laws support these good recommendations?

8. A Frenchman, filling out a form, writes "78 kg" in the space marked "Poids" (weight). However, weight is a force and the kilogram is a mass unit. What do Frenchmen (among others) have in mind when they use a mass unit to report their weight? Why don't they report their weight in newtons? How many newtons does this Frenchman weigh? How many pounds?

9. Figure 5–12 shows comet Kohoutek as it appeared in 1973. Like all comets, it moves around the sun under the influence of the gravitational pull that the sun exerts on it. The nucleus of the comet is a relatively massive core at a position indicated by P. The tail of a comet is produced by the action of the solar wind, which consists of charged particles streaming outward from the sun. By inspection, what, if anything, can you say about the direction of the force that acts on the nucleus of the comet? What about the direction in which the nucleus is being accelerated? What about the direction in which the comet is moving?

10. What is your mass in slugs? Your weight in newtons?

11. Using force, length, and time as fundamental quantities, what are the dimensions of mass?

12. A horse is urged to pull a wagon. The horse refuses to try, citing Newton's third law as his defense: "'The pull of the horse on the wagon is equal but opposite to the pull of the wagon on the horse.' If I can never exert a greater force on the wagon that it exerts on me, how can I ever start the wagon moving?" asks the horse. How would you reply?

13. Comment on whether the following pairs of forces are examples of action-reaction: (*a*) The earth attracts a brick; the brick attracts the earth. (*b*) A propellered airplane pulls air in toward the plane; the air pushes the plane forward. (*c*) A horse pulls forward on a cart, accelerating it; the cart pulls backward on the horse. (*d*) A horse pulls forward on a cart without moving it; the cart pulls back on the horse. (*e*) A horse pulls forward on a cart without moving it; the earth exerts an equal and opposite force on the cart.

Figure 5–12 Question 9. Courtesy Hale Observatories.

14. Comment on the following statements about mass and weight taken from examination papers. (*a*) Mass and weight are the same physical quantities expressed in different units. (*b*) Mass is a property of one object alone whereas weight results from the interaction of two objects. (*c*) The weight of an object is proportional to its mass. (*d*) The mass of a body varies with changes in its local weight.

15. How many ways can you think of in which you could experience weightlessness, however briefly?

16. In Fig. 5–13, we show four forces that are equal in magnitude. What combination of three forces, acting together on the same body, might keep that body in translational equilibrium?

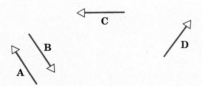

Figure 5–13 Question 16.

17. You are on the flight deck of the orbiting space shuttle *Discovery* and someone hands you two wooden balls, outwardly identical. One, however, has a lead core whereas the other does not. How many ways can you think of to tell them apart?

18. You are an astronaut in the lounge of an orbiting space station and you remove the cover from a long thin jar containing a single olive. How many ways can you think of—all taking advantage of the inertia of either the olive or the jar—to remove the olive from the jar?

19. A horizontal force acts on a body that is free to move. Can it produce an acceleration if the force is less than the weight of that body?

20. Does the acceleration of a freely falling body depend on the weight of the body?

21. What's the relation—if any—between the force acting on an object and the direction in which it is moving?

22. You shoot an arrow into the air and you keep your eye on it as it follows a parabolic flight path to the ground. You notice that the arrow turns in flight so that it is always tangent to its flight path. What makes it do that?

23. A bird alights on a stretched telegraph wire. Does this change the tension in the wire? If so, by an amount less than, equal to, or greater than the weight of the bird?

24. In November 1984 astronauts Joe Allen and Dale Gardner salvaged a Westar-6 communications satellite from a faulty orbit and manhandled it into the cargo bay of the space shuttle *Discovery;* see Fig. 5–14. Describing the experience, Joe Allen said of the satellite, "It's not heavy; it's massive." What did he mean?

25. Why do raindrops fall with constant speed during the latter stages of their descent?

26. In a tug of war, three men pull on a rope to the left at *A* and three men pull to the right at *B* with forces of equal magnitude. Now a 5-lb weight is hung vertically from the center of the rope. (*a*) Can the men get the rope *AB* to be horizontal? (*b*) If not, explain. If so, determine the magnitude of the forces required at *A* and *B* to do this.

27. The following statement is true; explain it. Two teams are having a tug of war; the team that pushes harder against the ground wins.

28. A massless rope is strung over a frictionless pulley. A monkey holds onto one end of the rope and a mirror, having the same weight as the monkey, is attached to the other end of the rope at the monkey's level. Can the monkey get away from his image seen in the mirror (*a*) by climbing up the rope, (*b*) by climbing down the rope, (*c*) by releasing the rope?

Figure 5–14   Question 24. NASA.

29. Two objects of equal mass rest on opposite pans of an equal-arm balance. Does the scale remain balanced when it is accelerated up or down in an elevator? When it is in free fall?

30. You stand on the large platform of a spring scale and note your weight. You then take a step on this platform and notice that the scale reads less than your weight at the beginning of the step and more than your weight at the end of the step. Explain.

31. Could you weigh yourself on a scale whose maximum reading is less than your weight? If so, how?

32. A weight is hung by a cord from the ceiling of an elevator. From the following conditions, choose the one in which the tension in the cord will be greatest . . . least. (*a*) elevator at rest; (*b*) elevator rising with uniform speed; (*c*) elevator descending with decreasing speed; (*d*) elevator descending with increasing speed.

33. A woman stands on a spring scale in an elevator. In which case below will the scale record the minimum reading . . . the maximum reading? (*a*) elevator stationary; (*b*) elevator cable breaks, free fall; (*c*) elevator accelerating upward; (*d*) elevator accelerating downward; (*e*) elevator moving at constant velocity.

34. Why do tires grip the road better on level ground than they do when going uphill or downhill?

35. In Fig. 5–15 a needle has been placed in each end of a broomstick, the tips of the needles resting on the edges of filled wine glasses. The "experimenter" strikes the broomstick a swift and sturdy blow with a stout rod. The broomstick breaks and falls to the floor but the wine glasses remain in place and no wine is spilled. This impressive parlor stunt was popular at the end of the last century. What is the physics behind it? (If you try it, practice first with empty soft drink cans. Come to think of it, you might ask your physics instructor to do it, as a lecture demonstration!)

Figure 5–15   Question 35. From *A Random Walk in Science* by Robert L. Weber, 1973, The Institute of Physics, Bristol and London.

36. Your car skids across the center line on an icy highway. Should you turn the front wheels in the direction of skid or in the opposite direction (*a*) when you want to avoid a collision with an oncoming car, (*b*) when no other car is near but you want to regain control of the steering? Assume rear-wheel drive.

37. What conclusion might a physicist draw if, while standing in an elevator, he observes that unequal masses hung over a pulley inside the elevator remain balanced, that is, that there is no tendency for the pulley to turn?

EXERCISES

Section 5–3   Newton's First Law

1. Suppose that gravity were suddenly turned off, so that the earth became a free object rather than being confined to orbit the sun. How long would it take for the earth to reach a distance from the sun equal to Pluto's present orbital radius? (Hint: You will find some of the data you need in Appendix C.)

Section 5–5   Mass; Newton's Second Law

2. An experimental rocket sled can be accelerated from rest to 1600 km/h in 1.8 s. What constant force would be required if the sled has a mass of 500 kg?

3. A neutron travels at a speed of about 1.4×10^7 m/s. Nuclear forces are of very short range, being essentially zero outside of a nucleus but very strong inside. If the neutron is captured by a nucleus whose diameter is 1.0×10^{-14} m, what is the minimum force acting on this neutron?

4. A 5.5-kg block is initially at rest on a frictionless horizontal surface. It is pulled with a constant horizontal force of 3.75 N. (*a*) What is its acceleration? (*b*) How long must it be pulled before its speed is 5.2 m/s? (*c*) How far does it move in this time?

5. In the Bohr model of the hydrogen atom, the electron revolves in a circular orbit around the nucleus. If the radius is 5.3×10^{-11} m and the electron makes 6.6×10^{15} rev/s, find (*a*) the acceleration (magnitude and direction) of the electron and (*b*) the centripetal force acting on the electron. (This force is due to the attraction between the positively charged nucleus and the negatively charged electron.) The mass of the electron is 9.1×10^{-31} kg.

Section 5–7   Newton's Third Law

6. Two blocks, mass m_1 and m_2, are connected by a light spring on a horizontal frictionless table. Find the ratio of their accelerations a_1 and a_2 after they have been pulled apart and then released.

7. (*a*) Two 10-lb weights are attached to a spring scale as shown in Fig. 5–16*a*. (*a*) What is the reading of the scale? (*b*) A single

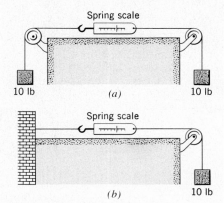

Spring scale

10 lb   (*a*)   10 lb

Spring scale

(*b*)   10 lb

Figure 5–16   Exercise 7.

10-lb weight is attached to a spring scale that itself is attached to a wall, as shown in Fig. 5–16*b*. What is the reading of the scale?

Section 5–9   Weight and Mass

8. What is the weight in newtons and the mass in kilograms of (*a*) a 5-lb bag of sugar, (*b*) a 240-lb fullback, and (*c*) a 1.8-ton automobile?

9. What are the mass and weight of (*a*) a 1400-lb snowmobile, and (*b*) a 400-kg heat pump?

10. A space traveler whose mass is 75 kg leaves Earth. Compute his weight (*a*) on Earth, (*b*) on Mars, where $g = 3.8$ m/s², and (*c*) in interplanetary space. (*d*) What is his mass at each of these locations?

11. A 400-lb motorcycle accelerates from rest to 50 mi/h in 6.0 s. (*a*) What average force acts on it? (*b*) What body provides the force?

12. A certain particle has a weight of 20 N at a point where the acceleration due to gravity is 9.8 m/s². (*a*) What are the weight and mass of the particle at a point where the acceleration due to gravity is 4.9 m/s²? (*b*) What are the weight and mass of the particle if it is moved to a point in space where the gravitational force is zero?

Section 5–11   Some Applications of Newton's Laws of Motion

13. What is the net force acting on a 3800-lb automobile accelerating at 12 ft/s²?

14. In a modified "tug-of-war" game, two people pull in opposite directions, not on a rope, but on a 25-kg block resting on a smooth surface. If the participants exert forces of 90 N and 92 N, what is the acceleration of the block?

15. A fireman weighing 160 lb slides down a vertical pole with an average acceleration of 10 ft/s². What is the average vertical force he exerts on the pole?

16. (*a*) Compute the initial upward acceleration of a rocket of mass 1.3×10^4 kg if the initial upward thrust of its engine is 2.6×10^5 N. (*b*) Can you neglect the weight of the rocket (the downward pull of the earth on it)?

17. A rocket and its payload have a total mass of 50,000 kg (weight $mg = 110,000$ lb). How large is the thrust of the rocket engine when (*a*) the rocket is "hovering" over the launch pad, just after ignition, and (*b*) when the rocket is accelerating upward at 20 m/s² (= 66 ft/s²)?

18. Determine the frictional force of the air on a body of mass 0.25 kg falling freely with an acceleration of 9.2 m/s².

PROBLEMS GROUPED BY SECTION

Section 5-5 Mass; Newton's Second Law

1. A car moving initially at a speed of 50 mi/h ($=80$ km/h) and weighing 3000 lb ($=13{,}000$ N) is brought to a stop in a distance of 200 ft ($=61$ m). Find (*a*) the braking force, and (*b*) the time required to stop. Assuming the same braking force, find (*c*) the distance, and (*d*) the time required to stop if the car was going 25 mi/h ($=40$ km/h) initially.

2. An electron is projected horizontally at a speed of 1.2×10^7 m/s into an electric field that exerts a constant vertical force of 4.5×10^{-16} N on it. The mass of the electron is 9.1×10^{-31} kg. Determine the vertical distance the electron is deflected during the time it has moved forward 30 mm horizontally.

3. An 8.5-kg object passes through the origin with a velocity of 30 m/s parallel to the *x*-axis. It experiences a constant 17-N force in the direction of the positive *y*-axis. (*a*) Describe the resulting motion. Calculate (*b*) the velocity and (*c*) the position of the particle after 15 s have elapsed.

4. In a prototype controlled thermonuclear fusion device, protons are caused to revolve in circular orbits of radius 3.7 m, making 10^5 revolutions each second. What is the net force on one such proton? A proton has a mass of 1.67×10^{-27} kg.

Section 5-7 Newton's Third Law

5. Two blocks are in contact on a frictionless table. A horizontal force is applied to one block, as shown in Fig. 5-17. (*a*) If $m_1 = 2.0$ kg, $m_2 = 1.0$ kg, and $F = 3.0$ N, find the force of contact between the two blocks. (*b*) Show that if the same force F is applied to m_2 rather than to m_1, the force of contact between the blocks is 2.0 N, which is not the same value derived in (*a*). Explain.

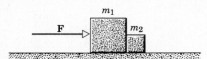

Figure 5-17 Problem 5.

Section 5-9 Weight and Mass

6. A body of mass 2.0 slugs is acted on by the downward force of gravity and a horizontal force of 130 lb. Find (*a*) its acceleration and (*b*) its velocity as functions of time, assuming that it starts from rest.

7. A landing craft approaches the surface of one of the moons of the planet Saturn. If an upward thrust of 3260 N is supplied by the rocket engine, the craft descends with constant speed. If the upward thrust is 2200 N, the craft accelerates downward at 0.82 m/s². (*a*) What is the weight of the landing craft in the vicinity of the moon's surface? (*b*) What is the mass of the craft? (*c*) What is the acceleration due to gravity near the moon's surface?

Section 5-11 Some Applications of Newton's Laws of Motion

8. *Sunjamming.* The sun yacht *Diana,* designed to navigate in the solar system using the pressure of sunlight, has a sail area of 3 km² and a mass of 900 kg. Near the earth's orbit the sun could exert a radiation force of 20 N on its sail. (*a*) What acceleration would such a force impart to the craft? (*b*) A small acceleration can produce large effects if it acts steadily for a long enough time. Starting from rest, then, how far would the craft have moved after one day under these conditions? (*c*) What would then be its speed? (See "The Wind from the Sun," a fascinating science fiction account by Arthur C. Clarke of a sun yacht race.)

9. An elevator and its load have a combined mass of 1600 kg. Find the tension in the supporting cable when the elevator, originally moving downward at 20 m/s, is brought to rest with constant acceleration in a distance of 50 m.

10. An object is hung from a spring balance attached to the ceiling of an elevator. The balance reads 60 lb when the elevator is standing still. (*a*) What is the reading when the elevator is moving upward with a constant speed of 24 ft/s? (*b*) What is the reading of the balance when the elevator is accelerating downward at 24 ft/s²?

11. Refer to Fig. 5-8. Let $m_1 = 4.0$ slugs and $m_2 = 2.0$ slugs. Find (*a*) the tension in the string and (*b*) the acceleration of the two blocks.

12. A charged sphere of mass 3.0×10^{-4} kg is suspended from a string. An electric force acts horizontally on the sphere so that the string makes an angle of 37° with the vertical when at rest. Find (*a*) the magnitude of the electric force and (*b*) the tension in the string.

13. A man of mass 80 kg (weight $mg = 176$ lb) jumps down to a concrete patio from a window ledge only 0.50 m ($=1.6$ ft) above the ground. He neglects to bend his knees on landing, so that his motion is arrested in a distance of about 2.0 cm ($=0.79$ in.). (*a*) What is the average acceleration of the man from the time his feet first touch the patio to the time he is brought fully to rest? (*b*) With what average force does this jump jar his bone structure?

14. Three blocks are connected, as shown in Fig. 5-18, on a horizontal frictionless table and pulled to the right with a force $T_3 = 60$ N. If $m_1 = 10$ kg, $m_2 = 20$ kg, and $m_3 = 30$ kg, find the tensions T_1 and T_2. Draw an analogy to bodies being pulled in tandem, such as an engine pulling a train of coupled cars.

15. A chain consisting of five links, each of mass 0.10 kg, is lifted vertically with a constant acceleration of 2.5 m/s², as shown in Fig. 5-19. Find (*a*) the forces acting between adjacent links, (*b*) the force **F** exerted on the top link by the agent lifting the chain, and (*c*) the *net* force acting on each link.

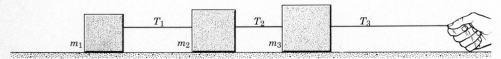

Figure 5-18 Problem 14.

Figure 5–19   Problem 15.

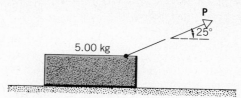

Figure 5–21   Problem 21.

16. A rocket of mass 3000 kg is fired from rest from the ground at an angle of elevation of 60°. The motor exerts a thrust of 6×10^4 N at a constant angle of 60° to the horizontal for 50 s and then cuts out. Ignore the mass of fuel consumed and neglect aerodynamic drag. Calculate (*a*) the altitude of the rocket at motor cut out and (*b*) the total horizontal distance from firing point to impact.

17. An elevator weighing 6000 lb is pulled upward by a cable with an acceleration of 4.0 ft/s^2. (*a*) What is the tension in the cable? (*b*) What is the tension when the elevator is accelerating downward at 4.0 ft/s^2, but is still moving upward?

18. An 80-kg man is parachuting and experiencing a downward acceleration of 2.5 m/s^2. The mass of the parachute is 5.0 kg. (*a*) What is the value of the upward force exerted on the parachute by the air? (*b*) What is the value of the downward force exerted by the man on the parachute?

19. A 100-kg man lowers himself to the ground from a height of 10 m by means of a rope passed over a frictionless pulley and attached to a 70-kg sandbag. (*a*) With what speed does the man hit the ground? (*b*) Is there anything he could do to reduce the speed with which he hits the ground?

20. Figure 5–20 shows a section of an alpine cable-car system. The maximum permitted mass of each car with occupants is 2800 kg. The cars, riding on a support cable, are pulled by a second cable attached to each pylon. What is the difference in tension between adjacent sections of pull cable if the cars are accelerated up the 35° incline at 0.81 m/s^2?

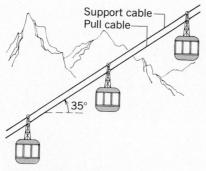

Figure 5–20   Problem 20.

21. A 5.0-kg block is pulled along a horizontal frictionless floor by a cord that exerts a force $P = 12$ N at an angle $\theta = 25°$ above the horizontal, as shown in Fig. 5–21. (*a*) What is the acceleration of the block: (*b*) The force P is slowly increased. What is the value of P just before the block is lifted off the floor? (*c*) What is the acceleration of the block just before it is lifted off the floor?

22. An elevator consists of the elevator cage (*A*), the counterweight (*B*), the driving mechanism (*C*), and the cable and pulleys as shown in Fig. 5–22. The mass of the cage is 1100 kg and the mass of the counterweight is 1000 kg. Neglect friction and the mass of the cable and pulleys. The elevator accelerates upward at 2.0 m/s^2 and the counterweight accelerates downward at the same rate. (*a*) What is the value of the tension T_1? (*b*) T_2? (*c*) What force is exerted on the cable by the driving mechanism?

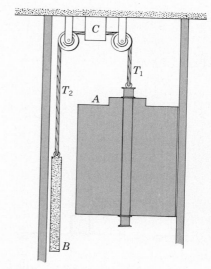

Figure 5–22   Problem 22.

*23. For the mass and pulley system shown in Fig. 5–23, assume that the surfaces and pulleys are frictionless, and neglect the masses of the cord and pulleys. (*a*) Find the acceleration of M. (*b*) What is the tension in the cord?

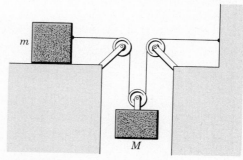

Figure 5–23   Problem 23.

ADDITIONAL PROBLEMS

24. An electron travels in a straight line from the cathode of a vacuum tube to its anode, which is exactly 1.0 cm away. It starts with zero speed and reaches the anode with a speed of 6.0×10^6 m/s. (a) Assume constant acceleration and compute the force on the electron. Take the electron's mass to be 9.1×10^{-31} kg. This force is electrical in origin. (b) Compare it with the gravitational force on the electron, which we neglected when we assumed straight line motion.

25. (a) Neglecting gravitational forces, what force would be required to accelerate a 1.0-kiloton spaceship from rest to one-tenth the speed of light in 3 days? In 2 months? (b) Assuming that the engines are shut down when this speed is reached, what would be the time required to complete a 5 light-month journey for each of these two cases?

26. Let the only forces acting on two bodies be their mutual interactions. If both bodies start from rest, show that the distance traveled by each is inversely proportional to the respective masses of the bodies.

27. A body of mass m is acted on by two forces F_1 and F_2, as shown in Fig. 5–24. If $m = 5.0$ kg, $F_1 = 3.0$ N, and $F_2 = 4.0$ N, find the vector acceleration of the body.

Figure 5–24 Problem 27.

28. An unconfined child is playing on the front seat of a car traveling in a residential neighborhood at 35 km/h. A small dog runs across the road and the driver applies the brakes, stopping the car and missing the dog. With what speed does the child strike the dashboard, presuming that the car stops before the child does so? Compare this speed with the world-record 100-m dash, about 10 s.

29. A small 150-g pebble is 9.0 km deep in the ocean and is falling with a constant terminal speed of 25 m/s. What force does the water exert on the falling pebble?

30. A 40-kg girl and an 8.4-kg sled are on the surface of a frozen lake, 15 m apart. By means of a rope the girl exerts a 5.2-N force on the sled, pulling it toward her. (a) What is the acceleration of the sled? (b) What is the acceleration of the girl? (c) How far from the girl's initial position do they meet?

31. A 60-kg woman stands in an elevator that is accelerating downward at 3.1 m/s². What force does she exert on the elevator floor?

32. What is the minimum acceleration with which a 50-kg girl can slide down a rope that can withstand a maximum tension of 425 N without breaking?

33. Refer to Fig. 5–7. Let the mass of the block be 2.0 slugs and the angle θ equal 30°. (a) Find the tension in the string and (b) the normal force acting on the block. (c) If the string is cut, find the acceleration of the block.

34. A block of mass $m_1 = 3.0$ slugs on a smooth inclined plane of angle 30° is connected by a cord over a small frictionless pulley to a second block of mass $m_2 = 2.0$ slugs hanging vertically (Fig. 5–25). (a) What is the acceleration of each body? (b) What is the tension in the cord?

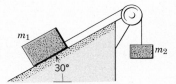

Figure 5–25 Problem 34.

35. How could a 100-lb object be lowered from a roof using a cord with a breaking strength of 87 lb without breaking the cord?

36. A block is projected up a frictionless inclined plane with a speed v_0. The angle of incline is θ. (a) How far up the plane does it go? (b) How long does it take to get there? (c) What is its speed when it gets back to the bottom? Find numerical answers for $\theta = 30°$ and $v_0 = 8.0$ ft/s.

37. A lamp hangs vertically from a cord in a descending elevator. The elevator has a deceleration of 8.0 ft/s² (=2.4 m/s²) before coming to a stop. (a) If the tension in the cord is 20 lb (=89 N), what is the mass of the lamp? (b) What is the tension in the cord when the elevator ascends with an acceleration of 8.0 ft/s² (=2.4 m/s²)?

Figure 5–26 Problem 38. P. H. Salesi/U.S. Navy.

38. A new 26-ton Navy jet requires an airspeed of 280 ft/s for lift-off. Its own engine develops a thrust of 24,000 lb. The jet is to take off from an aircraft carrier with a 300-ft flight deck. What force must be exerted by the catapult of the carrier? Assume that the catapult and the jet's engine each exert a constant force over the 300-ft take-off distance.

39. A 100-kg crate is pushed at constant speed up the smooth 30° ramp shown in Fig. 5–27. (a) What horizontal force **F** is required? (b) What force is exerted by the ramp on the crate?

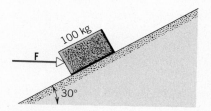

Figure 5–27   Problem 39.

40. A 10-kg monkey is climbing a massless rope attached to a 15-kg mass over a (frictionless!) tree limb. (a) Explain quantitatively how the monkey can climb up the rope so that he can raise the 15-kg mass off the ground. If, after the mass has been raised off the ground, the monkey stops climbing and holds on to the rope, what will now be (b) his acceleration and (c) the tension in the rope?

41. A research balloon of total mass M is descending vertically with downward acceleration a. (See Fig. 5–28.) How much ballast must be thrown from the car to give the balloon an *upward* acceleration a?

Figure 5–28   Problem 41.

42. A block of mass M is pulled along a horizontal frictionless surface by a rope of mass m, as shown in Fig. 5–29. A horizontal force **P** is applied to one end of the rope. (a) Show that the rope

must sag, even if only by an imperceptible amount. Then, assuming that the sag is negligible, find (b) the acceleration of rope and block, (c) the force that the rope exerts on the block M, and (d) the tension in the rope at its midpoint.

Figure 5–29   Problem 42.

43. A man sits in a bosun's chair supported by a light rope passing over a pulley, as shown in Fig. 5–30. The man pulls on the free end of the rope in order to lift himself. (a) If the mass of the man and chair together is 95 kg, with what force must he pull to raise himself at constant speed? (b) With what force must he pull if he desires an upward acceleration of 1.3 m/s²? Ignore friction and the mass of the pulley.

Figure 5–30   Problem 43.

***44.** Two blocks with masses m_1 and m_2 are connected by a string, as shown in Fig. 5–31. Ignore friction and assume that the pulley has zero mass. Derive an expression for the acceleration of m_2.

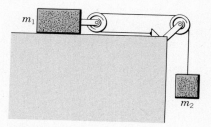

Figure 5–31   Problem 44.

6

Particle Dynamics—II

6–1 Introduction

In Chapter 5 we considered the accelerations produced by forces that were constant in both magnitude and direction. Their nature is illustrated by the forces that appear in Fig. 5–9 of Example 8; that is, they are forces that are exerted by the earth ($m_1\mathbf{g}$ and $m_2\mathbf{g}$), by taut cords ($\mathbf{T}$), or by contact with other bodies ($\mathbf{N}$). The first kind of force is gravitational in nature. The last two, at the level of the atoms that make up the bodies, are electromagnetic. In this chapter we introduce another kind of force, that due to friction. It too, at the atomic level, is electromagnetic in nature, although we shall treat it, not at this level, but at the level of the measureable properties of gross objects.

We shall also consider the dynamics of uniform circular motion, in which the force, though constant in magnitude, varies uniformly in direction with time.

6–2 Frictional Forces

We shall now consider the important forces that are classified under friction. Because a simple and relatively general law describes many frictional forces, we shall be able to solve many problems without necessarily understanding the origin of frictional forces themselves. Before considering applications of the frictional force law, however, we shall first consider briefly some ideas concerning the origin and nature of these forces.

If we project a block of mass m with initial velocity $\mathbf{v}_0$ along a long horizontal table, it eventually comes to rest. This means that, while it is moving, it experiences an average acceleration $\bar{\mathbf{a}}$ that points in the direction opposite to its motion. If we see that a body is being accelerated, we always associate a force, defined from Newton's second law, with the motion. In this case we declare that the table exerts a *force of friction*, whose average value is $m\bar{\mathbf{a}}$, on the sliding block.

The force of friction

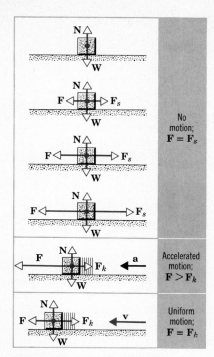

Figure 6–1 A block being put into motion as applied force **F** overcomes frictional forces. In the first four drawings the applied force is gradually increased from zero to magnitude $\mu_s N$. No motion occurs until this point because the frictional force always just balances the applied force. The instant F becomes greater than $\mu_s N$, the block goes into motion, as is shown in the fifth drawing. In general, $\mu_k N < \mu_s N$; this leaves an unbalanced force to the left and the block accelerates. In the last drawing F has been reduced to equal $\mu_k N$. The net force is zero, and the block continues with constant velocity. For clarity the frictional force is shown acting at the center of the block, even though it really acts on the bottom of the block.

Actually, whenever the surface of one body slides over that of another, each body exerts a frictional force on the other, parallel to the surfaces. The frictional force on each body is in a direction opposite to its motion relative to the other body. Even when there is no relative motion, frictional forces may exist between surfaces.

Although we have ignored its effects up to now, friction is very important in our daily lives. Left to act alone it brings every rotating shaft to a halt. In an automobile, about 20% of the engine power is used to counteract internal frictional forces. On the other hand, without friction we could not walk; we could not hold a pencil, and if we could it would not write; wheeled transport as we know it would not be possible.

We want to know how to express frictional forces in terms of the properties of the body and its environment; that is, we want to know the force law for frictional forces. In what follows we consider the sliding (not rolling) of one dry (unlubricated) surface over another. As we shall see later, friction, viewed at the microscopic level, is a very complicated phenomenon,* and the force laws for dry, sliding friction are empirical in character and approximate in their predictions. They do not have the elegant simplicity and accuracy that we find for the gravitational force law (Chapter 15) or for the electrostatic force law (Chapter 23). It is remarkable, however, considering the enormous diversity of surfaces one encounters, that many aspects of frictional behavior can be understood qualitatively on the basis of a few simple mechanisms.

Consider a block at rest on a horizontal table as in Fig. 6–1. Attach a spring scale to it to measure the force required to set the block in motion. We find that the block will not move even though we apply a small force. We say that our applied force is balanced by an opposite frictional force exerted on the block by the table, acting along the surface of contact. As we increase the applied force we find some definite force at which the block just begins to move. Once motion has started, this same force produces accelerated motion. By reducing the force once motion has started, we find that it is possible to keep the block in uniform motion without acceleration; this force may be small, but it is never zero.

The frictional forces acting between surfaces at rest with respect to each other are called forces of *static friction*. The maximum force of static friction will be the same as the smallest force necessary to start motion. Once motion is started, the frictional forces acting between the surfaces usually decrease so that a smaller force is necessary to maintain uniform motion. Forces acting between surfaces in relative motion are called forces of *kinetic friction*.

The maximum force of static friction between any pair of dry unlubricated surfaces follows these two empirical laws. (1) It is approximately independent of the area of contact, over wide limits and (2) it is proportional to the normal force. The normal force, sometimes called the loading force, is the one that either body exerts on the other at right angles to their mutual interface. It arises from the elastic deformation of the bodies in contact, such bodies never really being entirely rigid. For a block resting on a horizontal table or sliding along it, the normal force is equal in magnitude to the weight of the block. Because the block has no vertical acceleration, the table must be exerting a force on the block that is directed upward and is equal in magnitude to the downward pull of the earth on the block, that is, equal to the block's weight.

The ratio of the magnitude of the maximum force of static friction to the magnitude of the normal force is called the *coefficient of static friction* for the surfaces

---

* See, for example, "Stick and Slip," by Ernest Rabinowicz, *Scientific American*, May 1956.

involved. If F_s represents the magnitude of the force of static friction, we can write

Law of static friction

$$F_s \leq \mu_s N, \qquad (6-1)$$

where μ_s is the coefficient of static friction and N is the magnitude of the normal force. The equality sign holds only when F_s has its maximum value.

Law of kinetic friction

The force of kinetic friction F_k between dry, unlubricated surfaces follows the same two laws as those of static friction. (1) It is approximately independent of the area of contact over wide limits and (2) it is proportional to the normal force. The force of kinetic friction is also reasonably independent of the relative speed with which the surfaces move over each other.

The ratio of the magnitude of the force of kinetic friction to the magnitude of this normal force is called the *coefficient of kinetic friction*. If F_k represents the magnitude of the force of kinetic friction, then

$$F_k = \mu_k N, \qquad (6-2)$$

where μ_k is the coefficient of kinetic friction.

Both μ_s and μ_k are dimensionless constants, each being the ratio of (the magnitudes of) two forces. Usually, for a given pair of surfaces $\mu_s > \mu_k$. The actual values of μ_s and μ_k depend on the nature of both the surfaces in contact. Both μ_s and μ_k can exceed unity, although commonly they are less than one. Notice that Eqs. 6–1 and 6–2 are relations between the magnitudes only of the normal and frictional forces. These forces are always directed perpendicularly to one another.

The mechanism of friction

On the atomic scale even the most finely polished surface is far from plane. Figure 6–2, for example, shows an actual profile, highly magnified, of a steel surface that would be considered to be highly polished. When two bodies are placed in contact, the actual microscopic area of contact is much less than the apparent macroscopic area of contact; in a particular case these areas can be easily in the ratio of 1 to 10^4.

The actual (microscopic) area of contact is proportional to the normal force, because the contact points deform plastically under the great stresses that develop at these points. The actual contact area remains the same even when the apparent contact area is reduced because increased normal force per unit actual area produces further plastic deformation. Many contact points actually become "cold-welded" together. This phenomenon, called *surface adhesion,* occurs because at the contact point the molecules on opposite sides of the surface are so close together that they exert strong intermolecular forces on each other.

When one body is pulled across another, the frictional resistance is associated with the rupturing of these thousands of tiny welds, which continually reform as new chance contacts are made (see Fig. 6–3).

Figure 6–2 A highly magnified view of a section of a finely polished steel surface. The section was cut at an angle so that vertical distances are exaggerated by a factor of ten with respect to horizontal distances. The surface irregularities are several thousand atomic diameters high. From *Friction and Lubrication of Solids,* by F. P. Bowden and D. Tabor, Clarendon Press, Oxford, 1950.

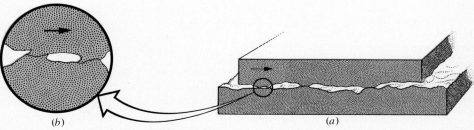

(b)   (a)

Figure 6–3 Details of sliding friction. (*a*) The upper body is sliding to the right over the lower body in this enlarged diagram (*b*) A further enlarged view showing two spots where surface adhesion has occurred. Force is required to break these welds apart and maintain the motion.

The coefficient of friction depends on many variables, such as the nature of the materials, surface finish, surface films, temperature, and extent of contamination. For example, if two carefully cleaned metal surfaces are placed in a highly evacuated chamber so that surface oxide films do not form, the coefficient of friction rises to enormous values and the surfaces actually become firmly "welded" together. The admission of a small amount of air to the chamber so that oxide films may form on the opposing surfaces reduces the coefficient of friction to its "normal" value.

Rolling friction

The frictional force that opposes one body rolling over another is much less than that for a sliding motion and this, indeed, is the advantage of the wheel over the sledge. This reduced friction is due in large part to the fact that, in rolling, the microscopic contact welds are "peeled" apart rather than "sheared" apart as in sliding friction. This will reduce the frictional force by a large factor.

Examples of the application of the empirical force law for friction follow. The coefficients of friction given are assumed to be constant. Actually, μ_k can be regarded as a good average value that is not greatly different from the value at any particular speed in the range.

Example 1 *A block on an inclined plane*   A block is at rest on an inclined plane making an angle θ with the horizontal, as in Fig. 6–4a. As the angle of incline is raised, it is found that slipping just begins at an angle of inclination θ_s. What is the coefficient of static friction between block and incline?

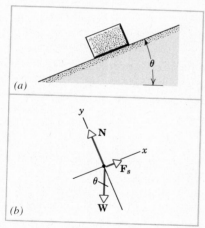

(a)

(b)

Figure 6–4   Example 1. (a) A block at rest on a rough inclined plane. (b) A free-body force diagram for the block.

The forces acting on the block, considered to be a particle, are shown in Fig. 6–4b. **W** is the weight of the block, **N** the normal force exerted on the block by the inclined surface, and **F**$_s$ the tangential force of friction exerted by the inclined surface on the block. Notice that the resultant force exerted by the inclined surface on the block, **N** + **F**$_s$, is no longer perpendicular to the surface of contact, as was true for frictionless surfaces (**F**$_s$ = 0). The block is at rest, so that

$$\mathbf{N} + \mathbf{F}_s + \mathbf{W} = 0.$$

Resolving our forces into x- and y-components, along the plane and the normal to the plane, respectively, we obtain

$$N - W \cos \theta = 0$$
$$F_s - W \sin \theta = 0. \qquad (6\text{--}3)$$

However, $F_s \leq \mu_s N$. If we increase the angle of incline slowly until slipping just begins, then for that angle, $\theta = \theta_s$ and we can use $F_s = \mu_s N$. Substituting this into Eqs. 6–3, we obtain

$$N = W \cos \theta_s$$

and

$$\mu_s N = W \sin \theta_s,$$

so that

$$\mu_s = \tan \theta_s.$$

Hence measurement of the angle of inclination at which slipping just starts provides a simple experimental method for determining the coefficient of static friction between two surfaces.

You can use similar arguments to show that the angle of inclination θ_k required to maintain a *constant speed* for the block as it slides down the plane, once it has been started by tapping, is given by

$$\mu_k = \tan \theta_k,$$

where $\theta_k < \theta_s$. With the aid of a ruler you can now determine μ_s and μ_k for a coin sliding down your textbook.

Example 2 *A braking automobile*   Consider an automobile moving along a straight horizontal road with a speed v_0. If the coefficient of static friction between the tires and the road is μ_s, what is the shortest distance in which the automobile can be stopped?

The forces acting on the automobile, considered to be a particle, are shown in Fig. 6–5. The car is assumed to be moving in the positive x-direction. If we assume that F_s is a constant force, we have uniformly decelerated motion.

From the relation (see Eq. 3–16)

$$v^2 = v_0{}^2 + 2ax,$$

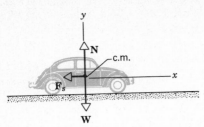

Figure 6–5 Example 2. The forces on a decelerating automobile. The frictional force $\mathbf{F}_s$, shown acting at the center of mass, is actually exerted by the road on the bottoms of the four tires.

with the final speed $v = 0$, we obtain

$$x = -\frac{v_0^2}{2a},$$

where the minus sign means that $\mathbf{a}$ points in the negative x-direction.

To determine a, apply the second law of motion to the x-component of the motion:

$$-F_s = ma = \left(\frac{W}{g}\right)a \quad \text{or} \quad a = -g\left(\frac{F_s}{W}\right).$$

From the y-components we obtain

$$N - W = 0 \quad \text{or} \quad N = W,$$

so that

$$\mu_s = \frac{F_s}{N} = \frac{F_s}{W}$$

and

$$a = -\mu_s g.$$

Then the distance of stopping is

$$x = -\frac{v_0^2}{2a} = \frac{v_0^2}{2g\mu_s}. \tag{6-4}$$

The greater the initial speed, the longer the distance required to come to a stop; in fact, this distance varies as the square of the initial velocity. Also, the greater the coefficient of static friction between the surfaces, the less the distance required to come to a stop.

We have used the coefficient of static friction in this problem, rather than the coefficient of sliding friction, because we assume there is no sliding between the tires and the road. Furthermore, we have assumed that the maximum force of static friction ($F_s = \mu_s N$) operates because the problem seeks the *shortest* distance for stopping. With a smaller static frictional force the distance for stopping would obviously be greater. The correct braking technique required here is to keep the car just on the verge of skidding. If the surface is smooth and the brakes are fully applied, skidding may occur. In this case μ_k replaces μ_s, and the distance required to stop is seen to increase.

As a specific example, if $v_0 = 55$ mi/h (=81 ft/s = 89 km/h) and $\mu_s = 0.60$ (a typical value), we obtain

$$x = \frac{v_0^2}{2\mu_s g} = \frac{(81 \text{ ft/s})^2}{2(0.60)(32 \text{ ft/s}^2)} = 170 \text{ ft } (=52 \text{ m}).$$

Notice that the mass of the car does not appear in Eq. 6–4. How can you explain the practice of "weighing down" a car in order to increase safety in driving on icy roads? (Hint: See Problem 8.)

Example 3   A boy pulls a 10-lb sled 30 ft along a horizontal surface at a *constant speed*. What force $\mathbf{T}$ does he exert on the sled if the coefficient of kinetic friction is 0.20 and his pull makes an angle of 45° with the horizontal?

The situation is shown in Fig. 6–6a, and the forces acting on the sled are shown in Fig. 6–6b. $\mathbf{T}$ is the boy's pull, $\mathbf{W}$ the sled's weight, $\mathbf{F}_k$ the frictional force, and $\mathbf{N}$ the normal force exerted by the surface on the sled.

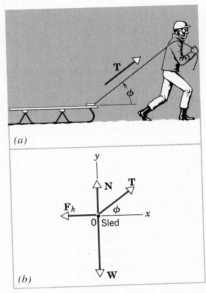

Figure 6–6   Example 3. (a) A boy exerts a force $\mathbf{T}$ on a sled, pulling it. (b) A free-body diagram for the sled.

The sled is unaccelerated, so that from the second law of motion we obtain

$$\Sigma F_x = 0, \quad \text{or} \quad T \cos \phi - F_k = 0,$$

and

$$\Sigma F_y = 0, \quad \text{or} \quad T \sin \phi + N - W = 0.$$

We know that F_k and N are related by

$$F_k = \mu_k N.$$

These three equations contain three unknown quantities, T, F_s, and N. To find T we eliminate F_k and N from these equations and solve the remaining equation for T. Verify that

$$T = \frac{\mu_k W}{\cos \phi + \mu_k \sin \phi}.$$

With $\mu_k = 0.20$, $W = 10$ lb, and $\phi = 45°$ we obtain

$$T = \frac{(0.20)(10 \text{ lb})}{0.707 + 0.141} = 2.4 \text{ lb}.$$

6–3 The Dynamics of Uniform Circular Motion

In Section 4–4, which deals with the kinematics of uniform circular motion, we pointed out that if a body is moving at uniform speed v in a circle of radius r, it experiences a centripetal acceleration **a** whose magnitude is v^2/r. The direction of **a** is always radially inward toward the center of rotation. Thus **a** is a variable vector because, even though its magnitude remains constant, its direction changes continuously as the motion progresses.

In general, there is no fixed relation between the directions of the acceleration **a** and the velocity **v** of a particle, as Fig. 4–7 shows. As it happens, for a particle in uniform circular motion the acceleration **a** and velocity **v** are always at right angles to each other.

Every accelerated body must have a force **F** acting on it, defined by Newton's second law (**F** = m**a**). Thus (assuming that we are in an inertial frame), if we see a body undergoing uniform circular motion, we can be certain that a net force **F**, given in magnitude by

$$F = ma = mv^2/r \qquad (6\text{--}5)$$

must be acting on the body; the body is *not* in equilibrium. The direction of **F** at any instant must be the direction of **a** at that instant, namely, radially inward. We must always be able to account for this force by pointing to a particular object in the environment that is exerting the force on the circulating, accelerating body.

If the body in uniform circular motion is a disk on the end of a string moving in a circle on a frictionless horizontal table as in Fig. 6–7, the force **F** on the disk is provided by the tension **T** in the string. This force **T** is the net force acting on the disk. It accelerates the disk by constantly changing the direction of its velocity so that the disk moves in a circle. **T** is always directed toward the pin at the center, and its magnitude is mv^2/R. If the string were to be cut where it joins the disk, there would suddenly be no net force exerted on the disk. The disk would then move with constant speed in a straight line along the direction of the tangent to the circle at the point at which the string was cut. Hence, to keep the disk moving in a circle, a force must be supplied to it pulling it *inward* toward the center.

Forces responsible for uniform circular motion are called *centripetal* forces because they are directed "toward the center" of the circular motion. To label a force as "centripetal," however, simply means that it always points radially inward; the name tells us nothing about the nature of the force or about the body that is exerting it. Thus, for the revolving disk of Fig. 6–7, the centripetal force is an elastic force provided by the string; for the moon revolving around the earth the centripetal force is the gravitational pull of the earth on the moon; for an electron circulating about an atomic nucleus the centripetal force is electrostatic. A centripetal force is not a new kind of force but simply a way of describing the behavior with time of forces that are attributable to specific bodies in the environment. Thus a force can be centripetal *and* elastic, centripetal *and* gravitational, or centripetal *and* electrostatic, among other possibilities.

Let us consider some examples of forces that act centripetally.

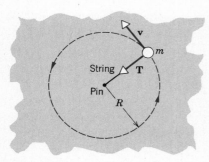

Figure 6–7 A disk of mass m moves with constant speed in a circular path on a horizontal frictionless air table. The only horizontal force acting on m is the centripetal force **T** with which the string pulls on the body.

Centripetal force

Example 4 *The Conical Pendulum* Figure 6–8a shows a small body of mass m revolving in a horizontal circle with constant speed v at the end of a string of length L. As the body swings around, the string sweeps over the surface of a cone.

This device is called a *conical pendulum*. Find the time required for one complete revolution of the body.

If the string makes an angle θ with the vertical, the radius of the circular path is $R = L \sin \theta$. The forces acting on the

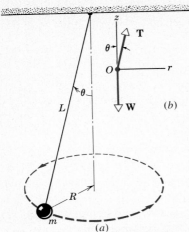

Figure 6-8   Example 4. (a) A mass m suspended from a string of length L swings so as to describe a circle. The string describes a right circular cone of semiangle θ. (b) A free-body force diagram for m.

body of mass m are **W**, its weight, and **T**, the pull of the string, as shown in Fig. 6-8b. It is clear that $\mathbf{T} + \mathbf{W} \neq 0$. Hence, the resultant force acting on the body is nonzero, which is as it should be because a force is required to keep the body moving in a circle with constant speed.

We can resolve **T** at any instant into a radial and a vertical component

$$T_r = T \sin \theta \quad \text{and} \quad T_z = T \cos \theta.$$

Since the body has no vertical acceleration,

$$\Sigma F_z = 0, \quad \text{or} \quad T_z - W = 0.$$

But

$$T_z = T \cos \theta \quad \text{and} \quad W = mg,$$

so that

$$T \cos \theta = mg.$$

The radial acceleration is v^2/R. This acceleration is supplied by T_r, the radial component of **T**, which is the centripetal force acting on m. Hence

$$T_r = T \sin \theta = \frac{mv^2}{R}.$$

Dividing this equation by the preceding one, we obtain

$$\tan \theta = \frac{v^2}{Rg}, \quad \text{or} \quad v^2 = Rg \tan \theta,$$

which gives the constant speed of the bob. If we let τ represent the time for one complete revolution of the body, then

$$v = \frac{2\pi R}{\tau} = \sqrt{Rg \tan \theta}$$

or

$$\tau = \frac{2\pi R}{v} = \frac{2\pi R}{\sqrt{Rg \tan \theta}} = 2\pi \sqrt{\frac{R}{g \tan \theta}}.$$

But $R = L \sin \theta$, so that

$$\tau = 2\pi \sqrt{\frac{L \cos \theta}{g}}.$$

This equation gives the relation between τ, L, and θ. Notice that τ, called the *period* of motion, does not depend on m.

If $L = 1.0$ m and $\theta = 30°$, what is the period of the motion? We have

$$\tau = 2\pi \sqrt{\frac{(1.0 \text{ m})(0.866)}{9.8 \text{ m/s}^2}} = 1.9 \text{ s}.$$

Example 5 *An unbanked roadway*   Let the block in Fig. 6-9 represent an automobile moving at constant speed v on a curved, unbanked roadway having a radius of curvature R. What must be the minimum coefficient of friction if the car is to remain on the roadway?

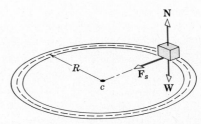

Figure 6-9   Example 5. A car on an unbanked roadway.

Two vertical forces act on the car, its weight **W** and a vertical normal force **N**. Since the car is not accelerating in the vertical direction, these forces must add vectorially to zero, so that $N = W = mg$.

The car, however, is accelerating radially toward the center of the turn. This acceleration must be provided by a centripetal force in the same direction; this can only be the sideways frictional force exerted by the roadway on the tires. The force in question is a force of *static* friction because there is no relative motion between the tire tread and the roadway in the radial direction.

The magnitude of the acceleration is v^2/R, so the magnitude of the required centripetal force must be $F = mv^2/R$. If this force is to be provided by friction, the required minimum coefficient of friction follows from Eq. 6-1, or

$$\mu_s = \frac{F}{N} = \frac{mv^2/R}{mg} = \frac{v^2}{gR}. \tag{6-6}$$

Note that the mass of the car does not enter. If a car, whether a Toyota or a Cadillac, travels at 50 km/h (=31 mi/h = 14 m/s) around an unbanked curve of 120-m radius, the coefficient of static friction between the tires and the road must be *at least*

$$\mu_s = \frac{v^2}{gR} = \frac{(14 \text{ m/s})^2}{(9.8 \text{ m/s}^2)(120 \text{ m})} = 0.17,$$

or the car will leave the road.

Example 6 *A banked roadway* The sideways frictional force in Example 5 cannot be safely relied upon under all circumstances, an icy road for example, to permit a car to negotiate a curve. Reliance on friction can be avoided by banking the roadbed on curves, as shown in Fig. 6-10. Derive an expression for θ, the correct angle of bank.

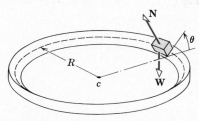

Figure 6-10   Example 6. A car on a banked roadway.

There is no acceleration in the vertical direction in Fig. 6-10, so

$$N \cos \theta = W \, (= mg).$$

Since we have assumed that frictional forces are not involved, the necessary centripetal force F must be provided by $N \sin \theta$, the radially inward component of **N**. This force must cause a centripetal acceleration whose magnitude a is v^2/R. Equation 6-5 then takes the form

$$F = N \sin \theta = \frac{mv^2}{R}.$$

Dividing this last equation by the preceding one yields

$$\tan \theta = \frac{v^2}{gR}. \qquad (6-7)$$

Notice that the proper angle of banking depends on the speed of the car and the radius of curvature of the road. For a given curvature, the road is banked at an angle corresponding to an expected average speed. Often curves are marked by signs giving the design speed for which the curve was banked. In driving a car around a properly banked curve, at the proper speed, the driver is unaware of *any* sideways force, frictional or otherwise, acting on the car or, for that matter, on the driver.

Check Eq. 6-7 for the limiting cases of $v = 0$, $R \to \infty$, $v \to \infty$, $g = 0$, and so on. Note that Eq. 6-7 is identical with the equation for the angle of a conical pendulum, discussed in Example 4. Review the two derivations side by side to see why this should be so. Note also that the quantity v^2/gR in Eq. 6-7 is identical with the expression for μ_s derived in Example 5. Why should this be so? Does the result of Example 1 throw any light on this equality?

Example 7 *The Rotor* In many amusement parks* we find a device called the *rotor*. The rotor is a hollow cylindrical room which can be set rotating about the central vertical axis of the cylinder. A person enters the rotor, closes the door, and stands up against the wall. The rotor gradually increases its rotational speed from rest until, at a predetermined speed, the floor below the person is opened downward, revealing a deep pit. The person does not fall but remains "pinned up" against the wall of the rotor. Find the coefficient of friction necessary to prevent falling.

The forces acting on the person are shown in Fig. 6-11. **W** is the person's weight, F_s is the force of static friction between person and rotor wall, and **N** is the centripetal force exerted by the wall on the person necessary to keep her moving in a circle. Let the radius of the rotor be R and the final speed of the passenger be v. Since the person does not move vertically, but experiences a radial acceleration v^2/R at any instant, we have

$$F_s - W = 0$$

and

$$N(=ma) = \left(\frac{W}{g}\right)\left(\frac{v^2}{R}\right).$$

If μ_s is the coefficient of static friction between person and wall necessary to prevent slipping, then $F_s = \mu_s N$ and

$$F_s = W = \mu_s N$$

or

$$\mu_s = \frac{W}{N} = \frac{gR}{v^2}.$$

This equation gives the minimum coefficient of friction necessary to prevent slipping for a rotor of radius R when a particle on its wall has a speed v. Notice that the result does not depend on the person's weight.

As a practical matter the coefficient of friction between the textile material of clothing and a typical rotor wall (canvas) is about 0.40. For a typical rotor the radius is 2.0 m, so that v must be about 7.0 m/s ($= 16$ mi/h) or more.

Note that the quantity v^2/gR occurs once more in the solution of this example, as an inverse ratio. Analyze the problems of the conical pendulum (Example 4), the banked roadway (Example 6), and the rotor (this example) to learn what elements they have in common.

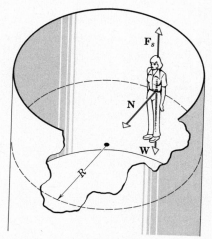

Figure 6-11   Example 7. The forces on a person in a "rotor" of radius R.

---

* See "Physics and the Amusement Park," by John L. Roeder, *The Physics Teacher*, September 1975.

REVIEW GUIDE AND SUMMARY

Static friction

Consider a horizontal force **F** applied to a block resting on a rough surface as in Fig. 6–1. As F is slowly increased from zero an opposing force F_s of equal magnitude automatically appears, associated with microscopic interlocking irregularities between the block and the surface at their areas of contact. When the applied force reaches a certain critical value the microscopic bonds break and the block suddenly moves. This is the phenomenon of static friction.

Law of static friction

Experimentally, the force of static friction turns out to be largely independent of the area of contact and proportional to the normal force acting between the body and the surface. The law of static friction is

$$F_s \leq \mu_s N, \qquad [6–1]$$

where μ_s is the coefficient of static friction, an essentially constant dimensionless quantity. See Examples 1 and 2.

Law of kinetic friction

Once the block has been set into motion, the force F_k needed to keep it in motion with a constant velocity—that is, to maintain it in equilibrium—is less than the critical force needed to get the motion started. The law of kinetic friction is

$$F_k = \mu_k N, \qquad [6–2]$$

where μ_k is the coefficient of kinetic friction. See Example 3.

Centripetal force

A body that moves at constant speed v in a circle of radius R is said to be in uniform circular motion. The kinematics of such motion (see Section 4–4) require that there be a centripetal acceleration **a**, directed toward the center of the circle and given in magnitude by v^2/R. Newton's second law yields, for the corresponding centripetal force,

$$F = ma = \frac{mv^2}{R}. \qquad [6–5]$$

Such a force may be provided, among other ways, by a taut string, as in a conical pendulum (Example 4); by a radially directed frictional force, as when a car drives around a curve on an unbanked roadway (Example 5); and by a contact force exerted by another body, such as a curved banked roadway (Example 6) or the wall of an amusement park rotor (Example 7).

QUESTIONS

1. There is a limit beyond which further polishing of a surface *increases* rather than decreases frictional resistance. Can you explain this?

2. Can the coefficient of static friction have a value greater than one? What about the coefficient of kinetic friction?

3. What are some examples of the beneficial uses of friction?

4. A crate, heavier than you are, rests on a rough floor. The coefficient of static friction between the crate and the floor is the same as that between the soles of your shoes and the floor. Can you push the crate across the floor? (See Fig. 6–12.)

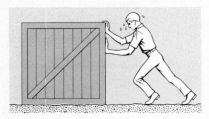

Figure 6–12 Question 4.

5. Often a base runner can get to a base quicker by running than by sliding. Explain why this is so. Why slide, then?

6. Which raindrops, if either, fall faster—small ones or large ones?

7. A downhill skier can move faster than a skydiver. (The 1985 world's record for skiers is 125 mi/h; the terminal speed for skydivers is somewhat less than this.) How can this be?

8. How could a person who is at rest on completely frictionless ice covering a pond reach shore? Could she do this by walking, rolling, swinging her arms, or kicking her feet? How could a person be placed in such a position in the first place?

9. If you want to stop the car in the shortest distance on an icy road, should you (a) push hard on the brakes to lock the wheels, (b) push just hard enough to prevent slipping, or (c) "pump" the brakes?

10. Why are train roadbeds and highways banked on curves?

11. How does the earth's rotation affect the apparent weight of a body at the equator?

12. Why is it that racing drivers actually speed up when traversing a curve?

13. You are flying a plane at constant altitude and you wish to make a 90° turn. Why must you bank in order to do so?

14. When a wet dog shakes himself, people standing nearby tend to get wet. Why does the water fly outward from the dog in this way?

15. You are driving a station wagon at uniform speed along a straight highway. A beachball rests at the center of the wagon bed and a helium-filled balloon floats above it, touching the roof of the wagon. What happens to each if you (*a*) turn a corner or (*b*) apply the brakes?

16. You must have noticed (Einstein did) that when you stir a cup of tea, the floating tea leaves collect at the center of the cup rather than at the outer rim. Can you explain this? (Einstein could.)

17. Explain why a plumb bob will not hang exactly in the direction of the earth's gravitational attraction at most latitudes.

18. Suppose that you need to measure whether a table top in a train is truly horizontal. If you use a spirit level, can you determine this when the train is moving down or up a grade? When the train is moving along a curve? (Hint: There are two horizontal components.)

19. In the conical pendulum of Example 4, what happens to the period τ and the speed v when $\theta = 90°$? Why is this angle not achievable physically? Discuss the case for $\theta = 0°$.

20. A coin is put on a phonograph turntable. The motor is started, but before the final speed of rotation is reached, the coin flies off. Explain.

21. A car is riding on a country road that resembles a roller coaster track. If the car travels with uniform speed, compare the force it exerts on a horizontal section of the road to the force it exerts on the road at the top of a hill and at the bottom of a hill. Explain.

22. A passenger in the front seat of a car finds himself sliding toward the door as the driver makes a sudden left turn. Describe the forces on the passenger and on the car at this instant if (*a*) the motion is viewed from a reference frame attached to the earth and (*b*) if attached to the car.

23. Your car skids across the center line on an icy highway. Should you turn the front wheels in the direction of skid or in the opposite direction (*a*) when you want to avoid a collision with an oncoming car, (*b*) when no other car is near but you want to regain control of the steering? Assume rear-wheel drive.

EXERCISES

Section 6–2 Frictional Forces

1. A 72-kg fireman ($W = mg = 160$ lb) slides down a vertical pole with an average acceleration of 3.0 m/s². What is the average vertical force he exerts on the pole?

2. A bedroom bureau with a mass of 45 kg, including drawers and clothing, rests on the floor. (*a*) If the coefficient of static friction between the bureau and the floor is 0.45, what is the minimum horizontal force a person must apply to start the bureau moving? (*b*) If the drawers and clothing, with 7.0 kg mass, are removed before the bureau is pushed, what is the minimum horizontal force required to start it moving?

3. A 35-kg crate is at rest on the floor. A man attempts to push it across the floor by applying a 100-N force horizontally. (*a*) Take the coefficient of static friction between the crate and floor to be 0.37 and show that the crate does not move. (*b*) A second man helps by pulling up on the crate. What minimum vertical force must he apply so that the crate starts to move across the floor? (*c*) If the second man applies a horizontal rather than a vertical force, what minimum force, in addition to the 100-N force of the first man, must he exert to get the crate started?

4. A horizontal force F of 12 lb pushes a block weighing 5 lb against a vertical wall (Fig. 6–13). The coefficient of static fric-

tion between the wall and the block is 0.6, and the coefficient of kinetic friction is 0.4. Assume that the block is not moving initially. (*a*) Will the block start moving? (*b*) What is the force exerted on the block by the wall?

5. A 70-kg mountain climber spans a rock chimney as shown in Fig. 6–14. If the effective coefficient of static friction between his hands and boots and the rock is 3.4, with what minimum force must he press with each hand and foot in order to avoid falling?

Figure 6–14   Exercise 5.

6. A 100-N force is applied at the angle θ above the horizontal to a 25-kg sofa resting on the floor. (*a*) For each of the following angles θ, calculate the normal force of the floor on the sofa and the

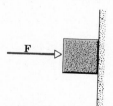

Figure 6–13   Exercise 4.

force of friction that the floor must exert on the sofa if it is to remain at rest: (*i*) 0°, (*ii*) 30°, (*iii*) 60°. (*b*) Take the coefficient of static friction between the sofa and the floor to be 0.42 and, for each of the above directions of the applied force, decide if the sofa does indeed remain at rest.

7. The coefficient of static friction between the tires of a car and a dry road is 0.60. If the mass of the car is 1500 kg, what maximum braking force is obtainable?

8. Suppose that only the rear wheels of an automobile can accelerate it, and that one-half the total weight of the automobile is supported by those wheels. (*a*) What is the maximum acceleration attainable if the coefficient of static friction between tires and road is μ_s? (*b*) Take $\mu_s = 0.35$ and get a numerical value for this acceleration.

9. What is the greatest acceleration that can be generated by a runner if the coefficient of static friction between her shoes and the road is 0.95?

10. A person applies a 220-N horizontal force to a 55-kg crate to push it across a level floor. If the coefficient of kinetic friction is 0.37, what is the acceleration of the crate?

11. A 50-lb (=220-N) chair rests on the floor. The coefficient of static friction between the chair and the floor is 0.41, while the coefficient of kinetic friction is 0.32. (*a*) What is the minimum horizontal force with which a person must push on the chair to start it moving? (*b*) Once moving, what horizontal force must the person apply to keep it moving with constant velocity? (*c*) If the person continued to push with the force used to start the motion, what would be the acceleration of the chair?

12. A 125-lb (=556-N) filing cabinet rests on the floor. The coefficient of static friction between it and the floor is 0.68, while the coefficient of kinetic friction is 0.56. In four different attempts to move it, it is pushed with horizontal forces of (*a*) 50.0 lb (=222 N), (*b*) 75.0 lb (=334 N), (*c*) 100 lb (=445 N), and (*d*) 125 lb (=556 N). For each attempt, tell if the cabinet moves and calculate the force of friction the floor exerts on it. The cabinet is initially at rest for each attempt.

13. A 110-g hockey puck slides on the ice for 15 m before it stops. (*a*) If its initial speed was 6.0 m/s, what is the force of friction between puck and ice? (*b*) What is the coefficient of kinetic friction?

Section 6-3 The Dynamics of Uniform Circular Motion
14. During an Olympic bobsled run, a European team takes a turn of radius 25 ft at a speed of 60 mi/h. How many *g*'s do the riders experience?

15. A bicylist travels in a circle of radius 25 m at a constant speed of 9.0 m/s. The combined mass of the bicycle and rider is 85 kg. Calculate (*a*) the force of friction exerted by the road and (*b*) the total force exerted by the road. See Fig. 6-15.

Figure 6-15   Exercise 15. EPU/Heine Pederson/Woodfin Camp & Associates.

16. A 2400-lb (=1.07×10^4-N) car traveling at 30.0 mi/h (=13.4 m/s) attempts to round an unbanked curve with a radius of 200 ft (=61.0 m). (*a*) What force of friction is required to keep the car on its circular path? (*b*) If the coefficient of static friction for the tires and road is 0.35, is the attempt successful?

17. If the coefficient of static friction for tires on a road is 0.25, at what greatest speed can a car round a level 150-ft (=47.5-m) radius curve without slipping?

18. What is the smallest radius of an unbanked curve around which a bicyclist can travel if her speed is 18 mi/h and the coefficient of static friction between the tires and the road is 0.32?

19. A child places a picnic basket on the outer rim of a merry-go-round that has a radius of 15 ft (=4.6 m and revolves once every 30 s. How large must the coefficient of static friction be for the basket to stay on the merry-go-round?

20. Show that the periods of two conical pendulums of different lengths that are hung from a ceiling and rotate with their bobs an equal distance below the ceiling are equal.

21. A conical pendulum is formed by attaching a 50-g mass to a 1.2-m string. The mass swings around a circle of radius 25 cm. (*a*) What is the speed of the mass? (*b*) What is the magnitude of the acceleration of the mass? (*c*) What is the tension in the string?

PROBLEMS GROUPED BY SECTION

Section 6-2 Frictional Forces
1. A man drags a 150-lb crate across a floor by pulling on a rope inclined 15° above the horizontal. (*a*) If the coefficient of static friction is 0.50, what tension in the rope is required to start the crate moving? (*b*) If $\mu_k = 0.35$, what is the initial acceleration of the crate?

2. A 3.5-kg block is pushed along a horizontal floor by a force $P = 15$ N that makes an angle $\theta = 40°$ with the horizontal, as shown in Fig. 6–16. The coefficient of kinetic friction between the block and floor is 0.25. Calculate (*a*) the frictional force exerted on the block and (*b*) the acceleration of the block.

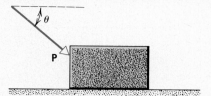

Figure 6–16   Problem 2.

3. The coefficient of kinetic friction in Fig. 6–17 is 0.2. What is the acceleration of the block?

Figure 6–17   Problem 3.

4. A 5.0-kg block on an inclined plane is acted upon by a horizontal force of 50 N (Fig. 6–18). The coefficient of friction between block and plane is 0.3°. (*a*) What is the acceleration of the block if it is moving up the plane? (*b*) How far up the plane will the block go if it has an initial upward speed of 4.0 m/s? (*c*) What happens to the block after it reaches the highest point?

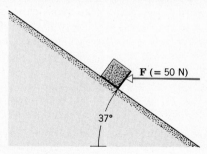

Figure 6–18   Problem 4.

5. A piece of ice slides down a 35° incline in twice the time it takes to slide down a frictionless 35° incline. What is the coefficient of kinetic friction between the ice and the incline?

6. In Fig. 6–19, *A* is a 10-lb (=44-N) block and *B* is a 5.0-lb

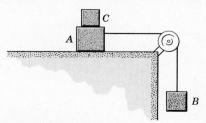

Figure 6–19   Problem 6.

(=22-N) block. (*a*) Determine the minimum weight (block *C*) that must be placed on *A* to keep it from sliding, if μ_s between *A* and the table is 0.20. (*b*) The block *C* suddenly lifted off *A*. What is the acceleration of block *A* if μ_k between *A* and the table is 0.20?

7. A 40-kg slab rests on a frictionless floor. A 10-kg block rests on top of the slab (Fig. 6–20). The coefficient of static friction between the block and the slab is 0.60, whereas the kinetic coefficient is 0.40. The 10-kg block is acted upon by a horizontal force of 100 N. What are the resulting accelerations of (*a*) the block, and (*b*) the slab?

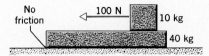

Figure 6–20   Problem 7.

8. A 4.0-kg block is put on top of a 5.0-kg block. In order to cause the top block to slip on the bottom one, held fixed, a horizontal force of 12 N must be applied to the top block. The assembly of blocks is now placed on a horizontal, frictionless table (Fig. 6–21). Find (*a*) the maximum horizontal force *F* that can be applied to the lower block so that the blocks will move together, and (*b*) the resulting acceleration of the blocks.

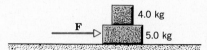

Figure 6–21   Problem 8.

9. Block *B* in Fig. 6–22 weighs 160 lb (=710 N). The coefficient of static friction between block and table is 0.25. Find the maximum weight of block *A* for which the system will be in equilibrium.

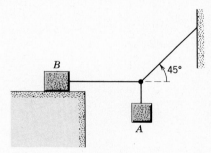

Figure 6–22   Problem 9.

10. Body *B* weighs 100 lb and body *A* weighs 32 lb (Fig. 6–23). Given $\mu_s = 0.56$ and $\mu_k = 0.25$, (*a*) find the acceleration of the

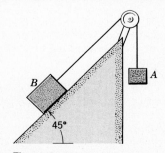

Figure 6–23   Problem 10.

system if B is initially at rest, and (b) find the acceleration if B is moving initially.

11. Two masses, $m_1 = 1.65$ kg and $m_2 = 3.30$ kg, attached by a massless rod parallel to the incline on which both slide, as shown in Fig. 6–24, travel down along the plane with m_1 trailing m_2. The angle of incline is $\theta = 30°$. The coefficient of kinetic friction between m_1 and the incline is $\mu_1 = 0.226$; between m_2 and the incline the corresponding coefficient is $\mu_2 = 0.113$. Compute (a) the tension in the rod linking m_1 and m_2 and (b) the common acceleration of the two masses. (c) Would the answers to (a) and (b) be changed if m_2 trails m_1?

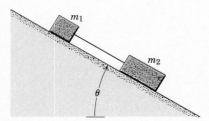

Figure 6–24   Problem 11.

12. The handle of a floor mop of mass m makes an angle θ with the vertical direction. Let μ_k be the coefficient of kinetic friction between mop and floor, and μ_s be the coefficient of static friction between mop and floor. Neglect the mass of the handle. (a) Find the magnitude of the force F directed along the handle required to slide the mop with uniform velocity across the floor. (b) Show that if θ is smaller than a certain angle θ_0, the mop cannot be made to slide across the floor no matter how great a force is directed along the handle. (c) What is the angle θ_0?

13. The two blocks shown in Fig. 6–25 are free to move. The coefficient of friction between the blocks is μ_s, but the surface beneath M is frictionless. What is the minimum horizontal force F required to hold m against M?

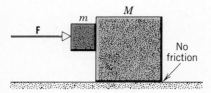

Figure 6–25   Problem 13.

14. A wire will break under tensions exceeding 1000 N ($= 230$ lb). (a) If the wire, not necessarily horizontal, is used to drag a box across the floor, what is the greatest weight that can be moved if the coefficient of static friction is 0.35? (b) If the wire is used to lift a box, what is the greatest weight that can be lifted with an upward acceleration of 0.92 m/s²?

Section 6–3   The Dynamics of Uniform Circular Motion

15. A mass m on a frictionless table is attached to a hanging mass M by a cord through a hole in the table (Fig. 6–26). Find the speed with which m must spin for M to stay at rest.

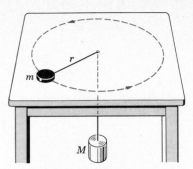

Figure 6–26   Problem 15.

16. A small coin is placed on a flat, horizontal turntable. The turntable is observed to make three revolutions in 3.14 s. (a) What is the speed of the coin when it rides without slipping at a distance 5.0 cm from the center of the turntable? (b) What is the acceleration (magnitude and direction) of the coin in part (a)? (c) What is the frictional force acting on the coin in part (a) if the coin has a mass of 2.0 g? (d) What is the coefficient of static friction between the coin and the turntable if the coin is observed to slide off the turntable when it is more than 10 cm from the center of the turntable?

17. A circular curve of highway is designed for traffic moving at 60 km/h ($= 37$ mi/h). (a) If the radius of the curve is 150 m ($= 490$ ft), what is the correct angle of banking of the road? (b) If the curve is not banked, what is the minimum coefficient of friction between tires and road that would keep traffic from skidding at this speed?

18. A banked circular highway curve is designed for traffic moving at 60 km/h. The radius of the curve is 200 m. Traffic is moving along the highway at 40 km/h on a stormy day. What is the minimum coefficient of friction between tires and road that will allow cars to negotiate the turn without sliding off the road?

19. A 150-lb student on a steadily rotating Ferris wheel has an apparent weight of 125 lb at the highest point. (a) What is his apparent weight at the lowest point? (b) What would be his apparent weight at the highest point if the speed of the Ferris wheel were doubled?

20. A block of mass m at the end of a string is whirled around in a vertical circle of radius R. Find the critical speed below which the string would become slack at the highest point.

21. A certain string can withstand a maximum tension of 9.0 lb without breaking. A child ties a 0.82-lb stone to one end. Holding the other end, he whirls the stone in a vertical circle of radius 3.0 ft, slowly increasing the speed until the string breaks. (a) Where is the stone on its path when the string breaks? (b) What is the speed of the stone as the string breaks?

22. Because of the rotation of the earth, a plumb bob may not hang exactly along the direction of the earth's gravitational pull (its weight) but deviate slightly from that direction. Calculate the deviation (a) at 40° latitude, (b) at the poles, and (c) at the equator.

23. An airplane is flying in a horizontal circle at a speed of 480 km/h. If the wings of the plane are tilted 40° to the vertical, what is the radius of the circle the plane is flying? See Fig. 6–27.

Figure 6–27 Problem 23. Courtesy Grumman Corporation.

24. A 1.34-kg ball is attached to a rigid vertical rod by means of two massless strings each 1.7 m long. The strings are attached to the rod at points 1.7 m apart. The system is rotating about the axis of the rod, both strings being taut and forming an equilateral triangle with the rod, as shown in Fig. 6–28. The tension in the upper string is 25 N. (*a*) Draw the free-body diagram for the ball. (*b*) What is the tension in the lower string? (*c*) What is the net force on the ball at the instant shown in the figure? (*d*) What is the speed of the ball?

Figure 6–28   Problem 24.

ADDITIONAL PROBLEMS

25. A student wants to determine the coefficients of static friction and kinetic friction between a box and a plank. He places the box on the plank and gradually raises the plank. When the angle of inclination with the horizontal reaches 30°, the box starts to slip and slides 2.5 m down the plank in 4.0 s. What are the coefficients of friction?

26. A frigate bird is soaring in a circular path. Her bank angle is estimated to be 25° and it takes 13 s for her to complete one circle. (*a*) How fast is she flying? (*b*) What is the radius of the circle? (See *Scientific American,* The Amateur Scientist, March 1985.)

27. A block weighing 80 N rests on a plane inclined at 20° to the horizontal, as shown in Fig. 6–29. The coefficient of static friction is 0.25, while the coefficient of kinetic friction is 0.15. (*a*) What is the minimum force *F*, parallel to the plane, that will

prevent the block slipping down the plane? (*b*) What is the minimum force *F* that will start the block moving up the plane? (*c*) What force *F* is required to move the block up the plane at constant velocity?

28. Frictional heat generated by the moving ski is the chief factor promoting sliding in skiing. The ski sticks at the start, but once in motion will melt the snow beneath it. Waxing the ski makes it water repellent and reduces friction with the film of water. A magazine reports that a new type of plastic ski is even more water repellent, and that on a gentle 200-m slope in the Alps, a skier reduced his time from 61 to 42 s with the new skis. (*a*) Determine the average acceleration for each pair of skis. (*b*) Assuming a 3° slope, compute the coefficient of kinetic friction for each case.

29. A driver's manual states that a driver traveling at 48 km/h and desiring to stop as quickly as possible travels 10 m before his foot reaches the brake. He travels an additional 21 m before coming to rest. (*a*) What coefficient of friction is assumed in these calculations? (*b*) What is the minimum radius for turning a corner at 48 km/h without skidding?

30. A man is driving a car at a speed of 85 km/h (= 53 mi/h) when he notices a barrier across the road exactly 60 m (= 200 ft) ahead. (*a*) What is the minimum static coefficient of friction be-

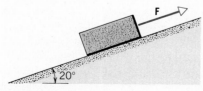

Figure 6–29   Problem 27.

tween his tires and the road that will allow him to stop before striking the barrier? (b) Suppose that he is driving on a large, empty parking lot. What is the minimum coefficient of static friction that would allow him to turn the car in a 60-m-radius circle and, in this way, avoid collision with the barrier?

31. A railroad flatcar is loaded with crates having a coefficient of static friction 0.25 with the floor. If the train is moving at 48 km/h, in how short a distance can the train be stopped without letting the crates slide?

32. A block slides down an inclined plane of slope angle θ with constant velocity. It is then projected up the same plane with an initial speed v_0. (a) How far up the incline will it move before coming to rest? (b) Will it slide down again?

33. A small object is placed 10 cm from the center of a phonograph turntable. It is observed to remain on the table when it rotates at $33\frac{1}{3}$ rev/min but slides off when it rotates at 45 rev/min. What are the limits of the coefficient of static friction between the object and the surface of the turntable?

34. A model airplane of mass 0.75 kg is flying at constant speed in a horizontal circle at one end of a 30-m cord and at a height of 18 m. The other end of the cord is tethered to the ground. The airplane makes 4.4 rev/min and there are no sideways aerodynamic forces on the airplane. (a) What is the acceleration of the plane? (b) What is the tension in the cord? (c) What is the lift produced by the plane's wings?

35. Two blocks are connected over a massless pulley as shown in Fig. 6–30. The mass of block A is 10 kg and the coefficient of kinetic friction is 0.20. Block A slides down the incline at constant speed. What is the mass of block B?

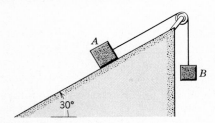

Figure 6–30   Problem 35.

36. Assume that the standard kilogram would weigh exactly 9.80 N at sea level on the earth's equator if the earth did not rotate. Then take into account the fact that the earth does rotate, so that this mass moves in a circle of radius 6.40×10^6 m (the earth's radius) at a constant speed of 465 m/s. (a) Determine the centripetal force needed to keep the standard moving in its circular path. (b) Determine the force exerted by the standard kilogram on a spring balance from which it is suspended at the equator (its apparent weight).

37. A stunt man drives a car over the top of a hill, the cross section of which can be approximated by a circle of radius 250 m, as in Fig. 6–31. What is the greatest speed at which he can drive without the car leaving the road at the top of the hill?

Figure 6–31   Problem 37.

38. Block m_1 in Fig. 6–32 has a mass of 4.0 kg and m_2 has a mass of 2.0 kg. The coefficient of friction between m_2 and the horizontal plane is 0.50. The inclined plane is smooth. Find (a) the tension in the string and (b) the acceleration of the blocks.

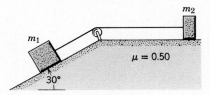

Figure 6–32   Problem 38.

39. A 10-kg block of steel is at rest on a horizontal table. The coefficient of static friction between block and table is 0.50. (a) What is the magnitude of the horizontal force that will just start the block moving? (b) What is the magnitude of a force acting upward 60° from the horizontal that will just start the block moving? (c) If the force acts down at 60° from the horizontal, how large can it be without causing the block to move?

40. A locomotive accelerates a 25-car train along a level track. Each car has a mass of 50 metric tons and is subject to a frictional force $f = 250v$, where the speed v is in meters per second and the force f is in newtons. At the instant when the velocity of the train is 30 km/h, the acceleration is 0.20 m/s². (a) What is the tension in the coupling between the first car and the locomotive? (b) If this tension is the maximum force the locomotive can exert on the train, what is the steepest grade up which the locomotive can pull the train at 30 km/h? (A metric ton equals 10^3 kg.)

41. An 8.0-lb block and a 16-lb block connected together by a string slide down a 30° inclined plane. The coefficient of kinetic friction between the 8.0-lb block and the plane is 0.10; between the 16-lb block and the plane it is 0.20. Find (a) the acceleration of the blocks and (b) the tension in the string, assuming that the 8.0-lb block leads. (c) Describe the motion if the blocks are reversed.

42. An old streetcar rounds a corner on unbanked tracks. If the radius of the tracks is 30 ft and the car's speed is 10 mi/h, what angle with the vertical will be made by the loosely hanging hand straps?

*43. A block of mass m slides in an inclined right-angled trough as in Fig. 6–33. If the coefficient of kinetic friction between the block and the material composing the trough is μ_k, find the acceleration of the block.

Figure 6–33   Problem 43.

*44. A 1000-kg boat is traveling at 90 km/h when it shuts off its engine. The force of friction F with the water is proportional to the speed v of the boat: $F = 70v$, where v is in meters per second and F is in newtons. Find the time required for the boat to slow down to 45 km/h.

7 Work and Energy

7–1 Introduction

A fundamental problem of particle dynamics is to find how a particle will move when we know the forces that act on it. By "how a particle will move" we mean how its position varies with time. If the motion is one-dimensional, the problem is to find x as a function of time, $x(t)$. In previous chapters we solved this problem for the special case of a constant force. The method used is this: We find the resultant force $\mathbf{F}$ acting on the particle from the appropriate force law. We then substitute $\mathbf{F}$ and the particle mass m into Newton's second law of motion. This gives us the acceleration $\mathbf{a}$ of the particle; or

$$\mathbf{a} = \frac{\mathbf{F}}{m}.$$

If the force $\mathbf{F}$ and the mass m are constant, the acceleration $\mathbf{a}$ must be constant and the kinematic equations of Table 3–1 apply. Let us choose the x-axis to be along the direction of this constant acceleration. We can then find the speed of the particle from Eq. 3–12,

$$v = v_0 + at,$$

and the position of the particle from Eq. 3–15 (with $x_0 = 0$), or

$$x = v_0 t + \tfrac{1}{2} at^2;$$

note that, for simplicity and convenience, we have dropped the subscript x in these equations. The last equation gives us directly what we usually want to know, namely, $x(t)$, the position of the particle as a function of time.

The problem is more difficult, however, when the force acting on a particle is *not constant*. In such a case we still obtain the acceleration of the particle, as before, from Newton's second law of motion. However, in order to get the speed or position of the particle, we can no longer use the formulas previously developed for constant acceleration because the acceleration now is *not* constant. To solve such problems, we use the mathematical process of integration, which we consider in this chapter.

We confine our attention to forces that depend only on the position of the particle in its environment. This type of force is common in physics. One example is the gravitational force between bodies, such as the sun and the earth or the earth and a projectile. Another is the force exerted by a stretched spring on a body to which it is attached. The procedure used to determine the motion of a particle subject to such a force leads us to the concepts of work and kinetic energy and to the development of the work-energy theorem, which is the central feature of this chapter. In Chapter 8 we consider a broader view of energy, embodied in the law of conservation of energy, a concept that has played a major role in the development of physics.

7–2 Work Done by a Constant Force

Consider a particle acted on by a force. In the simplest case the force **F** is constant and the motion takes place in a straight line in the direction of the force. In such a situation we define the work done *by* the force *on* the particle as the product of the magnitude of the force F and the distance d through which the particle moves. We write this as

$$W = Fd.$$

However, the constant force acting on a particle may not act in the direction in which the particle moves. In this case we define the work done by the force on the particle as the product of the component of the force along the line of motion by the distance d the body moves along that line. In Fig. 7–1 a constant force **F** makes an angle ϕ with the x-axis and acts on a particle whose displacement along the x-axis is **d**. If W represents the work done by **F** during this displacement, then according to our definition,

$$W = (F \cos \phi)d. \tag{7–1}$$

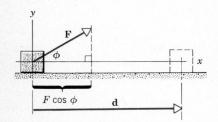

Figure 7–1 A force **F** makes the block undergo a displacement **d**. The component of **F** that does the work has magnitude $F \cos \phi$; the work done is $Fd \cos \phi$ $(=\mathbf{F} \cdot \mathbf{d})$.

Of course, other forces must act on a particle that moves in this way (its weight and the frictional force exerted by the plane, to name two). A particle acted on by only a single force may have a displacement in a direction other than that of this single force, as in projectile motion. But it cannot move in a straight line unless the line has the same direction as that of the single force applied to it. *Equation 7–1 refers only to the work done on the particle by the particular force* **F**. *The work done on the particle by the other forces must be calculated separately*. The total work done on the particle is the sum of the works done by the separate forces.

When ϕ is zero, the work done by **F** is simply Fd, in agreement with our previous equation. Thus, when a horizontal force draws a body horizontally, or when a vertical force lifts a body vertically, the work done by the force is the product of the magnitude of the force by the distance moved. When ϕ is 90°, the force has no component in the direction of motion. That force then does no work on the body. For instance, the vertical force holding a body a fixed distance off the ground does no work on the body, even if the body is moved horizontally over the ground. Also, the centripetal force acting on a body in motion does no work on that body because the force is always at right angles to the direction in which the body is moving. Of course, a force does no work on a body that does not move, for its displacement is then zero.

Notice that we can write Eq. 7–1 either as $(F \cos \phi)d$ or $F(d \cos \phi)$. This suggests that the work can be calculated in two different ways: Either we multiply the magnitude of the displacement by the component of the force in the direction of the displacement or we multiply the magnitude of the force by the component of the displacement in the direction of the force. These two methods always give the same result.

Work is a *scalar,* although the two quantities involved in its definition, force and displacement, are vectors. In Section 2–4 we defined the *scalar product* of two vectors as the scalar quantity that we find when we multiply the magnitude of one vector by the component of a second vector along the direction of the first. We promised in that section that we would soon run across physical quantities that behave like scalar products. Equation 7–1 shows that work is such a quantity. In the terminology of vector algebra we can write this equation as

Work by a constant force

$$W = \mathbf{F} \cdot \mathbf{d}, \tag{7-2}$$

where the dot indicates a scalar (or dot) product. Equation 7–2 for $\mathbf{F}$ and $\mathbf{d}$ corresponds to Eq. 2–11 for $\mathbf{a}$ and $\mathbf{b}$.

As Example 1 shows, work can be either positive or negative. If the particle on which a force acts has a displacement opposite to the direction of the force, the work done by that force is negative. This corresponds to an obtuse angle between the force and displacement vectors. For example, when a person lowers an object to the floor, the work done on the object by the upward force of his hand holding the object is negative. In this case ϕ is 180°, for $\mathbf{F}$ points up and $\mathbf{d}$ points down.

Work as we have defined it (Eq. 7–2) proves to be a very useful concept in physics. Our special definition of the word "work" does not correspond to the colloquial usage of the term. This may be confusing. As Fig. 7–2 suggests, a person holding a heavy weight at rest in the air may say that he is doing hard work—and he may work hard in the physiological sense—but from the point of view of physics we say that he is not doing any work on the weight. We say this because the ap-

(a)

(b)

Figure 7–2 (a) No force; no displacement; no work. (Sepp Seitz/Woodfin Camp.)
(b) Large force but no displacement so still no work. (Ed Goldfarb/Black Star.)

plied force causes no displacement. The word *work* is used only in the strict sense of Eq. 7–2. In many scientific fields words are borrowed from our everyday language and are used to name a very specific concept.

The *unit* of work is the work done by a unit force in moving a body a unit distance in the direction of the force. In SI units the unit of work is 1 *newton-meter*, called 1 *joule* (abbreviation J). In the British engineering system the unit of work is the *foot-pound*. In cgs systems the unit of work is 1 *dyne-centimeter*, called 1 *erg*. Using the relations between the newton, the dyne, and the pound, and the meter, the centimeter, and foot, we obtain

$$1 \text{ joule} = 10^7 \text{ ergs} = 0.738 \text{ foot-pound}$$

A convenient unit of work in atomic, nuclear, and particle physics is the *electron volt* (abbreviation eV) and its multiples (the keV, MeV, GeV, and TeV; see Table 1–2). We have

$$1 \text{ electron volt} = 1.60 \times 10^{-19} \text{ joule}.$$

Example 1 *Positive and negative work* A person lifts a stone of mass m from the floor a vertical distance d and puts it on a shelf. Discuss the work done on the stone. For simplicity, assume that the stone is lifted without acceleration.

As shown in Fig. 7–3, two forces act on the stone, the upward force F_p exerted by the person and the downward gravitational force, the weight mg, exerted by the earth. Because the stone is in equilibrium while it is being lifted, these forces must be oppositely directed and equal in magnitude, or

$$F_p = mg.$$

From Eq. 7–2, the work done on the stone by each of these forces is

$$W_p = \mathbf{F}_p \cdot \mathbf{d} = F_p d \cos 0° = +mgd$$

and

$$W_g = (m\mathbf{g}) \cdot \mathbf{d} = (mg)d \cos 180° = -mgd.$$

Thus the force exerted by the person does positive work on the stone but the gravitational force exerted by the earth does negative work. The *net* work $W_p + W_g$ done on the stone is zero, as we expect, because the net force on the stone is zero. This is not, of course, to say that it takes no work to lift a stone through a vertical height d. In such a context we do not refer to the *net* work but to the work done *by the person* and this, as we have just seen, is not zero but is $+mgd$.

Figure 7–3 Example 1.

Example 2 A block of mass 10.0 kg is to be raised from the bottom to the top of an incline 5.00 m long and 3.00 m off the ground at the top. Assuming frictionless surfaces, how much work must be done by a force **F** parallel to the incline pushing the block up at constant speed at a place where $g = 9.80 \text{ m/s}^2$?

The situation is shown in Fig. 7–4*a*. The forces acting on the block are shown in Fig. 7–4*b*. We must first find F, the magnitude of the force pushing the block up the incline. Because the motion is not accelerated, the resultant force must be zero; this means that the resultant force parallel to the plane must be zero. Thus,

$$F - mg \sin \theta = 0,$$

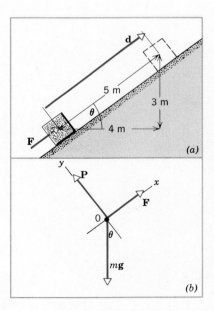

Figure 7–4 Example 2. (*a*) A force **F** displaces a block a distance **d** up an inclined plane which makes an angle θ with the horizontal. (*b*) A free-body force diagram for the block.

or

$$F = mg \sin \theta = (10.0 \text{ kg})(9.80 \text{ m/s}^2)(\tfrac{3}{5}) = 58.8 \text{ N}.$$

Then the work done by **F**, from Eq. 7–2 with $\phi = 0°$, is

$$W = \mathbf{F} \cdot \mathbf{d} = Fd \cos 0° = Fd = (58.8 \text{ N})(5.00 \text{ m}) = 294 \text{ J}.$$

If a man were to raise the block vertically without using the incline, the work he would do would be the vertical force mg times the vertical distance or

$$(98.0 \text{ N})(3.00 \text{ m}) = 294 \text{ J},$$

the same as before. The only difference is that with the incline he could apply a smaller force ($F = 58.8$ N) to raise the block than is required without the incline ($mg = 98.0$ N); on the other hand, he had to push the block a greater distance (5.00 m) up the incline than he had to raise the block directly (3.00 m).

Example 3 In Example 3 of Chapter 6 a boy pulls a sled along a rough horizontal surface at a constant speed. (*a*) What work is done by the tension force **T** in the rope (supplied by the boy) if the distance moved is 30 ft ($= 8.6$ m)? (*b*) What work is done by the frictional force **f** exerted by the ground on the sled? (*c*) What is the total work done on the sled by all the forces that act on it?

(*a*) From Eq. 7–2 we find that the work done by the boy on the sled, through the tension in the rope, is

$$W_T = \mathbf{T} \cdot \mathbf{d} = Td \cos \phi,$$

where **d** is the displacement of the sled. In the earlier example we found that the tension in the rope required to maintain a constant velocity is 2.4 lb. Since the angle ϕ is 45°, the work done is

$$\begin{aligned} W_T &= (2.4 \text{ lb})(30 \text{ ft})(\cos 45°) \\ &= + 51 \text{ ft} \cdot \text{lb} \ (= +69 \text{ J}). \end{aligned}$$

(*b*) To find the work done by friction, let us again apply Eq. 7–2, obtaining

$$W_f = \mathbf{f} \cdot \mathbf{d} = -fd.$$

The minus sign arises because **f** and **d** are oppositely directed. In Example 3 of Chapter 6 we saw that the force of friction is related to the tension in the rope by

$$T \cos \phi - f = 0,$$

so that the work is

$$W_f = -Td \cos \phi = -51 \text{ ft} \cdot \text{lb} \ (= -69 \text{ J}).$$

This is just $-W_T$, as we might expect.

(*c*) The four forces that act on the body are shown in Fig. 6–6*b*. Note that the work done by the normal force **N** and the weight **w** are zero because these forces act at right angles to the displacement **d**. The total work done by all four forces is then

$$\begin{aligned} W_{\text{total}} &= W_T + W_f + W_N + W_w \\ &= +51 \text{ ft} \cdot \text{lb} - 51 \text{ ft} \cdot \text{lb} + 0 + 0 \\ &= 0. \end{aligned}$$

We expect this result because the four forces add to zero as long as the sled is not accelerating. The resultant force, being zero, can do no work.

7–3 Work Done by a Variable Force—One-Dimensional Case

Let us now consider the work done by a force that is not constant. We consider first a force that varies in magnitude only. Let the force be given as a function of position $F(x)$ and assume that the force acts in the *x*-direction. Suppose that a body is moved along the *x*-direction by this force. What is the work done by this variable force in moving the body from x_1 to x_2?

In Fig. 7–5 we plot F versus x. Let us divide the total displacement into a large number of small equal intervals Δx (Fig. 7–5*a*). Consider the small displacement Δx from x_1 to $x_1 + \Delta x$. During this small displacement the force F has a nearly constant value and the work it does, ΔW, is approximately

$$\Delta W = F \Delta x, \tag{7–3}$$

where F is the average value of the force for this displacement. Likewise, during the small displacement from $x_1 + \Delta x$ to $x_1 + 2\Delta x$, the force F also has a nearly constant value and the work it does is approximately $\Delta W = F \Delta x$, where F is the average value of the force for *this* displacement. The total work done by F in displacing the body from x_1 to x_2, W_{12}, is approximately the sum of a large number of terms like that of Eq. 7–3, in which F has a different value of each term. Hence

$$W_{12} = \sum_{x_1}^{x_2} F \Delta x, \tag{7–4}$$

where the Greek letter sigma (Σ) stands for the sum over all intervals from x_1 to x_2.

To make a better approximation we can divide the total displacement from x_1 to x_2 into a larger number of equal intervals, as in Fig. 7–5*b*, so that Δx is smaller

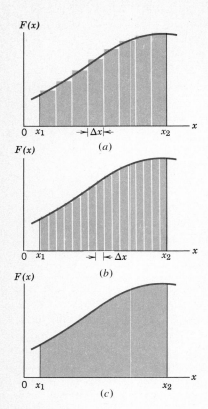

$F(x)$

$0 \quad x_1 \quad \rightarrow|\Delta x|\leftarrow \quad x_2$

(a)

$F(x)$

$0 \quad x_1 \quad \rightarrow| \leftarrow \Delta x \quad x_2$

(b)

$F(x)$

$0 \quad x_1 \qquad\qquad x_2$

(c)

Figure 7–5 Work done by a variable force—one dimension.

Hooke's law

and F is more nearly constant within each interval. It is clear that we can obtain better and better approximations by taking Δx smaller and smaller so as to have a larger and larger number of intervals. We can obtain an exact result for the work done by F if we let Δx go to zero and the number of intervals go to infinity. Hence the exact result is

$$W_{12} = \lim_{\Delta x \to 0} \sum_{x_1}^{x_2} F\,\Delta x. \tag{7–5}$$

The relation

$$\lim_{\Delta x \to 0} \sum_{x_1}^{x_2} F\,\Delta x = \int_{x_1}^{x_2} F\,dx,$$

as you may have learned in a calculus course, *defines the integral of F with respect to x from x_1 to x_2*. Numerically, this quantity is exactly equal to the area between the force curve and the x-axis between the limits x_1 and x_2 (Fig. 7–5c). Hence, graphically an integral can be interpreted as an area. The symbol $\int$ is a distorted S (for *sum*) and symbolizes the integration process. We can write the total work done by F in displacing a body from x_1 to x_2 as

$$W_{12} = \int_{x_1}^{x_2} F(x)\,dx. \tag{7–6}$$

As an example, consider a spring attached to a wall. Let the (horizontal) axis of the spring be chosen as an x-axis, and let the origin, $x = 0$, coincide with the endpoint of the spring in its normal, unstretched state. We assume that the positive x-direction points away from the wall. In what follows we imagine that we stretch the spring so slowly that it is essentially in equilibrium at all times ($\mathbf{a} = 0$).

If we stretch the spring so that its end point moves to a position x, the spring will exert a force on the agent doing the stretching given to a good approximation by

$$F = -kx, \tag{7–7}$$

where k is a constant called the *force constant* of the spring. Equation 7–7 is the *force law* for springs. The direction of the force is always opposite to the displacement of the end point from the origin. When the spring is stretched, $x > 0$ and F is negative; when the spring is compressed, $x < 0$ and F is positive. The force exerted by the spring is a *restoring force* in that it always points toward the origin. Real springs will obey Eq. 7–7, known as *Hooke's law*, if we do not stretch them beyond a limited range. We can think of k as the magnitude of the force per unit elongation. Thus very stiff springs have large values of k.

To stretch a spring we must exert a force F' on it equal but opposite to the force F exerted by the spring on us. The applied force is therefore $F' = kx$ and the work done by the applied force in stretching the spring so that its end point moves from x_1 to x_2 is*

$$W_{12} = \int_{x_1}^{x_2} F'(x)\,dx = \int_{x_1}^{x_2} (kx)\,dx = \tfrac{1}{2}kx_2{}^2 - \tfrac{1}{2}kx_1{}^2.$$

If we let $x_1 = 0$ and $x_2 = x$, we obtain

$$W = \int_0^x (kx)\,dx = \tfrac{1}{2}kx^2. \tag{7–8}$$

---

* The student just becoming familiar with calculus should consult the list of integrals in Appendix H.

This is the work done in stretching a spring so that its end point moves from its unstretched position to x. Note that the work to *compress* a spring by x is the same as that to stretch it by x because the displacement x is squared in Eq. 7–8; either sign for x gives a positive value for W.

We can also evaluate this integral by computing the area under the force-displacement curve and the x-axis from $x = 0$ to $x = x$. This is drawn as the shaded area in Fig. 7–6. The area is a triangle of base x and altitude kx. The shaded area is therefore

$$\tfrac{1}{2}(x)(kx) = \tfrac{1}{2}kx^2.$$

in agreement with Eq. 7–8.

Work to stretch a spring

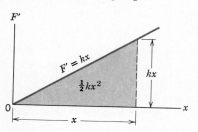

Figure 7–6   The force exerted in stretching a spring is $F' = kx$. The area under the force curve is the work done in stretching the spring a distance x and can be found by integrating or by using the formula for the area of a triangle.

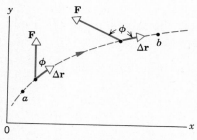

Figure 7–7   Showing how F and ϕ might vary along a path in a particular two-dimensional case.

7–4   Work Done by a Variable Force—General Case

The force **F** acting on a particle may vary in direction as well as in magnitude, and the particle may move along a curved path in three dimensions. To compute the work in this general case we divide the path into a large number of small displacements $\Delta\mathbf{r}$, each pointing along the path in the direction of motion. Figure 7–7 shows two selected displacements for a particular situation; it also shows the value of **F** and the angle ϕ between **F** and $\Delta\mathbf{r}$ at each location. We can find the amount of work done on the particle during a displacement $\Delta\mathbf{r}$ from

$$\Delta W = \mathbf{F} \cdot \Delta\mathbf{r} = F \cos \phi \, \Delta r \qquad (7\text{–}9)$$

The work done by the variable force **F** on the particle as the particle moves, say, from a to b in Fig. 7–7 is found very closely by adding up (summing) the elements of work done over each of the line segments that make it up. As the line segments $\Delta\mathbf{r}$ become smaller they may be replaced by differentials $d\mathbf{r}$ and the sum over the line segments may be replaced by an integral, as in Eq. 7–6. The work is then found from

$$W_{ab} = \int_a^b \mathbf{F} \cdot d\mathbf{r} = \int_a^b F \cos \phi \, dr. \qquad (7\text{–}10)$$

We cannot evaluate this integral until we are able to say how F and ϕ in Eq. 7–10 vary from point to point along the path; both are functions of the x- and y-coordinates of the particle in Fig. 7–7.

7–5   Kinetic Energy and The Work-Energy Theorem

In our previous examples of work done by forces, we dealt with unaccelerated objects. In such cases the resultant force acting on the object is zero. Let us suppose now that the resultant force acting on an object is not zero, so that the object is accelerated. The conditions are the same in all respects as those that exist when a single unbalanced force acts on the object.

The simplest situation to consider is that of a constant resultant force **F**. Such a force, acting on a particle of mass m, will produce a constant acceleration **a**. Let us choose the x-axis to be in the common direction of **F** and **a**. What is the work done by this force on the particle in causing a displacement x? We have (for constant acceleration) the relations

$$a = \frac{v - v_0}{t}$$

and

$$x = \frac{v + v_0}{2} \cdot t,$$

which are Eqs. 3–12 and 3–14, respectively (in which we have dropped the subscript x, for convenience, and chosen $x_0 = 0$ in the last equation). Here v_0 is the particle's speed at $t = 0$ and v its speed at the time t. Then the work done is

$$W = Fx = max$$
$$= m \left(\frac{v - v_0}{t}\right)\left(\frac{v + v_0}{2}\right) t = \frac{1}{2} mv^2 - \frac{1}{2} mv_0{}^2. \qquad (7\text{–}11)$$

We call one-half the product of the mass of a body and the square of its speed the *kinetic energy* of the body. If we represent kinetic energy by the symbol K, then

$$K = \tfrac{1}{2}mv^2. \qquad (7\text{–}12)$$

Kinetic energy

We may then state Eq. 7–11 in this way: The work done by the resultant force acting on a particle is equal to the change in the kinetic energy of the particle.

Although we have proved this result for a constant force only, it holds whether the resultant force is constant or variable. Let the resultant force vary in magnitude (but not in direction), for example. Take the displacement to be in the direction of the force. Let this direction be the x-axis. The work done by the resultant force in displacing the particle from x_0 to x is

$$W = \int \mathbf{F} \cdot d\mathbf{r} = \int_{x_0}^{x} F \, dx.$$

But from Newton's second law we have $F = ma$, and we can write the acceleration a as

$$a = \frac{dv}{dt} = \frac{dv}{dx} \cdot \frac{dx}{dt} = \frac{dv}{dx} v = v \frac{dv}{dx}.$$

Hence

$$W = \int_{x_0}^{x} F \, dx = \int_{x_0}^{x} mv \frac{dv}{dx} \, dx = \int_{v_0}^{v} mv \, dv = \frac{1}{2} mv^2 - \frac{1}{2} mv_0{}^2. \quad (7\text{–}13)$$

Even in the general case, in which the resultant force is variable in both magnitude *and* direction, Eq. 7–13 still holds: The work done on a particle by the resultant force acting on it is equal to the change in kinetic energy of the particle, or

$$W \text{ (of the resultant force)} = K - K_0 = \Delta K. \qquad (7\text{–}14)$$

Work-energy theorem

Equation 7–14 is known as the *work-energy theorem*. In Problem 11 you are asked to prove this theorem for the most general case.

Notice that when the speed of the particle is constant, there is no change in kinetic energy and the work done by the resultant force is zero. With uniform circular motion, for example, the speed of the particle is constant and the centripetal force does no work on the particle. A force at right angles to the direction of motion merely changes the direction of the velocity and not its magnitude.

Example 1 is another case in which the work-energy theorem applies in a particularly simple way. The kinetic energy of the stone is zero both initially (stone on the floor) and finally (stone on the shelf) so that $\Delta K = 0$. The work-energy theorem then predicts that the work done on the stone between these two states by the resultant force acting on the stone must be zero. Example 1 shows that this is indeed true.

If the kinetic energy of a particle decreases, the work done on it by the resultant force is negative. The displacement and the component of the resultant force along the line of motion are oppositely directed. The work done *on* the particle by the force is the negative of the work done *by* the particle on whatever produced the force. This is a consequence of Newton's third law of motion. Hence Eq. 7–14 can be interpreted to say that the kinetic energy of a particle *decreases* by an amount just equal to the amount of work that the particle *does*. A body is said to have energy stored in it because of its motion; as it does work it slows down and loses some of this energy. Therefore, *the kinetic energy of a body in motion is equal to the work it can do in being brought to rest*. This result holds whether the applied forces are constant or variable.

The units of kinetic energy and of work are the same. Kinetic energy, like work, is a scalar quantity. The kinetic energy of a group of particles is simply the (scalar) sum of the kinetic energies of the individual particles in the group.

Example 4   A neutron, one of the constituents of a nucleus, is found to pass two points 6.0 m apart in a time interval of 180 μs. Assuming that its speed is constant, find its kinetic energy. The mass of a neutron is 1.7×10^{-27} kg.

We find the speed from

$$v = \frac{d}{t} = \frac{6.0 \text{ m}}{180 \times 10^{-6} \text{ s}} = 3.3 \times 10^4 \text{ m/s}.$$

The kinetic energy is

$$K = \tfrac{1}{2}mv^2 = (\tfrac{1}{2})(1.7 \times 10^{-27} \text{ kg}) \times$$
$$(3.3 \times 10^4 \text{ m/s})^2 = 9.3 \times 10^{-19} \text{ J}.$$

For purposes of nuclear physics the joule is a very large energy unit. As we have seen, a unit more commonly used is the electron volt (eV), which is equal to 1.60×10^{-19} J. The kinetic energy of the neutron in our example can then be expressed as

$$K = (9.3 \times 10^{-19} \text{ J}) \left(\frac{1 \text{ eV}}{1.60 \times 10^{-19} \text{ J}} \right) = 5.8 \text{ eV}.$$

Example 5   A body is dropped from rest at a height h above the earth's surface. What will its kinetic energy be just before it strikes the ground?

The work-energy theorem (Eq. 7–14) states that the gain in kinetic energy is equal to the work done by the resultant force, which here is the force of gravity. This force is nearly constant as long as the distance fallen is small. It is directed along the line of motion, so that the work done by gravity is

$$W = \mathbf{F} \cdot \mathbf{d} = mgh.$$

Initially the body has a speed $v_0 = 0$ and finally a speed v. The gain in kinetic energy of the body is

$$K - K_0 = \tfrac{1}{2}mv^2 - \tfrac{1}{2}mv_0^2 = \tfrac{1}{2}mv^2 - 0.$$

Equating these two equivalent terms we obtain

$$K = \tfrac{1}{2}mv^2 = mgh$$

as the kinetic energy of the body just before it strikes the ground. The speed of the body is then

$$v = \sqrt{2gh}.$$

Show, also from the work-energy theorem, that in falling from a height h_1 to a height h_2 a body will increase its kinetic energy from $\tfrac{1}{2}mv_1^2$ to $\tfrac{1}{2}mv_2^2$, where

$$\tfrac{1}{2}mv_2^2 - \tfrac{1}{2}mv_1^2 = mg(h_1 - h_2).$$

Show how these results, obtained from the work-energy theorem, can also be obtained directly from the laws of motion for uniformly accelerated bodies.

Example 6   A 5.0-kg block slides on a horizontal frictionless table with a speed of 1.2 m/s. It is brought to rest in compressing a spring in its path. By how much is the spring compressed if its force constant k is 1.5×10^3 N/m?

If we apply the work-energy theorem to the motion of the block, we note that the change in kinetic energy, which appears on the right side of Eq. 7–14, is negative. The left side of this equation, which is the work done by the spring on the block, is also negative. The work-energy theorem thus yields

$$-\tfrac{1}{2}kx^2 = -\tfrac{1}{2}mv_0^2,$$

or

$$x = \sqrt{\frac{m}{k}}\, v_0$$

$$= \sqrt{\frac{5.0 \text{ kg}}{1.5 \times 10^3 \text{ N/m}}}\,(1.2 \text{ m/s})$$

$$= 6.9 \times 10^{-2} \text{ m} = 6.9 \text{ cm}.$$

7–6 Significance of
the Work-Energy Theorem

The work-energy theorem does not represent a new, independent law of classical mechanics. We have simply defined work and kinetic energy and derived the relation between them directly from Newton's second law. The work-energy theorem is useful, however, for solving problems in which the work done by the resultant force is easily computed and in which we are interested in finding the particle's speed at certain positions. Of greater significance, perhaps, is the fact that the work-energy theorem is the starting point for a sweeping generalization in physics. It has been emphasized that the work-energy theorem is valid when W is interpreted as the work done by the *resultant* force acting on the particle. However, it is helpful in many problems to compute separately the work done by certain types of force and give a special name to the work done by each type. This leads to the concepts of different types of energy and the principle of the conservation of energy, which is the subject of the next chapter.

Note that everything we have said in this chapter, and indeed in all previous chapters, dealt only with bodies that could be treated as a single particle. A block sliding down an inclined plane is an example. In more complicated systems, such as an accelerating automobile, we are dealing with systems of many particles rather than with a single particle. In the case of the automobile there are internal motions and forces associated with its engine so that internal work is done. In Chapter 9 we extend our treatment of mechanics to such many-particle systems. In particular, in Section 9–3 we shall see how to extend the work-energy theorem in such cases.

7–7 Power

Let us now consider the time involved in doing work. The same amount of work is done in raising a given body through a given height whether it takes one second or one year to do so but the *rate at which work is done* is quite different.

We define *power* as the time rate at which work is done. The average power delivered by an agent is the total work done by the agent divided by the total time interval, or

$$\overline{P} = W/t.$$

The instantaneous power delivered by an agent is

Power

$$P = dW/dt. \tag{7–15}$$

If the power is constant in time, then $P = \overline{P}$ and

$$W = Pt.$$

Units of power

In the International System of units, the unit of power is 1 J/s, which is called 1 *watt* (abbreviation W). This unit of power is named in honor of James Watt, who made major improvements to the steam engines of his day which pointed the way toward today's more efficient engines. In the British engineering system, the unit of power is 1 ft · lb/s. Because this unit is quite small for practical purposes, a larger unit, called the *horsepower* (abbreviation hp), has been adopted. Actually Watt himself suggested as a unit of power the power delivered by a horse as an engine. One horsepower was chosen to equal 550 ft · lb/s. One horsepower is equal to about 746 W or about three-fourths of a kilowatt.

Work can also be expressed in units of power × time. This is the origin of the term *kilowatt-hour*, for example. One kilowatt-hour is the work done in 1 hour by an agent working at a constant rate of 1 kilowatt. Hence 1 kilowatt-hour = $(1 \times 10^3 \text{ W})(3600 \text{ s}) = 3.6 \times 10^6$ J.

We can also express the power delivered to a body by a force **F** that acts on it by

$$P = \mathbf{F} \cdot \mathbf{v}, \tag{7-16}$$

where **v** is the velocity of the body. This equation holds at any instant of time provided only that **F** and **v** are specified at that instant. To convince ourselves that Eq. 7–16 is valid let us recall (Eq. 4–2) that **v** is defined as $d\mathbf{r}/dt$. With this substitution we have

$$P = \mathbf{F} \cdot \mathbf{v} = \mathbf{F} \cdot \left(\frac{d\mathbf{r}}{dt}\right) = \frac{\mathbf{F} \cdot d\mathbf{r}}{dt} = \frac{dW}{dt},$$

which agrees with Eq. 7–15.

Example 7 How fast would a 1500-lb horse have to run up a 30° incline to do work against gravity at a steady rate of 1.0 hp?

If the horse runs a distance x along the incline, in time t, he raises himself vertically by $x \sin \theta$, where $\theta = 30°$. The work W done against gravity is

$$W = \mathbf{F} \cdot \mathbf{d} = (mg)(x \sin \theta)$$

and the power P ($= 1.0$ hp) is

$$P = \frac{W}{t} = \frac{(mg)(x \sin \theta)}{t} = mgv \sin \theta.$$

The speed is

$$v = \frac{P}{mg \sin \theta} = \frac{1.0 \text{ hp}}{(1500 \text{ lb})(\sin 30°)} \left(\frac{550 \text{ ft} \cdot \text{lb/s}}{1.0 \text{ hp}}\right)$$
$$= 0.73 \text{ ft/s} = 0.50 \text{ mi/h}.$$

This is not a very demanding speed.

Example 8 An automobile uses 75 kW ($= 100$ hp) of power and moves at a uniform speed of 100 km/h ($= 28$ m/s $= 62$ mi/h). What is the forward thrust caused by the engine? The forward thrust **F** is in the direction of motion given by **v**, so that Eq. 7–16 becomes

$$P = Fv,$$

and

$$F = \frac{P}{v} = \frac{75 \times 10^3 \text{ W}}{28 \text{ m/s}} = 2.7 \times 10^3 \text{ N}(= 610 \text{ lb}).$$

Why doesn't the car accelerate?

REVIEW GUIDE AND SUMMARY

Work by a constant force

If a constant force **F** acts on a body that undergoes a displacement **d**, we say that the force does work on the body given by

$$W = \mathbf{F} \cdot \mathbf{d} = Fd \cos \phi, \tag{7-2}$$

where ϕ is the angle between **F** and **d**. W may be positive, zero, or negative, depending on the value of ϕ; see Example 1. Work is a scalar and its SI unit is the joule ($= 1$ newton-meter $= 0.735$ ft · lb). Example 3 shows a typical work situation involving constant forces.

Work by a variable force

Suppose that both force and displacement are constrained to the x-axis but the magnitude of the force depends on x. The work done by the force as the body moves from x_1 to x_2 is given, by extension of Eq. 7-2, as

$$W_{12} = \int_{x_1}^{x_2} F(x) \, dx \tag{7-6}$$

Hooke's law

An application is the work done to stretch or compress a spring by a distance x from its relaxed length. If the spring obeys Hooke's law,

$$F(x) = -kx, \qquad [7\text{-}7]$$

in which k is a constant characteristic of the spring, then Eq. 7–6 reduces to

Work to stretch a spring

$$W = \tfrac{1}{2}kx^2. \qquad [7\text{-}8]$$

Work is calculated from Eq. 7–2 or 7–6 with no consideration for the time involved. Power is the rate at which work is done and is given by,

Power

$$P = dW/dt. \qquad [7\text{-}15]$$

or by

$$P = \mathbf{F} \cdot \mathbf{v}, \qquad [7\text{-}16]$$

in which $\mathbf{F}$ is a force that acts on a body moving with velocity $\mathbf{v}$. Like work, power is a scalar. Its SI unit is the watt (W) ($= 1$ J/s). In the British system the horsepower (hp) is also used, with 1 hp $= 746$ W $= 550$ ft $\cdot$ lb/s.

Kinetic energy

If a body of mass m moves with speed v, we assign it a kinetic energy K, given by

$$K = \tfrac{1}{2}mv^2. \qquad [7\text{-}12]$$

This is defined as the work needed to accelerate the body from rest to speed v. Kinetic energy is a positive scalar and has the same units and dimensions as work.

The work-energy theorem,

Work-energy theorem

$$W = K - K_0 = \Delta K, \qquad [7\text{-}14]$$

states that the work W done by the resultant force acting on a body is equal to the change in the kinetic energy of the body. This theorem, which is derived from Newton's laws of motion, provides an alternative and often simpler way of solving problems. Examples 5 and 6 show applications of this theorem.

QUESTIONS

1. Can you think of other words like "work" whose colloquial meanings are often different from their scientific meanings?

2. The inclined plane (Example 2) is a simple "machine" that enables us to do work with the application of a smaller force than is otherwise necessary. The same statement applies to a wedge, a lever, a screw, a gear wheel, and a pulley combination. Far from saving us work, such machines in practice require that we do a little more work with them than without them. Why is this so? Why do we use such machines?

3. In a tug of war one team is slowly giving way to the other. What work is being done and by whom?

4. Why is it easier to push a wheelbarrow than to pull it?

5. Give an example in which positive work is done by a frictional force.

6. The displacement of a body depends on the reference frame of the observer. It follows that the work done a body should also depend on the observer's reference frame. You drag a crate across a rough floor by pulling on it with a rope. Identify reference frames in which the work done on the crate by the tension in the rope would be (a) positive, (b) zero, and (c) negative.

7. Why can you so much more easily ride a bicycle for a mile on level ground than run the same distance? In each case you transport your own weight for a mile; in the first case you must also transport the bicycle and, moreover, do so in a shorter time! (See *The Physics Teacher*, March 1981, p. 194.)

8. You slowly lift a bowling ball from the floor and put it on a table. Two forces act on the ball: its weight $-m\mathbf{g}$, and your upward force $+m\mathbf{g}$. These two forces cancel each other so that it would seem that no work is done. On the other hand, you know that you have done some work. What is wrong?

9. You cut a spring in half. What is the relationship of the spring constant k for the original spring to that for either of the half-springs?

10. Springs A and B are identical except that A is stiffer than B; that is, $k_A > k_B$. On which spring is more work expended if (a) they are stretched by the same amount, (b) they are stretched by the same force?

11. Does kinetic energy depend on the direction of the motion involved? Can it be negative? Does its value depend on the reference frame of the observer?

12. You throw a ball vertically in the air and catch it. What does the work-energy theorem say qualitatively about the free flight during this round trip? Answer the question first neglecting air resistance and then taking it into account.

13. Sally and Yuri are flying jet planes at the same speed on parallel low-altitude tracks. Suddenly Sally lands at a convenient air strip. Consider how this looks from the reference frame of Yuri, who keeps on flying. (a) Would he say that Sally's plane had gained or lost kinetic energy? (b) Would he say that work was done *on* her plane or *by* her plane as she landed? (c) Would he conclude that the work-energy theorem holds? (*d*) Answer these questions from the reference frame of Chang, who is in the control tower.

14. Why can a car so easily pass a loaded truck when going uphill? The truck is heavier, of course, but its engine is more powerful in proportion (or is it?). What considerations enter into choosing the design power of a truck engine? . . . of a car engine?

15. Does the power needed to raise a box onto a platform depend on how fast it is raised?

16. You lift some books from a library shelf to a higher shelf in time *t*. Does the work that you do depend on (*a*) the mass of the books, (*b*) the weight of the books, (*c*) the height of the upper shelf above the floor, (*d*) the time *t*, and (*e*) whether you lift the books sideways or directly upward?

17. We hear a lot about the "energy crisis." Would it be more accurate to speak of a "power crisis"?

EXERCISES

Section 7–2   Work Done by a Constant Force

1. To push a 50-kg crate across the floor, a man applies a force of 200 N, 20° above the horizontal. The floor exerts a 175-N force of friction on the crate. As the crate moves 3.0 m, what work is done by (*a*) the man, (*b*) the force of friction, (*c*) the force of gravity, and (*d*) the normal force of the floor on the crate? (*e*) What is the total work done on the crate?

2. To push a 25-kg crate up a 25° incline, a man applies a force of 200 N, parallel to the incline. The incline exerts a 96-N force of friction on the crate. As the crate slides 1.5 m, what work is done by (*a*) the man, (*b*) the force of friction, (*c*) the force of gravity, and (*d*) the normal force of the incline on the crate? (*e*) What is the total work done on the crate?

3. A man pushes a 30-kg crate with constant speed along a horizontal floor. First he pushes it 3.0 m east, then 4.0 m north. The coefficient of kinetic friction between crate and floor is 0.275. Calculate the total work done by the man.

4. A cord is used to lower vertically a block of mass *M* a distance *d* at a constant downward acceleration of *g*/4. (*a*) Find the work done by the cord on the block. (*b*) Find the work done by the force of gravity.

5. A block of mass $m = 3.57$ kg is drawn at constant speed a distance $d = 4.06$ m along a horizontal floor by a rope exerting a constant force of magnitude $F = 7.68$ N, at an angle $\theta = 15.0°$ above the horizontal. Compute (*a*) the work done by the rope on the block, (*b*) the work done by friction on the block, (*c*) the total work done on the block, and (*d*) the coefficient of kinetic friction between the block and the floor.

Section 7–3   Work Done by a Variable Force—One-Dimensional Case

6. A single force acts on a body in rectilinear motion. A plot of velocity versus time for the body is shown in Fig. 7–8. Find the sign (positive or negative) of the work done *by* the force *on* the body in each of the intervals *AB*, *BC*, *CD*, and *DE*.

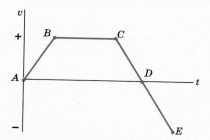

Figure 7–8   Exercise 6.

7. A 10-kg mass moves along the *x*-axis. Its acceleration as a function of its position is shown in Fig. 7–9. What is the net work performed on the mass as it moves from $x = 0$ to $x = 8$ m?

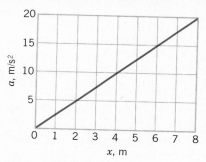

Figure 7–9   Exercise 7.

Section 7–5   Kinetic Energy and the Work-Energy Theorem

8. If a 2.9×10^5-kg Saturn V rocket with an Apollo spacecraft attached must achieve an escape velocity of 11.2 km/s (= 25,000 mi/h) near the surface of the earth, how much energy must the fuel contain? Would the system actually need as much or would it need more or less? Why?

9. From what height would an automobile have to fall to gain the kinetic energy equivalent to what it would have when going 55 mi/h (= 89 km/h)?

10. A 1000-kg car is traveling at 60 km/h on a level road. The brakes are applied long enough to do 5.0×10^4 J of work. (*a*) What is the final speed of the car? (*b*) How much more work must be done by the brakes to stop the car?

11. A 30-g bullet, initially traveling at 500 m/s, penetrates 12 cm into a wooden block, rigidly attached to a wall. (*a*) What work is done by the block in stopping the bullet? (*b*) Assume that the force of the block on the bullet is constant and calculate its value.

12. An outfielder throws a baseball with an initial speed of 60 ft/s (= 18.3 m/s). Just before an infielder catches the ball at the same level, its speed is reduced to 40 ft/s (= 12.2 m/s). What work was done by the force of the air on the ball? The weight of a baseball is 9.0 oz ($m = 255$ g).

13. A 240-g rubber ball is moving west at 15 m/s. It strikes a north-south wall and bounces east at 12 m/s. What work was done on the ball by the wall?

14. If the velocity components of a 0.5-kg object moving in the *x-y* plane change from $v_x = 3$ m/s, $v_y = 5$ m/s to $v_x = 0$ m/s, $v_y = 7$ m/s in 3 s, what is the work done on the object?

Section 7-7 Power

15. Your car averages 30 mi/gal of gasoline. (*a*) How far could you travel on 1 kW · h of energy consumed? (*b*) If you are driving at 55 mi/h, at what rate are you expending energy? The heat of combustion of gasoline is 140 MJ/gal.

16. A 45,000-kg satellite rocket acquires a speed of 6400 km/h in 1 min after launching. (*a*) What is its kinetic energy at the end of the first minute? (*b*) What is the average power expended during this time, neglecting frictional and gravitational forces?

17. A boy whose mass is 51 kg climbs, with constant speed, a vertical rope 6.0 m long in 10 s. (*a*) How much work does the boy perform? (*b*) What is the boy's power output during the climb?

18. A 55-kg woman ($W = mg = 120$ lb) runs with constant speed up a flight of stairs having a rise of 4.5 m (= 15 ft) in 3.5 s. What power must she supply?

19. In a 100-person ski lift, a machine raises passengers averaging 150 lb (= 667 N) a height of 500 ft (= 152 m) in 60 s, at constant speed. Find the power output of the motor, assuming no frictional loses.

PROBLEMS GROUPED BY SECTION

Section 7-2 Work Done by a Constant Force

1. A 100-kg object is initially moving in a straight line with a speed of 50 m/s. (*a*) If it is brought to a stop with a deceleration of 2.0 m/s², what force is required, what distance does it travel, and what work is done by the force? (*b*) Answer the same questions if its deceleration is 5.0 m/s².

2. A man pushed a 60-lb block ($m = W/g = 27$ kg) 30 ft (= 9.1 m) along a level floor at constant speed with a force directed 32° below the horizontal. If the coefficient of kinetic friction is 0.20, how much work did the man do on the block?

3. A 50-kg trunk is pushed 6.0 m at constant speed up a 30° incline by a constant horizontal force. The coefficient of sliding friction between the trunk and the incline is 0.20. Calculate the work done by (*a*) the applied force, (*b*) the frictional force, and (*c*) the gravitational force.

4. A crate weighing 500 lb ($m = W/g = 230$ kg) is suspended from the end of a rope 40 ft (= 12 m) long. By applying a horizontal force the crate is then pushed aside 4.0 ft (= 1.2 m) from the vertical and held there. (*a*) What is the force needed to keep the crate in this position? (*b*) How much work is being done to keep the crate in this position? (*c*) What work was done by gravity when the crate was moved aside? (*d*) How much work was done by the horizontal force? (*e*) Is the answer to (*d*) equal to the answer to (*a*) times 4.0 ft (= 1.2 m)? Explain why or why not. (*f*) How much work is done by the tension in the rope? (*g*) What was the total work done in moving the crate?

Section 7-3 Work Done by a Variable Force — One-Dimensional Case

5. (*a*) Estimate the work done by the force shown on the graph (Fig. 7-10) in displacing a particle from $x = 1$ to $x = 3$ m. Refine your method to see how close you can come to the exact answer of 6 J. (*b*) The curve is given analytically by $F = a/x^2$, where $a =$

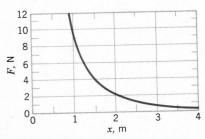

Figure 7-10 Problem 5.

9 N·m². Show how to calculate the work by the rules of integration.

6. The force exerted on an object is $F = F_0(x/x_0 - 1)$. Find the work done in moving the object from $x = 0$ to $x = 2x_0$, evaluating the integral both analytically and also by plotting $F(x)$ and finding the area under the curve.

Section 7-4 Work Done by a Variable Force — General Case

7. When the force **F** varies both in direction and magnitude and the motion is along a curved path, the work done by **F** is obtained from $dW = \mathbf{F} \cdot d\mathbf{r}$, the subsequent integration being taken along the curved path. Notice that both F and ϕ, the angle between **F** and $d\mathbf{r}$, may vary from point to point (See Fig. 7-7). Show that for two-dimensional motion,

$$W = \tfrac{1}{2}mv^2 - \tfrac{1}{2}mv_0^2.$$

where v is the final speed and v_0 is the initial speed.

8. The earth circles the sun once a year. How much work would have to be done on the earth to bring it to rest relative to the sun? See Appendix C for numerical data and ignore the rotation of the earth.

Section 7-5 Kinetic Energy and the Work-Energy Theorem

9. A proton (nucleus of the hydrogen atom) is being accelerated in a linear accelerator. In each stage of such an accelerator the proton is accelerated along a straight line at 3.6×10^{15} m/s². If a proton enters such a stage moving initially with a speed of 2.4×10^7 m/s and the stage is 3.5 cm long, compute (*a*) its speed at the end of the stage and (*b*) the gain in kinetic energy resulting from the acceleration. The mass of the proton is 1.67×10^{-27} kg. Express the energy in electron volts (eV); 1 eV = 1.6×10^{-19} J.

10. A helicopter is used to lift a 160-lb astronaut ($m = W/g = 72$ kg) 50 ft (= 15 m) vertically from the ocean by means of a cable. The acceleration of the astronaut is $g/10$. (*a*) How much work is done by the helicopter on the astronaut? (*b*) How much work is done by the gravitational force on the astronaut? What are (*c*) the kinetic energy and (*d*) the speed of the astronaut just before he reaches the helicopter?

11. A 60-lb (= 267-N) girl slides down a 20-ft (= 6.1-m) playground slide that makes an angle of 20° with the horizontal. The coefficient of kinetic friction is 0.10. (*a*) Find the work done by the force of gravity. (*b*) Find the work done by the force of friction. (*c*) If she starts at the top with a speed of 1.5 ft/s (= 0.457 m/s), what is her speed at the bottom?

12. A 0.55-kg projectile is launched from the edge of a cliff with an initial kinetic energy of 1550 J and at its highest point is 140 m above the launch point. (a) What is the horizontal component of its velocity? (b) What was the vertical component of its velocity just after launch? (c) At one instant during its flight the vertical component of its velocity is found to be 65 m/s. At that time, how far is it above or below the launch point?

13. A 250-g block is dropped onto a vertical spring with spring constant $k = 2.5$ N/cm (Fig. 7–11). The block becomes attached to the spring and the spring compresses 12 cm before coming momentarily to rest. (a) While the spring is being compressed, what work is done by the force of gravity? (b) By the spring? (c) What was the speed of the block just before it hit the spring? (d) If the initial speed of the block were doubled, what would be the maximum compression of the spring? Ignore friction.

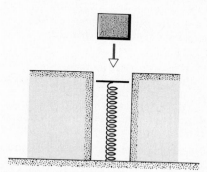

Figure 7–11 Problem 13.

Section 7–7 Power

14. A 100-kg block is pulled at a constant speed of 5.0 m/s across a horizontal floor by an applied force of 122 N, 37° above the horizontal. (a) At what rate is the applied force doing work? (b) At what rate is the force of friction doing work?

15. A body of mass m accelerates uniformly from rest to a speed v_f in time t_f. (a) Show that the work done on the body as a function of time t, in terms of v_f and t_f, is

$$\frac{1}{2} m \frac{v_f^2}{t_f^2} t^2.$$

(b) As a function of time t, what is the instantaneous power delivered to the body? (c) What is the instantaneous power at the end of 10 s delivered to a 1500-kg body that accelerates to 90 km/h in 10 s?

16. The loaded cab of an elevator has a mass of 3.0×10^3 kg and moves 200 m up the shaft in 20 s, at constant speed. At what rate does the cable do work on the cab?

17. A fully-loaded, slow-moving freight elevator has a total mass of 1200 kg. It is required to travel upward 54 m in 3.0 min. The counterweight has mass of 950 kg. Find the power output, in horsepower, of the elevator motor. Ignore the work required to start and stop the elevator; that is, assume that it travels at constant speed.

18. A 3200-lb automobile ($m = W/g = 1500$ kg) starts from rest on a horizontal road and gains a speed of 45 mi/h ($= 72$ km/h) in 30 s. (a) What is the kinetic energy of the auto at the end of the 30 s? (b) What is the average net power delivered to the car during the 30-s interval? (c) What is the instantaneous power at the end of the 30-s interval assuming that the acceleration was constant?

*19. At full power, a 1.5-MW railroad locomotive accelerates a train from a speed of 10 m/s to 25 m/s in 6.0 min. (a) Neglecting friction, calculate the mass of the train. (b) Find the speed of the train as a function of time in seconds during the interval. (c) Find the force accelerating the train as a function of time during the interval. (d) Find the distance moved by the train during the interval.

ADDITIONAL PROBLEMS

20. A 100-lb block of ice ($m = W/g = 45$ kg) slides down an incline 5.0 ft ($= 1.5$ m) long and 3.0 ft ($= 0.91$ m) high. A man pushes up on the ice parallel to the incline so that it slides down at constant speed. The coefficient of friction between the ice and the incline is 0.10. Find (a) the force exerted by the man, (b) the work done by the man on the block, (c) the work done by gravity on the block, (d) the work done by the surface of the incline on the block, and (e) the work done by the resultant force on the block.

21. A proton starting from rest is accelerated in a cyclotron to a final speed of 3.0×10^7 m/s (about one-tenth the speed of light). How much work, in electron volts (eV), is done on the proton by the electrical force of the cyclotron that accelerated it? 1 eV $= 1.6 \times 10^{-19}$ J.

22. A 0.63-kg ball is thrown straight up into the air with an initial speed of 14 m/s. It reaches a height of 8.1 m, then falls back down. Assume that the only forces acting are those of gravity and air resistance and calculate the work done during the ascent by the force of air resistance.

23. A running man has half the kinetic energy that a boy of half his mass has. The man speeds up by 1.0 m/s and then has the same kinetic energy as the boy. What were the original speeds of man and boy?

24. A 2.0-kg block is forced against a horizontal spring of negligible mass, compressing the spring by 15 cm. When the block is released from the compressed spring, it moves 60 cm across a horizontal tabletop before coming to rest. The force constant of the spring is 2.0 N/cm. What is the coefficient of sliding friction between the block and the table?

25. A horse pulls a cart with a force of 40 lb at an angle of 30° with the horizontal and moves along at a speed of 6.0 mi/h. (a) How much work does the horse do in 10 min? (b) What is the power output of the horse?

26. A 200-lb ($= 890$-N) running back starts from rest and runs 40 yd ($= 36.6$ m) in 4.4 s, accelerating uniformly throughout the run. (a) What is his final speed? (b) What average power does he generate over the course of the run?

27. A 2.0-kg object accelerates uniformly from rest to a speed of 10 m/s in 3.0 s. (a) How much work is done on the object? (b) What is the instantaneous power delivered to the body 1.5 s after it starts?

28. A 5.0-kg block moves in a straight line on a horizontal frictionless surface under the influence of a force that varies with position as shown in Fig. 7–12. How much work is done by the force as the block moves from the origin to $x = 8.0$ m?

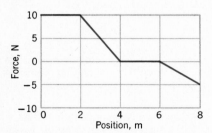

Figure 7–12 Problem 28.

29. A comet having a mass 8.38×10^{11} kg strikes the earth at a relative speed of 30 km/s. (*a*) Compute the kinetic energy of the comet in "megatons of TNT"; the detonation of one million tons of TNT releases 4.2×10^{15} J of energy. (*b*) The diameter of the crater blasted by a large explosion is proportional to the one-third power of the explosive energy released, with a one megaton of TNT explosion producing a crater about 1 km in diameter. What is the diameter of the crater produced by the impact of the comet?

30. A net force of 5.0 N acts on a 15-kg body initially at rest. Compute (*a*) the work done by the force in the first, second, and third seconds; and (*b*) the instantaneous power exerted by the force at the end of the third second.

31. A force acts on a 3.0-kg particle in such a way that the position of the particle as a function of time is given by $x = 3t - 4t^2 + t^3$, where x is in meters and t is in seconds. Find the work done by the force during the first 4.0 s.

32. A 1400-kg block of granite is pulled up a plane at a constant speed of 1.34 m/s by a steam winch (Fig. 7–13). The coefficient of kinetic friction between the block and plane is 0.40. (*a*) How much work is done by each of the forces that act on the block as it moves 9.0 m up the incline? (*b*) How much power must be supplied by the winch?

33. What power is developed by a grinding machine whose wheel has a radius of 20 cm and runs at 2.5 rev/s when the tool to be sharpened is held against the wheel with a force of 180 N (= 40 lb)? The coefficient of friction between the tool and the wheel is 0.32.

34. 1200 m³ of water passes each second over a waterfall 100 m high. Assuming that three-fourths of the kinetic energy gained by the water in falling is converted to electrical energy by a hydroelectric generator, what is the power output of the generator?

35. The force required to tow a boat at constant velocity is proportional to the speed. If it takes 10 hp to tow a certain boat at a speed of 2.5 mi/h, how much horsepower does it take to tow it at a speed of 7.5 mi/h?

*36. A truck can move up a road having a grade of 1.0-ft rise every 50 ft with a speed of 15 mi/h. The resisting force is equal to one twenty-fifth the weight of the truck. If its power output is the same, how fast will the same truck move down the hill? Assume that the resisting force remains unchanged.

*37. The resistance to motion of an automobile depends on road friction, which is almost independent of speed, and on aerodynamic drag, which is proportional to speed squared. For a 12,000-N car, then, the total resistant force F is given by $F = 300 + 1.8v^2$, where F is in newtons and v is in meters per second. Calculate the power required from the motor to accelerate the car at 0.92 m/s² when the speed is 80 km/h.

*38. A *governor* consists of two 200-g masses attached by light, rigid 10-cm rods to a vertical, rotating axle. The rods are hinged so that the masses swing out from the axle as they rotate with it. However, when the angle θ is 45°, the masses encounter the wall of the cylinder in which the governor is rotating; see Fig. 7–14. (*a*) What is the minimum rate of rotation, in revolutions per minute, required for the masses to touch the wall? (*b*) If the coefficient of kinetic friction between the masses and wall is 0.35, what power is dissipated due to the masses rubbing against the wall when the mechanism is rotating at 300 rev/min?

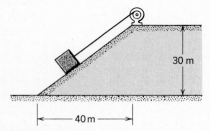

Figure 7–13 Problem 32.

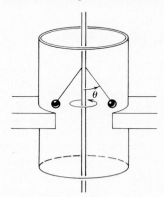

Figure 7–14 Problem 38.

The Conservation of Energy

8–1 Introduction

In Chapter 7 we derived the work-energy theorem from Newton's second law of motion. This theorem says that the work W done by the resultant force $\mathbf{F}$ acting on a particle as it moves from one point to another is equal to the change ΔK in the kinetic energy of the particle, or

Work-energy theorem

$$W = \Delta K. \tag{8–1}$$

Often several forces act on a particle, the resultant force $\mathbf{F}$ being their vector sum; that is, $\mathbf{F} = \mathbf{F}_1 + \mathbf{F}_2 + \cdots + \mathbf{F}_n$, in which we assume that n forces act. The work done by the resultant force $\mathbf{F}$ is the algebraic sum of the work done by these individual forces, or $W = W_1 + W_2 + \cdots + W_n$. Thus we can write the work-energy theorem (Eq. 8–1) as

$$W_1 + W_2 + \cdots + W_n = \Delta K. \tag{8–2}$$

In this chapter we shall consider systems in which a particle is acted upon by various kinds of forces and we shall compute W_1, W_2, and so on, for these forces; this will lead us to define different kinds of energy such as potential energy and internal energy. The process culminates in the formulation of one of the great principles of science, the *conservation of energy principle*.

8–2 Conservative Forces

Let us first distinguish between two types of forces, *conservative* and *nonconservative*. We shall consider an example of each type and then discuss each example from several different, but related, points of view.

Imagine a spring fastened at one end to a rigid wall as in Fig. 8–1. Let us slide a block of mass m with velocity $\mathbf{v}$ directly toward the spring; we assume that the horizontal plane is frictionless and that the spring is ideal, that is, that it obeys Hooke's law (Eq. 7–7),

$$F = -kx, \qquad (8-3)$$

where F is the magnitude of the force exerted by the spring when its free end is displaced through a distance x.

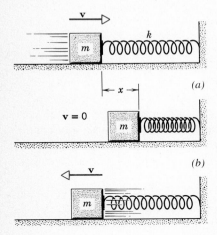

After the block touches the spring, the speed and hence the kinetic energy of the block decrease until finally the block is brought to rest by the action of the spring force, as in Fig. 8–1b. The block now reverses its motion as the compressed spring expands. It gains speed and kinetic energy and, when it comes once again to its position of initial contact with the spring, we find that it has the same speed and kinetic energy as it had originally; only the direction of motion has changed. The block loses kinetic energy during one part of its motion but gains it all back during the other part of its motion as it returns to its starting point (Fig. 8–1c).

The kinetic energy of a body may be interpreted as its ability to do work by virtue of its motion. It is clear that at the completion of a round trip the ability of the block in Fig. 8–1 to do work remains the same; it has been *conserved*. The elastic force exerted by an ideal spring, and other forces that act in this same way, are called *conservative*. The force of gravity is also conservative; if we throw a ball vertically upward, it will (if the air resistance is negligible) return to our hand with the same kinetic energy that it had when it left our hand.

If, however, a particle on which one or more forces act returns to its initial position with either more or less kinetic energy than it had initially, then in a round trip its ability to do work has been changed. In this case the ability to do work has not been conserved and at least one of the forces acting is labeled *nonconservative*.

Figure 8–1 (a) A block of mass m is projected with speed v against a spring. (b) The block is brought to rest by the action of the spring force. (c) The block has regained its initial speed v as it returns to its starting point.

To illustrate a nonconservative force let us assume that the surfaces of the block and the plane in Fig. 8–1 are not frictionless but rather that a force of friction **f** is exerted by the plane on the block. The frictional force opposes the motion of the block no matter which way the block is moving, and we find that the block returns to its starting point with less kinetic energy than it had initially. Since we showed in our first experiment that the spring force was conservative, we must attribute this new result to the action of the friction force. We say that this force, and other forces that act in this same way, are nonconservative. The induction force in a betatron (Section 32–6) is also a nonconservative force. Instead of dissipating kinetic energy, however, it generates it, so that an electron moving in the circular betatron orbit will return to its initial position with *more* kinetic energy than it had there originally. In a round trip the electron gains kinetic energy, as it must do if the betatron is to be effective.

To sum up: *A force is conservative if the kinetic energy of a particle on which it acts returns to its initial value after any round trip. A force is nonconservative if the kinetic energy of the particle changes after a round trip.* In this definition we assume that the force in question is the only one that does work on the particle. If more than one force does work we assume that the effects attributable to each such force can be analyzed separately.

We can define conservative force from another point of view, that of the work done by the force on the particle. In our first example above, the work done by the elastic spring force on the block while the spring was being compressed was negative, because the force exerted on the block by the spring (to the left in Fig. 8–1a) was directed opposite to the displacement of the block (to the right in Fig. 8–1a). While the spring was being extended the work that the spring force did on the block was positive (force and displacement in the same direction). In our first example, then, the net work done on the block by the spring force during a complete cycle or round trip, is zero.

In our second example we considered the effect of the frictional force. The work done on the block by this force was negative for each portion of the cycle because the frictional force always opposed the motion. Hence the work done by friction in a round trip cannot be zero. In general, then: *A force is conservative if the work done by the force on a particle that moves through any round trip is zero. A force is nonconservative if the work done by the force on a particle that moves through any round trip is not zero.*

The work-energy theorem shows that this second way of defining conservative and nonconservative forces is fully equivalent to our first definition. If there is no change in the kinetic energy of a particle that has moved through any round trip, then $\Delta K = 0$ and, from Eq. 8–1, $W = 0$ and the resultant force acting must be conservative. Similarly, if $\Delta K \neq 0$, then, from Eq. 8–1, $W \neq 0$ and at least one of the forces acting must be nonconservative.

We can consider the difference between conservative and nonconservative forces in still a third way. Suppose that a particle goes from a to b along path 1 and back from b to a along path 2 as in Fig. 8–2a. Several forces may act on the particle during this round trip; we consider each force separately. If the force being considered is conservative, the work done on the particle by that particular force for the round trip is zero, or

$$W_{ab,1} + W_{ba,2} = 0,$$

which we can write as

$$W_{ab,1} = -W_{ba,2}.$$

That is, the work in going from a to b along path 1 is the negative of the work in going from b to a along path 2. However, if we cause the particle to go from a to b along path 2, as shown in Fig. 8–2b, we merely reverse the direction of the previous motion along 2, so that

$$W_{ab,2} = -W_{ba,2}.$$

Hence

$$W_{ab,1} = W_{ab,2},$$

which tells us that the work done on the particle by a conservative force in going from a to b is the same for either path.

Paths 1 and 2 can be any paths at all as long as they go from a to b; and a and b can be chosen to be any two points at all. We always find the same result if the force is conservative. Hence, we have another equivalent definition of conservative and nonconservative forces: *A force is conservative if the work done by it on a particle that moves between two points depends only on these points and not on the path followed. A force is nonconservative if the work done by that force on a particle that moves between two points depends on the path taken between those points.*

To illustrate the third (equivalent) definition of a conservative force, consider the gravitational force. Let us pick up a stone of mass m at point a in Fig. 8–3a, move it horizontally to point c, and then raise it, without acceleration, through a vertical height h to point b. We label the path acb as path 1, and we state that the work done by the gravitational force $m\mathbf{g}$ (*not* the work done by the person moving the stone) is $-mgh$. For the horizontal distance ac the work is zero because the force $m\mathbf{g}$ and the displacement are at right angles; for the vertical distance cb we saw in Example 1 of Chapter 7 that the work is $-mgh$.

Conservative force—second definition

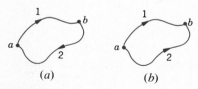

Figure 8–2

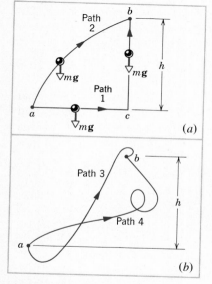

Figure 8–3 If a stone of mass m is made to move from point a to point b, the work done by the gravitational force is $-mgh$ for the four paths shown.

Conservative force—third definition

Consider now path 2 in Fig. 8–3a, which is a second, completely arbitrary path connecting points a and b. We assert that the work done by the gravitational force m**g** as the stone moves between points a and b along *this* path is also $-mgh$ and thus, by our definition, the gravitational force is conservative. To prove this assertion let us approximate path 2 by a series of alternate horizontal and vertical steps. We can have as many steps as we wish so that we can make this broken path arbitrarily close to path 2. Now, on the horizontal segments of this path no work will be done. The sum of the vertical segments is just h, so that the work on the broken path, and thus on path 2, is indeed $-mgh$. Convince yourself that the work would still be $-mgh$ for paths such as 3 and 4 shown in Fig. 8–3b.

For a nonconservative force, such as friction, the work done is not independent of the path taken between two fixed points. We need only point out that as we push a block over a (rough) table between any two points a and b by various paths, the distance traversed varies and so does the work done by the frictional force. It depends on the path.

The definitions of conservative force that we have given are equivalent to one another. Which one we use depends only on convenience or personal choice. The round-trip approach shows clearly that kinetic energy is conserved when conservative forces act. If we wish to develop the idea of potential energy, however, the path independence statement is preferable.

8–3   Potential Energy

In this section we shall focus attention not on the moving block of Fig. 8–1 but rather on the isolated system (block + spring). Instead of saying that the block is moving we prefer, from this point of view, to say that the configuration of the system is changing. We measure both the position of the block and the configuration of the system at any instant by the same parameter x, namely, the displacement of the free end of the spring from its normal position, corresponding to an unstretched spring. The kinetic energy of the system is the same as that of the block because we assume the spring to be massless.

We have seen that the kinetic energy of the system of Fig. 8–1 decreases during the first half of the motion, becomes zero, and then increases during the second half of the motion. If there is no friction, the kinetic energy of the system when it has regained its initial configuration returns to its initial value.

Potential energy

Under these circumstances (conservative forces acting) it is useful to introduce the concept of *energy of configuration,* or *potential energy U,* and to say that if K for the system changes by ΔK as the configuration of the system changes (that is, as the block moves in the system of Fig. 8–1), then U for the system must change by an equal but opposite amount so that the sum of the two changes is zero, or

$$\Delta K + \Delta U = 0. \tag{8–4a}$$

Alternatively, we can say that any change in kinetic energy K of the system is compensated for by an equal but opposite change in the potential energy U of the system so that their sum remains constant throughout the motion, or

$$K + U = \text{a constant.} \tag{8–4b}$$

The potential energy of a system represents a form of stored energy that can be fully recovered and converted into kinetic energy. We cannot associate a potential energy with a nonconservative force such as the force of friction because the kinetic energy of a system in which such forces act does not return to its initial value when the system returns to its initial configuration.

Equations 8–4 apply to a closed system of interacting objects, such as the (block + spring) system of Fig. 8–1. In this example, because we have taken the spring to be effectively massless, the kinetic energy may be associated with the moving block alone. The block slows down (or speeds up) because a force is exerted on it by the spring; it is appropriate, then, to associate the potential energy of the system with this force, that is to say, with the spring. No potential energy is associated with the block because we assume it to be an idealized rigid body, devoid of "springiness." Thus in this simple case we say that kinetic energy, localized in the block, decreases during the first part of the motion while potential energy, localized in the spring, increases during this same time.

Equations 8–4 are essentially bookkeeping statements about energy. However, they, and the concept of potential energy, have no real value until we have shown how to calculate U as a function of the configuration of the system within which the conservative forces act; in the example of Fig. 8–1 this means that we must be able to calculate $U(x)$, where x is the displacement of the end of the spring.

To refine our concept of potential energy U, let us consider the work-energy theorem, $W = \Delta K$, in which W is the work done by the resultant force on a particle as it moves from a to b. For simplicity let us assume that only a single force $\mathbf{F}$ acts on the particle; this is effectively true in the system of Fig. 8–1. If $\mathbf{F}$ is conservative, we can combine the work-energy theorem (Eq. 8–1) with Eq. 8–4a, obtaining

$$W = \Delta K = -\Delta U. \qquad (8\text{–}5a)$$

The work W done by a conservative force depends only on the starting and the end points of the motion and not on the path followed between them. Such a force can depend only on the position of a particle; it does not depend on the velocity of the particle or on the time, for example.

For motion in one dimension, Eq. 8–5a becomes

$$\Delta U = -W = -\int_{x_0}^{x} F(x)\,dx, \qquad (8\text{–}5b)$$

the particle moving from x_0 to x. Equation 8–5b shows how to calculate the change in potential energy ΔU when a particle, acted on by a conservative force $F(x)$, moves from point a, described by x_0, to point b, described by x. The equation shows that we can only calculate ΔU if the force $\mathbf{F}$ depends only on the position of the particle (that is, on the system configuration), which is equivalent to saying that potential energy has meaning only for conservative forces.

Now that we know that the potential energy U depends on the position of the particle only, we can write Eq. 8–4b as

Mechanical energy

$$K + U = \tfrac{1}{2}mv^2 + U(x) = E \qquad \text{(one-dimension)}, \qquad (8\text{–}6a)$$

in which E, which remains constant as the particle is moving, is called the *mechanical energy*. Suppose that the particle moves from point a (where its position is x_0 and its speed is v_0) to point b (where its position is x and its speed is v); the total mechanical energy E must be the same for each system configuration when the force is conservative, or, from Eq. 8–6a,

Conservation of mechanical energy

$$\tfrac{1}{2}mv^2 + U(x) = \tfrac{1}{2}mv_0^2 + U(x_0). \qquad (8\text{–}6b)$$

The quantity on the right depends only on the initial position x_0 and the initial speed v_0, which have definite values; it is, therefore, constant during the motion.

This is the constant total mechanical energy E. Notice that force and acceleration do not appear in this equation, only position and speed. Equations 8–6 are often called the *law of conservation of mechanical energy* for conservative forces.

In many problems we find that although some of the individual forces are not conservative, they are so small that we can neglect them. In such cases we can use Eqs. 8–6 as a good approximation. For example, air resistance may be present in projectile motion but may have so little effect that we can ignore it.

Notice that, instead of starting with Newton's laws, we can simplify problem solving when conservative forces alone are involved by starting with Eq. 8–6. This relation is derived from Newton's laws, of course, but it is one step closer to the solution. We often solve problems without analyzing the forces or writing down Newton's laws by looking instead for something in the motion that is constant; here the mechanical energy is constant and we can write down Eqs. 8–6 as the first step.

For one-dimensional motion we can also write the relation between force and potential energy (Eq. 8–5b) as

Force and potential energy—one dimension

$$F(x) = -\frac{dU(x)}{dx}. \tag{8-7}$$

To show this, substitute this expression for $F(x)$ into Eq. 8–5b and observe that you get an identity. Equation 8–7 gives us another way of looking at potential energy. The potential energy is a function of position whose negative derivative gives the force.

You may have noticed that we wrote down the quantity $U(x)$ in Eqs. 8–6 although we are only able to calculate changes in U (from Eq. 8–5b) and not U itself. Let us imagine that a particle moves from a to b along the x-axis and that a single conservative force $F(x)$ acts on it. To assign a value to U_b, the potential energy at point b, let us write

$$\Delta U = U_b - U_a,$$

or (see Eq. 8–5b),

$$U_b = \Delta U + U_a = -\int_{x_a}^{x_b} F(x)\,dx + U_a. \tag{8-8}$$

We cannot assign a value to U_b until we have assigned one to U_a. If point b is any arbitrary position x, so that $U_b = U(x)$, we give meaning to $U(x)$ by choosing point a to be some convenient reference position, described by $x_a = x_0$, and by arbitrarily assigning a value to the potential energy $U_a = U(x_0)$ when the body is at that point. Thus Eq. 8–8 becomes

Potential energy and force—one dimension

$$U(x) = -\int_{x_0}^{x} F(x)\,dx + U(x_0). \tag{8-9}$$

The potential energy when the body is at the reference position, that is, $U(x_0)$, is usually given the arbitrary value zero.

It is often convenient to choose the reference position x_0 to be that at which the force acting on the particle is zero. Thus the force exerted by a spring is zero when the spring has its normal unstretched length; we usually say that the potential energy is also zero for this condition. Also, the attraction of the earth on a body decreases as the body moves away from the earth, becoming zero at an infinite distance. We usually take infinity as our reference position and assign the value zero to the potential energy associated with the gravitational force at that position (see Chapter 15). So far, however, we have been more concerned with the gravitational pull on bodies such as baseballs, and so on, which, in comparison to the earth's radius, never move very far from the earth's surface. Here the gravitational force ($=m\mathbf{g}$) is essentially constant and we find it convenient to take the zero of potential energy to be, not at infinity, but at the surface of the earth.

The effect of changing the coordinate of the standard reference position x_0, or of the arbitrary value assigned to $U(x_0)$, is simply to change the value of $U(x)$ by an added constant. The presence of an arbitrary added constant in the potential energy expression (Eq. 8–9) makes no difference to the equations that we have written so far. This simply adds the same constant term to each side of Eq. 8–6b, for example, leaving that equation unchanged. Furthermore, changing $U(x)$ by an added constant does not change the force calculated from Eq. 8–7 because the derivative of a constant is zero. All this simply means that the choice of a reference point for potential energy is immaterial because we are always concerned with differences in potential energy, rather than with any absolute value of potential energy at a given point.

There is a certain arbitrariness in specifying kinetic energy also. In order to determine speed, and hence kinetic energy, we must specify a reference frame. The speed of a man sitting on a train is zero if we take the train as a reference frame, but it is not zero to an observer on the ground who sees the man move by with uniform velocity. The value of the kinetic energy depends on the reference frame used by the observer. Hence the important thing about mechanical energy E, which is the sum of the kinetic and the potential energies, is not its actual value during a given motion (this depends on the observer) but the fact that this value does not change during the motion for any particular observer when the forces are conservative.

8–4   One-Dimensional Conservative Systems

Let us now calculate the potential energy in one-dimensional motion for two examples of conservative forces, the force of gravity for motions near the earth's surface and the elastic restoring force of an (ideal) stretched spring.

The gravitational force

For the force of gravity we take the one-dimensional motion to be vertical, along the y-axis. We take the positive direction of the y-axis to be upward; the force of gravity is then in the negative y-direction, or downward. We have $F(y) = -mg$, a constant. The potential energy at position y is found from Eq. 8–9, or

$$U(y) = -\int_0^y F(y)\,dy + U(0) = -\int_0^y (-mg)\,dy + U(0) = mgy + U(0).$$

The potential energy can be taken as zero where $y = 0$, so that $U(0) = 0$, and

$$U(y) = mgy. \tag{8–10}$$

The gravitational potential energy is then mgy. The relation $F(y) = -dU/dy$ (Eq. 8–7) is satisfied, for $-d(mgy)/dy = -mg$. We choose $y = 0$ to be at the surface of the earth for convenience, so that the gravitational potential energy is zero at the earth's surface and increases linearly with altitude y.

If we compare points y and $y = 0$, the conservation of mechanical energy, Eq. 8–6b, gives us the relation

$$\tfrac{1}{2}mv^2 + mgy = \tfrac{1}{2}mv_0^2.$$

This is equivalent mathematically to the well-known result (see Eq. 3–17),

$$v^2 = v_0^2 - 2gy.$$

If our particle moves from a height h_1 to a height h_2, we can use Eq. 8–6b to obtain

$$\tfrac{1}{2}mv_1{}^2 + mgh_1 = \tfrac{1}{2}mv_2{}^2 + mgh_2.$$

This result is equivalent to the result of Example 5, Chapter 7. The mechanical energy E is constant and is conserved during the motion, even though the kinetic energy and the potential energy vary as the configuration of the system (particle + earth) changes.

The force due to an ideal spring

A second example of a conservative force is that exerted by an elastic spring on a block of mass m attached to it moving on a horizontal frictionless surface. If we take $x_0 = 0$ as the position of the end of the spring when unextended, the force exerted on the block when the spring is stretched a distance x from its unextended length is $F = -kx$. The potential energy is obtained from Eq. 8–9,

$$U(x) = -\int_0^x F(x)\,dx + U(0) = -\int_0^x (-kx)\,dx + U(0).$$

If we choose $U(0) = 0$, the potential energy, as well as the force, is zero when the spring is unextended, and

$$U(x) = -\int_0^x (-kx)\,dx = \tfrac{1}{2}kx^2.$$

The result is the same whether we stretch or compress the spring, that is, whether x is plus or minus.

The relation $F(x) = -dU/dx$ (Eq. 8–7) is satisfied, for $-d(\tfrac{1}{2}kx^2)/dx = -kx$. The elastic potential energy of the spring is then

$$U(x) = \tfrac{1}{2}kx^2. \tag{8–11}$$

The block of mass m will undergo a motion in which mechanical energy E is conserved (Fig. 8–4). From Eq. 8–6b we have

$$\tfrac{1}{2}mv^2 + \tfrac{1}{2}kx^2 = \tfrac{1}{2}mv_0{}^2.$$

Here v_0 is the speed of the block for $x = 0$. Physically we achieve such a result by stretching the spring with an applied force to some position, x_m, and then releasing the spring. Notice that at $x = 0$ the energy of the system (block + spring) is all kinetic. At $x = x_m$ (the maximum value of x), v must be zero, so that here the system energy is all potential. At $x = x_m$, we have

$$\tfrac{1}{2}kx_m{}^2 = \tfrac{1}{2}mv_0{}^2$$

or

$$x_m = \sqrt{m/k}\;v_0.$$

For positions x_1 and x_2 between 0 and x_m, Eq. 8–6b gives

$$\tfrac{1}{2}kx_1{}^2 + \tfrac{1}{2}mv_1{}^2 = \tfrac{1}{2}kx_2{}^2 + \tfrac{1}{2}mv_2{}^2.$$

We have seen that the kinetic energy of a body is the work that a body can do by virtue of its motion. We express the kinetic energy by the formula $K = \tfrac{1}{2}mv^2$. We cannot give a similar universal formula by which potential energy can be expressed. The potential energy of a system of bodies is the work that the system of bodies can do by virtue of the relative position of its parts, that is, by virtue of its configuration. In each case we must determine how much work the system can do in passing from one configuration to another and then take this as the difference in potential energy of the system between these two configurations.

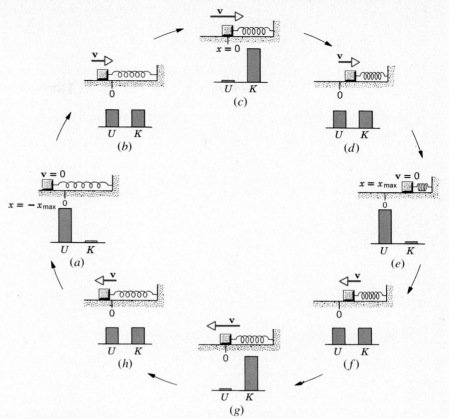

Figure 8–4 A block of mass m attached to a spring slides back and forth on a frictionless surface. The system is called a harmonic oscillator. The motion of the block through one cycle is illustrated. Starting at position (a) the block is in its extreme left position and momentarily at rest: $K = 0$. The spring is extended to its maximum length: $U = U_{max}$. (K and U are illustrated in the bar graphs below each sketch.) An eighth-cycle later (b) the block has gained kinetic energy, but the spring is no longer so elongated; K and U have here the same value, $K = U = U_{max}/2$. At (c) the spring is neither elongated nor compressed and the speed is a maximum: $U = 0$, $K = K_{max} = U_{max}$. The cycle continues, with the total energy $E = K + U$ always the same: $E = K_{max} = U_{max}$. The harmonic oscillator will be analyzed more closely in Chapter 14.

The potential energy of the spring depends on the relative position of the parts of the spring. Work can be obtained by allowing the spring to return from its extended to its unextended length, during which time it exerts a force through a distance. If a block is attached to the spring, as in our example, the block will be accelerated by this force and the potential energy will be converted to kinetic energy. In the gravitational case an object occupies a position with respect to the earth. The potential energy is a property of the object and the earth, considered as a system of bodies. It is the relative position of the parts of this system that determines its potential energy. The potential energy is greater when the parts are far apart than when they are close together. The loss of potential energy is equal to the work done by gravity in this process. This work is converted into kinetic energy of the bodies. In our example we ignored the kinetic energy acquired by the earth itself as an object fell toward it. In principle, this object exerts a force on the earth and causes it to acquire an acceleration, relative to some inertial frame. The resulting change in speed, however, is extremely small, and in spite of the enormous mass of the earth, its additional kinetic energy is negligible compared to that acquired by the falling object. This will be proved in a later chapter. In other cases, such as in planetary motion where the masses of the objects in our system may be comparable, we cannot ignore any part of the system. In general, potential energy is not assigned to either body separately but is considered a joint property of the system.

Example 1   What is the change in gravitational energy when a 1600-lb ($m = W/g = 720$ kg) elevator moves from street level to the top of the Empire State Building, 1250 ft ($= 380$ m) above street level?

The gravitational potential energy of the system elevator + earth) is $U = mgy$. Then

$$\Delta U = U_2 - U_1 = mg(y_2 - y_1).$$

But

$$mg = W = 1600 \text{ lb} \quad \text{and} \quad y_2 - y_1 = 1250 \text{ ft,}$$

so that

$$\Delta U = 1600 \times 1250 \text{ ft} \cdot \text{lb} = 2.00 \times 10^6 \text{ ft} \cdot \text{lb} = 2.7 \times 10^6 \text{ J.}$$

Example 2   As an example of the simplicity and usefulness of the energy method of solving dynamical problems, consider the problem illustrated in Fig. 8–5. A block of mass m slides down a curved frictionless surface. The force exerted by the surface on the block is always normal to the surface and to the direction of the motion of the block, so that this force does no work. Only the gravitational force does work on the block and that force is conservative. The mechanical energy E is, therefore, conserved and we write at once

$$\tfrac{1}{2}mv_1^2 + mgy_1 = \tfrac{1}{2}mv_2^2 + mgy_2.$$

This gives

$$v_2^2 = v_1^2 + 2g(y_1 - y_2).$$

The time the block takes to reach the bottom of the curved surface depends in the shape of the curve but the speed at the bottom does not. It depends only on the initial speed and the change in vertical height. In fact, if the block is initially at rest at $y_1 = h$, and if we set $y_2 = 0$, we obtain

$$v_2 = \sqrt{2gh}.$$

At this point you should recall the independence of path feature of work done by conservative forces and be able to justify applying the ideas developed for one-dimensional motion to this two-dimensional example.

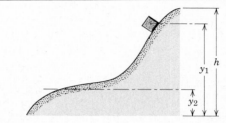

Figure 8–5   Example 2. A block sliding down a frictionless curved surface.

In this problem the value of the force depends on the slope of the surface at each point. Hence, the acceleration is not constant but is a function of position. To obtain the speed by starting with Newton's laws we would first have to determine the acceleration at each point and then integrate the acceleration over the path. We avoid all this labor by starting at once from the fact that the mechanical energy is constant throughout the motion.

Example 3   The spring in a spring gun has a force constant of 7.0 N/cm. It is compressed 3.0 cm from its natural length, and a 12-g ball is put into the barrel against it. Assuming no friction and a horizontal gun barrel, with what speed will the ball leave the gun when released?

The force is conservative so that mechanical energy is conserved in the process. The initial mechanical energy is the elastic potential energy of the spring, $\tfrac{1}{2}kx^2$, and the final mechanical energy is the kinetic energy of the ball, $\tfrac{1}{2}mv^2$. Hence,

$$\tfrac{1}{2}kx^2 = \tfrac{1}{2}mv^2$$

or

$$v = \sqrt{\frac{k}{m}}\,x = \sqrt{\frac{(7.0 \text{ N/cm})(100 \text{ cm/m})}{(12 \text{ g})(10^{-3} \text{ kg/g})}}\,(3.0 \times 10^{-2} \text{ m})$$

$$= 7.2 \text{ m/s.}$$

8–5   Mechanical Energy and the Potential Energy Curve

A large amount of information can be found simply from a qualitative consideration of a system by the use of the potential energy curve and the conservation of mechanical energy. Generally, one would like to solve the equations of motion that follow from the definition of mechanical energy,

$$\tfrac{1}{2}mv^2 + U(x) = E. \qquad\qquad [8\text{–}6a]$$

A complete solution would give the position of the particle as a function of time, $x(t)$ in the one-dimensional case.

For most systems, however, this solution is impossible to find without using difficult computer techniques. This is because either the mechanical energy, E, does not have a constant value with time (conservation of mechanical energy) or the potential function, $U(x)$, is not sufficiently simple.

However, even when $x(t)$ cannot be found explicitly, if E is constant a knowledge of $U(x)$ and the value of E can yield a great amount of useful information through the consideration of Eq. 8–6a.

For example, for a given mechanical energy E, Eq. 8–6a tells us that the particle is restricted to those regions on the x-axis where $E > U(x)$. Physically we cannot have a negative kinetic energy because this would require a negative mass or an imaginary velocity. Therefore $E - U(x)$ must be zero or positive. Furthermore, we can obtain a good qualitative description of the types of motion possible by plotting $U(x)$ versus x. This description depends on the fact that the speed is proportional to the square root of the difference between E and U.

To take a specific example, consider the potential energy function shown in Fig. 8–6. This could be thought of as an actual profile of a frictionless roller coaster, but in general it represents the potential energy of a particle constrained to move along the x-axis, acted on by forces that produce the potential energy variation shown. Since we must have $E \geqq U(x)$ for real motion, the lowest value of E possible is E_0. At this value of mechanical energy, $E_0 = U$ and the kinetic energy must be zero. The particle must be at rest at the point x_0. At a slightly higher energy E_1 the particle can move between x_1 and x_2 only. As it moves from x_0 its speed decreases on approaching either x_1 or x_2. At x_1 or x_2 the particle stops and reverses its direction. These points x_1 and x_2 are, therefore, called *turning points* of the motion. At an energy like E_2 there are four turning points, and the particle (constrained, as always, to the x-axis) can oscillate in either one of the two potential valleys. At an energy like E_3 there is only one turning point of the motion, at x_3. If the particle is initially moving in the negative x-direction, it will stop at x_3 and then move in the positive x-direction. It will speed up as U decreases and slow down as U increases. At energies above E_4 there are no turning points, and the particle will not reverse direction. Its speed will change according to the value of the potential at each point.

At a point where $U(x)$ has a minimum value, such as at $x = x_0$, the slope of the curve is zero so that the force is zero; that is, $F(x_0) = -(dU/dx)_{x=x_0} = 0$. A

Turning points

Stable equilibrium

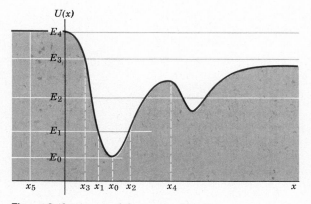

U(x)

E_4

E_3

E_2

E_1

E_0

x_5 x_3 x_1 x_0 x_2 x_4 x

Figure 8–6 A potential energy curve.

particle at rest at this point will remain at rest. Furthermore, if the particle is displaced slightly in either direction, the force, $F(x) = -dU/dx$, will tend to return it, and it will oscillate about the equilibrium point. This equilibrium point is, therefore, called a point of *stable equilibrium*.

At a point where $U(x)$ has a maximum value, such as at $x = x_4$, the slope of the curve is zero so that the force is again zero; that is, $F(x_4) = -(dU/dx)_{x=x_4} = 0$. A particle at rest at this point will remain at rest. However, if the particle is displaced even the slightest distance from this point, the force, $F(x) = -dU/dx$, will tend to push it farther away from the equilibrium position. Such an equilibrium point is, therefore, called a point of *unstable equilibrium*.

In an interval in which $U(x)$ is constant, such as near $x = x_5$, the slope of the curve is zero so that the force is zero; that is, $F(x_5) = -(dU/dx)_{x=x_5} = 0$. Such an interval is called one of *neutral equilibrium*, since a particle can be displaced slightly without experiencing either a repelling or a restoring force.

From this it is clear that if we know the potential energy function for the region of x in which the body moves, we know a great deal about the motion of the body.

8–6 Two- and Three-Dimensional Conservative Systems

So far we have discussed potential energy and energy conservation for one-dimensional systems in which the force was directed along the line of motion. We can easily generalize the discussion to three-dimensional motion.

If the work done by the force **F** depends only on the end points of the motion and is independent of the path taken between these points, the force is conservative. We define the potential energy U by analogy with the one-dimensional system and find that it is a function of three space coordinates, that is, $U = U(x, y, z)$. Again we obtain an expression for the conservation of mechanical energy.

The generalization of Eq. 8–5b to motion in three dimensions is

$$\Delta U = -\int_{\mathbf{r_0}}^{\mathbf{r}} \mathbf{F}(\mathbf{r}) \cdot d\mathbf{r} \tag{8-5c}$$

in which ΔU is the change in potential energy for the system as the particle moves from the point (x_0, y_0, z_0), described by the position vector $\mathbf{r_0}$, to the point (x, y, z), described by the position vector $\mathbf{r}$. F_x, F_y, and F_z are the components of the conservative force $\mathbf{F}(\mathbf{r}) = \mathbf{F}(x, y, z)$.

The generalization of Eq. 8–6b to three-dimensional motion is

$$\tfrac{1}{2}mv^2 + U(x, y, z) = \tfrac{1}{2}mv_0^2 + U(x_0, y_0, z_0). \tag{8-6c}$$

Likewise Eq. 8–6a becomes

$$\tfrac{1}{2}mv^2 + U(x, y, z) = E \tag{8-6d}$$

in three dimensions, E being the constant mechanical energy.

Finally, the generalization of Eq. 8–7 to three dimensions is*

$$\mathbf{F}(\mathbf{r}) = -\mathbf{i}\,\frac{\partial U}{\partial x} - \mathbf{j}\,\frac{\partial U}{\partial y} - \mathbf{k}\,\frac{\partial U}{\partial z}$$

---

* In general $U(x, y, z)$ is a function of the three variables, x, y, and z. The notation $\partial U/\partial x$, called a *partial derivative*, instructs us to take the derivative of $U(x, y, z)$ with respect to x *as if y and z were constants*. The meanings of $\partial U/\partial y$ and $\partial U/\partial z$ are similar.

If we substitute this expression for **F** into Eq. 8–5c we again obtain an identity. In vector language, the conservative force **F** is said to be the negative of the *gradient* of the potential energy $U(x, y, z)$.

Show that all these expressions reduce to the correct one-dimensional equation for motion along the x-axis.

Example 4 *The simple pendulum* An important two-dimensional conservative system that will be encountered several times in this text is the simple pendulum as shown in Fig. 8–7. The forces on the bob are $m\mathbf{g}$, the gravitational attraction of the earth, and the tension **T** in the supporting cord. In the absence of friction there are no dissipative forces, so we can define a potential energy. The mechanical energy will be constant.

If the cord does not stretch, the motion of the bob will always be at right angles to the direction of the tension so that this force does no work. After the bob is released only the gravitational force exerted on the bob by the earth does work on it. Since the gravitational force is conservative, we can use the equation of energy conservation in two dimensions,

$$\tfrac{1}{2}mv^2 + U(x, y) = E.$$

We previously showed that the gravitational potential energy in one dimension is $U(y) = mgy$ (see Eq. 8–10). Since the force of gravity is entirely in the y-direction, it will do no work by virtue of motion in the x-direction. Therefore, when the pendulum moves the gravitation potential will depend only on the y-coordinate; specifically $U(x, y)$ equals mgy, where y is taken as zero at the lowest point of the arc ($\phi = 0°$). Then

$$\tfrac{1}{2}mv^2 + mgy = E.$$

The bob is pulled through an angle ϕ_0 before being released. The potential energy there is mgh, corresponding to a height $y = h$ above the reference point. At the release point ($\phi = \phi_0$) the speed and the kinetic energy are zero so that the potential energy equals the total mechanical energy at that point.

Hence,

$$E = mgh$$

and

$$\tfrac{1}{2}mv^2 + mgy = mgh,$$

or

$$\tfrac{1}{2}mv^2 = mg(h - y).$$

Notice that (a) the maximum speed occurs at $y = 0$, where

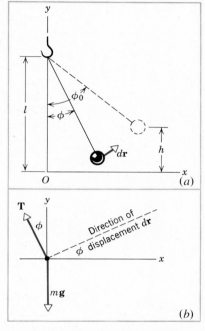

Figure 8–7   (a) A simple pendulum. The bob, imagined to be a point mass m, is suspended on a string of length l. Its maximum displacement is ϕ_0. (b) A free-body force diagram for the mass.

$v = \sqrt{2gh}$; (b) the minimum speed occurs at $y = h$, where $v = 0$; (c) At $y = 0$ the energy is entirely kinetic, the potential energy being zero; (d) At $y = h$ the energy is entirely potential, the kinetic energy being zero; (e). At intermediate positions the energy is partly kinetic and partly potential; and (f) $U \leqq E$ at all points of the motion—the pendulum bob cannot rise higher than $y = h$, its initial release point.

8–7   Nonconservative Forces

So far we have considered only the action of a single conservative force on a body. Starting from the work-energy theorem, or

$$W_1 + W_2 + \cdots + W_n = \Delta K \qquad [8\text{–}2]$$

we saw that, if only one force, say $\mathbf{F}_1$, was acting and if it was conservative, then we could represent the work W_1 that it did on the body as a change in potential energy ΔU_1 of the system of the opposite sign (see Eq. 8–5a), or

$$W_1 = -\Delta U_1.$$

Combining this with Eq. 8–2 yielded

$$\Delta K + \Delta U_1 = 0.$$

If several conservative forces such as gravity, an elastic spring force, an electrostatic force, and so on, are acting, we can easily extend these two equations to

$$\Sigma \, W_C = - \Sigma \, \Delta U \qquad (8\text{--}12)$$

and

$$\Delta K + \Sigma \, \Delta U = 0 \qquad (8\text{--}13)$$

in which $\Sigma \, W_C$ is the sum of the work done by the various conservative forces and the ΔU's are the changes in the potential energy of the system associated with these forces. The quantity on the left of Eq. 8–13 is simply ΔE, the change in the total mechanical energy, for the case in which several conservative forces are acting on a body. We can write this equation then as

$$\Delta E = 0 \qquad \text{(conservative forces)}, \qquad (8\text{--}14)$$

which tells us that, as the system configuration changes the total mechanical energy E for the system remains constant.

Let us now suppose that, besides one or more conservative forces, a single nonconservative force—perhaps due to friction—acts on the body. Let ΣW_c be the total work done on the body by conservative forces and let W_{nc} be the work done on the body by this single nonconservative force. We can then write Eq. 8–2 as

$$W_{nc} + \Sigma W_c = \Delta K, \qquad (8\text{--}15)$$

which we can recast (see Eq. 8–12) as

$$W_{nc} = \Delta K + \Sigma U. \qquad (8\text{--}16)$$

In turn we can express this equation as

$$W_{nc} = \Delta E. \qquad (8\text{--}17)$$

Thus the mechanical energy E is *not* constant when a nonconservative force acts; it changes by the amount of work done by this force. If there are no nonconservative forces, then $W_{nc} = 0$ and E is a constant, just as we expect; see Eq. 8–14. Hence, only when no nonconservative forces act, or when we can neglect the work that they do, can we assume that mechanical energy is conserved.

For a concrete example consider a table, firmly fastened to the laboratory floor. A block, projected horizontally onto its top surface, is brought to rest by the action of the frictional force exerted by this surface on the block. The work W_{nc} done by the (nonconservative) frictional force on the block is negative because the direction of this force is opposite to the displacement of the block as we watch it come to rest in the laboratory. It follows from Eq. 8–17 then that the final mechanical energy of the system (block) will be less than its initial mechanical energy E_0 by the magnitude of the work done. Potential energy plays no role in this example, however, so that changes in mechanical energy represent changes in kinetic energy. Thus, from Eq. 8–17 with ΔK substituted for ΔE, we see that the work (negative) done on the block by the frictional force is equal to the change (decrease) in its kinetic energy. This is simply a statement of the work-energy theorem as it applies to this example.

The work done by a frictional force can be positive as well as negative. Imagine, for example, that you are standing next to a moving conveyer belt, holding a package in your hands. The package has no kinetic energy in your reference frame. Suppose that you drop the package onto the moving belt and watch it pick up speed until it has come to rest on the belt and is moving away from you at the constant speed of the belt. The package (again, in your reference frame) now has kinetic energy. A force must have acted on it in the direction of its motion; that force can only be the frictional force exerted by the belt on the package. This force does positive work on the package because it is in the same direction as the displacement of the package. Thus, from Eq. 8–17, we see that the work (positive) done by the frictional force on the package is equal to the change (increase) in the kinetic energy of the package.

Example 5 A ball bearing is projected downward with an initial velocity v_0 of 14 m/s from a height of 240 m and buries itself in 20 cm of sand. The mass of the body is 5.0 g. Find the average resistive force exerted by the sand on the body. Neglect air resistance and solve the problem by considerations of work and energy.

The kinetic energy of the ball bearing just as it enters the sand is

$$K = \tfrac{1}{2}mv_0^2 + mgh,$$

where m is its mass and h is the height of fall.

Also, from the work-energy principle, we have (approximately)

$$K = \overline{F} s,$$

where $\overline{F}$ is the average resistive force and s is the distance of penetration into the ground.

Equating and solving for $\overline{F}$ gives

$$\overline{F} = \frac{mv_0^2}{2 s} + \frac{mgh}{s}$$

$$= \frac{(5.0 \times 10^{-3} \text{ kg})(14 \text{ m/s})^2}{2(0.20 \text{ m})}$$

$$+ \frac{(5.0 \times 10^{-3} \text{ kg})(9.8 \text{ m/s}^2)(240 \text{ m})}{(0.20 \text{ m})}$$

$$= 2.5 \text{ N} + 58.8 \text{ N} = 61.3 \text{ N}$$

What error do we make by neglecting (in comparison to h) the extra distance of fall s before the ball bearing is brought to rest? Show that this is equivalent to neglecting mg in comparison to $\overline{F}$ in arriving at the resultant force to be used in the work-energy theorem. Such terms are not always negligible in practice (see Problem 35, for example).

Example 6 A 4.5-kg block is projected up a 30° inclined plane with an initial speed of 5.0 m/s. It is found to travel 1.5 m up the plane, stop, and slide back to the bottom. Compute the force of friction **f** (assumed to have a constant magnitude) acting on the block and find the speed v of the block when it returns to the bottom of the inclined plane.

Consider first the upward motion. At the top, where this motion ends,

$$E = K + U = 0 + (4.5 \text{ kg})(9.8 \text{ m/s}^2)(1.5 \text{ m})(\sin 30°) = 33 \text{ J}.$$

At the bottom, where this motion begins,

$$E_0 = K_0 + U_0 = \tfrac{1}{2}(4.5 \text{ kg})(5.0 \text{ m/s})^2 + 0 = 56 \text{ J}.$$

But

$$W_f = -fs = -f(1.5 \text{ m})$$

and

$$E - E_0 = W_f,$$

so that

$$33 \text{ J} - 56 \text{ J} = -f(1.5 \text{ m})$$

and

$$f = 15 \text{ N}.$$

Now consider the downward motion. The block returns to the bottom of the inclined plane with a speed v. Then, at the bottom, where this motion ends,

$$E' = K' + U' = \tfrac{1}{2}(4.5 \text{ kg}) v^2 + 0 = (2.25 \text{ kg}) v_1^2$$

At the top, where this motion begins,

$$E'_0 = K'_0 + U'_0$$
$$= 0 + (4.5 \text{ kg})(9.8 \text{ m/s}^2)(1.5 \text{ m})(\sin 30°) = 33 \text{ J}.$$

But

$$W'_f = -(15 \text{ N})(1.5 \text{ m}) = -23 \text{ J}$$

and

$$E' - E'_0 = W'_f,$$

so that

$$(2.25 \text{ kg}) v^2 - 33 \text{ J} = -23 \text{ J}$$

and

$$v = 2.1 \text{ m/s},$$

only 42% of the initial speed of 5.0 m/s.

8–8 The Conservation of Energy

If the forces that act on a body are all conservative its kinetic energy K and its potential energy U may change as the body moves but their sum, the mechanical energy E, will remain constant. We have called this assertion (see Eqs. 8–6) the *law of the conservation of mechanical energy.* It is, as we have seen, a very useful relationship. If one or more nonconservative forces acts on the body Eq. 8–17 reminds us that this conservation law no longer holds. However, we can always broaden our definition of energy so that a *law of conservation of total energy* is valid.

Consider again a block projected horizontally onto a table top and brought to rest by frictional forces. The mechanical energy of the block (which in this case is entirely kinetic) drops to zero as the block comes to rest. What has happened to this "missing" energy? We find a clue in noting that both the block and the portion of the table top over which it moves have become slightly warmer as the block comes to rest. It is as if the kinetic energy associated with the directed motion of the block were transformed completely into kinetic energy of the disordered, random motions of the atoms that make up the block and the table top. We call such energy *thermal energy* or—more generally—*internal energy*—and we represent it by U_{int}. For blocks sliding over rough surfaces then, whether the surfaces are horizontal or inclined and whether they are plane or curved, we can broaden the definition of energy to include both mechanical energy E ($= K + U$) and internal energy U_{int} and we can write

$$\Delta K + \Delta U + \Delta U_{int} = 0$$

as a statement of the conservation of energy for these special situations. When we use this equation we must be sure to choose as our system—not the block alone—but the larger system made up of the block and the surface over which it slides. This is necessary because the internal energy U_{int} is shared between these two objects in a way that is not easy to predict.

It turns out that in new situations we can *always* identify new quantities like U_{int} that permit us to expand the definition of energy and to retain, in a more general form, the law of conservation of energy. Thus we can write

$$0 = \Delta K + \Sigma \Delta U + \Delta U_{int} + \text{(change in other forms of energy)}. \quad (8-18)$$

In other words, the total energy—kinetic plus potential plus internal plus other forms—does not change. *Energy may be transformed from one kind to another, but it cannot be created or destroyed; the total energy is constant.*

Conservation of energy

This statement is a generalization from our experience, so far not contradicted by observation of nature. It is called the *principle of the conservation of energy*. Often in the history of physics this principle seemed to fail. But its apparent failure stimulated the search for the reasons. Experimentalists searched for physical phenomena besides motion that accompany the forces that act between bodies. Such phenomena have always been found. With work done against friction, internal energy is produced; in other interactions energy in the form of sound, light, electricity, and so on, may be produced. Hence the concept of energy was generalized to include forms other than kinetic and potential energy of directly observable bodies. This procedure, which relates the mechanics of bodies observed to be in motion to phenomena that are not mechanical or in which motion is not directly detected, has linked mechanics to all other areas of physics. The energy concept now permeates all of physical science and has become one of the unifying ideas of physics.

In subsequent chapters we shall study various transformations of energy—from mechanical to internal, mechanical to electrical, nuclear to internal, and so on. It is during such transformations that we measure the energy changes in terms of work, for it is during these transformations that forces arise and do work.

Although the principle of the conservation of kinetic plus potential energy is often useful, we see that it is a restricted case of the more general principle of the

Figure 8–8 (*a*) Going up! Internal energy changes into gravitational potential energy. (*b*) Coming down! Gravitational potential energy changes into thermal energy—the rope and the carabiner get hot because of frictional work. Jim Elder/Image Bank.

(*a*)   (*b*)

conservation of energy. Kinetic plus potential energy is conserved only when conservative forces act. Total energy is always conserved.

Figure 8–8 provides an example of energy conservation in which nonconservative forces act. As the climber ascends the rock face in Fig. 8–8*a* he is gaining gravitational potential energy at the expense of his own store of biochemical internal energy. In Fig. 8–8*b* the climber descends the face at roughly constant speed, sliding along a rope that is led through a metal link (carabiner) and around his shoulders. The gravitational potential energy that he loses in descending appears as thermal internal energy—heat—in the rope and in the metal link. The purpose of the rope-link arrangement is to prevent the gravitational potential energy from transforming into kinetic energy of a falling climber!

8–9 Mass and Energy

For many years the laws of the conservation of mass and of energy were viewed as valid but quite independent laws. The science of chemistry was built up in large part by applying these two laws—always separately—to chemical reactions. Einstein however showed that mass and energy can be converted into each other so that these two laws are but two aspects of a single, deeper law, the law of conservation of mass-energy.

It turns out that in chemical reactions the amount of mass that is turned into energy (or vice versa) is so small a fraction of the total mass involved that, within the limits of measurement, it cannot be detected. Mass and energy truly *seem* to be separately conserved. In nuclear reactions however, in which the energy released per event is typically a million times greater than in chemical reactions, the change in mass can be easily measured. In such cases the interconvertibility of mass and energy must be fully taken into account.

In Einstein's papers introducing the theory of relativity he was led to take a new and broader view of the natures of mass, length, and time. In particular he showed that, if certain physical laws are to be retained, the mass of a particle must be redefined as

$$m = \frac{m_0}{\sqrt{1 - (v/c)^2}}. \qquad (8-19)$$

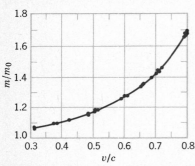

The variation of mass with speed

Figure 8-9 The solid curve is a plot of Eq. 8-19, and the dots are experimental points. As $v/c \to 0$, $m/m_0 \to$ unity; as $v/c \to$ unity, $m/m_0 \to \infty$.

In this view the mass of a particle is not constant but varies with v, the speed of the particle with respect to the observer. If $v = 0$ we have $m = m_0$ so that m_0, called the *rest mass,* is the mass of the particle when it is at rest with respect to the observer. The quantity c is the speed of light ($= 3 \times 10^8$ m/s). Equation 8-19 has been verified experimentally by deflecting high-speed electrons in a magnetic field and measuring the radius of curvature of their path. Figure 8-9 shows the results.

For gross objects at "ordinary" speeds the difference between m and m_0 in Eq. 8-19 is too small to be detected. Electrons and similar particles however can be easily accelerated to speeds high enough so that m becomes much larger than m_0. In the two-mile long linear electron accelerator at Stanford, for example, electrons with $v/c = 0.999\ 999\ 999\ 7$ can be generated; for such electrons $m/m_0 = 40,000$.

We turn now from considerations of mass to those of kinetic energy. In relativity theory we define kinetic energy just as we did in classical mechanics, namely (see Section 7–5), that it is equal to the work required to set the particle into motion at speed v. In our derivation of an expression for K, however, we use for the mass of the particle the quantity m defined by Eq. 8-19. This leads to

$$K = (m - m_0)c^2. \qquad (8-20)$$

Relativistic kinetic energy

When the velocity is small this approaches the classical expression, $\frac{1}{2}m_0 v^2$, and when $v = 0$, $m = m_0$, and $K = 0$, as expected.

The basic idea that energy is equivalent to mass can be extended to include energies other than kinetic. For example, when we compress a spring and give it elastic potential energy, U, its mass increases from m_0 to $m_0 + U/c^2$. When we add internal energy in amount U_{int} to an object, its mass also increases from m_0 to $m_0 + U_{\text{int}}/c^2$. Equation 8-20 can be modified to include these types of energy,

$$K + U + U_{\text{int}} + \cdots = (m - m_0)c^2, \qquad (8-21)$$

where m now is increased over that in Eq. 8-19 by the amount $(U + U_{\text{int}} + \cdots)/c^2$.

One might think that the total energy could now be defined as equal to $(m - m_0)c^2$. Classically, in a closed system (one acted on by no external forces and having no loss or gain of mass or energy through the sides) the sum of the energies on the left of Eq. 8-21 would indeed be a constant, the total energy. However, by introducing expressions containing mass where the classical total energy appeared, Einstein raised the possibility of converting rest-mass energy itself into other forms, just as kinetic and potential energies may be converted from one to the other. If this is the case, then the conserved quantity could be mc^2, which would then be quite properly called the total energy. It would certainly reduce to that in the classical limit. We would then have

Relativistic total energy

$$K + \Sigma U + U_{\text{int}} + \cdots + m_0 c^2 = mc^2 = E_{\text{tot}}. \qquad (8-22a)$$

If we consider the possibility of changes in these, recalling that E_{tot} remains constant, we have

$$\Delta K + \Sigma \, \Delta U + \Delta U_{\text{int}} + \cdots + \Delta m_0 c^2 = 0. \qquad (8\text{--}22b)$$

Equations 8–22 are statements of the law of *conservation of mass-energy*.

The study of energy exchanges between rest-mass form ($m_0 c^2$) and other forms are a workaday tool of physicists who study nuclei and subnuclear particles. The matter was brought dramatically to public attention when, in 1939, Lise Meitner and O. R. Frisch, interpreting some crucial experiments of Otto Hahn and F. Strassmann, discovered nuclear fission. This was followed only three years later by the development of the nuclear reactor.

In the fission process uranium nuclei break up into two major fragments of roughly equal mass plus a small number of neutrons. The sum of the rest masses of the particles so produced is less than that of the parent nucleus. The difference in rest mass Δm_0 appears as energy $\mathscr{E}$ of various forms, initially as kinetic energy of the fragments. The total energy E_T for this isolated system remains constant during the fission process so that, from Eq. 8–22b,

Mass-energy equivalence

$$\mathscr{E} = -\Delta m_0 c^2. \qquad (8\text{--}23)$$

The minus sign shows that if $\mathscr{E}$ is to increase when fission occurs, the rest mass must decrease, that is, Δm_0 must be negative. Equation 8–23 is Einstein's famous "$E = mc^2$" equation, written in a slightly more explicit form.

Example 7 *The binding energy of the deuteron* The rest masses of the proton and the neutron, which are two of the important particles of physics (see Appendix E), are 1.00728 u and 1.00867 u, respectively. Here u ($= 1.661 \times 10^{-27}$ kg) is an abbreviation for the *unified atomic mass unit*, which is defined in Section 1–4. A deuteron, the nucleus of a heavy hydrogen atom, consists of a proton and a neutron and has a measured mass of 2.01355 u. This is *less than* the combined masses of the proton and the neutron by 0.00240 u. The discrepancy is equivalent in energy to

$$\mathscr{E} = \Delta m_0 c^2 = (0.00240 \text{ u})(1.66 \times 10^{-27} \text{ kg/u})(3.00 \times 10^8 \text{ m/s})^2$$
$$= 3.58 \times 10^{-13} \text{ J} = 2.23 \text{ MeV},$$

in which the MeV ($= 10^6$ electron volts) is a convenient unit for measuring energy in nuclear physics. When a neutron and a proton combine to form a deuteron, this exact amount of energy is given off in the form of γ radiation. Similarly, it is found that this amount of energy must be added to the deuteron to break it up into a proton and a neutron. This energy is therefore called the *binding energy* of the deuteron.

REVIEW GUIDE AND SUMMARY

Conservative and nonconservative forces

A force acting on a particle is conservative if any one of the three equivalent conditions outlined in Section 8–2 is met. The forces of gravity and of an ideal spring are conservative. The force of friction, which meets none of the three requirements, is nonconservative.

Potential energy

Every conservative force has a potential energy U associated with it. When a system changes its configuration, the change in potential energy ΔU is defined as the negative of the work done, or

$$\Delta U = -W = -\int_{x_0}^{x} F(x) \, dx \qquad \text{(one dimension).} \qquad [8\text{--}5b]$$

Nonconservative forces do not have potential energies.

If only conservative forces act, the sum of the kinetic and potential energies of the particle, called the mechanical energy $E \, (= K + U)$ remains constant, or (see Eq. 8–4a)

Conservation of mechanical energy

$$\Delta K + \Delta U = 0.$$

See Examples 1, 2, and 3 and also Fig. 8–4 for illustrations of the conservation of mechanical energy.

Turning points, etc.

From the form of $U(x)$ alone, assuming no friction, we can learn a lot about the motion of the particle by simple inspection. Guided by the conservation of mechanical energy principle we can identify turning points, regions and points of equilibrium, forbidden regions and speed variations, as Fig. 8–6 shows.

Effect of friction

If a frictional force also acts on the particle, mechanical energy is not conserved but steadily decreases. We can retain a conservation of energy principle by associating the work done by the frictional force with an internal (thermal) energy U_{int} and by writing $K + U + U_{int} = $ a constant, or

$$\Delta K + \Delta U + \Delta U_{int} = 0. \qquad [8-16]$$

See Examples 5 and 6 for applications.

Conservation of energy principle

If several forces act, whether they are conservative or not, we can *always* associate some kind of energy with the work done by each of these forces. We always find that the sum of the energies so defined remains constant as the configuration of the system changes, and we write this conservation of energy principle as

$$\Delta K + \Sigma\ \Delta U + \Delta U_{int} + \text{(changes in other forms of energy)} = 0. \qquad [8-18]$$

Rest mass energy

The theory of special relativity allows us to extend this conservation principle to its ultimate form, the conservation of total relativistic energy, E_{tot}. We do so by recognizing a new kind of energy, rest mass energy m_0c^2, and by putting

$$\Delta K + \Sigma\ \Delta U + \Delta U_{int} + \cdots + \Delta m_0c^2 = 0. \qquad [8-22b]$$

The mass-energy equivalence formula ($\mathscr{E} = -\Delta m_0c^2$; Eq. 8–23) shows how this kind of energy can be changed to other forms; see Example 7.

QUESTIONS

1. An automobile is moving along a highway. The driver jams on the brakes and the car skids to a halt. In what form does the lost kinetic energy of the car appear?

2. In the above question, assume that the driver "pumps" the brakes in such a way that there is no skidding or sliding. In this case, in what form does the lost kinetic energy of the car appear?

3. You drop an "object" and observe that it bounces to one and one-half times its original height. What conclusions can you draw?

4. What happens to the potential energy an elevator loses in coming down from the top of a building to a stop at the ground floor?

5. Air bags greatly reduce the chance of injury in a car accident. Explain how they do so, in terms of energy transfers.

6. Pole vaulting was transformed when the wooden pole was replaced by the fiberglass pole. Explain why.

7. A ball dropped to earth cannot rebound higher than its release point. However, spray from the bottom of a waterfall can sometimes rise higher than the top of the falls. How come?

8. Figure 8–10 shows a circular glass tube fastened to a vertical wall. The tube is filled with water except for an air bubble that is temporarily at rest at the bottom of the tube. Discuss the subsequent motion of the bubble in terms of energy transfers. Do so both neglecting viscous and frictional forces and also taking them fully into account.

9. In Example 2 (see Fig. 8–5) we asserted that the speed at the bottom does not depend at all on the shape of the surface. Would this still be true if friction were present?

10. Give physical examples of unstable equilibrium. Of neutral equilibrium. Of stable equilibrium.

11. In an article called "Energy and the Automobile," which appeared in the October 1980 issue of *The Physics Teacher,* the author (Gene Waring) states: "It is interesting to note that *all* the fuel input energy is eventually transformed to thermal energy and strung out along the car's path." Analyze the various mechanisms by which this might come about. Consider, for example, road friction, air resistance, braking, the car radio, the headlamps, the battery, internal engine and drive train losses, the horn, and so on. Assume a straight and level roadway.

12. Trace back to the sun as many of our present energy sources as you can. Can you think of any that can't be so traced?

13. Explain, using work and energy ideas, how a child pumps a swing up to large amplitudes from a rest position. (See "How to Make a Swing Go," by R. V. Hesheth, *Physics Education,* July 1975.)

14. A swinging pendulum eventually comes to rest. Is this a violation of the law of conservation of energy?

15. Two disks are connected by a stiff spring. Can you press the upper disk down enough so that when it is released it will spring back and raise the lower disk off the table (see Fig. 8–11)? Can mechanical energy be conserved in such a case?

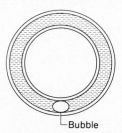

Bubble

Figure 8–10 Question 8.

Figure 8–11 Question 15.

16. Discuss the words "energy conservation" as used (*a*) in this chapter, and (*b*) in connection with the looming energy crisis (turning off lights, etc). How do these two usages differ?

17. The electric power for a small town is provided by a hydroelectric plant at a nearby river. If you turn off a light bulb in this closed energy system, conservation of energy requires that an equal amount of energy, perhaps in another form, appears some-where else in the system. Where and in what form does this energy appear?

18. The driver of an automobile suddenly sees a brick wall directly in front of him. To avoid crashing, is it better for him to slam on the brakes or to turn the car sharply away from the wall? (Hint: Consider the force required in each case.)

19. A spring is kept compressed by tying its ends together tightly. It is then placed in acid and dissolves. What happened to its stored potential energy?

20. The expression $E = mc^2$ tells us that perfectly ordinary objects such as coins or pebbles contain enormous amounts of energy. Why did these large stores of energy go unnoticed for so long?

21. "Nuclear explosions—weight for weight—release about a million times more energy than do chemical explosions because nuclear explosions are based on Einstein's $E = mc^2$ relation." What do you think of this statement?

EXERCISES

Section 8–3 Potential Energy

1. The summit of Mount Everest, the highest mountain on earth, is 8850 m above sea level. (*a*) How much energy would a 90-kg climber expend against gravity in climbing to the summit from sea level? (*b*) How many Mars bars, at 300 Calories per bar, would supply an energy equivalent to this? Your answer should suggest that work done against gravity is a very small part of the energy expended in climbing a mountain. [Note: 1 Calorie (nutritional) = 4187 J.]

2. A chain is held on a frictionless table with one-fourth of its length hanging over the edge, as shown in Fig. 8–12. If the chain has length *l* and a mass *m*, how much work is required to pull the hanging part back on the table?

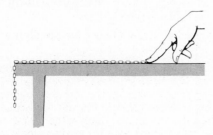

Figure 8–12 Exercise 2.

Section 8–4 One-Dimensional Conservative Systems

3. A 8.0-kg mortar shell is fired straight up with an initial speed of 100 m/s. What is the shell's potential energy at the top of its trajectory? Take the zero of potential energy to be at ground level.

4. Approximately 5.5×10^6 kg of water drops 50 m over Niagara Falls every second. (*a*) How much potential energy is lost every second by the falling water? (*b*) What would be the power output of a hydroelectric plant that could convert *all* of the water's potential energy to electrical energy? (*c*) If the utility company sold this energy at an industrial rate of 1 cent per kilowatt-hour, what would be their yearly income from this source?

5. You drop a physics textbook, whose mass is 2.0 kg, to a friend who is standing on the ground 10 m below, as in Fig. 8–13. (*a*) What is the potential energy of the book just before you release

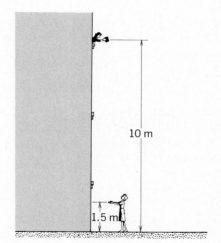

Figure 8–13 Exercise 5.

it? (*b*) What is its kinetic energy just before your friend catches it in her outstretched hands, which are 1.5 m above ground level? (*c*) How fast is the book moving as it is caught? Take the zero of potential energy to be ground level.

6. A 5.0-g marble is fired vertically upward, using a spring gun. The spring must be compressed at least 8.0 cm if the marble is to hit a target 20 m above it. What is the force constant of the spring?

7. Figure 8–14 shows an 8.0-kg stone resting on a spring. Call this position of the center of the stone $y = 0$. The spring is compressed 10 cm by the stone. (*a*) What is the force constant of the

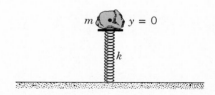

Figure 8–14 Exercise 7.

spring? (b) If the stone is pushed down an additional 30 cm and released, how high above the $y = 0$ position will it rise?

Section 8–5 Mechanical Energy and the Potential Energy Curve

8. A particle moves along the x-axis through a region in which its potential energy $U(x)$ varies as in Fig. 8–15. (a) Make a quantitative plot of the force $F(x)$ that acts on the particle, using the same x-axis scale as in Fig. 8–15. (b) The particle has a (constant) mechanical energy E of 4 J. Sketch a plot of its kinetic energy $K(x)$ directly on Fig. 8–15.

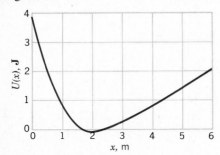

Figure 8–15 Exercise 8.

Section 8–6 Two- and Three-Dimensional Conservative Systems

9. An ice cube is released from the edge of a hemispherical frictionless bowl whose radius is 20 cm; see Fig. 8–16. How fast is the cube moving at the bottom of the bowl?

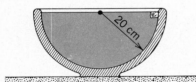

Figure 8–16 Exercise 9.

10. A particle has a potential energy given by $U(x, y, z) = ax^2 + bxy + cz$, in which a, b, and c are constants. What force acts on the particle?

11. A thin rod whose length is l and whose mass is negligible is pivoted at one end so that it can rotate in a vertical circle. The rod is pulled aside through an angle θ and then released, as shown in Fig. 8–17. How fast is the lead ball moving at its lowest point? Assume that $l = 2.0$ m and $\theta = 30°$.

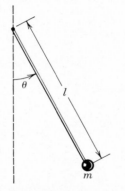

Figure 8–17 Exercise 11.

12. A projectile is fired from a cliff 125 m high with an initial velocity of 150 m/s, directed 41° above the horizontal. (a) Find the speed of the projectile just before it strikes the ground. Use the methods developed in Section 4–3 (Projectile Motion). (b) Now find the same answer using energy conservation methods. Compare your two solutions carefully and note how much simpler the energy method is in this case.

13. A runaway truck with failed brakes is barreling down grade at 80 mi/h. Fortunately, there is an emergency escape ramp at the bottom of the hill. The inclination of the ramp is 15°. How long must it be to make certain of bringing the truck to rest, at least momentarily?

Section 8–7 Nonconservative Forces

14. A 68-kg skydiver falls at a constant terminal speed of 59 m/s. At what rate is gravitational potential energy being removed from the earth-skydiver system? What happens to this energy?

15. A 25-kg bear slides, from rest, 10 m down a lodgepole pine tree, moving with a speed of 3.0 m/s at the bottom. What is the average frictional force that acts on the bear?

16. A 200-lb man jumps out a window into a fire net 30 ft below. The net stretches 9.0 ft before bringing him to rest and tossing him back into the air. What is the potential energy of the stretched net if no energy is dissipated by nonconservative forces?

17. Two blocks are connected by a string, as shown in Fig. 8–18. They are released from rest. Show that, after they have moved a distance L, their common speed is given by

$$V = \left[\frac{2(m_2 - \mu m_1)gL}{m_1 + m_2} \right]^{1/2}$$

in which μ is the coefficient of kinetic friction between the upper block and the surface.

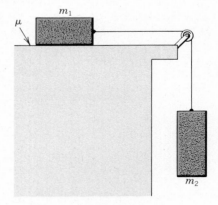

Figure 8–18 Exercise 17.

18. As Fig. 8–19 shows, a 3.5-kg block is released from a compressed spring whose force constant is 640 N/m. After leaving the spring, the block travels over a horizontal surface, with a coefficient of kinetic friction of 0.25, for a distance of 7.8 m before

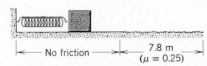

Figure 8–19 Exercise 18.

coming to rest. (a) What is the maximum kinetic energy of the block? (b) How far was the spring compressed before the block was released?

19. During a rockslide, a 520-kg rock slides from rest down a hillslope that is 500 m long and 300 m high. The coefficient of kinetic friction between the rock and the hill surface is 0.25. (a) What is the kinetic energy of the rock as it reaches the bottom of the hill? (b) How much work was done on the rock by frictional forces during the slide?

Section 8–9   Mass and Energy
20. The rest mass of a ball bearing is 10 g. (a) What is its mass when it is moving at 90% of the speed of light? What is the kinetic

energy of the ball bearing at this speed calculated according to (b) the (exact) relativistic formula, and (c) the (approximate) classical formula?

21. A 5.0-g micrometeorite is in the direct path of a spaceship, their relative speed being $0.99c$. From the point of view of the spaceship commander, what is the kinetic energy of this tiny particle? Express your answer in SI units and also in terms of the equivalent mass of TNT. (The explosion of 1 kg of TNT releases 4.6×10^6 J of energy.)

22. What is the ratio of the speed of an electron to that of light if the kinetic energy of the electron is (a) 1 keV? (b) 1 MeV? (c) 1 GeV?

PROBLEMS GROUPED BY SECTION

Section 8–3   Potential Energy
1. The magnitude of the gravitational force of attraction between a particle of mass m_1 and one of mass m_2 is given by

$$F(x) = G\frac{m_1 m_2}{x^2}$$

where G is a constant and x is the distance between the particles. (a) What is the potential energy function $U(x)$? [Assume that $U(x) \to 0$ as $x \to \infty$.] (b) How much work is required to increase the separation of the particles from $x = x_1$ to $x = x_1 + d$?

Section 8–4   One-Dimensional Conservative Systems
2. It is claimed that large trees can evaporate as much as 900 kg of water per day. (a) Evaporation takes place from the leaves. To get there the water must be raised from the roots of the tree. Assuming the average rise of water to be 9.0 m from the ground, how much energy must be supplied to do this? (b) What is the average power if the evaporation is assumed to occur during 12 h of the day?

3. A 2.0-kg block is dropped from a height of 40 cm onto a spring of force constant 19.6 N/cm; see Fig. 8–20. Find the maximum distance the spring will be compressed.

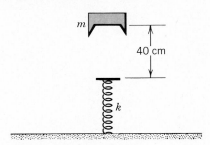

Figure 8–20   Problem 3.

4. In the 1984 Olympic Games the West German high jumper Ulrike Meyfarth set a women's Olympic record for this event with a jump of 2.02 m; see Fig. 8–21. Other things being equal, how high might she have jumped on the moon, where the surface gravity is only 0.17 that on earth? (Hint: The height that "counts" is the vertical distance that her center of gravity rose after her feet left the ground. Estimate that, at the instant her feet lost contact, her center of gravity was 110 cm above ground level. Assume also

Figure 8–21   Problem 4. © Duomo, photo by Paul J. Sutton.

that, as she clears the bar, her center of gravity is at the same height as the bar.)

5. A certain spring is found *not* to conform to Hooke's law. The force (in newtons) it exerts when stretched a distance x (in meters) is found to have magnitude $52.8x + 38.4x^2$ in the direction opposing the stretch. (a) Compute the total work required to stretch the spring from $x = 0.50$ m to $x = 1.00$ m. (b) With one end of the spring fixed, a particle of mass 2.17 kg is attached to the other end of the spring when it is extended by an amount $x = 1.00$ m. If the particle is then released from rest, compute its speed at the instant the spring has returned to the configuration in which the extension is $x = 0.50$ m. (c) Is the force exerted by spring conservative or nonconservative? Explain.

6. Figure 8–22a shows an atom of mass m at a distance r from a resting atom of mass M, where $m \ll M$. Figure 8–22b shows the potential energy function $U(r)$ for various positions of the lighter atom. Describe the motion of this atom if (a) the total mechanical energy is greater than zero, as at E_1 and (b) if it is less than zero, as at E_2. For $E_1 = 1 \times 10^{-19}$ J and $r = 0.3$ nm, find (c) the potential energy, (d) the kinetic energy, and (e) the force (magnitude and direction) acting on the moving atom.

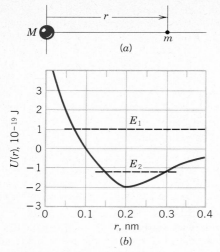

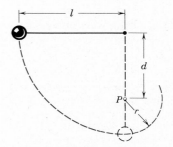

Figure 8-24 Problem 8.

9. The string in Fig. 8–25 is 120 cm long and the distance d to the fixed peg is 75 cm. When the ball is released from rest in the position shown, it will swing along the dotted arc. How fast will it be going (*a*) when it reaches the lowest point in its swing, and (*b*) when it reaches its highest point, after the string catches on the peg?

Figure 8-22 Problem 6.

Section 8–5 Mechanical Energy and the Potential Energy Curve

7. A particle of mass 2 kg moves along the x-axis through a region in which its potential energy $U(x)$ varies as shown in Fig. 8–23. When the particle is at $x = 2$ m, its velocity is -2 m/s. (*a*) What force acts on it at this position? (*b*) Between what limits does the motion take place? (*c*) How fast is it moving when it is at $x = 7$ m?

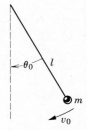

Figure 8-25 Problems 9 and 13.

10. Figure 8–26 shows a simple pendulum of length l. Its bob is observed to have a speed v_0 when the cord makes an angle θ_0 with the vertical. (*a*) Derive an expression for the speed of the bob when it is in its lowest position. (*b*) What is the least value that v_0 can have if the cord is to swing up to a horizontal position? (*c*) To a vertical position?

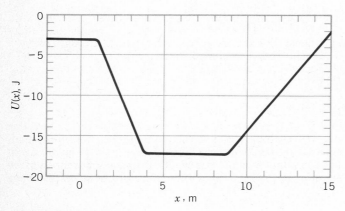

Figure 8-23 Problem 7.

Section 8–6 Two- and Three-Dimensional Conservative Systems

8. A ball is attached to the end of a very light rod. The other end of the rod is pivoted so that the ball can move in a vertical circle. The rod is pulled aside to the horizontal and given a downward push as shown in Fig. 8–24 so that the rod swings down and just reaches the vertically upward position. What initial speed was imparted to the ball?

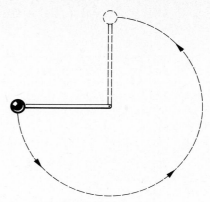

Figure 8-26 Problem 10.

11. A frictionless roller coaster car starts at point A in Fig. 8–27 with an initial speed v_0. (*a*) What will be the speed of the car at point B? (*b*) At point C? (*c*) What constant deceleration is required to stop the car at point E if the brakes are applied at point D? Assume that the car can be considered a particle and that it always remains on the track.

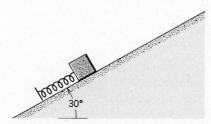

Figure 8–27   Problem 11.

12. A 2.0-kg block is placed against a compressed spring on a frictionless incline (Fig. 8–28). The spring, whose force constant is 19.6 N/cm is compressed 20 cm, after which the block is released. How far up the incline will it go before coming to rest? Measure the final position of the block with respect to its position just before being released.

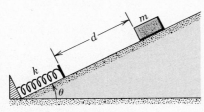

Figure 8–28   Problem 12.

13. In Fig. 8–25, show that, if the pendulum bob is to swing completely around the fixed peg, then $d > 3l/5$. (Hint: The bob must be moving at the top of its swing; otherwise the string will collapse.)

***14.** An escalator joins one floor with another one 8.0 m above. The escalator is 12 m long and moves along its length at 60 cm/s. (a) What power must its motor deliver if it is required to carry a maximum of 100 persons per minute, of average mass 75 kg? (b) An 80-kg man walks up the escalator in 10 s. How much work does the motor do on him? (c) If this man turned around at the middle and walked down the escalator so as to stay at the same level in space, would the motor do work on him? If so, what power does it deliver for this purpose? (d) Is there any (other?) way the man could walk along the escalator without consuming power from the motor?

***15.** A 3.2-kg block starts at rest and slides a distance d down a smooth 30° incline where it runs into a spring of negligible mass; see Fig. 8–29. The block slides an additional 21 cm before it is brought to rest momentarily by compressing the spring, whose force constant is 430 N/m. (a) What is d? (b) The speed of the block continues to increase for a certain interval after the block makes contact with the spring. What additional distance does the block slide before it reaches its maximum speed and begins to slow down?

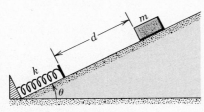

Figure 8–29   Problem 15.

Section 8–7   Nonconservative Forces

16. A factory worker accidentally releases a 400-lb crate that was being held at rest on a 12-ft-long ramp inclined at 39° to the horizontal. The coefficient of kinetic friction between crate and ramp, and also between the crate and the factory floor, is 0.28. (a) How fast is the crate moving as it reaches the bottom of the ramp? (b) How far will it subsequently slide across the factory floor?

17. Two snow-covered peaks are at elevations of 850 m and 750 m above the valley between them. A ski run extends from the top of the higher peak to the top of the lower one, with a total length of 3.2 km and average slope of 30°. (a) A skier starts from rest on the higher peak. At what speed will he arrive at the top of the lower peak if he goes as fast as possible, never trying to slow down? Ignore friction. (b) How large a coefficient of friction with the snow could be tolerated without preventing him from reaching the lower peak?

18. A 4.0-kg block starts up a 30° incline with 128 J of kinetic energy. How far will it slide up the plane if the coefficient of friction is 0.30?

19. A small particle slides along a track with elevated ends and a flat central part, as shown in Fig. 8–30. The flat part has a length $l = 2.0$ m. The curved portions of the track are frictionless. For the flat part the coefficient of kinetic friction is $\mu_k = 0.20$. The particle is released at point A, a height $h = 1.0$ m above the flat part of the track. Where does the particle finally come to rest?

Figure 8–30   Problem 19.

20. A block of mass m is projected up an inclined plane with an initial speed v_i. It reaches a maximum height h and then slides back down the plane, its speed at the bottom of the plane being v_f. The coefficients of friction μ_s and μ_k are not known. The angle of the inclined plane is θ. (a) Express h as a function of v_i, v_f, and g. (b) Express μ_k as a function of θ, v_i, and v_f. (c) What limits can be put on μ_s?

***21.** The cable of a 4000-lb elevator in Fig. 8–31 snaps when the elevator is at rest at the first floor so that the bottom is a distance $d = 12$ ft above a cushioning spring whose spring constant is $k = 10,000$ lb/ft. A safety device clamps the guide rails so that a constant friction force of 1000 lb opposes the motion of the elevator. (a) Find the speed of the elevator just before it hits the

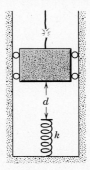

Figure 8–31   Problem 21.

spring. (b) Find the distance x that the spring is compressed. (c) Find the distance that the elevator will "bounce" back up the shaft. (d) Using the conservation of energy principle, find approximately the total distance that the elevator will move before coming to rest. Why is the answer not exact?

*22. While a 1700-kg automobile is moving at a constant speed of 15 m/s, the motor supplies 16 kW of power to overcome friction, wind resistance, etc. (a) What is the effective retarding force associated with all the frictional forces combined? (b) What power must the motor supply if the car is to move up an 8% grade (8 m vertically for each 100 m horizontally) at 15 m/s? (c) At what downgrade, expressed in percentage terms, would the car coast at 15 m/s?

Section 8-9   Mass and Energy

23. An electron has a total relativistic energy of 5.812×10^{-13} J. (a) What is its speed, in terms of the speed of light? (b) What is the ratio of its kinetic energy computed from the (approximate) classical formula to its kinetic energy computed from the correct relativistic formula?

24. The United States generated 2.31×10^{12} kW·h of electrical energy in 1983. How many kilograms of matter have to be destroyed to generate this amount of energy?

25. An aspirin tablet has a mass of 320 mg. For how many miles would the energy equivalent of this mass, in the form of gasoline, power a car? Assume 30 mi/gal and a heat of combustion of gasoline of 130 MJ/gal. Express your answer in terms of the equatorial circumference of the earth.

26. A spaceship, rest mass 2.0×10^4 kg, is to be accelerated from rest to a speed of $0.99c$. If energy costs one cent per kilowatt-hour, what will be the cost of the required energy? Express your answer in units of GNP83, the United States gross national product for 1983 ($= \$3.30 \times 10^{12}$).

ADDITIONAL PROBLEMS

27. A 5.0-kg mortar shell is fired upward at an angle of 34° with an initial speed of 100 m/s. What is its potential energy at the top of its trajectory? Neglect air resistance and assume the zero of potential energy to be at the ground level.

28. A nuclear power plant in Oregon supplies 1030 MW of useful power steadily for a year. In addition, 2100 MW of power is discharged as thermal energy to the Columbia river. How much mass is converted to energy at this plant in one year?

29. A particle moving along the x-axis is acted on by a force $F(x) = kx$, in which k is a positive constant. (a) Find the potential energy function $U(x)$ for this force. (Assume that $x = 0$ is the reference position at which the potential energy is taken as zero.) (b) Show that the motion is just what you would expect if the position $x = 0$ is a point of unstable equilibrium.

30. A volcanic ash flow is moving across horizontal ground when it encounters a 10° upslope. It is observed to travel 920 m on the upslope before coming to rest. The volcanic ash contains trapped gas, so the force of friction with the ground is very small and can be ignored. At what speed was the ash flow moving just before encountering the upslope?

31. One end of a vertical spring is fastened to the ceiling. A weight is attached to the other end and slowly lowered to its equilibrium position. Show that the loss of gravitational potential energy of the weight equals one-half the gain in elastic potential energy. (Why are these two quantities not equal?)

32. A 1.0-kg object is acted upon by a conservative force given by $F = -3.0x - 5.0x^2$, where F is in newtons if x is in meters. (a) What is the potential energy of the object at $x = 2.0$ m? (b) If the object has a speed of 4.0 m/s in the negative x-direction when it is at $x = 5.0$ m, describe the subsequent motion.

33. A 50-g ball is thrown from a window with an initial velocity of 8.0 m/s at an angle of 30° above the horizontal. Using energy methods determine (a) the kinetic energy of the ball at the top of its flight and (b) its speed when it is 3.0 m below the window.

34. How much matter must be converted into energy to accelerate a 1820-ton spaceship from rest to one-tenth of the speed of light?

35. An ideal massless spring can be compressed 2.0 cm by a force of 270 N. A block whose mass is 12 kg is released from rest at the top of an inclined plane as shown in Fig. 8-32, the angle of the plane being 30°. The block comes to rest momentarily after it has compressed this spring by 5.5 cm. (a) How far has the block moved down the plane at this moment? (b) What is the speed of the block just as it touches the spring?

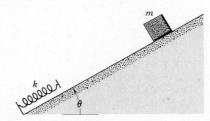

Figure 8-32   Problem 35.

36. A simple pendulum is made by tying a 2.0-kg stone to a string 4.0 m long. The stone is projected perpendicular to the string, away from the ground, with the string at an angle of 60° with the vertical. It is observed to have a speed of 8.0 m/s when it passes its lowest point. (a) What was the speed of the stone at the moment of release? (b) What is the largest angle with the vertical that the string will reach during the stone's motion? (c) Using the lowest point of the swing as the zero of gravitational potential energy, what is the total mechanical energy of the system?

37. Two children are playing a game in which they try to hit a small box on the floor with a marble fired from a spring-loaded gun that is mounted on a table. The target box is 2.2 m horizon-

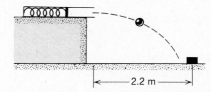

Figure 8-33   Problem 37.

tally from the edge of the table; see Fig. 8–33. Bobby compresses the spring by 1.10 cm, but the marble falls 27 cm short. How far should Rhoda compress the spring to score a hit?

38. Tarzan, who weighs 180 lb, swings from a cliff at the end of a convenient 50-ft grapevine; see Fig. 8–34. From the top of the cliff to the bottom of the swing, Tarzan's center of gravity would fall by 8.5 ft. The grapevine has a breaking strength of 250 lb. Will the grapevine break?

Figure 8–34   Problem 38.

39. A light rigid rod of length *l* has a ball with mass *m* attached to its end, forming a simple pendulum. It is inverted and then released. What are (*a*) its speed at the lowest point and (*b*) the tension in the suspension at that instant? (*c*) The same pendulum is next put in a horizontal position and released from rest. At what angle from the vertical will the tension in the rod just equal the weight of the ball in magnitude?

40. A small block of mass *m* slides along the frictionless loop-the-loop track shown in Fig. 8–35. (*a*) The block is released from rest at point *P*. What is the resultant force acting on it at point *Q*? (*b*) At what height above the bottom of the loop should the block be released so that it is on the verge of losing contact with the track at the top of the loop?

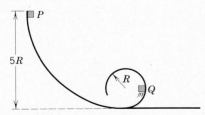

Figure 8–35   Problem 40.

41. The spring of a spring gun has a force constant of 4.0 lb/in. When the gun is inclined at an angle of 30°, a 2.0-oz ball is projected to a height of 6.0 ft above the muzzle of the gun. By how much must the spring have been compressed initially?

42. The magnitude of the force of attraction between the positively charged nucleus and the negatively charged electron in the hydrogen atom is given by

$$F = k\frac{e^2}{r^2},$$

where *e* is the charge of the electron, *k* is a constant, and *r* is the separation between electron and nucleus. Assume that the nucleus is fixed. The electron, initially moving in a circle of radius r_1 about the nucleus, jumps suddenly into a circular orbit of smaller radius r_2; see Fig. 8–36. (*a*) Calculate the change in kinetic energy of the electron, using Newton's second law. (*b*) Using the relation between force and potential energy, calculate the change in potential energy of the atom. (*c*) Show by how much the total energy of the atom has changed in this process. (This energy is often given off in the form of radiation.)

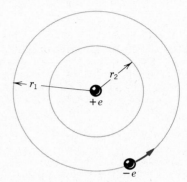

Figure 8–36   Problem 42.

43. A 2.5-kg block collides with a horizontal massless spring whose force constant is 320 N/m; see Fig. 8–37. The block compresses the spring a maximum of 8.5 cm from its rest position. How fast was the block going when it hit the spring? The coefficient of kinetic friction between the block and the horizontal surface is 0.25.

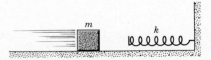

Figure 8–37   Problem 43.

44. A projectile whose mass is 9.4 kg is fired vertically upward. On its upward flight an energy of 68 kJ is dissipated because of air resistance. How much higher would the projectile have gone if air resistance had been made negligible (by streamlining the projectile, for example)?

45. A stone of weight *w* is thrown vertically upward into the air with an initial speed v_0. A constant force *f* due to air resistance acts on the stone throughout its flight. (*a*) Show that the maximum height reached by the stone is

$$h = \frac{v_0^2}{2g(1 + f/w)}.$$

(*b*) Show that the speed of the stone upon impact with the ground is

$$u = v_0 \left(\frac{w-f}{w+f}\right)^{1/2}.$$

46. A block is moving up a 40° inclined plane. At a point 1.8 ft from the bottom of the plane (measured along the plane), it has a speed of 4.5 ft/s. The coefficient of kinetic friction between block and plane is 0.15. (a) How much farther up the plane will the block move? (b) How fast will it be going when it slides back to the bottom of the plane?

47. The sun radiates energy at the rate of 4×10^{26} W. How many "tons of sunlight" does the earth intercept in one day?

48. A 40-lb body is pushed up a frictionless 20° inclined plane 10 ft long by a horizontal force **F**. (a) If the speed at the bottom is 2.0 ft/s and at the top is 10 ft/s, how much work is done by **F**? (b) Suppose that the plane is not frictionless and that $\mu_k = 0.15$. What work will this same force do? (c) How far up the plane does the body go?

49. A very light rigid rod whose length is L has a ball of mass m attached to one end (Fig. 8–38). The other end is pivoted frictionlessly in such a way that the ball moves in a vertical circle. The system is launched from the horizontal position A with downward initial velocity $\mathbf{v}_0$. The ball just reaches point D and then stops. (a) Derive an expression for v_0 in terms of L, m, and g. (b) What is the tension in the rod when the ball is at B? (c) A little sand is placed on the pivot, after which the ball just reaches C when launched from A with the same speed as before. How much work is done by friction during this motion? (d) How much total work is done by friction before the ball finally comes to rest at B after oscillating back and forth several times?

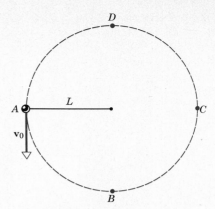

Figure 8–38 Problem 49.

*50. A boy is seated on the top of a hemispherical mound of ice (Fig. 8–39). He is given a very small push and starts sliding down the ice. Show that he leaves the ice at a point whose height is $2R/3$ if the ice is frictionless.

Figure 8–39 Problem 50.

9
The Dynamics
of Systems
of Particles

9–1 Center of Mass

Center of mass

So far we have treated objects as though they were particles, having mass but no size. In translational motion each point on a body experiences the same displacement as any other point as time goes on, so that the motion of one particle represents the motion of the whole body. But even when a body rotates or vibrates as it moves, there is one point on the body, called the *center of mass*, that moves in the same way that a single particle subject to the same external forces would move. Figure 9–1 shows the simple parabolic motion of the center of mass of an Indian club thrown from one performer to another; no other point in the club moves in such a simple way. Note that, if the club were moving in pure translation, then every point in it would experience the same displacements as does the center of mass in Fig. 9–1. For this reason the motion of the center of mass of a body is called the translational motion of the body.

When the system with which we deal is not a rigid body like the Indian club, a center of mass, whose motion can also be described in a relatively simple way, can be assigned, even though the particles that make up the system may be changing their positions with respect to each other in a relatively complicated way as the motion proceeds. In this section we define the center of mass and show how to calculate its position. In the next section we discuss the properties that make this concept useful for describing the motion of extended objects or systems of particles.

Consider first the simple case of a system of two particles m_1 and m_2 at distances x_1 and x_2, respectively, from some origin O as in Fig. 9–2. We define a point C, the center of mass of the system, at a distance x_{cm} from the origin O, where x_{cm} is defined by

$$x_{cm} = \frac{m_1 x_1 + m_2 x_2}{m_1 + m_2} \qquad (9–1)$$

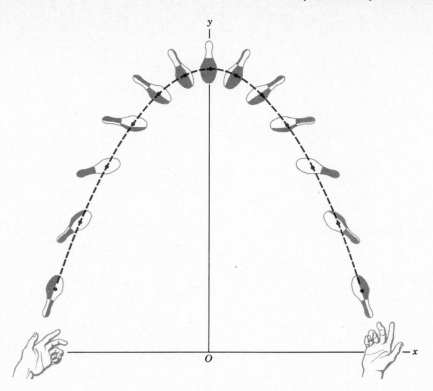

Figure 9-1 An Indian club is thrown from one performer to another. Even though it rotates and spins around its axis, as shown, there is one point on its axis, the center of mass, that follows a simple parabolic path.

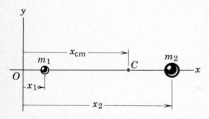

Figure 9-2 The center of mass of the two particles m_1 and m_2 lies on the line joining m_1 and m_2 at C, a distance x_{cm} from the origin.

The center of mass lies between the two particles and on the line joining them. Note that if all of the mass is concentrated at x_1 ($m_2 = 0$), then $x_{cm} = x_1$. If the masses are equal then the center of mass is equidistant from the two particles.

Although the usefulness of this definition will not become clear until later, we can get some feeling for the concept at this point. Figure 9–2 shows two particles and the center of mass at point C. Notice from Eq. 9–1 that the product of the total mass M ($= m_1 + m_2$) and the center of mass distance is the sum of similar products for each mass element of the system; that is,

$$Mx_{cm} = m_1 x_1 + m_2 x_2.$$

When we extend the definition to systems having more particles we shall simply add the products for the additional particles. Thus, knowing the center of mass will be useful where the mechanical equations involve adding for each mass element in a system the mass times the distance from some point (such as from an axis of rotation). The operation is then simplified greatly. Why such products arise will be discussed later.

If we have n particles, $m_1, m_2, \ldots, m_n$, *along a straight line,* by definition the center of mass of these particles relative to some origin is located at

$$x_{cm} = \frac{m_1 x_1 + m_2 x_2 + \cdots + m_n x_n}{m_1 + m_2 + \cdots + m_n} = \frac{\Sigma\, m_i x_i}{\Sigma\, m_i}, \qquad (9\text{--}2a)$$

where $x_1, x_2, \ldots, x_n$ are the distances of the masses from the origin from which x_{cm} is measured. The symbol Σ represents a summation operation, in this case over all n particles. The sum

$$\Sigma\, m_i = M$$

is the total mass of the system. We can then rewrite Eq. 9–2a in the form

$$x_{cm} = \frac{\Sigma m_i x_i}{M} \qquad (9\text{--}2b)$$

For a number of particles distributed in three dimensions, the center of mass is at x_{cm}, y_{cm}, z_{cm}, where

Center of mass—scalar form

$$x_{cm} = \frac{1}{M} \sum m_i x_i, \qquad y_{cm} = \frac{1}{M} \sum m_i y_i, \qquad z_{cm} = \frac{1}{M} \sum m_i z_i. \qquad (9\text{--}3a)$$

In vector notation each particle in the system can be described by a position vector $\mathbf{r}_i$ in a particular reference frame and the center of mass can be located by a position vector $\mathbf{r}_{cm}$. These vectors are related to x_i, y_i, z_i and x_{cm}, y_{cm}, z_{cm} in Eq. 9–3a by

$$\mathbf{r}_i = \mathbf{i} x_i + \mathbf{j} y_i + \mathbf{k} z_i$$

and

$$\mathbf{r}_{cm} = \mathbf{i} x_{cm} + \mathbf{j} y_{cm} + \mathbf{k} z_{cm}.$$

Thus the three scalar equations of Eq. 9–3a can be replaced by a single vector equation

Center of mass—vector form

$$\mathbf{r}_{cm} = \frac{1}{M} \sum m_i \mathbf{r}_i, \qquad (9\text{--}3b)$$

in which the sum is a vector sum. You can prove that Eq. 9–3b is true by substituting into it the expressions given for $\mathbf{r}_i$ and $\mathbf{r}_{cm}$ just above.

Equations 9–3 treat the most general case for a collection of particles. Equations 9–1 and 9–2 are special instances of this one. The location of the center of mass is independent of the reference frame used to locate it. The center of mass of a system of particles depends only on the masses of the particles and the positions of the particles relative to one another.

A rigid body, such as the Indian club, can be thought of as a system of closely packed particles; this would allow us to find a center of mass for it from the particle point of view (Eqs. 9–3). The number of particles (atoms) in the body is so large and their spacing so small, however, that we can treat such a body as though it had a continuous distribution of mass. To obtain the expression for the center of mass of a continuous body, we begin by subdividing the body into n small elements of mass Δm_i located approximately at the points x_i, y_i, z_i. The coordinates of the center of mass are then given approximately, by

$$x_{cm} \cong \frac{\sum \Delta m_i x_i}{\sum \Delta m_i}, \qquad y_{cm} \cong \frac{\sum \Delta m_i y_i}{\sum \Delta m_i}, \qquad z_{cm} \cong \frac{\sum \Delta m_i z_i}{\sum \Delta m_i}.$$

Now let the elements of mass be further subdivided so that the number of elements n tends to infinity. The points x_i, y_i, z_i will locate the mass elements more precisely as n is increased. We can now give the coordinates of the center of mass precisely as

Center of mass— integral form

$$x_{cm} = \lim_{\Delta m_i \to 0} \frac{\sum \Delta m_i x_i}{\sum \Delta m_i} = \frac{\int x \, dm}{\int dm} = \frac{1}{M} \int x \, dm,$$

$$y_{cm} = \lim_{\Delta m_i \to 0} \frac{\sum \Delta m_i y_i}{\sum \Delta m_i} = \frac{\int y \, dm}{\int dm} = \frac{1}{M} \int y \, dm, \qquad (9\text{--}4a)$$

$$z_{cm} = \lim_{\Delta m_i \to 0} \frac{\sum \Delta m_i z_i}{\sum \Delta m_i} = \frac{\int z \, dm}{\int dm} = \frac{1}{M} \int z \, dm.$$

In these expressions dm is the differential element of mass at the point x, y, z, and $\int dm$ equals M, where M is the total mass of the body. For a continuous body the summation of Eq. 9–3a is replaced by the integration of Eq. 9–4a.

The vector expression that is equivalent to the three scalar expressions of Eq. 9–4a is

$$\mathbf{r}_{cm} = \frac{1}{M} \int \mathbf{r} \; dm. \qquad (9\text{--}4b)$$

As before, the summation of Eq. 9–3b has been replaced by an integration.

Often we deal with homogeneous objects having a point, a line, or a plane of symmetry. Then the center of mass will lie at the point, on the line, or in the plane of symmetry. For example, the center of mass of a homogeneous sphere (which has a point of symmetry) will be at the center of the sphere, the center of mass of a cone (which has a line of symmetry) will be on the axis of the cone, and so on. We can understand that this is so because, from symmetry, an origin of coordinates can be found such that $\int \mathbf{r} \; dm$ in Eq. 9–4b is zero (at the center of the sphere, somewhere along the axis of the cone, etc.). From this it follows that $\mathbf{r}_{cm} = 0$ for this point, which identifies it as the center of mass.

Example 1   Locate the center of mass of three particles of mass $m_1 = 1.0$ kg, $m_2 = 2.0$ kg, and $m_3 = 3.0$ kg at the corners of an equilateral triangle 1.0 m on a side.

Choose the x-axis along one side of the triangle as shown in Fig. 9–3. Note that the coordinates of m_3 are then ($\frac{1}{2}$ m, $\sqrt{3}/2$ m). Then,

$$x_{cm} = \frac{\sum m_i x_i}{\sum m_i}$$

$$= \frac{(1.0 \text{ kg})(0) + (2.0 \text{ kg})(1.0 \text{ m}) + (3.0 \text{ kg})(\frac{1}{2} \text{ m})}{(1.0 + 2.0 + 3.0) \text{ kg}}$$

$$= \frac{7}{12} \text{ m} = 0.58 \text{ m},$$

$$y_{cm} = \frac{\sum m_i y_i}{\sum m_i}$$

$$= \frac{(1.0 \text{ kg})(0) + (2.0 \text{ kg})(0) + (3.0 \text{ kg})(\sqrt{3}/2 \text{ m})}{(1.0 + 2.0 + 3.0) \text{ kg}}$$

$$= \frac{\sqrt{3}}{4} \text{ m} = 0.43 \text{ m}.$$

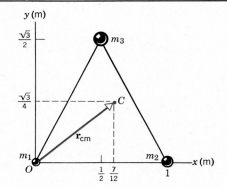

Figure 9–3   Example 1. Finding the center of mass C of three particles of unequal mass forming an equilateral triangle.

The center of mass C is shown in the figure. Why is it not at the geometric center of the triangle?

Example 2   Find the center of mass of the triangular plate of Fig. 9–4.

If we can divide a body into parts such that the center of mass of each part is known, we can usually find the center of mass of the body quite easily. The triangular plate may be divided into narrow strips parallel to one side. The center of mass of each strip lies on the line that joins the middle of that side to the opposite vertex. But we can divide up the triangle in three different ways, using this process for each of three sides. Hence the center of mass lies at the intersection of the three lines that join the middle of each side with the opposite vertices. This is the only point that is common to the three lines.

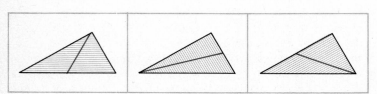

Figure 9–4   Example 2. Finding the center of mass C of a triangular plate.

9–2 Motion of the Center of Mass

Now we can discuss the physical importance of the center-of-mass concept. Consider the motion of a group of particles whose masses are $m_1, m_2, \ldots, m_n$ and whose total mass is M. For the time being we shall assume that mass neither enters nor leaves the system so that the total mass M of the system remains constant with time. In Section 9–7 we shall consider systems in which M is not constant; a familiar example is a rocket, which expels hot gases as its fuel burns, thus reducing its mass.

From Eq. 9–3b we have, for our fixed system of particles.

$$M\mathbf{r}_{cm} = m_1\mathbf{r}_1 + m_2\mathbf{r}_2 + \cdots + m_n\mathbf{r}_n,$$

where $\mathbf{r}_{cm}$ is the position vector identifying the center of mass in a particular reference frame. Differentiating this equation with respect to time, we obtain

$$M\mathbf{v}_{cm} = m_1\mathbf{v}_1 + m_2\mathbf{v}_2 + \cdots + m_n\mathbf{v}_n, \tag{9–5}$$

where $\mathbf{v}_1$ is the velocity of the first particle, and so on, and $d\mathbf{r}_{cm}/dt \ (=\mathbf{v}_{cm})$ is the velocity of the center of mass.

Differentiating Eq. 9–5 with respect to time, we obtain

$$M\frac{d\mathbf{v}_{cm}}{dt} = m_1\mathbf{a}_1 + m_2\mathbf{a}_2 + \cdots + m_n\mathbf{a}_n, \tag{9–6}$$

where $\mathbf{a}_1$ is the acceleration of the first particle, and so on, and $d\mathbf{v}_{cm}/dt \ (=\mathbf{a}_{cm})$ is the acceleration of the center of mass of the system. Now, from Newton's second law, the force $\mathbf{F}_1$ acting on the first particle is given by $\mathbf{F}_1 = m_1\mathbf{a}_1$. Likewise, $\mathbf{F}_2 = m_2\mathbf{a}_2$, and so on. We can then write Eq. 9–6 as

$$M\mathbf{a}_{cm} = \mathbf{F}_1 + \mathbf{F}_2 + \cdots + \mathbf{F}_n. \tag{9–7}$$

Hence *the total mass of the group of particles times the acceleration of its center of mass is equal to the vector sum of all the forces acting on the group of particles.*

Among all these forces will be internal forces exerted by the particles on each other. However, from Newton's third law, these internal forces will occur in equal and opposite pairs, so that they contribute nothing to the sum. Hence the internal forces can be removed from the problem. The right-hand side of Eq. 9–7 represents the sum of only the external forces acting on all the particles. We can then rewrite Eq. 9–7 as simply

$$M\mathbf{a}_{cm} = \mathbf{F}_{ext}. \tag{9–8}$$

Newton's second law for a system of particles

This states that *the center of mass of a system of particles moves as though all the mass of the system were concentrated at the center of mass and all the external forces were applied at that point.* From this perspective we can now see that Eq. 5–1 ($\mathbf{F} = m\mathbf{a}$) can be described as Newton's second law for a single particle, Eq. 9–8 being Newton's second law for a system of particles.

Notice that we obtain this simple result without specifying the nature of the system of particles. The system can be a rigid body in which the particles are in fixed positions with respect to one another, or it can be a collection of particles in which there may be all kinds of internal motion.

We have now found how to describe the translation motion of a system of particles and how to describe the translational motion of a body that may be rotating as well. In this chapter and the next we apply this result to the linear motion of a system of particles. In later chapters we shall see how it simplifies the analysis of rotational motion.

Example 3 Consider three particles of different masses acted on by external forces, as shown in Fig. 9–5. Find the acceleration of the center of mass of the system.

First we find the coordinates of the center of mass. From Eq. 9–3a,

$$x_{\text{cm}} = \frac{(8.0 \times 4) + (4.0 \times -2) + (4.0 \times 1)}{16}\ \text{m} = 1.75\ \text{m}.$$

$$y_{\text{cm}} = \frac{(8.0 \times 1) + (4.0 \times 2) + (4.0 \times -3)}{16}\ \text{m} = 0.25\ \text{m}.$$

These are shown as C in Fig. 9–5.

To obtain the acceleration of the center of mass, we first determine the resultant external force acting on the system consisting of the three particles. The x-component of this force is

$$F_x = 14\ \text{N} - 6.0\ \text{N} = 8.0\ \text{N},$$

and the y-component is

$$F_y = 16\ \text{N}.$$

Hence the resultant external force has a magnitude

$$F = \sqrt{(8.0)^2 + (16)^2}\ \text{N} = 18\ \text{N},$$

and makes an angle θ with the x-axis given by

$$\tan \theta = \frac{16\ \text{N}}{8.0\ \text{N}} = 2.0 \quad \text{or} \quad \theta = 63°.$$

Then, from Eq. 9–8, the magnitude of the acceleration of the center of mass is

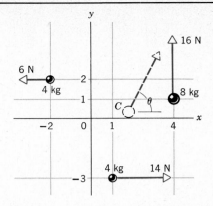

Figure 9–5 Example 3. Finding the motion of the center of mass of three particles, each subjected to a different force. The forces all lie in the plane defined by the particles. The distances indicated along the axes are in meters.

$$a_{\text{cm}} = \frac{F}{M} = \frac{18\ \text{N}}{16\ \text{kg}} = 1.1\ \text{m/s}^2,$$

making an angle of 63° with the x-axis.

Although the three particles will change their relative positions as time goes on, the center of mass will move, as shown, with this constant acceleration.

9–3 Internal Work and Kinetic Energy

In Section 7–5 we derived the work-energy theorem for a system that can be represented as a single particle. The theorem (Eq. 7–14) is

$$W = K - K_0 = \Delta K, \qquad (9\text{–}9)$$

Work-energy theorem for a single particle

and it tells us that the work W done on the particle by the resultant force acting on it is equal to ΔK, the change in kinetic energy of the particle.

There can be situations, however, in which although there is a resultant external force that does no work ($W = 0$), there *is* a change in kinetic energy ($\Delta K \neq 0$). It would seem that the work-energy theorem does not hold. Not so, however, because this theorem was derived for a force acting on a single particle and the situations to which we refer involve systems of many particles.

Figure 9–6 suggests an example. An astronaut clings by one hand to a support bracket on the outside of a spaceship that is floating in gravity-free space. We assume that the spaceship is so very much more massive than the astronaut that we can take it as a convenient inertial reference frame.

The astronaut gives a sharp push against the bracket, releases his grip, and finds himself moving away from the spaceship with a separation velocity **v**. While he is pushing on the bracket the spaceship exerts an external reaction force $\mathbf{F}_{\text{ext}}$ on him. This force does no work on him because, in our chosen reference frame, its point of application does not move. Thus $W = 0$ in Eq. 9–9. Yet the astronaut has acquired kinetic energy ($\Delta K > 0$ in Eq. 9–9). Where does it come from?

The answer is that the energy is provided by *internal work* done by the astronaut as he straightens his arm and pushes himself away from the spaceship. He

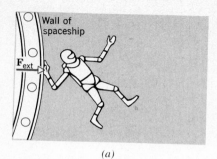

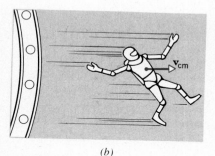

Figure 9–6   (a) An astronaut thrusts himself away from the side of a massive space ship. The ship exerts a reaction force $\mathbf{F}_{ext}$ on him. (b) The astronaut has released his grip and coasts away at constant velocity $\mathbf{v}_{cm}$.

Center-of-mass work

Work-energy theorem for a system of particles

would be very much aware of this muscular exertion. Equation 9–9 does not apply to the astronaut because it was derived for a single particle, for which the concept of internal work has no meaning. The astronaut must be treated as a system of particles, and the role of internal work is central.

Although we cannot easily analyze the internal work in the case of Fig. 9–6 in terms of detailed forces and displacements, we can still treat the problem quantitatively in terms of Newton's second law for a system of particles. In terms of magnitudes we can write for this law

$$F_{ext} = Ma_{cm}.$$

Let us choose the direction of $\mathbf{F}_{ext}$ in Fig. 9-6 as a convenient x-axis. Let the center of mass of the astronaut—or of any other system of particles—move a distance dx_{cm} along this axis. If we multiply each side of the above equation by this quantity we obtain

$$F_{ext}\, dx_{cm} = Ma_{cm}\, dx_{cm} = M\frac{dv_{cm}}{dt}\, dx_{cm}$$

$$= M\frac{dx_{cm}}{dt}\, dv_{cm} = Mv_{cm}\, dv_{cm}.$$

Let s_{cm} be the displacement of the center of mass of the system while the force $\mathbf{F}_{ext}$ acts. Integrating the above yields

$$\int_0^{s_{cm}} F_{ext}\, dx_{cm} = (\tfrac{1}{2}Mv_{cm}{}^2) - (\tfrac{1}{2}Mv_{cm_0}{}^2). \qquad (9-10a)$$

Let us represent the integral by the symbol W_{cm}; that is,

$$W_{cm} = \int_0^{s_{cm}} F_{ext}\, dx_{cm}. \qquad (9-10b)$$

We can then write Eq. 9–10a as

$$W_{cm} = K_{cm} - K_{cm_0} = \Delta K_{cm}. \qquad (9-10c)$$

in which K_{cm} $(=\tfrac{1}{2}Mv_{cm}{}^2)$ is the kinetic energy associated with the motion of the center of mass of the system. Kinetic energies associated with internal motions are *not* included in K_{cm}. Although we derived Eq. 9–10c by considering motion in one dimension only (the x-axis), it remains true for the general case of three-dimensional motion.

Examination of Eq. 9–10b shows that W_{cm} is the work that *would* be done by the external force F_{ext} *if* that force were applied at the center of mass of the system and moved with it. Note that W_{cm} is not the work actually done by the external force; for the case of Fig. 9–6 and similar cases this work is zero. We can call W_{cm} the *center-of-mass work*.

Equation 9–10c looks like the single-particle work-energy theorem of Eq. 9–9, but it is not. It is an extension of that equation to multiple-particle systems, in the same way that the relation $\mathbf{F}_{ext} = M\mathbf{a}_{cm}$ (Eq. 9–8) is an extension of the single-particle relation $\mathbf{F} = m\mathbf{a}$. We can use Eq. 9–10c to solve problems involving internal work in many-particle systems, even though we cannot analyze the details of the internal work.

Example 4   The suited astronaut of Fig. 9–6 has a mass M of 130 kg and, after pushing off from the space ship, finds himself moving with a constant velocity $\mathbf{v}_{cm}$ whose magnitude is 1.3 m/s. When he breaks contact with the ship, his center of mass has moved a distance s_{cm} of 12 cm. (a) What force F_{ext}, assumed constant, does the space ship exert on him while he is in contact with it? (b) How much internal work is done by the astronaut?

(a) For W_{cm}, Eq. 9–10b gives, for motion along the x-axis,

$$W_{cm} = \int_0^{s_{cm}} F_{ext}\, dx_{cm} = F_{ext} \int_0^{s_{cm}} dx_{cm} = +F_{ext}s_{cm},$$

where, because we have assumed F_{ext} to be constant, we have moved it outside the integral sign. Applying the theorem of Eq. 9–10c yields

$$W_{cm} = F_{ext}s_{cm} = K_{cm} - K_{cm_0} = \tfrac{1}{2}Mv_{cm}^2 - 0,$$

or

$$F_{ext} = \frac{Mv_{cm}^2}{2s_{cm}} = \frac{(130\text{ kg})(1.3\text{ m/s})^2}{(2)(0.12\text{ m})}$$
$$= 920\text{ N }(=210\text{ lb}).$$

This is a substantial push.

(b) Let us apply the conservation-of-energy principle to the situation of Fig. 9–6. No external work is done. We further assume that kinetic energy associated with the internal motions of the drifting astronaut (moving his legs, etc.) remain unchanged during the push-off process. Thus all of the kinetic energy K_{cm} of the astronaut must have its origin in the internal work W_{int} that he does, or

$$W_{int} = \tfrac{1}{2}Mv_{cm}^2 = \tfrac{1}{2}(130\text{ kg})(1.3\text{ m/s})^2 = +110\text{ J}.$$

Note, from Eq. 9–10c, that under the assumptions we have made we can also put $W_{int} = W_{cm}$. We leave to you the problem of getting the astronaut back to his space ship!

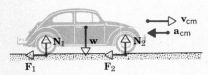

Figure 9–7   An automobile braking to a halt. An acceleration $\mathbf{a}_{cm}$, pointing to the left, is produced by the external horizontal force $\mathbf{F}_{ext}$ exerted by the road on the bottoms of the tires. Two of the components of $\mathbf{F}_{ext}$ are shown.

An accelerating or decelerating (braking) automobile provides another example of a system in which internal work plays a central role. The internal work may be positive—as when expanding gases in the cylinders of the engine provide power to turn the wheels during acceleration—or negative—as when friction-induced thermal energy heats up the braking surfaces when the brakes are applied.

Figure 9–7 shows the external forces acting on a car during braking. The four normal forces, exerted vertically upward by the road on the four tires, just cancel the weight and play no role in the horizontal motion of the car.* During accelerating or braking the road also exerts a horizontal force on the bottom of each tire. $\mathbf{F}_{ext}$, the resultant of these four forces, points opposite to the velocity of the car during braking (see Fig. 9–7) but in the same direction as the velocity during acceleration.

We assert that, just as for the astronaut of Fig. 9–6, the resultant external force $\mathbf{F}_{ext}$ acting on the car does no work on it. This is because the bottoms of the rotating tires at their points of contact with the roadway are momentarily at rest (always assuming no slipping or skidding). The points of application of the four forces that make up $\mathbf{F}_{ext}$ undergo no displacement relative to the roadway. In the absence of skidding the conditions of static friction prevail.

An accelerating car is very much like the astronaut of Fig. 9–6. Both must be treated as systems of particles. Positive internal work is done in each, by the engine or by muscles. An external force acts on each, exerted either by the space ship or by the roadway. In each case the force accelerates the center of mass of the system but does no external work on it.

Example 5 *A braking automobile.*   An automobile has a mass M, with driver, of 1500 kg. It brakes to rest from a speed v_{cm_0} of 80 km/h ($=22$ m/s $=50$ mi/h) over a distance s_{cm} of 180 m. What resultant frictional force F_{ext}, assumed constant, does the road exert on the bottoms of the tires? Assume that the driver manipulates the brakes so that the tires, though at the verge of skidding, never actually do so.

In this case the center of mass work W_{cm} is negative because $\mathbf{F}_{ext}$ and the displacement s_{cm} of the center of mass point in opposite directions. Thus, from Eq. 9–10b,

$$W_{cm} = -\int_0^{s_{cm}} F_{ext}\, dx_{cm} = -F_{ext}s_{cm},$$

where F_{ext} is the magnitude of the external force and the minus sign arises because $\mathbf{F}_{ext}$ and the displacement have opposite directions. Equation 9–10c yields

$$W_{cm} = K_{cm} - K_{cm_0},$$

or

$$-F_{ext}s_{cm} = 0 - \tfrac{1}{2}Mv_{cm_0}^2,$$

---

* The normal forces permit the horizontal frictional forces to act, of course.

which yields

$$F_{\text{ext}} = \frac{Mv_{\text{cmo}}^2}{2s} = \frac{(1500 \text{ kg})(22 \text{ m/s})^2}{(2)(180 \text{ m})}$$
$$= 2000 \text{ N} \ (=450 \text{ lb}).$$

Note how different the situation is if the driver simply jams on the brakes and skids to a halt with the wheels locked. Here the car behaves like a block projected onto a rough horizontal plane and brought to rest by frictional forces. The loss of kinetic energy is now associated with *external* frictional work done on the car by the road. The skid marks on the road and the elevated temperatures of the tires (and of the road) attest to the fact that the initial kinetic energy of the car appears at these sites and not, as previously, as thermal energy in the now-locked braking system. The car can be represented under these conditions as a single particle and the work-energy theorem in the original form of Eq. 9–9 applies.

In Example 2 of Chapter 6 we also analyzed a braking automobile but there from the point of view of force and acceleration rather than, as here, work and energy. Compare these two examples carefully.

9-4 Linear Momentum of a Particle

The *linear momentum* of a particle is a vector **p** defined as the product of its mass m and its velocity **v**. That is,

Linear momentum

$$\mathbf{p} = m\mathbf{v}. \tag{9-11}$$

Momentum, being the product of a scalar by a vector, is itself a vector. Because it is proportional to **v**, the momentum **p** of a particular particle depends on the reference frame of the observer; we must always specify this frame.

Newton expressed the second law of motion in terms of momentum (which he called "quantity of motion"). Expressed in modern terminology Newton's second law reads: *The rate of change of momentum of a body is proportional to the resultant force acting on the body and is in the direction of that force.* In equation form this becomes

Momentum form of Newton's second law

$$\mathbf{F} = \frac{d\mathbf{p}}{dt}. \tag{9-12}$$

If our system is a single particle of (constant) mass m, this formulation of the second law is equivalent to the form $\mathbf{F} = m\mathbf{a}$, which we have used up to now. That is, if m is a constant, then

$$\mathbf{F} = \frac{d\mathbf{p}}{dt} = \frac{d}{dt}(m\mathbf{v}) = m\frac{d\mathbf{v}}{dt} = m\mathbf{a}.$$

The relations $\mathbf{F} = m\mathbf{a}$ and $\mathbf{F} = d\mathbf{p}/dt$ for single particles are completely equivalent in classical mechanics.

In relativity theory the second law for a single particle in the form $\mathbf{F} = m\mathbf{a}$ is not valid. However, it turns out that Newton's second law in the form $\mathbf{F} = d\mathbf{p}/dt$ is still a valid law if the momentum **p** for a single particle is defined not as $m_0\mathbf{v}$ but as

Relativistic momentum

$$\mathbf{p} = \frac{m_0\mathbf{v}}{\sqrt{1 - v^2/c^2}}. \tag{9-13}$$

This result suggested a new definition of mass (compare Eqs. 9–11 and 9–13),

Relativistic mass

$$m = \frac{m_0}{\sqrt{1 - v^2/c^2}}, \tag{8-19}$$

so that the momentum could still be written as $\mathbf{p} = m\mathbf{v}$; see Section 8–9. In this equation v is the speed of the particle, c is the speed of light, and m_0 is the "rest mass'' of the body (its mass when $v = 0$). From this definition we must expect the mass of a particle to increase with its speed. Subatomic particles, such as electrons, protons, and so on, may acquire speeds comparable to the speed of light. This concept can be put to a direct test in such cases because the increase in mass over the rest mass for such particles is large enough to measure accurately. Results of all such experiments indicate that this effect is real and given exactly by the equation above. (See, for example, Fig. 8–9.)

9–5 Linear Momentum of a System of Particles

Suppose that instead of a single particle we have a system of n particles, with masses m_1, m_2, and so on. We shall continue to assume, as we did in Section 9–2, that no mass enters or leaves the system, so that the mass M $(= \Sigma\, m_i)$ of the system remains constant with time. The particles may interact with each other and external forces may act on them as well. Each particle will have a velocity and a momentum. Particle 1 of mass m_1 and velocity $\mathbf{v}_1$ will have a momentum $\mathbf{p}_1 = m_1\mathbf{v}_1$, for example. The system as a whole will have a total momentum $\mathbf{P}$ in a particular reference frame, which is defined to be simply the vector sum of the momenta of the individual particles in that same frame, or

$$\mathbf{P} = \mathbf{p}_1 + \mathbf{p}_2 + \cdots + \mathbf{p}_n \qquad (9-14)$$
$$= m_1\mathbf{v}_1 + m_2\mathbf{v}_2 + \cdots + m_n\mathbf{v}_n.$$

If we compare this relation with Eq. 9–5 we see at once that

Momentum of a system of particles

$$\mathbf{P} = M\mathbf{v}_{\mathrm{cm}}, \qquad (9-15)$$

which is an equivalent definition for the momentum of a system of particles. In words, Eq. 9–15 states: *The total momentum of a system of particles is equal to the product of the total mass of the system and the velocity of its center of mass.* Interestingly, we cannot write the *kinetic energy* of a system of particles as $\frac{1}{2}Mv_{\mathrm{cm}}^2$; see Question 16.

We have seen (Eq. 9–8) that Newton's second law for a system of particles can be written as

$$\mathbf{F}_{\mathrm{ext}} = M\mathbf{a}_{\mathrm{cm}}, \qquad [9-8]$$

in which $\mathbf{F}_{\mathrm{ext}}$ is the vector sum of all the external forces acting on the system; we recall that the internal forces acting between particles cancel in pairs because of Newton's third law. If we differentiate Eq. 9–15 with respect to time we obtain, for an assumed constant mass M,

$$\frac{d\mathbf{P}}{dt} = M\frac{d\mathbf{v}_{\mathrm{cm}}}{dt} = M\mathbf{a}_{\mathrm{cm}}. \qquad (9-16)$$

Comparison of Eqs. 9–8 and 9–16 allows us to write Newton's second law for a system of particles in the form

Momentum form of Newton's second law for a system of particles

$$\mathbf{F}_{\mathrm{ext}} = \frac{d\mathbf{P}}{dt}. \qquad (9-17)$$

This equation is the generalization of the single-particle equation $\mathbf{F} = d\mathbf{p}/dt$ (Eq. 9–12) to a system of many particles, no mass entering or leaving the system. Equation 9–17 reduces to Eq. 9–12 for the special case of a single particle, there being only external forces on a one-particle system.

9–6   **Conservation of Linear Momentum**

Conservation of linear momentum for a system of particles

Suppose that the sum of the external forces acting on a system is zero. Then, from Eq. 9–17,

$$\frac{d\mathbf{P}}{dt} = 0 \quad \text{or} \quad \mathbf{P} = \text{a constant.}$$

When the resultant external force acting on a system is zero, the total linear momentum of the system remains constant. This simple but quite general result is called *the principle of the conservation of linear momentum.* We shall see that it is applicable to many important situations.

The conservation of linear momentum principle is the second of the great conservation principles that we have met so far, the first being the conservation of energy principle. Later we shall meet several others, among them the conservation of electric charge and of angular momentum. Conservation principles are of theoretical and practical importance in physics because they are simple and universal. They are all cast in the form: While the system is changing there is one aspect of the system that remains unchanged. Different observers, each in his own reference frame, would all agree, if they watched the same changing system, that the conservation laws applied to the system. For the conservation of linear momentum, for example, observers in different reference frames would assign different values of $\mathbf{P}$ to the linear momentum of the system, but each would agree (assuming $\mathbf{F}_{ext} = 0$) that his own value of $\mathbf{P}$ remained unchanged as the particles that make up the system move about and interact with each other.

The total momentum of a system can only be changed by external forces acting on the system. The internal forces, being equal and opposite, produce equal and opposite changes in momentum which cancel each other. For a system of particles

$$\mathbf{p}_1 + \mathbf{p}_2 + \cdots + \mathbf{p}_n = \mathbf{P},$$

so that when the total momentum $\mathbf{P}$ is constant we have

$$\mathbf{p}_1 + \mathbf{p}_2 + \cdots + \mathbf{p}_n = \text{a constant} = \mathbf{P}_0. \qquad (9\text{--}18)$$

The momenta of the individual particles may change, but their sum remains constant if there is no net external force.

Momentum is a vector quantity. Equation 9–18 is therefore equivalent to three scalar equations, one for each coordinate direction. Hence the conservation of linear momentum supplies us with three conditions on the motion of a system to which it applies. The conservation of energy, on the other hand, supplies us with only one condition on the motion of a system to which it applies, because energy is a scalar.

The law of the conservation of linear momentum holds true even in atomic and nuclear physics, although Newtonian mechanics does not. Hence this conservation law must be even more fundamental than the Newtonian principles.

9–7   **Some Applications of the Momentum Principle**

Example 6 *An exploding projectile*   Consider first a problem in which an external force acts on a system of particles. Recall our previous discussion of projectile motion (Chapter 4). Now let us imagine that our projectile explodes while in flight. The path of the projectile is shown in Fig. 9–8. We assume that the air resistance is negligible. The system is the projectile, the earth is our reference frame, and the external force is that of gravity. At the point x_1 the projectile explodes and projectile

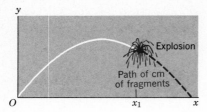

Figure 9-8 Example 6. A projectile, following the usual parabolic trajectory, bursts at x_1. The center of mass of the fragments continues along the same parabolic path.

fragments are blown in all directions. What can we say about the motion of this system thereafter?

The forces of the explosion are all internal forces; they are forces exerted by part of the system on other parts of the system. These forces may change the momenta of all the individual fragments from the values they had when they made up the projectile, but they cannot change the total vector momentum of the system. Only an external force can change the total momentum. The external force, however, is simply that due to gravity. Since a system of particles as a whole moves as though all its mass were concentrated at the center of mass with the external force applied there, the center of mass of the fragments will continue to move in the parabolic trajectory that the unexploded projectile would have followed. The change in the total momentum of the system attributable to gravity is the same whether the projectile explodes or not.

What can you say about the mechanical energy of the system before and after the explosion?

Example 7 Consider now two blocks A and B, of masses m_A and m_B, coupled by a spring and resting on a horizontal frictionless table. Let us pull the blocks apart and stretch the spring, as in Fig. 9-9, and then release the blocks. Describe the subsequent motion.

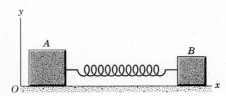

Figure 9-9 Example 7. Two blocks A and B, resting on a frictionless surface, are connected by a spring. If they are held apart and then released, the sum of their momenta remains zero.

If the system consists of the two blocks and spring, then after we have released the blocks there is no net external force acting on the system. We can therefore apply the conservation of linear momentum to the motion. The momentum of the system before the blocks were released was zero in the reference frame shown attached to the table, so the momentum must remain zero thereafter. The total momentum can be zero even though the blocks move because momentum is a vector quantity. One block will have positive momentum (A moves in the

$+x$-direction) and the other block will have negative momentum (B moves in the $-x$-direction). From the conservation of momentum we have

initial momentum = final momentum,

or

$$0 = m_B \mathbf{v}_B + m_A \mathbf{v}_A .$$

Therefore

$$m_B \mathbf{v}_B = - m_A \mathbf{v}_A$$

or

$$\mathbf{v}_A = - \frac{m_B}{m_A} \mathbf{v}_B .$$

For example, if m_A is 3.0 kg and m_B is 1.0 kg, then $\mathbf{v}_A$ will always be one-third $\mathbf{v}_B$ in magnitude and oppositely directed as the blocks move.

The kinetic energy of block A is $\frac{1}{2} m_A v_A^2$ and can be written as $(m_A v_A)^2/2m_A$ and that of block B is $\frac{1}{2} m_B v_B^2$ and can be written as $(m_B v_B)^2/2m_B$. But

$$\frac{K_A}{K_B} = \frac{2m_B(m_A v_A)^2}{2m_A(m_B v_B)^2} = \frac{m_B}{m_A},$$

in which $m_A v_A$ equals $m_B v_B$ because of momentum conservation. The kinetic energies of the blocks at any instant are *inversely* proportional to their respective masses. Because mechanical energy is conserved also, the blocks will continue to oscillate back and forth, the energy being partly kinetic and partly potential. What is the motion of the center of mass of this system?

If mechanical energy is not conserved, as would be true if friction were present, the motion will die out as the energy is dissipated. Can we apply the conservation of linear momentum in this case? Explain.

Example 8 As an example of recoil, consider radioactive decay. An α-particle (the nucleus of a helium atom) is emitted from a uranium-238 nucleus, originally at rest, with a speed of 1.4×10^7 m/s which corresponds to a kinetic energy of 4.1 MeV (million electron volts). Find the recoil speed of the residual nucleus (thorium-234).

We think of the system (thorium + α-particle) as initially bound and forming the uranium nucleus. The system then fragments into two separate parts. The momentum of the system before fragmentation is zero. In the absence of external forces, the momentum after fragmentation is also zero. Hence,

initial momentum = final momentum,

$$0 = M_\alpha \mathbf{v}_\alpha + M_{Th} \mathbf{v}_{Th},$$

$$\mathbf{v}_{Th} = - \frac{M_\alpha}{M_{Th}} \mathbf{v}_\alpha .$$

The ratio of the α-particle mass to the thorium nucleus mass, M_α/M_{Th}, is 4/234 and $v_\alpha = 1.4 \times 10^7$ m/s. Hence,

$$v_{Th} = -(4/234)(1.4 \times 10^7 \text{ m/s}) = -2.4 \times 10^5 \text{ m/s}.$$

The minus sign indicates that the residual thorium nucleus recoils in a direction opposite to the motion of the α-particle, so as to give a resultant vector momentum of zero.

How can we compute the kinetic energy of the recoiling nucleus (see the previous example)? Where does the energy of the fragments come from?

Example 9   A rocket   A rocket accelerates by ejecting mass from its nozzle at high speed. Thus its mass changes (decreases) as it accelerates. In our earlier examples, however, the mass of the system remained constant during the acceleration process. We deal with this new situation by applying Newton's second law, not to the rocket alone, but to the rocket and its ejected mass, taken together. The mass of *this* system does *not* change when the rocket engine is fired up.

Figure 9–10a shows the situation at $t = 0$. The rocket, whose mass is M, has been drifting in space with its engine turned off. Its speed, to an observer in an inertial reference frame coasting along with the rocket, is zero. The engine is now fired up and a "burn" begins.

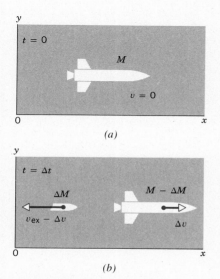

Figure 9–10   Example 9.

Figure 9–10b shows how things stand at $t = \Delta t$, the end of the burn. The mass of the rocket is now $M - \Delta M$, in which ΔM is the mass of fuel combustion products ejected from the nozzle, with an exhaust speed v_{ex} relative to the rocket. Our observer sees the rocket pulling ahead of him, with speed Δv. He also sees the mass ΔM moving with a forward speed of $\Delta v - v_{ex}$. (If $\Delta v < v_{ex}$, as in our example, this "forward" speed will be negative, which means that the inertial observer will actually see the ejected mass moving backwards; see Fig. 9–10b.)

Now let us apply Newton's second law in the form of Eq. 9–17 to the rocket plus its ejected mass. There are no external forces, such as gravity or air resistance, so that the Eq. 9–17 reduces to $dP/dt = 0$.

For the interval Δt we can then put

$$\frac{P_f - P_i}{\Delta t} = 0 \qquad (9\text{–}19a)$$

where $P_i (=0)$ is the initial system momentum and P_f is the final system momentum, which is (see Fig. 9–10b),

$$P_f = (M - \Delta M)(\Delta v) - (\Delta M)(v_{ex} - \Delta v)$$
$$= M(\Delta v) - (\Delta M)v_{ex}. \qquad (9\text{–}19b)$$

Combining Eqs. 9–19 and rearranging leads to

$$v_{ex}\left(\frac{\Delta M}{\Delta t}\right) = M\left(\frac{\Delta v}{\Delta t}\right).$$

In the differential limit this reduces to

$$v_{ex} R_{ex} = M(dv/dt), \qquad (9\text{–}20)$$

in which R_{ex} is the rate at which mass is expelled from the nozzle of the rocket engine and dv/dt is the acceleration of the rocket.

The left term of this equation has the dimensions of a force and depends only on the design characteristics of the rocket engine, namely, the rate at which it ejects mass and the speed with which that mass is ejected. We call this term the *thrust* of the rocket and represent it by T. To make the thrust as large as possible the rocket engine designer must arrange to eject mass as rapidly as possible (R_{ex} large) and with the highest possible relative speed (v_{ex} large). Newton's law emerges clearly if we write Eq. 9–20 as $T = Ma$ in which a is the acceleration of the rocket at a time when its mass is M.

Suppose that a rocket with a mass of 800 kg can eject mass during a burn at the rate of 2.0 kg/s. The speed of the exhaust gases relative to the rocket is 1800 m/s. The thrust of its engine is

$$T = v_{ex} R_{ex} = (1800 \text{ m/s})(2.0 \text{ kg/s}) = 3600 \text{ N} (=810 \text{ lb}).$$

Our rocket now performs a 60-s burn. Its acceleration at the beginning of the burn is

$$a = \frac{T}{M} = \frac{3600 \text{ N}}{800 \text{ kg}} = 4.5 \text{ m/s}^2.$$

At the end of the burn the mass of the rocket will have fallen to 680 kg and its acceleration will have risen to 5.3 m/s². Notice that this rocket, whose initial acceleration ($=4.5$ m/s²) is less than that of the earth's surface gravity ($=9.8$ m/s²), could not have been launched from the earth's surface. Put another way, its thrust ($=3600$ N) is less than its weight ($=800$ kg $\times$ 9.8 m/s² $= 7800$ N). We assume that our rocket was launched into space by another, more powerful, rocket.

Let us now find Δv, the increase in speed of the rocket during the burn. Because our rocket starts from rest, its increase in speed Δv is equal to its final speed v_f. From Eq. 9–20 we can write

$$dv = v_{ex}\frac{R_{ex}dt}{M} = -v_{ex}\frac{dM}{M}$$

in which $dM (= -R_{ex}dt)$ is the differential change in the mass of the rocket during the interval dt. Integrating leads to

$$v_f = \int_0^{v_f} dv = -v_{ex}\int_M^{M_f}\frac{dM}{M} = v_{ex}\ln\left(\frac{M}{M_f}\right)$$
$$= (1800 \text{ m/s}) \ln (800 \text{ kg}/680 \text{ kg}) = 290 \text{ m/s}.$$

Our rocket will reach its ultimate speed when it has burned all its fuel. Suppose that $M_f = 200$ kg (rather than the 680 kg assumed above) when this happens. Substituting 200 kg for M_f in the above yields 2500 m/s for the ultimate speed. We see the advantage of multistage rockets, in which M_f is kept small by discarding successive stages as their fuel is used up.

REVIEW GUIDE AND SUMMARY

Center of mass

The coordinates of the center of mass of a system of particles are given by

$$x_{cm} = \frac{1}{M} \sum m_i x_i, \qquad y_{cm} = \frac{1}{M} \sum m_i y_i, \qquad z_{cm} = \frac{1}{M} \sum m_i z_i; \qquad [9-3a]$$

see Examples 1 and 2. We use this concept to write Newton's second law for a system of particles, or

Newton's second law for a system of particles

$$\mathbf{F}_{ext} = M\mathbf{a}_{cm}, \qquad [9-8]$$

in which $\mathbf{F}_{ext}$ is the resultant external force acting on the system, M is the total mass of the system, and $\mathbf{a}_{cm}$ (see Example 3) is the acceleration of the center of mass of the system. Example 6 (an exploding projectile) shows that $\mathbf{a}_{cm}$ is quite independent of the actions of internal forces, depending only on $\mathbf{F}_{ext}$.

Kinetic energy and internal work

In some systems of particles an external force acts but does no work on the system. Nevertheless, the kinetic energy of the system may increase or decrease. The work associated with these changes in kinetic energy is internal work, and it may be positive (as for the astronaut in Example 4) or negative (as for the braking car in Example 5). The concept of center-of-mass work W_{cm} allows us to solve such problems without knowing the details of the internal work. W_{cm} is defined from

Center-of-mass work

$$W_{cm} = \int F_{ext} \, dx_{cm} \qquad [9-10b]$$

and is the work that *would* be done by the external force *if* it acted at the center of mass. W_{cm} is equal to the change in the kinetic energy associated with the motion of the center of mass, or

Work-energy theorem for a system of particles

$$W_{cm} = K_{cm} - K_{cmo} = \Delta K_{cm}. \qquad [9-10c]$$

For a single particle we define a quantity $\mathbf{p}$ called the linear momentum and then we express Newton's second law in its terms, or

Linear momentum and Newton's second law

$$\mathbf{p} = m\mathbf{v} \qquad \text{and} \qquad \mathbf{F} = \frac{d\mathbf{p}}{dt}. \qquad [9-11, 9-12]$$

For a system of particles these relations become

$$\mathbf{P} = M\mathbf{v}_{cm} \qquad \text{and} \qquad \mathbf{F}_{ext} = \frac{d\mathbf{P}}{dt}. \qquad [9-15, 9-17]$$

Conservation of linear momentum

The law of conservation of linear momentum tells us that if no external forces act on a system of particles, the linear momentum $\mathbf{P}$ of the system remains constant. Example 7 analyzes the conservation of linear momentum and energy for an isolated mass-spring system. Example 9 analyzes a system (a rocket) in which the mass of the system does not remain constant. The "trick" is to redefine the system, enlarging its boundaries until it encompasses a mass that *does* remain constant, and *then* to apply the laws of mechanics. For a rocket this means that the system must include not only the rocket proper but also the expelled gases.

QUESTIONS

1. Does the center of mass of a solid body necessarily lie within the body? If not, give examples.

2. Figure 9-11 shows (*a*) an isosceles triangle and (*b*) a right circular cone of the same width. The center of mass of the triangle is one-third of the way up from the base, but that of the cone is only one-fourth of the way up. Can you explain this difference?

Figure 9-11   Question 2.

3. How is the center of mass concept related to the concept of geographic center of the country? To the population center of the country? What can you conclude from the fact that the geographic center differs from the population center?

4. An amateur sculptor decides to portray a bird (Fig. 9–12). Luckily the final model is actually able to stand upright. The model is formed of a single sheet of metal of uniform thickness. Of the points shown, which is most likely to be the center of mass?

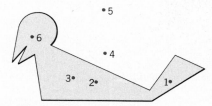

Figure 9–12 Question 4.

5. Someone claims that when a skillful high jumper clears the bar his center of mass actually goes *under* the bar. Is this possible?

6. If only an external force can change the state of motion of the center of mass of a body, how does it happen that the internal force of the brakes can bring a car to rest?

7. We say that a car is not accelerated by internal forces but rather by external forces exerted on it by the road. Why, then, do cars need engines?

8. Can the work done by internal forces decrease the kinetic energy of a body? . . . increase it?

9. If you do work on a system, does the system necessarily acquire kinetic energy? If a system acquires kinetic energy, does it necessarily mean that some external agent did work on it? Give examples. (By "kinetic energy" here we mean kinetic energy associated with the motion of the center of mass.)

10. A light and a heavy body have equal kinetic energies of translation. Which one has the large momentum?

11. A bird is in a wire cage hanging from a spring balance. Is the reading of the balance when the bird is flying about greater than, less than, or the same as that when the bird sits in the cage?

12. Can a sailboat be propelled by air blown at the sails from a fan attached to the boat?

13. When you run, do you transfer momentum to the earth?

14. A canoeist in a still pond can reach shore by jerking sharply on the rope attached to the bow of the canoe. How do you explain this? (yes she can! — it's true).

15. How might a person sitting at rest on a frictionless horizontal surface get altogether off of it?

16. A man stands still on a large sheet of slick ice; in his hand he holds a lighted firecracker. He throws the firecracker into the air. Describe briefly, but as exactly as you can, the motion of the center of mass of the firecracker and the motion of the center of mass of the system consisting of man and firecracker. It will be most convenient to describe each motion during each of the following periods: (a) after he throws the firecracker, but before it explodes: (b) between the explosion and the first piece of firecracker hitting the ice; (c) between the first fragment hitting the ice and the last fragment landing; (d) during the time when all fragments have landed but none has reached the edge of the ice.

17. Can you justify the following statement? "The law of conservation of linear momentum, as applied to a single particle, is equivalent to Newton's first law of motion."

18. In 1920 a prominent newspaper editorialized as follows about the pioneering rocket experiments of Robert H. Goddard, dismissing the notion that a rocket could operate in a vacuum: "That Professor Goddard, with his 'chair' in Clark College and the countenancing of the Smithsonian Institution, does not know the relation of action to reaction, and of the need to have something better than a vacuum against which to react — to say that would be absurd. Of course, he only seems to lack the knowledge ladled out daily in high schools." What is wrong with this argument?

19. Can a rocket reach a speed greater than the speed of the exhaust gases that propel it?

20. A ballet dancer doing a *grand jete* (great leap; see Fig. 9–13) seems to float horizontally in the central portion of her leap. Show how the dancer can maneuver her legs in flight so that, although her center of mass does indeed follow the expected parabolic trajectory, the top of her head moves more or less horizontally. (See "The Physics of Dance," by Kenneth Laws, *Physics Today*, February 1985.)

Figure 9–13 Question 20. Melinda Roy. Photo by Steven Caras, courtesy New York City Ballet.

EXERCISES

Section 9–1 Center of Mass

1. Show that the ratio of the distances of two particles from their center of mass is the inverse ratio of their masses.

2. Experiments using the diffraction of electrons show that the distance between the centers of the carbon (C) and oxygen (O) atoms in the carbon monoxide gas molecule is 1.130×10^{-10} m.

Locate the center of mass of a CO molecule relative to the carbon atom.

3. The mass of the moon is about 0.013 times the mass of the earth, and the distance from the center of the moon to the center of the earth is about 60 times the radius of the earth. How far is the center of mass of the earth-moon system from the center of the earth? Take the earth's radius to be 6400 km.

4. Where is the center of mass of the three particles shown in Fig. 9–14?

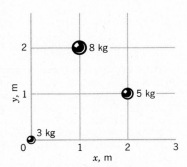

Figure 9–14 Exercise 4.

Section 9–2 Motion of the Center of Mass
5. Two blocks of masses 1.0 kg (weight $= mg = 2.2$ lb) and 3.0 kg (weight $= mg = 6.6$ lb) connected by a spring rest on a friction-less surface. If the two are given velocities such that the first travels at 1.7 m/s ($= 5.6$ ft/s) toward the center of mass, which remains at rest, what is the velocity of the second?

6. Two particles P and Q are initially at rest 1.0 m apart. P has a mass of 0.10 kg and Q a mass of 0.30 kg. P and Q attract each other with a constant force of 1.0×10^{-2} N. No external forces act on the system. (a) Describe the motion of the center of mass. (b) At what distance from P's original position do the particles collide?

7. A man of mass m clings to a rope ladder suspended below a balloon of mass M. The balloon is stationary with respect to the ground. (a) If the man begins to climb the ladder at a speed v (with respect to the ladder), in what direction and with what speed (with respect to the earth) will the balloon move? (b) What is the state of motion after the man stops climbing?

Section 9–3 Internal Work and Kinetic Energy
8. An automobile with passengers has weight 3680 lb ($= 16,400$ N) and is moving at 70 mi/h ($= 113$ km/h) when the driver brakes to a stop. The road exerts a force of 1850 lb ($= 8230$ N) on the wheels and there is no skidding. (a) What is the stopping distance? (b) Find the coefficient of static friction between tires and road.

Section 9–4 Linear Momentum of a Particle
9. What is the momentum of a 3000-lb automobile traveling at 55 mph?

10. How fast must an 816-kg Volkswagen travel (a) to have the same momentum as a 2650-kg Cadillac going 16 km/h? (b) To

have the same kinetic energy? (c) Make the same calculations using a 9080-kg truck instead of a Cadillac.

11. Find (a) the momentum and (b) the kinetic energy of a 10-g bullet with a speed of 760 m/s. (c) How fast must a 450-kg deer move to have the same momentum?

Section 9–5 Linear Momentum of a System of Particles
12. A 200-lb man standing on a surface of negligible friction kicks forward a 0.10-lb stone lying at his feet so that it acquires a speed of 10 ft/s. What velocity does the man acquire as a result?

13. A pellet gun fires ten 2-g pellets per second with a speed of 500 m/s. The pellets are stopped by a rigid wall. (a) What is the momentum of each pellet? (b) What is the kinetic energy of each pellet? (c) What is the average force exerted by the stream of pellets on the wall? (d) If each pellet is in contact with the wall for 0.6 ms, what is the average force exerted on the wall by each pellet while in contact? Why is this so different from (c)?

Section 9–6 Conservation of Linear Momentum
14. A stream of 40-g bullets, fired horizontally with a speed of 1000 m/s, strikes a 10-kg wooden block that is free to move on a horizontal frictionless tabletop. What is the speed of the block after it has absorbed 15 bullets?

15. A space vehicle is traveling at 4000 km/h with respect to the earth when the exhausted rocket motor is disengaged and sent backwards with a speed of 80 km/h with respect to the command module. The mass of the motor is four times the mass of the module. What is the speed of the command module after the separation?

Section 9–7 Some Applications of the Momentum Principle
16. A 6000-kg spaceprobe moving toward Jupiter at 100 m/s relative to the sun fires its rocket engine, ejecting 80 kg of exhaust at a speed of 250 m/s relative to the spaceprobe. What is the final velocity of the probe?

17. A 2000-kg railroad flatcar, which can move on the tracks with virtually no friction, is sitting motionless next to a station platform. A 100-kg football player is running along the platform parallel to the tracks at 10 m/s. He jumps onto the back of the flatcar. (a) What is the speed of the flatcar after he is aboard and at rest? (b) Now he starts to walk, at 0.5 m/s relative to the flatcar, to the front of the car. What is the speed of the flatcar as he walks?

18. A rocket is moving away from the solar system at a speed of 6.0×10^3 m/s. It fires its rocket engine, which ejects exhaust with a relative velocity of 3.0×10^3 m/s. The mass of the rocket at this time is 4.0×10^4 kg, and it experiences an acceleration of 2.0 m/s². (a) What is the velocity of the exhaust relative to the solar system? (b) At what rate was exhaust ejected during the firing?

19. Consider a rocket at rest in empty space. What must be its *mass ratio* (ratio of initial to final mass) in order that, after firing its engine, the rocket's speed is (a) equal to the exhaust speed? (b) equal to twice the exhaust speed?

PROBLEMS GROUPED BY SECTION

Section 9–1 Center of Mass
1. Three thin rods each of length L are arranged in an inverted "U," as shown in Fig. 9–15. The two rods on the arms of the U

each have mass M; the third rod has mass $3M$. Where is the center of mass of the assembly?

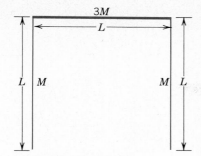

Figure 9–15   Problem 1.

2.   A box, open at the top, in the form of a cube of edge length 40 cm, is constructed from thin metal plate. Find the coordinates of the center of mass of the box with respect to the coordinate system shown in Fig. 9–16.

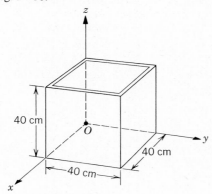

Figure 9–16   Problem 2.

3.   A precision-machined flat plate is formed from a uniform circular disk of radius R from which a circular section has been removed, as shown in Fig. 9–17. The diameter of the removed section equals the radius of the disk. The plate must rotate about an axle normal to its surface. For stability, this axle must pass through the center of mass of the plate. What must be the distance d between the axle and the "center" of the plate?

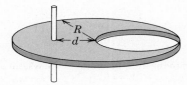

Figure 9–17   Problem 3.

*4.   Find the center of mass of a homogenous semicircular plate. Let a be the radius of the circle.

Section 9–2   Motion of the Center of Mass
5.   Ricardo, mass 80 kg, and Carmelita, who is lighter, are enjoying Lake Merced at dusk in a 30-kg canoe. When the canoe is at rest in the placid water they change seats, which are 3.0 m apart and symmetrically located with respect to the canoe's center. Ricardo notices that the canoe moved 40 cm relative to a submerged log, and calculates Carmelita's mass, which she has declined to tell him. What is it?

6.   A ball of mass m and radius R is placed inside a larger hollow sphere with the same mass and inside radius $2R$. The combination is at rest on a frictionless surface in the position shown in Fig.

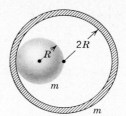

Figure 9–18   Problem 6.

9–18. The smaller ball is released, rolls around the inside of the hollow sphere, and finally comes to rest at the bottom. How far will the larger sphere have moved during this process?

Section 9–3   Internal Work and Kinetic Energy
7.   A 110-kg ice hockey player skates at 3.0 m/s toward a railing at the edge of the ice and stops himself by grasping the railing with his outstretched arms. During this stopping process his center of mass moves 30 cm toward the rail. (*a*) What average force must he exert on the rail? (*b*) How much internal work (magnitude and sign) does he do? List any assumptions that you make.

Section 9–4   Linear Momentum of a Particle
8.   A 5-kg object with a speed of 30 m/s strikes a steel plate at an angle of 40° and rebounds at the same speed and angle (Fig. 9–19). What is the change (magnitude and direction) of the linear momentum of the object?

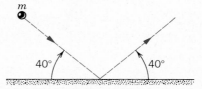

Figure 9–19   Problem 8.

9.   A 50-g ball is thrown into the air with an initial speed of 16 m/s at an angle of 30°. (*a*) What are the values of the kinetic energy of the ball initially and just before it hits the ground? (*b*) Find the corresponding values of the momentum (magnitude and direction). (*c*) Show that the change in momentum is just equal to the weight of the ball multiplied by the time of flight.

Section 9–5   Linear Momentum of a System of Particles
10.   A machine gun fires 50-g bullets at a speed of 1000 m/s. The gunner, holding the machine gun in his hands, can exert an average force of 180 N against the gun. Determine the maximum number of bullets he can fire per minute.

*11.   A very flexible uniform chain of mass M and length L is suspended from one end so that it hangs vertically, the lower end just touching the surface of a table. The upper end is suddenly released so that the chain falls onto the table and coils up in a small heap, each link coming to rest the instant it strikes the table. Find the force exerted by the table on the chain at any instant, in terms of the weight of chain already on the table at that moment.

Section 9–6   Conservation of Linear Momentum
12.   The last stage of a rocket is traveling at a speed of 7600 m/s. This last stage is made up of two parts which are clamped together, namely, a rocket case with a mass of 290 kg and a payload capsule with a mass of 150 kg. When the clamp is released, a compressed spring causes the two parts to separate with a relative speed of 910

m/s. (a) What are the speeds of the two parts after they have separated? Assume that all velocities are along the same line. (b) Find the total kinetic energy of the two parts before and after they separate and account for the difference, if any.

13. A radioactive nucleus, initially at rest, decays by emitting an electron and a neutrino at right angles to one another. The momentum of the electron is 1.2×10^{-22} kg·m/s and that of the neutrino is 6.4×10^{-23} kg·m/s. (a) Find the direction and magnitude of the momentum of the recoiling nucleus. (b) The mass of the residual nucleus is 5.8×10^{-26} kg. What is its kinetic energy of recoil? A *neutrino* is one of the fundamental particles of nature.

14. A shell is fired from a gun with a muzzle velocity of 1500 ft/s, at an angle of 60° with the horizontal. At the top of its trajectory the shell explodes into two fragments of equal mass. One fragment, whose speed immediately after the explosion is zero, falls vertically. How far from the gun does the other fragment land, assuming level terrain?

Section 9–7   Some Applications of the Momentum Principle
15. A railroad flatcar of weight W can roll without friction along a straight horizontal track. Initially a man of weight w is standing on the car, which is moving to the right with speed v_0. What is the change in velocity of the car if the man runs to the left (Fig. 9–20) so that his speed relative to the car is v_{rel} just before he jumps off at the left end?

Figure 9–20   Problems 15 and 16.

16. Assume that the car in Problem 15 is initially at rest. It holds n men each of weight w. If each man in succession runs with a relative velocity v_{rel} and jumps off the end, do they impart to the car a greater velocity than if they all run and jump at the same time?

17. A 6000-kg rocket is set for vertical firing. If the exhaust speed is 1000 m/s, how much gas must be ejected each second to supply the thrust needed (a) to overcome the weight of the rocket, and (b) to give the rocket an initial upward acceleration of 20 m/s²? Note that, in contrast to the situation described in Example 9, gravity is present here as an external force.

18. A rocket at rest in space, where there is virtually no gravity, has a mass of 2.55×10^5 kg, of which 1.81×10^5 kg is fuel. The engine consumes fuel at the rate 480 kg/s, and the exhaust speed is 3.27 km/s. The engine is fired for 250 s. (a) Find the thrust of the rocket engine. (b) What is the final speed attained?

19. A 1400-kg cannon, which fires a 70-kg shell with a muzzle speed of 556 m/s, is set at an elevation angle of 39° above the horizontal. The cannon is mounted on frictionless rails, so that it recoils freely. (a) What is the speed of the shell with respect to the earth? (b) At what angle with the ground is the shell projected? (Hint: The horizontal component of the momentum of the system remains unchanged as the gun is fired; see Section 4–5.)

20. A jet airplane is traveling 180 m/s (=600 ft/s). The engine takes in 68 m³ (=2400 ft³) of air making a mass of 70 kg (=4.8 slugs) each second. The air is used to burn 2.9 kg (=0.20 slugs) of fuel each second. The energy is used to compress the products of combustion and to eject them at the rear of the plane at 490 m/s (=1600 ft/s) relative to the plane. Find (a) the thrust of the jet engine and (b) the delivered power (horsepower).

ADDITIONAL PROBLEMS

21. An object is tracked by a radar station and found to have a position vector given by $\mathbf{r} = (3500 - 160t)\mathbf{i} + 2700\mathbf{j} + 300\mathbf{k}$, with $\mathbf{r}$ in meters and t in seconds. The radar station's x-axis points east, its y-axis north, and its z-axis vertically up. If the object is a 250-kg missile warhead, what are (a) its momentum and (b) its direction of motion?

22. A 55-kg woman leaps vertically into the air from a crouching position in which her center of mass is 40 cm above the ground. As her feet leave the floor her center of mass is 90 cm above the ground and rises to 120 cm at the top of her leap. (a) What upward force, assumed constant, does the ground exert on her? (b) What maximum speed does she attain?

23. A 2000-kg truck traveling north at 40 km/h turns east and accelerates to 50 km/h. (a) What is the change in kinetic energy of the truck? (b) What is the magnitude and direction of the change of the truck's momentum?

24. A vessel at rest explodes, breaking into three pieces. Two pieces, having equal mass, fly off perpendicular to one another with the same speed of 30 m/s. The third piece has three times the mass of each other piece. What is the direction and magnitude of its velocity immediately after the explosion?

25. Two 2-kg masses, A and B, collide. The velocities before the collision are $\mathbf{v}_A = 15\mathbf{i} + 30\mathbf{j}$ and $\mathbf{v}_B = -10\mathbf{i} + 5\mathbf{j}$. After the colli-

sion $\mathbf{v}'_A = -5\mathbf{i} + 20\mathbf{j}$. All speeds are given in meters per second. (a) What is the final velocity of B? (b) How much kinetic energy was gained or lost in the collision?

26. Many bicycles are braked by pressing a piece of hard rubber against each tire. Suppose that such a bicycle, initially traveling at 25.0 mi/h (=40.2 km/h), is stopped in 55 revolutions of its 14-in. (=35.6-cm) radius tires. The mass of bicycle plus rider is 3.50 slugs (=51.1 kg). Treat the front and rear brakes as identical and assume that the tires do not skid on the road. (a) What work is done by each brake? (b) Assume that the force of a brake against a tire is constant and calculate its value. (c) What is the acceleration of the bicycle? (d) What is the force of the road on each tire?

27. A 5.4-kg toboggan carrying 35 kg of sand slides from rest down an icy slope 90 m long, inclined 30° below the horizontal. The sand leaks from the back of the toboggan at the rate of 2.3 kg/s. How long does it take the toboggan to reach the bottom of the slope?

28. A body of mass 8 kg is traveling at 2 m/s with no external forces acting. At a certain instant an internal explosion occurs, splitting the body into two chunks of 4-kg mass each; 16 J of translational kinetic energy are imparted to the two-chunk system by the explosion. Neither chunk leaves the line of the original motion. Determine the speed and direction of motion of each of the chunks after the explosion.

29. A cannon and a supply of cannon balls are inside a sealed railroad car of length L, as in Fig. 9–21. The cannon fires to the right; the car recoils to the left. The cannon balls remain inside the car after hitting the far wall. (a) After all the cannon balls have been fired, what is the greatest distance the car can have moved from its original position? (b) What is the speed of the car after all the cannon balls have been fired?

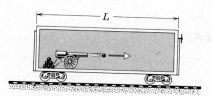

Figure 9–21 Problem 29.

30. In the ammonia (NH_3) molecule, the three hydrogen (H) atoms form an equilateral triangle, the distance between centers of the atoms being 16.28×10^{-11} m, so that the center of the triangle is 9.39×10^{-11} m from each hydrogen atom. The nitrogen (N) atom is at the apex of a pyramid, the three hydrogens constituting the base (see Fig. 9–22). The nitrogen-hydrogen distance is 10.14×10^{-11} m and the nitrogen-hydrogen atomic mass ratio is 13.9. Locate the center of mass relative to the nitrogen atom.

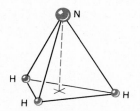

Figure 9–22 Problem 30.

31. Two long barges are floating in the same direction in still water, one with a speed of 10 km/h and the other with a speed of 20 km/h. While they are passing each other, coal is shoveled from the slower to the faster one at a rate of 1000 kg/min. How much additional force must be provided by the driving engines of each barge if neither is to change speed? Assume that the shoveling is always perfectly sideways and that the frictional forces between the barges and the water do not depend on the weight of the barges.

32. Each minute, a special game warden's machine gun fires 220 10-g rubber bullets with a muzzle velocity of 1200 m/s. How many bullets must be fired at an 85-kg animal charging toward the warden at 4.0 m/s in order to stop the animal in its tracks? (Assume that the bullets travel horizontally and drop to the ground after striking the target.)

33. During a violent thunderstorm, hail the size of marbles (diameter = 1.00 cm) falls at a speed of 25 m/s. There are estimated to be 120 hailstones per cubic meter of air. Ignoring the bounce of the hail on impact, what force is exerted by hail on a 10 m $\times$ 20 m flat roof at the height of the storm? Assume that, as for ice, 1 cm^3 of hail has a mass of 0.92 g.

34. A uniform square plate 6 m on a side has a square piece 2 m on a side cut out of it. The center of this notch is at $x = 2$ m, $y = 0$. The center of the square plate is at $x = y = 0$; see Fig. 9–23. Find the x- and y-coordinates of the center of mass of the remaining piece.

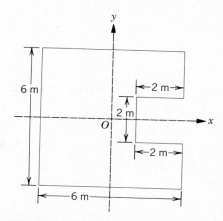

Figure 9–23 Problem 34.

35. Two bodies, each made up of weights from a set, are connected by a light cord which passes over a light, frictionless pulley with a diameter of 50 mm. The two bodies are at the same level. Each originally has a mass of 500 g. (a) Locate their center of mass. (b) Twenty grams are transferred from one body to the other, but the bodies are prevented from moving. Locate the center of mass. (c) The two bodies are now released. Describe the motion of the center of mass and determine its acceleration.

36. A dog, weighing 10 lb, is standing on a flatboat so that he is 20 ft from the shore. He walks 8.0 ft on the boat toward shore and then halts. The boat weighs 40 lb, and one can assume there is no friction between it and the water. How far is he from the shore at the end of this time? (Hint: The center of mass of boat + dog does not move. Why?) The shoreline is also to the left in Fig. 9–24.

Figure 9–24 Problem 36.

37. A cylindrical storage tank is initially filled with aviation gasoline. The tank is then drained through a valve on the bottom. See Fig. 9–25. (a) As the gasoline is withdrawn, describe qualitatively the motion of the center of mass of the tank and its remaining contents. (b) What is the depth x to which the tank is filled when the center of mass of the tank and its remaining contents reaches its lowest point? Express your answer in terms of H, the height of the tank; M, its mass; and m, the mass of gasoline it can hold.

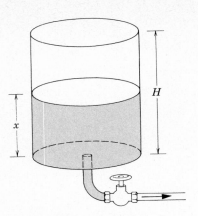

Figure 9-25 Problem 37.

*38. An 80-kg man is standing at the rear of a 400-kg iceboat that is moving at 4.0 m/s across ice that may be considered to be frictionless. He decides to walk to the front of the 18 m-long boat and does so at a speed of 2.0 m/s with respect to the boat. How far did the boat move across the ice while he was walking?

10 Collisions

10-1 What Is a Collision?

We can learn much about objects of all kinds by observing them as they collide with each other. As Fig. 10-1 shows, objects of interest range from subatomic particles, whose masses are $\sim 10^{-27}$ kg, to galaxies, for which typical masses are $\sim 10^{40}$ kg. Between these extremes lies a host of more directly familiar objects such as colliding billiard balls. In this chapter we confine ourselves to collisions between objects that can be represented as particles. As we shall see, our principal tools for analyzing collisions are the laws of conservation of energy and of momentum.

A collision defined

In a collision a relatively large force acts on each colliding particle for a relatively short time. The basic idea of a collision is that the motion of the colliding particles (or of at least one of them) changes rather abruptly and that we can make a relatively clean separation of times that are "before the collision" and those that are "after the collision."

When a bat strikes a baseball, for example the beginning and the end of the collision can be determined fairly precisely. The bat is in contact with the ball for an interval that is quite short in comparison to the time during which we are watching the ball. During the collision the bat exerts a large force on the ball (Fig. 10-2). This force varies with time in a complex way that we can measure only with difficulty. Both the ball and the bat are deformed during the collision.* Forces that act for a time that is short compared to the time of observation of the system are called *impulsive* forces.

---

* See "Batting the Ball," by P. Kirkpatrick, *American Journal of Physics*, August 1963.

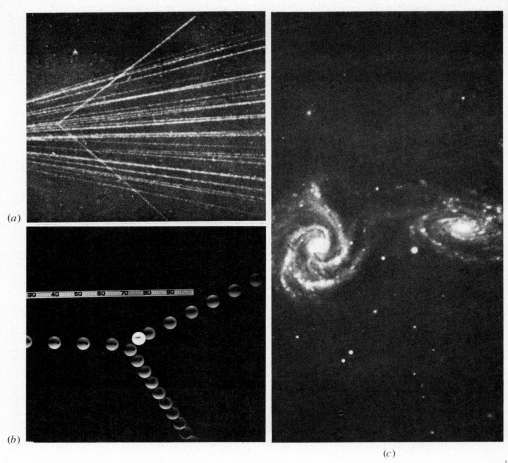

(a)

(b)

(c)

Figure 10–1 Collisions over a wide range of masses. (*a*) A cloud chamber photograph showing an energetic alpha particle (a nucleus of helium) coming from the left and colliding with a helium nucleus in the gas filling the chamber. *Typical Expansion Chamber Photographs,* by W. Gentner, Pergamon Press, London, 1954. (*b*) A multiflash photograph showing a billiard ball coming from the left and colliding with a resting billiard ball. (From *PSSC Physics,* 2nd ed., 1965, D.C. Heath and Company, Lexington, Mass. with Education Development Center, Newton, Mass.) (*c*) The near-collision of two spiral galaxies, photographed with the 200-in. telescope at Mount Palomar. Lick Observatories.

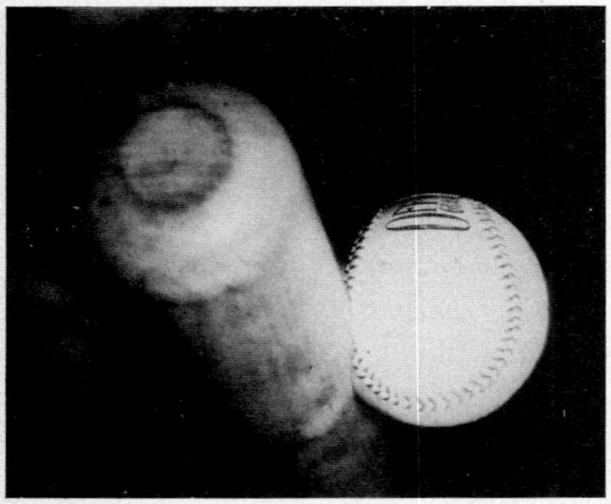

Figure 10–2 A high-speed flash photograph of a bat striking a baseball. Notice the deformation of the ball, indicating the enormous magnitude of the impulsive force at this instant. (Courtesy Harold E. Edgerton.)

When an α particle, (^{4}He) "collides" with a nucleus of gold (^{197}Au), the force acting between them may be the well-known repulsive electrostatic force associated with the positive charges of the particles. The particles may not "touch," but we still may speak of a "collision" because a relatively strong force, acting for a time that is short in comparison to the time that the α particle is under observation, has a marked effect on the motion of the α particle.

The concept of collision may be broadened to include events in which the identities of the interacting particles change during the event. Such collisions are called *reactions*. For instance, when a proton (^{1}H or p) with energy of 25 MeV ($= 4.0 \times 10^{-12}$ J) "collides" with a nucleus of a silver atom (^{107}Ag), the particles may in a sense actually "touch," the predominant force then acting between them being not the electrostatic repulsive force, but the strong, short-range, attractive nuclear force. The proton may enter the silver nucleus, forming a compound structure. A short time later—the collision interval may be 10^{-18} s—this compound structure may break up into two different particles, according to a scheme such as

$$p + {}^{107}Ag \rightarrow \text{compound nucleus} \rightarrow \alpha + {}^{104}Pd,$$

in which α ($= He^4$) is an α particle. The conservation principles apply to all such events.

We may, if we wish, broaden our definition of collision even further to include a reaction such as the spontaneous decay of a single particle into two or more other particles. An example is the decay of the elementary particle called the (negative) *sigma particle* (Σ-particle) into two other particles, the (negative) *pion* and the *neutron* or

$$\Sigma^- \rightarrow \pi^- + n.$$

Although two bodies do not come in contact in this process (unless we consider it in reverse), it has many features in common with collisions: (1) There is a clean distinction between "before the event" and "after the event," and (2) the laws of conservation of momentum and energy permit us to learn much about such processes by studying the before and after situations, even though we may know little about the force laws that operate during the event itself.

In studying collisions in this chapter our aim will be this: Given the initial motions of the colliding particles, what can we learn about their final motions from the principles of conservation of momentum and energy, assuming that we know nothing about the forces acting during the collision?

10-2 Impulse and Momentum

Let us assume that Fig. 10-3 shows the magnitude of the force exerted on a body during a collision. We assume that the force has a constant direction. The collision begins at time t_i and ends at time t_f, the force being zero before and after collision. From Eq. 9-12 we can write the change in momentum $d\mathbf{p}$ of a body in a time dt during which a force $\mathbf{F}$ acts on it as

$$d\mathbf{p} = \mathbf{F}\, dt. \tag{10-1}$$

Let us integrate each side of this equation over the time of the collision. We obtain

$$\int_{\mathbf{p}_i}^{\mathbf{p}_f} d\mathbf{p} = \int_{t_i}^{t_f} \mathbf{F}\, dt,$$

in which the subscripts i (*initial*) and f (*final*) refer to the times before and after the collision. The left side of this equation is just $\mathbf{p}_f - \mathbf{p}_i$, the (vector) change in the momentum of the body caused by the collision. The right side, which is a measure of the strength and duration of the collision force, is called the *impulse* $\mathbf{J}$, or

$$\mathbf{J} = \int_{t_i}^{t_f} \mathbf{F}\, dt \tag{10-2a}$$

Hence the change in momentum of a body acted upon by an impulsive force is equal to the impulse, or

$$\mathbf{p}_f - \mathbf{p}_i = \mathbf{J}. \tag{10-2b}$$

Both impulse and momentum are vectors and both have the same units and dimensions.

Spontaneous decay

Collisions and the conservation laws

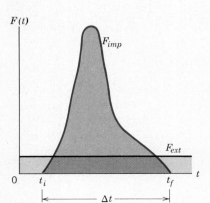

Figure 10-3 How an impulsive force $F(t)$ might vary with time during a collision starting at time t_i and ending at t_f. During a collision, the impulsive force F_{imp} is much greater than any external forces, F_{ext}, which may act on the system.

The impulse-momentum theorem

The impulsive force represented in Fig. 10–3 is assumed to have a constant direction. The impulse of this force, $\int_{t_i}^{t_f} \mathbf{F} \, dt$, is represented in magnitude by the area under the force-time curve.

Example 1 A baseball weighing 0.35 lb is struck by a bat while it is in horizontal flight with a speed of 90 ft/s. After leaving the bat the ball travels with a speed of 110 ft/s in a direction opposite to its original motion. Determine the impulse of the collision.

We cannot calculate the impulse from the definition $\mathbf{J} = \int \mathbf{F} \, dt$ because we do not know the force exerted on the ball as a function of time. However, we have seen (Eq. 10–2b) that the change in momentum of a particle acted on by an impulsive force is equal to the impulse. Hence

$$\mathbf{J} = \text{change in momentum}$$

$$= \mathbf{p}_f - \mathbf{p}_i = m\mathbf{v}_f - m\mathbf{v}_i = \left(\frac{W}{g}\right)(\mathbf{v}_f - \mathbf{v}_i).$$

Assuming arbitrarily that the direction of $\mathbf{v}_i$ is positive, the impulse is then

$$J = \left(\frac{0.35 \text{ lb}}{32 \text{ ft/s}^2}\right)(-110 \text{ ft/s} - 90 \text{ ft/s}) = -2.2 \text{ lb} \cdot \text{s}.$$

The minus sign shows that the direction of the impulse acting on the ball is opposite that of the original velocity of the ball.

We cannot determine the force of the collision from the data we are given. Actually, any force whose impulse is -2.2 lb · s will produce the same change in momentum. For example, if the bat and ball were in contact for 0.0010 s, the average force during this time would be

$$\overline{F} = \frac{\Delta p}{\Delta t} = \frac{-2.2 \text{ lb} \cdot \text{s}}{0.0010 \text{ s}} = -2200 \text{ lb}.$$

For shorter contact time the average force would be greater. The actual force would have a maximum value greater than this average value.

How far would gravity cause the baseball to fall during this collision time?

10–3 Conservation of Momentum During Collisions

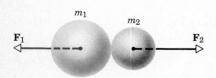

Figure 10–4 Two "particles" m_1 and m_2, in collision, experience equal and opposite forces along their line of centers, according to Newton's third law; $\mathbf{F}_2(t) = -\mathbf{F}_1(t)$.

Consider now a collision between two particles, such as those of masses m_1 and m_2, shown in Fig. 10–4. During the brief collision these particles exert large forces on one another. At any instant $\mathbf{F}_1$ is the force exerted on particle 1 by particle 2 and $\mathbf{F}_2$ is the force exerted on particle 2 by particle 1. By Newton's third law these forces at any instant are equal in magnitude but opposite in direction.

The change in momentum of particle 1 resulting from the collision is

$$\Delta\mathbf{p}_1 = \int_{t_i}^{t_f} \mathbf{F}_1 \, dt = \overline{\mathbf{F}}_1 \, \Delta t,$$

in which $\overline{\mathbf{F}}_1$ is the time average value of the force $\mathbf{F}_1$ during the time interval of the collision $\Delta t = t_f - t_i$.

The change in momentum of particle 2 resulting from the collision is

$$\Delta\mathbf{p}_2 = \int_{t_i}^{t_f} \mathbf{F}_2 \, dt = \overline{\mathbf{F}}_2 \, \Delta t,$$

in which $\overline{\mathbf{F}}_2$ is the time average value of the force $\mathbf{F}_2$ during the time interval of the collision $\Delta t = t_f - t_i$.

If no other forces act on the particles, then $\Delta\mathbf{p}_1$ and $\Delta\mathbf{p}_2$ give the total change in momentum for each particle. But we have seen that at each instant $\mathbf{F}_1 = -\mathbf{F}_2$, so that $\overline{\mathbf{F}}_1 = -\overline{\mathbf{F}}_2$, and therefore

$$\Delta\mathbf{p}_1 = -\Delta\mathbf{p}_2.$$

If we consider the two particles as an isolated system, the total momentum of the system is

$$\mathbf{P} = \mathbf{p}_1 + \mathbf{p}_2,$$

and the total change in momentum of the system as a result of the collision is zero; that is,

$$\Delta\mathbf{P} = \Delta\mathbf{p}_1 + \Delta\mathbf{p}_2 = 0.$$

Total momentum is conserved if no external forces act

Hence, if there are no external forces the total momentum of the system is not changed by the collision. The impulsive forces acting during the collision are internal forces which have no effect on the total momentum of the system.

We have defined a collision as an interaction that occurs in a time Δt that is very small compared to the time during which we are observing the system. We can also characterize a collision as an event in which the external forces that may act on the system are very small compared to the impulsive collision forces. When a bat strikes a baseball, a golf club strikes a golf ball, or one billiard ball strikes another, external forces act on the system. Gravity or friction exerts forces on these bodies, for example; these external forces may not be the same on each colliding body, nor are they necessarily cancelled by other external forces. Even so, it is quite safe to neglect these external forces during the collision and to assume momentum conservation provided, as is almost always true, that the external forces are much smaller than the impulsive forces of collision. As a result the change in momentum of a particle during a collision arising from an external force can be neglected in comparison with the change in momentum of that particle arising from the impulsive collisional force (Fig. 10–3).

For example, when a bat strikes a baseball, the collision lasts only a small fraction of a second. Since the change in momentum is large and the time of collision is small, it follows from

$$\Delta\mathbf{p} = \overline{\mathbf{F}}\,\Delta t$$

that the average impulsive force $\overline{\mathbf{F}}$ is relatively large. Compared to this force, the external force of gravity is negligible. During the collision we can safely ignore this external force in determining the change in motion of the ball; the shorter the duration of the collision the more likely this is to be true.

In practice, therefore, we can apply the principle of momentum conservation during collisions if the time of collision is small enough. We can then say that the momentum of a system of particles just before the particles collide is equal to the momentum of the system just after the particles collide.

10–4 Collisions in One Dimension

We can always calculate the motions of bodies after collision from their motions before collision if we know the forces that act during the collision, and if we can solve the equations of motion. Often we do not know these forces. Nevertheless, we may be able to predict the results of the collision by using the principle of conservation of momentum and the principle of conservation of total energy, both of which still hold.

Elastic and inelastic collisions

Collisions are usually classified according to whether or not *kinetic energy* is conserved in the collision. When kinetic energy is conserved, the collision is said to be *elastic*. Otherwise, the collision is said to be *inelastic*. Collisions between atoms, nuclei, and subatomic particles such as neutrons and protons are *sometimes* (but not always) elastic. These are, in fact, the only truly elastic collisions known. Collisions between gross bodies are always inelastic to some extent. We can often treat such collisions as approximately elastic, however, as, for example, collisions between ivory or steel balls. When two bodies stick together after collisions, the collision is said to be *completely inelastic*. For example, the collision between a bullet and a block of wood into which it is fired is completely inelastic when the bullet remains embedded in the block. The term completely inelastic does not mean that all the initial kinetic energy is lost; as we shall see, it means rather that the loss is as great as is consistent with momentum conservation.

When two macroscopic bodies undergo an inelastic collision, total energy is conserved even if kinetic energy is not. The "lost" kinetic energy appears at least in part as internal energy, manifesting itself as a small rise in temperature. Put another way, some of the energy associated with the initial directed motions of the colliding bodies is transformed into kinetic energy of the randomly-directed motions of the microscopic particles that make up those bodies. Beyond this, some of the initial kinetic energy may also appear after the collision as stored potential energy, associated with springlike deformations of the colliding bodies. Kinetic energy can also be gained in inelastic collisions. For example, one or both of the colliding bodies might contain an effective compressed spring, whose stored potential energy can be released on impact.

For the present we shall restrict our discussion to one-dimensional motion in which the relative motion after the collision is along the same line as the relative motion before the collision. Consider first an *elastic* one-dimensional collision. We can imagine two smooth nonrotating spheres moving initially along the line joining their centers, then colliding head-on and moving along the same straight line without rotation after collision (see Fig. 10–5). These bodies exert forces on each other during the collision that are along the initial line of motion, so that the final motion is also along this same line.

The masses of the spheres are m_1 and m_2, the (scalar) velocity components being v_{1i} and v_{2i} before collision and v_{1f} and v_{2f} after collision. We take the positive direction of the momentum and velocity to be to the right. We assume, unless we specify otherwise, that the speeds of the colliding particles are not so high as to require the use of the relativistic expressions for momentum and kinetic energy. Then, from conservation of momentum, we obtain

$$m_1 v_{1i} + m_2 v_{2i} = m_1 v_{1f} + m_2 v_{2f}.$$

Because we are considering an elastic collision, kinetic energy is conserved by definition and we obtain

$$\tfrac{1}{2} m_1 v_{1i}^2 + \tfrac{1}{2} m_2 v_{2i}^2 = \tfrac{1}{2} m_1 v_{1f}^2 + \tfrac{1}{2} m_2 v_{2f}^2.$$

It is clear at once that, if we know the masses and the initial velocities, we can calculate the two final velocities v_{1f} and v_{2f} from these two equations.

An elastic one-dimensional collision

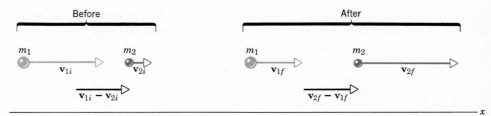

Figure 10–5 Two spheres before and after an elastic collision.
The velocity, $\mathbf{v}_{1i} - \mathbf{v}_{2i}$, of m_1 relative to m_2 before collision is
equal to the velocity, $\mathbf{v}_{2f} - \mathbf{v}_{1f}$, of m_2 relative to m_1 after collision.

The momentum equation can be written as

$$m_1(v_{1i} - v_{1f}) = m_2(v_{2f} - v_{2i}), \qquad (10-3)$$

and the energy equation can be written as

$$m_1(v_{1i}{}^2 - v_{1f}{}^2) = m_2(v_{2f}{}^2 - v_{2i}{}^2). \qquad (10-4)$$

Dividing Eq. 10–4 by Eq. 10–3, and assuming that $v_{2f} \neq v_{2i}$ and $v_{1f} \neq v_{1i}$ (see Question 4), we obtain

$$v_{1i} + v_{1f} = v_{2f} + v_{2i}$$

and, after rearrangement,

$$v_{1i} - v_{2i} = v_{2f} - v_{1f}. \qquad (10-5)$$

This tells us that in an elastic one-dimensional collision, the relative velocity of approach before collision is equal to the relative velocity of separation after collision.

To find the velocity components v_{1f} and v_{2f} after collision from the velocity components v_{1i} and v_{2i} before collision, we can use any two of the three previous numbered equations. Thus, from Eq. 10–5,

$$v_{2f} = v_{1i} + v_{1f} - v_{2i}.$$

Inserting this into Eq. 10–3 and solving for v_{1f}, we find that

$$v_{1f} = \left(\frac{m_1 - m_2}{m_1 + m_2}\right) v_{1i} + \left(\frac{2m_2}{m_1 + m_2}\right) v_{2i}. \qquad (10-6a)$$

Likewise, inserting $v_{1f} = v_{2f} + v_{2i} - v_{1i}$ (from Eq. 10–5) into Eq. 10–3 and solving for v_{2f}, we obtain

$$v_{2f} = \left(\frac{2m_1}{m_1 + m_2}\right) v_{1i} + \left(\frac{m_2 - m_1}{m_1 + m_2}\right) v_{2i}. \qquad (10-6b)$$

There are several cases of special interest. For example, when the colliding particles have the same mass, m_1 equals m_2 so that Eqs. 10–6 become simply

$$v_{1f} = v_{2i} \qquad \text{and} \qquad v_{2f} = v_{1i}.$$

That is, in a one-dimensional elastic collision of two particles of equal mass, the particles simply exchange velocities during collision.

Another case of interest is that in which one particle m_2 is initially at rest. Then v_{2i} equals zero and

$$v_{1f} = \left(\frac{m_1 - m_2}{m_1 + m_2}\right) v_{1i}, \qquad v_{2f} = \left(\frac{2m_1}{m_1 + m_2}\right) v_{1i}.$$

Of course, if $m_1 = m_2$ also, then $v_{1f} = 0$ and $v_{2f} = v_{1i}$ as we expect. The first particle is "stopped cold" and the second one "takes off" with the velocity the first one originally had. If, however, m_2 is very much greater than m_1, we obtain

$$v_{1f} \cong -v_{1i} \qquad \text{and} \qquad v_{2f} \cong 0.$$

That is, when a light particle collides with a very much more massive particle at rest, the velocity of the light particle is approximately reversed and the massive particle remains approximately at rest. For example, suppose that we drop a ball vertically onto a horizontal surface attached to the earth. This is in effect a collision between the ball and the earth. If the collision is elastic, the ball will rebound with a reversed velocity and will reach the same height from which it fell.

If, finally, m_2 is very much smaller than m_1, we obtain

$$v_{1f} \cong v_{1i}, \qquad v_{2f} \cong 2v_{1i}.$$

This means that the velocity of the massive incident particle is virtually unchanged by the collision with the light stationary particle, but that the light particle rebounds with approximately twice the velocity of the incident particle. The motion of a bowling ball is hardly affected by collision with an inflated beach ball of the same size, but the beach ball bounces away quickly.

If a collision is *inelastic* then, by definition, the kinetic energy is not conserved. However, the conservation of momentum still holds, as does the conservation of *total* energy. Let us consider now a *completely inelastic* collision. The two particles stick together after collision, so that there will be a common final velocity $\mathbf{v}_f$. It is not necessary to restrict the discussion to one-dimensional motion. Using only the conservation of momentum principle, we find

> A completely inelastic collision

$$m_1\mathbf{v}_{1i} + m_2\mathbf{v}_{2i} = (m_1 + m_2)\mathbf{v}_f.$$

This determines $\mathbf{v}_f$ when $\mathbf{v}_{1i}$ and $\mathbf{v}_{2i}$ are known.

Example 2 *A one-dimensional elastic collision* (*a*) By what fraction is the kinetic energy of a neutron (mass m_1) decreased in a head-on elastic collision with an atomic nucleus (mass m_2) initially at rest?

The initial kinetic energy of the neutron K_i is $\frac{1}{2}m_1v_{1i}^2$. Its final kinetic energy K_f is $\frac{1}{2}m_1v_{1f}^2$. The fractional decrease in kinetic energy is

$$\frac{K_i - K_f}{K_i} = \frac{v_{1i}^2 - v_{1f}^2}{v_{1i}^2} = 1 - \frac{v_{1f}^2}{v_{1i}^2}.$$

But, for such a collision, from Eq. 10–6a,

$$v_{1f} = \left(\frac{m_1 - m_2}{m_1 + m_2}\right) v_{1i},$$

so that

$$\frac{K_i - K_f}{K_i} = 1 - \left(\frac{m_1 - m_2}{m_1 + m_2}\right)^2 = \frac{4m_1m_2}{(m_1 + m_2)^2}.$$

(*b*) Find the fractional decrease in the kinetic energy of a neutron when it collides in this way with a lead nucleus, a carbon nucleus, and a hydrogen nucleus. The ratio of nuclear mass to neutron mass ($=m_2/m_1$) is 206 for lead, 12 for carbon, and 1 for hydrogen.

For lead, $m_2 = 206m_1$,

$$\frac{K_i - K_f}{K_i} = \frac{4 \times 206}{(207)^2} = 0.02 \qquad \text{or} \qquad 2\%.$$

For carbon, $m_2 = 12m_1$,

$$\frac{K_i - K_f}{K_i} = \frac{4 \times 12}{(13)^2} = 0.28 \qquad \text{or} \qquad 28\%.$$

For hydrogen, $m_2 = m_1$,

$$\frac{K_i - K_f}{K_i} = \frac{4 \times 1}{(2)^2} = 1 \qquad \text{or} \qquad 100\%.$$

These results explain why water, which of course contains hydrogen, is far more effective in slowing down neutrons than is lead. In nuclear reactors it is necessary to slow down the fast

neutrons produced during the fission process by elastic collisions with light nuclei. Otherwise the chain reaction could not be sustained.

Example 3 *The ballistic pendulum; a completely inelastic collision* The ballistic pendulum is used to measure bullet speeds. The pendulum is a large wooden block of mass M hanging vertically by two cords. A bullet of mass m, traveling with a horizontal speed v_i, strikes the pendulum and remains embedded in it (Fig. 10–6). If the collision time (the time required for the bullet to come to rest with respect to the block) is very small compared to the time of swing of the pendulum, the supporting cords remain approximately vertical during the collision. Therefore, no external horizontal force acts on the system (bullet + pendulum) during collision, and the horizontal component of momentum is conserved. The speed of the system after collision v_f is much less than that of the bullet before collision. This

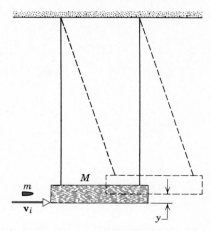

Figure 10–6 Example 3. A ballistic pendulum consisting of a large wooden block of mass M suspended by cords. When a bullet of mass m and velocity $\mathbf{v}_i$ is fired into it, the block swings, rising a maximum distance y.

final speed can be easily determined, so that the original speed of the bullet can be calculated from momentum conservation. The collision is completely inelastic.

The initial momentum of the system is that of the bullet mv_i, and the momentum of the system just after collision is $(m + M)v_f$, so that

$$mv_i = (m + M)v_f.$$

After the collision is over, the pendulum and bullet swing up to a maximum height y, where the kinetic energy left after impact is converted into gravitational potential energy. Then, using the conservation of mechanical energy for this part of the motion, we obtain

$$\tfrac{1}{2}(m + M)v_f{}^2 = (m + M)gy.$$

Solving these two equations for v_i, we obtain

$$v_i = \frac{m + M}{m}\sqrt{2gy}.$$

Hence, we can find the initial speed of the bullet by measuring m, M, and y.

The kinetic energy of the bullet initially is $\tfrac{1}{2}mv_i{}^2$ and the kinetic energy of the system (bullet + pendulum) just after collision is $\tfrac{1}{2}(m + M)v_f{}^2$. The ratio is

$$\frac{\tfrac{1}{2}(m + M)v_f{}^2}{\tfrac{1}{2}mv_i{}^2} = \frac{m}{m + M};$$

v_i and v_f can then be eliminated by using our first equation.

For example, if the bullet has a mass $m = 5$ g and the block has a mass $M = 2000$ g, only about one-fourth of 1% of the original kinetic energy remains; more than 99% is converted into internal energy, resulting in a local temperature rise in the block.

10–5 Collisions in Two and Three Dimensions

In two or three dimensions (except for a completely inelastic collision) the conservation laws alone cannot tell us the motion of particles after a collision if we know the motion before the collision. For example, for a two-dimensional elastic collision, which is the simplest case, we have four unknowns, namely, the two components of velocity for each of two particles after collision; but we have only three known relations between them, one for the conservation of kinetic energy and a conservation of momentum relation for each of the two dimensions. Hence we need more information than just the initial conditions. When we do not know the actual forces of interaction, as is often the case, the additional information must be obtained from experiment. It is usually simplest to measure the angle of recoil of one of the particles.

Let us consider what happens when one particle is projected at a target particle that is at rest. For example, much experimental work in nuclear physics involves projecting nuclear particles at a target that is stationary in the laboratory reference frame. In such collisions, because of momentum conservation, the motion is in a plane determined by the lines of recoil of the colliding particles. The initial motion need not be along the line joining the centers of the two particles. The force of interaction may be electromagnetic, gravitational (as for comets and spacecraft, for example), or nuclear. The particles need not "touch"; strong forces, which act at relatively close distances of approach and for a time short compared to the observation time, deflect the particles from their initial courses.

A typical situation is shown in Fig. 10–7. The distance b between the initial line of motion and a line parallel to it through the center of the target particle is called the *impact parameter*. This is a measure of the directness of the collision, $b = 0$, corresponding to a head-on collision. The direction of motion of the incident particle m_1 after collision makes an angle θ_1 with the initial direction, and the target projectile m_2, initially at rest, moves in a direction after collision making an angle θ_2 with the initial direction of the incident projectile. Applying the conservation of momentum, which is a vector relation, we obtain two scalar equations: for the x-component of motion we have

$$m_1 v_{1i} = m_1 v_{1f} \cos\,\theta_1 + m_2 v_{2f} \cos\,\theta_2, \tag{10–7a}$$

and for the y-component

$$0 = m_2 v_{2f} \sin\,\theta_2 - m_1 v_{1f} \sin\,\theta_1. \tag{10–7b}$$

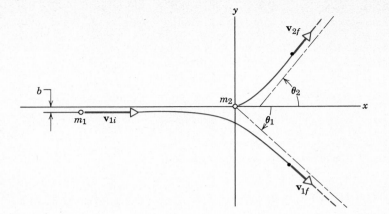

Figure 10–7 Two particles, m_1 and m_2, undergoing a collision. The open circles indicate their positions before collision, the solid ones after collision. Initially m_2 is at rest. The impact parameter b is the distance by which the collision misses being head-on.

Let us now assume that the collision is *elastic*. Here the conservation of kinetic energy applies and we obtain a third equation,

$$\tfrac{1}{2}m_1v_{1i}^2 = \tfrac{1}{2}m_1v_{1f}^2 + \tfrac{1}{2}m_2v_{2f}^2. \tag{10–7c}$$

If we know the initial conditions (m_1, m_2, and v_{1i}), we are left with four unknowns (v_{1f}, v_{2f}, θ_1, and θ_2) but only three equations relating them. The equations can be solved only if we have additional information, perhaps determined experimentally by measuring the deflection angle θ_1.

Example 4 *A completely inelastic collision in two dimensions*   Two skaters collide and embrace, as Fig. 10–8 suggests. One, whose mass m_1 is 80 kg ($W_1 = m_1 g = 180$ lb), is initially moving east at a speed v_1 of 6.0 km/h ($= 3.7$ mi/h). The other, whose mass m_2 is 50 kg ($W_2 = m_2 g = 110$ lb), is initially moving north at a speed v_2 of 8.0 km/h ($= 5.0$ mi/h). (a) What is the velocity of the couple immediately after the collision? (b) What fraction of the initial kinetic energy of the skaters is lost because of the collision?

(a) Figure 10–8 shows the initial and final situations. Because $\mathbf{P}_i = \mathbf{P}_f$ (the net external force is zero) we can write for the x-component of momentum

$$m_1v_1 = (m_1 + m_2)v\cos\theta.$$

and for the y-component of momentum

$$m_2v_2 = (m_1 + m_2)v\sin\theta.$$

Dividing the second equation by the first, we get

$$\tan\theta = \frac{m_2v_2}{m_1v_1} = \frac{(50\ \text{kg})(8.0\ \text{km/h})}{(80\ \text{kg})(6.0\ \text{km/h})}$$

$$= 0.83 \quad \text{or} \quad \theta = 40°$$

which gives the direction of the final velocity.
Then, from the y-component equation we have

$$v = \frac{m_2v_2}{(m_1 + m_2)\sin\theta}$$

$$= \frac{(50\ \text{kg})(8.0\ \text{km/h})}{(80\ \text{kg} + 50\ \text{kg})\sin 40°}$$

$$= 4.8\ \text{km/h}\ (= 3.0\ \text{mi/h}),$$

which gives the magnitude of the final velocity.

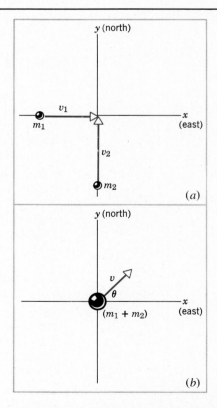

Figure 10–8 Example 4. (a) initial situation. (b) Situation just after the collision.

(b) The initial kinetic energy of the skaters is

$$K_i = \tfrac{1}{2}m_1v_1^2 + \tfrac{1}{2}m_2v_2^2$$
$$= \tfrac{1}{2}(80 \text{ kg})(6.0 \text{ km/h})^2 + \tfrac{1}{2}(50 \text{ kg})(8.0 \text{ km/h})^2$$
$$= \left(3040 \, \frac{\text{kg km}^2}{\text{h}^2}\right)\left(\frac{1 \text{ h}}{3600 \text{ s}}\right)^2\left(\frac{10^3 \text{ m}}{1 \text{ km}}\right)^2$$
$$= 230 \text{ J}.$$

The final kinetic energy of the couple is

$$K_f = \tfrac{1}{2}(m_1 + m_2)v^2$$
$$= \tfrac{1}{2}(80 \text{ kg} + 50 \text{ kg})(4.8 \text{ km/h})^2$$
$$= \left(1500 \, \frac{\text{kg km}^2}{\text{h}^2}\right)\left(\frac{1 \text{ h}}{3600 \text{ s}}\right)^2\left(\frac{10^3 \text{ m}}{1 \text{ km}}\right)^2$$
$$= 120 \text{ J}.$$

Hence,

$$\frac{K_i - K_f}{K_i} = \frac{230 \text{ J} - 120 \text{ J}}{230} = 0.48,$$

so that 48% of the initial kinetic energy is lost in the collision.

Example 5 *An elastic collision in two dimensions* A gas molecule having a speed of 300 m/s collides elastically with another molecule of the same mass which is initially at rest. After the collision the first molecule moves at an angle of 30° to its initial direction. Find the speed of each molecule after collision and the angle made with the incident direction by the recoiling target molecule.

This example corresponds exactly to the situation discussed in connection with Eqs. 10–7, with $m_1 = m_2$, $v_{1i} = 300$ m/s, and $\theta_1 = 30°$. Setting m_1 equal to m_2, Eqs. 10–7 become

$$v_{1i} = v_{1f} \cos \theta_1 + v_{2f} \cos \theta_2,$$
$$v_{1f} \sin \theta_1 = v_{2f} \sin \theta_2,$$

and

$$v_{1i}^2 = v_{1f}^2 + v_{2f}^2.$$

We must solve for v_{1f}, v_{2f}, and θ_2. To do this we square the first equation (rewriting it as $v_{1i} - v_{1f} \cos \theta_1 = v_{2f} \cos \theta_2$), and add this to the square of the second equation (noting that $\sin^2 \theta + \cos^2 \theta = 1$); we obtain

$$v_{1i}^2 + v_{1f}^2 - 2v_{1i}v_{1f} \cos \theta_1 = v_{2f}^2.$$

Combining this with the third equation, we obtain

$$2v_{1f}^2 = 2v_{1i}v_{1f} \cos \theta_1$$

or (since $v_{1f} \neq 0$)

$$v_{1f} = v_{1i} \cos \theta_1 = (300 \text{ m/s})(\cos 30°)$$

or

$$v_{1f} = 260 \text{ m/s}.$$

From the third equation

$$v_{2f}^2 = v_{1i}^2 - v_{1f}^2 = (300 \text{ m/s})^2 - (260 \text{ m/s})^2,$$

or

$$v_{2f} = 150 \text{ m/s}.$$

Finally, from the second equation

$$\sin \theta_2 = \left(\frac{v_{1f}}{v_{2f}}\right) \sin \theta_1$$
$$= \left(\frac{260 \text{ m/s}}{150 \text{ m/s}}\right)(\sin 30°) = 0.866$$

or

$$\theta_2 = 60°.$$

The two molecules move apart at right angles ($\theta_1 + \theta_2 = 90°$ in Fig. 10–6).

Show that in an elastic collision between particles of equal mass, one of which is initially at rest, the recoiling particles *always* move off at right angles to one another. This is evident in Fig. 10–1a (colliding helium nuclei) and in Fig. 10–1b (colliding billiard balls).

10-6 Reactions and Decay Processes

Conservation of total energy

We stated in Section 10–1 that reactions and spontaneous decay processes for atoms, nuclei, and elementary particles can be treated by the same methods used in collision studies, namely: We can apply the principles of conservation of linear momentum and energy to the (well-defined) periods "before the event" and "after the event." Total energy is conserved in these processes even though kinetic energy is not. In this section we consider only examples in which the speeds of the particles are negligible with respect to the speed of light. This means that we may use the classical expressions for momentum and energy and need not use the relativistic expressions.

Example 6 *A nuclear reaction* A thin film containing a fluorine (^{19}F) compound is bombarded by a beam of protons (p) which have been accelerated to an energy of 1.85 MeV (1 MeV = 1.60×10^{-13} J) in a Van de Graaff accelerator. Some of the protons interact with the fluorine nuclei to produce the following nuclear reaction:

$$^{19}\text{F} + p \rightarrow {}^{16}\text{O} + \alpha.$$

It is observed that the α particles (which are helium nuclei) that emerge at right angles to the incident proton beam (see Fig. 10–9) have speeds of 1.95×10^7 m/s. Our challenge is to learn as much as we can about the reaction by applying the laws of conservation of linear momentum and of total energy. The masses involved are, to a precision adequate for our purposes,

$$m_p = 1.01 \text{ u} \qquad m_O = 16.0 \text{ u}$$
$$m_F = 19.0 \text{ u} \qquad m_\alpha = 4.00 \text{ u,}$$

in which 1 u (*unified atomic mass unit*) = 1.66×10^{-27} kg.

The x- and y-components of linear momentum are conserved, which means that they have the same values before and after the reaction. In the laboratory reference frame of Fig. 10–9, then,

$$m_p v_p = m_O v_O \cos\theta \qquad (x\text{-component}) \qquad (10\text{–}8a)$$

and

$$m_O v_O \sin\theta = m_\alpha v_\alpha \qquad (y\text{-component}) \qquad (10\text{–}8b)$$

For the conservation of total energy we write

$$Q + \tfrac{1}{2}m_p v_p{}^2 = \tfrac{1}{2}m_O v_O{}^2 + \tfrac{1}{2}m_\alpha v_\alpha{}^2 \qquad (10\text{–}8c)$$

in which the energy balance Q is the amount by which the kinetic energy of the system after the reaction exceeds the kinetic energy before the reaction; it is exactly analogous to the heat of reaction for chemical reactions. Note that we have assumed that the particles are moving slowly enough so that we may use the classical expression for kinetic energy ($\tfrac{1}{2}mv^2$) rather than the relativistic one for total energy. If Q is positive, kinetic energy must be generated by the reaction and conversely.

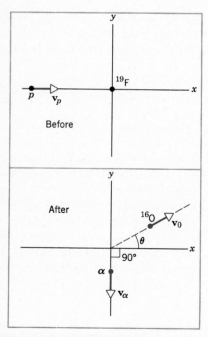

Figure 10–9 Example 6. The nuclear reaction $p + {}^{19}\text{F} \to \alpha + {}^{16}\text{O}$, showing the situation before and after the event in the laboratory reference frame.

The energy represented by Q must be associated with differences in the rest energies of the particles before or after the reaction, according to Einstein's well-known relation $E = \Delta mc^2$ (see Section 8–9) Thus (if Q is positive) we expect that the rest mass of the system after the reaction would be slightly less than its rest mass before the reaction and that Q would indeed be given by the Einstein relation

$$Q = -\Delta mc^2$$
$$= [(m_p + m_F) - (m_\alpha + m_O)]c^2. \qquad (10\text{–}9)$$

Note that Eqs. 10–8c and 10–9 are independent relations for Q, being connected through Einstein's mass-energy relation (see Problem 21).

The three conservation equations contain just three unknowns, v_O, θ, and Q. To find Q from them let us first eliminate θ between the first two equations by squaring and adding (recalling that $\cos^2\theta + \sin^2\theta = 1$). We obtain

$$m_p{}^2 v_p{}^2 + m_\alpha{}^2 v_\alpha{}^2 = m_O{}^2 v_O{}^2.$$

We can now eliminate v_O between this relation and Eq. 10–8c. After a little rearrangement, as you can easily show, we obtain

$$Q = K_\alpha \left(1 + \frac{m_\alpha}{m_O}\right) - K_p \left(1 - \frac{m_p}{m_O}\right). \qquad (10\text{–}10)$$

From the data given we know that K_p ($= \tfrac{1}{2}m_p v_p{}^2$) = 1.85 MeV and

$$
\begin{aligned}
K_\alpha &= \tfrac{1}{2}m_\alpha v_\alpha{}^2 \\
&= \tfrac{1}{2}(4.00 \text{ u})(1.66 \times 10^{-27} \text{ kg/u})(1.95 \times 10^7 \text{ m/s})^2 \\
&= (1.26 \times 10^{-12} \text{ J})\left(\frac{1 \text{ MeV}}{1.60 \times 10^{-13} \text{ J}}\right) \\
&= 7.89 \text{ MeV.}
\end{aligned}
$$

We may now calculate Q from Eq. 10–10 as

$$Q = (7.89 \text{ MeV})\left(1 + \frac{4.00}{16.00}\right) - (1.85 \text{ MeV})\left(1 - \frac{1.01}{16.0}\right)$$
$$= +8.13 \text{ MeV.}$$

Thus, by using the principles of conservation of linear momentum and total energy, we can calculate Q for the reaction without making any observations on the recoiling ${}^{16}\text{O}$ nucleus. If we want to know v_O and θ for this nucleus we can easily calculate them from Eqs. 10–8a and 10–8b.

The result $Q = 8.13$ MeV is an important bit of information about the reaction. From Eq. 10–9, which is a relation for Q independent of Eq. 10–8c, we can now calculate that the decrease in rest mass during the reaction is given by

$$
\begin{aligned}
-\Delta m &= \frac{Q}{c^2} \\
&= \frac{(8.13 \text{ MeV})(1.60 \times 10^{-13} \text{ J/MeV})}{(3.00 \times 10^8 \text{ m/s})^2} \\
&= (1.44 \times 10^{-29} \text{ kg})\left(\frac{1 \text{ u}}{1.66 \times 10^{-27} \text{ kg}}\right) \\
&= 0.00871 \text{ u.}
\end{aligned}
$$

We can verify this result by calculating Δm [$= (m_p + m_F) - (m_\alpha + m_O)$] from very precise mass-spectrometer measurements of the four separate masses. The excellent agreement that we get shows once again the validity of Einstein's mass-energy relationship (see Problem 21).

10–7 Cross Section

In many experimental situations, especially with particles of atomic or subatomic dimensions, we find that we cannot analyze the collisions individually because we do not have enough information about the situation just before collision. An example is the bombardment of nuclei in a thin target foil by a beam of particles accelerated in a cyclotron. We simply cannot aim individual projectile particles in the beam at individual target nuclei in the foil. In such cases we must deal statistically with averages over a large number of collisions.

The situation is much the same as if we were throwing stones at random (in the dark, say) at the side of a distant barn of area A on which someone had hung a number of small dinner plates, each of area σ, in random but not overlapping positions. If the number of plates is q and if the rate at which stones strike the barn is R_0, what is the rate R at which plates are broken?

The chance of hitting a plate with a single throw is $\sigma q/A$, since σq is the total area of the plates and we assume that every stone hits the barn wall somewhere. $\sigma q/A$ is just the fraction of the wall that is covered by plates. If this fraction is small (or if the plates are replaced every time one is hit), the total number of plates that we would expect to be broken after R_0 stones have been thrown would just be R_0 times the chance that a single stone would strike a plate. Thus the rate at which plates will be broken would be expected to be

$$R = R_0 \left(\sigma \frac{q}{A} \right). \tag{10–11}$$

We could in fact use this relation to measure σ, the geometrical area of a single plate. Solving for σ yields

Reaction cross section

$$\sigma = \frac{RA}{R_0 q}$$

which permits us to find σ from measured values of R, A, R_0, and q. We may call σ the *cross section* for the event consisting of the impact of a stone on a plate.

Similarly, in nuclear physics we often bombard targets with nuclear projectiles, measure the rate at which events of a selected type occur, and calculate a cross section for those events. Although the details of individual collisions remain unknown, much information is obtained that is independent of the number or density of target atoms or the flux of incident particles. For example, if we bombarded a thin gold foil (^{197}Au) with deuterons (^{2}H, or d) whose energy is 30 MeV, many events would occur. These would include elastic scattering of the deuterons, the nuclear reaction $d + {}^{197}\text{Au} \rightarrow p + {}^{198}\text{Au}$, and the reaction $d + {}^{197}\text{Au} \rightarrow n + {}^{198}\text{Hg}$, where d represents a deuteron, n a neutron, and p a proton. Each of these events (and many others) has its own cross section σ_x.

Let the area of the foil exposed to the beam be A and the thickness of the foil be d. If there are n target particles per unit volume in the foil, the total number of available target particles is nAd. If the effective area (that is, the cross section) for the event we are concerned with is σ_x, the total effective area of all the nuclei is $nA\sigma_x d$. If R_0 is the rate at which projectiles strike the target and R_x is the rate at which the events in which we are interested occur, we have, because of the random nature of the events (see Eq. 10–11),

Calculating reaction rates

$$\frac{R_x}{R_0} = \frac{(nA\sigma_x d)}{A}$$

or

$$R_x = R_0 n \sigma_x d. \tag{10–12}$$

Thus we can measure σ_x for the event by measuring R_x, R_0, n, and d and substituting into Eq. 10–12. Cross sections are commonly expressed in *barns* or submultiples thereof; one barn $= 10^{-28}$ m².

In atomic, nuclear, and particle physics the cross section only rarely has anything to do with the geometrical area of the target (whatever that may mean). It is an *effective* target area, measuring the probability that a given event will occur. Cross sections for particular events generally vary with the energy of the incident particle, often reaching very high values at sharply defined energies.

Example 7   If a slow neutron is absorbed by a uranium-235 nucleus, the nucleus so formed may, with high probability, break up into two roughly equal parts. The cross section σ_f for this *nuclear fission* process is 580 barns.

A slow-neutron beam of intensity R_0 ($=3.0 \times 10^{16}$ neutrons/m² · s) falls on a thin uranium foil whose thickness d is 5.0×10^{-4} cm. At what rate R_f will fission events occur? (The density of uranium is 18.9 g/cm³ and its atomic weight A is 238 g/mol. In uranium as found in nature only 0.72% of the uranium nuclei are ^{235}U, essentially all of the others being ^{238}U.)

We must first find n in Eq. 10–12, the number of target nuclei per unit volume. If N_0 is Avagadro's number, the number of atoms per mol, we have, for the number of uranium nuclei per unit volume,

$$n_U = \frac{N_0\, d}{A} = \frac{(6.0 \times 10^{23} \text{ nuclei/mol})(18.9 \text{ g/cm}^3)}{(238 \text{ g/mol})}$$
$$= 4.8 \times 10^{22} \text{ nuclei/cm}^3.$$

Of these only 0.72% are ^{235}U or

$$n = (0.0072)(4.8 \times 10^{22} \text{ nuclei/cm}^3) = 3.5 \times 10^{20} \text{ nuclei/cm}^3.$$

From Eq. 10–12, then,

$$
\begin{aligned}
R_f &= R_0 n \sigma_f d \\
&= (3.0 \times 10^{16}/\text{m}^2 \cdot \text{s})(3.5 \times 10^{20}/\text{cm}^3) \\
&\quad (580 \text{ barns})(5.0 \times 10^{-4} \text{ cm})(10^{-28} \text{ m}^2/\text{barn}) \\
&= 3.0 \times 10^8 \text{ fission events/cm}^2 \cdot \text{s}.
\end{aligned}
$$

The geometric cross-sectional area of a ^{235}U nucleus turns out to be 1.4 barns, much less than its cross section for fission by slow neutrons ($=580$ barns). This stresses again that reaction cross sections, although measured in units of area, are measures of reaction probability and are not related to the geometrical size of the nucleus. For example, ^{238}U, which has virtually the same geometrical cross-sectional area as ^{235}U, is much less susceptible to fission by slow neutrons, the ^{238}U fission cross section σ_f for such neutrons being only $\sim 5 \times 10^{-4}$ barn.

REVIEW GUIDE AND SUMMARY

Collisions

In a collision strong internal forces act between two particles for a short time. The definition may be broadened to include reactions, in which the identity of the particles changes, and spontaneous decays, which can be viewed as collisions in reverse. In all cases the challenge is to see what can be learned by applying the laws of conservation of momentum and energy to the "before" and "after" situations, assuming no knowledge of the processes or forces occuring during the "collision."

Impulse and momentum

Applying Newton's second law to either of the particles in a collision leads to

$$\mathbf{J} = \mathbf{p}_f - \mathbf{p}_i, \qquad\qquad [10\text{–}2b]$$

where $\mathbf{J}$ ($= \int_{t_i}^{t_f} \mathbf{F}\, dt$) is the impulse acting on that particle and $\mathbf{p}_f - \mathbf{p}_i$ is the change in its momentum; see Example 1. Note that the total momentum $\mathbf{P}$ for any system of two (or more) particles does not change during the collision, assuming that there are no external forces.

Elastic collision—one dimension

If kinetic energy is conserved during a collision, we call the collision elastic. Equations 10–6 give the final speeds for the two particles after a head-on, one-dimensional collision of this type; see Example 2. If the particle of mass m_2 is initially at rest, these equations become

$$v_{1f} = \left(\frac{m_1 - m_2}{m_1 + m_2}\right) v_{1i} \quad \text{and} \quad v_{2f} = \left(\frac{2m_1}{m_1 + m_2}\right) v_{1i}.$$

Completely inelastic collision

If the particles stick together after the collision, we call the collision completely inelastic. The common final velocity v_f follows from conservation of momentum alone, whether the collision is one-dimensional (as in Example 3) or two-dimensional (as in Example 4). Conservation of kinetic energy does *not* hold.

Elastic collision—two dimensions

Nuclear and subnuclear reactions

Reaction cross section

To solve elastic collisions in two dimensions (Example 5) we need not only conservation of momentum and kinetic energy but also one additional bit of information about the final state, typically the scattering angle of one of the particles.

In nuclear and subnuclear reactions (Example 6) we need the laws of conservation of momentum and of total energy, including the rest energy $m_0 c^2$. An important parameter is the energy balance or Q of the reaction, given by

$$Q = \Delta K \quad \text{or} \quad Q = -\Delta mc^2.$$

Here ΔK ($= K_f - K_i$) is the change in total kinetic energy during the reaction and Δm ($= m_f - m_i$) is the change in the rest mass resulting from the collision.

If we know the cross section σ for a specified nuclear or other reaction we can calculate, from Eq. 10–12, the rate at which such reactions will occur in a given experimental setup. Cross sections, which measure reaction probabilities and can be thought of as effective target areas, are given in barns; one barn = 10^{-28} m^2.

QUESTIONS

1. Explain how conservation of momentum applies to a handball bouncing off a wall.

2. Although the acceleration of a baseball after it has been hit does not depend on who hit it. Something about the baseball's flight must depend on the batter. What is it?

3. Many features on cars, such as collapsible steering wheels and padded dashboards, are meant to change more safely the momentum of the passengers during accidents. Explain their usefulness, using the impulse concept.

4. It is said that, during a 30-mi/h collision, a 10-lb child can exert a 300-lb force against a parent's grip. How can such a large force come about?

5. Taken from an exam paper: "The collision between two helium atoms is perfectly elastic, so that momentum is conserved." What do you think of this statement?

6. Susan is driving along a highway at 50 mi/h, followed by Ruth who is driving at the same speed. Susan slows to 40 mi/h, but Ruth does not, and there is a collision. What are the initial velocities of the colliding cars as seen from the reference frame of (*a*) Susan, (*b*) Ruth, and (*c*) trooper Paul, who is in a patrol car parked by the roadside? (*d*) Judge Sandra asks whether Ruth bumped into Susan or Susan bumped into Ruth. As a physicist, how would you answer?

7. It is obvious from inspection of Eqs. 10–3 and 10–4 that a valid solution to the problem of finding the final velocities of two particles in a one-dimensional elastic collision is $v_{1f} = v_{1i}$ and $v_{2f} = v_{2i}$. What does this mean physically?

8. Two identical cubical blocks, moving in the same direction with a common speed v, strike a third such block initially at rest on a horizontal frictionless surface. What is the motion of the blocks after the collision? Does it matter whether or not the two initially moving blocks were in contact or not? Does it matter whether these two blocks were glued together?

9. C. R. Daish has written that, for professional golfers, the initial speed of the ball off the clubhead is about 140 mi/h. He also says: (*a*) ". . . if the Empire State Building could be swung at the ball at the same speed as the clubhead, the initial ball velocity would only be increased by about 2% . . ."; and (*b*) that, once the golfer has started his downswing, camera clicking, sneezing, and so on, can have no effect on the motion of the ball. Can you give qualitative arguments to support these two statements?

10. Two clay balls of equal mass and speed strike each other head-on, stick together, and come to rest. Kinetic energy is certainly not conserved. What happened to it? Is momentum conserved?

11. A football player, momentarily at rest on the field, catches a football as he is tackled by a running player on the other team. This is certainly a collision (inelastic!) and momentum must be conserved. In the reference frame of the football field there is momentum before the collision but there seems to be none after the collision. Is linear momentum really conserved?

12. Consider a one-dimensional elastic collision between a given incoming body A and a body B initially at rest. How would you choose the mass of B, in comparison to the mass of A, in order that B should recoil with (*a*) the greatest speed, (*b*) the greatest momentum, and (*c*) the greatest kinetic energy?

13. An hourglass is being weighed on a sensitive balance, first when sand is dropping in a steady stream from the upper to the lower part and then again after the upper part is empty. Are the two weights the same or not? Explain your answer.

14. Give a plausible explanation for the breaking of wooden boards or of bricks by a karate punch (see "Karate Strikes," by Jearl D. Walker, *American Journal of Physics,* October 1975).

15. An evacuated box is at rest on a frictionless table. You punch a small hole in one face so that air can enter; see Fig. 10–10. How will the box move? What argument did you use to arrive at your answer?

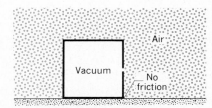

Figure 10–10 Question 15.

16. Mexican jumping beans have a hard rigid shell and contain a small live insect. If you put such beans on a hard table top they will occasionally jump. What might the insect be doing to make this happen?

EXERCISES

Section 10–2   Impulse and Momentum

1. A ball of mass m and speed v strikes a wall perpendicularly and rebounds with undiminished speed. If the time of collision is t, what is the average force exerted by the ball on the wall?

2. A stream of water impinges on a stationary "dished" turbine blade, as shown in Fig. 10–11. The speed of the water is u, both before and after it strikes the curved surface of the blade, and the mass of water striking the blade per unit time is constant at the value μ. Find the force exerted by the water on the blade.

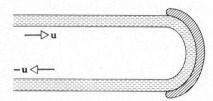

Figure 10–11   Exercise 2.

3. A cue strikes a pool ball, exerting an average force of 50 N over a time of 10 ms. If the ball has mass 0.20 kg, what speed does it have after impact?

4. A 150-g (weight = 5.3 oz) baseball pitched at 40 m/s (= 130 ft/s) is hit straight back to the pitcher at 60 m/s (= 200 ft/s). What average force was exerted by the bat if it was in contact with the ball for 5.0 ms?

5. A force that averages 1000 N is applied to a 0.40-kg steel ball moving at 14 m/s by a collision lasting 27 ms. If the force is in a direction opposite to the initial velocity of the ball, find the final speed of the ball.

Section 10–4   Collisions in One Dimension

6. A bullet of mass 10 g strikes a ballistic pendulum of mass 2.0 kg. The center of mass of the pendulum rises a vertical distance of 12 cm. Assuming that the bullet remains embedded in the pendulum, calculate its initial speed.

7. A 5.2-g bullet moving at 672 m/s strikes a 700-g wooden block at rest on a very smooth surface. The bullet emerges with its speed reduced to 428 m/s. Find the resulting speed of the block.

8. A 6.0-kg box sled is coasting across the ice at a speed of 9.0 m/s when a 12-kg package is dropped into it from above. What is the new speed of the sled?

9. An α particle (mass 4 u) experiences an elastic head-on collision with a gold nucleus (mass 197 u) that is originally at rest. What is the fractional loss of kinetic energy for the α particle?

10. An electron collides elastically with a hydrogen atom initially at rest. The initial and final motions are along the same straight line. What fraction of the electron's initial kinetic energy is transferred to the hydrogen atom? The mass of the hydrogen atom is 1840 times the mass of the electron.

11. A hovering fly is approached by an enraged elephant charging at 2 m/s. Assuming that the collision is elastic, at what speed does the fly rebound?

12. A 5.0-kg particle with a velocity of 3.0 m/s collides with a 10-kg particle that has a velocity of 2.0 m/s in the same direction.

After the collision, the 10-kg particle is observed to be traveling in the original direction with a speed of 4.0 m/s. (*a*) What is the velocity of the 5.0-kg particle immediately after the collision? (*b*) By how much does the total kinetic energy of the system of two particles change because of the collision?

13. Meteor Crater in Arizona (see Fig. 10–12) is thought to have been formed by the impact of a meteor with the earth some 20,000 years ago. The mass of the meteor is estimated to be 5×10^{10} kg, and its speed 7200 m/s. What speed would such a meteor impart to the earth in a head-on collision?

Figure 10–12   Exercise 13. Georg Gerster/Photo Researchers, Inc.

14. An object of mass m and speed v explodes into two pieces, one three times as massive as the other; the explosion takes place in gravity-free space. The lighter piece comes to rest. How much kinetic energy was added to the system?

Section 10–5   Collisions in Two and Three Dimensions

15. A 2500-kg unmanned spaceprobe is moving in a straight line at a constant speed of 300 m/s. A rocket engine on the spaceprobe executes a burn in which a thrust of 3000 N acts for 65 s. (*a*) What is the change in momentum (magnitude only) of the probe if the thrust is backward? . . . forward? . . . sideways? (*b*) What is the change in kinetic energy under the same three conditions? Assume that the mass of the ejected fuel is negligible compared to the mass of the spaceprobe.

16. In a game of billiards, the cue ball is given an initial speed V and strikes the pack of 15 stationary balls. All 16 balls then engage in numerous ball–ball and ball–cushion collisions. Sometime later it is observed that (by some accident) all 16 balls have the same speed v. Assuming that all collisions are elastic and ignoring the rotational aspect of the balls' motion, calculate v in terms of V.

17. Two vehicles A and B are traveling west and south, respectively, toward the same intersection, where they collide and lock together. Before the collision A (total weight, 2700 lb) is moving with a speed of 40 mi/h and B (total weight, 3600 lb) has a speed of

60 mi/h. Find the magnitude and direction of the velocity of the (interlocked) vehicles immediately after the collision.

18. A proton (atomic mass 1 u) with a speed of 500 m/s collides elastically with another proton at rest. The original proton is scattered 60° from its initial direction. (a) What are the speeds of the two protons after the collision? (b) What is the direction of the velocity of the target proton after the collision?

19. Two balls A and B, having different but unknown masses, collide. A is initially at rest when B has a speed v. After collision B has a speed $v/2$ and moves at right angles to its original motion. (a) Find the direction in which ball A moves after collision. (b) Can you determine the speed of A from the information given? Explain.

20. In a game of pool the cue ball strikes another ball initially at rest. After the collision, the cue ball moves at 3.50 m/s along a line making an angle of 65.0° with its original direction of motion. The

second ball acquires a speed of 6.75 m/s. (a) What is the angle between the direction of motion of the second ball and the original direction of motion of the cue ball? (b) What was the original speed of the cue ball?

Section 10–6 Reactions and Decay Processes

21. The precise masses in the reaction

$$p + {}^{19}F \rightarrow \alpha + {}^{16}O$$

have been determined by mass spectrometer measurements and are

$$m_p = 1.007825 \text{ u}, \qquad m_\alpha = 4.002603 \text{ u},$$
$$m_F = 18.998405 \text{ u}, \qquad m_O = 15.994915 \text{ u}.$$

Calculate the Q of the reaction from these data and compare with the Q calculated in Example 6 from reaction studies.

PROBLEMS GROUPED BY SECTION

Section 10–2 Impulse and Momentum

1. A 1.2-kg ball drops vertically onto the floor with a speed of 25 m/s. It rebounds with an initial speed of 10 m/s. (a) What impulse acts on the ball during contact? (b) If the ball is in contact for 0.020 s, what is the average force exerted on the floor?

2. A ball having a mass of 50 g strikes a wall with a speed of 5 m/s and rebounds with only 50% of its initial kinetic energy. (a) What is the final speed of the ball after rebounding? (b) What was the impulse delivered by the ball to the wall? (c) If the ball was in contact with the wall for 0.035 s, what was the average force exerted by the wall on the ball during this time interval?

3. A force exerts an impulse J on an object of mass m, changing its speed from v to u. The force and the object's motion are along the same straight line. Show that the work done by the force is $\frac{1}{2} J(u + v)$.

4. A golfer hits a golf ball, imparting to it an initial velocity of magnitude 50 m/s directed 30° above the horizontal. Assuming that the mass of the ball is 25 g and the club and ball are in contact for 0.01 s, find (a) the impulse imparted to the ball; (b) the impulse imparted to the club; (c) the average force exerted on the ball by the club; (d) the work done on the ball.

5. Two parts of a spacecraft are separated by detonating the explosive bolts that hold them together. The masses of the parts are 1200 kg and 1800 kg; the magnitude of the impulse delivered to each part is 300 N·s. What is the relative speed of recession of the two parts?

Section 10–4 Collisions in One Dimension

6. A railroad freight car weighing 32 tons and traveling 5.0 ft/s overtakes one weighing 24 tons traveling 3.0 ft/s in the same direction. (a) Find the speed of the cars after collision and the loss of kinetic energy during collision if the cars couple together. (b) If instead, as is very unlikely, the collision is elastic, find the speed of the cars after collision.

7. A bullet of mass 4.5 g is fired horizontally into a 2.4-kg wooden block at rest on a horizontal surface. The coefficient of kinetic friction between block and surface is 0.20. The bullet comes to rest in the block, which moves 1.8 m. Find the speed of the bullet.

8. A 3-ton weight falling through a distance of 6 ft drives a 0.5-ton pile one inch into the ground. Assuming that the weight–pile collision is completely inelastic, find the average force of resistance exerted by the ground.

9. A body of 2.0-kg mass makes an elastic collision with another body at rest and continues to move in the original direction but with one-fourth of its original speed. What is the mass of the struck body?

10. A block of mass m_1 is at rest on a long frictionless table, one end of which is terminated by a wall. Another block of mass m_2 is placed between the first block and the wall and set in motion to the left with constant speed v_{2i}, as in Fig. 10–13. Assuming that all collisions are completely elastic, find the value of m_2 for which both blocks move with the same velocity after m_2 has collided once with m_1 and once with the wall. Assume the wall to have infinite mass.

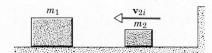

Figure 10–13 Problem 10.

11. A box is put on a scale that is adjusted to read zero when the box is empty. A stream of marbles is then poured into the box from a height h above its bottom at a rate of R (marbles per second). Each marble has a mass m. If the collisions between the marbles and the box are completely inelastic, find the scale reading at time t after marbles begin to fill the box. Determine a numerical answer when $R = 100 \text{ s}^{-1}$, $h = 25$ ft $(=7.6 \text{ m})$, $m = 3.125 \times 10^{-4}$ slug $(=4.5 \text{ g})$, and $t = 10$ s.

12. Two 50-lb (mass = 22.7 kg) ice sleds are placed a short distance apart, one directly behind the other, as shown in Fig. 10–14. An 8-lb (mass = 3.63 kg) cat, standing on one sled, jumps across to the other and immediately back to the first. Both jumps are made at a speed of 10 ft/s (= 3.05 m/s) relative to the ice. Find the final speeds of the two sleds.

Figure 10–14 Problem 12.

13. A 35-ton railroad freight car collides with a stationary caboose car. They couple together and 27% of the initial kinetic energy is dissipated as heat, sound, vibrations, etc. Find the weight of the caboose.

14. A block of mass $m_1 = 2$ kg slides along a frictionless table with a speed of 10 m/s. Directly in front of it, and moving in the same direction, is a block of mass $m_2 = 5$ kg moving at 3 m/s. A massless spring with a spring constant $k = 1120$ N/m is attached to the backside of m_2, as shown in Fig. 10–15. When the blocks collide, what is the maximum compression of the spring?

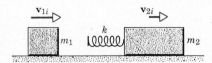

Figure 10–15 Problem 14.

*15. The two balls on the right of Fig. 10–16 are slightly separated and initially at rest; the third ball, on the left, is incident with speed v_0. Assuming head-on elastic collisions, (a) if $M \leq m$, show that there are two collisions and find all final velocities; (b) if $M > m$, show that there are three collisions and find all final velocities.

Figure 10–16 Problem 15.

Section 10–5   Collisions in Two and Three Dimensions
16. An α particle collides with an oxygen nucleus, initially at rest. The α particle is scattered at an angle of 64° above its initial direction of motion and the oxygen nucleus recoils at an angle of 51° below this initial direction. What is the ratio, α-particle to nucleus, of the final speeds of these particles? The mass of the oxygen nucleus is four times that of the α particle.

17. A 20-kg body is moving in the direction of the positive x-axis with a speed of 200 m/s when, due to an internal explosion, it

breaks into three parts. One part, whose mass is 10 kg, moves away from the point of explosion with a speed of 100 m/s along the positive y-axis. A second fragment, with a mass of 4 kg, moves along the negative x-axis with a speed of 500 m/s. (a) What is the velocity of the third (6-kg) fragment? (b) How much energy was released in the explosion? Ignore effects due to gravity.

18. A barge of mass 1.50×10^5 kg is proceeding downriver at 6.2 m/s in heavy fog when it collides broadside with a barge heading directly across the river; see Fig. 10–17. The second barge has mass 2.78×10^5 kg and was moving at 4.3 m/s. Immediately after impact the second barge finds its course deflected by 18° in the downriver direction and its speed increased to 5.1 m/s. The river current was practically zero at the time of the accident. (a) What is the speed and direction of motion of the first barge immediately after the collision? (b) How much kinetic energy was lost in the collision?

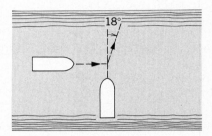

Figure 10–17 Problem 18.

19. After a totally inelastic collision, two objects of the same mass and initial speed are found to move away together at half their initial speed. Find the angle between the initial velocities of the objects.

*20. Show that, in the case of an elastic collision between a particle of mass m_1 with a particle of mass m_2 initially at rest, (a) the maximum angle θ_m through which m_1 can be deflected by the collision is given by $\cos^2\theta_m = 1 - m_2^2/m_1^2$, so that $0 \leq \theta_m \leq \pi/2$, when $m_1 > m_2$; (b) $\theta_1 + \theta_2 = \pi/2$, when $m_1 = m_2$; (c) θ_1 can take on all values between 0 and π, when $m_1 < m_2$.

Section 10–6   Reactions and Decay Processes
21. The Q of the reaction in which a ^{236}U nucleus at rest splits into just two fragments of masses 132 u and 98 u is 192 MeV. (a) How much energy was lost through radiation? (b) What is the speed of each fragment? (c) What is the kinetic energy of each fragment?

ADDITIONAL PROBLEMS

22. A 300-g ball with a speed of 6.0 m/s strikes a wall at an angle θ of 30° and then rebounds with the same speed and angle (Fig. 10–18). It is in contact with the wall for 10 ms. (a) What impulse was experienced by the ball? (b) What was the average force exerted by the ball on the wall?

23. The force on a 10-kg object increases uniformly from zero to 50 N in 4 s. What is the object's final speed if it started from rest?

24. Two titanium spheres approach each other head-on with the same speed and collide elastically. After the collision, one of the spheres, whose mass is 300 g, remains at rest. What is the mass of the other sphere?

25. A stream of water from a hose is sprayed on a wall. If the speed of the water is 5.0 m/s (= 16 ft/s) and the hose sprays 300 cm³/s (= 0.011 ft³/s), what is the average force exerted on the wall by the stream of water? Assume that the water does not spatter back appreciably. Each cubic centimeter of water has a mass of 1.0 g.

26. A billiard ball moving at a speed of 2.2 m/s strikes an identical stationary ball a glancing blow. After the collision one ball is found to be moving at a speed of 1.1 m/s in a direction making a 60° angle with the original line of motion. (a) Find the velocity of the other ball. (b) Can the collision be inelastic, given these data?

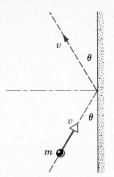

Figure 10–18 Problem 22.

the mass of the strings and any frictional effects. How high does the center of mass rise after the collision?

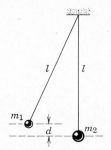

Figure 10–20 Problem 31.

27. A cart with mass 0.10 kg moving on a frictionless linear air track at an initial speed of 10 m/s strikes a second cart of unknown mass at rest. The collision between the carts is elastic. After the collision, the first cart continues in its original direction at 5.0 m/s. (*a*) What is the mass of the second cart? (*b*) What is its speed after impact?

28. A particle called Σ^- (sigma minus), at rest in a certain reference frame, decays spontaneously into two other particles according to

$$\Sigma^- \rightarrow \pi^- + n.$$

The masses are

$$m_\Sigma = 2340.5\ m_e,$$
$$m_\pi = 273.2\ m_e,$$
$$m_n = 1838.65\ m_e,$$

where m_e is the electron rest mass. (*a*) How much kinetic energy is generated in this process? (*b*) Which of the decay products (π^- and n) gets the larger share of this kinetic energy? Of the momentum?

29. An α-particle of kinetic energy 7.70 MeV strikes a ^{14}N nucleus at rest. A ^{17}O nucleus and a proton are produced, the proton emitted at 90° to the direction of the incident α particle and carrying kinetic energy of 4.44 MeV. (*a*) What is the Q of the reaction? (*b*) What is the kinetic energy of the oxygen nucleus?

30. A steel ball of mass 0.50 kg is fastened to a cord 70 cm long and is released when the cord is horizontal. At the bottom of its path the ball strikes a 2.5-kg steel block initially at rest on a frictionless surface (Fig. 10–19). The collision is elastic. Find (*a*) the speed of the ball and (*b*) the speed of the block just after the collision.

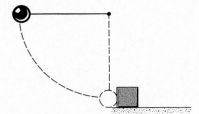

Figure 10–19 Problem 30.

31. Two pendulums each of length l are initially situated as in Fig. 10–20. The first pendulum is released and strikes the second. Assume that the collision is completely inelastic and neglect

32. Show that a slow neutron (called a *thermal* neutron) that is scattered through 90° in an elastic collision with a deuteron, initially at rest, loses two-thirds of its initial kinetic energy to the struck deuteron.

33. Two particles, one having twice the mass of the other, are held together with a compressed spring between them. The energy stored in the spring is 60 J. How much kinetic energy does each particle have after they are released? Assume that all the stored energy is transferred to the particles and that neither particle is attached to the spring after they are released.

34. A certain nucleus, at rest, spontaneously disintegrates into three particles. Two of them are detected; their masses and velocities are as shown in Fig. 10–21. (*a*) What is the momentum of the third particle, which is known to have a mass of 11.7×10^{-27} kg? (*b*) How much energy was involved in the disintegration process?

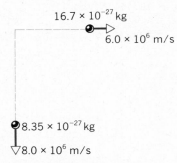

Figure 10–21 Problem 34.

35. A ball with an initial speed of 10 m/s collides elastically with two identical balls whose centers are on a line perpendicular to the initial velocity and that are initially in contact with each other (Fig. 10–22). The first ball is aimed directly at the contact point, and all the balls are frictionless. Find the velocities of all three balls after the collision.

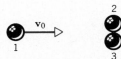

Figure 10–22 Problem 35.

36. It is well known that bullets and other missiles fired at Superman simply bounce off his chest as in Fig. 10–23. Suppose that a gangster sprays Superman's chest with 3-g bullets at the rate of 100 bullets/min, the speed of each bullet being 500 m/s. Suppose, too, that the bullets rebound straight back with no loss in speed. Show that the average force exerted by the stream of bullets on Superman's chest is only 5 N (=18 oz).

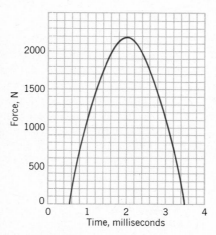

Figure 10–23 Problem 36. SUPERMAN is a registered trademark of DC Comics Inc. Copyright © 1963 and is used with permission.

Figure 10–24 Problem 37.

37. A croquet ball (mass 0.50 kg) is struck by a mallet, receiving the impulse shown in the graph (Fig. 10–24). What is the ball's velocity just after the force has become zero?

38. An elevator is moving up a shaft at 6.0 ft/s. At the instant the elevator is 60 ft from the top, a ball is dropped from the top of the shaft. The ball rebounds elastically from the elevator roof. (*a*) To what height can it rise relative to the top of the shaft? (*b*) Do the same problem assuming that the elevator is moving down at 6.0 ft/s.

39. A platform scale is calibrated to indicate the mass in kilograms of an object placed on it. Particles fall from a height of 3.5 m (=11.5 ft) and collide with the balance pan of the scale. The collisions are elastic; the particles rebound upward with the same speed they had before hitting the pan. If each particle has a mass of 110 g (=1/128 slug) and collisions occur at the rate of 42 s^{-1}, what is the scale reading?

40. The bumper of a 1200-kg car is designed so that it can just absorb all the energy when the car runs head-on into a solid stone wall at 5.0 km/h. The car is involved in a collision in which it runs at 70 km/h into the rear of a 900-kg car ahead moving at 60 km/h in the same direction. The 900-kg car is accelerated to 70 km/h as a result of the collision. (*a*) What is the speed of the 1200-kg car immediately after impact? (*b*) What is the ratio of the kinetic energy liberated in the collision to that which can be absorbed by the bumper of the 1200-kg car?

41. A ball of mass m is projected with speed v_i into the barrel of a spring-gun of mass M initially at rest on a frictionless surface; see Fig. 10–25. The ball sticks in the barrel at the point of maximum compression of the spring. No energy is lost in friction. What fraction of the initial kinetic energy of the ball is stored in the spring?

Figure 10–25 Problem 41.

42. An electron, mass m, collides head-on with an atom, mass M, initially at rest. As a result of the collision a characteristic amount of energy E is stored internally in the atom. What is the minimum initial speed v_0 that the electron must have? (Hint: Conservation principles lead to a quadratic equation for the final electron speed v and a quadratic equation for the final atom speed V. The minimum value v_0 follows from the requirement that the radical in the solutions for v and V be real.)

11 Rotational Kinematics

11-1 Rotational Motion

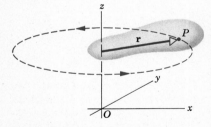

Figure 11-1 A rigid body rotating about the z-axis. Each point in the body, such as P, describes a circle about this axis.

Pure rotation

So far we have dealt mostly with the translational motion of single particles or of rigid bodies, that is, of bodies whose parts all have a fixed relationship to each other. No real body is truly rigid, but many bodies, such as molecules, steel beams, and planets, are rigid enough so that, in many problems, we can ignore the fact that they warp, bend or vibrate. We say that a rigid body moves in *pure translation* if each particle of the body undergoes the same displacement as every other particle in any given time interval.

In this chapter we are interested in the *rotation* of rigid bodies. We shall not consider such nonrigid rotational motions as those of the solar system or of water in a spinning beaker. We shall also deal only with rotation about axes that remain fixed in the reference frame in which we observe the rotation.

Figure 11-1 shows the rotational motion of a rigid body about a fixed axis, in this case the z-axis of our reference frame. Let P represent a particle in the rigid body, arbitrarily selected and described by the position vector **r**. We then say that: *A rigid body moves in pure rotation if every particle of the body* (such as P in Fig. 11-1) *moves in a circle, the center of which lies on a straight line called the axis of rotation* (the z-axis in Fig. 11-1). If we draw a perpendicular from any point in the body to the axis, each such line will sweep through the same angle in any given time interval as another such line. Thus we can describe the pure rotation of a rigid body by considering the motion of any one of the particles (such as P) that make it up. (We must rule out, however, particles that are on the axis of rotation. Why?)

The general motion of a rigid body is a combination of translation and rotation, however, rather than one of pure rotation. As we saw in Chapter 9, we can describe the translational motion of any system of particles—whether rigid or not, whether rotating or not—by imagining that all of the mass M of the body is concentrated at the center of mass and that $\mathbf{F}_{\text{ext}}$, the resultant of the external forces acting on the body, acts at this point. The acceleration of the center of mass is then given by Eq. 9–8 or $\mathbf{F}_{\text{ext}} = M\mathbf{a}_{\text{cm}}$. It is very helpful to be able to represent the translational motion of a rigid body by the motion of a single point—its center of mass; all that is left is to determine its rotational motion. To do this, we must first describe the rotational motion. We call this description *rotational kinematics;* we must define the variables of angular motion and relate them to each other, just as in particle kinematics we defined the variables of translational motion and related them to each other. The next part of our program is to relate the rotational motion of a body to the properties of the body and of its environment. This is rotational dynamics. In this chapter we study the kinematics of rotation. We develop the dynamics of rotation in the next chapter.

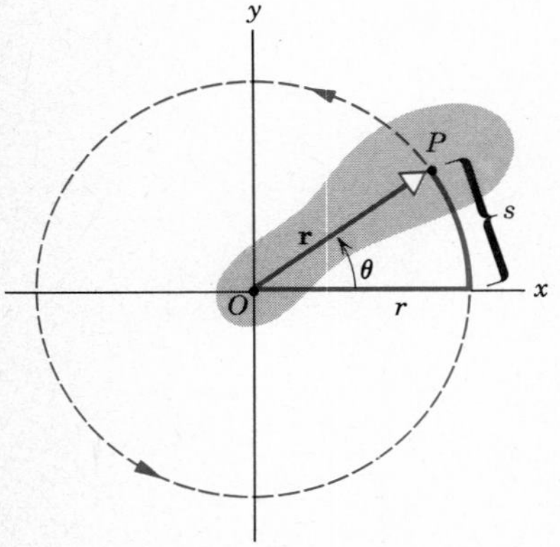

11–2 Rotational Kinematics—the Variables

In Fig. 11–1 let us pass a plane through P at right angles to the axis of rotation. This plane, which cuts through the rotating body, contains the circle in which particle P moves. Figure 11–2 shows this plane, as we look downward on it from above, along the z-axis in Fig. 11–1.

We can tell exactly where every point of the rotating body is in our reference frame if we know the location of any single particle (P) of the body in this frame. Thus, for the kinematics of this problem, we need consider only the (two-dimensional) motion of a particle in a circle. By convention we choose the positive sense of rotation in Fig. 11–2 to be counterclockwise, so that θ increases for counterclockwise rotation and decreases for clockwise rotation.

It is convenient to measure θ in radians rather than in degrees. By definition θ is given in radians by the relation

$$\theta = \frac{s}{r},$$

in which s is the arc length shown in Fig. 11–2.

Note that the radian is a pure number, having no physical dimension (in terms of our fundamental quantities M, L, and T) since it is the ratio of two lengths. Since the circumference of a circle of radius r is $2\pi r$, there are 2π radians in a complete circle (1 revolution). Therefore 1 revolution = 2π radians = 360°, and 1 radian = 57.3°.

Let the body of Fig. 11–2 be rotating counterclockwise. At time t_1 the angular position of P is θ_1 and at a later time t_2 its angular position is θ_2. This is shown in Fig. 11–3, which gives the positions of P and of the position vector $\mathbf{r}$ at these times; the outline of the body itself has been omitted in that figure for simplicity. The *angular displacement* of P will be $\theta_2 - \theta_1 = \Delta\theta$ during the time interval $t_2 - t_1 = \Delta t$. We define the *average angular speed* $\bar{\omega}$ of particle P in this time interval as

$$\bar{\omega} = \frac{\theta_2 - \theta_1}{t_2 - t_1} = \frac{\Delta\theta}{\Delta t}.$$

The *instantaneous angular speed* ω is then the limit approached by this ratio as Δt approaches zero:

$$\omega = \lim_{\Delta t \to 0} \frac{\Delta\theta}{\Delta t} = \frac{d\theta}{dt}. \tag{11–1}$$

Figure 11–2 A cross-sectional view of the rigid body of Fig. 11–1, showing point P and vector $\mathbf{r}$ of that figure. Point P, which is fixed in the rotating body, rotates counterclockwise about the origin in a circle of radius r. The angle θ in radians is defined as s/r.

Angular position

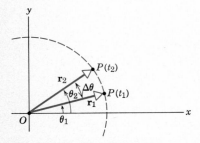

Figure 11–3 The reference line r ($= OP$), fixed in the body of Figs. 11–1 and 11–2, is displaced through angle $\Delta\theta (= \theta_2 - \theta_1)$ in time $\Delta t (= t_2 - t_1)$.

Angular speed

For a rigid body all radial lines fixed in it perpendicular to the axis of rotation rotate through the same angle in the same time, so that the angular speed ω about this axis is the same for each particle in the body. Thus ω is characteristic of the body as a whole. The units of angular speed are commonly taken to be radians per second (rad/s) or revolutions per second (rev/s). Since the angular units are dimensionless, angular speed has the dimension of inverse time (T^{-1}).

If the angular speed of P is not constant, then the particle has an angular acceleration. Let ω_1 and ω_2 be the instantaneous angular speeds at times t_1 and t_2, respectively; then the *average angular acceleration* $\overline{\alpha}$ of the particle P is defined as

Angular acceleration

$$\overline{\alpha} = \frac{\omega_2 - \omega_1}{t_2 - t_1} = \frac{\Delta\omega}{\Delta t}.$$

The *instantaneous angular acceleration* is the limit of this ratio as Δt approaches zero, or

$$\alpha = \lim_{\Delta t \to 0} \frac{\Delta\omega}{\Delta t} = \frac{d\omega}{dt}. \tag{11-2}$$

Because ω is the same for all particles in the rigid body, it follows from Eq. 11-2 that α must be the same for each particle and thus α, like ω, is a characteristic of the body as a whole. Angular acceleration has the dimensions of inverse time squared (T^{-2}); its units are commonly taken to be radians per second² (rad/s²) or revolutions per second² (rev/s²).

The rotation of a particle (or a rigid body) about a fixed axis has a formal correspondence to the translational motion of a particle (or a rigid body) along a fixed direction. The kinematical variables are θ, ω, and α in the first case and x, v, and a in the second. These quantities correspond in pairs: θ to x, ω to v, and α to a. Note that the angular quantities differ dimensionally from the corresponding linear quantities by a length factor. Note, too, that all six quantities may be treated as scalars in this special case. For example, a particle at any instant can be moving in one direction or the other along its straight-line motion, corresponding to a positive or a negative value for v; similarly a particle in a rigid body at any instant can be rotating in one direction or another about its fixed axis, corresponding to a positive or a negative value for ω.

11-3 Rotation with Constant Angular Acceleration

For translational motion of a particle or a rigid body along a fixed direction, such as the x-axis, we have seen (in Chapter 3) that the simplest type of motion is uniform linear motion, in which the acceleration a is zero. The next simplest type corresponds to $a =$ a constant (other than zero); for this motion we derived the equations of Table 3-1, which connect the kinematic variables x, v, a, and t in all possible combinations.

For the rotational motion of a particle or a rigid body around a fixed axis the simplest type of motion is uniform circular motion, in which the angular acceleration α is zero. The next simplest type of motion, in which $\alpha =$ a constant (other than zero), corresponds exactly to linear motion with $a =$ a constant (other than zero). As before, we can derive four equations linking the four kinematic variables θ, ω, α, and t in all possible combinations. We can either derive these angular equations by the methods used to derive the linear equations or we may write them down at once by substituting corresponding angular quantities for the linear quantities in the linear equations.

We list both sets of equations in Table 11–1, having chosen $x_0 = 0$ and $\theta_0 = 0$ in these relations for simplicity. Here ω_0 is the angular speed at the time $t = 0$. You might check these equations dimensionally before verifying them. Both sets of equations hold not only for particles but also for rigid bodies.

Table 11–1 The kinematic equations for motion with constant linear or angular acceleration

The kinematic equations for pure rotation with constant α

Equation	Translational motion (fixed direction)	Rotational motion (fixed axis)	Equation
(3–12)	$v = v_o + at$	$\omega = \omega_0 + \alpha t$	(11–3)
(3–14)	$x = \dfrac{v_0 + v}{2}\,t$	$\theta = \dfrac{\omega_0 + \omega}{2}\,t$	(11–4)
(3–15)	$x = v_0 t + \frac{1}{2}at^2$	$\theta = \omega_0 t + \frac{1}{2}\alpha t^2$	(11–5)
(3–16)	$v^2 = v_0^2 + 2ax$	$\omega^2 = \omega_0^2 + 2\alpha\theta$	(11–6)

For the angular quantities, we arbitrarily select one of the two possible directions of rotation about the fixed axis as the direction in which θ is increasing. From Eq. 11–1 ($\omega = d\theta/dt$) we see that if θ is increasing with time ω is positive. Similarly, from Eq. 11–2 ($\alpha = d\omega/dt$), we see that if ω is increasing with time α is positive. There are corresponding sign conventions for the linear quantities.

Example 1 A grindstone has a constant angular acceleration α of 3.0 rad/s². Starting from rest a line, such as OP in Fig. 11–4, is horizontal. Find (a) the angular displacement of the line OP (and hence of the grindstone) and (b) the angular speed of the grindstone 2.0 s later.

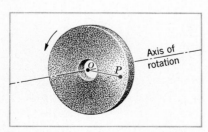

Figure 11–4 Example 1. The line OP is attached to a grindstone rotating as shown about an axis through O that is fixed in the reference frame of the observer.

(a) α and t are given; we wish to find θ. Hence, we use Eq. 11–5,

$$\theta = \omega_0 t + \tfrac{1}{2}\alpha t^2.$$

At $t = 0$, we have $\omega = \omega_0 = 0$ and $\alpha = 3.0$ rad/s². Therefore, after 2.0 s,

$$\theta = (0)(2.0\text{ s}) + \tfrac{1}{2}(3.0\text{ rad/s}^2)(2.0\text{ s})^2 = 6.0\text{ rad} = 0.96\text{ rev}.$$

(b) α and t are given; we wish to find ω. Hence we use Eq. 11–3,

$$\omega = \omega_0 + \alpha t,$$

and

$$\omega = 0 + (3.0\text{ rad/s}^2)(2.0\text{ s}) = 6.0\text{ rad/s}.$$

Using Eq. 11–6 as a check, we have

$$\omega^2 = \omega_0^2 + 2\alpha\theta,$$
$$\omega^2 = 0 + (2)(3.0\text{ rad/s}^2)(6.0\text{ rad}) = 36\text{ rad}^2/\text{s}^2,$$
$$\omega = 6.0\text{ rad/s}.$$

Example 2 For the grindstone of Example 1, assume again that the angular acceleration α has the constant value of $+3.0$ rad/s² and that at $t = 0$ the line OP is horizontal. However, we now put $\omega_0 = -6.0$ rad/s at $t = 0$, the angular acceleration acting to slow down this initial rotational rate.

(a) At what time t will the grindstone come momentarily to rest?

Solving Eq. 11-3 for t yields

$$t = \frac{\omega - \omega_0}{\alpha} = \frac{0 - (-6.0 \text{ rad/s})}{(+3.0 \text{ rad/s}^2)} = +2.0 \text{ s}.$$

(b) Through what angle θ will the line OP on the grindstone have rotated in this time?

From Eq. 11-5,

$$\theta = \omega_0 t + \tfrac{1}{2}\alpha t^2$$
$$= (-6.0 \text{ rad/s})(2.0 \text{ s}) + \tfrac{1}{2}(+3.0 \text{ rad/s}^2)(2.0 \text{ s})^2$$
$$= -6.0 \text{ rad} = -0.96 \text{ rev}.$$

The sign of θ is the same as the sign of ω_0, just as we expect.

(c) At what time t will the angle θ through which the line OP turns again be zero?

Putting $\theta = 0$ in Eq. 11-5 yields

$$0 = \omega_0 t + \tfrac{1}{2}\alpha t^2 = t\,(\omega_0 + \tfrac{1}{2}\alpha t).$$

One solution, which we expect, is $t = 0$. The other is

$$t = -\frac{2\omega_0}{\alpha} = -\frac{2(-6.0 \text{ rad/s})}{(+3.0 \text{ rad/s}^2)} = +4.0 \text{ s}.$$

Note that this is just twice the answer to (a) above; the grindstone moves in the direction of ω_0 for 2.0 s, slows to a momentary halt, and then reverses its direction. After an additional 2.0 s the line OP in Fig. 11-4 has returned to its initial horizontal position. Explore the analogy between the motion of the grindstone in this example and that of an upward-thrown stone. What similar analogy can be made to the angular motion of the grindstone as described in Example 1?

11-4   Relation Between Linear and Angular Kinematics for a Particle in Circular Motion

In section 4-4 we discussed the linear velocity and acceleration of a particle moving in a circle. When a rigid body rotates about a fixed axis, every particle in the body moves in a circle. Hence we can describe the motion of such a particle either in linear variables or in angular variables. The relation between the linear and angular variables enables us to pass back and forth from one description to another and is very useful.

Consider a particle at P in the rigid body, a distance r from the axis through O. This particle moves in a circle of radius r as the body rotates, as in Fig. 11-5a. The reference position is Ox. The particle moves through a distance s along the arc when the body rotates through an angle θ, such that

Linear and angular displacement

$$s = \theta r \tag{11-7}$$

where θ must be measured in radians.

Differentiating both sides of this equation with respect to the time, and noting that r is constant, we obtain

$$\frac{ds}{dt} = \frac{d\theta}{dt}\,r.$$

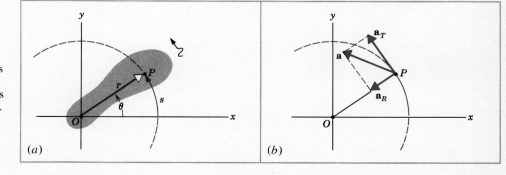

Figure 11-5   (a) A rigid body rotates about a fixed axis through O perpendicular to the page. The point P sweeps out an arc s which subtends an angle θ. (b) The acceleration **a** of point P has components a_T (tangential) where $a_T = \alpha r$ and a_R (radial) where $a_R = v^2/r = \omega^2 r$ (ω = angular speed).

But ds/dt is the linear speed of the particle at P and $d\theta/dt$ is the angular speed ω of the rotating body, so that

$$v = \omega r. \qquad (11\text{-}8)$$

This is a relation between the magnitudes of the linear velocity and the angular velocity; the linear speed of a particle in circular motion is the product of the angular speed and the distance r of the particle from the axis of rotation. If, in Eq. 11–8, we want to express v in m/s and r in m, then ω must be expressed in rad/s and not, for example, in degrees/s or revolutions/s.

Differentiating Eq. 11–8 with respect to the time, we have

$$\frac{dv}{dt} = \frac{d\omega}{dt}\, r.$$

dv/dt is the magnitude of the *tangential* component of acceleration of the particle and $d\omega/dt$ is the magnitude of the angular acceleration of the rotating body, so that

$$a_T = \alpha r. \qquad (11\text{-}9)$$

Hence the magnitude of the tangential component of the linear acceleration of a particle in circular motion is the product of the magnitude of the angular acceleration (in radians per second^2) and the distance r of the particle from the axis of rotation. If, in Eq. 11–9, we wish to express a_T in m/s^2 and r in m, then α must be expressed in rad/s^2.

We have seen that the radial component of acceleration is v^2/r for a particle moving in a circle. This can be expressed in terms of angular speed by use of Eq. 11–8. We have

$$a_R = \frac{v^2}{r} = \omega^2 r. \qquad (11\text{-}10)$$

The resultant acceleration of point P is shown in Fig. 11–5b.

Equations 11–7 through 11–10 enable us to describe the motion of one point on a rigid body rotating about a fixed axis either in angular variables or in linear variables. We might ask why we need the angular variables when we are already familiar with the equivalent linear variables. The answer is that the angular description offers a distinct advantage over the linear description when various points on the same rotating body must be considered. Different points on the same rotating body do not have the same linear displacement, speed, or acceleration, but all points on a rigid body rotating about a fixed axis do have the same angular displacement, speed, or acceleration at any instant. By the use of angular variables we can describe the motion of the body as a whole in a simple way.

Figure 11–6 shows an interesting application of the relation between angular and linear quantities. When a tall chimney is toppled by means of an explosive charge at its base it will often break near its middle, the rupture starting at the leading edge. The top part will then reach the ground later than the bottom part.

We note that as the chimney topples, it has at any instant an angular acceleration α about an axis through its base. The tangential acceleration a_T of its top is given by Eq. 11–9 ($a_T = \alpha r$). As the chimney leans more and more, the vertical component of a_T comes to exceed g, so that the bricks at the top are accelerating downward more than they would in free fall. This can happen only as long as the chimney is a rigid body. As the chimney continues to fall, internal tension stresses develop along its leading edge. In nearly all cases rupture occurs, thus relieving those stresses.

Figure 11–6 The toppling of a 49-m chimney in Louisville, Kentucky. Published by the courtesy of *The Courier-Journal* and *Louisville Times*. See "More on the Falling Chimney," by Albert A. Bartlett, *The Physics Teacher*, September 1976, for an account.

Example 3 If the radius of the grindstone of Example 1 is 0.50 m, calculate (*a*) the linear or tangential speed, (*b*) the tangential acceleration, and (*c*) the radial acceleration of a particle on the rim at the end of 2.0 s.

We have $\alpha = 3.0$ rad/s^2, $\omega = 6.0$ rad/s after 2.0 s, and $r = 0.50$ m. Then

(*a*)
$$v = \omega r$$
$$= (6.0 \text{ rad/s})(0.50 \text{ m})$$
$$= 3.0 \text{ m/s} \quad \text{(linear speed)};$$

(*b*)
$$a_T = \alpha r$$
$$= (3.0 \text{ rad/s}^2)(0.50 \text{ m})$$
$$= 1.5 \text{ m/s}^2 \quad \text{(tangential acceleration)};$$

(*c*)
$$a_R = \frac{v^2}{r} = \omega^2 r$$
$$= (6.0 \text{ rad/s})^2(0.50 \text{ m})$$
$$= 18 \text{ m/s}^2 \quad \text{(radial acceleration)}.$$

(*d*) Are the results the same for a particle halfway in from the rim, that is, at $r = 0.25$ m?

The angular variables are the same for this point as for a point on the rim. That is, once again,
$$\alpha = 3.0 \text{ rad/s}^2, \quad \omega = 6.0 \text{ rad/s}.$$

But now $r = 0.25$ m, so that for this particle
$$v = 1.5 \text{ m/s}, \quad a_T = 0.75 \text{ m/s}^2, \quad a_R = 9.0 \text{ m/s}^2.$$

REVIEW GUIDE AND SUMMARY

The angular variables

In this chapter we describe the rotation of a rigid body about a fixed axis. The variables, other than the time t, are the angular displacement θ about that axis, the angular speed ω, and the angular acceleration α. The latter two are given by

$$\omega = \frac{d\theta}{dt} \quad \text{and} \quad \alpha = \frac{d\omega}{dt}. \qquad [11-1, 11-2]$$

The kinematic equations for constant angular acceleration

An important special case is $\alpha = $ a constant. The appropriate kinematic equations, given in Table 11–1, are

$$\omega = \omega_0 + \alpha t; \qquad [11-3]$$
$$\theta = \frac{\omega_0 + \omega}{2} t; \qquad [11-4]$$
$$\theta = \omega_0 t + \tfrac{1}{2}\alpha t^2; \qquad [11-5]$$
$$\omega^2 = \omega_0^2 + 2\alpha\theta. \qquad [11-6]$$

Examples 1 and 2 show their application.

Linear and angular variables

Sometimes we wish to ask about the motion—not of the rotating rigid body as a whole—but of just one point P in it, distant r from the axis of rotation. We can describe the motion of P by three linear variables, the arc distance s ($= \theta r$), the velocity $\mathbf{v}$, and the acceleration $\mathbf{a}$. The tangential and radial components of the latter two are related to the angular variables by

$$v = v_T = \omega r, \quad v_R = 0; \qquad [11-8]$$
and
$$a_T = \alpha r, \quad a_R = \omega^2 r. \qquad [11-9, 11-10]$$

The angular quantities here must be expressed in radian measure, not in degrees or revolutions (1 radian = $57.3° = \frac{41}{2\pi}$ revolutions), if we want to express accelerations and velocities in m/s^2 and m/s and radii in m. Example 3 shows how the above equations may be used.

QUESTIONS

1. Could the angular quantities θ, ω, and α be expressed in terms of degrees instead of radians in the rotational equations of Table 11–1?

2. In what sense is the radian a "natural" measure of angle and the degree an "arbitrary" measure of that same quantity?

3. Taking the rotation and the revolution of the earth into account, does a tree move faster during the day or during the night? With respect to what reference frame is your answer given? (The earth's rotation and revolution are both in the same direction.)

4. What is the relationship between the angular velocities of a pair of coupled gears of different radii?

5. A golfer swings a golf club, making a long drive from the tee. Do all points on the club have the same angular velocity ω at any instant while the club is in motion?

6. If a car's speedometer is set to read a speed proportional to the rotational speed of its rear wheels, is it necessary to correct the reading when tires with larger outside diameter (such as snow tires) are used?

7. The frequency v of a rotating wheel is related to its angular frequency ω by $\omega = 2\pi v$. The accepted unit for v is the cycle per second or *hertz* (abbreviation Hz). It has been suggested that we also need a name for the radian per second, the accepted unit for ω. The word *avis* (abbreviation Av) has been put forward. What do you think of this idea?

8. When we say that a point on the equator has an angular speed of 2π rad/day, what reference frame do we have in mind?

9. Does the vector representing the angular velocity of a wheel rotating about a fixed axis necessarily have to lie along that axis?

10. A wheel is rotating about its axle. Consider a point on the rim. When the wheel rotates with constant angular velocity, does the point have a radial acceleration? A tangential acceleration? When the wheel rotates with constant angular acceleration, does the point have a radial acceleration? A tangential acceleration? Do the magnitudes of these accelerations change with time?

11. Suppose that you were asked to determine the distance traveled by a needle in playing a phonograph record. What information do you need? Discuss from the points of view of reference frames (*a*) fixed in the room, (*b*) fixed on the rotating record and (*c*) fixed on the arm of the record player.

EXERCISES

Section 11–2 Rotational Kinematics — the Variables
1. (*a*) What angle in radians is subtended from the center of a circle of radius 1.2 m by an arc of length 1.8 m? (*b*) Express the angle in (*a*) in degrees. (*c*) The angle between two radii of a circle is 0.62 rad. What arc length is subtended if the radius is 2.4 m? (*d*) Repeat (*c*) if the radius is 61 m.

2. What is the angular speed (*a*) of the second hand of a watch? (*b*) of the minute hand? (*c*) of the hour hand?

3. Our sun is 3×10^4 ly (light-years) from the center of our Milky Way galaxy, and is moving in a circle around this center at a speed of 250 km/s. How many revolutions has the sun completed since it was formed about 4.5×10^9 y ago?

4. A good baseball pitcher can throw a baseball toward home plate at 85 mi/h with a spin of 1800 rev/min. How many revolutions does the baseball make on its way to home plate? For simplicity, assume that the 60-ft trajectory is a straight line.

Section 11–3 Rotation with Constant Angular Acceleration
5. A heavy flywheel, rotating on its axis, is slowing down because of friction in its bearings. At the end of the first minute its angular velocity is 0.90 of its initial angular velocity ω_0. Assuming constant angular acceleration, find its angular velocity at the end of the second minute.

6. A disk, initially rotating at 120 rad/s, is slowed down with a constant angular acceleration of magnitude 4.0 rad/s^2. (*a*) How much time elapses before the disk stops? (*b*) Through what angle did the disk rotate in coming to rest?

7. A phonograph turntable rotating at 78 rev/min slows down and stops in 30 s after the motor is turned off. (*a*) Find its (uniform) angular acceleration. (*b*) How many revolutions did it make in this time?

8. The angular speed of an automobile engine is increased from 1200 rev/min to 3000 rev/min in 12 s. (*a*) What is its angular

acceleration, assuming it to be uniform? (*b*) How many revolutions does the engine make during this time?

9. The flywheel of an engine is rotating at 240 rev/min. When the engine is turned off, the flywheel decelerates at a constant rate and comes to rest after 20 s. Calculate (*a*) the angular acceleration (in rad/s^2) of the flywheel; (*b*) the angular velocity (in rad/s) of the flywheel 5.0 s before it comes to rest; (*c*) the angle (in rad) through which the flywheel rotates in coming to rest; (*d*) the number of revolutions made by the flywheel in coming to rest.

Section 11–4 Relation Between Linear and Angular Kinematics for a Particle in Circular Motion
10. What is the angular speed of a car rounding a circular turn of radius 110 m at 50 km/h?

11. A phonograph record on a turntable rotates at $33\frac{1}{3}$ rev/min. What is the linear speed of a point on the record at the needle at (*a*) the beginning and (*b*) the end of the recording? The distances of the needle from the turntable axis are 5.9 in. and 2.9 in., respectively, at these two positions.

12. What are (*a*) the angular speed, (*b*) the radial acceleration, and (*c*) the tangential acceleration of a spaceprobe negotiating a circular turn of radius 2000 mi ($=$ 3220 km) at a constant speed of 18,000 mi/h ($=$ 29,000 km/h)?

13. What is the acceleration of a point on the rim of a 12-in. ($=$ 30-cm) diameter record rotating at $33\frac{1}{3}$ rev/min?

14. A point on the rim of a 0.75-m diameter grinding wheel changes speed from 12 m/s to 18 m/s in 6.0 s. What is the average angular acceleration of the grinding wheel during this interval?

15. A 1.2-m diameter flywheel is rotating at 200 rev/min. (*a*) What is the angular velocity of the flywheel? (*b*) What is the linear velocity of a point on the rim of the flywheel? (*c*) What constant angular acceleration will cause the wheel to increase its speed to 1000 rev/min in 60 s? (*d*) How many revolutions does the wheel make during this 60-s interval?

PROBLEMS GROUPED BY SECTION

Section 11–2 Rotational Kinematics — the Variables
1. Starting from rest, a disk rotates about its axis with constant angular acceleration. After 5.0 s, it has rotated through 25 rad. (*a*)

What was the average angular acceleration during this time? (*b*) What was the average angular velocity? (*c*) What is the instantaneous angular velocity of the disk at the end of the 5.0 s? (*d*)

Assuming that the acceleration does not change, through what additional angle will the disk turn during the next 5.0 s?

2. The angular position of a point on the rim of a rotating wheel is described by $\theta = 4.0t - 3.0t^2 + t^3$, where θ is in radians if t is given in seconds. (a) What is the average angular acceleration for the time interval that begins at $t = 2.0$ s and ends at $t = 4.0$ s? (b) What is the instantaneous angular acceleration at the beginning and end of this time interval?

3. A diver makes 2.5 complete revolutions on her way from a 10-m platform to the water below. Assuming zero initial vertical velocity, calculate her average angular velocity for this dive.

4. A wheel has 36 spokes and a radius of 30 cm. It is mounted on a fixed axle and is spinning at 2.5 rev/s. You want to shoot a 20-cm arrow parallel to this axle and through the wheel without hitting any of the spokes. Assume that the arrow and the spokes are negligibly thin; see Fig. 11–7. (a) What minimum speed must the arrow have? (b) Does it matter where between the axle and rim of the wheel you aim? If so, where is the best location?

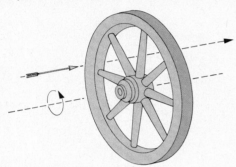

Figure 11–7 Problem 4.

Section 11–3 Rotation with Constant Angular Acceleration

5. A flywheel completes 40 rev as it slows from an angular speed of 1.5 rad/s to a complete stop. Assuming uniform acceleration, (a) what is the time required for it to come to rest? (b) What is the angular acceleration? (c) How much time is required for it to complete the first one-half of the 40 rev?

6. After leaving a helicopter, you notice that the rotor's motion changed from 300 rev/min to 225 rev/min in 1 min. (a) Find the average angular acceleration during the interval. (b) Assuming that this acceleration remains constant, calculate how long it will take for the rotor to stop. (c) How many revolutions will the rotor make after your second observation?

7. A wheel has a constant angular acceleration of 3.0 rad/s². In a 4.0-s interval, it turns through an angle of 120 rad. Assuming that the wheel started from rest, how long had it been in motion at the start of this 4.0-s interval?

8. Starting from rest at $t = 0$, a wheel undergoes a constant angular acceleration. When $t = 2.0$ s, the angular velocity of the wheel is 5.0 rad/s. The acceleration continues until $t = 20$ s, when it abruptly ceases. Through what angle does the wheel rotate in the interval $t = 0$ to $t = 40$ s?

9. A uniform disk rotates about a fixed axis starting from rest and accelerating with constant angular acceleration. At one time it is rotating at 10 rev/s. After completing 60 more complete revolutions, its angular speed is 15 rev/s. Calculate (a) the angular acceleration; (b) the time required to complete the 60 revolutions mentioned; (c) the time required to attain the 10 rev/s angular speed; and (d) the number of revolutions from rest until the time the disk attained the 10-rev/s angular speed.

Section 11–4 Relation Between Linear and Angular Kinematics for a Particle in Circular Motion

10. (a) What is the angular speed about the polar axis of a point on the earth's surface at a latitude of 40°N? (b) What is the linear speed? (c) How do these compare with the similar values for a point at the equator?

11. The earth's orbit about the sun is almost a circle. (a) What is the angular velocity of the earth (regarded as a particle) about the sun, and (b) what is its average linear speed in its orbit? (c) What is the acceleration of the earth with respect to the sun?

12. A gyroscope of radius 28.3 cm is accelerated from rest at 14.2 rad/s² until its angular speed is 2760 rev/min. (a) What is the tangential acceleration of a point on the rim of the gyroscope? (b) What is the radial acceleration of this point when the gyroscope is turning at full speed? (c) Through what distance does a point on the rim move during the acceleration?

13. An astronaut is being tested in a centrifuge. The centrifuge has a radius of 10 m, and rotates initially according to $\theta = 0.3t^2$, where t in seconds gives θ in radians. When $t = 5.0$ s, what are the astronaut's (a) tangential velocity, (b) tangential acceleration, (c) radial acceleration, and (d) resultant acceleration?

14. (a) What are the acceleration and velocity of a point on the top of a 66-cm diameter automobile tire if the automobile is traveling at 80 km/h on a level road? (b) What are the acceleration and velocity of a point on the bottom of the tire? (c) What are the acceleration and velocity of the center of the wheel? Calculate all quantities first as seen by a passenger in the car and then as seen by an observer standing on the side of the road as the car goes by.

15. A coin of mass M is placed a distance R from the center of a phonograph turntable. The coefficient of static friction is μ_g. The rotational speed ω of the turntable is slowly increased to a value ω_0, when the coin slides off. (a) Find ω_0 in terms of the quantities $M, R, g,$ and μ_s. (b) Make a sketch showing the path of the coin as it flies off the turntable.

ADDITIONAL PROBLEMS

16. A certain wheel turns through 90 rev in 15 s, its angular speed at the end of the period being 10 rev/s. (a) What was the angular speed of the wheel at the beginning of the 15-s interval, assuming constant angular acceleration? (b) How much time had elapsed between the time the wheel was at rest and the beginning of the 15-s interval?

17. A wheel, starting from rest, rotates with a constant angular acceleration of 2.0 rad/s². During a certain 3.0-s interval, it turns through 90 rad. (a) How long had the wheel been turning before the start of the 3.0-s interval? (b) What was the angular velocity of the wheel at the start of the 3.0-s interval?

18. The angle turned through by the flywheel of a generator during a time interval t is given by

$$\theta = at + bt^3 - ct^4,$$

where a, b, and c are constants. What is the expression for its angular acceleration?

19. If an airplane propeller of radius 5.0 ft ($= 1.5$ m) rotates at 2000 rev/min and the airplane is propelled at a ground speed of 300 mi/h ($= 480$ km/h), what is the speed of a point on the tip of the propeller, as seen by (a) the pilot and (b) an observer on the ground? Assume that the plane's velocity is parallel to the propeller's axis of rotation.

20. The flywheel of a steam engine runs with a constant angular speed of 150 rev/min. When steam is shut off, the friction of the bearings and of the air brings the wheel to rest in 2.2 h. (a) What is the average angular acceleration of the wheel? (b) How many rotations will the wheel make before coming to rest? (c) What is the tangential linear acceleration of a particle that is 50 cm from the axis of rotation when the flywheel is turning at 75 rev/min? (d) What is the magnitude of the total linear acceleration of the particle in part (c)?

21. A flywheel is accelerated from rest to 20π rad/s with a constant angular acceleration of $\pi/4$ rad/s^2. It then continues to rotate at 20π rad/s for 20 min. Then it slows down with a constant acceleration, taking 5.0 min to come to rest. How many revolutions did the flywheel make during each of these three phases of its motion?

22. The angular position of a rotating wheel is given by $\theta = 2 + 4t^2 + 2t^3$, where θ is in radians and t is in seconds. (a) What are the angular position and angular velocity at $t = 0$? (b) What is the angular velocity at $t = 4.0$ s? (c) Calculate the angular acceleration at $t = 2.0$ s. Is the angular acceleration constant?

23. A wheel rotates with an angular acceleration α given by

$$\alpha = 4at^3 - 3bt^2,$$

where t is the time and a and b are constants. If the wheel has an initial angular speed ω_0, write the equations for (a) the angular speed and (b) the angle turned through as functions of time.

24. The angle through which a flywheel of radius R rotates is given by $\theta = At + Bt^3 + Ct^4$, where A, B, and C are positive constants. (a) What are the angular velocity and linear velocity of a particle in the wheel a distance R from the axis? (b) Answer part (a) for a particle a distance $R/2$ from the axis. (c) What are the angular acceleration, tangential acceleration, and centripetal acceleration of the particle in part (a)?

25. An automobile traveling 80 km/h ($= 50$ mi/h) has tires of 75 cm ($= 30$ in.) diameter. (a) What is the angular speed of the tires about the axle? (b) If the car is brought to a stop uniformly in 30

turns of the tires (no skidding), what is the angular acceleration of the wheels? (c) How far does the car advance during this braking period?

26. A pulsar is a rapidly rotating neutron star which emits radio pulses with precise synchronization, there being one such pulse for each rotation of the star. The period T of rotation is found by measuring the time between pulses. At present, the pulsar in the central region of the crab nebula (see Fig. 11–8) has a period of rotation of $T = 0.033$ s, and this is observed to be increasing at the rate 1.26×10^{-5} s/y. (a) Show that the angular velocity ω of the star is related to the period of rotation by $\omega = 2\pi/T$. (b) What is the value of the angular acceleration in rad/s^2? (c) If its angular acceleration is constant, when will the pulsar stop rotating? (d) The pulsar originated in a supernova explosion in the year 1054 A.D. What was the period of rotation of the pulsar when it was "born"? (Assume constant angular acceleration.)

Figure 11–8 Problem 26. Photo by Dr. Brandt/Kitt Peak National Observatory.

27. An early method of measuring the speed of light makes use of a rotating toothed wheel. A beam of light passes through a slot at the outside edge of the wheel, as in Fig. 11–9, travels to a distant mirror, and returns to the wheel just in time to pass through the next slot in the wheel. One such toothed wheel has a radius of 5.0 cm and 500 teeth at its edge. Measurements taken when the mir-

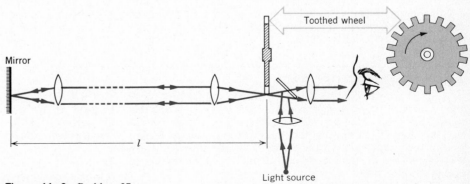

Figure 11–9 Problem 27.

ror was 500 m from the wheel indicated a speed of light of 3.0×10^5 km/s. (a) What was the (constant) angular speed of the wheel? (b) What was the linear speed of a point on its edge?

28. Wheel A of radius $r_A = 10$ cm is coupled by a belt B to wheel C of radius $r_C = 25$ cm, as shown in Fig. 11–10. Wheel A increases its angular speed from rest at a uniform rate of $\pi/2$ rad/s^2. Determine the time for wheel C to reach a rotational speed of 100 rev/min, assuming that the belt does not slip.

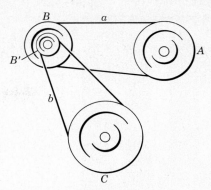

Figure 11–11 Problem 29.

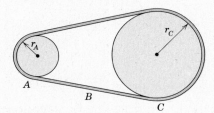

Figure 11–10 Problem 28.

29. Four pulleys (A, B, B', C) are connected by two belts (a, b) as shown in Fig. 11–11. Pulley A (radius 15 cm) is the drive pulley, and it rotates at 10 rad/s. Pulley B (radius 10 cm) is connected by belt a to pulley A. Pulley B' (radius 5 cm) is concentric with pulley B and is rigidly attached to it. Pulley C (radius 25 cm) is connected by belt b to pulley B'. Calculate (a) the linear speed of a point on belt a; (b) the angular speed of pulley B; (c) the angular speed of pulley B'; (d) the linear speed of a point on belt b; (e) the angular speed of pulley C.

30. A phonograph turntable is rotating at $33\frac{1}{3}$ rev/min. A small object is on the turntable 6.0 cm from the axis of rotation. (a) Calculate the acceleration of the object assuming that it does not slip. (b) What is the minimum value of the coefficient of static friction between object and turntable in (a)? (c) Suppose that the turntable achieved this angular velocity by starting from rest and undergoing a constant angular acceleration for 0.25 s. Calculate the minimum coefficient of static friction required for the object not to slip during the acceleration period.

12

Rotational Dynamics

12–1 Introduction

In Chapter 11 we considered the kinematics of rotation. In this chapter, following the pattern of our study of translational motion, we study the causes of rotation, a subject called *rotational dynamics*. Rotating systems are made up of particles and we have already learned how to apply the laws of classical mechanics to the motion of particles. For this reason rotational dynamics should contain no features that are fundamentally new. In the same way rotational kinematics contained no basic new features, the rotational parameters θ, ω, and α being related to corresponding translational parameters x, v, and a for the particles that make up the rotating system. As in Chapter 11, however, it is very useful to recast the concepts of translational motion into a new form, especially chosen for its convenience in describing rotating systems.

We restricted our kinematical studies in Chapter 11 to a single but important special case, the rotation of a rigid body about an axis that is fixed in the reference frame in which we make our measurements. In studying rotational dynamics we start from a more fundamental point of view, that of a single particle viewed from an inertial reference frame. Later we shall generalize to systems of many particles, including the special case of a rigid body rotating about a fixed axis.

Finally we shall consider systems on which no external torques act and shall introduce the important principle of *conservation of angular momentum*.

12–2 Torque

Definition of torque

In translational motion we associate a force with the linear acceleration of a body. In rotational motion, what quantity shall we associate with the angular acceleration of a body? It cannot be simply force because, as experiment with a heavy revolving door teaches us, a given force (vector) can produce various angular accel-

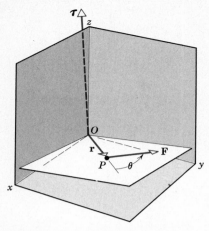

Figure 12–1 A force **F** is applied to a particle P, displaced **r** relative to the origin. The force vector makes an angle θ with the radius vector **r**. The torque $\boldsymbol{\tau}$ about O is shown. Its direction is perpendicular to the plane formed by **r** and **F** with the sense given by the right-hand rule.

erations of the door depending on where the force is applied and how it is directed; a force applied to the hinge line cannot produce any angular acceleration, whereas a force of given magnitude applied at right angles to the door at its outer edge produces a maximum acceleration.

Figure 12–1 shows a force **F** acting on a single particle located at point P in an inertial reference frame. We cannot give meaning to what we can loosely call the "turning effect" of this force until we specify a point with respect to which this effect is to be measured. Any point can be chosen, but we select the origin O in Fig. 12–1 as our reference point. To "turning effect" we give the formal name *torque*, and we define the torque $\boldsymbol{\tau}$ exerted by force **F** about point O in Fig. 12–1 as

$$\boldsymbol{\tau} = \mathbf{r} \times \mathbf{F} \qquad (12\text{–}1)$$

The vector **r** in this equation extends from reference point O to the point P at which the force acts.

Torque is a vector quantity.* Its magnitude is given by

$$\tau = rF \sin \theta, \qquad (12\text{–}2a)$$

where θ is the angle between **r** and **F**; its direction is normal to the plane formed by **r** and **F**. The sense is given by the right-hand rule for the vector product of two vectors, namely, we imagine swinging **r** into **F** through the smaller angle between them with the curled fingers of the right hand; the direction of the extended thumb then gives the direction of $\boldsymbol{\tau}$.

Torque has the same dimensions as force times distance, or in terms of our assumed fundamental dimensions, M, L, T, it has the dimensions ML^2T^{-2}. These are the same as the dimensions of work. However, torque and work are very different physical quantities. Torque is a vector and work is a scalar, for example. The unit of torque may be the newton-meter (N · m) or the pound-foot (lb · ft), among other possibilities.

Notice (Eq. 12–1) that the torque produced by a force depends not only on the magnitude and on the direction of the force but also on the point of application of the force relative to the chosen reference point O, that is, on the vector **r**. In particular, when particle P is at the origin, so that the line of action of **F** passes through the origin, **r** is zero and the torque $\boldsymbol{\tau}$ about the origin is zero.

We can also write the magnitude of $\boldsymbol{\tau}$ (Eq. 12–2a) either as

$$\tau = (r \sin \theta)F = Fr_\perp, \qquad (12\text{–}2b)$$

or as

$$\tau = r(F \sin \theta) = rF_\perp, \qquad (12\text{–}2c)$$

in which, as Fig. 12–2a shows, $r_\perp$ ($= r \sin \theta$) is the component of **r** at right angles to the line of action of **F**, and $F_\perp$ ($= F \sin \theta$) is the component of **F** at right angles to **r**. Torque is often called the *moment of force* and $r_\perp$ in Eq. 12–2b is called the *moment arm*. Equation 12–2c shows that only the component of **F** perpendicular to **r** contributes to the torque. In particular, when θ equals 0 or 180°, there is no perpendicular component ($F_\perp = F \sin \theta = 0$); then the line of action of the force passes through the origin and the moment arm $r_\perp$ about the origin is also zero. In this case both Eq. 12–2b and Eq. 12–2c show that the torque $\boldsymbol{\tau}$ is zero.

If, as in Fig. 12–2b, we reverse the direction of **F**, the magnitude of $\boldsymbol{\tau}$ remains unchanged but the direction of $\boldsymbol{\tau}$ is reversed. Similarly, if, as in Fig. 12–2c, we re-

---

* Equation 12–1 is our first application of a vector or cross product. You may wish to review this subject in Section 2–4.

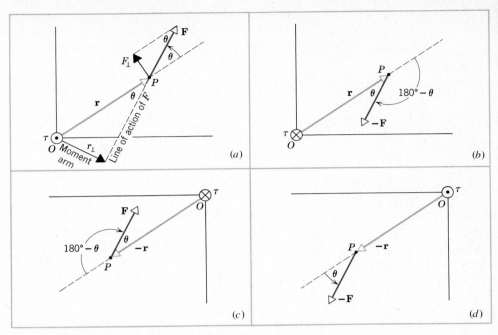

Figure 12–2   The plane shown is that defined by **r** and **F** in Fig. 12–1. (*a*) The magnitude of τ is given by $Fr_\perp$ (Eq. 12–2*b*) or by $rF_\perp$ (Eq. 12–2*c*). (*b*) Reversing **F** reverses the direction of τ but leaves its magnitude unchanged. (*c*) Reversing **r** also reverses the direction of τ but leaves its magnitude unchanged. (*d*) Reversing both **F** and **r** leaves both the direction and magnitude of τ unchanged. The directions of τ are represented by $\odot$ (perpendicularly out of the figure, the symbol representing the tip of an arrow) and by $\otimes$ (perpendicularly into the figure, the symbol representing the tail of an arrow).

verse **r**, thereby changing the point of application of **F**, the magnitude of τ remains unchanged but the direction of τ is again reversed.

If, as in Fig. 12–2*d*, we reverse both **r** and **F**, then both the magnitude and the direction of τ remain unchanged. These results follow formally from the facts that: (1) $\sin\theta = \sin(180° - \theta)$, so that Eq. 12–2*a* for the magnitude of τ is unaffected; (2) reversing the direction of one vector in a vector product (either **r** *or* **F**) reverses the direction of the product; and (3) reversing the directions of both vectors in a vector product (both **r** and **F**) leaves the direction of the product unchanged. Look at the directions of τ in Fig. 12–2 and verify them using the right-hand rule.

12–3   Angular Momentum

We have found linear momentum to be useful in dealing with the translational motion of single particles or of systems of particles (including rigid bodies). For example, linear momentum is conserved in collisions. For a single particle the linear momentum is $\mathbf{p} = m\mathbf{v}$ (Eq. 9–11); for a system of particles it is $\mathbf{P} = M\mathbf{v}_{cm}$ (Eq. 9–15), in which M is the total system mass and $\mathbf{v}_{cm}$ is the velocity of the center of mass. In rotational motion, what is the analog of linear momentum? We call it *angular momentum* and we define it below for the special case of a single particle. Later we shall broaden the definition to include systems of particles and shall show that angular momentum, as we define it, is as useful a concept in rotational motion as linear momentum is in translational motion.

Consider a particle of mass m and linear momentum $\mathbf{p}$ at a position $\mathbf{r}$ relative to the origin O of an inertial reference frame (Fig. 12–3). We define the *angular momentum* ℓ of the particle with respect to the origin O to be

$$\ell = \mathbf{r} \times \mathbf{p}. \tag{12–3}$$

Note that, as for torque, we cannot define angular momentum unless we specify a reference point as part of the definition. In each case we chose this point to be the origin O of an inertial reference frame. In each case the vector $\mathbf{r}$ locates the particle in question with respect to this reference point; compare Figs. 12–1 and 12–3.

Angular momentum is a vector. Its magnitude is given by

$$\ell = rp \sin \theta. \tag{12–4a}$$

where θ is the angle between $\mathbf{r}$ and $\mathbf{p}$; its direction is normal to the plane formed by $\mathbf{r}$ and $\mathbf{p}$. The sense is given by the right-hand rule, namely, we imagine swinging $\mathbf{r}$ into $\mathbf{p}$, through the smaller angle between them, with the curled fingers of the right hand; the extended right thumb then points in the direction of ℓ.

We can also write the magnitude of ℓ either as

$$\ell = (r \sin \theta)p = pr_\perp, \tag{12–4b}$$

or as

$$\ell = r(p \sin \theta) = rp_\perp, \tag{12–4c}$$

in which $r_\perp$ ($= r \sin \theta$) is the component of $\mathbf{r}$ at right angles to the line of action of $\mathbf{p}$ and $p_\perp$ ($= p \sin \theta$) is the component of $\mathbf{p}$ at right angles to $\mathbf{r}$. Angular momentum is often called *moment of* (linear) *momentum,* and $r_\perp$ in Eq. 12–4b is often called the *moment arm.* Equation 12–4c shows that only the component of $\mathbf{p}$ perpendicular to $\mathbf{r}$ contributes to the angular momentum. When the angle θ between $\mathbf{r}$ and $\mathbf{p}$ is zero or 180°, there is no perpendicular component ($p_\perp = p \sin \theta = 0$); then the line of action of $\mathbf{p}$ passes through the origin and $r_\perp$ is also zero. In this case both Eqs. 12–4b and 12–4c show that the angular momentum ℓ is zero.

We now derive an important relation between torque and angular momentum; it is Newton's second law for rotational motion. We have seen that Newton's second law, for a particle in translational motion, can be written as $\mathbf{F} = d\mathbf{p}/dt$ (Eq. 9–12). Let us take the vector product of $\mathbf{r}$ with both sides of this equation, obtaining

$$\mathbf{r} \times \mathbf{F} = \mathbf{r} \times \frac{d\mathbf{p}}{dt}.$$

But $\mathbf{r} \times \mathbf{F}$ is the torque, or moment of a force, about O. We can then write

$$\boldsymbol{\tau} = \mathbf{r} \times \frac{d\mathbf{p}}{dt}. \tag{12–5}$$

Next we differentiate Eq. 12–3 and obtain

$$\frac{d\ell}{dt} = \frac{d}{dt} (\mathbf{r} \times \mathbf{p}).$$

The derivative of a vector product such as this is taken in the same way as the derivative of an ordinary product, except that we must not change the order of the terms. We have

$$\frac{d\ell}{dt} = \frac{d\mathbf{r}}{dt} \times \mathbf{p} + \mathbf{r} \times \frac{d\mathbf{p}}{dt}.$$

Angular momentum

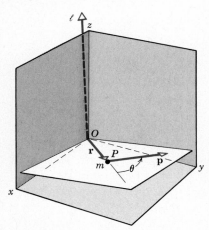

Figure 12–3 A particle of mass m is at point P displaced $\mathbf{r}$ relative to the origin, and has linear momentum $\mathbf{p}$. The vector $\mathbf{p}$ makes an angle θ with the radius vector $\mathbf{r}$. The angular momentum ℓ of the particle with respect to origin O is shown. Its direction is perpendicular to the plane formed by $\mathbf{r}$ and $\mathbf{p}$ with the sense given by the right-hand rule.

But $d\mathbf{r}/dt$ is the instantaneous velocity $\mathbf{v}$ of the particle. Also, $\mathbf{p}$ equals $m\mathbf{v}$, so that we can rewrite the equation as

$$\frac{d\boldsymbol{\ell}}{dt} = (\mathbf{v} \times m\mathbf{v}) + \mathbf{r} \times \frac{d\mathbf{p}}{dt}.$$

Now, $\mathbf{v} \times m\mathbf{v} = 0$, because the vector product of two parallel vectors is zero. Therefore,

$$\frac{d\boldsymbol{\ell}}{dt} = \mathbf{r} \times \frac{d\mathbf{p}}{dt}. \tag{12–6}$$

Inspection of Eqs. 12–5 and 12–6 shows that

Angular momentum change for a single particle

$$\boldsymbol{\tau} = \frac{d\boldsymbol{\ell}}{dt}, \tag{12–7}$$

which states that the time rate of change of the angular momentum of a particle is equal to the torque acting on it. This result is the rotational analog of Eq. 9–12, which stated that the time rate of change of the linear momentum of a particle is equal to the force acting on it, that is, that $\mathbf{F} = d\mathbf{p}/dt$.

Equation 12–7, like all vector equations, is equivalent to three scalar equations, namely,

$$\tau_x = \frac{d\ell_x}{dt}, \qquad \tau_y = \frac{d\ell_y}{dt}, \qquad \tau_z = \frac{d\ell_z}{dt}. \tag{12–8}$$

Hence, the x-component of the applied torque is given by the x-component of the change with time of the angular momentum. Similar results hold for the y- and z-directions.

Example 1 A particle of mass m is released from rest at point P in Fig. 12–4, falling parallel to the (vertical) y-axis. (a) Find the torque acting on m at any time t, with respect to origin O. (b) Find the angular momentum of m at any time t, with respect to this same origin. (c) Show that the relation $\boldsymbol{\tau} = d\boldsymbol{\ell}/dt$ (Eq. 12–7) yields a correct result when applied to this familiar problem.

(a) The torque is given by Eq. 12–1 or $\boldsymbol{\tau} = \mathbf{r} \times \mathbf{F}$, its magnitude being given by

$$\tau = rF \sin \theta.$$

In this example $r \sin \theta = x_0$ and $F = mg$, so that

$$\tau = mgx_0 = \text{a constant.}$$

Note that the torque is simply the product of the force (mg) times the moment arm (x_0). The right-hand rule shows that $\boldsymbol{\tau}$ is directed perpendicularly into the figure.

(b) The angular momentum is given by Eq. 12–3 or $\boldsymbol{\ell} = \mathbf{r} \times \mathbf{p}$, its magnitude being given by

$$\ell = rp \sin \theta.$$

In this example $r \sin \theta = x_0$ and $p = mv = m(gt)$, so that

$$\ell = mgx_0 t.$$

The right-hand rule shows that $\boldsymbol{\ell}$ is directed perpendicularly into the figure, which means that $\boldsymbol{\ell}$ and $\boldsymbol{\tau}$ are parallel vectors. The vector $\boldsymbol{\ell}$ changes with time in magnitude only, its direction always remaining the same in this case.

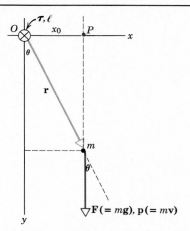

Figure 12–4 Example 1. A particle of mass m drops vertically from point P. The torque and the angular momentum about O are directed perpendicularly into the figure, as shown by the symbol $\otimes$ at O.

(c) Since $d\boldsymbol{\ell}$, the change in $\boldsymbol{\ell}$, and $\boldsymbol{\tau}$ are parallel, we can replace the vector relation $\boldsymbol{\tau} = d\boldsymbol{\ell}/dt$ by the scalar relation

$$\tau = \frac{d\ell}{dt}.$$

Using the expressions for τ and ℓ from (a) and (b) above, we have

$$mgx_0 = \frac{d}{dt}(mgx_0 t) = mgx_0,$$

which is an identity. Thus the relation $\tau = d\ell/dt$ yields correct results in this simple case. Indeed, if we cancel the constant x_0 out of the first two terms above and if we substitute for gt the equivalent quantity v, we have

$$mg = \frac{d}{dt}(mv).$$

Since $mg = F$ and $mv = p$, this is the familiar result $F = dp/dt$. Thus, as we indicated earlier, relations such as $\tau = d\ell/dt$, though often vastly useful, are not new basic postulates of classical mechanics but are rather the reformulation of the Newtonian laws for rotational motion.

Note that the values of τ and ℓ depend on our choice of origin, that is, on x_0. In particular, if $x_0 = 0$, then $\tau = 0$ and $\ell = 0$.

12–4   Systems of Particles

So far we have talked only about single particles. Let us now consider a system of many particles. To calculate the total angular momentum $\mathbf{L}$ of a system of particles about a given point, we must add vectorially the angular momenta of all the individual particles of the system about this same point. For a system containing n particles we have, then,

Angular momentum of a system of particles

$$\mathbf{L} = \ell_1 + \ell_2 + \cdots + \ell_n = \sum_{i=1}^{i=n} \ell_i, \qquad (12\text{–}9)$$

in which the (vector) sum is taken over all particles in the system.

As time goes on, the total angular momentum $\mathbf{L}$ of the system about a fixed reference point (which we choose, as in our basic definition of ℓ in Eq. 12–3, to be the origin of an inertial reference frame) may change. This change, $d\mathbf{L}/dt$, can arise from two sources: (1) torques exerted on the particles of the system by internal forces between the particles and (2) torques exerted on the particles of the system by external forces.

If the forces between any two particles are not only equal and opposite, as required by Newton's third law, but (as we assume) are also directed along the line joining the two particles, then the total internal torque is zero because the torque resulting from each internal action-reaction force pair is zero. (Forces that meet this condition are said to obey the *strong* form of Newton's third law.) The torques associated with internal forces cancel in pairs and only external torques, associated with external forces, remain.

The time derivative of Eq. 12–9 yields

$$\frac{d\mathbf{L}}{dt} = \sum_{i=1}^{i=n} \frac{d\ell_i}{dt},$$

or, using Eq. 12–7 for each particle,

$$\frac{d\mathbf{L}}{dt} = \sum_{i=1}^{i=n} \tau_i = \sum_{i=1}^{i=n} \tau_{i,\text{internal}} + \sum_{i=1}^{i=n} \tau_{i,\text{external}}.$$

Since the internal torques cancel in pairs we can write this as

Law of motion for a system of particles

$$\tau_{\text{ext}} = \frac{d\mathbf{L}}{dt}, \qquad (12\text{–}10)$$

where τ_{ext} stands for the sum of all the external torques acting on the system. In words, the time rate of change of the total angular momentum of a system of particles about the origin of an inertial reference frame is equal to the sum of the external torques acting on the system. Later, for convenience, in situations in which no confusion is likely to arise, we shall drop the subscript on τ_{ext}.

Equation 12–10 is the generalization of Eq. 12–7 to many particles. When we have only one particle, there are no internal forces or torques. This relation (Eq. 12–10) holds whether the particles that make up the system are in motion relative to each other or whether they have fixed spatial relationships, as in a rigid body.

Equation 12–10 is the rotational analog of Eq. 9–17,

$$\mathbf{F}_{ext} = \frac{d\mathbf{P}}{dt} \qquad [9–17]$$

which tells us that for a system of particles (rigid body or not) the resultant external force acting on the system equals the time rate of change of the linear momentum of the system.

As we have derived it, Eq. 12–10 holds when τ and $\mathbf{L}$ are measured with respect to the origin of an inertial reference frame. We may well ask whether it still holds if we measure these two vectors with respect to an arbitrary point (a particular particle, say) in the moving system. In general, such a point would move in a complicated way as the body or system of particles translated, tumbled, and changed its configuration and Eq. 12–10 would not apply to such a reference point. However, if the reference point is chosen to be the center of mass of the system, even though this point is not fixed in our reference frame, we state without proof that Eq. 12–10 *does* hold. This is another remarkable property of the center of mass. Thus we can separate the general motion of a system of particles into the translational motion of its center of mass (Eq. 9–17) and rotational motion about its center of mass (Eq. 12–10).

12–5 Kinetic Energy of Rotation and Rotational Inertia

We shall now confine our attention to an important special case of a system of particles, a rigid body. In a rigid body the particles in the system always maintain the same positions with respect to one another. In studying the rotation of a rigid body we shall restrict our attention to the special case, often encountered, in which the axis of rotation is fixed in an inertial reference frame. A grinding wheel and the rotor of a steam turbine are common examples.

Let us now imagine a rigid body rotating with angular speed ω about an axis that is fixed in a particular inertial frame, as in Fig. 11–1. Each particle in such a rotating body has a certain amount of kinetic energy. A particle of mass m at a perpendicular distance r from the axis of rotation moves in a circle of radius r with an angular speed ω about this axis and has a linear speed $v = \omega r$. Its kinetic energy therefore is $\frac{1}{2}mv^2 = \frac{1}{2}mr^2\omega^2$. The total kinetic energy of the body is the sum of the kinetic energies of its particles.

If the body is rigid, as we assume in this section, ω is the same for all particles. The radius r may be different for different particles. Hence the total kinetic energy K of the rotating body can be written as

$$K = \tfrac{1}{2}(m_1r_1^2 + m_2r_2^2 + \cdots)\omega^2 = \tfrac{1}{2}(\Sigma\ m_ir_i^2)\omega^2.$$

The term $\Sigma\ m_ir_i^2$ is the sum of the products of the masses of the particles by the squares of their respective distances from the axis of rotation. If we denote this quantity by I, then

Rotational inertia

$$I = \Sigma\ m_ir_i^2 \qquad (12–11)$$

is called the *rotational inertia,* or moment of inertia, of the body with respect to the particular axis of rotation.

Note that the rotational inertia of a body depends on the particular axis about which it is rotating as well as on the shape of the body and the manner in which its mass is distributed. Rotational inertia has the dimensions ML^2; its SI units are kg · m². In calculating I bear in mind that the term r_i in Eq. 12–11 is the perpendicular distance of the corresponding particle from the axis of rotation; it is not the radial distance of the particle from the origin of a coordinate system.

In terms of rotational inertia we can now write the kinetic energy of the rotating rigid body as

Kinetic energy of a rotating rigid body

$$K = \tfrac{1}{2}I\omega^2. \tag{12-12}$$

This is analogous to the expression for the kinetic energy of translation of a body, $K = \tfrac{1}{2}Mv^2$. We have already seen that the angular speed ω is analogous to the linear speed v. Now we see that the rotational inertia I is analogous to the mass, or the translational inertia M. Although the mass of a body does not depend on its location, the rotational inertia of a body does depend on the axis about which it is rotating.

We should understand that the rotational kinetic energy given by Eq. 12–12 is simply the sum of the ordinary translational kinetic energy of all the parts of the body and not a new kind of energy. Rotational kinetic energy is simply a convenient way of expressing the kinetic energy for a rotating rigid body.

Example 2   Consider a body consisting of two spherical masses of 5.0 kg each connected by a light rigid rod 1.0 m long (Fig. 12–5). Treat the spheres as point particles and neglect the mass of the rod. Determine the rotational inertia of the body (a) about an axis normal to it through its center at C, and (b) about an axis normal to it through the center of one sphere.

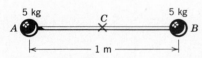

Figure 12–5   Example 2. Calculating the rotational inertia of a dumbbell.

(a) If the axis is normal to the page through C, we have
$$I_C = \Sigma\, m_i r_i^2 = m_a r_a^2 + m_b r_b^2$$
$$= (5.0\text{ kg})(0.50\text{ m})^2 + (5.0\text{ kg})(0.50\text{ m})^2 = 2.5\text{ kg} \cdot \text{m}^2.$$

(b) If the axis is normal to the page through A or B, we have
$$I_A = m_a r_a^2 + m_b r_b^2$$
$$= (5.0\text{ kg})(0\text{ m})^2 + (5.0\text{ kg})(1.0\text{ m})^2 = 5.0\text{ kg} \cdot \text{m}^2,$$
$$I_B = m_a r_a^2 + m_b r_b^2$$
$$= (5.0\text{ kg})(1.0\text{ m})^2 + (5.0\text{ kg})(0\text{ m})^2 = 5.0\text{ kg} \cdot \text{m}^2.$$

Hence the rotational inertia of this rigid dumbbell model is twice as great about an axis through an end as it is about an axis through the center.

For a body that is not composed of discrete point masses but is instead a continuous distribution of matter, the summation in $I = \Sigma\, m_i r_i^2$ becomes an integration. We imagine the body to be subdivided into infinitesimal elements, each of mass dm. We let r be the distance from such an element to the axis of rotation. Then the rotational inertia is obtained from

Rotational inertia—integral form

$$I = \int r^2\, dm, \tag{12-13}$$

where the integral is taken over the whole body. The procedure by which the summation Σ of a discrete distribution is replaced by the integral $\int$ for a continuous distribution is the same as that discussed for the center of mass in Section 9–1.

For bodies of irregular shape the integrals may be hard to evaluate. For bodies of simple geometrical shape the integrals are relatively easy when an axis of symmetry is chosen as the axis of rotation.

The rotational inertias about certain axes of some common solids (of uniform density) are shown in Table 12–1. Each of these results can be derived by integration. The total mass of the body is denoted by M in each equation.

There is a simple and very useful relation between the rotational inertia I of a body about any axis and its rotational inertia I_{cm} with respect to a parallel axis through the center of mass. If M is the total mass of the body and h the distance between the two axes, the relation is

Parallel axis theorem

$$I = I_{cm} + Mh^2. \tag{12-14}$$

Table 12–1 Some rotational inertias

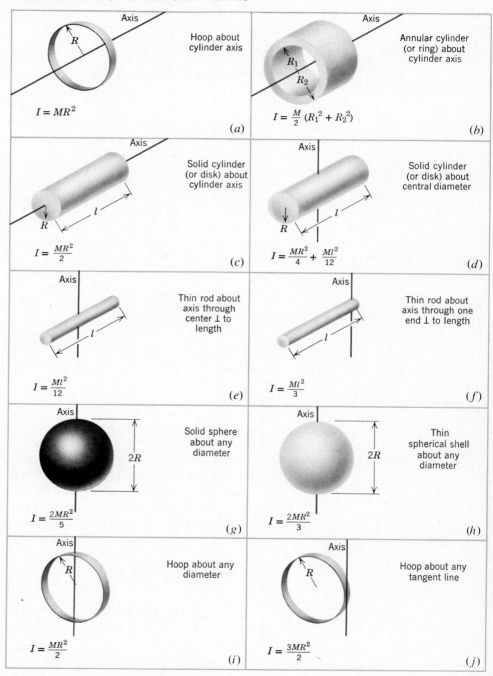

Hoop about cylinder axis $I = MR^2$ *(a)*	Annular cylinder (or ring) about cylinder axis $I = \frac{M}{2}(R_1^2 + R_2^2)$ *(b)*
Solid cylinder (or disk) about cylinder axis $I = \frac{MR^2}{2}$ *(c)*	Solid cylinder (or disk) about central diameter $I = \frac{MR^2}{4} + \frac{Ml^2}{12}$ *(d)*
Thin rod about axis through center ⊥ to length $I = \frac{Ml^2}{12}$ *(e)*	Thin rod about axis through one end ⊥ to length $I = \frac{Ml^2}{3}$ *(f)*
Solid sphere about any diameter $I = \frac{2MR^2}{5}$ *(g)*	Thin spherical shell about any diameter $I = \frac{2MR^2}{3}$ *(h)*
Hoop about any diameter $I = \frac{MR^2}{2}$ *(i)*	Hoop about any tangent line $I = \frac{3MR^2}{2}$ *(j)*

The proof of this relation, which is called the *parallel axis theorem,* follows. Let C be the center of mass of the arbitrarily shaped body whose cross section is shown in Fig. 12–6. The center of mass has coordinates x_{cm} and y_{cm}. We choose

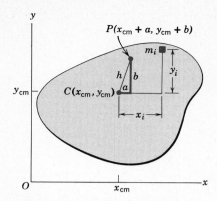

Figure 12–6 Derivation of the parallel axis theorem. Knowing the rotational inertia about an axis through C, we can find its value about a parallel axis through P.

the x-y plane to include C, so that z_{cm} equals zero. Consider an axis through C at right angles to the plane of the paper and another axis parallel to it through P at $(x_{cm} + a)$ and $(y_{cm} + b)$. The distance between the axes is $h = \sqrt{a^2 + b^2}$. Then the square of the distance of a particle from the axis through C is $x_i^2 + y_i^2$, where x_i and y_i measure the coordinates of a mass element m_i relative to the axis through C. The square of its distance from an axis through P is $(x_i - a)^2 + (y_i - b)^2$. Hence the rotational inertia about an axis through P is

$$I = \Sigma\, m_i[(x_i - a)^2 + (y_i - b)^2]$$
$$= \Sigma\, m_i(x_i^2 + y_i^2) - 2a\, \Sigma\, m_i x_i - 2b\, \Sigma\, m_i y_i + (a^2 + b^2)\, \Sigma\, m_i.$$

From the definition of center of mass,

$$\Sigma\, m_i x_i = \Sigma\, m_i y_i = 0,$$

so that the two middle terms are zero. The first term is simply the rotational inertia about an axis through the center of mass I_{cm} and the last term is Mh^2. Hence it follows that $I = I_{cm} + Mh^2$.

With the aid of this formula several of the results of Table 12–1 can be deduced from previous results. For example, (f) follows from (e), and (j) follows from (i) with the aid of Eq. 12–14.

12–6   Rotational Dynamics of a Rigid Body

In this section we continue to study the special case of a rigid body confined to rotate about an axis that is fixed in an inertial reference frame. First we shall review the concept of torque as applied to such a rigid body; then we shall show how the torque is related to the angular acceleration of the body about this axis.

Suppose that we apply a torque τ to one of the particles in a rigid body. Since all the particles of a truly rigid body maintain a fixed spatial relationship to all the other particles that make up the body, the torque may be said to act on the rigid body as a whole. In general, the vector τ will not lie along the axis around which the body is free to rotate. We are not concerned in this section with the actual torques applied to the body but only with the components of these torques that lie along the axis. Only these components can cause the body to rotate about this axis. Torque components perpendicular to the axis tend to turn the axis from its fixed direction.

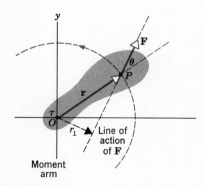

Figure 12–7 A force $\mathbf{F}$ acts on the particle P in a rigid body, exerting a torque $\tau = \mathbf{r} \times \mathbf{F}$ on the body, with respect to an axis through O at right angles to the plane of the figure. The moment arm $r_\perp$ is also shown, as is the torque τ, which is a vector emerging perpendicularly from the page.

In Fig. 12–7 (compare Fig. 11–2) we show a section through a rigid body that is free to rotate about the z-axis of an inertial reference frame. A force $\mathbf{F}$, taken for convenience to be in (or parallel to) the x-y plane of the section, acts on a particle at point P in the body, the position of P with respect to the rotational axis (the z-axis) being defined by the vector $\mathbf{r}$. The torque acting on the particle at P may be said to act on the rigid body as a whole and is given by Eq. 12–1, or

$$\tau = \mathbf{r} \times \mathbf{F}.$$

Because we have chosen $\mathbf{r}$ and $\mathbf{F}$ to lie in the x-y plane, the torque τ will point along the z-axis. The right-hand rule shows that it points perpendicularly out of the plane of Fig. 12–7. If $\mathbf{r}$ and $\mathbf{F}$ did not lie in the plane of the figure, τ would not be parallel to the z-axis and we would concern ourselves here only with the component of τ along this axis. The magnitude of τ is given by Eq. 12–2 or

$$\tau = rF \sin \theta$$

which, as we have seen, can also be written as $\tau = rF_\perp$ or $\tau = Fr_\perp$.

Example 3   A wagon wheel is free to rotate about a horizontal axis through O (Fig. 12–8). A force of 10 lb is applied to a spoke at the point P, 1.0 ft from the center. OP makes an angle of 30° with the horizontal (x-axis) and the force is in the plane of the wheel making an angle of 45° with the horizontal (x-axis). What is the torque on the wheel?

The angle between the displacement vector **r** from O to P and the applied force **F** (Fig. 12–8) is θ, where

$$\theta = 45° - 30° = 15°.$$

Then the magnitude of the torque is

$$\tau = rF \sin \theta$$
$$= (1.0 \text{ ft})(10 \text{ lb})(\sin 15°) = 2.6 \text{ lb} \cdot \text{ft}.$$

It is clear that we can obtain this same result from $\tau = rF_\perp$ or $\tau = Fr_\perp$ as well (see Eqs. 12–2). The torque ($\tau = \mathbf{r} \times \mathbf{F}$) is a vector pointing out $\odot$ along the axis through O having a magnitude 2.6 lb · ft.

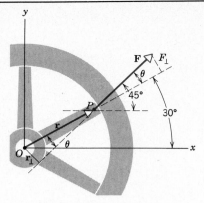

Figure 12–8   Example 3.

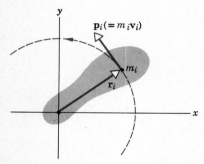

Figure 12–9   A rigid body rotating about the z-axis. The vector $\mathbf{r}_i$ locates one of the particles, of mass m_i, that make it up.

Angular momentum—rigid body with fixed axis

We now investigate the relationship between the torque applied to the rigid body of Fig. 12–7 and the angular acceleration of that body. Equation 12–10 ($\tau_{\text{ext}} = d\mathbf{L}/dt$) reminds us that torque is the time rate of change of angular momentum. We therefore start by finding an expression, in terms of the rotational inertia I, for the angular momentum L of the rotating rigid body. Guided by Fig. 12–9 and by Eqs. 12–9 and 4, we can write

$$L = \Sigma\, \ell_i = \Sigma\, r_i p_i = \Sigma\, r_i(m_i v_i).$$

Note that for every particle of mass m_i that makes up the rigid body, $\mathbf{r}_i$ and the linear momentum $\mathbf{p}_i$ are at right angles and lie in a plane perpendicular to the rotation axis. If we replace v_i in the above by ωr_i, we have

$$L = \Sigma\, m_i r_i^2 \omega = (\Sigma\, m_i r_i^2)\omega$$

or

$$L = I\omega. \tag{12–15}$$

We can remove the quantity ω from the sum because it has the same value for every particle in the rigid body. Finally, we recognize the remaining sum (see Eq. 12–11) as the rotational inertia of the rigid body about its rotation axis. Equation 12–15 is the rotational analog of the expression $P = Mv$, the linear momentum of a rigid body of mass M in translational motion with linear speed v.

The torque follows from Eq. 12–15 as

Equation of motion—rigid body with fixed axis

$$\tau = \frac{dL}{dt} = I\frac{d\omega}{dt}$$

or

$$\tau = I\alpha. \tag{12–16}$$

In deriving this relation we have simply transformed Newton's second law ($F = Ma$), written in scalar form suitable to describe rectilinear motion, into rotational terms. This suggests that just as we associate a force with the linear acceleration of a body, so we may associate a torque with the angular acceleration of a body about a given axis. The rotational inertia I is a measure of the resistance a body offers to having its rotational motion changed by a given torque just as the translational inertia, or mass, M is a measure of the resistance a body offers to having its translational motion changed by a given force.

The rotation of a rigid body about a fixed axis (to which $\tau = I\alpha$ applies) is not the most general kind of rotary motion; the body may not be rigid and the axis may not be fixed in an inertial reference frame. In this general case Eq. 12–10 or $\boldsymbol{\tau}_{\text{ext}} = d\mathbf{L}/dt$, applies. In the rest of this chapter, however, we consider only the case of a rigid body rotating about a fixed axis.

As the resultant torque τ continues to act on the rigid body of Fig. 12–7 (or 12–9), it changes (let us say, increases) the kinetic energy K of the body. We can find the rate P at which energy is delivered to the rotating body by the torque from (see Eq. 12–12)

$$ P = \frac{dK}{dt} = \frac{d}{dt}\,(\tfrac{1}{2}I\omega^2) = I\omega\frac{d\omega}{dt} = I\omega\alpha. $$

But $I\alpha$ is just the torque τ, so that the power P is

Power—rigid body with fixed axis

$$ P = \tau\omega. \tag{12–17} $$

This is the rotational analog of the relation $P = Fv$, which describes the rate at which kinetic energy is delivered to a rigid body moving in translational motion in a fixed direction.

Instead of the power we may want to know the work W done by the resultant torque as the rigid body rotates from an initial angular position θ_i to a final position θ_f. We find it from Eq. 12–17, or

$$ W = \int P\,dt = \int \tau\omega\,dt = \int \tau\left(\frac{d\theta}{dt}\right)dt $$

or

$$ W = \int_{\theta_i}^{\theta_f} \tau\,d\theta. \tag{12–18} $$

Work—rigid body with fixed axis

If the torque is constant, this relation reduces to

$$ W = \tau(\theta_f - \theta_i). \tag{12–19} $$

Equation 12–18 is the rotational analog of $W = \int F\,dx$, the work done by a force F acting on a rigid body in translational motion in a fixed direction.

In Table 12–2 we summarize the definitions and equations that we have developed here for the rotational motion of a rigid body about a fixed axis. Notice how exactly analogous they are to the corresponding linear equations for the translational motion of a rigid body in a fixed direction; see also Table 11–1.

Table 12–2 Some corresponding relations for translational and rotational motion

Translational motion (fixed direction)		Rotational motion (fixed axis)	
Displacement	x	Angular displacement	θ
Velocity	$v = \dfrac{dx}{dt}$	Angular velocity	$\omega = \dfrac{d\theta}{dt}$
Acceleration	$a = \dfrac{dv}{dt}$	Angular acceleration	$\alpha = \dfrac{d\omega}{dt}$
Mass (translational inertia)	M	Rotational inertia	I
Force	$F = Ma$	Torque	$\tau = I\alpha$
Work	$W = \int F\,dx$	Work	$W = \int \tau\,d\theta$
Kinetic energy	$\tfrac{1}{2}Mv^2$	Kinetic energy	$\tfrac{1}{2}I\omega^2$
Power	$P = Fv$	Power	$P = \tau\omega$
Linear momentum	Mv	Angular momentum	$I\omega$

Example 4 A uniform disk of radius R and mass M is mounted on an axle supported in fixed frictionless bearings, as in Fig. 12–10. A light cord is wrapped around the rim of the wheel and a steady downward pull **T** is exerted on the cord. Find the angular acceleration of the wheel and the tangential acceleration of a point on the rim.

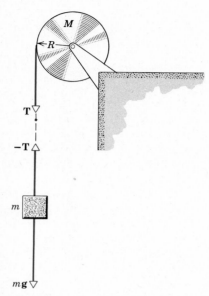

Figure 12–10 Example 4. A steady downward force **T** produces rotation of the disk. Example 5. Here **T** is supplied by the falling mass m.

The torque about the central axis is $\tau = TR$, and the rotational inertia of the disk about its central axis (see Table 12–1c) is $I = \frac{1}{2}MR^2$. From

$$\tau = I\alpha,$$

we have

$$TR = (\tfrac{1}{2}MR^2)\alpha,$$

or

$$\alpha = \frac{2T}{MR}.$$

If the mass of the disk is taken to be $M = 2.5$ kg, its radius $R = 20$ cm, and the force $T = 5.0$ N, then

$$\alpha = \frac{(2)(5.0 \text{ N})}{(2.5 \text{ kg})(0.20 \text{ m})} = 20 \text{ rad/s}^2.$$

The tangential acceleration of a point on the rim is given by

$$a = R\alpha = (20 \text{ rad/s}^2)(0.20 \text{ m}) = 4.0 \text{ m/s}^2.$$

Example 5 Suppose that we hang a body of mass m from the cord in the previous problem. Find the angular acceleration of the disk and the tangential acceleration of a point on the rim in this case.

Now, let T be the tension in the cord. Since the suspended body will accelerate downward, the magnitude of the downward pull of gravity on it, mg, must exceed the magnitude of the

upward pull of the cord on it, T. The acceleration a of the suspended body is the same as the tangential acceleration of a point on the rim of the disk. From Newton's second law,

$$mg - T = ma.$$

The resultant torque on the disk is TR and its rotational inertia is $\frac{1}{2}MR^2$, so that from

$$\tau = I\alpha$$

we obtain

$$TR = \tfrac{1}{2}MR^2\alpha.$$

Using the relation $a = R\alpha$, we can write this last equation as

$$2T = Ma.$$

Solving the first and last equations simultaneously leads to

$$a = \left(\frac{2m}{M + 2m}\right) g,$$

and

$$T = \left(\frac{Mm}{M + 2m}\right) g.$$

If now we let the disk have a mass $M = 2.5$ kg and a radius $R = 20$ cm as before, and we let the suspended body weigh 5.0 N (mass $= W/g = 510$ g), we obtain

$$a = \frac{2mg}{M + 2m} = \frac{(2)(5.0 \text{ N})}{(2.5 \text{ kg}) + 2(5.0/9.8) \text{ kg}} = 2.8 \text{ m/s}^2,$$

and

$$\alpha = \frac{a}{R} = \frac{(2.8 \text{ m/s}^2)}{0.20 \text{ m}} = 14 \text{ rad/s}^2.$$

Notice that the accelerations are less for a suspended 5.0-N body than they were for a steady 5.0-N pull on the string (Example 4). This corresponds to the fact that the tension in the string supplying the torque is now less than 5.0 N, namely,

$$T = \frac{Mmg}{M + 2m} = \frac{(2.5 \text{ kg})(5.0 \text{ N})}{(2.5 + 1.0) \text{ kg}} = 3.6 \text{ N}.$$

The tension in the string must be less than the weight of the suspended body if the body is to accelerate downward.

Example 6 Assuming that the disk of Example 5 starts from rest, compute the work done by the applied torque on the disk in 2.0 s. Compute also the increase in rotational kinetic energy of the disk.

Since the applied torque is constant, the resulting angular acceleration is constant. The total angular displacement in constant angular acceleration is obtained from Eq. 11–5,

$$\theta = \omega_0 t + \tfrac{1}{2}\alpha t^2,$$

in which

$$\omega_0 = 0, \qquad \alpha = 14 \text{ rad/s}^2, \qquad t = 2.0 \text{ s},$$

so that

$$\theta = 0 + (\tfrac{1}{2})(14 \text{ rad/s}^2)(2.0 \text{ s})^2 = 28 \text{ rad} = 4.5 \text{ rev}.$$

For constant torque the work done in a finite angular displacement is (Eq. 12-19)

$$W = \tau(\theta_2 - \theta_1),$$

in which

$$\tau = TR = (3.6 \text{ N})(0.20 \text{ m}) = 0.72 \text{ N} \cdot \text{m},$$

and

$$\theta_2 - \theta_1 = \theta = 28 \text{ rad}.$$

Therefore

$$W = (0.72 \text{ N} \cdot \text{m})(28 \text{ rad}) = 20 \text{ J}.$$

This work must result in an increase in rotational kinetic energy of the disk. Starting from rest the disk acquires an angular speed ω. The rotational energy is $\frac{1}{2}I\omega^2 = \frac{1}{2}(\frac{1}{2}MR^2)\omega^2$. To obtain ω we use Eq. 11-3,

$$\omega = \omega_0 + \alpha t,$$

in which

$$\omega_0 = 0, \qquad t = 2.0 \text{ s}, \qquad \alpha = 14 \text{ rad/s}^2,$$

so that

$$\omega = 0 + (14 \text{ rad/s}^2)(2.0 \text{ s}) = 28 \text{ rad/s}.$$

Then

$$\tfrac{1}{2}I\omega^2 = (\tfrac{1}{4})(2.5 \text{ kg})(0.20 \text{ m})^2(28 \text{ rad/s})^2 = 20 \text{ J},$$

as before. Hence the increase in kinetic energy of the disk is equal to the work done by the resultant force on the disk, as it must be.

Example 7   Show that the conservation of mechanical energy holds for the system of Example 5.

The resultant force acting on the system is the force of gravity on the suspended body. This is a conservative force. Viewing the system as a whole, we see that the suspended body loses potential energy U as it descends,

$$U = mgy,$$

where y is the vertical distance through which the block descends. At the same time the suspended body gains kinetic energy of translation and the disk gains kinetic energy of rotation. The total gain in kinetic energy is

$$\tfrac{1}{2}mv^2 + \tfrac{1}{2}I\omega^2,$$

where v is the linear speed of the suspended mass. We must show then that

$$mgy = \tfrac{1}{2}mv^2 + \tfrac{1}{2}I\omega^2.$$

For the linear motion starting from rest we have (see Eq. 3-16) $v^2 = 2ay$. From Example 5, we obtained $a = 2\,mg/(M + 2m)$. Hence

$$mgy = \frac{mgv^2}{2a} = \tfrac{1}{2}mv^2\left(\frac{g}{a}\right) = \tfrac{1}{2}mv^2\left(\frac{M + 2m}{2m}\right)$$
$$= \tfrac{1}{4}(M + 2m)v^2.$$

We also know that $\omega = v/R$ and $I = \frac{1}{2}MR^2$. Substituting these relations into the right-hand side of the conservation equation, we obtain

$$\tfrac{1}{2}mv^2 + \tfrac{1}{2}I\omega^2 = \tfrac{1}{2}mv^2 + \tfrac{1}{2}(\tfrac{1}{2}MR^2)(v^2/R^2) = \tfrac{1}{4}(M + 2m)v^2.$$

The mechanical energy is therefore conserved.

12-7   Rolling Bodies

Figure 12-11   rolling body may at any instant be thought of as rotating about a perpendicular axis through its point of contact P.

Up until now we have considered only bodies rotating about some fixed axis. If a body is rolling, however, it is rotating about an axis and also moving translationally. Therefore it would seem that the motion of rolling bodies must be treated as a combination of translational and rotational motion. It is also possible, however, to treat a rolling body as though its motion is one of pure rotation. We wish to illustrate the equivalence of the two approaches.

Consider, for example, a cylinder rolling along a level surface, as in Fig. 12-11. At any instant the bottom of the cylinder is at rest on the surface, since it does not slide. The axis normal to the diagram through the point of contact P is called the *instantaneous axis of rotation*. At that instant the linear velocity of every particle of the cylinder is directed at right angles to the line joining the particle and P and its magnitude is proportional to this distance. This is the same as saying that the cylinder is rotating about a fixed axis through P with a certain angular speed ω, *at that instant*. Hence, at a given instant the motion of the body is equivalent to a pure rotation. The total kinetic energy can therefore be written as

Kinetic energy in pure rotation

$$K = \tfrac{1}{2}I_P\omega^2, \tag{12-20}$$

where I_P is the rotational inertia about the axis through P.

Let us now use the parallel axis theorem, which tells us that

$$I_P = I_{\text{cm}} + MR^2,$$

where I_{cm} is the rotational inertia of the cylinder of mass M and radius R about a parallel axis through the center of mass. Equation 12–20 now becomes

$$K = \tfrac{1}{2}I_{\text{cm}}\omega^2 + \tfrac{1}{2}MR^2\omega^2. \qquad (12\text{--}21)$$

The quantity $R\omega$ is the speed with which the center of mass of the cylinder is moving with respect to the fixed point P. Let $R\omega = v_{\text{cm}}$. Equation 12–21 then becomes

$$K = \tfrac{1}{2}I_{\text{cm}}\omega^2 + \tfrac{1}{2}Mv_{\text{cm}}^2. \qquad (12\text{--}22)$$

Now notice that the speed of the center of mass with respect to P is the same as the speed of P with respect to the center of mass. Hence, the angular speed ω of the center of mass about P as seen by someone at P is the same as the angular speed of a particle at P about C as seen by someone at C (moving along with the cylinder). This is equivalent to saying that any reference line in the cylinder turns through the same angle in a given time whether it is observed from a reference frame fixed with respect to the surface on which the cylinder is rolling or from a frame moving translationally with respect to this fixed frame. We can therefore interpret Eq. 12–22, which was derived on the basis of a pure rotational motion, in another way; that is, the first term, $\tfrac{1}{2}I_{\text{cm}}\omega^2$, is the kinetic energy the cylinder would have if it were merely rotating about an axis through its center of mass, without translational motion; and the second term, $\tfrac{1}{2}Mv_{\text{cm}}^2$, is the kinetic energy the cylinder would have if it were moving translationally with the speed of its center of mass, without rotating. Notice that there is now no reference at all to the instantaneous axis of rotation. In fact, Eq. 12–22 applies to any body that is moving and rotating about an axis perpendicular to its motion whether or not it is rolling on a surface.

The combined effects of translation of the center of mass and rotation about an axis through the center of mass are equivalent to a pure rotation with the same angular speed about an axis through the point of contact of a rolling body.

To illustrate this result simply, let us consider the instantaneous speed of various points on the rolling cylinder. If the speed of the center of mass (as seen by an observer fixed with respect to the surface) is v_{cm}, the instantaneous angular speed about an axis through P is $\omega = v_{\text{cm}}/R$. A point Q at the top of the cylinder will therefore have a speed $\omega 2R = 2v_{\text{cm}}$ at that instant. The point of contact P is instantaneously at rest. Hence, from the point of view of pure rotation about P, the situation is as shown in Fig. 12–12.

Now let us regard the rolling as a combination of translation of the center of mass and rotation about the cylinder axis through C. If we consider translation only, all points on the cylinder have the same speed v_{cm}, the speed of the center of mass. This is shown in Fig. 12–13a. If we consider the rotation only, the center is at rest, whereas the point Q at the top has a speed ωR in the x-direction and the point P at the bottom of the cylinder has a speed ωR in the $-x$-direction. This is shown in Fig. 12–13b. Now let us combine these two results. Recalling that $\omega = v_{\text{cm}}/R$, we obtain

$$\text{for the point } Q \qquad v = v_{\text{cm}} + \omega R = v_{\text{cm}} + \frac{v_{\text{cm}}}{R}R = 2v_{\text{cm}},$$

$$\text{for the point } C \qquad v = v_{\text{cm}} + 0 = v_{\text{cm}},$$

$$\text{for the point } P \qquad v = v_{\text{cm}} - \omega R = v_{\text{cm}} - \frac{v_{\text{cm}}}{R}R = 0.$$

This result, shown in Fig. 12–13c, is exactly the same as that obtained from the purely rotational point of view, Fig. 12–12.

Kinetic energy in combined rotation and translation

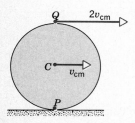

Figure 12–12 Since Q and C have the same angular velocity about P, therefore Q, being twice as far from P, moves with twice the linear velocity of C.

Figure 12–13 (a) For pure translation, all points move with the same velocity. (b) For pure rotation about C, opposite points move with opposite velocities. (c) Combined rotation and translation is obtained by adding together corresponding vectors in (a) and (b).

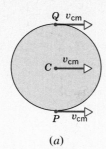

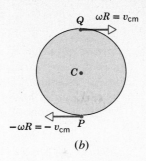

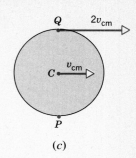

(a)   (b)   (c)

Example 8 Consider a solid cylinder of mass M and radius R rolling down an inclined plane without slipping. Find the speed of its center of mass when the cylinder reaches the bottom.

The situation is illustrated in Fig. 12–14. We can use the conservation of energy to solve this problem. The cylinder is initially at rest. In rolling down the incline the cylinder loses potential energy of an amount Mgh, where h is the height of the incline. It gains kinetic energy equal to

$$\tfrac{1}{2}I_{cm}\omega^2 + \tfrac{1}{2}Mv^2,$$

where v is the linear speed of the center of mass and ω is the angular speed about the center of mass at the bottom.

We have then the relation

$$Mgh = \tfrac{1}{2}I_{cm}\omega^2 + \tfrac{1}{2}Mv^2,$$

in which (see Table 12–1)

$$I_{cm} = \tfrac{1}{2}MR^2 \quad \text{and} \quad \omega = \frac{v}{R}.$$

Hence

$$Mgh = \tfrac{1}{2}(\tfrac{1}{2}MR^2)\left(\frac{v}{R}\right)^2 + \tfrac{1}{2}Mv^2 = (\tfrac{1}{4} + \tfrac{1}{2})Mv^2,$$

$$v^2 = \tfrac{4}{3}gh \quad \text{or} \quad v = \sqrt{\tfrac{4}{3}gh} = 1.15\sqrt{gh}.$$

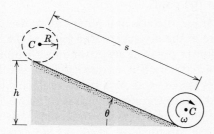

Figure 12–14 Example 8. A cylinder rolling down an incline.

The rotation in this case is caused by the friction between the cylinder and the incline. This provides the only force that exerts a torque around the cylinder's center of mass. If friction were not present, the cylinder would merely slide down the incline without rotating. Its speed at the bottom would then be $v = \sqrt{2\,gh} = 1.41\sqrt{gh}$. The speed of the rolling cylinder is therefore less than the speed of the sliding cylinder. For the rolling cylinder, part of the lost potential energy has been transformed into rotational kinetic energy, leaving less available for the translational part of the kinetic energy. Although the rolling cylinder arrives later at the bottom of the incline than an identical sliding cylinder started at the same time down a frictionless, but otherwise identical, incline, both arrive at the bottom with the same amount of energy; the rolling cylinder happens to be rotating as it moves, whereas the sliding one does not rotate as it moves.

Example 9 A sphere and a cylinder, having the same mass and radius, start from rest and roll down the same incline. Which body gets to the bottom first?

For a sphere, Table 12–1 shows that $I_{cm} = \tfrac{2}{5}MR^2$. If we solve Example 8 using a sphere instead of a cylinder we find, for the speed at the bottom of the incline,

$$v = \sqrt{\tfrac{10}{7}\,gh} = 1.20\sqrt{gh}.$$

Comparison with that example shows that the sphere is faster at all stages than the rolling cylinder ($= 1.15\sqrt{gh}$). Thus, in a race between a rolling sphere and a rolling cylinder, the sphere will always win. (But any object *sliding* down the incline without friction will beat them both.)

Which body has the greater rotational energy at the bottom? Which body has the greater translational energy at the bottom?

Note carefully that neither the mass nor the radius of the rolling object enters the previous results. How, then, would we expect the behavior of cylinders of different mass and radii to compare? How would we expect the behavior of spheres of different mass and radii to compare? How would the behavior of a cylinder and sphere having different masses and radii compare?

12–8 Conservation of Angular Momentum

In Section 12–4 we found that the time rate of change of the total angular momentum of a system of particles about a point fixed in an inertial reference frame (or about the center of mass) is equal to the sum of the external torques acting on the system; that is,

$$\tau_{\text{ext}} = \frac{d\mathbf{L}}{dt}. \qquad [12\text{–}10]$$

Suppose now that $\tau_{\text{ext}} = 0$; then $d\mathbf{L}/dt = 0$ so that $\mathbf{L} = $ a constant.

When the resultant external torque acting on a system is zero, the total vector angular momentum of the system remains constant. This is the *principle of the conservation of angular momentum.*

For a system of n particles, the total angular momentum $\mathbf{L}$ about some point is

$$\mathbf{L} = \boldsymbol{\ell}_1 + \boldsymbol{\ell}_2 + \cdots + \boldsymbol{\ell}_n.$$

When the resultant external torque on the system is zero, we have

$$\mathbf{L} = \text{a constant} = \mathbf{L}_0, \qquad (12\text{–}23)$$

where $\mathbf{L}_0$ is the constant total angular momentum vector. The angular momenta of the individual particles may change, but their vector sum $\mathbf{L}_0$ remains constant in the absence of a net external torque.

Angular momentum is a vector quantity so that Eq. 12–23 is equivalent to three scalar equations, one for each coordinate direction through the reference point. The conservation of angular momentum therefore supplies us with three conditions on the motion of a system to which it applies.

For a system consisting of a rigid body rotating about an axis (the z-axis, say) that is fixed in an inertial reference frame, we have

$$L_z = I\omega, \qquad (12\text{–}24)$$

where L_z is the component of the angular momentum along the rotation axis and I is the rotational inertia for this same axis. It is possible for the rotational inertia I of a rotating body to change by rearrangement of its parts. If no net external torque acts, then L_z must remain constant and, if I does change, there must be a compensating change in ω. The principle of conservation of angular momentum in this case is expressed as

$$I\omega = I_0\omega_0 = \text{a constant.} \qquad (12\text{–}25)$$

Equation 12–25 holds not only for rotation about a fixed axis but also about an axis through the center of mass of the system that moves so that it always remains parallel to itself.

Acrobats, divers, ballet dancers, ice skaters, and others often use this principle. Because I depends on the square of the distance of the parts of the body from the axis of rotation, a large variation is possible by extending or pulling in the limbs. Consider the diver* in Fig. 12–15. Let us assume that as she leaves the diving board she has a certain angular speed ω_0 about a horizontal axis through the center of mass, such that she would rotate through half a turn before she strikes water. If she wishes to make a one and one-half turn somersault instead, in the same time, she must triple her angular speed. Now there are no external forces acting on her except gravity, and gravity exerts no torque about her center of mass. Her angular momentum therefore remains constant, and $I_0\omega_0 = I\omega$. Since $\omega = 3\omega_0$, the diver must change her rotational inertia about the horizontal axis through the center of mass from the initial value I_0 to a value I, such that I equals $\frac{1}{3}I_0$. This she does by pulling in her arms and legs toward the center of her body. The greater her initial angular speed and the more she can reduce her rotational inertia, the greater the number of revolutions she can make in a given time.

Conservation of angular momentum

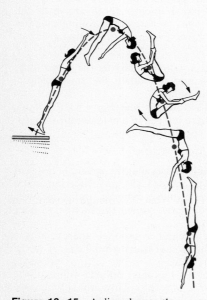

Figure 12–15 A diver leaves the diving board with arms and legs outstretched and with some initial angular velocity. Since no torques are exerted on her about her center of mass, $L(=I\omega)$ is constant while she is in the air. When she pulls her arms and legs in, since I decreases, ω increases. When she again extends her limbs, her angular velocity drops back to its initial value. Notice the parabolic motion of her center of mass, common to all two-dimensional motion under the influence of gravity.

---

* See "The Mechanics of Swimming and Diving," by R. L. Page, *The Physics Teacher,* February 1976, for an interesting biomechanical analysis.

We should notice that the rotational kinetic energy of the diver is not constant. In fact, in our example, since

$$I\omega = I_0\omega_0$$

and

$$I < I_0,$$

it follows that

$$\frac{1}{2}I\omega^2 = \frac{1}{2}\frac{(I\omega)^2}{I} > \frac{1}{2}\frac{(I_0\omega_0)^2}{I_0} = \frac{1}{2}I_0\omega_0^2$$

and the diver's rotational kinetic energy increases. This increase in energy is supplied by the diver, who does work when she pulls the parts of her body together.

In a similar way the ice skater and ballet dancer can increase or decrease the angular speed of a spin about a vertical axis.

Example 10 A small object of mass m is attached to a light string which passes through a hollow tube. The tube is held by one hand and the string by the other. The object is set into rotation in a circle of radius r_1 with a speed v_1. The string is then pulled down, shortening the radius of the path to r_2 (Fig. 12–16). Find the new linear speed v_2 and the new angular speed ω_2 of the object in terms of the initial values v_1 and ω_1 and the two radii.

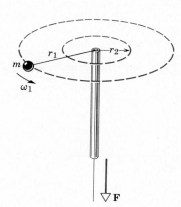

Figure 12–16 Example 10. An object of mass m at the end of a string moves in a circle of radius r_1 with angular speed ω_1. The string passes down through a tube. The tension in the string supplies the centripetal force.

The downward pull on the string is transmitted as a radial force on the object. Such a force exerts a zero torque on the object about the center of rotation. Since no torque acts on the object about its axis of rotation, its angular momentum in that direction is constant. Hence

initial angular momentum = final angular momentum,

$$mv_1r_1 = mv_2r_2,$$

and

$$v_2 = v_1\left(\frac{r_1}{r_2}\right).$$

Since $r_1 > r_2$, the object speeds up upon being pulled in.

In terms of angular speed, since v_1 equals ω_1r_1 and v_2 equals ω_2r_2,

$$mr_1^2\omega_1 = mr_2^2\omega_2$$

and

$$\omega_2 = \left(\frac{r_1}{r_2}\right)^2\omega_1,$$

so that there is an even greater increase in angular speed over the initial value. What effect does the force of gravity (the object's weight) have on this analysis?

Example 11 A student sits on a stool that is free to rotate about a vertical axis. He holds his arms extended horizontally with a 4.0-kg dumbbell ($W = mg = 8.8$ lb) in each hand. The instructor starts him rotating with an angular speed of 0.50 rev/s. Assume that friction is negligible and exerts no torque about the vertical axis of rotation. Assume also that the rotational inertia of the student remains constant at 5.0 kg · m² as he pulls his hands to his sides and that the change in rotational inertia is due only to pulling the dumbbells in. Take the original distance of the dumbbells from the axis of rotation to be 90 cm and their final distance 15 cm. Find the final angular speed of the student.

The only external force is gravity acting through the center of mass, and that exerts no torque about the axis of rotation. Hence the angular momentum is conserved about this axis and

initial angular momentum = final angular momentum,

$$I_0\omega_0 = I\omega.$$

We have

$$I = I_{\text{student}} + I_{\text{dumbbells}}$$

$$I_0 = 5.0 \text{ kg} \cdot \text{m}^2 + 2(4.0 \text{ kg})(0.90 \text{ m})^2 = 11.5 \text{ kg} \cdot \text{m}^2,$$

$$I = 5.0 \text{ kg} \cdot \text{m}^2 + 2(4.0 \text{ kg})(0.15 \text{ m})^2 = 5.2 \text{ kg} \cdot \text{m}^2,$$

$$\omega_0 = 0.50 \text{ rev/s}.$$

Therefore

$$\omega = \frac{I_0}{I}\omega_0 = \frac{(11.5 \text{ kg} \cdot \text{m}^2)}{(5.2 \text{ kg} \cdot \text{m}^2)}0.5 \text{ rev/s} = 1.1 \text{ rev/s}.$$

The final angular speed is approximately doubled. If we allowed for the decrease in I caused by the arms being pulled in, the final angular speed would have been greater.

What change would friction make? Is kinetic energy conserved as the student pulls in his arms and then puts them out again, assuming that there is no friction? Explain.

Example 12 A classroom demonstration that illustrates the vector nature of the law of conservation of angular momentum is worth considering.

A student stands in a platform that can rotate only about a vertical axis. In his hand he holds the axle of a rim-loaded bicycle wheel with its axis vertical. The wheel is spinning about this vertical axis with an angular speed ω_0, but the student and platform are at rest. The student tries to change the direction of rotation of the wheel. What happens?

Let us choose as the system the student plus platform plus wheel. The initial total angular momentum of this system is $I_0\omega_0$, arising from the spinning wheel alone, I_0 being the rotational inertia of the wheel about its axis. Figure 12–17a shows the initial condition.

The student next turns the axis of the wheel through an angle of 180° from the vertical (to do this he must supply a torque. This torque, however, is internal to the system as we have defined it). Since there is no external component of torque on the system about the vertical axis, the vertical component of angular momentum of the system must be conserved. The wheel, however, now contributes a vertical component of angular momentum of $-I_0\omega_0$ to the system. Hence the student and platform must supply the additional angular momentum about the vertical axis, and they begin to rotate. This extra vertical angular momentum $I_p\omega_p$ when added to $-I_0\omega_0$ must equal the initial vertical angular momentum of the system $I_0\omega_0$. That is,

$$-I_0\omega_0 + I_p\omega_p = I_0\omega_0$$

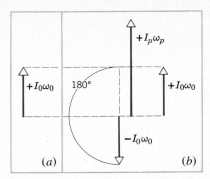

Figure 12–17 Example 12. (a) The initial angular momentum of the system is shown. In (b) the wheel has been turned through 180°. Its contribution to the vertical component of angular momentum is reduced by $2I_0\omega_0$. The deficit is made up by the student and platform.

This is shown in Fig. 12–17b. I_p is the rotational inertia of student and platform with respect to the vertical axis, and ω_p is their angular speed about this axis.

Thus, when the student turns the wheel through an angle $\theta = 180°$, the student and platform acquire a vertical angular momentum of $2I_0\omega_0$. The total vertical angular momentum of the system is still being conserved at the initial value $I_0\omega_0$.

Consider now the angular momentum of the wheel alone. As the student turns the axis of the wheel he exerts a torque on it which lasts for the time Δt that it takes to reorient the shaft. The vertical component of the reaction to this "torque-impulse" acts on the student and accounts for the vertical angular momentum acquired by him and the platform.

REVIEW GUIDE AND SUMMARY

Rotational motion—a single particle

Our understanding of rotational dynamics starts by considering a single particle of mass m whose location with respect to the origin O of an inertial reference frame is described by a position vector $\mathbf{r}$. Let a force $\mathbf{F}$ act on the particle and let its linear momentum be $\mathbf{p}$. Here are two definitions and a relation between them:

1. The torque τ about O acting on $\tau = \mathbf{r} \times \mathbf{F};$ [12–1]
   a particle due to the
   force $\mathbf{F}$,

2. The angular momentum ℓ $\ell = \mathbf{r} \times \mathbf{p};$ [12–3]
   of the particle about O,

3. Newton's second law of $\tau = \dfrac{d\ell}{dt}.$ [12–7]
   motion for a single par-
   ticle, in rotational form,

Rotational motion—a system of particles

Equation 12–7 above is the rotational analog of Eq. 9–12 ($\mathbf{F} = d\mathbf{p}/dt$), which is Newton's second law for single particle. Study Example 1 and review the material in Section 2–4 on vector products.

The above relations can be generalized to apply to a system of particles, or:

1'. The resultant external torque τ_{ext} on a system of particles about O,

$$\tau_{\text{ext}} = \Sigma\,\tau_i = \Sigma(\mathbf{r}_i \times \mathbf{F}_i);$$ [12–1]

2'. The resultant angular momentum $\mathbf{L}$ of a system of particles about O,

$$\mathbf{L} = \Sigma\,\ell_i = \Sigma(\mathbf{r}_i \times \mathbf{p}_i);$$ [12–3, 12–9]

3'. Newton's second law of motion for a system of particles, in rotational form,

$$\tau_{\text{ext}} = \frac{d\mathbf{L}}{dt}.$$ [12–10]

Equation 12–10 above is the rotational analog of Eq. 9–17 ($\mathbf{F}_{\text{ext}} = d\mathbf{P}/dt$), which is Newton's second law for a system of particles.

One important system of particles studied in this chapter is the special case of a rigid body free to rotate about a fixed axis. A central quantity here is the rotational inertia I of the body about that axis. I measures the resistance of the body to angular acceleration when a torque is applied, just as the mass M of a body measures its resistance to linear acceleration when a force is applied. For a hoop, a cylinder, and a sphere, I about the axis of axial symmetry is MR^2, $\frac{1}{2}MR^2$, and $\frac{2}{5}MR^2$, respectively; see Table 12–1 and Example 2.

A useful relation involving rotational inertia is the parallel axis theorem, or

$$I = I_{\text{cm}} + Mh^2.$$ [12–14]

This tells us how to find the rotational inertia of a body about *any* axis, provided we know its rotational inertia I_{cm} for a parallel axis through the center of mass; h in the above equation is the perpendicular distance between these two axes. Study the application of this theorem in Table 12–1.

The kinetic energy K and the angular momentum L of a rigid body rotating about a fixed axis involve I and are given by

$$K = \tfrac{1}{2}I\omega^2 \quad \text{and} \quad L = I\omega$$ [12–12, 12–15]

Finally, the law of motion, which is a special case of Eq. 12–10 ($\tau_{\text{ext}} = d\mathbf{L}/dt$) also involves I and is given by

$$\tau = I\alpha.$$ [12–16]

Here τ is the component of τ_{ext} along the fixed rotation axis, I is measured with respect to that axis, and the angular acceleration α is taken about that same axis. Examples 4, 5, 6, and 7, taken together, provide a detailed study, from several points of view, of rotational dynamics applied to a particular rigid body, the rotating wheel of Fig. 12–10.

A study of rolling hoops, cylinders, spheres, and so on, leads to the concept of an instantaneous axis of rotation. This is an axis lying in the rolling surface and passing through the point or line of contact between the rolling body and that surface. We can look at a rolling body in two quite equivalent ways:

1. As in pure rotation about the instantaneous axis. The kinetic energy is then

$$K = \tfrac{1}{2}I_P\,\omega^2,$$ [12–20]

where I_P is taken about this axis.

2. As in combined rotation (about an axis through the center of mass) and translation (of the center of mass). The kinetic energy in this case is

$$K = \tfrac{1}{2}I_{\text{cm}}\omega^2 + \tfrac{1}{2}Mv_{\text{cm}}^2,$$ [12–22]

where I_{cm} is taken about an axis through the center of mass.

Study carefully Fig. 12–13, which analyzes the combined rotation and translation of a wheel. Study also Examples 8 and 9, which compare the rolling and sliding of cylinders and spheres down a smooth inclined plane.

Rotational inertia

The parallel axis theorem

Rotational dynamics

Rolling bodies

Conservation of angular
momentum

The culmination of our study of rotational motion is the law of conservation of angular momentum. This states that the angular momentum **L** of a system remains constant if no external torques act on the system. This is true no matter what internal torques and rearrangements may take place. Give particular attention to the application of this law outlined in Example 12, in which a person sits on a stool that is free to rotate without friction and manipulates a spinning, rim-loaded bicycle wheel. This example illustrates the importance of the vector nature of angular momentum.

QUESTIONS

1. Can the mass of a body be considered as concentrated at its center of mass for purposes of computing its rotational inertia?

2. The language we use must be adapted to the case at hand. Ballet instructors, telling a dancer how to do a *pirouette,* may say, "Imagine that your body is being sucked up into a drinking straw." How would you give similar instructions in the langue of physics?

3. About what axis is the rotational inertia of your body the least?

4. If two circular disks of the same weight and thickness are made from metals having different densities, which disk, if either, will have the larger rotational inertia about its central axis?

5. The rotational inertia of a body of rather complicated shape is to be determined. The shape makes a mathematical calculation from $\int r^2 dm$ exceedingly difficult. Suggest ways in which the rotational inertia could be measured experimentally.

6. Five solids are shown in cross section (Fig. 12–18). The cross sections have equal heights and equal maximum widths. The solids have equal masses. Which one has the largest rotational inertia about a perpendicular axis through the center of mass? Which the smallest?

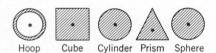

Hoop   Cube   Cylinder   Prism   Sphere

Figure 12–18   Question 6.

7. Why does the front of a car tip upward when you accelerate?

8. In Fig. 12–19*a* a meter stick, half of which is wood–the other half steel–is pivoted at the wooden end at *O* and a force is applied to the steel end at *a*. In Fig. 12–19*b* the stick is pivoted at the steel end at *O'* and the same force is applied at the wooden end at *a'*. Does one get the same angular acceleration in each case? Explain.

9. You can distinguish between a raw egg and a hardboiled one by spinning each one on the table. Explain how. Also, if you stop a spinning raw egg with your fingers and release it very quickly, it will resume spinning. Why?

10. Comment on each of these assertions about skiing. (*a*) In downhill racing one wants skis that do not turn easily. (*b*) In slalom racing, one wants skis that turn easily. (*c*) Therefore, the rotational inertia of downhill skis should be larger than that of slalom skis. (*d*) Considering that there is low friction between skis and snow and that the skier's center of mass is about over the center of the skis, how does a skier exert torques to turn or to stop a turn? (See "The Physics of Ski Turns," by J. I. Shonie and D. L. Nordick, *The Physics Teacher,* December 1972.)

11. A cylindrical drum, pushed along by a board from an initial position shown in Fig. 12–20, rolls forward on the ground a distance *l*/2, equal to half the length of the board. There is no slipping at any contact. Where is the board then? How far has the man walked?

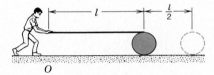

Figure 12–20   Question 11.

12. A solid wooden cylinder rolls down two different inclined planes of the same height but with different angles of inclination. Will it reach the bottom with the same speed in each case? Will it take longer to roll down one incline than the other? If your answer is yes to either question, explain why.

13. A solid brass cylinder and a solid wooden cylinder have the same radius and mass, the wooden cylinder being longer. You release them together at the top of an incline. Which will beat the other to the bottom? Suppose that the cylinders are now made to be the same length (and radius) and that the masses are made to be equal by boring a hole along the axis of the brass cylinder. Which cylinder will win the race now?

14. Two heavy disks are connected by a short rod of much smaller radius. The system is placed on an inclined plane so that the disks hang over the sides and the system rolls down on the rod without slipping (Fig. 12–21). Near the bottom of the incline the disks touch the horizontal table top and the system takes off with greatly increased translational speed. Explain carefully.

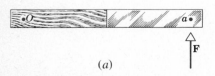

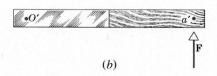

(*a*)   (*b*)

Figure 12–19   Question 8.

Figure 12-21   Question 14.

15. Consider a straight stick standing on end on (frictionless) ice. What would be the path of its center of mass if it falls?

16. A Yo-Yo is resting on a horizontal table and is free to roll (Fig. 12–22). If the string is pulled by a horizontal force such as $\mathbf{F}_1$, which way will the Yo-Yo roll? What happens when the force $\mathbf{F}_2$ is applied (its line of action passes through the point of contact of the Yo-Yo and table)? If the string is pulled vertically with the force $\mathbf{F}_3$, what happens?

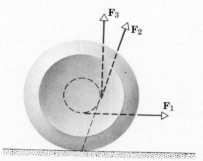

Figure 12-22   Question 16.

17. You are looking at the wheel of an automobile traveling at constant speed. Some one says to you: "The top of the wheel is moving twice as fast as the axle but the bottom is not moving at all." Can you accept this statement? Explain why or why not.

18. Ruth and Roger are cycling along a path at the same speed. The wheels of Ruth's bike are a little larger in diameter than the wheels of Roger's bike. How do the angular speeds of their wheels compare? What about the speeds of the top portions of their wheels?

19. Explain, in terms of angular momentum and rotational inertia, exactly how one "pumps up" a swing. (See "How Children Swing," by S. M. Curry, *American Journal of Physics,* October 1976.)

20. A man turns on a rotating table with an angular speed ω. He is holding two equal dumbells at arm's length. Without moving his arms, he drops the two dumbells. What change, if any, is there in his angular speed? Is the angular momentum conserved? Explain.

21. A helicopter flies off, its propellors rotating. Why doesn't the body of the helicopter rotate in the opposite direction?

22. If the entire population of the world moved to Antarctica, would it affect the length of the day? If so, in what way?

23. A circular turntable rotates at constant angular velocity about a vertical axis. There is no friction and no driving torque. A circular pan rests on the turntable and rotates with it: see Fig. 12–23. The bottom of the pan is covered with a layer of ice of uniform thickness, which is, of course, also rotating with the pan. The ice melts but none of the water escapes from the pan. Is the angular velocity now greater than, the same as, or less than the original velocity? Give reasons for your answer.

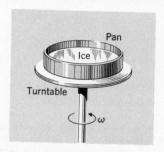

Figure 12-23   Question 23.

EXERCISES

Section 12-2   Torque
1. A 0.75-kg mass is attached to a 1.25-m massless rod and hung from a pivot. When the resulting pendulum is 30° from the vertical, what is the magnitude of the torque about the pivot?

2. (a) If the length of a bicycle pedal arm is 6.0 in. (=0.152 m) and a downward force of 25 lb (=111 N) is applied by the foot, what is the torque about the pivot point when the arm makes an angle of 30° with the vertical? (b) 90°? (c) 180°?

Section 12-3   Angular Momentum
3. A 1000-kg airplane is flying in a straight line at 80 m/s, 1.3 km above the ground. What is the magnitude of its angular momentum with respect to an observer on the ground directly under the path of the plane?

4. Using data in the Appendices, find the angular momentum of the earth due to its daily rotation about its own axis. Assume that the earth is a uniform sphere.

5. (a) Use the data given in the Appendices to compute the total angular momentum of all the planets due to their revolution about the sun. (b) What fraction of this is associated with the planet Jupiter?

6. Show that the angular momentum about any point of a single particle moving with constant velocity remains constant throughout the motion.

7. A particle P with mass 2.0 kg has position $\mathbf{r}$ and velocity $\mathbf{v}$ as shown in Fig. 12–24. It is acted on by the force $\mathbf{F}$. All three vectors lie in a common plane. Presume that $r = 3.0$ m, $v = 4.0$ m/s, and $F = 2.0$ N. Compute (a) the angular momentum of the particle and (b) the torque acting on the particle. What are the directions of these two vectors?

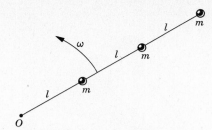

Figure 12-27 Exercises 11 and 12.

8. At a certain time the position vector in meters of a 0.25-kg object is $\mathbf{r} = 2.0\mathbf{i} - 2.0\mathbf{k}$. At that instant its velocity in meters per second is $\mathbf{v} = -5.0\mathbf{i} + 5.0\mathbf{k}$, and the force in newtons acting on it is $\mathbf{F} = 4.0\mathbf{j}$. (a) What is the angular momentum of the object about the origin? (b) What torque acts on it?

Section 12-4 Systems of Particles

9. Figure 12-25 shows the lines of action and the points of application of two forces about the origin O. Imagine these forces to be acting on a rigid body pivoted at O, all vectors being in the plane of the figure. (a) Find an expression for the magnitude of the resultant torque on the body. (b) If $r_1 = 1.30$ m, $r_2 = 2.15$ m, $F_1 = 4.20$ N, $F_2 = 4.90$ N, $\theta_1 = 75°$, and $\theta_2 = 60°$, what are the magnitude and direction of the resultant torque?

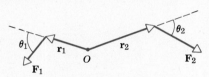

Figure 12-25 Exercise 9.

10. The body shown in Fig. 12-26 is pivoted at O. Three forces act on it in the directions shown on the figure: $F_A = 10$ N at point A, 8.0 m from O; $F_B = 16$ N at point B, 4.0 m from O; and $F_C = 19$ N at point C, 3.0 m from O. What are the magnitude and direction of the resultant torque about O?

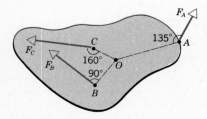

Figure 12-26 Exercise 10.

11. Three particles, each of mass m, are fastened to each other and to a rotation axis by three light strings each with length l as shown in Fig. 12-27. The combination rotates around the rotational axis with angular velocity ω in such a way that the particles remain in a straight line. (a) Calculate the rotational inertia of the combination about O. (b) What is the angular momentum of the middle particle? (c) What is the total angular momentum of the three particles? Express your answers in terms of m, l, and ω.

Section 12-5 Kinetic Energy of Rotation and Rotational Inertia

12. Presume that the strings in Exercise 11 are all replaced with uniform rods, each of mass M. (a) What is the total rotational inertia of the system about O? (b) What is the rotational kinetic energy of the system?

13. The masses and coordinates of four particles are as follows: 50 g, $x = 2.0$ cm, $y = 2.0$ cm; 25 g, $x = 0$, $y = 4.0$ cm; 25 g, $x = -3.0$ cm, $y = -3.0$ cm; 30 g, $x = -2.0$ cm, $y = 4.0$ cm. What is the rotational inertia of this collection with respect to the x-, y-, and z-axes?

14. Calculate the rotational inertia of a meter stick, weighing 5.5 N, about an axis perpendicular to the stick and located at the 20-cm mark.

15. Compare the kinetic energies of two uniform cylinders rotating about their axes of symmetry. They have the same mass, 1.25 kg, and rotate with the same angular velocity, 235 rad/s, but the first has a radius of 0.25 m and the second a radius of 0.75 m.

16. Assume the earth to be a sphere of uniform density. (a) What is its rotational kinetic energy? Take the radius of the earth to be 6.4×10^3 km and the mass of the earth to be 6.0×10^{24} kg. (b) Suppose that this energy could be harnessed for our use. For how long could the earth supply 1.0 kW of power to each of the 4.2×10^9 persons on earth?

17. Calculate the rotational inertia of a wheel that has a kinetic energy of 24,400 J when it is rotating at 600 rev/min.

18. Delivery trucks that operate by making use of energy stored in a rotating flywheel have been used in Europe. The trucks are "charged" by using an electric motor to get the flywheel up to its top speed of 200π rad/s. One such flywheel is a solid, homogenous cylinder with a mass of 500 kg and a radius of 1.0 m. (a) What is the kinetic energy of the flywheel after "charging"? (b) If the truck operates with an average power requirement of 8.0 kW, for how many minutes can it operate between "chargings"?

Section 12-6 Rotational Dynamics of a Rigid Body

19. A uniform rod rotates in a horizontal plane about a vertical axis through one end. The rod is 6.0 m long, weighs 10 N, and rotates at 240 rev/min clockwise when seen from above. Find the angular momentum of the rod.

20. When a torque of 32 N·m is applied to a certain wheel, it acquires an angular acceleration of 25 rad/s². What is the rotational inertia of the wheel?

21. Two uniform solid spheres have the same mass, 1.65 kg, but one has a radius of 0.226 m while the other has a radius of 0.854 m. (a) For each of the spheres, find the torque required to bring the sphere from rest to an angular velocity of 317 rad/s in 15.5 s. Each rotates about an axis through the center of the sphere. (b) For each sphere, what force applied tangentially at the equator would provide the needed torque?

22. A cylinder having a mass of 2.0 kg rotates about an axis through its center. Forces are applied as in Fig. 12–28. $F_1 = 6.0$ N, $F_2 = 4.0$ N, and $F_3 = 2.0$ N. Also, $R_1 = 5.0$ cm and $R_2 = 12$ cm. Find the magnitude and direction of the angular acceleration of the cylinder.

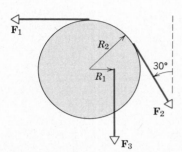

Figure 12–28 Exercise 22.

23. An automobile engine develops 100 hp (=74.6 kW) when rotating at a speed of 1800 rev/min. What torque does it deliver?

Section 12–7 Rolling Bodies

24. A hoop of radius 3.0 m has a mass of 150 kg. It rolls along a horizontal floor so that its center of mass has a speed of 0.15 m/s. How much work has to be done on the hoop to stop it?

25. A sphere rolls up an inclined plane of inclination angle 30°. At the bottom of the incline the center of mass of the sphere has a translational speed of 16 ft/s. How far does the sphere travel up the plane?

Section 12–8 Conservation of Angular Momentum

26. A man stands on a frictionless rotating platform which is rotating with an angular speed of 1.0 rev/s; his arms are outstretched and he holds a weight in each hand. With his hands in this position the total rotational inertia of the man, the weights, and the platform is 6.0 kg·m². If by moving the weights the man decreases the rotational inertia to 2.0 kg·m², (a) what is the resulting angular speed of the platform? (b) By how much is the kinetic energy increased?

27. A girl (mass M) stands on the edge of a frictionless merry-go-round (mass $10\,M$, radius R, rotational inertia I) that is not moving. She throws a rock (mass m) in a horizontal direction that is tangent to the outer edge of the merry-go-round. The speed of the rock, relative to the ground, is v. What are (a) the angular speed of the merry-go-round and (b) the linear speed of the girl after the rock is thrown?

PROBLEMS GROUPED BY SECTION

Section 12–2 Torque

1. (a) Given that $\mathbf{r} = \mathbf{i}x + \mathbf{j}y + \mathbf{k}z$ and $\mathbf{F} = \mathbf{i}F_x + \mathbf{j}F_y + \mathbf{k}F_z$, find the torque $\boldsymbol{\tau} = \mathbf{r} \times \mathbf{F}$. (b) Show that if $\mathbf{r}$ and $\mathbf{F}$ lie in a given plane, then τ has no component in that plane. (Hint: See Problem 18 in Chapter 2.)

2. A bicyclist of mass 70 kg puts all his weight on each pedal as he climbs up a steep road. Take the diameter of the circle in which the pedals rotate to be 0.40 m and determine the maximum torque he exerts in the process.

Section 12–3 Angular Momentum

3. If we are given r, p, and θ, we can calculate the angular momentum of a particle from Eq. 12–4a. Sometimes, however, we are given the components (x, y, z) of $\mathbf{r}$ and (p_x, p_y, p_z) of $\mathbf{p}$ instead. (a) Show that the components of l along the x-, y-, and z-axes are then given by

$$\ell_x = yp_z - zp_y,$$
$$\ell_y = zp_x - xp_z,$$
$$\ell_z = xp_y - yp_x.$$

(b) Show that if the particle moves only in the x-y plane, the resultant angular momentum vector has only a z-component. (Hint: See Problem 18 in Chapter 2.)

4. A 2.0-kg object moves in a plane with velocity components $v_x = 30$ m/s and $v_y = 60$ m/s as it passes through the point $(x, y) = (3.0, -4.0)$ m. (a) What is its angular momentum relative to the origin at this moment? (b) What is its angular momentum relative to the point $(-2.0, -2.0)$ m at this same moment? (Hint: See Problem 3.)

5. A 3.0-kg particle is at $x = 3.0$ m, $y = 8.0$ m with a velocity of $\mathbf{v} = 5\mathbf{i} - 6\mathbf{j}$ m/s. It is acted on by a 7.0-N force in the negative x-direction. (a) What is the angular momentum of the particle? (b) What torque acts on the particle? (c) At what rate is the angular momentum of the particle changing with time?

Section 12–4 Systems of Particles

6. Two particles, each of mass m and speed v, travel in opposite directions along parallel lines separated by a distance d. Show that the vector angular momentum of this system of particles is the same no matter what point is taken as the origin.

7. Starting from Newton's third law, prove that the resultant internal torque on a system of particles is zero. Assume that the force exerted by any particle on any other particle is directed along the line joining the particles.

Section 12–5 Kinetic Energy of Rotation and Rotational Inertia

8. (a) Show that a solid cylinder of mass M and radius R is equivalent to a thin hoop of mass M and radius $R/\sqrt{2}$, for rotation about a central axis. (b) The radial distance from a given axis at which the mass of a body could be concentrated without altering the rotational inertia of the body about that axis is called the *radius of gyration*. Let k represent radius of gyration and show that

$$k = \sqrt{\frac{I}{M}}.$$

This gives the radius of the "equivalent hoop" in the general case.

9. A thin rod of length l and mass m is suspended freely from one end. It is pulled aside and swung about a horizontal axis, passing through its lowest position with an angular speed ω. How high does its center of mass rise above its lowest position? Neglect friction and air resistance.

10. (*a*) Prove that the rotational inertia of a thin rod of length *l* about an axis through its center perpendicular to its length is $I = \frac{1}{12}Ml^2$. (See Table 12–1.) (*b*) Use the parallel axis theorem to show that $I = \frac{1}{3}Ml^2$ when the axis of rotation is through one end perpendicular to the length of the rod.

11. A rigid body is made of three identical thin rods fastened together in the form of a letter *H* (Fig. 12–29). The body is free to rotate about a horizontal axis that passes through one of the legs of the *H*. The body is allowed to fall from rest from a position in which the plane of the *H* is horizontal. What is the angular speed of the body when the plane of the *H* is vertical?

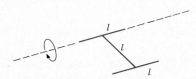

Figure 12–29 Problem 11.

12. A uniform spherical shell rotates about a vertical axis on frictionless bearings (Fig. 12–30). A light cord passes around the equator of the shell, over a pulley, and is attached to a small object that is otherwise free to fall under the influence of gravity. What is the speed of the object after it has fallen a distance *h* from rest? Use energy methods.

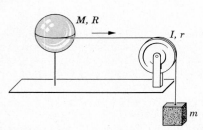

Figure 12–30 Problem 12.

Section 12–6   Rotational Dynamics of a Rigid Body
13. Calculate (*a*) the torque, (*b*) the energy, and (*c*) the average power required to accelerate the earth from rest to its present angular speed about its axis in one day.

14. The angular momentum of a flywheel having a rotational inertia of 0.125 kg·m² decreases from 3.0 to 2.0 kg·m²/s in a period of 1.5 s. (*a*) What is the average torque acting on the flywheel during this period? (*b*) Assuming a uniform angular acceleration, through how many revolutions will the flywheel have turned? (*c*) How much work was done on the wheel? (*d*) What was the average power supplied by the flywheel?

15. A pulley having a rotational inertia of 1.0×10^{-3} kg·m² and a radius of 10 cm is acted upon by a force, applied tangentially at its rim, that varies in time as $F = 0.50\,t + 0.30\,t^2$, where *F* is in newtons if *t* is given in seconds. If the pulley was initially at rest, find its angular velocity after 3.0 s.

16. A uniform steel rod of length 1.20 m and mass 6.40 kg has attached to each end a small ball of mass 1.06 kg. The rod is constrained to rotate in a horizontal plane about a vertical axis through its midpoint. At a certain instant it is observed to be rotating with an angular speed of 39.0 rev/s. Because of axle fric-

tion it comes to rest 32.0 s later. Compute, assuming a constant frictional torque, (*a*) the angular acceleration, (*b*) the retarding torque exerted by axle friction, (*c*) the total work done by the axle friction, and (*d*) the number of revolutions executed during the 32.0 s. (*e*) Now suppose that the frictional torque is known not to be constant. Which, if any, of the quantities (*a*), (*b*), (*c*) or (*d*) can still be computed without requiring any additional information? If such exists, give its value.

17. If $R = 12$ cm, $M = 400$ g, and $m = 50$ g in Fig. 12–10, find the speed of *m* after it has descended 50 cm starting from rest. Solve the problem using the results of Example 5 and by application of energy conservation principles.

Section 12–7   Rolling Bodies
18. A wheel of radius 0.25 m, moving initially at 43 m/s, rolls to a stop in 225 m. If its rotational inertia is 0.155 kg·m², calculate the torque exerted by rolling friction on the wheel.

19. A solid cylinder of radius 10 cm and mass 12 kg starts from rest and rolls without slipping a distance of 6.0 m down a house roof that is inclined at 30°. (*a*) What is the angular speed of the cylinder about its center as it leaves the house roof? (*b*) The outside wall of the house is 5.0 m high. How far from the wall does the cylinder hit the level ground? See Fig. 12–31.

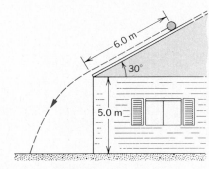

Figure 12–31 Problem 19.

20. A 1000-kg car has four 10-kg wheels. What fraction of the total kinetic energy of the car is due to rotation of the wheels about their axles? Assume that the wheels have the same rotational inertia as disks of the same mass and size. Explain why you do not need to know the radius of the wheels.

21. An automobile has a total mass of 1700 kg. It accelerates from rest to 40 km/h in 10 s. Each wheel has a mass of 32 kg and a radius of gyration (see Problem 8) of 30 cm. Find, for the end of the 10-s interval, (*a*) the rotational kinetic energy of each wheel about its axle, (*b*) the total kinetic energy of each wheel, and (*c*) the total kinetic energy of the automobile.

22. A homogeneous sphere starts from rest at the upper end of the track shown in Fig. 12–32 and rolls without slipping until it rolls off the right-hand end. If $H = 60$ m and $h = 20$ m and the track is horizontal at the right-hand end, determine the distance to the right of point *A* at which the ball strikes the horizontal base line.

23. A small sphere rolls without slipping on the inside of a large fixed hemisphere whose axis of symmetry is vertical. It starts at the top from rest. (*a*) What is its kinetic energy at the bottom? What fraction is rotational? What translational? (*b*) What normal force does the small sphere exert on the hemisphere at the bottom? Take

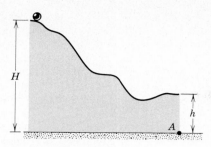

Figure 12-32 Problem 22.

the radius of the small sphere to be r, that of the hemisphere to be R, and let m be the mass of the sphere.

Section 12-8   Conservation of Angular Momentum

24. Two wheels, A and B, are connected by a belt as in Fig. 12-33. The radius of B is three times the radius of A. What would be the ratio of the rotational inertias I_A/I_B if (a) both wheels have the same angular momenta? (b) both wheels have the same rotational kinetic energy?

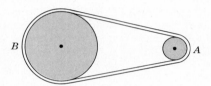

Figure 12-33 Problem 24.

25. Two skaters, each of mass 50 kg, approach each other along parallel paths separated by 3.0 m. They have equal and opposite velocities of 10 m/s. The first skater carries a long light pole, 3.0 m long, and the second skater grabs the end of it as he passes. (Assume frictionless ice.) (a) Describe quantitatively the motion of the skaters after they are connected by the pole. (b) By pulling on the pole, the skaters reduce their separation to 1.0 m. What is their motion then? (c) Compare the kinetic energy of the system in parts (a) and (b). Where does the change come from?

26. With center and spokes of negligible mass, a certain bicycle wheel has a thin rim of radius 1.14 ft and weight 8.36 lb; it can turn on its axle with negligible friction. A man holds the wheel above his head with the axis vertical while he stands on a turntable free to rotate without friction; the wheel rotates clockwise, as seen from above, with an angular speed 57.7 rad/s, and the turntable is initially at rest. The rotational inertia of wheel-plus-man-plus-turntable about the common axis of rotation is 1.54 slug·ft². The man's hand suddenly stops the rotation of the wheel (relative to the turntable). Determine the resulting angular velocity (magnitude and direction) of the system.

*27. Two cylinders having radii R_1 and R_2 and rotational inertias I_1 and I_2, respectively, are supported by axes perpendicular to the plane of Fig. 12-34. The large cylinder is initially rotating with angular velocity ω_0. The small cylinder is moved to the right until it touches the large cylinder and is caused to rotate by the frictional force between the two. Eventually, slipping ceases, and the two cylinders rotate at constant rates in opposite directions. (a) Find the final angular velocity ω_2 of the small cylinder in terms of I_1, I_2, R_1, R_2, and ω_0. (b) Is total angular momentum conserved in this case?

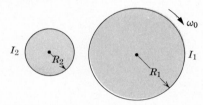

Figure 12-34 Problem 27.

ADDITIONAL PROBLEMS

28. (a) In Example 1, express **F** and **r** in terms of unit vectors and compute τ. Do the same in Example 3. (b) In Example 1, express **p** and **r** in unit vectors and compute l. (*Hint:* See Problems 1 and 3 above.)

29. The oxygen molecule, O_2, has a total mass of 5.30×10^{-26} kg and a rotational inertia of 1.94×10^{-46} kg·m² about an axis through the center perpendicular to the line joining the atoms. Suppose that such a molecule in a gas has a speed of 500 m/s and that its rotational kinetic energy is two-thirds of its translational kinetic energy. Find its angular velocity.

30. A projectile of mass m is fired from the ground with an initial speed v_0 and initial angle θ_0 above the horizontal. (a) Find an expression for its angular momentum about the firing point as a function of time. (b) Differentiate (a) to find the rate at which the angular momentum changes with time. (c) Evaluate **r** $\times$ **F** directly, and compare the result with (b). Why should the results be identical?

31. Suppose that the sun runs out of nuclear fuel and suddenly collapses to form a so-called white dwarf star, with a diameter about equal to that of the earth. Assuming no mass loss, what would then be the new rotation period of the sun, which currently is about 25 days?

32. In a playground there is a small merry-go-round of radius 1.2 m and mass 180 kg. The radius of gyration, see Problem 8, is 91 cm. A child of mass 44 kg runs at a speed of 3.0 m/s tangent to the rim of the merry-go-round when it is at rest and then jumps on. Neglect friction between the bearings and the shaft of the merry-go-round and find the angular velocity of the merry-go-round and child.

33. Show that the axis about which a given rigid body has its smallest rotational inertia must pass through its center of mass. (*Hint:* Use the parallel axis theorem.)

34. Figure 12-35 shows a top view of a uniform block of mass M. (a) Show that its rotational inertia about an axis perpendicular to the figure and passing through its center of mass is $\frac{1}{12}M(a^2 + b^2)$. (b) Show that the rotational inertia of the block about a parallel axis through point P is $\frac{1}{3}M(a^2 + b^2)$.

35. A body of radius R and mass m is rolling horizontally without slipping with speed v. It then rolls up a hill to a maximum height h. If $h = 3v^2/4g$, (a) what is the body's rotational inertia? (b) What might the body be?

36. The length of the day is increasing at the rate of about 1 ms/century. This is primarily due to frictional forces generated

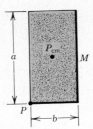

Figure 12-35 Problem 34.

by movement of water in the world's oceans and seas as a response to the tidal forces exerted by the sun and moon. (a) At what rate is the earth losing rotational kinetic energy? (b) What tangential force, exerted at (average) latitudes 60°N and 60°S, is applied by the oceans on the near-coastal seabed?

37. The rotor of an electric motor has a rotational inertia $I_m = 2 \times 10^{-3}$ kg·m² about its central axis. The motor is mounted parallel to the axis of a space probe having a rotational inertia $I_p = 12$ kg·m² about its axis. Calculate the number of revolutions of the rotor required to turn the probe through 30° about its axis.

38. A wheel is rotating with an angular speed of 800 rev/min on a shaft whose rotational inertia is negligible. A second wheel, initially at rest and with twice the rotational inertia of the first, is suddenly coupled to the same shaft. (a) What is the angular speed of the resultant combination of the shaft and two wheels? (b) Account for any changes in rotational kinetic energy experienced by this system.

39. Figure 12-36 shows two blocks of mass m suspended from the ends of a rigid weightless rod of length $l_1 + l_2$. The rod is held in the horizontal position shown in the figure and then released. (a) Derive algebraic expressions for the accelerations of the two blocks when they start to move. (b) Examine your expressions carefully and verify that they make sense in the following special cases: $g = 0$, $l_1 = 0$, $l_2 = 0$, $l_1 = l_2$. (c) Find the accelerations if $l_1 = 20$ cm and $l_2 = 80$ cm.

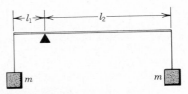

Figure 12-36 Problem 39.

40. In a lecture demonstration, a toy train track is mounted on a large wheel that is free to turn with negligible friction about a vertical axis. A toy electric train of mass m is placed on the track and, with the system initially at rest, the electrical power is turned on. The train reaches a steady speed v with respect to the track. What is the angular velocity ω of the wheel, if its mass is M and its radius R? (Neglect the mass of the spokes of the wheel.)

41. A cockroach, mass m, runs counterclockwise around the rim of a lazy Susan (a circular dish mounted on a vertical axle) of radius R and rotational inertia I with frictionless bearings. The cockroach's speed (relative to the earth) is v, whereas the lazy Susan turns clockwise with angular speed ω_0. The cockroach finds a bread crumb on the rim and, of course, stops, (a) What is the angular speed of the lazy Susan after the cockroach stops? (b) Is energy conserved?

42. A meter stick is held vertically with one end on the floor and is then allowed to fall. Find the speed of the other end when it hits the floor, assuming that the end on the floor does not slip.

43. Two 2.0-kg balls are attached to the end of a thin rod of negligible mass 50 cm long. The rod is free to rotate without friction about a horizontal axis through its center. A 50-g putty wad drops onto one of the balls with a speed of 3.0 m/s, and sticks to it. (See Fig. 12-37.) (a) What is the angular speed of the system just after the putty wad hits? (b) What is the ratio of the kinetic energy of the entire system after the collision to that of the putty wad just before? (c) Will the system subsequently rotate through a complete revolution? If not how far will it rotate?

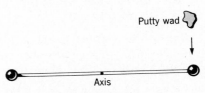

Figure 12-37 Problem 43.

44. Two girls, each of mass M, sit on opposite ends of a thin rod with length L and mass M (the same as the mass of each girl). The rod is pivoted at its center and is free to rotate in a horizontal circle without friction. (a) What is the rotational inertia of the rod plus the girls about a vertical axis through the center of the rod? (b) What is the angular momentum of the system if it is rotating with an angular speed ω_0 in a clockwise direction as seen from above? What is the direction of the angular momentum? (c) While the system is rotating, the girls pull themselves toward the center of the rod until they are half as far from the center as before. What is the resulting angular speed in terms of ω_0? (d) What is the change in kinetic energy of the system due to the girls changing their positions? How is energy conserved during this motion?

45. In an Atwood's machine (Fig. 5-9) one block has a mass of 500 g and the other a mass of 460 g. The pulley, which is mounted in horizontal frictionless bearings, has a radius of 5.0 cm. When released from rest the heavier block is observed to fall 75 cm in 5.0 s. What is the rotational inertia of the pulley?

46. Particle m in Fig. 12-38 slides down the frictionless surface and collides with the uniform vertical rod, sticking to it. The rod pivots about O and rotates through the angle θ before coming to rest. Find θ in terms of the other parameters given in the figure.

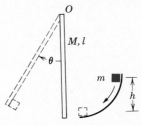

Figure 12-38 Problem 46.

47. Two identical blocks, each of mass M, are connected by a light string over a frictionless pulley of radius R and rotational inertia I (Fig. 12-39). The string does not slip on the pulley, and it is not known whether or not there is friction between the plane and the sliding block. When this system is released, it is found that the pulley turns through an angle θ in time t and the acceleration of the blocks is constant. (*a*) What is the angular acceleration of the pulley? (*b*) What is the acceleration of the two blocks? (*c*) What are the tensions in the upper and lower sections of the string. All answers are to be expressed in terms of M, I, R, θ, g, and t.

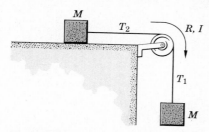

Figure 12–39 Problem 47.

48. A tall chimney cracks near its base and falls over. Express (*a*) the radial and (*b*) the tangential linear accelerations of the top of the chimney as a function of the angle θ made by the chimney with the vertical. (*c*) Can the resultant linear acceleration exceed g? (*d*) The chimney breaks during the fall. Explain how this can happen. (See "More on the Falling Chimney," by Albert A. Bartlett, *The Physics Teacher*, September 1976.)

49. A small solid marble of mass m and radius r rolls without slipping along the loop-the-loop track shown in Fig. 12–40, having been released from rest somewhere on the straight section of track. (*a*) From what minimum height above the bottom of the track must the marble be released in order that it just stay on the track at the top of the loop? (The radius of the loop-the-loop is R; assume $R \gg r$.) (*b*) If the marble is released from height $6R$ above the bottom of the track, what is the horizontal component of the force acting on it at point Q?

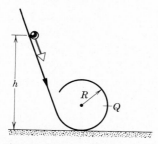

Figure 12–40 Problem 49.

tires is constructed as shown in Fig. 12–41. The tire is initially motionless and is held in a light framework which is freely pivoted

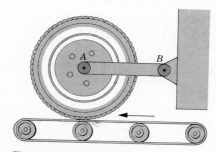

Figure 12–41 Problem 50.

at points A and B. The rotational inertia of the wheel about its axis A is 0.75 kg·m², its mass is 15 kg, and its radius is 0.30 m. The tire is placed on the surface of a conveyor belt that is moving with a surface velocity of 12 m/s, such that AB is horizontal. (*a*) If the coefficient of kinetic friction between the tire and the conveyor belt is 0.60, what time will be required for the wheel to achieve its final angular velocity? (*b*) What will be the length of the skid mark on the conveyor surface?

*51. A bullet of mass m is fired at a large block of wood of mass M that rests on a rough floor; see Fig. 12–42. What is the smallest speed of the bullet that will cause the block to tip over? Assume that $m = 35$ g, $M = 25$ kg, $a = 1.2$ m, $b = 0.40$ m, and $h = 1.0$ m. (*Hint:* This problem is an angular equivalent to the ballistic pendulum problem; see Example 3, Chapter 10; see the result of Problem 34 for the needed rotational inertia.)

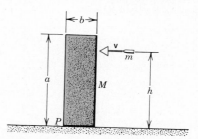

Figure 12–42 Problem 51.

*52. A car is fitted with an energy-conserving flywheel, which in operation is geared to the driveshaft so that it rotates at 240 rev/s when the car is traveling at 80 km/h. The total mass of the car is 800 kg, the flywheel weighs 200 N, and it is a uniform disk 1.1 m in diameter. The car descends a 1500-m long, 5° slope, from rest, with the flywheel engaged and no power supplied from the motor. Neglecting friction and the rotational inertia of the wheels, find (*a*) the speed of the car at the bottom of the slope; (*b*) the angular acceleration of the flywheel at the bottom of the slope; (*c*) the power being absorbed by the rotation of the flywheel at the bottom of the slope.

*53. In a circus act a clown rolls a hollow wooden cylinder down a sloping plank that makes an angle of 30° with the ground. As Fig. 12–43 shows, a small dog trots on top of the cylinder in such a way that it remains in place as the cylinder rolls downhill. The dog weighs three-tenths as much as the cylinder. How fast is the dog running when the cylinder has rolled 4.0 m down the plank? (*Hint:* Solve by energy considerations.)

Figure 12–43 Problem 53.

13 Equilibrium of Rigid Bodies

Figure 13–1 A balancing rock in the Chiricahua National Monument in Arizona. It is a spectacular example of an object in static equilibrium. (Alan Pitcairn/Grant Heilman.)

Static and dynamic equilibrium

13–1 A Rigid Body in Static Equilibrium

The feature of the balancing rock of Fig. 13–1 that attracts our attention is that the rock does not fall over but remains totally at rest as we stand in our inertial reference frame and watch it. It shares this property with more ordinary rocks and with bridges, power line towers, churches, tables, and all other presumably rigid bodies that are at rest in a chosen inertial reference frame. We say that such bodies are in *static equilibrium* and we note that (1) the linear acceleration a_{cm} of the center of mass of such bodies is zero, and (2) the angular acceleration α of such bodies about any point is also zero.

Actually a body may be moving in an inertial reference frame and still meet the conditions that $a_{cm} = 0$ and $\alpha = 0$. A spaceship drifting with constant velocity v_{cm} through gravity-free space and tumbling with constant angular velocity ω about an axis through its center of mass is an example. (A second spaceship can serve as a suitable inertial reference frame from which to observe the first.) Objects like the drifting and tumbling spaceship are said to be in *dynamic equilibrium*. We shall see that the restrictions imposed on the forces and torques that act on the rigid body are the same no matter whether the mechanical equilibrium is static or dynamic. Furthermore, we can transform any case of dynamic equilibrium to one of static equilibrium by choosing an appropriate new reference frame. In the rest of this chapter therefore we will consider only static equilibrium, in which the bodies are actually at rest.

The analysis of static equilibrium is very important in engineering practice. It requires the engineer to isolate and identify all the external forces and torques that act on the structure in question. By suitable design and choice of materials the engineer can then ensure that the structure will tolerate the external loads that bear on it. Such analyses are necessary to make certain that bridges do not collapse under their traffic and wind loads, that the landing gear of aircraft will survive rough landings, and so on.

The translational motion of a rigid body of mass M is governed by Eq. 9–8, or

$$\mathbf{F}_{ext} = M\mathbf{a}_{cm}, \qquad [9-8]$$

in which $\mathbf{F}_{ext}$ is the vector sum of all the external forces acting on the body. Because $\mathbf{a}_{cm}$ must be zero for equilibrium, the first condition of equilibrium (static or otherwise) is the following: *The vector sum of all the external forces acting on a body in equilibrium must be zero.*

We can write condition 1 as

The requirement for translational equilibrium

$$\mathbf{F} = \mathbf{F}_1 + \mathbf{F}_2 + \cdots = 0, \qquad (13-1)$$

in which we have dropped the subscript on $\mathbf{F}_{ext}$ for convenience. This vector equation leads to three scalar equations,

$$F_x = F_{1x} + F_{2x} + \cdots = 0,$$
$$F_y = F_{1y} + F_{2y} + \cdots = 0, \qquad (13-2)$$
$$F_z = F_{1z} + F_{2z} + \cdots = 0,$$

which state that the sum of the components of the forces along each of any three mutually perpendicular directions is zero.

The second requirement for equilibrium is that $\alpha = 0$ about any point. Since the angular acceleration of a rigid body is associated with torque—recall that $\tau = I\alpha$ for a fixed axis—we can state this second condition of equilibrium (static or otherwise) as follows: *The vector sum of all the external torques acting on a body in equilibrium must be zero.*

We can write condition 2 as

The requirement for rotational equilibrium

$$\boldsymbol{\tau} = \boldsymbol{\tau}_1 + \boldsymbol{\tau}_2 + \cdots = 0. \qquad (13-3)$$

This vector equation leads to three scalar equations,

$$\tau_x = \tau_{1x} + \tau_{2x} + \cdots = 0,$$
$$\tau_y = \tau_{1y} + \tau_{2y} + \cdots = 0, \qquad (13-4)$$
$$\tau_z = \tau_{1z} + \tau_{2z} + \cdots = 0,$$

which state that, at equilibrium, the sum of the components of the torques acting on a body, along each of any three mutually perpendicular directions, is zero.

The resultant torque $\boldsymbol{\tau}$ in Eq. 13–3, which must be zero for mechanical equilibrium, is defined with respect to a particular origin O. The quantities τ_x, τ_y, and τ_z in Eq. 13–4 are the scalar components of $\boldsymbol{\tau}$ and refer to any set of three mutually perpendicular axes whose origin is at O, no matter how these axes are oriented in space. This follows because, if a vector is zero, its scalar components must be zero no matter how we orient the axes of the reference frame. You may wonder whether the choice of a specific origin is essential. The answer—as we shall show below—is that it is not, because (for a body in translational equilibrium) if $\boldsymbol{\tau} = 0$ for any single origin O it is also zero for any other origin in the reference frame. The substance of this paragraph then is that condition 2 is satisfied for a body in translational equilibrium if we can show either that (*a*) $\boldsymbol{\tau} = 0$ with respect to any one point (Eq. 13–3) or that (*b*) the torque components along any three mutually perpendicular axes are zero (Eq. 13–4).

Hence we have six independent conditions on our forces for a body to be in equilibrium. These conditions are the six algebraic relations of Eqs. 13–2 and

13-4. These six conditions are a condition on each of the six degrees of freedom of a rigid body, three translational and three rotational.

Often we deal with problems in which all the forces lie in a plane. If we choose a reference frame so that this plane is the x-y plane, three of the equilibrium conditions are met automatically, leaving only three to be dealt with. There would be no z-components for any of the forces, so the third equation for translational equilibrium would be satisfied. Similarly, forces in the x-y plane can produce only torques that are in the direction of the z-axis. Thus the first and second of the rotational equilibrium conditions are satisfied. We then have only three conditions on our forces: The sum of their components must be zero for each of any two perpendicular directions in the plane, and the sum of their torques about any one axis perpendicular to the plane must be zero. These conditions correspond to the three degrees of freedom for motion in a plane, two of translation and one of rotation. We shall limit ourselves henceforth mostly to planar problems.

13-2 Center of Gravity

One of the forces encountered in rigid-body motions is the force of gravity. Actually, for an extended body, this is not just one force but the resultant of a great many forces. Each particle in the body is acted on by a gravitational force. If the body of mass M is imagined to be divided into a large number of particles, say, n, the gravitational force exerted by the earth on the ith particle of mass m_i is $m_i\mathbf{g}$. This force is directed down toward the earth. If the acceleration due to gravity $\mathbf{g}$ is the same everywhere in a region, we say that a uniform gravitational field exists there; that is, $\mathbf{g}$ has the same magnitude and direction everywhere in that region. For a rigid body in a uniform gravitational field, $\mathbf{g}$ must be the same for each particle in the body and the weight forces on the particles must be parallel to one another. If we assume that the earth's gravitational field is uniform, we can show that all the individual weight forces acting on a body can be replaced by a single force $M\mathbf{g}$ acting down at the center of mass of the body. This is equivalent to showing that the individual weight forces, acting downward, can be counteracted in their accelerating effects by a single force $\mathbf{F} (= -M\mathbf{g})$ acting upward, provided that this force $\mathbf{F}$ is applied at the center of mass of the body.

Figure 13-2 shows two typical particles or mass elements m_1 and m_2, selected from the n such elements into which the rigid body has been divided. An upward force $\mathbf{F} (= -M\mathbf{g})$ is applied at a certain point O. It remains to show that the body is in mechanical equilibrium if (and only if) point O is the center of mass. Condition 1 for equilibrium (Eq. 13-1) has already been satisfied by our choice of the magnitude and direction of $\mathbf{F}$. That is,

$$\mathbf{F} + m_1\mathbf{g} + m_2\mathbf{g} + \cdots + m_n\mathbf{g} = 0,$$

or

$$\mathbf{F} = -(m_1 + m_2 + \cdots + m_n)\mathbf{g} = -M\mathbf{g},$$

which corresponds to our assumption.

It remains to be proven that $\boldsymbol{\tau} = 0$ for any single point in the body, such as O. This is the second condition for equilibrium. By choosing O as our origin we ensure that the torque of $\mathbf{F}$ about this point is zero, because the moment arm of $\mathbf{F}$ is zero for this point. The torque about O due to the gravitational pull on the mass elements is

$$\boldsymbol{\tau} = \mathbf{r}_1 \times m_1\mathbf{g} + \mathbf{r}_2 \times m_2\mathbf{g} + \cdots + \mathbf{r}_n \times m_n\mathbf{g},$$

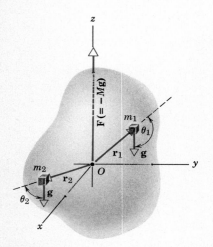

Figure 13-2 An irregular object is divided into n mass elements of which two typical elements m_1 and m_2 are shown. In the text we prove that if $\mathbf{g}$ is the same for all elements of the object, it can be held in translational and rotational equilibrium by a single force $\mathbf{F} (= -M\mathbf{g})$ directed upward and applied at its center of mass.

which (because m_1, m_2, etc., are scalars) we can write as

$$\boldsymbol{\tau} = m_1\mathbf{r}_1 \times \mathbf{g} + m_2\mathbf{r}_2 \times \mathbf{g} + \cdots + m_n\mathbf{r}_n \times \mathbf{g}.$$

Since $\mathbf{g}$ is the same in each term, we can factor it out, obtaining

$$\boldsymbol{\tau} = (m_1\mathbf{r}_1 + m_2\mathbf{r}_2 + \cdots + m_n\mathbf{r}_n) \times \mathbf{g}$$

$$= \left(\sum_{1}^{n} m_i\mathbf{r}_i \right) \times \mathbf{g},$$

in which the sum is taken over all the mass elements that make up the body.

Now if point O is the center of mass of the body, the sum above is zero. This follows from the definition of the center of mass (see Eq. 9–3b and the discussion following it). We conclude then that if point O is the center of mass, then $\boldsymbol{\tau} = 0$ and the second condition for mechanical equilibrium is satisfied. Actually, an object can be suspended from *any* point O, as in Fig. 13–3, and be in rotational equilibrium, with $\boldsymbol{\tau} = 0$. However, only if O is the center of mass do we have $\boldsymbol{\tau} = 0$ *for all orientations of the body* about its point of suspension.

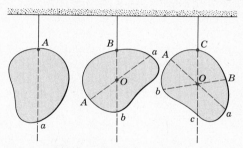

Figure 13–3   Since the center of mass O always hangs directly below the point of suspension, hanging a plate from two different points determines O.

Thus the gravitational forces acting on the individual mass elements that make up a rigid body are equivalent in their translational and rotational effects to a single force equal to $M\mathbf{g}$, the total weight of the body, acting at the center of mass. We can obtain the same result if the body is continuous and divided into an infinite number of particles. You should be able to do this by the methods of integral calculus (see Section 9–1). The point of application of the equivalent resultant gravitational force is often called the *center of gravity*.

Because almost all problems in mechanics involve objects having dimensions small compared to the distances over which $\mathbf{g}$ changes appreciably, we can usually assume that $\mathbf{g}$ is uniform over the body. The center of mass and the center of gravity can then be taken as the same point. In fact, we can use this fact to determine experimentally the center of mass in irregularly shaped objects. For example, let us locate the center of mass of a thin plate of irregular shape, as in Fig. 13–3. We suspend the body by a cord from some point A on its edge. When the body is at rest, the center of gravity must lie directly under the point of support somewhere on the line Aa, for only then can the torque resulting from the cord and the weight add to zero. We next suspend the body from another point B on its edge. Again, the center of gravity must lie somewhere on Bb. The only point common to the lines Aa and Bb is O, the point of intersection, so that this point must be the center of gravity. If now we suspend the body from any other point on its edge, as C, the vertical line Cc will pass through O. Since we have assumed a uniform field, the center of gravity coincides with the center of mass, which is therefore located at O.

Center of gravity

13-3   Examples of Equilibrium

In applying the conditions for equilibrium (zero resultant force and zero resultant torque about any axis), we can clarify and simplify the procedure by proceeding as follows.

First, we draw an imaginary boundary around the system under consideration. This assures that we see clearly just what body or system of bodies it is to which we are applying the laws of equilibrium. This process is called *isolating* the system.

Second, we draw vectors representing the magnitude, direction, and point of application of all external forces. An external force is one that acts from outside the boundary that was drawn earlier. Examples of external forces often encountered are gravitational forces and forces transmitted by strings, wires, rods, and beams which cross the boundary.

The diagram showing the isolated body together with the external forces that act upon it is called a *free-body diagram;* see Section 5–11. A question sometimes arises about the direction of a force. If you are in doubt, choose the direction arbitrarily. A negative value for a force in the solution means that the force acts in the direction opposite to that assumed. Note that only external forces acting on the system need be considered; all internal forces cancel one another in pairs.

Third, we choose a convenient reference frame along whose axes we resolve the external forces before applying the first condition of equilibrium (Eq. 13–2). The object here is to simplify the calculations. The preferable reference frame is usually obvious.

Fourth, we choose a convenient reference frame along whose axes we resolve the external torques before applying the second condition of equilibrium (Eq. 13–4). The object again is to simplify calculations, and we may use different reference frames in applying the two conditions for static equilibrium if this proves to be convenient. Suppose that an axis passes through the point at which the lines of action of two forces intersect and is at right angles to the plane formed by these forces; these forces will automatically have no torque component along (or about) this axis. The torque components resulting from all external forces must be zero about any axis for equilibrium. Internal torques will cancel in pairs and need not be considered.

Example 1   A uniform steel meter bar, weighing 4.0 lb, rests on two scales at its ends. A 6.0-lb block is placed at the 25-cm mark. Find the reading on the scales.

Our system is the bar and block. The forces acting on the bar are $\mathbf{W}$ and $\mathbf{w}$, the gravitational forces acting down at the centers of gravity of the bar and block, and $\mathbf{F_1}$ and $\mathbf{F_2}$, the forces exerted upward on the bar at its ends by the scales. These are shown in Fig. 13–4. By Newton's third law, the force exerted by a scale on the bar is equal and opposite to that exerted by the bar on the scale. Therefore, to obtain the readings on the scales, we must determine the magnitudes of $\mathbf{F_1}$ and $\mathbf{F_2}$.

For translational equilibrium (Eq. 13–1) the condition is

$$\mathbf{F_1} + \mathbf{F_2} + \mathbf{W} + \mathbf{w} = 0.$$

All forces act vertically, so that if we choose the y-axis to be vertical, no other axes need be considered. Then we get the scalar equation

$$F_1 + F_2 - W - w = 0.$$

For rotational equilibrium, the component of the resultant torque on the bar must be zero about any axis. We have seen

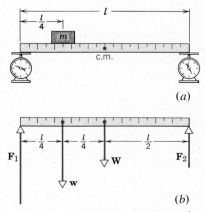

(a)

(b)

Figure 13–4   Example 1. (*a*) A weight is placed a quarter of the way from one end of a uniform steel bar resting on two spring scales. (*b*) The force diagram.

that it is enough to show that the torque components are zero for any set of three mutually perpendicular axes. These components are certainly zero for any two perpendicular axes that lie in the plane of Fig. 13–4 (Why?). It remains to require that the resultant torque is zero about any one axis at right angles to the plane of the figure. Let us choose an axis through the center of the bar. Then, taking clockwise rotation as positive and counterclockwise rotation as negative, the condition for rotational equilibrium (Eq. 13–4) is

$$F_1\left(\frac{l}{2}\right) - F_2\left(\frac{l}{2}\right) + W(0) - w\left(\frac{l}{4}\right) = 0,$$

or

$$F_1 - F_2 - \frac{w}{2} = 0.$$

Adding this equation to the one obtained for translational equilibrium and solving for F_1,

$$F_1 = \frac{W}{2} + \frac{3w}{4} = 6.5 \text{ lb}.$$

Similarly,

$$F_2 = F_1 - \frac{w}{2} = 3.5 \text{ lb}.$$

Notice, in the equation from which F_1 is obtained, if $w = 0$, $F_1 = W/2 = 2.0$ lb and consequently $F_2 = 2.0$ lb, also as we would have expected. If we had chosen an axis through one end of the bar, we would have obtained the same result.

Example 2 A 10-kg object M is held in the hand of a person whose upper arm is vertical and whose forearm is horizontal (Fig. 13–5a). What forces must be exerted on the forearm by the biceps muscle and by the upper arm? Assume that the forearm and hand together have a mass m of 2.0 kg with a center of mass 15 cm from the elbow. The center of the 10-kg object is 33 cm from the elbow and the biceps exerts a vertical force on the forearm at a point that is 4.0 cm from the elbow.

Figure 13–5b shows an appropriate free-body diagram of the forearm. The forearm itself is represented as a horizontal rod. Its own weight $m\mathbf{g}$ pulls downward at the center of gravity while the 10-kg object pushes downward on the hand with a force $M\mathbf{g}$ equal to its weight. The biceps pulls up on the forearm with a force $\mathbf{T}$. Muscles, like strings and cords, always pull. Their biological function is always to contract under appropriate stimulation.

We have shown the force $\mathbf{F}$ representing the force that the upper arm exerts at the elbow. We know that it is vertical because the forearm could not be in equilibrium if $\mathbf{F}$ had a horizontal component. We also chose the direction of $\mathbf{F}$ to be downward rather than upward. Do you see why? (There is a biological reason and an analytical one. Actually, it is not essential to choose the right direction for $\mathbf{F}$ at this point. Any error would be discovered in the analysis after the equilibrium conditions are applied.)

Figure 13–5b also shows the coordinate system we have chosen for the analysis. Since all the forces have only y-components, the first equilibrium condition becomes

$$F_y = T - F - mg - Mg = 0.$$

Calculating torques about O, with counterclockwise rotations taken as positive, reduces the second equilibrium condition to

$$\tau = F(0) + T(d) - mg(D) - Mg(L) = 0.$$

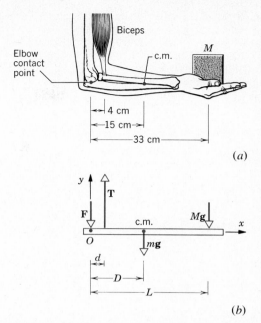

Figure 13–5 Example 2. (a) An object held with a horizontal forearm. (b) A free-body diagram of the forearm. Can you see, before doing any analysis, why we chose the force acting at the elbow contact to be a downward force rather than an upward force?

Since the coefficient of F in this equation is zero (we chose the origin at O so that this would happen), we can solve this equation for T.

$$T = \frac{mgD + MgL}{d}.$$

Inserting the numbers, we obtain

$$T = \frac{(2.0 \text{ kg})(9.8 \text{ m/s}^2)(0.15 \text{ m}) + (10 \text{ kg})(9.8 \text{ m/s}^2)(0.33 \text{ m})}{(0.04 \text{ m})}$$

$$= 880 \text{ N}.$$

Notice that the biceps must pull upwards on the forearm with a force T that is 9 times the weight (98 N) of the object being held. Knowing T, we can now use our first equation to calculate F.

$$F = T - mg - Mg$$
$$= 880 \text{ N} - (2.0 \text{ kg})(9.8 \text{ m/s}^2) - (10 \text{ kg})(9.8 \text{ m/s}^2)$$
$$= 760 \text{ N}.$$

F is also a strong force. The bone in the upper arm pushes down on the forearm, at the elbow, with a force F that is almost 8 times the weight of the object being held.

Example 3 (a) A 60-ft ladder weighing 100 lb rests against a wall at a point 48 ft above the ground. The center of gravity of the ladder is one-third the way up the ladder. A 160-lb man climbs halfway up the ladder. Assuming that the wall (but not the ground) is frictionless, find the forces exerted by the system on the ground and on the wall.

The forces acting on the ladder are shown in Fig. 13–6. **W** is the weight of the man standing on the ladder, and **w** is the weight of the ladder itself. A force $\mathbf{F}_1$ is exerted by the ground on the ladder. $\mathbf{F}_{1v}$ is the vertical component and $\mathbf{F}_{1h}$ is the horizontal component of this force (due to friction). The wall, being frictionless, can exert only a force normal to its surface, called $\mathbf{F}_2$. We are given the following data:

$$W = 160 \text{ lb}, \qquad w = 100 \text{ lb},$$

$$a = 48 \text{ ft}, \qquad c = 60 \text{ ft}.$$

From the geometry we conclude that $b = 36$ ft. The line of action of **W** intersects the ground at a distance $b/2$ from the wall and the line of action of **w** intersects the ground at a distance $2b/3$ from the wall.

We choose the x-axis to be along the ground and the y-axis along the wall. Then, the conditions on the forces for translational equilibrium (Eq. 13–2) are

$$F_2 - F_{1h} = 0,$$

$$F_{1v} - W - w = 0.$$

For rotational equilibrium (Eq. 13–4) choose an axis through the point of contact with the ground and obtain

$$F_2(a) - W\left(\frac{b}{2}\right) - w\left(\frac{b}{3}\right) = 0.$$

Using the data given, we obtain

$$F_2(48 \text{ ft}) - (160 \text{ lb})(18 \text{ ft}) - (100 \text{ lb})(12 \text{ ft}) = 0,$$

$$F_2 = 85 \text{ lb},$$

$$F_{1h} = F_2 = 85 \text{ lb},$$

$$F_{1v} = 160 \text{ lb} + 100 \text{ lb}$$
$$= 260 \text{ lb}.$$

By Newton's third law the forces exerted by the ground and the wall on the ladder are equal but opposite to the forces exerted by the ladder on the ground and the wall, respectively. Therefore, the normal force on the wall is 85 lb, and the force on the ground has components of 260 lb down and 85 lb to the right.

(b) If the coefficient of static friction between the ground and the ladder is $\mu_s = 0.40$, how high up the ladder can the man go before it starts to slip?

Let x be the fraction of the total length of the ladder the man can climb before slipping begins. Then our equilibrium conditions are

$$F_2 - F_{1h} = 0,$$

$$F_{1v} - W - w = 0,$$

and

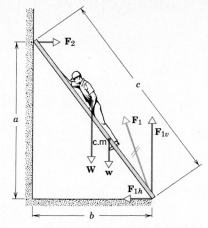

Figure 13–6   Example 3.

$$F_2 a - Wbx - w\left(\frac{b}{3}\right) = 0.$$

Now we obtain

$$F_2(48 \text{ ft}) = (160 \text{ lb})(36 \text{ ft})x + (100 \text{ lb})(12 \text{ ft}),$$

$$F_2 = (120x + 25) \text{ lb}.$$

Hence

$$F_{1h} = (120x + 25) \text{ lb},$$

and, as before,

$$F_{1v} = 260 \text{ lb}.$$

The maximum force of static friction is given by

$$F_{1h} = \mu_s F_{1v} = (0.40)(260 \text{ lb}) = 104 \text{ lb}.$$

Therefore,

$$F_{1h} = (120x + 25) \text{ lb} = 104 \text{ lb}$$

and

$$x = 79/120,$$

so that the man can climb up the ladder

$$60x \text{ ft} = 39.5 \text{ ft}$$

before slipping begins.

In this example the ladder is treated as a one-dimensional object, with only one point of contact at the wall and ground. You might reflect on how this limits consideration of the less artificial case of two contact points at each end.

Under-determined structures   In the preceding examples we have been careful to choose situations for which the number of unknown forces does not exceed the number of independent equations relating the forces. When all the forces act in a plane, we can have only three independent equations of equilibrium, one for rotational equilibrium about any axis normal to the plane and two others for translational equilibrium in the plane. However, we often have more than three unknown forces. For example, in the ladder problem of Example 3a, if we drop the artificial assumption of a frictionless wall, we have four unknown scalar quantities, namely, the horizontal and vertical components of the force acting on the ladder at the wall and the horizontal and vertical components of the force acting on the ladder at the ground. Since we have only three scalar equations, these forces cannot be determined. For any value assigned to one unknown force, the other three forces can be determined. But if we have no basis for assigning any particular value to an unknown force, there are an infinite number of solutions possible mathematically. We must therefore find another independent relation between the unknown forces if we hope to solve the problem uniquely.

Another simple example of such underdetermined structures is the automobile. In this case we wish to determine the forces exerted by the ground on each of the four tires when the car is at rest on a horizontal surface. If we assume that these forces are normal to the ground, we have four unknown scalar quantities. All other forces, such as the weight of the car and passengers, act normal to the ground. Therefore we have only three independent equations giving the equilibrium conditions, one for translational equilibrium in the single direction of all the forces and two for rotational equilibrium about the two axes perpendicular to each other in a horizontal plane. Again the solution of the problem is indeterminate, mathematically. A four-legged table with all its legs in contact with the floor is a similar example.

Of course, since there is actually a unique solution to any real physical problem, there must be a physical basis for the additional independent relation between the forces that enables us to solve the problem. The difficulty is removed when we realize that structures are never perfectly rigid, as we have tacitly assumed throughout. Actually our structures will be somewhat deformed. For example, the automobile tires and the ground will be deformed, as will the ladder and wall. The laws of elasticity and the elastic properties of the structure determine the nature of the deformation and will provide the necessary additional relation between the four forces. A complete analysis therefore requires not only the laws of rigid body mechanics but also the laws of elasticity. In courses in civil and mechanical engineering, many such problems are encountered and analyzed in this way. We shall not consider the matter further here.

REVIEW GUIDE AND SUMMARY

Definition of static equilibrium

A rigid body at rest in an inertial reference frame is said to be in *static equilibrium*. For such a body, (1) The linear acceleration $\mathbf{a}_{cm}$ of its center of mass is zero, and (2) its angular acceleration α about any point is also zero.

First condition for equilibrium

The first condition requires that the vector sum of all the external forces acting on the body add to zero, or

$$\mathbf{F} = \mathbf{F}_1 + \mathbf{F}_2 + \mathbf{F}_3 \cdots = 0 \qquad [13\text{–}1]$$

Internal forces cancel in pairs because of Newton's third law and do not appear in this sum. If all the forces lie in the x-y plane the vector equation above is equivalent to the two algebraic equations

$$F_x = F_{1x} + F_{2x} + F_{3x} \cdots \cdots = 0$$

and   [see 13–2]

$$F_y = F_{1y} + F_{2y} + F_{3y} \cdots \cdots = 0.$$

Second condition for equilibrium

The second condition requires that the vector sum of all the external torques acting on the body about any given point add to zero, or

$$\boldsymbol{\tau} = \boldsymbol{\tau}_1 + \boldsymbol{\tau}_2 + \boldsymbol{\tau}_3 \cdots = 0 \qquad [13\text{–}3]$$

As above, internal torques do not enter. If all the forces lie in the x-y plane, the vector equation above is equivalent to the single algebraic equation

$$\tau_z = \tau_{1z} + \tau_{2z} + \tau_{3z} \cdots = 0 \qquad \text{[see 13-4]}$$

in which the torques are about the z-axis or any axis parallel to it. Examples 1, 2, and 3 show how these conditions for equilibrium are applied.

Center of gravity

The gravitational force on a body acts diffusely throughout the body, exerting forces on all the mass elements that make it up. There is, however, one (and only one) point in the body at which we may imagine all these force contributions to be concentrated as far as their external effects are concerned. We can, then, replace the actual spread-out gravitational force by a single force **M**g acting at this *center of gravity*, where M is the total mass of the body. If the gravitational field **g** may be taken as uniform over the extent of the structure—which we assume in this chapter—the center of gravity coincides with the center of mass, which is defined in quite a different way; see Section 9-1.

QUESTIONS

1. Is a baseball in equilibrium at the instant it comes to rest at the top of a vertical pop fly?

2. In a simple pendulum, is the mass point in equilibrium at any point of its swing?

3. Are Eqs. 13-1 and 13-3 both necessary and sufficient conditions for static equilibrium?

4. A wheel rotating at constant angular velocity ω about a fixed axis is in mechanical equilibrium because no net external force or torque acts on it. However, the particles that make up the wheel undergo a centripetal acceleration **a** directed toward the axis. Since $\mathbf{a} \neq 0$, how can the wheel be said to be in equilibrium?

5. Give several examples of bodies that are not in equilibrium, even though the resultant of all the forces acting on it is zero.

6. Which is more likely to break in use, a hammock stretched tightly between two trees or one that sags quite a bit? Prove your answer.

7. A ladder is at rest with its upper end against a wall and the lower end on the ground. Is it more likely to slip when a man stands on it at the bottom or at the top? Explain.

8. In Example 3, if the wall were rough, would the empirical laws of friction supply us with the extra condition needed to determine the extra (vertical) force exerted by the wall on the ladder?

9. A picture hangs from a wall by two wires. What orientation should the wires have to be under minimum tension? Explain how equilibrium is possible with any number of orientations and tensions, even though the picture has a definite mass.

10. A book rests on a table. The table pushes up on the book with a force just equal to the weight of the book. Speaking loosely, just how does the table "know" what upward force it must provide? What is the mechanism by which this force comes into play? (See "The Smart Table," *The Physics Teacher,* December 1981.)

11. Sit in a straight-backed chair and try to stand up without leaning forward. Why can't you do it?

12. Why is it easier to balance a long rod on your finger tip than a short one?

13. Long balancing poles help a tightrope walker to keep her balance. How?

14. A composite block is made up of wood and metal. In which orientation of the two shown in Fig. 13-7 can you tip it over with the least force?

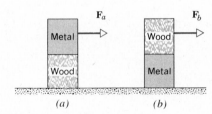

Figure 13-7   Question 14.

15. The Olympic gymnast Mary Lou Retton does some amazing things on the parallel bars. A friend tells you that careful analysis of films of her exploits shows that, no matter what she does, her center of mass is above her point(s) of support at all times, as required by the laws of physics. Comment on your friend's statement.

16. Do the center of mass and the center of gravity coincide for a building? For a lake? Under what conditions does the difference between the center of mass and the center of gravity of a body become significant?

17. If a rigid body is thrown into the air without spinning, it does not begin spinning during its flight, provided that air resistance can be neglected. What does this simple result imply about the location of the center of gravity?

18. Explain, using forces and torques, how a tree can maintain equilibrium in a high wind.

19. Is there such a thing as a truly rigid body?

20. You are sitting in the driver's seat of a parked automobile. You are told that the forces exerted upward by the ground on each of the four tries are different. Discuss the factors that enter into a consideration of whether this statement is true or false.

21. Consider this parlor stunt. You ask a friend to stand with his chest touching the edge of an open door, one foot being on each side of the door. He will not be able to stand on tip-toes. Explain, in terms of the conditions for equilibrium.

EXERCISES

Section 13-1   A Rigid Body in Static Equilibrium

1. A rigid square object of negligible weight is acted upon by three forces that pull on its corners as shown, to scale, in Fig. 13-8. (*a*) Is the first condition of equilibrium satisfied? (*b*) Is the second condition of equilibrium satisfied? (*c*) If either of the preceding answers is "No," could a fourth force restore the equilibrium of the object? If so, specify the magnitude, direction, and point of application of the needed force.

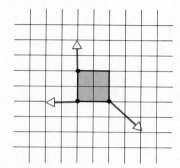

Figure 13-8   Exercise 1.

2. An eight-member family, whose weights in pounds are indicated, is balanced on a see-saw, as shown in Fig. 13-9. What is the number of the person who causes the largest torque, about the fulcrum, directed (*a*) out of the page, and (*b*) into the page?

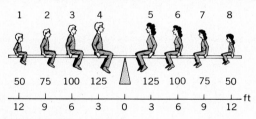

Figure 13-9   Exercise 2.

3. A 5-kg book rests on a table. Identify the forces on it.

4. A particle is acted upon by forces given, in newtons, by $\mathbf{F}_1 = 10\mathbf{i} - 4\mathbf{j}$ and $\mathbf{F}_2 = 17\mathbf{i} + 2\mathbf{j}$. (*a*) Find a force $\mathbf{F}_3$ that would keep it in equilibrium. (*b*) What direction does $\mathbf{F}_3$ have relative to the *x*-axis?

5. A uniform sphere of weight *w* and radius *r* is being held by a rope attached to a frictionless wall a distance *L* above the center of the sphere, as in Fig. 13-10. Find (*a*) the tension in the rope and (*b*) the force exerted on the sphere by the wall.

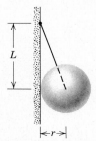

Figure 13-10   Exercise 5.

Section 13-9   Center of Gravity

6. A uniform thin wire is bent into the shape of an isosceles right triangle, each leg of which has length *L*. (*a*) What is the distance between its center of gravity and the right-angle vertex? (*b*) Is the answer the same for a uniform triangular plate of the same shape?

7. An automobile weighing 3000 lb (mass = 1360 kg) has a wheelbase of 120 in. (= 305 cm). Its center of gravity is located 70.0 in. (= 178 cm) behind the front axle. Determine (*a*) the force exerted on each of the front wheels (assumed the same) and (*b*) the force exerted on each of the back wheels (assumed the same) by the level ground.

Section 13-3   Examples of Equilibrium

8. A rope, assumed massless, is stretched horizontally between two supports that are 3.44 m apart. When an object of weight 3160 N is hung at the center of the rope, the rope is observed to sag by 35 cm. What is the tension in the rope?

9. In Fig. 13-11 a man is trying to get his car out of the mud on the shoulder of a road. He ties one end of a rope tightly around the front bumper and the other end tightly around a utility pole 60 ft away. He then pushes sideways on the rope at its midpoint with a force of 125 lb, displacing the center of the rope 1.0 ft from its previous position and the car barely moves. What force does the rope exert on the car? (The rope stretches somewhat under the tension.)

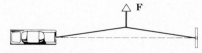

Figure 13-11   Exercise 9.

10. A 160-lb man is walking across a level bridge and stops three-fourths of the way from one end. The bridge is uniform and weighs 600 lb. What are the values of the vertical forces exerted on each end of the bridge by its supports?

11. A diver of weight 580 N stands at the end of a 4.5-m diving board of negligible weight. The board is attached by two pedestals 1.5 m apart, as shown in Fig. 13-12. Find the tension (or compression) in each of the two pedestals.

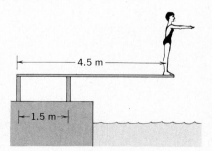

Figure 13-12   Exercise 11.

12. A meter stick balances on a knife edge at the 50.0-cm mark. When two nickels are stacked over the 12.0-cm mark, the loaded stick is found to balance at the 45.5-cm mark. A nickel has a mass of 5.0 g. What is the mass of the meter stick? Try this technique and check your answer experimentally.

13. A beam is carried by three men, one man at one end and the other two supporting the beam between them on a cross piece so placed that the load is equally divided among the three men. Find where the crosspiece is placed. Neglect the mass of the crosspiece.

14. An 1800-lb construction bucket is suspended by a cable that is attached at O to two other cables, these making angles of 66° and 51° with the horizontal. See Fig. 13–13. Find the tensions in the three cables. (*Hint:* To avoid solving two equations in two unknowns, use the "rotated axes" shown in the Figure.)

15. A uniform cubical crate is 0.75 m on each side and weighs 500 N. It rests on the floor with one edge against a very small, fixed obstruction. At what height above the floor must a horizontal force of 350 N be applied to just tip the crate?

16. A 50-kg uniform square sign, 2.0 m on a side, is hung from a 3.0-m bar of negligible mass. A cable is attached to the end of the bar and to a point on the wall 4.0 m above the point where the bar is fixed to the wall, as shown in Fig. 13–14. (*a*) What is the tension in the cable? (*b*) What are the horizontal and vertical components of the force exerted by the wall on the bar?

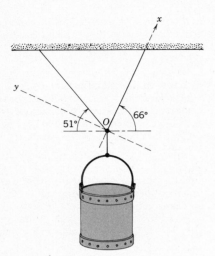

Figure 13–13   Exercise 14.

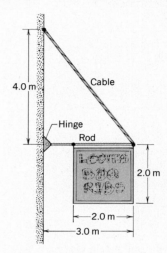

Figure 13–14   Exercise 16.

PROBLEMS GROUPED BY SECTION

Section 13–1   A Rigid Body in Static Equilibrium

1. The system in Fig. 13–15 is in equilibrium, but it begins to slip if any additional mass is added to the 5.0-kg object. What is the coefficient of static friction between the 10-kg block and the plane on which it rests?

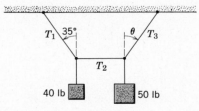

Figure 13–16   Problem 2.

3. Two identical uniform smooth spheres, each of weight W, rest as shown in Fig. 13–17 at the bottom of a fixed, rectangular container. Find, in terms of W, the forces acting on the spheres by (*a*) the container surfaces and (*b*) by one another, if the line of centers of the spheres makes an angle of 45° with the horizontal.

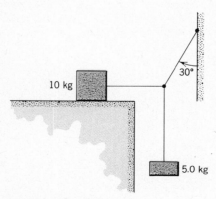

Figure 13–15   Problem 1.

2. The system in Fig. 13–16 is in equilibrium with the string in the center exactly horizontal. Find (*a*) the angle θ and (*b*) the tension in each string.

Figure 13–17   Problem 3.

Section 13-2   Center of Gravity

4.   The Leaning Tower of Pisa, see Fig. 13–18, is 55 m high and 7.0 m in diameter. The top of the tower is displaced 4.5 m from the vertical. Treating the tower as a uniform, circular cylinder, (a) what additional displacement, measured at the top, will bring the tower to the verge of toppling; and (b) what angle with the vertical will the tower make at that moment?

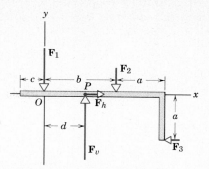

Figure 13–19   Problem 6.

8.   Figure 13–20 shows the anatomical structures in the lower leg and foot that are involved when the heel is raised off the floor so that the foot effectively contacts the floor at only one point, shown as P in the figure. Calculate the forces that must be exerted on the foot by the calf muscle and by the lower-leg bones when a 65.3-kg person stands tip-toe on one foot. Compare these forces to the person's weight. Assume that $a = 5.0$ cm and $b = 15$ cm.

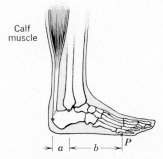

Figure 13–20   Problem 8.

9.   One end of a uniform beam weighing 50 lb and 3.0-ft long is attached to a wall with a hinge. The other end is supported by a wire (see Fig. 13–21). (a) Find the tension in the wire. (b) What are the horizontal and vertical components of the force at the hinge?

Figure 13–18   Problem 4. Barone/Monkmeyer Press.

5.   A circular section of radius r is cut out of a uniform disk of radius R, the center of the hole being $R/2$ from the center of the original disk. Locate the center of gravity of the resulting body.

Section 13-3   Examples of Equilibrium

6.   Forces $\mathbf{F}_1$, $\mathbf{F}_2$, and $\mathbf{F}_3$ act on the structure of Fig. 13–19 as shown. We wish to put the structure in equilibrium by applying a force, at a point such as P, whose vector components are $\mathbf{F}_h$ and $\mathbf{F}_v$. We are given that $a = 2.0$ m, $b = 3.0$ m, $c = 1.0$ m, $F_1 = 20$ N, $F_2 = 10$ N, and $F_3 = 5.0$ N. Find (a) F_h, (b) F_v, and (c) d.

7.   A trap door in a ceiling is 3.0 ft (=0.91 m) square, weighs 25 lb (mass = 11 kg), and is hinged along one side with a catch at the opposite side. If the center of gravity of the door is 4.0 in.(= 10 cm) from the door's center and closer to the hinged side, what forces must (a) the catch and (b) the hinge sustain?

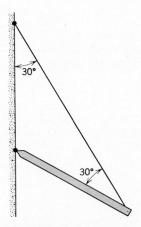

Figure 13–21   Problem 9.

10. The system shown in Fig. 13–22 is in equilibrium. The mass hanging from the end of the strut S weighs 500 lb (mass = 225 kg), and the strut itself weighs 100 lb (mass = 45 kg). Find (*a*) the tension T in the cable and (*b*) the force exerted on the strut by the pivot P.

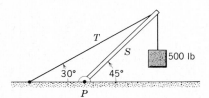

Figure 13–22   Problem 10.

11. A thin horizontal bar AB of negligible weight and length l is pinned to a vertical wall at A and supported at B by a thin wire BC that makes an angle θ with the horizontal. A weight W can be moved anywhere along the bar as defined by the distance x from the wall (Fig. 13–23). (*a*) Find the tension in the thin wire as a function of x. Find (*b*) the horizontal and (*c*) the vertical components of the force exerted on the bar by the pin at A.

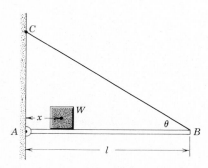

Figure 13–23   Problems 11 and 24.

12. Two uniform beams are attached to a wall with hinges and then loosely bolted together as in Fig. 13–24. Find (*a*) the forces on each hinge and (*b*) the force exerted by the bolt on each beam.

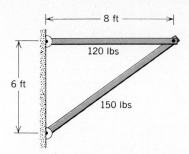

Figure 13–24   Problem 12.

13. Four identical bricks, each of length l, are put on top of one another (see Fig. (13–25) in such a way that part of each extends beyond the one beneath. Show that the largest equilibrium extensions are (*i*) top brick overhanging the one below by $l/2$, (*ii*) second brick from top overhanging the one below by $l/4$, and (*iii*) third brick from top overhanging the bottom one by $l/6$.

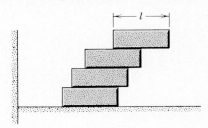

Figure 13–25   Problem 13.

14. A uniform cube of side length L rests on a horizontal floor. The coefficient of static friction between cube and floor is μ. A horizontal pull P is applied perpendicular to one of the faces of the cube, at a distance h above the floor on the vertical midline of the cube face. As P is slowly increased, the cube will either (*i*) begin to slide or (*ii*) begin to tip. What is the condition on μ for (*i*) to occur? For (*ii*)?

15. A crate in the form of a 4.0-ft cube contains a piece of machinery whose design is such that the center of gravity of the crate and its contents is located 1.0 ft above its geometrical center. The crate rests on a ramp that makes an angle θ with the horizontal. As θ is increased from zero, an angle will be reached at which the crate will either start to slide down the ramp or tip over. Which event will occur if the coefficient of static friction is (*a*) 0.60? (*b*) 0.70? In each case give the angle at which the event occurs.

ADDITIONAL PROBLEMS

16. A certain skydiver free falls until she reaches a speed of 70 ft/s. She then opens her parachute and experiences an initial upward acceleration of 30 ft/s². What will be her terminal speed (the speed with which she will fall with no acceleration) if the retarding force of her parachute is proportional to her speed of descent?

17. A 15-kg "weight" is being lifted by the pulley system shown in Fig. 13–26. The upper arm is vertical, whereas the forearm makes an angle of 30° with the horizontal. What forces are being exerted on the forearm by the triceps muscle and by the upper-arm bone (the humerus)? The forearm and hand together have a mass of 2.0 kg with a center of mass 15 cm (measured along the

arm) from the point where the two bones are in contact. The triceps pulls vertically upward at a point 2.5 cm behind the contact point.

18. A balance is made up of a rigid rod free to rotate about a point not at the center of the rod. It is balanced by unequal weights placed in the pans at each end of the rod. When an unknown mass m is placed in the left-hand pan, it is balanced by a mass m_1 placed in the right-hand pan, and similarly when the mass m is placed in the right-hand pan, it is balanced by a mass m_2 in the left-hand pan. Show that

$$m = \sqrt{m_1 m_2}.$$

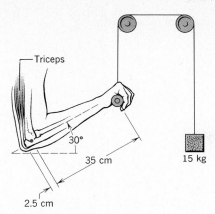

Figure 13–26 Problem 17.

19. Prove that when only three forces act on a body in equilibrium, they must be coplanar and their lines of action must meet at a point or at infinity.

20. What force F applied horizontally at the axle of the wheel in Fig. 13–27 is necessary to raise the wheel over an obstacle of height h? Take r as the radius of the wheel and W as its weight.

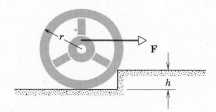

Figure 13–27 Problem 20.

21. A nonuniform bar of weight W is suspended at rest in a horizontal position by two light cords as shown in Fig. 13–28; the angle one cord makes with the vertical is $\theta = 36.9°$; the other makes the angle $\phi = 53.1°$ with the vertical. If the length l of the bar is 6.1 m, compute the distance x from the left-hand end to the center of gravity.

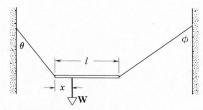

Figure 13–28 Problem 21.

22. A door 7.0 ft ($=2.1$ m) high and 3.0 ft ($=0.91$ m) wide weighs 60 lb (mass $=27$ kg). A hinge 1.0 ft ($=0.30$ m) from the top and another 1.0 ft ($=0.30$ m) from the bottom each support half the door's weight. Assume that the center of gravity is at the geometrical center of the door and determine the horizontal and vertical force components exerted by each hinge on the door.

23. Four similar, uniform bricks of length L are stacked on a table as shown in Fig. 13–29 (compare Problem 13). We seek to maximize the overhang distance h, measured from the edge of the

table. For each of the three arrangements, find the optimum indicated distances and calculate h. [See *Scientific American,* June 1985, for a discussion and an even better version of arrangement (*c*).]

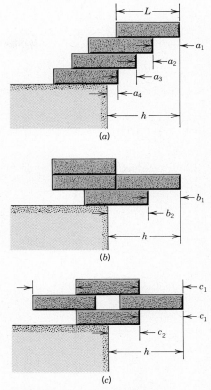

Figure 13–29 Problem 23.

24. In Fig. 13–23, the length of the bar is 3.0 m and its weight is 200 N. Also, $W = 300$ N and $\theta = 30°$. The wire can withstand a maximum tension of 500 N. (*a*) What is the maximum distance x possible before the wire breaks? (*b*) With W placed at this maximum x, what are the horizontal and vertical components of the force exerted on the plank by the pin?

25. A 75-kg window cleaner uses a 10-kg ladder that is 5.0 m long. He places one end down 2.5 m from a wall and rests the upper end against a cracked window and climbs the ladder. He climbs 3.0 m up the ladder when the window breaks. Neglecting friction between the ladder and window and assuming that the base of the ladder did not slip, find (*a*) the force exerted on the window by the ladder just before the window breaks, and (*b*) the magnitude and direction of the force exerted on the ladder by the ground just before the window breaks.

26. A 100-lb plank, of length $l = 20$ ft, rests on the ground and on a frictionless roller (not shown) at the top of a wall of height $h = 10$ ft (see Fig. 13–30). The center of gravity of the plank is at its center. The plank remains in equilibrium for any value of $\theta \geq 70°$, but slips if $\theta < 70°$. (*a*) Draw a diagram showing all forces acting on the plank. (*b*) Find the coefficient of static friction between the plank and the ground.

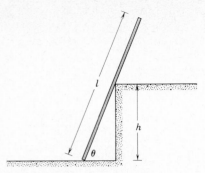

Figure 13–30 Problem 26.

27. In the stepladder shown in Fig. 13–31, *AC* and *CE* are 8.0 ft long and hinged at *C*. *BD* is a tie rod 2.5 ft long, halfway up. A man weighing 192 lb climbs 6.0 ft along the ladder. Assuming that the floor is frictionless and neglecting the weight of the ladder, find (*a*) the tension in the tie rod and (*b*) the forces exerted on the ladder by the floor. (*Hint:* It will help to isolate parts of the ladder in applying the equilibrium conditions.)

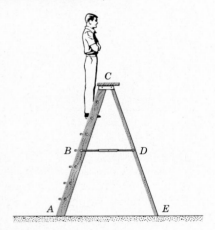

Figure 13–31 Problem 27.

28. By means of a turnbuckle *G*, a tension force *T* is produced in bar *AB* of the square frame *ABCD* in Fig. 13–32. Determine the forces produced in the other bars. The diagonals *AC* and *BD* pass each other freely at *E*. Symmetry considerations can lead to considerable simplification in this and similar problems.

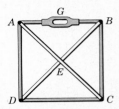

Figure 13–32 Problem 28.

29. A cubical box is filled with sand and weighs 200 lb (= 890 N). It is desired to "roll" the box by pushing horizontally on one of the upper edges. (*a*) What minimum force is required? (*b*) What minimum coefficient of static friction is required? (*c*) Is there a more efficient way to roll the box? If so, find the smallest possible force that would be required to be applied directly to the box.

14 Oscillations

14-1 Oscillations

Any motion that repeats itself in equal intervals of time is called *periodic* or *harmonic* motion. If a particle in periodic motion moves back and forth over the same path, we call the motion *oscillatory* or *vibratory*. The world is full of such motions. Some examples are the oscillations of the balance wheel of a watch, a violin string, a mass attached to a spring, atoms in molecules or in a solid lattice, and air molecules as a sound wave passes by.

Many oscillating bodies do not move back and forth between precisely fixed limits because frictional forces dissipate the energy of motion. Thus a violin string eventually stops vibrating and a pendulum stops swinging. We call such motions *damped* harmonic motions. Although we cannot eliminate friction from the periodic motions of gross objects, we can often cancel out its damping effect by feeding energy into the oscillating system so as to compensate for the energy dissipated by friction. The main spring of a watch and the falling weight in a pendulum clock supply external energy in this way, so that the oscillating system, that is, the balance wheel or the pendulum, moves as if it were undamped.

Not only mechanical systems can oscillate. Radio waves, microwaves, and visible light are oscillating magnetic and electric field vectors. Thus a tuned circuit in a radio and a closed metal cavity in which microwave energy is introduced can oscillate electromagnetically. The analogy is close, being based on the fact that mechanical and electromagnetic oscillations are described by the same basic mathematical equations. We will make the most of this analogy in later chapters.

The *period T* of a harmonic motion is the time required to complete one round trip of the motion, that is, one complete oscillation or *cycle*. The *frequency* of the motion ν is the number of oscillations (or cycles) per unit of time. The frequency is therefore the reciprocal of the period, or

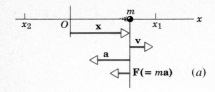

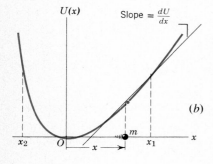

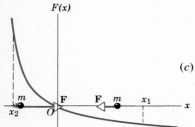

Figure 14–1 (*a*) A particle of mass *m* oscillates between points x_1 and x_2, *O* being the equilibrium position. (*b*) The potential energy of the particle as a function of position. The force acting on the particle at position *x* is given by $F = -dU/dx$. (*c*) The force acting on the particle as a function of position *x*; note that the force is always directed toward the equilibrium position.

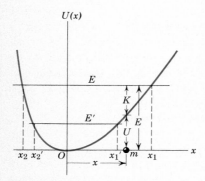

Figure 14–2 The mechanical energy *E* for the motion of Fig. 14–1 is shown. If the mechanical energy of the oscillating particle is reduced to *E'*, the limits of oscillation are reduced to x_1' and x_2' respectively.

$$\nu = \frac{1}{T}. \tag{14–1}$$

The SI unit of frequency is the cycle per second or *hertz* (abbr. Hz). The position at which no net force acts on the oscillating particle is called its *equilibrium* position. The *displacement* (linear or angular) is the distance (linear or angular) of the oscillating particle from its equilibrium position at any instant.

Let us focus attention on a particle oscillating back and forth along a straight line between fixed limits. Its displacement **x** changes periodically in both magnitude and direction. Its velocity **v** and acceleration **a** also vary periodically in magnitude and direction and, in view of the relation **F** = *m***a**, so does the force **F** acting on the particle.

In terms of energy, we can say that a particle undergoing harmonic motion passes back and forth through a point (its equilibrium position) at which its potential energy is a minimum. A swinging pendulum is a good example, its potential energy being a minimum at the bottom of the swing, that is, at the equilibrium position. Figure 14–1*a* shows the generalized case of a particle oscillating between the limits x_1 and x_2, *O* being the equilibrium position. Figure 14–1*b* shows the corresponding potential energy curve, which has a minimum value at that position. The force acting on the particle at any position is derivable from the potential energy function; it is given by Eq. 8–7,

$$F = -dU/dx, \tag{8–7}$$

and is displayed in Fig. 14–1*c*. The force is zero at the equilibrium position *O*, points to the right (that is, has a positive value) when the particle is to the left of *O*, and points to the left (that is, has a negative value) when the particle is to the right of *O*. The force is a *restoring force* because it always acts to accelerate the particle in the direction of its equilibrium position. Hence in harmonic motion the position of equilibrium is always one of stable equilibrium.

The mechanical energy *E* for an oscillating particle is the sum of its kinetic energy and its potential energy, or

$$E = K + U \tag{14–2}$$

in which *E* remains constant if no nonconservative forces, such as the force of friction, are acting. Figure 14–2 shows *E* for the motion of Fig. 14–1. Note how Eq. 14–2 is satisfied for the particle in the typical position shown. The particle cannot move outside the limits x_1 and x_2 because, in these regions, *U* exceeds *E*. This, as Eq. 14–2 shows, would require the kinetic energy to be negative, an impossibility.

For a given environment, that is, for a given function *U(x)*, an oscillating particle can have various total energies, depending on how we set it into motion initially. Thus the total energy may be *E'*, rather than *E*, in which case the limits of oscillation would be x_1' and x_2', as Fig. 14–2 shows, rather than x_1 and x_2.

14–2 The Simple Harmonic Oscillator

Let us consider an oscillating particle (Fig. 14–3*a*) moving back and forth about an equilibrium position through a potential that varies as

$$U(x) = \tfrac{1}{2}kx^2 \tag{14–3}$$

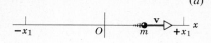

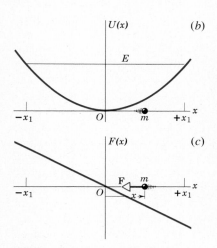

Figure 14-3 (a) A particle of mass m oscillates with simple harmonic motion between points $+x_1$ and $-x_1$, O being the equilibrium position. (b) The potential energy U(x) and the mechanical energy E. (c) The force acting on the particle. Compare this figure carefully with Fig. 14-1, which illustrates the general case of harmonic motion.

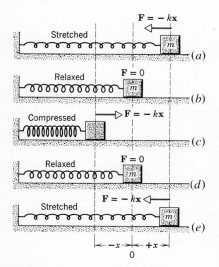

Figure 14-4 A block of mass m in simple harmonic motion. The force exerted by the spring on the block is shown in each case.

in which k is a positive constant; see Fig. 14-3b. The force acting on the particle is given by Eq. 8-7, or

$$F(x) = -\frac{dU}{dx} = -\frac{d(\frac{1}{2}kx^2)}{dx} = -kx; \quad (14-4)$$

see Fig. 14-3c. Such an oscillating particle is called a *simple harmonic oscillator* and its motion is called *simple harmonic motion* (SHM). In such motion, as Eq. 14-3 shows, the potential energy curve varies as the square of the displacement, and, as Eq. 14-4 shows, the force acting on the particle is proportional to the displacement but is opposite to it in direction. In simple harmonic motion the limits of oscillation are equally spaced about the equilibrium position. This is not true for the more general motion of Fig. 14-1 which, although harmonic, is not simple harmonic. The magnitude of the maximum displacement, the quantity x_1 in Fig. 14-3, always taken as positive, is called the *amplitude* of the simple harmonic motion.

You might have recognized Eq. 14-3 $[U(x) = \frac{1}{2}kx^2]$ as the expression for the potential energy of an "ideal" spring, compressed or extended by a distance x; see Section 8-4. In this same section an ideal spring was defined as one in which the force exerted by the stretched or compressed spring is given by $F(x) = -kx$ (see Eq. 14-4), k being called the *force constant*.

Hence, a body of mass m attached to an ideal spring of force constant k and free to move over a frictionless horizontal surface is an example of a simple harmonic oscillator (see Fig. 14-4). Note that there is a position (the equilibrium position; see Fig. 14-4b,d) in which the spring exerts no force on the body. If the body is displaced to the right (as in Fig. 14-4a), the force exerted by the spring on the body points to the left and is given by $F = -kx$. If the body is displaced to the left (as in Fig. 14-4c), the force points to the right and is also given by $F = -kx$. In each case the force is a restoring force. The motion of the oscillating mass is simple harmonic motion, one full cycle of motion being shown in the figure.

Let us apply Newton's second law, $F = ma$, to the motion of Fig. 14-4. For F we substitute $-kx$ (from Eq. 14-4) and for the acceleration a we put in d^2x/dt^2 ($= dv/dt$). This gives us

$$-kx = m\frac{d^2x}{dt^2}$$

or

$$\frac{d^2x}{dt^2} + \frac{k}{m}x = 0. \quad (14-5)$$

This equation involves derivatives and is called a *differential equation*. To solve this equation means to determine how the displacement x of the particle must depend on the time t in order that the equation be satisfied. When we know how x depends on time, we know the motion of the particle; thus, Eq. 14-5 is called the *equation of motion* of a simple harmonic oscillator. We shall solve this equation and describe the motion in detail in the next section.

The simple harmonic oscillator problem is important for two reasons. First, most problems involving mechanical vibrations reduce to that of the simple harmonic oscillator at small amplitudes of vibration, or to a combination of such vibrations. This is equivalent to saying that if we consider a small enough portion of the restoring force curve of Fig. 14-1c (around the origin), it becomes arbitrarily close to a straight line which, as Fig. 14-3c shows, is characteristic of simple harmonic motion. Or, in other words, the potential energy curve of Fig. 14-1b for general oscillatory motion reduces to that of Fig. 14-3b for simple harmonic oscillation when the amplitude of vibration is made sufficiently small about the equilibrium position O.

Second, as we have indicated, equations like Eq. 14–5 turn up in many physical problems in acoustics, in optics, in mechanics, in electrical circuits, and even in atomic physics. The simple harmonic oscillator exhibits features common to many physical systems.

Equation 14–4 ($F = -kx$) is an empirical relation known as *Hooke's law*. It is a special case of a more general relation, dealing with the deformation of elastic bodies, discovered by Robert Hooke (1635–1703). It is obeyed by springs and other elastic bodies provided the deformation is not too great. If the solid is deformed beyond a certain point, called its *elastic limit,* it will not even return to its original shape when the applied force is removed (Fig. 14–5). It turns out that Hooke's law holds almost up to the elastic limit for many common materials. The range of applied forces over which Hooke's law is valid is called the *proportional region.* Beyond the elastic limit, the force can no longer be specified by a potential energy function, because the force then depends on many factors including the speed of deformation and the previous history of the solid.

The restoring force and potential energy function of the simple harmonic oscillator are the same as that of a solid deformed in one dimension in the proportional region. If the deformed solid is released, it will vibrate, just as the simple harmonic oscillator does. Therefore, as long as the amplitude of the vibration is small enough, that is, as long as the deformation remains in the proportional region, mechanical vibrations behave exactly like simple harmonic oscillators. It is easy to generalize this discussion to show that any problem involving mechanical vibrations of small amplitude in three dimensions reduces to a combination of simple harmonic oscillators.

The vibrating string or membrane, sound vibrations, the vibrations of atoms in solids, and electrical or acoustical oscillations in a cavity can all be described in a form which is identical mathematically to a system of harmonic oscillators. The analogy enables us to solve problems in one area by using the techniques developed in other areas.

Hooke's law

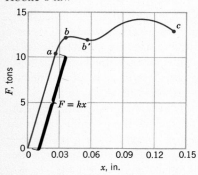

Figure 14–5 Typical graph of applied force F versus resulting elongation of an aluminum bar under tension. The sample was a foot long and a square inch in cross section. Notice that we may write $F = kx$ only for the portion Oa, since beyond this point the slope is no longer constant but varies in a complicated way with x. At some point such as b (the *elastic limit*) the sample does not return to its original length when the applied force is removed. Between b and b' the elongation increases, even though the force is held constant; the material flows like a viscous fluid. At c, the sample can be stretched no farther; any increase in elongation results in the sample's breaking in two. The applied force is equal in magnitude to the restoring force but is opposite in direction, so that no minus sign appears in the relation $F = kx$.

Equation of motion for a simple harmonic oscillator

14–3 Simple Harmonic Motion

Let us now solve the equation of motion of the simple harmonic oscillator,

$$\frac{d^2x}{dt^2} + \frac{k}{m} x = 0. \tag{14-6}$$

Recall that any system of mass m upon which a force $F = -kx$ acts will be governed by this equation. In the case of a spring, the proportionality constant k is the force constant of the spring, which is a measure of its stiffness. In other oscillating systems the proportionality constant k may be related to other physical features of the system, as we shall see later. We can use the oscillating mass-spring system as our prototype.

Equation 14–6 gives a relation between a function of the time $x(t)$ and its second time derivative d^2x/dt^2. To find the position of the particle as a function of the time, we must find a function $x(t)$ that satisfies this relation.

We can rewrite Eq. 14–6 as

$$\frac{d^2x}{dt^2} = -\left(\frac{k}{m}\right) x. \tag{14-7}$$

Equation 14–7 then requires that $x(t)$ be some function whose second derivative is the negative of the function itself, except for a constant factor k/m. We know from the calculus, however, that the sine function or the cosine function has this property. For example,

$$\frac{d}{dt} \cos t = -\sin t \qquad \text{and} \qquad \frac{d^2}{dt^2} \cos t = -\frac{d}{dt} \sin t = -\cos t.$$

This property is not affected if we multiply the cosine function by a constant A. We can allow for the fact that the sine function will do as well, and for the fact that Eq. 14–7 contains a constant factor, by writing as a tentative solution of Eq. 14–7,

Displacement equation

$$x = A \cos (\omega t + \phi). \tag{14–8}$$

Here since

$$\cos (\omega t + \phi) = \cos \phi \cos \omega t - \sin \phi \sin \omega t = a \cos \omega t + b \sin \omega t,$$

the constant ϕ allows for any combination of sine and cosine solutions. Hence, with the (as yet) unknown constants A, ω, and ϕ, we have written as general a solution to Eq. 14–7 as we can. In order to determine these constants such that Eq. 14–8 is actually the solution of Eq. 14–7, we differentiate Eq. 14–8 twice with respect to the time. We have

$$\frac{dx}{dt} = -\omega A \sin (\omega t + \phi)$$

and

$$\frac{d^2x}{dt^2} = -\omega^2 A \cos (\omega t + \phi).$$

Putting this into Eq. 14–7, we obtain

$$-\omega^2 A \cos (\omega t + \phi) = -\frac{k}{m} A \cos (\omega t + \phi).$$

Therefore, if we choose the constant ω such that

$$\omega^2 = \frac{k}{m}, \tag{14–9}$$

then

$$x = A \cos (\omega t + \phi)$$

is in fact a solution of the equation of a simple harmonic oscillator.

The constants A and ϕ are still undetermined and, therefore, still completely arbitrary. This means that any choice of A and ϕ whatsoever will satisfy Eq. 14–7, so that a large variety of motions is possible for the oscillator. In this case ω is common to all the allowed motions, but A and ϕ may differ among them. We shall see later that A and ϕ are determined for a particular harmonic motion by how the motion starts.

Let us find the physical significance of the constant ω. If we increase the time t in Eq. 14–8 by $2\pi/\omega$, the function becomes

$$x = A \cos \left[\omega \left(t + \frac{2\pi}{\omega} \right) + \phi \right],$$
$$= A \cos (\omega t + 2\pi + \phi),$$
$$= A \cos (\omega t + \phi).$$

That is, the function merely repeats itself after a time $2\pi/\omega$. Therefore, $2\pi/\omega$ is the *period* of the motion T. Since $\omega^2 = k/m$, we have

Period of SHM

$$T = \frac{2\pi}{\omega} = 2\pi\sqrt{\frac{m}{k}}. \tag{14–10}$$

Hence, all motions given by Eq. 14—7 have the same period of oscillation, and this is determined only by the mass m of the oscillating particle and the force constant k of the spring. The *frequency* ν of the oscillator is the number of complete vibrations per unit time and is given by

Frequency of SHM

$$\nu = \frac{1}{T} = \frac{\omega}{2\pi} = \frac{1}{2\pi}\sqrt{\frac{k}{m}}. \tag{14–11}$$

Hence,

Angular frequency of SHM

$$\omega = 2\pi\nu = \frac{2\pi}{T} = \sqrt{\frac{k}{m}}. \tag{14–12}$$

The quantity ω is called the *angular frequency;* it differs from the frequency ν by a factor 2π. It has the dimension of reciprocal time (the same as angular speed) and its unit is the radian per second. In Section 14–6 we shall give a geometric meaning to this angular frequency.

Amplitude of SHM

The constant A has a simple physical meaning. The cosine function takes on values from -1 to 1. The *displacement* x from the central equilibrium position $x = 0$, therefore, has a maximum value of A; see Eq. 14–8. We call A $(= x_{max})$ the *amplitude* of the motion. Because A is not fixed by our differential equation, motions of various amplitudes are possible, but all have the same frequency and period. *The frequency of a simple harmonic motion is independent of the amplitude of the motion.*

The quantity $(\omega t + \phi)$ is called the *phase* of the motion. The constant ϕ is called the *phase constant*. Two motions may have the same amplitude and frequency but differ in phase. If $\phi = -\pi/2$, for example,

$$x = A\cos(\omega t + \phi) = A\cos(\omega t - 90°)$$
$$= A\sin\omega t,$$

so that the displacement is zero at the time $t = 0$. When $\phi = 0$, the displacement $x = A\cos\omega t$ is a maximum at the time $t = 0$. Other initial displacements correspond to other phase constants.

The amplitude A and the phase constant ϕ of the oscillation are determined by the initial position and speed of the particle. These two initial conditions will specify A and ϕ exactly. Of course, ϕ may be increased by any integral multiple of 2π (or 360°) without altering the motion. Once the motion has started, the particle will continue to oscillate with the same amplitude and phase constant unless other forces disturb the system.

In Fig. 14–6 we plot the displacement x versus the time t for several simple harmonic motions described by Eq. 14–8. Three comparisons are made. In Fig. 14–6a, I and II have the same amplitude and frequency but differ in phase by $\phi = \pi/4$ or 45°. In Fig. 14–6b, I and III have the same frequency and phase constant but differ in amplitude by a factor of 2. In Fig. 14–6c, I and IV have the same amplitude and phase constant but differ in frequency by a factor $\frac{1}{2}$ or in period by a factor 2. Study these curves carefully to become familiar with the terminology used in simple harmonic motion.

Another distinctive feature of simple harmonic motion is the relation between the displacement, the velocity, and the acceleration of the oscillating particle. Let

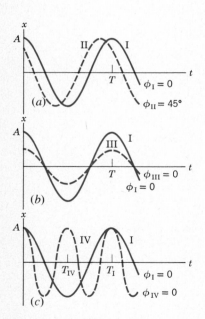

Figure 14–6 Several solutions of the simple harmonic oscillator equation. (*a*) Both solutions have the same amplitude and period but differ in phase by 45°. (*b*) Both have the same period and phase constant but differ in amplitude by a factor of 2. (*c*) Both have the same phase constant and amplitude but differ in period by a factor of 2.

us compare these quantities for curve I of Fig. 14–6, which is typical. In Fig. 14–7 we plot separately the displacement x versus the time t, the velocity $v = dx/dt$ versus the time t, and the acceleration $a = dv/dt = d^2x/dt^2$ versus the time t. The equations of these curves are

$$x = A \cos (\omega t + \phi),$$

$$v = \frac{dx}{dt} = -\omega A \sin (\omega t + \phi),$$ (14–13)

$$a = \frac{dv}{dt} = -\omega^2 A \cos (\omega t + \phi).$$

For the case plotted we have taken $\phi = 0$. The units and scale of displacement, velocity, and acceleration are omitted for simplicity of comparison. Notice that (see Eq. 14–13) the maximum displacement is A, the maximum speed is ωA, and the maximum acceleration is $\omega^2 A$.

When the displacement is a maximum in either direction, the speed is zero because the velocity must now change its direction. The acceleration at this instant, like the restoring force, has a maximum value but is directed opposite to the displacement. When the displacement is zero, the speed of the particle is a maximum and the acceleration is zero, corresponding to a zero restoring force. The speed increases as the particle moves toward the equilibrium position and then decreases as it moves out to the maximum displacement, just as for a pendulum bob.

In Fig. 14–8 we show the instantaneous values of **x**, **v**, and **a** at four instants in the motion of a particle oscillating at the end of a spring.

Displacement, velocity, and acceleration in SHM

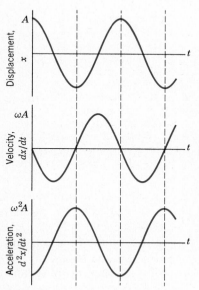

Figure 14–7 The relations between displacement, velocity, and acceleration in simple harmonic motion. The phase constant ϕ is zero in this particular case since the displacement is maximum at $t = 0$; see Eq. 14-8.

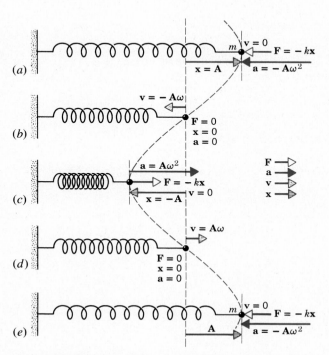

Figure 14–8 The force acting on, and the acceleration, velocity and displacement of a block of mass m undergoing simple harmonic motion. For simplicity, the block is represented by a particle. Compare carefully with Figs. 14–4 and 14–7.

Example 1 The horizontal spring of Fig. 14–4 is found to be stretched 3.3 in. from its equilibrium position when a force of 0.75 lb acts on it. Then a 1.5-lb body is attached to the end of the spring and is pulled 4.5 in. along a horizontal frictionless table from the equilibrium position. The body is then released and executes simple harmonic motion.

(a) What is the force constant of the spring?

A force of 0.75 lb *on* the spring produces a displacement of 0.28 ft. Hence,

$$k = \frac{F}{x} = \frac{0.75 \text{ lb}}{0.28 \text{ ft}} = 2.7 \text{ lb/ft}.$$

Why didn't we use $k = -F/x$ here?

(b) What is the force exerted by the spring on the 1.5-lb body just before it is released?

The spring is stretched 4.5 in. or 0.38 ft. Hence, the force exerted *by* the spring is

$$F = -kx = -(2.7 \text{ lb/ft})(0.38 \text{ ft}) = -1.0 \text{ lb}.$$

The minus sign indicates that the force is directed opposite to the displacement.

(c) What is the period of oscillation after release?

$$T = 2\pi \sqrt{\frac{m}{k}} = 2\pi \sqrt{\frac{(1.5 \text{ lb})/(32 \text{ ft/s}^2)}{(2.7 \text{ lb/ft})}} = 0.83 \text{ s}.$$

This corresponds to a frequency $\nu (= 1/T)$ of 1.2 Hz and to an angular frequency $\omega (= 2\pi\nu)$ of 7.6 rad/s.

(d) What is the amplitude of the motion?

The maximum displacement corresponds to zero kinetic energy and a maximum potential energy. This is the initial condition before release, so that the amplitude is the initial displacement of 4.5 in. Hence, $A = 0.38$ ft.

(e) What is the maximum speed of the vibrating body?

From Eq. 14–13,

$$v_{\max} = \omega A = (7.6 \text{ rad/s})(0.38 \text{ ft}) = 2.9 \text{ ft/s}.$$

The maximum speed occurs at the equilibrium position, where $x = 0$. This value is achieved twice in each period, the velocity being -2.9 ft/s when the body passes through $x = 0$ after release and $+2.9$ ft/s when the body passes through $x = 0$ on the return trip.

(f) What is the maximum acceleration of the body?

From Eq. 14–13,

$$a_{\max} = \omega^2 A = (7.6 \text{ rad/s})^2(0.38 \text{ ft}) = 20 \text{ ft/s}^2.$$

The maximum acceleration occurs at the ends of the path where $x = \pm A$ and $v = 0$. Hence, $a = -20 \text{ ft/s}^2$ at $x = +A$

and $a = +20 \text{ ft/s}^2$ at $x = -A$, the acceleration and displacement being oppositely directed.

(g) Express the displacement x as a function of time.

From Eq. 14–13 the general form is

$$x = A \cos(\omega t + \phi),$$

for which we have already found $A (= 0.38 \text{ ft})$ and $\omega (= 7.6 \text{ rad/s})$. It remains to find ϕ. From the conditions of release we must have $x = +A$ at $t = 0$, so $\cos \phi = +1$. Thus $\phi = 0$ and the displacement equation is

$$x = (0.38 \text{ ft}) \cos [(7.6 \text{ rad/s})t].$$

Example 2 An object of mass m is attached to a spring as in Fig. 14–4 and is oscillating with simple harmonic motion. We observe it at an arbitrary time $t = 0$ and note that its displacement x_0 from its center of oscillation, its velocity v_0, and its acceleration a_0 have the values -8.0 cm, -0.90 m/s, and $+50$ m/s^2, respectively. Find (a) the frequency ν, (b) the phase constant ϕ, and (c) the amplitude A for the motion.

We note that, because the velocity is not zero, the amplitude must be more than 8.0 cm. Putting $t = 0$ in Eq. 14–13 leads to

$$x_0 = +A \cos \phi,$$
$$v_0 = -\omega A \sin \phi,$$

and

$$a_0 = -\omega^2 A \cos \phi.$$

We see that we have three unknowns, ω, ϕ, and A, but with three equations we should be able to solve for them.

(a) If we divide the third equation by the first, we find that

$$\omega = \sqrt{-\frac{a_0}{x_0}} = \sqrt{-\frac{(+50 \text{ m/s}^2)}{(-0.080 \text{ m})}} = 25 \text{ rad/s}.$$

This corresponds to a frequency $\nu (= \omega/2\pi)$ of 4.0 Hz and a period $T (= 1/\nu)$ of 0.25 s.

(b) If we divide the second equation by the first we have

$$\tan \phi = -\frac{v_0}{\omega x_0} = -\frac{(-0.90 \text{ m/s})}{(25 \text{ rad/s})(-0.080 \text{ m})} = -0.45,$$

so that

$$\phi = \tan^{-1}(-0.45) = +156°.$$

(c) From the first equation we find that

$$A = \frac{x_0}{\cos \phi} = \frac{(-8.0 \text{ cm})}{\cos 156°} = 8.8 \text{ cm}.$$

14–4 Energy Considerations in Simple Harmonic Motion

Equation 14–2 tells us that for harmonic motion, including simple harmonic motion, in which no dissipative forces act, the total mechanical energy $E (= K + U)$ is conserved (remains constant). We can now study this in more detail for the special case of simple harmonic motion, for which the displacement is given by

$$x = A \cos(\omega t + \phi). \qquad [14\text{–}8]$$

The potential energy U at any instant is given by (Eq. 14-3)

Potential energy in SHM

$$U = \tfrac{1}{2}kx^2$$
$$= \tfrac{1}{2}kA^2 \cos^2(\omega t + \phi). \qquad (14-14)$$

The potential energy has a maximum value of $\tfrac{1}{2}kA^2$. During the motion the potential energy varies between zero and this maximum value, as the curves in Fig. 14-9a and 14-9b show.

The kinetic energy K at any instant is $\tfrac{1}{2}mv^2$. Using the relations

$$v = \frac{dx}{dt} = -\omega A \sin(\omega t + \phi)$$

and

$$\omega^2 = \frac{k}{m},$$

we obtain

Kinetic energy in SHM

$$K = \tfrac{1}{2}mv^2$$
$$= \tfrac{1}{2}m\omega^2 A^2 \sin^2(\omega t + \phi)$$
$$= \tfrac{1}{2}kA^2 \sin^2(\omega t + \phi). \qquad (14-15)$$

The kinetic energy, therefore, has a maximum value of $\tfrac{1}{2}kA^2$ or $\tfrac{1}{2}m(\omega A)^2$, in agreement with the maximum speed ωA noted earlier. During the motion the kinetic energy varies between zero and this maximum value, as shown by the curves in Fig. 14-9a and 14-9b.

Total mechanical energy

The total mechanical energy is the sum of the kinetic and the potential energy. Using Eqs. 14-14 and 14-15, we obtain

$$E = K + U = \tfrac{1}{2}kA^2 \sin^2(\omega t + \phi) + \tfrac{1}{2}kA^2 \cos^2(\omega t + \phi) = \tfrac{1}{2}kA^2. \quad (14-16)$$

We see that the total mechanical energy is constant, as we expect, and has the value $\tfrac{1}{2}kA^2$. At the maximum displacement the kinetic energy is zero, but the potential energy has the value $\tfrac{1}{2}kA^2$. At the equilibrium position the potential energy is zero, but the kinetic energy has the value $\tfrac{1}{2}kA^2$. At other positions the kinetic and potential energies each contribute energy whose sum is always $\tfrac{1}{2}kA^2$. This constant total energy is shown in Fig. 14-9a and 14-9b. It is clear from Fig. 14-9a that the *average* kinetic energy for the motion during one period is exactly equal to the *average* potential energy and that each of these average quantities is $\tfrac{1}{4}kA^2$.

Equation 14-16 can be written quite generally as

$$K + U = \tfrac{1}{2}mv^2 + \tfrac{1}{2}kx^2 = \tfrac{1}{2}kA^2. \qquad (14-17)$$

From this relation we obtain $v^2 = (k/m)(A^2 - x^2)$ or

$$v = \frac{dx}{dt} = \pm\sqrt{\frac{k}{m}(A^2 - x^2)}. \qquad (14-18)$$

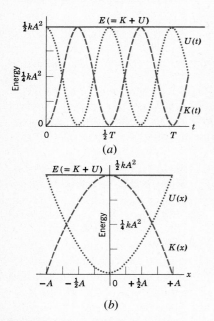

Figure 14-9 The potential energy U (.........), the kinetic energy K, (———), and the total energy E (———) of a simple harmonic oscillator (a) as a function of time (compare with Fig. 8-4); (b) as a function of displacement from the equilibrium position.

This relation shows clearly that the speed is a maximum at the equilibrium position $x = 0$ and zero at the maximum displacement $x = A$. In fact, we can start from the conservation of energy principle, Eq. 14-17 (in which $\tfrac{1}{2}kA^2 = E$), and by integration of Eq. 14-18 obtain the displacement as a function of time. The result is identical with Eq. 14-8, which we deduced from the differential equation of the motion, Eq. 14-6. (See Problem 33.)

Example 3 Consider further the mass-spring system of Example 1, oscillating in simple harmonic motion. Recall that its displacement from equilibrium as a function of time is given there as

$$x = (0.38 \text{ ft}) \cos [(7.6 \text{ rad/s})t].$$

(a) Find the velocity when the body has moved half-way from its release point toward the central equilibrium position.
Its position then will be given by

$$x = \tfrac{1}{2}A = \tfrac{1}{2}(0.38 \text{ ft}) = 0.19 \text{ ft}.$$

Let us find the time at which the body is in this position. From the above displacement equation we have

$$0.19 \text{ ft} = (0.38 \text{ ft}) \cos [(7.6 \text{ rad/s})t]$$

or

$$(7.6 \text{ rad/s})t = \cos^{-1} 0.50 = 60°,$$

so that

$$t = \left(\frac{60°}{7.6 \text{ rad/s}}\right)\left(\frac{1 \text{ rad}}{57.3°}\right) = 0.14 \ s.$$

To find the velocity let us take the time derivative of the displacement equation, or

$$v = dx/dt = (-2.9 \text{ ft/s}) \sin [(7.6 \text{ rad/s})t].$$

For $t = 0.14$ s (or, which is the same thing, for 7.6 $t = 60°$) we have

$$v = (-2.9 \text{ ft/s}) \sin 60° = -2.5 \text{ ft/s}.$$

(b) Compute the kinetic and the potential energies for $x = \tfrac{1}{2}A = 0.19$ ft.
We have

$$K = \tfrac{1}{2}mv^2 = \tfrac{1}{2}\left(\frac{1.5 \text{ lb}}{32 \text{ ft/s}^2}\right)(-2.5 \text{ ft/s})^2 = 0.15 \text{ ft} \cdot \text{lb}$$

and

$$U = \tfrac{1}{2}kx^2 = \tfrac{1}{2}(2.7 \text{ lb/ft})(0.19 \text{ ft})^2 = 0.049 \text{ ft} \cdot \text{lb}.$$

in which the values of m ($= W/g$) and of k are given in Example 1.

(c) Compute the total energy E of the oscillating system.
Since the total energy is conserved we can compute it at any stage of the motion. Using previous results we do so at three different displacements:

$$x = 0: E = K_{max} + 0 = \tfrac{1}{2}mv_{max}^2 = \tfrac{1}{2}\left(\frac{1.5 \text{ lb}}{32 \text{ ft/s}^2}\right)(2.9 \text{ ft/s})^2$$
$$= 0.20 \text{ ft} \cdot \text{lb}$$

$$x = \tfrac{1}{2}A: E = K + U = 0.15 \text{ ft} \cdot \text{lb} + 0.049 \text{ ft} \cdot \text{lb} = 0.20 \text{ ft} \cdot \text{lb}$$

$$x = A: E = 0 + U_{max} = \tfrac{1}{2}kA^2 = \tfrac{1}{2}(2.7 \text{ lb/ft})(0.38 \text{ ft})^2$$
$$= 0.20 \text{ ft} \cdot \text{lb}.$$

As expected, all three calculations yield the same result.

14–5 Applications of Simple Harmonic Motion

We have discussed at some length the simple harmonic motion of a mass-spring system. Two additional physical systems that move in this same manner will now be considered. Others will be discussed from time to time throughout the text.

The Simple Pendulum. A simple pendulum is an idealized body consisting of a point mass suspended by a light inextensible cord. When pulled to one side of its equilibrium position and released, the pendulum swings in a vertical plane under the influence of gravity. The motion is periodic and oscillatory. We wish to determine the period of the motion.

Figure 14–10 shows a pendulum of length l, particle mass m, making an angle θ with the vertical. The forces acting on m are $m\mathbf{g}$, the gravitational force, and $\mathbf{T}$, the tension in the cord. Choose axes tangent to the circle of motion and along the radius. Resolve $m\mathbf{g}$ into a radial component of magnitude $mg \cos \theta$ and a tangential component of magnitude $mg \sin \theta$. The radial components of the forces supply the necessary centripetal acceleration to keep the particle moving on a circular arc. The tangential component is the restoring force acting on m tending to return it to the equilibrium position. Hence, the restoring force is

$$F = -mg \sin \theta.$$

Notice that the restoring force is not proportional to the angular displacement θ but to $\sin \theta$ instead. The resulting motion is, therefore, not simple harmonic.

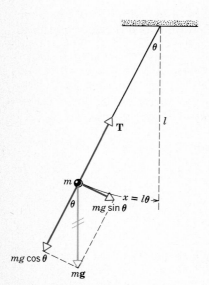

Figure 14–10 The forces acting on a simple pendulum are the tension $\mathbf{T}$ in the string and the weight $m\mathbf{g}$ of mass m. The magnitudes of the radial and tangential components of $m\mathbf{g}$ are labeled.

However, if the angle θ is small, $\sin \theta$ is very nearly equal to θ in radians. For example, if $\theta = 5° = 0.0873$ radian, $\sin \theta = 0.0872$; even at $\theta = 15°$ the two figures differ only by 1.1%. The displacement along the arc is $x = l\theta$, and for small angles this is nearly straight-line motion. Hence, assuming

$$\sin \theta \cong \theta,$$

we obtain

$$F = -mg\theta = -mg\,\frac{x}{l} = -\frac{mg}{l}\,x.$$

For small displacements, therefore, the restoring force is proportional to the displacement and is oppositely directed. This is exactly the criterion for simple harmonic motion. The constant mg/l represents the constant k in $F = -kx$. Check the dimensions of k and mg/l. The period of a simple pendulum when its amplitude is small is

$$T = 2\pi \sqrt{\frac{m}{k}} = 2\pi \sqrt{\frac{m}{mg/l}} \quad \text{or} \quad T = 2\pi \sqrt{\frac{l}{g}}. \qquad (14-19)$$

Notice that the period is independent of the mass of the suspended particle.

The Torsional Pendulum. In Fig. 14–11 we show a disk suspended by a wire attached to the center of mass of the disk. The wire is securely fixed to a solid support and to the disk. At the equilibrium position of the disk a radial line is drawn from its center to P, as shown. If the disk is rotated in a horizontal plane to the radial position Q, the wire will be twisted. The twisted wire will exert a torque on the disk tending to return it to the position P. This is a restoring torque. For small twists the restoring torque is found to be proportional to the amount of twist, or the angular displacement (Hooke's law), so that

$$\tau = -\kappa\theta. \qquad (14-20)$$

Here κ is a constant that depends on the properties of the wire and is called the *torsional constant*. The minus sign shows that the torque is directed opposite to the angular displacement θ. Equation 14–20 is the condition for *angular simple harmonic motion*.

The equation of motion for such a system is

$$\tau = I\alpha = I\,\frac{d\omega}{dt} = I\,\frac{d^2\theta}{dt^2},$$

so that, upon using Eq. 14–20, we obtain

$$-\kappa\theta = I\,\frac{d^2\theta}{dt^2}$$

or

$$\frac{d^2\theta}{dt^2} = -\left(\frac{\kappa}{I}\right)\theta. \qquad (14-21)$$

Notice the similarity between Eq. 14–21 for simple angular harmonic motion and Eq. 14–7 for simple linear harmonic motion. In fact, the equations are mathematically identical. We have simply substituted angular displacement θ for linear displacement x, rotational inertia I for mass m, and torsional constant κ for force constant k. By substituting these correspondences, we find the solution of Eq. 14–21, therefore, to be a simple harmonic oscillation in the angular coordinate θ, namely,

$$\theta = \theta_m \cos (\omega t + \phi). \qquad (14-22)$$

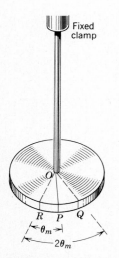

Figure 14–11 The torsional pendulum. The line drawn from the center to P oscillates between Q and R, sweeping out an angle $2\theta_m$, where θ_m is the (angular) amplitude of the motion.

Here, θ_m is the maximum angular displacement, that is, the amplitude of the angular oscillation. In Fig. 14–11 the disk oscillates about the equilibrium position $\theta = 0$ (line OP), the total angular range being $2\theta_m$ (from OQ to OR).

The period of the oscillation by analogy with Eq. 14–10 is

Period of a torsional pendulum

$$T = 2\pi \sqrt{\frac{I}{\kappa}}. \tag{14–23}$$

If κ is known and T is measured, the rotational inertia I about the axis of rotation of any oscillating rigid body can be determined. If I is known and T is measured, the torsional constant κ of any sample of wire can be determined.

Many laboratory instruments involve torsional oscillations, notably the galvanometer. The Cavendish balance is a torsional pendulum (Chapter 15). The balance wheel of a watch is another example of angular harmonic motion, the restoring torque here being supplied by a spiral hairspring.

Example 4 A thin rod of mass 0.10 kg and length 0.10 m is suspended by a wire that passes through its center and is perpendicular to its length. The wire is twisted and the rod set oscillating. The period is found to be 2.0 s. When a flat body in the shape of an equilateral triangle is suspended similarly through its center of mass, the period is found to be 6.0 s. Find the rotational inertia of the triangle about this axis.

The rotational inertia of the rod is $Ml^2/12$ (see Table 12–1). Hence

$$I_{\text{rod}} = \frac{(0.10\text{ kg})(0.10\text{ m})^2}{12} = 8.3 \times 10^{-5}\text{ kg} \cdot \text{m}^2.$$

From Eq. 14–23,

$$\frac{T_{\text{rod}}}{T_{\text{triangle}}} = \left(\frac{I_{\text{rod}}}{I_{\text{triangle}}}\right)^{1/2} \quad \text{or} \quad I_{\text{triangle}} = I_{\text{rod}} \left(\frac{T_t}{T_r}\right)^2,$$

so that

$$I_{\text{triangle}} = (8.3 \times 10^{-5}\text{ kg} \cdot \text{m}^2)\left(\frac{6.0\text{ s}}{2.0\text{ s}}\right)^2 = 7.5 \times 10^{-4}\text{ kg} \cdot \text{m}^2.$$

Does the amplitude of the oscillation affect the period in these cases?

14–6 Simple Harmonic Motion and Uniform Circular Motion

Let us consider the relation between simple harmonic motion along a straight line and uniform circular motion. This relation is useful in describing many features of simple harmonic motion. It also gives a simple geometric meaning to the angular frequency ω and the phase constant ϕ. Uniform circular motion is also an example of a combination of simple harmonic motions, a phenomenon we deal with rather often in wave motion.

In Fig. 14–12 Q is the point moving around a circle of radius A with a constant angular speed of ω, expressed, say, in radians per second. P is the perpendicular projection of Q on the horizontal diameter, along the x-axis. Let us call Q the *reference point* and the circle on which it moves the *reference circle*. As the reference point revolves, the projected point P moves back and forth along the horizontal diameter. The x-component of Q's displacement is always the same as the displacement of P; the x-component of the velocity of Q is always the same as the velocity of P; and the x-component of the acceleration of Q is always the same as the acceleration of P.

Let the angle between the radius OQ and the x-axis at the time $t = 0$ be called ϕ. At any later time t, the angle between OQ and the x-axis is $(\omega t + \phi)$, the point Q moving with constant angular speed. The x-coordinate of Q at any time is, therefore,

$$x = A \cos (\omega t + \phi). \tag{14–24}$$

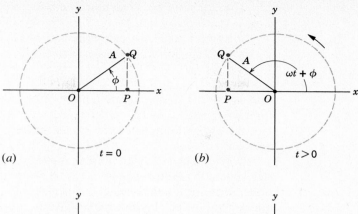

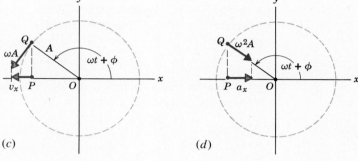

Figure 14-12 The relation of simple harmonic motion to uniform circular motion. Q moves in uniform circular motion and P in simple harmonic motion. Q has angular speed ω, P angular frequency ω. (a, b) The x-component of Q's displacement is always equal to P's displacement. (c) The x-component of Q's acceleration is always equal to P's acceleration.

Hence, the projected point P moves with simple harmonic motion along the x-axis. Therefore, *simple harmonic motion can be described as the projection along a diameter of uniform circular motion.*

The angular frequency ω of simple harmonic motion is the same as the angular speed of the reference point. The frequency of the simple harmonic motion is the same as the number of revolutions per unit time of the reference point. Hence, $\nu = \omega/2\pi$ or $\omega = 2\pi\nu$. The time for a complete revolution of the reference point is the same as the period T of the simple harmonic motion. Hence, $T = 2\pi/\omega$ or $\omega = 2\pi/T$. The phase of the simple harmonic motion, $\omega t + \phi$, is the angle that OQ makes with the x-axis at any time t (Fig. 14–12b, c, d). The angle that OQ makes with the x-axis at the time $t = 0$ (Fig. 14–12a) is ϕ, the phase constant or initial phase of the motion. The amplitude of the simple harmonic motion is the same as the radius of the reference circle.

The tangential velocity of the reference point Q has a magnitude of ωA. Hence, the x-component of this velocity (Fig. 14–12c) is

$$v_x = -\omega A \sin(\omega t + \phi).$$

This relation gives a negative v_x when Q and P are moving to the left and a positive v_x when Q and P are moving to the right. Notice that v_x is zero at the end points of the simple harmonic motion, where $\omega t + \phi$ is zero and π, as required.

The acceleration of point Q in uniform circular motion is directed radially inward and has a magnitude of $\omega^2 A$. The acceleration of the projected point P is the x-component of the acceleration of the reference point Q (Fig. 14–12d). Hence,

$$a_x = -\omega^2 A \cos(\omega t + \phi)$$

gives the acceleration of the point executing simple harmonic motion. Notice that a_x is zero at the midpoints of the simple harmonic motion, where $\omega t + \phi = \pi/2$ or $3\pi/2$, as required.

These results are all identical with those of simple harmonic motion along the x-axis; see Eqs. 14–13.

If we had taken the perpendicular projection of the reference point onto the y-axis, instead, we would have obtained for the motion of the y-projected point

$$y = A \sin (\omega t + \phi). \tag{14–25}$$

This is again a simple harmonic motion. It differs only in phase from Eq. 14–24, for if we replace ϕ by $\phi - \pi/2$, then $\cos (\omega t + \phi)$ becomes $\sin (\omega t + \phi)$. It is clear that the projection of uniform circular motion along *any* diameter gives a simple harmonic motion.

Conversely, uniform circular motion can be described as a combination of two simple harmonic motions. It is that combination of two simple harmonic motions, occuring along perpendicular lines, which have the same amplitude and frequency but differ in phase by 90°. When one component is at the maximum displacement, the other component is at the equilibrium point. If we combine these components (Eqs. 14–24 and 14–25), we obtain at once the relation

$$r = \sqrt{x^2 + y^2} = A.$$

By writing the relations for v_y and a_y (you should do this) and combining corresponding quantities, we obtain also the relations

$$v = \sqrt{v_x{}^2 + v_y{}^2} = \omega A,$$
$$a = \sqrt{a_x{}^2 + a_y{}^2} = \omega^2 A.$$

These relations correspond respectively to the magnitudes of the displacement, the velocity, and the acceleration in uniform circular motion.

It will be possible for us to analyze many complicated motions as combinations of individual simple harmonic motions. Circular motion is a particularly simple combination. In the next section we shall consider other combinations of simple harmonic motions.

Example 5 In Examples 1 and 3 we considered a body executing a horizontal simple harmonic motion. The equation of that motion (units?) was

$$x = (0.38 \text{ ft}) \cos [(7.6 \text{ rad/s})t].$$

This motion can also be represented as the projection of uniform circular motion along a horizontal diameter.

(a) Give the properties of the corresponding uniform circular motion.

The x-component of the circular motion is given by

$$x = A \cos (\omega t + \phi).$$

Therefore, the reference circle must have a radius $A = 0.38$ ft, the initial phase or phase constant must be $\phi = 0$, and the angular velocity must be $\omega = 7.6$ rad/s, in order to obtain the equation $x = (0.38 \text{ ft}) \cos [(7.6 \text{ rad/s})t]$ for the horizontal projection.

(b) From the motion of the reference point determine the time required for the body to come halfway in toward the center of motion from its initial position.

As the body moves halfway in, the reference point moves through an angle of $\omega t = 60°$ (Fig. 14–13). The angular velocity is constant at 8 rad/s, so the time required to move through 60° is

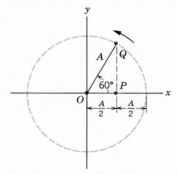

Figure 14–13 Example 5. The particles Q and P of Fig. 14–12 are shown for $\omega t = 60°$. Since ω is known, t may be found.

$$t = \frac{60°}{\omega} = \frac{\pi/3 \text{ rad}}{7.6 \text{ rad/s}} = 0.14 \text{ s}.$$

This is precisely the time we calculated directly from the equation of motion in Example 3a.

14–7 Combinations of Harmonic Motions

Often two linear simple harmonic motions at *right angles* are combined. The resulting motion is the sum of two independent oscillations. Consider first the case in which the frequencies of the vibrations are the same, such as

$$x = A_x \cos (\omega t + \phi_x),$$
$$y = A_y \cos (\omega t + \phi_y). \qquad (14\text{–}26)$$

The x- and y-motions have different amplitudes and different phase constants, however.

If the phase constants are the same so that $\phi_x = \phi_y = \phi$, the resulting motion is a straight line. This can be shown analytically, for when we eliminate t from the equations

$$x = A_x \cos (\omega t + \phi), \qquad y = A_y \cos (\omega t + \phi),$$

we obtain

$$y = \left(\frac{A_y}{A_x} \right) x.$$

This is the equation of a straight line, whose slope is A_y/A_x. In Fig. 14–14a and b we show the resultant motion for two cases, $A_y/A_x = 1$ and $A_y/A_x = 2$. In these cases both the x- and y-displacements reach a maximum at the same time and reach a minimum at the same time. They are in phase.

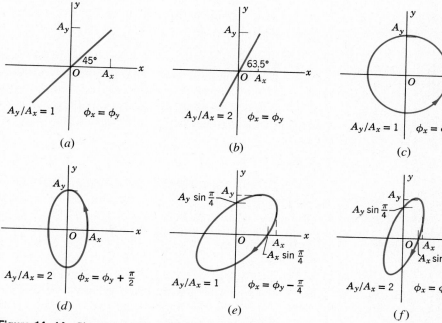

Figure 14–14 Simple harmonic motions in two dimensions. (a) The amplitudes of x and y (namely, A_x and A_y) are the same, as are their phase constants. (b) The amplitude of y is twice that of x, but the phase constants are the same. (c) The amplitudes are equal, but x leads y in phase by 90°. (d) Same as (c) except that the amplitude of y is twice that of x. (e) Equal amplitudes, but x lags y in phase by 45°. (f) Same as (e) except that the amplitude of y is twice that of x.

If the phase constants are different, the resulting motion will not be a straight line. For example, if the phase constants differ by $\pi/2$, the maximum x-displacement occurs when the y-displacement is zero and vice versa. When the amplitudes are equal, the resulting motion is circular; when the amplitudes are unequal, the resulting motion is elliptical. Two cases, $A_y/A_x = 1$ and $A_y/A_x = 2$, are shown in Fig. 14–14c and d, for $\phi_x = \phi_y + \pi/2$. The cases $A_y/A_x = 1$ and $A_y/A_x = 2$, for $\phi_x = \phi_y - \pi/4$, are shown in Fig. 14–14e and f.

All possible combinations of two simple harmonic motions at right angles having the same frequency correspond to *elliptical* paths, the circle and straight line being special cases of an ellipse. This can be shown analytically by combining Eqs. 14–26 and eliminating the time; you can show that the resulting equation is that of an ellipse. The shape of the ellipse depends only on the ratio of the amplitudes, A_y/A_x, and the *difference* in phase between the two oscillations, $\phi_x - \phi_y$. The actual motion can be either clockwise or counterclockwise, depending on which component leads in phase.

A simple way to produce such patterns is by means of an oscilloscope. In this, electrons are deflected by each of two electric fields at right angles to one another. The field strengths alternate sinusoidally with the same frequency, but their phases and amplitudes can be varied. In this way the electrons can be made to trace out the various patterns discussed above on a fluorescent screen. We can also produce these patterns mechanically by means of a pendulum swinging with small amplitude but not confined to one vertical plane. Such combinations of two simple harmonic motions at right angles having the same frequency are particularly important in the study of polarized light and alternating current circuits.

Combinations of simple harmonic motions of the same frequency in the *same direction*, but with different amplitudes and phases, are of special interest in the study of diffraction and interference of light, sound, and electromagnetic radiation. This will be discussed later in the text.

If two oscillations of *different frequencies* are combined at right angles, the resulting motion is more complicated. The motion is not even periodic unless the two component frequencies ω_1 and ω_2 are the ratio of two integers (see Problem 57). Oscillations of different frequencies in the same direction may also be combined. The treatment of this motion is particularly important in the case of sound vibrations and will be discussed in Chapter 18.

14–8 Damped Harmonic Motion

Up to this point we have assumed that no frictional forces act on the oscillator. If this assumption held strictly, a pendulum or a weight on a spring would oscillate indefinitely. Actually, the amplitude of the oscillation gradually decreases to zero as a result of friction. The motion is said to be damped by friction and is called *damped harmonic motion*. Often the friction arises from air resistance or internal forces. The magnitude of the frictional force usually depends on the speed. In most cases of interest the frictional force is proportional to the velocity of the body but directed opposite to it.

Figure 14–15 shows an idealization of a damped harmonic oscillator. We assume that all of the potential energy is concentrated in the ideal spring, which is taken to be massless and to have no internal frictional losses. The kinetic energy is taken to be concentrated in the oscillating mass and all of the internal thermal energy associated with the viscous or frictional losses is localized in the idealized "dash pot," as it is called.

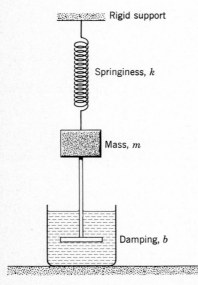

Figure 14–15 An idealized damped harmonic oscillator. A disk is attached to the mass and immersed in a fluid that exerts a damping force $-b\,dx/dt$. The elastic restoring force is $-kx$.

Rigid support

Springiness, k

Mass, m

Damping, b

The equation of motion of the damped simple harmonic oscillator is given by the second law of motion, $F = ma$, in which F is the sum of the restoring force $-kx$ and the damping force $-b\,dx/dt$. Here b, a positive constant, is a measure of the strength of the damping force. We obtain

$$F = ma,$$

or

$$-kx - b\frac{dx}{dt} = m\frac{d^2x}{dt^2}$$

or

The differential equation for damped harmonic motion

$$m\frac{d^2x}{dt^2} + b\frac{dx}{dt} + kx = 0. \qquad (14\text{–}27)$$

We consider only the case in which the damping is relatively small. The solution of Eq. 14–27 then is

Damped harmonic motion— the displacement

$$x = Ae^{-bt/2m}\cos{(\omega't + \phi)}, \qquad (14\text{–}28)$$

which is plotted in Fig. 14–16. We see that it is simply a harmonic oscillation whose amplitude $Ae^{-bt/2m}$ decreases exponentially with time. In the absence of damping, that is, if $b = 0$ in Eq. 14–28, this amplitude reduces to the constant A, as we expect. The frequency ω' in Eq. 14–28 is only slightly less than the undamped frequency $\omega\ (=\sqrt{k/m})$. For the small damping that we assume, we can put $\omega' = \omega$ with negligible error.

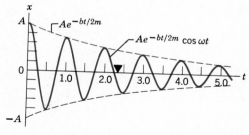

Figure 14–16 A plot of $x(t)$ for damped harmonic motion; see Eq. 14–28. For the initial conditions shown ($x = A$ for $t = 0$) $\phi = 0$ in that equation. The unit of time is the period T of the motion.

In damped harmonic motion the mechanical energy E of the oscillating system approaches zero as time increases, being transformed into internal thermal energy associated with the damping mechanism.

Example 6 For the damped harmonic oscillator of Fig. 14–15 we assume $m = 250$ g, $k = 2.0$ N/m, and a damping constant $b = 0.070$ kg/s. Assume that, as is indeed the case here, the damping is small enough so that we can take ω' in Eq. 14–28 to be given by $\sqrt{k/m}$, the frequency of the undamped motion.

(a) What is the period T of the motion?

We have

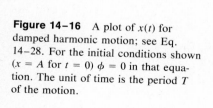

$$T = 2\pi\sqrt{\frac{m}{k}} = 2\pi\sqrt{\frac{0.25\text{ kg}}{2.0\text{ N/m}}} = 2.2\text{ s.}$$

(b) What time τ does it take for the amplitude of the motion to drop to half its initial value? How many periods of oscillation does this correspond to?

The amplitude factor in Eq. 14–28 is $Ae^{-bt/2m}$ and its initial value, corresponding to $t = 0$, is just A. We can find τ from

$$Ae^{-b\tau/2m} = \tfrac{1}{2}A.$$

Taking natural logarithms of each side gives, after rearrangement,

$$\tau = \frac{2m\ln 2}{b} = \frac{(2)(0.25\text{ kg})(0.693)}{(0.070\text{ kg/s})} = 5.0\text{ s.}$$

This is (5.0/2.2) or 2.3 periods of oscillation. The curve of Fig. 14–16 is drawn to the specifications of this Example, the half-amplitude time τ being marked by the filled triangle on the horizontal axis.

14–9 Forced Oscillations and Resonance

Forced oscillations

Up to this point we have discussed only the *natural oscillations* of a system, those that occur, for instance, when the system mass is displaced and then released. A different situation arises if the mass is subject to an oscillating external force. The system then responds with *forced oscillations*. As examples, a bridge vibrates under the influence of marching soldiers, the housing of a motor vibrates owing to periodic impulses from an irregularity in the shaft, and a tuning fork vibrates when exposed to the periodic force of a sound wave.

The differential equation for forced oscillations

To describe such motion the force equation again must be modified. For a mass-spring system we add to Eq. 14–27 a term describing the external force; for simplicity let this be given by $F_m \cos \omega t$. Here F_m is the maximum value of the external force and ω is its angular frequency. The equation of motion becomes

$$m \frac{d^2x}{dt^2} + b \frac{dx}{dt} + kx = F_m \cos \omega t \qquad (14-29)$$

The system responds to the driving force by oscillating with a constant amplitude and with the same frequency as the driving force, or

Forced oscillations—the displacement

$$x = x_m \sin (\omega t + \phi). \qquad (14-30)$$

The amplitude of the motion, x_m, depends on both the natural frequency of the system (which we now label ω_0) and on the frequency ω of the driving force.

Resonance

There is a characteristic value of the driving frequency ω at which the amplitude of oscillation is a maximum. This is called the *resonant frequency* and the corresponding state of the system is called *resonance*. As the damping gets smaller, the resonant frequency approaches the natural undamped frequency ω_0 and the amplitude can become very large. In effect, the driving force finds it easier to establish the oscillations if its frequency is close to ω_0 and also if there is little damping.

Figure 14–17 shows three resonance curves representing the amplitudes of vibration of the damped harmonic oscillator of Fig. 14–15. We imagine the rigid support shown in that figure replaced by a vibrating support whose angular driving frequency ω can be varied at will throughout a range of values. The amplitude of the vibrations induced in the oscillating body of mass m is plotted as a function of

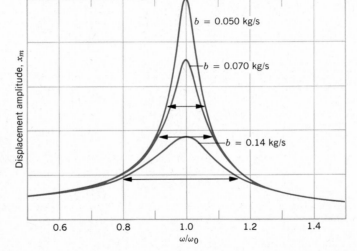

Figure 14–17 The amplitudes of forced vibrations for the mass-spring oscillator of Example 6, for three levels of damping. Notice that as the damping decreases (that is, as $b \to 0$) the amplitude of the resonance curve increases and the curve becomes sharper. The sharpness of resonance is measured by the horizontal arrow spanning each curve at its half-maximum point. Note that the curves are not symmetrical.

ω/ω_0, where ω_0 $(=\sqrt{k/m})$ is the natural frequency of oscillation of the undamped system.

Effect of damping

The intermediate curve of Fig. 14–17 is drawn for the numerical parameters assumed in Example 6; the upper and lower curves correspond to the oscillator of that example but with (a) less damping ($b = 0.050$ kg/s) and (b) more damping ($b = 0.14$ kg/s), respectively. Notice how the amplitude at resonance depends on the amount of damping, with the resonance amplitude increasing as the damping decreases.

If an oscillator had no damping at all ($b = 0$) the resonant amplitude of its oscillations would become infinitely great. Some friction is always present in real oscillations, however, so that the resonant amplitude will always remain finite but may become exceedingly large; often it will exceed the elastic limit and sometimes even reach the breaking point of an element of the system. That happened in the case of the Tacoma Bridge disaster (see Fig. 14–18). Of course these complex oscillating systems are much more difficult to analyze than our simple mass-spring system. An airplane, for example, has many natural frequencies of vibration; for safe travel these must not correspond to the frequencies of the engines that cause the forced vibrations. Vibrations in real systems is a complex and important study in mechanical engineering.

Figure 14–18 On July 1, 1940, the Tacoma Narrows Bridge at Puget Sound, Washington, was completed and opened to traffic. Just four months later a mild gale blowing across the bridge produced a fluctuating force in resonance with the natural frequency of the structure. This set the bridge oscillating with such a large amplitude that the structure rapidly weakened. In just a few hours the main span broke up, ripping loose from the cables and crashing into the water below. Many other bridges were subsequently redesigned to make them aerodynamically stable.

REVIEW GUIDE AND SUMMARY

Hooke's law and simple harmonic motion

If a particle moves under the influence of a restoring force given by

$$F = -kx, \qquad [14\text{–}4]$$

we say that the particle exhibits *simple harmonic motion*. Here x is the displacement of the particle from its equilibrium position and k is called the force constant.

The displacement equation

From Newton's second law we show that the displacement $x(t)$ is given by

$$x = A \cos(\omega t + \phi), \qquad [14\text{–}8]$$

in which A is the amplitude, ω is the angular frequency, and ϕ is the phase constant. Figure 14–6 shows how various choices for these three constants affect the motion.

In deriving the displacement equation above we discovered that the angular frequency ω ($= 2\pi\nu = 2\pi/T$) is related to the mass m of the particle and the force constant k by

The angular frequency

$$\omega = \sqrt{\frac{k}{m}}. \qquad [14\text{–}9]$$

Note that ω is measured in radians per second, whereas the frequency ν is in cycles per second or hertz.

The velocity and the acceleration

Differentiating Eq. 14–8 leads to the expressions for the velocity $v(t)$ and the acceleration $a(t)$ displayed in Eq. 14–13 and Fig. 14–17. Examples 1 and 2 analyze in detail the simple harmonic motion of a body of mass m attached to a spring of force constant k.

The energy

A particle undergoing simple harmonic motion has, at any given point in its cycle, both a kinetic energy $K (= \frac{1}{2}mv^2)$ and a potential energy $U (= \frac{1}{2}kx^2)$. If no friction is present the *mechanical energy* $E (= K + U)$ will remain constant, even though K and U vary throughout the cycle. We showed that

$$E = K_{\max} = U_{\max} = \tfrac{1}{2}kA^2.$$

Figure 14–9 and Example 3 analyze the situation.

Pendulums

Besides the oscillating mass-spring system of Fig. 14–4, we also analyzed the *simple pendulum* of Fig. 14–10 and the *torsion pendulum* of Fig. 14–11. Their periods of oscillation were shown to be

$$T = 2\pi \sqrt{\frac{l}{g}} \quad \text{and} \quad T = 2\pi \sqrt{\frac{I}{\kappa}}, \qquad [14\text{–}19,\ 14\text{–}23]$$

respectively.

Simple harmonic motion and uniform circular motion

Consider a particle moving with constant angular speed ω in a circle of radius A. The motion of the *projection* of this particle on any diameter of the circle turns out to be simple harmonic. Figure 14–12 shows that not only the displacement x but also the velocity v and the acceleration a of the particle moving in the reference circle project to the values expected for simple harmonic motion.

Two simple harmonic motions combined

We may ask how a particle moves if it is subjected to two simple harmonic motions simultaneously. The text considers only cases in which the separate motions (1) have the same angular frequency ω and (2) act at right angles to each other. As Fig. 14–14 shows, the resultant motion is an ellipse, which degenerates into a straight line if the two phase constants are equal and into a circle if (1) the amplitudes are equal and also (2) the phase constants differ by $\pm 90°$.

Damped harmonic oscillation

The mechanical energy E in real oscillating systems is drained away steadily, as time goes on, into internal thermal energy. This *damping* is caused by the action of frictionlike forces, which we assume are proportional to the velocity in magnitude but opposite in direction. The displacement $x(t)$ is given by Eq. 14–28 (see also Fig. 14–16 and Example 6), in which the constant b is a measure of the strength of the damping force.

Forced oscillations and resonance

It is possible to impose on the damped harmonic oscillator an external oscillatory force $F_m \cos \omega t$, where ω may have any chosen value. The system responds at this same angular frequency ω. Its amplitude of oscillation increases as the damping is made smaller (that is, as $b \to 0$) and as the external frequency is brought closer to the natural oscillation frequency of the system which, in this context, we label ω_0. Figure 14–17 displays this phenomenon of resonance for the oscillator of Example 6.

QUESTIONS

1. Give some examples of motions that are approximately simple harmonic. Why are motions that are exactly simple harmonic rare?

2. Is Hooke's law obeyed, even approximately, by a diving board? A trampoline? A coiled spring made of lead wire?

3. A spring has a force constant k, and a mass m is suspended from it. The spring is cut in half and the same mass is suspended from one of the halves. How are the frequencies of oscillation, before and after the spring is cut, related?

4. An unstressed spring has a force constant k. It is stretched by a weight hung from it to an equilibrium length well within the elastic limit. Does the spring have the same force constant k for displacements from this new equilibrium position?

5. Suppose that we have a block of unknown mass and a spring of unknown force constant. Show how we can predict the period of oscillation of this block-spring system simply by measuring the extension of the spring produced by attaching the block to it.

6. Will the frequency of oscillation of a torsional pendulum change if you take it to the moon? . . . a simple pendulum? . . . a mass-spring oscillator? . . . a physical pendulum, such as a wooden plank suspended between knife-edge supports?

7. How can an astronaut in an orbiting space shuttle measure his weight? After all, he is "weightless."

8. Any real spring has mass. If this mass is taken into account, explain qualitatively how this will affect the period of oscillation of a mass-spring system.

9. How are each of the following properties of a simple harmonic oscillator affected by doubling the amplitude: period, force constant, total mechanical energy, maximum velocity, maximum acceleration?

10. What changes could you make in a harmonic oscillator that would double the maximum speed of the oscillating mass?

11. Could we ever construct a true simple pendulum?

12. Could standards of mass, length, and time be based on properties of a pendulum? Explain.

13. What would happen to the motion of an oscillating system if the sign of the force term, $-kx$ in Eq. 14-4, were changed?

14. Predict by qualitative arguments whether a pendulum oscillating with large amplitude will have a period longer or shorter than the period for oscillations with small amplitude. (Consider extreme cases.)

15. How is the period of a pendulum affected when its point of suspension is (a) moved horizontally in the plane of the oscillation with acceleration a; (b) moved vertically upward with acceleration a; (c) moved vertically downward with acceleration $a < g$. Which case, if any, applies to a pendulum mounted on a cart rolling down an inclined plane?

16. How can a pendulum be used so as to trace out a sinusoidal curve?

17. What component simple harmonic motions would give a figure 8 as the resultant motion?

18. Why are damping devices often used on machinery? Give an example.

19. Give some examples of common phenomena in which resonance plays an important role.

20. In forced damped harmonic motion, is the damping ever useful?

21. In Fig. 14-17, what value does the amplitude of the forced oscillations approach as the driving frequency ω approaches zero? . . . as it approaches infinity? (Hint: see Problem 30.)

EXERCISES

Section 14-1 Oscillations

1. A particle moving along the x-axis is under the influence of the potential energy function $U(x) = 10x^2 - 40x + 60$, where U is in joules and x is in meters. (a) Make a sketch of $U(x)$ as a function of x. (b) About what value(s) of x is oscillation possible? (c) What is the minimum possible value for the total energy?

Section 14-2 The Simple Harmonic Oscillator

2. A (hypothetical) large slingshot is stretched 1.0 m to launch a 100-g projectile with speed sufficient to escape from the earth (11.2 km/s). (a) What is the force constant of the device, if all of the potential energy is converted to kinetic energy? (b) Assume that an average person can exert a force of 200 N. How many people are required to stretch the slingshot?

Section 14-3 Simple Harmonic Motion

3. An oscillating mass-spring system takes 0.75 s to begin repeating its motion. Find (a) the period, (b) the frequency in Hz, and (c) the angular frequency in radians per second.

4. An oscillator consists of a block of mass 0.50 kg connected to a spring. When set into oscillation with amplitude 35 cm, it is observed to repeat its motion every 0.50 s. Find (a) the period, (b) the frequency, (c) the angular frequency, (d) the spring constant, (e) the maximum speed, and (f) the maximum force exerted on the block.

5. A 4.0-kg block extends a spring 16 cm from its unstretched position. The block is removed and a 0.50-kg body is hung from the same spring. If the spring is then stretched and released, what is its period of oscillation?

6. A 20-N weight is hung from the bottom of a vertical spring, causing the spring to stretch 20 cm. This spring is now placed horizontally on a frictionless table. One end of it is held fixed and the other end is attached to a 5.0-N weight. The weight is then moved, stretching the spring an additional 10 cm, and released from rest. (a) What is the spring constant? (b) What is the period of oscillation?

7. A 2.0-kg block hangs from a spring. A 300-g body hung below the block stretches the spring 2.0 cm farther. If the 300-g body is removed and the block is set into oscillation, find the period of motion.

8. Find the maximum displacement of a 1.0×10^{-20}-kg particle vibrating with simple harmonic motion with a period of 1.0×10^{-5} s and a maximum speed of 1.0×10^3 m/s.

9. A 50-g mass is attached to the bottom of a vertical spring and set vibrating. If the maximum speed of the mass is 15 cm/s and the period is 0.50 s, find (a) the spring constant k of the spring, (b) the amplitude of the motion, and (c) the frequency of oscillation.

10. A particle executes linear harmonic motion about the point $x = 0$. At $t = 0$ it has displacement $x = 0.37$ cm and zero velocity. The frequency of the motion is 0.25 Hz. Determine (a) the period, (b) the angular frequency, (c) the amplitude, (d) the displacement at time t, (e) the velocity at time t, (f) the maximum speed, (g) the maximum acceleration, (h) the displacement at $t = 3.0$ s, and (i) the speed at $t = 3.0$ s.

11. At a certain harbor, the tides cause the ocean surface to rise and fall in simple harmonic motion, with a period of 12.5 h. How long does it take for the water to fall from its maximum height to one-half its maximum height above its average (equilibrium) level?

12. A speaker diaphragm is vibrating in simple harmonic motion with a frequency of 440 Hz and a maximum displacement of 0.75 mm. What are the maximum (a) speed and (b) acceleration of this diaphragm?

13. In an electric shaver, the blade moves back and forth over a distance of 2.0 mm. The motion is simple harmonic, with frequency 120 Hz. Find (a) the amplitude, (b) the maximum blade speed, and (c) the maximum blade acceleration.

Section 14-4 Energy Considerations in Simple Harmonic Motion

14. Find the mechanical energy of a mass-spring system having a force constant of 1.3 N/cm and an amplitude of 2.4 cm.

15. An oscillating mass-spring system has a mechanical energy of 1.0 J, amplitude of 0.10 m, and maximum speed of 1.2 m/s. Find (a) the force constant of the spring, (b) the mass, and (c) the frequency of oscillation.

16. A massless spring of force constant 19 N/m hangs vertically. A body of mass 0.20 kg is attached to its free end and then released. Assume that the spring was unstretched before the body was released. Find (a) how far below the initial position the body descends, (b) the frequency, and (c) the amplitude of the resulting motion, assumed to be simple harmonic.

17. A vertical spring stretches 9.81 cm when a 1.0-kg block is hung from its end. This block is then displaced an additional 5.0 cm downward and released from rest. Find (a) the period, (b) the frequency, (c) the amplitude, (d) the total energy, and (e) the maximum speed.

Section 14-5 Applications of Simple Harmonic Motion

18. What is the length of a simple pendulum whose period is 1.00 s at a point where $g = 32.2$ ft/s²?

19. A simple pendulum of length 1.00 m makes 100 complete oscillations in 204 s at a certain location. What is the acceleration of gravity at this point?

20. An 8.0-kg solid sphere with a 15-cm radius is suspended by a vertical wire attached to the ceiling of a room. A torque of 0.20 N·m is required to twist the sphere through an angle of 1.0 rad. What is the period of oscillation when the sphere is released from this position?

21. A flat uniform circular disk has a mass of 3.0 kg and a radius of 0.70 m. It is suspended in a horizontal plane by a vertical wire attached to its center. If the disk is rotated 2.5 rad about the wire, a torque of 0.060 N·m is required to maintain the disk in this position. Calculate (a) the rotational inertia of the disk about the wire, (b) the torsional constant κ, and (c) the angular frequency of this torsional pendulum.

22. An engineer wants to find the rotational inertia of an odd-shaped object of mass 10 kg about an axis through its center of mass. He supports the object with a wire through its center of mass and along the desired axis. The wire has a torsional constant $\kappa = 0.50$ N·m. The engineer observes that this torsional pendulum oscillates through 20 complete cycles in 50 s. What value of rotational inertia does he calculate?

Section 14-7 Combinations of Harmonic Motions

23. Sketch the path of a particle that moves in the x-y plane according to the equations

$$x = A \cos\left(\omega t - \frac{\pi}{2}\right), \qquad y = 2A \cos \omega t.$$

Section 14-8 Damped Harmonic Motion

24. In Example 6, what is the amplitude of the damped oscillations after 20 full cycles have elapsed?

PROBLEMS GROUPED BY SECTION

Section 14-1 Oscillations

1. A particle moving along the x-axis is under the influence of the potential energy function $U(x) = A(b^2x - x^3)$, where A and b are positive constants. (a) What are the SI units for A and b? (b) Make a sketch of $U(x)$ as a function of x. (c) About what equilibrium value(s) of x is oscillation possible? (d) For each such equilibrium value of x, what is the maximum possible total energy such that oscillation can occur? (e) What happens if the total energy of the particle exceeds that for which oscillation is possible?

Section 14-2 The Simple Harmonic Oscillator

2. Figure 14-19 shows a graph of applied force vs. elongation for a sample of quartzite. What are the sample's (a) elastic limit and (b) force constant?

Section 14-3 Simple Harmonic Motion

3. The scale of a spring balance reading from 0 to 32 lb is 4.0 in. long. A package suspended from the balance is found to oscillate vertically with a frequency of 2.0 Hz. How much does the package weigh?

4. The piston in the cylinder head of a locomotive has a stroke of 0.76 m (=2.5 ft). What is the maximum speed of the piston if the drive wheels make 180 rev/min and the piston moves with simple harmonic motion?

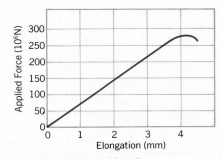

Figure 14-19 Problem 2.

5. An automobile can be considered to be mounted on four springs, as far as vertical oscillations are concerned. The springs of a certain car are adjusted so that the vibrations have a frequency of 3.0 Hz. (a) What is the force constant of each of the four springs (assumed identical) if the mass of the car is 1450 kg ($W = 3200$ lb)? (b) What will the vibration frequency be if five passengers, averaging 73 kg ($W = 160$ lb) each, ride in the car?

6. A 0.10-kg block slides back and forth along a straight line on a smooth horizontal surface. Its displacement from the origin is given by

$$x = (10 \text{ cm}) \cos \left[(10 \text{ rad/s}) \, t + \frac{\pi}{2} \text{ rad} \right].$$

(a) What is the oscillation frequency? (b) What is the maximum speed acquired by the block? At what value of x does this occur? (c) What is the maximum acceleration of the block? At what value of x does this occur? (d) What force must be applied to the block to give it this motion?

7. The end of one of the prongs of a tuning fork that executes simple harmonic motion of frequency 1000 Hz has an amplitude of 0.40 mm. Find (a) the maximum acceleration and maximum speed of the end of the prong and (b) the acceleration and the speed of the end of the prong when it has a displacement 0.20 mm.

8. The vibration frequencies of atoms in solids at normal temperatures are of the order 10^{13} Hz. Imagine the atoms to be connected to one another by "springs." Suppose that a single silver atom vibrates with this frequency and that all the other atoms are at rest. Compute the force constant of a single spring. One mole of silver has a mass of 108 g and contains 6.02×10^{23} atoms.

9. The end point of a spring vibrates with a period of 2.0 s when a mass m is attached to it. When this mass is increased by 2.0 kg the period is found to be 3.0 s. Find the value of m.

10. Two particles execute simple harmonic motion of the same amplitude and frequency along the same straight line. They pass one another when going in opposite directions each time their displacement is half their amplitude. What is the phase difference between them?

11. A block is on a horizontal surface which is moving horizontally with simple harmonic motion of frequency 2.0 Hz. The coefficient of static friction between block and plane is 0.50. How great can the amplitude be if the block does not slip along the surface?

12. Two blocks ($m = 1.0$ kg and $M = 10$ kg) and a spring ($k = 200$ N/m) are arranged on a horizontal, frictionless surface as shown in Fig. 14–20. The coefficient of static friction between the two blocks is 0.40. What is the maximum possible amplitude of the simple harmonic motion if no slippage is to occur between the blocks?

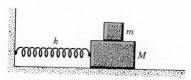

Figure 14–20   Problem 12.

Section 14–4   Energy Considerations in Simple Harmonic Motion

13. (a) When the displacement is one-half the amplitude A, what fraction of the total energy is kinetic and what fraction is potential in simple harmonic motion? (b) At what displacement is the energy half kinetic and half potential?

14. A 1.0×10^{-2}-kg particle is undergoing simple harmonic motion with an amplitude of 2.0×10^{-3} m. The maximum acceleration experienced by the particle is 8.0×10^{3} m/s². (a) Write an equation for the force on the particle as a function of time. (b) What is the period of the motion? (c) What is the maximum speed of the particle? (d) What is the total mechanical energy of this simple harmonic oscillator?

15. A 5.0-kg object moves on a horizontal frictionless surface under the influence of a spring with force constant 1.0×10^{3} N/m. The object is displaced 50 cm and given an initial velocity of 10 m/s back toward the equilibrium position. (a) What is the frequency of the motion? (b) How much total energy is associated with the motion? (c) What is the amplitude of the oscillation?

Section 14–5   Applications of Simple Harmonic Motion

16. For a simple pendulum, find the angular amplitude at which the deviation of the restoring torque from that required for simple harmonic motion is 1.0%.

17. An oscillator consists of a block attached to a spring ($k = 400$ N/m). At some time t, the position (measured from the equilibrium location), velocity, and acceleration of the block are $x = 0.10$ m, $v = -13.6$ m/s, $a = -123$ m/s². Calculate (a) the frequency, (b) the mass of the block, and (c) the amplitude of oscillation.

18. A long uniform rod of length l and mass m is free to rotate in a horizontal plane about a vertical axis through its center. A spring with force constant k is connected horizontally between the end of the rod and a fixed wall as shown in Fig. 14–21. What is the period of the small oscillations that result when the rod is pushed slightly to one side and released?

Figure 14–21   Problem 18.

19. A block weighing 14 N, which slides without friction on a plane inclined 40° to the horizontal, is connected to the top of the plane by a light spring of unstretched length 0.45 m and force constant 120 N/m, as shown in Fig. 14–22. (a) How far from the top of the plane does the block rest in equilibrium? (b) If the block is displaced slightly down the plane, what is the period of the ensuing oscillations?

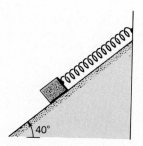

Figure 14–22   Problem 19.

20. A bicycle wheel is free to rotate about its fixed axle. A spring is attached to one of its spokes a distance r from the axle, as shown in Fig. 14–23. Assuming that the wheel is a hoop of mass m and radius R, obtain the angular frequency of small oscillations of this system in terms of m, R, r, and the spring constant k. Discuss the special cases $r = R$ and $r = 0$.

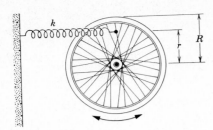

Figure 14-23 Problem 20.

21. A block of mass M, at rest on a horizontal frictionless table, is attached to a rigid support by a spring of constant k. A bullet of mass m and velocity v strikes the block as shown in Fig. 14–24. The bullet remains embedded in the block. Determine (a) the velocity of the block immediately after the collision and (b) the amplitude of the resulting simple harmonic motion.

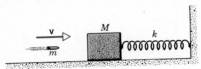

Figure 14-24 Problem 21.

22. Figure 14–25 shows a *physical pendulum*. An object of mass M is pivoted about a fixed point P; the distance from P to the center of gravity (CG) of the object is d. The object can then oscillate through small angles θ. (a) Show that simple harmonic motion occurs with period $T = 2\pi\sqrt{I/(Mgd)}$, where I is the rotational inertia of the object about the axis of rotation through P. (b) How would you apply this result to obtain the period of a simple pendulum?

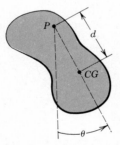

Figure 14-25 Problem 22.

23. A pendulum is formed by pivoting a long thin rod of length l and mass m about a point on the rod that is a distance d above the center of the rod. Find the small-amplitude period of this pendulum in terms of d, l, m, and g. (*Hint:* see Problem 22)

24. A uniform rod of mass M and length L is suspended from one end so as to form a physical pendulum; see Problem 22. (a) Find the period of small oscillations. (b) Is there another possible point of suspension that yields the same period? If so, where is it?

Section 14-7 Combinations of Harmonic Motions
25. Electrons in an oscilloscope are deflected by two mutually perpendicular electric fields in such a way that at any time t the displacement is given by

$$x = A \cos \omega t, \qquad y = A \cos (\omega t + \alpha).$$

Describe the path of the electrons and determine its equation (a) when $\alpha = 0°$, (b) when $\alpha = 30°$, (c) when $\alpha = 90°$.

26. Sketch the motion of a mass attached to two identical springs aligned along the x- and y-axes when at rest if the amplitude of the x-motion is twice that of the y, and the mass reaches maximum x-position and minimum y-position simultaneously. Find the phase difference $(\delta - \alpha)$ in Eq. 14–26. Why should you assume small vibrations?

Section 14-8 Damped Harmonic Motion
27. A damped harmonic oscillator involves a block ($m = 2.0$ kg), a spring ($k = 10$ N/m), and a damping force $F = -bv$. Initially it oscillates with an amplitude of 25 cm; due to the damping, the amplitude falls to three-fourths of this initial value after four complete cycles. (a) What is the value of b? (b) How much energy has been "lost" during these four cycles?

28. Assume that you are designing the suspension system for a 2000-kg automobile. The suspension is to "sag" 10 cm when the weight of the entire automobile is placed on it. In addition, the amplitude of oscillation is to decrease by 50% during one complete oscillation. Estimate the values of k and b for the spring and shock absorber system of each wheel (state any simplifying assumptions that you make).

29. For the system shown in Fig. 14–15, the block has a mass of 1.5 kg and the spring constant is 8.0 N/m. Suppose that the block is pulled down a distance of 12 cm and released. If the friction force is given by $-b \, dx/dt$, where $b = 0.23$ kg/s, find the number of oscillations made by the block during the time interval required for the amplitude to fall to one-third of its initial value.

Section 14-9 Forced Oscillations and Resonance
30. In Eq. 14–30, the amplitude x_m is given by

$$x_m = \frac{F_m}{[m^2 \, (\omega^2 - \omega_0^2)^2 + b^2\omega^2]^{1/2}},$$

where F_m is the (constant) amplitude of the externally imposed oscillating force. At resonance, what are (a) the amplitude and (b) the maximum speed of the oscillating object.

ADDITIONAL PROBLEMS

31. A body oscillates with simple harmonic motion according to the equation

$$x = (6.0 \text{ m}) \cos \left[(3\pi \text{ rad/s})t + \frac{\pi}{3} \text{ rad} \right].$$

What are (a) the displacement, (b) the velocity, (c) the acceleration, and (d) the phase at the time $t = 2.0$ s? Find also (e) the frequency and (f) the period of the motion.

32. (a) Show that the general relations for the period and frequency of any simple harmonic motion are

$$T = 2\pi \sqrt{-\frac{x}{a}} \quad \text{and} \quad v = \frac{1}{2\pi} \sqrt{-\frac{a}{x}}.$$

(b) Show that the general relations for the period and frequency of any simple angular harmonic motion are

$$T = 2\pi \sqrt{-\frac{\theta}{\alpha}} \quad \text{and} \quad v = \frac{1}{2\pi} \sqrt{-\frac{\alpha}{\theta}}.$$

33. A 3.0-kg particle is in simple harmonic motion in one dimension and moves according to the equation

$$x = (5.0 \text{ m}) \cos\left[\left(\frac{\pi}{3} \text{ rad/s}\right)t - \frac{\pi}{4} \text{ rad}\right].$$

(a) At what value of x is the potential energy equal to half the total energy? (b) How long does it take the particle to move to this position from the equilibrium position?

34. A loudspeaker produces a musical sound by the oscillation of a diaphram. If the amplitude of oscillation is limited to 1.0×10^{-3} mm, what frequencies will result in the acceleration of the diaphram exceeding g?

35. A small body of mass 0.10 kg is undergoing simple harmonic motion of amplitude 10 cm and period 0.20 s. (a) What is the maximum value of the force acting on it? (b) If the oscillations are produced by a spring, what is the force constant of the spring?

36. Start from Eq. 14–17 for the conservation of energy (with $\frac{1}{2}kA^2 = E$) and obtain the displacement as a function of the time by integration of Eq. 14–18. Compare with Eq. 14–8.

37. Show that the average values, over a complete cycle, of both the kinetic energy and the potential energy of a simple harmonic oscillator equal half the maximum value.

38. In Example 6, find the ratio of the maximum frictional force $(F_f = -b\,dx/dt)$ to the maximum spring force $(F_k = -kx)$ during the first cycle of the damped oscillation. Does this ratio change appreciably during later cycles?

39. A massless spring hangs from the ceiling with a small object attached to its lower end. The object is initially held at rest in such a position that the spring is not stretched. The object is then released from this position and oscillates up and down with its lowest position being 10 cm below the initial position. (a) What is the frequency of the oscillation? (b) What is the speed of the object when it is 8.0 cm below the initial position? (c) An object of mass 300 g is attached to the first object, after which the system oscillates with half the original frequency. What is the mass of the first object? (d) Where is the new equilibrium position with both objects attached to the spring?

40. A block is on a piston that is moving vertically with simple harmonic motion of period 1.0 s. (a) At what amplitude of motion will the block and piston separate? (b) If the piston has an amplitude of 5.0 cm, what is the maximum frequency for which the block and piston will be in contact continuously?

41. A 2200-lb car carrying four 180-lb people is traveling over a rough "washboarded" dirt road. The corrugations in the road are 13 ft apart. The car is observed to bounce with maximum amplitude when its speed is 10 mi/h. The car now stops and the four people get out. By how much does the car body rise on its suspension due to this decrease in weight?

42. Two oscillating systems that you have studied are the mass-spring and the simple pendulum. There is an interesting relationship between them. Suppose that you hang a weight on the end of a spring, and when the weight is in equilibrium, the spring is stretched a distance h. Show that the frequency of this mass-spring system is the same as that of a simple pendulum whose length is h. See Fig. 14–26.

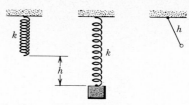

Figure 14–26 Problem 42.

43. Two identical springs are attached to a block of mass m and to fixed supports as shown in Fig. 14–27. Show that the frequency of oscillation on the frictionless surface is

$$v = \frac{1}{2\pi} \sqrt{\frac{2k}{m}}.$$

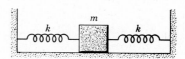

Figure 14–27 Problem 43.

44. The balance wheel of a watch vibrates with an angular amplitude of π rad and a period of 0.50 s. Find (a) the maximum angular speed of the wheel, (b) the angular speed of the wheel when its displacement is $\pi/2$ rad, and (c) the angular acceleration of the wheel when its displacement is $\pi/4$ rad.

45. Two particles oscillate in simple harmonic motion along a common straight line segment of length A. Each particle has a period of 1.5 s, but they differ in phase by 30°. (a) How far apart are they (in terms of A) 0.50 s after the lagging particle leaves one end of the path? (b) Are they moving in the same direction, toward each other, or away from each other at this time?

46. A physical pendulum (see Problem 22) consists of a uniform solid disk (mass M, radius R) supported in a vertical plane by a pivot located a distance d from the center of the disk, as shown in Fig. 14–28. The disk is displaced by a small angle and released. Find an expression for the period of the resulting simple harmonic motion.

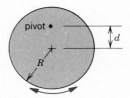

Figure 14–28 Problem 46.

47. A physical pendulum (see Problem 22) consists of a meter stick that is pivoted at a small hole drilled through the stick a distance x from the 50-cm mark. The period of oscillation is observed to be 2.5 s. Find the distance x.

48. Two springs are joined and connected to a mass m as shown in Fig. 14–29. The surfaces are frictionless. If the springs each have force constant k, show that the frequency of oscillation of m is

$$v = \frac{1}{2\pi}\sqrt{\frac{k}{2m}}.$$

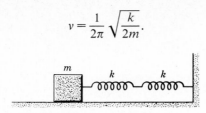

Figure 14–29   Problem 48.

49.   (*a*) What is the frequency of a simple pendulum 2.0 m long? (*b*) Assuming small amplitudes, what would its frequency be in an elevator accelerating upward at a rate of 2.0 m/s²? (*c*) What would its frequency be in free fall?

50.   A simple pendulum of length *l* and mass *m* is suspended in a car that is traveling with a constant speed *v* around a circle of radius *R*. If the pendulum undergoes small oscillations in a radial direction about its equilibrium position, what will its frequency of oscillation be?

51.   The bob on a simple pendulum of length *R* moves in an arc of a circle. By considering that the acceleration of the bob as it moves through its equilibrium position is that for uniform circular motion (mv^2/R), show that the tension in the string at that position is $mg(1 + \theta_m^2)$ if the amplitude is small. Would the tension at other positions of the bob be larger, smaller, or the same?

52.   A 4.0-kg block is suspended from a spring with a force constant of 500 N/m. A 50-g bullet is fired into the block from below with a speed of 150 m/s and comes to rest in the block. (*a*) Find the amplitude of the resulting simple harmonic motion. (*b*) What fraction of the original kinetic energy of the bullet is stored in the harmonic oscillator?

53.   A uniform spring whose unstressed length is *l* has a force constant *k*. The spring is cut into two pieces of unstressed lengths l_1 and l_2, with $l_1 = nl_2$. What are the corresponding force constants k_1 and k_2 in terms of *n* and *k*? Does your result seem reasonable for $n = 1$?

54.   Three 10,000-kg ore cars are held at rest on a 30° incline on a mine railway using a cable that is parallel to the incline (Fig. 14–30). The cable is observed to stretch 15 cm just before a coupling breaks, detaching one of the cars. Find (*a*) the frequency of the resulting oscillations of the remaining two cars; (*b*) the amplitude of the oscillations.

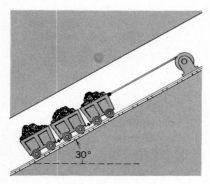

Figure 14–30   Problem 54.

55.   A simple harmonic oscillator consists of a block of mass 2.00 kg attached to a spring of force constant 100 N/m. When *t* =

1.00 s, the position and velocity of the block are $x = 0.129$ m and $v = 3.415$ m/s. (*a*) What is the amplitude of the oscillation? (*b*) What were the position and velocity at $t = 0.0$ s?

*56.   A solid cylinder is attached to a horizontal massless spring so that it can roll without slipping along a horizontal surface, as in Fig. 14–31. The force constant *k* of the spring is 3.0 N/m. If the system is released from rest at a position in which the spring is stretched by 0.25 m, find (*a*) the translational kinetic energy and (*b*) the rotational kinetic energy of the cylinder as it passes through the equilibrium position. (*c*) Show that under these conditions the center of mass of the cylinder executes simple harmonic motion with a period

$$T = 2\pi\sqrt{\frac{3M}{2k}},$$

where *M* is the mass of the cylinder.

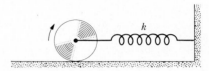

Figure 14–31   Problem 56.

*57.   *Lissajous figures.* When oscillations at right angles are combined, the frequencies for the motion of the particle in the *x*- and *y*-directions need not be equal, so that in the general case Eqs. 14–26 become

$$x = A_x \cos(\omega_x t + \delta) \qquad \text{and} \qquad y = A_y \cos(\omega_y t + \alpha).$$

The path of the particle is no longer an ellipse but is called a *Lissajous curve,* after Jules Antoine Lissajous, who first demonstrated such curves in 1857. (*a*) If ω_x/ω_y is a rational number, so that the angular frequencies ω_x and ω_y are "commensurable," then the curve is closed and the motion repeats itself at regular intervals of time. Assume that $A_x = A_y$ and $\delta = \alpha$ and draw the Lissajous curve for $\omega_x/\omega_y = \frac{1}{2}, \frac{1}{3}$, and $\frac{2}{3}$. (*b*) Let ω_x/ω_y be a rational number, either $\frac{1}{2}, \frac{1}{3}$, or $\frac{2}{3}$, say, and show that the shape of the Lissajous curve depends on the phase difference $\alpha - \delta$. Draw curves for $\alpha - \delta = 0, \pi/4$, and $\pi/2$ rad. (*c*) If ω_x/ω_y is not a rational number, then the curve is "open." Convince yourself that after a long time the curve will have passed through every point lying in the rectangle bounded by $x = \pm A_x$ and $y = \pm A_y$, the particle never passing twice through a given point with the same velocity.

58.   The Lissajous figure shown in Fig. 14–32 is the result of combining the two simple harmonic motions $x = A_x \cos \omega_x t$ and $y = A_y \cos(\omega_y t + \delta)$. (*a*) What is the value of A_x/A_y? (*b*) What is the value of ω_x/ω_y? (*c*) What is the value of δ?

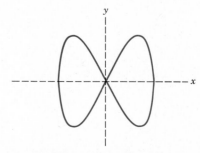

Figure 14–32   Problem 58.

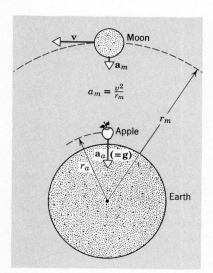

Figure 15–1 Both the moon and the apple are accelerated toward the center of the earth.

15 Gravitation

15–1 The Law of Universal Gravitation

From at least the time of the Greeks two problems were the subjects of searching inquiry: (1) the tendency of objects such as stones to fall to earth when released, and (2) the motions of the planets, including the sun and the moon, which were classified with the planets in those times. In early days these problems were thought of as completely separate. It is one of Isaac Newton's achievements that, building on the work of his predecessors, he saw them clearly as aspects of a single problem and subject to the same laws.

In 1665 the 23-year-old Newton was driven from Cambridge to Lincolnshire when the college was dismissed because of the plague. About 50 years later he wrote, "in the same year (1665) I began to think of gravity extending to the orb of the Moon . . . and having thereby compared the force requisite to keep the Moon in her orb with the force of gravity at the surface of the earth, and found them to answer pretty nearly."

Newton's young friend William Stukeley wrote of having tea with Newton under some apple trees when Newton said that the setting was the same as when he got the idea of gravitation. "It was occasion'd by the fall of an apple, as he sat in a contemplative mood . . . and thus by degrees he began to apply this property of gravitation to the motion of the earth and the heavenly bodies" (see Fig. 15–1.)

We can compute the acceleration of the moon toward the earth from its period of revolution and the radius of its orbit. We obtain 0.0027 m/s^2 (see Example 3, Chapter 4). This value is about 3600 times smaller than g, the acceleration due to gravity at the surface of the earth. Newton, guided as he says by Kepler's third law (see below and see Problem 14), sought to account for this difference by assuming that the acceleration of a falling body is inversely proportional to the square of its distance from the earth.

The question of what we mean by "distance from the earth" immediately arises. Newton eventually came to regard every particle of the earth as contributing to the gravitational attraction it had on other bodies. He made the daring assumption that the mass of the earth could be treated as if it were all concentrated at its center. (See Section 15–4.)

We can treat the earth as a particle with respect to the sun, for example. It is not obvious, however, that we can treat the earth as a particle with respect to an apple located only a few feet above its surface. If we make this assumption, however, a falling body near the earth's surface is a distance of one earth radius from the effective center of attraction of the earth, or 4000 mi. The moon is about 240,000 mi away. The square of the ratio of these distances is $(240,000/4000)^2 = 3600$, in agreement with the ratio of the accelerations of the moon and the apple. In Newton's words, quoted above, it does indeed "answer pretty nearly."

Newton's law of universal gravitation, which he developed from these early studies, may be expressed as follows:

The force between any two particles having masses m_1 and m_2 separated by a distance r is an attraction acting along the line joining the particles and has the magnitude

$$F = G\frac{m_1 m_2}{r^2}, \qquad (15\text{--}1)$$

where G is a universal constant having the same value for all pairs of particles.

The gravitational forces between two particles are an action-reaction pair. The first particle exerts a force on the second particle that is directed toward the first particle along the line joining the two. Likewise, the second particle exerts a force on the first particle that is directed toward the second particle along the line joining the two. These forces are equal in magnitude but oppositely directed. Note carefully that, even though the masses differ greatly, the forces are the same. Thus, if an apple drops toward the earth, the magnitude of the gravitational force of the earth acting on the apple equals that of the apple acting on the earth. What differs is the acceleration, the lighter body having the greater acceleration toward the common center of mass.

The universal constant G must not be confused with the **g** that is the acceleration of a body arising from the earth's gravitational pull on it. The constant G has the dimensions L^3/MT^2 and is a scalar; **g** has the dimensions L/T^2 is a vector, and is neither universal nor constant.

Notice that Newton's law of universal gravitation is not a defining equation for any of the physical quantities (force, mass, or length) contained in it. According to our program for classical mechanics in Chapter 5, force is defined from Newton's second law, $\mathbf{F} = m\mathbf{a}$. The essence of this law, however, is the assumption that the force on a particle, so defined, can be related in a simple way to measurable properties of the particle and of its environment; that is, the existence of simple force laws is assumed. The law of universal gravitation is such a simple law. The constant G must be found from experiment. Once G is determined for a given pair of bodies, we can use that value in the law of gravitation to determine the gravitational forces between any other pairs of bodies.

Notice also that Eq. 15–1 expresses the force between mass particles. If we want to determine the force between extended bodies, as for example the earth and the moon, we must regard each body as decomposed into particles. Then the interaction between all particles must be computed. Integral calculus makes such a calculation possible. Newton's motive in developing the calculus arose in part from a desire to solve such problems. In general, it is incorrect to assume that all the mass of a body can be concentrated at its center of mass for gravitational pur-

Newton's law of gravitation

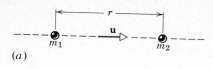

(a)

(b)

Figure 15-2 (a) A unit vector **u** is defined to point from m_1 to m_2. (b) The gravitational forces $\mathbf{F}_{12}$ and $\mathbf{F}_{21}$ point in the directions $+\mathbf{u}$ and $-\mathbf{u}$, respectively.

Vector form of the gravitational force law

The gravitational constant

poses. This assumption is correct for uniform spheres, however, a result that we shall use often and shall prove in Section 15-4.

Implicit in the law of universal gravitation is the idea that the gravitational force between two particles is independent of the presence of other bodies or the properties of the intervening space. This is consistent with all experimental evidence and has been used by some to rule out the possible existence of "gravity screens."

We can express the law of universal gravitation in vector form. Let a unit vector **u** be defined to point from the particle of mass m_1 to the particle of mass m_2 in Fig. 15-2a. A unit vector, as we have seen, is a vector with unit magnitude and no dimensions; its function is to specify a direction. The gravitational force $\mathbf{F}_{21}$, exerted on m_2 by m_1, is given in both magnitude and direction by

$$\mathbf{F}_{21} = - G \frac{m_1 m_2}{r^2} \mathbf{u}; \qquad (15\text{-}2a)$$

see Fig. 15-2b. Similarly the gravitational force $\mathbf{F}_{12}$, exerted on m_1 by m_2, is given in both magnitude and direction by

$$\mathbf{F}_{12} = + G \frac{m_1 m_2}{r^2} \mathbf{u}. \qquad (15\text{-}2b)$$

As we expect, $\mathbf{F}_{21} = -\mathbf{F}_{12}$; that is, the gravitational forces acting on the two bodies form an action-reaction pair.

15-2 The Constant of Universal Gravitation

To determine the value of G it is necessary to measure the force of attraction between two bodies of known mass. The first accurate measurement was made by Lord Cavendish in 1798. As of 1982 the accepted value is

$$G = 6.6726 \times 10^{-11} \text{ m}^3/\text{kg} \cdot \text{s}^2,$$

accurate to ± 5 in the last place of decimals.

The gravitational constant is a measure of the absolute strength of the gravitational force. If we hold an apple in each hand we are totally unaware of the tiny gravitational attraction between them. Yet just in holding them—or in releasing them—we are quite aware of the earth's gravitational attraction for each. Our sense of gravity as an important and ever-present force in our daily lives comes about simply because the earth is so massive. Someone has said: "I can fall off a cliff and break my leg but it takes a whole planet to do it."

From another point of view let us compare the electrostatic force between two protons (see p. 461) to the gravitational force that also acts between them; the gravitational force is weaker by a factor of $\sim 10^{39}$! However—and just because of the strong attractive electrostatic forces between charges of opposite sign— matter in bulk is electrically neutral to a remarkable extent. In the interactions among the stars of our galaxy, or between our galaxy and its galactic neighbors, the masses involved are so great and the electrical neutrality so complete that gravity controls the situation totally.

Consider also the formation of stars by condensation from cosmic dust and their evolution into incredibly compact white dwarfs, neutron stars (pulsars), or black holes, with central densities exceeding $\sim 10^{15}$ g/cm³. The story here is one of the increasing dominance of gravity over all other forces or influences of nature; it sweeps all before it.

We can measure the gravitational constant *G* using a Cavendish balance as illustrated in Fig. 15–3. Two small lead balls, each of mass *m*, are attached to the ends of a light rod. This rigid "dumbbell" is suspended, with its axis horizontal, by a fine vertical fiber. Two large lead balls, each of mass *M*, are placed near the ends of the dumbbell on opposite sides. When the large balls are in the positions *A* and *B*, the small balls are attracted by a gravitational force and a torque is exerted on the dumbbell rotating it counterclockwise, as viewed from above. When the large balls are in the positions *A'* and *B'*, the dumbbell rotates clockwise. The fiber opposes these torques as it is twisted. The angle θ through which the fiber is twisted when the balls are moved from one position to the other is measured by observing the deflection of a beam of light reflected from the small mirror attached to it. If the masses of the balls and their distances of separation and the torsional constant of the fiber are known, we can calculate *G* from the measured angle of twist. The force of attraction is very small, so the fiber must have an extremely small torsion constant if we are to obtain a detectable twist. In Example 1 at the end of this section some data are given from which *G* can be calculated.

The masses in the Cavendish balance of Fig. 15–3 are, of course, not particles but extended objects. Since they are uniform spheres, however, they act gravitationally as though all their mass were concentrated at their centers (Section 15–4).

Because *G* is so small, the gravitational forces between bodies on the earth's surface are extremely small and can be neglected for ordinary purposes. For ex-

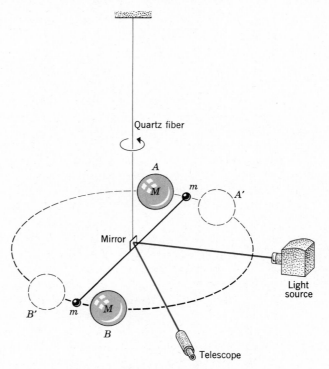

Figure 15–3 The Cavendish balance used for experimental verification of Newton's law of universal gravitation. Masses *m*, *m* are suspended from a fiber. Masses *M*, *M* can rotate on a stationary support. An image of the lamp filament is reflected by the mirror attached to *m*, *m* onto the scale so that any rotation of *m*, *m* can be measured.

ample, two spherical objects each having a mass of 100 kg (about 220 lb weight) and separated by 1.0 m at their centers attract each other with a force

$$F = \frac{Gm_1 m_2}{r^2} = \frac{(6.67 \times 10^{-11} \text{ m}^3/\text{kg} \cdot \text{s}^2) \times (100 \text{ kg}) \times (100 \text{ kg})}{(1.0 \text{ m})^2}.$$

$$= 6.7 \times 10^{-7} \text{ N}.$$

This is about the weight of a grain of sand ~0.4 mm in diameter! The Cavendish experiment must be a very delicate one indeed. Even so, it is often performed as an experiment in an introductory physics laboratory.

We can determine the mass of the earth from the law of universal gravitation and the value of G calculated from the Cavendish experiment. For this reason Cavendish is said to have been the first person to "weigh" the earth. Consider the earth, mass M_e, and an object on its surface of mass m. The force of attraction is given by both

$$F = mg$$

and

$$F = \frac{GmM_e}{R_e^2}.$$

Here R_e is the radius of the earth, which is the separation of the two bodies, and g is the acceleration due to gravity at the earth's surface. Combining these equations, we obtain

$$M_e = \frac{gR_e^2}{G} = \frac{(9.80 \text{ m/s}^2)(6.37 \times 10^6 \text{ m})^2}{6.67 \times 10^{-11} \text{ m}^3/\text{kg} \cdot \text{s}^2} = 5.97 \times 10^{24} \text{ kg}$$

or

$$6.6 \times 10^{21} \text{ tons} \qquad \text{"weight."}$$

If we were to divide the total mass of the earth by its total volume, we would obtain the average density of the earth. This turns out to be 5.5 g/cm³, or about 5.5 times the density of water. The average density of the rock on the earth's surface is much less than this value. We conclude that the interior of the earth contains material of density greater than 5.5 g/cm³. From the Cavendish experiment we have obtained information about the nature of the earth's core.

Example 1 Let the small lead balls of Fig. 15–3 each have a mass of 10.0 g and let the light rod connecting them be 50.0 cm long. The period of torsional oscillation of this system is found to be 769 s. Then two large fixed lead balls each of mass 10.0 kg are placed near each suspended ball so as to produce the maximum torsion. The angular deflection of the suspended rod is then 3.96×10^{-3} rad and the distance between centers of the large and small balls is 10.0 cm. Calculate the universal constant of gravitation G from these data.

The period of torsional oscillation is given by Eq. 14–23,

$$T = 2\pi \sqrt{\frac{I}{\kappa}}.$$

For the rigid dumbbell, if we neglect the contribution of the light rod,

$$I = \Sigma \, mr^2 = (10.0 \text{ g})(25.0 \text{ cm})^2 + (10.0 \text{ g})(25.0 \text{ cm})^2$$

or

$$I = 1.25 \times 10^{-3} \text{ kg} \cdot \text{m}^2.$$

With $T = 769$ s, we can obtain the torsional constant κ as

$$\kappa = \frac{4\pi^2 I}{T^2} = \frac{(4\pi^2)(1.25 \times 10^{-3} \text{ kg} \cdot \text{m}^2)}{(769 \text{ s})^2}$$

$$= 8.34 \times 10^{-8} \text{ kg} \cdot \text{m}^2/\text{s}^2.$$

The relation between the applied torque and the angle of twist is $\tau = \kappa\theta$. We now know κ and the value of θ at maximum deflection.

The torque arises from the gravitational forces exerted by the large lead balls on the small ones. This torque will be a maximum for a given separation when the line joining the centers of these balls is at right angles to the rod. The force on *each* small ball is

$$F = \frac{GMm}{r^2},$$

and the moment arm for each force is half the length of the rod ($l/2$). Then

$$\text{torque} = \text{force} \times \text{moment arm}$$

or

$$\tau = 2\frac{GMm}{r^2}\frac{l}{2}.$$

Combining this with

$$\tau = \kappa\theta,$$

we obtain

$$G = \frac{\kappa\theta r^2}{Mml}$$

$$= \frac{(8.34 \times 10^{-8}\text{ kg}\cdot\text{m}^2/\text{s}^2)(3.96 \times 10^{-3}\text{ rad})(0.100\text{ m})^2}{(10.0\text{ kg})(0.0100\text{ kg})(0.500\text{ m})}$$

$$= 6.63 \times 10^{-11}\text{ m}^3/\text{kg}\cdot\text{s}^2.$$

This result is about 1% lower than the accepted value. What have we neglected in this calculation that might account for this difference?

15-3   Inertial and Gravitational Mass

The gravitational force acting on a body is proportional to its mass, as Eq. 15–1 shows. Because of this proportionality we can measure a mass by measuring the gravitational force that acts on it. In practice we would use an equal-arm balance to do this. In this way we compare the mass of a body with that of a standard body (ultimately the standard kilogram) using the earth's gravitational field as a sensitive indicator of mass differences.

In Section 5–5, however, we described a totally different method of mass measurement. Here we apply a known force to the body, measure its acceleration, and compute its mass from $\mathbf{F} = m\mathbf{a}$. Gravity is in no way involved.

The question arises whether these two methods really measure the same property. The word "mass" has been used in two quite different experimental situations. For example, if we try to push a block that is at rest on a horizontal frictionless surface, we notice that it requires some effort to move it. The block seems to be inert and tends to stay at rest, or if it is moving it tends to keep moving. Gravity does not enter here at all. It would take the same effort to accelerate the block in a region of space free of gravity. It is the mass of the block that makes it necessary to exert a force to change its motion. This is the mass occurring in $\mathbf{F} = m\mathbf{a}$ in our original experiments in dynamics. We call this kind of mass m the *inertial mass*.

Inertial mass

Now, there is a different situation which involves the mass of the block. For example, it requires effort just to hold the block up in the air at rest above the earth. If we do not support it, the block will fall to the earth with accelerated motion. The force required to hold up the block is equal in magnitude to the force of gravitational attraction between it and the earth. Here inertia plays no role whatever; the property of material bodies, that they are attracted to other objects such as the earth, does play a role. The force is given by

$$F = G\frac{m'M_e}{R_e^2},$$

Gravitational mass

where we now call m' the *gravitational mass* of the block. Are the gravitational mass m' and the inertial mass m of the block really the same?

Newton devised an experiment to test directly the apparent equivalence of inertial and gravitational mass. If we go back (Section 14–5) and look up the derivation of the period of a simple pendulum, we find that the period (for small angles) should really have been written as

$$T = 2\pi\sqrt{\frac{ml}{m'g}},$$

where m in the numerator refers to the inertial mass of the pendulum bob and m' in the denominator is the gravitational mass of the pendulum bob, such that $m'g$

gives the gravitational pull on the bob. Only if we assume that m equals m', as we did there implicitly, do we obtain the expression

$$T = 2\pi \sqrt{\frac{l}{g}}$$

for the period. Newton made a pendulum bob in the form of a thin shell. Into this hollow bob he put different substances, being careful always to have the same weight of substance as determined by a balance. Hence, in all cases the force on the pendulum was the same at the same angle. Because the external shape of the bob was always the same, even the air resistance on the moving pendulum was the same. As one substance replaced another inside the bob, any difference in acceleration could be due only to a difference in the inertial mass. Such a difference would show up by a change in the period of the pendulum. But in all cases Newton found the period of the pendulum to be the same, always given by $T = 2\pi\sqrt{l/g}$. Hence, he concluded that $m = m'$ and that inertial and gravitational masses are equivalent.

In 1909, the Hungarian physicist Baron von Eötvös devised an apparatus that could detect a difference of 5 parts in 10^9 in gravitational force. He found that equal inertial masses always experienced equal gravitational forces within the accuracy of his apparatus. A refined version of the Eötvös experiment was reported in 1964 by R. H. Dicke and his collaborators at Princeton who improved the accuracy of the original experiment by a factor of several hundred.

In classical physics the equivalence of gravitational and inertial mass was looked upon as a remarkable accident. In modern physics, however, this equivalence is regarded as a clue leading to a deeper understanding of gravitation. This was, in fact, an important clue leading to the development of Einstein's general theory of relativity.

Principle of equivalence

Consider two reference frames: (1) a nonaccelerating (inertial) reference frame in which there is a uniform gravitational field and (2) a reference frame that is accelerating uniformly with respect to an inertial frame but in which there is no gravitational field. In his general theory of relativity, Albert Einstein showed that two such frames are exactly equivalent physically. That is, experiments carried out under the same conditions in these two frames should give the same results. This is the *principle of equivalence*.

Suppose that you were standing in an elevator at rest and you suddenly felt the sensation of being pulled toward the floor. You would probably conclude that the elevator was accelerating upward. However, could you not as well conclude that the gravitational force acting on you had increased? Two such reference frames would be equivalent. If the cable broke, the free-fall acceleration could be interpreted as a reduction of the gravitational force to zero.

The principle of equivalence says that no physical measurement can detect a difference between a situation where an observer is accelerating relative to an inertial frame in a region having no gravitational field and a situation in which one is not accelerating in an inertial frame where a uniform gravitational field exists. Of course, if it were found that gravitational mass and inertial mass were measurably different, then a way would indeed exist for telling the difference between the two frames and the principle of equivalence would be invalid.

15-4 Gravitational Effect of a Spherical Distribution of Mass

We have already used the fact that a large sphere attracts particles outside it just as though the mass of the sphere were concentrated at its center. Let us now prove this result.

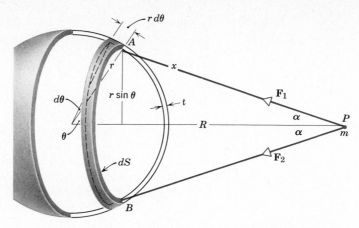

Figure 15–4 Gravitational attraction of a section dS of a spherical shell of matter on a particle of mass m.

Consider a uniformly dense shell whose thickness t is small compared to its radius r (Fig. 15–4). We seek the gravitational force it exerts on an external particle P of mass m. We assume that each particle of the shell exerts on P a force that is proportional to the mass of the small part, inversely proportional to the square of the distance between that part of the shell and P, and directed along the line joining them. We must then obtain the resultant force on P attributable to all parts of the spherical shell.

The small part of the shell at A attracts m with a force $\mathbf{F}_1$. A small part of equal mass at B, equally far from m but diametrically opposite A, attracts m with a force $\mathbf{F}_2$. The resultant of these two forces on m is $\mathbf{F}_1 + \mathbf{F}_2$. Notice, however, that the vertical components of these two forces cancel one another and that the horizontal components, $F_1 \cos \alpha$ and $F_2 \cos \alpha$, are equal. By dividing the spherical shell into pairs of particles like these, we can see at once that all transverse forces on m cancel in pairs. A small mass in the upper hemisphere exerts a force having an upward component of m that will annul the downward component of force exerted on m by an equal symmetrically located mass in the lower hemisphere of the shell. To find the resultant force on m arising from the shell, we need consider only horizontal components.

Let us take as our element of mass of the shell a circular strip labeled dS in the figure. Its length is $2\pi(r \sin \theta)$, its width is $r\, d\theta$, and its thickness is t. Hence, it has a volume

$$dV = 2\pi t r^2 \sin \theta \, d\theta.$$

Let us call the density ρ, so that the mass within the strip is

$$dM = \rho \, dV = 2\pi t \rho r^2 \sin \theta \, d\theta.$$

The force exerted by dM on the particle of mass m at P is horizontal and has the value

$$dF = G \frac{m \, dM}{x^2} \cos \alpha$$

$$= 2\pi G t \rho m r^2 \frac{\sin \theta \, d\theta}{x^2} \cos \alpha. \tag{15–3}$$

The variables x, α, and θ are related. From the figure we see that

$$\cos \alpha = \frac{R - r \cos \theta}{x}. \tag{15–4}$$

Since, by the law of cosines,

$$x^2 = R^2 + r^2 - 2Rr \cos \theta, \tag{15-5}$$

we have

$$r \cos \theta = \frac{R^2 + r^2 - x^2}{2R}. \tag{15-6}$$

On differentiating Eq. 15-5, we obtain

$$2x \, dx = 2Rr \sin \theta \, d\theta$$

or

$$\sin \theta \, d\theta = \frac{x}{Rr} \, dx. \tag{15-7}$$

We now put Eq. 15-6 into Eq. 15-4 and then put Eqs. 15-4 and 15-7 into Eq. 15-3. As a result we eliminate θ and α and obtain

$$dF = \frac{\pi Gt\rho mr}{R^2} \left(\frac{R^2 - r^2}{x^2} + 1 \right) dx.$$

This is the force exerted by the circular strip dS on the particle m.

We must now consider every element of mass in the shell and sum up over all the circular strips in the entire shell. This operation is an integration over the shell with respect to the variable x. But x ranges from a minimum value of $R - r$ to a maximum value $R + r$.

Since

$$\int_{R-r}^{R+r} \left(\frac{R^2 - r^2}{x^2} + 1 \right) dx = 4r,$$

we obtain the resultant force

A uniform spherical shell behaves like a mass point

$$F = \int_{R-r}^{R+r} dF = G \frac{(4\pi r^2 \rho t)m}{R^2} = G \frac{Mm}{R^2} \tag{15-8}$$

where

$$M = (4\pi r^2 t \rho)$$

is the total mass of the shell. This is exactly the same result we would obtain for the force between particles of mass M and m separated a distance R. We have proved, therefore, that *a uniformly dense spherical shell attracts an external mass point as if all its mass were concentrated at its center.*

A solid sphere can be regarded as composed of a large number of concentric shells. If each spherical shell has a uniform density, even though different shells may have different densities, the same result applies to the solid sphere. Hence, to the extent that they are such spheres, bodies like the earth, the moon, or the sun may be regarded gravitationally as point particles to bodies outside them. Notice that our proof applies only to spheres and then only when the density is constant over the sphere or is a function of radius alone.

A uniform spherical shell exerts no gravitational force at interior points

An interesting result of some significance is the force exerted by a spherical shell on a particle inside it. This force is zero. In this case R is now smaller than r. The limits of our integration over x are now $r - R$ to $R + r$. But

$$\int_{r-R}^{R+r} \left(\frac{R^2 - r^2}{x^2} + 1 \right) dx = 0,$$

so that $F = 0$.

This last result, although not obvious, is plausible because the mass elements of the shell on opposite sides of m now exert forces of opposite directions on m inside. The total annulment depends on the fact that the force varies precisely as an inverse square of the separation distance of two particles; see Problem 31. Important consequences of this result will be discussed in the chapters on electricity. There we shall see that the electrical force between charged particles also depends inversely on the square of the distance between them.

If the density of the earth were uniform—which it is not—the gravitational force exerted by the earth on a particle would be a maximum at the earth's surface. It would, as we expect, decrease as we move outward. If we were to move inward, the gravitational force would respond to two opposing influences. (1) It would tend to increase because we are moving closer to the earth's center. (2) It would tend to decrease because shells of the earth's crust external to the particle's position would not exert any force on the particle. For an earth of assumed constant density the second influence would prevail, so the gravitational force would decrease as we penetrated the earth.

For the real earth, however, the density is *not* constant everywhere; the outer crust ($\rho \cong 3$ g/cm³) is so much less dense than the inner core ($\rho \cong 13$ g/cm³) that influence 2 above is greatly weakened and no longer prevails. The gravitational force actually *increases* as we descend a mine shaft, eventually reaching a maximum and then decreasing to zero at the earth's center.

Example 2 Suppose that a tunnel could be dug through the earth from one side to the other along a diameter, as shown in Fig. 15–5.

(a) Show that the motion of a particle dropped into the tunnel is simple harmonic motion. Neglect all frictional forces and assume that the earth has a uniform density.

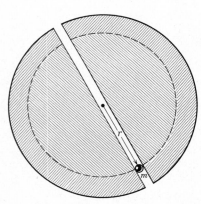

Figure 15–5   Example 2. Particle moving in a tunnel through the earth.

The gravitational attraction of the earth for the particle at a distance r from the center of the earth arises entirely from that portion of matter of the earth in shells internal to the position of the particle. The external shells exert no force on the particle. Let us assume that the earth's density is uniform at the value ρ. Then the mass inside a sphere of radius r is

$$M' = \rho V' = \rho \frac{4\pi r^3}{3}.$$

This mass can be treated as though it were concentrated at the center of the earth for gravitational purposes. Hence, the force on the particle of mass m is

$$F = \frac{-GM'm}{r^2}.$$

The minus sign is used to indicate that the force is attractive and directed toward the center of the earth.

Substituting for M', we obtain

$$F = -G \frac{(\rho 4\pi r^3)m}{3r^2} = -\left(G\rho \frac{4\pi m}{3}\right) r = -kr.$$

Here $G\rho 4\pi m/3$ is a constant, which we have called k. The force is, therefore, proportional to the displacement r but oppositely directed. This is exactly the criterion for simple harmonic motion.

(b) If mail were delivered through this chute, how much time would elapse between deposit at one end and delivery at the other end?

The period of this simple harmonic motion is

$$T = 2\pi \sqrt{\frac{m}{k}} = 2\pi \sqrt{\frac{3m}{G\rho 4\pi m}} = \sqrt{\frac{3\pi}{G\rho}}.$$

For $\rho = 5.51$ g/cm³ we have

$$T = \sqrt{\frac{3\pi}{G\rho}} = \sqrt{\frac{3\pi}{(6.67 \times 10^{-11} \text{ m}^3/\text{kg} \cdot \text{s}^2)(5.51 \times 10^3 \text{ kg/m}^3)}}$$
$$= 5050 \text{ s} = 84.2 \text{ min.}$$

The time for delivery is one-half period, or about 42 min. Notice that this time is independent of the mass of the mail and, indeed, of the radius of the earth (assuming that the density is fixed).

The earth does not really have a uniform density. Suppose that ρ were some function of r, rather than a constant. What effect would this have on our problem?

15-5   Gravitational Acceleration

In previous chapters we have taken the magnitude of the acceleration due to gravity **g** to be a constant, however, even on the surface of the earth it varies measurably depending on one's position. We can see why by explicitly calculating g from the law of gravitation. Consider, as we did in Section 15–2, the force exerted by the earth (mass M_e) on a small object of mass m as defined by Newton's second law of motion and also that given by the law of gravitation.

$$F = ma$$
$$= \frac{GmM_e}{r^2}.$$

We neglect, for the time being, the effects on the measured value of **g** caused by the earth's rotation. In addition, we not only assume that the earth is spherical but also that its density is constant for all the concentric spherical shells into which we can divide it, the value of the density depending only on the radius of the shell.

When the acceleration of the object is that caused by gravitational attraction alone, we define $a = g$. Equating the forces above and solving for g, we obtain

g for a spherical earth

$$g = \frac{GM_e}{r^2}. \tag{15–9}$$

This equation holds for points on or above the earth's surface and shows that g must depend on r, the distance from the earth's center. Table 15–1 shows, in fact, how g varies with altitude at a latitude of 45°. Note that at 1000 km ($= 620$ mi), a typical altitude for an orbiting satellite, g has dropped to 7.41 m/s², about 76% of its sea-level value.

Table 15–1   Variation of g with altitude at 45° latitude

Altitude, km	g, m/s²	Altitude, km	g, m/s²
0	9.806	32	9.71
1	9.803	100	9.60
4	9.794	500	8.53
8	9.782	1,000[a]	7.41
16	9.757	380,000[b]	0.00271

[a] Typical satellite orbit altitude ($= 620$ mi).
[b] Radius of moon's orbit ($= 240,000$ mi).

We can show from Eq. 15–9 that a given fractional change in radial distance, dr/r, produces a fractional change dg/g in the gravitational acceleration given by

$$\frac{dg}{g} = -2\frac{dr}{r}. \tag{15–10}$$

Thus if, at the earth's surface, we ascend 100 m, then $dr/r \cong (100 \text{ m})/(6.4 \times 10^6 \text{ m}) = 1.6 \times 10^{-5}$, and g changes by about $(-2)(1.6 \times 10^{-5}) \cong -3 \times 10^{-3}\%$. This is a measure of the extent to which we may take g to be constant near the earth's surface.

g for an ellipsoidal earth

To a good second approximation the earth is not a sphere but an ellipsoid of revolution, being flattened at the poles and bulging at the equator. The mean equatorial radius, in fact, exceeds the polar radius by 21 km. This flattening was caused by centrifugal effects in the rotating, plastic earth. The equatorial bulging means that there should be a steady increase in the measured value of g as one goes from the equator (latitude 0°) to the poles (latitude 90°); Table 15–2 verifies this expectation. Note that g increases by ~0.53% as one moves from the equator to the poles.

Table 15–2 Variation of g with latitude at sea level

Latitude	g, m/s²	Latitude	g, m/s²
0° (equator)	9.78039	50°	9.81071
10°	9.78195	60°	9.81918
20°	9.78641	70°	9.82608
30°	9.79329	80°	9.83059
40°	9.80171	90° (poles)	9.83217

The existence of oceans and continents alone tells us that the earth does not have a uniform surface density, as we assumed earlier. Because our concern is with gravity variations rather than the geometrical shape of the earth we focus attention, not on the reference ellipsoid above, but on a hypothetical closed surface called the *geoid*. It is defined so that the direction of a plumb line, that is, the direction of **g**, is everywhere at right angles to it. The geoid coincides with mean sea level over the oceans and with its extension through the continents. It lies somewhat outside the reference ellipsoid under the continents and somewhat inside over the oceans.

The geoid

In 1959 it was observed that the orbit of Vanguard, one of this country's early satellites, followed an orbit that did not agree exactly with that calculated using values of **g** based on a near-ellipsoidal geoid. It was concluded that the geoid is best approximated, not by an ellipsoid of revolution, but by a slightly pear-shaped figure, the small end of the "pear" being in the northern hemisphere and extending about 15 m above the reference ellipsoid. The motion of a satellite is governed at all times by the values of **g** along its orbit. Thus an earth satellite forms a useful probe to explore the values of **g** near the surface of the earth and from this to deduce information about the shape of the geoid.

The rotating earth

In all of the above we have ignored effects due to the rotation of the earth except that the values of g displayed in Tables 15–1 and 15–2, being actual measured values, necessarily take it into account. We analyze its effect in the following example.

Example 3 *Effect on g of the rotation of the earth*   Figure 15–6 is a schematic view of the earth looking down on the North Pole. In it we show an enlarged view of a body of mass m hanging from a spring balance at the equator. The forces on this body are the upward pull of the spring balance **w**, which is the apparent weight of the body, and the downward pull of the earth's gravitational attraction $F = GmM_e/R_e^2$. This body is not in equilibrium because it experiences a centripetal acceleration $\mathbf{a}_R$ as it rotates with the earth. There must, therefore, be a net force acting on the body toward the center of the earth. Consequently, the force **F** of gravitational attraction (the true weight of the body) must exceed the upward pull of the balance **w** (the apparent weight of the body).

From the second law of motion we obtain
$$F - w = ma_R,$$
$$\frac{GM_e m}{R_e^2} - mg = ma_R,$$
$$g = \frac{GM_e}{R_e^2} - a_R \qquad \text{at the equator.}$$

At the poles $a_R = 0$, so that
$$g = \frac{GM_e}{R_e^2} \qquad \text{at the poles.}$$

This is the value of g we would obtain anywhere (assuming a spherical earth) were the rotation of the earth to be neglected.

Actually, the centripetal acceleration is not directed in toward the center of the earth other than at the equator. It is directed perpendicularly in toward the earth's axis of rotation at any given latitude. The detailed analysis is, therefore, really a two-dimensional one. However, the extreme case is at the equator. There

$$a_R = \omega^2 R_e = \left(\frac{2\pi}{T}\right)^2 R_e = \frac{4\pi^2 R_e}{T^2},$$

in which ω is the angular speed of the earth's rotation, T is the period, and R_e is the radius of the earth. Using the values

$$R_e = 6.37 \times 10^6 \text{ m},$$

$$T = 8.64 \times 10^4 \text{ s},$$

we obtain

$$a_R = 0.0336 \text{ m/s}^2.$$

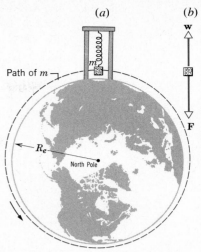

Figure 15-6 Example 3. Effect of the earth's rotation on the weight of a body as measured by a spring balance.

Any measurement of g as a function of latitude will reveal the combined effects of the earth's asphericity and its rotation. Table 15-2 shows the total variation. Comparing the data in this table with the results of Example 3 reveals that more than half (actually 65%) the difference between the observed values of g at low and high latitudes is due to the rotation effect, the remaining (35%) difference being associated with the earth's equatorial bulge.

15-6 The Gravitational Field

The central fact of gravitation is that two mass particles exert forces on one another. We can think of this as a direct interaction between the two mass particles, if we wish. This point of view is called *action at a distance,* the particles interacting even though they are not in contact. Another point of view is the field concept, which regards a mass particle as modifying the space around it in some way and setting up a *gravitational field.* This field then acts on any other mass particle in it, exerting the force of gravitational attraction on it. The field, therefore, plays an intermediate role in our thinking about the forces between mass particles. According to this view we have two separate parts to our problem. First, we must determine the field established by a given distribution of mass particles; and second, we must calculate the force that this field exerts on another mass particle placed in it.

For example, consider the earth as an isolated mass. If a body is now brought in the vicinity of the earth, a force is exerted on it. This force has a definite direction and magnitude at each point in space. The direction is radially in toward the center of the earth and the magnitude is mg. We can, therefore, associate with each point near the earth a vector $\mathbf{g}$, which is the acceleration that a body would experience if it were released at this point. We call $\mathbf{g}$ the *gravitational field* at the point in question. Since

The gravitational field

$$\mathbf{g} = \frac{\mathbf{F}}{m},$$ (15-11)

we may define the gravitational field at any point as the gravitational force per unit mass at that point. We calculate the force from the field simply by multiplying **g** by the mass *m* of the particle placed at any point.

The gravitational field is an example of a *vector field,* each point in this field having a vector associated with it. There are also scalar fields, such as the temperature field in a heat-conducting solid. The gravitational field arising from a fixed distribution of matter is also an example of a *stationary field,* because the value of the field at a given point does not change with time.

The field concept is particularly useful for understanding electromagnetic forces between moving electric charges. It has distinct advantages, both conceptually and in practice, over the action-at-a-distance concept. The field concept was not used in Newton's day. It was developed much later by Faraday for electromagnetism and only later applied to gravitation. Subsequently, this point of view was adopted for gravitation in the general theory of relativity. The chief purpose of introducing it here is to give you an early familiarity with a concept that proves to be important in the development of physical theory.

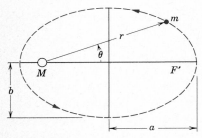

Figure 15–7 A planet of mass *m* moving in an elliptical orbit in the gravitational field of the sun, of mass *M*. The sun is at one focus of the ellipse; *F'* marks the other (empty) focus.

Kepler's laws of planetary motion

Kepler's first law

15–7 The Motions of Planets and Satellites

Through a lifelong study of the motions of bodies in the solar system Johannes Kepler (1571–1630) was able to derive empirically the three basic laws of planetary motion bearing his name. Tycho Brahe (1546–1601), who was the last and greatest astronomer to make observations without the use of a telescope, compiled the data. Kepler, who had been Brahe's assistant, after years of laborious calculations found these regularities, now known as Kepler's three laws of planetary motion.

1. All planets move in elliptical orbits having the sun as one focus (the law of orbits).

2. A line joining any planet to the sun sweeps out equal areas in equal times (the law of areas).

3. The square of the period of any planet is proportional to the cube of the semimajor axis of its orbit (the law of periods).

These laws, which were derived analytically by Isaac Newton from his laws of motion and his law of universal gravitation, apply not only to planets moving in the gravitational field of the sun but also to natural and artificial satellites moving in the gravitational field of a planet. Let us examine each of Kepler's laws in turn.

1. *The law of orbits.* Figure 15–7 shows a planet moving in an elliptical orbit around the sun. We assume that the mass *M* of the sun greatly exceeds the mass *m* of the planet so that the center of mass of the sun-planet system is virtually at the center of the sun. We thus take the sun to remain at rest at one focus of the elliptical orbit.

The orbit in Fig. 15–7 is specified by giving its semimajor axis *a* and its semiminor axis *b*. For planets in our solar system the percent difference between *a* and *b* ranges from ~0.002% (for the orbit of Venus) to ~3% (for the orbit of Pluto). For comparison, the orbit drawn in the figure has a percent difference between *a* and *b* of 35%.

The equation of the orbit, expressed in terms of the variables r and θ, is

$$\frac{1}{r} = \frac{1}{b^2} (a - \sqrt{a^2 - b^2} \cos \theta). \tag{15-12}$$

In the special case of $b = a$ the ellipse reduces to a circle. If we make this substitution in Eq. 15–12 we see that, as we expect, r always has the constant value a, no matter what the value of θ.

2.   *The law of areas.* Figure 15–8a again shows a planet moving around the sun. In a short time Δt the line joining these two bodies sweeps through an angle $\Delta\theta \ (= \omega \, \Delta t)$. The area ΔA of the long shaded wedge in the figure is approximately one-half its base ($\cong r \, \Delta\theta$) times its altitude ($\cong r$) or $\frac{1}{2} r^2 \, \Delta\theta$. This expression for ΔA becomes more exact in the limit as $\Delta t \to 0$. The instantaneous rate dA/dt at which area is being swept out is then found from

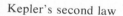

Kepler's second law

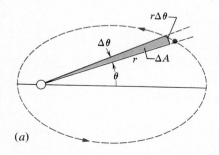

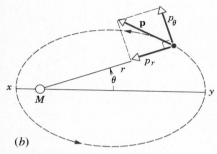

(a)

(b)

Figure 15–8   (*a*) In time Δt the line connecting the planet to the sun sweeps through an angle $\Delta\theta$. (*b*) Showing the linear momentum **p** of the planet; its angular momentum ℓ is equal to $p_\theta r$.

$$\frac{dA}{dt} = \lim_{\Delta t \to 0} \frac{\frac{1}{2} r^2 \, \Delta\theta}{\Delta t} = \frac{1}{2} r^2 \frac{d\theta}{dt} = \frac{1}{2} r^2 \, \omega. \tag{15-13a}$$

We turn now to Fig. 15–8b, which shows the linear momentum **p** of the orbiting planet, along with its scalar components $p_r \ (= mv_r)$ and $p_\theta \ (= mv_\theta)$. We can write for the angular momentum ℓ of the planet about the sun (see Eq. 12–3)

$$\ell = p_\theta r = (mv_\theta)(r) = (m\omega r)(r) = mr^2\omega. \tag{15-13b}$$

Combining Eqs. 15–13 gives

$$\frac{dA}{dt} = \ell/2m$$

But the law of conservation of angular momentum (see Section 12–8) tells us that for this isolated planet-sun system, on which no external torques act, the angular momentum ℓ must remain constant. Thus dA/dt must also remain constant, which is the content of Kepler's second law.

We see that Kepler's second law, which requires that the rate of sweeping out of area $\frac{1}{2}\omega r^2$ be constant, is entirely equivalent to the statement that the angular momentum of any planet about the sun remains constant. The angular momentum of the planet about the sun cannot be changed by a force on it directed toward the sun. Kepler's second law would, therefore, be valid for any *central force*, that is, any force directed toward the sun. The exact nature of this force—how it depends on distance of separation or other properties of the bodies—is not revealed by this law.

Kepler's first law, however, requires not only that the gravitational force be central but also that it depend exactly on the inverse square of the distance between two bodies, that is, on $1/r^2$. Only such a force, it turns out, can yield planetary orbits which are elliptical with the sun at one focus.

3.   *The law of periods.* We start by considering only the special case of a circular orbit of radius a, as shown in Fig. 15–9; later we will extend the analysis to the general case of elliptical orbits. As before we assume that $M \gg m$.

The gravitational force acting on the planet produces a centripetal acceleration of $\omega^2 a$ or, from Newton's second law,

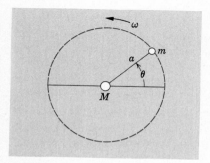

Figure 15–9   A planet of mass m moving in a circular orbit of radius a about the sun, of mass M.

$$\frac{GMm}{a^2} = (m)(\omega^2 a) \tag{15-14a}$$

If we replace ω by its equal ($2\pi/T$, where T is the period of the motion) we obtain after rearrangement,

Kepler's third law

$$T^2 = \left(\frac{4\pi^2}{GM}\right) a^3. \tag{15-14b}$$

This relation has been derived only for a circular orbit. We assert without proof however that it also holds for elliptical orbits if we take a to be the semimajor axis of the ellipse, as in Fig. 15–7. With this interpretation of a, Eq. 15–14 becomes a statement of Kepler's third law. Note that the mass m of the planet does not appear and that the proportionality constant $(4\pi^2/GM)$ is the same for all planets.

Example 4 From the period and radius of revolution of the earth about the sun, find the mass of the sun.

The earth's period is

$$T = 365 \text{ days} = 3.15 \times 10^7 \text{ s},$$

and its mean orbital radius is

$$a = 1.50 \times 10^{11} \text{ m}.$$

Hence, from Eq. 15–14,

$$M_s = \frac{4\pi^2 a^3}{GT^2} = \frac{(4\pi^2)(1.50 \times 10^{11} \text{ m})^3}{(6.67 \times 10^{-11} \text{ m}^3/\text{kg} \cdot \text{s}^2)(3.15 \times 10^7 \text{ s})^2}$$
$$= 2.01 \times 10^{30} \text{ kg}.$$

The mass of the sun is thus 330,000 times the mass of the earth. In a similar manner we can determine the mass of the earth from the period and radius of the moon's orbit about the earth. (See Problem 29.)

15–8 Gravitational Potential Energy

In Chapter 8 we discussed the gravitational potential energy of a particle (mass m) and the earth (mass M). We considered only the special case in which the particle remains close to the earth so that we could assume the gravitational force acting on the particle to be constant for all positions of the particle. In this section we remove that restriction and consider particle-earth separations that may be appreciably greater than the earth's radius.

Equation 8–5b, which we may write as

$$\Delta U = U_b - U_a = -W_{ab}, \qquad [8-5b]$$

defines the change ΔU in the potential energy of any system, in which a conservative force (gravity, say) acts, as the system changes from configuration a to configuration b. W_{ab} is the work done by that conservative force as the system changes.

The potential energy of the system in any arbitrary configuration b is (see Eq. 8–5b)

$$U_b = -W_{ab} + U_a. \qquad (15-15)$$

To give a value to U_b we must (arbitrarily) choose configuration a to be some agreed-upon reference configuration and we must assign to U_a some (arbitrarily) agreed-upon value, usually zero.

In Chapter 8 we chose as a reference configuration for the earth-particle system that in which the particle is resting on the surface of the earth and we assigned to this configuration the potential energy $U_a = 0$. When the particle is at a height y above the surface of the earth, the potential energy $U(= U_b)$ is given from Eq. 15–15 as

$$U = -W_{ab} + 0 = -(-mg)(y) = mgy.$$

The conservative force in question, gravity, points down and has the value $(-mg)$; the displacement of the particle $(+y)$ points up from the reference level; hence the difference in sign for these quantities.

For the more general case, in which the restriction $y \ll R$ (in which R is the radius of the earth) is not imposed, we find it convenient to select a different reference configuration, namely, that in which the particle and the earth are infinitely far apart. We assign the value zero to the potential energy of the system in this configuration. Thus the zero-potential-energy configuration is also the zero-force configuration. We made a similar choice when we defined the zero-energy configuration of a spring to be its normal unstressed state, for which the restoring force is zero.

When the particle of mass m is a distance r from the center of the earth, the system potential energy is given by Eq. 15–15 as

$$U(r) = -W_{\infty r} + 0, \tag{15-16}$$

in which $W_{\infty r}$ is the work done by the conservative force (gravity) on the particle as the particle moves in from infinity to a distance r from the center of the earth. For simplicity we assume for the present that the particle moves toward the earth along a radial line. The gravitational force $F(r)$ acting on the particle (assuming $r \geq R$) will then be $-GMm/r^2$, the minus sign indicating an attractive force, that is, a force that pulls the particle toward the earth. We may then find $U(r)$ from Eq. 15–16 as

Gravitational potential energy

$$
\begin{aligned}
U(r) &= -W_{\infty r} \\
&= -\int_{\infty}^{r} F(r')\, dr' \\
&= -\int_{\infty}^{r} \left(-\frac{GMm}{r'^2}\right) dr' = -\frac{GMm}{r'}\bigg|_{\infty}^{r} \\
&= -\frac{GMm}{r}.
\end{aligned} \tag{15-17}
$$

The minus sign indicates that the potential energy is negative at any finite distance; that is, the potential energy is zero at infinity and decreases as the separation distance decreases. This corresponds to the fact that the gravitational force exerted on the particle by the earth is attractive. As the particle moves in from infinity, the work $W_{\infty r}$ done by this force on the particle is positive, which means, from Eq. 15–16, that $U(r)$ is negative.

Equation 15–17 holds no matter what path is followed by the particle in moving in from infinity to radius r. We can show this by breaking up any arbitrary path into infinitesimal steplike portions, which are drawn alternately along a radius and perpendicular to one (Fig. 15–10). No work is done along perpendicular segments, like AB, because along them the force is perpendicular to the displacement. But the work done along the radial parts of the path, such as BC, adds up to the work done in going directly along a radial path, such as AE. The work done in moving the particle between two points in a gravitational field is, therefore, independent of the actual path connecting these points. Hence, the gravitational force is a conservative force.

Equation 15–17 shows that the potential energy of the particles M and m is a characteristic of the system $M + m$. The potential energy is a property of the

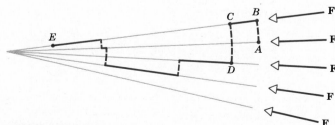

Figure 15–10 Work done in taking a mass from E to E is independent of the path.

system of bodies, rather than of either body alone. The potential energy changes whether M or m is displaced; each is in the gravitational field of the other. Nor does it make any sense to assign part of the potential energy to M and part of it to m. Often, however, we do speak of the potential energy of a body m (planet or stone, say) in the gravitational field of a much more massive body M (sun or earth, respectively). The justification for speaking as though the potential energy belongs to the planet or to the stone alone is this: When the potential energy of a system of two bodies changes into kinetic energy, the lighter body gets most of the kinetic energy. The sun is so much more massive than a planet that the sun receives hardly any of the kinetic energy; and the same is true for the earth in the earth-stone system.

We can derive the gravitational force from the potential energy. The relation for spherically symmetric potential energy functions is $F = -dU/dr$; (see Eq. 8–7). This relation is the converse of Eq. 15–17. From it we obtain

$$F = -\frac{dU}{dr} = -\frac{d}{dr}\left(-\frac{GMm}{r}\right) = -\frac{GMm}{r^2}. \qquad (15-18)$$

The minus sign here shows that the force is an attractive one, directed inward along a radius opposite to the radial displacement vector.

Example 5 *Escape velocity*   We can readily find the gravitational potential energy of a particle of mass m at the surface of the earth as (Eq. 15–17) $U(R) = -GM_e m/R_e$. The amount of work required to move a body from the surface of the earth to infinity is $GM_e m/R_e$, or about 6.0×10^7 J/kg. If we could give a projectile more than this energy at the surface of the earth, then, neglecting the resistance of the earth's atmosphere, it would escape from the earth never to return. As it proceeds outward its kinetic energy decreases and its potential energy increases, but its speed is never reduced to zero. The critical initial speed, called the escape speed v_0, such that the projectile does not return, is given by

$$\tfrac{1}{2}mv_0{}^2 = \frac{GM_e m}{R_e}$$

or

$$v_0 = \sqrt{2\,\frac{GM_e}{R_e}} = 7.0 \text{ mi/s} = 25{,}000 \text{ mi/h} = 11 \text{ km/s}.$$

Should a projectile be given this initial speed, it would escape from the earth. For initial speeds less than this the projectile will return. Its kinetic energy becomes zero at some finite distance from the earth and the projectile falls back to earth.

The lighter molecules in the earth's upper atmosphere can attain high enough speeds by thermal agitation to escape into outer space. Hydrogen gas, which must have been present in the earth's atmosphere a long time ago, has now disappeared from it. Helium gas escapes at a steady rate, much of it resupplied by radioactive decay from the earth's crust. The escape speed for the sun is much too great to allow hydrogen to escape from its atmosphere. On the other hand, for most gases the speed of escape on the moon is so small that it can hardly keep any atmosphere at all.

15–9   Potential Energy for Systems of Particles

If two particles are separated by a distance r, their potential energy is given from Eq. 15–16 as

$$U(r) = -W_{\infty r}, \qquad [15-16]$$

in which $W_{\infty r}$ is the work done by gravitational force as the particles move from an infinite separation to separation r. We now give another interpretation to $U(r)$.

Let us balance out the gravitational force by an external force applied by some external agent and let us arrange it so that, at all times, this external force is

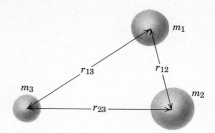

Figure 15-11   Three masses m_1, m_2, and m_3 brought together from infinity.

equal and opposite to the gravitational force for each particle. The work done by the external force as the particles move from an infinite separation to separation r is not $W_{\infty r}$ but $-W_{\infty r}$; this follows because the displacements are the same but the forces are equal and opposite. Thus we may interpret Eq. 15–16 as follows: *The potential energy of a system of particles is equal to the work that must be done by an external agent to assemble the system, starting from the standard reference configuration.*

Thus, if you lift a stone of mass m a distance of y above the earth's surface, you are the external agent (separating earth and stone) and the work you do in "assembling the system" is $+mgy$, which is also the potential energy. Similarly, the work done by an external agent as a body of mass m moves in from infinity to a distance r from the earth is negative because the agent must exert a restraining force on the body; this is in agreement with Eq. 15–16.

These considerations hold for systems that contain more than two particles. Consider three bodies of masses m_1, m_2, and m_3. Let them initially be infinitely far from one another. The problem is to compute the work done by an external agent to bring them into the positions shown in Fig. 15–11. Let us first bring m_2 in toward m_1 from an infinite separation to the separation r_{12}. The work done against the gravitational force exerted by m_1 on m_2 is $-Gm_1m_2/r_{12}$. Now let us bring m_3 in from infinity to the separation r_{13} from m_1 and r_{23} from m_2. The work done against the gravitational force exerted by m_1 on m_3 is $-Gm_1m_3/r_{13}$ and that against the gravitational force exerted by m_2 on m_3 is $-Gm_2m_3/r_{23}$. The total work done in assembling this system is the total potential energy of the system

$$-\left(\frac{Gm_1m_2}{r_{12}} + \frac{Gm_1m_3}{r_{13}} + \frac{Gm_2m_3}{r_{23}}\right).$$

Notice that no vector operations are needed in this procedure.

No matter how we assemble the system, that is, regardless of which bodies are moved or which paths are taken, we always find this same amount of work required to bring the bodies into the configuration of Fig. 15–11 from an initial infinite separation. The potential energy must, therefore, be associated with the system rather than with any one or two bodies. If we wanted to separate the system into three isolated masses once again, we would have to supply an amount of energy

$$+\left(\frac{Gm_1m_2}{r_{12}} + \frac{Gm_1m_3}{r_{13}} + \frac{Gm_2m_3}{r_{23}}\right)$$

This energy may be regarded as a sort of *binding energy* holding the mass particles together in the configuration shown. We define binding energy to be that amount of energy that must be added to a system in order to separate the particles of interest to infinite distances from one another. This is a useful concept in many areas of science such as molecular, atomic, and nuclear physics.

Just as we can associate elastic potential energy with the compressed or stretched configuration of a spring holding a mass particle, so we can associate gravitational potential energy with the configuration of a system of mass particles held together by gravitational forces. Similarly, if we want to think of the elastic potential energy of a particle as being stored in the spring, so we can think of the gravitational potential energy as being stored in the gravitational field of the system of particles. A change in either configuration results in a change of potential energy.

These concepts occur again when we meet forces of electric or magnetic origin, or, in fact, of nuclear origin. Their application is rather broad in physics. The advantage of the energy method over the dynamical method is derived from the fact that the energy method uses scalar quantities and scalar operations rather than vector quantities and vector operations. When the actual forces are not known, as is often the case in nuclear physics, the energy method is essential.

15–10 Energy Considerations in the Motions of Planets and Satellites

Consider again the motion of a body of mass m (planet or satellite, say) about a massive body of mass M (sun or earth, say). We shall consider M to be at rest in an inertial reference frame (a good approximation if $M \gg m$) with the body m moving about it in a circular orbit. The potential energy of the system is

$$U(r) = -\frac{GMm}{r_0},$$

where r_0 is the radius of the circular orbit. The kinetic energy of the system is approximately

$$K = \tfrac{1}{2}m\omega^2 r_0{}^2.$$

From Eq. 15–14a we obtain (with r_0 substituted for a)

$$\omega^2 r_0{}^2 = \frac{GM}{r_0},$$

so that

$$K = \frac{1}{2}\frac{GMm}{r_0}.$$

The mechanical energy is

$$E = K + U = \frac{1}{2}\frac{GMm}{r_0} - \frac{GMm}{r_0} = -\frac{GMm}{2r_0}. \tag{15–19}$$

This energy is constant and is negative. Now the kinetic energy can never be negative, but from Eq. 15–19 we see that it must go to zero as the separation goes to infinity. The potential energy is always negative, except for its zero value at infinite separation. The meaning of the negative mechanical energy, then, is that the system is a closed one, the planet m always being bound to the attracting solar center M and never escaping from it (Fig. 15–12).

Even when we consider elliptical orbits, in which r and ω vary, the mechanical energy is negative. It is also constant, corresponding to the fact that gravitational forces are conservative. Hence, both the mechanical energy and the total angular momentum are constant in planetary motion. These quantities are often called *constants of the motion*. We obtain the actual orbit of a planet with respect to the sun by starting with these conservation relations and eliminating the time variable by use of the laws of dynamics and gravitation. The result is that planetary orbits are elliptical.

In the earlier theories of the atom, as in the Bohr theory of the hydrogen atom, these identical mechanical relations are used in describing the motion of an electron about an attracting nuclear center. These same relations are used for open orbits (total energy positive) as in the experiments of Rutherford on the scattering of charged nuclear particles. Central forces, and particularly inverse square forces, are encountered often in physical systems.

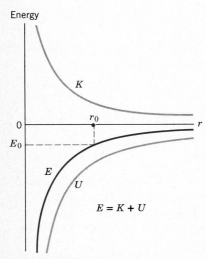

Figure 15–12 Kinetic energy K, potential energy U, and mechanical energy $E = U + K$ of a body in circular planetary motion. A planet with total energy $E_0 < 0$ will remain in an orbit of radius r_0. The farther the planet is from the sun, the greater (that is, less negative) its (constant) total energy E. To escape from the center of force and still have kinetic energy at infinity, it would need positive total energy.

Example 6 What is the binding energy of the earth-sun system? Neglect the presence of other planets or satellites.

For simplicity, assume that the earth's orbit about the sun is circular at a radius r_{es}. The work done against the gravitational force to bring the earth and sun from an infinite separation to a separation r_{es} is

$$-G\frac{M_s M_e}{r_{es}} \cong -5.0 \times 10^{33} \text{ J,}$$

where we take $M_s \cong 330,000 M_e$, $M_e = 6.0 \times 10^{24}$ kg, $r_{es} = 150 \times 10^9$ m. The minus sign indicates that the force is attrac-

tive, so that work is done by the gravitational force. It would take an equivalent amount of work by an outside agent to separate these bodies completely from rest. Because the kinetic energy of the earth in its orbit is half the magnitude of the potential energy of the earth-sun system, only half of this work is needed to break up the system, so that the effective binding energy, assuming that the earth-sun system is at rest after breakup, is about 2.5×10^{33} J.

What effect does the presence of the moon and other planets have on the energy binding the earth to the solar system?

Maneuvering a satellite

Let us now see how energy considerations are involved in the problem of maneuvering a manned satellite in earth orbit. The kinetic and potential energies of such a satellite in a circular orbit of radius r_0 are

$$K = +\frac{GMm}{2r_0} \quad \text{and} \quad U = -\frac{GMm}{r_0}. \qquad [15\text{--}19]$$

Suppose that our task is to increase the speed of the ship by firing a short burst from an on-board thruster. Just how should the astronaut do this?

To increase K the effect of the burst must be to move the satellite to an orbit of smaller radius, say $r - \Delta r$; we assume for simplicity in what follows that $\Delta r \ll r$. The resulting change in K is then

$$\Delta K = K_f - K_i$$
$$= \frac{GMm}{2}\left(\frac{1}{r - \Delta r} - \frac{1}{r}\right) = \frac{GMm}{2}\left[\frac{\Delta r}{r(r - \Delta r)}\right]$$
$$\cong +\frac{GMm\,\Delta r}{2r^2}.$$

In the same way we can show that the change in the potential energy U associated with moving to the new orbit is

$$\Delta U \cong -\frac{GMm\,\Delta r}{r^2},$$

which shows us that

$$\Delta K = \tfrac{1}{2}|\Delta U|.$$

Thus although the satellite loses a certain potential energy (let us say 400 J, for concreteness) in moving to the new orbit only half of this (200 J) appears as increased kinetic energy. This is in sharp contrast to simple free fall in which *all* of the "lost" potential energy is transformed into kinetic energy.

Thrusters ahead to gain speed

What happened to the "missing" 200 J? It represents work done *by* the satellite *on* the expelled gases during the burn (not the other way around). Thus the astronaut must fire his thruster in the *forward* direction if he wants to gain speed! If he fired it backwards the expelled gases would do work on the satellite, which would then slow down and move to an orbit of larger radius.

Example 7 Figure 15–13 shows two manned satellites in the same circular earth orbit, of radius r ($= 6800$ km). The astronauts, Sally and Igor, find themselves drifting around the orbit separated by a constant angle Φ ($= 4.5° = 0.0785$ rad). How can Sally catch up with Igor?

From Eq. 15–14a (with r substituted for a) the common angular speed of the two astronauts is

$$\omega = \frac{(GM)^{1/2}}{r^{3/2}} = \frac{(6.67 \times 10^{-11} \text{ m}^3/\text{kg} \cdot \text{s}^2)^{1/2}(5.98 \times 10^{24} \text{ kg})^{1/2}}{(6.80 \times 10^6 \text{ m})^{3/2}}$$
$$= 1.12 \times 10^{-3} \text{ rad/s}$$

To catch up Sally must increase her angular speed. She fires a (forward) burst in such a way as to take her ship to a circular orbit of smaller radius ($= r - \Delta r$), in which her angular speed will now be

$$\omega + \Delta\omega = \frac{(GM)^{1/2}}{(r - \Delta r)^{3/2}} = \frac{(GM)^{1/2}}{r^{3/2}}\left(1 - \frac{\Delta r}{r}\right)^{-3/2}$$

$$= \omega\left(1 - \frac{\Delta r}{r}\right)^{-3/2}.$$

If we assume that $\Delta r \ll r$ we can approximate this as

$$\omega + \Delta\omega \cong \omega\left(1 + \frac{3\Delta r}{2r}\right).$$

Thus Sally's angular speed now exceeds Igor's by

$$\Delta\omega = \frac{3\omega\Delta r}{2r}.$$

With this differential she will catch up in a time t_0 given by

$$t_0 = \frac{\Phi}{\Delta\omega} = \frac{2\Phi r}{3\omega\Delta r}.$$

If she adjusts the burn to give, say, $\Delta r = 80$ km (1.2% of r)

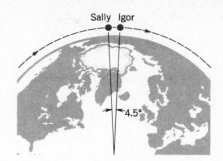

Figure 15–13 Example 7. How can S maneuver to catch up with I?

then

$$t_0 = \frac{(2)(0.0785 \text{ rad})(6.80 \times 10^6 \text{ m})}{(3)(1.12 \times 10^{-3} \text{ rad/s})(8.0 \times 10^4 \text{ m})} = 1.10 \text{ h.}$$

After this elapsed time Sally simply fires her thruster again (backwards this time!) and slows down to join Igor in the original orbit.

REVIEW GUIDE AND SUMMARY

Law of universal gravitation

Newton's law of universal gravitation says that any two mass particles attract each other with a force given by

$$F = G\frac{m_1 m_2}{r^2} \qquad [15\text{–}1]$$

where G, the *gravitational constant*, is 6.67×10^{-11} m³/kg · s².

Gravitational and inertial mass

The mass involved in this equation (the *gravitational mass*) is not necessarily the same property of a body as the mass in $\mathbf{F} = m\mathbf{a}$ (the *inertial mass*). The first need not involve inertia (lifting a bowling ball) and the second need not involve gravity (kicking a bowling ball). Experiment shows, however, that these masses are identical, an assumption called the *principle of equivalence*.

The gravitational behavior of uniform spherical shells

Equation 15–1 holds strictly for mass points. Consider, however, an object of finite extent, a solid spherical body whose density $\rho(r)$ is either constant or varies with r only. We show that for external points such a body behaves gravitationally as if all its mass were concentrated at its center. For internal points that part of the sphere beyond the radius in question exerts no gravitational effect at all; only the core of the sphere that lies within r is gravitationally effective; see Section 15–4.

Variations of g near the earth

The acceleration of gravity g due to the earth, which we often take as constant ($= 9.80$ m/s²) for points near the earth's surface, varies with altitude, as Eq. 15–9 and Table 15–1 show. It also varies with latitude (Table 15–2), increasing from the equator to the poles. There are local variations depending on the presence or absence of land masses and oceans, the nature of the underlying geological formations, and so on. Finally, the rotation of the earth causes the measured value of g to be lower than it would be on a nonrotating earth (Example 3).

Planets and satellites

Gravitational attraction holds the solar system together and makes possible orbiting earth satellites, both natural (the moon) and artificial (Telstar, etc.). Such motions are governed by Kepler's three laws:

Kepler's three laws

1. *The law of orbits:* All planets move in elliptical orbits with the sun at one focus.

2. *The law of areas:* A line joining any planet to the sun sweeps out equal areas in equal times. This turns out to be a consequence of the law of conservation of angular momentum; see Fig. 15–8.

3. *The law of periods:* The square of the period of any planet about the sun is proportional to the cube of the semimajor axis.

Gravitational potential energy

The gravitational potential energy $U(r)$ of two objects of masses M and m is the work required to bring them in from infinity to a separation r. This work is negative, and we showed that

$$U(r) = -\frac{GMm}{r}.$$ [15–17]

Escape velocity

If M is the mass of the earth (radius R) and m that of a rocket resting on its surface, the rocket can escape from earth if its initial kinetic energy is equal to $U(R)$; the *escape velocity* for objects from earth proves to be 11 km/s; see Example 5.

Energies in planetary motion

If m represents a planet or satellite moving in a circular orbit of radius r_0, the potential energy U, the kinetic energy K, and the mechanical energy $E\ (= K + U)$ prove to be

$$U = -\frac{GMm}{r_0}, \qquad K = +\frac{GMm}{2r_0}, \qquad \text{and} \qquad E = -\frac{GMm}{2r_0}.$$ [15–17, 15–19]

The fact that E is negative means that m is bound to M. A positive *binding energy* equal to $-E$ must be provided from an external source to separate them; study Fig. 15–11 and Example 6.

QUESTIONS

1. If the force of gravity acts on all bodies in proportion to their masses, why doesn't a heavy body fall correspondingly faster than a light body?

2. How does the weight of a body vary en route from the earth to the moon? Would its mass change?

3. You can experience weightlessness—however briefly—by jumping from the top of one of the World Trade Center towers. Can you think of any way(s) that you can experience weightlessness while moving *up*?

4. How could you determine whether two objects have the same gravitational mass? . . . the same inertial mass? . . . the same weight?

5. At the earth's surface a freely suspended object is given a horizontal blow by a hammer. The object is taken to the moon, suspended freely, and given an equal horizontal blow with the same hammer. How is the resulting horizontal speed on the moon related to the horizontal speed on the earth?

6. Would we have more sugar to the pound at the pole or the equator? What about sugar to the kilogram?

7. What approximately is the *gravitational* force of attraction between Rhoda and Ronald if they are 10 m apart? When they are dancing?

8. Because the earth bulges near the equator, the source of the Mississippi River, although high above sea level, is nearer to the center of the earth than is its mouth. How can the river flow "uphill"?

9. How could you determine the mass of the moon?

10. One clock is based on an oscillating spring, the other on a pendulum. Both are taken to Mars. Will they keep the same time there that they kept on Earth? Will they agree with each other? Explain. Mars has a mass 0.1 that of Earth and a radius half as great.

11. From Kepler's second law and observations of the sun's motion as seen from the earth, we can conclude that the earth is closer to the sun during winter in the Northern Hemisphere than during summer. Why isn't it colder in summer than in winter?

12. Saturn is about six times farther from the sun than Mars. Which planet has the greater period of revolution? . . . the greater orbital speed? . . . the greater angular speed?

13. A satellite orbiting the earth around the Arctic Circle would be very useful in maintaining surveillance of this strategically important part of the globe. Why don't we put one up?

14. As a car speeds around a curve, the passengers tend to be thrown radially outward. Why are astronauts in a space shuttle not similarly affected as their shuttle speeds in orbit around the earth?

15. The gravitational force exerted by the sun on the moon is about twice as great as the gravitational force exerted by the earth on the moon. Why then doesn't the moon escape from the earth?

16. Explain why the following reasoning is wrong. "The sun attracts all bodies on the earth. At midnight, when the sun is directly below, it pulls on an object in the same direction as the pull of the earth on that object; at noon, when the sun is directly above, it pulls on an object in a direction opposite to the pull of the earth. Hence, all objects should be heavier at midnight (or night) than they are at noon (or day)."

17. The gravitational attraction of the sun and the moon on the earth produces tides. The sun's tidal effect is about half as great as that of the moon's. The direct pull of the sun on the earth, however, is about 175 times that of the moon. Why is it, then, that the moon causes the larger tides?

18. Particularly large tides, called *spring tides,* occur at full moon and at new moon, when the configurations of the sun, earth, and moon are as shown in Fig. 15–14. From the figure you might conclude (incorrectly!) that the tidal effects of the sun and of the moon tend to add at new moon but to cancel at full moon. Instead, they add at both of these configurations. Can you explain why?

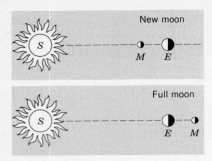

Figure 15–14   Question 18.

19. If lunar tides slow down the rotation of the earth (owing to friction), the angular momentum of the earth decreases. What happens to the motion of the moon as a consequence of the conservation of angular momentum? Does the sun (and solar tides) play a role here? (See "Tides and the Earth-Moon System" by Peter Goldreich, *Scientific American,* April 1972.)

20. A satellite in earth orbit experiences a small drag force as it starts to enter the earth's atmosphere. What happens to its speed? (Careful!)

21. Would you expect the total mechanical energy of the solar system to be constant? The total angular momentum? Explain your answers.

22. Does a rocket always need the escape speed of 25,000 mi/h to escape from the earth? What, then, does "escape speed" really mean?

23. Objects at rest on the earth's surface move in circular paths with a period of 24 h. Why are they not "in orbit" in the sense that an earth satellite is in orbit? What would the length of the "day" have to be to put such objects in true orbit?

24. Neglecting air friction and technical difficulties, can a satellite be put into an orbit by being fired from a huge cannon at the earth's surface? Explain.

25. Can a satellite coast in a stable orbit in a plane not passing through the earth's center? Explain.

26. As measured by an observer on earth, would there be any difference in the periods of two satellites, each in a circular orbit near the earth in an equatorial plane, but one moving eastward and the other westward?

27. After Sputnik I was put into orbit it was said that it would not return to earth but would burn up in its *descent.* Considering the fact that it did not burn up in its *ascent,* how is this possible?

28. An artificial satellite is in a circular orbit about the earth. How will its orbit change if one of its rockets is momentarily fired (*a*) toward the earth, (*b*) away from the earth, (*c*) in a forward direction, (*d*) in a backward direction, (*e*) at right angles to the plane of the orbit?

29. Inside a spaceship, what difficulties would you encounter in walking? In jumping? In drinking?

30. We have all seen TV transmissions from orbiting satellites and watched objects floating around in effective zero gravity. Suppose that an astronaut, bracing himself against the satellite frame, kicks a floating bowling ball. Will he stub his toe? Explain.

31. If a planet of given uniform density were made larger, its force of attraction for an object on its surface would increase because of the planet's greater mass but would decrease because of the greater distance from the object to the center of the planet. Which effect predominates?

32. The gravitational field associated with the earth is zero both at infinity and at the center of the earth. Is the gravitational potential also zero at each place? Is it indeed the same at each place? *Can* it be zero at either place? *Need* it be zero at either place?

33. Two identical cars travel at the same speed in opposite directions on an east–west highway. Which car presses down harder on the road?

34. What advantage does Florida have over California for launching (nonpolar) U.S. satellites?

35. You are a passenger on the *S.S. Arthur C. Clarke,* the first interstellar spaceship. The *Clarke* rotates about a central axis to simulate earth gravity. If you are in an enclosed cabin, how can you tell that you are not on earth?

36. Does it matter which way a rocket is pointed for it to escape from earth? Assume, of course, that it is pointed above the horizon, and neglect air resistance.

37. For a flight to Mars, a rocket is fired in the direction the earth is moving in its orbit. For a flight to Venus, it is fired backward along that orbit. Can you explain this?

38. It is said that Newton was inspired to think of his theory of gravitation by watching the fall of an apple. It would be amusing if an apple, in our modern SI nomenclature, turned out to weigh about 1 N. Is this within the range of the possible weights for apples?

EXERCISES

Section 15–1   The Law of Universal Gravitation

1. The sun and the moon each exert a gravitational force on the earth. What is the ratio of these two forces? See Appendix C for needed data.

2. What must the separation be between a 5.2-kg particle and a 2.4-kg particle in order for their gravitational attraction to be 2.3×10^{-12} N?

3. How far from the earth must a body be along a line toward the sun so that the sun's gravitational pull balances the earth's? The sun is 9.3×10^7 mi away and its mass is 2.0×10^{30} kg.

4. A spaceship is going from the earth to the moon in a trajectory along the line joining the centers of the two bodies. At what distance from the earth will the net gravitational force on the spaceship be zero? See Appendix C.

5. What is the percentage change in the acceleration of the earth toward the sun from an eclipse of the sun to an eclipse of the moon?

Section 15-4 Gravitational Effect of a Spherical Distribution of Mass

6. One of the *Echo* satellites consisted of an inflated aluminum balloon 30 m in diameter and of mass 20 kg. A meteor having a mass of 7.0 kg passes within 3.0 m of the surface of the satellite. If the effect of all bodies other than the meteor and satellite are ignored, what gravitational force does the meteor experience at closest approach to the satellite?

7. Two concentric shells of uniform density of mass M_1 and M_2 are situated as shown in Fig. 15-15. Find the force on a particle of mass m when the particle is located at (a) $r = a$, (b) $r = b$, and (c) $r = c$. The distance r is measured from the center of the shells.

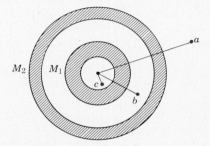

Figure 15-15 Exercise 7.

8. With what speed would mail pass through the center of the earth if it were delivered by the chute of Example 2?

Section 15-5 Gravitational Acceleration

9. What is the period of a "seconds pendulum" (period = 2 s on earth) on the surface of the moon? The moon's mass is 7.35×10^{22} kg and its radius is 1720 km.

10. At what altitude above the earth's surface would the acceleration of gravity be 4.9 m/s²? The mass of the earth is 6.0×10^{24} kg and its mean radius is 6.4×10^6 m.

11. If a pendulum has a period of exactly 1 s at the equator, what would be its period at the South Pole?

12. An object at rest on the earth's equator is accelerated (a) toward the center of the earth because the earth rotates, (b) toward the sun because the earth revolves around the sun in an almost circular orbit, and (c) toward the center of our galaxy; take the period of the sun's rotation about the galactic center to be 2.5×10^8 yr and the sun's distance from this center to be 2.4×10^{20} m. Compare these three accelerations.

Section 15-7 The Motions of Planets and Satellites

13. The year 1980 was the fiftieth anniversary of the discovery of the planet Pluto by the astronomer Clyde Tombaugh. How many revolutions has it made around the sun during that 50-year interval? See the Appendices for needed data.

14. (a) With what horizontal speed must a satellite be projected at 100 mi (= 160 km) above the surface of the earth so that it will have a circular orbit about the earth? Take the earth's radius as 4000 mi (= 6400 km). (b) What will be the period of revolution?

15. (a) Can a satellite be sent out to an orbit where it will revolve around the earth, in a circular orbit, with an angular velocity equal to that at which the earth rotates? (b) What will be the altitude of the orbit of such a "synchronous" earth satellite? (c) Show that, if the satellite's orbit lies in the equatorial plane of the earth, then the satellite will appear motionless—fixed in the sky—as seen from points on the earth's equator. (d) Will the satellite appear "fixed" from other locations on the earth?

16. A satellite is inserted into the geosynchronous orbit but, by colossal error, it is inserted in the direction *opposite* to the earth's rotation. How often would such a "wrong-way" satellite pass over any given equatorial point?

17. The mean distance of Mars from the sun is 1.52 times that of Earth from the sun. Find the number of years required for Mars to make one revolution about the sun.

Section 15-8 Gravitational Potential Energy

18. Mars has a mean diameter of 6.9×10^3 km; the earth's is 1.3×10^4 km. The mass of Mars is $0.11 M_e$. (a) How does the mean density of Mars compare with that of Earth? (b) What is the value of g on Mars? (c) What is the escape velocity on Mars?

19. Show that the velocity of escape from the sun at the earth's distance from the sun is $\sqrt{2}$ times the speed of the earth in its orbit, assumed to be a circle.

Section 15-9 Potential Energy for Systems of Particles

20. The masses and coordinates of three spheres are as follows: 20 kg, $x = 0.50$ m, $y = 1.0$ m; 40 kg, $x = -1.0$ m, $y = -1.0$ m; 60 kg, $x = 0.0$ m, $y = -0.50$ m. What are (a) the gravitational force and (b) the gravitational potential energy of a 20-kg sphere located at the origin?

21. An 800-kg particle and a 600-kg particle are separated by 25 m. (a) What is the gravitational field due to these particles at a point 20 m from the 800-kg particle and 15 m from the 600-kg particle? (b) What is the gravitational potential energy per unit mass at this point due to these same particles?

Section 15-10 Energy Considerations in the Motions of Planets and Satellites

22. Two earth satellites, A and B, each of mass m, are to be launched into circular orbits about the earth's center. Satellite A is to orbit at an altitude of 4000 mi. Satellite B is to orbit at an altitude of 12,000 mi. The radius of the earth R_e is 4000 mi (Fig. 15-16). (a) What is the ratio of the potential energy of satellite B to that of satellite A, in orbit? (Explain the result in terms of the work required to get each satellite from its orbit to infinity.) (b) What is the ratio of the kinetic energy of satellite B to that of satellite A, in orbit? (c) Which satellite has the greater total energy if each has a mass of 1.0 slug? By how much?

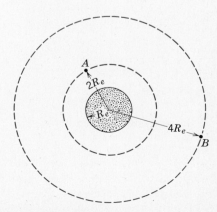

Figure 15-16 Exercise 22.

PROBLEMS GROUPED BY SECTION

Section 15–4   Gravitational Effect of a Spherical Distribution of Mass

1.   A typical "neutron star" may have a mass equal to that of the sun ($=2.0 \times 10^{30}$ kg) but a radius of only 10 km. (*a*) What is the acceleration due to gravity at the surface of such a star? (*b*) How fast would an object be moving if it fell from rest through a distance of 1.0 m on such a star?

2.   The following problem is from the 1946 "Olympic" examination of Moscow State University (see Fig. 15–17): A spherical hollow is made in a lead sphere of radius R, such that its surface touches the outside surface of the lead sphere and passes through its center. The mass of the sphere before hollowing was M. With what force, according to the law of universal gravitation, will the lead sphere attract a small sphere of mass m, which lies at a distance d from the center of the lead sphere on the straight line connecting the centers of the spheres and of the hollow?

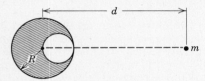

Figure 15–17   Problem 2.

*3.   Consider a mass particle at a point P anywhere inside a spherical shell of matter. Assume that the shell is of uniform thickness and density. Construct a narrow double cone with apex at P intercepting areas dA_1 and dA_2 on the shell (Fig. 15–18). (*a*) Show that the resultant gravitational force exerted on the particle at P by the intercepted mass elements is zero. (*b*) Show then that the resultant gravitational force of the entire shell on an internal particle is zero everywhere. This method was devised by Newton. (*Hint:* Equate the solid angles in the two cones.)

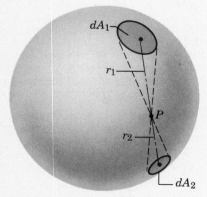

Figure 15–18   Problem 3.

Section 15–5   Gravitational Acceleration

4.   The fact that g varies from place to place over the earth's surface drew attention when Jean Richer in 1672 took a pendulum clock from Paris to Cayenne, French Guiana, and found that it lost 2.5 min/day. If $g = 9.81$ m/s^2 in Paris, what is g in Cayenne?

5.   (*a*) If g is to be determined by dropping an object through a distance of (exactly) 10 m, how accurately must the time be mea-sured in order to obtain a result good to 0.1% (*b*) Compare the accuracy of this method to that of timing a pendulum through a large number of swings.

6.   You weigh 120 lb at the sidewalk level outside the World Trade Center in New York City. Suppose that you ride from this level to the top of one of its 1350-ft towers. How much less will you weigh there because you are slightly farther away from the center of the earth?

7.   Show that, at the bottom of a vertical mine shaft dug to depth D, the measured value of g will be

$$g = g_s \left(1 - \frac{D}{R} \right),$$

g_s being the surface value. Assume that the earth is a uniform sphere of radius R.

8.   Several planets (Jupiter, Saturn, Uranus) possess nearly circular surrounding rings, perhaps composed of material that failed to form a satellite. In addition, many galaxies contain ring-like structures. Consider a homogenous ring of mass M and radius R. What gravitational attraction does it exert on a particle of mass m located a distance x from the center of the ring along its axis? See Fig. 15–19.

Figure 15–19   Problem 8.

9.   Sensitive meters that measure the vertical component of the local acceleration due to gravity g can be used to detect the presence of deposits of near-surface rocks of density significantly greater or less than that of the surroundings. Cavities such as caves and abandoned mine shafts can also be located. (*a*) Show that the vertical component of g a distance x from a point directly above the center of a spherical cave, see Fig. 15–20, is less than what would be expected assuming a uniform distribution of rock of density ρ, by the amount

$$\Delta g = \frac{4\pi}{3} R^3 G \rho \, \frac{d}{(d^2 + x^2)^{3/2}},$$

where R is the radius of the cave and d is the depth of its center. (*b*) These values of Δg, called *anomalies,* are usually very small and expressed in milligals, where 1 gal = 1 cm/s^2. Oil prospectors doing a gravity survey find Δg varying from 10 milligals to a maximum of 14 milligals over a 150-m distance. Assuming that the larger anomaly was recorded directly over the center of a spherical cave known to be in the region, find its radius and the depth to the roof of the cave at that point. Nearby rocks have a density of 2.8 g/cm^3. (*c*) Suppose that the cave, instead of being empty, is completely flooded with water. What do the gravity readings in (*b*) now indicate for its radius and depth?

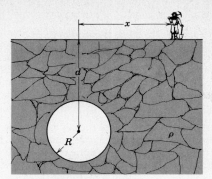

Figure 15–20 Problem 9.

10. A body is suspended on a spring balance in a ship sailing along the equator with a speed v. (a) Show that the scale reading will be very close to $W_0(1 \pm 2\omega v/g)$, where ω is the angular speed of the earth and W_0 is the scale reading when the ship is at rest. (b) Explain the plus or minus.

Section 15–7 The Motions of Planets and Satellites

11. The planet Mars has a satellite, Phobos, which travels in an orbit of radius 9.4×10^6 m with a period of 7 h 39 min. Calculate the mass of Mars from this information.

12. Spy satellites have been placed in the geosynchronous orbit above the earth's equator. What is the greatest latitude L from which the satellites are visible from the earth's surface? See Fig. 15–21.

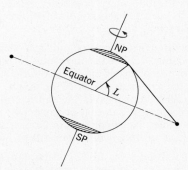

Figure 15–21 Problem 12.

13. One possibility for damaging a satellite in earth orbit is to launch a swarm of pellets in such a way that they move in the same orbit as the satellite but in the opposite direction. Consider a satellite in a circular orbit whose altitude above the earth's surface is 500 km. An on-board sensor detects a 100-g pellet approaching and determines that a head-on collision is inevitable. (a) What is the kinetic energy of the approaching pellet in the reference frame of the satellite? (b) How does this compare with the kinetic energy of a slug from an M15 rifle? Such a slug has a mass of 150 g and a muzzle velocity of 1000 m/s.

14. Show how, guided by Kepler's third law, Newton could deduce that the force holding the moon in its orbit, assumed circular, must vary as the inverse square of the distance from the center of the earth.

15. A certain triple-star system consists of two small stars, each of mass m, revolving about a larger central star, mass M, in the same circular orbit. The two small stars stay at opposite ends of a

diameter of the circular orbit; see Fig. 15–22. Derive an expression for the period of revolution of the small stars; the radius of the orbit is r.

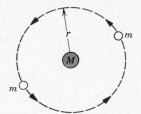

Figure 15–22 Problem 15.

16. A weather satellite is in a geosynchronous orbit, hovering over Nairobi. (a) What is its altitude above the earth's surface? (b) If its orbit radius is increased by 1.0 km, at what rate and in what direction would its reference spot, which was formerly stationary, move across the earth's surface?

17. Assume that a geosynchronous communications satellite is in orbit at the longitude of Chicago. You want to pick up its signals. In what direction should you point the axis of your parabolic antenna? (*Note:* The latitude of Chicago is 47.5°.)

*18. Three identical bodies of mass M are located at the vertices of an equilateral triangle with side L. At what speed must they move if they all revolve under the influence of one another's gravity in a circular orbit circumscribing the triangle while still preserving the equilateral triangle?

Section 15–8 Gravitational Potential Energy

19. A projectile is fired vertically from the earth's surface with an initial speed of 10 km/s. Neglecting atmospheric friction, how far above the surface of the earth would it go? Take the earth's radius as 6400 km.

20. Two identical stars are separated by a distance of 10^{10} km. They each have a mass of 10^{30} kg and a radius of 10^5 km. They are initially at rest with respect to one another. (a) How fast are they moving when their separation has decreased to one-half its initial value? (b) How fast are they moving just before they collide?

21. A rocket ship takes off from the Earth on a round trip to Jupiter with a stopover on the Moon. Assuming that the mass of the ship does not change, calculate the amount of energy required to escape from each of these two bodies relative to that required to escape from Earth.

22. Two particles of masses m and M are initially at rest an infinite distance apart. Show that at any instant their relative velocity of approach attributable to gravitational attraction is $\sqrt{2G(M + m)/d}$, where d is their separation at that instant.

Section 15–9 Potential Energy for Systems of Particles

23. In a double star, two stars of mass 3×10^{30} kg each rotate about their center of mass, 10^{11} m away. (a) What is their angular speed? (b) Suppose that a meteorite passes through this center of mass moving at right angles to the orbital plane of the stars. What must its speed be if it is to escape from the gravitational field of the double star?

24. What minimum initial speed (measured with respect to the earth) must be imparted to an object resting on the earth's surface if it is to escape not only from the gravitational field of the earth but also from that of the sun? Ignore the earth's rotation but not its orbital motion around the sun. (*Hint:* Note that for minimum speed the object must be projected in the direction of the earth's orbital motion. Treat the problem in two steps, escape from the sun following that from the earth. The earth's orbital speed, v_0, connects the two reference frames involved.)

Section 15–10 Energy Considerations in the Motions of Planets and Satellites

25. (*a*) Does it take more energy to get a satellite up to 1000 mi above the earth than to put it in orbit once it is there? (*b*) What about 2000 mi? (*c*) What about 3000 mi? Take the earth's radius to be 4000 miles.

26. A satellite travels initially in an approximately circular orbit 640 km above the surface of the earth; its mass is 220 kg. (*a*)

Determine its speed. (*b*) Determine its period. (*c*) For various reasons the satellite loses mechanical energy at the (average) rate of 1.4×10^5 J per complete orbital revolution. Adopting the reasonable approximation that the trajectory is a "circle of slowly diminishing radius," determine the distance from the surface of the earth, the speed, and the period of the satellite at the end of its 1500th orbital revolution. (*d*) What is the magnitude of the average retarding force? (*e*) Is angular momentum conserved?

*27. The orbit of the earth about the sun is only "almost" circular. The closest and farthest distances are 1.47×10^8 km and 1.52×10^8 km, respectively. Determine the maximum variations in (*a*) potential energy, (*b*) kinetic energy, (*c*) total energy, and (*d*) orbital speed that are due to the changing earth-sun distance in the course of one year. (*Hint:* Use conservation of energy and of angular momentum.)

ADDITIONAL PROBLEMS

28. Calculate the shortest possible period for an earth satellite from the value of G and the values of the earth's mass (6.0×10^{24} kg) and radius (6.4×10^6 m).

29. Determine the mass of the earth from the period T and the radius r of the moon's orbit about the earth: $T = 27.3$ days and $r = 2.39 \times 10^5$ mi ($= 3.85 \times 10^5$ km).

30. Certain neutron stars (extremely dense stars) are believed to be rotating at about one revolution per second. If such a star has a radius of 20 km, what must be its minimum mass so that objects on its surface will be attracted to the star and not "thrown off" by the rapid rotation?

31. An asteroid, whose mass is 2.0×10^{-4} times the mass of the earth, revolves in a circular orbit around the sun at a distance which is twice the earth's distance from the sun. (*a*) Calculate the period of revolution of the asteroid in earth years. (*b*) What is the ratio of the kinetic energy of the asteroid to that of the earth?

32. (*a*) Calculate the acceleration due to gravity on the surface of the moon if g_{earth} is 9.8 m/s², the moon's radius is 27% of the earth's radius, and the moon's mass is 1.2% that of the earth. (*b*) What will an object weigh on the moon's surface if it weighs 100 N on the earth's surface? (*c*) How many earth radii must this same object be from the center of the earth if it is to weigh the same as it does on the moon? (*d*) Is the gravitational force of the earth on the moon greater than, less than, or the same as the gravitational force of the moon on the earth?

33. The sun, mass 2.0×10^{30} kg, is revolving about the center of the Milky Way galaxy, which is 2.4×10^{20} m away. It completes one revolution every 2.5×10^8 yr. Estimate roughly the number of stars in the Milky Way. (*Hint:* Assume for simplicity that the stars are distributed with spherical symmetry about the galactic center and that our sun is essentially at the galactic edge.)

34. Assume that the earth expands to become a sphere twice its present radius but keeping the average mass density constant. In terms of the same quantities before the expansion, find the new (*a*) mass of a person, (*b*) value of gravitational acceleration, (*c*) weight of a person, and (*d*) speed of a satellite just skimming the surface.

35. Repeat the previous problem, assuming that the total mass of the earth remains constant and is uniformly distributed through the sphere.

36. The asteroid Eros, one of the many minor planets that orbit the sun in the region between Mars and Jupiter, has a radius of 7.0 km and a mass of 5.0×10^{15} kg. (*a*) If you were standing on Eros, could you lift a 2000-kg pickup truck? (*b*) Could you run fast enough to put yourself into orbit? Ignore effects due to the rotation of the asteroid. (*Note:* The Olympic records for the 400-m run corresponds to a speed of 9.1 m/s for men and 8.2 m/s for women.)

37. Consider an artificial satellite in a circular orbit about the earth. State how the following properties of the satellite vary with the radius r of its orbit: (*a*) period; (*b*) kinetic energy; (*c*) angular momentum; (*d*) speed.

38. It is conjectured that a "burned-out" star could collapse to a "gravitational radius," defined as the radius for which the work needed to remove an object of mass m_0 from the star's surface to infinity equals the rest energy $m_0 c^2$ of the object. Show that the gravitational radius of the sun is GM_s/c^2 and determine its value in terms of the sun's present radius. (For a review of this phenomenon see "Black Holes: New Horizons in Gravitational Theory," by Philip C. Peters, in *American Scientist*, Sept.–Oct. 1974.)

39. Masses of 200 and 800 g are 12 cm apart. (*a*) Find the gravitational force on a 1.0-g object situated at a point on the line joining the masses 4.0 cm from the 200-g mass. (*b*) Find the gravitational potential energy per unit mass at that point. (*c*) How much work is needed to move this object to a point 4.0 cm from the 800-g mass along the line of centers?

40. A spaceship is idling at the fringes of our galaxy, 80,000 light-years from the galactic center. What minimum speed must it have if it is to escape entirely from the gravitational attraction of the galaxy? The mass of the galaxy is 1.4×10^{11} times that of our sun. Assume, for simplicity, that the matter forming the galaxy is distributed with spherical symmetry.

41. Consider two satellites, A and B, of equal mass m, moving in the same circular orbit of radius r around the earth E but in opposite senses of rotation and therefore on a collision course (see Fig. 15–23). (*a*) In terms of G, M_e, m, and r, find the total mechanical energy $E_A + E_B$ of the two-satellite-plus-earth system before collision. (*b*) If the collision is completely inelastic so that the wreckage remains as one piece of tangled material (mass $= 2m$), find the total mechanical energy immediately after collision. (*c*) Describe the subsequent motion of the wreckage.

Figure 15-23   Problem 41.

42.   A sphere of matter, mass M, radius a, has a concentric cavity, radius b, as shown in cross section in Fig. 15-24. (a) Sketch the gravitational force F exerted by the sphere on a particle of mass m, located a distance r from the center of the sphere, as a function of r in the range $0 \leq r \leq \infty$. Consider points $r = 0$, b, a, and ∞ in particular. (b) Sketch the corresponding curve for the potential energy $U(r)$ of the system. (c) From these graphs, how would you obtain graphs of the gravitational field strength and the gravitational potential due to the sphere?

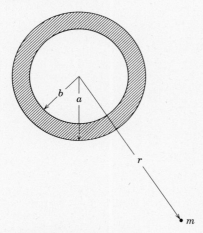

Figure 15-24   Problem 42.

43.   (a) What is the escape velocity on a hypothetical planet whose radius is 500 km and whose surface gravity is 3.0 m/s²? (b) How high will a particle rise if it leaves the surface of the planet with a vertical velocity of 1000 m/s? (c) With what speed will an object hit the planet if it is dropped from a height of 1000 km (1500 km from the center of the planet)?

44.   A rocket is accelerated to a speed of $v = 2\sqrt{gR_e}$ near the earth's surface and then coasts upward. (a) Show that it will escape from the earth. (b) Show that very far from the earth its speed is $v = \sqrt{2gR_e}$.

45.   (a) Write an expression for the potential energy of a body of mass m in the gravitational field of the earth and moon. Let M_e be the earth's mass, M_m the moon's mass, R the distance from the earth's center, and r the distance from the moon's center. (b) At what point between the earth and moon will the total gravitational field attributable to the earth and moon be zero? (c) What will be the gravitational potential and the gravitational field strength on the earth's surface? (d) Answer for the moon's surface.

46.   The fastest possible rate of rotation of a planet (or satellite, asteroid, etc.) is that for which the gravitational force on material at the equator barely provides the centripetal force needed for the rotation. (Why?) (a) Show, then, that the corresponding shortest period of rotation is given by

$$T = \left(\frac{3\pi}{G\rho}\right)^{1/2},$$

where ρ is the density of the planet, assumed to be homogenous. (b) Evaluate the rotation period assuming a density of 3.0 g/cm³, typical of many planets, satellites, and asteroids. No such object is found to be spinning with a period shorter than found by this analysis.

47.   Masses m, assumed equal, hang from strings of different lengths on a balance at the surface of the earth, as shown in Fig. 15-25. If the strings have negligible mass and differ in length by h, (a) show that the error in weighing, associated with the fact that W' is closer to the earth than W, is $W' - W = 8\pi G\rho mh/3$ in which ρ is the mean density of the earth (5.5 g/cm³). (b) Find the difference in length that will give an error of 1 part in a million.

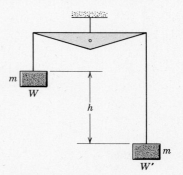

Figure 15-25   Problem 47.

48.   A scientist is making a precise measurement of g at a certain point in the Indian Ocean (on the equator) by timing the swings of a pendulum of accurately known construction. To provide a stable base the measurements are conducted in a submerged submarine. It is observed that a slightly different result for g is obtained when the submarine is moving eastward through the point than when it is moving westward, the speed in each case being 16 km/h. Account for this difference and calculate the fractional error $\Delta g/g$ in g.

49.   In the year 1610, Galileo made a telescope, turned it on Jupiter, and discovered four prominent moons. Their mean orbit radii a and periods T are

Name	a, 10⁸ m	T, days
Io	4.22	1.77
Europa	6.71	3.55
Ganymede	10.7	7.16
Callisto	18.8	16.7

(a) Plot log a (y-axis) against log T (x-axis) and show that you get a straight line. (b) Measure its slope and compare it with the value that you expect from Kepler's third law. (c) Find the mass of Jupiter from the intercept of this line with the y-axis. (*Note:* You may also use log-log graph paper.)

*50.   (a) Write an expression for the force exerted by the moon, mass M, on a particle of water, mass m, on the earth at A, directly under the moon, as shown in Fig. 15-26. The radius of the earth is R, and the center-to-center earth-moon distance is r. (b) Sup-

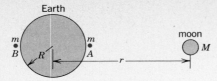

Figure 15-26   Problem 50.

pose that the particle of water was at the center of the earth. What force would the moon exert on it then? (c) Show that the difference in these forces is given by

$$F_T = \frac{2GMmR}{r^3},$$

and represents the *tidal force,* the force on water relative to the earth. What is the direction of the tidal force? (d) Repeat for a particle of water at B, on the far side of the earth from the moon. What is the direction of this tidal force? (e) Explain why there are two tidal bulges in the oceans (and solid earth), one pointing toward the moon and the other away from it?

*51.   (a) Show that in a frictionless chute through the earth along a chord line, rather than along a diameter, the motion of an object will be simple harmonic; assume a uniform earth density. (b) Find the period. (c) Will the object attain the same maximum speed along a chord as it does along a diameter?

*52.   Suppose in Problem 8 that the particle of mass m falls from rest due to the attraction of the ring of matter. Find an expression for the speed with which it passes through the center of the ring.

16 Fluid Mechanics

16–1 Fluids

We usually classify matter, viewed macroscopically, into solids and fluids. A *fluid* is something that can flow. Hence the term fluid includes liquids and gases. Such classifications are not always clearcut. Some fluids, such as glass or pitch, flow so slowly that they behave like solids for the time intervals that we usually work with them. Even the distinction between a liquid and a gas is not clearcut because, by changing the pressure and temperature in the right way, it is possible to change a liquid (water, say) into a gas (steam, say) without the appearance of a boundary layer between them and without boiling; the density and viscosity change in a continuous manner throughout the process.

In this chapter we shall apply Newton's laws of motion, including the appropriate force laws, to fluids. We shall find, just as we did for rigid bodies, that it will be useful to develop a special formulation of Newton's laws. In doing so, however, we introduce no new principles.

16–2 Fluid Pressure and Density

There is a difference in the way a surface force acts on a fluid and on a solid. For a solid there are no restrictions on the direction of such a force, but for a fluid at rest the surface force must always be directed at right angles to the surface. This is because a fluid at rest cannot sustain a tangential force; the fluid layers would simply slide over one another, resulting in fluid motion. Indeed, it is the inability of fluids to resist such tangential forces (or shearing stresses) that gives them their characteristic ability to change their shape or to flow.

It is convenient, therefore, to describe the force acting on a fluid by specifying the *pressure p*, which is defined as the magnitude of the *normal* force per

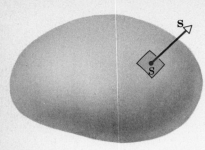

Figure 16-1 An element of surface S can be represented by a vector **S**, equal to its area in magnitude and normal to it in direction.

unit surface area. Pressure is transmitted to solid boundaries or across arbitrary sections of fluid *at right angles* to these boundaries or sections at every point. Pressure is a scalar quantity. The SI unit of pressure is the *pascal* (abbreviation Pa, 1 Pa = 1 N/m²). This unit is named after the French scientist Blaise Pascal (1623–1662) (see Section 16–4). Other units are the *bar* (1 bar = 10⁵ Pa), the *atmosphere* (1 atm = 14.7 lb/in.² = 1.01 × 10⁵ Pa), and the *millimeter of mercury* or *torr* (760 mm Hg = 760 torr = 1 atm).

A fluid under pressure exerts a force on any surface in contact with it. Consider a closed surface containing a fluid (Fig. 16–1). An element of the surface, assumed to be "small," can be represented by a vector **S** whose magnitude gives the area of the element and whose direction is taken to be the outward normal to the surface of the element. Then the force **F** exerted by the fluid against this surface element is

$$\mathbf{F} = p\mathbf{S}.$$

Since **F** and **S** have the same direction, the pressure p can be written as

$$p = \frac{F}{S}.$$

We assume that the element of area S is small enough so that the pressure p, defined as above, is independent of the size of the element S. The pressure may actually vary from point to point on the surface.

Density of a fluid

The density ρ of a homogeneous fluid (its mass divided by its volume) may depend on many factors, such as its temperature and the pressure to which it is subjected. For liquids the density varies very little over wide ranges in pressure and temperature, and we can safely treat it as a constant for our present purposes; see entries under "Water" in Table 16–1. The density of a gas, however, is very sensitive to changes in temperature and pressure; see entries under "Air" in Table 16–1. This table shows the range of densities that occur in nature. Note that the variation is by a factor of about 10³⁸.

Table 16-1 Densities of some materials and objects in kg/m³

Interstellar space	10^{-18}–10^{-21}
Best laboratory vacuum	$\sim 10^{-17}$
Hydrogen: at 0°C and 1.0 atm	9.0×10^{-2}
Air: at 0°C and 1.0 atm	1.3
at 100°C and 1.0 atm	0.95
at 0°C and 50 atm	65
Styrofoam	$\sim 1 \times 10^{2}$
Ice	0.92×10^{3}
Water: at 0°C and 1.0 atm	1.000×10^{3}
at 100°C and 1.0 atm	0.958×10^{3}
at 0°C and 50 atm	1.002×10^{3}
Aluminum	2.7×10^{3}
Mercury	1.36×10^{4}
Platinum	2.14×10^{4}
The earth: average density	5.52×10^{3}
density of core	9.5×10^{3}
density of crust	2.8×10^{3}
The sun: average density	1.4×10^{3}
density at center	1.60×10^{5}
White dwarf stars (central densities)	$\sim 10^{10}$
A uranium nucleus	$\sim 3 \times 10^{17}$
A neutron star (center)	$10^{17} - 10^{18}$
A black hole (one solar mass)	$> 10^{19}$

16–3   The Variation of Pressure in a Fluid at Rest

Consider a fluid, every part of which is at rest. Take a small volume element that is entirely submerged within the body of the fluid. Let this element be a thin disk a distance y above some reference level, as shown in Fig. 16–2a. The thickness of the disk is dy and each face has an area A. The mass of this element is $\rho A\, dy$. The forces exerted on the element are (1) its weight, $\rho g A\, dy$ and (2) those resulting from the pressure exerted on the disk by the surrounding fluid. The forces due to fluid pressure are perpendicular to the surface of the element at each point (Fig. 16–2b).

The fluid element is unaccelerated in the vertical direction, so that the result-ant vertical force on it must be zero. If we let p be the pressure on the lower face and $p + dp$ be the pressure on its upper face, the upward force due to the sur-rounding fluid is pA (exerted on the lower face) and the downward force (exerted on the upper face) is $-(p + dp)A$ plus the weight of the element $dw(= -\rho g A\, dy)$. Hence, for vertical equilibrium

$$pA - (p + dp)A - \rho g A\, dy = 0,$$

and, therefore,

$$\frac{dp}{dy} = -\rho g. \qquad (16\text{–}1)$$

Rate of change of pressure with height

This equation tells us how the pressure varies with elevation above some refer-ence level in a fluid in static equilibrium. As the elevation increases (dy positive), the pressure decreases (dp negative). The cause of this pressure variation is the weight per unit cross-sectional area of the layers of fluid lying between the points whose pressure difference is being measured. The quantity ρg is often called the *weight density* of the fluid. For water, for example, the weight density is 62.4 lb/ft³ (=9800 N/m²).

If p_1 is the pressure at elevation y_1 and p_2 is the pressure at elevation y_2 above some reference level, integration of Eq. 16–1 gives

$$\int_{p_1}^{p_2} dp = -\int_{y_1}^{y_2} \rho g\, dy$$

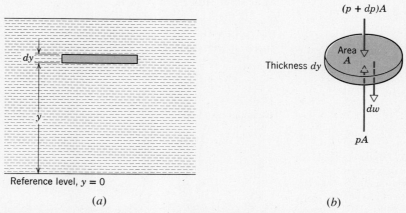

Reference level, $y = 0$

(a)

(b)

Figure 16–2 (a) A small volume element of fluid at rest. $(b$ The forces on the element.

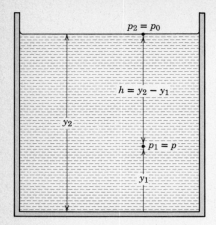

Figure 16–3 A liquid whose top surface is open to the atmosphere.

or

$$p_2 - p_1 = -\int_{y_1}^{y_2} \rho g \, dy. \qquad (16\text{–}2)$$

For liquids ρ is practically constant because liquids are nearly incompressible; and differences in level are rarely so great that any change in g need be considered. Hence, taking ρ and g as constants, we obtain

$$p_2 - p_1 = -\rho g(y_2 - y_1) \qquad (16\text{–}3)$$

for a homogeneous liquid.

If a liquid has a free surface such as the upper surface in Fig. 16–3, we may take y_2 to be the elevation of that surface, where the pressure is p_0, and we take the elevation y_1 to be at any level with pressure p, and $y_2 - y_1$ is the depth h (see Fig. 16–3). Equation 16–3 then becomes

$$p = p_0 + \rho g h. \qquad (16\text{–}4)$$

This shows that the pressure is the same at all points at the same depth. Often p_0 is the atmospheric pressure.

If we apply Eq. 16–2 to a gas, we find that $p_2 = p_1$ for many cases of interest. This is because the density ρ is so small it requires a very large height difference $y_2 - y_1$ to give a significant pressure change. In a vessel containing a gas we can therefore take the pressure as the same everywhere. Of course, this is not the case when $y_2 - y_1$ is very great, such as when we ascend to great heights in the atmosphere. In such cases the density ρ varies with altitude and we must know ρ (and also g) as a function of y before we can integrate Eq. 16–2. Although we may not know ρ, we can still equate p to the weight per unit area of a column of air from the point in question up to the "top" of the atmosphere. From Eq. 16–2 you can see that, since $\rho g \, dy$ is the weight per unit area of an element of air, an integration to the top of the atmosphere results in the total weight per unit area.

Equation 16–3 gives the relation between the pressures at any two points in a fluid, regardless of the shape of the containing vessel. For no matter what the shape of the containing vessel, two points in the fluid can be connected by a path made up of vertical and horizontal steps. For example, consider points A and B in the homogeneous liquid contained in the U-tube of Fig. 16–4a. Along the stepped path from A to B there is a difference in pressure $\rho g y'$ for each vertical segment of length y', whereas along each horizontal segment there is no change in pressure. Hence, the difference in pressure $p_B - p_A$ is ρg times the algebraic sum of the vertical segments from A to B, or $\rho g(y_2 - y_1)$.

If the U-tube contains two liquids that do not mix, perhaps a dense liquid in the right tube and a less dense one in the left tube, as Fig. 16–4b shows, the pressures at the same level on the two sides will be different. In the figure the liquid

Vertical pressure differences do not depend on the shape of the vessel

Figure 16–4 (a) The difference in pressure between two points A and B in a homogeneous liquid depends only on their difference in elevation $y_2 - y_1$. (b) Two points A and B at the same elevation can be at different pressures if the densities there differ.

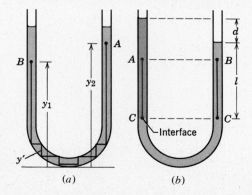

(a) (b)

surface is higher in the left tube than in the right. The pressure at A will be greater than at B. The pressure at C is the same on both sides, but the pressure falls less from C to A than from C to B, for a column of liquid of unit cross-sectional area connecting A and C will weigh less than a corresponding column connecting B and C.

Example 1 A U-tube is partly filled with water. Another liquid, which does not mix with water, is poured into one side until it stands a distance d above the water level on the other side (Fig. 16–4b). Find the density of the liquid relative to that of water.

In Fig. 16–4b points C are at the same pressure. Hence, the pressure drop from C to each surface is the same, for each surface is at atmospheric pressure.

The pressure drop on the water side is $\rho_w gl$; that on the other side is $\rho g(d + l)$, where ρ is the density of the unknown liquid. Hence,

and

$$\rho_w gl = \rho g(d + l)$$

$$\frac{\rho}{\rho_w} = \frac{l}{(l + d)}.$$

The ratio of the density of a substance to the density of water is called the *relative density* (or the *specific gravity*) of that substance.

16–4 Pascal's Principle and Archimedes' Principle

Figure 16–5 shows a liquid in a cylinder that is fitted with a piston to which we may apply an external pressure p_0. The pressure p at any arbitrary point P a distance h below the upper surface of the liquid is given by Eq. 16–4, which becomes

$$p = p_0 + \rho g h.$$

Let us increase the external pressure by an arbitrary amount Δp_0, which need not be small compared to p_0. Since liquids are nearly incompressible, the density ρ remains essentially constant. The equation shows that, to this extent, the change in pressure Δp at the arbitrary point P is equal to Δp_0. This result was stated by Blaise Pascal (1623–1662) and is called *Pascal's principle*. We usually say it this way: *Pressure applied to an enclosed fluid at rest is transmitted undiminished to every portion of the fluid and to the walls of the containing vessel.* This is a necessary consequence of the laws of mechanics, and not an independent principle.

Actually, slight changes in density do occur and provide the means by which a wave may propagate through a fluid (see Section 18–2). This means that a change of pressure applied to one portion of a liquid propagates through the liquid as a wave at the speed of sound in that liquid. Pascal's principle also holds for gases with slight complications of interpretation caused by the large volume changes that may occur when the pressure on a confined gas is changed.

Archimedes' principle is also a necessary consequence of the laws of fluid mechanics. When a body is wholly or partly immersed in a fluid (either a liquid or a gas) at rest, the fluid exerts pressure on every part of the body's surface in contact with the fluid (see Fig. 16–6). The pressure is greater on the parts immersed more deeply. The resultant or sum of all these fluid forces is an upward force **R** called the *buoyant force* on the immersed body. We can find the magnitude and direction of this resultant force as follows.

The pressure on each part of the surface of the body certainly does not depend on the material the body is made of. Suppose, then, that the body, or as much of it as is immersed, is replaced by more of the same fluid in which the body is immersed. This volume of fluid will experience the same pressures that acted on the immersed body (Fig. 16–6) and will be at rest. The sum of all forces (including

Pascal's principle

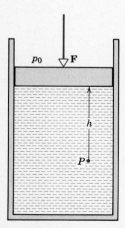

Figure 16–5 A fluid in a cylinder fitted with a movable piston. The pressure at any point P is due not only to the weight of the fluid above the level of P but also to the force exerted by the piston.

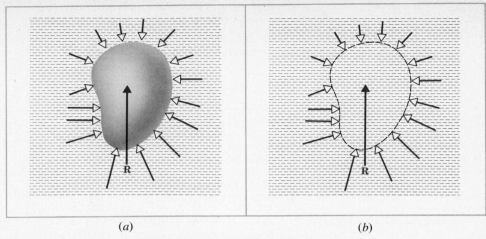

$$(a) \qquad\qquad (b)$$

Figure 16-6 If all of the fluid forces acting on a submerged body are added, the resultant **R** is the buoyant force as shown in (a). The same force would act if the body were replaced by an equal volume of fluid as shown in (b). Since the fluid volume is at rest, the total force on it must be zero. Therefore the magnitude of the buoyant force is equal to the weight of the displaced fluid.

gravity) acting on the fluid volume will be just sufficient to keep it in position without motion, which means that the total force must be zero. Hence, the buoyant force will just balance the weight of the fluid volume. From this follows *Archimedes' principle: A body wholly or partly immersed in a fluid is buoyed up with a force equal to the weight of the fluid displaced by the body.* The buoyant force acts vertically through the center of gravity of the fluid volume before its displacement. This is evident because the fluid is not rotating and a force directed anywhere but through this point would consitute a torque and result in rotation. The corresponding point in the immersed body is called its *center of buoyancy*.

Archimedes' principle

Example 2 What fraction of the volume of an iceberg is exposed? The density of ice is $\rho_i = 0.92$ g/cm³ and that of sea water is $\rho_w = 1.03$ g/cm³.

The weight of the iceberg is

$$W_i = \rho_i V_i g,$$

where V_i is the volume of the iceberg; the weight of the volume V_w of sea water displaced is the buoyant force

$$B = \rho_w V_w g.$$

But B equals W_i, for the iceberg is in equilibrium, so that

$$\rho_w V_w g = \rho_i V_i g,$$

and

$$\frac{V_w}{V_i} = \frac{\rho_i}{\rho_w} = \frac{0.92}{1.03} = 89\%.$$

The volume of water displaced, V_w, is the volume of the submerged portion of the iceberg, so that 11% of the iceberg is exposed.

16-5 Measurement of Pressure

The barometer

Evangelista Torricelli (1608–1647), for whom the pressure unit *torr* is named, invented the mercury barometer in 1643. This is a long glass tube that has been filled

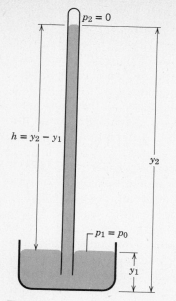

$p_2 = 0$

$h = y_2 - y_1$

y_2

$p_1 = p_0$

y_1

Figure 16–7 The Torricelli barometer.

with mercury and then inverted in a dish of mercury, as in Fig. 16–7. The space above the mercury column contains only mercury vapor, whose pressure is so small at ordinary temperatures that it can be neglected. You can easily show (see Eq. 16–3) that the atmospheric pressure p_0 is

$$p_0 = \rho gh,$$

where h is the height of mercury in the column above that in the dish.

We saw earlier that, if the air is at rest, the pressure of the atmosphere at any point is equal to the weight per unit area of a column of air extending from that point to the top of the atmosphere. The atmospheric pressure, therefore, decreases with altitude. There are variations in atmospheric pressure from day to day since the atmosphere is not static. The mercury column in the barometer will have a height of about 76 cm at sea level, varying with the atmospheric pressure. A pressure equivalent to that exerted by exactly 76 cm of mercury at exactly 0°C under standard gravity, $g = 32.174$ ft/s² $= 980.665$ cm/s², is defined to be *one atmosphere* (1 atm). The density of mercury at this temperature is 13.5950 g/cm³. Hence, one atmosphere is equal to

$$1 \text{ atm} = (13.5950 \text{ g/cm}^3)(980.665 \text{ cm/s}^2) \times (76.00 \text{ cm})$$
$$= 1.013 \times 10^5 \text{ N/m}^2 \ (\equiv 1.013 \times 10^5 \text{ Pa})$$
$$= 2116 \text{ lb/ft}^2$$
$$= 14.70 \text{ lb/in.}^2$$

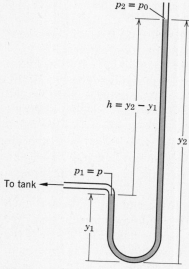

$p_2 = p_0$

$h = y_2 - y_1$

y_2

$p_1 = p$

To tank

y_1

Figure 16–8 The open-tube manometer, as used to measure the gauge pressure in a tank.

Often pressures are specified by giving the height of a mercury column, at 0°C under standard gravity, which exerts the same pressure. This is the origin of the expression "centimeters of mercury (cm Hg)" or "inches of mercury (in. Hg)" pressure. Pressure is the ratio of force to area, however, and not a length.

Many pressure gauges use atmospheric pressure as a reference level and measure the difference between the actual pressure and atmospheric pressure, called the *gauge pressure*. The actual pressure at a point in a fluid is called the *absolute pressure*. Gauge pressure is given either above or below atmospheric pressure.

The open-tube manometer (Fig. 16–8) measures gauge pressure. It consists of a U-shaped tube containing a liquid, one end of the tube being open to the atmosphere and the other end being connected to the system (tank) whose pressure p we want to measure. From Eq. 16–4 we obtain

$$p - p_0 = \rho gh.$$

Thus the gauge pressure, $p - p_0$, is proportional to the difference in height of the liquid columns in the U-tube.

Example 3 An open-tube mercury manometer (Fig. 16–8) is connected to a gas tank. The mercury is 39.0 cm higher on the right side than on the left when a barometer nearby reads 75.0 cm Hg. What is the absolute pressure of the gas?

The gas pressure is the pressure at the top of the left mercury column. This is the same as the pressure at the same horizontal level in the right column. The pressure at this level is the atmospheric pressure (75.0 cm Hg) plus the pressure exerted by the extra 39.0-cm column of Hg, or (assuming standard val-

ues of mercury density and gravity) a total of 114 cm Hg. Therefore, the *absolute* pressure of the gas is

$$114 \text{ cm Hg} = \tfrac{114}{76} \text{ atm} = 1.50 \text{ atm} = 1.52 \times 10^5 \text{ Pa}$$
$$= (1.50)(14.7) \text{ lb/in.}^2 = 22.1 \text{ lb/in.}^2.$$

The *gauge* pressure (that is, the difference between the pressures inside and outside the tank) of the gas is 39.0 cm Hg = 5.2×10^4 Pa.

16-6 Fluid Dynamics

Fluids in motion

Now we consider fluids that are flowing and on which definable forces act. One way of describing the motion of a fluid is to divide it into infinitesimal volume elements, which we may call fluid particles, and to follow the motion of each of these particles. This is a formidable task. We could give coordinates x, y, z to each such fluid particle and then specify these as functions of the time t and the initial position of the particle x_0, y_0, and z_0. This procedure is a direct generalization of the concepts of particle mechanics and was first developed by Joseph Louis Lagrange (1736–1813).

There is, however, a treatment, developed by Leonhard Euler (1707–1783), which is much more convenient. In it we give up the attempt to specify the history of each fluid particle and instead specify the density and the velocity of the fluid at each point in space at each instant of time. This is the method we shall follow here. We describe the motion of the fluid by specifying the density $\rho(x, y, z, t)$ and the velocity $\mathbf{v}(x, y, z, t)$ at the point (x, y, z) at the time t. We thus focus attention on what is happening at a particular point in space at a particular time, rather than on what is happening to a particular fluid particle. Any quantity used in describing the state of the fluid, for example, the pressure p, will have a definite value at each point in space and at each instant of time. Although this description of fluid motion focuses attention on a point in space rather than on a fluid particle, we cannot avoid following the fluid particles themselves, at least for short time intervals dt. For it is the particles, after all, and not the space points, to which the laws of mechanics apply.

Let us first consider some general characteristics of fluid flow and specify a number of restrictions we shall make in order to simplify the treatment.

Figure 16–9 We place a small free-floating paddle wheel in a flowing liquid. If it rotates, we call the flow *rotational*; if not, we call the flow *irrotational*.

Steady flow

1. Fluid flow can be *steady* or *nonsteady*. When the fluid velocity $\mathbf{v}$ at every point is constant in time, we say that the fluid motion is steady. That is, at any given point in a steady flow the velocity of each passing fluid particle is always the same. At any other point the velocity of the passing particles may be different from that at the first point. These conditions can be achieved at low flow speeds; a gently flowing stream is an example. In nonsteady flow, as in a tidal bore, the velocities $\mathbf{v}$ *are* a function of the time. In the case of turbulent flow, such as rapids or a waterfall, the velocities vary erratically from point to point as well as from time to time.

Irrotational flow

2. Fluid flow can be *rotational* or *irrotational*. If the element of fluid at each point has no net angular velocity about that point, the fluid flow is irrotational. We can imagine a small paddle wheel immersed in the moving fluid (Fig. 16–9). If the wheel moves without rotating, the motion is irrotational; otherwise it is rotational. Rotational flow includes vortex motion, such as whirlpools.

Incompressible flow

3. Fluid flow can be *compressible* or *incompressible*. Liquids can usually be considered as flowing incompressibly. But even a gas that is easily compressed may sometimes undergo unimportant changes in density. Its flow then can be considered incompressible. In flight at speeds much lower than the speed of sound in air (described by subsonic aerodynamics), the motion of the air relative to the wings is one of nearly incompressible flow. In such cases the density ρ is a constant, independent of x, y, z, and t, and the mathematical treatment of fluid flow is thereby greatly simplified.

Nonviscous flow

4. Finally, fluid flow can be *viscous* or *nonviscous*. Viscosity in fluid motion is the analog of friction in the motion of solids. In many cases, such as in lubrication problems, it is extremely important. Sometimes, however, it is negligible. Viscosity introduces tangential forces between layers of fluid in relative motion and results in dissipation of mechanical energy.

We shall confine our discussion of fluid dynamics for the most part to *steady, irrotational, incompressible, nonviscous* flow in order to simplify our treatment. We run the danger, however, of making so many simplifying assumptions that we are no longer talking about a meaningfully real fluid. Furthermore, it is sometimes difficult to decide whether a given property of a fluid—its viscosity, say—can be neglected in a particular situation. In spite of all this, the restricted analysis that we are going to give has wide application in practice.

16–7   Streamlines and the Equation of Continuity

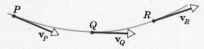

Figure 16–10   A particle passing through points P, Q, and R traces out a streamline, assuming steady flow. Any other particle passing through P must be traveling along the same streamline in steady flow.

A tube of flow

Figure 16–11   A tube of flow made up of a bundle of streamlines.

Figure 16–12   A tube of flow used in proving the equation of continuity.

Equation of Continuity

In steady flow the velocity **v** at a given point is constant in time. Consider the point P in Fig. 16–10. Since **v** at P does not change in time, every other particle arriving at P will pass on with the same speed in the same direction as the first. The same is true about the points Q and R. Hence, if we trace out the path of the first particle, as is done in the figure, that curve will be the path of every other particle arriving at P. Such a curve is called a *streamline*. A streamline is tangent to the velocity of the fluid particles at every point. No two streamlines can cross, for if they did, an oncoming fluid particle could go either one way or the other, and the flow could not be steady. In steady flow the pattern of streamlines in a flow is stationary with time.

In principle we can draw a streamline through every point in the fluid. Let us assume steady flow and select a finite number of streamlines to form a bundle, like the streamline pattern of Fig. 16–11. This tubular region is called a *tube of flow*. Since the boundary of such a tube consists of streamlines no fluid can cross the boundary of a tube of flow and the tube behaves somewhat like a pipe of the same shape. The fluid that enters at one end must leave at the other.

In Fig. 16–12 we have drawn a thin tube of flow. The velocity of the fluid inside will in general have different magnitudes at different points along the tube. Let the speed be v_1 for fluid particles at P and v_2 for fluid particles at Q. Let A_1 and A_2 be the cross-sectional areas of the tubes perpendicular to the streamlines at the points P and Q, respectively, and assume that the speed is essentially constant over each surface separately. In the time interval Δt a fluid element travels approximately the distance $v\,\Delta t$. Then the mass of fluid Δm_1 crossing A_1 in the time interval Δt is approximately

$$\Delta m_1 = \rho_1 A_1 v_1\,\Delta t$$

or the *mass flux*, $\Delta m_1/\Delta t$, is approximately $\rho_1 A_1 v_1$. We must take Δt small enough so that in this time interval neither v nor A varies appreciably over the distance the fluid travels. In the limit as $\Delta t \to 0$, we obtain the precise definitions:

$$\text{mass flux at } P = \rho_1 A_1 v_1, \quad \text{and} \quad \text{mass flux at } Q = \rho_2 A_2 v_2,$$

where ρ_1 and ρ_2 are the fluid densities at P and Q, respectively.

We shall assume that there are no "sources" or "sinks" wherein fluid is created or destroyed in the tube. Since in steady flow no fluid can leave through the walls nor can fluid particles temporarily bunch up, the mass crossing each section of the tube per unit time must be the same. In particular, the mass flux at P must equal that at Q:

$$\rho_1 A_1 v_1 = \rho_2 A_2 v_2,$$

or   (16–5)

$$\rho A v = \text{constant}.$$

This result (Eq. 16–5) is called the *equation of continuity* and expresses the law of conservation of mass in fluid dynamics.

Would you expect Eq. 16–5 to hold when the flow is (*a*) nonsteady, (*b*) rotational, (*c*) compressible, or (*d*) viscous?

If the fluid is incompressible, as we shall henceforth assume, $\rho_1 = \rho_2$ and Eq. 16–5 takes on the simpler form

Continuity equation for an incompressible fluid

$$A_1 v_1 = A_2 v_2,$$

or (16–6)

$$Av = \text{constant}.$$

The product Av gives the *volume flux* or flow rate, as it is often called. Its SI units are m³/s. Notice that it predicts that in steady incompressible flow the speed of flow varies inversely with the cross-sectional area, being larger in narrower parts of the tube. The fact that the product Av remains constant along a tube of flow allows us to interpret the streamline picture somewhat. In a narrow part of the tube the streamlines must crowd closer together than in a wide part. Hence, as the distance between streamlines decreases, the fluid speed must increase. Therefore, we conclude that widely spaced streamlines indicate regions of lower speed, and closely spaced streamlines indicate regions of higher speed.

We can obtain another interesting result by applying Newton's second law of motion to the flow of fluid. In moving from P to Q (Fig. 16–12) the fluid is decelerated from v_1 to the lower speed v_2. The deceleration can come about from a difference in pressure acting on the fluid particle flowing from P to Q or from the action of gravity. In a horizontal tube of flow the gravitational force does not change. Hence we can conclude that in steady horizontal flow the pressure is greatest where the speed is least. These results will be developed quantitatively in the next section.

16–8 Bernoulli's Equation

Bernoulli's equation is a fundamental relation in fluid mechanics. Like all equations in fluid mechanics it is not a new principle but is derivable from Newtonian mechanics. We shall derive it from the work-energy theorem (see Section 7–5), for it is essentially an application of this theorem to fluid flow.

Consider the nonviscous, steady, incompressible flow of a fluid through the pipeline or tube of flow in Fig. 16–13. The portion of pipe shown in the figure has a uniform cross section A_1 at the left. It is horizontal there at an elevation y_1 above some reference level. It gradually widens and rises until at the right it has a uniform cross section A_2. It is horizontal there also at an elevation y_2. Let us concentrate our attention on the shaded portion of fluid between A_1 and A_2 and call this fluid the "system." Consider then the motion of the system from the position shown in (*a*) to that in (*b*). At all points in the narrow part of the pipe the pressure is p_1 and the speed v_1; at all points in the wide portion the pressure is p_2 and the speed v_2.

The work energy theorem (see Eq. 7–14) states: The work done by the resultant force acting on a system is equal to the change in kinetic energy of the system assuming that there is no increase in internal energy. In Fig. 16–13 the forces that do work on the system, assuming that we can neglect viscous forces, are the pressure forces $p_1 A_1$ and $p_2 A_2$ that act on the left- and right-hand ends of the system, respectively, and the force of gravity. As fluid flows through the pipe the net effect, as a comparison of Figs. 16–13*a* and *b* shows, is to raise an amount

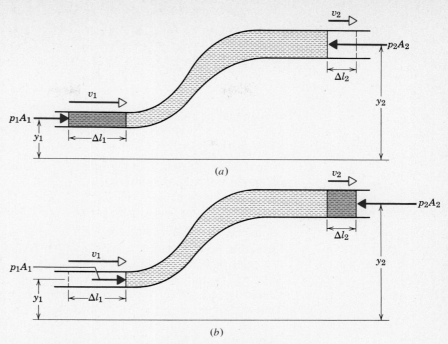

Figure 16–13 A portion of fluid (shown in color) moves through a section of pipeline from the position shown in (a) to that shown in (b).

of fluid represented by the cross-shaded area in Fig. 16–13a to the position shown in Fig. 16–13b. The amount of fluid represented by the horizontal shading is unchanged by the flow.

We can find the work W done on the system by the resultant force as follows:

1. The work done on the system by the pressure force p_1A_1 is $p_1A_1\Delta l_1$.

2. The work done on the system by the pressure force p_2A_2 is $-p_2A_2\Delta l_2$. Note that it is negative, which means that positive work is done by the system.

3. The work done on the system by gravity is associated with lifting the cross-shaded fluid from height y_1 to height y_2 and is $-mg(y_2 - y_1)$, in which m is the mass of fluid in either cross-shaded area. It too is negative because work is done by the system against the gravitational force.

$$W = p_1A_1\,\Delta l_1 - p_2A_2\,\Delta l_2 - mg(y_2 - y_1).$$

Now $A_1\,\Delta l_1\,(= A_2\,\Delta l_2)$ is the volume of the cross-shaded fluid element, which we can write as m/ρ, in which ρ is the (constant) fluid density. Recall that the two fluid elements have the same mass, so that in setting $A_1\,\Delta l_1 = A_2\,\Delta l_2$ we have assumed the fluid to be incompressible. With this assumption we have

$$W = (p_1 - p_2)\left(\frac{m}{\rho}\right) - mg(y_2 - y_1). \tag{16–7a}$$

The change in kinetic energy of the fluid element is

$$\Delta K = \tfrac{1}{2}mv_2{}^2 - \tfrac{1}{2}mv_1{}^2. \tag{16–7b}$$

From the work-energy theorem (Eq. 7–14) we then have

$$W = \Delta K$$

or

$$(p_1 - p_2)\left(\frac{m}{\rho}\right) - mg(y_2 - y_1) = \tfrac{1}{2}mv_2{}^2 - \tfrac{1}{2}mv_1{}^2, \qquad (16\text{–}8a)$$

which can be rearranged to read

$$p_1 + \tfrac{1}{2}\rho v_1{}^2 + \rho g y_1 = p_2 + \tfrac{1}{2}\rho v_2{}^2 + \rho g y_2. \qquad (16\text{–}8b)$$

Since the subscripts 1 and 2 refer to any two locations along the pipeline, we can drop the subscripts and write

Bernoulli's equation

$$p + \tfrac{1}{2}\rho v^2 + \rho g y = \text{constant.} \qquad (16\text{–}9)$$

Equation 16–9 is called *Bernoulli's equation* for steady, nonviscous, incompressible flow. It was first presented by Daniel Bernoulli in 1738.

Bernoulli's equation is strictly applicable only to steady flow, the quantities involved being evaluated along a streamline. In our figure the streamline used is along the axis of the pipeline. If the flow is irrotational, however, we can show that the constant in Eq. 16–9 is the same for all streamlines.

Just as the statics of a particle is a special case of particle dynamics, so fluid statics is a special case of fluid dynamics. It should come as no surprise, therefore, that the law of pressure change with height in a fluid at rest is included in Bernoulli's equation as a special case. For let the fluid be at rest; then $v_1 = v_2 = 0$ and Eq. 16–8b becomes

$$p_1 + \rho g y_1 = p_2 + \rho g y_2$$

or

$$p_2 - p_1 = -\rho g(y_2 - y_1),$$

which is the same as Eq. 16–3.

In Eq. 16–9 all terms have the dimension of a pressure (check this). The pressure $p + \rho g h$, which would be present even if there were no flow ($v = 0$), is called the *static pressure;* the term $\tfrac{1}{2}\rho v^2$ is called the *dynamic pressure.*

16–9 Applications of Bernoulli's Equation and the Equation of Continuity

Bernoulli's equation can be used to find fluid speeds by making pressure measurements. The principle generally used in such measuring devices is the following: The equation of continuity requires that the speed of the fluid at a constriction increase; Bernoulli's equation then shows that the pressure must fall there. That is, for a horizontal pipe $\tfrac{1}{2}\rho v^2 + p$ equals a constant; if v increases and the fluid is incompressible, p must decrease.

Measuring liquid flow rates

1. The Venturi Meter. (Fig. 16–14) is a gauge put in a flow pipe to measure the flow speed of a liquid. A liquid of density ρ flows through a pipe of cross-sectional area A. At the throat the area is reduced to a and a manometer tube is attached, as shown. Let the manometer liquid, such as mercury, have a density ρ'. By apply-

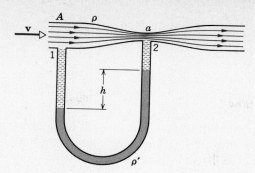

Figure 16–14 The Venturi meter, used to measure the speed of flow of a fluid.

ing Bernoulli's equation and the equation of continuity at points 1 and 2, you can show that the speed of flow at A (see Problem 43) is

$$v = a \sqrt{\frac{2(\rho' - \rho)gh}{\rho(A^2 - a^2)}}.$$

If we want the volume flux or flow rate R, which is the volume of liquid transported past any point per second, we simply compute

$$R = vA.$$

Measuring flow speeds in gases

2. The Pitot Tube. The Pitot tube (Fig. 16–15) is used to measure the flow speed of a gas. Consider the gas, say, air, flowing past the opening at a. These openings are parallel to the direction of flow and are set far enough back so that the velocity and pressure outside the openings have the free-stream values. The pressure in the left arm of the manometer, which is connected to these openings, is then the static pressure in the gas stream, p_a. The opening of the right arm of the manometer is at right angles to the stream. When the fluid in the manometer has come to rest the velocity is zero at b and the pressure is p_b.

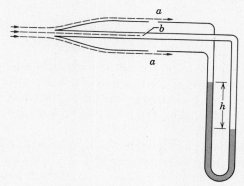

Figure 16–15 Cross-sectional diagram of a Pitot tube used to measure the flow speed in a gas.

Applying Bernoulli's equation to points a and b, we obtain

$$p_a + \tfrac{1}{2}\rho v^2 = p_b,$$

where, as the figure shows, p_b is greater than p_a. If h is the difference in height of the liquid in the manometer arms and ρ' is the density of the manometer liquid, then

$$p_a + \rho' gh = p_b.$$

Comparing these two equations, we find that

$$\tfrac{1}{2}\rho v^2 = \rho' gh$$

or

$$v = \sqrt{\frac{2gh\rho'}{\rho}},$$

which gives the gas speed. This device can be calibrated to read v directly and is then known as an air-speed indicator. You will see them on airplanes.

Accounting for dynamic lift

3. An Aircraft Wing.   The upward force on an aircraft wing, a hydrofoil, or a helicoptor rotor, by virtue of its motion through a fluid, is called *dynamic lift*. We must distinguish it from *static lift*, which is the buoyant force that acts on a body such as a balloon or an iceberg in accord with Archimedes' principle (Section 16–4).

Figure 16–16 shows the streamlines about an airfoil (or wing cross section) attached to an aircraft. Let us choose the aircraft as our frame of reference, as in a wind-tunnel experiment, and let us assume that the air is moving past the wing from right to left.

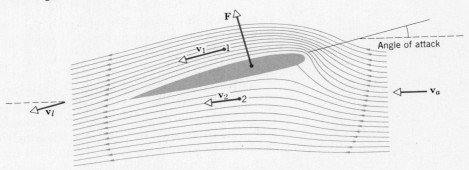

Figure 16–16   Streamlines about an airfoil. The velocity of the approaching air $\mathbf{v}_a$ is horizontal. That of the leaving air $\mathbf{v}_l$ has a downward component. Thus, because the airfoil has forced the air down, the air, from Newton's third law, must have forced the airfoil up.

The *angle of attack* of the wing causes air to be deflected downward. From Newton's third law the reaction of this downward force of the wing on the air is an upward force $\mathbf{F}$ exerted by the air on the wing. This force may be resolved into a vertical component, the lift $\mathbf{L}$, and a horizontal component $\mathbf{D}$, sometimes called the induced drag.

The pattern of streamlines is consistent. Above the wing (point 1) the streamlines are closer together than they are below the wing (point 2). Thus $v_1 > v_2$ and, from Bernoulli's principle, $p_1 < p_2$, which must be true if there is to be a lift.

Thrust of a rocket

4. A Rocket.   As our final example let us compute the thrust on a rocket produced by the escape of its exhaust gases. Consider a chamber (Fig. 16–17) of cross-sectional area A filled with a gas of density ρ at a pressure p. Let there be a small orifice of effective cross-sectional area A_0 at the bottom of the chamber.* We wish to find the speed v_0 with which the gas escapes through the orifice.

---

* The exiting jet converges for a short distance after leaving the hole so that the effective area A_0 for the jet is smaller than the actual area A_0' of the hole. That is, $A_0 = \epsilon A_0'$, where $\epsilon < 1$. For a round hole with a sharp edge $\epsilon = 0.62$. All exit hole or nozzle arrangements have a characteristic *efflux coefficient* ϵ.

Let us write Bernoulli's equation (Eq. 16–8b) as

$$p_1 - p_2 = \rho g(y_2 - y_1) + \tfrac{1}{2}\rho(v_2{}^2 - v_1{}^2).$$

For a gas the density is so small that we can neglect the variation in pressure with height in a chamber (*see* Section 16–3). Hence, if p represents the pressure p_1 in the chamber and p_0 represents the atmospheric pressure p_2 just outside the orifice, we have

$$p - p_0 = \tfrac{1}{2}\rho(v_0{}^2 - v^2)$$

or

$$v_0{}^2 = \frac{2(p - p_0)}{\rho} + v^2,$$

(16–10)

where v is the speed of the flowing gas inside the chamber and v_0 is the *speed of efflux* of the gas through the orifice. Although a gas is compressible and the flow may become turbulent, we can treat the flow as steady and incompressible for pressure and efflux speeds that are not too high.

Now let us assume continuity of mass flow (in a rocket engine this is achieved when the mass of escaping gas equals the mass of gas created by burning the fuel), so that (for an assumed constant density)

$$Av = A_0 v_0.$$

If the orifice is very small so that $A_0 \ll A$, then $v_0 \gg v$, and we can neglect v^2 compared to $v_0{}^2$ in Eq. 16–10. Hence, the speed of efflux is

$$v_0 = \sqrt{\frac{2(p - p_0)}{\rho}}.$$

(16–11)

If our chamber is the exhaust chamber of a rocket, the thrust F_{th} on the rocket (Section 9–6) is $v_0\, dM/dt$. But the mass of gas flowing out in time dt is $dM = \rho A_0 v_0\, dt$, so that

$$F_{\text{th}} = v_0 \frac{dM}{dt} = v_0 \rho A_0 v_0 = \rho A_0 v_0{}^2,$$

and from Eq. 16–11 the thrust is

$$F_{\text{th}} = 2A_0(p - p_0).$$

(16–12)

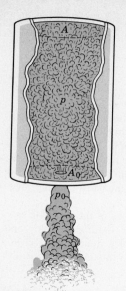

Figure 16–17 Fluid streaming out of a chamber.

REVIEW GUIDE AND SUMMARY

A fluid is a substance that can flow. Because of this it cannot sustain a static tangential or shearing force. It can, however, maintain a static compression force. This force results in a fluid pressure, defined as

Definition of pressure

$$p = \frac{F}{S},$$

where F is the force acting on a surface element of area S. The force due to a fluid pressure acts in a direction normal to the fluid surface.

Unit of pressure

The SI unit for pressure is the *pascal* ($= 1$ N/m²). Other units are: 1.00 atm $= 1.01 \times 10^5$ Pa $= 1.01$ bar $= 760$ mm-Hg $= 760$ torr $= 14.7$ lb/in.².

The pressure in a fluid at rest near the earth's surface depends on the depth h from the free surface where the pressure is p_0. The equation is

Pressure variation with height

$$p = p_0 + \rho g h,$$

[16–4]

where ρ is the fluid density, which is assumed to be constant. The pressure is the same at all points at the same level.

The above equation may be applied to the case of two fluids that are at rest, have different densities, and are in contact without mixing; see Example 1.

Pascal's principle

The equation for the depth dependence of pressure also leads to Pascal's principle, which states: *An external pressure applied to an enclosed fluid at rest is transmitted undiminished to every portion of the fluid and to the walls of the container.*

Archimedes' principle

The surface of an immersed body is acted upon by forces resulting from the fluid pressure. The sum of these forces is called the buoyant force. This force acts vertically up through the center of gravity of the displaced fluid. Archimedes' principle states that the magnitude of this force is equal to the weight of the fluid displaced by the body; see Example 2.

Various devices can be used to measure pressure. These include the barometer for measuring atmospheric pressure and the open-tube manometer for measuring gauge pressure (the deviation from atmospheric pressure) in a tank.

Fluid dynamics

In fluid dynamics, or the study of fluid flow, we divide the types of fluid motion into steady or nonsteady, rotational or irrotational, compressible or incompressible, and viscous or nonviscous. A *streamline* is the path a liquid particle takes in steady flow. A *tube of flow* is a bundle of streamlines. The conservation of mass applied to steady flow results in

Equation of continuity

$$\rho A v = \text{constant}, \qquad [16\text{–}5]$$

where A is the cross-sectional area of a tube of flow. If the flow is incompressible,

$$A v = \text{constant}. \qquad [16\text{–}6]$$

The application of Newton's laws to steady, incompressible, nonviscous flow leads to Bernoulli's equation:

Bernoulli's equation

$$p + \tfrac{1}{2}\rho v^2 + \rho g y = \text{constant}, \qquad [16\text{–}9]$$

where y is the fluid height above some reference point.

Four illustrations of Bernoulli's equation are given: (1) the Venturi meter, which measures the flow speed of a liquid; (2) the Pitot tube, which measures the flow speed of a gas; (3) dynamic lift, which results from the flow of a fluid past a body such as an airfoil; and (4) rocket thrust, which is the force on a rocket produced by the escape of its gases.

QUESTIONS

1. Make an estimate of the average density of your body. Explain a way in which you could get an accurate value using ideas in this chapter.

2. Two balls have the same shape and size, but one is denser than the other. Assuming that at a given speed the air resistance is the same on each, show that when they are released simultaneously from the same height the heavier body will get to the ground first.

3. Water is poured to the same level in each of the vessels shown, all of the same base area (Fig. 16–18). If the pressure is the same at the bottom of each vessel, the force experienced by the base of each vessel is the same. Why then do the three vessels have different weights when put on a scale? This apparently contradictory result is commonly known as the "hydrostatic paradox."

4. Does Archimedes' principle hold in a vessel in free fall? In a satellite moving in a circular orbit?

5. How does a suction cup work?

6. Is the buoyant force acting on a submerged submarine the same at all depths?

7. What, if anything, happens to the buoyant force acting on a helium-filled balloon if you replace the helium with hydrogen? (Hydrogen is less dense than helium.)

8. Does squeezing a tube of toothpaste illustrate Pascal's principle?

9. A block of wood floats in a pail of water in an elevator. When the elevator starts from rest and accelerates down, does the block float higher above the water surface?

10. Two identical buckets are filled to the brim with water, but one has a block of wood floating in the water. Which bucket, if either, is heavier?

11. "A hot air balloon rises because the heating causes the pressure of the air in the balloon envelope to rise above the external atmospheric pressure, thus providing the necessary lift." What do you think of this sentence?

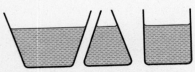

Figure 16–18 Question 3.

12. Can you sink an iron ship by siphoning sea water into it?

13. Scuba divers are warned not to hold their breath when swimming upward. Why?

14. Underwater swimmers often use a snorkel, a tube about a foot long, that, secured in the mouth, allows them to breath underwater. Why not extend this idea, providing the swimmer with a 50-ft length of hose so that he could breath through it at that depth?

15. A beaker is exactly full of liquid water at its freezing point and has an ice cube floating in it, also at the freezing point. As the cube melts, what happens to the water level in these three cases: (a) The cube is solid ice. (b) The cube contains some grains of sand. (c) The cube contains some bubbles.

16. Although parachutes are supposed to brake your fall, they are often designed with a hole at the top. Explain why.

17. In general, it is harder for a man to float on his back than for a woman to float on hers. Explain.

18. A ball floats on the surface of water in a container exposed to the atmosphere. Will the ball remain immersed at its former depth or will it sink or rise somewhat if (a) the container is covered and the air is removed, (b) the container is covered and the air is compressed?

19. Explain why an inflated balloon will rise to a definite height once it starts to rise, whereas a submarine will always sink to the bottom of the ocean once it starts to sink, if no changes are made.

20. A balloon weighs the same when empty as when filled with air at atmospheric pressure. Why? Would the weights be the same if measured in a vacuum?

21. During World War II, a damaged freighter that was barely able to float in the North Sea steamed up the Thames estuary toward the London docks. It sank before it arrived. Why?

22. Is it true that a floating object will only be in stable equilibrium if its center of buoyancy lies above its center of gravity? Illustrate with examples.

23. Very often a sinking ship will turn over as it becomes immersed in water. Why?

24. A barge filled with scrap iron is in a canal lock. If the iron is thrown overboard into the water, what happens to the water level in the lock? What if it is thrown onto the land beside the canal?

25. A bucket of water is suspended from a spring balance. Does the balance reading change when a piece of iron suspended from a string is immersed in the water? When a piece of cork is put in the water?

26. Why does a uniform wooden stick or log float horizontally? If enough iron is added to one end, it will float vertically. Explain this also.

27. Although there are practical difficulties it is possible in principle to float an ocean liner in a few barrels of water. How would you go about doing this?

28. A thin-walled pipe will burst more easily if, when there is a pressure differential between inside and outside, the excess pressure is on the outside. Explain.

29. Describe the forces acting on an element of fluid as it flows through a pipe of nonuniform cross section.

30. The height of the liquid in the standpipes indicates that the pressure drops along the channel, even though the channel has a uniform cross section and the flowing liquid is incompressible (Fig. 16–19). Explain.

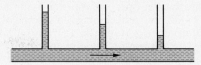

Figure 16–19 Question 30.

31. The taller the chimney, the better the draft taking the smoke out of the fireplace. Explain. Why doesn't the smoke pour into the room containing the fireplace?

32. In a lecture demonstration a Ping Pong ball is kept in midair by a vertical jet of air. Is the equilibrium stable, unstable, or neutral? Explain.

33. Two rowboats moving parallel to one another in the same direction are pulled toward one another. Two automobiles moving parallel are also pulled together. Explain such phenomena using Bernoulli's equation.

34. Can the action of a parachute in retarding free fall be explained by Bernoulli's equation?

35. Why does an object falling from a great height reach a steady terminal speed?

36. A stream of water from a faucet becomes narrower as it falls. Explain.

37. Sometimes people remove letters from envelopes by cutting a sliver from a narrow end, holding it firmly, and blowing toward it. Does Bernoulli's equation play a role in this enterprise? Explain.

38. On take-off, would it be better for an airplane to move into the wind or with the wind? On landing?

39. Does the difference in pressure between the lower and upper surfaces of an airplane wing depend on the altitude of the moving plane? Explain.

40. The accumulation of ice on an airplane wing may change its shape in such a way that its lift is greatly reduced. Explain.

41. How is an airplane able to fly upside down?

42. Why does the factor "2" appear in Eq. 16–12, rather than "1"? One might expect that the thrust would simply be the pressure difference times the area, that is, $A_0 (p - p_0)$.

43. The destructive effect of a tornado (twister) is greater near the center of the disturbance than near the edge. Explain.

44. In steady flow the velocity vector at any point is constant. Can there then be accelerated motion of the fluid particles? Discuss.

45. Explain why you can't remove the filter paper from the funnel of Fig. 16–20 by blowing into the narrow end.

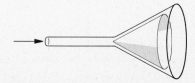

Figure 16–20 Question 45.

EXERCISES

Section 16–2   Fluid Pressure and Density

1. You inflate the front tires on your car to 28 psi. Later, you measure your blood pressure, obtaining a reading of 120/80, the readings being in mm-Hg. In metric countries (which is to say, most of the rest of the world), these quantities are customarily reported in kilopascals (kPa). What are (a) your tire pressure reading and (b) your blood pressure reading in these units?

2. Find the pressure increase in the fluid in a syringe when a nurse applies a force of 9.5 lb (= 42 N) to the syringe's piston of radius 1.1 cm.

3. An office window is 3.4 m by 2.1 m. Due to the passage of a storm, the outside air pressure drops to 0.92 atm, but inside the pressure is held at 1.0 atm. What net force pushes out on the window?

Section 16–3   The Variation of Pressure in a Fluid at Rest

4. The human lungs can operate against a pressure differential of up to one-twentieth of an atmosphere. If a diver uses a snorkel (long tube) for breathing, how far below the water level can he swim?

5. Find the total pressure, in pounds per square inch and in pascals, 500 ft (= 150 m) below the surface of the ocean. The relative density of sea water is 1.03 and the atmospheric pressure at sea level is 14.7 lb/in.^{2} (= 1.0×10^5 Pa).

6. A house is constructed on a slope 8.2 m below street level. If the sewer is 2.1 m below street level, find the minimum pressure that must be created by the sewage pump in order to transfer waste of average density 900 kg/m^3.

7. Crew members attempt to escape from a damaged submarine 100 m below the surface. What force must they apply to a hatch, which is 1.2 m by 0.60 m, in order to open it? The density of the ocean water is 1025 kg/m^3.

Section 16–4   Pascal's Principle and Archimedes' Principle

8. A piston of small cross-sectional area a is used in the hydraulic press to exert a small force f on the enclosed liquid. A connecting pipe leads to a larger piston of cross-sectional area A (Fig. 16–21). (a) What force F will the larger piston sustain? (b) If the small piston has a diameter of 1.5 in. and the large piston one of 21 in., what weight on the small piston will support 2.0 tons on the large piston?

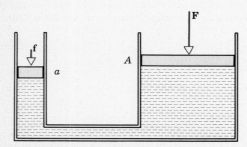

Figure 16–21   Exercise 8.

9. A cubical object of dimensions L (2.0 ft) on a side and weight W (1000 lb) in a vacuum is suspended by a rope in an open tank of liquid of density ρ (2.0 slugs/ft^3) as in Fig. 16–22. (a) Find the total downward force exerted by the liquid and the atmosphere on the top of the object of area A (4.0 ft^2). (b) Find the total upward force on the bottom of the object. (c) Find the tension in the rope.

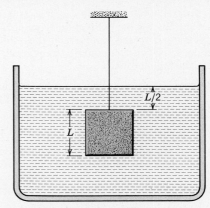

Figure 16–22   Exercise 9.

10. A block of wood floats in water with two-thirds of its volume submerged. In oil it has 0.90 of its volume submerged. Find the density of (a) the wood and (b) the oil.

11. Assume the density of brass weights to be 8.0 g/cm^3 and that of air to be 0.0012 g/cm^3. What percent error arises from neglecting the buoyancy of air in weighing an object of mass m and density ρ on a beam balance?

12. A physicist swimming in the Dead Sea notices that about one-third of the volume of her body will float above the water line. Assuming that the human body density is 0.98 g/cm^3, find the density of the water in the Dead Sea. Why is it so much greater than 1.0 g/cm^3?

13. An iron casting containing a number of cavities weighs 60 lb in air and 40 lb in water. What is the volume of the cavities in the casting? Assume the relative density of iron to be 7.8.

Section 16–7   Streamlines and the Equation of Continuity

14. A $\frac{3}{4}$-in. (inside diameter) water pipe is coupled to three $\frac{1}{2}$-in. pipes. (a) If the flow rates in the three pipes are 7.0, 5.0, and 3.0 gal/min, what is the flow rate in the $\frac{3}{4}$-in. pipe? (b) How will the speed of the liquid in the $\frac{3}{4}$-in. pipe compare with that carrying 7.0 gal/min?

15. A garden hose having an internal diameter of 0.75 in. is connected to a lawn sprinkler that consists merely of an enclosure with 24 holes, each 0.050 in. in diameter. If the water in the hose has a speed of 3.0 ft/s, at what speed does it leave the sprinkler holes?

Section 16–8   Bernoulli's Equation

16. How much work is done by pressure in forcing 50 ft^3 (= 1.4 m^3) of water through a 0.50-in. (= 13-mm)-internal diameter pipe if the difference in pressure at the two ends of the pipe is 1.0 atm (= 1.0×10^5 Pa)?

17. In a horizontal oil pipeline of constant cross-sectional area, the pressure decrease between two points 1000 ft apart is 5.0 lb/in.^{2}. What is the energy loss per cubic foot of oil per foot?

18. Water falls from a height of 20 m at the rate of 15 m^3/min and drives a water turbine. What is the maximum power that can be developed by this turbine?

Section 16–9 Applications of Bernoulli's Equation and the Equation of Continuity

19. A Pitot tube on a high-altitude aircraft measures a differential pressure of 180 Pa. What is the airspeed if the density of the air is 0.031 kg/m^3?

20. A Pitot tube is mounted on an airplane wing to determine the speed of the plane relative to the air, which has a density of 1.03×10^3 kg/m^3. The tube contains alcohol and indicates a level difference of 26 cm. What is the plane's speed relative to the air? The density of alcohol is 0.81×10^3 kg/m^3.

21. An 80-cm^2 plate of 500-g mass is hinged along one side. If air is blown over the upper surface only, what speed must the air have to hold the plate horizontal?

22. A Venturi meter has a pipe diameter of 10 in. and a throat diameter of 5.0 in. If the water pressure in the pipe is 8.0 lb/in.^{2} and in the throat is 6.0 lb/in.^{2}, determine the rate of flow of water in cubic feet per second (volume flux).

23. A hollow tube has a disk DD attached to its end (Fig. 16–23). When air is blown through the tube, the disk attracts the card CC. Let the area of the card be A and let v be the average airspeed between the card and the disk; calculate the resultant upward force on CC. Neglect the card's weight; assume that $v_0 \ll v$, where v_0 is the air speed in the hollow tube.

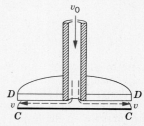

Figure 16–23 Exercise 23.

PROBLEMS GROUPED BY SECTION

Section 16–2 Fluid Pressure and Density

1. An airtight box having a lid with an area of 12 in.^{2} is partially evacuated. If a force of 108 lb is required to pull the lid off the box and the outside atmospheric pressure is 15 lb/in.^{2}, what was the pressure in the box?

Section 16–3 The Variation of Pressure in a Fluid at Rest

2. Estimate the hydrostatic difference in blood pressure between the brain and the foot in a person of height 1.83 m (=6.00 ft). The density of blood is 1.06×10^3 kg/m^3.

3. (a) Find the total weight of water on top of a nuclear submarine at a depth of 200 m, assuming that its (cross-sectional) hull area is 3000 m^2. (b) What water pressure would a diver experience at this depth? Express your answer in atmospheres. Do you think that occupants of a damaged submarine at this depth could escape without special equipment? The relative density of seawater is 1.03.

4. What would be the height of the atmosphere if the air density (a) were constant and (b) decreased linearly to zero with height? Assume a sea-level density of 1.3 kg/m^3.

5. Two identical cylindrical vessels with their bases at the same level each contain a liquid of density ρ. The area of either base is A, but in one vessel the liquid height is h_1 and in the other h_2. Find the work done by gravity in equalizing the levels when the two vessels are connected.

6. A U-tube is filled with a single homogeneous liquid. The liquid is temporarily depressed in one side by a piston. The piston is removed and the level of the liquid in each side oscillates. Show that the period of oscillation is $\pi \sqrt{2L/g}$, where L is the total length of the liquid in the tube.

Section 16–4 Pascal's Principle and Archimedes' Principle

7. (a) What is the minimum area of a block of ice 1.0 ft (0.3 m) thick floating on water that will hold up an automobile weighing 2500 lb (mass = 1100 kg)? (b) Does it matter where the car is placed on the block of ice?

8. A hollow sphere of inner radius 8.0 cm and outer radius 9.0 cm floats half submerged in a liquid of specific gravity 0.80. Calculate the density of the material of which the sphere is made.

9. Three children each of weight W (80 lb) make a log raft by lashing together logs of diameter D (1.0 ft) and length L (6.0 ft). How many logs will be needed to keep them afloat? Take the specific gravity of wood to be 0.80 (see Example 1).

10. (a) Consider a container of fluid subject to a *vertical upward* acceleration a. Show that the pressure variation with depth in the fluid is given by

$$p = \rho h(g + a),$$

where h is the depth and ρ is the density. (b) Show also that if the fluid as a whole undergoes a *vertical downward* acceleration a, the pressure at a depth h is given by

$$p = \rho h(g - a).$$

(c) What is the state of affairs in free fall?

11. It has been proposed to move natural gas from the North Sea gas fields in huge dirigibles, using the gas itself to provide lift. Calculate the force required to tether such an airship to the ground for off-loading when it is fully loaded with 1.0×10^6 m^3 of gas at a density of 0.80 kg/m^3. (The weight of the airship is negligible by comparison.)

12. A spherical, hydrogen-filled balloon has a radius of 12 m. The mass of the balloon plastic and support cables is 196 kg. What is the mass of the maximum load the balloon can carry? (Density of hydrogen = 0.09 kg/m^3; density of air = 1.25 kg/m^3.)

13. A car has a total mass of 1800 kg. The volume of air space in the passenger compartment is 5.0 m^3. The volume of the motor and front wheels is 0.75 m^3, and the volume of the rear wheels, gas tank, and luggage is 0.80 m³. Water cannot enter these areas. The car is parked on a hill; the handbrake cable snaps and the car rolls down the hill into a lake; see Fig. 16–24. (*a*) At first, no water enters the passenger compartment. How much of the car, in cubic meters, is below the water surface with the car floating as shown? (*b*) As water slowly enters, the car sinks. How many cubic meters of water are in the car as it disappears below the water surface?

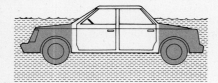

Figure 16–24   Problem 13.

*14. In analyzing certain geological features of the earth, it is often appropriate to assume that the pressure at some horizontal *level of compensation,* deep in the earth, is the same over a large region, and is equal to that exerted by the weight of the overlying material. That is, the pressure on the level of compensation is given by the hydrostatic (fluid) pressure formula. This requires, for example, that mountains have low density *roots;* see Fig. 16–25. Consider a mountain 6.0 km high. The continental rocks have a density of 2.9 g/cm³; beneath the continent is the mantle, with a density of 3.3 g/cm³. Calculate the depth *D* of the root. (*Hint:* Set the pressure at points *a* and *b* equal; the depth *y* of the level of compensation will cancel out.)

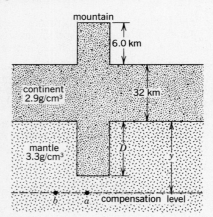

Figure 16–25   Problem 14.

Section 16–7   Streamlines and the Equation of Continuity

15. Water is pumped steadily out of a flooded basement at a speed of 5.0 m/s through a uniform hose of radius 1.0 cm. The hose passes out through a window 3.0 m above the water line. How much power is supplied by the pump?

16. A river 20 m wide and 4.0 m deep drains a 3000-km² land area in which the average precipitation is 48 cm per year. One-fourth of this subsequently returns to the atmosphere by evaporation, but the remainder ultimately drains into the river. What is the average speed of the river current?

Section 16–8   Bernoulli's Equation

17. If a person blows with a speed of 15 m/s across the top of one side of a U-tube containing water, what will be the difference between the water levels on the two sides?

18. Figure 16–26 shows liquid discharging from an orifice in a large tank at a distance *h* below the water level. (*a*) Apply Bernoulli's equation to a streamline connecting points 1 and 2 and show that the speed of efflux is

$$v = \sqrt{2gh}.$$

This is known as Torricelli's law. (*b*) If the orifice were curved directly upward, how high would the liquid stream rise? (*c*) How would viscosity or turbulence affect the analysis?

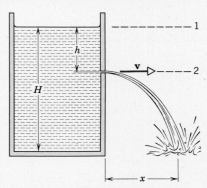

Figure 16–26   Problems 18 and 39.

19. A sniper fires a rifle bullet into a gasoline tank, making a hole 50 m below the surface. The tank was sealed and under 3.0 atm absolute pressure, as shown in Fig. 16–27. The stored gasoline has a density of 660 kg/m³. At what speed does the gasoline begin to shoot out of the hole?

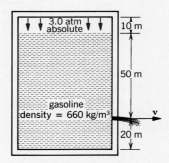

Figure 16–27   Problem 19.

Section 16–9   Applications of Bernoulli's Equation and the Equation of Continuity

20. A water pipe having a 1.0-in. inside diameter carries water into the basement of a house at a velocity of 3.0 ft/s at a pressure of 25 lb/in.². If the pipe tapers to ½ in. and rises to the second floor 25 ft above the input point, what are the (*a*) velocity and (*b*) water pressure there?

21. Consider the Venturi tube of Fig. 16–14 without the manometer. Let *A* equal 5*a*. Suppose that the pressure at *A* is 2.0 atm. (*a*) Compute the values of *v* at *A* and *v′* at *a* that would make the pressure *p′* at *a* equal to zero. (*b*) Compute the corresponding volume flow rate if the diameter at *A* is 5.0 cm. The phenomenon at *a* when *p′* falls to nearly zero is known as cavitation. The water vaporizes into small bubbles.

22. An airplane has a wing area (each wing) of 10 m². At a certain air speed, air flows over the upper wing surface at 48 m/s and over the lower wing surface at 40 m/s. What is the mass of the plane? Assume that the plane travels at constant velocity and that lift effects associated with the fuselage and tail assembly are small. Discuss the lift if the airplane, flying at the same air speed, is (*a*) in level flight, (*b*) climbing at 15°, and (*c*) descending at 15°.

23. A tank of large area is filled with water to a depth *D* (1.0 ft). A hole of cross section *A* (1.0 in.²) in the bottom of the tank allows water to drain out. (*a*) What is the rate at which water flows out in cubic feet per second? (*b*) At what distance below the bottom of the tank is the cross-sectional area of the stream equal to one-half the area of the hole?

24. Suppose that two tanks, 1 and 2, each with a large opening at the top, contain different liquids. A small hole is made in the side of each tank at the same depth *h* below the liquid surface, but the hole in tank 1 has half the cross-sectional area of the hole in tank 2. (*a*) What is the ratio ρ_1/ρ_2 of the densities of the fluids if it is observed that the mass flux is the same for the two holes? (*b*) What is the ratio of the flow rates (volume flux) from the two tanks? (*c*) To what height above the hole in the second tank should fluid be added or drained to equalize the flow rates?

ADDITIONAL PROBLEMS

25. In 1654 Otto von Guericke, burgomeister of Magdeburg and inventor of the air pump, gave a demonstration before the Imperial Diet in which two teams of eight horses could not pull apart two evacuated brass hemispheres. (*a*) Show that the force *F* required to pull apart the hemispheres is $F = \pi R^2 P$, where *R* is the (outside) radius of the hemispheres and *P* is the difference in pressure outside and inside the sphere (Fig. 16–28). (*b*) Taking *R* equal to 1.0 ft and the inside pressure as 0.10 atm, what force would the team of horses have had to exert to pull apart the hemispheres? (*c*) Why are two teams of horses used? Would not one team prove the point just as well?

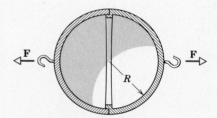

Figure 16–28   Problem 25.

26. A swimming pool has the dimensions 80 ft × 30 ft × 8.0 ft. (*a*) When it is filled with water, what is the force (due to the water alone) on the bottom? On the ends? On the sides? (*b*) If you are concerned with whether or not the concrete walls and floor will collapse, is it appropriate to take the atmospheric pressure into account?

27. Three liquids that will not mix are poured into a cylindrical container. The amounts and densities of the liquids are 0.50 liter, 2.6 g/cm³; 0.25 liter, 1.0 g/cm³; and 0.40 liter, 0.80 g/cm³. What is the total force acting on the bottom of the container? (Ignore the contribution due to the atmosphere.)

28. A simple U-tube contains mercury. When 11.2 cm of water is poured into the right arm, how high does the mercury rise in the left arm from its initial level?

29. Water stands at a depth *D* behind the vertical upstream face of a dam, as shown in Fig. 16–29. Let *W* be the width of the dam. (*a*) Find the resultant horizontal force exerted on the dam by the gauge pressure of the water and (*b*) the net torque due to the gauge pressure of the water exerted about a line through *O* parallel to the width of the dam. (*c*) What is the line of action of the equivalent resultant force?

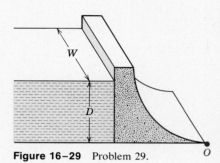

Figure 16–29   Problem 29.

30. A hollow spherical iron shell floats almost completely submerged in water. If the outer diameter is 60 cm and the relative density of iron is 7.80, find the inner diameter.

31. A block of wood has a mass of 3.67 kg and a relative density (see Example 1) of 0.60. It is to be loaded with lead so that it will float in water with 0.90 of its volume immersed. What mass of lead is needed (*a*) if the lead is on top of the wood? (*b*) if the lead is attached below the wood? The density of lead is 1.13 × 10⁴ kg/m³.

32. You place a glass beaker partially filled with water, in a sink (Fig. 16–20). It has a mass of 390 g and an interior volume of 500 cm³. You now start to fill the sink with water and you find, by experiment, that if the beaker is less than half full, it will float; but if it is more than half full, it remains on the bottom of the sink as the water rises to its rim. What is the density of the material of which the beaker is made?

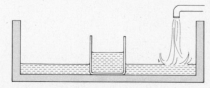

Figure 16–30   Problem 32.

33. The windows in an office building are 4.0 m by 5.0 m. On a stormy day, air is blowing at 30 m/s past a window on the 53rd floor. Calculate the net force on the window. The density of air is 1.23 kg/m³.

34. Water flows continuously from the outlet of a faucet of internal diameter *d* at an initial speed v_0. Determine the diameter of the stream in terms of the distance h below the outlet. (Neglect air resistance and assume droplets are not formed.)

35. A water intake at a pump storage reservoir (see Fig. 16–31) has a cross-sectional area of 8.0 ft². The water flows in at a speed of 1.33 ft/s. At the generator building 600 ft below the intake point, the cross-sectional area is 0.44 ft² and the water flows out at 31.0 ft/s. What is the difference in pressure, in pounds per square inch, between inlet and outlet?

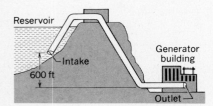

Figure 16–31 Problem 35.

36. The Goodyear blimp *Columbia* (see Fig. 16–32) is cruising slowly at low altitude, filled as usual with helium gas. Its maximum useful payload, including crew and cargo, is 1280. kg. (*a*) How much more payload could the *Columbia* carry if you replaced the helium with hydrogen? (*b*) What percent increase does this represent? (*c*) Why not do it? The volume of the helium-filled interior space is 5000 m³. The density of helium gas is two times the density of hydrogen at the same temperature and pressure. Assume a pressure of 1 atm and a temperature of 300 K.

Figure 16–32 Problem 36. Courtesy Goodyear Tire & Rubber.

37. A helium balloon is used to lift a 40-kg payload to an altitude of 27 km, where the air density is 0.035 kg/m³. The balloon has a mass of 15 kg and the density of the gas in the balloon is 0.0051 kg/m³. What is the volume of the balloon? (Neglect the volume of the payload.)

38. If the speed of flow past the lower surface of a wing is 110 m/s, what speed of flow over the upper surface will give a lift of 900 Pa (= 19 lb/ft²)? Take the density of air to be 1.3×10^{-3} g/cm³.

39. A tank is filled with water to a height H. A hole is punched in one of the walls at a depth h below the water surface (Fig. 16–26). (*a*) Show that the distance x from the foot of the wall at which the stream strikes the floor is given by $x = 2\sqrt{h(H - h)}$. (*b*) Could a hole be punched at another depth so that this second stream would have the same range? If so, at what depth? (*c*) At what depth should the hole be placed to make the emerging stream strike the ground at the maximum distance from the base of the tank?

40. Models of torpedoes are sometimes tested in a horizontal pipe of flowing water, much as a wind tunnel is used to test model airplanes. Consider a circular pipe of internal diameter 25 cm and a torpedo model, aligned along the axis of the pipe, with a diameter of 5.0 cm. The torpedo is to be tested with water flowing past it at 2.5 m/s. (*a*) With what speed must the water flow in the unconstricted part of the pipe? (*b*) What will the pressure difference be between the constricted and unconstricted parts of the pipe?

41. Water is moving with a speed of 5.0 m/s through a pipe with a cross-sectional area of 4.0 cm². The water gradually descends 10 m as the pipe increases in area to 8.0 cm². (*a*) What is the speed of flow at the lower level? (*b*) If the pressure at the upper level is 1.50×10^5 Pa what is the pressure at the lower level?

42. The fresh water behind the reservoir dam is 15 m deep. A horizontal pipe 4.0 cm in diameter passes through the dam 6.0 m below the water surface, as shown in Fig. 16–33. A plug secures the pipe opening. (*a*) Find the friction force between plug and pipe wall? (*b*) The plug is removed. What volume of water flows out of the pipe in 3.0 h?

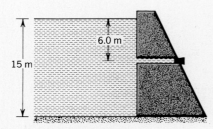

Figure 16–33 Problem 42.

43. By applying Bernoulli's equation and the equation of continuity to points 1 and 2 of Fig. 16–14, show that the speed of flow at the entrance is

$$v = a \sqrt{\frac{2(\rho' - \rho)gh}{\rho(A^2 - a^2)}}.$$

44. A siphon is a device for removing liquid from a container that cannot be tipped. It operates as shown in Fig. 16–34. The tube must initially be filled, but once this has been done the liquid will flow until its level drops below the tube opening at A. The liquid has density ρ and negligible viscosity. (*a*) With what speed does the liquid emerge from the tube at C? (*b*) What is the pressure in the liquid at the topmost point B? (*c*) What is the greatest possible height h_1 that a siphon may lift water?

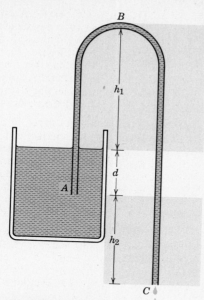

Figure 16–34 Problem 44.

45. A cylindrical wooden rod is loaded with lead at one end so that it floats upright in water as in Fig. 16–35. The length of the submerged portion is $l = 2.5$ m. The rod is set into vertical oscillation. (*a*) Show that the oscillation is simple harmonic. (*b*) Find the period of the oscillation. Neglect the fact that the water has a damping effect on the motion.

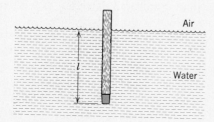

Figure 16–35 Problem 45.

*46. The tension in a string holding a solid block below the surface of a liquid (of density greater than the solid) is T_0 when the containing vessel (Fig. 16–36) is at rest. Show that the tension T, when the vessel has an upward vertical acceleration a, is given by $T_0(1 + a/g)$.

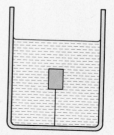

Figure 16–36 Problem 46.

*47. A jug contains 15 glasses of orange juice. When you open the tap at the bottom it takes 12 s to fill a glass with juice. If you leave the tap open, how long will it take to fill the remaining 14 glasses and thus empty the jug?

*48. In Fig. 16–37, the ocean is about to overrun the continent. Find the depth h of the ocean assuming isostatic compensation; see Problem 14.

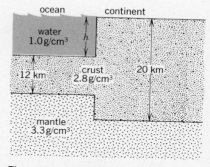

Figure 16–37 Problem 48.

17 Waves in Elastic Media

17-1 Mechanical Waves

Wave motion appears in almost every branch of physics. We are all familiar with water waves. There are also sound waves, as well as light waves, radio waves, and other electromagnetic waves. In Chapter 42 we shall see that it is even possible to treat a beam of electrons or other subatomic particles as a wave; we call such waves matter waves.

In this chapter and the next we confine our attention to waves in deformable or elastic media. These waves, among which ordinary sound waves in air are an example, are called *mechanical waves*. They originate in the displacement of some portion of an elastic medium from its normal position, causing it to oscillate about an equilibrium position. Because of the elastic forces on adjacent layers, the disturbance is transmitted from one layer to the next through the medium. The medium itself does not move as a whole; rather, the various parts oscillate in limited paths. For example, with surface waves in water, small floating objects like corks show that the actual motion of the water molecules is elliptical, slightly up and down and back and forth. Yet the water waves move steadily along the water. As they reach floating objects, they set them in motion, thus transferring energy to them. The energy in the waves is in the form of both kinetic and potential energy and its transmission comes about by its being passed from one part of the matter to the next, not by any long-range motion of the matter itself. Thus, mechanical waves are characterized by the transport of energy through matter by the motion of a disturbance in that matter without any corresponding bulk motion of the matter itself.

A material medium is necessary for the transmission of mechanical waves. Such a medium, however, is not needed to transmit electromagnetic waves. For example, light from stars comes to us through the near vacuum of space. These waves will be discussed later (Chapter 38).

17–2 Types of Waves

In listing water waves, light waves, and sound waves as examples of wave motion, we are classifying waves according to their broad physical properties. Waves can be classified in other ways.

We can distinguish different kinds of mechanical waves by considering how the motions of the particles of matter are related to the direction of propagation of the waves themselves. If the motions of the matter particles conveying the wave are perpendicular to the direction of propagation of the wave itself, we then have a *transverse* wave. For example, when a vertical string under tension is set oscillating back and forth at one end, a transverse wave travels down the string; the disturbance moves along the string but the string particles vibrate at right angles to the direction of propagation of the disturbance (Fig. 17–1a).

Although it is true that light waves are not mechanical, they are nevertheless transverse. Just as material particles move perpendicular to the direction of propagation in some mechanical waves, electric and magnetic fields are perpendicular to the direction of propagation of light waves.

If the motion of the particles conveying a mechanical wave is back and forth along the direction of propagation, we then have a *longitudinal* wave. For example, when a vertical spring under tension is set oscillating up and down at one end, a longitudinal wave travels along the spring; the coils vibrate back and forth in the direction in which the disturbance travels along the spring (Fig. 17–1b). Sound waves in a gas are longitudinal waves. We shall discuss them in greater detail in Chapter 18.

Some waves are neither purely longitudinal nor purely transverse. For example, in waves on the surface of water the particles of water move both up and down and back and forth, tracing out elliptical paths as the water waves move by.

Waves can also be classified as one-, two-, and three-dimensional waves, according to the number of dimensions in which they propagate energy. Waves moving along the string or the spring of Fig. 17–1 are one-dimensional. Surface waves or ripples on water, caused by dropping a pebble into a quiet pond, are two-dimensional. Sound waves and light waves that emanate radially from a small source are three-dimensional.

Waves may be classified further according to the behavior of a particle of the matter conveying the wave during the course of time the wave propagates. For example, we can produce a *pulse* traveling down a stretched string by applying a single sidewise movement at its end. Each particle remains at rest until the pulse reaches it, then it moves during a short time, and then it again remains at rest. If we continue to move the end of the string back and forth (Fig. 17–1a), we produce a *train of waves* traveling along the string. If our motion is periodic, we produce a *periodic train of waves* in which each particle of the string has a periodic motion.

The simplest special case of a periodic wave is a *simple harmonic wave,* which gives each particle a simple harmonic motion.

Consider a three-dimensional pulse. We can draw a surface through all points undergoing a similar disturbance at a given instant. As time goes on, this surface moves along showing how the pulse propagates. We can draw similar surfaces for subsequent pulses. For a periodic wave we can generalize the idea by drawing in surfaces, all of whose points are in the same phase of motion. These surfaces are

called *wavefronts*. If the medium is homogeneous and isotropic, the direction of propagation is always at right angles to the wavefront. A line normal to the wavefronts, indicating the direction of motion of the waves, is called a *ray*.

Wavefronts can have many shapes. If the disturbances are propagated in a single direction, the waves are called *plane waves*. At a given instant conditions

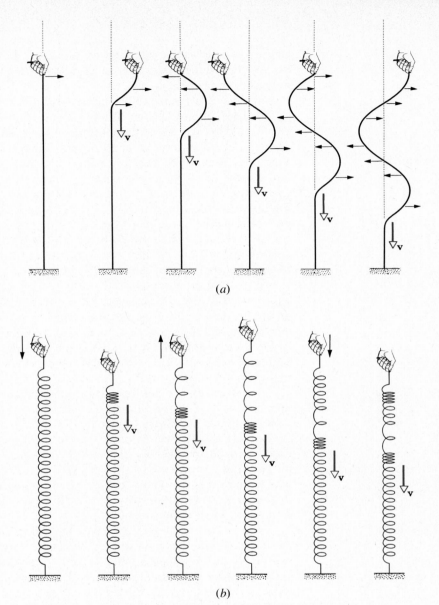

Figure 17–1 (*a*)In a transverse wave the particles of the medium (stretched string) vibrate at right angles to the direction in which the wave itself is propagated. (*b*) In a longitudinal wave the particles of the medium (stretched spring) vibrate in the same direction as that in which the wave itself is propagated. In both (*a*) and (*b*) we imagine that all the energy of the wave is absorbed by the device at the bottom. Thus we do not have to worry about waves reflected back up the string or spring.

are the same everywhere on any plane perpendicular to the direction of propagation. The wavefronts are planes and the rays are parallel straight lines (Fig. 17–2*a*). Another simple case is that of *spherical waves*. Here the disturbance is propagated out in all directions from a point source of waves. The wavefronts are spheres and the rays are radial lines leaving the point source in all directions (Fig. 17–2*b*). Far from the source the spherical wavefronts have very small curvature,

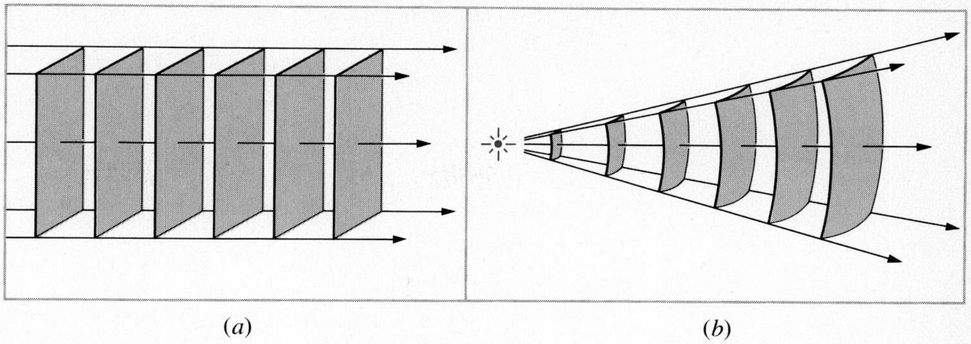

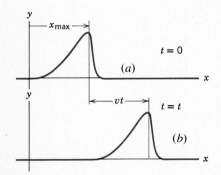

Figure 17-2 (*a*) A plane wave. The planes represent wave-fronts spaced a wavelength apart, and the arrows represent rays. (*b*) A spherical wave. The rays are radial and the wavefronts, spaced a wavelength apart, form spherical shells. Far out from the source, however, small portions of the wavefronts become nearly plane.

and over a limited region they can often be regarded as planes. Of course, there are many other possible shapes for wavefronts.

We shall refer to all these wave types as we progress through the wave phenomena of physics. In this chapter we often use the transverse wave in a string to illustrate the general properties of waves. In the next chapter we shall see the consequences of these properties for sound, a longitudinal mechanical wave. Later in the text we shall discuss the properties of nonmechanical waves such as light waves.

17-3 Traveling Waves

Let us consider a long string stretched in the *x*-direction along which a transverse wave is traveling. At some instant of time, say, $t = 0$, the shape of the string can be represented by

$$y = f(x) \qquad t = 0, \tag{17-1}$$

where *y* is the transverse displacement of the string at the position *x*. In Fig. 17–3*a* we show a possible waveform (a pulse) on the string at $t = 0$. Experiment shows that as time goes on such a wave travels along the string without changing its form, provided internal frictional losses are small enough. At some time *t* later the wave has traveled a distance *vt* to the right, where *v* is magnitude of the wave velocity, assumed constant. The equation of the curve at the time *t* is therefore

$$y = f(x - vt) \qquad \text{for all times } t \tag{17-2}$$

Figure 17-3 (*a*) The shape of a pulse in a stretched string at $t = 0$. The peak of the pulse is at $x = x_{max}$. (*b*) At a later time *t* the pulse has traveled to the right a distance *vt*.

A wave traveling in the + *x*-direction

This gives us the same waveform about the point $x = x_{max} + vt$ at time *t* as we had about $x = x_{max}$ at the time $t = 0$ (Fig. 17–3*b*). Equation 17–2 is the general equation representing a wave of any shape traveling to the right. To describe a particular shape we must specify exactly what the function *f* is. The variables *x* and *t*, of course, can only appear in the combination $x - vt$. For example, $\sin k(x - vt)$ and $(x - vt)^3$ are appropriate functions; $x^2 - v^2 t^2$ is not.

Let us look more carefully at this equation. If we wish to follow a particular part of the wave as time goes on, then in the equation we look at a particular value of *y* (say, the top of the pulse just described). Mathematically this means we look at how x changes with t when $(x - vt)$ has some particular fixed value. We see at once that as t increases x must increase in order to keep $(x - vt)$ fixed. Hence, Eq. 17–2 does in fact represent a wave traveling to the right (increasing x as time goes on). If we wished to represent a wave traveling to the left, we would write

$$y = f(x + vt), \tag{17-3}$$

for here the position x of some fixed value of $(x + vt)$ of the wave decreases as time goes on. The velocity of the waveform is easily obtained. For a wave traveling in the $+x$-direction we require that

$$x - vt = \text{constant}.$$

Then differentiation with respect to time gives

$$\frac{dx}{dt} - v = 0 \qquad \text{or} \qquad \frac{dx}{dt} = v. \tag{17-4}$$

v is called the *phase velocity* of the wave. For a wave traveling in the $-x$-direction we obtain $-v$, in the same way, as its phase velocity.

The general equation of a wave can be interpreted further. Note that for any fixed value of the time t the equation gives y as a function of x. This defines a curve, and this curve represents the actual shape of the string at this chosen time. It gives us a snapshot of the wave at this time. Suppose, on the other hand, that we wish to focus our attention on one point of the string, that is, a fixed value of x. Then the equation gives us y as a function of the time t. This describes how the transverse position of this point on the string changes with time.

For example, Fig. 17–4a (which is identical with Fig. 17–3a) shows y as a function of x for a particular time, namely $t = 0$. Figure 17–4b, on the other hand, shows y as a function of t for an observer stationed at a particular point along the string, namely at $x = x_0$ in Fig. 17–4a. Note that, for an observer stationed at this point and waiting for the pulse to reach him, the steep edge of the pulse arrives first, as indicated.

The argument just presented holds for longitudinal waves as well as for transverse waves. The analogous longitudinal example is that of a long straight tube of gas whose axis is taken as the x-axis, and the wave or pulse is a pressure change traveling along the tube. Then the same reasoning leads us to an equation, having the form of Eqs. 17–2 or 17–3, which gives the pressure variations with time at all points of the tube. (See Section 18–3.)

Let us now consider a particular waveform, whose importance will soon become clear. Suppose that at the time $t = 0$ we have a wavetrain along the string given by

$$y = y_m \sin \frac{2\pi}{\lambda} x. \tag{17-5}$$

The wave shape is a sine curve (Fig. 17–5). The maximum displacement y_m is the *amplitude* of the sine curve. The value of the transverse displacement y is the same at x as it is at $x + \lambda$, $x + 2\lambda$, and so on. The symbol λ is called the *wavelength* of the wavetrain and represents the distance between two adjacent points in the wave having the same phase. As time goes on let the wave travel to the right with a phase velocity v. Hence, the equation of the wave at the time t is

$$y = y_m \sin \frac{2\pi}{\lambda} (x - vt). \tag{17-6}$$

Notice that this has the form required for a traveling wave (Eq. 17–2).

A wave traveling in the $-x$-direction

Phase velocity

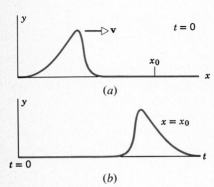

$t = 0$

(a)

$t = 0$

(b)

Figure 17–4 (*a*) Showing $y = f(x)$ at $t = 0$ for a pulse in a stretched string. (*b*) Showing $y = f(t)$ for an observer stationed at $x = x_0$ in (*a*) above.

Amplitude and wavelength

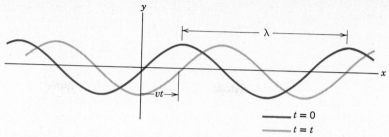

Figure 17–5 At $t = 0$, the string has a shape $y = y_m \sin 2\pi x/\lambda$. At a later time t the sine wave has moved to the right a distance $x = vt$, and the string has a shape given by $y = y_m \sin 2\pi(x - vt)/\lambda$.

Period

    The *period T* is the time required for the wave to travel a distance of one wavelength λ, so that

$$\lambda = vT. \tag{17–7}$$

Putting this relation into the equation of the wave, we obtain

$$y = y_m \sin 2\pi \left(\frac{x}{\lambda} - \frac{t}{T} \right). \tag{17–8}$$

From Eq. 17–8 it is clear that y, at any given time, has the same value at $x + \lambda$, $x + 2\lambda$, and so on, as it does at x, and that y, at any given position, has the same value at the time $t + T$, $t + 2T$, and so on, as it does at the time t.

Wave number, angular frequency

    To reduce Eq. 17–8 to a more compact form, we define two quantities, the *wave number k* and the *angular frequency* ω (see Eq. 14–12). They are given by

$$k = \frac{2\pi}{\lambda} \quad \text{and} \quad \omega = \frac{2\pi}{T}. \tag{17–9}$$

In terms of these quantities, the equation of a sine wave traveling to the right (positive x-direction) is

$$y = y_m \sin (kx - \omega t). \tag{17–10a}$$

For a sine wave traveling to the left (negative x-direction), we have

$$y = y_m \sin (kx + \omega t). \tag{17–10b}$$

Comparing Eqs. 17–7 and 17–9, we see that the phase velocity v of the wave is given by

$$v = \frac{\lambda}{T} = \frac{\omega}{k}. \tag{17–11}$$

    In the traveling waves of Eqs. 17–10a and 17–10b we have assumed that the displacement y is zero at the position $x = 0$ at the time $t = 0$. This, of course, need not be the case. The general expression for a sinusoidal wave traveling in the $+x$-direction is

$$y = y_m \sin (kx - \omega t - \phi),$$

Phase and the phase constant

where the quantity in parentheses is called the *phase* of the wave and ϕ is called the *phase constant*. For example, if $\phi = -90°$, the displacement y at $x = 0$ and $t = 0$ is y_m. This particular example is

$$y = y_m \cos (kx - \omega t),$$

since the cosine function is displaced by 90° from the sine function.

If we fix our attention on a given point of the string, say, $x = \pi/k$, the displacement y at that point can be written as

$$y = y_m \sin(\omega t + \phi).$$

This is similar to Eq. 14–25 for simple harmonic motion. Hence, any particular element of the string undergoes simple harmonic motion about its equilibrium position as this wavetrain travels along the string.

17–4 Wave Speed in a Stretched String

It is necessary to have a material medium to transmit mechanical waves. The properties of the medium that determine the speed of a wave through the medium are its inertia and its elasticity. All material media, including, say, air, water, and steel, possess these properties and can transmit mechanical waves. It is the elasticity that gives rise to the restoring forces on any part of the medium displaced from its equilibrium position; it is the inertia that tells us how this displaced portion of the medium will respond to these restoring forces. Together these two factors determine the wave speed.

For a stretched string the elasticity is measured by the tension F in the string; the greater the tension the greater will be the elastic restoring force on an element of the string that is pulled sideways. The inertia characteristic is measured by μ the mass per unit length of the string.

We can determine the general dependence of the wave speed v on F and μ by using dimensional analysis. In terms of mass M, length L, and time T, the dimensions of F are MLT^{-2} and the dimensions of μ are ML^{-1}. The only way these dimensions can be combined to get a velocity (which has the dimensions LT^{-1}) is to take the square root of F/μ. That is, F/μ has the dimensions L^2T^{-2} and $\sqrt{F/\mu}$ has the dimensions LT^{-1} of a velocity. Dimensional analysis cannot account for any dimensionless quantities, so that the result

$$v = \sqrt{\frac{F}{\mu}} \tag{17–12}$$

may or may not be complete. The most we can say is that the wave speed is equal to a dimensionless constant times $\sqrt{F/\mu}$.

We shall now derive the wave speed by a mechanical analysis and show that Eq. 17–12 is correct as it stands.

In Fig. 17–6 we show a wave pulse proceeding from right to left in the string with a speed v. For convenience in the derivation imagine the entire string to be moving from left to right with this same speed so that the wave pulse remains fixed in space, whereas the particles composing the string successively pass through the pulse. Instead of taking our reference frame to be the supports between which the string is stretched, we are choosing a reference frame that is in uniform motion with respect to the supports and at rest with respect to the pulse. Because Newton's laws involve only accelerations, which are the same in both frames, we can use them in either frame, but this one is more convenient.

Consider a small section of the pulse of length Δl to form an arc of a circle of radius R, as shown in the diagram. If μ is the mass per unit length of the string, the so-called linear density, then $\mu \Delta l$ is the mass of this element. The tension F in the string is a tangential pull at each end of this small segment of the string. The horizontal components cancel and the vertical components are each equal to $F \sin \theta$.

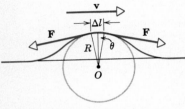

Figure 17–6 Derivation of wave speed by considering the forces on a section of string of length Δl.

Hence, the total vertical force is $2F \sin \theta$. Since θ is small, we can take $\sin \theta \cong \theta$ as we did in Section 14–5 and

$$2F \sin \theta \cong 2F\theta = 2F\frac{(\Delta l/2)}{R} = F\frac{\Delta l}{R}.$$

This gives the force supplying the centripetal acceleration of the string particles directed toward O. Now the centripetal force acting on a mass $\mu\,\Delta l$ moving in a circle of radius R with speed v is $\mu\,\Delta l\,v^2/R$; see Section 6–3. Notice that the tangential velocity v of this mass element along the top of the arc is horizontal and its magnitude is the same as the pulse phase speed. Combining the equivalent expressions just given we obtain

$$F\frac{\Delta l}{R} = \frac{\mu\,\Delta l\,v^2}{R}$$

or

Wave speed in a stretched string

$$v = \sqrt{\frac{F}{\mu}}, \qquad [17\text{–}12]$$

which is the result that we obtained earlier from dimensional analysis.

If the amplitude of the pulse were very large compared to the length of the string, we would not have been able to use the approximation $\sin \theta \cong \theta$. Furthermore, the tension F in the string would be changed by the presence of the pulse, whereas we assumed F to be unchanged from the original tension in the stretched string. Therefore, our result holds only for relatively small transverse displacements of the string—which case, however, is widely applicable in practice. Notice also that the wave speed is independent of the shape of the wave, for no particular assumption about the shape of the pulse was used in the proof.

The frequency of a wave is naturally determined by the frequency of the source. The speed with which the wave travels through a medium is determined by the properties of the medium, as illustrated before. Once the frequency ν and speed v of the wave are determined, the wavelength λ is fixed. In fact, from Eq. 17–7 and the relation, $\nu = 1/T$, we have

$$\lambda = \frac{v}{\nu}. \qquad (17\text{–}13)$$

Example 1   A transverse sinusoidal wave is generated at one end of a long horizontal string by a bar that moves the end up and down through a distance of 1.0 cm. The motion is continuous and is repeated regularly 120 times per second.

(a) If the string has a linear density of 120 g/m and is kept under a tension of 90 N, find the amplitude, frequency, period, speed, and wavelength of the wave motion.

i.   The end moves 0.50 cm away from the equilibrium position, first above it, then below it; therefore, the amplitude y_m is 0.50 cm.

ii.   The entire motion is repeated 120 times each second so that the frequency is 120 vibrations per second, or 120 Hz.

iii.   The period is given by

$$T = \frac{1}{\nu} = \frac{1}{120\ \text{Hz}} = 8.3\ \text{ms}$$

iv.   The wave speed is given by Eq. 17–12, or

$$v = \sqrt{\frac{F}{\mu}} = \sqrt{\frac{90\ \text{N}}{0.12\ \text{kg/m}}} = 27\ \text{m/s}.$$

v.   The wavelength is given by Eq. 17–13, or

$$\lambda = \frac{v}{\nu} = \frac{27\ \text{m/s}}{120\ \text{Hz}} = 23\ \text{cm}.$$

(b) Assuming the wave moves in the $+x$-direction and that, at $t = 0$, the end of the string described by $x = 0$ is in its equilibrium position $y = 0$, write the equation of the wave.

The general expression for a transverse sinusoidal wave moving in the $+x$-direction is

$$y = y_m \sin (kx - \omega t - \phi).$$

Requiring that $y = 0$ for the conditions $x = 0$ and $t = 0$ yields

$$0 = y_m \sin (-\phi),$$

which means that the phase constant ϕ may be taken to be zero. You should show that integral multiples of π yield the same final results. Hence for this wave

$$y = y_m \sin (kx - \omega t),$$

and with the values just found,

$$y_m = 0.50 \text{ cm},$$

$$\lambda = 23 \text{ cm} \quad \text{or} \quad k = \frac{2\pi}{\lambda} = \frac{2\pi}{23 \text{ cm}} = 0.27 \text{ rad cm}^{-1},$$

$v = 27 \text{ m/s} = 2700 \text{ cm/s} \quad \text{or} \quad \omega = (2700 \text{ cm/s})(0.27 \text{ cm}^{-1})$
$= 730 \text{ rad/s}$ we obtain as the equation for the wave

$$y = (0.50 \text{ cm}) \sin [(0.27 \text{ cm}^{-1})x - (730 \text{ s}^{-1})t]$$

where the quantity in brackets is an angle in radians.

(c) Find the displacement y for $x = 15$ cm and $t = 8.8$ ms. For these values we have

$y = (0.50 \text{ cm}) \sin [(0.27 \text{ cm}^{-1})x - (730\text{s}^{-1})t]$
$= (0.50 \text{ cm}) \sin [(0.27)(15) - (730)(8.8 \times 10^{-3})]$
$= (0.50 \text{ cm}) \sin (-2.37 \text{ rad})$
$= (-0.50 \text{ cm}) \sin 136° = -0.35 \text{ cm}$

Example 2   As this wave passes along the string, each particle of the string moves up and down at right angles to the direction of the wave motion. Find the velocity and acceleration of a particle 30 cm from the end.

The general form of this wave is

$$y = y_m \sin (kx - \omega t) = y_m \sin k(x - vt).$$

The v in this equation is the constant horizontal velocity of the wavetrain. What we are after now is the velocity of a particle in the string through which this wave moves; this particle velocity is neither horizontal nor constant. In fact, each particle moves vertically, that is, in the y-direction. In order to determine the particle velocity, which we shall designate by the symbol u, let us fix our attention on a particle at a particular position x—that is, x is now a constant in this equation—and ask how the particle displacement y changes with time. With x constant we obtain

$$u = \frac{\partial y}{\partial t} = -y_m \omega \cos (kx - \omega t),$$

in which the *partial derivative* $\partial y/\partial t$ tells us that although in general y is a function of both x and t, when we take the derivative we are to consider x a constant so that t is the only variable.* The acceleration a of the particle at this (constant) value of x is

$$a = \frac{\partial^2 y}{\partial t^2} = \frac{\partial u}{\partial t} = -y_m \omega^2 \sin (kx - \omega t) = -\omega^2 y.$$

This shows that for each particle through which this transverse sinusoidal wave passes we have precisely SHM (simple harmonic motion), for the acceleration a is proportional to the displacement y, but oppositely directed.

For a particle at $x = 30$ cm with the wave of Example 1, in which

$$y_m = 0.50 \text{ cm}, \quad k = 0.27 \text{ rad cm}^{-1} \text{ and } \omega = 730 \text{ rad/s}, \text{ we obtain}$$

$$u = -y_m \omega \cos (kx - \omega t)$$

$= -(0.50 \text{ cm})(730 \text{ s}^{-1}) \cos [(0.27 \text{ cm}^{-1})(30 \text{ cm}) - (730 \text{ s}^{-1})t]$
$= (-370 \text{ cm/s}) \cos [8.1 - (730 \text{ s}^{-1})t]$

and

$$a = -\omega^2 y$$

or

$a = -(730 \text{ s}^{-1})^2 (0.50 \text{ cm}) \sin [(0.27)(30) - (730 \text{ s}^{-1})t]$
$= (2.6 \times 10^5 \text{ cm/s}^2) \sin [8.1 - (730 \text{ s}^{-1})t]$

where all angles will be in radian measure. Show that for $t = 10^{-2}$ s, we have $y = +0.36$ cm, $u = -2.7$ m/s, and $a = -1900 \text{ m/s}^2$.

17–5 Power and Intensity in Wave Motion

An important quantity associated with one-dimensional waves is the amount of energy transferred past a given point per unit time. This is a measure of the power transfer. In the case of three-dimensional waves the corresponding quantity is the energy transferred across a given area per unit area per unit time or *intensity*.

We shall derive the power transfer in the case of a one-dimensional string. In Fig. 17–7 we draw an element of the stretched string at some position x and at a particular time t. The *transverse* component of the tension in the string exerted *by* the element to the left of x *on* the element of the right of x is

$$F_{\text{trans}} = -F \frac{\partial y}{\partial x}.*$$

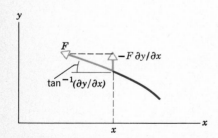

Figure 17–7   The transverse component of the tension in the string at each point x is $-F(\partial y/\partial x)$.

---

* The partial derivative notation here is to remind us that one independent variable in $y(x, t)$ is being held fixed while the derivative is taken with respect to the other. For example, imagine that we took a photograph (fixed time, t) of the string that showed the position of all string particles at one time; the expression $\partial y/\partial x$ represents the slope of that curve. On the other hand, we may consider a single string particle (fixed position, x); the expression $\partial y/\partial t$ represents the slope of the y versus t curve for that particle, which amounts to the particle's speed.

F is the tension in the string; $\partial y/\partial x$ gives the tangent of the angle made by the direction of F with the horizontal at the time t in question and, because we assume small displacements, this can be taken equal also to the sine of the angle. The transverse force is in the direction of increasing y; in the figure the slope is negative, so the transverse force is positive. The transverse velocity of the particle at x is $\partial y/\partial t$, which may be positive or negative. The power being expended by the force at x, or the energy passing through the position x per unit time in the positive x-direction (see Section 7–7), is

$$P = F_{\text{trans}}\, u = \left(-F\frac{\partial y}{\partial x}\right)\frac{\partial y}{\partial t}.$$

Suppose that the wave on the string is the simple sine wave

$$y = y_m \sin(kx - \omega t).$$

Then the magnitude of the slope at x is

$$\frac{\partial y}{\partial x} = ky_m \cos(kx - \omega t), \qquad [t = \text{constant}]$$

and the transverse force is

$$F_{\text{trans}} = -F\frac{\partial y}{\partial x} = -Fky_m \cos(kx - \omega t).$$

The transverse velocity of a particle of the string at x is

$$u = \frac{\partial y}{\partial t} = -\omega y_m \cos(kx - \omega t), \qquad [x = \text{constant}].$$

Hence, the power transmitted through x is

$$P = (-Fky_m)(-\omega y_m) \cos^2(kx - \omega t),$$
$$= y_m^2 k\omega F \cos^2(kx - \omega t).$$

Notice that the power or rate of flow of energy is not constant. The power is not constant because the power input oscillates. As the energy is passed along the string, it is stored in each element of string as a combination of kinetic energy of motion and potential energy of deformation. The power input to the string is often taken to be the *average* over one period of motion. The average power delivered is

$$\bar{P} = \frac{1}{T}\int_t^{t+T} P\, dt,$$

where T is the period. Using the fact that the average value of $\sin^2\theta$ or $\cos^2\theta$ over one cycle is $\frac{1}{2}$, we obtain for the string

$$\bar{P} = \tfrac{1}{2}y_m^2 k\omega F,$$

a result that does not depend on x or t. For the string, however, we can substitute $k = 2\pi/\lambda = 2\pi\nu/v$, $\omega = 2\pi\nu$, and (see Eq. 17–12) $F = v^2\mu$, so that

Wave power

$$\bar{P} = 2\pi^2 y_m^2 \nu^2 \mu v. \tag{17–14}$$

The fact that the rate of transfer of energy depends on the square of the wave amplitude and square of the wave frequency is true in general, holding for all types of waves.

Confirm that, if we had picked a wave traveling in the negative x-direction, we would have obtained the negative of this result. That is, the wave delivers power in the direction of wave propagation.

In a three-dimensional wave, such as a light wave or a sound wave from a

Wave intensity

point source, it is more significant to speak of the *intensity* of the wave. Intensity is defined as the power transmitted across a unit area normal to the direction in which the wave is traveling. Just as with power in the wave in a string, the intensity of a space wave is always proportional to the square of the amplitude.

As a wave progresses through space, its energy may be absorbed. For example, in a viscous medium, such as syrup or lead, mechanical waves rapidly decay in amplitude and disappear, owing to absorption of energy by internal friction. In most cases of interest to us, however, absorption will be negligible. Throughout this chapter we have assumed that there is no loss of energy in a given wave, no matter how far it travels.

Example 3 Spherical waves travel from a source of waves whose power output is P; see Fig. 17–8. Find how the wave intensity depends on the distance from the source. We assume that the medium is isotropic and that the source radiates uniformly in all directions, that is, that its emission is spherically symmetric.

The intensity of a three-dimensional wave is the power transmitted across a unit area normal to the direction of propagation. As the wavefront expands from a distance r_1 from the source at the center to a distance r_2, its surface area increases from $4\pi r_1^2$ to $4\pi r_2^2$. If there is no absorption of energy, the total energy transported per second by the wave remains constant at the value P, so that

$$P = 4\pi r_1^2 I_1 = 4\pi r_2^2 I_2,$$

where I_1 and I_2 are the wave intensities at r_1 and r_2, respectively. Hence,

$$\frac{I_1}{I_2} = \frac{r_2^2}{r_1^2},$$

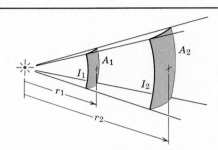

Figure 17–8 Example 3.

and the wave intensity varies inversely as the square of its distance from the source. Since the intensity is proportional to the square of the amplitude, the amplitude of the wave must vary inversely as the distance from the source.

17–6 The Superposition Principle

It is an experimental fact that, for many kinds of waves, two or more waves can traverse the same space independently of one another. The fact that waves act independently of one another means that the displacement of any particle at a given time is simply the sum of the displacements that the individual waves alone would give it. This process of vector addition of the displacements of a particle is called *superposition*. For example, radio waves of many frequencies pass through a radio antenna; the electric currents set up in the antenna by the superposed action of all these waves are very complex. Nevertheless, we can still tune to a particular station, the signal that we receive from it being in principle the same as that which we would receive if all other stations were to stop broadcasting. Likewise, in sound we can listen to notes played by individual instruments in an orchestra, even though the sound wave reaching our ears from the full orchestra is very complex.

For waves in deformable media the superposition principle holds whenever the mathematical relation between the deformation and the restoring force is one of simple proportionality. Such a relation is expressed mathematically by a linear equation. For electromagnetic waves the superposition principle holds because the mathematical relations between the electric and magnetic fields are linear.

For example, superposition will no longer hold in an elastic medium when the intensity of sound becomes so high that the elastic limit is exceeded and Hooke's

Superposition

law, $F = -kx$, no longer applies. A shock wave produced by a violent explosion is such a case for sound waves in air. Superposition also fails in a speaker that is being "driven too hard" by an amplifier and, as a result, sound distortion is introduced, which means that frequencies not present in the amplified signals are produced in the speakers.

The importance of the superposition principle physically is that, where it holds, it makes it possible to analyze a complicated wave motion as a combination of simple waves. In fact, as was shown by the French mathematician J. Fourier (1786–1830), all that we need to build up the most general form of periodic wave are simple harmonic waves. Fourier showed that any periodic motion of a particle can be represented as a combination of simple harmonic motions. The general expression of such a combination is called a Fourier series. An example of the analysis of a sawtooth wave as a combination of simple harmonic waves is shown in Fig. 17–9. Six terms of the Fourier series representing the sawtooth wave give a rather good fit.

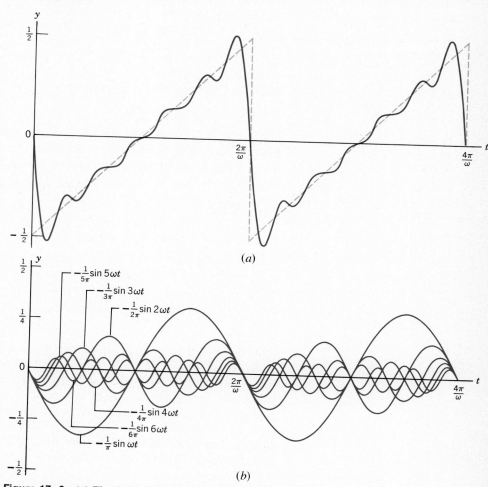

Figure 17–9 (a) The dashed line is a sawtooth "wave" commonly encountered in electronics. The Fourier series for this function is $y(t) = -\dfrac{1}{\pi}\sin \omega t - \dfrac{1}{2\pi}\sin 2\omega t - \dfrac{1}{3\pi}\sin 3\omega t - \cdots$. The solid line is the sum of the first six terms of this series and can be seen to approximate the sawtooth quite closely. (b) Here we show the first six terms of the Fourier series which, when added together, yield the solid curve in (a).

Dispersion

The elasticity of a medium might not obey Hooke's law exactly. Then a mechanical wave pulse produced at one end of a stretched string may change its shape as it travels along the string. Each of the component harmonic waves moves without changing its shape, but the speed of each is different; the speed, indeed depends on the frequency. This phenomenon is called *dispersion,* and the medium is said to be *dispersive* for the wave type in question. As a result the pulse shape changes as the wave moves along. Not only that, but the speed of the pulse itself is not the same for every pulse but depends on the details of its initial shape.

Light waves in a vacuum are nondispersive; however, light waves in glass are dispersive, as are water waves on the surface of a lake.

Another way in which the wave pulse may change its shape is by loss of mechanical energy to the medium or its surroundings, for example, by air resistance, viscosity, or internal friction. Then the amplitude of the wave decreases with time and the wave is said to be *attenuated.* The strength of the attenuation often depends on the frequency.

Attenuation

For the moment, we shall assume that the medium is nondispersive and that there is no dissipation of energy as the wave travels through the medium.

17–7   Interference of Waves

Interference refers to the physical effects of superimposing two or more wavetrains. Let us consider two waves of equal frequency and amplitude traveling with the same speed in the same direction $(+x)$ but with a phase difference ϕ between them. The equations of the two waves will be

$$y_1 = y_m \sin (kx - \omega t - \phi) \qquad (17-15)$$

and

$$y_2 = y_m \sin (kx - \omega t). \qquad (17-16)$$

We can rewrite the first equation in two equivalent forms

$$y_1 = y_m \sin \left[k \left(x - \frac{\phi}{k} \right) - \omega t \right] \qquad (17-15a)$$

or

$$y_1 = y_m \sin \left[kx - \omega \left(t + \frac{\phi}{\omega} \right) \right]. \qquad (17-15b)$$

Equations 17–15a and 17–16 suggest that if we take a "snapshot" of the two waves at any time t, we shall find them displaced from one another along the x-axis by the constant distance ϕ/k. Equations 17–15b and 17–16 suggest that if we station ourselves at any position x, the two waves will give rise to two simple harmonic motions having a constant time difference ϕ/ω. This gives some insight into the meaning of the phase difference ϕ.

Now let us find the resultant wave, which, on the assumption that superposition occurs, is the sum of Eqs. 17–15 and 17–16, or

$$y = y_1 + y_2 = y_m[\sin (kx - \omega t - \phi) + \sin (kx - \omega t)].$$

From the trigonometric equation for the sum of the sines of two angles

$$\sin B + \sin C = 2 \sin \tfrac{1}{2}(B + C) \cos \tfrac{1}{2}(C - B), \qquad (17-17)$$

we obtain

$$y = y_m \left[2 \sin \left(kx - \omega t - \frac{\phi}{2} \right) \cos \frac{\phi}{2} \right],$$

$$= \left(2 y_m \cos \frac{\phi}{2} \right) \sin \left(kx - \omega t - \frac{\phi}{2} \right). \qquad (17\text{–}18)$$

Resultant of two interfering waves differing only in phase

This resultant wave corresponds to a new wave having the same frequency but with an amplitude $2 y_m \cos (\phi/2)$. If ϕ is very small compared to 180°, the resultant amplitude will be nearly $2 y_m$. That is, when ϕ is very small, $\cos (\phi/2) \cong \cos 0° = 1$. When ϕ is zero, the two waves have the same phase everywhere. The crest of one corresponds to the crest of the other and likewise for the troughs. The waves are then said to interfere constructively. The resultant amplitude is just twice that of either wave alone. If ϕ is near 180°, on the other hand, the resultant amplitude will be nearly zero. That is, when $\phi \cong 180°$, $\cos (\phi/2) \cong \cos 90° = 0$. When ϕ is exactly 180°, the crest of one wave corresponds exactly to the trough of the other. The waves are then said to interfere destructively. The resultant amplitude is zero.

In Fig. 17–10a we show the superposition of two wavetrains almost in phase (ϕ small) and in Fig. 17–10b the superposition of two wavetrains almost 180° out of phase ($\phi \cong 180°$). Notice that in these figures the algebraic sum of the ordinates of the black (component) curves at any value of x equals the ordinate of the colored (resultant) curve. The sum of two waves can, therefore, have different values, depending on their phase relations.

The resultant wave will be a sine wave, even when the amplitudes of the component sine waves are unequal. Figure 17–11, for example, illustrates the addition of two sine waves of the same frequency and velocity but different amplitudes.

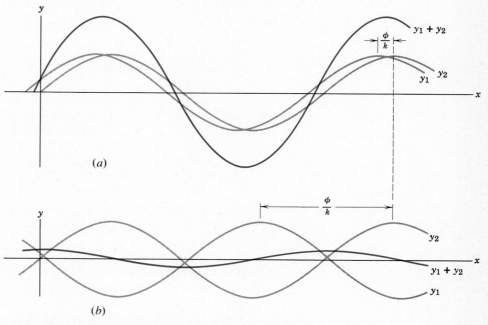

Figure 17–10 (a) The superposition of two waves of equal frequency and amplitude that are almost in phase results in a wave of almost twice the amplitude of either component. (b) The superposition of two waves of equal frequency and amplitude and almost 180° out of phase results in a wave whose amplitude is nearly zero. Note that in both the resultant frequency is unchanged.

Figure 17–11   The addition of two waves of same frequency and phase but differing amplitudes (light black lines) yields a third wave of the same frequency and phase (colored line).

The resultant amplitude depends on the phase difference, which is taken as zero in this figure. The resultant for other phase differences could be obtained by shifting one of the component waves sideways with respect to the other and would give a smaller resultant amplitude. The smallest resultant amplitude would be the difference in the amplitudes of the components, obtained when the phases differ by 180°. However, the resultant is always a sine wave. The addition of any number of sine waves having the same frequency and velocity gives a similar result. The resultant waveform will always have a constant amplitude because the component waves (and their resultant) all move with the same velocity and maintain the same relative position. The actual state of affairs can be pictured by having all the waves in Figs. 17–10 and 17–11 move toward the right with the same speed.

In practice, interference effects are obtained from wavetrains that originate in the same source (or in sources having a fixed phase relationship to one another) but that follow different paths to the point of interference. The phase difference ϕ between the waves arriving at a point can be calculated by finding the difference between the paths traversed by them from the source to the point of interference. The phase difference is $k\,\Delta L$ or $2\pi\,\Delta L/\lambda$ if the path difference is ΔL. When the path difference is 0, λ, 2λ, 3λ, etc., so that $\phi = 0, 2\pi, 4\pi$, etc., the two waves interfere constructively. For path differences of $\frac{1}{2}\lambda$, $\frac{3}{2}\lambda$, $\frac{5}{2}\lambda$, etc., ϕ is $\pi, 3\pi, 5\pi$, etc., and the waves interfere destructively. We shall return to these matters later in more detail.

17–8   Standing Waves

In a one-dimensional body of finite size, such as a taut string held by two clamps a distance l apart, traveling waves in the string are reflected from the boundaries of the body, that is, from the clamps. Each such reflection gives rise to a wave traveling in the string in the opposite direction. The reflected waves add to the incident waves according to the principle of superposition.

Consider two wavetrains of the same frequency, speed, and amplitude which are traveling in opposite directions along a string. Two such waves may be represented by the equations

$$y_1 = y_m \sin (kx - \omega t),$$
$$y_2 = y_m \sin (kx + \omega t).$$

We can write the resultant as

$$y = y_1 + y_2 = y_m \sin (kx - \omega t) + y_m \sin (kx + \omega t) \qquad (17\text{–}19a)$$

or, making use of the trigonometric relation of Eq. 17–17, as

$$y = 2y_m \sin kx \cos \omega t. \qquad (17\text{–}19b)$$

A standing wave

Equation 17–19b is the equation of a *standing wave*. Notice that a particle at any particular point x executes simple harmonic motion as time goes on, and that all particles vibrate with the same frequency. In a traveling wave each particle of the string vibrates with the same amplitude. Characteristic of a standing wave, however, is the fact that the amplitude is not the same for different particles but varies with the location x of the particle. In fact, the amplitude, $2y_m \sin kx$, has a maximum value of $2y_m$ at positions where

$$kx = \frac{\pi}{2}, \frac{3\pi}{2}, \frac{5\pi}{2}, \text{etc.}$$

Antinodes   or

$$x = \frac{\lambda}{4}, \frac{3\lambda}{4}, \frac{5\lambda}{4}, \text{etc.}$$

These points are called *antinodes* and are spaced one-half wavelength apart. The amplitude has a minimum value of zero at positions where

$$kx = \pi, 2\pi, 3\pi, \text{etc.}$$

Nodes   or

$$x = \frac{\lambda}{2}, \lambda, \frac{3\lambda}{2}, 2\lambda, \text{etc.}$$

These points are called *nodes* and are spaced one-half wavelength apart. The separation between a node and an adjacent antinode is one-quarter wavelength.

It is clear that energy is not transported along the string to the right or to the left, for energy cannot flow past the nodal points in the string, which are permanently at rest. Hence, the energy remains "standing" in the string, although it alternates between vibrational kinetic energy and elastic potential energy. We call the motion a wave motion because we can think of it as a superposition of waves traveling in opposite directions (Eq. 17–19a). We can equally well regard the motion as an oscillation of the string as a whole (Eq. 17–19b), each particle oscillating with simple harmonic motion of angular frequency ω and with an amplitude that depends on its location. Each small part of the string has inertia and elasticity and behaves like a tiny mass-spring system. The string as a whole can be viewed as an assembly of such systems, all coupled together. We note however that a single mass-spring has only one natural frequency, and a vibrating string has a large number of natural frequencies (Section 17–9).

In Fig. 17–12, in (a), (b), (c), and (d), we show a standing wave pattern separately at intervals of one-quarter of a period in the lower figures, 3. The traveling

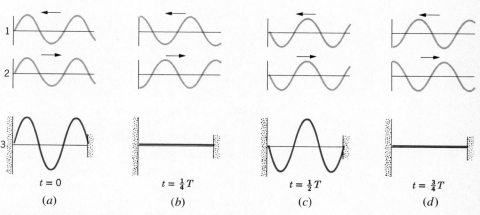

$t = 0$ $t = \frac{1}{4}T$ $t = \frac{1}{2}T$ $t = \frac{3}{4}T$

(a)   (b)   (c)   (d)

Figure 17–12 Standing waves as the superposition of left- and right-going waves; 1 and 2 are the components, 3 the resultant.

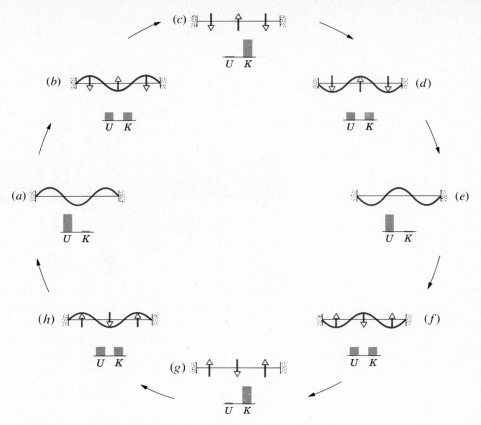

Figure 17–13 A standing wave in a stretched string, showing one cycle of oscillation. At (*a*) the string is momentarily at rest and the energy of the system is all potential energy of elastic deformation associated with the transverse displacement of the string. (*b*) An eighth-cycle later the displacement is reduced and the string is in motion. The three arrows show the velocities of the string particles at the positions shown. *K* and *U* have the same value. (*c*) The string is not displaced, but its particles have their maximum speeds; the energy is all kinetic. The motion continues until the initial condition (*a*) is reached after which the cycle continues to repeat itself.

waves, one moving in the positive *x*-direction and the other moving in the negative *x*-direction, whose superposition can be considered to give rise to the standing wave, are shown for the same quarter-period intervals in the upper figures 2 and 1. Standing waves can also be produced with electromagnetic waves and with sound waves.

In Fig. 17–13 we show how the energy associated with the oscillating string shifts back and forth between kinetic energy of motion *K* and potential energy of deformation *U* during one cycle. You should compare this with Fig. 8–4, which shows the same thing for a mass-spring oscillator. Oscillating strings often vibrate so rapidly that the eye perceives only a blur whose shape is that of the envelope of the motion; see Fig. 17–14.

The superposition of an incident wave and a reflected wave, being the sum of two waves traveling in opposite directions, will give rise to a standing wave. We shall now consider the process of reflection of a wave more closely. Suppose that

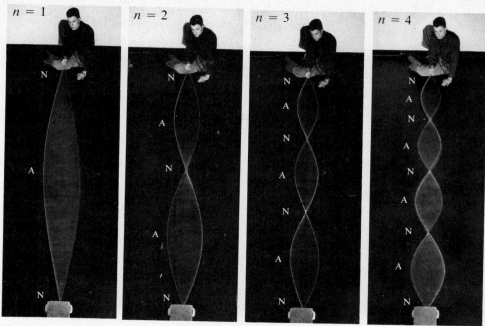

Figure 17-14 Standing waves. A long flexible rubber tube is rigidly attached at one end. If the experimenter wiggles the other end at the correct frequency, a standing wave is produced. Each particular pattern is associated with a definite frequency, which is given by Eq. 17-20 using the indicated values of n. In this situation the tension remains constant. N and A show nodes and antinodes, respectively. From Physical Science Study Committee, *Physics,* D. C. Heath, Lexington, Mass., 1960, with permission.

a pulse travels down a stretched string that is fixed at one end, as shown in Fig. 17-15a. When the pulse arrives at that end, it exerts an upward force on the support. The support is rigid, however, and does not move. By Newton's third law, the support exerts an equal but oppositely directed force on the string. This reaction force generates a pulse at the support, which travels back along the string in a direction opposite to that of the incident pulse. We say that the incident pulse has been *reflected* at the fixed end point of the string. Notice that the reflected pulse returns with its transverse displacement reversed. If a wavetrain is incident on the fixed end point, a reflected wavetrain is generated at that point in the same way. The displacement of any point along the string is the sum of the displacements caused by the incident and reflected wave. Since the end point is fixed, these two waves must always interfere destructively at that point so as to give zero displacement there. Hence, the reflected wave is always 180° out of phase with the incident wave at a fixed boundary. We say that on reflection from a fixed end a wave undergoes a phase change of 180°.

Let us now consider the reflection of a pulse at a free end of a stretched string, that is, at an end that is free to move transversely. This can be achieved by attaching the end to a very light ring free to slide without friction along a transverse rod, or to a long and very much lighter string. When the pulse arrives at the free end, it exerts a force on the element of string there. This element is accelerated and its inertia carries it past the equilibrium point; it "overshoots" and exerts a reaction force on the string. This generates a pulse which travels back along the string in a direction opposite to that of the incident pulse. Once again we get reflection, but now at a free end. The free end will obviously suffer the maximum displacement of the particles on the string; an incident and a reflected wavetrain

Reflection at a fixed end

Reflection at a free end

must interfere constructively at that point if we are to have a maximum there. Hence, the reflected wave is always in phase with the incident wave at that point (see Fig. 17–14b). We say that at a free end a wave is reflected without change of phase.

Hence, when we have a standing wave in a string, there will be a node at a fixed end and an antinode at a free end. Later, we shall apply these ideas to sound waves and electromagnetic waves.

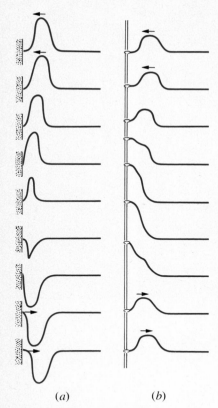

Figure 17–15 (a) Reflection of a pulse at the fixed end of a string. The drawings are spaced uniformly in time. The phase is changed by 180° on reflection. (b) Reflection of a pulse at an end free to move in a transverse direction. (The string is attached to a ring which slides vertically without friction.) The phase is unchanged on reflection.

17–9 Resonance

In general, whenever a system capable of oscillating is acted on by a periodic series of impulses having a frequency equal or nearly equal to one of the natural frequencies of oscillation of the system, the system is set into oscillation with a relatively large amplitude. This phenomenon is called *resonance,* and the system is said to resonate with the applied impulses.

Consider a string fixed at both ends. Oscillations or standing waves can be established in the string. The only requirement we have to satisfy is that the end points be nodes. There may be any number of nodes in between or none at all, so that the wavelength associated with the standing waves can take on many different values. The distance between adjacent nodes is $\lambda/2$, so that in a string of length l there must be exactly an integral number n of half wavelengths, $\lambda/2$. That is,

$$\frac{n\lambda}{2} = l$$

or

$$\lambda = \frac{2l}{n}, \qquad n = 1, 2, 3, \ldots .$$

But $\lambda = v/\nu$ and $v = \sqrt{F/\mu}$, so that the natural frequencies of oscillation of the system are

$$\nu = \frac{n}{2l}\sqrt{\frac{F}{\mu}}, \qquad n = 1, 2, 3, \ldots . \tag{17–20}$$

Resonant frequency of a stretched string

If the string is set vibrating and left to itself, the oscillations gradually die out. The motion is damped by dissipation of energy through the elastic supports at the ends and by the resistance of the air to the motion. We can pump energy into the system by applying a driving force. If the driving frequency is near that of any natural frequency, the string will vibrate at that frequency with a large amplitude. Because the string has a large number of natural frequencies, resonance can occur at many different frequencies. A mass-spring system, by contrast, has only one resonant frequency.

Resonance in a string is often demonstrated by attaching a string to a fixed end, by means of a weight attached to it over a pulley, and connecting the other end to a vibrator, as shown in Fig. 17–16. The transverse oscillations of the vibrator set up a traveling wave in the string which is reflected back from the fixed end. The frequency of the waves is that of the vibrator, and the wavelength is determined by $\lambda = v/\nu$. The fixed end P is a node, but the end Q vibrates and is not. If we now vary the tension in the string by changing the hanging weight, for example, we can change the wavelength. Changing the tension changes the wave velocity, and the wavelength changes in proportion to the velocity, the frequency being constant. Whenever the wavelength becomes nearly equal to $2l/n$, where l

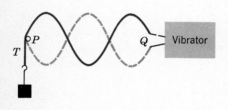

Figure 17–16 Standing waves in a driven string when the natural and driving frequencies are very nearly equal.

is the length of the string, we obtain standing waves of great amplitude. The string now vibrates in one of its natural modes and resonates with the vibrator. The vibrator does work on the string to maintain these oscillations against the losses due to damping. The amplitude builds up only to the point at which the vibrator expends all its energy input against damping losses. The point Q is almost a node because the amplitude of the vibrator is small compared to that of the string.

If the frequency of the vibrator is much different from a natural frequency of the system, as given by Eq. 17–20, the wave reflected at P on returning to Q may be out of phase with the vibrator, and it can do work on the vibrator. That is, the string can give up some energy to the vibrator just as well as receive energy from it. The "standing" wave pattern is not fixed in form but wiggles about. On the average the amplitude is small and not much different from that of the vibrator. This situation is analogous to the erratic motion of a swing being pushed periodically with a frequency other than its natural one. The displacement of the swing is rather small.

Hence, the string absorbs peak energy from the vibrator at resonance. Figure 17–14 shows a photograph of resonant standing waves produced in a stretched rubber tube. Tuning a radio is an analogous process. By tuning a dial the natural frequency of an alternating current in the receiving circuit is made equal to the frequency of the waves broadcast by the station desired. The circuit resonates with the transmitted signals and absorbs peak energy from the signal. We shall encounter resonance conditions again in sound, in electromagnetism, in optics, and in atomic and nuclear physics. In these areas, as in mechanics, the system will absorb peak energy from the source at resonance and relatively little energy off resonance.

Example 4 In a demonstration with the apparatus just described, the vibrator has a frequency $\nu = 20$ Hz and the string has a linear density $\mu = 7.5$ g/m and a length $l = 7.0$ m. The tension F is varied by pulling down on the end of the string over the pulley. If the demonstrator wants to show resonance, starting with one loop and then with two, three, and four loops, what force must he exert on the string?

At resonance,

$$\nu = \frac{n}{2l}\sqrt{\frac{F}{\mu}}.$$

Hence, the tension F is given by

$$F = \frac{4l^2\nu^2\mu}{n^2}.$$

For one loop, $n = 1$, so that

$$F_1 = 4l^2\nu^2\mu = 4(7.0\text{ m})^2(20\text{ s}^{-1})^2(7.5 \times 10^{-3}\text{ kg/m})$$
$$= 590\text{ N}(= 60\text{ kg wt})$$

For two loops, $n = 2$, and

$$F_2 = \frac{4l^2\nu^2\mu}{4} = \frac{F_1}{4} = 150\text{ N}.$$

Likewise, for three and four loops

$$F_3 = \frac{F_1}{(3)^2} = 66\text{ N},$$

and

$$F_4 = \frac{F_1}{(4)^2} = 37\text{ N}.$$

Hence, the demonstrator gradually relaxes the tension to obtain resonance with an increasing number of loops. Although the resonance frequency is always the same under these circumstances, the speed of propagation and the wavelength at resonance decrease proportionately.

If the tension were kept fixed, giving a definite wave speed, would we obtain more than one resonance condition by varying the frequency of the vibrator?

REVIEW GUIDE AND SUMMARY

Longitudinal and transverse waves

Mechanical waves can exist in any deformable, elastic medium such as air, water, or iron. The waves are called *longitudinal* if the motion of each individual particle of the medium is in the same direction as the wave propagation through that point. If the motion is perpendicular, the wave is called *transverse*.

Simple harmonic or sine wave

A wave is *periodic* if each particle has a motion that repeats itself. It is *simple harmonic* if the displacement can be described by a single sine function. A *wavefront* is a surface connecting medium particles all of which are moving in the same manner at any given moment. A wave is called a spherical wave or a plane wave, for instance, if the wavefront has the corresponding shape.

General one-dimensional wave equation

The general expression for the displacement y of a particle located at position x at time t in a one-dimensional traveling wave is

$$y = f(x \pm vt). \qquad [17-2]$$

The speed of the wave is v and f may represent any function. The $-$ sign is used when describing a wave traveling in the $+x$-direction and the $+$ sign for the $-x$-direction.

Simple harmonic traveling wave

The displacement of a particle located at position x and time t in a one-dimensional simple harmonic wave is given by

$$y = y_m \sin\left[2\pi\left(\frac{x}{\lambda} - vt\right) - \phi\right]. \qquad [\text{see } 17-8]$$

Here y_m is the *amplitude* of the wave, the quantity in parentheses is called the *phase* of the wave, and ϕ is the *phase constant*. The *wavelength* λ represents the shortest repetitive wave interval, and the *frequency* ν is given by $1/T$, where T is the *period* of the wave.

If we define the *wave number* k and the *angular frequency* ω from

$$k = \frac{2\pi}{\lambda} \qquad \text{and} \qquad \omega = 2\pi\nu = \frac{2\pi}{T}, \qquad [\text{see } 17-9]$$

we can write the above equation more compactly as

$$y = y_m \sin(kx - \omega t + \phi).$$

The *speed* v of such a wave is given by

$$v = \lambda\nu = \frac{\lambda}{T} = \frac{\omega}{k}. \qquad [\text{see } 17-11]$$

The speed of a wave depends on the properties of the medium. The speed of the wave in a stretched string, having a tension F and mass per unit length μ, was shown to be

$$v = \sqrt{\frac{F}{\mu}}. \qquad [17-12]$$

Power and intensity

The *power* of a wave is the energy transported past a given point per unit time. For three-dimensional waves a useful quantity is the *intensity*, which is the energy transferred across a given area per unit area per unit time. The power varies with time as the sine wave moves along, so the average power becomes a useful quantity for some purposes. Equation 17–14 gives the average power of a wave on a string. Example 3 develops the fact that the intensity in a spherical wave falls inversely with the square of the distance from the point source.

Two or more waves can traverse the same space independently of one another, and the displacement of any given particle at a given time is the sum of the displacements that the individual waves would give it. This type of addition is called *superposition*. The wavetrains involved are said to *interfere*.

Superposition and interference

Constructive and destructive interference

If two sine waves having the same frequency and amplitude y_m travel in the same direction but with a phase difference ϕ, superposition shows that the result is a sine wave having a phase constant $\phi/2$ and an amplitude $2y_m \cos \phi/2$; see Eq. 17–18. When $\phi = 0$ the interference is constructive and the amplitude is $2y_m$. When $\phi = \pi$ the interferrence is destructive and the amplitude is 0.

Standing Waves

The interference of two sine waves having the same frequency and amplitude but moving in opposite directions produces *standing waves* with the equation

$$y = 2y_m \sin kx \cos \omega t \qquad [17-19b]$$

Nodes and antinodes

Standing waves are characterized by fixed positions of zero displacement called *nodes* and positions of maximum displacement called *antinodes;* see Fig. 17–16. Each are separately spaced a half wavelength apart; a node is $\lambda/4$ from its adjacent antinodes.

Natural frequencies

When one applies the above theory to actual systems, the fact that strings, organ pipes, and so on, have finite sizes limits the frequencies of possible standing waves. These are called *natural frequencies.*

If a system is acted upon by some force that drives it at one of its natural frequencies, the condition called *resonance,* in which the system has a relatively large amplitude, occurs. Example 4 illustrates resonance in a string of finite length.

Resonance

QUESTIONS

1. How could you prove experimentally that energy is associated with a wave?

2. Energy can be transferred by particles as well as by waves. How can we experimentally distinguish between these methods of energy transfer?

3. Can a wave motion be generated in which the particles of the medium vibrate with angular simple harmonic motion? If so, explain how and describe the wave.

4. Are torsional waves transverse or longitudinal? Can they be considered as a superposition of two waves, which are either transverse or longitudinal?

5. How can one create plane waves? Spherical waves?

6. The following functions in which A is a constant are of the form $f(x \pm vt)$:

$$y = A(x - vt), \qquad y = A(x + vt)^2$$
$$y = A\sqrt{x - vt}, \qquad y = A\ln(x + vt).$$

Explain why these functions are not useful in wave motion.

7. Can one produce on a string a waveform which has a discontinuity in slope at a point, that is, one having a sharp corner? Explain.

8. How do the amplitude and the intensity of surface water waves vary with the distance from the source?

9. The inverse square law does not apply exactly to the decrease in intensity of sounds with distance. Why not?

10. When two waves interfere, does one alter the progress of the other?

11. When waves interfere, is there a loss of energy? Explain your answer.

12. Why don't we observe interference effects between the light beams emitted from two flashlights or between the sound waves emitted by two violins?

13. As Fig. 17–12 shows, twice during a cycle the configuration of standing waves in a stretched string is a straight line, exactly what it would be if the string were not vibrating at all. Discuss from the point of view of energy conservation.

14. If two waves differ only in amplitude and are propagated in opposite directions through a medium, will they produce standing waves? Is energy transported? Are there any nodes? (See Problem 34.)

15. Consider the standing waves in a string to be a superposition of traveling waves and explain, using superposition ideas, why there are no true nodes in the resonating string of Fig. 17–16, even at the "fixed" end. (*Hint:* Consider damping effects.)

16. In the discussion of transverse waves in a string, we have dealt only with displacements in a single plane, the x-y plane. If all displacements lie in one plane, the wave is said to be *plane polarized.* Can there be displacements in a plane other than the single plane dealt with? If so, can two differently plane-polarized waves be combined? What appearance would such a combined wave have?

17. A wave transmits energy. Does it transfer momentum? Can it transfer angular momentum: (See Question 16.) (See "Energy and Momentum Transport in String Waves," by D. W. Juenker, *American Journal of Physics,* January 1976.)

EXERCISES

Section 17–3 Traveling Waves

1. The speed of electromagnetic waves in vacuum is 3.0×10^8 m/s. (*a*) Wavelengths in the visible part of the spectrum (light) range from about 4.0×10^{-7} m in the violet to about 7.0×10^{-7} m in the red. What is the range of frequencies of light waves? (*b*) The range of frequencies for shortwave radio (for example, FM radio and VHF television) is 1.5 MHz (megahertz; see Table 1–2) to 300 MHz. What is the corresponding wavelength range? (*c*) X rays are also electromagnetic. Their wavelength range extends from about 5.0 nm (nanometers; see Table 1–2) to about 1.0×10^{-2} nm. What is the frequency range for x rays?

2. Show that $y = y_m \sin(kx - \omega t)$ may be written in the alternative forms

$$y = y_m \sin k(x - vt), \qquad y = y_m \sin 2\pi\left(\frac{x}{\lambda} - vt\right),$$
$$y = y_m \sin \omega\left(\frac{x}{v} - t\right), \qquad y = y_m \sin 2\pi\left(\frac{x}{\lambda} - \frac{t}{T}\right).$$

3. By rocking a boat, a boy produces surface water waves on a previously quiet lake. He observes that the boat performs 12 oscillations in 20 s, each oscillation producing a wave crest 15 cm above the undisturbed surface of the lake. He further observes that a given wave crest reaches the shore, 12 m away, in 6.0 s. What are (a) the period and (b) wavelength of this wave?

4. We know that $f(x - vt)$, where f is any function, represents a wave traveling in the positive x-direction. To illustrate this, consider the function shown in Fig. 17–17. (a) What is $f(0), f(1), f(2), f(3), f(4), f(5)$? (b) Plot $f(x - 5t)$ as a function of x for $0 \le x \le 20$ if $t = 0$. Here x is in centimeters and t is in seconds. (c) Repeat (b) for $t = 1$ s and $t = 2$ s. (d) From your graphs, what is the wave speed? (e) Plot $f(x - 5t)$ as a function of t for $0 \le t \le 2$ s if $x = 10$ cm.

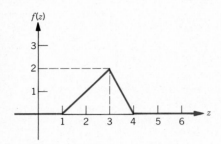

Figure 17–17 Exercise 4.

5. A traveling wave on a string is described by

$$y = 2 \sin \left[2\pi \left(\frac{t}{0.40} + \frac{x}{80} \right) \right],$$

where x and y are in centimeters and t is in seconds. (a) For $t = 0$, plot y as a function of x for $0 \le x \le 160$ cm. (b) Repeat (a) for $t = 0.05$ s and $t = 0.10$ s. (c) From your graphs, what is the wave speed, and in which direction ($\pm x$) is the wave traveling?

Section 17–4   Wave Speed in a Stretched String

6. What is the speed of a transverse wave in a rope of length 20 m ($= 6.6$ ft) and mass 0.060 kg ($= 0.0041$ slug) under a tension of 500 N ($= 110$ lb)?

7. The tension in a wire clamped at both ends is doubled without appreciably changing its length. What is the ratio of the new to the old wave speed for transverse waves in this wire?

8. The equation of a transverse wave on a string is

$$y = (2.0 \text{ mm}) \sin [(20 \text{ m}^{-1})x - (600 \text{ s}^{-1})t].$$

The tension in the string is 15 N. (a) What is the wave speed? (b) Find the linear density of this string in grams per meter.

9. The linear density of a vibrating string is 1.3×10^{-4} kg/m. A transverse wave is propagating on the string and is described by the equation $y = (0.021 \text{ m}) \sin [(1.0 \text{ m}^{-1})x + (30 \text{ s}^{-1})t]$, where x and y are measured in meters and t in seconds. What is the tension in the string?

Section 17–5   Power and Intensity in Wave Motion

10. Spherical waves are emitted from a 1.0-W source in an isotropic nonabsorbing medium. What is the wave intensity 1.0 m from the source?

11. A source emits spherical sound waves isotropically (i.e., with equal intensity in all directions). The intensity of the wave, 2.5 m from the source, is 1.91×10^{-4} W/m². What is the power output of this source?

Section 17–7   Interference of Waves

12. Determine the amplitude of the resultant motion when two sinusoidal motions having the same frequency and traveling in the same direction are combined, if their amplitudes are 3.0 cm and 4.0 cm and they differ in phase by $\pi/2$ rad.

13. Two waves are propagating on the same very long string. A generator at the left end of the string creates a wave given by $y = (6.0 \text{ cm}) \cos \pi/2 [(2.0 \text{ m}^{-1})x - (8.0 \text{ s}^{-1})t]$ and one at the right end of the string creates the wave $y = (6.0 \text{ cm}) \cos \pi/2 [(2.0 \text{ m}^{-1})x + (8.0 \text{ s}^{-1})t]$. (a) Calculate the frequency, wavelength, and speed of each wave. (b) Find the points at which there is no motion (the nodes). (c) At which points is the motion a maximum?

Section 17–8   Standing Waves

14. The equation of a transverse wave traveling in a string is given by

$$y = 0.10 \cos (0.79x - 13t - 0.89),$$

in which x and y are expressed in meters and t in seconds. Write down the equation of a wave that, when added to the given one, would produce standing waves on the rope.

15. A 15.0-cm violin string, fixed at both ends, is vibrating in its fundamental mode. The speed of waves in this wire is 250 m/s, and the speed of sound in air is 348 m/s. What are the frequency and wavelength of the emitted sound wave?

16. A 120-cm length of string is stretched between fixed supports. What are the three longest possible wavelengths for standing waves in this string? Sketch the corresponding standing waves.

17. One end of a 120-cm string is held fixed. The other end is attached to a weightless ring that can slide along a frictionless rod as shown in Fig. 17–18. What are the three longest possible wavelengths for standing waves in this string? Sketch the corresponding standing waves.

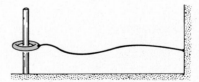

Figure 17–18 Exercise 17.

Section 17–9   Resonance

18. A 75-cm string is stretched between fixed supports. It is observed to have resonant frequencies of 420 Hz and 315 Hz and no other resonance frequency between these two. (a) What is the lowest resonance frequency for this string? (b) What is the wave speed for this string?

19. A 125-cm length of string has a mass of 2.0 g. It is stretched with a tension of 7.0 N between fixed supports. (a) What is the wave speed for this string? (b) What is the lowest resonant frequency of this string?

PROBLEMS GROUPED BY SECTION

Section 17-3   Traveling Waves

1. The equation of a transverse wave traveling along a very long string is given by $y = 6.0 \sin(0.020\pi x + 4.0\pi t)$, where x and y are expressed in centimeters and t in seconds. Calculate (a) the amplitude, (b) the wavelength, (c) the frequency, (d) the speed, (e) the direction of propagation of the wave, and (f) the maximum transverse speed of a particle in the string.

2. A sinusoidal wave travels along a string. If the time for a particular point to move from maximum displacement to zero displacement is 0.17 s, what are (a) the period, and (b) frequency? (c) If the wavelength is 1.4 m, what is the speed of the wave?

3. (a) Write an expression describing a transverse wave traveling on a cord in the $+y$-direction with a wave number of $6.0\,\text{cm}^{-1}$, a period of 0.20 s, and having an amplitude of 3.0 mm. Take the transverse direction to be the z-direction. (b) What is the maximum transverse velocity of a point on the cord?

4. (a) Write an expression describing a transverse wave traveling on a cord in the $+x$-direction with a wavelength of 10 cm, a frequency of 400 Hz, and an amplitude of 2.0 cm. (b) What is the maximum speed of a point on the cord? (c) What is the velocity of the wave?

Section 17-4   Wave Speed in a Stretched String

5. A stretched string has a mass per unit length of 5.0 g/cm and a tension 10 N. A wave on this string has an amplitude 0.12 mm, a frequency 100 Hz and is traveling in the negative x-direction. Write an equation for this wave.

6. A simple harmonic transverse wave is propagating along a string toward the left (or $-x$) direction. Figure 17-19 shows a plot of the displacement as a function of position at time $t = 0$. The string tension is 3.6 N and its linear density is 25 g/m. Calculate (a) the amplitude, (b) the wavelength, (c) the wave speed, (d) the period, and (e) the maximum speed of a particle in the string. (f) Write an equation describing the traveling wave.

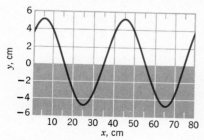

Figure 17-19   Problem 6.

7. Prove that the slope of a string at any point is numerically equal to the ratio of the particle speed to the wave speed at that point.

8. In Fig. 17-20, string #1 has a linear density of 3.0 g/m, and string #2 has a linear density of 5.0 g/m. They are under tension due to the hanging block of mass $M = 500$ g. (a) Calculate the wave speed in each string. (b) The block is now divided into two blocks (with $M_1 + M_2 = M$) and the apparatus rearranged as shown in Fig. 17-20b. Find M_1 and M_2 such that the wave speeds in the two strings are the same.

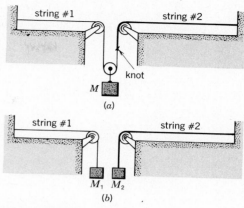

Figure 17-20   Problem 8.

***9.** A uniform circular hoop of string is rotating clockwise in the absence of gravity (see Fig. 17-21). The tangential speed is v_0. Find the speed of waves traveling on this string. (*Remark:* The answer is independent of the radius of the circle and the mass per unit length of the string!)

Figure 17-21   Problem 9.

Section 17-5   Power and Intensity in Wave Motion

10. A person stands at a distance D from an isotropic source of sound waves. He walks 50 m toward the source and observes that the intensity of these waves has doubled. Calculate the distance D.

11. (a) Show that the intensity I (the energy crossing unit area per unit time per unit area) is the product of the energy per unit volume u and the speed of propagation v of a wave disturbance. (b) Radio waves travel at a speed of 3.0×10^8 m/s. Find the energy density in a radio wave 480 km from a 50,000-W source, assuming the waves to be spherical and the propagation to be isotropic.

12. A line source emits a cylindrical expanding wave. Assuming that the medium absorbs no energy, find how (a) the intensity and (b) the amplitude of the wave depend on the distance from the source.

Section 17-7   Interference of Waves

13. A source S and a detector D of radio waves are a distance d apart on the ground. The direct wave from S is found to be in phase at D with the wave from S that is reflected from a horizontal layer at an altitude H (Fig. 17-22). The incident and reflected rays make the same angle with the reflecting layer. When the layer rises a distance h, no signal is detected at D. Neglect absorption in the atmosphere and find the relation between d, h, H, and the wavelength λ of the waves.

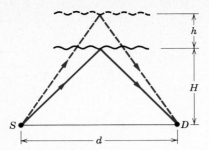

Figure 17–22 Problem 13.

14.   Three sinusoidal waves travel in the negative x-direction in the same string. All three waves have the same frequency. Their amplitudes are in the ratio $1 : \frac{1}{2} : \frac{1}{3}$ and their phase angles are 0, $\pi/2$, and π, respectively. Plot the resultant waveform when $t = 0$ and discuss its behavior as t increases.

Section 17–8 Standing Waves

15.   A string vibrates according to the equation

$$y = (0.5 \text{ cm}) \left[\sin\left(\frac{\pi}{3} \text{ cm}^{-1}\right)x \right] \cos(40\pi \text{ s}^{-1})t.$$

(*a*) What are the amplitude and velocity of the component waves whose superposition can give rise to this vibration? (*b*) What is the distance between nodes? (*c*) What is the velocity of a particle of the string at the position $x = 1.5$ cm when $t = \frac{9}{8}$ s?

16.   Two transverse sinusoidal waves travel in opposite directions along a string. Each has an amplitude of 0.30 cm and a wavelength of 6.0 cm. The speed of a transverse wave in the string is 1.5 m/s. Plot the shape of the string at each of the following times: $t = 0$ (arbitrary), $t = 5.0$, $t = 10$, $t = 15$, $t = 20$ ms.

17.   When played in a certain manner, the lowest frequency of vibration of a certain violin string is concert A (440 Hz). What two

higher frequencies could also be found on that string if the length isn't changed?

18.   Consider a standing wave that is the sum of two waves traveling in opposite directions but otherwise identical. Show that the energy in each loop of the standing wave is $2\pi^2\mu y_m^2 v v$.

Section 17–9 Resonance

19.   In a laboratory experiment on standing waves, a string 3.0 ft (=0.9 m) long is attached to the prong of an electrically driven tuning fork that vibrates perpendicular to the length of the string at a frequency of 60 vib/s (=60 Hz). The weight of the string is 0.096 lb (mass = 0.044 kg). (*a*) What tension must the string be under (weights are attached to the other end) if it is to vibrate in four loops? (*b*) What would happen if the tuning fork is turned so as to vibrate parallel to the length of the string?

20.   An aluminum wire of length $l_1 = 60.0$ cm and of cross-sectional area 1.00×10^{-2} cm² is connected to a steel wire of the same cross-sectional area. The compound wire, loaded with a block m of mass 10.0 kg, is arranged as shown in Fig. 17–23 so that the distance l_2 from the joint to the supporting pulley is 86.6 cm. Transverse waves are set up in the wire by using an external source of variable frequency. (*a*) Find the lowest frequency of excitation for which standing waves are observed such that the joint in the wire is a node. (*b*) What is the total number of nodes observed at this frequency, excluding the two at the ends of the wire? The density of aluminum is 2.60 g/cm³, and that of steel is 7.80 g/cm³.

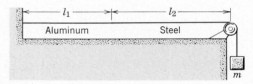

Figure 17–23 Problem 20.

ADDITIONAL PROBLEMS

21.   The equation of a transverse wave traveling in a string is given by

$$y = (2.0 \text{ mm}) \sin[(20 \text{ m}^{-1})x - (600 \text{ s}^{-1})t].$$

(*a*) Find the amplitude, frequency, velocity, and wavelength of the wave. (*b*) Find the maximum transverse speed of a particle in the string.

22.   Write the equation for a wave traveling in the negative direction along the x-axis and having an amplitude 0.010 m, a frequency 550 Hz, and a speed 330 m/s.

23.   (*a*) A continuous sinusoidal longitudinal wave is sent along a coiled spring from a vibrating source attached to it. The frequency of the source is 25 Hz, and the distance between successive rarefactions in the spring is 24 cm. Find the wave speed. (*b*) If the maximum longitudinal displacement of a particle in the spring is 0.30 cm and the wave moves in the $-x$-direction, write the equation for the wave. Let the source be at $x = 0$ and the displacement at $x = 0$ when $t = 0$ be zero.

24.   A continuous sinusoidal wave is traveling on a string with velocity 40 cm/s. The displacement of the particles of the string at $x = 10$ cm is found to vary with time according to the equation

$y = (5.0 \text{ cm}) \sin[1.0 - (4.0 \text{ s}^{-1})t]$. The linear density of the string is 4.0 g/cm. (*a*) What is the frequency of the wave? (*b*) What is the wavelength of the wave? (*c*) Write the general equation giving the transverse displacement of the particles of the string as a function of position and time. (*d*) Calculate the tension in the string.

25.   For a wave on a stretched cord, find the ratio of the maximum particle speed (the speed with which a single particle in the cord moves transverse to the wave) and the wave speed. If a wave having a certain frequency and amplitude is imposed on a cord, would this speed ratio depend on the material of which the cord is made, such as wire or nylon?

26.   Two pulses are traveling along a string in opposite directions, as shown in Fig. 17–24. (*a*) If the wave velocity is 2.0 m/s and the pulses are 6.0 cm apart, sketch the patterns after 5.0, 10, 15, 20, 25 ms. (*b*) What has happened to the energy at $t = 15$ ms?

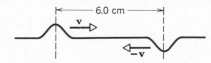

Figure 17–24 Problem 26.

27. Four sinusoidal waves travel in the positive x-direction in the same string. Their frequencies are in the ratio $1:2:3:4$ and their amplitudes are in the ratio $1:\frac{1}{2}:\frac{1}{3}:\frac{1}{4}$, respectively. When $t = 0$, at $x = 0$, the first and third are $180°$ out of phase with respect to the second and fourth. Plot the resultant waveform when $t = 0$ and discuss its behavior as t increases.

28. Two waves on a string are described by the equations

$$y_1 = (0.10 \text{ m}) \sin 2\pi [(0.50 \text{ m}^{-1})x + (20 \text{ s}^{-1})t]$$

and

$$y_2 = (0.20 \text{ m}) \sin 2\pi [(0.50 \text{ m}^{-1})x - (20 \text{ s}^{-1})t].$$

Sketch the total response for the point on the string at $x = 3.0$ m; that is, plot y vs. t for that value of x.

29. What are the three lowest frequencies for standing waves on a wire, 10-m long having a mass of 0.10 kg, which is stretched under a tension of 250 N?

30. A 3.0-m-long string is vibrating as a three-loop standing wave whose amplitude is 1.0 cm. The wave speed is 100 m/s. (a) What is the frequency? (b) Write equations for two waves that, when combined, will result in this standing wave.

31. Vibrations from a 600-Hz tuning fork sets up standing waves in a string clamped at both ends. The wave speed for the string is 400 m/s. The standing wave has four loops and an amplitude of 2.0 mm. (a) What is the length of the string? (b) Write an equation for the displacement of the string as a function of position and time.

32. A 1.0-m wire has a mass of 10 g and is held under a tension of 100 N. The wire is held rigidly at both ends and set into vibration. Calculate (a) the velocity of waves on the wire, (b) the wavelengths of the waves that produce one- and two-loop standing waves on the string, and (c) the frequencies of the waves that produce one- and two-loop standing waves.

33. The text discusses the addition of the two waves of the form $y_m \sin(kx \pm \omega t)$ to form a standing wave. Suppose that the two waves each have a different phase constant:

$$y_1 = y_m \sin (kx - \omega t - \phi_1),$$
$$y_2 = y_m \sin (kx + \omega t - \phi_2).$$

Show that we can always shift the spatial origin, $x' = x - x_0$ (shift to right by x_0), and shift the time origin, $t' = t - t_0$ (shift forward by t_0), so that (in terms of x' and t') the phase constants do not appear:

$$y_1 = y_m \sin (kx' - \omega t'),$$
$$y_2 = y_m \sin (kx' + \omega t').$$

Find x_0 and t_0. Discuss the implication of this problem for the production of standing waves.

34. *Standing wave ratio.* If an incident traveling wave is only partially reflected from a boundary, the resulting superposition of two waves having different amplitudes and traveling in opposite directions gives a standing wave pattern of waves whose envelope is shown in Fig. 17–25. The *standing wave ratio* (SWR) is defined

as $(A_i + A_r)/(A_i - A_r) = A_{max}/A_{min}$, and the percent reflection is defined as the ratio of the average power in the reflected wave to the average power in the incident wave, times 100. (a) Show that for 100% reflection SWR $= \infty$ and that for no reflection SWR $= 1$. (b) Show that a measurement of the SWR just before the boundary reveals the percent reflection occurring at the boundary according to the formula

$$\% \text{ reflection} = \frac{(\text{SWR} - 1)^2}{(\text{SWR} + 1)^2} \times 100.$$

35. A wave travels out uniformly in all directions from a point source. (a) Justify the following expression for the displacement y of the medium at any distance r from the source:

$$y = \frac{Y}{r} \sin k(r - vt).$$

Consider the speed, direction of propagation, periodicity, and intensity of the wave. (b) What are the dimensions of the constant Y?

36. For the traveling wave described quantitatively in Examples 1 and 2, find (a) the maximum value of the transverse speed u and (b) the maximum value of the transverse force F_{trans}. (c) The phase of this wave has been defined as $(kx - \omega t)$. Show that the two maximum values calculated above occur at the same phase values for the wave. What is the transverse displacement y of the string at these phases? (d) What is the maximum power transferred along the string? (e) What is the transverse displacement y for conditions under which this maximum power transfer occurs? (f) What is the minimum power transfer along the string? (g) What is the transverse displacement y for conditions under which this minimum power transfer occurs?

37. A wire 10 m long and having a mass of 100 g is stretched under a tension of 250 N. If two disturbances, separated in time by 0.030 s, are generated one at each end of the wire, where will the disturbances meet?

38. A wave of frequency 500 Hz has a velocity of 350 m/s. (a) How far apart are two points $60°$ out of phase? (b) What is the phase difference between two displacements at a certain point at times 1.0 ms apart?

39. The type of rubber band used inside some baseballs and golfballs obeys Hooke's law over a wide range of elongation of the band. The segment has an unstretched length l and a mass m. When a force F is applied, the band stretches an additional length Δl. (a) What is the speed (in terms of m, Δl, and the force constant k) of transverse waves on this rubber band? (b) Using your answer (a), show that the time required for a transverse pulse to travel the length of the rubber band is proportional to $1/(\Delta l)^{1/2}$ if $\Delta l \ll l$, and is constant if $\Delta l \gg l$.

*40. A uniform rope of mass m and length L hangs from a ceiling. (a) Show that the speed of a transverse wave in the rope is a function of y, the distance from the lower end, and is given by $v = \sqrt{gy}$. (b) Show that the time it takes a transverse wave to travel the length of the rope is given by $t = 2\sqrt{L/g}$. (c) Does the actual mass of the rope affect the results of (a) and (b)?

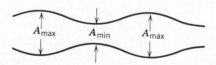

Figure 17–25 Problem 34.

18 Sound Waves

18-1 Audible, Ultrasonic, and Infrasonic Waves

Kinds of sound waves

Sound waves are longitudinal mechanical waves. They can be propagated in solids, liquids, and gases. The material particles transmitting such a wave oscillate in the direction of propagation of the wave itself. There is a large range of frequencies within which longitudinal mechanical waves can be generated, sound waves being confined to the frequency range that can stimulate the human ear and brain to the sensation of hearing. This range is from about 20 cycles/s (or Hz) to about 20,000 Hz and is called the *audible* range. A longitudinal mechanical wave whose frequency is below the audible range is called an *infrasonic* wave, and one whose frequency is above the audible range is called an *ultrasonic* wave.

Infrasonic waves of interest are usually generated by large sources, earthquake waves being an example. The high frequencies associated with ultrasonic waves may be produced by elastic vibrations of a crystal induced by resonance with an applied alternating electric field (piezoelectric effect). It is possible to produce ultrasonic frequencies as high as 6×10^8 Hz ($= 600$ MHz) in this way; the corresponding wavelength in air is about 5×10^{-5} cm, the same as the length of visible light waves.

Audible waves originate in vibrating strings (violin, piano, guitar), vibrating air columns (organ, clarinet), and vibrating plates and membranes (xylophone, loudspeaker, drum, human vocal cords). All of these vibrating elements alternately compress the surrounding air on a forward movement and rarefy it on a backward movement. The air transmits these disturbances outward from the source as a wave. Upon entering the ear, these waves produce the sensation of sound. Waveforms that are approximately periodic or consist of a small number of approximately periodic components often create a pleasant sensation (if the intensity is not too high), as, for example, musical sounds. Noise, on the other hand, is the unpleasant sound that may result, for instance, when the waveform is nonperiodic.

In this chapter we deal with the properties of longitudinal mechanical waves, using sound waves as the prototype.

18-2 Propagation and Speed of Longitudinal Waves

Sound waves, if unimpeded, will spread out in all directions from a source. It is simpler to deal with one-dimensional propagation, however, than with three-dimensional propagation, so we consider first the transmission of longitudinal waves in a tube.

Figure 18-1 shows a piston at one end of a long tube filled with a compressible medium. The vertical lines divide the compressional (fluid) medium into thin slices, each of which contains the same mass of fluid. Where the lines are relatively close together the fluid pressure and density are greater than they are in the normal undisturbed fluid, and conversely. We shall treat the fluid as a continuous medium and ignore for the time being the fact that it is made up of molecules that are in continual random motion.

If we push the piston of Fig. 18-1 forward, the fluid in front of it is compressed, the fluid pressure and density rising above their normal undisturbed values. The compressed fluid moves forward, compressing the fluid layers next to it, and a compressional pulse travels down the tube. If we then withdraw the piston, the fluid in front of it expands, its pressure and density falling below their normal undisturbed values; a pulse of rarefaction travels down the tube. These pulses are similar to transverse pulses traveling along a string, except that the oscillating fluid elements are displaced along the direction of propagation (longitudinal) instead of at right angles to this direction (transverse). If the piston oscillates back and forth, a continuous train of compressions and rarefactions will travel along the tube (Fig. 18-1). As for transverse waves in a string (see Section 17-4) we should be able, using Newton's laws of motion, to express the speed of propagation of this longitudinal wave in terms of an elastic and an inertial property of the medium. We now do so.

For the moment, let us assume that the tube is very long so that we can ignore reflections from the far end. As for the string of Fig. 17-6, we shall consider not an extended wave but a single (compressional) pulse that we might generate by giving the piston in Fig. 18-1 a short, rapid, inward stroke.

Figure 18-2 shows such a pulse (labeled "compressional zone") traveling at speed v along the tube from left to right. For simplicity we have assumed this pulse to have sharply defined leading and trailing edges and to have a uniform fluid

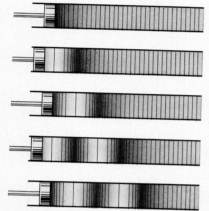

Figure 18-1 Sound waves generated in a tube by an oscillating piston. The vertical lines divide the compressible medium in the tube into layers of equal mass. We assume the tube to be infinitely long so that there is no confusion caused by reflection of the wave at the end of the tube.

Calculating the speed of sound

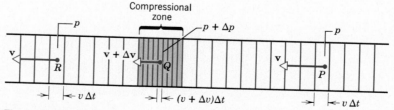

Figure 18-2 A compressional pulse travels along a gas-filled tube. In a reference frame in which the undisturbed gas is at rest the pulse moves from left to right with speed v. We view the pulse, however, from a reference frame in which the pulse is stationary; in such a frame the gas outside the pulse streams through the tube from right to left with speed v, as shown. Note that Δv is negative.

pressure and density in its interior. When we analyzed the motion of a transverse pulse in a string, we found it convenient to choose a reference frame in which the pulse remained stationary; we shall do this here also. In Fig. 18–2, then, the compressional zone remains stationary in our reference frame while the fluid moves through it from right to left with speed v, as shown.

Let us follow the motion of the element of fluid contained between the vertical lines at P in Fig. 18–2. This element moves forward at speed v until it strikes the compressional zone. While it is entering this zone it encounters a difference of pressure Δp between its leading and its trailing edges. The element is compressed and decelerated, moving with a lower speed $v + \Delta v$ within the zone, the quantity Δv being negative. The element eventually emerges from the left face of the zone where it expands to its original volume and the pressure differential Δp acts to accelerate it to its original speed v. The figure shows the element at point R, having passed through the compressional zone and moving again with speed v, as at P.

Let us apply Newton's laws to the fluid element while it is entering the compressional zone. The resultant force acting during entry points is to the right in Fig. 18–2 and has magnitude

$$F = (p + \Delta p)A - pA = \Delta pA$$

in which A is the cross-sectional area of the tube.

The length of the element outside the compressional zone (at P, say) is $v\,\Delta t$, where Δt is the time required for the element to move past any given point. The volume of the element is thus $vA\,\Delta t$ and its mass is $\rho_0 vA\,\Delta t$, where ρ_0 is the density of the fluid outside the compressional zone. The acceleration a experienced by the element as it enters the zone is $-\Delta v/\Delta t$; because Δv is inherently negative, a is positive, which means that, like the force ΔpA in Fig. 18–2, it points to the right. Thus Newton's second law,

$$F = ma,$$

yields

$$\Delta pA = (\rho_0 vA\,\Delta t)\frac{-\Delta v}{\Delta t},$$

which we may write as

$$\rho_0 v^2 = \frac{-\Delta p}{\Delta v/v}.$$

Now the fluid that would occupy a volume $V\,(= Av\,\Delta t)$ at P is compressed by an amount $A(\Delta v)\,\Delta t = \Delta V$ on entering the compressional zone. Hence,

$$\frac{\Delta V}{V} = \frac{A\,\Delta v\,\Delta t}{Av\,\Delta t} = \frac{\Delta v}{v},$$

and we obtain

$$\rho_0 v^2 = \frac{-\Delta p}{\Delta V/V}.$$

The ratio of the change in pressure on a body, Δp, to the resulting fractional change in volume, $-\Delta V/V$, is called the *bulk modulus of elasticity B* of the body. That is, $B = -V\,\Delta p/\Delta V$. B is positive because an increase in pressure causes a decrease in volume. In terms of B, the speed of the longitudinal pulse in the medium of Fig. 18–2 is

Speed of sound

$$v = \sqrt{\frac{B}{\rho_0}}. \qquad (18-1a)$$

A more extended analysis than given above shows that Eq. 18–1a applies not only to rectangular pulses of the type displayed in Fig. 18–2 but also to pulses of any shape and to extended wave trains. Notice that the speed of the wave is determined by the properties of the medium through which it propagates, and that an elastic property B and an inertial property ρ_0 are involved. Table 18–1 gives the speed of longitudinal (sound) waves in various media.

Table 18–1   Speed of sound

Medium[a]	Speed[b]	
	m/s	ft/s
Air (dry, at 0°C)	331	1,087
Air (dry, at 20°C)	343	1,130
Steam (at 134°C)	494	1,620
Hydrogen	1,330	4,360
Water (distilled)	1,486	4,876
Water (sea)	1,519	4,984
Lead	1,190	3,900
Copper	3,810	12,500
Aluminum	5,000	16,400
Glass (Pyrex)	5,170	17,000
Steel	5,200	17,100
Beryllium	12,900	42,300

[a] At 1.0 atm and 20°C.
[b] Values for solids measured in long thin rods.

If the medium is a gas, such as air, it is possible to express B in terms of the undisturbed gas pressure p_0. For a sound wave in a gas we obtain

Speed of sound in a gas

$$v = \sqrt{\frac{\gamma p_0}{\rho_0}}, \qquad (18-1b)$$

where γ is a constant called the ratio of specific heats for the gas (see Section 21–6).

If the medium is a solid, for a thin rod the bulk modulus is replaced by a stretch modulus (called Young's modulus). If the solid is extended, we must allow for the fact that, unlike a fluid, a solid offers elastic resistance to tangential or shearing forces and the speed of longitudinal waves will depend on the shear modulus as well as the bulk modulus.

18–3   Traveling Longitudinal Waves

Consider again the continuous train of compressions and rarefactions traveling down the tube of Fig. 18–1. As the wave advances along the tube, each small volume element of fluid oscillates about its equilibrium position. The displacement is to the right or left along the x-direction of propagation of the wave. For convenience let us represent the displacement of any such volume element (or layer of elements that move in the same way) from its equilibrium position at x by the letter y. You should understand that the displacement y is along the direction of propagation for a longitudinal wave, whereas for a transverse wave the displacement y is at right angles to the direction of propagation. In general, the equation of a longitudinal wave traveling to the right as in Fig. 18–1 may be written as

$$y = f(x - vt).$$

For the particular case of a simple harmonic wave we may have

$$y = y_m \cos \frac{2\pi}{\lambda} (x - vt).$$

In this equation v is the speed of the longitudinal wave, y_m is its amplitude, and λ is its wavelength; y gives the displacement of a particle at time t from its equilibrium position at x. As before, we may write this more compactly as

$$y = y_m \cos (kx - \omega t). \tag{18–2}$$

It is usually more convenient to deal with pressure variations in a sound wave than with the actual displacements of the particles conveying the wave. Let us therefore write the equation of the wave in terms of the pressure variation rather than in terms of the displacement.

From the definition of the bulk modulus of elasticity, given earlier,

$$B = -\frac{\Delta p}{\Delta V / V},$$

we have

$$\Delta p = -B \frac{\Delta V}{V}.$$

Just as we let y represent the displacement from the equilibrium position x, so we now let p represent the change from the undisturbed pressure p_0. Then p replaces Δp, and

$$p = -B \frac{\Delta V}{V}.$$

If a layer of fluid at pressure p_0 has a thickness Δx and cross-sectional area A, its volume is $V = A \, \Delta x$. When the pressure changes, its volume will change by $A \, \Delta y$, where Δy is the amount by which the thickness of the layer changes during compression or rarefaction. Hence,

$$p = -B \frac{\Delta V}{V} = -B \frac{A \, \Delta y}{A \, \Delta x}.$$

As we let $\Delta x \to 0$ so as to shrink the fluid layer to infinitesimal thickness, we obtain

$$p = -B \frac{\partial y}{\partial x}. \tag{18–3}$$

We have used partial derivative notation because (see Eq. 18–2) y is a function of both x and t and we take the latter quantity as constant in this discussion. If the particle displacement is simple harmonic, then, from Eq. 18–2, we obtain

$$\frac{\partial y}{\partial x} = -ky_m \sin (kx - \omega t),$$

and from Eq. 18–3,

$$p = Bky_m \sin (kx - \omega t). \tag{18–4}$$

Hence, the pressure variation at each position x is also simple harmonic.

Since $v = \sqrt{B/\rho_0}$, we can write Eq. 18–4 more conveniently as

A pressure wave

$$p = [k\rho_0 v^2 y_m] \sin (kx - \omega t).$$

Recall that p represents the change from standard pressure p_0. The term in brackets therefore represents the maximum change in pressure from equilibrium and is called the *pressure amplitude*. If we denote this by p_m, then

$$p = p_m \sin (kx - \omega t), \tag{18–5}$$

where

The pressure amplitude

$$p_m = k\rho_0 v^2 y_m. \tag{18–6}$$

Hence, a sound wave may be considered either as a displacement wave or as a pressure wave. If the former is written as a cosine function, the latter will be a sine function and vice versa. The displacement wave is thus 90° out of phase with the pressure wave. That is, when the displacement from equilibrium at a point is a maximum or a minimum, the excess pressure there is zero; when the displacement at a point is zero, the excess or deficiency of pressure there is a maximum. Equation 18–6 gives the relation between the pressure amplitude (maximum variation of pressure from equilibrium) and the displacement amplitude (maximum variation of position from equilibrium). Check the dimension of each side of Eq. 18–6 for consistency. What units may the pressure amplitude have?

Example 1   (*a*) The maximum pressure variation p_m that the ear can tolerate in loud sounds is about 28 N/m² (= 28 Pa). Normal atmospheric pressure is $\sim 1.0 \times 10^5$ Pa (= 1 bar). Find the corresponding maximum displacement for a sound wave in air at 20°C having a frequency of 1000 Hz.

From Eq. 18–6 we have

$$y_m = \frac{p_m}{k\rho_0 v^2}.$$

From Table 18–1, $v = 343$ m/s, so that

$$k = \frac{2\pi}{\lambda} = \frac{2\pi\nu}{v} = \frac{2\pi \times 10^3}{343} \text{ m}^{-1} = 18 \text{ m}^{-1}.$$

The density of air ρ_0 is 1.2 kg/m³. Hence, for $p_m = 28$ Pa we obtain

$$y_m = \frac{28}{(18)(1.2)(343)^2} \text{ m} = 1.1 \times 10^{-5} \text{ m}.$$

The displacement amplitudes for the *loudest* sounds are about 10^{-5} m, or about 0.01 mm, a very small value indeed.

(*b*) In the faintest sound that can be heard at 1000 Hz the pressure amplitude is about 2.8×10^{-5} Pa. Find the corresponding displacement amplitude.

From $y_m = p_m/k\rho_0 v^2$, using the same values for k, v, and ρ_0, we obtain, with $p_m = 2.8 \times 10^{-5}$ N/m²,

$$y_m \cong 10^{-11} \text{ m}.$$

This is smaller than the radius of an atom, which is about 10^{-10} m! How can it be that the ear responds to such a small displacement?

18–4   Sound Intensity

In Section 17–5 we developed the idea of wave power, which is the rate at which energy is transferred past a given point. For a three-dimensional wave we also defined intensity I to be the power transferred per unit area. We shall now find the average intensity of a sound wave.

The average power in a wave on a string was found to be

$$\overline{P} = 2\pi^2 y_m^2 \nu^2 \mu v \tag{17–14}$$

where y_m is the particle amplitude, ν is the frequency, v is the wave speed, and μ is the mass per unit length of the string. The same considerations that apply in this derivation also hold for sound waves if μ is interpreted as the mass per unit length in the direction of wave propagation. Thus if we consider an area A perpendicular to the direction of wave propagation $\mu = \rho_0 A$, where ρ_0 is the equilibrium density of the fluid (we assume that deviations from ρ_0 due to the wave motion will be small), the average intensity is

$$\bar{I} = \frac{\overline{P}}{A} = 2\pi^2 y_m{}^2 \nu^2 \rho_0 v. \tag{18-7}$$

In the last section we found the relationship between the pressure amplitude p_m and the particle amplitude y_m to be (Eq. 18–6) $p_m = k\rho_0 v^2 y_m$, where $k = 2\pi/\lambda = 2\pi\nu/v$, so that $p_m = 2\pi\rho_0 v y_m \nu$. Using this to eliminate y_m, we find an expression for average intensity as a function of the pressure amplitude

$$\bar{I} = \frac{1}{2} \frac{p_m{}^2}{v\rho_0}. \tag{18-8a}$$

Another useful form is obtained by applying Eq. 18–1 for the wave speed:

Average intensity for a sound wave

$$\bar{I} = \frac{1}{2} \frac{p_m{}^2}{\sqrt{B\rho_0}}. \tag{18-8b}$$

We saw in Example 1 that the pressure in sound waves acting on the human ear varies from 28 Pa for the loudest tolerable to 2.8×10^{-5} Pa for the faintest detectable, a ratio of 1×10^6. Because the intensity varies as the square of the pressure, the intensity ratio is 1×10^{12}, an impressive range.

We must distinguish carefully between the intensity I of a sound wave—a measureable objective quantity—and the subjective sensation of loudness produced in the mind of the hearer. A basic question is: When a sound wave of intensity I falls on the ear what change ΔI will cause the hearer to report a barely detectable change in the sensation of loudness? Our hearing system is so constructed that ΔI turns out to be proportional to I. That is, the more intense the sound the greater ΔI must be. It is, in fact, this feature of the human response that allows us to deal with sound intensities over such a wide range.

It is common to specify sounds not by their intensity I but by a related parameter β, called the *sound level* and defined from

The decibel,
a measure of sound level

$$\beta = 10 \log_{10} (I/I_0). \tag{18-9}$$

Here I_0 is a standard reference intensity ($= 10^{-12}$ W/m^2), chosen because it is near the lower limit of human audibility. The unit in which sound levels are expressed is the *decibel* (abbr. dB), named after Alexander Graham Bell. We see that if $I = I_0$ then $\beta = 10 \log 1 = 0$ so that our reference standard of intensity corresponds to zero decibels.

To investigate further the nature of β as a measure of the intensity of a sound wave let us take differentials of Eq. 18–9. We obtain

$$d\beta = 10(\log_{10} e) \frac{dI}{I} = 4.3 \frac{dI}{I}. \tag{18-10}$$

This shows us that to produce a specified change in the sound level β we must change I by an amount that is just proportional to I. As we have seen, this is just the way the human auditory system seems to operate! We conclude that sound levels measured in decibels should correlate at least reasonably well with our subjective sensations of loudness. This is indeed the case. In Table 18–2 we list approximate sound levels for several different sources.

Table 18–2   Sound Levels (β)*

Threshold of hearing	0 dB	Noisest spot at Niagara Falls	85 dB
Rustle of leaves	10 dB	Pneumatic drill (at 3 m)	90 dB
Average whisper (at 1 m)	20 dB	Hi-Fi phonograph, 10 W (at 3 m)	110 dB
City Street, no traffic	30 dB	Threshold of pain	120 dB
Office, classroom	50 dB	Jet engine (at 50 m)	130 dB
Normal conversation (at 1 m)	60 dB	Saturn rocket (at 50 m)	200 dB
City street, very busy traffic	70 dB		

* The sound levels are relative to $I_o = 10^{-12}$ W/m^2

The response of the ear is actually not the same at every frequency. Figure 18–3 shows how the thresholds for hearing (audibility) and for feeling (pain) depend on frequency. The audible range lies between these two limits.

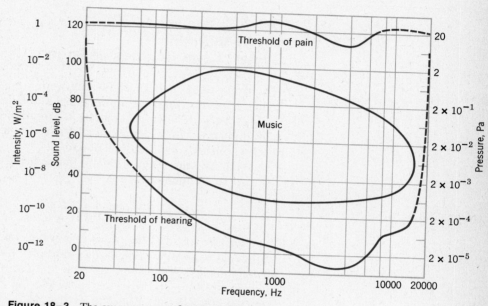

Figure 18–3   The average range of sound intensities for human hearing. Both the threshold of hearing and that of feeling of pain depend on the frequency. The approximate range of intensities occurring in music is also shown. (Courtesy Bell Telephone Laboratories.)

Example 2   (a) Suppose that you are given two intensities, I_1 and I_2. How are the sound levels, β_1 and β_2 related?

To find out, form the ratio I_2/I_1. Now,

$$10 \log_{10} \frac{I_2}{I_1} = 10 \log_{10} \left(\frac{I_2}{I_0} \cdot \frac{I_0}{I_1} \right) = 10 \log \frac{I_2}{I_0} + 10 \log \frac{I_0}{I_1}$$

$$= 10 \log \frac{I_2}{I_0} - 10 \log \frac{I_1}{I_0} = \beta_2 - \beta_1.$$

Thus we have

$$\beta_2 - \beta_1 = 10 \log_{10} \frac{I_2}{I_1},$$

so the ratio of the intensities results in the difference in decibel sound levels.

(b) Find the difference in sound levels for $I_2/I_1 = 100$, 1/100, and 5.2×10^5.

If $I_2/I_1 = 100$, $\beta_2 - \beta_1 = 10 \log_{10} 100 = 20$ dB.

If $I_2/I_1 = 1/100$, $\beta_2 - \beta_1 = 10 \log_{10} 1/100 = -20$ dB.

If $I_2/I_1 = 5.2 \times 10^5$, $\beta_2 - \beta_1 = 10 \log_{10} 5.2 \times 10^5 = 57$ dB.

(c) How has the intensity increased if the sound level has risen by 10 dB?

First find the exponential form of Eq. 18–9, $I_2/I_1 = 10^{(\beta_2 - \beta_1)/10}$, so in this case

$$\frac{I_2}{I_1} = 10^1 = 10.$$

Thus if you multiply the intensity by 10 you add 10 dB to the sound level. Verify that if you multiply the intensity by two you add 3.0 dB to the sound level.

Example 3 (a) How much more intense is an 80-dB shout than a 20-dB whisper?

Use the form

$$\frac{I_2}{I_1} = 10^{(\beta_2 - \beta_1)/10}.$$

Substituting the given values yields

$$\frac{I_s}{I_w} = 10^{(80-20)/10} = 10^6!$$

Would you have thought that a shout takes that much more energy than a whisper?

(b) How much energy is required to shout continuously for 5 min assuming that the sound level is measured 2.0 m from the source, which is considered to be a point?

To get the intensity of the shouting we use the exponential form of Eq. 18–9, remembering that $I_0 = 10^{-12}$ W/m².

$$I_{\text{shout}} = I_0 10^{0.10\beta} = 10^{-12}\,\frac{\text{W}}{\text{m}^2}\,10^{80/10} = 10^{-4}\ \text{W/m}^2.$$

Since intensity is energy per unit area per unit time and since the area may be considered to be that of a sphere with the point source at its center, the energy E is

$$E = 4\pi R^2 It = 4\pi (2\text{m})^2 \left(10^{-4}\,\frac{\text{W}}{\text{m}^2}\right) (5\ \text{min})\,\frac{60\ \text{s}}{\text{min}} \cdot \frac{\text{J}}{\text{W} \cdot \text{s}} = 1.5\ \text{J}.$$

You can show that this is enough energy to lift a penny (mass ≅ 3 g) about 50 m.

18–5 Vibrating Systems and Sources of Sound

Instruments with strings

If a string fixed at both ends is bowed, transverse vibrations travel along the string; these disturbances are reflected at the fixed ends, and a standing wave pattern is formed. The natural modes of vibration of the string are excited and these vibrations give rise to longitudinal waves in the surrounding air, which transmits them to our ears as a musical sound.

We have seen (Section 17–9) that a string of length l, fixed at both ends, can resonate at frequencies given by

The allowed frequencies

$$\nu_n = \frac{n}{2l}\,v = \frac{n}{2l}\sqrt{\frac{F}{\mu}}, \qquad n = 1, 2, 3, \ldots . \tag{18–11}$$

Here v is the speed of the transverse waves in the string whose superposition can be thought of as giving rise to the vibrations; the speed $v\,(=\sqrt{F/\mu})$ is the same for all frequencies. At any one of these frequencies the string will contain a whole number n of loops between its ends, and the condition that the ends be nodes is met (Fig. 18–4).

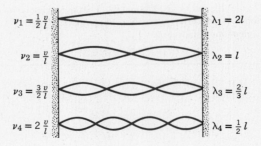

$$\nu_1 = \tfrac{1}{2}\frac{v}{l} \qquad \lambda_1 = 2l$$
$$\nu_2 = \frac{v}{l} \qquad \lambda_2 = l$$
$$\nu_3 = \tfrac{3}{2}\frac{v}{l} \qquad \lambda_3 = \tfrac{2}{3}l$$
$$\nu_4 = 2\,\frac{v}{l} \qquad \lambda_4 = \tfrac{1}{2}l$$

Figure 18–4 The first four modes of vibration of a string fixed at both ends. Note that $v_n\lambda_n = v = \sqrt{F/\mu}$.

Overtones and harmonics

The lowest frequency, $\sqrt{F/\mu}/2l$, is called the *fundamental* frequency ν_1, and the others are called *overtones*. Overtones whose frequencies are integral multiples of the fundamental are said to form a *harmonic series*. The fundamental is the first harmonic. The frequency $2\nu_1$ is the first overtone or the second harmonic, the frequency $3\nu_1$ is the second overtone or the third harmonic, and so on.

If the string is initially distorted so that its shape is the same as any one of the possible harmonics, it will vibrate at the frequency of that particular harmonic, when released. The initial conditions usually arise from striking or bowing the string, however, and in such cases not only the fundamental but many of the overtones are present in the resulting vibration. We have a superposition of several natural modes of oscillation. The actual displacement is the sum of the several harmonics with various amplitudes. The impulses that are sent through the air to the ear and brain give rise to one net effect that is characteristic of the particular stringed instrument. The quality of the sound of a particular note (fundamental frequency) played by an instrument is determined by the number of overtones present and their respective intensities. Figure 18–5 shows the sound spectra and corresponding waveforms for the violin and piano.*

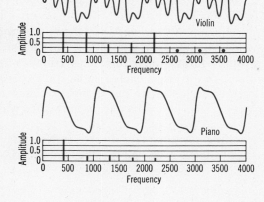

Figure 18-5 Waveform and sound spectrum for two stringed instruments, the violin and the piano. In each case the waveform is the curve which shows sound amplitude as a function of time. Below the waveform is the sound spectrum which shows the amplitude of the various harmonic components of the wave relative to the fundamental which, in both cases, is 440 Hz (concert A). These relative amplitudes are represented by vertical bars. Notice the presence of loud higher harmonics (especially the fifth) in the violin spectrum.

Standing longitudinal waves

Longitudinal waves traveling along a tube are reflected at the ends of the tube, just as transverse waves in a string are reflected at its ends. Interference between the waves traveling in opposite directions gives rise to standing longitudinal waves.

Displacement nodes and antinodes

If the end of the tube is closed by a rigid end wall, the displacement of small volume elements there must always be zero. Hence, a closed end is a displacement *node*. If the end of the tube is open, the fluid elements there are free to move. However, the nature of the reflection there depends on whether the tube is wide or narrow compared to the wavelength. If the tube is narrow compared to the wavelength, as in most musical instruments, the reflected wave has nearly the same phase as the incident wave. Then the open end is almost a displacement *antinode*. The exact antinode is usually somewhere near the opening, but the effective length of the air columns of a wind instrument, for example, is not as definite as the length of a string fixed at both ends.

Organ pipes

An organ pipe is a simple example of sound originating in a vibrating air column. If both ends of a pipe are open and a stream of air is directed against an edge, standing longitudinal waves can be set up in the tube. The air column will then resonate at its natural frequencies of vibration, given by

$$\nu_n = \frac{n}{2l} v, \qquad n = 1, 2, 3, \ldots .$$

---

* See "The Physics of the Piano," by E. Dunnell Blackham, *Scientific American*, December 1965, and "The Physics of the Violin," by Carleen M. Hutchins, *Scientific American*, November 1962.

Here v is the speed of the longitudinal waves in the column whose superposition can be thought of as giving rise to the vibrations, and n is the number of half wavelengths in the length l of the column. As with the bowed string, the fundamental and overtones are excited at the same time.

In an open pipe the fundamental frequency corresponds (approximately) to a displacement antinode at each end and a displacement node in the middle, as shown in Fig. 18–6a. The succeeding drawings of Fig. 18–6a show three of the overtones, the second, third, and fourth harmonics. Hence, in an open pipe the fundamental frequency is $v/2l$ and all harmonics are present.

In a closed pipe the closed end is a displacement node. Figure 18–6b shows the modes of vibration of a closed pipe. The fundamental frequency is $v/4l$ (approximately), which is one-half that of an open pipe of the same length. The only overtones present are those that give a displacement node at the closed end and an antinode (approximately) at the open end. Hence, as is shown in Fig. 18–6b, the second, fourth, etc., harmonics are missing. In a closed pipe the fundamental frequency is $v/4l$, and only the odd harmonics are present. The quality of the sounds from an open pipe is therefore different from that from a closed pipe.

Most musical wind instruments—trumpet, flute, oboe, etc.—are examples of open pipes. In the organ each pipe has only one tone, but in the instruments mentioned the effective length of the pipe is changed by opening and closing valves. The player selects a tone by blowing in such a way as to emphasize one mode over others, as well as by adjusting the valves for a particular pipe length.

Vibrating rods, plates, and stretched membranes also give rise to sound waves. Consider a stretched flexible membrane, such as a drumhead. If it is struck

Vibrating membranes, etc.

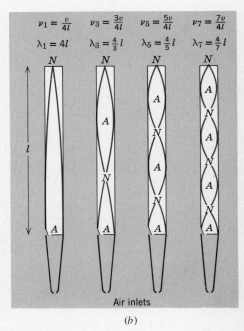

$$\nu_1 = \frac{v}{2l} \qquad \nu_2 = \frac{v}{l} \qquad \nu_3 = \frac{3v}{2l} \qquad \nu_4 = \frac{2v}{l}$$
$$\lambda_1 = 2l \qquad \lambda_2 = l \qquad \lambda_3 = \tfrac{2}{3}l \qquad \lambda_4 = \tfrac{1}{2}l$$

$$\nu_1 = \frac{v}{4l} \qquad \nu_3 = \frac{3v}{4l} \qquad \nu_5 = \frac{5v}{4l} \qquad \nu_7 = \frac{7v}{4l}$$
$$\lambda_1 = 4l \qquad \lambda_3 = \tfrac{4}{3}l \qquad \lambda_5 = \tfrac{4}{5}l \qquad \lambda_7 = \tfrac{4}{7}l$$

Air inlets (a) Air inlets (b)

Figure 18–6 (a) The first four modes of an open organ pipe. The distance from the center line of the pipe to the right lines drawn inside the pipe represent the displacement amplitude at each place. N and A mark the locations of the displacement nodes and antinodes. Note that both ends of the pipe are open. (b) The first four modes of vibration of a closed organ pipe. Notice that the evennumbered harmonics are absent and the upper end of the pipe is closed.

a blow, a two-dimensional pulse travels outward from the struck point and is reflected again and again at the boundary of the membrane. If some point of the membrane is forced to vibrate periodically, continuous trains of waves travel out along the membrane. Just as in the one-dimensional case of the string, so here too standing waves can be set up in the two-dimensional membrane. Each of these standing waves has a certain frequency natural to (or characteristic of) the membrane. Again the lowest frequency is called the fundamental and the others are overtones. Generally, a number of overtones are present along with the fundamental when the membrane is vibrating.

The nodes of a vibrating membrane are lines rather than points (as in a vibrating string) or planes (as in a pipe). Since the boundary of the membrane is fixed, it must be a nodal line. For a circular membrane fixed at its edge, possible modes of vibration along with their nodal lines are shown in Fig. 18–7. The natural frequency of each mode is given in terms of the fundamental ν_1. Notice that the frequencies of the overtones are not harmonics; that is, they are not integral multiples of ν_1. Vibrating rods also have a nonharmonic set of natural frequencies. Rods and plates have limited use as musical instruments for this reason.

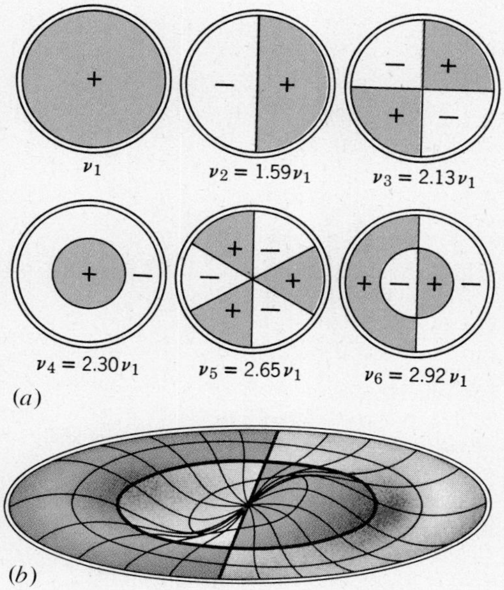

ν_1 $\nu_2 = 1.59\nu_1$ $\nu_3 = 2.13\nu_1$

$\nu_4 = 2.30\nu_1$ $\nu_5 = 2.65\nu_1$ $\nu_6 = 2.92\nu_1$

(a)

(b)

Figure 18–7 (a) The first modes of vibration of a circular drumhead clamped around its periphery. The lines represent nodes, the circumference being a node in every case. The + and − signs represent opposite displacements; at an instant when the + areas are raised, the − areas will be depressed. Note that the frequency of each mode is not an integral multiple of the fundamental ν_1 as is the case for strings and tubes. (b) A sketch of a drumhead vibrating in mode ν_6. The displacement shown here is exaggerated for clarity.

Example 4. A tuning fork with a frequency $\nu = 1080$ Hz sets up a resonance in a tube open on one end, closed on the other, and having a variable length (such as by using a tube partially filled with water). The shortest effective length l for which resonance occurs is 7.65 cm. Find the speed of sound in air.

Referring to Fig. 18–6b the relation between the wavelength λ of the sound and the effective length of the tube is

$$l = \frac{\lambda}{4}.$$

Knowing the frequency and the wavelength, we obtain the velocity as

$$v = \nu\lambda = (1080/\text{s})(4)(7.65 \text{ cm}) = 330 \text{ m/s}.$$

What series of measurements would you make to eliminate the edge effect? *Hint:* Consider measuring the distance of the water level from the actual physical edge for several different resonances using the same frequency. Then see whether or not you can infer the wavelength without using the position of the physical edge.

18–6   Beats

When two wavetrains of the same frequency travel along the same line in opposite directions, standing waves are formed in accord with the principle of superposition. We may characterize these waves by drawing a plot of the amplitude of oscillation as a function of distance, as in Fig. 18–4. This illustrates what we can call *interference in space*.

The same principle of superposition leads us to what we can call *interference in time*. It occurs when two wavetrains of slightly different frequency travel through the same region. With sound such a condition exists when, for example, two adjacent piano keys are struck simultaneously.

Consider some one point in space through which the waves are passing. In Fig. 18–8*a* we plot the displacements produced at such a point by the two waves separately as a function of time. For simplicity we have assumed that the two waves have equal amplitude, although this is not necessary. The resultant vibration at that point as a function of time is the sum of the individual vibrations and is plotted in Fig. 18–8*b*. We see that the amplitude of the resultant wave at the given point is not constant but varies with time. In the case of sound the varying amplitude gives rise to variations in loudness which are called *beats*. Two strings may be tuned to the same frequency by tightening one of them while sounding both until the beats disappear.

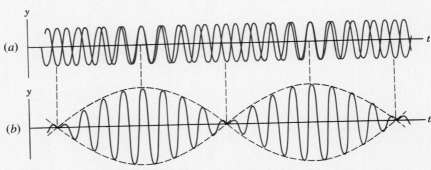

Figure 18–8 The beat phenomenon. Two waves of slightly different frequencies, shown in (*a*), combine in (*b*) to give a wave whose envelope (dashed line) varies periodically with time.

Let us represent the displacement at the point produced by one wave as

$$y_1 = y_m \cos 2\pi\nu_1 t,$$

and the displacement at the point produced by the other wave of equal amplitude as

$$y_2 = y_m \cos 2\pi\nu_2 t,$$

By the superposition principle, the resultant displacement is

$$y = y_1 + y_2 = y_m(\cos 2\pi\nu_1 t + \cos 2\pi\nu_2 t),$$

and since

$$\cos a + \cos b = 2 \cos \frac{a-b}{2} \cos \frac{a+b}{2},$$

this can be written as

$$y = \left[2y_m \cos 2\pi \left(\frac{\nu_1 - \nu_2}{2} \right) t \right] \cos 2\pi \left(\frac{\nu_1 + \nu_2}{2} \right) t. \qquad (18\text{–}12)$$

The resulting vibration may then be considered to have a frequency

$$\bar{\nu} = \frac{\nu_1 + \nu_2}{2},$$

which is the average frequency of the two waves, and an amplitude given by the expression in brackets. Hence, the amplitude itself varies with time with a frequency

$$\frac{\nu_1 - \nu_2}{2}.$$

If ν_1 and ν_2 are nearly equal, this term is small and the amplitude fluctuates slowly. This phenomenon is a form of amplitude modulation which has a counterpart (side bands) in AM radio receivers.

A beat, that is, a maximum of amplitude, will occur whenever

$$\cos 2\pi \left(\frac{\nu_1 - \nu_2}{2} \right) t$$

equals 1 or -1. Since each of these values occurs once in each cycle, the number of beats per second is twice the frequency displayed above or

The beat frequency

$$\nu_{\text{beat}} = \nu_1 - \nu_2.$$

Hence, the number of beats per second equals the difference of the frequencies of the component waves. Beats between two tones can be detected by the ear up to a frequency of about seven per second. At higher frequencies individual beats cannot be distinguished in the sound produced.

18–7   The Doppler Effect

When a listener is in motion toward a stationary source of sound, the pitch (frequency) of the sound heard is higher than when he is at rest. If the listener is in motion away from the stationary source, he hears a lower pitch than when he is at rest. We obtain similar results when the source is in motion toward or away from a stationary listener. The pitch of the whistle of the locomotive is higher when the source is approaching the hearer than when it has passed and is receding.

Christian Johann Doppler (1803–1853), an Austrian, in a paper of 1842, called attention to the fact that the color of a luminous body must be changed by relative motion of the body and the observer. This *Doppler effect,* as it is called, applies to waves in general. Doppler himself mentions the application of his principle to sound waves. An experimental test was carried out in Holland in 1845 by Buys Ballot, "using a locomotive drawing an open car with several trumpeters."

We now consider the application of the Doppler effect to sound waves, treating only the special case in which the source and observer move along the line joining them. Let us adopt a reference frame at rest in the medium through which the sound travels. Figure 18–9 shows a source of sound S at rest in this frame and an observer O (note the ear) moving toward the source at a speed v_0. The circles represent wavefronts, spaced one wavelength apart, traveling through the medium. If the observer were at rest in the medium he would receive vt/λ waves in time t, where v is the speed of sound in the medium and λ is the wavelength. Because of his motion toward the source, however, he receives $v_0 t/\lambda$ additional

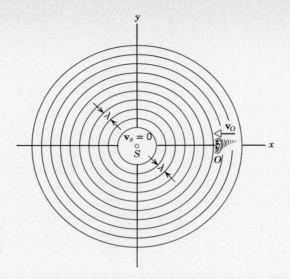

Figure 18–9 The Doppler effect due to motion of the observer (ear). The source is at rest.

waves in this same time t. The frequency ν' that he hears is the number of waves received per unit time or

$$\nu' = \frac{vt/\lambda + v_0 t/\lambda}{t} = \frac{v + v_0}{\lambda} = \frac{v + v_0}{v/\nu}.$$

That is,

$$\nu' = \nu\frac{v + v_0}{v} = \nu\left(1 + \frac{v_0}{v}\right). \tag{18–13a}$$

The frequency ν' heard by the observer is the ordinary frequency ν heard at rest plus the increase $\nu(v_0/v)$ arising from the motion of the observer. When the observer is in motion away from the stationary source, there is a decrease in frequency $\nu(v_0/v)$ corresponding to the waves that do not reach the observer each unit of time because of his receding motion. Then

$$\nu' = \nu\left(\frac{v - v_0}{v}\right) = \nu\left(1 - \frac{v_0}{v}\right). \tag{18–13b}$$

Hence, the general relation holding when the source is at rest with respect to the medium but the observer is moving through it is

Doppler equation for the source at rest

$$\nu' = \nu\left(\frac{v \pm v_0}{v}\right), \tag{18–13}$$

where the plus sign holds for motion toward the source and the minus sign holds for motion away from the source. Notice that the cause of the change here is the fact that the observer intercepts more or fewer waves each second because of his motion through the medium.

When the source is in motion toward a stationary observer, the effect is a shortening of the wavelength (see Fig. 18–10), for the source is following after the approaching waves and the crests therefore come closer together. If the frequency of the source is ν and its speed is v_s, then during each vibration it travels a distance v_s/ν and each wavelength is shortened by this amount. Hence, the wavelength of the sound arriving at the observer is not $\lambda = v/\nu$ but $\lambda' = v/\nu - v_s/\nu$. Therefore, the frequency of the sound heard by the observer is increased, being

$$\nu' = \frac{v}{\lambda'} = \frac{v}{(v - v_s)/\nu} = \nu\left(\frac{v}{v - v_s}\right). \tag{18–14a}$$

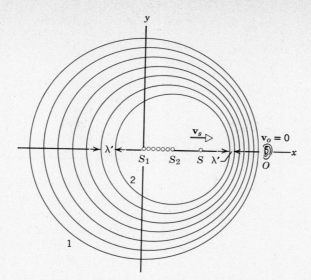

Figure 18-10 The Doppler effect due to motion of the source. The observer is at rest. Wavefront 1 was emitted by the source when it was at S_1, wavefront 2 was emitted when it was at S_2, etc. At the instant the "snapshot" was taken, the source was at S.

If the source moves away from the observer, the wavelength emitted is v_s/ν greater than λ, so that the observer hears a decreased frequency, namely,

$$\nu' = \frac{v}{(v + v_s)/\nu} = \nu \left(\frac{v}{v + v_s} \right). \tag{18-14b}$$

Hence, the general relation holding when the observer is at rest with respect to the medium but the source is moving through it is

Doppler equation for the observer at rest

$$\nu' = \nu \left(\frac{v}{v \mp v_s} \right), \tag{18-14}$$

where the minus sign holds for motion toward the observer and the plus sign holds for motion away from the observer. Notice that the cause of the change here is the fact that the motion of the source through the medium shortens or increases the wavelength transmitted through the medium.

If both source and observer move through the transmitting medium, you should be able to show that the observer hears a frequency

The general Doppler equation

$$\nu' = \nu \left(\frac{v \pm v_0}{v \mp v_s} \right). \tag{18-15}$$

There are four independent choices of sign in Eq. 18-15. You can keep them straight by associating, for both source and observer, the word "toward" with the words "frequency increase" and, of course, the converse. Thus if the observer moves toward the source (as in Fig. 18-9) you choose the plus sign in the numerator because that results in a frequency increase. Similarly, if the source moves toward the observer (as in Fig. 18-10), you choose the minus sign in the denominator because that also results in a frequency increase. Notice that Eq. 18-15 reduces to Eq. 18-13 when $v_s = 0$ and to Eq. 18-14 when $v_0 = 0$, as it must.

If a vibrating tuning fork on its resonating box is moved rapidly toward a wall, the observer will hear two notes of different frequency. One is the note heard directly from the receding fork and is lowered in pitch by the motion. The other note is due to the waves reflected from the wall, and this is raised in pitch. The superposition of these two wave trains produces beats.

Example 5 A submarine moving north with a speed of 75 km/h with respect to the ocean floor emits a sonar (sound waves in water used in ways similar to radar) signal of frequency 1000 Hz. If the ocean at that point has a current moving north with velocity 30 km/h with respect to the land, what frequency is observed by a ship north of the submarine that does not have its engines running?

All speeds in the Doppler equation must be taken with

respect to the medium, in this case water. Since the water and the submarine are both moving in the same direction, the speed of the water with respect to the ocean floor must be added to the submarine's speed with respect to the water v_{sw} to get the submarine's speed with respect to the ocean floor, $v_{sf} = v_{sw} + v_{wf}$.

Now, the source speed that is pertinent to the sound is that relative to the medium (water) or v_{sw}, which is therefore

$$v_{sw} = v_{sf} - v_{wf} = (75 - 30) \text{ km/h} = 45 \text{ km/h}.$$

The ship carrying the sonar detection equipment has a speed with respect to the water of zero since its engines are not running. The fact that the ship is being carried north with the ocean stream is of no consequence to the sound.

Now we apply the Doppler equation, Eq. 18–15, noting from Table 18–1 that the speed of sound in water is 1519 m/s (= 5470 km/h). The sign to be used in the denominator of the Doppler equation is negative because the source is moving through the medium toward the observer. The detected frequency is therefore

$$v_c = v_s \frac{v}{v - v_{sw}} = (1000 \text{ Hz}) \frac{5470 \text{ km/h}}{(5470 - 45) \text{ km/h}} = 1008 \text{ Hz}.$$

When v_o or v_s becomes comparable in magnitude to v, the formulas just given for the Doppler effect usually must be modified. One modification is required because the linear relation between restoring force and displacement assumed up until now may no longer hold in the medium. The speed of wave propagation is then no longer the normal phase velocity, and the wave shapes change in time. Components of the motion at right angles to the line joining source and observer also contribute to the Doppler effect at these high speeds. When v_o or v_s exceeds v, the Doppler formula does not apply; for example, if $v_s > v$, the source will get ahead of the wave in one direction; if $v_o > v$ and the observer moves away from the source, the wave will never catch up with the observer.

There are many instances in which the source moves through a medium at a speed greater than the speed of the wave in that medium. In such cases the wavefront takes the shape of a cone with the moving body at its apex. Some examples are the bow wave from a speedboat on the water and the "shock wave" from an airplane or projectile moving through the air at a speed greater than the velocity of sound in that medium (supersonic speeds). Cerenkov radiation consists of light waves emitted by charged particles that move through a medium with a speed greater than the speed of light in that medium.

In Fig. 18–11a we show the present positions of the spherical waves that originated at various positions of the source (assumed to be a tiny moving projectile)

Shock waves

Figure 18–11 (*a*) A representation of a group of spherical wavefronts associated with a tiny moving projectile. Note the conical envelope. (*b*) A shadow photograph of a delta winged aircraft model fired from a gun at approximately Mach 2. (Courtesy Exterior Ballistics Laboratory, Aberdeen Proving Ground, Maryland).

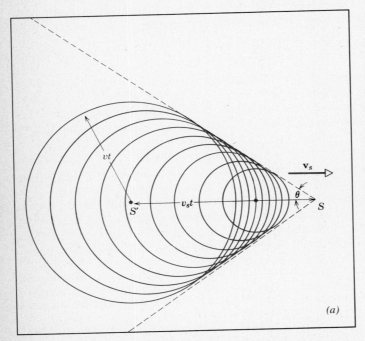

(a)

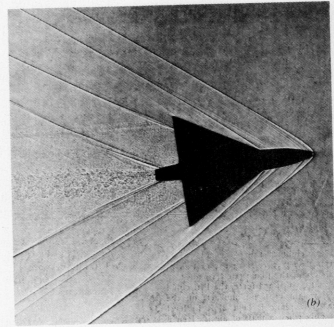

(b)

during its motion. The radius of each sphere at this time is the product of the wave speed v and the time t that has elapsed since the source was at its center. The envelope of these waves is a cone whose surface makes an angle θ with the direction of motion of the source. From the figure we obtain the result

$$\sin \theta = \frac{v}{v_s}.$$
(18–16)

For water waves the cone reduces to a pair of intersecting lines. In aerodynamics the ratio v_s/v is called the Mach number.

REVIEW GUIDE AND SUMMARY

Sound waves

Sound waves are longitudinal mechanical waves propagated in solids, liquids, or gases. The audible range is from about 20 Hz to about 20,000 Hz. Below this the waves are infrasonic, and above this they are ultrasonic. They originate in vibrating strings, air columns, plates, and membranes.

The speed v of the wave in a medium having a bulk modulus of elasticity B and density ρ_0 is

Wave speed

$$v = \sqrt{\frac{B}{\rho_0}}.$$
[18–1a]

If the medium is a gas having an undisturbed pressure p_0 and a ratio of specific heats γ, the speed becomes

Wave speed in a gas

$$v = \sqrt{\frac{\gamma p_0}{\rho_0}}.$$
[18–1b]

The equation describing a simple harmonic traveling wave is given in Chapter 17 as

Wave equation for displacement

$$y = y_m \cos (kx - \omega t),$$
[18–2]

where y_m is the maximum displacement from equilibrium, $k = 2\pi/\lambda$, and $\omega = 2\pi\nu$, λ and ν being the wavelength and frequency, respectively. In terms of the departure p of the pressure of the medium from equilibrium, the equation is

Wave equation for pressure

$$p = p_m \sin (kx - \omega t),$$
[18–5]

where the pressure amplitude is

$$p_m = k\rho_0 v^2 y_m.$$
[18–6]

Example 1 illustrates the range of pressures and displacements for sound waves in air.

The average intensity, which is the power per unit area, is found to be

Sound intensity

$$\bar{I} = \frac{1}{2}\frac{p_m^2}{v\rho_0} = \frac{1}{2}\frac{p_m^2}{\sqrt{B\rho_0}}$$
[18–8]

The range of intensities detectable by the human ear is $\sim 2 \times 10^{12}$.

The sound level β in decibels (dB) is defined to correspond to the human sensation of loudness. It is given by

Sound level in decibels

$$\beta = 10 \log_{10} \frac{I}{I_0}$$
[18–9]

I_0, usually taken as 10^{-12} W/m², is the reference intensity level to which I is compared. For every factor of 10 in intensity, you add 10 dB to the sound level. The range of the human ear is approximately 120 dB. Examples 2 and 3 illustrate calculations using intensity and sound level.

Modes of vibration

Most sources of sound can vibrate in many standing-wave modes each having a different frequency. The actual motion of the medium in a particular case is represented by the superposition of all the modes, each having a particular amplitude. The lowest fre-

Overtones and harmonics

quency is the *fundamental;* increasingly higher ones are called *overtones.* Those overtones that are integer multiples of the fundamental are called *harmonics.*

A vibrating string fixed at both ends or an organ pipe open at both ends can resonate at frequencies

Frequencies for fixed-fixed string or open-open pipe

$$\nu_n = \frac{n}{2l}\, v, \qquad n = 1, 2, 3, \ldots,$$ [18–11]

where v is the speed of the wave on the string. Example 4 illustrates a case of resonance in a pipe closed at one end and open at the other.

Beats arise when two waves having slightly different frequencies, ν_1 and ν_2, act together. The beat equation giving the particle displacement as a function of time is

The beat equation

$$y = \left[2y_m \cos 2\pi \left(\frac{\nu_1 - \nu_2}{2} \right) t \right] \cos 2\pi \left(\frac{\nu_1 + \nu_2}{2} \right) t.$$ [18–12]

The beat frequency is

The beat frequency

$$\nu_{\text{beat}} = \nu_1 - \nu_2.$$

The Doppler effect describes the change in the observed frequency when the source or the observer moves relative to the medium. The equation for the observed frequency ν' in terms of the frequency of the source ν is

Doppler equation

$$\nu' = \nu \left(\frac{v \pm v_0}{v \mp v_s} \right),$$ [18–15]

where v_0 is the speed of the observer relative to the medium, v_s is that of the source and v is the velocity of sound in the medium; the upper sign on v_s (or v_o) is used when the source (or observer) moves toward the observer (or source) while the lower sign is used when it moves away (four different cases). Example 5 illustrates the Doppler effect in water.

If the source speed relative to the medium exceeds the speed of sound in the medium, the Doppler equation no longer applies. In such a case shock waves result. The half angle θ of the wavefront is given by

Half angle of a shock wave

$$\sin \theta = \frac{v}{v_s}.$$ [18–16]

QUESTIONS

1. List some sources of infrasonic waves. Of ultransonic waves.

2. Ultrasonic waves can be used to reveal internal structures of the body. It can, for example, distingish between liquid and soft human tissues far better than can x rays. Why?

3. What experimental evidence is there for assuming that the speed of sound in air is the same for all wavelengths?

4. What is the meaning of zero decibels?

5. Could the reference intensity for audible sound be set so as to permit negative sound levels in decibels? If so, how?

6. How might you go about reducing the noise level in a machine shop?

7. Foghorns emit sounds of very low pitch. For what purpose?

8. Are longitudinal waves in air always audible as sound, regardless of frequency or intensity? What frequencies would give a person the greatest sensitivity, the greatest tolerance, and the greatest range?

9. What is the common purpose of the valves of a cornet and the slide of a trombone?

10. The bugle has no valves. How then can we sound different notes on it? To what notes is the bugler limited? Why?

11. The pitch of the wind instruments rises and that of the string instruments falls as an orchestra "warms up." Explain why.

12. Explain how a string instrument is "tuned."

13. Is resonance a desirable feature of every musical instrument?

14. When you strike one prong of a tuning fork, the other prong also vibrates, even if the bottom end of the fork is clamped firmly in a vise. How can this happen? That is, how does the second prong "get the word" that somebody has struck the first prong?

15. How can a sound wave travel down an organ pipe and be reflected at its open end? It would seem that there is nothing there to reflect it.

16. How can we experimentally locate the positions of nodes and antinodes in a string? In an air column? On a vibrating surface?

17. What physical properties of a sound wave correspond to the human sensation of pitch, of loudness, and of tone quality?

18. What is the difference between a violin note and the same note sung by a human voice that enables us to distinguish between them?

19. Does your singing really sound better in a shower? If so, are there physical reasons for this?

20. Explain the audible tone produced by drawing a wet finger around the rim of a wine glass.

21. Would a plucked violin string oscillate for a longer or shorter time if the violin had no sounding board? Explain.

22. A lightning flash dissipates an enormous amount of energy and is essentially instantaneous. How is that energy transformed into the sound waves of thunder? (See "Thunder," by Arthur A. Few, *Scientific American,* July 1975.)

23. Suppose that George blows a whistle and Gloria hears it. She will hear an increased frequency whether she is running toward George or George is running toward her. Are the increases in frequency the same in each case? Assume the same speeds of running.

24. Jenny, sitting on a bench, sees Lew, also sitting on a bench, across the campus. She blows a whistle to attract his attention. A steady wind is blowing from Jenny to Lew. How does the presence of the wind affect the frequency of the sound that Lew hears?

25. You are standing in the middle of the road and a bus is coming toward you at constant speed, with its horn sounding.

Because of the Doppler effect, is the pitch of the horn (*a*) rising, (*b*) falling, or (*c*) constant?

26. How might the Doppler effect be used in an instrument to detect the fetal heart beat? (Such measurements are routinely made; see "Ultrasound in Medical Diagnosis," by Gilbert B. Devey and Peter N. T. Wells, *Scientific American,* May 1978.)

27. Bats can examine the characteristics of objects—such as size, shape, distance, direction, motion—by sensing the way the high-frequency sounds they emit are reflected off the objects back to the bat. Discuss qualitatively each of these features. (See "Information Content of Bat Sonar Echoes," by J. A. Simmons, D. J. Howell, and N. Suga, *American Scientist,* March–April 1975.)

28. Two ships with steam whistles of the same pitch sound off in the harbor. Would you expect this to produce an interference pattern with regions of high and low intensity?

29. Suppose that, in the Doppler effect for sound, the source and receiver are at rest in some reference frame but the transmitting medium is moving with respect to this frame. Will there be a change in wavelength, or in frequency, received?

30. A satellite emits radio waves of constant frequency. These waves are picked up on the ground and made to beat against some standard frequency. The beat frequency is then sent through a loudspeaker and one "hears" the satellite signals. Describe how the sound changes as the satellite approaches, passes overhead, and recedes from the detector on the ground.

EXERCISES

Section 18–2 Propagation and Speed of Longitudinal Waves

1. The lowest pitch detectable as sound by the average human ear is about 20 Hz and the highest is about 20,000 Hz. What is the wavelength of each in air?

2. Bats emit ultrasonic waves. The shortest wavelength emitted in air by a bat is about 0.13 in. (= 3.3 mm). What is the highest frequency a bat can emit?

3. Diagnostic ultrasound of frequency 4.5 MHz is used to examine tumors in soft tissue. (*a*) What is the wavelength in air of such a sound wave? (*b*) If the speed of sound in tissue is 1500 m/s, what is the wavelength of this wave in tissue?

4. (*a*) A conical loudspeaker has a diameter of 6.0 in. At what frequency will the wavelength of the sound it emits in air be equal to its diameter? Be ten times its diameter? Be one-tenth its diameter? (*b*) Make the same calculations for a speaker of diameter 12 in. (*Note:* If the wavelength is large compared to the diameter of the speaker, the sound waves spread out almost uniformly in all directions from the speaker, but when the wavelength is small compared to the diameter of the speaker, the wave energy is propagated mostly in the forward direction.)

5. You are at a large outdoor concert, seated 300 m from the stage microphone. The concert is also being broadcast live, in stereo, around the world via satellite. Consider a listener 5000 km away. Who hears the music first and by what factor in time?

6. Figure 18–12 shows a remarkably detailed image, of a transistor in a microelectronic circuit, formed by an acoustic microscope. The sound waves have a frequency of 4.2 GHz. The speed

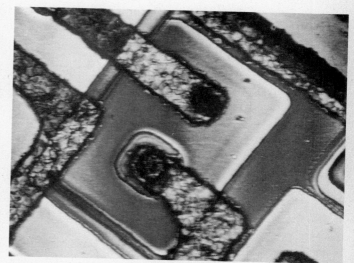

Figure 18–12 Exercise 6. Photo by C. F. Quate, courtesy John S. Foster, Stanford.

of such waves in the liquid helium in which the specimen is immersed is 240 m/s. (*a*) What is the wavelength of these ultrahigh frequency acoustic waves? (*b*) The ribbon-like conductors in the figure are ~2 μm wide. How many wavelengths does this correspond to?

Section 18-3 Traveling Longitudinal Waves

7. The pressure in a traveling sound wave is given by the equation

$$p = (1.5 \text{ Pa}) \sin \pi[(1.0 \text{ m}^{-1})x - (330 \text{ s}^{-1})t].$$

Find (a) the pressure amplitude, (b) the frequency, (c) the wavelength, and (d) the speed of the wave.

8. A note of frequency 300 Hz has an intensity of $1.0 \ \mu\text{W/m}^2$. What is the amplitude of the air vibrations caused by this sound?

Section 18-4 Sound Intensity

9. What is the intensity ratio of two sounds whose sound levels differ by 1.0 dB?

10. A certain sound level is increased by an additional 30 dB. Show that (a) its intensity increases by a factor of 1000 and (b) its pressure amplitude increases by a factor of 32.

11. A salesperson claimed that a stereo system would deliver 120 W of audio power. Testing the system with several speakers set up so as to simulate a point source, the consumer noticed that she could get as close as 1.2 m with the volume full on before the sound hurt her ears. Should she report the firm to the Consumer Protection Agency?

Section 18-5 Vibrating Systems and Sources of Sound

12. (a) Find the speed of waves on a 0.8-g violin string 22 cm long if the frequency of the fundamental is 920 Hz. (b) What is the tension in the string?

13. If a violin string is tuned to a certain note, by how much must the tension in the string be increased if it is to emit a note of double the original frequency (that is, a note one octave higher in pitch)?

14. A 3.0-m skip rope is used essentially in its fundamental mode of oscillation. If the rope has a mass of 1.0 kg and the children are pulling back with a force of 10 N, what is the frequency of oscillation?

15. A certain violin string is 50 cm long between its fixed ends and has a mass of 2.0 g. The string sounds an A note (440 Hz) when played without fingering. Where must one put one's finger to play a C (528 Hz)?

16. A sound wave in a fluid medium is reflected at a barrier so that a standing wave is formed. The distance between nodes is 3.8 cm and the speed of propagation is 1500 m/s. Find the frequency.

17. An organ is being designed for use in your living room. The organ pipes must be less than 1 m in length to keep the organ at a reasonable size. Would you use a pipe open at both ends, or a pipe open at just one end, to produce a note of 110 Hz when the pipe resonates at its fundamental frequency? Why? (Assume that the speed of sound in air is 343 m/s.)

Section 18-6 Beats

18. A tuning fork of unknown frequency makes three beats per second with a standard fork of frequency 384 Hz. The beat frequency decreases when a small piece of wax is put on a prong of the first fork. What is the frequency of this fork?

19. The A string of a violin is a little too taut. Four beats per second are heard when it is sounded together with a tuning fork that is vibrating accurately at the "concert pitch" of A (440 Hz). What is the period of the violin string vibration?

20. You are given four tuning forks. The fork with the lowest frequency vibrates at 500 Hz. By using two tuning forks at a time, the following beat frequencies are heard: 1, 2, 3, 5, 7, and 8 Hz. What are the possible frequencies of the other three tuning forks?

Section 18-7 The Doppler Effect

21. A whistle used to call a dog has a frequency of 30 kHz. The dog, however, ignores it. The owner of the dog, who cannot hear sounds above 20 kHz, wants to use the Doppler effect to make certain that the whistle is working. He asks a friend to blow the whistle from a moving car while the owner remains stationary and listens. (a) How fast must the car move, and in what direction? Is the experiment practical? (b) Repeat if the whistle frequency is 22 kHz instead of 30 kHz.

22. Trooper B is chasing speeder A along a straight stretch of road. Both are moving at a top speed of about 100 mi/h, which is about 150 ft/s. Trooper B, failing to catch up, sounds his siren again. Take the speed of sound in air to be 1100 ft/s and the frequency of the source to be 500 Hz. What, if any, will be the Doppler shift in the frequency heard by speeder A?

23. (a) Could you go through a red light fast enough to have it appear green? (b) If so, would you get a ticket for speeding? Take $\lambda = 620$ nm (= 620 nanometer = 620×10^{-9} m; see Table 1-2) for red light, $\lambda = 540$ nm for green light, and $c = 3.0 \times 10^8$ m/s as the speed of light.

24. The 16,000-Hz whine of the turbines in the jet engines of an aircraft moving with speed 200 m/s is heard at what frequency by the pilot of a second craft trying to overtake the first at a speed of 250 m/s?

25. A 2000-Hz siren and a civil defense official are both at rest with respect to the earth. (a) What frequency does the official hear if the wind is blowing at 12 m/s from source to observer? (b) From observer to source?

26. A bullet is fired with a speed of 2200 ft/s. Find the angle made by the shock wave with the line of motion of the bullet.

27. The speed of light in water is about three-fourths the speed of light in vacuum. A beam of high-speed electrons from a betatron emits Cerenkov radiation in water, the wavefront being a cone of half angle 60°. Find the speed of the electrons in the water.

PROBLEMS GROUPED BY SECTION

Section 18-2 Propagation and Speed of Longitudinal Waves

1. A rule for finding your distance from a lightning flash is to count seconds from the time you see the flash until you hear the thunder and then divide the count by 5. The result is supposed to give the distance in miles. Explain this rule and determine the percent error in it at standard conditions.

2. Two spectators at a soccer game in a large stadium see, and a moment later hear, the ball being kicked on the playing field. The time delay for one spectator is 0.90 s and for the other 0.60 s. The lines through each spectator and the player kicking the ball meet at an angle of 90°, (a) how far is each spectator from the player? (b) How far are the spectators from each other?

3. Earthquakes generate sound waves in the earth. Unlike in a gas, there are both transverse (S) and longitudinal (P) sound waves in a solid. Typically, the speed of S-waves is about 4.5 km/s and that of P-waves 8.0 km/s. A seismograph records P- and S-waves from an earthquake. The first P-waves arrive 3 min before the first S-waves; see Fig. 18–13. How far away did the earthquake occur?

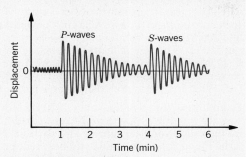

Figure 18–13 Problem 3.

4. A stone is dropped into a well. The sound of the splash is heard 3.0 s later. What is the depth of the well?

Section 18–3 Traveling Longitudinal Waves

5. Two waves give rise to pressure variations at a certain point in space given by

$$p_1 = p_m \sin \omega t$$
$$p_2 = p_m \sin (\omega t - \phi).$$

What is the pressure amplitude of the resultant wave at this point when $\phi = 0$, $\phi = \pi/2$, $\phi = \pi/3$, $\phi = \pi/4$? All ϕ's are measured in radians.

6. A certain loudspeaker produces a sound with a frequency of 2000 Hz and an intensity of 9.6×10^{-4} W/m² at a distance of 6.1 m. Presume that there are no reflections and that the loudspeaker emits the same in all directions. (a) What would be the intensity at 30 m? (b) What is the displacement amplitude at 6.1 m? (c) What is the pressure amplitude at 6.1 m?

7. Two stereo loudspeakers are separated by a distance of 6.0 ft. Assume that the amplitude of the sound from each speaker is approximately the same at the position of a listener who is 10 ft directly in front of one of the speakers. (a) For what frequencies in the audible range (20–20,000 Hz) will there be a minimum signal? (b) For what frequencies is the sound a maximum?

8. In Fig. 18–14 we show an acoustic interferometer, used to demonstrate the interference of sound waves. S is a diaphragm that vibrates under the influence of an electromagnet. D is a sound detector, such as the ear or a microphone. Path SBD can be varied in length, but path SAD is fixed. The interferometer contains air, and it is found that the sound intensity has a minimum value of 100 units at one position of B and continuously climbs to a maxi-

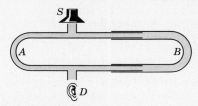

Figure 18–14 Problem 8.

mum value of 900 units at a second position 1.65 cm from the first. Find (a) the frequency of the sound emitted from the source, and (b) the relative amplitudes of the waves arriving at the detector for each of the two positions of B. (c) How can it happen that these waves have different amplitudes, considering that they originate at the same source?

9. A sound wave of 40-cm wavelength enters the tube shown in Fig. 18–15. What must be the smallest radius r such that a minimum will be heard at the detector?

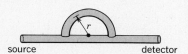

Figure 18–15 Problem 9.

Section 18–4 Sound Intensity

10. The source of a sound wave delivers $1.0 \, \mu$W of power. If it is a point source, (a) what is the intensity 3.0 m away? (b) What is the sound level in decibels?

11. Find the ratios of the intensities, the pressure amplitudes, and the particle-displacement amplitudes for two sounds whose intensity levels differ by 37 dB.

12. A certain loudspeaker (assumed to be a point source) emits 30 W of sound power. A small microphone of effective cross-sectional area 0.750 cm² is located 200 m from the loudspeaker. Calculate (a) the sound intensity at the microphone, and (b) the power incident on the microphone.

Section 18–5 Vibrating Systems and Sources of Sound

13. An open organ pipe has a fundamental frequency of 300 Hz. The first overtone of a closed organ pipe has the same frequency as the first overtone of the open pipe. How long is each pipe?

14. The strings of a cello have a length L. (a) By what length l must they be shortened by fingering to change the pitch by a frequency ratio r? (b) Find l, if $L = 0.80$ m and $r = 6/5, 5/4, 4/3$, and 3/2.

15. A tuning fork placed over an open vertical tube partly filled with water causes strong resonances when the water surface is 8 cm and 28 cm from the top of the tube and for no other positions. The speed of sound in the air in the room is 330 m/s. What is the frequency of the tuning fork?

16. In Fig. 18–16 a rod R is clamped at its center; a disk D at its end projects into a glass tube that has cork filings spread over its interior. A plunger P is provided at the other end of the tube. The rod is set into longitudinal vibration and the plunger is moved until the filings form a pattern of nodes and antinodes (the filings form well-defined ridges at the pressure antinodes). If we know the frequency v of the longitudinal vibrations in the rod, a measurement of the average distance d between successive antinodes determines the speed of sound v in the gas in the tube. Show that

$$v = 2vd.$$

This is Kundt's method for determining the speed of sound in various gases.

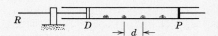

Figure 18–16 Problem 16.

17. *S* in Fig. 18–17 is a small loudspeaker driven by an audio oscillator and amplifier, adjustable in frequency from 1000 to 2000 Hz only. *D* is a piece of cylindrical sheetmetal pipe 18.0 in. long and open at both ends. (*a*) If the speed of sound in air is 1130 ft/s at the existing temperature, at what frequencies will resonance occur when the frequency emitted by the speaker is varied from 1000 to 2000 Hz? (*b*) Sketch the displacement nodes for each. Neglect end effects.

Figure 18–17 Problem 17.

Section 18–6 Beats
18. Two identical piano wires have a fundamental frequency of 600 Hz when kept under the same tension. What fractional increase in the tension of one wire will lead to the occurrence of 6 beats/s when both wires vibrate simultaneously?

19. You are given five tuning forks, each of which has a different frequency. By trying every pair of tuning forks, (*a*) what maximum number of *different* beat frequencies might be obtained? (*b*) What minimum number of *different* beat frequencies might be obtained?

Section 18–7 The Doppler Effect
20. In 1845 Buys Ballot first tested the Doppler effect for sound. He put a trumpet player on a flatcar drawn by a locomotive and another player near the tracks. If each player blows a 440-Hz note, and if there are 4.0 beats/s as they approach each other, what is the speed of the flatcar?

21. A man is pedaling his bike at 8.0 ft/s. An ambulance emitting a whine at 1600 Hz overtakes and passes him. After being passed, the man hears a frequency of 1590 Hz. How fast is the ambulance moving?

22. Two trains are traveling toward each other at 100 ft/s relative to the ground. One train is blowing a whistle at 500 Hz. (*a*) What frequency will be heard on the other train in still air? (*b*) What frequency will be heard on the other train if the wind is blowing at 100 ft/s toward the whistle and away from the listener? (*c*) What frequency will be heard if the wind direction is reversed?

23. A girl is sitting near the open window of a train that is moving at a velocity of 10 m/s to the east. The girl's uncle stands near the tracks and watches the train move away. The locomotive whistle vibrates at 500 Hz. The air is still. (*a*) What frequency does the uncle hear? (*b*) What frequency does the girl hear? A wind begins to blow from the east at 10 m/s. (*c*) What frequency does the uncle now hear? (*d*) What frequency does the girl now hear?

24. A siren emitting a sound of frequency 1000 Hz moves away from you toward a cliff at a speed of 10 m/s. (*a*) What is the frequency of the sound you hear coming directly from the siren? (*b*) What is the frequency of the sound you hear reflected off the cliff? (*c*) Could you hear the beat frequency? Take the speed of sound in air as 330 m/s.

25. A person on a railroad car blows a trumpet sounding at 440 Hz. The car is moving toward a wall at 20 m/s. Taking the speed of sound as 343 m/s, calculate (*a*) the frequency of the sound as received at the wall; (*b*) the frequency of the reflected sound arriving back at the source.

26. Two identical tuning forks can oscillate at 440 Hz. A person is located somewhere on the line between them. Calculate the beat frequency as measured by this individual if (*a*) she is standing still and the tuning forks both move to the right, say, at 30 m/s; (*b*) if the tuning forks are stationary and the listener moves to the right at 30 m/s. Take the speed of sound as 343 m/s.

27. A plane flies with $\frac{3}{4}$ the speed of sound. The sonic boom reaches a man on the ground exactly 1 min after the plane passed directly overhead. What is the altitude of the plane? Assume the speed of sound to be 330 m/s.

28. A jet plane passes overhead at a height of 5000 m and a speed of Mach 1.5 (1.5 times the speed of sound). (*a*) Find the angle made by the shock wave with the line of motion of the jet. (*b*) How long after the jet has passed directly overhead will the shock wave reach the ground?

ADDITIONAL PROBLEMS

29. The speed of sound in a certain metal is *V*. One end of a long pipe of that metal of length *l* is struck a hard blow. A listener at the other end hears two sounds, one from the wave that has traveled along the pipe and the other from the wave that has traveled through the air. (*a*) If *v* is the speed of sound in air, what time interval *t* elapses between the arrival of two sounds? (*b*) Suppose that *t* = 1.0 s and the metal is copper. Find the length *l*.

30. A man strikes a long steel girder at one end. Another man, at the other end with his ear close to the girder, hears the sound of the blow twice, with a 0.12-s interval between. How long is the girder?

31. An experimenter wishes to measure the speed of sound in an aluminum rod 10 cm long by measuring the time it takes for a sound pulse to travel the length of the rod. If results good to four significant figures are desired, how precisely must the length of the bar be known, and how closely must he be able to resolve time intervals?

32. (*a*) If two sound waves, one in air and one in water, are equal in intensity, what is the ratio of the pressure amplitude of the wave in water to that of the wave in air? (*b*) If the pressure amplitudes are equal instead, what is the ratio of the intensities of the waves?

33. Two loudspeakers are located 11 ft apart on the stage of an auditorium. A listener is seated 60 ft from one and 64 ft from the other. A signal generator drives the two speakers in phase with the same amplitude and frequency. The frequency is swept through the audio range. (*a*) What are the three lowest frequencies for which the listener will hear minimum intensity because of destructive interference? (*b*) What are the three lowest frequencies for which he will hear maximum intensity?

34. At 1.0 km a 100-Hz horn, assumed to be a point source, is barely audible. At what distance would it begin to cause pain?

35. A sound wave of frequency 1000 Hz propagating through air has a pressure amplitude of 10 Pa. What are the (a) wavelength, (b) particle displacement amplitude, and (c) maximum particle speed? (d) An open organ pipe has this frequency as a fundamental. How long is it?

36. What is the value for the bulk modulus of oxygen at standard temperature and pressure if one mole of oxygen (32 grams) occupies 22.4 liters under these conditions and the speed of sound in oxygen is 317 m/s?

37. The water level in a vertical glass tube 1.0 m long can be adjusted to any position in the tube. A tuning fork vibrating at 660 Hz is held just over the open top end of the tube. At what positions of the water level will there be resonance?

38. A tube 1.0 m long is closed at one end. A stretched wire is placed near the open end. The wire is 0.30 m long and has a mass of 0.010 kg. It is fixed at both ends and vibrates in its fundamental mode. It sets the air column in the tube into vibration at its fundamental frequency by resonance. Find (a) the frequency of oscillation of the air column and (b) the tension in the wire.

39. The period of a pulsating variable star may be estimated by considering the star to be executing radial longitudinal pulsations in the fundamental standing wave mode; that is, the radius varies periodically with the time, with a displacement antinode at the surface. (a) Would you expect the center of the star to be a displacement node or antinode? (b) By analogy with the open organ pipe, show that the period of pulsation T is given by

$$T = \frac{4R}{v_s},$$

where R is the equilibrium radius of the star and v_s is the average sound speed. (c) Typical white dwarf stars have pressures of 10^{22} Pa, densities of 10^{10} kg/m³, ratios of specific heats of 4/3, and radii 0.009 solar radii. What is the approximate pulsation period of a white dwarf? (See "Pulsating Stars," by John R. Percy, *Scientific American,* June 1975.)

40. Two "hunter-killer" submarines are on a collision course during maneuvers in the North Atlantic. The first sub is moving at 20 km/h and the second sub at 95 km/h. The first submarine sends out a sonar signal (sound wave in water) at 1000 Hz. Sonar waves travel at 5470 km/h. (a) The second sub picks up the signal. What frequency does the second sonar detector hear? (b) The first sub picks up the reflected signal. What frequency does the first sonar detector hear? See Fig. 18–18. The ocean is calm; assume no currents.

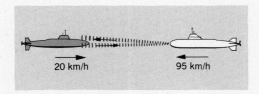

Figure 18–18 Problem 40.

41. Two sources of sound are separated by a distance of 5.0 m. They both emit sound at the same amplitude and frequency, 300 Hz, but they are 180° out of phase. At what points along the line between them will the sound intensity be the largest?

42. A hi-fi engineer has designed a speaker that is spherical in shape and emits sound isotropically (the same intensity in all

directions). The speaker emits 10 W of acoustic power into a room with completely absorbent walls, floor, and ceiling. (a) What is the intensity (W/m²) of the sound waves at 3.0 m from the center of the source? (b) How would the amplitude of the waves at 4.0 m compare with that 3.0 m from the center of the source? (*Hint:* see Example 3, Chapter 17.)

43. A whistle of frequency 540 Hz rotates in a circle of radius 2.0 ft at an angular speed of 15 rad/s. What is (a) the lowest and (b) the highest frequency heard by a listener a long distance away at rest with respect to the center of the circle?

44. A source of sound waves of frequency 1200 Hz moves to the right with a speed of 98 ft/s relative to the ground. To its right is a reflecting surface moving to the left with a speed of 216 ft/s relative to the ground. Take the speed of sound in air to be 1080 ft/s and find (a) the wavelength of the sound emitted in air by the source, (b) the number of waves per second arriving at the reflecting surface, (c) the speed of the reflected waves, and (d) the wavelength of the reflected waves.

45. Radar measurements in moving situations are relatively inaccurate compared to rest situations. (a) Consider a radar unit at rest and show that the difference Δv between the frequency reflected off a car moving at a speed V and the transmitted frequency v is given approximately by $\Delta v/v = 2V/c$. (b) Now consider the radar unit to be in a pursuing vehicle moving at a speed v and show that $\Delta v/v = 2(v - V)/c$. Discuss various cases and justify the first sentence; in particular, consider (c) the case in which the pursuer (police) is moving at the same speed as the speeder. What Doppler shift is observed in this case?

46. Estimate the speed of the projectile illustrated in the photograph in Fig. 18–11. Assume the speed of sound in the medium through which the projectile is traveling to be 380 m/s.

47. A 31.6-cm violin string with linear density 0.65 g/m is placed near a loudspeaker that is fed by an audio-oscillator of variable frequency. It is found that the string is set in oscillation only at the frequencies 880 and 1320 Hz as the frequency of the oscillator is varied continuously over the range 500 to 1500 Hz. What is the tension in the string?

48. In a discussion of Doppler shifts of ultrasonic waves used in medical diagnosis (see Question 26) the authors remark: "For every millimeter per second that a structure in the body moves, the frequency of the incident ultrasonic wave is shifted approximately 1.3 Hz/MHz." What speed of the ultrasonic waves in tissue do you deduce from this statement? (*Note:* When a wave is reflected from a moving boundary, it is as if the effective source of the wave is moving at twice the speed of the boundary; see Section 39–2.)

49. In a test, a jet fighter flies overhead at an altitude of 100 m. The sound intensity on the ground as the jet passes overhead is 150 dB. At what altitude should the plane fly so that the ground noise is no greater than 120 dB, the threshold of pain? Ignore the finite time required for the sound to reach the ground.

50. A spherical sound source is placed at P_1 near a reflecting wall AB and a microphone is located at point P_2, as shown in Fig. 18–19. The frequency of the sound source P_1 is variable. Find two different frequencies for which the sound intensity, as observed at P_2, will be a maximum. The speed of sound in air is 343 m/s. Assume the paths of the interfering waves to be parallel.

51. A bat is flittering about in a cave, navigating very effectively by the use of ultrasonic bleeps (short emissions lasting a millisec-

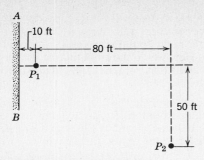

Figure 18–19 Problem 50.

ond or less and repeated several times a second). Assume that the sound emission frequency of the bat is 39,000 Hz. During one fast swoop directly toward a flat wall surface, the bat is moving at 1/40 of the speed of sound in air. What frequency does he hear reflected off the wall?

52. Microwaves, which travel with the speed of light, are reflected from a distant airplane approaching the wave source. It is found that when the reflected waves are beat against the waves radiating from the source the beat frequency is 990 Hz. If the microwaves are 0.10 m in wavelength, what is the approach speed of the airplane?

53. Two sound waves, from two different sources with the same frequency, 540 Hz, travel at a speed of 330 m/s. What is the phase difference of the waves at a point that is 4.40 m from one source and 4.00 m from the other if the sources are in phase?

54. A well with vertical sides and water at the bottom resonates at 7.0 Hz and at no lower frequency. The air in the well has a density of 1.1 kg/m^3, a pressure of 9.5×10^4 Pa, and a ratio of specific heats of 7/5. How deep is the well?

*55. Two loudspeakers, S_1 and S_2, are 7.0 m apart. Each emits sound of frequency 200 Hz uniformly in all directions. S_1 has an acoustic output of 1.2×10^{-3} W and S_2 one of 1.8×10^{-3} W. S_1 and S_2 vibrate in phase. Consider a point P that is 4.0 m from S_1 and 3.0 m from S_2. (*a*) How are the phases of the two waves arriving at P related? (*b*) What is the intensity of sound at P with both S_1 and S_2 on? (*c*) What is the intensity of sound at P if S_1 is turned off (S_2 on)? (*d*) What is the intensity of sound at P if S_2 is turned off (S_1 on)?

*56. A large parabolic reflector having a circular opening of radius 0.50 m is used to focus sound. If the energy is delivered from the focus to the ear of a listening detective through a tube of diameter 1.0 cm with 12% efficiency, how far away can he understand a whispered conversation? (Assume that the intensity level of a whisper is 20 dB at 1.0 m from the source, considered to be a point, and that the threshold for hearing is 0 dB.)

19 Temperature

19-1 Macroscopic and Microscopic Descriptions

In analyzing physical situations we usually focus our attention on some portion of matter that we separate, in our minds, from the environment external to it. We call such a portion the *system*. Everything outside the system that has a direct bearing on its behavior we call the *environment*. We then seek to determine the behavior of the system by finding how it interacts with its environment. For example, a ball can be the system and the environment can be the air and the earth. In free fall we seek to find how the air and the earth affect the motion of the ball. Or the gas in a container can be the system, and a movable piston and a Bunsen burner can be the environment. We seek to find how the behavior of the gas is affected by the action of the piston and burner. In all such cases we must choose suitable observable quantities to describe the behavior of the system. We classify these quantities, which are bulk properties of the system, as *macroscopic*. For processes in which heat is involved, the laws relating the appropriate macroscopic quantities (which include pressure, volume, temperature, internal energy, and entropy, among others) form the basis for the science of *thermodynamics*. Many of the macroscopic quantities (pressure, volume, and temperature, for example) are directly associated with our sense perceptions. We can also adopt a *microscopic* point of view. Here we consider quantities that describe the atoms and molecules that make up the system, their speeds, energies, masses, angular momenta, behavior during collisions, and so on. These quantities, or mathematical formulations based on them, form the basis for the science of *statistical mechanics*. The microscopic properties are not directly associated with our sense perceptions.

For any system the macroscopic and the microscopic quantities must be related because they are simply different ways of describing the same situation. In particular, we should be able to express the former in terms of the latter. The pressure of a gas, viewed macroscopically, may be measured using a manometer (Fig. 16–8). Viewed microscopically it is related to the average rate per unit area at which the molecules of the gas deliver momentum to the manometer fluid as they strike its surface. In Section 21–4 we shall make this microscopic definition of pressure quantitative. Similarly (see Section 21–5), the temperature of a gas may be related to the average kinetic energy of translation of the molecules.

If the macroscopic quantities can be expressed in terms of the microscopic quantities, we should be able to express the laws of thermodynamics quantitatively in the language of statistical mechanics. This has, in fact, been done. In the words of R. C. Tolman: "The explanation of the complete science of thermodynamics in terms of the more abstract science of statistical mechanics is one of the greatest achievements of physics. In addition, the more fundamental character of statistical mechanical considerations makes it possible to supplement the ordinary principles of thermodynamics to an important extent."

We begin our examination of heat phenomena in this chapter with a study of temperature. As we progress we shall try to gain a deeper understanding of these phenomena by interweaving the microscopic and the macroscopic description—statistical mechanics and thermodynamics. The interweaving of the microscopic and the macroscopic points of view is characteristic of present-day physics.

19–2   Thermal Equilibrium—
The Zeroth Law of Thermodynamics

The sense of touch is the simplest way to distinguish hot bodies from cold bodies. By touch we can arrange bodies in the order of their hotness, deciding that A is hotter than B, B than C, and so on. We speak of this as our *temperature* sense. This is a very subjective procedure for determining the temperature of a body and certainly not very useful for purposes of science. A simple experiment, suggested in 1690 by John Locke, shows the unreliability of this method. Let a person immerse his hands, one in hot water, the other in cold. Then let him put both hands in water of intermediate hotness. This will seem cooler to the first hand and warmer to the second hand. Our judgment of temperature can be rather misleading. Further, the range of our temperature sense is limited. What we need is an objective, numerical, measure of temperature.

To begin with, we should try to understand the meaning of temperature. Let an object A which feels cold to the hand and an identical object B which feels hot be placed in contact with each other. After a sufficient length of time, A and B give rise to the same temperature sensation. Then A and B are said to be in *thermal equilibrium* with each other. We can generalize the expression "two bodies are in thermal equilibrium" to mean that the two bodies are in states such that, if the two were connected, their condition would not change. The logical and operational test for thermal equilibrium is to use a third or test body, such as a thermometer. This is summarized in a postulate often called *the zeroth law of thermodynamics: If A and B are each in thermal equilibrium with a third body C (the "thermometer"), then A and B are in thermal equilibrium with each other.*

This discussion expresses the idea that the temperature of a system is a property that eventually attains the same value as that of other systems when all these systems are put in contact. This concept agrees with the everyday idea of temperature as the measure of the hotness or coldness of a system, because as far as our temperature sense can be trusted, the hotness of all objects becomes the same after they have been in contact long enough. The idea contained in the zeroth law,

Zeroth law of thermodynamics

although simple, is not obvious. For example, Jones and Smith each know Green, but they may or may not know each other. Two pieces of iron attract a magnet, but they may or may not attract each other.

19–3 Measuring Temperature

There are many measurable physical properties that vary as our perception of temperature varies. Among these are the volume of a liquid, the length of a rod, the electrical resistance of a wire, the pressure of a gas kept at constant volume, the volume of a gas kept at constant pressure, and the color of a lamp filament. Any of these properties can be used in the construction of a thermometer—that is, in the setting up of a particular "private" temperature scale. Such a temperature scale is established by choosing a particular thermometric substance and a particular thermometric property of this substance. We then define the temperature scale by an assumed continuous monotonic relation between the chosen thermometric property of our substance and the temperature as measured on our (private) scale. For example, the thermometric substance may be a liquid in a glass capillary tube and the thermometric property can be the length of the liquid column; or the thermometric substance may be a gas kept in a container at constant volume and the thermometric property can be the pressure of the gas; and so forth. We must realize that each choice of thermometric substance and property—along with the assumed relation between property and temperature—leads to an individual temperature scale whose measurements need not necessarily agree with measurements made on any other independently defined temperature scale unless some calibration is agreed upon.

This apparent chaos in the definition of temperature is removed by universal agreement, within the scientific community, on the use of a particular thermometric substance, a particular thermometric property, and a particular functional relation between measurements of that property and a universally accepted temperature scale. A private temperature scale defined in any other way can then always be calibrated against the universal scale. We describe such a universal scale in Section 19–4.

The *constant-volume gas thermometer* illustrates the techniques by which an arbitrary temperature scale might be defined. If the volume of a gas is kept constant, its pressure depends on the temperature and increases steadily with rising temperature. The constant-volume gas thermometer uses the pressure at constant volume as the thermometric property.

The thermometer is shown diagrammatically in Fig. 19–1. It consists of a bulb of glass, porcelain, quartz, platinum, or platinum-iridium (depending on the temperature range over which it is to be used), connected by a capillary tube to a mercury manometer. The bulb containing some gas is put into the bath or environment whose temperature is to be measured; by raising or lowering the mercury reservoir the mercury in the left branch of the U-tube can be made to coincide with a fixed reference mark, thus keeping the confined gas at a constant volume. Then we read the height of the mercury in the right branch. The pressure of the confined gas is the difference of the heights of the mercury columns (times ρg) plus the atmospheric pressure, as indicated by the barometer. In practice the apparatus is very elaborate, and we must make many corrections, for example, (1) to allow for the small volume change owing to slight contraction or expansion of the bulb and (2) to allow for the fact that not all the confined gas (such as that in the capillary) has been immersed in the bath. Assume that all corrections have been made, and let p be the corrected value of the pressure of the confined gas at

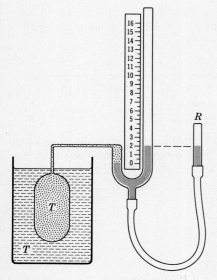

Figure 19–1 A representation of a constant-volume gas thermometer. As long as the mercury in the left manometer tube remains at the same position on the scale (zero), the volume of the confined gas will be constant. The meniscus can always be brought to the zero position by raising or lowering reservoir R.

the temperature of the bath. The temperature may be defined most simply to be linearly related to that pressure. Algebraically we write

$$T = ap, \qquad (19\text{--}1)$$

where a is a constant that may be defined arbitrarily.

The fixed point of the Kelvin temperature scale.

We next specify a *standard fixed point* at which all thermometers must give the same reading for temperature. This fixed point is chosen to be that at which ice, liquid water, and water vapor coexist in equilibrium and is called the *triple point of water*. This state can be achieved only at a water vapor pressure of 4.58 mm Hg and is unique. The temperature at this standard fixed point is chosen to be 273.16 kelvin* and is abbreviated as 273.16 K. The kelvin is a unit temperature interval.

Our constant-volume gas thermometer may now be calibrated (that is, the value of a may be determined) by inserting it into a bath of water maintained at the triple point and by measuring the pressure p_{tr} of the confined gas at the triple point temperature $T_{tr} = 273.16$ K. Inserting this value into Eq. 19–1 we obtain

$$T_{tr} = ap_{tr}$$
$$a = \frac{T_{tr}}{p_{tr}}$$
$$= 273.16 \text{ K}/p_{tr}.$$

This value of a is then inserted into Eq. 19–1. The temperature of the original bath, in which the thermometer gas pressure is p, is then provisionally (see below) defined to be

$$T_p = \left(\frac{p}{p_{tr}}\right) \times 273.16 \text{ K} \qquad (\text{constant } V). \qquad (19\text{--}2)$$

The constant-volume thermometer, used as described below, is the thermometer that serves to establish the temperature scale used universally in scientific work today.

Example 1   Suppose that we need to measure the temperature of a particular liquid inside a pressurized container maintained at a pressure of 3.00 atm. The bulb of a constant-volume gas thermometer is placed in the liquid where the gas pressure is measured to be 175 mm Hg. The thermometer bulb is then placed in a water bath which is maintained at the triple point. The thermometer gas pressure in this bath is 225 mm Hg. What is the indicated temperature of the high-pressure bath?

The indicated temperature is calculated from Eq. 19–2.

$$T = \frac{p}{p_{tr}} (273 \text{ K}) = \frac{(175 \text{ mm Hg})}{(225 \text{ mm Hg})} (273 \text{ K}) = 212 \text{ K}.$$

Notice that the pressures inside the thermometer bulb are used in this calculation, not the pressure of the liquid whose temperature is being measured.

19–4   Ideal Gas Temperature Scale

Let a certain amount of gas be put into the bulb of a constant-volume gas thermometer so that the bulb is surrounded by water at the triple point. The pressure p_{tr} is equal to a definite value, say 80 cm Hg. Now surround the bulb with steam condensing at 1 atm pressure and, with the volume kept constant at its previous value, measure the gas pressure p_s, the pressure at the steam point, in this case, p_{s80}. Then calculate the temperature provisionally from $T_{80} = (p_{s80}/80 \text{ cm Hg}) \times 273.16$ K. Next remove some of the gas so that p_{tr} has a smaller value, say 40 cm

---

* Adopted in 1967 at the Thirteenth General Conference on Weights and Measures in Paris.

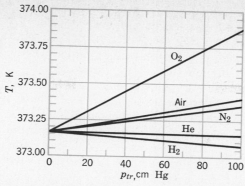

Figure 19–2 The readings of a constant-volume gas thermometer for the temperature T of condensing steam as a function of p_{tr}, when different gases are used. As the amount of gas in the thermometer is reduced its pressure p_{tr} at the triple point decreases. Note that at a particular p_{tr} the values of T given by different gas thermometers differ. The discrepancy is small but measurable, being about 0.2% in the most extreme cases shown (O_2 and H_2 at 100 cm Hg); note that the entire vertical axis covers only 1.00 K. Helium gives nearly the same T at all pressures (the curve is almost horizontal) so that its behaviour is the most similar to that of an ideal gas over the entire range shown.

Hg. Then measure the new value of p_s and calculate another provisional temperature from $T_{40} = (p_{s40}/40 \text{ cm Hg}) \, 273.16 \text{ K}$. Continue this same procedure, reducing the amount of gas in the bulb again, and at this new lower value of p_{tr} calculating the provisional temperature at the steam point. If we plot the provisional values of T against p_{tr} and have enough data, we can extrapolate the resulting curve to the intersection with the axis where $p_{tr} = 0$.

In Fig. 19–2, we plot curves obtained from such a procedure for constant-volume thermometers of some different gases. These curves show that the temperature readings of a constant-volume gas thermometer depend on the gas used at ordinary values of the reference pressure. However, as the reference pressure is decreased, the temperature readings of constant-volume gas thermometers using different gases approach the same value. Therefore, the extrapolated value of the temperature depends only on the general properties of gases and not on any particular gas. We therefore define an *ideal gas temperature scale* by the relation

Ideal gas temperature scale

$$T = \left[\lim_{p_{tr} \to 0} \left(\frac{p}{p_{tr}} \right) \right] \times 273.16 \text{ K} \qquad \text{(constant } V\text{).} \qquad (19\text{–}3)$$

Our standard thermometer is therefore chosen to be a constant-volume gas thermometer using a temperature scale defined by Eq. 19–3.

Although our temperature scale is independent of the properties of any one particular gas, it does depend on the properties of gases in general (that is, on the properties of a so-called ideal gas). Therefore, to measure a temperature, a gas must be used at that temperature. The lowest temperature that can be measured with any gas thermometer is about 1 K. To obtain this temperature we must use low-pressure helium, for helium becomes a liquid at a temperature lower than any other gas. Therefore we cannot give experimental meaning to temperatures below about 1 K by means of a gas thermometer.

We would like to define a temperature scale in a way that is independent of the properties of any particular substance. An *absolute thermodynamic temperature scale,* called the Kelvin scale, is such a scale. It can be shown that the ideal gas scale and the Kelvin scale are identical in the range of temperatures in which a gas thermometer may be used. For this reason we can write K after an ideal gas temperature, as we have already done.

It can also be shown that the Kelvin scale has an *absolute zero* of 0 K and that temperatures below this do not exist. The absolute zero of temperature has defied all attempts to reach it experimentally, although it is possible to come arbitrarily close. The existence of the absolute zero is inferred by extrapolation. You should not think of absolute zero as a state of zero energy and no motion. The conception that all molecular action would cease at absolute zero is incorrect. This notion assumes that the purely macroscopic concept of temperature is strictly connected to the microscopic concept of molecular motion. When we try to make such a connection, we find in fact that as we approach absolute zero the kinetic energy of the molecules approaches a finite value, the so-called zero-point energy. The molecular energy is a minimum, but not zero, at absolute zero.

In Table 19–1 we list the temperatures, on the Kelvin scale, of various bodies and processes.

Table 19–1 Some temperatures^a (K)

Plasma temperatures (laboratory)	5×10^7
Solar interior	10^7
Solar corona	10^6
Shock wave in air at Mach 20	2.5×10^4
Luminous nebulae	10^4
Solar surface	6×10^3
Tungsten melts	3.6×10^3
Lead melts	6.0×10^2
Water freezes	2.7×10^2
Oxygen boils (1 atm)	9.0×10^1
Hydrogen boils (1 atm)	2.0×10^1
Helium (^{4}He) boils at 1 atm	4.2
^{3}He boils at attainable low pressure	3.0×10^{-1}
Adiabatic demagnetization of paramagnetic salts	10^{-3}
Adiabatic demagnetization of nuclei	10^{-6}

^{a} See *Scientific American*, September 1954; special issue on heat.

19–5 The Celsius and Fahrenheit Scales

Two temperature scales in common use are the Celsius (formerly called "centigrade") and the Fahrenheit scales. These are defined in terms of the Kelvin scale, which is the fundamental temperature scale in science.

The Celsius temperature scale uses the unit "degree Celsius" (symbol °C), equal to the unit "kelvin." Temperature *intervals* thus have the same numerical value on the Celsius and Kelvin scales. The origin of the Celsius scale is just offset to what, for some, is a more convenient value. If we let T_C represent the Celsius temperature, then

Celsius temperature scale

$$T_c = T - 273.15° \tag{19–4}$$

relates the Celsius temperature T_C (°C) and the Kelvin temperature T (K). We see that the triple point of water ($= 273.16$ K by definition) corresponds to 0.01°C. The Celsius scale is defined so that the temperature at which ice and air-saturated water are in equilibrium at atmospheric pressure—the so-called ice point—is 0.00°C and the temperature at which steam and liquid water are in equilibrium at 1 atm pressure—the so-called steam point—is 100.00°C.

The Fahrenheit scale, still in use in some English-speaking countries (England itself adopted the Celsius scale for commercial and civil use in 1964) is usually not used in scientific work. The relationship between the Fahrenheit and Celsius scales is defined to be

$$T_F = 32 + \tfrac{9}{5}T_C.$$

From this relation we can conclude that the ice point (0.00°C) equals 32.0°F, that the steam point (100.0°C) equals 212.0°F, and that one Fahrenheit degree is exactly $\tfrac{5}{9}$ as large as one Celsius degree. In Fig. 19–3 we compare the Kelvin, Celsius, and Fahrenheit scales.

19–6 The International Practical Temperature Scale

Let us now summarize the ideas of the last few sections. The standard fixed point in thermometry is the triple point of water, which is arbitrarily assigned a value 273.16 K. The constant-volume gas thermometer is the standard thermometer. The extrapolated gas scale is used to define the ideal gas temperature from

$$T = \left[\lim_{n_r \to 0} \left(\frac{p}{p_{tr}} \right) \right] \times 273.16 \text{ K}.$$

This scale is identical with the (absolute thermodynamic) Kelvin scale in the range in which a gas thermometer can be used.

By using the standard thermometer in this way, we can experimentally determine other reference points for temperature measurements, called fixed points. We list the basic fixed points adopted for experimental reference in Table 19–2. The temperatures can be expressed on the Celsius scale, with the use of Eq. 19–4, once the Kelvin temperature is determined.

Fahrenheit temperature scale

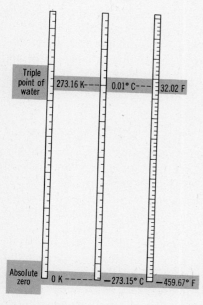

Figure 19–3 The Kelvin, Celsius, and Fahrenheit temperature scales.

Table 19–2 Fixed points on the international practical temperature scale[a]

Substance	State	Temperature K	°C
Hydrogen	Triple point	13.81	−259.34
Hydrogen	Boiling point[b]	17.042	−256.108
Hydrogen	Boiling point	20.28	−252.87
Neon	Boiling point	27.102	−246.048
Oxygen	Triple point	54.361	−218.789
Oxygen	Boiling point	90.188	−182.962
Water[c]	Triple point	273.16	0.01
Water[c]	Boiling point	373.15	100
Zinc	Freezing point	692.73	419.58
Silver	Freezing point	1235.08	961.93
Gold	Freezing point	1337.58	1064.43

[a] The so-called IPTS-68, adopted in 1968 by the International Committee on Weights and Measures.
[b] This boiling point is for a pressure of 25/76 atm. All other boiling points (and all freezing points) are for a pressure of 1 atm.
[c] The water used should have the isotopic composition of sea water.

Determining ideal gas temperatures is a painstaking job. It would not make sense to use this procedure to determine temperatures for all work. Hence, an International Practical Temperature Scale (IPTS) has been adopted (and is periodically revised) to provide a scale that can be used easily for practical purposes, such as for calibration of industrial or scientific instruments.* This scale consists of a set of recipes for providing in practice the best possible approximations to the Kelvin scale. A set of fixed points, the basic points in Table 19–2, is adopted, and a set of instruments is specified to be used in interpolating between these fixed points and in extrapolating beyond the highest fixed point. The IPTS-68 departs from the Kelvin scale at temperatures between the fixed points, but the difference is usually negligible. The IPTS-68 has become the legal standard in nearly all countries.

19–7 Thermal Expansion

Essentially all solids expand in volume when their temperature is raised. To see how this comes about let us take as a model of a crystalline solid an assembly of atoms held together in a regular three-dimensional cubic lattice by a system of spring-like interatomic forces. At any given temperature the atoms in the lattice will be vibrating about their equilibrium positions, the higher the temperature the greater being the amplitude of vibration.

Let x represent the distance between two nearest-neighbor atoms in the vibrating lattice. Figure 19–4 shows the potential energy function $U(x)$ associated with the interatomic force. It has a minimum at $x = x_0$ and this is the lattice spacing that the crystalline solid will have as the temperature approaches the absolute zero. We note at once that $U(x)$ is not a symmetrical function, rising more steeply when the atoms are pushed together ($x < x_0$) than when they are pulled apart ($x > x_0$). Put another way, the interatomic ''springs'' do not obey Hooke's law, as a comparison with Figs. 14–2 and 14–3 will make clear.

It is this lack of symmetry of the potential energy function that accounts for the thermal expansion of solids. To see this let us examine our vibrating atomic pair at an arbitrary temperature T, the corresponding total mechanical energy E being shown by the horizontal line in Fig. 19–4. Vibrations will occur between the two limits defined by setting $U(x)$ equal to E. The mid-point of the vibrations, labeled x_{av} in the figure, is a measure of the lattice spacing at temperature T. We see that $x_{av} > x_0$ and we see also that x_{av} continues to increase as the temperature is raised. In other words the lattice spacing, and thus the volume of the solid, increases with temperature. Note that if $U(x)$ were symmetrical about its minimum point then we would have $x_{av} = x_0$ for all temperatures; there would be no thermal expansion.

Thus we see that when the temperature is increased the average distance between atoms increases, which leads to an expansion of the whole solid body. The change in any linear dimension of the solid, such as its length, width, or thickness, is called a linear expansion. If the length of this linear dimension is l, the change in length, arising from a change in temperature ΔT, is Δl. We find from experiment that, if ΔT is small enough, this change in length Δl is proportional to the temperature change ΔT and to the original length l. Hence, we can write

$$\Delta l = \alpha l \, \Delta T, \qquad (19\text{–}5)$$

International practical temperature scale

An atomic model of thermal expansion

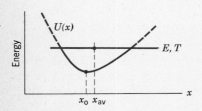

Figure 19–4 The average lattice spacing x_{av} for a crystalline solid increases with temperature.

Linear expansion

* See *Metrologia*, **5**, 35 (1969) for an English translation of the text on this scale.

where α, called the *coefficient of linear expansion,* has different values for different materials. Rewriting this formula we obtain

$$\alpha = \frac{1}{l}\frac{\Delta l}{\Delta T}, \qquad (19\text{–}6)$$

so that α has the meaning of a fractional change in length per degree temperature change.

Strictly speaking, the value of α depends on the actual temperature and the reference temperature chosen to determine l. However, its variation is usually negligible compared to the accuracy with which engineering measurements need to be made. We can often safely take it as a constant for a given material, independent of the temperature. In Table 19–3 we list the experimental values for the average coefficient of linear expansion $\bar\alpha$ of several common solids. For all the substances listed, the change in size consists of an expansion as the temperature rises, for $\bar\alpha$ is positive. The order of magnitude of the expansion is about 1 mm per meter length per 100°C.

Table 19–3 Some values^a of $\bar\alpha$

Substance	$\bar\alpha$ $10^{-6}/°C$	Substance	$\bar\alpha$ $10^{-6}/°C$
Aluminum	23	Lucite	60
Brass	19	Ice	51
Copper	17	Invar^b	0.7
Glass (ordinary)	9	Lead	29
Glass (Pyrex)	3.2	Steel	11

^{a} For the range 0°C to 100°C, except −10°C to 0°C for ice.
^{b} An iron-nickel-carbon alloy specifically designed to have a low value of $\bar\alpha$.

Example 2 A steel metric scale is to be ruled so that the millimeter intervals are accurate to within about 5×10^{-5} mm at a certain temperature. What is the maximum temperature variation allowable during the ruling?

From Eq. 19–5,

$$\Delta l = \alpha l\, \Delta T,$$

we have

$$5 \times 10^{-5}\text{ mm} = (11 \times 10^{-6}/°C)(1.0\text{ mm})\,\Delta T$$

in which we have used $\bar\alpha$ for steel, taken from Table 19–3. This yields $\Delta T \cong 5°C$. The temperature maintained during the ruling process must be maintained when the scale is being used and it must be held constant to within about 5°C.

Note (see Table 19–3) that if the alloy invar is used instead of steel, then for the same required tolerance one can permit a temperature variation of about 75°C; or for the same temperature variation ($\Delta T = 5°C$) the tolerance achieved would be more than an order of magnitude better.

For many solids, called *isotropic,* the percent change in length for a given temperature change is the same for all lines in the solid. The expansion is quite analogous to a photographic enlargement, except that a solid is three-dimensional. Thus, if you have a flat plate with a hole punched in it, $\Delta l/l\ (= \alpha\,\Delta T)$ for a given ΔT is the same for the length, the thickness, the face diagonal, the body diagonal, and the hole diameter. Every line, whether straight or curved, lengthens in the ratio α per degree temperature rise. If you scratch your name on the plate, the line representing your name has the same fractional change in length as any other line. The analogy to a photographic enlargement is shown in Fig. 19–5.

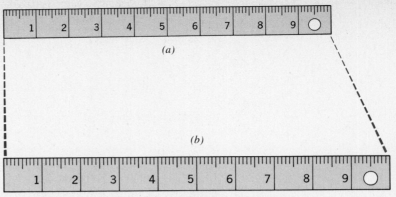

Figure 19-5 The same steel rule at two different temperatures. On expansion every dimension is increased by the same proportion; the scale, the numbers, the hole, and the thickness are all increased by the same factor. (The expansion shown, from (a) to (b), is obviously exaggerated, for it would correspond to an imaginary temperature rise of about 100,000°C.)

With these ideas in mind, you should be able to show (see Problems 12 and 14) that to a high degree of accuracy the fractional change in area A per degree temperature change for an isotropic solid is 2α, that is,

Area expansion

$$\Delta A = 2\alpha A \, \Delta T,$$

and the fractional change in volume V per degree temperature change for an isotropic solid is 3α, that is,

Volume expansion

$$\Delta V = 3\alpha V \, \Delta T.$$

Because the shape of a fluid is not definite, only the change in volume with temperature is significant. Gases respond strongly to temperature or pressure changes, whereas the change in volume of liquids with changes in temperature or pressure is much smaller. If we let β represent the coefficient of volume expansion for a liquid, that is,

Coefficient of volume expansion

$$\beta = \frac{1}{V} \frac{\Delta V}{\Delta T}, \tag{19-7}$$

we find that β is relatively independent of the temperature. Liquids typically expand with increasing temperature, their volume expansion being generally about ten times greater than that of solids.

The peculiar behavior of water

However the most common liquid, water, does not behave like other liquids. In Fig. 19-6 we show the expansion curve for water. Above ~4°C water expands as the temperature rises, although not linearly. Between 0°C and ~4°C, however, water contracts (see Fig. 19-6b) with increasing temperature instead of expanding as other common liquids do. The minimum specific volume of water occurs at 3.98°C and is 1.000 0250 cm³/g. The corresponding maximum density, which is just the inverse of the minimum specific volume, is 0.999 9750 g/cm³. At all other temperatures the density of water is less than this. This behavior of water is the reason why lakes freeze first at their upper surface. If water did not have a maximum density, lakes would freeze from the bottom up.

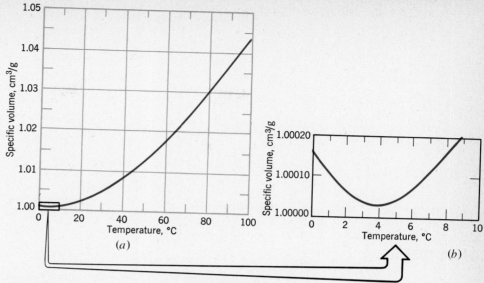

Figure 19-6 (*a*) The variation with temperature of the specific volume of water. (*b*) An expanded view of the lower left portion of the curve.

REVIEW GUIDE AND SUMMARY

Temperature; thermometers

Temperature is a quantitative measure of a macroscopic quantity related to our sense of hot and cold. It is measured by a thermometer, which contains a working substance with a measurable property, such as length or pressure, which changes in a regular way as the substance becomes hotter or colder. When a thermometer and some other object are placed in contact with each other, they eventually reach thermal equilibrium. The reading of the thermometer is then taken to be the temperature of the other object. The process is consistent because of the *zeroth law of thermodynamics:* If *A* and *B* are each in thermal equilibrium with a third body *C* (the "thermometer"), then *A* and *B* are in thermal equilibrium with each other.

The zeroth law of thermodynamics

The constant-volume gas thermometer

In a constant-volume gas thermometer, temperature is taken to be proportional to the pressure of a sample of gas whose volume is not allowed to change. The constant of proportionality is established by defining the numerical value of some reproducible condition, such as that in which water can exist with all three phases in equilibrium. The temperature of this triple point of water has been chosen, by international agreement, to be 273.16 K.

It is found experimentally that temperatures measured by constant-volume gas thermometers depend on the kind and amount of gas used. The discrepancies between different gases become smaller, however, if less gas is used. An ideal gas temperature can thus be defined as the limiting temperature approached by any constant-volume gas thermometer as repeated measurements are made with decreasing amounts of working gas. Algebraically,

Ideal gas temperature

$$T = \left[\lim_{p_{tr} \to 0} \left(\frac{p}{p_{tr}} \right) \right] \times 273.16 \text{ K (constant } V\text{)}. \qquad [19\text{--}3]$$

in which p is the pressure of the working gas at the measured temperature and p_{tr} is its pressure at the water triple-point temperature.

The Kelvin temperature scale

The ideal gas temperature scale is equivalent to the Kelvin scale, defined more carefully in thermodynamics (see Section 22–5). Two other temperature scales are in common use: the Celsius scale, defined by

$$T_C = T - 273.15°, \qquad [19\text{--}4]$$

The Celsius and Fahrenheit scales and the Fahrenheit scale, defined by

$$T_F = 32 + \frac{9}{5} T_C,$$

Thermal expansion All objects change size with changes in temperature. Such changes may be described in terms of a coefficient of linear expansion,

$$\alpha = \frac{1}{l} \frac{\Delta l}{\Delta T}, \qquad [19\text{--}6]$$

or a coefficient of volume expansion,

$$\beta = \frac{1}{V} \frac{\Delta V}{\Delta T}. \qquad [19\text{--}7]$$

Both β and α are properties of specific materials; see Table 19–3 for commonly used average values. It can be shown that $\beta \cong 3\alpha$.

QUESTIONS

1. Is temperature a microscopic or macroscopic concept?

2. Are there physical quantities other than temperature that tend to equalize if two different systems are joined?

3. Give a reasonable explanation for this: A piece of ice and a thermometer are suspended in an insulated evacuated enclosure so that they are not in contact and yet the thermometer reading decreases for a time.

4. A student, when told that the temperature at the center of the sun was thought to be about 1.5×10^7 degrees, asked whether that was on the Celsius or the Kelvin scale. How would you answer? How would you reply if he had asked whether it was on the Celsius or the Fahrenheit scale?

5. The editor-in-chief of a well-known business magazine, discussing possible warming effects associated with the increasing concentration of carbon dioxide in the earth's atmosphere, wrote: "The polar regions might be three times warmer than now. . . ." What do you suppose he meant? (See "Warmth and Temperature: A Comedy of Errors", by Albert A. Bartlett, *The Physics Teacher,* November 1984.)

6. Although the absolute zero of temperature seems to be experimentally unattainable, temperatures as low as 0.0000001 K have been achieved in the laboratory. Isn't this low enough for all practical purposes? Why would physicists (as indeed they do) strive to obtain still lower temperatures?

7. You put two uncovered pails of water, one containing hot water and one containing cold water, outside in below-freezing weather. The pail with the hot water will usually begin to freeze first. Why? What would happen if you covered the pails?

8. Can a temperature be assigned to a vacuum?

9. Does our "temperature sense" have a built-in sense of direction; that is, does hotter necessarily mean higher temperature, or is this just an arbitrary convention? Celsius, by the way, originally chose the steam point as 0°C and the ice point as 100°C.

10. How would you suggest measuring the temperature of (*a*) the sun, (*b*) the earth's upper atmosphere, (*c*) an insect, (*d*) the moon, (*e*) the ocean floor, and (*f*) liquid helium?

11. Is one gas any better than another for purposes of a standard constant-volume gas thermometer? What properties are desirable in a gas for such purposes?

12. State some objections to using water-in-glass as a thermometer. Is mercury-in-glass an improvement?

13. Can you explain why the column of mercury first descends and then rises when a mercury-in-glass thermometer is put in a flame?

14. What do the Celsius and Fahrenheit temperature conventions have in common?

15. What are the dimensions of α, the coefficient of linear expansion? Does the value of α depend on the unit of length used? When °F are used instead of °C as a unit of temperature change, does the numerical value of α change? If so, how?

16. A metal ball can pass through a metal ring. When the ball is heated, however, it gets stuck in the ring. What would happen if the ring, rather than the ball, were heated?

17. A bimetallic strip, consisting of two different metal strips riveted together, is used as a control element in the common thermostat. Explain how it works.

18. Two strips, one of iron and one of zinc, are riveted together side by side to form a straight bar which curves when heated. Why is it that the iron is on the inside of the curve?

19. Explain how the period of a pendulum clock can be kept constant with temperature by attaching vertical tubes of mercury to the bottom of the pendulum.

20. You can loosen a tight metal cap on a bottle by holding it briefly in hot water. Explain.

21. Why should a chimney be free standing, that is, not part of the structural support of the house?

22. Water expands when it freezes. Can we define a coefficient of volume expansion for the freezing process?

23. Explain why the apparent expansion of a liquid in a bulb does not give the true expansion of the liquid.

24. Does the change in volume of a body when its temperature is raised depend on whether the body has cavities inside, other things being equal?

25. What difficulties would arise if you defined temperature in terms of the density of water?

26. Explain why lakes freeze first at the surface.

27. What is it that causes water pipes to burst in the winter?

28. What can you conclude about how the melting point of ice depends on pressure from the fact that ice floats on water?

EXERCISES

Section 19–3 Measuring Temperature

1. To measure temperatures, physicists and astronomers often use the variation of intensity of electromagnetic radiation radiated by an object. The wavelength at which the intensity is greatest is given by the equation

$$\lambda_{max}T = 0.2898 \text{ cm} \cdot \text{K},$$

where λ_{max} is the wavelength of greatest intensity and T is the absolute temperature of the object. In 1965, microwave radiation peaked at $\lambda_{max} = 0.107$ cm was discovered coming in all directions from space. To what temperature does this correspond? The interpretation of this background radiation is that it is left over from the Big Bang some 15 billion years ago, when the universe began rapidly expanding and cooling.

2. A thermocouple is formed from two different metals, joined at two points in such a way that a small voltage is produced when the two junctions are at different temperatures. In a particular iron/constantan thermocouple, with one junction held at 0°C, the output voltage varies linearly from 0 to 28 mV as the temperature of the other junction is raised from 0°C to 510°C. Find the temperature of the variable junction when the thermocouple output is 10.2 mV.

3. The amplification or "gain" of a transistor amplifier may depend on the temperature. The gain for a certain amplifier at room temperature (20°C) is 30.0, whereas at 55°C it is 35.2. What would the gain be at 30°C if the gain depends linearly on temperature over this limited range?

Section 19–4 Ideal Gas Temperature Scale

4. If the ideal gas temperature at the steam point is 373.15 K, what is the limiting value of the ratio of the pressures of a gas at the steam point and at the triple point of water when the gas is kept at constant volume?

Section 19–5 The Celsius and Fahrenheit Scales

5. At what temperature is the Fahrenheit scale reading equal to (a) twice that of the Celsius? (b) half that of the Celsius?

6. If your doctor tells you that your temperature is 310° above absolute zero, should you worry? Explain.

7. At what temperature do the following pairs of scales give the same reading? (a) Fahrenheit and Celsius. (b) Fahrenheit and Kelvin. (c) Celsius and Kelvin.

8. (a) The temperature of the surface of the sun is about 6000 K. Express this on the Fahrenheit scale. (b) Express normal human body temperature, 98.6°F, on the Celsius scale. (c) In the continental United States, the highest officially recorded temperature is 134°F at Death Valley, California, and the lowest is −70°F at Rogers Pass, Montana. Express these extremes on the Celsius scale. (d) Express the normal boiling point of oxygen, −183°C, on the Fahrenheit scale. (e) At what Celsius temperature would you find a room to be uncomfortably warm?

Section 19–7 Thermal Expansion

9. A steel rod has a length of exactly 20 cm at 30°C. How much longer is it at 50°C?

10. Steel railroad tracks are laid when the temperature is 0°C. A standard section of rail is then 12.0 m long. What gap should be left between rail sections so that there is no compression when the temperature gets as high as 42°C?

11. The Pyrex glass mirror in the telescope at the Mount Palomar Observatory has a diameter of 200 in. The temperature ranges from −10°C to 50°C on Mount Palomar. Determine the maximum change in the diameter of the mirror.

12. A circular hole in an aluminum plate is 1.073 in. (= 2.725 cm) in diameter at 0°C. What is its diameter when the temperature of the plate is raised to 100°C?

13. An aluminum-alloy rod has a length of 10.000 cm at room temperature (20°C) and a length of 10.015 cm at the boiling point of water. (a) What is the length of the rod at the freezing point of water? (b) What is the temperature if the length of the rod is 10.009 cm?

14. A glass window is exactly 20 cm (= 7.9 in.) by 30 cm (= 11.8 in.) at 10°C. By how much has its area increased when its temperature is 40°C?

15. Find the change in volume of an aluminum sphere of 10.0-cm (= 3.94-in.) radius when it is heated from 0° to 100°C?

16. What is the volume of a lead ball at 30°C if its volume at 60°C is (exactly) 50 cm^3?

PROBLEMS GROUPED BY SECTION

Section 19–3 Measuring Temperature

1. A *resistance thermometer* is a thermometer in which the thermometric property is electrical resistance. We are free to define temperatures measured by such a thermometer in kelvins to be directly proportional to the resistance R, measured in ohms (Ω). A certain resistance thermometer is found to have a resistance R of 90.35 Ω when its bulb is placed in water at the triple-point temperature (273.16 K). What temperature is indicated by the thermometer if the bulb is placed in an environment such that its resistance is 96.28 Ω?

2. It is an everyday observation that hot and cold objects cool down or warm up to the temperature of their surroundings. If the temperature difference ΔT between an object and its surroundings ($\Delta T = T_{obj} - T_{sur}$) is not too great, the rate of cooling or warming of the object is proportional, approximately, to this temperature difference; that is,

$$\frac{d\,\Delta T}{dt} = -A(\Delta T),$$

where A is a constant. The minus sign appears because ΔT decreases with time if ΔT is positive and increases if ΔT is negative. This is known as *Newton's law of cooling*. (a) On what factors does A depend? What are its dimensions? (b) If at some instant $t = 0$ the temperature difference is ΔT_0, show that it is

$$\Delta T = \Delta T_0 e^{-At}$$

a time t later.

3. A *thermistor* is a semiconductor device with a temperature-dependent electrical resistance. It is commonly used in medical thermometers and to sense overheating in electronic equipment. Over a limited range of temperature, the resistance is given by

$$R = R_a e^{B(1/T - 1/T_a)},$$

where R is the resistance of the thermistor at temperature T, and R_a is the resistance at temperature T_a; B is a constant that depends on the particular semiconductor used. For one type of thermistor, $B = 4689$ K and the resistance at 273 K is $1.0 \times 10^4\ \Omega$. What temperature is the thermistor measuring when its resistance is 100 Ω?

Section 19–4 Ideal Gas Temperature Scale

4. Let p_{tr} be the pressure in the bulb of a constant-volume gas thermometer when the bulb is at the triple-point temperature of 273.16 K and p the pressure when the bulb is at room temperature. Given three constant-volume gas thermometers: For No. 1 the gas is oxygen, and $p_{tr} = 20$ cm Hg; for No. 2 the gas is also oxygen, but $p_{tr} = 40$ cm Hg; for No. 3 the gas is hydrogen, and $p_{tr} = 30$ cm Hg. The measured values of p for the three thermometers are p_1, p_2, and p_3. (a) An approximate value of the room temperature T can be obtained with each of the thermometers using

$$T_1 = 273.16\ \text{K}\ \frac{p_1}{20\ \text{cm Hg}}; \qquad T_2 = 273.16\ \text{K}\ \frac{p_2}{40\ \text{cm Hg}};$$

$$T_3 = 273.16\ \text{K}\ \frac{p_3}{30\ \text{cm Hg}}.$$

Mark each of the following statements "true" or "false": (1) With the method described, all three thermometers will give the same value of T. (2) The two oxygen thermometers will agree with each other but not with the hydrogen thermometer. (3) Each of the three will give a different value of T. (b) In the event that there is disagreement among the three thermometers, explain how you would change the method of using them to cause all three to give the same value of T.

5. Two constant-volume gas thermometers are assembled, one using oxygen as the working gas and the other using hydrogen. Both contain enough gas so that $p_{tr} = 80$ mm Hg. What is the difference between the pressures in the two thermometers if both are inserted into a water bath at the boiling point? Which pressure is the higher of the two?

Section 19–6 The International Practical Temperature Scale

6. In the interval between 0 and 700°C, a platinum resistance thermometer of definite specifications is used for interpolating temperatures on the International Practical Temperature Scale.

The Celsius temperature T_c is given by a formula for the variation of resistance with temperature:

$$R = R_0\,(1 + AT_c + BT_c^2).$$

R_0, A, and B are constants determined by measurements at the ice point, the steam point, and the zinc point. (a) If R equals 10.000 Ω at the ice point, 13.946 Ω at the steam point, and 24.172 Ω at the zinc point, find R_0, A, and B. (b) Plot R versus T_c in the temperature range from 0 to 700°C.

Section 19–7 Thermal Expansion

7. On a hot day in Las Vegas an oil trucker loaded 10,000 gal of diesel fuel from a tanker car and headed north. He soon encountered cold weather and, after traveling 375 miles to Payson, Utah, he delivered his entire load, measured at 9745 gal. Assume that the coefficient of linear expansion of his steel truck tank is $11 \times 10^{-6}/°$C, and that the coefficient of volume expansion of the diesel fuel is $9.5 \times 10^{-4}/°$C, and calculate by how many Fahrenheit degrees the fuel cooled during the trip.

8. A rod is measured to be exactly 20.0 cm long using a steel ruler at room temperature (20°C). Both the rod and the ruler are placed in an oven at 270°C, where the rod now measures 20.1 cm using the same ruler. What is the coefficient of thermal expansion for the material of which the rod is made?

9. A steel rod is 3.000 cm in diameter at 25°C. A brass ring has an interior diameter of 2.992 cm at 25°C. At what common temperature will the ring just slide onto the rod?

10. A 1.28-m-long vertical glass tube is half-filled with a liquid at 20°C. How much will the height of the liquid column change when the tube is heated to 30°C? Take $\alpha_{\text{glass}} = 1 \times 10^{-5}/°$C and $\beta_{\text{liquid}} = 4 \times 10^{-5}/°$C.

11. Imagine an aluminum cup of 0.10 liter capacity filled with glycerin at 22°C. How much glycerin, if any, will spill out of the cup if the temperature of the cup and glycerin is raised to 28°C? (The coefficient of volume expansion of glycerin is $5.1 \times 10^{-4}/°$C.)

12. The area A of a rectangular plate is ab. Its coefficient of linear expansion is α. After a temperature rise ΔT, side a is longer by Δa and side b is longer by Δb. Show that if we neglect the small quantity $\Delta a \Delta b/ab$ (see Fig. 19–7), then $\Delta A = 2\alpha A\Delta T$.

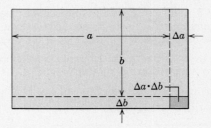

Figure 19–7 Problem 12.

13. Density is mass divided by volume. If the volume V is temperature dependent, so is the density ρ. Show that the change in density $\Delta\rho$ with change in temperature ΔT is given by

$$\Delta\rho = -\beta\rho\Delta T,$$

where β is the volume coefficient of expansion. Explain the minus sign.

14. Prove that, if we neglect extremely small quantities, the

change in volume of a solid upon expansion through a temperature rise ΔT is given by $\Delta V = 3\alpha V \Delta T$, where α is the coefficient of linear expansion.

15. When the temperature of a "copper" penny is raised by 100°C, its diameter increases by 0.18%. To two significant figures, give the percent increase in (a) the area of a face, (b) the thickness, (c) the volume, and (d) the mass of the penny. (e) What is the coefficient of linear expansion?

16. A pendulum clock with a pendulum made of brass is designed to keep accurate time at 20°C. What will be the error, in seconds per hour, if the clock operates at 0°C?

*17. Three equal-length straight rods, of aluminum, invar, and steel, all at 20°C, form an equilateral triangle with hinge pins at the vertices. At what temperature will the angle opposite the invar rod be 59.95°?

*18. The distance between the towers of the main span of the Golden Gate Bridge near San Francisco is 4200 ft (Fig. 19–8). The sag of the cable halfway between the towers at 50°F is 470 ft. Take $\alpha = 6.5 \times 10^{-6}/°F$ for the cable and compute (a) the change in length of the cable and (b) the change in sag for a temperature change from 10°F to 90°F. Assume no bending or movement of the towers and a parabolic shape for the cable.

Figure 19–8 Problem 18. Palmer/Monkmeyer Press.

ADDITIONAL PROBLEMS

19. (a) Express the coefficient of linear expansion of aluminum using the Fahrenheit temperature scale. (b) Now calculate the change in length of a 20-ft aluminum rod if it is heated from 40°F to 95°F.

20. In a certain experiment it was necessary to be able to move a small radioactive source at selected, extremely slow speeds. This was accomplished by fastening the source to one end of an aluminum rod and heating the central section of the rod in a controlled way. If the effective heated section of the rod in Fig. 19–9 is 2.0 cm, at what constant rate must the temperature of the rod be made to change if the source is to move at a constant speed of 100 nm/s?

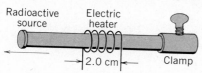

Radioactive source   Electric heater   Clamp
2.0 cm

Figure 19–9 Problem 20.

21. The heater of a house breaks down one day when the outside temperature is 7.0°C. As a result, the inside temperature drops from 22°C to 18°C in 1.0 h. The owner fixes the heater and adds insulation to the house. Now he finds that, on a similar day, the house takes twice as long to drop from 22°C to 18°C when the heater is not operating. What is the ratio of the constant A in Newton's law of cooling (see Problem 2) after the insulation is added to the value before?

22. When the temperature of a metal cylinder is raised from 0°C to 100°C, its length increases by 0.23%. (a) Find the percentage change in density. (b) What is the metal?

23. The timing of a certain electric watch is governed by a small tuning fork. The frequency of the fork is inversely proportional to the square root of the length of the fork. What is the percentage gain or loss in time for a steel tuning fork 8.0 mm long at (a) −40°F and (b) +120°F if it keeps perfect time at 25°C?

24. A mercury-in-glass thermometer is placed in boiling water for a few minutes and then removed. The temperature readings at various times after removal are as shown in the table. Plot A as a function of time, assuming Newton's law of cooling to apply (see

t, s	T, °C	t, s	T, °C
0	98.4	100	50.3
5	76.1	150	43.7
10	71.1	200	38.8
15	67.7	300	32.7
20	66.4	500	27.8
25	65.1	700	26.5
30	63.9	1000	26.1
40	61.6	1400	26.0
50	59.4	2000	26.0
70	55.4	3000	26.0

Problem 2). To what extent are you justified in assuming that Newton's law of cooling applies here?

25. A particular gas thermometer is constructed of two gas-containing bulbs, each of which is put into a water bath, as shown in Fig. 19–10. The pressure difference between the two bulbs is measured by a mercury manometer as shown in the figure. Appropriate reservoirs, not shown in the diagram, maintain constant gas volume in the two bulbs. There is no difference in pressure when both baths are at the triple point of water. The pressure difference is 120 mm Hg when one bath is at the triple point and the other is at the boiling point of water. Finally, the pressure difference is 90 mm Hg when one bath is at the triple point and the other is at an unknown temperature to be measured. What is the unknown temperature?

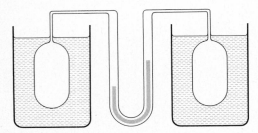

Figure 19–10   Problem 25.

26. A brass cube has an edge length of 30 cm. What is the increase in its surface area when it is heated from 20°C to 75°C?

27. As a result of a temperature rise of 32°C, a bar with a crack at its center buckles upward, as shown in Fig. 19–11. If the fixed distance $L_0 = 3.77$ m and the coefficient of linear expansion is $25 \times 10^{-6}/°C$, find x, the distance to which the center rises.

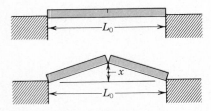

Figure 19–11   Problem 27.

28. Show that when the temperature of a liquid in a barometer changes by ΔT and the pressure is constant, the height h changes by $\Delta h = \beta h \Delta T$, where β is the coefficient of volume expansion.

29. A clock pendulum made of Invar has a period of 0.500 s, and is accurate at 20°C. If the clock is used in a climate where the temperature averages 30°C, what correction (approximately) is necessary at the end of 30 days to the time given by the clock?

30. (a) Show that if the lengths of two rods of different solids are inversely proportional to their respective coefficients of linear expansion at some initial temperature, the difference in length between them will be constant at all temperatures. (b) What should be the lengths of a steel and a brass rod at 0°C so that at all temperatures their difference in length is 0.30 m?

31. A composite bar of length $L = L_1 + L_2$ is made from a bar of material 1 and length L_1 attached to a bar of material 2 of length L_2, as shown in Fig. 19–12. (a) Show that the effective coefficient of linear expansion α for this bar is given by $\alpha = (\alpha_1 L_1 + \alpha_2 L_2)/L$. (b) Using steel and brass, design such a composite bar whose length is 52.4 cm and whose effective coefficient of linear expansion is $13 \times 10^{-6}/°C$.

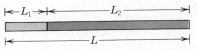

Figure 19–12   Problem 31.

32. A steel boiler in an electric power station is suspended from the ceiling of the boiler room by steel cables; this is to allow for thermal expansion of the boiler as it heats; see Fig. 19–13a. At 10°C, a 35.0-m tall boiler is suspended by 2.13-m steel cables and clears the floor by 1.82 m. What are the ceiling and floor clearances when the boiler is at 550°C? Assume that the cables reach 185°C. The variation of the coefficient of linear expansion of steel with temperature is shown in Fig. 19–13b.

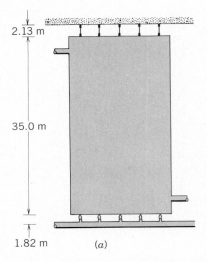

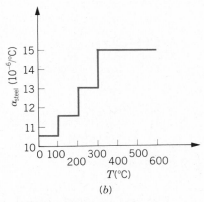

Figure 19–13   Problem 32.

*33. An aluminum cube 20 cm on an edge floats on mercury. How much farther will the block sink down when the temperature rises from 270 to 320 K? (The coefficient of volume expansion of mercury is $1.8 \times 10^{-4}/°C$.)

20 Heat and the First Law of Thermodynamics

20-1 Heat, a Form of Energy

When two systems at different temperatures are placed together, the final temperature reached by both systems is somewhere between the two starting temperatures. This is a common observation. Until the beginning of the nineteenth century, this and related phenomena were explained by postulating that a material substance, *caloric,* existed in every body. It was believed that a body at high temperature contained more caloric than one at a low temperature. When the two bodies were put together, the body rich in caloric lost some to the other until both bodies reached the same temperature. The caloric theory was able to describe many processes, such as heat conduction or the mixing of substances in a calorimeter, in a satisfactory way. However, the concept of heat as a substance, whose total amount remained constant, eventually could not stand the test of experiment. Nevertheless, we still describe many common temperature changes as the transfer of "something" from one body at a higher temperature to one at the lower, and this "something" we call heat. A common definition, is the following:

Heat is that which is transferred between a system and its surroundings as a result of temperature differences only.

Eventually it became generally understood that heat is a form of energy rather than a substance. The first conclusive evidence that heat could not be a substance was given by Benjamin Thompson (1753–1814), an American who later became Count Rumford of Bavaria.

Rumford made his discovery while supervising the boring of cannon for the Bavarian government. To prevent overheating, the bore of the cannon was kept full of water. The water was replenished as it boiled away during the boring process. It was accepted that caloric had to be supplied to water to boil it. The continuous production of caloric was explained by assuming that when a substance was more finely subdivided, as in boring, its capacity for retaining caloric became smaller, and that the caloric released in this way was what caused the water to boil. Rumford observed, however, that the water boiled away even when

Definition of heat

his boring tools became so dull that they were no longer cutting or subdividing matter. He was able to rule out, by experiment, all possible caloric interpretations and to conclude that the mechanical motion of the boring process produced the same effect on the water (a temperature rise and then boiling) as would be produced by putting the water near a flame.

Here we have the germ of the idea that doing mechanical work on a system, such as a can of water, and adding heat to it from an external source may be *equivalent* in their effects, both work and heat being forms of energy.

Although the concept of energy and its conservation is now well established, it was a novel idea as late as the 1850s and had eluded such men as Galileo and Newton. Throughout the subsequent history of physics this conservation idea led to new discoveries. Its early history was remarkable in many ways. Several thinkers arrived at this great concept at about the same time; at first, all of them either met with a cold reception or were ignored. The principle of the conservation of energy was established independently by Julius Mayer (1814–1878) in Germany, James Joule (1818–1889) in England, Hermann von Helmholtz (1821–1894) in Germany, and L. A. Colding (1815–1888) in Denmark.

Joule showed by experiment that a given amount of mechanical work done on a system was *quantitatively equivalent* in its effects to a specific quantity of heat added to the system from an external source at a higher temperature. Thus, the equivalence of heat and mechanical work as two forms of energy was definitely established.

Helmholtz first expressed clearly the idea that not only heat and mechanical energy but all forms of energy are equivalent, and that a given amount of one form cannot disappear without an equal amount appearing in some of the other forms.

20–2 Quantity of Heat and Specific Heat

The unit of heat Q used to be defined quantitatively by specifying that the temperature of one kilogram of water is raised from 14.5 to 15.5°C by one *kilocalorie* (kcal) of heat. The *calorie* ($= 10^{-3}$ kcal) is also used as a heat unit. (Incidentally, the "calorie" used to measure the energy content of foods is actually a kilocalorie.) In the engineering system the unit of heat is the *British thermal unit* (Btu), which is defined as the heat necessary to raise the temperature of one pound of water from 63 to 64°F.

The reference temperatures are stated because, near room temperature, there is a slight variation in the heat needed for a one-degree temperature rise for different initial temperatures. We shall neglect this variation for most practical purposes. The heat units are related as follows:

Heat units

$$1.000 \text{ kcal} = 1000 \text{ cal} = 3.968 \text{ Btu}.$$

Substances differ from one another in the quantity of heat needed to produce a given rise of temperature in a given mass. The ratio of the amount of heat energy Q supplied to a body to its corresponding temperature rise ΔT is called the *heat capacity* C of the body; that is,

Heat capacity

$$C = \text{heat capacity} = \frac{Q}{\Delta T}.$$

The word "capacity" may be misleading because it suggests the essentially meaningless statement, "the amount of heat a body can hold," whereas what is meant is simply the heat added per unit temperature rise.

The heat capacity per unit mass of a body, called *specific heat,* is characteristic of the material of which the body is composed:

Specific heat

$$c = \frac{\text{heat capacity}}{\text{mass}} = \frac{Q}{m\,\Delta T}. \qquad (20-1)$$

We properly speak of the heat capacity of a copper penny but of the specific heat of copper. The water contained in a swimming pool and in a drinking glass would have the same specific heat but quite different heat capacities.

Neither the heat capacity of a body nor the specific heat of a material is constant but depends on the location of the temperature interval. At ordinary temperatures and over ordinary temperature intervals, however, specific heats can be considered to be constants. For example, the specific heat of water varies less than 1% from its value of 1.000 cal/g °C over the temperature range from 0 to 100°C.

Equation 20-1 does not define specific heat uniquely. We must also specify the conditions under which the heat Q is added to the specimen. We have implied that the condition is that the specimen remain at normal (constant) atmospheric pressure while we add the heat. This is a common condition, but there are many other possibilities, each leading, in general, to a different value for c. To obtain a unique value for c we must specify the conditions, such as specific heat at constant pressure c_p, specific heat at constant volume c_v, and so on.

Table 20-1 (second column) shows the specific heats at constant pressure of some solid elements; we shall discuss the specific heats of gases later. You should realize from the way the calorie and the Btu are defined that 1 cal/g °C = 1 kcal/kg °C = 1 Btu/lb °F, exactly. Note that the specific heat of water, equal to 1.00 cal/g °C, is large compared to that of most substances.

Table 20-1   Values for c_p for some solids[a]

Substance	Specific heat cal/g °C	Specific heat J/g °C	Molecular weight g/mol[b]	Molar heat capacity cal/mol °C	Molar heat capacity J/mol °C
Aluminum	0.215	0.900	27.0	5.82	24.4
Carbon	0.121	0.507	12.0	1.46	6.11
Copper	0.0923	0.386	63.5	5.85	24.5
Silver	0.0564	0.236	108	6.09	25.5
Tungsten	0.0321	0.134	184	5.92	24.8
Lead	0.0305	0.128	207	6.32	26.5

[a] At 20°C and 1.0 atm.
[b] The molecular weight and the mole are defined on page 375.

Example 1   A 75-g block of copper, taken from a furnace, is dropped into a 300-g glass beaker containing 200 g of water. The temperature of the water rises from 12 to 27°C. What was the temperature of the furnace?

This is an example of two systems originally at different temperatures reaching thermal equilibrium after contact. No mechanical energy is involved, only heat exchange. Hence,

heat from copper = heat to (beaker + water),

$$m_C c_C (T_C - T_e) = (m_G c_G + m_W c_W)(T_e - T_W).$$

The subscript C stands for copper, G for glass, and W for water. The initial copper temperature is T_C, the initial beaker water temperature is T_W, and T_e is the final equilibrium temperature. Substituting the given values, with $c_C = 0.092$ cal/g °C, $c_G = 0.12$ cal/g °C, and $c_W = 1.0$ cal/g °C, we obtain

$$(75\text{ g})(0.092\text{ cal/g °C})(T_C - 27\text{°C}) = [(300\text{ g})(0.12\text{ cal/g °C})$$
$$+ (200\text{ g})(1.0\text{ cal/g °C})](27\text{°C} - 12\text{°C})$$

or, solving for T_C,

$$T_C = 530\text{°C}.$$

We have seen that energy must be added to a material to raise its temperature. Energy changes also occur when a material changes from one state to another. Melting (changing from solid to liquid), vaporization (from liquid to gas), or sublimation (changing directly from solid to gas) all require the addition of energy. The reverse processes involve the loss of energy from the material.

Heats of transformation

The amount of energy per unit mass that must be transferred as heat when any of these processes take place is called the *specific heat of transformation* or, sometimes, the *specific latent heat*. It is designated by the symbol *l*. The amount of heat required for a specific change is sometimes given a special name. For example, the *heat of vaporization* of water refers to the heat required to vaporize one unit mass of water at atmospheric pressure. Its numerical value is 539 cal/g or 539 kcal/kg. The *heat of fusion* of water is 80 kcal/kg and represents the amount of heat that must leave water when it freezes at atmospheric pressure. As with specific heats, heats of transformation vary with the external conditions under which the phase change takes place (see Example 3).

20–3 Heat Conduction

The transfer of energy arising from the temperature difference between adjacent parts of a body is called *heat conduction*. Consider a slab of material of cross-sectional area A and thickness Δx, whose faces are kept at different temperatures. We measure the heat Q that flows perpendicular to the faces in a time Δt. Experiment shows that Q is proportional to Δt and to the cross-sectional area A for a given temperature difference ΔT, and that Q is proportional to $\Delta T/\Delta x$ for a given Δt and A, providing both ΔT and Δx are small. That is,

$$\frac{Q}{\Delta t} \propto A \frac{\Delta T}{\Delta x} \qquad \text{approximately.} \qquad (20\text{–}2a)$$

In the limit of a slab of infinitesimal thickness dx, across which there is a temperature difference dT, we obtain the fundamental law of heat conduction in which the heat flow rate H is given by

Heat conduction equation

$$H = -kA \frac{dT}{dx}. \qquad (20\text{–}2b)$$

Here H (measured, say, in calories per second; see Eq. 20–2a) is the time rate of heat transfer across the area A, dT/dx is called the *temperature gradient,* and k is a constant of proportionality called the *thermal conductivity*. Heat flows in the direction in which the temperature decreases and we choose this direction to be that of increasing x. We must then put a minus sign in Eq. 20–2b, so that H will be positive when dT/dx is negative.

A substance with a large thermal conductivity k is a good heat conductor; one with a small thermal conductivity k is a poor heat conductor, or a good thermal insulator. The value of k depends on the temperature, increasing slightly with increasing temperature, but k can be taken to be practically constant throughout a substance if the temperature difference between its parts is not too great.

Let us apply Eq. 20–2b to a rod of length L and constant cross-sectional area A in which a steady state has been reached (Fig. 20–1). In a steady state the temperature at each point is constant in time. Hence, H is the same at all cross sections. (Why?) But $H = -kA(dT/dx)$, so that, for a constant k and A, the tempera-

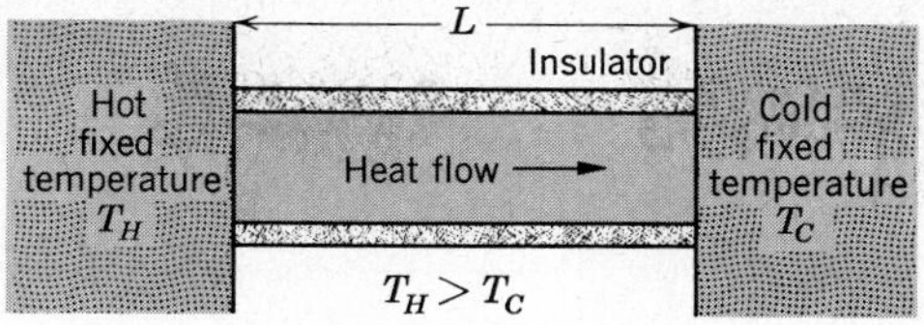

Figure 20–1 Conduction of heat through an insulated conducting rod.

ture gradient dT/dx is the same at all cross sections. Hence, T decreases linearly along the rod so that $-dT/dx = (T_H - T_C)/L$. Therefore, the time rate of transfer of heat energy is

Heat conduction for a uniform rod

$$H = kA \frac{T_H - T_C}{L}. \tag{20–3}$$

The phenomenon of heat conduction also shows that the concepts of heat and temperature are distinctly different. Different rods, having the same temperature difference between their ends, may transfer entirely different quantities of heat in the same time.

Persons interested in insulating their houses are more concerned with poor heat conductors than with good ones. For this reason the concept of thermal resistance R has been introduced into engineering practice. The R-value of a slab of material of thickness L is defined from

R-value of an insulating slab

$$R = L/k. \tag{20–4a}$$

The units for R—implied even if not commonly explicitly used, are ft² · °F · h/Btu. From Eqs. 20–3 and 20–4a we find

$$H = A \frac{T_H - T_C}{R}. \tag{20–4b}$$

In reasonably severe climates it is recommended that the ceilings of single-family dwellings be insulated to the level of $R30$. From Eq. 20–4b we see that this means that an average square foot of such a ceiling would loose heat by conduction at a rate of (1/30) Btu/hr for every 1 °F difference in temperature. As Example 2 shows, when materials are added in layers to form a composite ceiling or wall their R-values add.

Table 20–2 shows the thermal conductivities and the R-values of various materials, the latter being calculated for one-inch slabs. The use of R-values is normally restricted to commercial building and insulating materials, but we show them for a wide range of materials, for completion and comparison. Examination of the table shows us why a steel spoon is better than a silver spoon for stirring hot coffee. Among the gases helium and hydrogen are seen to be relatively good conductors, the former often being used both in experimental research and in industrial practice when it is desired to transfer heat efficiently into or out of a system. Still dry air has an insulating value (R-value) as great as that of any of the building materials shown. In fact, many commercial thermal insulating materials owe their efficacy to their ability to entrap isolated pockets of air. You can deduce from the table that to build a barrier with an insulating value of $R30$ you could use either 5.1 inches of polyurethane foam, 23 inches of white pine, 18 feet of window glass or 1.4 miles of silver!

Table 20–2 Some thermal conductivities and R-values[a]

Material	Conductivity, k	R-value for a one-inch slab
	cal/(m · °C · s)	(ft² · °F · h)/Btu
Metals		
Lead	8.4	0.0040
Steel	11	0.0030
Aluminum	48	0.00070
Copper	93	0.00036
Silver	98	0.00034
Gases		
Air (dry)	0.0057	5.9
Helium	0.033	1.0
Hydrogen	0.041	0.82
Building materials		
Polyurethane foam	0.0057	5.9
Rock wool	0.010	3.3
Fiber glass	0.011	3.0
White pine	0.026	1.3
Window glass	0.24	0.14

[a] For gases at 0 °C; other materials at room temperature.

Example 2 Consider a compound slab, consisting of two materials having different thicknesses, L_1 and L_2, and different thermal conductivities, k_1 and k_2. If the temperatures of the outer surfaces are T_H and T_c, find the rate of heat transfer through the compound slab (Fig. 20–2) in a steady state.

Let T_x be the temperature at the interface between the two materials. Then

$$H_2 = \frac{k_2 A(T_H - T_x)}{L_2}$$

and

$$H_1 = \frac{k_1 A(T_x - T_C)}{L_1}.$$

In a steady state $H_2 = H_1 = H$, so that

$$\frac{k_2 A(T_H - T_x)}{L_2} = \frac{k_1 A(T_x - T_C)}{L_1}.$$

Let H be the rate of heat transfer (the same for all sections). Then, solving for T_x and substituting into either of these equations, we obtain

$$H = \frac{A(T_H - T_C)}{(L_1/k_1) + (L_2/k_2)}$$

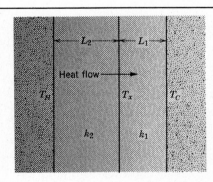

Figure 20–2 Example. 2. Conduction of heat through two slabs with different thermal conductivities.

The extension to any number of sections in series is obviously (see Eqs. 20–4)

$$H = \frac{A(T_H - T_C)}{\Sigma(L_i/k_i)} = \frac{A(T_H - T_C)}{\Sigma R_i}.$$

Note how the individual R-values add to form the R-value for the composite slab.

Convection

In addition to conduction, *convection* and *radiation* are important processes by which heat is transferred. Convection occurs when objects at different temperatures are both in contact with a fluid (Fig. 20–3). Fluid in contact with the warm object gains energy and, in most cases, expands. Since it is now less dense than

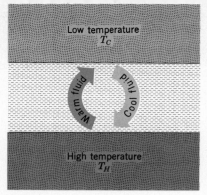

Figure 20–3   Convection in a fluid between objects at different temperatures.

Radiation

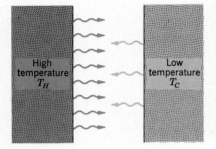

Figure 20–4   Both surfaces emit electromagnetic radiation and absorb some of the radiation emitted by the other. If both surfaces are made of the same material, the one with the higher temperature emits more radiation per unit area than the other.

Joule's experiment

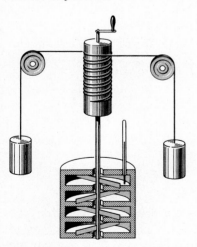

Figure 20–5   Joule's arrangement for measuring the mechanical equivalent of heat. The falling weights turn paddles which stir the water in the container, thus raising its temperature.

the surrounding cooler fluid, it rises because of the buoyant forces acting upon it. Its place is taken by cooler fluid which, in turn, gains energy from the warm object and rises in the same way. At the same time, fluid near the cool object loses energy, becomes more dense, and falls. The net effect is a continuous flow of warm fluid to the cool object and cool fluid to the warm object, a process that transfers energy from the warm object to the cooler one. Such processes are responsible for much of the variability in our weather pattern, for the transfer of significant amounts of energy within the oceans, and even for the flow of semifluid materials within the earth itself. The kind of convection described in Fig. 20–3 is *free* or *natural* convection. Convection can also be *forced*, as when a furnace blower causes circulation to heat the rooms of a home.

Radiation is represented in Fig. 20–4. All objects radiate electromagnetic radiation (see Chapter 38), the amount and character of the radiation being determined by the temperature and surface area of the object and the material of which it is made. In general, the rate of energy emission increases with the fourth power of the object's temperature T. Thus, a hot object radiates energy, some of which is absorbed by other objects it encounters. The cooler object also emits radiation, but may emit less than it absorbs because of its lower temperature. The net effect is a transfer of energy from the warmer object to the cooler one. Since electromagnetic radiation travels through empty space, radiation does not require physical contact for the transfer of energy. Thus energy can be transferred from the sun to the earth, by radiation, even though there is virtually no matter in the intervening space.

20–4   The Mechanical Equivalent of Heat

If heat is just another form of energy, any energy unit could be a heat unit. The calorie and Btu originated before it was generally accepted that heat is energy. Joule first carefully measured the mechanical energy equivalent of heat energy, that is, the number of joules equivalent to 1 cal, or the number of foot-pounds equivalent to 1 Btu.

The relative size of the "heat units" and the "mechanical units" can be found from experiments in which a measured quantity of mechanical energy—in joules, say—is added to a system such as a bucket of water; from the measured temperature rise we calculate, using the definition of Q in section 20–2, how much heat—in calories, say—we would have to add to the water sample to produce this same effect. Thus we can calculate the ratio of the joule to the calorie, the so-called *mechanical equivalent of heat*. Joule originally used an apparatus in which falling weights rotated a set of paddles in a water container (Fig. 20–5). The loss of mechanical energy was computed from a knowledge of the weights and the heights through which they fell and the equivalent heat energy by determining the mass of water and its rise in temperature. Joule wanted to show that the same amount of heat energy would be obtained from a given expenditure of work regardless of the method used to produce the work. He did so by stirring mercury, by rubbing together iron rings in a mercury bath, by adding electrical energy using a hot wire immersed in water, and in other ways. Always the constant of proportionality between the equivalent heat and work performed agreed within his experimental error of 5%. Joule did not have at his disposal the accurately standardized thermometers of today, nor could he make such reliable corrections for heat losses from the system as are possible now. His pioneer experiments are noteworthy not only for the skill and ingenuity he showed but also for the influence they had in convincing scientists everywhere of the correctness of the concept that heat is a form of energy.

The experimental results, are

$$1 \text{ kcal} = 1000 \text{ cal} = 4187 \text{ J},$$

$$1 \text{ Btu} = 252.0 \text{ cal} = 777.9 \text{ ft lb};$$

that is, 4187 J of mechanical energy will raise the temperature of 1 kg of water from 14.5 to 15.5°C, just as will 1000 cal of heat.

20–5 Heat and Work

We have seen that heat is the energy that flows from one body to another because of a temperature difference between them. Heat is not a property of a body in the sense that we can assign a value to the amount of heat ''contained'' in the body.

Work is similar to heat in that it is a measure of energy being transmitted from one body to another. In fact, *work* may be defined as *energy that is transmitted from one system to another in such a way that a difference of temperature is not directly involved.* Work is not a specific property of a body in the sense that we can assign a value to the work ''contained'' within the body. In Joule's experiment we can imagine that a steady state has been achieved, in which we do work on the system at a constant rate and the temperature of the system rises to such a steady value that heat energy is lost to the (cooler) surroundings at this same constant rate. We speak of work and heat only as they enter or leave the system.

Work is a measure of energy transfer by mechanical means, such as by gravitational, electrical, or magnetic forces. Heat is a measure of energy transfer by means of temperature differences.

Thermodynamics is concerned with the energy transfers that occur when a system undergoes any thermodynamic process from one state of the system to another. It is also concerned with the relationship of these energy transfers to changes in the measurable properties of the system.

For example, suppose that we wish to discuss a system composed of a certain number of gas molecules contained in a cylinder with a movable piston, as in Fig. 20–6. The environment with which the system of molecules interacts would be the cylinder walls and the piston. The properties that describe the system would be the pressure, temperature, volume, and chemical state of the gas. A thermodynamic process in such a system might involve a process in which an amount of heat energy Q is transferred to the gas because of a difference in temperature between the system (the gas) and its environment (the cylinder walls). The gas might do work W on its environment (the piston) by causing the piston to move because of an imbalance between the forces on the two sides of the piston. During the process the gas might change from a configuration described by initial parameters p_i, V_i, and T_i to a final configuration with parameters p_f, V_f, and T_f. (Chemical reactions or changes of state might also be involved, and these would also be described by appropriate system parameters.)

If the changes in the system occur slowly enough that all parts of the gas are always at the same temperature (that is, the gas is in thermal equilibrium), the sample of gas is characterized by a specific temperature, pressure, and volume at each point in the process. Such a process is called *quasi-static*. The state of the gas at any time and the process itself can then be represented in a diagram of pressure versus volume (a *p-V diagram*) as shown for a special case in Fig. 20–7.

The laws of thermodynamics allow us to calculate certain relationships

The mechanical equivalent of heat

Definition of work

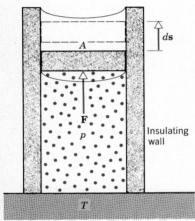

Figure 20–6 Work is done by the gas at pressure p as it expands against the piston. Heat may enter or leave the system from the reservoir on which the cylinder rests.

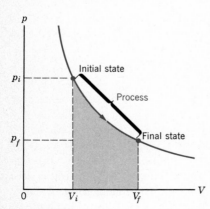

Figure 20–7 A *p-V* curve representing the slow expansion of a gas. The work done by a gas is equal to the area under the curve.

between the initial and final parameters of the system on the one hand and the energy transfers Q and W on the other.

Let us now compute W for a process of this type. In Fig. 20–6 we show the gas expanding against the piston. The work done by the gas in displacing the piston through an infinitesimal distance ds is

$$dW = \mathbf{F} \cdot d\mathbf{s} = pA \; ds = p \; dV,$$

where dV is the differential change in the volume of the gas. In general, the pressure will not be constant during a displacement. To obtain the total work W done on the piston by the gas in a large displacement, we must know how p varies with the displacement. Then we compute the integral

$$W = \int dW = \int_{V_i}^{V_f} p \; dV. \tag{20–5}$$

Work done by an expanding gas

This integral can be evaluated graphically as the area under the curve in a p-V diagram, as shown for a special case in Fig. 20–7.

There are many different ways in which the system can be taken from the initial state i to the final state f. For example (Fig. 20–8), the pressure may be kept constant from i to a and then the volume kept constant from a to f. Then the work done by the expanding gas is equal to the area under the line ia. Another possibility is the path ibf, in which case the work done by the gas is the area under the line bf. The continuous curve from i to f is another possible path, in which the work done by the gas is still different from the previous ones. We can see, therefore, that the work done by a system depends not only on the initial and final states but also on the intermediate states, that is, on the details of the process.

A similar result follows if we compute the flow of heat during the process. State i is characterized by a temperature T_i and state f by a temperature T_f. The heat flowing into the system, say, depends on how the system is heated. We can heat it at a constant pressure p_i, for example, until we reach the temperature T_f, and then change the pressure at constant temperature to the final value p_f. Or we can first lower the pressure to p_f and then heat it at that pressure to the final temperature T_f. Or we can follow many other paths. Each path gives a different result for the heat flowing into the system. Hence, the heat lost or gained by a system depends not only on the initial and final states but also on the intermediate states, that is, on the details of the process. This is an experimental fact.

Both heat and work "depend on the path" taken; neither one is independent of the path, and neither one can be conserved alone. There is, however, an important conservation law, *the first law of thermodynamics,* which involves Q and W taken together.

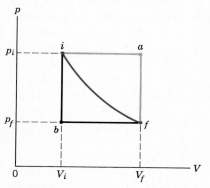

Figure 20–8 The work done by a system depends not only on the initial state (i) and the final state (f) but on the intermediate states as well.

20–6   The First Law of Thermodynamics

We can now tie all these ideas together. Let a system change from an initial equilibrium state i to a final equilibrium state f in a definite way, the heat absorbed by the system being Q and the work done by the system being W. Then we compute the value of $Q - W$. Now we start over and change the system from the same state i to the same state f, but this time by a different path. We do this over and over again, using different paths each time. We find that in every case the quantity $Q - W$ is the same. That is, although Q and W separately depend on the path taken, $Q - W$ does *not* depend at all on how we took the system from state i to state f but only on the initial and final (equilibrium) states.

You will remember from mechanics that when an object is moved from an initial point i to a final point f in a gravitational field in the absence of friction, the work done depends only on the positions of the two points and not at all on the path through which the body is moved. From this we concluded that there is a potential energy function of the space coordinates of the body whose final value minus its initial value equals the work done in displacing the body. Now, in thermodynamics we find from experiment that when a system has its state changed from state i to state f, the quantity $Q - W$ depends *only* on the initial and final coordinates and *not at all* on the path taken between these end points. We conclude that there is a function of the thermodynamic coordinates whose final value minus its initial value equals the change $Q - W$ in the process. We call this function the *internal energy function*.

Let us represent the internal energy function by the letter U. Then the internal energy of the system in state f, U_f, minus the internal energy of the system in state i, U_i, is simply the change in internal energy of the system, and this quantity has a definite value independent of how the system went from state i to state f. We have

The first law of thermodynamics

$$U_f - U_i = \Delta U = Q - W. \tag{20-6a}$$

Just as for potential energy, so for internal energy too it is the change that matters. If some arbitrary value is chosen for the internal energy in some standard reference state, its value in any other state can be given a definite value. Equation 20–6a is known as the *first law of thermodynamics*. In applying Eq. 20–6a Q is considered positive when heat *enters* the system and W is positive when work is done *by* the system. This convention can be traced to the fact that thermodynamics developed during the industrial revolution as engineers and scientists studied the efficiency of steam engines. In such engines you *add* heat (positive) and the engine *does* work (also positive).

The internal energy U

The internal energy function U may seem to be quite abstract to you at this point. Indeed, classical thermodynamics offered no explanation for it other than that it is a state function which changes in a predictable way. (By *state function* we mean just that its value depends only on the physical state of the material —its constitution, pressure, temperature, and volume.) In Chapter 21, however, we shall see that U just represents the microscopic kinetic and potential energy of the molecules and atoms of which the material is made. The first law of thermodynamics then becomes a statement of the law of conservation of energy for thermodynamic systems. The total energy of a system of particles (U) changes by exactly the amount that is added to the system less the amount that leaves.

If our system undergoes only an infinitesimal change in state, only an infinitesimal amount of heat dQ is absorbed and only an infinitesimal amount of work dW is done, so that the internal energy change dU is also infinitesimal. We may write the first law in differential form as*

$$dU = dQ - dW. \tag{20-6b}$$

We may express the first law in words by saying: *Every thermodynamic system in an equilibrium state possesses a state variable called the internal energy U whose change dU in a differential process is given by Eq. 20–6b.*

---

* Here dQ and dW, unlike dU, are not true differentials. There are, for example, no such functions as $Q(p, V)$ or $W(p, V)$ because Q and W refer to transitions between the states of a system and not to the states themselves. dQ and dW are called *inexact differentials* and are often represented by the symbols $đQ$ and $đW$. For our purposes we think of them simply as infinitesimally small energy transfers.

The first law of thermodynamics applies to every process in nature that starts in one equilibrium state and ends in another. We say that a system is in an equilibrium state when we can describe it by an appropriate set of constant system parameters such as pressure, volume, temperature, magnetic field, and so on. The first law still holds if the states through which the system passes from its initial (equilibrium) state to its final (equilibrium) state are not themselves equilibrium states. For example, we can apply the first law of thermodynamics to the explosion of a firecracker in a closed steel drum.

There are some important questions that the first law cannot answer. For example, although it tells us that energy is conserved in every process, it does not tell us if any particular process can actually occur. An entirely different generalization, called the second law of thermodynamics, gives us this information, and much of the subject matter of thermodynamics depends on this second law (Chapter 22).

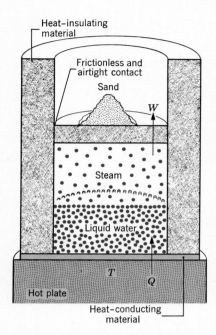

Figure 20–9 Water boiling at constant pressure (isobarically). The pressure is kept constant by the weight of the sand, the piston, and the external atmospheric pressure.

Work done during an isobaric process

20–7 Some Applications of the First Law of Thermodynamics

1. *Isobaric processes.* We have seen that when a gas expands the work it does on its environment is

$$W = \int p \, dV, \qquad [20-5]$$

where p is the pressure exerted by the gas and dV is the differential change in volume of the gas. Consider a special case in which the pressure remains constant while the volume changes by a finite amount, say, from V_i to V_f. Then

$$W = \int_{V_i}^{V_f} p \, dV = p \int_{V_i}^{V_f} dV = p(V_f - V_i) \qquad \text{(constant pressure).} \quad (20-7)$$

A process taking place at constant pressure is called an *isobaric* process. For example, water is heated in the boiler of a steam engine up to its boiling point and is vaporized to steam; then the steam is superheated, all processes proceeding at constant pressure.

In Fig. 20–9 we show an isobaric process. The system is H_2O in a cylindrical container. A frictionless airtight piston is loaded with sand to produce the desired pressure on the H_2O and to maintain it automatically. Heat can be transferred from the environment to the system by a "hot plate," whose temperature T can be readily adjusted. If the process continues long enough, the water boils and some is converted to steam; we assume that this occurs. The system may expand very slowly (quasi-statically) but the pressure it exerts on the piston is automatically always the same, for this pressure must be equal to the constant pressure that the piston exerts on the system and the expansion occurs slowly enough that the piston does not accelerate. If we wedged the piston so that it could not move, or if we added or took away some sand during the heating process, the process would not be isobaric.

Let us consider the boiling process. We know that substances will change their phase from liquid to vapor at a definite combination of values of pressure and temperature. Water will vaporize at 100°C and atmospheric pressure, for example. For a system to undergo a change of phase heat must be added to it, or taken from it, quite apart from the heat necessary to bring its temperature to the required value. Consider the change of phase of a mass m of liquid to a vapor occurring at constant temperature and pressure. Let V_l be the volume of liquid and V_v the vol-

The isobaric boiling of water

ume of vapor. The work done by this substance in expanding from V_l to V_v at constant pressure is given by Eq. 20–7;

$$W = p(V_v - V_l).$$

Let l represent the specific heat of vaporization, that is, the heat needed per unit mass to change a substance from liquid to vapor at constant temperature and pressure. Then the heat absorbed by the mass m during the change of state is

$$Q = ml.$$

From the first law of thermodynamics, we have

$$\Delta U = Q - W$$

so that

$$\Delta U = ml - p(V_v - V_l)$$

for this process.

Example 3 At atmospheric pressure 1.00 g of water, having a volume of 1.00 cm³, becomes 1671 cm³ of steam when boiled. The heat of vaporization of water is 539 cal/g at 1 atm. Hence, if $m = 1.00$ g,

$$Q = ml = 539 \text{ cal.}$$

This quantity, which represents heat added to the system from the environment, is positive.

$$W = p(V_v - V_l) = (1.013 \times 10^5 \text{ N/m}^2)[(1671 - 1) \times 10^{-6} \text{ m}^3]$$
$$= 169.5 \text{ J.}$$

This quantity, which represents work done by the system on the environment, is positive.

Since 1 cal equals 4.187 J, $W = 40$ cal. Then

$$\Delta U = U_v - U_l = ml - p(V_v - V_l) = (539 - 40) \text{ cal}$$
$$= 499 \text{ cal.}$$

This quantity is positive; the internal energy of the system increases during this process. Hence, of the 539 cal needed to boil 1 g of water (at 100°C and 1 atm), 40 cal go into external work of expansion and 499 cal go into internal energy added to the system. This energy represents the internal work done in overcoming the strong attraction of H_2O molecules for one another in the liquid state.

How would you expect the 80 cal that are needed to melt 1 g of ice to water (at 0°C and 1 atm) to be shared by the external work and the internal energy?

---

2. Adiabatic processes. A process that takes place in such a way that no heat flows into or out of the system is called an *adiabatic process*. Experimentally such processes are achieved either by sealing the system off from its surroundings with heat-insulating material or by performing the process quickly. Because the flow of heat takes time, any process can be made practically adiabatic if it is performed quickly enough.

Adiabatic processes

For an adiabatic process Q equals zero, so that from the first law we obtain

$$\Delta U = U_f - U_i = -W \qquad \text{(adiabatic process).}$$

Hence, the internal energy of a system increases exactly by the amount of work done on the system in an adiabatic process. If work is done by the system in an adiabatic process, the internal energy of the system decreases by exactly the amount of external work it performs. An increase of internal energy usually raises the system's temperature and conversely, a decrease of internal energy usually lowers the system's temperature. A gas that expands adiabatically does external work and its internal energy decreases; such a process is used to attain low temperatures. The increase of temperature during an adiabatic compression of air is well known from the heating of a bicycle pump.

In Fig. 20–10 we show a simple adiabatic process. The system is a gas inside

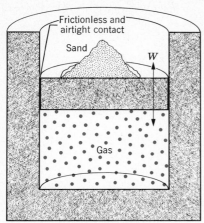

Figure 20–10 In an adiabatic process there is no flow of heat to or from the system. Here the walls are insulated and, as sand is removed or added, the volume of the gas changes adiabatically.

A free expansion

a cylinder made of heat-insulating material. Heat cannot enter the system from its environment or leave the system to the environment. Again we have a pile of sand on a frictionless airtight piston. The only interaction permitted between system and environment is through the performance of work. Such a process can occur when sand is added or removed from the piston, so that the gas can be compressed or can expand against the piston.

Among the many engineering examples of adiabatic processes are the expansion of steam in the cylinder of a steam engine, the expansion of hot gases in an internal combustion engine, and the compression of air in a Diesel engine or in an air compressor. These processes all take place rapidly enough so that only a very small amount of heat can enter or leave the system through its walls during that short time. The compressions and rarefactions in a sound wave are so rapid that the behavior of the transmitting gas is adiabatic.

Another important reason for studying adiabatic processes is that they are used in idealized, imaginary heat engines which may be analyzed to determine the theoretical limits to the operation and capabilities of real engines. We shall look further into this in Chapter 22.

A process of much theoretical interest is that of *free expansion*. This is an adiabatic process in which no work is performed on or by the system. Something like this can be achieved by connecting one vessel that contains a gas to another evacuated vessel with a stopcock connection, the whole system being enclosed with thermal insulation (Fig. 20–11). If the stopcock is suddenly opened, the gas rushes into the vacuum and expands freely. Because of the heat insulation this process is adiabatic, and because the walls of the vessels are rigid no external work is done on the system. Hence, in the first law we have $Q = 0$ and $W = 0$, so that $U_i = U_f$ for this process. The initial and final internal energies are equal in free expansion.

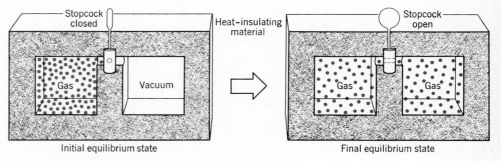

Figure 20–11 Free expansion. There is no change of internal energy U since there is no flow of heat Q and no external work W is done.

A free expansion differs from the other examples that we have given in that there is no way to carry it out very slowly (quasi-statically). After we open the stopcock we have no further control over the process. At intermediate states the pressure, volume and temperature do not have unique values characteristic of the system as a whole; that is, the system passes through nonequilibrium states so that we cannot plot the course of the process by a curve on a *p-V* diagram. We can plot the initial and final states as points on such diagrams because they are well-defined, equilibrium states. The free expansion is a good example of an irreversible process; see Section 22–2.

REVIEW GUIDE AND SUMMARY

Heat

Heat is the energy that flows from one object to another because of a temperature difference between them. It can be measured in kilocalories (kcal) or British thermal units (Btu), with

$$1.000 \text{ kcal} = 1000 \text{ cal} = 3.968 \text{ Btu.}$$

Heat capacity and specific heat

The heat capacity of an object is a measure of the amount of energy needed to raise its temperature. Specifically, it is the amount of heat energy Q supplied to the body divided by the corresponding temperature rise ΔT.

$$C = \text{heat capacity} = \frac{Q}{\Delta T}.$$

The specific heat of a material is its heat capacity per unit mass. See Table 20–1 and Example 1.

Heats of transformation

Heat supplied to a material may also change its physical state, for example, from solid to liquid or from liquid to gas. The amount required per unit mass for a particular material is its specific heat of transformation. The specific heat of vaporization refers to the amount of energy per unit mass required to vaporize a liquid and has the value 539 kcal/kg for water at atmospheric pressure. The specific heat of fusion is the amount of heat energy per unit mass that leaves a liquid when it freezes. Its value for water at atmospheric pressure is 80 kcal/kg.

The equivalence of heat and work

Since heat is a form of energy, it may also be measured in mechanical units. The equivalence between thermal units (kcal) and mechanical units (J) has been established by experiment to be

$$1 \text{ kcal} = 4187 \text{ J} \qquad \text{or} \qquad 1 \text{ Btu} = 778 \text{ ft lb.}$$

Heat conduction

Heat energy may be transferred between objects with differing temperatures by conduction, convection, or radiation. One-dimensional heat flow within a given material in which temperature varies with position obeys the law of heat conduction,

$$H = -kA \frac{dT}{dx}, \qquad [20-2b]$$

in which H represents the time rate of heat transfer across the area A, k is the thermal conductivity of the material, and dT/dx represents the temperature gradient within the material. When we deal with commercial building and insulating materials formed into slabs of thickness L, we often use the thermal resistance R ($= L/k$). See Table 20–2 and Example 2.

Convection

Convection occurs when temperature differences cause motion within a fluid (due to buoyant forces which vary within the fluid because of differences in density) which carries higher-temperature fluid from the hot object to the cooler one and lower-temperature fluid from the cool object to the warmer one. Radiation refers to the electromagnetic energy emitted by all objects, the amount increasing with increasing temperature.

Radiation

Work associated with volume change

Work is another process by which a material may exchange energy with its surroundings. The amount of energy as work W that a sample of material transmits to its surroundings as it expands (or contracts) from an initial volume V_i to a final volume V_f may be computed from

$$W = \int_{V_i}^{V_f} p \, dV. \qquad [20-5]$$

The integral calculation is necessary because the pressure p may vary during the volume change. W may also be computed as the area under the curve of p versus V representing the change (see Fig. 20–7).

The first law of thermodynamics

The conservation of energy for a sample of material exchanging energy by work and heat with its surroundings is expressed in the first law of thermodynamics. In mathematical form it assumes either of the forms

$$U_f - U_i = Q - W$$

or

$$dU = dQ - dW, \qquad [20-6b]$$

in which U represents the internal energy of the material, which depends only upon its state (temperature, pressure, volume, and constitution). Q represents heat added to the material and W represents the work done by the material on its surroundings.

Q and W are path-dependent; U is not

Q and W (as well as dQ and dW) will, in general, depend on the exact process through which the material goes from the initial to the final state. $U_f - U_i$ (or dU), however, does not, depending only on the nature of the initial and final states. U represents energy stored within the material in the form of microscopic kinetic and potential energy.

The first law applied to isobaric and adiabatic processes

Section 20–7 applies the first law of thermodynamics to isobaric (constant pressure) processes, using the boiling of water as an example; see Eq. 20–7 and Example 3. The law is also applied here to adiabatic ($Q = 0$) processes, using the free expansion of a gas ($Q = 0$; $W = 0$) as an example; see Figs. 20–10 and 20–11.

QUESTIONS

1. Temperature and heat are often confused, as in, "Bake in an oven at moderate heat." By example, distinguish between these two concepts as carefully as you can.

2. (*a*) Show how heat conduction and calorimetry could be explained by the caloric theory. (*b*) List some heat phenomena that cannot be explained by the caloric theory.

3. Give an example of a process in which no heat is transferred to or from a system but the temperature of the system changes.

4. Can heat be considered a form of stored (or potential) energy? Would such an interpretation contradict the concept of heat as energy in process of transfer because of a temperature difference?

5. Apply Eq. 20–1 to boiling water.

6. Can heat be added to a substance without causing the temperature of the substance to rise? If so, does this contradict the concept of heat as energy in process of transfer because of a temperature difference?

7. Why must heat energy be supplied to melt ice — the temperature doesn't change, after all?

8. Explain the fact that the presence of a large body of water nearby, such as a sea or ocean, tends to moderate the temperature extremes of the climate on adjacent land.

9. An electric fan not only does not cool the air but heats it slightly. How, then, can it cool you?

10. Both heat conduction and wave propagation involve the transfer of energy. Is there any difference in principle between these two phenomena?

11. When a hot body warms a cool one, are their temperature changes equal in magnitude? Give examples. Can one then say that temperature passes from one to the other?

12. A block of wood and a block of metal are at the *same* temperature. When the blocks feel cold the metal feels colder than the wood; when the blocks feel hot the metal feels hotter than the wood. Explain. At what temperature will the blocks feel equally cold or hot?

13. How can you best use a spoon to cool a cup of coffee? Stirring — which involves doing work — would seem to heat the coffee rather than to cool it.

14. How does a layer of snow protect plants during cold weather? During freezing spells, citrus growers in Florida often spray their fruit with water, hoping that it will freeze. How does that help?

15. Explain the wind-chill effect.

16. A thick glass is more likely to crack than a thin glass when you pour hot water into it. Why?

17. Why defrost a refrigerator?

18. You put your hand in a hot oven to remove a casserole and burn your fingers on the hot dish. However, the air in the oven is at the same temperature as the casserole dish but it does not burn your fingers. Why not?

19. Why is thicker insulation used in an attic than in the walls of a house?

20. Is ice always at 0°C? Can it be colder? Can it be warmer? What about an ice-water mixture?

21. Explain why your finger sticks to a metal ice tray just taken from the refrigerator.

22. On a winter day the temperature of the inside surface of a wall is much lower than room temperature and that of the outside surface is much higher than the outdoor temperature. Explain.

23. The physiological mechanisms that maintain a person's internal temperature operate in a limited range of external temperature. Explain how this range can be extended at each extreme by the use of clothes. (See "Heat, Cold, and Clothing," by James B. Kelley, *Scientific American,* February 1956.)

24. What requirements for thermal conductivity, specific heat capacity, and coefficient of expansion would you want a material to be used as a cooking utensil to satisfy?

25. Consider that heat can be transferred by convection and radiation, as well as by conduction, and explain why a thermos bottle is double-walled, evacuated, and silvered.

26. Explain, in terms of heat transfer by radiation, how the two surfaces in Fig. 20–4 may arrive at the same temperature. Does radiation continue thereafter?

27. A lake freezes first at its upper surface. Is convection involved? Conduction? Radiation?

28. Is the mechanical equivalent of heat, J, a physical quantity or merely a conversion factor for converting energy from heat units to mechanical units and vice versa?

29. Defend this statement: "In Joule's experiment on the mechanical equivalent of heat, described in Section 20–4, no heat is involved."

30. Is the temperature of an isolated system (no interaction with the environment) conserved?

31. Is heat the same as internal energy? If not, give an example in which a system's internal energy changes without a flow of heat across the system's boundary.

32. Can you tell whether the internal energy of a body was acquired by heat transfer or acquired by performance of work?

33. If the pressure and volume of a system are given, is the temperature always uniquely determined?

34. Does a gas do any work when it expands adiabatically? If so, what is the source of the energy needed to do this work?

35. A quantity of gas occupies an initial volume V_0 at a pressure p_0 and a temperature T_0. It expands to a volume V (*i*) at constant temperature and (*ii*) at constant pressure. In which case does the gas do more work?

36. Discuss the process of the freezing of water from the point of view of the first law of thermodynamics. Remember that ice occupies a greater volume than an equal mass of water.

37. A thermos bottle contains coffee. The thermos bottle is vigorously shaken. Consider the coffee as the system. (*a*) Does its temperature rise? (*b*) Has heat been added to it? (*c*) Has work been done on it? (*d*) Has its internal energy changed?

38. We have seen that "energy conservation" is a universal law of nature. At the same time national leaders urge "energy conservation" upon us (driving slower, etc.). Explain the two quite different meanings of these words.

EXERCISES

Section 20–2 Quantity of Heat and Specific Heat

1. A certain substance has a molecular weight of 50 g/mol. 75 cal of heat are added to a 30-g sample of this material and its temperature rises from 25 to 45°C. (*a*) What is the specific heat of this substance? (*b*) What is its molar heat capacity?

2. A diet doctor encourages dieting by drinking ice water. His theory is that the body must burn off enough fat to raise the temperature of the water from 0°C to body temperature. How many liters of ice water would have to be consumed to burn off 454 g ($w = mg = 1$ lb) of fat, assuming that this requires 3500 food calories? (Each food calorie is equal to 1 kcal.) Why is it not advisable to follow this "diet"?

Section 20–3 Heat Conduction

3. The thermal conductivity of Pyrex glass at 0°C is 2.9×10^{-3} cal/cm $\cdot$ °C $\cdot$ s. (*a*) Express this in W/m $\cdot$ °C and in Btu/ft $\cdot$ °F $\cdot$ h. (*b*) What is the R-value for a $\frac{1}{4}$-in. sheet of such glass?

4. (*a*) Calculate the rate at which body heat flows out through the clothing of a skier given the following data: the body surface area is 1.8 m² and the clothing is 1.0 cm thick; skin surface temperature is 33°C, whereas the outer surface of the clothing is at -5.0 C; the thermal conductivity of the clothing is 0.040 W/m $\cdot$ K. (*b*) How would the answer change if, after a fall, the skier's clothes become soaked with water? Assume that the thermal conductivity of water is 0.60 W/m $\cdot$ K and neglect ice formation.

5. Consider the rod shown in Fig. 20–1. Suppose that $L = 25$ cm, $A = 1.0$ cm², and the material is copper. If $T_H = 125$°C, $T_C = 0$°C, and a steady state is reached, find (*a*) the temperature gradient, (*b*) the rate of heat transfer, and (*c*) the temperature at a point in the rod 10 cm from the high-temperature end.

6. A cylindrical copper rod of length 1.2 m and cross-sectional area 4.8 cm² is insulated to prevent heat loss through its surface. The ends are maintained at a temperature difference of 100°C by having one end in a water–ice mixture and the other in boiling water and steam. (*a*) Find the rate at which heat is transferred along the rod. (*b*) Find the rate at which ice melts at the cold end.

7. (*a*) What is the rate of heat loss in watts per square meter through a glass window 3.0 mm thick if the outside temperature is -20°F and the inside temperature is $+72$°F? (*b*) A stormwindow is installed having the same thickness of glass but with an air gap of 7.5 cm between the two windows. What will be the corresponding rate of heat loss, presuming that conduction is the only important heat-loss mechanism?

Section 20–4 The Mechanical Equivalent of Heat

8. What quantity of butter (6000 cal/g) would supply the potential energy needed for a 160-lb man to ascend to the top of Mt. Everest, elevation 29,000 ft, from sea level?

9. A pickup truck whose mass is 2200 kg is speeding along the highway at 65 mi/h. (*a*) If you could use all this kinetic energy to boil water already at 100°C, how much water could you boil? (*b*) If you had to buy this amount of energy from your local utility company at 12¢/kW $\cdot$ h, how much would it cost you? Guess at the answers before you figure them out; you may be surprised.

10. In a Joule experiment, a mass of 6.0 kg falls through a height of 50 m and rotates a paddle wheel that stirs 0.60 kg of water. The water is initially at 15°C. By how much does its temperature rise?

11. An energetic athlete dissipates all the energy in a diet of 4000 kcal per day. If he were to release this energy at a steady rate, how would this output compare with the energy output of a 100-W bulb? (*Note:* The calorie of nutrition is really a kilocalorie, as we have defined it.)

12. Power is supplied at the rate of 0.40 hp for 2.0 min in drilling a hole in a 1.0-lb copper block. (*a*) How much heat is generated? (*b*) What is the rise in temperature of the copper if only 75% of the power warms the copper? (*c*) What happens to the other 25%?

13. (*a*) Compute the possible increase in temperature for water going over Niagara Falls, 162 ft high. (*b*) What factors would tend to prevent this possible rise?

Section 20-5   Heat and Work

14. A sample of gas expands from 1.0 m^3 to 4.0 m^3 while its pressure decreases from 40 N/m^2 to 10 N/m^2. How much work is done by the gas if its pressure changes with volume according to each of the three processes shown in the p-V diagram in Fig. 20-12?

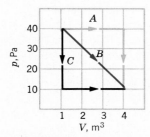

Figure 20-12   Exercise 14.

15. Suppose that a sample of gas expands from 1.0 m^3 to 4.0 m^3 along the path B in the p-V diagram shown in Fig. 20-13. It is then compressed back to 1.0 m^3 along either path A or path C. Compute the net work done by the gas for the complete cycle in each case.

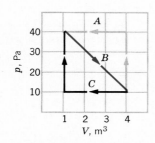

Figure 20-13   Exercise 15.

Section 20-6   The First Law of Thermodynamics

16. 200 J of work are done on a system and 70 cal of heat are extracted from the system. In the sense of the first law of thermodynamics, what are the values (including algebraic signs) of (a) W, (b) Q, and (c) ΔU?

Section 20-7   Some Applications of the First Law of Thermodynamics

17. A thermodynamic system is taken from an initial state A to another B and back again to A, via state C, as shown by the path A-B-C-A in the p-V diagram of Fig. 20-14a. (a) Complete the table in Fig. 20-14b by filling in appropriate + or −indications for the signs of the thermodynamic quantities associated with each process. (b) Calculate the numerical value of the work done by the system for the complete cycle A-B-C-A.

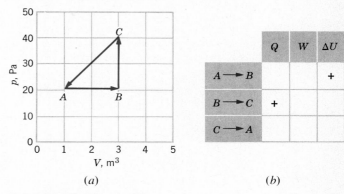

	Q	W	ΔU
$A \longrightarrow B$			+
$B \longrightarrow C$	+		
$C \longrightarrow A$			

(a)   (b)

Figure 20-14   Exercise 17.

18. Gas within a chamber passes through the cycle shown in Fig. 20-15. Determine the net heat added to the system during process CA if $Q_{AB} = 4.77$ cal, $Q_{BC} = 0$, and $W_{BCA} = 15.0$ J.

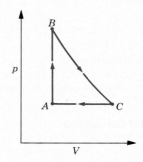

Figure 20-15   Exercise 18.

PROBLEMS GROUPED BY SECTION

Section 20-2   Quantity of Heat and Specific Heat

1. A 300-g copper object is dropped into a 150-g copper calorimeter containing 220 g of water at $20°$C. This causes the water to boil, 5.0 g being converted to steam. The heat of vaporization of water is 539 cal/g. What was the original temperature of the copper?

2. In a certain solar house, energy from the sun is stored in barrels filled with water. In a particular winter stretch of 5 cloudy days, 1.0×10^6 kcal are needed to maintain the inside of the house at $22°$C. Assuming that the water starts at $50°$C, what volume of water is required?

3. Calculate the specific heat of a metal from the following data. A container made of the metal weighs 8.0 lb (mass = 3.6 kg) and contains 30 lb (mass = 14 kg) of water. A 4.0-lb (mass = 1.8

kg) piece of the metal initially at a temperature of $350°$F ($= 180°$C) is dropped into the water. The container and water initially have a temperature of $60°$F ($= 16°$C), and the final temperature of the entire system is $65°$F ($= 18°$C).

4. A person makes a quantity of iced tea by mixing 500 g of the hot tea (essentially water) with an equal mass of ice at $0°$C. What is the final situation (temperature and mass of ice remaining) if the initial hot tea is at a temperature of (a) $90°$C; (b) $70°$C?

5. (a) Two 50-g ice cubes are dropped into 200 g of water in a glass. If the water was initially at a temperature of $25°$C, and if the ice came directly from a freezer operating at $-15°$C, what will be the final temperature of the drink? The specific heat of ice is approximately 0.5 cal/g·$°$C in this temperature range, and the heat required to melt ice to water is approximately 80 cal/g. (b)

Suppose that only one ice cube had been used in (a); what would be the final temperature of the drink?

6. The specific heat of a substance varies with temperature according to the formula $c = 0.20 + 0.14T + 0.023T^2$, with T in °C and c in cal/g·°C. Find the heat required to raise the temperature of 2.0 g of this substance from 5.0°C to 15°C.

Section 20–3   Heat Conduction
7. Two identical rectangular rods of metal are welded end to end as shown in Fig. 20–16a. Assuming that 10 cal of heat flows through the rods in 2.0 min, how long would it take for 10 cal to flow through the rods if they are welded as shown in Fig. 20–16b?

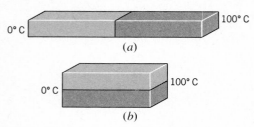

(a)

(b)

Figure 20–16   Problems 7 and 25.

8. Compare the heat flows through two storm doors 2.0 m high and 0.75 m wide. (a) One door is made with aluminum panels 1.5 mm thick and a 3.0-mm glass pane that covers 75% of its surface (the structural frame is considered to have a negligible area). (b) The second door is made entirely of wood averaging 2.5 cm in thickness. Take the temperature drop through the doors to be 33°C (=60°F). See Table 20–2.

9. A container of water has been outdoors in cold weather until a 5.0-cm-thick slab of ice has formed on its surface (Fig. 20–17). The air above the ice is at −10°C. Calculate the rate of formation of ice (in centimeters per hour) on the bottom surface of the ice slab. Take the thermal conductivity, density, and heat of fusion of ice to be 0.0040 cal/s·cm·°C, 0.92 g/cm³, and 80 cal/g, respectively. Assume that no heat enters or leaves the water through the walls of the container.

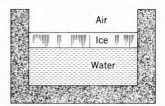

Figure 20–17   Problem 9.

10. Ice has formed on a shallow pond and equilibrium has been reached with the air above the ice at −5°C and the bottom of the pond at 4°C. If the total depth of ice + water is 1.0 m, how thick is the ice? (Assume that the thermal conductivities of ice and water are 0.40 and 0.12 cal/m·C·s, respectively.)

11. Assume that the thermal conductivity of copper is twice that of aluminum and four times that of brass. Three metal rods, made of copper, aluminum, and brass, are each 6.0 in. long and 1.0 in. in diameter. These rods are placed end to end, with the aluminum between the other two. The free ends of the copper and brass rods are maintained at 100°C and 0°C, respectively. Find the equilibrium temperatures of the copper-aluminum junction and the aluminum-brass junction.

Section 20–4   The Mechanical Equivalent of Heat
12. How long does it take a 2.0×10^5 Btu/h water heater to raise the temperature of a 40-gal tub from 70°F to 100°F?

13. An athlete needs to lose weight and he decides to do this by "pumping iron." (a) How many times does he have to raise an 80-kg "weight" a distance of 1 m in order to burn off 1 lb of fat, assuming that it takes 3500 food calories (=3500 kcal) to do this? (b) If he lifts the weight once every 2 s, how long does it take?

14. A 1500-kg Mercedes moving at 90 km/h brakes to rest, at uniform acceleration, over a distance of 80 m. At what average rate is thermal energy delivered to the brake system?

15. A chef, upon awakening one morning to find his stove out of order, decides to boil the water for his wife's coffee by shaking it in a Thermos flask. Suppose that he uses $\frac{1}{2}$ liter of tap water at 59°F, and that the water falls 1.0 ft each shake, the chef making 30 shakes each minute. Neglecting any loss of energy, how long must he shake the flask before the water boils?

Section 20–6   The First Law of Thermodynamics
16. Determine the value of J, the mechanical equivalent of heat, from the following data: 2000 cal (=7.936 Btu) of heat are supplied to a system; the system does 3350 J (=2471 ft·lb) of external work during that time; the increase of internal energy during the process is 5030 J (=3710 ft·lb).

Section 20–7   Some Applications of the First Law of Thermodynamics
17. Gas within a chamber undergoes the processes shown in the p-V diagram of Fig. 20–18. Calculate the net heat added to the system during one complete cycle.

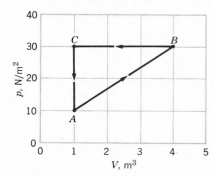

Figure 20–18   Problem 17.

18. Figure 20–19a shows a cylinder containing gas and closed by a movable piston. The cylinder is submerged in an ice-water mixture. The piston is *quickly* pushed down from position (1) to position (2). The piston is held at position (2) until the gas is again at 0°C and then is *slowly* raised back to position (1). Figure 20–19b is a p-V diagram for the process. If 100 g of ice are melted during the cycle, how much work has been done *on* the gas?

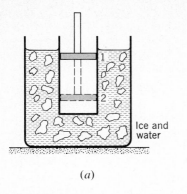
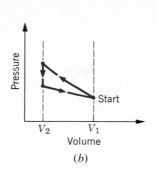

(a)

(b)

Figure 20–19 Problem 18.

*19. A cylinder has a well-fitted 2.0-kg metal piston whose cross-sectional area is 2.0 cm² (Fig. 20–20). The cylinder contains water and steam at constant temperature. The piston is observed to fall slowly at a rate of 0.30 cm/s because heat flows out of the cylinder through the cylinder walls. As this happens, some steam condenses in the chamber. The density of the steam inside the chamber is 6.0×10^{-4} g/cm³ and the atmospheric pressure is 1.0 atm. (a) Calculate the rate of condensation of steam. (b) At what rate is heat leaving the chamber? (c) What is the rate of change of internal energy of the steam and water inside the chamber?

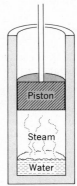

Figure 20–20 Problem 19.

ADDITIONAL PROBLEMS

20. A small electric immersion heater is used to boil 100 g of water for a cup of instant coffee. The heater is labeled "200 watts." Calculate the time required to bring this water from 23°C to the boiling point, ignoring any heat losses.

21. A block of ice at 0°C whose mass initially is 50.0 kg slides along a horizontal surface, starting at a speed of 5.38 m/s and finally coming to rest after traveling 28.3 m. Compute the mass of ice melted as a result of the friction between the block and the surface. (Assume that all the heat generated due to friction goes into the block of ice.)

22. Icebergs in the North Atlantic present hazards to shipping (see Fig. 20–21), causing shipping routes to increase by about 30% during the iceberg season. Attempts to destroy icebergs include planting explosives, bombing, torpedoing, shelling, ramming, and painting with lampblack. Suppose that direct melting of the iceberg, by placing heat sources in the ice, is tried. How much heat is required to melt 10% of a 200,000-ton iceberg?

Figure 20–21 Problem 22.

23. A large water tank with a bottom 1.67 m in diameter is made of iron boilerplate 5.2 mm thick. As the water is being heated, the gas burner underneath is able to maintain a temperature difference of 2°C between the surfaces of the iron. How much heat flows through to the water in 5 min? (Iron has a specific heat of 0.105 cal/g·°C and a thermal conductivity of 16 cal/m·°C·s.)

24. A thermometer of mass 0.055 kg and of specific heat 0.20 cal/g·°C reads 15.0°C. It is then completely immersed in 0.300 kg of water, and it comes to the same final temperature as the water. If the thermometer reads 44.4°C and is accurate, what was the temperature of the water before insertion of the thermometer, neglecting other heat losses?

25. Show that in a compound slab such as that in Fig. 20–16a, the temperature gradient in each portion in inversely proportional to the thermal conductivity.

26. Four square pieces of insulation of two different materials, all with the same thickness and area A, are available to cover an opening of area $2A$. This can be done in either of the two ways shown in Fig. 20–22. Which arrangement, (a) or (b), would give the lower heat flow if $k_2 \neq k_1$?

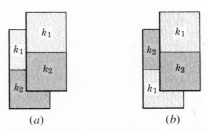

(a)   (b)

Figure 20–22 Problem 26.

27. In a solar water heater, energy from the sun is gathered by rooftop collectors, which circulate water through tubes in the collector. The solar radiation enters the collector through a transparent cover and warms up the water in the tubes; this water is pumped into a holding tank. Assuming that the efficiency of the overall system is 20% (i.e., 80% of the incident solar energy is lost from the system), what collector area is necessary to take water from a 200-liter tank and raise its temperature from 20°C to 40°C in 1.0 h? The intensity of incident sunlight is 700 W/m².

28. Show that the temperature T_x at the interface of a compound slab (see Example 2) is given by

$$T_x = \frac{R_1 T_H + R_2 T_C}{R_1 + R_2},$$

where T_H is the outer surface temperature of the slab with R-value $= R_2$, and T_C is the outer surface temperature of the slab with R-value $= R_1$.

29. An insulating cup contains 130 cm³ of hot coffee, at a temperature of 80°C. You put in a 12-g ice cube at 0°C to cool it. (*a*) By how many degrees will your coffee have cooled once the ice has melted? (*b*) What percent of this change is due to heating the ice water and what percent to melting the ice? The latent heat of fusion of water is 79.7 cal/g. Treat the coffee as though it were pure water.

30. A copper ring has a diameter of 1.00000 in. at its temperature of 0°C. An aluminum sphere has a diameter of 1.00200 in. at its temperature of 100°C. The sphere is placed on top of the ring (Fig. 20–23), and the two are allowed to come to thermal equilibrium, no heat being lost to the surroundings. The sphere just passes through the ring at the equilibrium temperature. What is the ratio of the mass of the sphere to the mass of the ring?

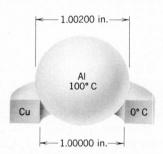

Figure 20–23   Problem 30.

31. When a system is taken from state *i* to state *f* along the path *iaf* in Fig. 20–24, it is found that $Q = 50$ cal and $W = 20$ cal. Along the path *ibf*, $Q = 36$ cal. (*a*) What is W along the path *ibf*? (*b*) If $W = -13$ cal for the curved return path *fi*, what is Q for this path? (*c*) Take $U_i = 10$ cal. What is U_f? (*d*) If $U_b = 22$ cal, what is Q for the process *ib*? for the process *bf*?

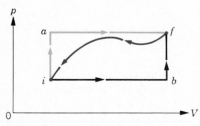

Figure 20–24   Problem 31.

32. A "flow calorimeter" is used to measure the specific heat of a liquid. Heat is added at a known rate to a stream of the liquid as it passes through the calorimeter at a known rate. Then a measurement of the resulting temperature difference between the inflow and the outflow points of the liquid stream enables us to compute the specific heat of the liquid. A liquid of density 0.85 g/cm³ flows through a calorimeter at the rate of 8.0 cm³/s. Heat is added by means of a 250-W electric heating coil, and a temperature difference of 15°C is established in steady-state conditions between the inflow and the outflow points. Find the specific heat of the liquid.

33. An iron ball is dropped onto a concrete floor from a height of 10 m. On the first rebound it rises to a height of 0.50 m. Assume that all the macroscopic mechanical energy lost in the collision with the floor goes into the ball. The specific heat of iron is 0.12 cal/g·°C. During the collision, (*a*) has heat been added to the ball? (*b*) Has work been done on it? (*c*) Has its internal energy changed? If so, by how much? (*d*) How much has the temperature of the ball risen after the first collision?

34. By means of a heating coil, energy is transferred at a constant rate to a substance in a thermally insulated container. The temperature of the substance is measured as a function of time. (*a*) Show how we can deduce from this information the way in which the heat capacity of the substance depends on the temperature. (*b*) Suppose that in a certain temperature range it is found that the temperature T is proportional to t^3, where t is the time. How does the heat capacity depend on T in this range?

35. A wall assembly consists of a 20 ft × 12 ft frame made of 16 two-by-four vertical studs, each 12 ft long and set with their center lines 16 in. apart. The outside of the wall is faced with ¼-in. plywood sheet ($R = 0.30$) and ¾-in. white pine siding ($R = 0.98$). The inside is faced with ¼-in. plasterboard ($R = 0.47$), and the space between the studs is filled with polyurethane foam ($R = 5.9$ for a 1-in. layer). A "two-by-four" is actually 1.75 in. × 3.75 in. in size. Assume that they are made of wood for which $R = 1.3$ for a 1-in. slab. (*a*) At what rate does heat flow through this wall for a 30°F temperature difference? (*b*) What is the R-value for the assembled wall? (*c*) What fraction of the wall area contains studs, as opposed to foam? (*d*) What fraction of the heat flow is through the studs, as opposed to the foam?

36. Two metal blocks are insulated from their surroundings. The first block, which has a mass $m_1 = 3.16$ kg and is at a temperature $T_1 = 17°C$, has a specific heat four times that of the second block. This second block is at a temperature $T_2 = 47°C$, and its coefficient of linear expansion is $15 \times 10^{-6}/°C$. When the two blocks are brought together and allowed to come to thermal equilibrium, the area of one face of the second block is found to have decreased by 0.030%. Find the mass m_2 of the second block.

37. A 2.0-g lead bullet moving at 187 m/s becomes embedded in a 215-g wooden block suspended as a pendulum bob (a ballistic pendulum). Calculate the rise in temperature of the bullet, assuming that all of the absorbed macroscopic mechanical energy raises the bullet's temperature.

38. When 1.0 liter of gasoline is burned completely, it releases 7.92×10^6 cal of thermal energy. If all that energy could be converted into kinetic energy for launching a 1.00-kg projectile vertically from the surface of the earth, what altitude would the projectile reach?

21 Kinetic Theory of Gases

21–1 Introduction

Thermodynamics, statistical mechanics, and kinetic theory

Thermodynamics deals only with macroscopic variables, such as pressure, temperature, and volume. Its basic laws, expressed in terms of such quantities, say nothing at all about the fact that matter is made up of atoms. *Statistical mechanics*, however, which deals with the same areas of science that thermodynamics does, presupposes the existence of atoms. Its basic laws are the laws of mechanics, which are applied to the atoms that make up the system.

No existing electronic computer could solve the problem of applying the laws of mechanics individually to every atom in a bottle of oxygen, for instance. Even if the problem could be solved, the results of such calculations would be too voluminous to be useful. Fortunately, the detailed life histories of individual atoms in a gas are not important if we want to calculate only the macroscopic behavior of the gas. We apply the laws of mechanics statistically, then, and we find that we are able to express all the thermodynamic variables as certain averages of atomic properties. For example, the pressure exerted by a gas on the wall of the containing vessel is the average rate per unit area at which the atoms of the gas transfer momentum to the wall as they collide with it. The number of atoms in a macroscopic system is usually so large that such averages are very sharply defined quantities indeed.

We can apply the laws of mechanics statistically to assemblies of atoms at two different levels. At the level called *kinetic theory* we proceed in a rather physical way, using relatively simple mathematical averaging techniques. In this chapter we shall use these methods to enlarge our understanding of pressure, temperature, specific heat, and internal energy at the atomic level. In this book we apply kinetic theory to gases only, because the interactions between atoms in gases are much weaker than in liquids and solids; this greatly simplifies the mathematical difficulties.

At another level, we can apply the laws of mechanics statistically, using techniques that are more formal and abstract than those of kinetic theory. This approach, developed by J. Willard Gibbs (1839–1903) and by Ludwig Boltzmann (1844–1906) among others, is called *statistical mechanics,* a term that includes kinetic theory as a sub-branch. Using these methods one can derive the laws of thermodynamics, thus establishing that science as a branch of mechanics. The fullest flowering of statistical mechanics (*quantum statistics*) involves the statistical application of the laws of quantum mechanics—rather than those of classical mechanics—to many-atom systems.

21–2   Ideal Gas—A Macroscopic Description

Let a certain amount of a gas be confined in a container of volume V. It is clear that we can reduce its density either by removing some gas from the container or by putting the gas in a larger container. We find from experiment that, at low enough densities, all gases tend to show a certain simple relationship among the thermodynamic variables p, V, and T. This suggests the concept of an *ideal gas,* one that would have the same simple behavior under all conditions of temperature and pressure. In this section we give a macroscopic or thermodynamic definition of an ideal gas. In Section 21–3 we shall define an ideal gas microscopically, from the standpoint of kinetic theory, and we shall see what we can learn by comparing these two approaches.

Given any gas in a state of thermal equilibrium we can measure its pressure p, its temperature T, and its volume V. For low enough values of the density experiment shows that (1) for a given mass of gas held at a constant temperature, the pressure is inversely proportional to the volume (Boyle's law), and (2) for a given mass of gas held at a constant pressure, the volume is directly proportional to the temperature (law of Charles and Gay-Lussac). We can summarize these two experimental results by the relation

$$\frac{pV}{T} = \text{a constant} \qquad \text{(for a fixed mass of gas).} \tag{21–1}$$

The volume occupied by a gas at a given pressure and temperature is proportional to its mass. Thus the constant in Eq. 21–1 must also be proportional to the mass of the gas. We therefore write the constant in Eq. 21–1 as nR, where n is the number of moles of gas in the sample* and R is a constant that must be determined for each gas by experiment. Experiment shows that, at low enough densities, R has the same value for all gases, namely,

$$R = 8.314 \text{ J/mol·K} = 0.08206 \text{ li·atm/mol·K.}$$

R is called the *universal gas constant.* We then write Eq. 21–1 as

Ideal gas equation of state

$$pV = nRT, \tag{21–2}$$

---

* A *mole* (abbr. mol) of any substance is the amount of the substance that contains a specified number of elementary entities, namely, 6.022045×10^{23}, called the Avogadro constant. This number is the result of the defining relation that one mole of carbon atoms (actually, of the isotope ^{12}C) shall have a mass of 12 g, exactly. The *gram molecular weight* M of a substance is the number of grams per mole of that substance. Thus the gram molecular weight of ordinary oxygen molecules is 32.0 g/mol. Although the mole is an amount of substance, we cannot translate it into mass, as grams, until we specify what the elementary entity is; it may be atoms, molecules, ions, electrons, other particles, or specified groups of such particles.

and we define an ideal gas as one that obeys this relation under all conditions. There is no such thing as a truly ideal gas, but it remains a useful and simple concept connected with reality by the fact that all real gases approach the ideal gas abstraction in their behavior if the density is low enough. Equation 21–2 is called the *equation of state* of an ideal gas.

If we could fill the bulb of an (ideal) constant-volume gas thermometer with an ideal gas, we see from Eq. 21–2 that we could define temperature in terms of its pressure readings; that is,

$$T = \left(\frac{p}{p_{tr}}\right) \times 273.16 \text{ K} \qquad \text{(ideal gas).}$$

Here p_{tr} is the gas pressure at the triple point of water, at which the temperature T_{tr} is 273.16 K by definition. In practice we must fill our thermometer with a real gas and measure the temperature by extrapolating to zero density using Eq. 19–3,

$$T = \left[\lim_{p_{tr} \to 0} \frac{p}{p_{tr}}\right] \times 273.16 \text{ K} \qquad \text{(real gas).} \qquad [19–3]$$

Example 1   A cylinder contains oxygen at a temperature of 20°C and a pressure of 15 atm in a volume of 100 liters. A piston is lowered into the cylinder, decreasing the volume occupied by the gas to 80 liters and raising the temperature to 25°C. Assuming oxygen to behave like an ideal gas under these conditions, what is the gas pressure?

From Eq. 21–1, since the mass of gas remains unchanged, we may write

$$\frac{p_i V_i}{T_i} = \frac{p_f V_f}{T_f}.$$

Our initial conditions are

$$p_i = 15 \text{ atm}, \qquad T_i = 293 \text{ K}, \qquad V_i = 100 \text{ liters}.$$

Our final conditions are

$$p_f = ?, \qquad T_f = 298 \text{ K}, \qquad V_f = 80 \text{ liters}.$$

Hence,

$$p_f = \left(\frac{T_f}{V_f}\right)\left(\frac{p_i V_i}{T_i}\right) = \left(\frac{298 \text{ K}}{80 \text{ liters}}\right)\left(\frac{15 \text{ atm} \times 100 \text{ liters}}{293 \text{ K}}\right) = 19 \text{ atm}.$$

Example 2   Calculate the work per mole done by an ideal gas that expands isothermally, that is, at constant temperature, from an initial volume V_i to a final volume V_f.

The work done may be represented as

$$W = \int_{V_i}^{V_f} p \, dV. \qquad [20–5]$$

From the ideal gas law we have

$$p = \frac{nRT}{V},$$

so that W/n, the work per mole, is

$$\frac{W}{n} = \int_{V_i}^{V_f} \frac{RT}{V} \, dV.$$

The temperature is constant so that

$$\frac{W}{n} = RT \int_{V_i}^{V_f} \frac{dV}{V} = RT \ln \frac{V_f}{V_i} \qquad \text{(ideal gas)}$$

is the work per mole done by an ideal gas in an isothermal expansion at temperature T from an initial volume V_i to a final volume V_f.

Notice that when the gas expands, so that $V_f > V_i$, the work done by the gas is positive; when the gas is compressed, so that $V_f < V_i$, the work done by the gas is negative. This is consistent with the sign convention adopted for W in the first law of thermodynamics. The work done is shown as the shaded area in Fig. 21–1. The curved line is an isotherm, that is, a curve giving the relation of p to V at a constant temperature.

In practice, how can we keep an expanding or contracting gas at constant temperature?

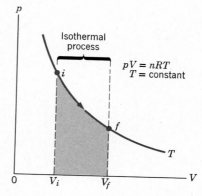

Figure 21–1   Example 2. The shaded area represents the work done by n moles of gas in expanding from V_i to V_f with the temperature held constant.

21–3 Ideal Gas—A Microscopic Description

From the microscopic point of view we define an ideal gas by making the following assumptions; it will then be our task to apply the laws of classical mechanics statistically to the gas atoms and to show that our microscopic definition is consistent with the macroscopic definition of the preceding section:

Description of an ideal gas

1. *A gas consists of particles, called molecules.* Depending on the gas, each molecule consists of one atom or a group of atoms. If the gas is an element or a compound and is in a stable state, we consider all its molecules to be identical.

2. *The molecules are in random motion and obey Newton's laws of motion.* The molecules move in all directions and with various speeds. In computing the properties of the motion, we assume that Newtonian mechanics works at the microscopic level. As for all our assumptions, this one will stand or fall depending on whether or not the experimental facts it predicts are correct.

3. *The total number of molecules is large.* The direction and speed of motion of any one molecule may change abruptly upon collision with the wall or another molecule. Any particular molecule will follow a zigzag path because of these collisions. However, because there are so many molecules we assume that the resulting large number of collisions maintains the overall distribution of molecular velocities and the randomness of the motion.

4. *The volume of the molecules is a negligibly small fraction of the volume occupied by the gas.* Even though there are many molecules, they are extremely small. We know that the volume occupied by a gas can be changed through a large range of values with little difficulty, and that when a gas condenses the volume occupied by the liquid may be thousands of times smaller than that of the gas. Hence, our assumption is plausible.

5. *No appreciable forces act on the molecules except during a collision.* To the extent that this is true a molecule moves with uniform velocity between collisions. Because we have assumed the molecules to be so small, the average distance between molecules is large compared to the size of a molecule. Hence, we assume that the range of molecular forces is comparable to the molecular size.

6. *Collisions are elastic and are of negligible duration.* Collisions between molecules and with the walls of the container conserve momentum and (we assume) kinetic energy. Because the collision time is negligible compared to the time spent by a molecule between collisions, the kinetic energy that is converted to potential energy during the collision is available again as kinetic energy after such a brief time that we can ignore this exchange entirely.

21–4 Kinetic Calculation of Pressure

Let us now calculate the pressure of an ideal gas from kinetic theory. To simplify matters, we consider a gas in a cubical vessel whose walls are perfectly elastic. Let each edge be of length l. Call the faces normal to the x-axis (Fig. 21–2) A_1 and A_2, each of area l^2. Consider a molecule that has a velocity $\mathbf{v}$. We can resolve $\mathbf{v}$ into components v_x, v_y, and v_z in the directions of the edges. Since each wall is perfectly elastic, this particle will rebound, after collision with A_1, with its x-

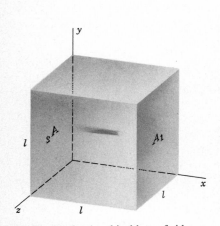

Figure 21–2 A cubical box of side l, containing an ideal gas. A molecule is shown moving toward A_1.

component of velocity exactly reversed. There will be no effect on v_y or v_z, so that the change Δp in the particle's momentum will be

$$\Delta p = p_f - p_i = (-mv_x) - (mv_x) = -2mv_x,$$

normal to A_1. Hence, the momentum imparted to A_1 will be $2mv_x$, since the total momentum is conserved.

Suppose that this particle reaches A_2 without striking any other particle on the way. The time required to cross the cube will be l/v_x. At A_2 it will again have its x-component of velocity reversed and will return to A_1. Assuming no collisions in between, the round trip will take a time $2l/v_x$. Hence, the number of collisions per unit time this particle makes with A_1 is $v_x/2l$, so that the rate at which it transfers momentum to A_1 is

$$2mv_x \frac{v_x}{2l} = \frac{mv_x^2}{l}.$$

To obtain the total force on A_1, that is, the rate at which momentum is imparted to A_1 by all the gas molecules, we must sum up mv_x^2/l for all the particles. Then, to find the pressure, we divide this force by the area of A_1, namely, l^2.

If m is the mass of each molecule, we have

$$p = \frac{m}{l^3}(v_{x1}^2 + v_{x2}^2 + \cdots),$$

where v_{x1} is the x-component of the velocity of particle 1, v_{x2} is that of particle 2, and so on. If N is the total number of particles in the container, we can write this as

$$p = \frac{mN}{V}\left(\frac{v_{x1}^2 + v_{x2}^2 + \cdots}{N}\right).$$

But mN is simply the total mass of the gas and mN/V is the mass per unit volume, that is, the density ρ. The quantity $(v_{x1}^2 + v_{x2}^2 + \cdots)/N$ is the average value of v_x^2 for all the particles in the container. Let us call this $\overline{v_x^2}$. Then

$$p = \rho \overline{v_x^2}.$$

For any particle $v^2 = v_x^2 + v_y^2 + v_z^2$. Because we have many particles and because they are moving entirely at random, the average values of v_x^2, v_y^2, and v_z^2 are equal and the value of each is exactly one-third the average value of v^2. There is no preference among the molecules for motion along any one of the three axes. Hence, $\overline{v_x^2} = \frac{1}{3}\overline{v^2}$, so that

Pressure and molecular speed

$$p = \rho \overline{v_x^2} = \tfrac{1}{3}\rho \overline{v^2}. \tag{21–3}$$

Although we derived this result by neglecting collisions between particles, the result is true even when we consider collisions. Because of the exchange of velocities in an elastic collision between identical particles, there will always be some one molecule that will collide with A_2 with momentum mv_x corresponding to the one that left A_1 with this same momentum. Also, the time spent during collisions is negligible compared to the time spent between collisions. Hence, our neglect of collisions is merely a convenient device for calculation. Likewise, we could have chosen a container of any shape —the cube merely simplifies the calculation. Although we have calculated the pressure exerted only on the side A_1, if we neglect the weight of the gas, it follows from Pascal's law that the pressure is the same on all sides and everywhere in the interior.

The square root of $\overline{v^2}$ is called the *root-mean-square* speed of the molecules and is a kind of average molecular speed (see Section 21–9). Using Eq. 21–3, we can calculate this root-mean-square speed from measured values of the pressure and density of the gas. Thus,

Speed of molecules

$$v_{\text{rms}} = \sqrt{\overline{v^2}} = \sqrt{\frac{3p}{\rho}}. \qquad (21\text{--}4a)$$

In Eq. 21–3 we relate a macroscopic quantity (the pressure p) to an average value of a microscopic quantity (that is, to $\overline{v^2}$ or v_{rms}^2). However, averages can be taken over short times or over long times, over small regions of space or large regions of space. The average computed in a small region for a short time might depend on the time or region chosen, so that the values obtained in this way may fluctuate. This could happen in a gas of very low density, for example. We can ignore fluctuations, however, when the number of particles in the system is large enough.

Example 3 Calculate the root-mean-square speed of hydrogen molecules at 0.00°C and 1.00 atm pressure, assuming hydrogen to be an ideal gas. Under these conditions hydrogen has a density ρ of 8.99×10^{-2} kg/m³. Then, since $p = 1.00$ atm $= 1.01 \times 10^5$ N/m²,

$$v_{\text{rms}} = \sqrt{\frac{3p}{\rho}} = \sqrt{\frac{3 \times 1.01 \times 10^5 \text{ N/m}^2}{8.99 \times 10^{-2} \text{ kg/m}^3}} = 1840 \text{ m/s}.$$

This is of the order of a mile per second, or 3600 mi/h.

Table 21–1 gives the results of similar calculations for some gases at 0°C. These molecular speeds are roughly of the same order as the speed of sound at the same temperature. For example, in air at 0°C, $v_{\text{rms}} = 485$ m/s and the speed of sound is 331 m/s, and in hydrogen $v_{\text{rms}} = 1838$ m/s and sound travels at 1286 m/s. These results are to be expected in terms of our model of a gas; see Problem 30. We must distinguish between the speeds of individual molecules, described by v_{rms}, and the very much lower speed at which one gas diffuses into another. Thus, if we open a bottle of ammonia in one corner of a room, we smell ammonia in the opposite corner only after an easily measurable time lag. The diffusion speed is slow because large numbers of collisions with air molecules greatly reduce the tendency of ammonia molecules to spread themselves uniformly throughout the room.

Table 21–1 Some molecular speeds

Gas	Molecular weight,[a] g/mol	v_{rms} (at 0°C), m/s	Translational kinetic energy per mole (at 0°C), $\frac{1}{2}Mv_{\text{rms}}^2$, J/mol
H_2	2.02	1838	3370
He	4.0	1311	3430
H_2O	18	615	3400
Ne	20.1	584	3420
N_2	28	493	3390
CO	28	493	3390
Air	28.8	485	3280
O_2	32	461	3400
CO_2	44	393	3400

[a] The molecular weight and the mole are defined on p. 375. We will discuss the last column in this table in the next section.

Example 4   Assuming that the speed of sound in a gas is the same as the root-mean-square speed of the molecules, show how the speed of sound for an ideal gas would depend on the temperature. (Actually this assumption is only crudely correct. Compare Eq. 21–4a and Example 6.)

The density of a gas is

$$\rho = \frac{nM}{V}$$

in which M is the molecular weight (grams per mole) and n is the number of moles. Combining this with the ideal gas law,

$$pV = nRT,$$

yields

$$\frac{p}{\rho} = \frac{RT}{M}.$$

We obtain, from Eq. 21–4a,

$$v_{rms} = \sqrt{\frac{3p}{\rho}} = \sqrt{\frac{3RT}{M}}, \qquad (21\text{–}4b)$$

so that the speed of sound v_1 at a temperature T_1 is related to the speed of sound v_2 in the same gas at a temperature T_2 by

$$\frac{v_1}{v_2} = \sqrt{\frac{T_1}{T_2}}.$$

For example, if the speed of sound at 273 K is 332 m/s in air, its speed in air at 300 K will be

$$\sqrt{\tfrac{300}{273}} \times 332 \text{ m/s} = 348 \text{ m/s}.$$

Would our result change if the speed of sound were proportional to, rather than equal to, the root-mean-square speed of the molecules of a gas? See Problem 30.

21–5 Kinetic Interpretation of Temperature

If we multiply each side of Eq. 21–3 by the volume V, we obtain

$$pV = \tfrac{1}{3}\rho V v_{rms}^2,$$

where ρV is simply the total mass of gas, ρ being the density. We can also write the mass of gas as nM, in which n is the number of moles and M is the molecular weight. Making this substitution yields

$$pV = \tfrac{1}{3}nM v_{rms}^2.$$

The quantity $\tfrac{1}{3}nM v_{rms}^2$ is two-thirds the total kinetic energy of translation of the molecules, that is, $\tfrac{2}{3}(\tfrac{1}{2}nM v_{rms}^2)$. We can then write

$$pV = \tfrac{2}{3}(\tfrac{1}{2}nM v_{rms}^2).$$

The equation of state of an ideal gas is

$$pV = nRT.$$

Combining these two expressions, we obtain

$$\tfrac{1}{2}M v_{rms}^2 = \tfrac{3}{2}RT. \qquad (21\text{–}5)$$

Kinetic energy per mole

That is, *the total translational kinetic energy per mole of the molecules of an ideal gas is proportional to the temperature*. We may say that this result, Eq. 21–5, is necessary to fit the kinetic theory to the equation of state of an ideal gas, or we may consider Eq. 21–5 as a definition of gas temperature on a kinetic theory or microscopic basis. In either case, we gain some insight into the meaning of temperature for gases.

The temperature of a gas is related to the total translational kinetic energy measured with respect to the center of mass of the gas. The kinetic energy associated with the motion of the center of mass of the gas has no bearing on the gas temperature. The temperature of a gas in a container does not increase when we put the container on a moving train.

Let us now divide each side of Eq. 21–5 by the Avogadro constant. N_A, which (see p. 408, footnote) is the number of molecules per mole of a gas. Thus M/N_A $(=m)$ is the mass of a single molecule and we have

$$\frac{1}{2}\left(\frac{M}{N_A}\right) v_{rms}^2 = \frac{1}{2}m v_{rms}^2 = \frac{3}{2}\left(\frac{R}{N_A}\right)T.$$

Now $\frac{1}{2}mv_{\text{rms}}^2$ is the average translational kinetic energy per molecule. The ratio R/N_A—which we call k, the *Boltzmann constant*—plays the role of the gas constant per molecule. We have

Kinetic energy per molecule

$$\frac{1}{2}mv_{\text{rms}}^2 = \frac{3}{2}kT \qquad (21\text{--}6)$$

in which

$$k = \frac{R}{N_A} = \frac{8.314 \text{ J/mol K}}{6.022 \times 10^{23} \text{ molecules/mol}} = 1.381 \times 10^{-23} \text{ J/molecule K}.$$

We shall return to the Boltzmann constant in Section 21–9.

In the last column of Table 21–1 we list calculated values of $\frac{1}{2}Mv_{\text{rms}}^2$. As Eq. 21–5 predicts, this quantity (the translational kinetic energy per mole) has (closely) the same value for all gases at the same temperatures, 0°C in this case. From Eq. 21–6 we conclude that at the same temperature T the ratio of the root-mean-square speeds of molecules of two different gases is equal to the square root of the inverse ratio of their masses. That is, from

$$T = \frac{2}{3k}\frac{m_1 v_{1\text{rms}}^2}{2} = \frac{2}{3k}\frac{m_2 v_{2\text{rms}}^2}{2}$$

we obtain

$$\frac{v_{1\text{rms}}}{v_{2\text{rms}}} = \sqrt{\frac{m_2}{m_1}}. \qquad (21\text{--}7)$$

We can apply Eq. 21–7 to the diffusion of two different gases in a container with porous walls placed in an evacuated space. The lighter gas, whose molecules move more rapidly on the average, will escape faster than the heavier one. The ratio of the numbers of molecules of the two gases that find their way through the porous walls for a short time interval is equal to the square root of the inverse ratio of their masses, $\sqrt{m_2/m_1}$. This diffusion process is one method of separating (readily fissionable) ^{235}U (0.7% abundance) from a normal sample of uranium containing mostly ^{238}U (99.3% abundance).

21–6   Specific Heats of an Ideal Gas

We picture the molecules in an ideal gas as hard elastic spheres; that is, we assume that there are no forces between the molecules except during collisions and that the molecules are not deformed by collisions. If this is so, there is no internal potential energy and the internal energy of an ideal gas is entirely kinetic. We have already found that the average translational kinetic energy per molecule is $\frac{3}{2}kT$, so that the internal energy U of an ideal monatomic gas containing N molecules is

Internal energy of an ideal monatomic gas

$$U = \frac{3}{2}NkT = \frac{3}{2}nRT. \qquad (21\text{--}8)$$

This prediction of kinetic theory says that *the internal energy of an ideal gas is proportional to the Kelvin temperature and depends only on the temperature,* being independent of pressure and volume. With this result we can now obtain information about the specific heats of an ideal gas.

The specific heat of a substance is the heat required per unit mass per unit temperature change. A convenient unit of mass is the mole. The corresponding specific heat is called the *molar heat capacity* and is represented by C. Only two varieties of molar heat capacity are important for gases, namely, that at constant volume, C_v, and that at constant pressure, C_p.

Let us confine a certain number of moles of an ideal gas in a piston-cylinder arrangement as in Fig. 21–3a. The cylinder rests on a heat reservoir whose tem-

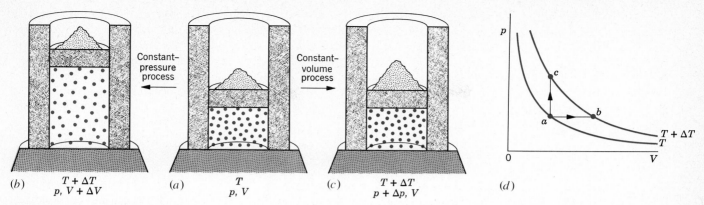

(b) $\quad T + \Delta T$
$\quad\quad p, V + \Delta V$

(a) $\quad\quad T$
$\quad\quad\quad p, V$

(c) $\quad\quad T + \Delta T$
$\quad\quad\quad p + \Delta p, V$

(d)

Figure 21–3 The temperature of a given mass of gas is raised by the same amount by a constant-pressure process ($a \rightarrow b$) and by a constant-volume process ($a \rightarrow c$).

perature can be raised or lowered at will, so that we may add heat to the system or remove it, as we wish. The gas has a pressure p such that its upward force on the (frictionless) piston just balances the weight of the piston and its sand load. The state of the system is represented by point a in the p-V diagram of Fig. 21–3d; this diagram shows two isothermal lines, all points on one corresponding to a temperature T and all points on the other to a (higher) temperature $T + \Delta T$.

Now let us raise the temperature of the system by ΔT, by slowly increasing the reservoir temperature. As we do this let us add sand to the piston so that the volume V does not change. This constant-volume process carries the system from the initial state of Fig. 21–3a to the final state of Fig. 21–3c. Equivalently, it goes from point a to point c in Fig. 21–3d. Let us apply the first law of thermodynamics,

$$\Delta U = Q - W,$$

to this process. By definition of C_v we have $Q = nC_v \, \Delta T$. Also, $W \,(= p \, \Delta V) = 0$ because $\Delta V = 0$. Thus

$$\Delta U = nC_v \, \Delta T. \tag{21-9}$$

Let us restore the system to its original state and again raise its temperature by ΔT, this time leaving the sand load undisturbed so that the pressure p does not change. This *constant-pressure process* carries the system from the initial state of Fig. 21–3a to the final state of Fig. 21–3b. Equivalently, it goes from point a to point b in Fig. 21–3d. Let us apply the first law to *this* process. By definition of C_p we have $Q = nC_p \, \Delta T$. Also, $W = p \, \Delta V$. Now for an ideal gas, U depends only on the temperature. Since processes $a \rightarrow b$ and $a \rightarrow c$ in Fig. 21–3 involve the same change ΔT in temperature, they must also involve the same change ΔU in internal energy, namely, that given by Eq. 21–9. Thus for the constant-pressure process the first law yields

$$nC_p \, \Delta T = nC_v \, \Delta T + p \, \Delta V.$$

Let us apply the equation of state $pV = nRT$ to the constant-pressure process $a \rightarrow b$. For p constant we have, by taking differences,

$$p \, \Delta V = nR \, \Delta T.$$

Combining these equations yields

$$C_p - C_v = R. \tag{21-10}$$

This shows that the molar heat capacity of an ideal gas at constant pressure is always larger than that at constant volume by an amount equal to the universal gas constant R ($= 8.31$ J/mol K or 1.99 cal/mol K). Although Eq. 21–10 is exact only for an ideal gas, it is nearly true for real gases at moderate pressure (see Table 21–2). Notice that in obtaining this result we did not use the specific relation $U = \frac{3}{2}nRT$, but only the fact that U depends on temperature alone.

If we can compute C_v, then Eq. 21–10 will give us C_p and vice versa. We *can* obtain C_v by combining Eq. 21–9 with the kinetic theory result for the internal energy of an ideal gas, $U = \frac{3}{2}nRT$ (Eq. 21–8). Thus, in the limit of differential changes,

C_v for an ideal monatomic gas

$$C_v = \frac{dU}{n\,dT} = \frac{d}{n\,dT}\,(\tfrac{3}{2}nRT) = \tfrac{3}{2}R. \qquad (21\text{--}11)$$

This result (2.98 cal/mol K) turns out to be rather good for monatomic gases. It is, however, in serious disagreement with values obtained for diatomic and polyatomic gases (see Table 21–2). This suggests that Eq. 21–8 is not generally correct. Since that relation followed directly from the kinetic theory model, we conclude that we must change the model if kinetic theory is to survive as a useful approximation to the behavior of real gases.

Example 5 Show that for an ideal gas undergoing an adiabatic process $pV^\gamma =$ a constant, where $\gamma = C_p/C_v$.

Let us apply the first law of thermodynamics,

$$Q = \Delta U + W.$$

For an adiabatic process $Q = 0$. For W we put $p\,\Delta V$. Since the gas is assumed to be ideal, U depends only on temperature and, from Eq. 21–9, $\Delta U = nC_v\,\Delta T$. With these substitutions we have

$$0 = nC_v\,\Delta T + p\,\Delta V$$

or

$$\Delta T = -\frac{p\,\Delta V}{nC_v}.$$

For an ideal gas $pV = nRT$, so that, if p, V, and T are allowed to take on small variations,

$$p\,\Delta V + V\,\Delta p = nR\,\Delta T$$

or

$$\Delta T = \frac{p\,\Delta V + V\,\Delta p}{nR}.$$

Equating these two expressions and using Eq. 21–10 ($C_p - C_v = R$), we obtain, after some rearrangement,

$$p\,\Delta V C_p + V\,\Delta p C_v = 0.$$

Dividing by pVC_v and recalling that, by definition, $C_p/C_v = \gamma$, we get

$$\frac{\Delta p}{p} + \gamma\frac{\Delta V}{V} = 0.$$

In the limiting case of differential changes this reduces to

$$\frac{dp}{p} + \gamma\frac{dV}{V} = 0,$$

which (assuming γ to be constant) we can integrate as

$$\ln p + \gamma \ln V = \text{a constant}$$

or

$$pV^\gamma = \text{a constant}. \qquad (21\text{--}12)$$

In Fig. 21–4 we compare the isothermal and adiabatic behaviors of an ideal gas.

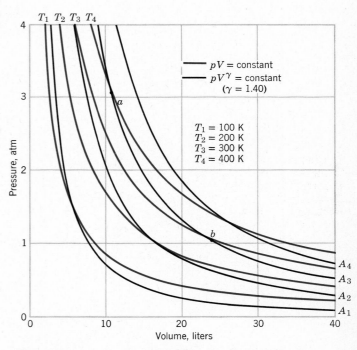

Figure 21–4 T_1, T_2, T_3, and T_4 show how the pressure of 1 mol of an ideal gas changes as its volume is changed, the temperature being held constant (isothermal process). A_1, A_2, A_3, and A_4 show how the pressure of an ideal gas changes as its volume is changed, no heat being allowed to flow to or from the gas (adiabatic process). An adiabatic increase in volume (for example, going from a to b along A_3) is always accompanied by a decrease in temperature, since at a, $T = 400$ K, whereas at b, $T = 300$ K.

Example 6 The compressions and rarefactions in a sound wave are practically adiabatic at audio frequencies. Show that in such a case the speed of sound in an ideal gas is given by

$$v = \sqrt{\frac{\gamma p}{\rho}}.$$

In Chapter 18 we showed the speed of sound to be $v = \sqrt{B/\rho}$, where ρ is the gas density and B is the bulk modulus of the gas, $B = -V(\Delta p/\Delta V)$. B will depend on the conditions that prevail as the pressure is changed. In a sound wave, the conditions are closely adiabatic. The appropriate bulk modulus is then

$$B_{\text{adiabatic}} = -V \left(\frac{dp}{dV}\right)_{\text{adiabatic}}.$$

In an adiabatic process for an ideal gas we have

$$pV^\gamma = \text{a constant}$$

or, by differentiating with respect to V,

$$p\gamma V^{\gamma-1} + V^\gamma \left(\frac{dp}{dV}\right)_{\text{adiabatic}} = 0.$$

Combining these equations yields

$$B_{\text{adiabatic}} = \gamma p$$

and, for the speed of sound,

$$v = \sqrt{\frac{B}{\rho}} = \sqrt{\frac{\gamma p}{\rho}}.$$

To understand why the compressions and rarefactions are adiabatic rather than isothermal, recall that compression of a gas causes a temperature rise and rarefaction a temperature fall unless heat energy is removed or added. Hence, in a gas through which sound propagates, the compressed regions are warmer than the rarefied ones. In principle, heat will be conducted from compression to rarefaction. The rate of heat conduction per unit area, however, depends (see Section 20–3) on the thermal conductivity of the gas and on the distance between compression and adjacent rarefaction, which is half a wavelength. The wavelength of audible sound is much too large for any significant rate of heat flow even in gases that are the best heat conductors. Hence, the conditions are essentially adiabatic in sound propagation. Actually, the condition for breakdown of the adiabatic approximation is that the wavelength of the wave be comparable with the mean free path of molecules in the gas, an extreme situation (see Section 21–8).

21–7 Equipartition of Energy

A modification of the kinetic theory model designed to explain the specific heats of gases was first suggested by Clausius in 1857. Recall that in our model we assumed a molecule to behave like a hard elastic sphere and we treated its kinetic energy as purely translational. The specific heat prediction was satisfactory for monatomic molecules. Further, because of the success of this simple model in other respects in predicting the correct behavior of gases of all kinds over wide temperature ranges, we feel confident that it is the average kinetic energy of translation that determines what we measure as the temperature of a gas.

However, in the case of specific heats we are concerned with all possible ways of absorbing energy and we must ask whether or not a molecule can store energy internally, that is, in a form other than kinetic energy of translation. This would certainly be so if we pictured a molecule, not as a rigid particle, but as an object with internal structure, for then a molecule could rotate and vibrate as well as move with translational motion. In collisions, the rotational and vibrational modes of motion could be excited, and this would contribute to the internal energy of the gas. Here, then, is a model that enables us to modify the kinetic theory formula for the internal energy of a gas.

Let us now find the total energy of a system containing a large number of such molecules, where each molecule is thought of as an object having internal structure. The energy will consist of kinetic energy of translation, kinetic energy of vibration of the atoms in a molecule, and potential energy of vibration of the atoms in a molecule. Although other kinds of energy contributions exist, such as magnetic, for gases we can describe the total energy quite accurately by terms such as these. We can show from statistical mechanics that when the number of particles is large and Newtonian mechanics holds, all these terms have the same average value, and this average value depends only on the temperature. In other words, the available energy depends only on the temperature and distributes itself in equal shares to each of the independent ways in which the molecules can absorb energy. This theorem, stated here without proof, is called the *equipartition of energy* and was deduced by James Clerk Maxwell. Each such independent way of absorbing energy is called a *degree of freedom*.

From Eq. 21–8 we know that the kinetic energy of translation per mole of gaseous molecules is $\frac{3}{2}RT$. The kinetic energy of translation per mole is the sum of three terms, however, namely, $\frac{1}{2}M\overline{v_x^2}$, $\frac{1}{2}M\overline{v_y^2}$, and $\frac{1}{2}M\overline{v_z^2}$. The theorem of equipartition requires that each such term contribute the same amount to the total energy per mole, or $\frac{1}{2}RT$ per degree of freedom.

For monatomic gases (gases for which each molecule is a single atom), the molecules have only translational motion (no internal structure in kinetic theory), so that $U = \frac{3}{2}nRT$. It follows from Eq. 21–11 that $C_v = \frac{3}{2}R \cong 3$ cal/mol K. Then from Eq. 21–10, $C_p = \frac{5}{2}R$, and the ratio of specific heats is

Specific heat of an ideal monatomic gas

$$\gamma = \frac{C_p}{C_v} = \frac{5}{3} = 1.67.$$

For a diatomic (two atoms per molecule) gas we can think of each molecule as having a dumbbell shape (two spheres joined by a rigid rod). Such a molecule can rotate about any one of three mutually perpendicular axes. However, the rotational inertia about an axis along the rigid rod should be negligible compared to that about axes perpendicular to the rod. This is consistent with our assumption that a monatomic molecule could not rotate. Implicitly, we have adopted a point mass model of the atom. The rotational energy, therefore, should consist of only two terms, such as $\frac{1}{2}I\omega_y^2$ and $\frac{1}{2}I\omega_z^2$. Each rotational degree of freedom is required by equipartition to contribute the same energy as each translational degree, so that for a diatomic gas having both rotational and translational motion,

$$U = 3n(\tfrac{1}{2}RT) + 2n(\tfrac{1}{2}RT) = \tfrac{5}{2}nRT,$$

or

Specific heat of an ideal diatomic gas

$$C_v = \frac{dU}{n\,dT} = \frac{5}{2}R \cong 5 \text{ cal/mol K}$$

and

$$C_p = C_v + R = \tfrac{7}{2}R,$$

or

$$\gamma = \frac{C_p}{C_v} = \frac{7}{5} = 1.40.$$

For polyatomic gases, each molecule contains three or more spheres (atoms) joined together by rods in our model, so that the molecule is capable of rotating energetically about each of three mutually perpendicular axes. Hence, for a polyatomic gas having both rotational and translational motion,

$$U = 3n(\tfrac{1}{2}RT) + 3n(\tfrac{1}{2}RT) = 3nRT,$$

or

Specific heat of an ideal polyatomic gas

$$C_v = \frac{dU}{n\,dT} = 3R = 5.96 \text{ cal/mol K},$$

and

$$C_p = 4R = 7.95 \text{ cal/mol K},$$

or

$$\gamma = \frac{C_p}{C_v} = 1.33.$$

Let us now turn to experiment to test these ideas. In Table 21–2 we list the experimentally determined molar heat capacities for common gases at 20°C and 1.0 atm. Notice that for monatomic and diatomic gases the values of C_v, C_p, and γ are close to the ideal gas predictions, which are shown by the numbers in parentheses. In some diatomic gases, such as chlorine, and in most polyatomic gases the experimental values are larger than the predicted values. This suggests that our model is not yet close enough to reality.

Table 21–2 Molar heat capacities—theory and experiment[a]

Type of molecule	Gas	C_p cal/mol K	C_v cal/mol K	$C_p - C_v$ cal/mol K	C_p/C_v $(= \gamma)$
Monatomic	Ideal	(4.97)	(2.98)	(1.99)	(1.67)
	He	4.97	2.99	1.98	1.66
	A	4.98	2.98	2.00	1.67
	Kr	4.99	2.95	2.04	1.69
	Xe	5.02	3.01	2.01	1.67
Diatomic	Ideal	(6.95)	(4.96)	(1.99)	(1.40)
	H_2	6.89	4.87	2.02	1.41
	N_2	6.96	4.97	1.99	1.40
	O_2	7.02	5.03	1.99	1.40
	Cl_2	8.15	5.99	2.16	1.36
Polyatomic	Ideal	(7.95)	(5.96)	(1.99)	(1.33)
	CO_2	8.74	6.75	1.99	1.30
	NH_3	8.91	6.80	2.11	1.31
	SO_2	9.42	7.43	1.99	1.27
	C_2H_6	12.34	10.35	1.99	1.19

[a] The numbers in parentheses are the predictions of kinetic theory, considerations of molecular vibrations being excluded.

We have not yet considered energy contributions from the vibrations of the atoms in diatomic and polyatomic molecules. That is, we can modify the dumbbell model and join the spheres instead by springs. This new model will greatly improve our results in some cases. However, instead of having a theoretical model for all gases, we now require an empirical model that differs from gas to gas. We can obtain a reasonably good picture of molecular behavior this way and the empirical model is therefore useful; however it ceases to be fundamental.

To see this more clearly, let us consider Fig. 21–5, which shows the variation of the molar heat capacity of hydrogen with temperature. The value of 5 cal/mol K, which is predicted for diatomic molecules by our model, is characteristic of hydrogen only in the temperature range from about 250 to 750 K. Above 750 K, C_v increases steadily to 7 cal/mol K; and below 250 K, C_v decreases steadily to 3 cal/mol K. Other gases show similar variations of molar heat with temperature.

Here is a possible explanation. At low temperatures apparently (see Example 7) the hydrogen molecule has translational energy only and, for some reason, cannot rotate. As the temperature rises, rotation becomes possible so that at "ordinary" temperatures a hydrogen molecule acts like our dumbbell model. At high temperatures the collisions between molecules cause the atoms in the molecule to vibrate, and the molecule ceases to behave as a rigid body. Different gases, because of their different molecular structure, may show these effects at different temperatures. Thus a chlorine molecule appears to vibrate at room temperature.

Variation of specific heat with temperature

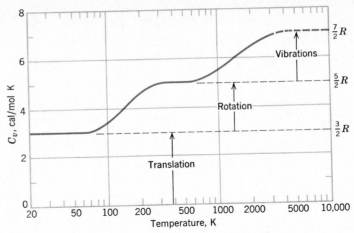

Figure 21–5 Variation of the molar heat C_v of hydrogen with temperature. Note that T is drawn on a logarithmic scale. Hydrogen dissociates above 3200 K and liquifies at 20 K. The dashed curve is for a hypothetical diatomic molecule that does not dissociate below 10,000 K.

Although this description is essentially correct, and we have obtained much insight into the behavior of molecules, this behavior contradicts classical kinetic theory. For kinetic theory is based on Newtonian mechanics applied to a large collection of particles, and the equipartition of energy is a necessary consequence of this classical statistical mechanics. But if equipartition of energy holds, then, no matter what happens to the total internal energy as the temperature changes, each part of the energy—translational, rotational, and vibrational—must share equally in the change. There is no classical mechanism for changing one mode of mechanical energy at a time in such a system. Kinetic theory requires that the specific heats of gases be independent of the temperature.

Hence, we have come to the limit of validity of classical mechanics when we try to explain the structure of the atom (or molecule). Just as Newtonian principles break down at very high speeds (near the speed of light), so here in the region of very small dimensions they also break down. Relativity theory modifies Newtonian ideas to account for the behavior of physical systems in the region of high speeds. It is quantum physics that modifies Newtonian ideas to account for the behavior of physical systems in the region of small dimensions. Both relativity theory and quantum physics are generalizations of classical theory in the sense that they give the (correct) Newtonian results in the regions in which Newtonian physics has accurately described experimental observations. We shall confine our attention to the very fruitful application of thermodynamics and the kinetic theory to "classical" systems.

Example 7 According to quantum theory the internal energy of an atom (or molecule) is "quantized"; that is, the atom cannot have any of a continuous set of internal energies but only certain discrete ones. After being raised from its lowest energy state to some higher one, the atom can give up this energy by emitting radiation whose energy equals the difference in energy between the upper and lower internal energy states of the atom.

When two atoms collide, some of their translational kinetic energy may be converted into internal energy of one or both of the atoms. In such a case the collision is inelastic, for translational kinetic energy is not conserved. In a gas, the average translational kinetic energy of an atom is $\frac{3}{2}kT$. When the temperature is raised to a value where $\frac{3}{2}kT$ is about equal to some allowed internal excitation energy of the atom, then an appreciable number of the atoms can absorb enough energy through inelastic collisions to be raised to this higher internal energy state. We can detect this because, after an interval, radiation corresponding to the absorbed energy will be emitted.

(a) Compute the average translational kinetic energy per molecule in a gas at room temperature.

We have, for $T = 300$ K,

$$\tfrac{3}{2}kT = \tfrac{3}{2}(1.38 \times 10^{-23} \text{ J/molecule K})(300 \text{ K})$$
$$= 6.21 \times 10^{-21} \text{ J/molecule}$$
$$= 3.88 \times 10^{-2} \text{ eV/molecule}.$$

This is about $\tfrac{1}{25}$ eV per molecule. Some molecules will have larger energies and some smaller energies than this average value.

(b) The first allowed (internal) excited state of a hydrogen atom is 10.2 eV above its lowest (ground) state. What temperature is needed to excite a "large" number of hydrogen atoms to emit radiation of this energy?

We require

$$\tfrac{3}{2}kT = 10.2 \text{ eV},$$

and we have from above

$$\tfrac{3}{2}k(300 \text{ K}) = \tfrac{1}{25} \text{ eV}.$$

Hence

$$T = 300 \text{ K} \times \frac{10.2}{(\tfrac{1}{25})} \simeq 7.5 \times 10^{4} \text{ K}.$$

Actually, because many molecules have energies much greater than the average value, appreciable excitation may occur at somewhat lower temperatures.

We can now appreciate why the kinetic theory assumption, that molecules can be regarded as having no internal structure and collide elastically with one another, holds true at ordinary temperatures. Only at temperatures high enough to give the molecules an average translational kinetic energy comparable to the energy difference between the lowest and the first allowed excited state of the molecule will the internal structure of the molecule change and the collisions become inelastic. Indeed, in retrospect one may say that early evidence that the internal energy of an atom is quantized existed in experiments with gas collisions and that the seeds of quantum theory lay in the kinetic theory of gases.

21-8 Mean Free Path

Between successive collisions a molecule in a gas moves with constant speed along a straight line. The average distance between such successive collisions is called the *mean free path* (Fig. 21–6). If molecules were points, they would not collide at all and the mean free path would be infinite. Molecules, however, are not points and hence collisions occur. If they were so numerous that they completely filled the space available to them, leaving no room for translational motion, the mean free path would be zero. Thus the mean free path is related to the size of the molecules and to their number per unit volume.

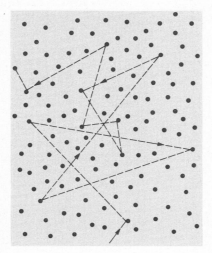

Figure 21–6 A molecule traveling through a gas, colliding with other molecules in its path. Of course, all the other molecules are moving in a similar fashion.

Consider the molecules of a gas to be spheres of diameter d. The effective cross section for a collision is then πd^2. That is, a collision will take place when the centers of two molecules approach within a distance d of one another. An equivalent

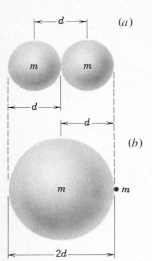

Figure 21-7 (a) If a collision occurs when two molecules of mass m come within a distance d of each other, the process can be treated equivalently (b) by thinking of one molecule as having an effective diameter $2d$ and the other as being a point.

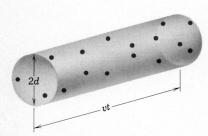

Figure 21-8 A molecule of equivalent diameter $2d$ traveling with speed v sweeps out a cylinder of base πd^2 and length vt in a time t. It suffers a collision with every other molecule whose center lies within this cylinder.

description of collisions made by any one molecule is to regard that molecule as having a diameter $2d$ and all other molecules as point particles (see Fig. 21-7).

Imagine a typical molecule of equivalent diameter $2d$ moving with speed v through a gas of equivalent point particles and let us assume, for the time being, that the molecule and the point particles exert no forces on each other. In time t our molecule will sweep out a cylinder of cross-sectional area πd^2 and length vt. If n_v is the number of molecules per unit volume, the cylinder will contain $(\pi d^2 vt) n_v$ particles (see Fig. 21-8). Since our molecule and the point particles do exert forces on each other, this will be the number of collisions experienced by the molecule in time t. The cylinder of Fig. 21-8 will, in fact, be a broken one, changing direction with every collision and v will not be a constant.

The mean free path $\bar{l}$ is the average distance between successive collisions. Hence, $\bar{l}$ is the total distance vt covered in time t divided by the number of collisions that take place in this time, or

$$\bar{l} = \frac{vt}{\pi\,d^2 n_v vt} = \frac{1}{\pi n_v d^2}.$$

This equation is based on the picture of a molecule hitting stationary targets. Actually the molecule hits moving targets. When the target molecules are moving, the two v's in the first equation above are not the same. The one in the numerator $(=\bar{v})$ is the mean molecular speed measured with respect to the container. The one in the denominator $(=\bar{v}_{rel})$ is the mean relative speed with respect to other molecules; it is this relative speed that determines the collision rate.

We can see qualitatively that $\bar{v}_{rel} > \bar{v}$. Thus two molecules of speed v moving toward each other have a relative speed of $2v$ ($>v$); two molecules with speed v moving at right angles on a collision course have a relative speed of $\sqrt{2}v$ (also $> v$); two molecules moving with speed v in the same direction have a relative speed of zero ($<v$). Thus molecules arriving from all of the forward hemisphere and part of the backward hemisphere have $\bar{v}_{rel} > \bar{v}$. The molecules arriving from the rest of the backward hemisphere have $\bar{v}_{rel} < \bar{v}$ but, since their numbers are smaller, the average over both hemispheres yields $\bar{v}_{rel} > \bar{v}$. A quantitative calculation, taking into account the actual speed distribution of the molecules, gives $\bar{v}_{rel} = \sqrt{2}\bar{v}$. With this change, the mean free path is reduced to

$$\bar{l} = \frac{1}{\sqrt{2}\pi n_v d^2}. \qquad (21-13)$$

Example 8 Let us calculate the magnitude of the mean free path and the collision frequency for air molecules at 0°C and 1 atm pressure.

We take 2×10^{-8} cm as an effective molecular diameter d. For the conditions stated, the average speed of air molecules is about 1×10^5 cm/s and there are about 3×10^{19} molecules/cm³. The mean free path is then

$$\bar{l} = \frac{1}{\pi\sqrt{2}\,n_v d^2} = \frac{1}{\pi\sqrt{2}(3 \times 10^{19}/\text{cm}^3)(2 \times 10^{-8}\text{ cm})^2}$$
$$= 2 \times 10^{-5}\text{ cm}.$$

This is about a thousand molecular diameters.

The corresponding collision frequency is

$$\frac{\bar{v}}{\bar{l}} = (1 \times 10^5\text{ cm/s})/(2 \times 10^{-5}\text{ cm})$$
$$= 5 \times 10^9/\text{s}.$$

Thus, on the average, each molecule makes 5 billion collisions per second!

In the earth's atmosphere the mean free path of air molecules at sea level (760 mm Hg) is, as we have seen, 2×10^{-5} cm. At 100 km above the earth (10^{-3} mm Hg) the mean free path is 2 mm. At 300 km (10^{-6} mm Hg) it is 15 cm, and yet there are about 10^8 molecules/cm³ in this region. This emphasizes that molecules are indeed small. At great enough heights the mean free path concept fails because the upward-directed molecules follow ballistic paths and may escape from the atmosphere.

In the laboratory the mean free path concept is useful in situations such as that of Example 8. In even modest laboratory vacuums, however, it loses some of its meaning because nearly all the collisions are with the wall of the containing vessel rather than with other molecules. Consider a box 10 cm on edge containing air at 10^{-6} mm Hg pressure. The mean free path (see above) is 15 cm, so that collisions between molecules are rare indeed. And yet this box contains about 10^{12} molecules!

Even in a finite "box," however, there are some conditions in which particles can travel great distances without striking the walls. In a typical proton synchrotron, used to accelerate protons to the billion-electron-volt range of energies, the protons are constrained by a magnetic field to move in a circular path and may travel several hundred thousand miles during the acceleration process. Mean free path considerations are important if the accelerating protons are to have essentially no collisions with residual air molecules. In this case the effective collision cross section is very small because of two factors. The proton is much smaller than any molecule, and it effectively "collides" only with the nuclei of the residual air molecules. Electrons in the molecules, because of their small mass, do not appreciably affect the high-energy protons. Thus there is essentially no beam loss by proton scattering from gas molecules inside the vacuum chamber if the residual pressure is as low as 10^{-6} mm Hg.

21–9   Distribution of Molecular Speeds

In earlier sections we discussed the root-mean-square speed of the molecules of a gas. However, the speeds of individual molecules vary over a wide range of magnitude; there is a characteristic distribution of molecular speeds of a given gas which depends, as we shall see below, on the temperature. If all the molecules of a gas had the same speed v, this situation would not persist for very long because the molecular speeds would be changed by collisions. However, we do not expect many molecules to have speeds $\ll v_{rms}$ (that is, near zero) or $\gg v_{rms}$ because such extreme speeds would require an unlikely sequence of preferential collisions.

James Clerk Maxwell first solved the problem of the most probable distribution of speeds in a large number of molecules of a gas. His molecular speed distribution law, for a sample of gas containing N molecules, is

Maxwell distribution of molecular speeds

$$N(v) = 4\pi N \left(\frac{m}{2\pi kT}\right)^{3/2} v^2 e^{-mv^2/2kT}. \qquad (21–14)$$

In this equation $N(v)\, dv$ is the number of molecules in the gas sample having speeds between v and $v + dv$, T is the absolute temperature, k is the Boltzmann constant, and m is the mass of a molecule. Note that for a given gas the speed distribution depends only on the temperature. We find N, the total number of molecules in the sample, by adding up (that is, by integrating) the number present in each differential speed interval from zero to infinity, or

$$N = \int_0^\infty N(v)\ dv. \qquad\qquad (21-15)$$

The SI unit of $N(v)$ is molecules/(m/s).

In Fig. 21–9 we plot the Maxwell distribution of speeds for molecules of oxygen at two different temperatures. The number of molecules having a speed between v_1 and v_2 equals the area under the curve between the vertical lines at v_1 and v_2. As Eq. 21–15 shows, the area under the speed distribution curve, which is the integral in that equation, is equal to the total number of molecules in the sample. At any temperature the number of molecules in a given speed interval Δv increases as the speed increases up to a maximum (the most probable speed v_p) and then decreases asymptotically toward zero. The distribution curve is not symmetrical about the most probable speed because the lowest speed must be zero, whereas there is no classical limit to the upper speed a molecule can attain. In this case the average speed $\bar{v}$ is somewhat larger than the most probable value. The root-mean-square value, v_{rms}, being the square root of the average of the sum of the squares of the speeds, is still larger.

Figure 21–9 The Maxwellian distribution of speeds of 10^6 oxygen molecules at two different temperatures. The number of molecules within a certain range of speeds (say, 300 to 600 m/sec) is the area under this section of the curve. The complete area under either curve is the total number of molecules (equals 10^6); this area must be the same for each temperature if, as in this case, the curves refer to a given number of molecules.

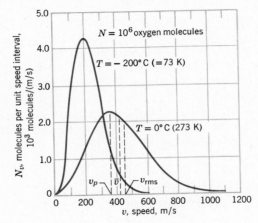

As the temperature increases, the root-mean-square speed v_{rms} (and $\bar{v}$ and v_p as well) increases, in accord with our microscopic interpretation of temperature. The range of typical speeds is now greater, so that the distribution broadens. Since the area under the distribution curve (which is the total number of molecules in the sample) remains the same, the distribution must also flatten as the temperature rises. Hence the number of molecules which have speeds greater than some given speed increases as the temperature increases (see Fig. 21–9). This explains many phenomena, such as the increase in the rates of chemical reactions with rising temperature.

The distribution of speeds of molecules in a liquid also resembles the curves of Fig. 21–9. This explains why some molecules in a liquid (the fast ones) can escape through the surface (evaporate) at temperatures well below the normal boiling point. Only these molecules can overcome the attraction of the molecules in the surface and escape by evaporation. The average kinetic energy of the remaining molecules drops correspondingly, leaving the liquid at a lower temperature. This explains why evaporation is a cooling process.

From Eq. 21–14 we see that the distribution of molecular speeds depends on the mass of the molecule as well as on the temperature. The smaller the mass, the larger the proportion of high-speed molecules at any given temperature. Hence hydrogen is more likely to escape from the atmosphere at high altitudes than oxygen or nitrogen.

Example 9 The speeds of ten particles in meters per second are 0, 1.0, 2.0, 3.0, 3.0, 3.0, 4.0, 4.0, 5.0, and 6.0. Find (a) the average speed, (b) the root-mean-square speed, and (c) the most probable speed of these particles.

(a) The average speed is

$$\bar{v} = \frac{0 + 1.0 + 2.0 + 3.0 + 3.0 + 3.0 + 4.0 + 4.0 + 5.0 + 6.0}{10}$$

$$= 3.1 \text{ m/s.}$$

(b) The mean-square speed is

$$\overline{v^2} = \frac{0 + (1.0)^2 + (2.0)^2 + (3.0)^2 + (3.0)^2 + (3.0)^2 + (4.0)^2 + (4.0)^2 + (5.0)^2 + (6.0)^2}{10}$$

$$= 12.5 \text{ m}^2/\text{s}^2,$$

and the root-mean-square speed is

$$v_{\text{rms}} = \sqrt{12.5 \text{ m}^2/\text{s}^2} = 3.5 \text{ m/s.}$$

(c) Of the ten particles three have speeds of 3.0 m/s, two have speeds of 4.0 m/s, and the other five each have a different speed. Hence, the most probable speed of a particle v_p is

$$v_p = 3.0 \text{ m/s.}$$

Example 10 Use Eq. 21–14 to determine the average speed $\bar{v}$, the root-mean-square speed v_{rms}, and the most probable speed v_p of the molecules in a gas in terms of the gas parameters.

The quantity $N(v)\, dv$ is the number of particles in the sample having a speed between v and $v + dv$, $N(v)$ being given by Eq. 21–14. We find the average speed $\bar{v}$ in the usual way: We multiply the number of particles in each speed interval by a speed v characteristic of that interval, we sum these products over all speed intervals, and we divide by the total number of particles. Replacing the summation by an integration, we obtain

$$\bar{v} = \frac{\int_0^\infty v N(v)\, dv}{N}.$$

When Eq. 21–14 is substituted for $N(v)$, the resulting integral is one that is found in most integral tables.* Upon integrating we obtain, as you should show,

$$\bar{v} = \sqrt{\frac{8}{\pi} \frac{kT}{m}} = 1.59 \sqrt{\frac{kT}{m}} \qquad \text{(average speed).}$$

The mean-square speed is given by

$$\overline{v^2} = \frac{\int_0^\infty v^2 N(v)\, dv}{N}.$$

Again the integral is easily evaluated using a standard tabulated integral, with the result

$$v_{\text{rms}} = \sqrt{\overline{v^2}} = \sqrt{\frac{3kT}{m}}$$

$$= 1.73 \sqrt{\frac{kT}{m}} \qquad \text{(root-mean-square speed).}$$

The most probable speed v_p is the speed at which $N(v)$ has its maximum value. It is given by requiring that

$$\frac{dN(v)}{dv} = 0.$$

By substituting Eq. 21–14 for $N(v)$ we obtain, as you should show,

$$v_p = \sqrt{\frac{2kT}{m}} = 1.41 \sqrt{\frac{kT}{m}} \qquad \text{(most probable speed).}$$

In Fig. 21–9 we show v_p, $\bar{v}$, and v_{rms} at 0°C for a molecular speed distribution in oxygen.

* Let $\lambda = m/2kT$. From tables of integrals,

$$\int_0^\infty v^2 e^{-\lambda v^2}\, dv = \frac{1}{4} \sqrt{\frac{\pi}{\lambda^3}}; \qquad \int_0^\infty v^3 e^{-\lambda v^2}\, dv = \frac{1}{2\lambda^2};$$

$$\int_0^\infty v^4 e^{-\lambda v^2}\, dv = \frac{3}{8} \sqrt{\frac{\pi}{\lambda^5}}.$$

REVIEW GUIDE AND SUMMARY

The kinetic theory of gases

The kinetic theory of gases relates the *macroscopic* properties of gases (pressure, temperature, etc.) to the *microscopic* properties of the gas molecules (average speed, molecular diameter, etc.). We deal largely with an *ideal gas*.

The macroscopic properties of an ideal gas are related by

$$pV = nRT. \qquad [21\text{--}2]$$

Ideal gas—a macroscopic description

Here n is the number of moles of the gas and R (= 8.31 J/mol K) is the *universal gas constant*. All real gases approach the ideal state at low enough densities. Examples 1 and 2 show applications of Eq. 21–2.

Ideal gas—a microscopic description

Microscopically, an ideal gas is made up of molecules whose aggregate volume is much less than that of the container in which the gas is confined. The molecules are in random motion and make only elastic collisions with each other and with the container walls.

In Section 21–4 we show that the pressure of an ideal gas is related to the speeds of its molecules by

Pressure—a kinetic theory interpretation

$$p = \tfrac{1}{3}\rho\overline{v^2} = \tfrac{1}{3}\rho v_{\text{rms}}^2. \qquad [21\text{-}3]$$

Here ρ is the gas density and v_{rms} is the square root of the average of the squares of the molecular speeds; see Example 3.

Temperature—a kinetic theory interpretation

In Section 21–5 we show that the total translational kinetic energy per mole of the molecules of an ideal gas is proportional to temperature and is $\tfrac{3}{2}RT$ ($=3400$ J/mol at 273 K; see Table 21–1). Similarly, the average translational kinetic energy for a single molecule is $\tfrac{3}{2}kT$. Here k ($=R/N_A = 1.38 \times 10^{-23}$ J/K) is called *Boltzmann's constant*.

Molar heats of an ideal gas

The above suggests that the internal energy U of an ideal gas depends only on its temperature and not, for example, on its pressure. Combining this fact with the first law of thermodynamics leads to

$$C_p - C_v = R, \qquad [21\text{-}10]$$

where C_p and C_v are the *molar heat capacities* at constant pressure and at constant volume, respectively; their SI units are the J/mol K. Table 21–2 shows that Eq. 21–10 holds well for many real gases.

The equipartition theorem

To find C_v and C_p separately we invoke the *equipartition of energy* theorem, which states that every *degree of freedom* of a molecule (that is, every independent way it can store energy) has the same associated energy, namely $\tfrac{1}{2}kT$. If ν is the number of degrees of freedom we find:

$$U = \left(\frac{\nu}{2}\right) nRT \qquad \text{and} \qquad C_v = \frac{1}{n}\frac{dU}{dT} = \left(\frac{\nu}{2}\right) R.$$

From Eq. 21–10 we have, then,

$$C_p = C_v + R = \left(1 + \frac{\nu}{2}\right) R \qquad \text{and} \qquad \gamma = \frac{C_p}{C_v} = 1 + \frac{2}{\nu}.$$

For monatomic gases $\nu = 3$ (all translational) and for diatomic gases $\nu = 5$ (three translational and two rotational). Table 21–2 shows that the above relations hold well for gases of these two types.

Mean free path

The *mean free path* of a gas molecule (that is, its average path length between collisions; see Fig. 21–6) is given by

$$\bar{l} = \frac{1}{\sqrt{2}\pi\eta_v d^2}. \qquad [21\text{-}13]$$

where η_v is the number of molecules per unit volume and d is the molecular diameter; see Example 8.

The distribution of molecular speeds

The *molecular speed distribution* function $N(v)$ is a quantity such that $N(v)\,dv$ gives the number of molecules with speeds between v and $v + dv$. As Eq. 21–14 shows, $N(v)$ depends only on the mass of the molecule and on the absolute temperature. Figure 21–9 shows $N(v)$ for oxygen at two different temperatures. This figure and Example 9 illustrate the definitions of the *most probable* speed v_p, the *average* speed $\bar{v}$, and the *root-mean-square* speed v_{rms}.

QUESTIONS

1. In discussing the fact that it is impossible to apply the laws of mechanics individually to atoms in a macroscopic system, Mayer and Mayer state: "The very complexity of the problem [that is, the fact that the number of atoms is large] is the secret of its solution." Discuss this sentence.

2. Is there any such thing as a truly continuous body of matter?

3. In kinetic theory we assume that there are a large number of molecules in a gas. Real gases behave like an ideal gas at low densities. Are these statements contradictory? If not, what conclusion can you draw from them?

4. We have assumed that the walls of the container are elastic for molecular collisions. Actually, the walls may be inelastic. In practice this makes no difference as long as the walls are at the same temperature as the gas. Explain.

5. In large-scale inelastic collisions mechanical energy is lost through internal friction resulting in a rise of temperature owing to increased internal molecular agitation. Is there a loss of mechanical energy in an inelastic collision between molecules?

6. What justification is there in neglecting the change in gravitational potential energy of molecules in a gas?

7. We have assumed that the force exerted by molecules on the wall of a container is steady in time. How is this justified?

8. The average velocity of the molecules in a gas must be zero if the gas as a whole and the container are not in translational motion. Explain how it can be that the average *speed* is not zero.

9. Consider a hot, stationary golf ball sitting on a tee and a cold golf ball just moving off the tee after being hit. Can the numerical value of the total kinetic energy of the molecules' motion relative to the tee be the same in each case? If so, what is the difference between the two cases?

10. Is it possible for a gas to consist of molecules that all have the same speed?

11. Justify the fact that the pressure of a gas depends on the *square* of the speed of its particles by explaining the dependence of pressure on the collision frequency and the momentum transfer of the particles.

12. Why does the boiling temperature of a liquid increase with pressure?

13. Pails of hot and cold water are set out in freezing weather. Explain (*a*) if the pails have lids, the cold water will freeze first but (*b*) if the pails do not have lids, it is possible for the hot water to freeze first.

14. How is the speed of sound related to the gas variables in the kinetic theory model?

15. Far above the earth's surface the gas kinetic temperature (see Eq. 21–5) is reported to be the order of 1000 K. However, a person placed in such an environment would freeze to death rather than vaporize. Explain.

16. Why doesn't the earth's atmosphere leak away? At the top of the atmosphere, molecules will occasionally be headed out with a speed exceeding the escape speed. Isn't it just a matter of time?

17. Titan, one of Saturn's many moons, has an atmosphere, but our own moon does not. What might the explanation be?

18. As ice is heated, it melts and then it boils. However, as solid carbon dioxide is heated, it goes directly to the vapor state—we say that it *sublimes*—without passing through a liquid state. Can you imagine any way that you could produce liquid carbon dioxide?

19. Does the concept of temperature apply to a vacuum? Consider interplanetary space, for example.

20. What direct evidence do you have for the existence of atoms? . . . What indirect evidence?

21. How, if at all, would you expect the composition of the air to change with altitude?

22. Is there a difference between a gas and a vapor?

23. We often say that we see the steam emerging from the spout of a kettle in which water is boiling. Can you really see steam? What is it that you see?

24. Why does smoke rise, rather than fall, from a lighted cigarette?

25. Would a gas whose molecules were true geometric points obey the ideal gas law?

26. If you fill a saucer with water at room temperature, the water, under normal conditions, will evaporate completely. It is easy to believe that some of the more energetic molecules can escape from the water surface, but how can *all* of them eventually escape? Many of them—in fact, the vast majority—do not have enough energy to do so.

27. Explain why the temperature of a gas drops in an adiabatic expansion.

28. If hot air rises, why is it cooler at the top of a mountain than near sea level?

29. Comment on this statement: "There are two ways to carry out an adiabatic process. One is to do it quickly and the other is to do it in an insulated box."

30. A sealed rubber balloon contains a very light gas. The balloon is released and it rises high into the atmosphere. Describe and explain the thermal and mechanical behavior of the balloon.

31. Explain why the specific heat at constant pressure is greater than the specific heat at constant volume.

32. Consider the case in which the mean free path is greater than the longest straight line in a vessel. Is this a perfect vacuum for a molecule in this vessel?

33. List effective ways of increasing the number of molecular collisions per unit time in a gas.

34. Consider Archimedes' principle applied to a gas. Isn't it true that once we accept a kinetic theory model of a gas, we need a new explanation for this principle? For example, suppose that the mean free path of a gas molecule is comparable to the depth of the body immersed in the gas, or greater; what is the origin of the buoyant force then? (See "Archimedes' Principle in Gases," by Alan J. Walton, *Contemporary Physics*, March 1969.)

35. The two opposite walls of a container of gas are kept at different temperatures. Describe the mechanism of heat conduction through the gas.

36. A gas can transmit only those sound waves whose wavelength is long compared with the mean free path. Can you explain this? Where might this limitation arise?

37. Justify qualitatively the statement that, in a mixture of molecules of different kinds in complete equilibrium, each kind of molecule has the same Maxwellian distribution in speed that it would have if the other kinds were not present.

38. What observation is good evidence that not all molecules of a body are moving with the same speed at a given temperature?

39. The fraction of molecules within a given range Δv of the root-mean-square speed decreases as the temperature of a gas rises. Explain why.

40. (*a*) Do half the molecules in a gas in thermal equilibrium have speeds greater than v_p? Than $\bar{v}$? Than v_{rms}? (*b*) Which speed, v_p, $\bar{v}$, or v_{rms}, corresponds to a molecule having average kinetic energy?

41. Keeping in mind that internal energy of a body consists of kinetic energy and potential energy of its particles, how would you distinguish between the internal energy of a body and its temperature?

EXERCISES

Section 21-2   Ideal Gas — A Macroscopic Description

1. (*a*) What is the volume occupied by 1.0 mol of an ideal gas at standard conditions, that is, pressure of 1.0 atm and temperature of 0°C? (*b*) Show that the number of molecules per cubic centimeter (Loschmidt number) at standard conditions is 2.687×10^{19}.

2. Compute the number of molecules in a gas contained in a volume of 1.00 cm³ at a pressure of 1.00×10^{-3} atm and a temperature of 200 K.

3. The best vacuum that can be attained in the laboratory corresponds to a pressure of about 10^{-18} atm, or about 10^{-14} mm Hg. How many molecules are there per cubic centimeter in such a "vacuum" at room temperature?

4. A quantity of ideal gas at 10°C and a pressure of 100 kPa occupies a volume of 2.5 m³. (*a*) How many moles of the gas are present? If the pressure is now raised to 300 kPa and the temperature is raised to 30°C, (*b*) how much volume will the gas now occupy, and (*c*) how many moles of the gas will now be present?

5. Oxygen gas having a volume of 1.0 liter at 40°C and a pressure of 76 cm Hg expands until its volume is 1.5 liters and its pressure is 80 cm Hg. Find (*a*) the amount of oxygen in the system and (*b*) its final temperature.

6. An automobile tire has a volume of 1000 in.³ and contains air at a gauge pressure of 24 lb/in.² when the temperature is 0°C. What is the gauge pressure of the air in the tires when its temperature rises to 27°C and its volume increases to 1020 in.³?

7. If the water molecules in 1.0 g of water were distributed uniformly over the surface of the earth, how many such molecules would there be on 1.0 cm² of the earth's surface?

Section 21-4   Kinetic Calculation of Pressure

8. The mass of the H_2 molecule is 3.3×10^{-24} g. If 10^{23} hydrogen molecules per second strike 2.0 cm² of wall at an angle of 55° with the normal when moving with a speed of 1.0×10^5 cm/s, what pressure do they exert on the wall?

Section 21-5   Kinetic Interpretation of Temperature

9. Calculate the root-mean-square speed of helium atoms at 1000 K.

10. What is the mean kinetic energy of individual nitrogen molecules at 1500 K (*a*) in joules; (*b*) in electron volts (eV)?

11. The lowest possible temperature in outer space is about 3 K. What is the average speed of hydrogen atoms at this temperature?

12. (*a*) Determine the average value of the kinetic energy of the particles of an ideal gas at 0.0°C and at 100°C. (*b*) What is the kinetic energy per mole of an ideal gas at these temperatures?

13. At what temperature do the atoms of helium gas have the same average speed as the molecules of hydrogen gas at 20°C?

14. Find the root-mean-square speed of (*a*) helium and (*b*) argon molecules at 40°C from that of oxygen molecules (40 m/s at 0.00°C). The molecular weight of oxygen is 32 g/mol, of argon 40 g/mol, of helium 4.0 g/mol.

Section 21-6   Specific Heats of an Ideal Gas

15. (*a*) What is the internal energy of 1.0 mol of an ideal gas at 273 K? (*b*) Does it depend on volume or pressure? Does it depend on the nature of the gas?

16. Air at 0.00°C and 1.00 atm pressure has a density of 1.291×10^{-3} g/cm³. The speed of sound is 332 m/s at that temperature. Compute the ratio of specific heats of air.

17. Show that the speed of sound in an ideal gas is independent of the pressure and density.

18. Show that the speed of sound in air increases about 0.61 m/s for each Celsius degree rise in temperature near 0°C.

19. We know that $pV^\gamma = $ "*constant*" for an adiabatic process. Evaluate the "*constant*" for an adiabatic process involving exactly 2 mol of an ideal gas passing through the state having exactly $p = 1$ atm and $T = 300$ K. Assume a diatomic gas.

20. One mole of an ideal gas expands adiabatically from an initial temperature T_1 to a final temperature T_2. Prove that the work done by the gas is $C_v(T_1 - T_2)$, where C_v is the molar heat capacity.

Section 21-7   Equipartition of Energy

21. One mole of oxygen is heated at constant pressure starting at 0°C. How much heat must be added to the gas to double its volume?

Section 21-8   Mean Free Path

22. The mean free path of nitrogen molecules at 0°C and 1.0 atm is 0.80×10^{-5} cm. At this temperature and pressure there are 2.7×10^{19} molecules/cm³. What is the molecular diameter?

23. Derive an expression, in terms of n_v, $\bar{v}$, and d, for the frequency of collisions suffered by a gas atom or molecule.

24. What is the mean free path for 15 spherical jelly beans in a bag that is vigorously shaken? Take the volume of the bag to be 1.0 liter and the diameter of a jelly bean to be 1.0 cm.

Section 21-9   Distribution of Molecular Speeds

25. The speeds of a group of 10 molecules are 2, 3, 4, 5, 6, 7, 8, 9, 10, 11 km/s. (*a*) What is the average speed of the group? (*b*) What is the root-mean-square speed for the group?

26. You are given the following group of particles (N_i represents the number of particles that have a speed v_i):

N_i	v_i (cm/s)
2	1
4	2
6	3
8	4
2	5

(*a*) Compute the average speed $\bar{v}$. (*b*) Compute the root-mean-square speed v_{rms}. (*c*) Among the five speeds shown, which is the most probable speed v_p for the entire group?

27. (*a*) Ten particles are moving with the following speeds: four at 200 m/s, two at 500 m/s, and four at 600 m/s. Calculate the average and root-mean-square speeds. Is $v_{rms} > \bar{v}$? (*b*) Make up your own speed distribution for the 10 particles and show that $v_{rms} > \bar{v}$ for your distribution. (*c*) Under what condition (if any) does $v_{rms} = \bar{v}$?

PROBLEMS GROUPED BY SECTION

Section 21–2   Ideal Gas—A Macroscopic Description

1.   What is the coefficient of volume expansion for an ideal gas at constant pressure and at 100°C? (See Section 19–7.)

2.   The equation of state for a certain material is given by

$$p = \frac{AT - BT^2}{V}.$$

Find an expression for the work done by the material if the temperature changes from T_1 to T_2 while the pressure remains constant at $p = p_0$.

3.   Calculate the work done in compressing 1.00 mol of oxygen from a volume of 22.4 liters at 0°C and 1.00 atm pressure to 16.8 liters at the same temperature.

4.   Air that occupies 5.0 ft³ ($= 0.14$ m³) at 15 lb/in.² ($= 1.03 \times 10^5$ Pa) gauge pressure is expanded isothermally to atmospheric pressure and then cooled at constant pressure until it reaches its initial volume. Compute the work done by the gas.

5.   Consider a given mass of an ideal gas. Compare curves representing constant-pressure, constant-volume, and isothermal processes on (a) a p-V diagram, (b) a p-T diagram, and (c) a V-T diagram. (d) How do these curves depend on the mass of gas chosen?

Section 21–3   Ideal Gas—A Microscopic Description

6.   A _____ of water contains about as many molecules as there are _____ s of water in all the oceans. What single word best fits the two blank spaces: drop, teaspoon, tablespoon, cup, quart, barrel, or ton? The oceans cover 75% of the earth's surface and have an average depth of about 5 km. (After Edward M. Purcell.)

Section 21–4   Kinetic Calculation of Pressure

7.   A container encloses two ideal gases. Two moles of the first gas are present, with molecular mass M_1. Molecules of the second gas have a molecular mass $M_2 = 3M_1$, and 0.50 mol of this gas is present. What fraction of the total pressure on the container wall is attributable to the second gas? (The kinetic theory explanation of pressure leads to the experimentally discovered law of partial pressures for a mixture of gases that do not react chemically: *The total pressure exerted by the mixture is equal to the sum of the pressures which the several gases would exert separately if each were to occupy the vessel alone.*)

Section 21–5   Kinetic Interpretation of Temperature

8.   Oxygen gas at 273 K and 1.00 atm pressure is confined to a cubical container 10 cm on a side. Compare the change in gravitational potential energy of an oxygen molecule falling the height of the box with its mean translational kinetic energy.

9.   At 273 K and 1.00×10^{-2} atm, the density of a gas is 1.24×10^{-5} g/cm³. (a) Find v_{rms} for the gas molecules. (b) Find the molecular weight of the gas and identify it.

10.   At what temperature is the average translational energy of a molecule equal to the kinetic energy of an electron accelerated from rest through a potential difference of 1.0 V (that is, an energy of 1.0 eV)?

11.   (a) Compute the root-mean-square speed of an argon atom at room temperature (20°C). (b) At what temperature will the root-mean-square speed be half that value? Twice that value?

12.   Water standing in the open at 27°C evaporates due to the escape of some of the surface molecules. The heat of vaporization (540 cal/g) may be found approximately from ϵn, where ϵ is the average energy of the escaping molecules and n is the number of molecules per gram. (a) Find ϵ. (b) How many times greater is ϵ than the average kinetic energy of H_2O molecules, assuming that the kinetic energy is related to temperature in the same way as it is for gases?

Section 21–6   Specific Heats of an Ideal Gas

13.   One mole of an ideal gas undergoes an isothermal expansion. Find the heat flow into the gas in terms of the initial and final volumes and the temperature.

14.   A mass of gas occupies a volume of 4.0 liters at a pressure of 1.0 atm and a temperature of 300 K. It is compressed adiabatically to a volume of 1.0 liter. Determine (a) the final pressure and (b) the final temperature, assuming it to be an ideal gas for which $\gamma = 1.5$.

15.   Prove that when an ideal gas undergoes an adiabatic expansion from volume V_1 to volume V_2, the initial and final temperatures T_1 and T_2 are related by

$$T_2 = T_1 \left(\frac{V_1}{V_2}\right)^{\gamma - 1}.$$

16.   Let 5.0 cal of heat be added to a particular substance. As a result, its volume changes from 50 to 100 cm³ while the pressure remains constant at 1.0 atm. (a) By how much did the internal energy of the substance charge? (b) If the quantity of the substance is 2.0×10^{-3} mol, find the molar heat capacity at constant volume. (c) Find the molar heat capacity at constant pressure.

17.   A quantity of ideal monatomic gas consists of n moles initially at temperature T_1. The pressure and volume are then slowly doubled in such a manner as to trace out a straight line on the p-V diagram. In terms of n, R, and T_1, what are (a) W, (b) ΔU, and (c) Q? (d) If one were to define an equivalent specific heat for this process, what would be its value?

Section 21–7   Equipartition of Energy

18.   How would you explain the observed value of $C_v = 7.50$ cal/mol·K for gaseous SO_2 at 15.0°C and 1.00 atm?

19.   Ten grams of oxygen are heated at constant atmospheric pressure from 27 to 127°C. (a) How much heat is transferred to the oxygen? (b) What fraction of the heat is used to raise the internal energy of the oxygen?

20.   An ideal diatomic gas (4.0 mol) at high temperature experiences a temperature increase of 60 K under constant pressure conditions. (a) How much heat was added to the gas? (b) By how much did the internal energy of the gas increase? (c) How much work was done by the gas? (d) By how much did the internal translational kinetic energy of the gas increase?

21.   Calculate the mechanical equivalent of heat from the value of R and the values of C_v and γ for oxygen from Table 21–2.

Section 21-8   Mean Free Path

22. In a certain particle accelerator the protons travel around a circular path of diameter 23 m in a chamber of 10^{-6} mm Hg pressure and 273 K temperature. (a) Estimate the number of gas molecules per cubic centimeter at this pressure. (b) What is the mean free path of the gas molecules under these conditions if the molecular diameter is 2.0×10^{-8} cm?

23. At what frequency would the wavelength of sound be of the order of the mean free path in oxygen at 1.0 atm pressure and 0°C? Take the diameter of the oxygen molecule to be 3.0×10^{-8} cm.

24. (a) What is the molar volume (the volume per mole) of an ideal gas at standard conditions (0°C, 1.0 atm)? (b) Calculate the ratio of the root-mean-square speeds of helium and neon under these conditions. (c) What would be the mean free path of helium atoms under these conditions? Assume molecular diameters d to be 1.0×10^{-8} cm. (d) What would be the mean free path of neon atoms under these conditions? (e) Comment on the results of parts (c) and (d) in view of the fact that the helium atoms are traveling faster than the neon atoms.

25. At 2500 km above the earth's surface, the density is about 1 molecule/cm³. (a) What mean free path is predicted by Eq. 21–13, and (b) what is its significance under these conditions?

Section 21-9   Distribution of Molecular Speeds

26. Consider the distribution of speeds shown in Fig. 21–10. (a) List v_{rms}, $\bar{v}$, and v_p in the order of increasing speed. (b) How does this compare with the Maxwellian distribution?

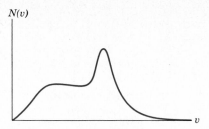

$N(v)$

Figure 21-10   Problem 26.

27. It is found that the most probable speed of molecules in a gas at equilibrium temperature T_2 is the same as the root-mean-square speed of the molecules in this gas when its equilibrium temperature is T_1. Find T_2/T_1.

28. For the hypothetical gas speed distribution of N particles shown in Fig. 21–11, $[N(v) = Cv^2, 0 < v < v_0, N(v) = 0, v > v_0]$, find (a) an expression for c in terms of N and v_0; (b) the average speed of the particles; (c) the root-mean-square speed of the particles.

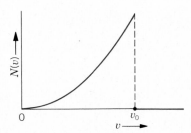

Figure 21-11   Problem 28.

ADDITIONAL PROBLEMS

29. In a gas of uranium hexafluoride there are isotopes $^{235}UF_6$ and $^{238}UF_6$ having molecular weights 349 and 352, respectively. (a) What is the ratio of the rms speeds of these two molecular isotopes? (b) How could this fact be used to separate the isotopes?

30. Show that the speed of sound in a gas v_s and the molecular speed v_{rms} are related by

$$\frac{v_{rms}}{v_s} = \left(\frac{3}{\gamma}\right)^{1/2}.$$

Test this prediction using the data given following Example 3; see also Section 18–2.

31. Two containers are at the same temperature. The first contains gas at pressure p_1 whose molecules have mass m_1 with a root-mean-square speed v_{rms1}. The second contains molecules of mass m_2 at a pressure $2p_1$ that have an average speed $\bar{v}_2 = 2v_{rms1}$. Find the ratio m_1/m_2 of the masses of their molecules.

32. (a) Compute the temperature at which the root-mean-square speed is equal to the speed of escape from the surface of the earth for molecular hydrogen; for molecular oxygen. (b) Do the same for the moon, assuming gravity on its surface to be $0.16g$. (c) The temperature high in the earth's upper atmosphere is about 1000 K. Would you expect to find much hydrogen there? Much oxygen?

33. An air bubble of 20 cm³ volume is at the bottom of a lake 40 m deep where the temperature is 4°C. The bubble rises to the surface, which is at a temperature of 20°C. Take the temperature of the bubble to be the same as that of the surrounding water and find its volume just before it reaches the surface.

34. The speed of sound in different gases at the same temperature depends on the molecular weight of the gas. Show that $v_1/v_2 = (M_2/M_1)^{1/2}$ (constant T), where v_1 is the speed of sound in the gas of molecular weight M_1 and v_2 is the speed of sound in the gas of molecular weight M_2.

35. Take the mass of a helium atom to be 6.66×10^{-27} kg. Compute the specific heat at constant volume for helium gas.

36. A container holds a mixture of three nonreacting gases: n_1 mol of the first gas with specific heat at constant volume C_1, and so on. Find the molar specific heat at constant volume of the mixture, in terms of the molar specific heats and quantities of the three separate gases.

37. The mass of a gas molecule can be computed from the specific heat at constant volume. Take $c_v = 0.075$ kcal/kg·K for argon and calculate (a) the mass of an argon atom and (b) the atomic weight of argon.

38. *Avogadro's law* states that under the same condition of temperature and pressure, equal volumes of gas contain equal numbers of molecules. Derive this law from kinetic theory using Eq. 21–3 and the equipartition of energy assumption.

39. The mean free path $\bar{l}$ of the molecules of a gas may be determined from measurements (e.g., from measurements of the viscosity of the gas). At 20°C and 75 cm Hg pressure, such measurements yield values $\bar{l}_A$ (argon) $= 9.9 \times 10^{-6}$ cm and $\bar{l}_{N_2}$ (nitrogen) $= 27.5 \times 10^{-6}$ cm. (a) Find the ratio of the effective cross-section diameters of argon and nitrogen. (b) What would the value be of the mean free path of argon at 20°C and 15 cm Hg? (c) What would the value be of the mean free path of argon at −40°C and 75 cm Hg?

40. (a) A liter of gas with $\gamma = 1.3$ is at 273 K and 1.0 atm pressure. It is suddenly compressed to half its original volume. Find its final pressure and temperature. (b) The gas is now cooled back to 0°C at constant pressure. What is its final volume?

41. The envelope and basket of a hot-air balloon have a combined weight of 550 lb, and the envelope has a capacity of 77,000 ft³. When fully inflated, what should be the temperature of the enclosed air to give the balloon a lifting capacity of 600 lb (in addition to its own weight)? Assume that the ambient air, at 20°C, has a weight density of 7.56×10^{-2} lb/ft³.

42. A quantity of ideal gas occupies an initial volume V_0 at a pressure p_0 and a temperature T_0. It expands to a volume V_1 (a) at constant pressure, (b) at constant temperature, (c) adiabatically. Graph each case on a p-V diagram. In which case is Q greatest? Least? In which case is W greatest? In which case is ΔU greatest? Least?

43. A reversible heat engine carries 1.00 mol of an ideal monatomic gas around the cycle shown in Fig. 21–12. Process $1 \rightarrow 2$ takes place at constant volume, process $2 \rightarrow 3$ is adiabatic, and process $3 \rightarrow 1$ takes place at a constant pressure. (a) Compute the heat Q, the change in internal energy ΔU, and the work done W, for each of the three processes and for the cycle as a whole. (b) If the initial pressure at point 1 is 1.00 atm, find the pressure and the volume at points 2 and 3.

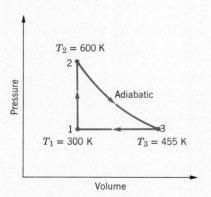

Figure 21–12 Problem 43.

44. From the knowledge that C_v, the molar heat capacity at constant volume, for a gas in a container is 5R, what can you conclude about the ratio of the speed of sound in that gas to the root-mean-square speed of its molecules at a temperature T?

45. The atomic weight of iodine is 127. A standing wave in a tube filled with iodine gas at 400 K has nodes that are 6.77 cm apart when the frequency is 1000 Hz. Is iodine gas monatomic or diatomic?

46. A steel tank contains 300 g of ammonia gas (NH_3) at an absolute pressure of 1.35×10^6 Pa and temperature 77°C. (a) What is the volume of the tank? (b) The tank is checked later when the temperature has dropped to 22°C and the absolute pressure has fallen to 8.7×10^5 Pa. How many grams of gas leaked out of the tank?

47. A room of volume V is filled with diatomic ideal gas (air) at temperature T_1 and pressure p_0. The air is heated to a higher temperature T_2, the pressure remaining constant at p_0 because the walls of the room are not airtight. Show that the internal energy content of the air remaining in the room is the same at T_1 and T_2, and that the energy supplied by the furnace to heat the air has all gone to heat the air *outside* the room. If we add no energy to the air, why bother to light the furnace? (Ignore the furnace energy used to raise the temperature of the walls, and consider only the energy used to raise the air temperature.)

48. An ideal gas experiences an adiabatic compression from $p = 1.0$ atm, $V = 1.0 \times 10^6$ liters, $T = 0°C$ to $p = 1.0 \times 10^5$ atm, $V = 1.0 \times 10^3$ liters. (a) Is this a monatomic, a diatomic, or a polyatomic gas? (b) What is the final temperature? (c) How many moles of the gas are present? (d) What is the total translational kinetic energy per mole before and after the compression? (e) What is the ratio of the squares of the rms speeds before and after the compression?

49. C_v for a certain ideal gas is 6.00 cal/mol·K. The temperature of 3.0 mol of the gas is raised 50 K by each of three different processes: at constant volume, at constant pressure, and by an adiabatic compression. Complete the accompanying table, showing for each process the heat added (or subtracted), the work done by the gas, the change in internal energy of the gas, and the change in total translational kinetic energy of the gas.

Process	Heat Added	Work Done by Gas	Change in Internal Energy	Change in Kinetic Energy
Constant volume	_____	_____	_____	_____
Constant pressure	_____	_____	_____	_____
Adiabatic	_____	_____	_____	_____

50. (a) An ideal gas, initially at pressure p_0, undergoes a free expansion (adiabatic, no external work) until its final volume is 3.0 times its initial volume. What is the pressure of the gas after the free expansion? (b) The gas is then slowly and adiabatically compressed back to its original volume. The pressure after compression is $(3.0)^{1/3} p_0$. Determine whether the gas is monatomic, diatomic, or polyatomic. (c) How does the average kinetic energy per molecule in this final state compare with that of the initial state?

51. A molecule of hydrogen (diameter 1.0×10^{-8} cm) escapes from a furnace (T = 4000 K) with the root-mean-square speed into a chamber containing atoms of cold argon (diameter 3.0×10^{-8} cm) at a density of 4.0×10^{19} atoms/cm³. (a) What is the speed of the hydrogen molecule? (b) If the molecule and an argon atom collide, what is the closest distance between their centers, considering each as spherical? (c) What is the initial number of collisions per unit time experienced by the hydrogen molecule?

52. A hypothetical gas of N particles has the speed distribution shown in Fig. 21-13. [$N(v) = 0$ for $v > 2v_0$.] (*a*) Evaluate a in terms of N and v_0. (*b*) Find the number of particles with speeds between $1.5v_0$ and $2.0v_0$. (*c*) Find the average speed of the particles. (*d*) Find v_{rms}.

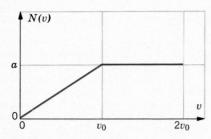

Figure 21-13   Problem 52.

53. In a motorcycle engine, after combustion occurs in the top of the cylinder, the piston is forced down as the mixture of gaseous products undergoes an adiabatic expansion. Find the average power involved in this expansion when the engine is running at 4000 rpm, assuming that the gauge pressure immediately after combustion is 15 atm, the initial volume is 50 cm³, and the volume of the mixture at the bottom of the stroke is 250 cm³. Assume that the gases are diatomic, and that the time involved in the expansion is one-half that of the total cycle. Express your answer in watts and horsepower.

*54. For a gas in which all molecules travel with the same speed $\bar{v}$, show that $\bar{v}_{\text{rel}} = \frac{4}{3}\bar{v}$ rather than $\sqrt{2}\,\bar{v}$ (which is the result obtained when we consider the actual distribution of molecular speeds).

22

Entropy and the Second Law of Thermodynamics

22–1 Introduction

The first law of thermodynamics states that energy is conserved. However, we can think of many thermodynamic processes that conserve energy but that actually never occur. For example, when a hot body and a cold body are put into contact, it simply does not happen that the hot body gets hotter and the cold body colder. Or again, a pond does not suddenly freeze on a hot summer day by giving up heat to its environment. And yet neither of these processes violates the first law of thermodynamics. Similarly, the first law does not restrict our ability to convert work into heat or heat into work, except that energy must be conserved in the process. And yet in practice, although we can convert a given quantity of work completely into heat, we have never been able to find a scheme that converts a given amount of heat completely into work. The second law of thermodynamics deals with this question of whether processes, assumed to be consistent with the first law, do or do not occur in nature. Although the ideas contained in the second law may seem subtle or abstract, in application they prove to be extremely practical.

22–2 Reversible and Irreversible Processes

Consider a typical system in thermodynamic equilibrium, say, a sample of a (real) gas confined in a cylinder-piston arrangement of volume V, the gas having a pressure p and a temperature T. In an equilibrium state these thermodynamic variables remain constant with time. Suppose that the cylinder, whose walls are an (ideal) heat insulator but whose base is an (ideal) heat conductor is placed on a large heat reservoir maintained at this same temperature T, as in Fig. 20–6. Now let us change the system to another equilibrium state in which the temperature T is the same but the volume V is reduced by one-half. Of the many ways in which we could do this we discuss two extreme cases.

I.   We depress the piston very rapidly; we then wait for equilibrium with the reservoir to be reestablished. During this process the gas is turbulent and its pressure and temperature are not well defined; we cannot plot the process as a continuous line on a p-V diagram because we would not know what value of pressure (or temperature) to associate with a given volume. The system passes from one equilibrium state i to another f through a series of nonequilibrium states (Fig. 22–1a).

II.   We depress the piston (assumed to be frictionless) exceedingly slowly — perhaps by adding sand to the top of the piston — so that the pressure, volume, and temperature of the gas are, at all times, well-defined quantities. We first drop a few grains of sand on the piston. This will reduce the volume of the system a little and the temperature will tend to rise; the system will depart from equilibrium, but only slightly. A small amount of heat will be transferred to the reservoir and in a short time the system will reach a new equilibrium state, its temperature again being that of the reservoir. Then we drop a few more grains of sand on the piston, reducing the volume further. Again we wait for a new equilibrium state to be established, and so forth. By many repetitions of this procedure we finally reduce the volume by one-half. During this entire process the system is never in a state differing much from an equilibrium state. If we imagine carrying out this procedure with still smaller successive increases in pressure, the intermediate states will depart from equilibrium even less. By indefinitely increasing the number of changes and correspondingly decreasing the size of each change, we arrive at an ideal process in which the system passes through a continuous succession of equilibrium states, which we can plot as a continuous line or path on a p-V diagram (Fig. 22–1b). During this process a certain amount of heat Q is transferred from the system to the reservoir.

Processes of type I are called *irreversible* and those of type II are called *reversible. A reversible process is one that, by a differential change in the environment, can be made to retrace its path.* Thus as we cause the piston to move slowly downward, in II, the external pressure on the piston exceeds the pressure exerted on it by the gas by only a differential amount dp. If at any instant we reduce the external pressure ever so slightly (by removing a few sand grains), so that it is less than the internal gas pressure by dp, the gas will expand instead of contracting and the system will retrace the equilibrium states through which it has just passed. In practice all processes are irreversible, but we can approach reversibility arbitrarily closely by making appropriate experimental refinements. The strictly reversible process is a simple and useful abstraction that bears a similar relation to real processes that the ideal gas abstraction does to real gases.

The process described in II is not only reversible but *isothermal,* because we have assumed that the temperature of the gas differs at all times by only a differential amount dT from the (constant) temperature of the reservoir on which the cylinder rests.

We could also reduce the volume *adiabatically* by removing the cylinder from the heat reservoir and putting it on a nonconducting stand. In an adiabatic process no heat is allowed to enter or to leave the system. An adiabatic process can be either reversible or irreversible — the definition does not exclude either. In a reversible adiabatic process we move the piston exceedingly slowly — perhaps using the sand-loading technique; in an irreversible adiabatic process we shove the piston down quickly.

An irreversible process

A reversible process

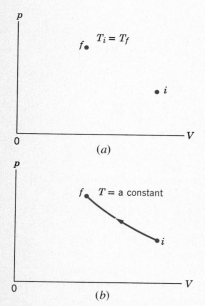

Figure 22–1   We cause a real gas to go from an initial equilibrium state i described by p_i, V_i, T_i to a final equilibrium state f described by p_f, $V_f(= \frac{1}{2}V_i)$, and $T_f(= T_i)$. We carry out the process (a) irreversibly, and (b) reversibly.

The temperature of the gas will rise during an adiabatic compression because, from the first law, with $Q = 0$, the work W done in pushing down the piston must appear as an increase ΔU in the internal energy of the system. W will have different values for different rates of pushing down the piston, being given by $\int p\, dV$—that is, by the area under a curve on a p-V diagram—only for reversible processes (for which p has a well-defined value). Thus ΔU and the corresponding temperature change ΔT will not be the same for reversible and irreversible adiabatic processes.

22–3 The Carnot Cycle

Suppose that we have a system (a gas, say) in an equilibrium state in a cylinder-piston arrangement. We can carry out, at our pleasure, a wide variety of processes. We can let the gas expand or we can compress it; we can add or subtract heat; we can do these things irreversibly or reversibly. We can also choose to carry out a sequence of processes such that the system returns to its original equilibrium state; we call this a *cycle*. If the processes are all reversible, we call it a *reversible cycle*.

Figure 22–2 shows a reversible cycle on a p-V diagram. Along the curve *abc* we allow the system to expand, and the area under this curve represents the work done by the system during the expansion. Along the curve *cda*, which returns the system to its original state, we compress the system, and the area under this curve represents the work we must do on the system during the compression. Hence, the *net* work done by the system is represented by the area enclosed by the curve and is positive. If we decided to traverse the cycle in the opposite sense, that is, expanding along *adc* and compressing along *cba*, the net work done by the system would be the negative of that of the previous case.

An important reversible cycle is the *Carnot cycle,* described by Sadi Carnot in 1824. We shall see later that this cycle will determine the limit of our ability to convert heat into work. The system consists of a "working substance," which we take to be an ideal gas, and the cycle is made up of two isothermal and two adiabatic reversible processes. The working substance is contained in a cylinder with a heat-conducting base and nonconducting walls and piston. We also provide a heat reservoir in the form of a body of large heat capacity at a temperature T_H, another reservoir at a colder temperature T_C and a nonconducting stand. We carry out the Carnot cycle in four steps, as shown in Figs. 22–3, 4.

The quantities Q and W that we shall use in describing the Carnot cycle can be either positive or negative, depending on the circumstances. However, it will be easier to understand what follows if we deal always with positive quantities. For this reason we will, in this and the following section, deal only with the absolute values of Q and W, which we represent by $|Q|$ and $|W|$ respectively. Thus, for both $Q = +10$ J and $Q = -10$ J we have $|Q| = 10$ J, and similarly for W. We will always make it explicitly clear in the context whether heat is being added to the system or withdrawn from it (and whether work is done on the system or by it).

Here are the four steps that make up the Carnot cycle:

Step 1.   The gas is in an initial equilibrium state represented by point *a* in Fig. 22–4. We put the cylinder on the heat reservoir at temperature T_H, and allow the gas to expand slowly to point *b* in Fig. 22–4. During the process heat $|Q_H|$ is absorbed by the gas by conduction through the base. The expansion is isothermal at T_H and the gas does work in raising the piston and its load.

A reversible cycle

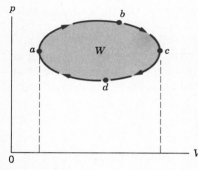

Figure 22–2 A *p-V* diagram of a gas undergoing a reversible cycle. The shaded area *W* represents the net work done by the gas in the cycle.

The Carnot cycle

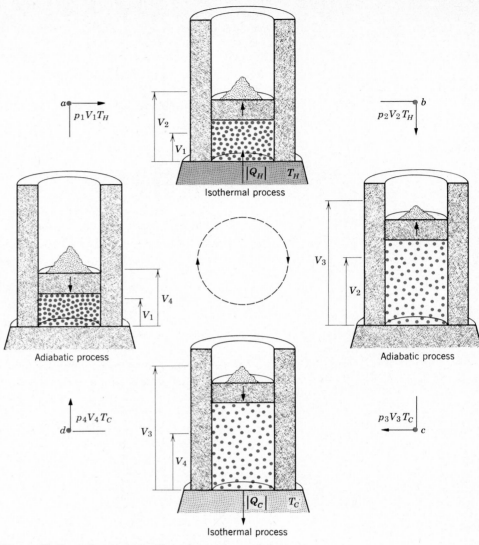

Figure 22–3   A Carnot cycle. The points a, b, c and d correspond to the points so labelled in Fig. 22–4. The cylinder-piston arrangements show intermediate steps in the processes that connect adjacent points. The arrows on the pistons suggest expansions (caused by removing sand) and compressions (caused by adding sand).

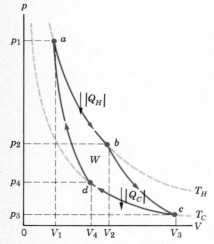

Figure 22–4   The Carnot cycle illustrated in the previous figure, plotted on a p-V diagram for an ideal gas as the working substance.

Step 2.   We put the cylinder on the nonconducting stand and, by reducing the piston load, allow the gas to further expand slowly to point c in Fig. 22–4. The expansion is adiabatic because no heat can enter or leave the system. The gas does work in raising the piston and its temperature falls to T_C.

Step 3.   We put the cylinder on the (colder) heat reservoir T_C and compress the gas slowly to point d in Fig. 22–4. During the process heat $|Q_C|$ is transferred from the gas to the reservoir by conduction through the base. The compression is isothermal at T_C and work is done on the gas by the piston, and its load.

Step 4.   We put the cylinder on the nonconducting stand and compress the gas slowly to the initial condition, point a of Fig. 22–4. The compression is adiabatic because no heat can enter or leave the system. Work is done on the gas and its temperature rises to T_H.

The net work $|W|$ done by the system during one cycle is represented by the area enclosed by the path *abcda* of Fig. 22–4. The net amount of heat that is absorbed by the system per cycle is $|Q_H| - |Q_C|$. The system returns to its initial state at the end of each cycle so that there is no net change in the internal energy U of the system. Hence, from the first law of thermodynamics,

$$|W| = |Q_H| - |Q_C|. \tag{22-1}$$

The result of the cycle is that heat has been converted into work. Any required amount of work can be obtained by simply repeating the cycle. Hence, the system is a *heat engine*.

Heat engine

We have used an ideal gas as an example of a working substance. The working substance can be anything at all, although the p-V diagrams for other substances would be different. Common heat engines use steam or a mixture of fuel and air, or fuel and oxygen as their working substance. Heat may be obtained from the combustion of a fuel such as gasoline or coal, or from fission processes in nuclear reactors. Heat may be discharged through an exhaust or to a condenser. Although real heat engines do not operate on a reversible cycle, the Carnot cycle, which is reversible, is especially important because, as we shall see, it sets an upper limit to the performance of real engines and thereby gives us a goal to work toward.

The efficiency e of a heat engine is the ratio of the net work done by the engine during one cycle to the heat taken in from the high temperature source in one cycle. Hence,

$$e = \frac{|W|}{|Q_H|} = \frac{|Q_H| - |Q_C|}{|Q_H|} = 1 - \frac{|Q_C|}{|Q_H|} \tag{22-2}$$

Equation 22–2 shows that the efficiency of a heat engine is less than one (100%) so long as the heat $|Q_C|$ delivered to the exhaust is not zero. Experience shows that every heat engine rejects some heat during the exhaust stroke. This represents the heat absorbed by the engine that is not converted to work in the process.

The Carnot refrigerator

We may choose to carry out the Carnot cycle by starting at any point, such as *a* in Fig. 22–4, and traversing each process in a direction opposite to that of the arrowheads in that figure. Then an amount of heat $|Q_C|$ is removed from the lower temperature reservoir at T_C, and an amount of heat $|Q_H|$ is delivered to the higher temperature reservoir at T_H; work $|W|$ must be done on the system by an outside agency. The system acts like a *refrigerator*, transferring heat from a body at a lower temperature (the freezing compartment) to one at a higher temperature (the room) by means of work supplied to it (the electric power input).

Example 1 *The efficiency of a Carnot engine* Show that the efficiency of a Carnot engine using an ideal gas as the working substance is $e = (T_H - T_C)/T_H$.

Along the isothermal path *ab*, the temperature, and hence the internal energy of the ideal gas, remains constant. From the first law, the heat $|Q_H|$ absorbed by the gas in its expansion must be equal to the work $|W_H|$ done in this expansion. From Example 2, Chapter 21, we have

$$|Q_H| = |W_H| = nRT_H \ln\left(\frac{V_2}{V_1}\right).$$

Likewise, in the isothermal compression along the path *cd*, we have

$$|Q_C| = |W_C| = nRT_C \ln\left(\frac{V_3}{V_4}\right).$$

On dividing the first equation by the second, we obtain

$$\frac{|Q_H|}{|Q_C|} = \frac{T_H}{T_C} \frac{\ln(V_2/V_1)}{\ln(V_3/V_4)}.$$

From the equation describing an isothermal process for an ideal gas we obtain for the paths ab and cd

$$p_1V_1 = p_2V_2,$$
$$p_3V_3 = p_4V_4.$$

From Eq. 21–12 describing an adiabatic process for an ideal gas we have for paths bc and da

$$p_2V_2^\gamma = p_3V_3^\gamma,$$
$$p_4V_4^\gamma = p_1V_1^\gamma.$$

Multiplying these four equations together and canceling the factor $p_1p_2p_3p_4$ appearing on both sides, we obtain

$$V_1V_2^\gamma V_3V_4^\gamma = V_2V_3^\gamma V_4V_1^\gamma,$$

from which

$$(V_2V_4)^{\gamma-1} = (V_3V_1)^{\gamma-1}$$

and

$$\frac{V_2}{V_1} = \frac{V_3}{V_4}.$$

Using this result in our expression for $|Q_H|/|Q_C|$, we see that

$$\frac{|Q_H|}{|Q_C|} = \frac{T_H}{T_C},$$

so that

$$e = 1 - \frac{|Q_C|}{|Q_H|} = 1 - \frac{T_C}{T_H}$$

or

$$e = \frac{|Q_H| - |Q_C|}{|Q_H|} = \frac{T_H - T_C}{T_H}. \tag{22–3}$$

The temperatures T_H and T_C are those measured on the ideal gas scale described in Chapter 19.

22–4 The Second Law of Thermodynamics

The first heat engines constructed were very inefficient devices. Only a small fraction of the heat absorbed at the high-temperature source could be converted to useful work. Even as engineering design improved, a sizable fraction of the absorbed heat was still discharged at the lower-temperature exhaust of the engine, remaining unconverted to mechanical energy. It remained a hope to devise an engine that could take heat from an abundant reservoir, like the ocean, and convert it completely into useful work. Then it would not be necessary to provide a source of heat at a higher temperature than the outside environment by burning fuels (Fig. 22–5). Likewise, we might hope to be able to devise a refrigerator that simply transfers heat from a cold body to a hot body, without requiring the expense of outside work (Fig. 22–6). Neither of these hopeful ambitions violates the first law of thermodynamics. The heat engine would simply convert heat energy completely into mechanical energy, the total energy being conserved in the process. In the refrigerator, the heat energy would simply be transferred from cold body to hot body without any loss of energy in the process. Nevertheless neither of these ambitions has ever been achieved, and there is reason to believe they never will be.

Figure 22–5 In an actual heat engine, some of the heat $|Q_H|$ taken in by the engine is converted into work $|W|$, but the rest is rejected as heat $|Q_C|$. In a "perfect" heat engine all the heat input would be converted into work output.

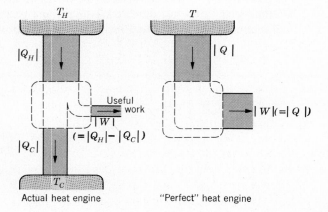

Actual heat engine   "Perfect" heat engine

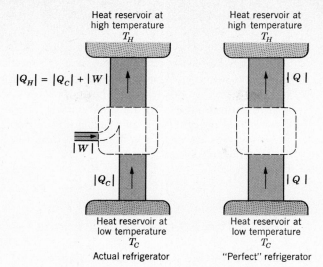

$|Q_H| = |Q_C| + |W|$

Heat reservoir at high temperature T_H

Heat reservoir at high temperature T_H

$|W|$

$|Q_C|$

$|Q|$

$|Q|$

Heat reservoir at low temperature T_C

Heat reservoir at low temperature T_C

Actual refrigerator

"Perfect" refrigerator

Figure 22–6 In an actual refrigerator, work $|W|$ is needed to transfer heat from a low-temperature to a high-temperature reservoir. In a "perfect" refrigerator, heat would flow from the low-temperature to the high-temperature reservoir without any work being done on the engine.

The second law of thermodynamics —Clausius statement

The *second law of thermodynamics,* which is a generalization of experience, is an assertion that such devices do not exist. There have been many statements of the second law, each emphasizing another facet of the law, but all can be shown to be equivalent to one another. Clausius stated it as follows: *It is not possible for any cyclical machine to convey heat continuously from one body to another at a higher temperature without, at the same time, producing some other (compensating) effect.* This statement rules out our ambitious refrigerator, for it implies that to convey heat continuously from a cold to a hot object it is necessary to supply work by an outside agent. We know from experience that when two bodies are in contact, heat energy flows from the hot body to the cold body. The second law rules out the possibility of heat energy flowing from cold to hot body in such a case and so determines the direction of transfer of heat. The direction can be reversed only by an expenditure of work.

Kelvin-Planck statement

Kelvin (with Planck) stated the second law in words equivalent to these: *A transformation whose only final result is to transform into work heat extracted from a source that is at the same temperature throughout is impossible.* This statement rules out our ambitious heat engine, for it implies that we cannot produce mechanical work by extracting heat from a single reservoir without returning any heat to a reservoir at a lower temperature.

To show that the two statements are equivalent we need to show that, if either statement is false, the other statement must be false also. Suppose that Clausius' statement were false, so that we could have a refrigerator operating without needing a work input. We could use an ordinary engine to remove heat from a hot body, to do work and to return part of the heat to a cold body. But by connecting our "perfect" refrigerator into the system, this heat would be returned to the hot body without expenditure of work and would become available again for use by the heat engine. Hence, the combination of an ordinary engine and the "perfect" refrigerator would constitute a heat engine that violates the Kelvin-Planck statement. Or we can reverse the argument. If the Kelvin-Planck statement were incorrect, we could have a heat engine that simply takes heat from a source and converts it completely into work. By connecting this "perfect" heat engine to an ordinary refrigerator, we could extract heat from the hot body, convert it completely to work, use this work to run the ordinary refrigerator, extract heat from the cold body, and deliver it plus the work converted to heat by the refrigerator to the hot body. The net result is a transfer of heat from cold to hot body without expenditure of work, and this violates Clausius' statement.

The second law tells us that many processes are irreversible. For example, Clausius' statement specifically rules out a simple reversal of the process of heat transfer from hot body to cold body. Not only will some processes not run backward by themselves, but no combination of processes can undo the effect of an irreversible process without causing another corresponding change elsewhere. In later sections we shall develop these ideas more fully and formulate the second law quantitatively.

22-5 The Efficiency of Engines

Carnot first wrote scientifically on the theory of heat engines. In 1824 he published *Reflections on the Motive Power of Heat*. By then the steam engine was commonly used in industry. Carnot wrote:

> In spite of labor of all sorts expended on the steam engine, and in spite of the perfection to which it has been brought, its theory is very little advanced. . . .
>
> The production of motion in the steam engine is always accompanied by a circumstance which we should particularly notice. This circumstance is the passage of caloric from one body where the temperature is more or less elevated to another where it is lower. . . .
>
> The motive power of heat is independent of the agents employed to develop it; its quantity is determined solely by the temperature of the bodies between which, in the final result, the transfer of the caloric occurs.

Hence, Carnot directed attention to the facts that the difference in temperature was the real source of "motive power," that the transfer of heat played a significant role, and that the choice of working substance was of no theoretical importance.

Carnot's achievement was remarkable when we recall that the mechanical equivalence of heat and the conservation of energy principle were not known in 1824. In his later papers, published posthumously in 1872, it became clear that Carnot had foreseen the principle of the conservation of energy and had made an accurate determination of the mechanical equivalent of heat. He had planned a program of research that included all the important developments in the field made by other investigators during the following several decades. However, he died during a cholera epidemic in 1832 at the age of 36, leaving it to others to extend his work. It was William Thomson (later Lord Kelvin) who modified Carnot's reasoning to bring it into accord with the mechanical theory of heat, and who, together with Clausius, successfully developed the science of thermodynamics.

All reversible engines have the same efficiency

Carnot developed the concept of a reversible engine and the reversible cycle named after him. He stated a theorem of great practical importance: *The efficiency of all reversible engines operating between the same two temperatures is the same, and no irreversible engine working between the same two temperatures can have a greater efficiency than this.* Clausius and Kelvin showed that this theorem was a necessary consequence of the second law of thermodynamics. Notice that nothing is said about the working substance, so that the efficiency of a reversible engine is independent of the working substance and depends only on the temperatures. Furthermore, a reversible engine operates at the maximum efficiency pos-

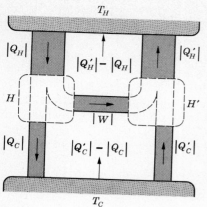

Figure 22–7 Proof of Carnot's theorem.

sible for any engine working between the same two temperature limits. The proof of this theorem follows.

Let us call the two reversible engines H and H'. They operate between the temperatures T_H and T_C, where $T_H > T_C$. They may differ, say, in their working substance or in their initial pressures and lengths of stroke. We choose H to run forward and H' to run backward (as a refrigerator). The forward-running engine H takes in heat energy $|Q_H|$ at T_H and gives out heat energy $|Q_C|$ at T_C. The backward-running engine (refrigerator) H' takes in heat $|Q_C'|$ at T_C and gives out heat $|Q_H'|$ at T_H. We now connect the engines mechanically and adjust the stroke lengths so that the work done per cycle by H is just sufficient to operate H' (Fig. 22–7). Suppose that the efficiency e of H were greater than the efficiency e' of H'. Then

$$e > e' \qquad \text{(assumption)},$$

or

$$\frac{|W|}{|Q_H|} > \frac{|W'|}{|Q_H'|}.$$

However, because we have arranged it so that $|W| = |W'|$, this inequality can only be true if

$$|Q_H'| > |Q_H|.$$

This means that $|Q_H'| - |Q_H|$ (see Fig. 22–7) is a positive quantity.

We can also write the work equality as

$$|Q_H| - |Q_C| = |Q_H'| - |Q_C'|$$

or

$$|Q_H'| - |Q_H| = |Q_C'| - |Q_C|.$$

Thus, for each cycle of operation of the combined system (see again Fig. 22–7), an amount of heat given by $|Q_C'| - |Q_C|$ leaves the cool reservoir and exactly equal amount of heat given by $|Q_H'| - |Q_H|$ enters the hot reservoir, no work being done in the process. Thus we have transferred heat from a body at one temperature to a body at a higher temperature without performing work—in direct contradiction to Clausius' statement of the second law. Hence, we conclude that e cannot be greater than e'. Likewise, by reversing the engines we can use the same reasoning to prove that e' cannot be greater than e. Hence,

$$e = e',$$

No engine can be more efficient than a reversible engine

proving the first part of Carnot's theorem.

Now suppose that H is an irreversible engine. Then by the exact same procedure we can prove that e_{irr} cannot be greater than e'. But H cannot be reversed, so we cannot prove that e' cannot be greater than e_{irr}. Therefore e_{irr} is either equal to or less than e'. Since $e' = e = e_{\text{rev}}$ we have

$$e_{\text{irr}} \le e_{\text{rev}},$$

thus proving the second part of Carnot's theorem.

Example 2 A steam engine takes steam from the boiler at 200°C (225 lb/in.² pressure) and exhausts directly into the air (14 lb/in.² pressure) at 100°C. What is its maximum possible efficiency?

Using the result of Example 1 (which applies to this case by virtue of Carnot's theorem, which we have just proved), we have

$$e_{max} = e_{reversible} = \frac{T_H - T_C}{T_H} = \frac{473 \text{ K} - 373 \text{ K}}{473 \text{ K}} = 0.21 \text{ or } 21\%.$$

Actual efficiencies of about 15% are usually realized. Energy is lost by friction, turbulence, and heat conduction. Lower exhaust temperatures on more complicated steam engines may raise the maximum possible efficiency to 35% and the actual efficiency to 20%. The efficiency of an ordinary automobile engine is about 22% and that of a large diesel engine about 40%.

22–6 Entropy—Reversible Processes

The zeroth law of thermodynamics is related to the concept of temperature T and the first law is related to the concept of internal energy U. In this and the following sections we show that the second law of thermodynamics is related to a thermodynamic variable called *entropy, S*, and that we can express the second law quantitatively in terms of this variable. We start by considering a Carnot cycle. For such a cycle we have seen (Eq. 22–3) that

$$\frac{|Q_H|}{T_H} = \frac{|Q_C|}{T_C},$$

Now let us discard the absolute value notation, recognizing in the process that, whether the Carnot cycle is operated (forwards) as an engine or (backwards) as a refrigerator, Q_H and Q_C have opposite algebraic signs. With this understanding we can rewrite this relation as

$$\frac{Q_H}{T_H} + \frac{Q_C}{T_C} = 0.$$

This equation states that the sum of the algebraic quantities Q/T is zero for a Carnot cycle.

As a next step, we assert that any reversible cycle is equivalent, to as close an approximation as we wish, to an assembly of Carnot cycles. Figure 22–8a shows an arbitrary reversible cycle superimposed on a family of isotherms. We can approximate the actual cycle by connecting the isotherms by suitably chosen adiabatic lines (Fig. 22–8b), thus forming an assembly of Carnot cycles. You should convince yourself that traversing the individual Carnot cycles in Fig. 22–8b is exactly equivalent, in terms of heat transferred and work done, to traversing the jagged sequence of isotherms and adiabatic lines that approximates the actual cycle. This is so because adjacent Carnot cycles have a common isotherm and the two traversals, in opposite directions, cancel each other in the region of overlap as far as heat transfer and work done are concerned. By making the temperature interval between the isotherms in Fig. 22–8b small enough, we can approximate the actual cycle as closely as we wish by an alternating sequence of isotherms and adiabatic lines.

We can write, then, for the isothermal-adiabatic sequence of lines in Fig. 22–8b,

$$\sum \frac{Q}{T} = 0,$$

or, in the limit of infinitesimal temperature differences between the isotherms of Fig. 22–8b,

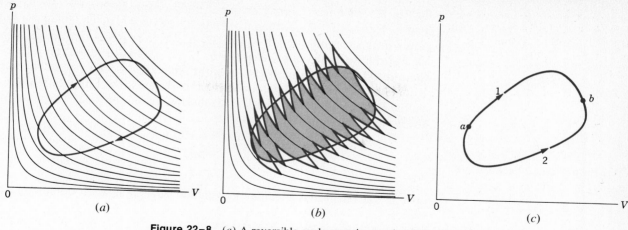

Figure 22-8 (a) A reversible cycle superimposed on a family of isotherms. (b) The isotherms are connected by adiabatic lines, forming an assembly of Carnot cycles that approximates the given cycle. (c) a and b are two arbitrary points on the cycle and 1 and 2 are reversible paths connecting them.

$$\oint \frac{dQ}{T} = 0, \tag{22-4}$$

in which $\oint$ indicates that the integral is evaluated for a complete traversal of the cycle, starting (and ending) at any arbitrary point of the cycle.

If the integral of a quantity around any closed path is zero, the quantity is called a *state variable*; that is, it has a value that is characteristic only of the state of the system, regardless of how that state was arrived at. We call the variable in this case the *entropy S* and we have, from Eq. 22-4,

State variable

Entropy; the entropy change for a closed cycle is zero

$$dS = \frac{dQ}{T} \quad \text{and} \quad \oint dS = 0. \tag{22-5}$$

Common units for entropy are joules per kelvin or calories per kelvin.

Gravitational potential energy U_g, internal energy U, pressure p, and temperature T are other state variables and equations of the form $\oint dX = 0$ hold for each of them, where for X we substitute the appropriate symbol. Heat Q and work W are not state variables and we know that, in general $\oint dQ \neq 0$ and $\oint dW \neq 0$, as you can easily show for the special case of a Carnot cycle.

The property of a state variable expressed by $\oint dX = 0$ can also be expressed by saying that $\int dX$ between any two equilibrium states has the same value for all (reversible) paths connecting those states. Let us prove this for the state variable called entropy. We can write Eq. 22-5 (see Fig. 22-8c) as

$$_1\int_a^b dS + {}_2\int_b^a dS = 0 \tag{22-6}$$

where a and b are arbitrary points and 1 and 2 describe the paths connecting these points. Since the cycle is reversible, we can write Eq. 22-6 as

$$_1\int_a^b dS - {}_2\int_a^b dS = 0$$

or

$$\tag{22-7}$$

$$_1\int_a^b dS = {}_2\int_a^b dS$$

In Eq. 22–7 we have simply decided to traverse path 2 in the opposite direction, that is, from a to b rather than from b to a. We do this by changing the order of the limits in the second integral of Eq. 22–6, which requires that we also change the sign of the integral, thus yielding Eq. 22–7. This latter equation tells us that the quantity $\int_a^b dS$ between any two equilibrium states of the system, such as a and b, is independent of the path connecting those states, for 1 and 2 are quite arbitrary paths. Recall our almost identical discussion in Section 8–2, where we introduced the concept of a conservative force.

The change in entropy between a and b in Fig. 22–8c is, then

The entropy change between two states is independent of path

$$S_b - S_a = \int_a^b dS = \int_a^b \frac{dQ}{T} \qquad \text{(reversible process)}, \qquad (22\text{–}8)$$

where the integral is evaluated over any reversible path connecting these two states.

22–7 Entropy—Irreversible Processes

In Section 22–6 we spoke only of reversible processes. However, entropy, like all state variables, depends only on the state of the system, and we must be able to calculate the change in entropy for irreversible processes, provided only that they begin and end in equilibrium states. Let us consider two examples.

1. Free Expansion. As in Section 20–7 (see Fig. 20–11), let a gas double its volume by expanding into an evacuated enclosure. Since no work is done against the vacuum, $W = 0$ and, since the gas is enclosed by nonconducting walls, $Q = 0$. From the first law, then $\Delta U = 0$ or

$$U_i = U_f, \qquad (22\text{–}9)$$

where i and f refer to the initial and final (equilibrium) states. If the gas is an ideal gas, then U depends on temperature alone and not on the pressure or the volume so that Eq. 22–9 implies $T_i = T_f$.

The free expansion is certainly irreversible because we lose control of the environment once we turn the stopcock in Fig. 20–11. There is, however, an entropy difference $S_f - S_i$ between the initial and final equilibrium states, but we cannot calculate it from Eq. 22–8 because that relation applies only to reversible paths; if we tried to use that equation we would have the immediate difficulty that $Q = 0$ for the free expansion and—further—we would not know how to assign meaningful values of T to the intermediate, nonequilibrium states.

How, then, do we calculate $S_f - S_i$ for these two states? We do so by finding a reversible path (any reversible path) that connects the states i and f and we calculate the entropy change for that path. In the free expansion a convenient reversible path (assuming an ideal gas) is an isothermal expansion from V_i to $V_f (= 2V_i)$. This corresponds to the isothermal expansion carried out between the points a and b of the Carnot cycle of Fig. 22–4. It represents quite a different set of operations from the free expansion and has in common with it only the fact that it connects the same set of equilibrium states, i and f. From Eq. 22–8 and Example 1 we have

Entropy change in a free expansion

$$S_f - S_i = \int_i^f \frac{dQ}{T} = nR \ln \left(\frac{V_f}{V_i} \right)$$

$$= nR \ln 2.$$

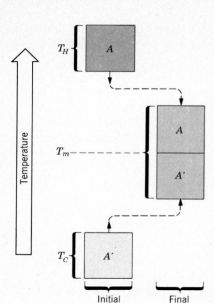

Figure 22–9 Two identical bodies held initially at different temperatures are put in thermal contact and eventually come to an equilibrium state at an intermediate temperature T_m. The entropy of the system increases during this irreversible process.

This is positive so that the entropy of the system increases in this irreversible, adiabatic process. Note that a free expansion is a process that, in nature, proceeds of its own accord once started. We cannot realistically conceive of its opposite, a free compression, in which the gas in a closed insulated vessel would spontaneously compress itself so that it occupied only half of the volume freely available to it. All our experience tells us that the first process is inevitable and the second is virtually unthinkable.

2. Irreversible Heat Transfer. For another example consider two bodies that are similiar in every respect except that one is at a temperature T_H and the other at temperature T_C, where $T_H > T_C$. If we put both objects in contact inside a box with nonconducting walls, they will eventually reach a common temperature T_m, somewhere between T_H and T_C; see Fig. 22–9. Like the free expansion, the process is irreversible because we lose control of the environment once we put the two bodies in the box. Like the free expansion this process is also (irreversibly) adiabatic because no heat enters or leaves the system during the process.

To calculate the entropy change for the system during this process we must again find a reversible process connecting the same initial and final states and calculate the system entropy change by applying Eq. 22–8 to that process. We can do so if we imagine that we have at our disposal a heat reservoir of large heat capacity whose temperature T is at our control, by turning a knob, say. We first adjust the reservoir temperature to T_H and put the first (hotter) object in contact with the reservoir. We then slowly (reversibly) lower the reservoir temperature from T_H to T_m, extracting heat from the hot body as we do so. The hot body loses entropy in this process, the entropy change being

$$\Delta S_H = -\frac{|Q|}{T_1}$$

Here T_1 is a suitably chosen average temperature that lies between T_H and T_m (see Fig. 22–9), and $|Q|$ is the heat extracted.

We then adjust our reservoir temperature to T_C and place it in contact with the second (cooler) body. We then slowly (reversibly) raise the reservoir temperature from T_C to T_m, adding heat to the cool body as we do so. The cool body gains entropy in this process, the entropy change being

$$\Delta S_C = +\frac{|Q|}{T_2}$$

Here T_2 is a suitably chosen average temperature that lies between T_C and T_m (see Fig. 22–9) and $|Q|$ is the heat added. The heat $|Q|$ added to the cold body is the same as the heat $|Q|$ extracted from the hot body.

The two bodies are now at the same temperature T_m and the system is in its final equilibrium state. The change of entropy for the complete system is

$$S_f - S_i = \Delta S_H + \Delta S_C$$
$$= -\frac{|Q|}{T_1} + \frac{|Q|}{T_2}$$

Entropy increases during an irreversible heat transfer

Since $T_1 > T_2$ (see Fig. 22–9) we have $S_f > S_i$. Again, as for a free expansion, the entropy of the system has increased in this reversible, adiabatic process.

Note that, like the free expansion, our heat conduction example is a process that in nature proceeds of its own accord once started. We cannot realistically conceive of the opposite process, in which for example a metal rod in thermal equilibrium at room temperature would spontaneously adjust itself so that one end grew hotter and the other colder. Once again, nature has a totally overwhelming preference that a process should go in one particular direction and not in the opposite direction.

In each of these examples we must distinguish carefully between the actual (irreversible) process (free expansion or heat transfer) and the reversible process that we introduce just so that we can calculate the entropy change in the actual process. We can choose any reversible process, as long as it connects the same initial and final states as the actual process; all such reversible processes will yield the same entropy change because this depends only on the initial and final states and not on the process connecting them—be it reversible or irreversible.

22-8   Entropy and the Second Law

We are now ready to formulate the second law of thermodynamics in terms of entropy. Since this law is a generalization from experience we cannot *prove* it. We can, however, show that this statement is in agreement with experiment and is equivalent to other formulations of the second law that we have given earlier. In this spirit we assert that the second law is as follows: *Any process that starts in one equilibrium state and ends in another will go in the direction that causes the entropy of the system plus environment to increase.*

Entropy statement of the second law

The two experiments of Section 22–7 (free expansion and heat transfer) are consistent with the second law. The entropy of the system increased in each of these irreversible processes. Note that the entropy of the environment in these two cases remains unchanged because, both being carried out in adiabatic enclosures, there was no interchange of heat with the environment. Thus, as required by our statement of the second law, the entropy of the system plus environment increased for each of these (natural) processes.

In the form that we have written it the second law applies only to irreversible processes because only such processes have a "natural direction." Indeed (see Section 22–1), the understanding of the natural direction of such processes is the main concern of the second law. Reversible processes can go equally well in either direction, however, and *for reversible processes the entropy of the system plus environment remains unchanged.* This is so because if heat dQ is transferred from the environment to the system the entropy of the environment *decreases* by dQ/T, whereas that of the system *increases* by dQ/T, the net change for the system plus environment being zero. The fact that the process is reversible means that the environment and the system can differ in temperature by only a differential amount dT when the heat transfer takes place; this is in sharp contrast to our (irreversible) heat transfer problem of the previous section, in which the temperature difference of the two bodies placed in contact was large.

Another class of processes of particular interest are adiabatic processes (reversible or irreversible); they involve no transfer of heat with the environment, so the only entropy change possible is that of the system. From our statement of the second law and from our remarks about reversible processes in the paragraph above, we conclude that

Entropy and adiabatic processes

$$S_f = S_i \quad \text{(reversible adiabatic process)}$$

and

$$S_f > S_i \quad \text{(irreversible adiabatic process)},$$

where S_f and S_i are the final and initial entropies of the system.

Our statement of the second law is consistent with the Clausius statement (p. 439), which declares that there is no such thing as a "perfect" refrigerator (see Fig. 22–6). If there were, the entropy of the lower temperature reservoir would decrease by Q/T_C; that of the upper temperature reservoir would increase by Q/T_H; that of the system would remain unchanged because the system traverses a cycle, returning to its starting point. Thus the net change in the entropy of the system plus environment is a decrease, because $T_C < T_H$. This violates the statement of the second law that we have just given and, if we wish to retain the statement, we must conclude (with Clausius) that there is no such thing as a "perfect" refrigerator.

Our statement of the second law is consistent with the Kelvin-Planck statement (p. 439), which declares that there is no such thing as a "perfect" heat engine (see Fig. 22–5). If there were, the entropy of the reservoir at temperature T would decrease by Q/T; that of the system would remain unchanged because the system traverses a cycle, returning to its starting point. Thus the net change of entropy of the system plus environment is a decrease. This violates the statement of the second law that we have just given and, if we wish to retain the statement, we must conclude (with Kelvin) that there is no such thing as a "perfect" heat engine.

Example 3 Compute the entropy change of a system consisting of 1.00 kg of ice at 0°C which melts (reversibly) to water at that same temperature. The latent heat of melting is 79.6 cal/g.

The requirement that we melt the ice reversibly means that we must put it in contact with a heat reservoir whose temperature exceeds 0°C by only a differential amount. (If we lower the reservoir temperature until it is a differential amount below 0°C, the melted ice will begin to freeze.) Since the process is reversible, we can use Eq. 22–8 to compute the entropy change of the system. The temperature remains constant at 273 K. Therefore,

$$S_{\text{water}} - S_{\text{ice}} = \int \frac{dQ}{T} = \frac{1}{T} \int dQ = \frac{Q}{T}.$$

But

$$Q = 1.00 \times 10^3 \text{ g} \times 79.6 \text{ cal/g} = 7.96 \times 10^4 \text{ cal}$$

or

$$S_{\text{water}} - S_{\text{ice}} = \frac{7.96 \times 10^4}{273} \text{ cal/K} = 292 \text{ cal/K} = 1220 \text{ J/K}.$$

In this example of reversible melting the entropy change of the system plus environment is zero, as it must be for all reversible processes. The entropy change calculated above is the increase in entropy of the system; there is an exactly equal decrease in entropy of the environment (-1220 J/K) associated with the heat that leaves the reservoir (environment), at 273 K, to melt the ice.

In practice, melting is likely to be irreversible, as when we put an ice cube in a glass of water at room temperature. This process has only one natural direction—the ice will melt. The entropy of the system plus environment will increase in this process as required by the second law. The (irreversible) heat transfer example of the previous section should make this understandable.

Example 4 Calculate the entropy change that an ideal gas undergoes in a reversible isothermal expansion from a volume V_i to a volume V_f.

From the first law,

$$dU = dQ - p \, dV.$$

But $dU = 0$, since U depends only on temperature for an ideal gas and the temperature is constant. Hence,

$$dQ = p \, dV$$

and

$$dS = \frac{dQ}{T} = \frac{p \, dV}{T}.$$

But

$$pV = nRT,$$

so that

$$dS = nR \frac{dV}{V}$$

and

$$S_f - S_i = \int_{V_i}^{V_f} nR \frac{dV}{V} = nR \ln \frac{V_f}{V_i}. \qquad (22–10)$$

Since $V_f > V_i$, $S_f > S_i$ and the entropy of the gas increases.

In order to carry out this process we must have a reservoir at temperature T that is in contact with the system and supplies the heat to the gas. Hence, the entropy of the reservoir decreases by $\int dQ/T \, [= nR \ln (V_f/V_i)]$, so that in this process the entropy of system plus environment does not change. As in the previous example, this is characteristic of a reversible process.

You should also convince yourself that overall entropy increases and that this is an irreversible expansion if the reservoir temperature is slightly higher than that of the gas.

REVIEW GUIDE AND SUMMARY

The role of the second law of thermodynamics

Many processes do not occur in nature even though their occurence would not violate the first law of thermodynamics. The spontaneous flow of heat from a cold to a hot body is an example. The *second law of thermodynamics* sets up a criterion for identifying such "forbidden" processes.

Reversible and irreversible processes

Thermodynamic processes may carry a system from an initial state i to a final state f either *reversibly* or *irreversibly*. In the first case (Fig. 22–1b) the system passes through a continuous sequence of equilibrium states; it can be made to reverse its path at any stage by making a suitable infinitesimal change in the environment. In the second case (Fig. 22–1a) the intermediate states are not equilibrium states; path reversal as defined above cannot occur. All real processes are irreversible but, by using suitable techniques, they can often be made to approximate reversibility closely.

The Carnot cycle — an idealized heat engine

The *Carnot cycle* describes the action of an ideal (that is, reversible) heat engine. As Figs. 22–3 and 22–4 show, it consists of two *adiabatic processes* (for which $Q = 0$) alternating with two *isothermal processes* (for which T = a constant). In each cycle heat $|Q_H|$ is withdrawn from a high-temperature reservoir T_H, work $|W|$ is done, and heat $|Q_C|$ is discharged to a lower-temperature reservoir T_C. The *efficiency* of the Carnot engine is

$$e = \frac{\text{Work output}}{\text{Heat input}} = \frac{|W|}{|Q_H|} = \frac{|Q_H| - |Q_C|}{|Q_H|} = 1 - \frac{|Q_C|}{|Q_H|}. \qquad [22\text{–}2]$$

If the working substance of a Carnot engine is an ideal gas, the efficiency can be shown (Example 1) to be

A limit on the efficiency of real engines

$$e = 1 - \frac{T_C}{T_H}. \qquad [22\text{–}3]$$

In Section 22–5 we show from the second law of thermodynamics that (1) All reversible engines, no matter what their working substance, have this same efficiency, and (2) No real engine can have a higher efficiency.

The second law of thermodynamics — three formulations

There are three equivalent formulations of the second law of thermodynamics. The *Kelvin-Planck statement* (Fig. 22–5 and text) tells us that a "perfect" heat engine is impossible. The *Clausius statement* (Fig. 22–6 and text) tells us that a "perfect" refrigerator is impossible. The *entropy statement* is as follows: If a system goes from a state i to a state f, the entropy of the *system plus environment* will either remain constant (if the process is reversible) or will increase (if the process is irreversible). Example 3 illustrates this for the melting of ice in water, both reversibly and irreversibly.

Calculating entropy changes for reversible processes — a defining equation

Entropy, S, like pressure, volume, and so on, is a property of a system in equilibrium. The change in S for a system that goes *reversibly* from state i to state f is *defined* from

$$S_f - S_i = \int_i^f dS = \int_i^f \frac{dQ}{T} \qquad \text{(reversible processes only)} \qquad [22\text{–}8]$$

where dQ is an increment of heat transferred at temperature T. $S_f - S_i$ depends only on the initial and final states and not in any way on the nature of the path connecting them. The SI unit for entropy is the joule per kelvin.

Calculating entropy changes for irreversible processes — a stratagem

If a system goes *irreversibly* from i to f, Eq. 22–8 cannot be applied and this stratagem must be used: (1) Find a *reversible* path that connects the states i and f. (2) Calculate $S_f - S_i$ for *this* path, using Eq. 22–8. The answer you get holds also for the actual irreversible process because both processes connect the same two states. Section 22–7 illustrates this for the (irreversible) *free expansion* of a gas and for the (irreversible) *transfer of heat* between two bodies.

QUESTIONS

1. What requirements should a system meet in order to be in thermodynamic equilibrium?

2. Are any of these phenomena reversible? (*a*) breaking an empty soda bottle; (*b*) mixing a cocktail; (*c*) melting an ice cube in a glass of ice tea; (*d*) burning a log of firewood; (*e*) puncturing an automobile tire; (*f*) finishing the "Unfinished Symphony"; (*g*) writing this book.

3. Give some examples of irreversible processes in nature.

4. Are there any natural processes that are reversible?

5. Is there an upper limit to temperature? . . . a lower limit?

6. In the irreversible process of Fig. 22–1a, can we calculate the work done in terms of an area on a p-V diagram? Is any work done?

7. Can a given amount of mechanical energy be converted completely into heat energy? If so, give an example.

8. Can you suggest a reversible process whereby heat can be added to a system? Would adding heat by means of a Bunsen burner be a reversible process?

9. Give a qualitative explanation of how frictional forces between moving surfaces produce heat energy. Does the reverse process (heat energy producing relative motion of these surfaces) occur? Can you give a plausible explanation?

10. A block returns to its initial position after dissipating mechanical energy to heat through friction. Is this process thermodynamically reversible?

11. To carry out a Carnot cycle we need not start at point a in Fig. 22–4, but may equally well start at points b, c, or d, or indeed any intermediate point. Explain.

12. If a Carnot engine is independent of the working substance, then perhaps real engines should be similarly independent, to a certain extent. Why then, for real engines, are we so concerned to find suitable fuels such as coal, gasoline, or fissionable material? Why not use stones as a fuel?

13. Couldn't we just as well define the efficiency of an engine as $e = W/Q_C$ rather than as $e = W/Q_H$? Why don't we?

14. Under what conditions would an ideal heat engine be 100% efficient?

15. What factors reduce the efficiency of a heat engine from its ideal value?

16. In order to increase the efficiency of a Carnot engine most effectively, would you increase T_H keeping T_C constant, or would you decrease T_C, keeping T_H constant?

17. Can a kitchen be cooled by leaving the door of an electric refrigerator open? Explain.

18. Is a heat engine operating between the warm surface water of a tropical ocean and the cooler water beneath the surface a possible concept? Is the idea practical? (See "Solar Sea Power," by Clarence Zener, *Physics Today,* January 1973.)

19. Why do you get poorer gasoline mileage from your car in winter than in summer?

20. From time to time inventors will claim to have perfected a device that does useful work but consumes no (or very little) fuel. What do you think is most likely true in such cases: (*i*) The claimants are right; (*ii*) the claimants are mistaken in their measurements; or (*iii*) the claimants are swindlers? Do you think that each such claim should be examined closely by a panel of scientists and engineers? In your opinion, would the time and effort be justified?

21. We have seen that real engines always discard substantial amounts of heat to their low-temperature reservoirs. It seems a shame to throw this heat energy away. Why not use this heat to run a second engine, the low-temperature reservoir of the first engine serving as the high-temperature reservoir of the second?

22. A *heat pump* can warm your house by operating as a refrigerator, withdrawing heat from outside (the cold reservoir), doing work, and discharging heat inside (the warm reservoir). Why is this method of heating intrinsically more effective—practical considerations aside—than heating your house with electric heating units?

23. Give examples in which the entropy of a system decreases and explain why the second law of thermodynamics is not violated.

24. Do living things violate the second law of thermodynamics? As a chicken grows from an egg, for example, it becomes more and more ordered and organized. Increasing entropy, however, calls for disorder and decay. Is the entropy of a chicken actually decreasing as it grows?

25. Is there a change in entropy in purely mechanical motions?

26. Two samples of a gas initially at the same temperature and pressure are compressed from a volume V to a volume (V/2), one isothermally, the other adiabatically. In which sample is the final pressure greater? Does the entropy of the gas change in either process?

27. Suppose that we had chosen to represent the state of a system by its entropy and its absolute temperature rather than by its pressure and volume. (*a*) What would a Carnot cycle look like on a T-S diagram? (*b*) What physical significance, if any, can be attached to the area under a curve on a T-S diagram?

28. Show that the total entropy increases when work is converted into heat by friction between sliding surfaces. Describe the increase in disorder.

29. Heat energy flows from the sun to the earth. Show that the entropy of the earth-sun system increases during this process.

30. Is it true that the heat energy of the universe is steadily growing less available? If so, why?

31. Discuss the following comment of Panofsky and Phillips: "From the standpoint of formal physics there is only one concept which is asymmetric in the time, namely entropy. But this makes it reasonable to assume that the second law of thermodynamics can be used to ascertain the sense of time independently in any frame of reference; that is, we shall take the positive direction of time to be that of statistically increasing disorder, or increasing entropy. . . ." (See, in this connection, "The Arrow of Time," by David Layzer, *Scientific American,* December 1975.)

EXERCISES

Section 22–3 The Carnot Cycle

1. An ideal gas heat engine operates in a Carnot cycle between 227°C and 127°C. It absorbs 6.0×10^4 cal per cycle at the higher temperature. (*a*) How much work per cycle is this engine capable of performing? (*b*) What is the efficiency of the engine?

2. (*a*) How much work must be done to extract 1.0 J of heat from a reservoir at 7.0°C and transfer it to one at 27°C by means of a refrigerator using a Carnot cycle? (*b*) From one at −73°C to one at 27°C? (*c*) From one at −173°C to one at 27°C? (*d*) From one at −223°C to one at 27°C?

3. In a Carnot cycle, the isothermal expansion of an ideal gas takes place at 400 K and the isothermal compression at 300 K. During the expansion, 500 cal of heat energy are transferred to the gas. Determine (a) the work performed by the gas during the isothermal expansion, (b) the heat rejected from the gas during the isothermal compression, (c) the work done on the gas during the isothermal compression.

4. For the Carnot cycle illustrated in Fig. 22–3, show that the work done by the gas during process bc (Step 2) has the same absolute value as the work done on the gas during process da (Step 4).

Section 22–5   The Efficiency of Engines
5. In a hypothetical nuclear fusion reactor, the fuel is deuterium (D) gas at a temperature of about 7×10^8 K. If this gas could be used to operate an ideal heat engine with $T_C = 100°C$, what would be its efficiency?

6. A Carnot engine has an efficiency of 22%. It operates between heat reservoirs differing in temperature by 75°C. What are the temperatures of the reservoirs?

7. Apparatus that liquefies helium is in a room at 300 K. If the helium in the apparatus is at 5.0 K, what is the minimum ratio of heat energy delivered to the room to the heat energy removed from the helium?

8. An air conditioner takes heat from a room at 70°F and transfers it to the outdoors, which is at 96°F. For each joule of electrical energy required to operate the air conditioner, how many joules of heat are removed from the room?

9. An inventor claims to have invented four engines, each of which operates between heat reservoirs at 400 K and 300 K. Data on each engine, per cycle of operation, are as follows: Engine (a): $Q_H = 200$ J, $Q_C = -175$ J, $W = 40$ J; Engine (b): $Q_H = 500$ J, $Q_C = -200$ J, $W = 400$ J; Engine (c): $Q_H = 600$ J, $Q_C = -200$ J, $W = 400$ J; Engine (d): $Q_H = 100$ J, $Q_C = -90$ J, $W = 10$ J. Which of the first and second laws of thermodynamics (if any) does each engine violate?

Section 22–6   Entropy—Reversible Processes
10. In Fig. 22–8c, suppose that the change in entropy of the system in passing from state a to state b along path 1 is +0.602 cal/K. (a) What is the entropy change in passing from state a to b along path 2? (b) From state b to a along path 2?

11. Find (a) the heat absorbed and (b) the change in entropy of a 1.0-kg block of copper whose temperature is increased reversibly from 25 to 100°C. The specific heat for copper is 9.2×10^{-2} cal/g·°C.

Section 22–7   Entropy—Irreversible Processes
12. Heat can be removed from water at 0°C and atmospheric pressure without causing the water to freeze, if done with little disturbance of the water. Suppose that the water is cooled to $-5.0°C$ before ice begins to form. What is the change in entropy per unit mass occurring during the sudden freezing that then takes place?

13. In a specific heat experiment, 200 g of aluminum ($c_p = 0.215$ cal/g·°C) at 100°C is mixed with 50 g of water at 20°C. Find the difference in entropy of the system at the end from its value before mixing.

Section 22–8   Entropy and the Second Law
14. (a) Show that when a substance of mass m having a constant specific heat c is heated from T_1 to T_2, the entropy change is

$$S_2 - S_1 = mc \ln \frac{T_2}{T_1}.$$

(b) Does the entropy of the substance decrease on cooling? (c) If so, does the total entropy of the universe decrease in such a process?

15. A brass rod is in thermal contact with a heat reservoir at 127°C at one end and a heat reservoir at 27°C at the other end. (a) Compute the total change in the entropy arising from the process of conduction of 1200 cal of heat through the rod. (b) Does the entropy of the rod change in the process?

PROBLEMS GROUPED BY SECTION

Section 22–3   The Carnot Cycle
1. If the Carnot cycle is run backward, we have an ideal refrigerator. A quantity of heat Q_C is taken in at the lower temperature T_C and a quantity of heat Q_H is given out at the higher temperature T_H. The difference is the work W that must be supplied to run the refrigerator. (a) Show that

$$|W| = |Q_C| \frac{T_H - T_C}{T_C}.$$

(b) The coefficient of performance K of a refrigerator is defined as the ratio of the heat extracted from the cold source to the work needed to run the cycle. Show that, ideally,

$$K = \frac{T_C}{T_H - T_C}.$$

In actual refrigerators, K has a value of 5 or 6. (c) In a mechanical refrigerator the low-temperature coils are at a temperature of $-13°C$, and the compressed gas in the condenser has a temperature of 27°C. What is the theoretical coefficient of performance?

2. How is the efficiency of a reversible ideal heat engine related to the coefficient of performance of the reversible refrigerator obtained by running the engine backward?

3. (a) A Carnot engine operates between a hot reservoir at 320 K and a cold reservoir at 260 K. If it absorbs 500 J of heat per cycle at the hot reservoir, how much work per cycle does it deliver? (b) If the same engine, working in reverse, functions as a refrigerator between the same two reservoirs, how much work per cycle must be supplied to remove 1000 J of heat from the cold reservoir?

Section 22–5   The Efficiency of Engines
4. A combination mercury-steam turbine takes saturated mercury vapor from a boiler at 876°F and exhausts it to heat a steam boiler at 460°F. The steam turbine receives steam at this temperature and exhausts it to a condenser at 100°F. What is the maximum efficiency of the combination?

5. A Carnot engine produces a power of 500 W. It operates between heat reservoirs at 100°C and 60°C. Calculate (a) the rate of heat input and (b) the rate of exhaust heat output, in kilocalories per second.

6. The motor in a refrigerator has a power output of 200 W. If the freezing compartment is at 270 K and the outside air is at 300 K, assuming ideal efficiency, what is the maximum amount of heat that can be extracted from the freezing compartment in 10 min?

7. Suppose that a deep shaft were drilled in the earth's crust near one of the poles where the surface temperature is $-40\,°C$ to a depth where the temperature is $800\,°C$. (a) What is the theoretical limit to the efficiency of an engine operating between these temperatures? (b) If all of the heat released into the low-temperature reservoir were used to melt ice that was initially at $-40\,°C$, at what rate could water at $0\,°C$ be produced by a power plant having an output of 100 MW? The specific heat of ice is 0.50 cal/g·°C; its heat of fusion is 80 cal/g. (Notice that the engine can operate only between $0\,°C$ and $800\,°C$ in this case. Energy exhausted at $-40\,°C$ cannot be used to raise the temperature of anything else above $-40\,°C$.)

Section 22–6 Entropy—Reversible Processes

8. (a) Show that a Carnot cycle, plotted on an absolute temperature vs. entropy (T-S) diagram, graphs as a rectangle. For the Carnot cycle shown in Fig. 22–10, calculate (b) the heat gained and (c) the work done by the system.

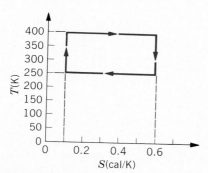

Figure 22–10 Problem 8.

9. A mole of a monatomic ideal gas is taken from an initial state of pressure p and volume V to a final state of pressure $2p$ and volume $2V$ by two different processes. (I) It expands isothermally until its volume is doubled, and then its pressure is increased at constant volume to the final state. (II) It is compressed isothermally until its pressure is doubled, and then its volume is increased at constant pressure to the final state.

Show the path of each process on a p-V diagram. For each process, calculate in terms of p and V: (a) the heat absorbed by the gas in each part of the process; (b) the work done by the gas in each part of the process; (c) the change in internal energy of the gas $U_f - U_i$; (d) the change in entropy of the gas $S_f - S_i$.

10. An ideal diatomic gas is caused to pass through the cycle shown on the p-V diagram in Fig. 22–11, where $V_2 = 3V_1$. Determine, in terms of p_1, V_1, T_1, and R: (a) p_2, p_3, and T_3; (b) W, Q, ΔU, and ΔS, per mole, for all three processes.

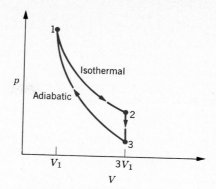

Figure 22–11 Problem 10.

Section 22–7 Entropy—Irreversible Processes

11. A 10-g ice cube at $-10\,°C$ is placed in a lake whose temperature is $+15\,°C$. Calculate the change in entropy of the system as the ice cube comes to thermal equilibrium with the lake. The specific heat of ice is 0.50 cal/g·°C.

12. An 8.0-g ice cube at $-10\,°C$ is dropped into a Thermos flask containing 100 cm³ of water at $20\,°C$. What is the change in entropy of the system when a final equilibrium state is reached? The specific heat of ice is 0.50 cal/g·°C.

Section 22–8 Entropy and the Second Law

13. Four moles of an ideal gas are caused to expand from a volume V_1 to a volume $V_2 = 2V_1$. (a) If the expansion is isothermal at the temperature $T = 400$ K, find the work done by the expanding gas. (b) Find the change in entropy, if any. (c) If the expansion were reversibly adiabatic instead of isothermal, would the change in entropy be positive, negative, or zero?

14. A 50-g block of copper having a temperature of 400 K is placed in an insulating box with a 100-g block of lead having a temperature of 200 K. (a) What is the equilibrium temperature of this two-block system? (b) What is the change in the internal energy of the two-block system as it changes from the initial condition to the equilibrium condition? (c) What is the change in the entropy of the two-block system? (See Table 20–1.)

15. An object of constant heat capacity C is heated from an initial temperature T_i to a final temperature T_f, by being placed in contact with a heat reservoir at T_f. Represent the process on a graph of C/T vs. T and (a) show graphically that the total change in entropy ΔS (object plus reservoir) is positive, and (b) show how the use of heat reservoirs at intermediate temperatures would allow the process to be carried out in a way that makes ΔS as small as desired.

16. A round silver rod 15 cm long, with a diameter of 1.0 cm, has its ends in contact with heat reservoirs at $60\,°C$ and $20\,°C$ and steady-state heat flow has been established. What will be the initial rate of change of the entropy of the rod if (a) the hot end is suddenly insulated from the $60\,°C$ reservoir, or (b) if the entire rod is suddenly insulated?

ADDITIONAL PROBLEMS

17. In a mechanical refrigerator, the low-temperature coils are at a temperature of $-13\,°C$, and the compressed gas in the condenser has a temperature of $27\,°C$. What is the theoretical maximum possible coefficient of performance? (See Problem 1.)

18. At very low temperatures, the molar specific heat of many solids is (approximately) proportional to T^3; that is, $C_v = AT^3$, where K depends on the particular substance. For aluminum, $K = 7.53 \times 10^{-6}$ cal/mol·K^4. Find the entropy change of 4.0 mol of aluminum when its temperature is raised from 5.0 K to 10 K.

19. A mixture of 1773 g of water and 227 g of ice at 0°C is, in a reversible process, brought to a final equilibrium state where the water-ice ratio, by mass, is 1:1 at 0°C. (a) Calculate the entropy change of the system during this process. (The heat of fusion for water is approximately 80 cal/g.) (b) The system is then returned to the first equilibrium state, but in an irreversible way (by using a Bunsen burner, for instance).Calculate the entropy change of the system during this process. (c) Is your answer consistent with the second law of thermodynamics?

20. Two pieces of molding clay of equal mass are moving in opposite directions with equal speeds. They strike and stick together. Treat the two pieces as a single system and state whether each of the following quantities is positive, negative, or zero for this process: ΔU, W, Q, and ΔS. Justify your answers.

21. In a two-stage Carnot heat engine, a quantity of heat Q_1 is absorbed at a temperature T_1, work W_1 is done, and a quantity of heat Q_2 is expelled at a lower temperature T_2 by the first stage. The second stage absorbs the heat expelled by the first, does work W_2, and expels a quantity of heat Q_3 at a lower temperature T_3. Prove that the efficiency of the combination engine is $(T_1 - T_3)/T_1$.

22. (a) Plot accurately a Carnot cycle on a p-V diagram for 1.0 mol of an ideal gas. Let point a correspond to $p = 1.0$ atm, $T = 300$ K, and let b correspond to $p = 0.50$ atm, $T = 300$ K; take the low-temperature reservoir to be at 100 K. Let $\gamma = 1.5$. (b) Compute graphically the work done in this cycle. (c) Compute the work analytically.

23. A Carnot engine works between temperatures T_1 and T_2. It drives a Carnot refrigerator that works between two different temperatures T_3 and T_4 (see Fig. 22–12). Find the ratio $|Q_3|/|Q_1|$ in terms of the four temperatures.

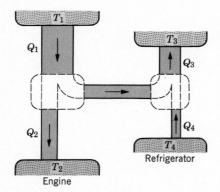

Figure 22–12 Problem 23.

24. One mole of a monatomic ideal gas initially at a volume of 10 liters and a temperature of 300 K is heated at constant volume to a temperature of 600 K, allowed to expand isothermally to its initial pressure, and finally compressed isobarically to its original volume, pressure, and temperature. (a) Compute the heat input to the system during one cycle. (b) What is the net work done by the gas during one cycle? (c) What is the efficiency of this cycle?

25. Two identical objects with different temperatures T_H and T_C are placed inside an insulated box and allowed to come to equilibrium. Presuming that their heat capacities are independent of temperature, show that the entropy increase in this process is

$$\Delta S = \frac{2Q}{T_H - T_C} \ln \frac{(T_H + T_C)^2}{4 T_H T_C},$$

in which Q represents the total heat transferred from the hot object to the cooler one.

26. An air conditioner operating between 93°F and 70°F is rated at 4000 Btu/h cooling capacity. Its coefficient of performance (see Problem 1) is 27% of a Carnot refrigerator operating between the same two temperatures. What is the required horsepower of the motor?

27. One mole of a monatomic ideal gas is caused to go through the cycle shown in Fig. 22–13. Process bc is a reversible adiabatic expansion. $p_b = 10$ atm, $V_b = 1.0$ m³, and $V_c = 8.0$ m³. Calculate (a) the heat added to the gas, (b) the heat leaving the gas, (c) the net work done by the gas, and (d) the efficiency of the cycle.

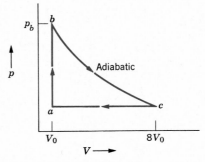

Figure 22–13 Problem 27.

28. One mole of an ideal monatomic gas is used as the working substance of an engine that operates on the cycle shown in Fig. 22–14. Calculate (a) the work done per cycle, (b) the heat added per cycle during the expansion stroke abc, and (c) the engine efficiency. (d) What is the Carnot efficiency of an engine operating between the highest and lowest temperatures present in the cycle? How does this compare to the efficiency calculated in (c)? Assume that $p = 2p_0$, $V = 2V_0$, $p_0 = 1.01 \times 10^5$ Pa, and $V_0 = 0.0225$ m³.

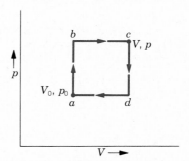

Figure 22–14 Problem 28.

29. In a heat pump, heat Q_C is extracted from the outside atmosphere at T_C and a larger quantity of heat Q_H is delivered to the inside of the house at T_H, with the performance of work W. (a) Draw a schematic diagram of a heat pump. (b) How does it differ in principle from a refrigerator? In practical use? (c) How are Q_H, Q_C, and W related to one another? (d) Can a heat pump be reversed for use in summer? Explain. (*e*) What advantages does such a pump have over other heating devices?

30.   In a heat pump, heat from the outdoors at $-5.0°C$ is transferred to a room at $17°C$, energy being supplied by an electric motor. How many joules of heat will be delivered to the room for each joule of electrical energy consumed? Assume an ideal heat pump and see Problem 29.

31.   Compute the efficiency of the cycle shown in Fig. 21–12. See Problem 43 in Chapter 21. In this case, the heat input will not be at a fixed temperature, as it is in the Carnot cycle.

32.   One mole of an ideal monatomic gas is caused to go through the cycle shown in Fig. 22–15. (*a*) How much work is done in expanding the gas from *a* to *c* along path *abc*? (*b*) What is the change in internal energy and entropy in going from *b* to *c*? (*c*) What is the change in internal energy and entropy in going through one complete cycle? Express all answers in terms of the pressure p_0, volume V_0, and temperature T_0 at point *a* in the diagram.

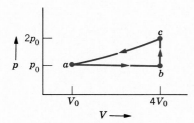

Figure 22–15   Problem 32.

***33.**   A gasoline internal combustion engine can be approximated by the cycle shown in Fig. 22–16. Assume an ideal gas and use a compression ratio of 4 : 1 ($V_4 = 4V_1$). Assume that $p_2 = 3p_1$. (*a*) Determine the pressure and temperature of each of the vertex points of the *p-V* diagram in terms of p_1, T_1, and the ratio of specific heats of the gas. (*b*) What is the efficiency of the cycle?

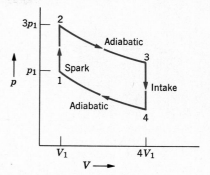

Figure 22–16   Problem 33.

23 Charge and Matter

23–1 Electromagnetism

The science of electricity has its roots in the observation, known to Thales of Miletus in 600 B.C., that a rubbed piece of amber will attract bits of straw. The study of magnetism goes back to the observation that naturally occurring "stones" (that is, magnetite) will attract iron. These two sciences developed quite separately until 1820, when Hans Christian Oersted observed a connection between them, namely, that an *electric* current in a wire can affect a *magnetic* compass needle.

The new science of electromagnetism was developed further by many workers, of whom one of the most important was Michael Faraday* (1791–1867). It fell to James Clerk Maxwell† (1831–1879) to put the laws of electromagnetism in essentially the form in which we know them today. These laws, called *Maxwell's equations*, are displayed in Table 37–2, which you may want to examine at this time. These laws play the same role in electromagnetism that Newton's laws of motion and of gravitation do in mechanics.

Although Maxwell's synthesis of electromagnetism rests heavily on the work of his predecessors, his own contribution was central and vital. Maxwell deduced that light is electromagnetic in nature and that its speed can be found by making purely electric and magnetic measurements. Thus the science of optics was intimately connected with those of electricity and of magnetism. The scope of Maxwell's equations is remarkable, including as it does the fundamental principles of all large-scale electromagnetic and optical devices such as motors, radio, television, radar, microscopes, and telescopes.

The English physicist Oliver Heaviside (1850–1925) and especially the Dutch physicist H. A. Lorentz (1853–1928) contributed substantially to the development

Maxwell's equations

---

* Michael Faraday," by Herbert Kondo, *Scientific American,* October 1953.
† "James Clerk Maxwell," by James R. Newman, *Scientific American* June 1955.

and clarification of Maxwell's theory. Heinrich Hertz (1857–1894)* took a great step forward when, more than twenty years after Maxwell set up his theory, he produced in the laboratory electromagnetic ''Maxwellian waves'' of a kind that we would now call short radio waves. It remained for Marconi and others to exploit the practical application of the electromagnetic waves of Maxwell and Hertz. At the level of engineering applications Maxwell's equations are used constantly and universally in the solution of a wide variety of practical problems.

The formulation of electromagnetism based on Maxwell's equations is called *classical electromagnetism*. It gives correct answers for all problems involving electric and magnetic fields down to dimensions of about 10^{-12} m. This is much smaller than an atom and is well into the region in which Newtonian mechanics must be replaced by quantum mechanics. A blending of classical electromagnetism, quantum mechanics, and relativity theory, called *quantum electrodynamics*, gives correct answers down to extremely small dimensions. F. J. Dyson has said of this theory: ''It is the only field in which we can choose a hypothetical experiment and predict the results to five places of decimals, confident that the theory takes into account all the factors that are involved.'' This book deals only with classical electromagnetism.

23-2 Electric Charge

Figure 23–1 is a particularly striking example of a kind of electrical charge transfer that is familiar to all. A similar but less spectacular phenomenon, well

Figure 23–1 A lightning display over the Kitt Peak National Observatory in Arizona. It represents the transfer of electric charge between clouds and ground. (Used by permission of KPNO, photo by Gary Ladd.)

---

* Heinrich Hertz,'' by Philip and Emily Morrison, *Scientific American*, December 1957.

known to those who live in dry climates, is the possibility of the generation of sparks by the friction involved in walking across a carpet. Such are the roots in common experience of the concept of electric charge.

We can show that there are *two kinds* of charge by rubbing a glass rod with silk and hanging it from a long thread as in Fig. 23–2. If a second rod is rubbed with silk and held near the rubbed end of the first rod, the rods will repel each other. On the other hand, a rod of plastic rubbed with fur will attract the glass rod. Two plastic rods rubbed with fur will repel each other. We explain these facts by saying that rubbing a rod gives it an electric charge and that the charges on the two rods exert forces on each other. Clearly the charges on the glass and on the plastic must be different in nature.

Benjamin Franklin (1706–1790), among his other achievements, was the first American physicist. He named the kind of electricity that appears on the glass positive and the kind that appears on the plastic (sealing wax or shell-lac in Franklin's day) negative; these names have remained to this day. We can sum up these experiments by saying that *like charges repel and unlike charges attract*.

Electric effects are not limited to glass rubbed with silk or to plastic rubbed with fur. Any substance rubbed with any other under suitable conditions will become charged to some extent; by comparing the unknown charge with a glass rod that had been rubbed with silk or a plastic rod that had been rubbed with fur, it can be labeled as either positive or negative.

The modern view of bulk matter is that, in its normal or neutral state, it contains equal amounts of positive and negative electricity. If two bodies like glass and silk are rubbed together, a small amount of charge is transferred from one to the other, upsetting the electric neutrality of each. In this case the glass would become positive, the silk negative.

Positive and negative charge

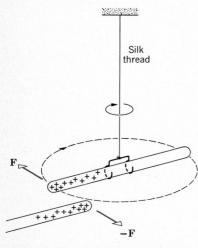

Figure 23–2 Two positively charged glass rods repel each other.

Conductors and insulators

Semiconductors

23–3   Conductors and Insulators

A metal rod held in the hand and rubbed with fur will not seem to develop a charge. It is possible to charge such a rod, however, if it is furnished with a glass or plastic handle and if the metal is not touched, say with the hands, while rubbing it. The explanation is that metals, the human body, and the earth are conductors of electricity and that glass, plastics, and so on are insulators.

In conductors electric charges are free to move through the material, whereas in insulators they are not. Although there are no perfect insulators, the insulating ability of fused quartz is about 10^{25} times as great as that of copper, so that for many practical purposes some materials behave as if they were perfect insulators.

In metals a fairly subtle experiment called the Hall effect (see Section 30–5) shows that only negative charge is free to move. Positive charge is as immobile in a metal as it is in glass or in any other insulator. The actual charge carriers in metals are the *free electrons*. When isolated atoms are combined to form a metallic solid, the outer electrons of the atom do not remain attached to individual atoms but become free to move throughout the volume of the solid.

Semiconductors are intermediate between conductors and insulators in their ability to conduct electricity. Among the elements, silicon and germanium are well-known examples. In semiconductors the electrical conductivity can often be greatly increased by adding small amounts of other elements; traces of arsenic or boron are often added to silicon for this purpose. Semiconductors have many practical applications, among which is their use in the construction of transistors. A detailed model explaining the properties of semiconductors cannot be described adequately without some understanding of quantum physics.

23-4 Coulomb's Law

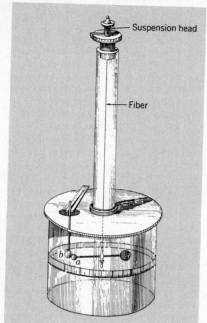

- Suspension head

- Fiber

b a

Figure 23-3 Coulomb's torsion balance, from his 1785 memoir to the French Academy of Sciences.

Charles Augustin Coulomb (1736–1806) measured electrical attractions and repulsions quantitatively and confirmed the law that governs them. His apparatus, shown in Fig. 23–3, resembles the hanging rod of Fig. 23–2, except that the charges in Fig. 23–3 are confined to small spheres a and b.

If a and b are charged, the electric force on a will tend to twist the suspension fiber. Coulomb canceled out this twisting effect by turning the suspension head through the angle θ needed to keep the two charges at the particular distance apart in which he was interested. The angle θ is then a relative measure of the electric force acting on charge a. The device of Fig. 23–3 is called a torsion balance; a similar arrangement was used later by Cavendish to measure gravitational attractions (Section 15–2).

Coulomb's first experimental results can be represented by

$$F \propto \frac{1}{r^2}.$$

Here F is the magnitude of the interaction force that acts on each of the two charges a and b; r is their distance apart. These forces, as Newton's third law requires, act along the line joining the charges but point in opposite directions. Note that the magnitude of the force on each charge is the same, even though the charges may be different.

The force between charges depends also on the magnitude of the charges. Specifically, it is experimentally found to be proportional to their product. Thus we arrive at

$$F \propto \frac{q_1 q_2}{r^2}, \tag{23-1}$$

Coulomb's law

where q_1 and q_2 are relative measures of the charges on spheres a and b. Equation 23–1, which is called *Coulomb's law,* holds only for charged objects whose sizes are much smaller than the distance between them. We often say that it holds only for point charges.

Coulomb's law resembles the inverse square law of gravitation, which was already more than 100 years old at the time of Coulomb's experiments; q plays the role of m in that law. In gravity, however, the forces are always attractive. There are two kinds of electricity but (apparently) only one kind of mass.

Our belief in Coulomb's law does not rest quantitatively on Coulomb's experiments. Torsion balance measurements are difficult to make to an accuracy of better than a few percent. Such measurements could not, for example, convince us that the exponent in Eq. 23–1 is exactly 2 and not, say, 2.01. In Section 25–7 we show that Coulomb's law can also be deduced from an indirect experiment which shows that the exponent in Eq. 23–1, if not 2, differs from 2 by a number which is less than 3×10^{-16}.

Although we have established the physical concept of electric charge, we have not yet defined a unit in which it may be measured. It is possible to do so operationally by putting equal charges q on the spheres of a torsion balance and by measuring the magnitude F of the force that acts on each when the charges are a measured distance r apart. One could then define q to have a unit value if a unit force acts on each charge when the charges are separated by a unit distance and one can give a name to the unit of charge so defined. This scheme is the basis for the definition of the unit of charge called the *statcoulomb*. However, we do not use this unit or the systems of units of which it is a part in this book.

For practical reasons having to do with the accuracy of measurements, the SI unit of charge is not defined using a torsion balance but is derived from the unit of electric current. If the ends of a long wire are connected to the terminals of a battery, it is common knowledge that an *electric current i* is set up in the wire. We visualize this current as a flow of charge. The SI unit of current is the *ampere* (abbr. A). In Section 31–4 we describe the laboratory procedures in terms of which the ampere is defined.

The SI unit of charge is the *coulomb* (abbr. C). *A coulomb is defined as the amount of charge that flows through any cross section of a wire in 1 s if there is a steady current of 1 A in the wire.* In symbols

$$q = it, \tag{23-2}$$

where q is in coulombs if i is in amperes and t is in seconds. Thus, if a wire is connected to an insulated metal sphere, a charge of 10^{-9} C can be put on the sphere if a current of 1.0 mA exists in the wire for 1.0 μs.

The ampere

The coulomb

Example 1 A copper penny has a mass of 3.1 g. Being electrically neutral, it contains equal amounts of positive and negative electricity. What is the magnitude q of these charges? A copper atom has a positive nuclear charge of 4.6×10^{-18} C and a negative electronic charge of equal magnitude.

The number N of copper atoms in a penny is found from the ratio

$$\frac{N}{N_A} = \frac{m}{M},$$

where N_A is the Avogadro number, m the mass of the coin, and M the atomic weight of copper. This yields, solving for N,

$$N = \frac{(6.0 \times 10^{23} \text{ atoms/mol})(3.1 \text{ g})}{64 \text{ g/mol}} = 2.9 \times 10^{22} \text{ atoms.}$$

The charge q is

$$q = (4.6 \times 10^{-18} \text{ C/atom})(2.9 \times 10^{22} \text{ atoms}) = 1.3 \times 10^5 \text{ C.}$$

The current in a 100-W, 110-V light bulb is 0.91 A. Verify that it would take 40 h for a charge of this amount to pass through this bulb.

Equation 23–1 can be written as an equality by inserting a constant of proportionality. Instead of writing this simply as, say, k, it is usually written in a more complex way as $1/4\pi\epsilon_0$ or

$$F = \frac{1}{4\pi\epsilon_0} \frac{q_1 q_2}{r^2}. \tag{23-3}$$

Certain equations that are derived from Eq. 23–3, but are used more often than it is, will be simpler in form if we do this.

To five significant figures the value of the *permittivity constant* ϵ_0, as it is called, is*

$$\epsilon_0 = 8.8542 \times 10^{-12} \text{ C}^2/\text{N}\cdot\text{m}^2.$$

In this book the value 8.85×10^{-12} C²/N·m² will be accurate enough for most problems. For direct application of Coulomb's law or in any problem in which the quantity $1/4\pi\epsilon_0$ occurs we may use, with sufficient accuracy for these problems,

$$1/4\pi\epsilon_0 = 8.99 + 10^9 \text{ N}\cdot\text{m}^2/\text{C}^2.$$

Using Example 1, Eq. 23–3, and this value for $1/4\pi\epsilon_0$, you should convince yourself that the force of attraction between the positive and negative charges of a

The SI form of Coulomb's law

The permittivity constant ϵ_0

---

* A consequence of the fact that the speed of light now has an assigned value, exact by definition (see Section 1–3), is that the permittivity constant ϵ_0 also has a defined constant value; see Section 38–3 and Appendix B.

copper penny would be 1.5×10^{16} N ($= 1.7 \times 10^{12}$ tons!) if they could be separated by 100 m. With 1 cm separation the force would be 1.5×10^{24} N. This suggests that it is not possible to upset the electrical neutrality of gross objects by any very large amount.

Multiple point charges

If more than two (point) charges are present, Eq. 23–3 holds for every pair of such charges. Let the charges be q_1, q_2, q_3, etc. We calculate the force exerted on any one (say q_1) by all the others from the vector equation

$$\mathbf{F}_1 = \mathbf{F}_{12} + \mathbf{F}_{13} + \mathbf{F}_{14} + \cdots, \tag{23–4}$$

where $\mathbf{F}_{12}$, for example, is the force exerted on q_1 by q_2.

Example 2 Figure 23–4 shows three fixed charges q_1, q_2, and q_3. What force acts on q_1? Assume that $q_1 = -1.0 \times 10^{-6}$ C, $q_2 = +3.0 \times 10^{-6}$ C, $q_3 = -2.0 \times 10^{-6}$ C, $r_{12} = 15$ cm, $r_{13} = 10$ cm and $\theta = 30°$.

From Eq. 23–3, ignoring the signs of the charges, since we are interested only in the magnitudes of the forces,

$$F_{12} = \frac{1}{4\pi\epsilon_0} \frac{q_1 q_2}{r_{12}{}^2}$$
$$= \frac{(9.0 \times 10^9 \ \text{N·m}^2/\text{C}^2)(1.0 \times 10^{-6} \ \text{C})(3.0 \times 10^{-6} \ \text{C})}{(1.5 \times 10^{-1} \ \text{m})^2}$$
$$= 1.2 \ \text{N}$$

and

$$F_{13} = \frac{(9.0 \times 10^9 \ \text{N·m}^2/\text{C}^2)(1.0 \times 10^{-6} \ \text{C})(2.0 \times 10^{-6} \ \text{C})}{(1.0 \times 10^{-1} \ \text{m})^2}$$
$$= 1.8 \ \text{N}.$$

Considering the signs of the charges, we see that q_1 is repelled by q_3 and attracted by q_2. Therefore the directions of $\mathbf{F}_{12}$ and $\mathbf{F}_{13}$ are as shown in Figure 23–4.

The components of the resultant force $\mathbf{F}_1$ acting on q_1 (see Eq. 23–4) are

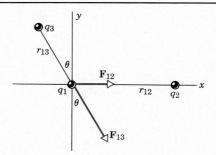

Figure 23–4 Example 2. Showing the forces exerted on q_1 by q_2 and q_3.

$$F_{1x} = F_{12x} + F_{13x} = F_{12} + F_{13} \sin\theta$$
$$= 1.2 \ \text{N} + (1.8 \ \text{N})(\sin 30°) = 2.1 \ \text{N}$$

and

$$F_{1y} = F_{12y} + F_{13y} = 0 - F_{13} \cos\theta$$
$$= -(1.8 \ \text{N})(\cos 30°) = -1.6 \ \text{N}.$$

Show that the magnitude of $\mathbf{F}_1$ is 2.6 N and that it makes an angle of $-37°$ with the x–axis.

23–5 Charge is Quantized

In Franklin's day electric charge was thought of as a continuous fluid, an idea that was useful for many purposes. The atomic theory of matter, however, has shown that fluids themselves, such as water and air, are not continuous but are made up of atoms. Experiment shows that the "electric fluid" is not continuous either but that it is made up of integral multiples of a certain minimum electric charge. This fundamental charge, to which we give the symbol e, has the approximate magnitude 1.602×10^{-19} C. Any physically existing charge q, no matter what its origin, can be written as ne when n is a positive or a negative integer.*

The quantized nature of electric charge

When a physical property such as charge exists in discrete "packets" rather than in continuous amounts, the property is said to be *quantized*. Later you will

---

* The results of modern particle physics research strongly suggest that protons, neutrons, and certain other particles may be made of even simpler particles called *quarks*. These are postulated to have fractional electric charge, quantized with values of either $\pm e/3$ or $\pm 2e/3$.

learn that several other properties prove to be quantized when suitably examined on the atomic scale; among them are energy and angular momentum.

The quantum of charge e is so small that the "graininess" of electricity does not show up in large-scale experiments, just as we do not realize that the air we breathe is made up of atoms. In an ordinary 110-V, 100-W light bulb, for example, 6×10^{18} elementary charges enter and leave the bulb every second.

23-6   Charge and Matter

Matter as we ordinarily experience it can be regarded as composed of three kinds of particles. Table 23–1 shows their masses and charges. Note that the masses of neutrons and protons are approximately equal but that electrons are less massive by a factor of about 1840.

Properties of protons, neutrons and electrons

Table 23-1   Some properties of three particles

Particle	Symbol	Charge	Mass
Proton	p	$+e$	1.6726×10^{-27} kg
Neutron	n	0	1.6750×10^{-27} kg
Electron	e^-	$-e$	9.1100×10^{-31} kg

Structure of atoms

Atoms are made up of a dense, positively charged *nucleus*, surrounded by a cloud of electrons; see Fig. 23–5. The nucleus varies in radius from about 1×10^{-15} m for hydrogen to about 7×10^{-15} m for the heaviest atoms. The outer diameter of the electron cloud, that is, the diameter of the atom, lies in the range $1–3 \times 10^{-10}$ m, about 10^5 times larger than the nuclear diameter.

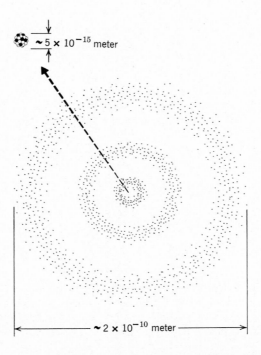

$\sim 5 \times 10^{-15}$ meter

$\sim 2 \times 10^{-10}$ meter

Figure 23-5   Our model of an atom, suggesting the electron cloud and, above, an enlarged view of the nucleus.

Example 3 The distance r between the electron and the proton in the hydrogen atom is about 5.3×10^{-11} m. What are the magnitudes of (a) the electrical force and (b) the gravitational force between these two particles?

From Coulomb's law,

$$F_e = \frac{1}{4\pi\epsilon_0} \frac{q_1 q_2}{r^2}$$
$$= \frac{(9.0 \times 10^9 \ \text{N·m}^2/\text{C}^2)(1.6 \times 10^{-19} \ \text{C})^2}{(5.3 \times 10^{-11} \ \text{m})^2}$$
$$= 8.2 \times 10^{-8} \ \text{N}.$$

The gravitational force is given by Eq. 15-1,

$$F_g = G \frac{m_1 m_2}{r^2}$$
$$= \frac{(6.7 \times 10^{-11} \ \text{N·m}^2/\text{kg}^2)(9.1 \times 10^{-31} \ \text{kg})(1.7 \times 10^{-27} \ \text{kg})}{(5.3 \times 10^{-11} \ \text{m})^2}$$
$$= 3.7 \times 10^{-47} \ \text{N}.$$

Thus the electrical force is about 10^{39} times stronger than the gravitational force in this case. How do the electrical and gravitational forces between two electrons compare?

Electric forces in bulk matter

The significance of Coulomb's law goes far beyond the description of the forces acting between charged spheres. This law, when incorporated into the structure of quantum physics, correctly describes (a) the electric forces that bind the electrons of an atom to its nucleus, (b) the forces that bind atoms together to form molecules, and (c) the forces that bind atoms or molecules together to form solids or liquids. Thus most of the forces of our daily experience that are not gravitational in nature are electrical. A force transmitted by a steel cable is basically an electrical force because, if we pass an imaginary plane through the cable at right angles to it, it is only the attractive electrical interatomic forces acting between atoms on opposite sides of the plane that keep the cable from parting. We ourselves are an assembly of nuclei and electrons bound together in a stable configuration by Coulomb forces.

The strong force

In the atomic nucleus we encounter a new force which is apparently neither gravitational nor electrical in nature. This strong attractive force, which binds together the protons and neutrons that make up the nucleus, is called simply *the strong force*. If this force were not present, the nucleus would fly apart at once because of the strong Coulomb repulsion force that acts between its protons.

Example 4 What repulsive Coulomb force exists between two protons in a nucleus of iron? Assume a separation of 4.0×10^{-15} m.

From Coulomb's law,

$$F = \frac{1}{4\pi\epsilon_0} \frac{q_1 q_2}{r^2}$$
$$= \frac{(9.0 \times 10^9 \ \text{N·m}^2/\text{C}^2)(1.6 \times 10^{-19} \ \text{C})^2}{(4.0 \times 10^{-15} \ \text{m})^2}$$
$$= 14 \ \text{N}.$$

This enormous repulsive force (3.2 lb acting on a single proton!) must be more than compensated for by the strong attractive nuclear force. This example, combined with Example 3, shows that nuclear binding forces are much stronger than atomic binding forces. Atomic binding forces are, in turn, much stronger than gravitational forces for the same particles separated by the same distance.

23-7 Charge is Conserved

Conservation of charge

When a glass rod is rubbed with silk, a positive charge appears on the rod. Measurement shows that a negative charge of equal magnitude appears on the silk. This suggests that rubbing does not create charge but merely transfers it from one object to another, disturbing slightly the electrical neutrality of each. This hypothesis of the conservation of charge, first proclaimed by Franklin, has stood up under close experimental scrutiny both for large-scale events and at the atomic and nuclear level; no exceptions have ever been found.

An interesting example of charge conservation comes about when an electron (charge $= -e$) and a positron (charge $= +e$) are brought close to each other. The two particles may simply disappear, converting all their rest mass into electromagnetic energy according to the well-known $E = mc^2$ relationship. The energy appears in the form of two oppositely directed gamma rays, which are similar in character to x rays; thus:

$$e^- + e^+ \rightarrow \gamma + \gamma. \tag{23-5}$$

The net charge is zero both before and after the event so that charge is conserved. Rest mass is *not* conserved, being turned completely into electromagnetic energy.

Another example of charge conservation is found in radioactive decay, of which the following process is typical:

$$^{238}\text{U} \rightarrow {}^{234}\text{Th} + {}^{4}\text{He}. \tag{23-6}$$

The radioactive "parent" nucleus, ^{238}U, contains 92 protons (that is, its atomic number Z is 92). It disintegrates spontaneously by emitting an α-particle (^{4}He; $Z = 2$) transmuting itself into the nucleus ^{234}Th, with $Z = 90$. Thus the amount of charge present before disintegration ($+92e$) is the same as that present after the disintegration.

Still another example of charge conservation is found in nuclear reactions, of which the bombardment of ^{44}Ca with cyclotron-accelerated protons is typical. In a particular collision a neutron may emerge from the nucleus, leaving ^{44}Sc as a "residual" nucleus:

$$^{44}\text{Ca} + p \rightarrow {}^{44}\text{Sc} + n.$$

The sum of the atomic numbers before the reaction ($20 + 1$) is exactly equal to the sum of the atomic numbers after the reaction ($21 + 0$). Again charge is conserved.

REVIEW GUIDE AND SUMMARY

Electromagnetism

Electromagnetism is a description of the interactions involving electric charge. Classical electromagnetism, summarized by Maxwell's equations, includes the phenomena of electricity, magnetism, electromagnetic induction (electric generators), and electromagnetic radiation, which includes all optical phenomena.

The strength of a particle's electromagnetic interaction is partly determined by its electric charge, which can be either positive or negative. Like charges repel and unlike charges attract each other. Objects with equal amounts of the two kinds of charge are electrically neutral, whereas those with an imbalance are electrically charged.

Electric charge

Conductors, insulators, and semiconductors

Conductors are materials in which a significant number of charged particles (electrons in metals) are free to move. The charged particles in insulators are not free. Semiconductors are intermediate between conductors and insulators in this respect.

The coulomb and ampere

The SI unit of charge is the coulomb (C). It is defined in terms of the unit of current, the ampere (A) as the charge passing a particular point in one second when a current of one ampere is flowing; see Eq. 23-2. Example 1 illustrates the large amount of electric charge in ordinary samples of matter.

Coulomb's law

Coulomb's law describes the forces between small (point) electric charges at rest. In SI form,

$$F = \frac{1}{4\pi\epsilon_0} \frac{q_1 q_2}{r^2}. \tag{23-3}$$

ϵ_0 is experimentally determined to be 8.854×10^{-12} C^2/N $\cdot$ m^2 or, with sufficient accuracy for the purposes of this chapter,

$$1/4\pi\epsilon_0 = 8.99 \times 10^9 \text{ N·m}^2/\text{C}^2.$$

The force of attraction or repulsion between point charges at rest acts along the line joining the two charges. If more than two charges are present, Eq. 23–3 holds for the forces between each pair. The resultant force on each charge is then found, using the superposition principle, as the vector sum of the forces exerted on it by all the others. Example 2 shows how such forces can be calculated.

Electric charge is quantized. This means that any charge found in nature can be written as ne, where n is a positive or negative integer and e is a constant of nature called the *electronic charge;* its value is approximately 1.602×10^{-19} C.

Matter as we ordinarily encounter it can be regarded as composed of protons, neutrons, and electrons whose properties are summarized in Table 23–1. Atoms contain a small, dense, positively charged nucleus (composed of neutrons and protons) surrounded by a cloud of electrons attracted to the nucleus by the electrical force. Most forces we ordinarily deal with, such as friction, fluid pressure, contact forces between solid surfaces, and elastic forces are all manifestations of the electrical interactions between the charged particles of which atoms are made. Examples 3 and 4 show that the electrical force is much stronger than gravity when both interactions occur simultaneously, and that the nuclear force is still stronger.

Electric charge is conserved. This means that the (algebraic) net charge of an isolated system of charges does not change, no matter what interactions occur within the system. Section 23–7 describes three illustrations of this important conservation law.

QUESTIONS

1. You are given two metal spheres mounted on portable insulating supports. Find a way to give them equal and opposite charges. You may use a glass rod rubbed with silk but may not touch it to the spheres. Do the spheres have to be of equal size for your method to work?

2. In Question 1, find a way to give the spheres equal charges of the same sign. Again, do the spheres need to be of equal size for your method to work?

3. A charged rod attracts bits of dry cork dust which, after touching the rod, often jump violently away from it. Explain.

4. The experiments described in Section 23–2 could be explained by postulating four kinds of charge, that is, on glass, silk, plastic, and fur. What is the argument against this?

5. If you rub a coin briskly between your fingers, it will not seem to become charged by friction. Why?

6. If you walk briskly across a carpet, you often experience a spark upon touching a door knob. (*a*) What causes this? (*b*) How might it be prevented?

7. Why do electrostatic experiments not work well on humid days?

8. An insulated rod is said to carry an electric charge. How could you verify this and determine the sign of the charge?

9. If a charged glass rod is held near one end of an insulated uncharged metal rod as in Fig. 23–6, electrons are drawn to one end, as shown. Why does the flow of electrons cease? After all, there is an almost inexhaustible supply of them in the metal rod.

10. In Fig. 23–6 does any net electric force act on the metal rod? Explain.

11. A person standing on an insulated stool touches a charged, insulated conductor. Is the conductor discharged completely?

12. (*a*) A positively charged glass rod attracts a suspended object.

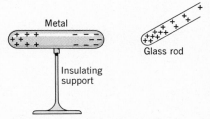

Figure 23–6   Questions 9 and 10.

Can we conclude that the object is negatively charged? (*b*) A positively charged glass rod repels a suspended object. Can we conclude that the object is positively charged?

13. Is the electric force that one charge exerts on another changed if other charges are brought nearby?

14. A solution of copper sulfate is a conductor. What particles serve as the charge carriers in this case?

15. If the electrons in a metal such as copper are free to move about, they must often find themselves headed toward the metal surface. Why don't they keep on going and leave the metal?

16. Coulomb's law predicts that the force exerted by one point charge on another is proportional to the product of the two charges. How might you go about testing this aspect of the law in the laboratory?

17. An electron (charge $= -e$) circulates around a helium nucleus (charge $= +2e$) in a helium atom. Which particle exerts the larger force on the other?

18. The charge of a particle is a true characteristic of the particle, independent of its state of motion. Explain how you can test this statement by making a rigorous experimental check of whether the hydrogen atom is truly electrically neutral.

19. Earnshaw's theorem says that no particle can be in stable equilibrium under the action of electrostatic forces alone. Consider, however, point P at the center of a square of four equal positive charges, as in Fig. 23–7. If you put a positive test charge there, it might seem to be in equilibrium. Every one of the four external charges pushes it toward P. Yet Earnshaw's theorem holds. Can you explain how?

20. The quantum of charge is 1.602×10^{-19} C. Is there a corresponding single quantum of mass?

21. What does it mean to say that a physical quantity is (a) quantized or (b) conserved? Give some examples.

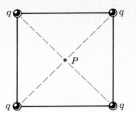

Figure 23–7   Question 19.

EXERCISES

Section 23–4   Coulomb's Law
1. What is the force of attraction between two 1.0-C charges separated by a distance of (a) 1.0 m? (b) 1.0 mile?

2. Figure 23–8a shows two charges, q_1 and q_2, held a fixed distance d apart. (a) What is the electric force that acts on q_1 because of the presence of q_2? (b) What is the the *total* force that acts on q_1? (c) What are the answers to these same two questions if a third charge q_3 is brought in from a large distance to a fixed position, as in Fig. 23–8b? Assume that $q_1 = 20$ μC, $q_2 = -40$ μC, $q_3 = 850$ μC, and $d = 1.5$ m.

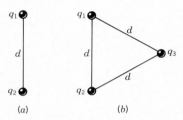

(a)   (b)

Figure 23–8   Exercise 2.

3. A point charge of $+3.0 \times 10^{-6}$ C is 12 cm distant from a second point charge of -1.5×10^{-6} C. Calculate the magnitude and direction of the force on each charge.

4. How far apart must two protons be if the electrical repulsive force acting on either one is equal to its weight at the earth's surface? The mass of a proton is 1.67×10^{-27} kg.

5. In the return stroke of a typical lightning bolt (see Fig. 23–1), a current of 2.0×10^4 A flows for 20×10^{-6} s. How much charge is transferred in this event?

6. Recent advances in the theory of elementary particles suggest that neutrons, protons, and other "heavy" particles are composed of more elementary particles called *quarks,* which have electrical charge less than e (where e is the charge of the electron).

A neutron is thought to be composed of one "up" quark of charge $+\frac{2}{3}e$, and two "down" quarks each having charge $-\frac{1}{3}e$. If the quarks are confined to a region of a size no greater than 2.64×10^{-15} m, what is the minimum repulsive electrical force between the two down quarks?

Section 23–6   Charge and Matter
7. The electrostatic force between two identical ions that are separated by a distance of 5.0×10^{-10} m is 3.7×10^{-9} N. (a) What is the charge on each ion? (b) How many electrons are missing from each ion?

8. (a) How many electrons would have to be removed from a penny to leave it with a charge of $+1.0 \times 10^{-7}$ C? (b) To what fraction of the electrons in the penny does this correspond? See Example 1.

9. Assuming that ions in a crystal can be thought of as point charges, find the force of attraction between a singly charged sodium ion and an adjacent chlorine ion in a salt crystal if their separation is 2.82×10^{-10} m.

10. Two small water droplets in air are 1.0 cm apart. Each has acquired a net charge of 1.0×10^{-16} C. (a) Find the magnitude of the electric force on each droplet. (b) How many excess electrons are on each droplet?

Section 23–7   Charge Is Conserved
11. When an electron and a positron interact, they may annihilate each other, thereby producing two uncharged photons. For this to occur, what must be the charge on the positron?

12. In beta decay a heavy fundamental particle changes to another heavy particle and either an electron (charge $-e$) or a positron (charge $+e$) is emitted. (a) If a proton becomes a neutron, which particle is emitted? (b) If a neutron becomes a proton, which particle is emitted?

PROBLEMS GROUPED BY SECTION

Section 23–4   Coulomb's Law
1. An electron is in a vacuum near the surface of the earth. Where should a second electron be placed so that the net force on the first electron, due to the other electron and to gravity, is zero?

2. Three charged particles lie on a straight line and are separated by a distance d as shown in Fig. 23–9. Charges q_1 and q_2 are held fixed. Charge q_3, which is free to move, is found to be in equilibrium under the action of the electric forces. What can you

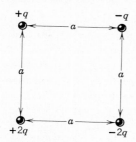

Figure 23-9 Problem 2.

say about the relative magnitudes and the signs of the three charges?

3. Charges q_1 and q_2 lie on the x-axis at points $x = -a$ and $x = +a$, respectively. (a) How must q_1 and q_2 be related for the net force on charge $+Q$, placed at $x = +a/2$, to be zero? (b) Answer the same question if the $+Q$ charge is placed at $x = +3a/2$.

4. Two equal positive charges, Q, are held fixed at a distance $2a$ apart. The force on a small positive test charge q midway between them is zero. If the test charge is displaced a short distance, either (a) toward one of the charges or (b) at right angles to the line joining the charges, find the direction of the force on it. Is the equilibrium stable or unstable in each case?

5. A 100-W lamp operated on a 120-V circuit has a current (assumed steady) of 0.833 A in its filament. How long does it take for one mole of electrons to pass through the lamp?

6. In Fig. 23-10, what is the resultant force on the charge in the lower left corner of the square? Assume that $q = 1.0 \times 10^{-7}$ C and $a = 5.0$ cm. The charges are fixed in position.

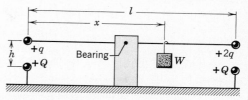

Figure 23-10 Problem 6.

7. Each of two small spheres is charged positively, the combined charge being 5.0×10^{-5} C. If each sphere is repelled from the other by a force of 1.0 N when the spheres are 2.0 m apart, how is the total charge distributed between the spheres?

8. Two identical conducting spheres, having charges of opposite sign, attract each other with a force of 0.108 N when separated by 50.0 cm. The spheres are connected by a conducting wire, which is then removed, and thereafter repel each other with a force of 0.0360 N. What were the initial charges on the spheres?

9. Figure 23-11 shows a long, insulating, massless rod of length l, pivoted at its center and balanced with a weight W at a distance x from the left end. At the left and right ends of the rod are attached positive charges q and $2q$, respectively. A distance h directly beneath each of these charges is a fixed positive charge Q. (a) Find the distance x for the position of the weight when the rod is balanced. (b) What value should h have so that the rod exerts no vertical force on the bearing when balanced? Neglect the interaction between charges at the opposite ends of the rod.

Figure 23-11 Problem 9.

Section 23-6 Charge and Matter

10. Estimate *roughly* the number of coulombs of positive charge in a glass of water. Assume the volume of the water to be 250 cm³.

11. (a) What equal positive charges would have to be placed on the earth and on the moon to neutralize their gravitational attraction? (b) Do you need to know the lunar distance to solve this problem? (Why or why not?) (c) How many tons of hydrogen would be needed to provide the positive charge calculated in part (a)?

12. We know that, within the limits of measurement, the magnitudes of the negative charge on the electron and the positive charge on the proton are equal. Suppose, however, that these magnitudes differed from each other by as little as 0.00010%. With what force would two copper pennies, placed 1.0 m apart, then repel each other? What do you conclude? (*Hint:* Use the result of Example 1.)

Section 23-7 Charge Is Conserved

13. In the radioactive decay of ^{238}U (see Eq. 23-6), the center of the emerging α particle is, at a certain instant, 9.0×10^{-15} m from the center of the residual nucleus ^{234}Th. At this instant, (a) what is the force on the α-particle, and (b) what is its acceleration?

14. Identify X in the following nuclear reactions:
(a) ^{1}H + ^{9}Be → X + n;
(b) ^{12}C + ^{1}H → X;
(c) ^{15}N + ^{1}H → X + ^{4}He.

ADDITIONAL PROBLEMS

15. The charges of an electron and a positron are $-e$ and $+e$, where $e = 1.60 \times 10^{-19}$ C. The mass of each is 9.11×10^{-31} kg. Calculate the ratio of the electrical force to the gravitational force between an electron and a positron. Why do you not need to know the distance between them?

16. Charge q_1 is fixed on the negative y-axis a distance a from the origin, and q_2 is fixed on the positive y-axis the same distance from the origin. What must be true of the two charges if the net force on a positive charge Q a distance b from the origin on the positive x-axis is (a) in the positive x-direction? (b) in the negative x-direction? (c) in the negative y-direction?

17. Two fixed charges, $+1.0 \times 10^{-6}$ C and -3.0×10^{-6} C, are 10 cm apart. (a) Where may a third charge be located so that no force acts on it? (b) Is the equilibrium of this third charge stable or unstable?

18. The charges and coordinates of two charged particles held fixed in the x-y plane are $q_1 = +3.0 \times 10^{-6}$ C; $x = 3.5$ cm, $y = 0.50$ cm, and $q_2 = -4.0 \times 10^{-6}$ C; $x = -2.0$ cm, $y = 1.5$ cm. (a) Find the magnitude and direction of the force on q_2. (b) Where could you locate a third charge $q_3 = +4.0 \times 10^{-6}$ C such that the total force on q_2 is zero?

19. Two *free* point charges $+q$ and $+4q$ are a distance l apart. A third charge is so placed that the entire system is in equilibrium. Find the location, magnitude, and sign of the third charge. Is the equilibrium stable?

20. Two equally charged particles, held 3.2×10^{-3} m apart, are released from rest. The initial acceleration of the first particle is observed to be 7.0 m/s^2 and that of the second to be 9.0 m/s^2. If the mass of the first particle is 6.3×10^{-7} kg, what are (a) the mass of the second particle and (b) the common charge?

21. Two engineering students (John at 200 lb and Mary at 100 lb) are 100 ft apart. Let each have a 0.01% imbalance in their amount of positive and negative charge, one student being positive and the other negative. Estimate *roughly* the electrostatic force of attraction between them. (*Hint:* Replace the students by equivalent spheres of water.)

22. Protons in cosmic rays strike the earth's upper atmosphere at a rate, averaged over the earth's surface, of 1500 protons/m$^2 \cdot$s. What total current does the earth receive from beyond its atmosphere in the form of incident cosmic-ray protons? The earth's radius is 6.4×10^6 m.

23. A charge Q is fixed at each of two opposite corners of a square. A charge q is placed at each of the other two corners. (a) If the resultant electrical force of Q is zero, how are Q and q related? (b) Could q be chosen to make the resultant force on *every* charge zero? Explain your answer.

24. A certain charge Q is to be divided into two parts, q and $Q - q$. What is the relationship of Q to q if the two parts, placed a given distance apart, are to have a maximum Coulomb repulsion?

25. Two similar conducting balls of mass m are hung from silk threads of length l and carry similar charges q as in Fig. 23–12. Assume that θ is so small that tan θ can be replaced by its approxi-mate equal, sin θ. To this approximation, (a) show that, for equilibrium,

$$x = \left(\frac{q^2 l}{2\pi\epsilon_0 mg} \right)^{1/3},$$

where x is the separation between the balls. (b) If $l = 120$ cm, $m = 10$ g, and $x = 5.0$ cm, what is q?

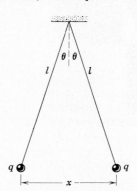

Figure 23–12 Problems 25 and 26.

26. If the balls of Fig. 23–12 are conducting, (a) what happens to them after one is discharged? Explain your answer. (b) Find the new equilibrium separation.

*27. An electron is projected with an initial speed of 3.2×10^5 m/s directly toward a proton that is essentially at rest. If the electron is initially a great distance from the proton, at what distance from the proton is its speed instantaneously equal to twice its initial value? (*Hint:* Use the work-energy theorem.)

The Electric Field

24-1 The Electric Field

Fields

Many phenomena are most conveniently described in terms of the *field* concept. A field is any physical quantity which can be specified simultaneously for all points within a given region of interest. We have already studied several fields without identifying them as such. One of the earliest occurred in Chapter 16 in which we showed that pressure can sometimes be predicted throughout the volume of a fluid. In this case pressure is the field variable. Another familiar example is the distribution of temperature in a lake or throughout the atmosphere. The field variable may or may not change with time.

A flow field

A field variable may be a scalar, as in the examples above, or a vector. The flow of water in a river is an example of a vector field, in this case called a *flow field*. Every point in the water has associated with it a vector quantity, the velocity **v** with which the water flows past the point. The field is called stationary if **v** at a particular point does not change with time. If we were to follow a particular small sample of water, we might find that its velocity changes as the sample moves from place to place. The field variable **v**, however, represents the velocity of water passing a particular point and does not change with time for steady-flow conditions.

The gravitational field

With every point in space near the earth we can associate a *gravitational field* vector **g** (see Eq. 15–11). This is the gravitational acceleration that a test body, placed at that point and released, would experience. If m is the mass of the body and **F** the gravitational force acting on it, **g** is given by

$$\mathbf{g} = \mathbf{F}/m. \tag{24-1}$$

This is another example of a vector field. For points near the surface of the earth the field is often taken as uniform; that is, **g** is the same for all points.

The electric field

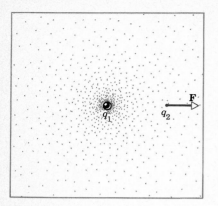

Figure 24–1 Charge q_1 sets up a field that exerts a force **F** on charge q_2.

If we place a test charge in the space near a charged rod, an electrostatic force will act on the charge. We speak of an *electric field* in this space. In a similar way we speak of a *magnetic field* in the space around a bar magnet. In the classical theory of electromagnetism the electric and magnetic fields are central concepts. In this chapter we deal with electric fields associated with charges viewed from a reference frame in which they are at rest, that is, with *electrostatics*.

Before Faraday's time, the force acting between charged particles was thought of as a direct and instantaneous interaction between the two particles. This action-at-a-distance view was also held for magnetic and for gravitational forces. Today we prefer to think in terms of electric fields as follows:

1. Charge q_1 in Fig. 24–1 sets up an electric field in the space around itself. This field is suggested by the shading in the figure.

2. The field acts on charge q_2; this shows up in the force **F** that q_2 experiences.

The field plays an intermediary role in the forces between charges. There are two separate problems: (*a*) calculating the fields that are set up by given distributions of charge and (*b*) calculating the forces that given fields will exert on charges placed in them. We think in terms of

$$\text{charge} \rightleftharpoons \text{field} \rightleftharpoons \text{charge}$$

and not, as in the action-at-a-distance point of view, in terms of

$$\text{charge} \rightleftharpoons \text{charge}.$$

The field vs. the
action-at-a-distance viewpoints

In Fig. 24–1 we can also imagine that q_2 sets up a field and that this field acts on q_1, producing a force $-\mathbf{F}$ on it in accord with Newton's third law. The situation is completely symmetrical, each charge being immersed in a field associated with the other charge.

If the only problem in electromagnetism was that of the forces between stationary charges, the field and the action-at-a-distance points of view would be equivalent. Suppose, however, that q_1 in Fig. 24–1 suddenly accelerates to the right. How quickly does the charge q_2 learn that q_1 has moved and that the force which it (q_2) experiences must increase? Electromagnetic theory predicts that q_2 learns about q_1's motion by a field disturbance that emanates from q_1, traveling with the speed of light. The action-at-a-distance point of view seems to imply that information about q_1's acceleration must be communicated instantaneously to q_2; this is not in accord with experiment. Accelerating electrons in the antenna of a radio transmitter influence electrons in a distant receiving antenna only after a time l/c where l is the separation of the antennas and c is the speed of light.

24–2 The Electric Field E

To define the electric field in laboratory terms, we imagine placing a small test charge q_0 (assumed positive for convenience) at the point in space that is to be examined, and we measure the electrical force **F** (if any) that acts on this body. The electric field **E** at the point is defined as

Definition of the electric field

$$\mathbf{E} = \mathbf{F}/q_0. \tag{24–2}$$

Here **E** is a vector because **F** is one, q_0 being a scalar. The direction of **E** is the direction of **F**, that is, it is the direction in which a resting positive charge placed at the point would tend to move.

The definition of gravitational field **g** is much like that of electric field, except that the mass of the test body rather than its charge is the property of interest. Although the units of **g** are usually written as m/s², they could also be written as N/kg (Eq. 24–1); those for **E** are N/C (Eq. 24–2). Thus both **g** and **E** are expressed as a force divided by a property (mass or charge) of the test body.

Example 1 What is the magnitude of the electric field **E** such that an electron, placed in the field, would experience an electrical force equal to its weight?

From Eq. 24–2, replacing q_0 by e and F by mg, where m is the electron mass, we have

$$E = \frac{F}{q_0} = \frac{mg}{e}$$

$$= \frac{(9.1 \times 10^{-31} \text{ kg})(9.8 \text{ m/s}^2)}{1.6 \times 10^{-19} \text{ C}} = 5.6 \times 10^{-11} \text{ N/C}.$$

This is a very weak electric field. Which way will **E** have to point if the electric force is to cancel the gravitational force?

In applying Eq. 24–2 we must use a test charge as small as possible. A large test charge might disturb the primary charges that are responsible for the field, thus changing the very quantity that we are trying to measure. Equation 24–2 should, strictly, be replaced by

$$\mathbf{E} = \lim_{q_0 \to 0} \frac{\mathbf{F}}{q_0}. \tag{24–3}$$

This equation instructs us to use a smaller and smaller test charge q_0, evaluating the ratio $\mathbf{F}/q_0$ at every step. The electric field **E** is then the limit of this ratio as the size of the test charge becomes very small. Because charge is quantized, however, we cannot contemplate free test charges whose magnitude is smaller than the electronic charge e ($= 1.6 \times 10^{-19}$ C).

24–3 Lines of Force

The concept of the electric field as a vector was not appreciated by Michael Faraday, who always thought in terms of lines of force. The lines of force still form a convenient way of visualizing electric-field patterns. We shall use them for this purpose but we shall not employ them quantitatively.

The relationship between the lines of force and the electric field vector is this:

1. The tangent to a line of force at any point gives the direction of **E** at that point.

2. The lines of force are drawn so that the number of lines per unit cross-sectional area (perpendicular to the lines) is proportional to the magnitude of **E**. Where the lines are close together E is large and where they are far apart E is small.

It is not obvious that it is possible to draw a continuous set of lines to meet these requirements. Indeed, it turns out that if Coulomb's law were not true, it would not be possible to do so; see Problem 2.

Figure 24–2 shows the lines of force for a uniform sheet of positive charge. We assume that the sheet is infinitely large, which, for a sheet of finite dimensions, is equivalent to considering only those points whose distance from the sheet is small compared to the distance to the nearest edge of the sheet. A positive test

Lines of force and the electric field

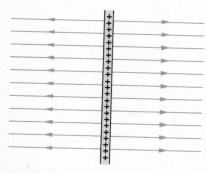

Figure 24–2 Lines of force for a section of an infinitely large sheet of positive charge.

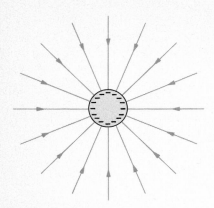

Figure 24-3 Lines of force for a negatively charged sphere.

charge, released in front of such a sheet, would move away from the sheet along a perpendicular line. Thus the electric field vector at any point near the sheet must be at right angles to the sheet. The lines of force are uniformly spaced, which means that **E** has the same magnitude for all points near the sheet.

Figure 24-3 shows the lines of force for a negatively charged sphere. The lines must lie along radii and point inward because a free positive charge would be accelerated in this direction. The electric field E is not constant but decreases with increasing distance from the charge. This is evident in the lines of force, which are farther apart at greater distances. From symmetry, E is the same for all points that lie a given distance from the center of the charge.

Example 2 In Fig. 24-3 how does E vary with the distance r from the center of the charged sphere?

Suppose that N lines terminate on the sphere. Draw an imaginary concentric sphere of radius r; the number of lines per unit cross-sectional area at every point on this sphere is $N/4\pi r^2$. Since E is proportional to this, we can write that

$$E \propto 1/r^2.$$

We derive an exact relationship in Section 24-4. How would you expect E to vary with distance from an infinitely long uniform cylinder of charge?

Figures 24-4 and 24-5 show the lines of force for two equal like charges and for two equal unlike charges, respectively. Michael Faraday, as we have said, used lines of force a great deal in his thinking. They were more real for him than they are for most scientists and engineers today. It is possible to sympathize with Faraday's point of view. Can we not almost "see" the charges being pushed apart in Fig. 24-4 and pulled together in Fig. 24-5 by the lines of force?

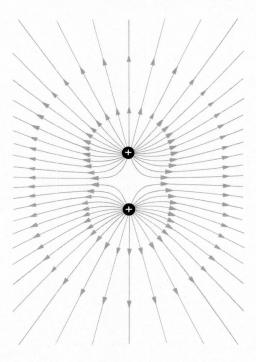

Figure 24-4 Lines of force for two equal positive charges.

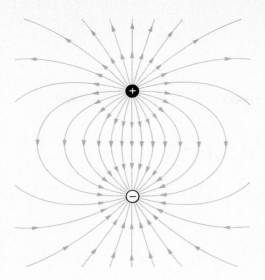

Figure 24-5 Lines of force for equal but opposite charges.

24-4 Calculation of E

In this section we deal with the charge-field interaction by showing how we may calculate **E** for various points near given charge distributions, starting with the simple case of a point charge q.

Let a test charge q_0 be placed a distance r from a point charge q. The magnitude of the force acting on q_0 is given by Coulomb's law, or

$$F = \frac{1}{4\pi\epsilon_0} \frac{qq_0}{r^2}.$$

The magnitude of the electric field at the site of the test charge is given by Eq. 24-2, or

E for a point charge

$$E = \frac{F}{q_0} = \frac{1}{4\pi\epsilon_0} \frac{q}{r^2}. \qquad (24-4)$$

The direction of **E** is on a radial line from q, pointing outward if q is positive and inward if q is negative.

To find **E** for a group of point charges: (a) Calculate $\mathbf{E}_n$ due to each charge at the given point as if it were the only charge present. (b) Add these separately calculated fields vectorially to find the resultant field **E** at the point. In equation form,

The superposition principle

$$\mathbf{E} = \mathbf{E}_1 + \mathbf{E}_2 + \mathbf{E}_3 + \cdots = \Sigma\mathbf{E}_n \qquad n = 1, 2, 3, \ldots \qquad (24-5)$$

The sum is a vector sum, taken over all the charges. Equation 24-5 (like Eq. 23-4) is an example of the *principle of superposition* (see Sec. 17-6), which states, in this context, that at a given point the electric fields due to separate charge distributions simply superimpose independently.

If the charge distribution is a continuous one, the field it sets up at any point P can be computed by dividing the charge into infinitesimal elements dq. The field $d\mathbf{E}$ due to each element at the point in question is then calculated, treating the elements as point charges. The magnitude of $d\mathbf{E}$ (see Eq. 24-4) is given by

$$dE = \frac{1}{4\pi\epsilon_0} \frac{dq}{r^2}, \qquad (24-6)$$

where r is the distance from the charge element dq to the point P. The resultant field at P is then found from the superposition principle by adding (that is, integrating) the field contributions due to all the charge elements, or

E for a continuous charge distribution

$$\mathbf{E} = \int d\mathbf{E}. \qquad (24-7)$$

The integration, like the sum in Eq. 24–5, is a vector operation; in Example 5 we will see how such an integral is handled in a simple case.

Example 3 *An electric dipole*   Figure 24–6 shows a positive and a negative charge of equal magnitude q placed a distance $2a$ apart, a configuration called an electric dipole. The pattern of lines of force is that of Fig. 24–5, which also shows an electric dipole. What is the field **E** due to these charges at point P, a distance r along the perpendicular bisector of the line joining the charges? Assume $r \gg a$.

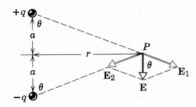

Figure 24–6   Example 3.

Equation 24–5 gives the vector equation

$$\mathbf{E} = \mathbf{E}_1 + \mathbf{E}_2,$$

where, from Eq. 24–4,*

$$E_1 = E_2 = \frac{1}{4\pi\epsilon_0} \frac{q}{a^2 + r^2}.$$

The vector sum of $\mathbf{E}_1$ and $\mathbf{E}_2$ points vertically downward and has the magnitude

$$E = 2E_1 \cos\theta.$$

From the figure we see that

$$\cos\theta = \frac{a}{\sqrt{a^2 + r^2}}.$$

Substituting the expressions for E_1 and for $\cos\theta$ into that for E yields

$$E = \frac{2}{4\pi\epsilon_0} \frac{q}{(a^2 + r^2)} \frac{a}{\sqrt{a^2 + r^2}} = \frac{1}{4\pi\epsilon_0} \frac{2aq}{(a^2 + r^2)^{3/2}}.$$

If $r \gg a$, we can neglect a in the denominator; this equation then reduces to

$$E \cong \frac{1}{4\pi\epsilon_0} \frac{(2a)(q)}{r^3}. \qquad (24-8a)$$

The essential properties of the charge distribution in Fig. 24–6, the magnitude of the charge q and the separation $2a$ between the charges, enter Eq. 24–8a only as a product. This means that, if we measure $\mathbf{E}$ at various distances from the elec-

tric dipole (assuming $r \gg a$), we can never deduce q and $2a$ separately but only the product $2aq$; if q were doubled and a simultaneously cut in half, the electric field at large distances from the dipole would not change.

The product $2aq$ is called the electric dipole moment p. Thus we can rewrite this equation for E, for distant points along the perpendicular bisector, as

$$E = \frac{1}{4\pi\epsilon_0} \frac{p}{r^3}. \qquad (24-8b)$$

The result for distant points along the dipole axis (see Problem 9) and the general result for any distant point (see Problem 33) also contain the quantities $2a$ and q only as the product $2aq$ ($= p$). The variation of E with r in the general result for distant points is also as $1/r^3$, as in Eq. 24–8b.

The dipole of Fig. 24–6 is two equal and opposite charges placed close to each other so that their separate fields at distant points almost, but not quite, cancel. On this point of view it is easy to understand that $E(r)$ for a dipole varies at large distances as $1/r^3$ (Eq. 24–8b), whereas for a point charge $E(r)$ drops off more slowly, namely as $1/r^2$ (Eq. 24–4).

Example 4   Figure 24–7 shows a charge q_1 ($= +1.0 \times 10^{-6}$ C) 10 cm from a charge q_2 ($= +2.0 \times 10^{-6}$ C). At what point on the line passing through the two charges is the electric field zero?

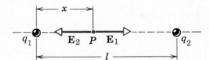

Figure 24–7   Example 4.

The point must lie between the charges because only here do the forces exerted by q_1 and q_2 on a test charge (no matter whether it is positive or negative) oppose each other. If $\mathbf{E}_1$ is the electric field due to q_1 and $\mathbf{E}_2$ that due to q_2, we must have

$$E_1 = E_2$$

or (see Eq. 24–4)

$$\frac{1}{4\pi\epsilon_0} \frac{q_1}{x^2} = \frac{1}{4\pi\epsilon_0} \frac{q_2}{(l - x)^2}$$

where x is the distance from q_1 and l equals 10 cm. Solving for x yields

$$x = \frac{l}{1 + \sqrt{q_2/q_1}} = \frac{10 \text{ cm}}{1 + \sqrt{2}} = 4.1 \text{ cm}.$$

---

* Note that the r's in Eq. 24–4 and in this equation have different meanings.

You should supply the missing steps. On what basis was the second root of the resulting quadratic equation discarded?

Example 5 *Ring of charge*   Figure 24–8 shows a uniform ring of charge q and radius a. Calculate $\mathbf{E}$ for points on the axis of the ring a distance x from its center.

Consider a differential element of the ring of length ds, located at the top of the ring in Fig. 24–8. It contains an element of charge given by

$$dq = q\,\frac{ds}{2\pi a}$$

where $2\pi a$ is the circumference of the ring. This element sets up a differential electric field $d\mathbf{E}$ at point P.

The resultant field $\mathbf{E}$ at P is found by adding, vectorially, the effects of all the elements that make up the ring. This looks like a hard job at first because the vectors $d\mathbf{E}$ all have components that are both parallel and perpendicular to the axis of the ring. Then we notice that the perpendicular components add to zero because every element of the ring can be paired to another that is directly opposite and produces a perpendicular component of $d\mathbf{E}$ that exactly cancels its own. Thus, all the components of $d\mathbf{E}$ that are perpendicular to the axis may be neglected and the resultant field may be obtained by adding only those components of $d\mathbf{E}$ that lie along the axis. These are all in the same direction, so the vector addition becomes a scalar addition of parallel axial components.

The quantity dE follows from Eq. 24–6, or

$$dE = \frac{1}{4\pi\epsilon_0}\frac{dq}{r^2} = \frac{1}{4\pi\epsilon_0}\frac{dq}{(a^2 + x^2)}.$$

From Fig. 24–8 we have

$$\cos\theta = \frac{x}{\sqrt{a^2 + x^2}}.$$

The component of $d\mathbf{E}$ parallel to the axis is thus

$$dE\cos\theta = \frac{dq}{4\pi\epsilon_0}\left[\frac{x}{(a^2 + x^2)^{3/2}}\right].$$

To add the various contributions, we need only add the dq's, since all the other factors in $dE\cos\theta$ above are the same for all dq's. Thus, the addition yields

$$E = \frac{1}{4\pi\epsilon_0}\left[\frac{qx}{(a^2 + x^2)^{3/2}}\right].$$

Verify that this expression for E reduces to an expected result for $x = 0$.

For $x \gg a$ we can neglect a in the denominator of this equation, yielding

$$E \cong \frac{1}{4\pi\epsilon_0}\frac{q}{x^2}.$$

This is an expected result (compare Eq. 24–4) because at great enough distances the ring appears to be a point charge q.

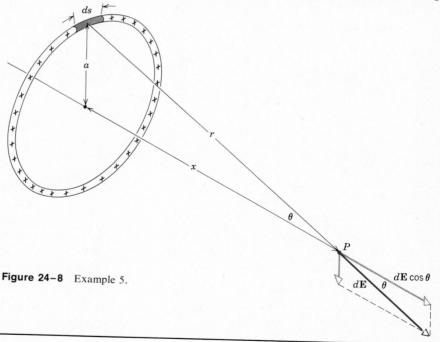

Figure 24–8   Example 5.

24–5   A Point Charge in an Electric Field

Here and in the following section, in contrast with Section 24–4, we investigate the other half of the charge-field interaction, namely, if we are given a field $\mathbf{E}$, what forces and torques will act on a charge configuration placed in it? We start with the simple case of a point charge in a uniform electric field.

An electric field will exert a force on a charged particle given by (Eq. 24–2)

$$\mathbf{F} = \mathbf{E}q.$$

Force and acceleration in an electric field

This force, acting alone, will produce an acceleration

$$\mathbf{a} = \mathbf{F}/m,$$

where m is the mass of the particle, We will consider two examples of the acceleration of a charged particle in a uniform electric field. Such a field can be produced by connecting the terminals of a battery of two parallel metal plates which are otherwise insulated from each other. If the spacing between the plates is small compared with the dimensions of the plates, the field between them will be fairly uniform except near the edges. Note that in calculating the motion of a particle in a field set up by external charges the field due to the particle itself (that is, its self-field) is ignored. In the same way, the earth's gravitational field can exert no net force on the earth itself but only on a second object, say a stone, placed in that field.

Example 6   A particle of mass m and charge q is placed at rest in a uniform electric field (Fig. 24–9) and released. Describe its motion.

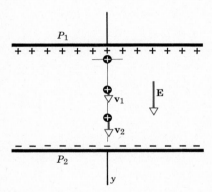

Figure 24–9   Example 6. A charge is released from rest in a uniform electric field set up between two oppositely charged metal plates P_1 and P_2.

The motion resembles that of a body falling in the earth's gravitational field. The (constant) acceleration is given by

$$a = \frac{F}{m} = \frac{qE}{m}.$$

The equations for uniformly accelerated motion (Eqs. 3–17) then apply. With subscripts dropped and with $v_0 = 0$, they are

$$v = at = \frac{qEt}{m},$$

$$y = \tfrac{1}{2}at^2 = \frac{qEt^2}{2m},$$

and

$$v^2 = 2ay = \frac{2qEy}{m}.$$

The kinetic energy attained after moving a distance y is found from

$$K = \tfrac{1}{2}mv^2 = \tfrac{1}{2}m\left(\frac{2qEy}{m}\right) = qEy.$$

This result also follows directly from the work-energy theorem because a constant force qE acts over a distance y.

Example 7   *Deflecting an electron beam*   Figure 24–10 shows an electron of mass m and charge e projected with speed v_0 at right angles to a uniform field $\mathbf{E}$. Describe its motion.

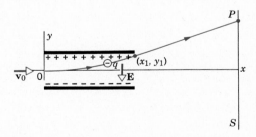

Figure 24–10   Example 7. An electron is projected into a uniform electric field.

The motion is like that of a projectile fired horizontally in the earth's gravitational field. The considerations of Section 4–3 apply, the horizontal (x) and vertical (y) motions being given by

$$x = v_0 t$$

and

$$y = \tfrac{1}{2}at^2 = \frac{eE}{2m}t^2.$$

Eliminating t yields

$$y = \frac{eE}{2mv_0^2}\,x^2 \tag{24-9}$$

for the equation of the trajectory.

When the electron emerges from the plates in Fig. 24–10 it travels (neglecting gravity) in a straight line tangent to the parabola of Eq. 24–9 at the exit point. We can let it fall on a fluorescent screen S placed some distance beyond the plates. Together with other electrons following the same path, it will then make itself visible as a small luminous spot; this is the principle of the electrostatic cathode-ray oscilloscope.

Example 8   The electric field between the plates of a cathode-ray oscilloscope is 1.2×10^4 N/C. What deflection will an electron experience if it enters at right angles to the field with

a kinetic energy of 2.0 keV ($=3.2 \times 10^{-16}$ J), a typical value? The deflecting assembly is 1.5 cm long.

Recalling that $K_0 = \frac{1}{2}mv_0^2$, we can rewrite Eq. 24–9 as

$$y = \frac{eEx^2}{4K_0}.$$

If x_1 is the horizontal position of the far edge of the plate, y_1 will be the corresponding deflection (see Fig. 24–10), or

$$
\begin{aligned}
y_1 &= \frac{eEx_1^2}{4K_0} \\
&= \frac{(1.6 \times 10^{-19}\ \text{C})(1.2 \times 10^4\ \text{N/C})(1.5 \times 10^{-2}\ \text{m})^2}{(4)(3.2 \times 10^{-16}\ \text{J})} \\
&= 3.4 \times 10^{-4}\ \text{m} = 0.34\ \text{mm}.
\end{aligned}
$$

The deflection measured at the fluorescent screen is much larger.

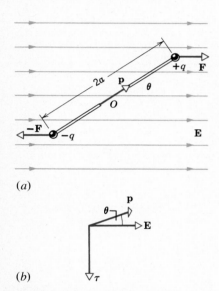

(a)

(b)

Figure 24–11   *(a)* An electric dipole in a uniform external field. *(b)* illustrating $\boldsymbol{\tau} = \mathbf{p} \times \mathbf{E}$.

Torque on an electric dipole

24–6   A Dipole in an Electric Field

An electric dipole moment can be regarded as a vector $\mathbf{p}$ whose magnitude p, for a dipole like that described in Example 3, is the product $2aq$ of the magnitude of either charge q and the distance $2a$ between the charges. The direction of $\mathbf{p}$ for such a dipole is from the negative to the positive charge. The vector nature of the electric dipole moment permits us to cast many expressions involving electric dipoles into concise form, as we shall see.

Figure 24–11a shows an electric dipole formed by placing two charges $+q$ and $-q$ a fixed distance $2a$ apart. The arrangement is placed in a uniform external electric field $\mathbf{E}$, its dipole moment $\mathbf{p}$ making an angle θ with this field. Two equal and opposite forces $\mathbf{F}$ and $-\mathbf{F}$ act as shown, where

$$F = qE.$$

The net force is clearly zero, but there is a net torque about an axis through O (see Eq. 12–2) given by

$$\tau = 2aF \sin \theta.$$

Combining these two equations and recalling that $p = (2a)(q)$, we obtain

$$\tau = 2aqE \sin \theta = pE \sin \theta. \tag{24-10}$$

Thus an electric dipole placed in an external electric field $\mathbf{E}$ experiences a torque tending to align it with the field. Equation 24–10 can be extended to vector form as

$$\boldsymbol{\tau} = \mathbf{p} \times \mathbf{E}, \tag{24-11}$$

the appropriate vectors being shown in Fig. 24–11b.

Work (positive or negative) must be done by an external agent to change the orientation of an electric dipole in an external field. This work is stored as potential energy U in the system consisting of the dipole and the arrangement used to set up the external field. If θ in Fig. 24–11a has the initial value θ_0, the work required to turn the dipole axis to an angle θ is (see Table 12–2)

$$W = \int dW = \int_{\theta_0}^{\theta} \tau\, d\theta = U,$$

where τ is the torque exerted by the agent that does the work. Combining this equation with Eq. 24–10 yields

$$U = \int_{\theta_0}^{\theta} pE \sin \theta \, d\theta = pE \int_{\theta_0}^{\theta} \sin \theta \, d\theta$$

$$= pE \left(-\cos \theta \right) \Big|_{\theta_0}^{\theta}.$$

Since we are interested only in changes in potential energy, we can choose the reference orientation θ_0 to have any convenient value, in this case 90°. This means we measure the potential energy with respect to a zero value in which the dipole is broadside to the external field. This gives

Potential energy of an electric dipole

$$U = -pE \cos \theta \qquad (24-12)$$

or, in vector terms,

$$U = -\mathbf{p} \cdot \mathbf{E}. \qquad (24-13)$$

Example 9 An electric dipole consists of two opposite charges of magnitude $q = 1.0 \times 10^{-6}$ C separated by $d = 2.0$ cm. The dipole is placed in an external field of 1.0×10^5 N/C.

(a) What maximum torque does the field exert on the dipole? The maximum torque is found by putting $\theta = 90°$ in Eq. 24–10 or

$$\tau = pE \sin \theta = qdE \sin \theta$$
$$= (1.0 \times 10^{-6} \text{ C})(0.020 \text{ m})(1.0 \times 10^5 \text{ N/C})(\sin 90°)$$
$$= 2.0 \times 10^{-3} \text{ N·m}.$$

(b) How much work must an external agent do to turn the dipole end for end, starting from a position of alignment ($\theta = 0$)? The work is the difference in potential energy U between the positions $\theta = 180°$ and $\theta = 0$. From Eq. 24–12,

$$W = U_{180} - U_0 = (-pE \cos 180°) - (-pE \cos 0)$$
$$= 2pE = 2qdE$$
$$= (2)(1.0 \times 10^{-6} \text{ C})(0.020 \text{ m})(1.0 \times 10^5 \text{ N/C})$$
$$= 4.0 \times 10^{-3} \text{ J}.$$

REVIEW GUIDE AND SUMMARY

The field interpretation of electric force

One way to explain the electric force between charges is to presume that each charge sets up an electric field in the space surrounding itself. Each charge then interacts with the resulting field at its own location.

We define the electric field at a point in space in terms of the electric force exerted on a small test charge placed at that point. The defining equation is

Definition of electric field

$$\mathbf{E} = \mathbf{F}/q_0. \qquad [24-2]$$

The field is defined as the limit of this ratio as q_0 approaches zero if the value obtained by using Eq. 24–2 is different for different test charges.

Lines of force

Lines of force are a convenient representation of an electric field. They are drawn so that the tangent to a given line is in the direction of the field and so that the number of lines per unit cross-sectional area perpendicular to the lines is proportional to the strength of the field. The number of lines leaving any charged particle is also proportional to its charge. Section 24–3 shows field lines for a large sheet of charge, a charged sphere, two equal charges and two equal but opposite charges.

The electric field due to a point charge

Any electric field can be calculated from Coulomb's law by knowing the location and charge of the charged particles that cause it. The field due to a point charge has magnitude

$$E = \frac{1}{4\pi\epsilon_0} \frac{q}{r^2}. \qquad [24-4]$$

The field points directly away from a positive charge and directly toward a negative one. The field due to a collection of point charges is obtained by adding, vectorially, the contributions from the individual charges. This is called the principle of superposition. The addition becomes an integration if the field-causing charge is distributed continuously rather than discretely. Examples 3, 4, and 5 illustrate both kinds of computations.

A point charge in an electric field $\mathbf{E}$ experiences an electric force

Force and acceleration in an electric field

$$\mathbf{F} = \mathbf{E}q$$

and an acceleration in accord with Newton's second law. Examples 6, 7, and 8 illustrate the application of these principles for a charged particle moving in a uniform electric field.

Electric dipoles

An assembly of two equal but opposite charges $\pm q$ separated by a distance $2a$ is called an *electric dipole*. It may be described in terms of a vector dipole moment $\mathbf{p}$ of magnitude $2aq$ and direction parallel to a line from the negative to the positive charge. The field due to an electric dipole is described in Example 3 and Problem 33. A dipole in a uniform electric field experiences no net force but does experience a torque given by

Torque on an electric dipole

$$\boldsymbol{\tau} = \mathbf{p} \times \mathbf{E}. \qquad [24\text{--}11]$$

Potential energy of an electric dipole

This torque tends to align the dipole in a direction parallel to the field. The potential energy of a dipole in an electric field may be conveniently taken to be

$$U = -\mathbf{p} \cdot \mathbf{E}. \qquad [24\text{--}13]$$

QUESTIONS

1. Name as many scalar fields and vector fields as you can.

2. In the gravitational attraction between the earth and a stone, can we say that the earth lies in the gravitational field of the stone? How is the gravitational field due to the stone related to that due to the earth?

3. A positively charged ball hangs from a long silk thread. We wish to measure $\mathbf{E}$ at a point in the same horizontal plane as that of the hanging charge. To do so, we put a positive test charge q_0 at the point and measure $\mathbf{F}/q_0$. Will F/q_0 be less than, equal to, or greater than E at the point in question?

4. Taking into account the quantization of electric charge, how can we justify the procedure suggested by Eq. 24–3?

5. In exploring electric fields with a test charge we have often assumed, for convenience, that the test charge was positive. Does this really make any difference in determining the field? Illustrate in a simple case of your own devising.

6. Electric lines of force never cross. Why?

7. In Fig. 24–4 why do the lines of force around the edge of the figure appear, when extended backward, to radiate uniformly from the center of the figure?

8. Figure 24–2 shows that $\mathbf{E}$ has the same value for all points in front of an infinite uniformly charged sheet. Is this reasonable? One might think that the field should be stronger near the sheet because the charges are so much closer.

9. If a point charge q of mass m is released from rest in a nonuniform field, will it follow a line of force?

10. A point charge is moving in an electric field at right angles to the lines of force. Does any force act on it?

11. Two point charges of unknown magnitude and sign are a distance d apart. The electric field is zero at one point between them, on the line joining them. What can you conclude about the charges?

12. Compare the way E varies with r for (a) a point charge (Eq. 24–4), (b) a dipole (Eq. 24–8a), and (c) a quadrupole (Problem 10).

13. Electric fields are very useful in removing dust and fly-ash from the air. How might such an electrostatic precipitator operate?

14. What is the origin of "static cling," a phenomenon that sometimes affects clothes as the are removed from a dryer?

15. In Example 5, give symmetry arguments to show that (a) the electric field for points along the axis of the ring must lie along that axis; (b) the electric field at the center of the ring must be zero; and (c) the magnitude of the electric field must pass through a maximum somewhere along the axis.

16. In Example 4 a charge placed at point P in Fig. 24–7 is in equilibrium because no force acts on it. Is the equilibrium stable for displacements along the line joining the charges? . . . for displacements at right angles to this axis?

17. Two point charges of unknown sign and magnitude are fixed a distance l apart. Can we have $\mathbf{E} = 0$ for off-axis points (excluding ∞)? Explain.

18. In Fig. 24–5 the force on the lower charge points up and is finite. The crowding of the lines of force, however, suggests that E is infinitely great at the site of this (point) charge. A charge immersed in an infintely great field should have an infinitely great force acting on it. What is the solution to this dilemma?

19. A positive and a negative charge of the same magnitude lie on a long straight line. What is the direction of $\mathbf{E}$ for points on this line that lie (a) between the charges, (b) outside the charges in the direction of the positive charge, (c) outside the charges in the direction of the negative charges, and (d) off the line but in the median plane of the charges?

20. In the median plane of an electric dipole, is the electric field parallel or antiparallel to the electric dipole moment $\mathbf{p}$?

21. If you turn an electric dipole end for end in a uniform electric field, does the work you do depend on the initial orientation of the dipole with respect to the field?

22. For what orientation of an electric dipole in a uniform electric field is the potential energy of the dipole the greatest? . . . the least?

23. An electric dipole is placed in a nonuniform electric field. Is there a net force on it?

24. An electric dipole is placed at rest in a uniform external electric field, as in Fig. 24–11a, and released. Discuss its motion.

EXERCISES

Section 24-2   The Electric Field E

1. Humid air breaks down (its molecules are ionized) in an electric field of about 10^6 N/C. What is the magnitude of the electric force on (a) an electron and (b) an ion (with a single electron missing) in this field?

2. An electron is released from rest in a uniform electric field of magnitude 2.0×10^4 N/C. Ignore gravity and find the acceleration of the electron.

3. An α particle, the nucleus of a helium atom, has a mass of 6.7×10^{-27} kg and a charge of $+2e$. What is the magnitude and direction of an electric field that will balance its weight?

4. In a uniform electric field near the surface of the earth, a particle having a charge of -2.0×10^{-9} C is acted on by a downward electric force of 3.0×10^{-6} N. (a) What is the magnitude of the electric field? (b) What is the magnitude and direction of the electric force exerted on a proton placed in this field? (c) What is the gravitational force on the proton? (d) What is the ratio of the electric to the gravitational forces in this case?

5. An electric field **E** with an average magnitude of about 150 N/C points downward in the earth's atmosphere. We wish to "float" a sulfur sphere weighing 1.0 lb ($=4.45$ N) in this field by charging it. (a) What charge (sign and magnitude) must be used? (b) Why is the experiment not practical? Give a qualitative reason supported by a very rough numerical calculation to prove your point.

6. In Fig. 24-12, charges are placed at the vertices of an equilateral triangle. For what value of Q (both sign and magnitude) does the total electric field vanish at C, the center of the triangle?

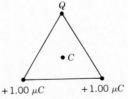

Figure 24-12   Exercise 6.

Section 24-3   Lines of Force

7. Figure 24-13 shows a field line representation of an electric field. (a) If the magnitude of the field at A is 40 N/C, what force does an electron at that point experience? (b) At B, does the field have a magnitude greater than, equal to, or less than its magnitude at A?

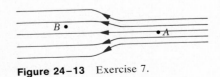

Figure 24-13   Exercise 7.

8. Sketch qualitatively the lines of force associated with two separated point charges $+q$ and $-2q$.

9. Three charges are arranged in an equilateral triangle as in Fig. 24-14. Consider the lines of force due to $+Q$ and $-Q$, and from them identify the direction of the force that acts on $+q$ because of the presence of the other two charges.

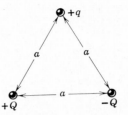

Figure 24-14   Exercise 9.

10. Sketch qualitatively the lines of force between two concentric conducting spherical shells, charge $+q_1$ being placed on the inner sphere and $-q_2$ on the outer. Consider the cases $q_1 > q_2$, $q_1 = q_2$, $q_1 < q_2$.

11. Sketch qualitatively the lines of force associated with a thin circular uniformly-charged disk of radius R. (*Hint:* Consider as limiting cases points very close to the disk, where the electric field is perpendicular to the surface, and points very far from it, where the electric field is like that of a point charge.)

Section 24-4   Calculation of E

12. What is the magnitude of a point charge chosen so that the electric field 50 cm away has a magnitude of 2.0 N/C?

13. Two point charges of magnitude $+2.0 \times 10^{-7}$ C and $+8.5 \times 10^{-8}$ C are 12 cm apart. (a) What electric field does each produce at the site of the other? (b) What force acts on each?

14. Two equal and opposite charges of magnitude 2.0×10^{-7} C are held 15 cm apart. (a) What are the magnitude and direction of **E** at a point midway between the charges? (b) What force (magnitude and direction) would act on an electron placed there?

Section 24-5   A Point Charge in an Electric Field

15. An electric field in a copper wire causes electrons to accelerate until they collide with copper atoms and thereby lose their kinetic energy. They are then accelerated again. Consider a 1.00-m-long wire plugged into a 120-V direct current power supply so that a 120-N/C electric field acts in the wire. What is the acceleration of the electrons between collisions?

16. One defensive weapon being considered for the Strategic Defense Initiative (Star Wars) uses particle beams. For example, a proton beam striking an enemy missile could render it harmless. Such beams can be produced in "guns" using electric fields to accelerate the charged particles. (a) What acceleration would a proton experience if the electric field is 2.0×10^4 N/C? (b) What speed would the proton attain if the field acts over a distance of 10 cm?

PROBLEMS GROUPED BY SECTION

Section 24–3   Lines of Force
1. (*a*) Sketch qualitatively the lines of force associated with three long parallel lines of charge, in a perpendicular plane. Assume that the intersections of the lines of charge with such a plane form an equilateral triangle (Fig. 24–15) and that each line of charge has the same linear charge density λ. (*b*) Discuss the nature of the equilibrium of a test charge placed on the central axis of the charge assembly.

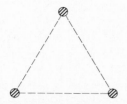

Figure 24–15   Problem 1.

2. Assume that the exponent in Coulomb's law is not "two" but *n*. Show that for $n \neq 2$ it is impossible to construct lines that will have the properties listed for lines of force in Section 24–3. For simplicity, treat an isolated point charge.

Section 24–4   Calculation of E
3. (*a*) In Fig. 24–16, locate the point (or points) at which the electric field is zero. (*b*) Sketch qualitatively the lines of force.

Figure 24–16   Problem 3.

4. Charges $+q$ and $-2q$ are fixed a distance *d* apart as in Fig. 24–17. (*a*) Find **E** at points *A, B,* and *C.* (*b*) Sketch roughly the lines of force.

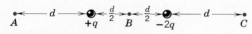

Figure 24–17   Problem 4.

5. In Fig. 24–6, assume that both charges are positive. (*a*) Show that *E* at point *P* in that figure, assuming that $r \gg a$, is given by

$$E = \frac{1}{4\pi\epsilon_0} \frac{2q}{r^2}.$$

(*b*) What is the direction of **E**? (*c*) Is it reasonable that *E* should vary as r^{-2} here and as r^{-3} for the dipole of Fig. 24–6?

6. Make a quantitative plot of the electric field on the axis of a charged ring having a diameter of 6.0 cm if the ring carries a uniform charge of 1.0×10^{-8} C. (See Example 5.)

7. What is **E** in magnitude and direction at the center of the square of Fig. 24–18? Assume that $q = 1.0 \times 10^{-8}$ C and $d = 5.0$ cm.

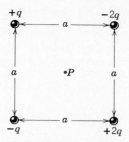

Figure 24–18   Problem 7.

8. A thin glass rod is bent into a semicircle of radius *r*. A charge $+Q$ is uniformly distributed along the upper half and a charge $-Q$ is uniformly distributed along the lower half, as shown in Fig. 24–19. Find the electric field **E** at *P*, the center of the semicircle.

Figure 24–19   Problem 8.

9. *Axial field due to an electric dipole.* In Fig. 24–6, consider a point a distance *r* from the center of the dipole along its axis. (*a*) Show that, at large values of *r*, the electric field is

$$E = \frac{1}{2\pi\epsilon_0} \frac{p}{r^3},$$

which is twice the value given for the conditions of Example 3. (*b*) What is the direction of **E**?

*10. *Electric quadrupole.* Figure 24–20 shows a typical electric quadrupole. It consists of two dipoles whose effects at external points do not quite cancel. Show that the value of *E* on the axis of the quadrupole for points distance *r* from its center (assume that $r \gg a$) is given by

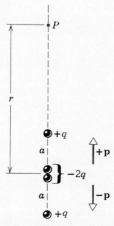

Figure 24–20   Problem 10.

$$E = \frac{3Q}{4\pi\epsilon_0 r^4},$$

where Q $(=2qa^2)$ is called the *quadrupole moment* of the charge distribution.

Section 24–5   A Point Charge in an Electric Field

11. (a) What is the acceleration of an electron in a uniform electric field of 1.0×10^6 N/C? (b) How long would it take for the electron, starting from rest, to attain one-tenth the speed of light?

12. An object having a mass of 10 g and a charge of $+8.0 \times 10^{-5}$ C is placed in a uniform electric field defined by $E_x = 3.0 \times 10^3$ N/C, $E_y = -600$ N/C, and $E_z = 0$. (a) What are the magnitude and direction of the force on the object? (b) If the object starts from rest at the origin, what will be its coordinates after 3.0 s?

13. A uniform vertical field $\mathbf{E}$ is established in the space between two large parallel plates. In this field one suspends a small conducting sphere of mass m from a string of length l. Find the period of this pendulum when the sphere is given a charge $+q$ if the lower plate (a) is charged positively; (b) is charged negatively.

14. *Millikan's oil drop experiment.* R. A. Millikan set up an apparatus (Fig. 24–21) in which a tiny, charged oil drop, placed in an electric field $\mathbf{E}$, could be "balanced" by adjusting E until the electric force on the drop was equal and opposite to its weight. If the radius of the drop is 1.64 μm and E at balance is 1.92×10^5

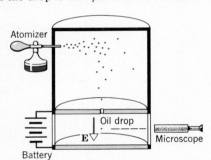

Atomizer

Oil drop

Microscope

Battery

Figure 24–21   Problem 14.

N/C, (a) what charge is on the drop in terms of the electronic charge e? (b) Why did Millikan not try to balance electrons in his apparatus instead of oil drops? The density of the oil is 851 kg/m³. (Millikan first measured the electronic charge in this way. He measured the drop radius by observing the limiting speed that the drops attained when they fell in air with the electric field turned off. He charged the oil drops by irradiating them with bursts of x rays.)

15. In a particular early run (1911), Millikan (see Problem 14) observed that the following measured charges, among others, appeared at different times on a single drop:

6.563×10^{-19} C	13.13×10^{-19} C	19.71×10^{-19} C
8.204×10^{-19} C	16.48×10^{-19} C	22.89×10^{-19} C
$11.50 \ \times 10^{-19}$ C	18.08×10^{-19} C	26.13×10^{-19} C

What value for the elementary charge e can be deduced from these data?

16. An electron is projected as in Fig. 24–22 at a speed of 6.0×10^6 m/s and at an angle θ of 45°; $\mathbf{E} = 2.0 \times 10^3$ N/C (directed upward), $d = 2.0$ cm, and $l = 10$ cm. (a) Will the electron strike either of the plates? (b) If it strikes a plate, where does it do so?

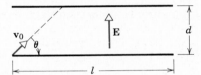

Figure 24–22   Problem 16.

*17. A positive charge q is located on the positive y-axis, 5.0 m from the origin, and an identical charge is located on the negative y-axis, also 5.0 m from the origin. At what position does an electron traveling along the x-axis have (a) the greatest acceleration? (b) the greatest deceleration?

Section 24–6   A Dipole in an Electric Field

18. Find the frequency of oscillation of an electric dipole, of moment p and rotational inertia I, for small amplitudes of oscillation about its equilibrium position in a uniform electric field E.

ADDITIONAL PROBLEMS

19. Two charges, q_1 $(=2.1 \times 10^{-8}$ C) and q_2 $(=-4q_1)$, are placed 50 cm apart. Find the point along the straight line joining the two charges at which the electric field is zero.

20. Two point charges of unknown magnitude and sign are placed a distance d apart. (a) If it is possible to have $\mathbf{E} = 0$ at any point *not* between the charges but on the line joining them, what are the necessary conditions, and where is the point located? (b) Is it possible, for any arrangement of two point charges, to find *two* points (neither at infinity) at which $\mathbf{E} = 0$; if so, under what conditions? Prove your answer.

21. Two point charges are fixed at a distance d apart (Fig. 24–23). Plot $E(x)$, assuming that $x = 0$ at the left-hand charge. Consider both positive and negative values of x. Plot E as positive if $\mathbf{E}$ points to the right and negative if $\mathbf{E}$ points to the left. Assume that $q_1 = +1.0 \times 10^{-6}$ C, $q_2 = +3.0 \times 10^{-6}$ C, and $d = 10$ cm.

22. A charge $q = 3.0 \times 10^{-6}$ C is 30 cm from a small dipole along its perpendicular bisector. The magnitude of the force on the charge is 5.0×10^{-6} N. Show on a diagram (a) the direction of

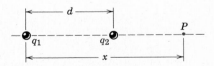

Figure 24–23   Problem 21.

the force on the charge and (b) the direction of the force on the dipole. (c) Determine the magnitude of the force on the dipole.

23. A clock face has negative point charges $-q, -2q, -3q, \ldots, -12q$ fixed at the positions of the corresponding numerals. The clock hands do not perturb the field. At what time does the hour hand point in the same direction as the electric field at the center of the dial? (*Hint:* Consider diametrically opposite charges.)

24. An electron is placed at each corner of an equilateral triangle having sides 20 cm long. (a) What is the electric field at the midpoint of one of the sides? (b) What force would another electron placed there experience?

25. Calculate **E** (direction and magnitude) at point P in Fig. 24–24.

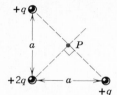

Figure 24–24   Problem 25.

26. An electron moving with a speed of 5.0×10^8 cm/s is shot parallel to an electric field of strength 1.0×10^3 N/C arranged so as to retard its motion. (a) How far will the electron travel in the field before coming (momentarily) to rest, and (b) how much time will elapse? (c) If the electric field ends abruptly after 0.8 cm, what fraction of its initial kinetic energy will the electron lose in traversing it?

27. At some instant the velocity components of an electron moving between two charged parallel plates are $v_x = 1.5 \times 10^5$ m/s and $v_y = 0.30 \times 10^4$ m/s. If the electric field between the plates is given by $\mathbf{E} = (1.2 \times 10^4 \text{ N/C})\mathbf{j}$, (a) what is the acceleration of the electron? (b) When the x-coordinate of the electron has changed by 2.0 cm, what will be the velocity of the electron?

28. Two large parallel copper plates are 5.0 cm apart and have a uniform electric field between them as depicted in Fig. 24–25. An electron is released from the negative plate at the same time that a proton is released from the positive plate. Neglect the forces of the particles on each other and find their distance from the positive plate when they pass each other. Does it surprise you that you don't need to know the electric field to solve this problem? (See Table 23–1 for needed data about the electron and proton.)

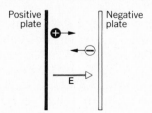

Figure 24–25   Problem 28.

29. A uniform electric field exists in a region between two oppositely charged plates. An electron is released from rest at the surface of the negatively charged plate and strikes the surface of the opposite plate, 2.0 cm away, in a time 1.5×10^{-8} s. (a) What is the speed of the electron as it strikes the second plate? (b) What is the magnitude of the electric field **E**?

30. An electron is constrained to move along the axis of the ring of charge in Example 5. Show that the electron can perform oscillations whose frequency is given by

$$\omega = \left(\frac{eq}{4\pi\epsilon_0 ma^3} \right)^{1/2}.$$

31. A thin nonconducting rod of finite length l carries a total charge q, spread uniformly along it. Show that E at point P on the perpendicular bisector in Fig. 24–26 is given by

$$E = \frac{q}{2\pi\epsilon_0 y} \frac{1}{(l^2 + 4y^2)^{1/2}}.$$

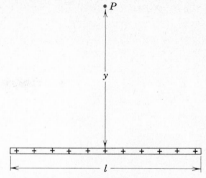

Figure 24–26   Problem 31.

32. An insulating rod of length L has charge $-q$ uniformly distributed along its length, as shown in Fig. 24–27. (a) What is the linear charge density of the rod? (b) What is the electric field at point P a distance a from the end of the rod? (c) If P were very far from the rod compared to L, the rod would look like a point charge. Show that your answer to (b) reduces to the electric field of a point charge for $a \gg L$.

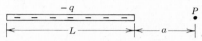

Figure 24–27   Problem 32.

***33.** *Field due to an electric dipole.* Show that the components of **E** due to a dipole are given, at distant points, by

$$E_x = \frac{1}{4\pi\epsilon_0} \frac{3pxy}{(x^2 + y^2)^{5/2}},$$

$$E_y = \frac{1}{4\pi\epsilon_0} \frac{p(2y^2 - x^2)}{(x^2 + y^2)^{5/2}},$$

where x and y are coordinates of a point in Fig. 24–28. Show that this general result includes the special results of Eq. 24–8b and of Problem 9.

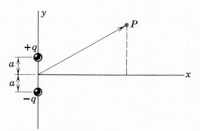

Figure 24–28   Problem 33.

***34.** A "semi-infinite" insulating rod (Fig. 24–29) carries a uniform charge per unit length of λ. Show that the electric field at the point P makes an angle of $45°$ with the rod and that this result is independent of the distance R.

Figure 24–29   Problem 34.

Gauss's Law

25–1 Introduction

In the preceding chapter we saw how we could use Coulomb's law to calculate **E** at various points if we knew enough about the distribution of charges that set up the field. This method always works. It is straight-forward but, except in the simplest cases, laborious. Given a versatile enough computer, however, we can always find the answer to any such problem, no matter how complicated.

We can express Coulomb's law in an another form, called Gauss's law.* If we use this formulation, the calculations are not laborious. However, the number of problems that we can solve by the Gauss law formulation is small. Those that we can solve are solved with grace and elegance but, by and large, the Gauss law formulation is more useful for the insights that it gives rather than for practical problem solving.

Before we discuss Gauss's law we must develop a new concept, that of the flux of a vector field.

25–2 Flux

Flux

Flux (symbol Φ) is a property of all vector fields. We are concerned in this chapter with the flux Φ_E of the electric field **E**. By way of introduction however, let us discuss the more familiar flux of fluid flow Φ_v. The word *flux* is derived from the Latin word *fluere* (to flow).

Figure 25–1 shows a stationary, uniform field of fluid flow (water, say) characterized by a constant flow vector **v**, the constant velocity of the fluid at any given point. Figure 25–1*a* suggests, in cross section, a hypothetical plane surface,

---

* See "Gauss," by Ian Stewart, *Scientific American*, July 1977.

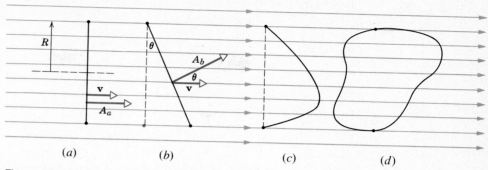

Figure 25-1 Showing four hypothetical surfaces immersed in a stationary, uniform flow field of an incompressible fluid (water, say) characterized by a constant field vector **v**, the velocity of the fluid at any given point. The horizontal lines are stream lines. R, in all four cases, is the radius of a circle at right angles to the stream lines.

Mass flux in fluid flow

a circle of radius R and area $\mathbf{A}_a$, immersed in the flow field at right angles to **v**. Notice that we have represented the surface area as a vector with magnitude A_a and direction perpendicular to the surface. The mass flux $\Phi_{v,a}$ (for example, kg/s) through this surface is given by

$$\Phi_{v,a} = \rho \, v \, A_a \qquad (25\text{-}1a)$$

in which ρ is the fluid density (for example, kg/m³). It represents the mass of fluid passing through the surface per unit time. Check that the dimensions are correct. We may also write this equation in vector notation as

$$\Phi_{v,a} = \rho \mathbf{v} \cdot \mathbf{A}_a. \qquad (25\text{-}1b)$$

Note that flux is a scalar.

Figure 25-1b suggests a plane surface whose projected area ($A_b \cos \theta$) is equal to A_a. It seems clear that the mass flux $\Phi_{v,b}$ through surface b must be the same as that through surface a. To gain some insight we can write

$$\Phi_{v,b} = \Phi_{v,a} = \rho v A_a = \rho v (A_b \cos \theta)$$
$$= \rho \mathbf{v} \cdot \mathbf{A}_b. \qquad (25\text{-}2)$$

Figure 25-1c suggests a curved hypothetical surface whose projected area is also equal to A_a. Once more it seems clear that $\Phi_{v,c} = \Phi_{v,a}$.

Figure 25-1d suggests a closed surface, the preceding three having been open. We assert that the flux $\Phi_{v,d}$ for this closed surface in this flow field is zero and justify it by noting that the amount of fluid that enters the left portion of the surface per unit time also leaves through the right portion. In this case the fluid (assumed incompressible) neither builds up nor disappears within the surface. We say that there happen to be no sources or sinks of fluid within the surface. Every stream line that enters on the left leaves on the right.

After these preliminaries we now turn our attention from Φ_v to Φ_E, the flux of the electric field. It may seem that in the latter case nothing is flowing. In a formal sense, however, Eqs. 25-1b and 25-2 concern themselves only with the (constant in this case) field vector **v**. If, in Fig. 25-1 we replace **v** by **E** and regard the stream lines as lines of force, all the discussion of this section remains true.

In this chapter we are most interested in closed surfaces immersed in the **E** field. This is because we are concerned here with Gauss's law, which is expressed only in terms of closed surfaces.

In the flow of incompressible fluids it is not true in general that, as in the special case of Fig. 25–1d, $\Phi_v = 0$ for *all* closed surfaces. There may be sources where fluid enters the system or sinks where fluid leaves within the surface. In such cases $\Phi_v \neq 0$.

In the same way it is not true that $\Phi_E = 0$ for every closed surface. There are sources (positive charges; here $\Phi_E > 0$) and sinks (negative charges; here $\Phi_E < 0$) of **E** that may be located within the hypothetical closed surface immersed in the **E** field.

25–3 Flux of the Electric Field

For closed surfaces in an electric field we shall see below that Φ_E is positive if the lines of force point outward everywhere and negative if they point inward. Figure 25–2 shows two equal and opposite charges and their lines of force. Curves S_1, S_2, S_3, and S_4 are the intersections with the plane of the figure of four hypothetical closed surfaces. From the statement just given, Φ_E is positive for surface S_1 and negative for S_2. Φ_E for surface S_3 (compare Fig. 25–1d) is zero. We shall discuss Φ_E for surface S_4 in Section 25–4. The flux of the electric field is important because Gauss's law, one of the four basic equations of electromagnetism (see Table 37–2), is expressed in terms of it.

To define Φ_E precisely, consider Fig. 25–3, which shows an arbitrary closed surface immersed in a nonuniform electric field. Let the surface be divided into elementary squares ΔS, each of which is small enough so that it may be consid-

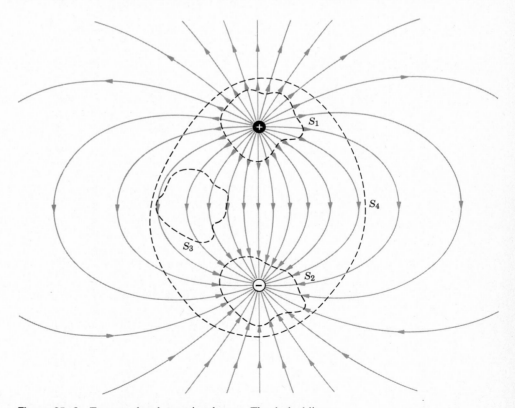

Figure 25–2 Two equal and opposite charges. The dashed lines represent the intersection with the plane of the figure of hypothetical closed surfaces.

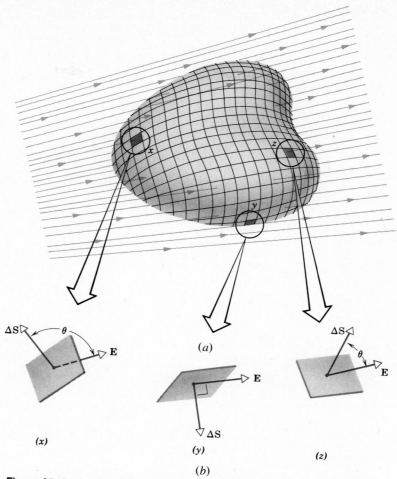

Figure 25–3 (a) A hypothetical surface immersed in a nonuniform electric field. (b) Three elements of area on this surface. shown enlarged.

ered to be plane. Such an element of area can be represented as a vector $\Delta\mathbf{S}$, whose magnitude is the area ΔS; the direction of $\Delta\mathbf{S}$ is taken as the outward-drawn normal to the surface.

At every square in Fig. 25–3 we can also construct an electric field vector $\mathbf{E}$. Since the squares have been taken to be arbitrarily small, $\mathbf{E}$ may be taken as constant for all points in a given square.

The vectors $\mathbf{E}$ and $\Delta\mathbf{S}$ that characterize each square make an angle θ with each other. Figure 25–3b shows an enlarged view of the three squares on the surface of Fig. 25–3a marked x, y, and z. Note that at x, $\theta > 90°$ ($\mathbf{E}$ points in); at y, $\theta = 90°$ ($\mathbf{E}$ is parallel to the surface); and at z, $\theta < 90°$ ($\mathbf{E}$ points out).

An approximate definition of flux is

$$\Phi_E \cong \Sigma\mathbf{E} \cdot \Delta\mathbf{S}, \tag{25–3}$$

which instructs us to add up the scalar quantity $\mathbf{E} \cdot \Delta\mathbf{S}$ for all elements of area into which the surface has been divided. For points such as x in Fig. 25–3 the contribution to the flux is negative; at y it is zero and at z it is positive. Thus if $\mathbf{E}$ is everywhere outward, $\theta < 90°$, $\mathbf{E} \cdot \Delta\mathbf{S}$ will be positive, and Φ_E for the entire surface will be positive; see Fig. 25–2, surface S_1. If $\mathbf{E}$ is everywhere inward, $\theta > 90°$, $\mathbf{E} \cdot \Delta\mathbf{S}$

will be negative, and Φ_E for the surface will be negative; see Fig. 25–2, surface S_2. From Eq. 25–3 we see that the appropriate simplest-form SI unit for Φ_E is the newton meter²/coulomb (N·m²/C).

The exact definition of electric flux is found in the differential limit of Eq. 25–3. Replacing the sum over the surface by an integral over the surface yields

Flux of the electric field

$$\Phi_E = \int \mathbf{E} \cdot d\mathbf{S}. \qquad (25-4)$$

This surface integral indicates that the surface in question is to be divided into infinitesimal elements of area $d\mathbf{S}$ and that the scalar quantity $\mathbf{E} \cdot d\mathbf{S}$ is to be evaluated for each element and the sum taken for the entire surface.

Flux may be evaluated in this way for any surface in the field whether it is open or closed. When the surface is closed, this fact is sometimes emphasized by placing a circle on the integral sign, as in, for example, Eq. 25–6 or the first equation in Example 1.

Example 1   Figure 25–4 shows a hypothetical cylinder of radius R immersed in a uniform electric field $\mathbf{E}$, the cylinder axis being parallel to the field. What is Φ_E for this closed surface?

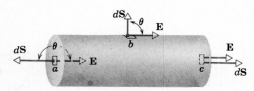

Figure 25–4   Example 1. A cylindrical surface immersed in a uniform field $\mathbf{E}$ parallel to its axis. The lines of force, which are uniformly spaced and lie parallel to the axis, are not shown.

The flux Φ_E can be written as the sum of three terms, an integral over (a) the left cylinder cap, (b) the cylindrical surface, and (c) the right cap. Thus

$$\Phi_E = \oint \mathbf{E} \cdot d\mathbf{S}$$
$$= \int_{(a)} \mathbf{E} \cdot d\mathbf{S} + \int_{(b)} \mathbf{E} \cdot d\mathbf{S} + \int_{(c)} \mathbf{E} \cdot d\mathbf{S}.$$

For the left cap the angle θ for all points is 180°, $\mathbf{E}$ has a constant value, and the vectors $d\mathbf{S}$ are all parallel. Thus

$$\mathbf{E} \cdot d\mathbf{S} = E \cos 180° \, dS$$
$$= -E \, dS$$

for all elements dS of surface a. Thus

$$\int_{(a)} \mathbf{E} \cdot d\mathbf{S} = -ES$$

where $S \, (= \pi R^2)$ is the cap area. Similarly, for the right cap,

$$\int_{(c)} \mathbf{E} \cdot d\mathbf{S} = +ES,$$

the angle θ for all points being zero here. Finally, for the cylinder wall,

$$\int_{(b)} \mathbf{E} \cdot d\mathbf{S} = 0.$$

because $\theta = 90°$, hence $\mathbf{E} \cdot d\mathbf{S} = 0$ for all points on the cylindrical surface. Thus

$$\Phi_E = -ES + 0 + ES = 0.$$

As we shall see in Section 25–4 we expect this because there are no sources or sinks of $\mathbf{E}$, that is, charges, within the closed surface of Fig. 25–4. Lines of (constant) $\mathbf{E}$ enter at the left and emerge at the right, just as in Fig. 25–1d.

25–4   Gauss's Law

Gauss's law, which applies to any closed hypothetical surface (called a Gaussian surface), gives a connection between Φ_E for the surface and the net charge q enclosed by the surface. It is

$$\epsilon_0 \Phi_E = q \qquad (25-5)$$

or, using Eq. 25–4,

Gauss's law

$$\epsilon_0 \oint \mathbf{E} \cdot d\mathbf{S} = q. \qquad (25-6)$$

The fact that Φ_E proves to be zero in Example 1 is predicted by Gauss's law because no charge is enclosed by the Gaussian surface in Fig. 25–4 ($q = 0$).

Note that q in Eq. 25–5 (or in Eq. 25–6) is the *net* charge, taking its algebraic sign into account. If a surface encloses equal and opposite charges, the flux Φ_E is zero. Charge outside the surface makes no contribution to the value of q, nor does the exact location of the inside charges affect this value.

Gauss's law can be used to evaluate **E** only if the charge distribution is so symmetric that by proper choice of a Gaussian surface we can easily evaluate the integral in Eq. 25–6. Conversely, if **E** is known for all points on a given closed surface, Gauss's law can be used to compute the charge inside. If **E** has an outward component for every point on a closed surface, Φ_E, as Eq. 25–4 shows, will be positive and, from Eq. 25–6, there must be a net positive charge within the surface (see Fig. 25–2, surface S_1). If **E** has an inward component for every point on a closed surface, there must be a net negative charge within the surface (see Fig. 25–2, surface S_2). Surface S_3 in Fig. 25–2 encloses no charge, so that Gauss's law predicts that $\Phi_E = 0$. This is consistent with the fact that lines of **E** pass directly through surface S_3, the contribution to the integral on one side canceling that on the other. For surface S_4 in Fig. 25–2 $\Phi_E = 0$ because the algebraic sum of the charges within the surface is zero. Put another way, as for surface S_3, as many lines of force leave the surface as enter it.

25–5   Gauss's Law and Coulomb's Law

Coulomb's law can be deduced from Gauss's law and symmetry considerations. To do so, let us apply Gauss's law to an isolated point charge q_1 as in Fig. 25–5. Although Gauss's law holds for any surface whatever, information can most readily be extracted for a spherical surface of radius r centered on the charge. The advantage of this surface is that, from symmetry, **E** must be normal to it and must have the same (as yet unknown) magnitude for all points on the surface. (See Question 11.)

In Fig. 25–5 both **E** and $d\mathbf{S}$ at any point on the Gaussian surface are directed radially outward. The angle between them is zero and the quantity $\mathbf{E} \cdot d\mathbf{S}$ becomes simply $E\,dS$. Gauss's law (Eq. 25–6) thus reduces to

$$\epsilon_0 \oint \mathbf{E} \cdot d\mathbf{S} = \epsilon_0 \oint E\,dS = q_1.$$

Because E is constant for all points on the sphere, it can be factored from inside the integral sign, leaving

$$\epsilon_0 E \oint dS = q_1,$$

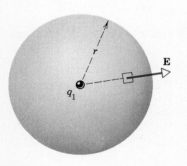

Figure 25–5   A spherical Gaussian surface of radius r surrounding a point charge q_1.

where the integral is simply the area of the sphere. This equation gives

$$\epsilon_0 E(4\pi r^2) = q_1$$

or

$$E = \frac{1}{4\pi\epsilon_0}\frac{q_1}{r^2}. \qquad (25\text{–}7)$$

Equation 25–7 gives the magnitude of the electric field **E** at any point a distance r from an isolated point charge q_1; compare Eq. 24–4. The direction of **E** is already known from symmetry.

Let us put a second point charge q_2 at the point at which **E** is calculated. The magnitude of the force that acts on it (see Eq. 24–2) is

$$F = Eq_2.$$

Combining with Eq. 25–7 gives

$$F = \frac{1}{4\pi\epsilon_0}\frac{q_1\,q_2}{r^2},$$ [23–3]

which is precisely Coulomb's law. Thus we have deduced Coulomb's law from Gauss's law. Note how vital symmetry arguments are in this derivation.

Gauss's law is one of the fundamental equations of electromagnetic theory and is displayed in Table 37–2 as one of Maxwell's equations, which summarize electromagnetic theory in its most elegant and simple form. Coulomb's law is not listed in that table because, as we have just proved, it can be deduced from Gauss's law and from simple assumptions about the symmetry of **E** due to a point charge. The usefulness of Gauss's law for determining **E** depends on our ability to find a surface over which, from symmetry, both E and θ (see Fig. 25–5) have constant values. Then $E \cos \theta$ can be factored out of the integral and E can be found simply, as in this example.

25–6   An Insulated Conductor

Gauss's law can be used to make an important prediction, namely: *An excess charge on an insulated conductor is, in equilibrium, entirely on its outer surface.* This hypothesis was shown to be true by experiment (see Section 25–7) before either Gauss's law or Coulomb's law was advanced. Indeed, the experimental proof of the hypothesis is the experimental foundation upon which both laws rest: We have already pointed out that Coulomb's torsion balance experiments, although direct and convincing, are not capable of great accuracy. In showing that the italicized hypothesis is predicted by Gauss's law, we are simply reversing the historical situation.

Figure 25–6 is a cross section of an insulated conductor of arbitrary shape carrying an excess charge q. The dashed lines show a Gaussian surface that lies a small distance below the actual surface of the conductor. Although the Gaussian surface can be as close to the actual surface as we wish, it is important to keep in mind that the Gaussian surface is inside the conductor.

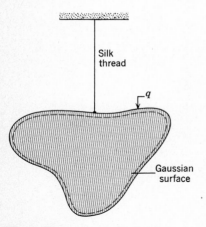

Figure 25–6   An insulated metallic conductor.

An initially uncharged insulated conductor must have **E** = 0 for all internal points. If this were not so, electrical forces would act on the charge carriers, producing internal currents. There are, in fact, no such currents in, say, a coin resting on a table top. Electrostatic conditions prevail. We must also have **E** = 0 for all internal points even if we put a charge on the conductor. It is true that random electric fields will appear inside a conductor at the instant that a charge is placed on it. However, these fields will not last long. Internal currents will act to redistribute the added charge until the electric fields vanish and we again have electrostatic conditions. What can we say about the distribution of the excess charge when such electrostatic conditions have been achieved?

If, at electrostatic equilibrium, **E** is zero everywhere inside the conductor, it must be zero for every point on the Gaussian surface. This means that the flux Φ_E for this surface must be zero. Gauss's law then predicts (see Eq. 25–5) that there must be no net charge inside the Gaussian surface. If the excess charge q is not inside this surface, it can only be outside it, that is, it must be on the actual surface of the conductor.

25–7 Experimental Proof of Gauss's and Coulomb's Laws

Let us turn to the experiments that prove that the hypothesis of Section 25–6 is true. For a simple test, charge a metal ball and lower it with a silk thread deep into a metal can as in Fig. 25–7. Touch the ball to the inside of the can; when the ball is removed from the can, all its charge will have vanished. When the metal ball touches the can, the ball and can together form an "insulated conductor" to which the hypothesis of Section 25–6 applies. That the charge moves entirely to the outside surface of the can can be shown by touching a small insulated metal object to the can; only on the outside of the can will it be possible to pick up a charge.

Benjamin Franklin* seems to have been the first to notice, in about 1755, that there can be no charge inside an insulated metal can. He recommended this "singular fact" to the attention of Joseph Priestley (1733–1804) who checked Franklin's observation and realized that the inverse square law followed from it. Many others, including Coulomb, continued to test the hypothesis. Modern experiments have confirmed Priestley's hypothesis to remarkable accuracy.

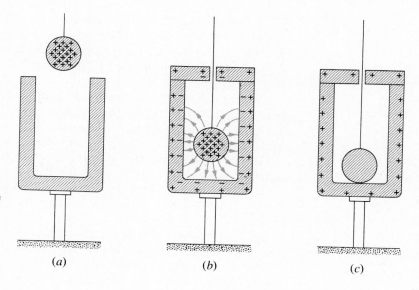

Figure 25–7 The entire charge of the ball is transferred to the outside surface of the can. After the ball is inserted a conducting lid is put in place, as shown in (b) and (c). If this is not done the "outside surface of the can" cannot be clearly defined.

(a)   (b)   (c)

For many reasons it is important to know whether or not the exponent in Coulomb's law is exactly "2"; experiments based on Gauss's law can help to determine this. Let us write Coulomb's law as

$$F = \frac{1}{4\pi\epsilon_0} \frac{q_1 q_2}{r^{2+\delta}} \qquad (25\text{–}8)$$

in which $\delta = 0$ yields an exact inverse square law. Table 25–1 shows the progress made in determining how close δ in Eq. 25–8 approaches zero.

---

* See "In Defense of Benjamin Franklin," by I. Bernard Cohen, *Scientific American*, August 1948.

Table 25-1 Test of Coulomb's inverse square law[a]

Experimenters	Date	δ (Eq. 25-8)
Benjamin Franklin	1755	—
Joseph Priestley	1767	". . . according to the squares of the distance . . ."
John Robison	1769	≤ 0.06
Henry Cavendish	1773	≤ 0.02
Charles A. Coulomb	1785	a few percent at most
James Clerk Maxwell	1873	$\leq 5 \times 10^{-5}$
Samuel J. Plimpton and Willard E. Lawton	1936	$\leq 2 \times 10^{-9}$
Edwin R. Williams, James E. Faller, and Henry A. Hill	1971	$\leq 3 \times 10^{-16}$

[a] Robison's and Coulomb's results were direct tests of the inverse square law; all others were "Gauss's law" experiments, in the spirit of Fig. 25-7.

A sensitive test of the inverse square law

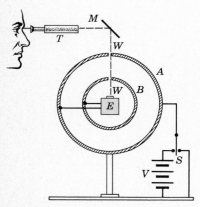

Figure 25-8 A representation of the apparatus of Plimpton and Lawton.

Figure 25-8 is an idealized sketch of the apparatus of Plimpton and Lawton. It consists in principle of two concentric metal shells, A and B, the former being 5 ft in diameter. The inner shell contains a sensitive electrometer E connected so that it will indicate whether any charge moves between shells A and B. If the shells are connected electrically, any charge placed on the shell assembly should reside entirely on shell A (see Section 25-6) if Gauss's law—and thus Coulomb's law—are correct as stated.

By throwing switch S to the left, a substantial charge can be placed on the sphere assembly. If any of this charge moves to shell B, it will have to pass through the electrometer and will cause a deflection, which can be observed optically using telescope T, mirror M, and windows W.

However, when the switch S is thrown alternately from left to right, thus connecting the shell assembly either to the battery or to the ground, no effect is observed. Knowing the sensitivity of their electrometer, Plimpton and Lawton calculated that δ in Eq. 25-8 has the upper limit shown in Table 25-1.

25-8 Gauss's Law—Some Applications

Gauss's law can be used to calculate **E** if the symmetry of the charge distribution is high. One example of this, the calculation of **E** for a point charge, has already been discussed (Eq. 25-7). Here we present other examples.

Example 2 *Spherically symmetric charge distribution* Figure 25-9 shows a spherical distribution of charge of radius R. The *charge density* ρ (that is, the charge per unit volume) at any point depends only on the distance of the point from the center and not on the direction, a condition called spherical symmetry. Find an expression for E for points (a) outside and (b) inside the charge distribution. Note that the object in Fig. 25-9 cannot be a conductor or, as we have seen, the excess charge would reside on its surface.

Applying Gauss's law to a spherical Gaussian surface of radius r in Fig. 25-9a (see Section 25-5) leads exactly to Eq. 25-7, or

$$E = \frac{1}{4\pi\epsilon_0} \frac{q}{r^2}, \qquad [25-7]$$

where q is the total charge. Thus for points outside a spherically symmetric distribution of charge, the electric field has the value that it would have if the charge were concentrated at its center.

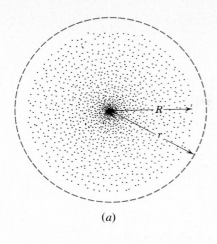

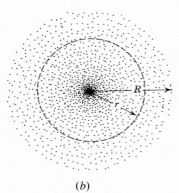

(a)

(b)

Figure 25–9 Example 2. A spherically symmetric charge distribution, showing two Gaussian surfaces. The density of charge, as the dots suggests, varies with distance from the center but not with direction.

This reminds us that a spherically symmetric distribution of mass m behaves gravitationally, for outside points, as if the mass were concentrated at its center. At the root of this similarity lies the fact that both Coulomb's law and the law of gravitation are inverse square laws. The gravitational case was proved in detail in Section 15–4; the proof using Gauss's law in the electrostatic case is certainly much simpler.

Figure 25–9b shows a spherical Gaussian surface of radius r drawn inside the charge distribution. Gauss's law (Eq. 25–6) gives

$$\epsilon_0 \oint \mathbf{E} \cdot d\mathbf{S} = \epsilon_0 E(4\pi r^2) = q'$$

or

$$E = \frac{1}{4\pi\epsilon_0} \frac{q'}{r^2},$$

in which q' is that part of q contained within the sphere of radius r. The part of q that lies outside this sphere makes no contribution to $\mathbf{E}$ at radius r. This corresponds, in the gravitational case (Section 15–4), to the fact that a spherically symmetric shell of matter exerts no gravitational force on a body inside it.

A special case of a spherically symmetric charge distribution is a uniform sphere of charge. For such a sphere, which

would be suggested by a uniform density of dots in Fig. 25–9, the charge density ρ would have a constant value for all points within a sphere of radius R and would be zero for all points outside this sphere. For points inside such a uniform sphere of charge we can put

$$q' = q \frac{\frac{4}{3}\pi r^3}{\frac{4}{3}\pi R^3}$$

or

$$q' = q \left(\frac{r}{R}\right)^3$$

where $\frac{4}{3}\pi R^3$ is the volume of the spherical charge distribution. The expression for E then becomes

$$E = \frac{1}{4\pi\epsilon_0} \frac{qr}{R^3}. \qquad (25\text{–}9)$$

This equation becomes zero, as it should, for $r = 0$. Note that Eqs. 25–7 and 25–9 give the same result, as they must, for points on the surface of the charge distribution (that is, if $r = R$). Note that Eq. 25–9 does not apply to the charge distribution of Fig. 25–9b because the charge density, suggested by the density of dots, is not constant in that case.

Figure 25–10 shows $E(r)$ for a uniform sphere of charge, using Eq. 25–7 for points outside the sphere and Eq. 25–9 for internal points. The "sphere" actually represents the nucleus of a gold atom, with $R = 6.2 \times 10^{-15}$ m and $q = Ze = (79)(1.6 \times 10^{-19}$ C$) = 1.3 \times 10^{-17}$ C.

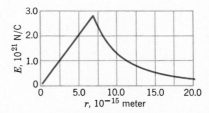

Figure 25–10 Example 2. The electric field both inside and outside a gold nucleus, due to its positive charge ($= +79e$).

Example 3 *An infinite line of charge* Figure 25–11 shows a section of an infinite line of charge, the linear charge density λ (that is, the charge per unit length) being constant for all points on the line. Find an expression for E at a distance r from the line.

From symmetry, $\mathbf{E}$ due to a uniform linear charge can only be radially directed. (See Question 21.) As a Gaussian surface we choose a circular cylinder of radius r and length h, closed at each end by plane caps normal to the axis. E is constant over the cylindrical surface and the flux of $\mathbf{E}$ through this surface is $E(2\pi rh)$ where $2\pi rh$ is the area of the surface. There is no flux through the circular caps because $\mathbf{E}$ here lies in the surface at every point.

The charge enclosed by the Gaussian surface of Fig. 25–11 is λh. Gauss's law (Eq. 25–6),

$$\epsilon_0 \oint \mathbf{E} \cdot d\mathbf{S} = q,$$

then becomes

$$\epsilon_0 E(2\pi rh) = \lambda h,$$

whence

$$E = \frac{\lambda}{2\pi\epsilon_0 r}. \qquad (25-10)$$

The direction of **E** is radially outward for a line of positive charge.

Note that the solution using Gauss's law is possible only if we choose our Gaussian surface to take full advantage of the cylindrical symmetry of the electric field set up by a long line of charge. We are free to choose any surface, such as a cube or a sphere, for a Gaussian surface. Even though Gauss's law holds for all such surfaces, they are not all useful for the problem at hand; only the cylindrical surface of Fig. 25–11 is appropriate in this case.

Gauss's law has the property that it provides a useful technique for calculation only in problems that have a certain degree of symmetry, but in these problems the solutions are strikingly simple.

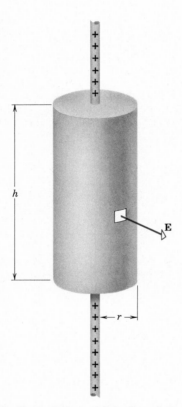

Figure 25–11 Example 3. An infinite line of charge, showing a closed, coaxial cylindrical Gaussian surface.

Example 4 *An infinite sheet of charge* Figure 25–12 shows a portion of a thin, infinite sheet of charge, the surface charge density σ (that is, the charge per unit area) being constant. A sheet of Saran wrap, uniformly charged by friction, can serve as a prototype. What is **E** at a distance r in front of the sheet?

A convenient Gaussian surface is a closed cylinder of cross-sectional area A and height $2r$, arranged to pierce the plane as shown. From symmetry, **E** points at right angles to the end caps and away from the plane. (See Question 22.) Since **E**

does not pierce the cylindrical surface, there is no contribution to the flux from this source. Thus Gauss's law,

$$\epsilon_0 \oint \mathbf{E} \cdot d\mathbf{S} = q$$

becomes

$$\epsilon_0(EA + EA) = \sigma A$$

where σA is the enclosed charge. This gives

$$E = \frac{\sigma}{2\epsilon_0}. \qquad (25-11)$$

Note that E is the same for all points on each side of the plane; compare Fig. 24–2. Although an infinite sheet of charge cannot exist physically, this derivation is still useful in that Eq. 25–11 yields substantially correct results for real (not infinite) charge sheets if we consider only points not near the edges whose distance from the sheet is small compared to the dimensions of the sheet.

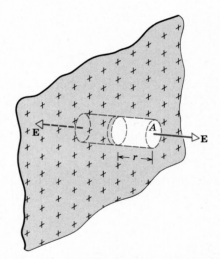

Figure 25–12 Example 4. An infinite sheet of charge pierced by a cylindrical Gaussian surface.

Example 5 *A charged conductor* Figure 25–13 shows a conductor carrying on its surface a charge whose surface charge density at any point is σ; in general σ will vary from point to point. What is **E** for points a short distance above the surface?

The direction of **E** for points close to the surface is at right angles to the surface, pointing away from the surface if the charge is positive. If **E** were not normal to the surface, it would have a component lying in the surface. Such a component would act on the charge carriers in the conductor and set up surface currents. Since there are no such currents under the assumed electrostatic conditions, **E** must be normal to the surface.

The magnitude of **E** can be found from Gauss's law using a small flat closed cylinder of cross section A as a Gaussian surface. Since **E** equals zero everywhere inside the conductor (see Section 25–6), the only contribution to Φ_E is through the plane cap of area A that lies outside the conductor. Gauss's law

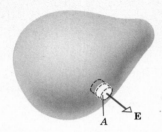

Figure 25–13 Example 5. A charged insulated conductor, showing a Gaussian surface.

$$\epsilon_0 \oint \mathbf{E} \cdot d\mathbf{S} = q$$

becomes

$$\epsilon_0(EA) = \sigma A$$

where σA is the net charge within the Gaussian surface. This yields

$$E = \frac{\sigma}{\epsilon_0}. \qquad (25-12)$$

Comparison with Eq. 25–11 shows that the electric field is twice as great near a conductor carrying a charge whose surface charge density is σ as that near an infinite sheet with the same charge density. To clarify this apparent paradox let us realize that we are comparing charge arrangements that have different geometries. Equation 25–11 (and Fig. 25–12) refer to a thin sheet of infinite extent; Eq. 25–12 (and Fig. 25–13) refer to a conductor of arbitrary shape. Let us therefore examine a conductor that is also in the form of a thin sheet of infinite extent. Example 6 clarifies the comparison.

Example 6 *A charged conducting sheet* An infinite conducting plate is charged with surface charge density σ on *both* surfaces (Figure 25–14*a*). What is the value of **E** both inside and outside the conductor? Solve the problem first using the results of Example 5 and then using Example 4.

Using the results of Example 5, E inside the plate is zero whereas that outside the plate points directly away from the plate (for a positive surface charge) and, as Eq. 25–12 shows, has magnitude σ/ϵ_0.

Viewed from the perspective of Example 4, the charges causing the field may be considered to be due to two sheets of charge, one on each surface of the plate, as shown in Figure 25–14*b*. Each of these sheets can be treated as though it were the single charged sheet of Fig. 25–12. The fields caused by the individual sheets of charge, as Eq. 25–11 shows, have magnitude $\sigma/2\epsilon_0$ and point directly away from each sheet, as indicated in Figures 25–14*c* and *d*. The total field, from superposition, is the sum of the two fields caused by the two charge sheets. This is shown in Figure 25–14*e*. These add to zero between the sheets, since the fields are in opposite directions in this region. They supplement each other outside the plate, since here they are in the same direction, and result in a total field whose magnitude is σ/ϵ_0, in agreement with the calculation using Example 5.

Figure 25–14 Example 6. (*a*) An infinite conducting plate with charge density σ on each surface. (*b*) The surface charges represented as infinite sheets of charge, each one resembling the charge sheet of Fig. 25–12. (*c*) The field due to the left-hand sheet alone. (*d*) The field due to the right-hand sheet alone. (*e*) The superposition of the fields in (*c*) and (*d*).

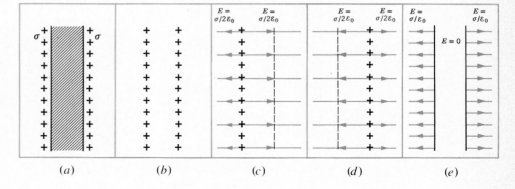

$(a) \qquad (b) \qquad (c) \qquad (d) \qquad (e)$

REVIEW GUIDE AND SUMMARY

Gauss's law and Coulomb's law, although expressed in different form, are equivalent ways of describing the relationship between charge and the electric field. Gauss's law is

Gauss's law

$$\epsilon_0 \Phi_E = q \qquad [25-5]$$

in which Φ_E is the flux of the electric field through any imaginary closed surface that may be drawn in this field. The charge q above is the charge inside this Gaussian surface only; charge outside does not contribute to the flux and is not counted in applications of Eq. 25–5.

The flux Φ_E in Eq. 25–5 is calculated from

The flux of the electric field

$$\Phi_E = \oint \mathbf{E} \cdot d\mathbf{S}. \qquad [25\text{–}4]$$

The discussion related to Fig. 25–3 shows how this calculation is to be carried out in the general case and Example 1 examines a simple special case. Figure 25–2 illustrates qualitatively four simple cases of the application of Gauss's law.

Coulomb's law can readily be derived from Gauss's law, the details being given in Section 25–5. The careful experimental verification of Gauss's law—and thus of the equivalent Coulomb's law—is described in Section 25–7. As Table 25–1 shows, the exponent of r in Coulomb's law is now known to be exactly "2" within an experimental uncertainty of less than 3×10^{-16}.

Coulomb's law and Gauss's law

Using Gauss's law and, in some cases, symmetry arguments, several important results may be derived. Among these are:

Charge on a conductor

a. An excess charge on an insulated conductor is, in equilibrium, entirely on its outer surface. (Section 25–6)

b. The electric field outside any spherically symmetrical distribution of charge is directed radially and has magnitude

Field outside a spherical charge

$$E = \frac{1}{4\pi\epsilon_0}\frac{q}{r^2}, \qquad [25\text{–}7]$$

in which q is the total charge.

c. The electric field inside any spherically symmetrical distribution of charge is directed radially and has magnitude

Field inside a spherical charge

$$E = \frac{1}{4\pi\epsilon_0}\frac{q'}{r^2}$$

in which q' is that part of q contained within the sphere of radius r centered at the center of symmetry.

d. The electric field inside a uniform sphere of charge is directed radially and has magnitude

Field inside a uniform spherical charge

$$E = \frac{1}{4\pi\epsilon_0}\frac{qr}{R^3} \qquad [25\text{–}9]$$

e. The electric field due to an infinite line of charge is in a direction perpendicular to the line and has magnitude

Field due to an infinite line of charge

$$E = \frac{\lambda}{2\pi\epsilon_0 r} \qquad [25\text{–}10]$$

f. The electric field due to an infinite sheet of charge is perpendicular to the plane of the sheet and has magnitude

Field due to an infinite sheet of charge

$$E = \frac{\sigma}{2\epsilon_0} \qquad [25\text{–}11]$$

g. The electric field near the surface of a charged conductor whose charges are in equilibrium is perpendicular to the surface and has magnitude

Field near a charged conductor

$$E = \frac{\sigma}{\epsilon_0} \qquad [25\text{–}12]$$

QUESTIONS

1. By analogy with Φ_E, how would you define the flux Φ_g of a gravitational field? What is the flux of the earth's gravitational field through the boundaries of a room, assumed to contain no matter?

2. A point charge is placed at the center of a spherical Gaussian surface. Is Φ_E changed (*a*) if the surface is replaced by a cube of the same volume, (*b*) if the sphere is replaced by a cube of one-tenth the volume, (*c*) if the charge is moved off-center in the original sphere, still remaining inside, (*d*) if the charge is moved just outside the original sphere, (*e*) if a second charge is placed near, and outside, the original sphere, and (*f*) if a second charge is placed inside the Gaussian surface?

3. In Gauss's law,

$$\epsilon_0 \oint \mathbf{E} \cdot d\mathbf{S} = q,$$

is **E** the electric field attributable to the charge *q*?

4. A surface encloses an electric dipole. What can you say about Φ_E for this surface?

5. Suppose that a Gaussian surface encloses no net charge. Does Gauss's law require that **E** equal zero for all points on the surface? Is the converse of this statement true, that is, if **E** equals zero everywhere on the surface, does Gauss's law require that there be no net charge inside?

6. Is Gauss's law useful in calculating the field due to three equal charges located at the corners of an equilateral triangle? Explain.

7. Is **E** necessarily zero inside a charged rubber balloon if the balloon is (*a*) spherical or (*b*) sausage-shaped? For each shape assume the charge to be distributed uniformly over the surface.

8. A spherical rubber balloon carries a charge that is uniformly distributed over its surface. As the balloon is blown up, how does *E* vary for points (*a*) inside the balloon, (*b*) at the surface of the balloon, and (*c*) outside the balloon?

9. We have seen that there are sources (positive charges) and sinks (negative charges) for the **E** field. Are there sources and/or sinks for the gravitational field **g**?

10. In Section 25–5 we have seen that Coulomb's law can be derived from Gauss's law. Does this necessarily mean that Gauss's law can be derived from Coulomb's law?

11. Explain why the spherical symmetry of Fig. 25–5 restricts us to a consideration of **E** that has only a radial component at any point. (Hint: Imagine other components, perhaps along the equivalent of longitude or latitude lines on the earth's surface. Spherical symmetry requires that these look the same from any perspective. Can you invent such field lines which satisfy this criterion?)

12. A large, insulated, hollow conductor carries a positive charge. A small metal ball carrying a negative charge of the same magnitude is lowered by a thread through a small opening in the top of the conductor, allowed to touch the inner surface, and then withdrawn. What is then the charge on (*a*) the conductor and (*b*) the ball?

13. Can we deduce from the argument of Section 25–6 that the electrons in the wires of a house wiring system move along the surfaces of those wires. If not, why not?

14. Does Gauss's law, as applied in Section 25–6, require that all the conduction electrons in an insulated conductor reside on the surface?

15. Suppose that you have a Gaussian surface in the shape of a donut and that there is a single point charge inside. Does Gauss's law hold? If not, why not? If so, is there enough symmetry in the situation to apply it usefully?

16. Does Gauss's law hold for the flow of water? Consider various Gaussian surfaces intersecting a fountain or a waterfall in different ways.

17. Two hollow metal spheres are identical except that one carries a charge and the other does not. Would they float at the same or at different levels in an oil bath?

18. A positive point charge *q* is located at the center of a hollow metal sphere. What charges appear on (*a*) the inner surface and (*b*) the outer surface of the sphere? (*c*) If you bring an (uncharged) metal object near the sphere, will it change your answer in (*a*) or (*b*) above? Will it change the way charge is distributed over the sphere?

19. As you penetrate a uniform sphere of charge, *E* should decrease because less charge lies inside a sphere drawn through the observation point. On the other hand, *E* should increase because you are closer to the center of this charge. Which effect predominates and why?

20. Given a spherically symmetric charge distribution (not of uniform density of charge radially), is *E* necessarily a maximum at the surface? Comment on various possibilities.

21. Explain why the symmetry of Fig. 25–11 restricts us to a consideration of **E** which has only a radial component at any point. Remember, in this case, that the field must not only look the same at any point along the line but that the field must also look the same if the figure is turned end for end.

22. Explain why the symmetry of Fig. 25–12 restricts us to a consideration of **E** which has only a component directed away from the sheet. Why, for example, could **E** not have components parallel to the sheet? Remember, in this case, that the field must not only look the same at any point along the sheet in any direction, but the field must also look the same if the sheet is rotated about any line perpendicular to the sheet.

EXERCISES

Section 25–2   Flux

1. Water in an irrigation ditch of width *w* = 3.22 m and depth *d* = 1.04 m flows with speed 0.207 m/s. Find the mass flux through the following surfaces: (*a*) a surface of area *wd*, entirely in the water, perpendicular to the flow; (*b*) a surface with area 3*wd*/2, of which *wd* is in the water, perpendicular to the flow; (*c*) a surface of area *wd*/2, entirely in the water, perpendicular to the flow; (*d*) a surface of area *wd*, half in the water and half out, perpendicular to the flow; (*e*) a surface of area *wd*, entirely in the water, with its normal 60° from the direction of flow.

Section 25–3   Flux of the Electric Field

2. Calculate Φ_E through a hemisphere of radius *R*. The field **E** is uniform and is parallel to the axis of the hemisphere.

3. A butterfly net is in a uniform electric field as shown in Fig.

25–15. The rim, a circle of radius a, is aligned perpendicular to the field. Find the electric flux through the netting.

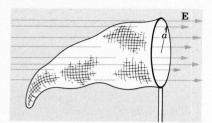

Figure 25–15   Exercise 3.

4.   A cube with 1.35-m edges is oriented as shown in Fig. 25–16 in a region of uniform electric field. Find the electric flux through the right face if the electric field in newtons per coulomb is given by (a) $6\mathbf{i}$; (b) $-2\mathbf{j}$; (c) $-3\mathbf{i} + 4\mathbf{k}$. (d) What is the total flux through the cube for each of these fields?

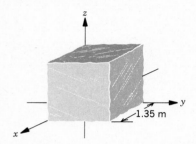

Figure 25–16   Exercise 4 and Problem 19.

5.   A point charge $+q$ is a distance $d/2$ from a square surface of side d and is directly above the center of the square as shown in Fig. 25–17. What is the electric flux through the square? (*Hint:* Think of the square as one face of a cube with edge d.)

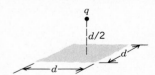

Figure 25–17   Exercise 5.

Section 25–4   Gauss's Law
6.   Consider a Gaussian surface that surrounds part of the charge distribution shown in Fig. 25–18. (a) Which of the charges contribute to the electric field at point P? (b) Would the value obtained for the flux through the surface, calculated using only the electric field due to q_1 and q_2, be greater than, equal to, or less than that obtained using the total field?

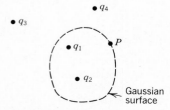

Figure 25–18   Exercise 6.

7.   Four charges, $2q$, q, $-2q$, and $-q$, are arranged at the corners of a square as shown in Fig. 25–19. If possible, describe a closed surface that encloses the charge $+2q$ and through which the net electric flux is (a) 0; (b) $+3q/\epsilon_0$; (c) $-2q/\epsilon_0$.

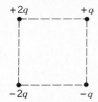

Figure 25–19   Exercise 7.

8.   Charge on an originally uncharged insulated conductor is separated by holding a positively charged rod very closely nearby, as in Fig. 25–20. What can you learn from Gauss's law about the flux for the five Gaussian surfaces shown? Assume that the induced negative charge on the conductor is equal to the positive charge q on the rod.

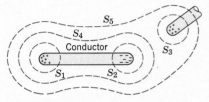

Figure 25–20   Exercise 8.

9.   The intensity of the earth's electric field near its surface is ~ 130 N/C, pointing down. What is the earth's net charge, assuming that this field is caused by it?

10.   A uniformly charged conducting sphere of 1.0 m diameter has a surface charge density of 8.0 C/m². What is the total electric flux leaving the surface of the sphere?

11.   A point charge of 1.0×10^{-6} C is at the center of a cubical Gaussian surface 50 cm on edge. What is Φ_E for the surface?

Section 25–5   Gauss's Law and Coulomb's Law
12.   A conducting sphere of radius 10 cm carries an unknown net charge. If the electric field 15 cm from its center is 3.0×10^3 N/C and points radially inward, what is the net charge on the sphere?

13.   A point charge at the origin causes an electric flux of -750 N·m²/C to pass through a spherical Gaussian surface of 10-cm radius centered at the origin. (a) If the radius of the Gaussian surface is doubled, how much flux would then pass through the surface? (b) What is the value of the point charge?

Section 25–6   An Insulated Conductor
14.   A conducting sphere carrying charge Q is surrounded by a spherical conducting shell. (a) What is the net charge on the inner surface of the shell? (b) Another charge q is placed outside the shell. Now what is the net charge on the inner surface of the shell? (c) If q is moved to a position between the shell and the sphere, what is the net charge on the inner surface of the shell? (d) Are your answers valid if the sphere and shell are not concentric?

Section 25–8 Gauss's Law—Some Applications

15. In units of 10^3 N·m²/C, the net electric flux through each face of a die (singular of dice) has magnitude equal to the number of spots on the face (1 through 6). The flux is inward for N odd and outward for N even. What is the net charge inside the die?

16. An infinite line of charge produces a field of 4.5×10^4 N/C at a distance of 2.0 m. Calculate the linear charge density.

17. A metal plate 8.0 cm on a side carries a total charge of 6.0×10^{-6} C. (a) Estimate the electric field 0.50 cm above the surface of the plate near the plate's center. (b) Estimate the field at a distance of 3.0 m.

18. Two charged concentric spheres have radii of 10 cm and 15 cm. The charge on the inner sphere is 4.0×10^{-8} C, and that on the outer sphere 2.0×10^{-8} C. Find the electric field (a) at $r = 12$ cm and (b) at $r = 20$ cm.

PROBLEMS GROUPED BY SECTION

Section 25–3 Flux of the Electric Field

1. In Example 1, compute Φ_E for the cylinder if it is turned so that its axis is perpendicular to the electric field. Calculate the flux directly, without using Gauss's law.

Section 25–4 Gauss's Law

2. Suppose that an electric field in some region is found to have a constant direction but to be decreasing in strength in that direction. What do you conclude about the charge in the region? Sketch the lines of force.

3. A point charge q is placed at one corner of a cube of edge a. What is the flux through each of the cube faces? (*Hint:* Use Gauss's law and symmetry arguments.)

4. "Gauss's law for gravitation" is

$$\frac{1}{4\pi G}\,\Phi_g = \frac{1}{4\pi G}\oint \mathbf{g}\cdot d\mathbf{S} = -m,$$

where m is the enclosed mass and G is the universal gravitation constant (Section 15–6). Derive Newton's law of gravitation from this. What is the significance of the minus sign?

Section 25–8 Gauss's Law—Some Applications

5. An insulated conductor of arbitrary shape carries a net charge of $+10 \times 10^{-6}$ C. Inside the conductor is a hollow cavity, within which is a point charge q of $+3.0 \times 10^{-6}$ C. What is the charge (a) on the cavity wall and (b) on the outer surface of the conductor?

6. Figure 25–21 shows a spherical shell of charge of uniform density ρ. Plot E for distances r from the center of the shell ranging from 0 to 30 cm. Assume that $\rho = 1.0 \times 10^{-6}$ C/m³, $a = 10$ cm, and $b = 20$ cm.

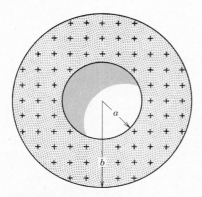

Figure 25–21 Problem 6.

7. A thin-walled metal sphere has a radius of 25 cm and carries a charge of 2.0×10^{-7} C. Find E for a point (a) inside the sphere, (b) just outside the sphere, and (c) 3.0 m from the center of the sphere.

8. A long conducting cylinder (length l) carrying a total charge $+q$ is surrounded by a conducting cylindrical shell, also of length l, with total charge $-2q$, as shown in cross section in Fig. 25–22. Use Gauss's law to find (a) the electric field at points outside the conducting shell, (b) the distribution of the charge on the conducting shell, and (c) the electric field in the region between the cylinders.

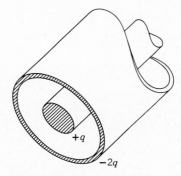

Figure 25–22 Problem 8.

9. Figure 25–23 shows a section through two long thin concentric cylinders of radii a and b. The cylinders carry equal and opposite charges per unit length λ. Using Gauss's law, prove (a) that $E = 0$ for $r < a$, and (b) that between the cylinders E is given by

$$E = \frac{1}{2\pi\epsilon_0}\frac{\lambda}{r}.$$

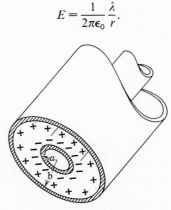

Figure 25–23 Problems 9 and 11.

`10.   An irregularly shaped conductor has an irregularly shaped cavity inside. A charge $+q$ is placed on the conductor, but there is no charge inside the cavity. Show that (a) $E = 0$ inside the cavity and (b) there is no charge on the cavity walls.

11.   In Problem 9, a positive electron revolves in a circular path of radius r between and concentric with the cylinders. What must be its kinetic energy K? Assume that $a = 2.0$ cm, $b = 3.0$ cm, and $\lambda = 3.0 \times 10^{-8}$ C/m.

12.   Equation 25–12 ($E = \sigma/\epsilon_0$) gives the electric field at points near a charged conducting surface. Apply this equation to a conducting sphere of radius r carrying charge q on its surface, and show that the electric field outside the sphere is the same as the field of a point charge at the position of the sphere center.

13.   Two large metal plates of area 1.0 m² face each other. They are 5.0 cm apart and carry equal and opposite charges on their inner surfaces. If E between the plates is 55 N/C, what is the charge on the plates? Neglect edge effects.

14.   Two large sheets of positive charge face each other as in Fig. 25–24. What is E at points (a) to the left of the sheets, (b) between them, and (c) to the right of the sheets? Assume the same surface charge density σ for each sheet. Consider only points not near the edges whose distance from the sheets is small compared to the dimensions of the sheet. (*Hint:* E at any point is the vector sum of the separate electric fields set up by each sheet.)

15.   Charge is distributed uniformly throughout an infinitely-long cylinder of radius R. (a) Show that E at a distance r from the cylinder axis ($r < R$) is given by

$$E = \frac{\rho r}{2\epsilon_0},$$

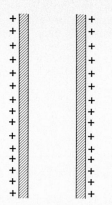

Figure 25–24   Problem 14.

where ρ is the density of charge. (b) What result do you expect for $r > R$?

*16.   The spherical region $a < r < b$ carries a charge per unit volume of $\rho = A/r$, where A is constant. At the center ($r = 0$) of the enclosed cavity is a point charge q. What should the value of A be so that the electric field in the region $a < r < b$ has constant magnitude?

*17.   A plane slab of thickness d has a uniform volume charge density ρ. Find the magnitude of the electric field at all points in space both (a) inside and (b) outside the slab.

ADDITIONAL PROBLEMS

18.   It is found experimentally that the electric field in a certain region of the earth's atmosphere is directed vertically down. At an altitude of 300 m the field is 60 N/C, and at an altitude of 200 m it is 100 N/C. Find the net amount of charge contained in a cube 100 m on edge located between 200 and 300 m altitude. Neglect the curvature of the earth.

19.   Find the net flux through the cube of Exercise 25–4 and Fig. 25–15 if the electric field is given by (a) $\mathbf{E} = 3y\mathbf{j}$ and by (b) $\mathbf{E} = -4\mathbf{i} + (6 + 3y)\mathbf{j}$. E is in N/C if y is in meters. (c) In each case, how much charge is inside the cube?

20.   An uncharged, spherical, thin, metallic shell has a point charge q at its center. Give expressions for the electric field (a) inside the shell and (b) outside the shell, using Gauss's law. (c) Has the shell any effect on the field due to q? (d) Has the presence of q any effect on the shell? (e) If a second point charge is held outside the shell, does this outside charge experience a force? (f) Does the inside charge experience a force? (g) Is there a contradiction with Newton's third law here? Why or why not?

21.   Figure 25–25 shows a charge $+q$ arranged as a uniform conducting sphere of radius a and placed at the center of a spherical conducting shell of inner radius b and outer radius c. The outer shell carries a charge of $-q$. Find $E(r)$, (a) within the sphere ($r < a$), (b) between the sphere and the shell ($a < r < b$), (c) inside the shell ($b < r < c$), and (d) outside the shell ($r > c$). (e) What charges appear on the inner and outer surfaces of the shell?

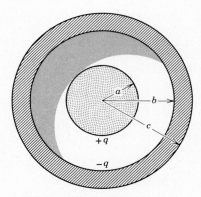

Figure 25–25   Problem 21.

22.   A thin, metallic, spherical shell of radius a carries a charge q_a. Concentric with it is another thin, metallic, spherical shell of radius b ($b > a$) carrying a charge q_b. Use Gauss's law to find the electric field at radial points r where (a) $r < a$; (b) $a < r < b$; (c) $r > b$. (d) Discuss the criterion one would use to determine how the charges are distributed on the inner and outer surfaces of each shell.

23.   Figure 25–26 shows a point charge of 1.0×10^{-7} C at the center of a spherical cavity of radius 3.0 cm in a piece of metal. Use

Gauss's law to find the electric field (a) at point P_1, halfway from the center to the surface and (b) at point P_2.

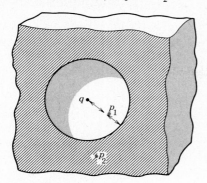

Figure 25-26 Problem 23.

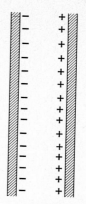

Figure 25-28 Problem 27.

24. Two charged concentric cylindrical shells have radii of 3.0 cm and 6.0 cm. The charge per unit length on the inner cylinder is 5.0×10^{-6} C/m and that on the outer cylinder is -7.0×10^{-6} C/m. Find the electric field at (a) $r = 4.0$ cm and (b) at $r = 8.0$ cm.

25. Figure 25-27 shows a section through a long, thin-walled metal tube of radius R, carrying a charge per unit length λ on its surface. Derive expressions for E for various distances r from the tube axis, considering both (a) $r > R$ and (b) $r < R$. Plot your results for the range $r = 0$ to $r = 5.0$ cm, assuming that $\lambda = 2.0 \times 10^{-8}$ C/m and $R = 3.0$ cm.

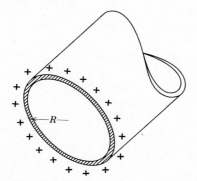

Figure 25-27 Problem 25.

26. A proton orbits with a speed v ($= 3.0 \times 10^5$ m/s) just outside a charged sphere of radius r ($= 10^{-2}$ m). What is the charge on the sphere?

27. Two large metal plates face each other as in Fig. 25-28 and carry charges with surface charge density $+\sigma$ and $-\sigma$, respectively, on their inner surfaces. What is **E** at points (a) to the left of the sheets, (b) between them, and (c) to the right of the sheets? Consider only points not near the edges whose distance from the sheets is small compared to the dimensions of the sheet.

28. An electron remains stationary in an electric field directed downward in the earth's gravitational field. If the electric field is due to charge on two large parallel conducting plates, oppositely charged and separated by 2.3 cm, what is the surface charge density on the plates?

29. A 100-eV electron is fired directly toward a large metal plate that has a surface charge density of -2.0×10^{-6} C/m². From

what distance must the electron be fired if it is to just fail to strike the plate?

30. A small sphere whose mass m is 1.0 mg carries a charge q of 2.0×10^{-8} C. It hangs in the earth's gravitational field from a silk thread that makes an angle θ of 30° with a large, uniformly charged nonconducting sheet as in Fig. 25-29. Calculate the charge density σ for the sheet.

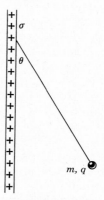

Figure 25-29 Problem 30.

31. In a 1911 paper, Ernest Rutherford said, "In order to form some idea of the forces required to deflect an α-particle through a large angle, consider an atom containing a point positive charge Ze at its centre and surrounded by a distribution of negative electricity, $-Ze$ uniformly distributed within a sphere of radius R. The electric field E . . . at a distance r from the center for a point *inside* the atom [is]

$$E = \frac{Ze}{4\pi\epsilon_0} \left(\frac{1}{r^2} - \frac{r}{R^3} \right) \; \ldots \text{"}$$

Prove the correctness of this equation.

*32. A spherical region carries a uniform charge per unit volume ρ. Let **r** be the vector from the center of the sphere to a general point P within the sphere. (a) Show that the electric field at P is given by $\mathbf{E} = \rho\mathbf{r}/3\epsilon_0$. (b) A spherical cavity is created in the above sphere, as shown in Fig. 25-30. Using superposition concepts, show that the electric field at all points within the cavity is $\mathbf{E} = \rho\mathbf{a}/3\epsilon_0$ (uniform field), where **a** is the vector connecting the center of the sphere with the center of the cavity. Note that both these results are independent of the radii of the sphere and the cavity.

Figure 25-30 Problem 32.

*33. Show that stable equilibrium under the action of electrostatic forces alone is impossible. (*Hint:* Assume that at a certain point *P* in an **E** field a charge +*q* would be in stable equilibrium if it were placed there. Draw a spherical Gaussian surface about *P*, imagine how **E** must point on this surface, and apply Gauss's law to show that the assumption leads to a contradiction.)

26 Electric Potential

26–1 Electric Potential

The electric field around a charged rod can be described not only by a (vector) electric field **E** but also by a scalar quantity, the *electric potential V*. These quantities are intimately related, and often it is only a matter of convenience which is used in a given problem.

To find the *electric potential difference* between an arbitrary point *A* and a second arbitrary point *B* in an electric field, let us move a (positive) test charge q_0 from *A* to *B*, keeping it in equilibrium at all times. As we do so the force $q_0\mathbf{E}$ exerted on the charge by the electric field will do a certain amount of work W_{AB} on the charge. (Note that, because the charge is in equilibrium at all times, an external agent must do work equal to the negative of this, or $-W_{AB}$, on the test charge; our concern, however, is with the work done by the electric field force.) We then define the electric potential difference between *A* and *B* from

$$V_B - V_A = -\frac{W_{AB}}{q_0}. \qquad (26-1)$$

The work W_{AB} done by the electric field force on the (positive) test charge as it moves from *A* to *B* may be (*a*) positive, (*b*) negative or (*c*) zero. Correspondingly the electric potential at point *B* will then be (*a*) lower than, (*b*) higher than, or (*c*) the same as, the electric potential at point *A*.

The SI unit of potential difference that follows from Eq. 26–1 is the joule/coulomb. This combination occurs so often that a special unit, the *volt* (abbr. V), is used to represent it; that is,

$$1 \text{ volt} = 1 \text{ joule/coulomb}.$$

Point *A* is often chosen to be at a large (strictly an infinite) distance from all charges, and the electric potential V_A at this infinite distance is arbitrarily taken as zero. This allows us to define the *electric potential at a point*. Putting $V_A = 0$ in Eq. 26–1 and dropping the subscripts leads to

$$V = -\frac{W}{q_0}, \qquad (26-2)$$

where W is the work that the electric field force $q_0\mathbf{E}$ does as the test charge q_0 is moved from infinity to the point in question. Keep in mind that *potential differences* are of fundamental concern and that Eq. 26–2 depends on the arbitrary assignment of the value zero to the potential V_A at the reference position (infinity); this reference potential could equally well have been chosen as any other value, say − 100 V. Similarly, any other agreed-upon point could be chosen as a reference position. In many problems the earth is taken as a reference of potential and assigned the value zero.

Equation 26–2 tells us that the potential V at all points near an isolated positive charge is positive. To see this, imagine that we push a test charge q_0 in from infinity toward such an isolated positive charge. The test charge will be repelled at all stages and the work W done by the electric field force on the test charge as it moves in will be negative. Equation 26–2 then tells us that the potential V will be positive. In the same way, the potential at points near an isolated negative charge is negative. Equation 26–2 also tells us that electric potential is a scalar because both W and q_0 in that equation are scalars.

Both W_{AB} and $V_B - V_A$ in Eq. 26–1 are independent of the path followed in moving the test charge from point A to point B. Such path independence of the work holds for all conservative fields, as discussed in Section 8–2. If this were not so, point B would not have a unique electric potential (with respect to point A as a defined reference position) and the concept of potential would not be useful.

Let us now show that the potential difference is path independent for the field due to the point charge of Fig. 26–1. Then we shall show that path independence holds in *all* electrostatic situations. Figure 26–1 shows two points A and B set up in the field of a point charge q, assumed positive.

Potential differences are independent of path

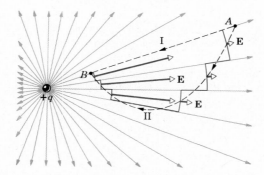

Figure 26-1 A test charge q_0 is moved from A to B in the field of charge q along either of two paths. The open arrows show **E** at six points on path II.

Let us imagine a positive test charge q_0 being moved by an external agent from point A to point B by two different paths. Path I is a simple radial line. Path II is a completely arbitrary path between the two points.

We may approximate path II by a broken path made up of alternating elements of circular arc and of radius. Since these elements can be arbitrarily small, the broken path can be made arbitrarily close to the actual path. On path II the electric field force does work only along the radial segments, because along the arcs the force $q_0\mathbf{E}$ and the displacement $d\boldsymbol{\ell}$ are at right angles, $q_0\mathbf{E} \cdot d\boldsymbol{\ell}$ being zero in such cases. The sum of the work done on the radial segments that make up path II is the same as the work done on path I because each path has the same array of radial segments. Since path II is arbitrary, we have proved that the work done is the same for all paths connecting A and B.

This proof holds only for the field due to a point charge. However, any charge distribution (discrete or continuous) can be considered as made up of an assembly of point charges or differential charge elements. Hence, from the principle of superposition, we conclude that path independence for the electrostatic potential holds for all electrostatic charge configurations.

Equipotential surfaces

The locus of points, all of which have the same electric potential, is called an *equipotential surface*. A family of equipotential surfaces, each surface corresponding to a different value of the potential, can be used to give a general description of the electric field in a certain region of space. We have seen earlier (Section 24–3) that electric lines of force can also be used for this purpose; in later sections (see, for example, Fig. 26–15) we explore the intimate connection between these two ways of describing the electric field.

No work is done by the electric field force as a test charge is moved between any two points on an equipotential surface. This follows from Eq. 26–1,

$$V_B - V_A = -\frac{W_{AB}}{q_0}.$$

because W_{AB} must be zero if $V_A = V_B$. This is true, because of the path independence of potential difference, even if the path connecting A and B does not lie entirely in the equipotential surface.

Figure 26–2 shows an arbitrary family of equipotential surfaces. The work done by the electric field force as a test charge is moved from one end point to the other along paths I and II is zero because both these paths begin and end on the same equipotential surface. The work done as a charge is moved from one end point to the other along paths I′ and II′ is not zero but is the same for each path because the initial and the final potentials are identical; paths I′ and II′ connect the same pair of equipotential surfaces.

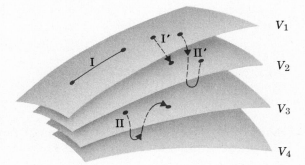

Figure 26–2 Portions of four equipotential surfaces. The heavy lines show four typical paths along which a test charge is moved.

From symmetry, the equipotential surfaces for a spherical charge are a family of concentric spheres. For a uniform field they are a family of planes at right angles to the field. In all cases (including these two examples) the equipotential surfaces are at right angles to the lines of force and thus to **E** (see Fig. 26–15). If **E** were *not* at right angles to the equipotential surface, it would have a component lying in that surface. Then work would be done by the electric field force as a test charge is moved about on the surface. Work cannot be done if the surface is an equipotential, so **E** must be at right angles to the surface.

26–2 Potential and the Electric Field

Let A and B in Fig. 26–3 be two points in a uniform electric field **E** where B is a distance l in the field direction from A. The field **E** has been set up by some arrangement of charges not shown. Now, assume that a positive test charge q_0 is moved,

Figure 26–3 A positive test charge q_0 is moved from A to B in a uniform electric field **E**.

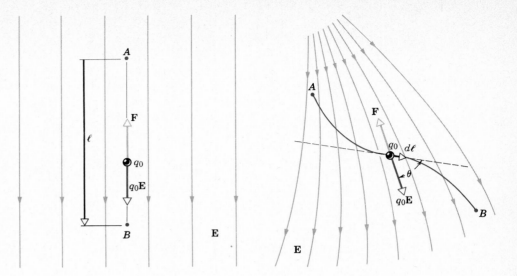

Figure 26–4 A positive test charge q_0 is moved from A to B in a non-uniform electric field by an external agent that exerts a force **F** on it.

without acceleration, from A to B along the straight line connecting them. An external agent provides the necessary balancing force **F**. With q_0 replaced by m and **E** by **g**, Fig. 26–3 could be used as it stands to illustrate the lowering of a stone from A to B in the uniform gravitational field near the earth's surface.

The electric force on the charge in Fig. 26–3 is $q_0\mathbf{E}$ and points down. The work W done by this force, as the test charge moves through displacement $\boldsymbol{\ell}$, is

$$W_{AB} = (q_0\mathbf{E})\cdot\boldsymbol{\ell} = q_0E \cos 0° \, l = q_0El. \tag{26–3}$$

Potential differences in a uniform electric field

Substituting this into Eq. 26–1 yields

$$V_B - V_A = -\frac{W_{AB}}{q_0} = -El, \tag{26–4}$$

showing that B is at a lower potential than A. Equation 26–4 gives the connection between potential difference and field strength for a simple special case. Note from this equation that another SI unit for **E** is the volt/meter. You may wish to prove that a volt/meter is identical to a newton/coulomb, the unit first presented for **E** in Section 24–2.

What is the connection between V and **E** in the more general case in which the field is *not* uniform and in which the test body is moved along a path that is *not* straight, as in Fig. 26–4? The electric field exerts a force $q_0\mathbf{E}$ on the test charge, as shown. To keep the test charge from accelerating, an external agent must again apply a force **F** chosen to be exactly equal to $-q_0\mathbf{E}$ for all positions of the test body.

If the test body moves through a displacement $d\boldsymbol{\ell}$ along the path from A to B, the element of work done by the electric field force is $(q_0\mathbf{E})\cdot d\boldsymbol{\ell}$. To find the total work W_{AB} done as the test charge moves from A to B, we add up (that is, integrate) the work contributions for all the infinitesimal segments into which the path is divided. This leads to

$$W_{AB} = q_0 \int_A^B \mathbf{E}\cdot d\boldsymbol{\ell}.$$

Such an integral is called a *line integral*.

Substituting this expression for W_{AB} into Eq. 26–1 leads to

$$V_B - V_A = -\frac{W_{AB}}{q_0} = -\int_A^B \mathbf{E}\cdot d\boldsymbol{\ell}. \tag{26–5}$$

If point A is taken to be infinitely distant and the potential V_A at infinity is taken to be zero, this equation gives the potential V at point B, or

$$V_B = -\int_\infty^B \mathbf{E}\cdot d\boldsymbol{\ell}. \tag{26–6}$$

Potential and the electric field

These two equations allow us to calculate the potential difference between any two points (or the potential at any point) if $\mathbf{E}$ is known at various points in the field.

Example 1 In Fig. 26–3 calculate $V_B - V_A$ using Eq. 26–5. Compare the result with that obtained by direct analysis of this special case (Eq. 26–4).

In moving the test charge the element of path $d\boldsymbol{\ell}$ always points in the direction of motion; this is down in Fig. 26–3. The electric field $\mathbf{E}$ in this figure also points down so that the angle θ between $\mathbf{E}$ and $d\boldsymbol{\ell}$ is 0°.

Equation 26–5 then becomes

$$V_B - V_A = -\int_A^B \mathbf{E}\cdot d\boldsymbol{\ell} = -\int_A^B E\cos 0°\, dl = -\int_A^B E\, dl.$$

E is constant for all parts of the path in this problem and can thus be taken outside the integral sign, giving

$$V_B - V_A = -E\int_A^B dl = -El,$$

which agrees with Eq. 26–4, as it must. Equipotential surfaces for this case are illustrated in Fig. 26–15b.

Example 2 In Fig. 26–5 let a test charge q_0 be moved without acceleration from A to B over the path ACB. Compute the potential difference between A and B.

Points A and C have the same potential because no work is done in moving a charge between them, $\mathbf{E}$ and $d\boldsymbol{\ell}$ being at right angles for all points on the line AC. In other words, A and C lie on the same equipotential surface.

For the path CB we have $\theta = 45°$ and, from Eq. 26–5,

$$V_B - V_C = -\int_C^B \mathbf{E}\cdot d\boldsymbol{\ell} = -\int_C^B E\cos 45°\, dl = -\frac{E}{\sqrt{2}}\int_C^B dl.$$

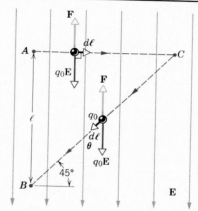

Figure 26–5 Example 2. A test charge q_0 is moved along path ACB in a uniform electric field.

The integral is the length of the line CB which is $\sqrt{2}l$. Thus

$$V_B - V_C = -\frac{E}{\sqrt{2}}(\sqrt{2}l) = -El.$$

Finally

$$V_B - V_A = V_B - V_C = -El$$

This is the same value derived for a direct path connecting A and B, a result to be expected because the potential difference between two points is path independent.

26-3 Potential Due to a Point Charge

Figure 26–6 shows two points A and B near an isolated positive point charge q. For simplicity we assume that A, B, and q lie on a straight line. Let us compute the potential difference between points A and B, assuming that a test charge q_0 is moved without acceleration along a radial line from A to B.

In Fig. 26–6 $\mathbf{E}$ points to the right and $d\boldsymbol{\ell}$, which is always in the direction of motion, points to the left. Therefore, in Eq. 26–5,

$$\mathbf{E}\cdot d\boldsymbol{\ell} = E\cos 180°\, dl = -E\, dl.$$

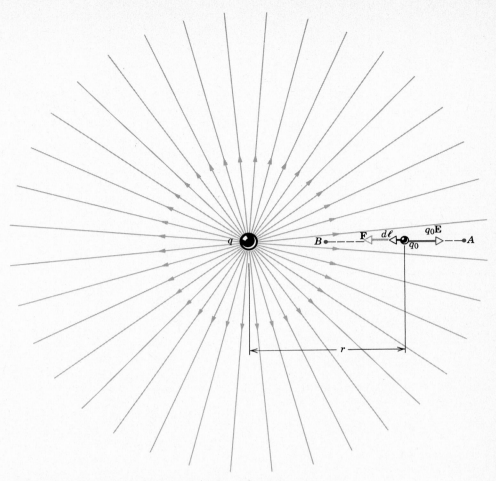

Figure 26-6 In the field set up by a positive point charge q, a test charge q_0 is moved from A to B by an external agent that exerts a force **F** on it.

However, as we move a distance dl to the left, we are moving in the direction of decreasing r because r is measured from q as an origin. Thus

$$dl = -dr.$$

Combining yields

$$\mathbf{E} \cdot d\boldsymbol{\ell} = E \, dr.$$

Substituting this into Eq. 26–5 gives

$$V_B - V_A = -\int_A^B \mathbf{E} \cdot d\boldsymbol{\ell} = -\int_{r_A}^{r_B} E \, dr.$$

Combining with Eq. 24–4,

$$E = \frac{1}{4\pi\epsilon_0} \frac{q}{r^2}$$

leads to

$$V_B - V_A = -\frac{q}{4\pi\epsilon_0} \int_{r_A}^{r_B} \frac{dr}{r^2} = \frac{q}{4\pi\epsilon_0} \left(\frac{1}{r_B} - \frac{1}{r_A} \right). \tag{26-7}$$

Choosing reference position A to be at infinity (that is, letting $r_A \rightarrow \infty$), choosing $V_A = 0$ at this position, and dropping the subscript B leads to

Potential due to a point charge

$$V = \frac{1}{4\pi\epsilon_0} \frac{q}{r}. \qquad (26\text{–}8)$$

This equation shows clearly that equipotential surfaces for an isolated point charge are spheres concentric with the point charge (see Fig. 26–15a). A study of the derivation will show that this relation also holds for points external to spherically symmetric charge distributions.

Example 3 What must the magnitude of an isolated positive point charge be for the electric potential at 10 cm from the charge to be $+100$ V?

Solving Eq. 26–8 for q yields

$q = V4\pi\epsilon_0 r = (100 \text{ V})(4\pi)(8.9 \times 10^{-12} \text{ C}^2/\text{N}\cdot\text{m}^2)(0.10 \text{ m})$
$= 1.1 \times 10^{-9}$ C.

This charge is comparable to charges that can easily be produced by friction.

Example 4 What is the electric potential at the surface of a gold nucleus? The radius is 6.2×10^{-15} m and the atomic number Z is 79.

The nucleus, assumed spherically symmetrical, behaves electrically for external points as if it were a point charge. Thus we can use Eq. 26–8, or, recalling that the proton charge is 1.6×10^{-19} C,

$$V = \frac{1}{4\pi\epsilon_0} \frac{q}{r} = \frac{(1/4\pi\epsilon_0)(Ze)}{r}$$

$$= \frac{(9.0 \times 10^9 \text{ N}\cdot\text{m}^2/\text{C}^2)(79)(1.6 \times 10^{-19} \text{ C})}{6.2 \times 10^{-15} \text{ m}}$$

$$= 1.8 \times 10^7 \text{ V} = 18 \text{ MV}.$$

26–4   A Group of Point Charges

The potential at any point due to a group of point charges is found by (a) calculating the potential V_n due to each charge, as if the other charges were not present and (b) adding the quantities so obtained, or (see Eq. 26–8)

Potential due to a group of point charges

$$V = \sum_n V_n = \frac{1}{4\pi\epsilon_0} \sum_n \frac{q_n}{r_n}, \qquad (26\text{–}9)$$

where q_n is the value of the nth charge and r_n is the distance of this charge from the point in question. The sum used to calculate V is an algebraic sum and not a vector sum like the one used to calculate $\mathbf{E}$ for a group of point charges (see Eq. 24–5). Herein lies an important computational advantage of potential over electric field.

If the charge distribution is continuous, rather than being a collection of points, the sum in Eq. 26–9 must be replaced by an integral, or

Potential due to a continuous charge distribution

$$V = \int dV = \frac{1}{4\pi\epsilon_0} \int \frac{dq}{r}, \qquad (26\text{–}10)$$

where dq is a differential element of the charge distribution, r is its distance from the point at which V is to be calculated, and dV is the potential it establishes at that point.

Example 5 What is the potential at the center of the square of Fig. 26–7? Assume that $q_1 = +1.0 \times 10^{-8}$ C, $q_2 = -2.0 \times 10^{-8}$ C, $q_3 = +3.0 \times 10^{-8}$ C, $q_4 = +2.0 \times 10^{-8}$ C, and $a = 1.0$ m.

The distance r of each charge from P is $a/\sqrt{2}$ or 0.71 m. From Eq. 26–9

$$V = \sum_n V_n = \frac{1}{4\pi\epsilon_0} \frac{q_1 + q_2 + q_3 + q_4}{r}$$

$$V = \frac{(9.0 \times 10^9 \text{ N}\cdot\text{m}^2/\text{C}^2)(1.0 - 2.0 + 3.0 + 2.0) \times 10^{-8} \text{ C}}{0.71 \text{ m}}$$

$$= 510 \text{ V}.$$

Is the potential constant within the square? Does any point inside have a negative potential?

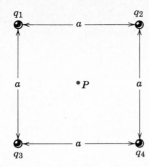

Figure 26-7   Example 5.

Example 6 *A charged disk*   Find the electric potential for points on the axis of a uniformly charged circular disk whose surface charge density is σ (see Fig. 26-8).

Consider a charge element dq consisting of a flat circular strip of radius y and width dy. We have

$$dq = \sigma(2\pi y)(dy),$$

where $(2\pi y)(dy)$ is the area of the strip. All parts of this charge element are the same distance r' $(=\sqrt{y^2 + r^2})$ from axial point P so that their contribution dV to the electric potential at P is given by Eq. 26-8, or

$$dV = \frac{1}{4\pi\epsilon_0}\frac{dq}{r'} = \frac{1}{4\pi\epsilon_0}\frac{\sigma 2\pi y\,dy}{\sqrt{y^2 + r^2}}.$$

The potential V is found by integrating over all the strips into which the disk can be divided (Eq. 26-10), or

$$V = \int dV = \frac{\sigma}{2\epsilon_0}\int_0^a (y^2 + r^2)^{-1/2}y\,dy = \frac{\sigma}{2\epsilon_0}(\sqrt{a^2 + r^2} - r).$$

This general result is valid for all values of r. In the special case of $r \gg a$ the quantity $\sqrt{a^2 + r^2}$ can be approximated as

$$\sqrt{a^2 + r^2} = r\left(1 + \frac{a^2}{r^2}\right)^{1/2} = r\left(1 + \frac{1}{2}\frac{a^2}{r^2} + \cdots\right) \cong r + \frac{a^2}{2r},$$

in which the quantity in parentheses in the second member of this equation has been expanded by the binomial theorem (see Appendix G). This equation means that V becomes

$$V \cong \frac{\sigma}{2\epsilon_0}\left(r + \frac{a^2}{2r} - r\right) = \frac{\sigma\pi a^2}{4\pi\epsilon_0 r} = \frac{1}{4\pi\epsilon_0}\frac{q}{r},$$

where q $(= \sigma\pi a^2)$ is the total charge on the disk. This limiting result is expected because the disk behaves like a point charge for $r \gg a$.

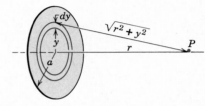

Figure 26-8   Example 6. A point P on the axis of a uniformly charged circular disk of radius a.

26-5   Potential Due to a Dipole

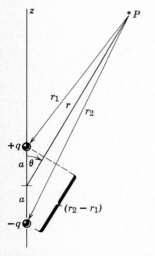

Figure 26-9   A point P in the field of an electric dipole.

Two equal charges of opposite sign, $\pm q$, separated by a distance $2a$, constitute an electric dipole; see Example 3, Chapter 24. The electric dipole moment **p** has the magnitude $2aq$ and points from the negative charge to the positive charge. Here we derive an expression for the electric potential V at any point of space due to a dipole, provided only that the point is not too close to the dipole ($r \gg a$).

A point P is specified by giving the quantities r and θ in Fig. 26-9. From symmetry, it is clear that the potential will not change as point P rotates about the z axis, r and θ being fixed. Thus we need only find $V(r, \theta)$ for any plane containing this axis; the plane of Fig. 26-9 is such a plane. Applying Eq. 26-9 gives,

$$V = \sum_n V_n = V_1 + V_2 = \frac{1}{4\pi\epsilon_0}\left(\frac{q}{r_1} - \frac{q}{r_2}\right) = \frac{q}{4\pi\epsilon_0}\frac{r_2 - r_1}{r_1 r_2},$$

which is an exact relationship.

We now limit consideration to points such that $r \gg 2a$. These approximate relations then follow from Fig. 26-9:

$$r_2 - r_1 \cong 2a\cos\theta \qquad \text{and} \qquad r_1 r_2 \cong r^2,$$

and the potential reduces to

$$V = \frac{q}{4\pi\epsilon_0}\frac{2a\cos\theta}{r^2} = \frac{1}{4\pi\epsilon_0}\frac{p\cos\theta}{r^2}, \tag{26-11}$$

Potential due to a dipole

in which $p \, (= 2aq)$ is the dipole moment. Note that V vanishes everywhere in the equatorial plane ($\theta = 90°$). This reflects the fact that it takes no work to bring a test charge in from infinity along the perpendicular bisector of the dipole. For a given radius, V has its greatest positive value for $\theta = 0$ and its greatest negative value for $\theta = 180°$. Note that the potential does not depend separately on q and $2a$ but only on their product p. See Fig. 26–15c for a representation of electric field lines and equipotential surfaces for an electric dipole.

Some dipoles

It is convenient to call any assembly of charges, for which V at distant points is given by Eq. 26–11, an *electric dipole*. Two point charges separated by a small distance behave this way, as we have just proved. However, other charge configurations also obey Eq. 26–11. Suppose that by measurement at points outside an imaginary box (Fig. 26–10) we find a pattern of lines of force that can be described quantitatively by Eq. 26–11. We then declare that the object inside the box is an electric dipole, that its axis is the line zz', and that its dipole moment **p** points vertically upward.

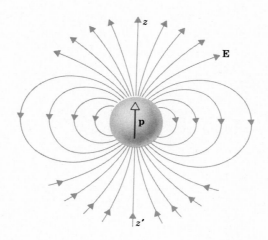

Figure 26–10 If an object inside the spherical box sets up the electric field shown (described quantitatively by Eq. 26–11), the object is an electric dipole. Compare this with Fig. 26–15c.

Many molecules have electric dipole moments. That for H_2O in its vapor state is 6.2×10^{-30} C·m. Figure 26–11 is a representation of this molecule, showing the three nuclei and the surrounding electron cloud. The dipole moment **p** is represented by the arrow on the axis of symmetry of the molecule. In this molecule the effective center of positive charge does not coincide with the effective center of negative charge. It is precisely because of this separation that the dipole moment exists.

Atoms, and many molecules, do not have permanent dipole moments. However, dipole moments may be induced by placing any atom or molecule in an external electric field. The action of the field (Fig. 26–12) is to separate the centers of positive and of negative charge. We say that the atom becomes polarized and acquires an induced electric dipole moment. Induced dipole moments disappear when the electric field is removed.

Electric dipoles are important in situations other than atomic and molecular ones. Radio and TV antennas are often in the form of a metal wire or rod in which electrons surge back and forth periodically. At a certain time one end of the wire or rod will be negative and the other end positive. Half a cycle later the polarity of the ends is exactly reversed. This is an *oscillating* electric dipole. It is so named because its dipole moment changes in a periodic way with time.

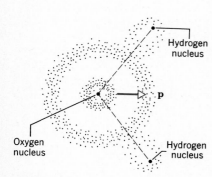

Figure 26–11 A schematic representation of a water molecule, showing the three nuclei, the electron cloud, and the orientation of the dipole moment.

26-6 Electric Potential Energy

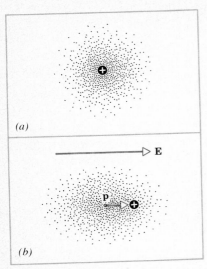

(a)

(b)

Figure 26-12 (*a*) An atom, showing the nucleus and the electron cloud. The center of negative charge coincides with the center of positive charge, that is, with the nucleus. (*b*) If an external field **E** is applied, the electron cloud is distorted so that the center of negative charge, marked by the dot, and the center of positive charge no longer coincide. An electric dipole appears. The distortion is greatly exaggerated.

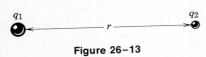

Figure 26-13

Potential energy of two point charges

If we raise a stone from the earth's surface, the work that we do against the earth's gravitational attraction is stored as potential energy in the system earth + stone. If we release the stone, the stored potential energy changes steadily into kinetic energy as the stone drops. After the stone comes to rest on the earth, this kinetic energy, equal in magnitude just before the time of contact to the originally stored potential energy, is transformed into thermal energy in the system earth + stone.

A similar situation exists in electrostatics. Consider two charges q_1 and q_2 a distance r apart, as in Fig. 26-13. If we increase the separation between them, an external agent must do work that will be positive if the charges are opposite in sign and negative if the charges are the same. The energy represented by this work can be thought of as stored in the system $q_1 + q_2$ as *electric potential energy*. This energy, like all varieties of potential energy, can be transformed into other forms. If q_1 and q_2, for example, are charges of opposite sign and we release them, they will accelerate toward each other, transforming the stored potential energy into kinetic energy of the accelerating masses. The analogy to the earth + stone system is exact, save for the fact that electric forces may be either attractive or repulsive whereas gravitational forces are always attractive.

We define the electric potential energy of a system of point charges as the work required to assemble this system of charges by bringing them in from an infinite distance. We assume that the charges are all at rest (no kinetic energy) in both their initial and final positions.

In Fig. 26-13 let us imagine q_2 removed to infinity and at rest. The electric potential at the original site of q_2, caused by q_1, is given by Eq. 26-8, or

$$V = \frac{1}{4\pi\epsilon_0}\frac{q_1}{r}.$$

If q_2 is moved in from infinity to the original distance r, the work required is, from the definition of electric potential, that is, from Eq. 26-2,

$$W = Vq_2. \qquad (26-12)$$

Combining these two equations and recalling that this work W is precisely the *electric potential energy* U of the system $q_1 + q_2$ yields

$$U\,(=W) = \frac{1}{4\pi\epsilon_0}\frac{q_1 q_2}{r_{12}}. \qquad (26-13)$$

The subscript on r emphasizes that the distance involved is that between the point charges q_1 and q_2. Potential (measured in volts) and potential energy (measured in joules) are quite different quantities and must not be confused.

For systems containing more than two charges the procedure is to compute the potential energy for every pair of charges separately and to add the results algebraically. This procedure rests on a physical picture in which (*a*) charge q_1 is brought into position, (*b*) q_2 is brought from infinity to its position near q_1, (*c*) q_3 is brought from infinity to its position near q_1 and q_2, etc.

Example 7 Two protons in a nucleus of ^{238}U are 6.0×10^{-15} m apart. What is their mutual electric potential energy expressed in joules and also in electron volts (see p. 515)?
From Eq. 26-13

$$U = \frac{1}{4\pi\epsilon_0}\frac{q_1 q_2}{r} = \frac{(9.0 \times 10^9\ \text{N·m}^2/\text{C}^2)(1.6 \times 10^{-19}\ \text{C})^2}{6.0 \times 10^{-15}\ \text{m}}$$

$$= 3.8 \times 10^{-14}\ \text{J} = 2.4 \times 10^5\ \text{eV} = 0.24\ \text{MeV}.$$

Example 8 Three charges are held fixed as in Fig. 26–14. What is their mutual electric potential energy? Assume that $q = 1.0 \times 10^{-7}$ C and $a = 10$ cm.

The total energy of the configuration is the sum of the energies of each pair of particles. From Eq. 26–13,

$$U = U_{12} + U_{13} + U_{23}$$

$$= \frac{1}{4\pi\epsilon_0}\left[\frac{(+q)(-4q)}{a} + \frac{(+q)(+2q)}{a} + \frac{(-4q)(+2q)}{a}\right]$$

$$= -\frac{10}{4\pi\epsilon_0}\frac{q^2}{a}$$

$$= -\frac{(9.0 \times 10^9 \text{ N·m}^2/\text{C}^2)(10)(1.0 \times 10^{-7} \text{ C})^2}{0.10 \text{ m}}$$

$$= -9.0 \times 10^{-3} \text{ J}.$$

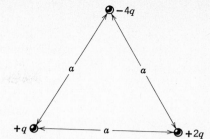

Figure 26–14 Example 8. Three charges are fixed, as shown, by external forces.

The fact that the total energy is negative means that negative work would have to be done to assemble this structure, starting with the three charges separated and at rest at infinity. Expressed otherwise, 9.0×10^{-3} J of work must be done to dismantle this structure, removing the charges to an infinite separation from one another.

26–7 Calculation of **E** from **V**

We have stated that V and **E** are equivalent descriptions of electric fields, and have determined (Eq. 26–6) how to calculate V from **E**. Let us now consider how to calculate **E** if we know V throughout a certain region.

This problem has already been solved graphically. If **E** is known at every point in space, the lines of force can be drawn; then a family of equipotentials can be sketched in by drawing surfaces at right angles. These equipotentials describe the behavior of V. Conversely, if V is given as a function of position, a set of equipotential surfaces can be drawn. The lines of force can then be found by drawing lines at right angles, thus describing the behavior of **E**. It is the mathematical equivalent of this second graphical process that we seek here. Figure 26–15 shows

Figure 26–15 Equipotential surfaces (dashed lines) and lines of force (solid lines) for (a) a point charge, (b) a uniform electric field set up by charges not shown, and (c) an electric dipole. In (a) and (c) the dashed lines represent intersections with the plane of the figure of closed surfaces; in (b) the dashed lines represent infinite sheets. In all figures there is a constant difference of potential ΔV between adjacent equipotential surfaces.

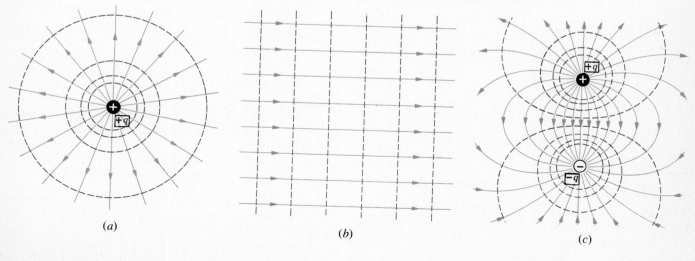

(a) (b) (c)

three examples of lines of force and of the corresponding equipotential surfaces.

Figure 26–16 shows the intersection with the plane of the figure of a family of equipotential surfaces. The figure shows that $\mathbf{E}$ at a typical point P is at right angles to the equipotential surface through P and points in the direction of decreasing potential, as it must do.

Let us move a test charge q_0 from P along the path marked $\Delta\ell$ to the equipotential surface marked $V - \Delta V$. The work ΔW that must be done by the electric field force (see Eq. 26–1) is $q_0 \Delta V$.

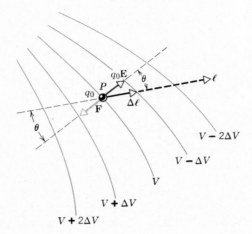

Figure 26–16 A test charge q_0 is moved from one equipotential surface to another along an arbitrarily selected direction marked ℓ.

From another point of view we can calculate the work from*

$$\Delta W = (q_0 \mathbf{E}) \cdot \Delta\boldsymbol{\ell}.$$

Thus

$$\Delta W = q_0 \mathbf{E} \cdot \Delta\boldsymbol{\ell} = q_0 E \cos\theta\, \Delta l.$$

These two expressions for the work must be equal, which gives

$$q_0\, \Delta V = q_0 E \cos\theta\, \Delta l$$

or

$$E \cos\theta = \frac{\Delta V}{\Delta l}. \tag{26–14}$$

Now $E \cos\theta$ is the component of $\mathbf{E}$ in the direction ℓ in Fig. 26–16; we call it E_l. In the differential limit Eq. 26–14 can then be written as

$$E_l = -\frac{dV}{dl}. \tag{26–15}$$

Electric field from potential

In words, Eq. 26–15 says: If we travel through an electric field along a straight line and measure V as we go, the rate of change of V with distance that we observe, when changed in sign, is the component of $\mathbf{E}$ in that direction. The minus sign tells us that $\mathbf{E}$ points in the direction of decreasing V, as in Fig. 26–16. It is clear from Eq. 26–15 that appropriate units for $\mathbf{E}$ are volt/meter (V/m).

---

* We assume that the equipotentials are so close together that $\mathbf{F}$ is constant for all parts of the path $\Delta\boldsymbol{\ell}$. In the limit of a differential path ($d\boldsymbol{\ell}$) there will be no difficulty.

There will be one direction ℓ for which the quantity $-dV/dl$ is a maximum. From Eq. 26–15, E_l will also be a maximum for this direction and will in fact be E itself. Thus

$$E = -\left(\frac{dV}{dl}\right)_{\text{max}}. \tag{26–16}$$

The maximum value of dV/dl at a given point is called the *potential gradient* at that point. The direction ℓ for which dV/dl has its maximum value is always at right angles to the equipotential surface, corresponding to $\theta = 0$ in Fig. 26–16.

If we take the direction ℓ to be, in turn, the directions of the x, y, and z axes, we can find the three components of **E** at any point, from Eq. 26–15.

The electric field components

$$E_x = -\frac{\partial V}{\partial x}; \qquad E_y = -\frac{\partial V}{\partial y}; \qquad E_z = -\frac{\partial V}{\partial z}. \tag{26–17}$$

Thus if V is known for all points of space, that is, if the function $V(x, y, z)$ is known, the components of **E**, and thus **E** itself, can be found by taking derivatives.*

Example 9   Calculate $E(r)$ for a point charge q, using Eq. 26–16 and assuming that $V(r)$ is given as (see Eq. 26–8)

$$V = \frac{1}{4\pi\epsilon_0}\frac{q}{r}.$$

From symmetry, **E** must be directed radially outward for a (positive) point charge. Consider a point P in the field a distance r from the charge. It is clear that $-dV/dl$ at P has its greatest value if the direction l is identified with that of r. Thus, from Eq. 26–16,

$$E = -\frac{dV}{dr} = -\frac{d}{dr}\left(\frac{1}{4\pi\epsilon_0}\frac{q}{r}\right)$$

$$= -\frac{q}{4\pi\epsilon_0}\frac{d}{dr}\left(\frac{1}{r}\right) = \frac{1}{4\pi\epsilon_0}\frac{q}{r^2}.$$

This result agrees exactly with Eq. 24–4, as it must.

26-8   An Insulated Conductor

We proved in Section 25–6, using Gauss's law, that after a steady state is reached an excess charge q placed on an insulated conductor (see Fig. 25–6) will move to its outer surface. We now assert that this charge q will distribute itself on this surface so that, in equilibrium, all points of the conductor, including those on the surface and those inside, have the same potential.

Consider any two points A and B in or on the conductor. If they were not at the same potential, the electronic charge carriers in the conductor near the point of lower potential would tend to move toward the point of higher potential. We have assumed, however, that a steady-state situation, in which such currents do not exist, has been reached; thus all points, both on the surface and inside it, must have the same potential. Since the surface of the conductor is an equipotential surface, **E** for points on the surface must be at right angles to the surface.

Potential and the electric field for a conductor

We saw in Section 25–6 that a charge placed on an insulated conductor will spread over the surface until **E** equals zero for all points inside. We now have an alternative way of saying the same thing; the charge will move until all points of the conductor (surface points and interior points) are brought to the same potential, for if V is constant in the conductor, then **E** is zero everywhere in the conductor ($E_l = -dV/dl$).

---

* Remember that the partial derivative $\partial V/\partial x$ instructs us to take the derivative of $V(x, y, z)$ considering x to be the variable and y and z to be constants. Similarly for $\partial V/\partial y$ and $\partial V/\partial z$.

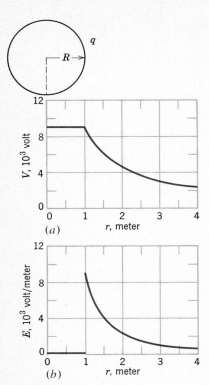

Figure 26-17 (a) The potential and (b) the electric field, for points near a conducting spherical shell of 1.0-m radius carrying a charge of +1.0 μC.

Electric field near sharp points

Corona discharge

Figure 26–17a is a plot of potential against radial distance for an isolated thin spherical conducting shell of 1.0-m radius carrying a charge of 1.0 μC. For points outside the shell $V(r)$ can be calculated from Eq. 26–8 because the charge q behaves, for such points, as if it were concentrated at the center of the sphere. Equation 26–8 is correct right up to the surface of the shell. Now let us push the test charge through the surface, assuming that there is a small hole, and into the interior. No extra work is needed because no electrical forces act on the test charge once it is inside the shell. Thus the potential everywhere inside is the same as that on the surface, as Fig. 26–17a shows.

Figure 26–17b shows the electric field for this same sphere. Note that E equals zero inside. The lower of these curves can be derived from the upper by differentation with respect to r, using Eq. 26–16; the upper can be derived from the lower by integration with respect to r, using Eq. 26–6.

Figure 26–17 holds without change if the conductor is a solid sphere rather than a spherical shell as we have assumed. It is instructive to compare Fig. 26–17b (conducting shell or sphere) with Fig. 25–10, which holds for a uniform spherical charge distribution. Try to understand the difference between these two figures, bearing in mind that in the first the charge lies entirely on the surface, whereas in the second it was assumed to have been spread uniformly throughout the volume of the sphere.

Finally we note that, as a general rule, the charge density σ tends to be high on isolated conducting surfaces whose radii of curvature are small, and conversely. For example, the charge density tends to be relatively high on sharp points and relatively low on plane regions on a conducting surface. The electric field E at points immediately above a charged surface is proportional to the charge density σ so that E may also reach very high values near sharp points. Glow discharges from sharp points during thunderstorms are a familiar example. The fact that σ, and thus E, can become very large near sharp points is important in the design of high-voltage equipment. Corona discharge can result from such points if the conducting object is raised to high potential and surrounded by air. Normally air is thought of as a nonconductor. However, it contains a small number of ions produced, for example, by the cosmic rays. A positively charged conductor will attract negative ions from the surrounding air and thus will slowly neutralize itself.

If the charged conductor has sharp points, the value of E in the air near the points can be very high. If the value is high enough, the ions, as they are drawn toward the conductor, will receive such large accelerations that, by collision with air molecules, they will produce vast additional numbers of ions. The air is thus made much more conducting, and the discharge of the conductor by this corona discharge may be very rapid indeed. The air surrounding sharp conducting points may even glow visibly because of light emitted from the air molecules during these collisions. This glowing can often be seen near pointed conductors during thunderstorms. Lightning rods are designed to have sharp points and a conducting wire to ground; the large electric fields near the points facilitate a flow of current between the air and ground thus neutralizing charged clouds and preventing lightning strokes.

26-9 The Electrostatic Generator

The electrostatic generator was conceived by Lord Kelvin in 1890 and put into useful practice in its modern form by R. J. Van de Graaff in 1931. It is a device for producing electric potential differences of the order of several millions of volts. Its chief application in physics is the use of this potential difference to accelerate charged particles to high energies. Beams of energetic particles made in this way can be used in many different "atom-smashing" experiments. The technique is to let a charged particle "fall" through a potential difference V, gaining kinetic energy as it does so.

Let a particle of (positive) charge q move in a vacuum under the influence of an electric field from one position A to another position B whose electric potential is lower by V. The electric potential energy of the system is reduced by qV because this is the work that an external agent would have to do to restore the system to its original condition. This decrease in potential energy appears as kinetic energy of the particle, or

$$K = qV. \tag{26–18}$$

K is in joules if q is in coulombs and V in volts. If the particle is an electron or a proton, q will be the quantum of charge e.

If we adopt the quantum of charge e as a unit in place of the coulomb, we arrive at another unit for energy, the *electron volt* (abbr. eV), which is used extensively in atomic and nuclear physics. By substituting into Eq. 26–18,

The electron volt, a unit of energy

$$\begin{aligned}
1 \text{ electron volt} &= (1 \text{ quantum of charge})(1 \text{ volt}) \\
&= (1.60 \times 10^{-19} \text{ coulomb})(1.00 \text{ volt}) \\
&= 1.60 \times 10^{-19} \text{ joule.}
\end{aligned}$$

The electron volt can be used interchangeably with any other energy unit. Thus a 10-g object moving at 10 m/s can be said to have a kinetic energy of 3.1×10^{18} eV. Most physicists would prefer to express this result as 0.50 J, the electron volt being inconveniently small. In atomic, nuclear, and particle physics, however, the electron volt (eV) and its multiples the keV ($= 10^3$ eV), MeV ($= 10^6$ eV), and the GeV ($= 10^9$ eV) are the usual units of choice.

Example 10   *The electrostatic or Van de Graaff generator*
Figure 26–18, which illustrates the basic operating principle of the electrostatic generator, shows a small sphere of radius r placed inside a large spherical shell of radius R. The two spheres carry charges q and Q, respectively. Calculate their potential difference.

The potential of the large sphere is caused in part by its own charge and in part because it lies in the field set up by the charge q on the small sphere. From Eq. 26–8,

$$V_R = \frac{1}{4\pi\epsilon_0}\left(\frac{Q}{R} + \frac{q}{R}\right).$$

The potential of the small sphere is caused in part by its own charge and in part because it is inside the large sphere; see Fig. 26–17a. From Eq. 26–8,

$$V_r = \frac{1}{4\pi\epsilon_0}\left(\frac{q}{r} + \frac{Q}{R}\right).$$

The potential difference is

$$V_r - V_R = \frac{q}{4\pi\epsilon_0}\left(\frac{1}{r} - \frac{1}{R}\right).$$

Note that the potential difference is independent of Q. Thus, assuming q is positive, the inner sphere will always be higher in potential than the outer sphere. If the spheres are connected by a fine wire, the charge q will flow *entirely* to the outer sphere, regardless of the charge Q that may already be present.

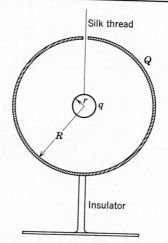

Figure 26–18   Example 10. A small charged sphere of radius r is suspended inside a charged spherical shell whose outer surface has radius R.

From another point of view, we note that since the spheres when electrically connected form a single conductor at electrostatic equilibrium there can be only a single potential. This means that $V_r - V_R = 0$, which can occur only if $q = 0$.

In actual electrostatic generators charge is carried into the shell on rapidly moving belts made of insulating material. Charge is "sprayed" onto the belts outside the shell by corona discharge from a series of sharp metallic points connected to a source of moderately high potential difference. Charge is removed from the belts inside the shell by a similar series of points connected to the shell. Electrostatic generators can be built commercially to accelerate protons to energies up to 10 MeV, using a single acceleration. Generators can also be built in which the accelerated particles are subject to two or three successive accelerations.

REVIEW GUIDE AND SUMMARY

Electric potential difference

If the electric field force does work W_{AB} in moving a test charge q_0 from point A to point B in an electric field the electric potential difference between these two points is given by

$$V_B - V_A = -\frac{W_{AB}}{q_0}. \qquad [26-1]$$

W_{AB} (and thus $V_B - V_A$) are independent of the path taken; see Fig. 26–1.

The electric potential V; the volt

Point A above is often treated as a universal reference point, located at infinity, and V_A is assigned the arbitrary value of zero. Point B is then a generalized field point, at which the electric potential is V_B (or simply V; see Eq. 26–2.). The SI unit of electric potential is the volt, defined as 1 joule/coulomb.

Equipotential surfaces

An equipotential surface is a locus of points all of which have the same electric potential. The work done in carrying a test charge from one such surface to another does not depend on the locations of the initial and terminal points on these surfaces or on the nature of the path that joins them; see Fig. 26–2. The lines of force associated with an electric field are always at right angles to the family of equipotential surfaces; see Fig. 26–15.

Calculating V from E

If the electric field $\mathbf{E}$ is known throughout a given region the potential difference between any two points in that region is given by

$$V_B - V_A = -\int_A^B \mathbf{E} \cdot d\boldsymbol{\ell}. \qquad [26-5]$$

Here the line integral is taken over any path connecting A and B. Examples 1 and 2 illustrate the application of this equation.

Calculating E from V

Conversely, if the potential $V(x, y, z)$ is known throughout a given region the components of $\mathbf{E}$ can be found by differentiation, or

$$E_x = -\frac{\partial V}{\partial x}, \qquad E_y = -\frac{\partial V}{\partial y}, \qquad E_z = -\frac{\partial V}{\partial z}, \qquad [26-17]$$

Equations 26–5 and 26–17 are reciprocal relations connecting $\mathbf{E}$ and V.

Potential due to a point charge or a distribution of charge

If point A in Eq. 26–5 is taken to be at infinity and if we put $V_A = 0$ the potential due to a single point charge can be shown (from Eqs. 26–5 and 24–4) to be

$$V(= V_B) = \frac{1}{4\pi\epsilon_0} \frac{q}{r}. \qquad [26-8]$$

For a continuous distribution of charge this can be extended to

$$V = \frac{1}{4\pi\epsilon_0} \int \frac{dq}{r} \qquad [26-10]$$

where the integral is taken over the entire charge. See Examples 3, 4, 5, and 6.

Electric dipoles

Two equal but opposite charges $\pm q$ separated by a distance $2a$ constitute an electric dipole. Its dipole moment $\mathbf{p}$ points from the negative to the positive charge and has a magnitude $2qa$. The potential due to a dipole is

$$V(r, \theta) = \frac{1}{4\pi\epsilon_0} \frac{p \cos \theta}{r^2} \qquad \text{(for } r \gg 2a\text{)} \qquad [26-11]$$

where r and θ are defined from Fig. 26–9. Many molecules have intrinsic dipole moments (Fig. 26–11). As Fig. 26–12 shows, a dipole moment can also be induced by the action of an external electric field.

Electric potential energy

The electric potential energy of a system of point charges is defined as the work needed to assemble the system, starting with the charges initially at rest and infinitely distant from each other. For two charges we have

$$U(=W) = \frac{1}{4\pi\epsilon_0} \frac{q_1 q_2}{r}.$$ [26–13]

Examples 7 and 8 show applications.

A charged conductor

We saw earlier (Sec. 25–6) that an excess charge placed on a conductor will, in equilibrium, be on its outer surface. From another point of view we say that in so distributing itself this charge brings the entire conductor, including both surface and interior points, to a uniform potential. Figure 26–17 shows $V(r)$—and also $E(r)$—for a charged conducting sphere or spherical shell. Such considerations explain the action of an electrostatic generator.

The electrostatic generator

Example 10 and Fig. 26–18 illustrate its working principle. It should be clear that if a connection is made between the two conducors in that figure the charge q on the inner sphere will move *entirely* to the outer shell, no matter what charge Q is already on that shell.

QUESTIONS

1. Are we free to call the potential of the earth $+ 100$ V instead of zero? What effect would such an assumption have on measured values of (a) potentials and (b) potential differences?

2. What would happen to a person on an insulated stand if his potential was increased by 10 kV with respect to the earth?

3. Do electrons tend to go to regions of high potential or of low potential?

4. Suppose that the earth has a net charge that is not zero. Is it still possible to adopt the earth as a standard reference point of potential and to assign the potential $V = 0$ to it?

5. Does the potential of a positively charged insulated conductor have to be positive? Give an example to prove your point.

6. Can two different equipotential surfaces intersect?

7. An electrical worker was accidentally electrocuted and a newspaper account reported: "He accidentally touched a high voltage cable and 20,000 V of electricity surged through his body." Criticize this statement.

8. Advice to mountaineers caught in lightning and thunderstorms is (a) get rapidly off peaks and ridges and (b) put both feet together and crouch in the open, only the feet touching the ground. What is the basis for this good advice?

9. If E equals zero at a given point, must V equal zero for that point? Give some examples to prove your answer.

10. If you know E at a given point, can you calculate V at that point? If not, what further information do you need?

11. In Fig. 26–2 is the electric field E greater at the left or at the right side of the figure?

12. Is the uniformly charged, nonconducting disk of Example 6 a surface of constant potential? Explain.

13. We have seen that, inside a hollow conductor, you are shielded from the fields of outside charges. If you are *outside* a hollow conductor that contains charges, are you shielded from the fields of these charges?

14. Distinguish between potential difference and difference of potential energy. Give examples of statements in which each term is used properly.

15. If the surface of a charged conductor is an equipotential, does that mean that charge is distributed uniformly over that surface? If the electric field is constant in magnitude over the surface of a charged conductor, does *that* mean that the charge is distributed uniformly?

16. In Example 10 we learned that charge delivered to the *inside* of an isolated conductor is transferred *entirely* to the outer surface of the conductor, no matter how much charge is already there. Can you keep this up forever? If not, what stops you?

17. Ions and electrons act like condensation centers; that is, water droplets form around them in air. Explain why.

18. If V equals a constant throughout a given region of space, what can you say about E in that region?

19. In Section 15–4 we saw that the gravitational field strength is zero inside a spherical shell of matter. The electrical field strength is zero not only inside an isolated charged spherical conductor but inside an isolated conductor of any shape. Is the gravitational field strength inside, say, a cubical shell of matter zero? If not, in what respect is the analogy not complete?

20. How can you insure that the electric potential in a given region of space will have a constant value?

21. We have seen (Fig. 26–17a) that the potential inside a conductor in the shape of a spherical shell is the same as that on its surface. (a) What if the conductor is a solid sphere? (b) What if the conductor is solid but irregularly shaped? (c) What if it is irregularly shaped and has an irregularly shaped cavity inside? (d) What if the cavity has a small "worm hole" connecting it to the outside? (e) What if the cavity is closed but has a point charge suspended within it? Discuss the potential within the conducting material and at different points within the cavities for these cases.

22. An isolated conducting spherical shell carries a negative charge. What will happen if a positively charged metal object is placed in contact with the shell interior? Assume that the positive charge is (a) less than, (b) equal to, and (c) greater than the negative charge in magnitude.

EXERCISES

Section 26-1   Electric Potential

1. In a typical lightning flash, the potential difference between discharge points is about 10^9 V and the quantity of charge transferred is about 30 C. (*a*) If all the energy released could be used to accelerate a 1000-kg automobile from rest, what would be its final speed? (*b*) If it could be used to melt ice, how much would it melt at 0°C?

Section 26-2   Potential and the Electric Field

2. In moving from A to B along an electric field line, the electric field does 3.94×10^{-19} J of work on an electron in the field illustrated in Fig. 26-19. (*a*) What is the difference in the electric potential $V_B - V_A$? (*b*) $V_C - V_A$? (*c*) $V_C - V_B$?

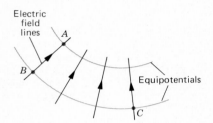

Figure 26-19   Exercise 2.

3. In the Millikan oil drop experiment (see Fig. 24-21), an electric field of 1.92×10^5 N/C is maintained at balance across two plates separated by 1.50 cm. Find the potential difference between the plates.

4. Two large parallel metal plates are 1.5 cm apart and carry equal but opposite charges on their facing surfaces. The negative plate is grounded and its potential is taken to be zero. If the potential halfway between the plates is $+5.0$ V, what is the electric field in this region?

5. Two large parallel conducting plates are 10 cm apart and carry equal but opposite charges on their facing surfaces. An electron placed midway between the two plates experiences a force of 1.6×10^{-15} N. What is the potential difference between the plates?

6. An infinite charged sheet has a charge density σ of 0.10 μC/m^2. How far apart are equipotential surfaces whose potentials differ by 50 V?

7. Two line charges are parallel to the z-axis. One, of charge per unit length $+\lambda$, is a distance a to the right of this axis. The other, of charge per unit length $-\lambda$, is a distance a to the left of this axis (the lines and the z-axis being in the same plane). Sketch some of the equipotential surfaces.

Section 26-3   Potential Due to a Point Charge

8. A point charge has $q = +1.0\,\mu$C. Consider point A, which is 2.0 m distant, and point B, which is 1.0 m distant in a direction diametrically opposite, as in Fig. 26-20*a*. (*a*) What is the potential difference $V_A - V_B$? (*b*) Repeat if points A and B are located as in Fig. 26-20*b*.

9. Consider a point charge with $q = 1.5 \times 10^{-8}$ C. (*a*) What is the radius of an equipotential surface having a potential of 30 V?

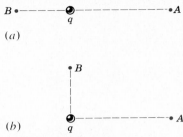

Figure 26-20   Exercise 8.

(*b*) Are surfaces whose potentials differ by a constant amount (1.0 V, say) evenly spaced in radius?

10. A charge of 10^{-8} C can be produced by simple rubbing. To what potential would such a charge raise an insulated conducting sphere of 10-cm radius?

11. We saw in Example 1, Chapter 23, that a single copper penny contains a positive charge of about 1.4×10^5 C and a negative charge of the same magnitude. Suppose that the negative charge were removed to a very large distance from the earth — perhaps to a distant galaxy — and that the positive charge were distributed uniformly over the earth's surface. By how much would the electric potential at the surface of the earth change?

Figure 26-21   Exercise 13. NASA.

12. As a space shuttle moves through the dilute ionized gas of the earth's outer atmosphere, its potential is typically changed by -1.0 V. By assuming that the shuttle is a sphere of radius 10 m, estimate the amount of charge it collects.

13. Much of the material comprising Saturn's rings, see Fig 26–21, is in the form of tiny dust particles having radii on the order of 10^{-6} m. These grains are in a region containing a dilute ionized gas, and they pick up excess electrons. If the electric potential at the surface of a grain is -400 V, how many excess electrons has it picked up?

Section 26–4   A Group of Point Charges
14. In Fig. 26–22, sketch qualitatively (a) the lines of force and (b) the intersections of the equipotential surfaces with the plane of the figure. (*Hint:* Consider the behavior close to each point charge and at considerable distances from the pair of charges.)

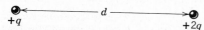

$+q$ $\qquad\qquad d \qquad\qquad$ $+2q$

Figure 26–22   Exercise 14.

15. Repeat the procedure explained in Exercise 14 for Fig. 26–23.

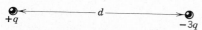

$+q$ $\qquad\qquad d \qquad\qquad$ $-3q$

Figure 26–23   Exercise 15.

Section 26–5   Potential Due to a Dipole
16. In the text the value for the electric dipole moment of a water molecule is given as 6.2×10^{-30} C·m. Estimate the maximum electric potential at the site of a neighboring water molecule, assuming that the molecular spacing is 3.0×10^{-8} cm.

Section 26–6   Electric Potential Energy
17. The electric potential difference between discharge points during a thunderstorm may be as large as 10^9 V (see Exercise 1). What is the magnitude of the change in the electrical potential energy of an electron that moves between these points?

18. A good 12-V car battery is rated to deliver a charge of 84 A·h (ampere·hours). (a) How many coulombs of charge does this represent? (b) If this entire charge is delivered at 12 V, how much energy is available?

Section 26–7   Calculation of E from V
19. An electric field of approximately 100 V/m is often observed near the surface of the earth. If this field were the same over the entire surface, what would be the electric potential of a point on the surface? Assume that the earth, being spherical, may be treated as a point charge, and use the results of Example 26–9.

20. Examine Fig. 26–15 and, in an approximate way, apply Eq. 26–16 to see whether or not the electric field magnitudes are consistent with the spacing of the equipotentials, assuming that the potential differences between every pair of adjacent equipotentials are equal in magnitude.

21. Problem 25–31 deals with Rutherford's calculation of the electric field a distance r from the center of an atom. He also gave the electric potential as

$$V = \frac{Ze}{4\pi\epsilon_0}\left(\frac{1}{r} - \frac{3}{2R} + \frac{r^2}{2R^3}\right).$$

Show that the expression for the electric field given in Problem 25–31 follows from the above expression for V.

Section 26–8   An Insulated Conductor
22. Can a conducting sphere 10 cm in radius hold a charge of 4 μC in air without breakdown? The dielectric strength (minimum field required to produce breakdown) of air at 1 atm is 3 MV/m.

Section 26–9   The Electrostatic Generator
23. Let the potential difference between the shell of an electrostatic generator and the point at which charges are sprayed onto the moving belt be 3.0 MV. If the belt transfers charge to the shell at the rate of 3.0 mC/s, what power must be provided to drive the belt, considering only electric forces?

24. (a) How much charge is required to raise an isolated metallic sphere of 1.0-m radius to a potential of 1.0 MV? Repeat for a sphere of 1.0-cm radius. (b) Why use a large sphere in an electrostatic generator when the same potential can be achieved using a smaller charge with a small sphere?

25. An α particle (which consists of two protons and two neutrons) is accelerated through a potential difference of one million volts in an electrostatic generator. (a) What kinetic energy does it acquire? (b) What kinetic energy would a proton acquire under these same circumstances? (c) Which particle would acquire the greater speed, starting from rest?

PROBLEMS GROUPED BY SECTION

Section 26–2   Potential and the Electric Field
*1. A charge q is distributed uniformly throughout a spherical volume of radius R. Show that the potential a distance r from the center, where $r < R$, is given by

$$V = \frac{q(3R^2 - r^2)}{8\pi\epsilon_0 R^3},$$

where the zero of potential is taken at $r = \infty$.

Section 26–3   Potential Due to a Point Charge
2. What is the charge density on the surface of a conducting sphere of radius 0.15 m whose potential is 200 V?

3. A spherical drop of water carrying a charge of 3×10^{-11} C

has a potential of 500 V at its surface. (a) What is the radius of the drop? (b) If two such drops of the same charge and radius combine to form a single spherical drop, what is the potential at the surface of the new drop so formed?

4. A thin conducting spherical shell of outer radius 20 cm carries a charge of $+3.0$ μC. Sketch (a) the magnitude of the electric field **E** and (b) the potential V, versus the distance r from the center of the shell.

Section 26–4   A Group of Point Charges
5. In Fig. 26–24, locate the points (a) where $V = 0$ and (b) where $\mathbf{E} = 0$. Consider only points on the axis, and choose $d = 1.0$ m.

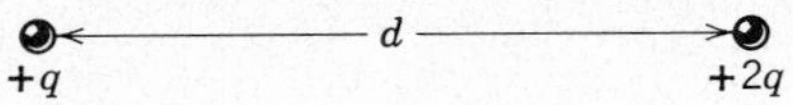

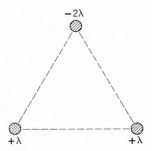

+q $\xleftarrow{\hspace{1.5cm} d \hspace{1.5cm}}$ +2q

Figure 26–24 Problem 5.

6. Three long parallel lines of charge have the relative linear charge densities shown in Fig. 26–25. Sketch the lines of force and the intersections of the equipotential surfaces with the plane of this figure.

−2λ

+λ +λ

Figure 26–25 Problem 6.

Section 26–5 Potential Due to a Dipole
7. The H_2O molecule (see Fig. 26–11) has an electric dipole moment of 6.2×10^{-30} C·m, an OH bond length of 0.96×10^{-10} m, and an angle of 104° between the two OH bonds. The oxygen nucleus has a charge of $+8e$, and the two hydrogen nuclei have charges of $+e$ each, where e ($= 1.60 \times 10^{-19}$ C) is the electronic charge. (a) Locate the center of positive charge for the molecule by giving its distance from the oxygen nucleus, measured along the symmetry axis. (b) Use the measured value of the electric dipole moment, given above, to locate the center of negative charge along this same axis. Is it about where you thought it would be?

Section 26–6 Electric Potential Energy
8. The charges and coordinates of two point charges located in the x-y plane are $q_1 = +3.0 \times 10^{-6}$ C; $x = 3.5$ cm, $y = +0.50$ cm, and $q_2 = -4.0 \times 10^{-6}$ C; $x = -2.0$ cm, $y = +1.5$ cm. (a) Find the electric potential at the origin. (b) How much work must be done to locate these charges at their given positions, starting from infinity?

9. A particle of charge q is kept in a fixed position at a point P and a second particle of mass m, having the same charge q, is initially held at rest a distance r_1 from P. The second particle is then released and is repelled from the first one. Determine its speed at the instant it is a distance r_2 from P. Let $q = 3.1 \, \mu C$, $m = 20$ mg, $r_1 = 0.90$ mm, and $r_2 = 2.5$ mm.

10. Three (large) charges of $+0.10$ C each are placed on the corners of an equilateral triangle, 1.0 m on a side. If energy is supplied at the rate of 1.0 kW, how many days would be required to move one of the charges onto the midpoint of the line joining the other two?

11. An electric charge of -9.0 nC is uniformly distributed around a ring of radius 1.5 m that lies in the y-z plane with its center at the origin. A point charge -6.0 pC is located on the x-axis at $x = 3.0$ m. Calculate the work done in moving the point charge to the origin.

12. Two small metal spheres of mass m_1 ($= 5.0$ g) and mass m_2 ($= 10$ g) carry equal positive charges q ($= 5.0 \, \mu C$). The spheres are connected by a massless string of length d ($= 1.0$ m), which is much greater than the sphere radius. (a) What is the electrostatic potential energy of the system? (b) You cut the string. At that instant, what is the acceleration of each of the spheres? (c) A long time after you cut the string, what is the velocity of each sphere?

13. A gold nucleus contains a positive charge equal to that of 79 protons. An α particle (which consists of two protons and two neutrons) has a kinetic energy K at points far from this charge and is traveling directly toward the charge. The particle just touches the surface of the charge (assumed spherical) and is reversed in direction. (a) Calculate K, assuming a nuclear radius of 6.2×10^{-15} m. (b) The actual α-particle energy used in the experiment of Rutherford and his collaborators that led to the discovery of the concept of the atomic nucleus was 5.0 MeV. What do you conclude?

14. A copper sphere whose radius is 1.00 cm has a very thin surface coating of nickel. Some of the nickel atoms are radioactive, each atom emitting an electron as it decays. Half of these electrons enter the copper sphere, each depositing 100 keV of energy there. The other half of the electrons escape, each carrying away a charge of $-e$. The nickel coating has an activity of 10 mCi ($= 10$ millicuries $= 3.70 \times 10^8$ radioactive decays per second). The sphere is hung from a long, nonconducting string and is insulated from its surroundings. (a) How long will it take for the potential of the sphere to increase by 1000 V? (b) How long will it take for the temperature of the sphere to increase by 5.00°C? (The heat capacity of the sphere is 14.3 J/°C.)

15. A decade before Einstein published his theory of relativity, J. J. Thomson proposed that the electron might be made up of small parts and that its mass results from the electrical interaction of the parts. Furthermore, he suggested that the energy equals mc^2. Make a rough estimate of the electron mass in the following way: Assume that the electron is composed of three identical parts that are brought in from infinity and placed at the vertices of an equilateral triangle having sides equal to the classical radius of the electron, 2.82×10^{-15} m. (a) Find the total electrical potential energy of this arrangement. (b) Divide by c^2 and compare your result to the accepted electron mass (9.11×10^{-31} kg). The result improves if more parts are assumed.

Section 26–7 Calculation of E from V
16. The electric potential varies along the x-axis as shown in the graph of Fig. 26–26. For each of the intervals shown (ignore the behavior at the end points of the intervals), determine the x-component of the electric field and plot E_x vs. x.

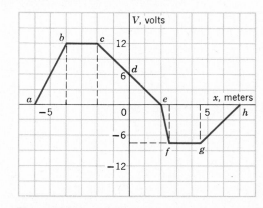

Figure 26–26 Problem 16.

17. What is the potential gradient at the surface of a gold nucleus? See Problem 13.

18. In Example 6 the potential at an axial point for a charged disk was shown to be

$$V = \frac{\sigma}{2\epsilon_0}(\sqrt{a^2 + r^2} - r).$$

(*a*) From this result, show that E for axial points is given by

$$E = \frac{\sigma}{2\epsilon_0}\left(1 - \frac{r}{\sqrt{a^2 + r^2}}\right).$$

(*b*) Does this expression for E reduce to an expected result for $r \gg a$ and for $r = 0$?

19. A charge per unit length λ is distributed uniformly along a straight-line segment of length L. (*a*) Determine the potential (chosen to be zero at infinity) at a point P a distance y from one end of the charged segment and in line with it (see Fig. 26–27). (*b*) Use the result of (*a*) to compute the component of the electric field at P in the y-direction (along the line). (*c*) Determine the component of the electric field at P in a direction perpendicular to the straight line.

Figure 26–27 Problem 19.

*20. On a thin rod of length L lying along the x-axis with one end at the origin ($x = 0$), as in Fig. 26–28, there is distributed a charge per unit length given by $\lambda = kx$, where k is a constant. (*a*) Taking the electrostatic potential at infinity to be zero, find V at the point P on the y-axis. (*b*) Determine the vertical component, E_y, of the electric field intensity at P from the result of part (*a*) and also by direct calculation. (*c*) Why cannot E_x, the horizontal component of the electric field at P, be found using the result of part (*a*)?

Section 26–8 An Insulated Conductor

21. What is the charge density σ on the surface of a conducting sphere of radius 15 cm whose potential is 200 V?

22. The metal object in Fig. 26–29 is a figure of revolution about the horizontal axis. If it is charged negatively, sketch roughly a few equipotentials and lines of force. Use physical reasoning rather than mathematical analysis.

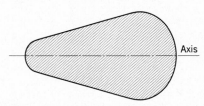

Figure 26–29 Problem 22.

23. Two identical conducting spheres of radius $r = 0.15$ m are separated by a distance $a = 10$ m. What is the charge on each sphere if the potential of one is $+1500$ V and if the other is -1500 V? What assumptions have you made?

24. Two thin, insulated, concentric conducting spheres of radii R_1 and R_2 carry charges q_1 and q_2. Derive expressions for $e(r)$ and $V(r)$, where r is the distance from the center of the spheres. Plot $E(r)$ and $V(r)$ from $r = 0$ to $r = 4.0$ m for $R_1 = 0.50$ m, $R_2 = 1.0$ m, $q_1 = +2.0$ μC, and $q_2 = +1.0$ μC. Compare with Fig. 26–17.

Section 26–9 The Electrostatic Generator

25. The high-voltage electrode of an electrostatic generator is a charged spherical metal shell having a potential $V (= +9.0$ MV). (*a*) Electrical breakdown occurs in the gas in this machine at a field $E (= 100$ MV/m). In order to prevent such breakdown, what restriction must be made on the radius r of the shell? (*b*) A long, moving rubber belt transfers charge to the shell at 300 μC/s, the potential of the shell remaining constant because of leakage. What minimum power is required to transfer the charge? (*c*) The belt is of width $w (= 0.50$ m) and travels at speed $v (= 30$ m/s). What is the surface charge density on the belt?

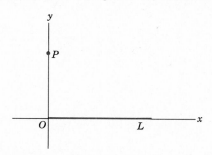

Figure 26–28 Problem 20.

ADDITIONAL PROBLEMS

26. In the quark model of fundamental particles, a proton is composed of three quarks: two "up" quarks, each having charge $+\frac{2}{3}e$, and one "down" quark, having charge $-\frac{1}{3}e$. Suppose that the three quarks are equidistant from each other. Take the distance to be 1.32×10^{-15} m and calculate (*a*) the potential energy of the interaction between the two "up" quarks and (*b*) the total electrical potential energy of the system.

27. In the rectangle shown in Fig. 26–30, the sides have lengths 5.0 cm and 15 cm, $q_1 = -5.0$ μC, and $q_2 = +2.0$ μC. (*a*) What is the electric potential at corner B? at corner A? (*b*) How much work is required to move a third charge $q_3 = +3.0$ μC from B to A along a diagonal of the rectangle? (*c*) In this process, is the external work converted into electrostatic potential energy or vice versa? Explain.

Figure 26-30   Problem 27.

28. Devise an arrangement of three point charges, separated by finite distances, that has zero electric potential energy.

29. The electric field inside a nonconducting sphere of radius a, containing uniform charge density, is radially directed and has magnitude

$$E(r) = \frac{qr}{4\pi\epsilon_0 a^3},$$

where q is the total charge in the sphere and r is the distance from the sphere center. What is the difference in electric potential between a point on the surface and the sphere center? If q is positive, which point is at the higher potential?

30. Calculate (*a*) the electric potential established by the nucleus of a hydrogen atom at the mean distance of the circulating electron ($r = 5.3 \times 10^{-11}$ m), (*b*) the electric potential energy of the atom when the electron is at this radius, and (*c*) the kinetic energy of the electron, assuming it to be moving in a circular orbit of this radius centered on the nucleus. (*d*) How much energy is required to ionize the hydrogen atom? Express all energies in electron volts.

31. Compute the escape velocity for an electron from the surface of a uniformly charged sphere of radius 1.0 cm and total charge 1.6×10^{-15} C. Neglect gravitational forces.

32. A Geiger counter has a metal cylinder 2.0 cm in diameter along whose axis is stretched a wire 1.3×10^{-2} cm in diameter. If 850 V is applied between them, what is the electric field at the surface of (*a*) the wire, and (*b*) the cylinder? (*Hint:* Use the result of Example 3, Chapter 25.)

33. Derive an expression for the work required to put the four charges together as indicated in Fig. 26-31.

Figure 26-31   Problem 33.

34. If the earth had a net charge equivalent to 1 electron/m^2 of surface area (a very artificial assumption), (*a*) what would be the earth's potential? (*b*) What would be the electric field due to the earth just outside its surface?

35. What is the electric potential energy of the charge configuration of Fig. 26-7? Use the numerical values of Example 5.

36. A charged metal sphere of radius 15 cm contains a total charge of 3.0×10^{-8} C. (*a*) What is the electric field at the sphere's surface? (*b*) What is the electric potential at the sphere's surface?

(*c*) At what distance from the sphere's surface has the electric potential decreased by 500 V?

37. (*a*) Show that the electric potential at a point on the axis of a ring of charge of radius a, computed directly from Eq. 26-10, is given by

$$V = \frac{1}{4\pi\epsilon_0} \frac{q}{\sqrt{x^2 + a^2}}.$$

(*b*) From this result, derive an expression for E at axial points; compare with the direct calculation of E in Example 5, Chapter 24.

38. Two charges q $(=+2.0\ \mu C)$ are fixed in space a distance d $(=2.0$ cm) apart, as shown in Fig. 26-32. (*a*) What is the electric potential at point C? (*b*) You bring a third charge q $(=+2.0\ \mu C)$ slowly from infinity to C. How much work must you do? (*c*) What is the potential energy U of the configuration when the third charge is in place?

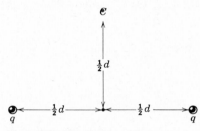

Figure 26-32   Problem 38.

39. Between two parallel, flat, conducting surfaces of spacing d $(=1.0$ cm) and potential difference V $(=10$ kV), an electron is projected from one plate directly toward the second. What is the initial velocity of the electron if it comes to rest just at the surface of the second plate?

40. Two electrons are fixed 2.0 cm apart. Another electron is shot from infinity and comes to rest midway between the two. What must be its initial velocity?

41. (*a*) For Fig. 26-33, derive an expression for $V_A - V_B$. (*b*) Does your result reduce to the expected answer when $d = 0$? When $a = 0$? When $q = 0$?

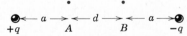

Figure 26-33   Problem 41.

42. Starting from Eq. 26-11, find E_r, the radial component of the electric field due to a dipole. Does the variation of E_r with θ seem reasonable?

43. A thin, spherical, conducting shell of radius R is mounted on an insulated support and charged to a potential $-V$. An electron is fired from point P a distance r from the center of the shell ($r \gg R$) with an initial speed v_0, directed radially inward. What is the value of v_0 chosen so that the electron will just reach the shell?

44. Two metal spheres are 3.0 cm in radius and carry charges of $+1.0 \times 10^{-8}$ C and -3.0×10^{-8} C, respectively, assumed to be uniformly distributed. If their centers are 2.0 m apart, calculate (*a*) the potential of the point halfway between their centers and (*b*) the potential of each sphere.

45. A particle of mass m, charge $q > 0$, and initial kinetic energy K is projected (from "infinity") toward a massive nucleus of charge Q, assumed to have a fixed position in our reference frame. If the aim is "perfect," how close to the center of the nucleus is the particle when it comes instantaneously to rest?

46. A particle of (positive) charge Q is assumed to have a fixed position at P. A second particle of mass m and (negative) charge $-q$ moves at constant speed in a circle of radius r_1, centered at P. Derive an expression for the work W that must be done by an external agent on the second particle in order to increase the radius of the circle of motion, centered at P, to r_2.

47. Near the center of our sun, a carbon-12 nucleus (charge $= 6e$) and a proton approach each other head-on. Each has an energy of 1300 eV, about the average thermal energy of a particle deep inside the sun. How close together will the centers of the two particles be when their mutual repulsion brings them momentarily to rest?

48. For the charge configuration of Fig. 26–34, show that $V(r)$ for points on the vertical axis, assuming $r \gg a$, is given by

$$V = \frac{1}{4\pi\epsilon_0} \frac{q}{r} \left(1 + \frac{2a}{r} \right).$$

Is this an expected result? (*Hint:* The charge configuration can be viewed as the sum of an isolated charge and a dipole.)

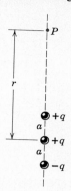

Figure 26–34 Problem 48.

*49. A thick spherical shell of charge of uniform charge density is bounded by radii r_1 and r_2, where $r_2 > r_1$. Find the electric potential V as a function of the distance r from the center of the distribution, considering the regions $(a)\, r > r_2$, $(b)\, r_2 > r > r_1$, and $(c)\, r < r_1$. (d) Do these solutions agree at $r = r_2$ and at $r = r_1$?

27 Capacitors and Dielectrics

27-1 Capacitance

Figure 27–1 shows a generalized *capacitor,* consisting of two insulated conductors, *a* and *b*, of arbitrary shape (later, regardless of their geometry, we will call them *plates*). We assume that they are totally isolated from objects in their surroundings and carry equal and opposite charges $+q$ and $-q$, respectively. Every line of force that originates on *a* terminates on *b*. We further assume, for the time being, that the conductors *a* and *b* exist in a vacuum.

The capacitor of Fig. 27–1 is characterized by q, the magnitude of the charge on either conductor, and by V, the positive potential difference between the con-

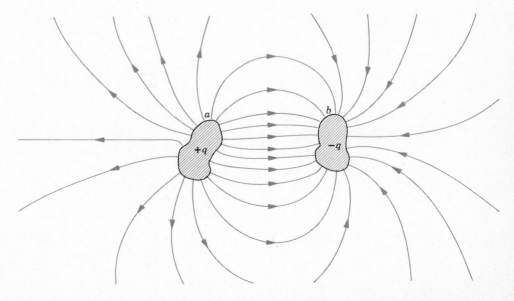

Figure 27–1 Two insulated conductors, totally isolated from their surroundings and carrying equal and opposite charges, form a capacitor.

ductors. Note (*a*) that *q* is *not* the net charge on the capacitor, which is zero, and (*b*) that *V* is *not* the potential of either conductor but the potential difference between them.

It is not hard to put equal and opposite charges on conductors such as those of Fig. 27–1. We need not charge them separately. All that we have to do is to connect the conductors momentarily to opposite poles of a battery. Equal and opposite charges ($\pm q$) will automatically appear.

For the moment we state without proof that *q* and *V* for a capacitor are proportional to each other, or

Definition of capacitance

$$q = CV \qquad (27\text{–}1)$$

in which *C*, the constant of proportionality, is called the *capacitance* of the capacitor. We state, again for the moment without proof, that *C* depends on the shapes and relative positions of the conductors. We will show in Section 27–2, for three important special cases, that *C* does indeed depend on these variables. *C* also depends on the medium in which the conductors are immersed but for the present we assume this to be a vacuum. Note that *C* is always positive since we use the positive magnitudes of both *q* and *V* in defining it.

The SI unit of capacitance that follows from Eq. 27–1 is the coulomb/volt. A special unit, the *farad* (abbr. F), is used to represent it. It is named in honor of Michael Faraday who, among other contributions, developed the concept of capacitance. Thus

The farad

$$1 \text{ farad} = 1 \text{ coulomb/volt.}$$

The submultiples of the farad, the *microfarad* (1 μF = 10^{-6} F) and the *picofarad* (1 pF = 10^{-12} F), are more convenient units in practice.

Capacitors are very useful devices, of great interest to engineers and physicists. For example:

1. In this book we stress the importance of fields to the understanding of natural phenomena. A capacitor can be used to establish desired electric field configurations for various purposes. In Section 24–5, for example, we described the deflection of an electron beam in a uniform field set up by a capacitor, although we did not use this term in that section.

2. A second concept stressed in this book is energy. By analyzing a charged capacitor we show that electric energy may be considered to be stored in the electric field between the plates and indeed in any electric field, however generated. Because capacitors can confine strong electric fields to small volumes, they can serve as useful devices for storing energy. In an electron synchrotron, which is a machine used to accelerate electrons to velocities near the speed of light, energy accumulated and stored in a large bank of capacitors over a relatively long period of time is made available intermittently to accelerate the electrons by discharging the capacitor in a much shorter time. Many researches and devices in plasma physics also make use of bursts of energy stored in this way.

Why study capacitors?

3. The electronic age could not exist without capacitors. They are used, in conjunction with other devices, to reduce voltage fluctuations in electronic power supplies, to transmit pulsed signals, to generate or detect electromagnetic oscillations at radio frequencies, to provide electronic time delays, and in many other ways. In most of these applications the potential difference between the plates will not be constant, as we assume in this chapter, but will vary with time, often in a sinusoidal or a pulsed fashion. In later chapters we consider some aspects of the capacitor used as a circuit element.

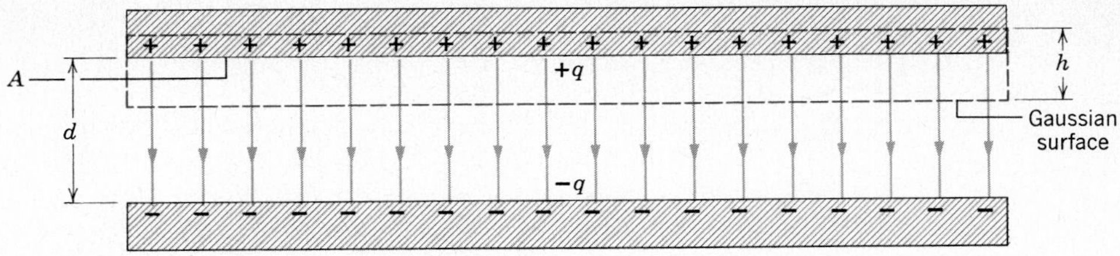

Figure 27–2 A parallel-plate capacitor with conductors (plates) of area A. The dashed line represents a Gaussian surface whose height is h and whose top and bottom caps are the same shape and size as the capacitor plates.

27–2 Calculating Capacitance

Figure 27–2 shows a *parallel-plate* capacitor in which the conductors of Fig. 27–1 take the form of two plane, parallel plates of area A separated by a distance d. If we connect each plate momentarily to the terminals of a battery, a charge $+q$ will automatically appear on one plate and a charge $-q$ on the other. If d is small compared with the plate dimensions, the electric field $\mathbf{E}$ between the plates will be uniform, which means that the lines of force will be parallel and evenly spaced. The laws of electromagnetism (see Problem 15, Chapter 32) require that there be some "fringing" of the lines at the edges of the plates, but for small enough d we can neglect it for our present purpose.

To calculate the capacitance we must relate V, the potential difference between the plates, to q, the capacitor charge. We can express V in terms of $\mathbf{E}$, the electric field between the plates and we can relate $\mathbf{E}$ to q by means of Gauss's law. Figure 27–2 shows (dashed lines) a Gaussian surface of height h closed by plane caps of area A that are the shape and size of the capacitor plates. The flux of $\mathbf{E}$ is zero for the part of the Gaussian surface that lies inside the top capacitor plate because the electric field inside a conductor carrying a static charge is zero. The flux of $\mathbf{E}$ through the vertical wall of the Gaussian surface is zero because, to the extent that we can neglect the fringing of the lines of force, $\mathbf{E}$ lies in the wall.

This leaves only the face of the Gaussian surface that lies between the plates. Here $\mathbf{E}$ is constant and the flux Φ_E is simply EA. Gauss's law (Eq. 25–5) gives

$$\epsilon_0 \Phi_E = \epsilon_0 EA = q. \tag{27–2}$$

The work required to carry a test charge q_0 from one plate to the other can be expressed either as $q_0 V$ (see Eq. 26–1) or as the product of a force $q_0 E$ times a distance d or $q_0 Ed$. These expressions must be equal, or

$$V = Ed. \tag{27–3}$$

Formally, Eq. 27–3 is a special case of the general relation (Eq. 26–5; see also Example 1, Chapter 26)

$$V = - \int \mathbf{E} \cdot d\boldsymbol{\ell},$$

where V is the difference in potential between the plates. The integral may be taken over any path that starts on one plate and ends on the other because each plate is an equipotential surface and the electrostatic force is path independent. Although the simplest path between the plates is a perpendicular straight line, Eq. 27–3 follows no matter what path of integration we choose.

If we substitute Eqs. 27–2 and 27–3 into the relation $C = q/V$, we obtain

A parallel-plate capacitor

$$C = \frac{q}{V} = \frac{\epsilon_0 E A}{E d} = \frac{\epsilon_0 A}{d}.$$ (27–4)

Equation 27–4 holds only for capacitors of the parallel-plate type; different formulas hold for capacitors of different geometry. This equation shows, for a particular case, that C does indeed depend on the geometry of the conductors (plates) as we pointed out in Section 27–1. Both A and d are geometrical factors.

In Section 23–4 we stated that ϵ_0, which we first met in connection with Coulomb's law, was not measured in terms of that law because of experimental difficulties. Equation 27–4 suggests that we might determine ϵ_0 by building a capacitor of accurately known plate area and plate spacing and measuring its capacitance experimentally by measuring q and V in the relation $C = q/V$. Thus we can solve Eq. 27–4 for ϵ_0 and find a numerical value in terms of the measured quantities A, d, and C; ϵ_0 has indeed been measured in this way.

Example 1 The parallel plates of an air-filled capacitor are everywhere 1.0 mm apart. What must the plate area be if the capacitance is to be 1.0 F?

From Eq. 27–4

$$A = \frac{dC}{\epsilon_0} = \frac{(1.0 \times 10^{-3} \text{ m})(1.0 \text{ F})}{8.9 \times 10^{-12} \text{ C}^2/\text{N·m}^2} = 1.1 \times 10^8 \text{ m}^2.$$

This is the area of a square sheet more than 6 miles on edge; the farad is indeed a large unit.

Example 2 *A cylindrical capacitor* A cylindrical capacitor consists of two coaxial cylinders (Fig. 27–3) of radii a and b and length l. What is the capacitance of this device? Assume that the capacitor is very long (that is, that $l \gg b$) so that we can ignore the fringing of the lines of force at the ends for the purpose of calculating the capacitance.

As a Gaussian surface construct a coaxial cylinder of radius r and length l, closed by plane caps. Gauss's law (Eq. 25–6)

$$\epsilon_0 \oint \mathbf{E} \cdot d\mathbf{S} = q$$

gives

$$\epsilon_0 E(2\pi r)(l) = q,$$

in which q is the magnitude of the enclosed charge. Note that the flux is entirely through the cylindrical surface and not through the end caps. Solving for E yields

$$E = \frac{q}{2\pi \epsilon_0 r l}.$$

The potential difference between the plates is given by Eq. 26–5 (note that $\mathbf{E}$ and $d\boldsymbol{\ell}\ (= d\mathbf{r})$ point in opposite directions) or

$$V = -\int_a^b \mathbf{E} \cdot d\boldsymbol{\ell} = \int_a^b E\, dr = \int_a^b \frac{q}{2\pi\epsilon_0 l} \frac{dr}{r} = \frac{q}{2\pi\epsilon_0 l} \ln \frac{b}{a}.$$

Finally, the capacitance is given by

$$C = \frac{q}{V} = \frac{2\pi\epsilon_0 l}{\ln (b/a)}.$$

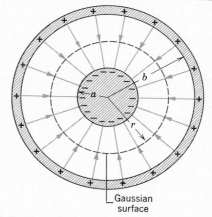

Figure 27–3 Example 2. A cross section of a cylindrical capacitor. The dashed circle is a cross-section of a closed cylindrical Gaussian surface of radius r and length l.

Like the relation for the parallel-plate capacitor (Eq. 27–4), this relation also depends only on geometrical factors, l, b, and a.

Example 3 *The capacitance of an isolated sphere* In Section 26–8 we showed that the potential of an isolated conducting sphere of radius R carrying a charge q is given by

$$V = \frac{1}{4\pi\epsilon_0} \frac{q}{R}.$$ (27–5)

We can regard this sphere as one plate of a capacitor, the other plate being a conducting sphere of infinite radius, with V chosen to be zero on the sphere at infinity.

The capacitance of the sphere of radius R is then given, from Eq. 27–5, by

$$C = \frac{q}{V} = 4\pi\epsilon_0 R.$$ (27–6)

Again, the only relevant geometric factor, the sphere radius R, appears.

In electrical circuits circumstances often arise where combinations of capacitors are used and one wishes to know the combined effect. The two most common combinations are capacitors in series and capacitors in parallel. To show that the effective capacitance of a number of capacitors hooked up in parallel equals the sum of the individual capacitances is left as an exercise (see Problem 9). The series case is discussed in Example 4 that follows.

Example 4 *Capacitors in series* Figure 27–4 shows three capacitors connected in series. What single capacitance C_{eq} is "equivalent" to this combination?

For capacitors connected in a series arrangement (Fig. 27–4) the magnitude q of the charge on each plate must be the same. This is true because the net charge on the part of the circuit enclosed by the dashed line in Fig. 27–4 must be zero; that is, the charge present on these plates initially is zero and connecting a battery between a and b will only produce a charge

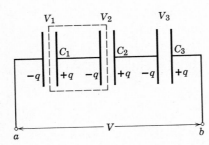

Figure 27–4 Example 4. Three capacitors in series.

separation, the *net* charge on these plates still being zero. Assuming that neither C_1 nor C_2 "sparks over," there is no way for charge to enter or leave the volume suggested by the dashed line.

Applying the relation $q = CV$ to each capacitor yields

$$V_1 = q/C_1; \qquad V_2 = q/C_2; \qquad \text{and} \qquad V_3 = q/C_3.$$

The potential difference for the series combination is

$$V = V_1 + V_2 + V_3$$
$$= q\left(\frac{1}{C_1} + \frac{1}{C_2} + \frac{1}{C_3}\right).$$

The equivalent capacitance is then

$$C_{eq} = \frac{q}{V} = \frac{1}{\dfrac{1}{C_1} + \dfrac{1}{C_2} + \dfrac{1}{C_3}},$$

or

$$\frac{1}{C_{eq}} = \frac{1}{C_1} + \frac{1}{C_2} + \frac{1}{C_3}.$$

The equivalent series capacitance is always less than the smallest capacitance

27–3 Energy Storage in an Electric Field

In Section 26–6 we saw that all charge configurations have a certain electric potential energy U, equal to the work W (which may be positive, zero, or negative) that must be done to assemble them from their individual components, originally assumed to be infinitely far apart and at rest. This potential energy reminds us of the potential energy stored in a compressed spring or the gravitational potential energy stored in, say, the earth-moon system.

For a simple example, work must be done to separate two equal and opposite charges. This energy is stored in the system and can be recovered if the charges are allowed to come together again. Similarly, a charged capacitor has stored in it an electrical potential energy U equal to the work W required to charge it. This energy can be recovered if the capacitor is allowed to discharge. We can visualize the work of charging by imagining that an external agent pulls electrons from the positive plate and pushes them onto the negative plate, thus bringing about the charge separation; normally the work of charging is done by a battery, at the expense of its store of chemical energy.

Suppose that at a time t a charge $q'(t)$ has been transferred from one plate to the other. The potential difference $V(t)$ between the plates at that moment will be

$q'(t)/C$. If an extra increment of charge dq' is transferred, the small amount of additional work needed will be

$$dW = V\, dq' = \left(\frac{q'}{C}\right) dq'.$$

If this process is continued until a total charge q has been transferred, the total work will be found from

$$W = \int dW = \int_0^q \frac{q'}{C}\, dq' = \frac{1}{2}\frac{q^2}{C}. \tag{27-7}$$

From the relation $q = CV$ we can also write this as

Energy stored in a capacitor

$$W\,(=U) = \tfrac{1}{2}CV^2. \tag{27-8}$$

It is reasonable to suppose that the energy stored in a capacitor resides in the electric field. As q or V in Eqs. 27-7 and 27-8 increase, for example, so does the electric field E; when q and V are zero, so is E.

In a parallel-plate capacitor, neglecting fringing, the electric field has the same value for all points between the plates. Thus the *energy density u,* which is the stored energy per unit volume, should also be uniform; u (see Eq. 27-8) is given by

$$u = \frac{U}{Ad} = \frac{\tfrac{1}{2}CV^2}{Ad},$$

where Ad is the volume between the plates. Substituting the relation $C = \epsilon_0 A/d$ (Eq. 27-4) leads to

$$u = \frac{\epsilon_0}{2}\left(\frac{V}{d}\right)^2.$$

However, V/d is the electric field E, so that

Energy density in an electric field

$$u = \tfrac{1}{2}\epsilon_0 E^2. \tag{27-9}$$

Although we derived this equation for the special case of a parallel-plate capacitor, it is true in general. If an electric field $\mathbf{E}$ exists at any point in space (a vacuum), we can think of that point as the site of stored energy in amount, per unit volume, of $\tfrac{1}{2}\epsilon_0 E^2$.

Example 5 A capacitor C_1 is charged to a potential difference V_0. The charging battery is then removed and the capacitor is connected as in Fig. 27-5 to an uncharged capacitor C_2.

(*a*) What is the final potential difference V across the combination?

The original charge q_0 is now shared by the two capacitors. Thus

$$q_0 = q_1 + q_2.$$

Applying the relation $q = CV$ to each of these terms yields

$$C_1 V_0 = C_1 V + C_2 V$$

or

$$V = V_0 \frac{C_1}{C_1 + C_2}.$$

This suggests a way to measure an unknown capacitance (C_2, say) in terms of a known one.

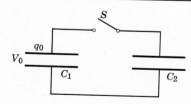

Figure 27-5 Example 5. C_1 is charged to a potential difference V_0 and then connected to C_2 by closing switch S.

(*b*) What is the stored energy before and after the switch in Fig. 27-5 is thrown?

The initial stored energy is

$$U_0 = \tfrac{1}{2}C_1 V_0^2.$$

The final stored energy is

$$U = \tfrac{1}{2}C_1 V^2 + \tfrac{1}{2}C_2 V^2 = \tfrac{1}{2}(C_1 + C_2)\left(\frac{V_0 C_1}{C_1 + C_2}\right)^2$$

$$= \frac{1}{2}\left(\frac{C_1}{C_1 + C_2}\right)C_1 V_0^2 = \left(\frac{C_1}{C_1 + C_2}\right)U_0.$$

Thus U is less than U_0! For $C_1 = C_2$, in fact, $U = \tfrac{1}{2}U_0$.

This is not a violation of the principle of conservation of energy. The example is over-idealized in that it does not take into account the resistance (see Chapter 28) and inductance (see Chapter 33) of the connecting wires. The "missing" energy would, in fact, appear as thermal energy in the wires and/or as energy radiated away from the circuit as electromagnetic radiation.

Example 6 An isolated conducting sphere of radius R, in a vacuum, carries a charge q. (a) Compute the total electrostatic energy stored in the surrounding space.

At any radius r from the center of the sphere (assuming $r > R$) E is given by

$$E = \frac{1}{4\pi\epsilon_0}\frac{q}{r^2}.$$

The energy density at any radius r is found from Eq. 27–9, or

$$u = \tfrac{1}{2}\epsilon_0 E^2 = \frac{q^2}{32\pi^2\epsilon_0 r^4}.$$

The energy dU that lies in a spherical shell between the radii r and $r + dr$ is

$$dU = (4\pi r^2)(dr)u = \frac{q^2}{8\pi\epsilon_0}\frac{dr}{r^2},$$

where $(4\pi r^2)(dr)$ is the volume of the spherical shell. The total energy U is found by integration, or

$$U = \int dU = \frac{q^2}{8\pi\epsilon_0}\int_R^\infty \frac{dr}{r^2} = \frac{q^2}{8\pi\epsilon_0 R}.$$

Note that this relation follows at once from Eq. 27–7 ($U = q^2/2C$), where C (see Example 3) is the capacitance ($= 4\pi\epsilon_0 R$) of an isolated sphere of radius R.

(b) What is the radius R_0 of a spherical surface such that half the stored energy lies within it?

In the equation just given we put

$$\tfrac{1}{2}U = \frac{q^2}{8\pi\epsilon_0}\int_R^{R_0} \frac{dr}{r^2}$$

or

$$\frac{q^2}{16\pi\epsilon_0 R} = \frac{q^2}{8\pi\epsilon_0}\left(\frac{1}{R} - \frac{1}{R_0}\right),$$

which yields, after some rearrangement,

$$R_0 = 2R.$$

Thus, most of the energy is stored in space rather close to the sphere.

27–4 Parallel-Plate Capacitor with a Dielectric

Equation 27–4 holds only for a parallel-plate capacitor with its plates in a vacuum. In 1837 Michael Faraday, after whom the SI unit of capacitance is named, first investigated the effect of completely filling the space between the plates with a dielectric; Table 27–1 shows a sampling of dielectrics in common use today.

Faraday found that the effect of the dielectric filling is to increase the capacitance of the device by a factor κ. This factor, known as the *dielectric constant* of the material, is displayed in Table 27–1. For a parallel-plate capacitor we thus have

Dielectric constant

$$C = \frac{\kappa\epsilon_0 A}{d}. \tag{27–10}$$

Parallel-plate capacitor with a dielectric

Equation 27–4 is a special case of this relation found by putting $\kappa = 1$, which corresponds to a vacuum between the plates. For capacitors in general we have

$$C = \kappa\epsilon_0 L$$

where L depends on the geometry and has the dimensions of a length. For the parallel-plate capacitor $L = A/d$. For a cylindrical capacitor (see Example 2), we have $L = 2\pi l/\ln(b/a)$.

Figure 27–6 shows in more detail the effect of introducing a dielectric between the plates of a capacitor. In Fig. 27–6a we have two capacitors, identical except that one contains a dielectric and the other does not. They are maintained at the same potential difference V by battery B. The charge on the capacitor containing the dielectric is larger, by a factor κ. In Fig. 27–6b on the other hand, the

Table 27–1 Properties of some dielectrics[a]

Material	Dielectric Constant	Dielectric Strength[b] (kV/mm)
Vacuum	1.00000	—
Air	1.00054	3
Titanium dioxide	100	6
Water	78	—
Neoprene	6.9	12
Porcelain	6.5	4
Ruby mica	5.4	160
Pyrex glass	4.5	13
Transformer oil	4.5	12
Fused quartz	3.8	8
Paper	3.5	14
Polystyrene	2.6	25

[a] At room temperature and for static electric fields.

[b] The maximum electric field that may exist in the dielectric without electrical breakdown.

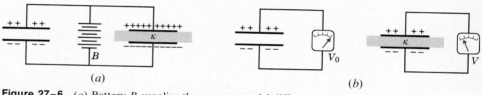

Figure 27–6 (a) Battery B supplies the same potential difference to each capacitor; the one on the right has the higher charge. (b) Both capacitors carry the same charge; the one on the right has the lower potential difference, as indicated by the meter readings.

capacitors are arranged to carry the same charge. The one with the dielectric then has the smaller potential difference, by the same factor. Thus

$$V = V_0/\kappa. \tag{27–11}$$

Both of these facts are consistent, by way of the relation $q = CV$, with the fact that filling a capacitor with a dielectric increases its capacitance by a factor κ.

Example 7 A parallel-plate capacitor has plates with area A and separation d. A battery charges the plates to a potential difference V_0. The battery is then disconnected, as in Fig. 27–6b, and a dielectric slab of thickness d is introduced. Calculate the stored energy both before and after the slab is introduced and account for any difference.

The energy U_0 before introducing the slab is

$$U_0 = \tfrac{1}{2}C_0V_0^2.$$

Because the charge remains constant while the slab is introduced, just as in Fig. 27–6b, we prefer to use the relation $q = CV$ to recast the expression for the energy as

$$U_0 = \tfrac{1}{2}q_0^2/C_0.$$

After the slab is in place we have $C = \kappa C_0$ and thus

$$U = \tfrac{1}{2}q_0^2/C = \tfrac{1}{2}q_0^2/\kappa C_0 = U_0/\kappa.$$

The energy after the slab is introduced is *less* by a factor $1/\kappa$. The "missing" energy would be apparent to the person who inserted the slab. He would feel a tug on the slab and would have to restrain it if he wished to insert the slab without acceleration. This means that he would have to do negative work on it, or, alternatively, that the capacitor + slab system would do positive work on him. This positive work is

$$W = U_0 - U = \tfrac{1}{2}C_0V_0^2 \left(1 - \frac{1}{\kappa}\right).$$

As expected, $W = 0$ for the case of $\kappa = 1$.

The following section will give some insight into how the tug referred to above arises, in terms of the attraction between what we will call "free" charges on the capacitor plates and "induced" charges on the dielectric.

Note from the relation $U = \frac{1}{2}CV^2$ that we can derive the energy density u, for a parallel-plate capacitor in which a dielectric slab is present, from

$$u = \frac{U}{(Ad)} = \left(\frac{1}{Ad}\right)(\tfrac{1}{2}CV^2) = \left(\frac{1}{Ad}\right)\left(\frac{1}{2}\right)\left(\frac{\epsilon_0 \kappa A}{d}\right)(V^2).$$

But $E = V/d$ so that we have

$$u = \tfrac{1}{2}\epsilon_0 \kappa E^2.$$

As for Eq. 27–9, this relation, although derived for a parallel-plate capacitor, holds in general; that is, at any point P in a dielectric of constant κ. As we expect, for $\kappa = 1$, this new relation reduces to Eq. 27–9.

27–5 Dielectrics—an Atomic View

We now seek to understand, in atomic terms, what happens when we place a dielectric in an electric field. There are two possibilities. The molecules of some dielectrics, like water (see Fig. 26–11), have permanent electric dipole moments. In such materials (called *polar*) the electric dipole moments **p** tend to align themselves with an external electric field, as in Fig. 27–7b; see also Section 24–6. Because the molecules are in constant thermal agitation, the degree of alignment will not be complete but will increase as the applied electric field is increased or as the temperature is decreased.

Induced dipoles

Whether or not the molecules have permanent electric dipole moments, they acquire them by induction when placed in an electric field. In Section 26–5 we saw that the external electric field tends to separate the negative and the positive charge in the atom or molecule. This induced electric dipole moment is present only when the electric field is present. It is proportional to the electric field (for moderate field strengths) and is created already lined up with the electric field as Fig. 26–12 suggests.

Figure 27–7 (a) Molecules with a permanent electric dipole moment, showing their random orientation in the absence of an external electric field. (b) An electric field is applied, producing partial alignment of the dipoles. Thermal agitation prevents complete alignment.

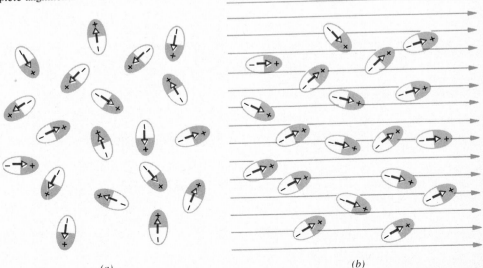

(a)   (b)

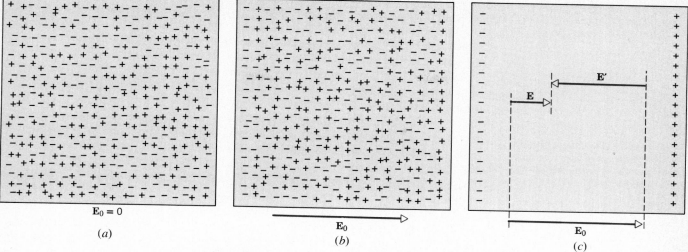

Figure 27–8 (a) A dielectric slab, showing a possible distribution of plus and minus charges. (b) An external field $\mathbf{E}_0$, established by putting the slab between the plates of a parallel-plate capacitor (not shown), separates the center of plus charge in the slab slightly from the center of minus charge, resulting in the appearance of surface charges. (c) The surface charges set up a field $\mathbf{E}'$ which opposes the external field $\mathbf{E}_0$ associated with the charges on the capacitor plates. The resultant field $\mathbf{E}$ ($= \mathbf{E}_0 + \mathbf{E}'$) in the dielectric is thus less than $\mathbf{E}_0$.

Let us use a parallel-plate capacitor, carrying a fixed charge q and not connected to a battery (see Fig. 27–6b), to provide a uniform external electric field $\mathbf{E}_0$ into which we place a dielectric slab. The over-all effect of alignment and induction is to separate the center of positive charge of the entire slab slightly from the center of negative charge. The slab, as a whole, although remaining electrically neutral, becomes polarized as Fig. 27–8b suggests. The net effect is a pile-up of positive charge on the right face of the slab and of negative charge on the left face; within the slab no excess charge appears in any given volume element. Since the slab as a whole remains neutral, the positive induced surface charge must be equal in magnitude to the negative induced surface charge. Note that in this process charges in the dielectric are displaced from their equilibrium positions by distances that are considerably less than an atomic diameter. There is no transfer of charge over macroscopic distances such as occurs when a current is set up in a conductor.

Figure 27–8c shows that the induced surface charges will always appear in such a way that the electric field set up *by them* ($\mathbf{E}'$) opposes the external electric field $\mathbf{E}_0$. The resultant field in the dielectric $\mathbf{E}$ is the vector sum of $\mathbf{E}_0$ and $\mathbf{E}'$. It points in the same direction as $\mathbf{E}_0$ but is smaller. If we place a dielectric in an electric field, induced surface charges appear that tend to weaken the original field within the dielectric.

This weakening of the electric field is quite consistent with the reduction in potential difference for a charged isolated capacitor that occurs when a dielectric slab is introduced. In connection with Fig. 27–6b we saw that

$$V = V_0/\kappa.$$

[27–11]

Induced surface charges

Because $E = V/d$ in this case we have directly

$$E = E_0/\kappa,$$ (27–12)

which describes the reduction in the strength of the electric field brought about by the action of the induced surface charges.

Induced surface charge is the explanation of the most elementary fact of static electricity, namely, that a charged rod will attract uncharged, nonconducting bits of paper, etc. Figure 27–9 shows a bit of paper in the field of a charged rod. Surface charges appear on the paper as shown. The negatively charged end of the paper will be pulled toward the rod and the positively charged end will be repelled. These two forces do not have the same magnitude because the negative end, being closer to the rod, is in a stronger field and experiences a stronger force. The net effect is an attraction. A dielectric body in a uniform electric field will not experience a net force.

In Example 7 we pointed out that, if we insert a dielectric slab into a parallel-plate capacitor carrying a fixed charge q, a force will act on the slab drawing it into the capacitor. This force is provided by the electrostatic attraction between the charges $\pm q$ on the capacitor plates and the induced surface charges $\mp q'$ on the dielectric slab. When the slab is only part way into the capacitor neither q nor q' will be uniformly distributed.

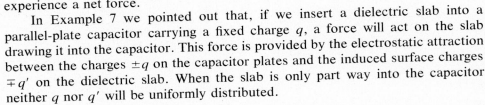

Figure 27–9 A charged rod attracts an uncharged piece of paper because unbalanced forces act on the induced surface charges.

27–6 Dielectrics and Gauss's Law

So far our use of Gauss's law has been confined to situations in which no dielectric was present. Now let us apply this law to a parallel-plate capacitor filled with a dielectric of dielectric constant κ.

Figure 27–10 shows the capacitor both with and without the dielectric. We assume that the charge q on the plates is the same in each case. Gaussian surfaces have been drawn after the fashion of Fig. 27–2.

If no dielectric is present (Fig. 27–10a), Gauss's law (see Eq. 27–2) gives

$$\epsilon_0 \oint \mathbf{E} \cdot d\mathbf{S} = \epsilon_0 E_0 A = q$$

or

$$E_0 = \frac{q}{\epsilon_0 A}.$$ (27–13)

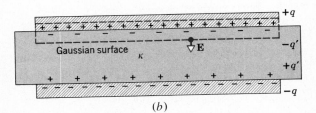

Figure 27–10 A parallel-plate capacitor (a) without and (b) with a dielectric. The charge q on the plates is assumed to be the same in each case.

If the dielectric is present (Fig. 27–10b), Gauss's law gives

$$\epsilon_0 \oint \mathbf{E} \cdot d\mathbf{S} = \epsilon_0 E A = q - q'$$

or

$$E = \frac{q}{\epsilon_0 A} - \frac{q'}{\epsilon_0 A}, \tag{27–14}$$

Free and induced charge

in which $-q'$, the induced surface charge, must be distinguished from q, the so-called free charge on the plates. These two charges, both of which lie within the Gaussian surface, are opposite in sign; $q - q'$ is the net charge within the Gaussian surface.

Combining Eqs. 27–12 and 27–13 yields

$$E = \frac{E_0}{\kappa} = \frac{q}{\kappa \epsilon_0 A}. \tag{27–15}$$

Inserting this in Eq. 27–14 yields

$$\frac{q}{\kappa \epsilon_0 A} = \frac{q}{\epsilon_0 A} - \frac{q'}{\epsilon_0 A} \tag{27–16a}$$

or

$$q' = q \left(1 - \frac{1}{\kappa} \right). \tag{27–16b}$$

This shows correctly that the induced surface charge q' is always less in magnitude than the free charge q and is equal to zero if no dielectric is present, that is, if $\kappa = 1$.

Now we write Gauss's law for the case of Fig. 27–10b in the form

Gauss's law when a dielectric is present

$$\epsilon_0 \oint \mathbf{E} \cdot d\mathbf{S} = q - q', \tag{27–17}$$

$q - q'$ again being the net charge within the Gaussian surface. Substituting from Eq. 27–16b for q' leads, after some rearrangement, to

$$\epsilon_0 \oint \kappa \mathbf{E} \cdot d\mathbf{S} = q. \tag{27–18}$$

This important relation, although derived for a parallel-plate capacitor, is true generally and is the form in which Gauss's law is usually written when dielectrics are present. Note the following:

1. The flux integral now deals not with $\mathbf{E}$ alone but with $\kappa \mathbf{E}$.

2. The charge q contained within the Gaussian surface is taken to be the free charge only. Induced surface charge is deliberately ignored on the right side of this equation, having been taken into account by the introduction of κ on the left side. Equations 27–17 and 27–18 are completely equivalent formulations.

Example 8 Figure 27–11 shows a dielectric slab of thickness b and dielectric constant κ placed between the plates of a parallel-plate capacitor of plate area A and separation d. A potential difference V_0 is applied with no dielectric present. The battery is then disconnected and the dielectric slab inserted. Assume that $A = 100 \text{ cm}^2$, $d = 1.0 \text{ cm}$, $b = 0.50 \text{ cm}$, $\kappa = 7.0$, and $V_0 = 100 \text{ V}$ and (a) calculate the capacitance C_0 before the slab is inserted.

From Eq. 27–4, C_0 is:

$$C_0 = \frac{\epsilon_0 A}{d} = \frac{(8.9 \times 10^{-12} \text{ C}^2/\text{N} \cdot \text{m}^2)(1.0 \times 10^{-2} \text{ m}^2)}{1.0 \times 10^{-2} \text{ m}}$$

$$= 8.9 \times 10^{-12} \text{ F} = 8.9 \text{ pF}.$$

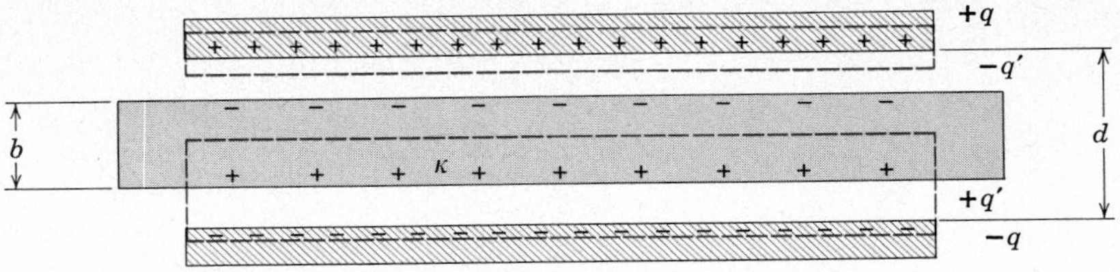

Figure 27–11 Example 8. A parallel-plate capacitor containing a dielectric slab.

(*b*) Calculate the free charge q.

From Eq. 27–1,

$q = C_0 V_0 = (8.9 \times 10^{-12} \text{ F})(100 \text{ V}) = 8.9 \times 10^{-10} \text{ C} = 890 \text{ pC}.$

Because of the technique used to charge the capacitor, the free charge remains unchanged as the slab is introduced. If the charging battery had not been disconnected, this would not be the case.

(*c*) Calculate the electric field E_0 in the gap.

Applying Gauss's law in the form given in Eq. 27–18 to the Gaussian surface of Fig. 27–11 (upper plate) yields

$$\epsilon_0 \oint \kappa \mathbf{E} \cdot d\mathbf{S} = \epsilon_0 E_0 A = q,$$

or

$$E_0 = \frac{q}{\epsilon_0 A} = \frac{8.9 \times 10^{-10} \text{ C}}{(8.9 \times 10^{-12} \text{ C}^2/\text{N·m}^2)(10^{-2} \text{ m}^2)} = 10 \text{ kV/m}.$$

Note that we put $\kappa = 1$ here because the surface over which we evaluate the flux integral does not pass through any dielectric. Note too that E_0 remains unchanged when the slab is introduced; this derivation takes no specific account of the presence of the dielectric.

(*d*) Calculate the electric field in the dielectric.

Applying Eq. 27–18 to the Gaussian surface of Fig. 27–11 (lower plate) yields

$$\epsilon_0 \oint \kappa \mathbf{E} \cdot d\mathbf{S} = \epsilon_0 \kappa E A = q.$$

Note that κ appears here because the surface cuts through the dielectric and that only the free charge q appears on the right.

Thus we have

$$E = \frac{q}{\kappa \epsilon_0 A} = \frac{E_0}{\kappa} = \frac{1.0 \times 10^4 \text{ V/m}}{7.0} = 1.4 \text{ kV/m}.$$

(*e*) Calculate the potential difference between the plates.

Applying Eq. 26–5 to a straight perpendicular path from the lower plate to the upper one yields

$$V = -\int_L^U \mathbf{E} \cdot d\boldsymbol{\ell} = -\int_L^U E \cos 180° \, dl$$

$$= \int_L^U E \, dl = E_0(d - b) + Eb.$$

Numerically

$V = (1.0 \times 10^4 \text{ V/m})(5.0 \times 10^{-3} \text{ m})$
$\qquad + (0.14 \times 10^4 \text{ V/m})(5.0 \times 10^{-3} \text{ m}) = 57 \text{ V}.$

This contrasts with the original applied potential difference of 100 V; compare Fig. 27–6*b*.

(*f*) Calculate the capacitance with the slab in place.

From Eq. 27–1,

$$C = \frac{q}{V} = \frac{8.9 \times 10^{-10} \text{ C}}{57 \text{ V}} = 16 \text{ pF}.$$

When the dielectric slab is introduced, the potential difference drops from 100 to 57 V and the capacitance rises from 8.9 to 16 pF, a factor of 1.8. If the dielectric slab had filled the capacitor, the capacitance would have risen by a factor of $\kappa (= 7.0)$ to 62 pF.

REVIEW GUIDE AND SUMMARY

A capacitor; capacitance

A capacitor consists of two isolated conductors (plates) carrying equal and opposite charges $+q$ and $-q$. The capacitance of the device is defined from

$$q = CV \qquad\qquad [27-1]$$

where V is the potential difference between the plates. The SI unit of capacitance is the farad (1 farad = 1 coulomb/volt).

Parallel-plate capacitor

A parallel-plate capacitor has plane parallel plates of area A and spacing d. In a vacuum its capacitance can be shown from Gauss's law to be

$$C = \frac{\epsilon_0 A}{d}; \qquad\qquad [27-4]$$

see Example 1.

Cylindrical capacitor

A cylindrical capacitor consists of two long coaxial cylinders of length l. If the inner and outer radii are a and b, respectively, the capacitance is

$$C = \frac{2\pi\epsilon_0 l}{\ln{(b/a)}},$$

as Example 2 shows.

Spherical capacitor

A spherical capacitor has plates of inner and outer radii a and b, respectively; its capacitance (see Problem 8) is

$$C = 4\pi\epsilon_0 \frac{ab}{b-a}.$$

If $b \rightarrow \infty$ and $a = R$, as in Example 3, we have the capacitance of an isolated sphere, or

$$C = 4\pi\epsilon_0 R. \qquad [27\text{-}6]$$

Capacitors in series and in parallel

The equivalent capacitances C_{eq} of combinations of individual capacitors arranged in series (see Example 4) and in parallel (see Problem 9) are

$$\text{Series: } \frac{1}{C_{eq}} = \frac{1}{C_1} + \frac{1}{C_2} + \frac{1}{C_3} \cdots,$$

$$\text{Parallel: } C_{eq} = C_1 + C_2 + C_3 \cdots.$$

Energy and energy density

The stored potential energy U of a capacitor, given by

$$U = \tfrac{1}{2}CV^2, \qquad [27\text{-}8]$$

is defined as the work required to charge it. This energy is conveniently thought of as stored in the electric field $\mathbf{E}$ associated with the capacitor. By extension we can associate stored energy with an electric field generally, no matter what its origin. The energy density u is given by

$$u = \tfrac{1}{2}\epsilon_0 E^2, \qquad [27\text{-}9]$$

in which it is assumed that the field $\mathbf{E}$ exists in a vacuum.

Capacitance with a dielectric

Figures 27–6a, b show the space between the plates of a capacitor being completely filled with a dielectric material. Experiment shows that this increases the capacitance C by a factor κ. This factor is characteristic of the material (see Table 27–1) and is called the dielectric constant.

Energy and energy density with a dielectric

For a fixed potential difference the increase of capacitance caused by adding a dielectric means (see Fig. 27–6a and Eq. 27–8) that the stored energy U of a capacitor also increases by a factor κ. By extension, Eq. 27–9 for the energy density must be generalized to

$$u = \tfrac{1}{2}\kappa\epsilon_0 E^2.$$

The effects of adding a dielectric can be understood physically in terms of the action of an electric field on the permanent or induced electric dipoles in the dielectric slab. As Fig. 27–8 shows, the result is the formation of induced surface charges, which results in a weakening of the field within the body of the dielectric.

When a dielectric is present Gauss's law may be generalized to

Gauss's law with a dielectric

$$\epsilon_0 \oint (\kappa\mathbf{E}) \cdot d\mathbf{S} = q. \qquad [27\text{-}18]$$

Here q includes only the free charge, the induced surface charge being accounted for by including κ inside the integral. Example 8 applies this relation in an important special case.

QUESTIONS

1. A capacitor is connected across a battery. (a) Why does each plate receive a charge of exactly the same magnitude? (b) Is this true even if the plates are of different sizes?

2. Can there be a potential difference between two adjacent conductors that carry the same amount of positive charge?

3. A sheet of aluminum foil of negligible thickness is placed between the plates of a capacitor as in Fig. 27–12. What effect has it on the capacitance if (a) the foil is electrically insulated and (b) the foil is connected to the upper plate?

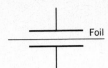

Figure 27–12 Question 3.

4. You are given two capacitors, C_1 and C_2, in which $C_1 > C_2$. Can C_1 always hold more charge than C_2? Explain.

5. In Fig. 27–1 suppose that a and b are nonconductors, the charge being distributed arbitrarily over their surfaces. (a) Would Eq. 27–1 ($q = CV$) hold, with C independent of the charge arrangements? (b) How would you define V in this case?

6. You are given a parallel-plate capacitor with square plates of area A and separation d, in a vacuum. What is the qualitative effect of each of the following on its capacitance? (a) Reduce d. (b) Put a slab of copper between the plates, touching neither plate. (c) Double the area of both plates. (d) Double the area of one plate only. (e) Slide the plates parallel to each other so that the area of overlap is, say, 50%. (f) Double the potential difference between the plates. (g) Tilt one plate so that the separation remains d at one end but is $\frac{1}{2}d$ at the other.

7. How can an isolated conductor have a capacitance? Where is the second plate?

8. You have two isolated conductors, each of which has a certain capacitance; see Fig. 27–13. If you join these conductors by a fine wire, how do you calculate the capacitance of the combination? In joining them with the wire, have you connected them in parallel or in series?

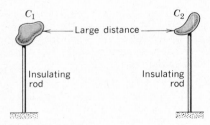

Figure 27–13 Question 8.

9. If you were not to neglect the fringing of the electric field lines in a parallel-plate capacitor, would you calculate a higher or a lower capacitance?

10. A capacitor is connected across the terminals of a battery. Does the charge that eventually appears on the capacitor plates depend on the value of the internal resistance of the battery?

11. Two circular copper disks are facing each other a certain distance apart. In what ways could you reduce the capacitance of this combination?

12. Would you expect the dielectric constant of a material to vary with temperature? If so, how? Does whether or not the molecules have permanent dipole moments matter here?

13. Discuss similarities and differences when (a) a dielectric slab and (b) a conducting slab are inserted between the plates of a parallel-plate capacitor. Assume the slab thicknesses to be one-half the plate separation.

14. An oil-filled, parallel-plate capacitor has been designed to have a capacitance C and to operate safely at or below a certain maximum potential difference V_m without arcing over. However, the designer did not do a good job and the capacitor occasionally arcs over. What can be done to redesign the capacitor, keeping C and V_m unchanged and using the same dielectric?

15. A dielectric body in a nonuniform electric field experiences a net force. Why won't it feel a force if the field is uniform?

16. A dielectric slab is inserted in one end of a charged parallel-plate capacitor (the plates being horizontal and the charging battery having been disconnected) and then released. Describe what happens. Neglect friction.

17. A capacitor is charged by using a battery, which is then disconnected. A dielectric slab is then slipped between the plates. Describe qualitatively what happens to the charge, the capacitance, the potential difference, the electric field, and the stored energy.

18. While a capacitor remains connected to a battery, a dielectric slab is slipped between the plates. Describe qualitatively what happens to the charge, the capacitance, the potential difference, the electric field, and the stored energy. Is work required to insert the slab?

19. Two identical capacitors are connected as shown in Fig. 27–14. A dielectric slab is slipped between the plates of one capacitor, the battery remaining connected. Describe qualitatively what happens to the charge, the capacitance, the potential difference, the electric field, and the stored energy for each capacitor.

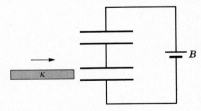

Figure 27–14 Question 19.

EXERCISES

Section 27–1 Capacitance
1. An electrometer is a device used to measure static charge. Unknown charge is placed on the plates of a capacitor and the potential difference is measured. What minimum charge can be measured by an electrometer with a capacitance of 50 pF and a voltage sensitivity of 0.15 V?

2. A storage capacitor on a random access memory chip (RAM) has a capacitance of 50 femtofarad (1 fF = 10^{-15} F). If it is

charged to 5.0 V, how many electrons are stored on its negative plate?

Section 27–2 Calculating Capacitance
3. If we solve Eq. 27–4 for ϵ_0, we see that its SI units are farad/meter. Show that these units are equivalent to those obtained earlier for ϵ_0, namely coulomb^2/newton·meter^2.

4. A parallel-plate capacitor has circular plates of 8.0-cm radius and 1.0-mm separation. What charge will appear on the plates if a potential difference of 100 V is applied?

5. You have two flat metal plates, each of area 1.0 m², with which to construct a parallel-plate capacitor. If its capacitance is to be 1.0 F, what must be the separation between the plates? Could this capacitor actually be constructed?

6. Calculate the capacitance of the earth, viewed as a spherical conductor of radius 6400 km.

7. The two metal objects in Fig. 27–15 have net charges of +70 pC and −70 pC, and this results in a 20-V potential difference between them. (a) What is the capacitance of the system? (b) If the charges are changed to +200 pC and −200 pC, what does the capacitance become? (c) What does the potential difference become?

Figure 27–15   Exercise 7.

8. The capacitor in Fig. 27–16 has a capacitance of 25 μF and is initialy uncharged. The battery supplies 120 V. After switch S has been closed for a long time, how much charge will have passed through the battery?

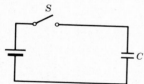

Figure 27–16   Exercise 8.

9. *Capacitors in parallel.* Figure 27–17 shows three capacitors connected in parallel. Show that the single capacitance C_{eq} that is equivalent to this combination is

$$C_{eq} = C_1 + C_2 + C_3.$$

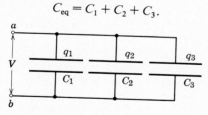

Figure 27–17   Exercise 9.

10. How many 1.0-μF capacitors must be connected in parallel to store a charge of 1.0 C with a potential of 300 V across them?

11. After you walk over a carpet on a dry day, your hand comes close to a metal doorknob and a 5-mm spark results. Such a spark means that there must have been a potential difference of possibly 15 kV between you and the doorknob. How much charge did you accumulate in walking over the carpet? For this extremely rough calculation, assume that your body can be represented by a uniformly charged conducting sphere 25 cm in radius and isolated from its surroundings. (*Hint:* See Example 3.)

Section 27–3   Energy Storage in an Electric Field
12. What capacitance is required to store an energy of 10 kW·h at a potential difference of 1000 V?

13. A parallel-plate air capacitor has a capacitance of 100 pF. (a) What is the stored energy if the applied potential difference is 50 V? (b) Can you calculate the energy density for points between the plates?

14. A parallel-plate air capacitor having area 40 cm² and spacing 1.0 mm is charged to a potential of 600 V. Find (a) the capacitance, (b) the magnitude of the charge on each plate, (c) the stored energy, (d) the electric field between the plates, and (e) the energy density between the plates.

15. Two capacitors (2.0 μF and 4.0 μF) are connected in parallel across a 300-V potential difference. Calculate the energy stored in the system.

16. Calculate the energy density of the electric field at distance r from the center of an electron at rest. If the electron is a point particle, what does this calculation yield for its energy density as you approach the electron?

17. A certain capacitor is charged to a potential V. If you wish to increase its stored energy by 10%, by what percentage should you increase V?

18. Attempts to build a controlled thermonuclear fusion reactor, which, if successful, could provide the world with a vast supply of energy from heavy hydrogen in sea water, usually involve huge electric currents for short periods of time in magnetic field windings. For example, ZT-40 at Los Alamos Scientific Laboratory has rooms full of capacitors to provide 350,000 A in the coils. The capacitor banks have 1.13 F capacitance at 40.0 kV and 81.6 F at 10.0 kV. (a) What total charge is stored? (b) How long could the current be sustained? (c) How much energy is stored?

Section 27–4   Parallel-Plate Capacitor with a Dielectric
19. Given a 7.40-pF air capacitor, you are asked to design a capacitor to store up to 1.69×10^{-5} C with a potential difference less than 600 V. What dielectric in Table 27–1 will you use if you do not allow for a margin of error?

20. For making a capacitor you have available two plates of copper, a sheet of mica (thickness = 0.10 mm, $\kappa = 5.4$), a sheet of glass (thickness = 2.0 mm, $\kappa = 7.0$), and a slab of paraffin (thickness = 1.0 cm, $\kappa = 2.0$). To obtain the largest capacitance, which sheet (or sheets) should you place between the copper plates?

Section 27–6   Dielectrics and Gauss's Law
21. A parallel-plate air capacitor has a capacitance of 50 pF. (a) If its plates each have an area of 0.35 m², what is their separation? (b) If the region between the plates is now filled with material having a dielectric constant of 5.6, what is the capacitance?

22. A coaxial cable used in a transmission line responds as a "distributed" capacitance to the circuit feeding it. Calculate the capacitance per unit length for a cable having an inner radius of 0.100 mm and an outer radius of 0.600 mm. Assume that the space between the conductors is filled with polystyrene.

23. A parallel-plate capacitor has a capacitance of 100 pF, a plate area of 100 cm², and mica dielectric ($\kappa = 5.4$). At 50-V potential difference, calculate (a) E in the mica, (b) the magnitude of the free charge on the plates, and (c) the magnitude of the induced surface charge.

24. In Example 7, how does the energy density between the plates compare, before and after the dielectric slab is introduced?

PROBLEMS GROUPED BY SECTION

Section 27–2 Calculating Capacitance

1. (a) Three capacitors are connected in parallel. Each has plate area A and plate spacing d. What must be the spacing of a single capacitor of plate area A if its capacitance equals that of the parallel combination? (b) What must be the spacing if the three capacitors are connected in series?

2. A 6.0-μF capacitor is connected in series with a 4.0-μF capacitor and a potential difference of 200 V is applied across the pair. (a) What is the charge on each capacitor? (b) What is the potential difference across each capacitor?

3. Repeat the previous problem for the same two capacitors connected in parallel.

4. In Fig. 27–18 a variable air capacitor of the type used in tuning radios is shown. Alternate plates are connected together, one group being fixed in position, the other group being capable of rotation. Consider a pile of n plates of alternate polarity, each having an area A and separated from adjacent plates by a distance d. Show that this capacitor has a maximum capacitance of

$$C = \frac{(n-1)\epsilon_0 A}{d}.$$

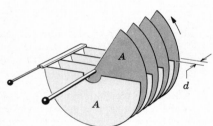

Figure 27–18 Problem 4.

5. In Fig. 27–19 find the equivalent capacitance of the combination. Assume that $C_1 = 10\ \mu$F, $C_2 = 5.0\ \mu$F, and $C_3 = 4.0\ \mu$F.

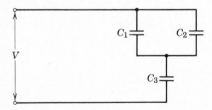

Figure 27–19 Problems 5, 33, and 40.

6. A potential difference of 300 V is applied to a 2.0-μF capacitor and an 8.0-μF capacitor connected in series. (a) What are the charge and the potential difference for each capacitor? (b) The charged capacitors are reconnected with their positive plates together and their negative plates together, no external voltage being applied. What are the charge and the potential differences for each? (c) The charged capacitors in (a) are reconnected with plates of *opposite* sign together. What are the charge and the potential difference for each?

7. If you have available several 2.0-μF capacitors, each capable of withstanding 200 V without breakdown, how would you assemble a combination having an equivalent capacitance of (a) 0.40 μF or of (b) 1.2 μF, each capable of withstanding 1000 V?

8. *A spherical capacitor.* A spherical capacitor consists of two concentric spherical shells of radii a and b, with $b > a$. (a) Show that its capacitance is

$$C = 4\pi\epsilon_0 \frac{ab}{b-a}.$$

(b) Does this reduce (with $a = R$) to the result of Example 3 as $b \to \infty$?

9. Suppose that the two spherical shells of a spherical capacitor have their radii approximately equal. Under these conditions the device approximates a parallel-plate capacitor with $b - a = d$. Show that the formula in Problem 8 does indeed reduce to Eq. 27–4 in this case.

*10. Find the equivalent capacitance between points x and y in Fig. 27–20. Assume that $C_2 = 10\ \mu$F and that the other capacitors are all 4.0 μF. (*Hint:* Apply a potential difference V between x and y and write down all the relationships that involve the charges and potential differences for the separate capacitors.)

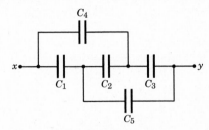

Figure 27–20 Problem 10.

Section 27–3 Energy Storage in an Electric Field

11. A parallel-connected bank of 2000 5.0-μF capacitors is used to store electric energy. What does it cost to charge this bank to 50,000 V, assuming a rate of 8.0¢/kW·h?

12. For the capacitors of Problem 6, compute the energy stored for the three different connections of parts (a), (b), and (c). Compare your answers and explain any differences.

13. A slab of copper of thickness b is thrust into a parallel-plate capacitor as shown in Fig. 27–21; it is exactly halfway between the plates. (a) What is the capacitance after the slab is introduced? (b) If a charge q is maintained on the plates, find the ratio of the stored energy before to that after the slab is inserted. (c) How much work is done on the slab? Is the slab sucked in, or do you have to push it in?

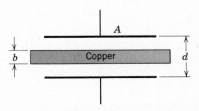

Figure 27–21 Problem 13.

14. Reconsider Problem 13 assuming that the potential difference rather than the charge is held constant.

15. A cylindrical capacitor has radii a and b as in Fig. 27–3. Show that half the stored electric potential energy lies within a cylinder whose radius is

$$r = (ab)^{1/2}.$$

16. Show that the plates of a parallel-plate capacitor attract each other with a force given by

$$F = \frac{q^2}{2\epsilon_0 A}.$$

Prove this by calculating the work necessary to increase the plate separation from x to $x + dx$.

17. Using the result of Problem 16, show that the force per unit area (the electrostatic stress) acting on either capacitor plate is given by $\frac{1}{2}\epsilon_0 E^2$. Actually, this result is true in general, for a conductor of *any* shape with an electric field **E** at its surface.

Section 27–4 Parallel-Plate Capacitor with a Dielectric
18. Two parallel plates of area 100 cm² are each given equal but opposite charges of 8.9×10^{-7} C. The electric field within the dielectric material filling the space between the plates is 1.4×10^6 V/m. (*a*) Find the dielectric constant of the material. (*b*) Determine the magnitude of the charge induced on each dielectric surface.

19. A parallel-plate capacitor is filled with two dielectrics as in Fig. 27–22. Show that the capacitance is given by

$$C = \frac{\epsilon_0 A}{d}\left(\frac{\kappa_1 + \kappa_2}{2}\right).$$

Check this formula for all the limiting cases that you can think of. (*Hint:* Can you justify regarding this arrangement as two capacitors in parallel?)

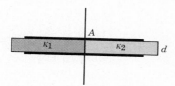

Figure 27–22 Problem 19.

20. What is the capacitance of the capacitor in Fig. 27–23?

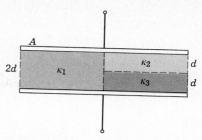

Figure 27–23 Problem 20.

21. Two parallel-plate capacitors have the same plate area A and separation d, but the dielectric constants of the materials between their plates are $\kappa + \Delta\kappa$ and $\kappa - \Delta\kappa$, respectively. (*a*) Find the equivalent capacitance when they are connected in parallel. (*b*) If the total charge on the parallel combination is Q, what is the charge on the capacitor with the largest capacitance?

Section 27–6 Dielectrics and Gauss's Law
22. A parallel-plate capacitor has plates of area 0.12 m² and a separation of 1.2 cm. A battery charges the plates to a potential difference of 120 V and is then disconnected. A dielectric slab of thickness 40 mm and dielectric constant 4.8 is then placed symmetrically between the plates. (*a*) Find the capacitance before the slab is inserted. (*b*) What is the capacitance with the slab in place? (*c*) What is the free charge q before and after the slab is inserted? (*d*) Determine the electric field in the space between the plates and dielectric. (*e*) What is the electric field in the dielectric? (*f*) With the slab in place, what is the potential difference across the plates? (*g*) How much external work is involved in the process of inserting the slab?

23. In the capacitor of Example 8 (Fig. 27–11), the dielectric slab fills half the space between the plates. (*a*) What percent of the energy is stored in the air gaps? (*b*) What percent is stored in the slab?

24. A dielectric slab of thickness b is inserted between the plates of a parallel-plate capacitor of plate separation d. Show that the capacitance is given by

$$C = \frac{\kappa \epsilon_0 A}{\kappa d - b(\kappa - 1)}.$$

(*Hint:* Derive the formula following the pattern of Example 8.) Does this formula correctly predict the numerical result of Example 8? Does the formula seem reasonable for the special cases of $b = 0$, $\kappa = 1$, and $b = d$?

ADDITIONAL PROBLEMS

25. In Fig. 27–24, find the equivalent capacitance of the combination. Assume that $C_1 = 10\ \mu\text{F}$, $C_2 = 5.0\ \mu\text{F}$, and $C_3 = 4.0\ \mu\text{F}$.

26. Figure 27–25 shows two capacitors in series, the rigid center section of length b being movable vertically. Show that the equivalent capacitance of the series combination is independent of the position of the center section and is given by

$$C = \frac{\epsilon_0 A}{a - b}.$$

27. Two sheets of aluminum foil have a separation of 1.0 mm, a capacitance of 10 pF, and are charged to 10 V. If the separation is decreased by 0.10 mm with the charge held constant, by how

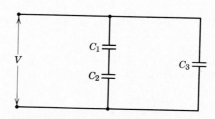

Figure 27–24 Problems 25 and 37.

much does the potential difference change? Explain how a microphone might be constructed using this principle.

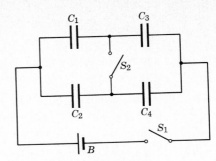

Figure 27–28   Problem 31.

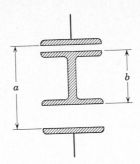

Figure 27–25   Problem 26.

28.   A 100-pF capacitor is charged to a potential difference of 50 V, the charging battery then being disconnected. The capacitor is then connected in parallel with a second (initially uncharged) capacitor. If the measured potential difference drops to 35 V, what is the capacitance of this second capacitor?

29.   In Figure 27–26 capacitors $C_1 (=1.0\ \mu F)$ and $C_2 (=3.0\ \mu F)$ are each charged to a potential $V (=100\ V)$ but with opposite polarity, so that points a and c are on the side of the respective positive plates of C_1 and C_2, and points b and d are on the side of the respective negative plates. Switches S_1 and S_2 are now closed. (a) What is the potential difference between points e and f? (b) What is the charge on C_1? (c) What is the charge on C_2?

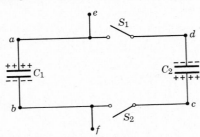

Figure 27–26   Problem 29.

30.   When switch S is thrown to the left in Fig. 27–27, the plates of the capacitor C_1 acquire a potential difference V_0. C_2 and C_3 are initially uncharged. The switch is now thrown to the right. What are the final charges q_1, q_2, q_3 on the corresponding capacitors?

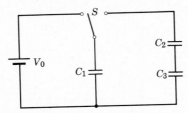

Figure 27–27   Problem 30.

31.   In Fig. 27–28 the battery B supplies 12 V. (a) Find the charge on each capacitor when switch S_1 is closed and (b) when (later) switch S_2 is also closed. Take $C_1 = 1.0\ \mu F$, $C_2 = 2.0\ \mu F$, $C_3 = 3.0\ \mu F$, and $C_4 = 4.0\ \mu F$.

32.   Figure 27–29 shows two identical capacitors C in a circuit with two (ideal) diodes D. A 100-V battery is connected to the input terminals, first with terminal a positive and later with terminal b positive. In each case, what is the potential difference across

the output terminals? (An ideal diode has the property that positive charge flows through it only in the direction of the arrow and negative charge flows through it only in the opposite direction.)

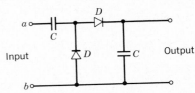

Figure 27–29   Problem 32.

33.   In Fig. 27–19, suppose that capacitor C_3 breaks down electrically, becoming equivalent to a conducting path. What *changes* in (a) the charge and (b) the potential difference occur for capacitor C_1? Assume that $V = 100$ V.

34.   In Example 2 the capacitance of a cylindrical capacitor was calculated. Using the approximation (see Appendix G) that $\ln(1 + x) \cong x$ when $x \ll 1$, show that the capacitance approaches that of a parallel-plate capacitor when the spacing between the two cylinders is small.

35.   A spherical drop of mercury of radius R has a capacitance given by $C = 4\pi\epsilon_0 R$ (see Example 3). If two such drops combine to form a single larger drop, what is its capacitance?

36.   Two spheres of radii 3.0 m and 1.0 m hang from long, non-conducting threads, far from each other. The larger sphere has a charge of 4.0 μC, while the smaller is uncharged. After the spheres are momentarily connected by a fine wire, what is the charge on the smaller sphere?

37.   In Fig. 27–24, find (a) the charge, (b) the potential difference, and (c) the stored energy for each capacitor. Assume the numerical values of Problem 25, with $V = 100$ V.

38.   A parallel-plate capacitor has plates of area A and separation d, and is charged to a potential difference V. The charging battery is then disconnected and the plates are pulled apart until their separation is $2d$. Derive expressions in terms of A, d, and V for (a) the new potential difference, (b) the initial and the final stored energy, and (c) the work required to separate the plates.

39.   One capacitor is charged until its stored energy is 4.0 J. A second uncharged capacitor is then connected to it in parallel. (a) If the charge distributes equally, what is now the total energy stored in the electric fields? (b) Where did the excess energy go?

40.   In Fig. 27–19, find (a) the charge, (b) the potential difference, and (c) the stored energy for each capacitor. Assume the numerical values of Problem 5, with $V = 100$ V.

41. An isolated metal sphere whose diameter is 10 cm has a potential of 8000 V. What is the energy density near the surface of the sphere?

42. A certain substance has a dielectric constant of 2.8 and a dielectric strength of 18 MV/m. If it is used as the dielectric material in a parallel-plate capacitor, what minimum area may the plates of the capacitor have in order that the capacitance be 7.0×10^{-2} μF and that the capacitor be able to withstand a potential difference of 4000 V?

43. You are asked to construct a capacitor having a capacitance near 1 nF and a breakdown potential in excess of 10,000 V. You think of using the sides of a tall drinking glass (Pyrex), lining inside and outside with aluminum foil (neglect the ends). Does this arrangement have about the right capacitance and breakdown potential? Assume a glass 15 cm tall with an inner radius of 3.6 cm and an outer radius of 3.8 cm.

44. A parallel-plate capacitor is filled with two dielectrics as in Fig. 27–30. Show that the capacitance is given by

$$C = \frac{2\epsilon_0 A}{d}\left(\frac{\kappa_1 \kappa_2}{\kappa_1 + \kappa_2}\right).$$

Check this formula for all the limiting cases that you can think of. (*Hint:* Can you justify regarding this arrangement as two capacitors in series?)

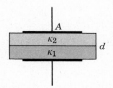

Figure 27–30 Problem 44.

45. In Example 8, suppose that the 100-V battery remains connected during the time that the dielectric slab is being introduced. Calculate (*a*) the charge on the capacitor plates, (*b*) the electric field in the gap, (*c*) the electric field in the slab, and (*d*) the capaci-

tance. For all of these quantities, give the numerical values before and after the slab is introduced. Contrast your results with those of Example 8 by constructing a tabular listing.

*46. Charges q_1, q_2, and q_3 are placed on capacitors C_1, C_2, and C_3, respectively, arranged in series as shown in Fig. 27–31. Switch S in that figure is then closed. (*a*) What are the final charges q'_1, q'_2, and q'_3 on the capacitors, each expressed in terms of the initial charges q_1, q_2, and q_3? (*b*) Show that the solution can also be written, more simply, as

$$q'_i = q_i - V_0 C_{eq} \qquad (i = 1, 2, 3).$$

Here V_0 is the original potential difference between the top and bottom plates and C_{eq} is the equivalent capacitance of the three capacitors in series. Can you give a simple qualitative interpretation for this form of solution? (The authors are indebted to Harold Froehling and Stanley M. Flatte of the University of California at Santa Cruz for pointing out this solution to us.)

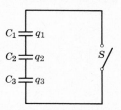

Figure 27–31 Problem 46.

*47. A soap bubble of radius R_0 is slowly given a charge q. Because of mutual repulsion of the surface charges, the radius increases slightly to R. The air pressure inside the bubble drops, because of the expansion, to $p(V_0/V)$, where p is the atmospheric pressure, V_0 is the initial volume, and V is the final volume. Show that

$$q^2 = 32\pi^2 \epsilon_0 pR(R^3 - R_0^3).$$

(*Hint:* Imagine that the bubble expands a further amount dR. Consider the energy changes associated with (*a*) the decrease in the stored electric field energy, (*b*) work ($=p\,dV$) done in pushing back the atmosphere, and (*c*) work done by the gas in the bubble. Apply conservation of energy. Neglect surface tension.)

28 Current and Resistance

28–1 Current and Current Density

The conduction electrons in an isolated metallic conductor, such as a length of copper wire, are in random motion like the molecules of a gas confined to a container. They have no net directed motion along the wire. If we pass a hypothetical plane through the wire perpendicular to its axis, the rate at which electrons pass through it from right to left is the same as the rate at which they pass through from left to right; the net rate is zero.

If the ends of the wire are connected to a battery, an electric field will be set up at every point within the wire. If the potential difference maintained by the battery is 10 V and if the wire (assumed uniform) is 5 m long, the strength of this field at every point will be 2 V/m. This field **E** will act on the conduction electrons and will give them a resultant motion in the direction of − **E**. We say that an *electric current i* is established. If a net charge q passes through any cross section of the conductor in time t, the current (assumed constant) is

$$i = q/t. \qquad (28-1)$$

The appropriate SI units are amperes (abbr. A) for i, coulombs for q, and seconds for t. Recall (Section 23–4) that Eq. 28–1 is the defining equation for the coulomb and that we have not yet given a definition of the ampere; we do so in Section 31–4.

If the rate of flow of charge with time is not constant, the current varies with time and is given by the differential limit of Eq. 28–1, or

$$i = dq/dt. \qquad (28-2)$$

Definition of current

In the rest of this chapter we consider only constant currents.

The current i is the same for all cross sections of a conductor, even though the cross-sectional area may be different at different points. In the same way the rate at which water (assumed incompressible) flows past any cross section of a pipe is the same even if the cross section varies. The water flows faster where the pipe is smaller and slower where it is larger, so that the volume rate, measured perhaps in liters/second, remains unchanged. This constancy of the electric current follows because charge must be conserved; it does not pile up steadily or drain away steadily from any point in the conductor under the assumed steady-state conditions. In the language of Section 16–7 there are no "sources" or "sinks" of charge.

The existence of an electric field inside a conductor does not contradict Section 25–6, in which we asserted that **E** equals zero inside a conductor. In that section, which dealt with a state in which all net motion of charge had stopped (electrostatics), we assumed that the conductor was insulated and that no potential difference was deliberately maintained between any two points on it, as by a battery. In this chapter, which deals with charges in motion, we relax this restriction.

Consider a length of copper wire as a typical conductor. It is a semirigid assembly of Cu^+ ions, coupled together by strong springlike forces of electromagnetic origin to form a three-dimensional, periodic lattice. The conduction electrons —one per atom for copper—are free to move about within this lattice. If an electric field is established within the copper wire, perhaps by using a battery, a force $-e\mathbf{E}$ acts on each such electron. The electrons accelerate in response to this force but only for successive short time intervals because they are continually bumping into the thermally jiggling ion cores that make up the lattice. The overall effect is to transfer kinetic energy from the accelerating conduction electrons into vibrational energy of the lattice. The electrons acquire a constant average *drift speed* v_d in the direction $-\mathbf{E}$. There is a close analogy to a ball bearing falling in a uniform gravitational field **g** at a constant terminal speed through a viscous oil. The gravitational force ($m\mathbf{g}$) acting on the ball as it falls does not go into increasing the ball's kinetic energy (which is constant) but is transferred to the fluid by molecular collisions, producing a small rise in temperature.

Current i has meaning only for a specified conductor such as the length of copper wire referred to above. Sometimes, however, we want to take a localized view and to focus attention on the flow of charge carriers at particular points within a conductor. For this we introduce the *current density* vector **j**. This vector is characteristic of a point inside a conductor and not of the conductor as a whole. If the current is distributed uniformly across a conductor of cross-sectional area A, as in Fig. 28–1b, the magnitude of the current density vector for all points on that cross section is

The current density

$$j = i/A. \tag{28-3}$$

The general relationship between **j** and i is that, for a particular surface (which need not be plane) that cuts across a conductor, i is the flux of the vector **j** over that surface, or

Current and current density

$$i = \int \mathbf{j} \cdot d\mathbf{S} \tag{28-4}$$

where $d\mathbf{S}$ is an element of surface area and the integral is taken over the surface in question. It is clear that Eq. 28–3 (written as $i = jA$) is a special case of this relationship in which the surface of integration is a plane cross section of the conductor and in which **j** is constant over this surface and at right angles to it. Equation 28–4 shows clearly that i is a scalar because the integrand $\mathbf{j} \cdot d\mathbf{S}$ is a scalar.

It remains to specify the direction of the current density vector **j**. This is complicated by the fact that in general the moving charges that form a current may be either positive or negative. In metallic conductors such as copper the carriers are the negative conduction electrons. In the liquid electrolyte of a storage battery or in the gaseous plasma of a fluorescent lamp both positive and negative ions may be present. As it happens, a positive charge moving in one direction is equivalent in nearly all its external effects to a negative charge of the same magnitude moving with the same speed in the opposite direction. We take advantage of this fact to treat all charge carriers *as if* they were positive; we then draw the arrows representing the current density **j** in the direction that such carriers would move.*

The direction of **j**—a sign convention for the charge carriers

Figure 28–1 illustrates the convention for **j**. In one case we have positive carriers drifting in the direction $+\mathbf{E}$ and in the other negative carriers drifting in the direction $-\mathbf{E}$. In each case we draw the current density vector **j** pointing in the direction the carriers would move *if* they were positive. The current *i*, being a scalar, does not have a direction as such. It does have a sense, however, related to the general direction of the current vectors that make it up. We represent the current sense by an ordinary arrow in Figs. 28–1*a, b*.

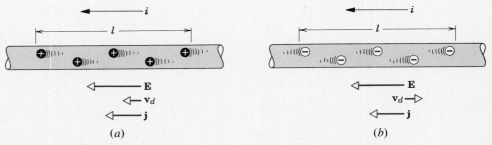

Figure 28–1 (*a*) Positive charge carriers drift in the direction of the applied electric field **E**. (*b*) Negative carriers drift in the opposite direction. However, the direction of the current density vector **j** and the sense of the current *i* are drawn, by convention, *as if* the carriers were positive, as in (*a*).

Calculation of drift speed

We can compute the drift speed v_d of charge carriers in a conductor from the current density *j*. Figure 28–1*b* shows the conduction electrons in a wire moving to the right at an assumed constant drift speed v_d. The number of conduction electrons in the wire is nAl where n is the number of conduction electrons per unit volume and Al is a specified volume of the wire. A charge of magnitude

$$q = (nAl)e$$

passes out of this volume, through its right end, in a time t given by

$$t = \frac{l}{v_d}.$$

The current *i* is given by

$$i = \frac{q}{t} = \frac{nAle}{l/v_d} = nAev_d.$$

Solving for v_d and recalling that $j = i/A$ (Eq. 28–3) yields

---

* In just one case in this book it *is* possible to distinguish the sign of the charge carriers by measuring external effects (see Section 30–5, the Hall Effect). In this case we scrap the convention and deal with the actual situation.

$$v_d = \frac{i}{nAe} = \frac{j}{ne}. \tag{28-5}$$

We can recast this equation in vector form as

$$\mathbf{j} = (ne)\,\mathbf{v}_d. \tag{28-6}$$

Here the product (ne), whose SI units are C/m³, is the charge density. Note that Eq. 28–6 is compatible with our convention for the sign of the current density vector $\mathbf{j}$. For positive charge carriers (which we always assume) the charge density (ne) is positive and Eq. 28–6 predicts that $\mathbf{j}$ and $\mathbf{v}_d$ point in the same direction, as Fig. 28–1a verifies.

Example 1 One end of an aluminum wire whose diameter is 2.5 mm is welded to one end of a copper wire which has a diameter of 1.6 mm. The composite wire carries a steady current of 10 A. What is the current density in each wire?

The current is distributed uniformly over the cross section of each conductor, except near the junction, which means that the current density may be taken as constant for all points within each wire. The cross-sectional area of the aluminum wire is 4.9 mm². Thus, from Eq. 28–3,

$$j_{Al} = \frac{i}{A} = \frac{10\ A}{4.9\ mm^2} = 2.0\ A/mm^2.$$

The cross-sectional area of the copper wire is 2.0 mm². Thus

$$j_{Cu} = \frac{i}{A} = \frac{10\ A}{2.0\ mm^2} = 5.0\ A/mm^2.$$

The fact that the wires are of different materials does not enter into consideration here.

Example 2 What is the drift speed v_d of the conduction electrons in the copper wire in Example 1?

To compute n we start from the assumption that there is one conduction electron per atom in copper. The number of

atoms per unit volume is $N_A d/M$ where d is the density, N_A is the Avogadro number, and M is the atomic weight. The number of conduction electrons per unit volume is then

$$
\begin{aligned}
n &= \frac{N_A d}{M} \\
&= \frac{(6.0 \times 10^{23}\ atoms/mol)(1\ electron/atom)(9.0\ g/cm^3)}{64\ g/mol} \\
&= 8.4 \times 10^{22}\ electrons/cm^3.
\end{aligned}
$$

Finally, v_d is, from Eq. 28–5,

$$
\begin{aligned}
v_d &= \frac{j}{ne} = \frac{(5.0\ A/mm^2)}{(8.4 \times 10^{22}\ electrons/cm^3)(1.6 \times 10^{-19}\ C/electron)} \\
&= 0.37\ mm/s.
\end{aligned}
$$

It takes 14 minutes for the electrons in this wire to drift one foot. Would you have guessed that v_d was so low? The drift speed of electrons must not be confused with the speed at which changes in the electric field configuration travel along wires, a speed that approaches that of light. When we apply a pressure to one end of a long water-filled garden hose, a pressure wave travels rapidly along the hose. The speed at which water moves through the hose is much lower, however.

28–2 Resistance and Resistivity

If we apply the same potential difference between the ends of geometrically similar rods of copper and of wood, very different currents result. The characteristic of the conductor that enters here is its *resistance*. We define the resistance of a conductor (often called a *resistor;* symbol ⌇⌇⌇⌇) between two points by applying a potential difference V between those points, measuring the current i, and dividing:

The resistance of a conductor

$$R = V/i. \tag{28-7}$$

If V is in volts and i in amperes, the resistance R will be in *ohms* (abbr. Ω).

The flow of charge through a conductor is often compared with the flow of water through a pipe, which occurs because there is a difference in pressure between the ends of the pipe, established perhaps by a pump. This pressure difference can be compared with the potential difference established between the ends of a resistor by a battery. The flow of water (liters/second, say) is compared with the current (coulombs/second, or amperes). The rate of flow of water for a given pressure difference is determined by the nature of the pipe. Is it long or short? Is it narrow or wide? Is it empty or filled, perhaps with gravel? These characteristics of the pipe are analogous to the resistance of a conductor.

Related to resistance is the *resistivity* ρ, which is a characteristic of a material rather than of a particular specimen of a material; it is defined, for isotropic materials, from

$$\rho = \frac{E}{j},\qquad (28-8)$$

which has the vector form

$$\mathbf{E} = \rho\mathbf{j}.\qquad (28-9)$$

The resistivity of copper is $1.7 \times 10^{-8}\ \Omega\cdot m$; that of fused quartz is about $10^{16}\ \Omega\cdot m$. Few physical properties are measurable over such a range of values; Table 28–1 lists some electrical properties for common materials.

Table 28–1 Properties of some conducting materials

Material	Resistivity[a] ρ $10^{-8}\ \Omega\cdot m$	Temperature Coefficient of Resistivity[a] α $10^{-3}\ K^{-1}$
Silver	1.6	3.8
Copper	1.7	3.9
Aluminum	2.8	3.9
Tungsten	5.6	4.5
Nickel	6.8	6.0
Iron	10	5.0
Manganin[b]	44	0.01
Carbon	3500	−0.50

[a] Measured at room temperature ($\cong 300$ K).
[b] An alloy specifically designed to have a small value of α.

Often we prefer to speak of the *conductivity* σ of a material rather than its resistivity. These are reciprocal quantities, related by

$$\sigma = 1/\rho.\qquad (28-10)$$

The SI units of σ are $(\Omega\cdot m)^{-1}$.

Consider a cylindrical conductor, of cross-sectional area A and length l, carrying a steady current i. Let us apply a potential difference V between its ends. If the cylinder cross sections at each end are equipotential surfaces, the electric field and the current density will be constant for all points in the cylinder and will have the values

$$E = \frac{V}{l}\qquad \text{and}\qquad j = \frac{i}{A}.$$

We may then write the resistivity ρ as

$$\rho = \frac{E}{j} = \frac{V/l}{i/A}.$$

But V/i is the resistance R, which leads to

$$R = \rho\frac{l}{A}.\qquad (28-11)$$

Example 3 A rectangular carbon block has dimensions $1.0 \times 1.0 \times 50$ cm. (a) What is the resistance measured between the two square ends and (b) between two opposing rectangular faces? The resistivity of carbon at 20°C is 3.5×10^{-5} Ω·m.

(a) The area of a square end is 1.0 cm² or 1.0×10^{-4} m². Equation 28–11 gives for the resistance between the square ends

$$R = \rho\,\frac{l}{A} = \frac{(3.5 \times 10^{-5}\ \Omega\cdot\text{m})(0.50\ \text{m})}{1.0 \times 10^{-4}\ \text{m}^2} = 0.18\ \Omega.$$

(b) For the resistance between opposing rectangular faces (area = 5.0×10^{-3} m²), we have

$$R = \rho\,\frac{l}{A} = \frac{(3.5 \times 10^{-5}\ \Omega\cdot\text{m})(10^{-2}\ \text{m})}{5.0 \times 10^{-3}\ \text{m}^2} = 7.0 \times 10^{-5}\ \Omega.$$

Thus a given conductor can have any number of resistances, depending on how the potential difference is applied to it. The ratio of resistances for these two cases is 2600. We assume in each that the potential difference is applied to the block in such a way that the surfaces between which the resistance is desired are equipotential. Otherwise Eq. 28–11 would not be valid.

The resistivity of all materials is temperature dependent. The solid curve of Fig. 28–2 shows, for example, how the resistivity of copper varies with temperature. For many such materials the relationship is nearly linear over a large temperature range and we can write, as an empirical approximation that is close enough for most purposes,

The variation of resistivity with temperature

$$\rho = \rho_0\,[1 + \alpha(T - T_0)]. \qquad (28\text{–}12)$$

Here ρ_0 is the resistivity of the material at the reference temperature T_0. The quantity α is a mean temperature coefficient of resistivity appropriate to the particular range of temperatures in question.

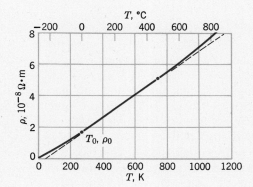

Figure 28–2 The resistivity of copper as a function of temperature. The dashed line is an approximation chosen to fit the curve at the two circled points. The point marked T_0, ρ_0 is chosen as a reference point.

Superconductivity

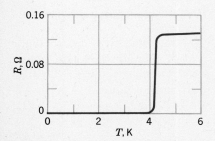

Figure 28–3 The resistivity of mercury disappears below about 4 K.

The curve of Fig. 28–2 does not go to zero at the absolute zero of temperature, even though it appears to do so, the residual resistivity depending on the level of impurities in the specimen. For many substances, however, the resistance does become zero at some low temperature. Figure 28–3 shows the resistance of a specimen of mercury for temperatures below 6 K. In the space of about 0.05 K the resistance drops abruptly to an immeasurably low value. This phenomenon, called *superconductivity*, was discovered by Kamerlingh Onnes in the Netherlands in 1911. The resistance of materials in the superconducting state seems to be truly zero; currents, once established in closed superconducting circuits, persist for weeks without diminution, even though there is no battery in the circuit. If the temperature is raised slightly above the superconducting point, or if a large enough magnetic field is applied, the material becomes resistive and such currents drop rapidly to zero.

28-3 Ohm's Law

Let us apply a variable potential difference V between the ends of a 400-foot coil of #18 copper wire. For each applied potential difference, let us measure the current i and plot it against V as in Fig. 28-4. The straight line that results means that the resistance of this conductor is the same no matter what applied voltage we use to measure it. This important result, which holds for metallic conductors, is known as *Ohm's law*.

The negative values of voltage in Fig. 28-4 correspond to reversing the polarity of the potential difference applied to the coil of wire. When this is done we see that the current simply reverses direction, its magnitude remaining unchanged. We assume that the temperature of the conductor is essentially constant throughout all these measurements.

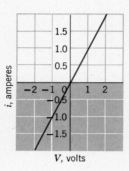

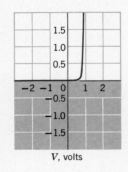

Figure 28-4 The current in a particular copper conductor as a function of the applied potential difference. This conductor obeys Ohm's law.

Figure 28-5 The current in a particular silicon *pn* junction diode as a function of the applied potential difference. This device does not obey Ohm's law.

Many conducting devices do not obey Ohm's law. Figure 28-5, for example, shows a V-i plot for a semiconducting diode, a so-called silicon-based *pn* junction; your pocket calculator contains many of them. The curve for the forward direction ($V > 0$ in the figure) is not a straight line. Also, the device exhibits a total lack of symmetry when the polarity of the applied potential difference is reversed. Modern electronics, and therefore much of the character of our present technological civilization, depends totally on devices such as semiconducting diodes and transistors that do *not* obey Ohm's law.

We stress that the relationship $V = iR$ is *not* by itself a statement of Ohm's law. A conductor obeys this law only if its V-i curve is linear, that is, if R is independent of V and i. The relationship $R = V/i$ remains as the definition of the resistance of a conductor whether or not the conductor obeys Ohm's law.

28-4 Ohm's Law—A Microscopic View

It is curious that many conductors (see Fig. 28-4) obey Ohm's law exceedingly well, whereas others (see Fig. 28-5) do not obey it at all. Our aim here is to understand why metals obey this law. Because our interest is in materials rather than in specific conducting bodies or devices we start, not from the relation $V = iR$ as we did in Section 28-3, but from Eq. 28-9

$$\mathbf{E} = \rho\mathbf{j} \qquad\qquad [28-9]$$

and we say that a material obeys Ohm's law if (assuming a constant temperature) its resistivity ρ is independent of the magnitude E of the applied electric field.

The free electron model

The effective speed

The mean free time τ

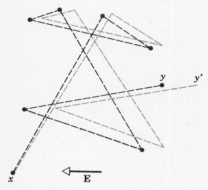

Figure 28-6 The colored lines show an electron moving from x to y, making six collisions. The black curves show what the electron path might have been in the presence of an electric field **E**. Note the steady drift in the direction of $-$**E**.

We adopt the so-called *free electron model,* in which we imagine the conduction electrons to move about within the volume of the metal much like the molecules of a gas confined to a container. We cannot push this analogy too far however. Although we can usefully describe the motions of gas molecules by classical mechanics—and indeed we did so in Chapter 21—it turns out that we cannot so easily do this for the "electron gas." Here the principles of quantum physics must be used from the beginning if we are to give a full description of the properties of the electron gas. However, we do not seek such a full description; we only want to understand why metals obey Ohm's law. It turns out that we can use classical ideas for this purpose if we are willing to accept one fact: The velocity distribution of the electrons that make up the electron gas can be usefully represented, over a wide range of temperatures, by a single effective speed v_{eff}. For copper it can be shown that $v_{\text{eff}} = 1.6 \times 10^6$ m/s.

We may also define a *mean free time* τ, the average time interval between the collisions experienced by any given electron as it moves through the lattice. In Example 4 we will shed some light on just what we mean by a "collision" in the case of the electron gas.

When we apply an electric field to a metal, the electrons modify their random motion in such a way that they drift slowly, in the opposite direction to that of the field, with an average drift speed v_d. This drift speed is very much less (by a factor of something like 10^{10}; see Example 2) than the effective speed v_{eff} mentioned above. Figure 28-6 suggests the relationship between these two speeds. The colored lines suggest a possible random path followed by an electron in the absence of an applied field; the electron proceeds from x to y, making six collisions on the way. The black curves show how this same event might have occurred if an electric field **E** had been applied. Note that the electron drifts steadily to the right, ending at y' rather than at y. In preparing Fig. 28-6, it has been assumed that the drift speed v_d is 0.02 v_{eff}; actually, it is more like $10^{-10}v_{\text{eff}}$, so that the "drift" exhibited in the figure is grossly exaggerated.

We can calculate the drift speed v_d in terms of the applied electric field E. When a field is applied to an electron of mass m it will experience a force eE which will impart to it an acceleration a given by Newton's second law,

$$a = \frac{eE}{m}.$$

Consider an electron that has just undergone a typical effective collision. Such a collision will, so to speak, completely destroy the electron's memory of its accumulated drift velocity. The electron will then start afresh, moving off in a truly random direction. In the average time interval τ to the next collision an average electron will change its velocity in the direction of $-$**E** by an amount $a\tau$. We identify this with the drift speed v_d and we write*

$$v_d = a\tau = \frac{eE\tau}{m}. \tag{28-13}$$

We may express v_d in terms of the current density (Eq. 28-5) and combine with Eq. 28-13 to obtain

$$v_d = \frac{j}{ne} = \frac{eE\tau}{m}.$$

---

* See "Drift Speed and Collision Time," Donald E. Tilley, *American Journal of Physics,* June 1976 for a more detailed analysis. See also *Electricity and Magnetism—Berkeley Physics Course,* Volume 2 (Section 4.4) by Edward M. Purcell, McGraw-Hill Book Company, 1965.

Combining this with Eq. 28-9 ($\rho = E/j$) leads finally to

$$\rho = \frac{m}{ne^2\tau}. \qquad (28-14)$$

The resistivity of a metal

Equation 28-14 can be taken as a statement that metals obey Ohm's law if we can show that τ does not depend on the applied electric field **E**. If this is true, then ρ will not depend on **E**, which is criterion that we have adopted for saying that a material obeys Ohm's law.

The mean free time τ is independent of **E**

The mean free time τ depends on the speed distribution of the conduction electrons. Figure 28-6 reminds us that this distribution is affected only very slightly by the application of even a relatively large electric field. The effect of the field is to superimpose a drift speed v_d, whose magnitude is commonly of the order of 10^{-4} m/s, on a speed distribution whose effective value v_{eff} is of the order of 10^6 m/s, greater by a factor of 10^{10}. We may be sure that whatever the value of τ is (for copper at 20°C, say) in the absence of a field it remains essentially unchanged when the field is applied.

Example 4 What are (a) the mean time τ between collisions and (b) the mean free path λ for the conduction electrons in copper at 300 K?

(a) From Eq. 28-14 (see also Example 2 and Table 28-1), we have

$$\tau = \frac{m}{ne^2\rho} = \frac{(9.1 \times 10^{-31}\ \text{kg})}{(8.4 \times 10^{28}/\text{m}^3)(1.6 \times 10^{-19}\ \text{C})^2(1.7 \times 10^{-8}\ \Omega\cdot\text{m})}$$

$$= 2.5 \times 10^{-14}\ \text{s}.$$

(b) We can define a mean free path λ as the product of τ and v_{eff}. Using the value of v_{eff} reported on page 510 yields

$$\lambda = \tau v_{eff} = (2.5 \times 10^{-14}\ \text{s})(1.6 \times 10^6\ \text{m/s}) = 4.0 \times 10^{-8}\ \text{m}.$$

This is about 150 times the distance between nearest neighbor ion cores in the copper lattice! Actually, a full quantum physics treatment reveals that we cannot view a "collision" as a direct interaction between an electron and an ion core. Rather it represents an interaction between an electron and (a) the thermal vibrations of the lattice, (b) a lattice imperfection, or (c) a lattice impurity atom.* For especially pure metals cooled to within a few degrees of the absolute zero, mean free paths as long as 10 cm have been measured.

28-5 Energy Transfers in an Electric Circuit

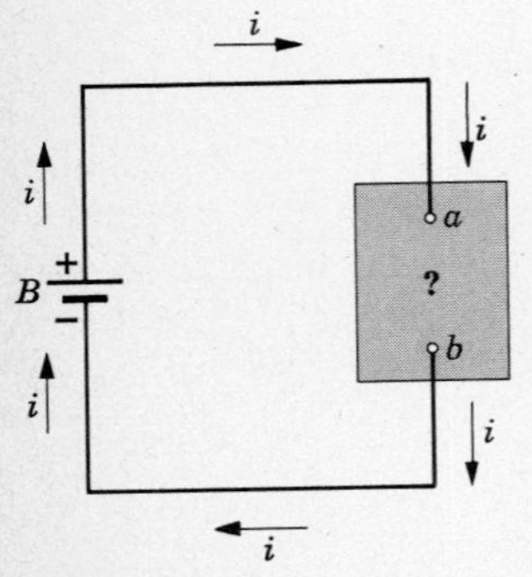

Figure 28-7 A battery B sets up a current in a circuit containing a "black box," that is, a box whose contents are not known to us.

Figure 28-7 shows a circuit consisting of a battery B connected to a "black box." A steady current i exists in the connecting wires and a steady potential difference V_{ab} exists between the terminals a and b. The box might contain a resistor, a motor, or a storage battery, among other things.

Terminal a, connected to the positive battery terminal, is at a higher potential than terminal b. If a charge dq moves through the box from a to b, this charge will decrease its electric potential energy by $dq\,V_{ab}$ (see Section 26-6). The conservation-of-energy principle tells us that this energy is transferred in the box from electric potential energy to some other form. What that other form will be depends on what is in the box. In a time dt the energy dU transferred inside the box is then

$$dU = dq\,V_{ab} = i\,dt\,V_{ab}.$$

We find the rate of energy transfer P by dividing by the time, or

$$P = \frac{dU}{dt} = iV_{ab}. \qquad (28-15)$$

Electrical power

If the device in the box is a motor connected to a mechanical load, the energy appears largely as mechanical work done by the motor; if the device is a storage bat-

---

* Direct electron-electron collisions play a negligible role.

tery that is being charged, the energy appears largely as stored chemical energy in this second battery.

If the device is a resistor, we assert that the energy appears as internal thermal energy in the resistor. To see this, consider a stone of mass m that falls through a height h. It decreases its gravitational potential energy by mgh. If the stone falls in a vacuum this energy is transformed into kinetic energy of the stone. If the stone falls in the ocean, however, its speed eventually becomes constant, which means that the kinetic energy no longer increases. The potential energy that is steadily being made available as the stone falls then appears as thermal energy in the stone and the surrounding water. It is the viscous, friction-like drag of the water on the surface of the stone that stops the stone from accelerating, and it is at this surface that thermal energy appears.

The course of the electrons through the resistor is much like that of the stone through water. The electrons travel with a constant drift speed v_d and thus do not gain kinetic energy. The electric potential energy that they lose is transferred to the resistor as thermal energy. On a microscopic scale we can understand this in that collisions between the electrons and the lattice increase the amplitude of the thermal vibrations of the lattice; on a macroscopic scale this corresponds to a temperature increase. There can be a flow of heat out of the resistor subsequently, if the environment is at a lower temperature than the resistor.

For a resistor we can combine Eqs. 28–7 ($R = V/i$) and 28–15 and obtain either

Joule's law

$$P = i^2 R \tag{28–16}$$

or

$$P = \frac{V^2}{R}. \tag{28–17}$$

Note that Eq. 28–15 applies to electrical energy transfer of all kinds; Eqs. 28–16 and 28–17 apply only to the transfer of electrical energy to thermal energy in a resistor. Equations 28–16 and 28–17 are known as *Joule's law*. This law is a particular way of writing the conservation-of-energy principle for the special case in which electrical energy is transferred into thermal energy (Joule energy).

The unit of power that follows from Eq. 28–15 is the volt-ampere. We can write it as

$$1 \text{ volt·ampere} = 1 \text{ volt·ampere} \left(\frac{1 \text{ joule}}{1 \text{ volt} \times 1 \text{ coulomb}}\right)\left(\frac{1 \text{ coulomb}}{1 \text{ ampere} \times 1 \text{ second}}\right)$$
$$= 1 \text{ joule/second} = 1 \text{ watt}.$$

The first conversion factor in parentheses comes from the definition of the volt (Eq. 26–1); the second comes from the definition of the coulomb. Recall that we introduced the watt ($= 1$ J/s) as a unit of power in Section 7–7, in connection with the performance of mechanical work.

Example 5 You are given a 4.0-m length of heating wire made of the special alloy Nichrome; it has a resistance of 24 Ω. Can you obtain more heat by winding one coil or by cutting the wire in two and winding two separate coils? In each case the coils are to be connected individually across a 110-V line.

The power P for the single coil is given by Eq. 28–17:

$$P = \frac{V^2}{R} = \frac{(110 \text{ V})^2}{24 \text{ }\Omega} = 0.50 \text{ kW}.$$

The power for a coil of half the length is given by

$$P' = \frac{(110 \text{ V})^2}{12 \text{ }\Omega} = 1.0 \text{ kW}.$$

There are two "half-coils," so that the total power obtained by cutting the wire in half is 2.0 kW, or four times that for the single coil. This would seem to suggest that we could buy an 0.50-kW heating coil, cut it in half, and rewind it to obtain 2.0 kW. Why is this not a practical idea?

REVIEW GUIDE AND SUMMARY

Current i

An electric current i in a conductor is defined by

$$i = dq/dt. \qquad [28-1]$$

Here dq is the amount of charge that passes in time dt through a hypothetical surface that cuts across the conductor. If the current is steady, it has the same value for all such cross sections.

Current (a scalar) is related to current density $\mathbf{j}$ (a vector) by

Current density $\mathbf{j}$

$$i = \int \mathbf{j} \cdot d\mathbf{S} \qquad [28-4]$$

where $d\mathbf{S}$ is an element of area and the integral is taken over any surface cutting across the conductor. The direction of $\mathbf{j}$ at any point is that in which a positive charge carrier would

The sign convention for charge carriers

move if placed at that point; see Fig. 28–1 and Example 1. We often label a current i with an arrow to remind us of the directions of the current density vectors that make it up.

The mean drift speed of the charge carriers

When an electric field $\mathbf{E}$ is established in a conductor the charge carriers (assumed positive) acquire a mean drift speed $\mathbf{v}_d$ in the direction of $\mathbf{E}$, related to the current density by

$$\mathbf{j} = (ne)\mathbf{v}_d \qquad [28-6]$$

where (ne) is the charge density

The resistance R between any two equipotential surfaces of a conductor is defined from

The resistance of a conductor

$$R = V/i \qquad [28-7]$$

where V is the potential difference between those surfaces and i is the current. The SI unit of R is the *ohm* ($=1$ V/A; abbr. Ω). A similar equation defines the resistivity ρ of a material. It is

The resistivity of a material

$$\rho = E/j \qquad [28-8]$$

where E is the applied electric field; see Table 28–1. For cylindrical conductors we can calculate R from

$$R = \rho l/A; \qquad [28-11]$$

see Example 3.

The change of ρ with temperature

The resistivity ρ for most materials changes with temperature. For many materials, including metals, the empirical linear relationship of Eq. 28–12 suffices. In it T_0 is a reference temperature, ρ_0 is the resistivity at T_0, and α is a mean temperature coefficient of resistivity; see Table 28–1 and Fig. 28–2.

Ohm's law for conductors

A given conductor obeys Ohm's law if its resistance R is independent of the applied potential difference V. This implies a linear relationship between V and i in the relation $V = iR$; compare Figs. 28–4 and 28–5.

Ohm's law for materials

A given *material* obeys Ohm's law if its resistivity is independent of the applied electric field $\mathbf{E}$. This implies a linear relationship between $\mathbf{E}$ and the current density $\mathbf{j}$ in the relation $\mathbf{E} = \rho \mathbf{j}$; see Eq. 28–9.

By treating the conduction electrons in a metal like the molecules of a gas it is possible to derive for the resistivity of a metal

Resistivity of a metal

$$\rho = \frac{m}{ne^2\tau}. \qquad [28-14]$$

Here n is the number of electrons per unit volume and τ is the mean time between the collisions of an electron with the ion cores of the lattice. The discussion based on Fig. 28–6 shows that τ is independent of E and thus accounts for the fact that metals obey Ohm's law; see Example 4.

Power

The power P or rate of energy transfer in an electrical device across which a potential difference V is maintained is

$$P = iV.$$ [28–15]

If the device is a resistor we can write this as

Joule's law

$$P = i^2R = V^2/R,$$ [28–16,17]

two equivalent forms of Joule's law; see Example 5. In a resistor electrical potential energy is transferred to the lattice by the drifting charge carriers, appearing as internal thermal energy.

QUESTIONS

1. What conclusions can you draw by applying Eq. 28–4 to a closed surface through which a number of wires pass in random directions, carrying steady currents of different magnitudes? Does Gauss's law hold?

2. In our convention for the direction of current arrows (a) would it have been more convenient, or even possible, to have assumed all charge carriers to be negative? (b) Would it have been more convenient, or even possible, to have labeled the electron as positive, the proton as negative, etc.?

3. Explain in your own words why we can have $\mathbf{E} \neq 0$ inside a conductor in this chapter but that we took $\mathbf{E} = 0$ for granted in Section 25–6.

4. Let a battery be connected to a copper cube at two corners defining a body diagonal. Pass a hypothetical plane completely through the cube, tilted at an arbitrary angle. (a) Is the current i through the plane independent of the position and orientation of the plane? (b) Is there any position and orientation of the plane for which $\mathbf{j}$ is a constant in magnitude, direction, or both? (c) Does Eq. 28–4 hold for all orientations of the plane? (d) Does Eq. 28–4 hold for a closed surface of arbitrary shape, which may or may not lie entirely within the cube?

5. A potential difference V is applied to a copper wire of diameter d and length l. What is the effect on the electron drift speed of (a) doubling V, (b) doubling l, and (c) doubling d?

6. If the drift speeds of the electrons in a conductor under ordinary circumstances are so slow (see Example 2), why do the lights in a room turn on so quickly after the switch is closed?

7. Can you think of a way to measure the drift speed for electrons by timing their travel along a conductor?

8. A potential difference V is applied to a circular cylinder of carbon by clamping it between circular copper electrodes, as in Fig. 28–8. Discuss the difficulty of calculating the resistance of the carbon cylinder using the relation $R = \rho l/A$.

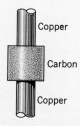

Figure 28–8 Question 8.

9. How would you measure the resistance of a pretzel-shaped conductor? Give specific details to clarify the concept.

10. Discuss the difficulties of testing whether the filament of a light bulb obeys Ohm's law.

11. Does the relation $V = iR$ apply to resistors that do not obey Ohm's law?

12. A cow and a man are standing in a meadow when lightning strikes the ground nearby. Why is the cow more likely to be killed than the man? The responsible phenomenon is called "step voltage".

13. The gray lines in Fig. 28–6 should be curved slightly. Why?

14. Explain the function and operation of a fuse.

15. Why does an incandescant light bulb grow dimmer with use?

16. The character and quality of our daily lives is influenced greatly by devices that do not obey Ohm's law. What can you say in support of this claim?

17. From a student's paper: "The relation $R = V/i$ tells us that the resistance of a conductor is directly proportional to the potential difference applied to it." What do you think of this proposition?

18. Table 28–1 shows us that carbon has a negative temperature coefficient of resistivity. This means that its resistivity drops as its temperature increases. Do you think that its resistivity would disappear entirely at some high enough temperature?

19. What special characteristics must (a) heating wire and (b) fuse wire have?

20. Equation 28–16 ($P = i^2R$) seems to suggest that the rate of increase of thermal energy in a resistor is reduced if the resistance is made less; Eq. 28–17 ($P = V^2/R$) seems to suggest just the opposite. How do you reconcile this apparent paradox?

21. Is the filament resistance lower or higher in a 500-W light bulb than in a 100-W bulb? Both bulbs are designed to operate on 110 V.

22. Five wires of the same length and diameter are connected in turn between two points maintained at constant potential difference. Will Joule energy be developed at the faster rate in the wire of (a) the smallest or (b) the largest resistance?

23. The windings of a motor (connected to a load) have a resistance of 1 Ω. If we apply a potential difference of 110 V to the motor, does it follow that the current through the motor will be 110 V/1 Ω = 110 A?

EXERCISES

Section 28–1   Current and Current Density

1. A current of 5.0 A exists in a 10-Ω resistor for 4.0 min. (a) How many coulombs and (b) how many electrons pass through any cross section of the resistor in this time?

2. The current in the electron beam of a typical video display terminal is 200 μA. How many electrons strike the screen each second?

3. We have 2.0×10^8 doubly charged positive ions per cubic centimeter, all moving north with a speed of 1.0×10^5 m/s. (a) What is the current density **j**, in magnitude and direction? (b) Can you calculate the total current i in this ion beam? If not, why?

4. A small but measurable current of 1.0×10^{-10} A exists in a copper wire whose diameter is 2.5 mm. Calculate the electron drift speed.

5. The (United States) National Electric Code, which sets maximum safe currents for rubber insulated copper wires of various diameters, is given (in part) below. Plot the safe current density as a function of diameter. What wire gauge has the maximum safe current density?

Gauge[a]	4	6	8	10	12	14	16	18
Diameter (mils)[b]	204	162	129	102	81	64	51	40
Safe current (A)	70	50	35	25	20	15	6	3

[a] A way of identifying the wire diameter.
[b] 1 mil = 10^{-3} in.

Section 28–2   Resistance and Resistivity

6. A coil is formed by winding 250 turns of insulated, gauge 16 wire (diameter = 1.3 mm) in a single layer on a cylindrical form whose radius is 12 cm. What is the resistance of the coil? Neglect the thickness of the insulation.

7. A steel trolley-car rail has a cross-sectional area of 7.1 in.² What is the resistance of 10 miles of single track? The resistivity of the steel is 3.0×10^{-7} Ω·m.

8. A conducting wire has a 1.0-mm diameter, a 2.0-m length, and a 50-mΩ resistance. What is the resistivity of the material?

9. A cylindrical copper rod is re-formed to twice its original length with no change in volume. If its resistance between its ends was R before the change, what is it after the change?

10. A wire 4.0 m long and 6.0 mm in diameter has a resistance of 96 Ω. If a potential difference of 20 V is applied between the ends, (a) What is the current in the wire? (b) What is the current density? (c) What is the resistivity of the wire material? Can you identify the material? See Table 28–1.

11. A wire of Nichrome (a nickel-chromium alloy commonly used in heating elements) is 1.0 m long and 1.0 mm² in cross-sectional area. It carries a current of 4.0 A when a 2.0-V potential difference is applied between its ends. What is the conductivity σ of Nichrome?

12. Nine copper wires of length l and diameter d are connected in parallel to form a single composite conductor of resistance R. What must be the diameter D of a single copper wire of length l if it is to have the same resistance?

13. (a) At what temperature would the resistance of a copper conductor be double its resistance at 0°C? (b) Does this same temperature hold for all copper conductors, regardless of shape or size?

14. The copper windings of a motor have a resistance of 50 Ω at 20°C when the motor is idle. After running for several hours the resistance rises to 58 Ω. What is the temperature of the windings?

Section 28–3   Ohm's Law

15. A human being can be electrocuted if a current as small as 50 mA passes in the vicinity of the heart. An electrician working with sweaty hands makes good contact with the two conductors he is holding. If his resistance is 2000 Ω what might the fatal voltage be?

16. Using data from Fig. 28–5, plot the resistance of the *pn* junction as a function of applied potential difference.

17. A 4.0-cm-long caterpillar crawls in the direction of electron drift along a 5.2-mm-diameter bare copper wire that carries a current of 12 A. (a) What is the potential difference between the two ends of the caterpillar? (b) Is his tail positive or negative compared to his head? (c) How much time could he take to crawl 1.0 cm and still keep up with the drifting electrons in the wire?

Section 28–5   Energy Transfers in an Electric Circuit

18. The headlights of a moving car draw about 10 A from the alternator, which is driven by the engine. Assume the alternator is 80 percent efficient and calculate the horsepower the engine must supply to run the lights.

19. An x-ray tube takes a current of 7.0 mA and operates at a potential difference of 80 kV. What power in watts is dissipated?

20. A space heater, operating from a 120-V line, has a hot resistance of 14 Ω. (a) At what rate is electrical energy transfered into heat? (b) At 5.0 ¢/kW·h, what does it cost to operate the device for 5.0 h?

21. Thermal energy is developed in a resistor at a rate of 100 W when the current is 3.0 A. What is the resistance in ohms?

22. A 500-W space heater operates from a 120-V line. (a) What is its (hot) resistance? (b) At what rate do electrons flow through any cross section of the filament?

PROBLEMS GROUPED BY SECTION

Section 28–1   Current and Current Density

1. A current is established in a gas discharge tube when a sufficiently high potential difference is applied across the two electrodes in the tube. The gas ionizes; electrons move toward the positive terminal and singly-charged positive ions move toward the negative terminal. What are the magnitude and sense of the current in a hydrogen discharge tube in which 3.1×10^{18} electrons and 1.1×10^{18} protons move past a cross-sectional area of the tube each second?

2. A *pn* junction is formed from two different semiconducting materials in the form of identical cylinders with radius 0.165 mm, as depicted in Fig. 28–9. In one application 3.50×10^{15} electrons/s flow across the junction from the *n* to the *p* side while 2.25×10^{15} holes/s flow from the *p* to the *n* side. (A hole acts like a particle with charge $+1.6 \times 10^{-19}$ C.) What is (*a*) the total current and (*b*) total current density?

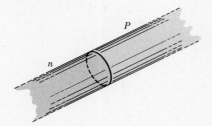

Figure 28–9 Problem 2.

3. How long does it take electrons to get from a car battery to the starting motor? Assume the current is 300 A and the electrons travel through a copper wire with cross-sectional area 0.21 cm² and length 1.0 m.

4. The belt of an electrostatic generator is 50 cm wide and travels at 30 m/s. The belt carries charge into the sphere at a rate corresponding to 100 μA. Compute the surface charge density on the belt. See Section 26–9.

5. You are given an isolated conducting sphere of 10-cm radius. One wire carries a current of 1.0000020 A into it. Another wire carries a current of 1.0000000 A out of it. How long would it take for the sphere to increase in potential by 1000 V?

Section 28–2 Resistance and Resistivity

6. Two conductors are made of the same material and have the same length. Conductor *A* is a solid wire of diameter 1.0 mm. Conductor *B* is a hollow tube of outside diameter 2.0 mm and inside diameter 1.0 mm. What is the resistance ratio, R_A/R_B, measured between their ends?

7. A certain wire has a resistance *R*. What is the resistance of a second wire, made of the same material, that is half as long and has half the diameter?

8. A copper wire and an iron wire of the same length have the same potential difference applied to them. (*a*) What must be the ratio of their radii if the current is to be the same? (*b*) Can the current density be made the same by suitable choices of the radii?

9. Copper and aluminum are being considered for a high-voltage transmission line that must carry a current of 60 A. The resistance per unit length is to be 0.15 Ω/km. Compute for each choice of cable material (*a*) the current density and (*b*) the mass per unit length of the cable. The densities of copper and aluminum are 9000 and 2700 kg/m³, respectively.

10. A wire with a resistance of 6.0 Ω is drawn out through a die so that its new length is three times its original length. Find the resistance of the longer wire, assuming that the resistivity and density of the material are not changed during the drawing process.

11. An electrical cable consists of 125 strands of fine wire, each having 2.65-μΩ resistance. The same potential difference is applied along the length of each strand and results in a total current of 0.75 A. (*a*) What is the current in each strand? (*b*) What is the

applied potential difference? (*c*) What is the resistance of the cable?

12. A common flashlight bulb is rated at 0.30 A and 2.90 V, the values of the current and voltage under operating conditions. If the resistance of the bulb filament when cold is 1.0 Ω, estimate the temperature of the filament when the bulb is on. The filament is made of tungsten.

13. Conductors *A* and *B*, having equal lengths of 40 m and a common diameter of 2.6 mm, are connected in series. A potential difference of 60 *V* is applied between the ends of the composite wire. The resistances of the wires are 0.13 and 0.75 Ω, respectively. Determine: (*a*) the current density in each wire; (*b*) the potential difference across each wire; (*c*) the magnitude of the electric field in each wire; (*d*) the resistivities of the two wires. Identify the wire materials.

14. A resistor is in the shape of a truncated right circular cone (Fig. 28–10). The end radii are *a* and *b*, the altitude is *l*. If the taper is small, we may assume that the current density is uniform across any cross section. (*a*) Calculate the resistance of this object. (*b*) Show that your answer reduces to $\rho(l/A)$ for the special case of zero taper ($a = b$).

Figure 28–10 Problem 14.

Section 28–3 Ohm's Law

15. List in tabular form similarities and differences between the flow of charge along a conductor, the flow of water through a horizontal pipe, and the conduction of heat through a slab. Consider such quantities as what causes the flow, what opposes it, what particles (if any) participate, and the units in which the flow may be measured.

Section 28–4 Ohm's Law: A Microscopic View

16. In the copper wire of Examples 1 and 2 the current density is 5.0 A/m² and the mean drift speed is 0.37 mm/s. (*a*) What electric field must exist within the wire? (*b*) What acceleration do the conduction electrons experience between collisions? (*c*) In this field how long does it take on the average for an electron to acquire the mean drift speed given above? Compare your answer with the result calculated somewhat more directly in part (*a*) of Example 4.

Section 28–5 Energy Transfers in an Electric Circuit

17. The National Board of Fire Underwriters has fixed safe current-carrying capacities for various sizes and types of wire. For #10 rubber-coated copper wire (diameter = 0.10 in.) the maximum safe current is 25 A. At this current, find (*a*) the current density, (*b*) the electric field, (*c*) the potential difference for 1000 ft of wire, and (d) the rate at which thermal energy is developed for 1000 ft of wire.

18. A potential difference of 1.0 V is applied to a 100-ft length of #18 copper wire (diameter = 0.040 in.). Calculate (a) the current, (b) the current density, (c) the electric field, and (d) the rate at which thermal energy is developed in the wire.

19. A cylindrical resistor of radius 5.0 mm and length 2.0 cm is made of material that has a resistivity of 3.5×10^{-5} $\Omega \cdot$m. What are (a) the current density and (b) potential difference when the power dissipation is 1.0 W?

20. Nichrome has a resistivity of 5.0×10^{-7} $\Omega \cdot$m. A heating element is made by maintaining a potential difference of 75 V along the length of a nichrome wire with a 2.6×10^{-6} m^2 cross-section. (a) If the element dissipates 5000 W, what is its length? (b) If a potential difference of 100 V is used to obtain the same power output, what should the length be?

21. A 400-W immersion heater is placed in a pot containing 2.0 liters of water at 20° C. (a) How long will it take to bring the water to boiling temperature, assuming that 80 percent of the available energy is absorbed by the water? (b) How much longer will it take to boil half the water away?

ADDITIONAL PROBLEMS

22. A beam of 16-MeV deuterons from a cyclotron falls on a copper block. The beam is equivalent to a current of 15 μA. (a) At what rate do deuterons strike the block? (b) At what rate is thermal energy produced in the block?

23. A square aluminum rod is 1.0 m long and 5.0 mm on edge. (a) What is the resistance between its ends? (b) What must be the diameter of a circular 1.0-m copper rod if its resistance is to be the same?

24. A 1250-W radiant heater is constructed to operate at 115 V. (a) What will be the current in the heater? (b) What is the resistance of the heating coil? (c) How much thermal energy is generated in one hour by the heater?

25. When a metal rod is heated, not only its resistance but also its length and its cross-sectional area change. The relation $R = \rho l/A$ suggests that all three factors should be taken into account in measuring ρ at various temperatures. (a) If the temperature changes by 1.0° C, what percent changes in R, l, and A occur for a copper conductor? (b) What conclusion do you draw? The coefficient of linear expansion is $1.7 \times 10^{-5}/$°C.

26. An electric immersion heater normally takes 100 min to bring cold water in a well-insulated container to a certain temperature, after which a thermostat switches the heater off. One day the line voltage is reduced by 6.0 percent because of a laboratory overload. How long will it now take to heat the water? Assume that the resistance of the heating element is the same for each of these two modes of operation.

27. If the gauge number of a wire is increased by 6 the diameter is halved; if a gauge number is increased by 1 the diameter decreases by the factor $2^{1/6}$ (see the table in Exercise 5). Knowing this, and also knowing that 1000 ft of #10 copper wire has a resistance of approximately 1.00 Ω, estimate the resistance of 25 ft of #22 copper wire.

28. When 115 V is applied across a 0.30-mm radius, 10-m-long wire, the current density is 1.4×10^4 A/m^2. Find the resistivity of the wire.

29. A block in the shape of a rectangular solid has a cross-sectional area of 3.50×10^{-4} m^2, a length of 15.8 cm, and a resistance of 935 Ω. The material of which the block is made has 5.33×10^{22} conduction electrons/m^3. A potential difference of 35.8 V is maintained along its length. (a) What is the current in the block? (b) If the current density is uniform, what is its value? (c) What is the drift velocity of the conduction electrons? (d) What is the electric field in the block?

30. A steady beam of alpha particles ($q = 2e$) traveling with constant kinetic energy, 20 MeV, carries a current 0.25 μA. (a) If the beam is directed perpendicular to a plane surface, how many alpha particles strike the surface in 3.0 s? (b) At any instant, how many alpha particles are there in a given 20-cm length of the beam? (c) Through what potential difference was it necessary to accelerate each alpha particle from rest to bring it to an energy of 20 MeV?

31. A 100-W light bulb is plugged into a standard 120-V outlet. (a) How much does it cost per month to leave the light turned on? Assume electric energy costs 5 cents/kW·h. (b) What is the resistance of the bulb? (c) What is the current in the bulb? (d) Is the resistance different when the bulb is turned off?

32. A rod of a certain metal is 1.00 m long and 5.50 mm in diameter. The resistance between its ends (at 20° C) is 2.87×10^{-3} Ω. A round disk is formed of this same material, 2.00 cm in diameter and 1.00 mm thick. (a) What is the resistance between the opposing round faces, assuming equipotential surfaces? (b) What is the material?

33. The resistance of an iron wire is 5.9 times that of a copper wire of the same dimensions. What must be the diameter of an iron wire if it is to have the same resistance as a copper wire 1.2-mm in diameter, both wires being the same length?

34. One end of a copper wire and one end of an iron wire of equal length l and diameter d are butted together and joined. A potential difference V is applied between the free ends of the composite wire. Calculate (a) the potential difference across each wire. Assume that $l = 10$ m, $d = 2.0$ mm, and $V = 100$ V. (b) Also calculate the current density in each wire, and (c) the electric field in each wire.

35. A potential difference V is applied to a wire of cross section A, length l, and resistivity ρ. You want to change the applied potential difference and draw out the wire so the power dissipated is increased by a factor of 30 and the current is increased by a factor of 4. What should the new values of l and A be?

36. A 500-W heating unit is designed to operate from a 115-V line. (a) By what percentage will its heat output drop if the line voltage drops to 110 V? Assume no change in resistance. (b) Taking the variation of resistance with temperature into account, would the actual heat output drop be larger or smaller than that calculated in (a)?

37. A Nichrome heater dissipates 500 W when the applied potential difference is 110 V and the wire temperature is 800° C.

How much power would it dissipate if the wire temperature were held at 200° C by immersion in a bath of cooling oil? The applied potential difference remains the same; α for Nichrome is about 4×10^{-4}/C°.

38. An electron linear accelerator produces a pulsed beam of electrons. The pulse current is 0.50 A and the pulse duration 0.10 μs. (a) How many electrons are accelerated per pulse? (b) What is the average current for a machine operating at 500 pulses/s? (c) If the electrons are accelerated to an energy of 50 MeV, what are the average and peak power outputs of the accelerator?

39. A 30-μF capacitor is connected across a programmed power supply. During the interval from $t = 0$ to $t = 3$ s the output voltage of the supply is given by $V(t) = 6 + 4t - 2t^2$ volts. At $t = 0.5$ s find (a) the charge on the capacitor, (b) the current into the capacitor and (c) the power output from the power supply.

40. It is desired to make a long cylindrical conductor whose temperature coefficient of resistance at 20° C is close to zero. (a) If such a conductor is made by assembling alternate disks of iron and carbon with the same diameter, what is the ratio of the thickness of an iron disk to the thickness of a carbon disk? Assume the temperature remains essentially the same in each disk. See Table 28–1. (b) What is the ratio of the heat dissipated in a carbon disk to the heat dissipated in an iron disk?

41. A temperature-stable resistor is made by connecting a resistor made from carbon in series with one made from iron. If the required total resistance is 1000 Ω at 20° C, what should the resistances of the two resistors be?

*42. A thin conducting spherical shell of radius r_1 is centered inside a similar larger shell of radius r_2 and the intervening space is filled with a uniform medium having resistivity ρ. Find the resistance between the shells. Assume total current i flows radially from the small to the large shell.

29

Electromotive Force and Circuits

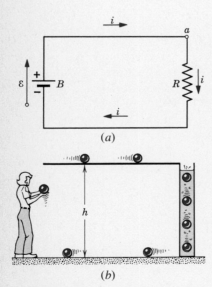

Figure 29–1 (*a*) A simple electric circuit and (*b*) its gravitational analog.

Definition of emf

29–1 Electromotive Force

Certain devices such as batteries and electric generators are able to maintain a potential difference between two points to which they are attached. We call such devices seats of *electromotive force* (symbol ε; abbr. emf). In this chapter we do not discuss their internal construction or detailed mode of action but confine ourselves to describing their gross electrical characteristics and to exploring their usefulness in electric circuits.

Figure 29–1*a* shows a seat of emf ε, represented by a battery, connected to a resistor R. The seat of emf maintains its upper terminal positive and its lower terminal negative, as shown by the + and − signs. In the circuit external to ε positive charge carriers would be driven in the direction shown by the arrows marked i. In other words, a clockwise current would be set up.

An emf is represented by an arrow that is placed next to the seat and points in the direction in which the seat, acting alone, would cause a positive charge carrier to move in the external circuit. We draw a small circle on the tail of an emf arrow so that we will not confuse it with a current arrow.

A seat of emf must be able to do work on charge carriers that enter it. In the circuit of Fig. 29–1*a*, for example, the seat acts to move positive charges from a point of low potential (the negative terminal) through the seat to a point of high potential (the positive terminal). This reminds us of a pump, which can cause water to move from a place of low gravitational potential to a place of high potential.

In Fig. 29–1*a* a charge dq passes through any cross section of the circuit in time dt. In particular, this charge enters the seat of emf ε at its low-potential end and leaves at its high-potential end. The seat must do an amount of work dW on the (positive) charge carriers to force them to go to the point of higher potential. The emf ε of the seat is defined from

$$\varepsilon = dW/dq. \tag{29–1}$$

The unit of emf is the joule/coulomb which is the *volt*. An ideal battery with an emf of one volt would maintain a one-volt potential difference between its terminals, regardless of the current passing through it. For actual batteries this condition holds only under certain conditions, which we describe in Section 29–4. Note also from Eq. 29–1 that the electromotive force is not actually a force, that is, we cannot measure it in newtons. The name is involved with the early history of the subject.

The volt as an emf unit

If a seat of emf does work on a charge carrier, energy must be transferred within the seat. In a battery, for example, chemical energy is transferred into electrical potential energy. Thus we can describe a seat of emf as a device in which chemical, mechanical, or some other form of energy is changed into electrical energy. The chemical energy provided by the battery in Fig. 29–1a is stored in the electric and the magnetic fields that surround the circuit. This stored energy does not increase because it is being drained away, by transfer to thermal or Joule energy in the resistor, at the same rate at which it is supplied. The electric and magnetic fields play an intermediary role in the energy transfer process, acting as a storage reservoir.

Emf and energy transfers

Figure 29–1b shows a gravitational analog of Fig. 29–1a. In the top figure the seat of emf does work on the charge carriers. This energy, stored in passage as electromagnetic field energy, appears eventually as thermal energy in resistor R. In the lower figure the person, in lifting the bowling balls from the floor to the shelf, does work on them. This energy is stored, in passage, as gravitational field energy. The balls roll slowly and uniformly along the shelf, dropping from the right end into a cylinder full of viscous oil. They sink to the bottom at an essentially constant speed, are removed by a trapdoor mechanism not shown, and roll back along the floor to the left. The energy put into the system by the person appears eventually as thermal energy in the viscous fluid, resulting in a temperature rise. The energy supplied by the person comes from internal (chemical) energy. The circulation of charges in Fig. 29–1a will stop eventually if the battery is not recharged; the circulation of bowling balls in Fig. 29–1b will stop eventually if the person does not replenish her store of internal energy by eating.

Emf, a gravitational analog

Figure 29–2 shows a circuit containing two storage batteries, A and B, a resistor R, and an (ideal) electric motor employed in lifting a weight. The batteries are connected so that they tend to send charges around the circuit in opposite directions; the actual direction of the current is determined by B, which supplies the larger potential difference. The energy transfers in this circuit are

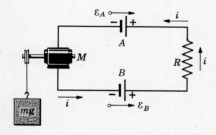

Figure 29–2 Two batteries, a resistor, and a motor, connected in a single-loop circuit. It is given that $\varepsilon_B > \varepsilon_A$.

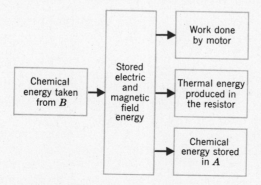

The chemical energy in B is steadily depleted, the energy appearing in the three forms shown on the right. Battery A is being charged while battery B is being discharged. Again, the electric and magnetic fields that surround the circuit act as an intermediary.

29-2 Calculating the Current

Joule's law (Eq. 28–16) tells us that in a time dt an amount of energy given by $i^2R\, dt$ will appear in the resistor of Fig. 29–1a as thermal energy. During this same time a charge $dq\,(=i\,dt)$ will have moved through the seat of emf, and the seat will have done work on this charge (see Eq. 29–1) given by

$$dW = \varepsilon\, dq = \varepsilon i\, dt.$$

From the conservation of energy principle, the work done by the seat must equal the thermal energy, or

$$\varepsilon i\, dt = i^2R\, dt.$$

Solving for i, we obtain

The current in a simple circuit

$$i = \varepsilon/R. \qquad\qquad (29\text{–}2)$$

We can also derive Eq. 29–2 by considering that, if electric potential is to have any meaning, a given point can have only one value of potential at any given time. If we start at any point in the circuit of Fig. 29–1a and, in imagination, go around the circuit in either direction, adding up algebraically the changes in potential that we encounter, we must arrive at the same potential when we return to our starting point. In other words, *the algebraic sum of the changes in potential encountered in a complete traversal of the circuit must be zero.*

The loop theorem

In a gravitational analogue this is no more than saying that any point on the side of a mountain can have only one value of gravitational potential. Expressed otherwise, every point must have a unique altitude above sea level. If you start from any such point and return to it after walking around the mountain the algebraic sum of the changes in altitude that you encounter on this round trip must be zero.

In Fig. 29–1a let us start at point a, whose potential is V_a, and traverse the circuit clockwise. In going through the resistor, there is a change in potential of $-iR$. The minus sign shows that the top of the resistor is higher in potential than the bottom; this is true because positive charge carriers lose potential energy in moving from high to low potential. As we traverse the battery from bottom to top, there is an increase of potential $+\varepsilon$ because the battery does (positive) work on the charge carriers, that is, it moves them from a point of low potential to one of high potential. Adding the algebraic sum of the changes in potential to the initial potential V_a must yield the identical value V_a, or

$$V_a - iR + \varepsilon = V_a.$$

We write this as

$$-iR + \varepsilon = 0,$$

which is independent of the value of V_a and which asserts explicitly that the algebraic sum of the potential changes for a complete circuit traversal is zero. This relation leads directly to Eq. 29–2.

These two ways to find the current in single-loop circuits, based on the conservation of energy and on the concept of potential, are completely equivalent because potential differences are defined in terms of work and energy (see Section 26–1). The statement that the sum of the changes in potential encountered in making a complete loop is zero is called the *loop theorem*. Always bear in mind that this theorem is simply a particular way of stating the law of conservation of energy for electric circuits.

To prepare for the study of more complex circuits, let us examine the rules for finding potential differences; these rules follow from the previous discussion. They are not meant to be memorized but rather to be so thoroughly understood that it becomes trivial to re-derive them on each application.

1. If a resistor is traversed in the direction of the current, the change in potential is $-iR$; in the opposite direction it is $+iR$.

2. If a seat of emf is traversed in the direction of the emf, the change in potential is $+\varepsilon$; in the opposite direction it is $-\varepsilon$.

29-3   Other Single-Loop Circuits

Figure 29–3a shows a circuit which emphasizes that all seats of emf have an intrinsic internal resistance r. This resistance cannot be removed—although we would usually like to do so—because it is an inherent part of the device. The figure shows the internal resistance r and the emf separately, although, actually, they occupy the same region of space.

If we apply the loop theorem, starting at b and going around clockwise, we obtain

$$V_b + \varepsilon - ir - iR = V_b$$

or

$$+\varepsilon - ir - iR = 0.$$

Compare these equations with Fig. 29–3b, which shows the changes in potential graphically. It will be helpful to imagine Fig. 29–3b folded into a cylinder, to suggest the continuity of Fig. 29–3a.

In writing the above equations note that we traversed r and R in the direction of the current and ε in the direction of the emf. The same equation follows if we

Figure 29-3   (a) A single-loop circuit. The rectangular block is a seat of emf with internal resistance r. (b) The same circuit is drawn for convenience as a straight line. Directly below are shown the changes in potential that one encounters in traversing the circuit clockwise from point b. Note that the two points marked "b" at the top of this figure are the same point.

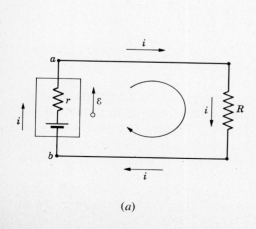

(a)

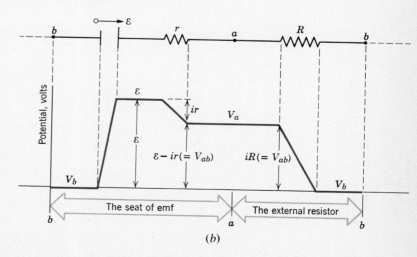

(b)

start at any other point in the circuit or if we traverse the circuit in a counterclock-wise direction. Solving for i gives

$$i = \frac{\mathcal{E}}{R + r}. \tag{29-3}$$

Example 1 *Resistors in series* Resistors in series are connected so that there is only one conducting path through them, as in Fig. 29–4. What is the equivalent resistance R_{eq} of this series combination? The equivalent resistance is the single re-

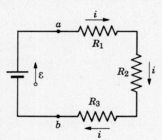

Figure 29–4 Example 1. Three resistors are connected in series between terminals a and b.

sistance R_{eq} which, substituted for the series combination between the terminals ab, will leave the current i unchanged.

Applying the loop theorem (going clockwise from a) yields

$$-iR_1 - iR_2 - iR_3 + \mathcal{E} = 0$$

or

$$i = \frac{\mathcal{E}}{R_1 + R_2 + R_3}.$$

For the equivalent resistance R_{eq}

$$i = \frac{\mathcal{E}}{R_{eq}}$$

or

$$R_{eq} = R_1 + R_2 + R_3. \tag{29-4}$$

The extension to more than three resistors is clear.

29–4 Potential Differences

We often want to compute the potential difference between two points in a circuit. In Fig. 29–3a, for example, what is the relationship between the potential difference $V_{ab}\,(= V_a - V_b)$ between points b and a and the fixed circuit parameters $\mathcal{E}$, r, and R? To find this relationship, let us start from point b and traverse the circuit to point a, passing through resistor R against the current. If V_b and V_a are the potentials at b and a, respectively, we have

$$V_b + iR = V_a$$

because we experience an increase in potential in traversing a resistor against the current arrow. We rewrite this relation as

$$V_{ab} = V_a - V_b = +iR.$$

which tells us that V_{ab}, the potential difference between points a and b, has the magnitude iR and that point a is more positive than point b. Combining this last equation with Eq. 29–3 yields

$$V_{ab} = \mathcal{E}\,\frac{R}{R + r}. \tag{29-5}$$

To sum up: To find the potential difference between any two points in a circuit start at one point and traverse the circuit to the other, following any path and add up algebraically the potential changes encountered. This algebraic sum will be the potential difference. This procedure is similar to that for finding the current in a closed loop, except that here the potential differences are added up over part of a loop and not over the whole loop.

The potential difference between any two points can have only one value; thus we must obtain the same answer for all paths that connect these points. In

Fig. 29–3a let us also calculate V_{ab}, using a path passing through the seat of emf. We have

$$V_b + \varepsilon - ir = V_a$$

or (see also Fig. 29–3b)

$$V_{ab} = V_a - V_b = +\varepsilon - ir.$$

Again, combining with Eq. 29–3 leads to Eq. 29–5.

A battery: the emf and the terminal potential difference

The terminal potential difference of the battery V_{ab}, as Eq. 29–5 shows, is less than ε unless the battery has no internal resistance ($r = 0$) or if it is on open circuit ($R = \infty$); then V_{ab} is equal to ε. Thus the emf of a device is equal to its terminal potential difference when on open circuit.

Example 2 In Fig. 29–5a let ε_1 and ε_2 be 2.0 and 4.0 V, respectively; let the resistances r_1, r_2, and R be 1.0, 2.0, and 5.0 Ω, respectively. What is the current?

Emfs ε_1 and ε_2 oppose each other, but because ε_2 is larger it controls the direction of the current. Thus i will be counterclockwise. The loop theorem, going clockwise from a, yields

$$-\varepsilon_2 + ir_2 + iR + ir_1 + \varepsilon_1 = 0.$$

Check that the same result is obtained by going around counterclockwise. Also, compare this equation carefully with Fig. 29–5b, which shows the potential changes graphically.

Solving for i yields

$$i = \frac{\varepsilon_2 - \varepsilon_1}{R + r_1 + r_2} = \frac{4.0\ \text{V} - 2.0\ \text{V}}{5.0\ \Omega + 1.0\ \Omega + 2.0\ \Omega} = 0.25\ \text{A}.$$

It is not necessary to know in advance what the actual direction of the current is. To show this, let us assume that the current in Fig. 29–5a is clockwise, an assumption that we know is incorrect. The loop theorem then yields (going clockwise from a)

$$-\varepsilon_2 - ir_2 - iR - ir_1 + \varepsilon_1 = 0$$

or

$$i = \frac{\varepsilon_1 - \varepsilon_2}{R + r_1 + r_2}.$$

Substituting numerical values (see above) yields -0.25 A for the current. The minus sign tells us that the current is in the opposite direction from the one we have assumed.

In more complex circuit problems involving many loops and branches it is often impossible to know in advance the correct directions for the currents in all parts of the circuit. The procedure is to assume directions for the currents at random. Those currents for which positive numerical values are obtained will have the correct directions; those for which negative values are obtained will be exactly opposite to the assumed directions. In all cases the numerical values will be correct.

Example 3 What is the potential difference (a) between points b and a in Fig. 29–5a, and (b) between points a and c?

(a) For points a and b we start at b and traverse the circuit to a, obtaining

$$V_{ab}(= V_a - V_b) = -ir_2 + \varepsilon_2 = -(0.25\ \text{A})(2.0\ \Omega) + 4.0\ \text{V}$$
$$= +3.5\ \text{V}.$$

Figure 29–5 Examples 2, 3. (a) A single-loop circuit. (b) The same circuit is shown schematically as a straight line, the potential differences encountered in traversing the circuit clockwise from point a being displayed directly below. In the lower figure the potential of point a was assumed to be zero for convenience.

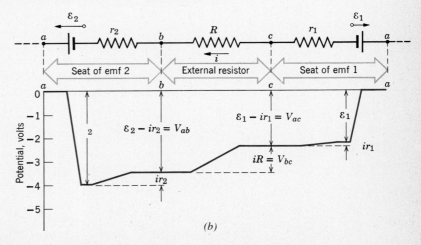

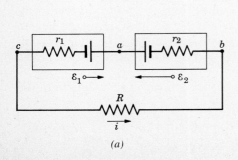

(a)

(b)

Thus a is more positive than b and the potential difference ($= 3.5$ V) is less than the emf ($= 4.0$ V); see Fig. 29–5b.

(b) For points c and a, we start at c and traverse the circuit to a, obtaining

$$V_{ac} \; (= V_a - V_c) = +\varepsilon_1 + ir_1 = +2.0 \text{ V} + (0.25 \text{ A})(1.0 \; \Omega)$$
$$= +2.25 \text{ V}.$$

This tells us that a is at a higher potential than c. The terminal potential difference of ε_1 ($=2.25$ V) is larger than the emf ($=2.0$ V); see Fig. 29–5b. Charge is being forced through ε_1 in a direction opposite to the one in which it would send charge if it were acting by itself; if ε_1 is a storage battery, it is being charged at the expense of ε_2.

Let us test the first result by proceeding from b to a along a different path, namely, through R, r_1, and ε_1. We have

$$V_{ab} = iR + ir_1 + \varepsilon_1 = (0.25 \text{ A})(5.0 \; \Omega) + (0.25 \text{ A})(1.0 \; \Omega) + 2.0 \text{ V}$$
$$= +3.5 \text{ V}.$$

which is the same as the earlier result.

29–5   Multiloop Circuits

Junctions and branches

Figure 29–6 shows a circuit containing two loops. For simplicity, we have neglected the internal resistances of the batteries. There are two junctions, b and d, and three branches connecting these junctions. The branches are the left branch *bad*, the right branch *bcd*, and the central branch *bd*. If the emfs and the resistances are given, what are the currents in the various branches?

We label the currents in the branches as i_1, i_2, and i_3, as shown. Current i_1 has the same value for any cross section of the left branch from b to d. Similarly, i_2 has the same value everywhere in the right branch and i_3 in the central branch. The directions of the currents may be chosen arbitrarily. You may already have noted that the current in the third branch must point up, not down as we have drawn it. We have deliberately assigned it this way to show how the formal mathematical procedure still gives the correct answer in that i_3 turns out to be negative. It doesn't matter how you assign the currents initially.

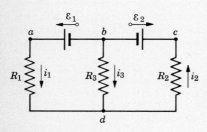

Figure 29–6   A multiloop circuit.

The three currents i_1, i_2, and i_3 carry charge either toward the junction d or away from it. Charge does not accumulate at junction d, nor does it drain away from this junction because the circuit is in a steady-state condition. Thus charge must be removed from the junction by the currents at the same rate that it is brought into it. If we arbitrarily call a current approaching the junction positive and one leaving the junction negative, then

$$i_1 + i_3 - i_2 = 0.$$

The junction theorem

This equation suggests a general principle, known as the *junction theorem*, for the solution of multiloop circuits: *At any junction the algebraic sum of the currents must be zero.* Note that it is simply a statement of the conservation of charge. Thus our basic tools for solving circuits are (a) the conservation of energy (see p. 562) and (b) the conservation of charge.

For the circuit of Fig. 29–6, the junction theorem yields only one relationship among the three unknowns. Applying the theorem at junction b leads to exactly the same equation, as you can easily verify. To solve for the three unknowns, we need two more independent equations; they can be found from the loop theorem.

In single-loop circuits there is only one conducting loop around which to apply the loop theorem and the current is the same in all parts of this loop. In multiloop circuits there is more than one loop and the current in general will not be the same in all parts of any given loop.

If we traverse the left loop of Fig. 29–6 in a counterclockwise direction, the loop theorem gives

$$\varepsilon_1 - i_1 R_1 + i_3 R_3 = 0. \tag{29–6}$$

The right loop gives

$$-i_3R_3 - i_2R_2 - \varepsilon_2 = 0. \qquad (29\text{–}7)$$

These two equations, together with the relation derived earlier with the junction theorem, are the three simultaneous equations needed to solve for the unknowns i_1, i_2, and i_3. Doing so yields

$$i_1 = \frac{\varepsilon_1(R_2 + R_3) - \varepsilon_2R_3}{R_1R_2 + R_2R_3 + R_1R_3}, \qquad (29\text{–}8a)$$

$$i_2 = \frac{\varepsilon_1R_3 - \varepsilon_2(R_1 + R_3)}{R_1R_2 + R_2R_3 + R_1R_3}, \qquad (29\text{–}8b)$$

and

$$i_3 = -\frac{\varepsilon_1R_2 + \varepsilon_2R_1}{R_1R_2 + R_2R_3 + R_1R_3}. \qquad (29\text{–}8c)$$

Supply the missing steps. Equation 29–8c shows that no matter what numerical values are given to the emfs and to the resistances, the current i_3 will have a negative value as long as the directions of the emfs don't change. This means that i_3 will point up in Fig. 29–6 rather than down, as we deliberately assumed. The currents i_1 and i_2 may be in either direction, depending on the particular numerical values of the resistances and emfs.

Verify that Eqs. 29–8 reduce to sensible conclusions in special cases. For $R_3 = \infty$, for example—which corresponds to clipping this resistor out of the circuit—we find

$$i_1 = i_2 = \frac{\varepsilon_1 - \varepsilon_2}{R_1 + R_2} \qquad \text{and} \qquad i_3 = 0.$$

What do these equations reduce to for $R_2 = \infty$?

The loop theorem can be applied to a large loop consisting of the entire circuit $abcda$ of Fig. 29–6. This fact might suggest that there are more equations than we need, for there are only three unknowns and we already have three equations written in terms of them. However, the loop theorem yields for this loop

$$-\varepsilon_1 + \varepsilon_2 + i_2R_2 + i_1R_1 = 0$$

which is nothing more than the sum of Eqs. 29–6 and 29–7. Thus this large loop does not yield another independent equation. It will never be found in solving multiloop circuits that there are more independent equations than variables.

Example 4 *Resistors in parallel* Figure 29–7 shows three resistors connected across the same seat of emf. Resistances across which the identical potential difference is applied are said to be in parallel. What is the equivalent resistance R_{eq} of this parallel combination? By *equivalent resistance* we mean that single resistance which, substituted for the parallel combination between terminals ab, would leave the current i unchanged.

The currents in the three branches are

$$i_1 = \frac{V}{R_1}, \qquad i_2 = \frac{V}{R_2}, \qquad \text{and} \qquad i_3 = \frac{V}{R_3},$$

where V is the potential difference that appears between points a and b. The total current i is found by applying the junction theorem to junction a, or

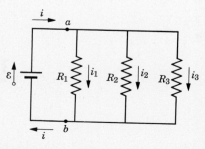

Figure 29–7 Example 4. Three resistors are connected in parallel between terminals a and b.

$$i = i_1 + i_2 + i_3 = V\left(\frac{1}{R_1} + \frac{1}{R_2} + \frac{1}{R_3}\right).$$

If the equivalent resistance is used instead of the parallel combination, we have

$$i = \frac{V}{R_{\text{eq}}}.$$

Combining these two equations gives

$$\frac{1}{R_{\text{eq}}} = \frac{1}{R_1} + \frac{1}{R_2} + \frac{1}{R_3}. \qquad (29-9)$$

This formula can easily be extended to more than three resistances. Note that the equivalent resistance of a parallel combination is less than any of the resistances that make it up.

29–6 Electrical Measuring Instruments

Several electrical measuring instruments involve circuits that can be studied by the methods of this chapter. In this section are simplified descriptions of the ammeter, the voltmeter, and the potentiometer.

1. The Ammeter. A meter to measure currents is called an *ammeter* (or a *milliammeter, microammeter,* etc., depending on the size of the current to be measured). To determine the current in a wire, it is necessary to break or cut the wire and to insert the ammeter, so that the current to be measured passes through the meter (see Fig. 29–8).

It is essential that the resistance R_A of the ammeter be small compared to other resistances in the circuit. Otherwise the presence of the meter will in itself change the current to be measured. An ideal ammeter would have zero resistance and zero power would be dissipated in it. In the circuit of Fig. 29–8 the required condition, assuming that the voltmeter is not connected, is

$$R_A \ll r + R_1 + R_2.$$

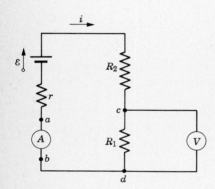

Figure 29–8 An ammeter (A) is connected to read the current in a circuit, and a voltmeter (V) is connected to read the potential difference across resistor R_1.

2. The Voltmeter. A meter to measure potential differences is called a *voltmeter* (or a *millivoltmeter, microvoltmeter,* etc.). To find the potential difference between two points in a circuit, it is necessary to connect one of the voltmeter terminals to each of the circuit points, without breaking the circuit (see Fig. 29–8).

It is essential that the resistance of the voltmeter R_V be large compared to any circuit resistance across which the voltmeter is connected. Otherwise the meter will itself constitute an important circuit element and will alter the circuit current and the potential difference to be measured. An ideal voltmeter would have an infinite resistance and, again, zero power would be dissipated in it. In Fig. 29–8 the required condition is

$$R_V \gg R_1.$$

The ultimate, sensitive element for both the ammeter and the voltmeter is the *galvanometer;* its mode of operation is described in Section 30–4. To form an ammeter from a galvanometer a low resistance is put in parallel with it, to insure that R_A will be sufficiently small. To form a voltmeter from a galvanometer a large resistance is put in series with it, to insure that R_V is sufficiently large.

Commonly a single galvanometer is boxed, provided with suitable scales, and arranged so that by suitable external switching it can serve as both an ammeter and a voltmeter, usually with several choices of scale for each mode. Often switching arrangements for measuring resistance (an *ohm-meter* mode) are also included. Such a versatile and packaged unit is called a *multimeter.*

3. The Potentiometer. Figure 29–9 shows the rudiments of a *potentiometer,* which is a device for measuring an unknown emf ε_x, under zero current conditions, by comparing it with a known standard emf ε_s.

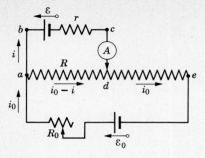

Figure 29-9 A potentiometer

When using the instrument $\mathcal{E}_x$ is first placed in position $\mathcal{E}$ and the precision variable resistor R is adjusted (to value R_x) until current i is zero as noted on the sensitive ammeter A. Then the process is repeated for $\mathcal{E}_s$, noting its zero-current value for R, namely R_s. We will now see how this can give a precise determination of $\mathcal{E}_x$.

Applying the loop theorem to loop $abcd$ yields

$$-\mathcal{E} - ir + (i_0 - i)R = 0,$$

where, by application of the junction theorem at a, $i_0 - i$ is the current in resistor R. Solving for i yields

$$i = \frac{i_0 R - \mathcal{E}}{R + r}$$

If R is adjusted to have the value

$$R = \mathcal{E}/i_0 \qquad (29-10)$$

the current i in the branch $abcd$ becomes zero. This can be done very accurately by changing R until A reads zero.

Repeating this procedure for the unknown as well as the standard emf's we get, from Eq. 29-10, $R_x = \mathcal{E}_x/i_0$ and $R_s = \mathcal{E}_s/i_0$. These equations can be reduced to a single one because i_0 is the same for both cases. This is because $i = 0$, which means that no current flows in the upper loop. The current in the lower loop then is completely determined by the circuit elements in that loop alone and therefore i_0 is independent of $\mathcal{E}$ and r. By eliminating i_0 from the two equations we obtain

$$\mathcal{E}_x = \mathcal{E}_s \frac{R_x}{R_s}. \qquad (29-11)$$

Thus the unknown emf can be obtained from the standard by making two determinations of the precision resistor. In practice, potentiometers are conveniently packaged units, containing a standard cell which, after calibration at the National Bureau of Standards or elsewhere, serves as a convenient known standard seat of emf $\mathcal{E}_s$. Switching arrangements for replacing the unknown emf by the standard and arrangements for ascertaining that the current i_0 remains constant are also incorporated.

29-7 RC Circuits

The preceding sections dealt with circuits in which the circuit elements were resistors and in which the currents did not vary with time. Here we introduce the capacitor as a circuit element, which will lead us to the concept of time-varying currents. In Fig. 29-10 let switch S be thrown to position a. What current is set up in the single-loop circuit so formed? Let us apply conservation of energy principles.

In time dt a charge dq ($= i\,dt$) moves through any cross section of the circuit. The work done by the seat of emf ($= \mathcal{E}\,dq$; see Eq. 29-1) must equal the energy that appears as thermal energy in the resistor during time dt ($= i^2 R\,dt$) plus the increase in the amount of energy U that is stored in the capacitor [$= dU = d(q^2/2C)$; see Eq. 27-7]. In equation form

$$\mathcal{E}\,dq = i^2 R\,dt + d\left(\frac{q^2}{2C}\right)$$

or

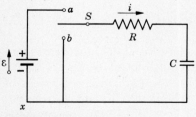

Figure 29-10 An RC circuit.

$$\varepsilon \, dq = i^2 R \, dt + \frac{q}{C} \, dq.$$

Dividing by dt yields

$$\varepsilon \frac{dq}{dt} = i^2 R + \frac{q}{C} \frac{dq}{dt}.$$

But dq/dt is simply i, so that this equation becomes

$$\varepsilon = iR + \frac{q}{C}. \tag{29-12}$$

This equation also follows from the loop theorem, as it must, since the loop theorem was derived from the conservation of energy principle. Starting from point x and traversing the circuit clockwise, we experience an increase in potential in going through the seat of emf and decreases in potential in traversing the resistor and the capacitor, or

$$\varepsilon - iR - \frac{q}{C} = 0,$$

which is identical to Eq. 29–12.

We cannot immediately solve Eq. 29–12 because it contains two variables, q and i, which, however, are related by

$$i = \frac{dq}{dt}. \tag{29-13}$$

Substituting for i into Eq. 29–12 gives

The differential equation for RC charging

$$\varepsilon = R \frac{dq}{dt} + \frac{q}{C}. \tag{29-14}$$

Our task now is to find the function $q(t)$ that satisfies this differential equation. Although this particular equation is not hard to solve, we choose to avoid mathematical complexity by simply presenting the solution, which is

RC charging, $q(t)$

$$q = C\varepsilon(1 - e^{-t/RC}). \tag{29-15}$$

We can easily test whether this function $q(t)$ is really a solution of Eq. 29–14 by substituting it into that equation and seeing whether an identity results. Differentiating Eq. 29–15 with respect to time yields

RC charging, $i(t)$

$$i = \frac{dq}{dt} = \frac{\varepsilon}{R} e^{-t/RC}. \tag{29-16}$$

Substituting q (Eq. 29–15) and dq/dt (Eq. 29–16) into Eq. 29–14 yields an identity, as you should verify. Thus Eq. 29–15 is indeed a solution of Eq. 29–14.

Figure 29–11 shows the variation with time during the charging process of V_C, the potential difference across the capacitor of Fig. 29–10, and V_R, the potential difference across the resistor. V_C $(= q/C)$ is a measure of the charge on the capacitor, as described by Eq. 29–15. V_R $(= iR)$ is a measure of the current through the resistor, as described by Eq. 29–16.

The time constant, RC

The quantity RC in Eqs. 29–15 and 29–16 has the dimensions of time (since the exponent must be dimensionless) and is called the *capacitive time constant* of the circuit. It is the time at which the charge on the capacitor has increased to within a factor of $(1 - e^{-1})$ $(\cong 63\%)$ of its equilibrium value. To show this, we put $t = RC$ in Eq. 29–15 to obtain

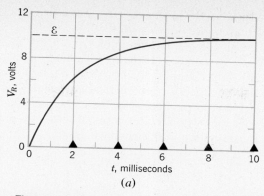

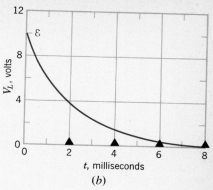

(a)

(b)

Figure 29–11 *RC charging.* The variation with time of (a) $V_C(=q/C)$ and (b) $V_R(=iR)$ after switch S is closed on a in Fig. 29–10. The filled triangles represent time constants ($RC = 2.0$ ms). For $R = 2000\ \Omega$, $C = 1.0\ \mu\text{F}$, and $\varepsilon = 10V$.

$$q = C\varepsilon(1 - e^{-1}) = 0.63\,C\varepsilon.$$

Since $C\varepsilon$ is the equilibrium charge on the capacitor, corresponding to $t \to \infty$, the foregoing statement follows.

Example 5 After how many time constants will the energy stored in the capacitor in Fig. 29–10 reach one-half its equilibrium value?

The energy is given by Eq. 27–7, or

$$U = \frac{1}{2C}\, q^2,$$

the equilibrium energy U_∞ being $(1/2C)(C\varepsilon)^2$. From Eq. 29–15, we can write the energy as

$$U = \frac{1}{2C}\,(C\varepsilon)^2(1 - e^{-t/RC})^2$$

or

$$U = U_\infty(1 - e^{-t/RC})^2.$$

Putting $U = \frac{1}{2}U_\infty$ yields

$$\tfrac{1}{2} = (1 - e^{-t/RC})^2$$

and solving this relation for t yields finally

$$t = 1.22\,RC = 1.22 \text{ time constants.}$$

If R in Fig. 29–10 were zero the capacitor charge q would jump to its equilibrium value of $C\varepsilon$ immediately after switch S in that figure was closed on a. The effect of introducing the resistance is to delay this precipitous rise of q, as Fig. 29–11a shows. Although we have proved this formally in our derivation of Eq. 29–15 it is also important to understand physically just how this delaying action comes about.

A physical feeling for the *RC* charging process

When switch S in Fig. 29–10 is closed on a, instantaneously the resistor experiences an applied potential difference of ε and an initial current of ε/R is set up. Initially, the capacitor experiences no potential difference because its initial charge is zero, the potential difference always being given by q/C. The flow of charge through the resistor starts to charge the capacitor, which has several effects. First, the existence of a capacitor charge means that there must now be a potential difference ($= q/C$) across the capacitor; this, in turn, means that the potential difference across the resistor must decrease by this amount, since the sum of the two potential differences must always equal ε. This decrease in the potential difference across R means that the charging current is reduced. Thus the charge of the capacitor builds up and the charging current decreases until the capacitor is fully charged. At this point the full emf ε is applied to the capacitor, there being no potential drop ($iR = 0$) across the resistor. This is precisely the reverse of the initial situation. Review the derivations of Eqs. 29–15 and 29–16 and study Fig. 29–11 with the qualitative arguments of this paragraph in mind.

Now let us consider the RC discharging process. Assume that the switch S in Fig. 29–10 has been in position a for a time t such that $t \gg RC$. The capacitor is then fully charged for all practical purposes. The switch S is then thrown to position b. How do the charge of the capacitor and the current now vary with time?

With the switch S closed on b, there is no emf in the circuit and Eq. 29–12 for the circuit, with $\varepsilon = 0$, becomes simply

$$iR + \frac{q}{C} = 0. \qquad (29\text{–}17)$$

Putting $i = dq/dt$ allows us to write, as the differential equation of the circuit (compare Eq. 29–14),

The differential equation for RC discharging

$$R\frac{dq}{dt} + \frac{q}{C} = 0. \qquad (29\text{–}18)$$

The solution is

RC discharging, $q(t)$

$$q = q_0 e^{-t/RC}, \qquad (29\text{–}19)$$

as you may readily verify by substitution, q_0 being the initial charge on the capacitor. The capacitive time constant RC appears in this expression for capacitor discharge as well as in that for the charging process (Eq. 29–15). We see that at a time $t = RC$ the capacitor charge is reduced to $q_0 e^{-1}$, which is about 37% of the initial charge q_0.

The current during discharge follows from differentiating Eq. 29–19, or

$$i = \frac{dq}{dt} = -\frac{q_0}{RC} e^{-t/RC}. \qquad (29\text{–}20)$$

The negative sign shows that the current is in the direction opposite to that shown in Fig. 29–10. This is as it should be, since the capacitor is discharging rather than charging. Since $q_0 = C\varepsilon$, we can write Eq. 29–20 as

RC discharging, $i(t)$

$$i = -\frac{\varepsilon}{R} e^{-t/RC}, \qquad (29\text{–}21)$$

in which ε/R appears as the initial current, corresponding to $t = 0$. This is reasonable because the initial potential difference for the fully charged capacitor is ε.

REVIEW GUIDE AND SUMMARY

Electromotive force

A *seat of electromotive force* is a device (a battery or a generator, say) that supplies energy (chemical or mechanical, say) to maintain a potential difference between its output terminals. If dW is the energy the seat provides to force positive charge dq through the seat from the negative to the positive terminal then

$$\varepsilon = dW/dq. \qquad [29\text{–}1]$$

The volt is the SI unit of emf as well as of potential difference. Study Figs. 29–1 and 29–2.

The current in a single-loop circuit

To find the current in a single-loop circuit we use the loop theorem, which follows from the conservation of energy and states that the algebraic sum of all the changes in potential encountered in traversing a circuit loop is zero. That is

The loop theorem

$$\Delta V_1 + \Delta V_2 + \Delta V_3 \ldots = 0 \qquad \text{any complete circuit loop.}$$

Study the sign conventions for ΔV given on p. 563; see Example 2.

The currents in a multi-loop circuit

To find the currents in a multi-loop circuit (see Fig. 29–6) we can apply the loop theorem to all closed circuit loops. We can also apply the *junction theorem* at all circuit junctions. This theorem, which follows from the conservation of charge, states that the algebraic sum of all currents approaching a junction is zero. That is

The junction theorem

$$i_1 + i_2 + i_3 \ldots = 0 \qquad \text{any circuit junction.}$$

Currents approaching a junction are taken as positive, those leaving being negative. In applying the loop and junction theorems it is never possible to obtain more independent equations than there are unknown currents. See Example 4.

To find the potential difference V_{ab} between any two points in either a single- or a multi-loop circuit we use

Potential Difference

$$V_{ab} = \sum_b^a \Delta V \qquad \text{any circuit path from } b \text{ to } a.$$

See Example 3.

The equivalent resistance of a number of resistors connected in series (see Example 1) is

Resistors in series

$$R_{eq} = R_1 + R_2 + R_3 \ldots \qquad [29\text{--}4]$$

For resistors in parallel (see Example 4) we have

Resistors in parallel

$$\frac{1}{R_{eq}} = \frac{1}{R_1} + \frac{1}{R_2} + \frac{1}{R_3} \ldots$$

Four electrical measuring instruments that involve circuits are analyzed by the methods of this chapter. The ammeter has a very small resistance so it can be introduced directly into the circuit to measure the current without unduly changing the operation of the circuit. The voltmeter is connected between two points in a circuit to measure the potential difference between those points. It has a high resistance so the little current that flows through it won't disturb the circuit. The potentiometer is an instrument that accurately compares an unknown potential difference with a standard potential difference; study Fig. 29–9. For the potentiometer two adjustments, R_x and R_s, of a precision resistor R are made yielding

Ammeter

Voltmeter

Potentiometer

$$\varepsilon_x = \varepsilon_s \frac{R_x}{R_s} \qquad [29\text{--}11]$$

Wheatstone bridge

In the Wheatstone bridge (See Problem 18 and Fig. 29–27.) a comparison of known precision resistances, R_1, R_2, and R_s results in a determination of an unknown resistance R_x through the equation

$$R_x = R_s \frac{R_2}{R_1}$$

A single-loop *RC* circuit

This table summarizes the discussion of the charging and the discharging of the *RC* circuit of Fig. 29–10. The product *RC* has the dimensions of time and is called the *capacitative time constant* of the circuit. See Example 5.

	RC **charging**	*RC* **discharging**
The differential equation	$\varepsilon = R\dfrac{dq}{dt} + \dfrac{q}{C}$ Eq. 29–14	$0 = R\dfrac{dq}{dt} + \dfrac{q}{C}$ Eq. 29–18
The charge $q(t)$ in the capacitor	$q = (C\varepsilon)(1 - e^{-t/RC})$ Eq. 29–15; see Fig. 29–11a	$q = (C\varepsilon)e^{-t/RC}$ Eq. 29–19
The current $i(t)$ in the resistor	$i = +(\varepsilon/R)e^{-t/RC}$ Eq. 29–16; see Fig. 29–11b	$i = -(\varepsilon/R)e^{-t/RC}$ Eq. 29–21

QUESTIONS

1. Does the direction of the emf provided by a battery depend on the direction of current flow through the battery?

2. In Fig. 29-2 discuss what changes would occur if we increased the mass m by such an amount that the "motor" reversed direction and became a "generator", that is, a seat of emf.

3. Discuss in detail the statement that the energy method and the loop theorem method for solving circuits are perfectly equivalent.

4. Devise a method for measuring the emf and the internal resistance of a battery.

5. The current passing through a battery of emf ε and internal resistance r is decreased by some external means. Does the potential difference between the terminals of the battery necessarily decrease or increase? Explain.

6. In calculating V_{ab} in Fig. 29-3a, is it permissible to follow a path from a to b that does not lie in the conducting circuit?

7. A 25-W, 110-V bulb glows at normal brightness when connected across a bank of batteries. A 500-W, 110-V bulb glows only dimly when connected across the same bank. Explain.

8. Under what circumstances can the terminal potential difference of a battery exceed its emf?

9. Automobiles generally use a 12-V electrical system. Years ago a 6-V system was used. Why the change? Why not 24 V?

10. Do the junction theorem and the loop theorem apply in a circuit containing a capacitor?

11. The loop theorem is based on the conservation of energy principle and the junction theorem on the conservation of charge principle. Explain just how these theorems are based on these principles.

12. Under what circumstances would you want to connect batteries in parallel? . . . in series?

13. Under what circumstances would you want to connect resistors in parallel? . . . in series?

14. What is the difference between an emf and a potential difference?

15. Compare and contrast the formulas for the equivalent values of (a) capacitors and (b) resistors, in series and in parallel.

16. Referring to Fig. 29-6 use a qualitative physical argument to convince yourself that i_3 is drawn in the wrong direction.

17. Explain in your own words why the resistance of an ammeter should be very small whereas that of a voltmeter should be very large.

18. Does the time required for the charge on a capacitor in an RC circuit to build up to a given fraction of its equilibrium value depend on the value of the applied emf? Does the time required for the charge to change by a given amount depend on the applied emf?

19. Devise a method whereby an RC circuit can be used to measure very high resistances.

20. In Fig. 29-10 suppose that switch S is closed on a. Explain why, in view of the fact that the negative terminal of the battery is not connected to resistance R, the initial current in R should be ε/R, as Eq. 29-16 predicts.

21. Show that the product RC in Eqs. 29-15 and 29-16 has the dimensions of time, that is, that 1 second = 1 ohm $\times$ 1 farad.

EXERCISES

Section 29-1 Electromotive Force

1. (a) How much work does a 12-V seat of emf do on an electron as it passes through from the positive to the negative terminal? (b) If 3.4×10^{18} electrons pass through each second, what is the power output of the seat?

2. A 5.0-A current is set up in an external circuit by a 6.0-V storage battery for 6.0 min. By how much is the chemical energy of the battery reduced?

3. A certain 12-V car battery carries an initial charge of 120 A·h. Assuming that the potential across the terminals stays constant until the battery is completely discharged, for how many hours can it deliver energy at the rate of 100 W?

Section 29-2 Calculating the Current

4. In Fig. 29-12 $\varepsilon_1 = 12$ V and $\varepsilon_2 = 8$ V. What is the sense of the current in the resistor? Which emf is doing positive work? Which point, A or B, is at the higher potential?

5. The current in a single-loop circuit is 5.0 A. When an additional resistance of 2.0 Ω is inserted, the current drops to 4.0 A. What was the resistance in the original circuit?

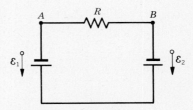

Figure 29-12 Exercise 4.

Section 29-3 Other Single-Loop Circuits

6. A wire of resistance 5.0 Ω is connected to a battery whose emf ε is 2.0 V and whose internal resistance is 1.0 Ω. In 2.0 min (a) how much energy is transferred from chemical to electrical form? (b) How much energy appears in the wire as thermal energy? (c) Account for the difference between (a) and (b).

7. In Fig. 29-3a put $\varepsilon = 2.0$ V and $r = 100$ Ω. Plot (a) the current and (b) the potential difference across R, as functions of r over the range 0 to 500 Ω. Make both plots on the same graph. (c) Make a third plot by multiplying together, for each value of R, the two curves plotted. What is the physical significance of this plot?

8. Assume that the batteries in Fig. 29–13 have negligible internal resistance. Find: (a) the current in the circuit, (b) the power dissipated in each resistor, and (c) the power supplied or absorbed by each emf.

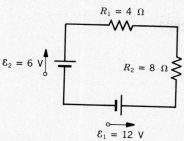

Figure 29–13 Exercise 8.

9. A gasoline gauge for an automobile is shown schematically in Fig. 29–14. The indicator (on the dashboard) has a resistance of 10 Ω. The tank unit is simply a float connected to a linear variable resistor that has a resistance of 140 Ω when the tank is empty and 20 Ω when it is full. Assume the indicator reading changes linearly with the current and find the current in the circuit when the tank is half full.

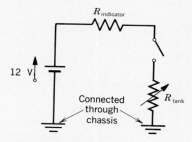

Figure 29–14 Exercise 9.

10. A 12-V emf car battery with an internal resistance of 0.04 Ω is being charged with a current of 50 A. (a) What is the potential difference across its terminals? (b) At what rate is heat being dissipated in the battery? (c) At what rate is electric energy being converted to chemical energy? (d) What are the answers to (a) and (c) if the same battery is used to supply 50 A to the starter motor?

Section 29–4 Potential Differences

11. In Fig. 29–15 the potential at point P is 100 V. What is the potential at point Q?

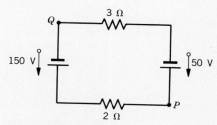

Figure 29–15 Exercise 11.

12. The section of circuit AB (see Fig. 29–16) absorbs 50 W of power when a current $i = 1.0$ A passes through it in the indicated direction. (a) What is the potential difference between A and B?

(b) If the element C does not have internal resistance, what is its emf? (c) What is its polarity?

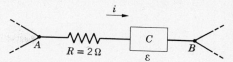

Figure 29–16 Exercise 12.

13. In Fig. 29–4 calculate the potential difference across R_2, using two different paths. Assume $\varepsilon = 12$ V, $R_1 = 3.0$ Ω, $R_2 = 4.0$ Ω, and $R_3 = 5.0$ Ω.

14. In Fig. 29–5a calculate the potential difference between a and c by considering a path that contains R and ε_2. (See Example 3.)

15. In the sketches of Fig. 29–17 find the equivalent resistance between the points (a) A and B, (b) A and C, (c) B and C, (d) D and E, (e) F and H, (f) F and G.

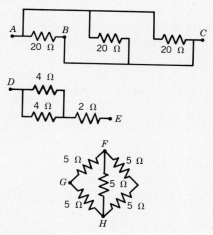

Figure 29–17 Exercise 15.

Section 29–5 Multiloop Circuits

16. By using only two resistance coils—singly, in series, or in parallel—you are able to obtain resistances of 3, 4, 12, and 16 Ω. What are the separate resistances of the coils?

17. In Fig. 29–18 find the current in each resistor and the potential difference between a and b. Put $\varepsilon_1 = 6.0$ V, $\varepsilon_2 = 5.0$ V, $\varepsilon_3 = 4.0$ V, $R_1 = 100$ Ω, and $R_2 = 50$ Ω.

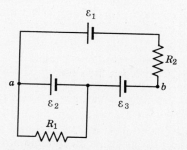

Figure 29–18 Exercise 17.

18. Fig. 29–19 shows a circuit containing an ammeter and three switches, labeled S_1, S_2, and S_3. Find the readings of the ammeter for all possible combinations of switch settings. Put $\varepsilon = 120$ V, $R_1 = 20\ \Omega$, and $R_2 = 10\ \Omega$.

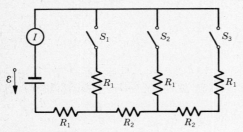

Figure 29–19 Exercise 18.

19. Two light bulbs, one of resistance R_1 and the other of resistance R_2 ($<R_1$) are connected (*a*) in parallel and (*b*) in series. Which bulb is brighter in each case?

20. For manual control of the current in a circuit, you can use a parallel combination of variable resistors of the sliding contact type, as in Fig. 29–20, with $R_1 = 20R_2$. (*a*) What procedure is used to adjust the current to the desired value? (*b*) Why is the parallel combination better than a single-variable resistor?

Section 29–6 Electrical Measuring Instruments
21. A simple ohmmeter is made by connecting a 1.5-V flashlight battery in series with a resistor R and a 1-mA ammeter, as shown in Fig. 29–21. R is adjusted so that when the circuit terminals are shorted together the meter deflects to its full-scale value of 1.0 mA. What external resistance across the terminals results in a deflection of (*a*) 10 percent, (*b*) 50 percent, and (*c*) 90 percent of

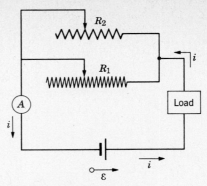

Figure 29–20 Exercise 20.

full scale? (*d*) If the ammeter has a resistance of 20 Ω and the internal resistance of the battery is negligible, what is R?

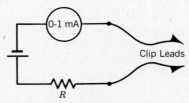

Figure 29–21 Exercise 21.

Section 29–7 RC Circuits
22. How many time constants must elapse before a capacitor in and RC circuit is charged to within 1.0 percent of its equilibrium charge?

PROBLEMS GROUPED BY SECTION

Section 29–1 Electromotive Force
1. A standard flashlight battery can deliver about 2.0 W·h of energy before it runs down. (*a*) If a battery costs 80 cents, what is the cost of operating a 100-W lamp for 8.0 h? (*b*) What is the cost if power provided by an electric utility company, at 12 cents per kW·h, is used?

Section 29–2 Calculating the Current
2. A battery of emf ε (=2.0 V) and internal resistance r (=0.5 Ω) is driving a motor. The motor is lifting a 2.0-kg mass at constant speed v (=0.50 m/s). Assuming no power losses, find (*a*) the current i in the circuit and (*b*) the potential difference V across the terminals of the motor. (*c*) Discuss the fact that there are two solutions to this problem.

Section 29–3 Other Single-Loop Circuits
3. (*a*) In Fig. 29–22 what value must R have if the current in the circuit is to be 1.0 mA? Take $\varepsilon_1 = 2.0$ V, $\varepsilon_2 = 3.0$ V, and $r_1 = r_2 = 3.0\ \Omega$. (*b*) What is the rate at which thermal energy appears in R?

4. You are given a number of 10-Ω resistors, each capable of dissipating only 1.0 W. What is the minimum number of such resistors that you need to combine in series or parallel combinations to make a 10-Ω resistor capable of dissipating at least 5.0 W?

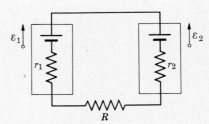

Figure 29–22 Problem 3.

5. (*a*) In the circuit of Fig. 29–3*a* show that the power delivered to R as thermal energy is a maximum when R is equal to the internal resistance r of the battery. (*b*) Show that this maximum power is $P = \varepsilon^2/4r$.

6. Thermal energy is to be generated in a 0.10-Ω resistor at the rate of 10 W by connecting it to a battery whose emf is 1.5 V. (*a*) What is the internal resistance of the battery? (*b*) What potential difference exists across the resistor?

Section 29–4 Potential Differences
7. Two batteries having the same emf ε but diffent internal resistances r_1 and r_2 are connected in series to an external resistance R. Find the value of R that makes the potential difference zero between the terminals of one battery. Which battery is it?

Section 29-5 Multiloop Circuits

8. Two batteries of emf ε and internal resistance r are connected in parallel across a resistor R, as in Fig. 29–23b. (a) For what value of R is the thermal energy delivered to the resistor a maximum? (b) What is the maximum energy dissipation rate?

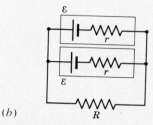

(a)

(b)

Figure 29–23 Problems 8 and 11.

9. In Fig. 29–6 calculate the potential difference between points c and d by as many paths as possible. Assume that $\varepsilon_1 = 4.0$ V, $\varepsilon_2 = 1.0$ V, $R_1 = R_2 = 10$ Ω, and $R_3 = 5.0$ Ω.

10. In Fig. 29–24 imagine an ammeter inserted in the branch containing R_3. (a) What will it read, assuming $\varepsilon = 5.0$ V, $R_1 = 2.0$ Ω, $R_2 = 4.0$ Ω, and $R_3 = 6.0$ Ω? (b) The ammeter and the source of emf are now physically interchanged. Show that the ammeter reading remains unchanged.

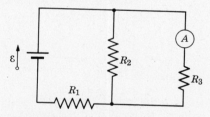

Figure 29–24 Problem 10.

11. You are given two batteries of emf ε and internal resistance r. They may be connected either in series or in parallel and are used to establish a current in a resistor R, as in Fig. 29–23. (a) Derive expressions for the current in R for both methods of connection. (b) Which connection yields the larger current if $R > r$ and if $R < r$?

12. (a) Find the three currents in Fig. 29–25. (b) Find V_{ab}. Assume that $R_1 = 1.0$ Ω, $R_2 = 2.0$ Ω, $\varepsilon_1 = 2.0$ V, and $\varepsilon_2 = \varepsilon_3 = 4.0$ V.

Section 29-6 Electrical Measuring Instruments

13. In Fig. 29–8 assume that $\varepsilon = 3.0$ V, $r = 100$ Ω, $R_1 = 250$ Ω, and $R_2 = 300$ Ω. If $R_V = 5.0$ kΩ, what percent error is made in reading the potential difference across R_1? Ignore the presence of the ammeter.

14. In Fig. 29–8 assume that $\varepsilon = 5.0$ V, $r = 2.0$ Ω, $R_1 = 5.0$ Ω, and $R_2 = 4.0$ Ω. If $R_A = 0.10$ Ω, what percent error is made in reading the current? Assume that the voltmeter is not present.

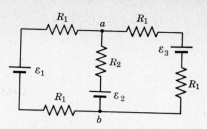

Figure 29–25 Problem 12.

15. *Resistance measurement.* A voltmeter (resistance R_V) and an ammeter (resistance R_A) are connected to measure a resistance R, as in Fig. 29–26a. The resistance is given by $R = V/i$, where V is the voltmeter reading and i is the current in the resistor R. Some of the current registered by the ammeter (i') goes through the voltmeter so that the ratio of the meter readings ($= V/i'$) gives only an *apparent* resistance reading R'. Show that R and R' are related by

$$\frac{1}{R} = \frac{1}{R'} - \frac{1}{R_V}.$$

Note that as $R_V \to \infty$, $R' \to R$.

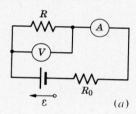

(a)

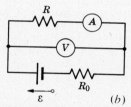

(b)

Figure 29–26 Problems 15, 16, and 17.

16. *Resistance measurement.* If meters are used to measure resistance, they may also be connected as they are in Fig. 29–26b. Again the ratio of the meter readings gives only an apparent resistance R'. Show that R' is related to R by

$$R = R' - R_A,$$

in which R_A is the ammeter resistance. Note that as $R_A \to 0$, $R' \to R$.

17. In Fig. 29–26 the ammeter and voltmeter resistances are 3.0 Ω and 300 Ω, respectively. If $R = 85$ Ω, (a) What will the meters read for the two different connections? (b) What apparent resistance R' will be computed in each case? Take $\varepsilon = 12$ V and $R_0 = 100$ Ω.

18. *The Wheatstone bridge.* In Fig. 29–27 R_s is to be adjusted in value until points a and b are brought to exactly the same potential. (One tests for this condition by momentarily connecting a sensitive ammeter between a and b; if these points are at the same potential, the ammeter will not deflect.) Show that when this adjustment is made, the following relation holds:

$$R_x = R_s (R_2/R_1).$$

An unknown resistor (R_x) can be measured in terms of a standard (R_s) using this device, which is called a Wheatstone bridge.

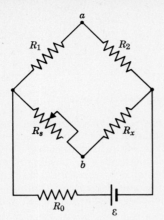

Figure 29–27   Problems 18 and 42.

Section 29–7   RC Circuits
19.   A 10-kΩ resistor and a capacitor are connected in series and a 10-V potential is suddenly applied. If the potential across the capacitor rises to 5.0 V in 1.0 μs, what is the capacitance of the capacitor?

20.   An RC circuit is discharged by closing a switch at time $t = 0$. The initial potential difference across the capacitor is 100 V. If the potential difference has decreased to 1.0 V after 10 s, (a) what will the potential difference be 20 s after $t = 0$? (b) What is the time constant of the circuit?

21.   A 1.0-μF capacitor with an initial stored energy of 0.50 J is discharged through a 1.0 MΩ resistor. (a) What is the initial charge on the capacitor? (b) What is the current through the resistor when the discharge starts? (c) Determine V_C, the voltage across the capacitor, and V_R, the voltage across the resistor, as a function of time. (d) Express the rate of generation of thermal energy in the resistor as a function of time.

22.   An initially uncharged capacitor C is fully charged by a constant emf ε in series with a resistor R. (a) Show that the final energy stored in the capacitor is half the energy supplied by the emf. (b) By direct integration of i^2R over the charging time show that the thermal energy dissipated by the resistor is also half the energy supplied by the emf.

ADDITIONAL PROBLEMS

23.   A solar cell generates a potential difference of 0.10 V when a 500-Ω resistor is connected across it and a potential difference of 0.15 V when a 1000-Ω resistor is substituted. (a) What is the internal resistance of the solar cell? (b) The area of the cell is 5.0 cm^2 and the intensity of light striking it is 2.0 mW/cm^2. What is the efficiency of the cell for converting light energy to heat in the 1000-Ω external resistor?

24.   Power is supplied by an emf ε to a transmission line with resistance R. Find the ratio of the power dissipated in the line for $\varepsilon = 110{,}000$ V to that dissipated for $\varepsilon = 110$ V, assuming the power supplied is the same for the two cases.

25.   The starting motor of an automobile is turning slowly and the mechanic has to decide whether to replace the motor, the cable, or the battery. The manufacturer's manual says that the battery can have no more than 0.020 Ω internal resistance, the motor no more than 0.200 Ω resistance, and the cable no more than 0.040 Ω resistance. The mechanic turns on the motor and measures 11.4 V across the battery, 3.0 V across the cable, and a current of 50 A. Which part is defective?

26.   A copper wire of radius a (=0.25 mm) has an aluminum jacket of outside radius b (=0.38 mm). (a) If there is a current i (=2.0 A) in the composite wire, find the current in each material. (b) What is the wire length if potential difference V (= 12 V) maintains the current?

27.   What current, in terms of ε and R, does the ammeter A in Fig. 29–28 read? Assume that A has zero resistance.

28.   When the lights of an automobile are switched on, an ammeter in series with them reads 10 A and a voltmeter connected across them reads 12 V. See Fig. 29–29. When the electric start-

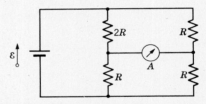

Figure 29–28   Problem 27.

ing motor is turned on the ammeter reading drops to 8.0 A and the lights dim somewhat. If the internal resistance of the battery is 0.050 Ω and that of the ammeter is negligible, what is (a) the emf of the battery and (b) the current through the starting motor when the lights are on?

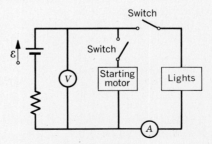

Figure 29–29   Problem 28.

29.   In the circuit of Fig. 29–30, ε and r have constant values but R can be varied. Find the value of R which results in the maximum heating in the resistor.

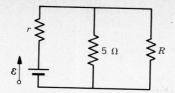

Figure 29-30 Problem 29.

30. (*a*) In Fig. 29-31 what is the equivalent resistance of the network shown? (*b*) What are the currents in each resistor? Put $R_1 = 100 \ \Omega$, $R_2 = R_3 = 50 \ \Omega$, $R_4 = 75 \ \Omega$, and $\varepsilon = 6.0V$.

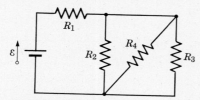

Figure 29-31 Problem 30.

31. What is the equivalent resistance between the terminal points x and y of the circuits shown in (*a*) Fig. 29-32*a*, (*b*) Fig. 29-32*b*, and (*c*) Fig. 29-32*c*? Assume that the resistance of each resistor is $10 \ \Omega$. Do you detect any similarities between Figs. 29-32*a* and *c*?

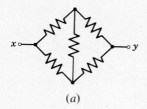

(*a*)

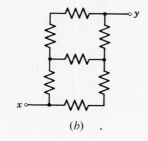

(*b*)

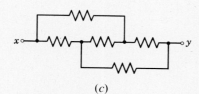

(*c*)

Figure 29-32 Problem 31.

32. *N* identical batteries of emf ε and internal resistance r may be connected all in series or all in parallel. Show that each arrangement will give the same current in an external resistor R, if $R = r$.

33. Fig. 29-33 shows a battery connected across a uniform resistor R_0. A sliding contact can move across the resistor from $x = 0$ at the left to $x = 10$ cm at the right. Find an expression for the power dissipated in the resistor R as a function of x. Plot the function for $\varepsilon = 50$ V, $R = 2000 \ \Omega$, and $r_0 = 100 \ \Omega$.

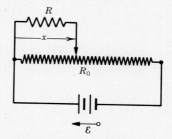

Figure 29-33 Problem 33.

34. (*a*) In Fig. 29-34 what power appears as thermal energy in R_1? In R_2? In R_3? (*b*) What power is supplied by ε_1? by ε_2? (*c*) Discuss the energy balance in this circuit. Assume that $\varepsilon_1 = 3.0$ V, $\varepsilon_2 = 1.0$ V, $R_1 = 5.0 \ \Omega$, $R_2 = 2.0 \ \Omega$, and $R_3 = 4.0 \ \Omega$.

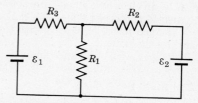

Figure 29-34 Problem 34.

35. Two resistors, R_1 and R_2 may be connected either in series or parallel across a (resistanceless) battery with emf ε. We desire the thermal energy transfer rate for the parallel combination to be five times that for the series combination. If $R_1 = 100 \ \Omega$, what is R_2?

36. (*a*) In the circuit of Fig. 29-35, for what value of R will the battery deliver energy to the circuit at a rate of 60 W? (*b*) What is the maximum power the battery can deliver to the circuit? (*c*) What is the minimum power the battery can deliver?

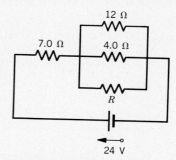

Figure 29-35 Problem 36.

37. A 3.0-MΩ resistor and a 1.0-μF capacitor are connected in a single-loop circuit with a seat of emf with $\varepsilon = 4.0$ V. At 1.0 s after the connection is made, what are the rates at which (*a*) the charge of the capacitor is increasing, (*b*) energy is being stored in the capacitor, (*c*) thermal energy is appearing in the resistor, and (*d*) energy is being delivered by the seat of emf?

38. The circuit of Fig. 29-36 shows a capacitor C, two batteries, two resistors, and a switch S. Initially S has been open for a long time. It is then closed for a long time. By how much does the charge on the capacitor change over this period? Assume $C = 10$ μF, $\varepsilon_1 = 1.0$ V, $\varepsilon_2 = 3.0$ V, $R_1 = 0.20$ Ω, and $R_2 = 0.40$ Ω.

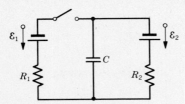

Figure 29-36   Problem 38.

39.   The potential difference between the plates of a leaky 2.0-μF capacitor drops to one-fourth its initial value in 2.0 s. What is the equivalent resistance between the capacitor plates?

40.   Prove that when switch S in Fig. 29–10 is thrown from a to b all the energy stored in the capacitor is transformed into thermal energy in the resistor. Assume that the capacitor is fully charged before the switch is thrown.

41.   A controller on an electronic arcade game consists of a variable resistor connected across the plates of a 0.22-μF capacitor. The capacitor is charged to 5.0 V, then discharged through the resistor. The time for the potential difference across the plates to decrease to 0.80 V is measured by an internal clock. If the range of discharge times that can be handled effectively is from 10 μs to 6.0 ms, what should the range of the resistor be?

42.   If points a and b in Fig. 29–27 are connected by a wire of resistance r, show that the current in the wire is

$$i = \frac{\varepsilon(R_s - R_x)}{(R + 2r)(R_s + R_x) + 2R_s R_x},$$

where ε is the emf of the battery. Assume that R_1 and R_2 are equal ($R_1 = R_2 = R$) and that R_0 equals zero. Is this formula consistent with the result of Problem 18?

*43.   In the circuit of Fig. 29–37, $\varepsilon = 1.2$ kV, $C = 6.5$ μF, $R_1 = R_2 = R_3 = 0.73$ MΩ. With C completely uncharged, the switch S is suddenly closed ($t = 0$). (a) Determine the currents through each resistor for $t = 0$ and $t = \infty$. (b) Draw qualitatively a graph of the potential drop V_2 across R_2 from $t = 0$ to $t = \infty$. (c) What are the numerical values of V_2 at $t = 0$ and $t = \infty$? (d) Give the physical meaning of "$t = \infty$" in this case.

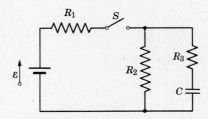

Figure 29-37   Problem 43.

*44.   Twelve resistors, each of resistance R ohms, form a cube (see Fig. 29–38). (a) Find R_{AB}, the equivalent resistance of an edge. (b) Find R_{BC}, the equivalent resistance of a face diagonal. (c) Find R_{AC}, the equivalent resistance of a body diagonal.

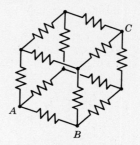

Figure 29-38   Problem 44.

30 The Magnetic Field

Figure 30–1 A research-type electromagnet showing iron frame F, pole faces P, and coils C. The pole faces are 12 in. in diameter. (Courtesy Varian Associates.)

The magnetic field **B**

Magnetic field lines

30–1 The Magnetic Field

The science of magnetism grew from the observation that certain "stones" (magnetite) would attract bits of iron. The word *magnetism* comes from the district of Magnesia in Asia Minor, which is one of the places at which the stones were found. The bar and horseshoe magnets with which we are all familiar are the lineal descendants of these natural magnets. Another "natural magnet" is the earth itself, whose orienting action on a magnetic compass needle has been known since ancient times.

In 1820 Oersted discovered that a current in a wire can also produce magnetic effects, namely, that it can change the orientation of a compass needle. We pointed out in Section 23–1 how this important discovery linked the then separate sciences of magnetism and electricity. We can intensify the magnetic effect of a current in a wire by forming the wire into a coil of many turns and by providing an iron core. Figure 30–1 shows how this is done in a large electromagnet of a type commonly used for research involving magnetism.

We define the space around a magnet or a current-carrying conductor as the site of a *magnetic field,* just as we defined the space near a charged rod as the site of an electric field. The basic magnetic field vector **B**, which we define in the following section, can be represented by field lines just as we represented the electric field by lines of force. As for the electric field (see Section 24–3), the magnetic field vector is related to its field lines in this way:

1. The tangent to a field line at any point gives the *direction* of **B** at that point.

2. The field lines are drawn so that the number of lines per unit cross-sectional area (perpendicular to the lines) is proportional to the *magnitude* of **B**. Where the lines are close together B is large and where they are far apart B is small.

Magnetic flux

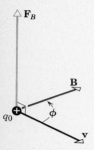

Figure 30–2 Illustrating $\mathbf{F}_B =$ $q_0\mathbf{v} \times \mathbf{B}$ (Eq. 30–1). Compare with Fig. 2–13 which shows the general case for a cross product.

Defining equation for **B**

The *flux* Φ_B for a magnetic field can be defined in exact analogy with the flux Φ_E for the electric field, namely $\Phi_B = \int \mathbf{B} \cdot d\mathbf{S}$, in which the integral is taken over the surface (closed or open) for which Φ_B is defined.

30–2 The Definition of B

We seek to define the magnetic field **B** in terms of procedures carried out in the laboratory. As in defining the electric field **E**, we take a small positive charge q_0 as a test body. Let us place it, at rest, at the point P for which we wish to determine the magnetic field. If no force acts on it—and this is what we assume—we take it that no electric field is present. Next we project our test charge q_0 with velocity **v** through the field point P. If a sideways deflecting force $\mathbf{F}_B$ acts on the test body we say that a magnetic field is present. By "sideways" we mean that $\mathbf{F}_B$ is at right angles to **v**, tending to make the test body move in a circle or a spiral.

We now define the magnetic field vector **B** in terms of the measurable quantities q_0, **v**, and $\mathbf{F}_B$ and we say that **B** is the vector that satisfies the relation*

$$\mathbf{F}_B = q_0\mathbf{v} \times \mathbf{B}. \tag{30–1}$$

Figure 30–2 shows these three vectors. Note that $\mathbf{F}_B$ is always at right angles to the plane formed by **v** and **B** and thus it is always at right angles to **v** (and also to **B**); as we said, it is a sideways force. The magnitude of the magnetic deflecting force, according to the rules for vector products, is

$$F_B = q_0 v B \sin \phi, \tag{30–2}$$

in which ϕ, as Fig. 30–2 shows, is the angle between **v** and **B**. Equation 30–1 tells us that (*a*) the magnetic force vanishes as $v \to 0$, (*b*) the magnetic force vanishes if **v** is either parallel or antiparallel to the direction of **B** (in these cases $\phi = 0$ or 180° and $\mathbf{v} \times \mathbf{B} = 0$), and (*c*) if **v** is at right angles to **B** ($\phi = 90°$), the deflecting force has its maximum value, given by $q_0 v B$.

Our definition of **B** is similar in spirit, although more complex, than the definition of the electric field **E**, which we can cast into this form: If we place a positive test charge q_0 at rest at point P and if a force $\mathbf{F}_E$ acts on it we say that an electric field **E** is present at P, where **E** is the vector satisfying the relation

$$\mathbf{F}_E = q_0\mathbf{E}, \tag{30–3}$$

q_0 and $\mathbf{F}_E$ being measured quantities. In defining **E**, the only characteristic direction to appear is that of the electric force $\mathbf{F}_E$ which acts on the positive test body; the direction of **E** is taken to be that of $\mathbf{F}_E$. In defining **B**, *two* characteristic directions appear, those of **v** and of the magnetic force $\mathbf{F}_B$; they prove always to be at right angles. Although we can easily solve the above equation for **E**, we cannot solve Eq. 30–1 for **B**. Why not?

To develop a feeling for Eq. 30–1 consider Fig. 30–3, which shows some tracks left by fast charged particles moving through a bubble chamber. At point P a positive and a negative electron are created from the energy of a gamma ray; the process is called *pair production* and the gamma ray, carrying no charge, leaves no track in the chamber.†

The chamber is immersed in a strong uniform magnetic field, with **B** pointing out of the plane of the figure, as suggested by the symbol $\odot$.‡ The relation $\mathbf{F}_B =$

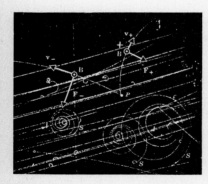

Figure 30–3 A *bubble chamber* is a device for rendering visible, by means of small bubbles, the tracks of charged particles. The figure is a photograph taken with such a chamber immersed in a magnetic field **B** and exposed to radiations from a large cyclotron-like accelerator. The track in the shape of a curved V at point P is formed by a positive and a negative electron, which deflect in opposite directions in the magnetic field. (Courtesy E. O. Lawrence Radiation Laboratory, University of California.)

---

* You may wish to review Section 2–4, which deals with vector products.
† See also the cover of this book.
‡ Remember that the symbol $\otimes$ indicates a vector into the page, the $\times$ being thought of as the tail of an arrow; $\odot$ indicates a vector out of the page, the dot being thought of as the tip of an arrow.

q_0 **v** $\times$ **B** (Eq. 30–1) predicts correctly that deflecting forces $\mathbf{F}_+$ and $\mathbf{F}_-$ will act on the particles, bending them away from a straight track as shown. The three spirals labeled S are the tracks of three low-energy electrons.

The unit of **B** that follows from Eq. 30–1 or 30–2 is the (newton/coulomb) (meter/second)$^{-1}$. This is given the SI name *tesla* (abbr. T). Recalling that a coulomb/second is an ampere, we have

The tesla

$$1 \text{ tesla} = 1 \text{ newton/(ampere·meter)}.$$

An earlier unit for **B**, still in common use, is the *gauss;* the relationship is

$$1 \text{ tesla} = 10^4 \text{ gauss.}$$

The fact that the magnetic force is always at right angles to the direction of motion means that (for steady magnetic fields at least) the work done by this force on the particle is zero. For an element of the path of the particle of length $d\ell$, this work dW is $\mathbf{F}_B \cdot d\ell$; dW is zero because $\mathbf{F}_B$ and $d\ell$ are always at right angles. Thus a static magnetic field cannot change the kinetic energy of a moving charge; it can only deflect it sideways.

If a charged particle moves through a region in which both an electric field and a magnetic field are present, the resultant force is found by combining Eqs. 30–3 and 30–1, or

The Lorentz force law

$$\mathbf{F} = q_0\mathbf{E} + q_0\mathbf{v} \times \mathbf{B}. \tag{30–4}$$

This is sometimes called the *Lorentz force law* in tribute to the Dutch physicist H. A. Lorentz who did so much to develop and clarify the concepts of the electric and magnetic fields.

Example 1 A uniform magnetic field **B** points horizontally from south to north; its magnitude is 1.5 T. If a 5.0-MeV proton moves vertically downward through this field, what force will act on it?

The kinetic energy of the proton is

$$K = (5.0 \times 10^6 \text{ eV})(1.6 \times 10^{-19} \text{ J/eV}) = 8.0 \times 10^{-13} \text{ J}.$$

We can find its speed from the relation $K = \frac{1}{2}mv^2$, or

$$v = \sqrt{\frac{2K}{m}} = \sqrt{\frac{(2)(8.0 \times 10^{-13} \text{ J})}{1.7 \times 10^{-27} \text{ kg}}} = 3.1 \times 10^7 \text{ m/s.}$$

Equation 30–2 gives

$$F = qvB \sin \phi = (1.6 \times 10^{-19} \text{ C})(3.1 \times 10^7 \text{ m/s})(1.5 \text{ T})(\sin 90°)$$
$$= 7.4 \times 10^{-12} \text{ N.}$$

You can show that this force is about 4×10^{14} times greater than the weight of the proton.

The relation $\mathbf{F} = q\mathbf{v} \times \mathbf{B}$ shows that the direction of the deflecting force is to the east. If the particle had been negatively charged, the deflection would have been to the west. This is predicted automatically by Eq. 30–1 if we substitute $-e$ for q_0.

30-3 Magnetic Force on a Current

A current is an assembly of moving charges. Because a magnetic field exerts a sideways force on a moving charge, we expect that it will also exert a sideways force on a wire carrying a current. Figure 30–4 shows a length l of wire carrying a current i and placed in a magnetic field **B**. For simplicity we have oriented the wire so that **B** is at right angles to it.

The current i in a metal wire is carried by the conduction electrons, n being the number of such electrons per unit volume of the wire. The magnitude of the average force on one such electron is given by Eq. 30–2, or, since $\phi = 90°$,

$$F' = q_0 vB \sin \phi = ev_d B$$

where v_d is the drift speed. From the relation $v_d = j/ne$ (Eq. 28–5),

$$F' = e\left(\frac{j}{ne}\right)B = \frac{jB}{n}.$$

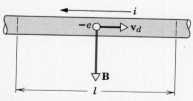

Figure 30-4 A wire carrying a current i is placed at right angles to a uniform magnetic field **B**. A typical conduction electron drifts to the right at speed v_d. The magnetic force on the electrons, and thus on the wire, points out of the page.

The length l of the wire contains nAl free electrons, Al being the volume of the section of wire of cross section A that we are considering. The total force on the free electrons in the wire, and thus on the wire itself, is

$$F = (nAl)F' = nAl\frac{jB}{n}.$$

Since jA is the current i in the wire, we have

$$F = ilB. \tag{30-5}$$

The negative charges, which move to the right in the wire of Fig. 30–4, are equivalent to positive charges moving to the left, that is, in the direction of the current arrow. For such positive charges the drift velocity $\mathbf{v}_d$ would point to the left and the force on the wire, given by Eq. 30–1 ($\mathbf{F}_B = q_0\mathbf{v} \times \mathbf{B}$) points up, out of the page. This same conclusion follows if we consider the actual negative charge carriers for which $\mathbf{v}_d$ points to the right but $q_0\,(= -e)$ has a negative sign. Thus by measuring the sideways magnetic force on a wire carrying a current and placed in a magnetic field, we cannot tell whether the current carriers are negative charges moving in a given direction or positive charges moving in the opposite direction.

Equation 30–5 holds only if the wire is at right angles to $\mathbf{B}$. We can express the more general situation in vector form as

Force on a straight wire segment

$$\mathbf{F} = i\boldsymbol{\ell} \times \mathbf{B}, \tag{30-6a}$$

where $\boldsymbol{\ell}$ is a vector whose magnitude is the length of the wire and which points along the (straight) wire in the direction of the current. Equation 30–6a is equivalent to the relation $\mathbf{F} = q_0\mathbf{v} \times \mathbf{B}$; either can be taken as a defining equation for $\mathbf{B}$. Note that the vector $\boldsymbol{\ell}$ in Fig. 30–4 points to the left and that the magnetic force $\mathbf{F}\,(= i\boldsymbol{\ell} \times \mathbf{B})$ points up, out of the page. This agrees with the conclusion obtained by analyzing the forces that act on the individual charge carriers.

If we consider a differential element of a conductor of length $d\boldsymbol{\ell}$, we can find the force $d\mathbf{F}$ acting on it, by analogy with Eq. 30–6a, from

Force on a current element

$$d\mathbf{F} = i\,d\boldsymbol{\ell} \times \mathbf{B}. \tag{30-6b}$$

By integrating this formula in an appropriate way we can find the force $\mathbf{F}$ on a conductor which is not straight.

Example 2 A wire bent as shown in Fig. 30–5 carries a current i and is placed in a uniform magnetic field $\mathbf{B}$ that emerges from the plane of the figure. Calculate the force acting on the wire. The magnetic field is represented by field lines, shown emerging from the page. The dots show that the sense of $\mathbf{B}$ is up, out of the page.

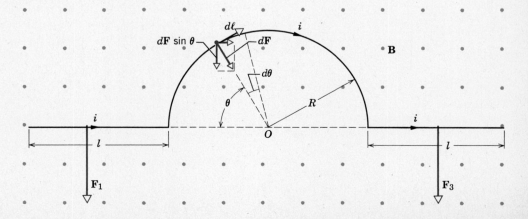

Figure 30–5 Example 2.

The force on each straight section, from Eq. 30–6a , has the magnitude

$$F_1 = F_3 = ilB$$

and points down as shown by the arrows in the figure.

A segment of wire of length $d\ell$ on the arc has a force $d\mathbf{F}$ on it whose magnitude is

$$dF = iB \, dl = iB(R \, d\theta)$$

and whose direction is radially toward O, the center of the arc. Only the downward component of this force is effective, the horizontal component being canceled by an oppositely directed component associated with the corresponding arc segment on the other side of O. Thus the total force on the semicircle of wire about O points down and is

$$F_2 = \int_0^\pi dF \sin \theta = \int_0^\pi (iBR \, d\theta) \sin \theta$$
$$= iBR \int_0^\pi \sin \theta \, d\theta = 2iBR.$$

The resultant force on the whole wire is

$$F = F_1 + F_2 + F_3 = 2ilB + 2iBR = 2iB(l + R).$$

Notice that this force is the same as that acting on a straight wire of length $2l + 2R$.

30–4 Torque on a Current Loop

Figure 30–6 shows a rectangular loop of wire of length a and width b placed in a uniform magnetic field $\mathbf{B}$, with sides 1 and 3 always normal to the field direction. The normal nn' to the plane of the loop makes an angle θ with the direction of $\mathbf{B}$. Assume the current to be as shown in the figure. Wires to lead the current into and out of the loop must be provided as well as a suspension system to support the loop while allowing it to turn.

The net force on the loop of Fig. 30–6 is the resultant of the forces on the four sides of the loop. On side 2 the vector ℓ points in the direction of the current and has the magnitude b. The angle between ℓ and $\mathbf{B}$ for side 2 (see Fig. 30–6b) is $90° - \theta$. Thus the magnitude of the force on this side is

$$F_2 = ibB \sin (90° - \theta) = ibB \cos \theta.$$

From the relation $\mathbf{F} = i\ell \times \mathbf{B}$ (Eq. 30–6a), we find the direction of $\mathbf{F}_2$ to be out of the plane of Fig. 30–6b. You can show that the force $\mathbf{F}_4$ on side 4 has the same magnitude as $\mathbf{F}_2$ but points in the opposite direction. Thus $\mathbf{F}_2$ and $\mathbf{F}_4$, taken together, have no effect on the motion of the loop. The net force they provide is zero, and, since they have the same line of action, the net torque due to these forces is also zero.

The common magnitude of $\mathbf{F}_1$ and $\mathbf{F}_3$ is iaB. These forces, too, are oppositely directed so that they do not tend to move the coil bodily. As Fig. 30–6b shows, however, they do not have the same line of action if the coil is in the position shown; there is a net torque, which tends to rotate the coil clockwise about the line xx'. The coil can be supported on a rigid axis that lies along xx', with no loss

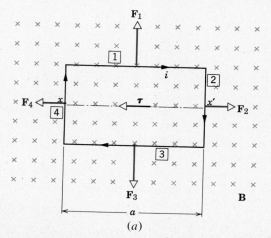
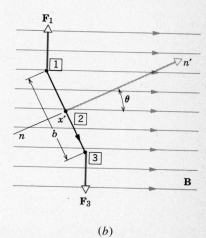

Figure 30–6 A rectangular coil carrying a current i is placed in a uniform external magnetic field.

(a)

(b)

of its freedom of motion. This torque can be represented in Fig. 30–6b by a vector pointing into the figure at point x' or in Fig. 30–6a by a vector pointing along the xx' axis from right to left.

The magnitude of the torque τ' is found by calculating the torque caused by $\mathbf{F}_1$ about axis xx' and doubling it, for $\mathbf{F}_3$ exerts the same torque about this axis that $\mathbf{F}_1$ does. Thus

$$\tau' = 2(iaB)\left(\frac{b}{2}\right)(\sin\theta) = iabB\sin\theta.$$

This torque acts on every turn of the coil. If there are N turns, the torque on the entire coil is

$$\tau = N\tau' = NiabB\sin\theta = NiAB\sin\theta, \tag{30–7}$$

Torque on a current loop

in which A, the area of the coil, is substituted for ab.

This equation can be shown to hold for all plane loops of area A, whether they are rectangular or not (see Problem 8). A torque on a current loop is the basic operating principle of the electric motor and of many electric meters used for measuring current or potential difference.

Example 3 A *galvanometer*. Figure 30–7 shows the rudiments of a galvanometer, which is the basic operating mechanism of some ammeters and voltmeters. The coil is 2.0 cm high and 1.0 cm wide; it has 250 turns and is mounted so that it can rotate about a vertical axis in a uniform radial magnetic field with $B = 0.20$ T ($=2000$ gauss). A spring Sp provides a countertorque that cancels out the magnetic torque, resulting in a steady angular deflection ϕ corresponding to a given current i in the coil. If a current of 0.10 mA produces an angular deflection of 30°, what is the torsional constant κ of the spring (see Eq. 14–20)?

Equating the magnitude of the magnetic torque to that of the torque caused by the spring (see Eq. 30–7) yields

$$\tau = NiAB\sin\theta = \kappa\phi$$

Note that the normal to the plane of the coil (that is, the pointer P) is always at right angles to the (radial) magnetic field so that $\theta = 90°$.

Solving for κ we obtain

$$
\begin{aligned}
\kappa &= \frac{NiAB\sin\theta}{\phi} \\
&= \frac{(250)(1.0\times10^{-4}\text{ A})(2.0\times10^{-4}\text{ m}^2)(0.20\text{ T})(\sin 90°)}{30°} \\
&= 3.3\times10^{-8}\text{ N·m/degree.}
\end{aligned}
$$

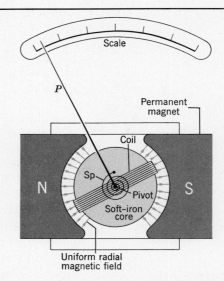

Figure 30–7 Example 3. The elements of a galvanometer, showing the coil, the helical spring Sp, and pointer P.

Magnetic dipoles

A current loop orienting itself in an external magnetic field reminds us of the action of a compass needle in such a field. One face of the loop behaves like the north pole of the needle; the other face behaves like the south pole. Compass needles, bar magnets, and current loops can all be regarded as magnetic dipoles. We show this here for the current loop, reasoning entirely by analogy with electric dipoles.

A structure is called an electric dipole if (a) when placed in an external electric field it experiences a torque given by Eq. 24–11,

$$\tau = \mathbf{p} \times \mathbf{E}, \tag{30–8}$$

where **p** is the electric dipole moment, and (*b*) it sets up a field of its own at distant points, described qualitatively by the lines of force of Fig. 26–10 and quantitatively by Eq. 26–11. These two requirements are not independent; if one is fulfilled, the other follows automatically.

The magnitude of the torque described by Eq. 30–8 is

$$\tau = pE \sin \theta, \tag{30–9}$$

where θ is the angle between **p** and **E**. Let us compare this with Eq. 30–7, the expression for the torque on a current loop:

$$\tau = (NiA)B \sin \theta. \tag{30–7}$$

In each case the appropriate field (*E* or *B*) appears, as does a term $\sin \theta$. Comparison suggests that *NiA* in Eq. 30–7 can be taken as the *magnetic dipole moment* μ, corresponding to *p* in Eq. 30–9, or

$$\mu = NiA. \tag{30–10}$$

Equation 30–7 suggests that we write the torque on a current loop in vector form, in analogy with Eq. 30–8, or

$$\boldsymbol{\tau} = \boldsymbol{\mu} \times \mathbf{B}. \tag{30–11}$$

Torque on a current loop, vector formulation

The magnetic dipole moment of the loop $\boldsymbol{\mu}$ must be taken to lie along an axis perpendicular to the plane of the loop; its direction is given by the following rule: Let the fingers of the right hand curl around the loop in the direction of the current; the extended right thumb will then point in the direction of $\boldsymbol{\mu}$. If $\boldsymbol{\mu}$ is defined by this rule and Eq. 30–10, check carefully that Eq. 30–11 correctly describes in every detail the torque acting on a current loop in an external field (see Fig. 30–6).

Since a torque acts on a current loop, or other magnetic dipole, when it is placed in an external magnetic field, it follows that work (positive or negative) must be done by an external agent to change the orientation of such a dipole. Thus a magnetic dipole has potential energy associated with its orientation in an external magnetic field. This energy may be taken to be zero for any arbitrary position of the dipole. By analogy with the assumption made for electric dipoles in Section 24–6, we assume that the magnetic energy U is zero when $\boldsymbol{\mu}$ and **B** are at right angles, that is, when $\theta = 90°$. This choice of a zero-energy configuration for U is arbitrary because we are interested only in the changes in energy that occur when the dipole is rotated.

The magnetic potential energy in any position θ is defined as the work that an external agent must do to turn the dipole from its zero-energy position ($\theta = 90°$) to the given position θ. Thus

$$U = \int_{90°}^{\theta} \tau \, d\theta = \int_{90°}^{\theta} NiAB \sin \theta \, d\theta = \mu B \int_{90°}^{\theta} \sin \theta \, d\theta = - \mu B \cos \theta,$$

in which Eq. 30–7 is used to substitute for τ. In vector symbolism we can write this relation as

Energy of a dipole

$$U = - \boldsymbol{\mu} \cdot \mathbf{B}, \tag{30–12}$$

which is in perfect correspondence with Eq. 24–13, the expression for the energy of an electric dipole in an external electric field,

$$U = - \mathbf{p} \cdot \mathbf{E}. \tag{24–13}$$

Example 4 A circular coil of N turns has an effective radius a and carries a current i. How much work is required to turn it in an external magnetic field **B** from a position in which θ equals zero to one in which θ equals 180°? Assume that $N = 100$, $a = 5.0$ cm, $i = 0.10$ A, and $B = 1.5$ T.

The work required is the difference in energy between the two positions, or, from Eq. 30–12,

$$W = U_{\theta=180°} - U_{\theta=0} = (-\mu B \cos 180°) - (-\mu B \cos 0) = 2\mu B.$$

But $\mu = NiA$, so that

$$W = 2NiAB = 2Ni(\pi a^2)B$$
$$= (2)(100)(0.10 \text{ A})(\pi)(5 \times 10^{-2} \text{ m})^2(1.5 \text{ T}) = 0.24 \text{ J}.$$

30–5 The Hall Effect

In 1879 E. H. Hall, at the Johns Hopkins University, reported an experiment that gives the sign of the charge carriers in a conductor. Figure 30–8 shows a flat conducting strip carrying a current i in the direction shown. The direction of the current arrow, labeled i, is the direction in which the charge carriers would move if they were positive. The current arrow can represent either positive charges moving down (as in Fig. 30–8a) or negative charges moving up (as in Fig. 30–8b). The Hall effect can be used to decide between these two possibilities.

A magnetic field **B** is set up at right angles to the strip by placing the strip between the polefaces of an electromagnet. This field exerts a deflecting force $\mathbf{F}_B$ on the strip (given by $i\boldsymbol{\ell} \times \mathbf{B}$), which points to the right in the figure. Since the sideways force on the strip is due to the sideways forces on the charge carriers (given by $q\mathbf{v} \times \mathbf{B}$), it follows that these carriers, whether they are positive or negative, will tend to drift toward the right in Fig. 30–8 as they drift along the strip, producing a *transverse Hall potential difference* V_{xy} between points such as x and y. The sign of the charge carriers is determined by the sign of this Hall potential difference. If the carriers are positive, y will be at a higher potential than x; if they are negative, y will be at a lower potential than x. Experiment shows that in metals the charge carriers are negative.

The Hall effect

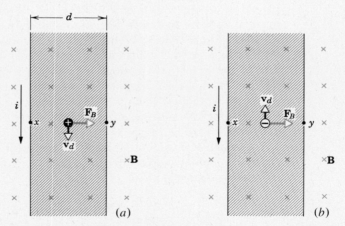

Figure 30–8 A current i is set up in a conducting strip placed in a magnetic field **B**, assuming (a) positive carriers and (b) negative carriers.

Hall effect studies give valuable information about the electrical conduction process in metals, semiconductors, and other conductive materials. In Problem 11 it is suggested how such studies can give not only the sign of the charge carriers but also their number per unit volume in the conductor.

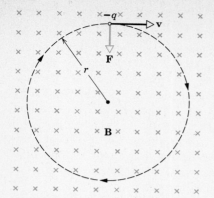

Figure 30–9 A charge $-q$ circulates at right angles to a uniform magnetic field.

Radius of path

The cyclotron frequency of a circulating ion

30–6 A Circulating Charge

Figure 30–9 shows a negatively charged particle introduced with velocity **v** into a uniform magnetic field **B**. We assume that **v** is at right angles to **B** and thus lies entirely in the plane of the figure. The relation $\mathbf{F} = q\mathbf{v} \times \mathbf{B}$ (Eq. 30–1) shows that the particle will experience a sideways deflecting force of magnitude qvB. This force will lie in the plane of the figure, which means that the particle cannot leave this plane.

This reminds us of a stone held by a rope and whirled in a horizontal circle on a smooth surface. Here, too, a force of constant magnitude, the tension in the rope, acts in a plane and at right angles to the velocity. The charged particle, like the stone, also moves with constant speed in a circular path. From Newton's second law we have

$$qvB = \frac{mv^2}{r} \quad \text{or} \quad r = \frac{mv}{qB}, \tag{30–13}$$

which gives the radius of the path. The three spirals in Fig. 30–3 show relatively low-energy electrons in a bubble chamber. The paths are not circles because the electrons lose energy by collisions in the chamber as they move.

The angular velocity ω is given by v/r or, from Eq. 30–13,

$$\omega = \frac{v}{r} = \frac{qB}{m}.$$

The frequency ν is given by

$$\nu = \frac{\omega}{2\pi} = \frac{qB}{2\pi m}. \tag{30–14}$$

Note that ν does not depend on the speed of the particle. Fast particles move in large circles (Eq. 30–13) and slow ones in small circles, but all require the same time T (the period) to complete one revolution in the field.

The frequency ν is a characteristic frequency for the charged particle in the field and may be compared to the characteristic frequency of a swinging pendulum in the earth's gravitational field or to the characteristic frequency of an oscillating mass-spring system. It is sometimes called the *cyclotron frequency* of the particle in the field because particles circulate at this frequency in a cyclotron.

Example 5 A 10-eV electron is circulating in a plane at right angles to a uniform magnetic field of 1.0×10^{-4} T ($= 1.0$ gauss).

(a) What is its orbit radius?

The velocity of an electron whose kinetic energy is K can be found from

$$v = \sqrt{\frac{2K}{m}}.$$

Verify that this yields 1.9×10^6 m/s for v. Then, from Eq. 30–13,

$$r = \frac{mv}{qB} = \frac{(9.1 \times 10^{-31} \text{ kg})(1.9 \times 10^6 \text{ m/s})}{(1.6 \times 10^{-19} \text{ C})(1.0 \times 10^{-4} \text{ T})} = 0.11 \text{ m} = 11 \text{ cm}.$$

(b) What is the cyclotron frequency? From Eq. 30–14,

$$\nu = \frac{qB}{2\pi m} = \frac{(1.6 \times 10^{-19} \text{ C})(1.0 \times 10^{-4} \text{ T})}{(2\pi)(9.1 \times 10^{-31} \text{ kg})}$$

$$= 2.8 \times 10^6 \text{ Hz} = 2.8 \text{ MHz}.$$

(c) What is the period of revolution T?

$$T = \frac{1}{\nu} = \frac{1}{2.8 \times 10^6 \text{ Hz}} = 3.6 \times 10^{-7} \text{ s} = 0.36 \text{ } \mu\text{s}.$$

(d) What is the direction of circulation as viewed by an observer sighting along the field?

In Fig. 30–9 the magnetic force $q\mathbf{v} \times \mathbf{B}$ must point radially inward, since it provides the centripetal force. Since **B** points into the plane of the paper, **v** would have to point to the left at the position shown in the figure if the charge q were positive. However, the charge is an electron, with $q = -e$, which means that **v** must point to the right. Thus the charge circulates clockwise as viewed by an observer sighting in the direction of **B**.

30-7 Cyclotrons and Synchrotrons

The cyclotron

The cyclotron accelerates charged particles such as hydrogen nuclei (protons) and heavy hydrogen nuclei (deuterons) to high energies so that they can be used in atom-smashing experiments and for the production of radioactive isotopes. Although conventional cyclotrons of the original design have been superseded by other accelerators we describe them for two reasons: (a) they provide an excellent framework within which to discuss the action of magnetic and electric fields on charged particles and (b) the cyclotron has led to the production of several generations of improved accelerators. These later devices provide even more opportunities for studying the interactions of charged particles with magnetic and electric fields although that, of course, is not their primary purpose.

In an *ion source* at the center of the cyclotron molecules of deuterium (heavy hydrogen) are bombarded with electrons whose energy is high enough (say 100 eV) so that many positive ions are formed during the collisions. Many of these ions are free deuterons, which enter the cyclotron proper through a small hole in the wall of the ion source and are available to be accelerated.

The cyclotron uses a modest potential difference for accelerating (say 100 kV), but it requires the ion to pass through this potential difference a number of times. To reach 10 MeV with a 100-kV accelerating potential requires 100 passages. A magnetic field is used to bend the ions around so that they may pass again and again through the same accelerating potential.

Figure 30–10 is a top view of the region of the cyclotron in which the particles are accelerated. The two D-shaped objects, called *dees,* are made of copper sheet and form part of an electric oscillator, which establishes an accelerating potential difference across the gap between the dees. The direction of this potential difference is made to change sign some millions of times per second.

The dees are immersed in a magnetic field ($B \cong 1.6$ T) whose direction is out of the plane of Fig. 30–10. The field is set up by a large electromagnet. Finally, the space in which the ions move is evacuated to a pressure of approximately 10^{-6} mm Hg. If this were not done, the ions would continually collide with air molecules.

Suppose that a deuteron, emerging from the ion source, finds the dee that it is facing to be negative; it will accelerate toward this dee and will enter it. Once inside, it is screened from electric fields by the metal walls of the dees. The magnetic field is not screened by the dees so that the ion bends in a circular path whose radius, which depends on the velocity, is given by Eq. 30–13, or

$$r = \frac{mv}{qB}.$$
[30–13]

After a time t_0 the ion emerges from the dee on the other side of the ion source. Let us assume that the accelerating potential has now changed sign. Thus the ion again faces a negative dee, is further accelerated, and again describes a semicircle, of somewhat larger radius (see Eq. 30–13), in the dee. The time of passage through this dee, however, is still t_0. This follows because the period of revolution T of an ion circulating in a magnetic field does not depend on the speed of the ion; see Eq. 30–14. This process goes on until the ion reaches the outer edge of one dee where it is pulled out of the system by a negatively charged deflector plate.

The key to the operation of the cyclotron is that the characteristic frequency ν at which the ion circulates in the field must be equal to the fixed frequency ν_0 of the electric oscillator, or

$$\nu = \nu_0.$$

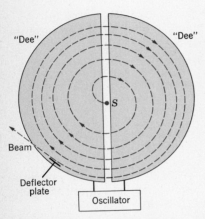

Figure 30–10 The elements of a cyclotron showing the ion source S and the dees. The deflector plate, held at a suitable negative potential, deflects the particles out of the dee system.

This *resonance condition* says that if the energy of the circulating ion is to increase, energy must be fed to it at a frequency ν_0 that is equal to the natural frequency ν at which the ion circulates in the field. In the same way we feed energy to a swing by pushing it at a frequency equal to the natural frequency of oscillation of the swing.

From Eq. 30–14, we can write the resonance equation as

The cyclotron resonance condition

$$\frac{qB}{2\pi m} = \nu_0. \tag{30-15}$$

Once we have selected an ion to be accelerated, q/m is fixed; usually the oscillator is designed to work at a single frequency ν_0. We then "tune" the cyclotron by varying B until Eq. 30–15 is satisfied and an accelerated beam appears.

The energy of the particles produced in the cyclotron depends on the radius R of the dees. From Eq. 30–13 ($r = mv/qB$), the velocity of a particle circulating at this radius is given by

$$v = \frac{qBR}{m}.$$

The kinetic energy is then

$$K = \tfrac{1}{2}mv^2 = \frac{q^2 B^2 R^2}{2m}. \tag{30-16}$$

Example 6 A cyclotron formerly in use at the University of Pittsburgh had an oscillator frequency of 12 MHz and a dee radius of 53 cm.

(*a*) What value of B was needed to accelerate deuterons? From Eq. 30–15,

$$B = \frac{2\pi\nu_0 m}{q} = \frac{(2\pi)(12 \times 10^6 \text{ Hz})(3.3 \times 10^{-27} \text{ kg})}{1.6 \times 10^{-19} \text{ C}} = 1.6 \text{ T}.$$

Note that the deuteron has the same charge as the proton but (very closely) twice the mass.

(*b*) What deuteron energy results? From Eq. 30–16,

$$K = \frac{q^2 B^2 R^2}{2m} = \frac{(1.6 \times 10^{-19} \text{ C})^2(1.6 \text{ T})^2(0.53 \text{ m})^2}{(2)(3.3 \times 10^{-27} \text{ kg})}$$

$$= (2.8 \times 10^{-12} \text{ J}) \left(\frac{1 \text{ eV}}{1.6 \times 10^{-19} \text{ J}} \right) = 17 \text{ MeV}.$$

At energies above ~50 MeV (for protons) the conventional cyclotron begins to fail because one of its assumptions, that the frequency of rotation of an ion circulating in a magnetic field is independent of its speed, is true only for speeds much less than that of light. As the particle speed increases, we must use the *relativistic mass m* in Eq. 30–14. The relativistic mass increases with velocity (Eq. 8–19) so that at high enough speeds the frequency of revolution of the ion decreases with velocity. Thus the ions get out of step with the oscillator—whose frequency remains fixed at ν_0—and eventually the energy of the circulating ion stops increasing.

Difficulties with the conventional cyclotron

A second difficulty associated with the acceleration of particles to high energies is that the size of the magnet that would be required to guide such particles in a circular path is very large. For a 500-GeV proton, for example, in a field of 1.5 T the radius of curvature is 1.1 km. A magnet of the cyclotron type of this size (about 1.4 miles in diameter!) would be prohibitively expensive.

The synchrotron

The *synchrotron* is a cyclotronlike accelerator designed to meet the two difficulties outlined above. The magnetic field B and the oscillator frequency ν_0, instead of having fixed values as in the conventional cyclotron, are caused to vary with time, each quantity sweeping repetitively through its design range of values. The design flexibility so introduced not only permits the cyclotron frequency

Figure 30–11 An aerial view of the Fermi National Accelerator
Laboratory, showing the main ring and the experimental areas.
Courtesy FNAL.

of the accelerating ions to remain in phase with the oscillator frequency (Eq.
30–15) but also maintains the radius of the path of the circulating ions at a constant value (Eq. 30–13). This constant-radius feature permits the use of an annular
or ring-shaped magnet rather than a circular one, at enormous saving in cost.

As of 1980 a proton synchrotron at the Fermi National Accelerator Laboratory (Fermilab) at Batavia, Illinois has accelerated protons up to ~500 GeV. Figure 30–11 shows the accelerator complex. The main ring consists of 954 separate
magnets arranged in a circle over four miles in circumference. The protons travel
~200,000 times around the ring while being accelerated to their full energy. They
are then extracted from the ring and enter the experimental area shown in the
lower left of Fig. 30–11, emerging as periodic pulses of ~10^{13} protons each.

30–8 Measuring e/m for the Electron

This crucial experiment, performed in 1897 by J. J. Thomson in the Cavendish
Laboratory in Cambridge, England, was a measurement of the ratio of charge e
to mass m of the electron by observing its deflection in combined magnetic and
electric fields. It amounted to the discovery of the electron as a fundamental particle and we discuss it here as another practical example of the action of magnetic
and electric fields on charged particles. Thomson received the Nobel Prize in 1906.

In Fig. 30–12, which is a modernized version of Thomson's apparatus, electrons are emitted from hot filament F and accelerated by an applied potential difference V. They then enter a region in which they move at right angles to an elec-
tric field $\mathbf{E}$ and a magnetic field $\mathbf{B}$; $\mathbf{E}$ and $\mathbf{B}$ are at right angles to each other. The
beam is made visible as a spot of light when it strikes fluorescent screen S. The
entire region in which the electrons move is highly evacuated so that collisions
with air molecules will not occur.

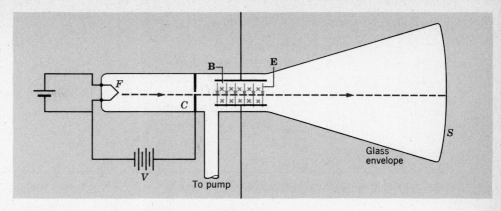

Figure 30–12   Electrons from the heated filament F are accelerated by a potential difference V and pass through a hole in the screen C. After passing through a region in which perpendicular electric and magnetic fields are present, they strike the fluorescent screen S.

The resultant force on a charged particle moving through an electric and a magnetic field is given by Eq. 30–4, or

$$\mathbf{F} = q_0\mathbf{E} + q_0\mathbf{v} \times \mathbf{B}.$$

Study of Fig. 30–12 shows that the electric field deflects the particle upward and the magnetic field deflects it downward. If these deflecting forces are to cancel (that is, if $\mathbf{F} = 0$), this equation, for this problem, reduces to

$$eE = evB$$

or

$$E = vB. \tag{30–17}$$

Thus for a given electron speed v the condition for zero deflection can be satisfied by adjusting either E or B.

Thomson's procedure was (a) to note the position of the undeflected beam spot, with $\mathbf{E}$ and $\mathbf{B}$ both equal to zero; (b) to apply a fixed electric field $\mathbf{E}$, measuring on the fluorescent screen the deflection so caused; and (c) to apply a magnetic field and adjust its value until the beam deflection is restored to zero.

In Section 24–5 we saw that the deflection y of an electron in a purely electric field (step b), measured at the far edge of the deflecting plates, is given by Eq. 24–9, or, with small changes in notation,

$$y = \frac{eEl^2}{2mv^2},$$

where v is the electron speed and l is the length of the deflecting plates; y is not measurable directly, but it may be calculated from the measured displacement of the spot on the screen if the geometry of the apparatus is known. Thus y, E, and l are known; the ratio e/m and the velocity v are unknown. We cannot calculate e/m until we have found the velocity, which is the purpose of step c.

If (step c) the electric force is set equal and opposite to the magnetic force, the net force is zero and we can write (Eq. 30–17)

$$v = \frac{E}{B}.$$

Substituting this equation into the equation for y and solving for the ratio e/m leads to

e/m for the electron

$$\frac{e}{m} = \frac{2yE}{B^2 l^2}, \tag{30-18}$$

in which all the quantities on the right can be measured. Thomson's value for e/m, expressed in SI units, was 1.7×10^{11} C/kg, in agreement with the 1973 value of 1.7588047×10^{11} C/kg.

REVIEW GUIDE AND SUMMARY

The magnetic field **B** defined

The magnetic field **B** is defined in terms of the sideways force $\mathbf{F}_B$ acting on a test particle with charge q_0 and moving with velocity **v**,

$$\mathbf{F}_B = q_0 \mathbf{v} \times \mathbf{B}; \tag{30-1}$$

study Fig. 30–2. The SI unit for **B** is the *tesla* (abbr. T) where $1\,\text{T} = 1\,\text{N}/(\text{A·m}) = 10^4$ gauss.

A wire carrying a current experiences a sideways force

The sideways force acting on a current element $i\,d\ell$ in a magnetic field is

$$d\mathbf{F} = i\,d\ell \times \mathbf{B}; \tag{30-6b}$$

The direction of the length element $d\ell$ is that of the current density vector **j**. If a straight wire carries a current i in a uniform magnetic field, Eq. 30–6b becomes

$$\mathbf{F} = i\ell \times \mathbf{B}; \tag{30-6a}$$

see Example 2.

A loop carrying a current experiences a torque

A current loop of area A in a uniform magnetic field will experience a torque τ given by

$$\boldsymbol{\tau} = \boldsymbol{\mu} \times \mathbf{B}. \tag{30-11}$$

Here $\boldsymbol{\mu}$ is the *magnetic dipole moment*. It has a magnitude $\mu = NiA$ where N is the number of turns. Curl the fingers of the right hand around the loop in the direction of the current; the extended thumb gives the direction of $\boldsymbol{\mu}$. See Example 3. The potential energy of orientation of a magnetic dipole in a magnetic field is

Orientation energy of a magnetic dipole

$$U = -\boldsymbol{\mu} \cdot \mathbf{B}. \tag{30-12}$$

See Example 4.

The Hall effect

When a conducting strip carrying a current i is placed in a magnetic field the charge carriers drift in a direction given by Eq. 30–6a. A potential difference V_{xy} builds up across the strip, as detailed in Fig. 30–8. The polarity of V_{xy} gives the sign of the charge carriers.

A charged particle circulating in a magnetic field

A charge moving with velocity **v** perpendicular to a magnetic field **B** will travel in a circle of radius

$$r = mv/qB \tag{30-13}$$

where m and q are the mass and charge of the particle. Its frequency of revolution in the field (the *cyclotron frequency*) is

$$\nu = qB/2\pi m; \tag{30-14}$$

see Example 5.

The cyclotron

A cyclotron is a particle accelerator that uses a magnetic field to hold a charged particle in a circular orbit so that a modest accelerating potential may act on it repeatedly, resulting in high energies. Because the particle mass increases relativistically with velocity there is an upper limit to the energy attainable with the cyclotron. A synchrotron avoids

The synchrotron

this difficulty. Here both B and the oscillator frequency ν_0 are programmed to change cyclically so that the particle can not only go to high energies but can do so at a constant orbital radius; this allows the use of a ring magnet rather than a solid magnet, at great saving in cost.

e/m for the electron

J. J. Thomson, in his 1897 discovery of the electron, used both magnetic and electric fields (see Fig. 30–12) to determine its charge to mass ratio.

QUESTIONS

1. Of the three vectors in the equation $\mathbf{F}_B = q\mathbf{v} \times \mathbf{B}$, which pairs are always at right angles? Which may have any angle between them?

2. Why do we not simply define the magnetic field **B** to point in the direction of the magnetic force that acts on the moving charge?

3. Imagine that you are sitting in a room with your back to one wall and that an electron beam, traveling horizontally from the back wall toward the front wall, is deflected to your right. What is the direction of the uniform magnetic field that exists in the room?

4. If an electron is not deflected in passing through a certain region of space, can we be sure that there is no magnetic field in that region?

5. If a moving electron is deflected sideways in passing through a certain region of space, can we be sure that a magnetic field exists in that region?

6. A conductor, even though it is carrying a current, has zero net charge. Why, then, does a magnetic field exert a force on it?

7. A beam of electrons can be deflected by either an electric field or by a magnetic field. Is one method better than the other? . . . in any sense easier?

8. A charged particle passes through a magnetic field and is deflected. This means that a force acted on it and changed its momentum. Where there is a force there must be a reaction force. On what body does it act?

9. A bare copper wire emerges from one wall of a room, crosses the room, and diappears into the opposite wall. You are told that there is a steady current in the wire. How can you find its direction? Describe as many ways as you can think of. You may use any reasonable piece of equipment, but you may not cut the wire.

10. In Example 2 (see Fig. 30–5) we saw that the magnetic force was the same as if the semicircular arc had been replaced by a straight wire of length $2R$. Would this same conclusion hold if we replaced the semicircular arc by a curve of irregular shape?

11. In Section 30–3 we state that a magnetic field **B** exerts a sideways force on the conduction electrons in, say, a copper wire carrying a current i. We have tacitly assumed that this same force acts on the conductor itself. Are there some missing steps in this argument?

12. Equation 30–11 ($\tau = \mu \times \mathbf{B}$) shows that there is no torque on a current loop in an external magnetic field if the angle between the axis of the loop and the field is (a) 0° or (b) 180°. Discuss the nature of the equilibrium (that is, is it stable, neutral, or unstable?) for these two positions.

13. In Example 4 we showed that the work required to turn a current loop end-for-end in an external magnetic field is $2\mu B$. Does this hold no matter what the original orientation of the loop was?

14. Imagine that the room in which you are seated is filled with a uniform magnetic field with **B** pointing vertically upward. A circular loop of wire has its plane horizontal. For what direction of current in the loop, as viewed from above, will the loop be in stable equilibrium with respect to forces and torques of magnetic origin?

15. A rectangular current loop is in an arbitrary orientation in an external magnetic field. Is any work required to rotate the loop about an axis perpendicular to its plane?

16. You wish to modify a galvanometer (see Example 3) to make it into (a) an ammeter and (b) a voltmeter. What do you need to do in each case?

17. Imagine the room in which you are seated to be filled with a uniform magnetic field with **B** pointing vertically downward. At the center of the room two electrons are suddenly projected horizontally with the same speed but in opposite directions. (a) Discuss their motions. (b) Discuss their motions if one particle is an electron and one a positron, that is, a positively charged electron.

18. In Fig. 30–3 why are the low-energy electron tracks spirals? That is, why does the radius of curvature change in the constant magnetic field in which the chamber is immersed?

19. What are the primary functions of (a) the electric field and (b) the magnetic field in the cyclotron?

20. What central fact makes the operation of a conventional cyclotron possible? Ignore relativistic considerations.

21. The arrangement of crossed electric and magnetic fields shown in the center part of Fig. 30–12 is sometimes called a *velocity filter*. How can this name be justified?

EXERCISES

Section 30–2   The Definition of B

1. Express magnetic field B and magnetic flux Φ_B in terms of the dimensions M, L, T, and Q (mass, length, time, and charge).

2. Four particles follow the paths shown in Fig. 30–13 as they pass through the magnetic field there. What can one conclude about each particle?

3. An airplane is flying west in level flight over Massachusetts, where the earth's magnetic field is directed downward at an angle of 73° below the horizontal in a northerly direction. As a result of the magnetic force on the free electrons in its wings, one of its wingtips will have more electrons than the other. Which one (right or left) is it? Will the answer be different if the plane is flying east?

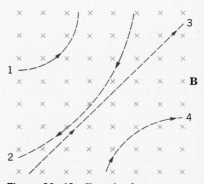

Figure 30–13   Exercise 2.

Section 30–3   Magnetic Force on a Current

4.   Figure 30–14 shows a magnet and a straight wire in which electrons are flowing out of the page at right angles to it. In which case will there be a force on the wire that points toward the top of the page? Lines of **B** are directed from the north poles to the south poles.

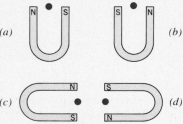

Figure 30–14   Exercise 4.

5.   A horizontal conductor in a power line carries a steady current of 5000 A from south to north. The earth's magnetic field (60 μT) is directed toward the north and is inclined downward at 70° to the horizontal. Find the magnitude and direction of the force on 100 m of the conductor due to the earth's field.

6.   A wire 1.0 m long carries a current of 10 A and makes an angle of 30° with a uniform magnetic field with $B = 1.5$ T. Calculate the magnitude and direction of the force on the wire.

Section 30–4   Torque on a Current Loop

7.   Figure 30–15 shows a rectangular, 20-turn loop of wire, 10 cm by 5.0 cm. It carries a current of 0.10 A and is hinged at one side. What torque about the hinge line acts on the loop if it is mounted with its plane at an angle of 30° to the direction of a uniform magnetic field of 0.50 T?

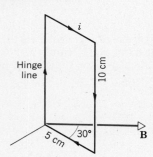

Figure 30–15   Exercise 7.

8.   A circular wire loop whose radius is 15 cm carries a current of 2.0 A. It is placed so that the normal to its plane makes an angle of 30° with a uniform magnetic field of 10 T. What torque acts on the loop?

9.   A stationary, circular wall clock has a face with a radius of 15 cm. Six turns of wire are wound around its perimeter; the wire carries a current 2.0 A in the clockwise sense. The clock is located where there is a constant, uniform external magnetic field of 70 mT (but the clock still keeps perfect time). At exactly 1:00 P.M., the hour hand of the clock points in the direction of the external magnetic field. (a) After how many minutes will the minute hand point in the direction of the torque on the winding due to the magnetic field? (b) What is the magnitude of this torque?

10.   A single-turn current loop, carrying a current of 5.0 A, is in the shape of a right triangle with sides 30, 40, and 50 cm. The loop

is in a uniform magnetic field of magnitude 80 mT whose direction is parallel to the current in the 50-cm side of the loop. Find the magnitude of (a) the magnetic dipole moment of the loop and (b) the torque on the loop.

Section 30–5   The Hall Effect

11.   A certain sample of material has both positive and negative charge carriers. In a Hall effect experiment, it exhibits zero Hall voltage. Comment on the following three possible conclusions: (a) The number of positive and negative charge carriers (per unit volume) must be the same. (b) The drift speed of the positive charge carriers must be the same as the drift speed of the negative charge carriers. (c) The positive and negative charge carriers must carry equal but opposite currents. (d) The positive and negative charge carriers must carry equal currents in the same direction.

12.   Figure 30–16 shows the cross section of a sample carrying a current directed out of the page. (a) Which pair of the four terminals (a, b, c, d) should be used to measure the Hall voltage if the magnetic field is in the +x-direction and the charge carriers are negative? What is the expected polarity of this voltage? (b) Repeat if the magnetic field is in the −y-direction and the charge carriers are positive. (c) Discuss the situation if the magnetic field is in the +z-direction.

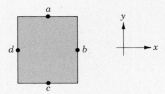

Figure 30–16   Exercise 12.

13.   A copper conductor with a cross section as shown in Fig. 30–17 carries current in the +z-direction in the presence of an external magnetic field $\mathbf{B} = 300\mathbf{i} - 300\mathbf{j} - 100\mathbf{k}$ mT. Of the indicated points on the surface of the wire, which point number will be at the location of the highest potential on the conductor surface?

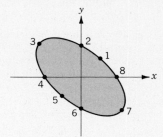

Figure 30–17   Exercise 13.

Section 30–6   Circulating Charges

14.   Magnetic fields are often used to bend a beam of electrons in physics experiments. What uniform magnetic field, applied perpendicular to a beam of electrons moving at 1.3×10^6 m/s, is required to make the electrons travel in a circular arc of radius 0.35 m?

15.   (a) In a magnetic field with $B = 0.50$ T, for what path radius will an electron circulate at 0.10 the speed of light? (b) What will its kinetic energy be?

16.   What uniform magnetic field must be set up in space to permit a proton of speed 1.0×10^7 m/s to move in a circle the size of the earth's equator?

17. A 10-keV electron is circulating in a plane at right angles to a uniform magnetic field. The orbit radius is 25 cm. Find (a) the magnetic field, (b) the cyclotron frequency, and (c) the period of the motion.

18. An electron is accelerated from rest by a potential difference of 350 kV. It then enters a uniform magnetic field of magnitude 200 mT, its velocity being at right angles to this field. What is the radius of curvature of its path in the magnetic field?

Section 30–7 Cyclotrons and Synchrotrons
19. In a certain cyclotron a proton moves in a circle of radius $r = 0.50$ m. The magnitude of the **B** field is 1.2 T. (a) What is the cyclotron frequency? (b) What is the kinetic energy of the proton?

20. A physicist is designing a cyclotron to accelerate protons to one-tenth the speed of light. The magnet used will produce a field of 1.4 T. Calculate (a) the radius of the cyclotron and (b) the corresponding oscillator frequency.

Section 30–8 Measuring e/m for the electron
21. A typical cathode-ray oscilloscope employs a cathode-ray tube in which electric fields are used for both horizontal and vertical deflections of the electron beam, but which is otherwise similar to the tube shown in Fig. 30–12. Figure 30–18 shows the face of such a tube. The solid straight line results when the electron beam is repeatedly swept left to right by a time-varying electric field. If a uniform magnetic field is applied perpendicularly inward through the face of the tube, one might expect the horizontal line to be shifted or tilted. Which of the four dashed lines in the figure will be the line that will result?

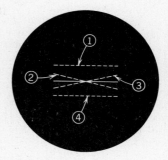

Figure 30–18 Exercise 21.

22. A 10-keV electron moving horizontally enters a region of space in which there is a downward-directed electric field of magnitude 10 kV/m. (a) What are the magnitude and direction of the (smallest) magnetic field that will allow the electron to continue to move horizontally? Ignore the gravitational force, which is rather small. (b) Is it possible for a proton to pass through this combination of fields undeflected? If so, under what circumstances?

PROBLEMS GROUPED BY SECTION

Section 30–2 The Definition of B
1. An electron in a uniform magnetic field has a velocity $\mathbf{v} = 40\mathbf{i} + 35\mathbf{j}$ km/s. It experiences a force $\mathbf{F} = -4.2\mathbf{i} + 4.8\mathbf{j}$ fN. If $B_x = 0$, find the magnetic field.

2. The electrons in the beam of a television tube have an energy of 12 keV. The tube is oriented so that the electrons move horizontally from south to north. The earth's magnetic field points down and has $B = 55$ μT. (a) In what direction will the beam deflect? (b) What is the acceleration of a given electron due to the magnetic field? (c) How far will the beam deflect in moving 20 cm through the television tube?

Section 30–3 Magnetic Force on a Current
3. A metal wire of mass m slides without friction on two horizontal rails spaced a distance d apart, as in Fig. 30–19. The track lies in a uniform vertical magnetic field **B**. A constant current i flows from generator G along one rail, across the wire, and back down the other rail. Find the velocity (speed and direction) of the wire as a function of time, assuming it to be at rest at $t = 0$.

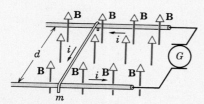

Figure 30–19 Problem 3.

4. Figure 30–20 shows a wire of arbitrary shape carrying a current i between points a and b. The wire lies in a plane at right angles to a uniform magnetic field **B**. Prove that the force on the wire is the same as that on a straight wire carrying a current i directly from a to b. (Hint: Replace the wire by a series of "steps" parallel and perpendicular to the straight line joining a and b.)

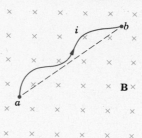

Figure 30–20 Problem 4.

5. A long, rigid conductor, lying along the x-axis, carries a current of 5.0 A in the $-x$-direction. A magnetic field **B** is present, given by $\mathbf{B} = 3\mathbf{i} + 8x^2\mathbf{j}$, with x in m and **B** in mT. Calculate the force on the 2.0-m segment of the conductor that lies between $x = 1.0$ m and $x = 3.0$ m.

Section 30–4 Torque on a Current Loop
6. A circular loop of wire having a radius of 8.0 cm carries a current of 0.20 A. A unit vector parallel to the dipole moment μ of the loop is given by $0.60\mathbf{i} - 0.80\mathbf{j}$. If the loop is located in a magnetic field given in T by $\mathbf{B} = 0.25\mathbf{i} + 0.30\mathbf{k}$, find (a) the magnitude

and direction of the torque on the loop and (b) the magnetic potential energy of the loop. Assume the same zero-energy configuration that we assumed in Section 30–4.

7. A certain galvanometer has a resistance of 75.3 Ω; its needle experiences a full-scale deflection when a current of 1.62 mA passes through its coil. (a) Determine the value of the auxiliary resistance required to convert the galvanometer into a voltmeter that reads 1.00 V at full-scale deflection. How is it to be connected? (b) Determine the value of the auxiliary resistance required to convert the galvanometer into an ammeter that reads 50.0 mA at full-scale deflection. How is it to be connected?

8. Prove that the relation $\tau = NAiB \sin \theta$ holds for closed loops of arbitrary shape and not only for rectangular loops as in Fig. 30–6. (*Hint:* Replace the loop of arbitrary shape by an assembly of adjacent long, thin — approximately rectangular — loops which are nearly equivalent to it as far as the distribution of current is concerned.)

9. Figure 30–21 shows a current loop *ABCDEFA* carrying a current $i = 5.0$ A. The sides of the loop are parallel to the coordinate axes, with $AB = 20$ cm, $BC = 30$ cm, and $FA = 10$ cm. Calculate the magnitude and direction of the magnetic dipole moment of this loop. (*Hint:* Imagine equal and opposite currents i in the line segment *AD*, then treat the two rectangular loops *ABCDA* and *ADEFA*.)

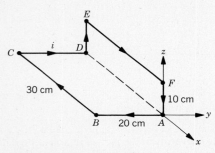

Figure 30–21 Problem 9.

10. Figure 30–22 shows a wooden cylinder with a mass m of 0.25 kg and a length l of 0.10 m with N equal to ten turns of wire wrapped around it longitudinally, so that the plane of the wire loop contains the axis of the cylinder. What is the least current through the loop that will prevent the cylinder from rolling down an inclined plane whose surface is inclined at an angle θ to the horizontal, in the presence of a vertical, uniform magnetic field of 0.50 T, if the plane of the windings is parallel to the incline.

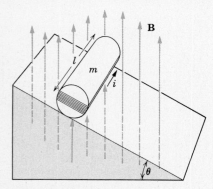

Figure 30–22 Problem 10.

Section 30–5 The Hall Effect

11. Referring to Fig. 30–8a, assume that the right face of the copper strip is charged with $+q$, the left face receiving $-q$. This sets up an electric field $\mathbf{E}_H$ and a potential difference $V_{xy} = E_H d$, much like in a capacitor. The total force on charges drifting with velocity $\mathbf{v}_d$ results from the magnetic field $\mathbf{B}$ and the Hall field $\mathbf{E}_H$. When equilibrium is reached the force on drifting charges carrying the current goes to zero. (a) Show that the Hall field is given by $\mathbf{E}_H = -\mathbf{v}_d \times \mathbf{B}$. (b) Using Eq. 28–5 for the drift velocity of charge carriers, show that the number of charge carriers per unit volume is

$$n = \frac{jB}{eE_H},$$

where j is the current density and e is the electronic charge.

12. In a Hall effect experiment, a current of 3.0 A lengthwise in a conductor 1.0 cm wide, 4.0 cm long, and 10 μm thick produced a transverse Hall voltage (across the width) of 10 μV when a magnetic field of 1.5 T passed perpendicularly through the thin conductor. From these data, find (a) the drift velocity of the charge carriers and (b) the number density of charge carriers. (c) Show on a diagram the polarity of the Hall voltage with a given current and magnetic field direction, assuming the charge carriers are (negative) electrons. See Problem 11.

13. Show that the ratio of the Hall electric field E_H to the electric field E responsible for the current is

$$\frac{E_H}{E} = \frac{B}{ne\rho},$$

where ρ is the resistivity of the material. See Problem 11.

Section 30–6 Circulating Charges

14. *Time-of-flight spectrometer.* S. A. Goudsmit devised a method for measuring accurately the masses of heavy ions by timing their period of circulation in a known magnetic field. A singly charged ion of iodine makes 7.00 rev in a field of 45.0 mT in about 1.29 ms. What (approximately) is its mass in kilograms? Actually, the mass measurements are carried out to much greater accuracy than these approximate data suggest.

15. A proton, a deuteron, and an α particle, accelerated through the same potential difference, enter a region of uniform magnetic field, moving at right angles to $\mathbf{B}$. (a) Compare their kinetic energies. If the radius of the proton's circular path is 10 cm, what are the radii of (a) the deuteron and (b) the α-particle paths?

16. A proton, a deuteron, and an α-particle with the same kinetic energies enter a region of uniform magnetic field, moving at right angles to $\mathbf{B}$. Compare the radii of their circular paths.

17. *Mass spectrometer.* Figure 30–23 shows an arrangement used to measure the masses of ions. An ion of mass m and charge $+q$ is produced essentially at rest in source S, a chamber in which a gas discharge is taking place. The ion is accelerated by potential difference V and allowed to enter a magnetic field $\mathbf{B}$. In the field it moves in a semicircle, striking a photographic plate at distance x from the entry slit. Show that the mass m is given by

$$m = \frac{B^2 q}{8V} x^2.$$

18. Two types of singly ionized atoms having the same charge q and mass differing by a small amount Δm are introduced into the mass spectrometer described in Problem 17. (a) Calculate the

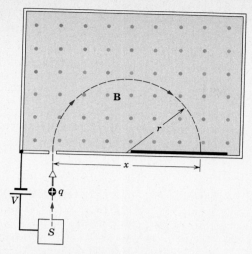

Figure 30–23 Problems 17 and 39.

difference in mass in terms of V, q, m (of either), B, and the distance Δx between the spots on the photographic plate. (b) Calculate Δx for a beam of singly ionized chlorine atoms of masses 35 and 37 u if $V = 7.3$ kV and $B = 0.50$ T.

Section 30–7   Cyclotrons and Synchrotrons

19. The cyclotron of Example 6 was normally adjusted to accelerate deuterons. (a) What energy of protons could it produce, using the same oscillator frequency as that used for deuterons? (b) What magnetic field would be required? (c) What energy of protons could be produced if the magnetic field was left at the value used for deuterons? (d) What oscillator frequency would then be required? (e) Answer the same questions for α particles.

20. A deuteron in a cyclotron is moving in a magnetic field with $B = 1.5$ T and an orbit radius of 50 cm. Because of a grazing collision with a target, the deuteron breaks up, with a negligible loss of kinetic energy, into a proton and a neutron. Discuss the subsequent motions of each. Assume that the deuteron energy is shared equally by the proton and neutron at breakup.

Section 30–8   Measuring e/m for the Electron

21. An electron is accelerated through a potential difference of 1.0 kV and directed into a region between two parallel plates separated by 20 mm with a potential difference of 100 V between them. If the electron enters moving perpendicular to the electric field between the plates, what magnetic field is necessary perpendicular to both the electron path and the electric field so that the electron travels in a straight line?

22. An electric field of 1.5 kV/m and a magnetic field of 0.40 T act on a moving electron to produce no force. (a) Calculate the minimum electron speed v. (b) Draw the vectors **E**, **B**, and **v**.

ADDITIONAL PROBLEMS

23. An electron has a velocity given in m/s by $\mathbf{v} = 2.0 \times 10^6\mathbf{i} + 3.0 \times 10^6\mathbf{j}$. It enters a magnetic field given in T by $\mathbf{B} = 0.030\mathbf{i} - 0.15\mathbf{j}$. (a) Find the magnitude and direction of the force on the electron. (b) Repeat your calculation for a proton having the same velocity.

24. A wire of 60 cm length and mass 10 g is suspended by a pair of flexible leads in a magnetic field of 0.40 T. What are the magnitude and direction of the current required to remove the tension in the supporting leads? See Fig. 30–24.

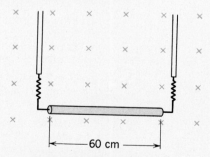

Figure 30–24 Problem 24.

25. A wire 50 cm long lying along the x-axis carries a current of 0.50 A in the positive x-direction. A magnetic field is present that is given in T by $\mathbf{B} = 0.0030\mathbf{j} + 0.010\mathbf{k}$. Find the components of the force on the wire.

26. A length l of wire carries a current i. Show that if the wire is formed into a circular coil, the maximum torque in a given magnetic field is developed when the coil has one turn only and the maximum torque has the value

$$\tau = \frac{1}{4\pi} l^2 iB.$$

27. (a) What is the cyclotron frequency of an electron with an energy of 100 eV in the earth's magnetic field of 100 μT? (b) What is the radius of curvature of the path of this electron if its velocity is perpendicular to the magnetic field?

28. In a nuclear experiment a 1.0-MeV proton moves in a uniform magnetic field in a circular path. What energy must (a) an α particle and (b) a deuteron have if they are to circulate in the same orbit?

29. An α particle travels in a circular path of radius 4.5 cm in a magnetic field with $B = 1.2$ T. Calculate (a) its speed, (b) its period of revolution, (c) its kinetic energy, and (d) the potential difference through which it would have to be accelerated to achieve this energy.

30. An ion source is producing ions of ^{6}Li (mass 1.0×10^{-26} kg) and ^{7}Li (mass 1.17×10^{-26} kg), each carrying a single positive elementary charge ($+e$). The ions are accelerated by a potential difference of 10 kV and pass through a region in which there is a magnetic field $B = 1.2$ T, perpendicular to the flight path. (a) Make a sketch showing the direction in which the ^{7}Li ions are deflected. (b) What electric field must be provided over the same region for the ^{6}Li ions to pass through undeflected?

31. An electron has an initial velocity $12\mathbf{j} + 15\mathbf{k}$ km/s and a constant acceleration of $2.0 \times 10^{12}\,\mathbf{i}$ m/s^2 in a region in which uniform electric and magnetic fields are present. If $\mathbf{B} = 400\mathbf{i}\,\mu$T, find the electric field **E**.

32. A beam of electrons whose kinetic energy is K emerges from a thin-foil "window" at the end of an accelerator tube. There is a metal plate a distance d from this window and at right angles to the direction of the emerging beam. Show that we can stop the beam from hitting the plate if we apply a magnetic field B such that

$$B \geq \left(\frac{2mK}{e^2d^2}\right)^{1/2},$$

in which m and e are the electron mass and charge. How should **B** be oriented?

33. A single-turn current loop, carrying a current of 4.0 A, is in the shape of a right triangle with sides 50 cm, 120 cm, and 130 cm. The loop is in a uniform magnetic field of magnitude 75 mT whose direction is parallel to the current in the 130-cm side of the loop. (*a*) Find the magnetic force on each of the three sides of the loop. (*b*) Show that the total magnetic force on the loop is zero.

34. A closed loop of wire carries a current i. The loop is in a uniform magnetic field **B**. Starting from the formula $d\mathbf{F} = id\boldsymbol{\ell} \times \mathbf{B}$, show that the total magnetic force on the loop is zero. Does your proof also hold for a nonuniform magnetic field?

35. A neutral particle is at rest in a uniform magnetic field of magnitude B. At time $t = 0$ it decays into two charged particles each of mass m. (*a*) If the charge of one of the particles is $+ q$, what is the charge of the other? The two particles move off in separate paths both of which lie in the plane perpendicular to **B**. (*b*) At a later time the particles collide. Express the time from decay until collision, t, in terms of m, B, and q.

36. Figure 30–25 shows a wire ring of radius a at right angles to the general direction of a radially symmetric diverging magnetic field. The magnetic field at the ring is everywhere of the same magnitude B, and its direction at the ring is everywhere at an angle θ with a normal to the plane of the ring. The twisted lead wires have no effect on the problem. Find the magnitude and direction of the force the field exerts on the ring if the ring carries a current i as shown in the figure.

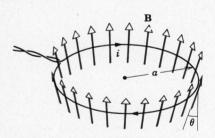

Figure 30–25 Problem 36.

37. Two concentric circular loops (radii 20 cm and 30 cm) in the $x - y$ plane each carry a clockwise current of 7.0 A, as shown in Fig. 30–26. (*a*) What is the net magnetic moment of this system? (*b*) Repeat if the current in the inner loop is reversed.

38. Consider the possibility of a new design for an electric train. The engine is driven by the force due to the vertical component of the earth's magnetic field on a conducting axle. Current is passed down one rail, into a conducting wheel, through the axle, through another conducting wheel, and then back to the source via the other rail. (*a*) What current is needed to provide a modest 10 kN force? Take the vertical component of **B** to be $10\,\mu$T and the length of the axle to be 3.0 m. (*b*) How much power would be lost for each ohm of resistance in the rails? (*c*) Is such a train totally unrealistic or marginally unrealistic?

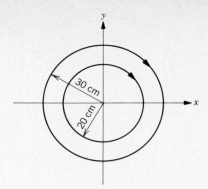

Figure 30–26 Problem 37.

39. In a mass spectrometer (see Problem 17) used for commercial purposes, uranium ions of mass 3.92×10^{-25} kg and charge 3.2×10^{-19} C are separated from related species. The ions are first accelerated through a potential difference of 100 kV and then pass into a magnetic field where they are bent in a path of radius 1.0 m. After traveling through 180°, they are collected in a cup after passing through a slit of width 1.0 mm and height 1.0 cm. (*a*) What is the magnitude of the (perpendicular) magnetic field in the separator? If the machine is designed to separate out 100 mg of material per hour, calculate (*b*) the current of the desired ions in the machine and (*c*) the heat dissipated in the cup in an hour.

40. Bainbridge's mass spectrometer, shown in Fig. 30–27, separates ions having the same velocity. The ions, after entering through slits S_1 and S_2, pass through a velocity selector composed of an electric field produced by the charged plates P and P', and a magnetic field **B** perpendicular to the electric field and the ion path. Those ions that pass undeviated through the crossed **E** and **B** fields enter into a region where a second magnetic field **B**′ exists, and are bent into circular orbits. A photographic plate registers their arrival. Show that $q/m = E/(rBB')$, where r is the radius of the circular orbit.

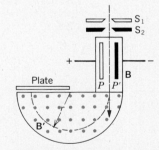

Figure 30–27 Problem 40.

41. Estimate the total path length traversed by a deuteron in the cyclotron of Example 6 during the acceleration process. Assume an accelerating potential between the dees of 80 kV.

42. A 2.0-keV positron is projected into a uniform magnetic field **B** of 0.10 T with its velocity vector making an angle of 89° with **B**. (*a*) Convince yourself that the path will be a helix, its axis being the direction of **B**. Find (*b*) the period, (*c*) the pitch p, and (*d*) the radius r of the helix; see Fig. 30–28.

43. A 1.0-kg copper rod rests on two horizontal rails 1.0 m apart and carries a current of 50 A from one rail to the other. The

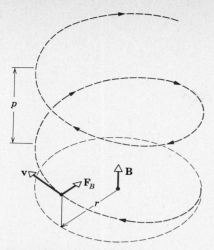

Figure 30-28 Problem 42.

coefficient of static friction is 0.60. What is the smallest magnetic field (not necessarily vertical) that would cause the bar to slide?

44. (a) What speed would a proton need to circle the earth at the equator, if the earth's magnetic field is everywhere horizontal there and directed along longitudinal lines? Relativistic effects must be taken into account. Take the magnitude of the earth's magnetic field to be 41 μT at the equator. (b) Draw the velocity and magnetic field vectors corresponding to this situation.

Ampere's Law

31-1 Magnetic Fields and Currents

A basic fact of electrostatics is that two charges exert forces on each other. Earlier (p. 468) we wrote

$$\text{charge} \rightleftharpoons \text{electric field} \rightleftharpoons \text{charge}$$

in which we introduced the electric field **E** as an intermediary in this interaction. This relationship suggests (*a*) that charges generate electric fields and (*b*) that electric fields exert forces on charges.

A basic fact of magnetism, one that we will explore in detail in Section 31–4, is that two parallel wires carrying currents also exert forces on each other. By analogy to the above we can write

$$\text{current} \rightleftharpoons \text{magnetic field} \rightleftharpoons \text{current}$$

in which we introduce the magnetic field **B** as an intermediary in this interaction. This relationship suggests (*a*) that currents generate magnetic fields and (*b*) that magnetic fields exert forces on currents. As Eq. 30–6*a* (**F** = $i\ell$ × **B**) reminds us, we dealt with (*b*) above in Section 30–3. In this chapter we deal with (*a*) above, namely: How do currents generate magnetic fields?

In 1820 Hans Christian Oerstead noted that establishing a current in a long straight wire could deflect a nearby compass needle. This was a landmark experiment in the history of physics because it marked the first experimental connection between the sciences of electricity (the current in the wire) and magnetism (the deflection of the needle). Up to this time these two sciences had developed as quite separate subjects.

Figure 31–1, which shows a wire surrounded by a number of small magnets, suggests a modification of Oersted's experiment. If there is no current in the wire, all the magnets are aligned with the horizontal component of the earth's magnetic field. When a strong current is present, the magnets point so as to suggest that the

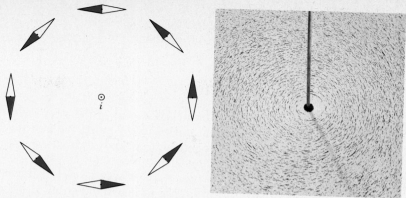

Figure 31–1   An array of compass needles surrounding a vertical wire that carries a large current. The colored ends of the needles mark their north poles.

Figure 31–2   Iron filings around a wire carrying a strong current. PSSC PHYSICS, 2nd ed., 1965, D. C. Heath & Co. with Education Development Center, Inc., Newton, MA 02160

magnetic field lines form closed circles around the wire. This view is strengthened by the experiment in Fig. 31–2, which shows iron filings on a horizontal glass plate, through the center of which a current-carrying conductor passes.

31–2   Calculating B; the Biot-Savart Law

We wish to learn how to calculate the magnetic field set up at any point P near a current-carrying conductor of arbitrary shape. Let us first recall how we dealt with the analogous problem in electrostatics, namely, how did we calculate the electric field set up at any point P near a charge distribution of arbitrary shape?

*Calculating **E** from q*

In that case we divided the charge distribution into differential charge elements dq and we calculated the magnitude of the differential field contribution $d\mathbf{E}$ set up by such an element from

Coulomb's law for the electric field

$$dE = \left(\frac{1}{4\pi\epsilon_0}\right)\frac{dq}{r^2}. \tag{31–1}$$

The direction of $d\mathbf{E}$ is that of $\mathbf{r}$, where $\mathbf{r}$ is a vector that points from the (assumed positive) charge element dq to the field point P. We find the resultant field $\mathbf{E}$ at point P by adding up (integrating) the contributions $d\mathbf{E}$ from all the charge elements that make up the distribution. Example 5 of Chapter 24 shows an application of this process to a charge distribution in the form of a uniform ring.

*Calculating **B** from i*

In the magnetic case we proceed by analogy. Figure 31–3 shows a wire of arbitrary shape carrying a current i. What is the magnetic field $\mathbf{B}$ at various points P near this wire? We first break up the wire into differential current elements $i\,d\boldsymbol{\ell}$, one of which is shown in the figure. Here $d\boldsymbol{\ell}$ is a differential element of the length of the wire, pointing along the tangent to the wire (see dashed lines) in the direction of i.

The differential current element

We notice already a complexity in our analogy to the electrostatic case. The differential charge element dq is a scalar but the differential current element $i\,d\boldsymbol{\ell}$ is a vector. Also, a current element cannot exist as an isolated entity because a way must be provided to lead the current into the element at one end and out of it at the other. Nevertheless, we can think of the actual circuit as made up of a large number of current elements placed end to end.

The Biot-Savart law

The permeability constant μ_0

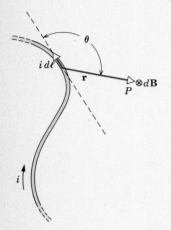

Figure 31-3   The current element $i\,d\ell$ sets up a differential magnetic field contribution $d\mathbf{B}$ at point P. For simplicity the wire is taken to lie in the plane of the figure, although this need not be the case.

In Fig. 31–3 the vector **r** extends from the current element $i\,d\ell$ to the field point P and θ is the angle between **r** and the direction of this element. The magnitude of the differential field contribution $d\mathbf{B}$ set up at P by the current element shown is given by

$$dB = \left(\frac{\mu_0}{4\pi}\right)\frac{i\,dl\,\sin\theta}{r^2}. \tag{31-2}$$

The direction of $d\mathbf{B}$ is that of the vector $d\ell \times \mathbf{r}$, namely, into the plane of Fig. 31–3 at right angles to the page. We find the resultant field **B** at point P by integrating the contributions $d\mathbf{B}$ from all the current elements $i\,d\ell$ that make up the arbitrary current distribution.

The constant μ_0 that appears as a factor in Eq. 31–2 is called the *permeability constant;* it has the assigned value, in SI units, of (exactly)

$$\mu_0 = 4\pi \times 10^{-7}\ \mathrm{T\cdot m/A}.$$

This constant plays a role in magnetic problems similar to the role played by the permittivity constant ϵ_0 in electrostatic problems. The permeability constant μ_0 must not be confused with magnetic dipole moment μ; the two quantities are unrelated.

Equation 31–1, which we can appropriately call Coulomb's law for the electric field, is our basic tool for calculating the electric field set up by any given charge distribution. We can view the analogous Eq. 31–2, called the *Biot-Savart law,* as the magnetic equivalent of Coulomb's law. It is our basic tool for calculating the magnetic field set up by any given distribution of currents.

The central features of the Biot-Savart law were established experimentally in Paris by these two scientists within a month or so of the time that news of Oersted's discovery reached that city. In Example 1 we use the Biot-Savart law to calculate the magnetic field set up at points near a long straight wire carrying a current i.

Example 1 B *near a long straight wire*   Use the Biot-Savart law (Eq. 31–2) to find **B** due to a current i in a long straight wire.

Figure 31–4, which is just like Fig. 31–3 except that the wire is straight, illustrates the problem. The differential magnetic field contribution $d\mathbf{B}$ set up at point P by the current element $i\,d\ell$ is given in magnitude by Eq. 31–2, or

$$dB = \frac{\mu_0 i}{4\pi}\frac{dl\,\sin\theta}{r^2}. \tag{31-2}$$

The direction of $d\mathbf{B}$ in Fig. 31–4 is that of the vector $d\ell \times \mathbf{r}$, namely, into the plane of the figure at right angles. We note that $d\mathbf{B}$ for every current element has this same direction. Thus, to find B at point P we simply integrate Eq. 31–2 above. This scalar operation yields

$$B = \int dB = \frac{\mu_0 i}{4\pi}\int_{l=-\infty}^{l=+\infty}\frac{\sin\theta\,dl}{r^2}.$$

Now, l, θ, and r are not independent, being related (see Fig. 31–4) by

$$r = \sqrt{l^2 + R^2}$$

and

$$\sin\theta\,[=\sin(\pi-\theta)] = \frac{R}{\sqrt{l^2 + R^2}},$$

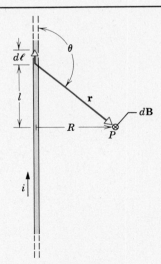

Figure 31-4   Example 1.

so that the expression for B becomes

$$B = \frac{\mu_0 i}{4\pi}\int_{-\infty}^{+\infty}\frac{R\,dl}{(l^2 + R^2)^{3/2}} = \frac{\mu_0 i}{4\pi R}\left|\frac{l}{(l^2 + R^2)^{1/2}}\right|_{l=-\infty}^{l=+\infty} = \frac{\mu_0}{2\pi}\frac{i}{R}.$$

We make a small change in notation and write this as

$$B(r) = \frac{\mu_0 i}{2\pi r} \qquad (31\text{-}3)$$

where r is now the perpendicular distance from the wire. For $i = 80$ A and $r = 2.0$ cm, Eq. 31–3 yields

$$B = \frac{\mu_0 i}{2\pi r} = \frac{(4\pi \times 10^{-7}\ \text{T·m/A})(80\ \text{A})}{(2\pi)(2.0 \times 10^{-2}\ \text{m})} = 800\ \mu\text{T}\ (= 8.0\ \text{gauss}).$$

For comparison, the magnitude of the average horizontal component of the earth's magnetic field in the central United States, is about 20 μT ($= 0.2$ gauss).

The lines of **B** for a current in a wire.

Experiment suggests that the lines of **B** for a current i in a long straight wire form concentric circles around the wire, as in Fig. 31–5, which shows the wire of Fig. 31–4 in cross section. Figure 31–5 is also consistent with the qualitative results of Figs. 31–1 and 31–2. Note the increase in spacing of the lines with increasing distance from the wire. This represents the $1/r$ decrease in B predicted by Eq. 31–3.

The right-hand rule

The direction of the lines of **B** in Fig. 31–5 follows from the Biot-Savart law. In Fig. 31–4 **B** at point P points into the page at right angles. This means that as you sight along the wire in the direction of the current the lines of **B** should be clockwise, just as Fig. 31–5 shows. A simple rule for finding the direction of **B** in such cases is this: Grasp the wire in your right hand with your extended right thumb pointing in the direction of the current; your fingers will then curl around the wire in the direction of **B**. Verify that this rule gives the correct direction for **B** in Figs. 31–4 and 31–5.

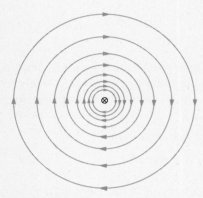

Figure 31–5   A cross-section of the wire of Fig. 31–4, sighting along the direction of the current. The lines of **B** are clockwise.

31–3   The Magnetic Force on a Current; a Second Look

We recall from Section 30–3 that a section of a long straight wire of length l, carrying a current i and placed in an external magnetic field, experiences a sideways deflecting force given by

$$\mathbf{F} = i\,\boldsymbol{\ell} \times \mathbf{B}_{\text{ext}}. \qquad [30\text{-}6a]$$

We introduce the subscript in $\mathbf{B}_{\text{ext}}$ to remind us that the magnetic field in this force equation must be set up by an external agent, an electromagnet of the type shown in Fig. 30–1 being one possibility.

The magnetic force law

In particular the field that appears in the force equation above must be carefully distinguished from the intrinsic field $\mathbf{B}_{\text{int}}$—as we may call it—set up by the current in the wire itself. Figure 31–5 shows this intrinsic field. This field cannot exert a force on its own source (the current-carrying wire) any more than can the gravitational field of the earth exert a force on the earth itself. In Fig. 31–5 there is no magnetic force on the wire because no *external* magnetic field is present.

$\mathbf{B}_{\text{ext}}$, $\mathbf{B}_{\text{int}}$, and their resultant

Figure 31–6 shows the resultant magnetic field lines associated with a current in a wire that is oriented at right angles to a uniform external magnetic field $\mathbf{B}_{\text{ext}}$. At any point the resultant magnetic field **B** will be the vector sum of $\mathbf{B}_{\text{ext}}$ and $\mathbf{B}_{\text{int}}$. These two fields tend to cancel each other above the wire and to reinforce each other below it. At point P in Fig. 31–6 $\mathbf{B}_{\text{ext}}$ and $\mathbf{B}_{\text{int}}$ cancel each other exactly. Very near the wire the field is represented closely by circular lines and is essentially the $\mathbf{B}_{\text{int}}$ of Fig. 31–5. Far from the wire the field is represented closely by uniformly spaced parallel lines and is essentially $\mathbf{B}_{\text{ext}}$.

Michael Faraday, who originated the concept of lines of force, endowed them with more reality than they are currently given. He imagined that, like stretched

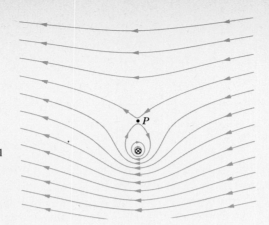

Figure 31–6 A wire carrying a current i is immersed in a uniform external magnetic field $\mathbf{B}_{ext}$ that extends from right to left. The magnetic field lines represent the resultant field $\mathbf{B}$ ($=\mathbf{B}_{ext} + \mathbf{B}_{int}$), where $\mathbf{B}_{int}$ is represented in Fig. 31–5.

rubber bands, they represent the site of mechanical forces. On this picture can we not readily visualize that the wire in Fig. 31–6 will be pushed up?

31–4 Two Parallel Conductors

It is a well-established experimental fact that two long parallel wires carrying currents exert forces on each other. Here we examine the situation more closely, paying particular attention to the role of the magnetic field as an intermediary in the relationship

$$\text{current} \rightleftharpoons \text{magnetic field} \rightleftharpoons \text{current.}$$

Figure 31–7 shows two such wires, separated by a distance d and carrying currents i_a and i_b.

Wire a in Fig. 31–7 will produce a magnetic field $\mathbf{B}_a$ at all nearby points. The magnitude of $\mathbf{B}_a$, due to the current i_a, at the site of the second wire is, from Eq. 31–3,

$$B_a = \frac{\mu_0 i_a}{2\pi d}.$$

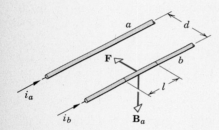

Figure 31–7 Two parallel wires that carry parallel currents attract each other.

The right-hand rule shows that the direction of $\mathbf{B}_a$ at wire b is down, as shown in the figure.

Wire b, which carries a current i_b, finds itself immersed in an external magnetic field $\mathbf{B}_a$. A length l of this wire will experience a sideways magnetic force ($=i\boldsymbol{\ell} \times \mathbf{B}_a$) whose magnitude is

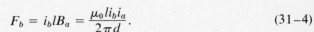

The force between parallel currents

$$F_b = i_b l B_a = \frac{\mu_0 l i_b i_a}{2\pi d}. \tag{31–4}$$

The rule for vector products tells us that $\mathbf{F}_b$ lies in the plane of the wires and points to the left in Fig. 31–7.

We could have started with wire b, computed the magnetic field which it produces at the site of wire a, and then computed the force on wire a. The force on wire a would, for parallel currents, point to the right. The forces that the two wires exert on each other are equal and opposite, as they must be according to Newton's law of action and reaction. For antiparallel currents the two wires repel each other.

The attraction between long parallel wires is used to define the ampere. Suppose that the wires are one meter apart ($d = 1.0$ m, exactly) and that the two currents are equal ($i_a = i_b = i$). If this common current is adjusted until, by measurement, the force of attraction per unit length between the wires is 2×10^{-7} N/m exactly, the current is defined to be one ampere. From Eq. 31–4,

$$\frac{F}{l} = \frac{\mu_0 i^2}{2\pi d} = \frac{(4\pi \times 10^{-7} \text{ T·m/A})(1 \text{ A})^2}{(2\pi)(1 \text{ m})}$$

$$= 2 \times 10^{-7} \text{ N/m}$$

as expected.

Example 2 Two parallel wires a distance d apart carry equal currents i in opposite directions, as shown in Fig. 31–8. Find the magnetic field **B** for points between the wires at a distance x from the midpoint.

Study of the figure and use of the right-hand rule show that $\mathbf{B}_a$ due to the current in wire a and $\mathbf{B}_b$ due to the current in wire b point in the same direction (up) at point P. Each is given in magnitude by Eq. 31–3 so that

$$B(x) = B_a + B_b = \frac{\mu_0 i}{2\pi \left(\frac{d}{2} + x\right)} + \frac{\mu_0 i}{2\pi \left(\frac{d}{2} - x\right)}$$

$$= \frac{2\mu_0 i d}{\pi (d^2 - 4x^2)}.$$

Note that B is symmetrical about the midpoint ($x = 0$) and has its minimum value ($= 2\mu_0 i / \pi d$) at that point; see Problem 8. For $x = \pm d/2$ point P in Fig. 31–8 is at the center lines of the wires. At these locations we see that $B \to +\infty$. However, in our derivation of Eq. 31–3 ($B = \mu_0 i / 2\pi r$) in Example 1 we considered only field points that were outside the wires. Our formula above therefore holds only up to the wire surfaces. See Example 3 for the variation of B with distance inside a wire.

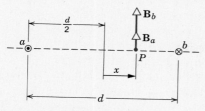

Figure 31–8 Example 2.

31–5 Ampere's Law

In electrostatics, as we have already reminded ourselves, a central problem is to calculate the electric field set up at various points by a given charge distribution. We can always solve such problems by direct integration methods based on Coulomb's law for the electric field (Eq. 31–1). Example 5 of Chapter 24 illustrates the method. The computational problems can easily become difficult, but we can always use a computer to obtain a numerical solution.

Some electrostatic problems, however, involve enough symmetry so that we can solve them, usually with ease and elegance, using Gauss's law (Eq. 25–6). Problems involving charge distributions with the symmetry of a point (Example 2, Chapter 25), a line (Example 3, Chapter 25), or a plane (Example 4, Chapter 25) come to mind. We stress that Gauss's law can be applied to *any* electrostatic charge distribution and will always yield a true result; only in problems of high symmetry however will such true results also prove to be useful results.

The situation in magnetism is just the same. We can in principle calculate the magnetic field set up by any current distribution by integration methods based on the Biot-Savart law (Eq. 31–2). We did this, for example, for the problem of a current in a long straight wire (Example 1). Once more we can easily find ourselves in great computational difficulties but, again, it is always possible to turn to a computer.

Is there, in magnetism, a relationship equivalent to the Biot-Savart law that will permit us to solve certain magnetic problems of high symmetry with ease and elegance, just as we did in electrostatics using Gauss's law? *Ampere's law* is such a formulation. Our plan is to display the law, to define its terms, and to apply it to some appropriate magnetic problems.

Ampere's law is

Ampere's law

$$\oint \mathbf{B} \cdot d\boldsymbol{\ell} = \mu_0 i. \qquad (31\text{-}5)$$

This relationship is to be applied to an arbitrary hypothetical closed loop drawn in a region where, in general, magnetic fields and current distributions exist. The integral on the left of Eq. 31–5 depends—as we shall see—on the way $\mathbf{B}$ varies in magnitude and direction as we traverse the loop. The quantity i on the right of Eq. 31–5 is—again as we shall see—the net current that pierces the loop.

In Fig. 31–9 we examine Ampere's law for a somewhat restricted two-dimensional case. The figure shows the cross sections of three long straight randomly arranged wires that pierce the plane of the page at right angles to it. The wires carry currents of arbitrary magnitudes that point in the directions indicated. The loop to which we intend to apply Ampere's law is taken to lie in the plane of the page. It threads arbitrarily among the wires, enclosing two of them but not the third.

We divide the loop of Fig. 31–9 into differential line elements $d\boldsymbol{\ell}$, one of which is shown. At this line element the magnetic field will have some particular value $\mathbf{B}$. Because the problem is two-dimensional $\mathbf{B}$ will lie in the plane of the figure, making an angle θ with the direction of the line element $d\boldsymbol{\ell}$.

The quantity $\mathbf{B} \cdot d\boldsymbol{\ell}$ in Eq. 31–5 is a scalar product and has the value $B \cos \theta \, dl$. The integral in that equation thus becomes

$$\oint \mathbf{B} \cdot d\boldsymbol{\ell} = \oint B \cos \theta \, dl.$$

This *line integral,* as it is called, instructs us to traverse the loop of Fig. 31–9, adding up (that is, integrating) the quantity $B \cos \theta \, dl$ as we go. We choose (arbitrarily) to traverse the loop in a counter-clockwise sense.

So much for the left side of Eq. 31–5. The term i on the right side is the *net* current that pierces the loop, the currents being added algebraically. For the special case of Fig. 31–9 we have

$$i = i_1 - i_2$$

For a counter-clockwise traversal of the loop, currents pointing upward out of the loop are taken as positive, those pointing downward through the loop being negative. This convention readily adapts to the right-hand rule that we enunciated on p. 605. Note that i_3 in Fig. 31–9 is not included on the right side of Eq. 31–5; it is outside the loop and does not pierce it.

The application of Ampere's law to the situation of Fig. 31–9 yields a statement that is true but not helpful or interesting, namely,

$$\oint B \cos \theta \, dl = \mu_0(i_1 - i_2).$$

Ampere's law; current in a long wire

The symmetry is simply not high enough to go further. We turn then to a familiar special case, that of a single long straight wire carrying a current i. As Fig. 31–10 shows we take our loop to be a concentric circle of radius r. This choice permits us to take full advantage of the symmetry of the problem. Because of this symmetry we conclude that $\mathbf{B}$ has the same magnitude B at every point on the circular loop and that $\mathbf{B}$ is everywhere tangent to the loop. Thus $\mathbf{B}$ and $d\boldsymbol{\ell}$ always point in the same direction, the angle θ between them being zero.

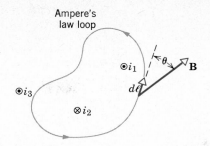

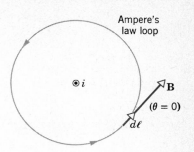

Figure 31–9   A two-dimensional problem. The three wires carrying the currents are at right angles to the page. The Ampere's law loop lies in the plane of the page.

Figure 31–10   A long straight wire carrying a current i. The Ampere's law loop is chosen to be a concentric circle.

For the case of Fig. 31–10 the left side of Ampere's law becomes

$$\oint \mathbf{B} \cdot d\boldsymbol{\ell} = \oint B \cos \theta \, dl = B \oint dl = B(2\pi r).$$

Note that $\oint dl$ above is simply the circumference of the circular loop. The right side of Ampere's law is simply $\mu_0 i$. Equating the two sides gives

$$B(2\pi r) = \mu_0 i$$

or

$$B = \frac{\mu_0 i}{2\pi r}. \qquad [31\text{–}3]$$

This is precisely the result that we derived using the Biot-Savart law, with considerable more effort, in Example 1. The advantage in simplicity of Ampere's law in this problem is clear.

Example 3   Derive an expression for $\mathbf{B}$ at a distance r from the center of a long cylindrical wire of radius R, where $r < R$. The wire carries a current i_0, distributed uniformly over the cross section of the wire.

Figure 31–11 shows a circular path of integration inside the wire. Symmetry suggests that $\mathbf{B}$ is tangent to the path as shown. Ampère's law,

$$\oint \mathbf{B} \cdot d\boldsymbol{\ell} = \mu_0 i,$$

gives

$$(B)(2\pi r) = \mu_0 i_0 \frac{\pi r^2}{\pi R^2},$$

since only the fraction of the current that passes through the path of integration is included in the factor i on the right. Solving for B and dropping the subscript on the current yields

$$B = \frac{\mu_0 i r}{2\pi R^2}.$$

At the surface of the wire ($r = R$) this equation reduces to the same expression as that found by putting $r = R$ in Eq. 31–3 ($B = \mu_0 i/2\pi R$).

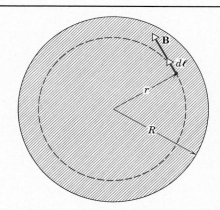

Figure 31–11   Example 3. A circular path of integration inside a wire. A current i_0, distributed uniformly over the cross section of the wire, emerges from the page.

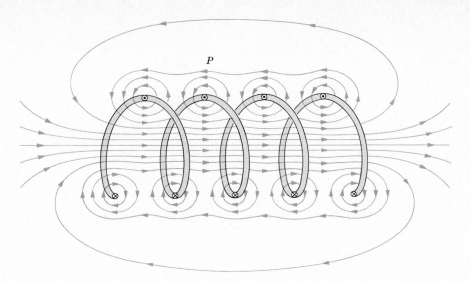

Figure 31-12   A loosely wound solenoid.

31-6   Solenoids and Toroids

Ampere's law; the solenoid

We now turn our attention to another problem of high symmetry in which Ampere's law will prove useful. It is that of finding the magnetic field set up by a current in the windings of a solenoid. A *solenoid* is a long wire wound in a close-packed helix and carrying a current *i*. We assume that the helix is very long compared with its diameter.

For points close to a single turn of the solenoid, the observer is not aware that the wire is bent in an arc. The wire behaves magnetically almost like a long straight wire, and the lines of **B** due to this single turn are almost concentric circles.

The solenoid field is the vector sum of the fields set up by all the turns that make up the solenoid. Figure 31–12, which shows a solenoid with widely spaced turns, suggests that the fields tend to cancel between the wires. It also suggests that, at points inside the solenoid and reasonably far from the wires, **B** is parallel to the solenoid axis. In the limiting case of adjacent square tightly packed wires, the solenoid becomes essentially a cylindrical current sheet and the requirements of symmetry then make the statement just given necessarily true. We assume that it is true in what follows.

For points such as *P* in Fig. 31–12 the field set up by the upper part of the solenoid turns (marked ⊙) points to the left and tends to cancel the field set up by the lower part of the solenoid turns (marked ⊗), which points to the right. As the solenoid becomes more and more ideal, that is, as it approaches the configuration of an infinitely long cylindrical current sheet, the field **B** at outside points approaches zero. Taking the external field to be zero is not a bad assumption for a practical solenoid if its length is much greater than its diameter and if we consider only external points near the central region of the solenoid, that is, away from the ends. Figure 31–13 shows the lines of **B** for a real solenoid, which is far from ideal in that the length is not much greater than the diameter. Even here the spacing of the lines of **B** in the central plane shows that the external field is much weaker than the internal field.

Let us apply Ampère's law,

$$\oint \mathbf{B} \cdot d\boldsymbol{\ell} = \mu_0 i, \qquad\qquad [31\text{--}5]$$

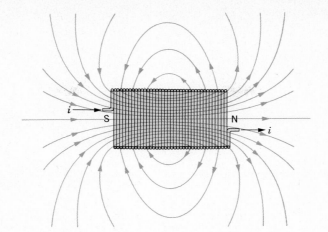

Figure 31–13 A solenoid of finite length. The right end, from which lines of **B** emerge, behaves like the north pole of a compass needle. The left end behaves like the south pole.

to the rectangular path *abcd* in the ideal solenoid of Fig. 31–14. We write the integral $\oint \mathbf{B} \cdot d\boldsymbol{\ell}$ as the sum of four integrals, one for each path segment:

$$\oint \mathbf{B} \cdot d\boldsymbol{\ell} = \int_a^b \mathbf{B} \cdot d\boldsymbol{\ell} + \int_b^c \mathbf{B} \cdot d\boldsymbol{\ell} + \int_c^d \mathbf{B} \cdot d\boldsymbol{\ell} + \int_d^a \mathbf{B} \cdot d\boldsymbol{\ell}.$$

The first integral on the right is Bh, where B is the magnitude of **B** inside the solenoid and h is the arbitrary length of the path from a to b. Note that path *ab*, though parallel to the solenoid axis, need not coincide with it.

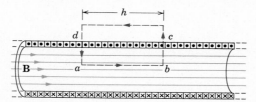

Figure 31–14 A section of an ideal solenoid, made of adjacent square turns, equivalent to an infinitely long cylindrical current sheet.

The second and fourth integrals are zero because for every element of these paths **B** is at right angles to the path. This makes $\mathbf{B} \cdot d\boldsymbol{\ell}$ zero and thus the integrals are zero. The third integral, which includes the part of the rectangle that lies outside the solenoid, is zero because we have taken **B** as zero for all external points for an ideal solenoid.

Thus $\oint \mathbf{B} \cdot d\boldsymbol{\ell}$ for the entire rectangular path has the value Bh. The net current i that passes through the area bounded by the path of integration is not the same as the current i_0 in the solenoid because the path of integration encloses more than one turn. Let n be the number of turns per unit length; then

$$i = i_0(nh).$$

Ampère's law then becomes

$$Bh = \mu_0 i_0 nh$$

or

B for a solenoid

$$B = \mu_0 i_0 n. \tag{31–6}$$

Although we derived Eq. 31–6 for an infinitely long ideal solenoid, it holds quite well for actual solenoids for internal points near the center of the solenoid. It shows that B does not depend on the diameter or the length of the solenoid and that B is constant over the solenoid cross section. A solenoid is a practical way to set up a known uniform magnetic field for experimentation, just as a parallel-plate capacitor is a practical way to set up a known uniform electric field.

Example 4 A solenoid is 1.0 m long and 3.0 cm in inner diameter. It has five layers of windings of 850 turns each and carries a current of 5.0 A. What is B at its center?

From Eq. 31–6,

$$B = \mu_0 i_0 n = (4\pi \times 10^{-7} \text{ T·m/A})(5.0 \text{ A})(5 \times 850 \text{ turns/m})$$
$$= 2.7 \times 10^{-2} \text{ T} = 27 \text{ mT·} (= 270 \text{ gauss}).$$

We can use Eq. 31–6 even if the solenoid has more than one layer of windings because the diameter of the windings does not enter the equation for B.

Example 5 *A toroid* Figure 31–15 shows a toroid, which we may describe as a solenoid bent into the shape of a doughnut. Calculate **B** at interior points.

From symmetry the lines of **B** form concentric circles inside the toroid, as shown in the figure. Let us apply Ampère's law to a concentric circular loop of radius r:

$$\oint \mathbf{B} \cdot d\boldsymbol{\ell} = \mu_0 i$$

or

$$(B)(2\pi r) = \mu_0 i_0 N,$$

where i_0 is the current in the toroid windings and N is the total number of turns. This gives

$$B = \frac{\mu_0}{2\pi} \frac{i_0 N}{r}. \qquad (31\text{–}7)$$

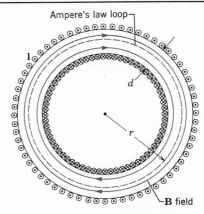

Figure 31–15 Example 5. A toroid.

In contrast to the solenoid, B is not constant over the cross section of a toroid. Show from Ampère's law that B equals zero for points outside an ideal toroid.

There is great current interest in toroids because they form the central feature of the tokamak, a device showing promise as the basis for a fusion power reactor. The toroidal space in a tokamak is filled with hot gaseous plasma—confined there by magnetic fields—within which the energy-producing fusion events take place.

31–7 Current Loop as a Magnetic Dipole

A current loop in an external field

So far we have studied magnetic fields set up by long straight wires, by solenoids, and by toroids. We turn our attention here to current loops. We learned in Section 30–4 that such a loop behaves like a magnetic dipole in that if we place it in an external magnetic field a torque $\boldsymbol{\tau}$, given by,

$$\boldsymbol{\tau} = \boldsymbol{\mu} \times \mathbf{B} \qquad [30\text{–}11]$$

acts on it. Here $\boldsymbol{\mu}$ is the magnetic dipole moment of the loop, given in magnitude by NiA (see Eq. 30–10) where N is the number of turns, i is the current, and A is the area enclosed by the loop.

The field set up by a current loop

We turn here to the other aspect of the current loop as a magnetic dipole: What magnetic field does it set up at nearby points? The problem does not have enough symmetry to make Ampère's law useful. To see this, try to locate an Ampère's law loop in Fig. 31–16 in such a way that the integral in Eq. 31–5 can be evaluated. We turn then to the Biot-Savart law and we limit consideration to points on the axis of the loop. Example 6 gives the derivation.

Example 6 *A circular current loop* Figure 31–16 shows a circular loop of radius R carrying a current i. Calculate **B** for points on the axis.

The vector $d\boldsymbol{\ell}$ for a current element at the top of the loop points perpendicularly out of the page. The angle θ between $d\boldsymbol{\ell}$ and **r** is 90°, and the plane formed by $d\boldsymbol{\ell}$ and **r** is normal to the

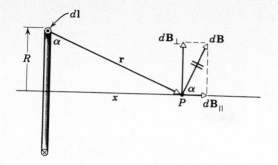

Figure 31–16 Example 6. A ring of radius R carrying a current i.

page. The vector $d\mathbf{B}$ for this element is at right angles to this plane and thus lies in the plane of the figure and at right angles to $\mathbf{r}$, as the figure shows.

Let us resolve $d\mathbf{B}$ into two components, one, $d\mathbf{B}_{\|}$, along the axis of the loop and another, $d\mathbf{B}_{\perp}$, at right angles to the axis. Only $d\mathbf{B}_{\|}$ contributes to the total magnetic field $\mathbf{B}$ at point P. This follows because the components $d\mathbf{B}_{\|}$ for all current elements lie on the axis and add directly; however, the components $d\mathbf{B}_{\perp}$ point in different directions perpendicular to the axis, and their resultant for the complete loop is zero, from symmetry. Thus

$$B = \int dB_{\|},$$

where the integral is a simple scalar integration over the current elements.

For the current element shown in Fig. 31–16 we have, from the Biot-Savart law (Eq. 31–2),

$$dB = \frac{\mu_0 i}{4\pi} \frac{dl \sin 90°}{r^2}.$$

We also have

$$dB_{\|} = dB \cos \alpha.$$

Combining gives

$$dB_{\|} = \frac{\mu_0 i \cos \alpha \, dl}{4\pi r^2}.$$

Figure 31–16 shows that r and α are not independent of each other. Let us express each in terms of a new variable x, the distance from the center of the loop to the point P. The relationships are

$$r = \sqrt{R^2 + x^2}$$

and

$$\cos \alpha = \frac{R}{r} = \frac{R}{\sqrt{R^2 + x^2}}.$$

Substituting these values into the expression for $dB_{\|}$ gives

$$dB_{\|} = \frac{\mu_0 i R}{4\pi (R^2 + x^2)^{3/2}} \, dl.$$

Note that i, R, and x have the same values for all current elements. Integrating this equation, noting that $\int dl$ is simply the circumference of the loop ($= 2\pi R$), yields

$$B = \int dB_{\|} = \frac{\mu_0 i R}{4\pi (R^2 + x^2)^{3/2}} \int dl$$

or

$$B(x) = \frac{\mu_0 i R^2}{2(R^2 + x^2)^{3/2}}. \tag{31–8}$$

If we put $x \gg R$ in Example 6 so that points close to the loop are not considered, Eq. 31–8 reduces to

$$B(x) = \frac{\mu_0 i R^2}{2x^3}.$$

Recalling that πR^2 is the area A of the loop and considering loops with N turns, we can write this equation as

$$B(x) = \frac{\mu_0}{2\pi} \frac{(NiA)}{x^3} = \frac{\mu_0}{2\pi} \frac{\mu}{x^3}, \tag{31–9}$$

where μ is the magnetic dipole moment of the current loop. This reminds us of the result derived in Problem 9, Chapter 24 $[E = (1/2\pi\epsilon_0)(p/x^3)]$, which is the formula for the electric field on the axis of an electric dipole.

A current loop as a magnetic dipole

Thus we have shown in two ways that we can regard a current loop as a magnetic dipole: It experiences a torque given by $\boldsymbol{\tau} = \boldsymbol{\mu} \times \mathbf{B}$ when we place it in an external magnetic field; it generates its own magnetic field given, for points on the axis, by the equation just developed.

Table 31–1 is a summary of the properties of electric and magnetic dipoles. The symmetry of the two sets of equations is striking.

Table 31–1   Some dipole equations

Property	Dipole Type	Equation	Reference
Torque in an external field	electric	$\tau = \mathbf{p} \times \mathbf{E}$	Eq. 24–11
	magnetic	$\tau = \mu \times \mathbf{B}$	Eq. 30–11
Energy in an external field	electric	$U = -\mathbf{p} \cdot \mathbf{E}$	Eq. 24–13
	magnetic	$U = -\mu \cdot \mathbf{B}$	Eq. 30–12
Field at distant points along axis	electric	$E(x) = \dfrac{1}{2\pi\epsilon_0}\dfrac{p}{x^3}$	Problem 9, Chapter 24
	magnetic	$B(x) = \dfrac{\mu_0}{2\pi}\dfrac{\mu}{x^3}$	Eq. 31–9
Field at distant points in the median plane	electric	$E(x) = \dfrac{1}{4\pi\epsilon_0}\dfrac{p}{x^3}$	Example 3, Chapter 24
	magnetic	$B(x) = \dfrac{\mu_0}{4\pi}\dfrac{\mu}{x^3}$	—

REVIEW GUIDE AND SUMMARY

The Biot-Savart law

The magnetic field set up by a current-carrying conductor can be found from the Biot-Savart law. This asserts that the contribution $d\mathbf{B}$ to the field set up by a current element $i\,d\ell$ at a point distant $\mathbf{r}$ from the current element has a magnitude given by

$$dB = \frac{\mu_0 i}{4\pi}\frac{\sin\theta\,dl}{r^2}.$$ [31–2]

The permeability constant μ_0

The direction of $d\mathbf{B}$ is that of the vector $d\ell \times \mathbf{r}$. Study Fig. 31–3 carefully. The quantity μ_0 is called the permeability constant and has the value $4\pi \times 10^{-7}$ T·m/A.

For a long straight wire carrying a current i the Biot-Savart law gives, for the magnetic field at a distance r from the wire,

A long, straight wire

$$B(r) = \frac{\mu_0 i}{2\pi r};$$ [31–3]

see Example 1. Figure 31–5 shows the concentric field lines. The right-hand rule described on p. 605 is helpful in relating the directions of $\mathbf{B}$ and $i\,d\ell$ in this and other magnetic problems.

If a wire carrying a current is placed in an external magnetic field $\mathbf{B}_{ext}$ a force given by Eq. 30–6a will act on it. $\mathbf{B}_{ext}$ must not be confused with $\mathbf{B}_{int}$, the intrinsic magnetic field set up by the current in the wire itself. Figure 31–6 shows field lines representing the vector sum of $\mathbf{B}_{ext}$ and $\mathbf{B}_{int}$ for a particular case.

$\mathbf{B}_{ext}$, $\mathbf{B}_{int}$, and their resultant

Parallel wires carrying currents in the same (opposite) direction attract (repel) each other. For a length l of the wire the magnitude of the force per unit length on either wire is

The force between parallel wires

$$\frac{F}{l} = \frac{\mu_0 i_a i_b}{2\pi d},$$ [31–4]

where d is the wire separation. The ampere is defined from this relation; Example 2.

For current distributions of high symmetry Ampere's law,

Ampere's law

$$\oint \mathbf{B} \cdot d\ell = \mu_0 i$$ [31–5]

can be used in place of the Biot-Savart law to calculate the magnetic field. One first chooses a closed *Ampere's law loop* around which to apply the law. Study the fairly general case of Fig. 31–9 to learn how to evaluate the line integral $\oint \mathbf{B} \cdot d\boldsymbol{\ell}$ and to find i.

B in a long solenoid at points near its center is given by

A solenoid

$$B = \mu_0 n i_0 \qquad [31-6]$$

where n is the number of turns per unit length; see Examples 4 and 5.

The magnetic field set up by a current loop of radius R at points a distance x along its axis is given by

A current loop

$$B(x) = \frac{\mu_0}{2} \frac{iR^2}{(R^2 + x^2)^{3/2}}. \qquad [31-8]$$

For $x \gg R$ the axial field is that of a magnetic dipole whose magnetic moment $\boldsymbol{\mu}$ is given in magnitude by NiA (Eq. 30–10). Table 31–1 summarizes and compares the impressively symmetric properties of both electric and magnetic dipoles.

QUESTIONS

1. A beam of 20-MeV protons emerges from a cyclotron. Do these particles cause a magnetic field?

2. Discuss analogies and differences between Coulomb's law and the Biot-Savart law.

3. Is **B** constant in magnitude for points that lie on a given magnetic field line?

4. In electronics, wires that carry equal but opposite currents are often twisted together to reduce their magnetic effect at distant points. Why is this effective?

5. Drifting electrons constitute the current in a wire and a magnetic field is associated with this current. Would the magnetic field disappear for an observer moving along with the drifting electrons?

6. Is there any way to set up a magnetic field other than by causing charges to move?

7. Like currents attract and unlike currents repel. Like stationary charges repel and unlike stationary charges attract. Is this a paradox?

8. You are given two pieces of wire of the same length. You bend the first into a circular loop of one turn and the second into a (smaller) circular loop of two turns. Identical currents are set up in each of the loops. How do the magnitudes of the magnetic fields set up at the centers of the two loops compare?

9. Figure 31–17 shows four vertical wires carrying equal currents in the same direction. What is the direction of the force on the left-hand wire, caused by the currents in the other three wires?

10. Two long parallel conductors carry equal currents i in the same direction. Sketch roughly the resultant lines of **B** due to the action of both currents. Does your figure suggest an attraction between the wires?

11. A current is sent through a vertical spring from whose lower end a weight is hanging. What will happen?

12. Two long straight wires pass near one another at right angles. If the wires are free to move, describe what happens when currents are sent through both of them.

13. Can the path of integration around which we apply Ampère's law pass through a conductor?

14. Suppose we set up a path of integration around a cable that contains twelve wires with different currents (some in opposite directions) in each wire. How do we calculate i in Ampère's law in such a case?

15. Apply Ampère's law qualitatively to the three paths shown in Fig. 31–18.

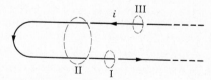

Figure 31–18   Question 15.

16. Discuss and compare Gauss's law and Ampère's law.

17. A steady longitudinal current is set up in a long copper tube. Is there a magnetic field (*a*) inside and (*b*) outside the tube?

18. Does it follow from symmetry arguments alone that the lines of **B** around a long straight wire carrying a current i must be concentric circles? Are other arrangements possible?

19. A long straight wire of radius R carries a steady current i. Does the magnetic field generated by this current depend on R? Consider points both outside and inside the wire.

Figure 31–17   Question 9.

20. A long straight wire carries a constant current i. Does Ampère's law hold for (a) a loop that encloses the wire but is not circular, (b) a loop that does not enclose the wire, and (c) a loop that encloses the wire but does not all lie in one plane? Discuss.

21. Two long solenoids are nested on the same axis, as Fig. 31–19 shows. They carry identical currents but in oposite directions. If there is no magnetic field inside the inner solenoid, what can you say about n, the number of turns per unit length, for the two solenoids. Which one, if either, has the larger value?

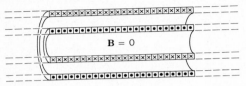

Figure 31–19 Question 21.

22. In Example 6 we learned (see Eq. 31–8) that the magnetic field at the center of a circular current loop has the value $\mu_0 i/2R$. However, the *electric* field at the center of a ring of charge is *zero*. Why this difference?

23. A steady current is set up in a cubical network of resistive wires, as in Fig. 31–20. Use symmetry arguments to show that the magnetic field at the center of the cube is zero.

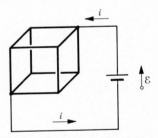

Figure 31–20 Question 23.

24. Does Eq. 31–6 ($B = \mu_0 in$) hold for a solenoid of square cross section?

25. Give details of three ways in which you can measure the magnetic field **B** at a point P, a perpendicular distance r from a long straight wire carrying a constant current i. Base them on: (a) projecting a particle of charge q through point P with velocity **v**, parallel to the wire; (b) measuring the force per unit length exerted on a second wire, parallel to the first wire and carrying a current i'; (c) measuring the torque exerted on a small magnetic dipole located a perpendicular distance r from the wire.

26. How might you measure the magnetic dipole moment of a compass needle?

27. A circular loop of wire lies on the floor of the room in which you are sitting and it carries a constant current i in a clockwise direction, as viewed from above. What is the direction of the magnetic dipole moment of this current loop?

28. In a circular loop of wire carrying a current i, is **B** uniform for all points within the loop?

EXERCISES

Section 31–2 Calculating B; the Biot-Savart Law

1. Which of the following are correct units for the permeability constant μ_0: (a) tesla, (b) newton/ampere^2, (c) tesla·meter/ampere, (d) weber/meter, (e) kilogram·meter/ampere?

2. Two infinitely long wires carry equal currents, i. Each follows a 90° arc on the circumference of the same circle of radius R in the configuration shown in Figure 31–21. Show, without doing a detailed calculation, that **B** at the center of the circle is the same

as the **B**-field a distance R below an infinite straight wire carrying a current i to the left.

3. (a) Show that the Biot-Savart Law, Eq. 31–2, can be correctly written in vector form as

$$dB = \left(\frac{\mu_0}{4\pi}\right)\frac{i\, d\ell \times r}{r^3}.$$

(b) What is the analogous vector form corresponding to Eq. (31–1)?

4. Figure 31–22 shows a 3.0-cm segment of wire, centered at the origin, carrying a current of 2.0 A in the $+y$-direction. (Of course this segment must be part of some complete circuit.) To calculate the **B**-field at a point P several meters from the origin one may use the Biot-Savart law in the form $B = (\mu_0/4\pi)i\,\Delta l\,\sin\theta/r^2$ with $\Delta l = 3.0$ cm. This is because r and θ are essentially constant over the segment of wire. Calculate **B** at the following (x, y, z) locations: (a) $(0, 0, 5\text{ m})$, (b) $(0, 6\text{ m}, 0)$, (c) $(7\text{ m}, 7\text{ m}, 0)$, (d) $(-3\text{ m}, -4\text{ m}, 0)$.

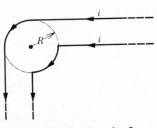

Figure 31–21 Exercise 2.

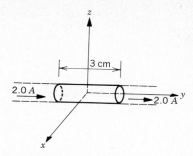

Figure 31–22   Exercise 4.

5.   A surveyor is using a magnetic compass 20 ft below a power line in which there is a steady current of 100 A. Will this interfere seriously with the compass reading? The horizontal component of the earth's magnetic field at the site is 20 μT (=0.20 gauss).

6.   A positive point charge of magnitude q is a distance d from the axis of a long straight wire carrying a current i and is traveling with speed v perpendicular to the axis of the wire. What are the direction and magnitude of the force acting on it if the charge is moving (a) toward, or (b) away from the wire?

7.   A long straight wire carries a current of 50 A. An electron, traveling at 1.0×10^7 m/s, is 5.0 cm from the wire. What force acts on the electron if the electron velocity is directed (a) toward the wire, (b) parallel to the wire, and (c) at right angles to the directions defined by (a) and (b)?

Section 31–3   The Magnetic Force on a Current; a Second Look

8.   A long wire carrying a current of 100 A is placed in a uniform external magnetic field of 5.0 mT (= 50 gauss). The wire is at right angles to this magnetic field. Locate the points at which the resultant magnetic field is zero.

9.   At a position in the Philippines the earth's magnetic field of 39 μT is horizontal and due north. The net field is zero exactly 8.0 cm above a long straight horizontal wire that carries a constant current. (a) What is the magnitude of the current and (b) what is the direction of the current?

10.   Figure 31–23 shows five long parallel wires in the $x - y$ plane. Each wire carries a current $i = 3.0$ A in the positive x-direction. The separation between adjacent wires is $d = 8.0$ cm. Find the magnetic force per length exerted on each of these five wires.

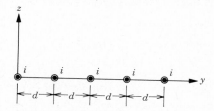

Figure 31–23   Exercise 10.

Section 31–4   Two Parallel Conductors

11.   Two long parallel wires are 8.0 cm apart. What equal currents must flow in the wires if the magnetic field half way between them is to have a magnitude of 300 μT? Consider both (a) parallel and (b) antiparallel currents.

12.   Two long straight parallel wires, separated by 0.75 cm, are

perpendicular to the plane of the page as shown in Fig. 31–24. Wire W_1 carries a current of 6.5 A into the page. What must be the current (magnitude and direction) in wire W_2 for the resultant magnetic field at point P to be zero?

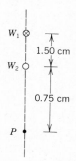

Figure 31–24   Exercise 12.

13.   Two long parallel wires carry currents of i and $3i$ in the same direction. Locate the point or points at which their magnetic fields cancel.

14.   Figure 31–25a shows an electrical circuit consisting of two "black boxes," X and Y, connected by two parallel wires. One of the boxes contains a battery (Fig. 31–25b) and the other contains a light bulb (Fig. 31–25c). The circuit is connected so that the battery is the energy source for the light bulb. In the vicinity of the point P there is an **E**-field (due to the charged parallel wires) and a **B**-field (due to the current in the parallel wires). Show that regardless of which black box contains the battery and regardless of the polarity of the connection of the battery, the direction of $\mathbf{E} \times \mathbf{B}$ in the vicinity of point P is from the battery to the light bulb, in the direction of the power flow.

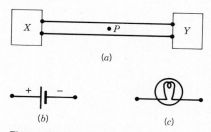

Figure 31–25   Exercise 14.

Section 31–5   Ampere's Law

15.   Each of the indicated eight conductors in Figure 31–26 carries 2.0 A of current into or out of the page. Two paths are indicated for the line integral $\oint \mathbf{B} \cdot d\boldsymbol{\ell}$. What is the value of the integral for (a) the dotted path, (b) the dashed path?

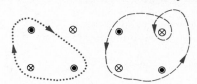

Figure 31–26   Exercise 15.

16.   Eight wires cut the page perpendicularly at the points shown in Fig. 31–27. A wire labeled with the integer $k (k = 1, 2, \ldots, 8)$ carries the current $k i_0$. For those with odd k, the current is out of the page; for those with even k it is into the page. Evaluate $\oint \mathbf{B} \cdot d\mathbf{l}$ along the closed path shown.

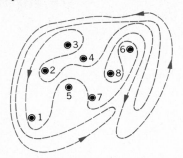

Figure 31–27   Exercise 16.

17. Two square conducting loops carry currents of 5.0 A and 3.0 A as shown in Fig. 31–28. What is the value of the line integral $\oint \mathbf{B} \cdot d\boldsymbol{\ell}$ for each of the two closed paths shown?

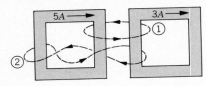

Figure 31–28   Exercise 17.

18. Figure 31–29 shows a cross section of a long cylindrical conductor of radius a, carrying a uniformly distributed current i. Assume $a = 2.0$ cm and $i = 100$ A and plot $B(r)$ over the range $0 < r < 6.0$ cm.

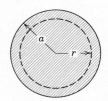

Figure 31–29   Exercise 18.

19. In a certain region there is a uniform current density of 15 A/m² in the $+z$-direction. What is the value of $\oint \mathbf{B} \cdot d\mathbf{l}$ when the line integral is taken along the four straight-line segments from $(4d,0,0)$ to $(4d,3d,0)$ to $(0,0,0)$ to $(0,0,4d)$ to $(4d,0,0)$ where $d = 20$ cm?

Section 31–6   Solenoids and Toroids
20. A 200-turn solenoid having a length of 25 cm and diameter of 10 cm carries a current of 0.30 A. Calculate the magnitude of the magnetic field **B** near the center of the solenoid.

21. A toroid having a square cross section, 5.0 cm on edge, and an inner radius of 15 cm has 500 turns and carries a current of 0.80 A. What is the magnetic field inside the toroid at (a) the inner radius and (b) the outer radius of the toroid?

Section 31–7   A Current Loop as a Magnetic Dipole
22. What is the magnetic dipole moment, μ, of the solenoid described in Exercise 20?

23. Figure 31–30a shows a length of wire carrying a current i and bent into a circular coil of one turn. In Fig. 31–30b the same length of wire has been bent more sharply, to give a double loop of smaller radius. (a) If B_a and B_b are the magnitudes of the magnetic fields at the centers of the two loops, what is the ratio B_b/B_a? (b) What is the ratio of their dipole moments, μ_b/μ_a?

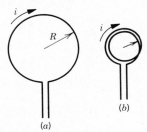

Figure 31–30   Exercise 23 and Problem 23.

24. A student makes an electromagnet by winding 300 turns of wire around a wooden cylinder of diameter $d = 5.0$ cm. She connects this to a battery producing a current of 4.0 A in the wire. (a) What is the magnetic moment of this device? (b) At what axial distance $R \gg d$ will the magnetic field of this dipole be 5.0 μT (approximately one-tenth that of the earth's magnetic field)?

PROBLEMS GROUPED BY SECTION

Section 31–2   Calculating B; the Biot-Savart Law
1. Two long straight parallel wires 10 cm apart each carry a current of 100 A. Figure 31–31 shows a cross section, with the wires running perpendicular to the page and point P lying on the perpendicular bisector of d. Find the magnitude and direction of the magnetic field at P when the current in the left-hand wire is out

of the page and the current in the right-hand wire is (a) out of the page and (b) into the page.

2. Use the Biot-Savart law to calculate the magnetic field **B** at C, the common center of the semicircular arcs AD and HJ, of radii R_2 and R_1, respectively, forming part of the circuit $ADJHA$ carrying current i, as shown in Fig. 31–32.

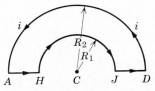

Figure 31–32   Problem 2.

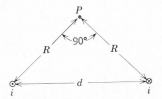

Figure 31–31   Problem 1.

3. A long hairpin is formed by bending a piece of wire as shown in Fig. 31–33. If the wire carries a current of 10 A, what are the direction and magnitude of **B** (*a*) at point *a*? (*b*) at point *b*? Take $R = 5.0$ mm.

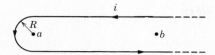

Figure 31–33 Problem 3.

4. A straight wire segment of length *l* carries a current *i*. Show that the magnetic field *B* associated with this segment, at *P* a distance *R* from the segment along a perpendicular bisector (see Fig. 31–34), is given in magnitude by

$$B = \frac{\mu_0 i}{2\pi R} \frac{l}{(l^2 + 4R^2)^{1/2}}.$$

Show that this expression reduces to an expected result as $l \to \infty$.

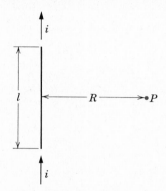

Figure 31–34 Problem 4.

5. A square loop of wire of edge *a* carries a current *i*. Show that the value of *B* at the center of the square is given by

$$B = \frac{2\sqrt{2}\ \mu_0 i}{\pi a}.$$

6. A square loop of wire of edge *a* carries a current *i*. (*a*) Show that *B* for a point on the axis of the loop and a distance *x* from its center is given by

$$B(x) = \frac{4\mu_0 i a^2}{\pi(4x^2 + a^2)(4x^2 + 2a^2)^{1/2}}.$$

Show that the result is consistent with the result of Problem 5.

7. You are given a length *l* of wire in which a current *i* may be established. The wire may be formed into a circle or a square. Show that the square yields the greater value for *B* at the central point.

Section 31–4 Two Parallel Conductors

8. Show that the expression derived in Example 2 for the magnetic field $B(x)$ between the wires of Fig. 31–8 also holds outside of the wires, that is, to the left of wire *a* and to the right of wire *b*. Assume that $i = 10$ A and $d = 4.0$ cm in Fig. 31–8 and plot $B(x)$ for the range -6 cm $< x < +6$ cm. Assume that the wire diameters are negligible.

9. Suppose, in Fig. 31–35, that the currents are all out of the page. (*a*) What is the force per unit length (magnitude and direction) on any one wire? (*b*) Repeat if the currents are all into the page. In the analogous case of parallel motion of charged particles in a plasma, this is known as the pinch effect.

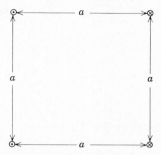

Figure 31–35 Problems 9, 11, and 30.

10. In Fig. 31–8 assume that both currents are in the same direction, out of the plane of the figure. Show that the magnetic field in the plane defined by the wires is

$$B(x) = \frac{4\mu_0 i x}{\pi(4x^2 - d^2)}.$$

Assume $i = 10$ A and $d = 4.0$ cm in Fig. 31–8 and plot $B(x)$ for the range -4 cm $< x < +4$ cm. Assume that the wire diameters are negligible.

11. In Fig. 31–35 what is the force per unit length acting on the lower left wire, in magnitude and direction?

Section 31–5 Ampere's Law

12. Figure 31–36 shows a cross section of a hollow cylindrical conductor of radii *a* and *b*, carrying a uniformly distributed current *i*. (*a*) Show that $B(r)$ for the range $b < r < a$ is given by

$$B(r) = \frac{\mu_0 i}{2\pi(a^2 - b^2)}\left(\frac{r^2 - b^2}{r}\right).$$

(*b*) Test this formula for the special cases of $r = a$, $r = b$, and $b = 0$. (*c*) Assume $a = 2.0$ cm, $b = 1.8$ cm, and $i = 100$ A and plot $B(r)$ for the range $0 < r < 6$ cm.

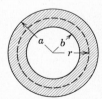

Figure 31–36 Problem 12.

13. Figure 31–37 shows a cross section of a long conductor of a type called a coaxial cable. Its radii (*a,b,c*) are shown in the figure. Equal but opposite currents *i* exist in the two conductors. Derive expressions for $B(r)$ in the ranges (*a*) $r < c$, (*b*) $c < r < b$, (*c*) $b < r < a$, and (*d*) $r > a$. (*e*) Test these expressions for all the special cases that occur to you. (*f*) Assume $a = 2.0$ cm, $b = 1.8$ cm, $c = 0.40$ cm, and $i = 100$ A and plot $B(r)$ over the range $0 < r < 6$ cm.

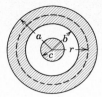

Figure 31-37 Problem 13.

14. Show that a uniform magnetic field **B** cannot drop abruptly to zero as one moves at right angles to it, as suggested by the horizontal arrow through point a in Fig. 31–38. (*Hint:* Apply Ampere's law to the rectangular path shown by the dotted lines.) In actual magnets "fringing" of the lines of **B** always occurs, which means that **B** approaches zero in a gradual manner. Modify the **B** lines in the figure to indicate a more realistic situation.

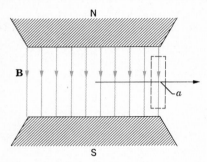

Figure 31-38 Problem 14.

15. The current density inside a long, solid, cylindrical wire of radius a is in the direction of the axis and varies linearly with radial distance r from the axis according to $j = j_0 r/a$. Find the magnetic field inside the wire.

16. Figure 31–39 shows a cross section of a long cylindrical conductor of radius a containing a long cylindrical hole of radius b. The axes of the two cylinders are parallel and are a distance h apart. A current i is uniformly distributed over the cross-hatched area in the figure. (*a*) Use superposition ideas to show that the magnetic field at the center of the hole is

$$B = \frac{\mu_0 i h}{2\pi(a^2 - b^2)}$$

(*b*) Discuss the two special cases $b = 0$ and $h = 0$. (*c*) Can you use Ampere's law to show that the magnetic field in the hole is uniform: (*Hint:* Regard the cylindrical hole as filled with two equal currents moving in opposite directions, thus canceling each other. Each of these currents must have the same current density as that in the actual conductor. Thus we superimpose the fields due to two complete cylinders of current, of radii a and b, each cylinder having the same current density.)

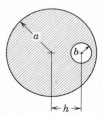

Figure 31-39 Problem 16.

Section 31-6   Solenoids and Toroids

17. An interesting (and agonizing) effect occurs when one attempts to confine a collection of electrons and positive ions (a plasma) in the magnetic field of a toroid. Particles whose motion is perpendicular to the **B**-field will not execute circular paths because the field strength varies with radial distance from the axis of the toroid. This effect which is shown (exaggerated) in Fig. 31–40, causes particles of opposite sign to drift in opposite directions parallel to the axis of the toroid. (*a*) What is the sign of the charge on the particle whose path is sketched in the figure? (*b*) If the particle path has a radius of curvature of 11 cm when its radial distance from the axis of the toroid is 125 cm, what will be the radius of curvature when the particle is 110 cm from the axis?

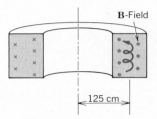

Figure 31-40 Problem 17.

18. Treat a solenoid as a continuous cylindrical current sheet, whose current per unit length, measured parallel to the cylinder axis, is λ. B inside a solenoid is given by Eq. 31–6 ($B = \mu_0 in$). Show that this can also be written as $B = \mu_0$. This is the value of the *change* in B that you encounter as you move from inside the solenoid to outside, through the solenoid wall. Show that this same change occurs as you move through an infinite plane current sheet such as that of Fig. 31–53 (see Problem 42). Does this equality surprise you?

19. In Example 5 we showed that the magnetic field at any radius r inside a toroid is given by

$$B = \frac{\mu_0 i_0 N}{2\pi r}. \qquad [31\text{-}7]$$

Show that as you move from a point just inside a toroid to a point just outside the magnitude of the *change* in B that you encounter —at any radius r—is just $\mu_0 \lambda$. Here λ is the current per unit length for any circumference ($= 2\pi r$) of the toroid. Compare the similar result found in Problem 18. Is the equality surprising?

Section 31-7   A Current Loop as a Magnetic Dipole

20. You are given a closed circuit with radii a and b (as shown in Fig. 31–41) carrying a current i. (*a*) What are the magnitude and direction of **B** at point P? (*b*) Find the dipole moment of the circuit.

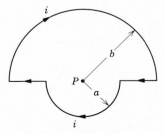

Figure 31-41 Problem 20.

21. *Helmholtz coils.* Two 300-turn coils each carry a current i. They are arranged a distance apart equal to their radius, as in Fig. 31–42. For $R = 5.0$ cm and $i = 50$ A, plot B as a function of distance x along the common axis over the range $x = -5$ cm to $x = +5$ cm, taking $x = 0$ at the midpoint P. (Such coils provide an especially uniform field of **B** near point P.)

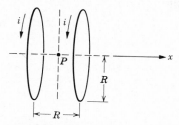

Figure 31–42   Problem 21.

22. In problem 21 let the separation of the coils be s (not necessarily equal to the coil radius R). (a) Show that the first derivative of the magnetic field (dB/dx) vanishes at the midpoint P regardless of the value of s. Why would you expect this to be true from symmetry? (b) Show that the second derivative of the magnetic field (d^2B/dx^2) also vanishes at P provided $s = R$. This accounts for the uniformity of B near P for this particular coil separation.

*23. Derive the solenoid equation (Eq. 31–6) starting from the expression for the field on the axis of a circular loop (Eq. 31–8). (*Hint:* Subdivide the solenoid into a series of current loops of infinitesimal thickness and integrate.) (See Fig. 31–16.)

*24. A plastic disk of radius R has a charge q uniformly distributed over its surface. If the disk is rotated at an angular frequency ω about its axis, show that (a) the magnetic field at the center of the disk is

$$B = \frac{\mu_0 \omega q}{2\pi R}$$

and (b) the magnetic dipole moment of the disk is

$$\mu = \frac{\omega q R^2}{4}.$$

(*Hint:* The rotating disk is equivalent to an array of current loops.)

ADDITIONAL PROBLEMS

25. A circular loop of radius 10 cm carries a current of 15 A. A second loop of radius 1.0 cm, having 50 turns and a current of 1.0 A, is at the center of the first loop. (a) What magnetic field B does the large loop set up at its center? (b) What torque acts on the small loop? Assume that the planes of the two loops are at right angles and that B due to the large loop is essentially uniform throughout the volume occupied by the small loop.

26. The wire shown in Fig. 31–43 carries a current i. What is the magnetic field B at the center C of the semicircle arising from (a) each straight segment of length l, (b) the semicircular segment of radius R, and (c) the entire wire?

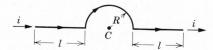

Figure 31–43   Problem 26.

27. A straight conductor carrying a current i is split into identical semicircular turns as shown in Fig. 31–44. What is the magnetic field at the center C of the circular loop so formed?

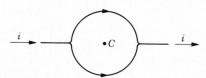

Figure 31–44   Problem 27.

28. Consider the circuit of Fig. 31–45. The curved segments are arcs of circles of radii a and b. The straight segments are along the radii. Find the magnetic field B at P, assuming a current i in the circuit.

29. A #10 bare copper wire (2.5 mm in diameter) can carry a current of 50 A without overheating. For this current, what is B at the surface of the wire?

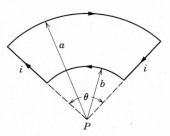

Figure 31–45   Problem 28.

30. Four long copper wires are parallel to each other, their cross section forming a square 20 cm on edge. A 20-A current is set up in each wire in the direction shown in Fig. 31–35. What are the magnitude and direction of **B** at the center of the square?

31. Figure 31–46 shows a long wire carrying a current of 30 A. The rectangular loop carries a current of 20 A. Calculate the resultant force acting on the loop. Assume that $a = 1.0$ cm, $b = 8.0$ cm, and $l = 30$ cm.

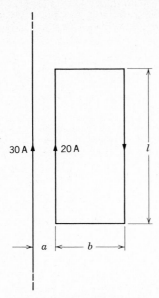

Figure 31–46   Problem 31.

32. A wire carrying current i has the configuration shown in Figure 31–47 in which two semi-infinite straight sections, each tangent to the same circle, are connected by a circular arc, of angle θ, along the circumference of the circle, with all sections lying in the same plane. What must θ be in order for B to be zero at the center of the circle?

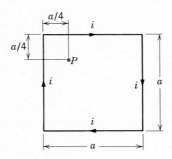

Figure 31–47   Problem 32.

33. Calculate B at point P in Fig. 31–48. Assume that $i = 10$ A and $a = 8.0$ cm.

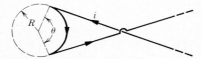

Figure 31–48   Problem 33.

34. Show that B at the center of a rectangular loop of wire of length l and width d, carrying a current i, is given by

$$B = \frac{2\mu_0 i}{\pi} \frac{(l^2 + d^2)^{1/2}}{ld}.$$

Show that this reduces to a result consistent with Example 2 for $l \gg d$.

35. A long solenoid with 10 turns/cm and a radius of 7.0 cm

carries a current of 20 mA. A current of 6.0 A flows in a straight conductor along the axis of the solenoid. At what radial distance from the axis will the direction of the resulting **B** field be at 45° from the axial direction? What is the **B** field magnitude there?

36. A 10-keV electron moves in a circular orbit in a 800-mT uniform magnetic field. (*a*) Calculate the magnetic dipole moment due to this circular motion of the electron. (*b*) Is this dipole moment parallel or antiparallel to the **B** field? (*c*) Does the answer to (*b*) depend on the fact that the electron has a negative charge?

37. A long circular pipe, with an outside radius of R, carries a (uniformly distributed) current of i_0 (into the paper as shown in Fig. 31–49). A wire runs parallel to the pipe at a distance of $3R$ from center to center. Calculate the magnitude and direction of the current in the wire that would cause the resultant magnetic field at the point P to have the same magnitude, but the opposite direction, as the resultant field at the center of the pipe.

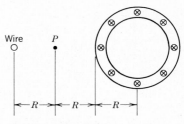

Figure 31–49   Problem 37.

38. Find the magnetic field B at point P in Fig. 31–50.

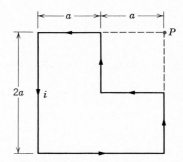

Figure 31–50   Problem 38.

39. (*a*) A long wire is bent into the shape shown in Fig. 31–51, without cross contact at P. The radius of the circular section is R. Determine the magnitude and direction of **B** at the center C of the circular portion when the current i is as indicated. (*b*) The circular part of the wire is rotated without distortion about its (dashed) diameter perpendicular to the straight portion of the wire. The magnetic moment associated with the circular loop is now in the direction of the current in the straight part of the wire. Determine **B** at C in this case.

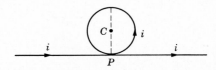

Figure 31–51   Problem 39.

40. A long solenoid has 100 turns per centimeter and carries a current i. An electron moves within the solenoid in a circle of radius 2.30 cm perpendicular to the solenoid axis. The speed of the electron is $0.046c$ (c = speed of light). Find the current i in the solenoid.

41. Two long wires a distance d apart carry equal antiparallel currents i, as in Fig. 31–52. (a) Show that B at point P, which is equidistant from the wires, is given by

$$B = \frac{2\mu_0 i d}{\pi(4R^2 + d^2)}.$$

(b) In what direction does **B** point?

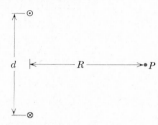

Figure 31–52   Problem 41.

42. Figure 31–53 shows a cross section of an infinite conducting sheet with a current per unit x-length λ emerging from the page at right angles. (a) Use the right-hand rule and symmetry arguments to convince yourself that the magnetic field **B** is constant for all points P above the sheet (and for all points P' below it) and is directed as shown. (b) Use Ampere's law to prove that

$$B = \tfrac{1}{2}\mu_0 \lambda.$$

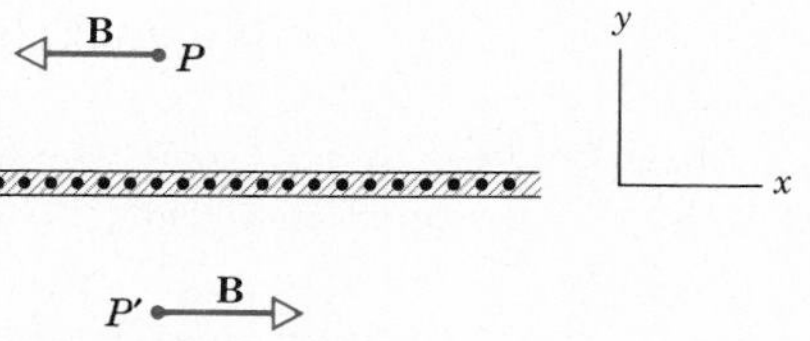

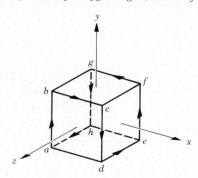

Figure 31–53   Problems 18 and 42.

43. A conductor carries a current of 6.0 A along the closed path $abcdefgha$ involving 8 of the 12 edges of a cube of side 10 cm as shown in Fig. 31–54. (a) Why can one regard this as the superposition of three square loops: $bcfgb$, $abgha$, and $cdefc$? (b) Find the

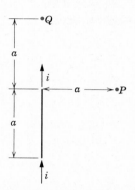

Figure 31–54   Problem 43.

magnetic dipole moment μ (magnitude and direction) of this current. (c) Using Table 31–1, calculate **B** at the points $(x, y, z) = (0, 5 \text{ m}, 0)$ and $(5 \text{ m}, 0, 0)$.

44. The magnetic field $B(x)$ for various points on the axis of a square current loop of side a is given in Problem 6. Show that the axial field for this loop for $x \gg a$ is that of a magnetic dipole (see Eq. 31–9). What is the magnetic dipole moment of this loop?

45. A current i flows in a straight wire segment of length a, as in Fig. 31–55. Show that the magnetic field at point Q in that figure is zero and that the field at P is given in magnitude by

$$B = \frac{\sqrt{2}\mu_0 i}{8\pi a}.$$

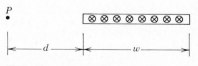

Figure 31–55   Problem 45.

*46. Derive the result $B = \tfrac{1}{2}\mu_0 \lambda$ given in Problem 42 by dividing the current sheet of Fig. 31–53 into infinitely long strips of width dx each carrying a current λdx. Find B at point P by integrating over all such strips.

*47. In a certain region there is a magnetic field given by $\mathbf{B} = 3i + 8(x^2/d^2)\mathbf{j}$ mT where x is the x-coordinate distance in meters and d is a constant, with units of length. Some current must be flowing in the region to cause the specified **B** field. (a) Evaluate the integral $\int \mathbf{B} \cdot d\boldsymbol{\ell}$ along the straight path from $(d,0,0)$ to $(d,d,0)$. (b) Let $d = 0.5$ m in the expression for **B** and apply Ampere's law to determine what current flows through a square of side length d ($=0.5$ m) that lies in the first quadrant of the $x - y$ plane with one corner at the origin. (c) Is this current in the $\mathbf{k}$ or $-\mathbf{k}$ direction?

*48. Figure 31–56 shows a cross section of a long, thin ribbon of width w that is carrying a uniformly-distributed total current I into the page. Calculate the magnitude and direction of the magnetic field, **B**, at a point P in the plane of the ribbon at a distance d from its edge.

Figure 31–56   Problem 48.

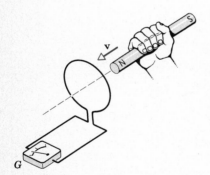

32 Faraday's Law of Induction

32–1 Two Experiments

Faraday's law of induction, which is one of the basic equations of electromagnetism was deduced to explain a number of simple experiments carried out by Michael Faraday in England in 1831 and by Joseph Henry in the United States at about the same time. We describe two of those experiments.

Experiment One. Figure 32–1 shows the terminals of a coil connected to a current-measuring device, perhaps a galvanometer such as that described in Section 30–4. Normally we would not expect this instrument to deflect because there seems to be no electromotive force in this circuit; but if we push a bar magnet toward the coil, with its north pole facing the coil, a remarkable thing happens. The galvanometer deflects, while the magnet is moving, showing that a current has been set up in the coil. If we hold the magnet stationary with respect to the coil, the galvanometer does not deflect. If we move the magnet *away* from the coil, the galvanometer again deflects, but in the opposite direction, which means that the current in the coil is in the opposite direction. If we use the south pole end of a magnet instead of the north pole end, the experiment works as described but the deflections are reversed. Further experimentation shows that what matters is the relative motion of the magnet and the coil. It makes no difference whether we move the magnet toward the coil or the coil toward the magnet.

The current that appears in this experiment is called an *induced current* and is said to be set up by an *induced electromotive force*. Note that there are no batteries anywhere in the circuit. Faraday was able to deduce from experiments like this the law that gives the magnitudes and directions of the induced currents and electromotive forces. Such emfs are very important in practice. The chances are good that the lights in the room in which you are reading this book are operated from an induced emf produced in a commercial electric generator.

Figure 32–1 Galvanometer *G* deflects while the magnet is moving with respect to the coil.

Induced emf

Experiment Two.   Here the apparatus of Fig. 32–2 is used. The coils are placed close together but at rest with respect to each other. When we close the switch *S*, thus setting up a steady current in the right-hand coil, the galvanometer deflects momentarily and then returns to zero; when we open the switch, thus interrupting this current, the galvanometer again deflects momentarily, but in the opposite direction. No gross objects are moving in this experiment.

Experiment shows that there will be an induced emf in the left coil of Fig. 32–2 whenever the current in the right coil is changing. It is the rate at which the current is changing and not the size of the current that is significant.

32-2   Faraday's Law of Induction

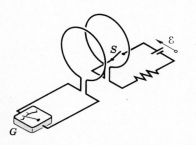

Figure 32–2   Galvanometer *G* deflects momentarily when switch *S* is closed or opened. No motion is involved.

Faraday had the insight to see that the induced emf in the two preceding experiments appeared only when the number of lines of **B** through the left-hand loop in Figs. 32–1 and 32–2 was changing. In the experiment of Fig. 32–1 the lines of **B** originate in the magnet and their number increases when we bring the magnet closer and decreases when we withdraw it. In the experiment of Fig. 32–2 the lines of **B** originate in the current set up in the right-hand circuit. The number of lines through the left-hand loop increases (from zero) when we close switch *S* and decreases (to zero) when we open it. The actual number of lines linking the left-hand loop at any moment is of no concern; it is the rate at which this number is changing that is related to the appearance of the induced emf.

We measure the number of magnetic lines that passes through a given surface by the magnetic flux Φ_B for that surface, where Φ_B is defined as

Magnetic flux

$$\Phi_B = \int \mathbf{B} \cdot d\mathbf{S}. \qquad (32–1a)$$

The integral is taken over the surface in question. If the magnetic field has a constant magnitude *B* and is everywhere at right angles to a plane surface of area *A* then Eq. 32–1*a* becomes simply

$$\Phi_B = BA. \qquad (32–1b)$$

This definition of the flux of the magnetic field **B** is precisely analagous to that of the flux of the electric field **E** given by Eq. 25–4.

From Eq. 32–1 we see that the SI unit for magnetic flux is the tesla·meter², to which we give the name *weber* (abbr. Wb). That is,

The weber

$$1 \text{ Wb} = 1 \text{ T·m}^2.$$

Faraday's law of induction says that the induced emf ε in a circuit is equal (except for a minus sign) to the rate at which the flux through the circuit is changing. If the rate of change of flux is in webers/second, the emf ε will be in volts. In equation form

Faraday's law of induction

$$\varepsilon = -\frac{d\Phi_B}{dt}. \qquad (32–2a)$$

This is *Faraday's law of induction*. The minus sign is an indication of the direction of the induced emf, a matter we discuss in Section 32–3.

If we apply Eq. 32–2*a* to a coil of *N* turns, an emf appears in every turn and these emfs are to be added. If the coil is so tightly wound that each turn can be said to occupy the same region of space, the flux through each turn will then be the same. The flux through each turn is also the same for (ideal) toroids and solenoids (see Section 31–6). The induced emf in all such devices is given by

$$\mathcal{E} = -N\frac{d\Phi_B}{dt} = -\frac{d(N\Phi_B)}{dt}, \qquad (32\text{-}2b)$$

where $N\Phi_B$ measures the so-called *flux linkages* in the device.

Figures 32–1 and 32–2 suggest that there are at least two ways in which we can make the flux through a circuit change and thus induce an emf in that circuit. The current through the galvanometer is the same for both experiments; the current measures only the change in magnetic flux through the coil no matter how the change occurs. The flux through a circuit can also be changed by changing its shape, that is, by squeezing or stretching it.

Example 1 The long solenoid in Fig. 32–3 has 200 turns/cm and carries a current of 1.5 A; its diameter is 3.0 cm. At its center we place a 100-turn, close-packed coil C of diameter 2.0 cm. This coil is arranged so that **B** at the center of the solenoid is parallel to its axis. The current in the solenoid is reduced to zero and then raised to 1.5 A in the other direction at a steady rate over a period of 50 ms. What induced emf appears in the coil while the current is being changed?

The field B at the center of the solenoid is given by Eq. 31–6, or

$B = \mu_0 ni = (4\pi \times 10^{-7}\,\text{T·m/A})(200 \times 10^2\,\text{turns/m})(1.5\,\text{A})$
$= 38\,\text{mT}.$

The area of the coil (not of the solenoid) is $3.1 \times 10^{-4}\,\text{m}^2$. The initial flux Φ_B through each turn of the coil is given by

$\Phi_B = BA = (38 \times 10^{-3}\,\text{T})(3.1 \times 10^{-4}\,\text{m}^2) = 12\,\mu\text{Wb}.$

The flux goes from an initial value of $+12\,\mu\text{Wb}$ to a final value of $-12\,\mu\text{Wb}$. The *change* in flux $\Delta\Phi_B$ for each turn of the

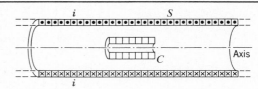

Figure 32–3 Example 1. A coil C inside a solenoid S, shown in cross section.

coil during the 50-ms period is thus twice the initial value. The induced emf is given by

$$\mathcal{E} = -\frac{N\Delta\Phi_B}{\Delta t} = -\frac{(100)(2 \times 12 \times 10^{-6}\,\text{Wb})}{50 \times 10^{-3}\,\text{s}}$$
$$= -4.8 \times 10^{-2}\,\text{V} = -48\,\text{mV}.$$

The minus sign deals with the direction of the emf, as we explain below.

32–3 Lenz's Law

So far we have not specified the directions of the induced emfs. We will now do so by using the conservation-of-energy principle which, in this context, takes the form of Lenz's law, deduced by Heinrich Friedrich Lenz (1804–1865) in 1834: *The induced current will appear in such a direction that it opposes the change that produced it.* The minus sign in our statement of Faraday's law suggests this opposition. In mechanics the energy principle often allows us to draw conclusions about mechanical systems without analyzing them in detail. We use the same approach here.

Lenz's law

Lenz's law refers to induced currents, which means that it applies only to closed conducting circuits. If the circuit is open, we can usually think in terms of what would happen if it were closed and in this way find the direction of the induced emf.

Consider the first of Faraday's experiments described in Section 32–1. Figure 32–4 shows the north pole of a magnet and a cross section of a nearby conducting loop. As we push the magnet toward the loop (or the loop toward the magnet) an induced current is set up in the loop. What is its direction?

A current loop sets up a magnetic field at distant points like that of a magnetic dipole, one side of the loop being a north pole, the opposite side being a south pole. The north pole, as for bar magnets, is that side from which the lines of **B** emerge. If, as Lenz's law predicts, the loop in Fig. 32–4 is to oppose the motion of the magnet toward it, the side of the loop toward the magnet must become a north pole. The two north poles—one of the current loop and one of the magnet—will repel each other. The right-hand rule shows that for the magnetic field set up by the loop to emerge from the right side of the loop, the induced current must be as shown. The current will be counterclockwise as we sight along the magnet toward the loop.

When we push the magnet toward the loop (or the loop toward the magnet), an induced current appears. In terms of Lenz's law this pushing is the "change" that produces the induced current, and, according to this law, the induced current will oppose the "push." If we pull the magnet away from the coil, the induced current will oppose the "pull" by creating a south pole on the right-hand side of the loop of Fig. 32–4. To make the right-hand side a south pole, the current must be opposite to that shown in Fig. 32–4. Whether we pull or push the magnet, its motion will always be automatically opposed.

If the current in Fig. 32–4 were in the opposite direction to that shown, the side of the loop toward the magnet would be a south pole, which would pull the bar magnet toward the loop. We would only need to push the magnet slightly to start the process and then the action would be self-perpetuating. The magnet would accelerate toward the loop, increasing its kinetic energy all the time. At the same time thermal energy would appear in the loop at a rate that would increase with time. This would indeed be a something-for-nothing situation! Needless to say, it does not occur.

The agent that causes the magnet to move, either toward the coil or away from it, will always experience a resisting force and will thus be required to do work. From the conservation-of-energy principle this work done on the system must appear as some other form of energy in the system. In this case it must be exactly equal to the internal or thermal energy produced in the coil, since these are the only two energy transfers that occur in the system. If we move the magnet more rapidly, we will have to do work at a greater rate and the rate of production of thermal energy will increase correspondingly. If we cut the loop and then perform the experiment, there will be no induced current, no thermal energy, no force on the magnet, and no work required to move it. There will still be an emf in the loop, but, like a battery connected to an open circuit, it will not set up a current.

Let us apply Lenz's law to Fig. 32–4 in a different way. Figure 32–5 shows the lines of **B** for the bar magnet.* On this point of view the "change" is the in-

Lenz's law and conservation of energy

Figure 32–4 If we move the magnet toward the loop, the induced current points as shown, setting up a magnetic field that opposes the motion of the magnet.

Figure 32–5 If we move the magnet toward the loop, we increase Φ_B through the loop.

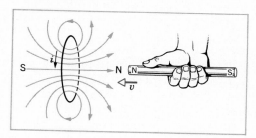
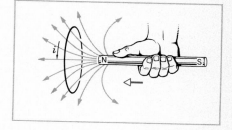

* There are two fields of **B** in this problem—one connected with the current loop and one with the bar magnet. You must always be certain which one is meant.

crease in Φ_B through the loop caused by bringing the magnet nearer. The induced current opposes this change by setting up a field that tends to oppose the increase in flux caused by the moving magnet. Thus the field due to the induced current must point from left to right through the plane of the coil, in agreement with our earlier conclusion.

It is not significant here that the induced field opposes the field of the magnet but rather that it opposes the change which in this case is the increase in Φ_B through the loop. If we withdraw the magnet, we reduce Φ_B through the loop. The induced field will now oppose this decrease in Φ_B (that is, the change) by reenforcing the magnet field. In each case the induced field opposes the change that gives rise to it.

32–4 Induction—A Quantitative Study

The example of Fig. 32–5, although easy to understand qualitatively, does not lend itself to quantitative calculations. Consider then Fig. 32–6, which shows a rectangular loop of wire of width l, one end of which is in a uniform field **B** pointing at right angles to the plane of the loop. This field **B** may be produced in the gap of a large electromagnet like that of Fig. 30–1. The dashed lines show the assumed limits of the magnetic field. The experiment consists in pulling the loop to the right at a constant speed v.

Note that the situation described by Fig. 32–6 does not differ in any essential particular from that of Fig. 32–5. In each case a conducting loop and a magnet are in relative motion; in each case the flux of the field of the magnet through the loop is being caused to change with time.

The flux Φ_B enclosed by the loop in Fig. 32–6 is

$$\Phi_B = Blx,$$

where lx is the area of that part of the loop in which B is not zero. We find the emf ε from Faraday's law, or

$$\varepsilon = -\frac{d\Phi_B}{dt} = -\frac{d}{dt}(Blx) = -Bl\frac{dx}{dt} = Blv, \tag{32–3}$$

where we have set $-dx/dt$ equal to the speed v at which the loop is pulled out of the magnetic field. Note that the only dimension of the loop that enters into Eq. 32–3 is the length l of the left end conductor.

Figure 32–6 A rectangular loop is pulled out of a magnetic field with velocity **v**.

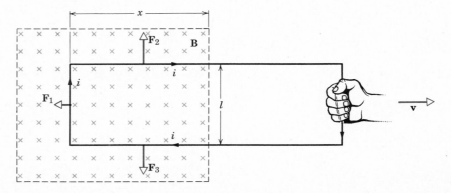

The emf Blv sets up a current in the loop given by

$$i = \frac{\varepsilon}{R} = \frac{Blv}{R}, \qquad (32\text{–}4)$$

where R is the loop resistance. From Lenz's law, this current (and thus ε) must be clockwise in Fig. 32–6; it opposes the "change" (the decrease in Φ_B) by setting up a field that is parallel to the external field within the loop.

The current in the loop will cause forces $\mathbf{F}_1$, $\mathbf{F}_2$, and $\mathbf{F}_3$ to act on the three conductors, in accord with Eq. 30–6a, or

$$\mathbf{F} = i\boldsymbol{\ell} \times \mathbf{B}. \qquad (32\text{–}5)$$

Because $\mathbf{F}_2$ and $\mathbf{F}_3$ are equal and opposite, they cancel each other; $\mathbf{F}_1$, which is the force that opposes our effort to move the loop, is given in magnitude from Eqs. 32–4 and 32–5 as

$$F_1 = ilB \sin 90° = \frac{B^2 l^2 v}{R}.$$

The agent that pulls the loop must do work at the steady rate of

$$P = F_1 v = \frac{B^2 l^2 v^2}{R}. \qquad (32\text{–}6)$$

While the agent is pulling the loop, a current given by Eq. 32–4 is set up. Thus thermal energy must appear in the loop because of the Joule effect described in Section 28–5. We can calculate the power dissipated in this way from

$$P = i^2 R. \qquad [28\text{–}16]$$

Inserting the value of i from Eq. 32–4 yields

$$P = \left(\frac{Blv}{R}\right)^2 R = \frac{B^2 l^2 v^2}{R},$$

which is just the same as the power supplied by the external agent; compare Eq. 32–6. Thus mechanical energy (the work done by the agent) is transformed into electrical energy (associated with the induced emf) and finally into thermal energy (manifested by a small temperature rise of the loop).

Figure 32–7 shows a side view of the loop in the field. In Fig. 32–7a the loop is stationary; in Fig. 32–7b we are moving it to the right; in Fig. 32–7c we are moving it to the left. The lines of $\mathbf{B}$ in these figures represent the resultant field produced by the vector addition of the field $\mathbf{B}_0$ due to the magnet and the field $\mathbf{B}_i$ due to the induced current, if any, in the loop. These lines suggest that the agent moving the loop always experiences an opposing force.

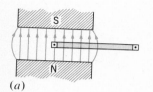

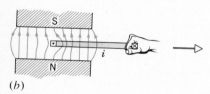

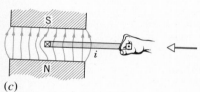

Figure 32–7 Side view of a rectangular loop in a magnetic field showing the loop (a) at rest, (b) being pulled out, and (c) being pushed in.

Example 2 Figure 32–8a shows a rectangular loop of resistance R, width l, and length a being pulled at constant velocity $\mathbf{v}$ through a region of thickness d in which a uniform magnetic field $\mathbf{B}$ is set up by a magnet.

(a) Plot the flux Φ_B through the loop as a function of the loop position x. Assume that $l = 40$ mm, $a = 10$ cm, $d = 15$ cm, $R = 16\ \Omega$, $B = 2.0$ T, and $v = 1.0$ m/s.

The flux Φ_B is zero when the loop is not in the field; it is Bla when the loop is entirely in the field; it is Blx when the loop is entering the field and $Bl[a - (x - d)]$ when the loop is leaving the field. These conclusions, which you should verify, are shown graphically in Fig. 32–8b.

(b) Plot the induced emf ε.

The induced emf ε is given by $\varepsilon = -d\Phi_B/dt$, which we can write as

$$\varepsilon = -\frac{d\Phi_B}{dt} = -\frac{d\Phi_B}{dx}\frac{dx}{dt} = -\frac{d\Phi_B}{dx}v,$$

where $d\Phi_B/dx$ is the slope of the curve of Fig. 32–8b. $\varepsilon(x)$ is plotted in Fig. 32–8c. Lenz's law, from the same type of rea-

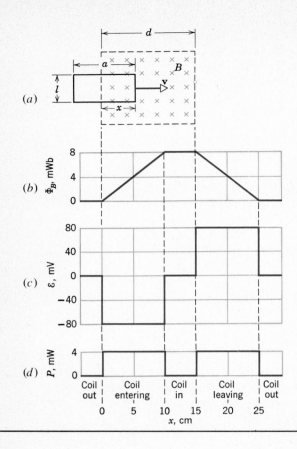

(a)

(b)

(c)

(d)

Coil out | Coil entering | Coil in | Coil leaving | Coil out

Figure 32–8 Example 2.

soning as that used for Fig. 32–6, shows that when the coil is entering the field, the emf ε acts counterclockwise as seen from above. Note that there is no emf when the coil is entirely in the magnetic field because the flux Φ_B through the coil is not changing with time, as Fig. 32–8b shows.

(c) Plot the rate P of thermal energy production in the loop.

This is given by $P = \varepsilon^2/R$. It may be calculated by squaring the ordinate of the curve of Fig. 32–8c and dividing by R. The result is plotted in Fig. 32–8d.

If the fringing of the magnetic field, which cannot be avoided in practice (see Problem 14, Chapter 31), is taken into account, the sharp bends and corners in the curves of Fig. 32–8 will be replaced by smooth curves. What changes would occur in the curves of Fig. 32–8 if the coil were open circuited?

32–5 Induced Electric Fields

Electric field due to a changing magnetic field

So far we have considered emfs induced by the relative motion of magnets and coils, as in the first experiment of Section 32–1. In this section we assume that there is no physical motion of gross objects but that the magnetic field may vary with time, much as in the second experiment of that section. If we place a conducting loop in such a time-varying field, the flux through the loop will change and an induced emf will appear in the loop. This emf will set the charge carriers in motion, that is, it will induce a current.

We can say, equally well, that the changing flux of **B** sets up an induced electric field **E** around the loop. This induced electric field is just as real as electric fields set up by static charges and will exert a force **F** on a test charge q_0 given by $\mathbf{F} = q_0\mathbf{E}$. Thus we can restate Faraday's law of induction in a loose but informative way as: *A changing magnetic field produces an electric field.*

To fix these ideas, consider Fig. 32–9, which shows a uniform magnetic field **B** at right angles to the plane of the page. We assume that **B** is increasing in magnitude at the same constant rate dB/dt at every point. This could be done by causing the current in the windings of the electromagnet that establishes the field to increase with time in the proper way.

The circle of arbitrary radius r, concentric with the circular pole face of the magnet as shown in Fig. 32–9, encloses, at any instant, a flux Φ_B. Because this flux is changing with time, an induced emf given by $\varepsilon = -d\Phi_B/dt$ will appear around the loop. The electric fields **E** induced at various points of the loop must,

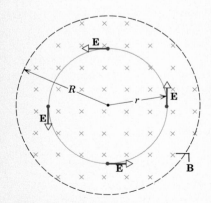

Figure 32–9 The induced electric field at four points produced by an increasing magnetic field. The magnetic field is produced by an electromagnet whose pole faces are circular, of effective radius R.

from symmetry, be tangent to the loop. Thus the electric lines of force that are set up by the changing magnetic field are in this case concentric circles.

If we consider a test charge q_0 moving around the circle of Fig. 32-9, the work W done on it per revolution is, in terms of our definition of an emf, simply εq_0. From another point of view it is $(q_0 E)(2\pi r)$, where $q_0 E$ is the force that acts on the charge and $2\pi r$ is the distance over which the force acts. Setting the two expressions for W equal and canceling q_0 yields

$$\varepsilon = E2\pi r. \tag{32-7}$$

In a more general case than that of Fig. 32-9 we must write

$$\varepsilon = \oint \mathbf{E} \cdot d\boldsymbol{\ell}. \tag{32-8}$$

If this integral is evaluated for the conditions of Fig. 32-9, we obtain Eq. 32-7 at once. If Eq. 32-8 is combined with Eq. 32-2 ($\varepsilon = -d\Phi_b/dt$), we can write Faraday's law of induction as

An integral form of Faraday's law

$$\oint \mathbf{E} \cdot d\boldsymbol{\ell} = -\frac{d\Phi_B}{dt}, \tag{32-9}$$

which is the form in which this law is expressed in Table 37-2.

Example 3 Let B in Fig. 32-9 be increasing at the rate dB/dt. Let R be the effective radius of the cylindrical region in which the magnetic field is assumed to exist.* What is the magnitude of the electric field $\mathbf{E}$ at any radius r? Assume that $dB/dt = 0.10$ T/s and $R = 10$ cm.

(a) For $r < R$, the flux Φ_B through the loop is

$$\Phi_B = B(\pi r^2).$$

Substituting into Faraday's law (Eq. 32-9), yields

$$(E)(2\pi r) = -\frac{d\Phi_B}{dt} = -(\pi r^2)\frac{dB}{dt}.$$

Solving for E yields

$$E = -\tfrac{1}{2}r\frac{dB}{dt}.$$

The minus sign is retained to suggest that the induced electric field $\mathbf{E}$ acts to oppose the change of the magnetic field. Note that $E(r)$ depends on dB/dt and not on B. Substituting numerical values, assuming $r = 5.0$ cm, yields, for the magnitude of E,

$$E = \tfrac{1}{2}r\frac{dB}{dt} = (\tfrac{1}{2})(50 \times 10^{-3} \text{ m})(0.10 \text{ T/s}) = 2.5 \text{ mV/m}.$$

(b) For $r > R$ the flux through the loop is

$$\Phi_B = \int \mathbf{B} \cdot d\mathbf{S} = B(\pi R^2).$$

This equation is true because $\mathbf{B} \cdot d\mathbf{S}$ is zero for those points of the loop that lie outside the effective boundary of the magnetic field.

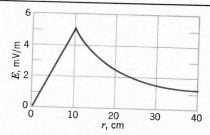

Figure 32-10 Example 3. If the fringing of the field in Fig. 32-9 were to be taken into account, the result would be a rounding of the sharp cusp at $r = R$ (= 10 cm).

From Faraday's law (Eq. 32-9),

$$(E)(2\pi r) = -\frac{d\Phi_B}{dt} = -(\pi R^2)\frac{dB}{dt}.$$

Solving for E yields

$$E = -\frac{1}{2}\frac{R^2}{r}\frac{dB}{dt}.$$

These two expressions for $E(r)$ yield the same result, as they must, for $r = R$. Figure 32-10 is a plot of the magnitude of $\mathbf{E}(r)$ for the numerical values given.

---

* The magnetic field in Fig. 32-9 cannot drop from a uniform value precipitously to zero but must do so gradually; see Problem 36, Chapter 31. The figure is idealized in that R represents an effective radius for the limit of the magnetic field.

In applying Lenz's law to the situation depicted in Fig. 32–9, imagine that a circular conducting loop is placed concentrically in the field. Since Φ_B through this loop is increasing, the induced current in the loop will tend to oppose this "change" by setting up a magnetic field of its own that points up within the loop. Thus the induced current i must be counterclockwise, which means that the lines of the induced electric field $\mathbf{E}$, which is responsible for the current, must also be counterclockwise. If the magnetic field in Fig. 32–9 were decreasing with time, the induced current and the lines of force of the induced electric field $\mathbf{E}$ would be clockwise, again opposing the change in Φ_B.

Figure 32–11 shows several of the lines of $\mathbf{E}$ in Fig. 32–9 together with four of many possible loops to which Faraday's law may be applied. For loops 1 and 2, the induced emf ε is the same because these loops lie entirely within the changing magnetic field and, having the same area, thus have the same value of $d\Phi_B/dt$. Note that even though the emf ε ($=\oint \mathbf{E}\cdot d\boldsymbol{\ell}$) is the same for these two loops, the distribution of electric fields $\mathbf{E}$ around the perimeter of each loop, as indicated by the electric lines of force, is different. For loop 3 the emf is less because Φ_B and thus $d\Phi_B/dt$ for this loop are less, and for loop 4 the induced emf is zero.

The induced electric fields that are set up by the induction process are not associated with static charges but with a changing magnetic flux. Although both kinds of electric fields exert forces on charges, there is a difference between them. The simplest manifestation of this difference is that lines of $\mathbf{E}$ associated with a changing magnetic flux can form closed loops (see Fig. 32–11); lines of $\mathbf{E}$ associated with static charges cannot form closed loops but can always be drawn to start on a positive charge and end on a negative charge.

Equation 26–5, which defined the potential difference between two points a and b, is

$$V_b - V_a = \frac{W_{ab}}{q_0} = -\int_a^b \mathbf{E}\cdot d\boldsymbol{\ell}.$$

We have insisted that if potential is to have any useful meaning this integral (and W_{ab}) must have the same value for every path connecting a and b. This proved to be true for every case examined in earlier chapters. If a and b are the same point the path connecting them becomes a closed loop; V_a must be identical with V_b and this equation reduces to

$$\oint \mathbf{E}\cdot d\boldsymbol{\ell} = 0. \qquad (32\text{–}10)$$

However, when changing magnetic flux is present, $\oint \mathbf{E}\cdot d\boldsymbol{\ell}$ is not zero but is, according to Faraday's law (see Eq. 32–9), $-d\Phi_B/dt$. Electric fields associated with stationary charges are conservative but those associated with changing magnetic fields are nonconservative; see Section 8–2. Electric potential, which can be defined only for a conservative force, has no meaning for electric fields produced by induction.

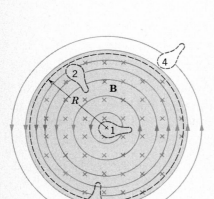

Figure 32–11 Showing the circular lines of $\mathbf{E}$ from an increasing magnetic field. The four loops are imaginary paths around which an emf can be calculated.

The nonconservative nature of induced electric fields

32–6 The Betatron

The betatron is a device used to accelerate circulating electrons to high energies by allowing them to be acted upon by electric fields induced by a changing central magnetic flux. Although the betatron is not widely used today, we describe it largely because it is an ideal example of the "reality" of induced electric fields. Also, like the cyclotron described in Section 30–7, it played a vital role as the forerunner of today's large accelerators such as the Fermilab proton synchrotron pictured on p. 592.

Figure 32–12 shows a vertical cross section through the central part of a betatron. The (time-varying) magnetic field that is depicted there has several functions: (a) it guides the electrons in a circular path; (b) the changing magnetic field generates an electric field that accelerates the electrons in this path; (c) it keeps the radius of the orbit in which the electrons are moving essentially constant; (d) it introduces the electrons into the orbit initially and removes them from the orbit after they have reached full energy; and finally (e) it provides a restoring force that resists any tendency for the electrons to leave their orbit, either vertically or radially. It is remarkable that it is possible to do all these things by proper shaping and control of the magnetic field.

The object marked D in Fig. 32–12 is an evacuated ceramic "doughnut," inside which the electrons circulate and are accelerated. Their orbit is a circle of constant radius R, its plane being at right angles to the page. In the figure the electrons are circulating counter clockwise as viewed from above. Thus we see them emerging from the figure on the left ($\cdot$) and entering it ($\times$) on the right.

The alternating magnetic field shown in Fig. 32–12 is produced by establishing an alternating current in coils (not shown) around the iron pole pieces. The field at the orbit position $\mathbf{B}_{\mathrm{orb}}$ serves to guide the electrons in a stable orbit. The field in the area encompassed by the orbit, whose average value is $\overline{B}$, contributes to the central flux Φ_B. It is the time variation of this flux that induces the electric fields that act on the electron and accelerate it.

It can be shown that, if the radius of the orbit is to remain constant while the electrons accelerate, the magnet must be designed so that $B = 2\,B_{\mathrm{orb}}$ at all times during the acceleration process. That is, the average value of the magnetic field encompassed by the orbit must always be twice as large as the magnetic field at the orbit; see Problem 16.

Figure 32–12 The central region of a betatron, showing the orbit of the accelerating electrons and a "snap shot" of the time-varying magnetic field at an instant during the accelerating interval.

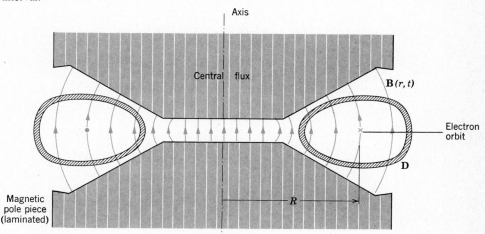

Example 4 In a 100-MeV betatron built at the General Electric Company the orbit radius R is 84 cm. The magnetic field in the central region encompassed by the orbit rises periodically (60 times per second) from zero to a maximum average value $\bar{B}_{max}$ of 0.80 T in an accelerating interval of 4.2 ms. How much energy does the electron gain in one trip around its orbit in this changing flux?

The central flux rises during the accelerating interval from zero to a maximum value of

$$\Phi_{B,max} = (\bar{B}_{max})(\pi R^2) = (0.80 \text{ T})(\pi)(0.84\text{m})^2 = 1.8 \text{ Wb}.$$

The average value of $d\Phi_B/dt$ during the accelerating interval is then

$$\left(\frac{d\Phi_B}{dt}\right)_{av} = \frac{1.8 \text{ Wb}}{4.2 \times 10^{-3} \text{ s}} = 430 \text{ Wb/s}.$$

From Faraday's law (Eq. 32–2a) this is also the average emf in volts. Thus the electron increases its energy by an average of 430 eV per revolution in this changing flux. To achieve its full final energy of 100 MeV the electron must therefore make about 230,000 revolutions during the acceleration process, a total path length of about 1200 km.

REVIEW GUIDE AND SUMMARY

Definition of magnetic flux

The flux Φ_B for a given surface immersed in a magnetic field **B** is defined from

$$\Phi_B = \int \mathbf{B} \cdot d\mathbf{S} \qquad [32\text{–}1a,b]$$

where the integral is taken over the surface. Review Section 25–2, which deals with the corresponding flux Φ_E for a surface immersed in an electric field **E**. The SI unit of magnetic flux is the weber, where 1 Wb = 1 T·m².

Faraday's law of induction, in its somewhat restricted form, states that if Φ_B for a surface bounded by a closed loop changes with time an emf given by

Faraday's law of induction

$$\varepsilon = -N \frac{d\Phi_B}{dt} \qquad [32\text{–}2a, b]$$

appears in the loop. If the loop is a conductor, a current will be set up. The two experiments of Section 32–1 and Examples 1 and 2 illustrate this law.

An induced emf will be present even if the loop through which the magnetic flux is changing is not a physical conductor but a mathematical construct. The changing flux induces an electric field **E** at every point of such a loop, the emf being related to **E** by

The emf and the induced electric field

$$\varepsilon = \oint \mathbf{E} \cdot d\boldsymbol{\ell}. \qquad [32\text{–}8]$$

The integral is taken around the loop.

Combining Eqs. 32–2 and 32–8 lets us write Faraday's law in its most general form, or

Faraday's law of induction—integral form

$$\oint \mathbf{E} \cdot d\boldsymbol{\ell} = -\frac{d\Phi_B}{dt}. \qquad [32\text{–}9]$$

The essence of this law is that *a changing magnetic field* $(d\Phi_B/dt)$ *induces an electric field* (**E**). Example 3 illustrates the law in this form; Figs. 32–10 and 32–11 show the induced electric fields.

Lenz's law

Lenz's law specifies the direction of the current induced in a closed conducting loop by a changing magnetic flux. The law states: *An induced current will flow in a direction such that it opposes the change that produced it.* If no conducting loop is present it is possible to imagine one and thus to deduce the direction of the induced electric fields around the actual loop. Study the applications of this law illustrated by Figs. 32–4 and 32–5 and by Examples 2 and 3.

Lenz's law and the conservation of energy

Lenz's law is a consequence of the conservation of energy principle. Section 32–4 for example shows that work is needed to pull a closed conducting loop out of a magnetic field and that this energy is accounted for as thermal energy of the loop material. If the direction of the induced current were opposite to that predicted by Lenz's law, work would be done *by* the loop and energy would not be conserved.

Induced electric fields

Electric fields induced by changing magnetic fields differ from those associated with static charges in that their lines of force form closed loops, as in Fig. 32–11. Electric potential has no meaning for such induced electric fields; the forces associated with them are nonconservative (see Section 8–2).

The betatron

The betatron (Fig. 32–12) uses induced electric fields to accelerate electrons to high energy. Example 4 shows how to use Faraday's law of induction to calculate the average energy gain per revolution of the orbit for a particular machine.

QUESTIONS

1. Are induced emfs and currents different in any way from emfs and currents provided by a battery connected to a conducting loop?

2. Can you explain in your own words the difference between a magnetic field **B** and the flux of a magnetic field Φ_B? Are they vectors or scalars? In what units may each be expressed? How are these units related? Are either or both (or neither) properties of a given point in space?

3. Can a charged particle at rest be set in motion by the action of a magnetic field? If not, why not? If so, how?

4. You drop a bar magnet along the axis of a long copper tube. Describe the motion of the magnet and the energy interchanges involved. Neglect air resistance.

5. You are playing with a metal loop, moving it back and forth in a magnetic field, as in Fig. 32–7. How can you tell, without detailed inspection, whether or not the loop has a narrow saw cut across it, rendering it nonconducting?

6. Figure 32–13 shows an inclined wooden track that passes, for part of its length, through a strong magnetic field. You roll a copper disk down the track. Describe the motion of the disk as it rolls from the top of the track to the bottom.

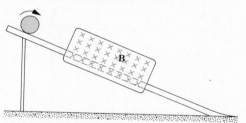

Figure 32–13 Question 6.

7. Figure 32–14 shows a copper ring, hung from the ceiling by two threads. Describe in detail how you might most effectively use a bar magnet to get this ring to swing back and forth.

Figure 32–14 Question 7.

8. Two conducting loops face each other a distance d apart (Fig. 32–15). An observer sights along their common axis from left to right. If a clockwise current i is suddenly established in the larger loop, by a battery not shown, (a) what is the direction of the induced current in the smaller loop? (b) What is the direction of the force (if any) that acts on the smaller loop?

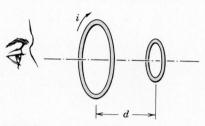

Figure 32–15 Question 8.

9. What is the direction of the induced emf in coil Y of Fig. 32–16 (a) when coil Y is moved toward coil X? (b) When the current in coil X is decreased, without any change in the relative positions of the coils?

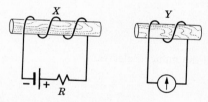

Figure 32–16 Question 9.

10. The north pole of a magnet is moved away from a copper ring, as in Fig. 32–17. In the part of the ring farthest from the reader, which way does the current point? Lines of **B** emerge from a magnetic north pole.

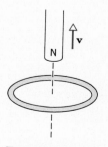

Figure 32–17 Question 10.

11. A short solenoid carrying a steady current is moving toward a conducting loop as in Fig. 32–18. What is the direction of the induced current in the loop as one sights toward it as shown?

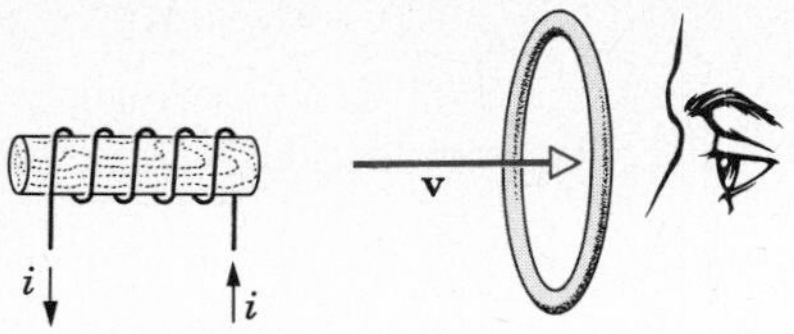

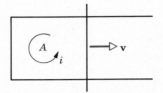

Figure 32–18   Question 11.

12. The resistance R in the left-hand circuit of Fig. 32–19 is being increased at a steady rate. What is the direction of the induced current in the right-hand circuit?

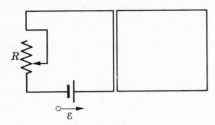

Figure 32–19   Question 12.

13. What is the direction of the induced current through resistor R in Fig. 32–20 (*a*) immediately after switch S is closed, (*b*) some time after switch S was closed, and (*c*) immediately after switch S is opened. (*d*) When switch S is held closed, from which end of the longer coil do field lines emerge? This is the effective north pole of the coil. (*e*) How do the conduction electrons in the coil containing R know about the flux within the long coil? What really gets them moving?

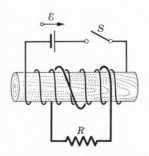

Figure 32–20   Question 13.

14. In Faraday's law of induction, does the induced emf depend on the resistance in the circuit?

15. Suppose that the direction of induced emfs was governed by what we can call the Antilenz law: The induced current will appear in such a direction that it aids the change that produced it. Design a machine based on this law that would make a lot of money for you. (Alas, the Antilenz law is *false.*)

16. In Fig. 32–21 the straight movable wire segment is moving to the right with a constant velocity **v**. An induced current appears in the direction shown. What is the direction of the uniform magnetic field (assumed constant and perpendicular to the page) in region A?

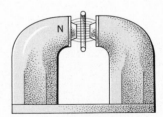

Figure 32–21   Question 16.

17. A conducting loop, shown in Fig. 32–22, is removed from the permanent magnet by pulling it vertically upward. (*a*) What is the direction of the induced current? (*b*) Is a force required to remove the loop? (*c*) Does the total amount of thermal energy produced in removing the loop depend on the time taken to remove it?

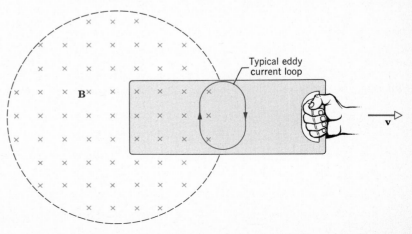

Figure 32–22   Question 17.

18. A plane closed loop is placed in a uniform magnetic field. In what ways can the loop be moved without inducing an emf? Consider motions of both translation and rotation.

19. *Eddy currents.* A sheet of copper is placed in a magnetic field as shown in Fig. 32–23. If we attempt to pull it out of the field or

Figure 32–23   Question 19.

push it further in, a resisting force automatically appears. Explain its origin. (*Hint:* Currents, called eddy currents, are induced in the sheet in such a way as to oppose the motion.)

20. *Magnetic damping.* A strip of copper is mounted as a pendulum about *O* in Fig. 32–24. It is free to swing through a magnetic field normal to the page. If the strip has slots cut in it as shown, it can swing freely through the field. If a strip without slots is substituted, the vibratory motion is strongly damped. Explain. (*Hint:* Use Lenz's law; consider the paths that the charge carriers in the strip must follow if they are to oppose the motion.)

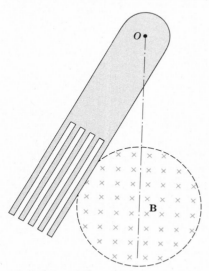

Figure 32–24   Question 20.

21. *Electromagnetic shielding.* Consider a conducting sheet lying in a plane perpendicular to a magnetic field **B**, as shown in Fig. 32–25. (*a*) If **B** suddenly changes, the full change in **B** is not immediately detected at points near *P*. Explain. (*b*) If the resistivity of the sheet is zero, the change is not ever detected at *P*. Explain. (*c*) If **B** changes periodically at high frequency and the conductor is made of a material of low resistivity, the region near *P* is almost completely shielded from the changes in flux. Explain. (*d*) Is such a conductor useful as a shield from static magnetic fields? Explain.

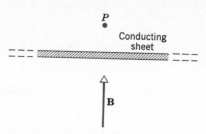

Figure 32–25   Question 21.

22. (*a*) In Fig. 32–9, need the circle of radius *r* be a conducting loop in order that **E** and ε be present? (*b*) If the circle of radius *r* were not concentric (moved slightly to the left, say), would ε change? Would the configuration of **E** around the circle change? (*c*) For a concentric circle of radius *r*, with $r > R$, does an emf exist? Do electric fields exist?

23. A copper ring and a wooden ring of the same dimensions are placed so that there is the same changing magnetic flux through each. How do the induced electric fields in each ring compare?

24. In Fig. 32–11 how can the induced emfs around paths 1 and 2 be identical? The induced electric fields are much weaker near path 1 than near path 2, as the spacing of the lines of force shows. See also Fig. 32–10.

25. In the betatron of Fig. 32–12 are the directions of the lines of **B** correctly drawn to be consistent with the direction of circulation shown for the electrons?

26. In the betatron of Fig. 32–12 must the magnetic field associated with the central flux be increasing or decreasing if the electrons are to be accelerated rather than decelerated?

27. In the betatron of Fig. 32–12 you want to increase the orbit radius by suddenly imposing an additional central flux $\Delta\Phi_B$ (set up by suddenly establishing a current in an auxilliary coil not shown). Should the lines of **B** associated with this flux increment be in the same direction as the lines shown in the figure or in the opposite direction? Assume that the magnetic field at the orbit position remains relatively unchanged by this flux increment.

28. In the betatron of Fig. 32–12 why is the iron core of the magnet made of laminated sheets rather than of solid metal as for the cyclotron of Section 30–7? (*Hint:* Consider the implications of Question 19.)

EXERCISES

Section 32–2   Faraday's Law of Induction

1. Express the dimensions of (*a*) the magnetic field **B** and (*b*) the magnetic flux Φ_B in terms of mass, length, time, and current (M, L, T, I).

2. A small loop of area *A* is inside of, and has its axis in the same direction as, a long solenoid of *n* turns per unit length and current *i*. If $i = i_o \sin \omega t$, find the emf in the loop.

3. A circular UHF television antenna has a diameter of 11.0 cm. The magnetic field of a TV signal is normal to the plane of the loop and, at one instant of time, its magnitude is changing at the rate 0.145 T/s. The field is uniform. What is the emf in the antenna?

4. A uniform magnetic field **B** is perpendicular to the plane of a circular wire loop of radius *r*. The magnitude of the field varies with time according to $B = B_o e^{-t/\tau}$ where B_o and τ are constants. Find the emf in the loop as a function of time.

5. A circular loop of wire 10 cm in diameter is placed with its normal making an angle of 30° with the direction of a uniform 0.50-T magnetic field. The loop is "wobbled" so that its normal rotates in a cone about the field direction at the constant rate of 100 rev/min; the angle between the normal and the field direction ($=30°$) remains unchanged during the process. What emf appears in the loop?

Section 32-3 Lenz's Law

6. (a) What would be the direction of the current induced in the loop of Fig. 32–26 if the current i in the long wire is increased from zero to 90 A in 15 ms? (b) What would be the direction if the current i in the long wire were then reduced back to zero during the next 30 ms?

Figure 32–26 Exercise 6.

7. A short bar magnet is pulled at constant speed through a conducting loop, along its axis. Assume that the north pole of the magnet enters the loop first. Sketch qualitatively (a) the induced current and (b) the rate of thermal energy production as a function of the position of the center of the magnet. Plot the induced current as positive if it is clockwise as viewed along the path of the magnet.

8. The magnetic field through a one-turn loop of wire 10 cm in radius and 10 Ω in resistance changes with time as shown in Fig. 32–27. Plot (a) the emf in the loop and (b) the rate of thermal energy production in the loop, each as a function of time. The (uniform) magnetic field is at right angles to the plane of the loop.

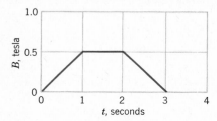

Figure 32–27 Exercise 8.

9. A circular loop moves with constant velocity through regions where uniform magnetic fields of the same magnitude are directed into or out of the plane of the page, as indicated in Fig. 32–28. At which of the seven indicated positions will the emf be (a) clockwise? (b) counterclockwise? (c) zero?

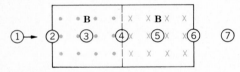

Figure 32–28 Exercise 9.

10. The loop of wire shown in Fig. 32–29 rotates with constant angular speed about the x-axis. A uniform magnetic field **B**, whose direction is that of the positive y-axis is present. For what portions of the rotation is the induced current in the loop (a) from P to Q,

(b) from Q to P, (c) zero? (d) Repeat if the sense of rotation is reversed from that shown in the figure.

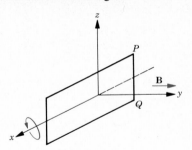

Figure 32–29 Exercise 10.

Section 32-4 Induction: A Quantitative Study

11. A metal rod moves with constant velocity along two parallel metal rails, connected with a strip of metal at one end, as shown in Fig. 32–30. A magnetic field $B = 0.35$ T points out of the page. (a) If the rails are separated by 25 cm and the speed of the rod is 55 cm/s, what emf is generated? (b) If the rod has a resistance of 18 Ω and the rails have negligible resistance, what is the current in the rod?

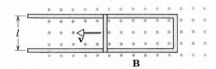

Figure 32–30 Exercises 11 and 13.

12. An airliner is cruising in level flight over Alaska, where the earth's magnetic field has a large downward component. Which of its wingtips (right or left) has more electrons than the other?

13. Figure 32–30 shows a conducting rod of length l being pulled along horizontal, frictionless, conducting rails at a constant velocity v. A uniform vertical magnetic field B fills the region in which the rod moves. Assume that $l = 10$ cm, $v = 5.0$ m/s, and $B = 1.2$ T. (a) What is the induced emf in the rod? (b) What is the current in the conducting loop? Assume that the resistance of the rod is 0.40 Ω and that the resistance of the rails is negligibly small. (c) At what rate is thermal energy being generated in the rod? (d) What force must be applied to the rod by an external agent to maintain its motion? (e) At what rate does this external agent do work on the rod? Compare this answer with the answer to (c).

Section 32-5 Induced Electric Fields

14. A long solenoid has a diameter of 12 cm. When a current i is passed through its windings, a uniform magnetic field $B = 30$ mT is produced in its interior. By decreasing i, the field is caused to decrease at the rate 6.5 mT/s. Calculate the magnitude of the induced electric field 2.0 cm from the axis of the solenoid.

15. Figure 32–31 shows two circular regions R_1 and R_2 with radii $r_1 = 20$ cm and $r_2 = 30$ cm, respectively. In R_1 there is a uniform magnetic field $B_1 = 50$ mT into the page and in R_2 there is a uniform magnetic field $B_2 = 75$ mT out of the page (ignore any fringing of these fields). Both fields are decreasing at the rate 8.5 mT/s. Calculate the integral $\oint \mathbf{E} \cdot d\ell$ for each of the three indicated paths.

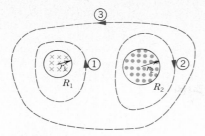

Figure 32–31   Exercise 15.

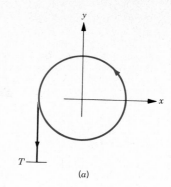

(a)

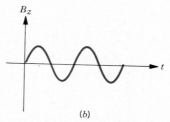

(b)

Figure 32–32   Exercise 16.

Section 32–6   The Betatron

16. Figure 32–32a shows a top view of the electron orbit in a betatron. Electrons are accelerated in a circular orbit in the x-y plane and then withdrawn to strike the target T. The magnetic field **B** is along the z-axis (the positive z-axis is out of the page). The magnetic field B_z along this axis varies sinusoidally as shown in Fig. 32–32b. Recall that the magnetic field must (i) guide the electrons in their circular path, and (ii) generate the electric field which accelerates the electrons. Which quarter cycle(s) in Fig. 32–32b are suitable (a) according to (i), (b) according to (ii), (c) for operation of the betatron?

PROBLEMS GROUPED BY SECTION

Section 32–2   Faraday's Law of Induction

1. A toroid having a 5.0-cm square cross section and an inside radius of 15 cm has 500 turns of wire and carries a current of 0.80 A. What is the magnetic flux through the cross section? (*Hint:* See Section 31–6.)

2. The current in the solenoid of Example 1 changes, not as in that example, but according to $i = 3.0t + 1.0t^2$, where i is in amperes and t is in seconds. (a) Plot the induced emf in the coil from $t = 0$ to $t = 4$ s. (b) The resistance of the coil is 0.15 Ω. What is the current in the coil at $t = 2.0$ s?

3. In Fig. 32–33 a 100-turn coil C of resistance 5.0 Ω is placed outside a solenoid S like that of Example 1. If the current in the solenoid is changed as in that example, (a) what current appears in the loop while the solenoid current is being changed? (b) How do the conduction electrons in the loop "get the message" from the solenoid that they should move to establish a current? After all, the magnetic field is entirely confined to the interior of the solenoid.

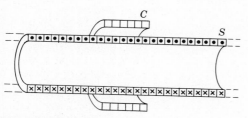

Figure 32–33   Problem 3.

4. A long solenoid with a radius of 25 mm has 100 turns/cm. A single loop of wire of radius 5.0 cm is placed around the solenoid,

the axis of the loop and the solenoid coinciding. The current in the solenoid is reduced from 1.0 to 0.50 A at a uniform rate over a time interval of 10 ms. What emf appears in the loop?

Section 32–3   Lenz's Law

5. In Fig. 32–34 the magnetic flux through the loop shown increases according to the relation

$$\Phi_B = 6t^2 + 7t,$$

where Φ_B is in milliwebers and t is in seconds. (a) What is the magnitude of the emf induced in the loop when $t = 2.0$ s? (b) What is the direction of the current through R?

Figure 32–34   Problems 5 and 8.

6. A closed loop of wire consists of a pair of equal semicircles, radius 3.7 cm, lying in mutually perpendicular planes. The loop was formed by folding a circular loop along a diameter until the two halves became perpendicular. A uniform magnetic field **B** of magnitude 76 mT is directed perpendicular to the fold diameter and makes equal angles (=45°) with the planes of the semicircles as shown in Fig. 32–35. The magnetic field is reduced at a uniform rate to zero during a time interval 4.5 ms. Determine the magnitude of the induced emf and the sense of the induced current in the loop during this interval.

Figure 32–35   Problem 6.

7.   A circular loop 10 cm in radius is placed with its plane at right angles to a uniform 0.80-T magnetic field. By a mechanism not specified the radius of the loop is caused to shrink at an instantaneous rate of 80 cm/s. What emf is induced in the loop at that instant?

8.   In Fig. 32–34 let the flux for the loop be $\Phi_B(0)$ at time $t = 0$. Then let the magnetic field **B** vary in a continuous but unspecified way, in both magnitude and direction, so that at time t the flux is represented by $\Phi_B(t)$. (*a*) Show that the net charge $q(t)$ that has passed through resistor R at time t is

$$q(t) = \frac{1}{R}[\Phi_B(0) - \Phi_B(t)],$$

independent of the way **B** has changed. (*b*) If $\Phi_B(t) = \Phi_B(0)$ in a particular case we have $q(t) = 0$. Is the induced current necessarily zero throughout the interval $0 \rightarrow t$?

9.   Figure 32–36 shows two parallel loops of wire having a common axis. The smaller loop (radius r) is above the larger loop (radius R) by a distance $x \gg R$. Consequently the magnetic field, due to the current i in the larger loop, is nearly constant throughout the smaller loop. Suppose that x is increasing at the constant rate $dx/dt = v$. (*a*) Determine the magnetic flux across the area bounded by the smaller loop as a function of x. (*b*) Compute the emf generated in the smaller loop. (*c*) Determine the direction of the induced current in the smaller loop. (*Hint:* See Example 6, Chapter 31.)

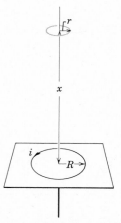

Figure 32–36   Problem 9.

Section 32–4   Induction: A Quantitative Study

10.   Two straight conducting rails form a right angle where their ends are joined. A conducting bar in contact with the rails starts at the vertex at time $t = 0$ and moves with a constant velocity of 5.2 m/s to the right, as shown in Fig. 32–37. A 0.35 T magnetic field points out of the page. Calculate (*a*) the flux through the triangle formed by the rails and bar at $t = 3.0$ s and (*b*) the emf around the triangle at that time. (*c*) In what manner does the emf around the triangle vary with time?

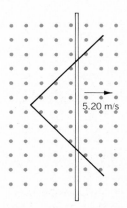

Figure 32–37   Problem 10.

11.   In Fig. 32–38 a conducting rod of mass m and length l slides without friction on two long horizontal rails. A uniform vertical magnetic field **B** fills the region in which the rod is free to move. The generator G supplies a constant current i that flows down one rail, across the rod, and back to the generator along the other rail. Find the velocity of the rod as a function of time, assuming it to be at rest at $t = 0$.

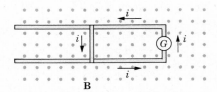

Figure 32–38   Problems 11 and 22.

*12.   A rod of length l, mass m, and resistance R slides without friction down parallel conducting rails of negligible resistance, as in Fig. 32–39. The rails are connected together at the bottom as shown, forming a conducting loop with the rod as the top member. The plane of the rails makes an angle θ with horizontal and a uniform vertical magnetic field **B** exists throughout the

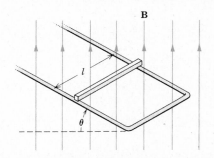

Figure 32–39   Problem 12.

region. (a) Show that the rod acquires a steady-state terminal velocity whose magnitude is

$$v = \frac{mgR}{B^2 l^2} \frac{\sin \theta}{\cos^2 \theta}.$$

(b) Show that the rate at which thermal energy is being generated in the rod is equal to the rate at which the rod is losing gravitational potential energy. (c) Discuss the situation if **B** were directed down instead of up.

Section 32–5 Induced Electric Fields

13. Early in 1981 the Francis Bitter National Magnet Laboratory at M.I.T. commenced operation of a 3.3-cm diameter cylindrical magnet, which produces a 30 T field, then the world's largest steady-state field. The field can be varied sinusoidally between the limits of 29.6 T and 30 T at a frequency of 15 Hz. When this is done, what is the maximum value of the induced electric field at a radial distance of 1.6 cm from the axis? This magnet is described in *Physics Today*, August 1984.

14. Figure 32–40 shows a uniform magnetic field **B** confined to a cylindrical volume of radius R. **B** is decreasing in magnitude at a constant rate of 10 mT/s. What is the instantaneous acceleration

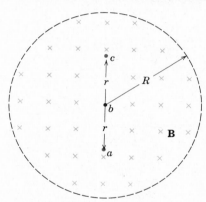

Figure 32–40 Problem 14.

(direction and magnitude) experienced by an electron placed at *a*, at *b*, and at *c*? Assume r = 5.0 cm. (The necessary fringing of the field beyond R will not change your answer as long as there is axial symmetry about the perpendicular axis through *b*.)

15. Prove that the electric field **E** in a charged parallel-plate capacitor cannot drop abruptly to zero as one moves at right angles to it, as suggested by the arrow in Fig. 32–41 (see point *a*). In actual capacitors fringing of the lines of force always occurs, which means that **E** approaches zero in a continuous and gradual way; compare with Problem 14, Chapter 31. (*Hint:* Apply Faraday's law to the rectangular path shown by the dashed lines.)

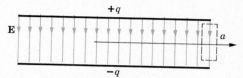

Figure 32–41 Problem 15.

Section 32–6 The Betatron

16. Some measurements of the maximum magnetic field as a function of radius for a betatron are

r, cm	B, tesla	r, cm	B, tesla
0	0.400	81.2	0.409
10.2	0.950	83.7	0.400
68.2	0.950	88.9	0.381
73.2	0.528	91.4	0.372
75.2	0.451	93.5	0.360
77.3	0.428	95.5	0.340

Show by graphical analysis that the relation $\overline{B} = 2B_{orb}$ mentioned on page 633 as essential to betatron operation is satisfied at the orbit radius, R = 84 cm. (*Hint:* Note that

$$\overline{B} = \frac{1}{\pi R^2} \int_0^R B(r)(2\pi r)\, dr$$

and evaluate the integral graphically.)

ADDITIONAL PROBLEMS

17. A uniform magnetic field is normal to the plane of a circular loop 10 cm in diameter made of copper wire (diameter = 2.5 mm). At what rate must the magnetic field change with time if an induced current of 10 A is to appear in the loop? The resistivity of copper is $1.7 \times 10^{-8}\ \Omega \cdot m$.

18. You are given 50 cm of copper wire (diameter = 1.0 mm). It is formed into circular loop and placed at right angles to a uniform magnetic field that is increasing with time at the constant rate of 10 mT/s. At what rate is thermal energy generated in the loop?

19. A generator consists of 100 turns of wire formed into a rectangular loop 50 cm by 30 cm, placed entirely in a uniform magnetic field with magnitude B = 3.5 T. What is the maximum value of the emf produced when the loop is spun at 1000 revolutions per minute about an axis perpendicular to **B**?

20. A stiff wire bent into a semicircle of radius *a* is rotated with a frequency *v* in a uniform magnetic field, as suggested in Fig. 32–

42. What are (a) the frequency and (b) the amplitude of the emf induced in the loop?

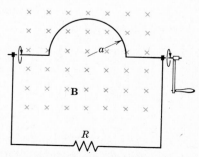

Figure 32–42 Problem 20.

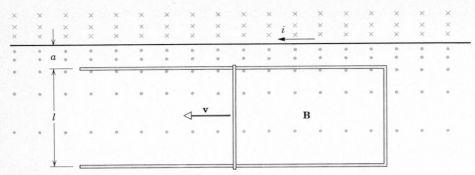

Figure 32-43   Problem 21.

21. Figure 32-43 shows a rod of length l caused to move at constant velocity v along horizontal conducting rails. In this case the magnetic field in which the rod moves is not uniform but is provided by a current i in a long parallel wire. Assume that $v = 5.0$ m/s, $a = 10$ mm, $l = 10$ cm, and $i = 100$ A. (a) Calculate the emf induced in the rod. (b) What is the current in the conducting loop? Assume that the resistance of the rod is 0.40 Ω and that the resistance of the rails is negligible. (c) At what rate is thermal energy being generated in the rod? (d) What force must be applied to the rod by an external agent to maintain its motion? (e) At what rate does this external agent do work on the rod? Compare this answer to (c).

22. In Problem 11 (see Fig. 32-38) the constant-current generator G is replaced by a battery that supplies a constant emf ε. (a) Show that the velocity of the rod now approaches a constant terminal value v and give its magnitude and direction. (b) What is the current in the rod when this terminal velocity is reached? (c) Analyze both this situation and that of Problem 11 from the point of view of energy transfers.

23. For the situation shown in Fig. 32-44, $a = 12$ cm, $b = 16$ cm. The current in the long straight wire is given by $i = 4.5t^2 - 10t$, where i is in amperes and t is in seconds. (a) Find the emf in the square loop at $t = 3.0$ s. Ignore magnetic fields due to the induced current. (b) What is the sense of the induced current in the loop?

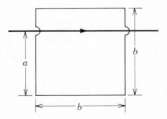

Figure 32-44   Problem 23.

24. In Fig. 32-45, the square has sides of length 2.0 cm. A magnetic field points out of the page; its magnitude is given by $B = 4t^2y$, where B is in tesla, t is in seconds, and y is in meters. Determine the emf around the square at $t = 2.5$ s and give its sense.

25. A hundred turns of insulated copper wire are wrapped around a wooden cylindrical core of cross-sectional area 1.0×10^{-3} m². The two terminals are connected to a resistor. The total resistance in the circuit is 10 Ω. If an externally applied uniform

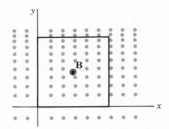

Figure 32-45   Problem 24.

longitudinal magnetic field in the core changes from 1.0 T in one direction to 1.0 T in the opposite direction, how much charge flows through the circuit?

26.   *Alternating current generator.* A rectangular loop of N turns and of length a and width b is rotated at a frequency v in a uniform magnetic field **B**, as in Fig. 32-46. (a) Show that an induced emf given by

$$\varepsilon = 2\pi vNabB \sin 2\pi vt = \varepsilon_0 \sin 2\pi vt$$

appears in the loop. This is the principle of the commercial alternating-current generator. (b) Design a loop that will produce an emf with $\varepsilon_0 = 150$ V when rotated at 60 rev/s in a magnetic field of 0.50 T.

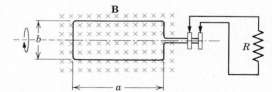

Figure 32-46   Problem 26.

27. At a certain place, the earth's magnetic field has magnitude $B = 0.59$ gauss and is inclined downwards at an angle of 70.0° to the horizontal. A flat horizontal circular coil of wire with a radius of 10 cm has 1000 turns and a total resistance of 85 Ω. It is connected to a galvanometer with 140 Ω resistance. The coil is flipped through a half revolution about a diameter, so it is again horizontal. How much charge flows through the galvanometer during the flip?

28. A square wire loop with 2.0-m sides is perpendicular to a uniform magnetic field, with half the area of the loop in the field, as shown in Fig. 32-47. The loop contains a 20-V battery with

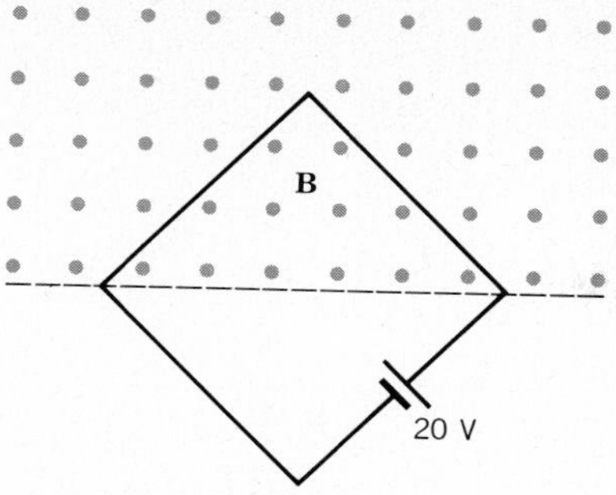

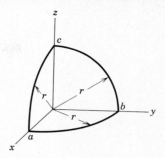

Figure 32–47 Problem 28.

negligible internal resistance. If the magnitude of the field varies with time according to $B = 0.042 - 0.87t$, with B in tesla and t in seconds, (a) what is the total emf in the circuit? (b) What is the direction of the current through the battery?

29. A rectangular loop of wire with length a, width b, and resistance R is placed near an infinitely long wire carrying current i, as shown in Fig. 32–48. The distance from the long wire to the center of the loop is r. Ignore the flux produced by current in the loop. Find (a) the magnitude of the magnetic flux through the loop and (b) the current in the loop as it moves away from the long wire with speed v. Your answers should be in terms of v, R, r, and i.

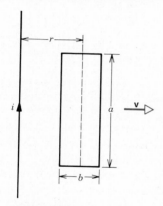

Figure 32–48 Problem 29.

30. A wire is bent into three circular segments of radius r ($= 10$ cm) as shown in Fig. 32–49. Each segment is a quadrant of a circle, ab lying in the x-y plane, bc lying in the y-z plane, and ca lying in the z-x plane. (a) If a uniform magnetic field **B** points in

the positive x-direction, what is the magnitude of the emf developed in the wire when B increases at the rate of 3.0 mT/s? (b) What is the direction of the current in the segment bc?

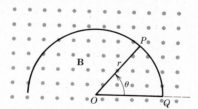

Figure 32–49 Problem 30.

31. Two long, parallel copper wires (diameter $= 2.5$ mm) carry currents of 10 A in opposite directions. (a) If their centers are 20 mm apart, calculate the flux per unit length of wire that exists in the space between the axes of the wires. (b) What fraction of this flux lies inside the wires? (c) Repeat the calculation of (a) for parallel currents. (*Hint:* See Section 31–4).

32. A wire whose cross-sectional area is 1.2 mm² and whose resistivity is 1.7×10^{-8} $\Omega \cdot$m is bent into a circular arc of radius $r = 24$, cm as shown in Fig. 32–50. An additional straight length of this wire, OP, is free to pivot about O and makes sliding contact with the arc at P. Finally, another straight length of this wire, OQ, completes the circuit. The entire arrangement is located in a magnetic field $B = 0.15$ T directed out of the plane of the figure. The straight wire OP starts from rest with $\theta = 0$ and has a constant angular acceleration of 12 rad/s². (a) Find the resistance of the loop $OPQO$ in terms of the angle θ. (b) Find the magnetic flux through the loop as a function of θ (neglect any magnetic field due to the induced current in the loop). (c) For what value of θ is the induced current in the loop a maximum? (d) What is the maximum value of the induced current in the loop?

Figure 32–50 Problem 32.

33 Inductance

33-1 Self Induction

Self induction

If two coils are near each other, a current i in one coil will set up a flux Φ_B through the second coil. If this flux is changed by changing the current, an induced emf will appear in the second coil according to Faraday's law; see Fig. 32–2. An induced emf also appears in a coil if the current in that same coil is changed. This is called *self induction* and the electromotive force produced is called a *self-induced emf*. It obeys Faraday's law of induction just as other induced emfs do.

Consider first a "close-packed" flat coil, a toroid, or the central section of a long solenoid. In all three cases the flux Φ_B set up in each turn by a current i is essentially the same for every turn. Faraday's law for such coils

$$\varepsilon = -\frac{d(N\Phi_B)}{dt} \qquad [32-2b]$$

shows that the number of *flux linkages* $N\Phi_B$ (N being the number of turns) is the important characteristic quantity for induction. For a given coil, provided no magnetic materials such as iron are nearby, this quantity is proportional to the current i, or

$$N\Phi_B = Li, \qquad (33-1)$$

Inductance

in which L, the proportionality constant, is called the inductance of the device.

From Faraday's law above we can write the induced emf as

$$\varepsilon = -\frac{d(N\Phi_B)}{dt} = -L\frac{di}{dt}. \qquad (33-2)$$

Written in the form

$$L = -\frac{\varepsilon}{di/dt}, \qquad (33-3)$$

this relation may be taken as the defining equation for *inductance* for coils of all shapes and sizes, whether or not they are close-packed and whether or not iron or other magnetic material is nearby. It is analogous to the defining relation for capacitance, namely,

$$C = \frac{q}{V}. \qquad [27-1]$$

Inductors

If no iron or similar materials are nearby, L depends only on the geometry of the device. In an *inductor* (symbol $\sim\!\!\sim\!\!\sim$) the presence of a magnetic field is the significant feature, corresponding to the presence of an electric field in a capacitor.

The SI unit of inductance, from Eq. 33–3, is the volt·second/ampere. A special name, the henry (abbr. H), has been given to this combination of units, or

The henry

$$1 \text{ henry} = 1 \text{ volt·second/ampere.}$$

The SI unit of inductance is named after Joseph Henry (1797–1878), an American physicist and a contemporary of Faraday.

You can find the direction of a self-induced emf from Lenz's law. Suppose that, by means of a battery, you set up a steady current i in a coil. By a switching arrangement you suddenly decrease the battery emf. The current starts to decrease at once. In the language of Lenz's law this decrease is the "change" that the self-induction must oppose. To do so the self-induced emf must point in the same direction as the current, as Fig. 33–1a shows. This conclusion is also predicted by Eq. 33–2. For a falling current di/dt is negative so that equation predicts a positive sign for the self-induced emf. This means that the emf must point in the direction of the current.

However, if you suddenly increase the battery emf the current will start to increase. To oppose *this* change the self-induced emf must point in the opposite direction to the current, as Fig. 33–1b shows. For an increasing current di/dt is positive so that Eq. 33–2 predicts a negative sign for the self-induced emf. This means that the emf must point in a direction opposite to the current.

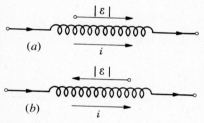

Figure 33–1 The currents in the two coils have the same instantaneous values but in (a) the current is decreasing ($di/dt < 0$) and in (b) it is increasing ($di/dt > 0$). In each case the self-induced emf opposes the change. The label on each emf arrow represents the magnitude (or absolute value) of the emf, a positive quantity.

33–2 Calculation of Inductance

We were able to make a direct calculation of capacitance in terms of geometrical factors for a few special cases, such as the parallel-plate capacitor. In the same way, we can calculate the self-inductance L for a few special cases.

For a close-packed coil with no iron nearby, we have, from Eq. 33–1,

$$L = \frac{N\Phi_B}{i}. \qquad (33-4)$$

Let us apply this equation to calculate L for a section of length l near the center of a long solenoid. The number of flux linkages in the length l of the solenoid is

$$N\Phi_B = (nl)(BA),$$

where n is the number of turns per unit length, B is the magnetic field inside the solenoid, and A is the cross-sectional area. B for a solenoid is given by Eq. 31–6 ($B = \mu_0 ni$) so that

$$N\Phi_B = \mu_0 n^2 liA.$$

Finally, from Eq. 33–4, the inductance is

Inductance of a long solenoid

$$L = \frac{N\Phi_B}{i} = \mu_0 n^2 lA. \qquad (33-5)$$

The inductance of a length l of a solenoid is proportional to its volume (lA) and to the square of the number of turns per unit length. Note that it depends on geometrical factors only. The proportionality to n^2 is expected. If we double the number of turns per unit length, not only is the total number of turns N doubled but the flux Φ_B through each turn is also doubled, an overall factor of four for the flux linkages $N\Phi_B$, hence also a factor of four for the inductance (Eq. 33–4).

Example 1 Derive an expression for the inductance of a toroid of rectangular cross section as shown in Fig. 33–2. Evaluate for $N = 10^3$, $a = 50$ mm, $b = 10$ cm, and $h = 10$ mm.

The lines of **B** for the toroid are concentric circles. Applying Ampère's law,

$$\oint \mathbf{B} \cdot d\boldsymbol{\ell} = \mu_0 i, \qquad [31-5]$$

to a circular path of radius r yields

$$(B)(2\pi r) = \mu_0(i_0 N),$$

where N is the number of turns and i_0 is the current in the toroid windings; recall that i in Ampère's law is the total current that passes through the path of integration. Solving for B yields

$$B = \frac{\mu_0 i_0 N}{2\pi r}.$$

The flux Φ_B for the cross section of the toroid is

$$\Phi_B = \int \mathbf{B} \cdot d\mathbf{S} = \int_a^b (B)(h\,dr) = \int_a^b \frac{\mu_0 i_0 N}{2\pi r} h\,dr$$

$$= \frac{\mu_0 i_0 Nh}{2\pi} \int_a^b \frac{dr}{r} = \frac{\mu_0 i_0 Nh}{2\pi} \ln\frac{b}{a},$$

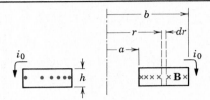

Figure 33–2 Example 1. A cross section of a toroid, showing the current in the windings and the magnetic field.

where $h\,dr$ is the area of the elementary strip shown between the dashed lines in the figure.

The inductance follows from Eq. 33–4, or

$$L = \frac{N\Phi_B}{i_0} = \frac{\mu_0 N^2 h}{2\pi} \ln\frac{b}{a}.$$

Substituting numerical values yields

$$L = \frac{(4\pi \times 10^{-7}\ \text{Wb/A·m})(10^3)^2(1.0 \times 10^{-2}\ \text{m})}{2\pi} \times$$

$$\ln\left(\frac{10 \times 10^{-2}\ \text{m}}{5 \times 10^{-2}\ \text{m}}\right) = 1.4 \times 10^{-3}\ \text{Wb/A} = 1.4\ \text{mH}.$$

33–3 An LR Circuit

In Section 29–7 we saw that if we suddenly introduce an emf ε, perhaps by using a battery, into a single loop circuit containing a resistor R and a capacitor C, the charge does not build up immediately to its final equilibrium value ($= C\varepsilon$) but approaches it in an exponential fashion described by Eq. 29–15, or

$$q = C\varepsilon\,(1 - e^{-t/\tau_C}). \qquad (33-6)$$

The delay in the rise of the charge is described by the capacitive time constant τ_C, defined from

$$\tau_C = RC. \qquad (33-7)$$

If in this same circuit the battery emf ε is suddenly removed, the charge does not immediately fall to zero but approaches zero in an exponential fashion, described by Eq. 29–19, or

$$q = C\varepsilon\,e^{-t/\tau}. \qquad (33-8)$$

The same time constant τ_C describes the fall of the charge as well as its rise.

An analogous delay in the rise (or fall) of the current occurs if we suddenly introduce an emf ε into (or remove it from) a single-loop circuit containing a resistor R and an inductor L. When the switch S in Fig. 33–3 is closed on a, for example, the current in the resistor starts to rise. If the inductor were not present, the current would rise rapidly to a steady value ε/R. Because of the inductor, however, a

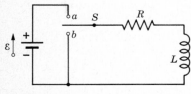

Figure 33–3 An *LR* circuit. Compare with the *RC* circuit of Fig. 29–10.

Current in an *LR* circuit —qualitative analysis

self-induced emf, which we label ε_L, appears in the circuit and, from Lenz's law, this emf opposes the rise of current, which means that it opposes the battery emf ε in polarity. Thus the resistor responds to the difference between two emfs, a constant one ε due to the battery and a variable one $\varepsilon_L\ (=-L\ di/dt)$ due to self-induction. As long as this second emf is present, the current in the resistor will be less than ε/R.

As time goes on, the rate at which the current increases becomes less rapid and the magnitude of the self-induced emf, which is proportional to di/dt, becomes smaller. Thus a time delay is introduced, and the current in the circuit approaches the value ε/R asymptotically.

When the switch S in Fig. 33–3 is thrown to a, the circuit reduces to that of Fig. 33–4. Let us apply the loop theorem, starting at x in this figure and going clockwise around the loop. For the direction of current shown, x will be higher in potential than y, which means that we encounter a drop in potential of $-iR$ as we traverse the resistor. Point y is higher in potential than point z because, for an increasing current, the induced emf will oppose the rise of the current by pointing as shown. Thus as we traverse the inductor from y to z we encounter a drop in potential of $-L(di/dt)$. We encounter a rise in potential of $+\varepsilon$ in traversing the battery from z to x. The loop theorem thus gives

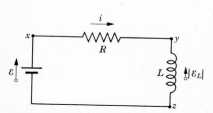

Figure 33–4 The circuit of Fig. 33–3 after switch S is closed on a. The self-induced emf opposes the battery emf.

$$-iR - L\frac{di}{dt} + \varepsilon = 0$$

or

$$L\frac{di}{dt} + iR = \varepsilon. \tag{33–9}$$

Equation 33–9 is a differential equation involving the variable i and its first derivative di/dt. We seek the function $i(t)$ such that when it and its first derivative are substituted in Eq. 33–9 the equation is satisfied.

Although there are formal rules for solving various classes of differential equations (and Eq. 33–9 can, in fact, be easily solved by direct integration, after rearrangement) we often find it simpler to guess at the solution, guided by physical reasoning and by previous experience. We can test any proposed solution by substituting it in the differential equation and seeing whether this equation reduces to an identity.

The solution to Eq. 33–9 is, we claim,

Rise of current in an *LR* circuit

$$i = \frac{\varepsilon}{R}(1 - e^{-Rt/L}). \tag{33–10}$$

To test this solution by substitution, we find the derivative di/dt, which is

$$\frac{di}{dt} = \frac{\varepsilon}{L}e^{-Rt/L}. \tag{33–11}$$

Substituting i and di/dt into Eq. 33–9 leads to an identity, as you can easily verify. Thus Eq. 33–10 is a solution of Eq. 33–9. Figure 33–5 shows how the potential difference V_R across the resistor ($=iR$) and V_L across the inductor ($=L\ di/dt$) vary with time for particular values of ε, L, and R. Compare this figure carefully with the corresponding figure for an RC circuit (Fig. 29–11).

We can rewrite Eq. 33–10 as

$$i = \frac{\varepsilon}{R}(1 - e^{-t/\tau_L}), \tag{33–12}$$

Inductive time constant

in which τ_L, the inductive time constant, is given by

$$\tau_L = L/R. \tag{33–13}$$

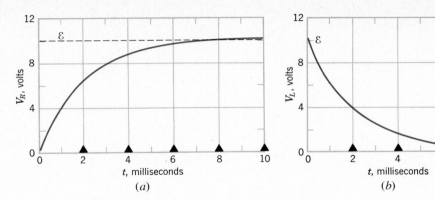

Figure 33–5 The variation with time of (a) V_R, the potential difference across the resistor, and (b) V_L, the potential difference across the inductor, after switch S is closed on a in Fig. 33–3. The filled triangles represent successive time intervals (τ_L = 2.0 ms). For R = 2000 Ω, L = 4.0 H, and ε = 10 V. Compare this figure carefully with Fig. 29–11, the corresponding figure for an RC circuit.

Note the correspondence between Eqs. 33–6 and 33–12.

To show that the quantity τ_L ($=L/R$) has the dimensions of time, we put

$$\frac{1\ \text{henry}}{\text{ohm}} = \frac{1\ \text{henry}}{\text{ohm}}\left(\frac{1\ \text{volt·second}}{1\ \text{henry·ampere}}\right)\left(\frac{1\ \text{ohm·ampere}}{1\ \text{volt}}\right) = 1\ \text{second}.$$

The first quantity in parentheses is a conversion factor based on the defining equation for inductance $[L = -\varepsilon/(di/dt)$; see Eq. 33–3]. The second conversion factor is based on the relation $V = iR$.

The physical significance of the time constant follows from Eq. 33–12. If we put $t = \tau_L = L/R$ in this equation, it reduces to

$$i = \frac{\varepsilon}{R}(1 - e^{-1}) = (1 - 0.37)\frac{\varepsilon}{R} = 0.63\frac{\varepsilon}{R}.$$

Thus the time constant τ_L is that time at which the current in the circuit will reach a value within $1/e$ (about 37%) of its final equilibrium value (see Fig. 33–5).

If the switch S in Fig. 33–3, having been left in position a long enough for the equilibrium current ε/R to be established, is thrown to b,* the effect is to remove the battery from the circuit. The differential equation that governs the subsequent decay of the current in the circuit can be found by putting $\varepsilon = 0$ in Eq. 33–9, or

$$L\frac{di}{dt} + iR = 0. \qquad (33-14)$$

Decay of current in an LR circuit

You can show by the test of substitution that the solution of this differential equation is

$$i = \frac{\varepsilon}{R}e^{-t/\tau_L}. \qquad (33-15)$$

Example 2 A solenoid has an inductance of 50 H and a resistance of 30 Ω. If it is connected to a 100-V battery, how long will it take for the current to reach one-half its final equilibrium value?

The equilibrium value of the current is reached as $t \to \infty$; from Eq. 33–12 it is ε/R. If the current has half this value at a particular time t_0, this equation becomes

$$\frac{1}{2}\frac{\varepsilon}{R} = \frac{\varepsilon}{R}(1 - e^{-t_0/\tau_L}).$$

Solving for t_0 yields

$$t_0 = \tau_L \ln 2 = 0.69\frac{L}{R}.$$

Using the values given, this reduces to

$$t_0 = 0.69\tau_L = 0.69\left(\frac{50\ \text{H}}{30\ \Omega}\right) = 1.2\ \text{s}.$$

---

* The connection to b must be made before the connection to a is broken. A switch which does this is called a ''make before break'' switch.

33-4 Energy and the Magnetic Field

When we lift a stone we do work, which we can get back by lowering the stone. It is convenient to think of the energy being stored temporarily in the gravitational field between the earth and the lifted stone, and being withdrawn from this field when we lower the stone.

When we pull two unlike charges apart we like to say that the resulting electric potential energy is stored in the electric field between the charges. We can get it back from the field by letting the charges move closer together again.

In the same way we can consider energy to be stored in a magnetic field. For example, two long, rigid, parallel wires carrying current in the same direction attract each other and we must do work to pull them apart. We can get this stored energy back at any time by letting the wires move back to their original positions.

To derive a quantitative expression for the energy stored in the magnetic field, consider Fig. 33-4, which shows a source of emf ε connected to a resistor R and an inductor L.

$$\varepsilon = iR + L\frac{di}{dt},$$

[33-9]

is the differential equation that describes the growth of current in this circuit. We stress that this equation follows immediately from the loop theorem and that the loop theorem in turn is an expression of the principle of conservation of energy for single-loop circuits. If we multiply each side of Eq. 33-9 by i, we obtain

$$\varepsilon i = i^2 R + Li\frac{di}{dt},$$

(33-16)

which has the following physical interpretation in terms of work and energy:

Energy transfer in an *LR* circuit

1. If a charge dq passes through the seat of emf ε in Fig. 33-4 in time dt, the seat does work on it in amount $\varepsilon\,dq$. The rate of doing work is $(\varepsilon\,dq)/dt$, or εi. Thus the left term in Eq. 33-16 is the rate at which the seat of emf delivers energy to the circuit.

2. The second term in Eq. 33-16 is the rate at which energy appears as thermal energy in the resistor.

3. Energy that does not appear as thermal energy must, by our hypothesis, be stored in the magnetic field. Since Eq. 33-16 represents a statement of the conservation of energy for *LR* circuits, the last term must represent the rate dU_B/dt at which energy is stored in the magnetic field, or

$$\frac{dU_B}{dt} = Li\frac{di}{dt}.$$

(33-17)

We can write this as

$$dU_B = Li\,di.$$

Integrating yields

Magnetic energy stored by an inductor

$$U_B = \int_0^{U_B} dU_B = \int_0^i Li\,di = \tfrac{1}{2}Li^2,$$

(33-18)

which represents the total magnetic energy stored by an inductor L carrying a current i.

We can compare this relation with the expression for the energy associated with a capacitor C carrying a charge q, namely,

$$U_E = \frac{1}{2}\frac{q^2}{C}. \qquad\qquad [27\text{-}7]$$

Here the energy is stored in an electric field. In each case the expression for the stored energy was derived by setting it equal to the work that must be done to set up the field.

Example 3 A coil has an inductance of 5.0 H and a resistance of 20 Ω. If a 100-V emf is applied, how much energy is stored in the magnetic field after the current has built up to its maximum value ε/R?

The maximum current is given by

$$i = \frac{\varepsilon}{R} = \frac{100\text{ V}}{20\ \Omega} = 5.0\text{ A}.$$

The stored energy is given by Eq. 33–18:

$$U_B = \tfrac{1}{2}Li^2 = \tfrac{1}{2}(5.0\text{ H})(5.0\text{ A})^2 = 63\text{ J}.$$

Note that the time constant for this coil $(= L/R)$ is 0.25 s. After how many time constants will half of this equilibrium energy be stored in the field?

Example 4 A 3.00-H inductor is placed in series with a 10.0-Ω resistor, an emf of 3.00 V being suddenly applied to the combination. At 0.30 s (which is one inductive time constant) after the contact is made, (a) what is the rate at which energy is being delivered by the battery?

The current is given by Eq. 33–12, or

$$i = \frac{\varepsilon}{R}(1 - e^{-t/\tau_L}),$$

which at $t = 0.300$ s $(= \tau_L)$ has the value

$$i = \left(\frac{3.00\text{ V}}{10.0\ \Omega}\right)(1 - e^{-1}) = 0.189\text{ A}.$$

The rate P at which energy is delivered by the battery is

$$P_\varepsilon = \varepsilon i = (3.0\text{ V})(0.189\text{ A}) = 0.567\text{ W}.$$

(b) At what rate does energy appear as thermal energy in the resistor? This is given by

$$P_J = i^2 R = (0.189\text{ A})^2(10.0\ \Omega) = 0.357\text{ W}.$$

(c) At what rate P_B is energy being stored in the magnetic field?

This is given by Eq. 33–17, which requires that we know di/dt. Differentiating Eq. 33–12 yields

$$\frac{di}{dt} = \left(\frac{\varepsilon}{R}\right)\left(\frac{R}{L}\right)e^{-t/\tau_L} = \frac{\varepsilon}{L}e^{-t/\tau_L}.$$

At $t = \tau_L$ we have

$$\frac{di}{dt} = \left(\frac{3.00\text{ V}}{3.00\text{ H}}\right)e^{-1} = 0.370\text{ A/s}.$$

From Eq. 33–17, the desired rate is

$$P_B = \frac{dU_B}{dt} = Li\frac{di}{dt} = (3.00\text{ H})(0.189\text{ A})(0.370\text{ A/s}) = 0.210\text{ W}.$$

Note that as required by the principle of conservation of energy (see Eq. 33–16)

$$P_\varepsilon = P_J + P_B,$$

or

$$0.567\text{ W} = 0.357\text{ W} + 0.210\text{ W} = 0.567\text{ W}.$$

33–5 Energy Density and the Magnetic Field

We now derive an expression for the density of energy u_B in a magnetic field. Consider a length l near the center of a very long solenoid; Al is the volume associated with this length. The stored energy must lie entirely within this volume because the magnetic field outside such a solenoid is essentially zero. Moreover, the stored energy must be uniformly distributed throughout the volume of the solenoid because the magnetic field is uniform everywhere inside. Thus, we can write

$$u_B = \frac{U_B}{Al}$$

or, since

$$U_B = \tfrac{1}{2}Li^2,$$

we have

$$u_B = \frac{\tfrac{1}{2}Li^2}{Al}.$$

To express this in terms of the magnetic field, we can substitute for L in this equation, using the relation $L = \mu_0 n^2 lA$ (Eq. 33–5). Also we can solve Eq. 31–6 ($B = \mu_0 in$) for i and substitute in this equation. Doing so yields finally

Energy density in a magnetic field

$$u_B = \frac{1}{2}\frac{B^2}{\mu_0}. \tag{33–19}$$

This equation gives the energy density stored at any point (in a vacuum or in a nonmagnetic substance) where the magnetic field is B. The equation is true for all magnetic field configurations, even though we derived it by considering a special case, the solenoid. Equation 33–19 is to be compared with Eq. 27–9,

$$u_E = \tfrac{1}{2}\epsilon_0 E^2, \tag{27–9}$$

which gives the energy density (in a vacuum) at any point in an electric field. Note that both u_B and u_E are proportional to the square of the appropriate field quantity, B or E.

The solenoid plays a role with relationship to magnetic fields similar to the role the parallel-plate capacitor plays with respect to electric fields. In each case we have a simple device that can be used for setting up a uniform field throughout a well-defined region of space and for deducing, in a simple way, some properties of these fields.

Example 5 A long coaxial cable (Fig. 33–6) consists of two concentric cylinders with radii a and b. Its central conductor carries a steady current i, the outer conductor providing the return path. (a) Calculate the energy stored in the magnetic field between the conductors for a length l of such a cable.

In the space between the two conductors Ampère's law,

$$\oint \mathbf{B} \cdot d\boldsymbol{\ell} = \mu_0 i,$$

leads to

$$(B)(2\pi r) = \mu_0 i$$

or

$$B = \frac{\mu_0 i}{2\pi r}.$$

The energy density for points in this region, from Eq. 33–19, is

$$u_B = \frac{1}{2}\frac{B^2}{\mu_0} = \frac{1}{2\mu_0}\left(\frac{\mu_0 i}{2\pi r}\right)^2 = \frac{\mu_0 i^2}{8\pi^2 r^2}.$$

Consider a volume element dV consisting of a cylindrical shell whose radii are r and $r + dr$ and whose length is l. The energy dU contained in it is

$$dU = u_B dV = \frac{\mu_0 i^2}{8\pi^2 r^2}(2\pi rl)(dr) = \frac{\mu_0 i^2 l}{4\pi}\frac{dr}{r}.$$

The total energy stored in the magnetic field in the space between the conductors is found by integration to be

$$U = \int dU = \frac{\mu_0 i^2 l}{4\pi}\int_a^b \frac{dr}{r} = \frac{\mu_0 i^2 l}{4\pi}\ln\frac{b}{a}. \tag{33–20}$$

No energy is stored in the space outside the outer conductor because the magnetic field is zero there, as you can easily show from Ampère's law.

(b) What is the inductance of a length l of coaxial cable?

We can find the inductance L from Eq. 33–18 ($U = \frac{1}{2}Li^2$) which leads to

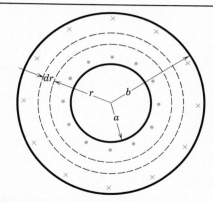

Figure 33–6 Example 5. A cross section of a coaxial cable consisting of two thin coaxial cylinders. Equal but opposite currents, represented by dots and crosses, are shown for the central and outer conductors.

$$L = \frac{2U}{i^2} = \frac{\mu_0 l}{2\pi}\ln\frac{b}{a}. \tag{33–21}$$

As an exercise calculate L also from Eq. 33–1 ($L = N\Phi_B/i$).

Example 6 Compare the energy required to set up, in a cube 10 cm on edge (a) a uniform electric field of 0.10 MV/m and (b) a uniform magnetic field of 1.0 T. Both these fields would be judged reasonably large but readily available in the laboratory.

(a) In the electric case we have, where V_0 is the volume of the cube,

$$\begin{aligned}
U_E &= u_E V_0 = \tfrac{1}{2}\epsilon_0 E^2 V_0\\
&= (0.5)(8.9 \times 10^{-12}\ \text{C}^2/\text{N·m}^2)(10^5\ \text{V/m})^2(0.1\ \text{m})^3\\
&= 4.5 \times 10^{-5}\ \text{J} = 45\ \mu\text{J}.
\end{aligned}$$

(b) In the magnetic case, from Eq. 33–19, we have

$$U_B = u_B \, V_0 = \frac{1}{2}\frac{B^2}{\mu_0} \, V_0 = \frac{(1.0 \text{ T})^2(0.1 \text{ m})^3}{(2)(4\pi \times 10^{-7} \text{ T·m/A})}$$

$$= 400 \text{ J}.$$

In terms of fields normally available in the laboratory, much larger amounts of energy can be stored in a magnetic field than in an electric one, the ratio being about 10^7 in this example. Conversely, much more energy is required to set up a magnetic field of "reasonable" laboratory magnitude than is required to set up an electric field of similarly reasonable magnitude.

33–6 Mutual Induction

In Section 32–1 we saw that if two coils are close together, as in Fig. 32–2, a steady current i in one coil will set up a magnetic flux Φ linking the other coil. If we change i with time, an emf ε given by Faraday's law appears in the second coil; we called this process induction. We could better have called it mutual induction, to suggest the mutual interaction of the two coils and to distinguish it from self-induction, in which only one coil is involved.

Let us look a little more quantitatively at mutual induction. Figure 33–7a shows two circular close-packed coils near each other and sharing a common axis. There is a steady current i_1 in coil 1, set up by the battery in the external circuit. This current produces a magnetic field suggested by the lines of $\mathbf{B}_1$ in the figure. Coil 2 is connected to a sensitive galvanometer G but contains no battery; a magnetic flux Φ_{21} is linked by its N_2 turns.

We define the *mutual inductance* M_{21} of coil 2 with respect to coil 1 as

$$M_{21} = \frac{N_2 \Phi_{21}}{i_1}. \qquad (33\text{–}22)$$

Compare this with Eq. 33–4 ($L = N\Phi/i$), the definition of (self-) inductance. We can recast Eq. 33–22 as

$$M_{21} i_1 = N_2 \Phi_{21}.$$

If, by external means, we cause i_1 to vary with time, we have

$$M_{21}\frac{di_1}{dt} = N_2 \frac{d\Phi_{21}}{dt}.$$

The right side of this equation, from Faraday's law, is, apart from a change in sign, just the emf ε_2 appearing in coil 2 due to the changing current in coil 1, or

$$\varepsilon_2 = -M_{21}\frac{di_1}{dt} \qquad (33\text{–}23)$$

which you should compare with Eq. 33–2 ($\varepsilon = -L \, di/dt$) for self-inductance.

Let us now interchange the roles of coils 1 and 2, as in Fig. 33–7b. That is, we set up a current i_2 in coil 2, by means of a battery, and this produces a magnetic flux Φ_{12} that links coil 1. If we change i_2 with time, we have, by the same argument given above,

$$\varepsilon_1 = -M_{12}\frac{di_2}{dt}; \qquad (33\text{–}24)$$

compare Eq. 33–23.

Thus we see that the emf induced in either coil is proportional to the rate of change of current in the other coil. The proportionality constants M_{21} and M_{12} seem to be different. We assert, without proof, that they are in fact the same so that no subscripts are needed. This conclusion is true but is in no way obvious. Thus we have

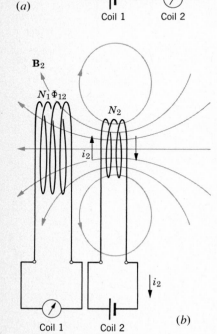

Figure 33–7 Mutual induction. (a) If the current in coil 1 changes an emf ε_2 will be induced in coil 2. (b) If the current in coil 2 changes an emf ε_1 will be induced in coil 1.

$$M_{21} = M_{12} = M \qquad (33-25)$$

and we can rewrite Eqs. 33–23 and 33–24 as

Induced emf due to mutual induction

$$\mathcal{E}_2 = -M \, di_1/dt \quad \text{and} \quad \mathcal{E}_1 = -M \, di_2/dt. \qquad (33-26)$$

The induction is indeed mutual. The SI unit for M (as for L) is the henry.

The calculation of M, like that of L, depends on the geometry of the system. The simplest case is that in which all of the flux from one coil links the other coil.

Example 7 Figure 33–8 shows two circular close-packed coils, the smaller (radius R_2) being coaxial with the larger (radius R_1) and in the same plane. Find the coefficient of mutual inductance M for this arrangement of these two coils. Assume that $R_1 \gg R_2$.

As the figure suggests we imagine that we establish a current i_1 in the larger coil and we note the magnetic field $\mathbf{B}_1$ that it sets up. The value of B_1 at the center of this coil is (see Eq. 31–8)

$$B_1 = \frac{\mu_0 i_1 N_1}{2R_1}.$$

Because we have assumed that $R_1 \gg R_2$ we may take B_1 to be the magnetic field at all points within the boundary of the smaller coil. The flux linkage for the smaller coil is then

$$N_2 \Phi_{21} = N_2(B_1)(\pi R_2{}^2) = \frac{\pi \mu_0 i_1 N_1 N_2 R_2{}^2}{2R_1}.$$

From Eq. 33–22 we then have

$$M = \frac{N_2 \Phi_{21}}{i_1} = \frac{\pi \mu_0 N_1 N_2 R_2{}^2}{2R_1}. \qquad (33-27)$$

For $N_1 = N_2 = 10^3$ turns, $R_2 = 1.0$ cm and $R_1 = 15$ cm this equation yields $M = 1.3$ mH.

Consider the situation if we reverse the roles of the two coils in Fig. 33–8, that is, if we set up a current i_2 in the smaller coil and try to calculate M from

$$M = \frac{N_1 \Phi_{12}}{i_2}.$$

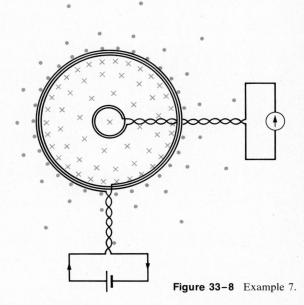

Figure 33–8 Example 7.

The exact calculation of Φ_{12} (the flux of the smaller coil's magnetic field encompassed by the larger coil) is not a simple matter. If we calculate it numerically using a computer we get the same value for M given by Eq. 33–27! This should make us appreciate even more the fact that the equality of Eq. 33–26 ($M_{12} = M_{21} = M$) is not obvious.

REVIEW GUIDE AND SUMMARY

If a current i in a coil changes with time, an emf described by Faraday's law (Eq. 32–2) is induced in the coil itself. This self-induced emf is given by

Self-induction

$$\mathcal{E} = -L \frac{di}{dt} \qquad [33-2]$$

where L, which is a constant in the absence of iron or similar magnetic materials, is called the inductance of the coil. The direction of $\mathcal{E}$ is found from Lenz's law, as in Section 32–3. The unit of L is the henry, where 1 H = 1 V·s/A.

Equation 33–2 is the basic definition of L. An equivalent definition that holds for **The inductance of close-packed coils** close-packed coils, solenoids and toroids, is

$$L = \frac{N \Phi_B}{i} \qquad [33-1]$$

Equation 33–2 follows if we introduce Eq. 33–1 into Faraday's law. Using Eq. 33–1 as a starting point we derived expressions for the inductance of a length l of a long solenoid (Eq. 33–5) and for a toroid (Example 1).

The rise and fall of current in an LR circuit

If a constant emf ε is introduced into a single loop circuit containing a resistor R and an inductor L (see Fig. 33–3) the current rises to an equilibrium value of ε/R according to

$$i(t) = (\varepsilon/R)(1 - e^{-t/\tau_L}). \qquad [33-10]$$

Here τ_L $(=L/R)$ governs the rate of rise of the current and is called the inductive time constant of the circuit. Figure 33–5, which shows the potential differences across the resistor and the inductor while the current is rising, deserves careful study; see also Example 2. The decay of the current when the source of constant emf is removed is given by

$$i(t) = (\varepsilon/R)e^{-t/\tau_L}. \qquad [33-15]$$

By applying the conservation of energy principle to the rise of current in the LR circuit of Fig. 33–3 we deduce that if an inductor L carries a current i an energy given by

Storage of energy by an inductor

$$U = \tfrac{1}{2}Li^2 \qquad [33-18]$$

can be viewed as stored in its magnetic field. Example 4 analyzes the energy transfers during the period of current rise for a particular case.

Applying Eq. 33–18 to a section of a long solenoid leads us to deduce the general result that, if B is the magnetic field at any point, the density of stored energy at that point is

Energy storage in a magnetic field

$$u = \frac{B^2}{2\mu_0}. \qquad [33-19]$$

Example 5 analyzes energy storage in a coaxial cable from this starting point.

If two coils or similar conductors are near each other (see Fig. 33–7) a changing current in either coil can induce an emf in the other. This mutual induction phenomenon is described by (compare Eq. 33–2)

Mutual induction

$$\varepsilon_2 = -M\frac{di_1}{dt} \quad \text{and} \quad \varepsilon_1 = -M\frac{di_2}{dt}, \qquad [33-26]$$

where M (which is measured in henries) is the coefficient of mutual inductance for the coil arrangement.

The mutual inductance of close-packed coils

Although Eq. 33–26 is the basic definition of M we can also write, for close-packed coils, solenoids, and toroids (compare Eq. 33–1)

$$M = \frac{N_2\Phi_{21}}{i_1} \quad \text{or} \quad M = \frac{N_1\Phi_{12}}{i_2} \qquad [\text{see } 33-22]$$

Eqs. 33–26 follow if we introduce the above equations into Faraday's law. Example 7 is a calculation of M for two close-packed, coaxial, and coplaner coils (see Fig. 33–8), using the left-hand definition for M given immediately above for a starting point.

QUESTIONS

1. Can a long straight wire show self-induction effects? How would you go about looking for them?

2. If the flux passing through each turn of a coil is the same, the inductance of the coil may be computed from $L = N\Phi_B/i$ (Eq. 33–4). How might one compute L for a coil for which this assumption is not valid?

3. Show that the dimensions of the two expressions for L, $N\phi_B/i$ (Eq. 33–4) and $\varepsilon/(di/dt)$ (Eq. 33–3), are the same.

4. You want to wind a coil so that it has resistance but essentially no inductance. How would you do it?

5. Is the inductance per unit length for a solenoid near its center the same as, less than, or greater than the inductance per unit length near its ends?

6. A steady current is set up in a coil with a very large inductive time constant. When the current is interrupted with a switch, a heavy arc tends to appear at the switch blades. Explain. (*Note:* Interrupting currents in highly inductive circuits can be dangerous.)

7. In an LR circuit like that of Fig. 33–4, can the self-induced emf ever be larger than the battery emf?

8. In an LR circuit like that of Fig. 33–4, is the current in the resistor always the same as the current in the inductor?

9. In the circuit of Fig. 33–3 the self-induced emf is a maximum at the instant the switch is closed on *a*. How can this be since there is no current in the inductor at this instant?

10. The switch in Fig. 33–3, having been closed on *a* for a "long" time, is thrown to *b*. What happens to the energy that is stored in the inductor?

11. A coil has a (measured) inductance *L* and a (measured) resistance *R*. Is its inductive time constant given by Eq. 33–13? Bear in mind that we derived that equation (see Fig. 33–3) for a situation in which the inductive and resistive elements are physically separated. Discuss.

12. Figures 33–5*a* and 29–11*b* are plots of $V_R(t)$ for, respectively, an *LR* circuit and an *RC* circuit. Why are these two curves so different? Account for each in terms of physical processes going on in the appropriate circuit.

13. Two solenoids, *A* and *B*, have the same diameter and length and contain only one layer of copper windings, with adjacent turns touching, insulation thickness being negligible. Solenoid *A* contains many turns of fine wire and solenoid *B* contains fewer turns of heavier wire. (*a*) Which solenoid has the larger self-inductance? (*b*) Which solenoid has the larger inductive time constant?

14. Can you make an argument based on the manipulation of bar magnets to suggest that energy may be stored in a magnetic field?

15. Draw all the formal analogies that you can between a parallel-plate capacitor (for electric fields) and a long solenoid (for magnetic fields).

16. The current in a solenoid is reversed. What changes does this make in the magnetic field **B** and the energy density *u* at various points along the solenoid axis?

17. Commercial devices such as motors and generators that are involved in the transformation of energy between electrical and mechanical forms involve magnetic rather than electrostatic fields. Why should this be so?

18. In a case of mutual induction, as in Fig. 33–7, is self-induction also present? Discuss.

19. You are given two similar flat circular coils of *N* turns each. For what geometry will their mutual inductance *M* be the greatest? For what geometry will it be the least? Assume that the centers of the coils are maintained at a fixed distance apart.

20. A flat circular coil is placed completely outside a long solenoid, near its center, the axes of the coil and the solenoid being parallel. Is there a mutual induction effect?

21. A circular coil of *N* turns surrounds a long solenoid. Is the mutual inductance greater when the coil is near the center of the solenoid or when it is near one end? Explain.

22. Suppose that you connect an ideal (that is, essentially resistanceless) coil across an ideal (again, essentially resistanceless) battery. You might think that, because there is no resistance in the circuit, the current would jump at once to a very large value. On the other hand, you might think that, because the inductive time constant ($=L/R$) is extremely large, the current would rise very slowly, if at all. What actually happens?

23. A coil, connected to an ac household power outlet, carries an alternating current. Can you think of any reason that the amplitude of the current on the coil might change if you place a conducting metal object near it? Design a metal detector based on this principle.

EXERCISES

Section 33–1 Self-Induction

1. The inductance of a close-packed coil of 400 turns is 8.0 mH. What is the magnetic flux through the coil when the current is 5.0 mA?

2. Each item (*a*) coulomb·ohm·meter/weber, (*b*) volt·second, (*c*) coulomb·ampere/farad, (*d*) kilogram·volt·meter²/(henry·ampere)², (*e*) (henry/farad)$^{1/2}$ is equal to one of the items in the following list: meter, second, kilogram, dimensionless number, newton, joule, volt, ohm, watt, coulomb, ampere, weber, henry, farad. Give the equalities.

3. A 12-H inductor carries a steady current of 2.0 A. How can a 60-V self-induced emf be made to appear in the inductor?

4. At a given instant the current and the induced emf in an inductor are as indicated in Fig. 33–9. (*a*) Is the current increasing or decreasing? (*b*) If the emf is 17 V and the rate of change of the current is 25 kA/s, what is the value of the inductance?

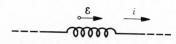

Figure 33–9 Exercise 4.

Section 33–2 Calculation of Inductance

5. A cylindrical solenoid with 100 turns per cm has a radius of 1.60 cm. Assume the magnetic field it produces is parallel to its axis and is uniform in its interior. (*a*) What is its inductance per (axis) meter? (*b*) If the current changes at the rate 12.5 A/s what emf is induced per meter?

6. A circular coil has a 10 cm radius and consists of 30 closely wound turns of wire. An externally produced magnetic field of 2.6 mT is perpendicular to the coil. (*a*) If no current is in the coil, what is the flux linkage? (*b*) When the current in the coil is 3.8 A in a certain direction, the net flux through the coil is found to vanish. What is the inductance of the coil?

Section 33–3 An LR Circuit

7. The current in an *LR* circuit builds up to one-third of its steady state value in 5.0 s. What is the inductive time constant?

8. How many "time constants" must we wait for the current in an *LR* circuit to build up to within 0.10 percent of its equilibrium value?

9. Suppose the emf of the battery in the circuit of Fig. 33–4 can be varied with time *t* so the current is given by $i(t) = 3.0 + 5.0t$, where i is in amperes and t is in seconds. Take $R = 4.0\ \Omega$, $L = 6.0$ H, and find an expression for the battery emf as a function of time.

Section 33–4 Energy and the Magnetic Field
10. The magnetic energy stored in a certain inductor is 25 mJ when the current is 60 mA. (*a*) What is the inductance? (*b*) What current is required for the magnetic energy to be four times as much?

11. For the circuit of Fig. 33–4, how long after the battery is connected will the energy stored in the magnetic field of the inductor be half its steady state value?

12. The inductive time constant for the circuit of Fig. 33–4 is 37 ms and the current in the circuit is zero at time $t = 0$. At what time does the rate at which energy is dissipated in the resistor equal the rate at which energy is being stored in the inductor?

Section 33–5 Energy Density and the Magnetic Field
13. A 90 mH toroidal inductor encloses a volume of 0.020 m³. If the average energy density in the toroid is 70 J/m³, what is the current?

14. The magnetic field in the interstellar space of our galaxy is thought to be about 10^{-10} T in magnitude. How much energy is stored in this field in a cube 10 light-years on edge? (For scale, note that the nearest star is 4.2 light-years distant and the radius of our galaxy is about 8×10^4 light-years.)

15. What must be the magnitude of a uniform electric field if it is to have the same energy density as that possessed by a 0.50-T magnetic field?

Section 33–6 Mutual Induction
16. Two coils are at fixed locations. When coil 1 has no current and the current in coil 2 increases at the rate 15 A/s, the emf in coil 1 is 25 mV. (*a*) What is their mutual inductance? (*b*) When coil 2 has no current and coil 1 has a current of 3.5 A, what is the flux linkage in coil 2?

17. Coil 1 has $L_1 = 25$ mH and $N_1 = 100$ turns. Coil 2 has $L_2 = 40$ mH and $N_2 = 200$ turns. The coils are rigidly positioned with respect to each other, their coefficient of mutual inductance M being 3.0 mH. A 6.0-mA current in coil 1 is changing at the rate of 4.0 A/s. (*a*) What flux Φ_{12} links coil 1 and what self-induced emf appears there? (*b*) What flux Φ_{21} links coil 2 and what mutually induced emf appears there?

PROBLEMS GROUPED BY SECTION

Section 33–1 Self-Induction
1. The flux linkage through a certain coil of negligible resistance is 25 mWb when there is a current of 5.5 A in it. If a 6.0-V battery is suddenly connected across the coil, how long will it take for the current to rise from 0 to 2.5 A?

2. The inductance of a closely wound N turn coil is such that an emf of 3.0 mV is induced when the current changes at the rate 5.0 A/s. A steady current of 5.0 A produces a magnetic flux of 40 μWb through each turn. (*a*) What is the inductance of the coil? (*b*) How many turns does the coil have?

Section 33–2 Calculation of Inductance
3. A solenoid is wound with a single layer of insulated copper wire (diameter, 2.5 mm). It is 4.0 cm in diameter and 2.0 m long. What is the inductance per unit length for the solenoid near its center? Assume that adjacent wires touch and that insulation thickness is negligible.

4. A long thin solenoid can be bent into a ring to form a toroid. Show that if the solenoid is long and thin enough, the equation for the inductance of a toroid (see Example 1) reduces to that for a solenoid of the appropriate length (Eq. 33–5).

5. A wide copper strip of width W is bent into a piece of slender tubing of radius R with two plane extensions, as shown in Fig. 33–10. A current i exists through the strip, distributed uniformly over its width. In this way a "one-turn solenoid" has been formed. (*a*) Find the magnitude of the magnetic field **B** in the tubular part (far away from the edges). (*Hint:* Assume that the field outside this one-turn solenoid is negligibly small.) (*b*) Find the inductance of this one-turn solenoid, neglecting the two plane extensions.

Section 33–3 An LR Circuit
6. A 50-V potential difference is suddenly applied to a coil with $L = 50$ mH and $R = 180\ \Omega$. At what rate is the current increasing after 1.0 ms?

7. A solenoid having an inductance of 6.0 μH is connected in series to a resistor. (*a*) If a 10-V battery is switched across the pair, how long will it take for the current through the resistor to reach 80 percent of its final value? (*b*) What is the current through the resistor after one time constant?

8. A wooden toroidal core with a square cross section has an inner radius of 10 cm and an outer radius of 12 cm. It is wound with one layer of wire (diameter 1.0 mm; "resistance" 0.02 Ω/m). What are (*a*) the inductance and (*b*) the inductive time constant? Ignore the thickness of the insulation.

*9. For the circuit shown in Fig. 33–11, switch S is closed at time $t = 0$. Thereafter the constant current source maintains a constant current i out of its upper terminal. (*a*) Derive an expression for the current through the inductor as a function of time.

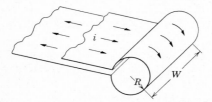

Figure 33–10 Problem 5.

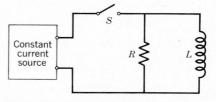

Figure 33–11 Problem 9.

(b) Show that the current through the resistor equals the current through the inductor at time $t = (L/R) \ln 2$.

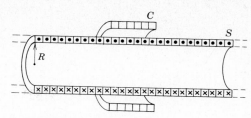

Figure 33–12   Problem 15.

Section 33–4   Energy and the Magnetic Field

10. A coil is connected in series with a 10-kΩ resistor. When a 50-V battery is applied to the two, the current reaches a value of 2.0 mA after 5.0 ms. (a) Find the inductance of the coil. (b) What is the energy stored in the inductance at this same moment?

11. A coil with an inductance of 2.0 H and a resistance of 10 Ω is suddenly connected to a resistanceless battery with $\varepsilon = 100$ V. At 0.10 s after the connection is made, what are the rates at which (a) energy is being stored in the magnetic field, (b) thermal energy is appearing, and (c) energy is being delivered by the battery?

12. For the circuit of Fig. 33–4, $\varepsilon = 10$ V, R = 6.7 Ω, and L = 5.5 H. The battery is connected at time $t = 0$. (a) How much energy is delivered by the battery during the first 2.0 s? (b) How much of this energy is stored in the magnetic field of the inductor? (c) How much has been dissipated in the resistor?

Section 33–5   Energy Density and the Magnetic Field

13. A circular loop of wire 50 mm in radius carries a current of 100 A. What is the energy density at the center of the loop?

14. (a) Find an expression for the energy density as a function of radial distance for the toroid of Example 1. (b) Integrating the energy density over the volume of the toroid, find the total energy stored in the field of the toroid; assume $i = 0.50$ A. (c) Using Eq. 33–18, evaluate the energy stored in the toroid directly from the inductance and compare with (b).

Section 33–6   Mutual Induction

15. A coil C of N turns is placed around a long solenoid S of radius R and n turns per unit length, as in Fig. 33–12. Show that the coefficient of mutual inductance for the coil-solenoid combination is given by

$$M = \mu_0 \pi R^2 nN.$$

Comment on the fact that M does not depend on the shape, size, or possible lack of close-packing of the coil.

16. Figure 33–13 shows a coil of N_2 turns linked as shown to a toroid of N_1 turns. The inner toroid radius is a, its outer radius is b, and its height is h, as in Fig. 33–2. Show that the coefficient of mutual inductance M for the toroid-coil combination is

$$M = \frac{\mu_0 N_1 N_2 h}{2\pi} \ln \frac{b}{a}.$$

Comment on the fact that M does not depend on the shape, size, or possible lack of close-packing of the coil.

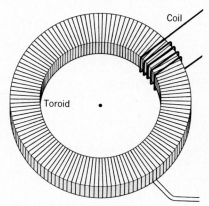

Figure 33–13   Problem 16.

17. A rectangular loop of N close-packed turns is positioned near a long straight wire as in Fig. 32–48. (a) What is the coefficient of mutual inductance M for the loop-wire combination? (b) Evaluate M for $N = 100$, $a = 1.0$ cm, $b = 8.0$ cm, and $l = 30$ cm.

ADDITIONAL PROBLEMS

18. Two long parallel wires whose centers are a distance d apart carry equal currents in opposite directions. Show that, neglecting the flux within the wires themselves, the inductance of a length l of such a pair of wires is given by

$$L = \frac{\mu_0 l}{\pi} \ln \frac{d-a}{a},$$

where a is the wire radius. See Example 2, Chapter 31.

19. *Inductors in series.* Two inductors L_1 and L_2 are connected in series and are separated by a large distance. (a) Show that the equivalent inductance is given by

$$L_{eq} = L_1 + L_2.$$

(b) Why must their separation be large for this relationship to hold? (c) What is the generalization of (a) for N inductors in series?

20. *Inductors in parallel.* Two inductors L_1 and L_2 are connected in parallel and separated by a large distance. (a) Show that the equivalent inductance is given by

$$\frac{1}{L_{eq}} = \frac{1}{L_1} + \frac{1}{L_2}.$$

(b) Why must their separation be large for this relationship to hold? (c) What is the generalization of (a) for N inductors in parallel?

21. How long would it take for the potential difference across the resistor in an LR circuit ($L = 2.0$ H, $R = 3.0$ Ω) to drop to 10 percent of its initial value?

22. In Fig. 33–14, $\varepsilon = 100$ V, $R_1 = 10$ Ω, $R_2 = 20$ Ω, $R_3 = 30$ Ω, and $L = 2.0$ H. Find the values of i_1 and i_2 (a) immediately after switch S is closed; (b) a long time later; (c) immediately after switch S is opened again; (d) a long time later.

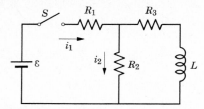

Figure 33-14 Problem 22.

23. The current is an LR circuit drops from 1.0 A at $t = 0$ to 10 mA one second later. If L is 10 H, find the resistance R in the circuit.

24. (a) In the LR circuit of Fig. 33–3, what is the self-induced emf ε_L when the switch has just been closed on a? (b) What is ε_L after two time constants? (c) After how many time constants will ε_L be just one-half of the battery emf ε?

25. A coil with an inductance of 2.0 H and a resistance of 10 Ω is suddenly connected to a resistanceless battery with $\varepsilon = 100$ V. (a) What is the equilibrium current? (b) How much energy is stored in the magnetic field when this current exists in the coil?

26. Prove that when switch S in Fig. 33–3 is thrown from a to b all the energy stored in the inductor appears as thermal energy in the resistor.

27. (a) What is the magnetic energy density of the earth's magnetic field of 50 μT? (b) Assuming this to be relatively constant over distances small compared with the earth's radius and neglecting the variations near the magnetic poles, how much energy would be stored in a shell between the earth's surface and 16 km above the surface?

28. A length of copper wire carries a current of 10 A, uniformly distributed. Calculate (a) the magnetic energy density and (b) the electric energy density at the surface of the wire. The wire diameter is 2.5 mm and its resistance per unit length is 3.3 Ω/km.

29. Figure 33–15 shows, in cross section, two coaxial solenoids. Show that the coefficient of mutual inductance M for a length l of this solenoid-solenoid combination is given by

$$M = \pi R_1^2 l \mu_0 n_1 n_2,$$

in which n_1 and n_2 are the respective numbers of turns per unit length and R_1 is the radius of the inner solenoid. Why does M depend on R_1 and not on R_2?

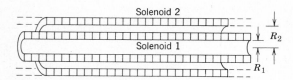

Figure 33-15 Problem 29.

30. In the circuit shown in Fig. 33–16, $\varepsilon = 10$ V, $R_1 = 5.0$ Ω, $R_2 = 10$ Ω, and $L = 5.0$ H. For the two separate conditions (I) switch S just closed and (II) switch S closed for a long time, calculate (a) the current i_1 through R_1, (b) the current i_2 through R_2, (c) the current i through the switch, (d) the potential difference across R_2, (e) the potential difference across L, and (f) di_2/dt.

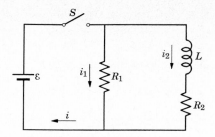

Figure 33-16 Problem 30.

31. In Fig. 33–17 two coils are connected as shown. The coils have self-inductances L_1 and L_2 and a coefficient of mutual inductance M. (a) Show that this combination can be replaced by a single coil of equivalent inductance given by

$$L_{eq} = L_1 + L_2 + 2M.$$

(b) How could the coils in Fig. 33–17 be reconnected to yield an equivalent inductance of

$$L_{eq} = L_1 + L_2 - 2M?$$

Note that this problem is an extension of Problem 19, in which the requirement that the coils be far apart is removed.

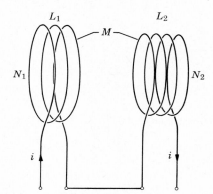

Figure 33-17 Problem 31.

32. A solenoid, with length 80 cm and radius 5.0 cm, consists of 3000 turns distributed uniformly over its length. Its total resistance is 10 Ω. 5.0 ms after it is connected to a 12-V battery (a) how much energy is stored in its magnetic field and (b) how much energy has been supplied by the battery up to that time?

33. In Fig. 33–18 the component in the upper branch is an ideal 3-A fuse. It has zero resistance as long as the current through it remains less than 3 A. If the current reaches 3 A, it blows and thereafter it has infinite resistance. Switch S is closed at time $t = 0$.

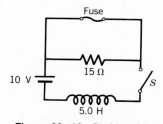

Figure 33-18 Problem 33.

(a) Find the time at which the fuse blows. (b) Sketch a graph of the current i through the inductor as a function of time. Mark the time at which the fuse blows.

*34. *Equality of M_{21} and M_{12}.* Figure 33–19 shows two LR circuits that are coupled by their mutual inductances M_{21} and M_{12}. (a) Show that the loop equations for this circuit are:

$$\varepsilon_1 = i_1 R_1 + L_1 di_1/dt + M_{12} di_2/dt,$$
$$\varepsilon_2 = i_1 R_1 + L_2 di_2/dt + M_{21} di_1/dt,$$

(The sign of the M terms depends on the relative orientation of the windings of the two inductors). (b) Combine the above two equations to obtain

$$\varepsilon_1 i_1 + \varepsilon_2 i_2 = i_1{}^2 R_1 + i_2{}^2 R_2 + L_1 i_1 di_1/dt + L_2 i_2 di_2/dt \\ + M_{12} i_1 di_2/dt + M_{21} i_2 di_1/dt.$$

(c) Following the discussion of Section 33–4, interpret the two terms in the left-hand side and the first two terms in the right-hand side of (b). (c) The remaining four terms in (b), i.e., those involving the self and mutual inductances, must be dU_B/dt. If $M_{12} = M_{21} = M$, concoct a function U_B (depending on L_1, L_2, M, i_1 and i_2) such that dU_B/dt equals these remaining four terms. (d) Now try to find a function U_B if $M_{12} \neq M_{21}$. Your inability to do this should convince you that these two mutual inductances must be equal.

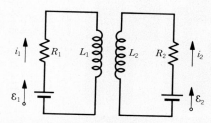

Figure 33–19   Problem 34.

34 Magnetic Properties of Matter

34-1 Poles and Dipoles

In electricity the isolated charge q is the simplest structure that can exist. If two such charges of opposite sign are placed near each other, they form an electric dipole, characterized by an electric dipole moment **p**. In magnetism isolated magnetic poles (usually called magnetic monopoles) which would correspond to isolated electric charges, apparently do not exist. The simplest magnetic structure is the magnetic dipole, characterized by a magnetic dipole moment $\boldsymbol{\mu}$. Table 31–1 summarizes some characteristics of electric and magnetic dipoles.

Current loops, bar magnets, and short solenoids all produce magnetic fields which, for points far enough from the source, are the same as the field from a simple idealized magnetic dipole. We sometimes refer to such field sources as magnetic dipoles, even though for points close to them the fields that they produce are more complicated than those of simple dipoles.

We can identify the north poles of such dipoles (that is, the regions from which lines of **B** emerge) by suspending them as compass needles and observing which end points north. We can find their magnetic dipole moments by placing the dipole in an external magnetic field **B**, measuring the torque $\boldsymbol{\tau}$ that acts on it, and computing $\boldsymbol{\mu}$ from

$$\boldsymbol{\tau} = \boldsymbol{\mu} \times \mathbf{B}. \tag{34-1}$$

Alternatively, we can measure **B** due to the dipole at a point along its axis a (large) distance r from its center and compute μ from the expression in Table 31–1, or

$$B = \frac{\mu_0}{2\pi} \frac{\mu}{r^3}. \tag{34-2}$$

Figure 34–1 A bar magnet is a magnetic dipole. The iron filings suggest the pattern of lines of **B** in Figure 34–4a. PSSC PHYSICS, 2nd ed., 1965, D. C. Heath & Co. with Education Development Center, Inc., Newton, MA 02160.

Example 1 Devise a method for measuring μ for a bar magnet.

(a) Place the magnet in a uniform external magnetic field **B**, with $\boldsymbol{\mu}$ making an angle θ with **B**. The magnitude of the torque acting on the magnet (see Eq. 34–1) is given by

$$\tau = \mu B \sin \theta.$$

Clearly we can find μ if we measure τ, B, and θ.

(b) A second technique is to suspend the magnet from its center of mass and to allow it to oscillate about its stable equilibrium position in the external field **B**. For small oscillations, $\sin \theta$ can be replaced by θ and the equation just given becomes

$$\tau = -(\mu B)\theta = -\kappa\theta,$$

where κ is a constant. The minus sign has been inserted to show that τ is a restoring torque. Since τ is proportional to θ, the condition for simple angular harmonic motion is met. The frequency ν is given by the reciprocal of Eq. 14–23, or

$$\nu = \frac{1}{2\pi}\sqrt{\frac{\kappa}{I}} = \frac{1}{2\pi}\sqrt{\frac{\mu B}{I}},$$

in which I is the rotational inertia. With this equation μ can be found from the measured quantities ν, B, and I.

Figure 34–1, which shows iron filings sprinkled on a sheet of paper under which there is a bar magnet, suggests that it might be viewed as two "poles" separated by a distance d. However, all attempts to isolate these poles fail. If we break the magnet, as in Fig. 34–2, the fragments prove to be dipoles and not isolated poles. Where one north pole and one south pole existed there are now three of each. If we break up a magnet into the electrons, protons, and neutrons that make up its atoms, we will find that even these subnuclear particles are magnetic dipoles.

Consider, for example, the electric and magnetic properties of the free electron, at rest in a convenient inertial reference frame. On the electric side we have given a complete description when we specify the electric charge, thus

$$e = -1.602 \times 10^{-19}\ \text{C}.$$

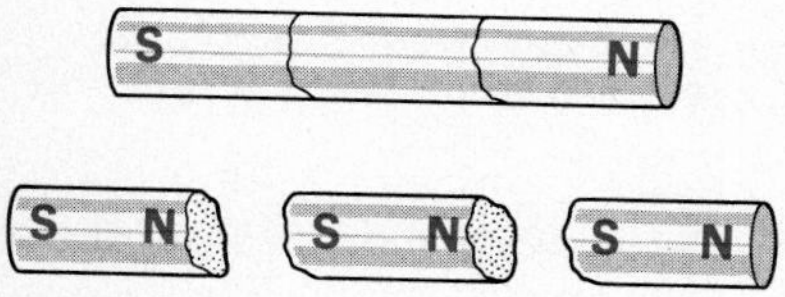

Figure 34–3a suggests the corresponding, spherically symmetric, lines of the electric field **E**.

On the magnetic side we learn that the electron has an intrinsic "spin" angular momentum **S**, whose direction establishes an axis of symmetry for the electron. There is also an associated intrinsic "spin" magnetic dipole moment $\boldsymbol{\mu_s}$, whose direction is opposite to that of **S**. The magnitudes of these two quantities are

$$S = 5.273 \times 10^{-35}\ \text{J}\cdot\text{s} \qquad \text{and} \qquad \mu_s = 9.285 \times 10^{-24}\ \text{J/T}.$$

Figure 34–3b suggests the corresponding, axially symmetric, lines of the magnetic field **B**.

It is tempting to account for the magnetic properties of the electron by viewing it as a tiny spinning sphere of charge, which we can treat classically as an assembly of infinitesimal current loops. Each such loop is a tiny magnetic dipole, its moment being given by (see Eq. 30–10)

$$\mu = iA, \tag{34–3}$$

Figure 34–2 If we break a bar magnet, each fragment becomes a dipole. There will always be equal numbers of north and south poles associated in pairs.

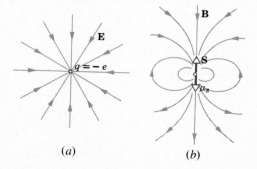

Figure 34–3 (a) The lines of **E** and (b) the lines of **B** for a resting free electron.

(a)   (b)

where i is the equivalent current in each infinitesimal loop and A is the loop area. The magnetic dipole moment of the spinning charge can be found by adding the moments of the infinitesimal current loops that make it up.

Although this model of the spinning electron is too mechanistic and is not in accord with modern quantum physics, it remains true that the magnetic dipole moments of elementary particles are closely connected with their intrinsic angular momenta, or spins. Those particles and nuclei whose spin angular momentum is zero (the α particle, the pion, the ^{16}O nucleus, etc.) have no magnetic dipole moment. We must distinguish the "intrinsic" or "spin" magnetic moment of the electron from any additional magnetic moment it may have because of its orbital motion in an atom; see Example 2.

Example 2 *The magnetic dipole moment of an electron due to its orbital motion*   An electron in a hydrogen atom circulating in an assumed circular orbit of radius r behaves like a tiny current loop and has an orbital magnetic dipole moment $\boldsymbol{\mu}_l$. Derive a connection between μ_l and the orbital angular momentum L.*

Newton's second law ($F = ma$) applied to the circulating electron yields, if we substitute Coulomb's law for F,

$$\frac{1}{4\pi\epsilon_0}\frac{e^2}{r^2} = ma = \frac{mv^2}{r},$$

or

$$v = \sqrt{\frac{e^2}{4\pi\epsilon_0 mr}}. \qquad (34-4)$$

The angular velocity ω is given by

$$\omega = \frac{v}{r} = \sqrt{\frac{e^2}{4\pi\epsilon_0 mr^3}}.$$

The current for the orbit is the rate at which charge passes any given point, or

$$i = ev = e\left(\frac{\omega}{2\pi}\right) = \sqrt{\frac{e^4}{16\pi^3\epsilon_0 mr^3}}.$$

The orbital dipole moment μ_l is given from Eq. 34–3 if we put $A = \pi r^2$;

$$\mu_l = iA = \sqrt{\frac{e^4}{16\pi^3\epsilon_0 mr^3}}(\pi r^2) = \frac{e^2}{4}\sqrt{\frac{r}{\pi\epsilon_0 m}}. \qquad (34-5)$$

The orbital angular momentum L is

$$L = (mv)r.$$

Combining with Eq. 34–4 leads to

$$L = \sqrt{\frac{e^2 mr}{4\pi\epsilon_0}}.$$

Finally, eliminating r between this equation and Eq. 34–5 yields

$$\mu_l = L\left(\frac{e}{2m}\right), \qquad (34-6)$$

which shows that the orbital magnetic moment of an electron is proportional to its orbital angular momentum. Convince yourself that the vectors $\boldsymbol{\mu}_l$ and $\mathbf{L}$ point in opposite directions.

Even though our derivation is too mechanistic in its approach and is not in the spirit of modern quantum physics it turns out that Eq. 34–6 is nevertheless a correct result. The lowest nonzero value of orbital angular momentum that the hydrogen atom can have is (we assert; see Section 41–8) 1.055×10^{-34} J·s. The corresponding orbital angular momentum is, from Eq. 34–6,

$$\mu_l = L\left(\frac{e}{2m}\right)$$
$$= \frac{(1.055 \times 10^{-34} \text{ J·s})(1.602 \times 10^{-19} \text{ C})}{(2)(9.110 \times 10^{-31} \text{ kg})}$$
$$= 9.276 \times 10^{-24} \text{ J/T}.$$

Note that this is very close to, but slightly less than, the value reported above for the intrinsic spin magnetic moment of the electron.

The apparent absence of free magnetic monopoles—the magnetic equivalents of free electric charges—represents a lack of symmetry in nature that many physicists find disturbing. This is especially true because in 1931 the physicist P. A. M. Dirac predicted, on the basis of theory, that magnetic monopoles should indeed exist. Physicists have sought for them constantly since that time but so far to no avail.†

---

* These two quantities must not be confused with μ_s and S (discussed above) which refer specifically to the spin motion of the electron. Our concern in this example is solely with the electron's orbital motion.

† See "Quest for the Magnetic Monopole" by Richard A. Carrigan, Jr., *The Physics Teacher*, October 1975.

The symmetry of nature has always been a fruitful guiding principle for physicists. In 1928, for example, this same P. A. M. Dirac developed a theory of the electron, which was later interpreted as predicting the existence of a positively charged counterpart to the well-known negatively charged electron. When positive electrons were discovered in 1933 by C. D. Anderson they fit in beautifully with the predictions of this totally symmetric theory. In the same way the existence of a (positive) proton suggested that there might also be a negative proton and a large accelerator was built at the University of California at Berkeley primarily to search for this particle; it was found. The discoveries of both the positive electron and the negative proton were recognized with Nobel prizes.

34-2 Gauss's Law for Magnetism

Gauss's law for magnetism, which is one of the basic equations of electromagnetism (see Table 37-2), is a formal way of stating a conclusion that seems to be forced on us by the facts of magnetism, namely, that isolated magnetic poles do not seem to exist. This equation asserts that the flux Φ_B through any closed Gaussian surface must be zero, or

Gauss's law for magnetism

$$\Phi_B = \oint \mathbf{B} \cdot d\mathbf{S} = 0, \tag{34-7}$$

where the integral is to be taken over the entire closed surface. We contrast this with Gauss's law for electricity, which is

$$\epsilon_0 \Phi_E = \epsilon_0 \oint \mathbf{E} \cdot d\mathbf{S} = q. \qquad \text{[see 25-6]}$$

The fact that a zero appears at the right of Eq. 34-7, but not at the right of Eq. 25-6, means that in magnetism there seems to be no counterpart to the free charge q in electricity.

Figure 34-4a suggests a Gaussian surface, marked I, enclosing one end of a short solenoid. Such a solenoid, as we have seen, sets up a magnetic dipole field at large distances; the end of the solenoid enclosed by surface I behaves, for such distant points, like a north magnetic pole. Note that the lines of the magnetic field $\mathbf{B}$ enter the Gaussian surface inside the solenoid and leave it outside the solenoid. No lines originate or terminate inside this surface; in other language, there are no sources or sinks of $\mathbf{B}$; in still other language, there are no free magnetic poles. Thus the total flux Φ_B for surface I in Fig. 34-4a is zero, as Gauss's law for magnetism (Eq. 34-7) requires.

No free magnetic poles

We also have $\Phi_B = 0$ for surface II in Fig. 34-4a and indeed for any closed surface that can be drawn in this figure. The situation is just the same if we replace the short solenoid by a short bar magnet, as in Fig. 34-4b. Here too $\Phi_B = 0$ for any closed surface that we can draw.

Figure 34-4c shows a close electrostatic analog to the two magnetic dipoles that we have been discussing. It consists of two oppositely charged circular disks facing each other as shown. The electric field $\mathbf{E}$ set up at distant points by this arrangement is also that of a dipole. In this case there *is* a net (outward) flux of the field lines for the Gaussian surface marked I; there *is* a source of the field, namely the positive charges enclosed by the surface. Of course, for a Gaussian surface such as that marked II in Fig. 34-4c we have $\Phi_E = 0$ because this surface happens to enclose no charge. For any closed Gaussian surface that we can draw Gauss's law for electrostatics holds.

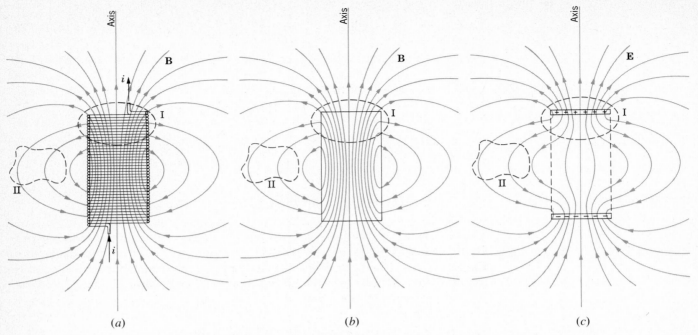

(a)   (b)   (c)

Figure 34-4 The lines of **B** for (a) a short solenoid and (b) a short bar magnet. In each case the top of the structure is a north pole, the bottom being a south pole. (c) The lines of **E** for two charged disks. At large distances all three fields vary like those for a dipole. The dashed curves represent closed Gaussian surfaces.

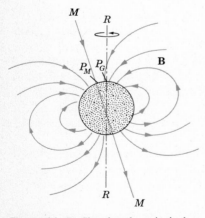

Figure 34-5 Showing the principal or dipole component of the earth's magnetic field. The earth's magnetic dipole axis *MM* makes an angle of 11.5° with its rotation axis *RR*. Points P_G and P_M are the geographic north and the geomagnetic south poles, respectively.

34-3 The Magnetism of the Earth

Our earth is associated with a magnetic field of considerable practical and theoretical interest. As Fig. 34-5 suggests this field can be represented to a good approximation as due to a small dipole (with $\mu = 8.0 \times 10^{22}$ J/T) located at the center of the earth. The dipole axis, shown as *MM* in Fig. 34-5, makes an angle of 11.5° with the earth's rotation axis, shown as *RR* in that figure. *MM* intersects the earth's surface at two points, defining the geomagnetic north pole (in northwest Greenland) and the geomagnetic south pole (in Antarctica). In general lines of **B** for the earth's field enter the earth in the northern hemisphere and leave it in the southern; thus the magnetic pole in the northern hemisphere is actually a south magnetic pole, similar to the bottom end of the bar magnet of Fig. 34-4b.

At any point on the earth's surface the observed magnetic field may differ appreciably from the idealized, best-fit, dipole field, both in magnitude and direction. At any location this observed field varies with time, by measureable amounts over a period of a few years and by substantial amounts over, say, 100 years. Thus between 1580 and 1820 the direction of the compass needle at London changed by 35°.

In spite of these variations the best-fit dipole field changes only slowly over such time periods, which are after all quite short compared to the age of the earth. The variation of the earth's field over much longer intervals can be studied by measuring the weak intrinsic magnetism of rocks, which preserve a ''frozen-in'' record of the earth's magnetic field at the time the rocks solidified and cooled.

Such studies tell us, among other things, that the earth's field completely changes its direction every million years or so.

There is at present no satisfactory detailed explanation for the origin of the earth's magnetic field. It seems certain that it arises in some way from circular current loops induced in the liquid and highly conducting outer region of the earth's core. The precise mode of action of this internal geomagnetic "dynamo" and the source of energy needed to keep it operating remain matters of continuing research interest.*

Several other planets in our solar system, Mercury and Jupiter among them, also have magnetic fields. So do the sun and many other stars. There is also a magnetic field associated with our Milky Way Galaxy. This field is quite weak (~ 2 pT, or 2×10^{-8} gauss) but also quite important because of its immense size.

Magnetic declination and inclination

Because of its practical applications in navigation, communication, and prospecting, the earth's magnetic field has been studied extensively for many years. The quantities of interest, as for any vector field, are the magnitude and direction of the field at different locations on the earth's surface and in the surrounding space. Field directions near the earth's surface are conveniently specified, with reference to the earth itself, in terms of the field declination (the angle between true geographic north and the horizontal component of the field) and the inclination (the angle between a horizontal plane and the field direction). There are a variety of commercially available magnetometers used to measure these quantities with high precision, but rough measurements can be made with a compass and a dip meter. The compass is just a magnet mounted for free rotation about a vertical axis. Its direction indicates the field declination. A dip meter is a similar magnet mounted for free rotation in a vertical plane. When aligned with its plane of rotation parallel to the compass direction, the angle between the needle and the horizontal is the inclination. The strength of the field can be estimated by measuring the frequency of the oscillations of either instrument after being displaced from its equilibrium position. The calculations are the same as we described earlier in Example 1.

As an example, the north pole of a compass needle in Tucson, Arizona pointed about 13° east of geographic north (the declination) in 1964. The magnitude of the horizontal component of the earth's field was 26 μT ($= 0.26$ gauss), and the north end of a dip needle pointed downward making an angle of about 59° (the inclination) with a horizontal plane. As we expect from Fig. 34–5, lines of **B** are entering the earth's surface at this point.

Example 3 From data given earlier in this section find (a) the vertical component B_v of the earth's magnetic field and (b) the magnitude of the resultant magnetic field B at Tucson.

Figure 34–6 shows the situation. We have

(a)
$$B_v = B_h \tan \varphi_i$$
$$= (26 \ \mu\text{T})(\tan 59°)$$
$$= 44 \ \mu\text{T} \ (= 0.44 \text{ gauss}),$$

and

(b)
$$B = \sqrt{B_h{}^2 + B_v{}^2}$$
$$= \sqrt{(26 \ \mu\text{T})^2 + (44 \ \mu\text{T})^2}$$
$$= 51 \ \mu\text{T} \ (= 0.51 \text{ gauss}).$$

Note that the magnetic declination at Tucson plays no role in this problem.

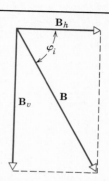

Figure 34–6 Example 3.

* "The Source of the Earth's Magnetic Field," by Charles R. Carrigan and David Gubbins, *Scientific American*, February 1979.

34–4   Paramagnetism

Magnetism as we know it in our daily experience is an important but special branch of the subject called *ferromagnetism;* we discuss this in Section 34–6. Here we discuss a weaker form of magnetism called *paramagnetism.*

For most atoms and ions, the magnetic effects of the electrons, including both their spins and orbital motions, exactly cancel so that the atom or ion is not magnetic. This is true for the rare gases such as neon and for ions such as Cu^+, which makes up ordinary copper.* For other atoms or ions the magnetic effects of the electrons do not cancel, so that the atom as a whole has a magnetic dipole moment **μ**. Examples are found among the transition elements, such as Mn^{++}; the rare earths, such as Gd^{+++}; and the actinide elements, such as U^{++++}.

If we place a sample of N atoms, each of which has a magnetic dipole moment **μ**, in a magnetic field, the elementary atomic dipoles tend to line up with the field. This tendency to align is called paramagnetism. For perfect alignment, the sample as a whole would have a magnetic dipole moment of $N\mu$. However, the aligning process is seriously interfered with by thermal agitation effects. The importance of thermal agitation may be measured by comparing two energies: one ($=\frac{3}{2}kT$) is the mean kinetic energy of translation of a gas atom at temperature T; the other ($=2\mu B$) is the difference in energy between an atom lined up with the magnetic field and one pointing in the opposite direction. As Example 4 shows, the effect of the collisions at ordinary temperatures and fields is very important. The sample acquires a magnetic moment when placed in an external magnetic field, but this moment is usually very much smaller than the maximum possible moment $N\mu$.

Paramagnetism

Alignment vs. thermal agitation

Example 4   A paramagnetic gas, whose atoms (see Example 2) have a magnetic dipole moment of about 10^{-23} J/T, is placed in an external magnetic field of magnitude 1.0 T ($=10^4$ gauss). At room temperature ($T = 300$ K) calculate and compare U_T, the mean kinetic energy of translation ($=\frac{3}{2}kT$), and U_B, the magnetic energy ($=2\mu B$):

$$U_T = \tfrac{3}{2}kT = (\tfrac{3}{2})(1.38 \times 10^{-23} \text{ J/K})(300 \text{ K}) = 6 \times 10^{-21} \text{ J},$$
$$U_B = 2\mu B = (2)(10^{-23} \text{ J/T})(1.0 \text{ T}) = 2.0 \times 10^{-23} \text{ J}.$$

Because U_T equals 300 U_B, we see that energy exchanges in collisions can interfere seriously with the alignment of the dipoles with the external field.

We can express the extent to which a given specimen of a material is magnetized by dividing its measured magnetic moment by its volume. This quantity, the magnetic moment per unit volume, is called the *magnetization M* of the material in question.

In 1895 Pierre Curie discovered experimentally that the magnetization M of a paramagnetic specimen is directly proportional to B, the effective magnetic field in which the specimen is placed, and inversely proportional to the temperature, or

Curie's law

$$M = C\frac{B}{T}, \tag{34–8}$$

in which C is a constant. This equation is known as Curie's law and the constant C is called Curie's constant. The law is physically reasonable in that increasing B tends to align the elementary dipoles in the specimen, that is, to increase M, whereas increasing T tends to interfere with this alignment, that is, to decrease M. Curie's law is well verified experimentally, provided that the ratio B/T does not become too large.

---

* Cu^+ indicates a copper atom from which one electron has been removed; Al^{+++} indicates an aluminum atom from which three electrons have been removed, etc.

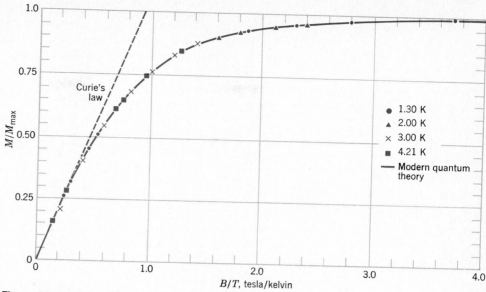

Figure 34–7 The ratio M/M_{max} for a paramagnetic salt, measured at various magnetic fields and at several low temperatures ($\leq$ 4.21 K). From W. E. Henry.

Paramagnetic saturation

M cannot increase without limit, as Curie's law implies, but must approach a value M_{max} ($= \mu N/V$) corresponding to the complete alignment of the N dipoles contained in the volume V of the specimen. Figure 34–7 shows this saturation effect for a sample of potassium chromium sulfate. The chromium ions are responsible for all the paramagnetism of this salt, all the other elements being paramagnetically inert. To achieve 99.5% saturation, it is necessary to use applied magnetic fields as high as 5.0 T ($=$ 50,000 gauss) and temperatures as low as 1.3 K. Note that for room temperature ($\cong$ 300 K) and for magnetic fields as strong as, say, 3.0 T ($=$ 30,000 gauss) the abscissa in Fig. 34–7 is only 0.01. This is well within the range that Curie's law, shown by the straight tangent line in the figure, is obeyed. The curve that passes through the experimental points in Fig. 34–7 is calculated from a theory based on modern quantum physics; we see that it is in excellent agreement with experiment.

34–5 Diamagnetism

If a sample of a paramagnetic material is placed near a pole of a strong magnet it will be weakly attracted. We can associate this with the fact that the atoms of a paramagnetic substance have permanent magnetic dipole moments; these intrinsic dipoles tend to line up with an external magnetic field, just like compass needles in the earth's field.

Substances repelled by magnetic poles are called diamagnetic

Some substances, called *diamagnetic*, are (weakly) repelled when placed near the pole of a strong magnet. The atoms of a diamagnetic substance have no intrinsic magnetic dipole moment. However, a magnetic dipole may be induced in such an otherwise magnetically inert atom by the action of an external field. These induced dipoles point in a direction opposite to that of the field that caused them.

Diamagnetism and Lenz's law

Diamagnetism (see Problem 16) is a manifestation of Lenz's law acting on the atomic scale. In a diamagnetic atom the electrons circulating around the nucleus behave like tiny current loops. Each has an orbital magnetic moment but for the atom as a whole the individual moments cancel, yielding a resultant moment of zero. An external magnetic field acts by induction to change the magnetic moments of the individual circulating electrons so that they no longer cancel. The magnetic moment so induced, as we shall see, points in a direction opposite to the external magnetic field and goes to zero when that field is removed.

Let us examine a particular case, qualitatively but in detail. Figures 34–8a and b show the orbits of two electrons in a single diamagnetic atom, circulating in the same plane but in opposite directions. An attractive electrostatic Coulomb force $\mathbf{F}_E$ acts as a centripetal force to hold them in their circular orbits. From Newton's second law we have

$$F_E = ma = mv^2/r.$$

Each of these two orbiting electrons is equivalent to a tiny current loop and has an associated magnetic moment of magnitude μ. In Fig. 34–8a this magnetic moment $(-\boldsymbol{\mu})$ points into the page at right angles; in Fig. 34–8b it points out $(+\boldsymbol{\mu})$, the net magnetic moment for the two orbits taken together being zero. This cancellation of moments is suggested in the upper right of Fig. 34–8.

Consider now the effect of applying an external magnetic field **B** to the diamagnetic atom. A new force, of magnetic origin, given by

$$\mathbf{F}_B = -e\,(\mathbf{v} \times \mathbf{B})$$

now acts on each orbiting electron. Convince yourself that in the case of Fig. 34–8c the resultant centripetal force is weakened but that in Fig. 34–8d it is strengthened. This means that the electron of Fig. 34–8a will slow down, to some speed $v - \Delta v$, but that of Fig. 34–8b will speed up to $v + \Delta v$. Thus the magnetic

Diamagnetism—analysis of a special case

Figure 34–8 (a, b) Two electrons orbit in opposite directions in the same diamagnetic atom; their two magnetic moments cancel. (c, d) When a magnetic field **B** is applied one electron is slowed down and the other is speeded up. The two magnetic moments no longer cancel, their resultant pointing in the direction of −**B**.

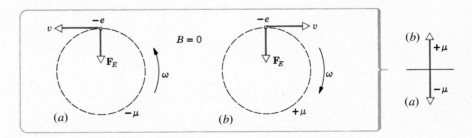

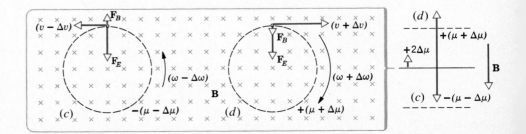

moment in the first case will be reduced in magnitude (to $\mu - \Delta\mu$) and in the second case it will be increased (to $\mu + \Delta\mu$). As the lower right portion of Fig. 34–8 shows the two magnetic moments no longer cancel. The resultant magnetic moment for the two electrons taken together is $+2\,\Delta\mu$ and points in a direction opposite to **B**, as we asserted above.

Diamagnetism is present in all atoms but if these atoms happen to have intrinsic magnetic dipole moments the diamagnetic effects are masked by the stronger paramagnetic or ferromagnetic behavior.

34–6 Ferromagnetism

When we speak of "magnetism" in everyday conversation we almost certainly have in mind the familiar behavior of bar magnets or electromagnets in attracting nails or other iron objects. We almost certainly do not have in mind the paramagnetic or diamagnetic effects discussed in the two preceding sections. These effects, although interesting and important for our understanding of the nature of matter, are weak by comparison.

It turns out that for iron and several other elements (most notably Co, Ni, Gd, and Dy) and for many alloys of these and other elements a special effect occurs that permits a specimen to achieve a high degree of magnetic alignment in spite of the randomizing tendency of the thermal motions of the atoms. In such materials, described as ferromagnetic, a special form of interaction called *exchange coupling* occurs between adjacent atoms and molecules, coupling their magnetic moments together in rigid parallelism. This is a purely quantum effect and cannot be explained in terms of classical physics. If the temperature is raised above a certain critical value, called the Curie temperature, the exchange coupling suddenly disappears and most materials become simply paramagnetic. For iron the Curie temperature is 1043 K (=770° C). Ferromagnetism is evidently a property not only of the individual atom or ion but also of the interaction of each atom or ion with its neighbors in the crystal lattice of the solid.

To study the magnetization of a ferromagnetic material such as iron it is convenient to form it into a toroidal or doughnut-shaped ring, as in Fig. 34–9. Primary coil P, containing n turns per unit length, is wrapped around it. This coil is essentially a long solenoid bent into a circle; if it carries a current i_p we can find the magnetic field within the toroidal space from Eq. 31–6 or

$$B_0 = \mu_0 n i_p \tag{34–9}$$

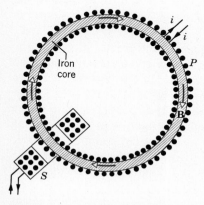

Figure 34–9 A specimen of iron whose ferromagnetic properties are being studied is formed into a Rowland ring. Secondary coil S is used to measure the magnetic field **B** within the specimen.

Note carefully that B_0 is the field that would be present within the toroid if the iron core were not in place. The arrangement of Fig. 34–9 is often called a *Rowland ring,* after the physicist H. A. Rowland (1848–1901) who devised it.

The actual magnetic field B in the toroidal space of the Rowland ring of Fig. 34–9, with the iron core in place, is greater than B_0, commonly by a large factor. We can write

$$B = B_0 + B_M \tag{34–10}$$

where B_M is the contribution of the iron core to the total magnetic field B. B_M is associated with the alignment of the elementary atomic dipoles in the iron and is proportional to the magnetization M of the iron, that is, to its magnetic moment per unit volume.

It is possible to measure B in the iron core by a method based on setting up an induced current in secondary coil S of Fig. 34–9. With the iron core initially unmagnetized set up a steady current i_P in the primary coil P. The magnetic flux

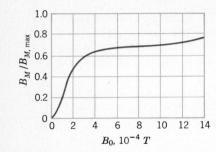

Figure 34–10 A magnetization curve for the core material of the Rowland ring of Fig. 34–9. On the vertical axis "1.0" corresponds to 100% alignment of the atomic magnetic dipoles within the specimen.

linking secondary coil S (of resistance R_S and N_S turns) will rise from zero to a value BA in a time Δt, A being the cross-sectional area of the iron core. From Faraday's law of induction (Eq. 32–2) an emf whose average value is

$$\overline{\mathcal{E}}_S = -N_S \frac{d\Phi}{dt} = -\frac{N_S BA}{\Delta t}$$

will appear in the secondary coil. The average current in this coil during the time that the flux is changing will then be

$$\bar{\imath}_S = \frac{\overline{\mathcal{E}}_S}{R_S} = \frac{N_S BA}{R_S \Delta t}.$$

We have then for B

$$B = \frac{(\bar{\imath}_S \Delta t) R_S}{N_S A} = \frac{\Delta q R_S}{N_S A}, \qquad (34\text{--}11)$$

in which $\Delta q\ (=\bar{\imath}_S \Delta t)$ is the charge that passes through the secondary coil during the time that the magnetic field in the iron core is changing. Thus a measurement of Δq (using an instrument called a ballistic galvanometer or in some other way) yields a measurement of the magnetic field B within the iron core.

Given the current i_P in the primary core we can easily calculate B_0 from Eq. 34–9. We can measure B as we have just described. Thus, from Eq. 34–10, we readily find B_M, a quantity that measures the extent to which the elementary dipoles in the iron core are aligned. B_M will have a maximum value $B_{M,\max}$ corresponding to complete alignment. It is instructive to plot, for any ferromagnetic specimen that can be formed into a Rowland ring, the ratio $B_M/B_{M,\max}$ as a function of B_0. Such a *magnetization curve*, as it is called, is displayed in Fig. 34–10. It is similar to the magnetization curve of Fig. 34–7 for a paramagnetic substance. Both are measures of the extent to which an applied magnetic field can succeed in aligning the elementary dipoles that make up the material in question.

A magnetization curve

For the ferromagnetic core material to which Fig. 34–10 refers the alignment of the dipoles is only about 75% effective for $B_0 = 14 \times 10^{-4}$ T ($= 14$ gauss). If B_0 were increased to 1.0 T ($= 10{,}000$ gauss) the fractional saturation of the specimen would still only be 99.7%. Such a value of B_0 would be ~ 120 ft to the right of the origin on the horizontal axis of Fig. 34–10; complete saturation of the dipoles is approached only with considerable difficulty.

Saturation effects

The use of iron in transformers, electromagnets, etc., greatly increases the strength of the magnetic field that can be generated by a given current in a given set of windings. That is, very often, $B_M \gg B_0$ in Eq. 34–10. However, in the design of electromagnets generating really intense magnetic fields ($3 - 25$ T, say) it proves poor practice to incorporate iron in the limited space available in the magnet core. The controlling design factor proves to be the rate at which cooling water can be pumped through this core to carry away the large amount of Joule heat generated in the energizing coils. A particular (iron-free) 25-T magnet at MIT, though only 36 in. in diameter, dissipates 16 MW and requires that cooling water course through its core at the rate of 2000 gal/min.

The magnetization curve for paramagnetism (Fig. 34–7) is explained in terms of the mutually opposing tendencies of alignment with the external field and of randomization because of the temperature motions. In ferromagnetism, however, we have assumed that adjacent atomic dipoles are locked in rigid parallelism. Why, then, does the magnetic moment of the specimen not reach its saturation value for very low—even zero—values of B_0? Why is not every iron nail a strong permanent magnet?

To understand this consider first a specimen of a ferromagnetic material such

Magnetic domains

as iron that is in the form of a single crystal. That is, the arrangement of atoms that make it up—its crystal lattice—extends with unbroken regularity throughout the volume of the specimen. Such a crystal will, in its normal unmagnetized state, be made up of a number of *magnetic domains*. These are regions of the crystal throughout which the alignment of the atomic dipoles is essentially perfect. For the crystal as a whole, however, the domains are so oriented that they cancel each other as far as their external magnetic effects are concerned.

Figure 34–11 is a photograph of such an assembly of domains in a single crystal of nickel. It was made by sprinkling a colloidal suspension of finely powdered iron oxide on a properly etched surface of the crystal. The domain boundaries, which are thin regions in which the alignment of the elementary dipoles changes from a certain orientation in one domain to a quite different orientation in the other, are the sites of intense, but highly localized and nonuniform magnetic fields. The suspended colloidal particles are attracted to these regions. Although the atomic dipoles in the individual domains are completely aligned, the crystal as a whole may have a very small resultant magnetic moment.

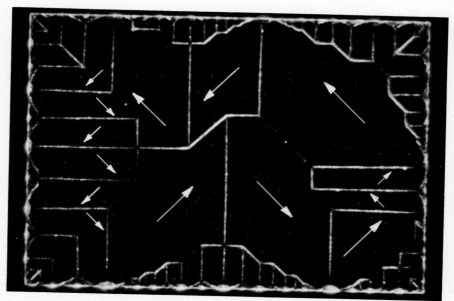

Figure 34–11 Domain patterns for a single crystal of nickel. The white lines show the boundaries between the domains. (Courtesy R. W. DeBlois.)

Actually a piece of iron as we ordinarily find it—an iron nail, say—is not a single crystal but is an assembly of many tiny crystals, randomly arranged; we call it a polycrystalline solid. Each tiny crystal, however, has its array of domains, just as in Fig. 34–11. Let us magnetize such a specimen by placing it in an external magnetic field of gradually increasing strength. Two effects take place, both contributing to the magnetization curve of Fig. 34–10. One is a growth in size of the domains that are favorably oriented at the expense of those that are not. Second, the orientation of the dipoles within a domain may swing around as a unit, becoming closer to the field direction.

Hysteresis

Magnetization curves for ferromagnetic materials do not retrace themselves as we increase and then decrease the external magnetic field B_0. Figure 34–12 shows the following operations with a Rowland ring: (1) starting with the iron un-

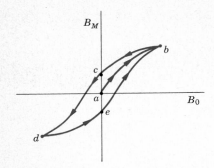

Figure 34–12 A magnetization curve (*ab*) for a specimen of iron and an associated hysteresis loop (*bcdeb*).

magnetized (point *a*), increase the toroid current until B_0 ($= \mu_0 ni$) has the value corresponding to point *b*; (2) reduce the current in the toroid winding back to zero (point *c*); (3) reverse the toroid current and increase it in magnitude until point *d* is reached; (4) reduce the current to zero again (point *e*); (5) reverse the current once more until point *b* is reached again. The lack of retraceability shown in Fig. 34–12 is called hysteresis. Note that at points *c* and *e* the iron core is magnetized, even though there is no current in the toroid windings; this is the familiar phenomenon of permanent magnetism.

Hysteresis can be understood on the basis of the magnetic domain concept outlined above. Evidently the motions of the domain boundaries and the reorientations of the domain directions are not totally reversible. When the applied magnetic field B_0 is increased and then decreased back to its initial value the domains do not return completely to their original configuration but retain some "memory" of the initial increase.

REVIEW GUIDE AND SUMMARY

Magnetic poles and dipoles

The simplest sources of the magnetic field are magnetic dipoles, associated either with the orbital motions of electrons in atoms (see Example 2) or with the intrinsic spin motions (see Fig. 34–3) of electrons, protons, and many other such particles. There is no convincing evidence that free magnetic poles (the magnetic equivalent of free electric charges) exist, either at the level of elementary particles or of gross objects; see Fig. 34–2.

Gauss's law for magnetism

Gauss's law for magnetism

$$\Phi_B = \oint \mathbf{B} \cdot d\mathbf{S} = 0 \qquad [34\text{--}7]$$

states that the magnetic flux through any closed Gaussian surface must be zero. See Fig. 34–4 for illustrations and for a comparison with Gauss's law for electrostatics. Equation 34–7 is simply a formal statement of the observation that there are no free magnetic poles.

The earth's magnetic field

The earth's magnetic field, approximately that of a dipole near the earth's center, finds its origin in current loops induced in the earth's liquid outer core by a mechanism not yet fully understood. The direction of the local magnetic field at any point is given by its angle of *declination* (in a horizontal plane) from true north and its angle of *inclination* (in a vertical plane) from the horizontal; see Example 3.

Diamagnetism

In the nature of their response to an external magnetic field materials may be broadly grouped as diamagnetic, paramagnetic, or ferromagnetic. Diamagnetic materials are (weakly) repelled by the pole of a strong magnet. The atoms of such materials do not have intrinsic magnetic dipole moments. A dipole moment may be induced, however, by an external magnetic field, its direction being opposite to that of the field. Figure 34–8 summarizes the induction mechanism.

Paramagnetism

Paramagnetic materials, which are (weakly) attracted by a magnetic pole, do have intrinsic magnetic dipole moments. The tendency of these dipoles to line up with an external magnetic field is interfered with (see Example 4) by thermal agitation. The magnetization M of a specimen, which is its magnetic moment per unit volume, is given approximately by Curie's law, or

$$M = C\frac{B}{T}. \qquad [34\text{--}8]$$

For strong enough fields or low enough temperatures this law breaks down as the atomic dipoles approach complete alignment; see the *magnetization curve* of Fig. 34–7.

Ferromagnetism

In ferromagnetic materials such as iron a special interaction between neighboring atoms locks the atomic dipoles in rigid parallelism in spite of the disordering tendency of thermal agitation. This interaction abruptly disappears at a well-defined Curie temperature, above which the material becomes simply paramagnetic.

The magnetization curve

Hysteresis

Figure 34–9 shows how a known external magnetic field B_0 can be applied to a ring-shaped ferromagnetic specimen. The resultant magnetic field can be measured by an induction method and a magnetization curve such as that of Fig. 34–10 can be plotted. The course of this curve as B_0 increases corresponds to the growth of favorably oriented magnetic domains in the material; see Fig. 34–11 and the related text.

As Fig. 34–12 shows, ferromagnetic magnetization curves do not retrace themselves, a phenomenon called hysteresis. Some alignment of dipoles remains even when the external magnetic field is completely removed; the result is the familiar "permanent" magnet.

QUESTIONS

1. Two iron bars are identical in appearance. One is a magnet and one is not. How can you tell them apart? You are not permitted to suspend either bar as a compass needle or to use any other apparatus.

2. Two iron bars always attract, no matter the combination in which their ends are brought near each other. Can you conclude that one of the bars must be unmagnetized?

3. The neutron, which has no charge, has a magnetic dipole moment. Is this possible on the basis of classical electromagnetism, or does this evidence alone indicate that classical electromagnetism has broken down?

4. Must all permanent magnets have identifiable north and south poles? Consider geometries other than the bar or horseshoe magnet.

5. A certain short iron rod is found, by test, to have a north pole at each end. You sprinkle iron filings over the rod. Where (in the simplest case) will they cling? Make a rough sketch of what the lines of **B** must look like, both inside and outside the rod.

6. Cosmic rays are charged particles that strike our atmosphere from some external source. We find that more low-energy cosmic rays reach the earth near the north and south magnetic poles than at the (magnetic) equator. Why is this so?

7. How might the magnetic dipole moment of the earth be measured?

8. Give three reasons for believing that the flux Φ_B of the earth's magnetic field is greater through the boundaries of Alaska than through those of Texas.

9. You are a manufacturer of compasses. (*a*) Describe ways in which you might magnetize the needles. (*b*) The end of a needle that points north is usually painted a characteristic color. Without suspending the needle in the earth's field, how might you find out which end of the needle to paint? (*c*) Is the painted end a north or a south magnetic pole?

10. Would you expect the magnetization at saturation for a paramagnetic substance to be very much different from that for a saturated ferromagnetic substance of about the same size? Why or why not?

11. The magnetization induced in a given diamagnetic sphere by a given external magnetic field does not vary with temperature, in sharp contrast to the situation in paramagnetism. Is this understandable in terms of the description that we have given of the origin of diamagnetism?

12. Explain why a magnet attracts an unmagnetized iron object such as a nail.

13. Does any net force or torque act on (*a*) an unmagnetized iron bar or (*b*) a permanent bar magnet when placed in a uniform magnetic field?

14. A nail is placed at rest on a smooth table top near a strong magnet. It is released and attracted to the magnet. What is the source of the kinetic energy it has just before it strikes the magnet?

15. Compare the magnetization curves for a paramagnetic substance (see Fig. 34–7) and for a ferromagnetic substance (see Fig. 34–10). What would a similar curve for a diamagnetic substance look like?

16. Why do iron filings line up with a magnetic field, as in Fig. 34–1? After all, they are not intrinsically magnetized.

17. The earth's magnetic field can be represented closely by that of a magnetic dipole located at or near the center of the earth. The earth's magnetic poles can be thought of as (*a*) the points where the axis of this dipole passes through the earth's surface or as (*b*) the points on the earth's surface where a dip needle would point vertically. Are these necessarily the same points?

18. A "friend" borrows your favorite compass and paints the entire needle red. When you discover this you are lost in a cave and have with you two flashlights, a few meters of wire, and (of course) this book. How might you discover which end of your compass needle is the north-seeking end?

19. How can you magnetize an iron bar if the earth is the only magnet around?

20. How would you go about shielding a certain volume of space from external magnetic fields? Can it be done?

EXERCISES

Section 34–1   Poles and Dipoles

1. A simple bar magnet hangs from a string as in Fig. 34–13. A uniform magnetic field **B** directed horizontally is then established. Sketch the resulting orientation of the string and the magnet.

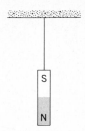

Figure 34-13 Exercise 1.

2. Using classical arguments show that (a) the magnetic moment set up by an orbiting electron in an atom points in a direction opposite to that of its (orbital) angular momentum and (b) the magnetic moment set up by a spinning, positively-charged proton points in the same direction as its (spin) angular momentum.

Section 34-2 Gauss's Law for Magnetism
3. A student rolls a sheet of paper into a cylinder. He then places a bar magnet near its end as shown in Fig. 34-14. (a) Sketch the **B** field lines as they cross the paper cylinder. (b) What can you say about the sign of $\mathbf{B} \cdot d\mathbf{S}$ for every $d\mathbf{S}$ of this paper cylinder? (c) Does this contradict Gauss's law for magnetism? Explain.

Figure 34-14 Exercise 3.

4. The magnetic flux through each of five faces of a die (singular of "dice") is given by $\Phi_B = \pm N$ Wb, where $N\,(= 1$ to 5) is the number of spots on the face. The flux is positive (outward) for N even and negative (inward) for N odd. What is the flux through the sixth face of the die?

Section 34-3 The Magnetism of the Earth
5. In New Hampshire the average horizontal component of the earth's magnetic field in 1912 was 16 μT and the average inclina-

tion of "dip" was 73°. What was the corresponding magnitude of the earth's magnetic field?

Section 34-4 Paramagnetism
6. A cylindrical rod magnet has a length of 5.0 cm and a diameter of 1.0 cm. It has a uniform magnetization of 5.3×10^3 A/m. What is its magnetic dipole moment?

7. A paramagnetic substance is (weakly) attracted to a pole of a magnet. Figure 34-15 shows a model of this phenomenon. The "paramagnetic substance" is a current loop L, which is placed on the axis of a bar magnet nearer to its north pole than its south pole. Because of the torque $\tau = \mu \times \mathbf{B}$ exerted on the loop by the **B** field of the bar magnet, the magnetic dipole moment μ of the loop will align itself to be parallel to **B**. (a) Make a sketch showing the **B** field lines due to the bar magnet. (b) Show the direction of the current i in the loop. (c) Using $d\mathbf{F} = i\,d\boldsymbol{\ell} \times \mathbf{B}$, show from (a) and (b) that the net force on L is toward the north pole of the bar magnet.

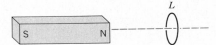

Figure 34-15 Exercises 7 and 9.

Section 34-5 Diamagnetism
8. In Fig. 34-8c verify that the magnetic force $\mathbf{F}_B$ is opposed to the electric force $\mathbf{F}_E$, as shown, and that the orbiting electron is slowed down. In Fig. 34-8d, verify that the magnetic force is aligned with the electric force and that the electron is speeded up.

9. A diamagnetic substance is (weakly) repelled by a pole of a magnet. Figure 34-15 shows a model of this phenomenon. The "diamagnetic substance" is a current loop L that is placed on the axis of a bar magnet nearer to its north pole than its south pole. Because the substance is diamagnetic the magnetic moment μ of the loop will align itself to be antiparallel to the **B** field of the bar magnet. (a) Make a sketch showing the **B** field lines due to the bar magnet. (b) Show the direction of the current i in the loop. (c) Using $d\mathbf{F} = i\,d\boldsymbol{\ell} \times \mathbf{B}$, show from (a) and (b) that the net force on L is away from the north pole of the bar magnet.

PROBLEMS GROUPED BY SECTION

Section 34-1 Poles and Dipoles
1. A flat current loop of 3.0-mm radius has a magnetic dipole moment of 2.0×10^{-4} J/T. (a) What is the current in the loop, assuming that it has five turns? (b) What magnetic field does the loop set up at a point on its axis 12 cm away from the plane of the loop?

2. In the lowest energy state of the hydrogen atom the most probable distance between the single orbiting electron and the central proton is 5.2×10^{-11} m. Calculate (a) the electric field and (b) the magnetic field set up by the proton at this distance, measured along the proton's axis of spin. The charge and magnetic moment of the proton are $+ 1.6 \times 10^{-19}$ C and 1.4×10^{-26} J/T, respectively.

Section 34-2 Gauss's Law for Magnetism
3. A Gaussian surface in the shape of a right circular cylinder has a radius of 12 cm and a length of 80 cm. Through one end

there is an inward magnetic flux of 25 μWb. At the other end there is a uniform magnetic field of 1.6 mT, normal to the surface and directed outward. What is the net magnetic flux through the curved surface?

*4. Two wires, parallel to the z-axis and a distance $4r$ apart, carry equal currents i in opposite directions, as shown in Fig. 34-16. A circular cylinder of radius r and length l has its axis on the z-axis, midway between the wires. Use Gauss's law for magnetism and the results of Example 2 of Chapter 31 to calculate the net outward magnetic flux through the half of the cylindrical surface above the x-axis. (Hint: find the flux through that portion of the x-z plane which is within the cylinder.)

Section 34-3 The Magnetism of the Earth
5. In Example 3 the vertical component of the earth's magnetic field in Tucson, Arizona was found to be 44 μT. Assume this is the average value for all of Arizona, which has an area of 114,000

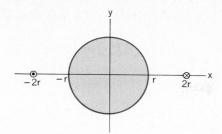

Figure 34-16 Problem 4.

square miles, and calculate the net magnetic flux through the rest of the earth's surface (the entire surface excluding Arizona). Is the flux outward or inward?

6. The earth has a magnetic dipole moment of 8.0×10^{22} J/T. (*a*) What current would have to be set up in a single turn of wire going around the earth at its magnetic equator if we wished to set up such a dipole? (*b*) Could such an arrangement be used to cancel out the earth's magnetism at points in space well above the earth's surface? (*c*) On the earth's surface?

Section 34-4 Paramagnetism

7. The paramagnetic salt to which the magnetization curve of Fig. 34-7 applies is to be tested to see whether it obeys Curie's law. The sample is placed in a 0.50-T magnetic field that remains constant throughout the experiment. The magnetization M is then measured at temperatures ranging from 10 to 300 K. Would it be found that Curie's law is valid under these conditions?

8. A 0.50-T magnetic field is applied to a paramagnetic gas whose atoms have an intrinsic magnetic dipole moment of 1.0×10^{-23} J/T. At what temperature will the mean kinetic energy of translation of the gas atoms be equal to the energy required to reverse such a dipole end-for-end in this magnetic field?

Section 34-5 Diamagnetism

9. An electron with kinetic energy K_e travels in a circular path that is perpendicular to a uniform magnetic field, subject only to the force of the field. (*a*) Show that the magnetic dipole moment due to its orbital motion has magnitude $\mu = K_e/B$ and that it is in the direction opposite to that of **B**. (*b*) What is the magnitude and direction of the magnetic dipole moment of a positive ion with kinetic energy K_i under the same circumstances? (*c*) An ionized gas consists of 5.3×10^{21} electrons/m³ and the same number of ions/m³. Take the average electron kinetic energy to be 6.2×10^{-20} J and the average ion kinetic energy to be 7.6×10^{-21} J. Calculate the magnetization of the gas for a magnetic field of 1.20 T.

10. Analyze qualitatively the appearance of induced magnetic dipole moments in diamagnetism from the point of view of Faraday's law of induction. [*Hint:* See Fig. 32-9. Also, note that for orbiting electrons the inductive effects (i.e., any change in the speed of the electron) persist after the magnetic field has stopped changing; they vanish only when the field is removed.]

Section 34-6 Ferromagnetism

11. The dipole moment associated with an atom of iron in an iron bar is 2.1×10^{-23} J/T. Assume that all the atoms in the bar, which is 5.0 cm long and has a cross-sectional area of 1.0 cm², have their dipole moments aligned. (*a*) What is the dipole moment of the bar? (*b*) What torque must be exerted to hold this magnet at right angles to an external field of 1.5 T? The density of iron is 7.9 g/cm³.

12. A Rowland ring is formed of ferromagnetic material. It is circular in cross section, with an inner radius of 5.0 cm and an outer radius of 6.0 cm and is wound with 400 turns of wire. (*a*) What current must be set up in the windings to attain $B_o = 2.0 \times 10^{-4}$ T in Fig. 34-10? (*b*) A secondary coil wound around the toroid has 50 turns and has a resistance of 8.0 Ω. If, for this value of B_o, we have $B_M = 800B_o$, how much charge moves through the secondary coil when the current in the toroid windings is turned on?

ADDITIONAL PROBLEMS

13. A charge q is distributed uniformly around a thin ring of radius r. The ring is rotating about an axis through its center and perpendicular to its plane at an angular speed ω. (*a*) Show that the magnetic moment due to the rotating charge is

$$\mu = \tfrac{1}{2}q\omega r^2.$$

(*b*) What is the direction of this moment if the charge is positive?

14. Figure 34-17 shows four arrangements of pairs of small

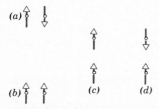

Figure 34-17 Problem 14.

compass needles, set up in a space in which there is no external magnetic field. Identify the equilibrium in each case as stable or unstable. For each pair consider only the torque acting on one needle due to the magnetic field set up by the other. Explain your answers.

15. A sample of the paramagnetic salt to which the magnetization curve of Fig. 34-7 applies is held at room temperature (300 K). At what applied magnetic field would the degree of magnetic saturation of the sample be (*a*) 50 percent? (*b*) 90 percent? (*c*) Comment on the magnitude of these fields.

16. A sample of the paramagnetic salt to which the magnetization curve of Fig. 34-7 applies is immersed in a magnetic field of 2.0 T. At what temperature would the degree of magnetic saturation of the sample be (*a*) 50 percent? (*b*) 90 percent? (*c*) Comment on the magnitude of these temperatures.

17. The magnetic dipole moment of the earth is 8.0×10^{22} J/T. (*a*) If the origin of this magnetism were a magnetized iron sphere at the center of the earth, what would be the needed radius? (*b*) What fraction of the volume of the earth would such a sphere occupy? Assume complete alignment of the dipoles. The density of the earth's inner core is 14 g/cm³. Take the magnetic dipole moment of an iron atom to be 2.1×10^{-23} J/T. (*Note:* The earth's inner core is in fact thought to be in both liquid and solid form and partly iron, but a permanent magnet as the source of the earth's magnetism has been ruled out by several considerations. For one, the temperature is almost certainly above the Curie point.)

18. *Dipole-dipole interaction.* The exchange coupling mentioned in Section 34–6 as being responsible for ferromagnetism is *not* the mutual magnetic interaction energy between two elementary magnetic dipoles. To show this, (*a*) compute *B* a distance *a* (= 10 nm) away from a dipole of moment μ (= 1.8×10^{-23} J/T); (*b*) compute the energy (= $2\mu B$) required to turn a second similar dipole end for end in this field. What do you conclude about the strength of this dipole-dipole interaction? Compare with the results of Example 4.

19. Figure 34–18 shows the apparatus used in a lecture demonstration of para- and diamagnetism. A sample of the magnetic material is suspended by a string ($L = 2$ m) in a region ($d = 2$ cm) between the two poles of a powerful electromagnet. Pole P_1 is sharply pointed and pole P_2 is rounded as indicated. Any deflection of the string from the vertical is visible to the audience by means of an optical projection system (not shown). (*a*) First a bismuth (highly diamagnetic) sample is used. When the electromagnet is turned on, the sample is observed to deflect slightly (about 1 mm) toward one of the poles. What is the direction of this deflection? (*b*) Next an aluminum (paramagnetic, conducting) sample is used. When the electromagnet is turned on, the sample is observed to deflect strongly (about 1 cm) toward one pole for about a second and then deflect moderately (a few mm) toward the other pole. Explain and indicate the direction of these deflections. (*Hint:* note that the sample is a conductor.) (*c*) What would happen if a ferromagnetic sample were used?

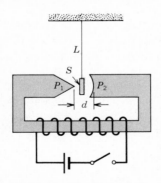

Figure 34–18 Problem 19.

20. Using the values of (spin) angular momentum *S* and (spin) magnetic moment μ_s given in Section 34–1 for the free electron show that

$$\mu_s = \left(\frac{e}{m}\right) S.$$

Verify that the units and dimensions are consistent. This result is a prediction of a relativistic theory of the electron advanced by P. A. M. Dirac in 1928.

21. Consider a solid containing *N* atoms per unit volume, each atom having a magnetic dipole moment μ. Suppose the direction of μ can be only parallel or antiparallel to an externally applied magnetic field **B** (this will be the case if μ is due to the spin of a single electron). According to statistical mechanics, it can be shown that the probability of an atom being in a state with energy *U* is proportional to $e^{-U/(kT)}$ where *T* is the temperature and *k* is Boltzmann's constant. Thus, since $U = -\mu \cdot \mathbf{B}$, the fraction of atoms whose dipole moment is parallel to **B** is proportional to $e^{(\mu B)/(kT)}$ and the fraction of atoms whose dipole moment is antiparallel to **B** is proportional to $e^{-(\mu B)/(kT)}$. (*a*) Show that the magnetization of this solid is $M = N\mu \tanh (\mu B/kT)$. Here tanh is the hyperbolic tangent function: $\tanh(x) = (e^x - e^{-x})/(e^x + e^{-x})$. (*b*) Show that (*a*) reduces to $M = N\mu^2 B/kT$ for $\mu B \ll kT$. (*c*) Show that (*a*) reduces to $M = N\mu$ for $\mu B \gg kT$. (*d*) Show that (*b*) and (*c*) agree qualitatively with Fig. 34–7.

22. The magnetic field of the earth can be approximated as a dipole magnetic field, with horizontal and vertical components, at a point a distance *r* from the earth's center, given by

$$B_h = \frac{\mu_0 \mu}{4\pi r^3} \cos \lambda_m, \qquad B_v = \frac{\mu_0 \mu}{2\pi r^3} \sin \lambda_m,$$

where λ_m is the *magnetic latitude* (latitude measured from the magnetic equator toward the north or south magnetic pole). The magnetic dipole moment $\mu = 7.94 \times 10^{22}$ A·m². What is the magnetic field, magnitude and direction, at (*a*) the north magnetic pole, and (*b*) the magnetic equator? (*c*) Show that the strength at latitude λ_m is given by

$$B = \frac{\mu_0 \mu}{4\pi r^3} (1 + 3 \sin^2 \lambda_m)^{1/2}.$$

(*d*) What is the predicted field strength at magnetic latitude 60°N at the surface of the earth? (*e*) Describe the field at the north *geographic* pole.

23. Find the altitude above the earth's surface where the earth's magnetic field has a magnitude one-half the surface value at the same magnetic latitude. (Use the dipole field approximation given in Problem 22.)

Electromagnetic Oscillations

35-1 *LC* Oscillations—Qualitative

Of the three circuit elements, resistance R, capacitance C, and inductance L, we have studied the combinations RC (Section 29–7) and RL (Section 33–3). These two circuit combinations are characterized by the exponential growth and decay of charge, current, and potential difference, as Figs. 29–11 and 33–5 attest. The quantity that most effectively describes the circuit behavior is the time constant τ.

In this chapter we examine the remaining circuit element combination LC. We will see that it is characterized by a sinusoidal variation of charge, current, and potential difference. The quantity that most effectively describes the circuit behavior in this case is the angular frequency ω of the oscillations. We first examine the situation from a physical but semiquantitative point of view.

Energy transfers in *LC* oscillations

Assume that initially the capacitor C in Fig. 35–1a carries a charge q_m and the current i in the inductor is zero. At this instant the energy stored in the capacitor is given by Eq. 27–8, or

$$U_E = \tfrac{1}{2} \frac{q_m^2}{C}. \tag{35-1}$$

The energy stored in the inductor, given by

$$U_B = \tfrac{1}{2} L i^2, \tag{35-2}$$

is zero because the current is zero. The capacitor now starts to discharge through the inductor, positive charge carriers moving counterclockwise, as shown in Fig. 35–1b. This means that a current i, given by dq/dt and pointing down in the inductor, is established.

As q decreases, the energy stored in the electric field in the capacitor also de-

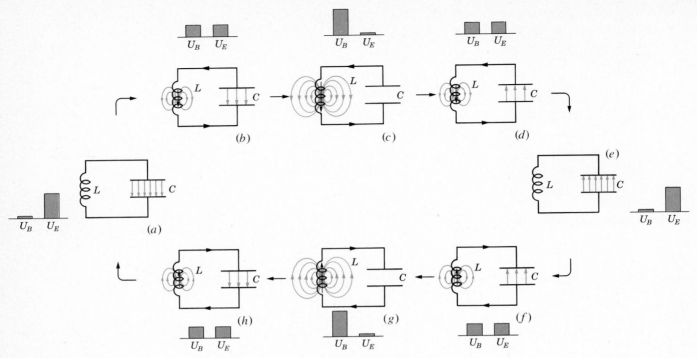

Figure 35-1   Showing eight stages in a cycle of oscillation of a resistanceless *LC* circuit. The bar graphs below each figure show the stored magnetic and electric energy. The vertical arrows on the inductor axis show the current.

creases. This energy is transferred to the magnetic field that appears around the inductor because of the current *i* that is building up there. Thus the electric field decreases, the magnetic field builds up, and energy is transferred from the former to the latter.

At a time corresponding to Fig. 35–1*c*, all the charge on the capacitor will have disappeared. The electric field in the capacitor will be zero, the energy stored there having been transferred entirely to the magnetic field of the inductor. According to Eq. 35–2, there must then be a current—and indeed one of maximum value—in the inductor. Note that even though *q* equals zero, the current (which is *dq/dt*) is not zero at this time.

The large current in the inductor in Fig. 35–1*c* continues to transport positive charge from the top plate of the capacitor to the bottom plate, as shown in Fig. 35–1*d*; energy now flows from the inductor back to the capacitor as the electric field builds up again. Eventually, the energy will have been transferred completely back to the capacitor, as in Fig. 35–1*e*. The situation of Fig. 35–1*e* is like the initial situation, except that the capacitor is charged oppositely.

The capacitor will start to discharge again, the current now being clockwise, as in Fig. 35–1*f*. Reasoning as before, we see that the circuit eventually returns to its initial situation and that the process continues at a definite frequency ν to which corresponds a definite angular frequency ω ($= 2\pi\nu$). Once started, such *LC* oscillations (in the ideal case described, in which the circuit contains no resistance) continue indefinitely, energy being shuttled back and forth between the electric field in the capacitor and the magnetic field in the inductor. Any configuration in Fig. 35–1 can be set up as an initial condition. The oscillations will then continue

Charge and current variations in
LC oscillations

from that point, proceeding clockwise around the figure. Compare these oscillations carefully with those of the mass-spring system described in Fig. 8–4.

To determine the charge q as a function of time, we can use a voltmeter to measure the variable potential difference $V_C(t)$ that exists across capacitor C. The relation

$$V_C = \left(\frac{1}{C}\right) q$$

shows that V_C is proportional to q. To measure the current we can insert a small resistance R in the circuit and measure the potential difference across it. This is proportional to i through the relation

$$V_R = Ri.$$

We assume here that R is so small that its effect on the behavior of the circuit is negligible. Both q and i, or more correctly V_C and V_R, which are proportional to them, can be displayed on a cathode-ray oscilloscope. This instrument can plot automatically, on its screen, graphs proportional to $q(t)$ and $i(t)$, as in Fig. 35–2.

Figure 35–2 A drawing of an oscilloscope screen showing potential differences proportional to (a) the charge and (b) the current, in the circuit of Fig. 35–1, as a function of time. The letters indicate corresponding phases of oscillation in that figure. Note that because $i = dq/dt$ the lower curve is proportional to the derivative of the upper.

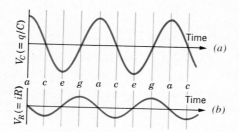

Example 1 A 1.0-μF capacitor is charged to 50 V. The charging battery is then disconnected and a 10-mH coil is connected across the capacitor, so that *LC* oscillations occur. What is the maximum current in the coil? Assume that the circuit contains no resistance.

The maximum stored energy in the capacitor must equal the maximum stored energy in the inductor, from the conservation-of-energy principle. This leads, from Eqs. 35–1 and 35–2, to

$$\tfrac{1}{2}\frac{q_m{}^2}{C} = \tfrac{1}{2}Li_m{}^2,$$

where i_m is the maximum current and q_m is the maximum charge. Note that the maximum current and the maximum charge do not occur at the same time but one-fourth of a cycle apart; see Figs. 35–1 and 35–2. Solving for i_m and substituting CV_0 for q_m gives

$$i_m = V_0 \sqrt{\frac{C}{L}} = (50\ \text{V}) \sqrt{\frac{1.0 \times 10^{-6}\ \text{F}}{10 \times 10^{-3}\ \text{H}}} = 0.50\ \text{A}.$$

Damping in *LC* Oscillations

In an actual *LC* circuit the oscillations will not continue indefinitely because there is always some resistance present that will drain energy from the electric and magnetic fields and dissipate it as thermal energy; the circuit will become warmer. The oscillations, once started, will die away as in Fig. 35–3. Compare this figure with Fig. 14–16, which shows the decay of the mechanical oscillations of a mass-spring system caused by frictional damping.

It is possible to have sustained electromagnetic oscillations if arrangements are made to supply, automatically and periodically (once a cycle, say), enough energy from an outside source to compensate for that dissipated as thermal energy. We are reminded of a clock escapement, which is a device for feeding energy from

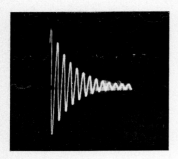

Figure 35-3 A photograph of an oscilloscope trace showing how the oscillations in an *RCL* circuit die away because energy is dissipated as thermal energy in the resistor. The figure is a plot of the potential difference across the resistor as a function of time.

Correspondence of Mechanical and Electrical Systems

a spring or a falling weight into an oscillating pendulum, thus compensating for frictional losses that would otherwise cause the oscillations to die away. *LC* oscillators, whose frequency may be varied between specified limits, are commercially available as packaged units over a wide range of frequencies, extending from low audio-frequencies (lower than 10 Hz) to microwave frequencies (higher than 10 GHz).

35-2 *LC* Oscillations—A Mechanical Analogy

The *LC* system of Fig. 35-1 resembles in many ways the mass-spring system of Fig. 8-4. Let us make the analogy explicit. Figure 8-4 shows that in a mass-spring system performing simple harmonic motion, as in an oscillating *LC* circuit, two kinds of energy occur. One is potential energy of the compressed or extended spring; the other is kinetic energy of the moving mass. These are given by the familiar formulas in the first column of Table 35-1. The table suggests that a capacitor is in some mathematical way like a spring and an inductor is like a mass and that certain electromagnetic quantities "correspond" to certain mechanical ones, namely,

$$q \text{ corresponds to } x,$$
$$i \text{ corresponds to } v,$$
$$C \text{ corresponds to } 1/k,$$
$$L \text{ corresponds to } m.$$

Table 35-1   Energy in oscillating systems

Mechanical (Fig. 8-4)		Electromagnetic (Fig. 35-1)	
spring	$U_P = \frac{1}{2}kx^2$	capacitor	$U_E = \frac{1}{2}\left(\frac{1}{C}\right)q^2$
mass	$U_K = \frac{1}{2}mv^2$	inductor	$U_B = \frac{1}{2}Li^2$
	$v = dx/dt$		$i = dq/dt$

Comparison of Fig. 35-1, which shows the oscillations of a resistanceless *LC* circuit, with Fig. 8-4, which shows the oscillations in a frictionless mass-spring system, indicates how close the correspondence is. Note how v and i correspond in the two figures; also x and q. Note, too, how in each case the energy alternates between two forms, magnetic and electric for the *LC* system, and kinetic and potential for the mass-spring system.

In Section 14-3 we saw that the natural angular frequency of oscillation of a (frictionless) mass-spring system is

$$\omega = \sqrt{\frac{k}{m}}.$$

The method of correspondences suggests that to find the natural frequency for a (resistanceless) *LC* circuit, k should be replaced by $1/C$ and m by L, obtaining

$$\omega = \frac{1}{\sqrt{LC}}. \tag{35-3}$$

This result is indeed correct, as we show in the next section.

35-3 *LC* Oscillations—Quantitative

We now derive an expression for the frequency of oscillation of a (resistanceless) *LC* circuit, our derivation being based on the conservation-of-energy principle. The total energy *U* present at any instant in an oscillating *LC* circuit is given by

$$U = U_B + U_E = \tfrac{1}{2}Li^2 + \tfrac{1}{2}\frac{q^2}{C},$$

which expresses the fact that at any arbitrary time the energy is stored partly in the magnetic field in the inductor and partly in the electric field in the capacitor. If we assume the circuit resistance to be zero, there is no energy transfer to thermal energy and *U* remains constant with time, even though *i* and *q* vary. In more formal language, dU/dt must be zero. This leads to

$$\frac{dU}{dt} = \frac{d}{dt}\left(\tfrac{1}{2}Li^2 + \tfrac{1}{2}\frac{q^2}{C}\right) = Li\frac{di}{dt} + \frac{q}{C}\frac{dq}{dt} = 0. \qquad (35\text{-}4)$$

Now, *q* and *i* are not independent variables, being related by

$$i = \frac{dq}{dt}.$$

Differentiating yields

$$\frac{di}{dt} = \frac{d^2q}{dt^2}.$$

Substituting these two expressions into Eq. 35-4 leads to

Differential equation for *LC* oscillations

$$L\frac{d^2q}{dt^2} + \frac{1}{C}q = 0. \qquad (35\text{-}5)$$

This is the differential equation that describes the oscillations of a (resistanceless) *LC* circuit. To solve it, note that Eq. 35-5 is mathematically of exactly the same form as Eq. 14-6,

$$m\frac{d^2x}{dt^2} + kx = 0, \qquad [14\text{-}6]$$

which is the differential equation for the mass-spring oscillations. Fundamentally, it is by comparing these two equations that the correspondences on p. 680 arise.

The solution of Eq. 14-6 is

$$x = A\cos(\omega t + \phi), \qquad [14\text{-}8]$$

where $A\,(=x_m)$ is the amplitude of the motion and ϕ is an arbitrary phase constant. Since *q* corresponds to *x*, we can write the solution of Eq. 35-5 as

Solution of the *LC* differential equation

$$q = q_m\cos(\omega t + \phi), \qquad (35\text{-}6)$$

where ω is the still unknown angular frequency of the electromagnetic oscillations.

We can test whether Eq. 35-6 is indeed a solution of Eq. 35-5 by substituting it and its second derivative in that equation. To find the second derivative, we write

$$\frac{dq}{dt} = i = -\omega q_m\sin(\omega t + \phi) \qquad (35\text{-}7)$$

and

$$\frac{d^2q}{dt^2} = -\omega^2 q_m \cos(\omega t + \phi). \qquad (35-8)$$

Substituting q and d^2q/dt^2 into Eq. 35–5 yields

$$-L\omega^2 q_m \cos(\omega t + \phi) + \frac{1}{C} q_m \cos(\omega t + \phi) = 0.$$

Canceling $q_m \cos(\omega t + \phi)$ and rearranging leads to

Angular frequency of
LC oscillations

$$\omega = \frac{1}{\sqrt{LC}}.$$

Thus, if ω is given the constant value $1/\sqrt{LC}$, Eq. 35–6 is indeed a solution of Eq. 35–5. This expression for ω agrees with Eq. 35–3, which we arrived at by the method of correspondences.

As in the mechanical analogy the phase constant ϕ in Eq. 35–6 is determined by the conditions that prevail at $t = 0$. If the initial condition is as represented by Fig. 35–1a, then we put $\phi = 0$ in order that Eq. 35–6 may predict $q = q_m$ at $t = 0$. What initial condition in Fig. 35–1 is implied if we select $\phi = 90°$?

Example 2 (a) In an oscillating LC circuit, what value of charge, expressed in terms of the maximum charge, is present on the capacitor when the energy is shared equally between the electric and the magnetic field? (b) How much time is required for this condition to arise, assuming the capacitor to be fully charged initially? Assume that $L = 10$ mH and $C = 1.0$ μF.

(a) The instantaneous and maximum stored energy in the capacitor are, respectively,

$$U_E = \frac{q^2}{2C} \quad \text{and} \quad U_{E,m} = \frac{q_m^2}{2C}.$$

Substituting $U_E = \tfrac{1}{2}U_{E,m}$ yields

$$\frac{q^2}{2C} = \tfrac{1}{2}\frac{q_m^2}{2C} \quad \text{or} \quad q = \frac{1}{\sqrt{2}} q_m = 0.71\, q_m$$

(b) To find the time, we write, setting $\phi = 0$ in Eq. 35–6,

$$q/q_m = \cos \omega t = \frac{1}{\sqrt{2}} \quad \text{or} \quad t = \frac{\pi}{4\omega}.$$

The angular frequency ω is found from Eq. 35–3, or

$$\omega = \frac{1}{\sqrt{LC}} = \frac{1}{\sqrt{(10 \times 10^{-3}\,\text{H})(1.0 \times 10^{-6}\,\text{F})}}$$
$$= 1.0 \times 10^4\ \text{rad/s}.$$

The time t is then

$$t = \frac{\pi}{4\omega} = \frac{\pi}{(4)(1.0 \times 10^4\ \text{rad/s})} = 7.9 \times 10^{-5}\ \text{s} = 79\ \mu\text{s}$$

Convince yourself that the frequency and the period of the oscillation are 1.6 kHz and 630 μs, respectively.

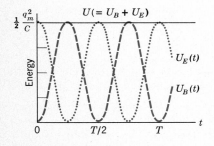

Figure 35–4 The stored magnetic ($---$) and electric ($\cdot\cdot\cdot\cdot$) energies in the circuit of Fig. 35–1. Note that their sum is a constant. T ($=1/\nu$) is the period of the oscillation.

The stored electric energy in the LC circuit, using Eq. 35–6, is

$$U_E = \tfrac{1}{2}\frac{q^2}{C} = \tfrac{1}{2}\frac{q_m^2}{C} \cos^2(\omega t + \phi), \qquad (35-9)$$

and the magnetic energy, using Eq. 35–7, is

$$U_B = \tfrac{1}{2}Li^2 = \tfrac{1}{2}L\omega^2 q_m^2 \sin^2(\omega t + \phi).$$

Substituting the expression for ω (see Eq. 35–3) into this last equation yields

$$U_B = \tfrac{1}{2}\frac{q_m^2}{C} \sin^2(\omega t + \phi). \qquad (35-10)$$

Figure 35–4 shows plots of $U_E(t)$ and $U_B(t)$ for the case of $\phi = 0$. Note that (a) the maximum values of U_E and U_B are the same ($= q_m^2/2C$); (b) at any instant the sum of U_E and U_B is a constant ($= q_m^2/2C$); and (c) when U_E has its maximum value, U_B is zero and conversely. This analysis supports the qualitative analysis of Section 35–1. Compare this discussion with that given in Section 14–4 for the energy transfers in a mass-spring system.

35-4 Damped *LC* Oscillations

If resistance R is present in an LC circuit the total electromagnetic energy E is no longer constant but decreases with time as it is transformed steadily to thermal energy in the resistor. As we shall see the analogy with the damped mass-spring oscillator of Section 14-8 is exact.

As before we have

$$E = U_B + U_E = \tfrac{1}{2} L i^2 + \tfrac{1}{2} \frac{q^2}{C}.$$

E is no longer constant but rather

$$\frac{dE}{dt} = -i^2 R,$$

the minus sign signifying that the stored energy E decreases with time, being converted to thermal energy at the rate $i^2 R$. Combining these two equations leads to

$$L i \frac{di}{dt} + \frac{q}{C} \frac{dq}{dt} = -i^2 R.$$

Substituting dq/dt for i and d^2q/dt^2 for di/dt and dividing by i leads to

$$L \frac{d^2 q}{dt^2} + R \frac{dq}{dt} + \frac{1}{C} q = 0, \tag{35-11}$$

which is the differential equation that describes the damped LC oscillations. If we put $R = 0$, this equation reduces, as expected, to Eq. 35-5.

Compare this differential equation for damped LC oscillations with Eq. 14-27, or

$$m \frac{d^2 x}{dt^2} + b \frac{dx}{dt} + kx = 0, \tag{14-27}$$

which describes damped mass-spring oscillations. Once again the equations are mathematically identical, the resistance R corresponding to the mechanical damping constant b and other quantities corresponding according to the list in Table 35-1.

The solution of Eq. 35-11 follows at once, by the method of correspondence, from the solution of Eq. 14-27. It is

Equation for damped oscillations

$$q = q_m e^{-Rt/2L} \cos (\omega' t + \phi). \tag{35-12}$$

Equation 35-12, which can be described as a cosine function with an exponentially decreasing amplitude, is the equation of the decay curve of Fig. 35-3. The frequency ω' is strictly less than the frequency $\omega \, (= 1/\sqrt{LC})$ of the undamped oscillations but here we will consider only those cases in which the resistance R is so small that we can put $\omega' = \omega$ with negligible error. Recall that we made a similar assumption of small damping in discussing the damping of mass-spring oscillations in Section 14-8.

Example 3 A circuit has $L = 10$ mH, $C = 1.0$ μF, and $R = 1.5$ Ω. (*a*) After what time t will the amplitude of the charge oscillations drop to one-half of its initial value? (*b*) To how many periods of oscillation does this correspond?

(*a*) This will occur when the amplitude factor $e^{-Rt/2L}$ in Eq. 35-12 has the value one-half, or

$$\tfrac{1}{2} = e^{-Rt/2L}.$$

We have

$$t = \frac{2L}{R} \ln 2 = \frac{(2)(10 \times 10^{-3} \text{ H})(0.69)}{1.5 \ \Omega} = 9.2 \text{ ms}.$$

(*b*) The angular frequency is

$$\omega = \frac{1}{\sqrt{LC}} = \frac{1}{\sqrt{(10 \times 10^{-3}\text{H})(1.0 \times 10^{-6}\text{F})}} = 1.0 \times 10^4 \text{ rad/s.}$$

From this we can easily show that the corresponding frequency

$\nu \, (= \omega/2\pi)$ is 1.6 kHz and the period $T \, (= 1/\nu)$ is 0.63 ms. Thus the amplitude drops to one-half after 9.2 ms/0.63 ms or about 15 cycles of oscillation. By comparison, the damping in this example is less severe than that shown in Fig. 35–3, where the amplitude drops to one-half in about three cycles.

35–5 Forced Oscillations and Resonance

We have discussed the so-called "free" oscillations of an *LC* circuit and also the "damped" oscillations, in which a resistive element *R* is present. If the damping is small enough—and we have assumed that it is—both kinds of oscillation proceed with an angular frequency given by

$$\omega_0 = \frac{1}{\sqrt{LC}} \qquad [35\text{--}3]$$

which we call the *natural frequency* of the oscillating system. To remind ourselves that, once the circuit parameters *L* and *C* are fixed, the natural frequency is a constant quantity we relabel it here as ω_0.

Suppose however that a time-varying emf given by

$$\mathcal{E} = \mathcal{E}_m \cos \omega t \qquad (35\text{--}13)$$

is impressed on the system by an external generator. Here ω, which can be varied at will, is the frequency imposed by this external source. We describe such oscillations as "forced." Whatever the (constant) natural frequency ω_0 may be, the oscillations of charge, current, or potential difference in the circuit must occur at the (variable) external or driving frequency ω.

Figure 35–5 shows the situation and makes a comparison with a corresponding mechanical system. A vibrator *V*, which imposes an external alternating force, corresponds to generator *G*, which imposes an external alternating emf. Other quantities "correspond" as before. Note incidentally that, although we use the same symbol for a spring and an inductor they are not corresponding elements. In the appropriate differential equations a spring is described mathematically like a capacitor and an inductor like a massive body.

Our starting point in deriving the differential equations for both free and damped *LC* oscillations was the conservation of energy principle. We could follow this course for the forced oscillations of the circuit of Fig. 35–5*a* but, for variety, we shall use the loop theorem instead. This theorem (see Section 29–2) is itself based on the conservation of energy principle so that our approach is not basically different.

The loop theorem tells us that

$$\Delta V_R + \Delta V_L + \Delta V_C + \Delta V_G = 0$$

where the four terms are the potential differences across the circuit elements of Fig. 35–5*a*. For concreteness let us apply this theorem at an instant when the current in the circuit is clockwise and is increasing. If we start from point *x* and traverse the circuit loop in a counter-clockwise direction, evaluating the potential differences as we go, we are led to

$$+iR + L\frac{di}{dt} + \frac{q}{C} - \mathcal{E}_m \cos \omega t = 0$$

Substituting $i = dq/dt$ and $di/dt = d^2q/dt^2$ and rearranging yields

$$L\frac{d^2q}{dt^2} + R\frac{dq}{dt} + \left(\frac{1}{C}\right) q = \mathcal{E}_m \cos \omega t \qquad (35\text{--}14)$$

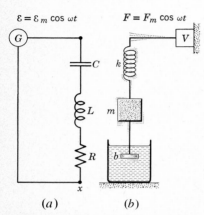

$\mathcal{E} = \mathcal{E}_m \cos \omega t \qquad F = F_m \cos \omega t$

(*a*)   (*b*)

Figure 35–5 Forced oscillations at angular frequency ω in (*a*) an electromagnetic oscillating system and (*b*) a corresponding mechanical oscillating system.

Free and damped oscillations

Forced oscillations

Deriving the differential equation

Forced RCL oscillations

as the desired differential equation. Compare this with the corresponding differential equation for the forced mechanical oscillations of the system of Fig. 35–5b, or

Forced mechanical oscillations

$$m\frac{d^2x}{dt^2} + b\frac{dx}{dt} + kx = F_m \cos \omega t.$$ [14–29]

We see that the two equations are mathematically identical.

The identity of these equations gives us confidence to write, as a solution of Eq. 35–14,

$$q(t) = q_m \sin (\omega t + \phi).$$ (35–15)

This corresponds exactly with Eq. 14–30, the solution of Eq. 14–29 above. Actually it is more customary to deal with the time-varying current in circuits like that of Fig. 35–5a rather than with the time-varying charge. Because $i = dq/dt$ we then have from Eq. 35–15

The current in forced oscillations

$$i(t) = i_m \cos (\omega t + \phi)$$ (35–16)

as the equation on which we will focus attention.

The current amplitude i_m in Eq. 35–16 is a measure of the response of the circuit of Fig. 35–5a to the impressed emf. It is reasonable to suppose, from experience in pushing swings if from no other source, that i_m will be larger the closer the driving frequency ω is to the natural frequency ω_0 of the system. In other words, we expect that a plot of i_m versus ω will exhibit a maximum when

Resonance condition

$$\omega = \omega_0,$$ (35–17)

a relationship described as the *resonance condition*.

Figure 35–6 shows three plots of i_m as a function of the ratio ω/ω_0, each plot

Figure 35–6 Resonance curves for forced oscillations in the circuit of Fig. 35–5a. The values of resistance R are as indicated. The horizontal arrows on each curve measure its width at the half-maximum level.

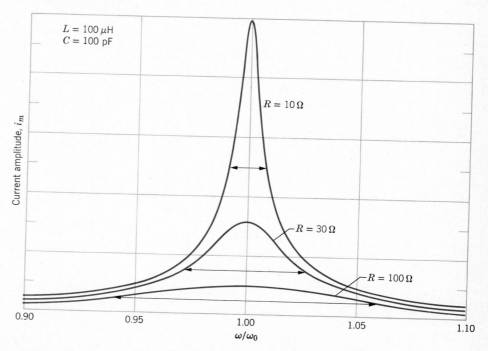

corresponding to a different value of the resistance R. We see that each of these *resonance peaks* does indeed have a maximum value when the resonance condition of Eq. 35–17 is satisfied. Note that as R is decreased the resonance peak becomes sharper, as shown by the three horizontal arrows drawn at the half-maximum level of each curve.

Figure 35–6 suggests to us the common experience of tuning a radio set. In turning the tuning knob we are adjusting the natural frequency ω_0 of an internal LC circuit to match the frequency ω of the signal transmitted by the antenna of the broadcasting station; we are seeking resonance. In a metropolitan area, where there are many signals whose frequencies are often close together, sharpness of tuning becomes important.

In general the response of circuits like that of Fig. 35–5a to impressed time-varying emfs is of great practical concern, not only in electronics generally but also in alternating current power distribution. Chapter 36, which follows, is devoted entirely to a more detailed study of the alternating currents generated by impressed alternating emfs.

We note finally that Fig. 35–6 finds a counterpart in Fig. 14–17, which shows resonance peaks for the forced oscillations of a mechanical oscillator such as that of Fig. 35–5b. In this case also the maximum response occurs when $\omega = \omega_0$ and the resonance peaks become sharper as the damping factor (the coefficient b) is reduced.

REVIEW GUIDE AND SUMMARY

Energy transfers

In a (resistanceless) oscillating LC circuit energy may be stored in the capacitor or in the inductor, according to

$$U_E = \tfrac{1}{2}q^2/C \quad \text{and} \quad U_B = \tfrac{1}{2}Li^2. \qquad [35–1, 2]$$

The total energy $E\ (= U_E + U_B)$ remains constant as energy oscillates back and forth between these elements, as in Fig. 35–1; see Example 1.

A mechanical analogy

Energy transfers in an oscillating circuit are analogous to those in an oscillating mass-spring system. The correspondences outlined in Table 35–1 allow us to predict that the angular frequency of the LC oscillations will be given by

$$\omega = \frac{1}{\sqrt{LC}}. \qquad [35–3]$$

A quantitative solution

The conservation of energy principle leads to

$$L\frac{d^2q}{dt^2} + \frac{q}{C} = 0 \qquad [35–5]$$

as the differential equation of free LC oscillations. From the method of correspondences (verified by trial) we find its solution to be

$$q = q_m \cos(\omega t + \phi) \qquad [35–6]$$

in which ω is indeed given by Eq. 35–3. The charge amplitude q_m and the phase constant ϕ are fixed by the initial conditions of the system. See Example 2.

$U_E(t)$ and $U_B(t)$

Given $q(t)$, and its derivative $i(t)$, we can find the energies $U_E(t)$ and $U_B(t)$ of Fig. 35–1. See Eqs. 35–9 and 10 and Fig. 35–4.

Damped oscillations

For damped LC oscillations the conservation of energy principle shows the differential equation to be

$$L\frac{d^2q}{dt^2} + R\frac{dq}{dt} + \frac{q}{C} = 0 \qquad [35–11]$$

Its solution, again found from the method of correspondences, is

$$q = q_m e^{-Rt/2L} \cos (\omega' t + \phi). \qquad [35-12]$$

We consider only light damping situations, in which ω' may be set equal to the ω of Eq. 35–3. See Example 3.

Forced oscillations

An LCR circuit such as that of Fig. 35–5a may be set into *forced oscillation* at angular frequency ω by an impressed emf such as

$$\mathcal{E} = \mathcal{E}_m \cos \omega t. \qquad [35-13]$$

Note that, when discussing forced oscillations, we relabel the (constant) natural frequency of the resonant system as ω_0, reserving ω for the (variable) frequency of the impressed emf. The differential equation proves to be

$$L \frac{d^2q}{dt^2} + R \frac{dq}{dt} + \frac{q}{C} = \mathcal{E}_m \cos \omega t \qquad [35-14]$$

and from its solution we find

$$i = i_m \cos (\omega t + \phi). \qquad [35-16]$$

Resonance

The current amplitude i_m has a maximum value when $\omega = \omega_0$, a condition called *resonance*. Figure 35–6 shows that resonance peaks becomes sharper as the circuit resistance R is decreased. Review the analogy to forced mechanical oscillations (see Fig. 35–5b) throughout.

QUESTIONS

1. Why doesn't the LC circuit of Fig. 35–1 simply stop oscillating when the capacitor has been completely discharged?

2. How might you start an LC circuit into oscillation with its initial condition being represented by Fig. 35–1c? Devise a switching scheme to bring this about.

3. In an oscillating LC circuit, assumed resistanceless, what determines (a) the frequency and (b) the amplitude of the oscillations?

4. In connection with Figs. 35–1c and g, explain how there can be a current in the inductor even though there is no charge on the capacitor.

5. Is it possible to have (a) an LC circuit without resistance, (b) an inductor without inherent capacitance, or (c) a capacitor without inherent inductance? Discuss the practical validity of the LC circuit of Fig. 35–1, in which each of the above possibilities is ignored.

6. In Fig. 35–1, what changes are called for if the oscillations are to proceed counterclockwise around the figure?

7. In Fig. 35–1, what phase constants ϕ in Eq. 35–6 would permit the eight circuit situations shown to serve in turn as initial conditions?

8. All practical LC circuits have to contain some resistance. However, one can buy a packaged audio oscillator in which the output maintains a constant amplitude indefinitely and does not decay, as it does in Fig. 35–3. How can this happen?

9. What constructional difficulties would you encounter if you tried to build an LC circuit of the type shown in Fig. 35–1 to oscillate (a) at 0.01 Hz, or (b) at 10^{10} Hz?

10. What is the difference between free, damped, and forced LC oscillations?

11. What would a resonance curve for $R = 0$ look like if plotted in Fig. 35–6?

12. Discuss the assertion that the resonance curves of Figs. 35–6 and 14–17 cannot truly be compared because the former is a plot of the current amplitude (i_m) and the latter of the displacement amplitude (x_m). Are these "corresponding" quantities? Does it make any difference if our purpose is only to exhibit the resonance phenomenon?

13. In comparing an electromagnetic oscillating system to a mechanical oscillating system, to what mechanical properties are the following electromagnetic properties analogous? . . . capacitance? . . . resistance? . . . charge? . . . electric field energy? . . . magnetic field energy? . . . inductance? . . . current?

14. Two identical springs are joined and connected to a body of mass m, the arrangement being free to oscillate on a horizontal frictionless surface as in Fig. 35–7. Sketch the electromagnetic analog of this mechanical oscillating system.

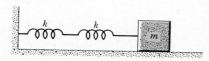

Figure 35–7   Question 14.

EXERCISES

Section 35–1   *LC* Oscillations: Qualitative

1. For a certain *LC* circuit the total energy is converted from electrical energy in the capacitor to magnetic energy in the inductor in 1.50×10^{-6} s. (*a*) What is the period of oscillation? (*b*) What is the frequency of oscillation? (*c*) How long after the magnetic energy is a maximum will it be a maximum again?

2. An *LC* circuit consists of a 75.0-mH inductor and a 3.55-μF capacitor. If the maximum charge on the capacitor is 2.95 μC, (*a*) what is the total energy in the circuit and (*b*) what is the maximum current?

Section 35–2   *LC* Oscillations: A Mechanical Analogy

3. Consider the list of four electrical quantities (q, i, C, L) and their corresponding mechanical quantities (x, v, $1/k$, m) as given in Section 35–2. (*a*) What is the mechanical quantity corresponding to potential difference V? (*b*) What are the SI units of the ratio of each of these five electrical quantities to the corresponding mechanical quantity?

4. Two capacitors are connected in parallel. (*a*) Complete the following sentence using either "charge" or "voltage": The sum of the _____ of the two capacitors is the _____ of the equivalent capacitor. (*b*) Using the correct mechanical analogies, rewrite (*a*) for a mechanical system. (*c*) Make a sketch showing the mechanical system in (*b*).

Section 35–3   *LC* Oscillations: Quantitative

5. What capacitance would you connect across a 1.0-mH inductor to make the resulting oscillator resonate at 3.5 kHz?

6. Consider the circuit shown in Fig. 35–8. With switch S_1

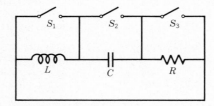

Figure 35–8   Exercise 6.

closed and the other two switches open, the circuit has a time constant τ_C (see Section 29–7). With switch S_2 closed and the other two switches open, the circuit has a time constant τ_L (see Section 33–3). With switch S_3 closed and the other two switches open, the circuit oscillates with a period T. Show that $T = 2\pi \sqrt{\tau_C \tau_L}$.

7. In an *LC* circuit with $L = 50$ mH and $C = 4.0$ μF the current is initially a maximum. How long will it take before the capacitor is fully charged for the first time?

Section 35–4   Damped *LC* Oscillations

8. Consider a damped *LC* circuit. (*a*) Show that the damping term $e^{-Rt/2L}$ (which involves L but not C) can be rewritten in the more symmetric manner (involving both L and C) as $e^{-\pi R \sqrt{C/L}(t/T)}$. Here T is the period of oscillation (neglecting resistance). (*b*) Using (*a*), show that the SI unit of $\sqrt{L/C}$ is "ohm." (*c*) Using (*a*), show that the condition that the fractional energy loss per cycle be small is $R \ll \sqrt{L/C}$.

9. A single loop consists of several inductors (L_1, L_2, . . .), several capacitors (C_1, C_2, . . .) and several resistors (R_1, R_2, . . .) connected in series as shown, e.g., in Fig. 35–9a. Show that, regardless of the sequence of these circuit elements in the loop, the behavior of this circuit is identical to that of a simple damped *LC* circuit shown in Fig. 35–9b. (*Hint:* Consider the loop equation.)

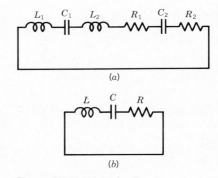

Figure 35–9   Exercise 9.

PROBLEMS GROUPED BY SECTION

Section 35–1   *LC* Oscillations: Qualitative

1. Find the capacitance of an *LC* circuit if the maximum charge on the capacitor is 1.0 μC and the total energy is 140 μJ.

2. A 1.5-mH inductor in an *LC* circuit stores a maximum energy of 10 μJ. What is the peak current?

3. In an oscillating *LC* circuit $L = 1.0$ mH and $C = 4.0$ μF. The maximum charge on C is 3.0 μC. Find the maximum current.

Section 35–2   *LC* Oscillations: A Mechanical Analogy

4. A 0.50-kg body oscillates on a spring that, when extended 2.0 mm from equilibrium, has a restoring force of 8.0 N. (*a*) What

is its period of oscillation? (*b*) What is the capacitance of the analogous *LC* system if *L* is chosen to be 5.0 H?

5. The energy in an *LC* circuit containing a 1.25-H inductor is 5.70×10^{-4} J. The maximum charge on the capacitor is 1.75×10^{-4} C. Find (*a*) the mass, (*b*) the spring constant, (*c*) the maximum displacement, and (*d*) the maximum speed for the analogous mechanical system.

Section 35–3   *LC* Oscillations: Quantitative

6. An *LC* circuit has an inductance of 3.0 mH and a capacitance of 10 μF. (*a*) Calculate the angular frequency of oscillation. (*b*) Find the period of the oscillation. (*c*) At time $t = 0$ the capacitor is charged to 200 μC, and the current is zero. Sketch roughly the charge on the capacitor as a function of time.

7. An inductor is connected across a capacitor whose capacitance can be varied by turning a knob. We wish to make the frequency of the LC oscillations vary linearly with the angle of rotation of the knob, going from 2×10^5 Hz to 4×10^5 Hz as the knob turns through $180°$. If $L = 1.0$ mH, plot C as a function of angle for the $180°$ rotation.

8. In an LC circuit in which $C = 4.0$ μF the maximum potential difference across the capacitor during the oscillations is 1.5 V and the maximum current through the inductor is 50 mA. (*a*) What is the inductance L? (*b*) What is the frequency of the oscillations? (*c*) How much time does it take for the charge on the capacitor to rise from zero to its maximum value?

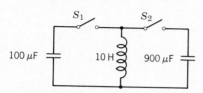

Figure 35–10 Problem 9.

9. In Fig. 35–10 the 900-μF capacitor is initially charged to 100 V and the 100-μF capacitor is uncharged. Describe in detail how one might charge the 100-μF capacitor to 300 V by manipulating switches S_1 and S_2.

10. Derive the differential equation for an LC circuit (Eq. 35–5) using the loop theorem.

Section 35–4 Damped *LC* Oscillations

11. A single loop circuit consists of a 5.50-Ω resistor, a 12.6-H inductor, and a 3.20-μF capacitor. Initially the capacitor has a charge of 6.25 μC and the current is zero. Calculate the charge on the capacitor N complete cycles later for $N = 5, 10, 100$.

*12. (*a*) By direct substitution of Eq. 35–12 into Eq. 35–11, show that $\omega' = \sqrt{(1/LC) - (R/2L)^2}$. (*b*) By what fraction does the frequency of oscillation shift when the resistance is increased from 0 to 100 Ω in a circuit with $L = 4.40$ H and $C = 7.35$ μF?

ADDITIONAL PROBLEMS

13. An oscillating LC circuit consisting of a 1.0-nF capacitor and a 3.0-mH coil carries a peak voltage of 3.0 V. (*a*) What is the maximum charge on the capacitor? (*b*) What is the peak current through the circuit? (*c*) What is the maximum energy stored in the magnetic field of the coil?

14. (*a*) In an oscillating LC circuit, in terms of the maximum charge on the capacitor, what value of charge is present when the energy in the electric field is one-half that in the magnetic field? (*b*) What fraction of a period must elapse following the time the capacitor is fully charged for this condition to arise?

15. An LC circuit oscillates at 10.4 kHz. (*a*) If the capacitance is 340 μF what is the inductance? (*b*) If the maximum current is 7.20 mA what is the total energy in the circuit?

16. You are given a 10-mH inductor and two capacitors, of 5.0- and 2.0-μF capacitance. List the resonant frequencies that can be generated by connecting these elements in various combinations.

17. A variable capacitor with a range from 10 to 365 pF is used with a coil to form a variable-frequency LC circuit to tune the input to a radio. (*a*) What ratio of maximum to minimum frequencies may be tuned with such a capacitor? (*b*) If this capacitor is to tune from 0.54 to 1.60 MHz, the ratio computed in (*a*) is too large. By adding a capacitor in parallel to the variable capacitor this range may be adjusted. How large should this capacitor be and what inductance should be chosen in order to tune the desired range of frequencies?

18. In an LC circuit $L = 25.0$ mH and $C = 7.65$ μF. At time $t = 0$ the current is 9.20 mA, the charge on the capacitor is 3.80 μC, and the capacitor is charging. (*a*) What is the total energy

in the circuit? (*b*) What is the maximum charge on the capacitor? (*c*) What is the maximum current? (*d*) If the charge on the capacitor is given by $q = q_m \cos(\omega t + \phi)$, what is the phase angle ϕ? (*e*) Suppose the data are the same, except that the capacitor is discharging at $t = 0$. What then is the phase angle ϕ?

19. In an oscillating LC circuit $L = 3.0$ mH and $C = 2.7$ μF. At $t = 0$ the charge on the capacitor is zero and the current is 2.0 A. (*a*) What is the maximum charge that will appear on the capacitor? (*b*) In terms of the period T of oscillation, how much time after $t = 0$ will elapse until the energy stored in the capacitor will be increasing at its greatest rate? (*c*) What is this greatest rate at which energy flows into the capacitor?

20. The resonant frequency of a series circuit containing inductance L_1 and capacitance C_1 is ω_0. A second series circuit, containing inductance L_2 and capacitance C_2, has the same resonant frequency. In terms of ω_0, what is the resonant frequency of a series circuit containing all four of these elements? Neglect resistance.

21. At some instant in an oscillating LC circuit, three-fourths of the total energy is stored in the magnetic field of the inductor. (*a*) In terms of the maximum charge on the capacitor, what is the charge on the capacitor at this instant? (*b*) In terms of the maximum current in the inductor, what is the current in the inductor at this instant?

22. In a damped LC circuit, find the time required for the maximum energy present in the capacitor during one oscillation to fall to one-half of its initial value. Assume $q = q_m$ at $t = 0$.

23. Three identical inductors L and two identical capacitors C are connected in a two-loop circuit as shown in Fig. 35–11. (*a*) Suppose the currents are as shown in Fig. 35–11*a*. What is the current in the middle inductor? Write down the loop equations and show that they are satisfied provided that the current oscillates with angular frequency $\omega = 1/\sqrt{LC}$. (*b*) Now suppose the currents are as shown in Fig. 35–11*b*. What is the current in the middle inductor? Write down the loop equations and show that they are satisfied provided the current oscillates with angular frequency $\omega = 1/\sqrt{3LC}$. (*c*) In view of the fact that the circuit can oscillate at two different frequencies, is it possible to replace this two-loop circuit by an equivalent single-loop LC circuit?

*24. In the damped LC circuit of Example 3 show that the fraction of the energy lost per cycle of oscillation, $\Delta U/U$, is given to a close approximation by $2\pi R/\omega L$. The quantity $\omega L/R$ is often called the Q of the circuit (for "quality"). A "high-Q" circuit has low resistance and a low fractional energy loss per cycle ($= 2\pi/Q$).

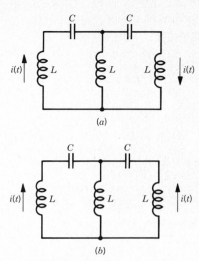

(*a*)

(*b*)

Figure 35–11 Problem 23.

36 Alternating Currents

36-1 Introduction

The applied emf

In Section 35–5 we saw that if an emf given by

$$\mathcal{E}(t) = \mathcal{E}_m \sin \omega t \qquad (36-1)$$

is applied to a circuit such as that of Fig. 36–1 an alternating current given by

The current

$$i(t) = i_m \sin(\omega t + \phi) \qquad (36-2)$$

is established. Alternating emfs and the alternating currents generated by them are key elements in power distribution systems, radio, television, satellite communications systems, computor systems, and much else that fashions our modern lifestyle. Because of this great technological importance we devote this entire chapter to the subject of alternating currents.

A change in notation

A small change in notation with respect to the usage of earlier chapters is embodied in the above two equations. We now represent the angular frequency of the emf and the current by ω rather than by ω_e, a step taken simply for convenience. In the sections that follow we are careful to indicate those few places where confusion with previous usage is possible.

The quantities of our concern in this chapter are the current amplitude i_m and the phase constant ϕ in Eq. 36–2. Our aim is to express them in terms of the characteristics of the generator ($\mathcal{E}$ and ω) and of the circuit (R, C, and L). Rather than attempt to solve the differential equation that applies to the circuit of Fig. 36–1 (that is, Eq. 35–14) we will use a geometrical method, introducing the concept of phasors.

36–2 *RCL* Elements, Considered Separately

Let us first simplify the problem suggested by Fig. 36–1 by considering three circuits, each containing the alternating current generator and only one other element, *R*, *C*, or *L*. We start with *R*.

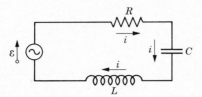

Figure 36–1 A single-loop circuit containing resistor, a capacitor, and an inductor. An alternating current generator supplies the emf of Eq. 36–1.

A resistor only

A Resistive Circuit. Fig. 36–2*a* shows a circuit containing a resistive element only, acted on by the alternating emf of Eq. 36–1. We can write

The voltage

$$V_R = V_{R,m} \sin \omega t \tag{36–3}$$

and, from the definition of resistance,

The current

$$i_R = \frac{V_R}{R} = \left(\frac{V_{R,m}}{R}\right) \sin \omega t = i_{R,m} \sin \omega t. \tag{36–4}$$

Comparison shows that the time-varying quantities V_R and i_R are in phase, which means that their corresponding maxima occur at the same time. Figure 36–2*b*, a plot of $V_R(t)$ and $i_R(t)$, illustrates this.

Figure 36–2 (*a*) A single-loop resistive circuit containing an ac generator. (*b*) The current and the potential difference across the resistor are in phase. (*c*) A phasor diagram shows the same thing.

Figure 36–3 (*a*) A single-loop capacitive circuit containing an ac generator. (*b*) The potential difference across the capacitor lags the current by 90°. (*c*) A phasor diagram shows the same thing.

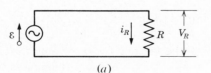

(*a*)

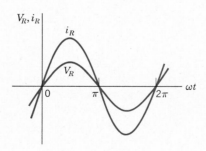

(*b*)

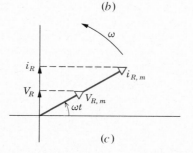

(*c*)

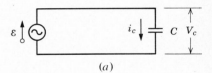

(*a*)

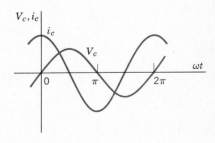

(*b*)

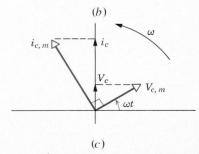

(*c*)

Phasor defined

Figure 36–2c shows another way of looking at the same situation. The two open arrows, called *phasors*, rotate counter clockwise about the origin with an angular frequency ω. The length of a phasor is proportional to the maximum value of the alternating quantity involved; thus $V_{R,m}$ and $i_{R,m}$. The projection of a phasor on the vertical axis is a measure of the instantaneous value of this alternating quantity; thus V_R and i_R. That V_R and i_R are in phase follows from the fact that their phasors lie along the same line in Fig. 36–2c. Follow the rotation of the phasors in this figure and convince yourself that it completely and correctly describes Eqs. 36–3 and 36–4.

A capacitor only

A Capacitive Circuit. Figure 36–3a shows a circuit containing a capacitive element only, acted on by the alternating emf of Eq. 36–1. We can write

The voltage

$$V_C = V_{C,m} \sin \omega t \qquad (36-5)$$

and, from the definition of capacitance,

$$q_C = CV_C = (CV_{C,m}) \sin \omega t.$$

Our concern, however, is with the current rather than the charge. Thus

$$i_C = \frac{dq_C}{dt} = \omega CV_{C,m} \cos \omega t \qquad (36-6a)$$

We now recast this equation in two aspects. First, for reasons of symmetry of notation, we introduce the quantity X_C, called the *capacitive reactance* of the capacitor at the frequency in question. It is defined from

The capacitive reactance

$$X_C = \frac{1}{\omega C}. \qquad (36-7)$$

Analysis shows that the SI unit of X_C is the ohm, just as for resistance R. Second, we replace cos ωt by its equal, or

$$\cos \omega t = \sin (\omega t + 90°)$$

This identity can be verified by expanding the right-hand side above according to the formula for sin $(\alpha + \beta)$ listed in Appendix G. Equation 36–6a then becomes

The current

$$i_C = \left(\frac{V_{C,m}}{X_C}\right) \sin (\omega t + 90°) = i_{C,m} \sin (\omega t + 90°). \qquad (36-6b)$$

Current leads voltage

Comparison of Eqs. 36–5 and 36–6b, or inspection of Fig. 36–3b, shows that the quantities V_C and i_C are 90° (= one-quarter cycle) out of phase. We see that V_C *lags* i_C, that is, as time goes on V_C reaches its maximum after i_C does. All this follows with equal clarity from the phasor diagram of Fig. 36–3c. As the phasors rotate counter clockwise we see that the phasor labeled $V_{C,m}$ does indeed lag behind that labeled $i_{C,m}$, and by an angle of 90°.

Example 1 In Figure 36–3a let $C = 15$ μF, $\nu = 60$ Hz, and $\mathcal{E}_m = V_{C,m} = 30$ V. Find (a) the capacitive reactance X_C and (b) the current amplitude $i_{C,m}$.
  (a) From Eq. 36–7

$$X_C = \frac{1}{\omega C} = \frac{1}{2\pi\nu C}$$

$$= \frac{1}{(2\pi)(60 \text{ Hz})(15 \times 10^{-6} \text{ F})} = 180 \ \Omega.$$

(b) From Eq. 36–6b

$$i_{C,m} = \frac{V_{C,m}}{X_C} = \frac{30 \text{ V}}{180 \ \Omega} = 0.17 \text{ A}.$$

How would the current amplitude change if you doubled the frequency? Does this seem intuitively reasonable?

An inductor only

The voltage

An Inductive Circuit. Figure 36–4a shows a circuit containing an inductive element only, acted on by the alternating emf of Eq. 36–1. We can write

$$V_L = V_{L,m} \sin \omega t \tag{36–8}$$

and, from the definition of inductance (see Eq. 33–3)

$$\frac{di_L}{dt} = \frac{V_L}{L} = \left(\frac{V_{L,m}}{L}\right) \sin \omega t.$$

Our concern however is with the current rather than with its time derivative. Thus

$$i_L = \int \left(\frac{di}{dt}\right) dt = \left(\frac{V_{L,m}}{\omega L}\right) \int \sin \omega t \; d(\omega t) = - \left(\frac{V_{L,m}}{\omega L}\right) \cos \omega t. \tag{36–9a}$$

Again, we recast this equation in two aspects. First, once more for reasons of symmetry of notation, we introduce the quantity X_L, called the *inductive reactance* of the inductor at the frequency in question. It is defined from

The inductive reactance

$$X_L = \omega L. \tag{36–10}$$

Analysis shows that the SI unit of X_L is the ohm, just as it is for X_C and for R. Second, we replace $-\cos \omega t$ by its equal, or

$$-\cos \omega t = \sin (\omega t - 90°).$$

This identity can be verified by expanding the right-hand side according to the formula in Appendix G. Equation 36–9a then becomes

The current

$$i_L = \left(\frac{V_{L,m}}{X_L}\right) \sin (\omega t - 90°) = i_{L,m} \sin (\omega t - 90°). \tag{36–9b}$$

Comparison of Eqs. 36–8 and 36–9b, or inspection of Fig. 36–4b, shows that the quantities V_L and i_L are 90° out of phase. In this case, however, V_L *leads* i_L, that is, as time goes on V_L reaches its maximum value before i_L does. The phasor diagram of Fig. 36–4c also contains this information. As the phasors rotate in that figure we see that the phasor labeled $V_{L,m}$ does indeed lead that labeled $i_{L,m}$, and by an angle of 90°.

Voltage leads current

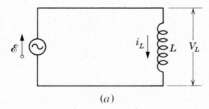

(a)

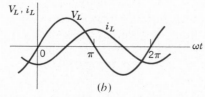

(b)

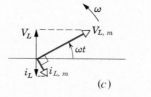

(c)

Figure 36–4 (a) A single-loop inductive circuit containing an ac generator. (b) The potential difference across the inductor leads the current by 90°. (c) A phasor diagram shows the same thing.

Example 2 In Fig. 36–4*a* let $L = 230$ mH, $\nu = 60$ Hz, and $\mathcal{E}_m = V_{L,m} = 30$ V. Find (*a*) the inductive reactance X_L and (*b*) the current amplitude $i_{L,m}$.

(*a*) From Eq. 36–10

$$X_L = \omega L = 2\pi\nu L = (2\pi)(60 \text{ Hz})(230 \times 10^{-3} \text{ H}) = 87\Omega.$$

(*b*) From Eq. 36–9*b*

$$i_{L,m} = \frac{V_{L,m}}{X_L} = \frac{30 \text{ V}}{87 \ \Omega} = 0.35 \text{ A}.$$

How would the current amplitude change if you doubled the frequency? Does your answer seem intuitively reasonable?

A summary

Table 36–1 summarizes what we have learned about the phase differences between the current i and the voltage V for each of the three kinds of circuit elements. It also shows the relationship between the maximum values of the current and the voltage in each of these cases.

Table 36–1 Phase and amplitude relationships for alternating currents and voltages

Circuit element	Phase relationship	Amplitude relationship
R	V_R and i are in phase	$V_{R,m} = i_m R$
C	V_C lags i by 90°	$V_{C,m} = i_m X_C$
L	V_L leads i by 90°	$V_{L,m} = i_m X_L$

36–3 The *RCL* Circuit

We are now ready to address the full problem posed by Fig. 36–1, in which the impressed alternating emf is

The applied emf

$$\mathcal{E} = \mathcal{E}_m \sin \omega t \tag{36–1}$$

and the resulting alternating current is

The current

$$i = i_m \sin (\omega t + \phi). \tag{36–2}$$

Our task, we recall, is to evaluate the current amplitude i_m and the phase constant ϕ. As a starting point we apply the loop theorem to the circuit of Fig. 36–1, obtaining

A phasor diagram—the voltages and the current

$$\mathcal{E} = V_R + V_C + V_L. \tag{36–11}$$

Each of these four quantities varies with time, this relationship among them remaining continually valid.

Consider now the phasor diagram of Fig. 36–5*a*. In constructing it we first drew a phasor representing, at an arbitrary instant of time, the as yet unknown alternating current. Its maximum value i_m, its phase $(\omega t + \phi)$, and its instantaneous value i are all shown. From the phase information summarized in Table 36–1 we can then draw the three phasors representing the voltages across the three circuit elements R, C, and L at that same instant. As we have learned, V_R is in phase with the current, V_C lags it by 90°, and V_L leads it by the same angle.

Note that the algebraic sum of the projections of the phasors $V_{R,m}$, $V_{C,m}$, and $V_{L,m}$ on the vertical axis is just the right-hand side of Eq. 36–11. This sum of projections must then equal the left-hand side of that equation, which is $\mathcal{E}$, the projection of the phasor $\mathcal{E}_m$. This equality continues to hold as the phasor assembly in Fig. 36–5*a* rotates about the origin with angular frequency ω.

In vector operations the (algebraic) sum of the projections of a set of vectors

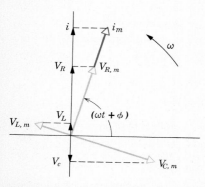

Figure 36–5*a* A phasor diagram showing the voltages and the current in the circuit of Fig. 36–1.

on a given axis is equal to the projection on that axis of the (vector) sum of those vectors. Equations 2–9 and 2–10 illustrate this for the case of two vectors projected on a coordinate axis. It follows that the phasor ε_m is equal to the (vector) sum of the phasors $V_{R,m}$ $V_{C,m}$, and $V_{L,m}$.

In Fig. 36–5b we have formed the phasor difference $V_{C,m} - V_{L,m}$ and we note that it is at right angles to $V_{R,m}$. We then see that

$$\varepsilon_m{}^2 = V_{R,m}{}^2 + (V_{C,m} - V_{L,m})^2$$

From the amplitude information displayed in Table 36–1 we can write this as

$$\varepsilon_m{}^2 = (i_m R)^2 + (i_m X_C - i_m X_L)^2$$

which can be rearranged in the form

$$i_m = \frac{\varepsilon_m}{\sqrt{R^2 + (X_C - X_L)^2}}. \tag{36–12}$$

The denominator above is called the *impedance* Z of the circuit for the frequency in question. Thus

$$Z = \sqrt{R^2 + (X_C - X_L)^2} \tag{36–13}$$

and

$$i_m = \frac{\varepsilon_m}{Z}. \tag{36–14}$$

If we substitute for X_C and X_L from Eqs. 36–7 and 36–10 we can write Eq. 36–12 more explicitly as

$$i_m = \frac{\varepsilon_m}{\sqrt{R^2 + \left(\dfrac{1}{\omega C} - \omega L\right)^2}}. \tag{36–15}$$

This relationship is half of the solution of our problem. It is an expression for the current amplitude i_m in terms of ε_m, ω, R, C, and L. It remains to find an equivalent expression for the phase angle ϕ.

From Fig. 36–5b and Table 36–1 we can write

$$\tan \phi = \frac{V_{C,m} - V_{L,m}}{V_{R,m}} = \frac{i_m X_C - i_m X_L}{i_m R} = \frac{X_C - X_L}{R}. \tag{36–16}$$

Thus we have solved the second half of the problem; we have expressed ϕ in terms of ω, R, C, and L. Note that ε_m is not involved in this case.

Equation 36–15 is essentially the equation of the resonance curves of Fig. 35–6. Inspection of Eq. 36–15 shows that the maximum value of i_m occurs when

$$\frac{1}{\omega C} = \omega L \qquad \text{or} \qquad \omega = \frac{1}{\sqrt{LC}}.$$

This is precisely the resonance condition that we discussed in connection with Fig. 35–6. The maximum value of i_m is, as we see from Eq. 36–15, just ε_m/R. Again this agrees with Fig. 35–6, in which we see that the resonance maximum increases as the circuit resistance decreases.

We drew Fig. 36–5b arbitrarily with $X_C > X_L$, that is, we assumed the circuit of Fig. 36–1 to be more capacitive than inductive. For this assumption ε_m lags i_m

A phasor diagram—the emf and the current

The current amplitudes

The impedance

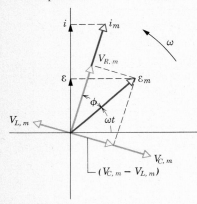

Figure 36–5b The same diagram as Fig. 36–5a with the impressed emf added. Note the phase angle ϕ between ε_m and i_m.

The phase angle

Resonance

A largely capacitive circuit

and the phase angle ϕ in Eqs. 36-2 and 16 is positive. In a limiting case we can put $R = X_L = 0$ in Eqs. 36-13 and 36-16, leading to $Z = X_C$ and $\tan \phi = +\infty$. The physical interpretation of this is that the circuit is now purely capacitive, as in Fig. 36-3a, and the phase angle ϕ is $+90°$, as in Fig. 36-3b, c.

A largely inductive circuit

If in drawing Fig. 36-5b we had chosen $X_L > X_C$ the circuit of Fig. 36-1 would then have been more inductive than capacitive. ε_m would then lead i_m and the phase angle ϕ in Eqs. 36-2 and 36-16 would be negative. In the limiting case we can put $R = X_C = 0$ in Eqs. 36-13 and 36-16, leading to $Z = X_L$ and $\tan \phi = -\infty$. This corresponds to a purely inductive circuit, as in Fig. 36-4a, with $\phi = -90°$, as in Figs. 36-4b,c.

Example 3   In Fig. 36-1 let $R = 160 \; \Omega$, $C = 15 \; \mu F$, $L = 230 \; mH$, $\nu = 60 \; Hz$, and $\varepsilon_m = 30 \; V$. Find (a) the capacitive reactance X_C, the (b) the inductive reactance X_L, (c) the impedance Z, (d) the current amplitude i_m, and (e) the phase angle ϕ.
  (a) $X_C = 180 \; \Omega$, as in Example 1.
  (b) $X_L = 87 \; \Omega$, as in Example 2.
  (c) From Eq. 36-13 we have

$$Z = \sqrt{R^2 + (X_C - X_L)^2}$$
$$= \sqrt{(160 \; \Omega)^2 + (180 \; \Omega - 87 \; \Omega)^2} = 190 \; \Omega.$$

Note that the circuit is more capacitive than inductive, that is, $X_C > X_L$, as in Fig. 36-5.

(d) From Eq. 36-14 we have

$$i_m = \frac{\varepsilon_m}{Z} = \frac{30 \; V}{190 \; \Omega} = 0.16 \; A.$$

(e) From Eq. 36-16 we have

$$\tan \phi = \frac{X_C - X_L}{R} = \frac{180 \; \Omega - 87 \; \Omega}{160 \; \Omega} = 0.58$$

or $\phi = +30°$. Because $X_C > X_L$, ϕ is positive and ε_m lags i_m, as in Fig. 36-5b. This figure is drawn to scale to correspond with the numerical values of this Example.

36-4   Power in Alternating Current Circuits

The flow of energy

In the *RCL* circuit of Fig. 36-1 the source of energy is the alternating current generator. Of the energy that it provides some is stored in the electric field in the capacitor, some in the magnetic field in the inductor, and some is dissipated as thermal energy in the resistor. We note that in steady-state operation—which we assume—the average energy stored in the capacitor and the inductor remains constant. Averaged over an integral number of cycles, energy is neither building up or leaking away in these elements. The net flow of energy is then from the generator to the resistor, where it is transformed from electromagnetic to thermal form.

With the aid of Eq. 36-2 we can write, for the *instantaneous* rate at which energy is transformed in the resistor,

$$P(t) = i^2 R = [i_m \sin (\omega t + \phi)]^2 R.$$

The *average* rate however, which is our real concern, is

$$P_{av} = \overline{P(t)} = i_m{}^2 R \; \overline{\sin^2 (\omega t + \phi)}.$$

Figure 36-6 shows us that the average value of $\sin^2 \theta$ (over any integral number of quarter-cycles) is just $\frac{1}{2}$, with θ being any angular quantity. Note how the parts of the curve that lie above the line "$\frac{1}{2}$" just fill in the gaps that lie below that line. Thus

$$P_{av} = \frac{1}{2} i_m{}^2 R = (i_m/\sqrt{2})^2 R.$$

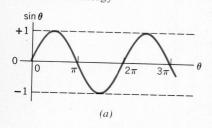

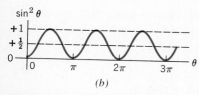

Figure 36-6   (a) A plot of $\sin \theta$ versus θ; its average value is zero. (b) A plot of $\sin^2 \theta$ versus θ; its average value is $\frac{1}{2}$. The angle θ represents any angular quantity.

We choose to call $i_m/\sqrt{2}$ the *root-mean-square* (rms) value of the current i and we write in place of the above equation

$$P_{av} = (i_{rms})^2\, R. \tag{36–17}$$

The rms notation is justified. First we *square* i_m, obtaining $i_m{}^2$. Then we average it (that is, we find its *mean* value), obtaining $i_m{}^2/2$, as in Fig. 36–6b. Finally we take its square *root*, obtaining $i_m/\sqrt{2}$ or i_{rms}.

Equation 36–17 looks much like Eq. 28–16 and the message is that if we use rms quantities for i, for V, and for ε, the *average* power dissipation, which is all that usually matters, will be the same for alternating current circuits as for direct current circuits with a constant emf. Alternating current instruments, such as ammeters and voltmeters, are deliberately calibrated to read i_{rms}, V_{rms}, and ε_{rms}. Thus, if you plug an alternating current voltmeter into a household electric outlet and it reads "120 V," the maximum value of the potential difference at the outlet is $\sqrt{2}$ (120 V) or 170 V. The sole reason for using rms values in alternating current circuits is to let us use the familiar direct current power relationships of Section 28–5. Thus, we summarize:

$$i_{rms} = i_m/\sqrt{2}, \; V_{rms} = V_m/\sqrt{2}, \text{ and } \varepsilon_{rms} = \varepsilon_m/\sqrt{2}. \tag{36–18}$$

Because the proportionality factor $(= 1/\sqrt{2})$ is the same in each case, we can write the important Eqs. 36–14 and 15 as

$$i_{rms} = \frac{\varepsilon_{rms}}{\sqrt{R^2 + (1/\omega C - \omega L)^2}} = \frac{\varepsilon_{rms}}{Z} \tag{36–19}$$

and, indeed, this is the form that we almost always use.

We can recast Eq. 36–17 in a useful equivalent way by combining it with the relationship $i_{rms} = \varepsilon_{rms}/Z$ (see Eq. 36–19) or

$$P_{av} = (\varepsilon_{rms}/Z)\, i_{rms}\, R = \varepsilon_{rms}\, i_{rms}\, (R/Z). \tag{36–20}$$

From Fig. 36–5b and Table 36–1, however, we can show that R/Z is just the cosine of the phase angle ϕ. Thus

$$\cos \phi = \frac{V_{R,m}}{\varepsilon_m} = \frac{i_m R}{i_m Z} = R/Z. \tag{36–21}$$

Equation 36–20 then becomes

$$P_{av} = \varepsilon_{rms}\, i_{rms} \cos \phi, \tag{36–22}$$

in which $\cos \phi$ is called the *power factor*.

Table 36–2 shows that Eq. 36–22 gives reasonable results in the three special cases in which the single circuit element present is a resistor (as in Fig. 36–2a), a capacitor (as in Fig. 36–3a), or an inductor (as in Fig. 36–4a).

Table 36–2 The average power in three special cases

Circuit element present	Impedance Z (Eq. 36–13)	Phase angle ϕ (Eq. 36–16)	Power factor $\cos \phi$	Power P_{av} (Eq. 36–22)
R	R	0	1	$\varepsilon_{rms} i_{rms}$
C	X_C	$+90°$	0	0
L	X_L	$-90°$	0	0

Example 4 Let us use the same parameters for Fig. 36–1 that we did in Example 3, namely $R = 160\ \Omega$, $C = 15\ \mu F$, $L = 230$ mH, $\nu = 60$ Hz, and $\varepsilon_m = 30$ V. Find (a) the emf ε_{rms}, (b) the current i_{rms}, (c) the power factor $\cos \phi$, and (d) the average power P_{av} dissipated in the resistor.

(a) $\varepsilon_{rms} = \varepsilon_m / \sqrt{2} = 30\ V / \sqrt{2} = 21$ V.

(b) In Example 3 we see that $i_m = 0.16$ A. We then have $i_{rms} = i_m / \sqrt{2} = 0.16\ A / \sqrt{2} = 0.12$ A.

(c) In Example 3 we see that $\phi = 30°$. The power factor is just $\cos \phi$ or 0.87.

(d) From Eq. 36–17 we have

$$P_{av} = (i_{rms})^2 R = (0.116\ A)^2 (160\ \Omega) = 2.1\ W.$$

Alternatively, Eq. 36–22 yields

$$P_{av} = \varepsilon_{rms} i_{rms} \cos \phi = (21.2\ V)(0.116\ A)(0.871) = 2.1\ W,$$

in full agreement. Note that to get agreement of the above two equations to two significant figures we have recalculated the values of i_{rms}, ε_{rms}, and $\cos \phi$ to three significant figures and used them in our calculations. These small calculational problems should not cause us to forget that Eqs. 36–17 and 36–22 are precisely equivalent expressions.

36–5 The Transformer

In dc circuits the power dissipation in a resistive load is given by the relation $P = iV$. This means that, for a given power requirement, we have a variety of choices ranging from a relatively large current i and a relatively small potential difference V or just the reverse, provided that their product remains constant. In the same way, for ac circuits the average power dissipation is given by Eq. 36–22 ($P_{av} = i_{rms} V_{rms}$) and we have the same choice as to the relative values of i_{rms} and V_{rms}.*

In electric power distribution systems it is clear that at both the generating end (the electric power plant) and the receiving end (the home or factory) it is desirable, both for reasons of safety and the efficient design of equipment, to deal with relatively low voltages. For example, no one wants an electric toaster or a child's electric train to operate at, say, 10 kV.

On the other hand, in the transmission of electric energy from the generating plant to the consumer, we want the lowest possible current (and thus the largest possible potential difference) so as to minimize the $i^2 R$ ohmic losses in the transmission line. $V_{rms} = 350$ kV is not uncommon. Thus there is a fundamental "mismatch" between the requirements for efficient transmission on the one hand and efficient and safe generation and consumption on the other. We need a device that can, as design considerations require, raise or lower the potential difference in a circuit, keeping the product iV essentially constant. The alternating current transformer of Fig. 36–7 is such a device. It has no direct-current counterpart of equivalent simplicity.

In Fig. 36–7 two coils are shown wound around a soft iron core. The *primary* winding, of N_1 turns, is connected to an alternating-current generator whose emf $\varepsilon_1(t)$ is given by $\varepsilon_1 = \varepsilon_m \sin \omega t$. The secondary winding, of N_2 turns, is on open circuit as long as switch S is open, which we assume for the present. Thus there is no secondary current. We assume further that the resistances of the primary and secondary windings and also the magnetic "losses" in the iron core are negligible. Actually, well-designed, high-capacity transformers can have energy losses as low as one percent so that our assumption of an ideal transformer is not unreasonable.

For the above conditions the primary winding is a pure inductance; compare Fig. 36–4a. Thus the (very small) primary current, called the *magnetizing current* $i_{mag}(t)$, lags the primary potential difference $V_1(t)$ by 90°; the power factor ($= \cos \phi$ in Eq. 36–22) is zero, and thus no power is delivered from the generator to the transformer.

The transformer

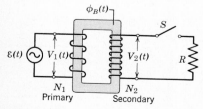

Figure 36–7 An ideal transformer, showing two coils wound on the same soft iron core.

---

* We assume a purely resistive load so that $\cos \phi = 1$ in Eq. 36–22.

The transformation of voltage

However, the small alternating primary current $i_{mag}(t)$ induces an alternating magnetic flux $\Phi_B(t)$ in the iron core and we assume that all this flux links the turns of the secondary windings. From Faraday's law of induction the emf per turn $\mathcal{E}_T$ is the same for both the primary and secondary windings. Thus, assuming that the symbols for time-varying quantities represent rms values, we can write

$$\mathcal{E}_T = -\frac{d\Phi_B}{dt} = \frac{V_1}{N_1} = \frac{V_2}{N_2}$$

or

$$V_2 = V_1(N_2/N_1). \tag{36-23}$$

If $N_2 > N_1$, we speak of a *step-up transformer;* if $N_2 < N_1$, we speak of a *step-down transformer.*

In all of the above we have assumed an open circuit secondary so that no power is transmitted through the transformer. If we now close switch S in Fig. 36–7, however, we have a more practical situation in which the secondary winding is connected with a resistive load R. In the general case the load would also contain inductive and capacitive elements but we confine ourselves to this special case.

Case of a resistive load

Several things happen when we close switch S. (1) A current $i_2(t)$ appears in the secondary circuit, with a corresponding power dissipation $i_2^2 R$ ($=V_2^2/R$) in the resistive load. (2) This current induces its own alternating magnetic flux in the iron core and this flux induces (from Faraday's law and Lenz's law) an opposing emf in the primary windings. The two windings appear now as a fully coupled mutual inductance. (3) $V_1(t)$, however, cannot change in response to this opposing emf because it must always equal $\mathcal{E}(t)$ as provided by the generator; closing switch S cannot change this fact. (4) For this reason a new resultant current $i_1(t)$ must appear in the primary windings, its magnitude and phase angle being just that needed to cancel the opposing emf generated in the primary windings by $i_2(t)$. In particular the phase angle ϕ between $i_1(t)$ and $\mathcal{E}(t)$ for an ideal transformer must approach 0°, so that the power factor in Eq. 36–22, $\cos \phi$, must approach unity.

Rather than analyze the above rather complex process in full detail we take advantage of the overall view provided by the conservation of energy principle. For an ideal transformer with a resistive load this tells us that

$$i_1 V_1 = i_2 V_2.$$

Because Eq. 36–23 holds whether or not the secondary circuit of Fig. 36–7 is closed (why?) we then have

The transformation of currents

$$i_2 = i_1 (N_1/N_2) \tag{36-24}$$

as the transformation relation for currents.

Example 5 A transformer on a utility pole operates at $V_1 = 8.0$ kV on the primary side and supplies electric energy to a number of nearby houses at $V_2 = 120$ V. (a) What is the turns ratio N_1/N_2? (b) If the average power consumption in the houses for a given time interval is 70 kW, what are the rms currents in the primary and secondary windings of the transformer? Assume an ideal transformer, a resistive load, and a power factor of unity. (c) What is the equivalent resistive load R in the secondary circuit?

(a) From Eq. 36–23,

$$N_1/N_2 = V_1/V_2 = 8.0 \text{ kV}/120 \text{ V} = 67.$$

(b) From Eq. 36–22 we have, with $\cos \phi = 1$

$$i_2 = P_{av}/[(V_2)(\cos \phi)] = (70 \text{ kW})/[(120 \text{ V})(1)] = 0.58 \text{ kA},$$

and also

$$i_1 = P_{av}/[(V_1)(\cos \phi)] = (70 \text{ kW})/[(8.0 \text{ kV})(1)] = 8.8 \text{ A}.$$

Note that, as required for an ideal transformer,

$$i_1 V_1 = i_2 V_2 = 70 \text{ kW}.$$

(c) Here we have

$$R = (V_2)^2/P_{av} = (120 \text{ V})^2/70 \text{ kW} = 0.21 \text{ }\Omega.$$

REVIEW GUIDE AND SUMMARY

The current amplitude and phase

The basic problem in alternating currents is to find expressions for the current amplitude i_m and the phase angle ϕ in

$$i = i_m \sin(\omega t + \phi) \qquad [36\text{--}2]$$

when an emf given by Eq. 36–1 is applied to the circuit of Fig. 36–1.

The problem subdivided

The circuit elements R, C, and L are first studied separately, using the phasor representation of alternating currents and voltages. Here the maximum value of any alternating quantity is represented by an open arrow rotating counter clockwise about the origin at the angular frequency ω of the generator. The projection of this arrow on the vertical axis gives the instantaneous value. See Figs. 36–2, 36–3, and 36–4.

Capacitive and inductive reactance

The useful quantities of capacitive reactance X_C and inductive reactance X_L, defined from

$$X_C = 1/\omega C \qquad \text{and} \qquad X_L = \omega L \qquad [36\text{--}7, 36\text{--}10]$$

are introduced. Their SI unit, like that of R, is the ohm. Table 36–1, which should be studied carefully, summarizes the results of the analysis of the separate current elements R, C, and L. See also Examples 1 and 2.

The problem solved

All three circuit elements are then considered together, as in Fig. 36–1. The phasor diagram of Fig. 36–5a, drawn with the help of Table 36–1, shows the relationship of the voltages across these three elements to the current in the circuit. By invoking the loop theorem (see Eq. 36–11) the emf phasor is then constructed, as in Fig. 36–5b. Derivations based on that figure yield

The current amplitude

$$i_m = \frac{\mathcal{E}_m}{\sqrt{R^2 + (X_C - X_L)^2}} \qquad [36\text{--}12]$$

and

The phase

$$\tan \phi = \frac{X_C - X_L}{R}. \qquad [36\text{--}16]$$

as the sought-for quantities. Introducing the impedance Z, defined from

The impedance

$$Z = \sqrt{R^2 + (X_C - X_L)^2} \qquad [36\text{--}13]$$

allows us to write Eq. 36–12 compactly as $i_m = \mathcal{E}_m/Z$. See Example 3.

Resonance in alternating current circuits

Equation 36–12, when combined with Eq. 36–7 and 10, yields the equation of the resonance curves of Fig. 35–6. Study of Eq. 36–12 shows that the peak current amplitude occurs when $X_C = X_L$ (the resonance condition; see Eq. 35–17) and has the value $\mathcal{E}_m/R$.

In the circuit of Fig. 36–1 the average power output P_{av} of the generator is delivered to the resistor, where it appears as thermal energy. Two equivalent expressions for P_{av} are

Power and rms notation

$$P_{av} = (i_{rms})^2 R \qquad \text{and} \qquad P_{av} = \mathcal{E}_{rms} i_{rms} \cos \phi. \qquad [36\text{--}17, 36\text{--}22]$$

Here "rms" stands for root-mean-square; rms quantities are related to maximum quantities by $i_{rms} = i_m/\sqrt{2}$ and $\mathcal{E}_{rms} = \mathcal{E}_m/\sqrt{2}$. These quantities are introduced so that the expressions for power in alternating and direct current circuits will have the same form. Alternating current voltmeters and ammeters have their scales adjusted to read rms values. The term $\cos \phi$ above is called the power factor. Table 36–2 examines P_{av} for the three special cases of Figs. 36–2, 36–3, and 36–4. See Example 4.

The transformer

A transformer (assumed "ideal;" see Fig. 36–7) is a soft-iron yoke on which are wound a primary coil of N_1 turns and a secondary coil of N_2 turns. If the primary is connected to an alternating current generator the primary and secondary voltages are related by

$$V_2 = V_1(N_2/N_1). \qquad [36\text{--}23]$$

If a resistive load R is connected across the secondary coil the power delivered by the generator ($= V_1 i_1$) must be equal to that dissipated in the load ($= V_2 i_2 = V_2{}^2/R$). The usefulness of a transformer is its ability to transform the magnitudes of alternating currents and voltages, maintaining their product (the power) constant. See Example 5.

QUESTIONS

1. In the relation $\omega = 2\pi v$ when using SI units we measure ω in radians/second and v in hertz or cycles/second. The radian is a measure of angle. What connection do angles have with alternating currents?

2. Problem 32–26 suggests how an alternating emf such as that described by Eq. 36–1 can be generated. If the output of such an ac generator is connected to an RCL circuit such as that of Fig. 36–1, what is the ultimate source of the power dissipated in the resistor?

3. Why would power distribution systems be less effective without alternating emfs?

4. In the circuit of Fig. 36–1, why is it safe to assume that (a) the alternating current of Eq. 36–2 has the same angular frequency ω as the alternating emf of Eq. 36–1, and (b) that the phase angle ϕ in Eq. 36–2 does not vary with time? What would happen if either of these (true) statements were false?

5. How does a phasor differ from a vector? We know, for example, that emfs, potential differences, and currents are not vectors. How then can we justify constructions such as Fig. 36–5?

6. Would any of the discussion of Section 36–2 be invalid if the phasor diagrams were to rotate in the clockwise direction, rather than the counter-clockwise direction which we assumed?

7. Does it seem intuitively reasonable that the capacitive reactance $(= 1/\omega C)$ should vary inversely with the angular frequency, whereas the inductive reactance $(= \omega L)$ varies directly with this quantity?

8. During World War II, at a large research laboratory in this country, an alternating current generator was located a mile or so from the laboratory building which it served. A technician increased the speed of the generator to compensate for what he called "the loss of frequency along the transmission line" connecting the generator with the laboratory building. Comment.

9. Discuss in your own words what it means to say that a potential difference "leads" or "lags" an alternating current.

10. If, as we stated in Section 36–3, a given circuit is "more inductive than capacitive," that is, that $X_L > X_C$, (a) does this mean, for a fixed angular frequency, that L is relatively "large" and C is relatively "small," or L and C are both relatively "large"? (b) For fixed values of L and C does this mean that ω is relatively "large" or relatively "small"?

11. What is wrong with this statement: "If $X_L > X_C$, then, regardless of the frequency, we must have $L > 1/C$."

12. Do the loop theorem (see Section 29–2) and the junction theorem (see Section 29–5) apply to multiloop ac circuits as well as to multiloop dc circuits?

13. In Example 4 what would be the effect on P_{av} if you increased (a) R, (b) C, and (c) L? How would ϕ in Eq. 36–22 change in these three cases?

14. Do commercial power station engineers like to have a low power factor or a high one, or does it make any difference to them? Between what values can the power factor range? What determines the power factor; is it characteristic of the generator, of the transmission line, of the circuit to which the transmission line is connected, or of some combination of these?

15. Can the instantaneous power delivered by a source of alternating current ever be negative? Can the power factor ever be negative? If so, explain the meaning of these negative values.

16. In a series RCL circuit (see Fig. 36–1), the voltage is leading the current for a particular frequency of operation. You now lower the frequency slightly. Does the total impedance of the circuit increase, decrease, or stay the same?

17. If you know the power factor $(= \cos \phi$ in Eq. 36–22) for a given RCL circuit, can you tell whether or not the applied alternating emf is leading or lagging the current? If so, how? If not, why not?

18. What is the permissible range of values of the phase angle ϕ in Eq. 36–2? Of the power factor in Eq. 36–22?

19. What is the usefulness of the rms notation for alternating currents and voltages?

20. If you want to reduce your electric bill, do you hope for a small or a large power factor or does it make any difference? If it does, is there anything that you can do about it? Discuss.

21. In Eq. 36–22 is ϕ the phase angle between $\varepsilon(t)$ and $i(t)$ or between ε_{rms} and i_{rms}? Discuss.

22. A doorbell transformer is designed for a primary rms input of 120 V and a secondary rms output of 6 V. What would happen if the primary and secondary connections were accidentally interchanged during installation? Would you have to wait for someone to push the doorbell to find out? Discuss.

23. You are given a transformer enclosed in a wooden box, its primary and secondary terminals being available at two opposite faces of the box. How could you find its turns ratio without opening the box?

24. In the transformer of Fig. 36–7, with the secondary on open circuit, what is the phase relationship between (a) the impressed emf and the primary current, (b) the impressed emf and the magnetic field in the transformer core, and (c) the primary current and the magnetic field in the transformer core?

25. What are some applications of a step-up transformer . . . a step-down transformer?

26. What determines which winding of a transformer is the primary and which the secondary? Can a transformer have a single primary and two secondaries? . . . a single secondary and two primaries?

EXERCISES

Section 36–1 Introduction

1. Let Eq. 36–1 describe the effective emf available at an ordinary 60-Hz ac outlet. What angular frequency ω does this correspond to? How does the utility company establish this frequency?

Section 36–2 *RCL* Elements, Considered Separately

2. A 1.5-μF capacitor is connected as in Fig. 36–3a to an ac generator with $\varepsilon_m = 30$ V. What is the amplitude of the resulting alternating current if the frequency of the emf is (a) 1.0 kHz, (b) 8.0 kHz?

3. A 5.0-mH inductor is connected as in Fig. 36–4a to an ac generator with $\varepsilon_m = 30$ V. What is the amplitude of the resulting alternating current if the frequency of the emf is (a) 1.0 kHz, (b) 8.0 kHz?

4. A 50-Ω resistor is connected as in Fig. 36–2a to an ac generator with $\varepsilon_m = 30$ V. What is the amplitude of the resulting alternating current if the frequency of the emf is (a) 1.0 kHz, (b) 8.0 kHz?

5. A 45-mH inductor has a reactance of 1.3 kΩ. (a) What is the frequency? (b) What is the capacitance of a capacitor with the same reactance at that frequency? (c) If the frequency is doubled, what are the reactances of the inductor and capacitor?

Section 36–3 The *RCL* Circuit

6. (a) Recalculate all the quantities asked for in Example 3 if the capacitor is removed from the circuit, all other parameters in that example remaining unchanged. (b) Draw to scale a phasor diagram like that of Fig. 36–5b for this new situation.

7. (a) Recalculate all the quantities asked for in Example 3 if the inductor is removed from the circuit, all other parameters in that example remaining unchanged. (b) Draw to scale a phasor diagram like that of Fig. 36–5b for this new situation.

8. (a) Recalculate all the quantities asked for in Example 3 for $C = 70$ μF, the other parameters in that example remaining unchanged. (b) Draw to scale a phasor diagram like that of Fig. 36–5b for this new situation and compare the two diagrams closely.

9. Consider the resonance curves of Fig. 35–6. (a) Show that for frequencies above resonance the circuit is predominantly inductive and for frequencies below resonance it is predominantly capacitive. (b) How does the circuit behave at resonance? (c) Sketch a phasor diagram like that of Fig. 36–5b for conditions at a frequency higher than resonance, at resonance and lower than resonance.

Section 36–4 Power in Alternating Current Circuits

10. Calculate the power dissipated in the circuits of Exercises 2, 3, 4, 6, and 7.

11. Show that the average power delivered to the circuit of Fig. 36–1 can also be written (see Eqs. 36–17 and 22) as

$$P_{av} = \varepsilon_{rms}^2 R/Z^2.$$

Does this expression give reasonable results for a purely resistive circuit? for an *RCL* circuit at resonance? for a purely capacitative circuit? for a purely inductive circuit?

Section 36–5 The Transformer

12. A transformer has 500 primary turns and 10 secondary turns. (a) If V_1 for the primary is 120 V (rms), what is V_2 for the secondary, assumed on open circuit? (b) If the secondary is now connected to a resistive load of 15 Ω, what are i_1 and i_2?

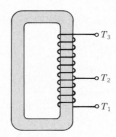

Figure 36–8 Exercise 13.

13. Figure 36–8 shows an "autotransformer." It consists of a single coil (with an iron core). Three "taps" are provided. Between taps T_1 and T_2 there are 200 turns and between taps T_2 and T_3 there are 800 turns. Any two taps can be considered the "primary terminals" and any two taps can be considered the "secondary terminals." List all the ratios by which the primary voltage may be changed to a secondary voltage.

PROBLEMS GROUPED BY SECTION

Section 36–2 *RCL* Elements, Considered Separately

1. The output of an ac generator is $\varepsilon = \varepsilon_m \sin \omega t$, with $\varepsilon_m = 25.0$ V and $\omega = 377$ rad/s. It is connected to a 12.7-H inductor. (a) What is the maximum value of the current? (b) When the current is a maximum, what is the emf of the generator? (c) When the emf of the generator is -12.5 V and increasing in magnitude, what is the current? (d) For the conditions of part (c), is the generator supplying energy to or taking energy from the rest of the circuit?

2. The ac generator of problem 1 is connected to a 4.15-μF capacitor. (a) What is the maximum value of the current? (b) When the current is a maximum, what is the emf of the generator? (c) When the emf of the generator is -12.5 V and increasing in magnitude, what is the current? (d) For the conditions of part (c), is the generator supplying energy to or taking energy from the rest of the circuit?

3. The output of an ac generator is given by $\varepsilon = \varepsilon_m \sin(\omega t - \pi/4)$, where $\varepsilon_m = 30$ V and $\omega = 350$ rad/s. The current is given by $i(t) = i_m \sin(\omega t - 3\pi/4)$, where $i_m = 620$ mA. (a) At what time, after $t = 0$, does the generator emf first reach a maximum? (b) At what time, after $t = 0$, does the current first reach a maximum? (c) The circuit contains a single element other than the generator. Is it a capacitor, and inductor, or a resistor? Justify your answer. (d) What is the value of the capacitance, inductance, or resistance, as the case may be?

4. The output of an ac generator is given by $\varepsilon = \varepsilon_m \sin(\omega t - \pi/4)$, where $\varepsilon_m = 30$ V and $\omega = 350$ rad/s. The current is given by $i(t) = i_m \sin(\omega t + \pi/4)$, where $i_m = 620$ mA. (a) At what time, after $t = 0$, does the generator emf first reach a maximum? (b) At what time, after $t = 0$, does the current first reach a maximum? (c) The circuit contains a single element other than the generator. Is it a capacitor, an inductor, or a resistor? Justify your answer.

(d) What is the value of the capacitance, inductance, or resistance, as the case may be?

Section 36–3   The RCL Circuit

5. When the generator emf in Example 3 is a maximum, what is the voltage across: (a) the generator, (b) the resistor, (c) the capacitor, and (d) the inductor? (e) By summing these with appropriate signs, verify that Kirchoff's loop rule is satisfied. (To obtain 2 significant figure agreement, carry three significant figures in the calculation. You must also recalculate the results of the example.)

6. An RCL circuit such as that of Fig. 36–1 has $R = 5.0\ \Omega$, $C = 20\ \mu F$, $L = 1.0$ H, and $\varepsilon_m = 30$ V. (a) At what angular frequency ω_0 will the current have its maximum value, as in the resonance curves of Fig. 35–6? (b) What is this maximum value? (c) At what two angular frequencies ω_1 and ω_2 will the current amplitude have one-half of this maximum value? (d) What is the fractional half-width $[=(\omega_1 - \omega_2)/\omega_0]$ of the resonance curve?

7. For a certain RCL circuit the maximum generator emf is 125 V and the maximum current is 3.20 A. If the current leads the generator emf by 0.982 rad, (a) what is the resistance and (b) what is the net reactance of the circuit? (c) Is the circuit predominantly capacitive or inductive?

*8. The ac generator in Fig. 36–9 supplies 120 V (rms) at 60 Hz. With the switch open as in the diagram, the resulting current leads the generator emf by 20°. With the switch in position 1 the current lags the generator emf by 10°. When the switch is in position 2 the rms current is 2.0 A. Find the values of R, L, and C.

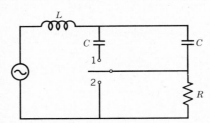

Figure 36–9   Problem 8.

Section 36–4   Power in Alternating Current Circuits

9. Show mathematically, rather than graphically as in Fig. 36–6b, that the average value of $\sin^2(\omega t + \phi)$ over an integral number of cycles is one-half.

10. In an RCL circuit such as that of Fig. 36–1, assume that $R = 5.0\ \Omega$, $L = 60$ mH, $v = 60$ Hz, and $\varepsilon_m = 30$ V. For what values of the capacitance would the average power dissipated in the resistor be (a) a maximum and (b) a minimum? (c) What are these maximum and minimum powers? (d) the corresponding phase angles? (e) the corresponding power factors?

11. In an RCL circuit, $R = 16.0\ \Omega$, $C = 31.2\ \mu F$, $L = 9.20$ mH, and $\varepsilon = \varepsilon_m \sin \omega t$, with $\varepsilon_m = 45.0$ V and $\omega = 3000$ rad/s. For

time $t = 0.442$ ms find (a) the rate at which energy is being supplied by the generator, (b) the rate at which energy is being stored in the capacitor, (c) the rate at which energy is being stored in the inductor, and (d) the rate at which energy is being dissipated in the resistor. (e) What is the meaning of a negative result for any of parts (a), (b), and (c)? (f) Show that the results of parts (b), (c), and (d) sum to the result of part (a).

12. Figure 36–10 shows an ac generator connected to a "black box" through a pair of terminals. The box could contain R's, C's and L's, possibly even in a multiloop circuit, whose elements and arrangements we do not know. Measurements outside the box reveal that

$$\varepsilon(t) = (75\ \text{V}) \sin \omega t$$

and

$$i(t) = (1.2\ \text{A}) \sin(\omega t + 42°).$$

(a) What is the power factor? (b) Does the emf lead or lag the current? (c) Is the behavior of the circuit in the box largely inductive or largely capacitive in nature? (d) Is the circuit in the box in resonance? (e) Must there be a capacitor in the box? an inductor? a resistor? (f) What power is delivered to the box by the generator? (g) Why don't you need to know the angular frequency ω to answer all these questions?

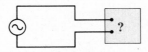

Figure 36–10   Problem 12.

Section 36–5   The Transformer

13. An ac generator delivers power to a resistive load in a remote factory over a two-cable transmission line. At the factory a step-down transformer reduces the voltage from its (rms) transmission value V_t to a much lower value, safe and convenient for use in the factory. The transmission line resistance is 0.30 Ω/cable and the power delivered to the factory is 250 kW. Calculate the voltage drop along the transmission line and the power dissipated in the line as thermal energy. Assume (a) $V_t = 80$ kV, (b) $V_t = 8.0$ kV, and (c) $V_t = 0.80$ kV and comment on the acceptibility of each choice.

14. In Fig. 36–7 (assuming switch S to be closed) show that the current in the primary circuit remains unchanged if a resistance R_1, given by

$$R_1 = R(N_1/N_2)^2,$$

is connected directly across the generator, the transformer and the secondary circuit being removed. In this sense we see that a transformer "transforms" not only voltages and currents but also resistances. (In the more general case, in which the secondary load in Fig. 36–7 contains capacitive and inductive elements also, we say that a transformer transforms impedances.)

ADDITIONAL PROBLEMS

15. A 1.5-μF capacitor has a capacitative reactance of 12 Ω. (a) What must be the frequency? (b) What will the capacitive reactance be if the frequency is doubled?

16. (a) At what frequency would a 6.0-mH inductor and a 10-μF capacitor have the same reactance? What would this reactance

be? (c) Show that this frequency would be equal to the natural frequency of free LC oscillations.

17. A typical "light dimmer" used to dim the stage lights in a theater consists of a variable inductor L connected in series with the light bulb B as shown in Fig. 36–11. The power supply is

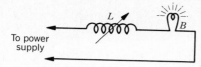

Figure 36-11 Problem 17.

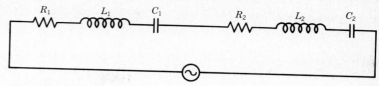

Figure 36-14 Problem 24.

120 V (rms) at 60 Hz; the light bulb is marked "120 V, 1000 W." (*a*) What maximum inductance L is required if the power in the light bulb is to be varied by a factor of five? Assume that the resistance of the light bulb is independent of its temperature. (*b*) Could one use a variable resistor instead of an inductor? If so, what maximum resistance is required? Why isn't this done?

18. In a certain RCL circuit, operating at 60 Hz, the maximum voltage across the inductor is twice the maximum voltage across the resistor, while the maximum voltage across the capacitor is the same as the maximum voltage across the resistor. (*a*) By what phase angle does the current lag the generator emf? (*b*) If the maximum generator emf is 30 V, what should the resistance of the circuit be to obtain a maximum current of 300 mA?

19. Can the amplitude of the voltage across an inductor be greater than the amplitude of the voltage across an inductor be greater than the amplitude of the generator emf in an RCL circuit? Consider an RCL circit with $\varepsilon_m = 10$ V, $R = 10\ \Omega$, $L = 1.0$ H, and $C = 1.0\ \mu$F. Find the amplitude of the voltage across the inductor at resonance.

20. Verify mathematically that the following geometrical construction correctly gives both the impedance Z and the phase constant ϕ. Referring to Fig. 36–12: (*i*) draw an arrow in the $+y$-direction of magnitude X_L, (*ii*) draw an arrow in the $-y$-direction of magnitude X_C, (*iii*) draw an arrow of magnitude R in the $+x$-direction. Then the magnitude of the "resultant" of these arrows is Z and the angle (measured below the $+x$-axis) of this resultant is ϕ.

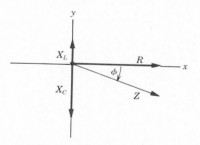

Figure 36-12 Problem 20.

21. The circuit of Example 3, to which the phasor diagram of Fig. 36–5*b* corresponds, is not in resonance. (*a*) How can you tell? (*b*) What capacitor would you combine, either in series or in parallel, with the capacitor already in the circuit to bring resonance about? (*c*) What would the current amplitude then be?

22. In Fig. 36–13, $R = 15\ \Omega$, $C = 4.7\ \mu$F, and $L = 25$ mH. The generator provides a sinusoidal voltage of amplitude $\varepsilon = 75$ V

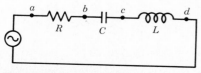

Figure 36-13 Problem 22.

(rms) and frequency $\nu = 550$ Hz. (*a*) Calculate the rms current amplitude. (*b*) Find the rms voltages V_{ab}, V_{bc}, V_{cd}, V_{bd}, V_{ad}. (*c*) What average power is dissipated by each of the three circuit elements?

23. An electric motor connected to a 120-V, 60-Hz power outlet does mechanical work at the same rate of 0.10 hp (1 hp = 746 W). If it draws an rms current of 0.65A, what is its effective resistance, in terms of power transfer? Would this be the same as the resistance of its coils, as measured with an ohm-meter with the motor disconnected from the power outlet?

24. A resistor-inductor-capacitor combination, R_1, L_1, C_1, has a resonant frequency that is just the same as that of a different combination, R_2, L_2, C_2. You now connect the two combinations in series, as in Fig. 36–14. Show that this new circuit also has the same resonant frequency as the separate individual circuits.

25. For an RCL circuit show that in one cycle with period T: (*a*) the energy stored in the capacitor does not change; (*b*) the energy stored in the inductor does not change; (*c*) the generator supplies energy $(\frac{1}{2}T)\varepsilon_m i_m \cos\phi$; and (*d*) the resistor dissipates energy $(\frac{1}{2}T)Ri_m^2$. (*e*) Show that the quantities found in (*c*) and (*d*) are equal.

26. Show that the fractional half-width of a resonance curve is given to a close approximation by

$$\frac{\Delta\omega}{\omega} = \sqrt{\frac{3C}{L}}\,R,$$

in which ω is the angular frequency at resonance and $\Delta\omega$ is the width of the resonance curve at half-amplitude. Note that $\Delta\omega/\omega$ decreases with R, as Fig. 35–5 shows. Use this formula to check the answer to part (*d*) of Problem 6.

27. In Fig. 36–15 show that the power dissipated in the resistor R is a maximum when $R = r$, in which r is the internal resistance of the ac generator. In the text we have tacitly assumed, up to this point, that $r = 0$.

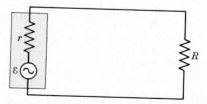

Figure 36-15 Problems 27 and 28.

28. *Impedance matching.* In Fig. 36–15 let the rectangular box on the left represent the (high-impedance) output of an audio amplifier, with $r = 1000\ \Omega$. Let $R\ (= 10\ \Omega)$ represent the (low-impedance) coil of a loudspeaker. In Problem 27 we learned that for maximum transfer of power to the load R we must have $R = r$ and that is not true in this case. However, we also learned, in Problem 27, that a transformer can be used to "transform" resistances, making them behave electrically as if they were larger or smaller than they actually are. Sketch the primary and secondary coils of a transformer to be introduced between the "amplifier" and the "speaker" in Fig. 36–15 to "match the impedances." What must be the turns ratio?

29. *Three-phase-power transmission.* A three-phase generator G produces electrical power that is transmitted by means of three wires as shown in Fig. 36–16. The potentials (relative to a common reference level) of these wires are: $V_1 = A \sin (\omega t)$, $V_2 = A \sin (\omega t - 120°)$, $V_3 = A \sin (\omega t - 240°)$. Some industrial equipment (e.g., motors) have three terminals and are designed to be connected directly to these three wires. To use a more conventional two terminal device (e.g., a light bulb), one connects it to any two

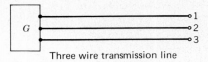

Three wire transmission line

Figure 36–16   Problem 29.

of the three wires. Show that the potential difference between *any two* of the wires (*i*) oscillates sinusoidally with angular frequency ω, and (*ii*) has amplitude $A\sqrt{3}$.

37 Maxwell's Equations

37–1 The Basic Equations of Electromagnetism

In classical mechanics and thermodynamics we sought to identify the smallest, most compact set of equations or laws that would define the subject as completely as possible. In mechanics we found this in Newton's three laws of motion and in the associated force laws, such as Newton's law of gravitation. In thermodynamics we found it in the three laws described in Sections 19–2, 20–6, and 22–4.

We have now reached the point in our studies of electromagnetism at which we can begin to assemble its basic equations. Table 37–1 shows a tentative set of them, pulled together from earlier sections of this book. After studying this table carefully we shall conclude from arguments of symmetry that there may be—and indeed is—an important missing term in one of them.

Table 37–1 The basic equations of electromagnetism—a tentative list

Symbol	Name	Equation	Text reference
I	Gauss's law for electricity	$\oint \mathbf{E} \cdot d\mathbf{S} = q/\epsilon_0$	Eq. 25–6
II	Gauss's law for magnetism	$\oint \mathbf{B} \cdot d\mathbf{S} = 0$	Eq. 34–7
III	Faraday's law of induction	$\oint \mathbf{E} \cdot d\ell = -\dfrac{d\Phi_B}{dt}$	Eq. 32–9
IV	Ampère's law	$\oint \mathbf{B} \cdot d\ell = \mu_o i$	Eq. 31–5

A review of the text will reveal no other candidates for inclusion in this list. Most of the many equations deal with special situations and thus are not "basic." Others, although basic, are equivalent to equations already listed; thus Coulomb's law (Eq. 23–3) is equivalent to I above and the law of Biot and Savart (Eq. 31–2) to IV.

The missing term to which we referred above will prove to be no trifling correction but will round out the complete description of electromagnetism and, beyond this, will establish optics as an integral part of electromagnetism. In particular it will allow us to prove that the speed of light c in free space is related to purely electric and magnetic quantities by

The speed of light in a vacuum

$$c = \frac{1}{\sqrt{\epsilon_0 \mu_0}}. \qquad (37-1)$$

It will also lead us to the concept of the electromagnetic spectrum, which lies behind the experimental discovery of radio waves.

We have seen how the principle of symmetry permeates physics and how it has often led to new insights or discoveries. For example, (*a*) if body A attracts body B with a force $\mathbf{F}$, then body B attracts body A with a force $-\mathbf{F}$ (it does), and (*b*) if there is a negative electron, there may well be a positive electron (there is), etc.

Symmetry and the electromagnetic equations

Let us examine Table 37–1 from this point of view. First we say that when we are dealing with symmetry considerations alone (that is, not making quantitative calculations) we can ignore ϵ_0 and μ_0. These constants result from our choice of unit systems and play no role in arguments of symmetry. There are in fact unit systems in which $\epsilon_0 = \mu_0 = 1$.

With this in mind we see that the left sides of the equations in Table 37–1 are completely symmetrical, in pairs. Equations I and II are surface integrals of $\mathbf{E}$ and $\mathbf{B}$, respectively, over closed surfaces. Equations III and IV are line integrals of $\mathbf{E}$ and $\mathbf{B}$, respectively, around closed loops.

On the right side of these equations, things are not symmetrical and, in fact, there are two kinds of asymmetries, which we shall discuss separately.

1. The first asymmetry, which is not really the concern of this chapter, deals with the apparent fact that although there are isolated centers of charge (electrons and protons, say) there seem not to be isolated centers of magnetism (magnetic monopoles; see Section 34–1). Thus we account for the "q" on the right of Eq. I and for the "0" on the right of Eq. II. In the same way the term $\mu_0 i (= \mu_0 dq/dt)$ appears on the right of Eq. IV but no similar term (a current of magnetic monopoles) appears on the right of Eq. III. The resolution of this asymmetry would also require the discovery of a magnetic monopole. Considerations of symmetry have motivated physicists to search for the magnetic monopole in great earnest and in many ways, yet none have been conclusively identified. It is as though nature were hinting and guiding physicists in their explorations.

2. The second asymmetry, with which this chapter deals, sticks out like a sore thumb. On the right side of Eq. III (Faraday's law of induction; see Eq. 32–9) we find the term $-d\Phi_B/dt$ and we interpret this law loosely by saying:

If you change a magnetic field ($d\Phi_B/dt$), you produce an electric field ($\oint \mathbf{E} \cdot d\boldsymbol{\ell}$).

We learned this in Section 32–1 where we showed that if you shove a bar magnet through a closed conducting loop, you do indeed induce an electric field, and thus a current, in that loop.

From the principle of symmetry we are entitled to suspect that the symmetrical relation holds, that is:

If you change an electric field ($d\Phi_E/dt$), you produce a magnetic field ($\oint \mathbf{B} \cdot d\boldsymbol{\ell}$).

This symmetry principle indeed meets the test of experiment and we discuss it fully in the next section. This supposition supplies us with the important "missing" term in Eq. IV in Table 37–1.

37–2 Induced Magnetic Fields

Here we discuss in detail the evidence for the supposition of the previous section, namely:

"A changing electric field induces a magnetic field."

Although we will be guided by considerations of symmetry alone, we will also point to direct experimental verification.

Figure 37–1a shows a uniform electric field **E** filling a cylindrical region of space. It might be produced by a circular parallel-plate capacitor, as suggested in Fig. 37–1b. We assume that E is increasing at a steady rate dE/dt, which means that charge must be supplied to the capacitor plates at a steady rate; to supply this charge requires a steady current i into the positive plate and an equal steady current i out of the negative plate.

Direct experiment shows (see Problem 11) that a magnetic field is set up by this changing electric field. Figure 37–1a shows **B** for four selected points. This figure suggests a beautiful example of the symmetry of nature. A changing magnetic field induces an electric field (Faraday's law); now we see that a changing electric field induces a magnetic field.

To describe this new effect quantitatively, we are guided by analogy with Faraday's law of induction,

$$\oint \mathbf{E} \cdot d\boldsymbol{\ell} = -\frac{d\Phi_B}{dt}, \qquad (37–2)$$

which asserts that an electric field (left term) is produced by a changing magnetic field (right term). For the symmetrical counterpart we might well venture to write

Figure 37–1 (a) Showing the induced magnetic fields **B** at four points, produced by a changing electric field **E**. The electric field is increasing in magnitude. Compare Fig. 32–9. (b) Such a changing electric field may be produced by charging a parallel-plate capacitor as shown.

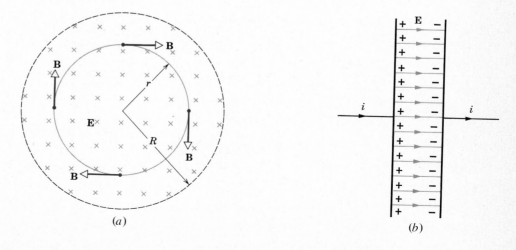

(a)

(b)

$$\oint \mathbf{B} \cdot d\ell = - \frac{d\Phi_E}{dt} \text{ (not correct).} \tag{37-3}$$

This is certainly symmetrical with Eq. 37–2, but there are two things wrong with it. The first is that experiment requires that we replace the minus sign by a plus sign. This in itself is a kind of symmetry and, in any case, nature requires it. The second difficulty with Eq. 37–3 is that it is not dimensionally correct. Our choice of a unit system requires that we insert a factor $\mu_0 \epsilon_0$ on the right-hand side. Thus the (correct) symmetrical counterpart of Eq. 37–2 is

$$\oint \mathbf{B} \cdot d\ell = + (\mu_0 \epsilon_0) \frac{d\Phi_E}{dt}. \tag{37-4}$$

To bring out the electromagnetic symmetries let us compare Fig. 37–1a carefully with Fig. 32–9. The former shows a magnetic field produced by a changing electric field; the latter shows the converse. In each figure the appropriate flux, Φ_E or Φ_B is increasing. However experiment requires that the lines of $\mathbf{E}$ in Fig. 32–9 be counterclockwise whereas those of $\mathbf{B}$ in Fig. 37–1a are clockwise. This is why Eqs. 37–2 and 37–4 differ by a minus sign.

In Section 31–1 we saw that a magnetic field can also be set up by a current in a wire. We described this quantitatively by Ampère's law:

$$\oint \mathbf{B} \cdot d\ell = \mu_0 i,$$

in which i is the conduction current passing through the loop around which the line integral is taken. Thus there are at least two ways of setting up a magnetic field: (a) by a changing electric field and (b) by a current. In general, both possibilities must be allowed for, or

Generalizing Ampère's Law

$$\oint \mathbf{B} \cdot d\ell = \mu_0 \epsilon_0 \frac{d\Phi_E}{dt} + \mu_0 i. \tag{37-5}$$

Maxwell is responsible for this important generalization of Ampère's law. It is a central and vital contribution, as we have pointed out earlier.

In Chapter 31 we assumed that no changing electric fields were present so that the term $d\Phi_E/dt$ in Eq. 37–5 was zero. In the discussion just given we assumed that there were no conduction currents in the space containing the electric field. Thus the term i in Eq. 37–5 is zero. We see now that each of these situations is a special case.

Note that we cannot claim to have derived Eq. 37–5 from deeper principles. Although our symmetry arguments should have made this equation at least reasonable it basically must stand or fall on whether or not its predictions agree with experiment. As we shall see in Chapter 38 this agreement is complete and impressive.

Example 1 A parallel-plate capacitor with circular plates is being charged as in Fig. 37–1. (a) Derive an expression for the induced magnetic field at various radii r. Consider both $r < R$ and $r > R$.

From Eq. 37–4,

$$\oint \mathbf{B} \cdot d\ell = \mu_0 \epsilon_0 \frac{d\Phi_E}{dt},$$

we can write, for $r \le R$,

$$(B)(2\pi r) = \mu_0 \epsilon_0 \frac{d}{dt}[(E)(\pi r^2)] = \mu_0 \epsilon_0 \pi r^2 \frac{dE}{dt}.$$

Solving for B yields

$$B(r) = \tfrac{1}{2} \mu_0 \epsilon_0 r \frac{dE}{dt} \qquad (r \le R).$$

For $r \ge R$, Eq. 37–4 yields

$$(B)(2\pi r) = \mu_0 \epsilon_0 \frac{d}{dt}[(E)(\pi R^2)] = \mu_0 \epsilon_0 \pi R^2 \left(\frac{dE}{dt}\right),$$

or

$$B(r) = \frac{1}{2} \frac{\mu_0 \epsilon_0 R^2}{r} \frac{dE}{dt} \qquad (r \ge R).$$

(b) Find B at $r = R$ for $dE/dt = 1.0 \times 10^{12}$ V/m·s and for $R = 50$ mm. At $r = R$ the two equations for B reduce to the same expression, or

$$B = \tfrac{1}{2}\mu_0\epsilon_0 R \frac{dE}{dt}$$

$$= (\tfrac{1}{2})(4\pi \times 10^{-7} \text{ T·m/A})(8.9 \times 10^{-12} \text{ C}^2/\text{N·m}^2)$$
$$\times (5.0 \times 10^{-2} \text{ m})(1.0 \times 10^{12} \text{ V/m·s})$$
$$= 0.28 \ \mu\text{T} = 0.0028 \text{ gauss}.$$

This shows that the induced magnetic fields in this example are so small that they can scarcely be measured with simple apparatus, in sharp contrast to induced electric fields (Faraday's law), which can be demonstrated easily. This experimental difference is in part due to the fact that induced emfs can easily be multiplied by using a coil of many turns. No technique of comparable simplicity exists for magnetic fields. In experiments involving oscillations at very high frequencies dE/dt above can be very large, resulting in significantly larger values of the induced magnetic field.

37–3 Displacement Current

Equation 37–5 shows that the term $\epsilon_0 \, d\Phi_E/dt$ has the dimensions of a current. Even though no motion of charge is involved, there are advantages in giving this term the name *displacement* current*. Thus we can say that a magnetic field can be set up either by a conduction current i or by a displacement current i_d ($= \epsilon_0 d\Phi_E/dt$), and we can rewrite Eq. 37–5 as

$$\oint \mathbf{B}\cdot d\boldsymbol{\ell} = \mu_0(i_d + i). \tag{37–6}$$

The concept of displacement current permits us to retain the notion that current is continuous, a principle established for steady conduction currents in Section 28–1. In Fig. 37–1b, for example, a current i enters the positive plate and leaves the negative plate. The conduction current is not continuous across the capacitor gap because no charge is transported across this gap. However, the displacement current i_d in the gap will prove to be exactly i, thus retaining the concept of the continuity of current.

To calculate the displacement current, recall (see Eq. 27–2) that E in the gap is given by

$$E = \frac{q}{\epsilon_0 A}.$$

Differentiating gives

$$\frac{dE}{dt} = \frac{1}{\epsilon_0 A}\frac{dq}{dt} = \frac{1}{\epsilon_0 A}\, i.$$

The displacement current i_d is by definition

Definition of Displacement Current

$$i_d = \epsilon_0 \frac{d\Phi_E}{dt} = \epsilon_0 \frac{d(EA)}{dt} = \epsilon_0 A \frac{dE}{dt}.$$

Eliminating dE/dt between these two equations leads to

$$i_d = (\epsilon_0 A)\left(\frac{1}{\epsilon_0 A}\, i\right) = i,$$

which shows that the displacement current in the gap is identical with the conduction current in the lead wires.

---

* The word "displacement" was introduced for historical reasons that need not concern us here.

Example 2   What is the displacement current for the situation of Example 1? From the definition of displacement current,

$$i_d = \epsilon_0 \frac{d\Phi_E}{dt} = \epsilon_0 \frac{d}{dt}[(E)(\pi R^2)] = \epsilon_0 \pi R^2 \frac{dE}{dt}$$

$$= (8.9 \times 10^{-12}\ \text{C}^2/\text{N·m}^2)(\pi)(5.0 \times 10^{-2}\ \text{m})^2$$
$$\times\ (1.0 \times 10^{12}\ \text{V/m·s})$$
$$= 70\ \text{mA}.$$

Even though this displacement current is reasonably large, it produces only a small magnetic field (see Example 1b) because it is spread out over a large area. In contrast, the conduction current i in the lead wires (also = 70 mA, because $i = i_d$) can produce magnetic effects close to the (thin) wires that are easily detectable by a compass needle. The discussion of Section 31–2 shows that the magnetic effect is greatest at the surface of the wire. The capacitor of Fig. 37–1 behaves like a "conductor" of radius 50 mm, carrying a (displacement) current of 70 mA. Its largest magnetic effect is at the capacitor edge, that is, at $r = 50$ mm.

37–4   Maxwell's Equations

Maxwell's equations of electromagnetism

Equation 37–5 completes our presentation of the basic equations of electromagnetism, called Maxwell's equations. They are summarized in Table 37–2, which rounds out the "tentative" Table 37–1 by supplying the "missing" term in Eq. IV of that table. All equations of physics that serve, as these do, to correlate experiments in a vast area and to predict new results have a certain beauty about them and can be appreciated, by those who understand them, on an aesthetic level. This is true for Newton's laws of motion, for the laws of thermodynamics, for the theory of relativity, and for the theories of quantum physics. As for Maxwell's equations, the physicist Ludwig Boltzmann (quoting a line from Goethe) wrote, "Was it a god who wrote these lines. . . ." In more recent times J. R. Pierce, in a book chapter entitled "Maxwell's Wonderful Equations" says, "To anyone who is motivated by anything beyond the most narrowly practical, it is worth while to understand Maxwell's equations simply for the good of his soul." The scope of these equations is remarkable, including as it does the fundamental operating principles of all large-scale electromagnetic devices such as motors, synchrotrons, television, and microwave radar.

Table 37–2   The basic equations of electromagnetism (Maxwell's equations)[a]

Number	Name	Equation	Describes	Crucial experiment	General text reference
I	Gauss's law for electricity	$\oint \mathbf{E} \cdot d\mathbf{S} = \dfrac{q}{\epsilon_0}$	Charge and the electric field	1. Like charges repel and unlike charges attract, as the inverse square of their separation. 1′. A charge on an insulated conductor moves to its outer surface.	Chapter 25
II	Gauss's law for magnetism	$\oint \mathbf{B} \cdot d\mathbf{S} = 0$	The magnetic field	2. It has thus far not been possible to verify the existence of a magnetic monopole.	Section 34–2
III	Faraday's law of induction	$\oint \mathbf{E} \cdot d\boldsymbol{\ell} = -\dfrac{d\Phi_B}{dt}$	The electrical effect of a changing magnetic field	3. A bar magnet, thrust through a closed loop of wire, will set up a current in the loop.	Chapter 32
IV	Ampère's law (as extended by Maxwell)	$\oint \mathbf{B} \cdot d\boldsymbol{\ell} = \mu_0 \epsilon_0 \dfrac{d\Phi_E}{dt} + \mu_0 i$	The magnetic effect of a changing electric field or of a current	4. The speed of light can be calculated from purely electro-magnetic measurements. 4′. A current in a wire sets up a magnetic field near the wire.	Section 38–7 Chapter 31

[a] Written on the assumption that no dielectric or magnetic material is present.

We suggested in Section 37–1 that Maxwell's equations (as they appear in Table 37–2) bear the same relationship to electromagnetism that Newton's laws of motion do to mechanics. There is, however, an important difference. Einstein presented his special theory of relativity in 1905, roughly 200 years after Newton's laws appeared and about 40 years after Maxwell's equations. As it turns out, Newton's laws had to be drastically modified in cases in which the relative speeds approached that of light. However, no changes whatever were required in Maxwell's equations.

REVIEW GUIDE AND SUMMARY

In Table 37–1 we pull together for study the basic equations of electromagnetism as they were presented in earlier chapters. In studying them for symmetry we come to see that to make Ampere's law symmetrical with Faraday's law we must write it as

Maxwell's extension of Ampere's law

$$\oint \mathbf{B} \cdot d\boldsymbol{\ell} = \mu_0 \epsilon_0 \frac{d\Phi_E}{dt} + \mu_0 i. \qquad [37\text{--}5]$$

The added first term on the right ("A changing electric field generates a magnetic field") is the symmetrical counterpart of Faraday's law ("A changing magnetic field generates an electric field"). See Example 1.

If we define displacement current as

Displacement current

$$i_d = \epsilon_0 \frac{d\Phi_E}{dt}$$

we can write Eq. 37–5 as

$$\oint \mathbf{B} \cdot d\boldsymbol{\ell} = \mu_0 (i_d + i). \qquad [37\text{--}6]$$

This concept allows us to retain the notion of continuity of current (conduction current + displacement current). Example 2 illustrates this for the charging capacitor of Fig. 37–1 and Example 1. Note that displacement current involves a changing electric field; it does *not* involve a transfer of charge.

Maxwell's equations

Maxwell's equations, displayed in Table 37–2, summarize all of electromagnetism and form its foundation. They will be used later and warrant careful study.

QUESTIONS

1. In your own words explain why Faraday's law of induction (see Table 37–2) can be interpreted by saying, "A changing magnetic field generates an electric field."

2. If a uniform flux Φ_E through a plane circular ring decreases with time, is the induced magnetic field (as viewed along the direction of **E**) clockwise or counterclockwise?

3. Why is it so easy to show that "a changing magnetic field produces an electric field" but so hard to show in a simple way that "a changing electric field produces a magnetic field"?

4. In Fig. 37–1a consider a circle with $r > R$. How can a magnetic field be induced around this circle, as Example 1 shows? After all, there is no electric field at the location of this circle and $dE/dt = 0$ here.

5. In Fig. 37–1a, **E** is into the figure and is increasing in magnitude. Find the direction of **B** if (a) **E** is into the figure and decreasing, (b) **E** is out of the figure and increasing, (c) **E** is out of the figure and decreasing, and (d) **E** remains constant.

6. In Fig. 35–1c a displacement current is needed to maintain continuity of current in the capacitor. How can one exist, considering that there is no charge on the capacitor?

7. In Figs. 37–1a,b what is the direction of the displacement current i_d? In this same figure, can you find a rule relating the directions of **B** and **E**? of **B** and dE/dt?

8. What advantages are there in calling the term $\epsilon_0 \, d\Phi_E/dt$ in Eq. IV, Table 37–2 a displacement current?

9. Why are the magnetic effects of conduction currents in wires so easy to detect but the magnetic effects of displacement currents in capacitors so hard to detect?

10. In Table 37–2 there are three kinds of apparent lack of symmetry in Maxwell's equations. (a) The quantities ϵ_0 and/or μ_0 appear in I and IV but not in II and III. (b) There is a minus sign in III but no minus sign in IV. (c) There are missing "magnetic pole terms" in II and III. Which of these represent genuine lack of symmetry? If magnetic poles were discovered, how would you rewrite these equations to include them? (Hint: Let p be the magnetic pole strength.)

EXERCISES

Section 37-1 The Basic Equations of Electromagnetism

1. Calculate the value of the speed of light from Eq. 37-1 and show that the equation is dimensionally correct. (See Appendix B.)

2. We know that $c = 1/\sqrt{\epsilon_0\mu_0}$. (a) Show that $\sqrt{\mu_0/\epsilon_0} = 377\ \Omega$ (called the "impedance of free space"). (b) Show that the angular frequency of ordinary 60 Hz ac is 377 s^{-1}. (c) Compare (a) with (b). Is this coincidence the reason that 60 Hz was originally chosen as the frequency for ac generators?

Section 37-3 Displacement Current

3. You are given a 1.0-μF parallel-plate capacitor. How would you establish an (instantaneous) displacement current of 1.0 A in the space between its plates?

4. Prove that the displacement current in a parallel-plate capacitor can be written as

$$i_d = C\,\frac{dV}{dt}.$$

Section 37-4 Maxwell's Equations

5. Indicate by an appropriate check which of Maxwell's equations is equivalent to or includes each of the phenomena listed in the table below. See Table 37-2.

6. Suppose that magnetic monopoles were shown to exist. (a) Which of the four Maxwell equations would have to be modified? (b) What is the general form of the above modification(s)? (Ignore constants of proportionality and signs.)

Phenomenon	Maxwell equation number			
	I	**II**	**III**	**IV**
1. Lines of **E** end only on electric charges.				
2. The displacement current.				
3. Under static conditions there cannot be any charge inside a conductor.				
4. A changing electric field generates a magnetic field.				
5. The net magnetic flux through a closed surface is zero.				
6. A changing magnetic field generates an electric field.				
7. Lines of **B** have no ends.				
8. The net electric flux through a closed surface may or may not be zero.				
9. An electric charge is always accompanied by an electric field.				
10. There are no free magnetic poles.				
11. An electric current is always accompanied by a magnetic field.				
12. Coulomb's law.				
13. The electrostatic field is conservative.				

Table for Exercise 5.

PROBLEMS GROUPED BY SECTION

Section 37–2   Induced Magnetic Fields

1. The capacitor of Example 1 has a radius R of 30 mm and a plate separation of 5.0 mm. A sinusoidal potential difference with a maximum value of 150 V and a frequency of 60 Hz is applied between the plates. Find $B_m(R)$, the maximum value of the induced magnetic field at $r = R$.

2. For the conditions of Problem 1, plot $B_m(r)$ for the range $0 < r < 10$ cm.

Section 37–3   Displacement Current

3. In Example 1 show that the *displacement current density* j_d is given, for $r < R$, by

$$j_d = \epsilon_0 \frac{dE}{dt}.$$

4. A parallel-plate capacitor has square plates 1.0 m on a side as in Fig. 37–2. There is a charging current of 2.0 A into (and out of) the capacitor. (*a*) What is the displacement current through the region between the plates? (*b*) What is dE/dt in this region? (*c*) What is the displacement current through the square dashed path between the plates? (*d*) What is $\int \mathbf{B} \cdot d\boldsymbol{\ell}$ around this square dashed path?

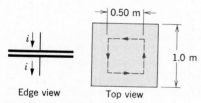

Figure 37–2   Problem 4.

5. The capacitor in Fig. 37–3 consisting of two circular plates with area $A = 0.10$ m² is connected to a source of emf/$\varepsilon = \varepsilon_m \sin \omega t$. where $\varepsilon_m = 200$ V and $\omega = 100$ rad/s. The maximum value of the displacement current is $i_d = 8.9$ μA. Neglect fringing of the electric field at the edges of the plates. (*a*) What is the maximum value of the current i? (*b*) What is the maximum value of $d\Phi_E/dt$, where Φ_E is the electric flux through the region between the plates? (*c*) What is the separation d between the plates? (*d*) Find the maximum value of the magnitude of $\mathbf{B}$ between the plates at a distance $R = 10$ cm from the center.

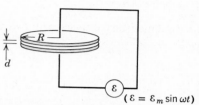

Figure 37–3   Problem 5.

*6. A long cylindrical conducting rod with radius R is centered on the x-axis as shown in Fig. 37–4. A narrow saw cut is made in the rod at $x = b$. A conduction current i, increasing with time and given by $i = \alpha t$, flows toward the right in the rod; α is a (positive)

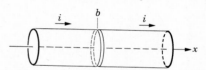

Figure 37–4   Problem 6.

Experimental situation	Field variation	
	$r \geq R$	$r \leq R$
1. A long wire of radius R carrying a current i. See Fig. 31–10. Find $B(r)$.	$\left(\dfrac{\mu_0 i}{2\pi}\right)\dfrac{1}{r}$ Section 31–5	$\left(\dfrac{\mu_0 i}{2\pi R^2}\right) r$ Example 3, Chapter 31
2. A long cylindrical region of radius R in which $\mathbf{E}$ is parallel to the axis and increasing at a rate dE/dt. See Fig. 37–1a. Find $B(r)$.	$\left(\dfrac{1}{2}\mu_0\epsilon_0 R^2 \dfrac{dE}{dt}\right)\dfrac{1}{r}$ or $\left(\dfrac{\mu_0 i_d}{2\pi}\right)\dfrac{1}{r}$	$\left(\dfrac{1}{2}\mu_0\epsilon_0 \dfrac{dE}{dt}\right) r$ or $\left(\dfrac{\mu_0 i_d}{2\pi R^2}\right) r.$
	Example 1, Chapter 37; Problem 12	
3. A long cylindrical region of radius R in which $\mathbf{B}$ is parallel to the axis and increasing at a rate dB/dt. See Fig. 32–9. Find $E(r)$.	$-\left(\dfrac{1}{2}R^2 \dfrac{dB}{dt}\right)\dfrac{1}{r}$	$-\left(\dfrac{1}{2}\dfrac{dB}{dt}\right) r$
	Example 3, Chapter 32	

Table for Problem 8.

proportionality constant. At $t = 0$ there is no charge on the cut faces near $x = b$. (a) Find the magnitude of the charge on these faces, as a function of time. (b) Use Eq. I in Table 37–2 to find E in the gap as a function of time. (c) Sketch the lines of **B** for $r < R$, where r is the distance from the x-axis. (d) Use Eq. IV in Table 37–2 to find $B(r)$ in the gap for $r < R$. (e) Compare the above answer with $B(r)$ in the rod for $r < R$.

Section 37–4   Maxwell's Equations

7. Suppose we agreed to reverse the arrowheads on all **B** field lines so that lines of **B** went from S-pole to N-pole outside a bar magnet. (a) How would the Lorentz force law, $\mathbf{F} = q\mathbf{E} + q\mathbf{v} \times \mathbf{B}$,

have to be modified? (b) How would the Maxwell equations have to be modified? (c) Is there any way that one could remove the asymmetry in signs in the Maxwell equations by changing the arrowhead directions for **E** or **B** or both?

8. In the table at the bottom of page 715 are listed three experimental situations of cylindrical symmetry, all of which have been previously discussed at the indicated text reference. (a) As a review and as an exercise in the use of Maxwell's equations, rederive each of the field variations shown from the appropriate Maxwell equation, chosen from Table 37–2. (b) Sketch the field lines, the conduction currents, and the displacement currents in each case, paying careful attention to their directions.

ADDITIONAL PROBLEMS

9. Figure 37–5 shows the plates P_1 and P_2 of a circular parallel-plate capacitor of radius R. They are connected as shown to long straight wires in which a constant conduction current i exists. A_1, A_2, and A_3 are hypothetical circles of radius r, two of them outside the capacitor and one between the plates. Show that the magnetic field at the circumference of each of these circles is given by

$$B(r) = \frac{\mu_0 i}{2\pi r}.$$

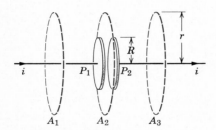

Figure 37–5   Problem 9.

10. A parallel-plate capacitor with circular plates 20 cm in diameter is being charged as in Fig. 37–1a. The displacement current density throughout the region is uniform, into the paper in the diagram, and has a value of 20 A/m². (a) Calculate the magnetic field B at a distance $r = 50$ mm from the axis of symmetry of the region. (b) Calculate dE/dt in this region.

11. In 1929 M. R. Van Cauwenberghe succeeded in measuring directly, for the first time, the displacement current i_d between the plates of a parallel-plate capacitor to which an alternating potential difference was applied, as suggested by Fig. 37–1. He used circular plates whose effective radius was 40 cm and whose capacitance was 1.0×10^{-10} F. The applied potential difference had a maximum value V_m of 174 kV at a frequency of 50 Hz. (a) What maximum displacement current was present between the plates? (b) Why was the applied potential difference chosen to be as high as it is? (The delicacy of these measurements is such that they were only performed in a direct manner more than 60 years after Maxwell enunciated the concept of displacement current! The experiment is described in *Journal de Physique* No. 8, 1929.)

12. In Example 1a show that the expressions derived for $B(r)$ can be written as

$$B(r) = \frac{\mu_0 i_d}{2\pi r} \qquad (r \geq R)$$

and

$$B(r) = \frac{\mu_0 i_d r}{2\pi R^2}. \qquad (r \leq R)$$

Note that these expressions are of just the same form as those derived in Section 31–5 except that the conduction current i has been replaced by the displacement current i_d.

13. *A self-consistency property of two of the Maxwell equations* (numbers III and IV in Table 37–2). Two adjacent closed paths *abefa* and *bcdeb* share the common edge *bc* as shown in Fig. 37–6. (a) We may apply $\oint \mathbf{E} \cdot d\boldsymbol{\ell} = -d\Phi_B/dt$ (III) to each of these two closed paths separately. Show that, from this alone, Eq. III is *automatically* satisfied for the composite closed path *abcdefa*. (b) Repeat using Eq. IV.

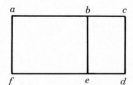

Figure 37–6   Problem 13.

14. *A self-consistency property of two of the Maxwell equations* (numbers I and II in Table 37–2). Two adjacent closed parallel-epipeds share a common face as shown in Fig. 37–7. (a) We may apply $\oint \mathbf{E} \cdot d\mathbf{S} = q/\epsilon_0$ (I) to each of these two closed surfaces separately. Show that, from this alone, Eq. I is *automatically* satisfied for the composite closed surface. (b) Repeat using Eq. II.

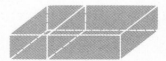

Figure 37–7   Problem 14.

38 Electromagnetic Waves

38–1 Maxwell and the Electromagnetic Spectrum

The crowning achievement of Maxwell's equations (see Table 37–2) is that we can use them, as Maxwell did, to predict that energy can be transported through free space by traveling configurations of electric and magnetic waves, that is, by traveling electromagnetic waves. We will show that the speed of these waves is predicted to be

$$c = \frac{1}{\sqrt{\epsilon_0 \, \mu_0}}. \tag{38–1}$$

The permeability constant μ_0 in this equation has a defined value of $4\pi \times 10^{-7}$ H/m; the permittivity constant ϵ_0 can be found experimentally by measuring, say, the capacitance of a parallel plate capacitor of known dimensions (see Eq., 27–4). Substituting these two values into Eq. 38–1 above yields 3.00×10^8 m/s, the speed of light in free space!* By such predictions Maxwell showed that optics, the study of visible light, is a branch of electromagnetism and falls within the scope of his equations.

Maxwell's equations put no limitation on the wavelengths of the electromagnetic waves whose existence they predict. As Fig. 38–1 shows there is an entire spectrum of them, refered to by one imaginative writer as "Maxwell's rainbow." The wavelength scale in Fig. 38–1 (and similarly the corresponding frequency scale) is drawn so that each scale marker represents a change in wavelength (and correspondingly in frequency) by a factor of ten. The scale is open-ended, the wavelengths of electromagnetic waves having no inherent upper or lower bounds.

Certain regions of the electromagnetic spectrum in Fig. 38–1 are identified by familiar labels, "x rays" and "microwaves" being examples. These labels denote roughly defined wavelength ranges within which certain kinds of sources and detectors of the radiations in question are in common use. Other regions of Fig.

---

* Since 1983 the speed of light has been assigned a precise numerical value (see Section 1–3). Equation 38–1, accepted as valid, is now used to assign a precise numerical value to the permittivity constant ϵ_0.

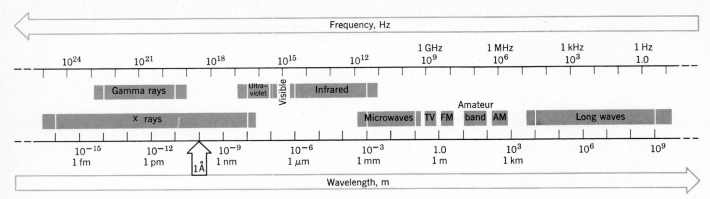

Figure 38–1 The electromagnetic spectrum. At the time that Maxwell advanced his theory of electromagnetism only the visible, the near-infrared, and the near ultraviolet portions of this spectrum were known. See ''The Allocation of the Radio Spectrum'' by Charles Lee Jackson, *Scientific American*, February 1980 for a description of the complex process of allocating frequencies in the range from ~10 kHz to ~300 GHz.

38–1, of which those labeled ''TV'' and ''AM'' are examples, represent specific wavelength bands assigned by law for certain commercial or other purposes. Only the most familiar of such allocated bands are shown.

There are no gaps in the electromagnetic spectrum. For example, we can produce electromagnetic radiation with wavelengths in the millimeter range either by microwave techniques (microwave oscillators) or by infrared techniques (heated sources). We stress that *all* electromagnetic waves, no matter where they lie in the spectrum, travel through free space with the same speed c.

It is interesting to ponder the extent to which we are bathed in electromagnetic radiation from various regions of the spectrum. The sun, whose radiations define the environment to which we as a species have adapted, is the dominant source. We are also criss-crossed by radio and TV signals. Microwaves from radar systems and from telephone relay systems may reach us. There are electromagnetic waves from light bulbs, from the heated engine blocks of automobiles, from x-ray machines, from lightning flashes, and from radioactive materials buried in the earth. Beyond this, radiation reaches us from stars and other objects in our galaxy and from other galaxies. We are even exposed, however weakly, to radiation ($\lambda \cong 2$ mm) from the primeval fireball, thought by many to be associated with the creation of our universe.

The visible region of the spectrum is of course of particular interest to us. Figure 38–2 shows the relative sensitivity of the eye of an assumed standard observer to radiations of various wavelengths. The center of the visible region is about 5.55×10^{-7} m or 555 nm; light of this wavelength produces the sensation of yellow-green.

The limits of the visible spectrum are not well defined because the eye sensitivity curve approaches the axis asymptotically at both long and short wavelengths. If we take the limits, arbitrarily, as the wavelengths at which the eye sensitivity has dropped to 1% of its maximum value, these limits are about 430 and 690 nm, less than a factor of two in wavelength. The eye can detect radiation beyond these limits if it is intense enough. In many experiments in physics one can

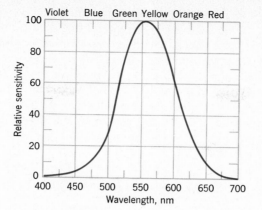

Figure 38-2 The relative eye sensitivity of an assumed standard observer at different wavelengths for a specified level of illumination.

use photographic plates or light-sensitive electronic detectors in place of the human eye.

38-2 Generating an Electromagnetic Wave

In this section we look more closely into the relationship between a traveling electromagnetic wave and its source. Some radiations such as x rays, gamma rays, and visible light come from sources that are of atomic or nuclear size. To simplify our treatment, we restrict ourselves here to that region of the spectrum ($\lambda \cong 1$ m, say) in which the source of the radiation (a shortwave radio antenna, say) is both macroscopic and of manageable dimensions.

Figure 38-3 shows, in broad outline, a generator of such waves. At its heart is an *LC* oscillator, which establishes an angular frequency $\omega(=1/\sqrt{LC})$. Charges and currents in this circuit vary sinusoidally at this frequency, as depicted in Fig. 35-1. An external source—possibly a battery—supplies the energy needed to compensate both for thermal losses in the circuit and for energy carried away by the radiated electromagnetic wave.

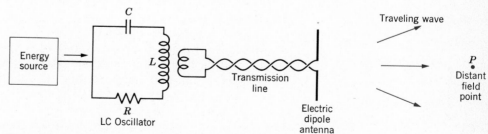

Figure 38-3 An arrangement for generating a traveling electromagnetic wave whose wavelength is in the range of short radio waves.

The *LC* oscillator of Fig. 38-3 is transformer-coupled to an antenna by a transmission line, which might be a coaxial cable among other possibilities. The two branches of the antenna alternate sinusoidally in potential at the angular frequency ω set by the oscillator, causing charges to accelerate back and forth along the antenna axis. The effect is that of an electric dipole whose electric dipole moment **p** is not constant but varies sinusoidally with time.

Figure 38–4 shows the traveling electromagnetic wave generated by the accelerating charges in the oscillating electric dipole antenna. The electric and magnetic field lines, which form a figure of revolution about the dipole axis, travel away from the antenna with speed c. The intensity of the traveling wave in any direction is proportional to $\sin^2 \theta$, being zero in the direction of the axis of the oscillating dipole ($\theta = 0, 180°$) and a maximum in the equatorial plane of the antenna ($\theta = 90°$). The field patterns close to the antenna (the so-called *near field*) are complicated and are not shown in Fig. 38–4. The field configurations that are shown in Fig. 38–4 (the so-called *radiation field*) hold for all radial distances from the antenna such that $r \gg \lambda$.

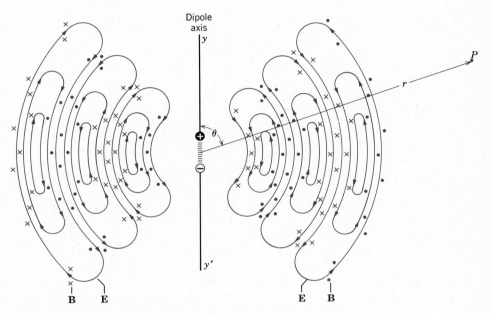

Figure 38–4   The electric and magnetic field lines associated with the traveling electromagnetic wave generated by an oscillating electric dipole. P is a distant field point.

Figure 38–5 shows "patches" of the wave fronts of the traveling wave as they would appear to an observer located at a distant point such as P. We assume that P is so far from the antenna that the observer can treat these local "patches," which sweep past him with a speed c, as defining a traveling plane wave. Figure 38–5 covers one complete cycle of oscillation and shows how the electric field **E** and the magnetic field **B** change with time as the wave goes by. Note that **E** and **B** are perpendicular to each other and to the direction of propagation of the wave.

38–3   Traveling Waves and Maxwell's Equations

Next we will analyze the sinusoidal wave that passes an observer at point P in Fig. 38–4. We will show that this wave is consistent with Maxwell's equations and prove that these equations predict that such an electromagnetic wave travels with a speed c, which turns out to be the speed of light. To simplify our analysis we assume that P is very far from the source so that the wavefronts near that point are

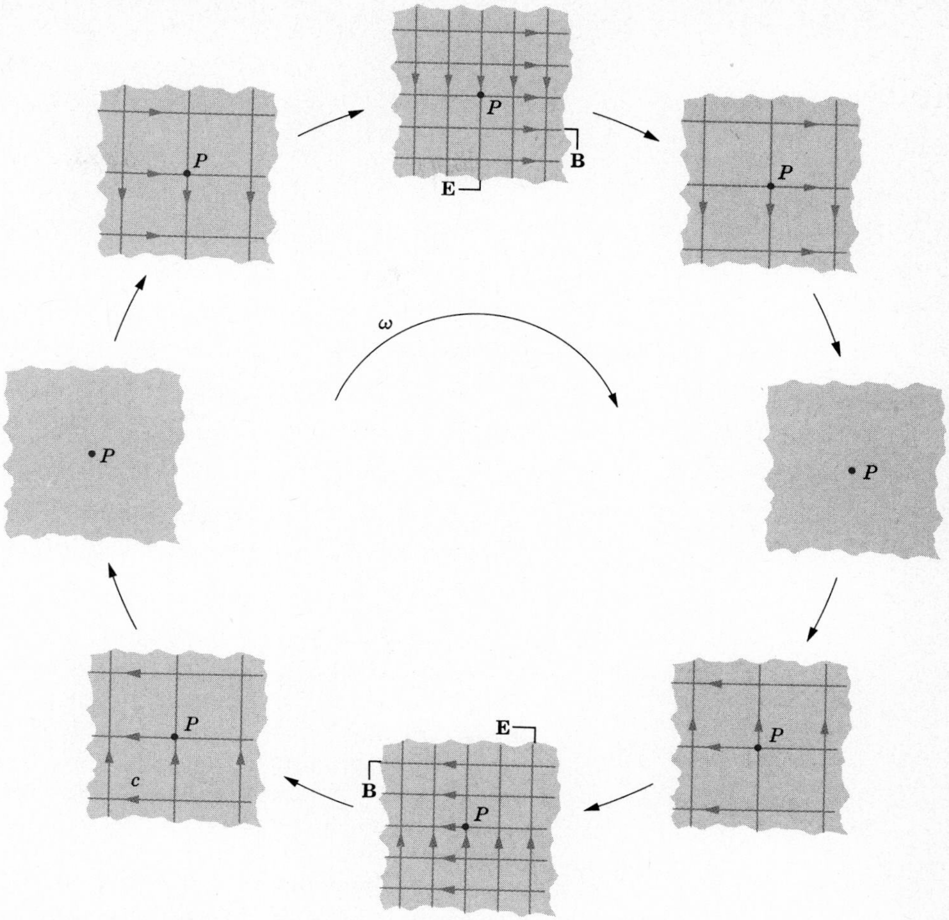

Figure 38-5 "Patches" of the wavefronts of the electromagnetic wave of Fig. 38-3 as it sweeps past an observer at point P in that figure.

planar; the observer at P will say that a plane wave (see Section 17-2) is rushing past him, as in Fig. 38-5.

Figure 38-6 shows a "snapshot" of such a plane wave traveling in the x-direction. The lines of **E** are chosen to be parallel to the y-axis and those of **B** to the z-axis. The values of **B** and **E** for this wave depend only on x and t (not on y or z). We postulate that they are given in magnitude by

A traveling electromagnetic wave

$$E = E_m \sin (kx - \omega t) \tag{38-2}$$

and

$$B = B_m \sin (kx - \omega t). \tag{38-3}$$

Figure 38-7 shows two sections through the three-dimensional diagram of Fig. 38-6. In Fig. 38-7a the plane of the page is the xy-plane and in Fig. 38-7b it is the xz-plane. Note that **E** and **B** are in phase, that is, at any point through which the wave is moving they reach their maximum values at the same time.

The shaded rectangle of dimensions dx and h in Fig. 38-7a is fixed at a particular point P on the x-axis. As the wave passes over it, the magnetic flux Φ_B through the rectangle will change, which will give rise to induced electric fields

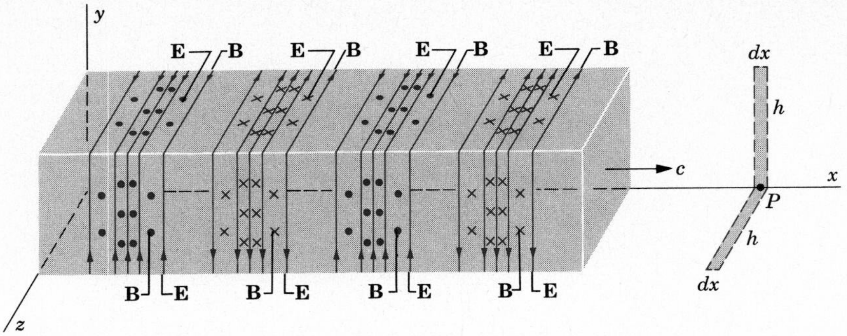

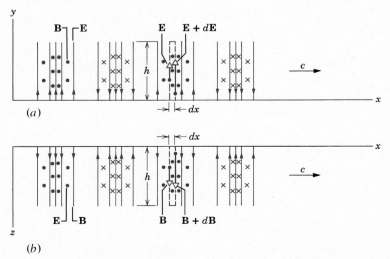

Figure 38–6 A "snapshot" of a plane electromagnetic wave traveling in the x-direction at speed c, sweeping past point P. Figure 38–5 shows another aspect. The shaded rectangles near P refer to Fig. 38–7.

around the rectangle, according to Faraday's law of induction. These induced electric fields are, in fact, simply the electric component of the traveling electromagnetic wave.

Let us apply Lenz's law. The flux Φ_B for the shaded rectangle of Fig. 38–7a is decreasing with time because the wave is moving through the rectangle to the right and a region of weaker magnetic fields is moving into the rectangle. The induced electric field will act to oppose this change, which means that if we imagine the boundary of the rectangle to be a conducting loop a counterclockwise induced current would appear in it. This current would produce a field of **B** that, within the rectangle, would point out of the page, thus opposing the decrease in Φ_B. There is, of course, no conducting loop, but the net induced electric field **E** does indeed act counterclockwise around the rectangle because $E + dE$, the field magnitude at the right edge of the rectangle is greater than E, the field magnitude at the left edge. Thus the electric field configuration is entirely consistent with the concept that it is induced by the changing magnetic field. The situation is similar to that of Fig. 32–9, which shows a counterclockwise configuration of **E** opposing an increase of Φ_B.

Figure 38–7 The wave of Fig. 38–6 viewed (a) in the xy-plane and (b) in the xz-plane.

For a more detailed analysis let us apply Faraday's law of induction, or

$$\oint \mathbf{E} \cdot d\boldsymbol{\ell} = -\frac{d\Phi_B}{dt}, \tag{38-4}$$

going counterclockwise around the shaded rectangle of Fig. 38-7a. There is no contribution to the integral from the top or bottom of the rectangle because $\mathbf{E}$ and $d\boldsymbol{\ell}$ are at right angles here. The integral then becomes

$$\oint \mathbf{E} \cdot d\boldsymbol{\ell} = [(E + dE)(h)] - [(E)(h)] = dE\, h.$$

The flux Φ_B for the rectangle is

$$\Phi_B = (B)(dx\, h),$$

where B is the magnitude of $\mathbf{B}$ at the rectangular strip and $dx\, h$ is the area of the strip. Differentiating gives

$$\frac{d\Phi_B}{dt} = h\, dx\, \frac{dB}{dt}.$$

From Eq. 38-4 we have then

$$dE\, h = -h\, dx\, \frac{dB}{dt},$$

or

$$\frac{dE}{dx} = -\frac{dB}{dt}. \tag{38-5}$$

Actually, both B and E are functions of x and t. In evaluating dE/dx, we assume that t is constant because Fig. 38-7a is an "instantaneous snapshot." Also, in evaluating dB/dt we assume that x is constant since what we want is the time rate of change of B at a particular place, the point P in Fig. 38-7a. The derivatives under these circumstances are called *partial derivatives,* and a special notation is used for them. In this notation Eq. 38-5 becomes

$$\frac{\partial E}{\partial x} = -\frac{\partial B}{\partial t}. \tag{38-6}$$

The minus sign in this equation is appropriate and necessary, for, although E is increasing with x at the site of the shaded rectangle in Fig. 38-7a, B is decreasing with t. Since $E(x, t)$ and $B(x, t)$ are known (see Eqs. 38-2 and 38-3), Eq. 38-6 reduces to

$$kE_m \cos (kx - \omega t) = \omega B_m \cos (kx - \omega t).$$

In Section 17-3 we saw that the ratio ω/k for a traveling wave is just its speed, which we choose here to call c. The above equation then becomes

$$E_m = cB_m. \tag{38-7}$$

Thus the speed of an electromagnetic wave c is the ratio of the amplitudes of the electric and the magnetic components of the wave.

From Eqs. 38-2 and 38-3 we see that the ratio of the amplitudes of $\mathbf{E}$ and $\mathbf{B}$ in an electromagnetic wave is the same as the ratio of instantaneous values, or

Relation between E and B in an electromagnetic wave

$$E = cB. \tag{38-8}$$

This result will be useful in Section 38-4.

We now turn to Fig. 38–7b, in which the flux Φ_E for the shaded rectangle is decreasing with time as the wave moves through it. According to Maxwell's fourth equation

$$\oint \mathbf{B}\cdot d\boldsymbol{\ell} = \mu_0\epsilon_0 \frac{d\Phi_E}{dt}, \tag{38–9}$$

where the conduction current term is zero because the wave is moving in a vacuum. The changing flux Φ_E will induce a magnetic field at points around the periphery of the rectangle. This induced magnetic field is simply the magnetic component of the electromagnetic wave. Thus the electric and the magnetic components of the wave are intimately connected, each depending on the time rate of change of the other. Although there are no conduction currents in the traveling wave we can associate $\oint \mathbf{B}\cdot d\boldsymbol{\ell}$ for the shaded rectangle of Fig. 38–7b with a displacement current. It points into the plane of the figure within the rectangle.

The integral in Eq. 38–9, evaluated by proceeding counterclockwise around the shaded rectangle of Fig. 38–7b, is

$$\oint \mathbf{B}\cdot d\boldsymbol{\ell} = [-(B + dB)(h)] + [(B)(h)] = -h\, dB,$$

where B is the magnitude of $\mathbf{B}$ at the left edge of the strip and $B + dB$ is its magnitude at the right edge.

The flux Φ_E through the rectangle of Fig. 38–7b is

$$\Phi_E = (E)(h\, dx).$$

Differentiating gives

$$\frac{d\Phi_E}{dt} = h\, dx\, \frac{dE}{dt}.$$

Equation 38–9 can thus be written

$$-h\, dB = \mu_0\epsilon_0 \left(h\, dx\, \frac{dE}{dt} \right)$$

or, substituting partial derivatives,

$$-\frac{\partial B}{\partial x} = \mu_0\epsilon_0 \frac{\partial E}{\partial t}. \tag{38–10}$$

Again, the minus sign in this equation is necessary, for, although B is increasing with x at point P in the shaded rectangle in Fig. 38–7b, E is decreasing with t.

Combining this equation with Eqs. 38–2 and 38–3 yields

$$-kB_m \cos(kx - \omega t) = -\mu_0\epsilon_0\omega E_m \cos(kx - \omega t),$$

which we can write as

$$E_m = \frac{1}{\epsilon_0\mu_0(\omega/k)} B_m = \frac{1}{\epsilon_0\mu_0 c} B_m. \tag{38–11}$$

Combining Eqs. 38–7 and 38–11 leads to

The speed of light

$$c = \frac{1}{\sqrt{\epsilon_0\mu_0}} \tag{38–1}$$

which, as we have seen, is equal to the measured speed of electromagnetic radiation in free space. Maxwell's theory of electromagnetism not only united optics with electricity and with magnetism but led directly to the discovery of radio waves. This was accomplished by Heinrich Hertz in a brilliant series of experiments he conducted during the period 1887–1890.

38-4 Energy Transport and the Poynting Vector

The Poynting vector

An electromagnetic wave can transport energy and—as sunbathers know—can deliver it to a body on which it falls. The rate of energy transport per unit area in such a wave is described by a vector **S**, called the Poynting vector after John Henry Poynting (1852–1914) who first pointed out its properties. **S** is defined from

$$\mathbf{S} = \frac{1}{\mu_0}\,\mathbf{E}\times\mathbf{B}, \tag{38-12}$$

its SI units being watts/meter². The direction of **S** at any point gives the direction of energy transport at that point.

For the traveling plane electromagnetic wave of Fig. 38–6 it is clear that **E** × **B**, and hence **S**, point in the direction of propagation, just as we expect. The magnitude of **S** for such a wave is

$$S = \frac{1}{\mu_0}\,EB. \tag{38-13}$$

We now show that Eq. 38–13 is indeed a correct result by deriving it from relations that we have dealt with earlier, without reference to Eq. 38–12. Figure 38–8 shows a traveling plane wave, along with a thin "box" of thickness dx and area A. The box, a mathematical construction, is fixed with respect to the axes while the wave moves through it. At any instant the energy stored in the box, from Eqs. 27–9 and 33–19, is

$$dU = dU_E + dU_B = (u_E + u_B)(A\,dx)$$
$$= \left(\tfrac{1}{2}\epsilon_0 E^2 + \tfrac{1}{2}\frac{1}{\mu_0}B^2\right) A\,dx, \tag{38-14}$$

where $A\,dx$ is the volume of the box and E and B are the instantaneous values of the field vectors in the box.

Using Eq. 38–8 ($E = cB$) to eliminate one of the E's in the first term in Eq. 38–14 and one of the B's in the second term leads to

Figure 38-8 A plane electromagnetic wave travels along the x-axis, transporting energy through an imaginary rectangular box.

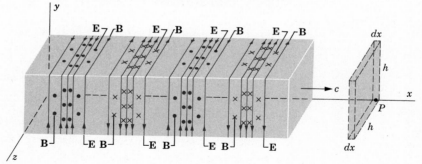

$$dU = \left[\tfrac{1}{2}\epsilon_0 E(cB) + \tfrac{1}{2}\frac{1}{\mu_0} B\left(\frac{E}{c}\right)\right] A \ dx$$

$$= \frac{(\mu_0\epsilon_0 c^2 + 1)(EBA \ dx)}{2\mu_0 c}.$$

From Eq. 38–1, however, $\mu_0\epsilon_0 c^2 = 1$, so that

$$dU = \frac{EBA \ dx}{\mu_0 c}.$$

This energy dU will pass through the right face of the box in a time dt equal to dx/c. Thus the energy per unit area per unit time, which is S, is given by

$$S = \frac{dU}{dt \ A} = \frac{EBA \ dx}{\mu_0 c (dx/c) A} = \frac{1}{\mu_0} EB, \qquad [38–13]$$

which (see Eq. 38–13) is what we set out to prove.

The quantity S as we have dealt with it so far is an instantaneous value, characteristic of a specified point in space at a particular instant of time. We are more often interested in $\overline{S}$, the time-averaged value of S at the point in question. Example 1 illustrates this point.

Example 1   An observer is 1.0 m from a point light source whose power output P_0 is 1.0 kW. Calculate the amplitudes E_m and B_m of the electric and the magnetic fields at that point. Assume that the source is monochromatic, that it radiates uniformly in all directions, and that at distant points it behaves like the traveling plane wave of Fig. 38–8.

The power that passes through a sphere of radius r is $(\overline{S})(4\pi r^2)$, where $\overline{S}$ is the time-averaged value of the Poynting vector at the surface of the sphere. This power must equal P_0, is

$$P_0 = \overline{S} 4\pi r^2.$$

From the definition of S (Eq. 38–13), we have

$$\overline{S} = \left(\frac{1}{\mu_0} \overline{EB}\right).$$

Using the relation $E = cB$ (Eq. 38–8) to eliminate B leads to

$$\overline{S} = \frac{1}{\mu_0 c} \overline{E^2}.$$

The average value of E^2 over one cycle is $\tfrac{1}{2} E_m^2$, since E varies sinusoidally (see Eq. 38–2). This leads to

$$P_0 = \left(\frac{E_m^2}{2\mu_0 c}\right)(4\pi r^2),$$

or

$$E_m = \frac{1}{r}\sqrt{\frac{P_0\mu_0 c}{2\pi}}.$$

For $P_0 = 1.0$ kW and $r = 1.0$ m this yields

$$E_m = \frac{1}{(1.0 \ \text{m})}$$

$$\times \sqrt{\frac{(1.0 \times 10^3 \ \text{W})(4\pi \times 10^{-7} \ \text{Wb/A·m})(3.0 \times 10^8 \ \text{m/s})}{2\pi}}$$

$$= 240 \ \text{V/m}.$$

The relationship $E_m = cB_m$ (Eq. 38–7) leads to

$$B_m = \frac{E_m}{c} = \frac{240 \ \text{V/m}}{3.0 \times 10^8 \ \text{m/s}} = 8.0 \times 10^{-7} \ \text{T}.$$

Note that E_m is appreciable as judged by ordinary laboratory standards but that B_m ($= 0.0080$ gauss) is quite small.

38–5   Radiation Pressure

The fact that energy is carried by electromagnetic waves is confirmed by everyday experience. Less familiar is the fact that electromagnetic waves may also transport linear momentum. It is possible to exert a pressure (a radiation pressure) on an object by shining a light on it. Such forces must be small in relation to forces of our daily experience because we do not ordinarily notice them.

Let a parallel beam of radiation, light for example, fall on an object for a time t and be entirely absorbed by the object. If energy U is absorbed during this time, the magnitude of the momentum p delivered to the object is given, according to Maxwell's prediction, by

$$p = \frac{U}{c} \qquad \text{(total absorption),} \qquad (38\text{–}15)$$

Momentum delivered
by radiation

where c is the speed of light. The direction of **p** is the direction of the incident beam. If the energy U is entirely reflected, the magnitude of the momentum delivered will be twice that given above, or

$$p = \frac{2U}{c} \qquad \text{(total reflection).} \qquad (38\text{–}16)$$

In the same way, twice as much momentum is delivered to an object when a perfectly elastic tennis ball is bounced from it as when it is struck by a perfectly inelastic ball (a lump of putty, say) of the same mass and velocity. If the light energy U is partly reflected and partly absorbed, the delivered momentum will lie between U/c and $2U/c$.

Example 2 A beam of laser light with an energy flux S of 10 W/cm² falls for 1.0 h on a perfectly reflecting plane mirror of 1.0-cm² area. (a) What momentum is delivered to the mirror in this time and (b) what force acts on the mirror?

(a) The energy that is reflected from the mirror is

$$U = (10 \text{ W/cm}^2)(1.0 \text{ cm}^2)(3600 \text{ s}) = 3.6 \times 10^4 \text{ J.}$$

The momentum delivered after one hour's illumination is, from Eq. 38–16

$$p = \frac{2U}{c} = \frac{(2)(3.6 \times 10^4 \text{ J})}{3 \times 10^8 \text{ m/s}} = 2.4 \times 10^{-4} \text{ kg·m/s.}$$

(b) From Newton's second law, the average force on the mirror is equal to the average rate at which momentum is delivered to the mirror, or

$$F = \frac{p}{t} = \frac{2.4 \times 10^{-4} \text{ kg·m/s}}{3600 \text{ s}} = 6.7 \times 10^{-8} \text{ N.}$$

This is a very small force, about equal to the weight of a small grain of table salt.

Measuring radiation pressure

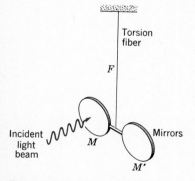

Figure 38–9 The arrangement of Nichols and Hull for measuring radiation pressure.

The first measurement of radiation pressure was made in 1901–1903 by Nichols and Hull at Dartmouth College and by Lebedev in Russia, about thirty years after the existence of such effects had been predicted theoretically by Maxwell.

Nichols and Hull measured radiation pressures and verified Eqs. 38–15 and 38–16, using a torsion balance technique. They allowed light to fall on mirror M as in Fig. 38–9; the radiation pressure caused the balance arm to turn through a measured angle θ, twisting the torsion fiber F. Assuming a suitable calibration for their torsion fiber, the experimenters could arrive at a numerical value for this pressure. Nichols and Hull measured the intensity of their light beam by allowing it to fall on a blackened metal disk of known absorptivity and by measuring the temperature rise of this disk. In a particular run these experimenters measured a radiation pressure of 7.01 μN/m²; the predicted value was 7.05 μN/m², in excellent agreement. Assuming a mirror area of 1 cm², this represents a force on the mirror of only 7×10^{-10} N, about 100 times smaller than the force calculated in Example 2.

The development of laser technology has permitted the achievement of radiation pressures much higher than those discussed so far in this section. This comes about because a beam of laser light—unlike a beam of light from, say, an incandescent filament—can be focused to a tiny spot of the order of magnitude of a wavelength in diameter. This permits the delivery of very large energy fluxes to small objects placed at the focal spot.*

---

* See "The Pressure of Laser Light" by Arthur Ashkin, *Scientific American*, February 1972.

38-6 Polarization

Electromagnetic waves are transverse waves, the directions of the alternating electric and magnetic field vectors being at right angles to the direction of propagation instead of parallel to it as is the case for a longitudinal wave. The transverse electromagnetic wave of Fig. 38-6 has the additional characteristic that it is *polarized* (more specifically, *plane polarized*). This means that the alternating electric field vectors are parallel to each other for all points in the wave. The magnetic field vectors are also parallel to each other but in dealing with polarization questions we focus our attention on the electric field, to which most detectors of electromagnetic radiation are sentitive.

Figure 38-10, which is simply another representation of the traveling electromagnetic wave of Fig. 38-6, is drawn to stress the polarization feature. The wave in this figure is traveling in the *x*-direction; it is said to be polarized in the *y*-direction because the electric field vectors are all parallel to this axis. The plane defined by the direction of propagation (the *x*-axis) and the direction of polarization (the *y*-axis) is called the *plane of vibration*.

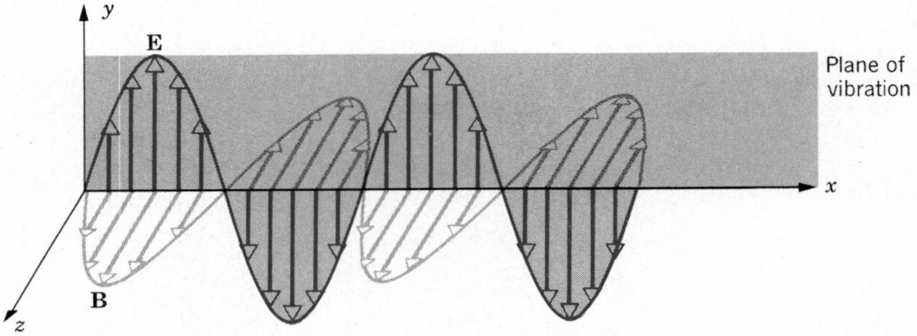

Figure 38-10 Another "snapshot" representation of the traveling electromagnetic wave of Figure 38-6, showing the alternating electric and magnetic field vectors at various points along the *x*-axis. The wave is polarized in the *y*-direction, the plane of vibration being the *xy*-plane.

Electromagnetic waves in the radio and microwave range readily exhibit polarization. Such a wave, generated by the surging of charge up and down in the dipole that forms the transmitting antenna of Fig. 38-11, has (for points along the horizontal axis) an electric field vector **E** that is parallel to the dipole axis. The plane of vibration of the transmitted wave is the plane of the figure. See also Fig. 38-12a, which shows a polarized beam emerging at right angles from the plane of the figure.

When the polarized wave of Fig. 38-11 falls on a second dipole connected to a microwave receiver, the alternating electric component of the wave will cause electrons to surge back and forth in the receiving antenna, producing a reading on the detector. If we turn the receiving antenna through 90° about the direction of propagation, the detector reading drops to zero. In this orientation the electric field vector is not able to cause charge to move along the dipole axis because it points at right angles to this axis. We can reproduce the experiment suggested by Fig. 38-11 by turning the receiving antenna of a television set (assumed an electric dipole type) through 90° about an axis that points toward the transmitting station. The intensity change should be evident.

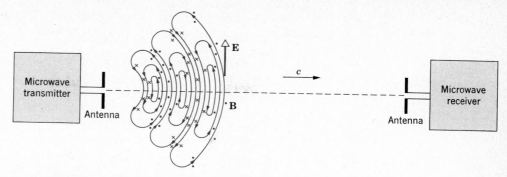

Figure 38–11 A polarized electromagnetic wave is generated by the transmitting antenna. The plane of vibration is the plane of the page.

Unpolarized light

In radio and microwave sources the elementary radiators, which are the electrons surging back and forth in the transmitting antenna, act in unison. In common light sources, however, such as the sun or a fluorescent lamp, the elementary radiators, which are the atoms that make up the source, act independently.* Because of this difference the light propagated from such sources in a given direction consists of many independent wavetrains whose planes of vibration are randomly oriented about the direction of propagation, as in Fig. 38–12b. Such light is unpolarized. Furthermore, the random orientation of the planes of vibration conceals the transverse nature of the waves. To study this transverse nature, we must find a way to unsort the different planes of vibration.

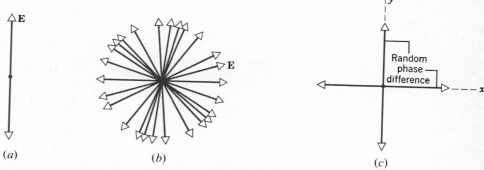

(a) (b) (c)

Figure 38–12 (a) A polarized wave moving out of the page, showing only the electric vector. (b) An unpolarized transverse wave viewed as a random superposition of many polarized wavetrains. (c) A second, completely equivalent, description of an unpolarized transverse wave; here we view the unpolarized wave as two plane-polarized waves with a random phase difference. The orientation of the x- and y-axes about the propagation direction is completely arbitrary.

Figure 38–13 shows unpolarized light falling on a sheet of commercial polarizing material called *Polaroid*. There exists in the sheet a certain characteristic polarizing direction, shown by the parallel lines. The sheet will transmit only those wavetrain components whose electric vectors vibrate parallel to this direction and will absorb those that vibrate at right angles to this direction. The emerging light will be polarized. This polarizing direction is established during the

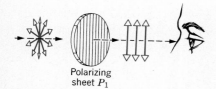

Figure 38–13 Unpolarized light is polarized by the action of a single polarizing sheet.

---

* This is not true for laser light. The atoms that generate such light act, not independently, but in unison. The emerging beam in most laser configurations is polarized.

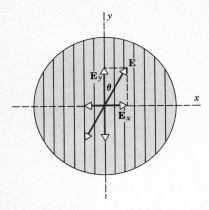

Figure 38–14 A wavetrain **E** is equivalent to two component wavetrains $\mathbf{E}_y$ and $\mathbf{E}_x$. Only the former is transmitted by the polarizer.

manufacturing process by embedding certain long-chain molecules in a flexible plastic sheet and then stretching the sheet so that the molecules are aligned parallel to each other.

In Fig. 38–14 the polarizing sheet or *polarizer* lies in the plane of the page and the direction of propagation is into the page. The arrow **E** shows the plane of vibration of a randomly selected wavetrain falling on the sheet. Two vector components, $\mathbf{E}_x$ (of magnitude $E \sin \theta$) and $\mathbf{E}_y$ (of magnitude $E \cos \theta$), can replace **E**, one parallel to the polarizing direction and one at right angles to it. Only $\mathbf{E}_y$ will be transmitted; $\mathbf{E}_x$ is absorbed within the sheet.

Let us place a second polarizing sheet P_2 (usually called, when so used, an *analyzer*) as in Fig. 38–15. If we rotate P_2 about the direction of propagation, there are two positions, 180° apart, at which the transmitted light intensity is almost zero; these are the positions in which the polarizing directions of P_1 and P_2 are at right angles.

If the amplitude of the polarized light falling on P_2 is E_m, the amplitude of the light that emerges is $E_m \cos \theta$, where θ is the angle between the polarizing directions of P_1 and P_2. Recalling that the intensity of the light beam is proportional to the square of the amplitude, we see that the transmitted intensity I varies with θ according to

$$I = I_m \cos^2 \theta, \qquad (38–17)$$

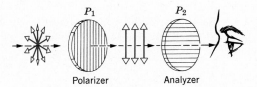

Polarizer   Analyzer

Figure 38–15 Unpolarized light is not transmitted by crossed polarizing sheets.

in which I_m is the maximum value of the transmitted intensity. It occurs when the polarizing directions of P_1 and P_2 are parallel, that is, when $\theta = 0$ or 180°. Figure 38–16a, in which two overlapping polarizing sheets are in the parallel position ($\theta = 0$ or 180° in Eq. 38–17) shows that the light transmitted through the region of overlap has its maximum value. In Fig. 38–16b one or the other of the sheets has been rotated through 90° so that θ in Eq. 38–17 has the value 90 or 270°; the light transmitted through the region of overlap is now a minimum.

Figure 38–16 Two sheets of Polaroid are laid over a drawing of the Luxembourg Palace in Paris. In (a) the axes of polarization of the two sheets are parallel so that light is transmitted through both sheets. In (b) one sheet has been rotated by 90° and no light passes through. (Malus discovered the phenomenon of polarization by reflection while looking at sunlight reflected from the windows of this building through a polarizing crystal.)

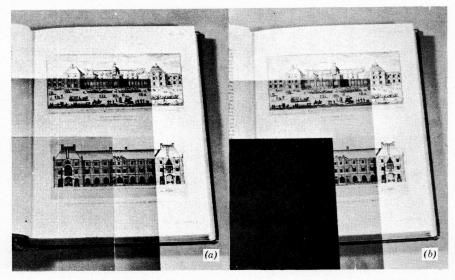

The law of Malus

Equation 38–17, called the law of Malus, was discovered by Étienne Louis Malus experimentally in 1809.

Example 3   Two polarizing sheets have their polarizing directions parallel so that the intensity I_m of the transmitted light is a maximum. Through what angle must either sheet be turned if the intensity is to drop by one-half?

From Eq. 38–17, since $I = \frac{1}{2}I_m$, we have

$$\tfrac{1}{2}I_m = I_m \cos^2 \theta$$

or

$$\theta = \cos^{-1}\left(\pm \frac{1}{\sqrt{2}}\right) = \pm 45°, \pm 135°.$$

The same effect is obtained no matter which sheet is rotated or in which direction.

Historically, polarization studies were made to investigate the nature of light. Today we reverse the procedure and deduce something about the nature of an object from the polarization state of the light emitted by or scattered from that object. For example, from studies of the polarization of light reflected from grains of cosmic dust present in our galaxy we can tell that they have been oriented in the weak galactic magnetic field ($\sim 10^{-8}$ T) so that their long dimension is parallel to this field. Polarization studies have suggested that Saturn's rings consist of ice crystals. We can find the size and shape of virus particles by the polarization of ultraviolet light scattered from them. We can learn a lot about the structure of atoms and nuclei from polarization studies of their emitted radiations. Thus we have a useful research technique for structures ranging in size from a galaxy ($\sim 10^{+20}$ m) to a nucleus ($\sim 10^{-14}$ m).

38–7   The Speed of Light

The speed of electromagnetic radiation in free space—usually called simply the speed of light—is the most investigated and one of the most accurately measured of all the fundamental constants of physics. Galileo was perhaps the first to attempt a measurement. In his chief work, *Two New Sciences,* published in 1638, is a conversation among three fictitious persons called Salviati, Simplicio, and Sagredo (who evidently is meant to be Galileo himself). Here is part of what they say about the speed of light.

Simplicio: Everyday experience shows that the propagation of light is instantaneous; for when we see a piece of artillery fired, at a great distance, the flash reaches our eyes without lapse of time; but the sound reaches the ear only after a noticeable interval.

Sagredo: Well, Simplicio, the only thing I am able to infer from this familiar bit of experience is that sound, in reaching our ear, travels more slowly than light; it does not inform me whether the coming of the light is instantaneous or whether, although extremely rapid, it still occupies time. . . .

Galileo's attempt to measure *c*

Sagredo then describes a possible method for measuring the speed of light. He and an assistant stand facing each other some distance apart, at night. Each carries a lantern which can be covered or uncovered at will. Galileo started the experiment by uncovering his lantern. When the light reached the assistant he uncovered his own lantern, whose light was then seen by Galileo. Galileo tried to measure the time between the instant at which he uncovered his own lantern and the instant at which the light from his assistant's lantern reached him. For a 1-mile separation we now know that the round trip travel time would be only 10^{-5} s. This is much less than human reaction times, so the method fails.

To measure such a large velocity directly, we must either deal with a small time interval or use a long base line. This latter idea suggests that astronomy, which involves great distances, might provide an experimental value for the speed of light.

In 1675 Ole Roemer, a Danish astronomer working in Paris, made some observations of the moons of Jupiter (see Problem 20) from which a speed of light of 2×10^8 m/s may be deduced. About fifty years later James Bradley, an English astronomer, made some observations of an entirely different kind, obtaining a value of 3.0×10^8 m/s. Actually, the speed of light is now so well known from terrestrial measurements that we use it to measure distances in the solar system. For example, laser pulses reflected from "mirrors" on the moon and microwave pulses reflected from Venus today provide the distances to these bodies.

Other early measurements of c

In 1849 Hippolyte Louis Fizeau, a French physicist, first measured the speed of light by a nonastronomical method, obtaining a value of 3.13×10^8 m/s. Figure 38–17 shows Fizeau's apparatus. Let us first ignore the toothed wheel. Light from source S is made to converge by lens L_1, is reflected from mirror M_1, and forms in space at F an image of the source. Mirror M_1 is a so-called "half-silvered mirror"; its reflecting coating is so thin that only half the light that falls on it is reflected, the other half being transmitted.

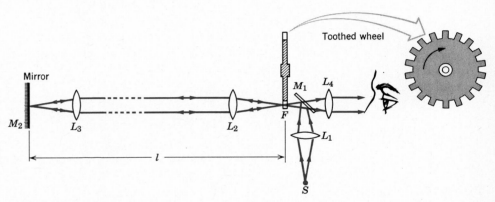

Figure 38–17   Fizeau's apparatus for measuring the speed of light. The distance between mirror M_2 and the toothed wheel is 8.63 km.

Light from the image at F enters lens L_2 and emerges as a parallel beam; after passing through lens L_3 it is reflected back along its original direction by mirror M_2. In Fizeau's experiment the distance l between M_2 and F was 8.63 km. When the light strikes mirror M_1 again, some will be transmitted, entering the eye of the observer through lens L_4.

The observer will see an image of the source formed by light that has traveled a distance $2l$ between the wheel and mirror M_2 and back again. To time the light beam a marker of some sort must be put on it. This is done by "chopping" it with a rapidly rotating toothed wheel. Suppose that during the round-trip travel time of $2l/c$ the wheel has turned just enough so that, when the light from a given "burst" returns to the wheel, point F is covered by a tooth. The light will hit the face of the tooth that is toward M_2 and will not reach the observer's eye.

If the speed of the wheel is exactly right, the observer will not see any of the bursts because each will be screened by a tooth. The observer measures c by increasing the angular speed ω of the wheel from zero until the image of source S disappears. Let θ be the angular distance from the center of a gap to the center of a tooth. The time needed for the wheel to rotate a distance θ is the round-trip travel time $2l/c$. In equation form,

$$\frac{\theta}{\omega} = \frac{2l}{c} \quad \text{or} \quad c = \frac{2\omega l}{\theta}. \tag{38–18}$$

Example 4 The wheel used by Fizeau had 720 teeth. What is the smallest angular speed at which the image of the source will vanish?

The angle θ is $(1/1440)$ rev; solving Eq. 38–18 for ω gives

$$\omega = \frac{c\theta}{2l} = \frac{(3.00 \times 10^8 \text{ m/s})[(1/1440) \text{ rev}]}{(2)(8630 \text{ m})} = 12.1 \text{ rev/s}.$$

The French physicist Foucault (1819–1868) greatly improved Fizeau's method by substituting a rotating mirror for the toothed wheel. The American physicist Albert A. Michelson (1852–1931) conducted an extensive series of measurements of c, extending over a fifty-year period, using this technique.

The terrestrial methods for measuring the speed of light that we have just described all involve timing the passage of a light pulse over a measured baseline. Another approach is to measure the frequency ν of radiation whose wavelength λ is known; the speed c can then be found from the relation

Modern measurements of c

$$c = \lambda\nu. \tag{38–19}$$

This method was first applied with precision in the 1950s by Essen and others, using radiation in the microwave region of the electromagnetic spectrum. With the development of lasers it has become possible to extend these methods to radiation in the visible or near-visible portion of the spectrum; our presently accepted value of the speed of light is measured in this way.

Table 38–1 shows some selected measurements that have been made of the speed of electromagnetic radiation since Galileo's day. It stands as a monument to human persistence and ingenuity. Note in the last column how the uncertainty in

Table 38–1 The speed of light—some selected measurements[a]

Date	Experimenter	Country	Method	Speed km/s	Uncertainty km/s
1600(?)	Galileo	Italy	Lanterns and shutters	"If not instantaneous, it is extraordinarily rapid"	—
1676	Roemer	France	Moons of Jupiter	214,000	—
1729	Bradley	England	Aberration of starlight	304,000	—
1849	Fizeau	France	Toothed wheel	315,300	—
1862	Foucault	France	Rotating mirror	298,000	±500
1879	Michelson	USA	Rotating mirror	299,910	±50
1906	Rosa and Dorsey	USA	From $c = 1/\sqrt{\epsilon_0\mu_0}$	299,781	±10
1927	Michelson	USA	Rotating mirror	299,798	±4
1950	Essen	England	Microwave cavity	299,792.5	±3
1950	Bergstrand	Sweden	Geodimeter	299,793.1	±0.25
1958	Froome	England	Microwave interferometer	299,792.5	±0.1
1965	Kolibuyev	USSR	Geodimeter	299,792.6	±0.06
1972	Bay et al.	USA	From $c = \lambda\nu$ (laser light)	299,792.462	±0.018
1973	Evenson et al.	USA	From $c = \lambda\nu$ (laser light)	299,792.4574	±0.0012
1974	Blaney et al.	England	From $c = \lambda\nu$ (laser light)	299,792.4590	±0.0008

[a] See "Some Recent Determinations of the Velocity of Light, III" by Joseph F. Mulligan, *American Journal of Physics*, October 1976.

the measurement has improved through the years. Note also the international character of the effort and the variety of methods.

The speed of light eventually became so well established by experiment that its precision was limited only by the uncertainties involved in the definition of the meter, then based on the wavelength of light emitted by atoms of krypton-86. To meet this situation, in 1983 the speed of light was given the *assigned* value of 299,792,458 m/s and the meter was redefined as described in Section 1–3. Today, if you send a light beam from one point to another and measure its transit time you are not measuring the speed of light; you are measuring the distance between the two points.

38–8 The Speed of Light
—Einstein's Postulate

When we say that the speed of sound in dry air at 0°C is 331.7 m/s we imply a reference frame fixed with respect to the air mass through which the sound wave is moving. When we say that the speed of electromagnetic radiation in free space is 2.997924590×10^8 m/s, what reference frame are we talking about? It cannot be the medium through which the light wave travels because, in contrast to sound, no medium is required.

The ether concept

Physicists of the nineteenth century, influenced as they then were by an analogy between light waves and sound waves or other purely mechanical disturbances did not accept the idea of a wave requiring no medium. They postulated the existence of an ether, which was a tenuous substance that filled all space and served as a medium of transmission for light.

Although it proved useful for many years, the ether concept did not survive the test of experiment. In particular, careful attempts to measure the speed of the earth through the ether always gave the result of zero. Physicists were not willing to believe that the earth was permanently at rest in the ether and that all other bodies in the universe were in motion through it.

Einstein's postulate

In 1905 Einstein resolved the dilemma by making a bold postulate: If a number of observers are moving (at uniform velocity) with respect to each other and to a source of electromagnetic radiation such as light and if each observer measures the speed of the radiation emerging from the source, they will all measure the same value. This is a fundamental assumption of Einstein's theory of relativity. It does away with the need for an ether by asserting that the speed of light is the same in *all* reference frames; none is singled out as fundamental. The theory of relativity, derived from this postulate, has been tested many times, and agreement with the predictions of theory has always emerged. These agreements, extending over half a century, lend strong support to Einstein's basic postulate about light propagation.

Figure 38–18 focuses specifically on the fundamental problem of light propagation. A source of light—perhaps a photographic flash bulb—emits a light pulse whose speed is measured by two observers. The first, S', is at rest with respect to the source and the second, S, is moving with speed u in the negative x'-direction thus seeing the source move with the same speed in the positive x-direction. The speed of light, v', measured by S' would be equal to c. Question: What speed, v,

Figure 38–18 A light source at the origin of reference frame S' sends out a pulse whose speed is measured by observers S and S'. The x and x' axes coincide but are shown as separate for clarity.

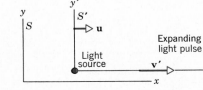

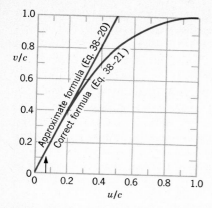

Figure 38-19 The speed of a particle P, as seen by an observer such as S in Fig. 38-18, for the special case of $v' = u$. All speeds are expressed as a ratio to c, the speed of light. The vertical arrow corresponds to 8×10^7 km/h, about 2500 times the speed of a typical earth satellite.

would observer S measure? Einstein's hypothesis asserts that each observer would measure the same speed, c, or that

$$v = v' = c.$$

This hypothesis contradicts the classical law of addition of velocities which asserts that

$$v = v' + u. \qquad \text{(true for low speeds only)} \qquad (38-20)$$

This law, which is so familiar that it seems (incorrectly) to be intuitively true, is in fact based on observations—not of light—but of gross moving objects in the world about us. Even the fastest of these—an earth satellite, say—is moving at a speed that is quite small compared to that of light. The body of experimental evidence that underlies Eq. 38-20 thus represents a severely restricted area of experience, namely, experiences in which $v' \ll c$ and $u \ll c$. If we assume that Eq. 38-20 holds for all particles regardless of speed, we are making a gross extrapolation. Einstein's theory of relativity predicts that this extrapolation is indeed not true and that Eq. 38-20 is a limiting case of a more general relationship that holds for light pulses and for material particles, whatever their speed, or

$$v = \frac{v' + u}{1 + v'u/c^2}. \qquad \text{(true for all speeds)} \qquad (38-21)$$

Equation 38-21 is quite indistinguishable from Eq. 38-20 at low speeds, that is, when $v' \ll c$ and $u \ll c$.

If we apply Eq. 38-21 to the case in which the moving object is a light pulse, and if we put $v' = c$, we obtain

$$v = \frac{c + u}{1 + cu/c^2} = c.$$

This is consistent, as it must be, with the fundamental assumption on which the derivation of Eq. 38-21 is based; it shows that *both* observers measure the same speed c for light. Equation 38-20 predicts (incorrectly) that the speed measured by S will be $c + u$. Figure 38-19 shows that we cannot tell the difference between the (correct) Eq. 38-21 and the (approximate) Eq. 38-20 at speeds that are small compared to the speed of light.

Example 5 Two electrons are ejected in opposite directions from radioactive atoms in a sample of radioactive material. Let each electron have a speed, as measured by a laboratory observer, of 0.60c (this corresponds to a kinetic energy of 130 keV). What is the speed of one electron as seen from the other?

Equation 38-20 gives

$$v = v' + u = 0.60c + 0.60c = 1.2c.$$

Equation 38-21 gives

$$v = \frac{v' + u}{1 + v'u/c^2} = \frac{0.60\,c + 0.60c}{1 + (0.60c)^2/c^2} = 0.88c.$$

This example shows that for speeds that are comparable to c, Eqs. 38-20 and 38-21 yield quite different results. A wealth of indirect experimental evidence points to the latter result as being correct.

38-9 The Doppler Effect for Light

We have seen that the same speed is measured for electromagnetic radiation no matter what the relative speeds of the light source and the observer are. The measured frequency and wavelength will change, however, but always in such a way that their product, which is the speed of light, remains constant. Such frequency shifts are called *Doppler shifts*, after Johann Doppler (1803-1853), who first predicted them. In what follows we will focus on visible light although our conclusions hold for all parts of the electromagnetic spectrum.

In Section 18–6 we showed that if a source of sound in moving away from an observer at a speed u, the frequency heard by the observer (see Eq. 18–10, which has been rearranged, with u substituted for v_s) is

Doppler effect for sound, I

$$\nu' = \nu \frac{1}{1 + u/v}. \qquad \begin{cases} 1. & \text{sound wave} \\ 2. & \text{observer fixed in medium} \\ 3. & \text{source receding from observer} \end{cases} \qquad (38\text{–}22)$$

In this equation ν is the frequency heard when the source is at rest and v is the speed of sound.

If the source is at rest in the transmitting medium but the observer is moving away from the source at speed u, the observed frequency (see Eq. 18–9b, in which u has been substituted for v_0) is

Doppler effect for sound, II

$$\nu' = \nu \left(1 - \frac{u}{v}\right). \qquad \begin{cases} 1. & \text{sound wave} \\ 2. & \text{source fixed in medium} \\ 3. & \text{observer receding from source} \end{cases} \qquad (38\text{–}23)$$

Even if the relative separation speeds u of the source and the observer are the same, the frequencies predicted by Eqs. 38–22 and 38–23 are different. This is not surprising, because a sound source moving through a medium in which the observer is at rest is simply not the same thing as an observer moving through that medium with the source at rest, as comparison of Figs. 18–9 and 18–10 shows.

We might be tempted to apply Eqs. 38–22 and 38–23 to light, substituting c, the speed of light, for v, the speed of sound. For light, as contrasted with sound, however, it has proved impossible to identify a medium of transmission relative to which the source and the observer are moving. This means that "source receding from observer" and "observer receding from source" are indistinguishable and must exhibit exactly the same Doppler frequency. As applied to light, either Eq. 38–22 or Eq. 38–23 or both must be incorrect. The Doppler frequency predicted by the theory of relativity is, in fact,

Doppler effect for light

$$\nu' = \nu \frac{\sqrt{1 - (u/c)^2}}{1 + u/c}. \qquad \begin{cases} 1. & \text{light wave} \\ 2. & \text{source and observer separating} \end{cases} \qquad (38\text{–}24)$$

In all three of the foregoing equations we obtain the appropriate relations for the source and the observer approaching each other if we replace u by $-u$.

If we replace v by c in Eqs. 38–22 and 38–23 we have the classical predictions for the Doppler effect of light waves, the "medium" in question being the (now-discredited) ether. They stand in contrast with Eq. 38–24, which is the prediction of the theory of relativity and does not require the ether concept.

Actually the above three equations are not as different as they seem to be. Let us expand all three by the binomial theorem (see Appendix G), first making the convenient substitution of β for the dimensionless ratio u/c. The results are:

$$\nu' = \nu(1 - \beta + \beta^2 - \ldots) \qquad \begin{cases} 1. & \text{classical theory} \\ 2. & \text{observer fixed in the "ether"} \\ 3. & \text{source receding from observer} \end{cases} \qquad (38\text{–}22')$$

$$\nu' = \nu(1 - \beta) \qquad \begin{cases} 1. & \text{classical theory} \\ 2. & \text{source fixed in the "ether"} \\ 3. & \text{observer receding from source} \end{cases} \qquad (38\text{–}23')$$

$$\nu' = \nu(1 - \beta + \tfrac{1}{2}\beta^2 - \ldots) \quad \begin{cases} 1. & \text{relativistic theory} \\ 2. & \text{source and observer separating} \quad (38\text{-}24') \end{cases}$$

For essentially all available monochromatic light sources, even those of atomic dimensions, the quantity β is much smaller than unity so that successive terms in these equations become small rapidly. If only the first two terms are retained all three equations yield identical results. If three terms are retained the (correct) relativistic equation is seen to give predictions that are just half-way between those of the two classical equations.

Under nearly all circumstances the differences between the above three equations are so small that it is of no practical concern which is used in calculations. It is important, however, to do at least one experiment of such sensitivity that it *is* possible to decide which equation agrees with experiment and which does not. To carry out such a test requires the use of a moving light source whose speed u is as large as possible compared with the speed of light. In 1938 H. E. Ives and G. R. Stilwell of the Bell Telephone Laboratories, following up a suggestion by Einstein, performed such an experiment, using moving hydrogen atoms ($\beta = u/c \cong 0.003$) as light sources. Their careful Doppler shift measurements showed beyond question that Eq. 38-24, which is based on the theory of relativity, agrees with experiment but that predictions based on classical theory and the ether concept do not.

The Doppler effect for light finds many applications in astronomy, where it is used to determine the speeds at which luminous heavenly bodies are moving toward us or receding from us. Such Doppler shifts measure only the radial or line-of-sight components of the relative velocity. Galaxies for which such measurements have been made (Fig. 38-20) appear to be receding from us, the recession velocity being greater for the more distant galaxies; these observations are the basis of the expanding-universe concept.

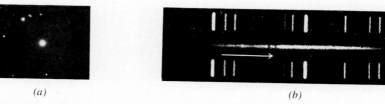

(a)   *(b)*

Figure 38-20   *(a)* The central spot is a nebula in the constellation Corona Borealis; it is 1.3×10^8 light years distant. *(b)* The central streak shows the distribution in wavelength of the light emitted from this nebula. The two vertical dark bands show the presence of calcium. The horizontal arrow shows that these calcium lines occur at longer wavelengths than those for terrestrial light sources containing calcium, the length of the arrow representing the wavelength shift. Measurement of this shift shows that the galaxy is receding from us at 2.2×10^4 km/s. The lines above and below the central streak represent light from a terrestrial source, used to establish a wavelength scale. (Courtesy Mount Wilson and Mount Palomar Observatories.)

Example 6   Certain characteristic wavelengths in the light from a galaxy in the constellation Virgo are observed to be increased in wavelength, as compared with terrestrial sources, by about 0.4%. What is the radial speed of this galaxy with respect to the earth? Is it approaching or receding?

If λ is the wavelength for a terrestrial source, then

$$\lambda' = 1.004\,\lambda.$$

Since we must have $\lambda'\nu' = \lambda\nu = c$, we can write this as

$$\nu' = 0.996\nu.$$

This frequency shift is so small that, in calculating the source velocity, it makes no practical difference whether we use Eq. 38–22, 38–23, or 38–24. Using Eq. 38–23 we obtain

$$\nu' = 0.996\nu = \nu\left(1 - \frac{u}{c}\right).$$

Solving yields $u/c = 0.0040$, or $u = (0.0040)(3.0 \times 10^8 \text{ m/s}) = 1.2 \times 10^6$ m/s. The galaxy is receding; had u turned out to be negative, the galaxy would have been moving toward us.

REVIEW GUIDE AND SUMMARY

Maxwell and the electromagnetic spectrum

Maxwell's equations predict the existence of a spectrum of electromagnetic waves (see Fig. 38–1) differing only in wavelength and traveling through free space with a speed given by

$$c = \frac{1}{\sqrt{\epsilon_0 \mu_0}} \quad (= 3.00 \times 10^8 \text{ m/s}). \qquad [38\text{–}1]$$

Generating a radio wave

Figure 38–3 shows an apparatus for generating an electromagnetic wave in the shortwave radio region of the spectrum. Figure 38–5 shows the cyclic variation of the electric and magnetic field components of this wave as it sweeps past a distant observer.

Analyzing an electromagnetic wave

Figure 38–6 shows a "snapshot" of the wave seen by observer P in Fig. 38–5. We assume that the electric and magnetic field components of this wave have the forms:

$$E(x, t) = E_m \sin (kx - \omega t) \qquad [38\text{–}2]$$

and

$$B(x, t) = B_m \sin (kx - \omega t). \qquad [38\text{–}3]$$

E and B in an electromagnetic wave

Figure 38–7 shows cuts through the wave of Fig. 38–6. By applying Maxwell's equations we can derive Eq. 38–1 above and can also show that the relation between the electric and magnetic field components in such a wave is given by Eq. 38–8, or $E = cB$.

In an electromagnetic wave the electric field component is generated by the alternating character of the magnetic field component (Faraday's law) and the magnetic field component is generated by the alternating character of the electric field component (Ampere's law).

The Poynting vector, defined from

Energy flow in an electromagnetic wave

$$\mathbf{S} = \frac{1}{\mu_0}\, \mathbf{E} \times \mathbf{B}, \qquad [38\text{–}12]$$

gives the energy flux (W/m²) in an electromagnetic wave. Figure 38–8 suggests a proof of this relationship for an important special case. Example 1 shows how the amplitudes of the electric and magnetic field components of a wave can be found if the energy flux is known.

Radiation pressure

Electromagnetic radiation falling on a surface exerts a radiation pressure on it. Given the energy flux S falling normally on an object, Example 2 shows how to calculate (a) the momentum delivered to the object in a time t along with (b) the radiation pressure and (c) the force acting on the object during this interval.

Polarization

An electromagnetic wave from an antenna like that of Fig. 38–3 is polarized, which means that its electric field vectors are all parallel. The direction of the electric field $\mathbf{E}$ is called the direction of polarization; the plane containing $\mathbf{E}$ and the direction of propagation is called the plane of vibration. See Figs. 38–10, 38–11, and 38–12a.

Unpolarized light

Light from an "ordinary" source such as the sun is unpolarized; its energy is emitted in wavetrains of finite length whose orientation about the propagation direction is random; see Fig. 38–12b.

Commercial polarizing sheets

A sheet of Polaroid transmits only electric field components parallel to a given direction, strongly absorbing the others. Thus it transforms unpolarized light into polarized light, as Fig. 38–13 shows.

The law of Malus

The intensity of polarized light passing through a polarizing sheet is reduced from I_m to I, where

$$I = I_m \cos^2 \theta. \qquad [38\text{--}17]$$

Here θ (see Fig. 38–14) is the angle between the plane of vibration and the polarizing direction of the sheet; see Figs. 38–15 and 38–16 and Example 3.

Einstein's postulate

Einstein postulated that no medium is required for the propagation of light. Thus as far as motion is concerned, only the relative speed of the source and the observer has physical meaning. The speed of light, given by Eq. 38–1, is independent of this relative speed; see Fig. 38–18.

The transformation of velocities and the speed of light

For a material particle (not light) whose speed in reference frame S' is v', the speed v in frame S is

$$v = \frac{v' + u}{1 + v'u/c^2}. \qquad [38\text{--}21]$$

The familiar Eq. 38–20 ($v = v' + u$) is an approximate result that holds only at low speeds; see Fig. 38–19 and Example 5. If we apply Eq. 38–21 to light by putting $v' = c$, the equation yields $v = c$, in accord with Einstein's postulate.

The Doppler effect

Even though c ($= \lambda \nu$) is independent of the relative speed of the source and the observer the wavelength λ and the frequency ν are not. If the source and the observer are separating, λ increases (to λ') and ν decreases (to ν') to compensate. This Doppler-shifted frequency is given by

$$\nu' = \nu \, \frac{\sqrt{1 - \beta^2}}{1 + \beta} \cong \nu(1 - \beta + \tfrac{1}{2}\beta^2 - \ldots). \qquad [38\text{--}24, 24']$$

Here $\beta = u/c$. Example 6 shows how Doppler shift measurements can be used to find the separation velocities of distant galaxies.

QUESTIONS

1. Electromagnetic waves reach us from the farthest depths of space. From the information they carry can we tell what the universe is like at the present moment? At any selected time in the past?

2. If you are asked on an examination what fraction of the electromagnetic spectrum lies in the visible range, what would you reply?

3. Project Seafarer was an ambitious program to construct an enormous antenna, buried underground on a site about 4000 square miles in area. Its purpose was to transmit signals to submarines while they were deeply submerged. If the effective wavelength was, say, about 10^4 earth radii, what would be the frequency and the period of the radiations emitted? Ordinarily electromagnetic radiations do not penetrate very far into conductors such as sea water. Can you think of any reason why such ELF (extremely low frequency) radiations should penetrate more effectively? Think of the limiting case of zero frequency. (Why not transmit signals at zero frequency?)

4. Why are danger signals in red when the eye is most sensitive to yellow-green?

5. Comment on this definition of the limits of the spectrum of visible light, given by a physiologist: "The limits of the visible spectrum occur when the eye is no better adapted than any other organ of the body to serve as a detector."

6. How might an eye-sensitivity curve like that of Fig. 38–2 be measured?

7. In connection with Fig. 38–2 (a) do you think it possible that the wavelength of maximum sensitivity could vary if the intensity of the light is changed? (b) What might the curve of Fig. 38–2 look like for a color blind person who could not, for example, distinguish red from green?

8. Suppose that human eyes were insensitive to visible light but were very sensitive to infrared light. What environmental changes would be needed if you were to (a) walk down a long corridor and (b) drive a car? Could the phenomenon of color exist? How would traffic lights have to modified?

9. Speaking loosely we can say that the electric and the magnetic components of a traveling electromagnetic wave "feed on each other." What does this mean?

10. "Displacement currents are present in a traveling electromagnetic wave and we may associate the magnetic field component of the wave with these currents." Is this statement true? Discuss it in detail.

11. H. G. Wells, in his novel, *The Invisible Man,* described a concoction developed by a "mad scientist" that would render the person who drank it invisible. Give arguments to prove that a truly invisible man would be blind.

12. Can an electromagnetic wave be deflected by a magnetic field? . . . by an electric field?

13. How does a microwave oven cook food? You can boil water in a plastic bag in such an oven. How can this happen?

14. Why is Maxwell's modification of Ampere's law (that is, the term $\epsilon_0 d\Phi_E/dt$; see Table 37–2) needed to understand the propagation of electromagnetic waves?

15. Can an object absorb light energy without having linear momentum transferred to it? If so, give an example. If not, explain why.

16. When you turn on a flashlight does it experience any force associated with the emission of the light?

17. As you recline in a beach chair in the sun why are you so conscious of the thermal energy delivered to you but totally unresponsive to the linear momentum delivered from the same source? Is it true that when you catch a hard-pitched baseball you are conscious of the energy delivered but not of the momentum?

18. When a parallel beam of light falls at right angles on an object the momentum transfers are given by Eqs. 38–15 and 38–16. Do these equations still hold if the light source is moving rapidly toward or away from the object, perhaps at a speed of $0.5\,c$?

19. If you were to calculate the Poynting vector for various points in and around a transformer, what would you expect the field pattern of these vectors to look like? Assume that an alternating potential difference has been applied to the primary windings and that a resistive load is connected across the secondary.

20. As we normally experience them, radio waves are almost always polarized and visible light is almost always unpolarized. Why should this be so?

21. Example 3 shows that, when the angle between the two polarizing directions is turned from 0 to 45°, the intensity of the transmitted beam drops to one-half its initial value. What happens to this "missing" energy?

22. You are given a number of polarizing sheets. Explain how you would use them to rotate the plane of polarization of a plane-polarized wave through any given angle.

23. What determines the desirable length and orientation of the rabbit ears antenna on a portable TV set?

24. Why do sunglasses made of polarizing materials have a marked advantage over those that simply depend on absorption effects?

25. Unpolarized light falls on two polarizing sheets so oriented that no light is transmitted. If a third polarizing sheet is placed between them, can light be transmitted?

26. Find a way to identify the polarizing direction of a sheet of Polaroid. No marks appear on the sheet.

27. When observing a clear sky through a polarizing sheet, you find that the intensity varies on rotating the sheet. This does not happen when viewing a cloud through the sheet. Why?

28. How could Galileo have tested experimentally that reaction times were an overwhelming source of error in his attempt to measure the speed of light, described in Section 38–7?

29. Can you think of any "every day" observation (that is, without experimental apparatus) to show that the speed of light is not infinite? Think of lightning flashes, possible discrepancies between the predicted time of sunrise and the observed time, radio communications between earth and astronauts in orbiting space ships, and so on.

30. Why is the rotating mirror method for measuring the speed of light better than the toothed wheel method of Fizeau? (See Fig. 38–17.)

31. A friend tells you that Einstein's postulate about the speed of light—namely, that it is not affected by the speed of the source or the observer—must be discarded because it violates common sense. How would you answer him?

32. Albert Einstein wrote about " . . . a paradox on which I already hit at the age of sixteen: If I pursue a beam of light with the velocity c, I should observe the beam as a nonprogressing spatially oscillating electromagnetic field." This makes no sense and the resolution of this paradox led him eventually to the special theory of relativity. What *is* the resolution of this paradox?

33. Atoms are mostly empty space. However, the speed of light passing through a transparent solid made up of such atoms is often considerably less than the speed of light in free space. How can this be?

34. In a vacuum, does the speed of light depend on (*a*) the wavelength, (*b*) the frequency, (*c*) the intensity, (*d*) the state of polarization, (*e*) the speed of the source, or (*f*) the speed of the observer?

35. Can a galaxy be so distant that its recession speed equals the speed of light? If so, how could we see it? Would its light ever reach us?

36. How and why do the Doppler effects for light and for sound differ? In what ways are they the same?

EXERCISES

Section 38–1 Maxwell and the Electromagnetic Spectrum

1. (*a*) How long does it take a radio signal to travel 150 km from a transmitter to a receiving antenna? (*b*) We see a full moon by reflected sunlight. How much earlier did the light that enters our eye leave the sun? The earth-moon and the earth-sun distances are 3.8×10^5 km and 1.5×10^8 km. (*c*) What is the round-trip travel time for light between Earth and a spaceship orbiting Saturn, 1.3×10^9 km distant? (*d*) The Crab nebula, which is about 6500 light-years away, is thought to be the result of a supernova explosion recorded by Chinese astronomers in 1054 A.D. In approximately what year did the explosion actually occur?

2. (*a*) The wavelength of the most energetic x rays produced when electrons accelerated to 18 GeV in the Stanford Linear Accelerator slam into a solid target is 0.067 fm. What is the frequency of these x rays? (*b*) A VLF (very low frequency) radio wave has a frequency of only 30 Hz. What is its wavelength?

Section 38-2 Generating an Electromagnetic Wave

3. Find the angle from the axis of an oscillating dipole at which the intensity of the radiation field is one-half its value in the equatorial plane.

Section 38-3 Traveling Waves and Maxwell's Equations

4. A plane electromagnetic wave has a maximum electric field of 3.2×10^{-4} V/m. Find the maximum magnetic field.

5. The electric field associated with a plane electromagnetic wave is given by: $E_x = 0$; $E_y = 0$; $E_z = 2\cos[\pi \times 10^{15} (t - x/c)]$, with $c = 3.0 \times 10^8$ m/s and all quantities in SI units. The wave is propagating in the $+x$-direction. Write expressions for the components of the magnetic field of the wave.

Section 38-4 Energy Transport and the Poynting Vector

6. Show, by finding the direction of the Poynting vector **S**, that the directions of the electric and magnetic fields at all points in Figs. 38-4, 38-5, 38-6, and 38-7 are consistent at all times with the assumed directions of propagation.

7. Currently operating Nd:glass lasers can provide 100 TW of power in 1.0-ns pulses at a wavelength of 0.26 μm. How much energy is contained in a single pulse?

8. Our closest stellar neighbor, α-Centauri, is 4.3 light years away. It has been suggested that TV programs from our planet have reached this star and may have been viewed by the hypothetical inhabitants of a hypothetical planet orbiting this star. A TV station on earth has a power output of 1.0 MW. What is the intensity of its signal at α-Centauri?

9. An electromagnetic wave is traveling in the negative y-direction. At a particular position and time, the electric field is along the positive z-axis and has a magnitude of 100 V/m. What are the direction and magnitude of the magnetic field at that position and at that time?

10. The earth's mean radius is 6.4×10^6 m and the mean earth-sun distance is 1.5×10^8 km. What fraction of the radiation emitted by the sun is intercepted by the disc of the earth?

11. The radiation emitted by a laser is not exactly a parallel beam; rather, the beam spreads out in the form of a cone with circular cross section. The angle θ of the cone, see Fig. 38-21, is called the *full-angle beam* divergence. An argon laser, radiating at 514.5 nm, is aimed at the moon in a ranging experiment. If the beam has a full-angle beam divergence of 0.88 μrad, what area on the moon's surface is illuminated by the laser?

Figure 38-21 Exercise 11.

Section 38-5 Radiation Pressure

12. High-power lasers are used to compress gas plasmas by radiation pressure. The reflectivity of a plasma is unity if the electron density is high enough. A laser generating pulses of radiation of peak power 1.5×10^3 MW is focused onto 1.0 mm² of high-electron-density plasma. Find the pressure exerted on the plasma.

13. Show that the force F exerted by a laser beam of intensity I on a perfectly reflecting object of area A normal to the beam is given by $F = 2IA/c$.

14. A helium-neon laser of the type often found in physics laboratories has a beam power output of 5.0 mW at a wavelength of 633 nm. The beam is focused by a lens to a circular spot whose effective diameter may be taken to be 2.0 wavelengths. Calculate (a) the intensity of the focused beam, (b) the radiation pressure exerted on a tiny perfectly-absorbing sphere whose diameter is that of the focal spot, (c) the force exerted on this body, and (d) the acceleration imparted to it. Assume a sphere density of 5.0×10^3 kg/m³.

15. The average intensity of the solar radiation that falls normally on a surface just outside the earth's atmosphere is 1.4 kW/m². (a) What radiation pressure is exerted on this surface, assuming complete absorption? (b) How does this pressure compare with the earth's sea-level atmospheric pressure, which is 1.0×10^5 N/m²?

16. What is the radiation pressure 1.0 m away from a 500-W light bulb? Assume that the surface on which the pressure is exerted faces the bulb and is perfectly absorbing and that the bulb radiates uniformly in all directions.

Section 38-6 Polarization

17. The magnetic field equations for an electromagnetic wave in free space are $B_x = B \sin (ky + \omega t)$, $B_y = B_z = 0$. (a) What is the direction of propagation? (b) Write the electric field equations. (c) Is the wave polarized? If so, in what direction?

18. Unpolarized light falls on two polarizing sheets placed one on top of the other. What must be the angle between the characteristic directions of the sheets if the intensity of the transmitted light is (a) one-third the maximum intensity of the transmitted beam or (b) one-third the intensity of the incident beam? Assume that the polarizing sheet is ideal, that is, that it reduces the intensity of unpolarized light by exactly 50%.

19. A beam of unpolarized light of intensity 0.01 W/m² falls at normal incidence upon a polarizing sheet. (a) Find the maximum value of the electric field of the transmitted beam. (b) What is the radiation pressure exerted on the polarizing sheet?

Section 38-7 The Speed of Light

20. How accurate a clock is required to measure the speed of light by a flight time measurement in a classroom of dimensions 10 m × 20 m × 5.0 m using one mirror.

Section 38-8 The Speed of Light—Einstein's Postulate

21. It is concluded from measurements of the red shift of the emitted light that quasar Q_1 is moving away from us at a speed of $0.80c$. Quasar Q_2, which lies in the same direction in space but is closer to us, is moving away from us at speed $0.40c$. What velocity for Q_2 would be measured by an observer on Q_1?

Section 38-9 The Doppler Effect for Light

22. A galaxy is receding from us at a velocity $v = 0.50c$. Find the observed wavelength of a hydrogen spectral line whose proper wavelength is $\lambda = 656$ nm.

23. A rocketship is receding from the earth at a speed of $0.20c$. A light in the rocketship appears blue to passengers on the ship. What color would it appear to be to an observer on the earth? See Fig. 38-2.

24. The period of rotation of the sun at its equator is 24.7 d; its radius is 7.0×10^8 m. What Doppler wavelength shift is expected for a characteristic wavelength in the vicinity of 550 nm emitted from the edge of the sun's disk?

PROBLEMS GROUPED BY SECTION

Section 38-1   Maxwell and the Electromagnetic Spectrum

1. In the electromagnetic spectrum displayed in Fig. 38-1 the wavelength scale markers represent successive powers of ten and are evenly spaced. Show that the frequency markers must also be evenly spaced, with the same interval.

2. The radiation from a certain HeNe laser, although centered on 632.8 nm, has a finite "linewidth" of 0.010 nm. Calculate the linewidth in frequency units.

Section 38-2   Generating an Electromagnetic Wave

3. Figure 38-22 shows an LC oscillator connected by a transmission line to an antenna of a so-called magnetic-dipole type. Compare with Fig. 38-3, which shows a similar arrangement but with an electric-dipole type of antenna. (a) What is the basis for the names of these two antenna types? (b) Draw figures corresponding to Figs. 38-4 and 38-5 to describe the electromagnetic wave that sweeps past the observer at point P in Fig. 38-22.

Section 38-3   Traveling Waves and Maxwell's Equations

4. (a) Show that Eqs. 38-2 and 38-3 satisfy the wave equations displayed in Problem 42. (b) Show that any expressions of the form

$$E = E_m f(kx \pm \omega t)$$

and

$$B = B_m f(kx \pm \omega t),$$

where $f(kx \pm \omega t)$ denotes an arbitrary function, also satisfy these wave equations.

5. Prove that, for any point in an electromagnetic wave such as that of Fig. 38-6, the time-averaged density of the energy stored in the electric field equals that of the energy stored in the magnetic field.

Section 38-4   Energy Transport and the Poynting Vector

6. Show that in a plane traveling electromagnetic wave the average intensity, that is, the average rate of energy transport per unit area, is given by

$$\overline{S} = \frac{E_m{}^2}{2\pi_0 c}.$$

7. What must be the average intensity of a plane traveling electromagnetic wave if B_m, the maximum value of its magnetic field component, is to be 1.0×10^{-4} T ($= 1.0$ gauss)?

8. An airplane flying at a distance of 10 km from a radio transmitter receives a signal of intensity 10 μW/m². Calculate (a) the amplitude of the electric field at the airplane due to this signal; (b) the amplitude of the magnetic field at the airplane; (c) the total

power radiated by the transmitter, assuming the transmitter to radiate uniformly in all directions.

9. A HeNe laser, radiating at 632.8 nm, has a power output of 3.0 mW and a full-angle beam divergence (see Exercise 11) of 0.17 mrad. (a) What is the intensity of the beam 40 m from the laser? (b) What would be the power output of an isotropic source that provides this same intensity at the same distance?

10. The intensity of direct solar radiation that was unabsorbed by the atmosphere on a particular summer day is 100 W/m². How close would you have to stand (in interstellar space) to a 1.0-kW electric heater to feel the same intensity? Assume that the heater radiates uniformly in all directions.

11. A copper wire (diameter 2.5 mm; resistance 1.0 Ω per 300 m) carries a current of 25 A. Calculate (a) **E**; (b) **B**; and (c) **S** for a point on the surface of the wire.

Section 38-5   Radiation Pressure

12. A laser has a power output of 4.6 W and a beam diameter of 2.0 mm. If it is aimed vertically upward, what is the diameter of a perfectly reflecting sphere that can be made to "hover" by the radiation pressure exerted by the beam? Assume that the density of the sphere is 1.0 g/cm³.

13. Radiation from the sun striking the earth has an intensity of 1400 W/m². (a) Assuming that the earth behaves like a flat disk at right angles to the sun's rays and that all the incident energy is absorbed, calculate the force on the earth due to radiation pressure. (b) Compare it with the force due to the sun's gravitational attraction.

14. A small spaceship whose mass, with occupant, is 1.5×10^3 kg is drifting in outer space, where no gravitational field exists. If the astronaut turns on a 10-kW laser beam, what speed would the ship attain in one day because of the reaction force associated with the momentum carried away by the beam?

15. Show that, for complete absorption of a parallel beam of light, the radiation pressure P on the absorbing object is given by $P = S/c$, where S is the magnitude of the Poynting vector and c is the speed of light in free space. (*Hint:* Consider the conservation of momentum for a system consisting of the light plus the absorber.)

16. A particle in the solar system is under the combined influence of the sun's gravitational attraction and the radiation force due to the sun's rays. Assume that the particle is a sphere of density 1.0×10^3 kg/m³ and that all of the incident light is absorbed. (a) Show that all particles with radius less than some critical radius, R_0, will be blown out of the solar system. (b) Calculate R_0. (c) Show that R_0 does not depend on the distance from the particle to the sun.

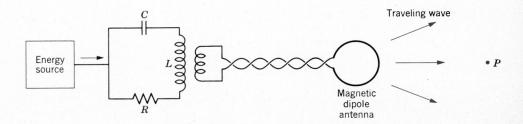

Figure 38-22   Problem 3.

Section 38–6   Polarization

17. An unpolarized beam of light is incident on a group of four polarizing sheets which are lined up so that the characteristic direction of each is rotated by 30° clockwise with respect to the preceding sheet. What fraction of the incident intensity is transmitted?

18. A beam of light is plane polarized in the vertical direction. The beam falls at normal incidence on a polarizing sheet with its polarizing direction at 70° to the vertical. The transmitted beam falls, also at normal incidence, on a second polarizing sheet with its polarizing direction horizontal. If the intensity of the original beam is 43 W/m², what is the intensity of the beam transmitted by the second sheet?

19. A beam of polarized light strikes two polarizing sheets. The characteristic direction of the second is 90° with respect to the incident light. The characteristic direction of the first is at angle θ with respect to the initial beam. Find angle θ for a transmitted beam intensity that is 0.10 times the incident beam intensity.

Section 38–7   The Speed of Light

20. Roemer's method for estimating the speed of light consisted in observing the apparent times of revolution of one of the moons of Jupiter. The true period of revolution is 42.5 h. (*a*) Taking into account the finite speed of light, how would you expect the apparent time of revolution to alter as the earth moves in its orbit from point *x* to point *y* in Fig. 38–23? (*b*) What observations would be needed to compute the speed of light? Neglect the motion of Jupiter in its orbit. Figure 38–23 is not drawn to scale.

Section 38–8   The Speed of Light—Einstein's Postulate

21. A rocket is moving at 0.90*c* to the right with respect to an observer. The rocket fires a projectile from its astrocannon at 0.80*c* to the left relative to the rocket. What is the speed of the projectile relative to the observer?

Section 38–9   The Doppler Effect for Light

22. The "red shift" of radiation from a distant nebula consists of the light H$_\gamma$, known to have a wavelength of 434 nm when observed in the laboratory, appearing to have a wavelength of 656 nm. (*a*) What is the speed of the nebula in the line of sight relative to the earth? (*b*) Is it approaching or receding?

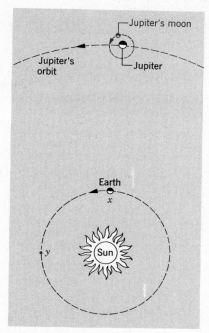

Figure 38–23   Problem 20.

23. The difference in wavelength between an incident microwave beam and one reflected from an approaching or receding car is used to determine automobile speeds on the highway. (*a*) Show that if *v* is the speed of the car and *ν* the frequency of the transmitter, the change of frequency is approximately $2v\nu/c$, where *c* is the speed of the electromagnetic radiation. (*b*) For microwaves of frequency 2450 MHz, what is the change of frequency per mi/h of speed?

24. Show that, for slow speeds, the Doppler shift can be written in the approximate form

$$\frac{\Delta\lambda}{\lambda} = \frac{u}{c},$$

where $\Delta\lambda$ is the change in wavelength.

ADDITIONAL PROBLEMS

25. (*a*) At what wavelengths does the eye of a standard observer have half its maximum sensitivity? (*b*) What are the wavelength, the frequency, and the period of the light for which the eye is the most sensitive?

26. What inductance is required with a 17 $\mu\mu$F capacitor in order to construct an oscillator capable of generating 550-nm (i.e., visible) electromagnetic waves. Comment on your answer.

27. Could you go through a red light fast enough to have it appear green? Calculate the minimum required speed. Would you rather get a ticket for speeding or for running the red light? Take $\lambda = 620$ nm for red light, $\lambda = 540$ nm for green light, and $c = 3 \times 10^8$ m/s as the speed of light.

28. Take for the speed of light the 1974 value displayed in Table 38–1, namely

$$299{,}792.4590 \pm 0.0008 \text{ km/s}.$$

A light beam traverses a baseline one mile long. (*a*) Assume that the length of the baseline has been measured so accurately that its uncertainty can be neglected. To what uncertainty in timing does the above uncertainty in the speed of light correspond? (*b*) Assume now that the time of travel over the baseline has been measured so accurately that its uncertainty can be neglected. To what uncertainty in the length of the baseline does the above uncertainty in the speed of light correspond?

29. Suppose in Problem 18 the incident beam was unpolarized. What now is the intensity of the beam transmitted by the second sheet?

30. A beam of light is a mixture of polarized light and unpolarized light. When it is sent through a Polaroid sheet, we find that the transmitted intensity can be varied by a factor of five depending on the orientation of the Polaroid. Find the relative intensities of these two components of the incident beam.

31. A plane electromagnetic wave, with wavelength 3.0 m, travels in free space in the +x-direction with its electric vector **E**, of amplitude 300 V/m, directed along the y-axis. (a) What is the frequency ν of the wave? (b) What are the direction and amplitude of the **B** field associated with the wave? (c) If $E = E_m \sin(kx - \omega t)$, what are the values of k and ω for this wave? (d) What is the time-averaged rate of energy flow per unit area associated with this wave? (e) If the wave falls upon a perfectly absorbing sheet of area 2.0 m², at what rate would momentum be delivered to the sheet and what is the radiation pressure exerted on the sheet?

32. Radiation of intensity I is incident on an object that absorbs a fraction f of it and reflects the rest. What would you expect the radiation pressure to be?

33. Frank D. Drake, an active investigator in the SETI (Search for Extra Terrestrial Intelligence) program, has said that the large radio telescope in Arecibo, Puerto Rico ". . . can detect a signal which lays down on the entire surface of the earth a power of only one picowatt." (a) What is the power actually received by the Arecibo antenna for such a signal? The antenna diameter is 1000 ft. (b) What would be the power output of a source at the center of our galaxy that could provide such a signal? The galactic center is about 3×10^4 ly away. Assume the source to be radiating uniformly in all directions.

34. In Figs. 38–5, 38–6, and 38–7 draw in lines to represent displacement currents. Pay particular attention to their directions and their spacings. (Hint: Review Section 37–3.)

35. Sunlight strikes the earth, outside its atmosphere, with an intensity of 1.4 kW/m². Calculate E_m and B_m for sunlight, assuming it to be a wave like that of Fig. 38–5.

36. Accelerated electrons from the Stanford Linear Accelerator have a kinetic energy, in a certain run, of 18 GeV. In a race (in a vacuum) with a light beam over a one-mile course which beam would win and by how many femtoseconds? (*Hint:* The electron speed can be calculated from Eqs. 8–19 and 8–20; for such energetic electrons $m_0 \ll m$ and can be neglected.)

37. The maximum electric field at a distance of 10 m from a point light source is 2.0 V/m. Calculate (a) the maximum value of the magnetic field, (b) the average intensity, and (c) the power output of the source.

38. In a plane radio wave the maximum value of the electric field component is 5.0 V/m. Calculate (a) the maximum value of the magnetic field component and (b) the wave intensity. (*Hint:* See Problem 6.)

39. Prove, for a stream of bullets striking a plane surface at normal incidence, that the "pressure" is twice the kinetic energy density in the stream above the surface; assume that the bullets are completely absorbed by the surface. Contrast this with the behavior of light (Problem 15).

40. It has been proposed that a spaceship might be propelled in

the solar system by radiation pressure, using a large sail made of foil. How large must the sail be if the radiation force is to be equal in magnitude to the sun's gravitational attraction? Assume that the mass of the ship plus sail is 1500 kg, that the sail is perfectly reflecting, and that the sail is oriented at right angles to the sun's rays. The sun's mass is 2.0×10^{30} kg.

41. A cube of edge a has its edges parallel to the x-, y-, and z-axes of a rectangular coordinate system. A uniform electric field **E** is parallel to the y-axis and a uniform magnetic field **B** is parallel to the x-axis. Calculate (a) the rate at which, according to the Poynting vector point of view, energy may be said to pass through each face of the cube and (b) the net rate at which the energy stored in the cube may be said to change.

42. Start from Eqs. 38–6 and 38–10 and show that $E(x, t)$ and $B(x,t)$, the electric and magnetic field components of a plane traveling electromagnetic wave, must satisfy the wave equations: tions":

$$\frac{\partial^2 E}{\partial t^2} = c^2 \frac{\partial^2 E}{\partial x^2}$$

and

$$\frac{\partial^2 B}{\partial t^2} = c^2 \frac{\partial^2 B}{\partial x^2}.$$

43. Prove, for a plane wave at normal incidence on a plane surface, that the radiation pressure on the surface is equal to the energy density in the beam outside the surface. This relation holds no matter what fraction of the incident energy is reflected.

44. It is desired to rotate the plane of vibration of a beam of polarized light by 90°. (a) How might this be done using only Polaroid sheets? (b) How many sheets are required in order that the total intensity loss is less than 40%?

45. A spaceship at rest in a certain reference frame S, is given a speed increment of $0.50c$. It is then given a further $0.50c$ increment in this new frame, and this process is continued until its speed with respect to the original frame is $0.999c$. How many increments are required?

46. Microwaves, which travel with the speed of light, are reflected from a distant airplane approaching the wave source. It is found that when the reflected waves are beat against the waves radiating from the source the beat frequency is 990 Hz. If the microwaves are 0.10 m in wavelength, what is the approach speed of the airplane?

47. An earth satellite, transmitting on a frequency of 40 MHz (exactly), passes directly over a radio receiving station at an altitude of 400 km and at a speed of 3.0×10^4 km/h. Plot the change in frequency, attributable to the Doppler effect, as a function of time, counting $t = 0$ as the instant the satellite is over the station. (*Hint:* The speed u in the Doppler formula is not the actual speed of the satellite but its component in the direction of the station. Use the nonrelativistic formula and neglect the curvature of the earth and of the satellite orbit.)

48. Figure 38–24 shows a cylindrical resistor of length l, radius a, and resistivity ρ carrying a current i. (a) Show that the Poynting vector **S** at the surface of the resistor is everywhere directed normal to the surface, as shown. (b) Show that the rate P at which energy flows into the resistor through its cylindrical surface, calculated by integrating the Poynting vector over this surface, is equal

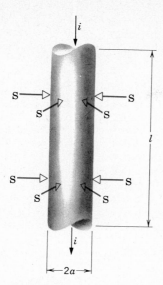

Figure 38–24   Problem 48.

to the rate at which thermal energy is produced; that is,

$$\int \mathbf{S} \cdot d\mathbf{A} = i^2 R,$$

where $d\mathbf{A}$ is an element of area of the cylindrical surface. This suggests that, according to the Poynting vector point of view, the energy that appears in a resistor as thermal energy does not enter it through the connecting wires but through the space around the wires and the resistor. (*Hint:* $\mathbf{E}$ is parallel to the axis of the cylinder, in the direction of the current; $\mathbf{B}$ forms concentric circles around the cylinder, in a direction given by the right-hand rule.)

39

Geometrical Optics

39-1 Geometrical Optics

In Chapter 38 we saw that light is an electromagnetic wave. It is a characteristic of waves of all kinds that, under most conditions, they do not travel in straight lines. We can hear sound waves around a corner. Ocean waves meeting an obstacle bend around it and meet on the other side. In a ripple tank, water waves that meet an opening in a barrier flare out as they pass through.

In certain circumstances, however, waves do travel in straight lines to a high degree of accuracy. We need only to agree not to place in the path of the wave any obstacle or aperture (mirror, lens, slit, baffle, etc.) unless its dimensions are much larger than the wavelength. Also, we must agree not to look too closely into what goes on at the edges of the obstacle or aperture. This special case, which we call *geometrical optics,* is the subject of this chapter. We will represent the straight lines in which light is traveling by *rays* which are lines perpendicular to the wave fronts. We can, in fact, ignore the wave nature of light completely and concentrate only on the rays. In later chapters we will remove these restrictions on the dimensions of obstacles and deal with the more general case of *wave optics.*

Geometrical optics

39-2 Reflection and Refraction— Plane Waves and Plane Surfaces

In Fig. 39–1a a plane light wave falls on a plane glass surface. The light beam is both reflected from the surface and bent (that is, *refracted*) as it enters the glass. We represent the incident beam in Fig. 39–1b by a line, the *incident ray,* parallel to the direction of propagation. We assume the incident wave in Fig. 39–1b to be a plane wave, the wavefronts being normal to the incident ray. The reflected and refracted waves are also represented by rays. The angles of *incidence* (θ_1), of *reflection* (θ_1'), and of *refraction* (θ_2) are measured between the normal to the surface and the appropriate ray, as shown in the figure.

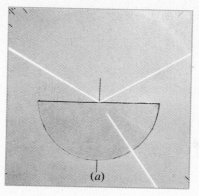

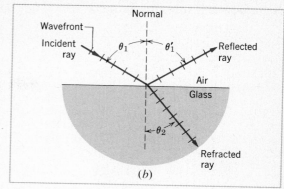

Figure 39–1 (a) A photograph showing reflection and refraction of plane waves at a (plane) air-glass interface. (b) A representation using rays. From *PSSC Physics*, 2nd ed., © 1965, D. C. Heath and Co., Lexington, Ma., with Education Development Center, Newton, Ma.

By experiment we find the laws governing reflection and refraction:

1. The reflected and the refracted rays lie in the plane (the *plane of incidence*) formed by the incident ray and the normal to the surface at the point of incidence, that is, the plane of Fig. 39–1b.

2. For reflection:

Law of reflection

$$\theta_1' = \theta_1. \tag{39–1}$$

3. For refraction:

Law of refraction (Snell's law)

$$\frac{\sin \theta_1}{\sin \theta_2} = n_{21}, \tag{39–2}$$

Index of refraction

where n_{21} is a constant called the *index of refraction* of medium 2 with respect to medium 1. Table 39–1 shows the indices of refraction for some common substances with respect to a vacuum for a particular wavelength.

Table 39–1 Some indices of refraction[a]

Medium	Index	Medium	Index
Air (STP)	1.00029	Typical crown glass	1.52
Water (20°C)	1.33	Sodium chloride	1.54
Sodium fluoride	1.33	Polystyrene	1.55
Acetone	1.36	Carbon disulfide	1.63
Ethyl alcohol	1.36	Heavy flint glass	1.65
Sugar solution (30%)	1.38	Sapphire	1.77
Fused quartz	1.46	Heaviest flint glass	1.89
Sugar solution (80%)	1.49	Diamond	2.42

[a] For λ = 589 nm (yellow sodium light) with respect to a vacuum.

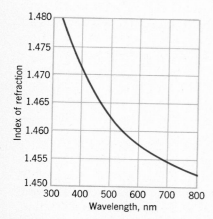

Figure 39–2 The index of refraction of fused quartz with respect to air at standard temperature and pressure.

The index of refraction of one medium with respect to another generally varies with wavelength, as Fig. 39–2 shows. Because of this fact refraction, unlike reflection, can be used to analyze light into its component wavelengths.

The law of reflection was known to Euclid. The law of refraction, usually called *Snell's law*, was discovered by Willebrod Snell (1591–1626) and deduced from an early corpuscular theory of light by René Descartes (1596–1650). The laws of reflection and refraction can be derived from Maxwell's equations. They hold for all regions of the electromagnetic spectrum.

Example 1   An incident ray from a mercury arc source falls, in air, on the plane surface of a block of quartz and makes an angle of 30.00° with the normal. This ray contains two wavelengths, 405 and 509 nm. The indices of refraction for quartz with respect to air (n_{qa}) at these wavelengths are 1.470 and 1.4629, respectively; see Fig. 39–2. What is the angle between the two refracted rays?

From Eq. 39–2 we have, for the 405-nm ray,

$$\sin \theta_1 = n_{qa} \sin \theta_2,$$

or

$$\sin 30.00° = (1.470) \sin \theta_2,$$

which leads to

$$\theta_2 = 19.89°.$$

For the 509-nm ray we have

$$\sin 30.00° = (1.463) \sin \theta_2',$$

or

$$\theta_2' = 19.98°.$$

The angle $\Delta\theta$ between the rays is 0.09°, the shorter wavelength component being bent through the larger angle, that is, having the smaller angle of refraction.

Example 2   An incident ray falls on one face of a glass prism in air as in Fig. 39–3. The angle θ is so chosen that the emerging ray also makes an angle θ with the normal to the other face. Derive an expression for the index of refraction of the prism material with respect to air.

Note that $\angle abc = \alpha$, the two angles having their sides mutually perpendicular. Therefore

$$\alpha = \tfrac{1}{2}\phi. \qquad (39-3)$$

The deviation angle ψ is the sum of the two opposite interior angles in triangle *aed*, or

$$\psi = 2(\theta - \alpha).$$

Substituting $\tfrac{1}{2}\phi$ for α and solving for θ yields

$$\theta = \tfrac{1}{2}(\psi + \phi). \qquad (39-4)$$

At point a, θ is the angle of incidence and α the angle of refraction. The law of refraction (see Eq. 39–2) is

$$\sin \theta = n_{ga} \sin \alpha,$$

in which n_{ga} is the index of refraction of the glass with respect to air.

From Eqs. 39–3 and 39–4 this yields

$$\sin \frac{\psi + \phi}{2} = n_{ga} \sin \frac{\phi}{2}$$

or

$$n_{ga} = \frac{\sin \tfrac{1}{2}(\psi + \phi)}{\sin (\phi/2)}, \qquad (39-5)$$

which is the desired relation. This equation holds only for θ so chosen that the light ray passes symmetrically through the prism. Under these conditions the deviation angle ψ has a minimum value (with respect to rotation of the prism) and is called the *angle of minimum deviation*.

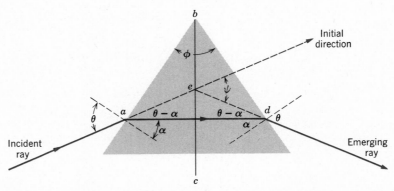

Figure 39–3   Example 2.

39–3   Huygens' Principle

The Dutch physicist Christian Huygens in 1678 proposed a wave theory for light, which, among other things, accounted for the laws of reflection and refraction. While much less comprehensive than the later theory of Maxwell, it was simpler mathematically and was useful for many years. It remains useful today for pedagogic and certain other practical purposes.

Huygens' theory simply assumes that light is a wave rather than, say, a stream of particles. It says nothing about the nature of the wave and, in particular—since Maxwell's theory of electromagnetism appeared only after the lapse of a century—gives no hint of the electromagnetic character of light. Huygens did not know whether light was a transverse wave or a longitudinal one; nor did he know the wavelengths of visible light.

The theory is based on a geometrical construction, called *Huygens' principle*, that allows us to tell where a given wavefront will be at any time in the future if we know its present position. Huygens' principle is:

All points on a wave front can be considered as point sources for the production of spherical secondary wavelets. After a time t the new position of the wave front will be the surface of tangency to these secondary wavelets.

We illustrate Huygens' principle by a trivial example: Given a wavefront, *ab* in Fig. 39-4, in a plane wave in free space, where will the wavefront be a time *t* later? Following Huygens' principle, we let several points on this plane (see dots) serve as centers for secondary spherical wavelets. In a time *t* the radius of these spherical waves is *ct*, where *c* is the speed of light in free space. We represent the plane of tangency to these spheres at time *t* by *de*. As we expect, it is parallel to plane *ab* and a perpendicular distance *ct* from it. Thus plane wavefronts are propagated as planes and with speed *c*. Note that the Huygens method involves a three-dimensional construction and that Fig. 39-4 is the intersection of this construction with the plane of the page.

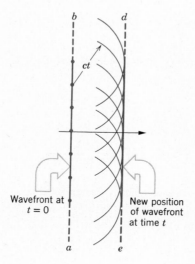

Figure 39-4 The propagation of a plane wave in free space is described by the Huygens construction. Note that the ray (horizontal arrow) representing the wave is perpendicular to the wavefronts.

39-4 The Law of Refraction

We shall now use Huygens' principle to derive the law of refraction, Eq. 39-2. Figure 39-5 shows four stages in the refraction of three wavefronts in a plane wave falling on a plane interface between air (medium 1) and glass (medium 2). We choose the wavefronts in the incident beam to be separated, arbitrarily, by λ_1, the wavelength in medium 1. Let the speed of light in air be v_1 and that in glass be v_2. We assume that $v_2 < v_1$, which happens to be true.

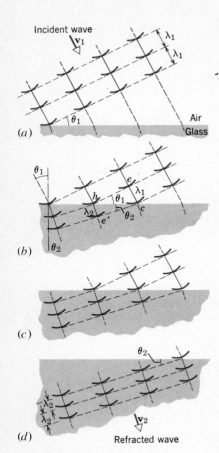

Incident wave

$\mathbf{v_1}$

λ_1

λ_1

θ_1

Air

Glass

(a)

θ_1

e

h

θ_1

λ_1

λ_2

e'

θ_2

c

(b)

θ_2

(c)

θ_2

λ_2

λ_2

$\mathbf{v_2}$

(d)

Refracted wave

Figure 39–5 The refraction of a plane wave as described by the Huygens construction; for simplicity we do not show the reflected wave. Note the change in wavelength on refraction.

Snell's law derived from Huygen's principle

Optical path length

The wavefronts in Fig. 39–5a are related in the same way as those in Fig. 39–4 which show the Huygens construction. The angle of incidence is θ_1. The time $(= \lambda_1/v_1)$ for a Huygens wavelet to expand from point e to include point c will equal the time $(= \lambda_2/v_2)$ for a wavelet in the glass to expand at the reduced speed v_2 from h to include e'. By equating these times we may obtain wavelength λ_2,

$$\lambda_2 = \lambda_1 \frac{v_2}{v_1} \qquad (39\text{--}6)$$

which, you can see, is less than that in air.

The refracted wavefront must be tangent to an arc of radius λ_2 centered on h. Since c lies on the new wavefront the tangent must pass through this point also. Note that θ_2, the angle between the refracted wavefront and the air-glass interface, is the same as the angle of refraction (that between the refracted ray and the normal to this interface).

For the right triangles hce and hce' we may write

$$\sin \theta_1 = \frac{\lambda_1}{hc} \qquad \text{(for } hce\text{)}$$

and

$$\sin \theta_2 = \frac{\lambda_2}{hc} \qquad \text{(for } hce'\text{)}.$$

Dividing and using Eq. 39–6 yields

$$\frac{\sin \theta_1}{\sin \theta_2} = \frac{\lambda_1}{\lambda_2} = \frac{v_1}{v_2}. \qquad (39\text{--}7)$$

Since v_1 and v_2 are constants, Eq. 39–7 is just the law of refraction (Eq. 39–2) provided we put

$$n_{21} = \frac{v_1}{v_2}. \qquad (39\text{--}8)$$

We now define the index of refraction n of a medium with respect to a vacuum, where v equals c. From Eq. 39–8 we have $n = c/v$. From Eq. 39–6 we may then express the wavelength in the medium as

$$\lambda_n = \frac{\lambda}{n} \qquad (39\text{--}9)$$

where λ is the wavelength in vacuum.

Multiplying the right side of Eq. 39–8 by c/c gives the relative index of refraction between any two media as $n_{21} = n_2/n_1$, and the law of refraction becomes

$$n_1 \sin \theta_1 = n_2 \sin \theta_2. \qquad (39\text{--}10)$$

Referring to Fig. 39–5b, we saw how two rays moving different distances, λ_1 in medium 1 and λ_2 in medium 2, nevertheless remained in phase; in other words the same number of wavelengths (in this case one) were contained in these distances. From Eq. 39–9 we can see that $n_1\lambda_1 = n_2\lambda_2 = \lambda$. This suggests that we define a quantity nl to be the *optical pathlength* of radiation traveling distance l in a medium of index of refraction n. This quantity, in effect, measures the number of wavelengths contained in distance l and is a useful quantity wherever phase differences between rays traveling in different media must be considered. The number of wavelengths contained in distance l is l/λ_n which, using Eq. 39–9 equals nl/λ. Thus, since λ is the fixed free-space wavelength, a condition that would insure that the same number of wavelengths are contained within distances l_1 and l_2 in two different media is

$$n_1 l_1 = n_2 l_2.$$ (39–11)

Equality of the optical pathlengths thus implies no relative change of phase.

39–5 Total Internal Reflection

Total internal reflection

Let rays in an optically dense medium (glass) fall on a plane surface on the other side of which is a less optically dense medium (air); see Fig. 39–6. As the angle of incidence θ is increased, we reach a situation (see ray e) at which the refracted ray points along the surface, the angle of refraction being 90°. For angles of incidence larger than this *critical angle* θ_c there is no refracted ray, and we speak of *total internal reflection*.

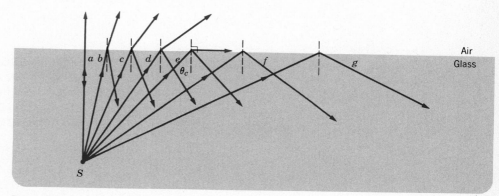

Figure 39–6 Showing the total internal reflection of light from a source S; the critical angle is θ_c.

We find the critical angle by putting $\theta_2 = 90°$ in the law of refraction (see Eq. 39–10):

$$n_1 \sin \theta_c = n_2 \sin 90°,$$

or

The critical angle for total internal reflection.

$$\sin \theta_c = \frac{n_2}{n_1}.$$ (39–12)

For glass with $n = 1.50$ and air, $\sin \theta_c = (1.00/1.50) = 0.667$, which yields $\theta_c = 41.8°$. Total internal reflection can not occur when light originates in the medium of lower index of refraction.

Example 3 Figure 39–7 shows a triangular prism of glass, a ray incident normal to one face being totally reflected. If θ_1 is 45°, what can you say about the index of refraction n of the glass?

The angle θ_1 must be equal to or greater than the critical angle θ_c where θ_c is given by Eq. 39–12:

$$\sin \theta_c = \frac{n_2}{n_1} = \frac{1}{n},$$

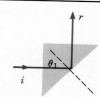

Figure 39–7 Example 3.

in which, for all practical purposes, the index of refraction of air ($= n_2$) is set equal to unity. Suppose that the index of refraction of the glass is such that total internal reflection just occurs, that is, that $\theta_c = 45°$. This would mean

$$n = \frac{1}{\sin 45°} = 1.41.$$

Thus the index of refraction of the glass must be equal to or larger than 1.41. If it were less, total internal reflection would not occur.

39–6 Brewster's Law

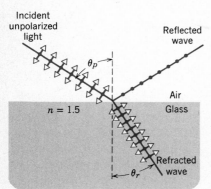

Figure 39–8 shows an unpolarized beam falling on a glass surface. The **E**-vector for each wavetrain in the beam can be resolved into two components, one perpendicular to the plane of incidence (the plane of Fig. 39–8), represented by dots, and one lying in this plane, represented by arrows. We shall call them the *perpendicular* and *parallel components*, respectively. For completely unpolarized incident light, these two components are of equal amplitude.

• Perpendicular component
⟷ Parallel component

Figure 39–8 For a particular angle of incidence θ_p, the reflected light is completely polarized, as shown. The transmitted light is partially polarized.

The laws of reflection and refraction give us information about the direction of reflected or refracted rays. They don't say anything about the intensities of these rays, except for the fact that the refracted beam intensity is zero for light striking at an angle equal to or greater than the critical angle. The complete intensity relationships can be derived from Maxwell's equations. The derivation is somewhat difficult; however, from it you will find that the reflection coefficient (the ratio of the reflected intensity to the incident intensity) not only depends upon the angle of incidence but also upon the direction of polarization of the incident beam. Similarly for the refracted beam.

Malus discovered in 1809 that light can be partially or completely polarized by reflection. Anyone who has watched the sun's reflection in water while wearing a pair of sunglasses made with polarizing lenses has probably noticed the effect. You need only to tilt your head from side to side, thus rotating the polarizers, to see that the intensity of the reflected sunlight passes through a minimum.

Experimentally, for glass or other dielectric materials, there is a particular angle of incidence, called the polarizing angle θ_p, at which the reflection coefficient for the parallel component is zero. This means that the beam reflected from the glass is polarized, with its plane of vibration at right angles to the plane of incidence. This polarization of the reflected beam can easily be verified by analyzing it with a polarizing sheet. The refracted beam at this incident angle still has both components, although they are not of equal amplitude.

Polarization by reflection

At the polarizing angle we find experimentally that the reflected and the refracted beams are at right angles, or (Fig. 39–8)

$$\theta_p + \theta_r = 90°.$$

From the law of refraction,

$$n_1 \sin \theta_p = n_2 \sin \theta_r.$$

Combining these equations leads to

$$n_1 \sin \theta_p = n_2 \sin (90° - \theta_p) = n_2 \cos \theta_p$$

or

$$\tan \theta_p = \frac{n_2}{n_1}, \qquad (39\text{–}13)$$

where the incident ray is in medium one and the refracted ray in medium two. We can write this as

$$\tan \theta_p = n, \qquad (39–14)$$

where $n \, (= n_2/n_1)$ is the index of refraction of medium two with respect to medium one.* Equation 39–14 is known as Brewster's law after Sir David Brewster, who deduced it empirically in 1812.

Example 4 We wish to use a plate of glass ($n = 1.50$) as a polarizer. What is the polarizing angle? What is the angle of refraction?

From Eq. 39–14

$$\theta_p = \tan^{-1} 1.50 = 56.3°.$$

The angle of refraction follows from Snell's law:

or

(1) $\sin \theta_p = n \sin \theta_r$

$$\sin \theta_r = \frac{\sin 56.3°}{1.50} = 0.555; \qquad \theta_r = 33.7°(= 90° - \theta_p).$$

39–7 Spherical Waves—Plane Mirror

In Section 39–2 we considered a special case of the reflection and refraction of light waves; that of plane waves falling upon plane surfaces. We will now proceed in stages to more complex situations. The next topic is spherical waves falling upon a plane mirror.

Figure 39–9 shows a point source of light O, the *object*, placed at a distance o in front of a plane mirror. The light falls on the mirror as a spherical wave represented in the figure by rays emanating from O. At the point in Fig. 39–9 at which each ray from O strikes the mirror we construct a reflected ray. If we extend the

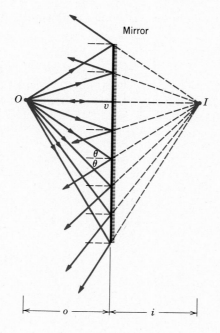

Figure 39–9 A point object O forms a virtual image I in a plane mirror. The rays appear to emanate from I, but actually light does not pass through this point.

---

* We will often drop the subscript on n_{21} if it is clear in which medium the incident ray travels.

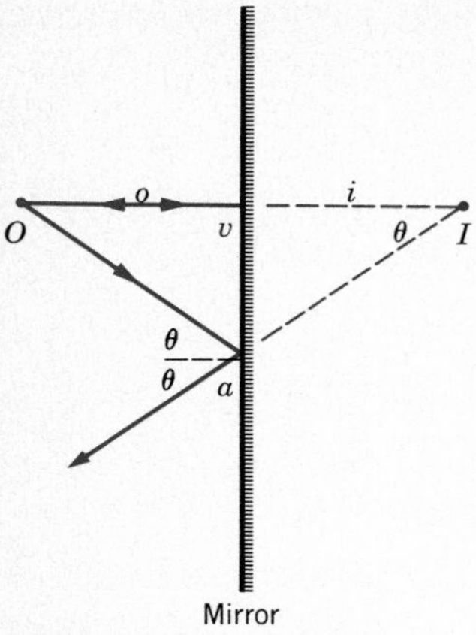

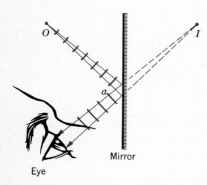

Figure 39–10 Two rays from Fig. 39–9; ray Oa makes an arbitrary angle θ with the normal.

reflected rays backward, they intersect in a point I which is the same distance behind the mirror that the object O is in front of it; I is called the *image* of O.

Images may be *real* or *virtual*. In a real image light actually passes through the image point; in a virtual image the light behaves as though it diverges from the image point even though it does not pass through this point; see Fig. 39–9. Images in plane mirrors are always virtual. We know from daily experience how "real" such a virtual image appears to be and how definite is its location in the space behind the mirror, even though this space may, in fact, be occupied by a brick wall.

Figure 39–10 shows two rays from Fig. 39–9. One strikes the mirror at v, along a perpendicular line. The other strikes it at an arbitrary point a, making an angle of incidence θ with the normal at that point. Elementary geometry shows that the right triangles $aOva$ and $aIva$ are congruent and thus

$$i = -o, \qquad (39-15)$$

in which we arbitrarily introduced the minus sign to signal that the image is virtual, a point we will expand upon below. Equation 39–15 does not involve θ, which means that all rays striking the mirror pass through I when extended backward, as we have claimed above. Beyond assuming that the mirror is truly plane and that the conditions for geometrical optics hold, we have made no approximations in deriving Eq. 39–15. A point object produces a point image in a plane mirror, with $i = -o$, no matter how large the angle θ in Fig. 39–10.

Because of the finite diameter of the pupil of the eye, only rays that lie fairly close together can enter the eye after reflection at a mirror. For the eye position shown in Fig. 39–11 only a small patch of the mirror near point a is used in forming the image. If we move our eye to another location, a different patch of the mirror will be used; the location of the virtual image I will remain unchanged, however, as long as the object remains fixed.

If the object is an extended source such as the head of a person, a virtual image is also formed. From Eq. 39–15, every point of the source has an image point that lies an equal distance directly behind the plane of the mirror. Thus the image reproduces the object point by point. Most of us prove this every day by looking in a mirror.

Images in plane mirrors differ from objects in that left and right are interchanged. The image of a printed page is different from the page itself. Similarly, if a top is made to spin clockwise, the image, viewed in a mirror, will seem to spin counterclockwise. Figure 39–12 shows an image of a left hand, constructed by using point-by-point application of Eq. 39–15; the image has the symmetry of a right hand.

Figure 39–11 A pencil of rays from O enters the eye after reflection at the mirror. Only a small portion of the mirror near a is effective. The small arcs represent portions of spherical wavefronts. The eye "thinks" that the point light source is at I.

Reversal of left-right symmetry due to reflection

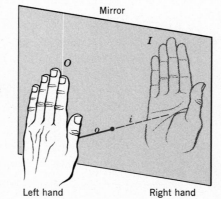

Figure 39–12 A plane mirror reverses right and left. The object O is a left hand; the image I is a right hand. Try it in a mirror.

Example 5   How tall must a vertical mirror be if a person 6.0 ft high is to be able to see his entire length? Assume that his eyes are 4.0 in. below the top of his head.

Figure 39–13 shows the paths followed by two of the many light rays leaving the top of the person's head and the tips of his toes. These rays, chosen so that they will enter the eye *e* after reflection, strike the vertical mirror at points *a* and *b*, respectively. The mirror need occupy only the region between these two points. Calculation shows that *b* is 2 ft, 10 in. and *a* is 5 ft, 10 in. above the floor. The length of the mirror is thus 3.0 ft, or half the height of the person. Note that this height is independent of the distance between the person and the mirror. Mirrors that extend below point *b* only add reflections of the floor between the person and the mirror.

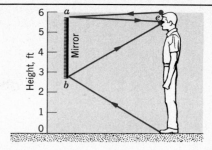

Figure 39–13   Example 5. A person can view his full-length image in a mirror that is only half his height.

39–8   Spherical Waves—Spherical Mirror

In Fig. 39–14 a spherical light wave from a point object *O* falls on a concave spherical mirror whose radius of curvature is *r*. (We will always judge whether a spherical mirror is concave ["caved in"] or convex by sighting along an incident ray.) A line through *O* and the center of curvature *C* makes a convenient reference axis.

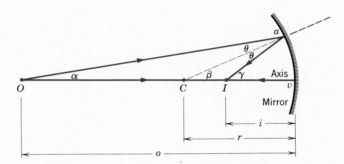

Figure 39–14   Two rays from *O* converge after reflection in a spherical concave mirror, forming a real image at *I*.

A ray from *O* that makes an arbitrary angle α with this axis intersects the axis at *I* after reflection from the mirror at *a*. A ray that leaves *O* along the axis will be reflected back along itself at *v* and will also pass through *I*. Thus, for these two rays at least, *I* is the image of *O*; it is a real image because light actually passes through *I*. Let us find the location of *I*.

A useful theorem is that the exterior angle of a triangle is equal to the sum of the two opposite interior angles. Applying this to triangles *OaCO* and *OaIO* in Fig. 39–14 yields

$$\beta = \alpha + \theta$$

and

$$\gamma = \alpha + 2\theta.$$

Eliminating θ between these equations leads to

$$\alpha + \gamma = 2\beta. \tag{39-16}$$

In radian measure we can write angles α, β, and γ as

$$\alpha \cong \frac{av}{vO} = \frac{av}{o}$$

$$\beta = \frac{av}{vC} = \frac{av}{r} \tag{39-17}$$

$$\gamma \cong \frac{av}{vI} = \frac{av}{i}.$$

Note that only the equation for β is exact, because the center of curvature of arc av is at C and not at O or I. However, the equations for α and γ are approximately correct if these angles are small enough. In all that follows we assume that the rays diverging from the object make only a small angle α with the axis of the mirror. We did not find it necessary to make such an assumption for plane mirrors. Substituting these equations into Eq. 39–16 and canceling av yields

The spherical mirror equation

$$\frac{1}{o} + \frac{1}{i} = \frac{2}{r}, \tag{39-18}$$

in which o is the object distance and i is the image distance. Both these distances are measured from the *vertex* of the mirror, which is the point v at which the axis intercepts the mirror.

Significantly, Eq. 39–18 does not contain α (or β, γ, or θ), so that it holds for all rays that strike the mirror provided that they make small enough angles with the axis. In an actual case we can insure this by putting a small enough circular diaphragm in front of the mirror, centered about the vertex v; this will impose a certain maximum value of α. As we let α in Fig. 39–14 become larger, it will become less true that a point object will form a point image; the image will become extended and fuzzy.

Although we derived Eq. 39–18 for the special case in which the object is located beyond the center of curvature of the mirror, it also holds if the object is located between the center of curvature and the mirror surface. It is also true for convex mirrors, as in Fig. 39–15. Finally, it remains true even if the incident light is caused to converge toward a (virtual) object, as in Fig. 39–16, rather than to diverge from a (real) object, as in Figs. 39–14 and 39–15.

In applying Eq. 39–18 to such general cases we must follow a consistent convention for the signs of o, i, and r. To develop such a convention we fix our minds

Definitions of *R*-side and *V*-side

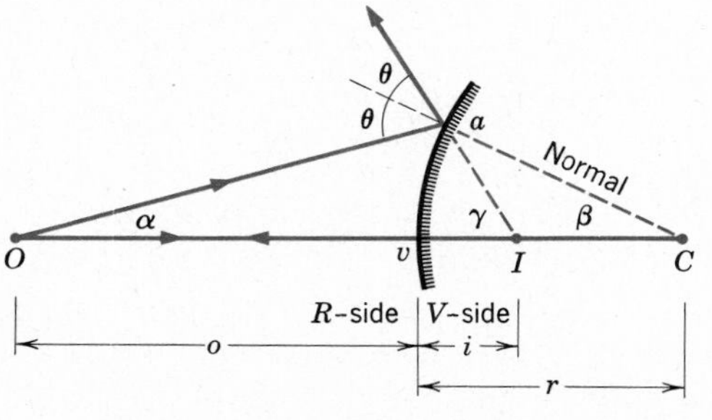

on the notion that if a real image is to be formed after reflection in a mirror it must lie on the side of the mirror from which the light comes; we call this the *R*-side (for real). We call the back of the mirror the *V*-side (for virtual) because images formed here must be virtual, no light actually being present.

Figure 39–15 Two rays from O diverge after reflection in a spherical convex mirror, forming a virtual image at I, the point from which they appear to originate.

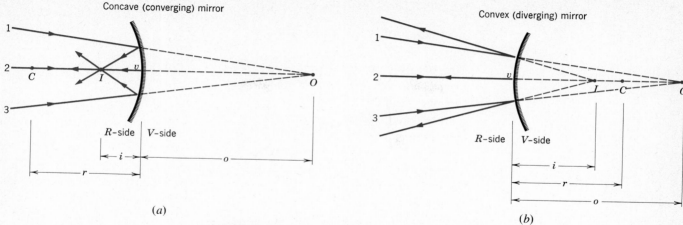

Figure 39-16 (a) Converging rays (see 1, 2, and 3) fall on a concave mirror. A virtual object O produces a real image. (b) Converging rays (see 1, 2, and 3) fall on a convex mirror. A virtual object O produces a virtual image I.

The conventions for the signs of o, i, and r are:

Rule 1. The quantities o and i are positive for real objects and images; they are negative for virtual objects and images.

The sign conventions

Rule 2. The quantity r is positive if the center of curvature C lies on the R-side; it is negative if it lies on the V-side.

Thus we see that in Fig. 39–14, o, i, and r are all positive. In Fig. 39–15 o is positive but i and r are negative. In Fig. 39–16a o is negative but i and r are positive. Finally, in Fig. 39–16b o, i, and r are all negative.

Example 6 A convex mirror has a radius of curvature of 20 cm. If a point source is placed 14 cm away from the mirror, as in Fig. 39–15, where is the image?

A rough graphical construction, applying the law of reflection at a in the figure, shows that the image will be on the V-side of the mirror and thus will be virtual. We may verify this from Eq. 39–18, noting that r is negative here because the center of curvature of the mirror is on its V-side, as it is for all convex mirrors. We have

$$\frac{1}{o} + \frac{1}{i} = \frac{2}{r}$$

or

$$\frac{1}{+14 \text{ cm}} + \frac{1}{i} = \frac{2}{-20 \text{ cm}},$$

which yields $i = -5.8$ cm, in agreement with the graphical prediction. The negative sign for i reminds us that the image is on the V-side of the mirror and thus is virtual.

Focal length for a mirror

When parallel light falls on a mirror (Fig. 39–17), the image point (real or virtual) is called the *focal point F* of the mirror. The focal length f is the distance between F and the vertex. If we put $o \to \infty$ in Eq. 39–18, thus insuring parallel incident light, we have

$$i = \tfrac{1}{2}r = f.$$

Equation 39–18 can then be rewritten

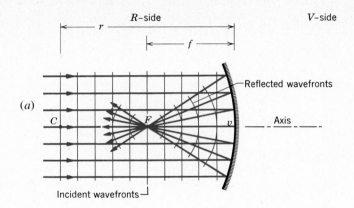

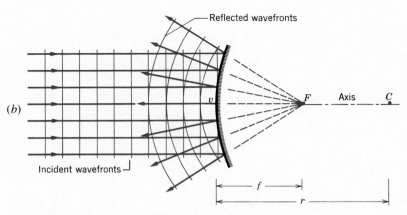

Figure 39–17 (*a*) The focal point for a concave spherical mirror, showing both the rays and the wavefronts. *F* and *C* lie on the *R*-side, the focal point is real, and the focal length *f* of the mirror is positive (as is *r*). (*b*) Same as (*a*) except that the mirror is convex; *F* and *C* lie on the *V*-side of the mirror. The focal point is virtual and the focal length *f* is negative (as is *r*).

The mirror equation in terms of focal length

$$\frac{1}{o} + \frac{1}{i} = \frac{1}{f}. \tag{39–19}$$

The sign convention for *f* is:

Rule 3. The quantity *f* is positive if incident parallel rays are brought to a real focus; it is negative if they are brought to a virtual focus.

Notice that all three of our sign convention rules associate the word ''positive'' with the word ''real'' and the word ''negative'' with the word ''virtual.''

We now consider objects that are not points. Figure 39–18 shows a luminous arrow in front of (*a*) a concave mirror and (*b*) a convex mirror. We choose to draw the mirror axis through the foot of the luminous arrow and, of course, through the center of curvature. We can find the image of any off-axis point, such as the tip of the luminous arrow, graphically by using the following facts:

Geometrical construction of mirror images

(*a*) A ray that strikes the mirror after passing (either directly or upon being extended) through the center of curvature *C* returns along itself (ray *x* in Fig. 39–18). Such rays strike the mirror at right angles.

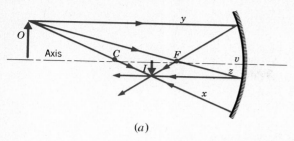

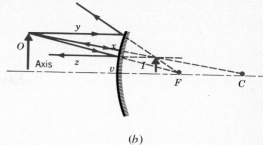

(a)

(b)

Figure 39–18   The image of an extended object in (a) a concave mirror and (b) a convex mirror is located graphically. Any two of the three special rays shown are sufficient. In (a) the image is *inverted;* in (b) it is *erect.*

(b)   A ray that strikes the mirror parallel to its axis passes (or will pass when extended) through the focal point (ray y).

(c)   A ray that strikes the mirror after passing (either directly or upon being extended) through the focal point emerges parallel to the axis (ray z).

Figure 39–19 shows a ray (dve) that originates on the tip of the object arrow of Fig. 39–18a, is reflected from the mirror at point v, and passes through the tip

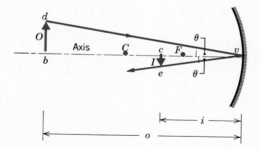

Figure 39–19   A particular ray for the arrangement of Fig. 39–18, used to illustrate lateral magnification.

of the image arrow. The law of reflection demands that this ray make equal angles θ with the mirror axis as shown. For the two similar right triangles in the figure we can write

$$\frac{ce}{bd} = \frac{vc}{vb}.$$

The quantity on the left (apart from a question of sign) is the lateral magnification m of the mirror. Since we want to represent an inverted image by a negative magnification, we arbitrarily define m for this case as $-(ce/bd)$. Since $vc = i$ and $vb = o$, we have at once

Lateral magnification for a mirror

$$m = -\frac{i}{o}. \tag{39–20}$$

This equation gives the magnification for spherical and plane mirrors under all circumstances. For a plane mirror, $o = -i$ and the predicted magnification is $+1$ which, in agreement with experience, indicates an erect image the same size as the object.

39–9 Spherical Waves— Spherical Refracting Surface

Figure 39–20 shows a point source O near a convex spherical refracting surface of radius of curvature r. The surface separates two media whose indices of refraction differ, that of the medium in which the incident light falls on the surface being n_1 and that on the other side of the surface being n_2.

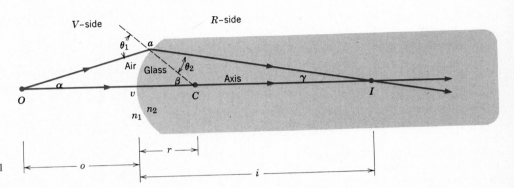

Figure 39–20 Two rays from O converge after refraction at a spherical surface, forming a real image at I.

From O we draw a line through the center of curvature C of the refracting surface, thus establishing a convenient axis which intercepts the surface at vertex v. From O we also draw a ray that makes a small but arbitrary angle α with the axis and strikes the refracting surface at a, being refracted according to

$$n_1 \sin \theta_1 = n_2 \sin \theta_2.$$

The refracted ray intersects the axis at I. A ray from O that travels along the axis will not be bent on entering the surface and will also pass through I. Thus, for these two rays at least, I is the image of O.

As in the derivation of the mirror equation, we use the theorem that the exterior angle of a triangle is equal to the sum of the two opposite interior angles. Applying this to triangles $COaC$ and $ICaI$ yields

$$\theta_1 = \alpha + \beta \tag{39–21}$$

and

$$\beta = \theta_2 + \gamma. \tag{39–22}$$

As α is made small, angles β, γ, θ_1, and θ_2 in Fig. 39–20 also become small. We at once assume that α, hence all these angles, are arbitrarily small. We also made this assumption for spherical mirrors. Replacing the sines of the angles by the angles themselves—since the angles are required to be small—permits us to write the law of refraction as

$$n_1 \theta_1 \cong n_2 \theta_2. \tag{39–23}$$

Combining Eqs. 39–22 and 39–23 leads to

$$\beta = \frac{n_1}{n_2} \theta_1 + \gamma.$$

Eliminating θ_1 between this equation and Eq. 39–21 leads, after rearrangement, to

$$n_1 \alpha + n_2 \gamma = (n_2 - n_1)\beta. \tag{39–24}$$

In radian measure the angles α, β, and γ in Fig. 39–20 are

$$\alpha \cong \frac{av}{o}; \ \beta = \frac{av}{r}; \ \gamma \cong \frac{av}{i}. \tag{39–25}$$

Only the second of these equations is exact. The other two are approximate because I and O are not the centers of circles of which av is an arc. However, for α small enough we can make the inaccuracies in Eqs. 39–25 as small as we wish.

Substituting Eqs. 39–25 into Eq. 39–24 leads readily to

$$\frac{n_1}{o} + \frac{n_2}{i} = \frac{n_2 - n_1}{r}. \tag{39–26}$$

This equation holds whenever light is refracted at a spherical surface, assuming only that the incident rays make a small enough angle α with the axis. In particular, Eq. 39–26 must hold whether the refracting surface is convex (as in Fig. 39–20), plane (which means $r \to \infty$), or concave (as in Fig. 39–21). It also holds whether $n_2 > n_1$ (as in Figs. 39–20 and 39–21) or $n_2 < n_1$ (as in Fig. 39–23).

Refraction at a single spherical surface

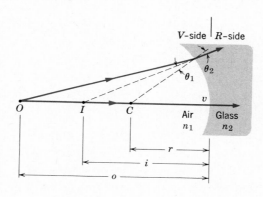

Figure 39–21 Two rays from real object O diverge after refraction at a spherical surface, forming a virtual image at I.

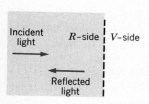

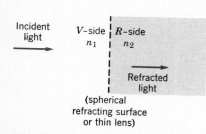

Figure 39–22 Real images are formed on the same side as the incident light for mirrors but on the opposite side for refracting surfaces and lenses.

Before stating the sign conventions for o, i, and r in Eq. 39–26, let us fix our attention on the side of the refracting surface from which the incident light falls on the surface. In contrast to mirrors, the light passes *through* a refracting surface to the other side, and if an image is formed on the far side, which we call the R-side, it must be a real image. The side from which the incident light comes is called the V-side because images formed here must be virtual. Figure 39–22 suggests this important distinction between reflection and refraction.

With the distinctions of Fig. 39–22 in mind we assert that the sign conventions for spherical refracting surfaces are the same as those for spherical mirrors.

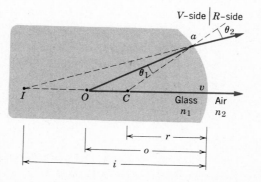

Figure 39–23 Two rays from O appear to originate from I (virtual image) after refraction at a spherical surface.

The three rules stated above hold without change. By applying them we see that in Fig. 39–20, o, i, and r are all positive. In Figs. 39–21 and 39–23, however, o is positive but i and r are negative.

Example 7   Locate the image for the geometry shown in Fig. 39–20, assuming the radius of curvature to be 10 cm, n_2 to be 2.0, and n_1 to be 1.0. Let the object be 20 cm to the left of v.

From Eq. 39–26,

$$\frac{n_1}{o} + \frac{n_2}{i} = \frac{n_2 - n_1}{r},$$

we have

$$\frac{1.0}{+20 \text{ cm}} + \frac{2.0}{i} = \frac{2.0 - 1.0}{+10 \text{ cm}}.$$

Note that r is positive because the center of curvature of the surface lies on the R-side. This relation yields $i = +40$ cm in agreement with the graphical construction. The light energy actually passes through I so that the image is real, as indicated by the positive sign for i.

Example 8   An object is immersed in a medium with $n_1 = 2.0$, being 15 cm from the spherical surface whose radius of curvature is -10 cm, as in Fig. 39–23; r is negative because C lies on the V-side. Locate the image.

Figure 39–23 shows a ray traced through the surface by applying the law of refraction at point a. A second ray from O along the axis emerges undeflected at v. The image I is found by extending these two rays backward; it is virtual.

From Eq. 39–26,

$$\frac{n_1}{o} + \frac{n_2}{i} = \frac{n_2 - n_1}{r},$$

we have

$$\frac{2.0}{+15 \text{ cm}} + \frac{1.0}{i} = \frac{1.0 - 2.0}{-10 \text{ cm}},$$

which yields $i = -30$ cm, in agreement with Fig. 39–23 and with the sign conventions. Remember that n_1 always refers to the medium on the side of the surface from which the light comes.

39–10   Thin Lenses

In most refraction situations there is more than one refracting surface. This is true even for a spectacle lens, the light passing from air into glass and then from glass into air. In microscopes, telescopes, cameras, etc., there are often many more than two surfaces.

Figure 39–24a shows a thick glass "lens" of length l whose surfaces are ground to radii r' and r''. A point object O' is placed near the left surface as shown. A ray leaving O' along the axis is not deflected on entering or leaving the lens because it falls on each surface along a normal.

A second ray leaving O', at an arbitrary angle α with the axis, strikes the surface at point a', is refracted, and strikes the second surface at point a''. The ray is again refracted and crosses the axis at I'', which, being the intersection of two rays from O'', is the image of point O', formed after refraction at two surfaces.

Figure 39–24b shows the first surface, which forms a virtual image of O' at I'. To locate I', we use Eq. 39–26,

$$\frac{n_1}{o} + \frac{n_2}{i} = \frac{n_2 - n_1}{r}.$$

Putting $n_1 = 1.0$ and $n_2 = n$ and bearing in mind that the image distance is negative (that is, $i = -i'$ in Fig. 39–24b), we obtain

$$\frac{1}{o'} - \frac{n}{i'} = \frac{n - 1}{r'}. \tag{39–27}$$

In this equation i' will be a positive number because we have arbitrarily introduced the minus sign appropriate to a virtual image.

Figure 39–24c shows the second surface. Unless an observer at point a'' were aware of the existence of the first surface, we would think that the light striking that point originated at point I' in Fig. 39–24b and that the region to the left of the surface was filled with glass. Thus the (virtual) image I' formed by the

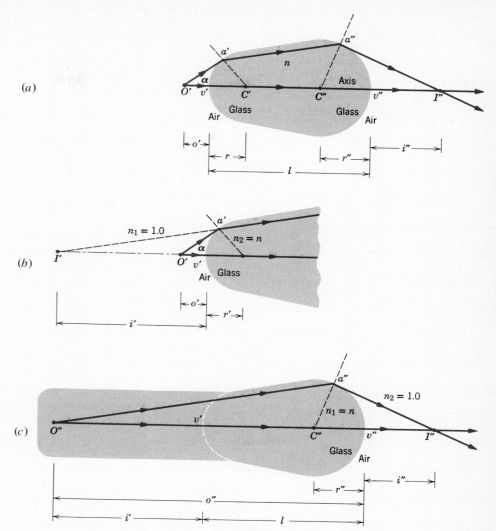

Figure 39-24 (*a*) Two rays from *O'* intersect at *I"* (real image) after refraction at two spherical surfaces. (*b*) The first surface and (*c*) the second surface shown separately. The quantities α and *n* have been exaggerated for clarity.

first surface serves as a real object *O"* for the second surface. The distance of this object from the second surface is

$$o'' = i' + l. \tag{39-28}$$

In applying Eq. 39-26 to the second surface, we insert $n_1 = n$ and $n_2 = 1$ because the object behaves as if it were imbedded in glass. If we use Eq. 39-28, Eq. 39-26 becomes

$$\frac{n}{i' + l} + \frac{1}{i''} = \frac{1 - n}{r''}. \tag{39-29}$$

Let us now assume that the thickness *l* of the "lens" in Fig. 39-24*a* is so small that we can neglect it in comparison with other linear quantities in this figure (such as *o'*, *i'*, *o"*, *i"*, *r'*, and *r"*). In all that follows we make this *thin-lens approximation*. Putting *l* = 0 in Eq. 39-29 leads to

$$\frac{n}{i'} + \frac{1}{i''} = -\frac{n - 1}{r''}. \tag{39-30}$$

Adding Eqs. 39–27 and 39–30 leads to

$$\frac{1}{o'} + \frac{1}{i''} = (n - 1) \left(\frac{1}{r'} - \frac{1}{r''} \right).$$

Finally, calling the original object distance simply o and the final image distance simply i leads to

Refraction by a thin lens

$$\frac{1}{o} + \frac{1}{i} = (n - 1) \left(\frac{1}{r'} - \frac{1}{r''} \right). \tag{39–31}$$

This equation holds only for rays that make small angles with the axis and only if the lens is so thin that it essentially makes no difference from which surface of the lens we measure the quantities o and i. In Eq. 39–31, r' refers to the first surface struck by the light as it traverses the lens and r'' to the second surface. If the lens is immersed in a medium with $n \neq 1$, Eq. 39–31 is still valid if we just replace n with the ratio n_{lens} / n_{medium}.

With the distinction of Fig. 39–22 in mind we assert that the sign conventions for thin lenses are the same as those for spherical mirrors and spherical refracting surfaces; the three rules given earlier hold. There is the complication however that there are now two radii of curvature to consider, r' always being that of the surface that the light first strikes. We see that in Fig. 39–25a o, i, and r' are positive, r'' being negative. In Fig. 39–25b o and r'' are positive, i and r' being negative. In Fig. 39–25c, i and r' are positive, o and r'' being negative. Finally, in Fig. 39–25d, r'' is positive, o, i, and r' being negative.

Figure 39–26a and b shows parallel light from a distant object falling on a thin lens. The image location is called the second focal point F_2 of the lens. The distance from F_2 to the lens is called the focal length f. The first focal point for a thin lens (F_1 in figure) is the object position for which the image is at infinity. For thin lenses the first and second focal points are on opposite sides of the lens and are equidistant from it.

Focal points of a thin lens

We can find the focal length from Eq. 39–31 by inserting $o \to \infty$ and $i = f$. This yields

The lens makers formula

$$\frac{1}{f} = (n - 1) \left(\frac{1}{r'} - \frac{1}{r''} \right). \tag{39–32}$$

This relation is called the lens maker's equation because it allows us to compute the focal length of a lens in terms of the radii of curvature and the index of refraction of the material. Combining Eqs. 39–31 and 39–32 allows us to write the thin-lens equation as

Thin lens equation

$$\frac{1}{o} + \frac{1}{i} = \frac{1}{f}. \tag{39–33}$$

Example 9 The lenses of Fig. 39–26 have radii of curvature of magnitude 40 cm and are made of glass with $n = 1.65$. Compute their focal lengths.

Since C' lies on the R-side of the lens in Fig. 39–26a, r' is positive ($= +40$ cm). Since C'' lies on the V-side, r'' is negative ($= -40$ cm). Substituting in Eq. 39–32 yields

$$\frac{1}{f} = (n - 1) \left(\frac{1}{r'} - \frac{1}{r''} \right) = (1.65 - 1) \left(\frac{1}{+40 \text{ cm}} - \frac{1}{-40 \text{ cm}} \right),$$

or

$$f = +31 \text{ cm}.$$

A positive focal length indicates that, in agreement with Fig. 39–26a, the focal point F_2 is on the R-side of the lens and parallel incident light converges after refraction to form a real image.

In Fig. 39–26b, C' lies on the V-side of the lens so that r' is negative ($= -40$ cm). Since r'' is positive ($= +40$ cm), Eq. 39–33 yields

$$f = -31 \text{ cm}.$$

A negative focal length indicates that, in agreement with Fig. 39–26b, the focal point F_2 is on the V-side of the lens and incident light diverges after refraction to form a virtual image.

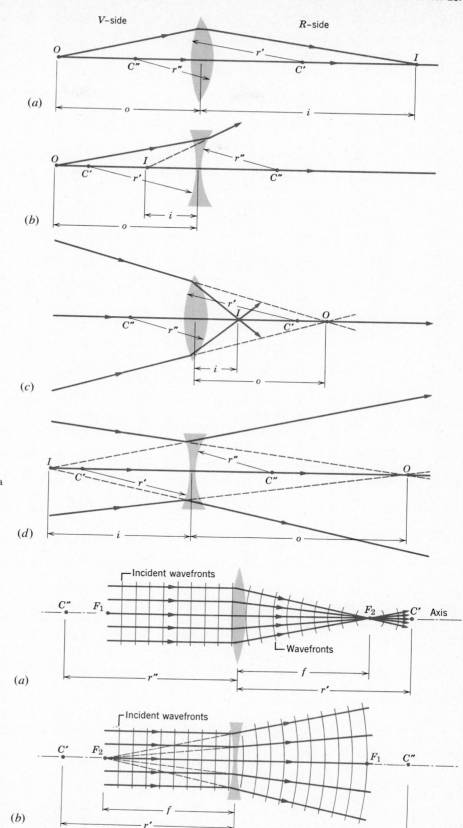

Figure 39-25 (a) A real object and a real image. (b) A real object and a virtual image. (c) A virtual object and a real image. (d) A virtual object and a virtual image.

Figure 39-26 (a) Parallel light passes through the second focal point F_2 of a converging lens. (b) Parallel light, passing through a diverging lens, seems to originate at the second focal point F_2. C' and C'' are centers of curvature for the lens surfaces; F_1 is the first focal point. A lens shaped like that in (a) is called *converging*; one shaped like that in (b) is called *diverging*; however, see Question 43.

Geometrical construction of images
from thin lens refraction

We can graphically locate the image of an extended object such as an illuminated arrow (Fig. 39–27) by using the following three facts:

(*a*)   A ray parallel to the axis and falling on the lens passes, either directly or
when extended, through the second focal point F_2 (ray *x* in Fig. 39–27).

(*b*)   A ray falling on a lens after passing, either directly or when extended,
through the first focal point F_1 will emerge from the lens parallel to the axis (ray *y*).

(*c*)   A ray falling on the lens at its center will pass through undeflected. There is
no deflection because the lens, near its center, behaves like a thin piece of glass
with parallel sides. The direction of the light rays is not changed and the sideways
displacement can be neglected because the lens thickness has been assumed to be
negligible (ray *z*; see also Problem 4).

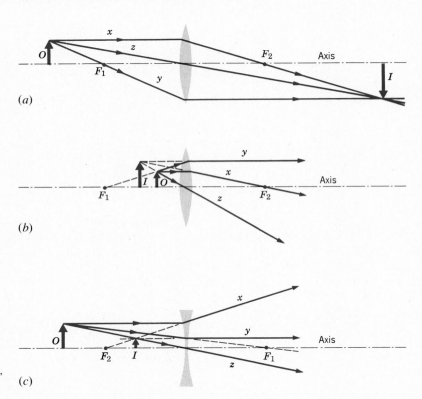

(*a*)

(*b*)

Figure 39–27   Showing the graphical
location of images for three thin
lenses. All three represent real objects,
with diverging rays falling on the lens.

(*c*)

Figure 39–28, which represents part of Fig. 39–27*a*, shows a ray passing
from the tip of the object through the center of curvature to the tip of the image.
For the similar triangles *abc* and *dec* we may write

$$\frac{de}{ab} = \frac{dc}{ac}.$$

The right side of this equation is i/o and the left side is $-m$, where *m* is the lateral
magnification. The minus sign is required because we wish *m* to be positive for an
erect image. This yields

Linear magnification of a thin lens

$$m = -\frac{i}{o},$$ (39–34)

which holds for all types of thin lenses and for all object distances.

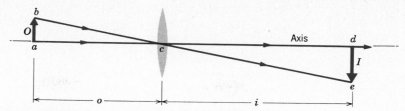

Figure 39–28   Two rays for the situation of Fig. 39–27a.

Example 10   A converging thin lens has a focal length of $+24$ cm. An object is placed 9.0 cm from the lens as in Fig. 39–27b; describe the image.

From Eq. 39–33,

$$\frac{1}{o} + \frac{1}{i} = \frac{1}{f},$$

we have

$$\frac{1}{+9.0 \text{ cm}} + \frac{1}{i} = \frac{1}{+24 \text{ cm}},$$

which yields $i = -14.4$ cm, in agreement with the figure. The minus sign means that the image is on the V-side of the lens and is thus virtual.

The lateral magnification is given by Eq. 39–34, or

$$m = -\frac{i}{o} = -\frac{-14.4 \text{ cm}}{+9.0 \text{ cm}} = +1.6,$$

again in agreement with the figure. The plus sign signifies an erect image.

In real-life optical devices, such as cameras or microscopes, we want the image to be a replica of the object, scaled up or down. Using the methods of this chapter, we can design a camera lens, say, only approximately because, among other things: (1) The object is not usually a plane but we want to record the image on a plane photographic film. (2) A point object produces a point image only approximately. (3) The refracting properties of the lens material vary with wavelength, as Fig. 39–2 shows. These imperfections are called *aberrations* and must be compensated for in the design of high-quality optical instruments. The design of a high-quality camera lens using the methods of geometrical optics calls for extensive ray-tracing calculations, using a computer. Such lenses contain several components, with different optical properties; the surfaces may not be spherical.

Even if we apply geometrical optics exactly (which we did not do here) we must remember that geometrical optics is itself an approximation because it assumes that light travels in straight lines, which it does not. To understand the ultimate performance of optical systems, we must take the wave nature of light into account.

Lens aberrations

39–11   Optical Instruments

The human eye is a remarkably effective organ but its range can be extended in many ways by a host of optical instruments such as spectacles, simple magnifiers, motion picture projectors, cameras (including TV cameras), microscopes, telescopes, etc. In many cases these devices extend the scope of our vision beyond the visible range; satellite-borne infrared cameras and x-ray microscopes are examples.

In almost all cases of modern sophisticated optical instruments the mirror and thin lens formulas (Eqs. 39–18 and 39–33) hold only as an approximation. The rays may not be paraxial, as anyone who has used a camera knows; in astronomical telescopes, however, the rays are indeed paraxial. In typical laboratory microscopes the lens cannot be considered "thin" in the sense in which that term was defined in Section 39–10. In most optical instruments lenses are compound, that is, they are made of several components, cemented together, the interfaces rarely being exactly spherical. This is done to improve image quality and brightness and to greatly relax the dependence on paraxial rays.

In what follows we describe three optical instruments, assuming, for simplicity of illustration only, that the thin lens formula applies.

The Simple Magnifier.   The normal human eye can focus a sharp image of an object on the retina if the object O is located anywhere from infinity, the stars, say, to a certain point called the near point P_n, which we take to be about 25 cm from the eye. If you move the object closer than the near point the perceived retinal image becomes fuzzy. The location of the near point normally varies with age. We have all heard stories about people who claim not to need glasses but who read their newspapers at arm's length; their near points are receding! Find your own near point by moving this page closer to your eyes, considered separately, until you reach a position at which the image begins to become indistinct.

Figure 39–29a shows an object O placed at the near point P_n. The size of the perceived image on the retina is measured by the angle θ. In Fig. 39–29b we insert a converging lens of focal length f just in front of the eye and move the object O to the first focal point F_1 of the lens. The eye now perceives an image at infinity, the angle of the image rays being θ', where $\theta' > \theta$. The angular magnification m_θ, not to be confused with the lateral magnification m given by Eq. 39–34, can be found from

$$m_\theta = \theta'/\theta$$

where

$$\theta \cong h/25 \text{ cm} \qquad \text{and} \qquad \theta' \cong h/f,$$

or

Angular magnification of a simple magnifier

$$m_\theta \cong 25 \text{ cm}/f. \tag{39–35}$$

Note that, as expected, if $f = 25$ cm, then $m_\theta = 1$, that is, $\theta' = \theta$. Lens aberrations limit the angular magnifications for a simple converging lens to something less than an order of magnitude. This is enough, however, for stamp collectors and for actors portraying Sherlock Holmes. More sophisticated magnifier designs have appreciably greater angular magnifications.

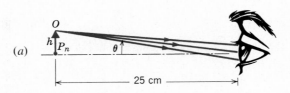

Figure 39–29   (a) An object O of height h is placed at the near point P_n of the human eye. If moved any closer, it would fail to form a clear image on the retina. (b) A converging lens (that is, a simple magnifier) is placed close to the eye and the object O is moved from P_n to F_1. Not drawn to scale.

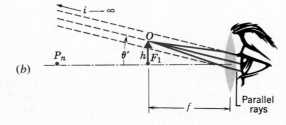

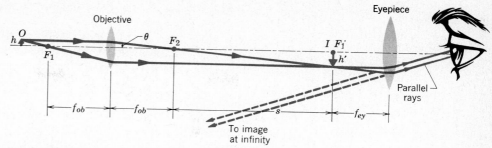

Figure 39–30 A simplified version of a compound microscope, using "thin" lenses. Some dimensions have been distorted for clarity. Two rays only are shown.

A Compound Microscope. Figure 39–30 shows a thin lens version of a compound microscope, used for viewing small objects that are very close to the instrument. The object O, of height h, is placed just outside the first focal point F_1 of the objective lens, whose focal length is f_{ob}. A real, inverted image I of height h' is formed by the objective, the lateral magnification being given by Eq. 39–34, or

$$m = \frac{h'}{h} = -\frac{s \tan \theta}{f_{ob} \tan \theta} = -\frac{s}{f_{ob}}. \tag{39–36}$$

Magnification of a microscope

As usual, the minus sign indicates an inverted image.

The distance s (sometimes called the tube length) is so chosen that the image I falls on the first focal point F_1' of the eyepiece, which then acts as a simple magnifier as described above. Parallel rays enter the eye and a final image I' forms at infinity. The final magnification M is given by the product of the linear magnification m for the objective lens, given by Eq. 39–36, and the angular magnification m_θ for the eyepiece, given by Eq. 39–35, or

$$M = m \times m_\theta = -\left(\frac{s}{f_{ob}}\right)\left(\frac{25 \text{ cm}}{f_{ey}}\right). \tag{39–37}$$

Refracting Telescope. Like microscopes, telescopes come in a large variety of forms. The form we describe here is the simple refracting telescope that consists of an objective lens and an eyepiece, both represented in Fig. 39–31 by thin lenses, although in practice, as for microscopes, they will each be compound lens systems.

Figure 39–31 A simplified version of an astronomical telescope, using "thin" lenses. Some dimensions have been distorted for clarity. Three rays only are shown.

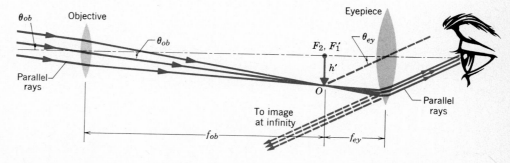

At first glance it may seem that the lens arrangements for telescopes (Fig. 39–31) and for microscopes (Fig. 39–30) are similar. However, telescopes are designed to view large objects, such as galaxies, stars, and planets, at large distances, whereas microscopes are designed for just the opposite purpose. Note also that in Fig. 39–31 the second focal point of the objective F_2 coincides with the first focal point of the eyepiece F_1', but in Fig. 39–30 these points are separated by the tube length s.

In Fig. 39–31 parallel rays from a distant object strike the objective lens, making an angle θ_{ob} with the telescope axis and forming a real, inverted image at the common focal point F_2, F_1'. This image acts as an object for the eyepiece and a (still inverted) virtual image is formed at infinity. The rays defining the image make an angle θ_{ey} with the telescope axis.

The angular magnification m_θ of the telescope is given by θ_{ey}/θ_{ob}. For paraxial rays we can write $\theta_{ob} = h'/f_{ob}$ and $\theta_{ey} = -h'/f_{ey}$ or

Angular magnification of a telescope

$$m_\theta = -\frac{f_{ob}}{f_{ey}}. \qquad (39-38)$$

Magnification is only one of the design factors of an astronomical telescope and is indeed easily achieved (How?). There is also *light gathering power,* which determines how bright the image is. This is important when viewing faint objects such as distant galaxies and is accomplished by making the objective lens diameter as large as possible. There is *field of view.* An instrument designed for galactic observation (narrow field of view) must be quite different from one designed for the observation of meteors (wide field of view). Also, there are lens and mirror aberrations including *spherical aberration* (that is, lenses and mirrors with truly spherical surfaces do not form sharp images) and *chromatic aberration* (that is, for simple lenses the focal length varies with wavelength so that fuzzy images are formed, displaying unnatural colors). There is also *resolving power,* which describes the ability of any optical instrument to distinguish between two objects (stars, say) whose angular separation is small. We will discuss this in more detail in Section 41–5. This by no means exhausts the design parameters of astronomical telescopes. We could also make a similar listing for compound microscopes and, indeed, for any high performance optical instrument.

Telescope design factors

Example 11 Figure 39–32 shows a simple astronomical telescope like that of Fig. 39–31 with the exceptions that (a) the incident parallel rays are parallel to the telescope axis, and (b) the incident rays fill the objective lens, whose diameter is d_{ob}, as they normally would. Find the diameter d of the *emergent pencil,* as it is called, which contains all the information that is available for entering the eye.

In Fig. 39–32 we have, from similar triangles,

$$\frac{d_{ob}/2}{f_{ob}} = \frac{d/2}{f_{ey}},$$

or (see Eq. 39–38)

$$d = d_{ob}\left(\frac{f_{ey}}{f_{ob}}\right) = -\frac{d_{ob}}{m_\theta}. \qquad (39-39)$$

Note that m_θ is inherently negative so that d is positive, as it must be.

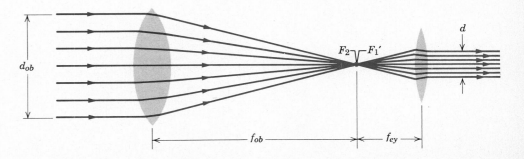

Figure 39–32   Example 11.

We must compare d with d_p, the diameter of the pupil of the eye. The pupil diameter is not constant but varies with illumination, between the limits of about 2 mm (sunlight) to 9 mm (darkness). The ideal condition is $d = d_p$. For example, if $d > d_p$ some light does not enter the eye and is wasted.

Actually these considerations are modified in practice. For most major optical instruments we use photographic recording.

Also, telescopes (and cameras) in orbiting satellites and space probes often send data to earth electronically, where it is processed by computor techniques into photographic form. The emergent pencil concept remains important however, not only for visual observation but for many lens systems design considerations.

REVIEW GUIDE AND SUMMARY

Geometrical optics

Light is most accurately described as an electromagnetic wave, whose speed and other properties are deriveable from Maxwell's equations. Geometrical optics is an approximate treatment in which the waves can be represented by straight-line rays; it is valid if the waves do not encounter obstacles comparable in size to the wavelength of the radiation.

When a light ray falls on a boundary between two media a reflected and a refracted ray generally appear. Both rays remain in the plane of incidence. The angle of reflection is equal to the angle of incidence and the angles of refraction and of incidence are related by

Reflection and refraction

$$n_1 \sin \theta_1 = n_2 \sin \theta_2 . \qquad [39\text{-}2, 10]$$

See Fig. 39–1 and Example 1.

Huygens' principle

Huygens' principle (see statement on p. 749) relates the rays of geometrical optics to a rudimentary wave theory and gives, for the speed and wavelength of a wave in a medium of index n,

$$v_n = c/n \quad \text{and} \quad \lambda_n = \lambda/n . \qquad [39\text{-}8, 9]$$

The laws of reflection and refraction (see Fig. 39–5) are deriveable from this principle.

A wave encountering a boundary for which a transmitted wave would have a higher speed will experience total internal reflection (see Fig. 39–6 and Example 3) if its angle of incidence is at least equal to θ_c, where

Total internal reflection.

$$\sin \theta_c = n_2/n_1 . \qquad [39\text{-}12]$$

A reflected wave will be polarized, with its **E** vector perpendicular to the plane of incidence, if it strikes a boundary at the polarizing angle θ_p, where

Polarization by reflection

$$\tan \theta_p = n_2/n_1 . \qquad [39\text{-}13]$$

See Fig. 39-8 and Example 4.

Rays diverging from a point object O can recombine to form an (approximately) point image I if they encounter a spherical mirror, a spherical refracting surface, or a thin lens. For rays sufficiently close to the axis we have the following:

1. Spherical mirror:

The spherical mirror equation

$$\frac{1}{o} + \frac{1}{i} = \frac{1}{f} = \frac{2}{r} . \qquad [39\text{-}18, 19]$$

See Figs. 39–14, 39–15, and 39–16 and also Example 6. A plane mirror is a special case for which $r \to \infty$, yielding $o = -i$.

2. Spherical refracting surface:

Refraction at a single spherical surface

$$\frac{n_1}{o} + \frac{n_2}{i} = \frac{n_2 - n_1}{r} . \qquad [39\text{-}26]$$

See Fig. 39–20 and Examples 7 and 8.

3. Thin lens:

Refraction by a thin lens

$$\frac{1}{o} + \frac{1}{i} = \frac{1}{f} = (n - 1) \left(\frac{1}{r'} - \frac{1}{r''} \right) . \qquad [39\text{-}31, 32, 33]$$

See Fig. 39–25 and Examples 9 and 10.

The same sign conventions hold for all three of the above situations. You can easily remember these conventions by associating the word "real" with the word "positive" (and the word "virtual" with the word "negative"). Thus, i and o are positives for *real* images and objects; f is positive if parallel rays are brought to a *real* focus; r, r' and r'' are positive if the corresponding center of curvature lies on the *R-side* of the surface in question.

Images of extended objects may be found graphically by ray tracing; see Fig. 39–18 for mirrors and Fig. 39–27 for thin lenses. The lateral magnification in these two cases is given by

Lateral magnification

$$m = -i/o, \qquad\qquad [39\text{–}20,\ 34]$$

a plus sign representing an erect image.

Three simplified treatments of optical instruments are given:

1. The simple magnifier (Fig. 39–29). The angular magnification is given by

$$m_\theta = (25\ \text{cm})/f. \qquad\qquad [39\text{–}35]$$

Optical instruments

2. The compound microscope (Fig. 39–30). The angular magnification is given by

$$M = -s(25\ \text{cm})/f_{ob}f_{ey}. \qquad\qquad [39\text{–}37]$$

3. The refracting telescope (Fig. 39–31). The overall angular magnification is given by

$$m_\theta = -f_{ob}/f_{ey}. \qquad\qquad [39\text{–}38]$$

QUESTIONS

1. Describe what your immediate environment would look like if all objects were totally absorbing. Sitting in a chair in a room, could you see anything? If a person entered the room could you see her?

2. Would you expect sound waves to obey the laws of reflection and of refraction obeyed by light waves? Does Huygens' principle apply to sound waves in air? If Huygens' principle predicts the laws of reflection and refraction, why is it necessary or desirable to view light as an electromagnetic wave, with all its attendant complexity?

3. A street light, viewed by reflection across a body of water in which there are ripples, appears very elongated. Explain.

4. By what percent does the speed of blue light in fused quartz differ from that of red light?

5. Can (*a*) reflection phenomena or (*b*) refraction phenomena be used to determine the wavelength of light?

6. How can one determine the indices of refraction of the media in Table 39–1 relative to water, given the data in that table?

7. Describe and explain what a fish sees as he looks in various directions above his "horizon."

8. Why does a diamond "sparkle" more than a glass imitation cut to the same shape?

9. Light has (*a*) a wavelength, (*b*) a frequency, and (*c*) a speed. Which, if any, of these quantities remains unchanged when light passes from a vacuum into a slab of glass?

10. Light, traveling through vacuum from a distant stationary source, strikes your eye. If the source starts to move rapidly toward you, how does this affect (*a*) the wavelength, (*b*) the frequency, and (*c*) the speed of the light?

11. Design a periscope, taking advantage of total internal reflection. What are the advantages compared with silvered mirrors?

12. For a plane mirror, what is the focal length? . . . the magnification?

13. At night, in a lighted room, you blow a smoke ring toward a window pane. If you focus your eyes on the ring as it approaches the pane, it will seem to go right through the glass into the darkness beyond. What is the explanation of this illusion?

14. How do our eyes adjust for seeing objects at different distances from us?

15. A machinist whose eyesight is failing finds that he can read his micrometer scale more easily if he squints at it through a tiny aperture formed by coiling his index finger into the base of his thumb. Although less bright, the image is sharper than when he looks at the scale directly. What is the explanation?

16. In driving a car you sometimes see vehicles such as ambulances with letters printed on them in such a way that they read in the normal fashion when you look at them through the rear-view mirror. Print your name so that it may be so read.

17. Brewster's law determines the polarizing angle on reflection from a material such as glass; see Fig. 39–8. A plausible interpretation for zero reflection of the parallel component at that angle is that the charges in the glass are caused to oscillate parallel to the reflected ray by this component and produce no radiation in this direction. Explain this and comment on the plausibility.

18. Can polarization by reflection occur if the light is incident on the interface from the side with the higher index of refraction (glass to air, for example)?

19. Can you think of a simple test or observation to prove that the law of reflection is the same for all wavelengths, under conditions in which geometrical optics prevails?

20. We all know that when we look into a mirror, right and left are reversed. Our right hand will seem to be a left hand; if we part our hair on the left it will seem to be parted on the right, etc. Can you think of a system of mirrors that would let us see ourselves as others see us? If so draw it and prove your point by sketching some typical rays.

21. If a mirror reverses right and left, why doesn't it reverse up and down?

22. Devise a system of plane mirrors that will let you see the back of your head. Trace the rays to prove your point.

23. If converging rays fall on a plane mirror, is the image virtual?

24. It is a bright sunny day and you want to create a rainbow in your back yard, using your garden hose. Exactly how do you go about it? Incidentally, why can't you walk under a rainbow?

25. Is it possible, by using one or more prisms, to recombine into white light the color spectrum formed when white light passes through a single prism?

26. How does atmospheric refraction affect the apparent time of sunset?

27. Stars twinkle but planets don't. Why?

28. The wavelength of, say, red light that we see in air is reduced when the light passes into water. Would that light then appear to be another color—blue, perhaps—if you viewed it from under the water surface?

29. In many city buses a convex mirror is suspended over the door, in full view of the driver. Why not a plane or a concave mirror?

30. What approximations were made in deriving the mirror equation (Eq. 39–18),

$$\frac{1}{o} + \frac{1}{i} = \frac{2}{r}?$$

31. Under what conditions will a spherical mirror, which may be concave or convex, form (a) a real image, (b) an inverted image, and (c) an image smaller than the object?

32. Can a virtual image be projected onto a screen? Photographed? If you put a piece of paper at the site of a virtual object (assuming a high-intensity light beam) will it ignite after sufficient exposure?

33. You are looking at a dog through a glass window pane. Where is the image of the dog? Is it real or virtual? Is it erect or inverted? What is the magnification? (*Hint:* Think of the window pane as the limiting case of a thin lens in which the radii of curvature have been allowed to become infinitely large.)

34. In some cars the right side mirror bears the notation: "Objects in the mirror are closer than they seem." What feature of the mirror requires this warning? What advantage does the mirror have to compensate for this disadvantage? Do cars viewed in this mirror seem to be moving faster or slower than they would be if viewed in a plane mirror?

35. We have all seen TV pictures of a baseball game shot from a camera located somewhere behind second base. The pitcher and the batter are about 60 ft apart, but they look much closer on the TV screen. Why are images viewed through a telephoto lens foreshortened in this way?

36. An unsymmetrical thin lens forms an image of a point object on its axis. Is the image location changed if the lens is reversed?

37. Why has a lens two focal points and a mirror only one?

38. Under what conditions will a thin lens, which may be converging or diverging, form (a) a real image, (b) an inverted image, and (c) an image smaller than the object?

39. A diver wants to use an air-filled plastic bag as a converging lens for underwater visibility. Sketch a suitable cross section for the bag.

40. What approximations were made in deriving the thin lens equation (Eq. 39–33),

$$\frac{1}{o} + \frac{1}{i} = \frac{1}{f}?$$

41. Under what conditions will a thin lens have a lateral magnification (a) of −1 and (b) of +1?

42. How does the focal length of a glass lens for blue light compare with that for red light, assuming the lens is (a) diverging and (b) converging?

43. Does the focal length of a lens depend on the medium in which the lens is immersed? Is it possible for a given lens to act as a converging lens in one medium and a diverging lens in another medium?

44. Are the following statements true for a glass lens in air? (a) A lens that is thicker at the center than at the edges is a converging lens for parallel light. (b) A lens that is thicker at the edges than at the center is a diverging lens for parallel light.

45. Under what conditions would the lateral magnification ($m = -i/o$) for lenses and mirrors become infinite? Is there any practical significance to such a condition?

46. Is the focal length of a spherical mirror affected by the medium in which it is immersed? . . . of a thin lens? . . . What's the difference?

47. Why is the magnification of a simple magnifier (see the derivation leading to Eq. 39–35) defined in terms of angles rather than image/object size?

48. Ordinary spectacles do not magnify but a simple magnifier does. What then, is the function of spectacles?

49. The "f-number" of a camera lens (see Problem 49) is its focal length divided by its aperture (effective diameter). Why is this useful to know in photography? How can the f-number of the lens be changed? How is exposure time related to f-number?

50. Does it matter whether (a) an astronomical telescope, (b) a compound microscope, (c) a simple magnifier, (d) a camera, including a TV camera, or (e) a projector, including a slide projector and a motion picture projector produce erect or inverted images? What about real or virtual images?

51. Why does chromatic aberration occur in simple lenses but not in mirrors?

52. The unaided human eye produces a real but inverted image on the retina. (a) Why then don't we perceive objects such as people and trees as upside down? (b) We don't, of course, but suppose that we wore special glasses so that we did. If you then turned this book upside down, could you read this question with the same facility that you do now?

EXERCISES

Section 39–2   Reflection and Refraction—Plane Waves and Plane Surfaces

1. The speed of yellow sodium light in a certain liquid is measured to be 1.92×10^8 m/s. What is the index of refraction of this liquid, with respect to air, for sodium light?

2. What is the speed in fused quartz of light of wavelength 550 nm? (See Fig. 39–2.)

3. The wavelength of yellow sodium light in air is 589 nm. (a) What is its frequency? (b) What is its wavelength in glass whose index of refraction is 1.52? (c) From the results of (a) and (b) find its speed in this glass.

4. The indices of refraction of water and glass, relative to vacuum, are 1.33 and 1.50, respectively. (a) What is the index of refraction of glass relative to water? (b) What is the index of refraction of water relative to glass?

Section 39–3   Huygen's Principle

5. Derive the law of reflection using Huygen's principle.

Section 39–4   The Law of Refraction

6. A laser beam travels along the axis of a straight section of pipeline, one mile long. The pipe normally contains air at standard temperature and pressure, but it may also be evacuated. (a) In which case would the travel time for the beam be greater and by how much? (b) Compare your answer with that for Problem 36, Chapter 38.

7. Light in vacuum is incident on the surface of a glass slab. In the vacuum the beam makes an angle of 32° with the normal to the surface, while in the glass it makes an angle of 21° with the normal. (a) What is the index of refraction of the glass? (b) What is the speed of light in the glass slab?

Section 39–5   Total Internal Reflection

8. A light ray falls on a square glass slab as in Fig. 39–33. What must the index of refraction of the glass be if total internal reflection occurs at the vertical face?

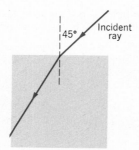

Figure 39–33   Exercise 8.

9. A ray of light is incident normally on the face ab of a glass prism ($n = 1.52$), as shown in Fig. 39–34. (a) Assuming that the prism is immersed in air, find the largest value for the angle ϕ so that the ray is totally reflected at face ac. (b) Find ϕ if the prism is immersed in water.

10. A fish is 2.00 m below the surface of a smooth lake. At what angle above the horizontal must it look to see the light from a

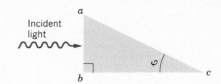

Figure 39–34   Exercise 9.

small fire burning at the water's edge 100 m away? Take the index of refraction for water to be 1.33.

Section 39–6   Brewster's Law

11. (a) At what angle of incidence will the light reflected from water be completely polarized? (b) Does this angle depend on the wavelength of the light?

Section 39–7   Spherical Waves—Plane Mirror

12. A small object is 10 cm in front of a plane mirror. If you stand behind the object, 30 cm from the mirror, and look at its image, for what distance must you focus your eyes?

13. Suppose you wished to photograph an object seen in a plane mirror. If the object is 5.0 m to your right and 1.0 m closer to the plane of the mirror than you, for what distance must you focus the lens of your camera?

14. You are standing in front of a large plane mirror, contemplating your image. If you move toward the mirror at speed v, at what speed does your image move toward you? Report this speed both (a) in your own reference frame and (b) in the reference frame of the room in which the mirror is at rest.

Section 39–8   Spherical Waves—Spherical Mirror

15. For clarity, the rays in figures like Fig. 39–14 are not drawn paraxial enough for Eq. 39–18 to hold with great accuracy. With a ruler, measure r and o in this figure and calculate, from Eq. 39–18, the predicted value of i. Compare this with the measured value of i.

Section 39–9   Spherical Waves—Spherical Refracting Surface

16. Define and locate the first and second focal points (see p. 764) for a single spherical refracting surface such as that of Fig. 39–20.

Section 39–10   Thin Lenses

17. You focus an image of the sun on a screen, using a thin lens whose focal length is 20 cm. What is the diameter of the image? (Note: The diameter of the sun is 1.39×10^9 m and its mean distance is 1.50×10^{11} m.)

18. An object is 20 cm to the left of a thin diverging lens having a 30 cm focal length. Where is the image formed? Calculate the position, then draw a ray diagram.

19. Two converging lenses, with focal lengths f_1 and f_2 are positioned a distance $f_1 + f_2$ apart, as shown in Fig. 39–35. Arrangements like this are called *beam expanders* and are often used to increase the diameters of light beams from lasers. (a) If W_1 is the incident beam width, show that the width of the emerging beam is $W_2 = (f_2/f_1)W_1$. (b) Show how a combination of one diverging

Figure 39–35 Exercise 19.

and one converging lens can also be arranged as a beam expander. Incident rays parallel to the axis should exit parallel to the axis.

20. A double-convex lens is to be made of glass with an index of refraction of 1.5. One surface is to have twice the radius of curvature of the other and the focal length is to be 60 mm. What are the radii?

21. A lens is made of glass having an index of refraction of 1.5. One side of the lens is flat and the other convex with a radius of curvature of 20 cm. (a) Find the focal length of the lens. (b) If an object is placed 40 cm to the left of the lens, where will the image be located?

22. Using the lensmaker's formula (Eq. 39–32), decide which of

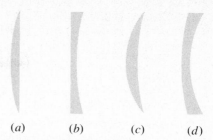

(a)   (b)   (c)   (d)

Figure 39–36 Exercise 22.

the thin lenses in Fig. 39–40 are converging and which diverging for parallel incident light.

23. Show that the focal length f for a thin lens whose index of refraction is n and which is immersed in a medium, water, say, whose index of refraction is n' is given by

$$\frac{1}{f} = \frac{n - n'}{n'}\left(\frac{1}{r'} - \frac{1}{r''}\right).$$

PROBLEMS GROUPED BY SECTION

Section 39–2 Reflection and Refraction: Plane Waves and Plane Surfaces

1. Prove that if a mirror is rotated through an angle α, the reflected beam is rotated through an angle 2α. Is this result reasonable for $\alpha = 45°$?

2. Figure 39–37 shows (top view) that Bernie B is walking directly toward the center of a vertical mirror. How close to the mirror will he be when Sarah S is just able to see him? Take $d = 3.0$ m.

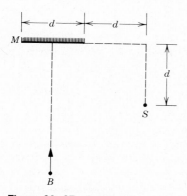

Figure 39–37 Problem 2.

3. A bottom-weighted vertical pole extends 2.0 m from the bottom of a swimming pool to a point 0.5 m above the water. Sunlight is incident at 45°. What is the length of the shadow of the pole on the bottom of the pool?

4. Prove that a ray of light incident on the surface of a sheet of plate glass of thickness t emerges from the opposite face parallel to its initial direction but displaced sideways, as in Fig. 39–38.

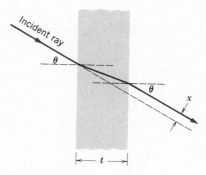

Figure 39–38 Problem 4.

Show that, for small angles of incidence θ, this displacement is given by

$$x = t\theta\, \frac{n - 1}{n}$$

where n is the index of refraction and θ is measured in radians.

Section 39–3 Huygen's Principle

5. One end of a stick is dragged through water at a speed v which is greater than the speed u of water waves. Applying Huygens' construction to the water waves, show that a conical wavefront is set up and that its half-angle α is given by

$$\sin \alpha = u/v.$$

This is familiar as the bow wave of a ship or the shock wave caused by an object moving through air with a speed exceeding that of sound, as in Fig. 18–11.

Section 39–4 The Law of Refraction

6. Light of free space wavelength 600 nm travels 1.6 μm in a medium of index of refraction 1.5. Find (*a*) the optical path length, (*b*) the wavelength in the medium, and (*c*) the phase difference after moving that distance, with respect to light traveling the same distance in free space.

7. Ocean waves moving at a speed of 4.0 m/s are approaching a beach at an angle of 30° to the normal, as shown in Fig. 39–39. Suppose the water depth changes abruptly and the wave speed drops to 3.0 m/s. Close to the beach, what is the angle θ between the direction of wave motion and the normal? Can you explain why most waves come in normal to a shore even though at large distances they approach at a variety of angles?

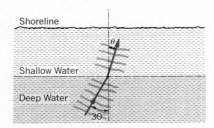

Figure 39–39 Problem 7.

Section 39–5 Total Internal Reflection

8. A given monochromatic light ray, initially in air, strikes the 90° prism at P (see Fig. 39–40) and is refracted there and at Q to such an extent that it just grazes the right-hand prism surface at Q. (*a*) Determine the index of refraction, relative to air, of the prism for this wavelength in terms of the angle of incidence θ_1 which gives rise to this situation. (*b*) Give a numerical upper bound for the index of refraction of the prism. Show, by ray diagrams, what happens if the angle of incidence at P is (*a*) slightly greater than θ_1 or (*b*), is slightly less than θ_1.

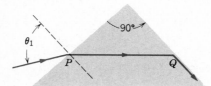

Figure 39–40 Problem 8.

9. A glass cube has a small spot at its center. (*a*) What parts of the cube face must be covered to prevent the spot from being seen, no matter what the direction of viewing? (*b*) What fraction of the cube surface must be so covered? Assume a cube edge of 10 mm and an index of refraction of 1.5. (Neglect the subsequent behavior of an internally reflected ray.)

10. A point source is 80 cm below the surface of a body of water. Find the diameter of the largest circle at the surface through which light can emerge from the water.

11. An optical fiber consists of a glass core (index of refraction n_1) surrounded by a coating (index of refraction $n_2 < n_1$). Suppose a beam of light enters the fiber from air at an angle θ with the fiber axis as shown in Fig. 39–41. (*a*) Show that the greatest possible value of θ for which a ray can be propagated down the fiber is given

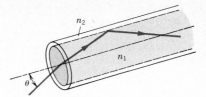

Figure 39–41 Problem 11.

by $\theta = \sin^{-1} (n_1{}^2 - n_2{}^2)^{1/2}$. (*b*) Assume the glass and coating indices of refraction are 1.58 and 1.53, respectively, and calculate the value of this angle.

12. In an optical fiber (see preceding problem), different rays travel different paths along the fiber, leading to different travel times. This causes a light pulse to spread out as it travels along the fiber, resulting in information loss. The delay time should be minimized in designing a fiber. Consider a ray that travels a distance L along a fiber axis and another that is reflected, at the critical angle, as it travels to the same destination as the first. (*a*) Show that the difference Δt in the times of arrival is given by

$$\Delta t = \frac{L}{c} \frac{n_1}{n_2} (n_1 - n_2),$$

where n_1 is the index of refraction of the glass core and n_2 is the index of refraction of the fiber coating. (*b*) Evaluate Δt for the fiber of the preceding problem, with $L = 300$ m.

Section 39–6 Brewster's Law

13. Calculate the range of polarizing angles for white light incident on fused quartz. Assume that the wavelength limits are 400 and 700 nm and use the dispersion curve of Fig. 39–2.

Section 39–7 Spherical Waves—Plane Mirror

14. A point object is 10 cm away from a mirror while the eye of an observer (pupil diameter 5.0 mm) is 20 cm away. Assuming both the eye and the point to be on the same line perpendicular to the surface, find the area of the mirror used in observing the reflection of the point.

15. Two plane mirrors make an angle of 90° with each other. What is the largest number of images of an object placed between them that can be seen by a properly placed eye? The object need not lie on the mirror bisector.

16. A small object O is placed one-third of the way between two parallel plane mirrors as in Fig. 39–42. Trace appropriate bundles of rays for viewing the four images that lie closest to the object.

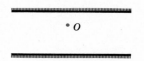

Figure 39–42 Problem 16.

***17.** Extend Fig. 39–48 to three dimensions by adding a mirror perpendicular to the common axis of the two mirrors shown. This forms a *corner reflector,* much used in optical, microwave, and other applications. It has the property that an incident ray is returned, after three reflections, with its direction exactly reversed. Prove this result.

Section 39–8 Spherical Waves—Spherical Mirror

18. Fill in the table below, each column of which refers to a spherical mirror and a real object. Check your results by graphical analysis. Distances are in centimeters; if a number has no plus or minus sign in front of it, find the correct sign.

Section 39–9 Spherical Waves—Spherical Refracting Surface

19. Fill in the table below, each column of which refers to a spherical surface separating two media with different indices of refraction. Distances are measured in centimeters. The object is real in all cases. Draw a figure for each situation and construct the appropriate rays graphically. Assume a point object.

20. A parallel beam of light from a laser falls on a solid transparent sphere of index of refraction n, as shown in Fig. 39–43. (a) Show that the beam cannot be brought to a focus at the back of the sphere unless the beam width is small compared with the radius of the sphere. (b) If the condition in (a) is satisfied, what is the index

of refraction of the sphere? (c) What index of refraction, if any, will focus the beam at the center of the sphere?

21. A narrow parallel incident beam falls on a solid glass sphere at normal incidence. Locate the image in terms of the index of refraction n and the sphere radius r.

Section 39–10 Thin Lenses

22. Show that the distance between a real object and its real image formed by a thin converging lens is always greater than or equal to four times the focal length of the lens.

23. To the extent possible, fill the table on page 778, each column of which refers to a thin lens. Distances are in centimeters. If a number (except in row n) has no plus sign in front of it, find the correct sign. Draw a figure for each situation and construct the appropriate rays graphically. The object is real in all cases.

24. A converging lens with a focal length of $+20$ cm is located 10 cm to the left of a diverging lens having a focal length of -15 cm. If a real object is located 40 cm to the left of the first lens, locate and describe completely the image formed.

25. An erect object is placed a distance in front of a converging lens equal to twice the focal length f_1 of the lens. On the other side of the lens is a converging mirror of focal length f_2 separated from the lens by a distance $2(f_1 + f_2)$. See Fig. 39–44. (a) Find the

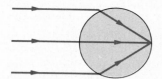

Figure 39–43 Problem 20.

Table for Problem 18.

	a	b	c	d	e	f	g	h
Type	Concave						Convex	
f	20		+20		20			
r				-40			40	
i				-10			4	
o	+10	+10	+30	+60				+24
m		+1		-0.5		+0.10		0.50
Real image?		no						
Erect image?								no

Table for Problem 19.

	a	b	c	d	e	f	g	h
n_1	1.0	1.0	1.0	1.0	1.5	1.5	1.5	1.5
n_2	1.5	1.5	1.5		1.0	1.0	1.0	
o	+10	+10		+20	+10		+70	+100
i		-13	+600	-20	-6	-7.5		+600
r	+30		+30	-20		-30	+30	-30
Real image?								

Table for Problem 23.

	a	b	c	d	e	f	g	h	i
Type	converging								
f	10	+10	10	10					
r'						+30	−30	−30	
r''						−30	+30	−60	
i									
o	+20		+5	+5	+5	+10	+10	+10	+10
n					1.5	1.5	1.5		
m			>1	<1				0.5	0.5
Real image?									yes
Erect image?								yes	

location, nature, and relative size of the final image, as seen by an eye looking toward the mirror through the lens. (b) Draw the appropriate ray diagram.

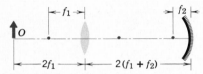

Figure 39–44 Problem 25.

26. An illuminated arrow forms a real inverted image of itself at a distance d of 40 cm, measured along the optic axis of a lens; see Fig. 39–45. The image is just half the size of the object. (a) What kind of lens must be used to produce this image? (b) How far from the object must the lens be placed? (c) What is the focal length of the lens?

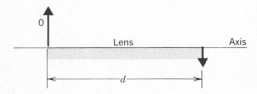

Figure 39–45 Problem 26.

27. An object is 20 cm to the left of a lens with a focal length of +10 cm. A second lens of focal length +12.5 cm is 30 cm to the right of the first lens. (a) Using the image formed by the first lens as the object for the second, find the location and relative size of the final image. (b) Verify your conclusions by drawing the lens system to scale and constructing a ray diagram. (c) Describe the final image.

Section 39–11 Optical Instruments
28. In connection with Fig. 39–29b, (a) show that if the object O is moved from the first focal point F_1 toward the eye, the image moves in from infinity and the angle θ' (and thus the angular

magnification m_θ) is increased. (b) If you continue this process, at what image location will m_θ have its maximum useable value? (c) Show that the maximum useable value of m_θ is $1 + (25\text{ cm})/f$. (d) Show that in this situation the angular magnification is equal to the linear magnification.

29. A microscope of the type shown in Fig. 39–30 has a focal length for the objective lens of 4.0 cm and for the eyepiece lens of 8.0 cm. The distance between the lenses is 25 cm. (a) What is the distance s in Fig. 39–30? (b) To reproduce the conditions of Fig. 39–30, how far beyond F_1 in that figure should the object be placed? (c) What is the lateral magnification m of the objective? (d) What is the angular magnification m_θ of the eyepiece? (e) What is the overall magnification M of the microscope?

30. *The eye—the basic optical instrument:* Figure 39–46a suggests a normal human eye. Parallel rays, entering a relaxed eye focused at infinity, produce a real, inverted image on the retina. The eye thus acts as a converging lens. Most of the refraction

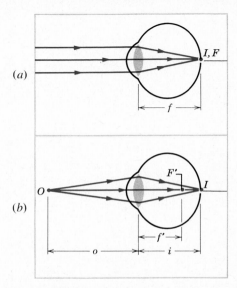

Figure 39–46 Problem 30.

occurs at the outer surface of the eye, the *cornea*. Assume a focal length f for the eye of 2.50 cm. In Fig. 39–46b the object is moved in to a distance o (= 40.0 cm) from the eye. To form an image on the retina the effective focal length of the eye must be reduced to f'. This is done by the action of the ciliary muscles that change the shape of the lens and thus the effective focal length of the eye. (*a*) Find f' from the above data. (*b*) Would the effective radii of curvature of the lens become larger or smaller in the transition from *a* to *b* in Fig. 39–46? (In the figure the structure of the eye is only roughly suggested and Fig. 39–46b is not to scale.)

31. The lens in a good camera can by no means be classified as "thin." If you want to measure its focal length by focusing a distant object on a screen, you are not sure what part of the lens to measure from. (*a*) Show that you can measure the focal length of a "thick" lens from the relation $f = I/\theta$, where I is the size of the image of a distant focussed object and θ is the angle that the object subtends. This way of measuring the focal length does not require any judgment of the lens position. (*b*) A distant tree subtends an angle of 3° at the lens position and the image of the tree is 2.0 cm tall. What is the focal length of the lens?

ADDITIONAL PROBLEMS

32. Claudius Ptolemy (c. 150 A.D.) gave the following measured values for the angle of incidence θ_1 and the angle of refraction θ_2 for a light beam passing from air to water:

θ_1	θ_2	θ_1	θ_2
10°	8°	50°	35°
20°	15°30′	60°	40°30′
30°	22°30′	70°	45°30′
40°	29°	80°	50°

Are these data consistent with Snell's law? If so, what index of refraction results? These data are interesting as perhaps the oldest recorded physical measurements.

33. A 60° prism is made of fused quartz. A ray of light falls on one face, making an angle of 35° with the normal. Trace the ray through the prism graphically with some care, showing the paths traversed by rays representing (*a*) blue light, (*b*) yellow-green light, and (*c*) red light. (See Fig. 39–2.)

34. You put a point source of light S a distance d in front of a screen A. How is the illumination at the center of the screen changed if you put a mirror M a distance d behind the source, as in Fig. 39–47?

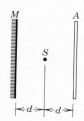

Figure 39–47   Problem 34.

35. *Cerenkov radiation.* When an electron moves through a medium at a speed exceeding the speed of light in that medium, it radiates electromagnetic energy (the Cerenkov effect). What minimum speed must an electron have in a liquid of refractive index 1.54 in order to radiate?

36. Figure 39–48 shows an incident ray i striking a plane mirror MM' at angle of incidence θ. Find the angle between i and r'. The two mirrors are at right angles.

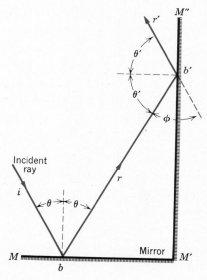

Figure 39–48   Problem 36.

37. A penny lies at the bottom of a pool with depth d and index of refraction n_1, as shown in Fig. 39–49. Take the index of refraction of air to be n_2 and prove that light rays that are close to the normal appear to come from a point $d_a = (n_2/n_1)d$ below the surface. This distance is the apparent depth of the pool.

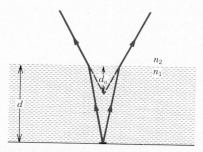

Figure 39–49   Problem 37.

38. A glass prism with an apex angle of 60° has $n = 1.60$. (*a*) What is the smallest angle of incidence for which a ray can enter one face of the prism and emerge from the other? (*b*) What angle of incidence would be required for the ray to pass through the prism symmetrically, as in Fig. 39–3?

39. A plane wave of white light traveling in fused quartz strikes a plane surface of the quartz, making an angle of incidence θ. Is it possible for the internally reflected beam to appear (*a*) bluish or (*b*) reddish? (*c*) Roughly what value of θ must be used? (*Hint:* White light will appear bluish if wavelengths corresponding to red are removed from the spectrum.)

40. You have a supply of flat glass disks ($n = 1.5$) and a lens grinding machine that can be set to grind radii of curvature of either 40 cm or 60 cm. You are asked to prepare a set of six lenses like those shown in Fig. 39–50. What will be the focal length of each lens? Will the lens form a real or a virtual image of the sun? (*Note:* Where you have a choice of radii of curvature, select the smaller one.)

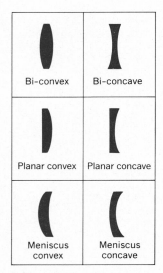

Figure 39–50   Problem 40.

41. When red light in vacuum is incident at the polarizing angle on a certain glass slab, the angle of refraction is 32°. What are (*a*) the index of refraction of the glass and (*b*) the polarizing angle?

42. Solve Problem 15 if the angle between the mirrors is (*a*) 45°, (*b*) 60°, (*c*) 120°, the object always being placed on the bisector of the mirrors.

43. In Fig. 39–11 you rotate the mirror 30° counterclockwise about its bottom edge, leaving the point object O in place. Is the image point displaced? If so, where is it? Can the eye still see the image without being moved? Sketch a figure showing the new situation.

44. A short linear object of length l lies along the axis of a spherical mirror, a distance o from the mirror. (*a*) Show that its image will have a length l' where

$$l' = l\left(\frac{f}{o-f}\right)^2.$$

(*b*) Show that the *longitudinal magnification* $m'\ (= l'/l)$ is equal to m^2 where m is the lateral magnification discussed in Section 39–8.

45. As an example of the importance of the paraxial-ray assumption, consider this problem. You place a coin at the bottom of a swimming pool filled with water ($n = 1.33$) to a depth of 2.4 m. What is the apparent depth of the coin below the surface when viewed (*a*) at near normal incidence (that is, by paraxial rays) and (*b*) by rays that leave the coin making an angle of 30° with the normal (that is, definitely not paraxial rays)?

46. The formula

$$\frac{1}{o} + \frac{1}{i} = \frac{1}{f}$$

is called the *Gaussian* form of the thin-lens formula. Another form of this formula, the *Newtonian* form, is obtained by considering the distance x from the object to the first focal point and the distance x' from the second focal point to the image. Show that

$$xx' = f^2.$$

47. Figure 39–51 shows a small light bulb suspended 250 cm above the surface of the water in a swimming pool. The water is 200 cm deep and the bottom of the pool is a large mirror. Where is the image of the light bulb? Consider only paraxial rays.

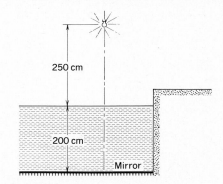

Figure 39–51   Problem 47.

48. In an eye that is *farsighted* the eye focuses parallel rays so that the image would form behind the retina, as in Fig. 39–52*a*. In an eye that is *nearsighted* the image is formed in front of the retina, as in Fig. 39–52*b*. (*a*) How would you design a corrective

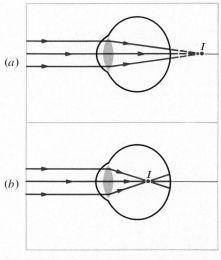

Figure 39–52   Problem 48.

lens for each eye defect? Make a ray diagram for each case. (*b*) If you need eyeglasses only for reading, are you nearsighted or far-sighted? (*c*) What is the function of bifocal eyeglasses, in which the upper parts and lower parts have different focal lengths?

49. *The camera.* Figure 39–53 shows an idealized camera focused on an object at infinity. A real, inverted image *I* is formed on the film, the image distance *i* being equal to the (fixed) focal length $f (= 5.0$ cm, say) of the lens system. In Fig. 39–53*b* the object *O* is closer to the camera, the object distance *o* being, say, 100 cm. To focus an image *I* on the film, we must extend the lens away from the camera (why?). (*a*) Find i' in Fig. 39–53*b*. (*b*) By how much must the lens be moved? Note that the camera differs from the eye (see Problem 30) in this respect. In the camera, f remains constant and the image distance *i* must be adjusted by moving the lens. For the eye the image distance *i* remains constant and the focal length f is adjusted by distorting the lens. Compare Figs. 39–46 and 39–53 carefully.

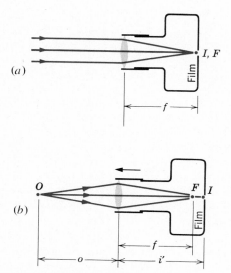

Figure 39–53 Problem 49.

50. *The reflecting telescope.* Isaac Newton, having convinced himself (erroneously as it turned out) that chromatic aberration was an inherent property of refracting telescopes, invented the reflecting telescope, shown schematically in Fig. 39–54. He presented his second model of this telescope, which has a magnifying power of 38, to the Royal Society, which still has it. In Fig. 39–54 incident light falls, closely parallel to the telescope axis, on the objective mirror *M*. After reflection from small mirror *M'* (the figure is not to scale), the rays form a real, inverted image in the focal plane through *F*. This image is then viewed through an eyepiece. (*a*) Show that the angular magnification m_θ is also given

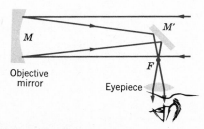

Figure 39–54 Problem 50.

by Eq. 39–38, or

$$m_\theta = -f_{ob}/f_{ey}$$

where f_{ob} is the focal length of the objective mirror and f_{ey} that of the eyepiece. (*b*) The 200-in. mirror of the reflecting telescope at Mt. Palomar in California has a focal length of 16.8 m. Estimate the size of the object formed in the focal plane of this mirror when the object is a meter stick 2.0 km away. Assume parallel incident rays. (*c*) The mirror of a reflecting astronomical telescope has an effective radius of curvature ("effective" because such mirrors are ground to a parabolic rather than a spherical shape, to eliminate spherical aberration defects) of 10 m. To give an angular magnification of 200, what must be the focal length of the eyepiece?

51. (*a*) A luminous point is moving at speed v_0 toward a spherical mirror, along its axis. Show that the speed at which the image of this point object is moving is given by

$$v_I = -\left(\frac{r}{2o - r}\right)^2 v_0.$$

(*Hint:* Start from Eq. 38–18.) (*b*) Assume that the mirror is concave, with $r = 15$ cm (and thus $f = 7.5$ cm) and that $v_0 = 5.0$ cm/s. Find the speed of the image if the object is far outside the focal point ($o = 30$ cm). (*c*) If it is close to the focal point ($o = 8.0$ cm). (*d*) If it is very close to the mirror ($o = 0.1$ cm).

52. A point source of light is placed a distance *h* below the surface of a large deep lake. (*a*) Show that the fraction *f* of the light energy that escapes directly from the water surface is independent of *h* and is given by

$$f = \tfrac{1}{2}(1 - \sqrt{1 - 1/n^2})$$

where *n* is the index of refraction of water. (*Note:* Absorption within the water and reflection at the surface — except where it is total — have been neglected.) (*b*) Evaluate this fraction for $n = 1.33$.

53. A laser beam travels through air at standard temperature and pressure, for which the index of refraction is 1.00029 (see Table 39–1). Nearby a collimated beam of accelerated electrons travels along the axis of a long straight evacuated pipeline. (*a*) What is the speed of the light forming the laser beam? (Its speed in a vacuum is given in Appendix B.) (*b*) What must be the kinetic energy of the electrons if they are to have this same speed?

54. Two perpendicular mirrors form the sides of a vessel filled with water, as shown in Fig. 39–55. A light ray is incident from above, normal to the water surface. (*a*) Show that the emerging ray is parallel to the incident ray. Assume that there are two reflections at the mirror surfaces. (*b*) Repeat the analysis for the case of oblique incidence, the ray lying in the plane of the figure.

Figure 39–55 Problem 54.

55. A layer of water ($n = 1.33$) 20 mm thick floats on carbon tetrachloride ($n = 1.46$) 40 mm thick. How far below the water surface, viewed at normal incidence, does the bottom of the tank seem to be?

56. An object is placed 1.0 m in front of a converging lens, of focal length 0.50 m, which is 2.0 m in front of a plane mirror. (*a*)

Where is the final image, measured from the lens, that would be seen by an eye looking toward the mirror through the lens? (b) Is the final image real or virtual? (c) Is the final image erect or inverted? (d) What is the lateral magnification?

57. Two thin lenses of focal lengths f_1 and f_2 are in contact. Show that they are equivalent to a single thin lens with a focal length given by

$$f = \frac{f_1 f_2}{f_1 + f_2}.$$

58. A luminous object and a screen are a fixed distance D apart. (a) Show that a converging lens of focal length f will form a real image on the screen for two positions that are separated by

$$d = \sqrt{D(D - 4f)}.$$

(b) Show that the ratio of the two image sizes for these two positions is

$$\left(\frac{D - d}{D + d}\right)^2.$$

*59. Sound waves generated in the earth by the detonation of a small amount of explosive obey the same laws of reflection, refraction, and total internal reflection as do light rays. Detectors, set up on the surface in a straight line from the detonation point S (see Fig. 39–56) detect the arrival of the sound waves. Suppose that a layer of soil in which the sound speed is v_1 covers solid bedrock in which the sound speed is v_2; suppose also, as is often the case, that $v_2 > v_1$. Waves arrive at a detector by two routes: (i) a direct surface wave; (ii) a wave striking the interface of soil and bedrock at the critical angle for total internal reflection; this wave travels along the boundary at speed v_2, generating waves that return to the surface, leaving the interface at the same angle as that of incidence. (Waves simply reflected from the interface are not considered.) (a) Show that the travel time of these critically reflected waves is given by

$$t_c = \frac{2D}{(1 - v_1^2/v_2^2)^{1/2}}\left(\frac{1}{v_1} - \frac{v_1}{v_2^2}\right) + \frac{x}{v_2},$$

where D is the thickness of the upper layer and x the distance from detonation S to the detector. (b) Show that beyond a certain distance x^* the critically reflected wave arrives before the direct wave, and that

$$x^* = 2D\left(\frac{1 + n}{1 - n}\right)^{1/2}, \quad n = v_1/v_2$$

and, therefore, by determining x^* the thickness D of the upper layer is determined. This method is widely employed in determining the suitability of land areas for construction purposes, to trace subsurface water-bearing zones, etc., and is called *seismic surveying*.

*60. A seismic surveying team is attempting to determine the depth to a horizontal bedrock layer. Detectors are placed 200 m apart. The data, shown graphically in Fig. 39–57, shows the time from the explosion to the arrival of the first sound waves at each detector plotted against the distance to the detector. Apply the theory of Problem 59 and find (a) the sound speeds v_1 and v_2, and (b) the depth to the bedrock layer.

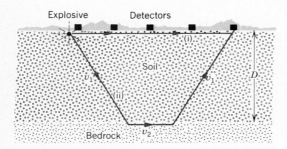

Figure 39–56 Problem 59.

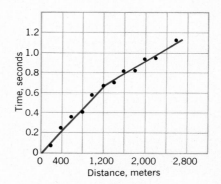

Figure 39–57 Problem 60.

40 Interference

40-1 Wave Optics

In Chapter 39 we made a special restriction, namely, that we would place no obstacle in the path of the incident light unless its dimensions were much greater than the wavelength of the light. Under these conditions we saw that light travels in straight lines and can be represented by rays. We called this special case geometrical optics.

In this chapter we remove this restriction and consider the general case of wave optics. In particular, we deal with slits and baffles whose dimensions are the same order as the wavelength of the incident light. Note that wave optics holds under all circumstances but geometrical optics is a special case.

Figure 40–1a shows schematically a plane wave of wavelength λ falling on a slit of width $a = 6.0\lambda$. We find that the light flares out into the geometrical shadow of the slit; in other words, the conditions for geometrical optics are not met. Figures 40–1b ($a = 3.0\lambda$) and 40–1c ($a = 1.5\lambda$) show that diffraction becomes more pronounced as $a/\lambda \to 0$ and that attempts to isolate a single ray from the incident plane wave are futile.

Figure 40–2 shows water waves in a shallow ripple tank, produced by tapping the water surface periodically with a flat stick. We see that the plane wave so generated flares out by diffraction when it encounters a gap in a barrier placed across it. Diffraction is characteristic of waves of all types. We can hear around corners, for example, because of the diffraction of sound waves.

The diffraction of waves at a slit (or at an obstacle such as a wire) is expected from Huygens' principle. Consider the portion of the wavefront that arrives at the slit in Fig. 40–1. Every point on it can be viewed as the site of an expanding spherical Huygens' wavelet (Section 39–3). The bending of light into the region of the

Diffraction

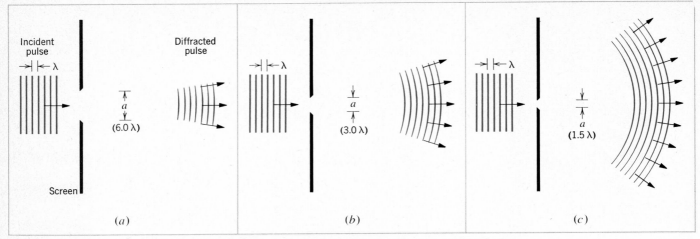

Figure 40-1 Suggesting how an incident plane light pulse is diffracted as it passes through a slit in an opaque screen. The smaller the ratio a/λ the more the wave flares into the geometrical shadow of the slit edges.

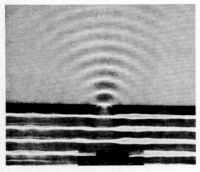

Figure 40-2 Diffraction of water waves at a slit in a ripple tank. Note that the slit width is about the same size as the wavelength. (Courtesy of Educational Services Incorporated.)

Young's two-slit interference experiment

geometrical shadow is associated with the blocking off of Huygens' wavelets from those parts of the incident wavefront that lie behind the slit edges. We shall discuss diffraction in more detail in Chapter 41.

40-2 Young's Experiment

Thomas Young in 1801 first established the wave theory of light on a firm experimental basis by demonstrating interference effects. He was even able to deduce the wavelength of light from his experiments, the first measurement of this important quantity.

We have already studied interference for sound waves in Chapter 18, where it led to an understanding of standing waves, beats, and other phenomena. In general, we would expect interference to occur whenever two or more waves of the same type and frequency, with a constant phase relationship between them, travel through the same region at the same time. Conversely, interference strongly indicates that the underlying phenomenon is a wave.

Young allowed sunlight to fall on a pinhole S_0 punched in a screen A. As represented in Fig. 40-3, the emerging light spread out by diffraction and fell on pinholes S_1 and S_2 punched into screen B. Diffraction occurred and two overlapping spherical waves expanded into the space to the right of screen B, where they could interfere with each other. The lines of dots show the directions in which the waves from S_1 and S_2 are mutually reenforcing, producing regions of maximum intensity on screen C. Minima appear on this screen between each pair of adjacent maxima. Figure 40-4 shows an actual pattern of maximum and minimum light intensity, generated in an apparatus like that of Fig. 40-3. In preparing this figure long narrow slits have been substituted in screens A and B for the pin holes originally used by Young.

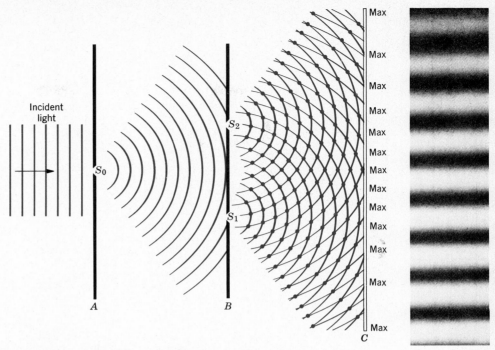

Incident
light

S_0

S_2

S_1

A

B

Max
Max
Max
Max
Max
Max
Max
Max
Max
Max
Max
Max
Max

C

Figure 40–3 Thomas Young produced an interference pattern on screen C by allowing diffracted waves from pinholes S_1 and S_2 to overlap in the region between screens B and C.

Figure 40–4 An interference pattern such as might be produced on screen C of Fig. 40–3. (From *Atlas of Optical Phenomena*, by M. Cagent, M. Francon, and J. C. Thrierr, Springer-Verlag, 1962.)

Let us analyze Young's experiment quantitatively, assuming that the incident light consists of a single wavelength only. In Fig. 40–5, P is an arbitrary point on the screen, a distance r_1 and r_2 from the narrow slits S_1 and S_2, respectively. Let us draw a line from S_2 to b in such a way that the lines PS_2 and Pb are equal. If d, the slit spacing, is much smaller than the distance D between the two screens (the ratio d/D in the figure has been exaggerated for clarity), S_2b is then almost perpendicular to both r_1 and r_2. This means that angle S_1S_2b is almost equal to angle PaO, both angles being marked θ in the figure. This is equivalent to saying that the lines r_1 and r_2 may be taken as parallel.

We often put a lens behind the two slits, as in Fig. 40–6, the screen C being in the focal plane of the lens. Under these conditions light focused at P must have struck the lens parallel to the line Px, drawn from P through the center of the (thin) lens. Under these conditions rays r_1 and r_2 are strictly parallel even though the requirement $D \gg d$ may not be met. The lens L may in practice be the lens and cornea of the eye, screen C being the retina.

The two rays arriving at P in Figs. 40–5 or 40–6 from S_1 and S_2 are in phase at the source slits, both being derived from the same wavefront in the incident plane wave. Because the rays contain different numbers of waves, they arrive at P with a phase difference. The number of wavelengths contained in S_1b, which is the path difference, determines the nature of the interference at P. If a lens is used as

The effect of a lens

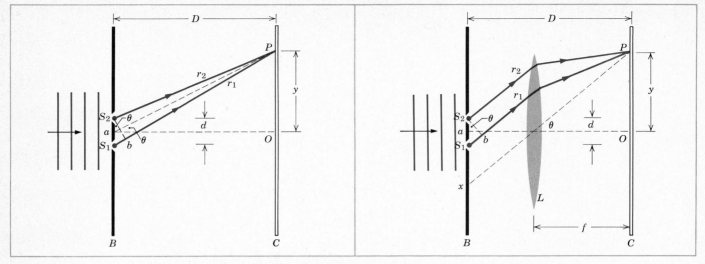

Figure 40–5 Rays from S_1 and S_2 combine at P. Actually, $D \gg d$, the figure being distorted for clarity. Point a is the midpoint of the slits.

Figure 40–6 A lens is normally used to produce interference fringes; compare with Fig. 40–5. The figure is again distorted for clarity in that $f \gg d$ in practice.

in Fig. 40–6, it may seem that a phase difference should develop between the rays beyond the plane represented by S_2b, the path lengths between this plane and P being clearly different. However, because of the presence of additional material of higher index of refraction in the path of the ray traveling the shortest distance, the optical path lengths are identical (see Section 39–4). Two rays with the same optical path lengths contain the same number of wavelengths, so that no phase difference will result because of the light passing through the lens.

Interference maxima and minima

To have a maximum at P, S_1b ($= d \sin \theta$) must contain an integral number of wavelengths, or

$$S_1b = m\lambda \qquad m = 0, 1, 2, \ldots,$$

which we can write as

$$d \sin \theta = m\lambda \qquad m = 0, 1, 2, \ldots \text{ (maxima)}. \qquad (40\text{–}1)$$

Note that each maximum above O in Figs. 40–5 and 40–6 has a symmetrically located maximum below O. There is a central maximum, described by $m = 0$.

For a minimum at P, S_1b ($= d \sin \theta$) must contain a half-integral number of wavelengths, or

$$d \sin \theta = (m + \tfrac{1}{2})\lambda \qquad m = 0, 1, 2, \ldots \text{ (minima)}. \qquad (40\text{–}2)$$

Example 1 What is the linear distance on screen C in Fig. 40–5 between adjacent maxima? The wavelength λ is 546 nm; the slit separation d as 0.10 mm, and the slit-screen separation D is 50 cm.

If θ is small enough, we can use the approximation

$$\sin \theta \cong \tan \theta \cong \theta.$$

From Fig. 40–5 we see that

$$\tan \theta = \frac{y}{D}.$$

Substituting this into Eq. 40–1 for $\sin \theta$ leads to

$$y = m \frac{\lambda D}{d} \qquad m = 0, 1, 2, \ldots \text{ (maxima)}.$$

The positions of any two adjacent maxima are given by

$$y_m = m \frac{\lambda D}{d}$$

and

$$y_{m+1} = (m + 1) \frac{\lambda D}{d}.$$

We find their separation Δy by subtracting:

$$\Delta y = y_{m+1} - y_m = \frac{\lambda D}{d}$$

$$= \frac{(546 \times 10^{-9} \text{ m})(50 \times 10^{-2} \text{ m})}{0.10 \times 10^{-3} \text{ m}} = 2.7 \text{ mm}.$$

As long as θ in Figs. 40–5 and 40–6 is small, the separation of the interference fringes is independent of m; that is, the fringes are evenly spaced. If the incident light contains more than one wavelength the separate interference patterns, which will have different fringe spacings, will be superimposed.

40–3 Coherence

Analysis of the derivation of Eqs. 40–1 and 40–2 shows that a fundamental requirement for the existence of well-defined interferences fringes on screen C in Fig. 40–3 is that the light waves that travel from S_1 and S_2 to any point on this screen must have a sharply defined phase difference ϕ that remains constant with time. If this condition is satisfied, a stable, well-defined fringe pattern will appear. At certain points on screen C in Fig. 40–3, ϕ will be given, independent of time, by $n\pi$ where $n = 1, 3, 5, \ldots$ so that the resultant intensity will be strictly zero and will remain so throughout the time of observation. At other points ϕ will be given by $n\pi$ where $n = 0, 2, 4, \ldots$ and the resultant intensity will be a maximum. The two beams emerging from slit S_1 and S_2 are said to be completely *coherent*, because their phase difference does not change with time.

Coherent waves

Imagine that the source in Fig. 40–3 is removed and that the slits S_1 and S_2 are replaced by two completely independent light sources, such as two fine incandescent wires placed side by side in a glass envelope. No interference fringes would appear on screen C but only a relatively uniform illumination. We can understand this if we assume that for completely independent light sources the phase difference between the two beams arriving at any point on screen C in Fig. 40–3 will vary with time in a random way. At a certain instant conditions may be right for cancellation and a short time later they may be right for reenforcement. This same random phase behavior holds for all points on screen C with the result that this screen is uniformly illuminated. The intensity at any point is equal to the sum of the intensities that each source S_1 and S_2 produces separately at that point. Under these conditions the two beams emerging from S_1 and S_2 are said to be completely incoherent.

Incoherent waves

Note that for completely coherent light beams we (1) combine the amplitudes vectorially, taking the (constant) phase difference properly into account, and then (2) square this resultant amplitude to obtain a quantity proportional to the resultant intensity. For completely incoherent light beams, on the other hand, we (1) square the individual amplitudes to obtain quantities proportional to the individual intensities and then (2) add the individual intensities to obtain the resultant intensity. This procedure is in agreement with the experimental fact that for completely independent light sources the resultant intensity at every point is always greater than the intensity produced at that point by either light source acting alone.

It is interesting to investigate the experimental conditions under which coherent or incoherent beams may be produced and to give an explanation for coherence in terms of the mode of production of the radiation. A coherent beam of radiowaves can be produced by an antenna connected by a transmission line to an LC oscillator as in Fig. 38–3. The oscillations are periodic with time and produce a periodic variation of **E** and **B** with time. The radiated wave at large enough distances from the antenna is well represented by Fig. 38–6. Note that (1) the wave has essentially infinite extent in time; see Fig. 40–7a. At any point, as the wave passes by, the wave disturbance (that is, **E** or **B**) varies with time in a perfectly periodic way. (2) The wavefronts at points far removed from the antenna are, for a small enough area of observation (see Fig. 38–5), parallel planes at right angles to the propagation direction. At any instant of time the wave disturbance varies with distance along the propagation direction in a perfectly periodic way.

Two beams generated from a single traveling wave like that of Fig. 38–6 will be completely coherent. One way to generate two such beams is to put an opaque screen containing two slits in the path of the beam. The waves emerging from the slits will always have a constant phase difference at any point in the region in which they overlap and interference fringes will be produced.

If we turn from radio sources to common sources of visible light, such as incandescent wires or an electric discharge passing through a gas, we become aware of a fundamental difference. In both of these sources the fundamental light emission processes occur in individual atoms and these atoms do not act together in a cooperative (i.e., coherent) way. The act of light emission by a single atom takes, in a typical case, about 10^{-8} s and the emitted light is properly described as a *wave-train* (Fig. 40–7b) rather than as a wave (Fig. 40–7a). For emission times such as these the wavetrains are a few meters long.

Interference effects from ordinary light sources may be produced by putting a narrow slit (S_0 in Fig. 40–3) directly in front of the source. This insures that the wavetrains that strike slits S_1 and S_2 in screen B in this figure originate from the same small region of the source. The diffracted beams emerging from S_1 and S_2 thus represent the same population of wavetrains and are coherent with respect to each other. If the phase of the light emitted from S_0 changes, this change is transmitted simultaneously to S_1 and S_2. Thus, at any point on screen C, a constant phase difference is maintained between the beams from these two slits and a stationary interference pattern occurs.

Lasers

The lack of coherence of the light from common sources such as the sun or incandescent lamp filaments is due to the fact that the emitting atoms in such sources act independently rather than cooperatively. Since 1960 it has been possible to make light sources in which the atoms *do* act cooperatively, the emitted light being highly coherent. Such devices are called lasers, a coined word derived from: *l*ight *a*mplification through the *s*timulated *e*mission of *r*adiation. Laser light is not only highly coherent but also extremely monochromatic, highly directional, and focusable to a spot whose dimensions are of the order of magnitude of one

Figure 40–7 (*a*) A section of an infinite wave and (*b*) a wavetrain of length *l*.

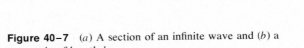

(*a*)

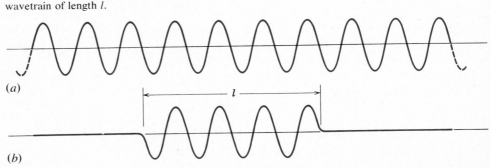

(*b*)

wavelength. This "light fantastic," as it has been called, has a bewildering variety of applications, among which are telephone communication over optical fibers, spotwelding detached retinas, hydrogen fusion research, measuring continental drift, and holography, a technique in which three-dimensional images can be produced from information stored in a two-dimensional matrix.

40–4 Intensity of Interfering Waves

Let us assume that the electric field components of the two waves in Fig. 40–5 vary with time at point P as

$$E_1 = E_0 \sin \omega t \tag{40–3}$$

and

$$E_2 = E_0 \sin (\omega t + \phi) \tag{40–4}$$

where ω (=$2\pi\nu$) is the angular frequency of the waves and ϕ is the phase difference between them. Note that ϕ depends on the location of point P, which, in turn, for a fixed geometrical arrangement, is described by the angle θ in Figs. 40–5 and 40–6. We assume that the slits are so narrow that the diffracted light from each slit illuminates the central portion of the screen uniformly. This means that near the center of the screen E_0 is independent of the position of P, that is, of the value of θ.

The resultant amplitude at point P is given by

$$E = E_1 + E_2. \tag{40–5}$$

This is a scalar and not a vector equation because $\mathbf{E}_1$ and $\mathbf{E}_2$ are essentially parallel at point P. We will combine E_1 and E_2 using an analytical, semigraphical method. The method will be especially useful later, when we want to combine a large number of wave disturbances with differing phases.

A sinusoidal wave disturbance E_1 (see Eq. 40–3) can be represented graphically, by a rotating vector. In Fig. 40–8a a vector of magnitude E_0 is allowed to rotate about the origin in a counterclockwise direction with an angular frequency ω. In our study of alternating currents (see Section 36–2) we learned to call such a rotating vector a *phasor*. The alternating wave disturbance E_1 is simply the projection of the phasor E_0 on the vertical axis.

A second wave disturbance E_2, which has the same amplitude E_0 but a phase difference ϕ with respect to E_1 (see Eq. 40–4) can be represented graphically (Fig. 40–8b) as the projection on the vertical axis of a second phasor of magnitude E_0, which makes a fixed angle ϕ with the first phasor. As this figure shows, the sum E of E_1 and E_2, which is the instantaneous amplitude of the resultant wave, is the sum of the projections of the two phasors on the vertical axis. This is revealed more clearly if we redraw the phasors, as in Fig. 40–8c, placing the foot of one arrow at the head of the other, maintaining the proper phase difference, and letting the whole assembly rotate counterclockwise about the origin.

In Fig. 40–8c E can also be regarded as the projection on the vertical axis of a phasor of length E_θ, which is the vector sum of the two phasors of magnitude E_0 and makes a phase angle β with respect to the phasor that generates E_1. Note that the (algebraic) sum of the projections of the two phasors is equal to the projection of the (vector) sum of the two phasors.

From the theorem that an exterior angle (ϕ) is equal to the sum of the two opposite interior angles ($\beta + \beta$) we see from Fig. 40–8c that $\beta = \frac{1}{2}\phi$. Thus we have

$$E = E_\theta \sin (\omega t + \beta), \tag{40–6a}$$

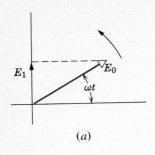

(a)

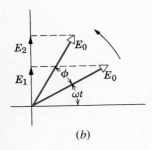

(b)

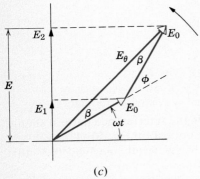

(c)

Figure 40–8 *(a)* A time-varying wave disturbance E_1 is represented by a rotating vector or *phasor*. *(b)* Two wave disturbances E_1 and E_2 with a phase difference ϕ between them are so represented. These two phasors can represent the two wave disturbances in the double-slit problem; see Eqs. 40–3 and 40–4. *(c)* Another way of drawing *(b)*.

Phase difference vs. path difference

where

$$\beta = \tfrac{1}{2}\phi \qquad (40\text{–}6b)$$

and

$$E_\theta = 2E_0 \cos \beta = E_m \cos \beta. \qquad (40\text{–}6c)$$

In most problems in optics we are concerned only with the amplitude E_θ of the resultant wave disturbance and not with its time variation. This is because the eye and most common measuring instruments respond to the resultant intensity of the light (that is, to the square of the amplitude) and cannot detect the rapid time variations that characterize visible light. For sodium light ($\lambda = 589$ nm), the frequency $\nu\,(= \omega/2\pi)$ is 5.1×10^{14} Hz. Usually, then, we need not consider the rotation of the phasors but can confine our attention to finding the magnitude of the resultant phasor.

The amplitude E_θ of the resultant wave disturbance, which determines the intensity of the interference fringes in Young's experiment, will turn out to depend strongly on the value of θ, that is, on the location of point P in Figs. 40–5 and 40–6.

In Section 17–5 we showed that the intensity of a wave I is proportional to the square of its amplitude. For the resultant wave then, leaving the proportionality constant unspecified,

$$I_\theta \propto E_\theta^2. \qquad (40\text{–}7)$$

This relationship seems reasonable if we recall (Eq. 27–9) that the energy density in an electric field is proportional to the square of the magnitude of the electric field. This is true for rapidly varying electric fields, such as those in a light wave, as well as for static fields.

The ratio of the intensities of two light waves is the ratio of the squares of the amplitudes of their electric fields. If I_θ is the intensity of the resultant wave at P and I_0 is the intensity that a single wave acting alone would produce, then

$$\frac{I_\theta}{I_0} = \left(\frac{E_\theta}{E_0}\right)^2. \qquad (40\text{–}8)$$

Combining with Eq. 40–6c leads to

$$I_\theta = 4I_0 \cos^2 \beta = I_m \cos^2 \beta. \qquad (40\text{–}9)$$

Note that the intensity of the resultant wave at any point P varies from zero (for a point at which $\cos \beta = 0$) to I_m, which is four times the intensity I_0 of each individual wave (for a point at which $\cos \beta = \pm 1$). Let us compute I_θ as a function of the angle θ in Figs. 40–5 or 40–6, that is, of the position of P on screen C in those figures.

The phase difference ϕ in Eq. 40–4 is associated with the path difference S_1b in Fig. 40–5 or 40–6. If S_1b is $\tfrac{1}{2}\lambda$, ϕ will be π; if S_1b is λ, ϕ will be 2π, etc. This suggests that

$$\frac{\text{phase difference}}{2\pi} = \frac{\text{path difference}}{\lambda},$$

$$\phi = \frac{2\pi}{\lambda} (d \sin \theta),$$

or, finally, from Eq. 40–6b,

$$\beta = \tfrac{1}{2}\phi = \frac{\pi d}{\lambda} \sin \theta. \qquad (40\text{–}10)$$

We can substitute this expression for β into Eq. 40–9 for I_θ, yielding the latter quantity as a function of θ. For convenience we collect here the expressions for the amplitude and the intensity in double-slit interference.

Intensity for two-slit interference

[Eq. 40–6c]	$E_\theta = E_m \cos \beta$	interference	(40–11a)
[Eq. 40–9]	$I_\theta = I_m \cos^2 \beta$	from narrow	(40–11b)
[Eq. 40-10]	$\beta(= \tfrac{1}{2}\phi) = \dfrac{\pi d}{\lambda} \sin \theta$	slits (that is, $a \ll \lambda$)	(40–11c)

To find the positions of the intensity maxima, we put

$$\beta = m\pi \qquad m = 0, 1, 2, \ldots$$

in Eq. 40–11b. From Eq. 40–11c this reduces to

$$d \sin \theta = m\lambda \qquad m = 0, 1, 2, \ldots \text{ (maxima)},$$

which is the equation derived in Section 40–2 (Eq. 40–1). To find the intensity minima we write

$$\frac{\pi d \sin \theta}{\lambda} = (m + \tfrac{1}{2})\pi \qquad m = 0, 1, 2, \ldots \text{ (minima)},$$

which reduces to the previously derived Eq. 40–2.

Figure 40–9, which is a representation of the central portion of the intensity pattern of Fig. 40–4, shows the intensity pattern for double-slit interference. The horizontal solid line is I_0; this describes the (uniform) intensity pattern on the screen if one of the slits is covered up. If the two sources were incoherent the intensity would be uniform over the screen and would be $2I_0$; see the horizontal dashed line in Fig. 40–9. For coherent sources we expect the energy to be merely redistributed over the screen, since energy is neither created nor destroyed by the interference process. Thus the average intensity in the interference pattern should be $2I_0$, as for incoherent sources. This follows at once if, in Eq. 40–11b, we substitute one-half for the cosine-squared term and if we recall (see Eq. 40–9) that $I_m = 4I_0$. We have seen several times that the average value of the square of a sine or a cosine term over one or more half-cycles is one-half.

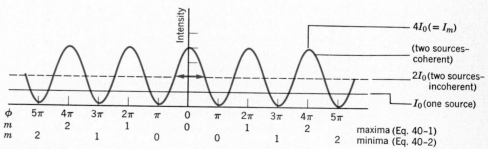

Figure 40–9 The intensity pattern for double-slit interference; Eq. 40–11b. The horizontal arrow in the central peak represents the half-width of the peak. This figure is constructed on the assumption that the two interfering waves each illuminate the central portion of the screen uniformly.

Addition of multiple phasors

In a more general case we might want to find the resultant of a number (>2) of sinusoidally varying wave disturbances. The general procedure is the following:

1. Construct a series of phasors representing the functions to be added. Draw them end to end, maintaining the proper phase relationships between adjacent phasors.

2. Construct the vector sum of this array. Its length gives the amplitude of the resultant. The angle between it and the first phasor is the phase of the resultant with respect to this first phasor. The projection of this phasor on the vertical axis gives the time variation of the resultant wave disturbance.

Example 2 Find graphically the resultant $E(t)$ of the following wave disturbances:

$$E_1 = E_0 \sin \omega t$$
$$E_2 = E_0 \sin (\omega t + 15°)$$
$$E_3 = E_0 \sin (\omega t + 30°)$$
$$E_4 = E_0 \sin (\omega t + 45°).$$

Figure 40–10 in which E_0 equals 10, shows the assembly of four phasors that represents these functions. Their vector sum, by graphical measurement, has an amplitude E_R of 38 and a phase ϕ_0 with respect to E_1 of 23°. In other words

$$E(t) = E_1 + E_2 + E_3 + E_4 = 38 \sin (\omega t + 23°).$$

Check this result by trigonometric calculation.

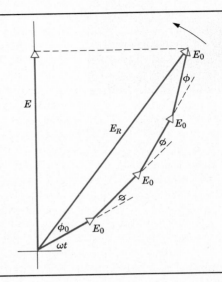

Figure 40–10 Example 2. Four wave disturbances are added graphically, using the method of phasors.

40–5 Interference from Thin Films

The colors that we see when sunlight falls on a soap bubble, on an oil slick, on a wet, dark pavement, or on a peacock feather are caused by the interference of light waves reflected from the front and back surfaces of a thin transparent film. The film thickness is typically of the order of magnitude of the wavelength of the light involved. Thin film technology, including the deposition of multiple-layered films, is highly developed and is widely used for the control of the reflection and/or transmission of light or radiant heat at optical surfaces. The familiar purple tint of a camera or a binocular lens shows that such films are in place, reducing unwanted reflections in the optical system.*

Figure 40–11 shows a transparent film of uniform thickness d illuminated by monochromatic light of wavelength λ from a point source S. The eye is so positioned that a particular incident ray i from the source enters the eye, as ray r_1, having been reflected from the front surface of the film at a. The incident ray also enters the film at a as a refracted ray and is reflected from the back surface of the film at b; it then reemerges from the front surface of the film at c and also enters the eye, as ray r_2. The geometry of Fig. 40–11 is such that rays r_1 and r_2 are parallel. Having originated in the same point source they are also coherent and are thus in a position to interfere. Because these two rays have traveled over paths of dif-

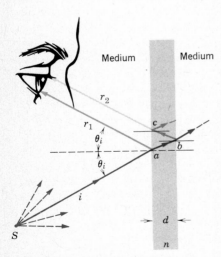

Figure 40–11 A thin film is viewed by light reflected from point source S. No account is taken of light transmitted through the film, as at b, or of light (weakly) reflected at c and beyond. The medium may be assumed to be air.

---

* See "Optical Interference Coatings" by Philip Baumeister and Gerald Pincus, *Scientific American*, December 1970.

ferent lengths, have traversed different media, and—as we shall see—have suffered different kinds of reflections at a and at b, there will be a phase difference between them. The intensity perceived by the eye as, focused to receive parallel light, it gazes through the region a,c of the film will be determined by this phase difference.

For near-normal incidence ($\theta_i \cong 0$ in Fig. 40–11) the geometrical path difference for the two rays from S will be close to $2d$. We might expect the resultant wave reflected from the film to be an interference maximum if the distance $2d$ is an integral number of wavelengths. This statement must be modified for two reasons.

First, the wavelength must refer to the wavelength of the light in the film λ_n and not to its wavelength in air λ; that is, we are concerned with optical pathlengths rather than geometrical path lengths. The wavelengths λ and λ_n (see Eq. 39–9) are related by

$$\lambda_n = \lambda/n. \qquad (40-12)$$

Figure 40–12 A soapy water film on a wire loop, viewed by reflected light. The black segment at the top is not a tear. It arises because the film, by drainage, is so thin here that there is destructive interference between the light reflected from its front surface and that reflected from its back surface. These two waves differ in phase by 180°.

Here n is the index of refraction of the film material with respect to the medium in which the film is immersed.

To bring out the second point, let us assume that the film is so thin that $2d$ is very much less than one wavelength. The phase difference between the two waves would be close to zero on our assumption, and we would expect such a film to appear bright on reflection. However, it appears dark. This is clear from Fig. 40–12, which shows a thin vertical film of soapy water, viewed by reflected monochromatic light. The action of gravity produces a wedge-shaped film, extremely thin at its top edge. As drainage continues, the dark area increases in size. To explain this and many similar phenomena, we assume that one or the other of the two rays of Fig. 40–11 suffers an abrupt phase change of 180° associated either with reflection at the air-film interface or transmission through it. The ray reflected from the front surface experiences this phase change. The other ray is not changed abruptly in phase, either on transmission through the front surface or on reflection at the back surface.

In Section 17–8 we discussed phase changes on reflection for transverse waves in strings. To extend these ideas, consider the composite string of Fig. 40–13, which consists of two parts with different masses per unit length, stretched to a given tension. If a pulse moves to the right in Fig. 40–13a, approaching the junction, there will be a reflected and a transmitted pulse, the reflected pulse being in phase with the incident pulse. In Fig. 40–13b the situation is reversed, the incident pulse now being in the lighter string. In this case the reflected pulse will differ in phase from the incident pulse by 180°. In each case the transmitted pulse will be in phase with the incident pulse.

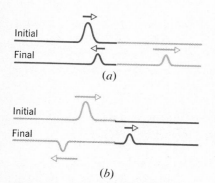

Figure 40–13 Phase changes on reflection at a junction between two stretched composite strings. (a) Incident pulse in heavy string and (b) incident pulse in light string.

Figure 40–13a suggests a light wave in glass, say, approaching a surface beyond which there is a less optically dense medium (one of lower index of refraction) such as air. Figure 40–13b suggests a light wave in air approaching glass. To sum up the optical situation, when reflection occurs from an interface beyond which the medium has a lower index of refraction, the reflected wave undergoes no phase change; when the medium beyond the interface has a higher index, there is a phase change of 180°. The transmitted wave does not experience a change of phase in either case.

We are now able to take into account both factors that determine the nature of the interference, namely, differences in optical path length and phase changes on reflection. For the two rays of Fig. 40–11 to combine to give a maximum intensity, assuming normal incidence, we must have

Thin-film interference—Maxima and minima

$$2d = (m + \tfrac{1}{2})\lambda_n \qquad m = 0, 1, 2, \ldots.$$

The term $\frac{1}{2}\lambda_n$ is introduced because of the phase change on reflection, a phase change of 180° being equivalent to half a wavelength. Substituting λ/n for λ_n yields finally

$$2nd = (m + \tfrac{1}{2})\lambda \qquad m = 0, 1, 2, \ldots \text{ (maxima)}. \qquad (40\text{-}13)$$

The condition for a minimum intensity is

$$2nd = m\lambda \qquad m = 0, 1, 2, \ldots \text{ (minima)}. \qquad (40\text{-}14)$$

These equations hold when the index of refraction of the film is either greater or less than the indices of the media on each side of the film. Only in these cases will there be a relative phase change of 180° for reflections at the two surfaces. A water film in air and an air film in the space between two glass plates provide examples of cases to which Eqs. 40–13 and 40–14 apply.

If the film thickness is not uniform, as in Fig. 40–12, where the film is wedge-shaped, constructive interference will occur in certain parts of the film and destructive interference will occur in others. Lines of maximum and of minimum intensity will appear—these are the interference fringes. They are called *fringes of constant thickness*, each fringe being the locus of points for which the film thickness d is a constant. If the film is illuminated with white light rather than monochromatic light, the light reflected from various parts of the film will be modified by the various constructive or destructive interferences that occur. This accounts for the brilliant colors of soap bubbles and oil slicks.

Example 3   A water film ($n = 1.33$) in air is 320 nm thick. If it is illuminated with white light at normal incidence, what color will it appear to be in reflected light?

By solving Eq. 40–13 for λ,

$$\lambda \approx \frac{2nd}{m + \tfrac{1}{2}} = \frac{(2)(320 \text{ nm})(1.33)}{m + \tfrac{1}{2}} = \frac{850 \text{ nm}}{m + \tfrac{1}{2}} \quad \text{(maxima)}.$$

From Eq. 40–14 the minima are given by

$$\lambda = \frac{850 \text{ nm}}{m} \quad \text{(minima)}.$$

Maxima and minima occur for the following wavelengths:

m	0 (max)	1 (min)	1 (max)	2 (min)	2 (max)
λ, nm	1700	850	570	425	340

Only the maximum corresponding to $m = 1$ lies in the visible region (see Fig. 38–2); light of this wavelength appears yellow-green. If white light is used to illuminate the film, the yellow-green component will be enhanced when viewed by reflection.

Example 4   *Nonreflecting glass*   Lenses are often coated with thin films of transparent substances like MgF$_2$ ($n = 1.38$) in order to reduce the reflection from the glass surface, using interference. How thick a coating is needed to produce a minimum reflection at the center of the visible spectrum (550 nm)?

We assume that the light strikes the lens at near-normal incidence (θ is exaggerated for clarity in Fig. 40–14), and we seek destructive interference between rays r_1 and r_2. Equation 40–14

does not apply because in this case a phase change of 180° is associated with each ray, for at both the upper and lower surfaces of the MgF$_2$ film the reflection is from a medium of greater index of refraction.

There is no net change in phase produced by the two reflections, which means that the optical path difference for destructive interference is $(m + \tfrac{1}{2})\lambda$ (compare Eq. 40–13), leading to

$$2nd = (m + \tfrac{1}{2})\lambda \qquad m = 0, 1, 2, \ldots \text{ (minima)}.$$

Solving for d and putting $m = 0$ yields

$$d = \frac{(m + \tfrac{1}{2})\lambda}{2n} = \frac{\lambda}{4n} = \frac{550 \text{ nm}}{(4)(1.38)} = 100 \text{ nm}.$$

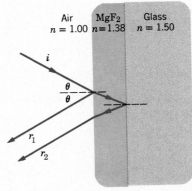

Figure 40-14   Example 4. Unwanted reflections from glass can be reduced by coating the glass with a thin transparent film.

40-6 Michelson's Interferometer

An interferometer is a device that can be used to measure lengths or changes in length with great accuracy by means of interference fringes. We describe the form originally devised and built by A. A. Michelson in 1881.

Consider light that leaves point P on extended source S (Fig. 40-15) and falls on half-silvered mirror (or "beam splitter") M. This mirror has a silver coating just thick enough to transmit half the incident light and to reflect half; in the figure we have assumed, for convenience, that this mirror possesses negligible thickness. At M the light divides into two waves. One proceeds by transmission toward mirror M_1; the other proceeds by reflection toward M_2. The waves are reflected at each of these mirrors and are sent back along their directions of incidence, each wave eventually entering the eye. Since the waves are coherent, being derived from the same point on the source, they will interfere.

If the mirrors M_1 and M_2 are exactly perpendicular to each other, the effect is that of light from an extended source S falling on a uniformly thick slab of air, between glass, whose thickness is equal to $d_2 - d_1$. Interference fringes appear, caused by small changes in the angle of incidence of the light from different points on the extended source as it strikes the equivalent air film. For thick films a path difference of one wavelength can be brought about by a very small change in the angle of incidence.

If M_2 is moved backward or forward, the effect is to change the thickness of the equivalent air film. Suppose that the center of the (circular) fringe pattern appears bright and that M_2 is moved just enough to cause the first bright circular fringe to move to the center of the pattern. The path of the light beam striking M_2 has been changed by one wavelength. This means (because the light passes twice through the equivalent air film) that the mirror must have moved one-half a wavelength. By such techniques the lengths of objects can be expressed in terms of the wavelength of light.

In Michelson's day the standard of length—the meter—was chosen by international agreement to be the distance between two fine scratches on a certain metal bar preserved at Sevres, near Paris. Michelson was able to show, using his interferometer, that the standard meter was equivalent to 1,553,163.5 wavelengths of a certain monochromatic red light emitted from a light source containing cadmium. For this careful measurement Michelson received the 1907 Nobel prize in physics. His work laid the foundation for the eventual abandonment (in 1961) of the meter bar as a standard of length and for the redefinition of the meter in terms of the wavelength of light, as described in Section 1-3.

Michelson's interferometer

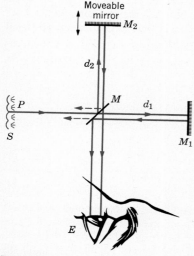

Figure 40-15 Michelson's interferometer, showing the path of a particular ray originating at point P of an extended source S.

The definition of the meter

REVIEW GUIDE AND SUMMARY

Diffraction and the failure of geometrical optics

Attempts to isolate a ray by forcing light through a narrow slit fail because of diffraction, the flaring out of the light into the geometrical shadow of the slit; see Figs. 40-1 and 40-2. If such slits are present the approximations of geometrical optics (Chapter 39) fail and the full treatment of wave optics must be used.

Interference

In Young's double-slit interference experiment light from a slit in screen A of Fig. 40-3 flares out (by diffraction!) and falls on the two slits in screen B. The light from these slits overlaps in the region beyond B and interference between the two sub-waves occurs. A fringe pattern like that of Fig. 40-4 is formed on screen C.

Locating the interference fringes

The light intensity at any point on screen C of Fig. 40-3 depends on the path difference between two rays drawn to that point from the two slits. If this difference is an integral number of wavelengths the waves will interfere constructively and an intensity maximum results. If it is a half-integral multiple there is destructive interference and an intensity minimum. Analysis shows:

$$d \sin \theta = \begin{cases} m\lambda & \text{maxima} & [40\text{-}1] \\ & m = 0, 1, 2, \dots \\ (m + \tfrac{1}{2})\lambda & \text{minima} & [40\text{-}2] \end{cases}$$

Coherence and incoherence

If two overlapping light waves are to interfere at a given point the phase difference between them must remain constant with time; the waves must be coherent. Very long wave-trains, as in Fig. 40-7a, must be involved. Laser light is coherent. Sunlight is incoherent, consisting of an assembly of short, overlapping wavetrains. Incoherent waves have random phase relationships and do not readily exhibit interference effects.

The intensity in Young's experiment

When two coherent waves overlap the resulting intensity may be found by the phasor method, as in Fig. 40-8. In this method the amplitude E_θ of the electric field vector of the resultant wave is calculated, taking the phase difference between the two combining waves properly into account. The intensity I_θ of the resultant wave is then taken to be proportional to $E_\theta{}^2$. As applied to Young's double-slit experiment we have:

$$I_\theta = I_m \cos^2 \beta, \qquad \text{where} \qquad \beta = (\pi d/\lambda) \sin \theta. \qquad [40\text{-}11b, c]$$

Equations 40-1 and 40-2, which identify the positions of the fringe maxima and minima, are contained within this relation. Figure 40-9 should be reviewed with some care.

Thin-film interference

When light falls on a thin transparent film the light waves reflected from the front and the rear surfaces, as in Fig. 40-11, interfere. For near-normal incidence the conditions for a maximum or a minimum intensity of the light reflected from the film are:

$$2nd = \begin{cases} (m + \tfrac{1}{2})\lambda & \text{maxima} & [40\text{-}13] \\ & m = 0, 1, 2, \dots \\ m\lambda & \text{minima} & [40\text{-}14] \end{cases}$$

Phase change on reflection

Here the medium in which the film is immersed is assumed to be the same on each side, n is the index of the film material with respect to this medium, and λ is the wavelength in the medium. The formula also assumes that one of the interfering waves has undergone a "hard" reflection (that is, from a higher index medium; compare Fig. 40-13b) and the other a "soft" reflection (that is, from a lower index medium; compare Fig. 40-13a). Hard reflections involve a phase change of 180°; soft reflections do not. Example 4 is a case in which some assumptions of the above formula do not hold and the problem must be solved from first principles.

The Michelson interferometer

In Michelson's interferometer (Fig. 40-15) a light beam is split into two sub-beams which, after traversing paths of different lengths, are recombined so that they interfere and form a fringe pattern. By varying the path length of one of the sub-beams distances can be accurately expressed in terms of wavelengths of light. Thus, in Fig. 40-15, as mirror M_2 is moved through the distance in question, one simply counts the number of fringes (wavelengths) that pass through the field of view of the observer.

QUESTIONS

1. Is Young's experiment an interference experiment or a diffraction experiment, or both?

2. Do interference effects occur for sound waves? Recall that sound is a longitudinal wave and that light is a transverse wave.

3. In Young's double-slit interference experiment, using a monochromatic laboratory light source, why is screen A in Fig. 40-3 necessary?

4. What changes occur in the pattern of interference fringes if the apparatus of Fig. 40-5 is placed under water?

5. If interference between light waves of different frequencies is possible, one should observe light beats, just as one obtains sound beats from two sources of sound with slightly different frequencies. Discuss how one might experimentally look for this possibility.

6. Why are parallel slits preferable to the pinholes that Young used in demonstrating interference?

7. Is coherence important in reflection and refraction?

8. If your source of light is a laser beam, you do not need the equivalent of screen A in Fig. 40–3. Why?

9. Describe the pattern of light intensity on screen C in Fig. 40–5 if one slit is covered with a red filter and the other with a blue filter, the incident light being white.

10. Define carefully, and distinguish between, the angles β, θ, and ϕ that appear in Eqs. 40–11.

11. If one slit in Fig. 40–5 is covered, what change would occur in the intensity of light in the center of the screen?

12. We are all bathed continuously in electromagnetic radiation, from the sun, from radio and TV signals, from the stars and other celestial objects. Why don't these waves interfere with each other?

13. Is polarization or interference a better test for identifying waves? Do they give the same information?

14. What causes the fluttering of a TV picture when an airplane flys overhead?

15. Is it possible to have coherence between light sources emitting light of different wavelengths?

16. Each slit in Fig. 40–5 is covered with a sheet of Polaroid, the polarizing directions of the two slits being at right angles. What is the pattern of light intensity on screen C? (The incident light is unpolarized.)

17. Suppose that the film coating in Fig. 40–14 had a refractive index greater than that of the glass. Could it still be nonreflecting? If so, what difference would it make?

18. What are the requirements for maximum intensity when viewing a thin film by *transmitted* light?

19. Why does a film (soap bubble, oil slick, etc.) have to be "thin" to display interference effects? Or does it? How thin is "thin"?

20. Why do coated lenses (see Example 4) look purple by reflected light?

21. A person wets his eyeglasses to clean them. As the water evaporates he notices that for a short time the glasses become markedly less reflecting. Explain.

22. A lens is coated to reduce reflection, as in Example 4. What happens to the energy that had previously been reflected? Is it absorbed by the coating?

23. Consider the following objects that produce colors when exposed to sunlight: (1) soap bubbles, (2) rose petals, (3) the inner surface of an oyster shell, (4) thin oil slicks, (5) nonreflecting coatings on camera lenses, and (6) peacock tail feathers. The colors displayed by all but one of these are purely interference phenomena, no pigments being involved. Which one is the exception? Why do the others seem to be "colored".

24. An automobile directs its headlights onto the side of a barn. Why are interference fringes not produced in the region in which light from the two beams overlaps?

25. A soap film on a wire loop held in air appears black at its thinnest portion when viewed by reflected light. On the other hand, a thin oil film floating on water appears bright at its thinnest portion when similarly viewed from the air above. Explain this phenomenon.

EXERCISES

Section 40–2 Young's Experiment

1. In an experiment using Young's arrangement to demonstrate the interference of light, the separation d of the two narrow slits is doubled. In order to maintain the same spacing of the fringes, how must the distance D of the screen from the slits be altered?

2. Young's experiment is performed with blue-green light of wavelength 500 nm. The slits are 1.0 mm apart and the screen is 5.0 m from the slits. How far apart are the bright fringes?

3. Design a double-slit arrangement that will produce interference fringes 1.0° apart on a distant screen. Assume sodium light ($\lambda = 589$ nm).

4. A double-slit arrangement produces interference fringes for sodium light ($\lambda = 589$ nm) that are 0.20° apart. For what wavelength would the angular separation be 10 percent greater?

5. In a double-slit arrangement the slits are separated by a distance equal to 100 times the wavelength of the light passing through the slits. (a) What is the angular separation between the central maximum and an adjacent maximum? (b) What is the linear distance between these maxima if the screen is 50 cm from the slits?

6. Referring to Fig. 40–6 in a double-slit experiment, $\lambda = 546$ nm, $d = 0.10$ mm, and $D = 20$ cm. What is the linear distance between the 5th maximum and 7th minimum?

7. In Young's experiment, how are Eqs. 40–1 and 40–2 changed if the light emitted from the two slits has a phase difference of 180°?

8. Sodium light ($\lambda = 589$ nm) falls on a double slit of separation $d = 0.20$ mm. A thin lens ($f = +1.0$ m) is placed near the slit as in Fig. 40–6. What is the linear fringe separation on a screen placed in the focal plane of the lens?

Section 40–3 Coherence

9. The coherence length of a wavetrain is the distance over which the phase constant is the same. (a) If an individual atom emits coherent light for 1×10^{-8} s, what is the coherence length of the wavetrain? (b) Suppose this wavetrain is separated into two parts with a partially reflecting mirror and later reunited after one beam travels 5.0 m and the other 10 m. Do the waves produce interference fringes observable by a human eye?

Section 40–4 Intensity of Interfering Waves

10. Light of wavelength 600 nm is incident normally on a system of two parallel narrow slits separated by 0.60 mm. Sketch the intensity pattern observed on a distant screen as a function of angle θ, as in Fig. 40-5, for the range of values $0 \leq \theta \leq 0.0040$ radians.

11. Add the following quantities graphically, using the phasor method:

$$y_1 = 10 \sin \omega t,$$
$$y_2 = 15 \sin (\omega t + 30°),$$
$$y_3 = 5 \sin (\omega t - 45°).$$

12. If a radiation source A leads source B by 90° and the distance r_A to a detector is greater than the distance r_B by 100 m, what is the phase difference at the detector? Both sources have a wavelength of 400 m.

13. A thin coating of refractive index 1.25 is applied to a glass camera lens to minimize the intensity of the light reflected from the lens. In terms of λ, the wavelength in air of the incident light, what is the smallest thickness of the coating which is needed?

Section 40-5   Interference from Thin Films

14. We wish to coat a flat slab of glass ($n = 1.50$) with a transparent material ($n = 1.25$) so that light of wavelength 600 nm (in vacuum) incident normally is not reflected. What thickness should the coating have?

15. A transparent liquid of refractive index 4/3 is allowed to displace the air from an air wedge that is formed from two glass plates touching each other along one edge. What happens, as a result, to the spacing of the dark lines caused by interference of the monochromatic light reflected back?

16. A thin film 0.41 μm thick is illuminated by white light normal to its surface. Its index of refraction is 1.5. What wavelengths within the visible spectrum will be intensified in the reflected beam?

17. A tanker leaks a large quantity of kerosene ($n = 1.2$) into the Gulf of Mexico, creating a large slick on top of the water ($n = 1.3$). (a) If you are looking straight down from an airplane onto a region of the slick where its thickness is 460 nm, for which wavelength(s) of visible light is the reflection the greatest? (b) If you are scuba diving directly under this same region of the slick, for which wavelengths of visible light is the transmitted intensity the strongest?

18. Light of wavelength 630 nm is incident normally on a thin wedge-shaped film of index of refraction 1.5. There are 10 bright and 9 dark fringes over the length of film. By how much does the film thickness change over this length?

19. Light of wavelength 585 nm is incident normally on a thin soapy film ($n = 1.33$) suspended in air. If the film is 0.00121 mm thick, determine whether it appears bright or dark when observed from a point near the light source.

Section 40-6   Michelson's Interferometer

20. If mirror M_2 in Michelson's interferometer is moved through 0.233 mm, 792 fringes are counted. What is the wavelength of the light?

21. A thin film with $n = 1.40$ for light of wavelength 589 nm is placed in one arm of a Michelson interferometer. If a shift of 7.0 fringes occurs, what is the film thickness?

PROBLEMS GROUPED BY SECTION

Section 40-2   Young's Experiment

1. In a double-slit experiment the distance between slits is 5.0 mm and the slits are 1.0 m from the screen. Two interference patterns can be seen on the screen, one due to light of wavelength 480 nm and the other due to light of wavelength 600 nm. What is the separation on the screen between the third-order interference fringes of the two different patterns?

2. In Young's interference experiment in a large ripple tank (see Fig. 40-3) the coherent vibrating sources are placed 120 mm apart. The distance between maxima 2.0 m away is 180 mm. If the speed of ripples is 25 cm/s, find the frequency of the vibrators.

3. If the distance between the first and tenth minima of a double-slit pattern is 2.4 mm and the slits are separated by 0.15 mm with the screen 50 cm from the slits, what is the wavelength of the light used?

4. As shown in Fig. 40-16, A and B are two identical radiators of waves that are in phase and of the same wavelength λ. The radiators are separated by distance 3.0 λ. Find the largest distance from A, along the line Ax, for which destructive interference occurs. Express this in terms of λ.

5. Two coherent radio point sources separated by 2.0 m are radiating in phase with $\lambda = 0.50$ m. A detector moved in a closed path around the two sources in a plane containing them will show how many maxima?

6. In Fig. 40-17, the source emits monochromatic light of wavelength λ. S is a narrow slit in an otherwise opaque screen I. A plane mirror, whose surface includes the axis of the lens shown, is located a distance h below S. Screen II is at the focal plane of the lens. (a) Find the condition for maximum and minimum brightness of fringes on screen II in terms of the usual angle θ, the wavelength λ, and the distance h. (b) Show that fringes appear only in region A (above the axis of the lens), but not in region B (below the axis of the lens) or in both regions A and B. (Hint: Consider the image of S formed by the mirror.)

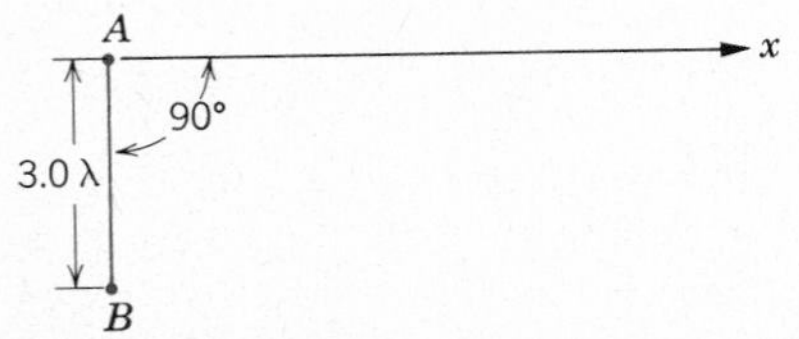

Figure 40-16   Problem 4.

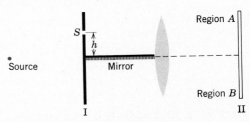

Figure 40-17   Problem 6.

7. A thin flake of mica ($n = 1.58$) is used to cover one slit of a double-slit arrangement. The central point on the screen is occupied by what used to be the seventh bright fringe. If $\lambda = 550$ nm, what is the thickness of the mica?

Section 40–4   Intensity of Interfering Waves

8. Prove Eq. 40–11a, and hence Eq. 40–11b, by trigonometry, without using phasors.

9. Two waves of the same frequency have amplitudes 1 and 2, respectively. They interfere at a point where their phase difference is 60°. What is the resultant amplitude?

10. Show that the half-width $\Delta\theta$ of the double-slit interference fringes (see horizontal arrow in Fig. 40–9) is given by

$$\Delta\theta = \frac{\lambda}{2d}$$

if θ is small enough so that $\sin\theta \cong \theta$.

11. S_1 and S_2 in Fig. 40–18 are effective point sources of radiation, excited by the same oscillator. They are coherent and in phase with each other. Placed 4.0 m apart, they emit equal amounts of power in the form of 1.0-m-wavelength electromagnetic waves. (a) Find the positions of the first (that is, the nearest), the second, and the third maxima of the received signal, as the detector is moved out along x. (b) Is the intensity at the nearest minimum equal to zero? Justify your answer.

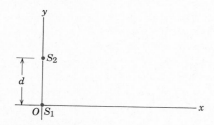

Figure 40–18   Problem 11.

12. Find the sum of the following quantities (a) by the phasor method and (b) analytically:

$$y_1 = 10 \sin \omega t.$$
$$y_2 = 8 \sin (\omega t + 30°).$$

13. A lens is coated with a thin transparent film to minimize reflection of the red component of white light. The index of refraction of the film is 1.30 and that of the lens is 1.65. In terms of λ_n, the wavelength of red light in the film, what minimum thickness of film is needed?

Section 40–5   Interference from Thin Films

14. A plane wave of monochromatic light falls normally on a uniformly-thin film of oil that covers a glass plate. The wavelength of the source can be varied continuously. Complete destructive interference of the reflected light is observed for wavelengths of 500 and 700 nm and for no other wavelengths in between. If the index of refraction of the oil is 1.3 and that of the glass is 1.5, find the thickness of the oil film.

15. White light reflected at perpendicular incidence from a soap film has, in the visible spectrum, an interference maximum at 600 nm and a minimum at 450 nm with no minimum in between. If $n = 1.33$ for the film, what is the film thickness, assumed uniform?

16. Two pieces of plate glass are held together in such a way that the air space between them forms a very thin wedge. Light of wavelength 4.8×10^{-5} cm strikes the upper surface perpendicularly and is reflected from the lower surface of the top glass and the upper surface of the bottom glass, thereby producing a series of interference fringes. How much thicker is the air wedge at the 16th fringe than it is at the 6th?

17. A sheet of glass having an index of refraction of 1.4 is to be coated with a film of material having a refractive index of 1.55 such that green light (wavelength = 525 nm) is preferentially transmitted. (a) What is the minimum thickness of the film that will achieve the result? (b) Why are other parts of the visible spectrum not also preferentially transmitted? (c) Will the transmission of any colors be sharply reduced?

18. A broad source of light ($\lambda = 680$ nm) illuminates normally two glass plates 120 mm long that touch at one end and are separated by a wire 0.048 mm in diameter at the other end (Fig. 40–19). How many bright fringes appear over the 120-mm distance?

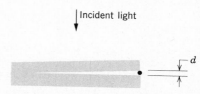

Figure 40–19   Problems 18 and 37.

19. A plane monochromatic light wave in air falls at normal incidence on a thin film of oil that covers a glass plate. The wavelength of the source may be varied continuously. Complete destructive interference in the reflected beam is observed for wavelengths of 500 and 700 nm and for no wavelength between. The index of refraction of glass is 1.5. Show that the index of refraction of the oil must be less than 1.5.

20. *Newton's rings.* Figure 40–20 shows a lens of radius of curvature R resting on an accurately-plane glass plate and illumi-

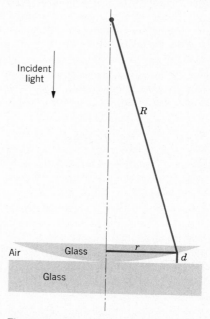

Figure 40–20   Problems 20, 39, 40 and 41.

Figure 40–21   Problem 20.

nated from above by light of wavelength λ. Figure 40–21 shows that circular interference fringes (Newton's rings) appear, associated with the variable thickness air film between the lens and the plate. Find the radii r of the circular interference maxima assuming that $r/R \ll 1$.

21. In a Newton's rings (see Problem 20) experiment the radius of curvature R of the lens is 5.0 m and its diameter is 20 mm. (a) How many rings are produced? (b) How many rings would be seen if the arrangement were immersed in water ($n = 1.33$)? Assume that $\lambda = 589$ nm.

Section 40–6   Michelson's Interferometer

22. A Michelson interferometer is used with a sodium discharge tube as a light source. The yellow sodium light consists of two wavelengths, 589.0 nm ($= 5890$ Å) and 589.6 nm ($= 5896$ Å). It is observed that the interference pattern disappears and reappears periodically as one moves mirror M_2 in Fig. 40–15. (a) Explain this effect. (b) Calculate the change in path difference between two successive reappearances of the interference pattern.

23. An air-tight chamber 5.0 cm long with glass windows is placed in one arm of a Michelson interferometer as indicated in Fig. 40–22. Light of $\lambda = 500$ nm is used. The air is slowly evacuated from the chamber using a vacuum pump. While the air is being removed, 60 fringes are observed to pass through the view. From these data, find the index of refraction of air at atmospheric pressure.

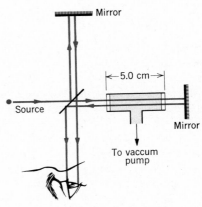

Figure 40–22   Problem 23.

ADDITIONAL PROBLEMS

24. Design a Young-type double-slit interference experiment (see Fig. 40–3) for sound waves in air. Assume a frequency of 500 Hz and a speed of 330 m/s. Discuss some of the design parameters such as the nature of the source, the width, height, and separation of the two slits, the "fringe" separation, the nature of the fringe detector, and so forth.

25. In the front of a lecture hall, a coherent beam of monochromatic light from a helium-neon laser ($\lambda = 632.8$ nm $= 6328$ Å) illuminates a double slit. From there it travels a distance d (20 m) to a mirror at the back of the hall, and returns the same distance to a screen. (a) In order that the distance between interference maxima be 10 cm what should be the distance between the two slits? (b) State briefly what you will see if the lecturer slips a thin sheet of cellophane over one of the slits. The optical pathlength through the cellophane is 2.5 wavelengths longer than the equivalent air pathlength.

26. Interference fringes are produced using white light with a double slit arrangement. A piece of parallel-sided mica of refractive index 1.6 is placed in front of one of the slits, as a result of which the center of the fringe system moves to the left a distance subsequently shown to accommodate 30 dark bands when light of wavelength 480 nm is used. What is the thickness of the mica?

27. In an air wedge formed by two plane glass plates, touching each other along one edge, there are 4001 dark lines observed when viewed by reflected monochromatic light. When the air between the plates is evacuated, only 4000 such lines are observed. What is the refractive index of the air?

28. One slit of a double-slit arrangement is covered by a thin glass plate of refractive index 1.4, and the other by a thin glass plate of refractive index 1.7. The point on the screen where the central maximum fell before the glass plates were inserted is now occupied by what had been the fifth bright fringe before. Assume $\lambda = 480$ nm and that the plates have the same thickness t. Find the value of t.

29. A double-slit arrangement produces interference fringes for sodium light ($\lambda = 589$ nm) that are 0.20° apart. What is the angular fringe separation if the entire arrangement is immersed in water?

30. Two point sources, S_1 and S_2, in Fig. 40–23 emit coherent waves. Show that curves, such as that given, over which the phase difference for rays r_1 and r_2 is a constant, are hyperbolas. (Hint: A constant phase difference implies a constant difference in length between r_1 and r_2.)

31. Sodium light ($\lambda = 589$ nm) falls on a double-slit of separation $d = 2.0$ mm. D in Fig. 40–5 is 40 mm. What percent error is made by using Eq. 40–1 to locate the tenth fringe on the screen?

32. A thin film of acetone (refractive index 1.25) is floated on a

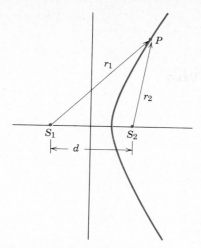

Figure 40-23 Problem 30.

thick glass plate (refractive index 1.50). Plane light waves of variable wavelengths are incident normal to the film. When one views the reflected wave it is noted that complete destructive interference occurs at 600 nm and constructive interference at 700 nm. Calculate the thickness of the acetone film.

33. In costume jewelry, rhinestones (made of glass with $n = 1.50$) are often coated with silicon monoxide ($n = 2.0$) to make them more reflective. How thick should the coating be to achieve strong reflection for 550 nm light, incident normally?

34. A soap film ($n = 1.33$) is illuminated by light of wavelength 624 nm in air, incident normally. What is (a) the smallest thickness, and (b) the second smallest thickness that results in no reflected light?

35. From a medium of index of refraction n_1, monochromatic light of wavelength λ falls normally on a thin film of uniform thickness and index of refraction n_2. The transmitted light travels in a medium of index of refraction n_3. Find expressions for the minimum film thickness (in terms of λ and the indices of refraction) for the following cases: (a) $n_1 < n_2 > n_3$—minimum reflected light; (b) $n_1 < n_2 > n_3$—maximum transmitted light; (c) $n_1 < n_2 < n_3$—minimum reflected light; (d) $n_1 < n_2 < n_3$—maximum transmitted light; and (e) $n_1 < n_2 < n_3$—maximum reflected light.

36. An oil drop ($n = 1.20$) floats on a water ($n = 1.33$) surface and is observed from above by reflected light (see Fig. 40-24). (a)

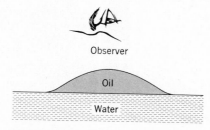

Observer

Figure 40-24 Problem 36.

Will the outer (thinnest) regions of the drop correspond to a bright or dark region? (b) Approximately how thick is the oil film where one observes the third blue region from the outside of the drop? (c)

Why do the colors gradually disappear as the oil thickness becomes larger?

37. In Fig. 40-19 white light is incident from above. (a) Observed from above, why does the region near the edge, where the two glass plates touch, appear black? (b) For what part of the visible spectrum does destructive interference next occur? (c) What color does an observer see where this destructive interference occurs?

38. A perfectly flat piece of glass ($n = 1.5$) is placed over a perfectly flat piece of black plastic ($n = 1.2$) as shown in Fig. 40-25a. They touch at A. Light of wavelength 600 nm is incident normally from above and the reflected pattern is shown in Fig. 40-25b. (a) How thick is the space between the glass and the plastic at B? (b) Water ($n = 1.33$) seeps into the region between the glass and plastic. How many dark lines are seen when all the air has been displaced by water?

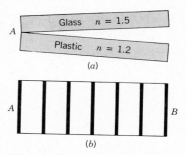

Figure 40-25 Problem 38.

39. In the Newton's rings experiment, use the result of Problem 20 to show that the difference in radius between adjacent rings (maxima) is given by

$$\Delta r = r_{m+1} - r_m \cong \tfrac{1}{2}(\lambda R/m)^{1/2},$$

assuming $m \gg 1$.

40. In the Newton's rings experiment, use the result of the preceding problem to show that the *area* between adjacent rings (maxima) is given by

$$A = \pi \lambda R,$$

assuming $m \gg 1$. Note that this area is independent of m.

41. A Newton's rings apparatus is used to determine the radius of a lens (see Fig. 40-20). The radii of the nth and $(n + 20)$th bright rings are measured and found to be 0.162 ± 0.005 cm and 0.368 ± 0.003 cm, respectively, in light of wavelength 546 nm. (a) Calculate the radius of curvature of the lower lens surface. (b) What is the possible error in the radius?

42. Write an expression for the intensity observed in Michelson's interferometer (Fig. 40-15) as a function of the position of the moveable mirror, measured from the point for which $d_1 = d_2$.

43. In Example 4, assume that there is zero reflection for light of wavelength 550 nm at normal incidence. Calculate the factor by which the reflection is diminished by the coating at 450 and at 650 nm.

*44. One of the slits of a double-slit system is wider than the other, so that the amplitude of the light reaching the central part of the screen from one slit, acting alone, is twice that from the other slit, acting alone. Derive an expression for I_θ in terms of θ, corresponding to Eqs. 40-11b and 40-11c.

41 Diffraction

Figure 41–1 The diffraction pattern formed when light falls on an IBM computing card in which rectangular holes have been punched. From *The Physics Teacher*. Vol. 7, #3, March 1969. Professor Roy Biser, Lamar University.

41–1 Diffraction—a Closer Look

In Chapter 40 we defined diffraction rather loosely as the flaring out of light as it emerges from the confines of a narrow slit. Figure 40–1 suggests the main feature of the phenomenon, namely, the narrower the slit, with respect to the wavelength of the light, the more the flaring out. Here we examine the diffraction phenomenon in more detail.

Figure 41–1 shows the diffraction pattern formed when light falls on the rectangular punched holes in a computer card. We see at once "flaring out" is far from a complete description of the process. There is a broad and intense central maximum—which was our principal concern in Chapter 40—but there are also secondary maxima and minima. Figure 41–2 shows the pattern for a single long narrow slit.

Figure 41–2 The diffraction pattern for a single slit. From *An Atlas of Optical Phenomena* by M. Cagnet, M. Francon, and J. C. Thrierr, Springer-Verlag, Berlin, 1962.

Figure 41–3 shows the generalized diffraction situation. Surface *A* is a wavefront that falls on *B*, which is an opaque screen containing an aperture of arbitrary shape; *C* is a screen or photographic film that receives the light that passes

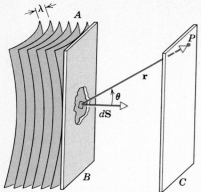

Figure 41–3 Coherent Light wavefronts such as *A* fall on an aperture in screen *B* and illuminate screen *C*. *P* is an arbitrary point on this screen at which we wish to find the light intensity.

Fresnel diffraction

Fraunhofer diffraction

through this aperture. We can calculate the pattern of light intensity on *C* by sub-dividing the wavefront into elementary areas *d***S**, each of which becomes a source of an expanding Huygens wavelet. The light intensity at an arbitrary point *P* is found by superimposing the wave disturbances (that is, the **E**-vectors) caused by the wavelets reaching *P* from all these elementary sources. Instead of a screen with a hole in it we could equally well use an opaque barrier such as a wire or a coin.

The wave disturbances reaching *P* differ in amplitude and in phase because (*a*) the elementary radiators are at varying distances from *P*, (*b*) the light leaves the radiators at various angles to the normal to the wavefront, and (*c*) some radiators are blocked by screen *B*; others are not. Diffraction calculations—simple in principle—may become difficult in practice. The calculation must be repeated for every point on screen *C* at which we wish to know the light intensity. We followed exactly this program in calculating the double-slit intensity pattern in Section 40–4. The calculation there was simple because we assumed only two radiators, the two narrow slits.

Figure 41–4*a* shows the general case of Fresnel diffraction, in which the light source and/or the screen on which the diffraction pattern is displayed are a finite distance from the diffracting aperture; the wavefronts that fall on the diffracting aperture in this case and that leave it to illuminate any point *P* of the diffusing screen are not planes; the corresponding rays are not parallel. As illustrated in Fig. 41–4*a*, Fresnel diffraction is a "close-in" situation.

A simplification results if source *S* and screen *C* are moved to a large distance from the diffraction aperture, as in Fig. 41–4*b*. This limiting "far-out" case is called Fraunhofer diffraction. The wavefronts arriving at the diffracting aperture from the distant source *S* are then planes, and the rays associated with these wave-fronts are parallel to each other. Fraunhofer conditions can be established in the laboratory by using two converging lenses, as in Fig. 41–4*c*. The first of these con-verts the diverging wave from the source into a plane wave. The second lens causes plane waves leaving the diffracting aperture to converge to point *P*. All rays that illuminate *P* will leave the diffracting aperture parallel to the dashed line *Px* drawn from *P* through the center of this second (thin) lens. We assumed Fraun-hofer conditions for Young's double-slit experiment in Section 40–2. Although Fraunhofer diffraction is a limiting case of the more general Fresnel diffraction, it is an important limiting case and is easier to handle mathematically. We shall treat only Fraunhofer diffraction.

41–2   Diffraction from a Single Slit—Locating the Minima

Figure 41–5 shows a plane wave falling at normal incidence on a long narrow slit of width *a*. Let us focus our attention on the central point P_0 of screen *C*. A set of horizontal, parallel rays (not shown in the figure) emerging from the slit will be fo-cused at P_0. These rays all have the same optical path lengths, as we saw in Sec-tion 39–4. Since they are in phase at the plane of the slit, they will still be in phase at P_0, and the central point of the diffraction pattern that appears on screen *C* has a maximum intensity.

We now consider another point on the screen. Light rays which reach P_1 in Fig. 41–5 leave the slit at an angle θ as shown. The ray xP_1 determines θ because it passes undeflected through the center of the lens. Ray r_1 originates at the top of the slit and ray r_2 at its center. If θ is chosen so that the distance bb' in the figure is

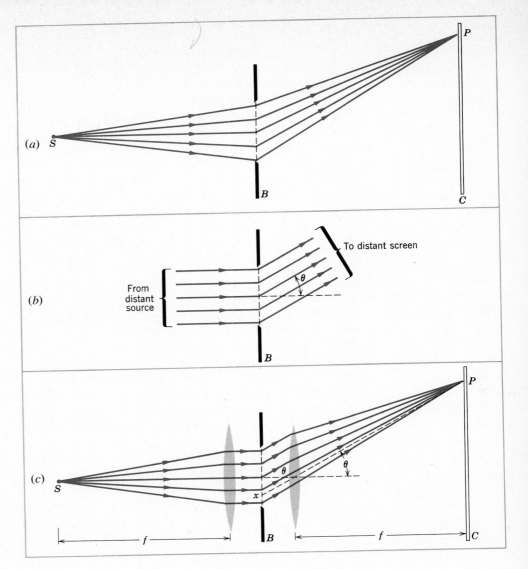

Figure 41-4 (a) Fresnel diffraction. (b) Source S and screen C are moved to a large distance, resulting in Fraunhofer diffraction, a limiting case of Fresnel diffraction. (c) Fraunhofer diffraction conditions produced by lenses, leaving source S and screen C on their original positions.

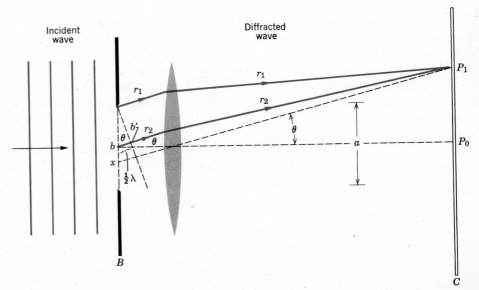

Figure 41-5 Conditions at the first minimum of the diffraction pattern.

one-half a wavelength, r_1 and r_2 will be out of phase and will produce no effect at P_1. In fact, every ray from the upper half of the slit will be canceled by a ray from the lower half, originating at a point $a/2$ below the first ray. The point P_1, the first minimum of the diffraction pattern, will have zero intensity.

The condition shown in Fig. 41–5 is

$$\frac{a}{2} \sin \theta = \frac{\lambda}{2},$$

or

$$a \sin \theta = \lambda. \tag{41-1}$$

Equation 41–1 shows that the central maximum becomes wider as the slit becomes narrower, that is, θ increases as a decreases, just as Fig. 40–1 suggests. For $a = \lambda$ we have $\theta = 90°$, which implies that the central maximum fills the entire forward hemisphere.

In Fig. 41–6 the slit is divided into four equal zones, with a ray shown leaving the top of each zone. Let θ be chosen so that the distance bb' is one-half a wavelength. Rays r_1 and r_2 will then cancel at P_2. Rays r_3 and r_4 will also be half a wavelength out of phase and will also cancel. Consider four other rays, emerging from the slit a given distance below the four rays above. The two rays below r_1 and r_2 will exactly cancel, as will the two rays below r_3 and r_4. We can proceed across the entire slit and conclude again that no light reaches P_2; we have located a second point of zero intensity.

Figure 41–6 Conditions at the second minimum of the diffraction pattern.

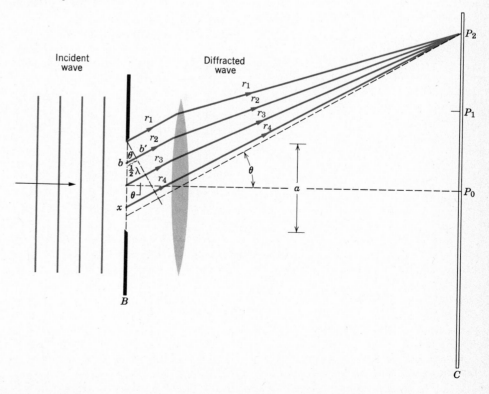

The condition described (see Fig. 41–6) requires that

$$\frac{a}{4} \sin \theta = \frac{\lambda}{2}, \quad \text{or} \quad a \sin \theta = 2\lambda.$$

By extension, the general formula for the minima in the diffraction pattern on screen C is

Location of minima in single slit diffraction

$$a \sin \theta = m\lambda \quad m = 1, 2, 3, \ldots \text{(minima).} \tag{41–2}$$

There is a maximum approximately halfway between each adjacent pair of minima.

Example 1   A slit of width a is illuminated by white light. For what value of a will the first minimum for red light ($\lambda = 650$ nm) fall at $\theta = 30°$?

At the first minimum we put $m = 1$ in Eq. 41–2. Doing so and solving for a yields

$$a = \frac{m\lambda}{\sin \theta} = \frac{(1)(650 \text{ nm})}{\sin 30°} = 1300 \text{ nm} = 1.3 \ \mu\text{m}.$$

Note that the slit width must be twice the wavelength in this case.

Example 2   In Example 1 what is the wavelength λ' of the light whose first diffraction maximum (not counting the central maximum) falls at $\theta = 30°$, thus coinciding with the first minimum for red light?

This maximum is about halfway between the first and second minima. We can find it without too much error by putting $m = 1.5$ in Eq. 41–2, or

$$a \sin \theta \cong 1.5 \ \lambda'.$$

From Example 1, however,

$$a \sin \theta = \lambda.$$

Dividing gives

$$\lambda' = \frac{\lambda}{1.5} = \frac{650 \text{ nm}}{1.5} = 430 \text{ nm}.$$

Light of this wavelength is violet. The second maximum for light of wavelength 430 nm will always coincide with the first minimum for light of wavelength 650 nm, no matter what the slit width. If the slit is relatively narrow, the angle θ at which this overlap occurs will be relatively large.

41–3   Diffraction from a Single Slit—Qualitative

We next turn our attention to the more general problem of calculating the intensity at any point P in a single-slit diffraction pattern. Figure 41–7 shows a slit of width a divided into N parallel strips of width Δx. Each strip acts as a radiator of Huygens' wavelets and produces a characteristic wave disturbance at point P, whose position on the screen, for a particular arrangement of apparatus, can be described by the angle θ.

If the strips are narrow enough ($\Delta x \ll \lambda$) all points on a given strip have essentially the same optical pathlength to P, and therefore all the light from the strip will have the same phase when it arrives at P. We may take the amplitudes ΔE_0 of the wave disturbances at P from the various strips as equal if θ in Fig. 41–7 is not too large.

The wave disturbances from adjacent strips have a constant phase difference $\Delta \phi$ between them at P given by

$$\frac{\text{phase difference}}{2\pi} = \frac{\text{path difference}}{\lambda},$$

or

$$\Delta \phi = \left(\frac{2\pi}{\lambda}\right) (\Delta x \sin \theta), \tag{41–3}$$

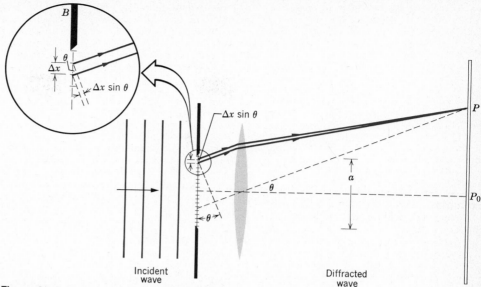

Figure 41–7   A slit of width a is divided into N strips of width Δx. The insert shows conditions at the second strip more clearly. In the differential limit the slit is divided into an infinite number of strips (that is, $N \to \infty$) of differential width dx. For clarity in this and the following figure, we take $N = 18$.

where $\Delta x \sin \theta$ is, as the figure insert shows, the path difference for rays originating at the top edges of adjacent strips. Thus, at P, N vectors with the same amplitude ΔE_0, the same frequency, and the same phase difference $\Delta \phi$ between adjacent members combine to produce a resultant disturbance. We ask, for various values of $\Delta \phi$ [that is, for various points P on the screen, corresponding to various values of θ (see Eq. 41–3)], what is the amplitude E_θ of the resultant wave disturbance? We find the answer by representing the individual wave disturbances ΔE_0 by phasors and calculating the resultant phasor amplitude, as described in Section 40–4.

At the center of the diffraction pattern θ equals zero, and the phase shift between adjacent strips (see Eq. 41–3) is also zero. As Fig. 41–8a shows, the phasor arrows in this case are laid end to end and the amplitude of the resultant has its maximum value E_m. This corresponds to the center of the central maximum.

As we move to a value of θ other than zero, $\Delta \phi$ assumes a definite nonzero value (again see Eq. 41–3), and the array of arrows is now as shown in Fig. 41–8b. The resultant amplitude E_θ is less than before. Note that the length of the "arc" of small arrows is the same for both figures and indeed for all figures of this series. As θ increases further, a situation is reached (Fig. 41–8c) in which the chain of arrows curls around through 360°, the tip of the last arrow touching the foot of the first arrow. This corresponds to $E_\theta = 0$, that is, to the first minimum.

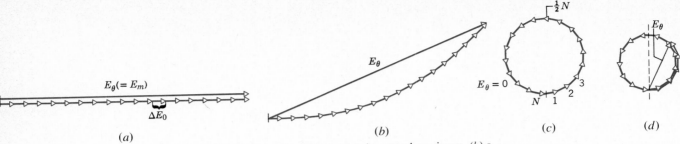

Figure 41–8 Conditions at (*a*) the central maximum, (*b*) a direction slightly removed from the central maximum, (*c*) the first minimum, and (*d*) the first maximum beyond the central maximum for single-slit diffraction. This figure corresponds to $N = 18$ in Fig. 41–7.

For this condition the ray from the top of the slit (1 in Fig. 41–8*c*) is 180° out of phase with the ray from the center of the slit ($\frac{1}{2}N$ in Fig. 41–8*c*). These phase relations are consistent with Fig. 41–5, which also represents the first minimum.

As θ increases further, the phase shift continues to increase, and the chain of arrows coils around through an angular distance greater than 360°, as in Fig. 41–8*d*, which corresponds to the first maximum beyond the central maximum. This maximum is much smaller than the central maximum. In making this comparison, recall that the arrows marked E_θ in Fig. 41–8 correspond to the amplitudes of the wave disturbance and not to the intensities. The amplitudes must be squared to obtain the corresponding relative intensities (see Eq. 40–7).

41–4 Diffraction from a Single Slit—Quantitative

We now derive an explicit expression for E_θ, in which the wavelength λ and the slit width a must appear as parameters. The square of E_θ will then give us a quantity proportional to I_θ, the intensity as a function of the diffraction angle θ.

The "arc" of small arrows in Fig. 41–9 shows the phasors representing, in amplitude and phase, the wave disturbances that reach an arbitrary point P on the screen of Fig. 41–7, corresponding to a particular angle θ. The resultant amplitude at P is E_θ. If we divide the slit of Fig. 41–7 into infinitesimal strips of width dx, the arc of arrows in Fig. 41–9 approaches the arc of a circle, its radius R being indicated in that figure. The length of the arc is E_m, the amplitude at the center of the diffraction pattern, for at the center of the pattern the wave disturbances are all in phase and this "arc" becomes a straight line as in Fig. 41–8*a*.

The angle ϕ in the lower part of Fig. 41–9 is revealed as the difference in phase between the infinitesimal vectors at the left and right ends of the arc E_m. This means that ϕ is the phase difference between rays from the top and the bottom of the slit of Fig. 41–7. From geometry we see that ϕ is also the angle between the two radii marked R in Fig. 41–9. From this figure we can write

$$E_\theta = 2R \sin \frac{\phi}{2}.$$

In radian measure ϕ, from the figure, is

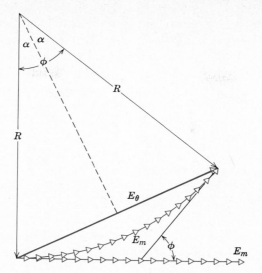

Figure 41–9 A construction used to calculate the intensity in single-slit diffraction. The situation corresponds to that of Fig. 41–8*b*.

$$\phi = \frac{E_m}{R}.$$

Combining yields

$$E_\theta = \frac{E_m}{\phi/2} \sin \frac{\phi}{2},$$

or

$$E_\theta = E_m \frac{\sin \alpha}{\alpha}, \tag{41–4}$$

in which

$$\alpha = \frac{\phi}{2}. \tag{41–5}$$

From Fig. 41–7, recalling that ϕ is the phase difference between rays from the top and the bottom of the slit and that the path difference for these rays is $a \sin \theta$, we have

$$\frac{\text{phase difference}}{2\pi} = \frac{\text{path difference}}{\lambda},$$

or

$$\phi = \left(\frac{2\pi}{\lambda}\right)(a \sin \theta).$$

Combining with Eq. 41–5 yields

$$\alpha = \frac{\phi}{2} = \frac{\pi a}{\lambda} \sin \theta. \tag{41–6}$$

Equation 41–4, taken together with Eq. 41–6, gives the amplitude of the wave disturbance for a single-slit diffraction pattern at any angle θ. The intensity I_θ for the pattern is proportional to the square of the amplitude, or

$$I_\theta = I_m \left(\frac{\sin \alpha}{\alpha}\right)^2. \tag{41–7}$$

For convenience we display together, and renumber, the formulas for the amplitude and the intensity in single-slit diffraction.

[Eq. 41-4]
$$E_\theta = E_m \frac{\sin \alpha}{\alpha}$$
(41-8a)

Intensity in single-slit diffraction [Eq. 41-7]
$$I_\theta = I_m \left(\frac{\sin \alpha}{\alpha} \right)^2 \quad \text{single-slit diffraction}$$
(41-8b)

[Eq. 41-6]
$$\alpha \left(= \tfrac{1}{2} \phi \right) = \frac{\pi a}{\lambda} \sin \theta$$
(41-8c)

Figure 41-10 shows plots of I_θ for several values of the ratio a/λ. Note that the pattern becomes narrower as we increase a/λ; compare this figure with Figs. 40-1 and 41-2.

Minima occur in Eq. 41-8b when

$$\alpha = m\pi \qquad m = 1, 2, 3, \ldots$$
(41-9)

Combining with Eq. 41-8c leads to

$$a \sin \theta = m\lambda \qquad m = 1, 2, 3, \ldots \text{(minima)},$$

which we derived earlier as Eq. 41-2. At that point however, we derived only this result, obtaining no quantitative information about the intensity of the diffraction pattern at places in which it was not zero. Here (Eqs. 41-8) we have complete intensity information.

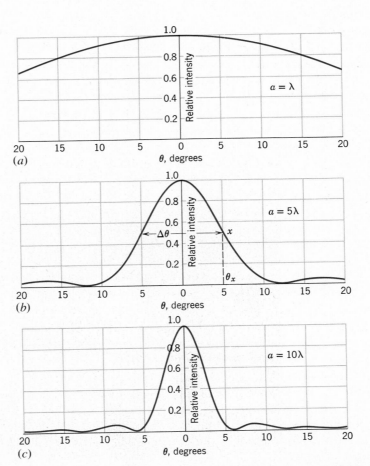

Figure 41-10 The relative intensity in single-slit diffraction for three values of the ratio a/λ. The arrow in (b) shows the half-width $\Delta\theta$ of the central maximum.

Example 3 *Intensities of the secondary diffraction maxima.* Calculate, approximately, the relative intensities of the secondary maxima in the single-slit Fraunhofer diffraction pattern.

The secondary maxima lie approximately halfway between the minima and are found (compare Eq. 41–9) from

$$\alpha \cong (m + \tfrac{1}{2})\pi \qquad m = 1, 2, 3, \ldots .$$

Substituting into Eq. 41–8*b* yields

$$I_\theta = I_m \left[\frac{\sin (m + \tfrac{1}{2})\pi}{(m + \tfrac{1}{2})\pi} \right]^2,$$

which reduces to

$$\frac{I_\theta}{I_m} = \frac{1}{(m + \tfrac{1}{2})^2 \pi^2}.$$

This yields, for $m = 1, 2, 3, \ldots$, $I_\theta/I_m = 4.5$, 1.6, 0.83%, etc. The successive maxima decrease rapidly in intensity. Figure 41–2 has been deliberately overexposed to reveal the secondary maxima.

41–5 Diffraction from a Circular Aperture

Here we consider diffraction by a circular aperture of diameter d, the aperture constituting the boundary of a circular converging lens.

Our previous treatment of lenses was based on geometrical optics, diffraction being specifically assumed not to occur. A rigorous analysis would be based from the beginning on wave optics, since geometrical optics is always an approximation, although often a very good one. Diffraction phenomena would emerge in a natural way from such a wave-optical analysis.

Figure 41–11 shows the image of a distant point source of light (a star, for instance) formed on a photographic film placed in the focal plane of a converging

Figure 41–11 The Fraunhofer diffraction pattern of a circular aperture. Note the central maximum and the circular secondary maxima. Compare Fig. 41–2. The central maximum is called the Airy disk, after Sir George Airy who first solved this problem mathematically in 1835. From *An Atlas of Optical Phenomena,* by M. Cagnet, M. Francon, and J. C. Thrierr, Springer-Verlag, Berlin, 1962.

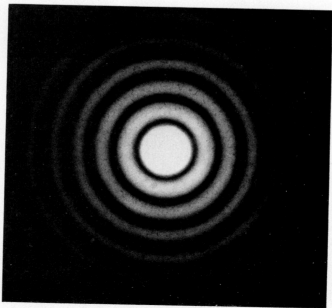

lens. It is not a point, as the (approximate) geometrical optics treatment suggests, but a circular disk surrounded by several progressively fainter secondary rings. Comparison with Fig. 41–2 leaves little doubt that we are dealing with a diffraction phenomenon in which, however, the aperture is a circle rather than a long narrow slit. The ratio d/λ, where d is the diameter of the lens (or of a circular aperture placed in front of the lens), determines the scale of the diffraction pattern, just as the ratio a/λ does for a slit.

Analysis shows that the first minimum for the diffraction pattern of a circular aperture of diameter d, assuming Fraunhofer conditions, is given by

Location of first minimum for circular diffraction

$$\sin \theta = 1.22 \frac{\lambda}{d}. \tag{41–10}$$

This is to be compared with Eq. 41–1, or

$$\sin \theta = \frac{\lambda}{a},$$

which locates the first minimum for a long narrow slit of width a. The factor 1.22 emerges from the mathematical analysis when we integrate over the elementary radiators into which the circular aperture may be divided.

The fact that lens images are diffraction patterns is important when we wish to distinguish two distant point objects whose angular separation is small. Figure 41–12 shows the visual appearances and the corresponding intensity patterns for two distant point objects (stars, say) with small angular separations. In (a) the objects are not resolved; that is, they cannot be distinguished from a single point object. In (b) they are barely resolved and in (c) they are fully resolved.

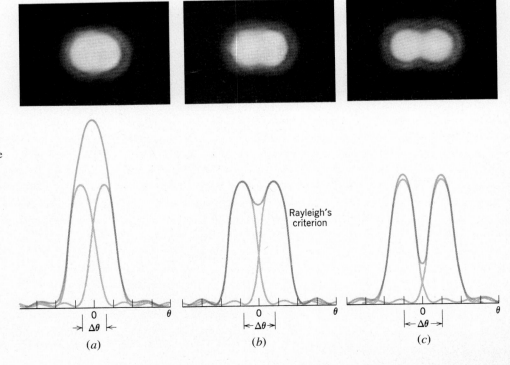

Figure 41–12 The images of two distant point objects (stars) are formed by a converging lens whose diameter (= 10 cm) is 200,000 times the effective wavelength (= 500 nm). Sketches of the images as they appear in the focal plane of the telescope objective lens are shown with the corresponding intensity plots below them. (a) The angular separation of the objects (see vertical ticks) is so small that the images are not resolved. (b) The objects are farther apart and meet Rayleigh's criterion for resolution. (c) The objects are still farther apart and the images are well resolved. The Rayleigh criterion is useful only if the two sources are of comparable brightness.

Rayleigh's criterion

In Fig. 41-12b the angular separation of the two point sources is such that the maximum of the diffraction pattern of one source falls on the first minimum of the diffraction pattern of the other. This is called *Rayleigh's criterion*. This criterion, though useful, is arbitrary. From Eq. 41-10, two objects that are barely resolvable by Rayleigh's criterion must have an angular separation θ_R of

$$\theta_R = \sin^{-1}\frac{1.22\lambda}{d}.$$

Since the angles involved are small, we can replace $\sin\theta_R$ by θ_R expressed in radians:

$$\theta_R = 1.22\frac{\lambda}{d}. \qquad (41-11)$$

If the angular separation θ between the objects is greater than θ_R, we can resolve the two objects; if it is significantly less, we cannot. The objects must be of comparable brightness for Rayleigh's criterion to be useful.

Example 4   A converging lens 30 mm in diameter has a focal length f of 20 cm. (a) What angular separation must two distant point objects have to satisfy Rayleigh's criterion? Assume that $\lambda = 550$ nm.

From Eq. 41-11,

$$\theta_R = 1.22\frac{\lambda}{d} = \frac{(1.22)(5.5\times10^{-7}\text{ m})}{3.0\times10^{-2}\text{ m}} = 2.2\times10^{-5}\text{ rad}.$$

(b) How far apart are the centers of the diffraction patterns in the focal plane of the lens? The linear separation is

$$x = f\theta = (20\text{ cm})(2.2\times10^{-5}\text{ radian}) = 4.4\ \mu\text{m}.$$

This is 8.0 wavelengths of the light employed.

When we wish to use a lens to resolve objects of small angular separation, it is desirable to make the central disk of the diffraction pattern as small as possible. This can be done (see Eq. 41-11) by increasing the lens diameter or by using a shorter wavelength. One reason for constructing large telescopes is to produce sharper images so that we can examine celestial objects in finer detail. The images are also brighter, because the energy is concentrated into a smaller diffraction disk and also because the larger lens collects more light. Thus we can view fainter objects such as distant galaxies.

To reduce diffraction effects in microscopes we often use ultraviolet light, which, because of its shorter wavelength, permits finer detail to be examined than would be possible for the same microscope operated with visible light. We shall see in Chapter 43 that beams of electrons behave like waves under some circumstances. In the electron microscope such beams may have an effective wavelength of 4 pm ($=4\times10^{-12}$ m), of the order of 10^5 times shorter than visible light. This permits the detailed examination of tiny objects like viruses. If a virus were examined with an optical microscope, its structure would be hopelessly concealed by diffraction.

41-6   Diffraction from a Double Slit

In Young's double-slit experiment (Section 40-2) we assume that the slits are arbitrarily narrow (that is, $a \ll \lambda$), which means that the central part of the diffusing screen was uniformly illuminated by the diffracted waves from each slit, as

Fig. 41–10a suggests. When such waves interfere, they produce fringes of uniform intensity, as in Fig. 40–9. This idealized situation cannot occur with actual slits because the condition $a \ll \lambda$ cannot usually be met. Waves from the two actual slits combining at different points of the screen will have intensities that are not uniform but are governed by the diffraction pattern of a single slit. The effect of relaxing the assumption that $a \ll \lambda$ in Young's experiment is to leave the fringes relatively unchanged in location but to alter their intensities. The net result is a combination of interference and diffraction.

The interference pattern for infinitesimally narrow slits is given by Eq. 40–11b and c or, with a small change in nomenclature,

$$I_{\theta,\text{int}} = I_{m,\text{int}} \cos^2 \beta, \tag{41–12}$$

where

$$\beta = \frac{\pi d}{\lambda} \sin \theta, \tag{41–13}$$

d being the distance between the centers of the slits.

The intensity for the diffracted wave from either slit is given by Eqs. 41–8b and c, or, with a small change in nomenclature,

$$I_{\theta,\text{dif}} = I_{m,\text{dif}} \left(\frac{\sin \alpha}{\alpha} \right)^2, \tag{41–14}$$

where

$$\alpha = \frac{\pi a}{\lambda} \sin \theta. \tag{41–15}$$

We find the combined effect by regarding $I_{m,\text{int}}$ in Eq. 41–12 as a variable amplitude, given in fact by $I_{\theta,\text{dif}}$ of Eq. 41–14. This assumption, for the combined pattern, leads to

Intensity for double-slit diffraction

$$I_\theta = I_m \, (\cos \beta)^2 \left(\frac{\sin \alpha}{\alpha} \right)^2, \tag{41–16}$$

in which we have dropped all subscripts referring separately to interference and diffraction.

Let us express this result in words. At any point on the screen the available light intensity from each slit, considered separately, is given by the diffraction pattern of that slit (Eq. 41–14). The diffraction patterns for the two slits, again considered separately, coincide because parallel rays in Fraunhofer diffraction are focused at the same spot (see Fig. 41–6). This pattern is often referred to as a diffraction envelope. Because the two diffracted waves are coherent, they will interfere.

The effect of interference is to redistribute the available energy over the screen, producing a set of fringes. In Section 40–2, where we assumed $a \ll \lambda$, the available energy was virtually the same at all points on the screen so that the interference fringes had virtually the same intensites (see Fig. 40–9). If we relax the assumption $a \ll \lambda$, the available energy is not uniform over the screen but is controlled by the diffraction pattern of a slit of width a. In this case the interference fringes will have intensities that are determined by the intensity of the diffraction pattern at the location of a particular fringe. Equation 41–16 is the mathematical expression of this argument.

Figure 41–13 is a plot of Eq. 41–16 for $d = 50\lambda$ and for $a = 5\lambda$. The interference factor (Fig. 41–13a), the diffraction factor (Fig. 41–13b), and their product

Figure 41–13 (*a*) The "interference factor" (due to double slit interference) and (*b*) the "diffraction factor" in Eq. 41–14 (due to diffraction from a single slit) and (*c*) their product (compare Figs. 40–9 and 41–10*b*). The figure is drawn for $a = 5\lambda$.

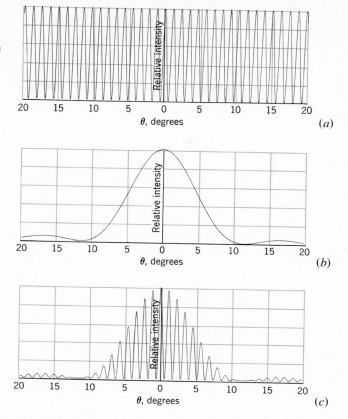

(Fig. 41–13*c*) are clearly shown. The interference fringes are contained within the diffraction envelope.

If we put $a = 0$ in Eq. 41–16, then (see Eq. 41–15) $\alpha = 0$ and $(\sin \alpha)/\alpha \underset{\alpha \to 0}{\cong} \alpha/\alpha = 1$. Thus this equation reduces, as it must, to the intensity equation for a pair of vanishingly narrow slits (Eq. 41–12). If we put $d = 0$ in Eq. 41–16, the two slits coalesce into a single slit of width a, as Fig. 41–26 shows; $d = 0$ implies $\beta = 0$ (see Eq. 41–13) and $\cos^2 \beta = 1$. Thus Eq. 41–16 reduces, as it must, to the diffraction equation for a single slit (Eq. 41–14).

Figure 41–14*a* shows a double-slit interference photograph. The uniformly

Figure 41–14 (*a*) Interference fringes for a double-slit system in which the slit width is not negligible in comparison to the wavelength. (*b*) If one of the slits is covered up, the fringes disappear and we see the diffraction pattern of a single slit, as in Fig. 41–2. (From *An Atlas of Optical Phenomena* by M. Cagnet, M. Francon, and J. C. Thrierr, Springer-Verlag, Berlin, 1962.)

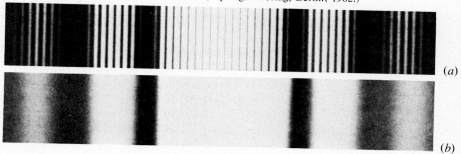

spaced interference fringes and their intensity modulation by the diffraction pattern of a single slit are clear. If one slit is covered up, as in Fig. 41–14*b*, the interference fringes disappear and we see the diffraction pattern of a single slit.

Example 5 In double-slit Fraunhofer diffraction what is the fringe spacing on a screen 50 cm away from the slits if they are illuminated with blue light ($\lambda = 480$ nm), if $d = 0.10$ mm, and if the slit width $a = 0.020$ mm? What is the linear distance from the central maximum to the first minimum of the fringe envelope?

The intensity pattern is given by Eq. 41–16, the fringe spacing being determined by the interference factor $\cos^2 \beta$. From Example 1, Chapter 40, we have

$$\Delta y = \frac{\lambda D}{d},$$

where D is the distance of the screen from the slits. Substituting yields

$$\Delta y = \frac{(480 \times 10^{-9} \text{ m})(50 \times 10^{-2} \text{ m})}{0.10 \times 10^{-3} \text{ m}} = 2.4 \text{ mm}.$$

The distance to the first minimum of the envelope is determined by the width of the diffraction pattern of each slit. From Eq. 41–1,

$$\sin \theta = \frac{\lambda}{a} = \frac{480 \times 10^{-9} \text{ m}}{0.020 \times 10^{-3} \text{ m}} = 0.024.$$

This is so small that we can assume that $\theta \cong \sin \theta \cong \tan \theta$, or

$$y = D \tan \theta \cong D \sin \theta = (50 \text{ cm})(0.024) = 12 \text{ mm}.$$

There are about ten fringes in the central peak of the diffraction envelope.

Example 6 What requirements must be met for the central maximum of the envelope of the double-slit Fraunhofer pattern to contain exactly eleven fringes?

The required condition will be met if the sixth minimum of the interference factor ($\cos^2 \beta$) in Eq. 41–16 coincides with the first minimum of the diffraction factor ($\sin \alpha/\alpha)^2$.

The sixth minimum of the interference factor occurs when

$$\beta = \tfrac{11}{2} \pi$$

in Eq. 41–12.

The first minimum in the diffraction term occurs for

$$\alpha = \pi.$$

Dividing (see Eqs. 41–13 and 41–15) yields

$$\frac{\beta}{\alpha} = \frac{d}{a} = \frac{11}{2}.$$

This condition depends only on the slit geometry and not on the wavelength. For long waves the pattern will be broader than for short waves, but there will always be eleven fringes in the central peak of the envelope.

The double-slit pattern as illustrated in Fig. 41–13*c* combines interference and diffraction in an intimate way. Both are superposition effects and depend on adding wave disturbances at a given point, taking phase differences properly into account. If the waves to be combined originate from a finite (and usually small) number of elementary coherent radiators, as in Young's double-slit experiment, we call the effect *interference*. If the waves to be combined originate by subdividing a wave into infinitesimal coherent radiators, as in our treatment of a single slit (Fig. 41–7), we call the effect *diffraction*. This distinction between interference and diffraction is convenient and useful. However, it should not cause us to lose sight of the fact that both are superposition effects and that often both are present simultaneously, as in Young's experiment.

Interference and diffraction

41–7 Multiple Slits

A logical extension of Young's double-slit interference experiment is to increase the number of slits from two to a larger number N. An arrangement like that of Fig. 41–15, usually involving many more slits (as many as 10^3/mm is not uncommon), is called a diffraction grating. As for a double slit, the intensity pattern that results when monochromatic light of wavelength λ falls on a grating consists of a series of interference fringes. The angular separation of these fringes are determined by the ratio λ/d, where d is the spacing between the centers of adjacent slits. The relative intensities of these fringes are determined by the diffraction pat-

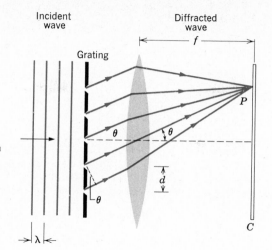

Figure 41–15 An idealized diffraction grating containing five slits. The slit width *a* is shown for convenience to be considerably smaller than λ, although this condition is not realized in practice. The figure is distorted in that *f* is much greater than *d* in practice.

tern of a single grating slit, which depends on the ratio λ/a, where *a* is the slit width.

Figure 41–16 compares the intensity patterns for $N = 2$ and for $N = 5$, showing only the central maximum of the single-slit diffraction envelope. It is clear that both patterns fade off at either end, because of the influence of this envelope. We also see that three secondary maxima appear between each of the primary maxima in the $N = 5$ pattern. What is not so clear—and this is only because the photograph has been deliberately overexposed to reveal these secondary maxima—is that the primary maxima for $N = 5$ are sharper than those for $N = 2$. As *N* increases, perhaps to 10^4 for a useful grating, the principal maxima become very sharp indeed and the secondary maxima, while increasing in number, become so reduced in intensity as to be negligible in their effects.

Location of principal maxima for multiple slits

A principal maximum in Fig. 41–15 will occur when the path difference between rays from adjacent slits ($= d \sin \theta$) is given by

$$d \sin \theta = m\lambda \qquad m = 0, 1, 2, \ldots \qquad \text{(principal maxima)}, \quad (41–17)$$

where *m* is called the *order number*. This equation is identical with Eq. 40–1, which locates the intensity maxima for a double slit. The locations of the (prin-

Figure 41–16 Intensity patterns for "gratings" with (*a*) $N = 2$ and (*b*) $N = 5$, showing only fringes well within the central diffraction maximum. In preparing (*b*) the film was deliberately overexposed to bring out the three faint secondary maxima that occur between each principal maximum. From *An Atlas of Optical Phenomena* by M. Cagnet, M. Francon, and J. C. Thrierr, Springer-Verlag, Berlin, 1962.

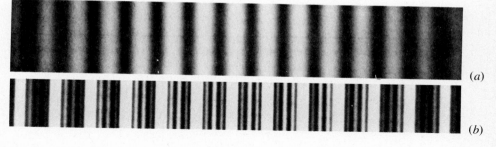

(*a*)

(*b*)

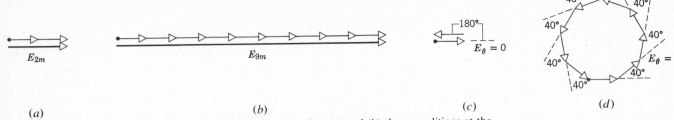

(a) $\qquad\qquad\qquad\qquad (b)$ $\qquad\qquad\qquad\qquad (c)$ $\qquad\qquad\qquad\qquad (d)$

Figure 41-17 Drawings (a) and (b) show conditions at the central principal maximum for a two-slit and a nine-slit grating respectively. Drawings (c) and (d) show conditions at the minimum of zero intensity that lies on either side of this central principal maximum.

cipal) maxima are thus determined only by the ratio λ/d and are independent of N. As for the double slit, the ratio a/λ determines the relative intensities of the principal maxima but does not alter their locations appreciably.

We can understand the sharpening of the principal maxima as N is increased by a graphical argument, using phasors. Figures 41–17a and b show conditions at any of the principal maxima for a two-slit and a nine-slit grating. The small arrows represent the amplitudes of the wave distrubances arriving at the screen at the position of each principal maximum. For simplicity we consider the central principal maximum only, for which $m = 0$, and thus $\theta = 0$, in Eq. 41–17.

Consider the angle $\Delta\theta_0$ corresponding to the position of zero intensity that lies on either side of the central principal maximum. Figures 41–17c and d show the phasors at this point. The phase difference between waves from adjacent slits, which is zero at the central principal maximum, must increase by an amount $\Delta\phi$ chosen so that the array of phasors just closes on itself, yielding zero resultant intensity. For $N = 2$, $\Delta\phi = 2\pi/2$ $(= 180°)$; for $N = 9$, $\Delta\phi = 2\pi/9$ $(= 40°)$. In the general case it is given by

$$\Delta\phi = \frac{2\pi}{N}.$$

This increase in phase difference for adjacent waves corresponds to an increase in the path difference Δl given by

$$\frac{\text{phase difference}}{2\pi} = \frac{\text{path difference}}{\lambda},$$

or

$$\Delta l = \left(\frac{\lambda}{2\pi}\right)\Delta\phi = \left(\frac{\lambda}{2\pi}\right)\left(\frac{2\pi}{N}\right) = \frac{\lambda}{N}.$$

From Fig. 41–15, however, the path difference Δl at the first minimum is also given by $d\sin\Delta\theta_0$, so that we can write

$$d\sin\Delta\theta_0 = \frac{\lambda}{N},$$

or

$$\sin\Delta\theta_0 = \frac{\lambda}{Nd}.$$

Since $N \gg 1$ for actual gratings, $\sin \Delta\theta_0$ will ordinarily be quite small (that is, the lines will be sharp), and we may replace it by $\Delta\theta_0$ to good approximation, or

$$\Delta\theta_0 = \frac{\lambda}{Nd} \qquad \text{(central principal maximum).} \qquad (41-18)$$

This equation shows specifically that if we increase N for a given λ and d, $\Delta\theta_0$ will decrease, which means that the central principal maximum becomes sharper.

We state without proof, and for later use, that for principal maxima other than the central one (that is, for $m \neq 0$) the angular distance between the position θ_m of the principal maximum of order m and the minimum that lies on either side is given by

$$\Delta\theta_m = \frac{\lambda}{Nd \cos \theta_m} \qquad \text{(any principal maximum).} \qquad (41-19)$$

For the central principal maximum we have $m = 0$, $\theta_m = 0$, and $\Delta\theta_m = \Delta\theta_0$, so that Eq. 41-19 reduces, as it must, to Eq. 41-18.

41-8 Diffraction Gratings

The grating spacing d for a typical grating that contains 12,000 "slits" distributed over a 1-in. width is 25.4 mm/12,000, or 2100 nm. Gratings are often used to measure wavelengths and to study the structure and intensity of spectrum lines.

Gratings are made by ruling equally spaced parallel grooves on a glass or a metal plate, using a diamond cutting point whose motion is automatically controlled by an elaborate ruling engine. Gratings ruled on metal are called *reflection gratings* because the interference effects are viewed in reflected rather than in transmitted light. Once such a master grating has been prepared, replicas can be formed by pouring a liquid plastic on the grating, allowing it to harden, and stripping it off. The stripped plastic, fastened to a flat piece of glass or other backing, forms a good grating.

Grating spectrographs

Figure 41-18 shows a simple grating spectroscope, used for viewing the spectrum of a light source, assumed to emit a number of discrete wavelengths, or *spectral lines*. The light from source S is focused by lens L_1 on a slit S_1 placed in the focal plane of lens L_2. The parallel light emerging from collimator C falls on grating G. Parallel rays associated with a particular interference maximum occurring at angle θ fall on lens L_3, being brought to a focus in plane F-F'. The image formed in this plane is examined, using a magnifying lens arrangement E, called an eyepiece. A symmetrical interference pattern is formed on the other side of the central position, as shown by the dotted lines. The entire spectrum can be viewed by rotating telescope T through various angles. Instruments used for scientific research or in industry are more complex than the simple arrangement of Fig. 41-18. They invariably employ photographic or photoelectric recording and are called *spectrographs*. Fig. 42-13 shows a small portion of the spectrum of iron, produced by examining the light produced in an arc struck between iron electrodes, using a research type spectrograph with photographic recording. Each line in the figure represents a different wavelength that is emitted from the source.

Wavelength measurement

Grating instruments can be used to make absolute measurements of wavelength, since the grating spacing d in Eq. 41-17 can be measured accurately with a microscope. Several spectra are normally produced in such instruments, corresponding to $m = \pm 1$, ± 2, etc., in Eq. 41-17 (see Fig. 41-19). This may cause some confusion if the spectra overlap. Further, this multiplicity of spectra reduces

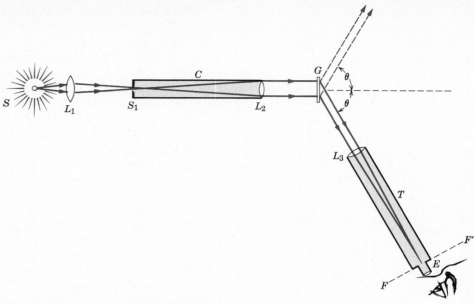

Figure 41–18   A simple type of grating spectroscope used to analyze the wavelengths of the light emitted by source S.

the recorded intensity of any given spectrum line because the available energy is divided among a number of spectra. However, by controlling the shape of the grating rulings, a large fraction of the energy can be concentrated in a particular order; this is called *blazing*.

Prism spectrographs

Light can also be analyzed into its component wavelengths if the grating in Fig. 41–18 is replaced by a prism. In a prism spectrograph each wavelength in the incident beam is deflected through a definite angle θ, determined by the index of refraction of the prism material for that wavelength. Curves such as Fig. 39–2, which gives the index of refraction of fused quartz as a function of wavelength, show that the shorter the wavelength, the larger the angle of deflection θ. Such curves vary from substance to substance and must be found by measurement.

Figure 41–19   The spectrum of white light as viewed in a grating instrument like that of Fig. 41–18. The different orders, identified by the order number m, are shown separated vertically for clarity. As actually viewed, they would not be so displaced. The central line in each order corresponds to $\lambda = 550$ nm, roughly the central point of the visible spectrum.

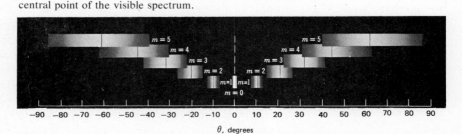

θ, degrees

Prism instruments are not adequate for accurate absolute measurements of wavelength because the index of refraction of the prism material at the wavelength in question is usually not known precisely enough. Both prism and grating instruments make accurate comparisons of wavelength, using a suitable comparison spectrum such as that shown in Fig. 42–13, in which careful absolute determinations have been made of the wavelengths of the spectrum lines.

Example 7 A diffraction grating has 10^4 rulings uniformly spaced over 25.4 mm. It is illuminated at normal incidence by yellow light from a sodium vapor lamp. This light contains two closely spaced lines (the well-known sodium doublet) of wavelengths 589.00 and 589.59 nm. (a) At what angle will the first-order maximum occur for the first of these wavelengths?

The grating spacing d is 2540 nm. The first-order maximum corresponds to $m = 1$ in Eq. 41–17. We thus have

$$\theta = \sin^{-1} \frac{m\lambda}{d} = \sin^{-1} \frac{(1)(589 \text{ nm})}{2540 \text{ nm}} = \sin^{-1} 0.232 = 13.3°.$$

(b) What is the angular separation between the first-order maxima for these lines?

The straightforward way to find this separation is to repeat this calculation for $\lambda = 589.59$ nm and to subtract the two angles. Alternatively, to calculate the difference in angular positions directly, let us write down Eq. 41–17, solved for $\sin \theta$, and differentiate it, treating θ and λ as variables:

$$\sin \theta = \frac{m\lambda}{d}$$

$$\cos \theta \, d\theta = \frac{m}{d} \, d\lambda.$$

If the wavelengths are close enough together, as in this case, $d\lambda$ can be replaced by $\Delta\lambda$, the actual wavelength difference; $d\theta$ then becomes $\Delta\theta$, the quantity we seek. This gives

$$\Delta\theta = \frac{m \, \Delta\lambda}{d \cos \theta} = \frac{(1)(0.59 \text{ nm})}{(2540 \text{ nm})(\cos 13.3°)} = 2.4 \times 10^{-4} \text{ radian}$$

$$= 0.014°.$$

Dispersion

The quantity $d\theta/d\lambda$, called the dispersion D of a grating, is a measure of the angular separation produced between two incident monochromatic waves whose wavelengths differ by a small wavelength interval. From this example we see that

$$D = \frac{d\theta}{d\lambda} = \frac{m}{d \cos \theta}. \qquad (41-20)$$

41–9 Resolving Power of a Grating

To distinguish light waves whose wavelengths are close together, the principal maxima of these wavelengths formed by the grating should be as narrow as possible. Expressed otherwise, the grating should have a high resolving power R, defined from

$$R = \frac{\lambda}{\Delta\lambda}. \qquad (41-21)$$

Resolving power definition

Here λ is the mean wavelength of two spectrum lines that can barely be recognized as separate and $\Delta\lambda$ is the wavelength difference between them. The smaller $\Delta\lambda$ is, the closer the lines can be and still be resolved; hence the greater the resolving power R of the grating. It is to achieve a high resolving power that gratings with many rulings are constructed.

The resolving power of a grating is usually determined by the same consideration (that is, the Rayleigh criterion) that we used in Section 41–5 to determine the resolving power of a lens. If two principal maxima are to be barely resolved, they must, according to this criterion, have an angular separation $\Delta\theta$ such that the maximum of one line coincides with the first minimum of the other; see Fig. 41–12. If we apply this criterion, we will show that

Resolving power of a grating

$$R = Nm, \qquad (41–22)$$

where N is the total number of rulings in the grating and m is the order. As expected, the resolving power is zero for the central principal maximum ($m = 0$), all wavelengths being undeflected in this order.

To derive Eq. 41–22 we note that the angular separation between two principal maxima whose wavelengths differ by $\Delta\lambda$ is found from Eq. 41–20, which we recast as

$$\Delta\theta = \frac{m\,\Delta\lambda}{d\cos\theta}. \qquad [41–20]$$

The Rayleigh criterion (Section 41–5) requires that this be equal to the angular separation between a principal maximum and its adjacent minimum. This is given from Eq. 41–19, dropping the subscript m in $\cos\theta_m$, as

$$\Delta\theta_m = \frac{\lambda}{Nd\cos\theta}. \qquad [41–19]$$

Equating Eqs. 41–20 and 41–19 above leads to

$$R(=\lambda/\Delta\lambda) = Nm,$$

which is the desired relation.

Example 8 How many rulings must a grating have if it is barely to resolve the sodium doublet in the third order? (See Example 7.)

From Eq. 41–21 the required resolving power is

$$R = \frac{\lambda}{\Delta\lambda} = \frac{589\ \text{nm}}{(589.00\ \text{nm} - 589.59\ \text{nm})} = 1000.$$

From Eq. 41–22 the number of rulings needed is

$$N = \frac{R}{m} = \frac{1000}{3} = 330.$$

This is a modest requirement.

The resolving power of a grating must not be confused with its dispersion. Table 41–1 shows the characteristics of three gratings, each illuminated with light of $\lambda = 589$ nm, the diffracted light being viewed in the first order ($m = 1$ in Eq. 41–17). You should verify that the values of D and R given in the table can be calculated from Eqs. 41–20 and 41–22, respectively.

Table 41–1 Some characteristics of three gratings
($\lambda = 589$ nm, $m = 1$)

Grating	N	d, nm	θ	R	D 10^{-2} degrees/nm
A	10,000	2540	13.3°	10,000	2.32
B	20,000	2540	13.3°	20,000	2.32
C	10,000	1370	25.5°	10,000	4.64

Resolving power versus dispersion

For the conditions of use noted in Table 41–1, gratings A and B have the same dispersion and A and C have the same resolving power. Figure 41–20 shows the intensity patterns that would be produced by these gratings for two incident waves of wavelengths λ_1 and λ_2, in the vicinity of $\lambda = 589$ nm. Grating B, which has high resolving power, has narrow intensity maxima and is inherently capable of distinguishing lines that are much closer together in wavelength than those in the figure. Grating C, which has high dispersion, produces twice the angular separation between rays λ_1 and λ_2 than does grating B.

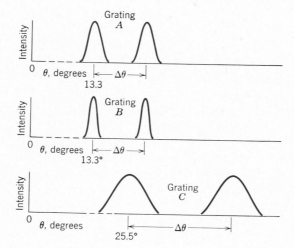

Figure 41–20 The intensity patterns for light of wavelengths λ_1 and λ_2 near 589 nm, incident on the gratings of Table 41–1. Grating B has the highest resolving power and grating C the highest dispersion.

Example 9   A grating has 8000 lines uniformly spaced over 25.4 mm and is illuminated by light from a mercury vapor discharge. (*a*) What is the expected dispersion, in the third order, in the vicinity of the intense green line ($\lambda = 546$ nm)? Noting that $d = 3170$ nm (25.4 mm/8000), we have, from Eq. 41–17,

$$\theta = \sin^{-1}\frac{m\lambda}{d} = \sin^{-1}\frac{(3)(546 \text{ nm})}{3170 \text{ nm}} = \sin^{-1} 0.517 = 31.1°.$$

From Eq. 41–20 we have

$$D = \frac{m}{d\cos\theta} = \frac{3}{(3170 \text{ nm})(\cos 31.1°)} = 1.1 \times 10^{-3} \text{ radian/nm}$$

$$= 6.3 \times 10^{-2} \text{ deg/nm}.$$

(*b*) What is the expected resolving power of this grating in the fifth order? Equation 41–22 gives

$$R = Nm = (8000)(5) = 40,000.$$

Thus near $\lambda = 546$ nm a wavelength difference $\Delta\lambda$ given by Eq. 41–21, or

$$\Delta\lambda = \frac{\lambda}{R} = \frac{546 \text{ nm}}{40,000} = 14 \text{ pm} \ (= 0.014 \text{ nm})$$

can be distinguished.

41–10   X-ray Diffraction

X rays are electromagnetic radiation whose wavelengths are of the order of 1 Å ($= 10^{-10}$ m).* Compare this with 550 nm ($= 5.5 \times 10^{-7}$ m) for the center of the visible spectrum. Figure 41–21 shows how x rays are produced when electrons from

---

* The angstrom (1 Å $= 10^{-10}$ m) is officially classified, by international agreement, as a unit "to be used with the International System for a limited time." Its usefulness stems from the fact that atomic dimensions are of about this size.

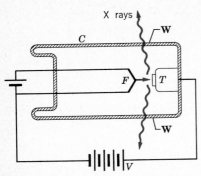

Figure 41–21 X rays are generated when electrons from heated filament F, accelerated by potential difference V, are brought to rest on striking metallic target T. W is a "window"—transparent to x rays—in the evacuated metal container C.

a heated filament F are accelerated by a potential difference V and strike a metal target T.

For such small wavelengths a standard optical diffraction grating, as normally employed, cannot be used to discriminate between different wavelengths. For $\lambda = 1.0$ Å $(=0.10$ nm) and $d = 3.0$ μm, for example, Eq. 41–17 shows that the first-order maximum occurs at

$$\theta = \sin^{-1}\frac{m\lambda}{d} = \sin^{-1}\frac{(1)(1.0\text{ Å})}{3.0\ \mu\text{m}} = \sin^{-1}0.33 \times 10^{-4} = 0.0020°.$$

This is too close to the central maximum to be practical. A grating with $d \cong \lambda$ is desirable, but, since x-ray wavelengths are about equal to atomic diameters, such gratings cannot be constructed mechanically.

In 1912 it occurred to Max von Laue that a crystalline solid, consisting as it does of a regular array of atoms, might form a natural three-dimensional "diffraction grating" for x rays. The idea is that in a crystal, such as sodium chloride (NaCl), there is a basic unit of atoms (called the unit cell) which repeats itself throughout the array. In NaCl four sodium ions and four chlorine ions are associated with each unit cell. Figure 41–22 represents a section through a crystal of NaCl and identifies this basic unit. This crystal has a cubic structure and the unit cell is thus itself a cube, measuring a_0 on the side.

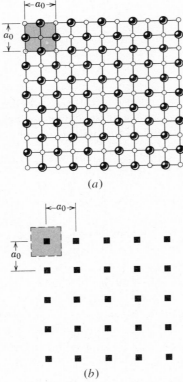

Figure 41–22 (a) A section through a crystal of sodium chloride, showing the sodium and chlorine ions. (b) The corresponding unit cells in this section, each cell being represented by a small black square.

Diffraction of electromagnetic radiation is accomplished by the electrons surrounding the atoms in the array. Each unit cell acts as a diffraction center just as a ruled groove acts as a two-dimensional diffraction line in a grating. The crystal, then, forms a three-dimensional array of diffraction centers. Just as in the two-

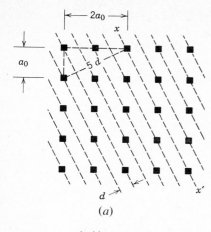

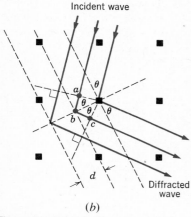

Figure 41–23 (a) A section through the NaCl unit cell lattice of Fig. 41–22b. The dashed sloping lines represent an arbitrary family of planes, with interplanar spacing d. (b) An incident wave falls on the entire family of planes shown in (a). A strong diffracted wave is formed.
Bragg's law

dimensional case, the intensity of any point outside the array is determined by the phase difference and intensity of radiation diffracted from each center to the point in question.

Bragg's law predicts the conditions under which diffracted x-ray beams from a crystal are possible. In deriving it, we ignore the structure of the unit cell, which is related only to the intensities of these beams. The dashed sloping lines in Fig. 41–23a represent the intersection with the plane of the figure of an arbitrary set of planes passing through the elementary diffracting centers. The perpendicular distance between adjacent planes is d. Many other such families of planes, with different interplanar spacings, can be defined.

Figure 41–23b shows an incident wave striking the family of planes. For a single plane, mirror-like "reflection" occurs for any incident angle as with diffraction gratings. To have a constructive interference in the beam diffracted from the entire family of planes in the direction θ, the rays from the separate planes must reinforce each other. In x-ray diffraction it is customary to specify the direction of a wave by giving the angle between the ray and the plane (the *glancing angle*) rather than the angle between the ray and the normal. This means that the path difference for rays from adjacent planes (*abc* in Fig. 41–23b) must be an integral number of wavelengths or

$$2d \sin \theta = m\lambda \qquad m = 1, 2, 3, \ldots . \tag{41–23}$$

This relation is called *Bragg's law* after W. L. Bragg who first derived it. The quantity d in this equation (the interplanar spacing) is the perpendicular distance between the planes. For the planes of Fig. 41–23a you can show that d is related to the unit cell dimension a_0 by

$$d = \frac{a_0}{\sqrt{5}}. \tag{41–24}$$

If an incident monochromatic x-ray beam falls at an arbitrary angle θ on a particular set of atomic planes, a diffracted beam will not result because Eq. 41–23 will not, in general, be satisfied. However if the incident x rays are continuous in wavelength, diffracted beams will result when wavelengths given by

$$\lambda = \frac{2d \sin \theta}{m} \qquad m = 1, 2, 3, \ldots$$

are present in the incident beam (see Eq. 41–23).

Example 10 At what angles must an x-ray beam with $\lambda = 1.10$ Å fall on the family of planes represented in Fig. 41–23b if a diffracted beam is to exist? Assume the material to be sodium chloride ($a_0 = 5.63$ Å).

The interplanar spacing d for these planes is given by Eq. 41–24 or

$$d = \frac{a_0}{\sqrt{5}} = \frac{5.63 \text{ A}}{2.24} = 2.52 \text{ A}.$$

Equation 41–23 gives

$$\sin \theta = \frac{m\lambda}{2d} = \frac{(m)(1.10 \text{ A})}{(2)(2.52 \text{ A})} = 0.218m.$$

Diffracted beams are possible at $\theta = 12.6°$ ($m = 1$), $\theta = 25.9°$ ($m = 2$), $\theta = 40.9°$ ($m = 3$), and $\theta = 60.7°$ ($m = 4$). Higher-order beams cannot exist because they require $\sin \theta$ to exceed unity. Actually, the odd-order beams ($m = 1, 3$) prove to have zero intensity because the unit cell in cubic crystals such as NaCl has diffracting properties such that the intensity of the light scattered in these orders is zero.

X-ray diffraction is a powerful tool for studying both x-ray spectra and the arrangements of atoms in crystals. To study spectra a particular set of crystal planes, having a known spacing d, is chosen. Diffraction from these planes disperses different wavelengths into different angles. A detector, then, which can discriminate one angle from another can be used to determine the wavelength of radiation reaching it. On the other hand, we can study the crystal itself, using a monochromatic x-ray beam, determining not only the spacings of various crystal planes but also the structure of the unit cell.

REVIEW GUIDE AND SUMMARY

Diffraction and the Huygens construction

Diffraction, which occurs when a wave encounters an obstacle or hole whose size is comparable to the wavelength of the wave, can be analyzed using the Huygens construction. The wave is divided at the obstruction into infinitesimal wavelets which then interfere with each other as they proceed (see Fig. 41–3). Fresnel diffraction is a description of diffraction effects near the obstruction while Fraunhofer diffraction, which is mathematically simpler, describes effects that occur a large distance from the obstruction. Fraunhofer patterns are usually studied experimentally by interposing a lens so that patterns which would otherwise occur at infinite distance are focused onto a screen at the lens focal plane. In this approximation, the only one treated here, all the waves which eventually interfere at a single point in the diffraction pattern leave the obstruction at the same angle θ, different angles corresponding to different points on the diffraction pattern.

Fresnel and Fraunhofer diffraction

Single-slit diffraction

Waves passing through a long narrow slit of width a produce a single-slit diffraction pattern with a central maximum together with minima corresponding to diffraction angles θ which satisfy

$$a \sin \theta = m\lambda \qquad m = 1, 2, 3 \ldots \text{(minima)}. \qquad [41\text{–}2]$$

Using phasor diagrams to add the Huygens wavelets, we show in Sec. 41–4 that the diffracted intensity for a given diffraction angle θ is

$$I_\theta = I_m \left(\frac{\sin \alpha}{\alpha}\right)^2 \qquad \text{where} \qquad \alpha = \frac{\pi a}{\lambda} \sin \theta. \qquad [41\text{–}8b,c]$$

Circular diffraction

Diffraction by a circular aperature or lens with diameter d also produces a central maximum with a first minimum at a diffraction angle θ given by

$$\sin \theta = 1.22 \frac{\lambda}{d}. \qquad [41\text{–}10]$$

See Fig. 41–11.

Rayleigh's criterion

Rayleigh's criterion suggests that two objects viewed through a telescope or microscope are on the verge of resolvability if the central diffraction maximum of one is no closer than the first minimum of the other. Their angular separation must be at least

$$\theta_R = 1.22 \frac{\lambda}{d}, \qquad [41\text{–}11]$$

in which d is the diameter of the objective lens.

Double-slit diffraction

Waves passing through two slits, each of width a, whose centers are distance d apart, display Fraunhofer diffraction patterns whose intensity I_θ at various diffraction angles θ is given by

$$I_\theta = I_m (\cos \beta)^2 \left(\frac{\sin \alpha}{\alpha}\right)^2, \qquad \text{with} \qquad \beta = \frac{\pi d}{\lambda} \sin \theta \qquad [41\text{–}16,13]$$

and α the same as for single-slit diffraction.

Multiple-slit diffraction

Diffraction by N multiple slits results in principle maxima whenever

$$d \sin \theta = m\lambda \qquad m = 0, 1, 2 \ldots \qquad [41\text{–}17]$$

with the angular width of the maxima given by

$$\Delta\theta_m = \frac{\lambda}{Nd \cos \theta_m}.$$ [41–19]

Diffraction gratings

A diffraction grating is a series of numerous "slits" used to separate an incident wave into different wavelength components whose principal diffraction maxima are directionally dispersed by the grating. A grating is characterized by two parameters, the dispersion D and the resolving power R.

$$D \equiv \frac{d\theta}{d\lambda} = \frac{m}{d \cos \theta}$$ [41–20]

Dispersion and resolving power

and

$$R \equiv \frac{\lambda}{\Delta\lambda} = Nm.$$ [41–21,22]

X-ray diffraction

The regular array of atoms in a crystal is a three-dimensional diffraction grating for short-wavelength waves such as x rays. The atoms can be visualized as being arranged in planes with characteristic inter-planar spacing d. Diffraction maxima (constructive interference) occur if the incident direction of the wave, measured from the surface of a plane of atoms, and the wavelength λ of the radiation satisfy Bragg's law:

Bragg's law

$$2d \sin \theta = m\lambda \qquad m = 1, 2, 3 \ldots$$ [41–23]

QUESTIONS

1. Why is the diffraction of sound waves more evident in daily experience than that of light waves?

2. Why do radio waves diffract around buildings, although light waves do not?

3. A person holds a single narrow vertical slit in front of the pupil of his eye and looks at a distant light source in the form of a long heated filament. Is the diffraction pattern that he sees a Fresnel or a Fraunhofer pattern?

4. *Poisson's bright spot.* If a coin or a ball bearing (or even a bowling ball) is suspended between a point source of light and a photographic film, a bright spot, called the Poisson bright spot, appears at the center of the geometrical shadow; see Fig. 41–24. Diffraction rings appear, both within the shadow and out of it. One might think that the center of the geometrical shadow, being most shielded from the light source, would be dark but just the reverse is true. Can you see the qualitative possibility that this experimental result can be consistent with Fresnel's diffraction theory?

5. A loud-speaker horn, used at a rock concert, has a rectangular aperture 1 m high and 30 cm wide. Will the pattern of sound intensity be broader in the horizontal plane or in the vertical?

6. A radar antenna is designed to give accurate measurements of the altitude of an aircraft but less accurate measurements of its direction in a horizontal plane. Must the height-to-width ratio of the radar antenna be less than, equal to, or greater than unity?

7. In single-slit Fraunhofer diffraction, what is the effect of increasing (*a*) the wavelength and (*b*) the slit width?

8. Why are the colors in the spectrum of some light sources called "lines"?

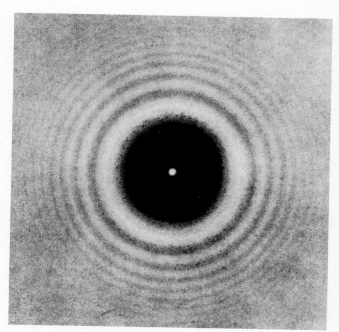

Figure 41–24 Question 4. From *An Atlas of Optical Phenomena* by M. Cagnet, M. Francon, and J. C. Thrierr, Springer-Verlag, Berlin, 1962.

9. While listening to the car radio, you may have noticed that the AM signal fades—but the FM signal doesn't—when you drive under a bridge. Could diffraction have anything to do with this?

10. In the formula for the first minimum of a single-slit diffraction pattern ($a \sin \theta = \lambda$), can a be less than λ? We can certainly find a wavelength that is greater than the slit width. What will the diffraction pattern look like in such a case?

11. What would the pattern on a screen formed by a double slit look like if the slits did not have the same width? Would the locations of the fringes be changed?

12. A crossed diffraction grating has lines ruled in two directions, at right angles to each other. What is the pattern of light intensity on the screen if light is sent through such a grating?

13. Sunlight falls on a single slit of width 1 μm. Describe qualitatively what the resulting diffraction pattern looks like.

14. In Fig. 41–6 rays r_1 and r_3 are in phase; so are r_2 and r_4. Why isn't there a maximum intensity at P_2 rather than a minimum?

15. When we speak of diffraction by a single slit we imply that the width of the slit must be much less than its height. Suppose that, in fact, the height was equal to twice the width. Make a rough guess at what the diffraction pattern would look like.

16. In connection with Fig. 41–5 we stated, correctly, that the optical pathlengths from the slit to point P_0 are all the same. Why?

17. Distinguish clearly between θ, α, and ϕ in Eq. 41–8.

18. If we were to redo our analysis of the properties of lenses in Section 39–10 by the methods of geometrical optics but without restricting our considerations to paraxial rays and to "thin" lenses, would diffraction phenomena, such as that of Fig. 41–11, emerge from the analysis? Discuss.

19. Give at least two reasons why the usefulness of large reflecting telescopes increases as we increase the mirror diameter.

20. We have seen that diffraction limits the resolving power of optical telescopes (see Fig. 41–12). Does it also do so for large radio telescopes?

21. Distinguish carefully between interference and diffraction in Young's double-slit experiment.

22. In what way are interference and diffraction similar? In what way are they different?

23. In double-slit interference patterns such as that of Fig. 41–14a we said that the interference fringes were modulated in intensity by the diffraction pattern of a single slit. Could we reverse this statement and say that the diffraction pattern of a single slit is intensity-modulated by the interference fringes? Discuss.

24. Discuss this statement: "A diffraction grating can just as well be called an interference grating."

25. Assume that the limits of the visible spectrum are 430 and 680 nm. Is it possible to design a grating, assuming that the incident light falls normally on it, such that the first-order spectrum barely overlaps the second-order spectrum?

26. For the simple spectroscope of Fig. 41–18 show (a) that θ increases with λ for a grating and (b) that θ decreases with λ for a prism.

27. You are given a photograph of a spectrum on which the angular positions and the wavelengths of the spectrum lines are marked. (a) How can you tell whether the spectrum was taken with a prism or a grating instrument? (b) What information could you gather about either the prism or the grating from studying such a spectrum?

28. Explain in your own words why increasing the number of slits, N, in a diffraction grating sharpens the maxima. Why does decreasing the wavelength do so? Why does increasing the slit spacing, d, do so?

29. How much information can you discover about the structure of a diffraction grating by analyzing the spectrum it forms of a monochromatic light source. Let $\lambda = 589$ nm, for an example.

30. (a) Why does a diffraction grating have closely spaced rulings? (b) Why does it have a large number of rulings?

31. Two nearly equal wavelengths are incident on a grating of N slits and are not quite resolvable. However, they become resolved if the number of slits is increased. Formulas aside, is the explanation of this that: (a) more light can get through the grating? (b) the principal maxima become more intense and hence resolvable? (c) the diffraction pattern is spread more and hence the wavelengths become resolved? (d) there are a large number of orders? or (e) the principal maxima become narrower and hence resolvable?

32. The relation $R = Nm$ suggests that the resolving power of a given grating can be made as large as desired by choosing an arbitrarily high order of diffraction. Discuss this possibility.

33. Show that at a given wavelength and a given angle of diffraction the resolving power of a grating depends only on its width W ($= Nd$).

34. How would you experimentally measure (a) the dispersion D and (b) the resolving power R for either a prism or a grating spectrograph.

35. For a given family of planes in a crystal, can the wavelength of incident x rays be (a) too large or (b) too small to form a diffracted beam?

36. If a parallel beam of x rays of wavelength λ is allowed to fall on a randomly oriented crystal of any material, generally no intense diffracted beams will occur. Such beams appear if (a) the x-ray beam consists of a continuous distribution of wavelengths rather than a single wavelength or (b) the specimen is not a single crystal but a finely divided powder. Explain each case.

EXERCISES

Section 41–2 Diffraction from a Single Slit— Locating the Minima

1. When monochromatic light is incident on a slit 0.02 mm wide, the first diffraction minimum is observed at an angle of 2.0° from the direction of the direct beam. What is the wavelength of the incident light?

2. Light of wavelength 633 nm is incident on a narrow slit. The angle between the first minimum on one side of the central maximum and the first minimum on the other side is 1.2°. What is the width of the slit?

3. Monochromatic light of wavelength 441 nm falls on a narrow slit. On a screen 2.0 m away, the distance between the second

minimum on one side of the central maximum and the second minimum on the other side is 3.0 cm. What is the width of the slit?

Section 41-3   Diffraction from a Single Slit—Qualitative

4. A 0.10-mm wide slit is illuminated by light with wavelength 589 nm. Consider rays for which $\theta = 30°$ in Fig. 41–4b and compare the phases at the screen of Huygens' wavelets from the top, midpoint, and bottom of the slit.

Section 41-4   Diffraction from a Single Slit—Quantitative

5. A single slit is illuminated by light whose wavelengths are λ_a and λ_b, so chosen that the first diffraction minimum of the λ_a component coincides with the second minimum of the λ_b component. (a) What relationship exists between the two wavelengths? (b) Do any other minima in the two patterns coincide?

6. Monochromatic light with wavelength 538 nm falls on a slit with width 0.025 mm. The distance from the slit to a screen is 3.5 m. Calculate the ratio of the intensity on the screen at a point 1.0 cm from the central diffraction maximum to the intensity at the central maximum.

Section 41-5   Diffraction from a Circular Aperture

7. Two headlights of an approaching automobile are 1.4 m apart. At what maximum distance will the eye resolve them? Assume a pupil diameter of 5.0 mm and a wavelength of 550 nm. Also assume that diffraction effects alone limit the resolution.

8. An astronaut in a satellite claims he can just barely resolve two point sources on the earth, 160 km below him. What is their separation, assuming ideal conditions? Take $\lambda = 550$ nm, and the pupil diameter of the astronaut's eye to be 5.0 mm.

9. Find the separation of two points on the moon's surface that can just be resolved by the 200-in. (= 5.1-m) telescope at Mount Palomar, assuming that this distance is determined by diffraction effects. The distance from the earth to the moon is 3.8×10^5 km. Assume $\lambda = 550$ nm.

Section 41-6   Diffraction from a Double Slit

10. Suppose that, as in Example 6, the envelope of the central peak contains 11 fringes. How many fringes lie between the first and second minima of the envelope?

11. For $d = 2a$ in Fig. 41–25, how many interference fringes lie in the central diffraction envelope?

Section 41-7   Multiple Slits

12. Show that in a grating with alternately transparent and opaque strips of equal width, all the even orders (except $m = 0$) are absent.

Section 41-8   Diffraction Gratings

13. A diffraction grating 20 mm wide has 6000 rulings. At what angles will maximum-intensity beams occur if the incident radiation has a wavelength of 589 nm?

14. A diffraction grating has 200 rulings/mm, and a strong diffracted beam is noted at $\theta = 30°$. (a) What are the possible wavelengths of the incident light? (b) What colors are they?

15. A grating has 315 rulings/mm. For what wavelengths in the visible spectrum can fifth-order diffraction be observed?

16. Given a grating with 400 lines/mm, how many orders of the entire visible spectrum (400–700 nm) can be produced?

17. A diffraction grating 30 mm wide produces a deviation of 30° in the second order with light of wavelength 600 nm. What is the total number of lines on the grating?

Section 41-9   Resolving Power of a Grating

18. The D line in the spectrum of sodium is a doublet with wavelengths 589.0 nm and 589.6 nm. What is the minimum number of lines in a grating needed to resolve this doublet in the second order spectrum?

19. A grating has 600 rulings/mm and is 5.0 mm wide. (a) What is the smallest wavelength interval that can be resolved in the third order at $\lambda = 500$ nm? (b) How many higher orders can be seen? Assume normal incidence of light on the grating.

20. A source containing a mixture of hydrogen and deuterium atoms emits light containing two closely spaced red colors at $\lambda = 656.3$ nm whose separation is 0.18 nm. Find the minimum number of lines needed in a diffraction grating that can resolve these lines in the first order.

21. How many rulings must a 4.0-cm wide diffraction grating have to resolve the wavelengths 415.496 nm and 415.487 nm in the second order? At what angle are the maxima found?

Section 41-10   X-ray Diffraction

22. The x-ray wavelength 1.0 Å is found to reflect in the second order from the face of a lithium fluoride crystal at a Bragg angle of 30°. What is the distance between adjacent crystal planes?

23. A beam of x rays of wavelength 3.0×10^{-11} m is incident on a calcite crystal of lattice spacing 0.30 nm. What is the smallest angle between the crystal planes and the x-ray beam that will result in constructive reflection of the x rays?

24. Monochromatic high energy x rays are incident on a crystal. If first order reflection is observed at Bragg angle 3.4°, at what angle would second order reflection be expected?

Figure 41–25 Exercise 11 and Problem 14.

PROBLEMS GROUPED BY SECTION

Section 41-1   Diffraction—A Closer Look

1. *Babinet's principle.* A monochromatic beam of parallel light is incident on a "collimating" hole of diameter $x \gg \lambda$. Point P lies in the geometrical shadow region on a distant screen, as shown in Fig. 41-26a. Two obstacles, shown in Fig. 41-26b, are placed in turn over the collimating hole. A is an opaque circle with a hole in it and B is the "photographic negative" of A. Using superposition concepts, show that the intensity at P is identical for each of the two diffracting objects A and B.*

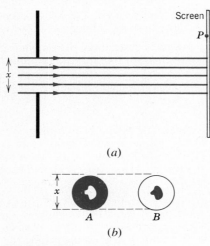

(a)

(b)

Figure 41-26   Problem 1.

Section 41-2   Diffraction from a Single Slit—Locating the Minima

2. If the distance between the first and fifth minima of a single slit pattern is 0.35 mm with the screen 40 cm away from the slit and using light having a wavelength of 550 nm, what is the width of the slit?

3. For a single slit the first minimum occurs at $\theta = 90°$, thus filling the forward hemisphere beyond the slit with light. What must be the ratio of the slit width to the wavelength for this to take place?

4. A slit 1.0 mm wide is illuminated by light of wavelength 589 nm. We see a diffraction pattern on a screen 3.0 m away. What is the distance between the first two diffraction minima on either side of the central diffraction maximum?

5. Manufacturers of wire (and other objects of small dimensions) sometimes use a laser to continually monitor the thickness of the product. The wire intercepts the laser beam, producing a diffraction pattern like that of a single slit of the same width as the wire diameter; see Fig. 41-27. Suppose a He-Ne laser, wavelength 632.8 nm, illuminates a wire, the diffraction pattern being projected onto a screen 2.6 m away. If the desired wire diameter is 1.37 mm, what should be the observed distance between the two 10th order minima on each side of the central maximum?

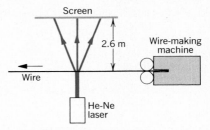

Figure 41-27   Problem 5.

Section 41-4   Diffraction from a Single Slit—Quantitative

6. If you double the width of a single slit the intensity of the central maximum of the diffraction pattern increases by a factor of four, even though the energy passing through the slit only doubles. Explain this quantitatively.

7. The half-width of the central diffraction maximum is defined as the angle between the two points in the pattern where the intensity is one-half that at the center of the pattern. (See Fig. 41-10b.) (a) Show that the intensity drops to one-half of the maximum value when $\sin^2\alpha = \alpha^2/2$. (b) Verify that $\alpha = 1.39$ radians (about 80°) is a solution to the transcendental equation of part (a). (c) Show that the half-width is $\Delta\theta = 2 \sin^{-1}(0.443 \lambda/a)$. (d) Calculate the half-width of the central maximum for slits whose widths are 1, 5, and 10 wavelengths.

8. (a) Show that the values of α at which intensity maxima for single-slit diffraction occur can be found exactly by differentiating Eq. 41-8b with respect to α and equating to zero, obtaining the condition

$$\tan \alpha = \alpha$$

(b) Find the values of α satisfying this relation by plotting graphically the curve $y = \tan \alpha$ and the straight line $y = \alpha$ and finding their intersections, or by using a pocket calculator to find an appropriate value of α by trial and error. (c) Find the (nonintegral) values of m corresponding to successive maxima in the single-slit pattern. Note that the secondary maxima do not lie exactly halfway between minima.

Section 41-5   Diffraction from a Circular Aperture

9. The wall of a large room is covered with acoustic tile in which small holes are drilled 5.0 mm from center to center. How far can a person be from such a tile and still distinguish the individual holes, assuming ideal conditions? Assume the diameter of the pupil of the observer's eye to be 4.0 mm and λ to be 550 nm.

10. The pupil of a person's eye has a diameter of 5.0 mm. What distance apart must two objects be if, when 250 mm from the eye, their images are just resolved when they are illuminated with light of wavelength 5.0×10^{-7} m?

11. Under ideal conditions, estimate the linear separation of two objects on the planet Mars that can just be resolved by an observer on earth (a) using the naked eye, and (b) using the 200-in. ($= 5.1$

---

* In this connection, it can be shown that the diffraction pattern of a wire is that of a slit of equal width. See "Measuring the Diameter of a Hair by Diffraction" by S. M. Curry and A. L. Schawlow, *American Journal of Physics,* May 1974.

m) Mt. Palomar telescope. Use the following data: distance to Mars $= 8.0 \times 10^7$ km; diameter of pupil $= 5.0$ mm; wavelength of light $= 550$ nm.

12. A "spy in the sky" satellite orbiting at 100 mi above the earth's surface, has a lens with a focal length of 8.0 ft. Its resolving power for objects on the ground is 1.2 ft; it could easily detect an automobile. What is the effective lens diameter, determined by diffraction consideration alone? Assume $\lambda = 550$ nm. Far more effective satellites are in operation today. See "Reconnaissance and Arms Control" by Ted Greenwood, *Scientific American,* February 1973.

13. A diffraction-limited laser aperture diameter d generates light of wavelength λ. If the beam is directed at the moon a distance D away, what is the approximate radius of the illuminated area on the moon, in terms of d, λ, and D?

Section 41–6   Diffraction from a Double Slit

14. If we put $d = a$ in Fig. 41–25, the two slits coalesce into a single slit of width $2a$. Show that Eq. 41–16 reduces to the diffraction pattern for such a slit.

15. Two slits of width a and separation d are illuminated by a coherent beam of light of wavelength λ. What is the separation of the interference fringes observed on a screen at a distance D along the perpendicular bisector of the line joining the slits?

16. (a) How many complete fringes appear between the first minima of the fringe envelope to either side of the central maximum for a double-slit pattern if $\lambda = 550$ nm, $d = 0.15$ mm, and $a = 0.030$ mm? (b) What is the ratio of the intensity of the third fringe to the side of the center to that of the central fringe?

Section 41–7   Multiple Slits

17. Derive Eq. 41–19, that is, the expression for $\Delta\theta_m$, the angular distance between a principal maximum of order m and either adjacent minimum.

18. The central intensity maximum formed by a grating, along with its subsidiary secondary maxima, can be viewed as the diffraction pattern of a single "slit" whose width is that of the entire grating. Treating the grating as a single wide slit, assuming that $m = 0$, and using the methods of Section 41–4, derive Eq. 41–18.

19. Assume that light is incident on a grating at an angle ψ as shown in Fig. 41–28. Show that the condition for a diffraction maximum is

$$d(\sin\psi + \sin\theta) = m\lambda \qquad m = 0, 1, 2, \ldots$$

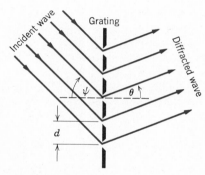

Figure 41–28   Problem 19.

Only the special case $\psi = 0$ has been treated in this chapter (compare Eq 41–17).

Section 41–8   Diffraction Gratings

20. A diffraction grating exactly one centimeter wide has 10,000 parallel slits. Monochromatic light which is incident normally is deviated through 30°. What is the wavelength of the light?

21. Light of wavelength 600 nm is incident normally on a diffraction grating. Two adjacent principal maxima occur at $\sin\theta = 0.2$ and $\sin\theta = 0.3$, respectively. The fourth order is missing. (a) What is the separation between adjacent slits? (b) What is the smallest possible individual slit width? (c) Name all orders actually appearing on the screen with the values chosen in (a) and (b).

22. Light containing a mixture of two wavelengths, 500 nm and 600 nm, is incident normally on a diffraction grating. It is desired (1) that the first and second principal maxima for each wavelength appear at $\theta \le 30°$, (2) that the dispersion be as high as possible, and (3) that the third order for 600 nm be a missing order. (a) What should be the separation between adjacent slits? (b) What is the smallest possible individual slit width? (c) Name all orders for 600 nm that actually appear on the screen with the values chosen in (a) and (b).

23. White light (400 nm $< \lambda <$ 700 nm) is incident on a grating. Show that, no matter what the value of the grating spacing d, the second- and third-order spectra overlap.

24. An optical grating with a spacing $d = 1500$ nm is used to analyze soft x rays of wavelength $\lambda = 0.50$ nm. The angle of incidence θ is $90° - \gamma$, where γ is a small angle. The first-order maximum is found at an angle $\theta = 90° - 2\beta$. Find the value of β.

Section 41–9   Resolving Power of a Grating

25. In a particular grating the sodium doublet (see Example 7) is viewed in third order at 10° to the normal and is barely resolved. Find (a) the grating spacing and (b) the total width of the rulings.

26. A diffraction grating has a resolving power $R = \lambda/\Delta\lambda = Nm$. (a) Show that the corresponding frequency range $\Delta\nu$ that can just be resolved is given by $\Delta\nu = c/Nm\lambda$. (b) From Fig. 41–15, show that the "times of flight" of the two extreme rays differ by an amount $\Delta t = Nd\sin\theta/c$. (c) Show that $(\Delta\nu)(\Delta t) = 1$, this relation being independent of the various grating parameters. Assume $N \gg 1$.

Section 41–10   X-ray Diffraction

27. Consider an infinite two-dimensional square lattice as in Fig. 41–22b. One interplanar spacing is obviously a_0 itself. (a) Calculate the next five smaller interplanar spacings by sketching figures similar to Fig. 41–23a. (b) Show that your answers obey the general formula

$$d = a_0/(h^2 + k^2)^{1/2}$$

where h and k are integers which have no common factor other than unity.

28. In comparing the wavelengths of two monochromatic X-ray lines, it is noted that line A gives a first order reflection maximum at a glancing angle of 30° to the smooth face of a crystal. Line B, known to have a wavelength of 0.97 Å, gives a third-order reflection maximum at an angle of 60° from the same face of the same crystal. Find the wavelength of line A.

29. Assume that the incident x-ray beam in Fig. 41–29 is not monochromatic but contains wavelengths in a band from 0.95 to 1.3 Å. Will diffracted beams, associated with the planes shown, occur? If so, what wavelength is diffracted? Assume $d = 0.275$ nm.

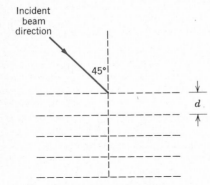

Figure 41–29 Problems 29 and 50.

30. First order Bragg scattering from a certain crystal occurs at an angle of incidence of 63.8°; see Fig. 41–30. The wavelength of the x rays is 0.26 nm. Assuming that the scattering is from the planes shown, find the unit cell size a_0.

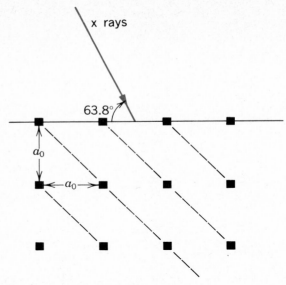

Figure 41–30 Problem 30.

ADDITIONAL PROBLEMS

31. A plane wave ($\lambda = 590$ nm) falls on a slit with $a = 0.40$ mm. A converging lens ($f = +70$ cm) is placed behind the slit and focuses the light on a screen. What is the linear distance on the screen from the center of the pattern to (a) the first minimum and (b) the second minimum?

32. (a) Design a double-slit system in which the fourth fringe, not counting the central maximum, is missing. (b) What other fringes, if any, are also missing?

33. In June 1985 a laser beam was fired from the Air Force Optical Station on Maui, Hawaii and reflected back from the shuttle *Discovery* as it sped by, 220 miles overhead. The diameter of the beam at the shuttle position was said to be 30 ft and the beam wavelength was 500 nm. What is the effective diameter of the laser at the Maui ground station? (*Hint:* A laser beam spreads because of diffraction; assume a circular exit aperture.)

34. Sound waves with frequency 3000 Hz diffract out of a speaker cabinet with a 0.30-m diameter opening into a large auditorium. Where does a listener standing against a wall 100 m from the speaker have the most difficulty hearing? Assume Fraunhofer diffraction and a sound speed of 343 m/s.

35. (a) A circular diaphragm 60 cm in diameter oscillates at a frequency of 25 kHz in an underwater source of sound used for submarine detection. Far from the source the sound intensity is distributed as a Fraunhofer diffraction pattern for a circular hole whose diameter equals that of the diaphragm. Take the speed of sound in water to be 1450 m/s and find the angle between the normal to the diaphragm and the direction of the first minimum. (b) Repeat for a source having an (audible) frequency of 1.0 kHz.

36. It can be shown that, except for $\theta = 0$, a circular obstacle produces the same diffraction pattern as a circular hole of the same diameter. Furthermore, if there are many such obstacles, as water droplets located randomly, then the interference effects vanish leaving only the diffraction associated with a single obstacle. (a) Explain why one sees a "ring" around the moon on a foggy night. Usually the ring is reddish in color; explain why. (b) Calculate the size of the water droplets in the air if the ring around the moon appears to have a diameter 1.5 times that of the moon. (c) At what distance from the moon might a bluish ring be seen? Sometimes the rings are white: explain. (d) The color arrangement is opposite to that in a rainbow—why should this be so?

37. The paintings of Georges Seurat consist of closely spaced small dots (≈ 2 mm in diameter) of pure pigment, as indicated in Fig. 41–31. The illusion of color mixing occurs because the

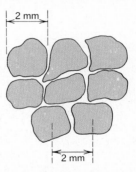

Figure 41–31 Problem 37.

pupils of the observer's eyes diffract light entering them. Calculate the minimum distance an observer must stand from such a painting to achieve the desired blending of color. Take the wavelength of the light to be 550 nm and the diameter of the pupil to be 1.5 mm.

38. (a) How small is the angular separation of two stars if their images are barely resolved by the Thaw refracting telescope at the Allegheny Observatory in Pittsburgh? The lens diameter is 76 cm and its focal length is 14 m. Assume $\lambda = 550$ nm. (b) Find the distance between these barely resolved stars if each of them is 10 light-years distant from the earth. (c) For the image of a single star in this telescope, find the diameter of the first dark ring in the diffraction pattern, as measured on a photographic plate placed at the focal plane. Assume that the star image structure is associated entirely with diffraction at the lens aperature and not with (small) lens "errors," etc.

39. A diffraction grating is made up of slits of width 300 nm with a 900 nm separation between their centers. The grating is illuminated by monochromatic plane waves, $\lambda = 600$ nm, the angle of incidence being zero. (a) How many diffraction maxima are there? (b) What is the angular width of the spectral lines observed, if the grating has 1000 slits? Angular width is defined to be the angle between the two positions of zero intensity on either side of the maximum.

40. If Superman really had x-ray vision at 0.10 nm and a 4.0 mm pupil diameter, at what maximum altitude could he distinguish villains from heroes assuming the minimum detail required was 5.0 cm?

41. An acoustic double-slit system (slit separation d, slit width a) is driven by two loudspeakers as shown in Fig. 41–32. By use of a variable delay line, the phase of one of the speakers may be varied. Describe in detail what changes occur in the intensity pattern at large distances as this phase difference is varied from zero to 2π. Take both interference and diffraction effects into account.

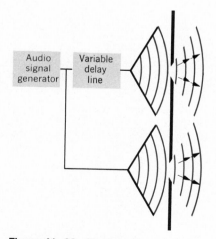

Figure 41–32 Problem 41.

42. In problem 39, how is the angular width of the spectral lines related to the resolving power of the grating?

43. Assume that the limits of the visible spectrum are arbitrarily chosen as 430 and 680 nm. Design a grating that will spread the first order spectrum through an angular range of 20°.

44. With light from a gaseous discharge tube incident normally on a grating with a distance 1.73 μm between adjacent slit centers, a green line appears with sharp maxima at measured transmission angles $\theta = \pm 17.6°$, $37.3°$, $-37.1°$, $65.2°$, and $-65.0°$. Compute the wavelength of the green line that best fits the data.

45. A transmission grating with $d = 1.50$ μm is illuminated at various angles of incidence by light of wavelength 600 nm. Plot as a function of angle of incidence (0 to 90°) the angular deviation of the first order diffracted beam from the incident direction. See Problem 19.

46. A grating has 350 rulings/mm and is illuminated at normal incidence by white light. A spectrum is formed on a screen 30 cm from the grating. If a 10-mm square hole is cut in the screen, its inner edge being 50 mm from the central maximum, what range of wavelengths passes through the hole?

47. A grating has 40,000 rulings spread over 76 mm. (a) What is its expected dispersion D for sodium light ($\lambda = 589$ nm) in the first three orders? (b) What is its resolving power in these orders?

48. Show that the dispersion of a grating can be written as

$$D = \frac{\tan \theta}{\lambda}.$$

49. Monochromatic x rays are incident on a set of NaCl crystal planes whose lattice spacing is 0.398 Å. When the beam is rotated 60° from the normal, first-order Bragg reflection is observed. What is the wavelength of the x rays?

50. Monochromatic x rays ($\lambda = 1.25$ Å) fall on a crystal of sodium chloride, making an angle of 45° with the reference line shown in Fig. 41–29. The planes shown are those of Fig. 41–23a, for which $d = 2.52$ Å. Through what angles must the crystal be turned to give a diffracted beam associated with the planes shown? Assume that the crystal is turned about an axis that is perpendicular to the plane of the page.

51. Prove that it is not possible to determine both wavelength of radiation and spacing of Bragg reflecting planes in a crystal by measuring the angles for Bragg reflection in several orders.

52. Two spectral lines have wavelengths λ and $\lambda + \Delta\lambda$, respectively, where $\Delta\lambda \ll \lambda$. Show that their angular separation $\Delta\theta$ in a grating spectrometer is given approximately by

$$\Delta\theta = \frac{\Delta\lambda}{[(d/m)^2 - \lambda^2]^{1/2}},$$

where d is the separation of adjacent slit centers and m is the order at which the lines are observed. Notice that the angular separation is greater in higher orders than in lower orders.

*53. In a Soviet-French experiment to monitor the moon's surface with a light beam, pulsed radiation from a ruby laser ($\lambda = 0.69$ μm) was directed to the moon through a reflecting telescope with a mirror radius of 1.3 m. A reflector on the moon behaved like a circular plane mirror with radius 10 cm, reflecting the light directly back toward the telescope on earth. The reflected light was then detected by a photometer after being brought to a focus by this telescope. What fraction of the original light energy was picked up by the detector? Assume that for each direction of travel all the energy is in the central diffraction circle.

*54. Derive this expression for the intensity pattern for a three-slit "grating":

$$I_\theta = \tfrac{1}{9}I_m (1 + 4 \cos \phi + 4 \cos^2 \phi),$$

where

$$\phi = \frac{2\pi d \sin \theta}{\lambda}.$$

Assume that $a \ll \lambda$ and be guided by the derivation of the corresponding double-slit formula (Eq. 41–7).

Light and Quantum Physics

42–1 Introduction

So far we have studied the behavior of light—by which we mean the entire electromagnetic spectrum—under the headings of propagation, reflection, refraction, interference, diffraction, and polarization. We saw that we could understand all these phenomena by assuming that light behaves as if it were a wave. Now we consider new experiments, involving light and its interaction with matter. We see that we can understand some of these only if we assume that light behaves as if it were a stream of particles.

These two views of the nature of light—wave and particle—are so different that they might seem to exclude each other. We shall face this apparent contradiction squarely in Section 43–5, where we seek to reconcile this wave-particle duality. First, however, we discuss the experimental evidence for the particle view of light. We begin by describing the emission of light by hot objects.

Spectral radiancy

The lower curve in Fig. 42–1 shows how the radiation emitted by a glowing ribbon of polished tungsten at 2000 K is distributed in wavelength. The quantity $\mathscr{R}_\lambda$, plotted on the vertical axis in Fig. 42–1, is called the *spectral radiancy*. It is defined so that $\mathscr{R}_\lambda \, d\lambda$ is the rate at which energy is radiated per unit area of surface for wavelengths lying in the interval λ to $\lambda + d\lambda$. Typical SI units for $\mathscr{R}_\lambda$ are $W/m^2 \cdot \mu m$; the corresponding units of $\mathscr{R}_\lambda \, d\lambda$ are W/m^2. In measuring $\mathscr{R}_\lambda$, all radiation emerging into the forward hemisphere is included.

Sometimes we wish to discuss the total radiated energy without regard to its wavelength. An appropriate quantity here is the *radiancy* $\mathscr{R}$, defined as the rate per unit surface area at which energy is radiated into the forward hemisphere, the SI units being W/m^2. We can find it by integrating the radiation present in all wavelength intervals:

Radiancy

$$\mathscr{R} = \int_0^\infty \mathscr{R}_\lambda \, d\lambda. \tag{42–1}$$

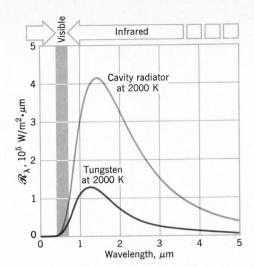

Figure 42-1 The full-colored curve shows the spectral radiancy of tungsten at 2000 K. The light-colored curve shows this quantity for a cavity radiator at the same temperature.

The radiancy $\mathscr{R}$ can be interpreted as the area under the plot of $\mathscr{R}_\lambda$ against λ. In Fig. 42-1, for tungsten at 2000 K, this area—and thus $\mathscr{R}$—is 2.4×10^5 W/m². Note the similarity between such curves and the Maxwell speed distribution curves of Section 21-9.

For every material there exists a family of curves much like the lower curve of Fig. 42-1, one curve for every temperature. If such families of curves are compared, no obvious regularities stand out. A quantitative understanding in terms of a basic theory presents serious difficulties. Fortunately, it is possible to work with an idealized heated solid, called a cavity radiator. Its light-emitting properties prove to be independent of any particular material and to vary in a simple way with temperature. In much the same way it proved convenient earlier to deal with an ideal gas rather than to analyze the properties of the infinite variety of real gases. The cavity radiator is the ideal solid as far as its light-emitting characteristics are concerned. The next two sections describe how the theoretical study of cavity radiation in 1900 by the German physicist Max Planck laid the foundations of modern quantum physics.

Cavity radiator

42-2 Cavity Radiation

A cavity radiator consists of a block of material having an internal cavity connected to the outside surface of the block by a small hole. Let the block be raised to a uniform temperature (say 2000 K) as determined by a suitable thermometer. The radiation emerging from the hole is called *cavity radiation**; its nature depends only on the temperature of the cavity walls and not at all on the shape or size of the cavity or on the material forming the cavity walls.

Figure 42-2 shows an actual cavity radiator, consisting of a hollow thin-walled cylinder of tungsten heated by passing an electric current through it. The cylinder is mounted in an evacuated glass bulb and a tiny hole is drilled through

Figure 42-2 An incandescent tungsten tube with a small hole drilled in its wall. The radiation emerging from the hole is cavity radiation.

---

* We can discuss a cavity as an absorber—as well as an emitter—of light. If our cavity block is held at room temperature and viewed by ambient light the small hole that penetrates to its interior appears black. Light that enters this hole is trapped within the cavity, which behaves like a perfect absorber of the incident light. It is on this basis that cavity radiation is often called *black body radiation*.

the cylinder wall. Consider that we form two additional similar cavities, with walls of tantalum and of molybdenum and that we heat all three cavities to the same temperature. Measurements of the radiancy $\mathcal{R}$ and the spectral radiancy $\mathcal{R}_\lambda$ for these three experimental cavities show the following:

1. At a given temperature the radiancy of the hole is identical for all three cavities; the radiancies of the three outer surfaces are all different and are smaller than the cavity radiancy which, at 2000 K, has the value 9.00×10^5 W/m².

2. The spectral radiancy for the cavity radiation varies with temperature in the way shown in Fig. 42–3. These curves depend only on the temperature and are quite independent of the material and of the size and shape of the cavity.

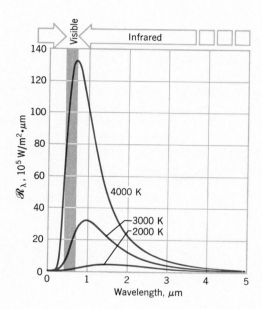

Figure 42–3 The spectral radiancy for cavity radiation at three different temperatures. As the temperature is raised the maximum of the radiation curve shifts to smaller wavelengths.

3. The cavity radiancy varies with temperature in a simple way, namely as

$$\mathcal{R} = \sigma T^4 \qquad \text{(cavity radiation)} \tag{42–2}$$

where σ is a universal constant called the Stefan-Boltzmann constant; its measured value is 5.67×10^{-8} W/m²·K⁴.

4. The radiancy of the outer surfaces varies with temperature in a more complicated way and is different for different materials. We often write it as

$$\mathcal{R} = e\sigma T^4, \tag{42–3}$$

Fourth-power law

where e, the *emissivity*, depends on the material and the temperature and is always less than unity. For our three cavity materials at 2000 K the measured emissivities are 0.259 (tungsten), 0.232 (tantalum), and 0.212 (molybdenum).

We can verify the universal character of cavity radiation by means of the arrangement of Fig. 42–4. It shows two cavities made of different materials, of arbitrary shapes, but with the same wall temperature T. Radiation, described by $\mathcal{R}_A$, goes from cavity A to cavity B and radiation described by $\mathcal{R}_B$ moves in the opposite direction. If these two rates of energy transfer are not equal, one end of the

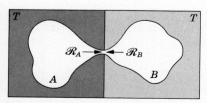

Figure 42–4 Two radiant cavities initially at the same temperature are placed together as shown.

composite block will start to heat up and the other end will start to cool down, which is a violation of the second law of thermodynamics. Thus we must have

$$\mathscr{R}_A = \mathscr{R}_B = \mathscr{R}_C, \tag{42-4}$$

where $\mathscr{R}_C$ describes the total radiation for *all* cavities.

Not only the total radiation but also the distribution of radiant energy with wavelength must be the same for each cavity in Fig. 42–4. We can show this by placing a filter between the two cavity openings, so chosen that it permits only a selected narrow band of wavelengths to pass. Applying the same argument, we can show that we must have

$$\mathscr{R}_{\lambda A} = \mathscr{R}_{\lambda B} = \mathscr{R}_{\lambda C}, \tag{42-5}$$

where $\mathscr{R}_{\lambda C}$ is a spectral radiancy characteristic of all cavities.

42–3 Planck's Radiation Law and the Origin of the Quantum

A theoretical explanation for the cavity radiation was one of the outstanding unsolved problems in physics during the years before the turn of the present century. A number of capable physicists advanced theories based on classical physics, which, however, had only limited success.

Wilhelm Wien derived a theoretical expression (see Problem 4) for the spectral radiancy of cavity radiation by applying the laws of thermodynamics to the radiation in the cavity. As we see from Fig. 42–5 Wien's law fits quite well at relatively short wavelengths. At longer wavelengths, however, it becomes increas-

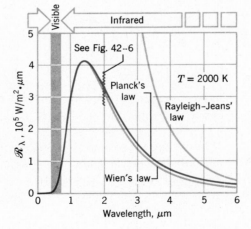

Figure 42–5 The full-colored curve, which can be shown to be in excellent agreement with experiment, is the Planck radiation law for cavity radiation at 2000 K. The curves predicted by the Wien and the Rayleigh-Jeans radiation laws for this temperature are also shown.

ingly in disagreement with experiment. This shows up not only in Fig. 42–5 but especially in Fig. 42–6, which shows the spectral radiancy (or rather its logarithm) for wavelengths up to 60 μm, far out in the tail of the curve of Fig. 42–5.

Lord Rayleigh also derived a radiation formula, later elaborated by Sir James Jeans. This Rayleigh-Jeans formula (see Problem 5) is the result of applying equipartition of energy ideas (see Section 21–7) to the radiation in the cavity. As we see from Fig. 42–5 their formula fails miserably for the range of wavelengths shown in that figure. In fact their curve does not even pass through a maximum

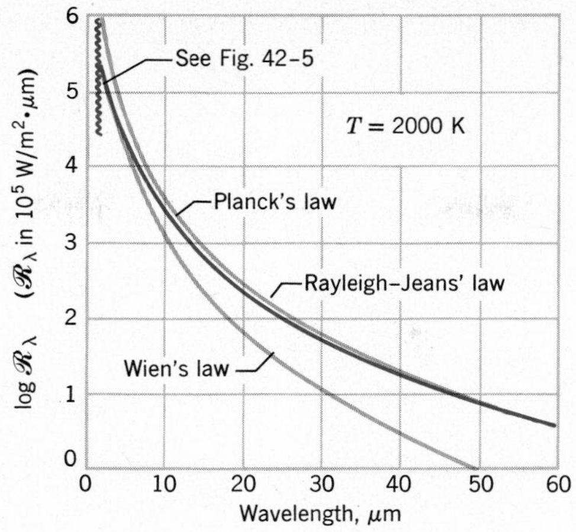

Figure 42–6 The three curves of Fig. 42–5 are extended to much longer wavelengths, starting from the vertical wavy line in that figure ($\lambda = 2\ \mu$m). Note that the Planck radiation law merges with Wien's law at short wavelengths and with the Rayleigh-Jeans law at long wavelengths. It is necessary to plot log $\mathscr{R}_\lambda$ rather than $\mathscr{R}_\lambda$ on the vertical axis because of the wide range of values involved.

but approachs infinity as the wavelength approachs zero! However, as Fig. 42–6 shows, the Rayleigh-Jeans formula fits quite well at very long wavelengths, precisely where the Wien formula fails.

The Wien and the Rayleigh-Jeans formulas represent the best that classical physics can offer as a solution to the cavity radiation problem. We see that one formula fits the experimental data just where the other formula fails, and conversely; neither formula holds over the complete range of wavelengths. Thus physicists at the turn of the century were faced with a well-formulated significant problem that the physics of their day could not explain. New ideas were needed.

Planck, seeking to reconcile the Wien and the Rayleigh-Jeans formulas, made a clever interpolation between them that turned out to fit the data extremely well at all wavelengths. Planck's formula, announced to the Berlin Physical Society on October 19, 1900, is

Planck's radiation formula

$$\mathscr{R}_\lambda = \frac{c_1}{\lambda^5} \frac{1}{e^{c_2/\lambda T} - 1}. \qquad (42-6)$$

in which c_1 and c_2 are empirical constants, chosen to give the best fit of Eq. 42–6 to the experimental data. This formula, although interesting and important, was still empirical at that stage and did not constitute a theory.

Planck sought such a theory in terms of a detailed model of the atomic processes taking place at the cavity walls. He assumed that the atoms that make up these walls behave like tiny electromagnetic oscillators, each with a characteristic frequency of oscillation. The oscillators emit electromagnetic energy into the cavity and absorb electromagnetic energy from it. Thus it should be possible to deduce the characteristics of the cavity radiation from those of the oscillators with which it is in equilibrium.

Planck was led to make two radical assumptions about the atomic oscillators. As eventually interpreted, these assumptions are the following:

1. An oscillator cannot have just any value of energy but only those values given by

Quantization of energy

$$E = nh\nu, \qquad (42-7)$$

where ν is the oscillator frequency, h is a constant (now called the *Planck constant*), and n is a number (now called a *quantum number*) that can take on only positive integral values. Equation 42–7 asserts that the oscillator energy is *quantized*.

2. The oscillators do not radiate energy continuously, but only in "jumps," or *quanta*. These quanta of energy are emitted when an oscillator changes from one to another of its quantized energy states. Thus, if *n* changes by one unit, Eq. 42–7 shows that an amount of energy given by

$$\Delta E = \Delta nh\nu = h\nu \qquad (42-8)$$

is radiated. As long as an oscillator remains in one of its quantized states, it neither emits nor absorbs energy.

These assumptions were radical ones and, indeed, Planck himself resisted accepting them for many years. In his words, "My futile attempts to fit the elementary quantum of action [that is, the quantity *h*] somehow into the classical theory continued for a number of years, and they cost me a great deal of effort."

Consider the application of Planck's hypotheses to a large-scale oscillator such as a mass-spring system or an *LC* circuit. It would be a stoutly defended common belief that oscillations in such systems could take place with *any* value of total energy and not with only certain discrete values. In the decay of such oscillations (by friction in the mass-spring system or by resistance and radiation in the *LC* circuit), it would seem that the mechanical or electromagnetic energy would decrease in a perfectly continuous way and not by "jumps." There is no basis in everyday experience, however, to dismiss Planck's assumptions as violations of "common sense," for the Planck constant proves to have a very small value, namely,

The Planck constant

$$h = 6.626 \times 10^{-34} \text{ joule·second.}$$

The following example makes this clear.

Example 1 A mass-spring system has a mass $m = 1.0$ kg and a spring constant $k = 20$ N/m and is oscillating with an amplitude of 10 mm. (*a*) If its energy is quantized according to Eq. 42–7, what is the quantum number *n*?

From Eq. 14–11 the frequency is

$$\nu = \frac{1}{2\pi}\sqrt{\frac{k}{m}} = \frac{1}{2\pi}\sqrt{\frac{20 \text{ N/m}}{1.0 \text{ kg}}} = 0.71 \text{ Hz.}$$

From Eq. 7–8 the mechanical energy is

$$E = \tfrac{1}{2}kx_{max}^2 = \tfrac{1}{2}(20 \text{ N/m})(1.0 \times 10^{-2} \text{ m})^2 = 1.0 \times 10^{-3} \text{ J.}$$

From Eq. 42–7 the quantum number is

$$n = \frac{E}{h\nu} = \frac{1.0 \times 10^{-3} \text{ J}}{(6.6 \times 10^{-34} \text{ J·s})(0.71 \text{ Hz})} = 2.1 \times 10^{30}.$$

(*b*) If *n* changes by unity, what fractional change in energy occurs?

The fractional change in energy is given by dividing Eq. 42–8 by Eq. 42–7, or

$$\frac{\Delta E}{E} = \frac{h\nu}{nh\nu} = \frac{1}{n} \cong 5 \times 10^{-31}.$$

Thus for large-scale oscillators the quantum numbers are enormous and the quantized nature of the energy of the oscillations will not be apparent. Similarly, in large-scale experiments we are not aware of the discrete nature of mass and the quantized nature of charge, that is, of the existence of atoms and electrons.

On the basis of his two assumptions, Planck was able to derive his radiation law (Eq. 42–6) entirely from theory. His theoretical expressions for the hitherto empirical constants c_1 and c_2 were

$$c_1 = 2\pi c^2 h \qquad \text{and} \qquad c_2 = \frac{hc}{k},$$

where k is the Boltzmann constant (see Section 21–5) and c is the speed of light. By inserting the experimental values for c_1 and c_2, Planck was able to derive the values of both h and k. Planck described his theory to the Berlin Physical Society on Friday, December 14, 1900. A new epoch in physics started on that date.

42–4   The Photoelectric Effect

We were led to the concept of the quantization of energy by considering the interaction between light and matter at the interior wall of a cavity radiator. Next we describe a second example of a light-matter interaction, the *photoelectric effect*. It involves the Planck constant in a central way and extends the concept of quantization to the very nature of light itself.

Figure 42–7 shows a typical laboratory photoelectric apparatus. Monochromatic light of suitably high frequency falls on a metal surface (emitter E) and ejects electrons, which we call *photoelectrons*, from it. These can be detected as a photoelectric current i if they are attracted to collector C by a suitable potential difference V established between E and C.*

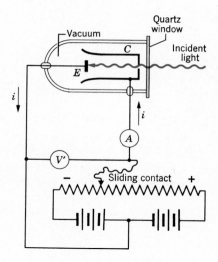

Figure 42–7   An apparatus used to study the photoelectric effect. The applied potential difference V' can be varied continuously from negative to positive values by moving the sliding contact. The collected photoelectrons, being negatively charged, move through ammeter A in a direction opposite to that shown by the conventionally-directed current arrow.

Figure 42–8 (curve a) shows the photoelectric current in an apparatus like that of Fig. 42–7 as a function of the potential difference V. We see that if V is positive and large enough the photoelectric current reaches a constant saturation value, at which all photoelectrons ejected from E are collected by C.

If we reverse V, the photoelectric current does not immediately drop to zero because the electrons are emitted from E with a finite velocity. Some will reach collector C even when the electric field opposes their motion. However, if this re-

---

* If the emitter and the collector in Fig. 42–7 are made of different materials—and this is usually the case—the actual potential difference V that acts between E and C is not the same as the voltage V' read on the meter in that figure. To V' must be added a "hidden" emf, the constant and measureable contact potential difference V_{EC} (a battery effect) that arises because E and C are made of different materials. In all that follows the symbol "V" will represent the algebraic sum of V' and V_{EC}.

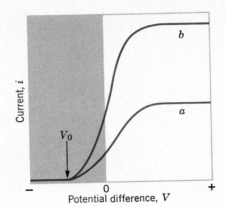

Figure 42-8 A representation (not to scale) of data taken with the apparatus of Fig. 42–7. The intensity of the incident light is twice as great for curve *b* as for curve *a*.

versed potential difference is made large enough, a value V_0 (the stopping potential) is reached at which the photoelectric current does drop to zero. This potential difference V_0, multiplied by the electron charge, measures the kinetic energy K_{max} of the fastest ejected photoelectron. In other words,

The stopping potential

$$K_{max} = eV_0. \qquad (42-9)$$

The important thing to note is that V_0, and thus K_{max}, is independent of the intensity of the incident light. This is shown by curve *b* in Fig. 42–8, in which the light intensity has been doubled; although the saturation current is also doubled the stopping potential V_0 remains unchanged.

Figure 42–9 is a plot of the measured stopping potential V_0 as a function of the frequency ν of the light falling on the emitting surface. We see by extrapolation that there is a sharp *cutoff frequency* ν_0, corresponding to a stopping potential of zero. For light of frequency lower than this no photoelectrons at all are emitted, no matter how intense the light; the photoelectric effect simply stops.

The cutoff frequency

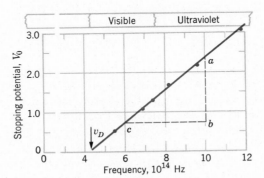

Figure 42-9 The stopping potential as a function of frequency for a sodium surface. Data taken by R. A. Millikan at the University of Chicago in 1916.

Three major features of the photoelectric effect cannot be explained in terms of the wave theory of light:

Failure of the wave theory

1. Wave theory suggests that the kinetic energy of the photoelectrons should increase as the light beam is made more intense. However, Fig. 42–8 shows that $K_{max} (= eV_0)$ is independent of the light intensity; this has been tested over a range of intensities of 10^7.

2. According to the wave theory, the photoelectric effect should occur for any frequency of the light, provided only that the light is intense enough. However, Fig. 42–9 shows that there exists, for each surface, a characteristic cutoff frequency ν_0. For frequencies less than this, the photoelectric emission disappears, no matter how intense the illumination.

3. If light were a wave, a particular electron in the metal would only slowly receive energy from the light because only a small portion of the wave fronts moving past would be intercepted by the electron. Because of this it would take time for the electron to "soak up" enough energy to be ejected. For light of sufficiently feeble intensity this time lag (see Example 2) should be measurable but no time lag has ever been detected.

Example 2 A zinc surface is placed 5.0 m from a weak monochromatic light source whose power output is 1.0 mW. Consider that a given ejected photoelectron may have collected its energy from a circular area taken to be as large as ten atomic diameters ($= 1.0 \times 10^{-9}$ m) in radius. The energy required to remove an electron from the surface is 4.2 eV. Assuming light to be a wave, how long would it take for such a "target" to soak up this much energy from the incident beam?

The target area is $\pi (1.0 \times 10^{-9}$ m$)^2$ or 3.1×10^{-18} m^2; the area of a 5.0-m sphere centered on the light source is $4\pi \times (5.0$ m$)^2 \cong 310$ m^2. Thus, if the light source radiates uniformly in all directions, the rate P at which energy falls on the target is given by

$$P = (1.0 \times 10^{-3} \text{ W}) \left(\frac{3.1 \times 10^{-18} \text{ m}^2}{310 \text{ m}^2} \right) = 1.0 \times 10^{-23} \text{ J/s.}$$

Assuming that all this power is absorbed, we may calculate the time required from

$$t = \left(\frac{4.2 \text{ eV}}{1.0 \times 10^{-23} \text{ J/s}} \right) \left(\frac{1.6 \times 10^{-19} \text{ J}}{1 \text{ eV}} \right) \cong 19 \text{ hours.}$$

Actually, no matter how feeble the light source, photoelectrons emerge within about 10^{-9} s after the light strikes the emitter!

42–5 Einstein's Photon Theory

In 1905 Einstein made a remarkable assumption about the nature of light, namely that, under some circumstances at least, it behaves as if its energy was concentrated into localized bundles, later called *photons*. The energy E of a single photon is given by

Energy of a photon

$$E = h\nu \tag{42–10}$$

where ν is the frequency of the light. This notion that a light beam behaves like a stream of particles is in sharp contrast to the notion that it behaves like a wave. In the wave theory of light the energy is *not* concentrated into bundles but is spread out uniformly over the wave fronts.

When Planck, in 1900, derived his radiation law and first introduced the quantity h into physics he made use of the relation $E = h\nu$. He applied it, however, not to the radiation within the cavity but to the atomic oscillators that made up its walls; see Eq. 42–7. Planck treated the cavity radiation on the basis of wave theory. Einstein however was later able to derive Planck's radiation law on the basis of his photon concept. His method was both clear and simple and avoided many of the special assumptions that Planck had found it necessary to make in his pioneering effort.

If we apply Einstein's photon concept to the photoelectric effect, we can write

Einstein's photoelectric equation

$$h\nu = \phi + K_{\text{max}} \tag{42–11}$$

where $h\nu$ is the energy of the photon. Equation 42–11 says that a photon carries an energy $h\nu$ into the surface where it is absorbed by a single electron. Part of this energy (ϕ, called the *work function* of the emitting surface) is used in causing the electron to escape from the metal surface. The excess energy ($h\nu - \phi$) becomes the electron kinetic energy; if the electron does not lose energy by internal collisions as it escapes from the metal, it will still have this much kinetic energy after it emerges. Thus K_{max} represents the maximum kinetic energy that the photoelectron can have outside the surface.

Consider how Einstein's photon hypothesis meets the three objections raised against the wave-theory interpretation of the photoelectric effect. As for objection 1 (the lack of dependence of K_{max} on the intensity of illumination), there is complete agreement of the photon theory with experiment. If we double the light intensity, we double the number of photons and thus double the photoelectric current; we do not change the energy ($= h\nu$) of the individual photons or the nature of the individual photoelectric processes described by Eq. 42–11.

Objection 2 (the existence of a cutoff frequency) follows from Eq. 42–11. If K_{max} equals zero, we have

$$h\nu_0 = \phi,$$

which asserts that the photon has just enough energy to eject the photoelectrons and none extra to appear as kinetic energy. If ν is reduced below ν_0, the individual photons, no matter how many of them there are (that is, no matter how intense the illumination), will not have enough energy to eject photoelectrons.

Objection 3 (the absence of a time lag) follows from the photon theory because the required energy is supplied in a concentrated bundle. It is not spread uniformly over the beam cross-section as in the wave theory.

Let us rewrite Einstein's photoelectric equation (Eq. 42–11) by substituting for K_{max} from Eq. 42–9. This yields, after rearrangement,

$$V_0 = (h/e)\nu - (\phi/e). \tag{42–12}$$

A linear relationship

Thus Einstein's theory predicts a linear relationship between V_0 and ν, in complete agreement with experiment; see Fig. 42–9. The slope of the experimental curve in this figure should be h/e, or

$$\frac{h}{e} = \frac{ab}{bc} = \frac{2.30\text{ V} - 0.68\text{ V}}{(10 \times 10^{14} - 6.0 \times 10^{14})\text{ Hz}} = 4.1 \times 10^{-15}\text{ V·s}.$$

We can find h by multiplying this ratio by the electron charge e,

$$h = (4.1 \times 10^{-15}\text{ V·s})(1.6 \times 10^{-19}\text{ C}) = 6.6 \times 10^{-34}\text{ J·s}.$$

From a more careful analysis of these and other data, including data taken with lithium surfaces, Millikan found the value $h = 6.57 \times 10^{-34}$ J·s with an accuracy of about 0.5%. This agreement with the value of h derived from Planck's radiation formula is a striking confirmation of Einstein's photon concept.

Example 3 Deduce the work function for sodium that follows from Fig. 42–9.

The intersection of the straight line in Fig. 42–9 with the horizontal axis is the cutoff frequency ν_0. Substituting yields

$$\phi = h\nu_0 = (6.6 \times 10^{-34}\text{ J·s})(4.4 \times 10^{14}\text{ Hz})$$
$$= 2.9 \times 10^{-19}\text{ J} = 1.8\text{ eV}.$$

The handbook value is 2.3 eV, in rough agreement.

Convince yourself that this determination of ϕ, which involves only the intercept of the straight line in Fig. 42–9, necessarily involves the contact potential difference correction outlined in the footnote to page 779. The determination of h in the text above, however, involves only the slope of this line and is independent of the contact potential difference correction.

For the 26-year old Einstein 1905 was a banner year. During that year he published, in sequence and in the same prestigious journal: (1) his theory of photons, (2) a theory of Brownian motion that contributed greatly to further understanding of the atomic nature of matter, (3) his special theory of relativity, and (4) the germ of his now-famous $E = mc^2$ relationship, derived from this theory. During all this time (and for four additional years), he was employed as a minor official in the Swiss patent office, working on physics privately and in his spare time!

When Einstein first advanced his photon theory of light the facts of photoelectricity were not nearly as well established experimentally as we have described. Precise photoelectric measurements are difficult and it was not until 1916 that the American physicist R. A. Millikan successfully subjected Einstein's photoelectric equation to rigorous experimental test. Although Millikan showed that this equation agreed with experiment in every detail he himself remained unconvinced that Einstein's light particles were real. He wrote of Einstein's "... bold, not to say reckless, hypothesis ..." and wrote further that Einstein's photon concept "... seems at present to be wholly untenable."

Planck, the very originator of the constant h, did not at once accept Einstein's photons either. In recommending Einstein for membership in the Royal Prussian Academy of Sciences in 1913 he wrote: "that he may sometimes have missed the target in his speculations, as for example in his theory of light quanta, cannot really be held against him." Far from being unusual it is almost commonplace that truly novel ideas are accepted only slowly, even by such men of genius as Millikan and Planck. It was, incidentally, for his photon theory that Einstein received the Nobel Prize in physics for 1921.

42–6 The Compton Effect

Cavity radiation was our first example of the interaction of light with matter; it deals largely with infrared light. The photoelectric effect, our second example, involves experiments with visible and ultraviolet light. We shall now describe the Compton effect, in which the key experiments occur in the x-ray and gamma-ray regions of the electromagnetic spectrum. This effect, explained in terms of billiard-ball-like collisions between photons and electrons, provides particularly convincing evidence of the particle-like nature of light. Its explanation requires that not only the energy but also the linear momentum of a light beam be concentrated in photons.

A. H. Compton*, in 1923, allowed a beam of x rays with sharply defined wavelength λ to fall on a graphite scatterer S, as in Fig. 42–10, and he measured, for

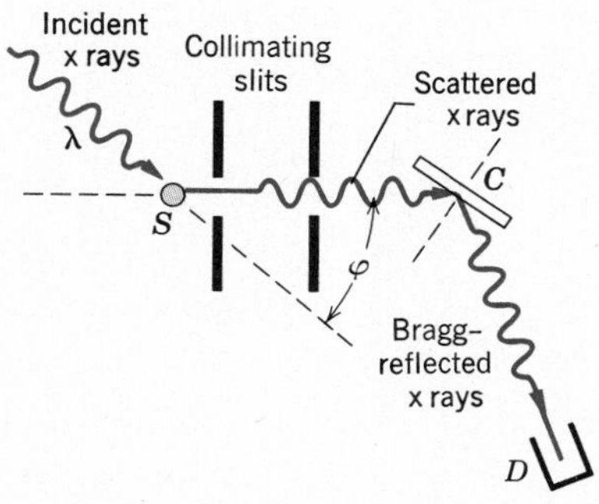

Compton's experiment

Figure 42–10 Compton's experimental arrangement. X rays of wavelength λ fall on graphite scatterer S. The wavelengths of the x rays scattered at selected angles φ are measured by Bragg reflection techniques (see Section 41–10) using crystal C. Detector D measures the intensities of the Bragg-reflected x rays.

* For an historical account of Compton's researches see "The Scattering of X-Rays as Particles" by A. H. Compton, *American Journal of Physics*, December 1961.

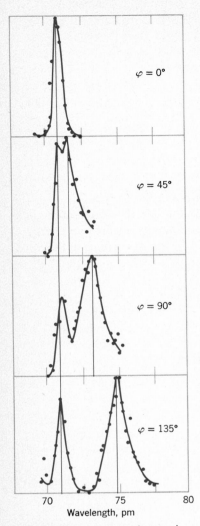

$\varphi = 0°$

$\varphi = 45°$

$\varphi = 90°$

$\varphi = 135°$

70   75   80
Wavelength, pm

Figure 42–11 Compton's experimental results for four different values of the scattering angle φ (1 pm = 1 picometer = 10^{-2} Å = 10^{-12} m).

various angles of scattering, the intensity of the scattered x rays as a function of their wavelength. Figure 42–11 shows his experimental results. We see that although the incident beam consists essentially of a single wavelength λ the scattered x rays have intensity peaks at two wavelengths; one of them is the same as the incident wavelength, the other, λ', being larger by an amount $\Delta\lambda$. This so-called *Compton shift* $\Delta\lambda$ varies with the angle at which the scattered x rays are observed.

The presence of a scattered wave of wavelength λ' cannot be understood if the incident x rays are regarded as an electromagnetic wave. On this picture the incident wave of frequency ν causes electrons in the scattering block to oscillate at that same frequency. These oscillating electrons, like charges surging back and forth in a small radio transmitting antenna, radiate electromagnetic waves that again have this same frequency ν. Thus, on the wave picture the scattered wave should have the same frequency ν and the same wavelength λ as the incident wave.

Compton explained his experimental results by postulating that the incoming x-ray beam was not a wave but an assembly of photons of energy E $(=h\nu)$ and that these photons experienced billiard-ball-like collisions with the free electrons in the scattering block. The "recoil" photons emerging from the block constitute, on this view, the scattered radiation. Since the incident photon transfers some of its energy to the electron with which it collides, the scattered photon must have a lower energy E'; it must therefore have a lower frequency ν' $(=E'/h)$, which implies a larger wavelength λ' $(=c/\nu')$. This point of view accounts, at least qualitatively, for the wavelength shift $\Delta\lambda$. Notice how different this particle model of x-ray scattering is from that based on the wave picture. Now let us analyze a single photon-electron collision quantitatively.

Figure 42–12 shows a collision between a photon and an electron, the electron assumed to be initially at rest and essentially free, that is, not bound to the atoms of the scatterer. Let us apply the law of conservation of energy to this collision. Since the recoil electrons may have a speed v that is comparable with that of

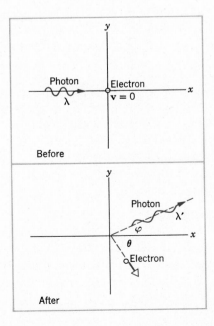

Figure 42–12 A photon of wavelength λ is incident on an electron at rest. On collision the photon is scattered at an angle φ with increased wavelength λ', whereas the electron moves off with speed v in direction θ.

Analyzing the Compton collision

light, we must use the relativistic expression for the kinetic energy of the electron. From the conservation of energy principle we may write

$$h\nu = h\nu' + (m - m_0)c^2,$$

in which the second term on the right (see Eq. 8–20) is the relativistic expression for the kinetic energy of the recoiling electron, m being the relativistic mass and m_0 the rest mass of that particle. Substituting c/λ for ν (and c/λ' for ν') and using Eq. 8–19 to eliminate the relativistic mass m leads us to

Energy conservation

$$\frac{hc}{\lambda} = \frac{hc}{\lambda'} + m_0c^2 \left(\frac{1}{\sqrt{1 - (v/c)^2}} - 1\right). \qquad (42–13)$$

Now let us apply the (vector) law of conservation of linear momentum to the collision of Fig. 42–12. We first need an expression for the momentum of a photon. In Section 38–5 we saw that if an object completely absorbs an energy U from a parallel light beam that falls on it, the light beam, according to the wave theory of light, will simultaneously transfer to the object a linear momentum given by U/c. On the photon picture we imagine this momentum to be carried along by the individual photons, each photon transporting linear momentum in amount $p = h\nu/c$, where $h\nu$ is the photon energy. Thus, if we substitute λ for c/ν, we can write

$$p = \frac{E}{c} = \frac{h\nu}{c} = \frac{h}{\lambda}. \qquad (42–14)$$

For the electron, the relativistic expression for the linear momentum is given by Eq. 9–13, or

$$\mathbf{p} = \frac{m_0\mathbf{v}}{\sqrt{1 - (v/c)^2}}.$$

Momentum conservation

We can then write for the conservation of the x-component of linear momentum

$$\frac{h}{\lambda} = \frac{h}{\lambda'} \cos \varphi + \frac{m_0v}{\sqrt{1 - (v/c)^2}} \cos \theta \qquad (42–15)$$

and for the y-component

$$0 = \frac{h}{\lambda'} \sin \varphi - \frac{m_0v}{\sqrt{1 - (v/c)^2}} \sin \theta. \qquad (42–16)$$

Our aim is to find $\Delta\lambda$ ($=\lambda' - \lambda$), the wavelength shift of the scattered photons, so that we may compare it with the experimental results of Fig. 42–11. Compton's experiment did not involve observations of the recoil electron in the scattering block. Of the five collision variables (λ, λ', v, φ, and θ) that appear in the three equations (42–13, 42–15, and 42–16) we may eliminate two. We chose to eliminate v and θ, which deal only with the electron, thereby reducing the three equations to a single relation among the variables.

Carrying out the necessary algebraic steps (see Problem 19) leads to this simple result:

The Compton shift

$$\Delta\lambda = \lambda' - \lambda = \frac{h}{m_0c} (1 - \cos \varphi). \qquad (42–17)$$

Thus the Compton shift $\Delta\lambda$ depends only on the scattering angle φ and not on the initial wavelength λ. Equation 42–17 predicts within experimental error the experimentally observed Compton shifts of Fig. 42–11. Note from the equation that $\Delta\lambda$ varies from zero (for $\varphi = 0$, corresponding to a "grazing" collision in Fig. 42–12, the incident photon being scarcely deflected) to $2h/m_0c$ (for $\varphi = 180°$, corresponding to a "head-on" collision, the incident photon being reversed in direction).

The Compton shift as a quantum effect

Remember that the Compton shift $\Delta\lambda$ is a purely quantum effect, not expected to occur on the basis of classical physics. Equation 42–17 supports this contention by showing that $\Delta\lambda \to 0$ as $h \to 0$. The device of letting the Planck constant approach zero is a formal way of testing quantum equations to see if they predict what would happen if the laws of classical physics applied not only to large objects but also to atoms and electrons.

The unshifted Compton peak

It remains to explain the presence of the peak in Fig. 42–11 for which the wavelength does not change on scattering. This peak results from collisions between photons and electrons which, instead of being free, are tightly bound in an ionic core in the scattering block. During photon collisions the bound electrons behave like very heavy free electrons. This is because the ionic core as a whole recoils during the collision. Thus the effective mass M for a carbon scatterer is approximately the mass of a carbon nucleus. Since this nucleus contains six protons and six neutrons, we have approximately, $M = 12 \times 1840m_0 = 22,000m_0$. If we replace m_0 by M in Eq. 42–17, we see that the Compton shift for collisions with tightly bound electrons is immeasurably small.

As in the cavity radiation problem (see Eq. 42–7) and the photoelectric effect (see Eq. 42–11), the Planck constant h is centrally involved in the Compton effect. The quantity h is the central constant of quantum physics. In a universe in which $h = 0$ there would be no quantum physics and classical physics would be valid in the subatomic domain. In particular, as Eq. 42–17 shows, there would be no Compton effect (that is, $\Delta\lambda = 0$) in such a universe.

Example 4 X rays with $\lambda = 1.00$ Å are scattered from a carbon block. The scattered radiation is viewed at 90° to the incident beam. (a) What is the Compton shift $\Delta\lambda$? (b) What kinetic energy is imparted to the recoiling electron?

(a) Putting $\varphi = 90°$ in Eq. 42–17, we have, for the Compton shift,

$$\Delta\lambda = \frac{h}{m_0c}(1 - \cos\varphi)$$
$$= \frac{6.63 \times 10^{-34} \text{ J·s}}{(9.11 \times 10^{-31} \text{ kg})(3.00 \times 10^8 \text{ m/s})}(1 - \cos 90°)$$
$$= 0.0243 \text{ Å}.$$

(b) If we put K for the kinetic energy of the electron, we can write Eq. 42–13 as

$$\frac{hc}{\lambda} = \frac{hc}{\lambda'} + K.$$

Since $\lambda' = \lambda + \Delta\lambda$, we obtain

$$\frac{hc}{\lambda} = \frac{hc}{\lambda + \Delta\lambda} + K,$$

which reduces to

$$K = \frac{hc\,\Delta\lambda}{\lambda(\lambda + \Delta\lambda)}$$
$$= \frac{(6.63 \times 10^{-34} \text{ J·s})(3.00 \times 10^8 \text{ m/s})(2.43 \times 10^{-12} \text{ m})}{(1.00 \times 10^{-10} \text{ m})(1.00 + 0.024) \times 10^{-10} \text{ m}}$$
$$= 4.72 \times 10^{-17} \text{ J} = 295 \text{ eV}.$$

You can show that the initial photon energy E in this case $(=h\nu = hc/\lambda)$ is 12.4 keV so that the photon lost about 2.4% of its energy in this collision. A photon whose energy was ten times as large (= 124 keV) can be shown to lose 20% of its energy in a similar collision. This is consistent with the fact that $\Delta\lambda$ does not depend on the initial wavelength. More energetic x rays, which have smaller wavelengths, will experience a larger percent increase in wavelength and thus a larger percent loss in energy.

42–7 Line Spectra

Continuous spectra and line spectra

We have seen how Planck successfully explained the nature of the radiation from heated solids, of which the cavity radiator formed the prototype. Such radiations form *continuous* spectra, Fig. 42–1 being an example. By contrast, Fig. 42–13

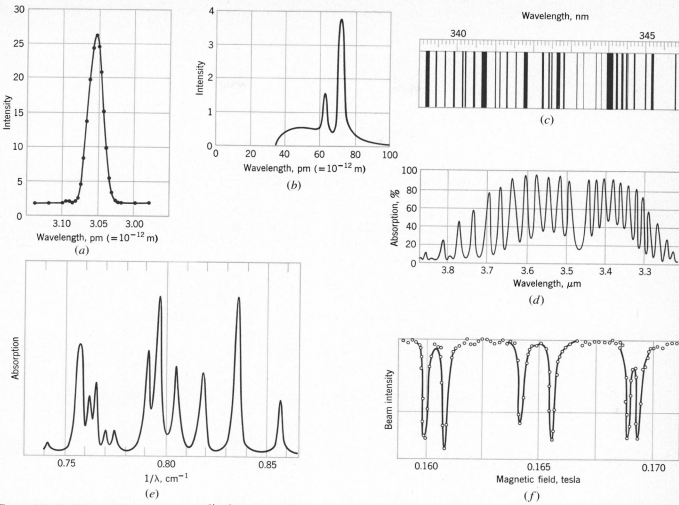

Figure 42-13 Selected lines from various regions of the electromagnetic spectrum; see Table 42-1. In (*f*) the spectrum lines are displayed by fixing the wavelength of the source (an electromagnetic oscillator) and varying the strength of the magnetic field through which the sample (a beam of hydrogen molecules) passes.

displays some *line spectra*, which are associated with the emission or the absorption of radiation by isolated atoms, molecules, and nuclei. These lines, described in Table 42-1, are selected from quite different regions of the electromagnetic

Table 42-1 Selected lines from various regions of the spectrum (see Figure 42-13)

Figure	Entity	Wavelength m	Spectrum region	Mode
a	^{198}Hg nucleus	$\sim 3 \times 10^{-12}$	Gamma ray	Emission
b	Mo atom	$\sim 6 \times 10^{-11}$	X-ray	Emission
c	Fe atom	$\sim 3 \times 10^{-7}$	Ultraviolet	Emission
d	HCl molecule	$\sim 3 \times 10^{-6}$	Infrared	Absorption
e	NH_3 molecule	$\sim 1 \times 10^{-2}$	Microwave	Absorption
f	H_2 molecule	~ 40	Radio	Absorption

spectrum, from gamma rays to radio waves. We shall see that the quantization idea, suitably extended, leads to an understanding of line spectra also. To help us understand the bewildering variety of line spectra, only hinted at in Fig. 42–13, we first study the spectrum of the hydrogen atom. Being the simplest atom it has the simplest spectrum.

42–8 Bohr's Theory of the Hydrogen Atom

The interaction between light and matter, which forms a theme for this chapter, appears in it simplest aspect when we study the emission or absorption of light by an isolated hydrogen atom. The wavelengths of many lines of the atomic hydrogen spectrum have been known with precision for many years and stand as a sensitive testing ground for any theory of atomic structure. Regularities in its spectrum are apparent on inspection, as Fig. 42–14 shows. This is not the case for more complex atoms such as iron; see Fig. 42–13c.

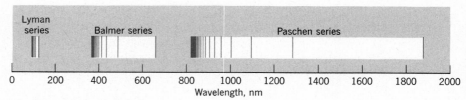

Figure 42–14 The spectrum of hydrogen. It consists of a number of series of lines, three of which are shown. Within each series the spectrum lines follow a regular pattern, approaching a so-called *series limit* at the short-wave end of the series.

It is tempting to identify the sharply defined frequency of a spectrum line as radiation emitted by an electron revolving about the central nucleus at that same angular frequency. Indeed, classical theory predicts that such an accelerating electron *will* radiate, and at its frequency of revolution. However, the radiation represents an energy loss and this same classical theory also predicts that the orbiting electron will quickly radiate away all its energy, its frequency changing continuously as the electron spirals into the nucleus. Thus the classical physics of Newton and Maxwell cannot even account for the very existence of line spectra, whose great variety and universality we have seen in the preceding section.

In 1913, just 2 years after Rutherford had proposed the nuclear model of the atom, Niels Bohr proposed a model for the hydrogen atom that, using no adjustable parameters whatever, predicted the wavelengths of its observed spectral lines with an accuracy of about 0.02%. Although wonderfully successful for hydrogen, and also for single-electron, hydrogen-like ions such as He^+ and Li^{++}, Bohr's theory is less useful for more complex atoms and has been largely replaced by modern quantum mechanics. We now regard Bohr's theory as an inspired and ingenious first step toward the more comprehensive theory that followed.

Bohr assumed that the hydrogen atom could exist in any one of a set of discrete stationary states, each with its sharply defined total energy E and that the atom would not radiate while in any of these states. Radiation occurs only when

Stationary states

the atom makes a transition from one state, with total energy E_k, to a second state, with lower total energy E_j. In equation form

Bohr's frequency condition

$$h\nu = E_k - E_j, \tag{42-18}$$

where $h\nu$ is the quantum of energy carried away by the single photon that is emitted from the atom during the transition. This *Bohr frequency condition* is quite general and applies to all atoms and also to molecules, solids, and nuclei.

To proceed beyond Eq. 42–18 it is necessary to describe the stationary states of the hydrogen atom and to derive an expression for their energies. Let us assume that the electron in the hydrogen atom moves in a circular orbit of radius r centered on its nucleus. This requires that the nucleus, which is a single proton, is so massive that the center of mass of the system is essentially at the position of the proton. Let us calculate the energy E of such an atom classically.

Writing Newton's second law for the motion of the electron, we have (using Coulomb's law)*

$$F = ma,$$

or

$$\frac{1}{4\pi\epsilon_0} \frac{(Ze)(e)}{r^2} = m\frac{v^2}{r}.$$

This allows us to calculate the kinetic energy of the electron, which is

$$K = \tfrac{1}{2}mv^2 = \frac{Ze^2}{8\pi\epsilon_0 r}. \tag{42-19}$$

The potential energy U of the nucleus-electron system is given by

$$U = V(-e) = -\frac{Ze^2}{4\pi\epsilon_0 r}, \tag{42-20}$$

where $V(=Ze/4\pi\epsilon_0 r)$ is the electric potential due to the nucleus at the radius of the electron.

The total energy E of the system is

E depends on r

$$E = K + U = -\frac{Ze^2}{8\pi\epsilon_0 r}. \tag{42-21}$$

Since the orbit radius can apparently take on any value, so can the energy E. The problem of quantizing E reduces to that of quantizing r.

Every property of the orbit is fixed if the radius is given. Equations 42–19, 42–20, and 42–21 show this specifically for the energies K, U, and E. From Eq. 42–19 we can show that the linear speed v for the electron is also given in terms of r by

$$v = \sqrt{\frac{Ze^2}{4\pi\epsilon_0 mr}}. \tag{42-22}$$

The frequency ν_0 of orbital revolution follows at once from

$$\nu_0 = \frac{v}{2\pi r} = \sqrt{\frac{Ze^2}{16\pi^3\epsilon_0 mr^3}}. \tag{42-23}$$

---

* We assume a nuclear charge of $+Ze$ for generality. For the hydrogen atom Z, the *atomic number*, is unity.

The linear momentum p follows from Eq. 42–22:

$$p = mv = \sqrt{\frac{Zme^2}{4\pi\epsilon_0 r}}. \tag{42–24}$$

The angular momentum L is given by

$$L = pr = \sqrt{\frac{Zme^2 r}{4\pi\epsilon_0}}. \tag{42–25}$$

Thus if r is known, the orbit parameters K, U, E, v, v_0, p, and L are also known. If any one of these quantities is quantized, all of them must be.

At this stage Bohr had no rules to guide him and so made (after some indirect reasoning which we do not reproduce) a bold hypothesis, namely, that the necessary quantization of the orbit parameters shows up most simply when applied to the angular momentum and that, specifically, L can take on only values given by

Bohr's quantization condition

$$L = n\frac{h}{2\pi} \qquad n = 1, 2, 3, \ldots. \tag{42–26}$$

The Planck constant appears again in a fundamental way; the integer n is a *quantum number*.

Combining Eqs. 42–25 and 42–26 leads to

$$r = n^2 \frac{h^2\epsilon_0}{Z\pi me^2} \qquad n = 1, 2, 3, \ldots, \tag{42–27}$$

which tells how r is quantized. Substituting Eq. 42–27 into Eq. 42–21 produces

The allowed energies

$$E = -\frac{Z^2 me^4}{8\epsilon_0^2 h^2}\frac{1}{n^2} \qquad n = 1, 2, 3, \ldots, \tag{42–28}$$

which gives directly the energy values of the allowed stationary states.

Figure 42–15 shows the energies of the stationary states and their associated quantum numbers. Equation 42–27 shows that the orbit radius increases as n^2.

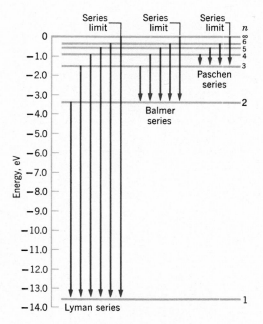

Figure 42–15 An energy level diagram for hydrogen showing the quantum number n for each level and some of the transitions that appear in the spectrum. An infinite number of levels is crowded in between the levels marked $n = 6$ and $n = \infty$. Compare this figure carefully with Fig. 42–14.

The upper level in Fig. 42–15, marked $n = \infty$, corresponds to a state in which the electron is completely removed from the atom (that is, $E = 0$ and $r = \infty$). Figure 42–15 also shows some of the quantum jumps that take place between the different stationary states.

Combining Eqs. 42–18 and 42–28 allows us to write a completely theoretical formula for the frequencies of the lines in the hydrogen spectrum. It is

The allowed frequencies

$$\nu = \frac{Z^2 m e^4}{8\epsilon_0^2 h^3}\left(\frac{1}{j^2} - \frac{1}{k^2}\right) \tag{42–29}$$

in which j and k are integers describing, respectively, the lower and the upper stationary states. The corresponding wavelengths can easily be found from $\lambda = c/\nu$. Table 42–2 shows some wavelengths so calculated; it should be compared carefully with Figs. 42–14 and 42–15.

Incidentally, because Einstein's equation for the energy of a photon, Eq. 42–10, was used in the derivation of the hydrogen emission wavelength formula, Eq. 42–29, its agreement with experiment helps confirm the particle model of light as well as Bohr's model of the atom.

Table 42–2 The hydrogen spectrum
(Some selected lines)

| Name of series | Quantum number | | Wavelength, nm |
	j (lower state)	k (upper state)	
Lyman	1	2	121.6
	1	3	102.6
	1	4	97.0
	1	5	94.9
	1	6	94.0
	1	∞ Series limit	91.2
Balmer	2	3	656.3
	2	4	486.1
	2	5	434.1
	2	6	410.2
	2	7	397.0
	2	∞ Series limit	365.0
Paschen	3	4	1875.1
	3	5	1281.8
	3	6	1093.8
	3	7	1005.0
	3	8	954.6
	3	∞ Series limit	822.0

Example 5 Calculate the *binding energy* of the hydrogen atom from Eq. 42–28. This is the energy required to remove the electron from the atom when the atom is in its lowest energy state.

Since the lowest possible energy for the electron after it leaves the atom is zero, the binding energy is equal to the absolute value of energy of the lowest state in Fig. 42–15. The largest negative value of E in Eq. 42–28 is found for $n = 1$. This yields, putting $Z = 1$,

$$E = -\frac{m e^4}{8\epsilon_0^2 h^2}$$

$$= -\frac{(9.11 \times 10^{-31}\ \text{kg})(1.60 \times 10^{-19}\ \text{C})^4}{(8)(8.85 \times 10^{-12}\ \text{C}^2/\text{N·m}^2)^2(6.63 \times 10^{-34}\ \text{J·s})^2}$$

$$= -2.17 \times 10^{-18}\ \text{J} = -13.6\ \text{eV},$$

which agrees with the experimentally observed binding energy for hydrogen.

42–9   The Correspondence Principle

Although all theories in physics have limitations, they usually do not break down abruptly but in a continuous way, yielding results that agree less and less well with experiment. Thus the predictions of Newtonian mechanics become less and less accurate as the speed approaches that of light. A similar relationship must exist between quantum physics and classical physics; it remains to find the circumstances under which the latter theory is revealed as a special case of the former.

The radius of the lowest energy state in hydrogen (the so-called *ground state*) is found by putting $n = 1$ in Eq. 42–27; it turns out to be 0.053 nm. If $n = 10,000$, however, the radius is $(10,000)^2$ times as large or 5.3 mm. This "atom" is so large that we suspect that its behavior should be accurately described by classical physics. Let us test this by computing the frequency of the emitted light on the basis of both classical and quantum assumptions. These calculations should differ at small quantum numbers but should agree at large quantum numbers. The fact that quantum physics reduces to classical physics at large quantum numbers is called the *correspondence principle*. This principle, due to Niels Bohr, was very useful during the years in which quantum physics was being developed. Bohr, in fact, based his theory of the hydrogen atom on correspondence principle arguments.

Bohr's correspondence principle

Classically, the fundamental frequency of the light emitted from an atom is ν_0, its frequency of revolution in its orbit. We can express this in terms of a quantum number n by combining Eqs. 42–23 and 42–27 to obtain, for $Z = 1$

A classical prediction

$$\nu_0 = \frac{me^4}{8\epsilon_0^2 h^3}\left[\frac{2}{n^3}\right]. \tag{42–30}$$

Quantum physics predicts that the frequency ν of the emitted light is given by Eq. 42–29. Considering a transition between an orbit with quantum number $k = n$ and one with $j = n - 1$ leads, with $Z = 1$, to

A quantum prediction

$$\nu = \frac{me^4}{8\epsilon_0^2 h^3}\left[\frac{1}{(n-1)^2} - \frac{1}{n^2}\right]$$

$$= \frac{me^4}{8\epsilon_0^2 h^3}\left[\frac{2n-1}{(n-1)^2 n^2}\right]. \tag{42–31}$$

As $n \to \infty$ the expression in the square brackets above approaches $2/n^3$ so that $\nu \to \nu_0$ as $n \to \infty$. Table 42–3 illustrates this example of the correspondence principle.

Table 42–3   The correspondence principle as applied to the hydrogen atom

Quantum number n	Frequency of revolution in orbit (Eq. 42–30) Hz	Frequency of transition to next lowest state (Eq. 42–31) Hz	Difference, %
2	8.20×10^{14}	24.6×10^{14}	67
5	5.26×10^{13}	7.38×10^{13}	29
10	6.57×10^{12}	7.72×10^{12}	14
50	5.25×10^{10}	5.42×10^{10}	3
100	6.578×10^{9}	6.677×10^{9}	1.5
1,000	6.5779×10^{6}	6.5878×10^{6}	0.15
10,000	6.5779×10^{3}	6.5789×10^{3}	0.015
25,000	4.2099×10^{2}	4.2102×10^{2}	0.007

REVIEW GUIDE AND SUMMARY

Photons

This chapter describes three experiments that lead ultimately to the concept that light is not wave-like but is made up of concentrated bundles of energy called photons. Each photon has an energy E and a linear momentum p, given by

$$E = h\nu \qquad \text{and} \qquad p = h/\lambda \qquad\qquad [42\text{-}10,14]$$

where h, the Planck constant, has the value 6.63×10^{-34} J·s. The fact that this constant, although small, is not zero is the determining feature of modern quantum physics. The experiments are:

Cavity radiation

1. Measurement of the distribution with wavelength of the radiation emerging from heated cavities; see Figs. 42–5 and 6 and Eq. 42–6. These studies introduced the concept of energy quantization (see Example 1) and caused the Planck constant h to appear in the equations of physics for the first time.

2. The photoelectric effect, in which electrons are ejected from a metal surface by incident light. Einstein's equation for this effect, based on his photon hypothesis, is

The photoelectric effect

$$h\nu = eV_0 + \phi. \qquad\qquad [42\text{-}12]$$

Here $h\nu$ is the energy of the photon that falls on the metal surface, knocking out an electron. The work function ϕ is the energy needed to remove this electron from the interior of the metal and eV_0 is the maximum kinetic energy of the electron once it has passed through the metal surface. Study Figs. 42–7, 8, and 9.

The Compton effect

3. The Compton effect, in which x-ray photons, scattered from free electrons as in Fig. 42–10, suffer a wavelength increase $\Delta\lambda$. This Compton shift (see Fig. 42–11) is given by

$$\Delta\lambda = \frac{h}{m_0 c}\,(1 - \cos\varphi). \qquad\qquad [42\text{-}17]$$

This equation follows from applying the laws of conservation of energy and momentum to a billiard-ball-like collision between a photon and a free electron, as in Fig. 42–12. Study Example 4.

Line spectra

Figure 42–13 and Table 42–1 show that the absorption and emission of radiation at sharply defined wavelengths is characteristic of atoms, molecules, and nuclei. Classical physics cannot explain this phenomenon. Attempts at understanding line spectra in quantum terms start with the spectrum of the simplest atom, hydrogen.

Bohr's theory of the hydrogen atom

Bohr's semiclassical theory of the hydrogen atom was a major step toward the development of a complete quantum theory of atomic structure. Bohr assumed that the atom emitted or absorbed radiation only in transitions between stationary states, the frequencies of the spectral lines being given by

The Bohr frequency condition

$$h\nu = E_k - E_j. \qquad\qquad [42\text{-}18]$$

Here E_k and E_j are the energies of the two states involved in the transition.

To find the energies of the stationary states Bohr assumed that the angular moment L of the orbiting electron can have only discrete quantized values, given by Eq. 42–26. This leads to

The energies of the stationary states

$$E_n = -\left(\frac{Z^2 me^4}{8\epsilon_0^2 h^2}\right)\frac{1}{n^2}, \qquad n = 1, 2, 3, \ldots \qquad\qquad [42\text{-}28]$$

for the energies of the allowed states. Combining Eqs. 42–18 and 42–28 leads to

The allowed frequencies

$$\nu = \left(\frac{Z^2 me^4}{8\epsilon_0^2 h^3}\right)\left(\frac{1}{j^2} - \frac{1}{k^2}\right) \qquad\qquad [42\text{-}29]$$

for the frequencies of the lines of the hydrogen spectrum. Study and compare Figs. 42–14 and 15, Table 42–2, and Example 5.

The correspondence principle

Bohr based his quantum theory of the hydrogen atom on the correspondence principle. This asserts that: (1) the fundamental description of any physical system (atom, baseball, etc.) must be based on quantum principles and (2) the familiar classical description of such systems is a limiting special case, becoming increasingly valid as the quantum number increases, that is, as $n \to \infty$. Table 42–3 provides an example for detailed study.

QUESTIONS

1. "Pockets" formed by the coals in a coal fire seem brighter than the coals themselves. Is the temperature in such pockets appreciably higher than the surface temperature of an exposed glowing coal? Explain this common observation.

2. The relation $\mathcal{R} = \sigma T^4$ (Eq. 42–2) is exact for true cavities and holds for all temperatures. Why don't we use this relation as the basis of a definition of temperature at, say, 100°C?

3. It is stated that if we look into a cavity whose walls are maintained at a constant temperature, no details of the interior are visible. Does this seem reasonable?

4. We claim that all objects radiate energy by virtue of their temperature and yet we cannot see all objects in the dark. Why?

5. Is charge quantized in physics? Is mass quantized? Describe some experiments that support your answers.

6. Show that the Planck constant has the dimensions of angular momentum. Does this necessarily mean that angular momentum is a quantized quantity?

7. For quantum effects to be "everyday" phenomena in our lives, what order of magnitude value would h need to have? (See G. Gamow, *Mr. Tompkins in Wonderland,* Cambridge University Press, Cambridge, 1957, for a delightful popularization of a world in which the physical constants c, G, and h make themselves obvious.)

8. An insulated metal plate yields photoelectrons when you first shine ultraviolet light on it; later it doesn't give up any more. Explain.

9. In Fig. 42–8, why doesn't the photoelectric current rise vertically to its maximum (saturation) value when the applied potential difference is just slightly more positive than V_0?

10. In the photoelectric effect, why does the existence of a cutoff frequency speak in favor of the photon theory and against the wave theory?

11. Why are photoelectric measurements so sensitive to the nature of the photoelectric surface?

12. How can a photon energy be given by $E = h\nu$ (Eq. 42–10) when the very presence of the frequency ν in the formula implies that light is a wave?

13. Does Einstein's theory of photoelectricity, in which light is postulated to be a stream of photons, invalidate Young's double-slit interference experiment?

14. What is the direction of a Compton-scattered electron with maximum kinetic energy, compared with the direction of the incident monochromatic photon beam?

15. In the Compton scattering picture (Fig. 42–12), why would you expect $\Delta\lambda$ to be independent of the material of which the scatterer is composed?

16. Why don't we observe a Compton effect with visible light?

17. Only a relatively small number of Balmer lines can be observed from laboratory discharge tubes, whereas a large number are observed in the spectra of stars. Explain this in terms of the small density, high temperature, and large volume of gases in stellar atmospheres.

18. In Bohr's theory for the hydrogen atom, what is the implication of the fact that the potential energy of the orbiting electron is negative and is greater in magnitude than the kinetic energy? Is this a result of quantum physics or is it true classically as well?

19. Why are some lines in the hydrogen spectrum brighter than others?

20. Radioastronomers observe lines in the hydrogen spectrum that originate in hydrogen atoms that are in a states with $n = 350$ or more. Why can't hydrogen atoms in states with such high quantum numbers be produced and studied in the laboratory?

21. When you put some salt (sodium chloride) in a colorless flame, you see a yellow line characteristic of the sodium but no color characteristic of the chlorine. What might be the explanation?

22. Given that $E = h\nu$ for a photon, the Doppler shift in frequency of radiation from a receding light source would seem to indicate a reduced energy for the emitted photons. Is this in fact true? If so, what happened to the conservation of energy principle?

23. (*a*) Can a hydrogen atom absorb a photon whose energy exceeds its binding energy (13.6 eV)? (*b*) What minimum energy must a photon have to initiate the photoelectric effect in hydrogen gas?

24. List and discuss the assumptions made by Planck in connection with the cavity radiation problem, by Einstein in connection with the photoelectric effect, by Compton in connection with the Compton effect, and by Bohr in connection with the hydrogen atom problem.

25. Describe several methods that can be used to experimentally determine the value of the Planck constant h.

26. Discuss Example 1 in terms of the correspondence principle.

27. According to classical mechanics, an electron moving in an orbit should be able to do so with any angular momentum whatever. According to Bohr's theory of the hydrogen atom, however, the angular momentum is quantized according to $L = nh/2\pi$. Reconcile these two statements, using the correspondence principle.

EXERCISES

Section 42-2 Cavity Radiation

1. Estimate the thermal power radiated from a fireplace assuming an emmisivity of 0.90, an effective radiating surface of 0.50 m², and an effective temperature of 500°C. Does your answer seem reasonable?

2. The radiancy of tungsten, in the form of an unoxidized polished ribbon at 2000 K, is 2.4×10^5 W/m². What is its emissivity at this temperature?

3. A 100-W incandescent lamp has a coiled tungsten filament whose diameter is 0.40 mm and whose extended length is 30 cm. The effective emissivity under operating conditions is 0.21. What is the operating temperature of the filament?

Section 42-3 Planck's Radiation Law and the Origin of the Quantum

4. The wavelength λ_{max} at which the spectral radiancy of a cavity radiator has its maximum value for a particular temperature T (see Fig. 42-3) is given by the Wien displacement law, or

$$\lambda_{max} T = \text{a constant} \ (= 2.898 \times 10^{-3} \ \text{m} \cdot \text{K}).$$

The effective surface temperature of the sun, calculated in Problem 2, is 5800 K. At what wavelength would you expect the sun to radiate most strongly? In what region of the spectrum is this? Why, then, does the sun appear yellow? Note that Wien's displacement law is not the same as the Wien radiation law referred to in Figs. 42-5 and 42-6 and Problem 42-4.

5. Calculate the wavelength of maximum spectral radiancy (see Exercise 4) and identify the region of the electromagnetic spectrum to which it belongs for each of the following: (a) the 3.0-K cosmic background radiation, a remnant of the primordial fireball; (b) your body, assuming a skin temperature of 20°C; (c) a tungsten lamp filament at 1800 K; (d) the sun, at an assumed surface temperature of 5800 K; (e) an exploding thermonuclear device, at an assumed fireball temperature of 10^7 K; (f) the universe immediately after the Big Bang, at an assumed temperature of 10^{38} K. Assume black-body conditions throughout.

6. Low-temperature physicists would not consider a temperature of 2.0 mK to be particularly low. (a) At what wavelength would a cavity whose walls were at this temperature radiate most copiously? (See Exercise 4) (b) To what region of the electromagnetic spectrum would this radiation belong? (c) What are some of the practical difficulties of operating a cavity radiator at such a low temperature?

7. At what temperature is a blackbody most visible to the human eye if the eye is most sensitive to green light (5500 Å)? Assume that the size of the black body source is adjusted so that the light intensity received by the eye does not change with temperature. Compare your answer to the temperature of the sun (see Exercise 4.)

Section 42-5 Einstein's Photon Theory

8. Show that the energy E of a photon (in electron volts) is related to its wavelength λ (in nanometers) by

$$E = \frac{1240}{\lambda}.$$

9. The energy required to remove an electron from sodium is 2.3 eV. Does sodium show a photoelectric effect for orange light, with $\lambda = 680$ nm?

10. An atom absorbs a photon having a wavelength of 375 nm and immediately emits another photon having a wavelength of 580 nm. How much energy was absorbed by the atom in this process? Ease the computation by using the result of Exercise 8.

11. Incident photons strike a sodium surface having a work function of 2.2 eV, causing photoelectric emission. When a stopping potential $V_0 = 5.0$ V is imposed, there is no photocurrent. What is the wavelength of the incident photons?

12. You wish to pick a substance for a photocell operable with visible light. Which of the following will do (work function in parentheses): tantalum (4.2 eV); tungsten (4.5 eV); aluminum (4.2 eV); barium (2.5 eV); lithium (2.3 eV)?

13. Find the maximum kinetic energy of photoelectrons if the work function of the material is 2.3 eV and the frequency of the radiation is 3.0×10^{15} Hz.

14. *The photographic process.* Consider monochromatic light falling on a photographic film. The incident photons will be recorded if they have enough energy to dissociate a AgBr molecule in the film. The minimum energy required to do this is about 0.6 eV. Find the cutoff wavelength greater than which the light will not be recorded. In what region of the spectrum does this wavelength fall?

15. To remove an inner, most tightly bound, electron from an atom of molybdenum requires an energy of 20 keV. If this is to be done by allowing a photon to strike the atom, (a) what must be the associated wavelength of the photon? (b) In what region of the spectrum does the photon lie? (c) Could this process be called a photoelectric effect? Discuss your answers.

Section 42-6 The Compton Effect

16. Photons of wavelength 2.40 pm are incident on a target containing free electrons. (a) Find the wavelength of a photon that is scattered at 30° from the incident direction and also the kinetic energy imparted to the recoil electron. (b) Do the same for a scattering angle of 120°.

17. A 0.510-MeV gamma-ray photon is Compton-scattered from a free electron in an aluminum block. (a) What is the wavelength of the incident photon? (b) What is the wavelength of the scattered photon? (c) What is the energy of the scattered photon? Assume a scattering angle of 90°.

18. Through what angle must a 200-keV photon be scattered by a free electron so that it loses 10% of its energy? (See Problem 16.)

19. Find the maximum wavelength shift for a Compton collision between a photon and a free *proton*.

Section 42-8 Bohr's Theory of the Hydrogen Atom

20. A line in the x-ray spectrum of gold has a wavelength of 18.5 pm ($= 18.5 \times 10^{-12}$ m). The emitted x-ray photons correspond to a transition of the gold atom between two stationary states, the upper one of which has the energy -13.7 keV. What is the energy of the lower stationary state?

21. What are (a) the energy, (b) the momentum, and (c) the wavelength of the photon that is emitted when a hydrogen atom undergoes a transition from the state $n = 3$ to $n = 1$?

22. (a) What are the wavelength intervals over which the Lyman, Balmer, and Paschen series extend? (Each interval extends from the longest wavelength to the series limit. See Fig. 42–14.) (b) What are the corresponding frequency intervals?

23. How much energy is required to remove the electron from a hydrogen atom in a state with $n = 8$? With $n = 25$?

24. Using Bohr's theory, calculate the energy required to re-move the electron from the ground state of singly ionized helium, that is, helium with one electron already removed.

25. (a) What are the first three energy levels of Li^{2+} according to the Bohr theory? (b) What is the ionization energy of Li^{2+}?

PROBLEMS GROUPED BY SECTION

Section 42–2   Cavity Radiation
1. An oven with an inside temperature $T_0 = 227°C$ is in a room having a temperature $T_r = 27°C$. There is a small opening of area 5.0 cm² in one side of the oven. How much net power is trans-ferred from the oven to the room? (*Hint:* Consider both oven and room as cavities.)

2. The sun, whose radius is 6.96×10^5 km, emits radiation at the rate of 3.90×10^{26} W. (a) What is the radiancy R of the solar surface? (b) What is the sun's effective surface temperature? As-sume that the emitted radiation is in thermal equilibrium with the material of the solar surface, that is, that the sun emits cavity radiation. (c) The mass of the moon is 7.4×10^{22} kg. What is the mass equivalent, measured in fractions of a moon-mass, of the energy radiated by the sun in one year?

3. A *thermograph* is a medical instrument used to measure radiation from the skin. For example, normal skin radiates at a temperature of about 34°C, and the skin over a tumor radiates at a slightly higher temperature. Assuming that the skin radiates with a constant emissivity, derive an expression for the fractional dif-ference in the radiancy between adjacent areas of the skin that are at slightly different temperatures. Evaluate the expression for a temperature difference of 1°C.

Section 42–3   Planck's Radiation Law and the Origin of the Quantum
4. *Wien's radiation law.* Show that this law, which is

$$\mathscr{R}_\lambda = \frac{c_1}{\lambda^5} e^{-c_2/\lambda T},$$

is a special case of Planck's radiation law (Eq. 42–6) holding for short wavelengths or low temperatures; see Figs. 42–5 and 42–6.

5. *The Rayleigh-Jeans radiation law.* Show that this law, which is

$$\mathscr{R}_\lambda = \frac{2\pi c k T}{\lambda^4},$$

is a special case of Planck's radiation law (Eq. 42–6) holding for long wavelengths or high temperatures; see Figs. 42–5 and 42–6.

Section 42–4   The Photoelectric Effect
6. In Example 2, suppose that the "target" is a simple gas atom of 0.10 nm radius and that the intensity of the light source is reduced to 1.0×10^{-5} W. If the binding energy of the most loosely bound electron in the atom is 2.0 eV, what time lag for the photo-electric effect is expected on the basis of the wave theory of light?

7. Photosensitive surfaces are not necessarily very efficient. Suppose that the fractional efficiency of a cesium surface (work function 1.8 eV) is 1.0×10^{-16}, that is, one photoelectron is pro-duced for every 10^{16} photons striking the surface. What would be the photocurrent from such a cesium surface if it is illuminated with 600-nm light from a 2.0-mW laser and all of the photoelec-trons produced take part in charge flow?

Section 42–5   Einstein's Photon Theory
8. In a photoelectric experiment in which a sodium surface is used, one finds a stopping potential of 1.85 V for a wavelength of 300 nm and a stopping potential of 0.82 V for a wavelength of 400 nm. From these data, find (a) a value for the Planck constant, (b) the work function for sodium, and (c) the cutoff wavelength for sodium.

9. Millikan's photoelectric data for lithium are as follows:

Wavelength, nm:	433.9	404.7	365.0	312.5	253.5
Stopping potential, V:	0.55	0.73	1.09	1.67	2.57

Make a plot like Fig. 42–9, which is for sodium, and find (a) the Planck constant, and (b) the work function for lithium.

10. A satellite in earth orbit maintains a panel of solar cells of 1.0-m² area at right angles to the direction of the sun's rays. (a) At what rate does solar energy strike the panel? (b) At what rate do solar photons strike the panel? Assume that the solar radiation is monochromatic with a wavelength of 550 nm. (c) How long would it take for a "mole of photons" to strike the panel? The solar constant (see Problem 36) is 1390 W/m².

11. What are (a) the frequency, (b) the wavelength, and (c) the momentum of a photon whose energy equals the rest energy of the electron?

12. An ultraviolet lightbulb, emitting at 400 nm, and an infrared lightbulb, emitting at 700 nm, each are rated at 400 W. (a) Which bulb radiates photons at the greater rate? (b) How many more photons does it generate per second than does the other bulb?

13. In the photon picture of radiation, show that if two parallel beams of light of different wavelengths are to have the same inten-sity, then the rates per unit area at which photons pass through any cross section of the beam are in the same ratio as the wavelengths.

14. Show, by analyzing a collision between a photon and a free electron (using relativistic mechanics), that it is impossible for a photon to give all of its energy to the free electron. In other words, the photoelectric effect cannot occur for completely free electrons; the electrons must be bound in a solid or in an atom.

Section 42–6   The Compton Effect
15. What is the maximum kinetic energy of the Compton-scat-tered electrons knocked out of a thin copper foil by an incident beam of 17.5-keV x rays?

16. *The fractional photon energy loss.* (a) Show that $\Delta E/E$, the fractional loss of energy of a photon during a Compton collision, is given by $(h\nu'/m_0c^2)(1 - \cos\phi)$. (b) Plot $\Delta E/E$ vs. ϕ and interpret the curve physically.

17. What fractional increase in wavelength leads to a 75% loss of photon energy in a Compton collision with a free electron?

18. The quantity h/m_0c in Eq. 42–17 is often called the *Compton wavelength* of the scattering particle, and that equation is written as

$$\Delta\lambda = \lambda_C (1 - \cos\phi).$$

(a) What is the Compton wavelength, λ_C, of an electron? of a proton? (b) What is the energy of a photon whose wavelength is equal to the Compton wavelength of the electron? of the proton? (c) Show that in general the energy of a photon whose wavelength is equal to the Compton wavelength of a particle is just the rest energy of that particle.

*19. Carry out the algebra needed to eliminate v and θ from Eqs. 42–13, 42–15, and 42–16 and thus to derive Eq. 42–17, the equation for the Compton shift.

Section 42–8 Bohr's Theory of the Hydrogen Atom
20. Light of wavelength 486.1 nm is emitted by a hydrogen atom. (a) What transition of the atom is responsible for this radiation? (b) To what series does this radiation belong?

21. (a) Using Bohr's formula for the frequencies of the lines of the hydrogen spectrum, calculate the three longest wavelengths of the Balmer series. (b) Between what wavelength limits does the Balmer series lie?

22. A hydrogen atom is excited from a state with $n = 1$ to one with $n = 4$. (a) Calculate the energy that must be absorbed by the atom. (b) Calculate and display on an energy-level diagram the different photon energies that may be emitted if the atom returns to the $n = 1$ state. (c) Calculate the recoil speed of the hydrogen atom, assumed initially at rest, if it makes the transition from $n = 4$ to $n = 1$ in a single quantum jump.

23. A neutron, with kinetic energy 6.0 eV, collides with a resting hydrogen atom in its ground state. Show that this collision must be elastic (that is, energy must be conserved). (*Hint:* Show that the atom cannot be raised to a higher excitation state as a result of the collision.)

24. From the energy-level diagram for hydrogen, explain the observation that the frequency of the second Lyman-series line is the sum of the frequencies of the first Lyman-series line and the first Balmer-series line. This is an example of the empirically discovered *Ritz combination principle.* Use the diagram to find some other valid combinations.

25. Use Bohr's theory to compare the spectrum of singly ionized helium with the spectrum of hydrogen.

26. Perhaps an atom could be formed by an electron and a neutron binding together by gravitational forces. Calculate the ground-state radius of an electron in such an atom by using a Bohr-type model in which the attractive electrical force is replaced by the attractive gravitational force.

*27. *Positronium.* Apply Bohr's theory to the positronium atom. This consists of a positive and a negative electron revolving around their center of mass, which lies halfway between them. (a) What relationship exists between this spectrum and the hydrogen spectrum? (b) What is the radius of the ground-state orbit? (*Hint:* It will be necessary to analyze this problem from first principles, because this "atom" has no nucleus. Both particles revolve about a point halfway between them.) See "Exotic Atoms" by E. H. S. Burhop, *Contemporary Physics,* July 1970.

Section 42–9 The Correspondence Principle
28. In Table 42–3, show that the quantity in the last column is given by

$$\frac{100 (\nu - \nu_0)}{\nu} \cong \frac{150}{n}$$

for large quantum numbers.

29. *Radio waves from a hydrogen atom.* (a) Show that the smallest quantum number of the levels in hydrogen between which transitions giving rise to radio waves are possible is given by $n = (2R_H \lambda)^{1/3}$, where λ is the wavelength of the radio wave and $R_H = Z^2 m e^4/8\epsilon_0^2 h^3$. (b) If the 21-cm radio emission from interstellar hydrogen were due to such a transition (it isn't), what would be the value of n?

ADDITIONAL PROBLEMS

30. (a) If the work function for a metal is 1.8 eV, what would be the stopping potential for light having a wavelength of 400 nm? (b) What would be the maximum velocity of the emitted photoelectrons at the metal's surface?

31. An x-ray photon of wavelength 0.10 Å strikes an electron head-on ($\phi = 180°$). Determine (a) the change in wavelength of the photon, (b) the change in energy of the photon, and (c) the kinetic energy imparted to the electron.

32. A 6.2-keV x-ray photon falling on a carbon block is scattered by a Compton collision and its frequency is shifted by 0.010%. (a) Through what angle is the photon scattered? (b) What kinetic energy is imparted to the electron?

33. Show, on an energy-level diagram for hydrogen, the quantum numbers corresponding to a transition in which the wavelength of the emitted photon is 121.6 nm.

34. Light of wavelength 200 nm falls on an aluminum surface. In aluminum, 4.2 eV are required to remove an electron. What is the kinetic energy of (a) the fastest and (b) the slowest emitted photoelectrons? (c) What is the stopping potential? (d) What is the cutoff wavelength for aluminum?

35. The absolute temperature of the sun's surface is roughly twice that of Barnard's star. (a) Assuming that both of these stars are perfect cavity radiators, find the ratio of the sun's radiancy to that of Barnard's star. (b) How large would the radius of Barnard's star have to be for it to produce 1% as much radiant power as the sun?

36. The *solar constant* ($= 1390$ W/m²) is the rate per unit area at which the sun delivers energy to the earth; it is measured just outside the earth's atmosphere and at right angles to the sun's rays. (a) At what average rate per unit area, measured along a normal to the earth's surface, does the earth reradiate this energy back into space? (b) To what average surface temperature for the earth does this correspond? Assume that the earth radiates like a cavity radiator, that is, that its emissivity is unity.

37. Calculate the percent change in photon energy for a Compton collision with ϕ in Fig. 42–12 equal to 90° for radiation in (a) the microwave range, with $\lambda = 3.0$ cm, (b) the visible range, with $\lambda = 500$ nm, (c) the x-ray range, with $\lambda = 25$ pm, and (d) the gamma-ray range, the energy of the gamma-ray photons being 1.0 MeV. What are your conclusions about the importance of the Compton effect in these various regions of the electromagnetic spectrum, judged solely by the criterion of energy loss in a single Compton encounter?

38. In the ground state of the hydrogen atom, according to Bohr's theory, what are (a) the quantum number, (b) the orbit radius, (c) the angular momentum, (d) the linear momentum, (e) the angular velocity, (f) the linear speed, (g) the force on the electron, (h) the acceleration of the electron, (i) the kinetic energy, (j) the potential energy, and (k) the total energy?

39. (a) If the angular momentum of the earth due to its motion around the sun were quantized according to Bohr's relation $L = nh/2\pi$, what would the quantum number be? (b) Could such quantization be detected if it existed?

40. A hydrogen atom in a state having a *binding energy* (the energy required to remove an electron) of 0.85 eV makes a transition to a state with an *excitation energy* (the difference in energy between the state and the ground state) of 10.2 eV. (a) Find the energy of the emitted photon. (b) Show this transition on an energy-level diagram for hydrogen, labeling the appropriate quantum numbers.

41. Assume that a 100-W sodium-vapor lamp radiates its energy uniformly in all directions in the form of photons with an associated wavelength of 589 nm. (a) At what rate are photons emitted from the lamp? (b) At what distance from the lamp will the average flux of photons be 1.0 photons/cm²·s? (c) At what distance from the lamp will the average density of photons be 1.0 photons/cm³? (d) What are the photon flux and the photon density 2.0 m from the lamp?

42. *Muonic atoms.* Apply Bohr's theory to a muonic atom, which consists of a nucleus of charge Ze with a negative muon (an elementary particle with a charge of $-e$ and a mass m that is 207 times as large as the electron mass) circulating about it. Calculate (a) the radius of the first Bohr orbit, (b) the ionization energy, and (c) the wavelength of the most energetic photon that can be emit-

ted. Assume that the muon is circulating about a hydrogen nucleus ($Z = 1$). See "The Muonium Atom," by Vernon W. Hughes, *Scientific American,* April 1966.

43. The star Alpha Centauri has an effective surface temperature of 5800 K and lies at a distance of 4.1×10^{16} m from Earth. Its light intensity at the top of the earth's atmosphere is 2.52×10^{-16} W/m². Assuming that Alpha Centauri radiates as a black body, estimate its radius. Repeat for the star Naos, which lies some 2.47×10^{19} from Earth, has a surface temperature of 35,000 K. Naos produces a light intensity of 2.3×10^{-19} W/m² just above the top of Earth's atmosphere.

44. Under ideal conditions the normal human eye will record a visual sensation at 550 nm, its most sensitive wavelength, if incident photons are absorbed at a rate as low as 100 s⁻¹. To what power level does this correspond?

45. *The photoelectric effect for tightly bound electrons.* X rays with a wavelength of 0.071 nm eject photoelectrons from a gold foil, the electrons originating from deep within the gold atoms. The ejected electrons move in circular paths of radius r in a region of uniform magnetic field **B**. Experiment shows that $rB = 1.88 \times 10^{-4}$ T·m. Find (a) the maximum kinetic energy of the photoelectrons and (b) the work done in removing the electrons from the gold atoms that make up the foil.

46. Show that when a photon of energy E scatters from a free electron, the maximum recoil kinetic energy of the electron is given by

$$K_{max} = \frac{E^2}{E + m_0c^2/2}.$$

*47. A diatomic gas molecule consists of two atoms of mass m separated by a fixed distance d rotating about an axis as indicated in Fig. 42–16. Assuming that its angular momentum is quantized as in the Bohr atom, determine (a) the possible angular velocities and (b) the possible quantized rotational energies.

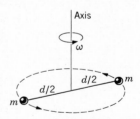

Figure 42–16 Problem 47.

43 Waves and Particles

43–1 Matter Waves

In 1924 Louis de Broglie of France reasoned that (*a*) nature is strikingly symmetrical; (*b*) our observable universe is composed entirely of radiation (encompassing all regions of the electromagnetic spectrum) and matter; (*c*) if radiation has a dual, wave-particle nature, perhaps matter has also. Since matter was at that time regarded as being composed of particles, de Broglie's reasoning suggested that it might be fruitful to search for a wave-like behavior for matter.

De Broglie's suggestion might not have received serious attention had he not predicted what the expected wavelength of the so-called matter waves would be. We recall that about 1680 Huygens put forward a wave theory of light that did not receive general acceptance, not only because it seemed to contradict Newton's particle theory but also because Huygens was not able to state what the wavelength of the light might be. When Thomas Young rectified this defect in 1800, the wave theory of light started on its way to acceptance.

For radiation, the frequency ν and the wavelength λ describe its familiar wave aspect. The photon energy E and momentum p describe its particle nature. We have seen that these latter are given, respectively, by Eq. 42–10 ($E = h\nu$) and Eq. 42–14 ($p = h/\lambda$). De Broglie boldly suggested that these same relations apply to matter as well as to radiation and that electrons, for example, have both wave and particle properties. The particle aspect of matter (energy E and momentum p) was most familiar at that time, of course. De Broglie suggested that the wave and particle properties are related in the same way as for radiation. Thus the frequency of a particle wave would be given by

$$\nu = E/h \tag{43–1}$$

and the wavelength (now called the de Broglie wavelength) by

The de Broglie wavelength

$$\lambda = h/p. \tag{43–2}$$

Note that the Planck constant h seems to be the essential link connecting the wave-particle nature of both matter and radiation.

Example 1 What wavelength is predicted by Eq. 43–2 for a beam of electrons whose kinetic energy is 100 eV?

We can find the speed of these relatively slow electrons from $K = \frac{1}{2}mv^2$, or

$$v = \sqrt{\frac{2K}{m}} = \sqrt{\frac{(2)(100\ \text{eV})(1.6 \times 10^{-19}\ \text{J/eV})}{9.1 \times 10^{-31}\ \text{kg}}}$$

$$= 5.9 \times 10^6\ \text{m/s}.$$

The momentum follows from

$$p = mv = (9.1 \times 10^{-31}\ \text{kg})(5.9 \times 10^6\ \text{m/s})$$

$$= 5.4 \times 10^{-24}\ \text{kg·m/s}.$$

The de Broglie wavelength is found from Eq. 43–2 or

$$\lambda = \frac{h}{p} = \frac{6.6 \times 10^{-34}\ \text{J·s}}{5.4 \times 10^{-24}\ \text{kg·m/s}}$$

$$= 1.2 \times 10^{-10}\ \text{m} = 120\ \text{pm} = 1.2\ \text{Å}.$$

This is the same order of magnitude as the size of an atom or the spacing between adjacent planes of atoms in a solid. Problem 5 gives a simple formula for making calculations of this kind.

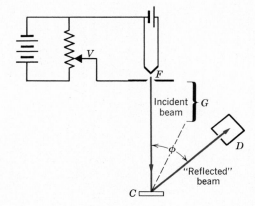

Figure 43–1 The apparatus of Davisson and Germer. Electrons from the hot filament F are accelerated by a variable potential difference V. After "reflection" from crystal C they are collected by detector D.

The Davisson-Germer experiment

In 1925 W. M. Elsasser pointed out that the wave nature of matter might be tested in the same way that the wave nature of x rays was first tested, namely, by allowing a beam of electrons of the appropriate energy to fall on a crystalline solid. The atoms of the crystal serve as a three-dimensional array of diffracting centers for the electron "wave"; we should look for strong diffracted peaks in certain characteristic directions, just as for x-ray diffraction.

This idea was tested by C. J. Davisson and L. H. Germer in this country and by G. P. Thomson in Scotland.* Figure 43–1 shows the apparatus of Davisson and Germer. Electrons from a heated filament are accelerated by a variable potential difference V and emerge from the "electron gun" G with kinetic energy eV. This electron beam is allowed to fall at normal incidence on a single crystal of nickel at C. Detector D is set at a particular angle ϕ and readings of the intensity of the "reflected" beam are taken at various values of the accelerating potential V. Figure 43–2 shows that a strong beam occurs at $\phi = 50°$ for $V = 54$ V.

All such strong reflected beams can be accounted for by assuming that the electron beam has a wavelength, given by $\lambda = h/p$ (Eq. 43–2), and that "Bragg re-

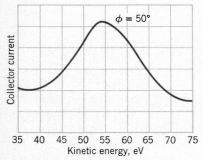

Figure 43–2 The collector current in detector D in Fig. 43–1 as a function of the kinetic energy of the incident electrons, showing a diffraction maximum.

---

* Sir G. P. Thomson, whose experiments verified the wave nature of the electron, was the son of Sir J. J. Thomson, whose discovery of the electron in 1897 amounted to a demonstration of its particle character! They each received Nobel prizes, 31 years apart.

flections'' occur from certain families of atomic planes precisely as described for x rays in Section 41–10.

Figure 43–3 shows such a Bragg reflection, obeying the Bragg relationship

$$m\lambda = 2d \sin \theta \qquad m = 1, 2, 3, \ldots . \qquad (43-3)$$

For the conditions of Fig. 43–3 the effective interplanar spacing d can be shown by x ray analysis to be 0.91 Å. Since ϕ equals 50°, it follows that θ equals $90° - (\frac{1}{2} \times 50°)$ or 65°. The wavelength to be calculated from Eq. 43–3, if we assume $m = 1$, is

$$\lambda = 2d \sin \theta = 2(0.91 \text{ Å})(\sin 65°) = 1.65 \text{ Å}.$$

The wavelength calculated from the de Broglie relationship $\lambda = h/p$ is, for 54 eV electrons (see Example 1), 1.66 Å. This excellent agreement, combined with much similar evidence, is a convincing argument for believing that electrons are wave-like in some circumstances.

Figure 43–4a shows an experimental arrangement, similar to that used by G. P. Thomson, in which the wave nature of electrons can be clearly shown by comparing the diffraction of a beam of electrons with that of a beam of x rays ($\lambda = 0.71$ A). The electron energy was chosen so that the predicted de Broglie wavelength of the electrons is approximately the same as the x-ray wavelength. The collimated beam, of either type, comes from the left and falls on the thin aluminum foil target, which is made up of large numbers of tiny randomly oriented crystallites. Diffraction—a wave phenomenon—occurs in the foil and the diffracted beams form a number of discrete circular rings on the photographic film. The diffraction patterns, which are characteristic of the atomic crystal structure of aluminum, are shown in Fig. 43–4b for the x rays and in Fig. 43–4c for the electrons. There can be little doubt that electrons and x rays behave in the same wave-like manner in this experimental situation.

Bragg's law

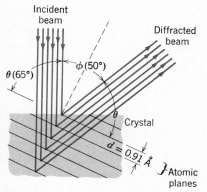

Figure 43–3 The strong diffracted beam at $\phi = 50°$ and $V = 54$ V arises from wavelike reflection from the family of atomic planes shown, for $d = 0.91$ Å. The Bragg angle θ is 65°. For simplicity, refraction of the wave as it leaves the surface is ignored.

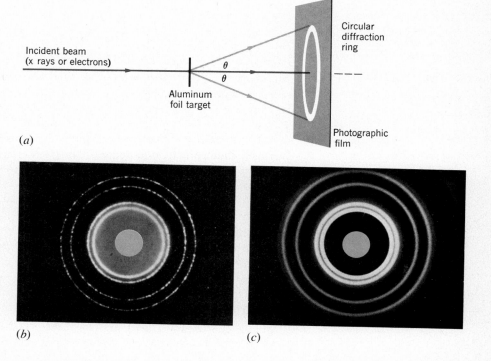

(a)

(b)

(c)

Figure 43–4 (a) An experimental arrangement for producing a diffraction pattern characteristic of an aluminum target. (b) Pattern for an incident x-ray beam. (c) Pattern for an incident electron beam. The electron energy has been chosen so that the de Broglie wavelength of the electron beam is approximately equal to the wavelength of the x rays employed. The central spot, identifying the directly transmitted beam, has been deleted in each case. (b and c) From the PSSC film *Matter Waves*, Production #68 Education Development Center, Newton, Ma.

Not only electrons but all other particles, charged or uncharged, show wave-like characteristics. Beams of slow neutrons from nuclear reactors are routinely used to investigate the atomic structure of solids. It has even been possible to build a Michelson-type interferometer (see Section 40–6) for neutron beams in which interference fringes are produced between two such beams.

The evidence for the existence of matter waves with wavelengths given by Eq. 43–2 is strong indeed. Nevertheless, the evidence that matter is composed of particles remains equally strong; see Fig. 10–1a. Thus, for matter as for radiation, we must face up to the existence of a dual character; matter behaves in some circumstances like a collection of particles and in others like a wave.

Another problem of considerable interest is that of an electron bound in a hydrogen atom. Here the electron is not free as it would be in an electron beam, but is held in the atom by its attraction to the positively charged nucleus. De Broglie made a significant start on this problem by applying his matter-wave concept to the Bohr atom.

Figure 43–5 shows a Bohr orbit of radius r. If we are to represent the orbiting electron by a wave, that wave must join on to itself in the same phase after going once around the orbit. If this does not occur there would be a mismatch and the circulating wave would gradually destroy itself by interference. Put another way, the wavelength must precisely fit around the circumference of the orbit an integral number of times, or

$$n\lambda = 2\pi r \qquad n = 1, 2, \ldots . \qquad (43\text{–}4)$$

Substituting h/p for λ (Eq. 43–2) allows us to predict that the orbital angular momentum L of the electron must be

$$L = pr = n\frac{h}{2\pi} \qquad n = 1, 2, \ldots . \qquad [42\text{–}26]$$

This is exactly Bohr's quantization assumption! This model of a hydrogen atom is now known to be oversimplified but it does suggest that we are headed in the right direction.

43–2 Wave Mechanics

So far we have seen that we can view a beam of electrons as a wave because we can measure its wavelength. We might also ask about the nature of the quantity whose variation with space and time make up the wave. In short, what feature of the electron is waving?

We have answered this question for other kinds of waves. In every case the wave is described by some time-varying field quantity. For waves on strings it is the transverse displacement y; for sound waves it is the differential pressure p; for electromagnetic waves it is the electric field **E**. For matter waves we represent the wave disturbance by a field quantity Ψ (spelled "psi" and pronounced "sigh") called a *wave function*. To describe a problem involving matter—the motion of the electron in the hydrogen atom, for example—we must know the value of Ψ for that problem as a function of both space and time. We can call this matter more closely to our attention by writing Ψ as Ψ (**r**, t), where **r** is a vector identifying the point we are talking about and t is the time.

The wave function Ψ, representing the wave nature of matter, is related to the particle nature of matter—individual electrons, for example—in almost the same way that you might imagine electromagnetic waves to be related to the individual photons. We have seen in Section 38–4 that the energy density u in an electromag-

Matter waves

Matter waves and the Bohr atom

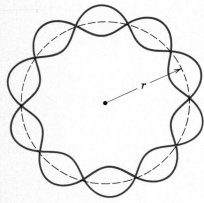

Figure 43–5 Showing how an electron wave can be adjusted in wavelength to fit an integral number of times around the circumference of a Bohr orbit of radius r.

The wave function

netic wave is proportional to the square of the field vector **E** (or **B**). This means that E^2 must be proportional to the density of photons in a particular region of space. If we were dealing with a wave containing only a few photons, perhaps only a single one, E^2 could be used to calculate the probability per unit volume that a photon is present.

Ψ in wave mechanics is analogous to **E** in electromagnetic waves except that Ψ itself cannot be directly determined by experiment. However Ψ^2 can be measured experimentally and has a simple physical interpretation.* When many particles are present Ψ^2 at any point represents the density of particles at that point. When only one particle is present—such as the single electron in the hydrogen atom problem—Ψ^2 at any point represents the probability per unit volume that the particle in question will be found at that point. We will return to this discussion of the physical significance of Ψ in Section 43–3.

Waves on strings and sound waves, as we have seen, are governed by the laws of Newtonian mechanics. Electromagnetic waves are predicted and described by Maxwell's equations. What corresponding general relationship governs the wave functions Ψ that represent matter waves?

In 1926 Erwin Schrödinger, inspired by de Broglie's matter-wave concept, pondered the observation that geometrical optics, with its rays and its rectilinear propagation, is a special case of wave optics. Is it possible that Newtonian mechanics, with *its* rays (particle trajectories) and *its* rectilinear propagation (of free particles) is a special case of some more general, as yet unknown, *wave mechanics?*

Schrödinger developed a successful theory by presenting a fundamental equation, now known as *Schrödinger's equation,* which governs the variation with space and time of the wave function Ψ for a great variety of physical problems. The resulting wave mechanics has proved almost incredibly successful in predicting the results of experiments involving objects the size of atoms or smaller. Classical Newtonian mechanics, on the other hand, works very well for larger objects such as viruses, baseballs, space-probes, and galaxies but fails completely when applied to atom-sized objects.

As an introduction to wave mechanics we shall first discuss an idealized one-dimensional problem, that of the possible motions of a particle of mass m—an electron, say—confined to move back and forth between rigid walls of separation l as in Fig. 43–6c. The formalism of wave mechanics suggests that all possible information about such motion is contained in the wave function Ψ. The solution to the problem, then, is to find and display a wave function that is consistent with Schrödinger's equation and with the limitations imposed by the rigid walls.

The required function is closely analogous to those that describe two classical situations also shown in Fig. 43–6. The first is the standing wave oscillations of a stretched string clamped at each end between massive supports, which are a distance l apart, as in Fig. 43–6a. The second analogous problem is that of electromagnetic radiation, let us say in the microwave region of the spectrum, trapped between two perfectly reflecting mirrors, as suggested by Fig. 43–6b. The standing electromagnetic waves that occur in this case are called *cavity oscillations.* They occur in a tuned microwave cavity used to produce or to detect microwaves with a specific wavelength. Figure 43–6b could also represent a tuned

The meaning of Ψ^2

Wave mechanics

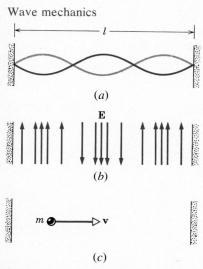

Figure 43–6 (a) A string stretched between massive clamps; a problem for Newtonian mechanics. (b) Electromagnetic radiation trapped between perfectly reflecting mirrors; a problem for Maxwell's equations. (c) A particle (electron) trapped between rigid walls; a problem for Schrödinger's wave mechanics.

---

* For those who know about complex numbers we state that Ψ is usually a complex quantity. By Ψ^2 (more properly written as $|\Psi|^2$) we mean the square of the absolute value of Ψ. This is a real quantity and is the focus of our attention in all that follows.

optical cavity, which can be used to restrict the operation of a laser to a specified wavelength.

We found in Section 17–9 that the constraints at the ends of a vibrating string produce nodes at these points and that this limits the possible wavelengths of the standing waves in the string to discrete (quantized) values given by

The quantization of wavelengths

$$\lambda = \frac{2l}{n}, \qquad n = 1, 2, 3, \ldots . \qquad (43\text{–}5)$$

If the string of Fig. 43–6a is not clamped but is infinitely long, waves of any length may be propagated. The quantization of wavelength comes about because of the localization of the wave to a length l by the clamps. In the same way a beam of electromagnetic radiation of any wavelength may be propagated in free space. The quantization suggested by Fig. 43–6b comes about because the radiation is confined within a cavity. Finally, we have seen that the localization of a matter wave by fitting it in the proper way around a Bohr orbit also led to quantization of its wavelength, as Eq. 43–4 and Fig. 43–5 show.

We saw in Section 17–8 that each point on the stretched string of Fig. 43–6a oscillates with simple harmonic motion. All the points have the same frequency of oscillation, but their amplitudes depend on position. The amplitude of a point a distance x from one of the ends was shown in Eq. 17–19b to be, with a small change of notation,

Amplitude function for a stretched string

$$y(x) = y_{max} \sin kx, \qquad (43\text{–}6)$$

where k is the wave number, defined from $k = 2\pi/\lambda$, and y_{max}, the "amplitude of amplitudes," is a constant representing the maximum amplitude anywhere along the string.

If we insert the quantized wavelength values from Eq. 43–5 into Eq. 43–6 we can write the amplitude function for the standing waves in the stretched string as

$$y_n(x) = y_{max} \sin \frac{n\pi x}{l} \qquad n = 1, 2, 3, \ldots . \qquad (43\text{–}7)$$

This shows that there are nodes at $x = 0$ and at $x = l$ for all values of n, as the boundary conditions require. Figure 43–7 shows plots of $y_n(x)$ for the first three modes of a vibrating string and for one higher mode. Note that Eq. 43–7 describes the space-dependent part only of the motion of the vibrating string. The variation with time, as we have seen, is harmonic, each oscillation mode in Fig. 43–7 having its characteristic frequency.

In the same way we can write, for the amplitude of the electric field component of the standing electromagnetic wave pattern of Fig. 43–6b,

$$E_n(x) = E_{max} \sin \frac{n\pi x}{l} \qquad n = 1, 2, 3, \ldots . \qquad (43\text{–}8)$$

As for the vibrating string there are electric field nodes at $x = 0$ and at $x = l$ for all values of n. These nodes exist because our "perfect mirrors" are also "perfect conductors," for which the transverse component of the electric field at the mirror surface must be zero. (Why?) Figure 43–7 serves equally well to show $E_n(x)$ for the first three modes and for one higher mode of the standing electromagnetic wave of Fig. 43–6b. Just as for the vibrating string, only the space-dependent part of the electric field variation is pictured.

We come now to Fig. 43–6c, the case of the particle trapped between rigid walls. If the walls are truly rigid—which we assume—the probability that the particle can penetrate them is zero. If the wave function describing the particle is a continuous function—which we also assume—it must have a node at each wall,

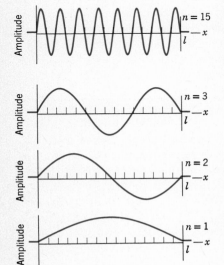

Figure 43–7 This figure represents equally well $y_n(x)$ of Eq. 43–7, $E_n(x)$ of Eq. 43–8, and $\psi_n(x)$ of Eq. 43–9.

just as for the two classical examples of Figs. 43–6a and b. Thus we can write

The wave function for a trapped particle

$$\psi_n(x) = \psi_{max} \sin \frac{n\pi x}{l}, \qquad n = 1, 2, 3, \ldots \qquad (43\text{--}9)$$

in exact analogy with Eqs. 43–7 and 43–8. Here $\psi_n(x)$ represents the space-dependent part only of the total wave function $\Psi_n(x, t)$, that is

$$\Psi_n(x, t) = \psi_n(x) f_n(t) \qquad (43\text{--}10)$$

where $f_n(t)$ describes the time-dependence of the wave function.* Figure 43–7 thus describes not only $y_n(x)$ for the stretched string (Eq. 43–7) and $E_n(x)$ for the electromagnetic standing wave (Eq. 43–8) but also $\psi_n(x)$ for the trapped particle of Fig. 43–6c.

The quantization of the wavelength led naturally, in the classical examples of the stretched string and the standing electromagnetic wave, to a quantization of frequency, each mode of oscillation of Eqs. 43–7 and 43–8 having a frequency given by

The quantization of frequency

$$\nu = \frac{v}{\lambda} = n \left(\frac{v}{2l} \right) \qquad n = 1, 2, 3, \ldots \qquad (43\text{--}11)$$

For the stretched string v in Eq. 43–11 is simply the speed with which a wave disturbance is propagated along the string. For the standing electromagnetic wave of Fig. 43–6b, $v = c$, the speed of electromagnetic radiation in free space.

For the particle trapped between rigid walls the quantization of wavelength leads to the quantization of the kinetic energy of the particle. Thus (See Eq. 43–2)

$$E_n = \frac{p^2}{2m} = \frac{(h/\lambda)^2}{2m}.$$

Putting $\lambda = 2l/n$ from Eq. 43–5 leads to

The quantization of energy

$$E_n = n^2 \frac{h^2}{8ml^2} \qquad n = 1, 2, 3, \ldots \qquad (43\text{--}12)$$

The particle cannot have any energy, as we would expect classically, but only the energies given by Eq. 43–12. Note especially that $E = 0$ is not permitted, that is, the particle simply cannot be at rest between the rigid walls. We will return to this important point, which is at variance with what Newtonian mechanics would lead us to believe, in Section 43–4.

Example 2 Consider an electron ($m = 9.1 \times 10^{-31}$ kg) confined by electrical forces to move between two rigid "walls" separated by 1.0 nm, which is about five atomic diameters. Find the quantized energy values for the three lowest stationary states.

From Eq. 43–12, for $n = 1$, we have

$$E = n^2 \frac{h^2}{8ml^2} = (1)^2 \frac{(6.6 \times 10^{-34} \text{ J·s})^2}{(8)(9.1 \times 10^{-31} \text{ kg})(1.0 \times 10^{-9} \text{ m})^2}$$
$$= 6.0 \times 10^{-20} \text{ J} = 0.38 \text{ eV}.$$

The energies for the next two states ($n = 2$ and $n = 3$) are $2^2 \times 0.38$ eV = 1.5 eV and $3^2 \times 0.38$ eV = 3.4 eV.

Example 3 Consider a grain of dust ($m = 1 \mu g = 1 \times 10^{-9}$ kg) confined to move between two rigid walls separated by 0.1 mm. Its speed is only 1×10^{-6} m/s, so that it requires 100 s to cross the gap. What quantum number describes this motion?

The energy is

$$E (= K) = \tfrac{1}{2}mv^2 = \tfrac{1}{2}(1 \times 10^{-9} \text{ kg})(1 \times 10^{-6} \text{ m/s})^2$$
$$= 5 \times 10^{-22} \text{ J}.$$

---

* It turns out (see the footnote on p. 865) that $f(t)$ is a complex quantity and that its square, defined as in that footnote, is unity. Thus ψ^2 is identical with Ψ^2. Each represents the probability per unit volume that the particle in question will be at a particular point in space.

Solving Eq. 43–12 for n yields

$$n = \sqrt{8mE}\,\frac{l}{h} = \sqrt{(8)(10^{-9}\text{ kg})(5 \times 10^{-22}\text{ J})}\left(\frac{1 \times 10^{-4}\text{ m}}{6.6 \times 10^{-34}\text{ J·s}}\right)$$

$$= 3 \times 10^{14}.$$

Even in these extreme conditions the quantized nature of the motion would never be apparent; we cannot distinguish experimentally between $n = 3 \times 10^{14}$ and $n = (3 \times 10^{14}) + 1$. Classical physics, which fails completely for the problem of Example 2, works extremely well for this problem. This example, together with the previous example, illustrates very well the correspondence principle of Section 42–9. As the quantum numbers become very large (that is, as $n \to \infty$) quantization phenomena become experimentally undetectable; quantum physics reduces to the familiar classical physics of gross objects.

43–3 The Physical Meaning of the Wave Function

We now return to the relationship between a wave function ψ and its interpretation in terms of possible experimental results. Max Born (1882–1970) first suggested that Ψ^2 at any point is a measure of the probability that the particle will be near that point. More exactly, if we imagine a volume element dV around that point, the probability that the particle will be found in that volume element is measured by $\Psi^2\,dV$. This interpretation of the wave function provides a statistical connection between the wave function and its associated particle. We are told where the particle is likely to be, not where it is.

The case of the particle trapped between rigid walls is one-dimensional so that the "volume" element dV is simply the length element dx. The probability that the particle will be found between two planes that are distance x and $x + dx$ from one wall is then $\Psi^2\,dx$ or, from Eq. 43–10 and Eq. 43–9,

$\psi^2(x)$ for particle between rigid walls

$$\psi_n^2(x)\,dx = \psi_{max}^2 \sin^2\frac{n\pi x}{l}\,dx \qquad n = 1, 2, 3, \ldots \qquad (43–13)$$

Figure 43–8 shows $\psi_n^2(x)$, which has the meaning of a probability per unit length, for the three modes corresponding to $n = 1$, 2, and 3 and for one higher mode. Note that for $n = 1$ the particle is much more likely to be found near the center than the ends. This correct prediction is in sharp contradiction to the prediction of classical physics, according to which the particle has the same probability of being located anywhere between the walls, as shown by the horizontal lines in Fig. 43–8.

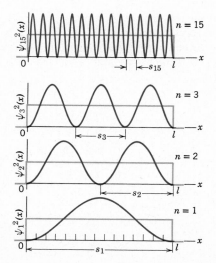

Figure 43–8 The curves show, for four values of the quantum number n, the quantity $\psi_n^2(x)$ for a particle trapped between rigid walls. Compare Fig. 43–7, which shows the wave functions $\psi_n(x)$ for the same four cases. The light lines show the classical probability expectations.

In Fig. 43–8 we see that as n increases the predictions of wave mechanics (curves) approach those of classical mechanics (light lines), just as we expect from the correspondence principle. It is perhaps troublesome that no matter how large n becomes there are still values of x for which the probability of finding the particle is zero, a fact that doesn't seem to make much sense classically. We will see in Section 43–4, however, that there is an inherent limitation to the precision with which the location of a particle can be measured. As applied to this problem this turns out to mean (see Problem 17) that the oscillations of $\psi_n^2(x)$ in Fig. 43–8 cannot be resolved experimentally. It is only the localized average behavior of $\psi_n^2(x)$—which coincides with the classical expectation for large quantum numbers—that has any physical meaning.

Let us now consider a problem that is more physically realizable than the particle trapped between rigid walls, the electron moving in a hydrogen atom. Here we examine the motion of the electron in the ground state only, defined by putting $n = 1$ in Eq. 42–28. We state without proof that the ground state wave function is given by

Hydrogen ground state wave function

$$\psi(r) = \frac{1}{\sqrt{\pi r_B^3}} e^{-r/r_B} \qquad (43-14a)$$

where

The Bohr radius

$$r_B = \frac{h^2 \epsilon_0}{\pi m e^2} = 5.29 \times 10^{-11} \text{ m} = 0.529 \text{ Å}.$$

Putting $n = 1$ in Eq. 42–27 shows that r_B above is the radius of the ground state orbit in Bohr's theory. This special interpretation has little meaning in wave mechanics, to which the very notion of orbits is foreign; r_B, taken here merely as a convenient unit of length for dealing with atomic problems, is called the *Bohr radius*.

The probability per unit volume that the electron will be found inside a volume element dV which is a radial distance r from the atomic center is given from Eq. 43–14a by

The probability density function for hydrogen

$$\psi^2(r) = \frac{1}{\pi r_B^3} e^{-2r/r_B} \qquad (43-14b)$$

which is suggested in Fig. 43–9a; the denser the distribution of dots at any point in this figure the more likely it is that the electron will be found near that point. This wave-mechanical "probability cloud" has replaced the Bohr orbit as our way of thinking about how the electron moves in the ground state of the hydrogen atom.

Another way to represent the electron distribution is to give the *radial probability density function* $P(r)$. This is defined so that $P(r)\,dr$ is the probability that the electron will be found between two spherical shells of radii r and $r + dr$. The volume dV between these shells is

$$dV = (4\pi r^2)\,(dr)$$

where $4\pi r^2$ is the surface area of a sphere of radius r. We can write this probability in two ways: (a) as $\psi^2(r)\,dV$ where dV is given above, and (b) as $P(r)\,dr$. These expressions must be equal, or

$$P(r)\,dr = \psi^2(r)\,dV = 4\pi r^2 \psi^2(r)\,dr$$

which gives us, from Eq. 43–14b,

$$P(r) = 4\pi r^2 \psi^2(r)$$

or

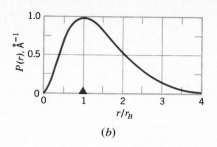

Figure 43–9 (b)

Figure 43–9 Two representations of the electron distribution in the ground state of the hydrogen atom. (a) The density of dots suggests the *probability density function* $\psi^2(r)$. The barred horizontal line represents the Bohr radius and shows the scale. (b) The *radial probability density function* $P(r)$. The filled triangle at $r = r_B$ indicates the maximum of the curve.

Radial probability density function for hydrogen

$$P(r) = \frac{4}{r_B{}^3}\, r^2 e^{-2r/r_B}. \qquad (43-14c)$$

Figure 43–9b shows a plot of $P(r)$. Note that the most probable radial coordinate for the electron corresponds to the Bohr radius. Thus in wave mechanics we do not say that the electron in the $n = 1$ state in hydrogen goes around the nucleus in a circular orbit of radius 0.529 Å, but only that the electron is more likely to be found at this distance from the nucleus than at any other distance.

Example 4 Verify that the maximum of the probability curve of Fig. 43–9b occurs for $r = r_B$.

To find the maximum we differentiate Eq. 43–14c, using the rule for differentiating a product, or

$$\frac{dP}{dr} = \left(\frac{4r^2}{r_B{}^3}\right)\left(-\frac{2}{r_B}\, e^{-2r/r_B}\right) + (e^{-2r/r_B})\left(\frac{8r}{r_B{}^3}\right)$$

$$= \frac{8}{r_B{}^3}\, r(1 - r/r_B)\, e^{-2r/r_B}.$$

At the maximum we have $dP/dr = 0$ and this does indeed occur at $r = r_B$. We also have $dP/dr = 0$ at $r = 0$, and for $r \to \infty$ but these correspond not to maxima but to minima and are quite consistent with Fig. 43–9b.

Example 5 What is the probability P' that the electron in the hydrogen-atom ground state will be found to have a radial coordinate r that lies between 1.000 and 1.020 Å?

These radii are close enough together so that P(r) is essentially constant in the interval. Thus we can approximate dr by Δr ($= 0.020$ Å) and we have

$$P' \cong P(r)\, \Delta r$$

$$= \frac{4r^2}{r_B}\, e^{-2r/r_B}\, \Delta r$$

$$= \frac{(4)(1.000 \text{ Å})^2}{(0.529 \text{ Å})^3}\, e^{-2(1.000)/(0.529)}\, (0.020 \text{ Å})$$

$$= 0.012.$$

Thus the electron would be found between these two radial limits 1.2% of the time.

43–4 The Uncertainty Principle

Wave mechanics predicts far more experimentally confirmed observations than does Bohr theory. For this reason essentially all scientists have accepted it, looking on Bohr theory today as an ingenious stepping stone to a level at which deeper understanding is possible.

There is another and deeper reason that compels us to give up the comfortable, easily understandable concept of orbital motion and to replace it by the statistical, probabilistic kind of picture summarized by Fig. 43–9. It is that all attempts to experimentally verify the details of orbital motion result in failure. This is not because the measurements are difficult, or because they require instruments much finer than any we now possess, but because—we now believe—they are not even possible in principle.

We observe the moon traveling in its orbit around the earth by means of sunlight that it reflects in our direction. Incident light transfers linear momentum to an object from which it is reflected. In principle, this reflected light would disturb the course of the moon in its orbit. We do not have to think very long, however, to convince ourselves that the disturbing effect is totally negligible in this case. We are able to get continuous information about the moon's position as a function of time without significantly disturbing the motion we are trying to observe.

For electrons the situation is quite different. Here, too, we can hope to "see" the electron only if we reflect light, or another particle, from it. In this case the recoil that the electron experiences when the light (photon) bounces from it alters the electron's motion in a way that cannot be avoided or even corrected for.

It is not surprising that the probability curve of Fig. 43–9b is the most detailed information that we can hope to obtain, by measurement, about the distribution of negative charge in the hydrogen atom. If orbits such as those envisaged by Bohr existed, they would be altered significantly in our attempts to verify their existence. Under these circumstances, we prefer to say that the wave function Ψ and not the orbits represents physical reality.

The fact that we can't describe the motions of electrons in a classical way finds expression in the *uncertainty principle*, enunciated by Werner Heisenberg in 1927.* To formulate this principle, consider a beam of monoenergetic electrons of speed v_0 moving from left to right in Fig. 43–10. Let us set ourselves the task of measuring the position of a particular electron in the vertical (y) direction and also its velocity component v_y in this direction. If we succeed in carrying out these measurements with unlimited accuracy, we can then claim to have established the position and motion of the electron (or one component of it at least) with precision. However, we shall see that it is impossible to make these two measurements simultaneously with unlimited accuracy.

To measure y we block the beam with an absorbing screen A in which we put a slit of width Δy. If an electron gets through the slit, its vertical position must be known to this accuracy. By making the slit narrower, we can improve the accuracy of this vertical position measurement as much as we wish.

Since the electron is a wave, it will undergo diffraction at the slit, and a photographic plate placed at B in Fig. 43–10 will reveal a typical diffraction pattern. The existence of this diffraction pattern means that there is an uncertainty in the values of v_y possessed by the electrons emerging from the slit. Let v_{ya} be the value of v_y that corresponds to an electron landing at the first minimum on the screen, marked by point a and described by a characteristic angle θ_a. We take v_{ya} as a rough measure of the uncertainty Δv_y in v_y for electrons emerging from the slit.

The first minimum in the diffraction pattern is given by Eq. 41–2, or

$$\sin \theta_a = \frac{\lambda}{\Delta y}.$$

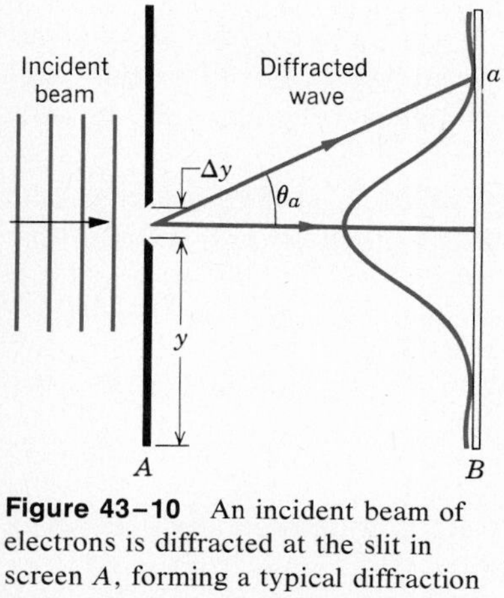

Figure 43–10 An incident beam of electrons is diffracted at the slit in screen A, forming a typical diffraction pattern on screen B. If the slit is made narrower, the pattern becomes wider.

* See "The Principle of Uncertainty" by George Gamow, *Scientific American*, January 1958.

If we assume that θ_a is small enough, we can write this equation as

$$\theta_a \cong \frac{\lambda}{\Delta y}. \tag{43-15}$$

To reach point a, $v_{ya}(=\Delta v_y)$ must be such that

$$\theta_a \cong \frac{\Delta v_y}{v_0}. \tag{43-16}$$

Combining Eqs. 43-15 and 43-16 leads to

$$\frac{\Delta v_y}{v_0} = \frac{\lambda}{\Delta y},$$

which we rewrite as

$$\Delta v_y \, \Delta y \cong \lambda v_0. \tag{43-17}$$

Now λ, the wavelength of the electron beam, is given by h/p or h/mv_0; substituting this into Eq. 43-17 yields

$$\Delta v_y \, \Delta y \cong \frac{hv_0}{mv_0}.$$

We rewrite this as

$$\Delta p_y \, \Delta y \cong h. \tag{43-18}$$

In Eq. 43-18 $\Delta p_y (= m\,\Delta v_y)$ is the uncertainty in our knowledge of the vertical momentum of the electrons; Δy is the uncertainty in our knowledge of their vertical position. The equation tells us that, since the product of these uncertainties is a constant, we cannot measure p_y and y simultaneously with unlimited accuracy.

If we want to improve our measurement of y (that is, if we want to reduce Δy), we use a finer slit. However, this will produce a wider diffraction pattern (see Eq. 43-15). A wider pattern means that our knowledge of the vertical momentum component of the electron has deteriorated, or, in other words, Δp_y has increased; this is exactly what Eq. 43-18 predicts.

The limits on measurement imposed by Eq. 43-18 have nothing to do with the crudity of our measuring instruments. We are permitted to postulate the existence of the finest conceivable measuring equipment. We are dealing with a fundamental limitation, imposed by nature.

Equation 43-18 is a statement, derived in a special case, of the general principle known as the *uncertainty principle*. As applied to position and momentum measurements, it asserts that

Heisenberg's uncertainty principle

$$\Delta p_x \, \Delta x \gtrsim h$$
$$\Delta p_y \, \Delta y \gtrsim h$$
$$\Delta p_z \, \Delta z \gtrsim h. \tag{43-19}$$

Thus no component of the motion of an electron, free or bound, can be described with unlimited precision. In Eq. 43-19 we have replaced the sign $\cong$ (of Eq. 43-18) by the sign $\gtrsim$ because in all cases involving actual measurements the products of the uncertainties will be greater than the irreducible minimum established by the uncertainty principle.

The Planck constant h probably has more significance in Eq. 43-19 than anywhere else we have seen it. If this product had been zero instead of h, the classical ideas about particles and orbits would be correct; it would then be possible to measure both momentum and position with unlimited precision. The fact that h appears instead of zero in Eq. 43–19 means that the classical ideas are wrong; the magnitude of h tells us under what circumstances these classical ideas must be replaced by quantum ideas. George Gamow* has speculated, in an interesting and readable fantasy, what our world would be like if the constant h were much larger than it is, so that nonclassical ideas would be apparent to our sense perceptions.

Example 6 A free electron whose kinetic energy is 10 eV has a corresponding speed of 1.9×10^6 m/s. Let us assume that this speed is known to 1.0% accuracy. With what irreducible accuracy can we measure the position of this electron?

The electron momentum is

$$p = mv = (9.1 \times 10^{-31} \text{ kg})(1.9 \times 10^6 \text{ m/s})$$
$$= 1.7 \times 10^{-24} \text{ kg·m/s}.$$

The uncertainty Δp in momentum is 1.0% of this or 1.7×10^{-26} kg·m/s. The uncertainty in position is then, from Eq. 43–19

$$\Delta x \cong \frac{h}{\Delta p} = \frac{6.6 \times 10^{-34} \text{ J·s}}{1.7 \times 10^{-24} \text{ kg·m/s}} = 39 \text{ nm},$$

or about 200 atomic diameters. If the electron momentum has really been determined by measurement to have the accuracy stated, there is no possibility that its position can be known more precisely than this.

Example 7 A golf ball has a mass of 45 g and a speed, measured to 1.0% accuracy, of 75 m/s. With what irreducible accuracy can we measure its position?

This example is exactly like Example 6 except that the golf ball is much more massive and much slower than the electron. The same calculation yields in this case

$$\Delta x = 1.9 \times 10^{-32} \text{ m}.$$

This is a very small number, about 10^{17} times smaller than the diameter of an atomic nucleus! In familiar cases such as this the uncertainty principle sets no practical limit whatever to the accuracy of our measuring procedures and classical Newtonian mechanics works just fine. A comparison with Example 6 shows once again that wave mechanics reduces to classical mechanics under the appropriate circumstances, just as predicted by the correspondence principle.

Example 8 *Zero point energy* Apply the uncertainty principle to the problem of a particle trapped between rigid walls as in Fig. 43–6c. Show that the minimum kinetic energy of the particle must be greater than zero. That is, show that the particle cannot be completely at rest between the walls.

All we know about the position of the particle is that it lies between the rigid walls and is not outside them. For the uncertainty in position we then put $\Delta x = l$.

As the particle bounces back and forth between the walls the x-component of its momentum alternates between $+p_x$ and $-p_x$. A measurement at any instant is just as likely to yield one value or the other. The magnitude of the difference between these two quantities is $2p_x$ and we can take this as a rough measure of the uncertainty in the x-component of the momentum. That is, $\Delta p_x = 2p_x$.

From the uncertainty relation we have

$$\Delta p_x \, \Delta x \gtrsim h \qquad \text{or} \qquad (2p_x)(l) \gtrsim h,$$

which yields

$$p_x \gtrsim h/2l \qquad \text{or} \qquad p_{x,\text{min}} = h/2l.$$

The corresponding minimum kinetic energy follows from

$$E_{\text{min}} = \frac{p^2}{2m} = \frac{h^2}{8ml^2}.$$

This happens to be exactly what we find for the ground state energy by putting $n = 1$ in Eq. 43–12! This exact agreement is accidental however, in part because of our rough assessment of Δp_x. Our concern, however, is only to show that the minimum allowed kinetic energy exceeds zero. The only way the kinetic energy can be zero, which corresponds to a resting particle, is to let the wall separation l become infinitely great.

The energy calculated above is called the *zero-point energy*. This example teaches us that whenever a particle is localized, whether by putting it between rigid walls, by binding it as an atom in a crystal lattice, or in other ways, it always has a finite minimum kinetic energy, even at the absolute zero of temperature. This is in sharp contrast to the classical assumption that all motion would stop at the absolute zero of temperature.

43–5 Wave-Particle Duality— The Complementarity Principle

In Table 43–1 we point to four of the strongest of the many possible experiments that underscore our belief that both matter and radiation have a dual character, emerging as wave-like in some situations and as particle-like in others.

Our mental images of both "wave" and "particle" are drawn from experiences with large objects such as ocean waves and moving balls. In a way we are

---

* *Mr. Tompkins in Wonderland,* by George Gamow, Macmillan, New York, 1940.

lucky that it is at all possible to extrapolate these concepts to the atomic domain, where we have no direct experience of seeing or touching. The difficulty, of course, is that the two concepts, waves and particles, are so very different that they seem to exclude each other.

Table 43–1 A selected experimental basis for the wave-particle duality of both matter and radiation

	Matter	**Radiation**
Wave nature	Davisson and Germer's electron diffraction experiments — Section 43–1	Young's double-slit interference experiment — Section 40–2
Particle nature	J. J. Thomson's measurement of e/m for the electron — Section 30–8	The Compton effect — Section 42–6

Niels Bohr's *complementarity principle* helps to resolve this conceptual problem. It tells us, as a distillation of accumulated experience:

The principle of complementarity

> The wave and particle natures of either matter or radiation complement each other. It is not possible to demonstrate an exclusive wave or particle nature of either matter or radiation. Both models are required.

A wave-particle duality experiment

As an example, consider the "thought experiment" illustrated in Fig. 43–11. A beam of electrons falls on a double-slit arrangement and produces interference fringes on a screen at C. As we saw in Section 43–1, this is convincing proof of the wave nature of the incident electron beam.

If we look carefully, however, we cannot explain everything in terms of waves. Suppose we replace screen C by a small electron detector, designed to generate and record a "click" every time an electron strikes it. We shall find that such clicks do indeed occur. By moving the detector up and down in Fig. 43–11 we shall, by plotting the click rate against the detector position, be able to trace out the interference pattern.

Here is the dilemma. The clicks suggest particles (electrons), falling like rain-drops on the detector. The interference pattern certainly suggests a wave. How are we to resolve the dilemma?

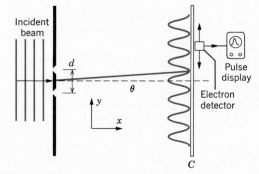

Figure 43–11 An electron beam falls on a double slit system, producing interference fringes in plane c. A small electron detector can be scanned across the plane in the y-direction.

Insight emerges when we realize that a mere click is not enough. The concept of "particle" is tied in closely with the concept of "trajectory." Our mental image is of a point following a path. As a minimum we would like to be able to say which of the two slits the electron passed through on its way from the source to the detector.

We might think of proceeding by putting an indicator—an atom, say—in front of the upper slit and then watching it with some kind of super microscope. (Remember, we said that this is a thought experiment!) If an electron goes through this slit it will strike the indicator atom and cause it to move. The electron's motion will also change in this collision and it will strike the plane C at a different place than it would have otherwise.

After many electrons have passed through the apparatus, the accumulated distribution will not show the interference pattern of Fig. 43–11, but rather a superposition of two overlapping, noninterfering diffraction patterns as shown in Figure 43–12. The interference pattern no longer appears. Whenever we succeed in tracing an electron through the upper slit to plane C we have indeed established it as a particle, but we shall then discover that we have lost the interference pattern we set out to explain.

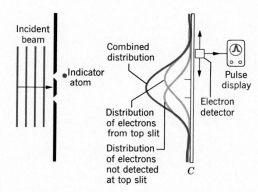

Figure 43–12 Distribution of electrons in a two-slit experiment when indicator atoms are used to identify each electron as having passed through one or the other slit.

We started above with an experiment in which the wave character (of matter or of radiation) was clearly evident and we tinkered with the experiment to bring out the particle character. To the extent to which we succeeded we destroyed the evidence for "wave". The converse is also true. If we start with an experiment in which the particle character is clearly evident and if we seek to develop evidence to support a wave concept we find that if we succeed in doing so we have destroyed the evidence for "particle". We cannot develop both wave and particle aspects simultaneously in a single experiment, any more than an ordinary tossed coin can display simultaneously both its "head" and its "tail" aspects.

A summary statement

REVIEW GUIDE AND SUMMARY

The wave nature of matter

Beams of electrons and other forms of matter exhibit wave properties, including interference and diffraction, with a *de Broglie wavelength* given by

$$\lambda = h/p \qquad [43\text{--}2]$$

These properties are most easily shown by diffraction, similar to x-ray diffraction, which occurs during reflection from atomic planes in crystals. (A moderate-energy 100-eV electron beam has a wavelength of only 1.2 Å, the same order of magnitude as atomic interplanar distances.)

The wave function

Wave mechanics

Matter waves are described by a wave function $\Psi(\mathbf{r}, t)$. The general theory of matter waves, successful in solving a host of problems in atomic physics and chemistry, is called wave mechanics. Classical mechanics is a limiting case, valid when applied to macroscopic particles. We focus attention here on the space-varying part of the wave function, designated $\psi(\mathbf{r})$.

A particle trapped between rigid walls

A simple one-dimensional introduction to wave mechanics is the study of the motion of a particle trapped between rigid walls. We can study the relevant wave function because of its close mathematical relationship to two classical problems: the standing-wave oscillations of a short constrained string and the electromagnetic oscillations inside a cavity with perfectly reflecting walls. The trapped particle's energy is shown to be limited to quantized values given by

$$E_n = n^2 \frac{h^2}{8ml^2} \qquad n = 1, 2, 3, \ldots. \qquad [43-12]$$

The statistical meaning of Ψ

The physical interpretation of $\Psi(\mathbf{r}, t)$ is that $\Psi^2 \, dV$ is the statistical probability that the particle will be found in a volume element dV placed at position $\mathbf{r}$. Figure 43–8 shows $\psi^2(x)$ for the confined particle of Fig. 43–6c. The distribution is not the same as would be predicted by classical mechanics, especially for low quantum numbers. Figure 43–9 represents an important three-dimensional situation, that of a ground-state electron in a hydrogen atom.

Heisenberg's uncertainty principle

The uncertainty principle suggests that the very concept of "particle" (as in "particle between rigid walls") is inherently fuzzy. In particular we cannot, even in principle, simultaneously measure both the position $\mathbf{r}$ and momentum $\mathbf{p}$ with arbitrary precision. In fact, the uncertainties associated with each component of $\mathbf{r}$ and $\mathbf{p}$ must obey a relationship of the form

$$\Delta p_x \Delta x \gtrsim h. \qquad [43-19]$$

Wave-particle duality and the principle of complementarity

The dual wave-particle nature of both matter and radiation (see Table 43–1) is summarized by Bohr's principle of complementarity. It states that exclusively wave or particle models are not adequate to describe either matter or radiation, but that both models must somehow be combined, in a complementary way, for a complete understanding. The thought experiment of Figs. 43–11 and 43–12 suggests the spirit of complementarity. Two-slit interference fringes show the wave nature of the incident beam, even though the detector responds to individual particles. Any attempt to demonstrate a more fundamental particle nature, perhaps by tracing such particles through the apparatus, causes the interference fringes to disappear.

QUESTIONS

1. How can the wavelength of an electron be given by $\lambda = h/p$? Doesn't the very presence of the momentum p in this formula implies that the electron is a particle?

2. In a repetition of Thomson's experiment for measuring e/m for the electron (see Section 30–8), a beam of electrons is collimated by passage through a slit. Why is the beamlike character of the emerging electrons not destroyed by diffraction of the electron wave at this slit?

3. Why is the wave nature of matter not more apparent to our daily observations?

4. Considering the wave behavior of electrons, we should expect to be able to construct an electron microscope. This, indeed, has been done. (a) How might an electron beam be focused? (b) What advantages might an electron microscope have over a light microscope? (c) Why not make a proton microscope? a neutron microscope?

5. How many experiments can you recall that support the wave theory of light? . . . the particle theory of light? . . . the wave theory of matter? . . . the particle theory of matter?

6. Discuss the analogy between (a) wave optics and geometrical optics and (b) wave mechanics and classical mechanics.

7. An electron and a neutron have the same kinetic energy. Which one has the shorter de Broglie wavelength?

8. Discuss similarities and differences between a matter wave and an electromagnetic wave.

9. Can the de Broglie wavelength of a particle be smaller than a the size of the particle? . . . Larger? Is there any relation necessarily between such quantities?

10. If, in the de Broglie formula $\lambda = h/mv$, we let $m \to \infty$, do we get the classical result for particles of matter?

11. A standing wave can be viewed as the superposition of two traveling waves. Can you apply this view to the problem of a particle confined between rigid walls, giving an interpretation in terms of the motion of the electron?

12. The allowed energies for a particle confined between rigid walls are given by Eq. 43–12. First, convince yourself that, as n increases, the energy levels become farther apart. How can this possibly be? The correspondence principle would seem to require that they move closer together as n increases, approaching a continuum.

13. How can the predictions of wave mechanics be so exact if the only information we have about the positions of the electrons is statistical?

14. In the $n = 1$ mode, for a particle confined between rigid walls, what is the probability that the particle will be found in a small volume element at the surface of either wall?

15. What are the dimensions of $\psi^2(x)$ in Fig. 43–8? What is the value of the classically expected probability density, represented by the horizontal lines? What values do the areas under the curves have? How does the area under any of these curves compare with the area under the horizontal line? All these questions can be answered by inspection of the figure; see Problem 10.

16. In Fig. 43–8 what do you imagine the curve for $\psi^2(x)$ for $n = 100$ looks like? Convince yourself that these curves approach classical expectations as $n \to \infty$.

17. (a) Give examples of how the process of measurement disturbs the system being measured. (b) Can the disturbances be taken into account ahead of time by suitable calculations?

18. Why does the concept of Bohr orbits violate the uncertainty principle?

19. Figure 43–8 shows that for $n = 3$ the probability function for a particle confined between rigid walls is zero at two points between the walls. How can the particle ever move across these positions? (Hint: Consider the implications of the uncertainty principle.)

20. Make up some numerical examples to show the difficulty of getting the uncertainty principle to reveal itself during experiments with an object whose mass is about one gram.

21. Several groups of experimenters are trying to detect gravity waves, perhaps coming from our galactic center, by measuring small distortions in a massive object through which the hypothesized waves pass. They seek to measure displacements as small as $\sim 10^{-21}$ m (the radius of a proton is $\sim 10^{-15}$ m, a million times larger!). Does the uncertainty principle put any restriction on the precision with which this measurement can be carried out?

22. State the complementarity principle in your own words and explain how the experiment described in connection with Figs. 43–11 and 43–12 illustrates it.

23. Is an electron a particle? Is it a wave? Explain your answer, citing relevant experimental evidence.

24. A laser projects a beam of light across a laboratory table. If you put a diffraction grating in the path of the beam and observe the spectrum you declare the beam to be a wave. If instead you put a clean metal surface in the path of the beam and observe the ejected photoelectrons you declare this same beam to be a stream of particles (photons). What can you say about the beam if you don't put anything in its path?

25. State and discuss (a) the correspondence principle, (b) the uncertainty principle, and (c) the complementarity principle.

26. In Figure 43–12, why would you expect the electrons from each slit to arrive at the screen over a range of positions? Shouldn't they all arrive at the same place? How does your answer relate to the complementarity principle?

EXERCISES

Section 43–1   Matter Waves

1. A bullet of mass 40 g travels at 1000 m/s. (a) What wavelength can we associate with it? (b) Why does the wave nature of the bullet not reveal itself through diffraction effects?

2. Show that, using the classical relation between momentum and kinetic energy, the de Broglie wavelength of an electron can be written as

$$\lambda = \frac{1.226 \text{ nm}}{\sqrt{K}}, \qquad (K \text{ in eV})$$

in which K is the kinetic energy in electron volts.

3. The 50-GeV electron accelerator at Stanford (1 GeV = 10^9 eV) provides an electron beam of small wavelength, suitable for probing the fine details of nuclear structure by scattering experiments. What will this wavelength be, and how does it compare with the size of an average nucleus? [At these energies, it is suffi-cient to use the extreme relativistic relationship between momentum and energy, namely, $p = E/c$. This is the same relationship used for light (Eq. 38–15) and is justified when the kinetic energy of a particle is much greater than its rest energy m_0c^2, as in this case.]

4. In an ordinary color television set, electrons are accelerated through a potential difference of 25 kV. What is the de Broglie wavelength of such electrons? (Hint: See Exercise 2.)

5. Compare the wavelength of (a) a 1-keV electron, (b) a 1-keV photon, and (c) a 1-keV neutron. Use the relation of Exercise 2 for the electron.

6. An electron and a photon each have a wavelength of 0.20 nm. What are their (a) momenta and (b) energies? Use the result of Exercise 2 for the electron.

Section 43-2   Wave Mechanics

7. (a) A proton or (b) an electron is trapped in a one-dimensional box of 1.0 Å (= 10^{-10} m) length. What is the minimum energy these particles can have?

8. (a) Find, approximately, the smallest allowed energy of an electron confined to an atomic nucleus (diameter about 1.4×10^{-14} m). (b) Compare this with the several MeV of energy binding protons and neutrons inside the nucleus; on this basis, should we expect to find electrons inside nuclei?

9. If an electron moves from a state represented by $n = 3$ in Fig. 43-7 to one represented by $n = 2$ in that figure, emitting electromagnetic radiation in the process, what are (a) the energy of the single emitted photon and (b) the corresponding wavelength? See Example 2.

Section 43-3   The Physical Meaning of the Wave Function

10. In the ground state of the hydrogen atom, what is the probability that the electron will be found within a sphere whose radius is that of the first Bohr orbit?

11. Find the ratio of the probabilities of finding the electron in the hydrogen atom between r_B and $r_B + \Delta r$ at the Bohr radius to that of finding it between $2r_B$ and $2r_B + \Delta r$.

12. In atoms there is a finite, though very small, probability that, at some instant, an orbital electron will actually be found inside the nucleus. In fact, some unstable nuclei use this occasional appearance of the electron to decay by *electron capture*. Assuming that the proton itself is a sphere of radius 1.1×10^{-15} m and that the hydrogen atom electronic wave function holds all the way to the proton's center, use the ground state wave function to estimate the probability that the hydrogen atom electron is inside its nucleus. (*Hint:* when $x \ll 1$, $e^{-x} \approx 1$.)

Section 43-4   The Uncertainty Principle

13. A microscope using photons is employed to locate an electron in an atom to within a distance of 1.0×10^{-2} nm. What is the minimum uncertainty in the momentum of the electron located in this way?

14. The uncertainty in the position of an electron is given as 0.50 Å, which is about the radius of the first Bohr orbit in hydrogen. What is the uncertainty in the linear momentum of the electron?

15. Imagine playing baseball in a universe where Planck's constant was 0.600 J·s. What would be the uncertainty in the position of a 0.500-kg baseball moving at 20.0 m/s with an uncertainty of 1.0 m/s. Why would it be hard to catch such a ball?

PROBLEMS GROUPED BY SECTION

Section 43-1   Matter Waves

1. The wavelength of the yellow spectral emission of sodium is 590 nm. At what kinetic energy would an electron have the same de Broglie wavelength?

2. Singly charged sodium ions are accelerated through a potential difference of 300 V. (a) What is the momentum acquired by the ions? (b) What is their de Broglie wavelength?

3. Thermal neutrons have an average kinetic energy $\frac{3}{2}kT$, where T may be taken to be 300 K. Such neutrons are in thermal equilibrium with normal surroundings. (a) What is the average energy of a thermal neutron? (b) What is the corresponding de Broglie wavelength?

4. Derive the commonly memorized (nonrelativistic) relation between the de Broglie wavelength λ of an electron (in angstroms) and the corresponding accelerating potential V (in volts):

$$\lambda = \left(\frac{150}{V}\right)^{1/2}.$$

5. A neutron crystal spectrometer utilizes crystal planes of spacing $d = 0.732$ Å in a beryllium crystal. What must be the Bragg angle θ so that only neutrons of energy $K = 4.0$ eV are reflected? Consider only first-order reflections.

6. *Can we treat gas molecules as small particles?* Consider a balloon filled with (monatomic) helium gas at room temperature and pressure. (a) How does the average de Broglie wavelength of the helium atoms compare with the average distance between atoms under these conditions? The average kinetic energy of an atom is equal to $\frac{3}{2}kT$. (b) What is the answer to the question posed at the beginning of this problem?

7. In the experiment of Davisson and Germer, (a) at what angles would the second- and third-order diffracted beams corresponding to a strong maximum in Fig. 43-2 occur, provided that they are present. (b) At what angle would the first-order diffracted beam occur if the accelerating potential were changed from 54 to 60 V? The experimenter is free to rotate the crystal.

Section 43-2   Wave Mechanics

8. Calculate the fractional difference between two adjacent energy levels of a particle confined in a one-dimensional box.

9. (a) What is the separation in energy between the lowest two energy levels for a container 20 cm on a side containing argon atoms? (b) How does this compare with the thermal energy of the argon atoms at 300 K? (c) At what temperature does the thermal energy equal the spacing between these two energy levels? The atomic weight of argon is 39.9 g/mol.

Section 43-3   The Physical Meaning of the Wave Function

10. A probability of unity is certainty. Since the particle described by the wave function of Eq. 43-13 must be *somewhere* between the rigid walls, we can write

$$\int_0^l \psi_n^2(x)\, dx = 1,$$

and we must adjust the constant ψ_{max} until this relation is satisfied. This process is called *normalization* of the wave function. Find ψ_{max} by this method, using a table of integrals if necessary.

11. A particle is confined between rigid walls separated by a distance l. What is the probability that it will be found within a distance $l/3$ from one wall (a) for $n = 1$, (b) for $n = 2$, (c) for $n = 3$, and (d) according to classical physics?

12. For the ground state of the hydrogen atom, show that the probability P_r that the electron lies within a sphere of radius r is given by

$$P_r = 1 - e^{-2r/r_B}\left(\frac{2r^2}{r_B^2} + \frac{2r}{r_B} + 1\right).$$

Does this yield expected values for (a) $r = 0$ and (b) $r = \infty$? (c) State clearly the difference in meaning between this expression and that given in Eq. 43–14c.

*13. In wave mechanics the average value of a measurable quantity f is called the *expectation value,* denoted by $\langle f \rangle$ and can be written for one-dimensional problems as

$$\langle f \rangle = \int_{-\infty}^{+\infty} f(x)\psi^2(x)dx,$$

where ψ is the normalized wave function. For a particle in the state n in a one-dimensional box of length l, show that the expectation value of the position x is $l/2$. Would you have expected this result?

*14. The wave function of a simple harmonic oscillator in the ground state is given by

$$\psi = \sqrt{\frac{a}{\pi}}\, e^{-a^2x^2/2}, \qquad -\infty \le x \le +\infty$$

where $a = 2\pi\omega m/h$ and ω is the angular frequency of the oscillator. What is the expectation value for the position x? (See Problem 13.)

Section 43–4 The Uncertainty Principle
15. In Example 2, the electron's energy was determined exactly by the size of the box. How do you reconcile this with the uncertainty principle since the uncertainty in the location of the electron cannot exceed 1.0×10^{-9} m (the size of the box) if the uncertainty principle is to be obeyed?

16. Another aspect of the uncertainty principle states that a measurement of the energy of a system performed during the time Δt must be uncertain by an amount ΔE, where

$$\Delta E \Delta t \gtrsim h.$$

The lifetime of an electron in the state $n = 2$ in hydrogen is about 10^{-8} s. What is the uncertainty in the energy of the $n = 2$ state? Compare this with the energy of this state.

17. (a) Prove that, for a particle trapped between rigid walls, the distance s_n between adjacent minima in the quantity $\psi_n^2(x)$ is given by

$$s_n = \frac{l}{n};$$

see Fig. 43–8. (b) Prove that, for such a trapped particle, the uncertainty principle gives for the uncertainty Δx in the position of the particle,

$$\Delta x \gtrsim s_n.$$

(*Hint:* In applying this principle, take Δp_x, the momentum uncertainty, to be $2p_x$, using the argument advanced in Example 8.) (c) Use the above results to show that, as n increases, the position of the trapped particle can be located with increasing precision, but the points for which $\psi_n^2(x) = 0$ can never be resolved experimentally.

ADDITIONAL PROBLEMS

18. If the de Broglie wavelength of a proton is 1.0×10^{-13} m, (a) what is the speed of the proton, and (b) through what electrical potential would the proton have to be accelerated to acquire this speed?

19. Show that if the uncertainty in the location of a particle is equal to its de Broglie wavelength, the uncertainty in its velocity is equal to its velocity.

20. Where are the points of (a) maximum and (b) minimum probability for a particle trapped in a box of length l if the particle is in the state n?

21. If the wave function for an electron is $\psi(x) = Ae^{-bx}$ from $x = 0$ to $x = \infty$ and is zero for negative values of x, find the constant A in terms of the constant b. (*Hint:* See Problem 10.)

22. What accelerating voltage would be required for electrons in an electron microscope to obtain the same ultimate resolving power as that which could be obtained from a gamma-ray microscope using 100-keV gamma rays?

23. A particle is confined between rigid walls located at $x = 0$ and $x = l$ and is in the fourth allowed energy state. (a) Sketch the probability curve for the particle's location. Calculate the approximate probabilities of finding the particle within a $\Delta x = 0.0003l$ when (b) x is equal to $l/8$ and (c) x is equal to $3l/16$. Refer to your figure to see whether or not your results seem reasonable. (*Hint:* No integration is necessary.)

24. Suppose that we wished to test the possibility that electrons in atoms move in orbits by "viewing" them with photons having sufficiently short wavelength, say 0.1 Å or less. (a) What would be the energy of such photons? (b) How much energy would such a photon transfer to a free electron in a head-on Compton collision? (c) What does this tell you about the possibility of confirming orbital motion by "viewing" an atomic electron at two or more points along its path?

25. *Bragg reflections for electrons and x rays.* A potassium chloride crystal is cut so that the layers of atomic planes parallel to its surface have an interplanar spacing of 0.314 nm. An incident beam makes an angle θ with the surface and a first-order diffraction peak is generated. Find θ if the incident beam is (a) a 40-keV x-ray beam and (b) if it is a beam of electrons whose kinetic energy is 40 keV.

26. *de Broglie waves and Rutherford scattering.* The existence of the atomic nucleus was discovered in 1911 by Ernest Rutherford, who properly interpreted some experiments in which a beam of alpha particles was scattered from a foil of atoms such as gold. (a) If the alpha particles had a kinetic energy of 7.5 MeV, what was their de Broglie wavelength? (b) Should the wave nature of the incident alpha particles have been taken into account in interpreting these experiments? The mass of an alpha particle is 4.00 u, and its distance of closest approach to the nuclear center in these experiments was about 30 fm. (The wave nature of matter was not postulated until more than a decade after these crucial experiments were first performed.)

27. According to the correspondence principle, as $n \rightarrow \infty$ we expect classical results in the Bohr atom. Hence, the de Broglie wavelength associated with the electron (a quantum result) should get smaller compared with the radius of the orbit as n increases. Indeed, we expect that $\lambda/r \rightarrow 0$ as $n \rightarrow \infty$. Show that this is the case.

28. The highest achievable resolving power of a microscope is limited only by the wavelength used, that is, the smallest detail that can be separated is about equal to the wavelength. Suppose that one wishes to "see" inside an atom. Assuming the atom to have a diameter of 1.0 Å, this means that we wish to resolve detail of separation about 0.10 Å. (*a*) If an electron microscope is used, what minimum energy of electrons is needed? (*b*) If a light microscope is used, what minimum energy of photons is needed? (*c*) Which microscope seems more practical for this purpose? Why?

One-Electron Atoms

44–1   Introduction

Reasons for studying the hydrogen atom

Most of this chapter deals with the application of wave mechanics to the hydrogen atom. Hydrogen, with its single electron, is the simplest atom. Even though it is also a very important atom—being a constituent of water and of all organic compounds—it may nevertheless strike us as overdoing it to devote so much space to just one element of the hundred or so that make up the periodic table of the elements. This concern is misplaced, however, for three reasons.

The first reason.   The hydrogen atom, because of its stark simplicity, has long served as a meeting ground for theory and experiment in any proposed study of atomic structure. A theory that cannot succeed at this level should almost certainly go no further. From the early days of Bohr theory, through wave mechanics, and into the latest developments of modern relativistic quantum electrodynamics the hydrogen atom has served as our principal laboratory for testing in precise and quantitative terms our understanding of atomic structure and properties.

The hydrogen atom is important because it is simple

The second reason.   What we learn about the hydrogen atom applies also to singly ionized helium (He^+), doubly ionized lithium (Li^{++}), etc. All such ions contain a single electron and in this sense are hydrogen-like atoms.

Hydrogen is not the only one-electron atom

Beyond this there are multi-electron atoms in which all of the electrons but a single "extra" valence electron form a core of one or more filled subshells, surrounding the nucleus symmetrically. The external properties of such atoms are determined largely by the motion of this single "extra" electron and hydrogen atom theory is very useful in such cases. Sodium, with one extra electron outside a filled ten-electron core and silver, with one extra electron outside a filled 46-electron core, are two examples that we shall discuss later in this chapter.

Finally, there are so-called exotic atoms consisting—not of an electron moving around a nucleus—but of an elementary particle and its (oppositely charged) antiparticle moving around their common center of mass. Because of their symmetry such atoms can also be treated by the methods developed for the (one-electron) hydrogen atom. These exotic atoms can be formed from electrons (positronium; see Problem 27 of Chapter 42), charmed quarks (charmonium), and bottom quarks (bottonium), among other possibilities. Positronium, for example, consists of an electron (charge $-e$) and a positron (the electron antiparticle; charge $+e$) circulating about their common center of mass.

The hydrogen atom as key to the periodic table

The third reason. In applying wave mechanics to the hydrogen atom we will be able to identify the allowed stationary states of this atom and also the quantum numbers used to label these states. It turns out that we can use these same hydrogen atom quantum numbers to label the allowed states of a single electron in *any* atom (tungsten, for example), no matter how many electrons this atom contains. This fact is extremely useful and permits us, for example, to account completely for the structure of the periodic table of the elements on wave mechanical principles. The ability to do this is perhaps the greatest triumph of wave mechanics and a full understanding of the hydrogen atom is its first, essential step.

An admonition

In all that follows we would naturally prefer to apply wave mechanics in a rigorous way to the hydrogen atom, showing in full detail exactly how the various predictions of this theory emerge. Such a full-blown mathematical treatment however, is beyond our present scope*. We have chosen instead to present certain of the predictions of wave mechanics without rigorous proof, relying on their consistency and especially on their agreement with experiment to convince the reader of their validity. Our aim is to develop a sense of the nature and the power of wave mechanics, omitting the mathematical details. It remains true, of course, that a deeper understanding of the power and beauty of this theory is not possible without a full mathematical treatment.

44–2 The Hydrogen Atom Quantum Numbers

In wave mechanics the problem of the motion of an electron in a particular circumstance—trapped between rigid walls, for example—is determined by giving the potential energy function $U(\mathbf{r})$ for the region in which the electron moves. Here $\mathbf{r}$ is the position vector of the electron. For the hydrogen atom—our basic one-electron atomic system—the potential energy function is given by Eq. 26–13, or

The hydrogen atom potential energy function

$$U(r) = -\frac{e^2}{4\pi\epsilon_0}\frac{1}{r}, \qquad (44-1)$$

which is plotted in Fig. 44–1a. Although the hydrogen atom is three-dimensional, its potential energy depends only on a single variable, the radial distance r of the electron from the atomic center (nucleus).

The program of wave mechanics is to start from the expression for $U(r)$ and to derive the wave functions that describe the various stationary states in which the

---

* The well-motivated reader will find a detailed treatment in Chapter 7 of *Quantum Physics,* by Robert Eisberg and Robert Resnick, John Wiley and Sons, Inc., New York, 1974.

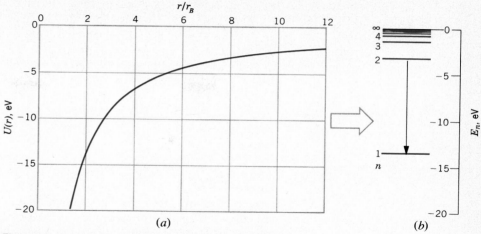

Figure 44-1   (a) The potential energy $U(r)$ of the hydrogen atom. The radial distance is expressed in terms of the Bohr radius r_B defined on p. 869. (b) The total energies E_n of the allowed states of the hydrogen atom. The vertical arrow shows a transition from the first excited state ($n = 2$) to the ground state ($n = 1$).

atom can exist. It is beyond our scope here to give the mathematical details and the physical arguments involved in the process of identifying the allowed stationary states and finding their wave functions.* Suffice it to say that in doing so the energies E_n corresponding to these allowed states emerge. For the hydrogen atom they are

Energies of the allowed hydrogen atom states

$$E_n = -\frac{me^4}{8\epsilon_0^2 h^2}\frac{1}{n^2} = -(13.6 \text{ eV})\frac{1}{n^2}, \; n = 1, 2, \ldots . \quad (44\text{-}2)$$

These allowed energies are plotted in Fig. 44-1b. A glance at Eq. 42-28 reminds us that Bohr theory also predicts this particular result.

In applying the machinery of wave mechanics to the hydrogen atom not only do the wave functions of the allowed states and the energies corresponding to them emerge. Quantum numbers, which identify the stationary state wave functions and are associated with various physical properties of the atom, also emerge.

In the (one-dimensional) problem of an electron trapped between rigid walls there was only one quantum number; see Eq. 43-12. For the (three-dimensional) problem of the hydrogen atom, however, there turn out to be three quantum numbers for each stationary state. This corresponds exactly to the three independent degrees of freedom available to the electron in such a problem. In addition to what we now call the *principal quantum number n*, there are also the *orbital quantum number l* and the *orbital magnetic quantum number m_l*. As Eq. 44-2 shows, n—and n alone—determines the energies of the allowed hydrogen atom states. We shall see in Section 44-4 what role l and m_l play in describing these states. Again, a detailed development of the process by which the quantum numbers emerge from the machinery of wave mechanics is beyond our scope here; see the reference on p. 882 for a full treatment.

The hydrogen atom quantum numbers

---

* See the footnote reference on p. 882.

Table 44–1 The hydrogen atom quantum numbers[a]

Name of quantum number	Symbol	Constant of the motion	Allowed values	Number of allowed values
Principal	n	Energy[b]	1, 2, 3, . . .	No limit
Orbital	l	Magnitude of the orbital angular momentum[c]	0, 1, 2, . . . , $(n-1)$	n
Orbital magnetic	m_l	z-Component of the orbital angular momentum[d]	$-l$, $-(l-1)$, . . . , 0, . . . , $+(l-1)$, $+l$	$2l + 1$

[a] The implications of electron spin are not included in this table. See Table 44–4, in which electron spin is taken fully into account. [b] See Eq. 44–2. [c] See Eq. 44–7. [d] See Eq. 44–8.

Relations among the quantum numbers

As the hydrogen atom quantum numbers are generated, relations among them and limitations on the values that they may have also appear. Table 44–1 summarizes them. For example, for $n = 3$, the quantum number l can have only the values 0, 1, and 2; the number of these allowed values (three) is just equal to n. For $l = 2$, the quantum number m_l can have the values -2, -1, 0, $+1$, and $+2$; the number of these allowed values (five) is just equal to $2l + 1$.

The importance of the quantum numbers

At this point we have asked you to take on faith alone both the existence of the three hydrogen atom quantum numbers and the interrelations among them displayed in Table 44–1. We have given neither a theoretical nor an experimental justification for the material we have presented. The motivated student, however, will find a full theoretical justification in the reference found on p. 882. As for an experimental verification, we shall describe, in Section 45–3, how the hydrogen atom quantum numbers and the rules of Table 44–1 that interconnect them play a direct and central role in the building up of the periodic table of the elements according to wave-mechanical principles. This achievement is one of the great triumphs of wave mechanics.

Classifying the hydrogen atom states

As a step toward examining the usefulness of the hydrogen atom quantum numbers, let us see how they are used to classify the various allowed stationary states of the hydrogen atom. Our first classification rule is that all states with the same principal quantum number n are said to form a *shell*. The various shells are identified, for historical reasons, by letter symbols according to the scheme

Shell notation

n	1	2	3	4	· · ·
Shell symbol	K	L	M	N	· · ·

All states that have the same values not only of n but also of l are said to form a *subshell*. Again for historical reasons, the value of l is identified by a letter symbol according to the scheme

Subshell notation

l	0	1	2	3	4	5	· · ·
Subshell symbol	s	p	d	f	g	h	· · ·

We can speak, for example, of a $3d$ state; it has the quantum numbers $n = 3$ and $l = 2$. A state such as $2f$ cannot exist; it would have quantum numbers $n = 2$ and $l = 3$, which violates the rule of Table 44–1 that the highest allowed value of l is $n - 1$, which is one in this case.

Example 1 *Counting states* Because the three hydrogen atom quantum numbers specify the various allowed hydrogen atom states, we can represent the wave functions corresponding to these states by the notation ψ_{nlm_l}. How many hydrogen atom states are there with $n = 3$? Put another way, how many allowed states are there in the M shell of an atom?

As we have seen, for $n = 3$, there are three allowed values of l, namely, 0, 1, and 2. For $l = 0$, Table 44–1 tells us that m_l can have only the value zero. For $l = 1$, we can have $m_l = -1$, 0, and +1. For $l = 2$, as already noted, we have $m_l = -2, -1, 0, +1$, and +2. This adds up to nine states, whose wave functions may be classified into subshells as follows:

Subshell	Wave function				
3s			ψ_{300}		
3p		$\psi_{31,-1}$	ψ_{310}	$\psi_{31,+1}$	
3d	$\psi_{32,-2}$	$\psi_{32,-1}$	ψ_{320}	$\psi_{32,+1}$	$\psi_{32,+2}$

Note that all of these states have the same energy because they all have the same principal quantum number, $n = 3$.

The spin magnetic quantum number m_s—a preview

At this point we note, by way of anticipation, that there are really four quantum numbers—not three—assigned to each state of the hydrogen atom. The fourth, or *spin magnetic quantum number*, has the symbol m_s.

This new quantum number arises because, although the electron seems to be a truly point particle, with no measurable lateral dimensions, it nevertheless seems to have a fourth degree of freedom, which we can choose to visualize classically as a motion of rotation, or *spin*, about an axis. This fourth variable of the electron's motion calls for a fourth quantum number to specify it.

As we shall see later, m_s can have only two values, $+\frac{1}{2}$ and $-\frac{1}{2}$. Its effect is to exactly double the number of states specified by the three quantum numbers n, l, and m_l. Thus there are really 18 states—and not nine—in Example 1. The numbers of states in the subshells 3s, 3p, and 3d in that example are really 2, 6, and 10, and not 1, 3, and 5.

We shall introduce the spin magnetic quantum number formally in Section 44–6, when we have established an experimental basis for it. Meanwhile we shall treat only properties that depend only on our three original quantum numbers, n, l, and m_l. Table 44–2 summarizes the allowed states of the hydrogen atom through $n = 3$.

Table 44–2 Hydrogen atom states through $n = 3$

n	1	2		3		
l	0	0	1	0	1	2
m_l	0	0	−1, 0, +1	0	−1, 0, +1	−2, −1, 0, +1, +2
Subshells Labels	1s	2s	2p	3s	3p	3d
Numbers of states[a] ($= 2l + 1$)	1	1	3	1	3	5
Shells Labels	K	L		M		
Numbers of states[a] ($= n^2$)	1	4		9		

[a] Our later introduction of the spin magnetic quantum number m_s will serve to double the numbers of states in these two rows; see Table 44–4.

44–3 The Hydrogen Atom Wave Functions

The $n = 1$ **State: A Review.** For $n = 1$ we must have $l = 0$, a $1s$ state. As Eq. 44–2 shows, $n = 1$ identifies the ground state of the hydrogen atom, that is, the state with the lowest energy. We discussed this state in some detail in the preceding chapter. Here we summarize the results and introduce the appropriate quantum number notation.

The ground state *wave function* for the hydrogen atom is

The $1s$ wave function

$$\psi_{1s}(r) = \frac{1}{\sqrt{\pi r_B{}^3}}\, e^{-r/r_B}, \qquad [43\text{–}14a]$$

in which r_B (= 0.529 Å) is the Bohr radius, our convenient measure of distance on the atomic scale. The notation ψ_{1s} is interchangeable with the notation ψ_{100}, in which the three quantum numbers are specifically displayed.

The square of the wave function, called the *probability density function*, has the property that $\psi^2(\mathbf{r})\, dV$ gives the probability of finding the electron in a volume element dV located at a position defined by the position vector $\mathbf{r}$. For the hydrogen atom in its ground state, we have

The $1s$ probability density function

$$\psi_{1s}{}^2(r) = \left(\frac{1}{\pi r_B{}^3}\right) e^{-2r/r_B}. \qquad [43\text{–}14b]$$

The distribution of dots in Fig. 43–9a suggests this quantity.

The *radial probability density function* $P(r)$ has the property that $P(r)\, dr$ gives the probability of finding the electron between two concentric spherical shells of radii r and $r + dr$. Thus we have (see p. 869)

$$P(r)\, dr = \psi^2(r)(4\pi r^2)\, dr, \qquad (44\text{–}3)$$

and we can write, for the hydrogen atom in its ground state,

The $1s$ radial probability density function

$$P_{1s}(r) = \left(\frac{4r^2}{r_B{}^3}\right) e^{-2r/r_B}. \qquad [43\text{–}14c]$$

Figure 43–9b is a plot of this quantity.

The $n = 2$ **States.** Following the pattern of Example 1, we can count four wave functions with $n = 2$. Thus,

Subshell	Wave function		
$2s$		ψ_{200}	
$2p$	$\psi_{21,-1}$	ψ_{210}	$\psi_{21,+1}$

All of these states correspond to an energy E_2 given by Eq. 44–2, or

$$E_2 = \frac{1}{2^2}\, E_1 = \tfrac{1}{4}(-13.6\text{ eV}) = -3.40\text{ eV}.$$

We can write ψ_{200} equally well as ψ_{2s}. Figure 44–2a suggests the behavior of $\psi_{2s}{}^2(r)$, the probability density function for the $2s$ state. Note that, like $\psi_{1s}{}^2(r)$ of Fig. 43–9a, it is spherically symmetrical. All s-states have this spherical symmetry.

Figure 44–2b shows the corresponding radial probability density function, which can be shown from wave mechanics to be

The $2s$ radial probability density function

$$P_{2s}(r) = \left(\frac{r^2}{8r_B{}^3}\right)\left(2 - \frac{r}{r_B}\right)^2 e^{-r/r_B}. \qquad (44\text{–}4)$$

Figure 44–2   Two representations of the hydrogen atom in a 2s state. (a) The density of dots suggests $\psi_{2s}^2(r)$. The barred horizontal line shows the scale. (b) The radial probability density function $P_{2s}(r)$.

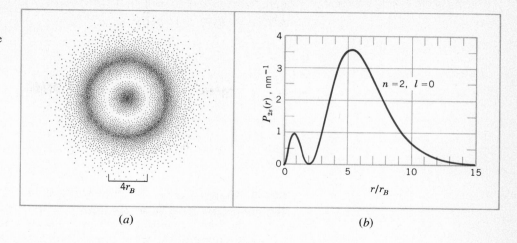

(a)　　　　　　　　　　　　　(b)

Inspection of this equation and of Fig. 44–2b shows that $P_{2s}(r)$ goes to zero at $r = 2r_B$.

The three 2p wave functions displayed in the above table are not spherically symmetrical; Fig. 44–3a, which shows the probability density functions for these states, makes this clear. We see that ψ^2 in these cases depends not only on r but also on the angular variable θ shown in that figure. All three distributions have rotational symmetry about the z-axis. Note that the patterns for ψ^2 with $m_l = -1$ and with $m_l = +1$ are identical, even though the wave functions themselves are different.

Although $\psi_{21,-1}^2(r, \theta)$, $\psi_{210}^2(r, \theta)$, and $\psi_{21,+1}^2(r, \theta)$ in Fig. 44–3a are certainly not spherically symmetrical, it turns out that if we add them together we obtain a quantity that *is* spherically symmetrical; all terms containing the variable θ add to zero, leaving only terms in the variable r. If each added quantity is suitably weighted by $\frac{1}{3}$, we identify the quantity so obtained as $\psi_{2p}^2(r)$, the radial probability density function for the 2p state. That is,

$$\psi_{2p}^2(r) = \tfrac{1}{3}\psi_{21,-1}^2(r, \theta) + \tfrac{1}{3}\psi_{210}^2(r, \theta) + \tfrac{1}{3}\psi_{21,+1}^2(r, \theta). \qquad (44\text{–}5)$$

The radial probability density function (see Eq. 44–3) turns out to be

The 2p radial probability density function

$$P_{2p}(r) = \left(\frac{r^4}{24r_B{}^5}\right) e^{-r/r_B}. \qquad (44\text{–}6)$$

Figure 44–3   Two representations of the hydrogen atom in a 2p state. (a) The density of dots suggests the probability density functions $\psi^2(r, \theta)$ for the three states whose quantum numbers are shown. (b) The radial probability density function $P_{2p}(r)$.

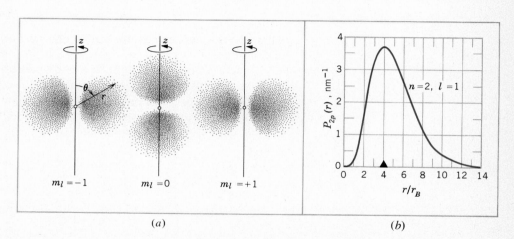

(a)　　　　　　　　　　　　　(b)

Figure 44–3b is a plot of this quantity.

This is a good place at which to pause and consider how very different are the descriptions of the hydrogen atom given by Bohr theory (Section 42–8) and by the theory of wave mechanics that superseded it. In Figs. 43–9 and 44–3, which reflect the wave mechanical viewpoint, there is no sign of orbits. Although both theories predict the same expression for the energy of the stationary states of the hydrogen atom (Eq. 44–2), they agree on little else. Where they differ, wave mechanics always proves to be correct.

Finally, the probability density patterns of Fig. 44–3a begin to suggest how atoms can form the directed bonds essential to the formation of molecules.

44–4 Angular Momentum and Magnetism

Bohr's theory of the hydrogen atom tells us that the stationary states in which the atom can exist have not only a quantized energy but also a quantized angular momentum. The Bohr orbits are tiny current loops, so there is also a quantized orbital magnetic moment associated with each state. In this section we shall learn how to deal with these two important atomic properties by the methods of wave mechanics.

The importance of angular momentum and magnetism

Angular momentum and magnetic moment are indeed important atomic properties. For example, when atoms interact with each other or when they absorb or emit radiation, angular momentum must be conserved. As for magnetism, the basis for the ferromagnetic, paramagnetic, or diamagnetic behavior of bulk matter (see Chapter 35) is found in the magnetism of the atomic electrons.

From wave mechanics we learn that the hydrogen atom states have an associated orbital angular momentum **L** whose magnitude L is given by

*The orbital angular momentum **L***

$$L = \sqrt{l(l + 1)}\, \hbar \qquad \text{(wave mechanics)}, \qquad (44–7)$$

where l is the orbital quantum number. We see now, incidentally, how this quantum number got its name; see Table 44–1. For convenience we have introduced the symbol $\hbar$ (pronounced "h-bar") as a shorthand notation for $h/2\pi$. Just as the quantum number n determines the energy of the hydrogen atom, the quantum number l determines the magnitude of its orbital angular momentum. Note that $L = 0$ is a possibility, corresponding to $l = 0$.

We note incidentally that the expression given by Bohr theory for the angular momentum of the hydrogen atom, namely,

Orbital angular momentum in Bohr theory

$$L = n\hbar \qquad \text{(Bohr theory)}, \qquad [42–26]$$

is not correct. For one thing, it predicts that the hydrogen atom would have one unit of angular momentum ($= \hbar$) in its ground state ($n = 1$). On the other hand, wave mechanics predicts that the angular momentum in the ground state ($n = 1$, $l = 0$) is zero, and experiment supports this prediction. Zero angular momentum on the Bohr picture would correspond to an electron oscillating back and forth on a line segment passing directly through the atomic nucleus.

Let us now ask what may seem to be a fairly straightforward question:

"What projections L_z along any specified direction in space, which we take to be the z-axis, can the orbital angular momentum vector **L** have?"

For concreteness we may imagine the z-axis to specify the direction of an imposed external magnetic field. We can choose this field to be very weak so that, while it clearly defines a direction in space, it does not measurably affect the behavior of atoms that are immersed in it.

The answer to our question would seem to be

"L_z can have any value between $+L$ and $-L$, depending on the angle θ that the orbital angular momentum vector makes with the z-axis."

This is certainly true in classical mechanics, but it is *not* true in wave mechanics. The answer given by wave mechanics is

"L_z can have only the discrete set of values given by

The allowed projections of **L**.

$$L_z = m_l \hbar, \tag{44–8}$$

where m_l is the orbital magnetic quantum number."

Here we begin to see where the orbital magnetic quantum number got its name; see Table 44–1. The maximum value of m_l is just l, so that the *maximum* allowed projection of **L** is just

The maximum value of L_z

$$L_{z,\text{max}} = l\hbar. \tag{44–9}$$

It turns out that it is L_z, and not the vector **L**, that is directly measurable experimentally and is thus, between the two, the quantity of practical interest. When we speak of the orbital angular momentum of an atomic system, we essentially always mean the maximum value of L_z, which, as Eq. 44–9 shows, is just $l\hbar$.

In spite of the failure of Bohr theory to account properly for the angular momentum of the hydrogen atom, it can help us greatly in understanding the relationship between the angular momentum **L** of an atom and the orbital magnetic dipole moment $\boldsymbol{\mu}$ associated with it. We used a Bohr orbit model in Example 2 of Chapter 34 to show that

Angular momentum and magnetism

$$\boldsymbol{\mu} = -\left(\frac{e}{2m}\right)\mathbf{L}. \tag{44–10}$$

The minus sign means that $\boldsymbol{\mu}$ and **L** point in opposite directions and comes about because the electron carries a negative charge.

Equation 44–10 is also valid in wave mechanics. If we restrict ourselves to the z-components of each quantity—which are the only quantities that we can measure—we find from Eqs. 44–10 and 44–8 that

The allowed projections of $\boldsymbol{\mu}$

$$\mu_z = -\left(\frac{e}{2m}\right)L_z = -\left(\frac{e}{2m}\right)(m_l\hbar) = -\left(\frac{e\hbar}{2m}\right)m_l$$
$$= -m_l\mu_B. \tag{44–11}$$

The quantity $e\hbar/2m$ above, which has the dimensions of a magnetic moment, is called the *Bohr magneton* and is represented by the symbol μ_B. Thus,

The Bohr magneton

$$\mu_B = \frac{e\hbar}{2m} = 5.79 \times 10^{-5} \text{ eV/T}. \tag{44–12}$$

The Bohr magneton, which turns out to be equal to the orbital magnetic moment of the electron circulating in the first Bohr orbit (see Problem 24) is a convenient unit for atomic magnetism.

We can visualize the relationships between the orbital angular momentum **L** and its projections L_z and between the orbital magnetic moment vector $\boldsymbol{\mu}$ and *its* projections μ_z, by constructing a vector model of the atom, as in Fig. 44–4.

A vector model of the atom

Space quantization

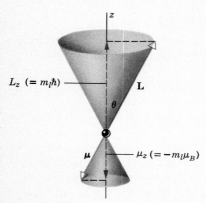

$L_z \ (= m_l \hbar)$ — **L**

θ

μ — $\mu_z \ (= -m_l \mu_B)$

Figure 44–4 A vector model of the atom, showing the orbital angular momentum vector **L** and the orbital magnetic moment vector μ, together with their projections L_z and μ_z.

Angular momentum and the uncertainty principle

Here the angular momentum vector **L** is confined to lie somewhere along the surface of a cone making an angle θ with the positive z-direction. The magnetic moment vector μ, which always points opposite to **L**, similarly lies somewhere on a cone, making the same angle θ with the negative z-direction. No matter where on these surfaces **L** and μ lie, their projections L_z and μ_z have opposite signs, as comparison of Eqs. 44–9 and 44–11 shows that they should.

The angle θ between the angular momentum vector and the z-axis is quantized. From the figure we see that its allowed values are given by

$$\theta = \cos^{-1} \frac{m_l}{\sqrt{l(l+1)}}. \tag{44-13}$$

The maximum value of m_l is just l, so that we see that θ can never be zero, as it can classically. The fact that θ has a minimum value corresponds to the fact that L_z is always less than L; that is, **L** can never lie exactly in the z-direction. The fact that θ is quantized is called *space quantization*.

The uncertainty principle throws considerable light on the concept of the quantization of angular momentum and on the related phenomenon of space quantization. In its angular form (compare Eq. 43–21), this principle is

$$\Delta L_z \, \Delta\phi \gtrsim h \qquad (z\text{-component}), \tag{44-14}$$

in which ϕ is the angle of rotation about the z-axis in, say, Fig. 44–4. Equation 44–8 ($L_z = m_l \hbar$) tells us that L_z is known precisely once the orbital magnetic quantum number m_l has been specified. It follows that ΔL_z, the uncertainty in L_z, must be zero, so that we must have $\Delta\phi \to \infty$ in Eq. 44–14. In other words, we have no information at all about the angular position about the z-axis of the precessing angular momentum vector **L** in Fig. 44–4. We know the magnitude of **L** (Eq. 44–7) and the angle θ that **L** makes with the z-axis (Eq. 44–13), and nothing else. Put another way, wave mechanics permits us to specify exactly the magnitude of **L** and of L_z but not of L_x and L_y. In this sense the vector model of the atom, which specifies a well-defined vector **L** precessing about the z-axis, must be viewed as a very useful but necessarily semiclassical construction.

Example 2 *Space quantization and the correspondence principle* Calculate $\theta_{\min}$ in Fig. 44–4 for $l = 1, 100, 1000,$ and $10,000$. Verify that $\theta_{\min} \to 0$ as $l \to \infty$.

The minimum value of θ comes when we put $m_l = l$ in Eq. 44–13. Doing so and rearranging leads to

$$\theta_{\min} = \cos^{-1} \frac{l}{\sqrt{l(l+1)}} = \cos^{-1} \left(1 + \frac{1}{l}\right)^{-1/2}.$$

We see clearly that as $l \to \infty$, $\theta_{\min} \to \cos^{-1} 1 = 0$. This is just what we expect from the correspondence principle.

Substitution above yields:

l	$\theta_{\min}$
1	45°
100	5.7°
1,000	1.8°
10,000	0.57°

For macroscopic spinning object such as a top, l would be very much larger indeed than 10,000 and $\theta_{\min}$ would be indistinguishable from zero beyond any hope of measurement. We see once again that the correspondence principle really works.

Example 3 Calculate, for a hydrogen atom in a state with $l = 2$, the allowed values of L_z, μ_z, and θ.

We find these values readily from Eqs. 44–9, 44–11, and 44–13, respectively. Note that these quantities depend only on l and are independent of the principal quantum number n:

m_l	+2	+1	0	−1	−2
L_z	$+2\hbar$	$+\hbar$	0	$-\hbar$	$-2\hbar$
μ_z	$-2\mu_B$	$-\mu_B$	0	$+\mu_B$	$+2\mu_B$
θ	35°	66°	90°	114°	145°

We see that, for the same value of m_l, the projections L_z and μ_z have opposite signs.

44–5   The Stern-Gerlach Experiment

Space quantization, that is, the fact that an atomic angular momentum vector **L** or an atomic magnetic moment vector $\boldsymbol{\mu}$ can take up only a finite number of discrete angular positions, was predicted (by Wolfgang Pauli) even before the development of wave mechanics. It is not an easy concept for the classically oriented mind to accept. Nevertheless, in 1922, Otto Stern and Walther Gerlach provided solid experimental evidence that this phenomenon really exists.

Figure 44–5 shows their apparatus. Silver is vaporized in an electrically heated "oven." A beam of neutral silver atoms emerges from a small hole in the oven wall, is collimated by a narrow slit, passes between the poles of an electromagnet, and finally deposits itself on a glass detector plate. The pole faces of the electromagnet are shaped so that they produce an extremely inhomogeneous magnetic field in the region close to the sharp V-shaped edge of the upper pole face.

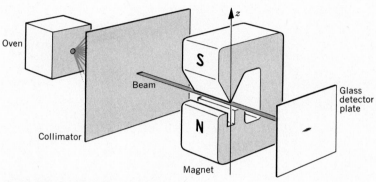

Figure 44–5   The apparatus of Stern and Gerlach with the magnet turned off, showing a beam of undeflected silver atoms. The beam was about 0.3 mm below the V-shaped ridge, and the path through the magnet was 3.5 cm long.

A magnetic dipole in a magnetic field

A neutral silver atom has an angular momentum **L** and also an associated magnetic moment $\boldsymbol{\mu}$. If we put such an atom in a magnetic field, the vector $\boldsymbol{\mu}$ will orient itself at one of a finite number of quantized angles θ to the magnetic field direction, as in Fig. 44–4. It will maintain a constant projection μ_z, which may be positive or negative, along this direction.

Precession of $\boldsymbol{\mu}$ about **B**

The magnetic moment vector $\boldsymbol{\mu}$ will also precess about the field direction, sweeping out a cone of half-angle θ. Physically, this precession comes about because the magnetic field exerts a torque $\boldsymbol{\tau}$ given by $\boldsymbol{\mu} \times \mathbf{B}$ (see Eq. 29–11) on the atom, tending to make $\boldsymbol{\mu}$ line up with **B**. It is because the atom has angular momentum that precession occurs, instead of simple oscillation about the field direction as would be the case if a simple compass needle were placed in the field.

In the same way, a torque exerted by the earth's gravitational field on a spinning top causes the axis of the top to precess about the vertical. If the top were not spinning, it would simply fall over, in an attempt to oscillate—pendulumlike—about the (downward) direction of the earth's gravitational field.

Precession of a spinning top about **g**

If the magnetic field is uniform the precession about **B**, mentioned above, is the only effect of the magnetic field on the magnetic dipole immersed in it. In particular, no net force of magnetic origin acts on the dipole.

Force on a dipole in an
inhomogeneous field—qualitative

If the magnetic field is nonuniform, however, there *is* a net magnetic force. Figure 44–6, which shows a small bar magnet placed in an inhomogeneous magnetic field, illustrates this qualitatively. Such a magnet has an effective north pole N and an effective south pole S. The magnetic moment vector μ, by convention, points from the latter to the former as indicated. The angle θ between μ and $\mathbf{B}$, the magnetic field at the site of the magnet, is also shown. For the orientation assumed in the figure, it is clear that the north pole is immersed in a stronger magnetic field than is the south pole, so that a net magnetic force, given by $\mathbf{F}_N - \mathbf{F}_S$, acts on the magnet, in the positive z-direction.

Figure 44–6   A small bar magnet placed in an inhomogeneous magnetic field. For the orientation shown there is a net upward force.

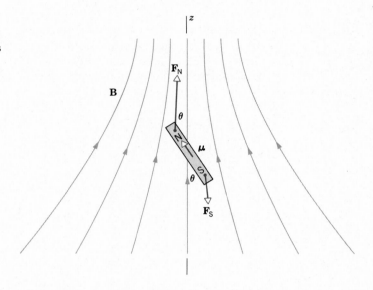

From energy considerations we can derive an expression for the force on a magnetic dipole placed in an inhomogeneous magnetic field. The magnetic potential energy of an atom of moment μ placed in a magnetic field $\mathbf{B}$ is given by Eq. 29–12, or

The potential energy of a magnetic
dipole

$$U = -\mu \cdot \mathbf{B} = -(\mu \cos \theta) B.$$

In our mind's eye let us follow a silver atom in the beam as it traverses the apparatus of Fig. 44–5 parallel to the knife edge; see also Fig. 44–6. We see that the magnetic field through which such atoms move has no x- or y-component. Thus $B = B_z$ and, from the above equation,

$$U = -\mu_z B_z. \tag{44–15}$$

From Eq. 8–7, the net force on the atom is

$$F_z = -\left(\frac{\partial U}{\partial z}\right)$$

or, from Eq. 44–15,

Force on a dipole in an
inhomogeneous field—quantitative

$$F_z = \mu_z \left(\frac{\partial B_z}{\partial z}\right). \tag{44–16}$$

Note that the deflecting force is determined by the field gradient $\partial B_z/\partial z$ and does not depend at all on the magnitude of the field itself.

In applying Eq. 44–16 to the bar magnet of Fig. 44–6, note that μ_z is directed along the positive direction of the z-axis and is therefore itself positive. The mag-

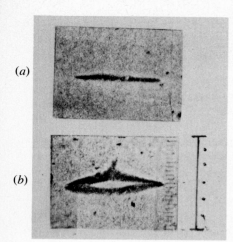

(a)

(b)

Figure 44-7 Silver deposits on the glass detector plate of the apparatus of Fig. 44-5 with the magnet (a) off and (b) on. For (b) the duration of a particular run in the original experiment was 8 h. The vertical line to the right represents 1.0 mm.

netic field gradient $\partial B_z/\partial z$ is also positive, because B_z increases as z increases. Therefore, F_z in Eq. 44-16 is positive—that is, the net magnetic force points upward in Fig. 44-6—as we know it must.

If the magnet of Fig. 44-5 is turned off, the beam will simply form an image of the defining slit on the detector plate. If space quantization exists—and the sole object of the experiment was to find out whether or not it does—then when the magnet is turned on the atoms in the beam will orient their precession angles θ so that μ_z has one or another of the discrete set of values allowed by the rules of space quantization. The subbeams so formed will undergo different vertical deflections in passing through the magnet and will become physically separated in space. They will form a discrete set of images on the detector plate.

This is exactly what happened! Figure 44-7 shows the result of the experiment (a) with the magnet turned off and (b) with the magnet turned on. When the magnet is turned on, the atomic beam splits cleanly into two separate beams. If space quantization did not exist, the silver atoms would have taken up random, continuously distributed orientations with respect to the magnetic field and the beam would simply have been broadened. Stern and Gerlach concluded the published report of their experiment by stating: "We view these results as direct experimental verifications of space quantization in a magnetic field." Physicists everywhere agree.

Example 4 In an experiment of the Stern-Gerlach type, the magnetic field gradient $\partial B_z/\partial z$ was 1.4 kT/m and the length h of the beam path through the magnet was 3.5 cm. The temperature of the small "oven" in which the silver was vaporized was adjusted so that the calculated value for the most probable speed was 750 m/s. The mass M of a silver atom is 1.8×10^{-25} kg, and its magnetic moment is 1.0 Bohr magneton (= 9.3×10^{-24} J/T). Find the total separation d between the two deflected beams, as in Fig. 44-7b.

The acceleration is given by (see Eq. 44-16)

$$a = \frac{F}{m} = \frac{\mu_z(\partial B_z/\partial z)}{M}.$$

The vertical deflection of one of the beams as it passes through the magnet is

$$\frac{d}{2} = \tfrac{1}{2}at^2 = \tfrac{1}{2}\left[\frac{\mu_z(\partial B_z/\partial z)}{M}\right]\left(\frac{h}{v}\right)^2,$$

so that

$$d = \frac{\mu_z(\partial B_z/\partial z)h^2}{Mv^2}$$

$$= \frac{(9.3 \times 10^{-24}\ \text{J/T})(1.4 \times 10^3\ \text{T/m})(3.5 \times 10^{-2}\ \text{m})^2}{(1.8 \times 10^{-25}\ \text{kg})(750\ \text{m/s})^2}$$

$$= 1.6 \times 10^{-4}\ \text{m} = 0.16\ \text{mm}.$$

This is the order of magnitude of the separation shown in Fig. 44-7b; note the scale to the right of the figure.

44-6 The Spinning Electron

A problem with the Stern-Gerlach experiment

A curious feature of the Stern-Gerlach experiment is that there are just two subbeams, corresponding to just two orientations of the silver atoms. We have seen that for a given orbital angular momentum l there are $2l + 1$ space-quantized orientations in a magnetic field, corresponding to the $2l + 1$ values of m_l. If we put $2l + 1 = 2$, we have $l = \tfrac{1}{2}$, which is not permitted because l must be an integer. Although the Stern-Gerlach experiment clearly demonstrates space quantization, it cannot account for it in terms of *orbital* angular momentum and magnetism.

There were yet other failures of agreement between theory and experiment occurring at the time of the Stern-Gerlach experiment. The spectral lines of hydrogen, for example, were shown not to be single, as we have tacitly assumed, but to be made up of a number of close components, the so-called *fine structure*. It

A fourth quantum number is needed

eventually became clear to the perceptive physicist Wolfgang Pauli that a fourth quantum number, in addition to n, l, and m_l, must be assigned to the electron to explain this and other spectral puzzles. The physical nature of this additional quantum number remained unclear, however.

In 1925, three years after the Stern-Gerlach experiment was performed, two young Dutch graduate students (George Uhlenbeck and Samuel Goudsmit), two years away from their Ph.D. degrees at the University of Leiden, proposed that the electron, in addition to its orbital motion, also has an intrinsic spin motion.* This "spin" has both an angular momentum and a magnetic moment associated with it and provides the basis for the anticipated fourth quantum number.

If the electron were a true geometrical point particle, the three degrees of freedom that we have assigned to it so far should suffice. We have seen that we may identify these with the energy of the state (specified by n), the angular momentum of the state (specified by l), and the orientation of the angular momentum vector in space (specified by m_l). Now we see that the electron cannot be treated so simply. Although it does indeed seem to have no measurable dimension, the electron behaves—to put it in terms of a classical picture—as if it had an axis about which it spins. Our fourth degree of freedom—and the fourth quantum number that we associate with it—must serve to define the direction of this spin axis in space.

Consider first the spin angular momentum vector, which we shall call **S**. By analogy with its orbital counterpart, we can write its quantized projections on the z-axis as

The spinning electron

The allowed projections of **S**

$$S_z = m_s \hbar. \tag{44-17}$$

We introduce here the *spin magnetic quantum number m_s*, which has only two allowed values, namely,

The spin magnetic quantum number

$$m_s = \pm \tfrac{1}{2}. \tag{44-18}$$

Table 44–3, an extension of Table 44–1 to include the implications of electron spin, summarizes the hydrogen atom quantum numbers.

Table 44–3   The hydrogen atom quantum numbers[a]

Name of quantum number	Symbol	Constant of the motion	Allowed values	Number of allowed values
Principal	n	Energy	1, 2, 3, . . .	No limit
Orbital	l	Magnitude of the orbital angular momentum	0, 1, 2, . . . , $(n - 1)$	n
Orbital magnetic	m_l	z-Component of the orbital angular momentum	$-l, -(l - 1), . . . , 0, . . . , +(l - 1), +l$	$2l + 1$
Spin[b]	s	Magnitude of the spin angular momentum	$\tfrac{1}{2}$	One
Spin magnetic	m_s	z-Component of the spin angular momentum	$\pm \tfrac{1}{2}$	Two

[a] Compare with Table 44–1, in which the implications of spin are not taken into account. [b] Because s has only one value, it is not really a quantum number. There are only four quantum numbers and not five, as this table suggests.

---

* See "Personal Reminiscences" by George E. Uhlenbeck and "It Might as Well be Spin" by Samuel A. Goudsmit, both in *Physics Today*, June 1976, for lively accounts of the subject on the fiftieth anniversary of the discovery of electron spin.

Now we turn our attention to the spin magnetic moment vector, which we call μ_s. To make sure that we do not confuse this with the orbital magnetic moment vector, which we have so far called simply μ, we shall from now on label the latter as μ_l. A host of experimental data, from atomic spectroscopy and from experiments of the Stern-Gerlach type, requires us to realize that the quantized projections of μ_s on the z-axis are given by

The allowed projections of μ_s

$$\mu_{s,z} = -2m_s\mu_B. \tag{44-19}$$

Compare this with the quantized projections of the orbital magnetic moment μ_l on the z-axis, which we rewrite, for clarity, as

$$\mu_{l,z} = -m_l\mu_B. \tag{[44-11]}$$

The factor 2

Notice that there is a difference of a factor of 2. This tells us that *spin angular momentum is twice as effective as orbital angular momentum in generating magnetism*. This factor of 2 was, in fact, conjectured as necessary by Uhlenbeck and Goudsmit when they first introduced the concept of electron spin.

The theory of the spinning electron

Schrödinger's wave mechanics says nothing at all about the electron spin or about the spin magnetic quantum number m_s. This is a result of the fact that Schrödinger's treatment of wave mechanics did not take into account the theory of relativity. When the English mathematical physicist P. A. M. Dirac developed a relativistic treatment of the electron the spin concept, and the factor 2, emerged in a natural way. Extremely precise measurements have shown that, as of 1980, the quantity "2" is actually 2.0023193044, with an uncertainty of one unit in the last digit. The latest refinements of theory predict a value that agrees with this within the precision of measurement. We may fairly say that the spinning electron is well understood, on the basis of both experiment and theory.

Doubling the number of states—a consequence of spin

We have already mentioned one consequence of the introduction of the spin magnetic quantum number m_s; the number of hydrogen atom states listed in Table 44-2 is doubled on its account. If this doubling did not occur, we would be totally unable to account for the structure of the periodic table in terms of wave mechanics, one of the principal triumphs of this theory.

For every n, l, m_l combination in Table 44-2, there are now two states, corresponding to the two possible values of m_s. Instead of the three $2p$ states shown in Fig. 44-3a and Example 1, for example, there are really six, with these quantum numbers:

n	l	m_l	m_s
2	1	-1	$-\frac{1}{2}$
2	1	-1	$+\frac{1}{2}$
2	1	0	$-\frac{1}{2}$
2	1	0	$+\frac{1}{2}$
2	1	+1	$-\frac{1}{2}$
2	1	+1	$+\frac{1}{2}$

Table 44-4 shows the hydrogen atom states, taking spin into account, that make up the various shells and subshells through $n = 4$.

Stern-Gerlach explained

The spinning electron allows us to understand fully the results of the Stern-Gerlach experiment. A silver atom has 47 electrons, 46 of which ($= 2 + 8 + 18 + 18$) form the closed K, L, and M shells and fill the first three subshells of the N shell; see Table 44-4. For these filled shells and subshells the magnetic effects, whether due to electronic orbital motion or to spin, cancel exactly to zero. The magnetic moment of the silver atom is thus due to the 47th electron only, which happens to be in a $4s$ state. According to our labeling scheme the s

Table 44–4 Hydrogen atom states through $n = 3^a$

n	1	2		3		
l	0	0	1	0	1	2
m_l	0	0	$-1, 0, +1$	0	$-1, 0, +1$	$-2, -1, 0, +1, +2$
m_s	$\pm\frac{1}{2}$	$\pm\frac{1}{2}$	$\pm\frac{1}{2}$	$\pm\frac{1}{2}$	$\pm\frac{1}{2}$	$\pm\frac{1}{2}$
Subshells Labels	$1s$	$2s$	$2p$	$3s$	$3p$	$3d$
Numbers of states $= 2(2l + 1)$	2	2	6	2	6	10
Shells Labels	K	L			M	
Number of states $(= 2n^2)$	2	8			18	

^{a} Compare with Table 44–2, in which the implications of electron spin are not taken into account.

tells us that this electron has zero orbital angular momentum. The entire magnetic effect of the silver atom is due to the spin of this "extra" electron. This accounts for the fact that the beam in Fig. 44–7 splits into just two components, corresponding to the alignments defined by $m_s = +\frac{1}{2}$ and $m_s = -\frac{1}{2}$.

44–7   The Total Angular Momentum J

We now see that for the hydrogen atom—and for atoms in general—there are two kinds of angular momentum, one (**L**) associated with the electron's orbital motion and another (**S**) associated with its intrinsic spin motion. It is natural to combine these momenta vectorially to form a total angular momentum **J** given by

Combining **L** and **S**

$$\mathbf{J} = \mathbf{L} + \mathbf{S}. \tag{44–20}$$

Combining the orbital and spin motions in this way is more than a mere formalism. There is a mechanism within the atom for linking these two motions physically. Later in this section we shall look into this *spin-orbit coupling* mechanism more fully, and we shall see that it produces observable effects in atomic spectra.

Just as for **L** and **S**, the z-axis projections of the total angular momentum **J** are quantized and are given by

$$J_z = m_j\hbar. \tag{44–21}$$

Recall the similar expressions for L_z and S_z. The quantum number m_j in Eq. 44–21 takes on the values

$$-j, -(j - 1), \ldots, + (j - 1), +j, \tag{44–22}$$

in which j—for one-electron atoms—takes on the positive half-integral values given by $l \pm \frac{1}{2}$. For every value of j there are $2j + 1$ values of m_j. We see that the maximum value of J_z is given by

$$J_{z,\text{max}} = j\hbar \quad \text{(maximum projected value)}. \tag{44–23}$$

Once again we have presented, without proof, some inter-related quantum numbers that are derivable from wave mechanics. Once again we appeal to their consistency and to their agreement with experiment (see Section 44–8) to convince the reader of their validity.

Let us apply what we have learned so far to the hydrogen atom. No state with $l = 0$ provides opportunity for new insights into this matter, because for such states $L = 0$ so that **J** is identical with **S**.

States with $l = 1$ (p states) have nonvanishing values of both **L** and **S**, so that we may usefully consider them. It does not matter whether they are $2p$, $3p$, $4p$, etc., states; the value of n is immaterial.

For states with $l = 1$, two values of j are possible, namely,

$$j = l + \tfrac{1}{2} = \tfrac{3}{2} \quad \text{and} \quad j = l - \tfrac{1}{2} = \tfrac{1}{2}. \tag{44–24}$$

We identify these two substates by using the j values as subscripts, and we call them the $p_{3/2}$ and the $p_{1/2}$ states. Figures 44–8a and 44–8b show how **L** and **S** are combined in these two cases.

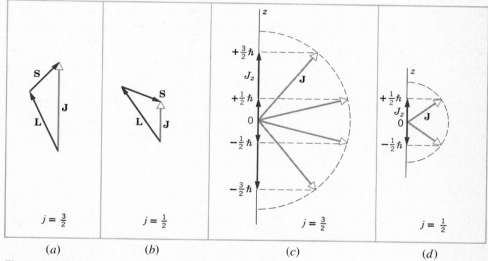

Figure 44–8 (a) **J** for a $p_{3/2}$ state; this is sometimes called the "stretch" mode. (b) **J** for a $p_{1/2}$ state; this is sometimes called the "jack-knife" mode. (c) The four values of J_z for the $p_{3/2}$ state. (d) The two values of J_z for the $p_{1/2}$ state.

Once having found **J**, we deal solely with it and not with **L** and **S** separately. For the $p_{3/2}$ state, **J** has four allowed projections, namely,

$$-\tfrac{3}{2}\hbar, \quad -\tfrac{1}{2}\hbar, \quad +\tfrac{1}{2}\hbar, \quad \text{and} \quad +\tfrac{3}{2}\hbar.$$

Figure 44–8c shows them. For the $p_{1/2}$ state, **J** has two allowed projections, namely,

$$-\tfrac{1}{2}\hbar \quad \text{and} \quad +\tfrac{1}{2}\hbar.$$

They are shown in Fig. 44–8d.

An analysis like the above could be carried out for any state with $l \gtrsim 1$, that is, for p states, d states, f states, and so on. All states except s states are divided into two substates in the process of combining **L** and **S** to form the total angular momentum **J**.

It may seem that in defining the quantum numbers j and m_j we have raised the total number of quantum numbers from four to six, but this is illusory. What we

Two systems of quantum numbers

have done is to provide two alternative sets of four quantum numbers, each capable of completely describing the hydrogen atom states. We would use n, l, m_l, and m_s if the orbital and spin motions were to act separately and independently, with no physical connection between them. We use n, l, j, and m_j if the orbital and spin motions are linked together and behave, as they do, as a single unit. This latter will indeed be the situation in all cases that we consider in this chapter.

Note that the quantum numbers n and l are common to both systems, so that we can speak of $1s$, $2d$, $3f$, etc., states without committing ourselves to either system. The populations of the shells and the subshells of Table 44–3 are also the same in each system, as Example 5 illustrates.

Example 5 *Assigning quantum numbers*   There are six hydrogen atom states in the $2p$ subshell, as Table 44–4 verifies. List their quantum numbers using each of the quantum number schemes outlined in the text.

Choose first the n, l, m_l, and m_s combination. There are $2l + 1$, or three, values of m_l, but, because m_s can have either of the two values $\pm\frac{1}{2}$, the number of states is doubled. Therefore the $2p$ subshell indeed contains six states. The quantum numbers identifying them are

Let us now choose the n, l, j, and m_j combination. From Eq. 44–24, the allowed values of j are $\frac{3}{2}$ and $\frac{1}{2}$. The number of allowed values of m_j is $2j + 1$, which yields four for $j = \frac{3}{2}$ and two for $j = \frac{1}{2}$. This again adds up to six states, just as above. The quantum numbers identifying them are shown below.

Actually, the number of states for *any* subshell must be the same no matter which system of quantum numbers we use to count them.

n	l	m_l	m_s
2	1	-1	$-\frac{1}{2}$
2	1	-1	$+\frac{1}{2}$
2	1	0	$-\frac{1}{2}$
2	1	0	$+\frac{1}{2}$
2	1	$+1$	$-\frac{1}{2}$
2	1	$+1$	$+\frac{1}{2}$

n	l	j	m_j
2	1	$\frac{3}{2}$	$-\frac{3}{2}$
2	1	$\frac{3}{2}$	$-\frac{1}{2}$
2	1	$\frac{3}{2}$	$+\frac{1}{2}$
2	1	$\frac{3}{2}$	$+\frac{3}{2}$
2	1	$\frac{1}{2}$	$-\frac{1}{2}$
2	1	$\frac{1}{2}$	$+\frac{1}{2}$

Now let us consider, classically and qualitatively, the *spin-orbit coupling* mechanism that actually binds together the orbital and the spin motions of the electron in the hydrogen (and other) atom. We focus on a p state, for which $l = 1$.

Consider an observer sitting on the electron in such an atom as it moves in its Bohr orbit around the atomic nucleus which, in the hydrogen atom, is simply a proton. This observer would see the (positively charged) proton circulating around him just as we on earth daily watch the apparent motion of the sun around us. The circulating proton forms a tiny current loop, which sets up an internal magnetic field $\mathbf{B}_{\text{int}}$ at its center, where the electron is. This internal magnetic field is surprisingly large, being about 0.4 tesla for the $n = 2$ state.

The mechanism of spin-orbit coupling

The electron, as we have seen, has a magnetic moment due to its spin, and this vector can be aligned either parallel to $\mathbf{B}_{\text{int}}$ or antiparallel to it. The parallel alignment, just as for a compass needle in the earth's magnetic field, gives the lower energy of the two.

Spin-orbit coupling in hydrogen

Thus the two states in Fig. 44–8 do not exactly have the same energy, the $p_{1/2}$ state being slightly lower than the $p_{3/2}$ state. For $n = 2$ the energy difference between these two states (derivable from Eq. 44–15) proves to be about 5×10^{-5} eV, much smaller (by a factor of $\sim 5 \times 10^5$) than the energy difference between the first excited state of the hydrogen atom ($n = 2$) and its ground state ($n = 1$), represented by the arrow in Fig. 44–1b. Although small, this level splitting causes a measurable "fine structure" in the wavelength of the spectral line joining these states.

Spin-orbit coupling in sodium

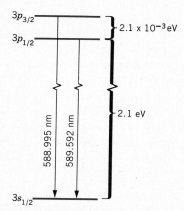

Figure 44–9 Showing the origin of the yellow sodium doublet. The upper two levels are split by spin-orbit coupling; the lower level (an *s* state) is not. Note the break in the vertical energy scale. A drawing of a photograph of the sodium doublet appears in Fig. 44–10*a*.

Effect of an external magnetic field—the Zeeman effect

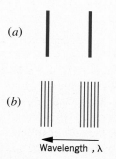

Figure 44–10 The Zeeman effect for sodium. (*a*) The two spectral lines of the normal sodium doublet. (*b*) When an external magnetic field is applied, one line splits into four components, the other into six.

The Zeeman effect as a test of wave mechanics

The spin-orbit level splitting is much larger in sodium than in hydrogen. Sodium has 11 electrons, ten of which (see Table 44–3) fill completely the $1s$, $2s$, and $2p$ subshells. For a filled subshell the net angular momentum and magnetic moment cancel to zero, so that these properties for the sodium atom are determined solely by the motion of the single remaining electron, called the *valence* electron. In fact, the atom can be pictured, to a good enough approximation for many purposes, as a single electron moving about a "nucleus" consisting of the actual nucleus and the core of non-valence electrons.

Normally the valence electron of sodium is in a $3s_{1/2}$ state; that is, it has $n = 3$, $l = 0$, and $j = \frac{1}{2}$. This state is *not* split by spin-orbit coupling because it has no orbital angular momentum. In such a case the angular momentum is due entirely to the spin of the valence electron.

As Fig. 44–9 shows, the $3p$ state *is* split by spin-orbit coupling, into $3p_{3/2}$ and $3p_{1/2}$ states. As in hydrogen, the $3p_{1/2}$ state is lower in energy. When valence electrons fall back from these states to the ground state, they generate the well-known sodium doublet ($\lambda_1 = 588.995$ nm and $\lambda_2 = 589.592$ nm). These so-called sodium D lines provide the yellow light characteristic of sodium vapor lamps.

44–8 The Atom in a Magnetic Field

In the last section we saw that the atom's *internal* magnetic field, involved as it is in the spin-orbit coupling mechanism, produces a small separation in energy of levels that would otherwise coincide. As Fig. 44–9 shows, this produces two spectral lines—the sodium doublet in this case—where there would otherwise have been but one.

In this section we wish to examine splitting of the spectrum lines caused by the application of an *external* magnetic field. The great English investigator Michael Faraday sought such effects, convinced that they should exist, but the apparatus available to him was not sufficiently sensitive. This search for an effect caused by a magnetic field on the light emitted from atoms in a flame was Faraday's last experiment, performed in 1862 when he was 71 years old. In 1897 the young Dutch physicist Pieter Zeeman, reasoning that if Faraday thought the experiment was worth doing, it probably was, repeated it with better equipment and found a positive result. The 1902 Nobel Prize, the second ever awarded, was shared by Zeeman and H. A. Lorentz, who provided a deeply perceptive interpretation of Zeeman's results.

Figure 44–10 shows the *Zeeman effect*, as it is called, for the two lines of the sodium doublet, as observed with modern equipment. The two lines in Fig. 44–10*a* are the normal doublet, no external magnetic field being present. Their separation, as we saw in connection with Fig. 44–9, is caused by the internal magnetic field associated with the spin-orbit coupling mechanism. When an external magnetic field is applied, as in Fig. 44–10*b*, one line ($\lambda = 589.592$ nm) splits into four components. The other line ($\lambda = 588.995$ nm) splits into six. The Zeeman effect is now so well understood that astronomers work it backwards, starting from the observed splittings of the spectral lines and deducing the strengths of the magnetic fields present over the solar disk (up to 0.4 T near sunspots), at the surfaces of certain stars (up to 10^4 T for certain white dwarfs), and in the interstellar spaces of our own galaxy ($\sim 10^{-9}$ to 10^{-10} T).

The most effective application of the Zeeman effect lies in the vast array of sharp and clear experimental data it provides as a testing ground for the validity of wave mechanics. It ranks high among the triumphs of this theory that the number of components into which a given line splits, their relative spacing as a function of the magnitude of the applied magnetic field, and indeed the nature of the polarization of the light associated with the various split components, all agree exactly with prediction.

The Zeeman effect is not confined to the optical spectra of atoms or ions, as in the example of Fig. 44-10. It is a useful tool in the analysis of the line spectra of molecules, which often occur in the microwave region of the electromagnetic spectrum. Beyond this it can be used—often in the radiofrequency region of the spectrum—to study the spins and the relatively small magnetic moments possessed by many atomic nuclei.

The splitting in wavelength of the spectral lines of the sodium doublet when the light source is immersed in an external magnetic field can be traced directly to a splitting in energy of the levels of the sodium atoms in the source. Both the levels from which the lines originate and those on which they terminate are split.

Figure 44-11 shows the splitting in energy of the 3p levels of sodium, the levels from which the sodium doublet lines originate; see Fig. 44-9. If there were no spin-orbit coupling mechanism or—put another way—if there were no internal magnetic field in the sodium atom, the 3p state would be represented by a single level as in Fig. 44-11a. There *is* an internal magnetic field in the sodium atom, however, so this figure does not correspond to reality. Figure 44-11b shows that, as we have seen, this internal field splits the 3p level into two components, labeled $3p_{3/2}$ and $3p_{1/2}$.

Level splitting by an external magnetic field—qualitative

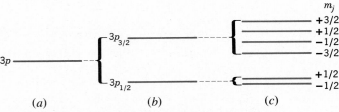

Figure 44-11   (a) The energies of the 3p levels for sodium would coincide if spin-orbit coupling did not exist (it does, though). (b) Spin-orbit coupling (that is, the action of an internal magnetic field) splits the 3p level into two levels. (c) An external magnetic field further splits these levels.

If now an external magnetic field is applied, it splits the levels further, splitting the upper level into four components and the lower level into two. This makes a total of six levels, as Fig. 44-11c illustrates. As Table 44-4 reminds us, there are six 3p states, and here we see them all spread out in energy by the combined actions of an internal and an external magnetic field. The single 3p level of Fig. 44-11a—if it existed—would be termed *sixfold degenerate*. In Fig. 44-11c this degeneracy has been (totally) removed by the magnetic field.

It is reasonable that an external magnetic field can cause the components of a degenerate level to split in energy. As Fig. 44-8c reminds us, the different values of m_j correspond to different orientations of the angular momentum vector **J** and—more to the point—of its associated magnetic moment vector $\boldsymbol{\mu}$. The magnetic potential energy is given by

Level splitting by an external magnetic field—quantitative

$$U = -\boldsymbol{\mu} \cdot \mathbf{B} = -\mu_z B. \qquad [44\text{-}15]$$

Note that if space quantization did not exist, that is, if *any* orientation of $\boldsymbol{\mu}$ with respect to **B** were possible, the level would simply broaden rather than split into discrete components. The fact that it unmistakably does the latter is strong support indeed for the wave mechanical concept of space quantization.

Figure 44–12, which should be viewed in connection with Fig. 44–10, shows in detail how the spectral line splitting of the sodium doublet lines comes about when an external magnetic field is applied. Note that the $3p_{3/2}$, $3p_{1/2}$, and $3s_{1/2}$ levels are all split by this field, the number of components being, as expected, just $2j + 1$; see Eq. 44–22.

The Zeeman effect in sodium

Figure 44–12 (a) The three levels of Fig. 44–9 as split by an external magnetic field. The transitions connecting the levels are shown. The two transitions indicated by dashed lines are predicted by wave mechanics not to occur; they indeed do not occur. (b) A wavelength plot of the predicted spectrum. (c) A wavelength plot of the spectrum with the magnetic field removed. Compare with Fig. 44–10.

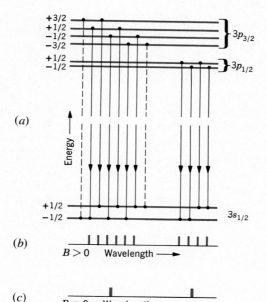

Equation 44–15 suggests that the level energy in the presence of a uniform external magnetic field **B** may be written as

$$E = E_0 - \mu_z B. \tag{44-25}$$

Here E_0 is the value of the energy when no external field is present and μ_z is the component of $\boldsymbol{\mu}$ in the field direction. We assume that **B** points in the $+z$-direction.

The g-factor

Now μ_z in Eq. 44–25 is related to the angular momentum of the state, and immediately a problem arises. Is it orbital angular momentum, spin angular momentum, or some combination of the two? We recall (p. 895) that spin angular momentum is twice as effective as orbital angular momentum in generating magnetism. Specifically, let us rewrite our earlier expressions for $\mu_{l,z}$ and $\mu_{s,z}$ as

$$\mu_{l,z} = -g_l m_l \mu_B \tag{44-11}$$

and

$$\mu_{s,z} = -g_s m_s \mu_B. \tag{44-19}$$

Experiment requires that we put $g_l = 1$ and $g_s = 2$ in the above.

When both orbital and spin motions are present, we can put

$$\mu_z = -g m_j \mu_B, \tag{44-26}$$

where g, the so-called *g-factor* is, in general, different for each state. Its numerical value depends on the quantum numbers that define the state, as Problem 17 reveals.

Substituting Eq. 44–26 into Eq. 44–25 yields

$$E = E_0 + gm_j\mu_B B, \tag{44-27}$$

which gives the energy of the state as a function of the external magnetic field **B**, once m_j and g are specified.

In Fig. 44–12 (see Problem 17) we have the following g-factors for the three states involved:

The g-factors for the sodium doublet levels

State	g-Factor
$3p_{3/2}$	$\tfrac{4}{3}$
$3p_{1/2}$	$\tfrac{2}{3}$
$3s_{1/2}$	2

The last result in this table is to be expected. An s-state has $l = 0$, which means that the angular momentum is determined entirely by the spin of the electron, so that Eq. 44–19 applies, with $g = 2$.

Example 6 Using data from the above table, find the energy interval between adjacent levels for the $3p_{3/2}$, the $3p_{1/2}$, and the $3s_{1/2}$ states of the sodium atom in an external magnetic field of 1.50 T; see Fig. 44–12.

For the two uppermost levels of the $3p_{3/2}$ state in Fig. 44–12, we can write, from Eq. 44–27,

$$E_{3/2} = E_0 + (\tfrac{4}{3})(\tfrac{3}{2})\mu_B B$$

and

$$E_{1/2} = E_0 + (\tfrac{4}{3})(\tfrac{1}{2})\mu_B B.$$

Subtracting yields

$$\begin{aligned}
\Delta E &= E_{3/2} - E_{1/2} \\
&= (\tfrac{4}{3})\mu_B B \\
&= (\tfrac{4}{3})(5.79 \times 10^{-5}\ \text{eV/T})(1.50\ \text{T}) \\
&= 1.16 \times 10^{-4}\ \text{eV}
\end{aligned}$$

Here $\tfrac{4}{3}$ is the g-factor for the $3p_{3/2}$ state, taken from the above table, and μ_B is the Bohr magneton.

In the same way, we can show that $\Delta E = 0.58 \times 10^{-4}$ eV for the $3p_{1/2}$ state (for which $g = \tfrac{2}{3}$) and $\Delta E = 1.74 \times 10^{-4}$ eV for the $3s_{1/2}$ state (for which $g = 2$).

We note that, as Fig. 44–12 suggests, the three level spacings above are all different. If they were the same, the Zeeman effect in sodium would differ unmistakably from that displayed in Fig. 44–10b; see Problem 20. This difference in level spacings can be traced directly to the difference in g-factors for each state.

REVIEW GUIDE AND SUMMARY

Wave functions and quantum numbers

The problem of the hydrogen atom is defined by introducing the potential energy function of Eq. 44–1 into the mathematical machinery of wave mechanics. What emerges is a set of wave functions ψ_{nlm} that describe the allowed states of the hydrogen atom. The energies of these states are

$$E_n = -\frac{me^4}{8\epsilon_0 h^2}\frac{1}{n^2}, \qquad n = 1, 2, 3, \ldots \tag{44-2}$$

The subscripts attached to the wave functions are the three quantum numbers n, l, and m_l, whose names and allowed values are shown in Table 44–1. In Table 44–2 the hydrogen atom states are conveniently grouped into shells (same n) and subshells (same n and l). The spin quantum number m_s, introduced later, serves to double the number of states assigned to each shell and subshell.

The $2s$ and $2p$ states of hydrogen

Figure 44–2 shows both ψ^2 and the radial distribution function $P(r)$ for the $2s$ state of hydrogen. Figure 44–3 shows these same quantities for the three $2p$ states of this atom.

Every hydrogen atom state has an orbital angular momentum **L** and an associated orbital magnetic moment **μ** (later labeled $\boldsymbol{\mu}_l$) whose magnitudes are fixed by the orbital quantum number l (see Eq. 44–7). Their allowed projections on the z-axis depend on the magnetic quantum number m_l and are given by

Space quantization

$$L_z = m_l \hbar \qquad \text{and} \qquad \mu_z = -m_l \mu_B. \qquad [44–8, 44–11]$$

The fact that only a discrete number ($= 2l + 1$) of these projections is allowed is called space quantization and is represented by the vector construction of Fig. 44–4. In atomic problems, angular momentum is measured in units of $\hbar$, where

$$\hbar = \frac{h}{2\pi} = 6.59 \times 10^{-16} \text{ eV·s},$$

and the magnetic moment in terms of the Bohr magneton μ_B, where

$$\mu_B = \frac{e\hbar}{2m} = 5.76 \times 10^{-5} \text{ eV/T}.$$

Proof of space quantization

The Stern-Gerlach experiment confirmed the reality of space quantization. It did so by splitting a beam of neutral silver atoms into two subbeams, each characterized by the orientation in space of the magnetic moments of the atoms that make it up.

The Stern-Gerlach experiment also called attention to the fact that the electron has a spin angular momentum $\mathbf{S}$ and a spin magnetic moment $\boldsymbol{\mu}_s$. Their projected values are given by

The spin of the electron

$$S_z = m_s \hbar \qquad \text{and} \qquad \mu_{s,z} = -2m_s \mu_B, \qquad [44–17, 14–19]$$

in which m_s, the spin quantum number, has only the two values $\pm \frac{1}{2}$. The introduction of spin doubles the number of allowed states in each shell and subshell, as Table 44–3 shows. The factor 2 in the equation for $\mu_{s,z}$ shows that spin is twice as effective as orbital motion in generating magnetism.

The total angular momentum $\mathbf{J}$

The orbital and spin motions of an atomic electron interact physically to generate a total angular momentum $\mathbf{J}$ for the atom. Its magnitude is fixed by a new quantum number j, given by

$$j = l \pm \tfrac{1}{2}, \qquad [44–24]$$

and its allowed projections are given by

$$J_z = m_j \hbar,$$

where m_j ranges by integral steps from $-j$ to $+j$. Figure 44–8 illustrates the case of $l = 1$. The quantum numbers n, l, j, m_j are a more useful set than n, l, m_l, m_s and are used henceforth.

Effect of an internal magnetic field (spin-orbit coupling)

The effect of spin-orbit coupling (see p. 898) is to split all the hydrogen atom levels except the s states into two. Figure 44–9 shows how the sodium doublet is accounted for on this basis, the $3p$ state being split into $3p_{3/2}$ and $3p_{1/2}$ substates. Spin-orbit coupling is equivalent to the action of an internally generated magnetic field.

An externally imposed magnetic field causes a level with a given value of j to split into $2j + 1$ sublevels, their energies being given by

$$E = E_0 + g m_j \mu_B B. \qquad [44–27]$$

Effect of an external magnetic field

This splitting of levels causes a splitting of the wavelengths of the transitions between levels. Example 6 and Figs. 44–10 and 54–12 explore this *Zeeman effect* for the sodium doublet lines of Fig. 44–9.

QUESTIONS

1. What are the dimensions and the SI units of a wave function?

2. Do the expressions for $\psi_{1s}^2(r)$ (Eq. 43–14b), for $P_{2s}(r)$ (Eq. 44–4), and for $P_{2p}(r)$ (Eq. 44–6) have the dimensions that you expect for them?

3. Figure 44–3a shows the three $3p$ wave functions for the hydrogen atom. What determines the direction in space that we choose for the z-axis?

4. How can you account for the fact that in the $2s$ state of the hydrogen atom $\psi_{2s}^2(r)$ is a maximum at $r = 0$ but the radial probability density function $P_{2s}(r)$ is zero there? See Fig. 44–2.

5. Define and distinguish among the terms wave function, probability density function, and radial probability density function.

6. If Bohr's theory and wave mechanics predict the same result for the energies of the hydrogen atom states (see Eq. 44–2), then why do we need wave mechanics with its greater complexity?

7. Compare Bohr theory and wave mechanics. In what respects do they agree? In what respects do they differ?

8. The angular momentum of the electron in the hydrogen atom is quantized. Is the magnitude of the linear momentum also quantized?

9. Consider the three $2p$ probability density functions of Fig. 44–3a, each of which is a figure of revolution about the z-axis. Do you see any connection between these figures and the semiclassical vector model of the atom (Fig. 44–4) for the case of $l = 1$?

10. In the laboratory how would you show that an atom has angular momentum? That it has a magnetic dipole moment?

11. A beam of neutral silver atoms is used in a Stern-Gerlach experiment. What is the origin of both the force and the torque that act on the atom? How is the atom affected by each?

12. Why don't we observe space quantization for a spinning top?

13. How do we arrive at the conclusion that the spin magnetic quantum number m_s can have only the values $\pm\frac{1}{2}$? What kinds of experiments support this conclusion?

14. What determines the number of subbeams into which a beam of neutral atoms is split in a Stern-Gerlach experiment?

15. How would the properties of helium differ if the electron had no spin, that is, if the only operative quantum numbers were n, l, and m_l?

16. Why is the magnetic moment of the electron directed opposite to its spin angular momentum?

17. An atom in a state with zero angular momentum has spherical symmetry as far as its interaction with other atoms is concerned. It is sometimes called a "billiard ball atom." Explain.

18. In this chapter we have defined seven quantum numbers, namely, n, l, m_l, s, m_s, j, and m_j. How can we say that only four of them are needed? Which four?

19. We assert that the number of quantum numbers needed for a complete description of the motion of the electron in the hydrogen atom is equal to the number of degrees of freedom which that electron possesses. What is this number? How can you justify it?

20. On p. 895 we note that, experimentally, spin angular momentum is twice as effective as orbital angular momentum in generating magnetism. How is this fact related to the number of Zeeman components induced in the sodium doublet by an external magnetic field? See Fig. 44–10.

21. If your own words, describe the mechanism of spin-orbit coupling.

22. How does spin-orbit coupling lead to a splitting of the spectral lines of sodium?

23. In Fig. 44–12, why do the $3p_{3/2}$, the $3p_{1/2}$, and the $3s_{1/2}$ levels have different energy splittings when an external magnetic field is applied?

EXERCISES

Section 44–2 The Hydrogen Atom Quantum Numbers
1. A hydrogen atom state is known to have the quantum number $l = 3$. What are the possible n, m_l and m_s quantum numbers?

2. Label as true or false these statements involving the quantum numbers n, l, and m_l. (a) One of these subshells cannot exist: $2p$, $4f$, $3d$, $1p$. (b) The number of values of m_l that are allowed depends only on l and not on n. (c) The N shell contains four subshells. (d) The smallest value of n that can go with a given l is $l + 1$. (e) All s states have $m_l = 0$, regardless of the value of n. (f) Every shell contains n subshells.

3. How many hydrogen atom states are there with $n = 4$? How are they distributed among the subshells? Use the quantum numbers n, l, and m_l as in Example 1, ignoring effects due to spin.

4. An electron in a hydrogen atom has a maximum m_l value of $+4$. Identify the shell and the subshell to which it belongs. How many states are in each? Ignore electron spin effects, as in Table 44–4.

Section 44–3 The Hydrogen Atom Wave Functions
5. What are the values of (a) $\psi_{1s}(r)$, (b) $\psi_{1s}^2(r)$ and (c) $P_{1s}(r)$ in the hydrogen atom at $r = r_B$? ($r_B = 0.529 \times 10^{-10}$ m).

6. Given the radial probability density function for the $2s$ state of the hydrogen atom (see Eq. 44–4), find the probability density function $\psi_{2s}^2(r)$.

Section 44–4 Angular Momentum and Magnetism
7. Calculate, for a hydrogen atom in a state with $l = 3$, the allowed values of L_z, μ_z, and θ. Find also the magnitudes of **L** and μ. Compare with Example 3.

8. Verify that $\mu_B = 5.79 \times 10^{-5}$ eV/T, as reported in Eq. 44–12.

Section 44–5 The Stern-Gerlach Experiment
9. What is the acceleration of a silver atom as it passes through the deflecting magnet in the Stern-Gerlach experiment of Example 4? Express your answer in terms of a ratio to the acceleration due to gravity.

Section 44–6 The Spinning Electron
10. Write down the quantum numbers for all the hydrogen atom states belonging to the $4f$ subshell.

11. What are the quantum numbers n, l, m_l, and m_s for the two electrons of the helium atom in its ground state?

Section 44–7 The Total Angular Momentum J
12. If a particular quantum system had quantum numbers $L = 2$ and $S = \frac{3}{2}$, what would be the possible values of the total angular momentum quantum number J?

13. An electron is in a $3p_{3/2}$ state. (*a*) What is the magnitude of the electron's total angular momentum? (*b*) What are the values of the *z*-components of the electron's total angular momentum?

14. What are the possible quantum numbers for an electron in a state with $n = 6$ and $l = 5$?

15. Which of the following levels of the hydrogen atom are split into two components by spin-orbit coupling and which are not: $1s$, $3p$, $4s$, $4d$, $2s$, $4p$? Give the reasons for your answer.

16. In a Stern-Gerlach experiment a beam of neutral nickel atoms splits into nine components. (*a*) What is the angular momentum of the ground state of the nickel atom? Give both its

magnitude and its maximum projected value. (*b*) In which quantum number do the atoms in the nine subbeams differ?

Section 44–8   The Atom in a Magnetic Field
17. A source emitting the potassium doublet (see Problem 14) is placed in a magnetic field with $B = 1.2$ T. (*a*) Into how many levels is each of the three states involved split? (*b*) For each state, what is the energy difference between adjacent levels when the magnetic field is imposed?

18. Verify that the *g*-factors given in the table preceding Example 6 are correct. Use the formula given in Problem 17.

PROBLEMS GROUPED BY SECTION

Section 44–2   The Hydrogen Atom Quantum Numbers
1. In connection with Table 44–2, show that the number of states in any shell is given by n^2. (Note that if the spin of the electron is taken into account, as in Table 44–3, this number is doubled, to $2n^2$.)

Section 44–3   The Hydrogen Atom Wave Functions
2. Show that the maximum of the radial probability density function $P_{2p}(r)$ of Fig. 44–3b occurs at $r = 4r_B$ and that this is the radius of the second Bohr orbit.

3. A small sphere of radius $0.10r_B$ is located a distance r_B from the nucleus of a hydrogen atom in a $1s$ state. What is the probability that the electron will be found inside this sphere?

4. Locate the two maxima for the radial probability density function $P_{2s}(r)$ of Fig. 44–2b.

5. For a hydrogen atom in a $2s$ state (see Eq. 44–4), what is the relative probability of finding the electron somewhere within the smaller of the two maxima of the radial probability density function? See Fig. 44–2b and the result of Problem 4. (*Hint:* Do not hesitate to use a table of integrals.)

6. The densities of dots in Fig. 44–3a represent these three functions:

$$\psi_{211}^2 (r, \theta) = \psi_{21,-1}^2 (r, \theta) = \left(\frac{r^2}{64\pi r_B^5}\right) e^{-r/r_B} \sin^2 \theta$$

and

$$\psi_{210}^2 (r, \theta) = \left(\frac{r^2}{32\pi r_B^5}\right) e^{-r/r_B} \cos^2 \theta.$$

If an electron moves randomly among these three states, all of which have the same energy, the square of its wave function ψ_{21}^2 ($=\psi_{2p}^2$) will be the equally weighted sum of the above three functions. Find an expression for this quantity, ψ_{21}^2, and show that it is spherically symmetric, that is, that it depends only on r and not on θ. See Eq. 44–5.

Section 44–4   Angular Momentum and Magnetism
7. Consider the relation

$$\cos \theta_{min} = \left(1 + \frac{1}{l}\right)^{-1/2}$$

in Example 2 for the limiting case of large l. (*a*) By expanding both sides of this equation in series form (see Appendix G), show that, to a good approximation for large l,

$$\theta_{min} \cong \frac{1}{\sqrt{l}},$$

where θ_{min} is to be expressed in radian measure. (*b*) Test the validity of this approximate formula for the four values of l given in Example 2. (*c*) Extend the table in Example 2 by finding θ_{min} for $l = 10^5$ and for $l = 10^6$. (*d*) What difficulties do you encounter if you do *not* use the above approximate formula to calculate θ_{min} for such large values of l?

8. Show, by reanalyzing Problem 42–46 for the diatomic molecule with the angular momentum quantized by Eq. 44–7, that the energy levels can be written as

$$E_l = \frac{h^2 l(l + 1)}{4\pi^2 md^2}.$$

Calculate the first three levels of the O_2 molecule for which the two atoms are spaced 0.20 nm apart.

* 9. Of the three scalar components of **L**, one, L_z, is quantized, according to Eq. 44–8. In view of the restrictions imposed by Eqs. 44–7 and 44–8, taken together, show that the most that can be said about the other two components of L is

$$[L_x^2 + L_y^2]^{1/2} = [l(l + 1) - m_l^2]^{1/2} \hbar.$$

Note that these two components are not separately quantized. Show that

$$\sqrt{l\hbar} \le [L_x^2 + Ly^2]^{1/2} \le [l(l + 1)]^{1/2} \hbar$$

and correlate this problem with Fig. 44–4.

Section 44–5   The Stern-Gerlach Experiment
10. A beam of chromium atoms is used in a Stern-Gerlach experiment, the apparatus being that of Example 4. The mass M of a chromium atom is 8.6×10^{-26} kg, and the "oven" temperature is adjusted so that the most probable speed of the chromium atoms in the beam is 1200 m/s. The beam forms seven evenly spaced images on the detector plate. The magnetic moment (i.e., the maximum projected value of μ) of the chromium atom is $6.0 \mu_B$. (*a*) To what projected values of μ do the seven beam components

correspond? (b) What is the spacing d between adjacent beam components at the detector plate?

Section 44–6   The Spinning Electron

11. Assume that in the Stern-Gerlach experiment described for neutral silver atoms, the magnetic field **B** has a magnitude of 0.500 T. (a) What is the energy difference between the orientations of the silver atoms in the two subbeams? (b) What is the frequency of the radiation that would induce a transition between these two states? (c) What is its wavelength, and to what part of the electromagnetic spectrum does it belong? Recall that the magnetic moment of a neutral silver atom is due entirely to the spin of its single valence electron.

Section 44–7   The Total Angular Momentum J

12. The wavelengths of the familiar yellow lines of the sodium doublet are $\lambda_1 = 588.995$ nm and $\lambda_2 = 589.592$ nm. The lines originate as shown in Fig. 44–9. From these data, find (a) the spin-orbit energy splitting of the 3p level and (b) the energy difference between the 3p and the 3s levels.

13. *Spin-orbit coupling.* In Problem 15 we shall see that the effective internal magnetic field "seen" by the spinning electron in a hydrogen atom in a 2p state is 0.39 T. The magnetic moment of this electron may line up either parallel or antiparallel to this internal field, forming the $2p_{1/2}$ and the $2p_{3/2}$ substates, respectively. Find the energy difference ΔE between these two substates. This quantity is a measure of the strength of the spin-orbit coupling interaction.

14. Potassium ($Z = 19$), like sodium ($Z = 11$), is an alkali metal, with its single valence electron outside a filled 18-electron, argon-like core. A potassium doublet originates in transitions from a $4p_{1/2}$ and a $4p_{3/2}$ state to the $4s_{1/2}$ ground state. The wavelengths of the doublet are $\lambda_1 = 764.5$ nm and $\lambda_2 = 769.9$ nm. Calculate (a) the energy splitting between the two 4p states due to spin-orbit coupling, and (b) the approximate energy difference between the 4p states (either of them) and the ground state.

15. From Eq. 31–8, with $x = 0$ and $R = r$, we learn that the magnetic field at the center of a circular loop of radius r carrying a current i is given by

$$B = \frac{\mu_0 i}{2r}.$$

Consider an observer seated on an electron in a hydrogen atom. He will see the proton (charge $= +e$) moving around him in a circle of radius r at a speed v that is equal to his own orbital speed as seen from the proton. (a) Show that, from Bohr theory (see Section 42–8), B due to the equivalent current loop formed by the proton is given by

$$B = \frac{\mu_0 \pi m^2 e^7}{8\varepsilon_0^3 h^5 n^5}$$

(b) Find B for $n = 2$.

16. Show, by considering in detail the mechanism of spin-orbit coupling discussed in the text, that a $p_{1/2}$ state lies lower in energy than a $p_{3/2}$ state.

Section 44–8   The Atom in a Magnetic Field

17. The g-factor of Eq. 44–26 is related to the quantum numbers by

$$g = 1 + \frac{j(j+1) + s(s+1) - l(l+1)}{2j(j+1)},$$

in which s, the spin quantum number, always has the value $\frac{1}{2}$; see Table 44–3. Show that if $l = 0$, then $g = 2$, as Eq. 44–19 requires.

18. What is the wavelength of a photon that will induce a transition of an electron spin from parallel to antiparallel orientation in a magnetic field of magnitude 0.20 T? You may assume that $l = 0$.

19. The magnetic moment μ of an electron corresponding to the total angular momentum **J** is given by

$$\mu = -\mu_B (\mathbf{L} + g_s \mathbf{S}),$$

where $g_s = 2$. Show that μ does not lie along the direction opposite of **J**.

20. Suppose that the energy intervals between adjacent levels for the three states shown in Fig. 44–12 were the same. How many Zeeman components would there then be for each of the two lines of the sodium doublet?

21. An electron in a hydrogen atom is in a 3d state. (a) List the allowed values of j, and for each of these values find (b) g, (c) J_z (maximum projected value), (d) μ_z (maximum projected value), and (e) the number of separate states generated when the atom is placed in an external magnetic field.

ADDITIONAL PROBLEMS

22. Write down the quantum numbers for all states in the 3d subshell, using both (a) the n, l, m_l, m_s, and (b) the n, l, j, m_j notations.

23. For a hydrogen atom in a 1s state, what is the probability of finding the electron between two spheres of radii $r = 1.000r_B$ and $r = 1.010r_B$?

24. Prove from Bohr theory that the magnetic moments of the electrons in the various Bohr orbits are given by

$$\mu = n\mu_B,$$

in which μ_B is the Bohr magneton and $n = 1, 2, 3, \ldots$.

25. The proton as well as the electron has spin $\frac{1}{2}$. In the hydrogen atom in its normal state with $n = 1$, $l = 0$ and $m_l = 0$, there are two energy levels depending on whether the electron and the proton spins are in the same direction or in opposite directions. The state with the spin in the same direction has the higher energy. If an atom is in this state and one of the spins "flips over," the small energy difference is released as a photon of wavelength 21 cm. This spin-flip process is rare; it takes about 10 million years for each atom. Radio astronomers observe this 21-cm radiation in space, where the density of hydrogen is so small that an atom can flip before being disturbed by collisions with other atoms. What is the effective magnetic field experienced by the electron?

26. Find all the blanks in the table for the five hydrogen atom states shown. See Problem 17.

State	$1s_{1/2}$	$2s_{1/2}$	$2p_{1/2}$	$2p_{3/2}$	$3d_{3/2}$
n					
l					
j					
g					
$J_z{}^a$					
$\mu_z{}^a$					

^{a} Give the maximum projected value.

27. An electron is one Bohr radius from a proton and has a total energy E of $+5.0$ eV. What are (a) its potential energy U and (b) its kinetic energy K? (c) What kinetic energy would this electron have at very large distances from the proton, assuming that its total energy E remains unchanged? (d) Describe this state of the hydrogen atom. Can a quantum number be assigned to it?

28. For the hydrogen atom in a $1s$ state, what is the radius of a sphere such that the probability of finding the electron inside it is just 0.50? (*Hint:* Use a table of integrals and solve the resulting equation by trial-and-error methods. It goes surprisingly rapidly with a hand calculator. See how closely you can guess the answer.)

29. Using Eq. 44–6, show that

$$\int_0^\infty P_{2p}(r)\,dr = 1.$$

What is the physical interpretation of this result? (*Hint:* Do not hesitate to use a table of integrals.)

30. The energy difference ΔE between the $3p_{1/2}$ and the $3p_{3/2}$ states of sodium is 2.14×10^{-3} eV as shown in Problem 12. Cal-culate the magnitude B_{int} of the internal magnetic field associated with this level splitting, through the spin-orbit coupling mechanism described on p. 896. Recall that the magnetic moment of the sodium atom is due entirely to the spin of its single valance electron.

31. What is the expectation value of r for hydrogen in the $1s$ state? (See Problem 13 in Chapter 43.)

32. Suppose that an electron (with spin) is trapped in a one-dimensional box of length l and is placed in a uniform magnetic field **B**. Give an expression for its energy levels.

33. Suppose that a hydrogen atom moves 0.80 m in a direction perpendicular to a magnetic field that has a gradient of 1.6×10^2 T/m. (a) What is the force on the atom due to the magnetic moment of the electron? (b) What is its vertical displacement if its speed is 1.2×10^5 m/s and the electron spin is parallel to the magnetic field?

34. Estimate (a) the quantum number l for the orbital motion of the earth around the sun and (b) the number of quantized orientations of the plane of the earth's orbit, according to the rules of space quantization. (c) Also find θ_{min}, the half-angle of the smallest cone that can be swept out by a perpendicular to the earth's orbit as the earth revolves around the sun. Discuss from the point of view of the correspondence principle. The earth's mass is $\sim 6 \times 10^{24}$ kg, and the mean earth–sun distance is $\sim 1.5 \times 10^{11}$ m.

35. Consider the transition of a hydrogen atom from its $2p$ state to its ground, or $1s$ state. The $2p$ state is split by spin-orbit coupling, the energy difference ΔE between the two split levels being that calculated in Problem 13 ($\sim 5 \times 10^{-5}$ eV). The transition is thus a doublet. Calculate (a) the approximate wavelength λ of the transition and (b) the approximate wavelength difference $\Delta\lambda$ between the components of the doublet. (c) Compare $\Delta\lambda/\lambda$ for this transition with that for the sodium doublet, whose origin is shown in Fig. 44–9. (d) Which doublet would be more difficult to resolve with a grating spectrometer?

Atomic Physics: Three Selected Topics

45–1 Introduction

In this chapter we deal with three selected topics in atomic physics: (1) x rays and the numbering of the elements; (2) the building up of the periodic table; and (3) the laser, its operating principles, and its applications. It is our plan in each case to build on the material of the preceding chapters, and it is our belief that careful study of these examples will suggest convincingly the broad applicability of wave mechanics in physics, chemistry, engineering, and industrial technology.

45–2 X rays and the Numbering of the Elements

The nature of x rays

So far we have dealt with the behavior of single electrons in atoms, either the lone electron of hydrogen or the single valence electron of sodium. We now shift our attention to the behavior of electrons deep within the atom. We move from a region of relatively low binding energy (~ 5 eV for the work required to remove the valence electron from sodium, for example) to a region of higher energy (~ 70 keV for the work required to remove an innermost electron from tungsten, for example). The radiations we deal with, though of course still part of the electromagnetic spectrum, differ drastically in wavelength, from $\sim 6 \times 10^{-7}$ m ($= 600$ nm $= 6000$ Å) for the sodium doublet lines, say to $\sim 2 \times 10^{-11}$ m ($= 20$ pm $= 0.2$ Å) for one of the tungsten characteristic radiations, say. We are now speaking of x rays.

X rays in science and technology

The usefulness of these penetrating radiations in medical and dental diagnostics and in therapy is well known, as are their many industrial applications, such as examining welded joints in pipe lines. In Section 41–10 we described how

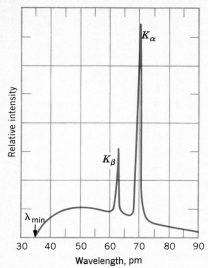

Figure 45-1 The distribution by wavelength of the x rays produced when 35-keV electrons strike a molybdenum target. After Ulrey. (1 pm = 10^{-12} m = 0.01 Å.)

The short wavelength cutoff

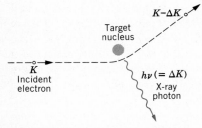

Figure 45-2 When an electron passes near the nucleus of a target atom, it may lose part of its kinetic energy by generating an x-ray photon.

x rays can be used to deduce the atomic structures of crystalline materials. The structures of such complex substances as insulin and DNA have been worked out by these methods. In astronomy, x-ray satellites such as Uhuru and the orbiting observatories Copernicus and Einstein have shown us an entirely new view of our universe, that provided by the x rays emitted by various objects in it.

Our concern in this section, however, is what x rays can teach us about the structure of the atoms that emit or absorb them. We focus on the work of the British physicist H. G. J. Moseley, who, by x-ray studies, developed the concept of atomic number and gave physical meaning in terms of atomic structure to the ordering of the elements in the periodic table.

We saw in Section 41–10 that x rays are produced when energetic electrons strike a solid target and are brought to rest in it. Figure 45–1 shows the wavelength spectrum of the x rays produced when 35-keV electrons strike a molybdenum target.

We first examine the continuous background spectrum of Fig. 45–1, from which the two prominent peaks arise. Consider an electron of kinetic energy K that happens to pass close to the nucleus of one of the molybdenum atoms in the target, as in Fig. 45–2. The electron may well lose a certain part ΔK of its kinetic energy, which will appear as the energy $h\nu$ of an x-ray photon that radiates away from the site of the encounter. Such *bremsstrahlung* processes (German: "braking radiation") account for the continuous x-ray spectrum.

If the target on which the incident electrons fall is sufficiently thick electrons in the kinetic energy range $0 < K < eV$ will be present, V being the accelerating potential and eV the initial kinetic energy. All electrons in this energy range can undergo bremsstrahlung processes, thus contributing x-ray photons to the continuous x-ray spectrum.

A prominent feature of the continuous spectrum of Fig. 45–1 is the sharply-defined cutoff wavelength λ_{min} below which the continuous spectrum does not exist. This minimum wavelength corresponds to a decelerating event in which one of the incident electrons (with initial kinetic energy eV) loses *all* of this energy in a single encounter, radiating it away as a single photon. Thus,

$$eV = h\nu_{max} = \frac{hc}{\lambda_{min}},$$

or

$$\lambda_{min} = \frac{hc}{eV}. \tag{45-1}$$

Equation 45–1 shows that if $h \to 0$, then $\lambda_{min} \to 0$, which is the prediction of classical theory. The existence of a minimum wavelength is a quantum phenomenon.

Example 1 Calculate the wavelength λ_{min} for the continuous spectrum of x rays emitted when 35-keV electrons fall on a molybdenum target, as in Fig. 45–1. From Eq. 45–1, we have

$$\lambda_{min} = \frac{hc}{eV} = \frac{(4.14 \times 10^{-15}\text{ eV·s})(3.00 \times 10^8\text{ m/s})}{(35.0 \times 10^3\text{ eV})}$$

$$= 3.54 \times 10^{-11}\text{ m} = 35.4\text{ pm}.$$

This is in agreement with the experimental result shown by the vertical arrow in Fig. 45–1. Note that Eq. 45–1 contains no reference to the target material. For a given accelerating potential all targets, no matter what they are made of, exhibit the same cutoff wavelength.

The characteristic x-ray spectrum

We now turn our attention to the two peaks of Fig. 45–1, labeled K_α and K_β. These peaks are characteristic of the target material and form part of the *charac-*

Origin of the characteristic x-ray spectrum

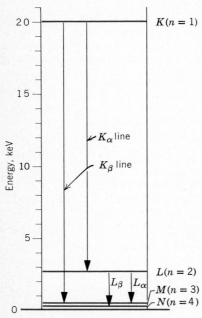

Figure 45-3 The x-ray atomic energy-level diagram for molybdenum. The lines marked K_α and K_β are those so labelled in Fig. 45-1, which also refers to molybdenum.

teristic x-ray spectrum of the element in question. Their origin is as follows: (1) An energetic incident electron strikes an atom in the target and knocks out one of its deep-lying orbital electrons. If it is a K electron, for which the principal quantum number n is 1, there remains a vacancy, or a "hole," as we shall call it, in the K shell. (2) One of the outer electrons moves in to fill this hole, and in the process it emits a characteristic x-ray photon. If the electron falls from the L shell ($n = 2$), we have the K_α line of Fig. 45-1; if it falls from the M shell ($n = 3$), we have the K_β line, and so on. Of course, such a transition will leave a hole in either the L or the M shell, but this will be filled by an electron from still farther out in the atom, thus emitting still another characteristic spectrum line.

Figure 45-3 shows an x-ray atomic energy-level diagram for molybdenum, the element to which Fig. 45-1 refers. The base line ($E = 0$) represents the energy of a neutral molybdenum atom in its ground state. The level marked K ($E = 20.0$ keV) represents the energy of a molybdenum atom with a hole in its K shell. Similarly, the level marked L ($E = 2.7$ keV) represents the energy of an atom with a hole in its L shell, and so on. Note that in representing the energy levels for the hydrogen atom (see Fig. 42-15), we chose a different base line. Rather than the atom in its ground state, we there selected the atom with its electron removed to infinity as our $E = 0$ configuration. Actually, the atomic configuration for which we choose to put $E = 0$ does not matter. Only differences in energy are physically significant, and these are the same no matter what our choice of an $E = 0$ base line is. Our actual choices, both for the hydrogen atom in Fig. 42-15 and for the x-ray spectrum of molybdenum in Fig. 45-3, correspond to convention.

Actually, the levels in Fig. 45-3, except for the K level, are not single as we have shown them. They are split, in part by the spin-orbit coupling mechanism, into a small number of close-lying components. For our purposes, however, we can safely ignore this relatively fine detail of the energy-level diagram.

The transitions K_α and K_β in Fig. 45-3 show the origin of the two peaks in Fig. 45-1. The K_α line, for example, originates when an electron from the L shell of molybdenum—moving *upward* on the energy-level diagram—fills the hole in the K shell. This is the same as saying that a hole—moving *downward* on the diagram—moves from the K shell to the L shell. It is easier to keep track of a single hole than of the 41 electrons in ionized molybdenum that are potentially available to fill it. We have drawn the arrows in Fig. 45-3 from the point of view of hole transitions.

In his investigation of the atomic number concept, Moseley* generated characteristic x rays by using as many elements as he could find—he found 38—as targets for electron bombardment in a special evacuated x-ray tube of his own design. He measured the wavelengths of a number of the lines of the characteristic x-ray spectrum by the crystal diffraction method described in Section 41-10. He then sought, and readily found, regularities in the spectra as he moved from element to element in the periodic table. In particular, he noted that if, for a given spectrum line such as K_α, he plotted the square root of the line frequency ($= \sqrt{\nu} = \sqrt{c/\lambda}$) against the position of the associated element in the periodic table, a straight line resulted. Figure 45-4 shows a portion of his data. We shall see below why it is logical to plot the data in this way and why a straight line is to be expected. Moseley's conclusion from the full body of his data was

---

* See *Moseley and the Numbering of the Elements,* by Bernard Jaffe, Doubleday & Company, New York, 1971, and also *H. G. J. Moseley,* by J. L. Heilbron, University of California Press, (Berkeley, 1974). Moseley was killed by enemy action in World War I at age 27.

Figure 45–4 A so-called Moseley plot of the K_α line of the characteristic x-ray spectra of 21 elements. The frequency ν $(=c/\lambda)$ is calculated from the measured wavelength λ.

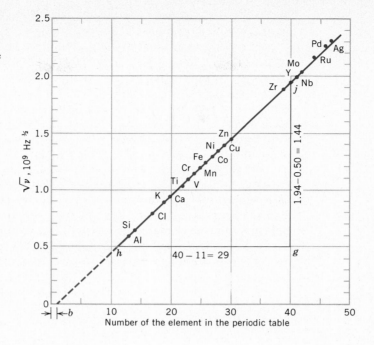

"We have here a proof that there is in the atom a fundamental quantity, which increases by regular steps as we pass from one element to the next. This quantity can only be the charge on the central atomic nucleus."

The atom in Moseley's time

Moseley's achievement can be appreciated all the more when we realize the status of understanding of atomic structure at that time (1913). The nuclear model of the atom had been proposed by Rutherford only 2 years earlier. Little was known about the magnitude of the nuclear charge or the arrangement of the atomic electrons; Bohr published his first paper on atomic structure only in that same year. An element's place in the periodic table was at that time assigned by atomic *weight,* although there were several cases in which it was necessary to invert this order to fit the demands of the chemical evidence. The table had several empty squares, and a surprisingly large number of claims for the discovery of new elements had been advanced; the rare earth elements, because of the problems caused by their similar chemical properties, had not yet been properly sorted out.

Due to Moseley's work, the characteristic x-ray spectrum became the universally accepted signature of an element. Through such studies it became possible to string the elements in a line, so to speak, and to assign consecutive numbers to them, all without the slightest need to know anything about their chemical properties.

X rays and the numbering of the elements

It is not hard to see why the characteristic x-ray spectrum shows such impressive regularities from element to element and the optical spectrum does not. The key to the identity of an element is the charge on its nucleus. This determines the number of its atomic electrons and thus its chemical and physical properties. Gold, for example, is what it is because its atoms have a nuclear charge of $+79e$. If it had just one more unit of charge, it would not be gold but mercury; if it had one fewer, it would be platinum. The K electrons, which play such a large role in the production of the characteristic x-ray spectrum, lie very close to the nucleus and are sensitive probes of its charge. The optical spectrum, on the other hand, is associated with transitions of the outermost or valence electrons. These are heavily screened from the nucleus by the remaining $Z = 1$ electrons of the atom, and are not sensitive probes of the nuclear charge.

Bohr theory and the Moseley plot
It is possible to account for the Moseley plot of Fig. 45–4 in terms of wave mechanics and, in fact, even in terms of Bohr's theory of atomic structure. It may seem surprising that this theory, which works so well for hydrogen but fails even for helium, the next element in sequence, should provide an excellent first approximation when applied to the inner electrons of heavy atoms. This proves to be the case, however, although its success seems to be, in part at least, a matter of the fortuitous cancellation of errors.*

Consider an electron in the L shell of an atom that is about to move into a hole in the K shell, emitting a K_α x-ray photon in the process. This electron "sees" the charge Ze of the atomic nucleus, but it sees it partially screened by the charge $-e$ of the single remaining K electron. In part because of this screening and in part because of readjustments that take place in the electron cloud as a whole the effective atomic number for the transition turns out to be $Z - b$, where $b \approx 1$.

Bohr's formula for the frequency of the radiation corresponding to a transition between any two atomic levels in hydrogenlike atoms is

$$\nu = \frac{me^4 Z^2}{8\epsilon_0^2 h^3}\left(\frac{1}{j^2} - \frac{1}{k^2}\right). \qquad [42\text{–}29]$$

For the K_α transition of Fig. 45–3, it is appropriate to replace Z by $Z - b$ and to substitute 1 for j and 2 for k. Doing so yields

$$\nu = \frac{me^4}{8\epsilon_0^2 h^3}\left(\frac{1}{1^2} - \frac{1}{2^2}\right)(Z - b)^2.$$

Taking the square root of each side leads to

$$\sqrt{\nu} = \left(\frac{3me^4}{32\epsilon_0^2 h^3}\right)^{1/2}(Z - b), \qquad (45\text{–}2a)$$

which we can write in the form

$$\sqrt{\nu} = aZ - ab, \qquad (45\text{–}2b)$$

where a is the indicated constant and $b \approx 1$.

Equation 45–2b represents a straight line, in full agreement with the experimental data of Fig. 45–4. If this plot is extended to higher atomic numbers, however, it turns out to be not quite straight but somewhat concave upwards. Nevertheless, the quantitative agreement with Bohr theory is surprisingly good, as Example 2 shows.

Example 2 Calculate the quantity a in Eq. 45–2b and compare it with the measured slope of the straight line of Fig. 45–4.

From Eq. 45–2 we see that

$$a = \left(\frac{3me^4}{32\epsilon_0^2 h^3}\right)^{1/2}$$

$$= \frac{\sqrt{3}\,(9.11 \times 10^{-31}\ \text{kg})^{1/2}(1.60 \times 10^{-19}\ \text{C})^2}{4\sqrt{2}\,(8.85 \times 10^{-12}\ \text{F/m})(6.63 \times 10^{-34}\ \text{J·s})^{3/2}}$$

$$= 4.95 \times 10^7\ \text{Hz}^{1/2}.$$

Careful measurement of Fig. 45–4, using the triangle hgj, yields

$$a = \frac{gj}{hg} = \frac{(1.94 - 0.50) \times 10^9\ \text{Hz}^{1/2}}{(40 - 11)} = 4.96 \times 10^7\ \text{Hz}^{1/2},$$

which is in agreement with the value predicted by Bohr theory within the uncertainty of the graphical measurement. Note also that the intercept in Fig. 45–4 is in fact ≈ 1, as expected from our shielding argument above.

The agreement with Bohr theory is not nearly as good for other lines in the x-ray spectrum, corresponding to the transitions of electrons farther from the nucleus; here one must rely on calculations based on wave mechanics.

---

* See "Reinterpretation of Moseley's Experiments Relating the K_α Line Frequencies and Atomic Number" by Arthur M. Lesk, *American Journal of Physics*, June 1980.

45–3 Atom Building
and the Periodic Table

Wave mechanics and the periodic table

In the preceding section we saw how it is possible to string the elements in a row and assign them to their proper places in the periodic table, that is, to give them an atomic number Z. These assignments depended on an experimental measurement of the wavelengths of the characteristic x-ray spectra of the elements in question.

Here we take a different approach. We try to see how far the principles of wave mechanics can take us in accounting for the periodic table of the elements. The attempt meets with essentially total success. Every detail of the table (see Appendix D) can be accounted for, including (1) the numbers of elements in the seven horizontal periods into which the table is divided, (2) the similarity of the chemical properties of the elements in the various vertical columns—the alkali metals and the noble gases, for example—and (3) the existence of the rare earth, or lanthanide, series of elements, all crammed into one square of the table. In short, wave mechanics, supplemented by certain guiding principles that we discuss below, accounts for every feature of this table and thus, essentially, for all of chemistry.

Atom building

Let us imagine that—in Tinker Toy fashion—we are going to construct a typical atom for each of the 100+ elements that make up the periodic table. Our starting materials will be a supply of nuclei, each characterized by a charge $+Ze$ and by an appropriate mass, with Z ranging by integers from 1 to 100+. We also need an ample supply of electrons. Our plan is to add Z electrons to each nucleus in such a way as to produce a neutral atom in its ground state.

Success follows only if we observe these three principles of atom building:

Labeling electrons

1. The Quantum Number Principle. The electron in a hydrogen atom may—to mention one possibility—be in a state described by the quantum numbers $n = 2$, $l = 1$, $j = \frac{3}{2}$, and $m_j = -\frac{1}{2}$. It turns out that a particular electron in an atom of, say, zirconium, may also be fully indentified by this same set of quantum numbers. That is not to say that the electrons in these two cases will move in the same way, because they will not. Put another way, although these two electrons may share the same set of quantum numbers, the potentials in which they move—and thus their wave functions—will be quite different. Specifically, the quantum number principle asserts:

The quantum number principle

The hydrogen atom quantum numbers can be used to describe electron states, and to assign electrons to shells and subshells, in any atom, no matter how many electrons it contains. Furthermore, the restrictions among the quantum numbers displayed in Table 44–3 remain in force.

2. The Pauli Exclusion Principle.* This powerful principle was put forward by the Austrian-born physicist Wolfgang Pauli in 1925. Speaking generally, it tells us that no two electrons can be in the same state of motion at the same time. More specifically, it asserts that:

The exclusion principle

In a multielectron atom there can never be more than one electron in any given quantum state.

If this principle did not hold, all atomic electrons would pile up in the K shell.

---

* See "The Exclusion Principle," by George Gamow, *Scientific American*, July 1959.

Chemistry as we know it could not exist, and you would not be here to read this sentence. Nor, in fact, would this sentence be here to be read!

3. The Minimum Energy Principle. As we fill subshells with electrons in the course of atom building, the question arises: In what order shall we fill them? The answer is: When one subshell is filled, put the next electron in which ever vacant subshell lies lowest in energy. To do otherwise would be to depart from our stated aim of building atoms in their ground states.

The lowest-energy subshell can be identified with the help of the following rule, which we first state and then try to make reasonable:

The minimum energy principle

> *For a given principal quantum number n in a multielectron atom, the order of increasing energy of the subshells is the order of increasing l.*

Table 45–1 helps to clarify this rule. Consider first a hydrogen atom whose single electron is in a state with $n = 4$. There are four allowed values of l, namely, 0, 1, 2, and 3. For electrons in true one-electron atoms—such as hydrogen—the energy does not depend on l at all but only on n, being given by

$$E = - \frac{me^4}{8\epsilon_0 h^2} \frac{Z^2}{n^2}, \qquad n = 1, 2, 3, \ldots \ . \qquad [42\text{–}28]$$

Recall that this relation is predicted not only by Bohr theory but also (see Eq. 44–2) by wave mechanics. Putting $Z = 1$ and $n = 4$ in this relation yields, for hydrogen, $E = -0.85$ eV, as Table 45–1 shows.

Table 45–1   Energy levels for electrons with $n = 4$, in three different atoms

State	Orbital quantum number l	Energy, eV		
		Hydrogen[a] $Z = 1$	"Lead"[b] $Z = 82$	Lead[c] $Z = 82$
$4s$	0	−0.85	−5720	−890
$4p$	1	−0.85	−5720	−710
$4d$	2	−0.85	−5720	−420
$4f$	3	−0.85	−5720	−140

[a] A neutral hydrogen atom; see Eq. 42–28.
[b] A hypothetical one-electron atom with $Z = 82$; see Eq. 42–28.
[c] An actual neutral lead atom ($Z = 82$); data from experiment.

A (phoney) lead "atom"

Consider now a lead nucleus ($Z = 82$) around which only a single electron circulates, again in a state with $n = 4$. Equation 42–28 also applies to this (admittedly rather unlikely) one-electron atom. For $Z = 82$ and $n = 4$, the table shows that we have $E = -5720$ eV, once more independent of l. It is not surprising that the single electron in this "atom" lies much lower in energy (that is, has a much higher binding energy) than the electron in hydrogen. It moves in the field of a nucleus with a charge of $+82e$; furthermore, it is drawn in very close to this nucleus, the equivalent Bohr orbit radius (see Eq. 42–27) being 82 times smaller than for hydrogen.

A real lead atom

Finally, let us construct a normal, neutral lead atom by "sprinkling in" the missing 81 electrons. The outermost or valence electrons in lead have $n = 6$, so that an electron with $n = 4$ would lie somewhere in the middle of the smeared-out electron cloud surrounding the lead nucleus.

Equation 42–28 no longer holds for this multielectron atom, but we can find the energies of the four $n = 4$ subshells experimentally from x-ray studies. Their approximate values are shown in the last column of Table 45–1. We see at once that they lie higher in energy (that is, the binding energies are smaller) than for our hypothetical one-electron lead "atom" and that they vary with l just as the minimum energy rule predicts.

That the electrons in lead become more loosely bound when the entire electron cloud is present follows because some substantial fraction of this cloud screens the nucleus electrically, following the principles of Gauss's law (see Example 2, Chapter 25). A typical $n = 4$ electron no longer "sees" the full positive nuclear charge, but rather sees this charge reduced by the negative charge of that part of the electron cloud that lies between the nucleus and the effective radius of the electron in question.

The variation of E with l

As for the variation of energy with l, let us ask ourselves what an $l = 0$ orbit would have to look like under the mechanical constraints of the Bohr picture. Truly to have no angular momentum, the electron would have to oscillate back and forth on a straight line segment passing directly through the nucleus. This does not happen, of course. The equivalent wave mechanical statement is that an electron with $l = 0$ must spend a larger fraction of its time near the nucleus than do electrons with higher values of l. Such electrons would then, on the average, "see" a higher effective nuclear charge and would be more tightly bound; they would lie lower in energy, just as the minimum energy principle and Table 45–1 predict. It is interesting to reexamine Fig. 44–2, which shows the $2s$ state of hydrogen. In this $l = 0$ state there is indeed a marked tendency for the electron to cluster near the nucleus—note the close-in secondary maximum—just as our qualitative argument suggests that it should.

Figure 45–5 embodies the three rules for atom building described above and shows us schematically how the periodic table can be constructed according to the principles of wave mechanics. Energy increases upward in this figure, and the horizontal boxes show the approximate relative energies of the subshells. States with increasing l have been displaced to the left for clarity and grouped into vertical columns according to their l value.

The dependence of E on l

The dependence of energy on l is a dominant feature of Fig. 45–5. Look, for example, at the sequence of states $4s$, $4p$, $4d$, and $4f$. They lie in energy in the order 0, 1, 2, and 3, just as the minimum energy rule requires. In fact, the $4f$ states lie so high that they are above the $5s$ and the $5p$ states in the next shell.

By starting with hydrogen ($Z = 1$) in Fig. 45–5 and following the dashed line, we can see how the seven horizontal periods of the periodic table are built up, each starting with an alkali metal (shaded ☐) and ending with a noble gas (shaded ■). Consider, for example, the long sixth horizontal period, which starts with the alkali metal cesium ($Z = 55$) and ends with the noble gas radon ($Z = 86$). The order in which the subshells are filled, as the dashed line indicates, is $6s$, $4f$, $5d$, and $6p$.

This sixth period contains a run of 15 elements, listed separately at the bottom of the periodic table of Appendix D and called the *lanthanides* or the *rare earths*. Their chemical properties are so similar that all are grouped into a single square of the table. This similarity of chemical properties arises because, while the $4f$ subshell is being filled deep within the electron cloud, an outer screen of one or more $6s$ valence electrons remains in place; it is these outermost or valence electrons that determine the chemical properties of an atom.

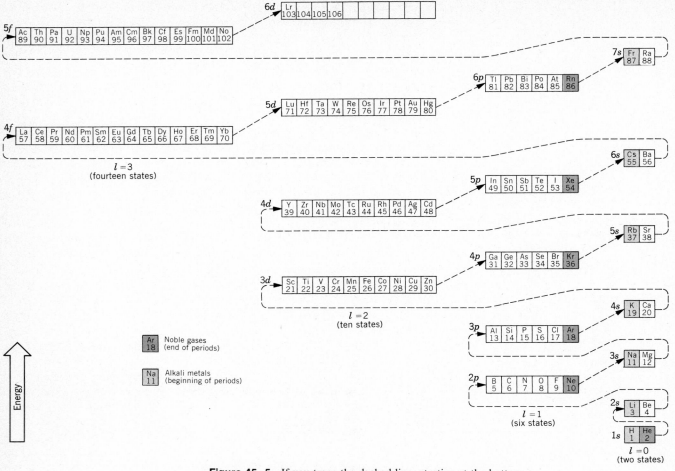

Figure 45–5 If you trace the dashed line, starting at the bottom with hydrogen ($Z = 1$), you will traverse consecutively the seven horizontal periods of the periodic table. Each period starts with an alkali metal and ends with a noble gas. Adapted with permission from *College Chemistry*, by Linus Pauling, W. H. Freeman & Company, San Francisco, 1955.

Example 3 (*a*) From Fig. 45–5, write down the subshells to which the various electrons in a xenon atom ($Z = 54$) in its ground state belong. Xenon, a noble gas, is the last element in the fifth horizontal period of the periodic table. (*b*) Do the same for the adjacent element cesium ($Z = 55$). Cesium, an alkali metal, is the first element in the sixth horizontal period.*

(*a*) Starting from hydrogen ($Z = 1$) and following the dashed line of Fig. 45–5 to xenon ($Z = 54$) allows us to write:

$$1s^2 2s^2 2p^6 3s^2 3p^6 4s^2 3d^{10} 4p^6 5s^2 4d^{10} 5p^6.$$

In this conventional representation the superscripts indicate the numbers of electrons in the corresponding subshell. Note that the superscripts add to 54, the atomic number of xenon.

An *s* state can accommodate two electrons, a *p* state six, and a *d* state ten; see Table 44–4. Thus all the subshells in xenon are filled. Xenon thus has a valence of zero and is chemically inert. Its ionization potential, that is, the energy required to pull an electron from its filled $5p$ shell (= 12.2 eV) is relatively large.

---

* Figure 45–5, which is intended to suggest only the broad sweep of the order of level filling, occasionally gives incorrect valence electron assignments for a few elements near the centers of the three long horizontal periods of the periodic table (copper, for example), for a few lanthanide elements (gadolinium, for example), and for a few actinide elements (curium, for example).

(b) For cesium ($Z = 55$) the electron configuration, from Fig. 45–5, is

$$1s^2 2s^2 2p^6 3s^2 3p^6 4s^2 3d^{10} 4p^6 5s^2 4d^{10} 5p^6 6s^1.$$

Here all the subshells are filled except $6s$, which contains only one electron. Because of this single, loosely bound valence electron, cesium is chemically active and readily enters into chemical combination with the atoms of many other elements. Its ionization potential (= 3.89 eV) is relatively small. The energy required to pull a *second* electron from a cesium atom, however (= 25.1 eV), is relatively large because this electron must be extracted from a filled $5p$ subshell.

Figure 45–6 shows the ionization potentials of the elements. The systematic variation of this property through five of the seven horizontal periods of the periodic table is clear. Note the values for xenon and for cesium.

Example 4 If the electron had no spin, and if the Pauli exclusion principle still held, how would the periodic table be affected? In particular, which of the present elements would be noble gases?

In the absence of spin there would be only three quantum numbers, the spin magnetic quantum number m_s ($= \pm \frac{1}{2}$) being missing. The effect of this would be to cut in half the populations of all the subshells listed in Fig. 45–5. Table 45–2 shows

what the situation would be for the noble gases. Only argon—and that by chance—would remain a noble gas. Hydrogen, one of the most chemically active elements, would be chemically inert, which raises the question of what would we make water out of. The electron simply must have spin!

Table 45–2 Electron spin and the noble gases

The electron spins			The electron does not spin	
Noble gases	Atomic number	Last filled Subshell[a]	Atomic number	Noble gases[b]
He	2	$1s$	1	H
Ne	10	$2p$	5	B
Ar	18	$3p$	9	F
Kr	36	$4p$	18	Ar
Xe	54	$5p$	27	Co
Ra	86	$6p$	43	Tc

[a] See Fig. 45–5.
[b] Other elements beyond $Z = 43$ would also be noble, but we choose not to speculate about subshells beyond $6p$.

Figure 45–6 The ionization energy of the elements as a function of their atomic number. The numbers in the alternately shaded strips are the numbers of elements in the horizontal periods two through five of the periodic table.

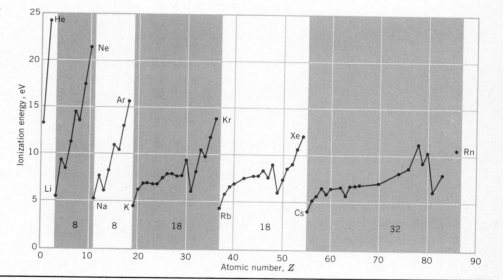

45–4 The Laser*

Quantum physics and technology

In the late 1940s and again in the early 1960s quantum physics made two enormous contributions to technology, the transistor and the laser. The first stimulated the growth of solid-state electronics, which deals with the interaction—at the quantum level—between *electrons* and bulk matter. The laser is leading the way in a new field which deals with the interaction—again at the quantum level—between *photons* and bulk matter.

---

* See ''Laser Light,'' by Arthur L. Schawlow, and ''Applications of Laser Light,'' by Donald H. Herriott, both in *Scientific American*, September 1968.

Let us describe the characteristics of laser light, comparing it with light from sources such as a tungsten filament lamp or a ^{86}Kr gas discharge lamp of the kind used to define the meter as a standard of length. We shall see that referring to laser light as "the light fantastic" goes far beyond whimsy.

Laser light is . . .

. . . sharply defined in wavelength

1. Laser Light Is Highly Monochromatic. Tungsten filament light, spread over a continuous spectrum, provides no basis for comparison. ^{86}Kr light, however, is conventionally taken as highly monochromatic, so much so that it was adopted as a length standard. Its wavelength can be established with a precision of about ± 0.001 nm. However, laser light of about the same wavelength can have a sharpness of definition about a thousand times greater, or $\pm 0.000\,001$ nm.

. . . stable in phase

2. Laser Light Is Highly Coherent. Wavetrains for laser light (see Fig. 40–7b) may easily be several hundred meters long. This means that in principle interference fringes can be set up by two subbeams with optical path lengths of this magnitude. The corresponding lengths for ^{86}Kr light and tungsten filament light are ~ 1 m and a few millimeters, respectively. Laser light wavefronts, which are surfaces over which the phase of the light wave remains constant, extend across the beam so that there is no need to limit the beam, by a pinhole aperture for example, to obtain a coherent source; the whole beam cross section can be used.

. . . virtually parallel

3. Laser Light Is Highly Directional. A laser beam departs from strict parallelism only because of diffraction effects, determined (see Section 41–5) by the wavelength and the diameter of the exit aperture. Light from other sources can be made into an approximately parallel beam by a lens or a mirror, but the beam divergence is much greater than for laser light. Each point on, say, a tungsten filament source forms its own separate beam, the angular divergence of the overall composite beam being determined—not by diffraction—but by the spatial extent of the filament.

. . . highly focusable

4. Laser Light Can Be Sharply Focused. This property is related to the parallelism of the laser beam. As for star light, the size of the focused spot for a laser beam is limited only by diffraction effects and not by the size of the source. Flux densities for focused laser light of $\sim 10^{15}$ W/cm^2 are readily achieved. An oxyacetylene flame, by contrast, has a flux density of only $\sim 10^3$ W/cm^2.

Lasers and their uses

The smallest lasers, used for telephone communication over optical fibers, have as their active medium a semiconducting gallium arsenide crystal about the size of a pinhead. The largest lasers, used for laser fusion research, fill a large building. They can generate pulses of laser light of $\sim 10^{-10}$ s duration which, during the pulse, have a power level of $\sim 10^{13}$ W. This is about ten times the total power-generating capacity of all the electric power stations on earth.

Other laser uses include spot-welding detached retinas, drilling tiny holes in diamonds for drawing fine wires, cutting cloth (50 layers at a time, with no frayed edges) in the garment industry, precision surveying, precise length measurements by interferometry, precise fluid-flow velocity measurements using the Doppler effect, and the generation of holograms, which are stabilized two-dimensional interference "gratings" that make it possible to reconstruct the images of objects in true three-dimensional perspective.

Figure 45–7 shows another example of laser technology, namely, a facility at the National Bureau of Standards used to measure the x-, y-, z-coordinates of a point, by laser interference techniques, with a precision of $\pm 2 \times 10^{-8}$ m ($= \pm 20$ nm). It is used for measuring the dimensions of special three-dimensional gauges which, in turn, are used in industry to check the dimensional accuracy of

Figure 45–7 A partial view of a Michelson interferometer arrangement for measuring the x, y, z-coordinates of a point in space to extreme precision. Note the many laser beam segments. Courtesy National Bureau of Standards.

complicated machined parts. A number of laser beams, visible by scattered light, appear in the figure.

Stimulated emission—the basis of laser operation

The word "laser" is an acronym for *l*ight *a*mplification by *s*timulated *e*mission of *r*adiation. The concept of *stimulated emission* was first introduced by Einstein in 1917.* Before defining it we look at two related but more familiar phenomena involving the interplay between matter and radiation, *absorption* and *spontaneous emission*.

Absorption. Figure 45–8*a* suggests an atomic system in the lower of two possible states, of energies E_1 and E_2. A continuous spectrum of radiation is present. Let a photon from this radiation field approach the two-level atom and interact with it and let the associated frequency ν of the photon be such that

$$h\nu = E_2 - E_1. \tag{45–3}$$

The result is that the photon vanishes and the atomic system moves to its upper energy state. We call this familiar process *absorption*.

Spontaneous emission; metastable states

Spontaneous Emission. In Fig. 45–8*b* the atomic system is in its upper state and there is no radiation nearby. After a mean time $\bar{t}$, this (isolated) atomic system

---

* Einstein not only introduced into physics the essential principle of the laser (stimulated emission) but, in his photon concept, he also introduced the essential principle of the photocell and all other photoelectric devices. Who says that pure theory may not be practical!

Figure 45–8 Showing the interaction between matter and radiation for the processes of (*a*) absorption, (*b*) spontaneous emission, and (*c*) stimulated emission.

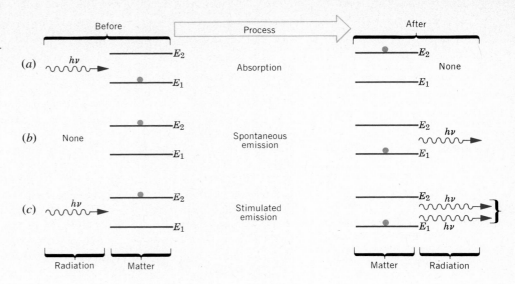

moves of its own accord to the state of lower energy, emitting a photon of energy $h\nu$ in the process. We call this familiar process *spontaneous emission,* in that no outside influence triggered the emission.

Normally the mean life $\bar{t}$ for spontaneous emission by excited atoms is $\sim 10^{-8}$ s. Occasionally, however, there are states for which $\bar{t}$ is much longer, perhaps $\sim 10^{-3}$ s. We call such states *metastable;* they play an essential role in laser operation.

The light from a glowing lamp filament is generated by spontaneous emission. Photons produced in this way are totally independent of each other. In particular, they have different directions and phases. Put another way, the light they produce has a low degree of coherence. To produce sharp interference fringes from the light from a tungsten filament, it is necessary to put a screen containing a pinhole or a narrow slit in front of the filament and to use only the light that emerges from this aperture.

Stimulated Emission. In Fig. 45–8*c* the atomic system is again in its upper state, but this time radiation of frequency given by Eq. 45–3 is present. As in absorption, a photon of energy $h\nu$ interacts with the system. The result is that the system is driven down to its lower state and there are now *two* photons where only one existed before.

The emitted photon in Fig. 45–8*c* is in every way identical with the "triggering" or "stimulating" photon. It has the same energy, direction, phase, and state of polarization. This is how laser light acquires its characteristics. We call the process of Fig. 45–8*c* *stimulated emission.* We can see how a chain reaction of similar processes could be triggered by one such event. This is the "*a*mplification" of the laser acronym.

Consider now a large number of two-level atomic systems. At thermal equilibrium most of them would be in the lower energy state, as in Fig. 45–9*a*. Only a few would be in the upper state, maintained there by thermal agitation of the assembly of atoms at their equilibrium temperature T.

If we expose a system like that of Fig. 45–9*a* to radiation, the dominant process, by sheer weight of numbers, will be absorption. However, if the level populations were inverted, as in Fig. 45–9*b*, the dominant process in the presence of radiation would be stimulated emission and with it the generation of laser light.

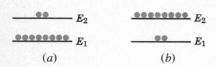

(a)   *(b)*

Figure 45–9 (*a*) The normal thermal equilibrium distribution of atomic systems between two states. If radiation is present, absorption is the dominant process. (*b*) An inverted population, achieved by special techniques. Stimulated emission is here the dominant process.

The three-level laser scheme

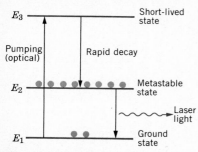

Figure 45-10 The basic three-level scheme for laser operation. Metastable state E_2 is populated more heavily than is the ground state E_1 by "pumping" atoms from this ground state through intermediate state E_3.

The He-Ne laser; the four-level laser scheme

A population inversion like that of Fig. 45-9b is a nonequilibrium situation, far outside normal experience, so that we must use clever tricks to bring it about.

Figure 45-10 shows schematically how a population inversion can be achieved so that laser action—or "lasing" as it is called—can occur. Atoms from ground state E_1 are "pumped" up to an excited state E_3 by some mechanism or other. One possibility, called *optical pumping*, is the absorption of light energy from an intense, continuous-spectrum source that is placed so that it surrounds the lasing material.

From E_3 the atoms decay rapidly to a state of energy E_2. For lasing to occur this state must be metastable, that is, it must have a relatively long mean life against decay by spontaneous emission. If conditions are right, state E_2 can then become more heavily populated than state E_1, thus providing the needed population inversion. A stray photon of the right energy can then trigger an avalanche of stimulated emission events, resulting in the production of laser light. A number of lasers using crystalline solids (such as ruby) as a lasing material operate in the three-level mode, whose essential features are exhibited in Fig. 45-10.

Figure 45-11 shows the elements of a type of laser that is often found in student laboratories. The glass discharge tube is filled with, say, an 80%–20% mixture of the noble gases helium and neon, the helium being the "pumping" medium and the neon the "lasing" medium. Figure 45-12 is a simplified version of the level structures for these two atoms. Note that four levels, labeled E_0, E_1, E_2, and E_3, are involved in this lasing scheme, rather than three levels as in Fig. 45-10.

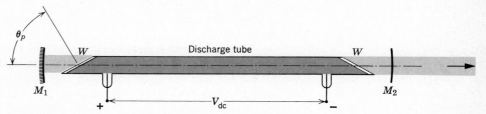

Figure 45-11 The elements of a helium-neon gas laser.

Function of mirrors M_1 and M_2

Pumping is accomplished by setting up an electrically induced gas discharge in the helium-neon mixture. Electrons and ions in this discharge, occasionally enough, collide with helium atoms, raising them to level E_3 in Fig. 45-12. This level is metastable, spontaneous emission to the ground state (level E_0) being very infrequent. Level E_3 in helium (= 20.61 eV) is, by chance, very close to level E_2 in neon (= 20.66 eV), so that, during collisions between helium and neon atoms, the excitation energy of the former can readily be transferred to the latter. In this way level E_2 in Fig. 45-12 can become more highly populated than level E_1 in that figure. This population inversion is maintained because (1) the metastability of level E_3 ensures a ready supply of neon atoms in level E_2 and (2) level E_1 decays rapidly (through intermediate stages not shown) to the neon ground state, E_0. Stimulated emission from level E_2 to level E_1 predominates and red laser light, of wavelength 632.8 nm, is generated.

Most stimulated emission photons initially produced in the discharge tube of Fig. 45-11 will not happen to be parallel to the tube axis, and will be quickly stopped at the walls. Stimulated emission photons that *are* parallel to the axis however can move back and forth through the discharge tube many times by successive reflections from mirrors M_1 and M_2 in Fig. 45-11. A chain reaction thus builds up rapidly in this direction and the inherent parallelism of the laser light finds its origin.

Figure 45–12 Showing four essential levels in helium and neon atoms in a He-Ne gas laser, leading to the production of red laser light of wavelength 632.8 nm.

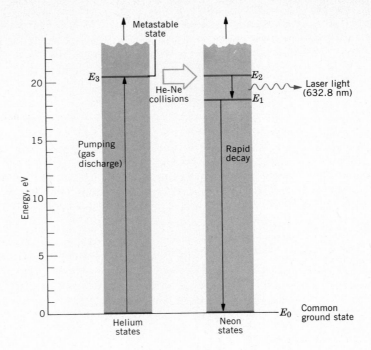

Rather than thinking in terms of the photons bouncing back and forth between the mirrors, it is perhaps more useful to think of the entire arrangement of Fig. 45–11 as an optical resonant cavity which, like an organ pipe for sound waves, can be tuned to be sharply resonant at one (or more) wavelengths.

The mirrors M_1 and M_2 are concave, with their focal points nearly coinciding at the center of the tube. Mirror M_1 is coated with a dielectric film whose thickness is carefully adjusted to make the mirror as close as possible to totally reflective at the wavelength of the laser light; see Example 4, Chapter 40. Mirror M_2, on the other hand, is coated so as to be slightly "leaky," so that a small fraction of the laser light can escape at each reflection to form the useful beam.

The windows W, W in Fig. 45–11, which close the ends of the discharge tube, are slanted so that their normals make an angle θ_p, the Brewster angle, with the tube axis, where

$$\tan \theta_p = n, \qquad [39\text{--}14]$$

Lossless transmission through windows W, W

n being the index of refraction of the glass for the laser light wavelength. In Section 39–6 we showed that such windows transmit light without loss by reflection, provided only that the light is polarized with its plane of polarization in the plane of Fig. 45–11. If the windows were normal to the tube ends, beam loss by reflection (~4% from each surface of each mirror) would make laser operation impossible.

Example 5 A pulsed ruby laser has as its active element a synthetic ruby crystal in the form of a cylinder 6.0 cm long and 1.0 cm in diameter. Ruby consists of Al_2O_3 in which—in this case—one aluminum ion in every 3500 has been replaced by a chromium ion, Cr^{3+}. It is in fact the optical absorption properties of this small chromium "impurity" that account for the characteristic color of ruby. These same ions also account for the lasing ability of ruby, which occurs—by the three-level mechanism of Fig. 45–10—at a wavelength of 694.4 nm.

Suppose that *all* of the Cr^{3+} ions are in a metastable state corresponding to state E_2 of Fig. 45–10 and that *none* are in the ground state E_1. How much energy is available for release in a single pulse of laser light if *all* of these ions revert to the ground state in a single stimulated emission chain reaction episode?

Our answer will be an upper limit only because the conditions postulated cannot be realized in practice. The density ρ of Al_2O_3 is 3700 kg/m³, and its molecular weight A is 0.102 kg/mol.

The number of Al^{3+} ions is

$$N_{Al} = \frac{2N_A M}{A} = \frac{2N_A \rho V}{A}$$

The volume V is given by

$$V = (\pi/4)(1.0 \times 10^{-2}\ \text{m})^2(6.0 \times 10^{-2}\ \text{m})$$
$$= 4.7 \times 10^{-6}\ \text{m}^3.$$

Thus

$$N_{Al} = \frac{(2)(6.0 \times 10^{23}/\text{mol})(3.7 \times 10^3\ \text{kg/m}^3)(4.7 \times 10^{-6}\ \text{m}^3)}{0.102\ \text{kg/mol}}$$
$$= 2.1 \times 10^{23}$$

Here M is the mass of the ruby cylinder and $V(= \pi d^2 h/4)$ is its volume. The factor 2 comes about because there are two aluminum ions in each "molecule" of Al_2O_3. The number of Cr^{3+} ions is then

$$N_{Cr} = \frac{N_{Al}}{3500} = 6.0 \times 10^{19}.$$

The quantum energy of the stimulated emission photon is

$$E = h\nu = \frac{hc}{\lambda} = \frac{(4.1 \times 10^{-15}\ \text{eV·s})(3.0 \times 10^8\ \text{m/s})}{(694 \times 10^{-9}\ \text{m})} = 1.8\ \text{eV},$$

and the total available energy per laser pulse is

$$U = N_{Cr}E = (6.0 \times 10^{19})(1.8\ \text{eV})(1.6 \times 10^{-19}\ \text{J/eV}) = 17\ \text{J}.$$

Such large pulse energies have indeed been achieved, but only by much more elaborate laser arrangements than that described here.

In this example we have postulated an ideal circumstance, namely, a total population inversion, in which the ground state remains virtually unpopulated. The actual population inversion in a working ruby laser will be very much less than total. For this and other reasons the pulse energy in practice will be very much less than the upper limit calculated above.

REVIEW GUIDE AND SUMMARY

The continuous x-ray spectrum

The smooth background curve in Fig. 45-1 is called the continuous x-ray spectrum. It arises (see Fig. 45-2) when energetic electrons are decelerated in the field of a nucleus. The minimum cutoff wavelength is given by

$$\lambda_{min} = \frac{hc}{eV}, \qquad\qquad [45-1]$$

where V, the only variable, is the accelerating potential of the electrons; see Example 1.

The characteristic x-ray spectrum

The sharp peaks in Fig. 45-1 belong to the characteristic x-ray spectrum of the target element. The peaks arise (see Fig. 45-3) from electron transitions deep within the atom. The frequency of any characteristic x-ray line (the K_α line, say) varies smoothly with the position of the element in the periodic table. This fact can be used to assign atomic numbers to the elements, by way of a Moseley plot such as that of Fig. 45-4. Example 2 suggests that Bohr theory gives a surprisingly good account of the Moseley plot, at least for x-ray lines associated with vacancies in the K shell.

Atom building and the periodic table

In building up a set of atoms to display as a periodic table, three rules must be followed: (1) Electron states in multielectron atoms are to be labeled with the hydrogen atom quantum numbers, and the rules for manipulating these numbers must be followed. (2) Only one electron may be assigned to each quantum state (the Pauli exclusion principle). (3) For a given n, the order of increasing energy is the order of increasing l; see Table 45-1. Figure 45-5, which should be compared in detail with the periodic table of Appendix D, suggests how this table can be completely accounted for on wave mechanical principles. Study Examples 3 and 4.

The laser, a stimulated emission device

Two requirements for laser operation

The special properties of laser light (see p. 918) are directly traceable to a chain reaction of stimulated emission events in a "lasing" medium. Figure 45-8c shows the basic process. Lasing always involves a pair of levels and, for stimulated emission to be the controlling process, there are two requirements: (1) The upper level (E_2) must be more heavily populated than the lower level (E_1); see Fig. 45-9b. (2) The upper level must be metastable, that is, it must have a relatively long mean lifetime against depletion by spontaneous emission (Fig. 45-8b). Figures 45-10 and 45-12 shows how the necessary population inversion can be brought about, by optical pumping in the first case and by gas collision excitation processes in the second. Study Example 5.

QUESTIONS

1. What is the origin of the cutoff wavelength λ_{min} of Fig. 45–1? Why is it an important clue to the photon nature of x rays?

2. What are the characteristic x rays of an element? How can they be used to determine the atomic number of the element?

3. Compare Figs. 45–1 and 45–3. How can you be sure that the two prominent peaks in the former figure do indeed correspond numerically with the two transitions similarly labeled in the latter figure?

4. Can atomic hydrogen be caused to emit x rays? If so, describe how. If not, why not?

5. How does the x-ray energy-level diagram of Fig. 45–3 differ from the energy-level diagram for hydrogen, displayed in Fig. 42–15? In what respects are the two diagrams similar?

6. When extended to higher atomic numbers, the Moseley plot of Fig. 45–4 is not a straight line but is slightly concave upwards. Does this affect the ability to assign atomic numbers to the elements?

7. Why is it that Bohr's theory, which otherwise does not work very well even for helium ($Z = 2$), gives such a good account of the characteristic x-ray spectra of the elements?

8. The periodic table of the elements was based originally on atomic weight, rather than on atomic number, the latter concept having not yet been developed. Why were such early tables as successful as they proved to be? In other words, why is the atomic weight of an element (roughly) proportional to its atomic number?

9. How does the structure of the periodic table support the need for a fourth quantum number, corresponding to electron spin?

10. If there were only three quantum numbers (that is, if the electron did not spin), how would the chemical properties of helium be different?

11. Why does it take more energy to remove an electron from neon ($Z = 10$) than from sodium ($Z = 11$)?

12. Does it make any sense to assign quantum numbers to a vacancy in an otherwise filled subshell?

13. Why do the lanthanide elements (see Appendix D) have such similar chemical properties? How can we justify putting them all into a single square of the periodic table? Why is it that, in spite of their similar chemical properties, they can be so easily sorted out by measuring their characteristic x-ray spectra?

14. In your own words, state the minimum energy principle for atom building and give a physical argument in support of it.

15. A beam of light emerges from an aperture in a "black box" and moves across your laboratory bench. How could you test this beam to find out the extent to which it is coherent over its cross section? How could you tell (without opening the box) whether or not the concealed light source is a laser?

16. Why is focused laser light inherently better than focused light from a tiny incandescent lamp filament for such delicate surgical jobs as spot-welding detached retinas?

17. Laser light forms an almost parallel beam. Does the intensity of such light fall off as the inverse square of the distance from the source?

18. In what ways are laser light and star light similar? In what ways are they different?

19. Arthur Schawlow, one of the laser pioneers, invented a typewriter eraser, based on focusing laser light on the unwanted character. What is its principle of operation.

20. In what ways do spontaneous emission and stimulated emission differ?

21. We have spontaneous emission and stimulated emission. From symmetry, why don't we also have spontaneous and stimulated absorption? Discuss in terms of Fig. 45–8.

22. Why is a population inversion between two atomic levels necessary for laser action to occur?

23. What is a metastable state? What role do such states play in the operation of a laser?

24. Comment on this statement: "Other things being equal, a four-level laser scheme such as that of Fig. 45–12 is preferable to a three-level scheme such as that of Fig. 45–10 because, in the latter scheme, half of the population of atoms in level E_1 must be moved to state E_2 before a population inversion can even begin to occur."

25. Comment on this statement: "In the laser of Fig. 45–11, only light whose plane of polarization lies in the plane of that figure is transmitted through the right-hand window W. Therefore half of the energy potentially available to the output beam is lost." (*Hint:* Is this second statement really true? Consider what happens to photons whose effective plane of polarization is at right angles to the plane of Fig. 45–11? Do such photons participate fully in the stimulated emission amplification process?)

EXERCISES

Section 45-2   X rays and the Numbering of the Elements

1. Determine Planck's constant from the fact that the minimum x-ray wavelength produced by 40.0-keV electrons is 31.1 pm.

2. Electrons bombard a molybdenum target, producing both continuous and characteristic x rays as in Fig. 45–1. In that figure the energy of the incident electrons is 35 keV. If the accelerating potential applied to the x-ray tube is increased to 50.0 kV, what new values of (a) λ_{min}, (b) λ_{K_α}, and (c) λ_{K_β} result?

3. Show that the short-wavelength cutoff in the continuous x-ray spectrum is given by

$$\lambda_{min} = 1240 \text{ pm}/V,$$

where V is the applied potential difference in kilovolts.

4. What is the minimum potential difference across an x-ray tube that will produce x rays with a wavelength of 1.0 Å?

5. In Fig. 45–1, the x rays shown are produced when 35.0-keV electrons fall on a molybdenum target. If the accelerating potential is maintained at 35.0 kV but a silver target ($Z = 47$) is substituted for the molybdenum target, what values of (a) λ_{min}, (b) λ_{K_β}, and (c) λ_{K_α} result? The K, L, and M atomic x-ray levels for silver (compare Fig. 45–3) are 25.51, ~3.56, and ~0.53 keV.

Section 45–3   Atom Building and the Periodic Table
6. Two electrons in lithium ($Z = 3$) have as their quantum numbers n, l, m_l, and m_s, the values 1, 0, 0, and $\pm\frac{1}{2}$, respectively. (a) What quantum numbers can the third electron have if the atom is to be in its ground state? (b) If the atom is to be in its first excited state?

7. What are the quantum numbers of the three electrons in lithium according to the n, l, j, m_j scheme? Assume the atom to be in its ground state.

8. Aluminum ($Z = 13$) in its ground state has, according to Fig. 45–5, filled $1s$, $2s$, $2p$, and $3s$ subshells and has a single electron in the $3p$ subshell. (a) What values of n, l, and j might this electron have? (b) How would you use a Stern-Gerlach type of experiment to decide between the possibilities?

Section 45–4   The Laser
9. Lasers have become very small as well as very large. The active volume of a laser constructed of the semiconductor GaAlAs has a volume of only 200 $(\mu m)^3$ (smaller than a grain of sand) and yet it can continuously deliver 5 mW of power at 0.8-μm wavelength. What is the production rate of photons?

10. A ruby laser emits light at wavelength 693.4 nm. If a laser pulse is emitted for 10^{-11} s and the energy release per pulse is 0.15 J, (a) what is the length of the pulse, and (b) how many photons are in each pulse?

11. Consider two single atomic levels with energies E_m and E_n, where $E_m > E_n$. If a system of such atoms is in thermal equilibrium at temperature T, the ratio of the populations of these two states is given by

$$\frac{N_m}{N_n} = e^{-(E_m - E_n)/kT},$$

in which k ($= 8.63 \times 10^{-5}$ eV/K) is Boltzmann's constant. In a particular atomic system, $E_m - E_n = 1.0$ eV. (a) Calculate N_m/N_n at room temperature ($T = 300$ K). The smallness of your answer shows the small role played by thermal agitation in populating the upper state. (b) At what temperature T would 0.01% of the atoms be in the state with the higher energy?

PROBLEMS GROUPED BY SECTION

Section 45–2   X rays and the Numbering of the Elements
1. Show that a moving electron cannot spontaneously change into an x-ray photon in free space. A third body (atom or nucleus) must be present. Why is it needed?

2. A tungsten target ($Z = 74$) is bombarded by electrons in an x-ray tube. (a) What is the minimum value of the accelerating potential that will permit the production of the characteristic K_β and K_α lines of tungsten? (b) For this same accelerating potential, what is λ_{min}? (c) What are λ_{K_β} and λ_{K_α}? The K, L, and M atomic x-ray levels for tungsten (see Fig. 45–3) are 69.5, ~11.3, and ~2.3 keV, respectively.

3. X rays from an x-ray tube with a cobalt target are composed of the strong K lines of cobalt and weak K line dues to impurities in the target. The measured wavelengths are 178.5 pm for the strong K_α line of cobalt, and 228.5 pm and 153.9 pm for the weak K_α lines of the impurities. Use Eq. 45–2, sometimes known as Moseley's law, to identify the impurities. In this equation assume that the constant a is unknown but that $b = 1$.

4. A molybdenum target ($Z = 42$) is bombarded with 35-keV electrons and the x-ray spectrum of Fig. 45–1 results. Here $\lambda_{K_\beta} = 63$ pm and $\lambda_{K_\alpha} = 71$ pm. (a) What are the corresponding photon energies? (b) It is desired to filter these radiations through a material that will absorb the K_β line much more strongly than it will absorb the K_α line. What substance would you use? The K ionization energies for molybdenum and for four neighboring elements are as follows:

Z	40	41	42	43	44
Element	Zr	Nb	Mo	Tc	Ru
E_K (keV)	18.00	18.99	20.00	21.04	22.12

(*Hint:* A substance will selectively absorb one of two x radiations more strongly if the photons of one have enough energy to eject a K-shell electron from the atoms of the substance but the photons of the other do not.)

5. The binding energies of K-shell and L-shell electrons in copper are 8.979 keV and 0.951 keV, respectively. A K_α x ray from copper is incident on a sodium chloride crystal and gives a first-order Bragg reflection at 15.9° when reflected from the alternating planes of the sodium atoms. What is the spacing between these planes?

Section 45–3   Atom Building and the Periodic Table
6. In the alkali metals there is one electron outside a closed shell. (a) Using Eq. 42–28, calculate the energy of this electron for each of the alkali metals through ^{55}Cs assuming, first, that $Z = 1$ in every case, and second, that Z has its actual value. (b) What logical support does each calculation have? (c) Which argument does the actual measured energies, being the negative of the ionization energies found in Fig. 45–6, seem to favor? The relevant quantum numbers are found in Fig. 45–5.

7. A chlorine atom ($Z = 17$) in its ground state has, according to Fig. 45–5, filled $1s$, $2s$, $2p$, and $3s$ subshells and has five $3p$ electrons, leaving one "hole" in that subshell. (a) What quantum numbers n, l, and j can you assign to the hole? Make use of the fact that in a given subshell, states with lower j are filled first. What are (b) the angular momentum (maximum projected value) and (c) the magnetic moment (maximum projected value) of the electrons in the entire chlorine atom?

8. Four identical, noninteracting particles are placed in a one-dimensional box of length 0.50 nm. Find the lowest energy level of the system and list the quantum states (a) if the exclusion principle did not apply and (b) taking into account the exclusion principle.

Section 45-4 The Laser

9. A high-powered laser beam ($\lambda = 600$ nm) with a beam diameter of 12 cm is aimed at the moon, 3.8×10^5 km distant. The spreading of the beam is caused only by diffraction effects. The angular location of the edge of the central diffraction disk (see Fig. 41-11) is given by

$$\sin\theta = \frac{1.22\lambda}{d}, \qquad [41-10]$$

where d is the diameter of the beam aperture. What is the diameter of the central diffraction disk at the moon's surface?

10. The beam from an argon laser ($\lambda = 515$ nm) has a diameter d of 3.0 mm and a continuous-wave power output of 5.0 W. The beam is focused onto a diffuse surface by a lens (assumed ideal) whose focal length f is 3.5 cm. A diffraction pattern such as that of Fig. 41-11 is formed, the radius of the central disk being given by

$$R = \frac{1.22\, f\lambda}{d}. \qquad [\text{see Eq. } 41-10]$$

The central disk can be shown to contain 84% of the incident power. Find (a) the radius R of the central disk, (b) the average power flux density in the incident beam, and (c) in the central disk.

11. The use of lasers for missile defense has been proposed as part of the Strategic Defense Initiative ("Star Wars"). A beam of intensity 10^8 W/m^2 would probably burn into and destroy a hardened missile in 1 s. (a) If the laser has 5-MW power, 3-μm wavelength and 4.0-m beam diameter (a very powerful laser indeed), would it destroy a missile at a distance of 3000 km? (b) If the wavelength could be changed, what maximum value would work? Use the equation for the central disk given in Problem 10 and take the focal length to be the distance to the target.

12. The active medium in a particular ruby laser ($\lambda \cong 694$ nm) is a synthetic ruby crystal 6.00 cm long and 1.00 cm in diameter. The crystal is silvered at one end and—to permit the formation of an external beam—only partially silvered at the other. (a) Treat the crystal as an optical resonant cavity in analogy to a closed organ pipe and calculate the number of standing-wave nodes there are along the crystal axis. (b) By what amount $\Delta\nu$ would the beam frequency have to shift to increase this number by unity? Show that $\Delta\nu$ is just the inverse of the travel time of light for one round trip back and forth along the crystal axis. (c) What is the corresponding fractional frequency shift $\Delta\nu/\nu$? The appropriate index of refraction is 1.75.

ADDITIONAL PROBLEMS

13. X rays are produced in an x-ray tube with a target potential of 50 kV. If an electron makes three collisions in the target before coming to rest and loses one-half of its remaining kinetic energy on each of the first two collisions, determine the wavelengths of the resulting photons. Neglect the recoil of the heavy target atoms.

14. Suppose that there are two electrons in the same system, both of which have $n = 2$ and $l = 1$. (a) If the exclusion principle did not apply, how many combinations of states are conceivably possible? (b) How many states does the exclusion principle forbid? Which ones are they?

15. From Fig. 45-1, calculate approximately the energy difference $E_L - E_M$ for the x-ray atomic energy levels of molybdenum. Compare with the result that may be found from Fig. 45-3.

16. Estimate the ratio of energies of photons from the K_α transitions from two atoms, Z and Z'. How much more energetic is a K_α x ray from uranium expected to be than from aluminum? From lithium?

17. It is entirely possible that techniques for modulating the frequency or amplitude of a laser beam will be developed so that such a beam can serve as a carrier for television signals, much as microwave beams do now. Also laser systems might be available whose wavelengths can be precisely "tuned" to any wavelength in the visible range, that is, in the range 400 nm $< \lambda <$ 700 nm. If a television channel occupies a bandwidth of 10 MHz, how many channels could be accommodated with this laser technology? Comment on the intrinsic superiority of visible light to microwaves as carriers of information.

18. Determine how close the theoretical K_α x-ray energies, obtained from Eq. 45-2, are to the measured energies in the light elements from lithium to magnesium. To do this you must first (a) determine the constant a^2h, in electron volts, to five significant figures using data in Appendix B. Next, (b) calculate the percentage deviation of the theoretical from the measured energies. (c) Plot the deviation and comment on the trend. The measured values of the K_α x rays are

Li	54.3 eV	0	524.9
Be	108.5	F	676.8
B	183.3	Ne	848.6
C	277	Na	1041
N	392.4	Mg	1254

(There is actually more than one K_α x ray because of splitting of the L energy level, but that effect is negligible in the elements considered.)

19. Show formally that for a filled subshell, both the resultant angular momentum and the resultant magnetic moment are zero.

20. The NOVA laser at Lawrence Livermore National Laboratory is used in fusion experiments (See Section 48-6). Its beam is designed to deliver between 0.1 and 1 MJ of energy in nanoseconds to a target of solid deuterium and tritium (isotopes of hydrogen) to heat the material and increase its density. (a) Assume that two 150-J laser pulses simultaneously impact from opposite directions on a sphere of 1.0-mm radius containing 4.0 μg of fuel. If the pulses last 3.0 ns and are completely absorbed, estimate the maximum radiation pressure on the sphere. [For simplicity, assume that each pulse is uniformly spread over an area $\pi(1.0$ mm$)^2$ and use Eq. 38-15.] (b) If 1% of the energy is effective in raising the temperature, how hot will the fuel get? Assume that the fuel is entirely deuterium with an atomic weight of 2.

46 Electrical Conductivity in Solids

46–1 Introduction

Wave mechanics applied to solids

Up to this point we have sought to demonstrate that wave mechanics gives a valid description, correct in essentially every detail, of the behavior of matter in the form of isolated atoms. Here we hope to show, by a single broad example, that this powerful theory is equally valid when applied to aggregates of atoms in the form of solids. In fact, one of the greatest technological advances in the human experience—the electronic computer—is based squarely on the insights gained in this way.

We treat only one aspect of solid state physics in this chapter, namely, the conduction of electricity in solids. With this in mind, we restrict our considerations to three types of solids: metals, semiconductors, and insulators. Copper, silicon, and carbon (diamond form) serve, respectively, as their prototypes.

46–2 Conduction Electrons—The Allowed States

The free-electron model

In the *free-electron model* of a metal, the conduction electrons are imagined to be free to move throughout the volume of the metal, much like the molecules of a gas confined to a container. In Section 28–4 (which should be reviewed carefully at this time) we used this model in a semiclassical way to derive Eq. 28–14 for the resistivity of a metal, and we interpreted that equation as a statement that metals obey Ohm's law. We now take a closer look at this model for the conduction electrons, from the point of view of wave mechanics. We start by reviewing a familiar, one-dimensional analog.

An electron between rigid walls—a review

In Section 43–2 we analyzed the one-dimensional problem of an electron confined between rigid walls of separation l. Our conclusion was that the energy of such a particle is quantized, the allowed energies being given by

$$E_n = n^2 \frac{h^2}{8ml^2}, \qquad n = 1, 2, 3, \ldots, \qquad [43-12]$$

in which n is a quantum number. Each such quantized energy characterizes a discrete state of the system, defined by the boundary condition that ψ^2 for the trapped electron must be zero at the (rigid) walls; see Figs. 43–7 and 43–8. Only those wave functions that can be "fitted" in this way between the walls are physically acceptable.

Confinement means quantization

An electron bouncing around in a three-dimensional "box" defined by the surface of a chunk of metal—a cube of copper, say—is just an extension to three dimensions of the one-dimensional problem of the electron trapped between rigid walls. Here too we must have $\psi^2 = 0$ at the walls of the box. We select only those discrete wave functions that can be fitted into the box with this boundary restriction. They will, as we expect, have discrete and quantized energies. Just as the quantized states for electrons in an atom determine the chemical properties of the atom, so too do the discrete quantized states for electrons in a metal determine the properties of the metal. Our particular interest, as we have indicated, is in the electrical conductivity.

For a string of truly infinite length, transverse waves of *any* wavelength are possible and there is no quantization of states. For a string of finite length, however, the phenomena of energy quantization and of discrete stationary states appear. In the same way, for a metallic specimen of truly infinite size, the allowed states for the conduction electrons would be distributed continuously in energy and quantization would be absent. For a metallic specimen of finite size, however, we do expect to find quantization of energy, manifested by the existence of discrete allowed states.

A statistical approach is called for

Most metallic specimens, while not of course of infinite extent, are nevertheless macroscopic. By this we mean that the volume of the "box" in which a conduction electron is free to move is enormously large in terms of atomic dimensions. The allowed states under these circumstances, although still discrete, are so numerous and lie so very close together in energy that we must use statistical methods to count them. In much the same way (see Fig. 21–9) we used a statistical approach to describe the distribution of speeds for the molecules of a gas.

For the conduction electron gas we define a *state distribution function* $n_s(E)$ such that the number of allowed conduction electron states whose energies lie between E and $E + dE$ is given by $n_s(E)\, dE$. The process of counting states that we referred to above yields, on a per unit volume basis,*

The state distribution function

$$n_s(E) = \left(\frac{35.5 m^{3/2}}{h^3} \right) E^{1/2}, \qquad (46-1)$$

in which m is the electron mass and 35.5 $(= 2^{7/2}\, \pi)$ is a numerical factor. The subscript s reminds us that we are at this stage counting states and not the electrons that may or may not occupy them. Figure 46–1a shows a plot of Eq. 46–1.

---

* See Chapter 11 of *Quantum Physics,* by Robert Eisberg and Robert Resnick, John Wiley & Sons, 1974, for a derivation of this and other results of this chapter.

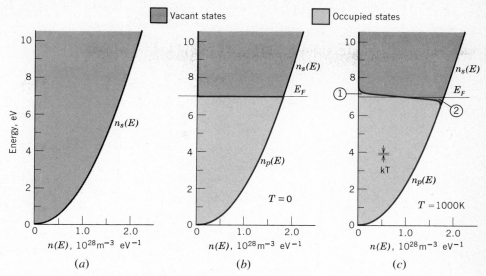

| Vacant states | Occupied states |

Figure 46-1 (a) Showing $n_s(E)$, the number of conduction electron states per unit volume and per unit energy interval as a function of energy. (b) The curve labeled $n_p(E)$ shows the distribution of electrons by energy at absolute zero. The curve labeled $n_s(E)$ shows the distribution of states. (c) The same for $T = 1000$ K.

Example 1 A tiny cube of copper is 0.10 mm on edge. How many conduction electron states are there whose energies lie in the range 5.000 to 5.025 eV?

These energy limits are so close together that we can say that the answer, on a per unit volume basis, is $n_s(E) \Delta E$, where $E = 5.00$ eV and $\Delta E = 0.025$ eV. From Eq. 46–1 we have

$n_s(E)$

$$= \frac{(35.5)(9.11 \times 10^{-31} \text{ kg})^{3/2}[(5.00 \text{ eV})(1.60 \times 10^{-19} \text{ J/eV})]^{1/2}}{(6.63 \times 10^{-34} \text{ J·s})^3}$$

$$= (9.48 \times 10^{46}) \text{m}^{-3} \text{ J}^{-1} = 1.52 \times 10^{28} \text{ m}^{-3} \text{ eV}^{-1}$$

This numerical result can readily be verified from Fig. 46–1a.

The total number of allowed states in the specified volume and the specified energy range is, then, if V is the volume of the copper cube,

$$N = n_s(E) \, \Delta E \, V$$
$$= [(1.52 \times 10^{28}) \text{m}^{-3} \text{ eV}^{-1}](0.025 \, eV)[(1.00 \times 10^{-4}) \text{m}]^3$$
$$= 3.8 \times 10^{14}.$$

Even in such a narrow energy band and for such a small—but still grossly macroscopic—volume, the number of allowed states is large. Small wonder that we must use statistical methods!

Note that neither Eq. 46–1 nor the result of this example depend in any way on the material of which the cube is made. Fitting and counting the allowed states is a purely geometrical problem.

46-3 Filling the Allowed States

The Pauli exclusion principle

Now that we have seen how many allowed conduction electron states there are, we can begin to fill them with electrons. We went through this process in Section 45–3 in connection with building up the periodic table. There we saw the central importance of the Pauli exclusion principle, which states that each allowed state may be occupied by no more than one electron.* This powerful principle is equally important in our present problem.

---

* We sometimes say that there can be two electrons per state provided that their spins point in opposite directions. This factor of 2, however, has already been absorbed into Eq. 46–1.

To produce the ground—or lowest-energy—state of a metal, corresponding to the absolute zero of temperature, we sprinkle the electrons, one at a time, into the lowest unoccupied levels. We proceed to fill up these vacant states until the last electron is in place. The energy of this highest-energy occupied state at the absolute zero of temperature is called the *Fermi energy* of the metal; it is marked E_F in Fig. 46–1b and for copper has the value of 7.05 eV. The colored curve marked $n_p(E)$ in Fig. 46–1b indicates the distribution of electrons that we have described. The subscript p reminds us that here we are counting *p*articles (electrons) or— what is the same thing—occupied states.

We can find the Fermi energy for a metal by adding up (integrating) the number of occupied states in Fig. 46–1b between zero energy and the Fermi energy. The result must be n_0, the number of conduction electrons per unit volume for that metal. In equation form,

$$n_0 = \int_0^{E_F} n(E) \, dE. \tag{46–2}$$

Note that n_0 is represented by the area shaded gray in Fig. 46–1b.

Example 2 Use Eq. 46–2 to find the Fermi energy for copper. Substituting Eq. 46–1 into Eq. 46–2 yields

$$n_0 = \left(\frac{35.5m^{3/2}}{h^3}\right) \int_0^{E_F} E^{1/2} \, dE = \left(\frac{35.5m^{3/2}}{h^3}\right) \left(\tfrac{2}{3} E_F^{3/2}\right)$$

or

$$E_F = \left(\frac{0.121h^2}{m}\right) n_0^{2/3}. \tag{46–3}$$

As in Example 2 of Chapter 28, we find that $n_0 = (8.49 \times 10^{28})\text{m}^{-3}$ for copper. Substituting this and other quantities into Eq. 46–3 yields

$$E_F = \frac{(0.121)(6.63 \times 10^{-34} \text{ J·s})^2 (8.49 \times 10^{28} \text{ m}^{-3})^{2/3}}{(9.11 \times 10^{-31} \text{ kg})}$$

$$= 1.13 \times 10^{-18} \text{ J} = 7.05 \text{ eV}$$

It is a popular misconception that, at the absolute zero of temperature, all motion ceases and all systems tend to a state of zero kinetic energy. Figure 46–1b, however, shows that, because of the Pauli exclusion principle, this statement is certainly not true for the conduction electrons in a metal. The relatively large average energy of the electrons at $T = 0$ is, in fact, the striking feature of this figure. The system is far from being in a state of no motion. Simple inspection shows that the average energy of the electrons is several electron volts. By contrast, we can easily show (see Example 7 of Chapter 21) that the mean translational kinetic energy for the molecules of an ideal gas is only ~ 0.040 eV at room temperature; it is only ~ 0.001 eV at $T = 1$ K, and approaches zero as $T \rightarrow 0$.

If, by transferring energy to a metal, we raise its temperature, some of the conduction electrons will be thermally excited to higher available energy states. Figure 46–1c shows the distribution of the conduction electrons over the allowed states for copper at 1000 K, a temperature at which the metal would glow brightly in a dark room. The striking feature of *this* figure is how little it differs from Fig. 46–1b, the distribution at the absolute zero. It is true that a few electrons have increased their energies from values slightly below the Fermi energy (region 2 in Fig. 46–1c) to values slightly above that energy (region 1). Thus the average electron energy at 1000 K is higher than that at the absolute zero, but it is not much higher. This is in striking contrast with the behavior of an (ideal) gas, for which the average kinetic energy of the molecules is directly proportional to the gas temperature, approaching zero as $T \rightarrow 0$; see Eq. 21–6.

It seems clear that the molecules of a gas and the conduction electrons in a metal behave in quite different ways. Formally, we say that the gas molecules,

Classical and quantum statistics

under ideal conditions, obey the (classical) Maxwell-Boltzmann statistics and that the conduction electrons in a metal obey the (quantum) Fermi-Dirac statistics. The word "statistics" here refers to the formal rules for counting the particles. In Maxwell-Boltzmann statistics, for example, we assume that we can tell identical particles apart, but in Fermi-Dirac statistics we assume that we cannot. Again, in Maxwell-Boltzmann statistics the Pauli exclusion principle plays no role, and all particles can pile up in the lowest energy state. In Fermi-Dirac statistics, however, the role of the Pauli principle, as we have seen, is vital; it dictates that only one particle can occupy any given state.

The relationship between the particle distribution function $n_p(E)$ (colored curves in Fig. 46-1) and the state distribution function $n_s(E)$ (black curve in that figure) for the conduction electrons in a metal is

$$n_p(E) = n_s(E)\, p(E), \tag{46-4}$$

where $n_s(E)$ is given by Eq. 46-1 and $p(E)$, whose values range between zero and one, must be interpreted as the probability that a state of energy E will be occupied.

From quantum theory it can be shown, following the rules of Fermi-Dirac statistics, that*

Fermi-Dirac probability function

$$p(E) = \frac{1}{e^{(E-E_F)/kT} + 1}, \tag{46-5}$$

in which E_F is the Fermi energy. Careful study of this equation shows that at $T = 0$ we have $p(E) = 0$ if $E > E_F$ and $p(E) = 1$ if $E < E_F$, just as Fig. 46-1b requires.

Combining Eqs. 46-1, 46-4, and 46-5 yields, finally, for the particle distribution function,

$$n_p(E) = C\frac{E^{1/2}}{e^{(E-E_F)/kT} + 1}, \tag{46-6}$$

where the constant C (see Exercise 3) has the value 6.78×10^{27} m^{-3} eV$^{-3/2}$.

Electrons whose energies are sufficiently close to the Fermi energy (see E_F in Fig. 46-1c) will find vacant levels above them in energy into which they can move by thermal excitation. The quantity kT, which has the value 0.086 eV at 1000 K, is a rough measure of what is meant by "sufficiently close." This quantity is displayed as an energy interval in Fig. 46-1c, and we see at once that, even at 1000 K, we have $kT \ll E_F$.

Electrons with $E = E_F$ have speeds close to v_F, the so-called *Fermi speed*, defined from

The Fermi speed

$$E_F = \tfrac{1}{2}mv_F^2. \tag{46-7}$$

For copper we have $E_F = 7.0$ eV, yielding $v_F = 1.6 \times 10^6$ m/s. Electrons with speeds close to this value are free to exchange thermal energy with the lattice because, as we have seen, they frequently find vacant levels slightly above them in energy. The remaining electrons are locked into their energies by the Pauli exclusion principle, because all suitably nearby states are already occupied.

Electrons whose energies are much less than the Fermi energy find the levels immediately above them occupied so that—because of the Pauli exclusion principle—no transitions are possible for them. One speaks—accurately enough and somewhat poetically—of "ripples on the surface of the Fermi sea."

---

* See footnote on p. 864.

Consider now the electrical conductivity of a metal. For concreteness let us apply a potential difference of 10 keV between the ends of a copper wire 1.0 m long, thus establishing an electric field of magnitude 1.0×10^4 V/m at all points within the wire. In Example 4 of Chapter 28 we showed that, for copper at room temperature, the mean free path of the conduction electrons is ~ 400 Å ($\approx 4.0 \times 10^{-8}$ m). The maximum kinetic energy that an electron can absorb from such a field while traversing such a distance is given by

$$K = qEl$$
$$= (1 \text{ electron})(1.0 \times 10^4 \text{ V/m})(4.0 \times 10^{-8} \text{ m})$$
$$= 4.0 \times 10^{-4} \text{ eV}.$$

This is so small that it defies representation on the scale of Fig. 46–1c. We see again that only the electrons with speeds close to the Fermi speed play a crucial role; the others—which are in fact the vast majority—play no role whatever in the conduction process, because all nearby levels into which they might move are already occupied. We anticipated this result in the semiclassical treatment of electrical conductivity given in Section 28–4. We there assumed, without explicit justification, an effective electron speed v_{eff} of 1.6×10^6 m/s. We now see that this is just the Fermi speed for copper.

Example 3 (a) For copper at 1000 K (see Fig. 46–1c), find the energy at which the probability $p(E)$ that a conduction electron state will be occupied is 0.90. Find also (b) the distribution of states function $n_s (E)$ and (c) the particle distribution function $n_p(E)$.

(a) Substitution into Eq. 46–5 yields

$$p(E) = \frac{1}{e^{\Delta E/kT} + 1} = 0.90,$$

where $\Delta E = E - E_F$. Manipulation leads readily to

$$\frac{\Delta E}{kT} = -2.20.$$

Thus

$$\Delta E = -2.20 \, kT = -(2.20)(8.63 \times 10^{-5} \text{ eV/K})(1000 \text{ K})$$
$$= -0.19 \text{ eV}.$$

The result up to this point is independent of the material consid-

ered. For copper, for which $E_F = 7.00$ eV, we finally have

$$E = E_F + \Delta E = 7.00 \text{ eV} - 0.19 \text{ eV} = 6.81 \text{ eV}.$$

(b) Carrying out a calculation like that of Example 1 leads, for $E = 6.81$ eV, to

$$n_s(E) = 1.77 \times 10^{28} \text{ m}^{-3} \text{ eV}^{-1}.$$

This result, which is independent of the material considered, can be verified by inspection of Fig. 46–1a. See also Problem 6.

(c) From Eq. 46–4 we have, finally, again for $E = 6.81$ eV,

$$n_p(E) = n_s(E)p(E)$$
$$= (1.77 \times 10^{28} \, m^{-3} \text{ eV}^{-1})(0.90)$$
$$= 1.59 \times 10^{28} \, m^{-3} \text{ eV}^{-1}.$$

This result can be verified from Fig. 46–1c, $E = 6.81$ eV, and $n_p(E) = (1.59 \times 10^{28})\text{m}^{-3} \text{ eV}^{-1}$ being a point on the curve marked $n_p(E)$ in that figure.

46–4 Conductors, Insulators, and Semiconductors

The potential for the free-electron model

In the free-electron model of the preceding section we tacitly assumed that the interior volume of the metal, throughout which the conduction electrons move, was a region of uniform potential. We assumed further that a repulsive electric potential, rising steeply to a large value at the metal surface, confined the electrons to their metallic "box." The curves of Fig. 46–1 result when these assumptions are introduced into the machinery of wave mechanics.

An improved potential assumption

It seems more likely, however, that the interior potential of a solid is not uniform but varies throughout the interior volume with the periodicity of the lattice. It is reasonable that this potential should exhibit maxima at the sites of the ion cores (which carry a net positive charge) and minima symmetrically between such sites.

Energy bands and gaps

Band theory—a second approach

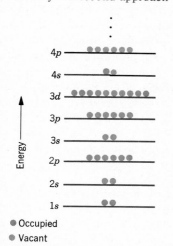

Figure 46-2 The electron configuration for a neutral isolated copper atom in its ground state. Not to scale.

Some band configurations

If we analyze the motions of the conduction electrons in a solid by the methods of wave mechanics and using such an improved periodic potential, the following important result emerges: The energies of the allowed states are not distributed smoothly and uniformly but are grouped into bands separated by energy gaps. Energies that fall within these gaps are forbidden to electrons.

Because it is based on a better assumption about the electric potential, this *band theory* model should give us more information about solids than does the free-electron model. This is indeed the case. For example, the free-electron model provides a good account of the nature of electrical conductivity in metals, but it does not help us to understand why some solids are conductors of electricity and some are not. The band theory provides such insights.

There is a second way to look at the existence of energy bands and gaps in solids. Consider, for example, a crystal of copper. The distance between nearest-neighbor ion cores is ~2.5 Å. Imagine the crystal to be expanded uniformly until this distance becomes, say, 25 Å. We are now no longer dealing with a solid but with an assembly of N isolated neutral copper atoms, the energies of the subshells for each atom being arranged as in Fig. 46-2.

If we begin to move the N copper atoms back together to again form a solid, nothing happens until the outermost atomic electrons begin to touch each other. In the language of wave mechanics, we say that at this point the wave functions of the 4s or valence electrons come to overlap. It can be shown that the effect of this overlap is to split the 4s level into N closely lying levels. Thus the 4s level for the isolated atom becomes the 4s band for the solid of N atoms. The 4s level contains a single valence electron for each atom, although it could accommodate two. In the same way, the 4s band for the solid contains N electrons although it could accommodate 2N; the band is said to be half-filled.

As we continue compressing the assembly of atoms toward its equilibrium nearest-neighbor separation of 2.5 Å, the 3d, the 3p, etc., wave functions come to overlap, and they too split and form bands. Thus every level of the isolated atom is associated with a band in the solid. The bands for the inner, more tightly bound electrons are relatively narrow because the overlap of their wavefunctions may be quite small.

Figure 46-3 is an idealized representation of the band configurations for three types of solids. Only bands corresponding to the outermost electrons are shown.

Figure 46-3*a* represents a conductor. Its essential feature is that the most energetic band that contains any electrons at all is only partially full. Just as in Figs. 46-1*b* and 46-1*c*, there are many vacant states lying just above the Fermi energy, so that electrical conduction can proceed as previously described. In

Figure 46-3 An idealized representation of the energy bands and gaps for (*a*, *d*) a conductor, (*b*) an insulator, and (*c*) a semiconductor. The horizontal coordinate has no physical meaning.

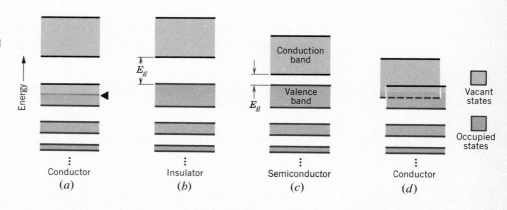

A conductor (copper)

An insulator (diamond)

A semiconductor (silicon)

actual conductors such as copper, the outermost bands often overlap, eliminating the gaps between them. In deciding whether or not a solid is a conductor, the presence or absence of such gaps is not significant; it is only necessary that vacant electron states be "accessible" to filled electron states. "Accessible" means that an electron can acquire on the order of kT of thermal energy to boost it to a formerly unfilled state.

In Fig. 46–3b we have a full band with a substantial energy gap immediately above it. If an electric field is applied to such a solid, the electrons will acquire extra energy only if there are empty levels within the range of energy that the strength of the applied electric field allows the electron to gain. If there are no nearby vacant states to go to, the electron will not be able to gain energy and the solid will be an insulator.

Carbon in its diamond form is an example of an insulator, the gap width E_g in this case being ~6 eV. Electrons cannot bridge such a wide energy gap, either by thermal excitation or by the action of even a rather strong applied electric field. The empty band just above the gap remains empty and no conduction occurs.

Figure 46–3c is just like Fig. 46–3b with the important difference that here the gap width E_g is small enough so that some thermal excitation of electrons across the gap *can* occur. This puts some electrons into the (nearly empty) band labeled "conduction band" in the figure and creates an equal number of vacant states in the (nearly filled) band labeled "valence band." Such solids—in which the width of the energy gap is small—are called *semiconductors*, and we discuss them in more detail in the following section. Our prototype is silicon, which has the same crystal structure as diamond but a gap width E_g of only ~1.1 eV. Note that kT, which determines the probability that an electron can be thermally excited across the gap, is only ~0.025 eV at room temperature. There are not many conduction electrons in semiconductors at room temperature and—indeed—this is why they are called *semi*conductors.

Figure 46–3d shows another type of conductor, zinc being an example. This element has two 4s electrons, which fill the 4s subshell of the isolated zinc atom. All other subshells of lower energy are also filled. We would normally conclude that the 4s band of a zinc specimen would be filled and that the material should be an insulator (Fig. 46–3b) or a semiconductor (Fig. 46–3c). However, we know zinc to be a metal and a reasonably good conductor of electricity. The dilemma is resolved when we realize that the next highest energy band (4p in this case) overlaps the 4s band in energy. Thus vacant states *are* available in the composite band of Fig. 46–3d, and zinc is indeed a conductor of electricity.

46–5   Semiconductors

Semiconductors are the basis for the revolution in microelectronics that characterizes our times.* Table 46–1 compares some electrical properties of copper, our prototype conductor, and silicon, our prototype semiconductor. We see at once how different they are. These differences can be understood from a study of Figs. 46–1a and 46–1c and of Eq. 28–14. This equation, which relates the resistivity ρ of a conductor to n_0, the density of its charge carriers, and to τ, the mean free time between collisions, is

$$\rho = \frac{m}{n_0 e^2 \tau}. \qquad [28\text{–}14]$$

---

* See *Scientific American*, September 1977, a special issue on microelectronics.

Table 46-1 Some electrical properties of copper (a conductor) and silicon (a semiconductor) at room temperature

	Density of charge carriers n_0	Resistivity ρ	Temperature coefficient of resistivity[a] α
Copper	9×10^{28} m^{-3}	2×10^{-8} Ω·m	$+4 \times 10^{-3}$ K^{-1}
Silicon	1×10^{16} m^{-3}[b]	3×10^{3} Ω·m	-70×10^{-3} K^{-1}

[a] Defined as $(1/\rho)(\partial\rho/\partial T)$. [b] Includes both electrons and holes.

Conductors and semiconductors compared

In a metallic conductor n_0 in the above equation is a fixed quantity, independent of the temperature. The mean free time τ decreases with temperature (because of a higher collision frequency at higher temperatures) so that Eq. 28-14 predicts that ρ will increase with temperature. This is consistent with the positive value for α, the temperature coefficient of resistivity, displayed in Table 46-1.

The density of charge carriers for semiconductors

In a semiconductor, only those few electrons that have been thermally excited from the valence band, through the energy gap, and into the conduction band contribute to n_0 in Eq. 28-14. As Table 46-1 shows, the density of charge carriers for silicon at 300 K is indeed much smaller than that for copper, by a factor of $\sim 10^{12}$. Moreover, n_0 for semiconductor is not constant, as it is for conductors, but increases rapidly with increasing temperature. In fact, the dependence of n_0 on temperature far outweighs the dependence of the mean free time τ on temperature and accounts for the fact that the resistivity of semiconductors *decreases* with temperature. As Table 46-1 shows, the temperature coefficient of resistivity α for a typical semiconductor is both negative and—by comparison with a typical conductor—large in its absolute value.

Conduction by holes

For every electron excited to the conduction band of a semiconductor a vacant electron state, called a *hole,* is created in the valence band. The holes permit some freedom of movement to the remaining electrons, within the ever-present restrictions of the Pauli exclusion principle. If an electric field **E** is established in a semiconductor the electrons in the valence band, being negatively charged, will drift in the direction of $-\mathbf{E}$. This causes the holes to drift in the direction of **E**. That is, the holes behave like positively-charged particles and are so treated. In much the same way a bubble rising in a pond behaves like a water droplet with a negative gravitational mass. It is easier to follow the simple motion of the bubble than to deal with the complicated motions of the liquid streaming down around it.

Doping of semiconductors

A semiconductor that contains no impurities—and that is what we have tacitly assumed so far—is called an *intrinsic semiconductor.* The density of charge carriers for such pure materials, however, is so small (see Table 46-1) that even as few as one impurity atom for every 10^5 to 10^{10} silicon atoms can have a controlling effect. In virtually all applications of semiconductors in modern technology, the silicon (or other) matrix is impregnated with a carefully controlled concentration of selected impurities, a process called *doping.* A doped semiconductor is called an *extrinsic* semiconductor. Table 46-2 shows some appropriate characteristics of silicon and of two commonly used "dopants."

Donor "impurities"

Silicon has a valence of four and, in a lattice of the pure element, it shares four two-electron bonds with each of its four nearest neighbors. Suppose that a silicon atom is removed from such a lattice and that an atom of phosphorus, which has a valence of five, is substituted. Four of the phosphorus valence electrons substitute directly for the four missing bonding electrons. The fifth or "extra" va-

Table 46-2 The properties of silicon and of two commonly used doping materials

Element	Number of valence electrons	Subshell assignments of the valence electrons[a]	Type of impurity	Type of semiconductor	Energy to create a charge carrier, meV[b]
Phosphorus (Z = 15)	5 (= 4 + 1)	$3p$ ● ● ● ● ● ● $3s$ ● ●	Donor	n-Type	45
Silicon (Z = 14)	4	$3p$ ● ● ● ● ○ ○ $3s$ ● ●	—	Intrinsic	1100
Aluminum (Z = 13)	3 (= 4 − 1)	$3p$ ● ● ● ○ ○ ○ $3s$ ● ●	Acceptor	p-Type	57

[a] The symbol ● represents an occupied state and ○ a vacant state.

[b] 1 meV = 10^{-3} eV; do not confuse with MeV (= 10^6 eV).

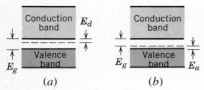

Figure 46-4 (a) An n-type semiconductor, showing a donor level that can contribute electrons to the conduction band. (b) A p-type semiconductor, showing an accepter level that can contribute holes to the valence band. See Table 46-2.

n-Type semiconductors

lence electron remains loosely bound to the phosphorus impurity atom. With an energy expenditure E_d of only ~ 45 meV,* however, this electron can be removed from its localized position in the lattice and "donated" to the conduction band. Phosphorus, then, is a *donor* impurity.

The fifth or "extra" electron in phosphorus moves in an orbit of very large radius (see Example 4) around this (singly charged) phosphorus ion. The orbit is so large that the Coulomb attraction of the phosphorus ion for the electron is appreciably shielded by the intervening silicon atoms, which form a dielectric. Hence this "extra" electron has a very small binding energy indeed to the phosphorus ion. It can be ionized, going into the conduction band, at a much lower temperature than would be needed for an electron from the valence band. In fact, at ordinary temperatures, essentially all such donor electrons are excited to the conduction band, where they are free to serve as carriers of the electric current.

Semiconductors in which donor impurities dominate are called *n-type semiconductors*, the n standing for the *n*egative sign of the charge carriers (electrons). Figure 46-4a shows the situation for such a semiconductor, the dashed horizontal line just below the conduction band representing a donor level in an impurity atom. Note that $E_d = 45$ meV and $E_g = 1100$ meV, so that $E_d \ll E_g$. In fact, at 300 K, E_d is not much larger than $kT(= 26$ meV), so that even a small concentration of donor impurities can account for a large fraction of the electrons present in the conduction band. The electrical conductivity of a semiconductor can be absolutely controlled by the amount of impurity used.

Acceptor impurities

If silicon is doped with aluminum (see Table 46-2), only three valence electrons are available to reestablish the four silicon bonds. There is a hole in one of the bonds that can be filled, with small expenditure of energy, by "accepting" an electron from the valence band. This creates a hole in this band and, as we have seen, such a hole behaves like a positive charge carrier. We speak of aluminum as an *acceptor* impurity. Semiconductors in which such impurities dominate are called *p-type semiconductors*, the p standing for the *p*ositive sign of the charge carriers (holes). Figure 46-4b shows the situation, the dashed line just above the valence band representing an acceptor level in an impurity atom. Note, again, that $E_a \ll E_g$.

p-Type semiconductors

Example 4 Consider a phosphorus impurity atom (Z = 15) substituting for a silicon atom (Z = 14) in a silicon lattice. Phosphorus (see Table 46-2) is a donor impurity, and its loosely bound "extra" electron may be imagined to move in a Bohr orbit about the central phosphorus ion core. Calculate (a) the binding energy and (b) the orbit radius for this electron.

---

* The symbol "meV" stands for "millielectron volts"; see Table 1-2. Thus 45 meV = 0.045 eV.

(a) The Bohr-theory expression for the binding energy is

$$E_b(=-E) = \frac{Z^2 m e^4}{8\epsilon_0^2 h^2} \frac{1}{n^2}. \qquad [42–28]$$

Here we put $n = 1$ because we are interested in the ground state of the donor atom and $Z = 1$ because the orbiting electron "sees" a net central charge of $+e$.

Equation 42–28 was derived by considering the hydrogen atom, whose orbiting electron moves in a vacuum. In this case, however, the orbiting electron moves through a silicon lattice. The effect of the silicon dielectric is to reduce the electric force by a factor of κ, the dielectric constant. Quantitatively, we must replace ϵ_0 in Coulomb's law (Eq. 23–3) by $\kappa\epsilon_0$. Making the same replacement in Eq. 42–28 leads to

$$E_b = \frac{1}{\kappa^2} \left(\frac{m e^4}{8\epsilon_0^2 h^2} \right),$$

in which the factor in parentheses (see Example 5 of Chapter 42) is just 13.6 eV. For silicon we have $\kappa = 12$, so that

$$E_b = \frac{13.6 \text{ eV}}{(12)^2} = 0.094 \text{ eV} = 94 \text{ meV}.$$

This is in very rough, order-of-magnitude agreement with the value of 45 meV listed in Table 46–2.*

(b) The orbit radius follows from Eq. 42–27. Substituting as above leads to

$$r = \kappa \left(\frac{h^2 \epsilon_0}{\pi m e^2} \right). \qquad [\text{see } 42–27]$$

The factor in parentheses is just the Bohr radius r_B ($= 0.531$ Å). Thus

$$r = (12)(0.531 \text{ Å}) \approx 6 \text{ Å}.$$

This is a large radius and justifies our treatment of the silicon lattice as a continuous medium.†

46–6 Semiconducting Devices

To understand the mode of operation of semiconducting devices such as diode rectifiers and transistors, it is necessary to be quite clear about the roles played by the charged particles in both n-type and p-type semiconductors. Table 46–3 summarizes the situation. Three points are worthy of emphasis:

1. The donor and acceptor impurity ions, although charged, are not charge carriers; they remain fixed in their lattice sites throughout the sample.

2. For n-type material the (majority) charge carriers are electrons in the conduction band; for p-type material they are holes in the valence band.

3. The concentrations of minority carriers (that is, of holes in n-type material and of electrons in p-type) are much smaller than those of the majority carriers, by factors in the range 10^3–10^6. This concentration difference is traceable to the fact that $E_d \ll E_g$ and $E_a \ll E_g$ in Fig. 46–4; that is, the majority carriers have much smaller excitation energies.

A *pn* junction

The Diode Rectifier. The diode rectifier is a two-terminal device that has a low resistance for one polarity of the potential difference applied between its terminals but a very high resistance if this polarity is reversed. A junction between a p-type and an n-type semiconductor, called a *pn junction*, has this property. Its rectifying characteristics are displayed in Fig. 28–5.

Figure 46–5 shows an idealized *pn* junction at the imagined moment of its formation. The shadings suggest the concentrations of the majority carriers, that is, of the electrons in the n-type material and of the holes in the p-type. However, a

---

* We should also replace the electron mass m in Eq. 42–28 by an effective electron mass m_{eff}, to take into account the fact that the silicon lattice exerts forces on the orbiting electron, forces not present in the hydrogen atom. There is evidence that for silicon we should put $m_{\text{eff}} \approx 0.2\, m$; this substitution would reduce the calculated value of E_b to ~ 20 meV.

---

† Making the effective mass substitution described in the preceding footnote would increase this value to ~ 30 Å.

Table 46–3 Charged particles involved in the semiconduction process

Particle	Charge	Origin	Remarks	Excitation energy[a]
		n-Type semiconductors		
1. Electrons	$-e$	Thermally excited by electron transitions from donor levels to the conduction band	Majority carriers	E_d
2. Holes	$+e$	Thermally excited by electron transitions from the valence band to the conduction band	Minority carriers	E_g
3. Ions	$+e$	Generated when donor atoms lose an electron	Not mobile	E_d
		p-Type semiconductors		
1'. Holes	$+e$	Thermally excited by electron transitions from the valence band to acceptor levels	Majority carriers	E_a
2'. Electrons	$-e$	Thermally excited by electron transitions from the valence band to the conduction band	Minority carriers	E_g
3'. Ions	$-e$	Generated when acceptor atoms gain an electron	Not mobile	E_a

[a] See Fig. 46–4.

A *pn* junction in electrical equilibrium

A barrier is established

A (small) diffusion current surmounts the barrier

The drift current

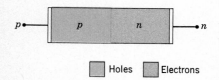

Figure 46–5 A *pn* junction at the imagined moment of its formation. The shadings suggest the majority carriers.

junction such as that of Fig. 46–5—if formed—would quickly assume the equilibrium configuration of Fig. 46–6b. The mechanism is outlined below.

In response to differences in concentration, electrons would diffuse from right to left across the boundary in Fig. 46–5. In the same way, holes would diffuse across this same boundary region from left to right. In a narrow zone of equilibrium thickness d_0, the diffusing majority carriers on each side of the junction would "uncover," so to speak, a few of the stationary donor and acceptor ions. Figure 46–6b displays them. These charges generate an electric field $\mathbf{E}_0$ at the junction and, as Fig. 46–6b' shows, a corresponding contact potential difference V_0. This potential difference serves as a barrier to the diffusion process, both for electrons and holes. Only those majority carriers with high enough energies will be able to surmount this barrier and continue to diffuse through the junction. We call the small current due to this process the *diffusion current* i_{df}, and we represent it in Fig. 46–6b''. As there displayed, it points in the conventional current direction and combines within itself both an electron and a hole component.

For a *pn* junction on open circuit—resting on a shelf, so to speak—it is simply not possible that there should be a steady equilibrium current flowing through it. The diffusion current i_{df} must therefore be cancelled by a suitable countercurrent. Such a countercurrent arises from the motions of the minority carriers.

Minority electron–hole pairs, for example, are created regularly by thermal agitation throughout the lattice, their excitation energy for creation being the intrinsic gap width E_g. In particular, those pairs that are created in or near the junction region are acted on by the field $\mathbf{E}_0$ that exists there and are caused to drift through the lattice under its influence. This drift occurs in an electrically downhill direction, holes drifting from right to left and electrons from left to right in Fig. 46–6b. Their combined motions constitute a *drift current* i_{dr} and we have, at equilibrium (see Fig. 46–6b'')

$$i_{df} = i_{dr}. \tag{46–8}$$

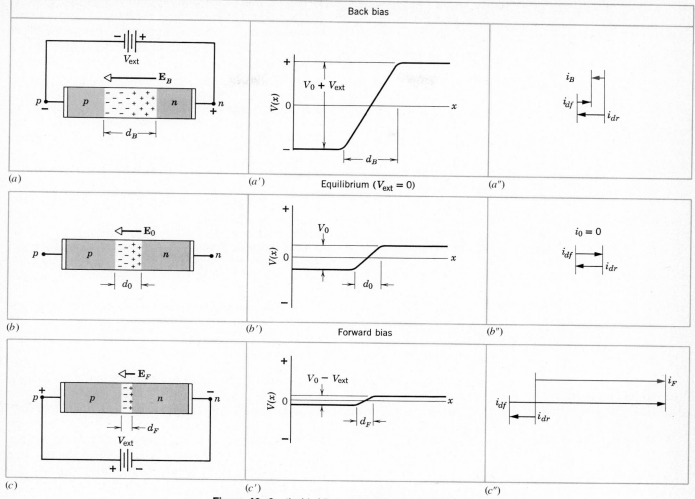

Figure 46–6 (b, b', b'') A pn junction in equilibrium, with no applied external potential difference and showing the depletion layer d_0, the electric field $\mathbf{E}_0$, the potential barrier V_0, and the junction currents i_{df} and i_{dr}. (a, a', a'') The same quantities for a back-biased pn junction. (c, c', c'') The same quantities for a forward-biased pn junction.

The resultant current i_0 is zero, as it must be in the absence of an external applied potential difference. The potential difference V_0 and the electric field $\mathbf{E}_0$ automatically adjust themselves so that the cancellation asserted by Eq. 46–8 comes about. For a typical silicon-based pn junction we may have $V_0 \approx 0.5$ V, $d_0 \approx 500$ nm (= 5000 Å), $E_0 \approx V_0/d_0 \approx 10^6$ V/m, and $i_{df} = i_{dr} \approx 10^{-8}$ A.

It is important to distinguish clearly between the diffusion current i_{df} of Eq. 46–8 and the drift current i_{dr}.

1. The diffusion current occurs in response to a concentration gradient. Put in simpler terms, there are more electrons in the right end of Fig. 46–5 than there are in the left end, so that—for this reason alone—electrons tend to diffuse from right to left. The diffusion current is sensitive to the height of the potential barrier (see Fig. 46–6b'); if the electrons (or holes) do not have an energy at least equal to

The diffusion current clarified

the barrier height, they cannot diffuse through it, no matter what the concentration gradient.

2. The drift current, on the other hand, occurs in response to the action of an electric field, just as in Section 28–4. This current is not sensitive to the barrier height, because no barrier exists for it; the journey through the barrier is a downhill trip. Note that the drift currents in Figs. 46–6a″, 46–6b″, and 46–6c″ are all the same length, even though the barriers have different heights.

The junction region, whose width is marked as d_0 in Fig. 46–6b, is called the *depletion layer* because it has been essentially depleted of charge carriers. Any electron or hole that appears there will, as we have seen, drift out of the region by the action of the electric field $\mathbf{E}_0$. The device as a whole resembles two conductors (the p and n regions) separated by a thin, somewhat leaky, dielectric slab (the depletion layer).

Let us now consider that an external potential difference V_{ext} is applied to the *pn* junction, using an ideal battery. We make the connections in such a way that the *n*-type material is positive with respect to the *p*-type material, as in Fig. 46–6a. The net effect of such *back biasing,* as it is called, is to increase E_0 (to E_B; see Fig. 46–6a), V_0 (to $V_0 + V_{ext}$; see Fig. 46–6a′), and d_0 (to d_B; see Fig. 46–6a′). The barrier is now larger, so fewer majority carriers have enough energy to surmount it. The diffusion current i_{df} drops.

The drift current i_{dr}, as we have seen, depends only on the rate at which minority carriers are generated and is independent of the applied potential difference; i_{dr} remains unchanged.

Figure 46–6a″ shows the currents involved in the back-biased *pn* junction, all current arrows being drawn in the conventional sense. The back current i_B, given by

$$i_B = i_{df} - i_{dr}, \tag{46–9}$$

is small indeed, being negligible for most purposes.

Now let an external potential difference V_{ext} be applied to the *pn* junction as in Fig. 46–6c, so that the *p*-type material is positive with respect to the *n*-type material. The effect of such *forward biasing,* as it is called, is to decrease E_0 (to E_F; see Fig. 46–6c), V_0 (to $V_0 - V_{ext}$; see Fig. 46–6c′), and d_0 (to d_F; see Fig. 46–6c′). The barrier is now smaller and many more carriers have enough energy to diffuse through it. The diffusion current i_{df} increases substantially. Again, the drift current i_{dr} remains unchanged. Figure 46–6c″ shows the currents involved. The forward current, given by

$$i_F = i_{dr} - i_{df} \tag{46–10}$$

and shown greatly abbreviated in the figure, is indeed substantial.

The Transistor. Figure 46–7 shows the physical arrangement of the type of transistor that we discuss here. It is called a MOSFET, the acronym standing for *m*etal-*o*xide-*s*emiconductor-*f*ield-*e*ffect-*t*ransistor. In this device the current i_{ds} passing between two of the terminals (the *source S* and the *drain D*) is sensitively controlled by the potential applied to a third terminal, the *gate G*.

In Fig. 46–7 a *p*-type substrate in the form of a thin wafer has embedded in it two "islands" of *n*-type material, the source *S* and the drain *D*. These are connected by a thin channel of *n*-type material, called the *n*-channel. An insulating layer of silicon dioxide (hence *o*xide in the acronym) is deposited on the substrate and penetrated by two metallic contacts (hence *m*etal) at *S* and at *D*, so that electrical contact can be made with the source and the drain. A thin metallic

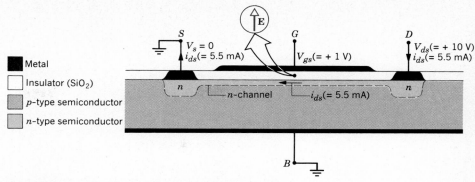

Figure 46-7 A typical transistor, of the MOSFET type. The potential differences and the current shown are particular values for a commercial MOSFET identified as 2N3797.

layer—the gate G—is deposited opposite to the n-channel, separated from it by the insulating silicon dioxide layer.

Suppose that the source and the substrate are grounded ($V_s = 0$) and that the drain is held at a positive potential with respect to the source ($V_{ds} = +10$ V, say). Assume further that the gate is also held at a (smaller) positive potential with respect to the source ($V_{gs} = +1.0$ V, say). Under these circumstances the n-channel will be conducting. Electrons will drift along it from left to right and a current ($i_{ds} = 5.5$ mA) will appear in the wire connected to the drain terminal.

Note that the potential of the n-channel varies from almost zero at its left end to almost $+10$ V at its right end, so that over essentially all of its length the gate ($V_{gs} = +1.0$ V) is at a lower potential than the channel. The electric field **E** in the insulating layer that separates the gate from the channel points upward in Fig. 46-7.

If now the gate potential is reduced (to $V_{gs} = 0.0$ V, say) the gate becomes more negative and the electric field **E** displayed in Fig. 46-7 will increase in magnitude. The effect of this is to repel the conduction electrons in the n-channel away from the gate, pushing some of them back toward or into the source and the drain. This reduces the number of charge carriers in the n-channel so that its resistance—measured along its length between the source and the drain—increases. This in turn means that the current i_{ds} must decrease, to 3.0 mA in our specific example.

Thus we see how a small change in V_{gs} can produce a significant change in i_{ds}. The gate in Fig. 46-7 totally controls the current that passes between the source and the drain, much as a valve controls the flow of fluid through a pipeline. Transistors like that of Fig. 46-7 are in common use in microelectronics, as amplifiers, as on–off switches or diodes, and in other ways.

The Integrated Circuit. The fabrication of a device such as the MOSFET of Fig. 46-7 is an incredible technological achievement. It involves: (1) producing a thin substrate wafer of suitably doped p-type silicon; (2) producing the source, drain, and channel regions by selective masking and doping procedures; and (3) superimposing the appropriate oxide layer and the necessary metallic gate and contact terminals.

Far beyond this lies the technology of manufacturing such devices in the thousands on a single semiconducting wafer and integrating them electrically to perform a specified function. Figure 46–8 shows such a silicon "chip," its dimensions suggested by the superimposed paper clip. The chip contains 30,000 circuit elements!

Figure 46–8 A "computer on a chip", containing 30,000 elements such as transistors and diodes. The standard paper clip shows the scale. Courtesy of Bell Laboratories.

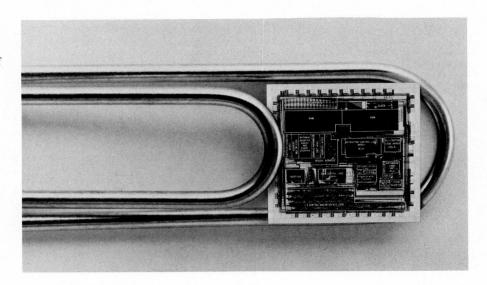

REVIEW GUIDE AND SUMMARY

The available states

The quantized states available to the conduction electrons in a metal are given by

$$n_s(E) = \left(\frac{35.5m^{3/2}}{h^3}\right) E^{1/2} = CE^{1/2}, \qquad [46\text{–}1]$$

where $n_s(E)\, dE$ is the number of states per unit volume that lie between E and $E + dE$. The constant C has the value $(6.78 \times 10^{27})\text{m}^{-3}\,\text{eV}^{-3/2}$. Figure 46–1a is a plot of the above function; see Example 1 and Exercise 3.

In filling the conduction electron states the Pauli exclusion principle must be observed. The particle distribution function is given by

The electron energy distribution

$$n_p(E) = C\,\frac{E^{1/2}}{e^{(E-E_F)/kT} + 1}, \qquad [46\text{–}6]$$

where $n_p(E)\, dE$ is the number of filled states between E and $E + dE$. Figures 46–1b and 46–1c show $n_p(E)$ for $T = 0$ and for $T = 1000$ K, respectively (for copper). See Example 3.

E_F in the above is the Fermi energy, that is, the energy of the highest filled state at $T = 0$. It is given by

The Fermi energy

$$E_F = \left(\frac{0.121h^2}{m}\right) n_0^{2/3} = An_0^{2/3}, \qquad [46\text{–}3]$$

where n_0 is the total number of conduction electrons per unit volume. The constant A has the value $(3.65 \times 10^{-19})\text{m}^2\,\text{eV}$. See Example 2 and Exercise 7.

The Fermi speed is defined from

The Fermi speed

$$v_F = \left(\frac{2E_F}{m}\right)^{1/2}. \qquad [46\text{–}7]$$

Only electrons with speeds near this value can contribute to the electrical conductivity of metals; the others are inhibited from changing their energies by the Pauli exclusion principle.

Energy band theory

If the potential $V(x, y, z)$ inside a metal is assumed to vary with the periodicity of the crystalline lattice, the phenomenon of energy bands and energy gaps appears. Figure 46–3 shows how to classify materials as conductors, insulators, and semiconductors on this basis.

Properties of a semiconductor

Table 46–1 contrasts the properties of a conductor and a semiconductor. In a semiconductor the resistivity is determined not so much by variations in the mean time τ between collisions, as for metals but by variations in n_0, the density of charge carriers; see Eq. 28–14. In a semiconductor both electrons in the conduction band and holes in the valence band can serve equally well as charge carriers.

Doping—donors and acceptors

In practice, semiconductors are "doped" with controlled levels of selected impurities; see Table 46–2. Donor impurities contribute electrons to the conduction band and produce n-type semiconductors. Acceptor impurities contribute holes to the valence band and produce p-type semiconductors. Study Fig. 46–4. Table 46–3 is very important to an understanding of semiconductor behavior.

A pn junction

A pn junction (see Fig. 46–6) can serve as a diode rectifier. For forward biasing (n positive with respect to p), the potential barrier is low, the junction is thin, and the forward current i_F is large. For back biasing, the potential barrier is high, the junction is thick, and the back current i_B is small, usually negligibly so. The junction region itself, regardless of the applied potential difference, is called a depletion layer. It is virtually free of charge carriers and behaves like a somewhat leaky insulating slab. Figure 28–5 shows the rectifying properties of a pn junction.

The depletion layer

MOSFETS and chips

Figure 46–7 shows a MOSFET-type transistor, in which a small variation of the potential difference V_{gs} between the gate G and the source S has a major controlling effect on the current i_{ds} between the drain D and the source. Figure 46–8 shows that thousands of such devices can be constructed on a relatively small silicon "chip."

QUESTIONS

1. What features of Fig. 46–1 make it specific for copper? What features are independent of the identity of the metal?

2. Why do the curves marked $n_p(E)$ in Figs. 46–1b and 46–1c differ so little from each other?

3. The conduction electrons in a metallic sphere occupy states of quantized energy. Does the average energy interval between adjacent states depend on (a) the material of which the sphere is made, (b) the radius of the sphere, (c) the energy of the state, and (d) the temperature of the sphere?

4. What role does the Pauli exclusion principle play in accounting for the electrical conductivity of a metal?

5. Distinguish carefully among (a) the state distribution function $n_s(E)$, (b) the particle distribution function $n_p(E)$, and the Fermi-Dirac probability function $p(E)$, all of which appear in Eq. 46–4.

6. In what ways do the free-electron model (see Section 28–4) and the Fermi-Dirac model (see Fig. 46–1) for the electrical conductivity of a metal differ?

7. In Section 21–6 we showed that the (molar) specific heat of an ideal monatomic gas is $\frac{3}{2}R$. If the conduction electrons in a metal behave like such a gas, we would expect them to make a contribution of about this amount to the measured specific heat of a metal. However, this measured specific heat can be accounted for quite well in terms of energy absorbed by the vibrations of the ion cores that form the metallic lattice. The electrons do not seem to absorb much energy as the temperature of the specimen is

increased. How does Fig. 46–1 provide an explanation of this prequantum-days puzzle?

8. If we compare the conduction electrons of a metal with the molecules of an ideal gas we are surprised to note (see Fig. 46–1b) that so much kinetic energy is locked into the conduction electron system at the absolute zero of temperature. Would it be better to compare the conduction electrons, not with the molecules of a gas, but with the inner electrons of a heavy atom? After all, a lot of kinetic energy is also locked up in this case, and we don't seem to find that surprising. Discuss.

9. Give a physical argument to account qualitatively for the existence of allowed and forbidden energy bands in solids. Why is it that no hint of bands appears in Fig. 46–1a?

10. Is the existence of a forbidden energy gap in an insulator any harder to accept than the existence of forbidden energies for an electron in, say, the hydrogen atom?

11. On the band theory picture, what are the *essential* requirements for a solid to be (a) a conductor, (b) an insulator, or (c) a semiconductor?

12. What can band theory tell us about solids that the free-electron model (see Section 28–4) cannot?

13. Why is it that, in a solid, the allowed bands become wider as one proceeds from the inner to the outer atomic electrons?

14. Do (intrinsic) semiconductors obey Ohm's law?

15. Consider these two statements: (a) At low enough temperatures silicon ceases to be a semiconductor and becomes a rather good insulator. (*b*) At high enough temperatures silicon ceases to become a semiconductor and becomes a rather good conductor. Discuss the extent to which each statement is either true or not true.

16. Can you tell by making Hall effect measurements (see Section 30–5) whether the current carriers in a semiconductor are electrons or holes?

17. What elements other than phosphorus are good candidates to use as donor impurities in silicon? What elements other than aluminum are good candidates to use as acceptor impurities? Consult the periodic table (Appendix D).

18. How do you account for the fact that the resistivity of metals increases with temperature but that of semiconductors decreases (see Table 46–1)?

19. Why does a *pn* junction, serving as a diode rectifier, rely so centrally on the concept of doping?

20. A semiconductor contains equal numbers of donor and acceptor impurities. Do they cancel each other in their electrical effects? If so, what is the mechanism? If not, why not?

21. Germanium and silicon are similar semiconducting materials whose principal distinction is that the gap width E_g (see Fig. 46–4) is 0.72 eV for the former and 1.09 eV for the latter. If you wished to construct a *pn* junction (see Fig. 46–5) in which the back current is to be kept as small as possible, which material would you choose and why?

22. Here are two possible techniques for fabricating a *pn* junction (see Fig. 46–5). (1) Prepare separately an *n*-type and a *p*-type sample and join them together, making sure that their abutting surfaces are plane and highly polished. (2) Prepare a single *n*-type sample and diffuse an excess acceptor impurity into it from one face, at high temperature. Which method is preferable and why?

23. In a *pn* junction (see Fig. 46–5) we have seen that electrons and holes may diffuse, in opposite directions, through the junction region. What is the eventual fate of each such particle as it diffuses into the material on the opposite side of the junction?

24. Explain in your own words how the MOSFET device of Fig. 46–7 works.

25. The acronym MOSFET stands for "metal oxide semiconductor field effect transistor." What is the significance of each of these terms as applied to the device shown in Fig. 46–7?

EXERCISES

Section 46–2   Conduction Electrons — The Allowed States

1. Gold is a monovalent metal with an atomic weight of 197 g/mol and a density of 1.93×10^4 kg/m³. What is the density of charge carriers?

2. At what pressure would an ideal gas have a density of molecules equal to that of the density of the conduction electrons in copper ($= 8.5 \times 10^{22}$ cm⁻³)? Assume that $T = 300$ K.

3. Show that Eq. 46–1 can be written as

$$n_s(E) = CE^{1/2},$$

where $C = 6.78 \times 10^{27}$ m⁻³·eV⁻³/². Use this relation to verify a calculation of Example 1, namely, that for $E = 5.00$ eV, $n_s(E) = 1.52 \times 10^{28}$ m⁻³·eV⁻¹.

4. Calculate the density $n_s(E)$ of conduction electron states in a metal for $E = 8.0$ eV and show that your result is consistent with the curve of Fig. 46–1*a*.

Section 46–3   Filling the Allowed States

5. Show that at $T = 0$, the Fermi-Dirac probability function is 1 for energies less than the Fermi energy, $\frac{1}{2}$ for energies equal to the Fermi energy and 0 for energies greater than the Fermi energy.

6. Calculate the result of Example 3, part (*c*), directly from Eq. 46–6. Put $E_F = 7.0$ eV.

7. Show that Eq. 46–3 can be written as

$$E_F = An_0^{2/3},$$

where the constant A has the value 3.65×10^{-19} m²·eV.

8. Figure 46–1*c* shows (colored curve) the energy distribution $n_p(E)$ of the conduction electrons in copper at 1000 K. Calculate

$n_p(E)$ for $E =$ 4.00, 6.75, 7.00, 7.25, and 9.00 eV and verify that the curve shown in the figure is correct at these points. Put $E_F = 7.00$ eV.

9. It can be shown that the conduction electrons in a metal behave like an ideal gas of the ordinary kind if the temperature is high enough. In particular, the temperature must be such that kT is much greater than the Fermi energy. What temperatures are required for copper ($E_F = 7.0$ eV) for this to be true? Study Fig. 46–1*c* in this connection and note that we have $kT \ll E_F$ for the conditions of that figure. This is just the reverse of the requirement cited above. Note also that copper boils at 2870 K.

Section 46–5   Semiconductors

10. In Fig. 46–9, two energy bands of a hypothetical solid are represented. The bands are filled to level E_e, which may be in either band 1 or band 2. There may be an impurity level at E_i. Fill in the table on the next page, indicating whether the solid is a

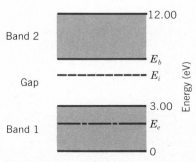

Figure 46–9   Exercise 10.

conductor, insulator, intrinsic semiconductor, or extrinsic semiconductor. The impurity type may be donor, acceptor, or none, and extrinsic semiconductors may be either p-type or n-type.

E_e	E_i (eV)	E_b	Type		
			Solid	Impurity	Extrinsic Semiconductor
3.00	—	9.00			
3.00	4.06	4.10			
3.00	—	4.10			
1.49	—	9.00			
4.40	—	4.10			
3.00	3.04	4.10			

11. In a semiconductor, electrons can be lifted by the action of light from the valence band to the conduction band. (a) What is the maximum wavelength that will produce photoconduction in diamond, which has a band gap of 6.1 eV? (b) In what part of the electromagnetic spectrum does this wavelength lie?

12. Identify the following as p-type or n-type semiconductors: (a) Sb in Si; (b) In in Ge; (c) Al in Ge; (d) P in Si.

Section 46–6   Semiconducting devices
13. When a photon enters the depletion region of a pn junction, electron-hole pairs can be created as electrons absorb part of the photon's energy and are excited from the valence band to the conduction band. These junctions are thus often used as detectors for photons, especially for x rays and nuclear gamma rays. When a 662-keV gamma-ray photon is totally absorbed by a semiconductor with an energy gap of 1.1 eV, on the average how many electron-hole pairs are created?

PROBLEMS GROUPED BY SECTION

Section 46–2   Conduction Electrons—The Allowed States
1. Calculate and compare the number of particles per unit volume for (a) the molecules of oxygen gas at standard temperature and pressure and (b) the conduction electrons in copper. (c) What is the ratio of these numbers? (d) What is the average distance between particles in each case? Assume that this distance is the edge of a cube whose volume is equal to the volume per particle.

2. Consider the small cube of copper analyzed in Example 1. (a) How many conduction electrons does it contain? (b) For the energy interval specified in that example, what is the energy separation between adjacent conduction electron states? (c) What would this separation be for a cube whose edge is 0.20 mm, double that of the cube of Example 1? (d) Which, if any, of the preceding results would differ if the cube were made of a material other than copper?

3. The Fermi energy of aluminum is 11.7 eV; its density is 2.70 g/cm³ and its atomic weight is 27.0 g/mol. From these data, determine the number of free electrons per atom.

Section 46–3   Filling the Allowed States
4. Silver is a monovalent metal with an atomic weight of 107.9 g/mol and a density of 1.05×10^4 kg/m³. Calculate (a) the number of conduction electrons per unit volume, (b) the Fermi energy E_F, (c) the Fermi speed v_F, and (d) the de Broglie wavelength corresponding to this speed.

5. A neutron star can be analyzed by techniques similar to those used for ordinary metals. In this case the neutrons (rather than electrons) obey the distribution function, Eq. 46–6. Consider a neutron star of 2.0 solar masses with a radius of 10 km. Calculate the Fermi energy of the neutrons.

6. Show that the distribution of states function given by Eq. 46–1 can be written in the form

$$n_s(E) = 1.5 n_0 E_F^{-3/2} E^{1/2}.$$

Explain how it can be that $n_s(E)$ is independent of material when the Fermi energy E_F (= 7.0 eV for copper, 9.4 eV for zinc, etc.) appears explicitly in this expression.

7. Consider the conduction electrons in copper as an electron gas. (a) Assume first that this gas behaves like an ideal gas of the familiar kind and calculate the root-mean-square speed v_{rms}. (b) Assume now that this gas follows the quantum rules exemplified by Fig. 46–1 and calculate the Fermi speed v_F. (c) What is the ratio v_s/v_{rms} of these two speeds? Note that assumption (b) is consistent with experiment, but assumption (a) is not.

8. Equation 21–14 which, with a small change of notation, is

$$n_p(v) = 4\pi n_0 \left(\frac{m}{2\pi kT} \right)^{3/2} v^2 e^{-mv^2/2kT},$$

gives the speed distribution for the molecules of an ideal gas at temperature T. Show that this can be rewritten in the form of an energy distribution, as

$$n_p(E) = \frac{1.13 n_0}{(kT)^{3/2}} E^{1/2} e^{-E/kT},$$

in which E $(=mv^2/2)$ is the kinetic energy. [Hint: Consider a group of particles whose speeds lie in a selected range dv, or in the corresponding kinetic energy range dE. The number of particles in this group can be written either as $n_p(v)\, dv$ or as $n_p(E)\, dE$, so that these two expressions can be set equal to each other.]

9. Convince yourself, from an inspection of Figs. 46–1b and 46–1c, that a rough but reasonable estimate of the fraction of the conduction electrons in a metal that are in excited states at any temperature T is

$$\frac{kT\, n_s(E_F)}{n_0},$$

where n_0 is the total number of conduction electrons per unit volume. This expression should be valid to within a factor of 2 or so. Evaluate for $T =$ (a) zero, (b) 300 K, and (c) 1000 K. For copper, $E_F = 7.0$ eV and $n_0 = 8.5 \times 10^{28}$ m⁻³.

***10.** Show that at the absolute zero of temperature, the average energy of the conduction electrons in a metal is equal to $\frac{3}{5}E_F$, where E_F is the Fermi energy. Evaluate for copper and sketch in this line in Fig. 46–1b.

Section 46-5 Semiconductors

11. The Fermi-Dirac distribution function can be applied to semiconductors as well as conductors. In semiconductors, E is the energy above the top of the valence band. The Fermi level for a intrinsic semiconductor is nearly midway between the top of the valence band and the bottom of the conduction band. For germanium these bands are separated by a gap of 0.67 eV. Calculate the probability that (a) a state at the bottom of the conduction band is occupied and (b) a state at the top of the valence band is unoccupied at 300 K. (c) For comparison, calculate the probability that a state at the Fermi level of a metal is occupied.

12. Doping changes the Fermi level of a semiconductor. Consider silicon, with a gap of 1.11 eV between the valence and conduction bands. At 300 K the Fermi level of the pure material is nearly at the midpoint of the gap. Suppose that it is doped with donor atoms, each of which has a state 0.15 eV below the bottom of the conduction band, and suppose further that doping raises the Fermi level to 0.11 eV below the bottom of that band. (a) For both the pure and doped silicon, calculate the probability that a state at the bottom of the conduction band is occupied. (b) Also calculate the probability that a donor state in the doped material is occupied. See Problem 11.

13. A silicon sample is doped with 5.65×10^{16} donor atoms/m³. Each atom contributes one state, with energy 0.11 eV below the bottom of the conduction band. (a) If each of these states are occupied with probability 5.00×10^{-5} at temperature 300 K, where is the Fermi level relative to the bottom of the conduction band? (b) What then is the probability that a state at the bottom of the conduction band is occupied? See Problem 11.

14. Consider a solid consisting of roughly 1 cm³ of material (about 10^{22} atoms). Estimate the number of conduction electrons at 300 K if the material is (a) a metal, (b) an intrinsic semiconductor ($E_g \approx 1$ eV), or (c) an insulator ($E_g \approx 6$ eV). (d) How would your estimates differ at 1000 K? At 4 K?

Section 46-6 Semiconducting Devices

15. For an ideal pn junction diode, with a sharp boundary between the two semiconducting materials, the current i is related to the potential difference V across the diode by

$$i = A(e^{eV/kT} - 1),$$

where A depends on the materials but not on the current or potential difference. V is positive if the junction is forward biased and negative if it is back biased. (a) Verify that this expression predicts the behavior expected of a diode by sketching i as a function of V over the range -0.12 V $< V < +0.12$ V. Take $T = 300$ K and $A = 5.00 \times 10^{-9}$ A. (b) For the same temperature, calculate the ratio of the current for a 0.50-V forward bias to the current for a 0.50-V back bias.

ADDITIONAL PROBLEMS

16. Carry out the calculation of Example 3 for the case of $p(E) = 0.10$. Compare your result and also the result of Example 3 carefully with Fig. 46-1c.

17. Pure silicon at room temperature has an electron density in the conduction band of approximately 1×10^{16}/m³ and an equal density of holes in the valence band. Suppose that one of every 10^7 silicon atoms is replaced by a phosphorus atom. (a) Which type will this doped semiconductor be, n or p? (b) What charge carrier density will the phosphorus add? (The density of silicon is 2.33×10^3 kg/m³, and its atomic weight is 28.1 g/mol.) (c) What is the ratio of the charge carrier density in the doped silicon to that for the pure silicon?

18. Zinc is a bivalent metal with an atomic weight of 65.37 g/mol and a density of 7.1×10^3 kg/m³. Calculate (a) the number of conduction electrons per unit volume, (b) the Fermi energy E_F, (c) the Fermi speed v_F, and (d) the de Broglie wavelength corresponding to this speed.

19. In Example 2 of Chapter 43 we considered the one-dimensional problem of an electron trapped between rigid walls. If the walls are 10 Å apart, we showed that the energy of the lowest level, corresponding to $n = 1$ in Eq. 43-12, is 0.38 eV. Suppose now that the walls are 10 cm apart. (a) What now is the quantum number n of the level for which $E = 0.38$ eV? (b) What is the density of levels at this energy? That is, how many levels per unit energy interval are there? (Note: The purpose of this problem is to show that as the "box" in which an electron is trapped is made larger, the density of levels increases.)

20. White dwarf stars represent a late stage in the evolution of stars like the sun. They become dense enough and hot enough that we can analyze their structure as a solid in which all Z electrons per atom are free. For a white dwarf with a mass equal to that of the sun and a radius equal to that of the earth, calculate the Fermi energy of the electrons. Assume the atomic structure to be represented by iron atoms.

21. In Example 2 of Chapter 43 we considered the one-dimensional problem of an electron trapped between rigid walls 10 Å apart. Consider that 100 electrons are to be placed in the region between the walls, two to a level, with opposite spins. What is the Fermi energy of the system? Compare with Fig. 46-1b.

22. Using the result of Problem 10, estimate how much energy would be released by the conduction electrons in a penny (assumed all copper; mass = 3.1 g) if we could suddenly turn off the Pauli exclusion principle. For how long would this amount of energy light a 100-W lamp? Would the other (bound) electrons in the penny also be affected? Note that there is no known way to turn off the Pauli principle.

23. Show that, if $E > E_F$, the particle distribution function $n_p(E)$ of Eq. 46-6 can be written as

$$n_p(E) \cong CE^{1/2}e^{-E/kT}.$$

Compare this result with that calculated for the Maxwell speed distribution in Problem 8. What do you conclude?

*24. In one model of a semiconductor, the actual energy spectrum is replaced by one for which there are N_v states in the valence band, all with energy E_v, and N_c states in the conduction band, all with energy E_c. For an intrinsic semiconductor the number of electrons in the conduction band equals the number of holes in the valence band. (a) Show that, when this condition is applied to the model,

$$\frac{N_c}{e^{(E_c-E_F)/kT}} + 1 = \frac{N_v}{e^{-(E_v-E_F)/kT}} + 1.$$

(b) If the Fermi level is in the gap between the two bands and is far from both bands compared to kT, then the exponentials dominate the denominators. Show that, under these conditions,

$$E_F = \tfrac{1}{2}(E_c + E_v) + \tfrac{1}{2}kT \ln\left(\frac{N_v}{N_c}\right).$$

47
Nuclear Physics: An Introduction

47-1 The Atomic Nucleus

The nucleus discovered

Ernest Rutherford was led, in 1911, to propose that the atomic nucleus existed when he interpreted some experiments carried out at the University of Manchester in England by his collaborators, Hans Geiger and Ernest Marsden.* These workers, at his suggestion, allowed a beam of α-particles to strike and be deflected by a thin gold foil. These α-particles, which are ~ 7300 times more massive than electrons, carry a charge of $+2e$ and are emitted spontaneously by many radioactive substances. They are now known to be nuclei of ^{4}He.

Geiger and Marsden counted the number of α-particles deflected through various scattering angles ϕ in an experimental arrangement like that of Fig. 47-1. Figure 47-2 shows the paths taken by typical α-particles as they scatter from atoms in the target foil. The angles ϕ through which the particles are deflected range from 0 to 180° as the character of the collision varies from "grazing" to "head-on."

The electrons in the atoms of the gold foil, being so light, have almost no effect on the motion of a fast incoming α-particle. The electrons in fact are themselves strongly deflected, somewhat as a swarm of insects would be deflected by a stone hurled through them. Any deflection of the α-particle must be caused by the repulsive action of the positive charge of the gold atom.

At the time of these experiments, most physicists believed in the so-called plum pudding model of the atom, which had been suggested by J. J. Thomson. In this view (see Exercise 1a), the positive charge of the atom was thought to be spread out through the whole volume of the atom, that is, through a spherical vol-

The scattering experiment

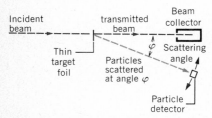

Figure 47-1 A generalized scattering arrangement. A beam of particles falls on a thin target foil. Most of the particles pass through the foil and are captured by the beam collector. Some particles however are scattered toward the detector, which may be set at any desired scattering angle.

---

* See "The Birth of the Nuclear Atom," by E. N. da C. Andrade, *Scientific American*, November 1956.

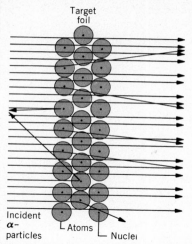

Figure 47-2 The deflection of the incident α-particles depends on the nature of the nuclear collision. (From "The Birth of the Nuclear Atom," by E. N. da C. Andrade, *Scientific American*, November 1956.)

Rutherford interprets the scattering experiments

Some nuclides

ume of radius about 10^{-10} m. The electrons were thought to vibrate about fixed centers within this sphere.

Rutherford showed that this model of the atom was not consistent with the α-scattering experiments and proposed instead the nuclear model of the atom that we now accept. Here the positive charge is confined to a very much smaller sphere whose radius is about 10^{-14} m (the *nucleus*). The electrons move around this nucleus and occupy a roughly spherical volume of radius about 10^{-10} m. This brilliant deduction by Rutherford laid the foundation for modern atomic and nuclear physics.

The feature of the α-scattering experiments that attracted Rutherford's attention was that a few such α-particles were deflected through very large angles, up to 180°. To scientists accustomed to thinking in terms of the plum pudding model, this was a very surprising result. In Rutherford's words: "It was quite the most incredible event that ever happened to me in my life. It was almost as incredible as if you had fired a 15-inch shell at a piece of tissue paper and it came back and hit you."

The α-particle must pass through a region in which the electric field is very high indeed in order to be deflected so strongly. Exercise 1a shows that, in Thomson's model, the maximum possible value of the electric field is $\sim 1 \times 10^{-13}$ N/C. Compare this with the value calculated in Exercise 1b for a point on the surface of a gold nucleus ($\sim 2 \times 10^{21}$ N/C). Thus the deflecting force acting on an α-particle can be up to 10^8 times as great if the positive charge of the atom is compressed into a small enough region (the nucleus) at the center of the atom. Rutherford made his hypothesis about the existence of nuclei only after a much more detailed mathematical analysis than that given here.

47-2 Some Nuclear Properties

The nucleus proves to have a structure every bit as complex as that of the atom for which it serves as the massive positively charged force center. Table 47-1 displays some properties of a few representative nuclear species, or *nuclides*, as they are called.

The Nuclear Force. The force that controls the electronic structure and properties of the atom is the familiar Coulomb force. To bind the nucleus together, however, there must be a strong attractive force of a totally new kind, acting between the

Table 47-1 Some properties of selected nuclides[a]

Nuclide	Z	N	A	Stability[b]	Atomic mass, u	Radius, fm	Binding energy, MeV/nucleon	Spin	Magnetic moment, μ_N
^{7}Li	3	4	7	92.5%	7.01604	2.1	5.60	$\frac{3}{2}$	+3.26
^{31}P	15	16	31	100%	30.97376	3.36	8.48	$\frac{1}{2}$	+1.13
^{81}Br	35	46	81	49.3%	80.91629	4.63	8.69	$\frac{3}{2}$	+2.27
^{120}Sn	50	70	120	32.4%	119.90221	5.28	8.50	0	0
^{157}Gd	64	93	157	15.7%	156.92397	5.77	8.21	$\frac{3}{2}$	−0.339
^{197}Au	79	118	197	100%	196.96655	6.23	7.91	$\frac{3}{2}$	+0.145
^{227}Ac	89	138	227	21.8 y	227.02777	6.53	7.65	$\frac{3}{2}$	+1.10
^{239}Pu	94	145	239	24,100 y	239.05217	6.64	7.56	$\frac{1}{2}$	+0.200

[a] The nuclides were chosen to lie along the stability line of Fig. 47-3.
[b] For stable nuclides the isotopic abundance is given; for radionuclides, the half-life.

Binding the nucleus together

neutrons and the protons that make up the nucleus. This force must be strong enough to overcome the repulsive Coulomb force between the (positively charged) protons and to bind both neutrons and protons into the tiny nuclear volume.

The strong force

Experiment suggests that this *strong force*, as it is simply called, has the same character between any pair of nuclear constituents, be they neutrons or protons. Because of this we refer to neutrons and protons collectively as *nucleons*, except when we specifically need to distinguish between them.

Nucleons

The strong force is a short-range force

The "strong force" has a short range. This means that the attractive force between pairs of nucleons drops rapidly to zero for nucleon separations greater than a certain critical value. This in turn means that, except in the smallest nuclei, a given nucleon cannot interact with all the other nucleons in the nucleus but only with a few of its nearest neighbors. By contrast, the Coulomb force is not a short-range force. A given proton in a nucleus exerts a Coulomb repulsion on all the other protons, no matter how large their separation; see Problem 35.

Z, N, and A

Nuclear Systematics. Nuclei are made up of protons and neutrons. The number of protons in the nucleus, as we have already seen, is called the *atomic number* and is represented by Z. The number of neutrons is called the *neutron number,* and we represent it by N. The total number of nucleons ($= Z + N$) is called the *mass number,* and we represent it by A.

We use A, the total number of nucleons, as an identifying superscript in labeling nuclides. In ^{81}Br, for example there are 81 nucleons. The symbol "Br" tells us that we are dealing with bromine, for which $Z = 35$. The remaining 46 nucleons are neutrons, so that, for this nuclide, $Z = 35$, $N = 46$, and $A = 81$.

The nuclidic chart

Figure 47–3 shows a chart of the nuclides. It is a plot of Z against N, each circle representing a nuclide. Filled black circles represent stable nuclides; filled colored circles represent radioactive nuclides—or radionuclides—that are relatively stable, having half-lives* greater than an arbitrarily chosen value of 100 y; open circles represent radionuclides with half-lives less than this value.

The line of stability

Note that there is a reasonably well-defined linear zone of stability in Fig. 47–3, traced only by the black and the colored circles. Relatively unstable radionuclides tend to lie on either side of the stability zone.

Figure 47–3 also shows that the lightest stable nuclides tend to lie on or close to the line $Z = N$. The heavier stable nuclides lie well below this line and thus typically have many more neutrons than protons. The tendency to an excess of neutrons at large mass numbers is a Coulomb repulsion effect. Because a given nucleon interacts with only a small number of its neighbors through the strong force, the amount of energy tied up in strong-force bonds between nucleons increases just in proportion to A. The energy tied up in Coulomb-force bonds between protons increases more rapidly than this because each proton interacts with every other proton in the nucleus. Thus the Coulomb energy becomes increasingly important at high mass numbers.

Neutron excess—a Coulomb effect

Consider a nucleus with 238 nucleons. If it were to lie on the $Z = N$ line it would have $Z = N = 119$. However, such a nucleus, if it could be assembled, would fly apart at once because of Coulomb repulsion. Relative stability is found only if we replace 27 of the protons by neutrons, thus greatly diluting the Coulomb repulsion effect. We then would have the nuclide ^{238}U, which has $Z = 92$ and $N = 146$, a neutron excess of 54.

Even in ^{238}U, Coulomb effects are evident in that (1) this nuclide is radioactive and emits α-particles and (2) it can easily break up (fission) into two fragments

---

* See p. 956 for a definition of half-life.

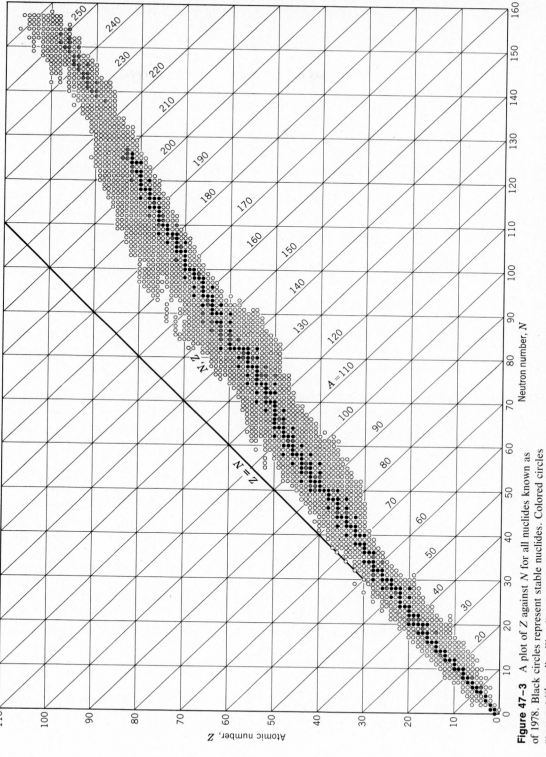

Figure 47-3 A plot of Z against N for all nuclides known as of 1978. Black circles represent stable nuclides. Colored circles represent "long-lived" radioactive nuclides (half-lives greater than 100 y). Open circles represent "short-lived" radioactive nuclides (half-lives less than 100 y).

if it absorbs a neutron. Both of these processes reduce the Coulomb energy more than they do the energy in the strong-force bonds.

Nuclear Radii. A convenient unit for reporting nuclear radii is the *fermi* (abbr. fm or F), defined by

$$1 \text{ fermi} = 10^{-15} \text{ m.}$$

The fermi

The fermi is a unit of convenience in nuclear physics, just as are the angstrom and the Bohr radius in atomic physics.

We can learn about the size and structure of nuclei by doing a scattering experiment, much as suggested by Fig. 47–1, using an incident beam of high-energy electrons. The energy of the incident electrons must be high enough (~200 MeV) so that their de Broglie wavelength will be small enough for them to act as structure-sensitive nuclear probes.

The distribution of nuclear matter

The distribution of nuclear matter for a typical nucleus is shown in Fig. 47–4. We see that the nucleus does not have a sharply defined surface. It does, however, have a characteristic mean radius R. The nuclear density $\rho(r)$ has a constant value

Figure 47–4 The variation with radial distance of the density of nuclear matter in the nucleus ^{197}Au.

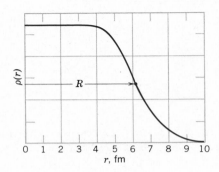

in the nuclear interior and drops to zero through the fuzzy surface zone. R is given by

The mean nuclear radius

$$R = R_0 A^{1/3}, \tag{47–1}$$

in which A is the mass number and R_0 is a constant with the value of 1.1 fm. For ^{63}Cu, for example,

$$R = (1.1 \text{ fm})(63)^{1/3} = 4.3 \text{ fm.}$$

By comparison, the mean radius of a copper ion in a lattice of solid copper is 0.96 Å, about 2×10^4 times larger.

Example 1 What is the approximate density of the "nuclear matter" that forms the nuclear interior?

We know that this number will be high, because virtually all of the mass of the atom resides in its tiny nucleus. The volume of the nucleus, approximated as a uniform sphere of radius R, is given by Eq. 47–1 as

$$V = \tfrac{4}{3}\pi R^3 = \tfrac{4}{3}\pi (R_0^3 A).$$

The density of nuclear matter, expressed in nucleons per unit volume, is then

$$\rho_0 = \frac{A}{V} = \frac{A}{(4\pi/3)\, R_0^3 A}$$

$$= \frac{1}{(4\pi/3)(1.1 \text{ fm})^3} = 0.18 \text{ nucleons/fm}^3.$$

The mass of a nucleon is 1.7×10^{-27} kg. In more familiar units, the density of nuclear matter is then

$$\rho_0 = (0.18 \text{ nucleons/fm}^3)$$
$$(1.7 \times 10^{-27} \text{ kg/nucleon})(1 \text{ fm}/10^{-15} \text{ m})^3$$
$$= 3.1 \times 10^{17} \text{ kg/m}^3,$$

or 3.1×10^{14} times greater than the density of water! Unlike the orbital electrons, the nucleons have a density nearly independent of their number. To some extent they are packed in like marbles in a bag.

Nuclear Masses and Binding Energies. Atomic masses can be measured with great precision using modern mass spectrometer and nuclear reaction techniques. We recall that such masses are measured in *unified atomic mass units* (abbr. u), chosen so that the atomic mass (*not* the nuclear mass) of ^{12}C is exactly 12 u. The relation of this unit to the SI mass standard is

The unified atomic mass unit

$$1 \text{ u} = 1.66 \times 10^{-27} \text{ kg}.$$

Note that the mass number (symbol A) identifying a nuclide is so named because this number is equal to the atomic mass of the nuclide, rounded to the nearest integer. Thus the mass number of the nuclide ^{137}Cs is 137; this nuclide contains 55 protons and 82 neutrons, a total of 137 particles; its atomic mass is 136.90707 u, which rounds off numerically to 137.

$E = \Delta m\, c^2$

In nuclear physics, as contrasted with atomic physics, the energy changes per event are commonly so great that Einstein's well-known mass-energy relation $E = \Delta m\, c^2$ is an indispensable work-a-day tool. See Fig. 47-5 for an example of the pervasiveness of this best-known of all equations in physics.

We shall often need to use the energy equivalent of 1 atomic mass unit, and we find it from

The mass-energy equivalence

$$E = \Delta m\, c^2 = (1.66 \times 10^{-27} \text{ kg})(3.00 \times 10^8 \text{ m/s})^2(1 \text{ MeV}/1.60 \times 10^{-13} \text{ J})$$
$$= 931 \text{ MeV}.$$

This means that we can write c^2 as 931 MeV/u and can thus easily find the energy equivalent (in MeV) of any mass or mass difference (in u), or conversely.

As an example, consider the deuteron, the nucleus of the heavy hydrogen atom. A deuteron consists of a proton and a neutron bound together by the strong interaction force. The energy E_B that we must add to the deuteron to tear it apart into its two constituent nucleons is called its *binding energy*. From the conservation of mass-energy we can write, for this pulling-apart process,

The deuteron binding energy

$$m_d c^2 + E_B = m_n c^2 + m_p c^2. \tag{47-2}$$

If we add $m_e c^2$, the energy equivalent of one electron mass, to each side of this equation, we have

$$(m_d + m_e)c^2 + E_B = m_n c^2 + (m_p + m_e)c^2,$$

Figure 47-5 Students of Shenandoah Junior High School, in Miami, Florida, honor Einstein in the year of the 100th anniversary of his birth by spelling out his famous formula with their bodies. Courtesy of Rocky Raisen; photo by Justo Alfonso.

or

$$m_{2H}c^2 + E_B = m_n c^2 + m_{1H}c^2. \tag{47-3}$$

Here m_{2H} and m_{1H} are the masses of the neutral *heavy* hydrogen atom and the neutral *ordinary* hydrogen atom, respectively. They are atomic masses, not nuclear masses. Solving Eq. 47–3 for E_B yields

$$E_B = (m_n + m_{1H} - m_{2H})c^2 = \Delta m c^2, \tag{47-4}$$

in which Δm is the change in atomic rest mass. In making calculations of this kind we always use atomic, rather than nuclear masses, as this is what is normally tabulated. As in this example, the electron masses always conveniently cancel.*

Use atomic—not nuclear—masses

For the deuteron problem the needed atomic masses are

Symbol	Atomic mass, u
¹H	1.00783
n	1.00867
²H	2.01410

Substituting into Eq. 47–4 and replacing c^2 by its equal, 931 MeV/u, yields for the binding energy

$$E_B = (1.00867\ u + 1.00783\ u - 2.01410\ u)(931\ \text{MeV/u})$$
$$= (0.00240\ u)(931\ \text{MeV/u}) = 2.23\ \text{MeV}.$$

Compare this with the binding energy of the hydrogen atom in its ground state, which is 13.6 eV, about five orders of magnitude smaller.

Example 2   Find E_B, the binding energy per nucleon required to pick apart a typical middle-mass nucleus such as ¹²⁰Sn into its constituent neutrons and protons. The atomic mass of ¹²⁰Sn (see Table 47–1) is 119.90221 u.

¹²⁰Sn contains 50 protons and $(120 - 50)$ or 70 neutrons. Their combined masses as free particles are, using the atomic mass data given above,

$$M = 50 \times 1.00783\ u + 70 \times 1.00867\ u$$
$$= 120.99840\ u.$$

This exceeds the atomic mass of ¹²⁰Sn by

$$\Delta m = 120.99840\ u - 119.90221\ u$$
$$\approx 1.10\ u.$$

Converting this to energy yields

$$E = \Delta m\ c^2 = (1.10\ u)(931\ \text{MeV/u}) = 1020\ \text{MeV}.$$

Finally, the binding energy per nucleon is

$$E_B = \frac{E}{A} = \frac{1020\ \text{MeV}}{120} = 8.50\ \text{MeV/nucleon}.$$

In Fig. 47–6 we plot E_B, the binding energy per nucleon, as a function of the mass number A. We see that the value calculated above is typical of middle-mass nuclides.

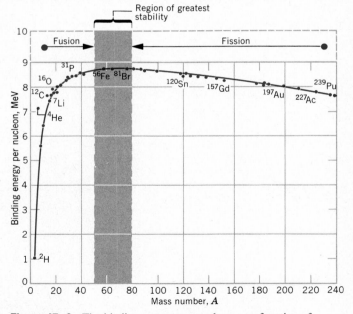

Figure 47–6   The binding energy per nucleon as a function of mass number for nuclides lying along the linear stability zone of Fig. 47–3. The nuclides in Table 47–1 are identified, along with a few other nuclides of interest.

* Not true for β^+ decay; see problem 40.

The fact that the binding energy curve of Fig. 47–6 is relatively low both for low and for high mass numbers has practical consequences of vast importance.

The "drooping" of the binding-energy at high mass numbers tells us that nucleons are more tightly bound when they are assembled into two middle-mass nuclei rather than into a single high-mass nucleus. In other words, energy can be released in the *nuclear fission* of a single massive nucleus into two smaller fragments.

The "drooping" of the binding-energy curve at low mass numbers, on the other hand, tells us that energy will be released if two nuclei of small mass number combine to form a single middle-mass nucleus. This process, the reverse of fission, is called *nuclear fusion*. It occurs inside our sun, and other stars, and is the mechanism by which the sun generates the energy it radiates to us.

Nuclear Spin and Magnetism. Nuclei, like atoms, have an intrinsic angular momentum whose maximum projected value is given by $I\hbar$. Here I is a quantum number, which may be integral or half-integral, called the *nuclear spin;* some values for selected nuclides are shown in Table 47–1.

Again as for atoms, a nuclear angular momentum has a nuclear magnetic moment associated with it. Recall that, in atomic magnetism the *Bohr magneton* μ_B, defined as

$$\mu_B = \frac{eh}{4\pi m_e} = 5.76 \times 10^{-5} \text{ eV/T},$$

is a unit of convenience.

In nuclear physics the corresponding unit of convenience is the *nuclear magneton,* defined as above except that the electron mass m_e is replaced by the proton mass m_p. That is,

$$\mu_N = \frac{eh}{4\pi m_p} = 3.14 \times 10^{-8} \text{ eV/T}.$$

Because the magnetic moment of the free electron is (very closely) one Bohr magneton, it might be supposed that the magnetic moment of the free proton would be (very closely) one nuclear magneton. It is not very close, however, the measured value being $+2.7929 \ \mu_N$. Nuclear magnetic moments are not yet as well understood as are electronic and atomic magnetic moments.

47–3 Radioactive Decay

Radionuclides normally decay by the spontaneous emission of a particle from the nucleus. It may be an alpha α-particle, in which case we refer to α-*decay*. Alternatively, the particle may be an electron, in which case we refer to β-*decay*.

All such decays are statistical in character. Consider, for example, a 1-g sample of uranium metal. It contains 2.5×10^{21} nuclei of the very long-lived α-emitter ^{238}U. During any given second about 12,000 of these nuclei will decay, but we have absolutely no way of predicting which of them will do so. Every ^{238}U nucleus in the sample has exactly the same probability of decay during the 1-s interval, namely, $(12,000)/(2.5 \times 10^{21})$, or 1 chance in 2.0×10^{17}.

In general, if a sample contains N radioactive nuclei, we can express the statistical character of the decay process by saying that the decay rate $R \ (= -dN/dt)$ is proportional to N, or

$$-\frac{dN}{dt} = \lambda N, \tag{47–5}$$

in which λ, the *distintegration constant,* has a characteristic value for every decay process. We can rewrite Eq. 47–5 as

$$\frac{dN}{N} = -\lambda \, dt,$$

which integrates readily to

Radioactive decay: $N(t)$

$$N = N_0 e^{-\lambda t}. \tag{47–6}$$

Here N_0 is the number of radioactive nuclei in the sample at $t = 0$. We see that the decrease of N with time follows a simple exponential law.

We are often more interested in the decay rate R ($= -dN/dt$) of the sample than we are in N. Differentiating Eq. 47–6 yields

Radioactive decay: $R(t)$

$$R = R_0 e^{-\lambda t}, \tag{47–7}$$

in which R_0 is the decay rate at $t = 0$. We assumed initially that R and N are proportional to each other, so we are not surprised to confirm that they both decrease with time according to the same exponential law.

A quantity of interest is the time $t_{1/2}$, called the *half-life,* after which both N and R are reduced to one-half of their initial values. Putting $R = \frac{1}{2}R_0$ in Eq. 47–7 gives

$$\tfrac{1}{2}R_0 = R_0 e^{-\lambda t_{1/2}}$$

which leads readily to

The half-life

$$t_{1/2} = \frac{0.693}{\lambda}, \tag{47–8}$$

a relationship between the half-life and the disintegration constant.

The following two examples show how λ can be measured for decay processes with relatively short half-lives and also with relatively long half-lives.

Example 3 *Short-lived decays* We define short-lived decays as cases in which it is possible to measure directly the decrease in the decay rate R with time. The following table gives some data for a sample of ^{128}I, a radionuclide often used medically as a tracer to measure the iodine uptake rate of the thyroid gland. Find (*a*) the disintegration constant λ and (*b*) the half-life $t_{1/2}$ from these data.

Time, min	R, counts/s	Time, min	R, counts/s
4	392.2	132	10.9
36	161.4	164	4.56
68	65.5	196	1.86
100	26.8		

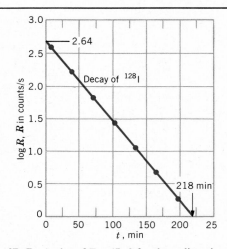

Figure 47–7 A plot of Eq. 47–9 for the radioactive decay of ^{128}I.

If we take the logarithm of each side of Eq. 47–7, we find that

$$\log R = \log R_0 - (\lambda \log e)t = \log R_0 + (-0.434\lambda)t. \tag{47–9}$$

Thus, if we plot $\log R$ against t, we should obtain a straight line whose slope is -0.434λ. Figure 47–7 shows such a plot. Equating the slope of the line to -0.434λ yields

$$-0.434\lambda = -\frac{(2.64 - 0)}{(218 \text{ min} - 0)},$$

or

$$\lambda = 0.0279 \text{ min}^{-1}.$$

Equation 47–8 yields for $t_{1/2}$:

$$t_{1/2} = \frac{0.693}{\lambda} = \frac{0.693}{0.0279 \text{ min}^{-1}} = 24.8 \text{ min}.$$

Example 4 *Long-lived decays* In the case of long-lived decays it is not possible to wait long enough to observe a measurable decrease in the decay rate R with time. We must find λ by measuring both N and $(-dN/dt)$ in Eq. 47–5.

A 1.00-g sample of pure KCl from the chemistry stock room is found to be radioactive and to decay at an absolute rate R of 1600 counts/s. The decays are traced to the element potassium and in particular to the isotope ^{40}K, which constitutes 1.18% of normal potassium. What is the half-life for this decay?

The molecular weight of KCl is 74.9 g/mol, so the number of potassium atoms in the sample is

$$N_K = \frac{(6.02 \times 10^{23} \text{ mol}^{-1})(1.00 \text{ g})}{(74.9 \text{ g/mol})} = 8.04 \times 10^{21}.$$

Of these the number of ^{40}K atoms is

$$N_{40} = (8.04 \times 10^{21})(1.18 \times 10^{-2}) = 9.49 \times 10^{19}.$$

From Eq. 47–5 we have

$$\lambda = \frac{-dN/dt}{N} = \frac{R}{N_{40}} = \frac{1600 \text{ s}^{-1}}{9.49 \times 10^{19}} = 1.69 \times 10^{-17} \text{ s}^{-1},$$

and the half-life, from Eq. 47–8, is

$$t_{1/2} = \frac{0.693}{\lambda} = \left(\frac{0.693}{1.69 \times 10^{-17} \text{ s}^{-1}}\right)\left(\frac{1 \text{ y}}{3.16 \times 10^7 \text{ s}}\right)$$
$$= 1.30 \times 10^9 \text{ y}.$$

This is almost equal to the age of the universe. No wonder we cannot see the decline of the decay rate with time!

47–4 Barrier Tunneling and α-Decay

A prototype for α-decay

We take as a prototype α-decay process

$$^{238}\text{U} \rightarrow {}^{234}\text{Th} + {}^4\text{He} \qquad (t_{1/2} = 4.47 \times 10^9 \text{ y}). \qquad (47–10)$$

α-Emission versus nucleon emission

We may wonder why it is energetically possible for ^{238}U to emit an α-particle spontaneously while spontaneous emission of a proton or a neutron is energetically forbidden. To start with, we verify in Example 5 that this statement is indeed true.

Example 5 Show that it is energetically favorable for ^{238}U to emit an α-particle but not a proton or a neutron. The needed atomic masses are

^{238}U	238.05081 u	^{4}He	4.00260 u
^{234}Th	234.04363 u	^{1}H	1.00783 u
^{237}Pa	237.05121 u	n	1.00867 u
^{237}U	237.04874 u		

In the α-decay process of Eq. 47–10 the atomic mass of the decay products (= 238.04623 u) is less than the atomic mass of ^{238}U by $\Delta m = 0.00458$ u, whose energy equivalent is

$$Q_\alpha = \Delta m \, c^2 = (0.00458 \text{ u})(931 \text{ MeV/u}) = +4.26 \text{ MeV}.$$

This *disintegration energy* is available to share as kinetic energy between the α-particle and the recoiling ^{234}Th atom.

If ^{238}U were to emit a proton, the decay process would be

$$^{238}\text{U} \rightarrow {}^{237}\text{Pa} + {}^1\text{H}.$$

In this case the mass of the decay products *exceeds* the mass of ^{238}U by $\Delta m = 0.00823$ u, the energy equivalent Q_p being -7.66 MeV. The minus sign tells us that ^{238}U is stable against spontaneous proton emission.

For neutron emission the process would be

$$^{238}\text{U} \rightarrow {}^{237}\text{U} + n,$$

and the energy release ($Q_n = -6.14$ MeV) again proves to be negative. Thus ^{238}U is also stable against spontaneous neutron emission.

Why is the α-decay of ^{238}U so slow?

We now ask the question: If energy can be released by the α-decay of ^{238}U, why did not such nuclides decay spontaneously when they were created? The creation process presumably occurred in the violent explosions of ancestral stars that predate the formation of our solar system. To answer this question we must study the physics of the α-decay process. We choose a model in which the α-particle is imagined to exist preformed in the nuclear interior before its escape.

Figure 47–8 shows an approximation to the potential energy function $U(r)$ for the α-particle and the residual ^{234}Th nucleus. It is a combination of a so-called po-

A model for α-decay

tential well associated with the attractive strong force in the nuclear interior and a Coulomb potential associated with the electrostatic repulsion between the α-particle and the residual ^{234}Th nucleus when the particle is outside the nucleus.

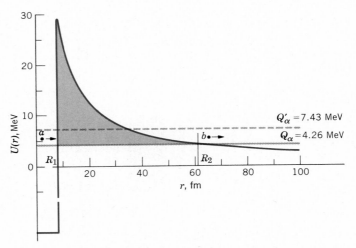

Figure 47–8   A potential energy function describing the emission of α-particles by ^{238}U. The shaded area represents a potential barrier that inhibits emission. The α-particle is shown inside the nucleus at a and also outside the potential barrier at b.

The potential barrier

Example 5 showed us that the disintegration energy Q_α for the α-decay of ^{238}U is 4.26 MeV. We show this in Fig. 47–8 as a horizontal line intersecting the curve of $U(r)$ at R_1 and at R_2. Now we see why the α-particle is not immediately emitted from the ^{238}U nucleus; there is a potential barrier of impressive height and thickness surrounding the nucleus, occupying a spherical shell between the radii R_1 and R_2.

Why is the α-decay of ^{238}U so rapid?

Indeed, we now shift the focus of our question and ask instead: How does the α-particle *ever* escape from the ^{238}U nucleus? It seems to be permanently trapped there by the barrier. Classically, the α-particle can exist inside this barrier, as at a in Fig. 47–8, or outside it as at b, but it cannot exist within the barrier itself because its kinetic energy would be negative in this region.

Barrier tunneling

According to wave mechanics, however, there is always a finite—although perhaps very small—chance that a particle can "leak through" a barrier such as that of Fig. 47–8 and suddenly appear on the other side. *Barrier tunneling* is a basic wave mechanical phenomenon of vast practical importance. Here are but two of many other possible examples: (1) The fusion reactions that account for the sun's radiated energy depend critically on this process. (2) More down to earth, the tunnel diode, a semiconducting device for the discovery of which Leo Esaki shared the 1973 Nobel Prize in physics, is based squarely on this phenomenon.

In wave mechanics we have seen that a particle cannot be represented as a geometric point. The concept of "position" simply loses meaning for distances less than the particle's de Broglie wavelength. If the width of a potential barrier were less than this wavelength, we could perhaps believe that such a barrier could be tunneled. Even for much wider barriers, however, it turns out that there is always a definite probability of tunneling—measured by the value of $\psi^2(r)$ on the opposite side of the barrier.

Barrier tunneling—a wave
mechanical digression

Figure 47–9 shows the tunneling of three one-dimensional rectangular barriers. In each case an incident matter wave (I) falls on the barrier from the left. It is partially reflected (R), the reflected wave combining with the incident wave to form a standing matter wave on the left side of the barrier. The incident wave is also partially transmitted (T); that is, barrier tunneling occurs. The arrows in each case suggest the amplitudes of these three waves. The figure shows that the amplitude of the transmitted wave is increased if the barrier is made either thinner, as in Fig. 47–9*b*, or lower, as in Fig. 47–9*c*.

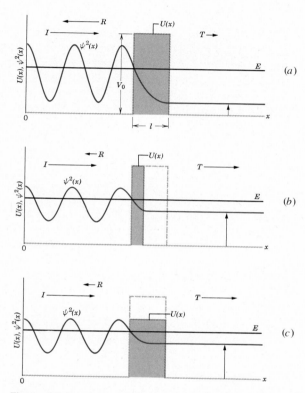

Figure 47–9 (*a*) An incident matter wave I falls on a potential barrier, generating a reflected wave R and a transmitted wave T.

Barrier tunneling in the α-decay of ^{238}U

Barrier tunneling is sensitive to the dimensions of the barrier

Table 47–2 The α-decay of ^{238}U and ^{228}U

Nuclide	Q_α	Half-life
^{238}U	4.26 MeV	4.9×10^9 y
^{228}U	6.81 MeV	550 s

For the long-lived ^{238}U decay of Fig. 47–8, the barrier penetrability is extremely small. We can show that the α-particle, presumed to be rattling back and forth within the nucleus, must present itself at the barrier surface $\sim 10^{38}$ times before it succeeds in tunneling through; this is $\sim 10^{20}$ times a second for $\sim 10^9$ years! Although the escape probability for the ^{238}U-decay is small, we note that we observe only those α-particles that *do* succeed in escaping.

If the α-disintegration energy were higher than the 4.26 MeV of our example, the barrier to α-decay would be both thinner and lower. α-Decay should then occur more readily, that is, with a shorter half-life. This prediction of wave mechanics is borne out quantitatively by experiment. To give just a tiny bit of the evidence, let us compare the α-decays of ^{238}U and ^{228}U. Table 47–2 shows the relevant data. We see that an increase in Q_α by a factor of only 1.6 produces a decrease in half-life (that is, an increase in the effectiveness of barrier tunneling) by a factor of 2.6×10^{14}. The probability of barrier tunneling is remarkably sensitive to small changes in the height and the thickness of the barrier.

47–5   β-Decay

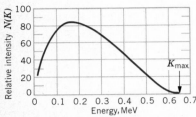

Figure 47–10   The distribution in energy of the positive electrons emitted by ⁶⁴Cu. $K_{max} = 0.653$ MeV.

The basic β-decay processes

The continuous β-spectrum

Here are two prototype examples of β-decay:

$$^{32}\text{P} \rightarrow {}^{32}\text{S} + e^- \qquad (t_{1/2} = 14.4 \text{ d}) \qquad (47\text{–}11a)$$

and

$$^{64}\text{Cu} \rightarrow {}^{64}\text{Ni} + e^+ \qquad (t_{1/2} = 12.7 \text{ h}). \qquad (47\text{–}11b)$$

Both of these processes are incomplete in an important way that we shall describe below but that need not concern us now. The symbol e^+ is a positively charged electron, sometimes called a positron.

It may seem surprising that the nucleus can emit an electron when we have said that it is made up of only protons and neutrons. However, we saw earlier that atoms emit photons, and we certainly do not say that atoms "contain" photons. We say that photons are *created* during the emission process.

So it is with electrons emitted from nuclei. Whether negative or positive, they are created within the nucleus during the emission process. A neutron transforms itself into a proton (or conversely) according to

$$n \rightarrow p + e^- \qquad (47\text{–}12a)$$

or

$$p \rightarrow n + e^+. \qquad (47\text{–}12b)$$

These are the elemental β-decay processes. Like those of Eqs. 47–11, they are incomplete in a way that we shall soon clarify.

In α-decay the emitted α-particles have sharply defined discrete energies. In β-decay, however, the emitted electrons are distributed in a continuous spectrum of energies, ranging from zero to a fixed upper limit K_{max}. Figure 47–10 shows the β-spectrum for the β⁺-decay of ⁶⁴Cu.

It is a matter of great interest to calculate the disintegration energy Q for either of the β-decay processes of Eqs. 47–11 and to compare it with the value of K_{max}. In Example 6 we do this for the β⁻-decay of ³²P.

Example 6   Calculate the disintegration energy Q in the β⁻-decay of ³²P. The needed atomic masses are 31.97391 u for ³²P and 31.97207 u for ³²S.

Because of the presence of the emitted electron, we must be especially careful to distinguish between nuclear and atomic masses. Let m'_P and m'_S represent the nuclear masses of ³²P and ³²S and let m_P and m_S be their atomic masses. We take the disintegration energy Q to be $\Delta m \, c^2$, where

$$\Delta m = m'_P - (m'_S + m_e),$$

m_e being the mass of the electron. If we add and subtract $15 m_e$ to the right-hand side, we have

$$\Delta m = (m'_P + 15 m_e) - (m'_S + 16 m_e).$$

The quantities in parentheses are the atomic masses. Thus we have

$$\Delta m = m_P - m_S.$$

If we subtract the atomic masses in this way, the mass of the emitted electron is automatically taken into account.*

The disintegration energy for the ³²P decay is then

$$Q = \Delta m \, c^2 = (39.97391 \text{ u} - 31.97207 \text{ u})(931 \text{ MeV/u})$$
$$= 1.71 \text{ MeV}.$$

This is just equal to the measured value of K_{max}! Thus although 1.71 MeV is released in every ³²P decay event, essentially all of the electrons have less energy than this. Where is the "missing" energy?

The neutrino; two problems solved

In 1927 Pauli proposed a solution of this puzzle (and also of an equally serious puzzle, involving "missing" angular momentum). He suggested that a second particle, dubbed later by Fermi the *neutrino* ("little neutral one"), is emitted simulta-

---

* This is not true for β⁺-decay; see Problems 40 and 41.

neously with the electron in β-decay and shares the available disintegration energy Q ($= K_{max}$) with it. If the electron has a kinetic energy K in a particular event the neutrino (symbol ν) will have a kinetic energy $K_{max} - K$ so that the sum of their energies is always K_{max}. If we further endow the neutrino with a spin of $\frac{1}{2}$, we are also able to satisfy the angular momentum conservation law, hinted at above. The neutrino must also have zero charge, because charge is already conserved in Eqs. 47–11 and 47–12. The neutrino has been thought to have zero rest mass but evidence adduced in 1980 suggests that this may not actually be the case. Research continues!

Equations 47–11 should then be rewritten as

β⁻-decay

$$^{32}P \rightarrow {}^{32}S + e^- + \nu \qquad (47\text{–}13a)$$

and

β⁺-decay

$$^{64}Cu \rightarrow {}^{64}Ni + e^+ + \nu. \qquad (47\text{–}13b)$$

Equations 47–12 should likewise be replaced by

$$n \rightarrow p + e^- + \nu \qquad (47\text{–}14a)$$

The basic β-decay processes

and

$$p \rightarrow n + e^+ + \nu. \qquad (47\text{–}14b)$$

A particle with such minimal properties should be hard to detect, and indeed this is the case. The mean free path of a neutrino in water, before it collides with a proton and induces a reaction, is ~ 3000 light years! In spite of the difficulties that such numbers suggest, experimental neutrino physics is now a well-developed branch of experimental physics, with avid practitioners at the Fermi National Accelerator Laboratory near Chicago (Fermilab) and at the Centre Europeén pour la Recherche Nucléaire (CERN) in Geneva, and elsewhere.

Detecting the neutrino

The valley of the nuclides

Our study of β- and α-decay permits us to look at the nuclidic chart of Fig. 47–3 in a new way. Let us construct a three-dimensional surface by plotting the mass of each nuclide in a direction at right angles to the N-Z plane of that figure.

The surface so formed gives a graphic representation of nuclear stability. As Fig. 47–11 shows, for the light nuclides, it describes a "valley of the nuclides,"

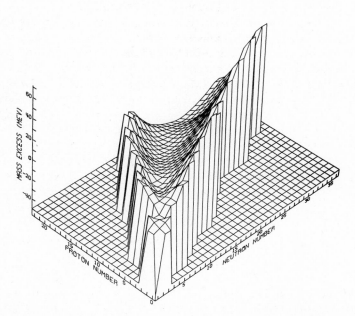

Figure 47–11 A portion of the valley of the nuclides, showing only the lightest nuclides. The quantity plotted on the vertical axis is the mass excess of the various nuclides, defined as $(m\text{-}A)\,c^2$, where m is the atomic mass. From "Exotic Light Nuclei" by Joseph Cerny and Arthur M. Poskanzer, *Scientific American*, June 1978. Computer drawing by Jef Poskanzer.

the stability zone of Fig. 47–3 running along its bottom. Nuclides on what climbers will recognize as the headwall of the valley decay into it largely by chains of α-decay and by spontaneous fission. Nuclides on the proton-rich side of the valley decay into it by chains of β^+-decay and those on the neutron-rich side do so by chains of β^--decay.

47–6   Nuclear Reactions

In this section we consider only those reactions in which the target (X) and the residual nuclides (Y) are at least moderately massive and the projectile (a) and the emerging particles (b) are reasonably light. Such reactions can be written as

$$X + a \rightarrow Y + b \tag{47-15a}$$

Reaction notation

or, in more compact notation, as

$$X(a, b)Y. \tag{47-15b}$$

We assume that the reaction is carried out in an arrangement like that of Fig. 47–1.

The *reaction energy Q* is defined as

$$Q = (m_X + m_a)c^2 - (m_Y + m_b)c^2 \tag{47-16a}$$

Reaction Q

or, equivalently, as

$$Q = (K_Y + K_b) - (K_X + K_a), \tag{47-16b}$$

in which the K's are kinetic energies. In Example 6 of Chapter 10 we studied the kinematics of reactions of the type described above, focusing particularly on the reaction $^{19}F(p, \alpha)^{16}O$, for which we found $Q = +8.13$ MeV. This example should be thoroughly reviewed at this point. A reaction with a positive Q is said to be *exothermic*. This means (see Eqs. 47–16) that the system loses rest mass during the reaction and gains a corresponding amount of kinetic energy. If Q is negative, the reaction is called *endothermic;* such reactions will not "go" unless a certain minimal kinetic energy is provided by the incoming particle.

If a and b are identical particles, which requires that X and Y also be identical, we describe the reaction as a *scattering* event. If the kinetic energy of the system is the same both before and after the event (which means that $Q = 0$), we have *elastic scattering*. If these energies are different ($Q \neq 0$), we have *inelastic scattering*.

Reaction systematics

Nuclear reactions can be systematized by plotting them on a nuclidic chart. Figure 47–12 shows a portion of such a chart, centered arbitrarily on the nuclide ^{197}Au. Stable nuclides are shaded in color and their isotopic abundances are shown. The squares without color represent radionuclides, their half-lives being indicated.

Figure 47–13 can best be viewed as a transparent overlay that may be placed at any location on a nuclidic chart whose scale is that of Fig. 47–12. If the colored central square is placed over a particular target nuclide on the chart, the residual nuclides resulting from the various reactions printed on the overlay are identified. Thus if ^{197}Au is chosen as a target, a (p, α) reaction will produce (stable) ^{194}Pt and either an (n, γ) or a (d, p) reaction will produce the radionuclide ^{198}Au ($t_{1/2} = 2.70$ d).

Figure 47–12 A portion of the nuclidic chart of Fig. 47–3, expanded and with more information added. The colored squares represent stable nuclides, their isotopic abundances being shown. The remaining squares represent radionuclides, their half-lives being indicated.

Atomic number, Z							
82	^{197}Pb 42 min	^{198}Pb 2.4 h	^{199}Pb 1.5 h	^{200}Pb 21.5 h	^{201}Pb 9.42 h	^{202}Pb 3×10^5 y	^{203}Pb 52.0 h
81	^{196}Tl 1.84 h	^{197}Tl 2.83 h	^{198}Tl 5.3 h	^{199}Tl 7.4 h	^{200}Tl 26.1 h	^{201}Tl 73.6 h	^{202}Tl 12.2 d
80	^{195}Hg 9.5 h	^{196}Hg 0.15%	^{197}Hg 64.1 h	^{198}Hg 10.0%	^{199}Hg 16.9%	^{200}Hg 23.1%	^{201}Hg 13.2%
79	^{194}Au 39.5 h	^{195}Au 183 d	^{196}Au 6.18 d	^{197}Au 100%	^{198}Au 2.70 d	^{199}Au 3.14 d	^{200}Au 48.4 min
78	^{193}Pt ~50 y	^{194}Pt 32.9%	^{195}Pt 33.8%	^{196}Pt 25.3%	^{197}Pt 18.3 h	^{198}Pt 7.2%	^{199}Pt 30.8 min
77	^{192}Ir 74.2 d	^{193}Ir 62.7%	^{194}Ir 19.2 h	^{195}Ir 2.5 h	^{196}Ir 52 s	^{197}Ir 9.85 min	^{198}Ir ~8 s
76	^{191}Os 15.4 d	^{192}Os 41.0%	^{193}Os 30.5 h	^{194}Os 6.0 h	^{195}Os 6.5 min	^{196}Os 35 min	—
	115	**116**	**117**	**118**	**119**	**120**	**121**

Neutron number, N

Figure 47–13 An "overlay" to be placed over a nuclidic chart such as that of Fig. 47–12. If the central square is placed over a target, the residual nuclides corresponding to various reactions can be read off.

			α,n	α,γ
	p,n	p,γ d,n	d,γ α,d	α,p
	γ,n p,d		n,γ d,p	
p,α	γ,d d,α	n,d γ,p	n,p	
γ,α	n,α			

Example 7 A gold foil is placed in a beam of slow neutrons ($E \ll 1$ eV) emerging from a nuclear reactor. Using Figs. 47–12 and 47–13, write down the various reactions that are possible.

If, in our mind, we place the shaded square of the overlay of Fig. 47–13 over the square occupied by ^{197}Au in Fig. 47–12, we see that, in principle, four neutron-induced reactions are possible:

Note that the n,p and the n,d reactions are endothermic and will not "go" for such slow incident neutrons.

After a ^{197}Au nucleus has absorbed a neutron from the incident beam, it is much more likely to get rid of its excitation energy by emitting a γ-ray than by emitting an α-particle, which must tunnel through a potential barrier, just as in Fig. 47–8. Thus, both the n,γ and the n,α reactions will occur, but the former will be much more probable.

Reaction	Residual nuclide	Half-life	Q, MeV (calculated)
n,γ	^{198}Au	2.70 d	+6.51
n,p	^{197}Pt	18.3 h	−0.74
n,d	^{196}Pt	Stable	−3.59
n,α	^{194}Ir	19.2 h	+7.02

Reactions and nuclear levels

Reaction studies can be used to investigate nuclear levels. Consider the reaction

$$^{27}\text{Al}(d,p)^{28}\text{Al}, \qquad Q = +5.49 \text{ MeV}, \qquad (47-17)$$

in which an aluminum target is bombarded with 2.10-MeV deuterons. The emerging protons are observed to come off with a number of well-defined discrete energies and to be accompanied by γ-rays. Figure 47–14 shows a portion of this proton spectrum.

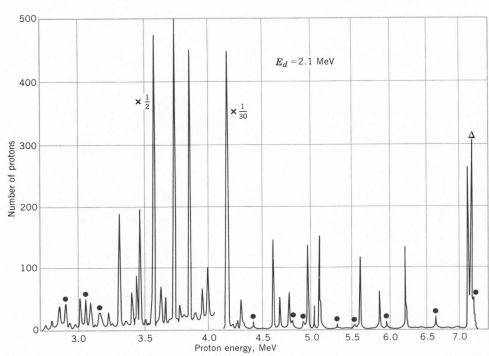

Figure 47–14 The distribution in energy of protons from the reaction $^{27}\text{Al}(d, p)^{28}\text{Al}$, emerging from the target at right angles to the incident deuteron beam. The symbol Δ identifies the proton peak of highest energy. The dots ● identify peaks due to impurities in the target foil. (From Enge, Buechner, and Sperduto.)

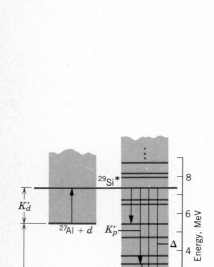

Figure 47–15 Energy levels of ^{28}Al. The vertical arrows correspond to the proton peaks of Fig. 47–14. The arrow marked Δ corresponds to the peak so labeled in that figure. (Target recoil effects must be taken into account in determining energy-level spacings from the energies of the proton peaks in Fig. 47–14.)

Let us rewrite Eq. 47–17 as

$$^{27}\text{Al} + d \rightarrow {}^{29}\text{Si}^* \rightarrow {}^{28}\text{Al} + p,$$

in which $^{29}\text{Si}^*$ represents an intermediate stage in the reaction. The asterisk indicates that this intermediate nuclide is in an excited state. It may get rid of its excitation energy by ejecting a proton, leaving ^{28}Al as a residual nucleus.

Figure 47–15 shows some of the levels involved. From the mass difference (Eq. 47–16a) we see that the ground states of the systems ($^{27}\text{Al} + d$) and ($^{28}\text{Al} + p$) differ by the reaction energy Q (= +5.49 MeV) as shown. The horizontal line that runs across the figure represents an excited level of the intermediate nuclide ^{29}Si. Protons are emitted from this nucleus with a variety of energies, each corresponding to a particular energy level in the residual nucleus ^{28}Al. Monoenergetic γ-ray photons—a typical one is shown—are also emitted as the excited states of ^{28}Al make transitions to the ground state.

47–7 Nuclear Models

Models

A "nuclear model" is a way of looking at the nucleus that permits us to understand a substantial, though perhaps limited, body of experimental nuclear data. We discuss two such models. Although based on assumptions that seem flatly to exclude each other, each accounts very well for a selected group of nuclear properties. We then show how these two models may be reconciled.

A nucleus is like a liquid drop

The Liquid Drop Model. In the liquid drop model, formulated by Niels Bohr, the nucleons are imagined to interact strongly with each other, like the molecules in a drop of liquid. A given nucleon collides frequently with the other nucleons in the nuclear interior, its mean free path as it moves about being substantially less than the nuclear diameter. This constant "jiggling around" reminds us of the thermal agitation of the molecules in a fluid. The liquid drop model has had impressive success in selected situations. It has been especially useful in explaining nuclear fission.

The liquid drop model and nuclear reactions

Consider how the liquid drop model may be applied to nuclear reactions of the type

$$X + a \rightarrow C^* \rightarrow Y + b. \qquad (47\text{–}18)$$

The compound nucleus

Here C^* represents an excited state of a so-called *compound nucleus* C. We imagine that projectile a enters target nucleus X, forming the compound nucleus C and conveying to it a certain excitation energy. The incoming particle a—let us say it is a neutron—is at once caught up in the random motions that characterize the nuclear interior. It quickly loses its identity so to speak, and the excitation energy that it has carried with it comes quickly to be shared more or less equally by all of the nucleons.

The above quasi-stable state, represented by C^* in Eq. 47–18, may endure for as long as, say, 10^{-16} s. This may seem to be a very short time, but it is a million times longer than the time it takes for a nucleon with a few MeV of kinetic energy to traverse a nuclear diameter.

Eventually, by a statistical fluctuation, one of the nucleons in C^*—perhaps a proton—will acquire enough energy to leave the compound nucleus, leaving as particle b.

Independent formation and decay of the compound nucleus

The central feature of the compound nucleus model is that the formation of the compound nucleus and its eventual decay are totally independent events. A given compound nucleus may be formed in a variety of ways and may also decay in a variety of ways. At the time of its decay, we assume, the nucleus has "forgotten" how it was formed. We show here three of the possible ways in which a compound nucleus $^{20}Ne^*$ might be formed and three different ways in which it might decay:

$$\left.\begin{matrix} ^{16}O + \alpha \\ (99.8\%) \\ ^{19}F + p \\ (100\%) \\ ^{20}Ne + \gamma \\ (90.5\%) \end{matrix}\right\} \rightarrow \begin{matrix} ^{20}Ne^* \\ \text{An excited} \\ \text{compound} \\ \text{nucleus} \end{matrix} \rightarrow \left\{\begin{matrix} ^{18}F + d \\ (110\text{ min}) \\ ^{19}Ne + n \\ (17.2\text{ s}) \\ ^{17}O + {}^{3}He \\ (0.038\%) \end{matrix}\right.$$

Three formation modes   Three decay modes

The stabilities of the target and of the residual nuclides are indicated by giving the isotopic abundances (for the stable nuclides) and the half-lives (for the radionuclides).

Example 8 *The compound nucleus and the uncertainty principle.* Consider the neutron-capture cross section reaction

$$^{109}\text{Ag} + n \rightarrow {}^{110}\text{Ag}^* \rightarrow {}^{110}\text{Ag} + \gamma \qquad (47-19)$$

Figure 47–16 shows its cross section (see Section 10–7) as a function of the energy of the incident neutron. Analyze this figure in terms of the compound nucleus concept and the uncertainty principle.

The cross-section curve of Fig. 47–16 is sharply peaked, reaching a maximum cross section of 12,500 barns. This "resonance peak" suggests that we are dealing with a single excited level in the compound nucleus $^{110}\text{Ag}^*$. When the available energy just matches the energy of this level above the ^{110}Ag ground state, we have "resonance," and the reaction really "goes."

However, the resonance peak is not infinitely sharp. From the figure we can measure that it has an approximate half-width at half maximum (that is, at 6250 barns) of 0.20 eV. We account for this by saying that the excited ^{110}Ag level is not sharply defined in energy; it is "fuzzy," with an energy uncertainty ΔE of ~0.20 eV.

We can use the uncertainty principle, written in the form

$$\Delta E \cdot \Delta T \gtrsim h \qquad (47-20)$$

to tell us something about any state of an atomic or nuclear system. We have seen that ΔE is a measure of the uncertainty of our knowledge of the energy of the state. The quantity Δt is interpreted as the time available to measure the energy of the state; it is in fact the mean life $\bar{t}$ of the state before it decays.

For the excited state $^{110}\text{Ag}^*$ we have, from Eq. 47–20,

$$\Delta t = \bar{t} = \frac{h}{\Delta E} \approx \frac{4.14 \times 10^{-15} \text{ eV} \cdot \text{s}}{0.20 \text{ eV}} \approx 2.1 \times 10^{-14} \text{ s}.$$

This is just the order of magnitude of the lifetime that we expect for a compound nucleus (see p. 965)! Thus we have convincing evidence for the compound nucleus concept.

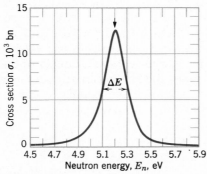

Figure 47–16 The cross section for the reaction ^{109}Ag (n, γ) ^{110}Ag for neutron energies near 5.2 eV. The halfwidth ΔE of the particular resonance peak that falls in this energy range is 0.20 eV.

Nucleon orbits

The Independent Particle Model. In the liquid drop model the nucleons were assumed to move about at random and to collide frequently with each other. The independent particle model, however, is based on an assumption just the opposite of this, namely, that each nucleon moves in a reasonably well-defined orbit within the nucleus and hardly makes any collisions at all! The potential in which each nucleon is assumed to move is determined by the smeared-out motions of all the other nucleons.

Nucleon levels

Just as for electrons in atoms, the concept of nucleon orbits leads to the concept of quantized nucleon states, each described by an appropriate set of quantum numbers. Like the electron, both the neutron and the proton have a spin quantum number of $\frac{1}{2}$. Because the Pauli exclusion principle applies to neutrons and protons as well as to electrons, we say that each neutron or proton state contains two particles with oppositely directed spins. In considering nucleon states, the neutrons and the protons are always dealt with separately, each having its own array of available quantized states.

Nucleon orbits and the Pauli principle

The fact that nucleons obey the Pauli principle helps us to understand the relative stability of the nucleon orbits. If two nucleons within a nucleus are to collide, the energy of each of them after the collision must correspond to one of the allowed quantized nucleon states. If that state happens to be filled already, the proposed collision simply cannot occur. If it did, there would then be a violation of the Pauli principle. In time any given nucleon will find it possible to collide, but meanwhile it will have made enough revolutions in its orbit to give meaning to the concept of a quantized state with a quantized total energy.

Atomic shells, a reminder

In atom building (see Fig. 45–5) we saw that as we form the elements by adding electrons one at a time, there are particular numbers of electrons that form atoms of exceptional stability, the noble gases. We learned to explain this from wave mechanics, in terms of filled shells and subshells. The atomic *"magic electron numbers,"* as we may call them, are the atomic numbers of the noble gases:

Atomic "magic" numbers

$$2, 10, 18, 36, 54, 86, \ldots$$

Nuclei also show shell effects, associated with certain *magic nucleon numbers:*

Nuclear "magic" numbers

$$2, 8, 20, 28, 50, 82, 126, \ldots$$

We shall see that any nuclide whose proton number Z or neutron number N has one of these values has a special stability that it makes evident in a variety of ways. Examples of "magic" nuclides are ^{18}O ($Z = 8$), ^{40}Ca ($Z = 20$, $N = 20$), ^{92}Mo ($N = 50$), and ^{208}Pb ($Z = 82$, $N = 126$). ^{40}Ca and ^{208}Pb are said to be "doubly magic" because they contain filled shells of *both* neutrons and protons.

The magic number "2" shows up in the exceptional stability of the α-particle (^4He), which has $Z = N = 2$. For example, the binding energy per nucleon for this nuclide stands well above that of its neighbors on the binding energy curve of Fig. 47–6. The α-particle is so tightly bound that it proves impossible to add another nucleon to it; there is no stable nuclide with $A = 5$.

Some of the evidence for the magic numbers

Figure 47–17 shows a small part of the evidence for $N = 20$, 28, 50, and 82 as magic neutron numbers. It displays the number of stable nuclides (isotones) that occur in nature for each neutron number N. The assumption is that, at the time the elements were formed, nuclides with especially stable neutron (or proton) configurations would have been formed in an especially large number of varieties. Unstable varieties would have broken up as quickly as they were formed.

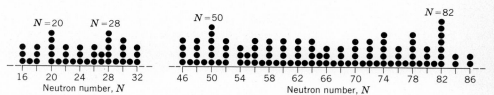

Figure 47–17 The number of stable nuclides for various neutron numbers N. The "magic" neutron numbers are identified. This figure may be verified by inspection of Fig. 47–3.

An interesting observation (the *pairing phenomenon*)

Just incidentally, Fig. 47–17 also shows a striking prevalence of stable nuclides with even neutron numbers. This can be traced to the apparent fact that nucleons have a *special stability* when they combine in pairs in such a way that the total angular momentum of the pair is zero.

Nucleus building

We build up nuclides from their constituent nucleons just as we built up atoms in Section 45–3 from their constituent electrons. Wave mechanics identifies the allowed nucleon states, allowing us to assign quantum numbers to them, and telling us how many neutrons or protons may occupy them. We then fill these states in the order of their increasing energy, being careful to follow the Pauli principle at all times.

If we make a crucial assumption (namely, that spin-orbit coupling effects are very large in nuclei), then it turns out that the nucleon levels form naturally into groups or shells, with an energy gap between each shell. In adding nucleons to fill these shells, the magic numbers are systematically revealed.

The nuclear levels

Table 47–3 shows a distribution of nucleon levels, with energy increasing upward. For simplicity we show the levels within the shells as evenly spaced, although in fact they are not.

Consider a typical level such as the second from the bottom in the uppermost shell. It is labeled "6 $h_{9/2}$." The "6" is the principal quantum number n, familiar from our study of the hydrogen atom. The "h" gives us the familiar orbital angular moment quantum number l, according to the code that we introduced for electrons in atoms on p. 884. Thus, for this level, $l = 5$, as displayed separately in Table 47–3.

The subscript "9/2" is the quantum number j, which is also separately displayed in the table. This quantum number identifies the total angular momentum of the nucleon, including both spin and orbital contributions. As for single electrons in atoms, j can have two values: $l - \frac{1}{2}$ ($= \frac{9}{2}$ in this case) and $l + \frac{1}{2}$ ($= \frac{11}{2}$; note that the 6 $h_{11/2}$ state belongs to the next lowest shell in Table 47–3.)

As for electrons in atoms, the level population is given by $2j + 1$ ($= 10$ for our 6 $h_{9/2}$ level). The total shell populations and the cumulative shell populations are also shown in the table. The latter are precisely the magic nucleon numbers!

Table 47–3 Nucleon levels in the shell model[a]

The levels	Quantum numbers			Level population	Shell population	Cumulative shell population
	Label	l	j			
⋮						
‾‾‾‾‾‾	7 $i_{13/2}$	6	$\frac{13}{2}$	14	44	126
‾‾‾‾‾‾	4 $p_{1/2}$	1	$\frac{1}{2}$	2		
‾‾‾‾‾‾	5 $f_{5/2}$	3	$\frac{5}{2}$	6		
‾‾‾‾‾‾	4 $p_{3/2}$	1	$\frac{3}{2}$	4		
‾‾‾‾‾‾	6 $h_{9/2}$	5	$\frac{9}{2}$	10		
‾‾‾‾‾‾	5 $f_{7/2}$	3	$\frac{7}{2}$	8		
‾‾‾‾‾‾	6 $h_{11/2}$	5	$\frac{11}{2}$	12	32	82
‾‾‾‾‾‾	4 $d_{3/2}$	2	$\frac{3}{2}$	4		
‾‾‾‾‾‾	3 $s_{1/2}$	0	$\frac{1}{2}$	2		
‾‾‾‾‾‾	5 $g_{7/2}$	4	$\frac{7}{2}$	8		
‾‾‾‾‾‾	4 $d_{5/2}$	2	$\frac{5}{2}$	6		
‾‾‾‾‾‾	5 $g_{9/2}$	4	$\frac{9}{2}$	10	22	50
‾‾‾‾‾‾	3 $p_{1/2}$	1	$\frac{1}{2}$	2		
‾‾‾‾‾‾	4 $f_{5/2}$	3	$\frac{5}{2}$	6		
‾‾‾‾‾‾	3 $p_{3/2}$	1	$\frac{3}{2}$	4		
‾‾‾‾‾‾	4 $f_{7/2}$	3	$\frac{7}{2}$	8	8	28
‾‾‾‾‾‾	3 $d_{3/2}$	2	$\frac{3}{2}$	4	12	20
‾‾‾‾‾‾	2 $s_{1/2}$	0	$\frac{1}{2}$	2		
‾‾‾‾‾‾	3 $d_{5/2}$	2	$\frac{5}{2}$	6		
‾‾‾‾‾‾	2 $p_{1/2}$	1	$\frac{1}{2}$	2	6	8
‾‾‾‾‾‾	2 $p_{3/2}$	1	$\frac{3}{2}$	4		
‾‾‾‾‾‾	1 $s_{1/2}$	0	$\frac{1}{2}$	2	2	2

Energy (increasing upward)

[a] The order of levels within the shells may differ from nuclide to nuclide and may depend on whether neutrons or protons are being considered. The integrity of the shells remains, however.

The magic numbers justified

Table 47–3 is just as much a triumph of wave mechanics as was the elucidation of the periodic table, symbolized by Fig. 45–5. Maria Mayer and Hans Jensen shared the 1963 Nobel Prize in physics for their (independent) derivations of the magic nucleon numbers from the principles of wave mechanics.

Example 9 According to the shell model, what should be the total angular momentum, as measured by the nuclear spin I, of the nuclide ^{91}Zr?

For ^{91}Zr we have $Z = 40$ (an even number) and $N = 91 - 40$ or 51 (an odd number). We mentioned earlier (see p. 967) that nucleons have a special stability when they cluster in pairs, each such pair having zero angular momentum. Thus we expect no angular momentum contribution to ^{91}Zr from its 40 protons.

As for the neutrons, we have one neutron outside a 50-neutron core. As above, we expect no angular momentum contribution from this core, 50 being an even number. Thus the total angular momentum of ^{91}Zr must be due to the single "extra" neutron outside the neutron core. This reminds us of the atomic case, in which the total angular momentum of the sodium atom ($Z = 11$) is determined by its single "extra" valence electron, moving outside a closed ten-electron, neonlike core.

Let us turn now to Table 47–3 and fill the neutron levels up to $N = 51$. We see that the last neutron, found just above the fifth filled neutron shell, is in a $4 d_{5/2}$ state. This tells us that j for this single "extra" neutron must be $\frac{5}{2}$ and that this must be the spin I of the entire ^{91}Zr nuclide. Measurement totally confirms this prediction of shell theory!

The two models combined

The Collective Model. The collective model of nuclear structure, developed largely by Aage Bohr (a son of Niels Bohr) and Ben Mottleson, succeeds in combining the seemingly irreconcilable points of view of the liquid drop and the independent particle models.

Consider a nucleus with a small number of neutrons (or protons) outside a filled core that contains a "magic" number of neutrons (or protons). The "extra" nucleons moved in quantized orbits, in a potential established by the central core, thus preserving the central feature of the independent particle model. These extra nucleons can also interact with the core, deforming it and setting up vibrational and rotational "tidal wave" motions in it. These "liquid drop" motions of the core preserve the central feature of that model. The collective model has had very great success indeed in many areas of nuclear physics.

REVIEW GUIDE AND SUMMARY

The nuclear force

The nuclidic chart of Fig. 47–3 shows the ~ 2000 nuclides that are known to exist. These nuclei are held together by an attractive *strong force* acting between the neutrons and the protons (*nucleons*) that make them up.

The nuclides, assumed spherical, have a mean radius given by

Nuclear radii

$$R = R_0 A^{1/3}, \qquad [47\text{–}1]$$

where $R_0 = 1.1$ fm (1 fermi $= 10^{-15}$ m).

Mass-energy exchanges

Nuclear masses, universally reported as the masses of the corresponding neutral atoms, are useful in calculating disintegration energies, binding energies, and so on. The energy equivalent of one mass unit (u) is 931 MeV. The curve of binding energies (see Fig. 47–6 and Example 2) shows that middle-mass nuclides are the most stable and that energy can be released both by fission of heavy nuclei and by fusion of light nuclei.

Radioactive decay

The basic assumption of radioactive decay is that the decay rate ($-dN/dt$) is proportional to the number N of radioactive atoms present, the proportionality constant being the *disintegration constant* λ; see Eq. 47–5. This leads to the law of exponential decay, or

$$R = R_0 e^{-\lambda t}. \qquad [47\text{–}7]$$

The half-life $t_{1/2}$ ($= 0.693/\lambda$) is the time after which the decay rate R (or the number N) has dropped to half of its initial value; see Examples 3 and 4.

α-Decay

For some nuclides the emission of an α-particle is energetically possible; see Example 5. Instantaneous decay is inhibited by a potential barrier, as in Fig. 47–8. Although such barriers cannot be penetrated according to classical mechanics, they can do so according to wave mechanics, as Fig. 47–9 suggests. The barrier penetrability—and thus the half-life for α-decay—are very sensitive to the dimensions of the barrier, as Table 47–2 shows.

β-Decay

In β-*decay* an electron is emitted, along with a neutrino. They share the available disintegration energy between them; see Eqs. 47–13 and 47–14 and Example 6. The emitted electrons have a continuous spectrum of energies up to a limit K_{max} (= Q); see Fig. 47–10.

The valley of the nuclides

New light is thrown on nuclear stability by constructing a "valley of the nuclides" as in Fig. 47–11, in which the nuclear mass is plotted vertically on a Z-N chart. The stability zone runs along the bottom of the valley and chains of β-decay processes descend its side walls.

Nuclear reactions

We study only those nuclear reactions in which relatively massive nuclei are bombarded by relatively light projectiles. Thus:

$$X + a \rightarrow Y + b \quad \text{or} \quad X(a,b)Y. \quad [47-15]$$

The reaction energy Q is given by

$$Q = (m_X + m_a)c^2 - (m_Y + m_b)c^2$$
$$= (K_Y + K_b) - (K_X + K_a). \quad [47-16]$$

Figures 47–12 and 47–13 and Example 7 analyze the systematics of nuclear reactions. Review also Example 6 of Chapter 10 and Section 10–7. Figures 47–14 and 47–15 show how nuclear levels can be studied by reaction methods.

The liquid drop model

In the liquid drop model of nuclear structure it is assumed that the nucleons collide constantly and that in nuclear reactions a long-lived *compound nucleus* is formed. Study Example 8 carefully.

The independent particle (or shell) model

In the independent particle model of nuclear structure it is assumed that each nucleon moves without collisions in a quantized orbit within the nucleus. Table 47–3 shows the predicted nucleon levels along with the predicted "magic nucleon numbers" (2, 8, 20, 28, 50, 82, 126, . . .) that define most stable nuclear configurations, as in Fig. 47–17. Study Example 9 carefully.

The collective model, currently accepted, combines the central features of both the liquid drop and the independent particle models; it is highly successful.

QUESTIONS

1. When a thin foil is bombarded with α particles, a few of them are scattered back toward the source. Rutherford concluded from this that the positive charge of the atom—and also most of its mass—must be concentrated in a very small "nucleus" within the atom. What was his line of reasoning?

2. In what ways do the so-called strong force and the electrostatic or Coulomb force differ?

3. Why does the relative importance of the Coulomb force compared to the strong nuclear force increase at large mass numbers?

4. In your body, are there more neutrons than protons? More protons than electrons? Discuss.

5. Why do nuclei tend to have more neutrons than protons at high mass numbers?

6. Why do we use atomic rather than nuclear masses in analyzing nuclear decay and reaction processes?

7. How might the equality 1.000 u = 1.660 × 10⁻²⁷ kg be arrived at in the laboratory?

8. The most stable nuclides have a mass number near 60 (see Fig. 47–6). Why don't *all* nuclides have mass numbers near 60?

9. If we neglect the very lightest nuclides, the binding energy per nucleon in Fig. 47–6 is roughly constant at 7 to 8 MeV/nucleon. Do you expect the mean electronic binding energy per electron in atoms also to be roughly constant throughout the periodic table?

10. Why is the binding energy (Fig. 47–6) low at low mass numbers? At high mass numbers?

11. In the binding-energy curve of Fig. 47–6, what is special or notable about the nuclides ^2H, ^4He, ^{56}Fe, and ^{239}Pu?

12. The magnetic moment of the neutron is −1.9131 nuclear magnetons. What is a nuclear magneton and how does it differ from a Bohr magneton? What does the minus sign mean? How can the neutron, which carries no charge, have a magnetic moment in the first place?

13. A particular ^{238}U nucleus was created in a massive stellar explosion, perhaps 10^{10} y ago. It suddenly decays by α-emission while we are observing it. After all those years, why did it decide to decay at this particular moment?

14. Can you justify this statement: "In measuring half-lives by the method of Example 3, it is not necessary to measure the absolute decay rate R; any quantity proportional to it will suffice. However, in the method of Example 4 an absolute rate *is* needed."

15. You are running longevity tests on light bulbs. Do you expect their "decay" to be exponential? What is the essential difference between the decay of light bulbs and of radionuclides?

16. The half-life of ^{238}U is 4.47×10^9 y, about the age of the solar system. How can such a long half-life be measured?

17. The half-life of ^{238}U is 4.47×10^9 years. How might measurements of its activity be used to determine the age of uranium-containing rocks? How do you get around the fact that you don't know how much ^{238}U was present in the rocks to begin with? (*Hint:* What is the ultimate decay product of ^{238}U?)

18. The half-life of ^{14}C is 5730 years. This isotope is produced in the upper atmosphere at an assumed constant rate by cosmic ray bombardment. How might measurements of its activity play a role in dating ancient carbon-containing specimens, such as wooden Egyptian artifacts or the remnants of ancient campfires?

19. Explain why, in α-decay, short half-lives correspond to large disintegration energies, and conversely.

20. Can you give a justification, even a partial one, for the barrier tunneling phenomenon in terms of basic ideas about the wave nature of matter?

21. In β-decay the emitted electrons form a continuous spectrum, but in α-decay they form a discrete spectrum. What difficulties did this cause in the explanation of β-decay, and how were these difficulties finally overcome?

22. How do neutrinos differ from photons? Each has zero charge and (presumably) zero rest mass and travels at the speed of light.

23. Which of these conservation laws apply to all nuclear reactions? Conservation of (*a*) charge, (*b*) relativistic mass, (*c*) relativistic total energy, (*d*) rest mass, (*e*) kinetic energy, (*f*) linear momentum, (*g*) angular momentum, and (*h*) total numbers of nucleons.

24. In the development of our understanding of the atom, did we use atomic models as we now use nuclear models? Is Bohr theory such an atomic model? Are models now used in atomic physics? What is the difference between a model and a theory?

25. What are the basic assumptions of the liquid drop and the independent particle models of nuclear structure? How do they differ? How does the collective model reconcile these differences?

26. What is so special ("magic") about the magic nucleon numbers?

27. Why aren't the magic nucleon numbers and the magic electron numbers the same? What accounts for each?

28. The numbers of stable (or very long-lived) isotopes of the noble gases are

He	Ne	A
2	3	3
Kr	Xe	Rn
5	9	0

an average of 3.7. The average number of stable isotones for the four magic neutron numbers shown in Fig. 47–17, however, is 5.8, considerably greater. If the noble gases are so stable, why were not more stable isotopes of them created when the elements were formed?

EXERCISES

Section 47–1 The Atomic Nucleus

1. (*a*) *The Thomson atom model.* At one time the positive charge of the atom was thought to be distributed uniformly throughout a sphere with a radius of about 1.0×10^{-10} m, that is, throughout the entire atom. Calculate the electric field at the surface of a gold atom on this (erroneous) assumption. Neglect the effect of the atomic electrons. (*b*) *The Rutherford, or nuclear, atom model.* Experiment shows that the positive charge of the atom is *not* spread uniformly throughout the volume of the atom, as (*a*) above suggests, but is concentrated in a small region (the nucleus) at the center of the atom. For gold the nuclear radius is 6.9×10^{-15} m. What is the electric field at the nuclear surface? Again, neglect effects associated with the atomic electrons.

Section 47–2 Some Nuclear Properties

2. Locate the nuclides displayed in Table 47–1 on the nuclidic chart of Fig. 47–3. Verify that, as intended, they lie along the stability zone.

3. Using the nuclidic chart of Fig. 47–3, write the symbols for (*a*) all stable nuclides (isotopes) with $Z = 60$, (*b*) all radioactive nuclides (isotones) with $N = 60$, and (*c*) all nuclides (isobars) with $A = 60$.

4. The atomic masses of ^{1}H, ^{12}C, and ^{238}U are 1.007831 u, 12.000000 u (by definition), and 238.051 u, respectively. (*a*) What would these masses be if the mass unit were defined in terms of ^{1}H rather than ^{12}C? (*b*) Use your result to suggest why this perhaps obvious choice was not made.

5. The electrostatic potential energy of a uniform sphere of charge Q and radius R is

$$U = \frac{3Q^2}{20\pi\epsilon_0 R}.$$

(*a*) Find the electrostatic potential energy for the nuclide ^{239}Pu, assumed spherical; see Table 47–1. (*b*) Compare its electrostatic potential energy per particle with its binding energy per nucleon of 7.57 MeV. (*c*) What do you conclude?

6. The strong neutron excess of heavy nuclei is illustrated by the fact that most heavy nuclides could never fission or break up into two stable nuclei without neutrons being left over. For example, consider the spontaneous fission of a $^{235}_{92}$U nucleus. If the two daughter nuclei had atomic numbers 39 and 53, and were stable, by referring to Fig. 47–3, determine the daughter nuclides and the number of neutrons left over.

7. *Mass excess.* To simplify calculations, atomic masses are sometimes tabulated, not as the actual atomic mass m, but as $(m - A)c^2$, where A is the mass number expressed in mass units. This quantity, usually reported in keV or MeV, is called the *mass excess,* symbol Δ. Using data from Example 2, find the mass excesses for (a) the proton, (b) the neutron, and (c) for ^{120}Sn.

Section 47–3   Radioactive Decay

8. *Alchemy?* A radioactive isotope of mercury, $^{197}_{80}$Hg, decays into gold, $^{197}_{79}$Au, with a disintegration constant of $0.0107/h$. (a) What is its half-life? (b) What fraction of the original amount will remain after 3 half-lives? (c) After 10 days? (The mode of decay is electron capture. For more on this see Exercise 18 and Problem 22.)

9. After long effort, in 1902, Marie and Pierre Curie succeeded in separating from uranium ore the first substantial quantity of radium, one decigram of pure $RaCl_2$. The radium was the radioactive isotope $^{226}_{88}$Ra, which has a decay half-life of 1602 y. (a) How many radium nuclei had they isolated? (b) What was the decay rate of their sample, in disintegrations per second? In curies? (The curie unit, Ci, was named in honor of the Curies, who received the 1903 Nobel Prize in physics for their work on radiation phenomena. One curie equals 3.7×10^{10} disintegrations/s.)

10. From data presented in the first paragraph of Section 47–3, deduce (a) the mean life t and (b) the half-life of ^{238}U.

11. The plutonium isotope $^{239}_{94}$Pu is produced as a byproduct in nuclear reactors (See Exercise 48–7) and hence is accumulating in our environment. It is radioactive, decaying by α-decay with a half-life of 2.44×10^4 y. But plutonium is also one of the most toxic chemicals known; as little as 2×10^{-3} g is lethal to a human. (a) How many nuclei constitute a chemically lethal dose? (b) What is the decay rate of this amount? If you were handling that quantity, would you fear being poisoned or suffering radiation sickness?

12. After a brief neutron irradiation of silver, two activities are present: ^{108}Ag ($t_{1/2} = 2.4$ min) with an initial decay rate of 3.1×10^5/s, and ^{110}Ag ($t_{1/2} = 24$ s) with an initial decay rate of 4.1×10^6/s. Make a semilog plot similar to Fig. 47–7 showing the total combined decay rate of the two isotopes as a function of time from $t = 0$ until $t = 10$ min. In Fig. 47–7, the extraction of the half-life for simple decays was illustrated. Given only the plot of total decay rate, can you suggest a way to analyze it in order to find the half-lives of both isotopes?

Section 47–4   Barrier Tunneling and α-Decay

13. Generally speaking, heavier nuclides tend to be more unstable to α-decay. For example, the most stable isotope of uranium, $^{238}_{92}$U, has an α-decay half-life of 4.5×10^9 y. The most stable isotope of plutonium is $^{244}_{94}$Pu with an 8.1×10^7-y half-life, and for curium we have $^{248}_{96}$Cm and 1.6×10^7 y. When half of an original sample of $^{238}_{92}$U has decayed, what fractions of the original isotopes of plutonium and curium are left?

14. Consider a ^{238}U nucleus to be made up of an α-particle (^{4}He) and a residual nucleus (^{234}Th). Plot the electrostatic potential energy $U(r)$, where r is the distance between these particles. Cover the range ~ 10 fm $< r < \sim 100$ fm and compare your plot with that of Fig. 47–8.

15. *Barrier tunneling.* For barriers like those of Fig. 47–9, wave mechanics tells us that the probability of transmission through the barrier is given (assuming that $T \ll 1$) by

$$T \cong 16 \left(\frac{E}{U_0}\right)\left(1 - \frac{E}{U_0}\right) e^{-2kl},$$

where

$$k = \left(\frac{2\pi}{h}\right)\left[2mU_0\left(1 - \frac{E}{U_0}\right)\right]^{1/2}.$$

Here m is the mass of the tunneling particle, E is the total energy, U_0 is the barrier height, and l is the barrier thickness. Calculate T for $m = 6.7 \times 10^{-27}$ kg (α particle), $E = 4.0$ MeV, $U_0 = 15$ MeV, and $l = 20$ fm. These are not unreasonable approximations for α-emission; compare Figs. 47–8 and 47–9.

Section 47–5 β-Decay

16. A certain stable nuclide, after absorbing a neutron, emits a negative electron and then splits spontaneously into two α-particles. Identify the nuclide.

17. Cancer cells are more vulnerable to x and γ radiation than are healthy cells. Though linear accelerators are now replacing it, in the past the standard source for radiation therapy has been radioactive $^{60}_{27}$Co, which β-decays into an excited nuclear state of $^{60}_{28}$Ni, which immediately drops into the ground state emitting two γ-ray photons, each of approximate energy 1.2 MeV. The controlling β-decay half-life is 5.26 y. How many radioactive ^{60}Co nuclei are present in a 6000-Ci source used in a hospital? (1 Ci = 3.7×10^{10} disintegrations/s.)

18. *Electron capture* (see Problem 22). Find the disintegration energy Q for the decay of ^{49}V by K-electron capture, as described in Problem 22. The needed data are $m_v = 48.94852$ u, $m_{T_i} = 48.94787$ u, and $E_K = 5.47$ keV.

19. $^{137}_{55}$Cs is present in the fallout from above-ground detonations of nuclear bombs. Because it β-decays with a slow 30.2-y half-life into $^{137}_{56}$Ba, releasing considerable energy in the process, it is an environmental concern. If the atomic masses of the parent and daughter are 136.9071 u and 136.9058 u, respectively, calculate the total energy released in each decay.

20. Heavy radionuclides, which may be either α- or β-emitters, belong to one of four decay chains, depending on whether their mass numbers A are of the form $4n$, $4n + 1$, $4n + 2$, or $4n + 3$, where n is a positive integer. (a) Justify this statement and show that if a nuclide belongs to one of these families, all its decay products will belong to the same family. (b) Classify these nuclides as to family: ^{235}U, ^{236}U, ^{238}U, ^{239}Pu, ^{240}Pu, ^{245}Cm, ^{246}Cm, ^{249}Cf, and ^{253}Fm.

Section 47–6   Nuclear Reactions

21. The radionuclide ^{60}Co ($t_{1/2} = 5.27$ y) is much used in cancer therapy. Tabulate possible reactions that might be used in preparing it. Limit the projectiles to neutrons, protons, and deutrons. Limit the targets to stable nuclides. The stable nuclides suitably close to ^{60}Co are ^{63}Cu, ^{60,61,62}Ni, ^{59}Co, and ^{57,58}Fe. [Commercially, ^{60}Co is made by bombarding elemental cobalt (^{59}Co; 100%) with neutrons in a reactor.]

22. A beam of deuterons falls on a copper target. Copper has two stable isotopes, ^{63}Cu (59.2%) and ^{65}Cu (30.8%). Tabulate the residual nuclides that can be produced by the reactions (d, n), (d, p), (d, α), and (d, γ). By inspection of Fig. 47–2, indicate which residual nuclides are stable and which are radioactive.

23. Making mental use of the "overlay" of Fig. 47–13, write down the reactions by which the radionuclide ^{197}Pt ($t_{1/2} = 18.3$ h) can be prepared, at least in principle. Except in special circumstances, only stable nuclides can serve as practical targets in bombardment arrangements such as that of Fig. 47-1.

24. A platinum target is bombarded with cyclotron-accelerated deuterons for several hours and then the element iridium ($Z = 77$) is separated chemically from it. What radioisotopes of iridium are present, and by what reactions are they formed? (*Note:* ^{190}Pt and ^{192}Pt, not shown in Fig. 47-12, are stable platinum isotopes, but their isotopic abundances are so small that we may ignore their presence.)

Section 47-7 Nuclear Models
25. An intermediate nucleus in a particular nuclear reaction decays within $\sim 10^{-22}$ s of its formation. Note that this is the characteristic nuclear time, defined and calculated in Problem 25. (*a*) What is the uncertainty ΔE in our knowledge of this intermediate state? (*b*) Can this state be called a compound nucleus?

26. A typical kinetic energy for a nucleon in a middle-mass nucleus may be taken as 5 MeV. To what effective nuclear temperature does this correspond, using the assumptions of the liquid drop model of nuclear structure? (*Hint:* See Eq. 21-6.)

27. From the following list of nuclides, identify (*a*) those with filled nucleon shells, (*b*) those with one nucleon outside a filled shell, and (*c*) those with one vacancy in an otherwise filled shell. Nuclides: ^{13}C, ^{18}O, ^{40}K, ^{49}Ti, ^{60}Ni, ^{91}Zr, ^{92}Mo, ^{121}Sb, ^{143}Nd, ^{144}Sm, ^{205}Tl, and ^{207}Pb.

28. Extend the illustration in Fig. 47-17 of greater stability for nuclides having a magic number of neutrons to include all the magic numbers, showing several neighbors for each magic number. Repeat the illustration for protons.

29. In Example 9 we mentioned that nucleons have a special stability when they associate in pairs, the angular momentum of the pair being zero. Use this phenomenon to account for: (*a*) the odd-even alternation of the abundances of stable isotones shown in Fig. 47-17; (*b*) the fact that nuclides with both Z and N even have zero spin; (*c*) the fact that almost all stable nuclides with even A have even N and even Z—almost none have odd N and odd Z, as they might.

PROBLEMS GROUPED BY SECTION

Section 47-2 Some Nuclear Properties
1. Arrange the 26 nuclides given below in squares as a nuclidic chart similar to Fig. 47-3 but showing greater detail. Draw in and label (*a*) all isobaric (constant A) lines and (*b*) all lines of constant neutron excess, defined as $N - Z$. Consider nuclides $^{118-122}$Te, $^{117-121}$Sb, $^{116-120}$Sn, $^{115-119}$In, and $^{114-118}$Cd.

2. Calculate and compare (*a*) the nuclear mass density ρ_m and (*b*) the nuclear charge density ρ_q for a fairly light nuclide such as ^{55}Mn and for a fairly heavy one such as ^{209}Bi. (*c*) Are the differences what you would expect?

3. How many particles make up the nuclide ^{7}Li on (*a*) the accepted neutron-proton theory of nuclear structure and (*b*) the long-since-abandoned proton-electron theory? (*c*) The nuclear spin of ^{7}Li is $\frac{3}{2}$. Can this fact be used to decide between these two theories? (*d*) The nuclear magnetic moment of ^{7}Li is $+3.26$ nuclear magnetons. Does this fact help to decide between the two theories?

4. Verify that the binding energy per particle given in Table 47-1 for ^{239}Pu is indeed 7.56 MeV/nucleon. The needed atomic masses are 239.0521 u (^{239}Pu), 1.00783 u (^{1}H), and 1.00867 u (neutron).

5. A penny has a mass of 3.0 g. Calculate the nuclear energy that would be required to separate all the neutrons and protons in this coin. Ignore the binding energy of the electrons. For simplicity, assume that the penny is made entirely of ^{64}Cu atoms (mass = 63.92976 u). The atomic masses of the proton and the neutron are 1.00783 u and 1.00867 u, respectively.

6. Show that an approximate formula for the mass M of an atom is

$$M = Am_p,$$

where A is the mass number and m_p is the proton mass. (*b*) What percent error is committed in using this formula to calculate the masses of the atoms in Table 47-1? The proton mass is 1.007276 u.

7. *Mass excess* (see Exercise 7). Show that the total binding energy of a nuclide can be written as

$$E = Z \Delta_p + N \Delta_n - \Delta,$$

where Δ_p, Δ_n, and Δ are the appropriate mass excesses. Using this method, calculate the binding energy per particle for ^{197}Au. Compare your result with the value listed in Table 47-1. The needed mass excesses, rounded to three significant figures, are $\Delta_p = +7.29$ MeV, $\Delta_n = +8.07$ MeV, and $\Delta_{197} = -31.2$ MeV. Note the economy of calculation that results when mass excesses are used in place of the actual masses.

8. As Table 47-1 shows, the nuclide ^{197}Au has a nuclear spin of $\frac{3}{2}$. (*a*) If we regard this nucleus as a spinning rigid sphere with a radius given by Eq. 47-1, what rotational frequency results? (*b*) What rotational kinetic energy? (*c*) What is the acceleration for a point on the equator of the spinning nucleus? Note that this picture is overly mechanistic.

Section 47-3 Radioactive Decay
9. The radionuclide ^{64}Cu has a half-life of 12.7 h. What fraction of the ^{64}Cu nuclei initially present will decay between the 14th and the 16th hours?

10. ^{238}U is an α-emitter with a half-life of 4.47×10^9 y. At what rate do ^{238}U α-decays occur in a 1.00-g sample of uranium metal? The isotopic abundance of ^{238}U in naturally occurring uranium is 99.3%; the molecular weight of such uranium is 238 g/mol.

11. The radioisotope ^{125}I is used routinely for medical research and diagnosis. The isotope is supplied in a compound (iodoestradiol) whose molecular weight is 362 g/mol and whose specific activity is 1.3×10^6 Ci/mol (1 Ci = 3.7×10^{10}/s). A sample whose activity is 10 μCi costs (1983 prices) $242. (*a*) What is the cost of an ounce of this substance? (b) The half-life of this isotope is 59.7 d. At what rate does a 1-oz sample decline in value?

12. The radionuclide ^{32}P ($t_{1/2} = 14.28$ d) is often used as a tracer to follow the course of biochemical reactions involving phosphorus. (a) If the counting rate in a particular experimental setup is 3050 counts/s, after what time will it fall to 170 counts/s? (b) A solution containing ^{32}P is fed to the root system of an experimental tomato plant and the ^{32}P activity in a leaf is measured 3.48 days later. By what factor must this reading be multiplied to correct for the decay that has occurred since the experiment began?

13. A source contains two phosphorus radionuclides, ^{32}P ($t_{1/2} = 14.3$ d) and ^{33}P ($t_{1/2} = 25.3$ d). Initially 10% of the decays come from ^{33}P. How long must one wait until 90% do so?

14. (a) Show that the *mean life* $\bar{t}$ of a radionuclide, that is, the average elapsed time before a nucleus decays, is $1/\lambda$. (b) Show that if it could be arranged — and there is no known way to do so — for the sample to continue to decay at its initial rate, the radioactivity would drop to zero after one mean life. (c) After how many mean lives will the radioactivity drop to one-half of its initial value?

15. ^{238}U decays to lead ^{206}Pb with a half-life of 4.47×10^9 y. Although the series has many individual steps, the first has by far the longest half-life; therefore one can often consider the decay to go directly to lead. That is,

$$^{238}\text{U} \rightarrow \,^{206}\text{Pb} + \text{various decay fragments.}$$

A rock is found to contain 4.2 mg of ^{238}U and 2.135 mg of ^{206}Pb. Assume that the rock contained no lead at formation, all of the lead now present arising from the decay of uranium. (a) How many atoms of ^{238}U and ^{206}Pb does the rock now contain? (b) How many atoms of ^{238}U did the rock contain at formation? (c) What is the age of the rock, in AE (1 AE $= 10^9$ y)?

Section 47–4 Barrier Tunneling and α-Decay

16. A ^{238}U nucleus emits an α particle of energy 4.196 MeV. What is the disintegration energy Q for this process, taking the recoil energy of the residual ^{234}Th nucleus into account?

17. Consider that a ^{238}U nucleus emits (a) an α particle or (b) a sequence of neutron, proton, neutron, proton. Calculate the energy released in each case. (c) Convince yourself both by reasoned argument and also by direct calculation that the difference between these two numbers is just the total binding energy of the α particle. Find the binding energy. Some needed atomic masses are

^{238}U: 238.05081 u,	^{4}He: 4.00260 u,
^{237}U: 237.04874 u,	^{1}H: 1.00783 u,
^{236}Pa: 236.04891 u,	n: 1.00867 u.
^{235}Pa: 235.04544 u,	
^{234}Th: 234.04363 u,	

18. *Barrier tunneling; a gravitational case* (see Exercise 15). A 2000-kg car moves at 80 km/h toward a hill that can be represented as a gravitational potential barrier (see Fig. 47–9) 50 m high and 80 m thick. From the relations given in Exercise 15, which hold for barriers for all kinds, calculate the probability T that the car, without changing speed, will suddenly appear on the far side of the hill, having tunneled through it quantum mechanically. Remember the correspondence principle, and don't hold your breath!

*19. Under certain circumstances, a nucleus can decay by emitting a particle heavier than an α particle. Such decays are very rare and have only recently been observed. Consider the decays

$$^{223}\text{Ra} \rightarrow \,^{209}\text{Pb} + \,^{14}\text{C},$$
$$^{223}\text{Ra} \rightarrow \,^{219}\text{Rn} + \,^4\text{He}.$$

(a) Calculate the Q-values for these decays and determine that both are energetically possible. (b) The coulomb barrier height for α particles in this decay is 30 MeV. What is the barrier height for ^{14}C decay? The needed masses are

^{223}Ra: 223.01850 u,	^{14}C: 14.00324 u,
^{209}Pb: 208.98108 u,	^{4}He: 4.00260 u.
^{219}Rn: 219.00948 u,	

Section 47–5 β-Decay

20. *Recoil during β-decay.* The radionuclide ^{32}P decays to ^{32}S as described by Eq. 47–13a. In a particular decay event a 1.71-MeV electron is emitted, the maximum possible value. What is the kinetic energy of the recoiling ^{32}S atom in this event? (*Hint:* For the electron it is necessary to use the relativistic expressions for the kinetic energy and the linear momentum. Newtonian mechanics may safely be used for the relatively slow-moving ^{32}S atom.)

21. *Q for positron decay.* The radionuclide ^{11}C decays according to

$$^{11}\text{C} \rightarrow \,^{11}\text{B} + e^+ + \nu, \qquad t_{1/2} = 20.3 \text{ min}$$

The maximum energy of the positron spectrum is 0.961 MeV. (a) Show that the disintegration energy Q for this process is given by

$$Q = (m_\text{C} - m_\text{B} - 2m_e)c^2,$$

where m_C and m_B are the *atomic* masses of ^{11}C and ^{11}B, respectively and m_e is the electron (positron) mass. (b) Given that $m_\text{C} = 11.011434$ u, $m_\text{B} = 11.009305$ u, and $m_e = 0.0005486$ u, calculate Q and compare it with the maximum energy of the positron spectrum, given above. (*Hint:* Let m'_C and m'_B be the nuclear masses and proceed as in Example 6 for β^--decay.)

22. *Electron capture.* Some radionuclides decay by capturing one of their own atomic electrons, a K-electron, say. An example is

$$^{49}\text{V} + e^- \rightarrow \,^{49}\text{Ti} + \nu \qquad (t_{1/2} = 331 \text{ d}).$$

Show that the disintegration energy Q for this process is given by

$$Q = (m_\text{V} - m_\text{Ti})c^2 - E_K,$$

where m_V and m_Ti are the atomic masses of ^{49}V and ^{49}Ti, respectively, and E_K is the binding energy of the vanadium K-electron. (*Hint:* Put m'_V and m'_Ti as the corresponding nuclear masses and proceed as in Example 6.)

23. A free neutron decays according to Eq. 47–14a. If the neutron-hydrogen atom mass difference is 840 μu, what is the maximum energy of the β-spectrum?

24. Two radioactive materials that are unstable to α-decay, $^{238}_{92}$U and $^{232}_{90}$Th, and one that is unstable to β-decay, $^{40}_{19}$K, are sufficiently abundant in granite to contribute significantly to the heating of the earth through the decay energy produced. The α-unstable isotopes give rise to decay chains that stop at stable lead isotopes. $^{40}_{19}$K has a single β-decay. Decay information follows:

Parent Nuclide	Decay Mode	Half-life, y	Stable Endpoint	Q, MeV	f, ppm
$^{238}_{92}$U	α	4.47×10^9	$^{206}_{82}$Pb	51.7	4
$^{232}_{90}$Th	α	1.41×10^{10}	$^{208}_{82}$Pb	42.7	13
$^{40}_{19}$K	β	1.28×10^9	$^{40}_{20}$Ca	1.31	4

Q is the total energy released in the decay of one parent nucleus to the final stable endpoint and f is the abundance of the isotope in kilograms per kilogram of granite; ppm means parts per million. (a) Show that these materials give rise to a total heat production of 9.8×10^{-10} W for each kilogram of granite. (b) Assuming that there are 2.7×10^{22} kg of granite in a 20-km-thick, spherical shell around the earth, estimate the power this will produce over the whole earth. (c) Compare this with the total solar power intercepted by the earth, 1.7×10^{17} W.

Section 47–6 Nuclear Reactions

25. *The characteristic nuclear time.* The characteristic nuclear time is a useful but loosely defined quantity, taken as equal to the time required for a nucleon with a few MeV of kinetic energy to travel a distance equal to the diameter of a middle-mass nuclide. What is this quantity for a 5-MeV neutron traversing a nuclear diameter of ^{197}Au; see Table 47–1?

26. Prepare an "overlay" like that of Fig. 47–13 in which that figure is extended to include reactions involving the light nuclides ^{3}H (tritium) and ^{3}He, considered both as projectiles and as emerging particles.

27. Prepare an "overlay" like that of Fig. 47–13 in which *two* nucleons or light nuclei may appear as emerging particles. The reaction ^{63}Cu(α, pn)^{65}Zn is an example. Consider the combinations nn, np, and pd as possibilities.

28. Consider the reaction $X(a, b)Y$, in which X is taken to be at rest in the laboratory reference frame. The initial kinetic energy in this frame is

$$K_{\text{lab}} = \tfrac{1}{2}m_a v_a^2.$$

(a) Show that the initial velocity of the center of mass of the system in the laboratory frame is

$$V = v_a \left(\frac{m_a}{m_X + m_a} \right).$$

Is this quantity changed by the reaction? (b) Show that the initial kinetic energy, viewed now in a reference frame attached to the center of mass of the two particles, is given by

$$K_{\text{cm}} = K_{\text{lab}} \left(\frac{m_X}{m_X + m_a} \right).$$

Is *this* quantity changed by the reaction? (c) In the reaction ^{27}Al$(d, p)^{28}$Al (see Fig. 47–15), the kinetic energy of the deuteron, measured in the laboratory frame, is 2.10 MeV. Find v_a $(=v_d)$, V, and K_{cm}.

*29. In the reaction ^{27}Al$(d, p)^{28}$Al, the kinetic energy K_d of the incident deuterons, as measured in the laboratory, is 2.10 MeV. The Q of the reaction is $+5.49$ MeV. Find (a) the energy K'_d delivered by the deuteron as excitation energy of the system $(^{27}\text{Al} + d)$, (b) the total excitation energy of the system $(^{28}\text{Al} + p)$, and (c) the kinetic energy of the most energetic emerging protons, as measured in the laboratory. See Figs. 47–14 and 47–15. (*Hint:* Apply the conservation of momentum principle both to the formation and the decay of the intermediate state ^{29}Si*.)

Section 47–7 Nuclear Models

30. Draw an energy-level diagram like that of Fig. 47–15 for the neutron-capture reaction ^{113}Cd$(n, \gamma)^{114}$Cd. Make the diagram as much to scale as possible. Take the energy of the incident neutron

to be 0.176 eV. Some needed atomic masses are

$$^{113}\text{Cd}:112.90440 \text{ u}, \qquad ^{114}\text{Cd}:113.90336 \text{ u}, \quad n:1.00867 \text{ u}.$$

31. Consider the three formation modes shown for the compound nucleus ^{20}Ne* on p. 965. What energy must (a) the α particle, (b) the proton, and (c) the X-ray photon have to provide 25.0 MeV of excitation energy to the compound nucleus? Some needed atomic masses are

$$\begin{aligned} ^{20}\text{Ne}: 19.99244 \text{ u}, & \qquad \alpha: 4.00260 \text{ u}, \\ ^{19}\text{F}: 18.99840 \text{ u}, & \qquad p: 1.00783 \text{ u}. \\ ^{16}\text{O}: 15.99491 \text{ u}, & \end{aligned}$$

32. Consider the three decay modes shown for the compound nucleus ^{20}Ne* on p. 965. If the compound nucleus is initially at rest and has an excitation energy of 25.0 MeV, what kinetic energy, measured in the laboratory, will (a) the deuteron, (b) the neutron, and (c) the ^{3}He nuclide, have when the nucleus decays? Some needed atomic masses are

$$\begin{aligned} ^{20}\text{Ne}: 19.99244 \text{ u}, & \qquad d: 2.01410 \text{ u}, \\ ^{19}\text{Ne}: 19.00188 \text{ u}, & \qquad n: 1.00867 \text{ u}, \\ ^{18}\text{F}: 18.00094 \text{ u}, & \qquad ^3\text{He}: 3.01603 \text{ u}. \\ ^{17}\text{O}: 16.99913 \text{ u}, & \end{aligned}$$

33. The nuclide ^{120}Sn has a filled proton shell, with $Z = 50$. The nuclide ^{121}Sb $(Z = 51)$ has an "extra" proton outside this shell. According to the shell concept, this extra proton should be easier to remove than a proton from the filled shell. Calculate the required energy in each case, using these mass data (The proton atomic mass is 1.00783 u.):

Nuclide	Z	N	Atomic Mass, u
^{121}Sb	51	70	120.9038
^{120}Sn	50	70	119.9022
^{119}In	49	70	118.9058

34. The nuclide ^{208}Pb is "doubly magic" in that both its proton number Z $(=82)$ and its neutron number N $(=126)$ represent filled nucleon shells. An additional proton would yield ^{209}Bi and an additional neutron ^{209}Pb. These "extra" nucleons should be easier to remove than a proton or a neutron from the filled shells of ^{208}Pb. (a) Calculate the energy required to move the "extra" proton from ^{209}Bi and compare it with the energy required to remove a proton from the filled proton shell of ^{208}Pb. (b) Calculate the energy required to remove the "extra" neutron from ^{209}Pb and compare it with the energy required to remove a neutron from the filled neutron shell of ^{208}Pb. Do your results agree with expectation? Use these atomic mass data (The masses of the proton and the neutron are 1.00783 u and 1.00867 u, respectively.):

Nuclide	Z	N	Atomic Mass, u
^{209}Bi	83	126	208.9804
^{208}Pb	82	126	207.9767
^{207}Tl	81	126	206.9774
^{209}Pb	82	127	208.9811
^{207}Pb	82	125	206.9759

ADDITIONAL PROBLEMS

35. (a) Convince yourself that the energy tied up in nuclear, or strong force, bonds is proportional to A, the mass number of the nucleus in question. (b) Convince yourself that the energy tied up in Coulomb force bonds between the protons is proportional to $Z(Z-1)$. (c) Convince yourself that, as we move to larger and larger nuclei (see Fig. 47–3), the importance of (b) increases more rapidly than does that of (a).

36. A 5.00-g charcoal sample from an ancient fire pit has a ^{14}C activity of 63.0 disintegrations/min. Carbon from a living tree has a specific activity of 15.3 disintegrations/g·min. The half-life of ^{14}C is 5730 y. How old is the charcoal sample?

37. Nuclear radii may be measured by scattering high-energy electrons from them, as in Fig. 47–1. (a) What is the de Broglie wavelength for 200-MeV electrons? (b) Are they suitable probes for this purpose?

38. A 1.00-g sample of samarium emits α particles at an absolute rate of 120 particles/s. ^{147}Sm, whose natural abundance in elemental samarium is 15.1%, is the responsible isotope. What is the half-life for the decay process?

39. Because a nucleon is confined to a nucleus, we can take its uncertainty in position to be approximately the nuclear radius R. What does the uncertainty principle say about the kinetic energy of a nucleon in a nucleus with, say, $A = 100$? (*Hint:* Take the uncertainty in momentum Δp to be the actual momentum p.)

40. Many radionuclides have two or more competing modes of decay. For ^{212}Bi, for example, we have

$$^{212}Bi \rightarrow {}^{208}Tl + \alpha, \quad \text{in 35.4\% of the cases,}$$

and

$$^{212}Bi \rightarrow {}^{212}Po + e^- + \nu, \quad \text{in 64.6\% of the cases.}$$

The half-life for the ^{212}Bi decay is 60.0 min. What would the half-life be if only the α-decay occurred? Although the answer is of interest, it must be realized that there is no known way to suppress the β-decay.

41. You are asked to pick apart an α particle (^4He) by removing, in sequence, a proton, a neutron, and a proton. Calculate (a) the work required for each step, (b) the total binding energy of the α particle, and (c) the binding energy per particle. Some needed atomic masses are

$$^4He: 4.00260 \text{ u}, \quad {}^2H: 2.01410 \text{ u},$$
$$^3H: 3.01605 \text{ u}, \quad {}^1H: 1.00783 \text{ u},$$
$$n: 1.00867 \text{ u}.$$

42. An electron is emitted from a middle-mass nuclide ($A = 150$, say) with a kinetic energy of 1.0 MeV. (a) What is its de Broglie wavelength? (b) What is the radius of the emitting nucleus? (c) Can such an electron be confined as a standing wave in a "box" of such dimensions? (d) Can you use these numbers to disprove the argument (long since abandoned) that electrons actually exist in nuclei?

43. Consider the reactions

$$^{19}F(p, \alpha)Y_1, \quad Q_1,$$

and

$$^{16}O(d, p)Y_2, \quad Q_2.$$

(a) Identify Y_1 and Y_2. (b) Identify a third reaction, of reaction energy Q_3, such that $Q_1 + Q_2 + Q_3 = 0$.

44. As of this writing there is speculation that the proton may not actually be a stable particle but may be radioactive, with a half-life of about 10^{31} y. If this turns out to be true — and measurements are underway — about how long would a person have to live to be sure that a single proton in his body had decayed? (*Hint:* Assume that you are made of water, which has 10 protons per molecule. Only an approximate answer is called for, of course.)

45. In the periodic table, the entry for magnesium is

| 12 |
| Mg |
| 24.312 |

There are three isotopes:

$$^{24}Mg, \text{ atomic mass} = 23.98504 \text{ u},$$
$$^{25}Mg, \text{ atomic mass} = 24.98584 \text{ u},$$
$$^{26}Mg, \text{ atomic mass} = 25.98259 \text{ u}.$$

The abundance of ^{24}Mg is 78.70% by weight. What are the abundances of the other two isotopes?

46. Heavy radionuclides emit α particles rather than other combinations of nucleons because the α-particle is such a stable, tightly-bound structure. To confirm this, calculate the disintegration energies for these hypothetical decay processes and discuss the meaning of your findings:

$$^{235}U \rightarrow {}^{232}Th + {}^3He + Q_3,$$
$$^{235}U \rightarrow {}^{231}Th + {}^4He + Q_4,$$
$$^{235}U \rightarrow {}^{230}Th + {}^5He + Q_5.$$

The needed atomic masses are

^{232}Th: 232.0381 u,	^3He: 3.0160 u,
^{231}Th: 231.0363 u,	^4He: 4.0026 u,
^{230}Th: 230.0332 u,	^5He: 5.0122 u,
^{235}U: 235.0439 u.	

47. From shell-theory considerations, what would be the nuclear spin of ^{87}Sr? (*Hint:* Convince yourself that the total angular momentum of a nucleon shell with one missing nucleon is equal to the angular momentum that nucleon would have were it present.)

48. In Example 8 we saw that ^{91}Zr ($Z = 40$, $N = 51$) has a single neutron outside a filled 50-neutron core. Because 50 is a magic number, this neutron should perhaps be especially loosely bound. (a) What is its binding energy? (b) What is the binding energy of the next neutron, which must be extracted from the filled core? (c) What is the average binding energy per particle for the entire nucleus? Compare these three numbers and discuss. Some needed atomic masses are

$$^{91}Zr: 90.90564 \text{ u}, \quad n: 1.00867 \text{ u},$$
$$^{90}Zr: 89.90471 \text{ u}, \quad p: 1.00783 \text{ u},$$
$$^{89}Zr: 88.90890 \text{ u}.$$

49. Because the neutron has no charge, its mass must be found in some way other than by using a mass spectrometer. When a resting neutron and a proton meet, they combine and form a deuteron, emitting a γ ray whose energy is 2.2244 MeV. The atomic masses of the proton and the deuteron are 1.00782522 u and 2.0141022 u, respectively. Find the mass of the neutron from these data, to as many significant figures as the data warrant. (A more precise value of the mass-energy conversion factor than that used in the text is 931.481 MeV/u.)

50. The spin and the magnetic moment of ^{81}Br in its ground state (see Table 47–1) are $\frac{3}{2}$ and $+2.27$ nuclear magnetons, respectively. A free ^{81}Br nucleus is placed in a magnetic field of 2.0 T. (*a*) Into how many levels will the ground state split because of space quantization? (*b*) What is the energy difference between adjacent pairs of levels? (*c*) What is the wavelength that corresponds to a transition between such a pair of levels? (*d*) In what region of the electromagnetic spectrum does this wavelength lie?

51. In an endothermic reaction ($Q < 0$), the interacting particles a and X must have a kinetic energy, measured in the center-of-mass reference frame, of at least $|Q|$ if the reaction is to "go." Show, using the result of Problem 28, that the so-called *threshhold energy* for particle a, measured in the laboratory reference frame (in which X is at rest), is

$$K_{\text{th}} = |Q| \frac{m_X + m_a}{m_X}.$$

Is it reasonable that K_{th} should be greater than $|Q|$?

52. One of the dangers of radioactive fallout from a nuclear bomb is ^{90}Sr, which β-decays with a long, 28-year half-life, emitting a 546-keV electron. Because it has chemical properties much like calcium, the strontium, if eaten by a cow, becomes concentrated in its milk and ends up in the bones of whoever drinks the milk. The energetic decay electrons damage the bone marrow and thus impair the production of red blood cells. A 1-megaton bomb produces approximately 0.4 kg of ^{90}Sr. If the fallout was spread uniformly over a 2000-km^2 area, what area would have radioactivity equal to the allowed bone burden for one person of 0.002 mCi? (1 Ci $= 3.7 \times 10^{10}$ disintegrations/s.)

*53. *Secular equilibrium.* A certain radionuclide is being manufactured, say, in a cyclotron, at a constant rate R. It is also decaying, with a disintegration constant λ. Let the production process continue for a time that is long compared to the half-life of the radionuclide. Convince yourself that the number of radioactive nuclei present at such times will be constant and will be given by $N = R/\lambda$. Convince yourself further that this result holds no matter how many of the radioactive nuclei were present initially. The nuclide is said to be in secular equilibrium with its source; in this state its decay rate is just equal to its production rate.

*54. *Secular equilibrium* (see Problem 53). The radionuclide ^{56}Mn has a half-life of 2.58 h and is produced in a cyclotron by bombarding a manganese target with deuterons. The target contains only the stable manganese isotope ^{55}Mn, and the reaction that produces ^{56}Mn is

$$^{55}\text{Mn} + d \rightarrow {}^{56}\text{Mn} + p.$$

After being bombarded for a time $\gg 2.58$ h, the activity of the target, due to ^{56}Mn, is 2.4 curie (Ci), where 1 Ci is 3.7×10^{10} disintegrations/s. (*a*) At what constant rate R are ^{56}Mn nuclei being produced in the cyclotron during the bombardment? (*b*) At what rate are they decaying (also during the bombardment)? (*c*) How many ^{56}Mn nuclei are present at the end of the bombardment? (*d*) What is their mass?

55. *Secular equilibrium* (see Problems 54 and 53). A radium source contains 1.00 mg of ^{226}Ra, which decays with a half-life of 1600 y to produce ^{222}Rn, a noble gas. This radon isotope in turn decays by α-emission with a half-life of 3.82 d. (*a*) What is the rate of disintegration of ^{226}Ra in the source? (*b*) How long does it take for the radon to come to secular equilibrium with its radium parent? (*c*) At what rate is the radon then decaying? (*d*) How much radon is in equilibrium with its radium parent?

48

Energy from the Nucleus

48-1 Introduction

Only during recent decades have we been able to turn the nucleus to the uses of technology. The application of radioisotopes in highly imaginative and useful ways permeates all branches of science, medicine, and industry. It is, however, the large-scale release of nuclear energy by fission or fusion—in reactors and in bombs—that captures our attention and forces major social and economic decisions of a global character upon us.

This chapter deals not so much with the details of the nuclear technologies themselves as with the basic fusion and fission processes from which they spring. Understanding the basic physics—apart from its intrinsic interest—can help us in arriving at an assessment of the issues involved.

To provide a benchmark for magnitudes, consider Table 48-1. It shows how much energy can be derived from 1.0 kg of matter by subjecting it to various processes. Instead of reporting the energy directly, we measure it by indicating how long the extracted energy could operate a 100-W lamp. Only processes represented in the first three rows have actually been carried out; the remaining three represent theoretical limits that are almost certainly not practically attainable.

Row 3 of Table 48-1 shows that, on a per unit mass of fuel basis, the energy derived from practical nuclear fission power is impressive. Row 4 suggests the performance of an imagined "ultimate" fission reactor—a seeming practical impossibility with today's technology. Row 5 represents the output of an imagined "ultimate" fusion reactor. If and when practical fusion power reactors are built, their outputs will certainly be much less. Row 6 represents the ultimate in the extraction of energy from matter, namely, the total transformation of matter into energy by the $E = mc^2$ process.

Table 48–1   Obtaining energy from 1.0 kg of matter[a]

Form of matter	Energy extraction process	Time
1. Water	A 50-m waterfall	5 s
2. Coal	Burning (2.7 kg of oxygen is also required)	8 h
3. UO_2 (enriched to 3% ^{235}U)	Fission in a typical reactor	680 y
4. ^{235}U	Complete fission	3×10^4 y
5. Hot deuterium gas	Complete fusion to $^3He + n$	3×10^4 y
6. Matter and antimatter (0.5 kg each)	Total mutual annihilation	3×10^7 y

[a] Measured by the length of time the energy generated could operate a 100-W lamp.

Incidentally, it must not be imagined that one can conclude by comparing rows 1 and 3 of Table 48–1 that nuclear fission power is in some way 300 y/5 s or ~10^9 times "better" than hydroelectric power. The comparisons in the table are made on a per unit mass basis; there is a lot of water in the Columbia river!

It is our plan to discuss first the physics of the fission process and also the basic physics—but not the technology—that underlies the design of fission reactors.

48–2   Nuclear Fission: The Basic Process

The neutron as a useful bombarding particle

In 1932 the English physicist James Chadwick discovered the neutron. It was soon learned that if various elements were bombarded by these new particles, new radioactive elements were produced. The neutron is a useful projectile for such experiments because, being uncharged, it experiences no repulsive electrostatic force when approaching a nuclear surface. There is no Coulomb barrier, so the slowest neutron can interact with even the most massive nucleus.

Thermal neutrons

Thermal neutrons, which are neutrons in equilibrium with matter at temperatures close to room temperature, are convenient and effective bombarding particles. At room temperature ($T = 300$ K, say) the mean kinetic energy of such neutrons is given from Eq. 21–6 as

$$E_{av} = \tfrac{3}{2}kT = \tfrac{3}{2}(8.6 \times 10^{-7} \text{ eV/K})(300 \text{ K}) \approx 0.04 \text{ eV}.$$

The discovery of fission

In 1939 the German chemists Otto Hahn and Fritz Strassman, following up work initiated by Enrico Fermi and his collaborators in Rome, bombarded uranium with such thermal neutrons. They found by chemical analysis that after the bombardment a number of new radioactive elements were present, among them one whose chemical properties were remarkably similar to those of barium. Repeated tests finally convinced these able chemists that this "new" element was not new at all; it really *was* barium. How could this middle-mass element ($Z = 56$) be produced by bombarding uranium ($Z = 92$) with neutrons?

The riddle was solved within a few weeks by the physicists Lise Meitner and her nephew Otto Frisch. They showed that a uranium nucleus, having absorbed a neutron, could split, with the release of energy, into two roughly equal parts, one of which might well be barium. They named this process *nuclear fission.** Figure 48–1 shows the tracks left in the gas of a cloud chamber by the two energetic fission fragments that result from a fission event occuring near the center of the chamber.

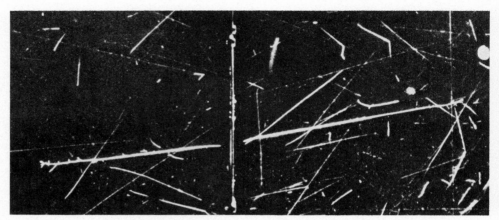

Figure 48–1   A cloud chamber is a device that makes visible the tracks of fast-moving charged particles by causing tiny drops of liquid to condense along them. The two (almost horizontal) tracks shown are of fission fragments, recoiling from a neutron-induced fission event that took place in the thin vertical uranium film. *An Atlas of Typical Expansion Chamber Photographs,* by W. Gentner, Pergamon Press, London, 1954.

We can represent the fission of ^{235}U by thermal neutrons—a process of great practical importance—by

$$^{235}\text{U} + n \rightarrow {}^{236}\text{U}^* \rightarrow \text{X} + \text{Y} + bn, \qquad (48\text{–}1)$$

Fission of ^{235}U by thermal neutrons—the general case

in which ^{236}U, as the asterisk indicates, is a compound nucleus. The factor b, which has the value 2.47 for fission events of this type, is the average number of neutrons released in such events.

Figure 48–2 shows the distribution by mass number of the fission fragments X and Y. We see that in only about 0.01% of the events will the fragments have equal mass. The most probable mass numbers, occurring in about 7% of the events, are $A = 140$ and $A = 95$. We can also tell from the difference in the length of the two fission fragment tracks in Fig. 48–1 that the two fragments in this particular fission event do not have the same mass.

^{236}U, which is the fissioning compound nucleus in Eq. 48–1, has 92 protons and $236 - 92$ or 144 neutrons, a neutron/proton ratio of ~ 1.6. The primary fragments formed immediately after fission will retain this same neutron/proton ratio. Study of the stability curve of Fig. 47–3, however, shows that stable nuclides in the middle-mass region ($75 < A < 150$) have a neutron/proton ratio of only 1.2 to 1.4. The primary fragments will thus be excessively neutron-rich and will "boil off" a small number of neutrons, 2.47 of them on the average for the reaction of Eq. 48–1. The fragments X and Y that remain are still too neutron-rich and will approach the stability line of Fig. 47–3 by a chain of successive β^--decays.

* See "The Discovery of Fission," by Otto Frisch and John Wheeler, *Physics Today,* November 1967, and also "The Discovery of Fission," by Otto Hahn, *Scientific American,* February 1958, for fascinating accounts of these early days of discovery.

Figure 48–2   The percent yield by mass number of the fragments resulting from the fission of ^{235}U by thermal neutrons; see Eq. 48–1. The yields add up to 200% because each fission event produces two fragments. Note that the vertical scale is logarithmic.

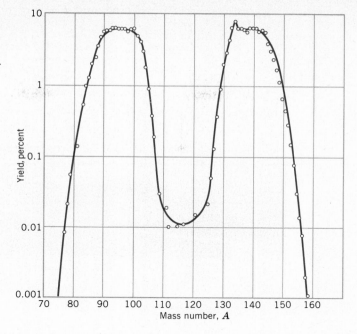

Equation 48–2 is a specific instance of the generalized fission process of Eq. 48–1, in which X and Y are taken as ^{140}Xe and ^{94}Sr, respectively, and in which just two neutrons are emitted. Thus

Fission of ^{235}U by thermal neutrons—a particular case

$$^{235}\text{U} + n \rightarrow {}^{236}\text{U}^* \rightarrow {}^{140}\text{Xe} + {}^{94}\text{Sr} + 2n. \qquad (48\text{–}2)$$

As Table 48–2 shows, the fission fragments in Eq. 48–2 decay until each reaches a so-called daughter nuclide that is stable. The decays are β^- events, the half-lives being indicated at each stage. As expected (see Section 47–5), the mass numbers (140 and 94) remain unchanged as the decay continues.

Table 48–2   The radioactive decay of the fission fragments produced in the fission event of Eq. 48–2

$^{140}\text{Xe} \xrightarrow{\beta^-}$	$^{140}\text{Cs} \xrightarrow{\beta^-}$	$^{140}\text{Ba} \xrightarrow{\beta^-}$	$^{140}\text{La} \xrightarrow{\beta^-}$	^{140}Ce
14 s	65 s	13 d	40 h	Stable

$^{94}\text{Sr} \xrightarrow{\beta^-}$	$^{94}\text{Y} \xrightarrow{\beta^-}$	^{94}Zr
75 s	18 min	Stable

Q for fission

From the first realization of fission, it was known that the disintegration energy Q would be very much larger than for chemical processes. We can support this by a rough calculation. From the binding-energy curve of Fig. 47–6 we see that for heavy nuclides ($A = 240$, say) the mean binding energy per nucleon is about 7.6 MeV. In the intermediate range ($A = 120$, say), it is about 8.5 MeV. The difference in total binding energy between a single large nucleus ($A = 240$) and two fragments (assumed equal) into which it may be split is then

$$Q = 2(8.5 \text{ MeV})\left(\frac{A}{2}\right) - (7.6 \text{ MeV})A \approx 200 \text{ MeV}.$$

The following example shows a more careful calculation.

Example 1   Calculate the disintegration energy Q for the fission process of Eq. 48–2, taking into account the decay of the fission fragments as outlined in Table 48–2. Needed atomic masses are

^{235}U	235.0439 u	^{140}Ce	139.9055 u
n	1.00866 u	^{94}Zr	93.9065 u

If we combine Eq. 48–2 and Table 48–2, we see that the overall transformation is

$$^{235}\text{U} \rightarrow {}^{140}\text{Ce} + {}^{94}\text{Zr} + n.$$

The single neutron comes about because the (initiating) neutron on the left side of Eq. 48–2 cancels one of the two neutrons on the right side of that equation.

The mass difference for the above relation is

$$\Delta m = (235.0439 \text{ u}) - (139.9055 \text{ u} + 93.9065 \text{ u} + 1.00866 \text{ u})$$
$$= 0.223 \text{ u},$$

and the corresponding disintegration energy is

$$Q = \Delta m\, c^2 = (0.223 \text{ u})(931 \text{ MeV/u}) = +208 \text{ MeV},$$

in good agreement with the rough estimate made above.

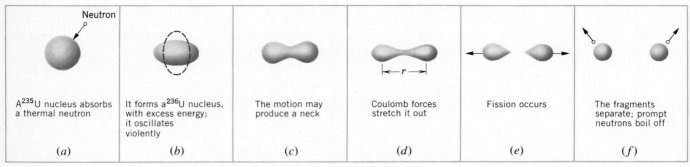

Figure 48–3   Stages in a typical fission process.

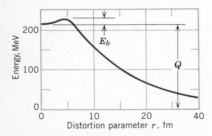

Figure 48–4   The potential energy at various stages of the fission process, showing the fission barrier E_b and the disintegration energy Q. The curve is only loosely to scale.

The fission barrier

The activation energy E_{act}

The excitation energy B_n

Soon after the discovery of fission, Niels Bohr and John Wheeler developed an understanding of its mechanism, based on an analogy between a nucleus and a charged drop of liquid (see p. 965). Figure 48–3 suggests the various stages of the fission process. Quantitatively, Bohr and Wheeler derived the curve of Fig. 48–4, which shows the potential energy of the system at each stage of the fission process. The disintegration energy Q is the energy difference between the initial state (a spherical heavy nucleus) and the final state (two separate fragments). The distortion parameter r, which is the horizontal coordinate in Fig. 48–4, has the value zero for the original spherical nucleus; in the limit of complete separation it is the distance between the centers of the recoiling fragments. At intermediate stages, r is a rough measure of the departure of the system from sphericity, as suggested by Fig. 48–3d.

Figure 48–4 reminds us of Fig. 47–8, the potential-energy curve for α-emission. In both processes a heavy nucleus breaks up into two energetic fragments. In both cases there is a potential barrier that must be surmounted or tunneled. If these barriers were not present, no heavy nuclide with $Z \gtrsim 85$ would exist long enough to be detected; all would disintegrate, either by α-emission or by spontaneous fission, as soon as they were formed.

The height of the fission barrier is shown as E_b in Fig. 48–4. For fission to occur we must supply to the nucleus an energy equal to or greater than a certain activation energy E_{act}. Because of barrier tunneling, E_{act} is somewhat less than the barrier height.

When a nucleus absorbs a slow neutron, as in Fig. 48–3a, that neutron falls into the potential well associated with the strong nuclear forces and transforms its potential energy into internal excitation energy, as suggested by Fig. 48–3b. The amount of excitation energy so provided is equal to B_n, the binding energy of the added or "last" neutron; we can easily calculate it if we know the masses involved.*

A gravitational analogy

We can make an analogy to a stone falling into a dry well. Here too the potential energy of the stone, associated now with gravitational forces, is transformed into (thermal) excitation energy at the bottom of the well. The amount of excitation energy so provided is just equal to the work required to pull the stone back out of the well, that is, to the "binding energy" B_s of the stone.

The criterion for fissionability

In the neutron case the question immediately arises: Is the excitation energy B_n large enough to permit the excited nucleus to surmount the fission barrier? Will fission occur? We see that fissionability depends critically on the relationship between B_n and E_{act}.

A test of fissionability

Table 48–3 shows a test of the fissionability by thermal neutrons of four heavy nuclides, chosen from dozens of candidates. For each nuclide both the activation energy E_{act} and the binding energy B_n of the added neutron are given. The former has been calculated from the theory of Bohr and Wheeler; the latter has been calculated from a knowledge of the masses involved.

Table 48–3 Test of the fissionability of four nuclides

Target nuclide	Nuclide being fissioned	E_{act}, MeV	B_n, MeV	$B_n - E_{act}$, MeV	Fission cross section, barns[a]
^{235}U	^{236}U	5.2	6.5	+1.3	500
^{238}U	^{239}U	5.7	4.8	−0.9	<0.5
^{239}Pu	^{240}Pu	4.8	6.4	+1.6	750
^{243}Am	^{244}Am	5.5	5.5	−0.3	<0.07

[a] The value given is for thermal neutrons. See Section 10–7 for the definition of cross section.

For ^{235}U and ^{239}Pu we see that $B_n > E_{act}$. This means that thermal neutron fission can occur for these nuclides, and this is confirmed by noting, in the table, the large measured values of the cross section for this process.

For the other two nuclides (^{238}U and ^{243}Am), we have $B_n < E_{act}$, so that there is not enough energy to tunnel through the fission barrier effectively. Table 48–3 shows, as we expect, that the cross sections for thermal neutron fission in these cases are exceedingly small. These nuclides can be made to fission, however, if they absorb, not a thermal neutron, but one that carries with it substantial kinetic energy. For ^{238}U, for example, the neutron must have at least ∼ 1.3 MeV of energy for fission to occur with meaningful probability.

48–3 Nuclear Reactors: The Basic Principles

Energy releases in individual nuclear events such as α-emission are roughly a million times larger than those of chemical events, calculated on a per-atom basis. To make large-scale use of nuclear energy, however, we must arrange it so that one nuclear event triggers off another nearby until the process spreads throughout

---

* Note that B_n is not the same as the binding energy B plotted in Fig. 47–6; this latter is the *mean* binding energy of *all* the nucleons in the nucleus.

bulk matter like a flame through a burning log. The fact that more neutrons are generated in fission than are consumed (see Eq. 48–1) raises just this possibility; the neutrons that are produced can cause fission in nearby nuclei and in this way a chain of fission events can propagate itself. Such a process is called a *chain reaction*. It can either be rapid and uncontrolled as in a nuclear bomb or controlled as in a nuclear reactor.

The chain reaction

Suppose that we wish to design a nuclear reactor based, as most present reactors are, on the fission of ^{235}U by slow neutrons. The fuel in such reactors is almost always artificially "enriched," so that ^{235}U makes up a few percent of the uranium nuclei rather than the 0.7% that occurs in natural uranium. Although on the average 2.47 neutrons are produced in ^{235}U fission for every thermal neutron consumed, there are serious difficulties in making a chain reaction "go." Here are three of the difficulties, together with their solutions:

Three reactor design problems

1. *The neutron leakage problem.* A certain percentage of the neutrons produced will simply leak out of the reactor core and be lost to the chain reaction. If too many do so, the reactor will not work. Leakage is a surface effect, its magnitude being proportional to the square of a typical reactor core dimension ($= 4\pi r^2$ for a sphere). Neutron production, however, is a volume effect, proportional to the cube of a typical dimension ($= \frac{4}{3}\pi r^3$ for a sphere). The fraction of neutrons lost by leakage can be made as small as we wish by making the reactor core large enough, thereby decreasing its surface to volume ratio ($= 3/r$ for a sphere).

Problem 1

2. *The neutron energy problem.* The neutrons produced by fission are fast neutrons, with kinetic energies of ~ 2 MeV. However, fission is induced most effectively by slow neutrons. The fast neutrons can be slowed down by mixing the uranium fuel with a substance that has these properties: (*a*) It is effective in causing neutrons to lose kinetic energy by elastic collisions and (*b*) it does not absorb neutrons excessively, thereby removing them from the fission chain. Such a substance is called a *moderator*. Most power reactors in this country today use water as a moderator, the hydrogen nuclei (protons) being the effective moderating element; see Example 2 of Chapter 10.

Problem 2

3. *The neutron capture problem.* Neutrons may be captured by nuclei in ways that do not result in fission, capture with the emission of a γ-ray being the most common possibility. In particular, as the fast (~ 2 MeV) neutrons generated in the fission process are slowed down in the moderator to thermal equilibrium (~ 0.04 eV), they must pass through an energy interval (~ 1 to ~ 100 eV) in which they are particularly susceptible to nonfission capture by ^{238}U.

Problem 3

To minimize such *resonance capture*, as it is called, the uranium fuel and the moderator (water, say) are not intimately mixed but are "clumped," remaining in close contact with each other but occupying different regions of the reactor volume. The hope is that a fast fission neutron, produced in a uranium "clump" (which might be a fuel rod), will with high probability find itself in the moderator as it passes through the "dangerous" resonance energy range. Once it has reached thermal energies, it will very likely wander back into a clump of fuel and produce a fission event. It seems clear that finding the optimum geometric arrangement of fuel and moderator is no simple problem.

Figure 48–5 shows the neutron balance in a typical power reactor operating with a steady power output. We start at the top with 1000 typical fast neutrons produced by fission. Eight of them produce fission in the fuel before they have had a chance to interact with the moderator (fast fission). Another 292 are permanently lost to the chain reaction by leakage (192) and by resonance capture (100); these two effects can be minimized, as we indicated above, but they cannot be eliminated. This leaves 700 fast neutrons available to be turned into slow neutrons by the moderator.

Neutron bookkeeping in a reactor

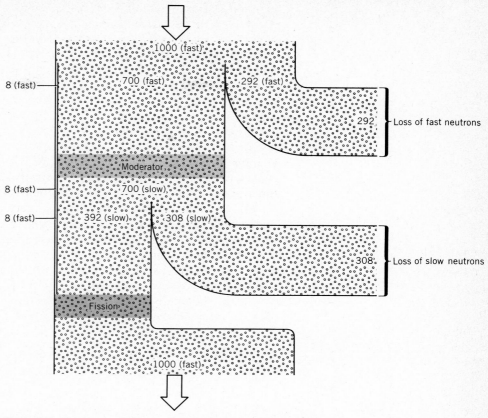

Figure 48-5 An accounting of one complete regeneration cycle for 1000 fast neutrons in a typical nuclear reactor using natural uranium as a fuel.

Of the 700 slow neutrons, some 308 are also lost to the chain reaction, by leakage (75) and by nonfission capture in the fuel (208), in the materials of which the reactor is made (2), and in the accumulated fission products (23). This leaves 392 slow neutrons free to cause fission. Together with the original 8 fast neutrons, they produce a new generation of 2.5 (392 + 8) or 1000 fast neutrons, and the cycle starts all over again. Note that we have assumed that b in Eq. 48-1 is 2.5.

The *multiplication factor k* is an important reactor parameter. It is the ratio of the number of neutrons produced in each fission cycle to the number present at the beginning of the cycle. If $k = 1$, as in Fig. 48-5, the reactor operation is said to be critical. If $k > 1$ it is supercritical, and if $k < 1$ it is subcritical. Reactors are designed with the inherent ability to go supercritical. This condition cannot last very long, because the rate at which fission processes occur, and thus the rate of energy production in the fuel, would increase exponentially. Criticality is achieved, at any desired power level in the design range of the reactor, by suitably adjusting the *control rods*. These rods, made of a material such as cadmium, which absorbs slow neutrons readily, may be moved mechanically in and out of the reactor volume.

A vital factor in the control of a reactor is its response time. If, for example, one of the control rods is suddenly partially withdrawn, how rapidly would the reactor power level increase? This time is controlled by the extremely fortunate circumstance that a small fraction of the neutrons generated in the overall fission

Reactor operation—critical, subcritical, and supercritical

Delayed neutron emission and reactor response time

process are not boiled off promptly from the newly formed fission fragments but are emitted from these fragments along the β-decay chain (see Table 48–2 in which, however, neutron emission does not happen to occur; in other decay chains it does occur). Of the 1000 neutrons analyzed in Fig. 48–5, about 16 are delayed, being emitted from the fragments following β-decays whose half-lives range from 0.2 s to 55 s. These delayed neutrons are few in number, but they serve an important function in slowing down reactor response times to within the range of human reaction times.

A nuclear power plant

Figure 48–6 shows the broad outlines of an electric power plant based on a so-called pressurized-water reactor (PWR), a type in common use in this country. In such a reactor, water is used both as the moderator and as the heat-transfer medium. In the primary loop, water at high temperature and pressure (possibly 600 K and 150 atm) circulates through the reactor vessel and transfers heat from the reactor core to the steam generator, which provides high-pressure steam to operate the turbine-generator combination. Low-pressure steam from the turbine is condensed to water and forced back into the steam generator by a pump, thus completing the secondary loop. To give some idea of scale, we note that a typical reactor pressure vessel for a 1000-MW(electric) plant may be 40 ft high and 15 ft in diameter, weighing 450 tons. Water flows through the primary loop at the rate of $\sim$ 300,000 gal/min.

Figure 48–6 A simplified layout of a nuclear power plant based on a reactor of the pressurized-water variety.

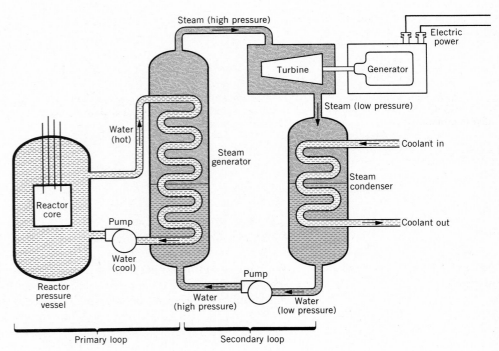

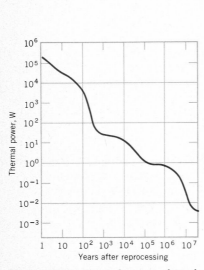

Figure 48–7 Thermal power released by the radioactive wastes generated by one year's operation of a typical large nuclear power plant. Note that both scales are logarithmic. Adapted from "High-Level Radioactive Waste from Light-Water Reactors," by Bernard L. Cohen, *Reviews of Modern Physics,* January 1977.

An unavoidable feature of nuclear power plants is the accumulation of radioactive wastes, including both fission products and heavy, "transuranic" nuclides such as plutonium and americium. One measure of their radioactivity is the rate at which they release energy in thermal form. Figure 48–7 shows the variation with time of the thermal power generated by the radioactive wastes from one year's operation of a typical large power plant. The curve is the sum of contributions from the many radioactive nuclides that make up the waste. Note that both scales are logarithmic. The total activity of the waste 10 years after processing is $\sim 3 \times 10^7$ Ci, where 1 Ci is 3.7×10^{10} disintegrations/s.

Example 2 A large electric power generating station operates on thermal energy derived from a nuclear reactor of the pressurized-water type. Here are some of its characteristics:

Electric power output of the plant: 1100 MW

Thermal power in the reactor core: 3400 MW

Fuel charge: 86,000 kg of uranium in the form of 110 tons of uranium oxide, distributed among 57,000 fuel rods. The uranium is enriched to 3.0% ^{235}U.

(*a*) What is the efficiency of the plant?

$$e = \frac{\text{Useful output}}{\text{Energy input}} = \frac{1100 \text{ MW(electric)}}{3400 \text{ MW(thermal)}} = 32\%.$$

The efficiency is limited by the strictures of the second law of thermodynamics; see Section 22-5. It is necessary to discharge 3400 MW − 1100 MW = 2300 MW as thermal energy to the environment.

(*b*) At what rate R do fission events occur in the reactor core?

Let us assume conditions at startup. If P (= 3400 MW) is the thermal power in the core and Q (= 200 MeV) is the energy released per fission event, we have

$$R = \frac{P}{Q} = \left(\frac{3.4 \times 10^9 \text{ W}}{200 \text{ MeV}}\right)\left(\frac{1 \text{ MeV}}{1.6 \times 10^{-13} \text{ J}}\right) = 1.1 \times 10^{20} \text{ s}^{-1}.$$

(*c*) At what initial rate (at startup) is ^{235}U disappearing?
^{235}U disappears by fission at the rate calculated in (*b*). It also is consumed by (n, γ) reactions at a rate about one-fourth as large. The total loss rate is then $(1.25)(1.1 \times 10^{20} \text{ s}^{-1})$ or $1.4 \times 10^{20} \text{ s}^{-1}$. We recast this as a mass rate by putting

$$\frac{dM}{dt} = (1.4 \times 10^{20} \text{ s}^{-1}) \left(\frac{235 \text{ g/mol}}{6.0 \times 10^{23} \text{ atoms/mol}}\right)\left(\frac{8.6 \times 10^4 \text{ s}}{1 \text{ d}}\right)$$
$$= 4.7 \text{ kg/d}.$$

The quantity in the second set of parentheses above is just the mass of a single ^{235}U atom.

(*d*) At this rate of fuel consumption, how long would the fuel supply last?

From the data given, we can calculate that, at startup, (0.030)(86,000 kg) or 2600 kg of ^{235}U were present. Thus a simplistic answer would be

$$T = \frac{2600 \text{ kg}}{4.7 \text{ kg/d}} = 550 \text{ d}.$$

However, complications associated with (1) the "poisoning" of the fuel by the buildup of fission products and (2) the buildup of (fissionable) ^{239}Pu render this simplistic result meaningless. In practice the fuel rods are replaced in batches—with their ^{235}U supply far from depleted—according to an efficiency-optimizing schedule.

(*e*) At what rate is mass being converted to energy in the reactor core?

From Einstein's $E = mc^2$ relationship, we can write

$$\frac{dM}{dt} = \frac{dE/dt}{c^2} = \frac{(3.4 \times 10^9 \text{ W})}{(3.0 \times 10^8 \text{ m/s})^2}$$
$$= 3.8 \times 10^{-8} \text{ kg/s} \quad \text{or} \quad 3.3 \text{ g/d}.$$

Note that this mass rate (conversion to energy) is quite a different physical quantity from the mass rate (loss of ^{235}U) calculated in part (*c*).

48-4 Thermonuclear Fusion: The Basic Process

Fusion hindered by the Coulomb barrier

We pointed out in connection with the binding-energy curve of Fig. 47-6 that energy can be released if light nuclei are combined to form nuclei of somewhat larger mass number, a process called *nuclear fusion*. However, this process is hindered by the mutual Coulomb repulsion that tends to prevent two such (positively) charged particles from coming within range of each other's attractive nuclear forces and "fusing." This reminds us of the potential barrier that inhibits nuclear fission (see Fig. 48-4) and also of the barrier that inhibits α-decay (see Fig. 47-8).

In the case of α-decay, two charged particles—the α-particle and the ^{234}Th residual nucleus for the situation represented in Fig. 47-8—are initially *inside* their mutual potential barrier. For α-decay to occur, the α-particle must leak through this barrier by the barrier tunneling process and appear on the *outside*. In nuclear fusion the situation is just reversed. Here the two particles must penetrate their mutual barrier from the *outside* if a nuclear interaction is to occur. Example 3 gives some indication of the magnitudes of such Coulomb barriers for interacting light nuclei.

Example 3 *The Coulomb barrier for fusion.* The deuteron (^{2}H) has a charge $+e$ and may be taken as a sphere whose effective radius R has been shown from scattering experiments to be 2.1 fm.* Two such particles are fired at each other with the same initial kinetic energy K. What must K be if the particles are brought to rest by their mutual Coulomb repulsion when the two assumed spheres are just "touching"? We shall take K as a measure of the height of the Coulomb barrier.

Because the two deuterons are momentarily at rest when they "touch" each other, their kinetic energy has all been transformed into electrostatic potential energy associated with the Coulomb repulsion between them. If we treat them as charged mass points separated by a distance $2R$, we have

$$2K = \frac{1}{4\pi\epsilon_0}\frac{q_1 q_2}{r} = \frac{1}{4\pi\epsilon_0}\frac{e^2}{2R},$$

which yields

$$K = \frac{e^2}{16\pi\epsilon_0 R}$$
$$= \frac{(1.6 \times 10^{-19}\ \text{C})^2}{(16\pi)(8.9 \times 10^{-12}\ \text{C}^2/\text{J·m})(2.1 \times 10^{-15}\ \text{m})}\left(\frac{1\ \text{keV}}{1.6 \times 10^{-16}\ \text{J}}\right)$$
$$\approx 200\ \text{keV}.$$

Although the calculation of Example 3 is somewhat simplistic, the result does indicate the order of magnitude of the Coulomb barrier for the deuteron–deuteron interaction. As we shall see later, this interaction is of particular importance in fusion research.

The barriers are high

The deuteron–deuteron barrier, calculated above, is actually a fairly modest one. The corresponding barrier for two interacting ^{3}He nuclei (charge $= +2e$), for example, is ~ 1 MeV. For more highly charged particles the barrier, of course, is correspondingly higher.

Thermonuclear fusion

One way to arrange for light nuclei to penetrate their mutual Coulomb barriers is to use one light particle as a target and to accelerate the other by means of a cyclotron or some such device. To generate power in a useful way from the fusion process, however, we must have the interaction of matter in bulk, just as in the combustion of coal. The cyclotron technique holds no promise in this direction. The best hope for obtaining fusion in bulk matter in a controlled fashion is to raise the temperature of the material so that the particles have sufficient energy to penetrate the barrier due to their thermal motions alone. This process is called *thermonuclear fusion*.

The mean translational kinetic energy of a particle

The mean thermal kinetic energy of a particle in equilibrium at a temperature T is given, as we have seen, by

$$\overline{K} = \tfrac{3}{2}kT, \tag{48–3}$$

where k ($= 8.63 \times 10^{-5}$ eV/K) is the Boltzmann constant. Substitution shows that at room temperature ($T \approx 300$ K) we have $K = 0.04$ eV, which is, of course, far too small for our purpose.†

Thermonuclear fusion in the sun

Even at the center of the sun, where $T = 1.5 \times 10^7$ K, the mean thermal kinetic energy calculated from Eq. 48–3 is only 1.9 keV. This still seems hopelessly small in view of the magnitude of the Coulomb barrier calculated in Example 3. Yet we know that thermonuclear fusion not only occurs in the solar interior but is the central and dominant feature of that body.

Two central points

The puzzle is solved with the realization that (1) The energy calculated from Eq. 48–3 is a *mean* kinetic energy; particles with energies much greater than this mean value constitute the high-energy "tails" of the Maxwellian speed distribution curves (see Fig. 21–9). Also, (2) the barrier heights that we have quoted above are just that; they represent the peaks of the barriers. Barrier tunneling can occur to a significant extent at energies well below these peaks.

---

* Note that Eq. 47–1 cannot be used to calculate the radii of such light nuclides.

† It is important at this point not to confuse k (the Boltzmann constant), keV (kiloelectron-volt), K (kinetic energy), and K (Kelvin temperature.)

The Maxwell energy distribution
$n(K)$

Figure 48–8 summarizes the situation by a quantitative example. The curve marked $n(K)$ in this figure is a Maxwell distribution curve drawn, for concreteness, to correspond to the sun's central temperature, 1.5×10^7 K. The vertical scale is arbitrary. Although the same curve holds no matter what particle is under consideration, we focus our attention on protons, bearing in mind that hydrogen forms about 35% of the mass of the sun's central core.

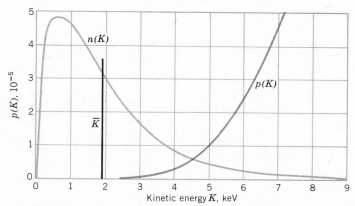

Figure 48–8   The curve marked $n(K)$ gives the Maxwellian energy distribution for particles at $T = 1.5 \times 10^7$ K, the temperature at the center of the sun. The vertical scale is arbitrary. The curve marked $p(K)$ gives the probability of Coulomb barrier penetration for colliding protons.

The distribution curve of Fig. 48–8 differs from the Maxwell distribution curves of Fig. 21–9 in that it is drawn in terms of energy and not speed. Specifically, $n(K)\, dK$ gives the probability per unit volume that a proton will have a kinetic energy lying between K and $K + dK$. See Problem 8 of Chapter 46 for a specific comparison of $n(K)$ and $n(v)$. For $T = 1.5 \times 10^7$ K, Eq. 48–3 yields $\bar{K} = 1.9$ keV, and this value is indicated by a vertical line in the figure. Note that there are many particles whose energies exceed this mean value.

The barrier penetration probability
$p(K)$

The curve marked $p(K)$ in Fig. 48–8 is the probability of barrier penetration for two colliding protons. At $K = 6$ keV, for example, we have $p = 2.4 \times 10^{-5}$. This is the probability that two colliding protons, each with $K = 6$ keV, will succeed in penetrating their mutual Coulomb barrier and coming within range of each other's strong nuclear forces. Put another way, on the average, one of every $\sim 42,000$ such encounters will succeed.

It turns out that the most probable energy for proton–proton fusion events to occur—at the sun's central temperature—is ~ 6 keV. If the energy is much higher, the barrier is transparent enough but there are too few protons in the Maxwellian "tail." If the energy is much lower, there are plenty of protons but the barrier is now too formidable.

48–5   Thermonuclear Fusion in the Sun and Other Stars

Here we consider in more detail the thermonuclear fusion processes that take place in our sun and in other stars. In the sun's deep interior, where its mass is concentrated and where most of the energy production takes place, the (central) temperature, as we have already seen, is $\sim 1.5 \times 10^7$ K and the central density is $\sim 1.5 \times 10^5$ kg/m³, about 13 times the density of lead. The central temperature is so high that, in spite of the high central pressure ($\sim 2 \times 10^{11}$ atm), the sun remains gaseous throughout. In fact, the ideal gas law holds as a good approximation.

The present composition of the sun's core is about 35% hydrogen by mass, about 65% helium, and about 1% other elements. At these temperatures the light elements are essentially totally ionized, so that our picture is one of an assembly of protons, electrons, α-particles, and photons in random motion.

The sun radiates at the rate of 3.9×10^{26} W and has been doing so for as long as the solar system has existed, which is about 4.5×10^9 y. It has been known since the early 1930s that thermonuclear fusion processes in the sun's interior account for its prodigious energy output. Before analyzing this further, however, let us dispose of two other possibilities that had been put forward earlier. Consider first chemical reactions such as simple burning. If the sun, whose mass is 2.0×10^{30} kg, were made of coal and oxygen in just the right proportions for burning, it would last only $\sim 10^3$ y, which of course is far too short. The sun, as we shall see, does not burn coal but hydrogen, and in a nuclear furnace, not an atomic or chemical one.

Another possibility is that, as the core of the sun cools and the pressure there drops, the sun will shrink under the action of its own strong gravitational forces. By transferring gravitational potential energy to thermal energy (just as we do when we drop a stone onto the earth's surface), the temperature of the sun's core will rise so that radiation may continue. Calculation shows, however, that the sun could radiate from this cause for only $\sim 10^7$ y, too short by a factor of about 500. This leaves only thermonuclear fusion.

The sun's energy is generated by the thermonuclear "burning" (that is, "fusing") of hydrogen to form helium. Table 48–4 shows the *proton–proton cycle* by means of which this is accomplished. Note that each reaction shown is a fusion reaction, in that one of the products, be it ²H, ³He, or ⁴He, has a higher mass number than any of the reacting particles that form it. As we expect for fusion reactions, the reaction energy Q, shown in the table, is positive. As we have seen, this characterizes an exothermic reaction, with the net production of energy.

The sun—age and energy output

Chemical processes ruled out

Gravitational heating ruled out

The proton–proton cycle

Table 48–4   The proton-proton cycle

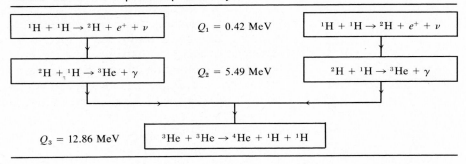

The cycle is initiated by the thermal collision of two protons (^{1}H + ^{1}H) to form a deuteron (^{2}H), with the simultaneous creation of a positive electron (e^+) and a neutrino (ν). In Table 48–4 we follow the consequences of two such events, as indicated in the top row of the table. Such events are extremely rare. In fact, only once in $\sim 10^{26}$ proton–proton collisions is a deuteron formed; in the vast majority of cases the colliding protons simply scatter from each other. It is the slowness of this process that regulates the rate of energy production and keeps the sun from exploding. Interestingly, however, in spite of this slowness, there are so very many protons in the huge volume of the sun's core that deuterium is produced there in this way at the rate of $\sim 10^{12}$ kg/s!

Once a deuteron has been produced it quickly (within a few seconds) collides with another proton and forms a ^{3}He nucleus, as the second row of Table 48–4 shows. Two such ^{3}He nuclei may then eventually (within $\sim 10^5$ y) collide, forming an α-particle (^{4}He) and two protons, as the third row of the table shows. There are other variations of the proton–proton cycle, involving other light elements, but we concentrate on the principal sequence as outlined above.

Taking an overall view of the proton–proton cycle, we see that it amounts to the combination of four protons to form an α-particle and two positive electrons with the release of energy Q' in the process. Note that although six protons are consumed in each cycle, two are generated, so that there is a net loss of four. Thus

Calculation of Q for the proton–proton cycle

$$Q' = (4m'_{1H} - m'_{4He} - 2m_e)c^2,$$

where the masses involved are here nuclear—not atomic—masses.

Consider now what happens to the two positive electrons that are produced in each cycle. Each will quickly encounter a negative electron and annihilate with it, turning their combined rest masses into γ-ray energy as described in Section 22–7. Thus the energy equivalent of the rest masses of four electrons (two positive and two negative) is also made available, or

$$Q'' = 4m_e c^2.$$

The total energy release for the cycle is then

$$\begin{aligned} Q &= Q' + Q'' \\ &= 4(m'_{1H} + m_e)c^2 - (m'_{4He} + 2m_e)c^2 \\ &= (4m_{1H} - m_{4He})c^2, \end{aligned}$$

in which the masses in the last line are now the customary atomic masses. Substitution yields

$$\begin{aligned} Q &= [(4)(1.00783 \text{ u}) - (4.00260 \text{ u})][931 \text{ MeV/u}] \\ &= 26.7 \text{ MeV.} \end{aligned}$$

Solar neutrinos

Not quite all of this energy is available as thermal energy inside the sun. About 0.5 MeV is associated with the two neutrinos that are produced in each cycle. Neutrinos are so penetrating that in essentially all cases they escape from the sun, carrying this energy with them. Some are intercepted by the earth, bringing us our only direct information about the sun's interior.

Subtracting the neutrino energy leaves 26.2 MeV per cycle available within the sun. It is not hard to show (see Example 4) that this corresponds to a "heat of combustion," as chemists call it, for the nuclear burning of hydrogen into helium of 6.3×10^{14} J/kg of hydrogen consumed. By comparison, the heat of combustion of coal is about 3.3×10^7 J/kg, some 20 million times lower, reflecting roughly the general ratio of energies in nuclear and chemical processes.

The burning of hydrogen in the sun's core to form helium is alchemy on a grand scale in the sense that one element is turned into another. The medieval alchemists, however, were more interested in changing lead into gold than in changing hydrogen into helium. In a sense they were on the right track, except that their furnaces were too cool. Instead of being at ~ 600 K, they should have been at $\sim 10^7$ K, the temperature of stellar interiors!

Example 4 At what rate is hydrogen being consumed in the core of the sun, assuming that all of the radiated energy is generated by the proton–proton cycle of Table 48–4?

We have seen that 26.2 MeV appear as thermal energy in the sun for every four protons consumed, a rate of 6.6 MeV/proton. We can express this as

$$6.6 \, \frac{\text{MeV}}{\text{proton}} \left(\frac{1 \text{ proton}}{1.67 \times 10^{-27} \text{ kg}} \right) \left(\frac{1.6 \times 10^{-13} \text{ J}}{1 \text{ MeV}} \right)$$

$$= 6.3 \times 10^{14} \text{ J/kg},$$

which tells us that the sun radiates away 6.3×10^{14} J for every kilogram of protons consumed. The hydrogen consumption rate is then just the output power (= 3.9×10^{26} W) divided by the above quantity, or

$$\frac{dm}{dt} = \frac{3.9 \times 10^{26} \text{ W}}{6.3 \times 10^{14} \text{ J/kg}} = 6.2 \times 10^{11} \text{ kg/s}.$$

To keep this number in perspective, we point out that the sun's mass is 2.0×10^{30} kg.

The sun—predictions for the future

We may ask how long the sun can continue to shine at its present rate before all the hydrogen in its core has been converted into helium. Hydrogen burning has been going on for about 4.5×10^9 y, and calculations show that there is enough available hydrogen left for about 5×10^9 y more. At that time major changes will begin to happen. The sun's core, which by then will be largely helium, will begin to collapse and to heat up while the outer envelope will expand greatly, perhaps so far as to encompass the earth's orbit. The sun will become what astronomers call a *red giant*.

Burning helium

If the core temperature heats up to $\sim 10^8$ K, energy can be produced once more by burning helium to make carbon. Helium does not burn readily, the only possible reaction being

$$^4\text{He} + {}^4\text{He} + {}^4\text{He} \rightarrow {}^{12}\text{C} + \gamma \qquad (Q = +7.3 \text{ MeV}).$$

Such a three-body collision of three α-particles must occur within $\sim 10^{-16}$ s if the reaction is to go. Nevertheless, if the density and temperature of the helium core are high enough, carbon will be manufactured by the burning of helium in this way.

Element building

As a star evolves and becomes still hotter, other elements can be formed by other fusion reactions. However, elements more massive than $A = 56$ (^{56}Fe, ^{56}Co, ^{56}Ni, etc.) cannot be manufactured by further fusion processes. $A = 56$ marks the peak of the binding-energy curve of Fig. 47–6, and fusion between nuclides beyond this point involves the consumption—and not the production—of energy.

If heavy elements cannot be produced by the fusion of middle-mass nuclides and if, as seems certain, they were not present in the primeval fireball that marked the creation of the universe, where then did they come from? The present view is that they were built up from medium-mass nuclides by the successive absorption of neutrons followed by β^--decay. A reaction such as

$$^{56}\text{Fe} + 3n \rightarrow {}^{59}\text{Fe},$$

$$^{59}\text{Fe} \rightarrow {}^{59}\text{Co} + e^- + \nu \qquad (t_{1/2} = 45 \text{ d}),$$

for example, produces ^{59}Co, a stable nuclide. Neutrons, being uncharged, experience no Coulomb barriers.

Supernovae

Conditions suitable for the buildup of the heaviest nuclides are thought to occur in *supernovae*, which are cataclysmic stellar explosions that occur in our galaxy, and also in other galaxies, with reasonable frequency. In such an event a very large amount of stellar matter is ejected violently into space where it mixes with, and becomes part of, the tenuous medium (~ 1 particle/cm^3) that fills the space between the stars. It is from this medium, continually enriched by debris from stellar explosions, that new stars form, by condensation under the influence of gravity.

The fact that elements heavier than hydrogen and helium are found on earth suggests to us that our solar system has condensed out of interstellar material that contained the remnants of such explosions. Thus all the elements around us— including those in our own bodies—were manufactured in the interiors of distant stars that have exploded long ago, returning their substance to the cosmic dust from which they evolved. As one scientist put it, "In truth we are the children of the Universe."

48–6 Controlled Thermonuclear Fusion

The fusion bomb

Thermonuclear reactions have been going on in the universe since its creation in the presumed cosmic "big bang" of some 7 to 15 billion years ago. Such reactions have taken place on earth, however, only since November 1952, when the first fusion (or hydrogen) bomb was exploded. The high temperatures needed to sustain the thermonuclear reactions in this case are provided by using a fission bomb as a trigger.

A controlled fusion power reactor

A sustained and controllable thermonuclear power source—a fusion reactor—is proving much more difficult to achieve. The goal, however, is being vigorously pursued because many look to the fusion reactor as the ultimate power source of the future, at least as far as the generation of electricity is concerned.

The $p–p$ interaction not suitable

The proton–proton interaction displayed in the top row of Table 48–4 is not suitable for use in a terrestrial fusion reactor because it is hopelessly slow. The reaction cross section is in fact so small that it cannot be measured in the laboratory. The reaction succeeds under the conditions that prevail in stellar interiors only because of the enormous number of protons available in the huge, high-density stellar cores.

The $d–d$ and $d–t$ reactions

The most attractive reactions for terrestrial use appear to be the deuteron–deuteron ($d–d$) and the deuteron–triton ($d–t$) reactions:

$$^2\text{H} + {}^2\text{H} \rightarrow {}^3\text{He} + n \qquad (d–d), \qquad Q = +3.27 \text{ MeV};$$
$$^2\text{H} + {}^2\text{H} \rightarrow {}^3\text{H} + {}^1\text{H} \qquad (d–d), \qquad Q = +4.03 \text{ MeV};$$

and

$$^2\text{H} + {}^3\text{H} \rightarrow {}^4\text{He} + n \qquad (d–t), \qquad Q = +17.59 \text{ MeV}.$$

As we see, *triton* is the name given to the nucleus of the hydrogen isotope ^3H. Note that each reaction above is indeed a fusion reaction and has a positive Q value. Deuterium, whose isotopic abundance in normal hydrogen compounds is 1/6500, is available in unlimited quantities as a component of sea water.

Three requirements for a fusion reactor

There are three basic requirements for the successful operation of a thermonuclear reactor.

First requirement

1. *A high particle density n.* The number of interacting particles (deuterons, say) per unit volume must be great enough to ensure a sufficiently high deuteron–deuteron collision rate. At the high temperatures required, the deuterium gas would be completely ionized into a neutral *plasma* consisting of deuterons and electrons.

Second requirement

2. *A high plasma temperature T.* The plasma must be hot. Otherwise the colliding deuterons will not be energetic enough to penetrate the mutual Coulomb barrier that tends to keep them apart; see Fig. 48–8. In fusion research, temperatures are often reported by giving the corresponding value of kT (not $\frac{3}{2}kT$). A plasma temperature of 6.5 keV, corresponding to 7.5×10^7 K, has been achieved in the laboratory. This is higher than the sun's central temperature (1.3 keV, or 1.5×10^7 K).

3. *A long confinement time* τ. A major problem is containing the hot plasma long enough to ensure that its density and temperature remain sufficiently high. It is clear that no actual solid container can withstand the high temperatures necessarily involved, so that special techniques, to be described later, must be employed. By use of one such technique, a confinement time as great as 100 ms has been achieved.

J. D. Lawson has shown that, for the successful operation of a thermonuclear reactor, it is necessary to have

$$n\tau \gtrsim 10^{14} \text{ s/cm}^3, \tag{48-4}$$

a condition called *Lawson's criterion.* The product $n\tau$ is called the *Lawson number* of any fusion device. Equation 48–4 tells us, loosely speaking, that we have a choice between confining a lot of particles for a relatively short time or confining fewer particles for a somewhat longer time. Beyond meeting this criterion, it is also necessary that the plasma temperature be sufficiently high. As of 1981 no experimental device has yet been able to satisfy Lawson's criterion.

Let us see how Lawson's criterion comes about. To raise a plasma to a suitably high temperature and to maintain it there against losses, energy must be added to the plasma at a rate per unit volume P_h, where the subscript stands for "heating." The heating may be done by passing an electric current through the plasma, by firing a beam of energetic neutral particles into it, or in other ways. The denser the plasma, the greater the heating power required, in direct proportion, or

$$P_h = C_h n, \tag{48-5a}$$

where C_h is a suitable constant.

If thermonuclear fusion occurs in the plasma, there will be a certain rate of energy generation per unit volume P_f, where the subscript now stands for "fusion." P_f is proportional to the confinement time τ. It is also proportional to n^2, where n is the particle density. To see this, suppose that we double the particle density. Not only will a given particle make twice as many collisions as it wanders through the plasma, but there will be twice as many wandering particles, an overall factor of four. Thus

$$P_f = C_f n^2 \tau. \tag{48-5b}$$

To have a net production of power, we must have

$$P_f > P_h$$

or, from Eqs. 48–5,

$$n\tau > \frac{C_h}{C_f},$$

which leads directly to Eq. 48–4 if the constants C_h and C_f are suitably evaluated. The condition in which $P_f = P_h$ is called *breakeven;* it has not yet (1980) been achieved by any laboratory fusion device.

We now turn our attention to plasma confinement techniques, of which we discuss two: (1) magnetic field confinement as manifested in the tokamak approach and (2) inertial confinement as manifested in the laser fusion approach.

"Tokamak" is a Russian-language acronym for a device, first developed in the USSR, in which the charged particles that make up the plasma are confined by a particular magnetic field configuration.* As Fig. 48–9 shows, the geometry is that of a doughnut or torus. The confining magnetic field is represented by helical lines of force—only one of which is shown in the figure—which spiral around the plasma "doughnut," pulling it away from contact with the walls of the vacuum chamber.

Origin of the tokamak confining field

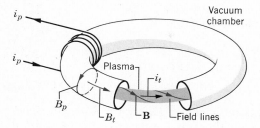

Figure 48–9 The toroidal vacuum chamber that forms the heart of a tokamak. Note the plasma, the helical magnetic field **B** that confines it, and the induced toroidal current i_t that heats it. B_p (generated by i_t) and B_t (generated by i_p) are the components of **B**. The transformer-magnet arrangement used to induce i_t is not shown.

The helical magnetic field is the resultant of two component fields, shown at the left in Fig. 48–9. (1) The *toroidal* field component B_t is generated by a current i_p in windings that wrap around the toroidal vacuum chamber just as for the toroid of Fig. 31–15; one such winding is shown at the left in Fig. 48–9. (2) The so-called *poloidal* field component B_p is generated by a toroidal current i_t (see figure) that is induced in the plasma itself by an alternating current in the windings of a transformer-magnet arrangement not shown. This induced toroidal current is also used to heat the plasma to the desired high temperature. As noted earlier, additional heating may be provided by firing one or more beams of energetic neutral particles into the plasma.

Figure 48–10 shows a tokamak, called the Princeton Large Torus (PLT), housed in the Princeton Plasma Physics Laboratory. It is with this machine that a then-record plasma temperature 7.5×10^7 K (or 6.5 keV) and a then-record confinement time of 100 ms were achieved in the summer of 1978. As of 1980 a considerably larger tokamak is under construction in this same laboratory. Tokamak research is also proceeding apace in many other laboratories around the world.

The PLT

Figure 48–10 A tokamak called the Princeton Large Torus, housed in the Princeton Plasma Physics Laboratory. The vapor clouds are from Dewar vessels being filled with liquid helium for the vacuum pumping system. The circle indicates a human figure to show the scale. Courtesy of the Princeton University Plasma Physics Laboratory.

* See "The Tokamak Approach in Fusion Research," by Bruno Coppi and Jan Rem, *Scientific American*, July 1972.

The "tokamak tree"

Figure 48–11, a plot of Lawson number versus plasma temperature, shows the steady progress being made in tokamak performance at various laboratories both in the United States and elsewhere. The "tokamak tree" is branching upward toward the breakeven region, but it has not yet reached it!

Figure 48–11 Showing the progress made in tokamak research in various laboratories. The energy breakeven region is being approached but has not yet been reached. Note that both scales are logarithmic. The short horizontal arrow on the left represents the sun's central temperature. From "Recent Progress in Tokamak Research," by Masanori Murakami and Harold P. Eubank, *Physics Today*, May 1979.

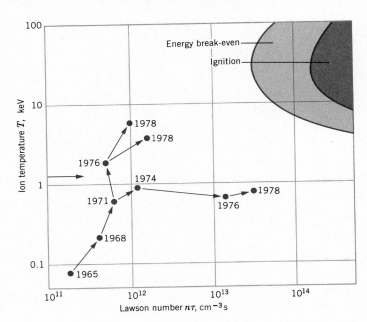

Example 5 The Princeton Large Torus (PLT) tokamak has achieved a confinement time of 0.10 s. (a) What must be the density of particles if Lawson's criterion is to be satisfied? (b) The Lawson number for this experiment was $\sim 10^{12}$ s/cm^3. By what factor must the particle density be increased to achieve breakeven performance?

(a) We must have

$$n = \frac{\sim 10^{14} \text{ s/cm}^3}{0.10 \text{ s}} \approx 10^{15} \text{ cm}^{-3}.$$

(b) The actual number density is

$$n' = \frac{\sim 10^{12} \text{ s/cm}^3}{0.10 \text{ s}} \approx 10^{13} \text{ cm}^{-3}.$$

Thus the particle density is ~ 100 times too small to achieve breakeven performance.

Inertial confinement

A second technique for confining plasma so that thermonuclear fusion can take place is called *inertial confinement*. In terms of Lawson's criterion (Eq. 48–4), it involves working with extremely high particle densities n for extremely short confinement times τ. These times are arranged to be so short that the fusion episode is over before the particles of the plasma have time to move appreciably from the positions they occupy at the onset of fusion. The interacting particles are confined by their own inertia. The fusion bomb operates on the same principle; after ignition the bomb explodes before the interacting particles have time to drift away.

Laser fusion

Laser fusion, which relies on the inertial confinement principle, is being investigated in many laboratories both in the United States and in other countries.* At

* See "Fusion Power by Laser Implosion," by John L. Emmett, John Nuckolls, and Lowell Wood, *Scientific American*, June 1974.

the Lawrence Livermore Laboratory, for example, in the SHIVA laser fusion arrangement (named for the many-armed Hindu god), liquid deuterium-tritium fuel pellets about 1 mm in diameter are to be "zapped" by 20 synchronized high-powered laser pulses, symmetrically arranged around the pellet. The laser pulses are designed to deliver in total some 200 kJ of energy to each fuel pellet in less than a nanosecond. This is a delivered power of $\sim 2 \times 10^{14}$ W during the pulse, which far exceeds the total electrical power generating capacity of the United States!

The laser pulse energy serves to heat the fuel pellet, ionizing it to a plasma and—it is hoped—raising its temperature to $\sim 10^8$ K. As the surface layers of the droplet evaporate at these high thermal speeds, the reaction force of the escaping particles compresses the core of the droplet, increasing its density by a factor of $\sim 10^4$. If all these things happened, then conditions would be right for thermonuclear fusion to occur in the core of the highly compressed droplet of plasma, the fusion reaction being

$$^2\text{H} + {}^3\text{H} \rightarrow {}^4\text{He} + n, \qquad Q = +17.59 \text{ MeV}. \qquad (48-6)$$

A projected laser fusion reactor

In an operating thermonuclear reactor of the laser fusion type, it is visualized that fuel pellets would be exploded, like miniature hydrogen bombs, at the rate of perhaps 10 to 100 per second. The energetic emerging particles of the fusion reaction ($^4\text{He}, n$) plus some of the reacting particles ($^2\text{H}, {}^3\text{H}$) escaping at thermal ($\sim 10^8$ K) energies might be absorbed in a "blanket" consisting of a moving stream of molten lithium, heating it up. Thermal energy would then be extracted from the lithium stream at another location and used to generated steam, just as in a fission reactor or a fossil-fuel power plant. Lithium would be a suitable choice for a heat-transfer medium because the energetic neutron of Eq. 48-6 would, with high probability, deliver up its energy to the "blanket" by the reaction

$$^6\text{Li} + n \rightarrow {}^4\text{He} + {}^3\text{H};$$

the two charged particles would be readily brought to rest in the lithium matrix. The feasibility of laser fusion as the basis of a fusion reactor has not been conclusively demonstrated as of 1980; the matter is being pursued with vigor in many laboratories.

Example 6 A 1.0-mg liquid deuterium-tritium (50% mixture) fuel pellet in a laser fusion device has a radius of 1 mm. Its density is increased by a factor of $\sim 10^4$ by the action of the laser pulses. (a) What is the density of the compressed fuel pellet? (b) How many particles per unit volume (either deuterons or tritons) does it contain? (c) At breakeven operation, what confinement time is required by Lawson's criterion?

(a) The density of the compressed pellet is

$$\rho = 10^4 \frac{m}{V} = \frac{10^4 m}{(\frac{4}{3})\pi r^3}$$

$$= \frac{(10^4)(1 \times 10^{-3} \text{ g})}{(4\pi/3)(1 \times 10^{-1} \text{ cm})^3} \approx 2400 \text{ g/cm}^3.$$

(b) Let N be the number of tritium nuclei, each of mass m_t and also the number of deuterium nuclei, each of mass m_d. If m is the mass of the pellet,

$$Nm_d + Nm_t = m.$$

Then, if A_t is the atomic weight of tritium and A_d that of deuterium,

$$n = \frac{2N}{V} = \frac{2m/(m_d + m_t)}{m/\rho} = \frac{2\rho}{m_d + m_t}$$

$$= \frac{2\rho}{(A_d/N_A) + (A_t/N_A)} = \frac{2\rho N_A}{A_d + A_t}$$

$$= \frac{2(2400 \text{ g/cm}^3)(6 \times 10^{23} \text{ particles/mol})}{2.0 \text{ g/mol} + 3.0 \text{ g/mol}}$$

$$\cong 6 \times 10^{26} \text{ cm}^{-3}.$$

(c) From the above we have, where L is the Lawson number,

$$\tau = \frac{L}{n} = \frac{\sim 10^{14} \text{ s·cm}^{-3}}{6 \times 10^{26} \text{ cm}^{-3}} \approx 2 \times 10^{-13} \text{ s}.$$

This time is short enough to permit inertial confinement.

REVIEW GUIDE AND SUMMARY

Energy from the nucleus

Table 48–1 suggests that, on a per unit mass basis, nuclear processes are about a million times more effective than chemical processes in transforming matter into energy.

The basic fission process

Equation 48–1 shows a generalized fission reaction that of ^{235}U by thermal neutrons. Equation 48–2 shows a particular example; see also Fig. 48–1. Table 48–2 shows the β^--decay chains of the primary fragments generated in the reaction of Eq. 48–2. Example 1 shows that $Q \approx 200$ MeV for such fission events.

A model for fission

Fission can be understood in terms of the model of a charged liquid drop carrying a certain excitation energy; see Fig. 48–3. A potential barrier (see Fig. 48–4) must be tunneled if fission is to occur. Table 48–3 shows that fissionability depends on the relationship between E_{act}, the activation energy for barrier tunneling, and B_n, the excitation energy.

Chain reactions and reactors

The fact that $\nu > 1$ in Eq. 48–1 makes a fission chain reaction possible. In a practical reactor the three problems raised starting on p. 984 must be satisfactorily solved. Figure 48–5 shows the neutron balance for a typical reactor. Figure 48–6 suggests the outlines of a complete nuclear power plant; see also Example 2.

Thermonuclear fusion and barrier tunneling

The release of energy by the fusion of two light nuclei is inhibited by their mutual Coulomb barrier; see Example 3. Fusion can occur in bulk matter only if the temperature is high enough (that is, if the particle energy is high enough) for appreciable barrier tunneling to occur. Figure 48–8 shows (1) the energy distribution $n(K)$ for protons at the central temperature of the sun, and (2) the proton–proton barrier penetrability factor $p(K)$. The figure deserves careful study.

The proton–proton cycle

Element building

The sun's energy arises mainly from the thermonuclear "burning" of hydrogen to form helium by the proton–proton cycle outlined in Table 48–4. The overall Q is 26.7 MeV per cycle; see Example 4. Elements up to $A = 56$ (the peak of the binding-energy curve of Fig. 47–6) can be built up by other thermonuclear processes once the hydrogen fuel supply of a star has been exhausted. Heavier elements are probably formed by successive neutron captures in supernovae explosions.

Controlled fusion

Controlled thermonuclear fusion for power generation has not yet been achieved, even on a laboratory scale. The d–d and the d–t reactions (p. 993) are contemplated. A successful fusion reactor must satisfy Lawson's criterion,

Lawson's criterion

$$n\tau \gtrsim 10^{14} \text{ s/cm}^3, \qquad [48\text{–}4]$$

and must have a plasma temperature T of $\sim 10^8$ K ($kT \approx 9$ keV). Confining such a hot plasma is a major problem. In the tokamak approach the plasma is generated as a torus and is confined by a helical magnetic field; see Fig. 48–9. Plasma heating is done by an induced toroidal current i_t. Figure 48–11 shows the progress made by tokamaks toward achieving breakeven performance.

The tokamak (magnetic field confinement)

Laser fusion (inertial confinement)

In laser fusion a deuterium-tritium fuel pellet is "zapped" with extremely intense laser pulses, compressing it so that the d–t fusion reaction (p. 993) occurs. The confinement time τ is very small so that the fusion episode is complete before the interacting particles can move appreciably, a principle called inertial confinement.

QUESTIONS

1. To which of the processes in Table 48–1 does the relationship $E = mc^2$ apply?

2. From Table 48–3 we can see that ^{238}U is not fissionable by thermal neutrons. What minimum neutron energy do you think would be necessary to induce fission in this nuclide?

3. The half-life for the decay of ^{235}U by α-emission is 7×10^8 y; by spontaneous fission, acting alone, it would be $\sim 3 \times 10^{17}$ y. Both are barrier tunneling processes, as Figs. 47–8 and 48–4

reveal. Why this enormous difference in barrier tunneling probability?

4. In Eq. 48–1, the generalized equation for the fission of ^{235}U by thermal neutrons, do you expect the Q of the reaction to depend on the identity of X and Y.

5. Is the fission fragment curve of Fig. 48–2 necessarily symmetrical about its central minimum?

6. In the chain decays of the primary fission fragments (see Table 48–2), why do no β^+-decays occur?

7. In connection with the stable isotone data displayed in Fig. 47–17 we saw evidence that pairs of nucleons (in the same state but with their angular momenta cancelling) exhibit a special stability. Can you discern any similar odd-even stability effects lurking in Table 48–3? (*Hint:* Write down the neutron numbers for the compound nuclides in the table.)

8. The half-life of ^{235}U is 7.0×10^8 y. Discuss the assertion that if it had turned out to be shorter by a factor of ten or so, there would not be any atomic bombs today.

9. Compare a nuclear reactor with a coal fire. In what sense does a chain reaction occur in each? What is the energy-releasing mechanism in each case?

10. Describe how to operate the control rods of a nuclear reactor (*a*) during initial startup, (*b*) to reduce the power level, and (*c*) on a long term basis, as fuel is consumed.

11. Not all neutrons produced in a reactor are destined to initiate a fission event. What happens to those which do not? See Fig. 48–5.

12. Why is a moderator needed in a fission reactor?

13. A reactor is operating at full power with its multiplication factor k adjusted to unity. If the reactor is now adjusted to operate stably at half power, what value must k now assume?

14. In Fig. 48–8, are you surprised that, as judged by the areas under the curve marked $n(K)$, the number of particles with $K > \bar{K}$ is smaller than the number with $K < \bar{K}$, where $\bar{K}$ is the average thermal energy?

15. The sun's energy is assumed to be generated by nuclear reactions such as the proton–proton cycle. What alternative ways of generating solar energy were proposed in the past, and why were they rejected?

16. If two protons, colliding head-on at the sun's center, succeed in tunneling through the Coulomb barrier that inhibits the collision, does that necessarily mean that the reaction (see Table 48–4)

$$^1H + {}^1H \rightarrow {}^2H + e^+ + \nu$$

will occur?

17. Can you think of any reason that the reaction referred to in the preceding question should have such a low probability of occurrence? (*Hint:* Think about the time required for β-decay processes to occur.)

18. Elements up to mass number 56 are thought to have been created by thermonuclear fusion in the cores of stars. Why stop here? Why are heavier elements not also created by this process?

19. The earth's core is thought to be made of iron because, during the formation of the earth, heavy elements such as iron would have sunk toward the earth's center and lighter elements, such as silicon, would have floated upward to form the earth's crust. However, iron is far from the heaviest element. Why isn't the earth's core made of uranium?

20. Heavy elements such as uranium are thought to have been created during the explosions of stars (supernova events). How, then, do such elements end up in the earth's crust?

21. Why does it take so long ($\sim 10^6$ y!) for γ-ray photons generated by nuclear reactions in the sun's central core to diffuse to the surface? What kinds of interactions do they have with the protons, α particles, and electrons that make up the core?

22. Does Lawson's criterion hold both for tokamaks and for laser fusion devices?

23. Which would generate more radioactive waste products, a fission reactor or a fusion reactor?

EXERCISES

Section 48–2   Nuclear Fission: The Basic Process

1. (*a*) How much energy is produced by the complete fissioning of 1.0 kg of ^{235}U? (*b*) For how long would this energy light a 100-W lamp? See Table 48–1.

2. Fill in the following table, which refers to the generalized fission reaction of Eq. 48–1.

X	Y	b
^{140}Xe	————	1
^{139}I	————	2
————	^{100}Zr	2
^{141}Cs	^{92}Rb	————

3. At what rate must ^{235}U nuclei undergo fission by thermal neutrons to generate 1.0 W? Assume that $Q = 200$ MeV.

4. ^{238}Np has an activation energy for fission of 4.2 MeV. To remove a neutron from this nuclide requires an energy expenditure of 5.0 MeV. Is ^{237}Np fissionable by thermal neutrons?

5. Very occasionally a ^{235}U nucleus, having absorbed a neutron, breaks up into *three* fragments. If two of these fragments are identified chemically as isotopes of chromium and gallium and if no prompt neutrons are involved, what is at least one possibility for the identity of the fragments? Consult a nuclidic chart or table.

Section 48–3   Nuclear Reactors: The Basic Principles

6. Neutrons in a reactor are in thermal equilibrium with the moderator at an effective temperature of 330 K. Find (*a*) the most probable energy and (*b*) the average energy. See Problem 32.

7. *The Breeder Reactor.* Many fear that helping additional nations develop nuclear power reactor technology will increase the likelihood of nuclear war because reactors can be used not only to produce energy but, as a by-product through neutron capture with inexpensive ^{238}U, to make ^{239}Pu, which is a "fuel" for nuclear bombs. What simple series of reactions involving neutron capture and β-decay would yield this plutonium isotope?

Section 48–4   Thermonuclear Fusion: The Basic Process

8. Calculate the height of the Coulomb barrier for the head-on collision of two protons. The effective radius of a proton may be taken to be 0.8 fm. See Example 3.

9. From information given in the text, collect and write down the approximate heights of the Coulomb barriers for (*a*) the α-

decay of ^{238}U, (b) the fission of ^{235}U by thermal neutrons, and (c) the head-on collision of two deuterons.

Section 48–5 Thermonuclear Fusion in the Sun and Other Stars

10. We have seen that Q for the overall proton-proton cycle is 26.7 MeV. How can you relate this number to the Q-values for the three reactions that make up this cycle, as displayed in Table 48–4?

11. The gravitational potential energy of a uniform sphere of radius r and mass m is

$$U = -\frac{3}{5}\frac{Gm^2}{r}.$$

(a) Use this to find the maximum energy that could be released by a uniform spherical mass, initially of infinite radius, in shrinking to the present size of the sun (use Appendix C for appropriate data). (b) Calculate the net energy the sun would radiate at the present rate (3.9×10^{26} W) for its lifetime (4.5×10^9 y). (c) What is the ratio of radiated to gravitational energy? (d) Does this calculation tend to verify the claim in Section 48–5? (e) Why is the answer to (c) not 500 as given in the text?

12. Show that the energy released when three α particles fuse to form ^{12}C is 7.3 MeV. Use $m(^4\text{He}) = 4.0026$ u.

13. *Helium burning in stars.* After much of a star's hydrogen has been fused into helium, its core will collapse and heat up through the release of gravitational energy. When the core temperature reaches 10^8K, the helium itself will begin to fuse. By the "triple alpha process" it will produce stable ^{12}C with a considerable re-lease of energy. (a) Except for possible γ rays produced, give the two reactions involved. (b) Why would there be no neutrinos involved as there are with hydrogen fusion? (c) Why would the reaction $^4\text{He} + {}^4\text{He} + {}^4\text{He} \rightarrow {}^{12}\text{C} + \gamma$ not be significant?

14. *Nucleosynthesis.* The process of fusing light nuclei into heavier and heavier nuclei with consequent releases of energy inside stars must ultimately stop. (a) Referring to Fig. 47–6, determine the region of nuclides where this happens. (b) Explain why. (c) One way to continue the buildup of heavy nuclei is through a net addition of energy for each reaction. This is thought to occur in heavy stars that suffer gigantic supernova explosions. Explain how such explosions could account for all the heavy nuclei.

Section 48–6 Controlled Thermonuclear Fusion

15. Methods other than heating the material have been suggested for overcoming the coulomb barrier for fusion. For example, one might consider using particle accelerators. If you were to use two of them to accelerate two beams of deuterons directly toward each other so as to collide "head-on," (a) what voltage would each require to overcome the coulomb barrier? (b) Would this voltage be difficult to achieve? (c) Why do you suppose this method is not presently used?

16. By the summer of 1985, the TFTR tokamac at the Princeton Plasma Physics Laboratory could be run consistently with plasma number densities of 3×10^{13}/cm^3, confinement times of 400 ms, and ion temperatures of 3 keV. Under special experimental conditions these same parameters are 6×10^{12}/cm^3, 100 ms, and 10 keV. Plot the two points representing these data on Fig. 48–11. Realizing that this represents just one machine while the figure represents many, what is your impression of progress being made in this type of fusion research?

PROBLEMS GROUPED BY SECTION

Section 48–2 Nuclear Fission: The Basic Process

1. In a particular fission event of ^{235}U by slow neutrons, it happens that no neutron is emitted ($b = 0$) and that one of the primary fission fragments is ^{83}Ge. (a) What is the other fragment? (b) How is the disintegration energy Q split between the two fragments? (c) If $Q = 170$ MeV, what is the velocity of each fragment?

2. ^{235}U decays by α-emission with a half-life of 7.0×10^8 y. It also decays (rarely) by spontaneous fission, and if the α-decay did not occur, its half-life due to this process alone would be $\sim 3 \times 10^{17}$ y. (a) At what rate do spontaneous fission decays occur in 1.0 g of ^{235}U? (b) How many α-decay events are there for every spontaneous fission event?

3. Consider the fission of ^{238}U by fast neutrons. In one fission event no neutrons were emitted and the final stable end products, after the β-decay of the primary fission fragments, were ^{140}Ce and ^{99}Ru. (a) How many β^--decay events were there in the two β-decay chains, considered together? (b) Calculate Q. The relevant atomic masses are

$$^{238}\text{U: } 238.05081 \text{ u}, \qquad ^{140}\text{Ce: } 139.90549 \text{ u},$$
$$n: \quad 1.00867 \text{ u}, \qquad ^{99}\text{Ru: } 98.90594 \text{ u}.$$

4. Calculate the disintegration energy Q for the fission of ^{52}Cr into two equal fragments. The needed masses are ^{52}Cr, 51.94051 u; and ^{26}Mg, 25.98260 u. Interpret your answer in terms of the binding-energy curve of Fig. 47–6.

5. Assume that just after the fission of ^{236}U* according to Eq. 48–2, the resulting ^{140}Xe and ^{94}Sr nuclei are just touching at their surfaces. (a) Assuming the nuclei to be spherical, calculate the coulomb potential energy (in MeV) of repulsion between the two fragments. (b) Compare this energy with the energy released in a typical fission process. In what form will this energy ultimately appear in the laboratory?

Section 48–3 Nuclear Reactors: The Basic Principles

6. *The neutron generation time t_{gen}* in a reactor is the average time for a fast neutron emitted in a fission event to be slowed down to thermal energies by the moderator and to initiate another fission event. If $t_{\text{gen}} = 1.0$ ms and if the control rods are adjusted so that $k = 1.0005$, how many neutron generations would it take for the power output of the reactor to double? How much actual time?

7. The neutron generation time t_{gen} (see Problem 6) in a particular reactor is 1.0 ms. If the reactor is operating at a power level of 500 MW, about how many free neutrons are present in the reactor at any moment?

8. *Radionuclide power sources* (see Problem 25). Among the many fission products that may be extracted chemically from the spent fuel of a nuclear power reactor is ^{90}Sr ($t_{1/2} = 29$ y). It is produced in typical large reactors (1000 MW, electric) at the rate of about 18 kg/y. By its radioactivity it generates thermal energy at the rate of 0.93 W/g. (a) What is the effective disintegration energy Q_{eff} associated with the decay of a ^{90}Sr nucleus? (Q_{eff} includes contributions from the decay of the ^{90}Sr daughter products in its decay chain but not from neutrinos, which escape totally from the sample.) (b) It is desired to construct a power source generating 150 W (electric) to use in operating electronic equipment in an underwater acoustic beacon. If the source is based on the thermal energy generated by ^{90}Sr and if the efficiency of the thermal-electric conversion process is 5.0%, how much ^{90}Sr is needed?

9. Verify that, as reported in Table 48–1, the fission of the ^{235}U in 1.0 kg of UO_2 (enriched so that ^{235}U is 3.0% of the total uranium) could keep a 100-W lamp burning for 680 y.

10. A 200-MW fission reactor consumes half its fuel in 3 years. How much ^{235}U did it contain initially? Assume that all the energy generated arises from the fission of ^{235}U and that this nuclide is consumed only by the fission process.

Section 48–4 Thermonuclear Fusion: The Basic Process

11. At the center of the sun the density is 1.5×10^5 kg/m^3 and the composition is essentially 35% hydrogen by mass and 65% helium. (a) What is the density of protons at the sun's center? (b) How much larger is this than the density of particles for an ideal gas at standard conditions of temperature and pressure?

12. The equation of the curve $n(K)$ in Fig. 48–8 is (see Problem 8 in Chapter 46)

$$n(K) = 1.13 n_0 \frac{K^{1/2}}{(kT)^{3/2}} e^{-K/kT},$$

where n_0 is the total density of particles. At the center of the sun the temperature is 1.5×10^7 K and the mean proton energy $\overline{K}$ is 1.94 keV. Find the ratio of the density of protons at 5.00 keV to that at the mean proton energy.

Section 48–5 Thermonuclear Fusion in the Sun and Other Stars

13. Verify the values of Q_1, Q_2, and Q_3 reported for the three reactions in Table 48–4. The needed atomic masses are

^{1}H: 1.007825 u,	^{3}He: 3.016029 u,
^{2}H: 2.014102 u,	^{4}He: 4.002603 u,
$e^\pm$: 0.0005486 u.	

(*Hint:* Distinguish carefully between atomic and nuclear masses, and take the positrons properly into account.)

14. Calculate and compare the energy released by (a) the fusion of 1.0 g of hydrogen deep within the sun and (b) the fission of 1.0 g of ^{235}U in a fission reactor.

15. The sun has a mass of 2.0×10^{30} kg and radiates energy at the rate of 3.9×10^{26} W. (a) At what rate does the sun transfer its mass into energy? (b) What fraction of its original mass has the sun lost in this way since it began to burn hydrogen, about 4.5×10^9 y ago?

16. Coal, assumed to be all carbon, burns according to

$$C + O_2 \rightarrow CO_2.$$

The heat of combustion is 3.3×10^7 J/kg of atomic carbon consumed. (a) Express this in terms of energy per carbon atom. (b) Express it in terms of energy per unit mass of the initial reactants, carbon and oxygen. (c) Suppose that the sun (mass $= 2.0 \times 10^{30}$ kg) were made of carbon and oxygen in combustible proportions and that it continued to radiate energy at its present rate of 3.9×10^{26} W. How long would it last?

17. In certain stars the so-called *carbon cycle* is more likely than the proton-proton cycle to be effective in generating energy. This cycle is

$^{12}C + {}^1H \rightarrow {}^{13}N + \gamma,$	$Q_1 = 1.95$ MeV,
$^{13}N \rightarrow {}^{13}C + \beta^+ + v,$	$Q_2 = 1.19,$
$^{13}C + {}^1H \rightarrow {}^{14}N + \gamma,$	$Q_3 = 7.55,$
$^{14}N + {}^1H \rightarrow {}^{15}O + \gamma,$	$Q_4 = 7.30,$
$^{15}O \rightarrow {}^{15}N + \beta^+ + v,$	$Q_5 = 1.73,$
$^{15}N + {}^1H \rightarrow {}^{12}C + {}^4He,$	$Q_6 = 4.97.$

(a) Show that this cycle of reactions is exactly equivalent in its overall effects to the proton-proton cycle of Table 48–4. (b) Verify that both cycles, as expected, have the same Q.

Section 48–6 Controlled Thermonuclear Fusion

18. Verify the Q values reported on p. 993 for the two (d, d) reactions and the (d, t) reaction. The needed masses are

^{1}H: 1.007825 u,	^{3}He: 3.016029 u,
^{2}H: 2.014102 u,	^{4}He: 4.002603 u,
^{3}H: 3.016049 u,	n: 1.008665 u.

19. Suppose that the compressed fuel pellet of Example 6 "burns" with an efficiency of 10%. That is, only 10% of the deuterons and 10% of the tritons participate in the fusion reaction of Eq. 48–6. (a) How much energy is released in each such microexplosion of a pellet? (b) To how much TNT is each such 1.0-mg pellet equivalent? The heat of combustion of TNT is 500 kcal/lb. (c) If a fusion reactor is constructed on the basis of 100 microexplosions per second, what power would be generated? (Note that part of this power must be used to operate the lasers.)

20. Ordinary water consists of roughly 0.015% of "heavy water," in which the two hydrogens are replaced with deuterium, ^{2}H. How much fusion power could be extracted if we "burned" all of the ^{2}H in 1 liter of water in one day through the reaction $^2H + {}^2H \rightarrow {}^3He + n$?

ADDITIONAL PROBLEMS

21. Calculate the disintegration energy Q for the fission of ^{98}Mo into two equal parts. The needed masses are ^{98}Mo, 97.90541 u; and ^{49}Sc, 48.95003 u. If Q turns out to be positive, discuss why this process does not occur spontaneously.

22. Verify that, as reported in Table 48–1, the fusion of 1.0 kg of

deuterium by the reaction

$$^2H + {}^2H \rightarrow {}^3He + n, \qquad Q = +3.27 \text{ MeV},$$

could keep a 100-W lamp burning for 2.5×10^4 y.

23. Estimate the Coulomb barrier height for two ^{7}Li nuclei, fired at each other with the same initial kinetic energy K. See Example 3.

24. Let us assume that the core of the sun includes half its mass and is compressed within a sphere whose radius is one-fourth of the solar radius. We assume further that the composition of the core is 35% hydrogen by mass and that essentially all of the sun's energy is generated there. If the sun continues to burn hydrogen at the rate calculated in Example 4, how long will it be before the hydrogen is entirely consumed? The sun's mass is 2.0×10^{30} kg.

25. *Radionuclide power sources.* The thermal energy generated when radiations from radionuclides are absorbed in matter can be used as the basis for a small power source for use in satellites, remote weather stations, and so on. Such radionuclides are manufactured in abundance in nuclear power reactors and may be separated chemically from the spent fuel. One suitable radionuclide is ^{238}Pu ($t_{1/2} = 87.7$ y), which is an α-emitter with $Q = 5.50$ MeV. At what rate per kilogram is thermal energy generated in this material?

26. A ^{236}U nucleus undergoes fission and breaks up into two middle-mass fragments, ^{140}Xe and ^{96}Sr. (*a*) By what percentage does the surface area of the ^{236}U nucleus change during this process? (*b*) By what percentage does its volume change? (*c*) By what percentage does its electrostatic potential energy change? The potential energy of a uniformly charged sphere of radius r is given by

$$U = \frac{3}{5}\left(\frac{Q^2}{4\pi\epsilon_0 r}\right).$$

27. In the deuteron-triton fusion reaction of Eq. 48–6, how is the reaction energy Q shared between the α particle and the neutron? Neglect the relatively small kinetic energies of the two combining particles.

28. The effective Q for the proton-proton cycle of Table 48–4 is 26.2 MeV. (*a*) Express this as energy per kilogram of hydrogen consumed. (*b*) The luminosity of the sun is 3.9×10^{26} W. If its energy derives from the proton-proton cycle, at what rate is it losing hydrogen? (*c*) At what rate is it losing mass? Account for the difference in the results for (*b*) and (*c*). (*d*) The sun's mass is 2.0×10^{30} kg. If it loses mass at the constant rate calculated in (*c*), how long will it take before it loses 0.10% of its mass?

29. Suppose that we had a quantity of N deuterons (^{2}H nuclei). (*a*) Which of the following procedures for fusing these N nuclei releases more energy? (*i*) $N/2$ fusion reactions of the type ^{2}H $+$ ^{2}H $\rightarrow$ ^{3}H $+ p$, or (*ii*) $N/3$ fusion reactions of the type ^{2}H $+ ^3$H $\rightarrow$ ^{4}He $+ n$, using $N/3$ nuclei of ^{3}H that are first made in $N/3$ reactions of type (*i*). (*b*) List the ultimate product nuclei resulting from the two procedures and the quantity of each.

30. Assume that a plasma temperature of 1×10^8 K is reached in a laser fusion device. (*a*) What is the most probable speed of a deuteron at this temperature? (See Example 10 of Chapter 21.) (*b*) How far would such a deuteron move in the confinement time calculated in Example 6? (*c*) What is the radius of the compressed fuel pellet in Example 6, and how does this compare with the result calculated in part (*b*)?

*31. *Energy loss in a moderator.* (*a*) A neutron with initial kinetic energy K makes a head-on elastic collision with a resting atom of mass m. Show that the fractional energy loss of the neutron is given by

$$\frac{\Delta K}{K} = \frac{4m_n m}{(m + m_n)^2},$$

in which m_n is the neutron mass. (*b*) Find $\Delta K/K$ if the resting atom is hydrogen ($m = m_n$), deuterium ($m = 2m_n$), carbon ($m = 12m_n$), or lead ($m = 208m_n$). (*c*) If $K = 1.00$ MeV initially, how many such collisions would it take to reduce the neutron energy to thermal values (0.025 eV) if the material is deuterium, a commonly used moderator?

*32. Expressions for the Maxwell speed and energy distributions for the molecules in a gas are given in Problem 8 of Chapter 46. (*a*) Show that the *most probable energy* is given by

$$K_p = \tfrac{1}{2}kT.$$

Verify this result from the energy distribution curve of Fig. 48–8, for which $T = 1.5 \times 10^7$ K. (*b*) Show that the *most probable speed* is given by

$$v_p = \left(\frac{2kT}{m}\right)^{1/2}.$$

Find its value for protons at $T = 1.5 \times 10^7$ K. (*c*) Show that *the energy corresponding to the most probable speed* (which is not the same as the most probable energy) is

$$K_{p}, K_{v,\,p} = kT.$$

Locate this quantity on the energy-distribution curve of Fig. 48–8.

*33. In addition to ^{238}U, uranium mined today contains 0.72% of fissionable ^{235}U, too little to make reactor fuel for slow neutrons. For this reason, the natural uranium must be enriched or concentrated in ^{235}U. Both ^{235}U ($t_{1/2} = 7.0 \times 10^8$ y) and ^{238}U ($t_{1/2} = 4.5 \times 10^9$ y) are radioactive. How far back in time would natural uranium have been a practical reactor fuel, with a ^{235}U/^{238}U ratio of 3%?

Supplementary Topic
Special Relativity—A Summary of Conclusions*

1 Introduction

Here we simply display in one place some conclusions drawn from the special theory of relativity (hereafter, SR) proposed by Einstein in 1905. We omit all proofs and make only a modest attempt to make the conclusions acceptable in terms of "common sense." We give examples of the application of the theory and, at the end, a set of questions and problems for discussion and practice.

2 The Postulates (RR, Section 1.9)

Einstein based his theory on two postulates, and *all* of the conclusions of SR derive from them.

The First Postulate. From the time of Galileo it was known that the laws of mechanics were the same in all inertial frames (see Fig. 1). This means that all inertial observers having relative motion, even though they may measure different values for the velocities, momenta, and so on, of the particles involved in a given experiment (a game of pool, perhaps), would nevertheless agree on the laws of mechanics involved (conservation of linear momentum, etc.) and on the outcome of the experiment (who won).

Einstein took the bold step of extending this invariance principle to *all* of physics, not only mechanics, including especially electromagnetism. Einstein's first postulate is

The laws of physics are the same in all inertial frames. No preferred inertial frame exists.

---

* For a fuller treatment, see *Introduction to Special Relativity*, by Robert Resnick, John Wiley & Sons, Inc., New York, 1968. References to this work will be in the style RR, p. 187; RR, Section 1.9, etc.

The Second Postulate.   In pre-SR days a vexing question was this: Given that the speed of light c is 2.998×10^8 m/s, with respect to what is this speed measured? For sound waves in air the answer is simple — it is with respect to the medium (air) through which the sound wave travels. Light, however, travels through a vacuum. Even so, is there a tenuous medium (the luminiferous, or light-carrying, ether) that plays the same role for light that air does for sound? If such an ether exists, can we detect it? Alternatively, should c be measured with respect to the source that emits the light?

All attempts to make experimental verifications along these lines failed completely (see RR, Sections 1.5 through 1.8). Einstein made a second bold postulate.

The speed of light is the same in all inertial frames.

Note that no ether is needed or involved. This second postulate means, for example, that if you consider three light sources, (1) one at rest with respect to you, (2) one moving toward you at speed $0.9c$, say, and (3) one moving away from you at speed $0.9c$, you would measure the *same* speed of light from all three sources.

This second postulate has been tested directly (see RR, p. 34) using as a moving "light" source neutral pions generated in a proton synchrotron at speeds of $0.99975c$. These elementary particles disintegrate by emitting γ-rays which, like light, are electromagnetic in character and travel with the same speed. The speed measured for the radiation emitted by these fast moving sources was, within experimental error, just c, as Einstein's second postulate predicts.

3 Special Relativity and Newtonian Mechanics (RR, Section 2.8)

Many of the conclusions of SR simply don't seem reasonable on the basis of everyday experience. Even Einstein's second postulate seems to violate common sense. If you catch a pitched baseball thrown by a pitcher (1) at rest with respect to you, (2) moving toward you (in a pick-up truck, say) at 30 mi/h, and (3) moving away from you at this same speed, you expect a different baseball speed in each case with respect to you. But if you extend this experience to a source (the pitcher) emitting light (photons), you would contradict Einstein's second postulate. And yet experiment shows that light does have the same speed in each case, in support of Einstein's postulate.

The solution to this dilemma comes about when we realize that the basis of our "common sense" experience is very limited indeed. It is restricted to speeds v such that $v \ll c$, where c is the speed of light. For example, the speed of a satellite in earth orbit may be about 8000 m/s, which seems fast to us, but in terms of the speed of light (3.0×10^8 m/s), it is only $0.000027c$. We simply have no personal experience in regions of high relative velocity.

As an example, to accelerate an average person (to say nothing of a spaceship) to $0.90c$ would require no less than 150 percent of the 1976 total electrical energy consumption in the United States. However, the particles of physics (electrons, mesons, protons, etc.) can readily be accelerated to high speeds. Electrons emerging from the 2-mi-long linear accelerator at Stanford University can have speeds as high as $0.9999999997c$, for example, corresponding to an electron energy of 22 GeV. In the arena of particle physics, SR is absolutely necessary for the solution of mechanical problems.

It turns out that in nature there is a certain finite speed that cannot be exceeded and that we call the limiting speed. This limiting speed is the speed of light, c, the greatest speed with which signals can be transmitted. Classical physics assumes that signals can be transmitted with infinite speed, but nature contradicts that assumption, and it really does seem fanciful that such a signal could exist. Experiment confirms c as the limiting speed, so that in a sense the speed of light plays the role in relativity that infinity does in classical physics. It is then not difficult to understand—in fact, it becomes very plausible—that the finite speed of the source of light cannot affect the measured value of the speed of an emitted signal already having the limiting value.

The world in which we live and develop our sense perceptions is a world of Newtonian mechanics, in which $v \ll c$. Newtonian mechanics is revealed as a special case of SR for the limit of low speeds. Indeed, a test of SR is to allow $c \to \infty$ (in which case $v \ll c$ always holds true) and see that the corresponding formulas of Newtonian mechanics emerge.

Newtonian mechanics, although a special case, is an all-important one. It describes the essential motions of our solar system, the tides, our space ventures, the behavior of baseballs and pinball machines, and so on. It works beautifully in the vastly important realm $v \ll c$. But it breaks down at speeds approaching that of light.

Few theories have been subject to more rigorous experimental tests than SR. Not the least among them is the fact that particle accelerators work. They are designed using SR at the level of engineering and technology. An accelerator designed on the basis of Newtonian mechanics simply would not work. Nuclear reactors and, alas, nuclear bombs, confirm that SR really works.

Einstein once said that no number of experiments could prove him right but a single experiment could prove him wrong. To date this single experiment has not been found.

4 The Transformation Equations (RR, Section 2.2)

The basic observation made in SR (or in Newtonian mechanics for that matter) is this. Consider observers to be in different inertial frames, S and S' (Fig. 1). The corresponding axes of S and S' are parallel, the x-x' axes being common, and S' moves to the right with speed v as seen by S; the two origins coincide at $t = t' = 0$. Each observer, S and S', records the same event, which might be the detonation of a tiny flashbulb, and assigns space and time coordinates to the event, namely, x, y, z, t and x', y', z', t'. What are the relations between these two sets of numbers written down in the observers' notebooks?

Before SR the accepted relations were

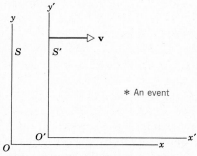

Figure 1 Two inertial frames with parallel axes, the $x - x'$ axis being common. S' moves to the right with speed v as seen by S. At $t = t' = O$ the two origins, O and O', coincide.

$$x' = x - vt, \qquad y' = y,$$
$$t' = t, \qquad z' = z, \tag{1}$$

called the *Galilean transformation equations* (RR, Section 1.2). Though impressively correct in the important region $v \ll c$, they nevertheless fail as $v \to c$.

The corresponding equations used in SR, called the *Lorentz transformation equations*, are (RR, Table 2-1)

$$x' = \frac{x - vt}{\sqrt{1 - (v/c)^2}}, \qquad y' = y,$$
$$t' = \frac{t - (v/c^2)x}{\sqrt{1 - (v/c)^2}}, \qquad z' = z. \tag{2}$$

Note certain things about these equations. (1) Space and time coordinates are thoroughly intertwined. In particular, time is not the same for each observer; t' depends on x as well as on t. (2) If we let $c \to \infty$, the Lorentz equations reduce to the Galilean equations, as promised! Finally, (3) We must have $v < c$ or else the quantities x' and t' become infinite ($v = c$) or imaginary ($v > c$). The speed of light is an upper limit for the speeds of material objects.

The Lorentz equations, like everything else in SR, can be derived from Einstein's two postulates (RR, Section 2.2).

5   Time Dilation and Length Contraction (RR, Sections 2.3 and 2.4)

Let S' observe two events that occur at the same place in this reference frame. They might be two successive positions of the hand of a clock located at a fixed position, x'. Let S' measure the time interval $\Delta t'$ between these events. S, for whom the clock appears to be moving, observes the same two events and measures a different time interval Δt, which is given by

$$\Delta t = \frac{\Delta t'}{\sqrt{1 - (v/c)^2}}. \tag{3}$$

This fact, that $\Delta t > \Delta t'$, is called *time dilation,* and we often verbalize it as "moving clocks run slow." Observer S records a longer time interval than shown to have transpired on the moving clock.

Equation 3 has been tested experimentally and found to be correct. In one test the "moving clocks" were fast particles called pions ($\pi^{\pm}$). Pions are radioactive, and their rate of radioactive decay is a measure of their time-keeping ability. See Example 2.

Now let us consider a rod, parallel to the x-x' axes, to be at rest in the S' frame. S' will measure a length $\Delta x'$ for it. S, however, measuring the same rod, which is moving with respect to him, would find a length Δx, which is given by

$$\Delta x = \sqrt{1 - (v/c)^2}\, \Delta x'. \tag{4}$$

This fact, that $\Delta x < \Delta x'$, is called *length contraction.*

The length contraction has been verified in the design of, say, the linear electron accelerator at Stanford University. At an exit speed of $v = 0.99999998c$, each meter of the accelerating tube seems like 0.20 mm to an observer moving with the electron. If these length-contraction considerations had not been taken into account, the machine simply would not work.

The simplest way to understand these results—the time dilation and the length contraction—is to note that one observer, S', is at rest with respect to what he or she is measuring (clock or rod), whereas for the other observer, S, the objects are in motion. Relativity therefore asserts that *motion affects measurement.* If we had interchanged frames, letting the clock and rod be at rest in S, for example, we would have found the observers again disagreeing on the measured values, but now we would have $\Delta x' < \Delta x$ and $\Delta t' > \Delta t$. So the results are reciprocal, neither observer being "absolutely" right or wrong.

What both observers *will* agree on however, is the *rest length* (also called *proper length*) of a given rod (they will both measure the rod to have the same length when the rod is at rest with respect to their measuring instruments) and the *proper time interval* of a given clock (they will both measure the successive positions of the hand of the clock to have taken the same elapsed time when the clock is at rest with respect to their measuring instruments).

That motion should affect measurement is not a strange idea, even in classical physics. For example, the measured frequency of sound or of light depends on the motion of the source with respect to the observer; we call this the Doppler effect and everyone is familiar with it. And in mechanics the measured values of the speed, the momentum, or kinetic energy of moving particles are different for observers on the ground than those on a moving train. However, in classical physics measurements of space intervals and time intervals *are* absolute, whereas in SR such measurements are relative to the observer. Not only does experiment contradict classical physics, but only by adopting the relativity of space and time do we arrive at the invariance (the absoluteness) of all of the laws of physics for all observers. Surely, giving up the absoluteness of the laws of physics (would they then be laws?), as classical notions of time and length require, would leave us with an arbitrary and complex world. By comparison, relativity is absolute and simple.

Example 1 The ratio v/c occurs often, so it simplifies the notation to let $v/c = \beta$. We see that the factor $\sqrt{1 - \beta^2}$ occurs in Eq. 4 and the factor $\gamma = 1/\sqrt{1 - \beta^2}$ in Eq. 3. Because they arise frequently in relativity, it is helpful to be able to estimate their values as a function of β. Compute $\sqrt{1 - \beta^2}$ and $\gamma = 1/\sqrt{1 - \beta^2}$ for $\beta = v/c = 0.100,\ 0.300,\ 0.600,\ 0.800,$ 0.900, 0.950, 990, and 0.999, and plot them as functions of β. We find that:

$\beta =$	0.100	0.300	0.600	0.800	0.900	0.950	0.990	0.999
$\sqrt{1 - \beta^2} =$	0.995	0.954	0.800	0.600	0.436	0.312	0.141	0.045
$\gamma = 1/\sqrt{1 - \beta^2} =$	1.005	1.048	1.250	1.667	2.294	3.205	7.092	22.37

These factors are plotted as a function of β in Fig. 2.

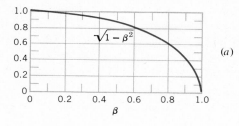

(a)

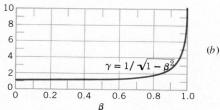

(b)

Figure 2 (a) A plot of $\sqrt{1 - \beta^2}$ as a function of β. (b) A plot of $\gamma = 1/\sqrt{1 - \beta^2}$ as a function of β.

Example 2 Among the particles of high-energy physics are charged pions, particles of mass between that of the electron and the proton and of positive or negative electronic charge. They can be produced by bombarding a suitable target in an accelerator with high-energy protons, the pions leaving the target with speeds close to that of light. It is found that the pions are radioactive and, when they are brought to rest, their half-life is measured to be 1.77×10^{-8} s. That is, half of the number present at any time have decayed 1.77×10^{-8} s later. A colli-

mated pion beam, leaving the accelerator target at a speed of $0.991c$, is found to drop to half its original intensity 39.2 m from the target.

(a) Are these results consistent?

If we take the half-life to be 1.77×10^{-8} s and the speed to be 2.97×10^8 m/s ($= 0.991c$), the distance traveled over which half the pions in the beam should decay is

$$d = vt = (2.97 \times 10^8 \text{ m/s})(1.77 \times 10^{-8} \text{ s}) = 5.3 \text{ m}.$$

This appears to contradict the direct measurement of 39.2 m.

(b) Show how the time dilation accounts for the measurements.

If the relativistic effects did not exist, then the half-life would be measured to be the same for pions at rest and pions in motion (an assumption we made in part a). In relativity, however, the nonproper and proper half-lives are related by

$$\Delta t = \frac{\Delta \tau}{\sqrt{1 - \beta^2}}.$$

The proper time $\Delta \tau$ in this case is 1.77×10^{-8} s, the time interval measured by a clock attached to the pion, that is, at one place in the rest frame of the pion. In the laboratory frame, however, the pions are moving at high speeds, and the time interval there (a nonproper one) will be measured to be larger (moving clocks appear to run slow). The nonproper half-life, Δt, measured by two different clocks in the laboratory frame, would then be

$$\Delta t = \frac{1.77 \times 10^{-8} \text{ s}}{\sqrt{1 - (0.991)^2}} = 1.32 \times 10^{-7} \text{ s}.$$

This is the half-life appropriate to the laboratory reference frame. Pions that live this long, traveling at a speed $0.991c$, would cover a distance

$$d = 0.991c \times \Delta t$$
$$= (2.97 \times 10^{-8} \text{ m/s})(1.32 \times 10^{-7} \text{ s})$$
$$= 39.2 \text{ m}$$

exactly as measured in the laboratory.

(*c*) Show how the length contraction accounts for the measurements.

In part *a* we used a length measurement (39.2 m) appropriate to the laboratory frame and a time measurement (1.77×10^{-8} s) appropriate to the pion frame and incorrectly combined them. In part *b* we used the length (39.2 m) and time (1.3×10^{-7} s) measurements appropriate to the laboratory frame. Here we use length and time measurements appropriate to the pion frame.

We already known the half-life in the pion frame, that is, the proper time 1.77×10^{-8} s. What is the distance covered by the pion beam during which its intensity falls to half its original value? If we were sitting on the pion, the laboratory distance of 39.2 m would appear much shorter to us because the laboratory moves at a speed 0.991c relative to us (the pion). In fact, we would measure the distance

$$d' = d \sqrt{1 - \beta^2} = 39.2 \sqrt{1 - (0.991)^2} \text{ m}.$$

The time elapsed in covering this distance is $d'/0.991c$ or

$$\Delta \tau = \frac{39.2 \text{ m} \sqrt{1 - (0.991)^2}}{0.991c} = 1.77 \times 10^{-8} \text{ s},$$

exactly the measured half-life in the pion frame.

Thus, depending on which frame we choose to make measurements in, this example illustrates the physical reality of either the time-dilation or the length-contraction predictions of relativity. Each pion carries its own clock, which determines the proper time τ of decay, but the decay time observed by a laboratory observer is much greater. Or, expressed equivalently, the moving pion sees the laboratory distances contracted and in its proper decay time can cover laboratory distances greater than those measured in its own frame.

Notice that in this region of $v \approx c$, the relativistic effects are large. There can be no doubt whether in our example the distance is 39.2 m or 5.3 m. If the proper time were applicable to the laboratory frame, the time (1.3×10^{-7} s) to travel 39.2 m would correspond to more than seven half-lives (that is, $(1.3 \times 10^{-7} \text{ s})/(1.8 \times 10^{-8} \text{ s}) \approx 7$). Instead of the beam being reduced to half its original intensity, it would be reduced to $(\frac{1}{2})^7$ or $\frac{1}{128}$ its original intensity in traveling 39.2 m. Such differences are very easily detectable.

This example is by no means an isolated result. All the kinematic (and dynamic) measurements in high-energy physics are consistent with the time-dilation and length-contraction results. The experiments and the accelerators themselves are designed to take relativistic effects into account. Indeed, relativity is a routine part of the everyday world of high-energy physics and engineering.

6 Relativistic Addition of Velocities and the Doppler Effect (Sections 38–8 and 38–9; RR, Sections 2.6 and 2.7)

Let *S* observe a particle moving with speed *u'* parallel to the *x'*-axis. What speed *u* would *S* measure? From the Galilean transformation equations (Eq. 1) we can easily show that

$$u = u' + v. \tag{5}$$

This relation, which seems to most of us to be intuitively obvious, is, alas, not true (except for the very important special case of $v \ll c$). The Lorentz transformation equations lead us to

$$u = \frac{u' + v}{1 + (u'v/c^2)}. \tag{6}$$

As we expect, for $c \to \infty$, Eq. 6 reduces to Eq. 5. You can prove that if $u' < c$ and $v < c$, then it must always be true that $u < c$ (see Problem 9). There is no way to generate speeds $\geq c$ by compounding velocities.

Using the relativistic velocity addition result (Eq. 6), we can deduce the Doppler effect for light. In relativity theory there is no difference between the two cases, which are different in classical theory (namely, source at rest—observer moving and observer at rest—source moving); only the relative motion *v* of source and observer counts. This fact and the result

$$\nu = \nu' \sqrt{\frac{c \pm v}{c \mp v}} \tag{7}$$

are in agreement with experiment. Here, ν' is the frequency of the source at rest in S' and ν is the frequency observed in frame S with respect to which the source moves at speed v; the upper signs refer to source and observer moving *toward* one another and the lower signs refer to source and observer moving *away* from one another. Equation 7 is called the *longitudinal* Doppler effect, and v refers to the relative velocity of source and observer along the line connecting them.

There is in relativity, however, an effect not predicted by classical physics, namely, a *transverse* Doppler effect; that is, if the relative velocity **v** is at right angles to the line connecting source and observer, we find that

$$\nu = \nu' \sqrt{1 - \frac{v^2}{c^2}}. \tag{8}$$

This result, confirmed by experiment, can be interpreted simply as a time dilation, moving clocks appearing to run slow.

Example 3 Two electrons are ejected in opposite directions from radioactive atoms in a sample of radioactive material at rest in the laboratory. Each electron has a speed $0.67c$ as measured by a laboratory observer. What is the speed of one electron as measured from the other, according to the classical velocity addition theorem?

Here, we may regard one electron as the S-frame, the laboratory as the S'-frame, and the other electron as the object whose speed in the S-frame is sought (see Fig. 3). In the S'-frame, the other electron's speed is $0.67c$, moving in the positive x'-direction, say, and the speed of the S-frame (one electron) is $0.67c$, moving in the negative x'-direction. Thus, $u'_x = +0.67c$ and $v = +0.67c$, so that the other electron's speed with respect to the S-frame is

$$u_x = u'_x + v = +0.67c + 0.67c = +1.34c,$$

according to the classical velocity addition theorem, Eq. 5.

Example 4 Repeat the previous example, to find the speed of one electron relative to the other, using the more general relativistic equation. Again, we may regard one electron as the S-frame, the sample as the S'-frame, and the other electron as the object whose speed in the S-frame we seek. Then

$$u' = 0.67c \qquad \text{and} \qquad v = 0.67c.$$

Because u' and v are *not* very small compared to c, we must use Eq. 6, obtaining

$$u = \frac{u' + v}{1 + u'v/c^2} \cdot \frac{(0.67 + 0.67)c}{1 + (0.67)^2} = \frac{1.34}{1.45} c = 0.92c.$$

Hence, the speed of one electron relative to the other is less than c, according to the relativistic velocity addition formula, Eq. 6. This is consistent with the role played by c as a limiting speed and with experimental measurements.

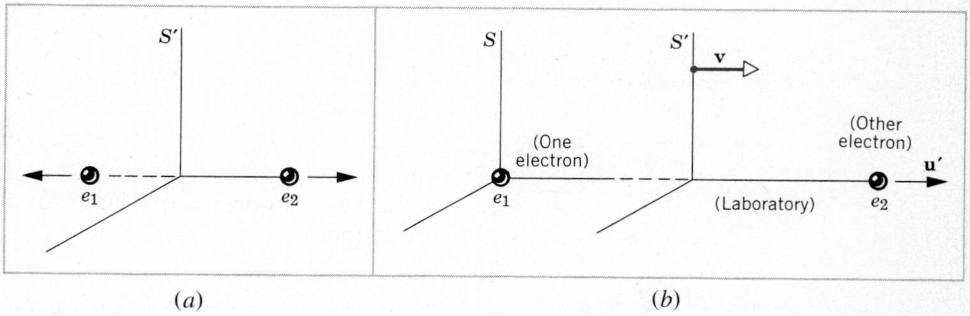

Figure 3 (*a*) In the laboratory frame, the electrons are observed to move in opposite directions at the same speed. (*b*) In the rest frame, S, of one electron, the laboratory moves at a velocity **v**. In the laboratory frame, S', the second electron has a velocity denoted by **u'**. What is the velocity of this second electron as seen by the first?

(*a*)   (*b*)

7 Mass, Momentum, and Kinetic Energy (Sections 8–9 and 9–4; RR, Sections 3.3 and 3.5)

We have seen that time and length measurements are functions of velocity v. Should mass be any different? SR tells us that the *relativistic mass* m of a particle moving at speed v with respect to the observer is

$$m = \frac{m_0}{\sqrt{1 - (v/c)^2}}, \tag{9a}$$

in which m_0 is the rest mass, that is, the mass measured when the particle is at rest ($v = 0$) with respect to the observer.

It is m and not m_0 that must be taken into account when designing magnets to bend charged particles in arcs of circles. By these techniques, Eq. 9 has been thoroughly tested. Incidentally, the ratio m/m_0 for electrons emerging from the Stanford University linear accelerator at $K = 20$ GeV is the order of 40,000.

To preserve the law of conservation of linear momentum in SR, we redefine the momentum of a particle of rest mass m_0 and speed v as

$$p = mv = \frac{m_0 v}{\sqrt{1 - (v/c)^2}}. \qquad (9b)$$

We made use of this expression for the recoil momentum of an electron in the derivation of the Compton effect formulas in Chapter 42.

As a result of the considerations above, in SR the kinetic energy of a particle is no longer given by $\frac{1}{2}m_0 v^2$ but by

$$K = mc^2 - m_0 c^2$$

$$= m_0 c^2 \left(\frac{1}{\sqrt{1 - (v/c)^2}} - 1 \right). \qquad (10)$$

You can show that $K \to \frac{1}{2}m_0 v^2$ as $c \to \infty$. (see Problem 20).

Example 5 For what value of v/c ($= \beta$) will the relativistic mass m of a particle exceed its rest mass m_0 by a given fraction f?

From Eq. 9a we have

$$f = \frac{m - m_0}{m_0} = \frac{m}{m_0} - 1 = \frac{1}{\sqrt{1 - \beta^2}} - 1,$$

which, solved for β, is

$$\beta = \frac{\sqrt{f(2 + f)}}{1 + f}.$$

The following table shows some computed values, which hold for all particles regardless of their rest mass.

f		β
0.001	(0.1%)	0.045
0.01		0.14
0.1		0.42
1	(100%)	0.87
10		0.996
100		0.99995
1000		0.99999950

Example 6 The kinetic energy acquired by a particle of charge q starting from rest in a uniform electric field when it falls through an electrostatic potential difference of V_0 volts is $K = qV_0$. Assume the particle to be an electron and the potential difference to be 10^4 V. Find the kinetic energy of the electron, its speed, and its mass at the end of the acceleration.

The charge on the electron is $e = -1.602 \times 10^{-19}$ C. The potential difference is now a rise, $V_i - V_f = -10^4$ V, a nega-

tive charge accelerating in a direction opposite to the electric field. Hence, the kinetic energy acquired is

$$K = qV_0 = (-1.602 \times 10^{-19} \text{ C})(-10^4 \text{ V}) = 1.602 \times 10^{-15} \text{ J.}$$

From Eq. 10, $K = mc^2 - m_0 c^2$, we obtain

$$\frac{K}{c^2} = (m - m_0)$$

or

$$m - m_0 = (1.602 \times 10^{-15} \text{ J})(3.00 \times 10^8 \text{ m/s})^2$$
$$= 1.78 \times 10^{-32} \text{ kg}$$

and, with $m_0 = 9.109 \times 10^{-31}$ kg, we find the mass of the moving electron to be

$$m = (9.109 + 0.178) \times 10^{-31} \text{ kg} = 9.287 \times 10^{-31} \text{ kg.}$$

Notice that $m/m_0 = 1.02$, so that the mass increase due to the motion is about 2% of the rest mass.

From Eq. 9a, $m = m_0/\sqrt{1 - v^2/c^2}$, we have

$$\frac{v^2}{c^2} = 1 - \left(\frac{m_0}{m} \right)^2 = 1 - \left(\frac{9.109}{9.287} \right)^2 = 0.0038$$

or

$$v = 0.195c = 5.85 \times 10^7 \text{ m/s.}$$

The electron acquires a speed of about one-fifth the speed of light.

These are the relativistic predictions. They are confirmed by direct experiment. If we had used the classical formula for kinetic energy, $K = \frac{1}{2}m_0 v^2$, instead of Eq. 10, we would have obtained a speed $v = 5.93 \times 10^7$ m/s and, of course, a mass $m = m_0$. The error in the classical predictions increases rapidly as the voltage (and the kinetic energy) increases. (See Problems 17 and 21.)

8 The Equivalence of Mass and Energy (Section 8–9; RR, Section 3.6)

The best known result of SR is the so-called mass-energy equivalence. That is, the conservation of total energy is equivalent to the conservation of relativistic mass. Mass and energy are equivalent; they form a single invariant that we can call mass-energy. The relation

$$E = mc^2 \qquad (11)$$

expresses the fact that mass-energy can be expressed in energy (E) units or equivalently in mass ($m = E/c^2$) units. In fact, it has become common practice to refer to masses in terms of electron-volts, such as saying that the rest mass of an electron is 0.51 MeV, for convenience in energy calculations. Likewise, entities of zero rest mass, such as photons, may be assigned an effective mass equivalent to their energy. We associate mass with each of the various forms of energy.

In classical physics we had two separate conservation principles: (1) the conservation of (classical) mass, as in chemical reactions, and (2) the conservation of energy. In relativity, these merge into one conservation principle, that of the conservation of mass-energy. The two classical laws may be viewed as special cases that would be expected to agree with experiment only if energy transfers into or out of the system are so small compared with the system's rest mass that the corresponding fractional change in rest mass of the system is too small to be measured.

For example, the rest mass of a hydrogen atom is 1.00797 u (= 938.8 MeV). If enough energy (13.58 eV) is added to ionize hydrogen, that is, to break it up into its constituent parts, a proton and an electron, the fractional change in rest mass of the system is

$$\frac{13.58 \text{ eV}}{938.8 \times 10^6 \text{ eV}} = 1.45 \times 10^{-8}$$

or 1.45×10^{-6} %, too small to measure. However, for a nucleus such as the deuteron, whose rest mass is 2.01360 u (= 1876.4 MeV), one must add an energy of 2.22 MeV to break it up into its constituent parts, a proton and a neutron. The fractional change in rest mass of the system is

$$\frac{2.22 \text{ MeV}}{1876.4 \text{ MeV}} = 1.18 \times 10^{-3}$$

or 0.12%, which is readily measureable. This is characteristic of the fractional rest-mass changes in nuclear reactions, so that one must use the relativistic law of conservation of mass-energy to get agreement between theory and experiment in nuclear reactions. The classical (rest) mass is *not* conserved, but total energy (mass-energy) is.

The most direct example of exchange of energy between rest mass and other forms is given by the phenomenon of pair annihilation or pair production. In this phenomenon an electron and a positron, elementary material particles differing only in the sign of their electric charge, can combine and literally disappear. In their place we find high-energy radiation, called γ-radiation, whose radiant energy is exactly equal to the rest mass plus kinetic energies of the disappearing particles. The process is reversible, so that a materialization of rest mass from radiant energy can occur when a high enough energy γ-ray, under proper conditions, disappears; in its place appears a positron-electron pair whose total energy (rest mass + kinetic) is equal to the radiant energy lost. Today we use the term *photon* to describe such radiant energy (see Section 42–5).

Example 7 Consider an electron and a positron, which are essentially at rest near one another, that unite and are annihilated. The initial momentum of the electron-positron system is zero, so that momentum conservation requires that the momentum be zero after the annihilation process. Because a single γ-ray photon cannot have zero momentum, we must have more than one photon created. The most probable process is the creation of two photons. Find the energy and wavelength of each photon in such a process.

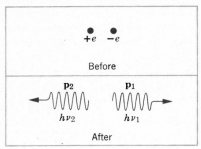

Figure 4 Pair annihilation producing two photons. Example 7.

In the two-photon process (see Fig. 4), momentum conservation gives

$$0 = \mathbf{P}_1 + \mathbf{P}_2$$

or

$$\mathbf{P}_1 = -\mathbf{P}_2,$$

so that the photon momenta are equal in magnitude but oppositely directed. Hence,

$$P_1 = P_2$$

or (see Section 42–6)

$$\frac{h\nu_1}{c} = \frac{h\nu_2}{c}$$

and

$$\nu_1 = \nu_2 = \nu,$$

so that each photon has the same frequency, ν.

Energy conservation then requires that

$$m_0 c^2 + m_0 c^2 = h\nu + h\nu,$$

the electron and the positron each having rest energy $m_0 c^2$ but no initial kinetic energy, and the two photons each having the same energy $h\nu$.

Hence, the energy of each photon is

$$h\nu = m_0 c^2 = 0.51 \text{ MeV},$$

and the wavelength λ of each photon, from $h\nu = hc/\lambda = m_0 c^2$, is

$$\lambda = \frac{h}{m_0 c} = 2.4 \text{ pm} \qquad \text{(or 0.024 Å)}.$$

If the initial pair of particles had some kinetic energy, then the photon energy would exceed 0.51 MeV and its wavelength could be less than 2.4 pm.

QUESTIONS

1. A friend asserts that Einstein's postulate (that the speed of light is not affected by the uniform motion of the source or the observer) must be discarded because it violates "common sense." How would you answer him?

2. In a vacuum, does the speed of light depend on (*a*) the wavelength, (*b*) the frequency, (*c*) the intensity, (*d*) the speed of the source, or (*e*) the speed of the observer?

3. Compare the results obtained for length- and time-interval measurements by observers in frames whose relative velocity is *c*. In what sense, from this point of view, does *c* become a limiting velocity?

4. Consider a spherical light wavefront spreading out from a source. As seen by the source, what is the *difference in velocity* of portions of the wavefront traveling in opposite directions? What is the *relative velocity* of one portion of the wavefront with respect to the other portion?

5. The sweep rate of the tail of a comet can exceed the speed of light. Explain this phenomenon and show that there is no contradiction with relativity.

6. If photons have a speed *c* in one reference frame, can they be found at rest in any other frame? Can photons have a speed other than *c*?

7. How would you expect the relativistic variation of mass to affect the performance of a cyclotron?

8. Can we simply substitute *m* for m_0 in classical equations to obtain the correct corresponding relativistic equation? Give examples.

9. Can a body be accelerated to the speed of light? Explain.

10. A hot metallic sphere cools off on a scale. Does the scale indicate a change in rest mass?

11. A spring is kept compressed by tying its ends together tightly. It is then placed in acid and dissolves. What happens to its stored potential energy?

12. Is the sum of the interior angles of a triangle equal to 180° on a spherical surface? On a plane surface? Under what circumstances does spherical geometry reduce to plane geometry? Draw an analogy to relativistic mechanics and classical mechanics.

PROBLEMS

Section 4

1. At what speed v will the Gallilean and Lorentz expressions for x' (see Eqs. 1 and 2) differ by 0.10%? By 1%? By 10%?

2. Two events, one at position x_1, y_1, z_1 and another at a different position x_2, y_2, z_2, occur at the *same time* t according to observer S. (*a*) Do these events appear to be simultaneous to an observer in S' who moves relative to S at speed v? (*b*) If not, what is the time interval he measures between occurrences of these events? (*c*) How is this time interval affected as $v \to 0$? As the separation between events goes to zero?

Section 5

3. The length of a spaceship is measured to be exactly half its rest length. (*a*) What is the speed of the spaceship relative to the observer's frame? (*b*) What is the dilation of the spaceship's unit time?

4. Two spaceships, each of rest length 100 m, pass near one another heading in opposite directions. If an astronaut at the front of one ship measures a time interval of 2.50 μs for the second ship to pass him, then (*a*) what is the relative velocity of the spaceships? (*b*) What time interval is measured on the first ship for the front of the second ship to pass from the front to the back of the first ship?

5. In the target area of an accelerator laboratory there is a straight evacuated tube 300 m long. A momentary burst of 1 million radioactive particles enters at one end of the tube moving at a speed of 0.80c. Half of them arrive at the other end without having decayed. (*a*) How long is the tube as measured in the rest frame of the particles? (*b*) What is the half-life of the particles (the time during which half the particles initially present have decayed) measured in this same frame? (*c*) With what speed is the tube measured to move in the rest frame of the particles?

6. The mean lifetime of elementary particles called muons stopped in a lead block in the laboratory is measured to be 2.2μs. The mean lifetime of high-speed muons in a burst of cosmic rays observed from the earth is measured to be 16 μs. Find the speed of these cosmic-ray muons.

7. Laboratory experiments on muons at rest show that they have a (proper) average lifetime of about 2.2 μs. Such muons are produced high in the earth's atmosphere by cosmic-ray reactions and travel at a speed 0.99c relative to the earth a distance of from 4 to 13 km after formation before decaying. (*a*) Show that the average distance a muon can travel before decaying is much less than even the shorter distance of 4 km, if its lifetime in flight is only 2.2 μs. (*b*) Explain the consistency of the observations on length traveled and lifetime by computing the lifetime of a muon in flight as measured by a ground observer. (*c*) Explain the consistency by computing the length traveled as seen by an observer at rest on the muon in its flight through the atmosphere.

8. A passenger walks forward along the aisle of a train at a speed of 2.2 mi/h as the train moves along a straight track at a constant speed of 57.5 mi/h with respect to the ground. What is the passenger's speed with respect to the ground? To the accuracy cited, do classical and relativistic predictions differ?

9. (*a*) Show, using Eq. 6, that if $u' < c$ and $v < c$, then it must always be true that $u < c$. (*b*) Suppose $u' = c$, as for example a photon in the S'-frame; what then is the observed value of u, the speed of the same photon as measured in the S-frame?

Section 6

10. An observer in an inertial system S reports that two missiles are moving parallel to one another on a straight-line path, one with a speed 0.90c and the other with a speed 0.70c. Find the speed of one missile with respect to the other.

11. One cosmic-ray particle approaches the earth along its axis with a velocity 0.80c toward the North Pole and another with a velocity 0.60c toward the South Pole. What is the relative speed of approach of one particle with respect to the other? (*Hint:* It is useful to consider the earth and one of the particles as the two inertial systems.)

12. A, on earth, signals with a flashlight every 6 min. B is on a space station that is stationary with respect to the earth. C is on a rocket traveling from A to B with a constant velocity of 0.6c relative to A (see Fig. 5). (*a*) At what intervals does B receive the signals from A? (*b*) At what intervals does C receive signals from A? (*c*) If C flashes a light using intervals equal to those she received from A, at what intervals does B receive C's flashes?

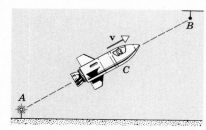

Figure 5. Problem 12.

13. In the spectrum of Quasar 3C9, some of the familiar hydrogen lines appear, but they are shifted so far toward the red that their frequencies are observed to be only one-third as large as that observed in a hydrogen rest frame. (*a*) Show that the classical Doppler equation (see Section 38–9) gives a velocity of recession greater than c. (*b*) Assuming that the relative motion of 3C9 and the earth is entirely one of recession, find the recession speed predicted by the relativistic Doppler equation.

14. A rocket ship moving away from the earth at a speed $v = \frac{12}{13}c$ reports back by sending waves of frequency 100 MHz as measured in the frame of the rocket ship. At what frequency are the signals received on earth?

15. Give the wavelength shift, if any, in the Doppler effect for the sodium D_2 line 589 nm emitted from a source moving in a circle with constant speed 0.10c measured by an observer fixed at the center of the circle.

Section 7

16. What is the speed of an electron whose kinetic energy equals its rest energy? Does the result depend on the mass of the electron?

17. Compute the speed of (a) electrons and (b) protons that fall through an electrostatic potential difference of 10 million volts. (c) What is the ratio of relativistic mass to rest mass in each case?

18. (a) What potential difference will accelerate electrons to the speed of light, according to classical physics? (b) With this potential difference, what speed would an electron acquire relativistically? (c) What would its mass be at this speed? Its kinetic energy?

19. If $m/m_0 = 40,000$ for electrons emerging from the Stanford linear accelerator, what is their laboratory speed? What is the value of β?

20. (a) Show that at small values of β, $K = \frac{1}{2}m_0v^2$—the classical result. (b) Show that if there were no limiting speed (that is, let $c \rightarrow \infty$), then $K = \frac{1}{2}m_0v^2$—again the classical result.

21. (a) Show that when $v/c < \frac{1}{10}$, then $K/m_0c^2 < \frac{1}{200}$ and the classical expressions for kinetic energy and momentum, that is, $K = \frac{1}{2}m_0v^2$ and $p = m_0v$, may be used with an error of less than 1%. (b) Show that when $v/c > 99/100$, then $K/m_0c^2 > 7$ and the relativistic relation $p = E/c$ for a zero rest-mass particle may be used for a particle of rest mass m_0 with an error of less than 1%.

22. (a) Show that a particle which travels at the speed of light must have a zero rest mass. (b) Show that for a particle of zero rest mass, $v = c$, $K = E$, and $p = E/c$.

Section 8
23. The "effective mass" of a photon (bundle of electromagnetic radiation of zero rest mass and energy $h\nu$) can be determined from the relation $m = E/c^2$. Compute the "effective mass" for a photon of wavelength 5000 Å (visible region), and for a photon of wavelength 1.0 Å (x-ray region).

24. (a) Prove that 1 u = 931.5 MeV/c^2. (b) Find the energy equivalent to the rest mass of the electron, and to the rest mass of the proton. Use accurate data from Appendix B.

25. (a) How much energy is released in the explosion of a fission bomb containing 3.0 kg of fissionable material? Assume that 0.10% of the rest mass is converted to released energy. (b) What mass of TNT would have to explode to provide the same energy release? Assume that each mole of TNT liberates 820,000 cal upon exploding. The molecular weight of TNT is 0.227 kg/mol. (c) For the same mass of explosive, how much more effective are fission explosions than TNT explosions? That is, compare the fractions of rest mass converted to released energy for the two cases.

26. The nucleus ^{12}C consists of six protons (^{1}H) and six neutrons (n) held in close association by strong nuclear forces. The atomic rest masses are

^{12}C	12.000000 u,
^{1}H	1.007825 u,
n	1.008665 u.

How much energy would be required to separate ^{12}C nucleus into its constituent protons and neutrons? This energy is called the binding energy of the ^{12}C nucleus. (The masses are really those of the neutral atoms, but the extranuclear electrons have relatively negligible binding energy and are of equal number before and after the breakup of ^{12}C.)

APPENDIX A

The International System of Units (SI)*

1. The SI Base Units

Quantity	Name	Symbol	Definition
length	meter	m	"... the length of the path traveled by light in vacuum in 1/299,792,458 of a second." (1983)
mass	kilogram	kg	"... this prototype [a certain platinum-iridium cylinder] shall henceforth be considered to be the unit of mass." (1889)
time	second	s	"... the duration of 9,192,631,770 periods of the radiation corresponding to the transition between the two hyperfine levels of the ground state of the cesium-133 atom." (1967)
electric current	ampere	A	"... that constant current which, if maintained in two straight parallel conductors of infinite length, of negligible circular cross section, and placed 1 meter apart in vacuum, would produce between these conductors a force equal to 2×10^{-7} newton per meter of length." (1946)
thermodynamic temperature	kelvin	K	"... the fraction 1/273.16 of the thermodynamic temperature of the triple point of water." (1967)
amount of substance	mole	mol	"... the amount of substance of a system which contains as many elementary entities as there are atoms in 0.012 kilogram of carbon 12." (1971)
luminous intensity	candela	cd	"... the luminous intensity, in the perpendicular direction, of a surface of 1/600,000 square meter of a blackbody at the temperature of freezing platinum under a pressure of 101,325 newton per square meter." (1967)

2. Some SI Derived Units

Quantity	Name of unit	Symbol	
area	square meter	m^2	
volume	cubic meter	m^3	
frequency	hertz	Hz	s^{-1}
mass density (density)	kilogram per cubic meter	kg/m^3	
speed, velocity	meter per second	m/s	
angular velocity	radian per second	rad/s	
acceleration	meter per second squared	m/s^2	
angular acceleration	radian per second squared	rad/s^2	
force	newton	N	$kg \cdot m/s^2$
pressure	pascal	Pa	N/m^2
work, energy, quantity of heat	joule	J	$N \cdot m$
power	watt	W	J/s
quantity of electricity	coulomb	C	$A \cdot s$
potential difference, electromotive force	volt	V	W/A
electric field strength	volt per meter	V/m	
electric resistance	ohm	Ω	V/A
capacitance	farad	F	$A \cdot s/V$
magnetic flux	weber	Wb	$V \cdot s$
inductance	henry	H	$V \cdot s/A$
magnetic flux density	tesla	T	Wb/m^2
magnetic field strength	ampere per meter	A/m	
entropy	joule per kelvin	J/K	
specific heat capacity	joule per kilogram kelvin	$J/(kg \cdot K)$	
thermal conductivity	watt per meter kelvin	$W/(m \cdot K)$	
radiant intensity	watt per steradian	W/sr	

3. The SI Supplementary Units

plane angle	radian	rad
solid angle	steradian	sr

* Adapted from "The International System of Units (SI)," National Bureau of Standards Special Publication 330, 1972 edition. The definitions above were adopted by the General Conference of Weights and Measures, an international body, on the dates shown. In this book we do not use the candela.

APPENDIX B

Some Fundamental Constants of Physics*

Constant	Symbol	Computational value	Best (1973) value	
			Value[a]	Uncertainty[b]
Speed of light in a vacuum	c	3.00×10^8 m/s	2.99792458	exact
Elementary charge	e	1.60×10^{-19} C	1.6021892	2.9
Electron rest mass	m_e	9.11×10^{-31} kg	9.109534	5.1
Permittivity constant	ϵ_0	8.85×10^{-12} F/m	$1/\mu_0 c^2$	exact
Permeability constant	μ_0	1.26×10^{-6} H/m	$0.4\,\pi$	exact
Electron rest mass[c]	m_e	5.49×10^{-4} u	5.4858026	0.38
Neutron rest mass[c]	m_n	1.0087 u	1.008665012	0.037
Hydrogen atom rest mass[c]	m_{1H}	1.0078 u	1.007825035	0.011
Deuterium atom rest mass[c]	m_{2H}	2.0141 u	2.0141019	0.053
Helium atom rest mass[c]	m_{4He}	4.0026 u	4.0026032	0.067
Electron charge to mass ratio	e/m_e	1.76×10^{11} C/kg	1.7588047	2.8
Proton rest mass	m_p	1.67×10^{-27} kg	1.6726485	5.1
Ratio of proton mass to electron mass	m_p/m_e	1840	1836.15152	0.38
Neutron rest mass	m_n	1.68×10^{-27} kg	1.6749543	5.1
Muon rest mass	m_μ	1.88×10^{-28} kg	1.883566	5.6
Planck constant	h	6.63×10^{-34} J·s	6.626176	5.4
Electron Compton wavelength	λ_c	2.43×10^{-12} m	2.4263089	1.6
Universal gas constant	R	8.31 J/mol·K	8.31441	31
Avogadro constant	N_A	6.02×10^{23} mol^{-1}	6.022045	5.1
Boltzmann constant	k	1.38×10^{-23} J/K	1.380662	32
Molar volume of ideal gas at STP[d]	V_m	2.24×10^{-2} m³/mol	2.241383	31
Faraday constant	F	9.65×10^4 C/mol	9.648456	2.8
Stefan–Boltzmann constant	σ	5.67×10^{-8} W/m²·K⁴	5.67032	125
Rydberg constant	R	1.10×10^7 m^{-1}	1.097373177	0.075
Gravitational constant	G	6.67×10^{-11} m³/s²·kg	6.6726	75
Bohr radius	r_B	5.29×10^{-11} m	5.2917706	0.82
Electron magnetic moment	μ_e	9.28×10^{-24} J/T	9.284832	3.9
Proton magnetic moment	μ_p	1.41×10^{-26} J/T	1.4106171	3.9
Bohr magneton	μ_B	9.27×10^{-24} J/T	9.274078	3.9
Nuclear magneton	μ_N	5.05×10^{-27} J/T	5.050824	3.9

[a] Same unit and power of ten as the computational value.
[b] Parts per million.
[c] Mass given in unified atomic mass units, where 1 u = 1.660565×10^{-27} kg.
[d] STP–standard temperature and pressure = 0°C and 1.0 atm.

* The values in this table were largely selected from a longer list developed by E. Richard Cohen and B. N. Taylor, *Journal of Physical and Chemical Reference Data*, vol. 2, no. 4 (1973).

APPENDIX C

Some Astronomical Data

Some distances from the earth

To the moon*	3.82×10^8 m
To the sun*	1.50×10^{11} m
To the nearest star (Proxima Centauri)	4.04×10^{16} m
To the center of our galaxy	$\sim 3 \times 10^{20}$ m
To the Andromeda Galaxy	2.1×10^{22} m
To the edge of the observable universe	$\sim 10^{26}$ m

* Mean distance.

The sun, the earth and the moon

Property	Unit	Sun[a]	Earth	Moon
Mass	kg	1.99×10^{30}	5.98×10^{24}	7.36×10^{22}
Mean radius	m	6.96×10^8	6.37×10^6	1.74×10^6
Mean density	kg/m³	1410	5520	3340
Surface gravity	m/s²	274	9.81	1.67
Escape velocity	km/s	618	11.2	2.38
Period of rotation[c]	—	37d—poles[b] 26d—equator	23.9h	27.3d

[a] The sun radiates energy at the rate of 3.90×10^{26} W; just outside the earth's atmosphere solar energy is received, assuming normal incidence, at the rate of 1340 W/m².
[b] The sun—a ball of gas—does not rotate as a rigid body.
[c] Measured with respect to the distant stars.

Some properties of the planets

	Mercury	Venus	Earth	Mars	Jupiter	Saturn	Uranus	Neptune	Pluto
Mean distance from sun, 10^6 km	57.9	108	150	228	778	1,430	2,870	4,500	5,900
Period of revolution, y	0.241	0.615	1.00	1.88	11.9	29.5	84.0	165	248
Period of rotation[a], d	58.7	−243[b]	0.997	1.03	0.409	0.426	−0.451[b]	0.658	6.39
Orbital speed, km/s	47.9	35.0	29.8	24.1	13.1	9.64	6.81	5.43	4.74
Inclination of axis to orbit	<28°	~3°	23.5°	24.0°	3.08°	26.7°	82.1°	28.8°	?
Inclination of orbit to earth's orbit	7.00°	3.39°	—	1.85°	1.30°	2.49°	0.77°	1.77°	17.2°
Eccentricity of orbit	0.206	0.0068	0.0167	0.0934	0.0485	0.0556	0.0472	0.0086	0.250
Equatorial diameter, km	4,880	12,100	12,800	6,790	143,000	120,000	51,800	49,500	6,000(?)
Mass (earth = 1)	0.0558	0.815	1.000	0.107	318	95.1	14.5	17.2	0.11(?)
Density (water = 1)	5.60	5.20	5.52	3.95	1.31	0.704	1.21	1.67	?
Surface gravity[c], m/s²	3.78	8.60	9.78	3.72	22.9	9.05	7.77	11.0	4.3(?)
Escape velocity[c], km/s	4.3	10.3	11.2	5.0	59.5	35.6	21.2	23.6	5.3(?)
Known satellites	0	0	1	2	14 + ring	15 + rings	5 + rings	2	0

[a] Measured with respect to the distant stars.
[b] The sense of rotation is opposite to that of the orbital motion
[c] Measured at the planet's equator.

The Periodic Table of the Elements

Actinium	Ac	89	Hafnium	Hf	72	Praseodymium	Pr	59
Aluminum	Al	13	Hahnium	Ha	105	Promethium	Pm	61
Americium	Am	95	Helium	He	2	Protactinium	Pa	91
Antimony	Sb	51	Holmium	Ho	67	Radium	Ra	88
Argon	Ar	18	Hydrogen	H	1	Radon	Rn	86
Arsenic	As	33	Indium	In	49	Rhenium	Re	75
Astatine	At	85	Iodine	I	53	Rhodium	Rh	45
Barium	Ba	56	Iridium	Ir	77	Rubidium	Rb	37
Berkelium	Bk	97	Iron	Fe	26	Ruthenium	Ru	44
Beryllium	Be	4	Krypton	Kr	36	Rutherfordium	Rf	104
Bismuth	Bi	83	Lanthanum	La	57	Samarium	Sm	62
Boron	B	5	Lawrencium	Lw	103	Scandium	Sc	21
Bromine	Br	35	Lead	Pb	82	Selenium	Se	34
Cadmium	Cd	48	Lithium	Li	3	Silicon	Si	14
Calcium	Ca	20	Lutetium	Lu	71	Silver	Ag	47
Californium	Cf	98	Magnesium	Mg	12	Sodium	Na	11
Carbon	C	6	Manganese	Mn	25	Strontium	Sr	38
Cerium	Ce	58	Mendelevium	Md	101	Sulfur	S	16
Cesium	Cs	55	Mercury	Hg	80	Tantalum	Ta	73
Chlorine	Cl	17	Molybdenum	Mo	42	Technetium	Tc	43
Chromium	Cr	24	Neodymium	Nd	60	Tellurium	Te	52
Cobalt	Co	27	Neon	Ne	10	Terbium	Tb	65
Copper	Cu	29	Neptunium	Np	93	Thallium	Tl	81
Curium	Cm	96	Nickel	Ni	28	Thorium	Th	90
Dysprosium	Dy	66	Niobium	Nb	41	Thulium	Tm	69
Einsteinium	Es	99	Nitrogen	N	7	Tin	Sn	50
Erbium	Er	68	Nobelium	No	102	Titanium	Ti	22
Europium	Eu	63	Osmium	Os	76	Tungsten	W	74
Fermium	Fm	100	Oxygen	O	8	Uranium	U	92
Fluorine	F	9	Palladium	Pd	46	Vanadium	V	23
Francium	Fr	87	Phosphorus	P	15	Xenon	Xe	54
Gadolinium	Gd	64	Platinum	Pt	78	Ytterbium	Yb	70
Gallium	Ga	31	Plutonium	Pu	94	Yttrium	Y	39
Germanium	Ge	32	Polonium	Po	84	Zinc	Zn	30
Gold	Au	79	Potassium	K	19	Zirconium	Zr	40

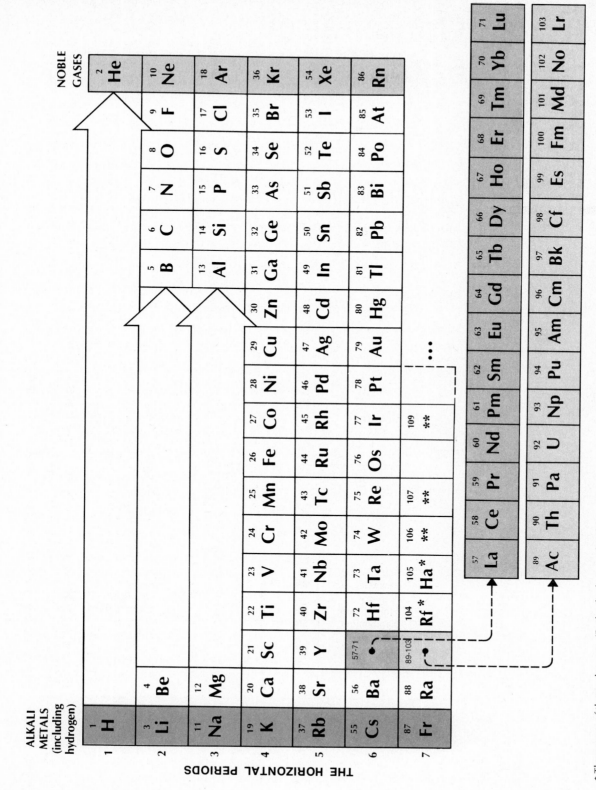

ALKALI METALS (including hydrogen)

NOBLE GASES

THE HORIZONTAL PERIODS

1	2	3	4	5	6	7

Period 1: 1 H — 2 He

Period 2: 3 Li, 4 Be — 5 B, 6 C, 7 N, 8 O, 9 F, 10 Ne

Period 3: 11 Na, 12 Mg — 13 Al, 14 Si, 15 P, 16 S, 17 Cl, 18 Ar

Period 4: 19 K, 20 Ca, 21 Sc, 22 Ti, 23 V, 24 Cr, 25 Mn, 26 Fe, 27 Co, 28 Ni, 29 Cu, 30 Zn, 31 Ga, 32 Ge, 33 As, 34 Se, 35 Br, 36 Kr

Period 5: 37 Rb, 38 Sr, 39 Y, 40 Zr, 41 Nb, 42 Mo, 43 Tc, 44 Ru, 45 Rh, 46 Pd, 47 Ag, 48 Cd, 49 In, 50 Sn, 51 Sb, 52 Te, 53 I, 54 Xe

Period 6: 55 Cs, 56 Ba, 57-71, 72 Hf, 73 Ta, 74 W, 75 Re, 76 Os, 77 Ir, 78 Pt, 79 Au, 80 Hg, 81 Tl, 82 Pb, 83 Bi, 84 Po, 85 At, 86 Rn

Period 7: 87 Fr, 88 Ra, 89-103, 104 Rf*, 105 Ha*, 106 **, 107 **, 109 **

Lanthanides: 57 La, 58 Ce, 59 Pr, 60 Nd, 61 Pm, 62 Sm, 63 Eu, 64 Gd, 65 Tb, 66 Dy, 67 Ho, 68 Er, 69 Tm, 70 Yb, 71 Lu

Actinides: 89 Ac, 90 Th, 91 Pa, 92 U, 93 Np, 94 Pu, 95 Am, 96 Cm, 97 Bk, 98 Cf, 99 Es, 100 Fm, 101 Md, 102 No, 103 Lr

* The names of these elements (Rutherfordium and Hahnium) have not been accepted because of conflicting claims of discovery. A group in the USSR has proposed the names Kurchatovium and Nielsbohrium.

** Discovery of these three elements has been reported but names for them have not yet been proposed.

Conversion Factors

Conversion factors may be read off directly from the tables. For example, 1 degree = 2.778×10^{-3} revolutions, so $16.7° = 16.7 \times 2.778 \times 10^{-3}$ rev. The SI quantities are capitalized. Adapted in part from G. Shortley and D. Williams, *Elements of Physics,* Prentice-Hall, Englewood Cliffs, N.J., 1971.

Plane angle

	°	′	″	**RADIAN**	**rev**
1 degree =	1	60	3600	1.745×10^{-2}	2.778×10^{-3}
1 minute =	1.667×10^{-2}	1	60	2.909×10^{-4}	4.630×10^{-5}
1 second =	2.778×10^{-4}	1.667×10^{-2}	1	4.848×10^{-6}	7.716×10^{-7}
1 RADIAN =	57.30	3438	2.063×10^{5}	1	0.1592
1 revolution =	360	2.16×10^{4}	1.296×10^{6}	6.283	1

Solid angle

1 sphere = 4π steradians = 12.57 steradians

Length

	cm	**METER**	**km**	**in.**	**ft**	**mi**
1 centimeter =	1	10^{-2}	10^{-5}	0.3937	3.281×10^{-2}	6.214×10^{-6}
1 METER =	100	1	10^{-3}	39.37	3.281	6.214×10^{-4}
1 kilometer =	10^{5}	1000	1	3.937×10^{4}	3281	0.6214
1 inch =	2.540	2.540×10^{-2}	2.540×10^{-5}	1	8.333×10^{-2}	1.578×10^{-5}
1 foot =	30.48	0.3048	3.048×10^{-4}	12	1	1.894×10^{-4}
1 mile =	1.609×10^{5}	1609	1.609	6.336×10^{4}	5280	1

1 angström = 10^{-10} m
1 nautical mile = 1852 m
   = 1.151 miles = 6076 ft
1 fermi = 10^{-15} m

1 light-year = 9.460×10^{12} km
1 parsec = 3.084×10^{13} km
1 fathom = 6 ft
1 Bohr radius = 5.292×10^{-11} m

1 yard = 3 ft
1 rod = 16.5 ft
1 mil = 10^{-3} in.

Area

	METER^2	cm^2	ft^2	in.^{2}
1 SQUARE METER =	1	10^4	10.76	1550
1 square centimeter =	10^{-4}	1	1.076 × 10^{-3}	0.1550
1 square foot =	9.290 × 10^{-2}	929.0	1	144
1 square inch =	6.452 × 10^{-4}	6.452	6.944 × 10^{-3}	1

1 square mile = 2.788 × 10^7 ft^2 = 640 acres   1 acre = 43,560 ft^2
1 barn = 10^{-28} m^2 1 hectare = 10^4 m^2 = 2.471 acre

Volume

	METER^3	cm^3	li	ft^3	in.^{3}
1 CUBIC METER =	1	10^6	1000	35.31	6.102 × 10^4
1 cubic centimeter =	10^{-6}	1	1.000 × 10^{-3}	3.531 × 10^{-5}	6.102 × 10^{-2}
1 liter =	1.000 × 10^{-3}	1000	1	3.531 × 10^{-2}	61.02
1 cubic foot =	2.832 × 10^{-2}	2.832 × 10^4	28.32	1	1728
1 cubic inch =	1.639 × 10^{-5}	16.39	1.639 × 10^{-2}	5.787 × 10^{-4}	1

= 0.264 gal

1 U.S. fluid gallon = 4 U.S. fluid quarts = 8 U.S. pints = 128 U.S. fluid ounces = 231 in.^{3} *⟹ 1 L. = 0.264 gal*
1 British imperial gallon = 277.4 in.^{3} 1 liter = 10^{-3} m^3.

Mass

Quantities in the colored areas are not mass units but are often used as such. When we write, for example, 1 kg "=" 2.205 lb this means that a kilogram is a *mass* that *weighs* 2.205 pounds under standard condition of gravity (g = 9.80665 m/s^2).

	g	KILOGRAM	slug	u	oz	lb	ton
1 gram =	1	0.001	6.852 × 10^{-5}	6.022 × 10^{23}	3.527 × 10^{-2}	2.205 × 10^{-3}	1.102 × 10^{-6}
1 KILOGRAM =	1000	1	6.852 × 10^{-2}	6.022 × 10^{26}	35.27	2.205	1.102 × 10^{-3}
1 slug =	1.459 × 10^4	14.59	1	8.786 × 10^{27}	514.8	32.17	1.609 × 10^{-2}
1 u =	1.661 × 10^{-24}	1.661 × 10^{-27}	1.138 × 10^{-28}	1	5.857 × 10^{-26}	3.662 × 10^{-27}	1.830 × 10^{-30}
1 ounce =	28.35	2.835 × 10^{-2}	1.943 × 10^{-3}	1.718 × 10^{25}	1	6.250 × 10^{-2}	3.125 × 10^{-5}
1 pound =	453.6	0.4536	3.108 × 10^{-2}	2.732 × 10^{26}	16	1	0.0005
1 ton =	9.072 × 10^5	907.2	62.16	5.463 × 10^{29}	3.2 × 10^4	2000	1

Density

Quantities in the colored areas are weight densities and, as such, are dimensionally different from mass densities. See note for mass table.

	slug/ft³	KILOGRAM/ METERS³	g/cm³	lb/ft³	lb/in.³
1 slug per ft³ =	1	515.4	0.5154	32.17	1.862×10^{-2}
1 KILOGRAM per METER³ =	1.940×10^{-3}	1	0.001	6.243×10^{-2}	3.613×10^{-5}
1 gram per cm³ =	1.940	1000	1	62.43	3.613×10^{-2}
1 pound per ft³ =	3.108×10^{-2}	16.02	1.602×10^{-2}	1	5.787×10^{-4}
1 pound per in.³ =	53.71	2.768×10^{4}	27.68	1728	1

Time

	yr	d	h	min	SECOND
1 year =	1	365.2	8.766×10^{3}	5.259×10^{5}	3.156×10^{7}
1 day =	2.738×10^{-3}	1	24	1440	8.640×10^{4}
1 hour =	1.141×10^{-4}	4.167×10^{-2}	1	60	3600
1 minute =	1.901×10^{-6}	6.944×10^{-4}	1.667×10^{-2}	1	60
1 SECOND =	3.169×10^{-8}	1.157×10^{-5}	2.778×10^{-4}	1.667×10^{-2}	1

Speed

	ft/s	km/h	METER/ SECOND	mi/h	cm/s
1 foot per second =	1	1.097	0.3048	0.6818	30.48
1 kilometer per hour =	0.9113	1	0.2778	0.6214	27.78
1 METER per SECOND =	3.281	3.6	1	2.237	100
1 mile per hour =	1.467	1.609	0.4470	1	44.70
1 centimeter per second =	3.281×10^{-2}	3.6×10^{-2}	0.01	2.237×10^{-2}	1

1 knot = 1 nautical mi/h = 6076 ft/h   1 mi/min = 88.00 ft/s = 60.00 mi/h

Force

Quantities in the colored areas are not force units but are often used as such. For instance, if we write 1 gram-force ''='' 980.7 dynes, we mean that a gram-mass experiences a force of 980.7 dynes under standard conditions of gravity (g = 9.80665 m/s²)

	dyne	NEWTON	lb	pdl	gf	kgf
1 dyne =	1	10^{-5}	2.248×10^{-6}	7.233×10^{-5}	1.020×10^{-3}	1.020×10^{-6}
1 NEWTON =	10^5	1	0.2248	7.233	102.0	0.1020
1 pound =	4.448×10^5	4.448	1	32.17	453.6	0.4536
1 poundal =	1.383×10^4	0.1383	3.108×10^{-2}	1	14.10	1.410×10^{-2}
1 gram-force =	980.7	9.807×10^{-3}	2.205×10^{-3}	7.093×10^{-2}	1	0.001
1 kilogram-force =	9.807×10^5	9.807	2.205	70.93	1000	1

Pressure

	atm	dyne/cm²	inch of water	cm Hg	PASCAL	lb/in.²	lb/ft²
1 atmosphere =	1	1.013×10^6	406.8	76	1.013×10^5	14.70	2116
1 dyne per cm² =	9.869×10^{-7}	1	4.015×10^{-4}	7.501×10^{-5}	0.1	1.405×10^{-5}	2.089×10^{-3}
1 inch of water[a] at 4° C =	2.458×10^{-3}	2491	1	0.1868	249.1	3.613×10^{-2}	5.202
1 centimeter of mercury[a] at 0° C =	1.316×10^{-2}	1.333×10^4	5.353	1	1333	0.1934	27.85
1 PASCAL =	9.869×10^{-6}	10	4.015×10^{-3}	7.501×10^{-4}	1	1.450×10^{-4}	2.089×10^{-2}
1 pound per in.² =	6.805×10^{-2}	6.895×10^4	27.68	5.171	6.895×10^3	1	144
1 pound per ft² =	4.725×10^{-4}	478.8	0.1922	3.591×10^{-2}	47.88	6.944×10^{-3}	1

[a] Where the acceleration of gravity has the standard value 9.80665 m/s².

1 bar = 10^6 dyne/cm² = 0.1 MPa   1 millibar = 10^3 dyne/cm² = 10^2 Pa   1 Torr = 1 millimeter of mercury

Energy, work, heat

Quantities in the colored areas are not properly energy units but are included for convenience. They arise from the relativistic mass-energy equivalence formula $E = mc^2$ and represent the energy released if a kilogram or unified atomic mass unit (u) is completely converted to energy.

	Btu	erg	ft · lb	hp · h	JOULE	cal	kW · h	eV	MeV	kg	u
1 British thermal unit =	1	1.055×10^{10}	777.9	3.929×10^{-4}	1055	252.0	2.930×10^{-4}	6.585×10^{21}	6.585×10^{15}	1.174×10^{-14}	7.070×10^{12}
1 erg =	9.481×10^{-11}	1	7.376×10^{-8}	3.725×10^{-14}	10^{-7}	2.388×10^{-8}	2.778×10^{-14}	6.242×10^{11}	6.242×10^{5}	1.113×10^{-24}	670.2
1 foot-pound =	1.285×10^{-3}	1.356×10^{7}	1	5.051×10^{-7}	1.356	0.3238	3.766×10^{-7}	8.464×10^{18}	8.464×10^{12}	1.509×10^{-17}	9.037×10^{9}
1 horsepower-hour =	2545	2.685×10^{13}	1.980×10^{6}	1	2.685×10^{6}	6.413×10^{5}	0.7457	1.676×10^{25}	1.676×10^{19}	2.988×10^{-11}	1.799×10^{16}
1 JOULE =	9.481×10^{-4}	10^{7}	0.7376	3.725×10^{-7}	1	0.2388	2.778×10^{-7}	6.242×10^{18}	6.242×10^{12}	1.113×10^{-17}	6.702×10^{9}
1 calorie =	3.969×10^{-3}	4.187×10^{7}	3.088	1.560×10^{-6}	4.187	1	1.163×10^{-6}	2.614×10^{19}	2.614×10^{13}	4.660×10^{-17}	2.806×10^{10}
1 kilowatt-hour =	3413	3.6×10^{13}	2.655×10^{6}	1.341	3.6×10^{6}	8.598×10^{5}	1	2.247×10^{25}	2.247×10^{19}	4.007×10^{-11}	2.413×10^{16}
1 electron volt =	1.519×10^{-22}	1.602×10^{-12}	1.182×10^{-19}	5.967×10^{-26}	1.602×10^{-19}	3.826×10^{-20}	4.450×10^{-26}	1	10^{-6}	1.783×10^{-36}	1.074×10^{-9}
1 million electron volts =	1.519×10^{-16}	1.602×10^{-6}	1.182×10^{-13}	5.967×10^{-20}	1.602×10^{-13}	3.826×10^{-14}	4.450×10^{-20}	10^{6}	1	1.783×10^{-30}	1.074×10^{-3}
1 kilogram =	8.521×10^{13}	8.987×10^{23}	6.629×10^{16}	3.348×10^{10}	8.987×10^{16}	2.146×10^{16}	2.497×10^{10}	5.610×10^{35}	5.610×10^{29}	1	6.022×10^{26}
1 unified atomic mass unit =	1.415×10^{-13}	1.492×10^{-3}	1.101×10^{-10}	5.559×10^{-17}	1.492×10^{-10}	3.564×10^{-11}	4.146×10^{-17}	9.32×10^{8}	932.0	1.661×10^{-27}	1

Power

	Btu/h	ft · lb/s	hp	cal/s	kW	WATT
1 British thermal unit per hour =	1	0.2161	3.929×10^{-4}	6.998×10^{-2}	2.930×10^{-4}	0.2930
1 foot-pound per second =	4.628	1	1.818×10^{-3}	0.3239	1.356×10^{-3}	1.356
1 horsepower =	2545	550	1	178.1	0.7457	745.7
1 calorie per second =	14.29	3.088	5.615×10^{-3}	1	4.187×10^{-3}	4.187
1 kilowatt =	3413	737.6	1.341	238.8	1	1000
1 WATT =	3.413	0.7376	1.341×10^{-3}	0.2388	0.001	1

Magnetic flux

	maxwell	WEBER
1 maxwell =	1	10^{-8}
1 WEBER =	10^8	1

Magnetic field

	gauss	TESLA	milligauss
1 gauss =	1	10^{-4}	1000
1 TESLA =	10^4	1	10^7
1 milligauss =	0.001	10^{-7}	1

1 tesla = 1 weber/meter^2

Mathematical Symbols and the Greek Alphabet

Mathematical Signs and Symbols

$=$ equals
$\cong$ equals approximately
$\neq$ is not equal to
$\equiv$ is identical to, is defined as
$>$ is greater than ($\gg$ is much greater than)
$<$ is less than ($\ll$ is much less than)
$\geq$ is greater than or equal to (or, is no less than)
$\leq$ is less than or equal to (or, is no more than)
$\pm$ plus or minus ($\sqrt{4} = \pm2$)
$\propto$ is proportional to
Σ the sum of
$\bar{x}$ the average value of x

The Greek Alphabet

Alpha	A	α	Nu	N	ν
Beta	B	β	Xi	Ξ	ξ
Gamma	Γ	γ	Omicron	O	o
Delta	Δ	δ	Pi	Π	π
Epsilon	E	ε	Rho	P	ρ
Zeta	Z	ζ	Sigma	Σ	σ
Eta	H	η	Tau	T	τ
Theta	Θ	θ	Upsilon	Y	υ
Iota	I	ι	Phi	Φ	ϕ, φ
Kappa	K	κ	Chi	X	χ
Lambda	Λ	λ	Psi	Ψ	ψ
Mu	M	μ	Omega	Ω	ω

APPENDIX G

Mathematical Formulas

Geometry

Circle of radius r: circumference $= 2\pi r$; area $= \pi r^2$.
Sphere of radius r: area $= 4\pi r^2$; volume $= \frac{4}{3}\pi r^3$.
Right circular cylinder of radius r and height h: area $= 2\pi r^2 + 2\pi rh$; volume $= \pi r^2 h$.
Triangle of base a and altitude h: area $= \frac{1}{2}ah$.

Quadratic Formula

$$\text{If } ax^2 + bx + c = 0, \text{ then } x = \frac{-b \pm \sqrt{b^2 - 4ac}}{2a}.$$

Trigonometric Functions of Angle θ

$\sin \theta = \dfrac{y}{r}$ $\cos \theta = \dfrac{x}{r}$

$\tan \theta = \dfrac{y}{x}$ $\cot \theta = \dfrac{x}{y}$

$\sec \theta = \dfrac{r}{x}$ $\csc \theta = \dfrac{r}{y}$

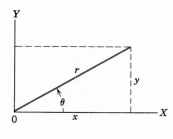

Pythagorean Theorem

$$x^2 + y^2 = r^2$$

Trigonometric Identities

$\sin^2 \theta + \cos^2 \theta = 1$ $\sec^2 \theta - \tan^2 \theta = 1$ $\csc^2 \theta - \cot^2 \theta = 1$

$\sin 2\theta = 2 \sin \theta \cos \theta$

$\cos 2\theta = \cos^2 \theta - \sin^2 \theta = 2 \cos^2 \theta - 1 = 1 - 2 \sin^2 \theta$

$\sin (\alpha \pm \beta) = \sin \alpha \cos \beta \pm \cos \alpha \sin \beta$

$\cos (\alpha \pm \beta) = \cos \alpha \cos \beta \mp \sin \alpha \sin \beta$

$\tan (\alpha \pm \beta) = \dfrac{\tan \alpha \pm \tan \beta}{1 \mp \tan \alpha \tan \beta}$

$\sin \alpha \pm \sin \beta = 2 \sin \frac{1}{2}(\alpha \pm \beta) \cos \frac{1}{2}(\alpha \mp \beta)$

Binomial Theorem

$$(1 + x)^n = 1 + \frac{nx}{1!} + \frac{n(n - 1)}{2!} x^2 + \cdots$$

Exponential Expansion

$$e^x = 1 + x + \frac{x^2}{2!} + \frac{x^3}{3!} + \cdots$$

Logarithmic Expansion

$$\ln (1 + x) = x - \tfrac{1}{2}x^2 + \tfrac{1}{3}x^3 - \cdots$$

Trigonometric Expansions (θ in radians)

$$\sin \theta = \theta - \frac{\theta^3}{3!} + \frac{\theta^5}{5!} - \cdots$$

$$\cos \theta = 1 - \frac{\theta^2}{2!} + \frac{\theta^4}{4!} - \cdots$$

Derivatives and Indefinite Integrals

In what follows, the letters u and v stand for any functions of x, and a and m are constants. To each of the integrals should be added an arbitrary constant of integration. The *Handbook of Chemistry and Physics* (Chemical Rubber Co.) gives a more extensive tabulation.

1. $\dfrac{dx}{dx} = 1$

2. $\dfrac{d}{dx}(au) = a\dfrac{du}{dx}$

3. $\dfrac{d}{dx}(u + v) = \dfrac{du}{dx} + \dfrac{dv}{dx}$

4. $\dfrac{d}{dx}x^m = mx^{m-1}$

5. $\dfrac{d}{dx}\ln x = \dfrac{1}{x}$

6. $\dfrac{d}{dx}(uv) = u\dfrac{dv}{dx} + v\dfrac{du}{dx}$

7. $\dfrac{d}{dx}e^x = e^x$

8. $\dfrac{d}{dx}\sin x = \cos x$

9. $\dfrac{d}{dx}\cos x = -\sin x$

10. $\dfrac{d}{dx}\tan x = \sec^2 x$

11. $\dfrac{d}{dx}\cot x = -\csc^2 x$

12. $\dfrac{d}{dx}\sec x = \tan x \sec x$

13. $\dfrac{d}{dx}\csc x = -\cot x \csc x$

1. $\displaystyle\int dx = x$

2. $\displaystyle\int au\, dx = a\int u\, dx$

3. $\displaystyle\int (u + v)\, dx = \int u\, dx + \int v\, dx$

4. $\displaystyle\int x^m\, dx = \dfrac{x^{m+1}}{m + 1}$ $(m \neq -1)$

5. $\displaystyle\int \dfrac{dx}{x} = \ln |x|$

6. $\displaystyle\int u\dfrac{dv}{dx}\, dx = uv - \int v\dfrac{du}{dx}\, dx$

7. $\displaystyle\int e^x\, dx = e^x$

8. $\displaystyle\int \sin x\, dx = -\cos x$

9. $\displaystyle\int \cos x\, dx = \sin x$

10. $\displaystyle\int \tan x\, dx = \ln |\sec x|$

11. $\displaystyle\int \cot x\, dx = \ln |\sin x|$

12. $\displaystyle\int \sec x\, dx = \ln |\sec x + \tan x|$

13. $\displaystyle\int \csc x\, dx = \ln |\csc x - \cot x|$

Vector Products

Let **i**, **j**, **k** be unit vectors in the x, y, z directions. Then

$$\mathbf{i} \cdot \mathbf{i} = \mathbf{j} \cdot \mathbf{j} = \mathbf{k} \cdot \mathbf{k} = 1, \qquad \mathbf{i} \cdot \mathbf{j} = \mathbf{j} \cdot \mathbf{k} = \mathbf{k} \cdot \mathbf{i} = 0,$$

$$\mathbf{i} \times \mathbf{i} = \mathbf{j} \times \mathbf{j} = \mathbf{k} \times \mathbf{k} = 0,$$

$$\mathbf{i} \times \mathbf{j} = \mathbf{k}, \qquad \mathbf{j} \times \mathbf{k} = \mathbf{i}, \qquad \mathbf{k} \times \mathbf{i} = \mathbf{j}.$$

Any vector **a** with components a_x, a_y, a_z along the x, y, z axes can be written

$$\mathbf{a} = a_x \mathbf{i} + a_y \mathbf{j} + a_z \mathbf{k}.$$

Let **a**, **b**, **c** be arbitrary vectors with magnitudes a, b, c. Then

$$\mathbf{a} \times (\mathbf{b} + \mathbf{c}) = (\mathbf{a} \times \mathbf{b}) + (\mathbf{a} \times \mathbf{c})$$

$$(s\mathbf{a}) \times \mathbf{b} = \mathbf{a} \times (s\mathbf{b}) = s(\mathbf{a} \times \mathbf{b}) \qquad (s = \text{a scalar}).$$

Let θ be the smaller of the two angles between **a** and **b**. Then

$$\mathbf{a} \cdot \mathbf{b} = \mathbf{b} \cdot \mathbf{a} = a_x b_x + a_y b_y + a_z b_z = ab \cos \theta$$

$$\mathbf{a} \times \mathbf{b} = -\mathbf{b} \times \mathbf{a} = \begin{vmatrix} \mathbf{i} & \mathbf{j} & \mathbf{k} \\ a_x & a_y & a_z \\ b_x & b_y & b_z \end{vmatrix} = (a_y b_z - b_y a_z)\mathbf{i}$$
$$+ (a_z b_x - b_z a_x)\mathbf{j} + (a_x b_y - b_x a_y)\mathbf{k}$$

$$|\mathbf{a} \times \mathbf{b}| = ab \sin \theta$$

$$\mathbf{a} \cdot (\mathbf{b} \times \mathbf{c}) = \mathbf{b} \cdot (\mathbf{c} \times \mathbf{a}) = \mathbf{c} \cdot (\mathbf{a} \times \mathbf{b})$$

$$\mathbf{a} \times (\mathbf{b} \times \mathbf{c}) = (\mathbf{a} \cdot \mathbf{c})\mathbf{b} - (\mathbf{a} \cdot \mathbf{b})\mathbf{c}$$

APPENDIX H

Winners of the Nobel Prize in Physics*

1901	Wilhelm Konrad Röntgen	1845–1923	for the discovery of the remarkable rays subsequently named after him
1902	Hendrik Antoon Lorentz	1853–1928	for their researches into the influence of magnetism upon radiation phenomena
	Pieter Zeeman	1865–1943	
1903	Antoine Henri Becquerel	1852–1908	for his discovery of spontaneous radioactivity
	Pierre Curie	1859–1906	for their joint researches on the radiation phenomena discovered by Professor Henri Becquerel
	Marie Sklowdowska-Curie	1867–1934	
1904	Lord Rayleigh (John William Strutt)	1842–1919	for his investigations of the densities of the most important gases and for his discovery of argon
1905	Philipp Eduard Anton von Lenard	1862–1947	for his work on cathode rays
1906	Joseph John Thomson	1856–1940	for his theoretical and experimental investigations on the conduction of electricity by gases
1907	Albert Abraham Michelson	1852–1931	for his optical precision instruments and metrological investigations carried out with their aid
1908	Gabriel Lippmann	1845–1921	for his method of reproducing colors photographically based on the phenomena of interference
1909	Guglielmo Marconi	1874–1937	for their contributions to the development of wireless telegraphy
	Carl Ferdinand Braun	1850–1918	
1910	Johannes Diderik van der Waals	1837–1932	for his work on the equation of state for gases and liquids
1911	Wilhelm Wien	1864–1928	for his discoveries regarding the laws governing the radiation of heat
1912	Nils Gustaf Dalén	1869–1937	for his invention of automatic regulators for use in conjunction with gas accumulators for illuminating lighthouses and buoys
1913	Heike Kamerlingh Onnes	1853–1926	for his investigations of the properties of matter at low temperatures which led, *inter alia,* to the production of liquid helium
1914	Max von Laue	1879–1960	for his discovery of the diffraction of Röntgen rays by crystals
1915	William Henry Bragg	1862–1942	for their services in the analysis of crystal structure by means of Röntgen rays
	William Lawrence Bragg	1890–1971	
1917	Charles Glover Barkla	1877–1944	for his discovery of the characteristic Röntgen radiation of the elements
1918	Max Planck	1858–1947	for his discovery of energy quanta
1919	Johannes Stark	1874–1957	for his discovery of the Doppler effect in canal rays and the splitting of spectral lines in electric fields
1920	Charles-Édouard Guillaume	1861–1938	for the service he has rendered to precision measurements in Physics by his discovery of anomalies in nickel steel alloys
1921	Albert Einstein	1879–1955	for his services to Theoretical Physics, and especially for his discovery of the law of the photoelectric effect
1922	Niels Bohr	1885–1962	for the investigation of the structure of atoms, and of the radiation emanating from them
1923	Robert Andrews Millikan	1868–1953	for his work on the elementary charge of electricity and on the photoelectric effect
1924	Karl Manne Georg Siegbahn	1888–1979	for his discoveries and research in the field of x-ray spectroscopy
1925	James Franck	1882–1964	for their discovery of the laws governing the impact of an electron upon an atom
	Gustav Hertz	1887–1975	
1926	Jean Baptiste Perrin	1870–1942	for his work on the discontinuous structure of matter, and especially for his discovery of sedimentation equilibrium
1927	Arthur Holly Compton	1892–1962	for his discovery of the effect named after him
	Charles Thomson Rees Wilson	1869–1959	for his method of making the paths of electrically charged particles visible by condensation of vapor

* See *Nobel Lectures, Physics,* 1901–1970, Elsevier Publishing Company for biographies of the awardees and for lectures given by them on receiving the prize.

1928	Owen Willans Richardson	1879–1959	for his work on the thermionic phenomenon and especially for the discovery of the law named after him
1929	Prince Louis-Victor de Broglie	1892–	for his discovery of the wave nature of electrons
1930	Sir Chandrasekhara Venkata Raman	1888–1970	for his work on the scattering of light and for the discovery of the effect named after him
1932	Werner Heisenberg	1901–1976	for the creation of quantum mechanics, the application of which has, among other things, led to the discovery of the allotropic forms of hydrogen
1933	Erwin Schrödinger	1887–1961	for the discovery of new productive forms of atomic theory
	Paul Adrien Maurice Dirac	1902–1984	
1935	James Chadwick	1891–1974	for his discovery of the neutron
1936	Victor Franz Hess	1883–1964	for the discovery of cosmic radiation
	Carl David Anderson	1905–1984	for his discovery of the positron
1937	Clinton Joseph Davisson	1881–1958	for their experimental discovery of the diffraction of electrons by crystals
	George Paget Thomson	1892–1975	
1938	Enrico Fermi	1901–1954	for his demonstrations of the existence of new radioactive elements produced by neutron irradiation, and for his related discovery of nuclear reactions brought about by slow neutrons
1939	Ernest Orlando Lawrence	1901–1958	for the invention and development of the cyclotron and for results obtained with it, especially with regard to artificial radioactive elements
1943	Otto Stern	1888–1969	for his contribution to the development of the molecular ray method and his discovery of the magnetic moment of the proton
1944	Isidor Isaac Rabi	1898–	for his resonance method for recording the magnetic properties of atomic nuclei
1945	Wolfgang Pauli	1900–1958	for the discovery of the Exclusion Principle, also called the Pauli Principle
1946	Percy Williams Bridgman	1882–1961	for the invention of an apparatus to produce extremely high pressures, and for the discoveries he made therewith in the field of high-pressure physics
1947	Sir Edward Victor Appleton	1892–1965	for his investigations of the physics of the upper atmosphere, especially for the discovery of the so-called Appleton layer
1948	Patrick Maynard Stuart Blackett	1897–1974	for his development of the Wilson cloud chamber method, and his discoveries therewith in the fields of nuclear physics and cosmic radiation
1949	Hideki Yukawa	1907–1981	for his prediction of the existence of mesons on the basis of theoretical work on nuclear forces
1950	Cecil Frank Powell	1903–1969	for his development of the photographic method of studying nuclear processes and his discoveries regarding mesons made with this method
1951	Sir John Douglas Cockcroft	1897–1967	for their pioneer work on the transmutation of atomic nuclei by artificially accelerated atomic particles
	Ernest Thomas Sinton Walton	1903–	
1952	Felix Bloch	1905–1983	for their development of new methods for nuclear magnetic precision methods and discoveries in connection therewith
	Edward Mills Purcell	1912–	
1953	Frits Zernike	1888–1966	for his demonstration of the phase-contrast method, especially for his invention of the phase-contrast microscope
1954	Max Born	1882–1970	for his fundamental research in quantum mechanics, especially for his statistical interpretation of the wave function
	Walther Bothe	1891–1957	for the coincidence method and his discoveries made therewith

1955	Willis Eugene Lamb	1913–	for his discoveries concerning the fine structure of the hydrogen spectrum
	Polykarp Kusch	1911–	for his precision determination of the magnetic moment of the electron
1956	William Shockley	1910–	for their researches on semiconductors and their discovery of the transistor effect
	John Bardeen	1908–	
	Walter Houser Brattain	1902–	
1957	Chen Ning Yang	1922–	for their penetrating investigation of the so-called parity laws which has led to important discoveries regarding the elementary particles
	Tsung Dao Lee	1926–	
1958	Pavel Aleksejevič Čerenkov	1904–	for the discovery and the interpretation of the Cerenkov effect
	Il' ja Michajlovič Frank	1908–	
	Igor' Evgen'evič Tamm	1895–1971	
1959	Emilio Gino Segrè	1905–	for their discovery of the antiproton
	Owen Chamberlain	1920–	
1960	Donald Arthur Glaser	1926–	for the invention of the bubble chamber
1961	Robert Hofstadter	1915–	for his pioneering studies of electron scattering in atomic nuclei and for his thereby achieved discoveries concerning the structure of the nucleons
	Rudolf Ludwig Mössbauer	1929–	for his researches concerning the resonances absorption of γ-radiation and his discovery in this connection of the effect which bears his name
1962	Lev Davidovič Landau	1908–1968	for his pioneering theories of condensed matter, especially liquid helium
1963	Eugene P. Wigner	1902–	for his contributions to the theory of the atomic nucleus and the elementary particles, particularly through the discovery and application of fundamental symmetry principles
	Maria Goeppert Mayer	1906–1972	for their discoveries concerning nuclear shell structure
	J. Hans D. Jensen	1907–1973	
1964	Charles H. Townes	1915–	for fundamental work in the field of quantum electronics which has led to the construction of oscillators and amplifiers based on the maser-laser principle
	Nikolai G. Basov	1922–	
	Alexander M. Prochorov	1916–	
1965	Sin-itiro Tomonaga	1906–1979	for their fundamental work in quantum electrodynamics, with deep-ploughing consequences for the physics of elementary particles
	Julian Schwinger	1918–	
	Richard P. Feynman	1918–	
1966	Alfred Kastler	1902–1984	for the discovery and development of optical methods for studying Hertzian resonance in atoms
1967	Hans Albrecht Bethe	1906–	for his contributions to the theory of nuclear reactions, especially his discoveries concerning the energy production in stars
1968	Luis W. Alvarez	1911–	for his decisive contribution to elementary particle physics, in particular the discovery of a large number of resonance states, made possible through his development of the technique of using hydrogen bubble chamber and data analysis
1969	Murray Gell-Mann	1929–	for his contributions and discoveries concerning the classification of elementary particles and their interactions
1970	Hannes Alvén	1908–	for fundamental work and discoveries in magneto-hydrodynamics with fruitful applications in different parts of plasma physics
	Louis Néel	1904–	for fundamental work and discoveries concerning antiferromagnetism and ferrimagnetism which have led to important applications in solid state physics

1971	Dennis Gabor	1900–1979	for his discovery of the principles of holography
1972	John Bardeen	1908–	for their development of a theory of super-
	Leon N. Cooper	1930–	conductivity
	J. Robert Schrieffer	1931–	
1973	Leo Esaki	1925–	for his discovery of tunneling in semiconductors
	Ivar Giaever	1929–	for his discovery of tunneling in superconductors
	Brian D. Josephson	1940–	for his theoretical prediction of the properties of a super-current through a tunnel barrier
1974	Antony Hewish	1924–	for the discovery of pulsars
	Sir Martin Ryle	1918–1984	for his pioneering work in radioastronomy
1975	Aage Bohr	1922–	for the discovery of the connection between
	Ben Mottelson	1926–	collective motion and particle motion and
	James Rainwater	1917–	the development of the theory of the structure of the atomic nucleus based on this connection
1976	Burton Richter	1931–	for their (independent) discovery of an im-
	Samuel Chao Chung Ting	1936–	portant fundamental particle.
1977	Philip Warren Anderson	1923–	for their fundamental theoretical investiga-
	Nevill Francis Mott	1905–	tions of the electronic structure of magnetic
	John Hasbrouck Van Vleck	1899–1980	and disordered systems
1978	Peter L. Kapitza	1894–1984	for his basic inventions and discoveries in the area of low-temperature physics
	Arno A. Penzias	1926–	for their discovery of cosmic microwave
	Robert Woodrow Wilson	1936–	background radiation
1979	Sheldon Lee Glashow	1932–	for their unified model of the action of the
	Abdus Salam	1926–	weak and electromagnetic forces and for
	Steven Weinberg	1933–	their prediction of the existence of neutral currents
1980	James W. Cronin	1931–	for the discovery of violations of fundamental
	Val L. Fitch	1923–	symmetry principles in the decay of neutral K mesons
1981	Nicolaas Bloembergen	1920–	for their contribution to the development of
	Arthur Leonard Schawlow	1921–	laser spectroscopy
	Kai M. Siegbahn	1918–	for his contribution to the development of high-resolution electron spectroscopy
1982	Kenneth Geddes Wilson	1936–	for his method of analyzing the critical phenomena inherent in the changes of matter under the influence of pressure and temperature
1983	Subrahmanyan Chandrasekhar	1910–	for his theoretical studies of the structure and evolution of stars
	William A. Fowler	1911–	for his studies of the formation of the chemical elements in the universe
1984	Carlo Rubbia	1934–	for their decisive contributions to the large
	Simon van der Meer	1925–	project, which led to the discovery of the field particles W and Z, communicators of the weak interaction
1985	Klaus von Klitzing	1943–	for his discovery of the quantized Hall resistance

Answers to Odd Numbered Exercises and Problems

CHAPTER 1

Exercises

3. 186 mi.   **5.** (a) 10^9.   (b) 10^{-4}.   (c) 9.14×10^5.   **7.** (a) 2.99×10^{-26} kg.   (b) 4.68×10^{46}.
9. 52.6 min; 5.2%.   **11.** 720 days.

Problems

1. 32.2 km.   **3.** (a) 251 ft^2.   (b) 23.3 m^2.   (c) 3060 ft^3.   (d) 86.6 m^3.   **5.** (a) 11.3 m^2/l.   (b) 1.13×10^4 m^{-1}.
7. (a) 4.85×10^{-6} pc; 1.58×10^{-5} ly.   (b) 5.87×10^{12} mi; 1.91×10^{13} mi.   **9.** 1.32×10^9 kg.   **11.** 0.12 AU/min.
13. 0.0012 s shorter.   **15.** 2 days 5 hours.   **17.** (a) 100 m exceeds 100 yd by 8.56 m or 28.1 ft.   (b) One mile exceeds one
metric mile by 109 m or 358 ft.   **19.** 3.80 mg/s.   **21.** (a) 1 in.^{2} = 6.45 cm^2.   (b) 1 mi^2 = 2.59 km^2.   (c) 1 m^3 = 10^6 cm^3.
(d) 1 ft^2 = 0.111 yd^2.   **23.** C, D, A, B, E (best to worst). The important criterion is the constancy of the daily variation,
not its magnitude.   **25.** 2.03 h.

CHAPTER 2

Exercises

1. The displacements should be (a) parallel;   (b) antiparallel;   (c) perpendicular.   **3.** (a) 370 m, 35.8° north of east.
(b) Displacement magnitude = 370 m; distance walked = 425 m.   **5.** (a) 47 units.   (b) 122°.   **7.** (a) 13 m; 7.5 m.
(b) 15 m; 0.   (c) 7.5 m; -13 m.   **9.** (a) $\mathbf{i} + 10\mathbf{j}$.   (b) 10.05, at 84.3° with the x-axis.   **11.** (a) 10 m, north.
(b) 7.5 m, south.   **13.** (a) The $+z$-direction.   (b) The $-z$-direction.   (c) The $+y$-direction for $d > 0$;
the $-y$-direction for $d < 0$.   (d) Zero.   **17.** (a) Up, unit magnitude.   (b) Zero.   (c) South, unit magnitude.   (d) 1.00.
(e) Zero.

Problems

1. 81 km, 40° north of east.   **3.** (a) 38 units at 320°.   (b) 130 units at 1.2°.   (c) 62 units at 130°.   **5.** (a) $a_x = -2.83$,
$a_y = -2.83$; $b_x = +5.00$, $b_y = 0$; $c_x = +3.00$, $c_y = 5.20$, m.   (b) $d_x = +5.17$, $d_y = 2.37$, m.   (c) 5.69 m, 24.6°
north of east.   (d) 5.69 m, 24.6° south of west.   **7.** 4.74 km.   **11.** 168 cm, 32.5° above the floor.   **13.** (a) -18.8.
(b) 26.9, $+z$-direction.   **19.** (a) 2.97.   (b) $1.51\mathbf{i} + 2.67\mathbf{j} - 1.36\mathbf{k}$.   (c) 48.6°.   **21.** 6.0 ft, 20.5° east of north.
23. (a) 5; 323°.   (b) 10; 53°.   (c) 11; 27°.   (d) 11; 80°.   (e) 11; 260°. The angles are with the $+x$-axis.
25. 3390 ft, horizontally.   **27.** $(\mathbf{A} + \mathbf{B})$ points 37° east of north and has a magnitude of 3.6 units; $(\mathbf{B} - \mathbf{A})$ points 70° west of
north and has a magnitude of 8.3 units.   **29.** (a) 57°.   (b) $c_x = \pm 2.2$; $c_y = \mp 4.5$.   **31.** 22.2°.   **39.** (b) 11,230 km.

CHAPTER 3

Exercises

1. 81 ft (24 m).   **3.** 1 h 13 min (1 h 14 min).   **5.** 3.50 m/s.   **7.** (a) 1.00 m/s; 0.33 m/s^2.   (b) 1.33 m/s; 0.33 m/s^2.
9. (a) v_x a_x (b) No.   **11.** Both accelerations are equal to 0.278 m/s^2.   **13.** (a) 25g.   (b) 400 m.

OA	+	$-$
AB	0	0
BC	+	+
CD	+	0

15. (a) 3.06×10^6 s.   (b) 4.59×10^{13} m.   **17.** (a) 8.36 m/s^2; $0.85g$.   (b) 3.21 s; $\approx 8T$.   **19.** (a) No.   (b) m^2/s^2; m/s^2.
21. (a) 47.7 m.   (b) 30.6 m/s.

Problems

1. 40 km/h.   **3.** (a) 6.79 km/h.   (b) $6.96°$.   **5.** 47.7 s.   **9.** (b) $-0.040, -0.035, -0.010, 0.0$ m/s.
(c) $-0.040, -0.020, 0.0, +0.020, +0.040, +0.060$ m/s.   (e) $+0.020$ m/s^2.   **11.** (a) 80 m/s.   (b) 110 m/s.   (c) 20 m/s^2.
13. 0.10 m.   **15.** 8.0×10^{14} m/s^2.   **17.** (a) 5.00 m/s.   (b) 1.67 m/s^2.   (c) 7.50 m.   **19.** (a) $x = 6.0$ m; $t = 2.0$ s.
(b) 3.0 m/s; -4.0 m/s.   **21.** (a) L/T^2, ft/s^2; L/T^3, ft/s^3.   (b) 2 s.   (c) 24 ft.   (d) -16 ft.   (e) $3, 0, -9, -24$ ft/s.
(f) $0, -6, -12, -18$ ft/s^2.   **23.** (a) 1.77 s.   (b) 0.73 s.   **25.** 3.96 m/s.   **27.** 857 m/s^2, up.   **29.** (a) 5.45 s.
(b) 41.4 m/s.   **31.** (a) 45.0 mi/h (72.4 km/h).   (b) 42.8 mi/h (68.8 km/h).   (c) 43.9 mi/h (70.6 km/h).
33. (a) 11.9 ft/s^2 (3.41 m/s^2).   (b) 3.73 ft/s (1.92 m/s).   **35.** (a) 8.85 m/s.   (b) 1.00 m.   **37.** (a) 73.5 mi/h.
(b) 49.0 s.   (c) 26.1 mi/h.   **39.** (a) 28.5 cm/s.   (b) 18.0 cm/s.   (c) 40.5 cm/s.   (d) 28.1 cm/s.   (e) 30.4 cm/s.
41. (a) 40 m (130 ft).   (b) 8.8 m/s (29 ft/s).   (c) 16 m/s (51 ft/s), up.   **43.** (a) 1.54 s.   (b) 27.1 m/s.   **45.** $96g$.
47. (a) 40.0 ft/s.   **49.** (a) 0.75 s.   (b) -20 ft/s^2.   **51.** (a) 76 m above the ground.   (b) 4.1 s.   **53.** 0.75 in. (1.54 cm).
55. (a) 0.707 s.   (b) 2.34 ft.

CHAPTER 4

Exercises

3. (a) $\mathbf{v} = 8t\mathbf{j} + \mathbf{k}$; $\mathbf{a} = 8\mathbf{j}$.   (b) A parabola.   **5.** 10 ft/s.   **7.** 800 ft/s.   **9.** (a) 3.03 s.   (b) 758 m.   (c) 29.7 m/s.
11. 1.92 in.   **13.** (a) 16.9 m; 8.21 m.   (b) 27.6 m; 7.26 m.   **15.** (a) 1.15 s.   (b) 12.0 m.   (c) 19.2 m/s; 4.80 m/s.
(d) No.   **17.** 2370 m/s^2.   **19.** 8×10^5 m/s^2.   **21.** 36 s.   **23.** The motorist is approaching the police car along the line
of sight at 100 km/h.

Problems

1. (a) 670 mi, $63°$ south of east.   (b) 300 mi/h, $63°$ south of east.   (c) 400 mi/h.   **3.** $-2.10\mathbf{i} + 2.81\mathbf{j}$, m/s^2.   **9.** (a) It will hit
the ground 54 ft from third base.   (b) $7.6°$.   (c) 1.03 s.   **11.** 78.3 ft/s (24.0 m/s) at $64.8°$ ($64.9°$) above the horizontal.
13. Approximately 40 m.   **15.** (b) His record would have been longer by about 1 cm.   **17.** (a) 18.5 m/s.   (b) 35.4.
19. 2.58 cm/s^2.   **21.** (a) 92.   (b) 9.6.   (c) $92 = (9.6)^2$.   **23.** 0.964 m/s, vertically up.   **25.** 185 km/h, $22°$ south of west.
27. (a) $46.8°$ east of north.   (b) 6 min 36 s.   **29.** $60°$.   **31.** (a) 2.0 mm.   (b) $v_{hor} = 1.0 \times 10^9$ cm/s;
$v_{vert} = 0.20 \times 10^9$ cm/s down.   **33.** (a) $x = -27$ m, $y = -27$ m; $\mathbf{r} = 38.2$ m at $225°$.   (b) 12.7 m/s at $225°$.
(c) 22.8 m/s at $203°$.   (d) 8.54 m/s^2 at $201°$.   (e) 10.0 m/s^2 at $143°$.   **35.** (a) $-18\mathbf{i}$, m/s^2 (b) 0.750 s.   (c) Never.
(d) 2.19 s.   **37.** (a) $-1.5\mathbf{j}$, m/s.   (b) $x = 4.5$ m, $y = -2.25$ m.   **39.** 19 ft/s.   **41.** 80.1 m/s.   **43.** (a) 4.2 m, $45°$;
5.5 m, $68°$; 6.0 m, $90°$.   (b) 4.2 m, $135°$.   (c) 0.85 m/s, $135°$.   (d) 0.94 m/s, $90°$; 0.94 m/s, $180°$.
(e) 0.27 m/s^2, $225°$. (f) 0.30 m/s^2, $180°$; 0.30 m/s^2, $270°$. Angles measured counterclockwise from the x-axis.   **45.** (a) $v_x = 0.36\pi$,
$v_y = -0.48\pi$, m/s.   (b) $a_x = -0.48\pi^2$, $a_y = -0.36\pi^2$, m/s^2.   **47.** $78.5°$.   **49.** Between the angles $31°$ and $63°$
above the horizontal.   **51.** (a) 38.4 knots, $1.53°$ east of north.   (b) 4.17 h.   (c) $1.53°$ west of south.   **53.** 5.66 s.

CHAPTER 5

Exercises

1. 6.3 years.   **3.** 16.4 N.   **5.** (a) 9.11×10^{22} m/s^2, toward the nucleus.   (b) 8.29×10^{-8} N, toward the nucleus.
7. (a) 10 lb.   (b) 10 lb.   **9.** (a) Mass $= 43.75$ slug; weight $= 1400$ lb.   (b) Mass $= 400$ kg; weight $= 3920$ N.
11. (a) 15 lb.   (b) The road.   **13.** 1425 lb.   **15.** 110 lb.   **17.** (a) 4.9×10^5 N (110,000 lb).
(b) 1.49×10^6 N (337,000 lb).

Problems

1. (a) 1300 lb (5400 N).   (b) 5.5 s.   (c) 50 ft (15 m).   (d) 2.7 s.   **3.** (a) $\mathbf{r} = 30t\mathbf{i} + t^2\mathbf{j}$, m; a parabola.
(b) $\mathbf{v} = 30\mathbf{i} + 30\mathbf{j}$, m/s.   (c) $\mathbf{r} = 450\mathbf{i} + 225\mathbf{j}$, m.   **5.** (a) 1.0 N.   **7.** (a) 3260 N.   (b) 1290 kg.   (c) 2.52 m/s^2.
9. 2.2×10^4 N.   **11.** (a) 42.7 lb.   (b) 10.7 ft/s^2.   **13.** (a) 250 m/s^2 (790 ft/s^2).   (b) 2.0×10^4 N (4300 lb).
15. (a) 1.23, 2.46, 3.69, 4.92 N.   (b) 6.15 N.   (c) 0.25 N.   **17.** (a) 6750 lb.   (b) 5250 lb.   **19.** (a) 5.9 m/s.
21. (a) 2.18 m/s^2.   (b) 116 N.   (c) 21.0 m/s^2.   **23.** (a) $\dfrac{M}{M + 4M}g$.   (b) $\dfrac{2mM}{M + 4m}g$.   **25.** (a) 1.0×10^8 N; 5.2×10^6 N.
(b) 4.2 y.; 4.3 y.   **27.** 1.0 m/s^2, $37°$ counterclockwise from the 4.0-N face.   **29.** 1.47 N.   **31.** 402 N.   **33.** (a) 32 lb.
(b) 55 lb.   (c) 16 ft/s^2.   **35.** Lower object with an acceleration ≥ 4.2 ft/s^2.   **37.** (a) 0.50 slug (7.3 kg).
(b) 20 lb (89 N).   **39.** (a) 566 N.   (b) 1130 N.   **41.** $2M \dfrac{a}{a + g}$.   **43.** (a) 466 N.   (b) 527 N.

CHAPTER 6

Exercises

1. 490 N, down. **3.** (b) 72.7 N. (c) 26.9 N. **5.** 50.4 N. **7.** 8820 N. **9.** 9.31 m/s². **11.** (a) 20.5 lb (90.2 N).
(b) 16.0 lb (70.4 N). (c) 2.88 ft/s² (0.882 m/s²). **13.** (a) 0.13 N. (b) 0.12. **15.** (a) 275 N. (b) 877 N.
17. 34.6 ft/s (10.8 m/s). **19.** 0.0206. **21.** (a) 0.722 m/s. (b) 2.09 m/s². (c) 0.501 N.

Problems

1. (a) 68 lb. (b) 4.2 ft/s². **3.** 7.51 m/s². **5.** 0.525. **7.** (a) 6.1 m/s². (b) 0.98 m/s². **9.** 40 lb (180 N).

11. (a) 1.05 N, in tension. (b) 3.62 m/s². (c) No, except that the rod is under compression. **13.** $\left(1 + \dfrac{m}{M}\right)mg/\mu_s$.

15. $(Mgr/m)^{1/2}$. **17.** (a) 11°. (b) 0.19. **19.** (a) 175 lb. (b) 50 lb. **21.** (a) At the bottom of the circle.
(b) 30.9 ft/s. **23.** 1.52 km. **25.** $\mu_s = 0.577$; $\mu_k = 0.541$. **27.** (a) 8.57 N. (b) 46.2 N. (c) 38.6 N. **29.** (a) 0.43.
(b) 42 m. **31.** 36 m. **33.** $0.124 < \mu_s < 0.227$. **35.** 3.27 kg. **37.** 178 km/h. **39.** (a) 5.0 lb. (b) 5.4 lb.
(c) 75 lb. **41.** (a) 11.4 ft/s². (b) 0.46 lb. (c) Blocks move independently unless the plane is so long that they ultimately collide.
43. $g(\sin\theta - \sqrt{2}\mu_k\cos\theta)$.

CHAPTER 7

Exercises

1. (a) 564 J. (b) −525 J. (c) Zero. (d) Zero. (e) 39 J. **3.** 566 J. **5.** (a) 30.1 J. (b) −30.1 J.
(c) Zero. (d) 0.225. **7.** 800 J. **9.** 102 ft (31.2 m). **11.** (a) −3750 J. (b) 3.13 × 10⁴ N. **13.** −9.72 J.
15. (a) 0.771 mi. (b) 71.3 kW. **17.** (a) 3000 J. (b) 300 W. **19.** 227 hp (1.69 × 10⁵ W).

Problems

1. (a) 200 N; 625 m; −1.25 × 10⁵ J. (b) 500 N; 250 m; −1.25 × 10⁵ J. **3.** (a) 2240 J. (b) −770 J. (c) −1470 J.
9. (a) 2.9 × 10⁷ m/s. (b) 1.3 MeV. **11.** (a) 410 ft·lb (557 J). (b) −113 ft·lb (−153 J). (c) 17.9 ft/s (5.46 m/s).
13. (a) 0.294 J. (b) −1.80 J. (c) 3.47 m/s. (d) 22.9 cm. **15.** (b) $m(v_f^2/t_f^2)t$. (c) 93750 W. **17.** 0.985 hp.
19. (a) 2.1 × 10⁶ kg. (b) $\sqrt{100 \times 1.46t}$, m/s. (c) $(1.5 \times 10^6)/\sqrt{100 \times 1.46t}$, N. (d) 6.69 km. **21.** 4.7 MeV.
23. Man, 2.4 m/s; boy, 4.8 m/s. **25.** (a) 1.8 × 10⁵ ft·lb. (b) 0.55 hp. **27.** (a) 100 J. (b) 33.3 W.
29. (a) 8.98 × 10⁴ "megatons of TNT". (b) 44.8 km. **31.** 528 J. **33.** 0.24 hp. **35.** 90 hp. **37.** 69.0 hp.

CHAPTER 8

Exercises

1. (a) 7.81 × 10⁶ J = 7.81 MJ. (b) 6.21. **3.** 4.00 × 10⁴ J. **5.** (a) 196 J. (b) 167 J. (c) 12.9 m/s.
7. (a) 7.84 N/cm. (b) 50.0 cm. **9.** 1.98 m/s. **11.** 2.29 m/s. **13.** 831 ft. **15.** 234 N, up. **19.** (a) 1.02 × 10⁶ J.
(b) −5.10 × 10⁵ J. **21.** 2.74 × 10¹⁵ J; 5.96 × 10⁸ kg of TNT.

Problems

1. (a) $U(x) = -Gm_1m_2/x^2$. (b) $Gm_1m_2d/x_1(x_1 + d)$. **3.** 10 cm. **5.** (a) 31.0 J. (b) 5.33 m/s. (c) Conservative.
7. (a) 4.67 N. (b) Between $x = 1.2$ m and $x = 14$ m. (c) 3.61 m/s. **9.** (a) 4.85 m/s. (b) 2.42 m/s.
11. (a) v_0. (b) $(v_0^2 + gh)^{1/2}$. (c) $(v_0^2 + 2gh)/2L$. **15.** (a) 39.5 cm. (b) 3.65 cm. **17.** (a) 44.3 m/s. (b) 0.036.
19. In the center of the flat part. **21.** (a) 3.00 ft. (b) 9.00 ft. (c) 48.8 ft. **23.** (a) 0.99c. (b) 0.0805.
25. 266 times the mean circumference of the earth. **27.** 7.82 kJ. **29.** (a) $U(x) = -1/2kx^2$. **33.** (a) 1.2 J. (b) 11 m/s.
35. (a) 34.7 cm. (b) 169 cm/s. **37.** 1.25 cm. **39.** (a) $2\sqrt{gl}$. (b) $5mg$. (c) 71°. **41.** 4.2 in. **43.** 1.16 m/s.
47. 191 tons. **49.** (a) $(2gl)^{1/2}$. (b) $5mg$. (c) $-mgl$. (d) $-2mgl$.

CHAPTER 9

Exercises

3. 4900 km. **5.** 0.57 m/s (1.9 ft/s), toward the center of mass. **7.** (a) Down; $mv/(m + M)$. (b) Balloon again stationary.
9. 7560 slug·ft/s. **11.** (a) 7.6 kg·m/s. (b) 2900 J. (c) 1.7 cm/s. **13.** (a) 1.0 kg·m/s. (b) 250 J. (c) 10 N.
(d) 1670 N. **15.** 4060 km/h. **17.** (a) 0.476 m/s. (b) 0.452 m/s. **19.** (a) 2.72. (b) 7.39.

Problems

1. 0.2L from the heavy rod, along the symmetry axis. **3.** $R/6$. **5.** 58 kg. **7.** (a) 1700 N. (b) 500 J. **9.** (a) 6.4 J.
(b) $P_i = 0.80$ kg·m/s, 30° above the horizontal; $P_f = 0.80$ kg·m/s, 30° below the horizontal. **11.** Three times the weight of chain
already on the table. **13.** (a) 1.4×10^{-22} kg·m/s, 150° from the electron track and 120° from the neutrino track.
(b) 1.0 eV (= 1.6×10^{-19} J). **15.** $wv_{rel}/(W + w)$. **17.** (a) 59 kg. (b) 180 kg. **19.** (a) 540 m/s. (b) 40.4°.
21. (a) 4.00×10^4 kg·m/s, west. (b) West. **23.** (a) 6.94×10^4 J. (b) 3.56×10^4 kg·m/s; 38.7° south of east.
25. (a) $10\mathbf{i} + 15\mathbf{j}$, m/s. (b) 500 J is lost. **27.** 6.1 s. **29.** (a) L. (b) Zero. **31.** Fast barge: 46 N more; slow barge:
no change. **33.** 7230 N. **35.** (a) Midway between them. (b) It moves 1.0 mm toward the heavier body. (c) 0.0016g, down.

37. (b) $(HM/m)\left(\sqrt{1 + \dfrac{m}{M}} - 1\right)$.

CHAPTER 10

Exercises

1. $2mv/t$. **3.** 2.5 m/s. **5.** 53.5 m/s. **7.** 1.81 m/s. **9.** 7.8%. **11.** 4 m/s. **13.** ~2 mm/y.
15. (a) 1.95×10^5 kg·m/s for each direction of thrust. (b) Backward: +66.1 MJ; forward: −50.9 MJ; sideways: +7.6 MJ.
17. 38.3 mi/h, 63.4° south of west. **19.** (a) 117° from the final direction of B. (b) No. **21.** 8.12 MeV.

Problems

1. (a) 42 N·s. (b) 2100 N. **5.** 41.7 cm/s. **7.** 1420 m/s. **9.** 1.2 kg. **11.** 11.25 lb (49.6 N). **13.** 12.9 tons.
15. (a) Left mass comes to rest; center mass: $\dfrac{m - M}{m + M} v_0$ to the right; right mass: $2\dfrac{m}{m + M} v_0$ to the right. (b) Left mass: $\dfrac{M - m}{M + m} v_0$ to

the left; center mass comes to rest; right mass: $2\dfrac{m}{m + M} v_0$ to the right. **17.** (a) 1010 m/s, 9.46° clockwise from the +x-axis.

(b) 3.23 MJ. **19.** 120°. **21.** (a) 5400 MeV. (b) 1.09×10^7 m/s; 1.47×10^7 m/s. (c) 81.7 MeV; 110 MeV.
23. 10 m/s. **25.** 1.5 N (0.33 lb). **27.** (a) 33.3 g. (b) 15.0 m/s. **29.** (a) −1.19 MeV. (b) 2.07 MeV.

31. $d\left(\dfrac{m_1}{m_1 + m_2}\right)^2$. **33.** 20 J for the heavy particle, 40 J for the light particle. **35.** $\mathbf{v}_2$ and $\mathbf{v}_3$ will be at 30° to $\mathbf{v}_0$ and will

have a magnitude of 6.9 m/s; $\mathbf{v}_1$ will be in the opposite direction to $\mathbf{v}_0$ and will have a magnitude of 2.0 m/s. **37.** 8.8 m/s.
39. 7.81 kg (weight = 17.8 lb). **41.** $M/(m + M)$.

CHAPTER 11

Exercises

1. (a) 1.50 rad. (b) 85.9°. (c) 1.49 m. (d) 37.8 m. **3.** 20. **5.** $0.80\omega_0$. **7.** (a) −0.27 rad/s². (b) 20.
9. (a) −1.26 rad/s². (b) 6.28 rad/s. (c) 251 rad. (d) 40.0. **11.** (a) 20.6 in./s. (b) 10.1 in./s.
13. 6.1 ft/s² (1.83 m/s²). **15.** (a) 20.9 rad/s. (b) 12.5 m/s. (c) 1.40 rad/s². (d) 600.

Problems

1. (a) 2.00 rad/s². (b) 5.00 rad/s. (c) 10.0 rad/s. (d) 75.0 rad. **3.** 11.0 rad/s. **5.** (a) 340 s.
(b) $−4.5 \times 10^{-3}$ rad/s². (c) 104 s. **7.** 8.00 s. **9.** (a) 1.04 rev/s². (b) 4.8 s. (c) 9.6 s. (d) 48.
11. (a) 2.0×10^{-7} rad/s. (b) 30 km/s. (c) 6.0 mm/s², toward the sun. **13.** (a) 30.0 m/s. (b) 6.00 m/s².
(c) 90.0 m/s². (d) 90.2 m/s². **15.** (a) $\sqrt{\mu_s g/R}$. **17.** (a) 13.5 s. (b) 27.0 rad/s. **19.** (a) 6.3×10^4 ft/min
(1.9×10^4 m/min). (b) 6.8×10^4 ft/min (2.1×10^4 m/min). **21.** 400; 12,000; 1500. **23.** (a) $\omega_0 + at^4 - bt^3$.
(b) $\theta_0 + \omega_0 t + at^5/5 - bt^4/4$. **25.** (a) 59.3 rad/s (58.7 rad/s). (b) −9.32 rad/s² (−9.14 rad/s²). (c) 70.7 m (236 ft).
27. (a) 3.8×10^3 rad/s. (b) 190 m/s. **29.** (a) 150 cm/s. (b) 15.0 rad/s. (c) 15.0 rad/s. (d) 75.0 cm/s.
(e) 3.00 rad/s.

CHAPTER 12

Exercises

1. 4.59 N·m. **3.** 1.04×10^8 kg·m²/s. **5.** (a) 3.15×10^{43} kg·m²/s. (b) 0.616. **7.** (a) 12 kg·m²/s, out of page.
(b) 3.0 N·m, out of page. **9.** (a) $r_2F_2\sin\theta_2 - r_1F_1\sin\theta_1$. (b) 3.85 N·m, into page. **11.** (a) $14ml^2$. (b) $4ml^2\omega$.
(c) $14ml^2\omega$. **13.** $I_{xx} = 1305$; $I_{yy} = 545$; $I_{zz} = 1850$, g·cm². **15.** First cylinder: 1079 J; second cylinder: 9708 J.
17. 12.4 kg·m². **19.** 308 kg·m²/s, down. **21.** (Solid spheres) (a) Small sphere, 0.689 N·m; large sphere, 9.84 N·m.

(b) Small sphere, 3.05 N; large sphere, 11.5 N. (Spherical shells)   (a) Small sphere, 1.15 N·m; large sphere, 16.4 N·m.
(b) Small sphere, 5.08 N; large sphere, 19.2 N.   **23.** 292 ft·lb (396 N·m).   **25.** 11.2 ft.   **27.** (a) $mvR/(I + MR^2)$.
(b) $mvR^2/(I + MR^2)$.

Problems

1. (a) $\mathbf{i}(yF_z - zF_y) + \mathbf{j}(zF_x - xF_z) + \mathbf{k}(xF_y - yF_x)$.   **5.** (a) $-170\mathbf{k}$, kg·m²/s.   (b) $56\mathbf{k}$, N·m.   (c) $56\mathbf{k}$, N·m.   **9.** $l^2\omega^2/6g$.
11. $\sqrt{9g/4l}$.   **13.** (a) 8.3×10^{28} N·m.   (b) 2.6×10^{29} J.   (c) 3.0×10^{21} kW.   **15.** 500 rad/s.   **17.** 140 cm/s.
19. (a) 62.6 rad/s.   (b) 4.01 m.   **21.** (a) 990 J.   (b) 3000 J.   (c) 1.1×10^5 J.   **23.** (a) $mg(R - r)$; $\dfrac{2}{7}$; $\dfrac{5}{7}$.

(b) $\dfrac{17}{7}\,mg$.   **25.** (a) Each revolves in a circle of radius 1.5 m at 6.7 rad/s.   (b) The angular speed increases to 60 rad/s.
(c) $K_b/K_a = 9.0$ **27.** (a) $\omega_0(R_1I_2/R_2I_1 + R_2/R_1)^{-1}$.   (b) No.   **29.** 6.75×10^{12} rad/s.   **31.** 3.0 min.   **35.** (a) ½mR^2.
(b) A solid circular cylinder.   **37.** 500.   **39.** (c) 1.72 m/s²; 6.92 m/s².   **41.** (a) $(I\omega_0 - mvR)/(I + mR^2)$.   (b) No.
43. (a) 0.148 rad/s.   (b) 0.0123.   (c) It will rotate through 181°.   **45.** 1.39×10^5 g·cm².   **47.** (a) $2\theta/t^2$.

(b) $2R\theta/t^2$.   (c) $T_2 = Mg - (2\theta/t^2)(MR + I/R)$; $T_1 = M(g - 2R\theta/t^2)$.   **49.** (a) 2.7R.   (b) $\dfrac{50}{7}\,mg$.

51. 416 m/s.   **53.** 4.53 m/s.

CHAPTER 13

Exercises

1. (a) Yes.   (b) No.   (c) No.   **3.** Its weight, 49 N; the upward force (= 49 N) exerted by the table.   **5.** (a) $(w/L)\sqrt{L^2 + r^2}$.
(b) wr/L.   **7.** (a) 625 lb (2780 N).   (b) 875 lb (3890 N).   **9.** 1900 lb.   **11.** Left pedestal: 1160 N (tension); right pedestal:
1740 N (compression).   **13.** Three-fourths the length of the beam from the man at the end.   **15.** 0.536 m.

Problems

1. 0.29.   **3.** (a) Bottom: 2W; sides: W.   (b) $W\sqrt{2}$.   **5.** Along a line from the center of the hole through the center of the disk,
beyond the latter point by a distance $Rr^2/2(R^2 - r^2)$.   **7.** (a) 9.7 lb (42 N).   (b) 15 lb (66 N).   **9.** (a) 43.30 lb.

(b) 21.65 lb; 12.50 lb.   **11.** (a) $Wx/l\sin\theta$.   (b) $Wx/l\tan\theta$.   (c) $W\left(1 - \dfrac{x}{l}\right)$.   **15.** (a) Slides at 31°.   (b) Tips at 34°.

17. Triceps: 1940 N, up; humerus: 2070 N, down.   **21.** 2.20 m.   **23.** (a) $a_1 = L/2$, $a_2 = L/4$, $a_3 = L/6$, $a_4 = L/8$; $h = 25L/24$.
(b) $b_1 = L/2$, $b_2 = 5L/8$; $h = 9L/8$.   (c) $c_1 = 2L/3$, $c_2 = L/2$; $h = 7L/6$.   **25.** (a) 283 N.   (b) 880 N,
71.2° above the horizontal.   **27.** (a) 47 lb.   (b) $F_A = 120$ lb, $F_E = 72$ lb.   **29.** (a) 445 N.   (b) 0.50.   (c) 315 N.

CHAPTER 14

Exercises

1. (a) $x = +2.0$ m.   (b) 20 J.   **3.** (a) 0.75 s.   (b) 1.33 Hz.   (c) 8.38 rad/s.   **5.** 0.28 s.   **7.** 0.73 s.
9. (a) 7900 dyne/cm.   (b) 1.19 cm.   (c) 2.00 Hz.   **11.** 2.08 h.   **13.** (a) 1.00 mm.   (b) 0.754 m/s.   (c) 568 m/s².
15. (a) 200 N/m.   (b) 1.39 kg.   (c) 1.91 Hz.   **17.** (a) 0.629 s.   (b) 1.59 Hz.   (c) 0.050 m.   (d) 0.125 J.
(e) 0.500 m/s.   **19.** 9.49 m/s².   **21.** (a) 0.735 kg·m².   (b) 0.024 N·m.   (c) 0.181 rad/s.

Problems

1. (a) J/m³; m.   (c) $-b/\sqrt{3}$.   (d) $\dfrac{2\sqrt{3}}{9}\,Ab^3$.   **3.** 19 lb.   **5.** (a) 1.29×10^5 N/m (8880 lb/ft).   (b) 2.68 Hz (2.68 Hz).

7. (a) 1.6×10^4 m/s²; 2.5 m/s.   (b) 7.9×10^3 m/s²; 2.2 m/s.   **9.** 1.60 kg.   **11.** 3.1 cm.   **13.** (a) $\dfrac{3}{4}$; $\dfrac{1}{4}$.   (b) $A/\sqrt{2}$.

15. (a) 2.25 Hz.   (b) 375 J.   (c) 86.6 cm.   **17.** (a) 5.58 Hz.   (b) 0.325 kg.   (c) 0.400 m.   **19.** (a) 0.525 m.

(b) 0.686 s.   **21.** (a) $mv/(m + M)$.   (b) $mv/\sqrt{k(m + M)}$.   **23.** $2\pi\left(\dfrac{l^2 + 12d^2}{12gd}\right)^{1/2}$.   **25.** (a) $x = y$; straight line.

(b) $x^2 - \sqrt{3}xy + y^2 = A^2/4$; ellipse.   (c) $x^2 + y^2 = A^2$; circle.   **27.** (a) 0.102 kg/s.   (b) 0.137 J.   **29.** 5.26.
31. (a) 3.0 m.   (b) -49 m/s.   (c) -270 m/s².   (d) 20 rad.   (e) 1.5 Hz.   (f) 0.67 s.   **33.** (a) 3.5 m.   (b) 0.75 s.
35. (a) 9.87 N.   (b) 98.7 N/m.   **39.** (a) 2.23 Hz.   (b) 56.0 cm/s.   (c) 100 g.   (d) 20.0 cm below the original position of
the first mass in (a).   **41.** 1.88 in.   **45.** (a) 0.183A.   (b) Same direction.   **47.** 5.57 cm.   **49.** (a) 0.35 Hz.   (b) 0.39 Hz.
(c) Zero.   **53.** $k_1 = (n + 1)k/n$; $k_2 = (n + 1)k$.   **55.** (a) 0.500 m.   (b) -0.251 m; 3.06 m/s.

CHAPTER 15

Exercises

1. $F_{sun}/F_{moon} = 175$.   **3.** 1.6×10^5 mi.   **5.** 0.10%.   **7.** (a) $G(M_1 + M_2)m/a^2$.   (b) GM_1m/b^2.   (c) Zero.   **9.** 4.9 s.
11. 0.9974 s.   **13.** Only 0.20 revolutions.   **15.** (a) Yes.   (b) 36,000 km.   (c) No.   **17.** 1.87 y.
21. (a) 2.22×10^{-10} N/kg.   (b) -5.34×10^{-9} J/kg.

Problems

1. (a) 1.33×10^{12} m/s^2.   (b) 1.63×10^6 m/s.   **5.** (a) To 0.2%, or to 2.8 ms.   **9.** (b) 250 m; 50 m.   (c) 288 m; 12.0 m.
11. 6.5×10^{23} kg.   **13.** (a) 1.16×10^7 J.   (b) The pellet energy is greater than the rifle slug energy by a factor of 155.
15. $2\pi r^{3/2}/G(M + \frac{1}{4}m)$.   **17.** South, at 35.4° above the horizon.   **19.** 2.60×10^4 km.   **21.** Energy ratios are 0.0453 and 28.5
for the moon and Jupiter.   **23.** (a) 2.24×10^{-7} rad/s.   (b) 8.95×10^4 m/s.   **25.** (a) No.   (b) Same.   (c) Yes.
27. (a) 1.79×10^{32} J.   (b) -1.79×10^{32} J.   (c) Zero.   (d) 987 m/s.   **29.** 4.1×10^{23} slug (6.1×10^{24} kg).
31. (a) 2.8 y.   (b) 1.0×10^{-4}.   **33.** 6.5×10^{10}.   **35.** (a) Same.   (b) One-fourth pre-expansion value.
(c) One-fourth pre-expansion value.   (d) $\sqrt{2}/2$ times pre-expansion value.   **37.** (a) $r^{3/2}$.   (b) r^{-1}.   (c) $\sqrt{r}$.   (d) $1/\sqrt{r}$.
39. (a) Zero.   (b) -1.00×10^{-9} J/kg.   (c) -5.00×10^{-13} J.   **41.** (a) $-GM_e m/r$.   (b) $-2GM_e m/r$.
(c) It falls vertically.   **43.** (a) 1700 m/s.   (b) 250 km.   (c) 1400 m/s.   **45.** (a) $-Gm(M_e/R + M_m/r)$.
(b) 3.4×10^8 m from earth.   (c) -6.3×10^7 J/kg; 9.8 m/s^2.   (d) -3.9×10^6 J/kg; 1.6 m/s^2.   **47.** (b) 3.2 m.
51. (b) 84 min.   (b) No.

CHAPTER 16

Exercises

1. (a) 192 kPa.   (b) 159/106.   **3.** 5.77×10^4 N.   **5.** (a) 230 lb/in.^{2} (1.6×10^6 Pa).   **7.** 7.23×10^5 N.   **9.** (a) 8720 lb.
(b) 9240 lb.   (c) 480 lb.   **11.** $0.12\left(\dfrac{1}{\rho} - \dfrac{1}{8}\right)$, with ρ in g/cm^3.   **13.** 0.20 ft^3.   **15.** 28 ft/s.   **17.** 0.72 ft·lb.   **19.** 108 m/s.
21. 31 m/s.   **23.** $\frac{1}{2}\rho v^2 A$, where ρ is the density of air.

Problems

1. 6.0 lb/in.^{2}.   **3.** (a) 6.06×10^9 N.   (b) 20 atm.   **5.** $\frac{1}{4}\rho g A(h_2 - h_1)^2$.   **7.** (a) 500 ft^2 (46 m^2).   **9.** Five.
11. 4.9×10^6 N.   **13.** (a) 1.80 m^3.   (b) 4.75 m^3.   **15.** 65.8 W.   **17.** 1.4 cm.   **19.** 39.9 m/s.   **21.** (a) $v = 4.1$ m/s;
$v' = 21$ m/s.   (b) 8.0×10^{-3} m^3/s.   **23.** (a) 5.5×10^{-2} ft^3/s.   (b) 3.0 ft.   **25.** (b) 6000 lb.   **27.** 18 N.
29. (a) $\frac{1}{2}\rho g W D^2$.   (b) $\dfrac{1}{6}\rho g W D^3$.   (c) $\dfrac{1}{3}D$.   **31.** (a) 1.8 kg.   (b) 2.0 kg.   **33.** 1.11×10^4 N.   **35.** 254 lb/in.^{2}.
37. 1840 m^3.   **39.** (b) $H - h$.   (c) $H/2$.   **41.** (a) 2.5 m/s.   (b) 2.6×10^5 Pa.   **45.** (b) 3.17 s.   **47.** 5 min 54 s.

CHAPTER 17

Exercises

1. (a) 400 THz to 800 THz (THz = terahertz; see Table 1-2).   (b) 1.0 m to 200 m.   (c) 6.0×10^4 THz to 3.0×10^7 THz.
3. (a) 1.67 s.   (b) 3.33 m.   **5.** (c) 200 cm/s; $-x$-direction.   **7.** $\sqrt{2}$.   **9.** 0.12 N.   **11.** 15.0 mW.
13. (a) 2.0 Hz; 200 cm; 400 cm/s.   (b) $x = 50$ cm, 150 cm, 250 cm, etc.   (c) $x = 0$, 100 cm, 200 cm, etc.   **15.** (a) 833 Hz.
(b) 0.418 m.   **17.** 480 cm; 160 cm; 96 cm.   **19.** (a) 66.1 m/s.   (b) 26.4 Hz.

Problems

1. (a) 6.0 cm.   (b) 100 cm.   (c) 2.0 Hz.   (d) 200 cm/s.   (e) $-x$-direction.   (f) 75 cm/s.   **3.** (a) $z = (3.0$ mm$)$
$\sin\{(6.0$ cm$^{-1})y - (10\pi$ s$^{-1})t\}$.   (b) 9.42 cm/s.   **5.** $y = (0.12$ mm$)\cos\{(141$ m$^{-1})x - (628$ s$^{-1})t\}$.   **9.** v_0.
11. (b) 5.76×10^{-17} J/m^3.   **13.** $\lambda = 2\{d^2 + 4(H + h)^2\}^{1/2} - 2(d^2 + 4H^2)^{1/2}$.   **15.** (a) 0.25 cm; 120 cm/s.   (b) 3.0 cm.
(c) Zero.   **17.** 880 Hz and 1320 Hz.   **19.** (a) 8.1 lb (36 N).   **21.** (a) 2.00 mm; 95.5 Hz; 30.0 m/s; 31.4 cm.
(b) 1.20 m/s.   **23.** (a) 6.00 m/s.   (b) $y = 0.30\sin(\pi x/12 + 50\pi t)$, with x and y in centimeters and t in seconds.
25. $2\pi A/\lambda$.   **29.** 7.91 Hz; 15.8 Hz; 23.7 Hz.   **31.** (a) 1.33 m.   (b) $y = (0.002)\sin 9.42x\cos 3770t$, with x and y in meters,
t in seconds.   **33.** $x_0 = (\phi_1 + \phi_2)/2k$; $t_0 = (\phi_1 - \phi_2)/2\omega$.   **35.** (b) (length)^2.   **37.** 2.63 m from the end of the
wire from which the later disturbance originates.   **39.** (a) $\{k(\Delta l)(l + \Delta l)/m\}^{1/2}$.

CHAPTER 18

Exercises

1. 17 m; 1.7 cm.   **3.** (*a*) 0.0762 mm.   (*b*) 0.333 mm.   **5.** The radio listener; the travel time factor is 52.4.   **7.** (*a*) 1.5 Pa.
(*b*) 165 Hz.   (*c*) 2.0 m.   (*d*) 330 m/s.   **9.** 1.26.   **11.** Yes.   **13.** By a factor of four.   **15.** 8.3 cm from one end.
17. A pipe open at just one end.   **19.** 2.25 ms.   **21.** (*a*) 384 mi/h, away from owner.   (*b*) 76.7 mi/h, away from owner.
23. (*a*) No.   **25.** (*a*) 2.0 kHz.   (*b*) 2.0 kHz.   **27.** 2.60×10^8 m/s.

Problems

1. About 3%.   **3.** 1850 km.   **5.** $2.00p_m$; $1.41p_m$; $1.73p_m$; $1.85p_m$.   **7.** (*a*) $343(1 + 2n)$ Hz, with n being an integer from 0 to 28.
(*b*) $686\,n$ Hz, with n being an integer from 1 to 29.   **9.** 17.5 cm.   **11.** 5000; 71; 71.   **13.** 57.2 cm; 42.9 cm.   **15.** 830 Hz.
17. (*a*) 1130 Hz; 1500 Hz; 1880 Hz.   **19.** (*a*) Ten.   (*b*) Four.   **21.** 15.1 ft/s.   **23.** (*a*) 485 Hz.   (*b*) 500 Hz.   (*c*) 485 Hz.
(*d*) 500 Hz.   **25.** (*a*) 467 Hz.   (*b*) 494 Hz.   **27.** 33.0 km.   **29.** (*a*) $l(V - v)/Vv$.   (*b*) 377 m.   **31.** If only the length is
uncertain, it must be known to within 10^{-4} cm. If only the time is imprecise, the uncertainty must be no more than one part in 10^5.
33. (*a*) 136; 408; 679 Hz.   (*b*) 272; 544; 815 Hz.   **35.** (*a*) 33 cm.   (*b*) 4.0 μm.   (*c*) 2.5 cm/s.   (*d*) 17 cm.
37. Water filled to a height of 7/8, 5/8, 3/8, 1/8 m.   **39.** (*a*) Node.   (*b*) 22 s.   **41.** ±0, 0.572, 1.14, 1.72, 2.29 m
from the midpoint.   **43.** (*a*) 526 Hz.   (*b*) 555 Hz.   **45.** (*c*) None.   **47.** 50 N.   **49.** 3.16 km.   **51.** 41 kHz.   **53.** 236°.
55. (*a*) Phases differ by 30°.   (*b*) 38.8 μW/m².   (*c*) 15.9 μW/m².   (*d*) 5.97 μW/m².

CHAPTER 19

Exercises

1. 2.71 K.   **3.** 31.5.   **5.** (*a*) 320°F.   (*b*) −12.3°F.   **7.** (*a*) −40°.   (*b*) 575°.   (*c*) Celsius and Kelvin cannot give
the same reading.   **9.** 4.4×10^{-3} cm.   **11.** 0.038 in.   **13.** (*a*) 9.996 cm.   (*b*) 68°C.   **15.** 29 cm³ (1.8 in.³).

Problems

1. 291.1 K.   **3.** 373 K.   **5.** 0.0185 mm Hg; oxygen.   **7.** −48.3°F.   **9.** 360°C.   **11.** 0.265 cm³.   **15.** (*a*) 0.36%.
(*b*) 0.18%.   (*c*) 0.54%.   (*d*) 0.00%.   (*e*) 1.8×10^{-5}/°C.   **17.** 66.4°C.   **19.** (*a*) 12.8×10^{-6}/°F.   (*b*) 0.169 in.
21. ½.   **23.** (*a*) +0.036%.   (*b*) −0.013%.   **25.** 348 K.   **27.** 7.5 cm.   **29.** 9.07 s, the clock running slow.
31. (*b*) Use 39.3 cm of steel and 13.1 cm of brass.   **33.** 0.266 mm.

CHAPTER 20

Exercises

1. (*a*) 0.13 cal/g·°C.   (*b*) 6.3 cal/mol·°C.   **3.** (*a*) 1.2 W/m·°C; 0.70 Btu/ft·°F·h.   (*b*) 0.0297 ft²·°F·h/Btu.   **5.** (*a*) 500°C/m.
(*b*) 4.6 cal/s.   (*c*) 75°C.   **7.** (*a*) 17.1 kW/m².   (*b*) 16 W/m².   **9.** (*a*) 411 g.   (*b*) 3.09¢.   **11.** 1.9 times as great.
13. (*a*) 0.12°C.   **15.** +45 J along path *BC* and −45 J along path *BA*.   **17.** (*a*) $A \rightarrow B$: + + +; $B \rightarrow C$: + 0 +;
$C \rightarrow A$: − − −.   (*b*) −20 J.

Problems

1. 873°C.   **3.** 0.13 Btu/lb·°F (0.13 cal/g·°C).   **5.** (*a*) 0°C.   (*b*) 2.50°C.   **7.** 0.5 min.   **9.** 0.39 cm/h.   **11.** Cu-Al: 86°C;
Al-Brass: 57°C.   **13.** (*a*) 18,700.   (*b*) 10.4 h.   **15.** 2.8 d.   **17.** −30 J.   **19.** (*a*) 0.360 mg/s.   (*b*) 0.814 J/s.
(*c*) −0.694 J/s.   **21.** 2.16 g.   **23.** 4.04×10^6 cal.   **27.** 33.2 m².   **29.** (*a*) 13.5°C.   (*b*) 45.5% heating; 54.5% melting.
31. (*a*) 6.0 cal.   (*b*) −43 cal.   (*c*) 40 cal.   (*d*) 18 cal; 18 cal.   **33.** (*d*) 0.185°C.   **35.** (*a*) 400 Btu/h.   (*b*) 18.
(*c*) 12%.   (*d*) 33%.   **37.** 136°C.

CHAPTER 21

Exercises

1. (*a*) 22.4 l.   **3.** 25.   **5.** (*a*) 0.039 mol.   (*b*) 220°C.   **7.** 6600.   **9.** 2.50 km/s.   **11.** 252 m/s.   **13.** 313°C.
15. (*a*) 3400 J.   **19.** 1500 N·m².²   **21.** 7940 J.   **23.** $\pi d^2 n_v v$.   **25.** (*a*) 6.50 km/s.   (*b*) 7.11 km/s.   **27.** (*a*) 420 m/s; 458 m/s.

Problems

1. 2.68×10^{-3}/K.   **3.** 648 J.   **7.** 1/5.   **9.** (*a*) 430 m/s.   (*b*) N_2.   **11.** (*a*) 430 m/s.   (*b*) 73 K; 1200 K.
13. $RT\ln(V_f/V_i)$.   **17.** (*a*) $1.5nRT_1$.   (*b*) $4.5nRT_1$.   (*c*) $6nRT_1$.   (*d*) $2R$.   **19.** (*a*) 920 J.   (*b*) 71%.   **21.** 4.13 J/cal.
23. 3500 MHz.   **25.** (*a*) 7×10^9 km.   **27.** 1.50.   **29.** (*a*) 1.0043.   **31.** 4.71.   **33.** 103 cm³.   **35.** 3.11×10^3 J/kg·°C.

37. (*a*) 6.6×10^{-26} kg.   (*b*) 40.   **39.** (*a*) 1.7.   (*b*) 5.0×10^{-5} cm.   (*c*) 7.9×10^{-6} cm.   **41.** 198°F.
43. (*a*) In joules: $1 \rightarrow 2$:   3740   3740   0
            $2 \rightarrow 3$:   0 -1810 1810
            $3 \rightarrow 1$: -3220 -1930 -1290
            Cycle: 520   0   520
(*b*) $V_2 = 0.0246$ m³; $p_2 = 2$ atm; $V_3 = 0.0373$ m³; $p_3 = 1$ atm.   **45.** Diatomic.   **49.** In cal: 900   0   900   450
                                                1200   300   900   450
                                                0   -900 900   450

51. (*a*) 7.1 km/s.   (*b*) 2.0×10^{-8} cm.   (*c*) 5.0×10^{10} s^{-1}.   **53.** 12.8 kW; 17.1 hp.

CHAPTER 22

Exercises

1. (*a*) 50.2 kJ.   (*b*) 20%.   **3.** (*a*) 2090 J.   (*b*) 1570 J.   (*c*) 1570 J.   **5.** 99.999947%.   **7.** 60.   **9.** (*a*) First.
(*b*) First and second.   (*c*) Second.   (*d*) Neither.   **11.** (*a*) 6900 cal.   (*b*) 21 cal/K.   **13.** 0.68 cal/K.
15. (*a*) $+1.0$ cal/K.   (*b*) No.

Problems

1. (*c*) 6.5.   **3.** (*a*) 94 J.   (*b*) 230 J.   **5.** (*a*) 1.115 kcal/s.   (*b*) 0.995 kcal/s.   **7.** (*a*) 78%.   (*b*) 81.5 kg/s.
9. (*a*) Path I: $Q_T = pV\ln2$, Path II: $Q_T = -pV\ln2$,

$$Q_V = \frac{9}{2} pV, \qquad Q_P = \frac{15}{2} pV.$$

(*b*) Path I: $W_T = pV\ln2$, Path II: $W_T = -pV\ln2$,
    $W_V = 0$, $W_p = 3pV$.

(*c*) $\frac{9}{2} pV$ for either case.   (*d*) $4R\ln2$ for either case.   **11.** 0.18 cal/K.   **13.** (*a*) 9220 J.   (*b*) 23.1 J/K.   (*c*) Zero.

17. 6.5.   **19.** (*a*) -226 cal/K.   (*b*) $+226$ cal/K.   **23.** $\{1 - (T_2/T_1)\}/\{1 - (T_4/T_3)\}$.   **27.** (*a*) 1.47×10^6 J.
(*b*) 5.54×10^5 J.   (*c*) 9.18×10^5 J.   (*d*) 0.624.   **31.** 0.139.   **33.** (*a*) $T_2 = 3T_1$; $T_3 = 3T_1/4^{\gamma-1}$; $T_4 = T_1/4^{\gamma-1}$; $p_2 = 3p_1$; $p_3 =$
$3p_1/4^\gamma$; $p_4 = p_1/4^\gamma$.   (*b*) $1 - \frac{1}{4^{\gamma-1}}$.

CHAPTER 23

Exercises

1. (*a*) 9×10^9 N.   (*b*) 3.5×10^3 N.   **3.** 2.8 N; attractive.   **5.** 0.40 C.   **7.** (*a*) 3.2×10^{-19} C.   (*b*) Two.
9. 2.9×10^{-9} N.   **11.** $+1.6 \times 10^{-19}$ C.

Problems

1. 5.1 m below the electron.   **3.** (*a*) $q_1 = 9q_2$.   (*b*) $q_1 = -25q_2$.   **5.** 1.34 d.   **7.** 1.2×10^{-5} C and 3.8×10^{-5} C.
9. (*a*) $\frac{1}{2}l\left(1 + \frac{1}{4\pi\epsilon_0}\frac{qQ}{Wh^2}\right)$.   (*b*) $\left(\frac{3}{4\pi\epsilon_0}\frac{qQ}{W}\right)^{1/2}$ **11.** (*a*) 5.7×10^{13} C.   (*b*) No.   (*c*) 660 tons.
13. (*a*) 510 N.   (*b*) 7.7×10^{28} m/s².   **15.** 4.16×10^{42}.   **17.** (*a*) 14 cm from the postive charge, 24 cm
from the negative charge.   (*b*) Unstable.   **19.** A charge $-\frac{4}{9}g$ must be located on the line segment joining the two positive charges,
a distance $l/3$ from the $+q$ charge. The equilibrium is unstable.   **21.** 10^{18} N.   **23.** (*a*) $Q = -2\sqrt{2}q$.   (*b*) No.
25. (*b*) $\pm 2.4 \times 10^{-8}$ C.   **27.** 1.6×10^{-9} m.

CHAPTER 24

Exercises

1. (*a*) 1.6×10^{-13} N.   (*b*) 1.6×10^{-13} N.   **3.** 2.1×10^{-7} N/C, up.   **5.** (*a*) -0.029 C.   (*b*) Sphere would blow up because
of mutual Coulomb repulsion.   **7.** (*a*) 6.4×10^{-18} N.   (*b*) Less.   **9.** To the right in the figure.
13. (*a*) The larger sphere produces a field of 1.3×10^5 N/C at the site of the smaller; the smaller sphere produces a field of 5.3×10^4
N/C at the site of the larger.   (*b*) 1.1×10^{-2} N, repulsive.   **15.** 2.11×10^{13} m/s².

Problems

1. (*b*) Equilibrium is unstable.   **3.** (*a*) $0.172a$ to the right of the $+2q$ charge.   **5.** (*b*) At right angles to the axis and away from it.

7. 1.02×10^5 N/C, upward.   **9.** (b) Parallel to **p**.   **11.** 0.171 ns.   **13.** (a) $2\pi\left(\dfrac{l}{|g - qE/m|}\right)^{1/2}$.   (b) $2\pi\left(\dfrac{l}{g + qE/m}\right)^{1/2}$.

15. 1.64×10^{-19} C ($\approx$3% high).   **17.** (a) $x = -3.54$ m.   (b) $x = 3.54$ m. These answers assume an electron travelling in the $+x$-direction.   **19.** 50 cm from q_1 and 100 cm from q_2.   **23.** 9:30.   **25.** $E = q/\pi\epsilon_0 a^2$, along the bisector, away from the triangle.   **27.** (a) $-\mathbf{j}(2.1 \times 10^{15}$ m/s$^2)$.   (b) $\mathbf{i}(1.5 \times 10^5$ m/s$) - \mathbf{j}(2.8 \times 10^8$ m/s$)$.   **29.** (a) 2.7×10^6 m/s.   (b) 1000 N/C.

CHAPTER 25

Exercises

1. (a) 693 kg/s.   (b) 693 kg/s.   (c) 347 kg/s.   (d) 347 kg/s.   (e) 347 kg/s.   **3.** $\pi a^2 E$.   **5.** $q/6\epsilon_0$.   **9.** -6×10^5 C.
11. 1.1×10^5 N·m^2/C.   **13.** (a) -750 N·m^2/C.   (b) -6.64 nC.   **15.** 26.6 fC.   **17.** (a) 5.3×10^7 N/C.
(b) 6.0×10^3 N/C.

Problems

1. Zero.   **3.** Through each of the faces meeting at q: zero; through each of the other three faces: $q/24\epsilon_0$.   **5.** (a) -3.0×10^{-6}C.
(b) 1.3×10^{-5}C.   **7.** (a) Zero.   (b) 2.9×10^4 N/C.   (c) 200 N/C.   **11.** 4.3×10^{-17} J.   **13.** $\pm4.9 \times 10^{-10}$ C.
15. (b) $\rho R^2/2\epsilon_0 r$.   **17.** (a) $E = \rho x/\epsilon_0$, x measured from the median plane of the slab.   (b) $\rho d/2\epsilon_0$.   **19.** (a) 7.38 N·m^2/C.
(b) 7.38 N·m^2/C.   (c) 6.53×10^{-11} C in each case.   **21.** (a) $E = (q/4\pi\epsilon_0 a^3)r$.   (b) $E = q/4\pi\epsilon_0 r^2$.
(c) Zero.   (d) Zero.   (e) Inner: $-q$; outer: zero.   **23.** (a) 4.0×10^6 N/C.   (b) Zero.   **25.** (a) $E = \lambda/2\pi\epsilon_0 r$.
(b) Zero.   **27.** (a) Zero.   (b) $E = \sigma/\epsilon_0$, to the left.   (c) Zero.   **29.** 0.44 mm.

CHAPTER 26

Exercises

1. (a) 7.75 km/s.   (b) 9.0×10^4 kg ($= 99$ tons).   **3.** 2.90 kV.   **5.** 1000 V.   **9.** (a) 4.5 m.   (b) No.   **11.** 198 MV.
13. 2.78×10^5.   **17.** 10^9 eV.   **19.** 637 MV.   **23.** 9.0 kW.   **25.** (a) 3.2×10^{-13} J or 2.0 MeV.
(b) 1.6×10^{-13} J or 1.0 MeV.   (c) The proton.

Problems

3. (a) 0.54 mm.   (b) 790 V.   **5.** (a) There are no points with $V = 0$, except at "infinity".   (b) 41.4 cm from $+q$,
between the charges.   **7.** (a) 0.12×10^{-10} m.   (b) 0.08×10^{-10} m.   **9.** 2.48 km/s.   **11.** 1.79×10^{-10} J.
13. (a) 37 MeV.   **15.** (a) 2.72×10^{-14} J.   (b) 3.02×10^{-31} kg, in error by a factor of about three.   **17.** 4.6×10^{21} V/m.
19. (a) $\dfrac{\lambda}{4\pi\epsilon_0}\ln\left(\dfrac{L + y}{y}\right)$.   (b) $\dfrac{\lambda}{4\pi\epsilon_0}\dfrac{L}{y(L + y)}$.   (c) Zero.   **21.** 1.2×10^{-8} C/m^2.   **23.** $\pm2.5 \times 10^{-8}$ C.
25. (a) $r > 9.0$ cm.   (b) 2.7 kW.   (c) 20 μC/m^2.   **27.** (a) -7.8×10^5 V; 0.6×10^5 V.   (b) 2.5 J.   (c) Work is converted
into potential energy.   **29.** $q/8\pi\epsilon_0 a$.   **31.** 1.12 km/s.   **33.** $-0.21q^2/\epsilon_0 a$.   **35.** -1.5×10^{-7} J.   **39.** 5.85×10^7 m/s.

41. (a) $qd/2\pi\epsilon_0 a(a + d)$.   **43.** $(2eV/m)^{1/2}$.   **45.** $qQ/4\pi\epsilon_0 K$.   **49.** (a) $Q/4\pi\epsilon_0 r$.   (b) $\dfrac{\rho}{3\epsilon_0}\left(\dfrac{3}{2}r_2^2 - \tfrac{1}{2}r^2 - r_1^3/r\right)$;

$\rho = Q/\left\{\dfrac{4\pi}{3}(r_2^3 - r_1^3)\right\}$.   (c) $\dfrac{\rho}{2\epsilon_0}(r_2^2 - r_1^2)$, with ρ as in (b).

CHAPTER 27

Exercises

1. 7.5×10^{-12} C.   **5.** 8.85×10^{-12} m.   **7.** (a) 3.5 pF.   (b) 3.5 pF.   (c) 57.1 V.   **11.** $\approx 4 \times 10^{-7}$ C.
13. (a) 1.3×10^{-7} J.   (b) No.   **15.** 0.27 J.   **17.** 4.88%.   **19.** Fused quartz.   **21.** (a) 6.20 cm.   (b) 280 pF.
23. (a) 10 kV/m.   (b) 5.0 nC.   (c) 4.1 nC.

Problems

1. (a) $d/3$.   (b) $3d$.   **3.** (a) $q_4 = 8.0 \times 10^{-4}$ C; $q_6 = 12 \times 10^{-4}$ C.   (b) 200 V.   **5.** 3.2 μF.   **7.** (a) Five in series.
(b) Three arrays as in (a) in parallel. There are other possibilities.   **11.** 27.8¢.   **13.** (a) $\epsilon_0 A/(d - b)$.   (b) $d/(d - b)$.
(c) $-q^2 b/2\epsilon_0 A$; sucked in.   **21.** (a) $2\kappa\epsilon_0 A/d$.   (b) $\frac{1}{2}Q(1 + \Delta\kappa/\kappa)$.   **23.** (a) 88%.   (b) 12%.   **25.** 7.3 μF.
27. 1.00 V.   **29.** (a) 50 V.   (b) 0.50×10^{-4} C.   (c) 1.5×10^{-4} C.   **31.** (a) $q_1 = 9.0$ μC; $q_2 = 16$ μC;
$q_3 = 9.0$ μC; $q_4 = 16$ μC.   (b) $q_1 = 8.4$ μC; $q_2 = 17$ μC; $q_3 = 11$ μC; $q_4 = 14$ μC.   **33.** (a) $+7.9 \times 10^{-4}$ C.   (b) $+79$ V.
35. $5.05\pi\epsilon_0 R$.   **37.** (a) $q_1 = q_2 = 0.33$ mC; $q_3 = 0.40$ mC.   (b) $V_1 = 33$ V; $V_2 = 67$ V; $V_3 = 100$ V.   (c) $U_1 = 5.4$ mJ;
$U_2 = 11$ mJ; $U_3 = 20$ mJ.   **39.** (a) 2.0 J.   **41.** 0.11 J/m^3.   **43.** Yes; the capacitance is near 0.7 nF and the breakdown

potential is 26 kV.   **45.** (a) 8.9×10^{-10} C (before); 1.6×10^{-9} C after.   (b) 1.0×10^4 V/m (before); 1.8×10^4 V/m (after).
(c) 2.5×10^3 V/m.   (d) 8.9 pF (before); 16 pF (after).

CHAPTER 28

Exercises

1. (a) 1200 C.   (b) 7.5×10^{21}.   **3.** (a) 6.4 A/m^2, north.   **5.** Gauge 14.   **7.** $1.1 \, \Omega$.   **9.** $4R$.   **11.** $2.0 \times 10^6 \, (\Omega \cdot \text{m})^{-1}$.
13. (a) $260°$C.   **15.** 100 V.   **17.** (a) 0.384 mV.   (b) Negative.   (c) 3 min 58 s.   **19.** 560 W.   **21.** $11.1 \, \Omega$.

Problems

1. 0.67 A, toward the negative terminal.   **3.** 941 s.   **5.** 5.6 ms.   **7.** $2R$.   **9.** (a) 5.29×10^5 A/m^2 for copper;
3.21×10^5 A/m^2 for aluminum.   (b) 1.02 kg/m for copper; 0.504 kg/m for aluminum.   **11.** (a) 6.00 mA.   (b) 1.59×10^{-8} V.
(c) $2.12 \times 10^{-8} \, \Omega$.   **13.** (a) 1.28×10^7 A/m^2 in each.   (b) $V_A = 8.86$ V; $V_B = 51.1$ V.   (c) $E_A = 0.222$ N/C; $E_B = 1.28$ N/C.
(d) A: copper; B: iron.   **17.** (a) 4.9 MA/m^2.   (b) 8.3×10^{-2} V/m.   (c) 25 V.   (d) 630 W.   **19.** (a) 1.3×10^5 A/m^2.
(b) 94 mV.   **21.** (a) 28 min.   (b) 1.6 h.   **23.** (a) 1.1 mΩ.   (b) 4.4 mm.   **25.** (a) 0.39%; 0.0017%; 0.0034%.
27. $0.40 \, \Omega$.   **29.** (a) 38.3 mA.   (b) 109 A/m^2.   (c) 1.28 cm/s.   (d) 227 V/m.   **31.** (a) \$3.72 for a 31-day month.
(b) $144 \, \Omega$.   (c) 0.833 A.   **33.** 2.9 mm.   **35.** New length $= 1.369l$; new area $= 0.730A$; new potential difference $= 7.5V$.
37. 620 W.   **39.** (a) 2.25×10^{-4} C.   (b) 6.00×10^{-5} A.   (c) 0.450 mW.   **41.** Carbon: $909 \, \Omega$; iron: $90.9 \, \Omega$.

CHAPTER 29

Exercises

1. (a) 12 eV.   (b) 6.53 W.   **3.** 14 h 24 min.   **5.** $8.0 \, \Omega$.   **9.** 0.240 A.   **11.** -10 V.   **13.** 4.0 V.   **15.** (a) $6.67 \, \Omega$.
(b) $6.67 \, \Omega$.   (c) Zero.   (d) $4.00 \, \Omega$.   (e) $2.50 \, \Omega$.   (f) $3.125 \, \Omega$.   **17.** $i_1 = 50$ mA; $i_2 = 60$ mA; $V_{ab} = 9.0$ V.
19. (a) R_2.   (b) R_1.   **21.** (a) 13.5 kΩ.   (b) $1500 \, \Omega$.   (c) $167 \, \Omega$.   (d) $1480 \, \Omega$.

Problems

1. (a) \$320.   (b) 9.6¢.   **3.** (a) $990 \, \Omega$.   (b) 9.4×10^{-4} W.   **7.** $r_1 - r_2$.   **9.** $V_d - V_c = +0.25$ V, by all paths.
11. (a) $2\varepsilon/(2r + R)$, series; $2\varepsilon/(r + 2R)$, parallel.   (b) Series if $r < R$; parallel if $R < r$.   **13.** 112%.   **17.** (a) Top arrangement:
70.9 mA; 4.91 V; bottom arrangement: 55.2 mA; 4.86 V.   (b) Top: $69.3 \, \Omega$; bottom: $88.0 \, \Omega$.   **19.** 1.4×10^{-10} F.
21. (a) 10^{-3} C.   (b) 10^{-3} A.   (c) $V_C = 10^3 e^{-t}$, $V_R = -10^3 e^{-t}$, volts.   (d) e^{-2t}, watts.   **23.** 0.225%.
25. The cable, assuming a 12-V battery.   **27.** $\varepsilon/7R$.   **29.** $5r/(5 + r)$.   **31.** (a) $10 \, \Omega$.   (b) $14 \, \Omega$.
(c) $10 \, \Omega$. Arrangements (a) and (c) are identical.   **33.** (50 kW) $\left(\dfrac{x}{2000 \times 10x - x^2} \right)^2$, x in cm.   **35.** $38 \, \Omega$ or $260 \, \Omega$.
37. (a) 0.955 μC/s.   (b) 1.08 μW.   (c) 2.74 μW.   (d) 3.82 μW.   **39.** 0.72 MΩ.   **41.** 24.8 Ω to 14.9 kΩ.
43. (a) At $t = 0$, $i_1 = 1.1$ mA, $i_2 = i_3 = 0.55$ mA; at $t = \infty$, $i_1 = i_2 = 0.82$ mA, $i_3 = 0$.   (c) At $t = 0$, $V_2 = 400$ V;
at $t = \infty$, $V_2 = 600$ V.   (d) After several time constants ($\tau = 7.1$ s) have elapsed.

CHAPTER 30

Exercises

1. M/QT; ML^2/QT.   **3.** Right; No.   **5.** 28.2 N, horizontally west.   **7.** 4.3×10^{-3} N·m; the torque vector is parallel to the long
side of the loop and points down.   **9.** (a) 20 min.   (b) 5.94×10^{-2} N·m.   **13.** Point #5.   **15.** (a) 0.34 mm.   (b) 2.6 keV.
17. (a) 1.4 mT.   (b) 38 MHz.   (c) 26 ns.   **19.** (a) 18 MHz.   (b) 17 MeV.   **21.** 2.

Problems

1. 0.75k, T.   **3.** $Bitd/m$, away from the generator.   **5.** -0.347k, N.   **7.** (a) $540 \, \Omega$, connected in series.   (b) $2.52 \, \Omega$,
connected in parallel.   **9.** 0.335 A·m^2, $297°$ counterclockwise from the $+y$-axis in the y, z-plane.   **15.** (a) $K_p = K_d = \frac{1}{2}K_\alpha$.
(b) $R_d = R_\alpha = 14$ cm.   **19.** (a) 8.5 MeV.   (b) 0.80 T.   (c) 34 MeV.   (d) 24 MHz.   (e) 34 MeV; 1.6 T;
34 MeV; 12 MHz.   **21.** 0.27 mT.   **23.** (a) 6.2×10^{-14}k, N.   (b) -6.2×10^{-14}k, N.   **25.** $(-2.5\mathbf{j} + 0.75\mathbf{k}) \times 10^{-3}$ N.
27. (a) 2.8 MHz.   (b) 0.34 m.   **29.** (a) 2.60×10^6 m/s.   (b) 2.17×10^{-7} s.   (c) 0.140 MeV.   (d) 70 kV.
31. $-11.4\mathbf{i} - 6.00\mathbf{j} + 4.8\mathbf{k}$, V/m.   **33.** (a) 0; 1.38 mN; 1.38 mN.   **35.** (a) $-q$.   (b) $\pi m/qB$.   **37.** (a) 2.86 A·m^2.
(b) 1.10 A·m^2.   **39.** (a) 495 mT.   (b) 22.7 mA.   (c) 8.17 MJ.   **41.** 240 m.   **43.** 0.10 T, at $31°$ from the vertical.

CHAPTER 31

Exercises

1. (b) and (c).   **5.** Yes.   **7.** (a) 3.2×10^{-16} N, parallel to the current.   (b) 3.2×10^{-16} N, radially outward if **v** is parallel to the current.   (c) Zero.   **9.** (a) 15.6 A.   (b) West to east.   **11.** (a) It is impossible to have other than $B = 0$ midway between them.   (b) 30.0 A.   **13.** At all points between the wires, on a line parallel to them, at a distance one-fourth of the wire separation, from the wire carrying current i.   **15.** (a) $-4\mu_0$.   (b) $2\mu_0$.   **19.** 4.52×10^{-6} T·m.   **21.** (a) 533 μT.   (b) 400 μT.   **23.** (a) 4.   (b) 1/2.

Problems

1. (a) 400 μT; to the left.   (b) 400 μT; up.   **3.** (a) 1.03 mT; out of figure.   (b) 0.80 mT; out of figure.
9. (a) $0.338\mu_0 i^2/a$, towards P.   (b) $0.338\mu_0 i^2/a$, away from P.   **11.** $0.791\mu_0 i^2/\pi a$, 117° counterclockwise from P.

13. (a) $\mu_0 ir/2\pi c^2$.   (b) $\mu_0 i/2\pi r$.   (c) $\dfrac{\mu_0 i}{2\pi(a^2 - b^2)}\left(\dfrac{a^2 - r^2}{r}\right)$.   (d) Zero.   **15.** $\mu_0 j_0 r^2/3a$.   **17.** (a) Negative.   (b) 9.68 cm.   **25.** (a) 94 μT.   (b) 1.5×10^{-6} N·m.   **27.** Zero.   **29.** 8.0 mT. (#10 wire should have a diameter of 2.6 mm; for this, $B = 7.7$ mT.)   **31.** 3.2 mN, toward the wire.   **33.** 200 μT, into the page.   **35.** 4.77 cm.   **37.** $3i_0/8$, into the page.

39. (a) $(\mu_0 i/2R)\left(1 + \dfrac{1}{\pi}\right)$, out of page.   (b) $(\mu_0 i/2\pi R)(1 + \pi^2)^{1/2}$, 18° out of page.   **41.** (b) To the right.

43. (a) **0.060j**, A·m².   (b) 9.6×10^{-11}**j**, T.   (c) 4.8×10^{-11}**i**, T.   **47.** (a) $8.0 \times 10^{-3}d$, in T, where d is in m.
(b) 3180 A.   (c) **k**.

CHAPTER 32

Exercises

1. (a) M/IT^2.   (b) ML^2/IT^2.   **3.** 1.38 mV.   **5.** Zero.   **9.** (a) 2, 6.   (b) 4.   (c) 1, 3, 5, 7.   **11.** (a) 48.1 mV.
(b) 2.67 mA.   **13.** (a) 600 mV, up.   (b) 1.5 A, clockwise.   (c) 0.90 W.   (d) 0.18 N.   (e) Same as (c).   **15.** 1: -1.07 mV;
2: -2.40 mV; 3: 1.33 mV.

Problems

1. 1.15×10^{-6} Wb.   **3.** (a) 21 mA.   **5.** (a) 31 mV.   (b) Right to left.   **7.** 0.40 V.   **9.** (a) $\mu_0 iR^2\pi r^2/2x^3$.
(b) $\dfrac{3}{2}\mu_0 i\pi R^2 r^2 v/x^4$.   (c) In the same direction as the current in the large loop.   **11.** $Blit/m$, away from G.
13. 0.151 V/m.   **17.** 1.39 T/s.   **19.** 5.50 kV.   **21.** (a) 240 μV.   (b) 0.60 mA.   (c) 0.144 μW.   (d) 2.88×10^{-8} N.
(e) Same as (c).   **23.** (a) 0.598 μV.   (b) Counterclockwise.   **25.** 20 mC.   **27.** 1.55×10^{-5} C.
29. (a) $(\mu_0 ia/2\pi) \ln\left(\dfrac{2r + b}{2r - b}\right)$.   (b) $2\mu_0 iabv/\pi R(4r^2 - b^2)$.   **31.** (a) 13 μWb/m.   (b) 17%.

CHAPTER 33

Exercises

1. 0.10 μWb.   **3.** Let the current change at 5.0 A/s.   **5.** (a) 0.101 H.   (b) 1.26 V.   **7.** 12 s.   **9.** $42 + 20t$, V.
11. $1.23\tau_L$.   **13.** 5.58 A.   **15.** 1.5×10^8 V/m.   **17.** (a) 1.50 μWb; 100 mV.   (b) 0.090 μWb; 12 mV.

Problems

1. 1.89 ms.   **3.** 2.53×10^{-4} H.   **5.** (a) $\mu_0 i/W$.   (b) $\pi\mu_0 R^2/W$.   **7.** (a) 9.7 ns.   (b) 6.3 mA. These answers assume a 1000-Ω resistor.   **9.** (a) $i(1 - e^{-Rt/L})$.   **11.** (a) 240 W.   (b) 150 W.   (c) 390 W.   **13.** 0.63 J/m³.
17. (a) $\mu_0 \dfrac{Nl}{2\pi} \ln\left(1 + \dfrac{b}{a}\right)$.   (b) 13.2 μH.   **21.** 1.54 s.   **23.** 46 Ω.   **25.** (a) 10 A.   (b) 100 J.
27. (a) 1.0×10^{-3} J/m³.   (b) 8.4×10^{15} J.   **33.** (a) 1.50 s.

CHAPTER 34

Exercises

5. 55 μT.

Problems

1. (a) 1.4 A.   (b) 23 nT, pointing along the axis.   **3.** 47.4 μWb, inward.   **5.** 13.0 × 10⁶ Wb, outward.   **7.** Yes.
9. (b) K_i/B, opposite to the field.   (c) 307 A/m.   **11.** (a) 8.9 A·m².   (b) 13 N·m.   **15.** (a) 160 T.   (b) 600 T.
17. (a) 1800 km.   (b) 2.3%.   **23.** 1660 km.

CHAPTER 35

Exercises

1. (a) 6.00 μs.   (b) 3.00 μs.   **3.** (a) Force.   (b) Coulomb/meter for q, i; (coulomb/meter)² for C; (meter/coulomb)² for L;
meter/coulomb for V.   **5.** 2.1 μF.   **7.** 7.02 × 10⁻⁴ s.

Problems

1. 3.6 nF.   **3.** 47 mA.   **5.** (a) 1.25 kg.   (b) 3.72 × 10⁴ N/m.   (c) 1.75 × 10⁻⁴ m.   (d) 3.02 × 10⁻² m/s.
9. Let T_2 = 0.596 s be the period of the inductor and 900-μF capacitor and T_1 = 0.199 s be the period of the inductor and 100-μF capacitor.
Close S_2, wait $T_2/4$; quickly close S_1, then open S_2; wait $T_1/4$ and then open S_1.   **11.** 5.98 μC; 5.73 μC; 2.61 μC.   **13.** (a) 3.0 nC.
(b) 1.7 mA.   (c) 4.5 nJ.   **15.** (a) 0.689 μH.   (b) 17.9 pJ.   **17.** (a) 6.0:1.   (b) 36 pF; 0.22 mH.   **19.** (a) 0.18 mC.
(b) $T/8$.   (c) 66.7 W.   **21.** $q_m/2$.   (b) $0.87i_m$.

CHAPTER 36

Exercises

1. 377 rad/s.   **3.** (a) 0.95 A.   (b) 0.12 A.   **5.** (a) 4.60 kHz.   (b) 26.6 pF.   (c) X_L = 2.6 kΩ; X_C = 0.65 kΩ.
7. (a) 180 Ω; zero; 240 Ω; 0.13 A; 48°.   **13.** Step up: 5.00, 4.00, 1.25; step down: 0.800, 0.250, 0.200.

Problems

1. (a) 5.22 mA.   (b) Zero.   (c) 4.51 mA.   (d) Taking energy.   **3.** (a) 6.73 ms.   (b) 11.2 ms.   (c) Inductor.
(d) 138 mH.   **5.** (a) 30 V.   (b) 22.8 V.   (c) 14.2 V.   (d) −6.95 V.   **7.** (a) 26.1 Ω.   (b) 39.1 Ω.   (c) Capacitive.
11. (a) 41.4 μW.   (b) − 17.0 μW.   (c) 44.1 μW.   (d) 14.4 μW.   **13.** (a) 1.9 V; 5.8 W.   (b) 19 V; 0.58 kW.
(c) 0.19 kV; 58 kW.   **15.** (a) 8.84 kHz.   (b) 6.00 Ω.   **17.** (a) 76.4 mH.   (b) 17.8 Ω.   **19.** 1000 V.   **21.** (b) Add a 16-μF
capacitor in parallel.   (c) 0.19 A.   **23.** 180 Ω.

CHAPTER 37

Exercises

3. Change the potential difference between the plates at the rate of 1.0 MV/s.

Problems

1. 1.88 pT.   **5.** (a) 8.9 μA.   (b) 1.0 MV·m/s.   (c) 2.0 mm.   (d) 5.6 pT.   **11.** (a) 5.47 mA.

CHAPTER 38

Exercises

1. (a) 0.50 ms.   (b) 8.4 min.   (c) 2.4 h.   (d) 5500 B.C.   **3.** 45°.   **5.** B_x = 0; B_y = −0.67 × 10⁻⁸cos{π × 10¹⁵
$(t − x/3 × 10^8)$}; B_z = 0; SI units in B_y.   **7.** (a) 10⁵ J.   (b) 1.31 × 10²³.   **9.** 0.33 μT, in the −x-direction.
11. 8.78 × 10⁻² km².   **15.** (a) 4.7 × 10⁻⁶ N/m².   (b) 2.1 × 10¹⁰ times smaller.   **17.** (a) −y-direction.
(b) E_z = − cBsin(ky + ωt); E_x = E_y = 0.   (c) Plane polarized with **E** along the z-axis.   **19.** (a) 1.94 V/m.
(b) 1.67 × 10⁻¹¹ N/m².   **21.** Receding at 0.59c.   **23.** Yellow-orange.

Problems

7. 1.2 MW/m².   **9.** (a) 82.6 W/m².   (b) 1.66 MW.   **11.** (a) 83.3 mV/m parallel to current.   (b) 4.00 mT, tangent to surface
and perpendicular to current.   (c) 265 W/m², radially inward.   **13.** (a) 6.0 × 10⁸ N.   (b) F_{grav} = 3.6 × 10²² N.   **17.** 27/128.
19. 20° or 70°.   **21.** 0.357c.   **23.** (b) 7.3 Hz.   **25.** (a) 510 nm; 610 nm.   (b) 555 nm; 5.4 × 10¹⁴ Hz; 1.9 × 10⁻¹⁵ s.
27. 0.137c.   **29.** 19.0 W/m².   **31.** (a) 100 MHz.   (b) 1.0 μT along the z-axis.   (c) 2.1 m⁻¹;

6.3×10^8 rad/s.   (d) 120 W/m².   (e) 8.0×10^{-7} N; 4.0×10^{-7} N/m².   **33.** (a) $\approx 6 \times 10^{-22}$ W.   (b) $\approx 8 \times 10^{15}$ W.
35. 1.03 kV/m; 3.43 μT.   **37.** (a) 6.7 nT.   (b) 5.3 mW/m².   (c) 6.7 W.   **41.** (a) $\pm EBa^2/\mu_0$ for faces parallel to
the x,y-plane; zero through each of the other four faces.   (b) Zero.   **45.** Seven.

CHAPTER 39

Exercises

1. 1.56.   **3.** (a) 5.1×10^{14} Hz.   (b) 388 nm.   (c) 1.98×10^8 m/s.   **7.** (a) 1.48.   (b) 2.03×10^8 m/s.   **9.** (a) 49°.
(b) 28°.   **11.** (a) 53°.   (b) Yes.   **13.** $(4D^2 - 4D + 26)^{1/2}$, in meters, where D = camera-mirror distance in meters.
17. 1.85 mm.   **19.** (b) Separate the lenses by a distance $f_2 - f_1$, where f_2 = focal length of the converging lens.
21. (a) +40 cm.   (b) Image to the right at infinity.

Problems

3. 1.44 m.   **7.** 22.0°.   **9.** (a) Cover the center of each face with an opaque disk of radius 4.5 mm.   (b) 63%.
11. (b) 23.2°.   **13.** 55.49° to 55.77°.   **15.** Three.   **18.** For alternate vertical columns: (a) +, +40, −20, +2, no, yes.
(c) Concave, +40, +60, −2, yes, no.   (e) Convex, −20, +20, +0.5, no, yes.   (g) −20, −, −, +5, +0.8, no, yes.
19. For alternate vertical columns: (a) −18, no. (c) +71, yes.   (e) +30, no.   (g) −26, no.   **21.** Assuming light incident

from the left, $i = \dfrac{(2 - n)r}{2(n - 1)}$, to the right of the right edge of the sphere, if $n < 2$, as it is for glass.   **23.** For alternate vertical

columns (an X means that the quantity cannot be found from the data given): (a) +, X, X, +20, X, −1, yes, no.
(c) Converging, +, X, X, −10, X, no, yes.   (e) Converging, +30, −15, +1.5, no, yes.   (g) Diverging, −120, −9.2,
+0.92, no, yes.   (i) Converging, +3.3, X, X, +5, X, −, no.   **25.** (a) The final image coincides in location with the object.
It is real, inverted and $m = -1.0$.   **27.** (a) Coincides in location with the original object and is enlarged 5.0 times.
(c) Virtual and inverted.   **29.** (a) 13 cm.   (b) 1.2 cm.   (c) −3.3.   (d) 3.1.   (e) −10.   **31.** (b) 38.2 cm.
35. 1.9×10^8 m/s.   **39.** (a) Yes.   (b) No.   (c) 43°.   **41.** (a) 1.60.   (b) 58.0°.   **45.** (a) 1.8 m.   (b) 1.2 m.
47. 401 cm beneath the mirror surface.   **49.** (a) 5.3 cm.   (b) 3.0 mm.   **51.** (b) 0.556 cm/s.   (c) 11.25 m/s.   (d) 5.14 cm/s.
53. (a) 2.99706×10^8 m/s.   (b) 21 MeV.   **55.** 42 mm.

CHAPTER 40

Exercises

1. The distance D must also be doubled.   **3.** Slit separation must be 34 μm.   **5.** (a) 0.57°.   (b) 5.0 mm.
7. Conditions for maxima and minima are interchanged.   **9.** (a) 3.0 m.   (b) No.   **11.** $y = 27\sin(\omega t + 8.5°)$.   **13.** $\lambda/5$.
15. Spacing decreases by a factor of 3/4.   **17.** (a) 552 nm.   (b) 442 nm.   **19.** Bright.   **21.** 5200 nm.

Problems

1. 0.072 mm.   **3.** 80 nm.   **5.** 32.   **7.** 6.6 μm.   **9.** 2.65.   **11.** (a) 1.17 m; 3.00 m; 7.50 m.   (b) No.   **13.** $\lambda_n/4$.
15. 338 nm.   **17.** (a) 169 nm.   (c) Blue-violet will be sharply reduced.   **21.** (a) 34.   (b) 46.   **23.** 1.003.
25. (a) 0.253 mm.   (b) In place of the central maximum you get a minimum.   **27.** 1.00025.   **29.** 0.15°.   **31.** 0.03%.
33. 137.5 nm.   **35.** (a) $\lambda/2n_2$.   (b) $\lambda/2n_2$.   (c) $\lambda/4n_2$.   (d) $\lambda/4n_2$.   (e) $\lambda/2n_2$.   **37.** (b) Violet.   (c) Red.
41. (a) 1.00 m.   (b) 4 cm.   **43.** The intensity is diminished by 88% at 456 nm and by 94% at 650 nm.

CHAPTER 41

Exercises

1. 698 nm.   **3.** 0.118 mm.   **5.** (a) $\lambda_a = 2\lambda_b$.   (b) Coincidences occur when $m_b = 2m_a$.   **7.** 10 km.   **9.** 51 m.
11. Three.   **13.** 0°; ±10°; ±21°; ±32°; ±45°; ±62°.   **15.** All wavelengths shorter than 635 nm.   **17.** 12,500.
19. (a) 55.6 pm.   (b) None.   **21.** 23,100; 28.7°.   **23.** 2.87°.

Problems

3. 1.00.   **5.** 24.0 mm.   **7.** (d) 53°; 10°; 5.1°.   **9.** 30 m.   **11.** (a) 1.1×10^4 km.   (b) 11 km.   **13.** $1.22\lambda D/d$.
15. $\lambda D/d$.   **21.** (a) 6.0 μm.   (b) 1.5 μm.   (c) $m = 0, 1, 2, 3, 5, 6, 7, 9$.   **25.** (a) 1.0×10^4 nm.   (b) 3.3 mm.
29. Yes; $m = 3$ for $\lambda = 0.124$ nm; $m = 4$ for $\lambda = 0.097$ nm.   **31.** (a) 1.03 mm.   (b) 2.06 mm.   **33.** 4.73 cm.   **35.** (a) 6.7°.
(b) Since $1.22\lambda > d$, there is no answer for 1.0 kHz.   **37.** 4.5 m.   **39.** (a) Two.   (b) 0.103°.   **43.** 1100 lines/mm.
47. (a) 0.032°/nm; 0.076°/nm; 0.241°/nm.   (b) 40,000; 80,000; 120,000.   **49.** 0.30 nm.   **53.** $\approx 10^{-12}$.

CHAPTER 42

Exercises

1. 9.1 kW.   **3.** 2170 K.   **5.** (a) 0.97 mm; microwave.   (b) 9.9 μm; infrared.   (c) 1.6 μm; infrared.   (d) 500 nm; visible.
(e) 0.29 nm; x ray.   (f) 2.9×10^{-41} m; hard gamma ray.   **7.** 5270 K; sun's "surface" temperature is 5800 K.   **9.** No.
11. 172 nm.   **13.** 10.1 eV.   **15.** (a) 62 pm.   (b) X-ray region.   **17.** (a) 2.43 pm.   (b) 4.86 pm.   (c) 0.255 MeV.
19. 2.65 fm.   **21.** (a) 12 eV.   (b) 6.5×10^{-27} kg·m/s.   (c) 103 nm.   **23.** 0.213 eV; 0.0218 eV.   **25.** (a) -122.4 eV;
-30.6 eV; -13.6 eV.   (b) 122.4 eV.

Problems

1. 1.54 W.   **3.** $4\triangle T/T$; 0.0130.   **7.** 9.68×10^{-20} A.   **11.** (a) 1.2×10^{20} Hz.   (b) 2.4 pm.   (c) 2.7×10^{-22} kg·m/s.
15. 1.12 keV.   **17.** 300%.   **21.** (a) 661 nm; 489 nm; 437 nm.   (b) 661 nm to 367 nm.   **25.** $\lambda_{He} = \frac{1}{4}\lambda_H$, for corresponding
spectral lines.   **27.** (a) Corresponding positron wavelengths are longer by a factor of two.   (b) Radius to center of mass is equal to
the corresponding radius for hydrogen.   **29.** 166.   **31.** (a) $+0.048$ Å.   (b) -41 keV.   (c) 41 keV.   **35.** (a) 16.
(b) $0.4R_{sun}$.   **37.** (a) 8.1×10^{-9}%.   (b) 4.9×10^{-4}%.   (c) 9.6%.   (d) 68%.   **39.** (a) 3×10^{74}.   (b) No.
41. (a) 2.96×10^{20} photons/s.   (b) 4.86×10^{4} km.   (c) 5.89×10^{18} photons/m²·s; 1.96×10^{10}/m³.   **43.** 81.3 km; 40.6 km.
45. (a) 3.1 keV.   (b) 14.4 keV.   **47.** (a) $nh/\pi md^2$.   (b) $n^2h^2/4\pi^2md^2$.

CHAPTER 43

Exercises

1. (a) 1.7×10^{-35} m.   **3.** 2.5×10^{-17} m; nuclear radii are approximately 7×10^{-15} m.   **5.** (a) 0.0387 nm.   (b) 1.24 nm.
(c) 9.04×10^{-4} nm.   **7.** (a) 2.05×10^{-2} eV.   (b) 37.7 eV.   **9.** (a) 1.9 eV.   (b) 660 nm.   **11.** 1.85.
13. 6.63×10^{-23} kg·m/s.   **15.** 1.2 m.

Problems

1. 4.3 μeV.   **3.** (a) 3.88×10^{-2} eV.   (b) 0.146 nm.   **5.** 5.6°.   **7.** (a) Higher orders cannot exist for this accelerating
potential and these planes.   (b) 59°.   **9.** (a) 3.9×10^{-22} eV.   (b) The thermal energy is about 10^{20} times as great.
(c) 3.0×10^{-18} K.   **11.** (a) 0.195.   (b) 0.402.   (c) 0.333.   (d) 0.333.   **21.** $\sqrt{2}b$.   **23.** (b) 6×10^{-4}.   (c) 3×10^{-4}.
25. (a) 2.83°.   (b) 0.55°.

CHAPTER 44

Exercises

1. $n > 3$; $m_l = +3, +2, +1, 0, -1, -2, -3$; $m_s = +\frac{1}{2}, -\frac{1}{2}$.   **3.** The 4s, 4p, 4d and 4f subshells have 1, 3, 5 and 7 electrons,
respectively, a total of 16 for the N shell.   **5.** (a) 5.40×10^{14} m$^{-3/2}$.   (b) 2.91×10^{29} m^{-3}.   (c) 1.02×10^{10} m^{-1}.
7. $L = \sqrt{12}\hbar$; $L_z = -3\hbar, -2\hbar, -\hbar, 0, \hbar, 2\hbar, 3\hbar$; $\mu = \sqrt{12}\mu_B$; $\mu_z = 3\mu_B, 2\mu_B, \mu_B, 0, -\mu_B, -2\mu_B, -3\mu_B$; $\theta = 30°, 55°, 73°, 90°,$
$107°, 125°, 150°$.   **9.** $a/g = 7.4 \times 10^3$.   **11.** 1, 0, 0, $\pm\frac{1}{2}$.   **13.** (a) $\frac{\sqrt{15}}{2}\hbar$.   (b) $\frac{3}{2}\hbar, \frac{1}{2}\hbar, -\frac{1}{2}\hbar, -\frac{3}{2}\hbar$.

15. 1s, 2s, 4s are not split since these states have zero orbital angular momentum.   **17.** (a) $4p_{1/2}$ and $4s_{1/2}$: two states each;
$4p_{3/2}$: four states.   (b) $4p_{3/2}$: 9.2×10^{-5} eV; $4p_{1/2}$: 4.6×10^{-5} eV; $4s_{1/2}$: 1.4×10^{-4} eV.

Problems

3. 1.8×10^{-4}.   **5.** 2.25×10^{-2}.   **7.** (c) 0.18°; 0.057°.   **11.** (a) 5.8×10^{-5} eV.   (b) 1.4×10^{10} Hz.   (c) 2.1 cm;
short radio-wave region.   **13.** 4.5×10^{-5} eV.   **15.** (b) 0.39 T.   **21.** (a) $\frac{3}{2}; \frac{5}{2}$.   (b) $\frac{4}{5}; \frac{6}{5}$.   (c) $\frac{3}{2}\hbar, \frac{5}{2}\hbar$.   (d) $\frac{6}{5}\mu_B; 3\mu_B$.
(e) 4; 6.   **23.** 5.4×10^{-3}.   **25.** 51.0 mT.   **27.** (a) -27.2 eV.   (b) 32.2 eV.   (c) 5.0 eV.   (d) Since $E > 0$, the
system is unbound; no quantum number can be assigned.   **31.** $1.5r_B$.   **33.** (a) 1.48×10^{-21} N.   (b) 0.02 mm.
35. (a) 122 nm.   (b) $\approx 6 \times 10^{-4}$ nm.   (c) $\approx 5 \times 10^{-5}$ for hydrogen; $\approx 1.0 \times 10^{-3}$ for sodium.   (d) The hydrogen doublet.

CHAPTER 45

Exercises

1. 6.63×10^{-34} J·s.   **5.** (a) 35.4 pm, as for molybdenum.   (b) 50 pm.   (c) 57 pm.   **7.** (1, 0, $\frac{1}{2}$, $\pm\frac{1}{2}$) and (2, 0, $\frac{1}{2}$, $\pm\frac{1}{2}$).
9. 2.02×10^{16} s^{-1}.   **11.** (a) 10^{-17}.   (b) 1260 K.

Problems

3. Chromium and copper. **5.** 0.282 nm. **7.** (a) $n = 3, l = 1, j = \frac{3}{2}$. (b) $\frac{3}{2}\hbar$. (c) $2\mu_B$. **9.** 4.8 km. **11.** (a) No.

(b) 0.138 μm. **13.** 49.6 pm; 99.2 pm; 99.2 pm. **15.** 2.2 keV. **17.** 2×10^7.

CHAPTER 46

Exercises

1. 5.90×10^{28} m^{-3}. **9.** $T \gg 10^5$ K. **11.** (a) 203 nm. (b) Ultraviolet. **13.** 6.02×10^5.

Problems

1. (a) 2.7×10^{25} m^{-3}. (b) 8.4×10^{28} m^{-3}. (c) 3100. (d) 3.3 nm for oxygen, 0.23 nm for the electrons. **3.** 3
5. 137 MeV. **7.** (a) 1.2×10^5 m/s. (b) 1.6×10^6 m/s. (c) 13. **9.** (a) Zero. (b) 0.5%. (c) 1.8%.
11. (a) 2.4×10^{-6}. (b) 2.4×10^{-6}. (c) 0.5. **13.** (a) 0.744 eV. (b) 7.14×10^{-7}. **15.** (b) -2.49×10^8.
17. (a) n type. (b) 5×10^{21} m^{-3}. (c) 2.5×10^5. **19.** (a) 10^8. (b) 1.3 eV^{-1} for 10 Å; 1.3×10^8 eV for 10 cm.
21. 0.95 keV.

CHAPTER 47

Exercises

1. (a) 1.1×10^{13} N/C. (b) 2.4×10^{21} N/C. **3.** (a) ^{142}Nd; ^{143}Nd; ^{146}Nd; ^{148}Nd; ^{150}Nd. (b) ^{97}Rb; ^{98}Sr; ^{99}Y; ^{100}Zr; ^{101}Nb; ^{102}Mo; ^{103}Tc;
^{105}Rh; ^{109}In; ^{110}Sn; ^{111}Sb; ^{112}Te. (c) ^{60}Zn; ^{60}Cu; ^{60}Ni; ^{60}Co; ^{60}Fe. **5.** (a) 1150 MeV. (b) 4.8 MeV/nucleon; 12 MeV/proton.
7. (a) $+7.29$ MeV. (b) $+8.07$ MeV. (c) -91.0 MeV. **9.** (a) 2.03×10^{20}. (b) 2.78×10^9/s $= 75$ mCi.
11. (a) 5.04×10^{18}. (b) 4.54×10^6/s. **13.** Pu: 1.9×10^{-17}; Cm: 2.2×10^{-85}. **15.** 1.6×10^{-25}. **17.** 5.32×10^{22}.
19. 1.21 MeV. **21.** ^{63}Cu(n,α); ^{60}Ni(n,p); ^{61}Ni(n,d); ^{62}Ni(d,α); ^{59}Co(n,γ); ^{58}Fe(d,γ). **25.** (a) 6.6 MeV. (b) No.
27. (a) ^{18}O; ^{60}Ni; ^{92}Mo; ^{144}Sm; ^{207}Pb. (b) ^{40}K; ^{91}Zr; ^{121}Sb; ^{143}Nd. (c) ^{13}C; ^{40}K; ^{49}Ti; ^{205}Tl; ^{207}Pb.

Problems

3. (a) $3p + 4n = 7$. (b) $7p + 4e = 11$. (c) No: both numbers are odd and consistent with $I = \frac{3}{2}$. (d) Not really: even
though the electron has a much larger magnetic moment, there are four of them and they could cancel. **5.** 1.58×10^{25} MeV.
7. 7.92 MeV. **9.** 4.82×10^{-2}. **13.** 209 d. **15.** (a) U: 1.062×10^{19}; Pb: 0.624×10^{19}. (b) 1.686×10^{19}. (c) 2.98 AE.
17. (a) 4.26 MeV. (b) -24.03 MeV. (c) 28.3 MeV. **19.** (a) 5.98 MeV; 31.8 MeV. (b) 78 MeV. **21.** (b) 0.961 MeV.
23. 0.782 MeV. **25.** 4×10^{-22} s. **29.** (a) 1.96 MeV. (b) 7.45 MeV. (c) 7.19 MeV. **31.** (a) 25.3 MeV.
(b) 12.8 MeV. (c) 25.0 MeV. **33.** Energy to remove the "extra" proton is 5.8 MeV. Energy to remove a filled-shell
proton is 10.6 MeV. **37.** (a) 6.2 fm. (b) Yes. **39.** $K \simeq 30$ MeV. **1.** (a) 19.8 MeV; 6.26 MeV; 2.22 MeV.
(b) 28.3 MeV. (c) 7.07 MeV. **43.** (a) $Y_1 = {}^{16}$O; $Y_2 = {}^{17}$O. (b) ^{17}O$(\alpha,d)^{19}$F. **45.** ^{25}Mg: 9.88%; ^{26}Mg: 11.42%. **47.** 9/2.
49. 1.0087 u. **55.** (a) 3.66×10^7/s. (b) $t \gg 3.82$ d. (c) Same as (a). (d) 6.42 ng.

CHAPTER 48

Exercises

1. (a) 8.2×10^{13} J. (b) 2.6×10^4 y. **3.** 3.1×10^{10} s^{-1}. **7.** ^{238}U $+ n \rightarrow {}^{239}$U $\rightarrow {}^{239}$Np $+ e$; ^{239}Np $\rightarrow {}^{239}$Pu $+ e$.
9. (a) 30 MeV. (b) Zero. (c) 200 keV. **11.** (a) 2.28×10^{41} J. (b) 5.54×10^{43} J. (c) 1260. **15.** (a) 100 keV.

Problems

1. (a) ^{153}Nd. (b) 110 MeV to ^{83}Ge; 60 MeV to ^{153}Nd. (c) 1.60×10^7 m/s for ^{83}Ge; 8.7×10^6 m/s for ^{153}Nd. **3.** (a) Ten.
(b) 231 MeV. **5.** (a) 276 MeV. (b) Typical fission energy $= 200$ MeV. **7.** 1.6×10^{17}. **11.** (a) 3.1×10^{31} protons/m^3.
(b) 1.2×10^6 times. **15.** (a) 4.3×10^9 kg/s. (b) 3.1×10^{-4}. **19.** (a) 35 MJ. (b) 17 lb. (c) 3500 MW.
21. $Q = +4.98$ MeV. **23.** $\simeq 1.5$ MeV. **25.** 560 W. **27.** $K_\alpha = 3.5$ MeV; $K_n = 14.1$ MeV.
29. (a) Procedure ii. **31.** (b) 1.0; 0.89; 0.28; 0.019. (c) 8. **33.** 1.7×10^9 y.

Index

Some Useful Numbers

$\sqrt{2} = 1.414$　　$\sqrt{3} = 1.732$　　　$\sqrt{10} = 3.162$　　　$\pi = 3.142$

$\pi^2 = 9.870$　　　$\sqrt{\pi} = 1.773$　　　$\log \pi = 0.4972$　　$4\pi = 12.57$

$e = 2.718$　　　$\ln 10 = 2.303$　　　$\log e = 0.4343$　　$\ln 2 = 0.6932$

$\sin 30° = \cos 60° = 0.5000$　　$\cot 30° = \tan 60° = 1.7321$

$\cos 30° = \sin 60° = 0.8660$　　$\sin 45° = \cos 45° = 0.7071$

$\tan 30° = \cot 60° = 0.5774$　　$\tan 45° = \cot 45° = 1.0000$

Change of Base

$\log x = \ln x / \ln 10 = 0.4343 \ln x$

$\ln x = \log x / \log e = 2.303 \log x$